ENCYCLOPEDIA *of* PLANETARY SCIENCES

JOIN US ON THE INTERNET VIA WWW, GOPHER, FTP OR EMAIL:

WWW: http://www.thomson.com
GOPHER: gopher.thomson.com
FTP: ftp.thomson.com
EMAIL: findit@kiosk.thomson.com

A service of I(T)P®

Chapman & Hall Encyclopedia Of Earth Sciences Series

ENCYCLOPEDIA OF PLANETARY SCIENCES

Aim of the series

The Chapman & Hall *Encyclopedia of Earth Sciences Series* provides comprehensive and authoritative coverage of all the main areas in the earth sciences. Each volume comprises a focused and carefully chosen collection of contributions from leading names in the subject, with copious illustrations and detailed reference lists.

These books represent one of the world's leading reference resources for the earth sciences community. Previous volumes are being updated and new works published so that the volumes will continue to be essential reading for all professional geologists, geophysicists, climatologists, oceanographers, for teachers, and for students.

Series Editor

Rhodes W. Fairbridge has edited more than two dozen encyclopedias in the *Encyclopedia of Earth Sciences Series* and 90 volumes of the Benchmark reprint series, as well as authoring over 300 other scientific publications. During his career he has worked as a petroleum geologist in the Middle East, been an intelligence officer in the SW Pacific, and led expeditions to the Sahara, Arctic Canada, Arctic Scandinavia, Brazil and New Guinea. He is now Emeritus Professor of Geology at Columbia University.

Volume Editor

James H. Shirley has authored scientific publications in lunar and terrestrial seismology, in solar physics, and in the atmospheric sciences. He is a member of the Near-Infrared Mapping Spectrometer Science team of the Galileo Mission to the planet Jupiter, based at the Jet Propulsion Laboratory in Pasadena, California.

Previous titles in series

Schwartz: *The Encyclopedia of Beaches and Coastal Environments*, 1982
Finkl: *The Encyclopedia of Applied Geology*, 1984
Oliver: *The Encyclopedia of Climatology*, 1987
Finkl: *The Encyclopedia of Field and General Geology*, 1988
Bowes: *The Encyclopedia of Igneous and Metamorphic Petrology*, 1989
James: *The Encyclopedia of Solid Earth Geophysics*, 1989

New and forthcoming volumes

Moores: *The Encyclopedia of European and Asian Regional Geology*
Herschy: *The Encyclopedia of Hydrology and Water Resources*
Finkl: *The Encyclopedia of Soil Science and Technology*
Gerrard: *The Encyclopedia of Geomorphology*
Alexander: *The Encyclopedia of Environmental Science*
Marshal: *The Encyclopedia of Geochemistry*
Gerrard: *The Encyclopedia of Quaternary Science*

Editorial Advisory Board

Daniel C. Boice
Southwest Research Institute
6220 Culebra Road
San Antonio, Texas 78228–0510, USA

Hans Haubold
United Nations Office for Outer Space
Vienna International Center
Post Office Box 500
A-1400 Vienna, Austria

Robert M. Nelson
Jet Propulsion Laboratory
4800 Oak Grove Drive
Pasadena, California 91109, USA

Christopher T. Russell
Institute of Geophysics and Planetary Physics
University of California
Los Angeles, California 90024, USA

Richard A. Simpson
Space, Telecommunications, and Radiosciences Laboratory
Stanford University
Stanford, California 94305–9515, USA

Conway W. Snyder
21206 Seep Willow Way
Canyon Country, California 91351, USA

Timothy D. Swindle
Lunar and Planetary Laboratory
University of Arizona
Tucson, Arizona 85721, USA

Published by Chapman & Hall, 2–6 Boundary Row, London SE1 8HN, UK

Chapman & Hall, 2–6 Boundary Row, London SE1 8HN, UK

Chapman & Hall GmbH, Pappelallee 3, 69469 Weinheim, Germany

Chapman & Hall USA, 115 Fifth Avenue, New York, NY 10003, USA

Chapman & Hall Japan, ITP-Japan, Kyowa Building, 3F,
2-2-1 Hirakawacho, Chiyoda-ku, Tokyo 102, Japan

Chapman & Hall Australia, 102 Dodds Street, South Melbourne,
Victoria 3205, Australia

Chapman & Hall India, R. Seshadri, 32 Second Main Road, CIT East,
Madras 600 035, India

First edition 1997

© 1997 Chapman & Hall

Typeset in 8/8½pt Times by Photoprint, Torquay, Devon

Printed and bound in Croatia by Zrinski d.d., Cakovec

ISBN 0 412 06951 2

Apart from any fair dealing for the purposes of research or private study, or criticism or review, as permitted under the UK Copyright Designs and Patents Act, 1988, this publication may not be reproduced, stored, or transmitted, in any form or by any means, without the prior permission in writing of the publishers, or in the case of reprographic reproduction only in accordance with the terms of the licences issued by the Copyright Licensing Agency in the UK, or in accordance with the terms of licences issued by the appropriate Reproduction Rights Organization outside the UK. Enquiries concerning reproduction outside the terms stated here should be sent to the publishers at the London address printed on this page.

The publisher makes no representation, express or implied, with regard to the accuracy of the information contained in this book and cannot accept any legal responsibility or liability for any errors or omissions that may be made.

A catalogue record for this book is available from the British Library

Library of Congress Catalog Card Number: 96–86590

∞ Printed on permanent acid-free text paper, manufactured in accordance with ANSI/NISO Z39.48–1992 and ANSI/NISO Z39.48–1984 (Permanence of Paper).

ENCYCLOPEDIA OF EARTH SCIENCES SERIES

ENCYCLOPEDIA *of* PLANETARY SCIENCES

edited by

JAMES H. SHIRLEY *and*
RHODES W. FAIRBRIDGE

CHAPMAN & HALL
London · Weinheim · New York · Tokyo · Melbourne · Madras

Contents

Colour plates appear between pages 514 and 515

List of contributors	xvii
Preface	xxvii
Absorption, absorption spectrum *Richard A. Simpson and Jon M. Jenkins*	1
Accretion *Tamara V. Ruzmaikina and G. Wuchterl*	1
Achondrite meteorites *David W. Mittlefehldt*	2
Adams, John Couch (1819–1892) *Hans J. Haubold*	3
Aerosol *Rhodes W. Fairbridge*	4
Airglow *Jane L. Fox*	4
Albedo *Rhodes W. Fairbridge*	10
Alfvén, Hannes Olof Gösta (1908–1995), and Alfvén wave *Richard A. Simpson, Christopher T. Russell and Rhodes W. Fairbridge*	11
Amor object *James H. Shirley*	11
Angular momentum *Jane B. Blizard*	12
Angular momentum cycle in planet Earth *Abraham A. Oort*	13
Antarctic meteorites *Marilyn M. Lindstrom*	19
Antoniadi, Eugenios (1870–1944) *Patrick Moore*	22
Aphelion *Curtis Wilson*	23
Apollo missions *James H. Shirley*	23
Apollo object *James H. Shirley*	25
Apsis, apsides *Rhodes W. Fairbridge*	26
Archeoastronomy *David S.P. Dearborn*	26
Aristarchus (c. 310–230 BC) *Patrick Moore*	27
Asteroid *Linda M. French*	27
Asteroid: compositional structure and taxonomy *Daniel T. Britt and Larry A. Lebofsky*	33
Asteroid: families *Giovanni B. Valsecchi*	35
Asteroid: lightcurve *Alan W. Harris*	38
Asteroid: photometry *Alan W. Harris*	38
Asteroid: resonance *Fergus J. Wood*	42
Asteroid: thermal infrared studies *Larry A. Lebofsky*	43
Asthenosphere *Rhodes W. Fairbridge*	45
Astrogeology *Don E. Wilhelms*	46
Astrometric observation *Jeremy B. Tatum*	46
Astronomical constants *James H. Shirley*	48
Astronomical unit *Gareth V. Williams*	48
Aten object *James H. Shirley*	51
Atmosphere *Anthony D. Del Genio*	51
Atmospheric thermal structure *Robert E. Samuelson*	54
Aurora: historical record *Rhodes W. Fairbridge*	57

Aurora, planetary *J. Hunter Waite Jr*	58	Charged particle observation *Andrew F. Cheng*	102
Barnard, Edward Emerson (1857–1923) *Patrick Moore*	62	Charon *James H. Shirley*	103
Barycenter *James H. Shirley and Rhodes W. Fairbridge*	62	Chemical element *Clare P. Marshall*	103
Basalt *John Longhi*	63	Chiron *Linda M. French*	104
Basaltic achondrite meteorites *David W. Mittlefehldt and John Longhi*	65	Chondrites, ordinary *Derek W.G. Sears*	105
Biosphere *Rhodes W. Fairbridge*	68	Chronology: meteorite *Frank A. Podosek*	110
Blackbody radiation *Imke de Pater and Richard A. Simpson*	69	Clementine mission *James H. Shirley*	112
Bradley, James (1693–1762) *Curtis Wilson*	69	Cold accretion theory *William K. Hartmann*	113
Brahe, Tycho (1546–1601) *Hans J. Haubold*	70	Collisions *William K. Hartmann*	114
Breccia *Rhodes W. Fairbridge*	71	Color *Patricia A. Jacobberger*	114
Brouwer, Dirk (1902–1966) *Patrick Moore*	71	Comet *Daniel C. Boice and Walter F. Huebner*	115
Brückner cycle *Rhodes W. Fairbridge*	71	Comet: dynamics *J.A. Fernández and W.-H. Ip*	119
Callisto *Paul M. Schenk and William B. McKinnon*	73	Comet: historical perspective *Patrick Moore*	125
Campbell, William Wallace (1862–1938) *Patrick Moore*	75	Comet: impacts on Earth *Christopher F. Chyba*	126
Capture mechanisms *Alan P. Boss*	76	Comet: impacts on Jupiter *James H. Shirley*	132
Carbon, carbon dioxide *Rhodes W. Fairbridge and Clare P. Marshall*	76	Comet: observation *Jacques Crovisier*	134
Carbonaceous chondrite *Michael Zolensky*	77	Comet: origin and evolution *Paul R. Weissman*	140
Carrington, Richard Christopher (1826–1875) *Rhodes W. Fairbridge*	79	Comet: structure and composition *Daniel C. Boice and Walter F. Huebner*	145
Cartography *R.M. Batson*	79	Commensurability *James H. Shirley*	153
Cassini, Jean-Dominique (1625–1712), and Cassini's laws *Rhodes W. Fairbridge and Hans J. Haubold*	83	Coordinate systems *Philip J. Stooke and Jeremy B. Tatum*	154
Cassini mission *Dennis L. Matson*	84	Copernicus, Nicolaus (1473–1543) *Rhodes W. Fairbridge and Hans J. Haubold*	159
Celestial mechanics *Morris S. Davis*	88	Core, terrestrial planetary *Yuan-Chong Zhang*	160
Center of mass, gravity and inertia *Yuan-Chong Zhang*	92	Coriolis force, geostrophic motion *Anthony D. Del Genio*	162
Ceres *Larry A. Lebofsky*	93	Corona (Venus) *R.R. Herrick*	163
Chamberlin, Thomas Chrowder (1843–1928); Chamberlin–Moulton planetesimal hypothesis *Rhodes W. Fairbridge and Norriss S. Hetherington*	93	Coronal mass ejections *J.T. Gosling*	164
Chandler wobble *Benjamin Fong Chao*	94	Corpuscular radiation *J.P. Rozelot*	165
Chaotic dynamics in the solar system *Jacques Laskar*	94	Cosmic rays *Arnold W. Wolfendale*	166
Charge-coupled device *Robert S. McMillan*	98	Cosmic ray exposure ages *Robert C. Reedy*	169

Cosmochemistry *Bruce Fegley*	169	Eccentric dipole *Conway W. Snyder*	228
Crater *Richard A.F. Grieve*	177	Eccentricity *Curtis Wilson*	228
Cretaceous–Tertiary (K–T) boundary *Alan Hildebrand*	178	Eclipse *Ewen Whitaker*	229
Croll, James (1821–1890) *Rhodes W. Fairbridge*	178	Ecliptic *Fergus J. Wood*	231
Crust *Rhodes W. Fairbridge*	179	Ejecta *Richard A.F. Grieve*	231
Cyclone, anticyclone *Anthony D. Del Genio*	179	Electromagnetic radiation *Richard A. Simpson*	232
Darwin, George Howard (1845–1912) *Patrick Moore*	181	Ellipsoid *John P. Snyder*	232
Dating methods *Frank A. Podosek*	181	Emissivity *Richard A. Simpson*	233
Dawes, William Rutter (1799–1868) *Patrick Moore*	183	Enceladus *Jonathan I. Lunine*	233
Deep Space Network *Nicholas A. Renzetti*	183	Encke, Johann Franz (1791–1865) *Patrick Moore*	233
Deimos *Peter C. Thomas*	184	Enstatite meteorites *Derek W.G. Sears*	234
Denning, William Frederick (1848–1931) *Patrick Moore*	186	Eolian transport *Steven H. Williams*	237
Determinism *Sheldon Ackley*	186	Ephemeris *X.X. Newhall*	238
Differentiation *Timothy D. Swindle*	188	Eratosthenes (276–195 BC) *Rhodes W. Fairbridge*	240
Diurnal variation *Anthony D. Del Genio*	188	Erosion *Rhodes W. Fairbridge*	241
Dome *Adrian F. Park*	189	Eucrite meteorites *John Longhi*	241
Dust *James H. Shirley*	189	Euler, Leonhard (1707–1783), and Eulerian motion *Hans J. Haubold and Benjamin Fong Chao*	242
Dynamo theory *James H. Shirley*	192	Europa *William B. McKinnon*	243
Early bombardment *Nadine G. Barlow*	193	European Space Agency *Rüdeger Reinhard*	246
Earth *Paul D. Lowman Jr*	194	Exosphere *James H. Shirley*	248
Earth: atmosphere *David H. Grinspoon*	198	Flood basalt *John Longhi*	249
Earth: geology, tectonics and seismicity *Paul D. Lowman Jr*	201	Forbush decrease *Rhodes W. Fairbridge*	251
Earth: magnetic field and magnetosphere *Christopher T. Russell and Janet G. Luhmann*	208	Forbush effect *Arnold W. Wolfendale*	252
Earth Observing System *Ralph Kahn and Daniel Wenkert*	211	Fracture, fault *Rhodes W. Fairbridge*	252
Earth rotation *Benjamin Fong Chao*	215	Fraunhofer line *Roger Ferlet*	253
Earth: rotational history *George E. Williams*	217	Galactic cycle *Michael R. Rampino*	255
Earth–Moon system: dynamics *J.M. Anthony Danby*	221	Galilean satellites *Deborah L. Domingue*	257
Earth–Moon system: origin *Alan P. Boss*	223	Galileo Galilei (1564–1642) *Hans J. Haubold*	258

Galileo mission *Karen Buxbaum*	258	Huygens, Christian (1629–1695) *Rhodes W. Fairbridge and Hans J. Haubold*	310
Ganymede *Jeffrey M. Moore and William B. McKinnon*	263	Huygens Titan atmospheric probe *Jean-Pierre Lebreton*	311
Gaspra *H. Herbert Breneman*	267	Hydrosphere *Rhodes W. Fairbridge*	313
Gauss, Carl Friedrich (1777–1855) *Rhodes W. Fairbridge*	267	Ice age *Rhodes W. Fairbridge*	315
Geoid *Kurt Lambeck*	268	Ida *Marcia Segura*	316
Geomagnetic polarity reversals and the geological record *William Lowrie*	269	Igneous rock *Peter A. Sabine*	317
Geomagnetic storm *Gary D. Parker*	272	Imaging science *Carolyn C. Porco and Patricia T. Eliason*	319
Gilbert, William (1540–1603) *Rhodes W. Fairbridge*	273	Impact cratering *H.J. Melosh*	326
Giotto mission *Rüdeger Reinhard*	274	Inertia, inertial frame *Yuan-Chong Zhang*	335
Global Positioning System *Timothy H. Dixon*	279	Infrared radiation *James H. Shirley*	336
Goddard, Robert Hutchings (1882–1945) *Rhodes W. Fairbridge*	283	Infrared spectroscopy *D.E. Jennings*	336
Gravimetry *William L. Sjogren*	283	Insolation *Yuan-Chong Zhang*	338
Gravitation *John D. Anderson*	283	International Astronomical Union *Brian G. Marsden*	339
Gravity-assist navigation *John D. Anderson*	287	Interplanetary magnetic field *J.T. Gosling*	342
Gravity fields of the terrestrial planets *Kurt Lambeck*	289	Interstellar grains *James H. Shirley*	343
Greenhouse effect *Robert E. Samuelson*	292	Interstellar medium *Bruce T. Draine*	344
Hadley circulation *Anthony D. Del Genio*	293	Io *Robert M. Nelson*	345
Hale, George Ellery (1868–1938), and Hale cycle *Rhodes W. Fairbridge*	293	Ion and neutral mass spectrometry *David T. Young*	351
Hall, Asaph (1829–1907) *Patrick Moore*	295	Ionosphere *Thomas E. Cravens*	354
Halley, Edmond (1656–1742), and Halley's comet *Daniel C. Boice and Rhodes W. Fairbridge*	295	Iridium anomaly *Alan Hildebrand*	359
Heliosphere *D. Venkatesan*	297	Iron *Peter A. Sabine*	360
Herschel, William (1738–1822) *Rhodes W. Fairbridge*	298	Iron meteorites *Alfred Kracher*	361
Hilda asteroids *Linda M. French*	299	Isostasy *Rhodes W. Fairbridge*	363
Hipparchus of Nicaea (190–120 BC) *Rhodes W. Fairbridge*	299	Jones, Harold Spencer (1890–1960) *Patrick Moore*	365
History of planetary science I: pre-space age *Patrick Moore*	300	Julian calendar, year and day *Rhodes W. Fairbridge*	365
History of planetary science II: space age *James A. Van Allen*	302	Jupiter *James H. Shirley*	366
Hohmann transfer orbit *John D. Anderson*	309	Jupiter: atmosphere *Timothy E. Dowling*	367
Hot spot tectonics *Robert R. Herrick*	310	Jupiter: interior structure *William B. Hubbard*	371

Jupiter: magnetic field and magnetosphere *Christopher T. Russell and Janet G. Luhmann*	372	Mantle *James H. Shirley*	415
Jupiter: ring system *Mark R. Showalter*	373	Mantle convection *Gerald Schubert*	416
Jupiter: satellite system *Anthony R. Dobrovolskis*	375	Map projections *John P. Snyder*	421
Kepler, Johannes (1571–1630) *Hans J. Haubold*	378	Maraldi, Giacomo Filippo (1665–1729) *Patrick Moore*	427
Kepler's laws *Hans J. Haubold*	379	Mariner missions *Conway W. Snyder*	427
Kirchhoff, Gustav Robert (1824–1887), and Kirchhoff's laws *Rhodes W. Fairbridge and Hans J. Haubold*	380	Mars *Nadine G. Barlow*	430
Kirkwood, Daniel (1814–1895) *Rhodes W. Fairbridge*	380	Mars: atmosphere *Robert M. Haberle*	432
Kuiper belt *Daniel C. Boice*	381	Mars: geology *Vivien Gornitz*	441
Kuiper, Gerard Peter (1905–1973) *Ewen A. Whitaker*	381	Mars: gravity *William L. Sjogren*	450
Lagrange, Joseph Louis (1736–1813) *Rhodes W. Fairbridge and Hans J. Haubold*	382	Mars: impact cratering *Nadine G. Barlow*	451
Lagrangian point *Fergus J. Wood*	382	Mars: magnetic field and magnetosphere *Janet G. Luhmann and Christopher T. Russell*	454
Laplace, Pierre Simon de (1749–1827) *Rhodes W. Fairbridge and Hans J. Haubold*	383	Mars Observer mission *Arden L. Albee and Frank D. Palluconi*	456
Laser ranging *James H. Shirley*	385	Mars: remote sensing *Ted L. Roush*	459
Length of day *Benjamin Fong Chao*	386	Mars: structural geology and tectonics *David H. Scott and James M. Dohm*	461
Leverrier, Urbain Jean Joseph (1811–1877) *Patrick Moore*	386	Mascon *William L. Sjogren*	463
Libration *Jeremy B. Tatum*	387	Mass extinction *Michael R. Rampino*	464
Life: origin *Christopher P. McKay*	387	Maunder, Edward Walter (1851–1928), and Maunder minimum *Patrick Moore and John E. Oliver*	467
Lithosphere *Rhodes W. Fairbridge*	392	Maxwell theory *Michael H. Farris and Christopher T. Russell*	468
Lomosomov, Mikhael Vasilyevich (1711–1765) *Rhodes W. Fairbridge*	393	Mercury *Faith Vilas*	469
Lowell, Percival (1855–1916) *Nadine G. Barlow and Norriss S. Hetherington*	393	Mercury: atmosphere *Ann L. Sprague*	471
Luna missions *Timothy D. Swindle*	394	Mercury: geology *Paul D. Spudis*	473
Lunar meteorites *Marilyn M. Lindstrom*	395	Mercury: magnetic field and magnetosphere *Christopher T. Russell and Janet G. Luhmann*	476
Lunar Orbiter missions *Conway W. Snyder*	397	Mesosphere *Wilfried Schröder*	478
Lyman alpha *Alfred Vidal-Madjar*	397	Meteor, meteoroid *James H. Shirley*	479
Magellan mission *R. Stephen Saunders*	399	Meteor shower, meteoroid stream *James H. Shirley*	480
Magnetism *Michael D. Fuller*	400	Meteorite *Timothy D. Swindle*	481
Magnetometry *Mario H. Acuña*	406	Meteorite parent bodies *Daniel T. Britt and Larry A. Lebofsky*	486
Magnetospheres of the outer planets *Andrew F. Cheng*	410	Micrometeorite *Michael Zolensky*	489

Microwave spectroscopy *Therese Encrenaz*	490
Milankovitch, Milutin (1879–1958), and Milankovitch theory *George Kukla and Rhodes W. Fairbridge*	494
Miller–Urey experiment *Christopher P. McKay*	495
Miranda *Paul Schenk*	495
Moon (Earth's Moon) *Donald R. Davis*	498
Moon: atmosphere *Thomas H. Morgan*	501
Moon: geology *Graham Ryder*	501
Moon: gravity *William L. Sjogren*	508
Moon: magnetism and interior *Lon L. Hood*	509
Moon: origin *William K. Hartmann*	512
Moon: seismicity *Yosio Nakamura*	513
NASA *Conway W. Snyder*	517
Near-Earth object *Richard P. Binzel*	518
Nebular hypothesis: Kant–Laplace *Norriss S. Hetherington*	519
Nemesis *John D. Anderson*	520
Neptune *William B. Hubbard*	521
Neptune: atmosphere *Kevin H. Baines*	523
Neptune: magnetic field and magnetosphere *Christopher T. Russell and Janet G. Luhmann*	532
Neptune: ring system *Cecile Ferrari*	532
Neptune: satellite system *Jeffrey S. Kargel*	534
Nereid *Anthony R. Dobrovolskis*	539
Newcomb, Simon (1835–1909) *Patrick Moore*	540
Newton, Sir Isaac (1642–1727), and Newton's laws *Hans J. Haubold and Rhodes W. Fairbridge*	540
Noble gas *Timothy D. Swindle*	542
Nomenclature *Joel F. Russell and R.M. Batson*	543
Oberon *Daniel M. Janes*	554
Obliquity *Jeremy B. Tatum*	554
Obliquity: terrestrial record *George E. Williams*	555
Occultation *Richard A. Simpson*	558
Olbers, Heinrich Wilhelm Matthäus (1758–1840), and Olbers' paradox *Rhodes W. Fairbridge*	559
Oort, Jan Hendrik (1900–1992), and Oort cloud *Daniel C. Boice and Rhodes W. Fairbridge*	559
Öpik, Ernst Julius (1893–1985) *Patrick Moore*	560
Opposition *Jeremy B. Tatum*	560
Opposition effect *Robert M. Nelson*	561
Optical depth *Richard A. Simpson*	561
Orbit *Jeremy B. Tatum*	562
Orbital commensurability and resonance *Rhodes W. Fairbridge*	564
Paleomagnetism *J.F. Harper*	572
Periapse, perihelion, perigee, peribac *Fergus J. Wood and Rhodes W. Fairbridge*	572
Phobos *Peter C. Thomas*	573
Phobos mission *Conway W. Snyder*	574
Photoclinometry *David G. Jankowski*	576
Photogrammetry, radargrammetry and planetary topographic mapping *Sherman S.C. Wu*	576
Photometry *James H. Shirley*	578
Piazzi, Giuseppe (1746–1826) *Rhodes W. Fairbridge*	579
Pickering, William Henry (1858–1938) *Patrick Moore*	579
Pioneer 10 and 11 missions *Lawrence E. Lasher*	579
Pioneer Venus missions *Lawrence Colin*	582
Planet *Andrea Carusi*	584
Planet: extrasolar *Robert S. McMillan*	588
Planet X *John D. Anderson*	590
Planetary Data System *Susan McMahon and Sandy Dueck*	591
Planetary dynamical astronomy *P.K. Seidelmann*	592
Planetary geodesy *Bruce G. Bills*	594
Planetary ice *William D. Smythe*	596

Planetary lightning *William J. Borucki*	598	Remote sensing *Patricia A. Jacobberger and Gerald P. Jellison*	689
Planetary ring *Linda J. Horn*	602	Resonance in Saturn's rings *Fergus J. Wood*	696
Planetary rotation *Jack J. Lissauer*	608	Resonance in the solar system *Jacques Henrard*	698
Planetary sampling: *in situ* analysis *Benton C. Clark*	613	Roche, Édouard Albert (1820–1883) *Rhodes W. Fairbridge*	701
Planetary torus *Renée Prangé*	616	Roche limit *Rhodes W. Fairbridge*	701
Planetesimal *Rhodes W. Fairbridge*	623	Sabine, Edward (1788–1883) *Rhodes W. Fairbridge*	704
Plasma *Fran Bagenal*	624	Sakigake and Suisei missions *Daniel C. Boice*	704
Plasma wave *Donald A. Gurnett*	630	Saros cycle *Rhodes W. Fairbridge*	705
Plate tectonics *J.F. Harper*	641	Satellite, natural *Bonnie J. Buratti*	705
Pluto *Steve Mueller and William B. McKinnon*	645	Saturn *Jonathan I. Lunine*	708
Polar cap *Rhodes W. Fairbridge*	650	Saturn: atmosphere *Reta Beebe*	714
Polarimetry *Audouin Dollfus*	650	Saturn: interior structure *William B. Hubbard*	717
Polarity reversals *William Lowrie*	654	Saturn: magnetic field and magnetosphere *Christopher T. Russell and Janet G. Luhmann*	718
Poynting–Robertson drag *Michael C. Nolan*	656	Saturn: ring system *Luke Dones*	719
Precession and nutation *Fergus J. Wood*	656	Saturn: satellite system *Peter C. Thomas*	723
Ptolemy (Claudius Ptolemaeus, AD c. 100–c. 170) *Rhodes W. Fairbridge*	658	Schiaparelli, Giovanni Virginio (1835–1910) *Rhodes W. Fairbridge*	725
Radar astronomy *Richard A. Simpson and Steven J. Ostro*	660	Seismicity *James H. Shirley*	726
Radiation belts *William S. Kurth*	663	Sharonov, Vsevolod Vasilievich (1901–1964) *Rhodes W. Fairbridge*	732
Radiative transfer in planetary atmospheres *Robert A. West*	664	Shepherd satellite *Carl D. Murray*	732
Radio astronomy *Imke de Pater*	667	Shock metamorphism *Richard A.F. Grieve*	733
Radio science *G. Leonard Tyler*	676	Shock waves *Charles P. Sonett*	734
Radiometry *James H. Shirley*	682	Sidereal period *Jeremy B. Tatum*	737
Ranger missions *Conway W. Snyder*	683	Silica, silicate *Rhodes W. Fairbridge*	737
Reflectance spectroscopy *Bruce W. Hapke*	684	Slipher, Earl Carl (1883–1964) *Nadine G. Barlow*	737
Reflectivity *Richard A. Simpson*	686	Small satellite *Carl D. Murray*	738
Regional Planetary Image Facilities *Leslie J. Pieri and Marian E. Rudnyk*	686	SNC meteorites *James L. Gooding*	739
Regolith *Rhodes W. Fairbridge*	687	Solar activity *James H. Shirley*	741
Relativistic cosmology *Wilfried Schröder and Hans-Jürgen Treder*	688	Solar constant *James H. Shirley*	745

Solar corona *Jacques-Clair Noëns*	746	Tektite *Billy P. Glass*	802
Solar day and time *Rhodes W. Fairbridge*	747	Temperature *James H. Shirley and Rhodes W. Fairbridge*	806
Solar flare *James H. Shirley*	747	Terrestrial planets *James H. Shirley*	806
Solar luminosity *Gary A. Chapman*	748	Tessera *Robert R. Herrick*	807
Solar motion *Ivanka Charvátová*	748	Thales of Miletus (624–548 BC) *Rhodes W. Fairbridge*	808
Solar nebula *Tamara V. Ruzmaikina*	751	Thermal evolution of planets and satellites *Gerald Schubert*	808
Solar neutrino *Hans J. Haubold*	751	Thermal plasma instrumentation *W.B. Hanson and R.A. Heelis*	814
Solar particle event *Robert C. Reedy*	754	Thermosphere *Stephen W. Bougher and Raymond G. Roble*	819
Solar photosphere *J.P. Rozelot*	754	Tidal friction *David P. Rubincam*	825
Solar system *James H. Shirley*	756	Tidal heating *Jonathan I. Lunine*	828
Solar system: origin *Tamara V. Ruzmaikina*	756	Tide-raising force *Yuan-Chong Zhang*	828
Solar system: stability *Jacques Laskar*	762	Titan *Cindy C. Cunningham*	831
Solar wind *Philip A. Isenberg*	766	Titania *Daniel M. Janes*	833
Soviet Mars missions *Conway W. Snyder*	772	Titius–Bode law *Rhodes W. Fairbridge*	834
Spacewatch *Tom Gehrels*	774	Tombaugh, Clyde William (1906–) *Patrick Moore*	835
Spectrophotometry *Robert M. Nelson*	775	Torque *Yuan-Chong Zhang*	835
Spectroscopy: atmospheres *Robert W. Carlson*	776	Transit *Jeremy B. Tatum*	835
Stony iron meteorites *David W. Mittlefehldt*	783	Triton *Lance A.M. Benner*	836
Stratosphere *Anthony D. Del Genio*	786	Trojan asteroids *Linda M. French*	841
Sun *Hans J. Haubold and A.M. Mathai*	786	Troposphere *Anthony D. Del Genio*	841
Surface gravity *James H. Shirley*	794	Trouvelot, Étienne Léopold (1827–1895) *Patrick Moore*	842
Surface pressure *James H. Shirley*	794	Tsiolkovsky, Konstantin Eduardovich (1857–1935) *Rhodes W. Fairbridge*	842
Surface processes *Steven H. Williams*	794	Ultraviolet radiation *James H. Shirley*	843
Surveyor missions *Conway W. Snyder and James H. Shirley*	796	Ultraviolet spectroscopy *Bill R. Sandel*	843
Synergetic tidal force *Fergus J. Wood*	797	Ulysses mission *D. Edgar Page*	852
Synodic period *Jeremy B. Tatum*	800	Umbriel *Daniel M. Janes*	855
Syzygy *Jeremy B. Tatum*	800	Uranus *William B. Hubbard*	856
Tectonics *Rhodes W. Fairbridge*	801	Uranus: atmosphere *Kevin H. Baines*	858

Uranus: magnetic field and magnetosphere *Christopher T. Russell and Janet G. Luhmann*	863	Volcanism in the solar system *Adrian F. Park*	915
Uranus: ring system *Philip D. Nicholson*	864	Voyager missions *Ellis D. Miner*	922
Uranus: satellite system *Daniel M. Janes*	866	Voyager planetary radio astronomy *James W. Warwick*	927
Ureilite meteorites *John L. Berkley*	868	Water *Rhodes W. Fairbridge*	933
Urey, Harold Clayton (1893–1981) *Rhodes W. Fairbridge*	874	Weathering *Rhodes W. Fairbridge*	935
Väisälä orbit *Gareth V. Williams*	875	Whistler *William S. Kurth*	936
Van Allen, James Alfred (1914–) *Bruce Randall*	876	Wolf, Max (1863–1932) *Rhodes W. Fairbridge*	936
Vega mission *M.I. Verigin*	876	Wolf, Rudolf (1816–1893), and Wolf number *Rhodes W. Fairbridge*	937
Venera missions *Oleg L. Vaisberg*	879	Yarkovsky effect *Michael C. Nolan*	938
Venus *R. Stephen Saunders*	887	Year *Rhodes W. Fairbridge*	938
Venus: atmosphere *Fred W. Taylor*	890	Zeeman effect *Alfred Vidal-Madjar*	940
Venus: geology and geophysics *Sean C. Solomon*	895	Zodiac *J.B. Tatum*	940
Venus: gravity *William L. Sjogren*	904	Zodiacal light *James H. Shirley*	941
Venus: magnetic field and magnetosphere *Janet G. Luhmann and Christopher T. Russell*	905	Zond *Conway W. Snyder*	941
Vesta *Timothy D. Swindle*	907	Zwicky, Fritz (1898–1974) *Rhodes W. Fairbridge*	941
Viking mission *Conway W. Snyder*	908	Appendix A Lists of entries by subject	942
		Appendix B The international system of units	947
Visible and near-infrared spectroscopy *James F. Bell III*	911	Author index	951
		Subject index	975

Contributors

Sheldon Ackley
322 West 57th Street
New York, NY 10019, USA

Mario H. Acuña
NASA Goddard Space Flight Center Code 695
Greenbelt, MD 20771, USA

Arden L. Albee
California Institute of Technology MC 02–31
Pasadena, CA 91125, USA

John D. Anderson
Jet Propulsion Laboratory MS 301–230
4800 Oak Grove Drive
Pasadena, CA 91109, USA

Fran Bagenal
Department of Astrophysical, Planetary and
  Atmospheric Sciences
Campus Box 391
University of Colorado
Boulder, CO 80309–0391, USA

Kevin H. Baines
Jet Propulsion Laboratory MS 169–237
4800 Oak Grove Dr.
Pasadena, CA 91109, USA

Nadine G. Barlow
Department of Physics
University of Central Florida
Orlando, FL 32816, USA

R.M. Batson
United States Geological Survey, Branch of
  Astrogeology
2255 North Gemini Drive
Flagstaff, AZ 86001, USA

Reta Beebe
Department of Astronomy
University of New Mexico
Box 30001/Department 4500
Las Cruces, NM 88003–0001, USA

James F. Bell III
Department of Astronomy
424 Space Sciences Building
Cornell University
Ithaca, NY 14853-6801, USA

Lance A.M. Benner
Jet Propulsion Laboratory MS 300–233
4800 Oak Grove Drive
Pasadena, CA 91109, USA

John L. Berkley
Department of Geosciences
State University of New York
Fredonia, NY 14063, USA

Bruce G. Bills
Goddard Space Flight Center Code 921
Greenbelt, MD 20771, USA

Richard P. Binzel
Department of Earth, Atmospheric and Planetary
  Sciences
Massachusetts Institute of Technology
Cambridge, MA 02139, USA

Jane B. Blizard
827 16th Street
Boulder, CO 80302, USA

Daniel C. Boice
Instrumentation and Space Research Division
Southwest Research Institute
San Antonio, TX 78228–0510, USA

William J. Borucki
NASA Ames Research Center MS 245-3
Moffett Field, CA 94035-1000, USA

Alan P. Boss
Carnegie Institute of Washington
5241 Broad Branch Road NW
Washington, DC 20015, USA

Steven W. Bougher
Lunar and Planetary Laboratory
University of Arizona
Tucson, AZ 85721, USA

H. Herbert Breneman
Jet Propulsion Laboratory MS 168-222
4800 Oak Grove Drive
Pasadena, CA 91109, USA

Daniel T. Britt
Lunar and Planetary Laboratory
University of Arizona
Tucson, AZ 85721, USA

Bonnie J. Buratti
Jet Propulsion Laboratory MS 183-501
4800 Oak Grove Drive
Pasedena, CA 91109, USA

Karen Buxbaum
Jet Propulsion Laboratory MS 264-765
4800 Oak Grove Drive
Pasadena, CA 91109, USA

Robert W. Carlson
Jet Propulsion Laboratory MS 183-601
4800 Oak Grove Drive
Pasadena, CA 91109, USA

Andrea Carusi
Reparto Planetologia
Instituto Astrofisica Spaziale
Viale Universita 11
1-00185 Roma, Italy

D.P. Cauffman
Lockheed Solar and Astrophysics Laboratory
0/91-30B/256 R&D Div.
3251 Hanover Street
Palo Alto, CA 94304, USA

Benjamin Fong Chao
Goddard Space Flight Center Code 621
Greenbelt, MD 20771, USA

Gary A. Chapman
Department of Physics and Astronomy
California State University
18111 Nordhoff Street
Northridge, CA 91330, USA

Ivanka Charvátóva
Vitezna 12
11800 Prague 1,
The Czech Republic

Andrew F. Cheng
Applied Physics Laboratory
Johns Hopkins University
Laurel, MD 20723, USA

Christopher F. Chyba
Department of Geosciences
Guyot Hall, Princeton University
Princeton, NJ 08544-1003, USA

Benton C. Clark
Martin-Marietta Corporation, MS 0560
P.O. Box 179
Denver, CO 80201, USA

Lawrence Colin
3913 Nelson Drive
Palo Alto, CA 94306, USA

Thomas E. Cravens
Department of Physics and Astronomy
University of Kansas
Lawrence, KS 66045-2151, USA

Jacques Crovisier
Observatoire de Paris-Meudon
5, place J. Janssen
F-92195 Meudon Cedex Principal
France

Cindy C. Cunningham
Space Astrophysics Laboratory
4850 Keele Street
North York
Ontario M3J 3K1, Canada

J.M. Anthony Danby
Department of Mathematics
Box 8205
North Carolina State University
Raleigh, NC 27695, USA

Donald R. Davis
Planetary Science Institute
620 N. 6th Ave.
Tucson, AZ 85705-8331, USA

Morris S. Davis
Department of Physics and Astronomy
University of North Carolina
CB# 3255, Phillips Hall
Chapel Hill, NC 27599, USA

David S.P. Dearborn
Lawrence Livermore National University
L-58
P.O. Box 808
Livermore, CA 94550, USA

Imke de Pater
Astronomy Department, RA O3
University of California
Berkeley, CA 94720, USA

Anthony D. Del Genio
NASA Goddard Institute for Space Studies
2880 Broadway
New York, NY 10025, USA

Timothy H. Dixon
Department of Marine Geology and Geophysics
University of Miami
4600 Rickenbacker Causeway
Miami, FL 33149-1098, USA

Anthony R. Dobrovolskis
NASA Ames Research Center MS 245-3
Moffett Field, CA 94035-1000, USA

James M. Dohm
United States Geology Survey Branch of Astrogeology
2255 N. Gemini Drive
Flagstaff, AZ 86001-1698, USA

Audouin Dollfus
Observatoire de Meudon
5, Place Jules Janssen
F-92195 Meudon Principal Cedex
France

Deborah L. Domingue
Lunar and Planetary Institute
3600 Bay Area Blvd.
Houston, TX 77058, USA

Luke Dones
NASA Ames Research Center MS 245-3
Moffett Field, CA 94035, USA

Timothy E. Dowling
Department of Earth, Atmospheric, and Planetary Sciences
Massachusetts Institute of Technology 54-420
Cambridge, MA 02139, USA

Bruce T. Draine
Department of Astrophysical Sciences
109 Peyton Hall, Princeton University
Princeton, NJ 08544-1001, USA

Sandy Dueck
Jet Propulsion Laboratory MS 168-514
4800 Oak Grove Drive
Pasadena, CA 91109, USA

Patricia T. Eliason
Department of Planetary Sciences
University of Arizona
Tucson, AZ 85721, USA

Therese Encrenaz
Observatoire de Meudon
5, Place Jules Janssen
F-92195 Meudon Principal Cedex
France

Rhodes W. Fairbridge
NASA Goddard Institute for Space Studies
2880 Broadway
New York, NY 10025, USA

Michael H. Farris
417 Galsworthy Street
Thousand Oaks, CA 91360-5316, USA

Bruce Fegley
Department of Earth and Planetary Sciences
Washington University
St Louis, MO 63130, USA

Roger Ferlet
Institut d'Astrophysique de Paris
98 bis Boulevard Arago
F-75014 Paris
France

J.A. Fernández
Departmento Di Astronomia
Facultad di Humanides y Ciencias
Tristan Narvaja 1674
Montevideo, Uruguay

Cecile Ferrari
Observatoire de Paris
92195 Meudon Principal Cedex
France

Jane L. Fox
Marine Sciences Research Center
State University of New York
Stony Brook, NY 11794, USA

Linda M. French
Department of Liberal Arts
Wheelock College
200 The Riverway
Boston, MA 02215, USA

Michael D. Fuller
Department of Geosciences
University of California
Santa Barbara, CA 93106, USA

Tom Gehrels
Lunar and Planetary Laboratory
University of Arizona
Tucson, AZ 85721, USA

Billy P. Glass
Department of Geology
University of Delaware
Newark, DE 19716–2544, USA

James L. Gooding
Johnson Space Center SN 2
Houston, TX 77058, USA

Vivien Gornitz
Goddard Institute for Space Studies
2880 Broadway
New York, NY 10025, USA

J.T. Gosling
Los Alamos National Laboratory MS D438
Los Alamos, NM 87545, USA

Richard A.F. Grieve
Geological Survey of Canada
1 Observatory Crescent
Ottawa, Ontario K1A OYA
Canada

David Grinspoon
Laboratory for Atmospheric and Space Physics
University of Colorado
Boulder, CO 80309, USA

Donald A. Gurnett
Department of Physics and Astronomy
University of Iowa
Iowa City, IA 52242, USA

Robert M. Haberle
NASA Ames Research Center MS 245-3
Moffett Field, CA 94035, USA

W.B. Hanson (deceased)
Center for Space Sciences
University of Texas at Dallas
Box 830688 MS FO22
Richardson, TX 75083, USA

Bruce W. Hapke
Department of Geological and Planetary Sciences
321 Old Engineering Hall
University of Pittsburgh
Pittsburgh, PA 15360, USA

J.F. Harper
Department of Mathematics
Victoria University
Wellington, New Zealand

Alan W. Harris
Jet Propulsion Laboratory MS 183–501
4800 Oak Grove Drive
Pasadena, CA 91109, USA

William K. Hartmann
Planetary Science Institute
620 N. 6th Ave.
Tucson, AZ 85705–8331, USA

Hans J. Haubold
Office for Outer Space
Building F-8
Vienna International Center
P.O. Box 500
A-1400 Vienna, Austria

R.A. Heelis
Center for Space Sciences
University of Texas at Dallas
Box 830688 MS F022
Richardson, TX 75083, USA

Jacques Henrard
Facultes Universitaires de Namur
Rue de Bruxelles 61
B-5000 Namur
Belgium

Robert R. Herrick
Lunar and Planetary Institute
3600 Bay Area Blvd.
Houston, TX 77058-1113, USA

Norriss S. Hetherington
470 Stephens Hall
University of California
Berkeley, CA 94720, USA

Alan Hildebrand
Geological Survey of Canada
1 Observatory Crescent
Ottawa, Ontario KIA OY3
Canada

Lon L. Hood
Lunar and Planetary Laboratory
University of Arizona
Tucson, AZ 85721, USA

Linda J. Horn
Jet Propulsion Laboratory MS 183–501
4800 Oak Grove Drive
Pasadena, CA 91109, USA

William B. Hubbard
Planetary Sciences Department
University of Arizona
Tucson, AZ 85721, USA

Walter F. Huebner
Southwest Research Institute
6220 Culebra Road
Post Office Drawer 28510
San Antonio, TX 78228–0510, USA

W.-H. Ip
Max Planck Institut fur Aeronomie
D-3411 Katlenberg-Lindau
Germany

Phillip A. Isenberg
Institute for the Study of Earth, Oceans and Space
University of New Hampshire
Durham, NH 03824, USA

Patricia A. Jacobberger
National Air and Space Museum
Washington, D.C. 20560, USA

Daniel M. Janes
420 Space Sciences Building
Cornell University
Ithaca, NY 14853, USA

David G. Jankowski
Department of Physics and Astronomy
University of Wisconsin
Eau Claire
WI 54702-4004, USA

Gerard P. Jellison
Mitre Corporation
7525 Colshire Drive
McLean, VA 22102, USA

Jon M. Jenkins
NASA Ames Research Center
Moffett Field, CA 94035, USA

Donald E. Jennings
Goddard Space Flight Center Code 693.2
Greenbelt, MD 20771, USA

Ralph Kahn
Jet Propulsion Laboratory MS 169-237
4800 Oak Grove Drive
Pasadena, CA 91109, USA

Lucas W. Kamp
Jet Propulsion Laboratory MS 168-414
4800 Oak Grove Drive
Pasadena, CA 91109, USA

Jeffrey S. Kargel
Lunar and Planetary Laboratory
University of Arizona
Tucson, AZ 85721, USA

Alfred Kracher
Department of Earth Sciences
Iowa State University
Ames, IA 50011, USA

George Kukla
Lamont-Doherty Earth Observatory
Palisades, NY 10964, USA

William S. Kurth
Department of Physics and Astronomy
University of Iowa
Iowa City, IA 52242, USA

Kurt Lambeck
Research School of Earth Sciences
Australian National University
GPO Box 4
Canberra, ACT 2601, Australia

Lawrence E. Lasher
NASA Ames Research Center MS 244-14
Moffett Field, CA 94035-1000, USA

Jacques Laskar
Bureau des Longitudes
77, Avenue Denfert-Rochereau
75014 Paris, France

Larry A. Lebofsky
Lunar and Planetary Laboratory
University of Arizona
Tucson, AZ 85721, USA

Jean-Pierre Lebreton
European Space Research and Technology Center
  Planetary and Space Science Division
Postbus 299
2200 AG Noordwijk
The Netherlands

Marilyn M. Lindstrom
Johnson Space Center SN 2
Houston, TX 77058, USA

Jack J. Lissauer
Department of Earth and Space Sciences
State University of New York
Stony Brook, NY 11794, USA

John Longhi
Lamont-Doherty Geological Observatory
Palisades, NY 10964, USA

Paul D. Lowman Jr.
12403 Shafer Lane
Bowie, MD 20720, USA

William Lowrie
Institut fur Geophysik
ETH-Honggerberg
8093 Zurich, Switzerland

Janet T. Luhmann
Space Sciences Laboratory
University of California
Berkeley, CA 94720, USA

Jonathan I. Lunine
Lunar and Planetary Laboratory
University of Arizona
Tucson, AZ 85721

Brian G. Marsden
Smithsonian Center for Astrophysics
60 Garden Street
Cambridge, MA 02138, USA

Clare P. Marshall
Associate Editor, Geochemistry
Box 280931
Lakewood, CO 80228, USA

A.M. Mathai
Department of Mathematics and Statistics
McGill University
H3A 2K6 Montreal
Canada

Dennis L. Matson
Jet Propulsion Laboratory MS 183–501
4800 Oak Grove Drive
Pasadena, CA 91109, USA

Alfred McEwen
United States Geology Survey
Branch of Astrogeology
2255 N. Gemini Drive
Flagstaff, AZ 86001, USA

Christopher P. McKay
NASA Ames Research Center MS 245–3
Moffett Field, CA 94035, USA

William B. McKinnon
Department of Earth and Planetary Sciences
Washington University
St. Louis, MO 63130, USA

Susan McMahon
Jet Propulsion Laboratory MS 525–3610
4800 Oak Grove Drive
Pasadena, CA 91109, USA

Robert S. McMillan
Lunar and Planetary Laboratory
University of Arizona
Tucson, AZ 85721, USA

H.J. Melosh
Lunar and Planetary Laboratory
University of Arizona
Tucson, AZ 87521, USA

Ellis Miner
Jet Propulsion Laboratory MS 264–411
4800 Oak Grove Drive
Pasadena, CA 91109, USA

David W. Mittlefehldt
Lockheed Engineering and Sciences Program, MC C23
2400 NASA Rd. 1
Houston, TX 77058, USA

Jeffrey M. Moore
NASA Ames Research Center MS 245–3
Moffett Field, CA 94035, USA

Patrick Moore
Farthings, West Street
Selsey, Sussex
UK

Thomas H. Morgan
NASA Headquarters Code SL
Washington, DC 20546, USA

Steve Mueller
Department of Earth and Planetary Sciences
Washington University
St. Louis, MD 63130, USA

Carl D. Murray
Queen Mary and Westfield College
Mile End Road, University of London
London E1 4NS
UK

Yosio Nakamura
Institute for Geophysics
University of Texas at Austin
Austin, TX 78759–8397, USA

Robert M. Nelson
Jet Propulsion Laboratory MS 183–501
4800 Oak Grove Drive
Pasadena, CA 91109, USA

X.X. Newhall
Jet Propulsion Laboratory MS 238–332
4800 Oak Grove Drive
Pasadena, CA 91109–8099

Phillip D. Nicholson
Space Sciences Building
Cornell University
Ithaca, NY 14853, USA

J.C. Noëns
Observatoire du Pic-du-Midi
65204 Bangeres-de-Bigorre Cedex
France

Michael C. Nolan
Arecibo Observatory
Cornell University
P.O. Box 995
Arecibo, PR 00613, USA

John E. Oliver
Department of Geography and Geology
Indiana State University
Terre Haute, IN 47809, USA

Abraham H. Oort
Geophysical Fluid Dynamics Laboratory
P.O. Box 308, Princeton University
Princeton, NJ 08542, USA

Steven J. Ostro
Jet Propulsion Laboratory MS 300–233
4800 Oak Grove Drive
Pasadena, CA 91109, USA

D. Edgar Page
Jet Propulsion Laboratory MS 264–318
4800 Oak Grove Drive
Pasadena, CA 91109, USA

Frank D. Palluconi
Jet Propulsion Laboratory MS 264–267
4800 Oak Grove Drive
Pasadena, CA 91109, USA

Adrian F. Park
Department of Geology
Concordia University
7141 Sherbrooke Street West
Montreal, PQ H4B 1R6, Canada

Gary D. Parker
Department of Physics
Norwich University
Northfield, VT 05663, USA

Leslie J. Pieri
Jet Propulsion Laboratory MS 168–522
4800 Oak Grove Drive
Pasadena, CA 91109, USA

Frank A. Podosek
Department of Earth and Planetary Sciences
Washington University
St. Louis, MO 63130, USA

Carolyn C. Porco
Lunar and Planetary Laboratory
University of Arizona
Tucson, AZ 85721, USA

Renée Prangé
Institut d'Astrophysique Spatiale
Batiment 121, Universite Paris XI
F-91405 Orsay Cedex, France

Michael Rampino
Earth Systems Group
New York University
26 Stuyvesant St.
New York, NY 10003, USA

Bruce Randall
Department of Physics and Astronomy
University of Iowa
Iowa City, IA 52242, USA

Robert C. Reedy
Los Alamos National Laboratory
SST-8, MS D-438
Los Alamos, NM 87545, USA

Rüdeger Reinhard
European Space Research and Technology Center
Postbus 299
2200 AG Noordwijk
The Netherlands

Nicholas A. Renzetti
Jet Propulsion Laboratory MS 303–401
4800 Oak Grove Drive
Pasadena, CA 91109

Raymond G. Roble
National Center for Atmospheric Research
P.O. Box 3000
Boulder, CO 80307, USA

Ted L. Roush
NASA Ames Research Center MS 245-3
Moffett Field, CA 94035

J.P. Rozelot
Observatoire de la Cote D'Azur
Dept. CERGA, Ave. Copernic
06130 Grasse
France

David P. Rubincam
NASA Goddard Space Flight Center C921
Greenbelt, MD 20771, USA

Marian E. Rudnyk
732 W. Hillcrest Blvd.
Monrovia, CA 91016, USA

Christopher T. Russell
Institute of Geophysics and Planetary Physics
405 Hilgard Avenue
Los Angeles, CA 90024–1567, USA

Joel F. Russell
United States Geological Survey Branch of
  Astrogeology
2255 N. Gemini Drive
Flagstaff, AZ 86001, USA

Tamara V. Ruzmaikina
Lunar and Planetary Laboratory
University of Arizona
Tucson, AZ 85721, USA

Graham Ryder
Lunar and Planetary Institute
3600 Bay Area Blvd.
Houston, TX 77058, USA

Peter A. Sabine
Lark Rise, Camp Road
Gerrards Cross
Buckinghamshire SL9 7PF
UK

Robert E. Samuelson
NASA Goddard Space Flight Center Code 693.2
Greenbelt, MD 20771, USA

Bill R. Sandel
Lunar and Planetary Laboratory West
University of Arizona
Tucson, AZ 85721, USA

R. Stephen Saunders
Jet Propulsion Laboratory MS 230–225
4800 Oak Grove Drive
Pasadena, CA 91109, USA

Paul M. Schenk
Lunar and Planetary Institute
3600 Bay Area Blvd
Houston, TX 77058, USA

Wilfried Schröder
Geophysical Station
Hechelstrasse 8
D-2820 Bremen-Roennebeck
Germany

Gerald Schubert
Department of Earth and Space Sciences
University of California
Los Angeles, CA 90024, USA

David H. Scott
United States Geological Survey Branch of Astrogeology
2255 North Gemini Drive
Flagstaff, AZ 86001–1698, USA

Derek W.G. Sears
Department of Chemistry and Biochemistry
University of Arkansas
Fayetteville, AR 72701, USA

Marcia Segura
Jet Propulsion Laboratory MS 264–744
4800 Oak Grove Drive
Pasadena, CA 91109, USA

P.K. Seidelmann
United States Naval Observatory
34th and Massachussetts Ave., NW
Washington, DC 20392–5100, USA

James H. Shirley
Jet Propulsion Laboratory MS 183–601
4800 Oak Grove Drive
Pasadena, CA 91109, USA

Mark R. Showalter
NASA Ames Research Center MS 245–3
Moffett Field, CA 94035, USA

Richard A. Simpson
Space, Telecommunications, and Radiosciences Laboratory
Stanford University
Stanford, CA 94305–9515, USA

William L. Sjogren
Jet Propulsion Laboratory MS 301–150
4800 Oak Grove Drive
Pasadena, CA 91109, USA

William D. Smythe
Jet Propulsion Laboratory MS 183–601
4800 Oak Grove Drive
Pasadena, CA 91109, USA

Conway W. Snyder
21206 Seep Willow Way
Canyon Country, CA 91351, USA

John P. Snyder
17212 Friends House Road
Sandy Spring, MD 20860–1200, USA

Sean C. Solomon
Department of Terrestrial Magnetism
5241 Broad Branch Road
Washington, DC 20015, USA

Charles P. Sonett
Lunar and Planetary Laboratory
University of Arizona
Tucson, AZ 85721, USA

Ann L. Sprague
Lunar and Planetary Laboratory
Bldg. 92
University of Arizona
Tucson, AZ 85721, USA

Paul D. Spudis
Lunar and Planetary Institute
3600 Bay Area Blvd.
Houston, TX 77058, USA

Philip J. Stooke
Department of Geology
University of W. Ontario
London, Ontario N6A 5C2
Canada

Timothy D. Swindle
Lunar and Planetary Laboratory
University of Arizona
Tucson, AZ 85721, USA

Jeremy B. Tatum
Department of Physics and Astronomy
University of Victoria
British Columbia V8W 3P6
Canada

Fred W. Taylor
Clarendon Laboratory
University of Oxford
Parks Road
Oxford OX1 3PU
UK

Peter C. Thomas
Space Sciences Building
Cornell University
Ithaca, NY 14853, USA

Hans-Jürgen Treder
Einstein Laboratorium
Rosa-Luxemburg-Strasse 17a
D-1590 Potsdam
Germany

G. Leonard Tyler
Starlab
Department of Electrical Engineering
Stanford University
Stanford, CA 94305-4055, USA

Oleg Vaisberg
Space Research Institute
Academy of Science
117810 GSP 1 Profsoyuznaya U1 84/32
Moscow
Russia

Giovanni B. Valsecchi
Instituto Di Astrofisica Spaziale Reparto Planetologia
Viale dell'Universita 11
00185 Roma
Italy

James A. Van Allen
Department of Physics and Astronomy
University of Iowa
Iowa City, IA 52242, USA

D. Venkatesan
Department of Physics
University of Calgary
Calgary, Alberta
Canada

M.I. Verigin
Space Research Institute
Academy of Science
117810 GSP 1 Profsoyuznaya U1 84/32
Moscow
Russia

Alfred Vidal-Madjar
Institut d'Astrophysique de Paris
98 bis Boulevard Arago
F-75014 Paris, France

Faith Vilas
Lyndon B. Johnson Space Center SN3
Houston, TX 77058, USA

J. Hunter Waite Jr.
Southwest Research Institute
6220 Culebra Road, P.O. Drawer 28510
San Antonio, TX 78284, USA

James W. Warwick
Radiophysics, Inc.
5475 Western Avenue
Boulder, CO 80301, USA

Paul R. Weissman
Jet Propulsion Laboratory MS 183–601
4800 Oak Grove Drive
Pasadena, CA 91109, USA

Daniel Wenkert
Jet Propulsion Laboratory MS 233–305
4800 Oak Grove Drive
Pasadena, CA 91109, USA

Robert A. West
Jet Propulsion Laboratory MS 169–237
4800 Oak Grove Drive
Pasadena, CA 91109, USA

Ewen A. Whitaker
4332 East 6th Street
Tucson, AZ 85711, USA

Don E. Wilhelms
P.O. Box 595
Monte Rio, CA 95462, USA

Gareth V. Williams
Minor Planet Center
60 Garden Street
Cambridge, MA 02138, USA

George E. Williams
Department of Geology and Geophysics
University of Adelaide
Adelaide SA 5005, Australia

Steven H. Williams
Department of Space Studies
University of North Dakota
P.O. Box 9008
Grand Forks, ND 58202-9008
USA

Curtis Wilson
St John's College
P.O. Box 2800
Annapolis, MD 21404
USA

Arnold W. Wolfendale
Department of Physics
Science Laboratories, South Road
Durham DH1 3LE
UK

Fergus J. Wood
3103 Casa Bonita Dr.
Bonita, CA 92002
USA

Sherman S.C. Wu
United States Geological Survey Branch of
  Astrogeology
2255 North Gemini Drive
Flagstaff, AZ 86001, USA

G. Wuchterl
Institut fur Theoretische Astrophysik
Universitat Heidelberg
Im Neuenheimer Feld 561
900 Heidelberg, Germany

David T. Young
Southwest Research Institute
6220 Culebra Road, P.O. Drawer 28510
San Antonio, TX 78284, USA

Yuan-Chong Zhang
Goddard Institute for Space Studies
2880 Broadway
New York, NY 10025, USA

Michael Zolensky
Johnson Space Center SN 2
Houston, TX 77058, USA

Preface

In September 1886 several fragments of a meteorite fell to Earth near the town of Karamzinka in the Gorkovski province of Russia. One of the pieces of the meteorite was recovered and preserved, and another fell into a bog and was lost. The peasants who recovered a third piece reportedly *ate* part of the meteorite. The results of this early planetary science experiment are not known.

Today we know that this meteorite, called Novo Urei, is one of a class of similar meteorites, the ureilites. The ureilites are fragments of a larger parent body that accreted during the early stages of formation of our solar system, some 4.55 billion years ago. The ureilites contain microscopic diamonds and show evidence of shock metamorphism; they also have flow features indicating at least partial melting. About 40 ureilite meteorites have been found.

Studies of meteorites like Novo Urei have yielded more information on the formation, composition, and early evolution of our solar system than any other source. But where do the meteorites come from? Most come from the asteroid belt, between the planets Mars and Jupiter. This connection is of more than academic interest, since asteroids seem to have played an important role in the evolution of life on Earth. Many investigators believe that the impact of a small asteroid on the Earth, 65 million years ago, rang down the curtain on the age of dinosaurs, and set the stage for the evolution of mammals.

Until recently, asteroids and meteorites were studied by different groups of scientists. In the 19th century asteroid studies were performed by astronomers, and meteorites were studied by specialists in the geological fields of mineralogy and petrology. Today's investigator needs to be knowledgeable in both fields. Considerable present research, for instance, focuses on identifying the particular asteroid parent bodies of particular groups of meteorites. Some progress has been made; both astronomical remote sensing evidence and laboratory evidence now suggest that the basaltic achondrite meteorites are pieces of the large main-belt asteroid Vesta.

The connection linking asteroid and meteorite studies serves to illustrate the increasingly transdisciplinary nature of modern scientific investigation in the planetary sciences. To an outsider, planetary science appears to consist of fragmentary bits drawn from many traditional disciplines. One may compartmentalize planetary science investigations in terms of astronomy, atmospheric sciences, celestial mechanics, planetary geology and so on; for this reason, we chose to use the plural, planetary sciences, in our title. (We use the singular, planetary science, to inclusively characterize the full range of investigations performed by planetary scientists in all disciplines).

The goal of planetary science is to understand the nature and evolution of the planets, satellites, and smaller bodies of the solar system. To understand the nature and evolution of planets and satellites it is necessary to understand the larger environment of our solar system. To understand the environment of the solar system we need to consider the physics of the Sun and the radiation, plasma and magnetic fields present between the planets. No discussion of the solar system could be complete without mention of the dynamical motions and rotations of solar system bodies. In addition, no discussion of solar system bodies would be complete without mention of geological and atmospheric processes.

Planetary science thus involves virtually every branch of the physical sciences. In practice, planetary science often requires the most sophisticated application of the tools and knowledge of the traditional disciplines of physics, chemistry, astronomy, geology, geophysics and biology. Planetary science is our current best approximation to a 'science of everything', pushing back the frontiers of knowledge in fundamental areas such as the origins of life, the origins of the Earth, and the origins of the solar system.

Planetary science offers a uniquely powerful means of understanding the Earth, through comparative studies of the other terrestrial planets. While we know that heat flow from within the Earth must be responsible for tectonic phenomena such as earthquakes and volcanic eruptions, our understanding is far from complete; comparisons between the Earth and the planets Venus and Mars are already providing new insights into the nature and origins of plate tectonics at home.

The reader will find that the knowledge gained from the great missions of the golden age of planetary exploration has been systematized and greatly expanded in recent years. The information collected here reflects the growing maturity of the planetary sciences; the first wave of exploration has visited every planet but Pluto, and the results have been well assimilated, leading to the formulation of new and sharper questions.

Level and content

Our objective in this encyclopedia is to provide comprehensive coverage of the planetary sciences in a form that is useful for professionals, for students, and for the interested layperson. It is a challenge to cover so broad a field in a single volume, and we have therefore limited the length of contributions in order to include a larger number of topics. It is not possible to cover a subject completely and in exhaustive detail in an encyclopedia article; in fact it is not even desirable to do so. The user of an encyclopedia needs an introduction to the main ideas, a sense of the level of current understanding and a good list of references as a basis for more in-depth study. The list of references may be the most useful component of the article. While current interpretations and the latest data may become stale with time, a knowledge of prior work is always a necessity for scientific investigation. Unfortunately, the editors of some volumes seem to have forgotten that the literature represents the fundamental resource of knowledge, providing only skimpy references. Thousands of references are included within this volume.

The articles making up this encyclopedia can be broken down into a variety of categories. There are several articles on each planet, covering the interiors, magnetic fields, atmospheres and surfaces. All of the major moons are described in one or more articles, and there are several entries on comets, on asteroids and on meteorites. The accompanying table gives a partial listing of the articles on asteroids: 25 articles are listed.

A breakdown like this can be a valuable aid for those exploring an unfamiliar subject. We have compiled 15 tables like this one, which appear in Appendix A of this volume.

Asteroids and related articles

Amor object
Apollo object
Asteroid
Asteroid: compositional structure and taxonomy
Asteroid: families
Asteroid: lightcurves
Asteroid: photometry
Asteroid: resonances
Asteroid: thermal infrared studies
Aten object
Ceres
Chaotic dynamics in the solar system
Collision
Dust
Gaspra
Hilda asteroids
Ida
Mass extinction
Meteorite parent bodies
Near-Earth objects
Planetesimal
Resonances in the solar system
Trojan asteroids
Väisälä orbit
Vesta

The categories include, in addition to Asteroids: Astronomy; Atmospheres; Biographies; Celestial mechanics and gravitation; Comets; Geology and Geophysics; Magnetic fields and the interplanetary environment; Meteorites; Missions to planets and the moon; Planets; Rings; Satellites; the Sun and solar physics and Techniques.

Although this list of article categories is useful it does not tell the whole story. We had a number of goals and objectives in constructing this volume. For instance, we wanted to give a sense of how planetary science is carried out. Thus under the category Techniques one will find nearly two dozen articles on remote sensing, including spectroscopy, polarimetry, radiometry, magnetometry, and so on. Articles on Cartography, the process of preparing maps of planets and satellites, and Imaging science, the process of generating images from digital data, are also included in this category.

When we began, we felt that the smaller bodies (asteroids, comets and meteorites) had not been adequately surveyed in an encyclopedia format elsewhere, and so we wanted to do a good job with these topics.

The subjects of magnetic fields and the interplanetary environment also lack wide exposure.

Many short biographical entries are included in this volume. The individuals profiled have made important contributions to our knowledge of the planets and the solar system; most are historic figures. We could not include a 'who's who' of contemporary scientists; such a project would have rapidly become unmanageable. Most of Earth's planetary scientists are alive and still active! There are a number of historical entries that trace the development of planetary science, and more than two dozen articles on past and current planetary and lunar exploration missions.

We also tried to ensure that important theories and processes were well represented. The articles Gravitation, Impact cratering, Mantle convection, Plate tectonics, Resonance in the solar system, Tidal heating, and Chaotic dynamics in the solar system fall into this category.

We have attempted to place the Earth on an equal footing with the other planets; thus there are articles on the atmosphere, surface geology and magnetic field of the Earth, corresponding to similar articles for Venus, Mercury and Mars. This approach cannot do justice to what is known about our own planet. Other encyclopedias in this series, such as the *Encyclopedia of Solid Earth Geophysics*, the *Encyclopedia of Structural Geology and Plate Tectonics*, and the *Encyclopedia of Climatology* may be consulted for more in-depth coverage of our own planet.

Lastly, a note about maps. The format of this volume does not permit us to include large-scale maps of the surfaces of the planets and moons in the solar system. However, detailed maps have been constructed for a number of solar system bodies other than the Earth. One source for such maps is the Map Distribution Office of the US Geological Survey, Box 25286, Denver Federal Center, Denver, Colorado 80225, USA.

How to use this Encyclopedia

Topics are presented in alphabetical order. This generally provides direct access to primary subject matter. If a subject does not appear as a separate article, the reader should next consult the index at the back of the book. As noted above, Appendix A provides lists of articles for major subject areas.

Cross references to other entries are included both within the text of articles and as a separate list at the end of each article. The abbreviation q.v. (*quod vide*) means that an entry with that title appears elsewhere in the volume. Added perspective may be gained by consulting apparently unrelated articles that are cross-referenced. For example, the article Radio science includes examples and data bearing on studies of planetary rings and planetary atmospheres.

References to cited works follow each entry and are designed to lead the interested reader to additional, more detailed information or to other summaries that contain extensive bibliographies. The journals most commonly cited in the articles in this volume, such as *Icarus*, the *Journal of Geophysical Research*, *Planetary and Space Science*, and so on, provide a starting point for obtaining the most recent information.

Other resources available to researchers include published abstracts and abstracting services, available at university libraries, and computerized databases, which may be searched using subject or author keywords, date of publication and other parameters.

Units and abbreviations

Essentially all of the articles in this volume employ Système International units. Appendix B lists the SI units together with a few additional tables.

CD-ROM

The CD-ROM included with this Encyclopedia contains more than 200 images from the archives of NASA's National Space Science Data Center. Images of all the planets visited by spacecraft are included, along with images of asteroids, comets, planetary rings, and satellites. These images complement those included as color plates in the Encyclopedia, offering additional views and higher resolution in some cases.

Viewing software for both Macintosh and IBM-PC compatible computers is included on the disk. Users should consult the 'Read Me First' file contained in the top-level directory of the CD for detailed instructions.

The National Space Science Data Center archives images and data products from many sources. For instance, images of Earth's Moon captured during the Lunar Orbiter, Surveyor, Luna, Zond, Ranger, Clementine, and Apollo missions are available. Researchers and interested individuals can obtain these directly from NSSCD. The CD-ROM includes instructions for assessing the resources of the Center. We are grateful to the National Aeronautics and Space Administration and to the National Space Science Data Center for enabling us to include the CD-ROM with this volume.

Acknowledgements

This Encyclopedia could not have been completed without the support of the planetary science community, and the contributing authors in particular, who took time from their own research and other commitments to write the articles included here. The members of the Editorial Board – Dan Boice, Hans Haubold, Bob Nelson, Chris Russell, Dick Simpson, Conway Snyder and Tim Swindle – have made special contributions and deserve special acknowledgement. They have suggested necessary topics, identified potential authors, written important articles and they have assisted in reviewing contributed articles. Their efforts have substantially improved the volume.

It is a time-consuming and sometimes unrewarding task to review articles submitted for publication. A competent review of each article is nonetheless essential in the production of a scientific encyclopedia. The following individuals, among others, have served as reviewers for the present volume: Nadine Barlow, Jane Blizard, Bonnie Buratti, Ben Chao, Gary Chapman, Tony Del Genio, Tim Dowling, Vivien Gornitz, Richard Grieve, Norriss Hetherington, Linda Horn, William Hubbard, Pat Jacobberger, Dan Janes, Jeff Kargel, Heinz Lowenstam, Patrick Moore, J.P. Rozelot, Paul Schenk, Gerry Schubert, Nick Short, Ann Sprague, Jeremy Tatum, Hunter Waite, Curtis Wilson, Arnold Wolfendale, Fergus Wood and Y.-C. Zhang.

Many individuals have contributed to this project in other ways, for instance by suggesting a potential author for a particular topic, or by drafting figures for articles without compensation. We warmly thank the following persons for their help: Raymond Arvidson, Roger Bonnet, James Burke, Charles Card, Bob Carlson, Ivanka Charvátóva, Barney Conrath, William Cox, Von Del Chamberlin, R.G. Currie, James A. Dunn, Richard H. Dunn, Cesare Emiliani, Hugues Faure, Ian Francis, David Hicks, Joe King, Theodor Kostiuk, George Kukla, Theodor Landscheidt, Lindsey Lee, John Longhi, Paul Lowman, Lucy-Ann McFadden, Eldredge Moores, Peter Mouginis-Mark, Marcia Neugebauer, Gary Parker, Tina Pauro, John Perras, Suzanne Peiffer, Robert Reedy, P.K. Seidelmann, Gerry Simila, Larry Soderblom, Charles Sonett, Sean Solomon, Richard Stothers, Peter Tucker, S.J. Wiedenschilling, Dave Williams, and Göran Windelius.

We are grateful for the patience and long-suffering support of family members, particularly Maureen Shirley and Dolores Fairbridge, during the five years that were required to complete this volume.

James H. Shirley and Rhodes W. Fairbridge

A

ABSORPTION, ABSORPTION SPECTRUM

Absorption is the conversion of electromagnetic energy to another form of energy through interaction with matter (Townes and Schawlow, 1955). The process can be viewed at several levels, but the end result is usually heat – radiation impinges on a body, and the body becomes warmer. An absorption spectrum indicates relative efficiency in the absorption process as a function of frequency or wavelength of the radiation. In planetary studies absorption and absorption spectra generally refer to radiation passing through material rather than being reflected by material (compare reflectance spectroscopy). For example, an absorption spectrum indicates what fraction of the radiation passing through a gas is absorbed at each wavelength.

Energy absorbed by an atom results in a change in state of the atom – one or more electrons move to higher energy levels (excitation) or are removed entirely (ionization). Energy absorbed by a molecule will cause it to rotate or vibrate more rapidly, or change structural configuration. The amount of energy corresponding to any particular degree of excitation (or for ionization) is limited to discrete levels according to quantum theory; the energy quanta ΔE have corresponding wavelengths λ given by $\Delta E = hc/\lambda$ where h is Planck's constant and c is the speed of light.

Energy absorption by molecules usually results from coupling between the electric or magnetic component of the wave and the respective dipole moment of the molecule. Vibrational and rotational resonances, for example, can lead to strong absorption at and near certain wavelengths, called 'absorption lines.' The number of 'lines' can be quite large for complex molecules.

Lines associated with excitation of electrons or ionization of atoms typically occur in the visible spectrum. Lines associated with transitions in rotational and vibrational states have energies in the infrared region of the spectrum. Structural transitions can split rotational bands, so that the energy associated with transitions between states in the split bands causes strong absorption at microwave frequencies (e.g. the 'inversion' lines of gaseous ammonia).

Bulk materials exhibit absorption characteristics which are composites of the absorption spectra of their component atoms and molecules, modified by the 'structures' within which they are arranged. Molecule–molecule collisions in a dense hot gas, for example, broaden absorption lines so that a range of wavelengths is absorbed rather than the narrow line predicted by simple theory. The amount of broadening depends on both the pressure and temperature in the gas. Absorption by a plasma results from interaction of the charged particles with the wave and collisions among the particles, including neutrals; the amount of absorption depends on the density of the particles.

The Earth's atmosphere preferentially absorbs at certain wavelengths. Over the optical and infrared range absorption results primarily from ozone, carbon dioxide and water vapor. Quantitative astronomical measurements from Earth must often be corrected for an assumed amount of water vapor along the observing path through the atmosphere; ozone and carbon dioxide are presumed much less variable and standard values may be applied for correction. Astronomical measurements at certain wavelengths are simply impractical because of the high opacity. On the other hand, carefully calibrated measurements of observed absorption may be used to infer water vapor content in the atmosphere.

In other contexts measurements of absorption have led not only to inference of concentration but also to identification of the species responsible. Unexpectedly high absorption during radiowave passage through Venus' atmosphere was eventually traced to sulfuric acid vapor below clouds composed of liquid sulfuric acid droplets at 48 km above the planet's surface (Steffes and Eshleman, 1982). High absorption during radio occultations by the giant outer planets can be explained by a combination of absorption by methane clouds and distributed ammonia.

Richard A. Simpson and Jon M. Jenkins

Bibliography

Steffes, P.G. and Eshleman V.R.(1982) Sulfuric acid vapor and other cloud-related gases in the Venus atmosphere: abundances inferred from observed radio opacity. *Icarus*, **51**, 322–33.

Townes, C.H. and Schawlow A.L. (1955) *Microwave Spectroscopy*. New York: McGraw Hill.

Cross references

Emissivity
Radio science
Reflectance spectroscopy
Spectrophotometry
Spectroscopy: atmosphere

ACCRETION

Accretion originates from Latin (*accretio – accretere*), and means a growth or increase in size by gradual external addition or accumulation. In astronomy and planetary physics, accretion assumes the increase in the mass of a celestial object by collection of the surrounding gas and objects (of smaller size) by gravity.

The extraction of gravitational potential energy from material that accretes on to the gravitating body is widely believed to be the principal source of energy in several types of evolved objects – evolved close binary stars, pulsars and black holes. It also could be the energy source for the active galactic nuclei and quasars (Frank, King and Raine, 1985).

In the 19th century, gravity was considered to be the only source of energy in celestial bodies, including stars. Further, it has been assumed that for the most of the lifetime of the Sun and other stars, the luminosity is sustained by nuclear fusion at the center. However the formation of both the Sun and planets involves stages when accretion of the dispersed material (gas and solid bodies) by the gravitating embryonic body is the dominant process (see Solar system: origin).

If non-interacting particles having a zero velocity at infinity accrete on to a gravitating body of mass M, then their velocity is equal to $v = -\sqrt{2GM/r}$, where G is a gravitational constant and r is the distance from the center of the gravitating body. For steady, spherically symmetric accretion, the accretion rate is $\dot{M} = 4\pi r^2 \rho (-v)$, where ρ is the density of accreting material. This rate is determined by the inward flux at infinity.

Actually, forces other than gravity are also important for accretion. The pressure gradient in the spherically symmetric, accreting gas restricts the rate of accretion by the value of inward flux at the radius $r_c = GM/(2c_s^2)$, which is the sonic radius, where the flow of the accreting gas changes from subsonic $v < c_s$ (for $r > r_c$) to supersonic $v > c_s$ (for $r < r_c$).

If the accreting material possesses a non-radial component of velocity at infinity, then the value of the specific angular momentum of the material with respect to the gravitating body is an important parameter. If the angular momentum is so small that the trajectories of the accreting bodies (for solids) or accretion flux lines (for gas) cross the surface of the gravitating body, then the character of accretion is similar in many respects to accretion with zero initial velocities; the accretion can impart spin to the gravitating body, provided that the angular momenta of different portions of the accreting materials do not perfectly compensate each other. In the case of larger angular momenta, the accreting gas or/and solid bodies have to lose kinetic energy and angular momentum to be accreted. Historically this type of accretion was first studied for a gravitating body moving through a homogeneous infinite medium by Hoyle and Lyttelton (1939) and Bondi and Hoyle (1944). In this model freely streaming accreting material loses most of its angular momentum in the accretion column behind the gravitating body, where flow lines of different patches of the accreting material cross.

In the context of physics of the early solar system a more important role is played by so-called disk accretion. It assumes that the accreting material (which has a remarkable average angular momentum) is captured first into closed orbits around the gravitating body, and then spirals toward the center due to dissipation of the rotational energy associated with the orbital motion. This type of accretion is of dominant importance for the formation of the Sun and the solar nebula, other stars with protoplanetary disks, and possibly for the formation of the giant planets. Disk accretion has been studied mainly for viscous disks in the thin disk approximation. For the steady-state disk model the accretion rate is equal to the mass inflow through the ring of any radius, which is equal to $\dot{M} = 2\pi R \Sigma (-v_R)$, where Σ is the surface density, which is the mass per unit surface area of the disk, and v_R is determined by the rate of the redistribution of the angular momentum, which is associated with the dissipation of the rotational energy for the most-studied case of the viscous disk.

For the solar nebula in the early stage the accretional process begins when the embryonic Sun and solar nebula form within the collapsing presolar cloud from which the solar system originated about 4.5 billion years ago.

The character of accretion was dependent on the value and distribution of the angular momentum within the cloud. It might have started as a nearly spherically symmetric accretion when gas was accreted directly onto the protosun, but probably continued by the formation of a disk (embryonic solar nebula), as more distant and higher angular momentum material was involved in the collapse. This process coincides with or is followed by the redistribution of mass and angular momentum within the solar nebula, and addition of the mass to the Sun in the regime of this disk accretion. The strong nonhomogeneity of the distribution of angular momentum in the present-day solar system reveals that the exchange of mass and angular momentum between the Sun and solar nebula was significant, and that the present-day distribution of mass, which is concentrated in the Sun, and angular momentum, which is mainly contained in the orbital motion of planets, resulted from disk accretion.

An accretional process is also associated with planetary formation. If a solar nebula had a mass of the same order of magnitude as the Sun, then the development of gravitational instability in it could result in formation of the massive gaseous protoplanets whose further evolution resembles in some details the process of formation of the Sun. However in the lower mass solar nebula the process of planetary formation comes about through the separation of the solids from the gas, growth of the solid bodies, their accumulation to form planetary embryos, and accretion of the remnant planetesimals and gas (in the case of giant planets). The process of formation of planets from solid bodies is discussed in some detail in Solar system: origins and Cold accretion theory. It is worth noting that the terrestrial planets and the cores of giant planets were plausibly formed mainly in the regime of direct accretion due to collision of bodies in crossing orbits, while the collisional cross-section could be enlarged by the gravity of large bodies. The disk accretion took place, possibly, in the circumplanetary disk, and was also related to both the formation of the regular satellites and gas accretion by the giant planets.

Planetary gas accretion in low mass solar nebulae follows solid-body accretion. The prior factor determining the gas accretion is the embryo's mass. As the solid embryo grows larger in mass a hydrostatic gaseous envelope forms. The envelope's mass is determined by the embryo's increasing gravity competing with the heating of the envelope gas due to dissipation of kinetic energy of the planetesimals. As the mass of the gaseous envelope becomes comparable to the solid embryo (typically at a few Earth masses), further gas accretion is increasingly determined by the ability of the envelope to radiate away the energy produced by its own gravitational contraction. If this is done sufficiently effectively, gas accretion will involve a brief phase of hydrodynamic inflow. When the mass of a planet increases to a few tens of Earth masses (for gas plus solids), the planet's gravity starts to modify the structure of the ambient solar nebula. This might inhibit any further accretion and was possibly responsible for the final masses of Jupiter and Saturn.

Tamara V. Ruzmaikina and G. Wuchterl

Bibliography

Bondi, H. and Hoyle, F. (1944) On the mechanism of accretion by stars. *Mon. Not. Roy. Astron. Soc.*, **104**, 273–81.
Frank, J., King, A.R. and Raine, D.J. (1985) *Accretion Power in Astrophysics*, Cambridge University Press, Cambridge.
Hoyle, F. and Lyttleton, R.A. (1939) The effect of interstellar matter on climatic variation, *Proc. Cambridge Phil. Soc.*, **35**, 405–15.

Cross references

Cold accretion theory
Collision
Earth–Moon system: origin
Solar system: origin

ACHONDRITE METEORITES

Achondrite meteorites are stony meteorites lacking chondritic texture; that is, they lack the spherical chondrules that are the defining characteristic of chondrites. As originally defined in the Rose–Tschermak–Brezina classification system, achondrites were also distinct from chondritic meteorites in composition. In particular, the original achondrite groups were poor in iron–nickel metal, and many had broadly basic compositions, rather than the ultrabasic compositions typical of chondrites. Hence, the chemical compositions of most achondrites are significantly different from that of material formed directly from the solar nebula. However, recent achondrite falls and finds have blurred the boundary between achondritic and chondritic compositions. A subset of the achondrites, often referred to as the primitive achondrites, are rocks with broadly chondritic compositions, but with achondritic textures. Almost all achondrites have ages of about 4.6 Ga, essentially identical to the ages of chondrites. Table A1 summarizes the major groups of achondritic meteorites.

Most achondrites are igneous rocks and were therefore formed by planetary processes. The most common types of igneous achondrites are basalts similar to basalts from the Earth and Moon. The largest group of achondrites is the basaltic achondrite group, commonly known as the HEDs for the three types; howardites, eucrites and diogenites (see Basaltic achondrite meteorites). The HED meteorites are basalts, gabbros and orthopyroxenites similar to those occurring

Table A1 The major achondrite meteorite groups

Name	Characteristics	Mode of origin	Abundance (%)*
Igneous achondrites			
Angrites	Olivine–anorthite-fassaite basalts	Igneous	1.7
Aubrites	Enstatite orthopyroxenites	Igneous	6.0
Basaltic achondrites			62.6
Diogenites	Hypersthene orthopyroxenites	Igneous	10.2
Eucrites	pigeonite-anorthite basalts	Igneous	34.0
Howardites	Hypersthene-pigeonite-anorthite breccias	Brecciated igneous	18.4
Lunar meteorites			3.9
Highlands rocks	Anorthosite dominated breccias	Igneous	2.2
Mare rocks	Pyroxene-plagioclase gabbros and breccias	Igneous	1.7
SNCs			4.3
ALH84001	Orthopyroxenite	Igneous	0.4
Chassigny	Dunite	Igneous	0.4
Nakhlites	Augite clinopyroxenites	Igneous	1.3
Shergottites	Pigeonite-andesine basalts	Igneous	2.2
Ureilites	Pigeonite-olivine dunites and wehrlites	Igneous	10.2
Primitive achondrites			
Acapulco-like	Olivine-orthopyroxene-diopside-oligoclase granulites	Metamorphic	2.2
Lodran-like	Olivine-orthopyroxene-diopside-oligoclase granulites	Metamorphic	3.9
Winona-like	Olivine-orthopyroxene-diopside-oligoclase granulites	Metamorphic	3.0
Brachina-like	Olivine-orthopyroxene-diopside-oligoclase granulites	Metamorphic? Igneous?	2.2

* Approximate abundance among achondrites. An estimated correction for pairing among the Antarctic achondrites has been made.

on Earth, plus breccias composed of these rock types. The next largest group of achondrites is the ureilites; rocks generally consisting of olivine and pigeonite (see Ureilite meteorites). The ureilites exhibit a curious mixture of nebular and planetary features. The mineralogy, texture and lithophile element composition appear to be those of either a magmatic cumulate or partial melting residue, while their oxygen isotopic compositions show variable anomalous enrichments in ^{16}O, which are typical of some nebular materials (Clayton and Mayeda, 1988). The aubrites are the third most abundant type of achondrite. They are highly reduced achondrites composed of coarse-grained, nearly pure enstatite. They are believed to be cumulates from a differentiated asteroid (Okada et al., 1988). Two small but diverse groups of igneous achondrites are distinguished from other achondrites by their younger ages (among other things). Lunar meteorites are meteorites from a known parent body. Their lunar origin was established by their similarity to samples returned by the Apollo and Luna missions. Most lunar meteorites are breccias of igneous rocks from the mare and/or highlands regions and have ages between 3–4 Ga (see Lunar meteorites). The other small, diverse group of achondrites is commonly referred to as the SNC group after the principal members; shergottites, nakhlites and the meteorite Chassigny (see SNC meteorites). This group includes basalts, clinopyroxenites, plagioclase-harzburgites and a dunite. The SNCs have ages < 1.3 Ga, and are widely believed to be fragments of the Martian crust. The smallest group of igneous achondrites is the angrites. These rocks are broadly equivalent to alkali-olivine basalts on Earth in that they are critically silica undersaturated. The angrites are very poor in the volatile elements, however, including the alkali elements (Mittlefehldt and Lindstrom, 1990). The angrites are composed of fassaitic pyroxene and Ca-rich olivine, with or without pure anorthite.

The primitive achondrites consist of Acapulco-like, Brachina-like, Lodran-like and Winona-like achondrites. All these achondrite types exhibit granulitic texture and have broadly chondritic composition. The Acapulco-like and Londran-like achondrites are closely related achondrites. They are distinct from either the Brachina-like or Winona-like achondrites in oxygen isotopic composition and in the amount of FeO in the silicates (e.g. Nehru et al., 1992). The Lodran-like achondrites were originally classified as stony-iron meteorites in the Rose–Tschermak–Brezina classification system because the only member known at the time, Lodran, contained about 25% metal plus troilite by volume. Recently recovered Lodran-like achondrites have metal contents that vary from about 3 to 30%, and petrologic and isotopic work on them suggest that they are closely related to the Acapulco-like achondrites (McCoy et al., 1992). Therefore, it is better to classify the Lodran-like meteorites with the achondrites, rather than the stony-irons. The Acapulco-like and Lodran-like achondrites are believe to be chondritic materials that have experienced extensive metamorphism, perhaps up to temperatures where melting initiated. This metamorphism totally recrystallized the rocks, completely wiping out their presumed original chondritic texture. The Brachina-like achondrites are dunitic–wehrlitic meteorites and may be either ultra-metamorphosed chondrites (Nehru et al., 1992) or igneous rocks (Warren and Kallemeyn, 1989). The Winona-like achondrites are compositionally similar to the chondritic silicates from IAB iron meteorites and appear to be lithic fragments of the same parent body (Bild, 1977; and see Iron meteorites).

David W. Mittlefehldt

Bibliography

Bild, R.W. (1977) Silicate inclusions in group IAB irons and a relation to the anomalous stones Winona and Mt Morris (Wis). *Geochim. Cosmochim. Acta*, **41**, 1439–56.

Clayton, R.N. and Mayeda T.K. (1988) Formation of ureilites by nebular processes. *Geochim. Cosmochim. Acta*, **52**, 1313–8.

McCoy, T.J., Keil, K., Mayeda T.K. and Clayton, R.N. (1992) Petrogenesis of the Lodranite-Acapulcoite parent body (abstract). *Meteoritics*, **27**, 258–9.

Mittlefehldt, D.W. and Lindstrom M.M. (1990) Geochemistry and genesis of the angrites. *Geochim. Cosmochim. Acta*, **54**, 3209–18.

Nehru, C.E., Prinz M., Weisberg M.K. et al. (1992) Brachinites: a new primitive achondrite group (abstract). *Meteoritics*, **27**, 267.

Okada, A., Keil, K. Taylor, G.J. and Newsom, H. (1988) Igneous history of the aubrite parent asteroid: evidence from the Norton County enstatite achondrite. *Meteoritics*, **23**, 59–74.

Warren, P.H. and Kallemeyn, G.W. (1989) Allan Hills 84025: the second brachinite, far more differentiated than Brachina, and an ultramafic achondritic clast from L chondrite Yamato 75097. *Proc. Lunar Planet. Sci. Conf.*, **1**, pp. 475–486.

Cross references

Basaltic achondrite meteorites
Chondrites, ordinary
Meteorite
Meteorite parent bodies

ADAMS, JOHN COUCH (1819–1892)

An English astronomer and mathematician, who was the first individual to predict the position of a planet beyond Uranus. Already

Figure A1 J.C. Adams. (Reproduced by permission of American Institute of Physics, Emilio Segre Visual Archives.)

in 1820 it was known by astronomers that the motion of the planet Uranus could not be explained exclusively by Newton's law of gravitation and the perturbations of the then-known planets on Uranus. Adams analysed the irregularities in the motion of Uranus' orbit and came to the conclusion that they were due to an undiscovered planet. He reported his discovery to the Astronomer Royal, G.B. Airy (1801–1892), who ignored Adams' results. Only when the French astronomer V.J.J. Leverrier (1811–1877) made the same prediction almost 1 year later, did astronomers start searching for the planet. Neptune was finally discovered in 1846 by the German astronomer J. Galle, who directly received from Leverrier a request to search for the planet. The discovery of the new planet Neptune was one of the biggest triumphs of celestial mechanics in the 19th century.

Adams was a very modest man and later became good friends with Leverrier. He was awarded the Copley Medal by the Royal Society, its highest honor, and twice served as president of the Royal Astronomical Society. After laborious calculations he refined the tables of the lunar parallax and in 1853 published a remarkable paper on the Moon's secular acceleration, showing that, contrary to Laplace, only about half this acceleration could be attributed to the secular diminution of the orbital eccentricity of the Earth. The remainder of the acceleration has since been ascribed to a deceleration in the Earth's rotation caused by tidal friction (see Darwin, George Howard; Solar day).

Hans J. Haubold

Bibliography

Berry, A. (1898) *A Short History of Astronomy*. London: J. Murray [Dover Publications, Inc., 1961]
Grosser, M. (1962) *The Discovery of Neptune*. Cambridge, Massachusetts: Harvard University Press.
Grosser, M. (1970) Adams, John Couch, *Dict. Sci. Biogr.*, Vol. 1, pp. 53–4.

Spencer Jones, A. (1947) *John Couch Adams and the Discovery of Neptune*. Cambridge: Cambridge University Press.

AEROSOL

In physical chemistry and atmospheric sciences an aerosol is a system of colloidal particles dispersed in a gas; examples include smoke and fog. (The term also has commercial use for both contents and canisters of sprays of deodorant, disinfectant or insecticide). The particles, that may be solid or liquid, have been defined (by Lodge, 1962) as 'larger than single molecules yet small enough to remain dispersed for a significant length of time.'

In atmospheres, aerosols are important in cloud formation, nucleation of condensation droplets, atmospheric electricity and radiation balance. They thus play a major role in climate control. Sources of aerosols on planet Earth include sea salts (from storm waves), biological products, desert dust, volcanic ash and human pollution. Artificial aerosols (such as silver iodide, acting as freezing nuclei) can be used in artificial rain production in suitable cloud formations. Extraterrestrial aerosols from interplanetary space and meteorites are sufficiently abundant that they are recognized in deep-sea 'red clay' deposits.

In the stratosphere, at 3–9 km above the tropopause, is a worldwide aerosol layer, principally of $(NH_4)_2SO_4$ that is believed to result from photodissociation of SO_2 to H_2SO_4. The source of SO_2 may be natural, from volcanic eruptions and biological activity and partly as an anthropogenic pollutant. The extreme variability of volcanic eruptions in this way constitutes an important forcing factor in climate events.

During ice ages (q.v.) an enormous increase of desert dust production (at least one order of magnitude), evidenced by loess deposits, has also been monitored in Antarctic ice cores. Due to widespread desiccation in the mid-latitudes, the dust sets up a positive feedback mechanism in helping to nucleate snowfall, thus amplifying the glacial growth.

Rhodes W. Fairbridge

Bibliography

Jaenicke, R. and Davies, C.N. (1976) The mathematical expression of the size distribution of atmospheric aerosols. *J. Aerosol Sci.*, 7, 255–9.
Junge, C.E. (1963) *Air Chemistry and Radioactivity*. New York: Academic Press.
Lodge, J.P. (1962) Identification of aerosols, *Adv. Geophys.*, 9, 67–130.
Oliver, J.E. and Fairbridge, R.W. (eds) (1987) *The Encyclopedia of Climatology*. New York: Van Nostrand Reinhold, 986pp.
Pasquill, F. (1974). *Atmospheric Diffusion*, 2nd edn, New York: Wiley.
Stern, A.C. (ed.) (1976). *Air Pollution*, Vol. 1, 3rd edn, New York: Academic Press.

AIRGLOW

Airglow is the luminosity of an atmosphere that arises from radiative transitions between internal energy states of atoms or molecules; the source of the emissions is ultimately the interaction of solar photons with the atmosphere. Emissions that are produced by the more or less direct interaction of solar radiation with atmospheric gases are classified as dayglow, as are those that are produced by the interaction of atmospheric species with photoelectrons. Nightglow emissions are those that arise from chemiluminescent reactions of fragments or ions that are produced during the day or transported from the dayside. The production mechanisms for twilight glow, which are emissions in sunlit regions of the atmosphere that are observed from the nightside near the terminator, are not conceptually different from dayglow. Airglow is usually distinguished from aurora by the source of the emissions. Auroral emissions are those that originate from excited states of atmospheric species that are produced by the impact of particles other than photoelectrons. The particles, which may be electrons, protons or heavier ions, originate outside the atmosphere, and usually precipitate into the atmosphere near the poles of the planet along magnetic field lines. Auroral emissions are thus usually more localized and exhibit more variability than airglow emissions. Although Venus has no intrinsic magnetic field, the diffuse and variable ultraviolet emissions of atomic oxygen at 1304 and 1356 Å that have been observed on the nightside have been called 'auroral' (e.g. Phillips, Stewart and Luhmann 1986). We exclude from our discussion both localized polar auroral emissions and the diffuse Venusian nightside emissions, but the interested reader is referred to the entry Aurora, planetary.

In general, measurements and models of the intensities of airglow emissions are important because they may be used to derive information about the composition and dynamics of an atmosphere, about its interaction with solar radiation, and about the mechanisms for production of the emissions. In particular, the intensities can be used to infer density profiles of the emitting species, or the strength of sources, such as photoelectron or solar fluxes. Measurements of emission lineshapes or Doppler shifts can be used to obtain information about winds. High-resolution spectra can be used to determine rotational distributions, which usually reflect the temperature of the emitting region. Vibrational distributions of molecular emissions can reflect the ambient temperature or the excitation mechanisms, depending on the effectiveness of quenching processes. In many cases, intensities of emissions have been used to determine cross-sections or reactions rates that are difficult to measure in the laboratory.

Measurement of airglow

Terrestrial airglow emissions have been recorded from a variety of vantage points, including rockets, satellites, aircraft, and the Space Shuttle. Only those emissions to which the atmosphere is transparent, including those in the visible and parts of the infrared, may be recorded from the ground. Electronic transitions usually appear in the ultraviolet or visible regions of the spectrum, but a few, such as the 1.27 μm band of O_2, which arises from the transition O_2 ($^1\Delta_g \rightarrow\, ^3\Sigma_g^-$), occur in the infrared.

Information about planetary airglow emissions is of necessity more limited than that about their terrestrial counterparts. Only a few emissions are sufficiently intense to be detected by Earth-based methods, including ground-based, rocket- and satellite-borne instruments. The instruments carried on spacecraft are subject to restrictions on size, weight and data rate, which results in limited sensitivity, spectral resolution and coverage and dynamic range. Orbiting spacecraft have circled Venus and Mars, but the outer planets have thus far been probed only by instruments on flyby missions. Venus ultraviolet airglow emissions have been detected by the US Mariner 5 and 10 flyby spacecraft (1967 and 1974, respectively), the Pioneer Venus orbiter (1978–1992) and by the Soviet flyby spacecraft Venera 4 (1967), and the Venera 11 and 12 (1978) orbiters. The Soviet orbiters Veneras 9 and 10 (1975) recorded the visible nightglow spectrum on Venus. Instruments that have observed Mars ultraviolet airglow include the US Mariners 6 and 7 flybys (1969) the Mariner 9 orbiter (1971–1972), and the Soviet spacecraft Mars 2 and 3 (1971-1972). The Soviet orbiter Mars 5 (1974), like Veneras 9 and 10, carried a Lyman alpha filter photometer and a visible spectrometer past Mars. Ultraviolet emissions from the atmospheres of Jupiter and its satellites were detected by the US Pioneer 10 and 11 ultraviolet photometers (1973) and the Voyager 1 and 2 ultraviolet spectrometers (1979). Pioneer 11 (1979) and Voyagers 1 and 2 (1980–1981) measured emissions from Saturn; Voyager 1 also recorded emissions from Titan (1980). Emissions from Uranus (1986), Neptune and its satellite Triton (1989) were detected by the ultraviolet spectrometer on Voyager 2. In addition, there have been many ultraviolet spectra measured from rockets and the International Ultraviolet Explorer (IUE) and Copernicus satellites. The advent of the Hubble Space Telescope has provided the capability to resolve the rotational structure of sufficiently intense planetary emission.

Mechanisms for production of airglow emissions

The major mechanisms for production of airglow include photoionization and excitation of an atomic or molecular atmospheric species X

$$X + h\nu \rightarrow X^{+*} + e \tag{A1}$$

and photoelectron impact ionization and excitation

$$X + e \rightarrow X^{+*} + 2e \tag{A2}$$

These processes are followed by the emission of a photon, as the species decays to the ground state or another lower energy state. Photoelectrons, which are represented here by the symbol e, are electrons with energies averaging about 20 eV, which are created in photoionization of atmospheric gases by solar ultraviolet radiation. In these equations, the superscript asterisk (*) represents electronic excitation. In photodissociative excitation of a molecule AB

$$AB + h\nu \rightarrow A^* + B \tag{A3}$$

and electron impact dissociative excitation

$$AB + e \rightarrow A^* + B + e \tag{A4}$$

fragments A and B are produced, either or both of which may be formed in excited states. Resonance scattering involves the absorption of a photon, causing a dipole allowed transition to an excited state of an atom A

$$A + h\nu \rightarrow A^* \tag{A5a}$$

followed by the re-emission of a photon of nearly the same wavelength, as the species decays back to the ground state:

$$A^* \rightarrow A + h\nu \tag{A5b}$$

In the analogous process for a molecule, fluorescent scattering, a photon is absorbed by a molecule AB in a vibrational state v producing an excited molecule in a vibrational state v':

$$AB(v) + h\nu \rightarrow AB^*(v') \tag{A6a}$$

which is then followed by emission to a lower state with vibrational quantum number v'':

$$AB^*(v') \rightarrow AB(v'') + h\nu' \tag{A6b}$$

Since the transition is usually to a range of lower vibrational states, the wavelength of the photon emitted may be longer than that of the photon absorbed. It should be noted here that we exclude from our discussion non-selective scattering processes, such as Rayleigh or aerosol scattering.

Electron impact excitation of an atom or molecule

$$X + e \rightarrow X^* + e \tag{A7}$$

may be followed by the emission of a photon. Unlike resonance or fluorescent scattering, the excited state produced in electron impact excitation can be connected to the ground state by a dipole forbidden transition.

A chemiluminescent reaction is one in which one or more of the products is formed in an excited state, which then radiates to a lower state. It can be represented symbolically by

$$X + Y \rightarrow Z^* + \text{products} \tag{A8a}$$

where X and Y are reactants, followed by emission of a photon from an excited state of the product Z:

$$Z^* \rightarrow Z + h\nu. \tag{A8b}$$

In practice, however, chemiluminescent reactions may involve two or three reactants; one or more of the products may be formed in excited states. Dayglow can result from chemiluminescent reactions of species that occur promptly, that is, close in time and space to their place of origin. Nightglow arises from delayed or persistent chemiluminescence, in which the reacting species are produced during the daytime or transported from the dayside.

Venus and Mars

The first planetary airglow emission to be observed from a spacecraft was the Lyman alpha line of atomic hydrogen, which was detected by the photometers aboard the Venus flyby spacecraft Mariner 5 and Venera 4 in October 1967 (Barth *et al.* 1967; Kurt, Dostavalow and Scheffer, 1968). The Lyman alpha line (q.v.) arises from the transition H($2p \rightarrow 1s$), and is produced mainly by resonance scattering in the extended hydrogen corona of Venus. The first planetary airglow spectra were taken of Venus and Jupiter in 1967 by a low-resolution spectrometer carried on a rocket (Moos *et al.*, 1969). Only Lyman alpha was detected on Jupiter. The features identified in the Venus dayglow spectrum included Lyman alpha and a broad feature that was interpreted as consisting of the O 1304 and 1356 Å

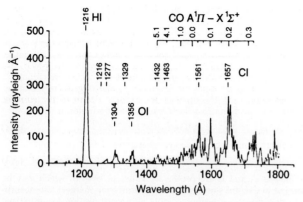

Figure A2 Martian dayglow spectrum (1100–1800 Å) recorded by Mariners 6 and 7 at 10 Å resolution. The spectrum is the sum of four limb observations with tangent altitudes between 140 and 160 km. (From Barth *et al.*, 1971.)

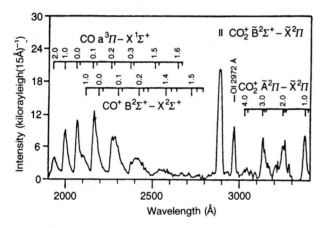

Figure A3 Martian airglow spectrum (1900–3400 Å) recorded by Mariner 9 at 15 Å resolution. This spectrum is the result of averaging 120 individual limb observations. (From Barth *et al.*, 1972.)

multiplets, which are produced by the $O(^3S^o \rightarrow {}^3P)$ and $O(^5S^o \rightarrow {}^3P)$ transitions, respectively. Later Rottman and Moos (1973) flew a medium-resolution spectrometer, which revealed also a few lines of atomic carbon, and some of the CO fourth positive (A $^1\Pi \rightarrow$ X $^1\Sigma$) bands. The first measurements of the Martian dayglow were made by Mariner 6 and 7 in 1969 and by Mariner 9 in 1971 (Barth *et al.*, 1969, 1971). Spectra from the short wavelength channel (1100–1800 Å) of Mariner 6 and 7 and the long wavelength channel (1900–3400 Å) of Mariner 9 are shown in Figures A2 and A3, respectively. The features detected include Lyman alpha, the 1304 and 1356 Å multiplets of atomic oxygen, the O 2972 Å line, the CO fourth positive bands, the atomic carbon lines at 1561 and 1657 Å, the CO Cameron bands (a $^3\Pi \rightarrow$ X $^1\Sigma$), and the CO_2^+ Fox–Duffendack–Barker bands (A $^2\Pi_u \rightarrow$ X $^2\Pi_g$) and the ultraviolet doublet (B $^2\Sigma_u^+ \rightarrow$ X $^2\Pi_g$).

The potential sources of excited O atoms include dissociative excitation of CO and CO_2, but the major sources of the 1304 and 1356 Å emissions are fluorescent scattering of solar radiation and electron impact on atomic oxygen respectively. The CO fourth positive bands are the only molecular features that appear in the far ultraviolet region of the dayglow spectra of Venus and Mars; they can be excited by electron impact dissociative excitation and photodissociative excitation of CO_2, by dissociative recombination of CO_2^+:

$$CO_2^+ + e \rightarrow CO(A\ ^1\Pi) + O \quad (A9)$$

as well as by fluorescent scattering of solar photons and electron impact excitation of CO. In equation (A9), the symbol e represents a thermal electron. The solar Lyman alpha line occurs at nearly the same wavelength as the (14, 0) transition of the CO fourth positive band. Fluorescent scattering of solar Lyman alpha therefore results in the production of the $v' = 14$ progression, i.e. the (14, v'') bands, and has been found to be the dominant source of fourth positive emission in the dayglow spectra of both Venus and Mars (Durrance, Barth and Stewart 1980; Fox and Dalgarno 1981).

Potential sources of the atomic carbon lines at 1657 and 1561 Å include dissociative excitation of CO and CO_2, electron impact excitation of atomic carbon and resonance scattering of solar radiation by atomic carbon, and the latter has been found to dominate the production.

The Cameron bands of CO, which occur in the 1800 to 2600 Å range, are the brightest features in the Martian ultraviolet dayglow, and are predicted to be so for Venus also, although no measurements have been published. With the exception of fluorescent scattering, which is not important for this dipole forbidden transition, the potential sources are the same as those of the fourth positive bands. The most important sources are electron impact dissociative excitation and photodissociative excitation of CO_2. The CO_2^+ Fox–Duffendack–Barker bands and the ultraviolet doublet are prominent features of the Martian dayglow, and are predicted to be important on Venus as well. The major excitation mechanism for both emissions is simultaneous photoionization and excitation of CO_2.

The red line of atomic O at 6300 Å and the green line at 5577 Å arise from the transitions $O(^1D \rightarrow {}^3P)$ and $O(^1S \rightarrow {}^1D)$ respectively. Significant intensities of these emissions have been predicted in the dayglows of both Mars and Venus, but they have been neither sought nor observed. Upper limits have been placed on the nightglow intensities on Venus by the Venera 9 and 10 and on Mars by the Mars 5 visible spectrometer measurements (Krasnopolsky, 1981). Although the transition probability is about 5% of that of the green line, $O(^1S)$ can also radiate to the ground $O(^3P)$ state, emitting a photon at 2972 Å. This emission has been observed in both the Martian and Venusian ultraviolet dayglow spectra. The major source of $O(^1S)$ on both planets is photodissociative excitation of CO_2.

N_2 comprises 2.5–4% of the atmospheres of both Venus and Mars. Emissions from excited states of N_2, mostly arising from electron impact excitation of states in the triplet manifold, have been predicted. Model calculations show that two band systems of N_2^+, the Meinel (A $^2\Pi_u \rightarrow$ X $^2\Sigma_g^+$) and first negative (B $^2\Sigma_u^+ \rightarrow$ X $^2\Sigma_g^+$) bands, are excited mostly by fluorescent scattering of sunlight, and are predicted to have significant total intensities. No bands of N_2 or N_2^+, however, have been definitively identified in the spectra of either planet.

Nightglow spectra taken by the Pioneer Venus orbiter ultraviolet spectrometer have revealed the presence of the nitric oxide gamma and delta band systems, produced by radiative association of N and O that are transported from the dayside:

$$N + O \rightarrow NO^* + h\nu \quad (A10)$$

A sample Pioneer Venus nightglow spectrum is shown in Figure A4. Spectra taken by the visible spectrometers on the Venera 9 and 10 spacecraft, presented here in Figure A5, showed the presence of the O_2 Herzberg II (c $^1\Sigma_u^- \rightarrow$ X $^3\Sigma_g^-$) bands, and three other O_2 band systems. These band systems arise from the three-body recombination of O atoms transported from the dayside:

$$O + O + M \rightarrow O_2^* + M \quad (A11)$$

Images of the Venus nightside at ultraviolet or visible wavelengths thus provide information about the day-to-night circulation of the atmosphere. Although the Mars 5 orbiter searched for comparable visible emissions, no nightglow emissions were detected in the Martian atmosphere.

The (0,0) band of the O_2 infrared atmospheric system at 1.27 μm is the brightest feature (in terms of photon flux) in the airglows of the terrestrial planets. Using a ground-based high resolution spectrometer, Connes *et al.* (1979) first detected the emission from both the dayside and nightside of Venus; Noxon *et al.* (1976) first measured 1.27 μm emission from Mars. The source of the emission in the Martian dayglow is photolysis of ozone. On Venus the intensities in the dayglow and nightglow are comparable, and the emission ultimately arises from recombination of oxygen atoms created in photolysis of CO_2, although it is uncertain whether the recombinations are direct reactions between O atoms or result from catalytic cycles involving chlorine compounds or other minor constituents.

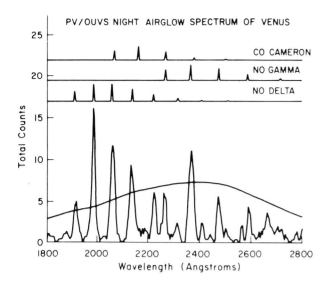

Figure A4 The far-ultraviolet (1800–2800 Å) nightglow spectrum of Venus obtained by the Pioneer Venus Orbiter ultraviolet spectrometer. The spectrum is the sum of 96 individual spectra, and has been smoothed by a three-point running average. The solid line represents the instrument sensitivity. The predicted relative responses for three band systems are also shown. (From Stewart and Barth, 1979.)

The outer planets

On their journeys past Jupiter and Saturn, Pioneer 10 and 11 carried two-channel photometers designed to detect the H Lyman alpha and He 584 Å lines, which are produced primarily by resonance scattering in the thermospheres of the outer planets. The ultraviolet spectrometers on Voyager 1 and 2 measured, in addition to Lyman alpha, emission in the Lyman (B $^1\Sigma_u^+ \to$ X $^1\Sigma_g^+$) and Werner (C $^1\Pi_u \to$ X $^1\Sigma_g^+$) bands of H_2. These emissions may be produced by fluorescent scattering of sunlight or electron impact on H_2. Two Jovian spectra taken by the Voyager 2 ultraviolet spectrometer are shown in Figure A6. A longitudinal asymmetry in the Lyman alpha intensity, which takes the form of a bright region between 50 and 100° magnetic longitude, has been observed by both the Voyager spacecraft and the IUE satellite. This so-called Lyman alpha bulge has been ascribed to increases in temperature and/or increases in the column density of H above the absorbing layer. Voyager 2 also measured Lyman alpha emission on the equatorial nightside of Jupiter, probably produced by resonance scattering of sky background emission.

The Pioneer 11 two-channel photometer measured Lyman alpha emission from Saturn, but failed to detect the He 584 Å line (Judge, Wu and Carlson, 1980). Both emissions and the H_2 Lyman and Werner bands were measured in the Saturnian dayglow by the ultraviolet spectrometers carried on Voyager 1 (Broadfoot et al., 1981). Weak Lyman alpha and He 584 Å emission were detected also from the nightside of Saturn. Lyman alpha emission from Uranus has been detected by the IUE spectrograph and by the Voyager 2 spectrometer (Broadfoot et al., 1986), but emission from Neptune has been recorded only during the Voyager 2 flyby (Broadfoot et al., 1989). The spectra taken by the Voyager 2 ultraviolet spectrometer at both planets are dominated by emissions of H and H_2, just as those of Jupiter and Saturn (Strobel et al. 1991). The Uranus spectrum exhibits anomalously large intensities between 1216 and 1500 Å, which have been ascribed to Rayleigh scattering of sunlight. In the thermosphere of Uranus there is an apparent dearth of the strongly absorbing molecule methane, relative to the abundances in the thermospheres of the other outer planets. The presence of a prominent feature at 1280 Å, identified as solar Lyman alpha that has been Raman scattered by H_2, tends to support this hypothesis (Yelle et al. 1987). The source of the H_2 band emissions in the dayglow spectra of the outer planets has been the subject of some controversy. Arguments have been advanced that the emissions are caused by impact of low energy electrons, possibly photoelectrons accelerated by parallel electric fields created by an atmospheric dynamo; others claim that the intensity produced by fluorescent scattering of solar radiation is sufficient to power the emissions. The arguments have been reviewed critically by Strobel et al. (1991a).

Titan and Triton

The ultraviolet spectrometer on Voyagers 1 recorded the spectrum of the Titan daylow in the range 500–1700 Å, and it was found to be dominated by emission features of N_2 and N (Strobel et al., 1991b). The most intense features of N_2 were the c'_4 $^1\Sigma_u^+ \to$ X $^1\Sigma_g^+$ Rydberg bands, but the Birge–Hopfield (b $^1\Pi_u \to$ X $^1\Sigma_g^+$) and Lyman–Birge–Hopfield (a $^1\Pi_g \to$ X $^1\Sigma_g^+$) bands were also present in the spectrum. The former two band systems are dipole allowed, and could in theory be excited by fluorescent scattering, but the predicted intensities are small. In fact, the source is probably electron impact excitation. The emissions were found to be brightest in the quadrant facing into the corotating magnetosphere of Saturn, and it was therefore suggested that the emissions could be produced by impact of low energy electrons from Saturn's magnetosphere. Later analyses, however, showed that the solar source was sufficient to produce most of the observed emissions. Other features identified include the 1134, 1200 and 1243 Å lines of N and the 1085 Å line of N^+. Voyager 2 measured ultraviolet emissions from Neptune's satellite Triton. As for Titan, features of N_2, N and N^+ dominate the spectrum, but for Triton interactions of atmospheric species with magnetospheric electrons may be important in producing the observed emissions. The strong emission at Lyman alpha in both spectra results from resonance scattering by thermospheric and exospheric atomic hydrogen.

Earth

The primary terrestrial dayglow emissions arise from the interactions of solar photons and photoelectrons with the major thermospheric species, N_2, O_2 and O. Figure A7 shows a spectrum of the terrestrial airglow for the 1200 to 9000 Å wavelength region that was taken by a spectrograph flown on the Space Shuttle (Broadfoot et al. 1992). The most important emission features of O are those at 1304, 1356 and 2972 Å in the ultraviolet and 5577 and 6300 Å in the visible. They are produced by direct excitation of O and to a lesser extent by dissociative excitation of O_2. The Lyman–Birge–Hopfield band system of N_2 (a $^1\Pi_g$–X $^1\Sigma_g^+$), which appears in the far ultraviolet

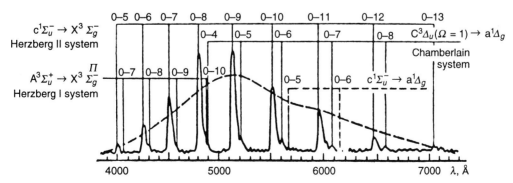

Figure A5 Visible nightglow spectrum of Venus measured by the spectrometers on Veneras 9 and 10. The dashed line is the instrument sensitivity. (From Krasnopolsky and Tomashova, 1980.)

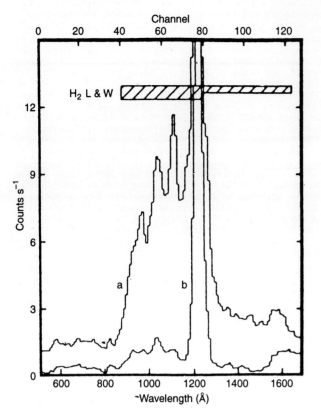

Figure A6 Jovian airglow spectrum recorded by the ultraviolet spectrometer on Voyager 2. Spectrum (a) was recorded when a portion of the auroral zone lay within the slit. Spectrum (b) was recorded near the equator. The zero for spectrum a is shifted upward by 1 count s^{-1}. (From Sandel et al., 1979.)

(1200–1700 Å), is dipole forbidden, and is therefore excited only by photoelectron impact on N_2. Most electronically excited states of N_2 in the singlet manifold, other than the a $^1\Pi_g$ state, have significant probabilities for predissociation or branching to states other than the ground state. Therefore, few emissions from these states are observed in dayglow spectra of the optically thick terrestrial atmosphere, although some bands of the Birge–Hopfield I (b $^1\Pi_u$–X $^1\Sigma_g^+$) and Birge–Hopfield II (b' $^1\Sigma_u^+$–X $^1\Sigma_g^+$) systems have been identified. Several emissions arising from triplet states of N_2 are important features of the dayglow, including the second positive (C $^3\Pi_u$–B $^3\Pi_g$) and Vegard–Kaplan (A $^3\Sigma_u^+$–X $^1\Sigma_g^+$) bands in the near ultraviolet and the first positive (B $^3\Pi_g$–A $^3\Sigma_u^+$) bands in the visible to near infrared. These emissions also are produced by photoelectron-impact excitation of N_2. Two prominent band systems of N_2^+, the first negative (B $^2\Sigma_u^+$–X $^2\Sigma_g^+$) in the visible and the Meinel (A $^2\Pi_u$–X $^2\Sigma_g^+$) bands in the infrared can be produced by fluorescent scattering of solar radiation by N_2^+ in the electronic ground state, and by photoionization and electron-impact ionization of N_2. Dissociative ionization of N_2 produces the prominent N^+ 1085 Å line. Many more multiplets of N, N^+, O and O^+, produced by direct excitation or ionization of O and N or dissociative excitation or ionization of O_2 or N_2, are observed in the EUV portion of the spectrum.

NO is a minor constituent in the thermosphere, but the the NO γ (A $^2\Sigma^+$–X $^2\Pi$) bands, which are excited by fluorescent scattering of sunlight, are important features of the dayglow. Because of diffusive separation above the homopause, light species become dominant at great heights, and extended emissions (called geocoronae) are observed at Lyman alpha (1216 Å) and 584 Å, which are produced by resonance scattering of the analogous solar lines by H and He respectively.

The terrestrial nightglow contains several emissions that are due to recombination of species produced during the day. Three-body recombination of O atoms results in emission primarily in the Herzberg I (A $^3\Sigma_u^+$–X $^3\Sigma_g^-$) and Chamberlain (A' $^3\Delta_u$–a $^1\Delta_g$) bands, although the Herzberg II bands have also been identified in the terrestrial nightglow. Just as on Venus and Mars, the (0,0) infrared atmospheric band O_2(a $^1\Delta_g$) at 1.27 μm is a strong feature of both the dayglow and the nightglow. It is produced in the dayglow by photolysis of ozone and in the nightglow by recombination of O atoms. Radiative association of N and O atoms produces emission in the $v' = 0$ progression of the NO γ and δ (C $^2\Pi$–X $^2\Pi$) bands. Continuum emission due to radiative recombination of O^+:

$$O^+ + e \rightarrow O + h\nu \qquad (A12)$$

can be seen at wavelengths shortward of 911, 1304 and 1356 Å in the terrestrial nightglow.

One of the first nightglow features ever recorded was the atomic oxygen green line. O atoms in the ^1S state are produced at high altitudes in the terrestrial ionosphere in dissociative recombination of O_2^+:

$$O_2^+ + e \rightarrow O(^1S, ^1D, ^3P) + O(^1D, ^3P) \qquad (A13)$$

At low altitudes $O(^1S)$ is produced in a two-step process, called the Barth mechanism, that begins with three-body recombination of O atoms to produce O_2 in an electronically excited state:

$$O + O + M \rightarrow O_2^* + M \qquad (A14)$$

Excitation transfer from this state to O produces $O(^1S)$:

$$O_2^* + O \rightarrow O_2 + O(^1S) \qquad (A15)$$

Among the most interesting features of the terrestrial nightglow are the OH Meinel bands, which appear longward of 5200 Å and arise from vibrational transitions within the ground $X^2\Pi_{3/2,1/2}$ state. Vibrationally excited OH is produced via the reaction

$$H + O_3 \rightarrow OH(v \leq 9) + O_2 \qquad (A16)$$

in the 85 to 90 km region. Wave structure has been observed in images of the Meinel emissions that has been utilized to obtain information about the occurrence morphology and sources of gravity waves in the upper mesosphere (Taylor and Hill, 1991).

Recommended reading

Further information and a more complete bibliography can be found in the following references. A pedagogical discussion of planetary ultraviolet spectroscopy has been presented by Barth (1969); (see also Ultraviolet spectroscopy). Luminosity from the atmosphere of Venus has been reviewed by Fox and Bougher (1991) and that from Venus and Mars by Fox (1992). Paxton and Anderson (1992) have dicussed ultraviolet remote sensing of Venus and Mars. Krasnopolsky (1983) has reviewed the visible night-glow measurements from Veneras 9 and 10. He has also summarized the understanding of the photochemistry of Venus and Mars gained from measurements, including airglow spectroscopy, and models (Krasnopolsky, 1986). Ultraviolet spectroscopy and other measurements of relevance to the upper atmosphere of Saturn have been discussed by Atreya et al. (1984), along with models of the thermosphere and ionosphere. Hunten et al. (1984) have reviewed measurements and models of the upper atmosphere of Titan. A critical review of studies of airglow and auroral emissions from the outer planets has been presented by Strobel et al. (1991). Fox (1986) has surveyed measurements and models of airglow and auroral emissions from the atmospheres of all the planets.

An excellent pedagogical discussion of terrestrial airglow features and excitation mechanisms can be found in Rees (1989). Meier (1991) has presented an extensive review of ultraviolet spectroscopy and remote sensing of the terrestrial atmosphere. Solomon (1991) and Meier (1987) have reviewed recent advances in optical aeronomy. Meriwether (1989) has discussed the photochemistry of some nightglow emissions from the mesopause.

Jane L. Fox

Bibliography

Atreya, S.K., Waite, J.H., Donahue, T.M. et al. (1984) Theory, measurements and models of the upper atmosphere and ionosphere of Saturn, in *Saturn* (eds T. Gehrels and M.S. Matthews). Tucson: Univ. of Arizona Press, pp. 239–80.

Barth, C.A. (1969) Planetary ultraviolet spectroscopy. *Appl. Optics*, **8**, 1295.

Barth, C.A., Pearce, J.B., Kelly, K.K. et al. (1967) Ultraviolet emissions observed near Venus from Mariner 5. *Science*, **158**, 1675.

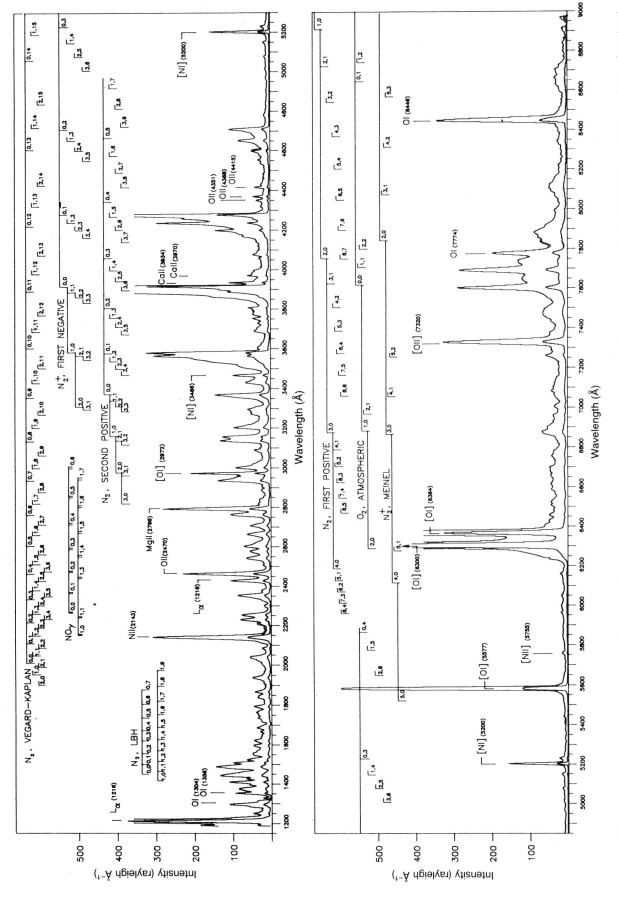

Figure A7 A terrestrial dayglow spectrum recorded in a single 32 s exposure by the Arizona Imager Spectrograph flown on the Space Shuttle STS-53. The spectrum is shown plotted with two scales, 1.0 (top curve in both parts of the figure) and 0.2 (bottom curve). The major vibrational transitions of the band systems are indicated at the tops of the spectra. (From A.L. Broadfoot, personal communication).

Barth, C.A., Fastie, W.G., Hord, C.W. et al. (1969) Mariner 6: ultraviolet spectrum of Mars upper atmosphere. *Science*, **165**, 1004.

Barth, C.A., Hord, C.W., Pearce, J.B. et al. (1971) Mariner 6 and 7 ultraviolet spectrometer experiment: upper atmospheric data. *J. Geophys. Res.*, **76**, 2213.

Barth, C.A., Hord, C.W., Stewart, A.I. and A.L. Lane (1972) Mariner 9 ultraviolet spectrometer experiment: Mars airglow spectroscopy and variations in Lyman alpha. *Science*, **175**, 309.

Broadfoot, A.L., Sandel, B.R., Shemansky, D.E. et al. (1981) Extreme ultraviolet observations from Voyager 1 encounter with Saturn. *Science*, **212**, 206.

Broadfoot, A.L., Herbert, F., Holberg, J.B. et al. (1986) Ultraviolet spectrometer observations of Uranus. *Science*, **233**, 74.

Broadfoot, A.L., Atreya, S.K., Bertaux, J.L. et al. (1989) Ultraviolet spectrometer observations of Neptune and Triton. *Science*, **246**, 1459.

Connes, P., Noxon, J.F., Traub, W.A. and Carleton, N.P. (1979) $O_2(^1\Delta)$ emission in the day and night airglow of Venus. *Astrophys. J.*, **233**, L29.

Durrance, S.T., Barth, C.A. and Stewart, A.I.F. (1980) Pioneer Venus observations of the Venus dayglow spectrum 1250–1430 Å. *Geophys. Res. Lett.*, **7**, 222.

Fox, J.L. (1986) Models for airglow and auroral emissions from other planetary atmospheres. *Can. J. Phys.*, **61**, 1631.

Fox, J.L. (1992) Airglow and aurora in the atmospheres of Venus and Mars, in *Venus and Mars: Atmospheres, Ionospheres and Solar Wind Interactions* (eds J.G. Luhmann, M. Tatrallyay and R. Pepin). Washington, DC AGU Press, pp. 191–224.

Fox, J.L. and Bougher, S.W. (1991) Structure, luminosity and dynamics of the Venus thermosphere. *Space Sci. Rev.*, **55**, 357.

Fox, J.L. and Dalgarno, A. (1981). Ionization, luminosity and heating of the upper atmosphere of Venus. *J. Geophys. Res.*, **86**, 629.

Hunten, D.M., Tomasko, M.G., Flasar, F.M. et al. (1984) Titan, in *Saturn* (eds T. Gehrels and M.S. Matthews). Tucson; University of Arizona Press, pp. 239–80.

Judge, D.L., Wu, F.-M. and Carlson, R.W. (1980) Ultraviolet photometer observations of the Saturnian system. *Science*, **207**, 431.

Krasnopolsky, V.A. (1981) Excitation of oxygen emissions in the night airglow of the terrestrial planets. *Planet. Space Sci.*, **29**, 925.

Krasnopolsky, V.A. (1983) Venus spectroscopy in the 3000–8000 Å region, in *Venus* (eds D.M. Hunten, L. Colin, T.M. Donahue, and V.I. Moroz). Tuscon: University of Arizona Press, pp. 459–483.

Krasnopolsky, V.A. (1986) *Photochemistry of the Atmospheres of Mars and Venus*. New York: Springer-Verlag.

Krasnopolsky, V.A. and Tomashova (1980) Venus nightglow variations. *Cosmic Res.*, **18**, 766.

Kurt, V.G., Dostavalow, S.B. and Scheffer, E.K. (1968) The Venus far ultraviolet observations with Venera 4. *J. Atmos. Sci.*, **25**, 574.

Meier, R.R. (1987) Thermospheric aurora and airglow. *Rev. Geophys.*, **25**, 471.

Meier, R.R. (1991) Ultraviolet spectroscopy and remote sensing of the upper atmosphere, *Space Sci. Rev.*, **58**, 1.

Meriwether, J.W. (1989) A review of the photochemistry of selected nightglow emissions from the mesopause. *J. Geophys. Res.*, **94**, 14629.

Moos, H.W., W.G. Fastie and M. Bottema (1969) Rocket measurement of ultraviolet spectra of Venus and Jupiter between 1200 and 1800 Å. *Astrophys. J.*, **155**, 887.

Noxon, J.L., Traub, W.A., Carleton, N.P. and Connes, P. (1976) Detection of O_2 dayglow emission from Mars and Martian ozone abundance. *Astrophys. J.*, **207**, 1025.

Paxton, L.J. and D.E. Anderson (1992) Far ultraviolet remote sensing of Venus and Mars, in *Venus and Mars: Atmospheres, Ionospheres and Solar Wind Interactions* (eds J.G. Luhmann, M. Tatrallyay and R. Pepin). Washington, DC: AGU Press, pp. 113–90.

Phillips, J.L., A.I.F. Stewart and J.G. Luhmann (1986) The Venus ultraviolet aurora: observations at 130.4 nm. *Geophys. Res. Lett.*, **13**, 1047.

Rees, M.H. (1989) *Physics and Chemistry of the Upper Atmosphere*. Cambridge, England: Cambridge University Press.

Rottman, G.J. and Moos, H.W. (1973) The ultraviolet (1200–1900 Å) spectrum of Venus. *J. Geophys. Res.*, **78**, 8033.

Sandel, B.R. et al. (1979) Extreme ultraviolet observations from Voyager 2 encounter with Jupiter. *Science*, **206**, 962.

Solomon, S.C. (1991) Optical aeronomy. *Rev. Geophys., Suppl.*, 1089.

Stewart, A.I.F. and Barth, C.A. (1979). Ultraviolet night airglow of Venus. *Science*, **205**, 59.

Strobel, D.F., Yelle, R.V., Shemansky, D.E. and Atreya, S.K. (1991a) The upper atmosphere of Uranus, in *Uranus* (eds J.T. Bergstralh, E.D. Miner and M.S. Matthews). Tucson: University of Arizona Press, pp. 65–109.

Strobel, D.F., Meier, R.R., Summers, M.E. and Strickland, D.L. (1991b) Nitrogen airglow sources: comparison of Titan, Titan and Earth. *Geophys. Res. Lett.*, **18**, 689–92.

Taylor, M.J. and Hill, M.J. (1991) Near infrared imaging of hydroxyl wave structure over an ocean site at low latitudes. *Geophys. Res. Lett.*, **18**, 1333.

Yelle, R.V., Doose, L.R., Tomasko, M.G. and Strobel, D.F. (1987) Analysis of Raman scattered Ly-α emissions from the atmosphere of Uranus. *Geophys. Res. Lett.*, **14**, 483.

Cross references

Absorption, absorption spectrum
Atmosphere
Aurora, planetary
Emissivity
Lyman alpha
Mariner missions
Mars: atmosphere
Pioneer 10/11 missions
Pioneer Venus mission
Radiative transfer
Spectroscopy: atmosphere
Titan
Triton
Ultraviolet spectroscopy
Venera missions
Venus: atmosphere
Voyager missions

ALBEDO

The term *albus* is Latin for 'white', and *albedo* means 'whiteness', but in astronomy and atmospheric sciences the 'normal albedo' refers to the normal reflectance (T_n) potential or brightness of any solid, liquid or gaseous surface when illuminated vertically. For planets it is specifically 'the proportion of the solar light incident upon an element of the surface of a planet which is again diffusely reflected from it' *Oxford English Dictionary*; *Mon. Not. Roy. Astr. Soc.*, 1859, **20**, p. 103. It may be expressed as $A = pq$, where p at the phase angle zero is the ratio of brightness of a substance to the brightness of a perfectly diffusing disk under the same conditions; and q is a factor representing the phase law. The sunlight not reflected will be absorbed and heat the surface concerned, whether a planet, asteroid or satellite. Thus the albedo is the ratio of reflected radiation to incoming electromagnetic radiation. If 50% of that radiation is reflected the major albedo is indicated as 0.5. A white substance is close to 1.0, while a non-reflecting black surface approaches zero.

A 'Bond albedo' (A_B) establishes the energy balance of a planet, determined over all wavelengths, being the fraction of incident solar energy that is reflected back into space. Approximate albedos of solar system bodies are: Mercury, 0.06; Venus, 0.76; Earth, 0.39; Moon, 0.07; Mars, 0.15, Jupiter, 0.51; Saturn, 0.50; Uranus, 0.66; and Neptune, 0.62. The rocky or desert-covered planets absorb most of the incoming radiation (0.06–0.15), whereas the cloud or gas-covered ones reflect greater proportions of it (0.15–0.76).

The four largest asteroids have albedos as follows: Ceres, 0.06; Pallas, 0.07; Juno, 0.12; and Vesta, 0.26. For small asteroids, their approximate size can be estimated from their luminosity assuming an albedo comparable to some reference (average) value (see Asteroid lightcurve). Measurement of the polarization of reflected sunlight from more than 20 asteroids by Veverka, Zellner and Bowell (cited in Milton, 1977) and comparison with laboratory-known albedos makes it possible to estimate their dimensions. Inasmuch as most of them are irregular in shape, the reflectance varies as they rotate, and thus the spin rate has been calculated in more than 50 examples.

On the surface of planet Earth different materials are found to possess a wide range of albedos. For example, snow is 0.4 to 0.85 and clouds are similar, 0.4 to 0.8. (Europa, one of Jupiter's moons, displays an albedo comparable to snow.) The total albedo of the Earth averages about 0.39 but is a constantly changing variable because of clouds, soil moisture, vegetation, ocean (sea state), sea ice, snow, glaciers and so on (Fairbridge, 1967). Although clouds are constantly changing, the mean cloudiness of the Earth is almost stable at about 50%. Thus snow cover, sea ice and vegetation are the three biggest variable factors in terrestrial albedo. The surface of water, with the Sun near zenith, has an albedo of about 0.02, i.e. 98% incident radiation is absorbed, but at a low angle of incidence it can be > 0.8. Forest is usually around 0.04 to 0.1, and green fields up to 0.15.

Albedo is thus easily changed on planet Earth by anthropogenic activity. Generally, such changes are inadvertent, but Bloch (1964, noted in Fairbridge, 1967, p. 13) cited examples in China and Peru where drought conditions were relieved by spreading coal dust on glacier surfaces, thereby inducing rapid melting. In paleoclimatology, volcanic ash showers over ice caps could trigger brief deglaciation events.

On the planet Mars the large orbital eccentricity favors a regular build-up and decay of snow-white polar caps (CO_2 ice), thus creating a fluctuating albedo and the consequent planetary climate state. Mercury, because of its highly eccentric and inclined orbit, displays a correspondingly variable albedo.

Rhodes W. Fairbridge

Bibliography

Fairbridge, R.W. (1967) Albedo and reflectivity, in *The Encyclopedia of Atmospheric Sciences and Astrogeology* (ed. R.W. Fairbridge). New York: Reinhold Publ. Corp., pp. 12–13.

Milton, S. (ed.) (1977) *Cambridge Encyclopedia of Astronomy*. Cambridge: University Press, 495 pp.

Veverka, J. (1987) Albedo, in *McGraw-Hill Encyclopedia of Science and Technology*, Vol. 1, pp. 309–10.

Cross references

Asteroid: lightcurve
Mars: remote sensing
Opposition effect

ALFVÉN, HANNES OLOF GÖSTA (1908–1995), AND ALFVÉN WAVE

Distinguished Swedish astronomer and specialist in plasma physics, Alfvén earned his PhD at the University of Uppsala in 1934 and was awarded the Nobel Prize for physics in 1970. He was long associated with the Royal Institute of Technology in Stockholm, as professor of electricity, then electronics and later (1963–1973) of plasma physics.

Key publications by Alfvén include his *Cosmical Electrodynamics* (1948), *On the Origin of the Solar System* (1956), *Atom, Man and the Universe* (1969) and *Cosmic Plasma* (1981). Jointly with Gustaf Arrhenius he prepared *Evolution of the Solar System*, which was issued by NASA as a special publication (No. SP-345, 1976).

Alfvén waves

At very low frequencies a plasma may be considered a low-density, perfectly conducting fluid. In the presence of a static magnetic field, physical displacement of the charged particles can launch magnetohydrodynamic waves. Mechanical motion of the particles sets up an electromotive force and currents; the currents, in turn, interact with the magnetic field, modifying the motion of the particles. When the background magnetic field strength B_0 is large and the density of the medium ρ_0 is low, the characteristic velocity of these waves $V_A = B_0/(4\pi\rho_0)$ is large enough that significant energy transport can result. Alfvén originally proposed that such waves might move energy from the Sun's interior to the photosphere and be partly responsible for sunspots. Alfvén waves may also be found in the solar wind and in planetary magnetospheres.

Simultaneous solution of Maxwell's equations and the continuity equation for a perfectly conducting medium yields a vector expression involving only the small amplitude velocity perturbation δv of the plasma particles. There are four solutions for δv. One solution describes a compressional wave that can travel both along and across the magnetic field. This wave is called the fast magnetosonic mode because it is the fastest of the four modes. In this wave the magnetic and density compression are in phase. The next mode is a transverse perturbation of the magnetic field that twists the field but compresses neither the field nor the plasma density. It is most properly called the Alfvén/ion cyclotron mode, but it is often referred to as the intermediate mode or shear Alfvén wave. The third mode is the slow mode/sound oscillation in which the density and the field compression are generally out of phase. In general it is the slowest of the three propagating modes. The fourth mode is referred to as the entropy wave in MHD and the mirror mode in kinetic analysis. This wave exhibits total pressure balance across the magnetic field and does not propagate.

Interaction of the transverse wave with the magnetic field may be likened to oscillations on an elastic string. The curvature in the magnetic field provides a restoring force. Figure A8 illustrates a fast magnetosonic wave moving perpendicular to the field and an Alfvén wave moving along it. In the former case both the magnetic field and the charged particles become compressed. In the latter the field oscillates and the wave moves parallel to the field but neither the field nor the particles are compressed. If the medium is not perfectly conducting, the oscillations will be damped. In media such as solar plasmas, field strengths are large and densities low, yielding high Alfvén velocities with relatively little damping.

If a disturbance propagates through a plasma at greater than the velocity of any of the three propagating waves, a collisionless shock may be formed. Whereas conventional acoustic shocks transfer energy largely through collisions of particles at the shock front, in the plasmas in which shocks associated with Alfvén waves occur, the density of the medium is very low and collisions are unimportant. Instead it is the electric and magnetic fields at the 'collisionless' shock that provide the dissipation required.

Richard A. Simpson, Christopher T. Russell and Rhodes W. Fairbridge

Bibliography

Alfvén, H. (1943) On the existence of electromagnetic-hydrodynamic waves. *Arkin for Matematik, Astronomi, och Fysik*, **29B**(2) 1–7.

Alfvén, H. (1956) *On the Origin of the Solar System*.

Alfvén, H. (1965) Origin of the Moon. *Science*, **148**, 476–7.

Alfvén, H. and Arrhenius G. (eds) (1976) *Evolution of the solar system*. Washington: NASA SP-345.

Fonkal, P.V. (1990) *Solar Astrophysics*. New York: Wiley Interscience. Jackson, J.D (1962) *Classical Electrodynamics*. New York: John Wiley and Sons.

ter Haar, D. and Cameron, A.G.W. (1967) Solar system: review of theories, in *The Encyclopedia of Atmospheric Sciences and Astrogeology* (ed. R.W. Fairbridge). New York: Reinhold Publ. Co. pp.890–9.

Cross references

Plasma
Plasma wave
Solar wind

AMOR OBJECT

Amor objects are a subset of near-Earth asteroids (see Near-Earth objects), distinguished from the Apollo objects (q.v.) and Aten objects (q.v.) by orbital characteristics. The Amor objects have semimajor axes greater than that of the Earth (i.e. somewhat greater than 1 AU). They are named for a member of the class, asteroid Amor, discovered in 1932 by E. Delporte. The best known member

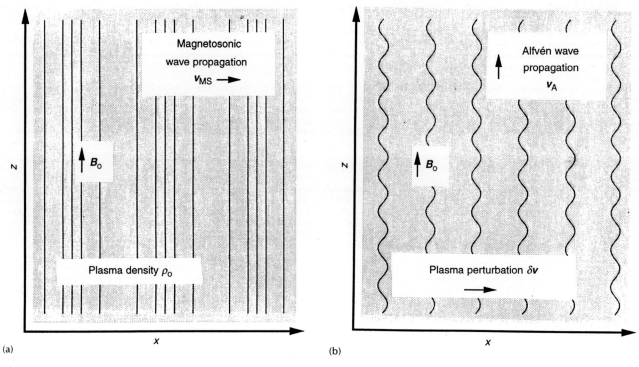

Figure A8 (a) Fast magnetosonic wave moving perpendicular to a magnetic field and (b) an Alfvén wave moving along it.

of the class, however, is the larger asteroid Eros. The Amors are the most populous of the near-Earth objects, probably numbering in the thousands (McFadden, Tholen and Veeder, 1989).

No clear compositional trends are present in the Aten objects; they appear to have originated in multiple source regions. S-class asteroids are most common, but this may be a selection effect, since these are brighter than most other asteroid classes (C in particular). (See Asteroid: compositional structure and taxonomy.)

Since the projected dynamical lifetimes of objects in near-Earth space are, at about 10^8 years, much shorter than the age of the solar system, they must be replenished from some source. The main asteroid belt is the most likely source region (Binzel et al., 1992). There are indications that some Amor objects are members of 'asteroid streams', analogous to meteor streams, in which a number of separate objects share very similar orbital characteristics (Drummond, 1991).

Active programs are underway to detect near-Earth objects (see Spacewatch). Since objects in near-Earth space have the potential to collide with the Earth, there is considerable interest in determining the population and orbital characteristics of these bodies (Matthews 1992). Although the Amor objects are not in Earth-crossing orbits at the present time, perturbations could in future reduce their perihelia to make them so.

James H. Shirley

Bibliography

Binzel, R.P., Xu, S., Bus, S.J. and Bowell, E. (1992) Origins for the near-Earth asteroids. *Science*, **257**, 779–81.
Drummond, J.D. (1991) Earth-approaching asteroid streams. *Icarus*, **89**, 14–25.
Matthews, R. (1992) A rocky watch for Earthbound asteroids. *Science*, **255**, 1204–5.
McFadden, L.-A., Tholen, D.J. and Veeder, G.J. (1989) Physical properties of Aten, Apollo, and Amor asteroids, in *Asteroids II* (eds R.P. Binzel, T. Gehrels and M. Matthews). Tucson: University of Arizona, pp. 442–67.

Cross references

Apollo object
Asteroid
Aten object
Mass extinction
Near-Earth object
Spacewatch

ANGULAR MOMENTUM

A vector quantity that is the product of moment of inertia and angular velocity. In rotational motion, angular velocity replaces linear velocity and moment of inertia replaces mass. Thus the angular momentum of a body with respect to a given axis of rotation is defined as the product of its moment of inertia by its angular velocity about that axis. The concept of angular momentum (and its conservation) followed from Galileo's law of inertia, but more directly from Newton's laws of motion and Kepler's second law (equal areas). The conservation of momentum was verified experimentally in the last half of the 17th century by Huygens, Wallis and Wren.

To calculate the total angular momentum of the Earth with respect to an axis passing through the Sun, we use the Earth's mass, 6.0×10^{24} kg, its mean radius: 6.4×10^6 m, and its mean distance from the Sun, 1.5×10^{11} m. There are two contributions to the Earth's angular momentum: the spin angular momentum of rotation about an axis through its center of mass, and its orbital angular momentum about the Sun or mass center of the solar system. The magnitude of the spin angular momentum is given by the moment of inertia of the Earth, taken as a (near) sphere of uniform density multiplied by the angular speed of rotation:

$$L_{spin} = I_{spin}\omega_{spin} = (\tfrac{2}{5}mr^2)\omega_{spin}$$

$$L_{spin} = \tfrac{2}{5}\,(6 \times 10^{24}\ \text{kg})\,(6.4 \times 10^6\ \text{m})^2\,(2\pi\ \text{rad day})^{-1}$$

$$L_{spin} = 6.9 \times 10^{33}\ \text{kg m}^2\ \text{s}^{-1}$$

To compute the Earth's orbital momentum about the Sun, the Earth is regarded as a particle moving in a circle at a distance R from the Sun:

$$L_{orb} = I_{orb}\omega_{orb} = (mR^2)\omega_{orb}$$

$$L_{orb} = (6 \times 10^{24}\ \text{kg})\,(1.5 \times 10^{11}\ \text{m})^2\,(2\pi\ \text{rad year})^{-1}$$
$$= 2.7 \times 10^{40}\ \text{kg m}^2\ \text{s}^{-1}$$

Compared with the orbital angular momentum, the Earth's spin angular momentum is negligible. So the total angular momentum of the Earth about the Sun is approximately 2.7×10^{40} kg m^2 s^{-1}.

The total angular momentum of the solar system about the mass center is comprised of the spin and orbital momenta of the several planets, together with the spin and orbital angular momentum of the Sun. Nearly all (98%) of the angular momentum arises from the orbital angular momenta of the planets, the remainder being spin angular momenta of the Sun and planets. The law of conservation of angular momentum holds for a system of bodies whenever the bodies can be treated as particles. When the individual bodies have rotation, the conservation of angular momentum is still valid, providing we include the angular momentum associated with this rotation. If we were to regard the Sun, planets and satellites as particles having no intrinsic spinning motion, the angular momentum of the solar system would not be constant. But these bodies do have rotation; in fact, tidal forces convert some of the spin angular momentum into orbital angular momentum of the planets and satellites. The conservation of angular momentum plays a key role in the evaluation of theories of the origin of the solar system.

Jane B. Blizard

Bibliography

Alfvén, H. and Arrhenius, G. (1976) *Evolution of the Solar System* Washington, DC: NASA SP-345.

Dole, S. (1970) *Habitable Planets for Man*, New York: Am. Elsevier Publ. Co.

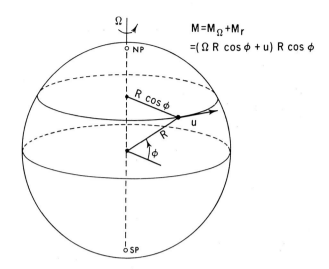

Figure A9 Schematic diagram of the atmospheric angular momentum around the Earth's axis of rotation, $M_{atm} = M_\Omega + M_r$ where $M_\Omega = \Omega R^2 \cos^2\phi$, $M_r = uR\cos\phi$, Ω = angular velocity of the Earth, R = mean radius of the Earth, ϕ = latitude and u is the eastward (zonal) wind component.

ANGULAR MOMENTUM CYCLE IN PLANET EARTH

The angular momentum budget of the Earth represents a beautiful and simple example of how the various climatic elements (atmosphere, oceans and solid Earth) work together and are united through a basic physical conservation law, despite enormous differences in their space and time scales and in their masses.

The angular momentum is a vector quantity that is the product of the moment of inertia and the angular velocity. Here we will consider only the component of the angular momentum vector that is parallel to the Earth's polar axis. In the case of the atmosphere it has two components, one connected with the solid rotation of the Earth, the Ω-angular momentum, and the other with the zonal component of the air flow with respect to the rotating Earth, the relative angular momentum: $M = M_\Omega + M_r$ (see Figure A9). Most of the temporal variability is found in the relative angular momentum.

Global atmospheric angular momentum and the length of day

Considering the Earth as a whole, the angular momentum integrated over all the masses of the solid Earth, oceans and atmosphere combined is conserved, assuming external torques (due to the Moon and Sun) to be negligible at the time scale of months to decades. Thus, the total angular momentum of the system does not vary with time, and if there were changes in one component, say the atmosphere, there would have to be compensating changes in the other components so that the total angular momentum is conserved:

$$dM/dt = 0$$

where

$$M = M_{atm} + M_{oceans} + M_{ice} + M_{crust} + M_{mantle} + M_{core}$$

It has been well known, for some time that substantial changes in the relative angular momentum of the global atmosphere are observed to occur at the time scale of days, months, years, decades and even longer. Recently it has been firmly established that there is a close negative correlation with the changes of the angular momentum of the solid Earth on time scales of days to years, and that the oceans and snow and ice do not play an important role in this. These changes in the solid Earth are readily observed as changes in the length of day (LOD); for example, an increase in angular momentum of the Earth corresponds to a decrease in the LOD.

Traditionally, the LOD data are obtained from astronomical measurements, but recently have also been obtained from lunar or satellite laser measurements. A comparison between a 13-year record of the LOD (thin solid curve) and the global mean angular momentum of the atmosphere (thick solid curve) is shown in Figure A10. The two independent datasets confirm the expected relationship at all time scales, except for a slow trend in the LOD data and occasional, still largely unexplained, differences. The slow negative decadal trend in LOD is opposite in direction to the expected very long positive trend of about 2 ms per century connected with tidal friction and the Earth–Moon torque (see Figure A11). The decadal trends are thought to be associated with the coupling between the Earth's crust and the underlying mantle and core. The occasional discrepancies between the two curves in Figure A10 may be caused by the neglected storage of angular momentum in the oceans, ice caps and snow masses, or may be due to inaccuracies in the determination of M_{atm} and LOD.

Because the changes in the Earth's rotation rate are exceedingly small, i.e. changes in the Earth's velocity at the equator are on the order of only a few micrometers per second compared to atmospheric velocities on the order of a few meters per second, the Earth cannot noticeably disturb the atmospheric flow. Thus the main point in the comparison shown in Figure A10 is to point out the remarkable fact that the atmosphere seems to be pulling and pushing the Earth around on time scales of days to years. This tiny, highly turbulent component of the climatic system with a mass of only 10^{-6} of the Earth is accelerating or slowing down the giant Earth; it seems that the tail of the elephant is driving the elephant around! The interaction is clearly a one-way street; atmosphere → Earth. The extension to more regional interactions between the atmosphere and the solid Earth will be discussed later.

It is of interest to quote a sentence from Taylor, Mayr and Kramer (1985) on core–mantle coupling: 'It is postulated that through core–mantle coupling, secular (decadal) changes (in the length of day) on the order of 0.5 ms may be driven from preceding changes in core rotation.' This sentence suggests a definite direction of forcing: core → mantle → crust. An even more extreme statement connecting changes in the core with climatic changes in the atmosphere is mentioned in Salstein and Rosen (1986): 'The hypothesis has also

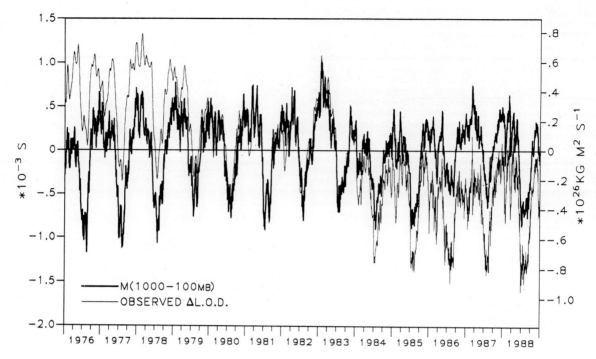

Figure A10 Time series of daily values of the relative westerly angular momentum M_r of the global atmosphere between 1000 and 100 mb based on NMC analyses (thick line), and 3-day means of the length of day (LOD; thin line) for the years 1976–88. The mean value of each series has been removed, as have solid body tidal terms from LOD. (After Rosen, 1988; see update in Rosen, Salstein and Wood, 1991).

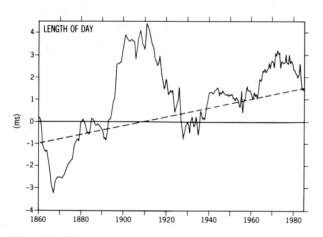

Figure A11 Time series of semiannual values of LOD during 1860–1985, taken from the work by McCarthy and Babcock (1986). A mean annual signal has been subtracted. The dashed line indicates the secular trend from tidal friction estimates. (Adapted from Salstein and Rosen, 1986).

been offered (Courtillot et al., 1982) that the decade changes in LOD originating in the Earth's core are responsible for at least some of the changes in climate, by affecting the angular momentum of the atmosphere and hence its circulation.'

However it seems inconceivable that the minute changes (10^{-8}) in the rotation rate Ω and in the Coriolis force could in any way affect the atmospheric motions, and thereby the general circulation and climate. In fact, the core-mantle forcing could be in the opposite direction so that

$$\text{atmosphere} \rightarrow \text{crust} \rightarrow \text{mantle} \rightarrow \text{core}$$

It is perhaps a more plausible hypothesis to assume that stochastic or white-noise forcing by the atmosphere would induce the long-period decadal variations in the rotation rate of the Earth's core, somewhat like the forcing mechanism proposed by Hasselmann (1976) to explain long-period variations in the oceans.

Transfer of angular momentum across the Earth's surface

In order to discuss the transfer of angular momentum at the interface between the atmosphere and the Earth's crust, it is instructive to consider first the normal seasonal cycle in the relative angular momentum of the global atmosphere given in Fig. A12. The curve shows that the atmosphere rotates faster in northern winter and slower in northern summer due mainly to the strengthening and weakening of the jet streams (superrotation) in the northern hemisphere. The derivative of this curve would show the weakening of the westerlies in April–June and strengthening in August–October. There is fair agreement with the observations of the LOD, with differences of about 1 ms in the LOD between January and July (Wahr and Oort, 1984).

The variations in the total angular momentum of the atmosphere and those in the LOD must be connected through a net transfer of angular momentum across the Earth's surface. This transfer of

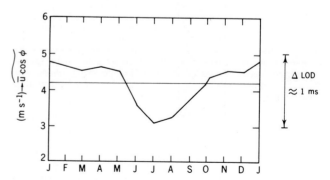

Figure A12 Normal annual cycle of the global atmospheric angular momentum M_r in terms of the mean zonal velocity, $u \cos\phi$, in units of m s^{-1} for the period 1963–73. The equivalent variation in LOD is shown on the right (in ms).

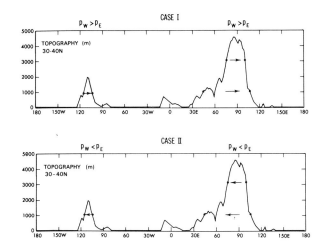

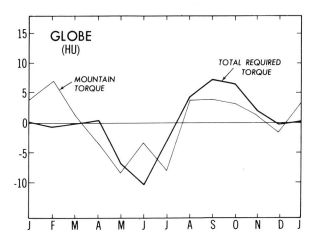

Figure A13 Zonal profile of the mountain heights averaged over the latitude belt 30–40°N. Schematic pictures of the atmosphere affecting the mountains for case I, $p_W > p_E$, and case II, $p_W < p_E$, are added.

Figure A14 Normal annual cycle of the required surface torque (thick line) and the large-scale mountain torque (thin line) in Hadley units (1 HU = 10^{18} kg m^2 s^{-2}) integrated over the globe. Only the annual and semiannual components are included in the required torque. The annual mean value of the mountain torque has been removed. (From Wahr and Oort, 1984).

angular momentum may occur by two processes, namely friction torques over oceans and land and pressure torques across mountains. We will assume here that the oceans rotate together with the land in solid rotation. When the Earth is speeding up in the northern spring there must be a net downward transfer of westerly angular momentum by an excess in eastward friction stress and/or by higher pressures on the west side than on the east side of the major mountain ranges (see Figure A13, case I). On the other hand, in northern fall there must be a net upward transfer across the Earth's surface and/or lower pressures on the west than on the east side of the mountain ranges (see Figure A13, case II). Taking the opposite view, i.e. the mountains decelerating or accelerating the atmospheric flow, would lead to a very implausible situation.

In principle the mountain torque is easy to obtain by measuring the surface pressure on both sides of the mountains (White, 1949). However, with a realistic topography of the mountains this becomes a major task and it has been accomplished so far only in a very coarse manner. Over time intervals from days to months the pressure torque seems to dominate the variability in the surface transfer (see e.g. the general circulation model results reported by Swinbank, 1985, and Boer, 1990). The data given in Figure A14 suggest that the large-scale pressure torques can explain an important fraction of the normal annual variation in the required surface torque. Thus it is apparently relatively easy for the atmosphere to transfer locally about 1–2% of the mass to affect a change, for example, of $\Delta p = 15$ mb across a mountain range of 1000 km length and 1 km height, which would be sufficient to lead to the observed seasonal changes in relative angular momentum of the atmosphere. Assuming that the large-scale mountain torque is perhaps the dominant term, we can investigate which mountain ranges are especially important. To do this we will study

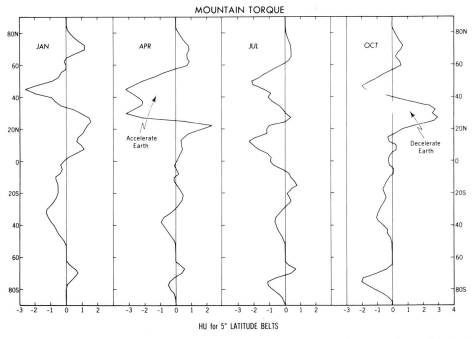

Figure A15 Meridional profiles of the mountain torque in Hadley units (1 HU = 10^{18} kg m^2 s^{-2}) integrated over 5° latitude belts for January, April, July and October 1963–73. Global integrals for the four months are −3.2, −9.8, −14.5 and −3.1 HU respectively, and for the year, −6.4 HU.

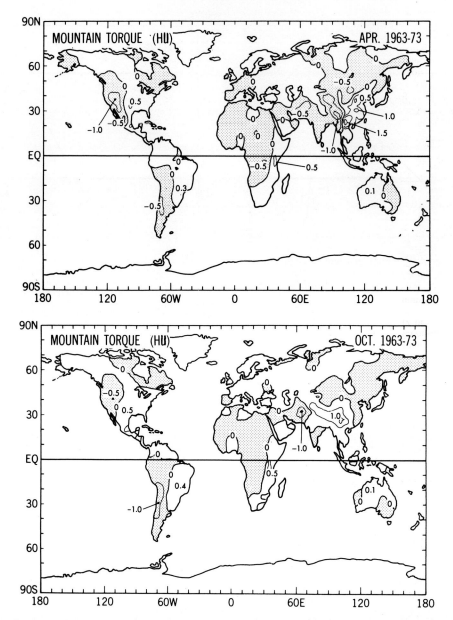

Figure A16 Global distribution of the mountain torque in Hadley units integrated over 5° latitude × 5° longitude boxes, for the transition months of April (top) and October (bottom) 1963–73. Positive values indicate source regions of westerly angular momentum for the atmosphere or, in other words, regions where the atmospheric winds exert a westward torque on the earth. Negative values shaded.

first some meridional profiles (Figure A15). A major imbalance in the northern spring seems to come from the northern mid-latitudes, suggesting the influence of the Rocky Mountains, and in northern fall from the latitude belt around 30°N, suggesting the influence of the Himalayas. To some extent, these ideas are confirmed by the global maps for April and October given in Figures A16a and A16b. These results hint at how and where the transfer may occur. In summary, we have found evidence of a clear link between the Earth and the atmosphere on the time scales up to 1 year in the direction atmosphere → Earth that could occur principally through mountain torques.

Angular momentum exchange between the tropics and mid-latitudes

Let us give a more detailed look at what happens within the various climatic belts of the Earth. Figure A17a gives a cross-section of the mean zonal circulation in the atmosphere. We notice near the surface the familiar westerlies in mid-latitudes of both hemispheres, and easterlies in low latitudes. Considering first the land and oceans to rotate like a solid sphere, friction and mountain drag will tend to slow down both the easterlies and westerlies so that the Earth's surface in the tropics will act as a source of westerly (eastward) momentum for the atmosphere and in the mid-latitudes as a sink (Figure A17b). The angular momentum is carried high into the atmosphere largely by the action of the mean meridional overturnings, as originally pointed out by Lorenz (1967). But at the level of the jet streams the angular momentum is transferred from low to mid-latitudes through tilted troughs and ridges (SW–NE in the northern hemisphere and SE–NW in the southern hemisphere) as first suggested on theoretical grounds by Jeffreys (1926) and later identified and documented using observed atmospheric winds by Starr (1948, 1968) and his coworkers (Figure A18). Of course, under steady conditions (no changes in LOD) the total surface torque integrated over the entire globe must vanish, requiring an approximate balance between the area covered by westerlies and easterlies.

The surface transfer in the various belts will again occur through friction and mountain torques. In the long-term (annual) mean the

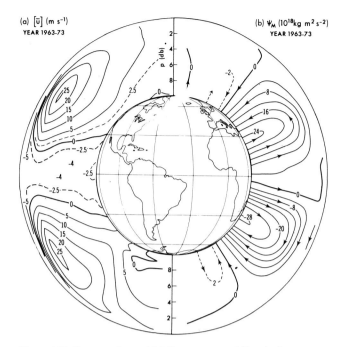
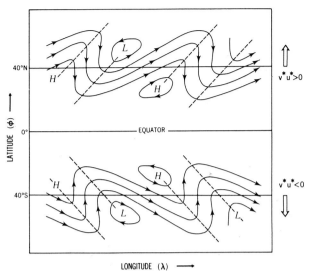

Figure A17 Cross sections of (a) the mean zonal flow in the atmosphere in units of m s^{-1}, and (b) the streamlines of relative angular momentum in Hadley units, both for annual mean conditions 1963–73 (after Oort and Peixoto, 1983). Note that the vertical scale of the atmosphere is grossly exaggerated.

Figure A18 Schematic diagram of the dominant mechanism of poleward transport of angular momentum by mid-latitude waves or eddies. In the northern hemisphere the waves are tilted from southwest to northeast, leading to a northward transport of relative angular momentum ($v^*u^* > 0$), and in the southern hemisphere from southeast to northwest, leading to a southward transport of relative angular momentum ($v^*u^* < 0$), where u and v are the eastward and northward components of the wind respectively, and the asterisks indicate departures from the zonal mean values.

friction term is now certainly an important factor. Since most of the Earth's surface is taken up by oceans where the surface winds are also strongest, much of the atmospheric angular momentum must be lost to the mid-latitude oceans and gained from the low-latitude oceans in the general cycle depicted in Figure A17b. How is the cycle closed at the surface? Does the return flow of angular momentum from mid-latitudes to the tropics take place within the oceans, or does it involve a transfer within the solid Earth? If we use typical values of the velocities in the oceans, any possible north–south transport turns out to be much too weak by one or two orders of magnitude. What appears to be happening instead is that the ocean wind stresses (see Fig. A19) lead to changes in sea level on the two sides of the oceans and to a transfer of the received angular momentum laterally to the continents through a 'continental' torque that is very similar to the mountain torque (Figure A20). This hypothesis is confirmed by independent estimates of the sea level from geopotential thickness calculations that are precisely of the right magnitude with ≈ 50–70 cm differences in sea level between the east and west sides of the continents (Figure A21).

Possible geological implications

In the preceding section we found that all torques are focused onto the continents with the tendency to rotate the continents clockwise in

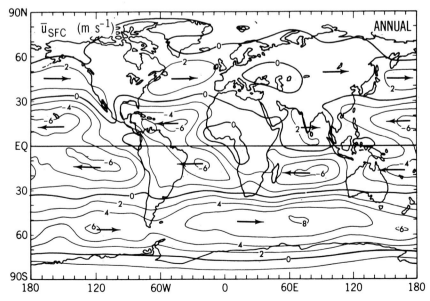

Figure A19 Global distribution of zonal wind component u at the surface for annual mean conditions in units of m s^{-1}.

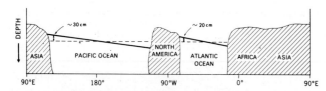

Figure A20 Schematic diagram of the east-west sloping of sea level along the 25° latitude circle. (After Oort, 1985).

the northern hemisphere and counterclockwise in the southern hemisphere. We should add that similar torques must have been working on the continents throughout the ages. Thus we can speculate about several hypothetical scenarios:

- the continents may actually move slowly under the influence of these torques in a continental drift pattern;
- the continents are so strongly anchored in the crust that they can withstand this continually acting torque; and
- the continents may release the stress intermittently by relative internal motions such as those along the San Andreas Fault.

The foregoing is speculative and tentative, but the possible implications of a real link between atmospheric, oceanic and geological processes are so important that they certainly justify further research.

Summary

There seems to be an undeniable link between the atmosphere and the solid Earth, two basic components of the climatic system, in the sense that the Earth follows the atmosphere, presumably mainly through east–west atmospheric pressure torques across the mountains and through east–west ocean pressure torques across the continental shelves. However, there is still considerable uncertainty and need for further research regarding whether differential motions may be set up in the Earth through regional pressure torque anomalies.

Abraham H. Oort

Reprinted in slightly modified form from: Oort, A.H. (1989) Angular momentum cycle in the atmosphere–ocean–solid Earth system. *Bull. Am. Meteor. Soc.* **70**, 1231–1241.

Bibliography

Boer, G.J. (1990) Earth–atmosphere exchange of angular momentum simulated in a general circulation model and implications for the length of day. *J. Geophys. Res.*, **95**(D5), 5511–31.
Courtillot, V., LeMouel, J.L., Ducruix, J. and Cazenave, A. (1982) Geomagnetic secular variation as a precursor of climatic change. *Nature*, **297**, 386–7.
Hasselmann, K. (1976) Stochastic climate models, Part I. Theory. *Tellus*, **28**, 473–85.
Jeffreys, H. (1926) On the dynamics of geostrophic winds. *Quart. J. Roy. Meteorol. Soc.*, **52**, 85–104.
Levitus, S. and Oort, A.H. (1977) Global analysis of oceanographic data. *Bull. Am. Meteorol. Soc.*, **58**, 1270–84.
Lorenz, E.N. (1967) The Nature and Theory of the General Circulation of the Atmosphere. WMO Publ. No. 218, T.P. 115, World Meteorological Organization, Geneva, Switzerland.
McCarthy, D.D. and Babcock, A.K. (1986) The length of day since 1656. *Phys. Earth Planet. Inter.*, **44**, 281–92.
Newton, C.W. (1971) Mountain torques in the global angular momentum balance. *J. Atmos. Sci.*, **28**, 623–8.
Oort, A.H. (1985) Balance conditions in the Earth's climate system. *Adv. Geophys.*, **28A**, 75–98.
Oort, A.H. (1989) Angular momentum cycle in the atmosphere–ocean–solid Earth system. *Bull. Am. Meteor. Soc.*, **70**, 1231–242.
Oort, A.H. and Peixoto, J.P. (1983) Global angular momentum and energy balance requirements from observations. *Adv. Geophys.*, **25**, 355–490.
Rosen, R.D. (1988) Recent developments in the study of the Earth–atmosphere angular momentum budget. Contribution No. 51, Institut d'Astronomie et de Géophysique, Université Catholique de Louvain, Louvain-la-Neuve, Belgium.
Rosen, R.D., Salstein, D.A. and Wood, T.M. (1991) Zonal contributions to global momentum variations on intraseasonal through interannual time scales. *J. Geophys. Res.*, **96**(D3), 5145–51.
Salstein, D.A. and Rosen, R.D. (1986) Earth rotation as a proxy for interannual variability in atmospheric circulation, 1860–present. *J. Climate Appl. Meteorol.*, **25**, 1870–7.
Starr, V.P. (1948) An essay on the general circulation of the Earth's atmosphere. *J. Meteor.*, **5**, 39–43.
Starr, V.P. (1953) Note concerning the nature of the large-scale eddies in the atmosphere. *Tellus*, **5**, 494–8.

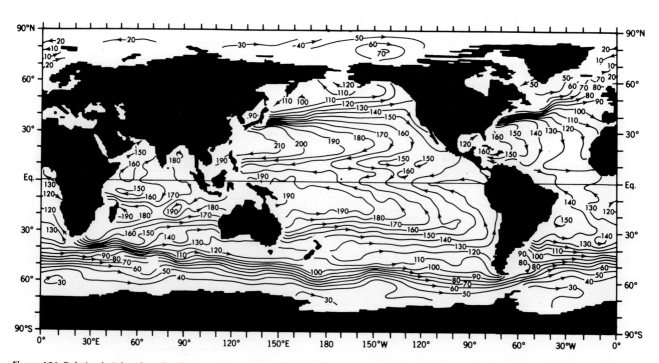

Figure A21 Relative height of sea level (in cm) for annual mean conditions as computed from global density data in the oceans, assuming a level of no motion at 1000 m depth. (From Levitus and Oort, 1977).

Starr, V.P. (1968) *Physics of Negative Viscosity Phenomena*. New York City: McGraw-Hill Book Co.

Swinbank, R. (1985) The global atmospheric angular momentum balance inferred from analyses made during the FGGE. *Quart. J. Roy. Meteorol. Soc.*, **111**, 977–92.

Taylor, H.A., Jr, Mayr, H.G. and Kramer, L. (1985) Contributions of high-altitude winds and atmospheric moment of inertia to the atmospheric angular momentum–earth rotation relationship. *J. Geophys. Res.*, **90**, 2889–96.

Wahr, J.M. and Oort, A.H. (1984) Friction and mountain-torque estimates from global atmospheric data. *J. Atmos. Sci.*, **41**, 190–204.

White, R.M. (1949) The role of the mountains in the angular momentum balance of the atmosphere. *J. Meteorol.*, **6**, 353–5.

Cross references

Chandler wobble
Earth: rotational history
Planetary rotation

ANTARCTIC METEORITES

Meteorites are rocks from space that have fallen on Earth. They are fragments of various bodies in the solar system which were sent toward Earth by impacts on asteroids, the Moon, and the planets. They are ancient rocks which provide evidence for the origin and evolution of the solar system.

More meteorite fragments have been recovered from Antarctica than from the rest of the world combined. As of 1993, over 15 000 meteorites have been collected in Antarctica compared to around 2600 in the rest of the world. Presumably, the probability of a meteorite falling on Antarctica is no greater than that of it falling on any other part of the world. Several factors contribute to making it easier to find meteorites in Antarctica. The first is that it is easier to see a dark meteorite, even a small one, on Antarctic ice than on bare ground or vegetation. The next is that the movement of Antarctic ice appears to concentrate meteorites and increase the number of specimens found. Meteorites fall on the ice and become entrained in the moving icefield. When the ice comes up against a barrier such as a mountain, ice movement is stopped and the ice is gradually ablated away. This exposes meteorites which fell at different times and locations on the ice. A third factor is that Antarctic meteorites have been on Earth for much longer times than those found in other parts of the world. It is not that there were no meteorites elsewhere long ago, but that weathering in warmer climates destroyed old meteorites before they could be found.

Futhermore, the large number of Antarctic meteorite recoveries is misleading in that several meteorite fragments may represent the same fall. Many meteorites break up during atmospheric entry and fall as showers of numerous fragments. When this happens at most places in the world all fragments of the same fall are given the same name and counted as one meteorite. Because there are so many meteorites collected in close proximity in Antarctica, it is difficult to tell which ones are part of the same meteorite, so each specimen is given a separate name. Scientists have attempted to group together identical meteorites that were collected from the same area. This pairing, if it could be done for all Antarctic meteorites, would reduce the number of meteorites by a factor of two to ten (Graham and Annexstad, 1989; Scott 1989; Ikeda and Kimura, 1992). Even with this reduction, the number of Antarctic meteorites is comparable to the number of meteorites from the rest of the world. Antarctica truly is a wonderful place to collect meteorites.

Antarctic meteorite programs

Of major importance in the large number of Antarctic meteorite recoveries are the systematic meteorite searches that have been conducted over the last 20 years by Japanese, American and European teams. Similar systematic searches of the Australian desert have also yielded significant numbers of meteorites (Bevan and Binns, 1989). Prior to 1969 only four meteorites had been found in Antarctica. The first Antarctic meteorite was collected on ice in Adelie Land by Sir Douglas Mawson's Australian expedition to search for the south magnetic pole in 1912. Meteorites were found in three other locations around Antarctica (Table A2 and Figure A22) before a Japanese geologic field team discovered a concentration of nine meteorites of four distinct types on ice in the Yamato mountains in 1969.

This apparent concentration of meteorites on ice led to systematic meteorite searches in Antarctica. The Japanese Antarctic Research Expeditions (JARE) led by K. Yanai of the National Institute for Polar Research (NIPR) recovered 985 meteorites in the Yamato mountains between 1973 and 1975. The US Antarctic Search for Meteorites (ANSMET) under W. Cassidy was funded by NSF in 1976 and Japanese–American teams conducted joint meteorite searches in the Allan Hills from 1976 to 1978. Since 1979 Japanese and American teams have conducted independent searches almost yearly from their national bases on opposite sides of the continent (Figure A23). Europeans also became involved in Antarctic meteorite collection. German teams collected meteorites in the Frontier Mountains in 1984 and Allan Hills in 1988. In 1990 the European Community funded EUROMET (European Meteorite Consortium) to collect and curate Antarctic meteorites. EUROMET's first expedition was to the Frontier Mountains in 1990. EUROMET also has a program of collecting micrometeorites from melted ice. As of 1993, a total of over 15 000 meteorites have been collected (Table A1) in various locations (Figure A22) around Antarctica (Cassidy *et al.*, 1992).

Meteorites are traditionally named after the nearest post office, town or country. Antarctic meteorites present a problem for this scheme because there are so many meteorites and so few human settlements in Antarctica. A system of alphanumeric names was developed in which the location of a local geologic feature is followed by a 3–7 digit number representing the year and sample number. Thus LEW85320, the meteorite shown in Figure A23, was collected at Lewis Cliff in 1985.

The frozen Antarctic environment preserves meteorites far better than temperate climates and is less contaminated by industrial pollution. In order to preserve the meteorites in as pristine condition as possible, great care is taken in their collection and curation. For example, US meteorites are collected with clean tools, packed in clean teflon bags with metal labels, and shipped frozen back to the US. All three international groups set up clean-room meteorite curation facilities: JARE at NIPR; ANSMET at NASA Johnson Space Center and the Smithsonian Natural History Museum; and EUROMET at the Open University, UK. All these facilities are involved in the classification, distribution and preservation of meteorites.

Distribution of meteorite types

Meteoriticists have known for some time (i.e. Mason, 1962) that the distribution of meteorite types among finds is very different from that among observed falls (Table A3). Iron meteorites are much more abundant among finds than falls because they are very dense and easier to distinguish from terrestrial rocks than are stony meteorites. Comparison of the distribution of meteorite types between Antarctic and non-Antarctic meteorites provides new insights (Dennison, Lingner and Lipschutz, 1986). All Antarctic meteorites are finds, and it is true that more irons are found by chance by field parties doing other types of Antarctic research (Cassidy *et al.*, 1992). However, Table A3 shows that the overall distribution of meteorite types among Antarctic meteorites is much more similar to that of falls than of finds. Ordinary chondrites are slightly more abundant among Antarctic meteorites, and achondrites, stony irons and irons are less abundant than among falls. A problem with this comparison is the uncertainty in the number of individual Antarctic meteorites. It is relatively simple to evaluate pairing among the less common types, but difficult for ordinary chondrites. Pairing ratios for the less common meteorites range from 1 to 8, but are typically 2–3. These meteorites are frequently small and the pairing ratio increases with sample size, so a higher pairing ratio is appropriate for larger chondrites. Table A3 gives a possible distribution of Antarctic meteorites that adjusts for pairing. It includes the known pairing for the five less common meteorite types and estimates the pairing for ordinary chondrites based on a moderate pairing ratio of five chondrite specimens per individual meteorite. This pairing-corrected distribution of Antarctic meteorites is remarkably similar to that of observed falls.

Another way to get around the pairing issue is to compare the total masses of meteorite types rather than the numbers of meteorites.

Table A2 Locations and numbers of Antarctic meteorites

Geographic name	Abbreviation	Number	Country	Latitude	Longitude
Adelie Land	(1912)	1	Australia	67 11 S	142 23 E
Lazarev	(1961)	2	Russia	71 57 S	11 30 E
Neptune Mountains	(1964)	1	USA	83 15 S	55 00 W
Thiel Mountains	(1962)	2	USA	85 15 S	91 00 W
Allan Hills	ALH	1753	USA[a]	76 43 S	159 40 E
Asuka	A	2479	Japan	75 50 S	24 30 E
Bates Nunataks	BTN	4	USA	80 15 S	153 30 E
Beckett Nunatak	BEC	2	USA	76 02 S	160 11 E
Belgica	B	5	Japan	72 35 S	31 15 E
Bowden Neve	BOW	1	USA	83 30 S	165 00 E
David Glacier	DAV	9	USA	75 19 S	162 00 E
Derrick Peak	DRP	25	N. Zealand[b]	80 04 S	156 23 E
Dominion Range	DOM	11	USA	85 20 S	166 30 E
Elephant Moraine	EET	1785	USA	76 11 S	157 10 E
Frontier Mountain	FRO	284	EUROMET	75 59 S	160 20 E
Geologists Range	GEO	2	USA	82 30 S	155 30 E
Grosvenor Mountains	GRO	19	USA	85 40 S	175 00 E
Inland Forts	ILD	1	USA	77 38 S	161 00 E
LaPaz Icefield	LAP	3	USA	86 22 S	70 00 W
Lewis Cliff	LEW	1751	USA	84 17 S	161 05 E
MacAlpine Hills	MAC	128	USA	84 13 S	160 30 E
MacKay Glacier	MCY	4	USA	76 58 S	162 00 E
Meteorite Hills	MET	28	USA	79 41 S	155 45 E
Mount Baldr	MBR	2	USA	77 35 S	160 34 E
Mount Howe	HOW	4	USA	87 22 S	149 30 W
Mount Wegener		1	EUROMET	80 42 S	23 35 W
Miller Range	MIL	1	USA	83 15 S	157 00 E
Outpost Nunatak	OTT	1	USA	75 50 S	158 12 E
Patuxent Range	PAT	52	USA	84 43 S	64 30 W
Pecora Escarpment	PCA	519	USA	85 38 S	68 42 W
Purgatory Peak	PGP	1	USA	77 20 S	162 18 E
Queen Alexandra Range	QUE	90	USA	84 00 S	168 00 E
Reckling Peak	RKP	135	USA	76 16 S	159 15 E
Stewart Hills	STE	1	USA	84 12 S	86 00 W
Taylor Glacier	TYR	1	USA	77 44 S	162 10 E
Thiel Mountains	TIL	41	USA	85 15 S	91 00 W
Wisconsin Range	WIS	33	USA	85 45 S	125 00 W
Yamato	Y	5940	Japan	71 30 S	35 40 E

The meteorites from the first four locations were found before formal meteorite searches were organized. The year of collection is listed instead of an abbreviation. The remaining meteorites were collected between 1969 and 1993 mostly as part of formal programs by Japan (JARE), USA (ANSMET) and Europe (EUROMET).
[a]The USA has most of Allan Hills samples (1555 meteorites); Japan has splits of the 578 meteorites collected with the USA in 1976–1978; EUROMET has 198 meteorites collected in 1988.
[b]New Zealand collected all 25 Derrick Peak meteorites and donated samples 1–9 to the USA and Japan. All 25 samples are the same iron meteorite.

This also shows that the distribution of Antarctic meteorites is very similar to that of observed falls (Cassidy and Harvey, 1991; Huss, 1991). Thus the systematic meteorite searches in the Antarctic recover a more representative suite of meteorites than the randomly collected world finds. As more meteorites are collected from Australia and other desert areas (Bevan and Binns, 1989) it will be interesting to see if the distribution of desert meteorites also resembles that of observed falls.

Rare meteorites

Although the overall distribution of Antarctic meteorite types does not differ much from that of observed falls, the abundances of rare and unusual types is distinctly higher (Koeberl and Cassidy, 1991). Antarctic iron meteorites include a very large proportion (39% compared to 7% for finds and 1% for falls) of anomalous or ungrouped specimens (Clarke, 1986; Wasson, 1990). Stony and stony iron meteorites have 15–25% rare meteorites compared to 1% for both finds and falls. Most of the previously known unique or anomalous meteorites have one or more counterparts in the Antarctic collections. These include achondrites similar to Acapulco, Angra dos Reis, Brachina, Shergotty and Winona; stony irons like Lodran; and chondrites similar to Carlisle Lakes, Karoonda, Renazzo and Shaw. There are also chondrites which have no counterparts in the world collection, ALH85085 and LEW85332. Recent finds of desert meteorites have also added to the ranks of rare meteorites. Most of these rare meteorites are small (< 30 g) and are unlikely to be found in other environments.

Surely the most surprising discovery was the identification of lunar and Martian meteorites in the Antarctic collections. In 1982 the first meteorite from the Moon, ALHA81005, was discovered on Antarctic ice (Marvin, 1983). Earlier studies of lunar samples made it possible for the first time to identify the parent body of a single meteorite. Since then ten more lunar meteorites were found in Antarctica and, in 1991, the first non-Antarctic lunar meteorite was found in the Australian desert (see **Lunar meteorites**).

The existence of lunar meteorites fueled speculation about the SNC (Shergottite, Nahklite, Chassignite) meteorites (q.v.), igneous meteorites which have very young crystallization ages. Four of the ten SNC meteorites were collected in Antarctica. One of these, EETA79001 (Figure A24) has glass veins and bubbles containing noble gases with the isotopic characteristics of the Martian atmosphere (Bogard, Nyquist and Johnson, 1984) as measured by the Viking spacecraft (see **SNC** meteorite). The Antarctic collections thus

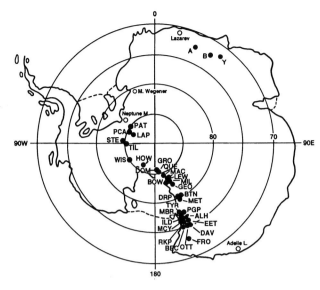

Figure A22 Antarctic meteorite locations. Locations of samples found since 1969 by meteorite collection teams are abbreviated. Sites of pre-1969 meteorites are spelled out except Thiel Mountains (TIL), which is also a site for program collection.

Table A3 Distributions of meteorite types. Percentages of meteorite types are compared for US Antarctic meteorites and non-Antarctic falls and finds (Graham *et al.*, 1985 data).

Meteorite type	Non-Antarctic falls	Non-Antarctic finds	US Antarctic finds	US Antarctic $P=5$
Ordinary chondrite	79.5	49.4	90.9	81.5
Carbonaceous chondrite	4.2	1.9	3.4	4.9
Enstatite chondrite	1.6	0.7	1.1	1.5
Achondrite	8.3	1.4	3.3	8.3
Stony iron	1.2	3.6	0.3	0.7
Iron	5.1	43.0	0.8	2.5
Total meteorites	830	1588	5537	1036

contain meteorites from both the Moon and Mars as well as the more common meteorites from asteroids.

Continued systematic searches for meteorites in Antarctica are expected to yield many more specimens, and especially more rare and unusual meteorites, which are so valuable for the study of the origin and evolution of the solar system.

Marilyn M. Lindstrom

Figure A23 Collection of LEW85320. This 110 kg ordinary chondrite is one of the largest Antarctic meteorites.

Figure A24 Martian meteorite EETA79001. The glass veins and bubbles contain gases similar to those in the Martian atmosphere.

Bibliography

Bevan, A.W.R. and Binns, R.A. (1989) Meteorites from the Nullarbor Plain: I. A review of past recoveries and a procedure for naming new finds. *Meteoritics*, **124**, 127–33.

Bogard, D.D., Nyquist, L.E. and Johnson, P. (1984) Noble gas contents of shergottites and implications for the Martian origin of SNC meteorites. *Geochim. Cosmochim. Acta*, **48**, 1723–39.

Cassidy, W.A. and Harvey, R.P. (1991) Are there real differences between Antarctic finds and modern falls meteorites. *Geochim. Cosmochim. Acta*, **55**, 99–104.

Cassidy, W., Harvey, R., Schutt, J. et al. (1992) The meteorite collection sites of Antarctica. *Meteoritics*, **27**, 490–525.

Clarke, R.S. (1986) Antarctic iron meteorites: An unexpectedly high proportion of falls of unusual interest, in *International Workshop on Antarctic Meteorites* (eds. J.O. Annexstad, L. Schultz and H. Wanke). Houston: Lunar and Planetary Institute, pp. 28–9.

Dennison, J., Lingner, D. and Lipschutz, M. (1986) Antarctic and non-Antarctic meteorites form different populations. *Nature*, **319**, 390–3.

Graham, A.L. and Annexstad, J.O. (1989) Antarctic meteorites. *Antarctic Sci*, **1**, 3–14.

Graham, A.L., Bevan, A.W.R. and Hutchison, R. (1985) *Catalogue of Meteorites*, 4th eds. London: British Museum (Natural History).

Huss, G. (1991) Meteorite mass distributions and differences between Antarctic and non-Antarctic meteorites. *Geochim. Cosmochim. Acta*, **55**, 105–11.

Ikeda, Y. and Kimura, M. (1992) Mass distribution of Antarctic ordinary chondrites and the estimation of the fall-to-specimen ratios. *Meteoritics*, **27**, 435–41.

Koeberl, C. and Cassidy, W. (1991) Differences between Antarctic and non-Antarctic meteorites. An assessment. *Geochim. Cosmochim. Acta*, **55**, 3–18.

Marvin, U. (1983) The discovery and initial characterization of Allan Hills 81005: the first lunar meteorite. *Geophys. Res. Lett.*, **10**, 775–8.

Mason, B. (1962) *Meteorites*. New York: John Wiley and Sons.

Scott, E.R.D. (1989) Pairing of meteorites from Victoria Land and the Thiel Mountains, Antarctica. *Smithsonian Contrib. Earth Sci.*, **28**, 103–11.

Wasson, J. (1990) Ungrouped iron meteorites in Antarctica: origin of anomalously high abundance. *Science*, **249**, 900–2.

Cross references

Lunar meteorites
Meteorite
SNC meteorites

ANTONIADI, EUGENIOS (1870–1944)

Antoniadi had a somewhat mixed career. He was of Greek origin, but was born at Constantinople, in Turkey, on 10 March 1870. In

1893 he went to France as assistant to Camile Flammarion at the Juvisy Observatory, and in 1909 became assistant astronomer at the Observatoire de Meudon, near Paris. He took French nationality and remained in France for the rest of his life.

Antoniadi was probably the best planetary observer of his time, and made full use of the 33-inch (83 cm) refractor at Meudon. He paid close attention to Mercury and Mars, and drew maps of their surfaces. His map of Mercury was inaccurate, though this was not Antoniadi's fault; he believed the rotation period to be synchronous with the revolution period (88 days), a conclusion now known to be wrong. His charts of Mars were excellent, and his nomenclature is still used, albeit in modified form; he was a strong opponent of the 'canal network' drawn by Lowell (q.v.) and others. For many years he was Director of the Mars Section of the British Astronomical Association. Antoniadi was also noted for his historical researches, and in 1934 published a book about Egyptian astronomy. He died in Occupied France on 10 February 1944.

Patrick Moore

Bibliography

Abetti, G. (1970) Antoniadi, Engène M. *Dict. Sci. Biogr.*, vol 1, p. 172.
Antoniadi, E.M. (1930) *La Planète Mars*. Paris: Gauthier-Villars. English translation by P. Moore and K. Reid, Shaldon, 1975.
Antoniadi, E.M. (1934) *La Planète Mercure et la Rotation des Satellites*. Paris: Gauthier-Vilalrs English translation by P. Moore and K. Reid, Shaldon, 1974.
Antoniadi, E.M. (1934) *L'Astronomie Egyptienne*. Paris: Gauthier-Villars.

APHELION

The aphelion of a circumsolar planet or periodic comet is the point of its orbit farthest from the Sun (in Figure A25 it is point A, ADPF being the orbit and S the Sun). It is in contrast with the perihelion (q.v.) or point closest to the Sun (P in Figure A25). These two terms were first introduced by Johannes Kepler in his *Mysterium Cosmographicum* of 1596. He formed them from the Greek roots $\alpha\phi = \alpha\pi o$, from, and $\pi\epsilon\rho\iota$, near + $\eta\lambda\iota o s$, Sun, thus constructing them in analogy with the terms *apogee* and *perigee* used by Ptolemy to denote the points on an epicycle or eccentric circle farthest from and closest to the Earth's center. When Copernicus set the Earth in motion about the Sun, he employed the Ptolemaic terms, contrary to their original meanings, to denote the points on a planetary orbit respectively farthest from and nearest to the mean Sun, or center of the Earth's circular orbit.

Kepler was the first heliocentrist to take as primary reference point the true and apparent Sun, and to stipulate that the line of apsides (through aphelion and perihelion) and line of nodes (through the planes of the orbit and the ecliptic) pass through the center of the Sun's body. With his discovery that the orbit was elliptical the aphelion and perihelion became the vertices of the major axis in the ellipse. Through his *Epitome Astronomiae Copernicanae* (1618) the new terms aphelion and perihelion became current among 17th century astronomers, most often in their latinate forms, *aphelium* and *perihelium*.

Throughout the 17th and 18th centuries, the aphelion served as the zero-point for reckoning the anomaly in a planet's motion. True anomaly v was the angle between the planet's radius vector and the line through Sun and aphelion: $v = \sphericalangle ASp$. Eccentric anomaly was an angle singled out by Kepler at the center of the circle circumscribing the elliptical orbit: $E = \sphericalangle ACq$ in Figure A25. The two angles v and E were related by the equation.

$$\tan \frac{E}{2} = \tan \frac{v}{2} \sqrt{\frac{1+e}{1-e}}$$

where e is the eccentricity or SC/CA. The mean anomaly, finally, was an angle proportional to the time, completing 360° when the planet returned to aphelion; it was related to the eccentric anomaly by 'Kepler's equation', which embodies Kepler's second or areal law:

$$M = E + e \sin E.$$

Assuming this second equation solved by approximation for M in terms of E, one could use the first equation to find the corresponding v, or orbital position of the planet for a given time interval since its passage through aphelion.

Nicolas-Louis Lacaille (1713–1762) in his *Leçons d'astronomie géométrique et physique* of 1746 (with later editions in 1755, 1761 and 1780) proposed that perihelion be substituted for aphelion as the zero-point of anomaly, in order that the rules of calculation used in establishing planetary orbits could be applied as well to cometary orbits, in which the aphelia are not observationally accessible. This proposal was adopted by the French Bureau des Longitudes shortly after its inception in 1795. Laplace followed it throughout his *Mécanique Céleste*, as did all subsequent writers on planetary astronomy. The formulas of the preceding paragraph are converted to the new zero-point simply by substituting $-e$ for $+e$.

The lines of apsides, and hence the aphelia and perihelia, of all the circumsolar planets slowly advance (move eastward), owing to mutual planetary attractions.

Curtis Wilson

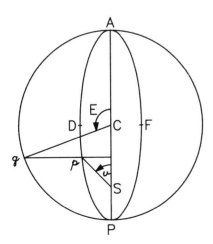

Figure A25 Aphelion and perihelion in an elliptical orbit. ADPF = orbit, S = Sun at focus, C = center, A = aphelion, P = perihelion, SC/CP = eccentricity (shown with large value appropriate to a comet). If p = place of planet or comet, then $\sphericalangle ASp = v$ = true anomaly from aphelion, and with AqP a circumscribing circle and qp perpendicular to ASP, $\sphericalangle ACq = E$ = eccentric anomaly from aphelion.

Bibliography

Delambre, J.B.J. (1827) *Histoire de l'astronomie au dix-huitième siècle*.
Delambre, J.B.J. (1806) *Tables du soleil*. In *Tables astronomiques publiée par le Bureau des Longitudes de France*, Première Partie.
Kepler, (1621) *Mysterium Cosmographicum*, 2d ed, footnote to Chap. 15. (Modern edition, 1981, transl. A.M. Duncan, p. 163. New York: Abaris Books).
Laplace, P.S. (1796) *Mécanique Céleste*, Vol.I, Book II, Chapter iii.
Moulton, F.R. (1914) *An Introduction to Celestial Mechanics*, pp. 352–55. New York: Macmillan (Dover edition, 1970).

APOLLO MISSIONS

The objective of the Apollo program was to accomplish the first manned exploration of Earth's Moon. The program was mounted by the United States of America during the 1960s and early 1970s, in a context of competition between the United States and the Soviet Union (the 'space race'). At that time the Soviet Union was actively pursuing a program of lunar exploration by unmanned probes (see Luna missions; Zond missions). The Apollo Program cost $24 billion and provided a significant stimulus to technological innovation in many areas. Twelve astronauts landed on the lunar surface and returned safely to Earth (Table A.4). This article can touch only briefly on the development of the program and the planetary science performed. Hallion and Crouch (1979) provide a more complete history with many references.

Table A4 Manned flights to the Moon

Mission	Date of start of mission	Participating astronauts[a]
Apollo 8	21 December 1968	F. Borman, J.A. Lovell, W.A. Anders
Apollo 10	18 May 1969	T.P. Stafford, J.W. Young, E.A. Cernan
Apollo 11	16 July 1969	N.A. Armstrong, M. Collins, E.E. Aldrin
Apollo 12	14 November 1969	Ch. Conrad, R.F. Gordon, A.L. Bean
Apollo 13	11 April 1970	J.A. Lovell, J.L. Swigert, F.W. Haise
Apollo 14	31 January 1971	A.B. Shepard, S.A. Roosa, E.D. Mitchell
Apollo 15	26 July 1971	D.R. Scott, A.M. Worden, J.B. Irwin
Apollo 16	16 April 1972	J.W. Young, T.K. Mattingly, Ch. M. Duke
Apollo 17	6 December 1972	E.A. Cernan, R.E. Evans, H.H. Schmitt

[a]The first name is that of the commander of the respective mission and the second that of the command module pilot (i.e. orbiting astronaut).
After Kopal, 1979, Table 5.

Table A5 Manned landings on the Moon

Mooncraft	Place of landing or region	Longitude	Latitude	Date of landing
Apollo 11	Mare Tranquillitatis	23°29′24″E	0°40′12″N	20 July 1969
Apollo 12	Oceanus Procellarum	23°20′23″W	2°27′0″S	18 November 1969
Apollo 14	Fra Mauro	17°27′55″W	3°40′24″S	5 February 1971
Apollo 15	Mare Imbrium	3°39′10″E	26°6′4″N	30 July 1971
Apollo 16	Descartes	15°30′47″E	8°59′34″S	21 April 1972
Apollo 17	Taurus–Littrow	30°45′26″E	20°9′41″N	11 December 1972

After Kopal, 1979, Table 4.

Early missions in the series were focused on stepwise development of capabilities and subsystems needed. The tragic deaths of astronauts Grissom, White and Chaffee as a result of a fire on the launch pad in 1967 took place during this phase of the program. Apollo 8 was the first mission to visit (and orbit) the Moon, in December 1968. The first landing was accomplished by Apollo 11 on 20 July 1969, and it was followed 4 months later by Apollo 12. On Apollo 13 an explosion in the fuel cell very nearly incapacitated the spacecraft, but 4 days later after a loop around the Moon, the crew was able to bring it home safely. Apollo missions 15, 16 and 17 included vehicles for travel on the lunar surface. The final mission to the Moon (Apollo 17) was launched on 6 December 1972. Four subsequent Skylab missions and the joint American–Soviet Apollo–Soyuz mission utilized Apollo equipment.

In a typical mission the travel time to the Moon was about 65–70 h. Once in orbit, the lunar landing module would detach from the orbiter; the pilot would remain in orbit, while his two companions would journey to the surface. Some time later the lander would return to the orbiter, uniting the crew, and the mission would return to Earth and (via parachute) splash down in a predetermined patch of ocean, where the astronauts would be recovered by the US Navy.

Apollo planetary science

A variety of observations and experiments were carried out during the six manned landings on the Moon (Table A5). Analysis of the data and samples obtained continues today. The *Lunar Sourcebook* (Heiken, Vaniman and French, 1991) provides a comprehensive overview of scientific investigations.

The astronauts took many photographs and collected a total of nearly 382 kg of lunar soil and rocks (Figure A26). In all more than 2000 samples were obtained from nine sites. Dating of these materials has shed light on the timing of important events in the solar system and on the history of the Moon itself (see Earth–Moon system: origin; Moon: origin; Plate 14). The lunar maria are composed of basaltic materials similar to terrestrial lavas, while the highlands are made of materials similar to those of the Earth's mantle. In addition to the many samples of breccias, regolith, glasses and rocks, Apollos 15, 16 and 17 recovered 'soil' cores up to 3 m in length (see Moon: geology). Heiken, Vaniman and French (1991) review the geology of the Apollo landing sites.

Each mission carried a set of scientific instruments that was emplaced on the lunar surface and left to continue acquiring data after the astronauts had left. Called Apollo Lunar Surface Experiment Packages (ALSEP), each set was slightly different. Altogether they included magnetometers, ion detectors to measure the solar wind and the lunar atmosphere, subsurface heat-flow detectors and seismometers. The magnetometers detected a weak field (about 300 gamma; for comparison, the Earth's field is about 50 000 gamma). (See Moon: magnetism and interior).

The astronauts also deployed an array of four seismometers, which detected many deep moonquakes and a few energetic shallow events (see Moon: seismicity; Plate 14). The data from these instruments helps to resolve features of the interior structure of the Moon. The seismometers operated for a period of about 7 years.

The astronauts left on the surface a number of corner-cube reflectors, which return laser light beams precisely back along the direction from which they came. Earth-based telescopes send light pulses to them and measure the elapsed time to acquisition of the reflected pulses. With the precisely known velocity of light, the distance between the telescope and the reflector is determined with a precision of a few centimeters. These experiments provide information on the dynamics of the Earth–Moon system (q.v.), including the recession of the Moon from the Earth (about 3 cm per year) and the temporal variations in the distance and rotation rate of the Moon.

The orbiters of later Apollo missions included a number of other instruments, such as a laser altimeter, x-ray and gamma-ray spectrometers, and a mass spectrometer to detect constituents of the atmosphere. These missions also deployed 'subsatellites', with instruments to measure solar wind structure and the lunar magnetic field.

Some of the major findings of the Apollo missions are summarized in Courtright (1975) and Lowman (1991).

The Apollo program represents nothing less than a watershed in human history; individuals of our species *homo sapiens* have now traveled beyond the confines of the Earth, to land upon and explore our nearest neighbor the Moon.

James H. Shirley

Bibliography

Courtright, E.M. (ed.) (1975) *Apollo Expeditions to the Moon*, Washington, DC.: NASASP-350.
Hallion, R.P. and Crouch, T.D. (eds) (1979) *Apollo: Ten Years Since Tranquillity Base*. Washington: Smithsonian Institution Press.

Figure A26 Collection of a tube of regolith material in the ejecta zone outside Van Serg crater, during the Apollo 17 mission. (AS-17–143–21837.)

Heiken, G., Vaniman, D. and French, B.M. (1991) *Lunar Sourcebook*. New York: Cambridge University Press.

Lowman, P.D., Jr (1991) Moon, space missions, in *The Astronomy and Astrophysics Encyclopedia*. New York: Van Nostrand Reinhold.

Tauber, G.E. (1979) *Man and the Cosmos*. New York: Crown Publishers.

Kopal, Z. (1979) *The Realm of the Terrestrial Planets*. New York: Halstead Press.

Cross references

Lunar meteorites
Moon (Earth's Moon)
Moon: geology

APOLLO OBJECT

A subset of the population of Earth-crossing asteroids, distinguished from the Aten objects (q.v.) and the Amor objects (q.v.) by orbital characteristics. The three groups are collectively termed near-Earth objects (q.v.). The Apollo objects have orbits with perihelia closer to the Sun than the Earth's aphelion; that is, they spend part of the time close to and inside the orbit of the Earth. They are named for a member of the class, Asteroid 1862 Apollo. Apollo is an unusual asteroid of taxonomic class Q (see Asteroid: compositional structure and taxonomy). Yeomans (1991) has suggested that it may in fact be an extinct comet nucleus. No clear compositional trends are evident among the near-Earth objects; they appear to have originated in multiple source regions. Since the projected dynamical lifetimes of objects in near-Earth space are, at about 10^8 years, much shorter than the age of the solar system, they must be replenished from some source. The main asteroid belt is the most likely source region (Binzel et al., 1992).

The number of Apollo objects is not known. McFadden, Tholen and Veeder (1989) quote an earlier estimate of the Apollo population by Shoemaker of 700 ± 300 objects, but that number will probably increase in coming years as current programs to observe these objects are implemented (see Spacewatch). The interest in observing and categorizing these objects derives from their potential to generate catastrophic impacts on the Earth. The 1.2 km diameter 'Meteor Crater', in Arizona, was excavated some 50 000 years ago by an object only about 50 m in diameter; some near-Earth objects are more than 20 km in diameter.

James H. Shirley

Bibliography

Binzel, R.P., Xu, S., Bus, S.J. and Bowell, E. (1992) Origins for the near-Earth asteroids. *Science*, **257**, 779–81.

McFadden, L.-A., Tholen, D.J. and Veeder, G.J. (1989) Physical properties of Aten, Apollo, and Amor asteroids in *Asteroids II* (eds R.P. Binzel, T. Gehrels and M. Matthews), Tucson: University of Arizona. pp. 442–67.

Yeomans, D.K. (1991) A comet among the near-Earth asteroids. *Astrophys. J.*, **101**, 1920–8.

Cross references

Amor objects
Asteroid
Aten object
Near-Earth object
Spacewatch

APSIS, APSIDES

Apsis is a term in celestial mechanics meaning 'one of the two points in the elliptic orbit of a planetary body at which it is respectively at its greatest and least distance from the body about which it revolves' (Oxford English Dictionary). The apsides is the plural of apsis and the 'line of apsides' is the major axis that in graphical representations joins those two points (Krogdahl, 1962). (Apsidal is an adjective form.)

Derived from Greek and Latin, apsis, meaning an arch or loop, in its anglicized form apse, refers to the architectural recess form in temples going back to Roman times, or later to the east end of Christian churches. (It is used in distinction to the 'basis' or 'base'.) In the 18th century the term apsis and apside (the Italian form) were adopted in astronomy in a loose way for identifying the major axis of a planetary orbit.

A series of related and analogous terms are applied to the two apices of any orbital major axis, with greatest and least differences in radius vectors, as follows:

- Earth–Moon: apogee; perigee (first used by Ptolemy); when coinciding with syzygy: exogee; proxigee (Wood, 1986).
- Sun–planets: aphelion; perihelion (terms introduced by Kepler).
- System barycenter–Sun: apobac; peribac (Fairbridge and Sanders, 1987).

In the orbital trajectories of artificial satellites, two comparable terms have been adopted which can be applied to any orbit around any celestial body, namely apoapsis and periapsis.

In the case of the Earth–Moon separation, the mean distance (body center to body center) is 384 318 km, but at extreme apogee (exogee) may exceed 406 691 km, and at extreme perigee (proxigee) may be less than 355 800 km (see Synergetic tidal forces). The Earth–Moon line of apsides displays a counterclockwise shift, or progression, over the course of an apsides cycle (8.849 years, a mean value, in tropical years; 3232 days). During this progression the line of apsides rotates 360° in the lunar orbit at a mean rate of 0.111404° per day (about 3° per anomalistic month). This shift was a fact known to Hipparchus in the 2nd century BC. It was Newton (in *Principia*, Book I, Prop. 66, th. 26, Con VII) who explained it in terms of the Sun's gravitational attraction. (This apsides progression is opposite in sense to the 18.6134 years lunar nodal revolution, a clockwise migration, or regression.)

At such times in Earth history as the date of perigee and that of perihelion very nearly coincide, and especially when augmented by other lunar periodicities (syzygy, etc.), the amplitude of ocean tides is maximized. During the 8.849 years of the apsides cycle, the Earth at the same time is moving in its own orbit, and so it is only at every third completion that the line of apsides returns to an approximate starting alignment with the Sun, a 26.54 years periodicity (Pettersson, 1930). The last peak was January 4 1912 AD (Wood, 1986). The principal recurrence interval is 558.12 year, approximately 21 × 26.547 years. At the co-alignment the tide-raising potential can be up to 40% higher than during the long intervals when the line of apsides is not oriented towards the Sun.

A secondary tidal and climatic cycle is set up at the apsides hemicycle, 4.42 years, when the perigee-syzygy aligns with aphelion. In terrestrial climate cycles a periodicity of either 4 or 5 years is commonly observed; however much the lunar cycle is represented, the 12-month solar cycle must always be dominant.

In Sun–Earth relations, the mean separation is taken as the definition of the Astronomical Unit or AU, currently calculated to be 149 572 827 km. The approximation recommended by IAU as a 'primary constant' is 149 600 000 km. At the epoch of perihelion it is 1.5% less than the mean; at aphelion (July) it is that amount greater.

Long-term commensurability relationships are found to exist between the lunar apsides cycle and the number of planetary relationships. The causal factors always seem to be the gravitational accelerations set up by the passage of other celestial bodies. Examples of these beat frequencies or resonance periods include:

- 17.3769 years (B.F. apsides/18.03 year Saros period);
- 34.7538 years (2 × 17.3768 years; see Brückner cycle);
- 115.047 years (13 × 8.849; 9 × 12.783 year Neptune–Jupiter lap; 72 × 1.5978 year Earth–Venus lap);
- 208.523 years (Earth-Moon/Uranus–Jupiter lap; 6 × 34.7538 years; 12 × 17.3769 years; 25 × 83.4 year mean solar flare period).
- 230.094 years (2 × 115.047 years Neptune–Jupiter/Earth–Venus/apsides/solar).

What is particularly interesting about the two last-named periods is that they correspond to the principal ^{14}C flux-rate periods measured in tree rings on planet Earth (Thomson, 1990), and are taken to be proxies of long-term solar activity. The complex linkage mechanisms between lunar, planetary periods and solar activity are still, of course, problems of ongoing research.

Rhodes W. Fairbridge

Bibliography

Fairbridge, R.W. and Sanders, J.E. (1987), in *History, Periodicity, Predictability*. New York: Van Nostrand Reinhold.
Harris, R.A. (1895–1908) *Manual of Tides*. Washington: US Government Printing Office, in 5 parts.
Krogdahl, W.S. (1962) *The Astronomical Universe*, 2nd edn. New York: Macmillan, 535 pp.
Petersson, O. (1930) The tidal force. *Geogr. Annaler*, **12**, 261–322.
Thomson, D.J. (1990) Time series analysis of Holocene climate data. *Phil. Trans. Roy. Soc. London*, **A330**, 601–16.
Wood, F.J. (1986) *Tidal Dynamics, Coastal Flooding and Cycles of Gravitational Force*. Dordrecht: Reidel, 712 pp.

ARCHEOASTRONOMY

Archeoastronomy is an interdisciplinary activity, investigating not just the mechanics of astronomical practices, but the uses to which astronomical knowledge was put in ancient societies. The synchronization of celestial motions with the seasons has resulted in the practical activities of time keeping and orientation by humanity. They have aided the development of mathematics, navigation, surveying, measurement and urban planning. The value given to studying and understanding the sky led to the incorporation of astronomy into social and religious activities of many cultures, involving folklore, mythology and the world view of indigenous peoples. Modern studies of the role of astronomy in regulating social activities are called ethnoastronomy.

Ancient observations range from precisely timed and measured eclipses, through monitoring the Sun's motion along the horizon, to beginning a season by watching for the first day that a particular star rises in the morning sky. Explanations of those observations range from mathematically precise geometrical models capable of predicting eclipses or planetary motion, to ambiguous mythological accounts.

Archeoastronomy has gradually evolved beyond collecting astronomical artifacts (e.g. testing megalithic sites, medieval churches or Mayan temples for astronomical orientations) to a collaboration between social archeologists and astronomers. The origins of astronomy are unclear. Marshack (1972, 1984) has shown that certain Paleolithic materials contain coded information, consistent with the markings of a lunar calendar. Megalithic (late Neolithic to early Bronze Age) sites like Newgrange and Stonehenge in Britain include solstitial alignments. While the existence of other astronomical functions for these sites is debated, it is evident that the resource expenditure necessary for constructing such monumental architecture indicates a developing social organization that places a high value on certain astronomical phenomena (cf. papers in volumes by Ruggles and Whittle, 1981; Ruggles, 1988).

Evidence of pre-existing knowledge emerges during the historical development of western astronomy, from the Greeks through Copernicus (Neugebauer, 1952). The extent of our debt to the Assyrians, Babylonians and Egyptians is tremendous. They clearly used mathematics to organize the systematic observations that they gathered, and their mathematical models enabled them to predict planetary positions and the periodic nature of eclipses. The quality of these records led to the discovery of such subtle effects as the Saros and the precession of the equinoxes (see Precession and nutation), and provided the foundation for the first physical models of the solar system.

Astronomical observation in China, Japan and Korea has left a rich data set for identifying former sunspot conditions and auroras (Schove, 1983), as well as transient celestial events like comets, novae and supernovae. The belief in a solar–terrestrial linkage resulted in an official program to gather observations of unexpected events.

Astronomy also developed in the Americas. The few surviving Mayan documents show a well-developed interest in astronomy with a mathematically based ability to predict planetary appearances and periods when eclipses could occur (or conversely those periods that were 'safe' from eclipses). Here, the astronomical regulation of social activities extended to warfare as well as agricultural ritual. The Inca legitimized their rule through a claim of descent from the Sun. Native Americans of the southwest also monitored the Sun and used it to regulate their ritual activities (McCluskey, 1990).

Resource materials for archeoastronomy include the *Journal for the History of Astronomy* which has an annual supplement on it. The journal *Archaeoastronomy*, and the quarterly newsletter *Archaeoastronomy and Ethnoastronomy News* are published by the Center for Archaeoastronomy. Finally, there are numerous conference proceedings, books and monographs (Hadingham, 1984; Krupp, 1984; Aveni, 1986) that have appeared in recent years.

David S.P. Dearborn

Bibliography

Aveni, A. (1986) *World Archaeoastronomy* (selected papers from the 2nd Oxford International Conference on Archaeoastronomy, Merida, Yucatan). New York: Cambridge University Press.

Clark, H. and Stephenson, F. (1977) *The Historical Supernovae*. Oxford: Pergamon Press.

Hadingham, E. (1984) *Early Man and the Cosmos*. New York: Walker and Co.

Krupp, E.C. (1984) *Archaeoastronomy and the roots of science*. Boulder, Colorado: Westview Press.

Marshack, A. (1972) *The Roots of Civilization: The Cognitive Beginnings of Man's First Art, Symbol, and Notation*. New York: McGraw-Hill.

Marshack, A. (1985) *Hierarchical Evolution of the Human Capacity: The Paleolithic Evidence*. American Museum of Natural History, New York.

McCluskey, S.C. (1990) Calendars and symbolism: functions of observation in hopi astronomy. *Archaeoastronomy*, 15 (*J. Hist. Astron.*, **21**, S1–16).

Neugebauer, O. (1952) *The Exact Sciences in Antiquity*, Princeton, NJ: Princeton University Press.

Ruggles, C.L.N. (1988) *Records in Stone: Papers in Memory of Alexander Thom*, New York: Cambridge University Press.

Ruggles, C.L.N. and Whittle, A.W.R. (1981) *Astronomy and Society in Britain During the Period 4000–1500 B.C.*. BAR British Series; 88, Oxford, England.

Schove, D.J. (ed.) (1983) *Sunspot Cycles*. Benchmark Papers in Geology, Vol. 68. Stroudsburg: Hutchinson Ross Publ. Co., 392 pp.

Cross references

Eclipse
History of planetary science I: pre-space age
Precession and nutation
Saros cycle

ARISTARCHUS (c. 310–230 BC)

Aristarchus of Samos ranks as one of the greatest astronomers of Ancient Greece. We know nothing about his character or career, and we are not even certain of the dates of his birth and death; however, he was certainly an older contemporary of Archimedes, who has in fact left us the only real accounts of his most important discoveries.

Aristarchus made two significant contributions. First, he was one of the earliest to maintain that the Earth is in orbit round the Sun instead of lying in the centre of the universe. Secondly, he made a noble attempt to measure the relative sizes and distances of the Sun and Moon. His method was perfectly sound in theory, and gave poor results only because the necessary measurements could not be made with sufficient precision.

Just why Aristarchus concluded that the Earth is in orbit round the Sun is not entirely clear. Archimedes, in his *Psammites* or Sand-reckoner, wrote as follows:

> But Aristarchus of Samos brought out a book consisting of certain hypotheses, in which the premises lead to the conclusion that the universe is many times greater than that now so called. His hypotheses are that the fixed stars and the Sun remain motionless, that the Earth revolves round the Sun in the circumference of a circle, the Sun lying in the middle of the orbit, and that the sphere of the fixed stars, situated about the same centre as the Sun, is so great that the circle in which he supposes the Earth to revolve bears such a proportion to the distances of the fixed stars as the centre of the sphere bears to its surface.

Aristarchus found few followers, partly because of the charge of impiety (according to Plutarch, Cleanthes the Stoic thought that Aristarchus ought to be indicted for 'putting in motion the hearth of the universe') and partly because he could give no proof of the correctness of his ideas.

His method of measuring the relative distances of the Sun and Moon depended upon timing the moment when the Moon appears at exact half-phase. He concluded that 'the distance of the Sun from the Earth is greater than 18 times, but less than 20 times, the distance of the Moon from the Earth'. In fact the real distance of the Sun is 400 times greater than that of the Moon, and Aristarchus' estimates of the relative sizes of the two bodies were equally wide of the mark, but there was nothing wrong with his reasoning, and at least he was well aware that the Sun is considerably larger than the Earth – an opinion not shared by all his contemporaries. The fact that his ideas were mainly ignored for many centuries after his death in no way diminishes the magnitude of his achievement.

Patrick Moore

Bibliography

Archimedes (ca. 250 BC) *Psammites* (Sand reckoner), chapter 1, 1–10.

Dreyer, J.L.E. (1953) *A History of Astronomy from Thales to Kepler*, New York: Dover Books, pp. 135–45. (Original edition Cambridge, 1906. The Dover reprint is unabridged.)

Heath, T.L. (1932) *Greek Astronomy*. London and Toronto: J.M. Dent.

Stahl, W.H. (1970) Aristarchus of Samos. *Dict. Sci. Biogr.*, Vol. 1, pp. 246–50.

ASTEROID

One of a group of small planetary bodies (also called minor planets or planetoids) orbiting the Sun, primarily between the orbits of Mars and Jupiter. The empirical Titius–Bode relation (q.v.) can be used to represent the distances of most of the planets from the Sun fairly accurately with one exception: it predicts a planet at 2.8 AU from the Sun where none exists. Late in the 18th century a search for the 'missing' planet was undertaken by six German astronomers who called themselves the 'celestial police'. Their effort had barely begun when they were scooped by the Sicilian astronomer Giuseppe Piazzi (q.v.), who was measuring stellar positions from his observatory near Palermo. On New Year's Day, 1801, Piazzi discovered the first asteroid, which he named Ceres after the patron Roman goddess of Sicily. Its motion showed that Ceres must be at a distance of 2.8 AU, just as the Titius–Bode relation predicts. For predicting the position of Ceres, C.F. Gauss (q.v.) developed the first method of orbit determination, and the asteroid was observed again in December 1801. Ceres was much fainter than any planet observed to that time, implying that it must be much smaller than the known planets. The celestial police persevered, discovering three more asteroids – Pallas, Juno and Vesta – in the next 6 years. The abundance of new objects prompted H. Olbers and others to speculate that they were fragments

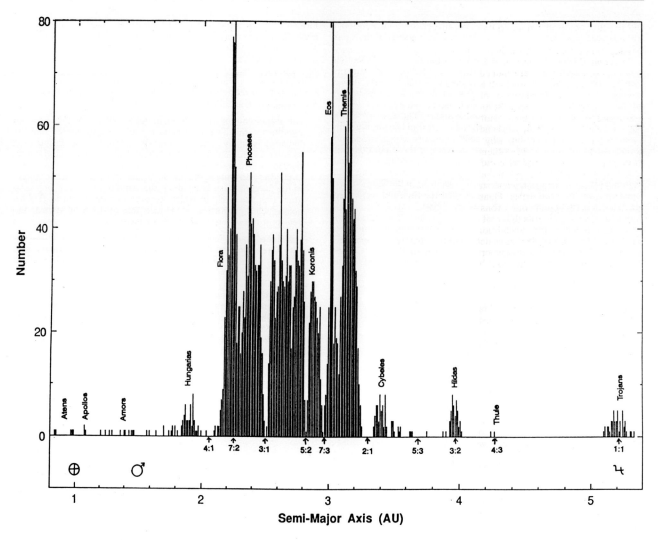

Figure A27 Distribution of orbital semimajor axes for approximately 4000 numbered asteroids. Major Jovian resonances and regions referred to in the text are labeled. (From Binzel, 1989.)

of an exploded or fragmented planet, an idea which has been discarded today.

The next asteroid discovery did not occur until 1845, but from that time on they were catalogued regularly as astronomical instrumentation improved. Today more than 5000 asteroids have been observed with sufficient accuracy to derive orbits. All asteroid discoveries are reported to the International Astronomical Union Minor Planet Centers in Cambridge, Massachusetts, USA, and in St Petersburg, Russia. These centers assign temporary designations, predict future positions and circulate the results to interested observers. After an asteroid has been observed successfully on another opposition it is assigned a permanent number and a name chosen by the discoverer. Thus the first four asteroids are now known as 1 Ceres, 2 Pallas, 3 Juno and 4 Vesta. Certain groups of asteroids with unusual orbits are named by special conventions. For example, the Trojan asteroids (q.v.), found in the stable L4 and L5 points of Jupiter's orbit, are named after Greek and Trojan heroes of the *Iliad* and the *Aenead*. Asteroids with semimajor axes beyond the orbit of Saturn, such as 2060 Chiron and 5145 Pholus, are named after 'good' centaurs from Greek mythology.

Orbits

Figure A27 shows the heliocentric distribution of semimajor axes for approximately 4000 numbered asteroids. The majority of asteroids have orbital semimajor axes between 2.1 and 3.3 AU. A few hundred asteroids have orbits with larger or smaller semimajor axes.

The innermost asteroids, with semimajor axes inside the Earth's orbit, are called the Atens (groups of asteroids with similar orbits are often designated by their first discovered member, in this case 2062 Aten). Apollo asteroids have orbits that cross the Earth's, while Amor asteroids have perihelia between the orbits of Earth and Mars. Together, these three groups are often referred to as planet-crossing asteroids, near-Earth asteroids or AAAO, for Aten-Apollo-Amor objects. At the edge of the main belt, the Cybele ($a = 3.2$ AU) and Hilda groups ($a = 4.0$ AU) are isolated from main belt asteroids. Trojan asteroids are more distant still at Jupiter's distance, 5.2 AU from the Sun. These extremely distant asteroids, difficult to detect and study due to their faintness, hold special interest because they are likely to be little altered since the early days of solar system formation. 2060 Chiron (q.v.), first discovered in 1977, was observed in 1989 to be undergoing comet-like outbursts and is now considered a 'hybrid' – part comet, part asteroid. The most distant asteroid discovered to date, 1992QB1, has a semimajor axis of at least 41 AU, putting it well outside the orbit of Pluto. Extremely distant asteroids and their possible relationship to comets will be discussed in detail below.

In general, asteroid orbits are more elliptical and more inclined to the plane of the ecliptic than are those of major planets. The average orbital eccentricity (e) of main belt asteroids is about 0.15 and the average inclination (i) about 10°. Occasionally these values exceed

0.5 and 30°, which are more typical of short period comets than the majority of asteroids.

Asteroid orbits are not randomly distributed in a, e and i. Figure A27 shows distinct breaks in the distribution of semimajor axes. These 'Kirkwood gaps' are named after Daniel Kirkwood who in 1867 noted depletions of asteroids at the positions of resonances with Jupiter. The mean orbital motion of an asteroid at a resonance would be an exact multiple of Jupiter's; for example, an asteroid at the 2 : 1 resonance would have a mean motion exactly twice that of Jupiter. The correspondence of the gaps and resonances have long indicated a connection with Jupiter, but why some resonances (e.g. 3 : 1 and 2 : 1) should clear gaps while others (e.g. 3 : 2; see Figure A27) have higher concentrations of asteroids was unclear until the 1980s. Wisdom (1983, 1987) has shown that chaotic orbit zones appear at the 3 : 1 and 2 : 1 resonances while stable, quasi-periodic solutions are found for the 3 : 2 resonance. Quasi-periodic orbits allow concentrations to occur while chaotic orbits lead to increased eccentricities, allowing asteroids to cross the orbit of Mars where they are subject to gravitational interactions or collisions with the inner planets. Wisdom (1985) and Wetherill (1985) have demonstrated that chaotic dynamics could provide a means to transport asteroids and meteorites to the inner solar system from the main belt.

Families

The first identification of asteroid families was made by K. Hirayama in 1918. He identified groups of asteroids with similar values of a, e and i and proposed that these objects were the result of the disruption of a single parent body from a collision with another asteroid. With relative impact velocities of $\cong 5$ km s^{-1}, most fragments from the 'target' asteroid would have relative velocities after such a collision of only a few hundred meters per second, small compared to their orbital speeds around the Sun. Thus, while each fragment would have its own orbit in the asteroid belt, the new orbits should be similar to that of the parent body. Studies of asteroid orbital elements have identified anywhere from 20 to more than 100 families, depending upon the selection criteria used. All researchers agree, however, on the existence of a few large and well-populated groups such as the Eos, Koronis and Themis families. If these objects are fragments of a single parent body they should have similar compositions, a prediction which is confirmed for the three largest families (Chapman et al., 1989). Other groupings of asteroids which are isolated by mean motion and/or secular resonances are not considered families, as they are not thought to be common descendents from a single parent body. These are called 'groups'; some examples include the Hungarias, Phocaeas, Cybeles, Hildas and Trojans.

Origin and collisional evolution

The idea that asteroids are fragments of a single exploded planet is no longer tenable. It is hard to imagine a mechanism for so thoroughly dispersing a full-sized planet against the force of its own gravity. The origin of the asteroids appears more consistent with a 'planetesimal' theory of formation, in which the planets accumulated gradually from smaller bodies (Safronov, 1979; Wetherill, 1989). In the region between Mars and Jupiter, it is thought that the accretion was interrupted by some mechanism which is not yet understood, but which was likely driven by nearby, massive Jupiter. The orbital eccentricities and inclinations of planetesimals in this region must have been increased, producing high-speed collisions and collisional grinding among the proto-asteroid planetesimals (Davis et al., 1989). Evidence for significant collisional evolution is provided by the asteroids which are apparently composed of metal (see 'composition' section below); these are widely regarded as the exposed cores of differentiated parent bodies. Collisions are the only viable mechanism for shattering and stripping away the mantles of the parent bodies. The existence of 4 Vesta (q.v.), however, puts an upper limit on the amount of collisional grinding that occurred; with its basaltic crust, Vesta must be a highly differentiated object. For it to have survived largely intact implies that the initial mass in the asteroid region must have been only a few times greater than the present mass (Davis et al., 1985).

Shapes and spin rates

When the brightnesses of most asteroids are measured, they vary in a regular, periodic manner (Figure A28). The lightcurves are usually

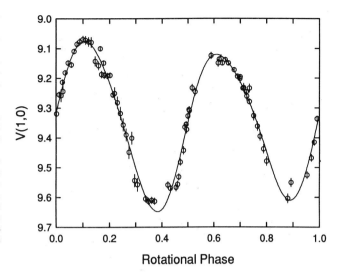

Figure A28 Typical asteroid lightcurve showing rotational light variation due primarily to irregular shape. Trojan asteroid 1173 Anchises has a rotation period of 11.60 h. Data points are from four different nights; smooth curve is a four-component Fourier fit. (From French, 1987.)

double peaked and are thought to be due to the changing surface area of a rotating, irregularly shaped object. The peak-to-peak variation of the lightcurve gives an estimate of the shape of the asteroid. If the asteroid is observed at many viewing angles (usually over several years) an estimate of the true shape and the orientation of the pole of the asteroid can be determined (Magnusson et al., 1989).

Why are most asteroids non-spherical? The largest, such as Ceres and Pallas, have sufficient self-gravity to achieve near-spherical shapes. Lesser bodies, below 125 km in diameter, are more irregular in shape as evidenced by their higher average lightcurve amplitudes (Lagerkvist, Harris and Zappala, 1989; Binzel et al., 1989). These objects are probably either 'rubble piles' consisting of collisional fragments with insufficient gravity to re-accumulate into a single body, or single collision fragments. Suspicions that some asteroids could have a binary nature have long been raised: first, by observations of occultations (see below) and then by theoretical considerations (Weidenschilling, Paolicchi and Zappala, 1989). Radar is a powerful tool for determining shapes of asteroids which approach the Earth. Radar spectra of 216 Kleopatra (Ostro, Campbell and Shapiro, 1986) and 4769 Castalia (Ostro, 1992) suggested a binary nature for both objects; the radar cross-section increased with distance from the asteroid spin axis. Only recently has the existence of a binary asteroid been confirmed by the spectacular radar observations of the near-Earth asteroid 4179 Toutatis. The images reveal Toutatis to be a contact binary, consisting of two objects about 1.6 and 4 km in average diameter and rotating once every 10 or 11 days. Craters are clearly visible on the surface of both components, indicating a complex collisional history. Whether the two components were once fragments of a single parent body or whether they formed independently and were joined by a relatively gentle collision is not known. Compositional study of each Toutatis component could help resolve this question.

Typical asteroid rotation periods are in the range 8–12 h, with extreme values ranging from ~ 2.8 h to more than 700 h. A rotation period of 2.5 h is thought to be the smallest possible for a single body at typical asteroid sizes and strengths: an object rotating faster would fly apart because centripetal acceleration would exceed gravity at the equator for reasonable material strengths. Because rotational angular momentum is gained or lost through collisions, a knowledge of asteroid rotation rates is essential for understanding their collisional evolution. Significant correlations are found between asteroid rotation rate and diameter, with distinct changes occurring around a diameter of 125 km. For asteroids larger than 125 km, a Maxwellian distribution of rotation rates is a good fit to the data, implying that their rotation properties have been determined by collisional evolution. Below a diameter of 125 km, however, asteroids do not show a

Maxwellian distribution, suggesting that other processes may be more important in modifying their rotation rates (Binzel et al., 1989). In particular, an excess number of slowly rotating small asteroids is seen. The most extreme cases known to date are those of 288 Glauke, with a period of more than 48 days (Harris, 1983) and 1220 Crocus, whose period is approximately 30 days (Binzel, 1985). Tidal braking due to a binary companion has been suggested as the most plausible explanation for such long rotation periods (Weidenschilling et al., 1989). It should be noted that the one confirmed binary asteroid, 4179 Toutatis, has a relatively long rotation period of 10–11 days (see above).

Sizes

Since the 1970s infrared observations have been used to determine the diameters of asteroids. At visual wavelengths asteroids shine by reflected sunlight, while at longer wavelengths (beyond 5 μm) the asteroids' own thermal emission dominates. A given visual magnitude could be matched by a large dark object or a bright small one, but the former will be warmer as well as larger than the latter, and will thus show greater infrared emission. Radiometric diameter determination involves finding a diameter and albedo that will simultaneously match both the observed reflected sunlight and infrared thermal emission from an object. Ground-based observations were the source of asteroid radiometric data for more than two decades (Lebofsky and Spencer, 1989). In 1983 the Infrared Astronomical Satellite (IRAS) was launched. Over 11 months IRAS surveyed the entire sky at four wavelengths between 12 and 100 μm. By applying the standard thermal model developed for ground-based asteroid radiometry and by using mean visual magnitudes for asteroids, diameter estimates of over 1800 asteroids were determined (IRAS Asteroid and Comet Survey, 1986: Matson et al., 1989).

The occultation of a star by an asteroid provides a direct measurement of the asteroid's cross-section at the time of the occultation. Since asteroid distances from the Earth and orbital velocities are well known, timing the duration of an occultation gives a measure of one chord across the asteroid. Observers in different locations along the path of the occultation observe different chords; with sufficient numbers of observers the apparent limb profile of the asteroid can be closely determined (Millis and Dunham, 1989). Diameter measurements of more than 30 asteroids have been made using occultations. The agreement with diameters derived using the radiometric method is generally satisfactory.

Masses and densities

The masses of asteroids are not easily measured, since their determination requires a measurable gravitational interaction with another planetary body. Vesta's mass has been derived from its perturbations on the orbit of 197 Arete, while the masses of Ceres and Pallas have been derived from their mutual perturbations and from the action of Vesta on Ceres (Schubart and Matson, 1979; Standish, 1989). The mass of 10 Hygeia was measured by its effect on the orbit of 829 Academia by Scholl, Schmadel and Roser (1987).

Density is a characteristic property which is diagnostic of composition and thus its measurement is extremely desirable. Asteroids for which densities have been determined to date are given in Table A6. The densities of 1 Ceres, 2 Pallas and 10 Hygeia are consistent with a composition similar to that of carbonaceous chondrites, while Vesta's higher density is consistent with a rocky composition.

Binary asteroids could also be used to determine masses and densities of asteroids. In principle, the mean density of a binary system can be determined from lightcurves of satellite transits and occultations, a technique which has been used by Tholen et al. (1985) for the Pluto–Charon system. If a binary system is resolved so that its separation can be measured, the total mass is given by Kepler's third law. Relative masses could be estimated by the relative brightnesses of the components or by astrometric measurements of the components about the center of mass of the system. The extremely irregular shapes of the components of 4179 Toutatis suggest that, in practice, density determinations will be limited by uncertainties in the volumes of the components.

Composition

Spectral analysis of the electromagnetic radiation received from asteroids provides information about their composition. Visible and near-infrared reflectance spectroscopy (q.v.) has been the most heavily used technique for characterizing asteroid surface materials. Because the asteroids reflect solar light, asteroid spectra are calibrated by observing a solar-type star through the same amount of atmosphere as the asteroid. The asteroid spectrum divided by the calibration spectrum then shows the reflectance spectrum of the material on the asteroid's surface.

It is currently believed that most meteorites originated in the asteroid belt. Although there is not always a sharp distinction between comets and asteroids (see below), significant compositional differences exist between cometary nuclei and known chondritic meteorites (Brownlee et al., 1987). Direct orbital evidence, however, supports an asteroidal origin for at least three chondrites. Pribram (L5), Lost City (H5) and Innisfree (LL5) were photographed between 1959 and 1977 by camera networks so that their orbits and fall directions could be determined. The aphelion of each meteorite's orbit lay in the asteroid belt. Modern studies of asteroid composition, therefore, make extensive use of terrestrial meteorite collections (Lipschutz, Gaffey and Pellas, 1989). No specific meteorite has yet been unequivocally linked to a specific asteroid, however.

The earliest attempts to classify asteroid spectral and albedo data utilized broadband UBV (ultraviolet, blue and visual) colors and albedos (Chapman, 1975; Bowell et al., 1978). Two major groups were recognized: low-albedo objects with neutral colors were designated 'C types' because of their spectral similarities to carbonaceous chondrites, while 'S types', objects with redder colors and more moderate albedos, had spectra resembling those of stony iron meteorites. As instrumental capabilities have increased, more detailed reflectance spectra of asteroids have become available and more complex taxonomic schemes have been devised (Tholen and Barucci, 1989). Approximately one-third of the asteroids still fall into the C group or its related subgroups. The surface material of C asteroids is still thought to resemble carbonaceous chondrites (q.v.), which have near-solar elemental abundances. The spectral signature of water of hydration has been detected on the surfaces of several C-type asteroids, including 1 Ceres, by infrared measurements (Lebofsky et al., 1981). The presence of water in the surface layers of asteroids means that they have been subjected to only moderate heating since the formation of the solar system. S asteroids, the most numerous, have moderate-albedo, reddish surfaces containing pyroxene and olivine silicates mixed with metallic iron. The meteorites which are most numerous on Earth, the ordinary chondrites, appear to have no analogs in the asteroid belt. The S asteroids have qualitatively the same mineralogy, but the relative proportions on the surfaces differ. Observationally the S asteroids appear more closely related to the stony iron meteorites. One Earth crosser, 1862 Apollo, appears to be mineralogically consistent with the ordinary chondrites, although this single small asteroid is too small to be the only source. Thus the parent bodies of the ordinary chondrites have not been definitively identified (Gaffey, Bell and Cruikshank, 1989, Bell et al., 1989).

It was clear from the earliest taxonomy studies that different types of asteroids predominate at different distances. The S objects are found predominantly in the inner part of the main asteroid belt. The M objects, with metal-like spectra, are located throughout the main belt with a peak near 2.6 AU. The C objects, with their solar abundances and surface ice, are predominant in the outer part of the main belt. Beyond the main belt, new classes appear: the P objects are found in the Cybele–Hilda region just beyond the main belt and the D objects are found predominantly in the Trojan regions of Jupiter's orbit (French et al., 1989). The D class have reddish spectra in the visible region and extremely low albedos; P-type spectra are intermediate in slope between C and D classes and also have low

Table A6 Asteroid densities (from Millis and Dunham, 1989)

Object	Mass (10^{-10} solar masses)	Density (g cm^{-3})
1 Ceres	5.9 ± 0.3	2.7 ± 0.14
2 Pallas	1.08 ± 0.22	2.6 ± 0.5
4 Vesta	1.38 ± 0.12	3.62 ± 0.35
		3.3 ± 1.5
		2.4 ± 0.3
10 Hygeia	0.47 ± 0.23	2.05 ± 1

albedos. No mechanism has been discovered for delivering samples to the inner solar system from these distant regions, consistent with the observation that no meteorite samples corresponding to P or D asteroids have been found. The surface composition of the D-class asteroids was first investigated by Gradie and Veverka (1980) who mixed organic polymers, clay silicates and an opaque to simulate the low albedo and red slope of D asteroid spectra. In contrast to the other primitive group of C asteroids (the C objects), no 3-µm water of hydration feature has been observed in P or D spectra. Jones *et al.* (1990) suggest that the C asteroids have undergone moderate solar heating, melting internal ice that has become bound to the surface material. P and D asteroids, at their greater solar distances, have not undergone such heating episodes. Ice was once common on these asteroid surfaces, but has sublimated away. Considerable water may be present inside the distant asteroids – up to 50% water ice content is possible, given the compositions of Jupiter's Galilean satellites. The red slope and low albedo of these distant objects is similar to spectra observed for Saturn's moon Iapetus (Bell, Cruikshank and Gaffey, 1985) and for some comet nuclei.

Bell *et al.* (1989) present a overview of asteroid compositional taxonomy. In their view, the spectral groups found in the inner part of the belt, primarily the S and M classes, are classed as 'igneous', signifying that their current surface mineralogy was formed from a melt. 'Metamorphic' objects have undergone some heating, while 'primitive' groups such as the C, P and D classes have undergone little or no heating.

Surface properties

The surface of a typical asteroid is thought to be covered with a regolith, a layer of dusty rubble caused by billions of years of meteorite bombardment. The evidence for asteroid regoliths is indirect but several types of observations are consistent with this hypothesis. Precise measurements of an asteroid's brightness at different solar phase angles (the phase angle is the angle between Sun and Earth, as seen from the asteroid) enable estimation of the importance of mutual shadowing between particles. Most asteroid surfaces show a strong surge in brightness near zero phase angle ('full asteroid') due to the disappearance of shadows. Radar and microwave observations of asteroids also are useful in determining their surface properties. Typical microwave estimates of regolith thickness range from 1–8 cm (Webster and Johnson, 1989). Ostro *et al.* (1985) presented observations of 19 Psyche which showed a porosity comparable to that of the lunar surface and a metallic composition, consistent with the observed M-type spectrum of Psyche. Recent upgrades of major radio facilities such as Arecibo and Goldstone will make possible further detailed studies of asteroid surfaces.

In a few cases spectral variations are observed on the surfaces of asteroids. The most notable example is 4 Vesta, whose surface is largely composed of plagioclase and pyroxene. Time-resolved photometry (Gaffey, 1983) indicates a localized region with an olivine-rich spectrum, indicating the presence of a crater sufficiently deep to reveal the mantle material of Vesta. When a spacecraft eventually returns pictures from Vesta, at least one large impact crater should be visible on the surface. Spectra of small asteroids belonging to Vesta's family are consistent with a plagioclase/pyroxene composition, indicating they are 'chips' off the larger object (Binzel *et al.*, 1993).

Connections between asteroids and comets

Both the asteroids and the comets are remnants of the planetesimals which are believed to have formed in the solar nebula some 4.5 billion years ago. Throughout human history they have been observed, identified and catalogued as different classes of objects because of their different orbits and observational properties. Recent evidence suggests, however, that the distinction between 'comets' and 'asteroids' is less than sharp in many cases. Observations of near-Earth asteroids, some of which may be extinct comet nuclei, have provided vital clues. Additionally, the last 5 years have brought more observations of small bodies outside the orbit of Jupiter, some displaying both cometary and asteroidal behavior. Dynamical mechanisms for transporting both comets and asteroids to the near-Earth environment have been identified, making the case for a connection between extinct comets and near-Earth asteroids.

Kresák (1985) has shown that Jupiter-crossing and Jupiter-approaching objects – asteroids or comets – are most easily captured into Earth-crossing orbits. Before 1980 only a handful of Jupiter-crossing asteroids were known to exist. Between 1981 and 1988, improved search techniques led to the discovery of 18 Jupiter crossers (Weissman *et al.*, 1989). Many of these objects have chaotic orbits which are not stable over long time periods, implying some source of replenishment. The passage of small bodies from the trans-Jovian region to the inner solar system is thus a more common occurrence than previously believed. At about the same time, theoretical dynamical studies showed that secular resonances and chaotic motion are viable mechanisms for transporting objects to the near-Earth region (Wetherill, 1988; Wisdom, 1987).

A new piece of evidence connecting near-Earth asteroids with comets is the recognition that some Earth-crossing asteroids appear to be connected with meteor showers. Most meteor showers are associated with active short-period comets. However, asteroid 3200 Phaethon, discovered in 1983 by the IRAS satellite, is in an identical orbit with the Geminid meteor stream, which gives rise to one of the brightest meteor showers every year.

For a comet to appear as an asteroid, it must lose its ability to generate a detectable coma, either through loss of its volatiles or through the development of a thick, non-volatile lag deposit on the surface which prevents the heating of the interior. Complete volatile loss is likely to be slow, taking on the order of 600 perihelion passages for a 1-km comet nucleus with zero albedo and a perihelion distance of 1 AU (Weissman *et al.*, 1989). The development of an insulating crust on the surface would hinder the process further. Observations of the surface of Comet Halley by the Vega and Giotto spacecraft showed a surface that was $\approx 70\%$ covered by an inactive crust. Comets may eventually cover themselves with non-volatile crusts, becoming dormant and asteroidal in appearance. The crust may fail catastrophically after gas pressure builds up within the core, returning the comet to an active state periodically (Kresák, 1985). Asteroid 4015, discovered in 1979 and as yet unnamed, was recently connected with Comet Wilson-Harrington, discovered in 1949. Between 1949 and 1979 the volatiles apparently sublimated from the surface, leaving a dormant, perhaps dead, nucleus. 2060 Chiron, the first asteroid to be discovered with a semimajor axis outside that of Saturn, began showing cometary activity in 1989 and is a focus of intense study as it nears perihelion in 1996.

The reflectance spectra of the most distant asteroids are similar to the few spectra of bare cometary nuclei obtained to date. Their featureless, red-sloping spectra are consistent with a surface composed of primitive hydrocarbons (Gradie and Veverka, 1980; Fink *et al.*, 1992; Mueller *et al.*, 1992; Binzel, 1992). Such spectra are consistent with those obtained in laboratory studies of low molecular weight hydrocarbons such as methane frosts following irradiation with ultraviolet light (Khare *et al.*, 1989). Although evidence of the presence of volatiles such as water and methane ice is not seen in the spectrum of trans-Jovian asteroids or in the Trojan asteroids, these asteroids could have volatile contents approaching 50% in their interiors if the lag deposit crusts were thick enough.

In late 1992 the most distant asteroid yet observed, 1992 QB_1, was discovered by D. Jewitt and J. Luu. The object's distance is so great and its daily motion so slow that several months' observations were needed to get even an approximate orbit. Currently the best fit orbit is nearly circular, with a semimajor axis of 41 AU, putting 1992QB1 just outside the orbit of Pluto. This is consistent with the existence of the Kuiper belt (q.v.), a 'deep freeze' of comets just beyond Pluto.

Linda M. French

Bibliography

Bell, J.F., Cruikshank, D.P. and Gaffey, M.J. (1985) The composition and origin of the Iapetus dark material. *Icarus*, **61**, 192–207.

Bell, J.F., Davis, D.R., Hartmann, W.K. and Gaffey M.J. (1989) Asteroids: the big picture, in *Asteroids II* (eds R.P. Binzel, T. Gehrels and M.S. Matthews). Tuscon. University of Arizona Press, pp. 921–45.

Binzel, R.P. (1985) Is 1220 Crocus a precessing, binary asteroid? *Icarus*, **72**, 99–108.

Binzel, R.P. (1992) 1991 Urey Prize Lecture: Physical evolution in the solar system – present observations as a key to the past. *Icarus*, **100**, 274–87.

Binzel, R.P. (1992) The optical spectrum of 5145 Pholus. *Icarus*, **99**, 238–40.

Binzel, R.P., Farinella, P., Zappala, V. and Cellino, A. (1989) Asteroid rotation rates: distributions and statistics, in *Asteroids II* (eds R.P. Binzel, T. Gehrels and M.S. Matthews). Tucson: University of Arizona Press, pp. 416–41.

Binzel, R.P., Xu, S., Bus, S.J. *et al.* (1993) Discovery of a main-belt asteroid resembling ordinary chondrite meteorites. *Science*, **262**, 1541–3.

Bowell, E., Chapman, C.R., Gradie, J.C. *et al.* (1978) Taxonomy of asteroids. *Icarus*, **35**, 313–35.

Brownlee, D.E., Wheelock, M.M., Temple, S. *et al.* (1987) A quantitative comparison of Comet Halley and carbonaceous chondrites at the submicron level. *Lunar Planet. Sci. XVIII*, pp. 133–4.

Chapman, C.R. (1975) The nature of asteroids. *Sci. Amer.*, **232**(1), 24–33.

Chapman, C.R., Morrison, D. and Zellner, B. (1975) Surface properties of asteroids: a synthesis of polarimetry, radiometry, and spectrophotometry. *Icarus*, **25**, 104–30.

Chapman, C.R., Paolicchi, P. Zappala, V. *et al.* (1989) Asteroid families: physical properties and evolution, in *Asteroids II* (eds R.P. Binzel, T. Gehrels and M.S. Matthews). Tucson: University of Arizona Press, pp. 386–415.

Davis, D.R., Chapman, C.R., Weidenschilling, S.J., and Greenberg, R. (1985) Collisional history of asteroids: evidence from the Vesta and the Hirayama families. *Icarus*, **63**, 30–53.

Davis, D.R., Weidenschilling, S.J., Farinella, P. *et al.* (1989) Asteroid collisional history: effects on sizes and spins, in *Asteroids II* (eds R.P. Binzel, T. Gehrels, and M.S. Matthews). Tucson: University of Arizona Press, pp. 805–26.

Fink, U., Hoffman, M., Grundy, W. *et al.* (1992) The steep red spectrum of 1992 AD: an asteroid covered with an organic material? *Icarus*, **97**, 145–149.

French, L.M. (1987) Rotation properties of four L5 Trojan asteroids from CCD photometry. *Icarus*, **72**, 325–341.

French, L.M., Vilas, F., Hartmann, W.K. and Tholen, D.J. (1989) Distant asteroids and Chiron, in *Asteroids II* (eds R.P. Binzel, T. Gehrels and M.S. Matthews). Tucson: University of Arizona Press, pp. 468–86.

Gaffey, M.J. (1983) The asteroid (4) Vesta: rotational spectral variations, surface material heterogeneity, and implications for the origin of the basaltic achondrites, in *Lunar Sci. I*, pp. 312–313 (abstract).

Gaffey, M.J., Bell, J.F. and Cruikshank, D.P. (1989) Reflectance spectroscopy and asteroid surface mineralogy, in *Asteroids II* (eds R.P. Binzel, T. Gehrels, and M.S. Matthews). Tucson: University of Arizona Press, pp. 98–127.

Gehrels, T. (1979) The asteroids: history, surveys, techniques, and future work, in *Asteroids* (eds T. Gehrels). Tucson: Univ. of Arizona Press, pp. 3–24.

Gradie, J. and Tedesco, E.F. (1982) The compositional structure of the asteroid belt. *Science*, **216**, 1405–7.

Gradie, J. and Veverka, J. (1980) The composition of the Trojan asteroids. *Nature*, **283**, 840–2.

Harris, A.W. (1983) Slowly rotating asteroids: evidence for binary asteroids? *Br. A. A. Soc.*, **15**, 828.

Harris, A.L. and Burns, J.A. (1979) Asteroid rotation I. Tabulations and analysis of rates, pole positions, and shapes. *Icarus*, **40**, 115–44.

IRAS Asteroid and Comet Survey (1986) Preprint Version No. 1. (ed. D.L. Matson), JPL Document No. D-3698.

Jet Propulsion Laboratory Press Release (January 4–1993) [Radar observations of 4179 Toutatis.]

Jones, T.D., Lebofsky, L.A., Lewis, J.S. and Marley, M.S. (1990) The composition and origin of the C, P, and D asteroids: water as a tracer of thermal evolution in the outer belt. *Icarus*, **88**, 172–92.

Khare, B.N., Thompson, W.R., Murray, B.G.J.P.T. *et al.* (1989) Solid organic residues produced by irradiation of hydrocarbon-containing H_2O and H_2O/NH_3 ices: infrared spectroscopy and astronomical implications. *Icarus*, **79**, 350–61.

Kresak, L. (1985) The aging and lifetimes of comets, in *Dynamics of Comets: Their Origin and Evolution* (eds. A. Carusi and G.B. Valsecchi). Dordrecht: D. Reidel, pp. 279–302.

Lagerkvist, C.-I., Harris, A.W. and Zappala, V. (1989) Asteroid lightcurve parameters in *Asteroids II* (eds R.P. Binzel, T. Gehrels and M.S. Matthews). Tucson: University of Arizona Press, pp. 1162–79.

Lebofsky, L.A., Feierberg, M.A., Tokunaga, A.T. *et al.* (1981) The 1.7- to 4.2-μm spectrum of asteroid 1 Ceres: evidence for structural water in clay minerals. *Icarus*, **48**, 453–59.

Lebofsky, L.A. and Spencer, J.R. (1989) Radiometry and thermal modeling of asteroids, in *Asteroids II* (eds R.P. Binzel, T. Gehrels and M.S. Matthews). Tucson: University of Arizona Press, pp. 128–47.

Lipschutz, M.E., Gaffey, M.J. and Pellas, P. (1989) Meteoritic parent bodies: nature, number, size, and relation to present-day asteroids, in *Asteroids II* (eds R.P. Binzel, T. Gehrels and M.S. Matthews). Tucson: University of Arizona Press, pp. 740–77.

Magnusson, P., Barucci, M.A., Drummond, J.D. *et al.* (1989) Determination of pole orientations and shapes of asteroids, in *Asteroids II* (eds R.P. Binzel, T. Gehrels and M.S. Matthews), Tucson: University of Arizona Press, pp. 66–97.

Matson, D.L., Veeder, G.J., Tedesco, E.F. and Lebofsky, L.A. (1989) The IRAS asteroid and comet survey, in *Asteroids II*, (eds R.P. Binzel, T. Gehrels and M.S. Matthews). Tucson: University of Arizona Press, pp. 269–81.

Millis, R.L. and Dunham, D.W. (1989) Precise measurements of asteroid sizes and shapes from occultations, in *Asteroids II* (eds R.P. Binzel, T. Gehrels and M.S. Matthews) Tucson: University of Arizona Press, pp. 148–67.

Mueller, B.E.A., Tholen, D.J., Hartmann, W.K. and Cruikshank, D.P. (1992) Extraordinary colors of asteroidal object (5145) 1992 AD. *Icarus*, **97**, 150–4.

Ostro, S.J. (1989) Radar observations of asteroids, in *Asteroids II* (eds R.P. Binzel, T. Gehrels and M.S. Matthews). Tucson: University of Arizona Press, pp. 192–212.

Ostro, S.J. (1992) Plans for radar investigation of asteroid 4179 Toutatis. *Bull. Am. Astron. Soc.*, **24**, 934.

Ostro, S.J., Campbell, D.B. and Shapiro, I.I. (1985) Mainbelt asteroids: dual-polarization radar observations. *Science*, **224**, 442–6.

Ostro, S.J., Campbell, D.B. and Shapiro, I.I. (1986) Radar detection of 12 asteroids from Arecibo. *Bull. Am. Astron. Soc.*, **18**, 796 (abstract).

Safronov, V.S. (1979) On the origin of asteroids, in *Asteroids* (ed. T. Gehrels). Tucson: University of Arizona Press, pp. 975–93.

Scholl, H., Schmadel, L.D. and Roser, S. (1987) The mass of the asteroid 10 Hygeia derived from observations of 829 Academia. *Astron. Astrophys.*, **179**, 311–6.

Schubart, J. and Matson, D.L. (1979) Masses and densities of asteroids, in *Asteroids* (ed. T. Gehrels). Tucson: University of Arizona Press, pp. 84–97.

Standish, E.M., Jr (1989) A determination of the masses of Ceres, Pallas and Vesta from their perturbations upon the orbit of Mars. *Icarus*, **46**, 124–6.

Tholen, D.J. and Barucci, M.A. (1989) Asteroid taxonomy, in *Asteroids II* (eds R.P. Binzel, T. Gehrels and M.S. Matthews). Tucson: University of Arizona Press, pp. 298–315.

Tholen, D.J., Buie, M., Binzel, R. and Frueh, M. (1987) Improved orbital and physical parameters for the Pluto–Charon system. *Science*, **237**, 512–4.

Webster, W.J., Jr and Johnston, K.J. (1989) Passive microwave observations of asteroids, in *Asteroids II* (eds R.P. Binzel, T. Gehrels and M.S. Matthews). Tucson: University of Arizona Press, pp. 213–27.

Weidenschilling, S.J., Chapman, C.R., Davis, D.R. *et al.* (1987) Photometric geodesy of main belt asteroids. I. Lightcurves of 26 large, rapid rotators. *Icarus*, **70**, 190–245.

Weidenschilling, S.J., Paolicchi, P. and Zappala, V. (1989) Do asteroids have satellites? in *Asteroids II* (eds R.P. Binzel, T. Gehrels, and M.S. Matthews). Tucson: University of Arizona Press, pp. 643–658.

Weissman, P.R. (1980) Physical loss of long-period comets. *Astron. Astrophys*, **85**, 191–6.

Weissman, P.R., A'Hearn, M.F., McFadden, L.A., and Rickman, H. (1989) Evolution of comets into asteroids, in *Asteroids II* (eds R.P. Binzel, T. Gehrels and M.S. Matthews). Tucson: University of Arizona Press, pp. 880–920.

Wetherill, G.W. (1985) Asteroidal source of ordinary chondrites. *Meteoritics*, **20**, 1–22.

Wetherill, G.W. (1988) Where do the Apollo asteroids come from? *Icarus*, **76**, 1–18.

Wetherill, G.W. (1989) Origin of the asteroid belt, in *Asteroids II* (eds R.P. Binzel, T. Gehrels and M.S. Matthews). Tucson: University of Arizona Press, pp. 661–80.

Wisdom, J. (1983) Chaotic behavior and the origin of the 3:1 Kirkwood gap. *Icarus*, **56**, 51–74.

Wisdom, J. (1985) Meteorites may follow a chaotic route to Earth. *Nature*, **315**, 731–3.
Wisdom, J. (1987) Urey Prize Lecture: chaotic dynamics in the solar system. *Icarus*, **72**, 241–75.

Cross references

Chaotic dynamics in the solar system
Lagrangian point
Meteorite parent bodies
Reflectance spectroscopy
Resonance in the solar system

ASTEROID: COMPOSITIONAL STRUCTURE AND TAXONOMY

Asteroid taxonomies are based on observable traits such as morphology or mineralogy. Observations of asteroids are strictly limited, in most cases, to ground-based telescopic remote sensing of the visible and near infrared spectra and radiometry (see Infrared Spectroscopy Radiometry). IR spectra and albedo (q.v.) provide most of our information on most asteroids and are the basis for all the current asteroid taxonomies.

There are currently three published asteroid taxonomic schemes (Tholen, 1984; Barucci *et al.*, 1987; Tedesco *et al.*, 1989). All of these classifications are 'spectral' in nature. Asteroids that have similar spectra and albedo characteristics are grouped together in a class denoted by a letter or group of letters. Asteroids in subgroups or asteroids with unusual spectra are often assigned multiple letters, the first letter denoting the dominant group and the succeeding letters denoting less prominent spectral affinities or subgroups. Although other workers have made mineralogical interpretations of the classes (see Meteorite parent bodies; Bell *et al.*, 1989; Gaffey *et al.*, 1989), none of the classifications presuppose specific mineralogies. In this paper we will focus primarily on the widely used Tholen (1984) taxonomy.

Brief history of asteroid taxonomy

Asteroid taxonomy has developed in tandem with the increase in the range and detail of asteroid observational data sets. Early observations were often limited in scope to the larger and brighter asteroids, and in wavelength range to filter sets used for stellar astronomy. As observations widened in scope, and more specialized filter sets and observational techniques have been applied to asteroids, our appreciation of the variety and complexity of asteroid spectra has also increased. The asteroid classification system has evolved to reflect this complexity, and the number of spectral classes has steadily increased. Since any asteroid scientist can propose a new class or an entire new classification, the growth of asteroid taxonomy can be anarchistic. Figure A29 is a schematic of the growth and evolution of the Tholen (1984) asteroid spectral classification system. This taxonomy was based on the Chapman, Morrison and Zellner (1975) taxonomy of only three classes. Additions, subtractions and splits have increased the number of classes to 19, including four added since 1984.

The original three classes were the C class, which showed neutral spectra and low albedos similar to carbonaceous chondrite meteorites; the S class, which had higher albedos and spectral absorption features indicting large amounts of the rock-forming minerals olivine and pyroxene, and the U class. The C and S characters were mnemonics for 'carbonaceous' and 'stony' asteroids respectively although the authors cautioned that the classifications were based on reflectance parameters and not on inferred surface mineralogy. The U class was for the unclassified or unusual spectra. The unique large asteroid 4 Vesta (q.v.) was previously recognized to have spectra analogous to basalt and the highly differentiated basaltic achondrite meteorites (q.v.), the eucrites, diogenites and howardites. This object eventually became the type asteroid for the V class.

Soon after the original taxonomy was published the high-albedo asteroid 44 Nysa was added as the E class. Once again, mnemonics for the presumed surface mineralogy dictated the choice of the class letter since Nysa was thought to be dominated by enstatite, and analogous to enstatite achondrites. In addition to Nysa, a group of

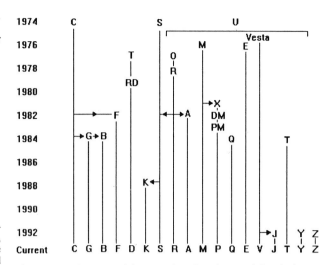

Figure A29 Schematic of the growth and evolution of the Tholen (1984) asteroid spectral classification system. This taxonomy was based on the Chapman, Morrison and Zellner (1975) taxonomy of only three classes. Additions, subtractions and splits have increased the number of classes to 19, including four added since 1984.

moderate-albedo objects were recognized as the M class at this time. Again, the letter was used to suggest mineralogy, in this case analogous to iron-nickel meteorites or 'metallic' asteroids. Zellner and Bowell (1977) published an expanded taxonomy that included objects with spectra similar to ordinary chondrites (the O class) and added the T class for the few Trojan asteroids that had been observed at that time. Both these classes were soon changed; the O class became the R class for high-albedo, red-sloped objects and the T class became the RD class when it was recognized that the IR colors of these asteroids were very red and their albedos were very low (RD = red-dark class). This class eventually became the D class.

Another group whose designation became an odyssey through the alphabet was first recognized in 1981. The objects were spectrally M class but with albedos much lower than the standard for M asteroids or their suggested metallic meteorite analogs. The first proposed name for this group was the X class. This was soon changed to the 'dark' M or DM class and changed again to the PM or pseudo-M class. The first results of the Eight-Color Asteroid Survey (ECAS) (Gradie and Tedesco, 1982) were used to simplify this group to the P class. This work also defined the F class as asteroids with low albedos and 'flat' spectra. Other workers divided up the R class with some objects going to the S class and some going into a new A class.

The evolving asteroid classification system was given an enormous boost by the publication of the ECAS survey. In all 535 asteroids were observed in the full ECAS system and this work remains by far the largest single survey of asteroid spectral characteristics. Based on this large and standardized data set, Tholen (1984) regularized the asteroid classification system by defining the classes in terms of their spectra in statistical parameter space. The C class was subdivided by the addition of the G and B classes to reflect variations in albedo and UV spectral characteristics. The R class was cut down to a single asteroid, the unique 349 Dembowska and the old R asteroids divided between the S and A classes. Other single-member classes were defined around 4 Vesta (V class) and 1862 Apollo (Q class). The T class was created for the handful of asteroids that fell in the statistical space between the S and D classes.

Since 1984 the classification system has been expanded by the addition of the K class for moderate albedo S class asteroids that are statistically between the bulk of the S population and the C class (hence the use of K which is alphabetically midway between C and S; Tedesco *et al.*, 1989). Howell, Merenyl and Lebofsky (1993) have proposed subdividing the remaining S class asteroids by adding lowercase letters to the S designation for some asteroids. The So class would be those S asteroids that have spectra indicating a mineralogy rich in the mineral olivine. The Sp class would be the pyroxene-rich subset. The remaining asteroids denoted by the classification S would indicate a silicate mineralogy that was a mix of olivine and pyroxene. Gaffey *et al.* (1993) have also subdivided the S class, but into seven

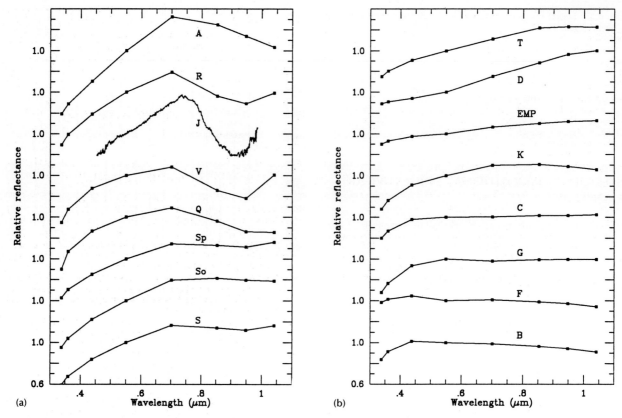

Figure A30 (a), (b) The scaled 0.3 to 1.1 μm reflectance spectra of the Tholen (1984) asteroid classes. All spectra are scaled to unity at 0.55 μm, offset vertically by 0.4 for clarity, and each tickmark on the vertical scale equals a reflectance value of 0.1. Except for the J class, all spectra are the average ECAS reflectance for that class. The J class has been observed only with CCD spectra and the spectra included here is of asteroid 3155 Lee. The S class spectra shown are the Howell, Merengi and Lebofsky (1993) subdivisions of the class into pyroxene rich (Sp), olivine-rich (So) and mixtures of olivine and pyroxene (S) subgroups.

subclasses. Binzel, after greatly increasing the number of identified Vesta-like asteroids, subdivided the V class by adding the J class to denote asteroids with orthopyroxene mineralogy similar to the diogenite Johnstown. Finally, Bell and Tholen have added the Z class for the unique and very red trans-Saturnian asteroid 5145 Pholus to complete the current evolution of this classification system.

Since Tholen (1984) two additional asteroid classification systems have been proposed. Barucci *et al.* (1987) added the newly available IRAS albedo data to the ECAS colors. This allowed direct classification on the basis of albedo as well as spectra, an advantage when dealing with spectrally similar classes such as the E, M and P classes that have strong albedo differences. Tedesco and coworkers used a subset of the ECAS data, the U, V and X filters, together with the IRAS albedos to classify (initially) 357 asteroids (Tedesco *et al.*, 1989). The advantage of this system was the modest data requirements that could incorporate data from other asteroid observations and potentially greatly expand the number of classified asteroids. The common thread of all these taxonomies is that they are based on visible and near-IR spectroscopy, and photometry (0.33–1.1 μm) and, to a varying extent, albedo.

Spectra of the asteroid classes

Shown in Figure A30 are the average 0.35 to 1.1 μm reflectance spectra of the Tholen (1984) asteroid classes. The E, M and P classes are shown as a single spectrum because of the strong similarities in their spectra; they are distinguished in the taxonomy by large differences in albedo. The new J class has only been observed by charge-coupled device (CCD) observations in the 0.43 to 1.0 μm wavelength range. Because of this relatively limited spectral coverage, the full resolution CCD spectrum is included. Also included are the three subclasses of the S class proposed by Howell, Merenyi and Lebofsky (1993).

Distribution of taxonomic classes

In addition to proposing the first asteroid taxonomy, Chapman, Morrison and Zellner (1975) noted that their new asteroid classes were not uniformly distributed through the asteroid belt. The S class seemed to dominate the inner asteroid belt while the C class was far more abundant in the outer asteroid belt. This relationship between heliocentric distance and taxonomic type was formalized by Gradie and Tedesco (1982) using preliminary results from ECAS and a radiometric survey. They observed that if abundance of taxonomic type is plotted against semimajor axis, the most populous taxonomic classes (the E, S, C, P and D classes) peak in abundance at different heliocentric distances. In addition, the distribution of each class is roughly Gaussian in shape and tends to range over about 1 AU. Shown in Figure A31 is the distribution of taxonomic classes from Bell *et al.* (1989). This plot is based on 584 asteroids classified by Tholen (1984) and is not corrected for observational bias.

What does this remarkable heliocentric distribution mean? If we assume that the bulk of the spectral and albedo differences between the asteroid classes reflect real differences in mineralogy and thermal evolution, then what we are seeing are rough compositional zones in the asteroid belt. These zones probably are the result of a combination of effects that include the primordial distribution of mineralogy, the subsequent thermal evolution of the asteroid belt and the sum of dynamical processes that scatter and pulverize asteroids.

The dominant cause of this taxonomic zonation may be an echo of the conditions in the early solar nebula (see Solar system: origin; Solar nebula). During the earliest period of its formation the solar system was a flattened whirlpool of hot gases and dust. As the nebula cooled the temperature, pressure and chemical state of the nebular gas controlled the mineralogy of the condensing grains. According to models of solar system condensation the high to moderate temperature silicate minerals would tend to dominate the inner solar system, while lower-temperature carbonaceous minerals would be common in the

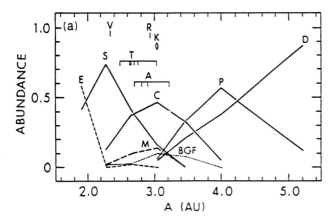

Figure A31 The distribution of taxonomic classes from Bell et al. (1989). This plot is based on 584 asteroids classified by Tholen (1984) and is not corrected for observational bias. The plot is dominated by the large S, C, P and D classes and the single-member V and R classes only show up as individual points in the distribution.

cooler, outer regions of the solar system. The transition between moderate and low temperature nebular condensates is apparently what we are seeing in the taxonomic zonation of the asteroid belt. The innermost major group of asteroids are the E class, which are thought to be composed of the iron-free silicate enstatite, indicating formation under relatively reducing conditions (see Meteorite parent bodies). This class peaks in abundance at about 2 AU. The next group out is the S class (including the subclasses So, Sp and S) which peaks in abundance at 2.3 AU. This class is thought to be rich in the moderate-temperature silicates olivine and pyroxene, and also to contain large amounts of free metal. The mix of mineralogies in the S class suggests more oxidizing conditions in this region of the solar nebula. The C class, which peaks in abundance at 3 AU, shows a major transition in asteroid mineralogy. Compared to the S class, the C asteroids probably contain less free metal, more oxidized silicates, important low-temperature carbon minerals, and significant amounts of volatiles such as water. The outer asteroid belt is dominated by the P asteroids which peak in abundance at about 4 AU, and the D asteroids, peaking at 5.2 AU. Although there are no direct meteorite analogs for these asteroid types, their spectra indicate mineralogies rich in low-temperature materials such as carbon compounds, complex organics and volatiles. The general trend of the taxonomic zonation supports the predictions of some solar system condensation models and may provide a guide to the chemical and mineralogical variations within the solar nebula. Geochemical studies of the different taxonomic classes may provide constraints on the temperature, pressure and chemical structure of the nebula during solar system formation.

The taxonomic imprint from the original condensation has probably been blurred by several post-accretion processes. Apparently a radially dependent thermal event heated much of the asteroid belt soon after accretion. Evidence from meteorites shows that some parent asteroids were completely melted and differentiated (basaltic achondrites, irons, stony irons), some asteroids were strongly metamorphosed (ordinary chondrites) and some were heated only enough to boil off volatiles and produce aqueous alteration (CI and CM carbonaceous chondrites). This event seems to have been much more intense in the inner asteroid belt and strongly affected the E, S, A, R, V and M class asteroids. The dynamical interaction of asteroids with each other and the planets, particularly Jupiter, has altered the original orbital distribution of the asteroids and cleared whole sections of the belt (see Asteroid: resonance). The net result probably has been to expand the original compositional zones and produce orbital overlaps of zones that were once distinct from each other.

Summary

Asteroid taxonomy has evolved along with the expansion of asteroid data sets. The widely used Tholen (1984) classification system has its roots in a classification that began with just three types in 1975 and now has expanded to a current total of 19 types. This is a reflection of the greater breath and detail of asteroid data sets and also a greater recognition of the complexity and variety of the asteroid belt. Similarly, the recognition of taxonomic zones suggests that the asteroid belt retains priceless information about the conditions of the primeval solar nebula as well as the history of many subsequent solar system processes.

Daniel T. Britt and Larry A. Lebofsky

Bibliography

Barucci, M.A., Capria, M.T., Coradini, A. and Fulchignoni, M. (1987) Classification of asteroids using G-mode analysis. *Icarus*, **72**, 304–24.

Bell, J.F., Davis, D.R., Hartmann, W.K. and Gaffey, M.J. (1989) Asteroids: the big picture in *Asteroids II* (eds R.P. Binzel, T. Gehrels and M.S. Matthews). Tucson: University of Arizona Press; pp. 921–45

Chapman, C.R., Morrison D. and Zellner, B. (1975) Surface properties of asteroids: a synthesis of polarimetry, radiometry, and spectrophotometry. *Icarus*, **25**, 104–30.

Gaffey, M.J., Bell, J.F. and Cruikshank D.P. (1989). Reflectance spectroscopy and asteroid surface mineralogy, in *Asteroids II* (eds R.P. Binzel, T. Gehrels and M.S. Matthews) Tucson: University of Arizona Press, pp. 98–127.

Gaffey, M.J., Bell, J.F., Brown, R.H. *et al.* (1993) Mineralogical variations within the S-type asteroid class. *Icarus*, **106**, 573–602.

Gradie, J.C. and Tedesco, E.F. (1982) Compositional structure of the asteroid belt. *Science*, **216**, 1405–7.

Howell, E.S., Merenyi, E. and Lebofsky, L.A. (1993) Classification of asteroid spectra using a neural network. *J. Geophy. Res.*, 99, 10847.

Tedesco, E.F., William, J.G., Matson, D.L. *et al.* (1989) A three-parameter asteroid taxonomy. *Astron. J.*, **97**, 580–606.

Tholen, D.J. (1984) Asteroid taxonomy from cluster analysis of photometry. PhD thesis, University of Arizona, Tucson.

Zellner, B. and Bowell, E. (1977) Asteroid compositional types and their distributions, in *Comets, Asteroids, Meteorites: Interrelations, Evolution and Origins* (ed. A.H. Delesemme. Toledo: University of Toledo Press, pp. 185–97.

Cross references

Meteorite
Meteorite parent bodies
Solar system: origin

ASTEROID: FAMILIES

The orbital elements of the main-belt population of asteroids tend to cluster about specific sets of values. Three of these groups (whose largest members are 24 Themis, 221 Eos and 158 Koronis) were discovered by Hirayama (1918a), who called them asteroid families. Hirayama hypothesized that the family members were in fact fragments of an originally larger parent asteroid, which had been disrupted by a collision with another smaller one.

Osculating and proper orbital elements

Asteroids move on elliptic orbits with the Sun located at one of the foci. Each of these orbits is described by five quantities, which are the semimajor axis a, the eccentricity e, the inclination i, the argument of perihelion ω and the longitude of node Ω, characterizing respectively the size, shape, inclination with respect to the ecliptic and (for the last two) the orientation in space of the ellipse. Together with the mean anomaly M (giving the instantaneous position of the body in its orbit) these quantities constitute the set of osculating orbital elements of an asteroid.

Hirayama realized that the distribution of osculating a, e and i of asteroidal orbits is not uniform, but rather shows the presence of clusters, not due to chance, close to some specific values. After some time he also realized that a better procedure to identify these

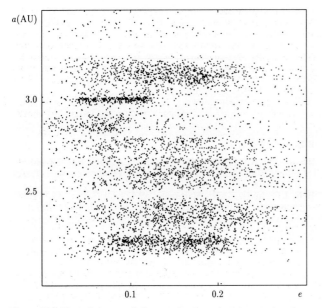

Figure A32 Plot of the osculating semimajor axis a versus the sine of osculating eccentricity e for asteroids up to No. 5383.

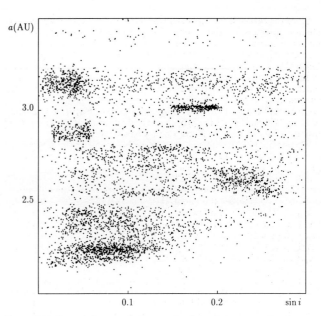

Figure A34 Plot of the osculating semimajor axis a versus the sine of osculating inclination i for the same asteroids of Figure A32.

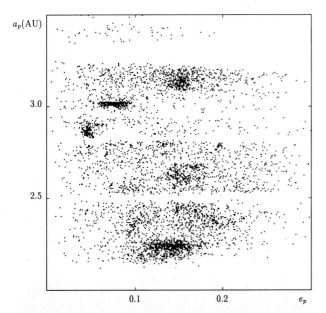

Figure A33 Plot of the proper semimajor axis a_p versus the proper eccentricity e_p for the same asteroids of Figure A32; the proper elements used here come from computations by Milani and Knežević, version 6.8. The three first Hirayama families Themis, Eos and Koronis are now recognizable, respectively at $a_p \sim 3.1$, $e_p \sim 0.15$, at $a_p \sim 3.0$, $e_p \sim 0.08$, and at $a_p \sim 2.9$, $e_p \sim 0.05$; Hirayama's Flora family is embedded in the conspicuous cluster at $a_p \sim 2.2$, $e_p \sim 0.15$, where unambiguous identification of individual families is not trivial, and Hirayama's small family Maria is located at $a_p \sim 2.55$, $e_p \sim 0.10$.

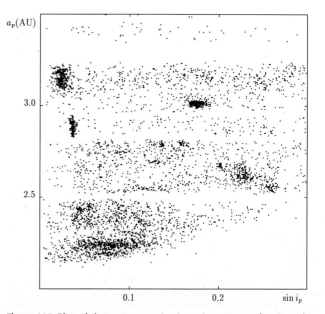

Figure A35 Plot of the proper semimajor axis a_p versus the sine of proper inclination i_p for the same asteroids of Figure A32; also in this plot the families Themis, Eos and Koronis are easily recognizable, respectively at $a_p \sim 3.1$, $\sin i_p \sim 0.02$, at $a_p \sim 3.0$, $\sin i_p \sim 0.17$, and at $a_p \sim 2.9$, $\sin i_p \sim 0.04$, and Hirayama's Flora family is embedded in the cluster at $a_p \sim 2.2$, $\sin i_p \sim 0.1$, whereas Maria family is recognizable at $a_p \sim 2.5$, $\sin i_p \sim 0.25$.

clusterings should be based not on the osculating, but rather on the proper orbital elements (Hirayama, 1923; see Orbit).

In fact, in the absence of planetary perturbations, all the osculating elements, except M, would remain unchanged, as in the classical two-body problem, and M would increase with time at a constant rate; however, the effect of planetary perturbations is to introduce small quasi-periodic variations of the values of a, e and i, and a slow rotation of ω and Ω.

Among the perturbations, those of short period depend on the instantaneous positions of the planets along their orbits, and thus have periods comparable to the revolution periods of the planets; it is possible to take into account these perturbations by averaging them over all the positions of the planets on their orbits, thus defining a new set of elements, the mean elements, such that the mean a, e and i vary with time much less than the osculating ones.

Since the orbits of the planets are also subject to perturbations, the mean elements still exhibit some variations over longer time scales; taking into account also long-period perturbations is a complex task that can be performed only to a certain limited degree of accuracy. Once this is done, the resulting proper elements are almost constant in time; actually, since the n-body problem is not integrable (i.e the positions and velocities of the bodies of such a system cannot be accurately computed for all times through analytical methods), approximate theories are used to model planetary perturbations, and as a result in most cases the proper a, e and i of an asteroidal orbit are found to vary in time by about one part in a thousand.

The idea behind Hirayama's interpretation of the origin of asteroid families is that if a catastrophic collision between two asteroids takes place, leading to the disappearance of the smaller (the 'projectile') and to the fragmentation of the larger one (the 'target'), there should be a range of conditions in which a recognizable clustering of proper a, e and i is produced; this should happen when the collision process is energetic enough to allow the escape of the fragments, while not being too energetic (Chapman et al., 1989). If the typical velocity v_{ej} of the fragments ejected from the target asteroid were less than, or nearly equal to, the escape velocity v_{esc} of its largest surviving fragment, re-accretion onto the latter of most of the small pieces would take place. On the other hand, if v_{ej} were much higher than v_{esc}, and comparable to the typical relative velocities among main-belt asteroids, then the fragments flying away from the largest remnant of the target would disperse too much, in the space of the elements a, e and i, among the background asteroids, becoming indistinguishable from them; but if v_{ej} were just slightly larger than v_{esc}, then the proper a, e and i of the fragments would remain, with respect to those of the largest remnant, closer than the proper a, e and i of the other asteroids in the same region of the belt.

Asteroid classifications in families

After Hirayama's pioneering work (Hirayama, 1918a, 1918b, 1919, 1920, 1923, 1928, 1933), aimed at identifying the main asteroid families and finding their members, no further work was done on the subject for about two decades, until Brouwer re-addressed the issue (Brouwer, 1951), using a more refined perturbation theory to compute the proper elements and a larger asteroid sample (this is a feature distinguishing practically every new classification from the previous ones, due to the ever increasing number of asteroids with well-studied orbits).

About another two decades then passed before a new investigation was carried out by Arnold (1969); this time, not only was the number of asteroids available for the investigation larger, but a computer-assisted method, instead of the visual inspection of tables and plots, was used to identify the groupings of asteroids in proper element space. Many more studies were performed in the next decade, using either automatic, computer-assisted techniques for the identification of groupings (Lindblad and Southworth, 1971; Carusi and Massaro, 1978), or improved perturbation theories for the computation of proper elements, with visual techniques for the identification (Kozai, 1979; Williams, 1979). In this last case the visual identification was coupled with a statistical criterion for the rejection of spurious groupings.

While the work of Hirayama led him to the conclusion that there are five main asteroid families, those associated with 24 Themis, 221 Eos, 158 Koronis, 170 Maria and 8 Flora, the search by Brouwer substantially raised this number, up to 28; after that, the number of families identified varied according to the different authors of the searches: 37 for Arnold, 34 for Lindblad and Southworth, only 15 for Carusi and Massaro, but 72 for Kozai and 104 for Williams.

In a comparison of the published family classifications Carusi and Valsecchi (1982) pointed out that they agreed very well with each other only for the three large families first identified by Hirayama, but on almost nothing else, and that there were many methodological reasons for that; in particular, the various different methods used for family identification seemed to have played a crucial role, and the large range in the total number of families found appeared to be related to this issue. To resolve the controversy, the best available theory for the computation of proper elements and several statistical techniques for identifying the groupings should be tried (Valsecchi et al., 1989); this should lead to more reliable results.

Zappalà et al. (1990) and Bendjoya, Slézak, and Froeschlé (1991) have used two automatic techniques, respectively hierarchical clustering and wavelet analysis for their classifications, and have both used the recent theory of proper elements by Milani and Knežević (1990) The number of families found in both investigations is 21 and, although there is little agreement for the families found in the innermost part of the asteroid belt (due to some shortcomings of the perturbation theory, which gives proper elements of lower quality in this zone) the overall level of agreement is better than in the past. Further refinements in proper element theory and identification techniques are expected in the near future.

Williams (1992) advocates a different point of view. According to him, visual identification of clusters, with a statistical test of the probability of their occurrence by chance, is a valid tool for the purpose. In addition, he observes that the lack of agreement among the results of different searches does not necessarily imply that the controversial families are not real, but just that they cannot be easily identified in a unique way.

Physical studies of family members

An important consequence of a confirmation of the reality of an asteroid family is that it becomes possible to obtain information on the interior of the parent body. Various attempts have been made to compare the surface properties of family members with those of asteroids belonging to the same region of the belt (Chapman et al., 1989); an important result is that the members of the two well-established and populous Themis and Eos families are clustered not only in orbital elements, but also in spectral properties, strengthening the plausibility of their genetic linkage. For other cases the results are less conclusive, and more observational data have to be gathered. The Galileo spacecraft (q.v.), on its route towards Jupiter, has passed close to 243 Ida, a member of the well-established Koronis family, discovering that it has a very peculiar shape, is heavily cratered and is accompanied by a small satellite, which has been named Dactyl.

Giovanni B. Valsecchi

Bibliography

Arnold, J.R. (1969) Asteroid families and 'jet streams'. Astron. J., **74**, 1235–42.

Bendjoya, Ph., Slézak, E. and Froeschlé, Cl. (1991) The wavelet transform: a new tool for asteroid family determination. Astron. Astrophys., **251**, 312–30.

Brouwer, D. (1951) Secular variations of the orbital elements of minor planets. Astron. J., **56**, 9–32.

Carusi, A. and Massaro, E. (1978) Statistics and mapping of asteroid concentrations in the proper elements' space. Astron. and Astrophysics Suppl. Ser., **34**, 81–90.

Carusi, A. and Valsecchi, G.B. (1982) On asteroid classifications in families. Astron. and Astrophys., **115**, 327–35.

Chapman, C.R., Paolicchi, P., Zappalà, V. et al. (1989) Asteroid families: physical properties and evolution, in Asteroids II (eds R.P. Binzel, T. Gehrels and M.S. Matthews). Tucson: University of Arizona Press, pp. 386–415.

Hirayama, K. (1918a) Groups of asteroids probably of common origin. Proc. Physico-Mathematical Soc. Jap., Ser. II, **9**, 354–61.

Hirayama, K. (1918b) Groups of asteroids probably of common origin. Astron. J., **31**, 185–8.

Hirayama, K. (1919) Further notes on the families of asteroids. Proc. Physico-Mathematical Soc. Jap., Ser. III, **1**, 52–9.

Hirayama, K. (1920) New asteroids belonging to new families. Proc. Physico-Mathematical Soc. Jap., Ser. III, **2**, 236–40.

Hirayama, K. 1923. Families of asteroids. Jap. J. Astron. Geophys., **1**, 55–93.

Hirayama, K. (1928) Families of asteroids. Second paper. Jap. J. Astron. and Geophys., **6**, 137–62.

Hirayama, K. (1933) Present state of the families of asteroids. Proc. Imp. Acad. Jap., **9**, 482–5.

Kozai, Y. (1979) The dynamical evolution of the Hirayama family, in Asteroids, (ed. T. Gehrels), Tucson: University of Arizona Press, pp. 334–58.

Lindblad, B.A. and Southworth, R.B. (1971) A study of asteroid families and streams by computer techniques, in Physical Studies of Minor Planets (ed. T. Gehrels). Washington, DC: NASA SP-267, US Govt Printing Office, pp. 337–52.

Milani, A. and Knežević, Z. (1990) Secular perturbation theory and computation of asteroid proper elements. *Celest. Mechan. Dynamical Astron.*, **49**, 347–411.
Valsecchi, G.B., Carusi, A., Knežević, Z. *et al.* 1989. Identification of asteroid dynamical families, in *Asteroids II* (eds R.P. Binzel, T. Gehrels and M.S. Matthews). Tucson: University of Arizona Press, pp. 368–85.
Williams, J.G. (1979) Proper elements and family memberships of the asteroids, in *Asteroids*, (ed. T. Gehrels) Tucson: University of Arizona Press, pp. 1040–63.
Williams, J.G. (1992) Asteroid families – an initial search. *Icarus*, **96**, 252–80.
Zappalà, V., Cellino, A., Farinella, P. and Knežević, Z. (1990) Asteroid families. Identification by hierarchical clustering and reliability assessment. *Astron. J.*, **100**, 2030–46.

Cross references

Dust
Meteorite parent bodies

ASTEROID: LIGHTCURVE

The brightness of an asteroid changes slowly with time due to the changing distances from the Earth and Sun, and the changing solar phase angle α, which is the angle between the lines of illumination and viewing. The brightness also changes more rapidly due to rotation. This is mostly due to the irregular shapes of asteroids, but also to some degree due to albedo variegation of the surface. This latter, periodic variation is called the asteroid's lightcurve. The first effect, due to changing distances, is trivial and can be removed by reducing observed magnitudes to values equivalent to the asteroid being at a standard distance of 1 AU from both the Earth and Sun. The lightcurve variation is periodic, with a fundamental period being that of the asteroid's rotation, and including higher harmonics. Since typical rotation periods are about 10 h, the lightcurve can be separated from the slower variation due to changing phase angle (the phase relation) as follows:

$$V(\alpha, t) = \overline{V}(\alpha) + \sum_{l=1}^{n} \left[A_l \sin \frac{2\pi l}{P}(t - t_0) + B_l \cos \frac{2\pi l}{P}(t - t_0) \right]$$

where α is the solar phase angle, $\overline{V}(\alpha)$ is the mean magnitude at α, and A_l and B_l are the amplitudes of periodic variation of harmonic l. P is the rotation period, and t_0 is a reference time, taken in practice to be somewhere near the span of observations. It is thus possible to fit a set of observations to the above relation. Generally data from a single night can be treated as if the phase angle α were constant; thus for each night of observations fitted to the above relation, a different value of $\overline{V}(\alpha)$ is obtained. Thus a least squares fit to many days' data results in values for $\overline{V}(\alpha)$ for each day, a single value of P and one set of Fourier coefficients. The optimum order n of the harmonic solution can be found by trying solutions of ever increasing order until no further improvement in fit results. Figure A36 is a lightcurve plot for a typical asteroid, which is a composite of observations on ten different nights, each plotted with a different symbol. The plotted points are the observed magnitudes (reduced) minus the mean level at α for the night, $V_{obs} - \overline{V}(\alpha)$. The curve is the solution Fourier fit from the above equation, $V(\alpha, t) - \overline{V}(\alpha)$. A plot of $\overline{V}(\alpha)$ vs. α is called the *phase curve*, or *phase relation*, of the asteroid. Figure A37 is the phase curve for the same asteroid as in Figure A36, with the same plot symbols used for each day.

Alan W. Harris

Bibliography

Harris, A.W. and Lupishko, D.F. (1989) Photometric lightcurve observations and reduction techniques, in *Asteroids II* (eds R.P. Binzel, T.Gehrels, and M.S. Matthews) Tucson: University of Arizona, pp. 39–53.

Cross references

Asteroid
Asteroid: photometry
Chiron
Hilda asteroid

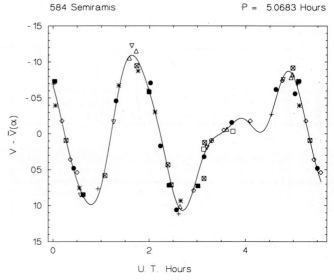

Figure A36 A typical asteroid lightcurve. Different symbols are used for observations on different nights. The curve is the computed Fourier fit to the observations.

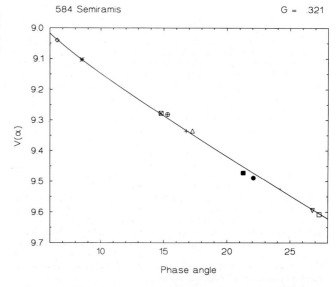

Figure A37 The phase relation derived from the same observations as plotted in Figure A36. The plot symbols used for each day are the same as in Figure A36.

ASTEROID: PHOTOMETRY

Photometric lightcurves of asteroids are used to determine periods of rotation, to estimate shapes and spin axis orientations, and to define phase relations, or brightness as a function of solar phase angle.

Observational techniques and data reduction

The brightness of an asteroid varies due to (1) the changing distances between it and the Earth and Sun, (2) the rotation of the asteroid, and (3) the solar phase angle, or angle between the lines of

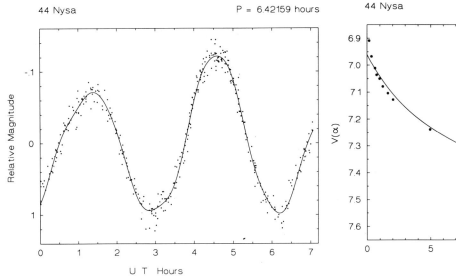

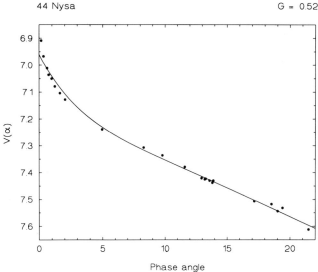

Figure A38 The composite lightcurve of the asteroid 44 Nysa. The observations were taken on 23 different nights over a span of nearly 5 months. A seventh-order Fourier analysis of the data yielded the above composite curve, with a solution period of 6.42159 ± 0.00005 h.

Figure A39 The phase relation of 44 Nysa, which consists of the zero-point offsets of the observed magnitudes each night to adjust the points to fall on the lightcurve of Figure A38.

illumination and viewing. The first effect is trivial and is removed by computing equivalent magnitudes as if observed at 1 AU from both the Earth and Sun. Such magnitudes are called *reduced magnitudes*. Usually the second and third components of variation can be separated because that due to rotation is periodic, with a fundamental frequency equal to the rotation period (typically ~ 10 h), while the variation due to changing phase angle generally has a much longer timescale. Thus we can represent the reduced magnitude of an asteroid as the sum of a mean magnitude $\overline{V}(\alpha)$ at solar phase angle α, plus a harmonic series in time with a fundamental period equal to the asteroid's rotation period P, and including as many harmonics as necessary to describe fully the structure of the curve. In practice, photometric data from many nights can be fit to yield a *composite lightcurve*, consisting of the harmonic component of variation (Figure A38), and a *phase relation*, which is a plot of the mean magnitude levels $\overline{V}(\alpha)$ as a function of α (Figure A39). The lightcurve in Figure A39 contains data from 23 nights. The Fourier fit (curved line) required harmonics up to the seventh order to fit the data fully. For most asteroids the rotational lightcurve is dominated by the second harmonic (hence two maxima and two minima per rotation cycle), which implies that most of the variation is due to shape rather than albedo variegation. In Figure A39 the mean magnitude levels $\overline{V}(\alpha)$ are plotted for each night. It should be noted that $\overline{V}(\alpha)$ is not the mean of the measurements for a night, but is instead the magnitude at phase angle α that corresponds to the zero level of the composite lightcurve on a given night.

Harris and Lupishko (1989) discuss methods of lightcurve observations and reductions more fully.

Rotation rate statistics

At present, periods of rotation have been determined from lightcurves for about 500 asteroids. Figure A40 is a plot of those periods, versus the estimated diameter of each object. Among the largest objects, $D > 100$ km, the dispersion in rotation rates is more or less Maxwellian about a mean of ~ 10 h. This is consistent with rotations arising through collisions among the asteroids, either in the course of formation (accretional growth) or subsequent fragmentation. Below $D \approx 100$ km, the dispersion in rotation rates becomes distinctly non-Maxwellian. Among the smallest asteroids, it appears that about 80% of the population obey more or less Maxwellian statistics, with a mean rotation rate somewhat faster than the mean for the larger bodies ($P \approx 5$ h), while the rest seem to rotate much slower, with a mean period of perhaps as much as 50 h. This result is not fully apparent in the figure, because several small asteroids which have

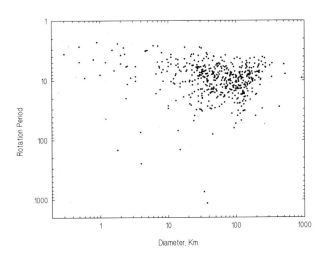

Figure A40 A plot of the rotation rates versus diameters for nearly 400 asteroids.

been observed only partially are not included in the plot. For these the exact rotation rates are not determined, but the data are sufficient to know reliably that the periods of rotation are tens of hours long. It has been speculated that these slow rotators are the result of tidal friction between binary pairs, like the Pluto–Charon system, or that they are extinct comets that have slowed down due to outgassing. Neither of these explanations appears viable. Among the very small objects, the rotation statistics appear the same for both near-Earth asteroids (some of which probably are extinct comets) and for main-belt asteroids, which are unlikely to contain a component of extinct comets. Tidal friction can only operate to slow the spin of a small, relatively rigid body to a period of about 5 days. Several asteroids have longer periods, notably 288 Glauke (~ 60 days), 1220 Crocus (~ 30 days) and 4179 Toutatis (~ 10 days). Both 288 Glauke and 4179 Toutatis have been observed with radar. These observations confirm that the 2-month period of Glauke is a solid body rotation period, and that Toutatis does not consist of two detached binary components. Thus tidal friction can be ruled out as the operator producing either of these rotation states. The explanation for the slow rotators remains a mystery.

Another application of the rotation statistics of asteroids is to investigate rotation rates versus taxonomic class. Dividing the various classes crudely into those which are considered to be undifferentiated, or primitive (C, G, B, F, D, P, T, X), differentiated (S, A, Q, E, V, R) and metallic (M), we find that the mean rotation rates differ at a significant level. The geometric mean rotation periods for each of the above groups are 10.4 ± 0.2, 9.7 ± 0.2, and 7.0 ± 0.3 h respectively. A theoretical model of collisional evolution of spins (Harris, 1979) predicts that the mean spin rate should be proportional to the square root of the mean density of a class of bodies. While that model is insufficient to make quantitative estimates of the mean densities, we can infer ratios of densities. The above mean rotation rates imply density ratios of the three classes of 2.6 : 3.0 : 5.9. The numbers have been normalized to 3.0 for the differentiated class, which is the expected density for such material, e.g. lunar rocks or ordinary chondritic meteorites. The other two densities implied are likewise consistent with their interpretations as undifferentiated material and iron-rich material, respectively.

A thorough review of asteroid rotation rate statistics can be found in Binzel *et al.* (1989).

Shapes and pole orientations

Lightcurves taken at different apparitions of an asteroid, that is, from different directions in the sky, usually exhibit varying forms, from which one might infer both the shape of the asteroid, and the direction in space of the spin axis. H.N. Russell (1906) first considered this problem. He showed that one could infer the pole direction from lightcurves taken at different aspects in the sky, but that any possible lightcurve could be reproduced by an infinite variety of albedo patterns on an arbitrarily shaped body. Thus he declared that the problem of inverting lightcurves to obtain shape and albedo patterns is intractable. This conclusion discouraged serious work on the problem for several decades. It has become apparent that asteroids exhibit very little albedo variegation, probably because they become covered with a fairly uniform layer of ejecta dust (or regolith) as a result of meteoroid bombardment. Thus one can consider the question of determining shape only from lightcurves, assuming that the albedo is constant over the surface.

Conceptually, there are two methods of determining the spin axis orientation. The 'epoch' method consists of comparing the times of arrival of a lightcurve feature (e.g. maximum brightness) from several aspects, from which one might infer the direction of rotation of a lighthouse beam from timings of the beam arrival in various directions. The 'magnitude–amplitude' method consists of determining the shape and orientation of the asteroid based on the intrinsic brightness and range of variation as seen from various aspects. The first method is insensitive to shape, as long as enough variation exists to define 'epochs', while the second method is insensitive to direction of rotation, or for that matter even rate of rotation. By combining the two methods, one can obtain a solution for all of those parameters. Figure A41 (Magnusson *et al.*, 1992) is an illustration of these methods applied to 951 Gaspra (q.v.), which was the target of the first asteroid flyby observations by the Galileo spacecraft. Gaspra was observed at four different aspects, in 1982, 1988, 1990 and 1991. The figure shows schematically Gaspra's spin axis orientation as determined from those observations, and sample lightcurves observed at each aspect. In 1982 Gaspra was observed relatively pole-on, thus the low-amplitude lightcurve was not very constraining for the epoch method, but provided a very strong constraint for the magnitude–amplitude method. Conversely, the nearly equatorial lightcurve obtained in 1988 was very powerful for the epoch method, but not particularly helpful for determining the latitude of the pole. Using a simultaneous solution employing both methods, the lightcurves from the four aspects sufficed to define the spin rate, pole orientation and rough shape of Gaspra, which agreed very well with the images returned by Galileo. The shape of Gaspra was modeled more completely by Barucci *et al.* (1992).

What can be said about the statistics of asteroid shapes and pole orientations? To date, about 30 asteroids have fairly reliably determined pole orientations. There may be a slight preference for prograde rather than retrograde rotation, and a tendency toward avoidance of very high obliquity (near 90°), but the statistics are not convincing, especially in the absence of a physical explanation of how it could be so. It seems much more likely that axis orientations are random. With regard to shapes, there are some minor trends with respect to subclasses of objects, but the only firmly established trend

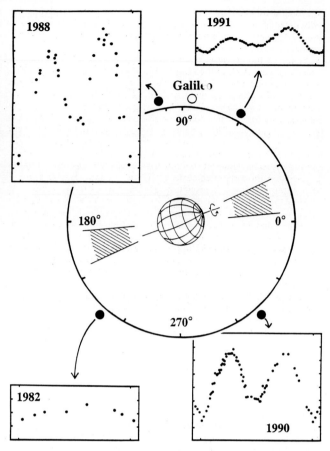

Figure A41 The asteroid 951 Gaspra was observed at four different aspects before the Galileo encounter. The four lightcurves are depicted with the celestial longitude of each indicated. In the center is a figure indicating the orientation and spin of Gaspra derived from the observations. (Reproduced from Magnusson *et al.* (1992), with permission of Academic Press.)

is that the largest asteroids have smaller amplitudes of variation, indicating less shape irregularity than smaller bodies, no doubt reflecting the effect of gravity tending to draw the largest bodies into a more spherical form. The onset of decreasing amplitude is at about 100 km diameter. Among these larger asteroids, there is a correlation between amplitude and spin rate – that is, the fast-spinning asteroids have a tendency to be more elongate. This has been interpreted as a tendency toward figures of rotational equilibrium, indicating that these bodies may be 'rubble piles' of loosely consolidated regolith. Among smaller asteroids, the mean lightcurve amplitude is consistent with shapes similar to the average irregularity of impact fragments, as determined in the laboratory.

Magnusson *et al.* (1989) provide a recent review of methods of shape and pole determinations; current results are presented by Magnusson (1986, 1990). A new method of shape determination including the higher order harmonics of the lightcurves has been developed by Kaasalainen *et al.* (1992), yielding the most accurate results to date.

Phase relations

The integrated brightness of an asteroid surface depends in a rather complex way on the solar phase angle, due to albedo, shadowing effects, surface roughness, the irregular shape of the body, and many other factors. It is of importance to try to model phase relations not only for physical interpretation of the asteroid surface, but also to allow comparison of observations taken at different phase angles, in particular for the determination of shapes and pole orientations, as discussed above. A phase model is also necessary in order to predict the brightness of an asteroid at any given time, as a part of the ephemeris of the asteroid.

Some of the essential features of a typical phase relation can be seen in Figure A39. In the range 6–30°, the slope of the phase relation (in magnitude units) is nearly constant. The slope is determined primarily by the large-scale roughness of the surface and by the albedo of the surface. These two parameters are highly correlated, so it is difficult to separate them from phase relations alone. In the range $\alpha < 6°$, the phase relation steepens, which is the so-called 'opposition effect' (q.v.). This has traditionally been ascribed to shadowing in a porous surface. As the phase angle approaches zero, microscopic shadows become hidden behind the foreground surfaces that cast them, thus the integrated surface brightness increases. According to this interpretation, the amplitude of the opposition effect should be greater for darker surfaces, since high-albedo surfaces tend to fill in shadows due to multiple reflections. The width of the opposition effect should be related to the porosity of the surface, since a more 'fluffy' surface should have more vertical relief and require a lower phase angle for shadow-hiding. At large phase angles, > 30° (beyond the range seen in Figure A38), the phase relation steepens again, this time mainly due to the geometric effect of decreasing illuminated surface area.

The H–G photometric model (Bowell et al., 1989) was developed primarily for the purpose of magnitude prediction, but also has some use for physical interpretations. The fundamental assumption of this model is that the total light returned at a given phase angle from any asteroid surface is the sum of the intensities of two phase functions, which are themselves not dependent on a specific asteroid. That is, the phase relation of any asteroid can be expressed as the linear sum (in intensity units) of two functions. The two parameters defining the phase relation for any object are thus H, the absolute brightness at zero phase angle, and G, the mixing ratio of the two functions. In concept, the two phase functions are the phase relations for singly scattered light and for multiply scattered light, but since the model is a linear sum of the two, any two well-separated, empirically determined phase relations can be used as a basis. The curve in Figure A39 is the fitted H–G phase relation for the asteroid 44 Nysa. About 100 asteroids have been observed in enough detail to determine at least crude phase relations. The principal variation among asteroids is in the slope of the phase curve. The two empirically determined phase curves of the H–G model differ mostly in slope, hence the mixing ratio G is usually referred to as the slope parameter.

An optical process that has only recently been recognized in asteroid phase curves is that of 'coherent backscatter' (Mishchenko, 1993). This process results from doubly scattered light, where the two reflecting surfaces are separated by at most a few wavelengths of light. Thus at low phase angle, two paths exist for any beam, which differ in pathlength by only a fraction of a wavelength (Figure A42). At zero phase angle the two beams interfere constructively, while at larger phase angle the interference becomes destructive, leading to a brightness surge at zero phase angle. It is this effect that is primarily responsible for the very bright radar albedos of the Galilean satellites and Saturn's rings, and indeed is likely the explanation for the 'opposition effect' of Saturn's rings in visible light. A similar very narrow 'opposition spike', was first observed for the asteroid 44 Nysa, easily apparent in Figure A39 as the misfit between the nominal H–G function plotted and the data points. This opposition spike has now been seen in the phase curves of other high albedo asteroids. These observations cast into doubt the model of shadowing in porous surfaces as the explanation of the opposition effect, at least among higher-albedo asteroids.

Recently phase relations of several asteroids have been obtained that violate the fundamental assumption of the H–G model, that is they cannot all be represented by linear combinations of two functions of phase angle. It is possible that the failure lies in not accounting for coherent backscatter explicitly. It is not yet clear whether a two-parameter (i.e. absolute magnitude and a 'slope parameter' of some sort) can successfully fit all phase curve data, or if a third parameter will become necessary.

Alan W. Harris

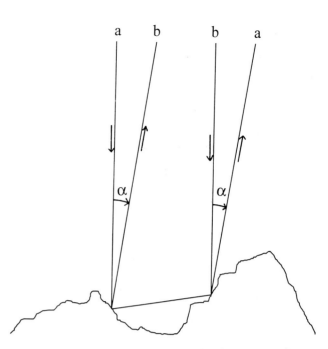

Figure A42 A schematic representation of the phenomenon of coherent backscatter. For any two scattering elements on a surface, a light ray can be doubly scattered along two paths (a and b) from the light source to the observer. At zero phase angle ($\alpha = 0$), the two paths have the same length, hence light waves traveling the two paths interfere constructively. As α increases, path a becomes shorter than b, so the interference becomes increasingly destructive. For a surface where scattering centers are typically only wavelengths apart, this leads to a narrow opposition spike near zero phase angle.

Bibliography

Barucci, M.A., Cellino, A., De Sanctis, C. et al. (1992) Ground-based Gaspra modeling: comparison with the first Galileo image. *Astron. Astrophys.*, **266**, 385–94.

Binzel, R.P., Farinella, P., Zappala, V. and Cellino, A. (1989) Asteroid rotation rates: distributions and statistics, in *Asteroids II* (eds R.P. Binzel, T. Gehrels and M.S. Matthews). Tucson: University of Arizona Press, pp. 416–41.

Bowell, E., Hapke, B., Domingue, D. et al. (1989) Application of photometric models to asteroids, in *Asteroids II* (eds R.P. Binzel, T. Gehrels and M.S. Matthews). Tucson: University of Arizona Press, pp. 524–56.

Harris, A.W. (1979). Asteroid rotation rates II. A theory for the collisional evolution of rotation rates. *Icarus*, **40**, 145–53.

Harris, A.W. and Lupishko, D.F. (1989) Photometric lightcurve observations and reduction techniques, in *Asteroids II* (eds R.P. Binzel, T. Gehrels and M.S. Matthews). Tucson: University of Arizona Press, pp. 39–53.

Kaasalainen, M., Lamberg, L., Lumme, K. and Bowell, E. (1992). Interpretation of lightcurves of atmosphereless bodies. I. General theory and new inversion schemes. *Astro. Astrophys.*, **259**, 318–32.

Magnusson, P. (1986) Distribution of spin axes and senses of rotation for 20 large asteroids. *Icarus*, **68**, 1–39.

Magnusson, P. (1990) Spin vectors of 22 large asteroids. *Icarus*, **85**, 229–40.

Magnusson, P., Barucci, M.A., Drummond, J.D. et al. (1989) Determination of pole orientations and shapes of asteroids, in *Asteroids II* (eds R.P. Binzel, T. Gehrels and M.S. Matthews). Tucson: University of Arizona Press. pp. 66–97.

Magnusson, P., Barucci, M.A., Binzel, R.P. et al. (1992) Asteroid 951 Gaspra: pre-Galileo physical model. *Icarus*, **97**, 124–9.

Mishchenko, M.I. and Dlugach, J.M. (1993) Coherent backscatter and the opposition effect for E type asteroids. *Planet. Space Sci.*, **41**, 173–81.

Russell, H.N. (1906) On the light-variation of asteroids and satellites. *Astrophys. J.*, **24**, 1–18.

Cross references

Asteroid
Asteroid: lightcurve

ASTEROID: RESONANCE

Resonance is a physical phenomenon that produces an amplification in the strength of an acting force, an increase in the acceleration and velocity of an object subject to this force, and a continuing influence building through these same factors toward a potentially maximized amplitude of displacement. The phenomenon is achieved by a process of synchronously timed force impulses, applied in rhythmic beat to an object constrained to harmonic motion or oscillation.

The familiar childhood example of resonance is that of a youngster on a swing who attains a surprising height by applying a propelling force repeatedly in synchronism with the oscillations of the swing, always at the same position in the arc of its ascent. The swing acquires a steadily greater amplitude of motion through the action of resonance reinforcement – in the same manner that a wave induced in a tilted bowl, when sloshed rhythmically back and forth, grows increasingly larger. The augmentation of the wave peaks continues, accompanied by a corresponding lowering in the wave troughs, with each coincidence of the extremes in a similar-phase relationship. Wherever, in nature, a vibrating body – or an oscillating particle or wave – possesses the same natural or induced frequency of vibration as another, a resonance reinforcement may occur.

In astrophysics and in its terrestrial relationships, the phenomenon of resonance is represented, for example, by (1) various actions affecting the transformation of radiant energy from one form to another (e.g. the establishment of the response frequency for Fraunhofer lines, resonance radiation and absorption band resonance); (2) a theory of nuclear disintegration; and (3) such other aspects as quantum mechanics resonance, the resonance Raman effect, etc.

In gravitational and dynamical astronomy, two additional examples of resonance influence exist. (1) A relatively small celestial body such as an asteroid revolves around the Sun as its central attracting force in accordance with the same Keplerian laws as the massive planet Jupiter. (2) Similarly, a small fragment of particulate matter which, in its compositional aggregate, forms one of Saturn's rings, revolves (along with the satellites of Saturn) around the planet at the same time the planet itself revolves around the Sun. Should the less massive body revolve around its primary in the same period – or in a period that is mathematically commensurable, in the respective cases, with (1) that of the planet Jupiter or (2) one of Saturn's satellites (see Resonance in Saturn's rings), resonance can also occur. A reinforcing impulse is imparted to the lesser of the two bodies by the more massive body, and both the velocity and displacement of the lesser body are increased.

In analogy to the 'pumping' action on the swing, the smaller orbiting body receives resonance-induced bursts of energy with each successive arrival at its minimum distance from the larger body of commensurable period. The result is an increase in the velocity of the small object or particle to one which exceeds the stability level for its distance from the central attracting source, and the object moves inward to a Keplerian orbit corresponding to this increased velocity.

Cumulatively, this action results in the physical displacement of asteroids from local regions of the asteroid belt in which the orbital velocities are commensurable with that of Jupiter. In this process a ratio of 1 : 2, 1 : 3, 2 : 5, or similar exact commensurability between the revolutionary *period* of an asteroid and that of Jupiter (at their respective distances from the Sun) is the basis for the resonant action produced.

The special case of the Trojan asteroids and the Lagrangian points (q.v.) represents a 1 : 1 commensurability of these asteroids with the period of Jupiter. The significance of other commensurable ratios has been established in many studies (e.g. Giffen, 1973; Franklin *et al.*, 1975; Milani and Nobili, 1984; Henrard and Lemaître, 1986; Murray, 1986; and Schubert, 1988). General summary volumes of papers describing recent discoveries in connection with the asteroids, made by NASA space probes and by flyby expeditions to the outer planets, are provided by Morrisson and Samz (1980) and Binzel, Gehrels and Matthews (1990).

The following analytic and quantitative discussion details the creation of five distinctive gaps within the principal asteroid belt by the phenomenon of resonance, as described above.

Kirkwood's theory of resonance gaps in the asteroid belt

According to this theory, during the formation of the primordial asteroid belt, an asteroid initially orbited the Sun with a period of revolution which represented a commensurable fraction of the period of Jupiter around the Sun. The asteroid thus revolved in fractional synchronism with Jupiter, cumulatively and after successive orbits reaching the closest position in its orbit to Jupiter, in phase with the motion of the planet. Resonance resulted. At each return to their positions of least separation, the object of smaller mass – the asteroid – received an additional burst of resonant energy, manifest in increased orbital velocity.

A situation therefore arose in which, from Kepler's third law (see below), the asteroid acquired an orbital velocity which was kinetically too large in relationship to its distance from the Sun; the body sought to compensate by an adjustment in position. The asteroid moved inward to a position closer to the Sun, abandoning its former orbit. A logical assumption can be made that, with the large number of planetoid fragments present in the belt, the above peculiarity applied to various cases of dynamically related fragments. These can be assumed to be moving originally with periods close to, or fractionally commensurable with, the period of the first example. Accordingly, this entire family of asteroids likewise developed periods commensurable with that of Jupiter.

By the evacuation of multiple asteroids from the same relatively narrow, annular zones, gaps were created within the asteroid belt at mean distances from the Sun approximating those of the asteroids which formerly occupied these zones. In consequence, after the formation of the principal asteroid belt, a number of 'zones of avoidance,' today known as Kirkwood's gaps, were established. These occur at distances from the Sun at which the corresponding periods of revolution of any random asteroids which might subsequently wander into these belts – being commensurable with the period of Jupiter – would likewise be ejected therefrom. The comprehensive theory of origin of these gaps, and numerous studies of their topology and other characteristics, are available (including papers by Kirkwood, 1866, 1869; Froeschlé and Scholl, 1976; and Wisdom, 1982).

Verification of existing discontinuities

Representative gaps in the asteroid belt are shown in Figure L2 of Lagrangian points. The mean distances of the more prominent gaps (adjusted slightly, in order to accommodate better the width of each gap) are 2.50, 2.82, 2.96, 3.28 and 3.65 AU. (The astronomical unit, (AU) defined as the mean distance of the Earth from the Sun in its annual orbit around the Sun, is equal to 149 572 827 km.) To test the validity of Kirkwood's theory, it is only necessary to convert these figures into the periods of time (i.e. sidereal periods of revolution) which bodies at each of these distances would take to orbit the Sun.

The revolution period of an object in orbit at a mean distance a from the Sun can be computed from the generalized form of Kepler's (third) harmonic law of motion under gravitation, namely: 'the squares of the periods of two celestial bodies around a central attracting source are proportional to the cubes of their distances from this source.' The planet Jupiter, having a major part in the explanation of this phenomenon, also provides a scaling factor for the interim analysis. A hypothetical object revolving at a distance from the Sun equal to one of the observed asteroid gaps is the third body in the system. Thus, in symbolic form:

$$P_A^2/P_J^2 = a_A^3/a_J^3$$

where the subscripts A and J refer respectively to the asteroid and Jupiter. Jupiter's mean distance from the sun is 5.203 AU, and its corresponding sidereal period of revolution around the sun is 11.862 years.

Substituting for a_A the first of the known gap values among the asteroid orbits, as well as the values of a_J and P_J, and solving for P_A:

$$P_A^2 = (11.862)^2 \times (2.50)^3/(5.203)^3 \text{ or } P_A = 3.951 \text{ years}$$

The results for the various gap distances may be grouped for comparison in a single table (Table A7).

Table A7 Resonance gaps in the asteroid belt, relative to Jupiter's period

Mean a_A (AU)	Mean P_A (years)	Equivalent commensurable fraction of Jupiter's period of 11.862 years	Corresponding value of this fractional portion of Jupiter's period (years; compare with col. 2)
2.50	3.951	1/3	3.954
2.82	4.733	2/5	4.745
2.96	5.090	3/7	5.083
3.28	5.937	1/2	5.931
3.65	6.970	7/12	6.920

It is evident that the period of revolution of an asteroid occurring within any of the zones of avoidance has a close approximation to a value which represents a commensurable fraction of Jupiter's sidereal revolution period.

Fergus J. Wood

Bibliography

Binzel, R.P., Gehrels, T. and Matthews, M.S. (eds) (1990) *Asteroids II*. Tucson: University of Arizona Press, 1258 pp.
Franklin, F.A. *et al.* (1975) Minor planets and comets in libration about the 2:1 resonance with Jupiter. *Astron. J.*, **80**, 729–46.
Froeschlé, C. and Scholl, H. (1976) On the dynamical topology of the Kirkwood gaps. *Astron. Astrophys.*, **48**, 389–96.
Giffen, R. (1973) A study of commensurable motion in the asteroid belt. *Astron. Astrophys.*, **23**, 387–403.
Henrard, J. and Lemaître, A. (1986) A perturbative treatment of the 2:1 Jovian resonance. *Icarus*, **69**, 266–79.
Kirkwood, D. (1866) (published 1867) On the theory of meteors (resonance between Jupiter and the asteroid belt). *Proc. Am. Assoc. Adv. Sci.*, 1866, 8–14.
Kirkwood, D. (1869) On the nebular hypothesis, and the approximate commensurability of the planetary periods. (An analytical development of the original theory of Kirkwood's gaps in the asteroid belt and in Saturn's rings.) *Mon. Not. Roy. Astron. Soc.*, **29**, 96–102.
Milani, A. and A.M. Nobili (1984) Resonant structure of the outer asteroid belt. *Celest. Mech.*, **34**, 343–55.
Morrison, D. and Samz, J. (1980) *Voyage to Jupiter*. Washington: NASA, SP-439, 211 pp.
Murray, C.D. (1986) Structure of the 2:1 and 3:2 Jovian resonances. *Icarus*, **65**, 70–82.
Schubert, J. (1988) Resonant asteroids between the main belt and Jupiter's orbit. *Celest. Mech.*, **43**, 309–17.
Wisdom, J. (1982) The origin of the Kirkwood gaps: A mapping for asteroidal motion near the 3:1 commensurability. *Astron. J.*, **87**, 577–93.

Cross references

Chaotic dynamics in the solar system
Commensurability
Resonance in the solar system

ASTEROID: THERMAL INFRARED STUDIES

Thermal infrared studies of asteroids afford planetary scientists the opportunity to study the physical and mineralogical properties of asteroids: size, albedo, surface roughness and composition. For the purposes of this article, we define the thermal infrared to be the spectral region beyond 5 μm. For asteroids, which are illuminated by the Sun, the spectral region beyond 5 μm is dominated by radiation thermally emitted from the asteroids themselves. Below 2.5 μm, their flux is dominated by reflected solar radiation, and between 2.5 and 5 μm is a transition region (Figure A43). The wavelength where this transition occurs is strongly dependent on the asteroid's albedo, solar distance and thermophysical properties.

Generally speaking, we can divide thermal studies into two types of observations, photometric and spectroscopic. Photometric observations are used for studying the physical properties of asteroids, while spectroscopic observations can be used to study the mineralogy of asteroids. We will describe these below.

Radiometric albedo and diameter determination

Thermal infrared observations of asteroids (5–20 μm) have been used for the determination of asteroid diameters and albedos for over two decades (Allen 1970). Radiometric diameter determinations involve finding a diameter and albedo that will simultaneously match the observed reflected sunlight and thermal emission from an object. A given visual magnitude can be matched by a large dark object or a small bright one, but the former will be warmer as well as larger than the latter and so will show much greater thermal emission. To make the technique quantitative, various assumptions are needed to determine a bolometric albedo (albedo averaged over all solar phase angles) from a visual magnitude and diameter, and a thermal model is required to predict the emission expected from a body of given size and bolometric albedo. Unfortunately, there are a number of assumptions that go into the thermal model calculation. These include microscopic and macroscopic surface roughness which affect the directionality of the thermal emission [similar to the visual opposition effect (q.v.)], pole orientation (along with rotation rate) and maturity of the surface regolith (rocky versus dusty).

Three basic models have been developed for the reduction of radiometric observations: the non-rotating or standard thermal model, the fast-rotating (rocky or isothermal latitude) thermal model, and the thermophysical model. As we discuss below, all three models assume a balance of solar insolation with re-emitted thermal radiation (see Lebofsky and Spencer, 1989, for more details). For simplicity, the asteroid is assumed to be spherical. This can be generalized in the form:

$$\pi R^2 (1 - A)S = \eta\epsilon\sigma R^2 \int_{-\pi}^{\pi}\int_{-\pi/2}^{\pi/2} T^4(\theta,\phi) \cos\phi \, d\phi d\theta \quad (A17)$$

where R is the radius of the asteroid, A is the bolometric Bond albedo, S is the solar flux at the distance of the asteroid, η is a normalization constant for adjusting the surface temperatures to compensate for the angular distribution of the thermal emission (infrared beaming) so that the correct flux is obtained at zero phase angle, ϵ is the wavelength-independent emissivity, σ is the Stefan–Boltzmann constant and $T(\theta,\phi)$ is the model temperature at longitude θ and latitude ϕ.

The relationship between the visual magnitude and the thermal flux is used to determine asteroid diameter and albedo. The visual magnitude is a function of the asteroid diameter and the visual geometric albedo, and the visual solar magnitude. As shown in equation (A17), the thermal flux is a function of asteroid diameter, the bolometric Bond albedo and the solar flux. The geometric and Bond albedos are related by the equation:

$$A = pq \quad (A18)$$

where p is the bolometric geometric albedo and q is the bolometric phase integral. Since the absolute magnitude at opposition (within the uncertainty of the lightcurve variation), total radiated flux of the Sun, and q are known or calculable, then the observable, the thermal flux, uniquely determines the diameter and bolometric geometric albedo of an asteroid.

The non-rotating and fast-rotating models are idealized end-members of the thermophysical model which assumes certain properties of the asteroid: rotation rate and direction, pole orientation and thermophysical properties of the surface material. A recent discussion of these models can be found in Lebofsky and Spencer (1989).

The standard thermal model (STM) for asteroids is a simplistic thermal model that has been evolving over the last 20 years. The model assumes the ideal situation of a non-rotating spherical asteroid in instantaneous equilibrium with solar insolation observed at opposition (Figure A44a). In reality, observations are never made *exactly* at opposition, so the observed fluxes must be adjusted to correct for this using a thermal phase coefficient. Lebofsky *et al.* (1986) observed several large asteroids and determined a mean thermal phase coefficient of about 0.01 mag deg^{-1}. Although this is an oversimplification, given the uncertainties of telescopic observations, and because of usually limited data sets, more sophisticated models have not been warranted for most asteroid observations.

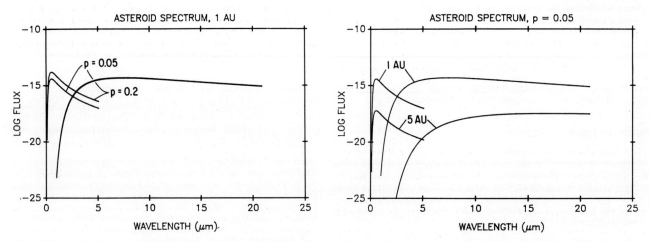

Figure A43 A comparison of the reflected and thermal flux from two 100-km radius asteroids. (a) Both asteroids at 1 AU from the Sun, one with a geometric albedo of 0.05 and one with a geometric albedo of 0.20, typical of C- and S-class asteroids respectively. While the reflected fluxes differ, the thermal fluxes are indisguishable at the scale of this figure. However, the crossover from reflected to thermal shifts about 0.4 μm. (b) One asteroid at 1 AU and the other at 5 AU, both with geometric albedos of 0.05. The crossover from reflected to thermal shifts by about 2.5 μm.

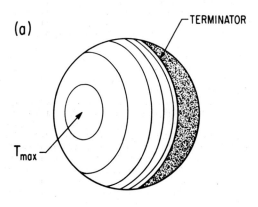

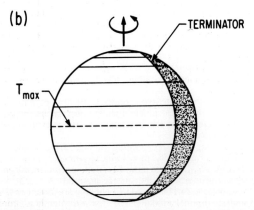

Figure A44 Illustration of the two endmember thermal models. (a) The standard thermal (non-rotating) model, with the temperature dependent only on the solar incidence angle. (b) The fast-rotating (isothermal latitude) model, with the temperature dependent only on the distance from the equator.

The STM does not accurately predict the thermal emission from solar system objects of known size, unless the beaming factor η is included. η is effectively an adjustment to the subsolar temperature, T_{ss}, and was first introduced in that form by Jones and Morrison (1974). η is needed for two reasons: no real rotating asteroid radiates all its heat on the dayside, as the STM (without η) assumes, and individual points on the surface radiate their heat preferentially in the Sunward direction, and not isotropically as the STM assumes.

Without a real understanding of physical causes of thermal anisotropy, the value of the beaming factor can still be estimated. It is possible to tie the thermal model diameter to actual asteroid diameters determined from stellar occultations. The most recent attempt to use occultation diameters to calibrate radiometric diameter determinations was that of Lebofsky et al. (1986). They determined the value of η that resulted in radiometric diameters for Ceres and Pallas in agreement with the observed occultation diameters.

More recently, attempts have been made to determine more realistic models for thermal emission from rough surfaces for determination of the anisotropy of asteroid surfaces (Spencer, 1990). However, these models still give surfaces that appear to be unrealistically rough.

For small, Earth-approaching asteroids, the standard thermal model sometimes gives albedos which imply compositions inconsistent with those inferred from visual spectroscopy. In all cases, the model albedos are high by as much as a factor of two over the albedo expected from the spectrally determined compositions. The derived albedos are more consistent with the other data if a fast-rotating thermal model is used.

The fast-rotating model differs from the standard thermal model in that it assumes the extreme case of a spherical asteroid whose surface is extremely rocky (high thermal inertia) and/or executes relatively rapid rotation, and/or is very cold. This results in a temperature distribution that is isothermal in longitude (temperature constant through the day and night) and depends only on latitude, so the model is also called the isothermal latitude model. If the Sun is in the equatorial plane, temperature decreases with latitude due to the decreased solar insolation but is independent of longitude (Figure A44b). There is assumed to be no beaming due to roughness and, in theory, the thermal flux does not decrease with increasing solar phase angle.

In reality, most asteroids probably lie somewhere between the two extreme cases just discussed. The Appendix of Spencer, Lebofsky and Sykes (1989) describes the simplest form of thermophysical model in detail. The model numerically solves the one-dimensional thermal diffusion equation with the surface boundary condition determined by the diurnal variations in insolation, and assumes constant thermal parameters with depth and temperature, and absorption of all insolation at a smooth surface with albedo independent of solar incidence angle. Relaxation of any of these assumptions produces more complex and potentially more realistic models.

Detailed studies of the the phase-dependent variation of the thermal flux of several of the largest asteroids have shown that these asteroids have regoliths similar to that of the Moon and rotate in a prograde direction (warmer afternoon side before opposition and cooler morning side after opposition); see Lebofsky *et al.* (1986) for a more detailed description. As noted above, the smaller Earth-approaching asteroids appear to have surfaces that are much rockier than the larger main-belt asteroids (main-belt asteroids of sizes comparable to the Earth-approaching asteroids are generally too faint, due to their greater distances, to be easily observable). This is consistent with the Earth-approaching asteroids being younger and thus having less mature surfaces than the main-belt asteroids.

Asteroid shapes

The visual lightcurves of asteroids can be caused by one of two effects (or a combination of both): non-spherical shape or non-uniform albedo. Thermal infrared measurements, when coupled with visual or near-infrared reflected measurements of the lightcurve, can be used to discriminate variations that are due to shape from those that are due to albedo. A spotted, spherical asteroid will have reflected and thermal lightcurves that are out of phase. At reflected wavelengths, maxima occur when high-albedo areas are visible, while minima occur when dark areas are in view. Conversely, in the thermal IR, maxima occur when warmer dark areas are visible. However if the lightcurve is due to non-spherical shape, both the reflected and thermal lightcurves will be in phase, peaking when the largest cross section is presented. To date there is little evidence for albedo spots on asteroids (Lebofsky *et al.*, 1988), with the possible exception of asteroid 4 Vesta. Recent observations by Redman *et al.* (1992) have shown that the thermal lightcurve of Vesta at submillimeter wavelengths is consistent with the presence of a dark spot on one side of Vesta.

Thermal spectroscopy

Thermal radiation carries compositional as well as temperature information in the form of discrete spectral features resulting from variations in emissivity with wavelength. The Moon shows emission features in the 10-μm region due to silicates (Murcray, Murcray and Williams, 1970), and extensive laboratory measurements have investigated the dependence of thermal emission spectra on composition, surface structure and surface temperature distribution. The recent work of Sprague (1993) suggests the possible existence of compositionally significant features near 10 μm. However, the interpretation of these features is still controversial.

Summary and conclusions

Thermal models of asteroids have come a long way over the past two decades. From the early models that ignored (or at least avoided) many of the physical properties of asteroids have come more sophisticated models that more closely represent physically realistic asteroid models. In many ways even these new models still represent simplifications, but more and more they are helping us understand the true nature of asteroids and their surfaces.

Larry A. Lebofsky

Bibliography

Allen, D.A. (1970) The infrared diameter of Vesta. *Nature*, **227**, 158–9.
Jones, T.J. and Morrison, D. (1974) Recalibration of the photometric method of determining asteroid sizes. *Astron. J.*, **79**, 892–5.
Lebofsky, L.A. and Spencer, J.R. (1989) Radiometry and thermal modeling of asteroids, in *Asteroids II* (eds R. Binzel, T. Gehrels and M.S. Mathews). Tucson: University of Arizona Press, pp. 128–47.
Lebofsky, L.A., Sykes, M.V., Tedesco, E.F. *et al.* (1986) A refined 'standard' thermal model for asteroids based on observations of 1 Ceres and 2 Pallas. *Icarus*, **68**, 239–51.
Lebofsky, L.A., Greenberg, R., Tedesco, E.F. and Veeder, G.J. (1988) Infrared lightcurves of asteroids 532 Herculina and 45 Eugenia: proof of the absence of significant albedo markings. *Icarus*, **75**, 518–26.
Murcray, F.H., Murcray, D.G. and Williams, W.J. (1970) Infrared emissivity of lunar surface features 1. Balloon-borne observations. *J. Geophys. Res.*, **75**, 2662–9.
Redman, R.O., Feldman, P.A., Mathers, H.E. *et al.* (1992) Millimeter and submillimeter observations of the asteroid 4 Vesta. *Astron. J.*, **104**, 405–11.
Spencer, J.R. (1990) A rough-surface thermophysical model for airless planets. *Icarus*, **83**, 27–38.
Spencer, J.R., Lebofsky, L.A. and Sykes, M.V. (1989) Systematic biases in radiometric diameter determinations. *Icarus*, **78**, 337–354.
Sprague, A.L. (1993) Mid-infrared (7.7–13.5 μm) spectroscopy of asteroids: the potential for mineralogic determination. *Publ. Astron. Soc. Pacific*, **41**, 41.

Cross references

Asteroid: lightcurve
Infrared spectroscopy
Radiometry

ASTHENOSPHERE

The layer or 'shell' of the planet Earth that is located in the upper mantle beneath the lithosphere (q.v.). It is marked by the strong attenuation of seismic waves (James, 1989) and consequently also termed the 'low velocity zone'. It is a structurally weak zone that behaves in a plastic manner, providing the accommodation medium for isostatic adjustment of the Earth's crust and horizontal motions for plate tectonics (q.v.). Its existence and name was proposed on theoretical grounds by Joseph Barrell (1914) based on the field evidence of crustal isostatic adjustment; the name is Greek-based meaning the 'layer of no strength'.

The idea of a plastic layer beneath the crust had been mooted by earlier writers, notably Wegener in a series of articles on 'continental drift' from 1909 onward (in German; first English translation, 1924; newest, from his fourth edition, 1929/1966; see also Marvin, 1973). Wegener had the impression that the layer consisted of basalt, on which the granitic continents floated like 'icebergs on the ocean'.

With the development of plate tectonics (q.v.) in the 1970s Wegener's theory had to be severely modified. It is true that the density of continental crust is distinctly lower than that of basalt (2.7 versus > 3.0), but unfortunately the melting point of basalt is far higher than that of granite, so there would be no way for the iceberg hypothesis to work in a basalt 'sea'. However, it now seems likely that the asthenosphere composition is in the nature of peridotite, an even denser material. At the appropriate depth the temperature–pressure relationships are close enough to its melting temperature to have a viscosity low enough to permit the 'drifting' movement of lithospheric plates. In contrast to Wegener's drift concept, the drifting plates consist of lithosphere (q.v.), having both granitic (continental) and basaltic (ocean) crust, riding on a dense strong cold layer of peridotite.

Rhodes W. Fairbridge

Bibliography

Barrell, J. (1914) The strength of the Earth's crust. *J. Geol.*, **22** (8 articles; also ibid. 1915, **23**, 3 articles).
Bowes, D.R. (ed.) (1989) *The Encyclopedia of Igneous and Metamorphic Petrology*. New York: Van Nostrand Reinhold, 666 pp.
Garfunkel, Z. (ed.) (1985) *Mantle Flow and Plate Theory*. New York: Van Nostrand Reinhold (Benchmark Papers in Geology, Vol. 84), 395 pp.
James, D.E. (ed.) (1989) *The Encyclopedia of Solid Earth Geophysics*. New York: Van Nostrand Reinhold, 1328 pp.
Marvin, U. (1973) *Continental Drift: The Evolution of a Concept*. Washington: Smithsonian Institution Press, 239 pp.
Wegener, A. (1919) *Die Enstehung der Kontinente und Ozeane*, 4th ed. Braunschweig: F. Vieweg & Sohn, (transl. by J. Biram, as *The Origin of Continents and Oceans*, New York: Dover Publ., 1966, 246 pp.

Acknowledgement

Draft kindly read and improved by Eldridge Moores (Davis, CA).

Cross references

Atmosphere
Crust
Hydrosphere
Lithosphere
Mantle

ASTROGEOLOGY

Astrogeology is the study of planetary surfaces by all applicable geologic means. *Astro* in this sense refers not to self-luminous stars as it usually does in English – for these have no geology – but rather to celestial bodies in general as *astro* or *astre* do in the romance languages. As in the original Greek γῆ *geo* refers here to land, ground or soil and not only to the planet Earth (Ronca, 1965). Astrogeology also includes investigations of meteorites found on Earth and the craters made by their impacts. Close synonyms include the rarely used 'space geology', the awkward 'extraterrestrial geology' and the now commonly used 'planetary geology' in its usual extension from true planets to satellites and asteroids. 'Planetology' is a broader term that includes astronomic and atmospheric disciplines.

In 1876 the Russian philosopher V.V. Lesevich coined the term astro-geology, assuming it would be based on meteoritics and telescopic spectroscopy (Milton, 1969). This prior mention was not known, however, when the age of lunar exploration approached and Eugene M. Shoemaker founded the Astrogeologic Studies Group in the US Geological Survey (USGS) at Menlo Park, California, in August 1960 (Wilhelms, 1993). In September 1961 the group acquired formal status as the USGS Branch of Astrogeology. The headquarters and part of the scientific activity of the branch were moved to Flagstaff, Arizona, in 1963. In 1967 the Astrogeology branch split into two branches, Surface Planetary Exploration (SPE) and Astrogeologic Studies, and the latter name survived when the two branches recombined in 1973 as the end of Apollo exploration of the Moon's surface eliminated the main reason for SPE's existence.

The principles of photogeology are central to astrogeology. Relative ages of geologic units can be detected on photographs simply from such geometric relations as superposition (overlap) and transection (cross-cutting). A completely developed crater ejecta blanket is younger than another blanket whose features it partly obscures. This holds true regardless of the origin of the craters. A geologic unit emplaced as a fluid of any kind will encroach on or flood an older unit of any kind. For instance, this kind of reasoning conclusively overturned an earlier confusion between lunar maria and basins by proving that a mare is much younger than the ringed basin it occupies. Local stratigraphic sequences determined from geometric relations and counts of craters superposed on neighboring geologic units are integrated into global stratigraphic schemes by correlating densities of superposed craters and by mapping widespread marker horizons. Extraterrestrial geologic units, like terrestrial ones, are three-dimensional. Their distributions, age relations and inferred depths are portrayed on geologic maps and cross-sections constructed by principles long applied to mapping Earth's rocks (Wilhelms, 1990). Structures or structural terrain units can similarly be placed in proper historical sequence on the basis of their geometric relations.

These photogeologic methods are actually easier to apply to airless and quiescent planets than to Earth. As Shoemaker recognized by his adoption of the term astrogeology, however, the findings of photogeology would remain speculative unless supplemented by terrestrial research in the field and laboratory, robotic orbital and surface probes and, eventually, sample collection and human fieldwork. This higher level of exploration is well advanced for the Moon. Between September 1959 and August 1976, 18 Soviet robotic probes, 15 American robotic probes and two American manned orbiters (Apollos 8 and 10) returned photographs, instrumental data and even samples (three Soviet Lunas). Lunar geologic fieldwork was carried out between July 1969 and December 1972 by six crews of two astronauts each who were trained, guided and supported during the missions by teams of astrogeologists drawn from the USGS, universities and NASA (Wilhelms, 1993). Pre-exploration astrogeologic scrutiny and planning greatly increased the productivity of the astronauts' fieldwork. For example the photogeologic identification and mapping of the ejecta blanket of the Imbrium basin enabled the astronauts of Apollo 14 (February 1971) and Apollo 15 (August 1971) to return samples from two points on the surface that represent a vast area of the lunar nearside. Absolute ages for much of the Moon's unsampled terrain can now be estimated because stratigraphic sequences older than the Imbrium basin are impinged upon by this relatively and absolutely dated marker horizon throughout its extent, whereas those younger than the basin are superposed on the ejecta blanket or are overlain by fewer craters than it is. A confluence of photogeology and terrestrial fieldwork on craters furthermore showed that the Imbrium ejecta would provide samples from considerable depth within the Moon.

So far, the Moon is the only planet for which the relative stratigraphic timescale has been calibrated by absolute ages of samples returned to Earth's laboratories. Meteorites provide other data points but not from a known geologic context. Two Viking landings have begun the *in situ* study of Mars. Mercury, Venus, the asteroid Gaspra, the two asteroid-like moons of Mars and 16 satellites of the four giant outer planets have been photographed at geologically useful scales by spacecraft flybys or orbiters. Astrogeologic techniques originally developed for the Moon have gathered a rich harvest of insight from this remote scrutiny and are available for further application as space exploration progresses.

Don E. Wilhelms

Bibliography

Milton, D.J. (1969) Astrogeology in the 19th century. *Geotimes*, **14**(6), 22.

Ronca, L.B. (1965) Selenology vs geology of the Moon etc. *Geotimes* **9**(9), 13.

Wilhelms, D.E. (1990) Geologic mapping, in *Planetary Mapping* (eds R. Greeley and R.M. Batson). Cambridge: Cambridge University Press, pp. 208–60.

Wilhelms, D.E. (1993) *To a Rocky Moon: A Geologist's History of Lunar Exploration.* Tucson: University of Arizona.

ASTROMETRIC OBSERVATION

Astrometry is the art and science of the precise measurement of the positions of the heavenly bodies. In stellar astronomy it has important applications in the measurement of stellar parallaxes and proper motions. This article is restricted to consideration of astrometry of solar system bodies such as comets and asteroids.

Around the middle of the 20th century, there was little incentive for carrying out large-scale precise astrometry of the more than 1000 minor planets and a few hundred comets known at the time. Not only was the computational effort required for astrometric measurement and subsequent orbit determination prohibitive, but there seemed to be little scientific purpose to this activity. Times have changed. Not only have observational techniques and computational power made it possible to track these numerous objects, but there has developed a parallel need from astrophysical observers for astrometric data. Many large telescopes used, for example, for photometry have very small fields of view. Radiometric observers working at infrared or submillimeter wavelengths or with radar-bounce techniques have need of accurate and up-to-date astrometric data. Diameters and shapes of asteroids can be determined by observations of occultations (q.v.) of stars by asteroids. Predictions of such events require unusually precise astrometry, usually performed when both the occulting asteroid and the star to be occulted can be photographed together on the same plate days before the event. And of course for spacecraft missions high-quality astrometric data are essential. Consequently there has been a vibrant resurgence of interest in astrometry, with innovative techniques emerging.

Transit instruments

The most fundamental method of astrometric observation is with the use of a transit instrument or meridian circle. This is a telescope which is mounted on a fixed east–west axis, and can move only in the plane of the meridian. The right ascension of an object is found by timing it as it passes a crosshair in an eyepiece, while the declination is found by reading a graduated circle. There are relatively few observers today who use a transit instrument, and the method is restricted to the brighter planets and asteroids. However, great precision, particularly in right ascension, can be achieved. Furthermore, the observations have the advantage of being fundamental, in the sense that positions are measured absolutely with respect to an inertial reference frame rather than by comparison with the 'fixed' stars. The Carslberg Automatic Meridian Circle in the Canary Islands is a modern example of one of the most advanced instruments of this type.

Photographic astrometry

Until recently, most astrometric observations were carried out photographically. At one time long focal length refractors were favored because of the large plate scale. Many observers, however, have had good success with relatively short focal length Schmidt cameras, in which the small plate scale is more than compensated for by the very sharp images and by the larger number of comparison stars available in the wide field.

For detection of an asteroid, two photographs are obtained about an hour apart, and the two are examined simultaneously and overlapping by means of a stereocomparator. Because of the motion between the two exposures, the asteroid image appears to stand up above or below the plane of the star field, depending on whether the motion is direct or retrograde. For fast-moving objects, the motion is predicted in advance, and the telescope is made to drive at the predicted speed and direction of the asteroid. The stars then appear as short streaks, and the asteroid as a sharp point. This allows the observer not only to recognize the asteroid, but also to photograph much fainter objects than if the image of the asteroid were allowed to trail across the plate.

Suppose the optical axis of a photographic telescope is directed towards a point in the sky whose right ascension and declination are (A, D), so that this point is at the center of the photograph. The photograph is a projection of a portion of the celestial sphere on a plane, or nearly plane, plate or film. It can be shown that the image of a star at (α, δ) will appear on the plate at a position whose linear coordinates measured east and north from the center are (ξ, η), given by

$$\xi = \frac{\sin (\alpha - A)}{\sin D \tan \delta + \cos D \cos (\alpha - A)} \quad (A19)$$

$$\eta = \frac{1 - \cot \delta \tan D \cos (\alpha - A)}{\tan D + \cot \delta \cos (\alpha - A)} \quad (A20)$$

in units of the focal length. The coordinates (ξ, η) are the standard coordinates of the star.

If the plate is placed on the stage of a measuring microscope, the measured coordinates (x, y) of a star can be obtained, but it is not easy to set up the microscope so that its origin of coordinates is at (A, D) or its x- and y-motions are exactly east and north. In general the (x, y) axes are displaced from the (ξ, η) axes by an unknown translation and rotation. Thus to a first approximation (for refinements, see below), the (x, y) and (ξ, η) bear a linear relation to each other:

$$\xi - x = ax + by + c \quad (A21)$$

$$\eta - y = dx + ey + f \quad (A22)$$

The constants a, b, c, d, and e, and f are the plate constants and can be obtained by solving equations (A21) and (A22) simultaneously for three comparison stars. The measured coordinates (x_0, y_0) of an object of interest, such as an asteroid, can then be converted to standard coordinates, and to right ascension and declination by inversion of equations (A19) and (A20).

The (α, δ) of the comparison stars will be found from a standard catalog, but must be corrected for proper motion between the catalog epoch and the time when the photograph was obtained. Additional corrections to the positions of the comparison stars need to be made for differential refraction differential aberration and pincushion or barrel distortion introduced by the optical system. These can be made either by the use of specific theoretical formulas for these effects, or by adding higher-order terms to equation (A21) and (A22). In the latter case, more than three comparison stars will be needed to solve for the additional plate constants so introduced.

In practice, more than the minimum number of comparison stars are used, and 'best' values for the plate constants are calculated from the criterion of least squares. The quotation from C.F. Gauss in the closing paragraph of the article on Orbits is equally pertinent here.

CCD astrometry – stare mode

Recent years have seen the development of astrometry with CCDs (charge-coupled devices; q.v.). These typically have rather small fields of view – a few arcminutes rather than the several degrees typical of photographic Schmidt telescopes. It is expected that this limitation may diminish as technological advances allow the construction of larger CCD arrays. Initially the use of CCDs for astrometry was impracticable because there were almost never enough comparison stars in the field of view. The preparation, for the Hubble Space Telescope, of the Guide Star Catalog (GSC) containing positions of more than 18 000 000 objects has to a large extent alleviated this difficulty, although an observer can still be frustrated to find only two identifiable stars on a frame. The GSC does not give proper motions of the stars. This is not important at present, because the stellar positions are given for a recent epoch. It will be necessary, however, to determine the proper motions during the coming decade. In order to find an asteroid with a CCD detector, its position must be predictable to within a few arcminutes. This is often not possible with a newly discovered object. The photographic Schmidt cameras are still useful for immediate follow-up of newly discovered objects.

In order to recognize a faint asteroid with a CCD, three or four images of the field are obtained a few minutes apart, and these images are displayed in rapid succession ('blinked') on the screen of a computer terminal, and the asteroid can easily be picked out as it moves across the screen. Successive frames can also be displaced electronically relative to each other in the direction of and at the speed of the asteroid's motion. The stellar images are then short streaks, while the successive images of a very faint asteroid build up from below detectability on a single frame to a measurable image on the co-added series. Alternatively, two exposures can be made, and one image can be subtracted from the other. All the stars in the field then vanish, but the asteroid, which has moved between exposures, remains as a pair of images, one positive, the other negative.

Except for discovery and follow-up of new objects, CCD astrometry has overwhelming advantages over photographic techniques. CCD astrometry can reach much fainter objects and can operate even in moonlight, which would fog a photographic emulsion. The positional measurements are more precise, and reduction of the observations can be performed almost instantaneously rather than by subsequent laborious measurement of a photographic plate.

The image of a star or asteroid typically spreads over several pixels in two dimensions, and each pixel might be a square of about 1 arcsec on a side. However, several image reduction software packages (such as IRAF – Image Reduction and Analysis Facility) are available to determine the centroid of the image to a fraction of a pixel. In a typical reduction, the operator will place the cursor on the screen of a computer terminal on top of a comparison star by means of a 'mouse'. The software package will count the number of photoelectrons held by each pixel and will calculate the position of the centroid by fitting, for example, a double gaussian profile. It will also identify the star and read its right ascension and declination from the GSC. Thus, in one 'click' of the mouse, the (x, y) coordinates and the right ascension and declination are found. This is done for each star and for the asteroid, and the subsequent computation of standard coordinates, plate constants and final position proceeds automatically with no further intervention from the observer. This can be done within minutes of an observation, and the position can be transmitted immediately by electronic mail to the Minor Planet Center in Cambridge, Massachusetts.

CCD astrometry – scan mode

The number of usable comparison stars and the chance of discovery of new objects are somewhat restricted by the small size of currently available CCDs. 'Scan' mode techniques, developed to a high degree of sophistication with the Spacewatch telescope (q.v.) at Kitt Peak, Arizona, considerably increase the area of the sky that can be covered, at the same time permitting very precise astrometry. Some aspects of the technique recall that of classical meridian circle astrometry.

In the scan mode the telescope drive is stopped, so that a narrow strip of sky drifts across the CCD detector for perhaps 30 min or so. The strip has a narrow range of declination, but a wide range of right ascension, thus covering an adequate supply of usable comparison stars. Relative right ascensions are found by timing, which is capable of high precision. As the image of the sky drifts across the CCD detector, the peripheral software and circuitry move the signal charges from one diode column to the next at a speed that exactly compensates for the Earth's rotation. Not only is accurate astrometry possible, but special techniques allow the detection of very faint moving objects, including some of the smallest extra terrestrial rocks, of a few meters in size, ever detected. The system is currently among the most prolific in the detection and real-time astrometry of near-Earth asteroids, and is described more fully by Gehrels (1991) and references therein.

Natural satellites

Astrometric measurements of the Earth's Moon are routinely obtained from observations of occultations (q.v.) of stars by the Moon.

Positions of the satellites of other planets are made by measuring the angular distance and position angle of the satellites with respect to their parent planets, in a manner that is in principle akin to the manner in which visual binary stars are measured. Such measurements may be made visually, photographically or by CCD, and are carried out by relatively few observatories and telescopes, such as the United States Naval Observatory, Pulkovo, Yunnan, the Kapteyn Observatory, the Carlsberg Meridian Circle and the Hubble Space Telescope. The Royal Greenwich Observatory maintains an enormous data bank of satellite observations, which includes more than 100 000 observations of 15 satellites of Saturn, Uranus and Neptune.

Hipparcos

The astrometric satellite Hipparcos was launched by the European Space Agency in August 1989. While it was intended for astrometric measurement of stars, rather than of solar system objects, with a view to obtaining stellar parallaxes and proper motions, it is of interest to solar system astrometrists because they of course make use of stellar astrometric positions for planetary and cometary work. Hipparcos was intended to be placed in a near-circular geostationary orbit, but the failure of the apogee booster motor meant that Hipparcos has remained in a highly elliptical transfer orbit. In spite of this misfortune, astrometric positions of some tens of thousands of stars have been obtained with an internal precision of the order of tens of milliarcseconds.

Jeremy B. Tatum

Bibliography

Gehrels, T. (1991) Scanning with charge-coupled devices. *Space Sci. Rev.*, **58**, 347–75.
Pravec, P., Tichý, M., Tichá, J. et al. (1994) CCD astrometry of asteroids and comets using the Guide Star Catalogue, *Planet. Space Sci.*, **42**, 345–8.
Tatum, J.B. (1982) The measurement of comet positions. *J. Roy. Astron. Soc. Can.*, **76**, 97–108.

Cross references

Celestial mechanics
Orbit
Planetary dynamical astronomy
Spacewatch

ASTRONOMICAL CONSTANTS

A set of quantities employed in the construction of theories of planetary motion and in the calculation of the motions and positions of the bodies. In order to process observational data, and to perform calculations involving planetary data, it is necessary to have an agreed system of numerical values of important quantities such as the speed of light and the constant of gravitation. If different investigators, working at different times, employ different values of these quantities, chaos will ensue. The results cannot easily be compared, and the correctness of one or the other approach cannot be readily determined.

Clemence (1965) notes that the concept of a system of astronomical units probably originated with Simon Newcomb (q.v.) sometime before 1877. The current system of astronomical constants was defined in 1976 and adopted by the International Astronomical Union (IAU) (q.v.) in 1984. This system is given in Table A8. The listing shows defining constants, primary constants, derived constants and other quantities. Some values were modified after 1976; the modified values appear here in brackets (see also Astronomical unit; Appendix 2: SI units).

James H. Shirley

Bibliography

Clemence, G.M. (1965) The system of astronomical constants. *Ann. Rev. Astron. Astrophys.*, **3**, 93–111.
Naval Observatory (1991) *American Ephemeris and Nautical Almanac*. Washington, DC: US Govt Printing Office.

ASTRONOMICAL UNIT

One astronomical unit is the distance at which a massless particle in an unperturbed circular orbit about the Sun would have a mean daily motion of k radians per day, where k is the Gaussian constant, $k \equiv 0.01720209895$. The oft-quoted 'definition' of an astronomical unit as 'the mean distance of the Earth from the Sun' is merely an approximation.

In the 1976 IAU System of Astronomical Constants, introduced for use in the national almanacs from 1984, the meter, kilogram and second are the units of length, mass and time respectively, in the International System of Units (SI). For astronomical applications it would be inconvenient to use these units: for example, astronomical distances are too large to be measured accurately and conveniently in meters.

The astronomical unit of time is the day, a time interval of 86 400 SI seconds. The astronomical unit of mass is the mass of the Sun, $M_\odot$. The astronomical unit of length is the length for which $k \equiv 0.01720209895$. This length is known as the astronomical unit (AU).

Carl Friedrich Gauss (1777–1855) determined the value of k from the motion of the Earth using Kepler's third law (q.v.):

$$k^2(1+m) = n^2 a^3$$

where m is the mass of the Earth (plus the Moon) in solar masses, n is the mean daily motion of the Earth ($n = 2\pi/P$, where P is the orbital period in days), a is the semimajor axis of the earth's orbit and k is a constant, whose value depends on the units chosen for m, n and a. Using a value of 365.2563835 days for the length of the sidereal year, a reciprocal mass of the Earth of 354 710 $M_\odot^{-1}$ and adopting the mean distance of the earth from the sun as the unit of distance, Gauss derived $k = 0.01720209895$.

Although Gauss used the mean distance of the Earth from the Sun as the unit of distance to derive k, he was aware that future determinations of the length of the year and of the mass of the Earth would alter the value of k. Also, the length of the year is not constant, but is slowly decreasing, so k would become time dependent. Rather than allowing the value of k to change, it seemed more reasonable to keep k fixed and to dispense with the notion that the semimajor axis of the Earth be exactly one AU. Using modern values for the length of the year (365.25636 days) and the reciprocal mass of the Earth-Moon system (328 900.5 $M_\odot^{-1}$), it is found that the semimajor axis of the Earth-Moon barycenter is $a = 1.00000003$ AU. Due to planetary perturbations the mean distance of the Earth from the Sun is 1.0000002 AU.

In principle, it is possible to determine the length of the astronomical unit by measuring the distance between two bodies for which the AU separation is known. The determination of the solar parallax at unit distance, $\pi_\odot$, was one of the fundamental problems in astronomy for

Table A8 International Astronomical Union System of astronomical constants

Defining constants
1. Gaussian gravitational constant — $k = 0.01720209895$
2. Speed of light — $c = 299\,792\,458$ m s^{-1}

Primary constants
3. Light-time for unit distance — $\tau_A = 499.004782$ s [$499.0047837\ldots$]
4. Equatorial radius for Earth — $a_e = 6378\,140$ m
   [IUGG value] — [$a_e = 6378\,137$ m]
5. Dynamical form factor for Earth — $J_2 = 0.00108263$
6. Geocentric gravitational constant — $GE = 3.986005 \times 10^{14}$ m^3 s^{-2} [$3.98600448\ldots \times 10^{14}$]
7. Constant of gravitation — $G = 6.672 \times 10^{-11}$ m^3 kg^{-1} s^{-2}
8. Ratio of mass of Moon to that of Earth — $\mu = 0.01230002$ [0.012300034]
9. General precession in longitude, per Julian century, at standard epoch 2000 — $\rho = 5029''.0966$
10. Obliquity of the ecliptic, at standard epoch 2000 — $\epsilon = 23°26'21''.448$ [$23°26'21''.4119$]

Derived constants
11. Constant of nutation, at standard epoch 2000 — $N = 9''.2025$
12. Unit distance — $c\tau_A = A = 1.49597870 \times 10^{11}$ m [$1.4959787066 \times 10^{11}$]
13. Solar parallax — $\arcsin(a_e/A) = \pi_\odot = 8''.794148$
14. Constant of aberration, for standard epoch 2000 — $\kappa = 20''.49552$
15. Flattening factor for the Earth — $f = 0.00335281$, $1{,}298.257$
16. Heliocentric gravitational constant — $A^3 k^2/D^2 = GS = 1.32712438 \times 10^{20}$ m^3 s^{-2} [$1.32712440\ldots \times 10^{20}$]
17. Ratio of mass of Sun to that of the Earth — $(GS)/(GE) = S/E = 332\,946.0$ [$332\,946.038\ldots$]
18. Ratio of mass of Sun to that of Earth + Moon — $(S/E)/(1+\mu) = 328\,900.5$ [$328\,900.55$]
19. Mass of the Sun — $(GS)/G = S = 1.9891 \times 10^{30}$ kg
20. System of planetary masses
    Ratios of mass of Sun to masses of the planets

Planet	Ratio	Planet	Ratio	
Mercury	6 023 600	Jupiter	1 047.355	[1 047.350]
Venus	408 523.5	Saturn	3 498.5	[3 498.0]
Earth	328 900.5	Uranus	22 869	[22 960]
Mars	3 098 710	Neptune	19 314	
		Pluto	3 000 000	[130 000 000]

Other quantities

21. Masses of minor planets

Minor planet	Mass in solar mass
(1) Ceres	5.9×10^{-10}
(2) Pallas	1.1×10^{-10} [1.081×10^{-10}]
(4) Vesta	1.2×10^{-10} [1.379×10^{-10}]

22. Masses of satellites

Planet	Satellite	Satellite/planet
Jupiter	Io	4.70×10^{-5}
	Europa	2.56×10^{-5}
	Ganymede	7.84×10^{-5}
	Callisto	5.6×10^{-5}
Saturn	Titan	2.41×10^{-4}
Neptune	Triton	2×10^{-3}

23. Equatorial radii in km

Mercury	2 439	Jupiter	71 398	Pluto	2 500
Venus	6 052	Saturn	60 000		
Earth	6 378.140	Uranus	25 400	Moon	1 738
Mars	3 397.2	Neptune	24 300	Sun	696 000

24. Gravity fields of planets

Planet	J_2	J_3	J_4
Earth	+0.00108263	-0.254×10^{-5}	-0.161×10^{-5}
Mars	+0.001964	$+0.36 \times 10^{-4}$	
Jupiter	+0.01475		-0.58×10^{-3}
Saturn	+0.01645		-0.10×10^{-2}
Uranus	+0.012		
Neptune	+0.004		

(Mars: $C_{22} = -0.000055$, $S_{22} = +0.000031$, $S_{31} = +0.000026$)

25. Gravity field of the Moon

$\gamma = (B - A)/C = 0.0002278$ $\quad C/MR^2 = 0.392$
$\beta = (C - A)/B = 0.0006313$ $\quad I = 5552''.7 = 1°32'32''.7$

$C_{20} = -0.0002027$	$C_{30} = -0.000006$	$C_{32} = +0.0000048$
$C_{22} = +0.0000223$	$C_{31} = +0.000029$	$S_{32} = +0.0000017$
	$S_{31} = +0.000004$	$C_{33} = +0.0000018$
		$S_{33} = -0.000001$

Table A9 Selected historical determinations of the length of the astronomical unit

Authority	Date	Method	$\pi_\odot$ (arcsec)		A ($\times 10^{11}$ m)
Parallax measurements					
G.D. Cassini, J. Richer	1672	Mars	9.5		1.38
J. Flamsteed	1672	Mars	10		1.3
J.F. Encke	1824	1761/9 transits of Venus	8.5776		1.53375
S. Newcomb	1867	Mars (1862)	8.855	± 0.020	1.4857
G. Airy	1877	1874 transit of Venus	8.754		1.5028
E.T. Stone	1877	1874 transit of Venus	8.884	± 0.037	1.4808
G.L. Tupman	1877	1874 transit of Venus	8.813	± 0.033	1.4928
D. Gill	1881	Mars (1877)	8.78	± 0.01	1.498
J.G. Galle	1878	(25) Phocæa (1872) and (8) Flora (1873)	8.87		1.483
M. Houzeau	1884	1882 transit of Venus	8.907	± 0.084	1.4770
D. Gill	1896	(12) Victoria (1889)	8.801	± 0.006	1.4948
D. Gill	1896	(7) Iris (1888)	8.798	± 0.011	1.4953
D. Gill	1896	(80) Sappho (1888)	8.812	± 0.009	1.4929
J. Hinks	1911	Eros (1900–1)	8.807	± 0.003	1.4938
H. Spencer Jones	1925	Mars (1924)	8.809	± 0.005	1.4935
H. Spencer Jones	1939	Eros (1931)	8.790	± 0.001	1.4967
Dynamical methods					
S. Newcomb	1867	Light time measures	8.860		1.4849
S. Newcomb	1867	Parallactic inequality	8.838	± 0.025	1.4886
S. Newcomb	1867	Lunar equation	8.809	± 0.054	1.4935
S. Newcomb	1867	Aberration constant	8.793	± 0.005	1.4962
H. Spencer Jones	1928	Radial velocities of stars	8.803	± 0.004	1.4945
E. Rabe	1950	Motion of Eros (1926–1945)	8.79835	± 0.00039	1.495264
D. Brouwer	1950	Parallactic inequality	8.798	± 0.003	1.4953
E. Rabe	1967	Motion of Eros (1926–1945)	8.7945	± 0.0002	1.49592
Radar ranging					
D.O. Muhleman *et al.*	1962	Venus (1961)	8.794491	± 0.000024	1.4959204
G.H. Pettingill *et al.*	1962	Venus (1961)	8.7940976	± 0.0000147	1.49598728
I.I. Shapiro	1963	Venus	8.79416	± 0.00002	1.495977
M.E. Ash *et al.*	1967	Venus and Mercury, plus optical			1.49597892

three centuries. The length of the AU, A, follows from $\sin \pi_\odot = r_e/A$, where r_e is the equatorial radius of the earth.

The first attempt involved parallax measurements of Mars at the 1672 opposition. In the 18th century, parallax measurements of Venus were made during the 1761 and 1769 transits. Although the parallax of Venus is greater than that of Mars, the reduction of the transit observations was hampered by the 'black-drop' effect caused by Venus' dense atmosphere, which made the precise timing of the transit very difficult. A problem with making parallax determinations for extended bodies, such as Venus and Mars, is the determination of the center of the disk, particularly when the bodies show phases. In the 19th century it was realized that parallax observations of minor planets, which did not show resolved disks, would be easier to make, even though the parallax of a main-belt object would be much smaller than that of Venus or Mars. The 1898 discovery of the Amor (q.v.) object (433) Eros, which has a perihelion distance of 1.13 AU, afforded the opportunity of observing a non-resolved object with a parallax that could be more than seven times larger than the solar parallax. Eros was well observed at its 1900–1901 and 1930–1931 oppositions.

Several other methods have been used to determine the solar parallax at unit distance, or the length of the astronomical unit:

1. The coefficient of the parallactic inequality in the motion of the Moon depends on the ratio of the distances of the Sun and the Moon, and on the masses of the Earth and the Moon. The value of the coefficient can be determined from observation, from which the solar parallax may be determined.
2. The coefficient of the lunar inequality in the motion of the Earth, due to the motion of the Earth about the Earth-Moon barycenter, can be combined with the parallax and mass of the Moon, and with observations of the Sun, planets or stars to derive the solar parallax.
3. The observed value of the constant of aberration, combined with values for the radius of the Earth, the speed of light and the length of the sidereal year, may be used to derive the solar parallax. The uncertainty in the observed value of the aberration constant is the limiting factor in the use of this method.
4. The length of the AU may also be obtained by multiplying the speed of light by the travel time of light from the Sun to the Earth. The limiting factor in using this method was obtaining the light travel time to a sufficient precision. The radar detection of Venus from 1959 allowed a very accurate determination of how long it took the radar waves to make the round trip from the Earth to Venus.

Details on the derivation of $\pi_\odot$ and A using the methods mentioned above may be found in Newcomb (1867) or Kulikov (1956).

The table gives selected historical determinations of the length of the AU. All the values of A, with the exception of the values from Ash, Shapiro and Smith (1967) and from Rabe (1967), were derived from the stated values of $\pi_\odot$ given by the various authorities, using the 1976 IAU value $r_e = 6\,378\,140$ m. Ash, Shapiro and Smith (1967) did not give a value for $\pi_\odot$, but A was obtained by multiplying the given value of τ_A by the given value for c. Rabe (1967) gave neither $\pi_\odot$ nor A, so $\pi_\odot$ was derived from the given mass of the Earth–Moon system.

The modern value for the length of the AU, $1.49597870 \times 10^{11}$ m, is derived from two constants in the 1976 (IAU) System of Astronomical Constants: c, the speed of light (a defining constant with the value 299 792 458 ms^{-1}); and τ_A, the light time for unit distance (a primary constant with the value 499.004782 s). The length of the astronomical unit follows from $A = c\tau_A$.

Gareth V. Williams

Bibliography

Airy, G. (1877) On the inferences for the value of the mean solar parallax . . . *Mon. Not. Roy. Astron. Soc.*, **38**, 11.
Ash, M.E., Shapiro, I.I. and Smith, W.B. (1967) Astronomical constants and planetary ephemerides deduced from radar and optical observations. *Astron. J.*, **72**, 338.

Astronomical Almanac (1984) HMSO and US Government Printing Office.
Brouwer, D. and Clemence, G.M. (1961) *Methods of Celestial Mechanics*. Academic Press.
Gauss, C.F. (1809) *Theoria Motus Corpurum Cœlestium in Sectionibus Conicis Solem Ambientium*. Translated into English by C.H. Davis (1857). Reprinted 1964 by Dover.
Gill, D. (1881) Account of a determination of the solar parallax from observations of Mars, made at Ascension in 1877. *Mem. Roy. Astron. Soc.*, **46**, 1.
Gill, D. (1896) *Ann. Cape Observatory*, **6**.
Gill, D. (1897) *Ann. Cape Observatory*, **7**.
Kulikov, K.A. (1956) *Fundamental Constants of Astronomy*. Translated from the Russian by the Israel Program for Scientific Translation, 1964.
Marsden, B.G. (1963) An attempt to reconcile the dynamical and radar determinations of the astronomical unit. *Bull. Astron.*, **25**, 225.
Muhleman, D.O., Holdridge, D.B. and Block, N. (1962) The astronomical unit determined by radar reflections from Venus. *Astron. J.*, **67**, 191.
Newcomb, S. (1867) *Washington Astronomical and Meteorological Observations 1865*.
Pettingill, G.H., Briscoe, H.W., Evans, J.V. et al. (1962) A radar investigation of Venus. *Astron. J.*, **67**, 181.
Rabe, E. (1950) Derivation of fundamental astronomical constants from the observations of Eros during 1926–1945. *Astron. J.*, **55**, 112.
Rabe, E. (1967) Corrected derivation of astronomical constants from the observations of Eros 1926–1945. *Astron. J.*, **72**, 852.
Shapiro, I.I. (1963) Radar determination of the astronomical unit. *Bull. Astron.*, **25**, 177.
Spencer Jones, H. and Halm, J. (1925). Determination of the solar parallax I. *Mon. Not. Roy. Astron. Soc.*, **85**, 832.
Spencer Jones, H. (1939) *Mem. Roy. Astron. Soc.*, **66**, part 2.
Stone, E.T. (1877) On the Telescopic observations of the transit of Venus 1874 . . . *Mon. Not. Roy. Astron. Soc.*, **38**, 279.
Tupman, G.L. (1877) Note on the mean solar parallax as derived from the observations of the recent transit of Venus. *Mon. Not. Roy. Astron. Soc.*, **38**, 334.

ATEN OBJECT

Aten objects are a subset of near-Earth asteroids (see near-Earth object), distinguished from the Apollo objects (q.v.) and Amor objects (q.v.) by orbital characteristics. The Aten objects have semimajor axes smaller than that of the Earth (i.e. less than 1 AU). Since most have significant orbital eccentricities, these are a class of Earth-crossing asteroids. They are named for a member of the class, asteroid 2062 Aten. No clear compositional trends are present in the Aten objects; they appear to have originated in multiple source regions. Since the projected dynamical lifetimes of objects in near-Earth space are, at about 10^8 years, much shorter than the age of the solar system, they must be replenished from some source. The main asteroid belt is the most likely source region (Binzel et al., 1992).

Fewer than 100 Aten objects are known (McFadden, Tholen and Veeder, 1989), although this number will increase in future as there are active programs under way to detect near-Earth objects (see Spacewatch). Since objects in near-Earth space have the potential to collide with the Earth, there is considerable interest in determining the population and orbital characteristics of these bodies (Matthews, 1992).

James H. Shirley

Bibliography

Binzel, R.P., Xu, S., Bus S.J and Bowell E. (1992) Origins for the near-Earth asteroids. *Science*, **257**, 779–81.
Matthews, R. (1992) A rocky watch for Earthbound asteroids. *Science*, **255**, 1204–5.
McFadden, L.-A., Tholen, D.J. and Veeder G.J. (1989) Physical properties of Aten, Apollo, and Amor asteroids, in *Asteroids II* (eds R.P. Binzel, T. Gehrels and M. Matthews) Tucson: University of Arizona, pp.442–67.

Cross references

Amor object
Apollo object
Asteroid
Asteroid: composition and taxonomy
Near-Earth object
Spacewatch

ATMOSPHERE

The gaseous envelopes surrounding the terrestrial planets make up a small fraction of their total mass but exert an inordinate influence on both their underlying surfaces and the energy balance between planet and Sun. The Jovian planets' atmospheres, on the other hand, represent a significant fraction of their total mass and dominate discussion of all aspects of their physical states. All planets have atmospheres, but on some of the smaller ones there is only a rarefied regional atmosphere produced either by interaction with the solar wind or by local surface or subsurface sources. We consider here only those objects with atmospheres thick enough to be hydrodynamic in nature: the planets Earth, Mars, Venus, Jupiter, Saturn, Uranus and Neptune, and the satellites Titan and Triton. Pluto may have a substantial atmosphere as well, but little is known about it at the present time.

Composition and clouds

The composition of present-day planetary atmospheres reflects the make-up of the primordial solar nebula from which the planets were formed, modulated to a greater or lesser extent by a variety of physical, geological, chemical and biological processes. All terrestrial planet and icy satellite atmospheres have molecular nitrogen (N_2) as either their primary or secondary atmospheric constituent. N_2 presumably derives from primordial organic material that was heated and subjected to chemical reactions during planetary accretion. Earth is unique among the planets in having a substantial inventory of molecular oxygen (O_2), a symptom of biological activity.

On Earth both carbon dioxide (CO_2) and water vapor (H_2O) are minor but important constituents, both regulated by much larger reservoirs in the oceans and the biosphere. CO_2 is supplied by volcanic outgassing, biogenic decay and recently by anthropogenic influences; it is dissolved in ocean water and ultimately removed by chemical weathering, limestone formation and photosynthesis. H_2O is kept within about a factor of two of the maximum amount that would be in equilibrium with the ocean surface by atmospheric motions which regulate precipitation and evaporation.

Venus, about the same size as Earth, has orders of magnitude more CO_2 and equally less H_2O by comparison. The best explanation is that the two planets started with similar compositions, but Venus, being closer to the Sun, was heated to a greater extent. This caused its water to evaporate, which trapped additional heat. Photodissociation stripped hydrogen from water molecules, and the light hydrogen atoms escaped to space via molecular collisions. Without an ocean or biosphere, and with its hot surface, there are no surface sinks for CO_2 on Venus, so most of it resides in the atmosphere. Mars is also dominated by CO_2 but has a very low total pressure. Its water is locked primarily in a polar cap. The presence of meandering channels in the Martian terrain has led to speculation that it once had a thicker, warmer atmosphere that supported running water. The current thinness of Mars' atmosphere is plausibly related to its small size in two ways. Thermal escape of molecules was more efficient in Mars' past. In addition, Mars' interior probably cooled after formation more quickly than Earth's, leading to a thicker, geologically more dormant lithosphere and less recent outgassing of CO_2 and other volatiles.

The Jovian planets, in contrast, are massive enough to have lost comparatively little of their primordial material, even the lightest gases. As a result, their compositions are all similar to that of the Sun, with molecular hydrogen (H_2) as the dominant species and secondary amounts of helium (He). Unlike the terrestrial planets, therefore, these are highly reducing atmospheres. Their carbon, nitrogen and oxygen exist primarily in trace amounts of methane (CH_4), ammonia (NH_3) and water. These gases may actually be somewhat overabundant relative to their solar proportions, especially CH_4 on Neptune.

The icy satellites Titan and Triton were probably formed in conditions dominated by the planets around which they revolve. Both have CH_4 as an important secondary constituent. Titan's CH_4 should have been long depleted by photochemical reactions in the stratosphere, leading to the idea that Titan's surface contains a source in the form of regional methane–ethane lakes. Triton's extremely thin atmosphere is presumably in vapor pressure equilibrium with its more substantial surface frost.

Gravity exerts a downward force on the gases in an atmosphere, which can only be balanced if the atmosphere compresses toward the surface to an extent that an equal and opposite pressure gradient force exists. This is the statement of hydrostatic equilibrium:

$$\frac{dp}{dz} = -\rho g \tag{A23}$$

where p is pressure, z altitude, ρ density and g the acceleration of gravity. The mass per unit area contained within an altitude range Δz is thus simply $\Delta p/g$. Invoking the ideal gas law, one can show that pressure decreases with height according to

$$p = p_0 \exp\left(-\int_0^z \frac{g}{RT} dz\right) = p_0 \exp\left(-\int_0^z \frac{dz}{H}\right) \tag{A24}$$

where p_0 is the surface pressure, R the universal gas constant divided by the mean molecular weight of the atmosphere, T the temperature and $H = RT/g$, called the scale height, is the height at which atmospheric mass is reduced to $1/e$ (about 37%) of its original value. Most atmospheric gases are well mixed up to very high altitudes, and their masses thus decrease with height according to equation (A24). However, when there is a local source, e.g. H_2O at the terrestrial ocean surface or ozone (O_3) in the photochemically active stratosphere, the concentration of that gas peaks near its source.

For gases which condense to form clouds at ambient temperatures and pressures, there is a local vapor sink. Earth is 50–60% covered by H_2O condensate clouds resulting primarily from lifting of moist air. Thin H_2O clouds also occur sporadically on Mars, but more important are the CO_2 ice clouds which form the Martian polar hood and supply the seasonal polar caps. Venus is too hot to sustain lifting condensate clouds, but a deep global haze of hydrated sulfuric acid (H_2SO_4) produced by photochemical reactions envelopes the planet. Titan too has a planetwide stratospheric haze, in this case the product of photochemically produced hydrocarbons. In addition, scattered CH_4 stratus and cumulus may exist in Titan's troposphere. Deep, dense H_2O clouds are thought to exist on all the Jovian planets, but cannot be directly observed. The visible cloud features on Jupiter and Saturn are primarily composed of NH_3 ice and ammonium hydrosulfide (NH_4SH), while those of colder Uranus and Neptune are thought to be composed of CH_4 and hydrogen sulfide (H_2S) with deeper, hidden NH_3 and NH_4SH layers.

Energy balance and thermal structure

In thermal equilibrium a planet emits as much energy as it receives from the Sun. The sunlight absorbed depends on the solar constant, i.e. the flux incident at Earth's distance from the Sun ($S_0 = 1367$ W m^{-2}), the planet's distance from the Sun in astronomical units (d), and the planetary albedo (A), representing the fraction of incident sunlight reflected by the planet. The emitted energy is described by the Stefan–Boltzmann law for a black body. As a result, an observer in space could describe a planet as a black body radiating at an effective temperature T_e given by

$$T_2 = \left(\frac{S_0(1-A)}{4\sigma d^2}\right)^{1/4} \tag{A25}$$

where $\sigma = 5.67 \times 10^{-8}$ W m^{-2} K^{-4} is the Stefan–Boltzmann constant. In the absence of an atmosphere, a planet's effective temperature would equal its surface temperature T_s.

The presence of an atmosphere changes the energy balance in two ways. Gases absorb and emit radiation selectively in certain wavelength ranges, determined by the quantization of rotational and vibrational levels of molecular excitation and by the broadening of the resulting spectral lines associated with molecular collisions and Doppler shifts of randomly moving molecules (see Spectroscopy: atmosphere). Consequently, atmospheres do not behave like simple black bodies. In addition, radiation leaving the planet's surface is not radiated directly to space. Instead, some of it is absorbed by the overlying atmosphere and re-emitted to space at a colder effective temperature. Part of the atmospheric radiation is re-emitted downward toward the surface. Since the surface must balance not only the incoming sunlight but also the downward component of atmospheric radiation, it must radiate at a higher temperature than T_e. The extent to which T_s exceeds T_e is a measure of the blanketing efficiency of the atmosphere, known as the greenhouse effect (Table A10). The most extreme example is the hot, thick, CO_2-dominated atmosphere of Venus.

The lapse rate, or vertical distribution of temperature, in an atmosphere is determined by a variety of factors. For thin atmospheres the horizontally averaged temperature may satisfy the condition that the net radiative flux does not vary with height, so that there is no radiative heating or cooling tendency. Such an atmosphere is said to be in radiative equilibrium. The radiative equilibrium temperature usually decreases with height, most steeply near the surface. A notable exception is terrestrial ozone, which forms at high altitudes by photodissociation of molecular oxygen. Radiative heating by this localized ozone layer causes the temperature to increase between about 15 and 50 km altitude and defines the stratosphere.

If an atmosphere is thick enough or close enough to the Sun, however, the radiative equilibrium lapse rate near the surface may be so steep that a parcel of air lifted from near the surface will be buoyant relative to its surroundings and will spontaneously rise. This occurs if the lapse rate $\Gamma = -dT/dz$ satisfies the condition

$$\Gamma > \frac{g}{c_p} = \Gamma_d, \tag{A26}$$

where c_p is the specific heat of the atmosphere at constant pressure. Γ_d is referred to as the dry adiabatic lapse rate. The consequence of a superadiabatic lapse rate is efficient vertical convective mixing of heat, which drives Γ toward Γ_d. In an atmosphere with condensible species, Γ_d must be modified to account for the release of latent heat, variations in molecular weight and the effect of condensate mass loading. Atmospheres in which radiative heating drives convective mixing are said to be in radiative–convective equilibrium.

The stability of the vertical thermal structure is quite sensitive to the fraction of incoming sunlight that reaches the surface. Earth is only partly cloud covered, and about 50% of the insolation is absorbed at the ground. The result is a turbulent convective bound-

Table A10 Selected meteorological parameters for planetary atmospheres: pressure (at surface for terrestrial planets and icy satellites, at base of H_2O cloud for Jovian planets), effective temperature, surface (or H_2O cloud base) temperature, emission/insolation ratio E/I, rotation period, revolution period and tropospheric radiative relaxation time.

Planet	p_s	T_e(K)	T_s(K)	E/I	P_{rot}	P_{rev}	$\tau_r(s)$
Earth	1.013 bar	255	288	1.0002	23.93 h	365.26 d	5×10^6
Mars	7 mbar	210	214	1.000	24.62 h	686.98 d	3×10^5
Triton	17 μbar	38	38	1.000	5.877 d	164.79 years	7×10^5
Venus	92.1 bar	229	731	1.000	243.01 d	224.70 d	1×10^9
Titan	1.5 bar	85	94	1.000	15.95 s	29.46 years	3×10^9
Jupiter	7 bar	124	290	1.7	9.93 h	11.86 years	2×10^8
Saturn	20 bar	95	300	1.8	10.66 h	29.46 years	9×10^8
Uranus	240 bar	59	350	<1.1	17.23 h	84.01 years	5×10^9
Neptune	330 bar	59	350	2.7	16.10 h	164.79 years	5×10^9

ary layer in many locations and a vigorously mixed troposphere in radiative–convective equilibrium, especially in the tropics where moist convection dominates. Under normal conditions the daytime Martian atmosphere behaves in a similar fashion. During dust storms, however, absorption of sunlight and upwelling infrared radiation by suspended dust significantly stabilizes the Martian lapse rate. Venus' thick planetwide cloud cover intercepts most of the incoming sunlight; only 2–3% reaches the surface. Partly as a result of the cloud, the vertical structure is complex, with isolated convective layers but extensive stable regions. Titan's planetwide stratospheric haze has an even more stabilizing influence. Except for a shallow boundary layer, Titan's troposphere is subadiabatic and overlain by a highly stable stratosphere. However, the possibility exists that methane-driven moist convection provides more efficient mixing of the troposphere. Triton's very thin atmosphere may vary greatly in structure between sunlit and dark sides and between locations covered with and bare of frost. One model of the vertical structure has a low-level thermal inversion topped by an adiabatic troposphere.

Not all planets are in thermal equilibrium. The Jovian planets, with the exception of Uranus, emit substantially more radiation than they absorb (Table A10). Their effective temperatures are therefore higher than would be predicted by equation (3). This internal heat flux is thought largely to be remnant cooling of the primordial heat with which these planets were formed. However, on at least one Jovian planet, Saturn, interior conditions may be cool enough for helium to be condensing and precipitating toward the center of the planet. The resulting conversion of gravitational potential energy ultimately to heat would provide an additional internal flux. Because of the internal heating, all the Jovian planets are thought to be convective to varying degrees throughout their interiors and adiabatic in their thermal structure up to the visible cloud level. Complicating factors are latent heating/cooling due to phase changes of trace gases, resulting molecular weight fluctuations and variations in the specific heat of H_2, all of which affect the adiabat followed by a rising parcel. Nearly neutral conditions extend through the visible cloud levels in Jovian atmospheres, but quickly give way to a more stable upper troposphere and capping stratosphere due to methane heating.

If the thermal structure is perturbed from equilibrium by radiative heating/cooling, the perturbation decays over a time scale τ_r known as the radiative relaxation time:

$$\tau_r = \frac{p \, c_p T}{g \, \sigma T_e^4} \quad (A27)$$

The linear dependence on p means that thicker atmospheres cool more slowly than thinner ones. As planets rotate about their axes and revolve around the Sun, diurnal and seasonal variations of insolation take place. The temperature responds on the time scale τ_r, varying with an amplitude of $T/(1 + \Omega^2\tau_r^2)^{1/2}$, where $\Omega = 2\pi/P$ and P is the length of the day or year. For Mars' thin atmosphere, τ_r is comparable to the length of its day, so nights are much cooler than days. In the thick lower atmosphere of Venus, on the other hand, τ_r is a few decades, and there is consequently no detectable difference between day and night.

The phase of the temperature variation relative to the solar heating is $\tan^{-1}\Omega\tau_r$. Thus on Earth, whose radiative relaxation time is several months, northern hemisphere winter solstice (minimum insolation) occurs in December while the coldest temperatures climatologically occur near the end of January. (The response time of the upper ocean layers is also a factor.) On the Jovian planets, which have extremely long τ_r, the phase lag approaches 90°, so that temperature extremes actually occur at the equinoxes.

Dynamics and general circulation

On all planets except Uranus (whose spin axis tilts slightly more than 90° away from the ecliptic), more sunlight falls on the equator than on the poles. As a result, latitudinal pressure gradients are created, driving winds that redistribute heat between the tropics and polar regions. The interaction of this thermally driven flow with the planet's rotation, and frictional interactions with the underlying surface determine the general circulation. On the Jovian planets the global radiation imbalance caused by the internal heat flux, and the concomitant need to transport heat vertically, may determine the style of dynamics instead.

As air moves poleward on a rotating planet, toward the axis of rotation, it conserves its angular momentum. The decreasing moment of inertia thus implies a corresponding increase in the air parcel's angular velocity about the axis. Viewed from the reference frame of the underlying planetary surface, the wind is apparently deflected at right angles to its initial direction of motion, in the same direction as that of the planet's rotation. The argument can be generalized to motion in any direction, the deflection always being to the right in the hemisphere of the positive rotation axis and to the left in the opposite hemisphere. This is called the Coriolis force, and is given by the formula

$$F_c = -2\rho\Omega \times v \quad (A28)$$

where Ω and v are the planetary rotation angular velocity and wind velocity vectors respectively and $\times$ denotes the cross-product. For large-scale motions on some planets the Coriolis force approximately offsets the pressure gradient force $-\nabla p$, a condition known as geostrophic balance. Two familiar examples are the cyclonic (counterclockwise in the northern hemisphere) flow of air around low pressure centers in terrestrial mid-latitudes and the general eastward flow of the jet stream around the Earth. In other situations, nonlinear terms in the momentum balance caused by transport of the momentum of the wind or by centrifugal forces may be important. If the centrifugal force dominates, the balance with $-\nabla p$ is called cyclostrophic balance, the best example of which is tornadoes.

The dimensionless ratio of the nonlinear terms to the Coriolis term in the momentum balance is known as the Rossby number, and is given by

$$Ro = \frac{U}{fL} \quad (A29)$$

where U is the characteristic zonal (east–west) wind speed, $f = 2\Omega \sin\theta$, θ is latitude and L is a typical horizontal length scale of weather systems. $Ro \ll 1$ is a necessary and sufficient condition for geostrophy, while $Ro \gg$ is required for cyclostrophy.

By combining the equations of geostrophic balance, hydrostatic equilibrium and the ideal gas law, one can relate the vertical gradient of the zonal wind to the latitudinal temperature gradient. The result is the thermal wind equation

$$H\frac{\Delta U}{\Delta z} = -\frac{R}{fa}\frac{\Delta T}{\Delta\theta} \quad (A30)$$

where a is the planetary radius. Thus, if temperature decreases toward the pole, the zonal wind increases with height. This is generally the case in atmospheres up to the tropopause, where peak winds associated with jet streams occur. A cyclostrophic version of the thermal wind equation can also be defined.

Another controlling factor for planetary dynamics is the vertical stability of the atmosphere to perturbations. Dynamic perturbations can draw on either the gravitational potential energy associated with buoyancy or the kinetic energy of a shear flow to grow in strength. A dimensionless measure of buoyancy versus shear effects on stability is the Richardson number, given by

$$Ri = \frac{gT^{-1}(\Gamma_d - \Gamma)}{(dU/dz)^2} \quad (A31)$$

Dynamical stability increases as Ri increases. For large-scale motions the vertical wind shear is related to the latitudinal temperature gradient by the thermal wind equation, and it is indicative of the potential energy caused by differential heating of equator and pole that is available for the growth of perturbations.

Based on the two dimensionless parameters Ro and Ri, the known planetary atmospheres can be grouped into three dynamic categories. The rapidly rotating terrestrial planets Earth and Mars are characterized by geostrophic dynamics ($Ro \approx 0.1$–0.5) and moderate to large stability ($Ri \approx 10$–100). On these planets warm subtropical air displaced poleward becomes buoyant and thus spontaneously rises upon entering a colder, denser environment, while cold air moving equatorward sinks in warmer, less dense surroundings. As a result, perturbations of the jet stream can grow. This process, known as baroclinic instability, is responsible for terrestrial mid-latitude storms and the wind and pressure fluctuations observed at the Mars Viking probe lander sites. The most rapidly growing disturbances have a horizontal length scale close to the Rossby radius of deformation

$$L_d = \left(\frac{g}{T}(\Gamma_d - \Gamma)\right)^{1/2}\frac{H}{f} \quad (A32)$$

For terrestrial midlatitudes $L_d \approx 1000$ km, slightly smaller than the observed scale of low pressure systems (due to nonlinear effects).

In the tropics of both planets, heat transport is supplemented by gentle, forced rising of warm air, poleward drift aloft, sinking of cool air at higher latitudes and equatorward drift near the surface, known as the Hadley cell. On Earth the Hadley cell is forced by convective latent heating, while on Mars it is affected by radiative heating of suspended dust from regional and occasionally global dust storms. On Earth ocean currents do about half the work of poleward heat transport. On Mars the primary atmospheric constituent, CO_2, condenses to ice in the winter polar region and enhances the latitudinal pressure gradient, noticeably impacting the dynamics. Triton's dynamics may also be affected by seasonal surface frost variations, in this case composed of N_2, CH_4 and CO_2. Its moderately slow rotation rate and small size may place Triton either in the geostrophic, baroclinic Earth/Mars class or the slowly rotating, cyclostrophic class to be discussed below.

The Jovian planets Jupiter, Saturn, Uranus and Neptune are also rapidly rotating and geostrophic in their dynamics ($Ro \approx 0.1$), but their internal heating and near-adiabatic lapse rates probably give them low dynamical stability ($Ri \approx 1$). The dominant cloud-level motions on Jupiter (Plates 22 and 24) and Saturn (Plate 28) consist of a series of closely spaced, alternating eastward and westward moving jets, with a strong superrotating (in the direction of planetary rotation) equatorial jet. Uranus and Neptune are characterized instead by equatorial subrotation and a single transition to superrotating flow at higher latitudes, somewhat like the terrestrial situation. Observations beneath the clouds are lacking, so we do not know how deep the jets extend. In one view the jets are the cloud-level manifestation of deep rotating convective cylinders and extend well into the interior, driven by the internal heating. Another possibility is that a shallow weather layer exists within and just below the visible clouds, in which dynamics is driven by latent heating by water clouds, orthopara hydrogen conversion (see Neptune: atmosphere) and perhaps indirectly by differential insolation. In either case the weak vertical stability of these planets, and the absence of a solid surface beneath the clouds, assures a different style of dynamics than that of the terrestrial baroclinic regime.

Venus and Titan differ from all the other planets by virtue of their extremely slow rotation (Table A10). Venus' general circulation is dominated by a planetwide superrotation, with the atmosphere moving some 50–60 times as fast as the solid planet (see Plate 8). Titan is thought to occupy a similar dynamical regime, although observations are not conclusive. The resulting dynamics is cyclostrophic ($Ro \approx 10$–100) in nature in most locations. The planetwide cloud cover on both Venus and Titan, and the resulting high static stability at most altitudes, produce a moderate to large $Ri \approx 10$–100. L_d on Venus and Titan exceeds the size of the planets owing to the slow rotation, so baroclinic instability is not an important heat transport mechanism. Instead, a broad Hadley cell driven by differential insolation does most of the work. The Coriolis torque on the poleward branch of the Hadley cell produces a high-latitude jet but cannot by itself explain the large equatorial superrotation. At deep levels a horizontal shear instability of the zonal flow, known as barotropic instability, may transport angular momentum from the jet latitudes toward the equator to maintain superrotation there. At higher levels on Venus, at least, significant diurnal variations in solar heating occur, driving waves known as thermal tides. The tides pump momentum both vertically and horizontally to maintain the cloud-level flow.

Anthony D. Del Genio

Bibliography

Allison, M. and Travis L.D. (1986) Astronomical, physical, and meteorological parameters for planetary atmospheres, in *The Jovian Atmospheres* (eds M. Allison and L.D. Travi). Washington, D.C.: NASA CP-2441, pp. 293–319.

Donahue, T.M. and Pollack J.B. (1983) Origin and evolution of the atmosphere of Venus, in *Venus* (eds D.M. Hunten, L. Colin, T.M. Donahue and V.I. Moroz). Tucson, AZ: University of Arizona Press, pp. 1003–36.

Goody, R.M. and Walke, J.C.G. (1972) *Atmospheres*. Englewood Cliffs, NJ: Prentice-Hall.

Hunten, D.M., Tomasko, M.G. Flasar, F.M. *et al*. (1984) Titan, in *Saturn* (eds T. Gehrels and M.S. Matthews). Tucson, AZ: University of Arizona Press, pp. 671–759.

Ingersoll, A.P. (1990a) Dynamics of Triton's atmosphere. *Nature*, **344**, 315–7.

Ingersoll, A.P. (1990b) Atmospheric dynamics of the outer planets. *Science*, **248**, 308–15.

Pollack, J.B (1981) Atmospheres of the terrestrial planets, in *The New Solar System* (eds J.K. Beatty, B. O'Leary and A. Chaikin). Cambridge, MA: Sky Publishing Corp., pp. 57–70.

Rossow, W.B. (1985) Atmospheric circulation of Venus. *Adv. Geophys.*, **28A**, 347–79.

Weidenschilling, S.J. and Lewis J.W. (1973) Atmospheric and cloud structures of the Jovian planets. *Icarus*, **20**, 465–76.

Zurek, R.W. (1982) Martian great dust storms: an update. *Icarus*, **50**, 288–310.

Cross references

Coriolis effect, geostrophic motion
Earth: atmosphere
Greenhouse effect
Hadley circulation
Insolation
Jupiter: atmosphere
Mars: atmosphere
Mercury: atmosphere
Moon: atmosphere
Neptune: atmosphere
Saturn: atmosphere
Solar constant
Stratosphere
Thermosphere
Titan
Triton
Troposphere
Uranus: atmosphere
Venus: atmosphere

ATMOSPHERIC THERMAL STRUCTURE

The thermal structure of any atmosphere depends on several factors: the vertical distributions of solar and thermal radiative energy deposition and radiative losses to space, the magnitude of the internal heat source and the dynamic response of the atmosphere to the various energy sources and sinks. Deep in the atmosphere of the giant planets there is negligible deposition of solar radiation, and the heat flux is provided by an internal energy reservoir. Energy transport is by thermal radiation and convection, and is spherically symmetric except for dynamical perturbations due to planetary rotation. Higher up, where solar energy deposition and infrared radiative losses to space become noticeable (but where thermal inertia is still large and the radiative time constant long compared with the season), the thermal structure remains time independent but becomes variable with latitude. Still higher, where the radiative time constant becomes short compared with the season, there is a latitudinal variation of thermal structure with time of year. Terrestrial planets, with their thinner atmospheres, tend to be dominated by the latter characteristics.

The large-scale thermal structure of planetary atmospheres depends primarily on radiative heating and cooling. However, because heat transport by mass motion can be extremely effective where such motions are allowed, dynamical activity is often responsible for decreasing large-scale thermal gradients established by radiative processes. For example, the establishment of a baroclinic eddy regime or a zonally symmetric meridional circulation system can lead to greatly reduced latitudinal temperature gradients, while turbulent convection can reduce the vertical lapse rate established by radiative processes.

If the absorption of solar radiation did not depend strongly on wavelength, we would expect solar heating to be restricted to levels covering no more than about one decade in pressure. However, in practice, many sharp absorption lines of different strengths of such gases as carbon dioxide (CO_2), water vapor (H_2O), ozone (O_3), and methane (CH_4) give rise to absorption covering several decades of pressure. Far-infrared bands of the above gases plus others such as those of ammonia (NH_3), acetylene (C_2H_2) and ethane (C_2H_6) are responsible for radiative cooling at these levels. In addition to sharp line bands, a general continuum also adds to the atmospheric opacity. Aerosols, far wings of lines, and collision-induced absorption of such gases as H_2, N_2 and CH_4 contribute in different amounts, depending on the planet.

Thus radiative processes influence the vertical thermal structure of atmospheres to pressures as high as many tens of bars. Very little solar radiation penetrates to higher pressures, where the vertical thermal structure is determined almost entirely by convection driven by a heat flux from the deep interior. At low pressures extremely high in the atmosphere, both diurnal and non-local thermodynamic equilibrium (non-LTE) processes become important, and it is no longer possible to associate the local temperature with Planckian emission. In this article we restrict our attention to the pressure range from about 10 bar to 0.1 mbar, covering approximately five decades of pressure (Venus is an exception, with a surface pressure ~ 100 bar). The regions of the atmosphere covered are the troposphere (middle to upper troposphere for the giant planets) and the stratosphere.

General analysis

Because radiative processes are dominantly responsible for the large-scale vertical thermal structure of planetary atmospheres, it is useful to consider the extreme case of an atmosphere in radiative equilibrium. Solar radiation is deposited at various levels, leading to radiative heating. This gives rise to an infrared radiation field that both heats and cools, depending on whether it is absorbed or emitted at these levels. When the radiative heating and cooling rates are equal in magnitude at every level, the atmosphere is said to be in a state of radiative equilibrium.

In practice many sharp absorption lines contribute to both heating and cooling. The strengths of these lines vary by many orders of magnitude, leading to heating and cooling over several decades of pressure. Strong lines in the visible part of the spectrum absorb solar radiation very high in the atmosphere, whereas radiation between these lines penetrates to large depths before being absorbed. Similarly, strong infrared vibration and rotation bands cause cooling at high levels, while a collision-induced absorption continuum gives rise to cooling from the deeper regions.

A physical understanding of the thermal structure resulting from this complex array of heating and cooling sources can be obtained from a much simpler model consisting of two channels of solar radiation and one infrared channel. Each channel represents a radiation field having associated with it a single absorption coefficient independent of wavelength. One solar channel consists of a 'visible' field of radiation (subscript/superscript v) characterized by a large absorption coefficient, thereby restricting its heating to the upper levels, or stratosphere. A second solar channel consists of a 'conservative' field of radiation (subscript/superscript c) characterized by no absorption. Radiation from this channel diffuses down to the surface where it is partially absorbed and leads to heating there. Because none of the radiation in this field is absorbed in the atmosphere, it does not contribute directly to atmospheric heating, but does contribute indirectly through a conversion to infrared radiation at the surface. The infrared channel has its own characteristic absorption coefficient, and is responsible for redistributing thermal energy in the atmosphere and cooling to space. In particular, emission from the heated surface, with subsequent reabsorption in the lower atmosphere, gives rise to the formation of a troposphere.

Two-stream approximations to the appropriate equations of radiative transfer yield an analytic solution for the vertical thermal structure of this simplified model. The Planck intensity can be expressed in the form

$$B(\tau) = \frac{1}{4}F_0 \{g_0 + g_1\tau + g_2 e^{-\sqrt{3}\kappa\beta\tau}[1 + g_3 e^{-2\sqrt{3}\kappa\beta(\tau_1 - \tau)}]\} \quad (A33)$$

where πF_0 is the solar constant (adjusted for the correct planetary distance from the Sun), τ is the thermal infrared optical depth, τ_1 is the infrared optical thickness of the atmosphere,

$$\beta = \frac{\chi_E^v}{\chi_E^{ir}} \quad (A34)$$

is the ratio of visible to infrared extinction (absorption plus scattering) cross-sections, and

$$\kappa = [(1 - \omega^v)(1 - \omega^v <\cos\theta>_v)]^{1/2} \quad (A35)$$

where ω^v and $<\cos\theta>_v$ are the single scattering albedo and asymmetry factor for the visible channel, respectively. The multipliers $g_0, ..., g_3$ are functions of the physical parameters defining the mode and are independent of optical depth.

The form of equation (A33) mimics the overall vertical thermal structures of real planetary atmospheres rather well. Limiting cases expose the basic physics involved. If an atmosphere is thin, then $\kappa\beta\tau_1 << 1$ and $\exp(-2\sqrt{3}\kappa\beta\tau_1) \sim 1 - 2\sqrt{3}\kappa\beta\tau_1$, leading to

$$B(0) < B(\tau_1) \quad (A36)$$

if $2g_2g_3 > g_2(1 + g_3) - g_1/\sqrt{3}\kappa\beta$, and conversely if the inequalities are reversed. Scattering tends to be unimportant in thin atmospheres due to a lack of clouds, and $\omega^v \sim 0$, leading to simplifications for κ, g_1, g_2 and g_3. A complete reduction demonstrates that equation (A36) remains valid if

$$\frac{(1-q)(1-A_c)}{q(1-A_v)} > \beta^2 - 1 \quad (A37)$$

where q and $(1 - q)$ are the fractional amounts of the incoming solar flux in the visible and conservative channel respectively, and A_v and A_c are the corresponding surface albedos. Clearly equation (A37) is true if $\beta < 1$; that is, if the visible channel is less opaque than the infrared channel [equation (A34)]. In practice, because only the strongest lines would be expected to contribute to absorption in thin atmospheres, $(1 - q)$ should be considerably larger than q, while A_c should be comparable to A_v. Hence, generally for thin atmospheres the surface temperature should be greater than the temperature at the top of the atmosphere, unless β is quite large.

Very thick atmospheres behave differently. Clouds tend to be ubiquitous and considerable scattering of solar radiation is to be expected, although scattering in the thermal infrared can generally be ignored in the first approximation. In this case $2\sqrt{3}\kappa\beta\tau_1 >> 1$ and, in the absence of an internal heat source, the limiting solutions to equation (A33) reduce to

$$B(0) = \frac{\sqrt{3}}{4}F(0)[1 + \kappa\beta] \quad (A38)$$

and

$$B(\infty) = \frac{\sqrt{3}}{4}F(0)\left[1 + \frac{1}{\kappa\beta}\right] + \frac{1}{2}(1-q)F_0 \frac{1}{\beta_c(1 - <\cos\theta>_c)} \quad (A39)$$

where $\pi F(0)$ is the outgoing infrared flux at the top of the atmosphere, and β_c and $<\cos\theta>_c$ are the conservative channel counterparts to β and $<\cos\theta>_v$. If there is a non-negligible internal heat source, an additional term proportional to the optical thickness is required on the right hand side of equation (A39), and $B(\tau_1)$ increases without bound as $\tau_1 \to \infty$.

In the absence of the conservative field, $B(\infty)$ can be either $< B(0)$ or $> B(0)$, depending on whether $\kappa\beta$ is >1 or < 1 respectively. In the former case a temperature inversion occurs, and the temperature increases with altitude. In the latter case a greenhouse effect is created, even though the visible radiation field is exponentially attenuated as it diffuses downward. In this case the atmosphere is effectively more transparent to the visible field than to the infrared field arising as a result of solar heating. The fraction of atmosphere above a given level acts as a thermal blanket that is more opaque to infrared radiation than to solar radiation, and the temperature at that level is elevated above what it would be in the absence of the overlying atmosphere.

When the variability of the absorption coefficient with wavelength is taken into account by adding the conservative channel, the second term on the right hand side of equation (A39) is also required. The greenhouse effect is considerably enhanced, especially if β_c is small and $<\cos\theta>_c$ is close to unity. Small β_c indicates that the atmosphere is relatively transparent to the conservative field, while $<\cos\theta>_c \sim 1$ implies that forward scattering is very strong – solar radiation originally directed downward will tend to remain directed in a downward direction even after many scattering processes, thereby allowing the conservative field to maintain a high intensity even after penetrating very deeply into the atmosphere.

In order for a temperature inversion and a greenhouse effect to be present simultaneously, heating from both above and below is required. The visible field will heat the uppermost layers, while the conservative field and/or the internal heat reservoir will heat from below. Differentiation of equation (A33) yields

$$\frac{\partial B(\tau)}{\partial \tau} = \frac{1}{4}F_0\{g_1 - \sqrt{3}\kappa\beta g_2 e^{-\sqrt{3}\kappa\beta\tau}[1 - g_3 e^{-2\sqrt{3}\kappa\beta(\tau_1 - \tau)}]\} \quad (A40)$$

Because $g_3 < 1$, the quantity in brackets is always positive. Hence $\partial B(\tau)/\partial \tau$ can vanish if g_2 is sufficiently large and a temperature minimum will occur at $\tau = \tau_m$, say. For thick atmospheres τ_m tends to be much smaller than τ_1, and $\kappa\beta\tau_1 \gg 1$. Hence $g_1 = \sqrt{3}\kappa\beta g_2 \exp(-\sqrt{3}\kappa\beta\tau_m)$ approximately, requiring $g_1 < \sqrt{3}\kappa\beta g_2$. Thus, if the ratio $\sqrt{3}\kappa\beta g_2/g_1 > 1$ a temperature minimum exists – otherwise it does not.

A complete reduction yields

$$\sqrt{3}\kappa\beta\frac{g_2}{g_1} = \frac{q(1 - a_v)}{(1 - q)(1 - a_c) + \sqrt{3}F_n/F_0}[(\beta\kappa)^2 - 1] \quad (A41)$$

where πF_n is the heat flux from the interior, and a_v and a_c are the albedos at the top of the atmosphere associated with the visible and conservative fields respectively. Clearly, if $\beta\kappa < 1$ no temperature minimum is possible. Even if $\beta\kappa$ is slightly greater than unity, no minimum will exist if the denominator, which is proportional to the net upward heat flux at the surface, is large compared with $q(1 - a_v)$. In this case heating from below dominates heating from the visible channel at all levels, and the temperature decreases with altitude everywhere.

Real atmospheres

Figures A45 and A46 show the vertical thermal structure of all the planets with substantial atmospheres (including Titan). Mars and Venus are the only two planets with dominantly CO_2 atmospheres. Because CO_2 absorbs only weakly in the visible and near infrared, but emits strongly in the ν_2 band at 15 μm, $\beta < 1$ for both planets. According to equations (A37) (for Mars) and (A41) (for Venus) and the related discussion, neither planet should have a stratospheric inversion; this is in accord with observation.

Venus also has a strong greenhouse effect, as illustrated by equations (A38) and (A39). Both $\kappa\beta$ and $\beta_c(1 - \langle\cos\theta\rangle_c)$ are small compared with unity, and

$$F(0) = \frac{1}{\sqrt{3}}F_0 q(1 - a_v) \quad (A42)$$

is small compared with $(1 - q)F_0$. This is partly because CO_2 at low pressures is relatively transparent at solar wavelengths, and partly because the sulfuric acid clouds in Venus' upper troposphere are both highly reflecting and forward scattering. Considerable solar radiation penetrates to pressures where pressure broadening and collision induced absorption by CO_2 greatly increases the thermal infrared opacity. According to Figure A45, a greenhouse effect appears to exist for all the other planets shown, although for Jupiter, Saturn and Neptune the effect is overwhelmed at depth by internal heat sources.

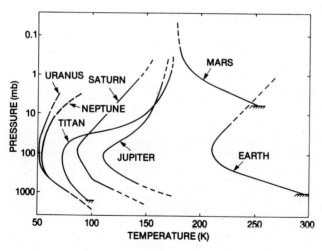

Figure A45 Vertical temperature profiles for various planets. All profiles were obtained by inversion of infrared spectral measurements from Michelson interferometers, except for that of Titan, obtained from Voyager 1 radio occultation data (Hanel et al., 1992).

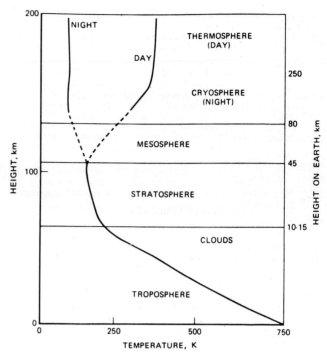

Figure A46 Typical temperature profiles of the Venus atmosphere. The discussion in this article has relevance only below ~ 100 km (Fimmel, Colin and Burgess, 1983).

With the exceptions of Venus and Mars, the planets also have temperature minima as a result of stratospheric inversions. Ozone is responsible for this inversion on Earth. It absorbs strongly in the ultraviolet below 0.3 μm, but emits only weakly at 9.6 μm in the infrared, yielding a value of β large enough to create the inversion. Methane plays a comparable role on the giant planets. Although many strong absorption bands occur in the near infrared, only the ν_4 band of CH_4 at 7.7 μm is available for emission. Even though this band is strong, it lies well outside the spectral range of maximum emission at the low temperatures associated with the outer planets, and β substantially exceeds unity.

Titan's atmosphere is exceptionally rich in organic gases. Photochemistry and charged particle bombardment from Saturn's magnetosphere have produced a dense, absorbing aerosol from these gases, which in turn provides most of the opacity in Titan's stratosphere. Particle diameters are less than a micron, leading to infrared cross-sections small compared with the geometric cross-section, though the optical and geometric cross-sections are comparable at solar wavelengths. According to equation (2) this should yield an exceptionally large value of β. Analysis yields a value ~ 10^3 for $[(\kappa\beta)^2 - 1]$ in equation (9), leading to the largest stratospheric temperature gradient of any of the planetary bodies.

Even though Titan's aerosol is dense, enough near-infrared radiation penetrates to heat the surface slightly and create a small greenhouse effect. Collision-induced absorption by N_2-CH_4, CH_4-CH_4 and H_2-N_2 molecular pairs is primarily responsible for the far-infrared opacity of the atmosphere. Because the amount of H_2 available in Titan's atmosphere is small, a fairly transparent window exists at about 18.5 μm, and this limits the magnitude of the greenhouse effect. The major planets all have considerably more H_2 than Titan, however, and for them the window is effectively closed by H_2-H_2 and H_2-He collision-induced absorption. This, along with strong internal heat sources, ensures high temperatures at depth for these planets.

Robert E. Samuelson

Bibliography

Conrath, B.J., Hanel, R.A. and Samuelson, R.E. (1989) Thermal structure and heat balance of the outer planets, in *Origin and Evolution of Planetary and Satellite Atmospheres* (eds S.K.

Atreya, J.B. Pollack and M.S. Matthews). Tucson: University of Arizona Press, pp. 513–38.
Fimmel, R.O., Colin, L. and Burgess, E. (1983) *Pioneer Venus*. NASA Special Publication SP-461.
Hanel, R.A., Conrath, B.J., Jennings, D.E. and Samuelson, R.E. (1992) *Exploration of the Solar System by Infrared Remote Sensing*. Cambridge, New York: Cambridge University Press.
Samuelson, R.E. (1983) Radiative equilibrium model of Titan's atmosphere. *Icarus*, **53**, 364–87.

Cross references

Earth: atmosphere
Greenhouse effect
Jupiter: atmosphere
Mars: atmosphere
Neptune: atmosphere
Radiative transfer in planetary atmospheres
Saturn: atmosphere
Spectroscopy: atmospheres
Titan
Uranus: atmosphere
Venus: atmosphere

AURORA: HISTORICAL RECORD

Auroras of planet Earth provide us with a proxy record of solar–terrestrial behavior spanning over 2400 years. The earliest description of the aurora borealis, together with a peak in sunspots and a passage of Halley's comet, appears to have been recorded about 467 BC (Stothers, 1979; Schove, 1983).

The auroras were commonly assumed to be divine messages, signifying omens for better or worse, but commonly indicating the wrath of the deity for humanity's sins, real or imagined. Accordingly, the auroral displays were probably better documented (although less understood) by untrained observers than any other celestial phenomena.

Early work on auroras, e.g. in 1710 by G.W. von Leibnitz (1646–1716), traced records back to the Saxon chronicles of the 10th century. De Mairan, in 1733 (revised 1754) collected the first comprehensive catalog going back to classical Greece and Rome (corrected by Stothers, 1979; see also Siscoe, 1980). In the 19th century the correlation between auroras and the sunspot cycle was noticed, and daily monitoring of solar behavior became widespread. An obvious clustering was seen, but the gaps or low points in both auroras and sunspot numbers established the periodicity more precisely because of the occurrence of multiple peaks (Fairbridge and Hameed, 1983; Schove, 1983).

Auroral frequency and strength are both reduced during the extended sunspot cycle 'minima'; dated approximately from low to low, these are: Sabine or Dalton (1796–1835), Maunder (1632–1712), Spörer (1394–1533), Wolf (1275–1353), Oort (1018–1065) and 'Dark Ages Minimum' (661–738 AD). As shown by Fairbridge and Shirley (1987) and Charvátová (1995), the low activity intervals correspond mainly to the most disordered periods of the Sun's circum-barycentric inertial orbit. A long-term 'Solar index' of amplitudes of the solar cycle comparable to the Wolf numbers, based on a combination of sunspot and auroral data, is provided by Bray (1980). An update with geomagnetic index is provided by Silverman (1992).

Solar and lunar correlations

Terrestrial auroras show a clear correlation between auroras and the sunspot cycle (SSC), which averages 11.12 years (± 5.5 years). The aurora peak is usually 1 to 2 years after the SSC.

In 1859 R.C. Carrington (and colleagues) observed a gigantic 'white-light' solar flare that was followed soon after by a violent magnetic storm and a brilliant, long lasting series of aurora borealis as well as aurora australis. The displays tended to follow one another in 26-day peaks, the mean period Carrington had discovered for the solar rotation.

More complete auroral (and sunspot) catalogs were gathered during the late 19th century which expanded and revised earlier ones (Fritz 1873; English translation 1928). The lists were expanded to Scandinavia and analyzed by Ekholm and Arrhenius (1898), who also recognized the 18.6 year lunar cycle. The presence of a lunar cycle provides a significant clue as to the modulation of the geomagnetic field by the nutation of the Earth's spin axis.

Ancient Chinese records (Schove, 1955, 1983, 1987; Stephenson and Wolfendale, 1988) disclose intensities that are usually higher than in Europe. Archeomagnetism brought the explanation: the north magnetic pole lay, for extended periods, in the hemisphere of the eastern Arctic (Bucha, 1979) and only shifted to the western hemisphere (in northern Canada) during the last three centuries (Siscoe and Verosub, 1983).

When the Earth's main dipole field weakens, auroral sightings are reported as far as 45° from the magnetic pole. When that pole is more than 10° from the geographic pole (as today, in northern Canada), rare sightings in North America may reach 35° latitude or still farther south. In medieval times when the magnetic pole lay in the eastern hemisphere there were increased sightings not only in China, but also in Constantinople, Baghdad and Cairo (Siscoe, 1980; Schove, 1983). In the five centuries BC, frequent sightings in classical Greece and Rome suggests that the magnetic pole was also favorably placed at that time.

Rhodes W. Fairbridge

Bibliography

Bray, J.R. (1980) A sunspot–auroral Solar Index from 522 B.C. to A.D. 1968. *New Zealand J. Sci.*, **23**, 99–106.
Bucha, V. (1979) Connections between geophysical and meteorological processes. *Studia Geophys. Geodet.*, **23**, 55–67, 102a–102d.
Chapman, S. (1964) Aurora and geomagnetic storms, in *Space Physics* (eds D.P. LeGalley and A. Rosen) New York: J. Wiley (752 pp.) pp. 226–269.
Charvátová, I. (1995) Solar–terrestrial and climatic variability during the last millennia. *J. Coastal Res.*, Spread Issue **17**, 343–354.
Charvátová-Jakabcová, Streštik, J. and Křivsky, L. (1988) The periodicity of aurorae in the years 1001–1900. *Studia Geophys. Geodet.*, **32**, 70–7.
Courtillot, V. and Lemouël, J.L. (1984) Geomagnetic secular variation impulses. *Nature*, **311**, 709–16.
Ekholm, N. and Arrhenius, S. (1898) Ueber den Einfluss des Mondes auf die Polarlichter und Gewitter. *Kongl. Svenska Vetenskap-sakademiens Handlingar*, **32**(2).
Fairbridge, R.W. (1987) Sunspots, in *The Encyclopedia of Climatology* (eds J.E. Oliver and R.W. Fairbridge). New York: Van Nostrand Reinhold, pp. 815–23.
Fairbridge, R.W. and Hameed, S. (1983) Phase coherence of solar cycle minima over two 178-year periods. *Astron. J.*, **88**(6), 867–9.
Fairbridge, R.W. and Shirley, J.H. (1987) Prolonged minima and the 179-yr cycle of the solar inertial motion. *Solar Phys.*, **110**, 191–220.
Fritz, H. (1873) *Verzeichniss Beobachter Polarlichter*. Wien: C. Gerold.
Fritz, H., 1928. The periods of solar and terrestrial phenomena. *Monthly Weather Rev.*, **56**, 401–407. [English translation of *Vierteljahrsch. Natf. Ges., Zürich*, **1**, 1893].
Gleissberg, W. (1965) The eighty-year cycle in auroral frequency numbers. *J. Br. Astron. Assoc.*, **75**, 227–231.
Schove, D.J. (1955) The sunspot cycle, 649 B.C. to A.D. 2000. *J. Geophys. Res.*, **60**, 127–146.
Schove, D.J. 1983. *Sunspot Cycles*. Benchmark Papers in Geology, vol. 68. Stroudsburg: Hutchinson Ross Publ. Co., 392 pp.
Schove, D.J. 1984. *Chronology of Eclipses and Comets*. Woodbridge, Suffolk (UK): Boydell Press, 356 pp.
Schove, D.J. 1987. Sunspot cycles and weather history, in *Climate: History, Periodicity and Predictability* (ed. M.R. Rampino). New York: Van Nostrand.
Silverman, S.M. (1992) Secular variation of the aurora for the past 500 years. *Rev. Geophys.*, **30**(4), 333–51.
Siscoe, G.L. (1980) Evidence in the auroral record for secular solar variability. *Rev. Geophys. Space Phys.*, **18**, 647–58.
Siscoe, G.L. and Verosub, K.L. (1983) High medieval auroral incidence over China and Japan: implications for the medieval site of the geomagnetic pole. *Geophys. Res. Letters*, **10**(4), 345–8.
Stephenson, F.R. and Wolfendale, A.W. (eds) (1988) *Secular Solar and Geomagnetic Variations in the Last 10,000 Years*. Dordrecht: Kluwer Academic, 510 pp.
Stothers, R. (1979) Ancient aurorae. *Isis*, **70**(251), 85–95.

Cross references

Coronal mass ejection
Geomagnetic storm
Precession and nutation
Solar activity
Solar flare

AURORA, PLANETARY

Auroral displays, with their beautiful colors and constantly changing patterns, have intrigued Earth-bound observers for centuries (Plates 3, 9). The earliest written observational records date back to the Old Testament, but it was not until the writings of Kepler and Galileo that a scientific inquiry into auroral phenomena began. The real advance in our understanding of the aurora, however, came with the advent of the space age, when it became possible for the first time to make measurements of the charged particles and electric and magnetic fields in the Earth's space environment. Later, spacecraft orbiting thousands of kilometres above the Earth obtained spectacular global images of the aurora, which gave us our first views of its large-scale behavior and morphology. Not only has the space age advanced our knowledge of the terrestrial aurora; Earth-orbiting observatories such as the International Ultraviolet Explorer (IUE) and planetary probes like Pioneer and Voyager have permitted us to discover and study auroral activity at other planets in the solar system as well. Auroral emissions of varying strengths have been detected at Venus, Saturn, Jupiter, Uranus and Neptune. Of these planetary auroras, the most powerful – and, next to Earth's, the most intensively investigated – is Jupiter's.

The aurora presents scientists with a host of questions about the complex interactions that occur between a planet's upper atmosphere and the plasma or charged-particle medium in which the planet is embedded. Even after years of extensive research and significant advances in our understanding of auroral processes, many questions remain to be answered, particularly about the less well-observed planetary auroras but also about the Earth's aurora. While the terrestrial and planetary auroras may differ from each other in certain critical respects, they all have in common three essential elements: emissions at a variety of wavelengths, a source of auroral power and particle acceleration mechanisms.

Auroral emissions

The colorful visible-light emissions of the Earth's aurora borealis and aurora australis are the best known auroral phenomenon. The familiar green, white and red emissions occur when charged particles are accelerated downward along magnetic field lines into Earth's upper atmosphere. There, at altitudes of 100 to 150 km, they collide with and excite molecular and atomic oxygen and molecular nitrogen, which then relax and emit photons at characteristic wavelengths (e.g. the green and red atomic oxygen emissions at 5577 Å and 6300 Å respectively, and blue nitrogen emissions at 4278 Å and 3914 Å).

Auroral emissions also occur in other regions of the electromagnetic spectrum, ranging from x-ray to infrared and radio wavelengths. At Earth, for example, a powerful (10^7 W) radio emission – the auroral kilometric radiation (AKR) – is generated along the auroral magnetic field lines by the resonance between a Doppler shift of the cycloidal motion of electrons in a magnetic field and naturally occurring fluctuations in the background plasma. This mechanism is known as the cyclotron maser. Similar auroral radio emissions have been observed at other planets.

The rapid deceleration of energetic auroral electrons as they pass close to atomic nuclei in the Earth's upper atmosphere produces radiation at x-ray wavelengths ('bremsstrahlung' or 'braking radiation'). Measurements of these auroral bremsstrahlung x-rays in the mid to late 1950s by balloon- and rocketborne instrumentation made it possible to establish the importance of energetic electrons (as opposed to energetic solar protons) in the generation of the terrestrial aurora. Indeed, today one of the best ways to measure the energy flux and distribution of auroral electrons is to image the bremsstrahlung x-ray emissions, an experiment that is routinely carried out on the NASA Upper Atmospheric Research Satellite (UARS).

In 1963 Yasha Feldstein of the USSR showed that terrestrial auroras appear along oval-shaped bands around both magnetic poles. The instantaneous distribution of the aurora is located near 67° magnetic latitude at local midnight but at about 78° local midday (for quiet solar conditions). During disturbed times, when a magnetospheric substorm is triggered by a southward turning of the solar interplanetary field and a reconfiguration of the Earth's magnetotail occurs, significant brightenings of the aurora occur near local midnight and the auroral oval extends to lower magnetic latitudes. The global morphology of the aurora can be easily seen in the high-altitude images obtained by the NASA Dynamics Explorer and the Scandinavian Viking spacecraft. Such high-altitude imaging has revealed previously unknown emission patterns, such as the theta aurora, that continue to challenge our understanding of auroral processes.

Power sources

Auroras can be roughly categorized in terms of their power sources as internally driven and externally driven auroras. The Earth's aurora is an externally driven aurora, powered by the interaction of the solar wind (the stream of charged particles and the embedded interplanetary magnetic field that originate at the Sun) with the Earth's magnetosphere (the 'magnetically shielded' region around the Earth where the behavior of the charged particles is controlled by the magnetic field). The nature of this interaction is not completely understood, but it appears to involve both a viscous-like process in which mass, momentum and energy are transferred as a result of the solar wind's 'rubbing' against the Earth's magnetosphere and a more impulsive process in which antiparallel magnetic fields from the Earth and interplanetary magnetic field 'merge' or 'reconnect' with a resulting release of energy. Jupiter's powerful aurora is an example of an internally driven aurora. While the solar wind interaction with Jupiter's giant magnetosphere may play a role in the production of the Jovian aurora, it is thought that the tapping of the planet's internal rotational energy is predominantly responsible for auroral processes.

Particle acceleration mechanisms

At Earth, auroral particles are accelerated downward within currents that map onto the auroral oval. In addition to the so-called region 1 currents, which are produced by the interaction of the solar wind with the Earth's magnetosphere, the neutral sheet current, which maintains the Earth's magnetotail, is partially diverted into the ionosphere during magnetospheric substorms, causing a brightening and poleward expansion of the auroral oval known as an auroral substorm. When these currents exceed certain well-defined thresholds, the auroral particles are further accelerated, forcing current continuity within the circuit. The regions where this acceleration occurs exist at altitudes of one to two Earth radii above the high-latitude atmosphere and are known as 'inverted-v' potential drops because of the characteristic energy versus latitude signature of the precipitating auroral electrons. These accelerated electrons produce the bright arcs, striations and other dynamic features that we associate most readily with the visible aurora.

Energetic ions also stimulate auroral emissions; these emissions are weaker and more diffuse, however, than those due to electron precipitation. These energetic ions form part of the ring current and associated radiation belts; their precipitation into the atmosphere is controlled by scattering from plasma waves present in the local equatorial magnetospheric environment near geosynchronous orbit (at an altitude of 5.6 Earth radii).

These three elements of auroral activity – multiple wavelength emissions, a power source and particle acceleration (and resulting electrical currents) – can be scientifically studied by both local and remote-sensing techniques. Local (or *in situ*) spacecraft measurements of charged particle densities, temperatures and motions, as well as of the local structure of electric and magnetic fields, are necessary to understand the microphysics of auroral phenomena and provide valuable information about auroral acceleration processes and the power sources that drive these processes. *In situ* measurements have contributed significantly to our understanding of the Earth's aurora and of the solar wind–magnetosphere interaction that drives it. Unfortunately, compared to the spacecraft missions to study the Earth's space environment, there have been relatively few missions to the other planets; moreover, the planetary missions that have been flown have left unsampled or inadequately sampled the

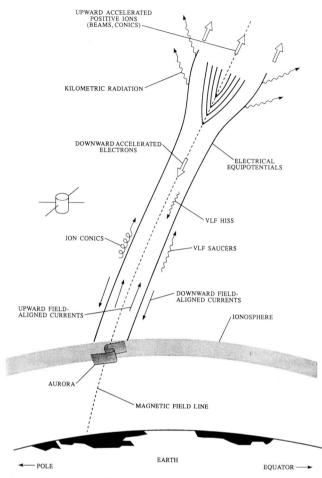

Figure A47 Schematic illustrating the terrestrial auroral process.

high-latitude regions of the planetary magnetospheres that are critical for an understanding of auroral processes. Extensive *in situ* data on the auroras at other planets are thus generally not available, and the investigation of these auroras must rely heavily on remote-sensing observations of auroral emissions at x-ray, ultraviolet, infrared and radio wavelengths.

Other terrestrial planets

Besides Earth, the only other inner or terrestrial planet where auroral emissions have been detected is Venus. The Venera and Pioneer Venus spacecraft have shown Venus to have little or no intrinsic magnetic field. The deflection of the solar wind is thus produced by direct interaction with the charged upper layer of the atmosphere, the ionosphere. Although this process is quite different from Earth's solar wind–magnetosphere interaction, which drives the terrestrial aurora, weak auroral emissions have been observed at Venus. In 1986 a team of researchers reported observations by the Pioneer Venus Orbiter of aurora-like ultraviolet emissions on the planet's nightside. The observed emissions, which vary in intensity and sometimes cover nearly the entire nightside, are due to the excitation of atomic oxygen in Venus' upper atmosphere and may be stimulated by precipitating electrons energized by the interaction of the solar wind and Venus' ionosphere. Attempts to correlate the emissions with solar wind parameters suggest that solar flares and interplanetary shocks may play a role in the generation of the Venusian aurora.

Like Venus, Mars has no or only a weak intrinsic magnetic field and thus would not be expected to have an Earth-like aurora; further, no aurora-like ultraviolet emissions like those observed by PVO at Venus have been detected at Mars. Spacecraft measurements at Mercury, on the other hand, have indicated that Mercury has a substantial magnetic field that forms a magnetosphere around the planet. Mercury's magnetosphere deflects the solar wind flow, in a manner quite analogous to that at Earth, and the resulting solar wind–magnetosphere interaction appears to be responsible for producing particle acceleration much like that encountered in a terrestrial auroral substorm. However, Mercury's atmosphere is only 10^{-15} that of Earth, and there is no evidence of any auroral emissions resulting from the particle acceleration that occurs in these substorm-like events.

Outer planets

The two most characteristic and reproducible outer planet auroral emissions are radio waves produced by accelerated particles and ultraviolet emissions produced by electron impact on molecular and atomic hydrogen, which are the major constituents of the upper atmospheres of Jupiter, Saturn, Uranus and Neptune. All the outer planets have revealed radio emissions to detectors aboard the Voyager spacecraft that appear to be associated with auroral processes. The first-order similarities of the emissions from Earth, Jupiter, Saturn, Uranus and Neptune suggest common processes for radio wave generation. The situation at Jupiter is rendered somewhat more complex than that at the other outer planets, however, because of the presence of additional radio emissions caused by the interaction of Io with the Jovian magnetosphere and frequency shifting associated with the intensity of Jupiter's magnetic field.

Ultraviolet auroral emissions have also been observed at every outer planet visited by the Voyager spacecraft, with varying emission intensities and different global emission morphologies. Weak emissions were seen on Neptune's nightside at high magnetic latitudes; researchers disagree about whether these auroral emissions are powered by the solar wind interaction or by the planet's rotational energy. Bright emissions were seen at Uranus, forming a polar cap at the nightside magnetic pole and a polar ring at the dayside magnetic pole. The asymmetries apparently are the result of the offset of the magnetic dipole moment towards northern polar latitudes. The ultraviolet auroral emissions observed at Saturn emanate from two ovals ringing either magnetic pole at $\approx 80°$ latitude. An interesting feature of Saturn's aurora is a longitudinally fixed brightness 'hot spot'. Closely correlated with the auroral emissions in both latitude and longitude is the radio wavelength Saturn kilometric radiation (SKR), which is confined in longitude near the noon meridian in both the northern and southern auroral regions. SKR has been shown to be correlated strongly with the solar wind pressure, which suggests that, as on Earth, the source of auroral power at Saturn is probably the solar wind. The longitudinal dependence of the radio emissions and of the auroral hot spot has not been satisfactorily explained but may be due to some anomaly in the planet's magnetic field.

The most extensively observed of the outer planet auroras – and the most powerful aurora in the solar system ($> 5 \times 10^{13}$ W) – is the Jovian aurora. Pioneer, Voyager and, most recently, Ulysses made *in situ* measurements of charged particles and of electric and magnetic fields that have provided clues to auroral processes at Jupiter. These measurements yield only a partial picture of Jovian magnetospheric processes, however, and must be supplemented by remote sensing data acquired at multiple wavelengths by both spacecraft-borne and ground-based observatories. Taken together with local plasma measurements, observations at x-ray, ultraviolet, infrared and radio wavelengths permit researchers to set constraints on the energies of the precipitating auroral particles and to make inferences about the identities of these particles and about the processes that energize them.

A distinctive feature of Jupiter's magnetospheric system is the interaction of the planet's inner magnetosphere with the volcanically active Jovian satellite Io. Sulfur and oxygen ions created from the sulfur dioxide emitted by Io's volcanoes form a plasma torus around Jupiter, which is a significant source of plasma for Jupiter's magnetosphere and which is thought to play an important role in generating the Jovian aurora. The torus is electrodynamically coupled with Jupiter's ionosphere and corotating inner magnetosphere and is thus able to tap the vast rotational energy of the planet to accelerate the newly created sulfur and oxygen ions, of which some are convected into the outer magnetosphere where they are further energized. These energetic ions return to the inner magnetosphere and are apparently precipitated into the auroral atmosphere by the interactions of the ions with plasma waves.

Table A11 Planetary auroras

Planet	Auroral input power (W)	Emission	Auroral zone morphology	Precipitating particles	Radio emissions	Energy source
Mercury	10^9–10^{10}	Surface IR emissions	?	Electrons	?	Solar wind magnetosphere interaction in magnetotail
Venus	?	OI1304Å~7R	Patchy, sometimes fills nightside, intense evening	Low energy electrons	?	Venus umbra suprathermal electrons
Earth	10^{11}–10^{12}	N_2 Bands OI1304Å	Auroral ring IL = 65–75° MLT = ~187–24 h	Energetic electrons (discrete arcs) energetic protons (diffuse aurora)	Auroral kilometric radiation, nightside 70°IL, 23 h MLT	Solar wind magnetosphere interaction in magnetotail
Mars	?	?	?	?	?	?
Jupiter	$> 2 \times 10^{13}$	Lyman α, H_2 bands, X-rays, C_N H_M and H_3^+ infrared	Auroral oval maps in latitude to magnetopause longitudinal variation: Max intensity North S_{III} ~ 180° South S_{III} ~ 0°	Energetic electrons (10–50 keV), sulfur and oxygen ions (<40 meV/nucleon)	Decametric Io related polar cap related	Planetary rotation, solar wind interaction (?)
Saturn	10^{11}–10^{12}	Lyman α, H_2 bands	78–81° latitude near noon local time	Energetic electrons (5–10 keV), nitrogen ions (?)	Saturn kilometric radiation dayside cusp	Solar wind magnetosphere interaction in cusp
Uranus	10^{11}	Lyman α, H_2 bands H_3^+ infrared	Dayside: 55–60° MLT ring Nightside: 20° polar cap circle	Energetic electrons (~10 keV)	Nightside, southern, polar cap/auroral zone	Solar wind magnetosphere interaction
Neptune	10^{10}	Lyman α, H_2 bands		?	Yes	?

The question of Io's role in the generation of Jupiter's aurora is far from settled, however, and is at the center of a controversy over which particles are responsible for the Jovian auroral emissions and which processes are responsible for their energization. Precipitation of Iogenic heavy sulfur and oxygen ions, downward acceleration of electrons within field-aligned currents, and a combination of both of these mechanisms have all been proposed to account for the observed auroral emissions. Unfortunately, remote sensing observations at ultraviolet and x-ray wavelengths and the available *in situ* data support conflicting interpretations and do not permit a clear-cut resolution of the controversy.

The soft x-ray emissions observed by the Einstein and Röntgensatellit (ROSAT) observatories, for example, could be either K-shell emissions caused by heavy ion precipitation or bremsstrahlung emissions excited by energetic electron precipitation. Calculations of the energy requirements for each mechanism and emissions present in the measured x-ray spectra consistent with the recombination lines of high-charge states of sulfur and oxygen appear to support the case for sulfur and oxygen ion precipitation. However, corresponding sulfur and oxygen emissions at extreme ultraviolet wavelengths, which would be expected to accompany the x-ray K-shell emissions, have not been verified in observations at ultraviolet wavelengths made by the Earth-orbiting International Ultraviolet Explorer (IUE).

While *in situ* plasma measurements by Voyager 1 and 2 provided evidence for heavy ion precipitation, more recent *in situ* measurements, made by Ulysses as it passed through Jupiter's high-latitude magnetosphere (a region not sampled by the Voyagers), identified both magnetic field signatures and electrical current flow (precipitating energetic electrons) reminiscent of those in the Earth's high-latitude magnetosphere. The Ulysses data suggest that electrons affected by the solar wind interaction with the Jovian magnetosphere also excite auroral emissions. This interpretation was reinforced by simultaneous remote measurements of the Jovian auroral ultraviolet emissions by the Hubble Space Telescope. Thus, Jupiter may exhibit Earth-like auroral phenomena as well as Io-generated heavy-ion auroral processes. The question remains open for now, although recent ground-based measurements of infrared emissions from Jupiter's auroral zones hold the promise of a significant advance toward finding an answer.

Conclusion

This article has focused on the emissions at various wavelengths that characterize auroras, the extrinsic (solar wind) and intrinsic (planetary rotation) sources of auroral power, and the mechanisms by which auroral particles are accelerated into the upper atmosphere (potential drops, wave-particle interactions). Table A11 provides a comparative summary of the present state of our knowledge of the aurora at Earth and other planets. It should be noted in conclusion, however, that the emission of photons at various wavelengths is not the only atmospheric effect of auroral particle precipitation. The energy deposited by auroral processes into the upper atmosphere also influences through Joule (resistive) heating the temperature structure and circulation patterns of the thermosphere. During equinox, for example, auroral heating of the Earth's high latitude thermosphere due to strong geomagnetic storms essentially reverses the equator-to-pole pattern characteristic of periods of quiet auroral activity. Auroral activity may also alter the compositional structure of the upper atmosphere through such mechanisms as the impact dissociation of molecular species. Auroral modification of atmospheric compositional structure is thought to be particularly important in the case of Jupiter, where dissociation of molecular hydrogen by auroral particle impact is believed to make a significant contribution to Jupiter's global atomic hydrogen budget.

J. Hunter Waite Jr

Bibliography

Akasofu, S.-L. and Kamide, Y. (1987) The aurora, in *The Solar Wind and the Earth*. (eds S.-I. Akasofu and Y. Kamide) Tokyo: Terra Scientific Publishing Company, pp. 143–59.

Bagenal, F. and Sullivan, J.D. (1981) Direct plasma measurements in the Io torus and inner magnetosphere of Jupiter. *J. Geophys. Res.*, **86**, 8447.

Barbosa, D.D. (1990) Auroral precipitation flux of ions and electrons in Saturn's outer magnetosphere. *Planet. Space Sci.*, **38**(10), 1295–304.

Broadfoot, A.L. Abreya, S.K., Bertaux, J.L. *et al.* (1989) Ultraviolet spectrometer observations of Neptune and Triton. *Science*, **246**, 1459–66.

Broadfoot, A.L., Herbert, F., Molberg, J.B. *et al.* (1986) Ultraviolet spectrometer observations of Uranus. *Science*, **223**, 74–9.

Burch, J.L. (1987) Plasma populations in the magnetosphere, in *The Solar Wind and the Earth* (eds S.-I. Akasofu and Y. Kamide). Tokyo: Terra Scientific Publishing Company, pp. 103–22.

Connerney, J.E.P. (1987) The magnetospheres of Jupiter, Saturn, and Uranus. *Rev. Geophys.*, **25**(3), 615–38.

Cheng, A.F. (1990) Triton torus and Neptune aurora. *Geophys. Res. Lett.*, **17**(10), 1669–72.

Desch, M.D. (1982) Evidence for solar wind control of Saturn radio emission. *J. Geophys. Res.*, **87**, 4549.

Eather, R.H. (1980) *Majestic Lights*. Washington, DC: American Geophysical Union.

Fox, J.L. and Stewart, A.L.F. (1991) The Venus ultraviolet aurora: a soft electron source. *J. Geophys. Res.*, **96**(A6), 9821–8.

Gehrels, N. and Stone, E.C. (1983) Energetic oxygen and sulfur ions in the Jovian magnetosphere and their contribution to the auroral excitation. *J. Geophys. Res.*, **88**(A7), 5537–50.

Kurth, W.S. (1991) Magnetospheric radio and plasma wave research: 1987–1990. *Rev. Geophys.*, Suppl., 1075–86.

Luhmann, J.G. (1991) Space plasma physics research progress 1987–1990: Mars, Venus, and Mercury. *Rev. Geophys.*, Supplement, 965–75.

McNutt, R.L., Jr(1990) The magnetospheres of the outer planets. *Rev. Geophys.*, Supplement, 985–997.

Metzger, A.E., Gilman, D.A., Luthey, J.L. *et al.* (1983) The detection of x-rays from Jupiter. *J. Geophys Res.*, **88**(A10), 7731–41.

Phillips, J.L. Stewart, A.I.F. and Luhmann, J.G. (1986) The Venus ultraviolet aurora: observations at 130.4 nm. *Geophys. Res. Lett.*, **13**(10), 1047–50.

Russell, C.T. (1987) The magnetosphere, in *The Solar Wind and the Earth* (eds S.-I. Akasofu and Y. Kamide). Tokyo: Terra Scientific Publishing Company, pp. 73–100.

Roble, R.G. (1987) The Earth's thermosphere, in *The Solar Wind and the Earth* (eds S.-I. Akasofu and Y. Kamide). Tokyo: Terra Scientific Publishing Company, pp. 245–64.

Sandel, B.R. and Broadfoot A.L. (1981) Morphology of Saturn's aurora. *Nature*, **292**, 679–82.

Sandel, B.R., Herbert, F. Dessler, A.J., and Hill, T.W. (1990) Aurora and airglow on the night side of Neptune. *Geophys. Res. Lett.*, **17**(10), 1693–6.

Siscoe, G.L., F.N. Ness and M.C. Yeates (1975) Substorms on Mercury? *J. Geophys. Res.*, **80**(31), 4359–63.

Waite, J.H., Jr, Cravens, T.E., Kozyn, J.U. *et al.* (1983) Electron precipitation and related aeronomy of the Jovian thermosphere and ionosphere. *J. Geophys. Res.*, **88**(A8), 6143–63.

Waite, J.H., Jr, Clarke, J.T., Cravens, T.E. and Hammond, C.M. (1988) The Jovian aurora: electron or ion precipitation? *J. Geophys. Res.*, **93**(A7), 7244–50.

Cross references

Charged particle observation
Geomagnetic storm
Jupiter: magnetic field and magnetosphere
Magnetospheres of the outer planets
Planetary torus
Plasma wave
Thermosphere
Voyager missions

B

BARNARD, EDWARD EMERSON (1857–1923)

Though Edward Emerson Barnard became one of the greatest astronomical observers of his time, his early life was far from promising. His family were very poor; the civil war swept over Nashville, Tennessee, where he had been born on 16 December 1857, and he had only two months' formal schooling. At the age of nine, he was put to work in a photographer's studio; young though he was, the experience he gained there proved to be very useful later on.

He had a natural interest in astronomy; his first telescope was made from an old lens and a cardboard tube. In 1877, after a meeting with Simon Newcomb, one of America's leading astronomers, he began to study mathematics. In this he was much helped by Rhoda Calvert, an English immigrant, whom he married in 1881.

He began hunting for comets and had his first success in 1881. Others followed; his final total was 16. A wealthy American had offered a prize of 200 dollars for each discovery, and Barnard found the money very useful; it went toward the building of his first home.

He then went to Vanderbilt University, and later obtained a post as assistant at the new Lick Observatory, where he quickly established a reputation as an excellent observer. He made systematic observations of the Moon and planets as well as continuing with his searches for comets; he was an independent discoverer of the Gegenschein or Counterglow, and in 1889 he took part in his first eclipse expedition. He used the 36-inch (90 cm) Lick refractor to study Mars; he was unable to see the 'canals' drawn by Lowell and others – which is to his credit, since they are now known not to exist! On several occasions he drew craters that he perceived on the Martian surface. He did not publish these sketches for fear of ridicule, and even tore the pages out of his observing logbook.

His most famous discovery came in 1892, when he identified Amalthea, the fifth satellite of Jupiter. This was in fact the last satellite discovery made visually.

In 1897 Barnard moved to the Yerkes Observatory. During his photographic survey of the sky he discovered many dark nebulæ, and in 1919 published a catalogue of them; he correctly claimed that they are due to obscuring matter rather than voids. In 1916 he measured the proper motion of a faint red dwarf star, now known as Barnard's Star or the 'Runaway' star; it has the greatest known proper motion, and is only 6 light years away. He received many honours, including the Gold Medal of the Royal Astronomical Society. Barnard died at Williams Bay, Wisconsin, on 6 February 1923.

Patrick Moore

Bibliography

Barnard, E.E. (1892) Jupiter's fifth satellite. *Astr. J.*, Nos. 275, 325, 367, 472. Also: *Observatory Mag.*, **15**, 25.
Gordon, R.W. (1892) *Craters on Mars and Mercury. 1983 Yearbook of Astronomy*, London: Sidgwick and Jackson. pp. 138–54.
Hardie, R.H. (1970) Barnard, Edward Emerson. *Dict. Sci. Biogr.*, vol. 1, pp. 463–467.

BARYCENTER

From the Greek for 'mass' and 'mid-point', barycenter is the term employed for the center of mass of a system of gravitationally interacting bodies. The barycenter has dynamical significance in that the motion of the barycenter is equivalent to the motion of the system as a whole.

Although one may define a barycenter for any system of interacting bodies, such as the planet and satellite system of Jupiter, the examples most often encountered in planetary science are the solar system barycenter and the Earth–Moon barycenter. The solar system barycenter is the origin of the solar system inertial frame (or 'local standard of rest'). This is the fundamental coordinate system of Newtonian dynamics. If the totality of the solar system angular momentum were represented as a vector, that vector would originate at the system barycenter and point in the direction of motion of the solar system in its orbit about the center of the galaxy.

For observational purposes it is customary to use heliocentric rather than barycentric coordinates, as the position of the Sun is easily determined. This introduces some mathematical complications in celestial mechanics. The complications arise due to the fact that the center of the Sun only rarely coincides with the barycenter of the solar system. The Sun in fact describes a looping orbit about the solar system barycenter; the center of the Sun is at times more than two solar radii from this origin (see Solar motion). This movement of the Sun gives rise to detectable changes in the motions of planets, which are termed 'indirect perturbations'.

Figure B1 illustrates how the displacement of the Sun relative to the solar system barycenter is determined by the distribution of the major planets in space. The principal interest in the solar motion derives from evidence that this motion may be involved in the generation of solar activity (q.v.) (Jose, 1965; Fairbridge and Shirley, 1987; Fairbridge and Sanders, 1987; Charvátová, 1990). The fundamental period of the solar motion is 19.859 years, which corresponds to the synodic period of Jupiter and Saturn. This has been characterized as the 'pulse' of the solar system (Fairbridge and Sanders, 1987).

The Earth and Moon are often characterized as a 'twin-planet' system, due to the large size of the satellite relative to the primary. Although it is often stated that the Moon orbits the Earth, in fact the two bodies each orbit the barycenter of this two-body system, which is located along the line connecting their centers. The shapes of the barycentric orbits of the Earth and Moon are similar, though that of the Moon is much larger. The barycenter is typically located at a distance of about 5000 km from the center of the Earth, which is about 1300 km beneath the Earth's surface.

Thus we may envision the translational inertial motion of the Earth as follows: Earth orbits the Earth–Moon barycenter, while this

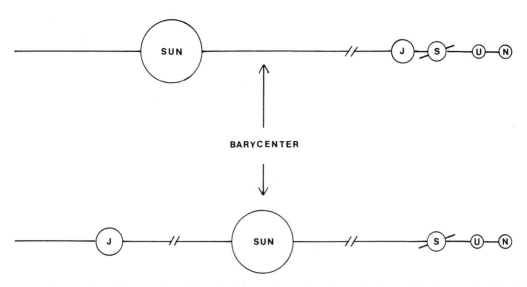

Figure B1 Diagram, not to scale, to illustrate the relationship between the major planets, the Sun and the barycenter (after Fairbridge and Sanders, 1987). When the major planets are aligned on one side of the Sun, the latter must swing away from the barycenter by more than 1.5×10^6 km. In the case where Jupiter is found on the opposite side of the Sun from the other major planets, the center of the Sun may approach the barycenter. The actual motion is complex (see Solar motion).

barycenter orbits the Sun, which in turn loops about the barycenter of the solar system.

James H. Shirley and Rhodes W. Fairbridge

Bibliography

Charvátová, I. (1990) The relations between solar Motion and solar variability. *Bull. Astron. Inst. Czech.*, **41**, 56–9.
Fairbridge, R.W. and Sanders, J.E. (1987) The sun's orbit, A.D. 750–2050: basis for new perspectives on planetary dynamics and the Earth–Moon linkage, in *Climate: History, Periodicity, Predictability*, (eds M.R. Rampino). New York: Van Nostrand Reinhold, pp. 446–71.
Fairbridge, R.W. and Shirley J.H. (1987) Prolonged minima and the 179-yr cycle of the solar inertial motion. *Solar Phys.*, **110**, 191–210.
Jose, P.D. (1965) Sun's motion and sunspots. *Astron. J.*, **70**, 193.

Cross references

Celestial mechanics
Earth–Moon system: dynamics
Solar activity
Solar motion

BASALT

Basalt is a general term given to fine-grained, dark volcanic rocks composed primarily of plagioclase and pyroxene in subequal proportions. Opaque minerals – usually ilmenite and/or a spinel – are ubiquitous accessory minerals; when present, olivine typically signifies a higher temperature basalt. Traditional definitions have restricted the compositional range to 45 to 52 wt% SiO_2 and have specified augitic (high-calcium) pyroxene (e.g. Williams, Turner and Gilbert, 1955). However, in recent years the usage of the term has become more liberal and there are several varieties of basalt – low-titanium lunar mare volcanics, eucrites and terrestrial boninites – in which a low-calcium pyroxene is the dominant ferromagnesian mineral. Also, it is now commonplace to refer to some rocks as basalts that have less than 45% SiO_2 and contain no plagioclase, e.g. melilite basalts, although strictly speaking such rocks are foidites, and are referred to by the major feldspathoidal mineral, e.g. melilitite (Streckeisen, 1978).

There is a consensus that most basalts are erupted melts of planetary mantles and as such may provide important information about planetary composition and evolution. However, there are proposals that some terrestrial basalts (high-alumina basalt) are melts of subducted oceanic crust (Marsh, 1979), while others (monzonorites) are melts of the lower continental crust (Duchesne, 1989). Isotope ratios of Sr, Nd and Hf in most terrestrial basalts bear evidence of earlier melting events that depleted their source regions in incompatible elements and led to the formation of the continents. Among terrestrial basalts there are two major classes: tholeiites and alkali basalts. The distinction is usually made on the basis of a set of calculated or normative minerals with tholeiites containing normative hypersthene (low-calcium pyroxene) and alkali basalts containing normative nepheline. Tholeiitic differentiation produces derivative magmas with progressively increasing SiO_2 that may crystallize low-calcium pyroxene and a silica mineral; derivatives of alkali basalts typically have progressively decreasing SiO_2 and may crystallize nepheline (or some other feldspathoid), but not low-calcium pyroxene or silica (Yoder and Tilley, 1962). The difference in crystallization paths is caused by a thermal divide at low pressure involving olivine, plagioclase and augite. However, there is a continuous range in composition between primitive tholeiites and alkali basalts that is produced by melting at high pressures where plagioclase is not stable and where the thermal divide is not present, so the distinction between the two classes is largely the result of differing extents of partial melting with alkali basalts representing the lesser amount of melting. Consequently, the distinction between alkalic and tholeiitic basalts is, to a certain degree, accidental. Both tholeiites and alkali basalts are found on intraplate oceanic volcanic islands such as Hawaii (hence ocean island basalts). Both types are also found along divergent plate margins, such as continental rifts and mid-ocean ridges; however, tholeiites are by far the dominant type of basalt along mid-ocean ridges (BVSP, 1981). Along convergent plate margins tholeiites are also dominant, although in this case a secondary distinction is made between tholeiitic and calc-alkaline series with the former characterized by a marked increase in Fe/Mg and constant or decreasing SiO_2 in derivative lavas, whereas the latter is characterized by only modest increases in Fe/Mg and marked increases in SiO_2. Calc-alkaline differentiation may produce a nearly continuous series of lavas in the sequence basalt → andesite → dacite → rhyolite. It is generally agreed that water and assimilation of continental crust play a role in this process (Grove and Kinzler, 1986), although the details are still subject to debate. Another major type is continental flood basalt. Flood basalts comprise huge accumulations of lava that are hundreds of meters thick and cover thousands of square kilometers. In detail, the floods are formed from the successive eruptions of dozens of flows that cover their predecessors. Flood basalts typically are monotonous in composition and generally tholeiitic in character, although the earliest stages may include alkali basalts and olivine-rich tholeiites called picrites. Despite the large volumes of lava, flood basalts appear to erupt over a relatively short period of time ($\sim 10^6$ years) and are thought to result from the ascension and partial melting of an especially large mantle plume under continental crust (Campbell and Griffiths, 1990).

In recent years a new concept of petrogenesis has developed that posits basalts as composites of melts aggregated over a range of depths from a decompressing source (Kline and Langmuir, 1987). Thus the chemical composition may reveal only an average depth of melting. However, studies of rare earth element (REE) abundance patterns in clinopyroxene from peridotites (Johnson, Dick and Shimizu, 1989) and studies of Hf and Nd isotope fractionations in mid-ocean ridge basalts both suggest that melting beneath the ridges begins at depths where garnet is stable in the mantle (i.e. ≥ 60 km). The manner in which melt segregates from the rising, residual mantle is the subject of widespread investigation.

Lunar basalts are usually separated into two categories: mare and KREEP basalts. Mare basalts are the dark volcanic infillings of the large lunar impact basins. Radiometric crystallization ages lie between 4.2×10^9 and 3.2×10^9 years, although photogeological studies have determined that younger basalts exist, perhaps as young as 1.1×10^9 years (BVSP, 1981). Chemically there is a nearly continuous variation in TiO_2 concentration from ~ 0.5 wt% to ~ 13 wt% among the mare basalts (BVSP, 1981). The Apollo missions returned subequal proportions of high- and low-Ti varieties and so there was a tendency to discuss mare basalts in terms of these endmembers; however, remote spectral studies of the lunar surface have shown that basalts with intermediate concentrations of TiO_2 are common (Pieters et al., 1980). Mineralogically, mare basalts differ from terrestrial basalts in several ways: (1) pyroxene, not plagioclase, usually crystallizes after olivine; (2) mare plagioclase is typically more calcic (85–95% of the anorthite component versus 65–80%); (3) magnetite spinel is absent and small amounts of iron metal are present; and (4) there are no hydrous minerals or even hydrous alterations. Among the low-Ti basalts pigeonite is the dominant pyroxene, whereas augite dominates among high-Ti basalts. Ilmenite is the primary carrier of titanium; however, in basalts with the highest TiO_2 armalcolite ($[Mg,Fe]Ti_2O_5$) crystallizes before ilmenite. Besides the large variation in Ti, the most prominent chemical feature of mare basalts is the high samarium (Sm) to europium (Eu) ratio relative to the ratios of these elements in primitive chondrites. This high ratio is diagnostic of plagioclase fractionation, but since experimental studies have shown that plagioclase could not have been an early crystallizing phase near the surface at low pressure or a residual phase in the mantle source region at high pressure (Green et al., 1971), it follows that high Sm/Eu was a feature of the source imparted to it by depletion of plagioclase during its formation. Given that the lunar highlands possess an apparent superabundance of plagioclase, the theory has developed that initially a substantial portion of the Moon was melted and that floating plagioclase accumulated at the top of this 'magma ocean' to form a plagioclase-enriched crust atop a complementary plagioclase-depleted interior; and that this interior subsequently remelted to yield the mare basalts (Taylor, 1982). Another apparent product of this global differentiation was a series of evolved rocks and residual liquids probably situated at the base of the lunar crust and highly enriched in incompatible elements such as potassium (K), rare earths and phosphorus (P); hence the name KREEP. There are no known samples of this primordial KREEP, but numerous pre-mare magmas are thought to have assimilated KREEP while intruding the crust (Warren, 1988), and KREEP basalts may represent volcanic expressions of this process.

There is compelling isotopic evidence of trapped Martian atmosphere (Bogard and Johnson, 1983) in a distinctive group of basaltic meteorites called SNC (shergottites, nakhlites, chassignites) (q.v.). Additionally, shergottite magmas have compositions very similar to the volatile-free portions of the Martian soil analyses derived from Viking lander measurements (McSween, 1985). The distinction between these three groups is based on mineral proportions: shergottites have subequal proportions of pigeonite and augite; nakhlites are augite-rich rocks with minor olivine; and the one chassignite is olivine-rich with minor augite. Unlike other meteorites, these rocks have relatively young crystallization ages ($\leq 1.3 \times 10^9$ years) and are mineralogically similar to terrestrial basaltic komatiites (Longhi and Pan, 1989). Unlike lunar basalts, ferric iron is present in solid solution and so are traces of hydrous minerals (Floran et al., 1978). However, oxygen isotopes clearly show that these meteorites are not terrestrial rocks (Clayton and Mayeda, 1983). As with the Earth and Moon, strontium (Sr) and neodymium (Nd) isotopes in the nakhlites and chassignites show clear evidence of remelting of a source region previously depleted, probably by a primordial crust-forming event (Nakamura et al., 1982). Interpetation of the shergottite isotopic data is hampered by shock effects which obscure the true crystallization ages, but the most likely scenario dates these rocks at $\sim 180 \times 10^6$ years and also requires a strongly depleted mantle (Jones, 1986). Patterns of incompatible trace element ratios in the parent magmas of the SNC meteorites are different from lunar and most terrestrial basalts (Longhi, 1991). The Sm/Eu ratio is chondritic so there is no evidence for a lunar-style differentiation (and plagioclase-rich crust) on Mars; however, there are marked fractionations of the Nd/Sr, Sm/Hf (hafnium), and La/Ta (lanthanum/tantalum) ratios in the SNC magmas from their chondritic values that are not typical of terrestrial ocean ridge or ocean island basalts. The greatest similarities exist between SNC meteorites and terrestrial convergent plate margin basalts for which differential transport of trace elements by fluids in the source region is often invoked to explain unusual elemental ratios (Kay, 1980). These similarities are surprising because there is no evidence of plate tectonics on Mars.

Eucrites (q.v.) are probably the simplest basaltic rocks in the solar system. They are igneous meteorites that are characterized by subequal proportions of low-calcium pyroxene (pigeonite) and anorthitic plagioclase. Mineralogically and chemically, eucrites are similar to mare basalts, e.g. in their dominance of pigeonitic pyroxene, presence of calcic plagioclase, a low oxidation state and low abundances of water and volatile elements (BVSP, 1981). However, the crystallization ages of the eucrites, $\sim 4.5 \times 10^9$ years, are much older than those of lunar basalts and there is no clear evidence from the patterns of incompatible trace element abundances in eucrites of a prior crust-forming event on their parent body.

John Longhi

Bibliography

Bogard, D. and Johnson, P. (1983) Martian gases in an Antarctic meteorite. *Science*, **221**, 651–4.
BVSP [Basaltic Volcanism Study Project] (1981) *Basaltic Volcanism on the Terrestrial Planets*, New York: Pergamon Press, 1286 pp.
Campbell, I.H. and Griffiths, R.W. (1990) Implications of mantle plume structure for the evolution of flood basalts. *Earth Planet. Sci. Lett.*, **99**, 79–93.
Clayton, R.N. and Mayeda, T.K. (1983) Oxygen isotopes in eucrites, shergottites, nakhlites, and chassignites. *Earth Planet. Sci. Lett.*, **62**, 1–6.
Duchesne, J.C. (1989) Origin and evolution of monzonorites related to anorthosites. *Schweiz. Mineral. Petrogr. Mitt.*, **70**, 189–98.
Floran, R.J., Prinz, M., Hlava, P.F. et al. (1978) The chassigny meteorite: a cumulate dunite with hydrous amphibole-bearing melt inclusions. *Geochim. Cosmochim. Acta*, **42**, 1213–29.
Green, D.H., Ware, N.G., Hibberson, W.O. and Major, A. (1971) Experimental petrology of Apollo 12 basalts: part 1, sample 12009. *Earth Planet. Sci. Lett.*, **13**, 85–96.
Grove, T.L. and Kinzler, R.J. (1986) Petrogenesis of andesites. *Ann. Rev. Earth Planet. Sci.*, **14**, 417–454.
Johnson, K.T.M., Dick, H.J.B. and Shimizu, N. (1990) Melting in the oceanic upper mantle: an ion microprobe study of diopsides in abyssal peridotites. *J. Geophys. Res.*, **95**, 2661–78.
Jones, J.H. (1986) A discussion of isotopic systematics and mineral zoning in the shergottites: Evidence for a 180 m.y. igneous crystallization age. *Geochim. Cosmochim. Acta*, **50**, 69–977.
Kay, R.W. (1980) Volcanic arc magmas: implications of a melting–mixing model for element recycling in the crust–upper mantle system. *J. Geol.*, **88**, 497–522.
Klein, E.M. and Langmuir, C.H. (1987) Global correlation of ocean ridge basalt chemistry with axial depth and crustal thickness. *J. Geophys. Res.*, **92**, 8089–115.
Longhi, J. (1991) Complex magmatic processes on Mars: inferences from the SNC meteorites, in *Proc. Lunar Planet. Sci. Conf.*, **21**, pp. 695–709.
Longhi, J. and Pan, V. (1989) The parent magmas of the SNC meteorites, in *Proc. Lunar Planet. Sci. Conf.*, **19**, pp. 451–64.
Marsh, B.D. (1979) Island-arc magmatism. *Am. Sci.*, **67**, 161–72.

McSween, H.Y. (1985) SNC meteorites: clues to Martian petrologic evolution. *Rev. Geophys.*, **23**, 391–416.
Nakamura, N., Unruh, D.M., Tatsumoto, M. and Hutchinson, R. (1982) Origin and evolution of the Nakhla meteorite inferred from Sm–Nd and U–Pb systematics, and REE, Ba, Sr, Rb, and K abundances. *Geochim. Cosmochim. Acta*, **46**, 1555–73.
Pieters, C.M., Head, J.W., Adams, J.B. et al. (1980) Late high-titanium basalts of the western Maria: geology of the Flamsteed region of Oceanus Procellarum. *J. Geophys. Res.*, **85**, 3913–38.
Salters, V.J.M. and Hart, S.R. (1989) The Hf-paradox and the role of garnet in the source of mid-ocean ridge basalts. *Nature*, **342**, 420–2.
Streckeisen, A.L. (1978) Classification and nomenclature of volcanic rocks, lamprophyres, carbonatites and melilitic rocks, *Neues Jahrb. Mineral. Abh.*, **134**, 1–14.
Taylor, S.R. (1982) *Planetary Science: A Lunar Perspective*. Houston: The Lunar and Planetary Institute, 481 pp.
Warren, P.W. (1988) The origin of pristine KREEP: effects of mixing between urKREEP and the magmas parental to the Mg-rich cumulates, in *Proc. Lunar Planet. Sci. Conf.*, **18**, pp. 233–41.
Williams, H., Turner, F.J. and Gilbert, C.M. (1955) *Petrography*. W.H. Freeman and Co., 406 pp.
Yoder, H.S. and Tilley, C.E. (1962) Origin of basalt: an experimental study of natural and synthetic rock systems. *J. Petrol.*, **3**, 342–532.

Cross references

Basaltic achondrite meteorites
Eucrite meteorites
Igneous rock
SNC meteorites

BASALTIC ACHONDRITE METEORITES

The largest group of achondritic meteorites is the basaltic achondrite group (see Achondrite meteorites). These achondrites are also referred to as the HED meteorites after the three types: howardites, eucrites and diogenites. Diogenites are not truly basaltic in composition, but are frequently referred to as members of the basaltic achondrite meteorite group because of their genetic relationship with eucrites and howardites. Together, these three igneous meteorite types form the most complete crustal sample available from any differentiated body, excluding the Earth and Moon. Most of the HED meteorites are either monomict breccias, composed of a single lithologic type, or polymict breccias, composed of a mixture of lithologies. The basaltic achondrites record the igneous and impact metamorphic evolution of one of the oldest planetary crusts we have available for study. The shergottites were formerly classified as basaltic achondrites, but they are now known to be closely related to the nakhlites and Chassigny and not the HED meteorites (see SNC meteorites).

Eucrites are basaltic rocks comparable to tholeiitic basalts on Earth and mare basalts from the Moon. Eucrites are composed of approximately 45 vol% anorthitic plagioclase, 50 vol% ferroan pigeonite, and minor tridymite or quartz, ilmenite, merrillite, chromite, troilite and Fe–Ni metal (Delaney, Prinz and Takeda, 1984). Most eucrites are brecciated, but nonetheless relict igneous textures show that there are two main textural varieties: basaltic and gabbroic. Basaltic eucrites are fine- to medium-grained and have relict ophitic, subophitic and/or intersertal textures indicating that they were crystallized in the near surface environment, either as basalt flows or shallow, diabasic intrusions. Two eucrites, Juvinas and Ibitira, contain vesicles, thus suggesting eruption onto the surface of the parent asteroid. Gabbroic eucrites are coarse-grained and have equigranular or hypidiomorphic–granular textures indicating slower cooling at depth in their parent asteroid's crust.

The distinction between basaltic and gabbroic eucrites is not just textural; their compositions are distinct as well, although the differences are not dramatic. The basaltic eucrites have more ferroan compositions, with low molar 100 MgO/(MgO + FeO) (hereafter mg#) in the range 42–32. Their incompatible refractory element (e.g. La, Hf, Ta, Th) contents are enriched by 10 to 20 times those of primitive CI chondrites and are generally present in chondritic ratios (Figure B2). In the most incompatible element-rich basaltic eucrites, Eu/Sm

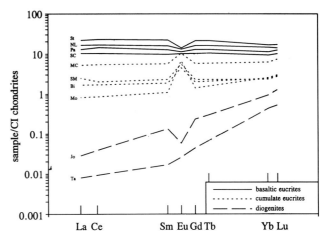

Figure B2 Rare earth element diagram for typical basaltic achondrites normalized to CI chondrites. The meteorite abbreviations are St – Stannern, NL – Nuevo Laredo, Pa – Pasamonte, SC – Sioux County, MC – Moore County, SM – Serra de Magé, Bi – Binda, Mo – Moama, Jo – Johnstown, Ta – Tatahouine. The data are means obtained from an extensive database of achondrite analyses.

and Lu/La ratios are lower than those of CI chondrites. The basaltic eucrites are depleted in the volatile elements, including the alkali elements (e.g. Mittlefehldt, 1987). Typical Na_2O contents are 0.4 to 0.5 wt% and K concentrations are usually 300 to 600 $\mu g\ g^{-1}$. In this the basaltic eucrites are similar to lunar mare basalts and distinct from terrestrial thoeliites. Basaltic eucrites are also very depleted in the siderophile elements; typical Ni and Co contents are 10^{-4} to 10^{-3} and 10^{-2} times CI respectively. All aspects of the compositions of the basaltic eucrites indicate that they represent melt compositions.

The gabbroic textured eucrites are more magnesian in composition, with mg# in the range 65 to 50, and they have lower abundances of the incompatible refractory elements in the range 0.8 to 5 times CI chondrites (Figure B2). In contrast with basaltic eucrites, the incompatible refractory elements in gabbroic eucrites are modestly fractionated relative to chondritic ratios. Their Lu/La ratios are frequently greater than those of CI chondrites, and all gabbroic eucrites have Eu/Sm ratios greater than those of CI chondrites (Figure B2). Like the basaltic eucrites, gabbroic eucrites are poor in volatile and siderophile elements. The texture and major and trace element compositions of the gabbroic eucrites indicate that they are crystal accumulations from a basaltic melt. For this reason, these eucrites are commonly referred to as the cumulate eucrites.

Diogenites, sometimes referred to as the hypersthene achondrites, are brecciated orthopyroxenites. Relict, coarse-grained textures are preserved in the largest clasts. Diogenites are similar to orthopyroxenite layers in terrestrial layered intrusions, such as the Stillwater. Diogenites are composed of 90–95 vol% orthopyroxene, with minor coarse-grained olivine and chromite. Finer-grained Fe–Ni metal, troilite and silica are common accessory minerals, while plagioclase and merrillite are trace phases. Very early on, the coarse-grained texture and nearly monomineralic composition led meteoriticists to posit that diogenites were cumulate orthopyroxenites (Mason, 1963).

Almost all diogenites are virtually identical in major element composition. All but a few are composed of orthopyroxene of composition $Wo_2En_{73}Fs_{25}$, and whole rock SiO_2, FeO, MnO, MgO and CaO are essentially those of the pyroxene. In contrast, the Cr_2O_3 content of diogenites varies with the amount on chromite present in the rock. There are significant differences in trace element contents, however, and the major element and trace element compositions of diogenites are decoupled relative to what one would expect for simple igneous processes. For example, the majority of diogenites have an mg# of ~ 75, yet the heavy rare earth element Yb can vary by a factor of ten (Mittlefehldt, 1994). The uniformity in mg# would suggest that the diogenites were formed from a single melt, or melts of similar composition, while the large variation in Yb suggests an extensive fractionation sequence. The most incompatible elements, such as La, Ta, Th, U, etc. are in very low concentrations in

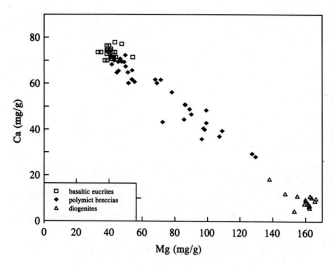

Figure B3 Ca versus Mg for basaltic achondrites. Monomict breccias, the basaltic eucrites and diogenites, form small clusters, while polymict breccias, the howardites and polymict eucrites form a mixing line between the two igneous endmembers. The data are means obtained from an extensive database of achondrite analyses.

diogenites, between 0.008 and 0.03 times those of CI chondrites (Figure B2). The incompatible elements are highly fractionated, and Lu/La ratios can be up to 50 times those of CI chondrites. These low abundances and fractionated ratios are compatible with an origin for the diogenites as cumulate orthopyroxenites.

Howardites are regolith breccias composed of mineral and lithic clasts in a fine-grained, fragmental matrix (Duke and Silver, 1967). The lithic clasts are most commonly composed of basalt fragments similar in texture, mineralogy and composition to basaltic eucrites. The larger mineral fragments are usually orthopyroxene grains similar in composition to diogenitic pyroxenes. However, howardites contain a wider range of materials than simply eucrite-like basalts and diogenite-like orthopyroxenes. In particular, chondritic fragments are a minor but important component in howardites (Zolensky et al., 1992).

Howardites are intermediate in bulk composition between basaltic eucrites and diogenites (Figure B3), and are heterogeneous; unlike basaltic eucrites and diogenites, individual analyses of howardites can be quite different. Howardites have mg# in the range 70–50, and contain incompatible trace element contents roughly half that of basaltic eucrites, although the trace element content varies considerably. In most compositional aspects, howardites can be considered to be simple mixtures of basaltic eucrites and diogenites (Jerome and Goles, 1971) to a first approximation. However, the siderophile element contents of howardites are enriched compared to those of eucrites and diogenites because about 1–3% chondritic debris is present in howardites (Chou et al., 1976).

Early models for the formation of howardites posited that they are phenocryst-enriched basalts (Mason, 1962). However, detailed petrographic study showed that howardites were composed of distinct materials (Duke and Silver, 1967). At about the same time, detailed compositional investigation of HED meteorites led to the hypothesis that howardites are brecciated mixtures of distinct igneous rock types and are not themselves a separate igneous lithology (Jerome and Goles, 1971). More recently it has been shown that some eucrites contain a small amount of diogenitic material, and they have been reclassified as polymict eucrites (Delaney et al., 1983). In reality there is a virtual continuum from monomict basaltic eucrites to monomict diogenites (Figure B3), and both polymict eucrites and polymict diogenites exist in meteorite collections (Delaney et al., 1983).

Eucrites, diogenites and howardites have the same oxygen isotopic composition. Other meteorites sharing this same oxygen isotopic composition are the mesosiderites, main-group pallasites, angrites and IIIAB irons (Clayton, Onuna and Mayeda, 1976; Clayton and Mayeda, 1978). This similarity indicates that these meteorite groups were formed within a limited region of the solar nebula. The silicates in mesosiderites are very similar to those of the HED meteorites, but numerous subtle differences between HED meteorites and mesosiderites indicate that these two meteorite groups formed on separate parent bodies (see Stony iron meteorites; Rubin and Mittlefehldt, 1993).

Most HED meteorites have undergone impact metamorphism which modified their original textures and mineral compositions. During impacts the rocks were brecciated, rapidly heated and then cooled slowly, either in an ejecta blanket or in the crater floor. During slow cooling the original igneous zoning in pigeonite was annealed out, leaving a mixture of homogeneous orthopyroxene and augite (Takeda and Graham, 1991). Textures were modified as well, and in the extreme case, eucrites and diogenites have granulitic texture as exhibited by the eucrite Ibitira (Steele and Smith, 1976) and the diogenites paired with Y-6902 (Takeda, Mori and Yanai, 1981). Because the HED crust is free of volatile species such as H_2O and CO_2, metamorphism on the HED asteroid was essentially isochemical on the scale of about a centimeter. Therefore, even highly metamorphosed eucrites faithfully retained their original bulk compositions.

The formation age of the HED asteroid crust is essentially identical to the ages determined for chondritic meteorites. Internal isochron ages using the Rb–Sr and Sm–Nd systems for little-modified igneous materials yield ages of about 4.6 Ga (Nyquist et al., 1986). Considering the most precise chondrite and basaltic achondrite ages, it is evident that the magmatic differentiation of the HED asteroid occurred within a few tens of million years of the formation of chondrites (Rubin and Mittlefehldt, 1993). Ages substantially younger than 4.6 Ga have been determined for several HED meteorites. For example, the basaltic eucrite Cachari has an Ar–Ar age of 3.0 Ga (Bogard et al., 1985), and Sioux County has a Rb–Sr age of 4.2 Ga (Birck and Allègre, 1978). These younger ages are the result of impact metamorphism of the HED crust; Cachari is heavily shock damaged and includes veins of shock produced glass (Bogard et al., 1985), and Sioux County is brecciated (Duke and Silver, 1967). There is no evidence for internal (as opposed to impact melting) igneous activity on the HED parent asteroid after about 4.4 Ga ago.

The origin of eucrites has been extensively studied, but nonetheless, there is no consensus as to whether they represent primary partial melts or residual liquids. The earliest petrogenetic model for eucrites had them as residual melts left over from crystallization of a totally molten asteroid (Mason, 1962). This model was prompted by consideration of the range of meteorite types in collections, early experimental work in the forsterite–anorthite–silica system by Bowen and analogy with terrestrial layered intrusions. Mason believed that once totally molten, an asteroid of chondritic composition would form an iron core, similar to iron meteorites, an olivine mantle, as suggested by the olivine–metal stony irons, the pallasites (see Stony iron meteorites), and as demonstrated by Bowen's phase diagrams, a layered crust of, from bottom up, orthopyroxene cumulates (the diogenites) through phenocryst-enriched basalts (the howardites) to residual basalts (the eucrites). This sequence was suggested by analogy with terrestrial layered intrusions and Bowen's phase diagrams.

It wasn't until Stolper performed melting experiments under controlled oxygen fugacity on natural eucrites that an alternative model was presented. Stolper (1977) showed that some eucrites are saturated on their liquidus with olivine, low-Ca pyroxene, plagioclase, spinel and metal. He also showed that eucrite compositions cluster strongly about a point on a silica–plagioclase–olivine pseudo-ternary diagram (Figure B4). He inferred that this point represented a peritectic, where only melting and not crystallization is likely to cluster silicate liquid compositions. As eucrites are saturated with appropriate phases for a chondritic body, Stolper concluded that the most primitive of the basaltic eucrites are primary partial melts of chondritic material under oxygen fugacities about one log unit below that of the iron–wüstite oxygen buffer. Stolper's experiments also showed that those eucrites with lower mg#, near 32, and higher incompatible element contents, are more evolved melts formed as residual magmas by crystallization of primitive eucrites. Stolper further suggested that there was no direct genetic link between the diogenites and eucrites, that is, they did not share a common parent melt. He posited that the diogenites were formed from a second generation melt of nearly orthopyroxenitic composition. Recent melting experiments on chondrites have shown that moderate degree partial melts, about 10–20%, of CM chondrites are remarkably similar to eucrites in composition, with the exception of low MnO content (Jurewicz, Mittlefehldt and Jones, 1993).

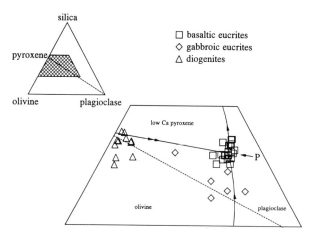

Figure B4 Silica – plagioclase – olivine pseudo-ternary diagram for basaltic achondrites, with schematic phase boundaries taken from Stolper (1977). Basaltic eucrites cluster tightly around point P, the inferred pseudo-ternary peritectic point. Gabbroic eucrites plot between the pyroxene – plagioclase join (dashed line) and the peritectic point, as would be expected for cumulates from basaltic eucrite melts. Diogenites plot near the olivine – silica boundary at the pyroxene point. The data are means obtained from an extensive database of achondrite analyses.

In spite of Stolper's detailed experimental work, several researchers have found evidence suggesting a residual liquid origin for even the most primitive basaltic eucrites. Dreibus et al. (1977) inferred from Stolper's experiments that the parent body for the eucrites would have a mg# of about 65. For their calculated eucrite parent body composition with a MnO/MgO ratio about 0.7 times CI chondrites, one would then expect that the FeO/MnO ratio of the bulk silicates would be about 88, versus 35 as observed in basaltic eucrites. (In fact, there are no chondrites with a high enough MnO/MgO ratio (Wasson and Kallemeyn, 1988) to solve the Dreibus et al. conundrum.) Dreibus et al. (1977) noted that Fe^{+2} and Mn^{+2} are geochemically very similar and are not significantly fractionated by igneous processes. Therefore, the FeO/MnO ratio of eucrites should be close to that of their parent body. Extensive fractional crystallization of a magnesian, ultramafic magma would yield a basaltic melt with low mg#, but FeO/MnO ratio similar to the parent melt. Dreibus et al. (1977) therefore concluded that eucrites must be residual liquids from an extensive fractional crystallization sequence, much like Mason's model.

Warren (1985) and Warren and Jerde (1987) have argued, based on major and trace element trends and mass balance constraints, that most of the eucrites are part of an extensive fractional crystallization sequence that first formed diogenitic cumulates. However, fractional crystallization scenarios pose additional questions. If the diogenite–eucrite association developed from crystallization of a totally molten asteroid, then how is it that distinct mineral composition trends, indicating different parent magmas are observed in polymict eucrites (Delaney, Prinz and Takeda, 1984), and why is it that anorthosites, which would have been buoyant when the asteroidal magma ocean reached eucritic composition, apparently did not form a stable crust as they did on the Moon? On the other hand, if the diogenite–eucrite association developed after partial melting of an asteroidal mantle, how is it that eucritic magmas, which are calculated to be denser than diogenitic parent magmas (Longhi and Pan, 1989), erupted to form fine-grained volcanic rock, but the less dense diogenitic magmas did not? Thus, the petrogenesis of eucrites remains an unresolved problem.

The parent body of the basaltic achondrites is unknown. However, the asteroid 4 Vesta, with a diameter of 550 km, has a reflectance spectrum similar to that of eucrites and has long been considered a potential parent body for the HED meteorites (McCord, Adams and Johnson, 1970). Recently, three Earth-approaching asteroids have been identified as having spectral reflectivities similar to those of HED meteorites. These asteroids, 1980 PA, 1983 RD and 1985 DO2, are between 1 and 3 km in diameter and are considered to be the likely immediate source of HED meteorites (Cruikshank et al., 1991). However, these small asteroids are undoubtedly fragments of a much larger, basaltic achondrite parent asteroid. Cruikshank et al. (1991) do not consider it likely that these three small asteroids are fragments of 4 Vesta because of the differences in their orbital elements. More recently, Binzel and Xu (1993) have demonstrated that there is a family of 20 small asteroids between 4 and 10 km diameter with reflectance spectra similar to those of eucrites and diogenites and to Vesta's. Of these, 12 have orbital elements similar to those of Vesta, while eight bridge the gap in orbital space between Vesta and the 3 : 1 resonance with Jupiter (Binzel and Xu, 1993). These authors have shown that the small asteroids can plausibly be impact spalls off of Vesta, and that some of their brethren could have reached the 3 : 1 resonance or the v_6 resonance and thence been perturbed into Earth-approaching orbits. Binzel and Xu (1993) believe that this is a plausible origin for the three Earth-approaching basaltic asteroids identified by Cruikshank et al. (1991).

The basaltic achondrite meteorites make up the largest group of meteorites from an igneously differentiated asteroid crust. Their compositions and textures record the igneous history of small bodies in the earliest stages of solar system evolution. Although intensive study has revealed much about the nature of the igneous processes that formed these meteorites, many fundamental questions remain unanswered.

David W. Mittlefehldt and John Longhi

Bibliography

Binzel, R.P. and Xu, S. (1993) Chips off of asteroid 4 Vesta: evidence for the parent body of basaltic achondrite meteorites. *Science*, **260**, 186–91.

Birck, J.L. and Allègre, C.J. (1978) Chronology and chemical history of the parent body of basaltic achondrites studied by the ^{87}Rb–^{87}Sr method. *Earth Planet. Sci. Lett.*, **39**, 37–51.

Bogard, D.D., Keil, K. Taylor, G.J. et al. (1985) Impact melting of the Cachari eucrite 3.0 Gy ago. *Geochim. Cosmochim. Acta*, **49**, 941–6.

Chou, C.-L., Boynton, W.V., Bild R.W. et al. (1976) Trace element evidence regarding a chondritic component in howardite meteorites, in *Proc. Lunar Sci. Conf.*, **7**, pp. 3501–18.

Clayton, R.N., Onuma, N. and Mayeda, T.K. (1976) A classification of meteorites based on oxygen isotopes. *Earth Planet. Sci. Lett.*, **30**, 10–8.

Clayton, R.N. and Mayeda, T.K. (1978) Genetic relations between iron and stony meteorites. *Earth Planet. Sci. Lett.*, **40**, 168–74.

Cruikshank, D.P., Tholen, D.J. Hartmann, W.K. et al. (1991) Three basaltic Earth-approaching asteroids and the source of the basaltic meteorites. *Icarus*, **89**, 1–13.

Delaney, J.S., Takeda, H., Prinz, M., et al. (1983) The nomenclature of polymict basaltic achondrites. *Meteoritics*, **18**, 103–11.

Delaney, J.S., Prinz, M., and Takeda, H. (1984) The polymict eucrites, in *Proc. Lunar Planet. Sci. Conf.*, **15**, Part 1, *Geophys. Res.*, Suppl., **89**, C251–C88.

Dreibus, G., Kruse, H., Spette, B., and Wänke, H. (1977) The bulk composition of the moon and the eucrite parent body, in *Proc. Lunar Sci. Conf.*, **8**, pp. 211–27.

Duke, M.B. and Silver, L.T. (1967) Petrology of eucrites, howardites and mesosiderites. *Geochim. Cosmochim. Acta*, **31**, 1637–65.

Jerome, D.Y. and Goles, G.G. (1971) A re-examination of relationships among pyroxene–plagioclase achondrites, in: *Activation Analysis in Geochemistry and Cosmochemistry* (eds A.O. Brunfelt and E. Steinnes). Oslo; Universitetsforlaget, pp. 261–6.

Jurewicz, A.J.G., Mittlefehldt, D.W. and Jones, J.H. (1993) Experimental partial melting of the Allende (CV) and Murchison (CM) chondrites and the origin of asteroidal basalts. *Geochim. Cosmochim. Acta*, **59**, 2123–39.

Longhi, J. and Pan, V. (1989) The parent magmas of the SNC meteorites. *Proc. Lunar Planet. Sci. Conf.*, **19**, 451–64.

Mason, B. (1962) *Meteorites*. New York; Wiley, 274 pp.

Mason, B. (1963) The hypersthene achondrites. *American Museum Novitates* **2155**, 13 pp.

McCord, T.B., J.B. Adams, and Johnson, T.V. (1970) Asteroid Vesta: spectral reflectivity and compositional implications. *Science*, **168**, 1445–7.

Mittlefehldt, D.W. (1987) Volatile degassing of basaltic achondrite parent bodies: evidence from alkali elements and phosphorus. *Geochim. Cosmochim. Acta*, **51**, 267–78.

Mittlefehldt, D.W. (1994) The genesis of diogenites and HED parent body petrogenesis. *Geochim. Cosmochim. Acta,* **58**, 1537–1552.

Nyquist, L.E., Takeda, H., Bansal, B.M. *et al.* (1986) Rb–Sr and Sm–Nd internal isochron ages of a subophitic basalt clast and a matrix sample from the Y75011 eucrite. *J. Geophys. Res.*, **91**, 8137–50.

Rubin, A.E. and Mittlefehldt, D.W. (1993) Evolutionary history of the mesosiderite asteroid: a chronologic and petrologic synthesis. *Icarus*, **101**, 201–12.

Steele, I.M. and Smith, J.V. (1976) Mineralogy of the Ibitira eucrite and comparison with other eucrites and lunar samples. *Earth Planet. Sci. Lett.*, **33**, 67–78.

Stolper, E. (1977) Experimental petrology of eucritic meteorites. *Geochim. Cosmochim. Acta*, **41**, 587–611.

Takeda, H. and Graham, A.L. (1991) Degree of equilibration of eucritic pyroxenes and thermal metamorphism of the earliest planetary crust. *Meteoritics*, **26**, 129–34.

Takeda, H., Mori, H., and Yanai, K. (1981) Mineralogy of the Yamato diogenites as possible pieces of a single fall, in *Proc. 6th Symp. Antarctic Meteorites, Mem. Natl Inst. Polar Research, Special Issue* **20**, pp. 81–99.

Warren, P.H. (1985) Origin of howardites, diogenites and eucrites: a mass balance constraint. *Geochim. Cosmochim. Acta*, **49**, 577–86.

Warren, P.H. and Jerde, E.A. (1987) Composition and origin of Nuevo Laredo Trend eucrites. *Geochim. Cosmochim. Acta*, **51**, 713–25.

Wasson, J.T. and Kallemeyn, G.W. (1988) Compositions of chondrites. *Phil. Trans. Roy. Soc. London*, **A325**, 535–544.

Zolensky, M.E., Hewins, R.H., Mittlefehldt, D.W. *et al.* (1992) Mineralogy, petrology and geochemistry of carbonaceous chondritic clasts in the LEW 85300 polymict eucrite. *Meteoritics*, **27**, 596–604.

Cross references

Eucrite meteorites
Meteorite

BIOSPHERE

A unique feature of planet Earth, the biosphere is that part of the Earth's crust, its natural waters and atmosphere where living organisms exist. The biota is the total living population in a region, seen on any scale. Essential parameters for the life process are substrate and media such as air or water that are favorable to gas exchange and energy supply. What makes planet Earth unique among the celestial bodies of the solar system is the widespread availability of liquid water.

Since the first emergence of primitive self-reproducing unicellular life on the planet about 3.9 billion years ago this favorable state has been maintained without interruption, even for one instant, in spite of hazardous events such as asteroid impacts and episodic cycles of unfavorable climatic and paleogeographic conditions. This continuity has made possible the uninterrupted, though episodically accelerated sequence of organic evolution throughout geologic time. It is known as the 'law of biologic continuity'. During Phanerozoic time (the last 560 Ma), which has seen most of the evolution of advanced multicellular life, no major phylum has ever become extinct, although at certain points in time large families or groups (e.g. trilobites, ammonites, dinosaurs) have indeed gone out of existence.

Thanks to its enormous range of physiographic landscapes and natural environments, planet Earth is uniquely favorable for the enormously varied biota that it supports. Ecologically, each life form has evolved into a degree of specialization that enables it to occupy a specific niche. The range of such environments extends from the deep sea (at extremely high pressure and temperatures approaching 0°C) at over 10 km depth, to the highest mountains (with very low atmospheric pressure and temperature) at over 8 km elevation. The biosphere thus occupies an irregular envelope relating to the topography of the lithosphere–hydrosphere–atmosphere boundary, and characterized by a very wide range of microclimates. 'The biosphere extends the interface upward as forest crowns receive sun and rain, downward as roots seek moisture' (Miller 1965, p. 180). Birds extend that interface still farther, and *Homo sapiens*, with protective clothing, life-support systems and technology, has vastly expanded the unaided ranges.

Products of the metabolic processes of the biosphere include such geochemical substances as coal, lignite and petroleum. This 'organic debris', by sedimentation and burial, is withdrawn from circulation in the active life cycles of the biosphere, and committed to deep, long-term storage. It may eventually be chemically altered and recycled through subduction and plate tectonic activity, or simply uplifted and eroded, or in the industrial age exploited by *Homo sapiens* as energy sources. Combustion of such hydrocarbons adds considerable quantities of CO_2 to the atmosphere, and is an important factor in the present problem of the greenhouse effect.

The biosphere is, in principle, a closed and self-regulating system. It is maintained by solar and chemical energy and characterized by major cycles of the life-sustaining elements, notably carbon, oxygen, nitrogen and phosphorus, together with potassium, magnesium, sodium, etc. (the 'biophile elements' of Goldschmidt, 1954). Metabolism, mediated for the planet world by photosynthesis, is balanced by decay or animal consumption. On a short term, the cyclical budgets would appear to be balanced, but on a long term this is not so, and the biota systematically modify their own environment. Over geologic time, as that environment has changed the organisms have adapted to it, following Darwinian principles. The Gaia hypothesis of Lovelock (1979) seized on this fact of biogeological history and made the useful comment that the Earth's surface behaves as an interactive physiological system.

As an evolving star the Sun's irradiance should increase through time, perhaps 30% in the last 4.5 Ga. However, a high level of CO_2 in the early atmosphere, generated by very active volcanism in the protoplanet, would have led to an important greenhouse warming. Using fossil and mineralogical proxies, it is possible to calculate approximate temperatures for geologic time since about 3.9 Ga when carbonate-preserved stromatolites and bacterial traces first appeared. Since then, the mean global temperatures have remained about 20 ± 5°C, notwithstanding climatic cycles ranging from ice ages (the 'icehouse' state) to the universal tropical (the 'hothouse' state).

Throughout this long and cyclical history there has been a quasi-secular decrease in the pCO_2 to its present 0.03% of the total atmosphere, paralleled by a progressive rise in available oxygen. In spite of fluctuating environmental stress, the biota has nevertheless managed to maintain itself, developing progressively greater diversity and evolving to systematically higher and more sophisticated forms of life, culminating in the primates during the last 50 Ma or so.

Evolution of the biosphere

The evolution of the biotic environment has not been a secular one, but has proceeded in a series of revolutionary steps, each achieved as a threshold phenomenon. Five such thresholds are recognizable (Fairbridge, 1964; Oliver and Fairbridge, 1987, p. 301).

1. First life, ca. 3.5–4.0 Ga, from an earliest atmosphere, when volcanic exhalation of H_2O vapor, CO_2, Cl_2 and SO_2 provided the inorganic building blocks for the self-reproducing life cells.
2. First photosynthesis, perhaps 2.9 Ga ago, made possible by the early life cells to synthesize chlorophyll. These were bacterial autotrophs, which took CO_2 and H_2O from their environment to synthesize sugars. Eventually the pO_2 availability reached a level that could support animal life, in the form of worms, jellyfish and other shell-less invertebrates, and their fossils are now widely known.
3. First carbonate shells (560 Ma ago) appeared in a truly revolutionary event that is marked worldwide by the sudden transference of the earlier shell-free forms to the new age of exoskeletons. It must have been triggered by a rise in marine pH (to 7.5–8.0) that made it possible for calcium carbonate to accumulate within the dermal tissues.
4. The great coal age (beginning about 300 Ma ago). Land plants had begun to exploit the weathered rock surfaces and coastal mudflats in increasing complexity and size. Coastal swamps in warm latitudes, comparable to present-day mangrove, were inundated and buried by sediment during eustatic oscillations, alternating with forest expansion during the negative intervals. Geochemically, a systematic removal of carbon took place from the global budget.
5. Calcareous plankton bloom (around 100 Ma ago). In the late Mesozoic a worldwide expansion of floating unicellular organisms.

plant and animal, began a systematic withdrawal of $CaCO_3$ (and therefore CO_2) from the world budget, as it became buried in deep-sea deposits. Because of plate tectonics it would become gradually recycled but by the Cenozoic Era (beginning 65 Ma) the ocean and atmosphere assumed approximately their present oxygen and CO_2 levels.

Rhodes W. Fairbridge

Bibliography

Fairbridge, R.W. (1964) The importance of limestone and its Ca/Mg content to paleoclimatology, in *Problems in Paleoclimatology*, (ed. Nairn, A.E.M.) New York: Wiley, pp. 431–530.
Goldschmidt, V.M. (1954) *Geochemistry*. Oxford: Clarendon Press, 730 pp.
Holland, H.D. (1984) *The Chemical Evolution of the Atmosphere and Oceans*. Princeton University Press, 582 pp.
Lovelock, J.E. (1979) *Gaia: a New Look at Life on Earth*. Oxford: Oxford University Press, 157 pp.
Miller, S.L. (1953) A production of amino acids under possible primitive Earth conditions. *Science*, 117, 528–9.
Miller, S.L. and Orgel, L.E. (1974) *The Origins of Life on Earth*. Englewood Cliffs, NJ: Prentice-Hall.
Oliver, J.E. and Fairbridge, R.W. (eds) (1987) *The Encyclopedia of Climatology*. New York: Van Nostrand Reinhold, 986 pp. [see p. 301].
Whitehead, W.D. and Bieger, I.A. (1963) Geochemistry of Petroleum, in *organic Geochemistry* (ed. I.E. Bieger). New York: Macmillan, pp. 248–332.

Cross references

Atmosphere
Carbon
Chemical element
Greenhouse effect
Hydrosphere
Life: origin
Lithosphere
Water

BLACKBODY RADIATION

Any object with a temperature above absolute zero emits a continuous spectrum of electromagnetic radiation. This emission is called thermal, or 'blackbody', radiation. A perfect black body is not defined by its emission, however. Instead it is an object which absorbs all radiation falling on it at all wavelengths and from all angles of incidence; none of the radiation is reflected.

All radiation absorbed by an ideal black body is re-emitted. The black body has a 'brightness' B_f which varies with the physical temperature T of the body and the frequency f at which it is observed. The brightness is given by Planck's law (Figure B5)

$$B_f(T) = \frac{2hf^3}{c^2(e^{hf/kT} - 1)} \text{ W m}^{-2} \text{ sr}^{-1} \text{ Hz}^{-1} \quad (B1)$$

where c is the velocity of light, k is Boltzmann's constant (1.38×10^{-23} J K^{-1}) and h is Planck's constant (6.63×10^{-34} J s). The radiation integrated over all frequencies is given by the Stefan–Boltzmann law

$$B = \int_0^\infty B_f(T) \, df = \frac{\sigma T^4}{\pi} \text{ W m}^{-2} \text{ sr}^{-1} \quad (B2)$$

where the Stefan–Boltzmann constant is $\sigma = 5.673 \times 10^{-8}$ W m^{-2} K^{-4} sr^{-1}. The brightness may also be given as a function of wavelength $\lambda = c/f$

$$B\lambda(T) = \frac{2\pi hc^2}{\lambda^5(e^{hc/\lambda kT} - 1)} \text{ W m}^{-2} \text{ sr}^{-1} \text{ m}^{-1} \quad (B3)$$

The wavelength λ_m at which the maximum brightness $B_\lambda(T)$ occurs can be determined by evaluating the expression $\frac{\partial B\lambda}{\partial \lambda} = 0$. The result is known as Wien's displacement law; the wavelength for peak brightness can be obtained from

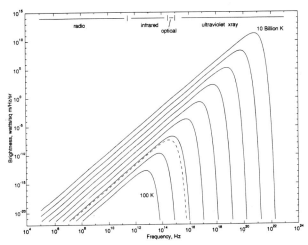

Figure B5 The radiation B_f emitted by an ideal black body is a function of temperature only, shown here increasing from 10^2 K to 10^{10} K in multiples of 10. Measured emissions from real objects with high emissivity ($e \sim 1$) can provide information about their temperatures. The Sun behaves as a black body at many wavelengths with a temperature of about 5700 K (dashed curve). Planets have temperatures up to a few hundred kelvin. Remnant radiation from the big bang is spread almost uniformly through the universe and appears to have a blackbody temperature of about 3 K.

$$\lambda_m T = 2.879 \times 10^{-3} \text{ m K} \quad (B4)$$

Historically it has been the wavelength λ_m that has been used to indicate peak brightness rather than the frequency f_m at which (B1) is maximum. Note that $\lambda_m \neq c/f_m$.

The brightness curve for a body like the Sun, with a surface temperature ~ 5700 K, peaks at optical wavelengths while those of most planets peak at infrared wavelengths. The brightness of most solar system objects can be approximated quite well by blackbody curves near their spectral peaks, especially after a scale factor (the emissivity or reflectivity, depending on the application) has been applied to 'correct' for the discrepancy from ideal blackbody behavior. At low frequencies (e.g. radio), the blackbody behavior of many objects (B1) may be approximated by the Rayleigh–Jeans law

$$B_f = \frac{2kT}{\lambda^2} \quad (B5)$$

This expression is widely used in radio studies of planetary surfaces, for which T rarely exceeds a few hundred kelvin.

Imke de Pater and Richard A. Simpson

Bibliography

Elachi, C. (1987) *Introduction to the Physics and Techniques of Remote Sensing*. New York: John Wiley and Sons.
Feynmann, R.P. et al. (1963–1965) *Lectures on Physics*. Reading, MA: Addison-Wesley Publ. Co.
Ulaby, F.T. et al. (1981) *Microwave Remote Sensing: Volume 1 – Fundamentals and Radiometry*. Reading, MA: Addison-Wesley.

Cross references

Electromagnetic radiation
Emissivity
Infrared radiation
Radio astronomy
Reflectivity
Remote sensing

BRADLEY, JAMES (1693–1762)

The astronomical discoveries of Bradley, together with the body of meridian observations that he initiated, mark the true beginning of the modern science of stellar positions.

Born in England, at Sherborne, Gloucestershire, to parents of limited income, Bradley was intended for the Church. An uncle, the Reverend James Pound, rector of Wanstead in Essex, and an able astronomical observer, instructed him in the observational art, and introduced him to Edmond Halley, for whom Bradley carried out a number of astronomical observations. With Halley's support Bradley was elected a fellow of the Royal Society in 1718, a year after receiving his MA from Oxford. By 1719 he had completed a study of the motions of the Galilean satellites of Jupiter, in which he detected the inequalities that would later (1766) be explained by Lagrange; his account of them, however, appeared only in 1749 after that of Pehr Wargentin.

In 1721 Bradley was appointed Savilian Professor of Astronomy at Oxford, a post he would retain till 1760. He had been ordained as a priest and granted livings in 1719, but he now resigned these in order to devote himself entirely to astronomy.

In 1725 Samuel Molyneux brought Bradley into his design of repeating the attempt made by Robert Hooke in 1669 to detect stellar parallax. Hooke, with a telescope fixed vertically, had claimed to find changes in the declination of γ Draconis, proving parallax. Using a specially constructed zenith sector, Molyneux and Bradley began observations in December 1725, but soon found that γ Draconis was still moving southward after the winter solstice, whereas a parallactic displacement would have reached its southerly limit at the solstice. In fact the star moved southward from autumn equinox to spring equinox, and northward in the rest of the year. The effect in other stars was similar but varied in quantity. A new zenith sector with wider aperture was set up at Wanstead in 1727 to study this second variation, which proved to be proportional to the sine of the star's latitude. The hypotheses first thought of to explain the motion, such as a nutation of the Earth's axis, proved incompatible with it.

Bradley continued the project alone after Molyneux, becoming a Lord of the Admiralty, had to abandon it. In the autumn of 1728, while on a pleasure trip on the Thames, Bradley watched the pennant's change of direction as the boat was put about, and formed the hypothesis that the stellar displacements were proportional, not as in parallax to the Earth's displacement, but to its velocity in a given direction. The 'aberration' thus discovered was the first direct evidence for the Earth's circumsolar motion. At maximum it amounted to over 20″ of displacement from the star's mean position.

The discovery of nutation followed soon as a spin-off: there were residual changes in declination, not accounted for by precession or aberration, and maximal for stars on the solstitial colure. These changes, Bradley hypothesized, were due to a nutation of the Earth's axis, consequent on the changing angle of the lunar orbit to the Earth's equatorial bulge. A full verification of the effect required observations over the 18.57-year cycle of revolution of the Moon's orbital nodes; Bradley announced the result in 1748. The maximum displacement due to nutation amounted to a little more than 9″.

After the death of Halley in 1742 Bradley was appointed Astronomer Royal to succeed him. An initial attempt at a rigorous program of observation showed the untrustworthiness of the Greenwich instruments as Halley had left them; a new set of instruments to Bradley's order was installed in 1750. The 60 000 or so observations of right ascension and declination that he and his assistants carried out in the next dozen years laid the basis for all later exact astronomy of the stars and planets.

Curtis Wilson

Bibliography

Alexander, A.F. O'D. (1971) Bradley, J., *Dict. Sci. Biogr.*, Vol. 2, pp. 387–9.
Pannekoek, A. (1961) *A History of Astronomy*. New York: Interscience, 521 pp.
Rigaud, S.P. (1833) *Miscellaneous Works and Correspondence of the Rev. James Bradley*, Oxford.

BRAHE, TYCHO (1546–1601)

Born to a Danish noble family with a military tradition, Tycho was both headstrong and autocratic, but happily possessed 'a passion for accuracy' (Thoren, 1990). He was to become the Danish astronomer who is considered to have made the most accurate astronomical observations of the pre-telescope era. In 1559 he started studying rhetoric and philosophy at the University of Copenhagen. However, the observation of a solar eclipse in 1560 turned his interest to astronomy, and he devoted the major part of the time in Copenhagen to mathematics and astronomy. In 1563 Brahe observed a close approach of Jupiter and Saturn, the time of which deviated a whole month from that predicted by the *Alfonsine Tables* and several days from the *Prussian Tables*. After 1565 he started travelling throughout Europe to study science at universities in Leipzig, Wittenberg, Rostock, Basel, Ingolstadt and Augsburg. In November 1572, while staying at a relative's manor at Herrevad Abbey, he observed a new star in Cassiopeia, the first nova to be visible to the unaided eye since 134 BC. By observing the nova for many nights, Tycho was able to determine that the new star (actually, an exploding star) must be beyond the moon's orbit and therefore in the celestial region, which was supposed to be unchanging. This observation provided evidence against the Aristotelian view of the cosmos and helped bring about a revision. Tycho published his observational results in his first work *De Nova Stella* (1573).

The discovery of the nova established Tycho's reputation, and with royal funding he built the world's first modern scientific observatory, which he called Uraniborg, on the Danish island of Hven. (Unfortunately, almost nothing remains of it today.) Here his observations were carried out with meticulous precision until 1596 when his funding was cut and he emigrated to Germany and eventually Prague.

From the wealth of observational data accumulated at Uraniborg, Tycho was able to refine the Ptolemaic and Copernican ideas about lunar motions. Empirically he discovered the speeding up ('variation') of the Moon through syzygies, as well as the oscillation of the inclination and the nodes. Thus he paved the way with an empirical base for Newton a century later.

From 1577 to 1596 Tycho made repeated observations of comets which were exact enough to convince him that comets were not generated in the Earth's atmosphere as taught in the universities at that time. His observations led him to believe that comets revolved about the Sun, at a distance from it greater than that of the planet Venus. This conclusion was strongly in opposition to the then common doctrine of the solid crystalline spheres. His work *De Mundi Aetherei Recentioribus Phaenomenis Liber Secundus* (Second Book About Recent Appearances in the Celestial World, 1588), dealing mainly with observations of a comet in 1577, also contains an account of Tycho's system of the celestial world, which was a compromise between those of Ptolemy and of Copernicus. The 'Tychonic System', in which the Earth is the fixed center of the celestial world, the Sun and Moon revolve around the Earth, and the other planets revolve around the Sun, proved to be a popular alternative for 17th century cosmologists.

Beginning in 1577, Tycho devoted the next 20 years to measuring the positions of 777 stars with unprecedented accuracy. In 1597 he left Denmark and settled in Prague in 1599 after almost three years of travelling through northern Europe. In 1600 Tycho acquired an assistant, a young and promising mathematician, Johannes Kepler (1571–1630), who had recently published his first important work, the *Mysterium Cosmographicum*. Two years later, in 1601, Tycho died, leaving Kepler to publish their star catalogue, as well as new planetary tables, the *Rudolphine Tables* of 1627. Kepler (q.v.) thus enjoyed a unique advantage in having Tycho's meticulous observational data before him in his planetary calculations.

Hans J. Haubold

Bibliography

Berry, A. (1898) *A Short History of Astronomy*. London: J. Murray (Dover Publications Inc., 1961).
Dreyer, J.L.E. (1890) *Tycho Brahe: A Picture of Scientific Life and Work in the Sixteenth Century*. Edinburgh: Adams and Charles Black.
Hellman, C.D. (1970) Brahe, Tycho. *Dict. Sci. Biogr.*, Vol. 2, 401–16.
Shackleford, J. (1993) Tycho Brahe, Laboratory Design and the Aim of Science. **Isis 84**, 211–230.
Thoren, V.E. (1990) *The Lord of Uraniborg: A Biography of Tycho Brahe*. New York: Cambridge University Press.
Weaver, J.H. (ed.) (1987) *The World of Physics*. New York: Simon and Schuster (Vol. 1).

BRECCIA

A rock consisting of a mass of angular fragments of pre-existing rocks, cemented together. The term (pronounced bretch-ia), derived from Italian (*brèche*, in French), has widespread use and applies equally well to lithological types of many different origins. Multiple components are indicated as polymictic; single-component types are monomictic. The various categories are sedimentary breccias (including cemented talus, collapse and solution breccias, intraclastic breccias, autoclastic breccias); igneous breccias (especially volcanic breccias or pyroclastic agglomerates); tectonic breccias (kataclastic, cataclastic or crush breccias, cataclasite, mylonite, the last marking a transition to recrystallized rock); and explosion breccias [cemented, meteorite impact debris; lithified ejecta (q.v.)]. More or less analogous explosion breccias are generated by nuclear bomb tests (Shoemaker 1963).

The expressions authigenic and allogenic (breccias) are used to indicate degree of displacement:

- authigenic: breccias produced by shattering approximately *in situ*;
- allogenic: breccias generated by major displacement and now either part of the 'throwout' (ejecta), or 'fallback' within the crater.

Explosion breccias often display shatter cones, high-pressure minerals like coesite and stishovite, and various other forms of shock metamorphism (q.v.; see also Craters; Ejecta).

On the Moon, loose, pre-existing lunar regolith may be converted by 'shock lithification' into 'instant rocks' (Short, 1975). In contrast, a little farther away from the impact point there is an 'annealed breccia', often polymict, probably transported by a base surge of hot gases, possibly in the manner of a terrestrial ignimbrite.

On planet Earth explosion breccias are limited to a peripheral belt in and around meteorite or astrobleme craters. On a waterless celestial body like the Moon, meteorite ejecta (q.v.) covers the entire surface (see also Regolith), and includes an appreciable fraction of impact breccia.

Rhodes W. Fairbridge

Bibliography

Chao, E.C.T., Boreman, J.A., Minkin, J.A. *et al.* (1970) Lunar glasses of impact origin: physical and chemical characteristics and geologic implications. *J. Geophys. Res.*, **75**, 7445–79.
Middlehurst, B.M. and Kuiper, G.P. (eds) (1963) *The Moon Meteorites and Comets* (*The Solar System*, Vol. 4). Chicago: University of Chicago Press.
Shoemaker, E.M. (1963) Impact mechanics at Meteor Crater, Arizona, in *The Moon Meteorites and Comets* (eds B.M. Middlehurst and G.P. Kuiper). Chicago: University of Chicago Press, pp. 301–36.
Short, N.M. (1975) *Planetary Geology*. Englewood Cliffs: Prentice-Hall, 361 pp.

Cross references

Crater
Ejecta
Impact cratering
Regolith

BROUWER, DIRK (1902–1966)

Dirk Brouwer was Dutch, but spent most of his working life in America. He was born in Rotterdam on September 1 1902 and was trained at Leiden University under the guidance of the great cosmologist Willem de Sitter. His first research program was an investigation of the movements of Jupiter's Galilean satellites, and his doctoral thesis in 1927 was a discussion of the long series of observations of the satellites carried out by RTA Innes.

In 1928 he went to the United States and joined the staff of Yale University, working with E.W. Brown on the theory of the movements of the Moon. At that time it was thought that small irregularities in the lunar motion could be explained by variations in the Earth's rate of rotation; to check this, Brown and Brouwer undertook an extensive analysis of observations of lunar occultations. Many of these observations were made by amateurs, and Brouwer became a very firm supporter of amateur astronomy. His interest in the Earth's rotation was maintained throughout his career.

At Yale he worked with the Director, Frank Schlesinger, in star cataloging and the determination of star distances; he continued this when he himself became Director in 1941. His solar system studies included theoretical research into asteroid families and the determination of asteroid positions (1933); later he wrote on Kirkwood's gaps (1963). For many years he was President of the IAU Commission 20, dealing with minor planets.

Brouwer was very active in the training of young astronomers. In 1959 he founded an annual summer institute in dynamical astronomy, and in 1963 founded a celestial mechanics research center at Yale. His death in New Haven, on January 31 1966, came when he was still at the height of his powers.

Patrick Moore

Bibliography

Brouwer, D. (1952) A study of changes in the rate of rotation of the Earth. *Astron. J.*, **57**, 125–46.
Brouwer, D. (1956) The motions of the outer planets. *Mon. Not. Roy. Astron. Soc.*, **115**, 221–35.
Brouwer, D. (1963) The problem of the Kirkwood gaps. *Astron. J.*, **68**, 152–9.
Brouwer, D., Eckert W.J. and Clemence, G.M (1951) Coordinates of the five outer planets, 1653–2060. *Astron. Papers for Am. Ephemeris*, **12**, Washington, DC.
Brouwer, D., and Clemence G.M. 1961. *Methods of Celestial Mechanics*. New York.
Danby, J.M.A. (1967) Obituary notice. *Quant. J. Roy. Astron. Soc.*, **8**, 84–8.
Dieke, S.H. (1970) Brouwer, Dirk. *Dict. Sci. Biogr.*, Vol. 2, pp. 551–12.

BRÜCKNER CYCLE

An ill-defined *cycle* of about 35 years, first noted by Sir Francis Bacon in 1625 from evidence of an alternation of cool – damp and warm – dry periods in Holland. In 1890 the cycle was rediscovered by E. Brückner, who regarded it as a worldwide phenomenon.

Brückner, a distinguished Swiss specialist in Quaternary Geology, is well known for his collaboration with Albrecht Penck, which established the distinctive cycles of Alpine glaciation that are now firmly locked in to the Milankovitch periodicities of astronomic forcing.

The Brückner cycle appears to be only one of many short-term periods that have no forecasting value, although they show up over certain periods in tree ring and rainfall records. The Brückner weather patterns for northwestern Europe suggest seasonal trends as follows:

Dry: 1756–1770, 1781–1805, 1826–1840, 1856–1870.
Cold: 1731–1745, 1756–1790, 1806–1820, 1836–1850, 1871–1885.

A dendrochronological analysis of wood from ancient Greece shows that during the 200 years prior to 440 BC (when the Acropolis was built), there were clear 34.5-year cycles as well as the solar cycle (Mariolopoulos 1962). It has also been cited in recent decades in Russian literature concerning fluctuations of the Caspian Sea, which has been rising (since 1978) after a long fall in water level. The anthropogenic effect is regarded as rather minor. A study by A.O. Selivanov (personal communication) shows it to oscillate over 100 m; the warm interglacial-type cycles (10^5 years) are matched by a major rise. During the Holocene, on a decadal to century scale, this is reversed: the water level rises during cool – humid cycles (mean global trends) as in the Little Ice Age and falls during warm – dry intervals. Secondary fluctuations around 33–35 years may be identified with the Brückner cycle.

Possible planetary periodicities and harmonics may be considered as potential forcing functions for the Brückner cycle, bearing in mind that climatologic linkage to a large catchment area such as that of European Russia will probably involve both solar emissions –

modulating the terrestrial heat budget and lunar tides (modulating the thermal exchanges in the shallow North Sea, Baltic, Barents Sea and White Sea). Tentatively, the long-term resonance concordium is 139.015 years (Saturn – Jupiter 19.859 year beat, Earth's 17.3769 year perigee-syzygy–apsides beat frequency, and the Earth/Moon–Venus 1.5978 year beat); one-fourth of the long-term resonance is 34.7538 years, which appeas to be a reasonable approximation of the climatic Brückner cycle.

Rhodes W. Fairbridge

Bibliography

Baur, F. (1951) Extended-range weather forecasting, in *Compendium of Meteorology*, (ed. T.F. Malone). Boston: American Meteorological Society, pp. 814–33.

Brooks, C.E.P. (1949) *Climate through the Ages*. London: Earnest Benn; New York: McGraw-Hill; 2nd edn (Dover reprint, 1970), 395 pp.

Brückner, E. (1890) Klimaschwankungen seit 1700 nebst Bemerkungen über die Klimaschwankungen der Diluvialzeit, *Geog. Abh.*, **4**(2).

Brückner, E. (1895) Der Einfluss der Klimaschwankungen auf die Ernteerträge und Getreidepreise. *Geog. Z.*, **39**, 100.

Brückner, E. (1901) Zur Frage der 35 jährigen Klimaschwankungen, *Petermanns Geog. Mitt.*, pp. 173–8.

Craig, R.A. and Willett H.C. (1951) Solar energy variations as a possible cause of anomalous weather changes, in *Compendium of Meteorology* (ed. T.F. Malonel). Boston: American Meteorological Society, pp. 379–90.

Mariolopoulos, E.G. (1962) Fluctuation of rainfall in Attica during the years of the erection of the Parthenon. *Geofis. Pura Appl.*, **51**, 243–50.

Schuster, A. (1914) On Newcomb's method of investigating periodicities and its application to Brückner's weather cycle. *Proc. Roy. Soc. London*, **90A**, 349–55.

Streiff, A. (1926) On the investigation of cycles and the relation of the Bruckner and solar cycle. *Mon. Weather Rev.*, **54**(7), 289–96.

Cross references

Orbital commensurability and resonance

C

CALLISTO

Callisto is the second largest and outermost of the four Galilean satellites of Jupiter (Plate 25). Discovered by Galileo in 1610, these large moons were crucial evidence that all worlds did not circle the Earth. Prior to the Voyager encounters in 1979, understanding of Callisto was limited. Subsequently, knowledge of the basic characteristics has significantly improved, and the first understanding of the geologic history of this satellite was gained. With a radius of 2400 km, Callisto is just smaller than the planet Mercury. It orbits Jupiter at a distance of 1.88 million kilometers, completing one revolution in 16.689 d. Callisto rotates synchronously with its revolution, keeping the same face toward Jupiter.

Callisto has a bulk density of only 1860 kg m^{-3} significantly less than that of any of the rocky, terrestrial planets. Water ice is the primary species identified on the surface spectroscopically, and as water ice is stable and abundant in the Jupiter region, the low density of Callisto indicates that it is composed of water ice and silicate rock in close to equal proportions by mass. Callisto is also very similar in size and density to Ganymede (which orbits Jupiter just interior to Callisto), but unlike its geologically complex neighbor, the surface of Callisto is dominated by impact craters (Figure C1) and lacks evidence for widespread resurfacing by bright icy material. This apparent lack of geologic activity in comparison with Ganymede is puzzling, and is perhaps the single most interesting aspect of Callisto.

Surface composition

Despite having the highest bulk ice content, Callisto has the lowest visible surface albedo of any of the Galilean satellites, close to 20% on average. This is approximately half as reflective as Ganymede but nearly twice as reflective as the Moon. At Callisto's distance from the Sun, this implies maximum temperatures of ~ 170 K at the equator, with diurnally averaged, subsurface temperatures of ~ 125 K at the equator, declining toward the poles (McKinnon and Parmentier, 1986).

Water-ice absorption bands have been identified in Callisto's spectra, but the absorptions are rather weak. This implies that there is a significant amount of rocky material in the visible surface (that is, the upper few millimeters to centimeters). The surface of Callisto is estimated to be as much as 55 to 80 wt% rocky non-ice material. Laboratory and theoretical spectra of ice mixed with serpentinite (a hydrated silicate) and opaque minerals such as magnetite produce a good (although not unique) match to Callisto's spectra (Roush et al., 1990; Calvin and Clark, 1991). These materials are commonly found in primitive carbonaceous chondritic meteorites.

Optical spectroscopy samples only the visible surface, but because cold ice is extraordinarily transparent to radar, such signals can probe to depths of many meters. Radar echoes have been returned from the

Figure C1 Global view of Callisto taken from a range of 1.1 million km. Callisto is very heavily cratered, with the ejecta of younger craters appearing brighter. The two brightest features mark the centers of the Asgard double multiringed system; the rings of the largest of the two (barely discernible in this image) are over 1600 km in diameter (see Schenk and McKinnon, 1987). (Voyager 2 image FDS 20583.21; north is at upper left.)

surface of Callisto since the late 1970s. The icy Galilean satellites are unusually reflective at radar wavelengths (Ostro et al., 1992). This reflectivity has been attributed to coherent backscatter in an icy regolith (fragmental surface layer), implying that the regolith of Callisto is also ice rich. The radar albedo of Callisto is approximately 50% lower than that of Ganymede but over four times greater than that of the Moon. Crude radar albedo maps of Callisto have been constructed, and reveal a large radar bright spot corresponding to the bright center of the Valhalla multiringed structure (discussed below); radar albedo may in general correlate with visual albedo. Although the crust of Callisto could be an undifferentiated 50–50 rock–ice mixture, a surficial rock-rich concentration developed from contamination of dark cometary debris and the preferential loss of water ice on the surface may be a more likely explanation for the high surface rock content but relatively icy regolith.

Water frost is concentrated locally on Callisto and is apparently mobile over geologic time. Bright frost or ice deposits have been identified on pole-facing crater walls in the high northern latitudes (the southern hemisphere, unfortunately, was not seen by Voyager). Average temperatures on these slopes are relatively low due to preferential shading, and these areas probably act as cold sinks for migrating frost molecules, given that solar-driven thermal segregation of rock and ice should occur on Callisto (Spencer, 1987). In comparison with Ganymede, however, Callisto does not have polar frost caps or shrouds. Nor is Callisto's trailing hemisphere (with respect to its orbit about Jupiter) darker or redder than the opposite, leading hemisphere, as is true for Ganymede and Europa. Apparently, Callisto's orbit is too far out in the Jovian magnetosphere to suffer much charged particle bombardment, which is thought to cause the darkening and reddening of the icy surfaces of the other satellites. In fact, Callisto's leading hemisphere is slightly darker.

Geology

The fundamental geological feature of Callisto's surface is the abundance of impact craters (Passey and Shoemaker, 1982). Many older craters on Callisto (and Ganymede) are anomalously shallow compared to fresh craters. This flattening has been attributed to slow viscous creep of the icy crust over geologic time, and virtually all topographic expression of an original crater can be removed over time if the thermal gradient is high enough. On the other hand, the depths of more recently formed, fresh craters on these two satellites are anomalously shallow relative to their lunar crater counterparts, to which they should be very similar due to the nearly identical surface gravity on all three bodies. These fresh craters have not had time to relax viscously. Crater morphology depends in large part on crustal properties, so as ice is mechanically very weak as a crustal material compared with most silicates these differences in crater morphology are consistent with an ice-rich composition for the upper crusts of Callisto and Ganymede (Schenk, 1991).

Among the multitude of craters on Callisto and Ganymede are some unusual landforms. These include central pit craters, which have small rimmed depressions in their centers, rather than conical peaks as in larger lunar craters. These depressions are also commonly filled by small rounded domes. One interpretation is that these pits and domes formed when warm, rheologically weak ice was uplifted from depths of a few kilometers during impact. Deep-seated material is usually found uplifted in the centers of large impact craters on Earth and the Moon, and a similar process may occur on Callisto and Ganymede (Schenk, 1993). In other cases, deeper material beneath larger craters may rise diapirically over time and eventually breach the surface.

Unique to Callisto and Ganymede are the mysterious palimpsests. Palimpsests are circular high-albedo patches several hundred kilometers across; some have faintly expressed concentric ridges near the center. They have very subdued topographic expressions and are in general very ancient. Secondary craters surround some palimpsests, indicating that the palimpsests are impact related. The bright material appears to be relatively thin, perhaps just a few hundred meters in the outer portions, but thicker in the center. Palimpsests could have formed from the viscous relaxation of old impact craters, or from impact into a very warm, soft (nearly strengthless) crust in the earliest epoch of Callisto's surface. In the former case, the origin of the high-albedo unit is unclear because younger craters do not have similar bright deposits. In the latter case, the bright material would result from excavation into and subsequent outflow of a warm, relatively rock-free lower crustal material (Thomas and Squyres, 1990).

The dominant geologic features on Callisto are the extensive concentric multiringed fracture systems. The largest, Valhalla, is almost 4000 km across and consists of several dozen individual ring arcs, which appear to be normal fault scarps and down-dropped graben (double-walled troughs) in an outer zone and graben or ridges in an inner zone (Figure C2). Both normal faults and graben indicate extensional deformation. Crater densities decrease toward the center of the ring system, consistent with burial by an ejecta deposit and with an impact origin for the ring system (an origin that is rather

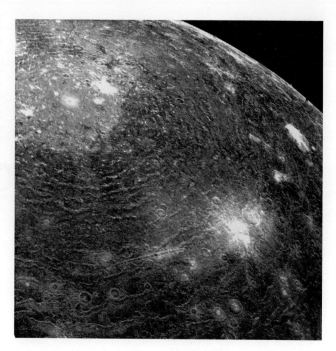

Figure C2 The Valhalla multiringed system on Callisto. Surrounding the central, high-albedo palimpsest (~ 600–800 km in diameter) are up to two dozen rings, with a maximum radius of ~ 1900 km from the center of the structure. The outermost rings in this image, covering the northern quadrant of the system, are best interpreted as outward-facing normal faults, while closer in the rings appear to be sinuous graben or troughs, possibly filled in with brighter, icier material or possibly ridges. Cutting across the foreground is a younger crater chain, ~ 340 km long. (Voyager 1 image FDS 16422.25; north is at lower right.)

obvious from its size and circular nature). The center of Valhalla is partially obscured by a ~ 600-km wide palimpsest, and the original crater rim is probably hidden within this palimpsest.

Multiringed structures observed on the rocky, inner planets consist only of between two and several rings. The large number of ring arcs in basins on Callisto has been interpreted as the result of impact into a relatively thin, brittle lithosphere overlying a rheologically soft viscous asthenosphere. As the large crater cavity collapses, the weaker asthenosphere flows inward, inducing drag and extension and fracturing of the overlying lithosphere (McKinnon and Melosh, 1980; Melosh, 1982). Because the lithosphere is thought to have been so thin on these large icy satellites at the time of impact, the surface fractured much more readily than on the inner planets. Thus ring morphologies on both Callisto and Ganymede are related to the weak rheology of water ice, compared with that of rock, in these worlds.

The dense cratering implies an ancient surface, dating back to an early period of heavy bombardment that affected the outer solar system. Estimated ages for different regions range from ~ 4.0 to 4.4 Ga (Passey and Shoemaker, 1982), but these are rather uncertain because the source of impactors has not been definitely established and the cratering timescale in the outer solar system has not been calibrated.

Despite the density of craters and the apparent lack of geologic activity, there is some evidence for possible volcanism very early in Callisto's history. The albedo of the surface is not uniform, irrespective of crater deposits. High-albedo, polygonal areas several hundred kilometers across can be recognized. Irregular wispy dark deposits and sporadic smooth intercrater deposits on scales of tens of kilometers can also be seen. Some craters are truncated by smooth deposits whose origin is unclear, because the resolution of the images is not much better than the scale of the features. Similar smooth deposits have been identified in the ancient terrains of Ganymede and have been interpreted as ancient dark volcanic units (Murchie, Head and Plescia, 1989). Callisto may have undergone a similar earlier episode of discrete volcanism, although there is no evidence that it experienced the global-scale clean-ice resurfacing that affected Ganymede.

Cratering history

As one of the most heavily cratered bodies in the outer solar system, Callisto serves as a key benchmark for comparing the cratering histories and impactor populations of the inner and outer solar systems. The crater size–frequency distribution (density of craters of a given size per unit area) on Callisto (and Ganymede) indicates a relative lack of craters larger than ~ 60 km in diameter on these bodies relative to the highlands of the terrestrial planets (Chapman and McKinnon, 1986), and there is also a near absence of craters larger than 150 km. The lack of large craters could be due to either preferential degradation and removal of large craters (in which case the same population of impactors may have swept through the entire solar system) or a real deficiency in large impactors in a differing Jovian cratering population. Viscous relaxation is most effective in larger craters and is the suggested mechanism for the removal of large craters from the geologic record. The formation of (subsequently relaxed) large craters would, however, tend to destroy pre-existing craters of all sizes in that area, creating circular 'scars' or gaps in the areal crater distribution, features that are not generally observed. Comparison of computer cratering simulations with the rather uniform areal crater distribution on Callisto indicates that removal or relaxation of a sufficient number of large craters has not occurred and implies a real difference in the impactor populations of the inner solar system and the Jovian system.

The source of this Jovian population is controversial. Comets dominate the present population of impactors there and may have been responsible for the early cratering record as well. Others advocate that this Jovian population may have been planetocentric debris orbiting Jupiter, but there are severe dynamical difficulties in having such a population last any significant fraction of solar system history.

There is some evidence that relatively young bright-rim and bright-ray craters larger than 30 km may be concentrated towards the leading hemisphere (Passey and Shoemaker, 1982). All four of the largest, most recent multiringed impact structures are also located on the leading hemisphere. This may reflect the preferential bombardment of the leading hemisphere of the synchronously orbiting Callisto (similar to the preferential impact of insects on the windshield of a moving automobile). The leading hemisphere of Callisto is also, as noted above, darker than the trailing hemisphere, and as determined from photometric studies (Buratti, 1991), appears to have a more porous visible surface. Both features may be related to the preferential bombardment of the leading hemisphere by dark, cometary micrometeorites, if such meteoritic debris accumulates.

As a final note on cratering history, the ancient surface of Callisto is remarkable for having 13 chains of aligned craters, or catenae, one of which can be seen in Figure C2. Catena craters range between ~ 10 and 30 km in diameter and the largest catena is ~ 620 km long. While these chains have been hypothesized to be composed of secondary craters from larger primary impacts, the source craters are not obvious. The catenae may instead be the result of collisions of comets tidally fragmented during a close pass by Jupiter, in a manner similar to the recently disrupted comet Shoemaker-Levy 9. The lengths of the chains and presence of 11 of the 13 on Callisto's Jupiter-facing hemisphere support this hypothesis (Melosh and Schenk, 1993).

Internal evolution

The divergent geologic histories of Callisto and Ganymede remain an outstanding paradox. It is unknown whether Callisto is completely or partially differentiated into an icy mantle and rocky core or if it is an undifferentiated mixture. Numerical and analytical simulations suggest that if Ganymede differentiated, which seems likely based on its extensive resurfacing, then Callisto may have also because of its very similar size and density (implying similar heating during accretion and from the decay of radioactive elements U, K and Th). This is not consistent with the lack of resurfacing on Callisto, however, unless such activity occurred very early in Callisto's history (McKinnon and Parmentier, 1986).

Recent work on the orbital evolution of the Galilean satellites suggests an early, significant resonant excitation of Ganymede's eccentricity, predating the current three-way Laplace resonance amongst Io, Europa and Ganymede (Malhotra, 1991). The higher orbital eccentricity (at the percent level) would have considerably enhanced heating within Ganymede because of the tidal dissipation induced. This tidal heating (q.v.) could have been comparable to Ganymede's radiogenic heating at the time, and if so would have been enough to cause water-ice melting within the satellite and perhaps trigger the global-scale resurfacing observed. In contrast, Callisto has never been affected by similar tidal interactions and so might not ever have experienced the global resurfacing that Ganymede did.

Paul M. Schenk and William B. McKinnon

Bibliography

Buratti, B.J. (1991) Ganymede and Callisto: surface textural dichotomies and photometric analysis. *Icarus*, **92**, 312–23.
Calvin, W.M. and Clark, R.N. (1991) Modeling the reflectance spectrum of Callisto 0.25 to 4.1 μm. *Icarus*, **89**, 305–317.
Chapman, C.R. and McKinnon W.B. (1986) Cratering of planetary satellites, in *Satellites* (eds J.A. Burns and M.S. Matthews). Tucson: University of Arizona Press, pp. 492–580.
Malhotra, R. (1991) Tidal origin of the Laplace resonance and the resurfacing of Ganymede. *Icarus*, **94**, 339–412.
McKinnon, W.B. and Melosh, H.J. (1980) Evolution of planetary lithospheres: evidence from multiringed basins on Ganymede and Callisto. *Icarus*, **44**, 454–71.
McKinnon, W.B and Parmentier E.M. (1986) Ganymede and Callisto, in Satellites (eds J.A. Burns and M.S. Matthews). Tucson: University of Arizona Press, pp. 718–763.
Melosh, H.J. (1982) A simple mechanical model of Valhalla Basin, Callisto. *Geophys. Res.*, **87**, 1880–90.
Melosh, H.J. and Schenk P. (1993) Split comets and the origin of crater chains on Ganymede and Callisto. *Nature*, **365**, 731–3.
Murchie, S.L., Head, J.W. and Plescia, J.B. (1989) Crater densities and crater ages of different terrain types on Ganymede. *Icarus*, **81**, 271–97.
Ostro, S.J., Campbell, D.B., Simpson. R.A. et al. (1992) Europa, Ganymede, and Callisto: new radar results from Arecibo and Goldstone. *J Geophys. Res.*, **97**, 18227–44.
Passey, Q.R. and Shoemaker, E.M. (1982) Craters and basins on Ganymede and Callisto: Morphological indicators of crustal evolution, in *Satellites of Jupiter* (ed. D. Morrison). Tucson: University of Arizona Press. pp. 379–434.
Roush, T.L., Pollack, J.B., Witteborn F.C. et al. (1990) Ice and minerals on Callisto: a reassessment of the reflectance spectra. *Icarus*, **86**, 355–82.
Schenk, P.M. (1991) Ganymede and Callisto: complex crater formation and planetary crusts. *J. Geophys. Res.*, **96**, 1563–64.
Schenk, P.M. (1993) Central pit and dome craters: exposing the interiors of Ganymede and Callisto. *J. Geophys. Res.*, **98**, 7475–98.
Schenk, P.M. and McKinnon, W.B. (1987) Ring geometry on Ganymede and Callisto. *Icarus*, **72**, 209–34.
Spencer, J.R. (1987) Thermal segregation of water ice on the Galilean satellites. *Icarus*, **69**, 297–313.
Thomas, P.J. and Squyres, S.W. (1990) Formation of crater palimpsests on Ganymede. *J. Geophys. Res.* **95**, 19161–74.

Cross references

Europa
Galilean satellite
Ganymede
Impact cratering
Io
Jupiter: satellite system
Spectrophotometry

CAMPBELL, WILLIAM WALLACE (1862–1938)

Campbell was born on 11 April 1862 at Hancock County, Ohio. He entered the University of Michigan in 1882, initially to study engineering, but soon changed over to astronomy, studying under J.M. Schaeberle. He graduated in 1886, and became Professor of Mathematics at the University of Colorado, though he returned to Michigan 2 years later to assume the post of Lecturer in Astronomy. In 1891 he joined the staff of Lick Observatory, becoming Director in 1901 – a post which he retained until his retirement in 1930. He was President of the University of California from 1923 to 1930; President of the United States National Academy of Sciences from 1931 to 1933, and President of the International Astronomical Union from 1922 to 1925. He was equally skilled as an observer, a theorist and an organizer.

His main work was in connection with spectroscopy. He determined the radial motions of 3000 stars and published a catalogue of them in 1928; during the course of his researches he discovered many spectroscopic binaries, including Polaris (1899) and Capella (1900). He made a long series of observations of spectroscopic changes in Nova Aurigæ 1892, and also designed new spectroscopic equipment.

He did not neglect the planets, and made spectroscopic observations of Jupiter and Saturn. He made visual observations of Mars, using the Lick 36-inch (90-cm) refractor, but here too his main work was spectroscopic. Up to that time it had been generally believed that the atmosphere of Mars was reasonably dense, and contained a considerable amount of water vapor, while the polar caps were made up of water ice and snow. Campbell disagreed. His method was ingenious; what he did was to compare the spectrum of Mars with that of the Moon, making the conditions of observation as similar as possible. He found that to all intents and purposes the spectra were identical – and, of course, it was known that the Moon has virtually no atmosphere. Campbell did not claim that Mars too was airless, but he did maintain that the atmosphere was extremely thin, lacking in water vapour; there could be no seas, and he was inclined to the view that the polar caps were mainly composed of carbon dioxide ice. The ochre tracts, he maintained, were genuinely coloured. It is now known from the space-probe results that his conclusions about the atmosphere were essentially correct.

Campbell's health declined in his latter years, and finally he decided to end his life to spare his relatives the trouble of looking after him. He died in San Francisco on 14 June 1938.

Patrick Moore

Bibliography

Abrams, J.W. (1971) Campbell, William Wallace. *Dict. Sci. Biogr.*, Vol. 3, pp. 35–7.
Campbell, W.W. (1894) The spectrum of Mars. *Publ. Astron. Soc. Astrophys.*, **16**, 273.
Campbell, W.W. (1884) The spectrum of Mars. *Astron. Astrophys.*, **13**, 752.
Campbell, W.W. (1895) A review of spectroscopic observations of Mars. *Astrophysics*, **2**, 28.
Flammarion, C. (1909) Campbell's observations of Mars, in *La Planète Mars*, vol. 2. Paris: Gauthier-Villars, pp. 144–59.

CAPTURE MECHANISMS

Capture mechanisms are dynamical mechanisms for turning two gravitationally unbound bodies into a gravitationally bound pair. In order to achieve capture, sufficient relative kinetic energy between the two bodies must be dissipated in order to produce a bound system. A variety of mechanisms for dissipating this energy have been proposed, such as tidal dissipation and disruption, gas drag, impacts and three-body interactions. Tidal dissipation relies on the energy dissipated in tidal flexing of the smaller body and the torques associated with the tidal forces between the two bodies to effectively remove relative kinetic energy; tidal disruption presupposes the tidal disintegration of the smaller body within the Roche limit of the larger body (see Earth–Moon system: origin). If one body is surrounded by a sufficiently dense gaseous envelope or disk, then the second body may be captured through the effects of aerodynamic drag forces with the gas or the gravitational effects from the wake in the gas produced by passage of the second body. Major impacts can result in the trapping in orbit of dispersed matter derived from the smaller body that could later accumulate into a companion object, or impacts can result in the direct orbital capture of a sizeable chunk of the smaller body (see Earth–Moon system: origin). If three bodies should experience a close mutual encounter, certain orbital configurations for the triplet can lead to the formation of a bound pair and an ejected third body that carries off the excess kinetic energy.

Capture has been suggested as an explanation for the origin of a number of planetary satellites, in particular the Earth's Moon, though capture is no longer thought to be the most likely mechanism for the Moon's origin (see Earth–Moon system: origin). Capture remains as a likely explanation for the formation of the outermost (irregular) satellites of Jupiter, Saturn's satellite Phoebe and Neptune's satellite Triton, largely because the orbits of all of these satellites are retrograde compared to the planets' rotations and to the orbits of most of those planets' satellites (see Triton; Neptune: satellite system). If the regular satellites of the giant planets formed in a rotating gas and dust disk surrounding the growing protoplanets, this disk might very well have been the agent responsible for capturing the irregular satellites.

Alan P. Boss

Bibliography

Boss, A.P. and Peale, S.J. (1986) Dynamical constraints on the origin of the Moon, in *Origin of the Moon* (eds W.K. Hartmann, R.J. Phillips and G.J. Taylor). Houston: Lunar and Planetary Institute, pp. 59–101.
Malcuit, R.J., Mehringer, D.M. and Winters, R.R. (1989) Numerical simulation of gravitational capture of a lunar-like body by Earth, in *Proc. Lunar Planet. Sci. Conf.*, **19** (eds V. Sharpton and G. Ryder). Houston: Lunar and Planetary Institute, pp. 581–91.

Cross references

Celestial mechanics
Earth–Moon system: dynamics
Earth–Moon system: origin
Moon: origin

CARBON, CARBON DIOXIDE

Carbon (C), the element with atomic number 6, is named for *carbo* meaning charcoal. Known since prehistoric time, it occurs in many forms, the most renowned being diamond. Carbon compounds are the basis for all life as we know it, and the study of carbon compounds is known as organic chemistry. Carbon forms covalent bonds, single, double or triple bonds, in complex arrangements that allow the molecules the variety and complexity to form the basis of life on Earth. The dioxide CO_2 plays an important role as a greenhouse gas in planetary atmospheres.

Carbon may be studied in an astrophysical context; in stars such as our Sun, energy is liberated by the conversion of hydrogen to helium through the intermediary actions of elemental carbon, nitrogen and oxygen. The isotope ^{12}C with ^1H converts to ^{13}N + energy, and $^{13}C + {}^1H$ converts to ^{14}N + energy (see Sun). Carbon is widely distributed in the universe (see Cosmochemistry). Organic compounds are one component of interstellar grains (q.v.); these may in turn be important in the context of theories of the origins and compositions of comets (see Comet: structure and composition).

Studies of organic materials in meteorites provide insights into conditions prevailing and processes occurring during the formation of the solar system some 4.5 Ga ago. These topics are well covered elsewhere in this volume (see Carbonaceous chondrite; Meteorite; Solar system: origin). Comets and meteorites may also have played an important role in the delivery of organic materials to the early Earth (see Cometary impacts on Earth). Arrhenius (q.v.) suggested as much in his 'Panspermic hypothesis'. These facts and hypotheses bear on the question of the origins of life on Earth (see Life: origin).

In the present solar system the gas carbon dioxide (CO_2) plays an important role in the atmospheres of the planets Venus, Mars and the Earth. CO_2 is the primary constituent of the Mars atmosphere; it cycles between the gaseous and solid forms during the seasonal cycle on Mars, forming polar caps of solid CO_2 during the local winter (see Mars: atmosphere). On Venus the high concentration of atmospheric

CO_2 has produced extremely high surface temperatures through the greenhouse effect (q.v.; see also Venus: atmosphere). On Earth carbon dioxide and greenhouse warming is an important area of scientific investigation due to the possible consequences for societies of irreversible climatic changes. Combustion of fossil fuels will double the pre-industrial level of atmospheric CO_2 some time in the next century. These topics are also covered elsewhere in this volume (see Greenhouse effect; Atmospheric thermal structure; Atmosphere; Earth: atmosphere).

In any planetary setting, some sort of carbosphere probably existed sometime in its cosmochemical history. A carbosphere is a celestial environment (as on Earth-like bodies and large asteroids) where carbonaceous complexes exist containing covalent C–C bonds. Associated in these complexes are C–H–O–S–N, in decreasing numeric abundance. Typical associations are found on carbonaceous chondrites (q.v.). On the outer planets (the great 'gas planets'), methane and ammonia (CH_4 and NH_3) predominate, as primary constituents on Jupiter and Saturn, while only methane is predominant on Uranus and Neptune (see Atmosphere, and separate entries). For the inner planets, carbospheres also exist, differing from one another by their individual histories. A common feature is the photodissociation and gravitational escape of hydrogen from CH_4, NH_3, H_2O, etc., as well as a concentration of CO_2 in the carbosphere. Large accumulations of O_2 are only possible on biogenic planets like Earth.

The origin of carbon on Earth is still under investigation. Two standard scenarios offer contrasting views of the immediate source of the Earth's carbon reservoirs: it is derived (1) via CO_2 outgassing from the primitive planet, originally devoid of atmosphere, immediately following its accretion from planetesimals; (2) via photodissociation of the primitive CH_4–NH_3–H_2O (methane, ammonia, water) atmosphere, allowing the hydrogen to escape from the Earth's gravitational field, leaving a largely nitrogen atmosphere with CO_2 supplied partly by the dissociation and partly by degassing. The loss of hydrogen and gain of CO_2 eventually led to the partial oxidation of the initially reducing environment. The scenario (2) was argued by Urey (1952, 1959) by analogy with the great 'gas planets' and Holland (1984) supported it, with evidence that about 0.5 Ga is needed for the process to convert the first atmosphere from a reducing state to a slightly oxidizing one.

Carbon cycle on Earth

The element carbon, while quantitatively low in abundance on the Earth, is one of the most important components of the Earth's outer spheres and is the most important component of the biosphere and all organic processes. The dioxide plays a vital ('greenhouse') role in modulating the climate in the atmosphere of this planet. CO_2 is transparent to visual light but not to infrared. Much of the solar energy reaching the Earth comes in the form of visible light, but much of the Earth's radiated energy leaves in the form of infrared. CO_2 allows the incoming radiation to pass but blocks the outgoing radiation.

The carbonate and bicarbonate ions play a key part in the hydrosphere, where $CaCO_3$ sedimentation acts as a long-term storage reservoir in deep sea deposits. Burial of carbonate sediments and organic debris (which become altered to petroleum and natural gas) represent a geologic sink for carbon.

Organic molecules, atmospheric CO_2 and oceanic CO_3^{-2} are continuously linked in the carbon cycle which involves complex exchanges between gaseous, liquid and solid states. The rates of such exchanges vary: in some cases, minute to minute and in others, over geologic time.

Rhodes W. Fairbridge and Clare P. Marshall

Bibliography

Berkner, L.V. and Marshall, L.C. (1972) Oxygen: evolution in the Earth's atmosphere, in *The Encyclopedia of Geochemistry* (ed. R.W. Fairbridge). New York: Van Nostrand Reinhold, pp. 849–61.

Bowen, H.J.M. (1966) *Trace Elements in Biochemistry*. New York: Academic Press.

Holland, H.D. (1984) *The Chemical Evolution of the Atmosphere and Oceans*. Princeton University Press, 582 pp.

Urey, H.D. (1952) *The Planets*. New Haven: Yale University Press, 240 pp.

Urey, H.C. (1959) The atmospheres of the planets, in *Astrophysics III, The Solar System* (ed. S. Flügge), *Hand buch des Physic*, Vol. 52. Berlin: Springer-Verlag, pp. 363–418.

Cross references

Arrhenius, Svante
Atmosphere
Atmospheric thermal structure
Carbonaceous chondrite
Cosmochemistry
Earth: atmosphere
Greenhouse effect
Life: origin
Mars: atmosphere
Meteorite
Solar system: origin

CARBONACEOUS CHONDRITE

Chondritic meteorites are chemically primitive in that their compositions approach that of the solar photosphere for all but the most volatile elements and noble gases; the solar photosphere is assumed to be compositionally similar to that of the early solar nebula. The most volatile-rich (and therefore primitive) chondrites have been assigned to a group termed carbonaceous (C) chondrites. Although the distinction between C and ordinary chondrites is not always clear-cut (Dodd, 1981), the former do contain a generally higher amount of carbon (0.5 to 5 wt%) than the other chondritic meteorites (only about 0.1 wt%). Although much of this carbon remains poorly characterized, graphite, diamond, and complex organic molecules (e.g. amino acids and hydrocarbons) are known to be present (Cronin, Pizzarello and Cruikshank, 1988). The C chondrites are unmetamorphosed to weakly metamorphosed, generally unequilibrated agglomerates of chondrules (spherical aggregates of silicates), unmelted to melted nebular aggregates, lithic fragments, individual crystals and volatile-rich, fine-grained matrix materials. They can best be visualized as mechanical mixtures of unrelated nebular materials which have, in some cases, experienced subsequent processing on small planetary bodies. The C chondrites are generally black in appearance, due to a high content of fine-grained iron–nickel sulfides and/or metal. They are physically weak and generally fragment during atmospheric entry, resulting in large meteorite strewn fields on the Earth. Two notable instances of C chondrite showers occurred in 1969 at Pueblito de Allende, Mexico, and Murchison, Victoria, Australia.

The C chondrites have been separated into individual groups based upon their bulk carbon, water and sulfur content. To date six groups have been recognized, designated CI, CM, CR, CV, CO and CK; with the former four being the most common and best characterized. Among C Chondrites the CI chondrites have the highest content of volatile elements, and CK at least. A complementary classification scheme is based upon the mineralogy of each chondrite, which is embodied in the concept of petrologic type. This scale was originally intended to indicate increasing levels of parent body metamorphism, in well-defined steps from petrologic types 1 to 6. However, recent research indicates that most type 3 chondrites experienced little or no processing following chondrite parent body accretion, that types 2 to 1 indicate increasing levels of low-temperature parent body aqueous alteration, and that types 4 and up indicate increasing levels of parent body thermal metamorphism (McSween, 1979). This re-evaluation of the concept of petrologic type is the subject of continuing controversy and research. Those C chondrites which have experienced the most extensive aqueous alteration are also host to a complex suite of organic compounds, at least some of which may have been processed or synthesized by the aqueous activity. One possible example of this process would be amino acid production via Strecker–cyanohydrin synthesis (Cronin, Pizzarello and Cruikshank, 1988).

The most primitive C chondrites (type 3) consist dominantly of ferromagnesian silicates (principally olivine), in grain sizes ranging from 0.5 mm down to a fraction of 1 micrometer, most often accompanied by lesser amounts of silicate glass, iron–nickel sulfides

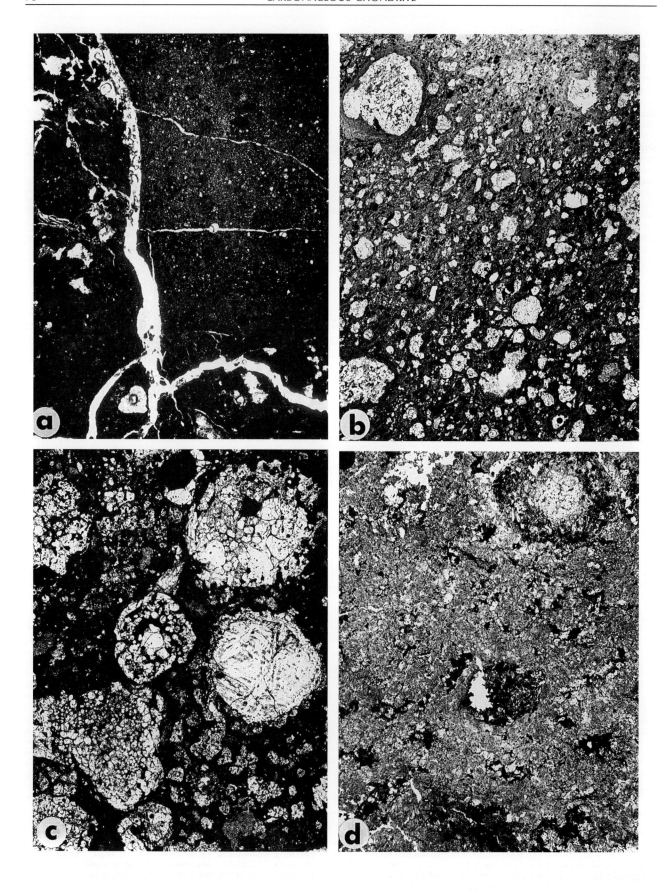

and metal. Many other minerals are also present in minor amounts. Chondrules, mineral aggregates, lithic fragments (sometimes from other meteorites) and inclusions rich in refractory elements calcium and aluminum (called CAIs) are present in varying abundance. The CAIs, in particular, record a complex history of nebular condensation, subsequent melting and partial vaporization, all important processes occurring in the early solar system. Recently, exciting discoveries have been made of micrometer-sized (and smaller) grains of diamonds, SiC, TiC and graphite located within the matrix of type 3 (as well as 1 and 2) C chondrites. Based upon isotopic investigations, these grains appear to be preserved interstellar material from the primordial solar nebula, having originated in the circumstellar clouds about neighboring (Anders, 1988). Study of these interstellar grains opens a window onto galactic history before the birth of the solar system.

In going from petrologic type 3 to 1 the original mineralogy of C chondrites is gradually replaced by hydrous ferromagnesian silicates, carbonates, sulfides and oxides as the bulk composition becomes increasingly enriched in volatile elements. Chondrules are entirely absent in type 1 C chondrites, which commonly exhibit veins filled with carbonates and sulfates. These features indicate progressive alteration by aqueous solutions, and reveal that liquid water was present on primitive solar system bodies. Isotopic studies indicate that this aqueous activity occurred within the initial 100 million years of solar system history. In passing from petrologic type 3 to 6, the fine-grained matrix of C chondrites is observed to coarsen, and aggregate and chondrule boundaries disappear as mineral grains recrystallize. Silicate glass is observed to crystallize into plagioclase feldspar. These effects indicate progressive thermal metamorphism to a peak temperature of approximately 950°C (Scott et al., 1989). These mineralogical and compositional factors are consistent with a model of type 1 to 2 C chondrites lying within or near the outer regolith of a parent body, and types 3–6 originating from progressively greater depths within the interior.

It is currently believed that the C chondrites were derived from hydrous asteroids (which could include inactive cometary nuclei), with diameters varying up to several hundred kilometers (McSween, 1987). Likely candidates are the C- and P-class asteroids lying in the main and outer regions of the asteroid belt.

Michael Zolensky

Bibliography

Anders, E. (1988) Circumstellar material in meteorites: noble gases, carbon and nitrogen, in *Meteorites and the Early Solar System* (eds J.F. Kerridge and M.S. Matthews). Tucson: University of Arizona Press, pp. 927–55.

Cronin, J.R., Pizzarello, S. and Cruikshank D.P. (1988) Organic matter in carbonaceous chondrites, planetary satellites, asteroids and comets, in *Meteorites and the Early Solar Sytem* (eds J.F. Kerridge and M.S. Matthews). Tucson: University of Arizona Press, pp. 819–57.

Dodd, R.T. (1981) *Meteorites: A Petrologic–Chemical Synthesis*. Cambridge: Cambridge University Press.

McSween, H.Y., Jr (1979) Are carbonaceous chondrites primitive or processed? A review. *Rev. Geophy. Space Phys.*, 17, 1059–78.

McSween, H.Y., Jr (1987) *Meteorites and Their Parent Planets*. Cambridge: Cambridge University Press.

Scott, E.R.D., Taylor, G.J., Newsom, H.E. et al. (1989) Chemical, thermal and impact processing of asteroids, in *Asteroids II* (eds R.P. Binzel, T. Gehrels and M.S. Matthews). Tucson: University of Arizona Press, pp. 701–39.

CARRINGTON, RICHARD CHRISTOPHER (1826–1875)

An English amateur astronomer best known for his extensive work on sunspots, especially 1853–1861, summarized in a paper to the Royal Society (1863) 'Observations of the spots on the Sun'. He discovered the latitudinal drift of the spots over the 11-year period from mid-latitudes to the equatorial belt, a phenomenon that became known as the 'law of zones' (often called Spörer's law after the man who studied it most thoroughly).

Carrington also devised a system for enumerating the solar rotations, beginning with no. 1 on November 9, 1853, a system which subsequently became known as the 'Carrington numbers'; for example, Rotation no. 1500 began October 19, 1965. The mean synodic period is 27.2753 d (approx. 0.075 year). Near the equator the sunspot rotation period is only about 25 d; it is here, in the low latitudes, that the spots (in each hemisphere) dissipate. The sunspot rotation period is greater (> 30 d) in the solar mid-latitudes (more than 35°) where they initiate. Thus, through his observations the difference in the rate of rotation of the different parts of the surface of the Sun (differential rotation) was clearly confirmed. Carrington also appears to have been the first person to observe a 'white-light' solar flare (of great magnitude) on September 1 1859.

Carrington was relatively well-to-do and had set up his own solar observatory at his home in Redhill, Surrey. Unfortunately his business, a brewery, suffered meanwhile and he was forced to discontinue detailed studies. Fortunately the observations were taken up by Spörer and to him credit is given for the latitudinal observation.

Rhodes W. Fairbridge

Bibliography

Anon. (1876) Obituary. *Monthly Notices, Roy. Astron. Soc.*, 36, 137–42.

Azimov, I. (1964) *Biographical Encyclopedia of Science and Technology*. Garden City, NY: Doubleday, 661 pp.

Forbes, E.G. (1971) Carrington, Richard Christopher, *Dict. Sci. Biogr.*, Vol. 3, pp. 92–4.

CARTOGRAPHY

The word 'map' calls to mind a scale drawing of a surface on which symbols are used to depict the distribution of features. Increasingly, maps are being prepared in computers and maintained as digital databases from which traditional printed maps can be designed and extracted on demand. Whereas political boundaries and cultural features usually dominate terrestrial maps, planetary maps show the forms and structures of natural surfaces. Terrestrial map data are usually presented as lines and graphics, whereas image (photographic) maps are the typical format in planetary cartography. Derivative geological maps showing stratigraphy, structure or other information are often prepared as overlays on both terrestrial and planetary base maps.

To provide a framework for their maps, cartographers rely on the associated science of geodesy, which seeks to define the shape, gravitational field and coordinate system of a planet. Where possible, photogrammetry (mapmaking by means of photographic images) is used to provide precise vertical dimensions or topographic contours.

Maps play a key role both in understanding the geology of other planets and in illustrating geologic interpretations. Relative ages of surfaces and the sequences of their disruption by meteoritic bombardment, volcanic activity, the flow of lava or water and tectonism

---

Figure C3 Views of C chondrite thin sections showing the effects of aqueous alteration (a, b) and metamorphism (d), compared to a pristine type 3 meteorite; all views measure 4 mm high. (a) The Orgueil CI1 meteorite showing fine-grained, dark matrix of hydrated ferromagnesian silicates (principally serpentine and saponite), sulfides, oxides and light carbonates, cut by carbonate and sulfate veins; (b) the Murray CM2 meteorite showing light chondrules and aggregates, surrounded by a dark matrix of partially hydrated ferromagnesian silicates (principally serpentine and olivine) and sulfides; (c) the Vigarano CV3 meteorite with large chondrules and aggregates in an opaque matrix of anhydrous ferromagnesian silicates (principally olivine); (d) the PCA 82500 C4 meteorite with a very coarsely recrystallized, almost transparent matrix of anhydrous ferromagnesian silicates containing nearly indistinguishable chondrules. (Photo credit: Michael Zolensky, NASA Johnson Space Center.)

can be deduced by studying the dimensions and distributions of features that appear on maps. Other questions are crucial to the mechanical success of a planetary mission. Is a targeted site smooth enough for a spacecraft landing? Can an exploration vehicle climb or cross a particular slope? Will the programmed descent trajectory of a spacecraft intersect the surface at the desired landing site, or is a mountain or ridge in the way?

Given the data at hand, it is rarely possible to compile precise topographic maps to answer all of these questions quantitatively, but many can be answered by inference from indirect evidence on image maps.

History

Although mapmaking dates from earliest recorded history, systematic mapping of planets and satellites began only with the space age (Greeley and Batson, 1990). The successes of the Magellan radar mapper at Venus are the latest in a history of extraterrestrial mapping that began with W. Gilbert's sketch of the Moon in 1600 (10 years before the appearance of the first primitive telescope). The first 360 years of that history saw the compilation of many sketches made from painstaking observing sessions at the telescope. Some were noteworthy primarily because of their subject matter and because considerable effort went into them, but others were highly credible cartographic accomplishments.

The advent of practical astronomical photography during the last decade of the 19th century strongly affected lunar cartography. It permitted the mapping of the Moon's surface details with previously unheard of ease and accuracy. Toward the end of the 1950s space exploration enthusiasts began to realize that the space age would require better maps of planetary bodies than had been made previously. A successful, continuing program of lunar mapping was begun in the fall of 1959 by the US Air Force Chart and Information Center (ACIC) (Kopal and Carder, 1974). The centuries-old difficulty of portraying lunar features realistically was overcome for the first ACIC lunar chart by drawing shaded relief maps with an airbrush. Relative elevations were computed by shadow measurements. Only a few of these 'lunar astronautic charts' have been superseded by new editions incorporating space-age image data.

No mountains or craters can be seen on Mercury or Mars with even the most powerful telescopes; only albedo markings can be resolved. This did not deter early workers from sketching their observations in map form, but their products are fewer and cruder than the early lunar maps. On Mars recognizable detail such as the dark area now known as Syrtis Major had been sketched in 1659 by C. Huygens (q.v.). In 1878 G.V. Schiaparelli (q.v.) made a map on which he named features; his naming scheme forms the basis of modern Martian nomenclature. Not all observers saw the famous 'canals' on Mars popularized by Percival Lowell (q.v.), and those who did seldom agreed on their positions. The solid surfaces of planets and satellites other than Mercury, the Moon and Mars cannot be resolved even with the best telescopes, and the few sketches that have been attempted cannot really be called maps.

Mapping the planets with spacecraft

In this century planetary cartography did not become professionally respectable until the advent of spaceflight, and the making of maps from pictures taken by cameras mounted on spacecraft emerged as a new discipline (Plate 4). Planetary mapping methods resemble those developed long ago for mapping the Earth, although the peculiarities of data available from telescopes and spacecraft have required modifications to the technology. Although terrestrial cartography is heavily dependent upon cultural and geopolitical surveys, photogrammetry (q.v.) and digital image processing are the fundamental technologies on which planetary cartography is based.

Only planets and moons with solid surfaces have been mapped in the traditional sense; the colors and patterns on the giant gaseous planets (Jupiter, Saturn, Uranus and Neptune) change hourly. Existing photomosaics of these bodies are weather maps, showing atmospheric phenomena only.

The television cameras used on spacecraft typically have an image format of roughly 1000 pixels (picture elements) on a side; the total number of pixels per image is thus less than 1% of that for a film camera. To obtain high spatial resolution with such small pixel arrays, spacecraft cameras are designed with extremely narrow fields of view. High-resolution coverage of large areas can therefore be achieved only by mosaicking many images taken from different locations along the spacecraft trajectory. Stereoscopic coverage is not obtained routinely as a result of overlap, as is the case with wide-angle film cameras. Even when stereocoverage is acquired, by design or by chance, the non-standard imaging geometry makes it difficult to produce relief measurements and contour maps from the data with conventional stereoplotting equipment. Until the advent of solid-state detectors, television cameras were also plagued by serious image distortions that had to be measured and corrected before mapping could be done.

These disadvantages are more than offset by the significant advantages of digital television over film cameras for planetary mapping. For example, subtle color and brightness variations can be recorded electronically much more accurately than on film. Electronic cameras are also more sensitive than film and hence perform better in the low light levels of the outer solar system. Most importantly, the number of images returned by a given mission is not limited by the film supply, but only by the duration of the mission and available power. It is not unusual for planetary spacecraft to function for many years and to return tens of thousands of pictures, more than offsetting the lesser amount of information in each frame compared with film. Electronic images are returned to Earth at the speed of light, rather than at the speed of a spacecraft. Voyager 1 and 2 images of the Saturn system were recovered 1.5 hours after they were acquired, whereas at least 2 years and a prohibitively complex mission design would have been required to recover them by spacecraft return.

Not all spacecraft imaging systems operate as framing cameras. The Magellan spacecraft has imaged most of Venus despite that planet's thick cloud cover by using side-looking imaging radar. On each orbit this instrument produced a nearly pole-to-pole image strip or 'noodle' $\approx$ 300 pixels wide but $\approx$ 200 000 pixels long. The planned Cassini mission to the Saturn system will image the surface of Titan with a similar radar system but less systematically; its image strips will be laid down criss-cross in a series of Titan flybys. Strip-imaging optical systems also exist (e.g. Landsat) and are planned for the near future. The unique geometric properties of these imaging systems present planetary cartographers with a challenge to develop innovative new mapping techniques.

Planetary mapping methods

Geodetic control surveys of the planets are based on a combination of data. These include astronomical measurements of positions and orientations (ephemerides) of the involved objects, spacecraft position measurements made on Earth by analysis of radio signals from the spacecraft, camera orientation angles transmitted with spacecraft telemetry signals and geometric analysis of images returned by spacecraft imaging systems. The photogrammetric triangulation of the locations of features is the final step in producing accurate map coordinates for a collection of features.

Making planetary maps involves digital and manual processing and mosaicking of images transmitted by spacecraft. In some cases a painting of the planetary surface is made on a map projection by the airbrush relief-shading technique used in lunar mapping. Maps (whether image mosaics or airbrush drawings) are commonly produced in more than one version because no single version can serve all scientific investigations. For example, some maps emphasize surface relief while others emphasize surface markings or albedo.

Photomosaics are arrays of overlapping pictures. They differ from true maps because photographic images are collections of light and dark patterns reflecting a complicated mixture of terrain and surface coloration projected from three-dimensional surfaces onto two-dimensional perspective views. Thus they cannot provide the precise schematic definitions normally expected from maps. As the only resources available, however, mosaics are often published as maps in their own right. They nearly always are the fundamental compilations upon which more elegant cartography is based (Figure C4).

Manual compilation of controlled mosaics was an iterative process in which picture scale was assumed, pictures were printed to scale and a trial mosaic was made. The inevitable errors were evaluated, a new scale selected and the pictures reprinted and remosaicked. Three or four reprintings of large numbers of frames were common in controlled mosaicking.

Digital mosaicking has now replaced manual mosaicking; computers and image processing software have replaced scissors and glue. With computers it is possible to correct the brightnesses and positions of

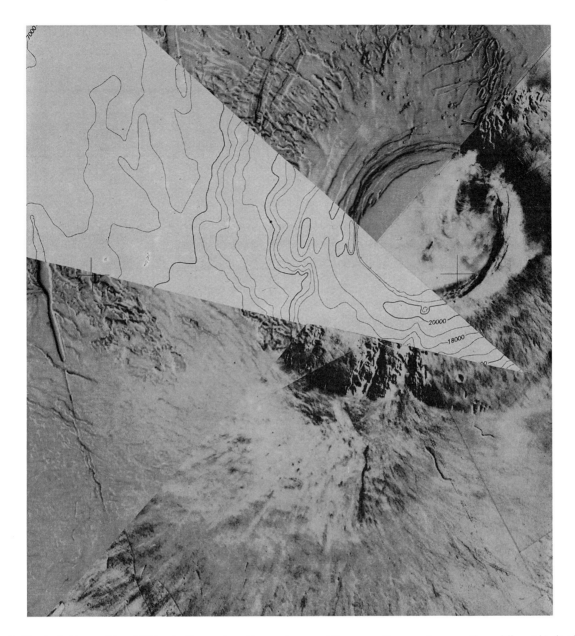

Figure C4 The Arsia Mons caldera (125 km in diameter) on Mars and some of its surroundings, as shown on three different kinds of maps. North is at the top. Airbrushed shaded relief is shown in the northwest half and part of a photomosaic in the southeast half; a wedge-shaped segment of a contour map (interval = 1 km) lies between them. (The contour compilation did not become available until long after the other two maps were made and could not, therefore, be incorporated in the earlier versions.)

individual image elements of digital images to make clear and accurate mosaics. Sophisticated algorithms have been developed that can be used to automate many aspects of the mosaicking process. Scaling and geometric transformation of images and the matching of tones and contrasts between image frames can be performed on the basis of known solar illumination conditions (Batson, 1987). Special composite products may also be made, such as mosaics of low-resolution color images coregistered with mosaics of high-resolution black and white images of the same area, for maximum utilization of all available image data (Figure C5). None of the processes is completely automated, however, and each requires human supervision.

Although there will always be a need for a basic set of printed maps, the time is near when most terrestrial and planetary maps will be made digitally and used with computer workstations. Comprehensive map datasets have become too large to manage in any other way. Maps of this kind, containing geophysical, geochemical and geological data as well as image mosaics, are already being distributed on magnetic media and on optical disks.

Shaded-relief ('hill-shaded') maps present the clearest pictures of surface morphology. Although image mosaics provide the control basis for shaded-relief mapping, the shading technique developed for planetary cartography requires the ability to visualize landforms by examining many different photographs. These usually have inconsistent scale, surface resolution and viewing geometry. The clouds and haze that obscure parts of many pictures must be identified and eliminated from the analysis. After evolving a mental image based on this examination, the airbrush cartographer must be able to draw the result so that it is clearly visible to others. The technique differs from the hill shading of terrestrial maps, which is based on existing topographic contour maps not available to the planetary cartographer. Airbrushing is so labor intensive and

Figure C5 Composite photomosaic of part of the Ophir Chasma area of Mars, made from 48 high-resolution (about 60 m per resolution element) black and white Viking Orbiter images and three low-resolution (about 1 km per resolution element) color images. North is at the top. The mosaic covers about 355 km on a side.

requires such specialized skills that its dominance in planetary cartography is diminishing. Computerized methods that emulate some aspects of the technique have been developed, but they require systematic and specialized data collection methods that have become practical only recently.

Topographic contour maps of Earth are made from stereoscopic aerial photographs by photogrammetry. The specialized cameras designed solely for photogrammetric work on Earth are not only too heavy to be used in planetary exploration, but they require that film be returned to the photogrammetrist. Film cameras were taken to the Moon on some Apollo (q.v.) and Zond missions (q.v.). For the most part, however, the planetary explorer must use photogrammetric techniques that have been modified extensively for use with electronic spacecraft cameras; results range from good to marginally useful. In addition, several topographic maps of the planets have been based on discrete altitude measurements. These include Earth- or spacecraft-based soundings made with radar or lasers, with atmospheric pressure measurements, and with radii determined by timing the occultation of radio signals from an orbiting spacecraft when it passes behind the planet. Like digital image maps, digital topographic maps are becoming increasingly important and are being reproduced on optical disks.

Maps of spherical or spheroidal planets are familiar and well understood, but the asteroids and small satellites that we have explored are not at all spherical. Because they are primordial bodies containing significant evidence relating to planetary evolution, exploring and mapping them is becoming increasingly important. Mapping an asteroid is like mapping a potato, except that the 'potatographic projection' has yet to be defined. Therefore, analytical work involving volumes, dimensions and spatial distributions of materials on these bodies is best done with digital maps, which consist of arrays of radii registered with images transmitted by spacecraft. The mapping process is analogous to projecting a set of pictures onto a physical model of the object and assembling a systematic collection of latitudes, longitudes, radii and image brightnesses. The initial model upon which image projections are made is derived from photogrammetric measurements of radii at conspicuous features on the object. Once the digital map has been made, it can be projected to any desired map view (Figure C6).

Summary

Maps of the planets are required to plan and execute their safe and productive exploration and to analyze the scientific results of that exploration. Although the technological elements of planetary mapping are the same as those of terrestrial cartography, the use of film is virtually precluded, and extensive modifications have been made to accommodate electronic imaging and measurement methods. The

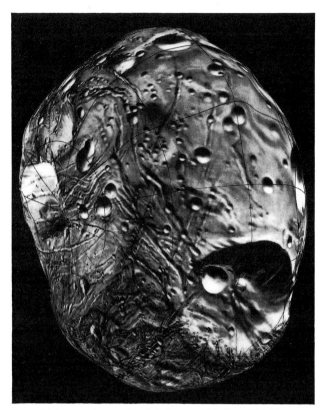

Figure C6 Image of part of a digital map of Phobos, Mars' largest moon (27 km maximum diameter). A shaded relief map was digitized, registered with a map of radii of Phobos, and transformed to this orthographic view. The large crater in the upper right is called 'Stickney', after the wife of astronomer Asaph Hall.

final maps are physiographic representations; only on Earth do cultural, political and hydrologic characterizations have relevance.

More than 1600 maps have been published of 22 planets and satellites other than Earth. Only two major bodies are yet unmapped: no spacecraft has yet been sent to Pluto, and Titan can be mapped only by imaging radar because its surface is obscured by clouds.

R.M. Batson

Bibliography

Batson, R.M. (1987) Digital cartography of the planets: new methods, its status, and its future. *Photogramm. Eng. Remote Sensing*, **53**(9), 1211–8.

Greeley, R. and Batson, R.M. (eds) (1990) *Planetary Mapping*. New York: Cambridge University Press.

Kopal, Z. and Carder, R.W. (1974) *Mapping of the Moon*. Dordrecht The Netherlands/Boston, USA.: D. Reidel.

Cross references

Coordinate systems
Map projection
Photoclinometry
Photogrammetry, radargrammetry and planetary topographic mapping

CASSINI, JEAN-DOMINIQUE (1625–1712) AND CASSINI'S LAWS

Born and raised near Nice (then in Italy) as Giovanni Domenico, Cassini became professor of astronomy at Bologna at the age of 25. In 1669 he was appointed to the new Paris Observatory by Louis XIV, and later assumed the French spelling of his name and French citizenship.

Between 1671 and 1674 Cassini was particularly interested in Saturn, discovering its second, third, fourth and fifth satellites [Iapetus (1671), Rhea (1672), Dione and Tethys (1684)]. In 1675 he observed the major gap ('Cassini's division') in Saturn's rings. The rings, he postulated, consisted of millions of small particles (satellites), an idea supported later on theoretical grounds by James Clerk Maxwell (in 1857), and spectroscopically by Keeler (in 1895). He derived the rotational periods of Jupiter, Mars and Venus, and tabulated the movement of the Jovian satellites discovered by Galileo. Later these results were used by Roemer to calculate the speed of light.

Cassini's laws

In 1693 Cassini published three laws describing his observations of lunar motion (Cassini, 1693). In modern terms, these laws are:

1. the Moon is in synchronous rotation with the Earth;
2. the spin axis of the Moon maintains a constant inclination of 1.5° to the ecliptic plane;
3. the lunar spin axis, the lunar orbit normal and the ecliptic normal remain coplanar at all times.

The first law results from tidal 'spin locking'. Tides raised on the body of the Moon by the Earth result in a torque which acts to alter the Moon's spin until it is in synchronous rotation with the Earth (for an elementary discussion see Munk and MacDonald, 1960). A large majority of the natural satellites in the solar system are in such a spin-locked state.

An explanation of the second and third laws is considerably more involved. Due to gravitational torques exerted by the Sun, the lunar orbit plane currently precesses with a period of 18.6 years. The lunar orbit normal thus sweeps out a cone, maintaining a constant inclination of 5.2° from the normal to the ecliptic plane. In satisfying Cassini's third law, the Moon's spin axis also precesses about the normal to the ecliptic plane with the same 18.6-year period. This latter precession is caused by torques exerted by the Earth on the non-spherical shape of the Moon.

Cassini's laws can be generalized to any spin-locked satellite undergoing orbital precession at a uniform rate. For the general case, the coprecession required by Cassini's third law is only possible for special values of the satellite's obliquity (angle between the spin axis and the orbit normal). These special obliquity values, which correspond to extrema of the system Hamiltonian, are known as 'Cassini states' (Colombo, 1966; Peale, 1969).

In general, a satellite will possess either four Cassini states, with two of them dynamically stable, or two Cassini states, with only one stable endpoint (Peale 1977). The Moon currently falls into the latter category and is observed to lie in its only stable state [although orbital evolution may have resulted in a very complicated history for the lunar spin axis (Ward, 1975)]. All spin-locked satellites observed to date have obliquities corresponding to a stable Cassini state.

Rhodes W. Fairbridge and Hans J. Haubold

Bibliography

Berry, A. (1898) *A Short History of Astronomy*. London: Murray. [also New York: Dover reprint, 1961.]

Cassini, G.D. (1693) *Traité de l'Origine de Progrès de l'Astronomie*. Paris.

Cohen, I.B. (1990) G.D. Cassini and the number of the planets: an example of 17th-century astro-numerological patronage, in *Nature, Experiment, and the Sciences: Essays on Galileo and the History of Science in Honor of Stillman Drake* (eds T.H. Levere and W.R. Shea). Dordrecht: Kluwer Academic.

Colombo, G. (1966) Cassini's second and third laws. *Astron. J.*, **71**, 891–6.

Marsden, B.G. and Cameron A.G.W. (1966) *The Earth–Moon System*. New York: Plenum, 288 pp.

Munk, W.H. and MacDonald, G.J.F. (1960) *The Rotation of the Earth*. London: Cambridge University Press.

Peale, S.J. (1969) Generalized Cassini's laws. *Astron. J.*, **74**, 483–9.

Peale, S.J. (1977) Rotational histories of the natural satellites, in *Planetary Satellites* (ed. J.A. Burns). Tucson: University of Arizona Press, pp. 87–112.

Taton, R. (1971) Cassini, Gian Domenico. *Dict. Sci. Biogr.*, Vol. 3, 100–104 (younger Cassinis: ibid. pp. 104–9).
Ward, W.R. (1975) Past orientation of the lunar spin axis. *Science*, **189**, 377–9.
Wolf, C. (1902) *Histoire de l'Observatoire de Paris*. Paris.

Cross reference

Saturn: ring system

CASSINI MISSION

Cassini is the name given to both an interplanetary spacecraft and the mission which it is designed to carry out. The mission and spacecraft are named in honor of the French/Italian astronomer Jean-Dominique Cassini (q.v.). In the period 1671–1685 he discovered four of Saturn's satellites as well as the large gap in the rings which is named after him. The Cassini spacecraft will be used for a detailed exploration of the planet Saturn and its rings, satellites and magnetosphere. It will also study the relationships which link the rings and the icy satellites, and the interaction of the magnetospheric plasma with the rings, the satellites and the ionosphere of Titan. Thus Cassini will study the whole system, which is more than the sum of the properties of each of the parts.

Scheduled for launch in October 1997, Cassini will take 7 years to reach Saturn. The path of its journey includes flybys of (and gravity assists from) Venus, Earth and Jupiter. In late 2004, a few weeks after the Cassini spacecraft has been placed in orbit around Saturn, an atmospheric probe, called Huygens (see Huygens Titan atmosphere probe), will be dropped into the atmosphere of Titan (q.v.), the largest of Saturn's satellites. Huygens will descend slowly, by parachute, through Titan's thick atmosphere until it reaches the surface. On the way down it will measure the properties of the atmosphere. The data acquired will be radioed to the main Cassini spacecraft which will, in turn, send the data back to the Earth. Cassini will then embark upon a 4 year orbital tour of the Saturn system, including a series of flybys of Titan and close encounters with several of the smaller, icy satellites. Instruments on board the spacecraft will obtain many images and detailed spectroscopic and other types of measurements in all parts of the Saturn system. These data will provide the basis for an in-depth study of this unique planetary system.

The Cassini mission is a joint undertaking by NASA (q.v.) and the European Space Agency (ESA) (q.v.). The Huygens probe is supplied by ESA while the Saturn Orbiter is provided by NASA. The scientific payloads on both the main spacecraft and the probe are provided by scientific groups supported by NASA and by the national funding agencies of the participating countries. The Italian Space Agency contributes to the instrumentation on both the probe and the orbiter and it also provides, through a bilateral agreement with NASA, part of the orbiter telecommunication subsystem, including the large high-gain antenna.

Cassini spacecraft

Cassini is a three-axis-stabilized spacecraft (Figure C7). The body of the spacecraft consists of six sections, all stacked on top of each other. From the bottom to the top they are the lower module, the propellant tanks and the engines, the upper equipment module, the twelve-bay electronics bus, and the high-gain antenna. Attached on one side is the Huygens probe and its support electronics. Science instruments are installed at various locations on the spacecraft. Most of them are on one of the two platforms attached to the side of the

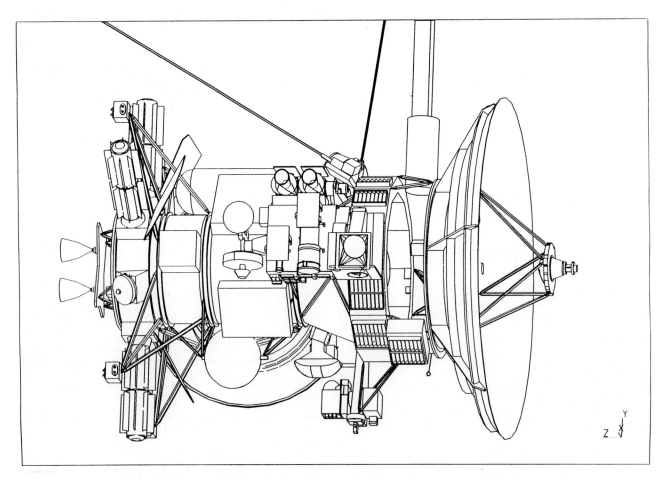

Figure C7 The Cassini spacecraft with the Huygens probe attached. The remote sensing instrument platform is in the foreground. The long, narrow, plasma wave antennas are shown beneath the magnetometer boom.

spacecraft. These are the remote sensing pallet and the particles and fields pallet. The magnetometer has two sensors. One is at the tip of an 8 m long boom. The other is located halfway out on the same boom. The magnetic and the electric antennae of the radio and plasma wave spectrometer are thin, rod-like structures, fixed to the body of the spacecraft. They extend away from the spacecraft as shown in Figure C7. Both the radar and the radio-science instruments use the high-gain antenna. At launch the mass of the fully fueled spacecraft will be about 5050 kg.

Electrical power for the Cassini spacecraft is supplied by three radioisotope thermoelectric generators. Cassini will communicate with the Deep Space Network (q.v.) of antennae on the Earth. The spacecraft will be able to receive X-band radio signals, using either the 4 m diameter high-gain antenna or one of the two low-gain antennae. It can also send X-band signals to the Earth. The attitude of the spacecraft can be adjusted by torquing the onboard reaction wheels, or by actuating a set of 0.5 N thrusters.

Since the instruments are fixed to the body of the spacecraft, the whole spacecraft must turn in order to point them. Thus the Cassini spacecraft will be turned frequently. The data obtained will usually be recorded on two state-of-the-art solid-state recorders. Each has a data storage capacity of about 2 Gbits (10^9 bits). Most of the scientific data will be acquired in two modes of operation. These modes are the remote sensing mode, and the fields and particles plus downlink mode. During remote sensing operations the spacecraft points to various targets in the Saturn system. The images, spectra, and other acquired data are stored on the recorder. Then, during the fields and particles plus downlink mode, the high-gain antenna is pointed at the Earth and the stored data are transmitted. While the spacecraft is in this mode it is set to rotate slowly around the axis of the high-gain antenna. In this way the fields and particles instruments can survey the sky. Other less frequently used modes are for the radar mapping of Titan, the measurement of occultations of Saturn, Titan and the rings, and for aeronometric measurement of Titan's upper atmosphere.

After the Huygens probe has deployed and the data has been stored, the Cassini spacecraft will be turned to point the cameras at Titan. Remote sensing observations will be made of Titan's atmosphere at and near the Huygens probe entry site. These global-scale remote sensing observations by Cassini will allow the local *in situ* observations obtained by Huygens to be placed in context. Also, the quasi-simultaneity of Huygens' and Cassini's observations will permit the probe data to provide a 'ground-truth' calibration for the orbiter's observations.

Payload and scientific objectives

Together, Cassini and Huygens carry 18 instruments which support some 28 different investigations. The tabulation (Table C1) identifies the instruments and lead investigators, and gives a brief indication of what will be measured or studied.

The exploration objectives of the Cassini mission encompass the whole of the Saturn system. The largest and most massive target is Saturn itself. Cassini's cameras and imaging spectrographs will take pictures of the details of the ever-changing cloud features. The records of these moving features, in turn, will help in characterizing Saturn's winds. Temperatures, cloud properties and atmospheric composition will be sensed remotely. The upper atmosphere and ionosphere will be explored by ultraviolet spectroscopy and radio sounding. These and other data will be used to infer the properties of the deeper atmosphere which lies beyond the reach of direct optical sensing. From these studies will emerge a better understanding of Saturn's present state and more accurate scenarios of how it was created as a planet and how it has subsequently evolved with time.

Special attention will be paid to the moon Titan, the largest of Saturn's satellites. With a diameter of 5 150 km, it is larger than the planets Mercury (4 878 km) and Pluto (2 300 km), but smaller than Mars (6 786 km). It has a thick atmosphere which obscures its surface, and is the only moon in the solar system with such an atmosphere. Cassini and the Huygens probe, together, will determine Titan's atmospheric composition and investigate the sources of energy which power chemical reactions in its atmosphere. Winds will be measured, global temperatures sensed and clouds and aerosols studied. The surface of Titan will be observed directly by Huygens as it nears its landing point. The surface will also be remotely sensed by Cassini using radar. Many of Cassini's close flybys will pass through the extreme upper atmosphere. During these passages, compositions will be measured directly by the ion and neutral mass spectrometer. On many occasions radio occultations will enable us to probe the ionosphere and deeper atmosphere. The Titan flybys will also allow an accurate determination of its gravitational field. The detailed shape of the gravitational field will give an indication of Titan's internal structure.

Many images of Saturn's famous ring system will be obtained. The configuration of the rings will be measured and the dynamical processes responsible for the ring structure will be inferred. The rings are thought to be composed mostly of water ice. Cassini's mapping spectrometers will chart the compositions of different rings. These are believed to vary due to different types of impurities in the ice. The size distributions of the ring particles will also be obtained. Interactions between the rings and the satellites, and between the magnetosphere and Saturn's ionosphere and atmosphere, will all be studied.

The icy satellites offer a number of new worlds to explore. Cassini will image and map their surfaces over the course of the orbital tour. The compositions and distributions of surface materials will be measured and observed. Special attention will be focused on defining the mechanisms of surface modifications whose action is evident in the contorted stratigraphy of satellites such as Enceladus. Other measurements obtained will further define the general characteristics of the satellites, placing constraints on bulk composition and internal structures.

Cassini will be within Saturn's magnetosphere during most of the mission. Here it will determine the configuration of the magnetic field, find the sources and sinks of magnetospheric charged particles, and investigate the interaction of the magnetosphere with the solar wind and with Titan's atmosphere and ionosphere.

Cassini mission

The plan calls for Cassini to be launched in October 1997 from Kennedy Space Center in Florida using staged Titan IV and Centaur rockets. The Cassini interplanetary trajectory includes gravity assists from Venus (in April 1998 and June 1999), Earth (in August 1999) and Jupiter (in December 2000). The October 1997 launch opportunity is the last in the present set of opportunities which permits Jupiter to be used for a gravitational assist (see Gravity-assist navigation). This allows Cassini to reach Saturn in about 7 years. Back-up launch opportunities (without a Jupiter flyby) exist in 1998 and 1999 but are not desirable because they take longer to fly to Saturn. This longer transit would result in less electrical power for the spacecraft when it reaches Saturn because the radioactive heat source which powers the thermal-electric generators becomes weaker with time.

The interplanetary trajectory is illustrated in Figure C8, together with several key dates. The arrival at Saturn is planned for June 2004. The most critical phase of the mission, after launch, is Saturn orbit insertion (SOI), which will take place in June 2004. Not only will it be a crucial maneuver, slowing the spacecraft and placing it in orbit about Saturn, but it will also be a period of unique science activity. The spacecraft will fly closer to the planet than at any other point in the mission and will provide the best close-up view of the rings.

After SOI Cassini will release the Huygens probe during the initial orbit around Saturn. This event occurs about 22 days before Titan encounter, as shown in Figure C9. Shortly after the probe release, Cassini will perform a deflection maneuver to avoid Titan and to be in a proper position for establishing the radio communication link with the probe during its descent. This maneuver also sets up the initial conditions for the start of the satellite tour which begins after completion of the probe mission.

Huygens' encounter with Titan will be on or about 11 November 2004. The celestial mechanics does not allow much adjustment for the arrival date at Saturn. However, the date on which Huygens encounters Titan is dictated by the duration of the initial orbit around Saturn, which is somewhat adjustable.

After the end of the probe mission phase, the Saturn Orbiter will begin a 4-year tour of the Saturn system. This tour is comprised of more than 50 Saturn-centered orbits, set up by either Titan gravity assist flybys or by propulsive maneuvers, or both. Within the constraints of the celestial mechanics, the choice of the size of these orbits, their orientation to the Sun–Saturn line, and their inclination to Saturn's equator are governed by the various scientific requirements on the mission. These requirements include Titan ground-track coverage, icy-satellite flybys, Saturn, Titan and ring occultations, orbit

Table C1 Cassini/Huygens mission science investigations[a]

Investigation/acronym	Scientist/affiliation	Brief objectives
Cassini Saturn Orbiter science investigations		
Cassini plasma spectrometer (CAPS)	D. Young (PI), Southwest Research Institute	*In situ* study of plasma within and near Saturn's magnetic field
Cosmic dust analyzer (CDA)	E. Grün (PI), Max Planck Inst für Kernphysik	*In situ* study of ice and dust grains in the Saturn system
Composite infrared spectrometer (CIRS)	V. Kunde (PI), NASA Goddard Space Flight Ctr	Spectral mapping to study temperature/composition of surfaces/atmospheres/rings within the Saturn system
Interdisciplinary scientist (IDS) – magnetosphere and plasma	M. Blanc (IDS), Obs. Midi-Pyrenees	Interdisciplinary study of plasma circulation and magnetosphere–ionosphere coupling
Interdisciplinary scientist (IDS) – rings and dust	J. Cuzzi (IDS), NASA Ames Research Ctr	Interdisciplinary study of rings and dust within the Saturn system
Interdisciplinary scientist (IDS) – magnetosphere and plasma	T. Gombosi (IDS), Univ. of Michigan	Interdisciplinary study of the plasma environment in Saturn's magnetosphere
Interdisciplinary scientist (IDS) – atmospheres	T. Owen (IDS), Inst. for Astronomy	Interdisciplinary study of the atmospheres of Titan and Saturn
Interdisciplinary scientist (IDS) – satellites	L. Soderblom (IDS), US Geological Survey	Interdisciplinary study of the satellites of Saturn
Interdisciplinary scientist (IDS) – aeronomy and solar wind interaction	D. Strobel (IDS), Johns Hopkins Univ.	Interdisciplinary study of aeronomy in the Titan and Saturn atmospheres
Ion and neutral mass spectrometer (INMS)	H. Waite (TL), Southwest Research Institute	*In situ* compositions of neutral and charged particles within the Saturn magnetosphere
Imaging science subsystem (ISS)	C. Porco (TL), Univ. of Arizona	Multispectral imaging of Saturn, Titan, rings and the icy satellites to observe their properties
Dual technique magnetometer (MAG)	D. Southwood (PI), Imperial College	Study Saturn's magnetic field and interactions with the solar wind
Magnetospheric imaging instrument (MIMI)	S. Krimigis (PI), Applied Physics Lab.	Global magnetospheric imaging and *in situ* measurements of Saturn's magnetosphere and solar wind interactions
Cassini radar (RADAR)	C. Elachi (TL), Jet Propulsion Lab.	Radar imaging, altimetry, radiometry and backscatter of Titan's surface
Radio and plasma wave science (RPWS)	D. Gurnett (PI), Univ. of Iowa	Study plasma waves, radio emissions and dust in the Saturn system
Radio science subsystem (RSS)	A Kliore (TL), Jet Propulsion Lab.	Study atmospheres and ionospheres of Saturn and Titan, rings and gravity fields of Saturn and its satellites
Ultraviolet imaging spectrograph (UVIS)	L. Esposito (PI), Univ. of Colorado	Spectra and low-resolution imaging of atmosphere and rings for structure, chemistry and composition
Visual and infrared mapping spectrometer (VIMS)	R. Brown (TL), Jet Propulsion Lab.	Spectral mapping to study composition and structure of surfaces, atmospheres and rings
Huygens Titan probe science investigations		
Aerosol collector pyrolyser (ACP)	G. Israel (PI), Cnrs, Service d'Aéronomie	*In situ* study of clouds and aerosols in the Titan atmosphere
Descent imager and spectral radiometer (DISR)	M. Tomasko (PI), Univ. of Arizona	Temperatures and images of Titan's atmospheric aerosols and surface
Doppler wind experiment (DWE)	M. Bird (PI), Univ. Bonn	Study of winds from their effect on the probe during Titan descent
Gas chromatograph and mass spectrometer (GCMS)	H. Niemann (PI), NASA Goddard Space Flight Ctr	*In situ* measurement of chemical composition of gases and aerosols in Titan's atmosphere
Huygens atmospheric structure instrument (HASI)	M. Fulchignoni (PI), Observatoire de Paris-Meudon	*In situ* study of Titan atmospheric physical and electrical properties
Interdisciplinary scientist (IDS) – Titan aeronomy	D. Gautier (IDS), Obs. de Paris-Meudon	Interdisciplinary study of the aeronomy of Titan's atmosphere
Interdisciplinary scientist (IDS) – Titan atmosphere–surface interactions	J. Lunine (IDS), Univ. of Arizona	Interdisciplinary study of Titan atmosphere–surface interactions
Interdisciplinary scientist (IDS) – Titan organic chemistry	F. Raulin (IDS), Univ. Paris Val de Marne	Interdisciplinary study of Titan's chemistry and exobiology
Surface science package (SSP)	J. Zarnecki (PI), Univ. of Kent	Measurement of the physical properties of Titan's surface

[a] IDS = Interdisciplinary scientist; no instrumentation is provided by an IDS, but data from several PI or TL investigations will be used.
PI = Principal investigatory; each PI proposed the team and is responsible for providing the instrumentation for the investigation.
TL = Team Leader; each TL utilizes a facilities instrument provided as part of the spacecraft systems; team members were individually selected by NASA.

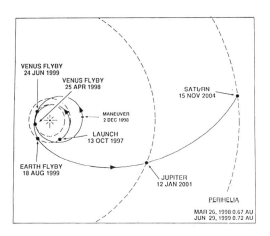

Figure C8 The Cassini interplanetary trajectory. Gravitational assists occur at Venus, Earth and Jupiter.

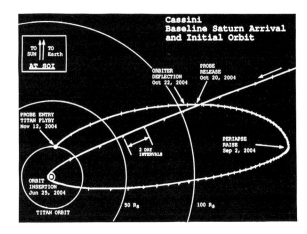

Figure C10 Cassini's orbital tour. This schematic diagram depicts the first 36 orbits of the 4-year tour. These orbits lie primarily in the equatorial plane of Saturn and involve close flybys of many of Saturn's icy satellites. The later orbits will place the spacecraft in a highly inclined orbit about Saturn's pole and provide excellent viewing of Saturn and its magnetosphere at high latitudes.

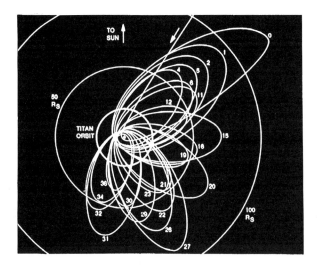

Figure C9 The arrival of Cassini at Saturn and the initial orbit. The spacecraft performs an orbit insertion burn on 25 June 2004 to slow Cassini, put it into orbit about Saturn and start a 4-year orbital mission. A periapsis raise maneuver is performed on 10 September 2004 to increase the minimum orbital distance and avoid Saturn's rings. The Huygens probe is released on 6 November 2004 and Cassini performs a deflection maneuver on 8 November 2004 so that it will not fly into Titan.

inclinations and ring-plane crossings. An example of one possible tour is illustrated in Figure C10.

Operations

The operation of the Cassini spacecraft and the mission flight activities will be carried out at the Jet Propulsion Laboratory in Pasadena, California. The 'uplink' operations organization produces spacecraft commands and transmits them to the spacecraft. The data downlinked from the spacecraft will be processed to assess spacecraft health and status. Data from the scientific instruments will be sent to the respective instrument teams for scientific analysis. The Cassini scientists will be interacting on a daily basis with the spacecraft operations teams at JPL. This will extend from the start of the Saturn encounter phase until the end of the orbital tour.

A Probe Operations Center (POC) will be established at the European Space Operations Center (ESOC) in Darmstadt, Germany. The Huygens-related mission activities (periodic checkouts and Probe mission phase), will be carried out at the POC. The Huygens telemetry data will be routed to the POC where they will be processed, distributed to the investigators' computers, displayed and archived. The Huygens post-flight products will be prepared and distributed by the POC.

Dennis L. Matson

Bibliography

Calcutt, S., Taylor, F., Ade, P. et al. (1992) The composite infrared spectrometer. *J. Br. Interplanet. Soc.*, **45**(9), 811–6.
Coates, A.J., Alsop, C., Coker, D.R. et al. (1992) The electron spectrometer for the Cassini spacecraft. *J. Br. Interplanet. Soc.*, **45**(9), 387–92.
Kolhase, C. (1993) Meeting with a majestic giant: the Cassini mission to Saturn. *Planet. Rep.*, **13**(4), 4–11.
Lebreton, J.-P. (1992) The Huygens probe, in *Proc. Symp. on Titan*, ESA SP-338, pp. 287–92.
Lebreton, J.-P. and Matson, D.L. (1990) Cassini – a mission to Saturn and Titan. *Proc. 24th ESLAB Symp. on the Formation of Stars and Planets, and the Evolution of the Solar System.* ESA SP-315.
Lebreton, J.-P. and Matson, D.L. (1992) An overview of the Cassini Mission. *Nuovo Cimento*, **15C**(6), 1137–47.
Matson, D.L. (1992) Cassini — a mission to Saturn and Titan, in *Proc. Symp. on Titan*, ESA SP-338, pp. 281–6.
Murray, C.D. (1992) The Cassini imaging science experiment. *J. Br. Interplanet. Soc.*, **45**(9), 359–64.
Ratcliff, P.R., McDonnell, J.A.M. Firth, J.G. and Gruen, E. (1992) The cosmic dust analyser. *J. Br. Interplanet. Soc.*, **45**(9), 375–80.
Sandford, M.C.W. (1992) The Cassini/Huygens mission and the scientific involvement of the United Kingdom. *J. Br. Interplanet. Soc.* **45**(9), 355–8.
Southwood, D.J., Blogh A. and Smith, E.J. (1992) Dual technique magnetometer experiment for the Cassini Orbiter spacecraft. *J. Br. Interplanet. Soc.*, **45**(9), 371–4.
Zarneki, J.C. (1992) Surface science package for the Huygens Titan probe. *J. Br. Interplanet. Soc.*, **45**(9), 365–70.

CELESTIAL MECHANICS

In contemporary parlance, celestial mechanics is a subfield of the more general field of dynamical astronomy, which may be defined as the study of the motions of all possible objects, or collections of objects, in the astronomical universe under the action of defined physical forces. On the macroscopic scale this would include the motions of clusters of galaxies, galaxies within clusters, objects such as stars within galaxies, star clusters, HI and HII regions, and dust clouds within galaxies, under the influence of gravitation and possibly other force fields. Historically, celestial mechanics has encompassed the motions of objects in the solar system (planets, asteroids, natural satellites, ring systems and comets), and questions of nongravitational forces. Celestial mechanics is employed to address a number of general problems, such as the stability of individual motions, the stability of systems, the origins of comets and the origin of the solar system. Celestial mechanics is also employed in the study of double star and other small, n-body systems. With the advent of the space age, which may be said to have begun in 1957 with the launching of the first Sputnik of the USSR, there was a renaissance in the field of celestial mechanics. The exigencies of the space age introduced many new, unforeseen problems, and led to further work in analytical and numerical planetary and satellite theories (to achieve the new orders of magnitude of precision required by new optical technologies and spacecraft exploration).

History

Celestial mechanics had its kinematic beginnings in attempts to understand the apparent motions of the Sun, Moon and planets by the ancient Greeks as far back as Pythagoras ($c.$ 500 BC). These efforts culminated in Ptolemy's ingenious, geocentric cosmology described in the 13-volume *Almagest* ($c.$ AD 140). In this system (see Figure C11) the Sun moves in an off-center, circular orbit around the Earth. The inferior planets, Mercury and Venus, move in circular orbits, called epicycles, whose centers are always on the line Earth–Sun, while the superior planets, Mars, Jupiter and Saturn, move in epicycles whose centers move on larger circular orbits called deferents, outside the Sun's orbit. The epicyclic radius vectors of the superior planets are parallel to the Earth–Sun radius vector. The Moon's complicated motion is explained by having it revolve in an epicycle with a period of 1 month, while the epicycle's center revolves in its deferent about the Earth, inside the Sun's orbit, also in 1 month. Tycho Brahe (ca. 1600) briefly improved this system by having all the planets revolve in circular orbits around the Sun, which in turn revolved in a circular, off-center orbit around the Earth. Ptolemy's system permitted the inferior planets, Mercury and Venus, to exhibit only crescent phases to an observer on Earth. This was exposed as a fatal flaw in this system when Galileo first observed the gibbous phase of Venus $c.$ 1610 through a telescope.

An important, revolutionary step forward was taken ($c.$ 1543) in the history of celestial mechanics when Copernicus moved the Earth from its central position in the universe into an orbit around the Sun, thus making the Sun the center of the solar system and producing a new heliocentric universe. This model was still based on circular deferents and epicycles and, while philosophically far more satisfying than the Ptolemaic system, did not improve its accuracy.

Tycho (q.v.) compiled a long series of observations of Mars, which were later employed in an extended and painstaking analysis by Kepler (q.v.), who then provided the foundation for the beginnings of modern celestial mechanics by establishing his three important empirical laws of planetary motion (1609–1619):

1. the orbit of each planet is an ellipse with the Sun located at one focus;
2. the radius vector of each planet sweeps out equal areas in equal times;
3. the squares of the periods of any two planets are to each other as the cubes of their semimajor axes.

In the *Principia Mathematica*, published in 1687, Newton (q.v.) stated his three well-known laws of motion. From Kepler's second law of areas, he deduced that the force between the Sun and a planet is a central force and, because the orbit is concave toward the Sun, it must also be attractive. From Kepler's first law he concluded that the acting force obeys an inverse square law. Kepler's third law may be written

$$\tau^2 = a^3$$

if the unit of time is the sidereal year and the astronomical unit is the unit of distance. Newton showed that the period of a planet, τ, is given more precisely by the expression in Figure C13 which contains the masses of the Sun and planet. When the period of any planet is compared with that of the Earth, in the same units as above,

$$\tau^2 = \frac{m_\odot}{m_\odot + m} a^3$$

where the mass of the Earth–Moon system $m_{(E+M)} = 0.0000030 \approx 0$, in units of the Sun's mass $m_\odot = 1$. For Jupiter, the most massive planet in the solar system, $m = 0.00095$, and the inclusion of the mass terms decreases the period calculated from Kepler's third law by 2.06 days in a period of 11.86 years, an error of 0.05%. Thus Kepler's third law is an excellent approximation. With the more accurate statement of this law, Newton concluded that the force between two objects is proportional to the product of their masses. Combining all of the above results, Newton was able to formulate the law of universal gravitation, thus establishing the principal foundation of celestial mechanics.

Two-body problem and orbit determination

A typical development of the subject matter begins with an investigation of motions taking place in a general, central force field and is followed by a thorough study of the most important elementary problem on which all others depend: the two-body problem. This naturally leads to orbit determination, the calculation of six constants of integration, usually the elements of the orbit. Figures C12 and C13 illustrate one of the simplest sets of elements and related quantities. Two early, general solutions for the orbit of a solar system object, given three observations of right ascension and declination and the time are (1) the Laplace method, which conceptually is a solution of the differential equations of motion, and (2) the Gauss method, which begins with the assumption that the orbit is a conic section. Laplace's method provides the best fit of the radius vector and velocity vector for the middle observation while Gauss' method provides the best fit of a conic passing through the three observational points. A multitude of other methods are based either on the Laplace method or the Gauss method, or some combination of both (Danby, 1988, p. 213). Gauss introduced his method in 1801 when he was 23 years old and applied it to the first asteroid, Ceres, discovered by Piazzi on January 1. Because of illness, Piazzi lost track of Ceres which was subsequently recovered as a result of Gauss' calculations.

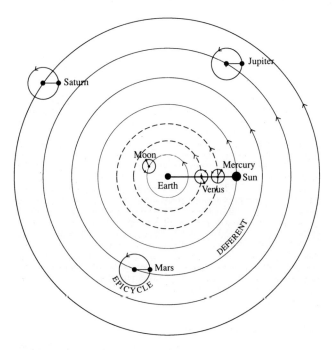

Figure C11 The Ptolemaic system.

CELESTIAL MECHANICS

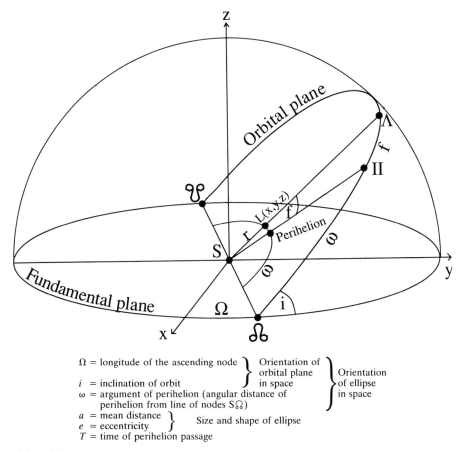

Ω = longitude of the ascending node } Orientation of orbital plane in space } Orientation of ellipse in space
i = inclination of orbit
ω = argument of perihelion (angular distance of perihelion from line of nodes S☊)
a = mean distance } Size and shape of ellipse
e = eccentricity
T = time of perihelion passage

Figure C12 Elements of the orbit.

Early applications treated solar system objects as though they were particles or point masses. This reasonable assumption follows because (1) the planets and the Sun have extremely small dimensions compared to their distances apart and (2) objects with spherical symmetry (true to a first approximation in the solar system) attract each other as though they were particles, a fact which Newton established in the *Principia*. When an object is not spherical, for example as is the case for Jupiter, and one considers the motion of a nearby satellite, such as J5 (Amalthea), it becomes necessary to determine the gravitational potential of the object, a measure of the force field. This, in turn, requires that assumptions be made regarding the distribution of density within the object. In lunar theory the non-sphericity of the Earth gives rise to a very small effect on account of the Moon's distance. In the case of near-Earth satellites, however, it is necessary to determine the Earth's potential to a high degree of accuracy (see Geoid).

Classical methods (before the 1950s) employed three telescopic observations, yielding a set of two angles (such as right ascension and declination) and the time for each observation. Contemporary observations use measurements in multiple ranges of the electromagnetic spectrum. In particular these include radar and Doppler radar (range and range-rate), telemetry and interferometry in radio, and image intensification and lasers in the optical range. Thus, modern orbit determination deals with a mix of observations which may be optimally combined to yield the elements. When more than three observations are available, least square methods are applied and, as additional observations accrue, differential corrections of the elements are made using least squares again, thus producing an improved orbit.

Perturbation theories

Of great interest in the further development of the subject are the celebrated three-body problem, Lagrange's elegant solutions of the restricted problem, the n-body problem and a discussion of integrability. The groundwork is thus laid for the exposition of perturbation theories, which consist of two fundamental types: general perturbation theory and special perturbation theory. The schematic in Figure C14 illustrates the broad categories.

General perturbation theory

General perturbation theory is an analytical treatment of a dynamical system, and may be further subdivided into two general kinds. In the method of variation of the elements (often called 'arbitrary constants'), a particular set of orbital elements is expressed as a function of the time. For example, using the elements described in Figure C12, in a first-order theory there would be expressions for da/dt, de/dt, di/dt, $d\omega/dt$, $d\Omega/dt$ and $d\sigma/dt$ (where $\sigma = nT$). At a given time t the variations of the elements may be determined and added to the elements of a reference ellipse to yield the elements at time t. Examples of perturbations in rectangular coordinates would be expressions for dx/dt, dy/dt, dz/dt, the choice made by Brouwer in his planetary theory [Brouwer and Clemence, 1961, p. 376]. Another set of perturbations in the coordinates would be expressions for dr/dt, dM/dt and $d\beta/dt$, the first two representing perturbations in the radius vector and the mean anomaly in the plane of the orbit. The third gives the perturbations in the inclination of the orbit, or the latitude (Brouwer and Clemence, 1961, p. 416). These perturbations were first used by Hansen in his lunar theory and later applied to planetary theory by others. In all of these theories a reference orbit, an auxiliary orbit, a mean orbit or an osculating orbit (in the case of a first-order theory) is utilized. General perturbation theories are useful in studying the intrinsic nature of the motion and its long-period behavior.

Special perturbation theory

Special Perturbation theory is concerned with numerical methods for obtaining solutions to the differential equations of motion of a given

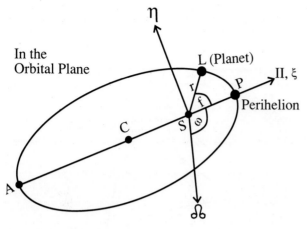

a = CP ≡ mean distance ≡ semimajor axis
e = CS/CP ≡ eccentricity
(T ≡ time of perihelion passage)
$\xi = r \cos f$ } rect. coordinates of planet
$\eta = r \sin f$ } in its orbital plane
 r ≡ radius vector
 f ≡ true anomaly
$L(x,y,z)$ ≡ rectangular coordinates of planet with respect to fundamental plane
$n = 2\pi/\tau$ ≡ mean motion
 where τ ≡ period
 $= \dfrac{2\pi a^{3/2}}{k(m_\odot + m)^{1/2}}$
[a] = astronomical units (AU)
k = Gaussian constant
 = 0.0172029895
$m_\odot$ = mass of Sun = 1
m = mass of planet

Figure C13 Relations and elements in the orbital plane.

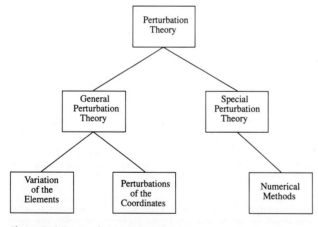

Figure C14 Types of pertubation theory.

dynamical system, or for finding solutions to any given analytical theories. These encompass many schemes for numerical integration of differential equations. In addition, there also exist hybrid solutions combining various analytical and numerical perturbation methods. Numerical methods are of most frequent use in the computations of trajectories of spacecraft and other complicated systems in the solar system, because all kinds of non-gravitational forces, such as air resistance for artificial satellites around the Earth, or solar radiation pressure, where it is sensible for solar spacecraft, can be incorporated in the equations.

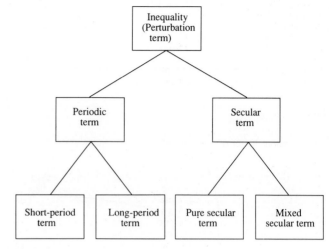

Figure C15 Types of perturbation terms.

Three types of perturbation

The solution of a general perturbation theory gives rise to three types of perturbations: periodic, secular and mixed secular. Individual terms of the solution are called inequalities; Figure C15 shows a schematic of these types. As an example, the first order solution for the motion of the node $d\Omega/dt$, in the simplest case where there is one disturbing body, has the form:

$$\Omega = \Omega_0 + At + \Sigma \left(B_p/(p_1 n_1 + p_2 n_2)\right) \sin\left[(p_1 n_1 + p_2 n_2)t + C\right]$$

where B_p is a function of various elements and the disturbing mass, C depends on several elements, p_1 and p_2 are integers and n_1 and n_2 are the mean motions of the perturbed and perturbing objects.

Short-period and long-period inequalities

Periodic inequalities are cyclic, trigonometric functions which fall into two classes: short-period inequalities, if there is no near-commensurability in the mean motions contained in the argument of the trigonometric function and, otherwise, long-period inequalities. In periodic inequalities the coefficient contains a denominator that is a linear function of the mean motions, such as $(p_1 n_1 + p_2 n_2)$ above. In long-period inequalities this denominator is small and hence has an important influence on the motion over a long interval of time. An example is the so-called great inequality in the motion of Jupiter and Saturn around the Sun in which $2n_J - 5n_S = 0.0011$ degrees per day ($\tau_J = 11.86$ years, $\tau_S = 29.46$ years, the sidereal periods of Jupiter and Saturn). This results in an inequality which causes the longitude of Saturn to change by 50′ over a period of 900 years, a puzzle to astronomers until Laplace discovered its cause.

Secular perturbations

Secular perturbations are inequalities that have the time as a factor. A pure secular term is a constant times some power of the time and, in the first order motion of the node above, is the term At. As t increases, the node regresses along the fundamental plane. In the case of the Moon the node regresses along the ecliptic with a period of 18 years, while the secular term in the motion of perigee causes it to advance with a period of 9 years. A mixed secular term is a periodic inequality which has a power of the time as a factor. When secular, or mixed secular, terms appear in the solutions of other elements, such as a, it places a limit on the usefulness of the theory beyond some point in time and raises questions about the stability of motion over a very long period of time. In the late 18th century Laplace and Lagrange showed that the first-order perturbations in the semi-major axes do not contain any secular terms. Poisson and others extended these results, and in 1976 Message proved that it was possible to derive explicit expressions for the orbital elements with no secular terms present at all, thus establishing the long-term stability of a solar system model in which gravitation is the only recognized force (Message, 1984).

One of the greatest of the early triumphs of celestial mechanics was the independent work of J.C. Adams in England and U.J.J. Leverrier in France leading to the discovery of the planet Neptune in 1846. They correctly assumed that Uranus was not in its predicted orbit because of perturbations arising from the gravitational attraction of an unknown planet. It was on the basis of those perturbations that they determined the elements of this planet and predicted its position, leading to the discovery of the planet now known as Neptune.

Lunar theory

An historically important problem in celestial mechanics has been the lunar theory. Some of the Moon's perturbations were known to ancient astronomers. The largest inequality in the Moon's motion is called the evection and its value of approximately 1° 15′ was already known to Hipparchus in the second century BC. It is a displacement of the Moon's position in geocentric longitude. In 1590 Tycho Brahe discovered two important lunar inequalities: the variation and the annual equation. The variation has a value of 40′ in a period of half a synodic month and is the difference between the true and mean position of the Moon. Its existence was explained by Newton. The annual equation has a value of 11′ with a period of 1 year and affects the length of the month in different parts of the Earth's orbit around the Sun. When Halley compared ancient eclipses with those of his time, he found that the mean motion of the Moon was slowly decreasing. In 1787, 100 years later, Laplace explained this secular acceleration in the motion of the Moon amounting to 6″ per century.

The lunar theory has played a major role in the development of celestial mechanics because it is a challenging problem of great complexity. Among the distinguished astronomers and mathematicians who worked on this problem were Newton, Laplace, de Pontécoulant, Hansen, Delaunay, Hill and Brown. Looked at simply, the Moon may be thought of as moving in an elliptical orbit about the Earth, which in turn revolves in an invariant ellipse about the Sun. If the Sun is considered the only disturbing influence, this formulation is called the main problem of the lunar theory. In the further development of this theory, direct and indirect planetary perturbations would be added, as well as the effects produced by the oblateness of the Earth. Direct planetary perturbations are produced by the attraction of relevant planets pulling the Moon out of its orbit about the Earth while indirect perturbations are the result of those planets pulling the Earth out of its (former invariant) ellipse, thus indirectly affecting the motion of the Moon. Of these the strongest perturbations are the solar perturbations calculated in the main problem, followed by the weaker indirect planetary perturbations which are in turn stronger than the direct perturbations. Hansen separated the perturbations in latitude from those in the orbit (radius vector and mean anomaly) by considering them as departures from an auxiliary ellipse contained in the osculating plane of the actual motion (Brouwer and Clemence, 1961, p. 416).

Artificial satellites

The beginning of the space age in 1957 immediately required solutions to hitherto unexplored problems, such as precision orbit determination for artificial satellites. An early approach to the solution of this problem was given by Brouwer (1946). He remarked that up till then the problem was largely ignored but that in the solar system it was useful in the motion of Jupiter's very close fifth satellite because of Jupiter's considerable oblateness of 0.062. In 1958, 1959 and 1960 the problem of artificial satellites was directly addressed by Brouwer, Garfinkel, Kozai and others.

The first-order solution by the von Zeipel method (Kovalevsky, 1967, p. 69) shows that there is a regression of the nodes which has a zero value in polar orbits and is maximum in equatorial orbits. Furthermore, it reveals that in the motion of perigee there exists a critical inclination i_o equal to approximately 63°, below which the perigee advances and above which it retrogresses. The importance of this result arises from the fact that the expression

$$(1 - 5\cos^2 i_o)$$

appears in the denominator of long-period terms in the motion of perigee, causing them to become infinite as the inclination approaches its critical value (Kovalevsky, 1967, p. 91). It thus becomes necessary to use a different formulation of the problem to avoid this embarrassing difficulty. Further study brings out the fact that under some conditions the perigee will oscillate about a mean position with a given amplitude and period, an example of a very common phenomenon found in the solar system called libration.

Librations

The simplest libration is the swinging pendulum of a clock. In the solar system librations come in a variety of species when the periods of long-period terms grow without bound as the result of particular initial conditions, and otherwise are symptomatic of fundamental resonances arising from the commensurability of arguments in the perturbations which lead to small divisors. Attention was brought to the subject of librations in Lagrange's solution of the restricted problem of three bodies, which consists of five stationary points in the rotating coordinate system. Two equilateral solutions, L_4 and L_5, are located at the apexes equidistant from the two finite masses. In the collinear solutions, L_1, L_2 and L_3, a third body of infinitesimal mass occupies a position between the two masses or on either side of the other two. A slight displacement of the point mass from an equilateral solution will cause it to librate in a variety of ways but in stable equilibrium to the first order. Motion about the collinear solutions are generally unstable.

The best-known example of these librations is the group of some two dozen Trojan asteroids in the three-body problem Sun–Jupiter–asteroid displaying a range of librations including so-called horseshoe orbits about L_4 and L_5, 60° ahead of and 60° behind Jupiter (and equidistant from Jupiter) and the Sun. No longer held is the view promulgated in the early years of this century that the gegenschein, a faint glow in the ecliptic in opposition to the Sun, is a collection of planetary dust at the opposition collinear point in the three-body problem Sun–Earth–particle (see Zodiacal light). In recent years it was recommended by some scientists that a space station be established at L_5.

Another well-known libration is the physical libration of the Moon, which is a triaxial ellipsoid. If the Moon were in a circular orbit, its longest equatorial axis would point towards the Earth, but because of the Moon's variable motion in its Keplerian ellipse, a libration ensues which manifests itself as an oscillation in longitude. This is not to be confused with the libration in longitude, an optical effect which is the consequence of the Moon's variable Keplerian motion but constant rotation. Both effects together permit more than 50% of the lunar surface to be seen in the course of its revolution about the Earth.

Modern developments in celestial mechanics

The robust progress of celestial mechanics in the second half of the 20th century is directly attributable to the development and application of high-speed computers, the demands of the new space age, new observational techniques of high accuracy and new mathematical techniques and disciplines. Celestial mechanicians were among the first to recognize that computers could be used to develop literal, mathematical manipulators to mechanize the complex, algebraic equations and substitutions which make up so large a part of general perturbation theory. Already in the works of Poincaré near the beginning of this century are the glimmerings of the new mathematical field of chaos born 50 years later. There he recognized that small changes in initial conditions may give rise to such large changes in the final results that prediction may become impossible. In 1964 Hénon applied some of the new methods to the study of motion in a globular cluster (Hénon and Heiles, 1964) and in 1976 he devised strange attractors (Hénon, 1976).

In addition to work involving chaos and attractors, a great deal of current research in the mathematical aspects of celestial mechanics is devoted to stability of motions, asymptotic behavior, periodicity, quasi-periodicity, ergodicity, and the structure of n-dimensional phase space. With regard to computer technology, advances are being made in the development of numerical and analytical algorithms, analysis of accuracy and accumulation of error in numerical integration. Using numerical techniques, greater attention is being paid to non-gravitational forces such as drag, radiation pressure, etc.

Other areas which are currently under intensive study are the dynamics of ring systems, the problem of shepherding satellites, resonances in the asteroid belt (known as the Kirkwood gaps) and Saturn's ring system, as well as the ring systems of the other Jovian planets. Of great interest nowadays is the progress in the origin and evolution of the solar system and the effect tidal forces play. Stability of the system remains a central problem.

Not until the space age arrived was much attention paid to the more precise foundation of celestial mechanics attainable on the basis of Einstein's general theory of relativity. The importance of this has finally been recognized with the appearance of textbooks in relativistic celestial mechanics (Brumberg, 1991) and astrometry (Murray, 1983; Kovalevsky, 1990) and the attention which national observatories have paid to more precise definitions of reference frames and timescales.

Unsolved problems in celestial mechanics

Among the interesting, still not completely solved, problems under consideration today are the unusual rotations of Mercury and Venus. Why are there still unexplained residuals in the motion of Uranus? What is the origin of our own Moon? What are the origins of the satellites of Neptune, and what of their possible connection with Pluto? What are the origins of the asteroids and of Oort's comet cloud? This is only a small part of the list of outstanding problems of celestial mechanics (Seidelmann, 1986; also Brumberg and Kovalevsky, 1986). The space age has opened a new Pandora's box of unanswered questions to provide a challenge for the celestial mechanicians of the 21st century.

Morris S. Davis

Bibliography

Brouwer, D. (1946) The motion of a particle of negligible mass under the attraction of a spheroid. *Astron.*, **51**, 223.
Brouwer, D. and Clemence G.M. (1961) *Methods of Celestial Mechanics*. New York: Academic Press.
Brown, E.W. (1896) *An Introductory Treatise on the Lunar Theory*. London: Cambridge University Press.
Brown, E.W. and Shook, C.A. (1933) *Planetary Theory*. London: Cambridge University Press.
Brumberg, V.A. (1991) *Essential Relativistic Celestial Mechanics*. New York: Adam Hilger.
Brumberg, V.A. and Kovalevsky, J. (1986) Unsolved, problems of celestial, mechanics. *Celest. Mech.*, **39**(2), 133.
Danby, J.M.A. (1988) *Fundamentals of Celestial Mechanics*. Richmond, VA: Willmann-Bell, Inc.
Fitzpatrick,. P M. (1970) *Principles of Celestial Mechanics*. New York: Academic Press.
Gleick, J. (1987) *Chaos*. New York: Penguin Books.
Hagihara, Y. (1957) *Stability in Celestial Mechanics*. Tokyo: Kasai Publishing and Printing Co.
Hagihara, Y. (1975) *Celestial Mechanics* (4 volumes). Tokyo: Japan Society for the Promotion of Science.
Hénon, M. (1976) A two-dimensional mapping with a strange attraction, *Commun. Math. Phys.*, **50**, 69.
Hénon, M. and C. Heiles (1964) The Application of the Third Integral: Some Numerical Experiments, *Astron. J.*, **60**, 73.
Herget, P. (1948) *The Computation of Orbits* (Published privately by the author).
Kovalevsky, J. (1967) *Introduction to Celestial Mechanics*. New York: Springer-Verlag.
Kovalevsky, J. (1990) Astrométrie Moderne, in *Lecture Notes in Physics*, **358**. Berlin: Springer-Verlag.
McCuskey, S.W. (1963) *Introduction to Celestial Mechanics*. Reading, Massachusetts: Addison-Wesley.
Message, P.J. (1984) The stability, of our solar system. *Celest. Mech.*, **34**(1–4), 155.
Moulton, F.R. (1914) *An Introduction to Celestial Mechanics*. New York: The Macmillan Company.
Murray, C.A. (1983) *Vectorial Astrometry*. Bristol: Adam Hilger Ltd.
Pollard, H. (1966) *Mathematical Introduction to Celestial Mechanics*. Englewood Cliffs, NJ: Prentice-Hall, Inc.
Roy, A.E. (1982) *Orbital Motion*. Bristol, England: Adam Hilger Ltd.
Seidelmann, P.K. 1986. Unsolved problems, of celestial mechanics – the solar system, *Cel. Mech.*, **39**(2), 141.
Smart, W.M. (1953) *Celestial Mechanics*. London: Longmans, Green and Co.
Szebehely, V. (1967) *Theory of Orbits: The Restricted Problem of Three Bodies*. New York: Academic Press.

An invaluable resource for current research in celestial mechanics is the journal *Celestial Mechanics* published by D.Reidel Publishing Co. in Dordrecht, Holland, beginning with Volume 1 in 1969. In 1989, the title changed to *Celestial Mechanics and Dynamical Astronomy*. The journal has been published since 1987 by Kluwer Academic Publishers of Dordrecht, New York and London.

Cross references

Chaotic dynamics in the solar system
Earth–Moon system: dynamics
Galileo Galilei
Kepler, Johannes
Kepler's laws
Lagrange, Louis de
Laplace, Pierre Simon de
Newton, Isaac
Newton's laws
Orbit
Planetary dynamical astronomy
Resonance in the solar system
Solar system: stability

CENTER OF MASS, GRAVITY AND INERTIA

In classical mechanics the center of mass of a system of bodies or discrete particles is an imaginary point where the total mass of the system may be thought to be concentrated, for the purpose of dealing with mechanical problems of the system. For a system of particles it is defined as a point determined by a position vector R with respect to an inertial reference frame:

$$R = \frac{\sum_i m_i r_i}{\sum_i m_i} = \frac{\sum_i m_i r_i}{M} \qquad (C1)$$

where m_i is mass of the ith particle of the system, r_i is its position vector and M is the total mass of all the particles of the system. If a body of continuous mass is involved, the above summations should be replaced by integrations over the body. For a simple continuous body of homogeneous material, the center of mass coincides with its centroid (geometrically symmetric center), e.g. the center of mass for a homogeneous sphere is at its geometric center.

Newton's three laws essentially apply to mass points or particles. When Newton applied his laws to extended bodies, he employed the concept of a 'center of motion' or occasionally 'center of gravity'; the latter actually coincides with the center of mass (Herivel, 1965). The introduction of the concept of the center of mass greatly simplifies treatments of mechanical problems for a system of discrete particles or continuous bodies (or both). With the definition (C1), the motion of a system of particles or bodies can always be conveniently separated into the translational motion of its center of mass and the rotational motion with respect to the center. The center of mass will move as if the total external forces were acting on the entire mass of the system concentrated at the center of mass and the time rate of change of the total moment of momentum about the center of mass is equal to the total torque of external forces about the center of mass. Purely internal forces within the system, as long as they obey Newton's third law, have no effect on the motion of the center of mass and the rotational motion about it. For example, an exploding shell moves in such a way that the center of mass of the fragments travels as if the shell were still in a single piece (neglecting air resistance). In the same fashion, the Earth–Moon system moves around the Sun as if the total mass of the Earth and the Moon were concentrated at their center of mass, i.e. Earth–Moon barycenter (q.v.), with the Sun's gravitational field acting solely at the center.

When particles or bodies of a system are acted on by parallel gravitational forces, all the acting parallel forces can then be replaced by a single resultant force at the center of mass. The center of mass in this particular case is generally named the center of gravity as first introduced by Archimedes when he dealt with the lever problem in the 3rd century BC (Mach, 1960). He defined the center of gravity as the point where the weight of a body (or bodies) can be conceived as being concentrated regardless of the position(s) of the body (bodies).

If the bodies or particles of a system are given a linear acceleration in one arbitrary direction without rotation, since the inertia of each particle is proportional to its mass and acts in the direction opposite to the acceleration, the resultant inertia of the whole system then acts in a line through the center of mass, which is therefore also called the center of inertia.

Yuan-Chong Zhang

Bibliography

Goldstein, H. (1980) *Classical Mechanics*, 2nd ed. London: Addison-Wesley Publishing Company.
Herivel, J. (1965) *The Background to Newton's Principia*. Oxford: Clarendon Press.
Lindsay, R.B. and Margenau, H. (1957) *Foundation of Physics*. New York: Dover Publications Inc.
Mach, E. (1960) *The Science of Mechanics*. LaSalle: Open Court Publishing Company.

CERES

Asteroid 1 Ceres was discovered on 1 January 1801 by Giuseppe Piazzi while searching for the missing planet, predicted by Bode's Law to exist between Mars and Jupiter. Though it is the largest asteroid with a diameter of about 935 km (Millis *et al.*, 1987), it is far smaller than any planet. With a density of 2.7 ± 0.13 gm cm^{-3} (Millis *et al.*, 1987), Ceres probably accounts for more than 25% of the mass of the entire asteroid belt. Its distance from the Sun is, coincidently, very close to that predicted by Bode's law (2.8 AU) for the hypothetical planet between Mars and Jupiter.

Visible and near-IR photometric and spectrophotometric observations

Spectral reflectance studies of Ceres span the 0.25- to 4.2-μm spectral region. Based on the occultation diameter given above, Ceres has a visual albedo of about 0.095. Lightcurve measurements indicate that Ceres has a rotation period of 9.075 h with an amplitude of only 0.05 magnitudes, implying that Ceres is nearly spherical in shape. Visible spectra (0.3 to 1.1 μm) have been compared to the laboratory spectra of meteorites. These studies place Ceres in the G taxonomic class of asteroids (Tholen, 1984), which implies a composition similar to that of carbonaceous chondritic meteorites. Spectral studies in the infrared (out to 4.2 μm) have shown the existence of a 3-μm absorption feature indicative of the presence of water in the form of water of hydration in clay minerals. These observations would indicate a surface composition similar to the aqueously altered CI carbonaceous chondrites.

At 3.1 μm, within the water of hydration feature, is another, weaker, spectral absorption feature. This feature was originally attributed to a very short pathlength of water ice (Lebofsky *et al.*, 1981). However, a more recent study attributes this feature to the presence of an ammonia compound in the surface minerals of Ceres (King *et al.*, 1992). Ultraviolet observations made with the International Ultraviolet Explorer (IUE) spacecraft (A'Hearn and Feldman, 1992) have shown the presence of an extremely thin atmosphere of water around Ceres. These observations would require that there be a continuous supply of water from the interior of Ceres; such a rate could also support the amount of ice necessary for the original identification of Lebofsky *et al.* (1981).

Radiometric studies

Thermal IR observations of Ceres are consistent with a surface whose thermophysical properties are similar to those of the Moon. Radiometric observations from 10 to 1000 μm over a range of solar phase angles are consistent with Ceres having a dusty surface regolith overlying a more rocky subsurface.

Larry A. Lebofsky

Bibliography

A'Hearn, M. and Feldman, P.D. (1992) Water vapor on Ceres. *Icarus*, **98**, 54–60.
King, T.V.V., Clark, R.N., Calvin, W.M. *et al.* (1992) Evidence for ammonium-bearing minerals on Ceres. *Science*, **225**, 1551–3.
Lebofsky, L.A., Feierberg, M.A., Tokunaga, A.T. *et al.* (1981) The 1.7- to 4.2-μm spectrum of asteroid 1 Ceres: evidence for structural water in clay minerals. *Icarus*, **48**, 453–459.
Millis, R.L. Wasserman, L.H., Franz, O.G. *et al.* (1987) The size, shape, density and albedo of Ceres from its occultation of BD + 8° 471. *Icarus*, **72**, 507–518.
Tholen, D.J. (1984) Asteroid taxonomy from cluster analysis of photometry. Ph D Thesis, University of Arizona.

Cross references

Asteroid
Asteroid: compositional structure and taxonomy
Carbonaceous chondrite
Meteorite parent bodies

CHAMBERLIN, THOMAS CHROWDER (1843–1928); CHAMBERLIN–MOULTON PLANETESIMAL HYPOTHESIS

Chamberlin was an outstanding geologist and planetary scientist who, with F.R. Moulton (a distinguished mathematician and specialist in celestial mechanics), jointly developed the *planetesimal hypothesis* for the accretionary origin of planets, asteroids and the natural satellites. The term 'planetesimal' (q.v.), coined by Chamberlin, described the small, cold particles or 'dirty snowballs' (Shapley's term) of the primeval dust cloud believed to have been the source of the solar system (sometimes called the 'dust-cloud hypothesis'). The nebular hypothesis of Kant–Laplace was vigorously attacked by Chamberlin and rejected. While the Kant–Laplace model required a very hot beginning, Chamberlin and Moulton's model permitted a cold accretion such as indicated by the geochemical evidence in the Earth's Precambrian rocks, which display salt deposits and glacial cycles. Chamberlin was also interested in the study of meteorites and proposed the 'chondrulitic hypothesis' for the origin of chondrulites, meteorites and comets.

Examining the nebular hypothesis critically, Chamberlin noted that photographs of the Andromeda nebula, originally hailed as supporting Laplace's theory, offered only a vague and general analogy. Saturn's rings, by 1897 known to be composed not of gas but of solid particles, would no longer furnish a helpful analogy for Laplace's ideas. Furthermore, if a molten earth had condensed from the gaseous mass, the temperature of the Earth's atmosphere would probably have been so great that water vapor would have evaporated and been lost into space, as shown by Forest R. Moulton, an assistant at the University of Chicago. Alternatively, water would have dissociated into hydrogen and oxygen, and the hydrogen would then have quickly escaped into space. Thus Laplace's hypothesis of a molten earth was untenable.

Other tests further weakened faith in the theory and served to forestall acceptance of even a modified form of the Laplacian hypothesis in which meteoroids slowly aggregated. A large portion of the solar system's momentum was observed to reside in the outer planets (> 98%) but the mass is heavily concentrated near the center, an unlikely occurrence within Laplace's nebular hypothesis. The unsymmetrical distribution of matter and momentum suggested to Chamberlin that the solar system had been formed by the collision between a small nebula with large momentum and the periphery of a large nebula with very little momentum. Spiral galaxies might be a result of such collisions, but little was then known of their dynamics. The peculiar form of spiral nebulae suggested to Chamberlin a combined outward and rotatory movement. He called for spectroscopic or photographic determination of the motions of their constituents as a test of his hypothesis, and accordingly spent the summer of 1915 at the Mount Wilson Observatory. There was disagreement between Chamberlin's theory and those of Percival Lowell, who favored Laplace, but this helped inspire Vesto M. Slipher's measurement at the Lowell Observatory of the radial velocity of the Andromeda nebula, a measurement at the roots of the later demonstration of the expansion of the universe.

The Chamberlin–Moulton hypothesis was eventually superseded by others in the 1920s. Spiral nebulae were too large to be the progenitors of planetary systems. Furthermore, regarding the origin of the solar system, the idea of a gaseous filament drawn out of the Sun's very hot interior also fell from favor in the late 1930s because it would rapidly dissipate into space before it could condense into planetesimals.

Chamberlin's professional life was extraordinarily varied. While teaching geology at Beloit College, Wisconsin (1873–1882), he also served on the Geological Survey of Wisconsin (chief geologist, 1876–1882), and published the *Geology of Wisconsin* (1879–82). He then became president of the University of Wisconsin and simultaneously joined the US Geological Survey as chief of its glacial division, becoming one of the foremost specialists on Pleistocene stratigraphy and glacial dynamics, which led to travels in Greenland in 1894 and a 'granular theory' for ice motion. He became fascinated by climate change, the role played by CO_2 in heating the atmosphere (a topic pioneered by S.A. Arrhenius) and especially appreciated the role of deep-sea circulation. His studies of glaciers eventually led him to consider the question of the Earth's origin.

In 1897 Lord Kelvin (q.v.) presented his astonishing deduction that the Sun could not have heated the Earth for more than 20–30 Ma, which generated a vigorous rejoinder from Chamberlin (1898), who pointed out the enormous potential energy of the atom, a concept ignored by Kelvin.

Chamberlin became president of the University of Chicago (1892–1918) and while there founded the *Journal of Geology* (1893), a journal which carried many of his major papers. Jointly with his former student, R.D. Salisbury (1858–1922), he published a three-volume compendium, *Geology* (1906), a subject which he saw in its broadest meaning as science of the Earth. Following this came *The Tidal and Other Problems* (1909), *The Origin of the Earth* (1916) and *The Two Solar Families, the Sun's Children* (1928).

With the late 20th century interest in the greenhouse problem, attention has recently returned to Chamberlin, who was one of the first (after Tyndall) to recognize the feedback role of water vapor. While rising CO_2 unquestionably contributes to planetary warming, increased evaporation, water vapor and cloudiness might appear to cap the warming trend, but there is a positive feedback whereby the warmer the atmosphere grows the more moisture it will hold, Chamberlin's 'dynamic pyramid' (Fleming, 1992). Nevertheless at a certain point a negative feedback comes into play because increasing cloud cover raises the albedo to reradiate the incoming solar energy. However, it seems that he failed to recognize the biological role in negative feedback, inasmuch as the metabolic rate rises with rising CO_2 and H_2O vapor, and with it there is a simultaneous long-term storage of CO_2 as $CaCO_3$ and woody material. Although he was interested in Croll's orbital model, he was convinced that the CO_2–H_2O feedback fluctuations were basic in controlling multiple glacial cycles.

Rhodes W. Fairbridge and Norriss S. Hetherington

Bibliography

Brush, S.G. (1978) A geologist among astronomers: the rise and fall of the Chamberlin–Moulton cosmogony. *J. Hist. Astron.*, **9**, 1–41, 77–104.

Chamberlin, R.T. (1934) Biographical memoir of Thomas Chrowder Chamberlin, 1843–1928. *Natl Acad. Sci. Biogr. Mem.*, **15**, pp. 305–407.

Chamberlin, T.C. (1897) A group of hypotheses bearing on climatic changes, *J. Geol.*, **5**, 653–83.

Fenton, C.L. and Fenton, M.A. (1952) *Giants of Geology*. Garden City, NY: Doubleday, 333 pp.

Fleming, J.R. (1992) T.C. Chamberlin and H_2O climate feedbacks: a voice from the past. *Eos*, **73**(47), 505–6.

Hetherington, N.S. (1992) Converting an hypothesis into a research program: T.C. Chamberlin, his planetesimal hypothesis, and its effect on research at the Mt Wilson Observatory, in *The Earth, the Heavens, and the Carnegie Institution of Washington; Historical Perspectives after Ninety Years* (ed. G.A. Good). Washington, DC: American Geophysical Union.

MacMillan, W.D. (1929) The field of cosmogony (development of Chamberlin's planetesimal hypothesis). *J. Geol.*, **27**, 341–56.

Mather, K.F. (1971). Chamberlin, Thomas Chrowder. *Dict. Sci. Biogr.*, Vol. 3, pp. 189–91.

Moulton, F.R. (1900) An attempt to test the nebular hypothesis by an appeal to the laws of motion. *Astrophys. J.*, **11**, 103–30.

Willis, B. (1929) Memorial of Thomas Chrowder Chamberlin. *Geol. Soc. Am. Bull.*, **40**, 23–45.

Cross references

Lowell, Percival
Nebular hypothesis (Kant–Laplace)
Planetesimal

CHANDLER WOBBLE

The term wobble pertains to a free mode of motion of a rotating object in the absence of external torques. The free rotation of an object will be uniform only when the rotation axis happens to coincide with one of the three principal axes of the object; in this case the object does not wobble. Otherwise it will wobble, or even 'tumble' if the wobble is too severe. An easy way to demonstrate wobble is to throw a plate (or a 'frisbee') spinning in the air. Unless thrown perfectly, the plate would wobble (for a plate the wobble period is about half of the spin period). The rotational dynamics of a rigid body were first established by Euler in the 1750s. As a result, especially in the early literature, wobble has been referred to as the '(free) Eulerian nutation' or '(free) Eulerian precession', although it is neither a nutation nor precession in the classical sense.

The Earth also wobbles, independently of external luni-solar tidal torques. To an observer on the Earth (analogous to an ant riding on the spinning plate), the wobble manifests itself as a circular motion, known as the polar motion, of the rotation axis in the counterclockwise sense around the Earth's mean polar principal axis (the nominal geographic North Pole), or as a periodic variation of latitude of any given point on Earth. The Eulerian theory predicts a period that equals the reciprocal of the Earth's ellipticity in days (about 300 days). For a long time, using astrometric records, astronomers looked for a 300-day wobble, but to no avail. This changed in 1891 when an American actuary and astronomer by the name of Seth Carlo Chandler (1846–1913) reported the finding of two wobbles in the polar motion: one with an annual period and the other with a period of about 14 months.

The amplitude of the polar motion is only several meters. The annual wobble is a wobble forced by seasonality in the atmospheric and hydrospheric fluctuations (through surface torques under the conservation of angular momentum). But the identity of the 14-month wobble, now called the Chandler wobble, was not explained until a year later by Simon Newcomb: the predicted 300-day period is for a rigid Earth, while in reality the elastic deformation in the Earth caused by the wobble itself feeds back to the system and lengthens the characteristic period by about 40%. Today we have a much more sophisticated Earth model, including additional contributions from the ocean and the fluid core, to explain the Chandler period, the modern determination of which is about 434 days.

The Chandler wobble is inevitably dissipative; its Q value has not been well determined nor is its energy dissipation mechanism fully understood. At any rate, the very existence of the wobble means it is being continually excited, presumably by geophysical processes indigenous to the Earth. Ironically, although the Chandler wobble has been studied for a century now and is one of the best-measured geophysical quantities today, geophysicists are just beginning to have a clue as to what are the major excitation sources: atmospheric mass movement and possibly ocean circulations. One other candidate has been studied in some detail: this is the mass redistribution in the solid Earth associated with large earthquake faulting. It does not seem to be energetically significant.

Benjamin Fong Chao

Cross references

Earth rotation
Euler, Leonhard, and Eulerian motion

CHAOTIC DYNAMICS IN THE SOLAR SYSTEM

The dynamics of solar system bodies has been long considered to be the paradigm of regularity. Since the work of Newton and Laplace, celestial mechanicians thought that knowledge of the initial conditions of the celestial bodies would allow one to predict their positions with great accuracy for an ever extending time span. The work of astronomers was thus reduced to the determination of these precise initial conditions. This view was destroyed by Henri Poincaré (1892), when he demonstrated that the three-body problem is not integrable and showed the very complicated behavior of the solutions of the differential equations. He demonstrated that in the vicinity of the separatrix of a resonance, very small variations of the initial conditions can lead to very large variations of the final solutions. He also showed that the perturbation methods used by the astronomers at the time were not converging. Later on the mathematicians Kolmogorov, Arnold and Moser demonstrated that under certain conditions, in conservative systems that are close to being integrable, with Hamiltonians of the form

$$H_o + \epsilon H_1$$

(where H_o is the integrable part of the Hamiltonian), there exist initial conditions for which the motion is quasi-periodic, and thus predictable for all time. For other initial conditions arbitrarily close, there exist zones of instability, or chaotic zones, where the dynamics is much more complicated (these results are known as the KAM theorem). For small values of the perturbation (i.e. when ϵ is small), the measure of the regular solution is large, and the system behaves over a finite time span (which can eventually be as large as the age of the universe) like a regular system. Since Poincaré's work and development of the KAM theorem, it is now known that chaotic behavior is everywhere present in the phase spaces of the motions of the bodies of the solar system. Nevertheless, this chaotic behavior is only of practical interest for understanding the dynamics of the solar system when it manifests itself on a macroscopic scale over a smaller time span than the age or lifetime of the solar system, that is about 4.5–10 Ga.

Chaotic behavior was first described by Poincaré with application to the three-body problem, and some of the first numerical experiments were applied to stellar dynamics (Hénon and Heiles, 1964). Chaotic dynamics has been recognized only recently in application to the solar system. Indeed, the timescale involved for the observation of this chaotic behavior is often very long, and many motions seem to be regular on the short timescales which were investigated at the beginning of computer experiments.

Chaotic motion of Hyperion

One of the first examples of chaotic behavior in the solar system to be studied was the chaotic tumbling of Hyperion, a small satellite of Saturn (Wisdom, Peale and Mignard, 1984). More generally, the same study applies to any satellite of irregular shape in the vicinity of spin-orbit resonance (Wisdom, 1987).

The equations of motion for the orientation of the satellite S orbiting around a planet P on a fixed elliptical orbit of semimajor axis a and eccentricity e are given by the Hamiltonian

$$H = \frac{y^2}{2} - \frac{3}{4} \frac{B-A}{C} \left(\frac{a}{r(t)}\right)^3 \cos 2(x - v(t))$$

where x gives the orientation of the satellite with respect to a fixed direction (here the direction of periapse), $y = dx/dt$ is its conjugate variable, v is the true anomaly of the satellite, and $A < B < C$ are the principal moments of inertia of the satellite. The associated equations of motion are

$$\frac{dy}{dt} = \frac{\partial H}{\partial x}; \quad \frac{dx}{dt} = -\frac{\partial H}{\partial y}$$

The unit of time is taken such that the mean motion $n = 1$. When expanding the Hamiltonian with respect to the eccentricity (e) and retaining only the terms of first order in eccentricity, one obtains

$$H = \frac{y^2}{2} - \frac{\alpha}{2}\cos 2(x-t) + \frac{\alpha e}{4}[\cos(2x-t) - 7\cos(2x-3t)] \quad (C2)$$

with $\alpha = 3(B-A)/2C$. If S has a rotational symmetry, $\alpha = 0$, and the Hamiltonian is reduced to $H_0 = y^2/2$. The satellite rotates with constant velocity $dx/dt = y_0$. When the orbit is circular, the problem is also integrable, and H_0 is reduced to the first two terms of equation (C2).

$$H_0 = \frac{y^2}{2} - \frac{\alpha}{2}\cos 2(x-t) \quad (C3)$$

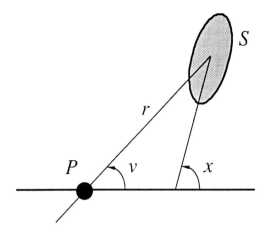

Figure C16 The position of the satellite S around the planet P is defined by the distance r and the angle v from the direction of periapse (true anomaly). The angle x defines the orientation of the satellite.

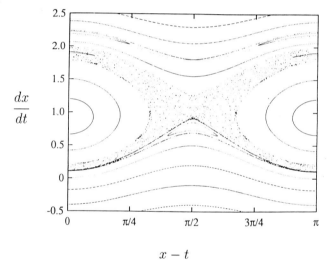

Figure C17 Surface of section in the phase space of Phobos. $x - t$ defines the orientation of the satellite, and dx/dt its rotational velocity. A small chaotic zone appears in the vicinity of the separatrix of the unperturbed problem ($e = 0$). (Wisdom, 1987).

This motion will be similar to simple pendulum motion, with the possibility of libration of the satellite around the direction of the planet (i.e. spin–orbit resonance occurs) or circulatory motion. In the general case, $\alpha e \neq 0$, and the Hamiltonian H_0 of (C3) is perturbed by the remaining terms of (C2). At the transition between librational motion and rotational motion of the satellites appears a small chaotic zone. This is what can be observed in a section of the phase space portrait of Phobos' motion (in orientation) when orbiting Mars (Figure C17). When the size of the perturbation αe increases, resonant zones corresponding to the various possible resonant terms $\cos 2(x-t)$, $\cos(2x-t)$, $\cos(2x-3t)$ will overlap, and give rise to large-scale chaotic motion, which is the case for Hyperion (Figure C18), where $\alpha e \approx 0.039$. The resulting effect is that the rotational motion of Hyperion is not regular, and it becomes impossible to adjust any periodic or quasi-periodic model to its lightcurve (Klavetter, 1989).

3 : 1 Kirkwood gap

The distribution of the asteroids, the minor planets whose orbits lie primarily between Mars and Jupiter, has puzzled astronomers for

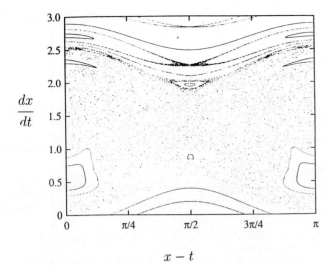

Figure C18 Surface of section in the phase space of Hyperion. When the perturbation (αe) is large, as in the case of Hyperion, many of the resonances overlap and give rise to a large chaotic zone where the motion is not regular. (Wisdom, 1987).

many decades, since Kirkwood observed that they are not randomly distributed. Indeed, when plotting the number of asteroids against their semimajor axes, one can observe gaps and accumulations. Kirkwood noticed that these gaps coincide with commensurabilities of mean motion with Jupiter, and that the extent of the gaps also coincides with the libration zones of the resonances (Dermott and Murray, 1983). It was thus thought that these gaps result from the effect of these resonances, but although the first numerical integrations of asteroids placed inside the resonance lead to some increase of their eccentricities (Froeschlé and Scholl, 1977), no satisfying explanation was given. Later on, Wisdom (1983, 1985) managed to integrate the orbits approximately over a much extended time to show that in the vicinity of the 3 : 1 resonance, there exists a chaotic zone which can be observed in a surface of section of the trajectories, corresponding to the successive intersections with a given plane (Figure C19). An orbit starting in this chaotic zone will have a moderate eccentricity for a long time, as long as it stays in the largest part of the chaotic zone of Figure C19, but after some time it could enter the narrow branch of this chaotic zone, which would lead to a large increase of the eccentricity, sufficient to cross Mars' orbit. A possible close encounter with this planet would then expel the asteroid from its primitive orbit. The location and extent of the chaotic zone related to the 3 : 1 resonance is in good agreement with the 3 : 1 asteroid gap. The understanding of this complex dynamics, which differs very much from the ordered motion of the integrable problems, thus allowed Wisdom to obtain a convincing explanation for one of the most famous problems of celestial mechanics. Since the work of Wisdom, which applied specifically to the 3 : 1 gap, many other studies have analysed the possible chaotic behaviors in the vicinity of other asteroidal resonances.

Chaotic motion of comets

The asteroids are not the only small bodies of the solar system which can be subject to chaotic motion. Indeed, many cometary orbits are chaotic.

When Halley's comet returned in 1985, several numerical integrations were carried out to retrace its orbit over the full time span of the observations, that is back to 163 BC, the date of the most ancient observation of this comet (Stephenson, Yau and Hunger, 1984). After such a long time all the different numerical integrations showed different behavior, and their accuracy was questioned. In fact these divergences were later explained by the analysis of Chirikov and Vecheslavov (1989), which demonstrated that the motion of Halley's comet could be chaotic, and may thus be practically unpredictable after 29 revolutions. Indeed, it can be shown that, due to the perturbation of Jupiter, there exists a large chaotic zone for nearly

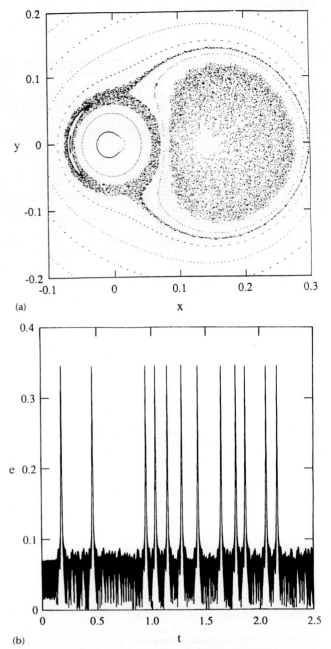

Figure C19 (a) Surface of section for asteroidal motion near the 3:1 resonance. The coordinates are $x = e \cos(\overline{\omega} - \overline{\omega}_J)$ and $y = e \sin(\overline{\omega} - \overline{\omega}_J)$, where $\overline{\omega}$ and $\overline{\omega}_J$ are the longitude of perihelion of the asteroid and Jupiter. (b) Example of an asteroidal orbit starting in the chaotic zone of (a) located around (0,0). For some time the asteroid stays in the broad ring around (0,0), and the eccentricity remains small, but from time to time the asteroid escapes in the narrow branch where it can gain a very large eccentricity, and eventually cross the orbit of Mars. (Wisdom, 1987).

parabolic cometary orbits which extend out to the Oort cloud (Petrosky, 1986; Sagdeev and Zaslavsky, 1987; Natenzon et al., 1990). This chaotic behavior of Halley's comet was later confirmed by direct numerical integration (Froeschlé and Gonczi, 1988).

Chaotic motion could also possibly drive short-period comets from a transneptunian belt towards the inner parts of the solar system, under planetary perturbations (Duncan, Quinn and Tremaine, 1989, Gladman and Duncan, 1990, Holman and Wisdom, 1993).

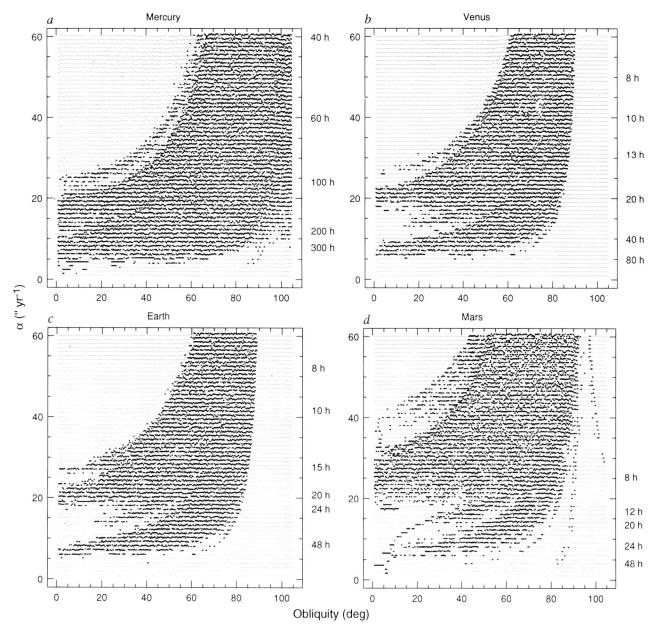

Figure C20 Zones of large-scale chaotic behavior for the obliquity of (a) Mercury, (b) Venus, (c) the Earth and (d) Mars, for a wide range of spin rate. The precession rate of the planet is given on the left in arcsec per year, and the estimated corresponding rotation period of the planet on the right, in hours. The regular solutions are represented by small dots, while large black dots represent solutions with large-scale chaotic behavior. (Laskar and Robutel, 1993).

Chaotic motion of the planets

The first examples of chaotic behavior which were found in the solar system concern small objects: small satellites, asteroids and comets. But contrary to what was expected until very recently, chaotic behavior affects also the motion of the major bodies of the solar system: the planets (see Solar system: stability). The chaotic behavior of Pluto was recognized by Sussman and Wisdom (1988) with a numerical integration of the outer planets of 875 Ma. Later on, Laskar (1989, 1990) showed that the whole solar system is chaotic, when integrating numerically the averaged equations of the planetary motions. The chaotic motion affects essentially the inner planets (Mercury, Venus, Earth and Mars), and results from the existence of secular resonances (Laskar, 1990, 1992; Sussman and Wisdom, 1992). The Lyapunov exponent of the planetary system is about $1/(5\ Ma)$, which means that an error of 15 m in the position of the Earth will become about 150 m after 10 Ma, but more than 150 million kilometers after 100 Ma. In this sense the chaotic behavior of the solar system precludes the construction of ephemerides for the motion of the inner planets over more than 10 to 20 Ma. The motions of the large planets (Jupiter, Saturn, Uranus and Neptune) are more stable and are closer to quasi-periodic motion, although small chaotic zones are present.

The chaotic behavior of the inner planets has a secondary effect of great importance: due to the planetary perturbations, the rotational motion of the inner planets would be largely chaotic and unstable, for many values of its rotational spin rate (Laskar and Robutel, 1993) (Figure C20). The rotational motion of the Earth is stable because of the presence of the Moon, but if this large satellite were not here, the tilt of the Earth would probably behave chaotically, with very large amplitude variations, ranging from 0 to about 85° (Laskar, Joutel and Robutel, 1993). Similar chaotic behavior could have also driven Venus' obliquity to nearly 90°, when dissipative effects would bring it to its present value of 178°. Mars' obliquity is chaotic, with a Lyapunov exponent of about 1/(5 Ma) (Laskar and Robutel, 1993; Touma and Wisdom, 1993). The extent of the chaotic zone for Mars' obliquity goes from 0 to about 60° (Laskar and Robutel, 1993).

Jacques Laskar

Bibliography

Chirikov, B.V. and Vecheslavov, V.V. (1989) Chaotic dynamics of comet Halley. *Astron. Astrophys.*, **221**, 146–54.
Dermott, S.F. and Murray, C.D. (1983) Nature of the Kirkwood gaps in the asteroidal belt. *Nature*, **301**, 201–5.
Duncan, M., Quinn, T. and Tremaine, S. (1988) The origin of short period comets. *Astrophys. J. Lett.*, **328**, L69–L73.
Duncan, M., Quinn, T. and Tremaine, S. (1989) The long-term evolution of orbits in the solar system: a mapping approach. *Icarus*, **82**, 402–18.
Froeschlé, C. and Scholl, H. (1977) A qualitative comparison between the circular and elliptic Sun–Jupiter–asteroid problem at commensurabilities. *Astron. Astrophys.*, **57**, 33–59.
Froeschlé, C. and Gonzci, R. (1988) On the stochasticity of Halley like comets, *Celes. Mech.*, **43**, 325–30.
Gladman, B. and Duncan, M. (1990) On the fates of minor bodies in the outer solar system. *Astron. J.*, **100**(5).
Hénon, M. and Heiles, C. (1964) The applicability of the third integral of motion: some numerical experiment. *Astron. J.*, **69**, 73–9.
Holman, M.J. and Wisdom, J. (1993) Dynamical stability in the outer solar system and the delivery of short period comets. *Astron. J.*, **105**(5), 1987–99.
Klavetter, J.J. (1989) *Astron. J.*, **97**(2), 570–9.
Laskar, J. (1989) A numerical experiment on the chaotic behaviour of the solar system. *Nature*, **338**, 237–38.
Laskar, J. (1990) The chaotic motion of the solar system. A numerical estimate of the size of the chaotic zones. *Icarus*, **88**, 266–91.
Laskar, J. (1992) A few points on the stability of the solar system, in *Chaos, Resonances, and Collective Dynamical Phenomena Symp.*, (ed. S. Ferraz-Mello) IAU 152, pp. 1–16.
Laskar, J. and Robutel, P. (1993) The chaotic obliquity of the planets. *Nature*, **361**, 608–12.
Laskar, J., Joutel, F. and Robutel, P. (1993) Stabilization of the Earth's obliquity by the Moon. *Nature*, **361**, 615–7.
Natenzon, M.Y., Neishtadt, A.I., Sagdeev, R.Z. *et al.* (1990) Chaos in the Kepler problem and long period comet dynamics. *Phys. Lett. A*, **145**, 255–63.
Petrosky, T.Y. (1986) Chaos and cometary clouds in the solar system, *Phys. Lett. A*, **117**, 328–332.
Poincaré, H. (1892–1899) *Les Méthodes Nouvelles de la Mécanique Céleste*, tomes I–III. Paris: Gauthier Villard, (reprinted by Blanchard, 1987).
Sagdeev, R.Z. and Zaslavsky, G.M. (1987) Stochasticity in the Kepler problem and a model of possible dynamics of comets in the Oort cloud, *Il Nuovo Cimento*, **97B**, 119–30
Stephenson, F.R., Yau, K.K.C. and Hunger, H. (1984) *Nature*, **314**, 587.
Sussman, G.J. and Wisdom, J. (1988) Numerical evidence that the motion of Pluto is chaotic. *Science*, **241**, 433–7.
Sussman, G.J. and Wisdom, J. (1992) Chaotic evolution of the solar system. *Science*, **257**, 56–62.
Torbett M.V. and Smoluchovski, R. (1990) Chaotic motion in a primordial comet disk beyond Neptune and comet influx to the solar system. *Nature*, **345**, 49–51.
Touma, J. and Wisdom, J. (1993) The chaotic obliquity of Mars. *Science*, **257**, 56–62.
Wisdom, J. (1983) Chaotic behavior and the origin of the 3/1 Kirkwood gap. *Icarus*, **56**, 51–74.
Wisdom, J. (1985) A perturbative treatment of motion near the 3/1 commensurability. *Icarus*, **63**, 272–89.
Wisdom, J. (1987) Urey Prize Lecture: chaotic dynamics in the solar system. *Icarus*, **72**, 241–75.
Wisdom, J. (1987) Rotational dynamics of irregularly shaped natural satellites. *Astron. J.*, **94**(5), 1350–60.
Wisdom, J., Peale, S.J. and Mignard F. (1984) The chaotic rotation of Hyperion. *Icarus*, **58**, 137–52.

Cross references

Celestial mechanics
Earth: rotational history
Planetary dynamical astronomy
Resonance in the solar system

CHARGE-COUPLED DEVICE

A charge-coupled device, or CCD, is one of several types of two-dimensional arrays of semiconductor photodetectors. CCDs are the detectors of choice in planetary science for detecting and imaging light of wavelengths less than 1100 nm, by virtue of their sensitivity, spatial resolution and other properties that allow accurate representation of the flux of photons from celestial objects imaged through telescopes, cameras and spectrographs. Popular-level introductory descriptions of CCDs as used in astronomy have been written by Eberhart (1977), Raloff (1978), Kristian and Blouke (1982) and Janesick and Blouke (1987). Janesick *et al.* (1987, 1989) and Janesick and Elliott (1992) provide comprehensive technical summaries of CCDs from the electronic engineer's point of view. Mackay (1986) and Tyson (1986, 1990) review the use and revolutionary capabilities of CCDs as applied to problems in astronomy. The history of CCD technology in astronomy is reviewed by Geary (1989). Rufener (1989) summarizes the literature on the calibration of CCDs. Janesick and Elliott (1992) provide an impressive list of current and future spacecraft that use CCDs for planetary remote sensing. Nowadays, most articles in the planetary science journal *Icarus* that are written about observations in visible light are based on the use of CCDs. Some recent examples are given below.

Description

A CCD is a large-scale microelectronic semiconductor circuit on a silicon substrate, comprising a two-dimensional rectangular array of light-sensitive picture elements (pixels). Each pixel senses light by converting photon energy into a separation of electrons and their complementary 'holes' in a silicon crystal lattice. In this article we use the term 'charge' to refer to the numbers of electron–hole pairs in the CCD pixels. Designed as metal–insulator–semiconductor (MIS) capacitors, the pixels maintain separation of electric charge locally and in proportion to the cumulative exposure to incident photons, until (say) a mechanical shutter is closed and the accumulated electronic picture is 'read'. It is the process of reading the picture off the array that gives the CCD its name. Since it is impractical to wire outputs from every pixel, the electronic image is read sequentially, pixel by pixel, by transferring the signal charges in the two spatial coordinates through the imaging array itself. Pixels in adjacent rows are said to be 'charge coupled' because their signal charges can be moved in parallel from row to row by the manipulation of potentials on electrode lines that define the boundaries of the rows. As each row of signals reaches the end of the CCD it is transferred into a serial register, a one-dimensional array of capacitors that are also charge coupled. The contents of this register are sequentially transferred, a pixel at a time, through an output gate where the electric charge in each pixel can be transmitted to additional electronics that amplify and measure the number of electrons in it. The signal amplitudes are converted to digital numbers and stored sequentially in a computer memory, where they can be quantitatively processed and displayed as a picture on a video monitor.

Capabilities

In use and function, CCDs are analogous to photographic emulsions because they permit long time exposures to accumulate signal on faint light sources and have comparable spatial resolution. The photographic analog to the readout of a CCD is the chemical development process. However, compared to photographic plates and other imaging detectors of the 1970s, CCDs are far more sensitive, linear, uniform spatially, stable with time, and quiet (Janesick and Elliott, 1992). CCDs as large as 55 mm square and CCDs with as many as 4096 × 4096 pixels have been developed (Blouke *et al.*, 1987; Bredthauer *et al.*, 1991; Janesick *et al.*, 1990). The numbers of pixels in such CCDs rival those in photographic

Figure C21 A CCD mounted on a printed circuit board, on the underside of which are electronic components to control the transfer and readout of the electronic image. The CCD was made in 1991 by Tektronix, Inc. of Beaverton, OR. It contains 2048 rows and 2048 columns of pixels, each 24 μm square. The image-sensing area appears black because very low reflectivity is required for high quantum efficiency. At the time of this writing this is the largest thinned, backside-illuminated CCD available: 49 mm square. The circuit board is made by Princeton Scientific Instruments, Inc. of Monmouth Jct., NJ. The housing, from which the front window has been removed for this photograph, is a liquid nitrogen cryostat built by Infrared Laboratories, Inc. of Tucson, AZ. Photograph by the author for the Spacewatch Project, Lunar and Planetary Laboratory, University of Arizona.

plates and the uniform and stable positions of the pixels provide better image fidelity than the random arrangement of developed silver grains in a photographic emulsion. Figure C21 illustrates a large, sensitive scientific CCD. Janesick et al. (1990) have reduced CCD read noise to arbitrarily low values, and sensitivity to incident light (the quantum efficiency) is approaching the theoretical limit at some wavelengths (Janesick and Elliott, 1992). With such high sensitivity, large area and low noise, CCDs can be as much as two orders of magnitude more efficient with telescope time than the detectors of the 1970s.

Types

Even a non-user of CCDs needs a rudimentary perspective of CCDs to understand scientific results obtained with them. CCDs for science differ considerably from those found in mass-produced portable video cameras. Scientific CCDs can be used for direct imagery of solar system objects at high spatial resolution, for surveying large areas of sky to find new objects, for differential or absolute photometry of object-integrated light, or to record spectra. Time resolution can be long (hours) or as short as milliseconds (Elliot et al., 1989). The variety of uses and the rapid advance of technology from 1980 to 1992 (Geary, 1989; Janesick and Elliott, 1992) mean that CCD systems can differ widely in several parameters: Array dimensions, pixel size, quantum efficiency, readout noise and operating temperature are among the most important. Quantum efficiency is the probability that a photon incident on the CCD will create an electron–hole pair in the solid state material. Mostly the quantum efficiency depends on whether the CCD is illuminated on the front side, where the pixel control electrodes are overlaid, or on the back side of the substrate that has been chemically etched ('thinned') to permit light to reach the pixels without obstruction by the gate structures. Thinned, back-illuminated CCDs are about twice as sensitive as front-illuminated ones but are much more expensive. Quantum efficiency versus wavelength also depends on which of a variety of coatings might be applied to the CCD.

Performance and calibration

The exceptional capabilities of CCDs have raised the standards of observational planetary science dramatically since the 1970s. Increasingly exacting demands have gradually uncovered the systematics of CCDs: they are not perfect detectors. Considerable care and skill is needed to build, operate, maintain and calibrate CCDs and the electronics used to control and 'read them out' (Djorgovsky, 1984; Djorgovsky and Dickinson, 1989). A CCD is a complex device that is not self-contained. Electronic circuits that are not on the chip (or even in its housing) play a key role in the operation of the device (Djorgovsky, 1984; Gudehus and Hegyi, 1985; Leach, 1987). To use a CCD for precise and stable measurements of image position or brightness requires precise control of all functions that could affect the CCD's electronic representation of the optical image formed on it. A CCD not only must interact with incoming photons, it must also collect and store the electronic charge accumulating during an astronomical exposure that lasts many minutes without letting the charge spread over the array or recombine into neutral atoms. After the exposure the CCD must transfer all the pixels' charges in two coordinates through hundreds or thousands of other pixels and register locations before the sequence of pixel signals can be 'read' as a representation of exposure to light. The transfers must not leave behind any significant amount of signal ('deferred charge') and the external readout electronics must be quieter than the detector. In a well designed, well built, well tuned CCD system these requirements can be met and re-verified as often as desired (Massey and Jacoby, 1992; Robinson et al., 1991; Tyson, 1990).

Signal-to-noise ratio

By 'signal' we mean the number of electron–hole pairs created in the CCD pixels of interest by incident photons. 'Noise' in CCDs has three major components: fundamental statistical fluctuations in the process of counting photons or electrons, noise introduced by the CCD or circuitry that processes the signal, and uncertainties in calibrating measured signal as photon flux. The ratio of signal to noise (S : N) must be made larger than the effect one is trying to measure. A S : N of 3 is considered a marginal detection; routine quantitative work ranges from 10 to 50, and rigorous exacting measurements may have to be as high as 1000 in this ratio. At low values of S : N, photon statistics and readout noise contribute about equally; intermediate S : N is mostly controlled by photon statistics, and above S : N = 50 rigorous calibrations of the properties of the CCD are necessary. Equations describing S : N in CCDs are given by Rabinowitz (1991).

Increasing the exposure time is the most straightforward way to improve S : N because more photons are detected. Other ways of improving S : N are refrigeration, slowing down the readout, integrating over several pixels, taking multiple observations, using less spectral resolution, using quieter readout electronics and (as a last resort) obtaining time on a larger telescope.

Readout noise

This is the fluctuation in pixel readings obtained from a short dark exposure. For a 'skipper' CCD of Janesick et al. (1990), the readout noise has been reduced to 0.5 electron–hole pairs per pixel per readout. Most contemporary scientific CCD systems have readout noise less than 10 electrons. In addition there is thermal noise from the CCD, which depends on temperature and exposure time. The latest CCD designs have reduced thermal background, permitting operation with less or even no refrigeration (Janesick and Elliott 1992).

Bias frame

This is the underlying instrumental 'picture' one obtains when the CCD is read following a dark exposure of negligible duration. Mostly it is a constant voltage set to prevent the analog-to-digital converter from sampling signals with algebraic sign opposite to that obtained on exposure to light. The bias frame also may show spatial dependences, 'features' and other defects associated with imperfections on the imaging area.

Cosmetic defects

A 'long' dark exposure (at least as long as the longest science exposure) should be made to show pixels and columns that generate spurious signals. These are usually fixed in position, so they can be masked off in the data reduction process or otherwise avoided when observing.

Flat field

In quantitative work the relative sensitivities of the CCD pixels must be measured at the wavelengths of the observations and compensated to bring the CCD picture to a uniform scale of spectral radiance. The techniques and pitfalls of measuring the map of pixel sensitivities, called the 'flat-field calibration', have been described by Djorgovsky and Dickinson (1989), Massey and Jacoby (1992), Stetson (1989), Tobin (1993) and Tyson (1990). Most of the systematic errors remaining after flat-field correction are caused by differences between the illumination by the calibrating source and that by the science targets (Kostkowski, 1979; Stetson, 1989).

Point spread function

This is the two-dimensional appearance of a point source imaged by the CCD. It represents the accuracy with which the device represents the shape of a real image. The PSF of the electronic image transferred out of the CCD depends on the voltages and relative timing of the waveforms applied to the CCD electrodes, the location of the image on the CCD array, the number, placement and type of physical imperfections in the imaging array and serial register, the flat-field map of the CCD, the linearity of the CCD, the physical flatness of the CCD chip and the temperature of the CCD (Gudehus and Hegyi, 1985; Janesick, Klaasen and Elliott, 1987; Epperson et al., 1987; Leach, 1987; Walker, 1993). If these parameters change with time, the shape of the PSF will also, introducing systematic errors if not corrected.

Linearity

To what degree is the raw CCD signal directly proportional to the cumulative spectral radiance (photons per telescope collecting area per pixel per spectral bandpass per exposure time)? The process of electron–hole pair generation in the CCD itself is intrinsically very linear, but the processes of charge collection, transfer, amplification and readout must also be linear if the CCD's linearity is to be preserved in the observer's data. If charge transfer efficiency is inadequate, or if charge tends to spread into adjacent pixels or the substrate during an exposure, significant nonlinearities could be present in the images of point sources such as stars or asteroids. Electronic readout clock voltages must be tuned and analog-to-digital converters must have the specified linearity. With good CCD systems and routine flat-fielding procedures, departures from linearity will be less than 1%, except near saturation and in some cases at very low light levels (Djorgovsky and Dickinson, 1989; Walker, 1993). Exacting observations, as always, must be collected in a context of relevant checks and calibrations. For the most exacting absolute work, calibration sources such as standard stars should be observed with the same exposure time, signal level and location on the CCD array as the target object. More expanded image scales and avoidance of extremely low or high signal levels tend to reduce systematic errors. Image retention should always be checked. It has not been established that CCDs used as single-channel detectors are more linear than photomultipliers, except at very low light levels (Young, 1974; 1992, personal communication).

Cosmic rays

CCDs detect cosmic ray secondaries. They appear as single-pixel or two-pixel signals at random times and locations on the array. They can be distinguished from optical images by avoiding an image scale that is too compressed. Telescopic 'seeing' disks, lines in spectra, etc., should occupy several pixels. Confusion by cosmic rays also can be minimized by repeating observations.

Afterimages

Some types of CCD show image retention after readout. This effect depends on the type of CCD, its temperature, the read-out clocking technique, the wavelength of illuminating light, recent exposure history, recent readout history and recent temperature cycling history (Djorgovsky, 1984; Janesick, Klaasen and Elliott, 1987; Janesick and Elliott, 1992). Such latent images can affect the intended images of later targets. In state-of-the-art CCD systems, afterimages can be erased electronically (Janesick and Elliott, 1992).

Full well

The full-well capacity of CCD pixels (the amount of signal charge that can be held in the potential well without spreading into adjacent pixels or the substrate) depends on the voltages of the potential barriers that define the CCD rows (Gudehus and Hegyi, 1985; Janesick, Klaasen and Elliott, 1987; Leach, 1987). A change in these voltages, controlled in external electronics, affects the high-end linearity of the CCD's representation of the number of collected photons. Depending on pixel size and other parameters, CCD full wells can be as small as 10 000 or as large as 1 000 000 electrons per pixel.

Applications

Many of the discoveries and measurements in optical observational planetary science since the 1970s have been made possible by CCDs. Below are some recent investigations selected to emphasize the variety of ways these revolutionary detectors are used. This list is not intended to prioritize planetary science, nor to claim that these observations would have been impossible without a CCD in every case. However, at the very least the enormously high efficiency of a CCD makes routine what would otherwise require a hopelessly long observing time.

Asteroids

Drift scanning of large areas of sky with CCDs has permitted discovery of the smallest natural objects ever observed outside the Earth's atmosphere: Earth-approaching asteroids as small as 9 m in diameter, as well as detection of larger Earth approaches when they are their farthest from the Sun (Gehrels, 1991; Scotti, Rabinowitz and Marsden, 1991; Rabinowitz, 1992a). Figure C22 illustrates a CCD observation of a very small, fast-moving asteroid. Lightcurves and compositional studies of faint, fast-moving asteroids (impossible with single-channel detectors) have been made possible with CCDs (Wisniewski and McMillan, 1987; French and Binzel, 1989; Wisniewski, 1991). The first close-up image of an asteroid (Gaspra) was obtained with a CCD on the Galileo spacecraft (Kerr, 1991; Veverka et al., 1992).

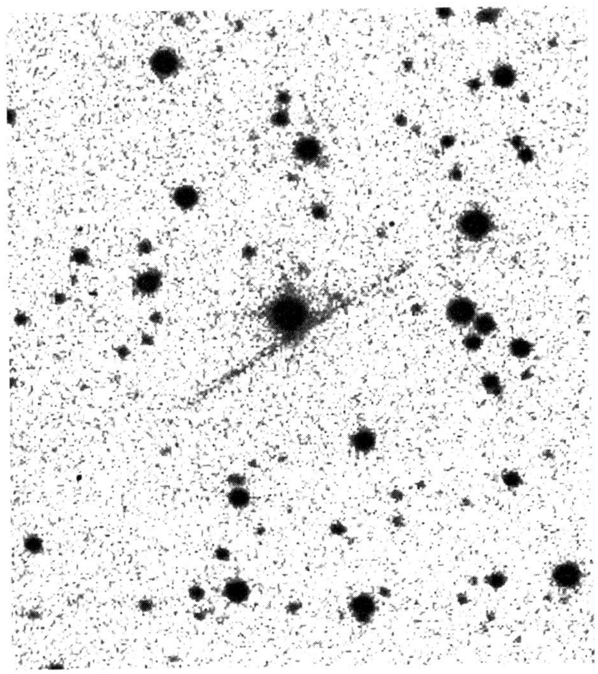

Distant members of the solar system

CCD surveys by Rabinowitz (1992b) and Jewitt and Luu (1992) have discovered objects greater than 100 km in diameter at great distances from the Sun. Such a class of objects was long suspected but not detectable before the advent of large, sensitive imaging detectors.

Planetary atmospheres

The exotic subject of atmospheres on Mercury and Pluto were investigated with CCDs by Sprague et al. (1992) and Elliot et al. (1989) respectively. Karkoschka (1992) and Moreno, Molina and Ortiz (1991) have observed the atmospheres of the giant planets and Titan with CCDs at high spectral resolving power. Rizk et al. (1991) observed water vapor in the atmosphere of Mars with a CCD.

Planet detection

Searches for planets orbiting other stars are being conducted with CCDs on high resolution spectrographs by Cochran, Hatzes and Hancock (1991), Marcy and Butler (1992) and McMillan et al. (1993). In these investigations the gravitational effect of planets orbiting other stars is sought via changes in the stars' Doppler shifts. There is no doubt that CCDs will introduce many surprises in the planetary science of the 21st century.

Robert S. McMillan

Figure C22 An observation made with the CCD in Figure C21, and which would not have been possible with earlier detector technology. The 2.6 arcmin long trail seen on this negative print was made by Earth-approaching asteroid 1993 KA2 on the night of 21 May 1993 UT. Nine hours earlier the object made a record closest approach of 0.001 AU from the Earth. This is the record smallest asteroid to date, less than 10 m in diameter (*IAU Circular* 5817). The exposure time was 157 s, and although the apparent magnitude of the object was 18.6 V at the time of this observation, it would not have registered on a photographic emulsion because of its rapid motion of a second of arc per second of time. Discovery observation made by T. Gehrels, Principal Investigator of the Spacewatch Project, Lunar and Planetary Laboratory, University of Arizona. Photograph composed digitally by J.V. Scotti of LPL.

Bibliography

Blouke, M.M. et al. (1987) Large-format, high resolution image sensors. *Optical Eng.*, **26**, 837–43.
Bredthauer, R.A. et al. (1991) Notch and large-area CCD imagers. *Proc. SPIE*, **1447**, 310–5.
Cochran, W.D., Hatzes, A.P. and Hancock, R.J. (1991). Constraints on the companion object to HD 114762. *Astrophys. J. Lett.*, **380**, L35–L38.
Djorgovsky, S. (1984). CCDs: their cause and cure. *NASA Conf. Proc.*, **2350**, 152–76.
Djorgovsky, S. and Dickinson, M. (1989) CCD data taking modes and flatfielding problems. *Highlights of Astron*, **8**, 645–50.
Eberhart, J. (1977) The CCD: new eye on the sky. *Sci. News*, **111**, 169–73.
Elliot, J.L., Dunham, E.W., Bosh, A.S. et al. (1989) Pluto's atmosphere. *Icarus*, **77**, 148–70.
Epperson, P.M., Sweedler, J.V., Denton, M.B. et al. (1987) Electro-optical characterization of the Tektronix Tk512M-011 charge-coupled device. *Optical Eng.* **26**, 715–24.
French, L.M. and Binzel, R.P. (1989) CCD photometry of asteroids, in *Asteroids II* (eds R.P. Binzel, T. Gehrels and M.S. Matthews). Tucson: University of Arizona Press, pp. 54–65.
Geary, J.C. (1989) CCD imagers for astronomy: past problems and future hopes. *Highlights of Astron*, **8**, 623–7.
Gehrels, T. (1991) Scanning with CCDs. *Space Sci. Rev.*, **58**, 347–75.
Gudehus, D.H. and Hegyi, D.J. (1985) The design and construction of a CCD imaging system. *Astron. J.*, **90**, 130–8.
Janesick, J. and Blouke, M. (1987) Sky on a chip: the fabulous CCD. *Sky and Telescope*, **74**, 238–42.
Janesick, J. and Elliott, T. (1992) History and advancement of large area array scientific CCD imagers. *Astron. Soc. Pacific Conf. Ser.* **23**, 1–68.
Janesick, J.R., Klaasen, K.P. and Elliott, T. (1987) CCD charge-collection efficiency and the photon-transfer technique. *Optical Eng.*, **26**, 972–980.
Janesick, J. R., Elliott, T., Collins, S. et al. (1987) Scientific CCD's. *Optical Eng.* **26**, 692–714.
Janesick, J., Elliott, T., Bredthauer, R. et al. (1989) Recent developments in large area scientific CCD image sensors. *Proc. SPIE*, **1071**, 115–133.
Janesick, J., Elliott, T., Dingizian, A. et al. (1990) New advancement in CCD technology – sub-electron noise and 4096 × 4096 pixel CCDs. *Astron. Soc. Pacific Conf. Ser.*, **8**, 18–39.

Jewitt, D. and Luu, J. (1992) 1992QB1. *IAU Circular* 5611.
Karkoschka, E. (1992) Diurnal variations on Jupiter and Saturn? *Icarus*, **97**, 182–6.
Kerr, R.A. (1991). Galileo hits its target. *Science*, **254**, 1109.
Kostkowski, H.J. (1979) Precise instruments and accurate standards do not insure accurate measurements. Reprinted in 1990 in *Selected Papers on Radiometry* (ed. I.J. Spiro). Bellingham, WA: SPIE Optical Engineering Press, pp. 184–9.
Kristian, J. and Blouke, M. (1982) Charge-coupled devices in astronomy. *Sci. Am.*, **247**, 67–74.
Leach, R.W. (1987) Optimizing CCD operation for optical astronomy. *Optical Eng.*, **26**, 1061–6.
Mackay, C. (1986) CCDs in astronomy. *Ann. Rev. Astron. Astrophys.*, **24**, 255–75.
Marcy, G.W. and Butler, R.P. (1992) Precision radial velocities with an iodine absorption cell. *Publ. Astron. Soc. Pacific*, **104** 270–7.
Massey, P. and Jacoby, G.H. (1992) CCD data: the good, the bad, and the ugly. *Astron. Soc. Pacific Conf. Ser.*, **23**, 240–58.
McMillan, R.S., Moore, T.L., Perry, M.L. et al. (1993) Radial velocity observations of the sun at night. *Astrophys. J.* **403**(2), 801–9.
Moreno, F., Molina, A. and Ortiz, J.L. (1991) CCD spectroscopic observations of Saturn, Uranus, Neptune, and Titan during the 1990 apparitions. *Icarus*, **93**, 88–95.
Rabinowitz, D.L. (1991) Detection of Earth-approaching asteroids in near real time. *Astron. J.*, **101**(4), 1518–29.
Rabinowitz, D.L. (1992a) The flux of small asteroids near the Earth, in *Asteroids, Comets, and Meteors IV*, (eds A. Harris and E. Bowell). Houston: Lunar and Planetary Institute, pp. 481–6.
Rabinowitz, D.L. (1992b) 1992AD. *IAU Circular* 5435.
Raloff, J. (1978) CCDs: astronomy's superchips. *Sci. News*, **114**, 146–7.
Rizk, B., Wells, W.K., Hunten, D.M. et al. (1991) Meridional Martian water abundance profiles during the 1988–1989 season. *Icarus*, **90**, 205–13.
Robinson, L.B., Brown, W. and Gilmore, K. (1991) Performance tests of large CCDs. *Proc. SPIE*, **1447**, 214–28.
Rufener, F. (1989) Introduction and basic references for stellar photometry with CCDs. *Highlights of Astron.*, **8**, 617–21.
Scotti, J.V., Rabinowitz, D.L. and Marsden, B.G. (1991) Near miss of the earth by a small asteroid. *Nature*, **354**, 287–9.
Sprague, A.L., Kozlowski, R.W.H., Hunten, D.M. et al. (1992) The sodium and potassium atmosphere of the Moon and its interaction with the surface. *Icarus*, **96**, 27–42.
Stetson, P. (1989) Some factors affecting the accuracy of stellar photometry with CCDs (and some ways of dealing with them). *Highlights of Astron.*, **8**, 635–44.
Tobin, W. (1993) Problems of CCD flat fielding, in *Stellar Photometry* (eds I. Elliott and J. Butler). Proc. IAU Colloquium 136. Cambridge: Cambridge University Press, pp. 304–10.
Tyson, J.A. (1986) Low-light-level CCD imaging in astronomy. *Optical Soc. Am.*, **A3**, 2131–8.
Tyson, J.A. (1990) Progress in low-light-level CCD imaging in astronomy. *J. Optical Soc. Am.*, **A7**, 1231–6.
Veverka, J., Belton, M., Chapman, C. and the Galileo Imaging Team (1992) Gaspra: overview of Galileo imaging results. *Bull. Am. Astron. Soc.*, **24** (3), 931–3.
Walker, A. (1993) Photometry with CCDs, in *Stellar Photometry*, (eds I. Elliott and J. Butler) Proc. IAU Colloquium 136. Cambridge: Cambridge University Press.
Wisniewski, W.Z. (1991) Physical studies of small asteroids. I. Lightcurves and taxonomy of 10 asteroids. *Icarus*, **90**, 117–22.
Wisniewski, W.Z. and McMillan, R.S. (1987) Differential CCD photometry of faint asteroids in crowded star fields and nonphotometric sky conditions. *Astron. J.*, **93**, 1264–7.
Young, A.T. (1974) Photomultipliers: their cause and cure. In *Astrophysics* (ed. N. Carleton) *Methods of Experimental Physics*, **12**, part A: New York: Academic Press, pp. 1–94.

Cross references

Astrometric observation
Comet: observation
Photometry
Polarimetry
Remote sensing
Spectroscopy: atmosphere
Visible and near-infared spectroscopy

CHARGED PARTICLE OBSERVATION

Charged particles are ubiquitous in the space environments of planets, their satellites and the solar wind. The particle energies of interest in planetary and space environments range from less than 1 eV to more than 1 GeV, but this article will focus on experimental techniques for observing charged particles above several keV. In the solar wind, a spacecraft detector typically measures particles of energies up to a few keV for the bulk of the solar wind population, with the distribution including a high-energy tail extending to above 1 MeV. The energetic particle populations are enhanced by orders of magnitude following solar activity such as solar flares and coronal mass ejections, and they are also greatly enhanced by shock waves in the solar wind which can be produced not only by supersonic collisions of solar wind plasma streams with each other but also by the solar wind collisions with planetary magnetospheres or ionospheres. Within planetary magnetospheres, charged particles are found across the full spectrum of particle energies, but the total plasma density is generally dominated by particles below a few tens of eV, while the total plasma pressure is typically dominated by energetic particles above several tens of keV. The total plasma pressure generally determines the distortion of the magnetic field away from a vacuum (zero current) configuration. Energetic particles measurements address issues of magnetospheric particle sources and losses, transport, acceleration mechanisms and dynamic processes.

Energetic charged particle measurements seek to determine one or more of the following: (1) species identification, meaning electron or ion, mass and charge state (number of charges) of an ion, and atomic or molecular nature of an ion; (2) differential intensity, which is defined as the number of particles of given species incident per unit area, per unit time, per unit solid angle in a specified direction per energy range; (3) the variation of differential intensity with energy, which is the spectrum; (4) the variation of this quantity with direction, which is the angular distribution; and (5) the variation with position and time within the object under study, which reflect the plasma structure and dynamics. In addition to characterizing incident charged particles, actual spacecraft instruments must also reject background sources (e.g. photons) and satisfy constraints on cost, size, mass and power consumption. Actual instruments typically attempt only a subset of the above measurements with carefully chosen energy and angular coverage.

Energetic charged particles are detected following their interaction with the matter in a detector. The interaction of charged particles with matter is conventionally described in terms of energy per nucleon, which is a function of velocity. In the energy range of interest above several keV per nucleon, charged particles lose energy mainly to ionization of detector material; at lower energies other energy loss mechanisms such as elastic collisions with atomic cores become important. The widely used silicon solid state detector measures total ionization energy loss of an incident charged particle. If the particle stops in the detector, this measurement allows determination of the incident energy. Solid state detectors are often used in combinations with thin foils or additional absorbers. Such combinations, with coincidence–anticoincidence techniques, measure both total energy and energy loss rate, and allow determination of particle species above a few hundred keV per nucleon. Another type of detector in widespread use is the microchannel plate, which converts the one or more secondary electrons from an incident particle impact into a large electron cloud that can trigger a counting circuit. Microchannel plate detectors can be combined with position-sensitive anodes so as to record both the position and the time of a particle impact, e.g. in the focal plane of an imager or particle spectrometer.

Additional techniques are used to determine particle species. Magnetic fields can provide unambiguous separation of electrons from ions up to some maximum energy determined by mass and volume constraints on the magnet and collimator. Electric fields can provide separation of incident particles according to their energy per charge, and time-of-flight measurements directly determine velocity. One or more of these measurements can be combined to characterize the incident particle. For example, the energy per charge, velocity and total energy of an ion can all be measured so that its mass, energy and charge state are all determined. Multiple coincidence and time-of-flight techniques serve the additional purpose of suppressing background counts from photons and cosmic rays.

Andrew F. Cheng

Bibliography

Gloeckler, G. et al. (1985). The charge–energy–mass (CHEM) spectrometer for 0.3–300 keV ions on the AMPTE CCE. *IEEE Trans. Geosci. Remote Sensing*, **23**, 234–40.

Krimigis, S.M. et al. (1977). The low energy charged particle (LECP) experiment on the Voyager spacecraft. *Space Sci. Rev.*, **21**, 329–54.

McEntire, R.W. et al. (1985). The medium energy particle analyzer (MEPA) on the AMPTE CCE spacecraft. *IEEE Trans. Geosci. Remote Sensing*, **23**, 230–4.

Hovestadt, D. et al. (1978). The nuclear and ionic charge distribution particle experiments on the ISEE-1 and ISEE-C spacecraft. *IEEE Trans. Geosci. Remote Sensing*, **16**, 166–75.

Williams, D. J. et al. (1992). The Galileo energetic particles detector. *Space Sci. Rev.*, **60**, 385–412.

Cross references

Ionosphere
Magnetospheres of the outer planets
Solar wind
Whistler

CHARON

Satellite of Pluto (q.v.), Charon was discovered in 1978 by James W. Christy of the US Naval Observatory (see Pluto). It orbits only about 19000 km from Pluto (less than 1 arcsec separation seen from Earth), with a period of 6.387 days. This is also the rotational period of Pluto. A series of rare eclipses within the Pluto–Charon pair were observed beginning in 1985 (Binzel et al., 1985). These enabled observers to better determine the masses, sizes and albedos of the two bodies. Charon has a radius of 600–640 km, and a density of about 1.3 kg m^{-3}. It appears to have no atmosphere, but the presence of water ices was confirmed. The latter fact contrasts with the observations for Pluto, where CH_4, N_2 and CO ices were detected. Water ice is structurally strong and may preserve features formed during the earliest period of planetary formation. The Pluto–Charon system represents the remaining unexplored planetary component of the solar system. The pair are currently near perihelion, offering a rare opportunity for direct exploration during the next two decades. Conditions will be unfavorable thereafter until the 2060s (Stern, 1993).

James H. Shirley

Bibliography

Binzel, R.P., Tholen, D.J. et al. (1985) The detection of eclipses in the Pluto–Charon system. *Science*, **228**, 1193–1195.

Stern, S.A. (1993) The Pluto reconnaissance flyby mission. *EOS*, **74**, 73–78.

Cross references

Pluto
Satellite (natural)

CHEMICAL ELEMENT

Chemical elements and their geochemical classification

A chemical element is a substance which cannot be decomposed or resolved into simpler substances by ordinary chemical means. Altogether 105 elements are known. Each element consists of one or more atoms with the same number of protons in the nucleus and electrons outside the nucleus. Not all atoms of the same element contain the same number of neutrons. Eighty-nine elements occur naturally within the Earth's surface layers; of these, 82 are stable. Some elements (16) have been synthesized experimentally on Earth and are believed to exist in the interiors of stars. Only about 30 elements occur in their free (uncombined) state at or near the Earth's surface.

Spectroscopic analysis of stars and nebulae and chemical analyses of meteorites yield information on the chemical composition of the universe (see Cosmochemistry). Hydrogen (H) and Helium (He), atomic numbers 1 and 2, are the most abundant elements known. Because the heavier elements probably formed by nuclear reactions involving H and He in stars, the abundance of elements is related to their nuclear properties, not their chemical properties. A general trend of decreasing natural abundance with increasing atomic number is observed. Also, those elements with an even atomic number are relatively more abundant than elements with odd numbers of protons.

Some elements behave in ways chemically similar to other elements. In ancient to medieval times, only four elements were recognized (by associated qualities): earth, air, fire and water. The modern concept of elements began with Robert Boyle in 1661 and the search for distinct elements and their quantitative attributes. As early as 1829, scientists started to look for a systematic arrangement of the elements in order to simplify their study. In the mid-19th-century the periodic law established that 'properties of elements are periodic functions of their atomic weights'. Numerous scientists in the nineteenth and twentieth centuries published versions of the Periodic Table, but in 1882 the Davey medal of the Royal Society of London was awarded jointly to the Russian D.I. Mendeleyev (1834–1907) and the German Lothar Meyer (1830–1895) for the development of the modern Periodic Table. Gaps left in the first periodic tables led to discoveries of many unknown elements. For more information on the history and structure of the Periodic Table including an extensive bibliography, see Mazurs (1974).

The size of the element and its ionization potential establishes its place in the geochemical classification of the elements originally published in 1937 by the Norwegian Victor M. Goldscmidt (1888–1947). The geochemical classification of the elements is a qualitative expression of the chemical behavior of the elements in and on the Earth. It shows chemical tendencies, not quantitative relationships. Elements can be classified in more than one group, depending on the relative abundance of iron, sulfur and oxygen in the environment. Goldschmidt used data on the affinity of various elements for oxygen and sulfur to establish his basic geochemical classification.

Originally, Goldschmidt used five chemical groups, atmophile, biophile, chalcophile, lithophile and siderophile (spelled, in German, without the final 'e'). To classify all elements, the categories 'artificial' elements and decay products, encompassing elements either unknown or not well researched in 1937, should be added to the list. For most geochemical purposes and for planetary science, chalcophile, lithophile and siderophile are the most accepted and useful classes (Ahrens, 1964). A diagram showing the Periodic Table and the geochemical classifications of the elements is shown in Figure C23.

- 'Artificial' (synthetic elements, elements which do not occur naturally except in stellar interiors). Artificial elements are radioactive elements, generally with very short half-lives. Sixteen artificial elements have been synthesized to date.
- Atmophile (Greek: steam or vapor-loving, elements which occur either in the uncombined state or as volatile compounds). Atmophile elements are prominent in the Earth's atmosphere and other natural gases. Examples of the latter are the noble gases such as argon and neon and the halogens such as chlorine, bromine and iodine.
- Biophile (Greek: life-loving, elements which are concentrated in or by living plants and animals). Biophile elements include carbon, hydrogen, oxygen, nitrogen, phosphorus, sulfur, chlorine and iodine. The more common, oxygen, carbon, hydrogen, sulfur and phosphorus are involved in what Goldschmidt called the 'biogeochemical cycle'.
- Chalcophile (Greek: copper-loving, elements like copper, tending to concentrate in sulfide ores). Chalcophile elements generally form covalent bonds with sulfur. Two groups of chalcophile elements are recognized, depending on the amount of sulfur and

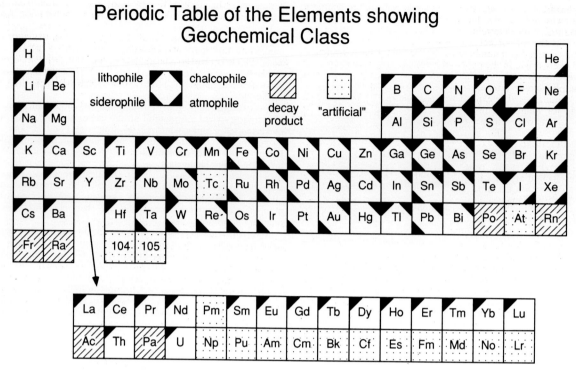

Figure C23 Periodic Table and geochemical classification of the elements. Elements can occur in more than one geochemical classification depending on the chemical environment. Biophile elements are not distinguished on this diagram.

oxygen in the environment of formation. When oxygen and sulfur compete for metals, as in meteorites or in the metallic core of the Earth, certain elements prefer the sulfide environment. A different distribution of elements prefer the sulfide phases when excess oxygen and sulfur are present, that is, when little or no uncombined metallic elements are present, as in the crust of the Earth. In the environment of excess metal, elements such as Mn, Cr and V are highly concentrated in sulfides but the sulfides include practically no Ni or Co. In the environment of excess sulfur and oxygen, Ni and Co are concentrated in the sulfide ores while Cr, V and Mn are present in the oxygen compounds.

- Decay products (those elements which occur in nature only as decay products of uranium and thorium). Elements which are decay products occur where the parent elements occur, regardless of their geochemical preference. They are often concentrated in soils, regolith and adjacent sediments.
- Lithophile (Greek: stone-loving, elements tending to concentrate in stony matter). Lithophile elements form ionic bonds with oxygen. They occur in silicates in both the crust of the Earth and in the stony portion of stony meteorites. Examples include the rare earth elements, the alkali and alkaline earth metals and the halogens.
- Siderophile (Greek: iron-loving, elements tending to concentrate in metallic iron). Siderophile elements are those which are not readily available for combination with other elements; they occur in their native state, although rarely in the Earth's crust. They tend to form metallic bonds. Siderophile elements are concentrated in the Earth's core and in the metallic portions of meteorites.

Clare P. Marshall

Bibliography

Ahrens, L.H. (1964) The significance of the chemical bond for controlling the geochemical distribution of the elements, in *Physics and Chemistry of the Earth*, Vol. 5, Part 1, 211–55.
Ahrens, L.H. (ed.) (1968) *Symposium on the Origin and Distribution of the Elements*. Permagon Press, 1178 pp.
Anders, E. and Grevesse, N. (1989) Abundances of the elements: meteoritic and solar. *Geochim. Cosmochim. Acta*, **53**, 197–214.
Brownlow, A.H. (1979) *Geochemistry*. Prentice-Hall, Inc. 498 pp.
Gill, R. (1989) *Chemical Fundamentals of Geology*. Cambridge: Cambridge University Press. 291 pp.
Goldschmidt, V.M. (1954) *Geochemistry*, translation from German, (ed. A. Muir). London: Oxford University Press, 730 pp.
Mazurs, E.G. (1974) *Graphic Representations of the Periodic System During One Hundred Years*. University of Alabama Press, 251 pp.
Suess, H.E. and Urey, H.C. (1956) Abundances of the elements, *Rev. Modern Phys.*, **28**(1), 53–74.

Cross references

Carbon, carbon dioxide
Cosmochemistry
Iron
Iridium anomaly
Water

CHIRON

2060 Chiron was the first asteroid discovered with a perihelion distance inside the orbit of Saturn and an aphelion distance beyond Uranus. The semimajor axis is 13.8 AU. The perihelion distance is 8.51 AU; there was a perihelion passage in 1996. Chiron was discovered in 1977 by C.T. Kowal as part of a survey searching for slow-moving outer solar system objects. The object was named after Chiron the centaur, son of Kronos (Saturn) and grandson of Uranus. Chiron's present orbit is chaotic and subject to strong perturbations by Saturn (Hahn and Rickman, 1985). Numerical integrations by Scholl (1979) suggest that Chiron is evolving inward from the Uranus–Neptune zone. Because the orbit is chaotic, it is not possible to trace the orbit of Chiron backward in time to determine its origin, or to specify how long it has been in its present orbit.

Physical studies of Chiron show a neutral reflectance spectrum, consistent with a low-albedo surface composed of dark carbonaceous material (Hartmann *et al.*, 1990). Radius estimates range from 65–200 km, depending on the value assumed for the albedo. In any case, Chiron is larger than all but a few asteroids. The rotation period is 5.92 ± 0.0001 h (Bus *et al.*, 1989); the lightcurve amplitude during the 1980s was only a few hundredths of a magnitude. Further monitoring will allow determination of Chiron's shape.

Chiron was the first object observed to display cometary activity after being classified and numbered as an asteroid. Chiron's unusual orbit and likely origin in the outer solar system suggested soon after its discovery that it might be related to comets. By 1989, Chiron was nearly one magnitude brighter than it had been in the early 1980s. Such brightening would be expected for a comet moving close enough to the Sun for volatiles in the surface layers to sublimate. In 1990 images of Chiron were finally obtained which showed a distinct coma (Meech and Belton, 1990). The solar distance at that point was 11.8 AU. CO_2 is the most likely candidate material for outgassing at that distance (Stern, 1989). Some materials, such as CO, CH_4, Ar, N_2 and $CH_4 \times H_2O$ clathrates, would be expected to show substantial activity throughout all or part of Chiron's orbit. Thus these materials cannot be present on the surface of Chiron in large quantities. The time for CO_2 to sublimate completely from the surface is estimated by Stern to be $< 10^8$ years. The sublimation time is 10^6–10^7 years for the other volatiles. Thus the presence of volatiles on the surface indicates either that the volatiles have been excavated recently or that Chiron has been inside 20 AU for less than approximately 10^9 years. More recently, CN emission from Chiron was detected (Bus *et al.* 1991) when Chiron was at a heliocentric distance of 11.26 AU. The parent molecules which give rise to CN, probably HCN and CN-bearing polymers, are thought to be carried along by outgassing driven by sublimation of CO_2 ice. Bus *et al.* (1991) observed outgassing activity from small, localized regions on the surface of Chiron, a behavior characteristic of comets which have made many passages close to the Sun.

Linda M. French

Bibliography

Bus, S.J., Bowell, E., Harris, A.W. and Hewitt, A.V. (1989) 2060 Chiron: CCD and electronographic photometry. *Icarus*, **77**, 223–38.
Bus, S.J., A'Hearn, M.F., Schleicher, D.G. and Bowell, E. (1991) Detection of CN emission from (2060) Chiron. *Science*, **251**, 774–7.
Hahn, G. and Rickman, H. (1985) Asteroids in cometary orbits. *Icarus*, **61**, 417–42.

Hartmann, W.K., Tholen, D.J., Meech, K.J. and Cruikshank, D.P. (1990) 2060 Chiron: calorimetry and cometary behavior. *Icarus*, **83**, 1–15.
Meech, K.J. and Belton, M.J.S. (1990) The atmosphere of 2060 Chiron. *Astron. J.*, **100**, 1323–38.
Scholl, H. (1979) History and evolution of Chiron's orbit. *Icarus*, **40**, 345–9.
Stern, S.A. (1989) Implications of volatile release from object 2060 Chiron. *Publ. Astr. Soc. Pacific*, **101**, 126–132.

Cross references

Asteroid
Chaotic dynamics in the solar system
Comet

CHONDRITES, ORDINARY

The chondrites constitute about 85% of the meteorites that fall to Earth. Their chief characteristics are (1) they have essentially the same composition as the solar photosphere, (2) they are comparable in age with the solar system, and (3) they have textures which indicate little or no alteration since their formation. They therefore provide a unique opportunity to investigate processes occurring during the earliest phases of solar system history (Kerridge and Matthews, 1988). The chondrites are divided into the carbonaceous (see Carbonaceous chondrite), enstatite (see Enstatite meteorite) and ordinary chondrites (Table C2), of which the last division is much the largest ($\sim 79\%$ of all meteorites).

H, L and LL classes and their significance

It has been known since the last century that iron exists in several forms in the chondrites, as metal, as silicates and as sulfides. In the carbonaceous chondrites very little iron exists as metal, while in enstatite chondrites virtually all the iron exists in metallic and sulfide forms with little in the silicates. The ordinary chondrites lie between these extremes. It was therefore assumed that the chondrites came from a common source that underwent progressive oxidation (or reduction).

After a major survey of compositional data, Urey and Craig (1952) found that the idea of a simple oxidation series was incorrect. An updated version of their plot is shown in Figure C24. Chondrites from

Table C2 Chondrite classes

Petrogic type Fa[a] (mol%)		Fe^o/Fe_t[b]	Fe/Si	Mg/Si	Ca/Si	Co in kam[c]	$\delta^{17}O$[d]	$\delta^{18}O$[d]	
			(atom/atom)						
Carbonaceous chondrites									
1–6	0–35	0	~ 0.8	~ 1.05	0.6–0.8	–	Well dispersed, generally below terrestrial fractionation line		
Ordinary chondrites									
H3–6	16–20	0.58	0.81	0.96	0.050	~ 6.5	2.9	4.1	Above
L3–6	22–25	0.29	0.57	0.93	0.046	~ 10	3.5	4.6	terrestrial fractionation
LL3–6	27–32	0.11	0.52	0.94	0.049	20–100	3.9	4.9	line
Enstatite chondrites									
3–6	<2	0.80	0.6–1.0	0.80	0.03	–	On terrestrial fractionation line		

[a] The major mineral, olivine, is a solid solution of fayalite (Fe_2SiO_4) and forsterite (Mg_2SiO_4). The above parameter refers to the concentration of fayalite (Fa) in the olivine, expressed as mole percent.
[b] Refers to the ratio of reduced Fe (metallic Fe) to the total Fe.
[c] This parameter, the Co content of the kamacite quoted above in units of mg g^{-1}, has only been discussed as a classification parameter for the ordinary chondrites.
[d] $\delta^{18}O$ and $\delta^{17}O$ are defined by the following, using $\delta^{18}O$ as an example:

$$\delta^{18}O = \left[\frac{(^{18}O/^{16}O)_{sample}}{(^{18}O/^{16}O)_{SMOW}} - 1 \right] \times 1000,$$

where SMOW is a standard, (Standard Mean Ocean Water). 'Above', 'on' and 'below' the terrestrial fractionation line refers to measurements on fairly large bulk samples. Certain components within at least one highly unequilibrated ordinary chondrite (Allan Hills 76004) plot below the terrestrial line.

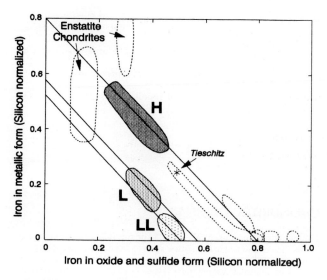

Figure C24 The ratio of Fe/Si for iron in the metallic state (Fe⁰) against Fe/Si for iron in the non-metallic (i.e. oxide, silicate and sulfide forms). Many classes of chondrite can be distinguished on this plot, most notably the H, L and LL classes of ordinary chondrite. Several type 3 chondrites (like Tieschitz) plot to the lower right of their respective class fields. The unlabeled fields refer to the carbonaceous chondrite classes.

Table C3 Major minerals in ordinary chondrites and amount by weight percent (modified after Van Schmus, 1969)[a]

		H	L	LL
Olivine	Fayalite, Fa, Fe_2SiO_4 Forsterite, Fo, Mg_2SiO_4	33–37	45–49	56–60
Low-Ca pyroxene	Ferrosilite, Fs, $FeSiO_3$ Enstatite, En, $MgSiO_3$	23–27	21–25	14–18
Ca-pyroxene	Ferrosilite, Fs, $FeSiO_3$ Enstatite, En, $MgSiO_3$ Wollastonite, Wo, $CaSiO_3$	4–5	4–5	4–5
Feldspar	Albite, Ab, $NaAlSi_3O_8$ Anorthite, An, $CaAl_2Si_2O_8$ Orthoclase, Or, $KAlSi_3O_8$	9–10	9–10	9–10
Troilite	FeS	5–6	5–6	5–6
Kamacite	FeNi (BCC)	15–17	6–8	1–2
Taenite	Fe,Ni (FCC)	2–3	2–3	2–4
Chromite	$FeCr_2O_4$	0.5	0.5	0.5
Whitlockite Chlorapatite	$Ca_2(PO_4)_3$ $Ca_5(PO_4)_3Cl$	0.6	0.6	0.6

[a] Olivine, the pyroxenes and feldspar are solid solutions of the minerals listed. Whitlockite and chlorapatite are different minerals which are difficult to distinguish optically and are here listed together.

a single oxidation series would plot along a diagonal line appropriate to the bulk iron–silicate ratio for the meteorites. What Urey and Craig observed was that the ordinary chondrites plotted along two diagonals of different Fe/Si ratio. We now know that there are three. The ratios of other siderophile elements (like cobalt, nickel and iridium which tend to concentrate in the metal phase, whenever possible) to silicon show similar trends. Apparently, the oxidation (or reduction) process experienced by the chondrites was accompanied by loss (or gain) of Fe. Ordinary chondrites are thus divided into three classes, termed H, L and LL, after their high, low and very low Fe/Si ratios.

Several other measurements which relate to the trends observed in the Urey–Craig plot may be used to classify ordinary chondrites. As the amount of metal decreases along the H–L–LL series, the amount of Fe in the silicates increases. Thus the amount of FeO in the major silicates, as measured by the amount of fayalite in the olivine or ferrosilite in the pyroxene (see Table C3), is a characteristic of each class (Fredriksson, Nelen and Frederiksson, 1968; Figure C25a). Similarly, the amount of Ni in the metal also increases along the series, as Fe is removed from the metal phase. Finally, the amount of Co in the metal (to be precise, the metal phase known as kamacite) increases along the H–L–LL series (Figure C25b; Sears and Axon, 1975).

Improvement in analytical techniques yielded other differences between the classes which precluded a simple oxidation relationship. In the 1960s it became clear that the abundance of refractory elements relative to the moderately volatile elements (like silicon) varied from class to class. While carbonaceous chondrites are close to the solar value in their Mg : Si ratio (1.05), the ordinary and enstatite chondrite values are slightly but significantly below this value and H chondrites plot in the upper part of the ordinary chondrite range (Figure C26). The Ca : Si ratios show similar effects. It is as if, during their formation, a component containing refractory material (represented by Ca and Mg) and a component containing moderately volatile elements (represented in this case by Si), became physically separated so that their Ca : Si ratios were changed.

Evidence that these compositional and oxidation state differences between the classes were due to processes occurring in the primordial solar nebula, and not on a planet-like body, was provided by measurements of the relative proportions of the three isotopes of oxygen (Clayton, Grossman and Mayeda, 1973; Figure C27). On a plot of $\delta^{17}O$ against $\delta^{18}O$, terrestrial and lunar samples lie along a line of slope 0.5. Such a slope indicates the influence of processes which are mass dependent, and this is true of all of the chemical and physical processes commonly encountered. The ordinary chondrites lie slightly but significantly above the terrestrial fractionation line, with the H, L and LL classes on a line with a slope equal to 1; $\delta^{17}O$ and $\delta^{18}O$ values increase along the H–L–LL series. For data to plot off the terrestrial line, with a slope of unity, requires the mixing of components of distinctly different nucleosynthetic origin than is normally found in terrestrial and lunar samples (e.g. an ^{16}O-rich solid and a relatively ^{16}O-poor gas). The process must have occurred prior to the isotopic homogenization of the Earth and Moon, presumably in the primordial solar nebula (see Accretion; Solar nebula; Solar system: origin).

Mineralogy

The ordinary chondrites consist primarily of the silicates olivine, pyroxene and plagioclase and of metal which exists as two alloys (kamacite and taenite). There is also a sulfide unique to meteorites, termed troilite (FeS). There are also a great many minor and trace minerals found in chondrites, such as chromite, whitlockite and chlorapatite. A recent count revealed over 100 minerals recorded in chondrites. Some of the trends in metal and silicate abundance discussed above are apparent in the data in Table C3.

Texture

Chondrites are also noteworthy for their texture; indeed their name derives from *chondros* (Greek, 'a grain') which refers to ubiquitous spherical or near-spherical objects present in the meteorites, which are now known as 'chondrules'. There is agreement among petrologists that the chondrules were formed by a flash-melting event, but although it must have occurred before the rock was assembled in its present form, the details of where and how the heating occurred are unknown. Located between the chondrules and the grains of metal and sulfide is a silicate-rich material rather loosely termed the 'matrix'.

Chondrules

Chondrules range from a few micrometers to over 200 μm in size and have a variety of internal textures (Gooding and Keil, 1979), which may be due to factors such as maximum temperature, number and type of nucleation centers and cooling rate (Hewins, 1988; Lofgren, 1989). They are often surrounded by rims of matrix-like material, often rich in fine-grained metal and sulfide, which are thought to be dust accreted onto the chondrule before it became part of the meteorite.

There is considerable compositional diversity among the chondrules with some being highly refractory (free of volatiles) and reduced

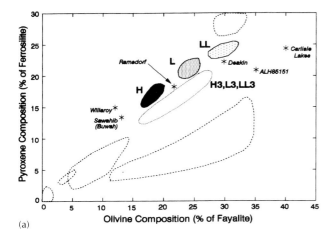

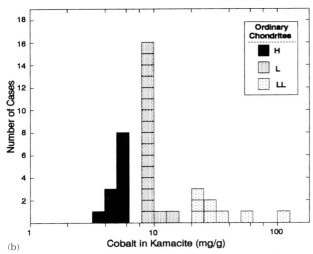

Figure C25 (a) Ferrosilite content of the pyroxene against the fayalite content of the olivine for ordinary chondrites. The three ordinary chondrite groups, corresponding to H, L and LL chondrites, are apparent in this plot. The field occupied by type 3 ordinary chondrites and data for several unusual ordinary chondrites are also plotted. The unlabeled fields refer to the carbonaceous chondrite classes. (Enstatite chondrites contain little or no olivine). (b) Cobalt content of the kamacite for ordinary chondrites. This parameter also enables the H, L and LL chondrites to be identified.

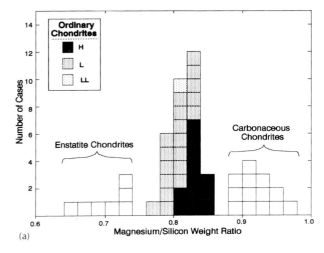

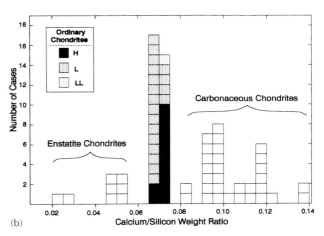

Figure C26 Histograms of (a) Mg/Si and (b) Ca/Si ratios in chondritic meteorites. These ratios permit subdivision of the chondrites into carbonaceous, ordinary and enstatite chondrite classes. Within the ordinary chondrites, H chondrites plot at the upper end of the ranges.

(their silicates are Fe free), while others contain volatile element abundances and silicates quite rich in iron (e.g. Sears *et al.*, 1992). Parent body processes (i.e. metamorphism, see below) caused these two groups to converge (in mineral composition) to average values. Significantly, there are a few chondrules of these average mineral compositions in the unmetamorphosed chondrites.

The main question concerning the origin of the chondrule groups is whether they are due to differences in the original material, or whether the differences were caused by the flash-heating event. Consistent with the latter, the amount of the volatile element Na and the amount of ^{16}O in the chondrules is related to their size. Both the loss of Na and exchange of ^{16}O would be size-dependent processes. Many chondrules contain fragments of other chondritic components such as fine-grained matrix, refractory inclusions and grains from other chondrule classes, and the narrow spread in their oxygen isotope ratios has been ascribed to multiple episodes of reheating (Clayton *et al.*, 1991).

Matrix

The matrix is a mixture of silicate minerals which may be either fine-grained or very coarse. Some authors have suggested that the matrix is primitive nebula dust (e.g. Scott *et al.*, 1984), while others have suggested that it was derived from the chondrules (e.g. Alexander, Hutchison and Barbes, 1989). The picture is complicated by brecciation, which adds fragments and chondrules to the matrix.

Sulfide and metal

These tend to be associated with each other and in the unmetamorphosed meteorites take a variety of textures, including spherules resembling chondrules. Some of the metal and sulfide exists as very finely dispersed grains in rims around chondrules. In metamorphosed meteorites these phases have increased in grain size and are intergrown with the silicates (Afiattalab and Wasson, 1981).

Metamorphism and other secondary processes

Metamorphism

After the formation of chondrites by the accretion of chondrules, matrix, sulfide and metal onto their parent bodies, most chondrites underwent a protracted period of heating which caused many solid state physical and chemical changes (metamorphism). This has lead to subdivision of the chondrites by Van Schmus and Wood (1967) into six 'petrographic types' using a variety of petrologic and compositional criteria (Table C4).

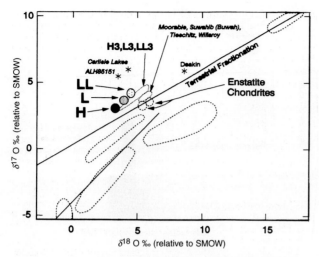

Figure C27 Plot of the oxygen isotope abundances in chondritic meteorites. The data are shown as the differences (in parts per thousand) in the ratio for ^{17}O and ^{18}O to ^{16}O compared with the laboratory standard SMOW (standard mean ocean water). The diagonal line is the terrestrial fractionation line, defined by terrestrial and lunar samples. The ordinary chondrites plot above this line, the H, L and LL chondrites occupying close but separate fields on a line with a slope of 1. For comparison, the enstatite chondrites plot on the terrestrial line while the carbonaceous chondrites (unlabeled fields) plot on or below the line and spread widely. Data for some unusual ordinary chondrites and the field occupied by type 3 ordinary chondrites are also plotted.

Types 1 and 2 are reserved for carbonaceous chondrites and largely describe the effects of aqueous alteration. The type 3 ordinary chondrites have experienced a wide range of metamorphic alteration and are subdivided into types 3.0 to 3.9 (Sears et al., 1980; Table C4). Among the best studied of the ordinary chondrites is Semarkona, which is the only known 3.0 ordinary chondrite. Types 4–6 (sometimes termed 'equilibrated') represent high levels of metamorphism, so that in some even the coarsest textures are blurred.

The most unequilibrated ordinary chondrites contain isotopically heavier oxygen than the equilibrated (types 4–6) chondrites. This probably reflects the loss of CO or CO_2 during open-system metamorphism (Clayton et al., 1991). Abundances of highly volatile elements, e.g. In, Tl, Bi and Pb, decrease by several orders of magnitude as petrographic type increases. Some authors also attribute this to evaporative loss during metamorphism, although others suggest that it may reflect accretion of the meteorites at different temperatures.

The lack of equilibrium between minerals makes it difficult to estimate metamorphic temperatures for type 3 chondrites, but thermoluminescence data suggest values of about 500°C for type 3.5 chondrites (Guimon, Keck and sears, 1985). For the higher petrographic types oxygen isotope data and mineral compositions suggest metamorphic temperatures of 600–700°C for type 4, 700–750°C for type 5 and 750–950°C for type 6 (McSween, Sears and Dodd, 1988). A few chondrites which underwent incipient melting during metamorphism are termed 'type 7'.

Shock heating

Some chondrites experienced short-term excursions to high temperature as a result of impact in space. Stöffler, Keil and Scott (1991) have systematized and summarized observations in a shock classification scheme. Shock causes the loss of volatile elements, especially the light rare gases; it causes feldspars and sulfides to melt and olivines and metal to assume distinctive optical properties.

Table C4 Petrographic types of chondritic meteorites (see Sears et al., 1991, for references. Data on water in type 3 chondrites are from Jarosewich, 1980)

Petrologic type	Characteristic[b]							
	1	2	3	4	5	6	7	8
1	n.a.[a]	0	n.a.[a]	0	>1.9	~3.5	~6	≥65
2	≥50	0	n.a.[a]	0	>1.9	1.5–2.8	3–11	≥65
3.0	≥50	<0.0046	>26.3	<10	>1.9	≥0.60	~2	≥65
3.1	≥50	0.0046–0.01	26.3–21.3	10–20	1.9–1.7	0.60–0.50	~2	65–55
3.2	≥50	0.010–0.022	21.3–16.3	10–20	1.7–1.6	0.50–0.43	~2	55–45
3.3	≥50	0.022–0.046	16.3–12.5	10–20	1.6–1.5	0.43–0.38	≤2	45–35
3.4	≥50	0.046–0.10	12.5–10.0	~20	1.5–1.4	0.38–0.33	≤2	35–27
3.5	41–50	0.10–0.22	10.0–7.5	~50	1.4–1.3	0.33–0.30	≤2	27–18
3.6	31–40	0.22–0.46	7.5–6.3	>60	1.3–1.2	0.30–0.27	≤2	18–13
3.7	21–30	0.46–1.0	6.3–3.1	>60	1.2–1.1	0.27–0.24	<2	13–8
3.8	11–20	1.0–2.2	3.1–1.9	>60	1.1–1.0	0.24–0.21	<2	8–4
3.9	5–10	2.2–4.6	<1.9	>60	1.0	≤0.21	<2	<4
4	<5	4.6–10	<1.9	>60	1.0	<0.2	<2	<4
5	<5	>10	<1.9	100	1.0	<0.2	<2	<4
6	<5	>10	<1.9	100	1.0	<0.2	<2	<4

[a] n.a. = not applicable.
[b] Characteristics:
1. Olivine (or pyroxene) heterogeneity (standard deviations of Fa or Fs, divided by the mean, expressed as a percentage).
2. Thermoluminescence sensitivity (on a scale of Dhajala equals 1.0).
3. Heterogeneity of Co in the kamacite (standard deviation divided by the mean, expressed as a percentage).
4. Percentage of the matrix which is recrystallized.
5. FeO/(FeO+MgO) value for the matrix divided by the same value for the bulk rock.
6. Bulk carbon content (wt%).
7. Bulk water content for observed falls (wt%).
8. Bulk ^{36}Ar content (in units of $10^{-8} cm^{-3} STP g^{-1}$).

Other (mainly descriptive) characteristics:
a. The bulk Ni content of metal in type 1 and 2 chondrites is <20% so that taenite is absent or minor in these meteorites.
b. The sulfides in types 1 and 2 contain significant Ni (>0.5%).
c. The structural state of the low-Ca pyroxene in types 2 and 3 is predominantly monoclinic, in type 4 it is >20% monoclinic, in type 5 it is < 20% monoclinic while in type 6 it is orthorhombic. Pyroxene is absent in type 1 chondrites.
d. Igneous glass is clear and isotropic in types 2 and 3, turbid in type 4 and absent in types 1 and 6.
e. Secondary feldspar is absent in types 1–3, present as <2 μm grains in type 4, as <50 μm grains in type 5 and as >50 μm grains in type 6.
f. Chondrules are absent in type 1, very sharply defined in types 2 and 3, well-defined in type 4, readily delineated in type 5 and poorly defined in type 6.

Aqueous alteration

Hutchison, Alexander and Barber (1987) reported hydrous silicates and calcite with a morphology indicative of deposition from aqueous fluids in the meteorites Semarkona (LL3.0) and, subsequently, Bishunpur (LL3.1). The relative importance of aqueous alteration and dry thermal metamorphism in the history of type 3.0–3.2 chondrites is unclear (Alexander, Hutchison and Barber, 1989; Guimon, Keck and Sears, 1985; McSween, Sears and Dodd 1988).

Unusual ordinary chondrites

There are several anomalous ordinary chondrites. Tieschitz and Chainpur appear to be intermediate between H and L chondrites. Carlisle Lakes and Allan Hills A80151 are type 3.7 chondrites with especially high Fa values, high amounts of matrix and unusual oxygen isotope properties (Rubin and Kallemeyn, 1989). Pecora Escarpment 91002 may be a similar meteorite. Deakin 001 is a highly unequilibrated LL3 chondrite with heavier oxygen than observed for other type 3 chondrites (Bevan and Binns, 1989). Moorabie, Willaroy and Suwahib (Buwah) are type 3 ordinary chondrites with low Fe olivines and heavy oxygen isotopes but whose olivine heterogeneity and thermoluminescence data suggest are type 3.5–3.9 (Scott, Clayton and Mayeda, 1985). It is likely that the history of these 'unusual' ordinary chondrites, and several others that have been described in the literature, will not be well understood until more samples are available.

History and origin

Early history

A variety of radiometric dating methods indicate formation ages of 4.6 Ga for the ordinary chondrites, comparable to the ages of the Moon, Earth and Sun (e.g. Tilton, 1988) (see Chronology: meteorite). The presence of the decay products of now extinct parent nuclides provides a means of discriminating small intervals of time near the beginning of solar system history. The data indicate that the chondrites formed soon after the end of nucleosynthesis and that they formed over a small time span of less than 100 million years (Swindle and Podosek, 1988).

Many low-petrographic type meteorites also contain unusual patterns of rare gas isotopes which are thought to indicate the presence of components which predate the formation of the solar system. These exotic mixtures of isotopes are found in diamond, silicon carbide and other carbon-bearing phases (Anders, 1988; Zinner, 1988). Their abundance decreases with petrographic type, presumably indicating destruction of the carrier phases during metamorphism (Huss, 1989).

Their essentially solar composition, their age and their texture indicate that the ordinary chondrites are aggregations of dust, chondrules, metal and sulfide which were present in the primordial solar nebula during planet formation. The small but significant differences in elemental and isotopic composition from which we derive the ordinary chondrite classes may be the result of (1) chemical reactions and related physical processes in the nebula, or (2) they may be associated with the ubiquitous, but poorly understood, chondrule formation process. Detailed thermodynamic models have been developed which are successful in explaining many features of the elemental and isotopic properties of the ordinary chondrite classes (e.g. Sears, 1988), but they may be applied equally well to processes occurring in the nebula and during chondrule formation.

Recent history

Most authorities accept that chondrites are pieces of asteroids. Spectral reflectivity data for several small meteorite classes closely resemble those of certain asteroids, but the largest chondrite classes do not have close matches in the main asteroid belt (Gaffey, 1976). There are, however, several asteroids currently on Earth-crossing orbits whose spectra match those of the ordinary chondrites (see Asteroid; Reflectance spectroscopy). Orbital calculations indicate that gravitational reasonances with Jupiter provide a mechanism for transferring asteroids to Earth-crossing orbits (Wisdom, 1985).

The rates of cooling following metamorphism can be estimated from nickel profiles in the metal grains and the from the abundance of tracks of radiation damage caused by Pu fission fragments (Pellas and Storzer, 1981). The results range from 1 to 100°C per million years, which are consistent with asteroid sized parent bodies, assuming that the heating was caused by long-lived radioactive elements (Wood, 1979). If the heat source was ^{26}Al, the objects could have been much smaller. Some authors claim that the parent object of the ordinary chondrites had a simple onion skin structure with the slowly cooled, highly metamorphosed meteorites at the center (Pellas and Storzer, 1981). The inverse correlation between Pb–Pb age and petrographic type recently reported by Göpel, Manhes and Allegre (1992) is consistent with such an idea, since presumably the first solids to form in the nebula would be buried most deeply. However, others have questioned the existence of a relationship between cooling rate and metamorphism and suggest instead that the parent bodies may have been random mixtures of materials, perhaps even produced by the reassembling of previously disrupted objects (Taylor et al., 1987).

A significant fraction of the L chondrites have K–Ar ages of about 500 Ma and show petrographic evidence for intense shock heating. It is widely assumed that the L chondrite parent body broke up in a very violent manner 500 Ma ago (Heymann, 1967).

More recently, ordinary and enstatite chondrites underwent further fragmentation, this time down to meter-sized objects which were small enough to undergo nuclear reactions with cosmic rays (Crabb and Schultz, 1981). The abundances of the isotopes produced by these reactions provide an estimate of the length of exposure to cosmic rays (see Cosmic ray exposure age). About one-third of the H chondrites have cosmic ray exposure ages of about 8 Ma, indicating a major impact event, or perhaps even break-up of the H chondrite parent body at that time. Others range from ≤ 1 up to about 100 Ma, consistent with the timescales for material coming from the asteroid belt to Earth via the resonance mechanisms. There is evidence that H chondrites from the 8 Ma impact have been coming to Earth over the last 1 Ma in a way that can provide new insights into parent body structure (Benoit and Sears, 1992).

Derek W.G. Sears

Bibliography

Afiattalab, F. and Wasson, J.T. (1981) Composition of the metal phases in ordinary chondrites: implications regarding classification and metamorphism. *Geochim. Cosmochim. Acta*, **44**, 431–46.

Alexander, C.M.O., Hutchison, R. and Barber, D.J. (1989) Origin of chondrule rims and interchondrule matrices in unequilibrated ordinary chondrites. *Earth Planet. Sci. Lett.*, **95**, 187–207.

Anders, E. (1988) Circumstellar material in meteorites: noble gases, carbon and nitrogen, in *Meteorites and the Early Solar System* (eds J.F. Kerridge and M.S. Matthews). Tucson: University of Arizona Press, pp. 927–56.

Benoit, P.H. and Sears, D.W.G. (1992) The break-up of a meteorite parent body and the delivery of meteorites to Earth. *Science*, **255**, 1685–7.

Bevan, A.W.R. and Binns, R.A. (1989) Meteorites from the Nullarbor Region, Western Australia: II. Recovery and classification of 34 new meteorite finds from the Mundrabilla, Forrest, Reid and Deakin areas. *Meteoritics*, **24**, 135–41.

Clayton, R.N., Grossman, L. and Mayeda, T.K. (1973) A component of primitive nuclear composition in carbonaceous meteorites. *Science*, **182**, 485–8.

Clayton, R.N., Mayeda, T.K., Goswami, J.N. and Olsen, E.J. (1991) Oxygen isotope studies of ordinary chondrites. *Geochim. Cosmochim. Acta*, **55**, 2317–37.

Crabb, J. and Schultz, L. (1981) Cosmic-ray exposure ages of the ordinary chondrites and their significance for parent body stratigraphy. *Geochim. Cosmochim. Acta*, **45**, 2151–60.

Fredriksson, K., Nelen, J. and Fredriksson, B.J. (1968) The LL-group chondrites, in *Origin and Distribution of the Elements* (ed. L.H. Ahrens). Pergamon, pp. 457–66.

Gaffey, M.J. (1976) Spectral reflectance characteristics of the meteorite classes. *J. Geophys. Res.*, **81**, 905–920.

Gooding, J.L. and Keil, K. (1979) Relative abundances of chondrule primary textural types in ordinary chondrites and their bearing on conditions of chondrule formation. *Meteoritics*, **16**, 17–43.

Göpel, C., Manhes, G. and Allegre, C.J. (1992) Constraints on the time of accretion and thermal evolution of chondrite parent bodies by precise U–Pb dating of phosphates. *Meteoritics*, **26**, 338.

Guimon, R.K., Keck, B.D. and Sears, D.W.G. (1985) Chemical and physical studies of type 3 chondrites – IV: annealing studies of a type 3.4 ordinary chondrite and the metamorphic history of meteorites. *Geochim. Cosmochim. Acta*, **19**, 1515–24.

Hewins, R.H. (1988) Experimental studies of chondrules, in *Meteorites and the Early Solar System* (eds. J.F. Kerridge and M.S. Matthews). Tucson: University of Arizona Press, pp. 660–79.

Heymann, D. (1967) On the origin of the hypersthene chondrites: ages and shock effects of black chondrites. *Icarus*, **6**, 189–221.

Huss, G.R. (1989) Ubiquitous interstellar diamond and SiC in primitive chondrites: abundances reflect metamorphism. *Nature*, **347**, 159–162.

Hutchison, R., Alexander, C.M.O. and Barber, D.J. (1987) The Semarkona meteorite: first recorded occurrence of smectite in an ordinary chondrite, and its implications. *Geochim. Cosmochim. Acta*, **51**, 1875–82.

Jarosewich, E. (1980) Chemical analysis of meteorites: a compilation of story and iron meteorite analyses. *Meteoritics*, **25**, 323–7.

Kerridge, J.F. and Matthews, M.S. (eds) (1988) *Meteorites and the Early Solar System*. Tucson: University of Arizona Press.

Lofgren, G.E. (1989) Dynamic crystallization of chondrule melts of porphyritic olivine composition: texture experimental and natural. *Geochim. Cosmochim. Acta*, **53**, 461–70.

McSween, H.Y., Sears, D.W.G. and Dodd, R.T. (1988) Thermal metamorphism, in *Meteorites and the Early Solar System* (eds J.F. Kerridge and M.S. Matthews). Tucson: University of Arizona Press, pp. 102–13.

Pellas, P. and D. Storzer (1981) ^{244}Pu fission track thermometry and its application to stony meteorites. *Proc. Roy. Soc. London*, **A374**, 253–70.

Rubin, A.E. and Kallemeyn, G.W. (1989) Carlisle Lakes and Allan Hills 85151: members of a new chondrite grouplet. *Geochim. Cosmochim. Acta*, **53**, 3035–44.

Scott, E.R.D., Clayton, R.N. and Mayeda, T.K. (1985) Properties and genesis of two anomalous type 3 chondrites, Suwahib (Buwah) and Willaroy. *Lunar Planet. Sci.*, **16**, pp. 749–50.

Scott, E.R.D., Rubin, A.E., Taylor, G.J. and Keil, K. (1984) Matrix material in type 3 chondrites – occurrence, heterogeneity and relationship with chondrules. *Geochim. Cosmochim. Acta.*, **48**, 1741–57.

Sears, D.W.G. (1988) Chemical processes in the early solar system: A discussion of meteorites and astrophysical processes. *Vistas Astron.*, **32**, 1–21.

Sears, D.W. and H.J. Axon (1975) Ni and Co content of chondritic metal. *Nature*, **260**, 34–35.

Sears, D.W.G., Lu Jie, Keck, B.D. and Batchelor, D.J. (1991) Metamorphism of CO and CO-like chondrites and comparisons with type 3 ordinary chondrites. *Proc. NIPR Symp. Antarct. Meteorit.*, **4**, 1745–1805.

Sears, D.W., Grossman, J.N., Melcher, C.L. et al. (1980) Measuring metamorphic history of unequilibrated ordinary chondrites. *Nature*, **287**, 791–5.

Sears, D.W.G., Jie, L., Benoit, P.H. et al. (1992) A compositional classification scheme for meteoritic chondrules. *Nature*, **357**, 207–10.

Stöffler, D., Keil, K. and Scott, E.R.D. (1991) Shock metamorphism of ordinary chondrites. *Geochim. Cosmochim. Acta*, **55**, 3845–3867.

Swindle, T.D. and Podosek, F.A. (1988) Nucleo-cosmochronology, in *Meteorites and the Early Solar System* (eds J.F. Kerridge and M.S. Matthews). Tucson: University of Arizona Press, pp. 1114–26.

Taylor, G.J., Maggiore, P., Scott, E.R.D. et al. (1987) Original structures and fragmentation and reassembling histories of asteroids: evidence from meteorites. *Icarus*, **69**, 1–13.

Tilton, G.R. (1988) Age of the solar system, in *Meteorites and the Early Solar System* (eds J.F. Kerridge and M.S. Matthews). Tucson: University of Arizona Press, pp. 259–275.

Urey, H.C. and Craig, H. (1952) The composition of the stone meteorites and the origin of the meteorites. *Geochim. Cosmochim. Acta*, **4**, 36–82.

Van Schmus, W.R. (1969) The mineralogy and petrology of chondritic meteorites. *Earth Sci. Rev.*, **5**, 145–84.

Van Schmus, W.R. and Wood, J.A. (1967) A chemical–petrologic classification for the chondritic meteorites. *Geochim. Cosmochim. Acta*, **31**, 747–65.

Wisdom, J. (1985) Meteorites may follow a chaotic route to Earth. *Nature*, **315**, 731–3.

Wood, J.A. (1979) Review of metallographic cooling rates of meteorites and a new model for the planetesimals in which they formed, in *Asteroids* (ed. T. Gehrels). Tucson: University of Arizona Press, pp. 849–91.

Zinner, E. (1988) Interstellar cloud material in meteorites, in *Meteorites and the Early Solar System* (eds J.F. Kerridge and M.S. Matthews). Tucson: University of Arizona Press, pp. 956–83.

Cross references

Carbonaceous chondrite
Meteorite
Solar system: origin

CHRONOLOGY: METEORITE

Meteorites are samples of other (than the Earth) planetary bodies in our solar system, and except for the collection of returned lunar materials they are the only such samples available for laboratory examination. Since nearly all absolute chronometric methodologies require laboratory observation, meteorites provide the major basis for planetary chronology outside the Earth-Moon system. The principal non-laboratory approach to planetary chronology is analysis of impact crater density on the surfaces of major bodies observed remotely; such ages are considered in the articles for these planetary bodies, while this entry concentrates on the information which can be gathered from meteorites.

With but a few exceptions meteorites are believed to be samples of asteroids, but it has proved very difficult to associate specific meteorite types with specific asteroids, and for some types even the asteroid association can be challenged. The meteorites themselves have much to say about the nature of their parent bodies, however, so that meteorite chronology provides important constraints on the formation and evolution of small bodies in the terrestrial planetary zone of the solar system. Moreover, the most abundant variety of meteorites, those classified as undifferentiated, are believed never to have experienced planetary differentiation but instead to be assemblages of materials originally formed in the solar nebula. Undifferentiated meteorites – also called chondrites – thus provide chronological information not only on planetary body evolution but also on the formation of the solar system itself.

Formation of the solar system

As a good first-order generalization, the age of chondritic (undifferentiated) meteoritic material, and of the formation of the solar system itself, can be taken to be 4.56 Ga. The first indications of a characteristic meteorite age around 4.5 Ga emerged as an upper limit of concordant U–He and K–Ar ages and has been adequately substantiated and refined as other dating methods – U–Pb, Rb–Sr, Sm–Nd, Re–Os – were developed and applied to meteorites. Presently the most precise and accurate absolute age for chondritic material is based on application of Pb–Pb (^{207}Pb/^{206}Pb) analysis to a class of objects generally called CAIs ('calcium–aluminium inclusions'). These are inclusions, commonly millimeter-sized, found in some classes of carbonaceous chondrites and characterized by high abundances of very refractory elements, notably Ca and Al; they are particularly suitable for precise U–Pb analysis because of their high ratios of (refractory) U to (volatile) Pb. Nominal analytical uncertainties for Pb–Pb ages of CAIs are usually around 1 Ma, but absolute ages reported by different laboratories differ by around 10 Ma, so that the 'best' absolute age for CAIs, and for the solar system itself, can be taken to be 4.56 Ga with a reasonable uncertainty of around 0.01 Ga (Chen and Wasserburg, 1981; Tilton, 1988).

That the age of chondritic materials, specifically including CAIs, is also the age of the solar system as a whole, can be argued on several grounds. The most compelling is that chondritic materials contained some very short-lived and now-extinct radionuclides (see Dating methods) which were produced in stellar nucleosynthesis, necessarily predating the solar system. The tightest limit comes from the shortest-lived such radionuclide believed to have been present in meteoritic solids: ^{26}Al (half-life 0.72 Ma). For ^{26}Al to survive, the interval between ejection of freshly synthesized ^{26}Al into the interstellar medium and formation of solids in the solar nebula which resulted from collapse of part of the interstellar medium cannot have exceeded more than a few half-lives of ^{26}Al, i.e. no more than a few Ma. Since ^{26}Al is prominently present in CAIs, the same class of objects for which we have a well-established absolute age, nothing made in the solar nebula can be very much older than CAIs. The absolute age of formation can thus be taken with considerable confidence to be 4.56 Ga, within a precision that seems adequate for currently posed questions.

CAIs are a very restricted class of materials, however. Whole rock chondrites and various other classes of their constituents have also been the subject of extensive geochronological studies by essentially all applicable methods. To the extent that the resultant ages can be inferred to represent the original formation of solids in the solar nebula, these ages agree with the 4.56-Ga figure noted above. In general, however, analytical uncertainties in the absolute ages of other kinds of chondritic materials are substantially greater than the CAI-based absolute age of the solar system as a whole, typically a few to several tens of Ma.

Theoretical expectations based on models for the evolution of the solar nebula suggest that the span of original (nebular) formation times of meteoritic solids probably does not exceed about 10 Ma, a span shorter than analytical uncertainties for most available absolute ages. It can also be argued, again on the basis of nebular theory, that the span of original formation times should be shorter still, about 1 Ma and perhaps significantly less. Although this question is of considerable interest it is extremely difficult to explore it empirically by absolute ages because of analytical limitations. Exploration of nebular ages on a sub-Ma scale is possible through relative chronologies based on extinct radionuclides (see dating methods). Although considerable data are available, however, the results remain inconclusive. Some results suggest nebular processes spanning ≈ 10 Ma, but alternative interpretations are possible: in some cases it can be argued that the ages reflect secondary (parent body) rather than primary (nebular) processing, and in other cases it can be argued that the relevant short-lived radionuclides were not uniformly distributed and so cannot be used as a basis for geochronology.

Chondrite parent bodies

All meteorites which fall to Earth were formerly parts of larger objects – parent bodies – in which they have resided through most of the age of the solar system. These bodies have their own diverse histories and their study is of interest in its own right as well as instructive in helping to distinguish between nebular and parent body effects in undifferentiated meteorites and in providing likely constraints on the formation and earliest evolution of the major planets.

Several different processes are involved in accreting initially dispersed nebular dust into macroscopic planetary objects (Weidenshilling, 1988). Theoretical expectations are that accretion to planetesimal or asteroidal size should be quite fast, of the order of 1 Ma, with further growth to the size of the major terrestrial planets somewhat slower, of the order of 10 to perhaps 100 Ma. It is, however, quite difficult to construct an empirical chronology, primarily because there is no unambiguous association of isotopic closure events with the accretion process itself.

All chondritic meteorites, as far as is known, have experienced thermal metamorphism or aqueous alteration in their parent bodies; some have been affected more than others, but none have escaped some degree of overprinting by parent body processing. The problem of parent body chronology has been attacked by a wide variety of experimental techniques, but despite the wealth of data there is still no generally accepted coherent picture of parent body histories. This may be due in part to inapplicability of some of the techniques employed and in part to model dependence of other techniques, but the major difficulty is probably that for the geological processes involved different chronometers record different epochs in a given sample's history.

Nevertheless, some generalizations seem secure. At one extreme, apparently robust dating of metamorphic minerals in some H-group ordinary chondrites indicates their formation some 5 Ma later than carbonaceous chondrite CAIs (Zinner and Göpel, 1992). This places a solid (if not very surprising) upper limit on the time taken to accrete this parent body and to generate and begin to relax metamorphic heating. More generally, several different chronometric systems, responding to lower post-metamorphic temperatures, indicate timescales in the range of a few tens to about 100 Ma after the formation of the solar system (Pellas and Störzer, 1981; Swindle and Podosek, 1988; Turner, 1988). Timescales for alteration in carbonaceous chondrites are harder to study but seem to be of the same order or shorter (MacDougall, Lugmair and Kerridge, 1984). The energy source for such relatively early parent body processing remains controversial (heating by short-lived radionuclides, solar electromagnetic fields, and by the accretion process itself have been proposed), but thermal modeling indicates that these timescales are appropriate for asteroid sized radii of the order of 100 km, which is one of the factors supporting an asteroid association for chondrites.

Beyond this pervasive early parent body processing there are datable events which occurred more recently, some very much more recently (Turner, 1988). It is generally believed that most if not all of these more recent events result from high-velocity impacts in the asteroid belt rather than intrinsic processes in a given parent body. A prominent example is the frequency of 0.5-Ga Ar retention ages among L-group ordinary chondrites, ascribed to a major impact on a single parent body, but there are also scattered instances of older and even younger events.

Differentiated meteorites

In contrast to undifferentiated (chondritic) meteorites, which are basically non-equilibrium mechanical assemblages of diverse constituents with diverse origins, differentiated meteorites are basically igneous rocks (or breccias thereof). Differentiated meteorites are samples of parent bodies which were at least partially melted and which correspondingly experienced large-scale chemical differentiation, similar in kind if not in degree to that which occurred in the Earth and the Moon. The diversity of types suggests that they represent many parent bodies. Some, e.g. basaltic achondrites (q.v.) are not radically different from some terrestrial or lunar rocks; others are quite different, e.g. irons and stony irons (pallasites and mesosiderites) (q.v.). One specific class of differentiated meteorites, the eucrites (q.v.) is the only case of plausibly proposed association with a specific asteroid, Vesta (q.v.).

In broad terms, most differentiated meteorites, like undifferentiated meteorites, are found to be approximately 4.5–4.6 Ga old. By itself, this observation indicates that at least one fast-acting energy source for planetary-scale melting was available early in solar system history; whether it was fundamentally the same source as that responsible for chondrite parent body metamorphism is unclear. The quality of chronological information available for the various types of differentiated meteorites is rather uneven. For some types it is difficult to say much more than that they formed at about the same time as chondrites, within a few to several tens of Ma. For other types much more extensive and precise data are available.

The best examples are the basaltic achondrites (q.v.) and angrites, which are amenable to several geochronologic techniques. Primarily on the basis of precise U–Pb/Pb–Pb absolute ages and models for the evolution of initial ^{87}Sr/^{86}Sr, it is widely considered that eucrites (q.v.) formed (in the sense of cooling of a magma) shortly but significantly after the formation of the solar system, by some 10 to 20 Ma (Tilton, 1988). For some time it has also been considered that angrites also formed significantly later than the solar system itself, by a few to several Ma, but also significantly before the eucrites (Tilton, 1988); more recent work (Lugmair and Galer, 1992) however suggests that eucrites and angrites may be contemporaneous. The earliest known parent body igneous activity occurred some 2 Ma after CAI formation, as inferred from ^{26}Al in an igneous lithic clast in an ordinary chondrite (Hutcheon and Hutchison, 1989). These observations indicate that igneous activity on asteroidal bodies began very quickly after formation of the solar system and evidently persisted for some 10–20 Ma.

There are also some well-established examples of chemical differentiation events occurring substantially later than the early formational epoch. These include the 3.8 Ga age for the iron meteorite Kodaikanal (Burnett and Wasserburg, 1967) and the 3.6–3.9 Ga age of the howardite Kapoeta (Papanastassiou et al., 1974). It is difficult to determine whether such events reflect intrinsic parent body evolution or external events such as high-velocity impacts. There are, however, two interesting classes of meteorites whose ages are widely believed to mark stages in the intrinsic evolution of major planetary bodies. One of these classes is lunar meteorites (q.v.), samples of the Earth's

Moon recently delivered to the Earth by impacts on the Moon. The other class comprises the so-called SNC meteorites (q.v.). A variety of techniques applied to various SNC meteorites provide evidence for discernible isotopic closure events at ages from 180 Ma to 4.6 Ga. Of particular note is the age of 1.3 Ga, firmly established by several techniques and several meteorites, for an igneous differentiation event. This age itself is taken as prima facie evidence that SNCs came from a major planetary body, since only large bodies would be expected to continue intrinsic igneous activity (presumably powered by the long-lived radioactive elements U, Th and K) for so long. A number of other lines of evidence strongly suggest that the parent body of SNC meteorites (q.v.) is Mars.

Presolar materials

The solar system was made from pre-existing interstellar gas and dust, so that in the broadest sense everything in the solar system is a sample of presolar materials. As far as we know, however, all samples of the Earth and Moon, and even most meteorite samples, are made from a thoroughly homogenized mix of presolar materials, so that their present chemical and isotopic identities were determined by processes acting within the solar system. There are some meteoritic samples (e.g. CAIs) which have modestly anomalous isotopic compositions which cannot have been generated by solar system processes acting on average solar system materials; these are believed to have formed from imperfectly mixed presolar materials, but are nevertheless not themselves considered presolar materials in that their present isotopic, chemical and mineralogical identities were established by solar system processes. There are, however, more restricted types of materials which can be isolated from meteorites and for which the terms 'presolar grains' or 'presolar materials' are usually reserved: grains believed to have existed in the interstellar medium from which the solar system was formed and to have survived, largely unmodified, through solar nebula processing and subsequent parent body history. Several types have been identified (Anders and Zinner, 1993): diamond, silicon carbide, graphite, titanium carbide and corundum (Al_2O_3). The most compelling argument that these materials are indeed presolar is that they contain elements with isotopic compositions radically different from the solar system mix.

Presolar grains are necessarily older than the solar system. Just how old they actually are is a question of keen interest for a variety of reasons, but thus far there have been no direct absolute age determinations. In part this is due to the very limited quantities of these materials so far available for study, and it is likely that such data will be forthcoming in the future. It is interesting to note that studies of radiometric geochronology on interstellar grains will pose intellectual as well as extremely exacting analytical challenges. To varying degrees all the radiometric methods involve implicit assumptions about the isotopic structure of both parent and daughter elements. For presolar materials these assumptions will almost surely be violated, so that geochronology will have to be part of general isotopic characterization.

For SiC, presolar age information may be obtained indirectly. SiC samples contain excess ^{21}Ne which is attributed to exposure to cosmic rays in the interstellar medium (see Cosmic ray exposure age). Nominal exposure ages for 'bulk' samples (very small, but still involving many grains) range from 13 to 133 Ma (Lewis, Amari and Anders, 1994). It is likely, however, that these effective average ages reflect interstellar medium processing as much or more than stellar formation events, and that individual grains had a wide range of ages at the time they were incorporated into the solar system.

Conclusions

The absolute age of formation of the solar system is known with considerable precision and confidence: 4.56 Ga. Much happened quickly, in only a small fraction of the solar system's age. Solid objects were formed within a few Ma of the time that some of their constituent atoms were made in other stars. Within no more than a few Ma after formation of the first solids, at least some asteroidal-size bodies had formed by accretion of nebular dust. At least one fast (no more than a few Ma) energy source (possibly short-lived radionuclides electromagnetic induction or accretional energy) was available to heat asteroid-sized bodies. Some were quickly melted and experienced igneous differentiation; such igneous activity evidently occurred over the first 10–20 Ma of solar system history, and it seems likely that at least some of the bodies accreting to form the major terrestrial planets were already differentiated when they arrived. Other parent bodies did not melt but experienced metamorphism or aqueous alteration, cooling on a 100-Ma timescale. Subsequent to this early epoch around 4.5 Ga, evolution of small bodies seems powered primarily by orbital interactions and mutual collisions. Most of this information is obtained from laboratory analysis of meteorites, our only samples of materials originating outside the Earth–Moon system.

A small proportion of meteorites have substantially younger ages, reflecting the effects of collisions or of continuing evolution of larger bodies due to longer-lasting internal energy sources, notably including the group of meteorites which many investigators believe come from Mars.

Frank A. Podosek

Bibliography

Anders, E. and Zinner, E. (1993) Interstellar grains in primitive meteorites: diamond, silicon carbide, and graphite. *Meteoritics*, **28**, 490–514.
Burnett, D.S. and Wasserburg, G.J. (1967) Evidence for the formation of an iron meteorite at 3.8×10^9 years. *Earth Planet. Sci. Lett.*, **2**, 137–47.
Chen, J.H. and Wasserburg, G.J. (1981) The isotopic composition of uranium and lead in Allende inclusions and meteorite phosphates. *Earth Planet. Sci. Lett.*, **52**, 1–15.
Hutcheon, I.D. and Hutchison, R. (1989) Evidence from the Semarkona ordinary chondrite for ^{26}Al heating of small planets. *Nature*, **337** 238–41.
Lewis, R.S., Amari, S. and Anders, E. (1993) Interstellar grains in meteorites. II. SiC and its noble gases. *Geochim. Cosmochim. Acta*, **58**, 471–94.
Lugmair, G.W. and Galer, S.J.G. (1992) Age and isotopic relationships among the angrites Lewis Cliff 86010 and Angra dos Reis. *Geochim. Cosmochim. Acta*, **56**, 1673–94.
MacDougall, J.D., Lugmair, G.W. and Kerridge, J.F. (1984) Early solar system aqueous activity: Sr isotopic evidence from the Orgueil CI meteorite. *Nature*, **307**, 249–51.
Papanastassiou, D.A., Rajan, R.J., Huneke, J.C. and Wasserburg, G.J. (1974) Rb–Sr ages and lunar analogs in a basaltic achondrite: implications for early solar system chronologies, in *Lunar Sci. V*, pp. 583–5.
Pellas, P. and Störzer, D. (1981) ^{244}Pu fission track thermometry and its application to stony meteorites. *Proc. Roy. Soc. London*, **A374**, 253–70.
Swindle, T.D. and Podosek, F.A. (1988) Iodine–xenon dating, in *Meteorites and the Early Solar System* (eds. J.F. Kerridge and M.S. Matthews). Tucson: University of Arizona Press, pp. 1127–46.
Tilton, G.R. (1988) Age of the solar system, in *Meteorites and the Early Solar System* (eds J.F. Kerridge and M.S. Mathews). Tucson: University of Arizona Press, pp. 259–75.
Turner, G. (1988) Dating of secondary events, in *Meteorites and the Early Solar System* (eds. J. F. Kerridge and M.S. Matthews). Tucson: University of Arizona Press, pp. 276–88.
Weidenshilling, S.J. (1988) Formation processes and time scales for meteorite parent bodies, in *Meteorites and the Early Solar System* (eds J.K. Kerridge and M.S. Matthews). Tucson: University of Arizona Press, pp. 348–71.
Zinner, E. and Göpel, C. (1992) Evidence for ^{26}Al in feldspars from the H4 chondrite Ste. Marguerite. *Meteoritics*, **27**, 311–2.

Cross references

Asteroid
Chondrite
Differentiation
Meteorite
Meteorite parent bodies
Solar system: origin

CLEMENTINE MISSION

The Clementine spacecraft was launched on 25 January 1994 and subsequently placed in a 400-km by 2940-km, 5-h orbit around the

Moon. Clementine observed the Moon from 19 February to 3 May 1994. The mission was originally designed to include an asteroid flyby, but a software error caused the depletion of spacecraft attitude control propellant, and the flyby portion of the mission was aborted.

The spacecraft instrumentation consisted of four cameras: an ultraviolet–visible (UVVIS) camera, a long-wave infrared (LWIR) camera, a high-resolution (HIRES) camera, and a near-infrared (NIR) camera. The high resolution camera incorporated a laser ranging (LIDAR) system. The spacecraft systems have been described by Nozette et al. (1994). Information on the gravity field of the Moon was obtained from spacecraft tracking data.

Clementine mapped the Moon with unprecedented resolution, returning nearly 1 million images in 11 visible and near-infrared colors. The average surface resolution was about 200 m per pixel. Mulitspectral image data enables the determination of the composition of lunar surface materials and the mapping of the distribution of rock and soil types on the Moon. [Plate 14a is a multispectral image showing lunar surface mineralogy, obtained at lower resolution by the Galileo spacecraft (q.v.)]. Preliminary results of the Clementine mission were presented in a special issue of the journal *Science* (volume 266, number 5192). Important results include the following.

A high-resolution global topographic data set for the Moon was obtained for the first time. The topographic data provide information about crustal structures and thickness. Impact crater structures were well resolved, and the presence of a number of ancient, partially buried and nearly erased impact basins was confirmed. The South Pole–Aitken basin was recognized for the first time to be the largest and deepest impact basin in the solar system. The diameter of the basin is 2500 km; its maximum depth is 8.2 km below the reference ellipsoid for the Moon. The highest topography, 8 km above the reference, is found adjacent to the basin in the farside highlands.

High-resolution gravity data reveal the Moon to be a more dynamic body than was previously appreciated. Large subsurface density variations occur beneath highlands and farside areas, as well as beneath basins (see Mascon). The basins are not isostatically compensated, as might be expected for an ancient, cold body. Much additional information about the state of stress and internal structure of the Moon may be derived from the Clementine data in future.

The compositional information derived from the multispectral imagery provides a wealth of new detail on the geological history of the Moon. Earlier geologic maps (see color plates) show considerable detail, based on superposition relationships, hand specimen analysis, and related information. The new data will, for instance, enable investigators to distinguish between successive lava eruptions in a basin, given small compositional differences, and can thus provide much more detailed information on the nature and timing of geologic events.

The Clementine mission was sponsored by the Ballistic Missile Defense Organization, with participation from NASA. The name was selected with reference to the traditional American ballad (or folk song) 'My Darlin' Clementine.' Clementine was a miner's daughter, and the spacecraft was to map the distribution of lunar minerals. Additionally, the spacecraft, like the miner's daughter, was eventually to be 'lost and gone forever.'

James H. Shirley

Bibliography

Nozette, S. et al. (1994). The Clementine mission to the Moon: scientific overview. *Science*, **266**, 1835–9. (This issue of the journal *Science* includes six other articles describing the scientific results of the mission.)

Cross references

Impact cratering
Mascon
Moon: geology
Moon: gravity
Reflectance spectroscopy

COLD ACCRETION THEORY

Cold accretion theory is a term that can be used to distinguish modern theories of planet growth from earlier theories. The mechanics of modern cold accretion are reviewed in the article Accretion.

Earlier theories visualized planets as forming out of hot blobs of solar plasma, or large, gravitationally unstable 'protoplanet' clouds which were relatively large segments of the solar nebula at any given solar distance. The planets were generally visualized as forming either from very hot matter, or (at least) forming in a molten state because of the high temperatures reached during gravitational collapse.

However, Urey (1952) and Shmidt (1958) theorized that the planets formed from innumerable small, solid (i.e. 'cold') bodies. Urey used the argument that planets are extremely deficient in inert gases. This means they could not have formed by gravitational contraction of solar or nebular clouds because those would have contained solar abundances of inert gases, since such gases are too heavy to have been lost by gravitational escape from larger planets. Urey argued that preplanetary dust grains accreted by collision (see Collision) into small, asteroid-sized bodies, or 'planetesimals,' which later accreted into planets. Because of the chemical inertness of the noble gases, they would not be trapped in minerals of these bodies, nor would they be accreted by them because the planetesimals' gravity was too slight. Urey's picture thus explains why planets are extremely deficient in inert gases.

The Russian researcher Otto Shmidt (1958) came to the same conclusion by considering solar system dynamics. He argued that the planets accumulated from innumerable small bodies, with statistical averaging of the impact effects, because this was the only way to explain their near-circular, low-inclination orbits and mostly prograde, low-obliquity rotations.

Shmidt's student, Russian dynamicist V.S. Safronov, made the pivotal study describing the collisional accretion of planetesimals in terms of orbits and collision mechanics. This work, published in Russia in 1969, reached Western scientists in a 1972 English translation (Safronov, 1972), spurring further numerical models. Computer models by Wetherill (1976, 1990), Greenberg et al. (1979) and others, incorporating results of experimental collisions (Hartmann, 1978), have attempted to start with nebular conditions and follow the growth of planetesimals from colliding dust particles all the way to planet scale. The latter work showed that an initial low velocity collision regime will result in accumulation of granular 'regolith debris' on at least some planetesimal surfaces, which restrains rebound and greatly facilitates accretion, and thus in turn triggers rapid accretionary growth without invoking any special 'sticking' mechanisms (which once seemed necessary to start growth). The models give some confidence that the planets grew in this way, although various problems are still being worked out.

Accumulation from 'cold' planetesimals does not mean that the planets did not become molten during their formation. Discovery of the anorthosite crust (of low-density minerals) on the Moon gave rise to the theory of magma oceans on accreting planets. Once the growing planet got large enough, impact velocities of the accreting bodies increased, and the kinetic energy of impact deposited enough heat to keep at least the outer layers of the planet molten.

William K. Hartmann

Bibliography

Greenberg, R. et al. (1978) Planetesimals to planets: numerical simulation of collisional evolution. *Icarus*, **35**, 1.
Hartmann, W.K. (1978) Planet formation: mechanism of early growth. *Icarus*, **24**, 504.
Safronov, V.S. (1972) *Evolution of the Protoplanetary Cloud and Formation of the Earth and the Planets*. Jerusalem: Israel Program for Scientific Translations.
Shmidt, O. (1958) A theory on the origin of the Earth. Moscow: Foreign Languages Publishing House.
Urey, H.C. (1952) *The Planets: Their Origin and Development*. New Haven: Yale University Press.
Wetherill, G.W. (1976) The role of large bodies in the formation of the earth and moon. *Proc. Lunar Sci. Conf.*, **7**, p. 3245.
Wetherill, G.W. (1990) Formation of the Earth. *Ann. Rev. Earth Plane. Sci.*, **18**, 205.

Cross references

Accretion
Earth–Moon system: origin

Moon: origin
Solar system: origin

COLLISIONS

Collisions play a role in planetary evolution that can hardly be overemphasized. Collisions between planetary bodies are now believed to be responsible for the growth of planets, the production of meteorites, the size distribution of the asteroids, the formation of Earth's Moon and the solar system's ubiquitous craters (see Impact cratering).

Early work by Piotrowski (1953), Öpik (1963) and others established that typical asteroids must have undergone many collisions at speeds around 5 km s^{-1} – a speed sufficient to shatter them. This led to the realization that collisions have shaped asteroidal and crater populations in the solar system, and that the resulting fragments, perturbed from the belt, are the main source of meteorites (Davis et al., 1979). Additional studies showed that when natural rocky bodies are fragmented, the resulting size distributions have approximately the power law form

$$N = m^{-b} \qquad (C4)$$

where N = cumulative number of bodies of mass exceeding m, b = power law exponent, or slope on a plot of log N versus log m. The exponent b was found to vary from roughly 0.7 for lower-energy collisions to about 1.0 to 1.2 for high energy collisions (Hartmann, 1969). In other words, the smaller the fragment, the more there are; and higher-energy collisions, such as cratering impacts on the Moon, eject the largest numbers of small fragments, relative to the large fragments.

This law accounts for many phenomena in planetary science. The smaller the asteroid or meteorite, the more there are. The smaller the particle size in surface regoliths (produced by collisional grinding), the more there are. The smaller the bolide considered for Earth impact, the more common they are. This is why atom bomb scale meteoritic explosions, such as the 1908 Siberia event, happen once every century or so, while species-extinguishing catastrophes like the Cretaceous–Tertiary impact happen only every hundred million years or so.

Asteroids, meteorites, and craters fit this general theoretical picture, supporting the theory of origin by collisional fragmentation and impact. The asteroid belt shows size distributions with exponents $b \sim 0.7$ or 0.8. Fresh primary impact craters (the craters formed from interplanetary debris), when deconvolved from crater size back to impactor size, generally indicate a similar mass distribution for the impactors, supporting their origin by impact of interplanetary degree. Secondary impact craters, however, (caused by ejecta from primary craters) have steeper size distributions with exponents $b \sim 1.0$ to 1.2.

In spite of this general understanding of the role of collisions in shaping interplanetary bodies, there are some puzzles. Asteroids of some compositional types probably have 'bumps' in their size distribution curve, departing from smooth power laws. Explanations advanced to account for the irregularities have ranged from temporally decreasing fragmentation of bodies with iron cores, to transitions from strength to gravity as the dominant cohesive force (Davis and Ryan, 1990). Comets apparently have irregular shapes that may depend more on ice sublimation than on collisional fragmentation (Hartmann and Tholen, 1990).

Research in the 1980s and 1990s has focused on the results of experiments performed at a range of velocities from 1 m s^{-1} up to several km s^{-1} to clarify the effects of collisions at different times and places in the solar system. In the early solar system, as well as in ring systems, nearly parallel circular orbits permitted low-velocity collisions. This favored accretion instead of fragmentation, leading to growth of planets (except that in ring systems inside Roche's limit, accretion is inhibited). Researchers continue to investigate partitioning of the kinetic energy into heat, kinetic energy of ejecta, and so on (Ahrens and O'Keefe, 1978); deviations from power law size distribution (Fujiwara, Kamimoto and Tsukamoto, 1977; Ryan, Hartmann, and Davis 1991); planet-disrupting events by rare giant collisions (Housen and Holsapple, 1990); regolith production (Housen et al., 1979) and cratering mechanics (Melosh, 1989) etc., all within the framework of collision theory.

William K. Hartmann

Bibliography

Ahrens, T.J. and O'Keefe J. (1978) Energy and mass distributions of impact ejecta blankets on the moon and Mercury, in *Impact and Explosion Cratering* (eds D. Roddy, R. Pepin and R. Merrill). New York: Pergamon Press.

Davis, D.R., Chapman, C., Greenberg, R. *et al.* (1979) Collisional evolution of asteroids: Populations, rotations, and velocities, in *Asteroids* (ed. T. Gehrels). Tucson: University of Arizona Press.

Davis, D.R. and Ryan, E.V. (1990) On collisional disruption: Experimental results and scaling laws. *Icarus*, **83**, 156–82.

Fujiwara, A., Kamimoto, G. and Tsukamoto, A. (1977) Destruction of basalt bodies by high-velocity impact. *Icarus*, **31**, 277–88.

Hartmann, W.K. (1969) Terrestrial, lunar, and interplanetary rock fragmentation. *Icarus*, **10**, 201.

Hartmann, W.K. and Tholen, D.J. (1990) Comet nuclei and Trojan asteroids: A new link and a possible mechanism for comet splittings. *Icarus*, **86**, 448–54.

Housen, K, Wilkening, L, Chapman, C. and Greenberg, R. (1979) Regolith development and evolution on asteroids and the moon, in *Asteroids* (ed. T. Gehrels). Tucson: University of Arizona Press.

Housen, K.R. and Holsapple, K.A. (1990) On the fragmentation of asteroids and planetary satellites. *Icarus*, **84**, 226–53.

Melosh, H.J. (1989) *Impact Cratering*. New York: Oxford University Press.

Öpik, E.J. (1963) The stray bodies in the solar system 1. Survival of cometary nuclei and the asteroids. *Ad. Astron. Astrophys.*, **2**, 219.

Piotrowski, S. (1953) The collisions of asteroids. *Acta Astron.*, **A5**, 115.

Ryan, E., Hartmann, W. and Davis, D. (1991) Impact experiments II. Catastrophic fragmentation of aggregate targets and relation to asteroids. *Icarus*, **94**, 283–98.

Cross references

Accretion
Impact cratering
Solar system: origin

COLOR

The color of an object (or substance) depends upon the object's material properties, the wavelengths of radiant energy incident upon it or emitted from it, and the ability of the human retina and associated neural structures to be stimulated by that energy. The material properties of an object, such as composition and texture, control its ability to absorb, transmit or reflect portions of the incident radiant energy, while the temperature of the material generally plays the central role in controlling emitted wavelengths. While color can thus be defined in terms of physical characteristics, the eye and neurologic response of the observer precisely controls the way in which color is seen; thus, perceived color exists at the interface between physical characteristics and the visual color response of the eye and neural system. The spectral responsiveness of the average human eye peaks at approximately 555 nm (close to the mean solar spectral irradiance); this means that humans can on average resolve subtler differences between shades of yellowish-green than they can for reds or blues.

Assuming that an object or substance is evenly illuminated by radiant energy, the color of that object can be defined via three tristimulus values, which represent the degree of stimulation of the eye's red-, green- and blue-perceiving receptors. For a spectral color of wavelength λ, the tristimulus values are given as $x(\lambda)$, $y(\lambda)$ and $z(\lambda)$. For non-spectral colors, the tristimulus values X, Y and Z are derived by integrating:

$$X = \int x(\lambda) E_C(\lambda) \rho(\lambda) \, d\lambda \qquad (C5)$$

$$Y = \int y(\lambda) E_C(\lambda) \rho(\lambda) \, d\lambda \qquad (C6)$$

$$Z = \int z(\lambda) E_c(\lambda) \rho(\lambda) \, d\lambda \qquad (C7)$$

where $E_C(\lambda)$ is the illumination of the object and $\rho(\lambda)$ is the spectral reflectance of the object. Each tristimulus value can then be ratioed

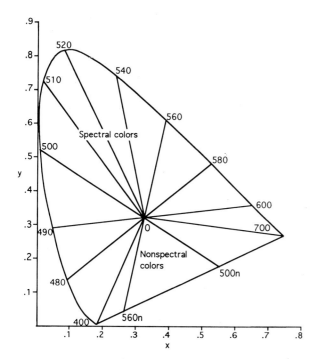

Figure C28 The Cie chromaticity diagram is a plot of the x and y trichromatic coefficients, thus representing the hue and saturation of a color. In a three-dimensional rendering, the z coefficient, representing color intensity, would be plotted in the z direction. The central point O represents white, or completely unsaturated color. The curved perimeter of the horseshoe on the diagram represents fully saturated spectral colors (blue through red), while the straight line segment that closes the horseshoe contains the fully saturated non-spectral colors (purple and magenta). The numbers along the perimeter of the horseshoe are the wavelengths in nanometers of corresponding colors.

to the sum of the three to form the trichromatic coefficients which are independent of light intensity:

$$x = X/(X + Y + Z) \quad \text{(C8)}$$
$$y = Y/(X + Y + Z) \quad \text{(C9)}$$

and

$$z = Z/(X + Y + Z) \quad \text{(C10)}$$

Note that $x + y + z = 1$ and that for an equal energy spectrum, $X = x$, $Y = y$ and $Z = z$. A plot of x versus y yields the CIE (Commission Internationale de l'Eclairage) chromaticity diagram, a standard graphic means of representing the hue and saturation of a color. In a three-dimensional diagram, color intensity, i.e. the achromatic aspect of color from white through gray to black, would be plotted in the z direction (Figure C28). The curved perimeter of the horseshoe on the diagram represents fully saturated spectral colors (blue through red), while the straight line segment that closes the horseshoe contains the fully saturated non-spectral colors (purple and magenta). The central point represents unsaturated color (white; Miller 1948).

The visual portrayal of color relies on either additive or subtractive techniques, depending on the medium involved. In each approach the primary colors are those that can be combined to form all other colors. Subtractive color is based on the tendency for materials to absorb or attenuate certain colors; hence, the mixing of artist's pigments or the use of absorptive filters (where the primary colors are cyan, magenta and yellow) as in color photography are subtractive processes. In additive color, multiple images of an object or scene are recorded, each preserving the object's spectral response in a given color or wavelength range. The hues, intensities and saturations represented by these spectral images are then recombined as light for display. In the additive process the primary colors are red, green and blue; thus, television screens and video monitors rely on the excitation of red, green and blue phosphors to produce additive color images.

Patricia A. Jacobberger

Bibliography

Miller, C.W. (1948) *Principles of Photographic Reproduction*, New York: MacMillan, pp. 234–47.

Cross references

Electromagnetic radiation
Remote sensing

COMET

The word 'comet' originates from the Greek, *kometes*, meaning long-haired, because of a comet's appearance as a hairy star or tail star. The large phenomenon 'comet' visible in the sky, is disproportionate to its small source, the nucleus. Thus the term 'comet' usually refers to the observable phenomenon, rather than the actual body. The proper body of a 'comet' is its solid nucleus. Cometary nuclei originated during the formation of the solar system, some 4.6 Ga ago (1 Ga = 10^9 years). When the orbit of a nucleus brings it into the inner solar system, the visible phenomenon known as a comet occurs.

Cometary nuclei are minor bodies of the solar system, composed of frozen gases (mostly H_2O but also CO, CO_2, H_2CO and small amounts of others) and organic- and silicate-based dust. When a nucleus is heated by sunlight in the inner solar system, a diffuse atmosphere or coma (surrounded by a very large hydrogen halo), a dust tail and a plasma tail develop from the sublimation (vaporization) of the icy component of the nucleus and from dust particles entrained by the escaping gases (Plates 20, 21). Comets are usually distinguished from minor planets (asteroids) by their diffuse appearance and by the character of their orbits, but cases are known in which the distinction is poorly defined. For example, the object 2060 Chiron (q.v.) was first thought to be an asteroid, until it was discovered to have a diffuse coma and tail.

The orbital eccentricity of a comet can range from nearly circular to almost parabolic to hyperbolic for comets strongly affected by planetary perturbations. One distinguishes between long-period and short-period comets; the limiting period is arbitrarily set at 200 years since comets with greater periods have been observed only once in modern times. Long-period comets move in highly elliptical, nearly parabolic orbits that bring them close to the edge of the solar system at aphelion (the farthest point from the Sun). This means that long-period comets spend most of their orbital period of several million years at an average distance of about 50 000 astronomical units (AU, the mean distance between the Earth and the Sun) from the Sun, or about one-fifth of the distance to the nearest star. Aside from some clustering caused by the passage of nearby stars, the distribution of orbital aphelia of the long-period comets is nearly isotropic, forming an almost spherical distribution around the Sun. Short-period comets are usually divided into two subgroups: the Halley-type comets with periods between 20 and 200 years, and the Jupiter-family comets with periods less than 20 years. The orbits of the Jupiter-type comets are restricted predominately to the plane of the ecliptic, and are unstable and chaotic because of frequent perturbations by Jupiter. A few short-period comets have orbits that resemble those of minor planets, making the distinction between them and minor planets difficult on the basis of orbital characteristics. More than 80% of known comets have long-period orbits; only about 17% belong to the Jupiter family, and about 2% are Halley-type comets (see Comet: dynamics).

Comets continually lose mass because they are gravitationally too weak to bind the gases evaporating and escaping from active areas on the nucleus. As a result, the typical lifetime of a comet is limited. Its volatiles may become depleted or its active areas may become mantled with dust, forming a regolith of refractory material and preventing further losses. In this respect, comets are related to a few asteroids (inactive comet nuclei) and meteor streams (debris released from comets) (see Meteor shower, Meteoroid stream).

A comet consists of five parts: a nucleus, coma (atmosphere), hydrogen halo, dust tail and plasma tail. Not all comets display both

Figure C29 Comet Mrkos (1957 V) photographed by J.A. Farrell on August 15 1957 with a Schmidt camera near Fort Worth, Texas. The head (consisting of the coma and the nucleus) is seen at the left of the photograph, the plasma tail is the narrow, straight feature pointing toward the upper right corner. The dust tail is bright near the comet head and fades as it curves toward the bottom right corner. (Courtesy J.A. Farrell, Los Alamos.)

types of tails. Associated with the orbit of some comets is a trail of particles. The nucleus is a solid body consisting of ices (frozen gases) and refractory components (dust) with a fluffy structure and fragile cohesive strength. The concept of the nucleus as a dirty 'snowball' was originally proposed by the American astronomer Fred L. Whipple in 1950. Whipple's model has been essentially confirmed by the detailed images returned by the European spacecraft, Giotto (q.v.), which was the first to clearly image the nucleus of a comet, P/Halley (Keller et al., 1986). The relative proportion of mass in ice (volatiles) and dust varies from comet to comet; in the case of comet P/Halley, McDonnell, Lamy and Pankiewicz (1991) found that the mass of dust detected in the coma was twice the mass of the gas. This value represents a lower limit for the dust-to-gas mass ratio of the nucleus as only dust particles with masses smaller than a limiting value can be lifted from the nucleus by the entraining gas. One of the surprising results of Giotto mission was the blackness of P/Halley's nucleus; it is one of the blackest objects in the solar system. The size of the nucleus is on the order of 10 km, its irregular shape approximates a triaxial ellipsoid, roughly like that of an avocado. The mass of a comet can be estimated from the non-gravitational acceleration in its orbital motion caused by recoil from anisotropic outgassing (Newton's third law). Hans Rickman (1989) performed this analysis for P/Halley and estimated its mass to be in the range from 1 to 3.5×10^{14} kg, corresponding to a range in bulk density from 200 to 700 kg m^{-3}.

The coma of a comet develops fully when the comet is near the Sun. Solar irradiation, mostly visible light, sublimates the volatiles (up to about 90% water), which escape into the near vacuum of space because of the low gravitational binding of the nucleus. From roughly a dozen different parent molecules, a plethora of chemical species is formed by solar photolytic processes, secondary electron processes and gas-phase chemical reactions. The photolytic processes depend mainly on solar ultraviolet light to form radicals and ions. The highly reactive radicals and ions interact mainly with the neutrals via fast ion–neutral reactions in the inner coma, where pressures are high enough to make molecular collisions frequent. The expanding coma ions and electrons eventually interact with the solar wind (a high-speed stream of charged particles from the Sun). On the sunward side the coma plasma is decelerated by the opposing solar wind and the flow of the plasma is diverted to the antisolar direction, forming the so-called ion (or plasma) tail. Other structures and boundaries may exist in the plasma environment, including a bow shock or wave, a magnetic field pile-up region, a region of plasma stagnation, a central magnetic field-free cavity close to the nucleus, and a boundary called the ionopause separating the magnetic field-free cavity from the plasma stagnation region. Dust particles released from the nucleus and entrained by the rapidly expanding gas decouple from the gas; some large particles eventually enter Keplerian orbits around the Sun, resulting in a dust trail. Other forces, primarily radiation pressure, act on the small particles to form the curved dust tail.

Depending on the dust-to-gas abundance ratio in the comet and the particle composition and size distribution, the dust tail may be spectacular or absent altogether. While many comets share the above similarities, each has its own unique peculiarities, as noted in Table C5 (see Comet: structure and composition).

The gradual erosion of a comet during its passage through the inner solar system leads to the formation of a trail of particles strewn throughout its orbit. If the trail's orbit crosses that of the Earth, cometary particles enter the atmosphere with high speeds and are heated to temperatures hot enough to vaporize the material, giving rise to a meteor shower. This association between comets and meteor showers is important for gaining further insight into the nature of comets. Observations of the meteors yield information on the composition of comet dust, cometary orbital evolution, the ages of present orbits and lower limits of the masses of comet nuclei. In general, meteor showers are not always observed to be the same magnitude each year, so the cometary debris must be distributed inhomogeneously in the orbit. Indeed, non-uniform trails were observed by the Infrared Astronomical Satellite (IRAS). Also detected by IRAS was a small asteroid (1983 TB) with an orbit that matched the Geminid meteor shower (December). This object is thought to be an extinct cometary nucleus, either depleted in its reservoir of ices or with a refractory regolith, preventing the release of volatiles from its interior. It is possible that some Earth-crossing asteroids may be defunct comets. Eventually, a comet trail disperses due to gravitational perturbations from the planets and its particles may become part of the sporadic (random) meteors that are continually seen throughout the year.

The fact that some cometary orbits intersect the Earth's orbit leads to the possibility of a collision. There is evidence that the Earth, like the other members of the inner solar system, has undergone many such collisions in its history. More than 100 impact craters caused by asteroid and comet bombardment have been found, and many more have undoubtedly been erased through erosion and tectonic activity. Impacts of this magnitude release cataclysmic amounts of energy. For example, a small stony asteroid about 60 m in diameter and travelling at 15 km s^{-1} has a kinetic energy equivalent of about 15 megatons of TNT. The most recent such impact occurred in central Siberia on 30 June 1908. Called the Tunguska event, it knocked down trees in a heavily forested area up to 30 km away from the center but no sizable crater was found. At present, the type of the impactor has not been firmly established; however, a recent investigation by Christopher Chyba and colleagues (1993) supports the hypothesis of a mid-air explosion of a large stony meteor. In 1983, comet IRAS-Araki-Alcock (1983 VII) came within 6×10^6 km of the Earth (a near miss in terms of the dimensions of the solar system). Brian Marsden at the Smithsonian Astrophysical Observatory found that comet P/Swift-Tuttle (1862 III) will make a close approach in August 2126, and may miss the Earth by only tens of millions of kilometers. As more information is gathered about near-Earth comets and asteroids, we will learn more about the frequency of life-threatening impacts on the Earth (see Near-Earth object; Mass extinction; Impact cratering).

Comets are designated by the International Astronomical Union (IAU) according to the following nomenclature. After a comet discovery (or recovery) is confirmed, a preliminary designation is assigned consisting of the names of the discoverers of the comet (up to three codiscoverers are allowed) and the year of discovery or recovery, followed by a lowercase letter indicating the order of discovery or recovery relative to other comets in that year. For example, comet Bowell (1980b) was discovered by Edward Bowell (Lowell Observatory) in 1980 and was the second comet discovered in that year. In the case of the recovery of a well-known comet, the recoverer's name is not used since the comet already bears the name of the original discoverer. When the orbit has been determined accurately, another designation is given consisting of the names of the discoverers (as in the preliminary designation) and the year followed by a Roman numeral indicating the order in that year in which the comet passed perihelion (the point in the orbit closest to the Sun). If the period of the orbit is less than 200 years, the prefix P/ is added to the comet's name to indicate a short-period comet. For example, comet P/Tempel 2 (1983 X) was the tenth comet to pass perihelion in 1983 and was originally discovered by William Tempel in 1873. It is a short-period comet and was the second comet to be named after him. Other examples are found in Table C5. A comet may be discovered in one year and pass perihelion in another year. Thus it may correctly be given two different year designations. The second, permanent designation supersedes the first when accurate orbital elements have

Table C5 Table of selected comets

Comet	Comment
Bowell (1982 I)	Dust activity at heliocentric distances > 4 AU
2060 Chiron	Very large comet, at first thought to be an asteroid, hence its designation as a minor planet
P/Encke (1984 VI)	Shortest period of only 3.3 years. Observed 55 perihelion passages, the most of any comet
P/Giacobini-Zinner (1985 XIII)	First comet encountered by a spacecraft, the International Cometary Explorer (ICE), in September 1985
P/Grigg-Skjellerup (1982 IV)	Second target of the Giotto spacecraft encountered in July 1992
P/Halley (1986 III)	First recorded observation 240 BC by Chinese, mean period of 76.1 years, target of space missions in March 1986, retrograde orbit far out of the ecliptic, outburst at 14 AU post-perihelion
Humason (1962 VIII)	Unusually strong activity of CO^+ far from the Sun
IRAS-Araki-Alcock (1983 VII)	Close approach to Earth in 1983, the only comet to have shown S_2
Kohoutek (1973 XII)	Predicted to be 'comet of the century' shortly after discovery in 1973, but didn't follow normal cometary brightening relations and fizzled in the public's eye!
P/Kopff (1983 XIII)	Potential target of a space mission because of short period (6.43 years), low orbital inclination, and reasonably strong activity
Kreutz Group	Group of Sun-grazing comets (perihelion < 0.0009 AU), includes the Great March Comet (1843 I), comet 1880 I, comet 1882 II, the Great Southern Comet of 1887 (1887 I), comet du Toit (1945 VII), comet Pereya (1963 V), the Day-Light Comet Ikeya-Seki (1965 VIII), and comet White-Ortiz-Bolelli (1970 VI), as well as several other likely members. They are comets that probably resulted from the fragmentation of a large comet nucleus as suggested by Heinrich Kreutz (1854–1907)
Morehouse (1908 III)	Unusually strong activity of CO^+
P/Neujmin 1 (1984 XIX)	Became inactive to give appearance of a minor planet
Schuster (1975 II)	Observed to very large heliocentric distance, also had a large perihelion distance (6.88 AU)
P/Schwassmann-Wachmann 1 (1974 II)	Almost circular orbit at 6 AU, outbursts
Stearns (1927 IV)	Observed to very large heliocentric distance (11.5 AU)
P/Swift-Tuttle (1862 III)	Predicted to have close approach with Earth in 2126
	Associated with the yearly Perseid meteor shower seen in August.
P/Tempel 2 (1983 X)	Potential target of a space mission because of short period (5.29 years), low orbital inclination and reasonably strong activity
West (1976 VI)	Split into at least four large pieces near perihelion

been derived and sufficient time has elapsed to ensure that new discoveries will not upset the order of perihelion passages determined within that year. Under this scheme a periodic comet will have several year and Roman numeral designations, one for each observed perihelion passage. Comets Halley and Encke are exceptions to the above naming convention. They were named in honor of extensive work done on cometary orbits by Edmond Halley and Johann Encke, respectively.

Besides the intrinsic interest that the physics and chemistry of comets hold for researchers, the study of comets is important from two other aspects. It is commonly held that comets are the icy remnants left over from the accretion of the outer planets. According to this scenario, comets were gravitationally scattered by the outer planets from the plane of the solar system to the Oort cloud, a remote region spherically distributed about the Sun and extending up to about one-third of the distance to the nearest star. Many comets remain there, in a frozen state of storage, until gravitational perturbations from passing stars, interstellar molecular clouds or the tidal forces of the galaxy bring them back into the inner solar system or eject them from the solar system. Many comets may also reside in the ecliptic region beyond the orbit of Neptune in a disk-like distribution named the Kuiper belt. In both cases, comets are probably the least modified objects in the solar system, preserved in a state similar to their formation and relatively unaltered by the processes that have changed the larger objects closer to the Sun. Therefore, cometary studies reveal important clues concerning the chemical composition of the solar nebula and the physical processes that occurred in the early stages of the formation of our solar system.

Another important aspect of cometary studies involves the question of the origins of life on Earth. The discovery of organic material in the dust of comet P/Halley by the Giotto spacecraft has raised the intriguing possibility that prebiotic chemical reactions may have taken place on the dust particles in comets, perhaps even in interstellar space before these particles were incorporated into cometary nuclei during their formation. Current studies suggest that the early atmosphere on Earth was inhospitable to the evolution of organic material necessary for life. A way out of this dilemma is the possibility that prebiotic molecules that may be abundant in the outer solar system were delivered to the Earth by impacting comets. Thus, important clues to understanding the origins of life on Earth may be found by investigating the organic material of comets (see Comet: impacts on Earth; Life: origin).

Brief historical summary

Throughout history, comets have been associated with mystery and superstition, sometimes seen as 'celestial daggers,' omens or harbingers of disasters. This was probably caused by the unpredictability of the appearance of a comet in the dark, ancient skies as a spectacular phenomenon that changed the otherwise near-perfect celestial heavens. To this day, the appearance of a comet is called an apparition (ghost).

Early records of comets are preserved in cuneiform tablets dating back to the Babylonian era. The Greeks (about 500–300 BC) began the debate about the true nature of comets. Two schools of thoughts emerged. One considered comets as atmospheric phenomena, while the other treated them as true celestial bodies, beyond the Earth's atmosphere. Aristotle (384–322 BC) was a strong proponent of the first school of thought and wrote extensively about the effects of comets on meteorology. The Roman philosopher Seneca (4 BC–AD 64) defended the other school of thought, arguing by logic that comets are in the realm of the stars. Observations of comets were not limited to the Middle East and Europe. From about 300 BC until the Renaissance, the Chinese kept the most reliable records of comets. Ancient Chinese annals contain observations of many comets, including several apparitions of comet Halley. In the Western hemisphere, appearances of comets are found in Aztec records.

With the onset of astronomy in Europe in the Middle Ages, questions about the nature and origin of comets led to extensive observations. European astronomers adopted the school of thought that comets are 'atmospheric exhalations.' From observations of the parallax of the bright comet of 1577, Tycho Brahe (1546–1601) (q.v.) determined that comets are not phenomena within the Earth's atmosphere but are farther away than the Moon, i.e. that comets are interplanetary bodies. His pupil Johannes Kepler (1571–1630) (q.v.)

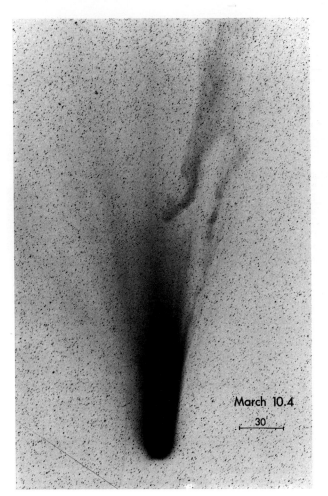

Figure C30 Comet P/Halley on March 10 1986, photographed with the 1-m Schmidt camera at the European Southern Observatory in La Silla, Chile. The comet was at a heliocentric distance of 0.85 AU (post-perihelion) and 1.05 AU from the Earth. This photograph, printed as a negative, was taken 1 day after the closest approach of the Vega 2 spacecraft. The narrow plasma tail, seen on the left, has a filamentary structure near the comet head. Just above the center of the figure a plasma condensation (or disconnection event) stands out. The diffuse appearance of the dust tail is seen curving to right. The scale bar corresponds to a distance at the comet of nearly 1.4 million kilometers. (Courtesy European Southern Observatory.)

2×10^{11} nuclei, based on an analysis of the almost parabolic orbits of some observed comets. These comets were supposedly formed in the planetary nebula and scattered by the planets into the Oort cloud. Passing stars and interstellar clouds randomize the orbits and eventually perturb some comets so that they pass again through the inner solar system while others escape. This is the traditional model of the source of new comets, many details of which are still under discussion (Duncan and Quinn, 1993).

From the similarity of the spectrum of a dusty comet to that of the Sun, it was recognized by Dominique Arago (1786–1853) in 1820 that the light from a comet is mostly scattered sunlight. Soon after, in 1836, Friedrich Bessel (1784–1846) developed a mechanical theory to explain the tail and its observed direction away from the Sun; he introduced the idea of a repulsive force opposing the force of solar gravity. The first detailed study of the structure of comet tails thus began with Bessel's investigation of the tail of comet 1835 III (P/Halley).

In 1866 Giovanni Schiaparelli (1835–1910) (q.v.) found that the orbit of the Perseid meteoroid stream, seen yearly as a meteor shower in August, was associated with the orbit of comet Swift-Tuttle (1862 III). This was the first definite connection to be established between comets and meteor showers. Since then over a dozen comets have been identified as sources of meteor showers. These include the associations of comet P/Halley with the η Aquarids (May) and Orionids (August), comet P/Tempel-Tuttle (1965 IV) with the Leonids (November), and comet P/Encke with the Taurids (October–December).

Around 1900 Fedor Bredikhin (1831–1904) refined the concept of solar radiation pressure as the repulsive force responsible for the formation of the dust tail. This force originates from the interaction of electromagnetic radiation (sunlight) with matter. It describes the momentum transfer from the radiation field, or photon stream, to the scattering and absorbing dust. Nearly seven decades later, Michael Finson and Ronald Probstein (1968) formulated the complete description of the dynamics of dust tails, which also allows comparison with observations. They showed that particles of different sizes moved with different speeds and are therefore separated in the dust tail. With this theory, it is now possible to characterize dust particles according to size and speed of emission from the sphere of influence in the coma.

In 1943, Karl Wurm published a review paper proposing that many highly reactive chemical species observed in cometary comae were derived by photodissociation of more stable chemical compounds released from the nucleus, such as H_2O, CH_4 and $(CN)_2$. About the same time that Oort developed his theory of the comet cloud in 1950, Fred Whipple built on Wurm's suggestion and described the nucleus of a comet as a 'dirty snowball,' i.e. as a solid body containing large amounts of frozen gases, such as H_2O, CO_2, CO, NH_3 and CH_4, and refractory grains with sizes ranging from dust to boulders (Whipple, 1950). This has come to be known as the icy conglomerate model. In 1951 Gerard Kuiper (1905–1973) (q.v.) proposed comet nuclei as building blocks of the solar system and suggested a belt of comets beyond the region of Neptune and Pluto. During the next two decades, variations on these ideas and competing hypotheses were developed by Ernst Öpik (1973), Victor Safronov (1977) and Alastair Cameron (1973).

Before comet plasma physics matured as an important discipline of space physics, theories were advanced by several pioneering researchers. In a historical context, the modern study of comet–solar wind interaction began with the statistical investigation of the pointing direction of cometary ion tails by the German astrophysicist Ludwig Biermann (1907–1986) in 1951. From the aberration angle of about 3° relative to the radial direction from the Sun, Biermann deduced that there must exist a solar corpuscular radiation in order to sweep away the cometary ions at high speed (Biermann, 1951). The radial velocity of these solar charged particles (i.e. the solar wind) was deduced to be on the order of a few hundred km s^{-1}. However, to facilitate the momentum coupling between the solar plasma and the cometary ions by Coulomb collisions (interactions between charged particles via the electrostatic force), a very high plasma number density, far exceeding the limit set by the coronal white light measurements, had to be invoked. To overcome this difficulty, Hannes Alfvén (1957) advocated the presence of a magnetic field in the interplanetary space so that the comet ionosphere would drape the field lines into a magnetic tail. In this classical picture the cometary ions are channeled into the antisunward direction by the magnetic field.

thought that comets followed straight lines and concluded in 1619 that they originate outside the solar system. Johannes Hevelius (1611–1687) suggested parabolic orbits in 1668. Isaac Newton (1642–1727) (q.v.) argued against a parabolic orbit of the comet of 1681 (calculated by John Flamsteed); he preferred the hypothesis of two independent comets, one for the inbound and one for the outbound leg. Comets were firmly established as part of the solar system by the analysis of Edmond Halley (1656–1742) (q.v.) that the comets of 1531, 1607 and 1682 were apparitions of the same comet and his prediction that it would return in 1758.

While Immanuel Kant (1724–1804) included comets along with planets in his nebular hypothesis (q.v.) for the formation of the solar system, Pierre Laplace (1749–1827) (q.v.) considered comets to be extrasolar in his cosmogony of the protosolar nebula. The question about the origin and formation of comets has been debated ever since. In 1932 Ernst Öpik (1893–1985) (q.v.) was the first to suggest that a cloud of comets gravitationally bound to the Sun surrounds the planetary system. This idea was given a solid basis by the Dutch astronomer Jan Oort (1900–1992) (q.v.) in 1950. He postulated a comet cloud with dimensions of about 10^5 AU containing more than

Ludwig Biermann and Eleonore Trefftz (1964) put forth the idea of the photodissociation of water vapor (assumed to be abundant in comets, based on Whipple's icy conglomerate model of the nucleus) by solar ultraviolet (UV) radiation to explain observations of atomic oxygen in the coma. They concluded that in active comets a halo or coma of atomic hydrogen should exist, larger than the visible coma, because atomic hydrogen is the fragment of the photodissociation of water that carries off most of the excess energy imparted by the UV photons. This hydrogen coma would best be detected by observing the Lyman α transition in atomic hydrogen at the ultraviolet wavelength, 1216 Å. However, this spectral region is inaccessible from the ground. One of the early successes of the space age was the confirmation of the existence of a hydrogen coma surrounding comet Tago-Sato-Kosaka (1969 IX) by Arthur Code and colleagues (1970) using the OAO-2 satellite.

Advances in astronomical instrumentation in the last two decades have opened new wavelength regions (the ultraviolet, infrared, microwave and submillimeter regions) for comet investigations. Rockets and the Space Shuttle have been utilized to make UV observations possible. Radar has been used to probe the nucleus and large particles surrounding it, and radio telescopes continue to be employed in studying comets as new limits of sensitivity and resolution are reached. In addition, technological advances in conventional methods of imaging and analyzing light from astronomical sources, such as the charge-coupled device (CCD) camera, have vastly improved our ability to observe sources with very low surface brightness, typical of cometary comae and comet nuclei at large heliocentric distances. All these new tools and techniques give us insights from many different perspectives and continue to advance our knowledge concerning the nature of comets.

On 11 September 1985 we entered a new era of cometary science with the encounter of comet P/Giacobini-Zinner (1985 XIII) by the International Cometary Explorer (ICE), a spacecraft originally launched to study the solar wind and the Earth's magnetosphere but cleverly diverted to the comet after completing its intended mission (Plate 20). This was the first time that scientific instruments were placed directly in the environment of a comet. The *in situ* measurements returned by ICE gave us a first-hand look at cometary plasma. Since that date other spacecraft encounters with comets have taken place and more are planned. In early March 1986 an international armada of five spacecraft encountered comet P/Halley. The European spacecraft, Giotto was partially damaged just prior to its closest approach but functioned well enough to fly by a second comet, P/Grigg-Skjellerup, in July 1992 (see Giotto mission; Suisei and Sakigake missions; Vega missions). Our knowledge of comets has been tremendously expanded by the results from these missions. Accompanying articles highlight some of the new results of the era of spacecraft observations of comets.

Daniel C. Boice and Walter F. Huebner

Bibliography

Alfvén, H. (1957) On the theory of comet tails. *Tellus*, **9**, 92–6.
Bailey, M.E., Clube, S.V.M. and Napier, W.M. (1990) *The Origin of Comets*. Oxford: Pergamon Press.
Biermann, L. (1951), Kometenschweife und solare Korpuskularstrahlung *Z. Astrophys.*, **29**, 274–86.
Biermann, L. and Trefftz, E. (1964) Über die Mechanismen der Ionisation und der Anregung in Kometenatmosphären. *Z. Astrophys.*, **59**, 1–28.
Brandt, J.C. and Chapman, R.D. (1992) *Rendezvous in Space: The Science of Comets*. New York: Freeman.
Cameron, A.G.K. (1973) Accumulation processes in the primitive solar nebula. *Icarus*, **18**, 407–49.
Code, A.D., Houck, T.E. and Lillie C.F. (1972) Ultraviolet observations of comets, in *The Scientific Results from the Orbiting Astronomical Observatory*, (ed. A.D. Code). Washington, DC: NASA SP-310, pp. 109–14.
Chyba, C.F., Thomas, P.J. and Zahnle, K.J. (1993) The 1908 Tunguska explosion: atmospheric disruption of a stony asteroid. *Nature*, **361**, 40–4.
Duncan, M.J. and Quinn, T. (1993) The long-term dynamical evolution of the solar system *Ann. Rev. Astron. Astrophys.*, **31**, 265–95.
Finson, M.L. and Probstein, R.F. (1968) Theory of dust comets. I. Model and equations. *Astrophys. J.*, **154**, 327–52.
Huebner, W.F. (ed.) (1990) *The Physics and Chemistry of Comets*. Berlin, Heidelberg: Springer-Verlag.
Keller, H.U., Arpigny, C.B. Bonnet, C.M. *et al.* (1986) First Halley multicolour camera imaging results from Giotto, *Nature*, **321**, 320–26.
McDonnell, J.A.M., Lamy, P.L. and Pankiewicz, G.S. (1991) Physical properties of cometary dust, in *Comets in the Post-Halley Era*, (eds R.L. Newburn, Jr, M.Neugebauer and J. Rahe). Dordrecht: Kluwer Academic.
Newburn, R.L., Jr, Neugebauer, M. and Rahe J. (eds) (1991) *Comets in the Post-Halley Era*, Dordrecht: Kluwer Academic
Oort, J.H. (1950) The structure of the cloud of comets surrounding the Solar System, and a hypothesis concerning its origin. *Bull. Astron. Inst. Neth.*, **11**, 91–10.
Öpik, E.J. (1973) Comets and the formation of planets. *Astrophys. Space Sci.*, **21**, 307–98.
Rickman, H. (1989) The nucleus of comet Halley: surface structure, mean density, gas and dust production, *Adv. Space Sci.*, **9**, 59–71.
Safronov, V.S. (1977) Oort's cometary cloud in the light of modern cosmogony, in *Comets, Asteroids, Meteorites*, (ed. A.H. Delsemme). Toledo: University of Toledo Press, pp. 483–84.
Whipple, F.L. (1950) A comet model I. The acceleration of comet Encke. *Astrophys. J.*, **III**, 375–94.
Whipple, F.L. (1985) *The Mystery of Comets*. Washington, DC: Smithsonian Institution Press.
Wilkening, L.L. (ed.) (1982) *Comets*. Tucson: University of Arizona Press.
Yeomans, D.K. (1991) *Comets: A Chronological History of Observations, Science, Myth, and Folklore*. New York: John Wiley and Sons.

Cross references

Chiron
Giotto mission
Halley, Edmond; Halley's comet
Kuiper belt
Life: origin
Oort, Jan Hendrik; Oort cloud
Sakigake and Suisei missions
Vega missions

COMET: DYNAMICS

Some orbital properties

Comets coming to the vicinity of the Sun are found to move on orbits with very different orbital periods (P), ranging from a few years to several million years. They are usually classified as periodic and non-periodic [or long-period (LP)]. The boundary is placed rather arbitrarily at 200 years, based on the fact that comets with $P > 200$ years have been observed only once. To a large extent this is an observational artifact due to the lack of systematic observations of comet apparitions for more than about two centuries. Periodic comets are sometimes divided into intermediate-period (IP), or Halley-type, and short-period (SP), or Jupiter-family comets, with the boundary at $P \approx 20$ years. The prefix P/ stands for periodic comet.

There are 810 comets listed in Marsden's (1989) catalogue of cometary orbits, of which 655 are LP comets, 20 are IP comets and 135 are SP comets. It is interesting to analyze the distributions of orbital inclinations (i) of the different dynamical classes. As Figure C31 shows, the orbital planes of LP comets are roughly randomly distributed; the observed distribution of i fits a sine law quite well, although showing a small excess of retrograde orbits (i.e. with $i > 90°$). By contrast, IP comets move predominantly in direct orbits and SP comets move mostly in near-ecliptic orbits.

Short-period comets move in very unstable, chaotic orbits subject to frequent perturbations by Jupiter. There are in fact some SP comets found to librate temporarily around mean motion resonances, for instance P/Tempel 1, though they will tend to resume their chaotic behavior after some time. Indeed, the fundamental dynamical role of Jupiter is reflected in a spatial concentration of SP comets in its vicinity at a given time (Tancredi and Lindgren, 1992). Furthermore,

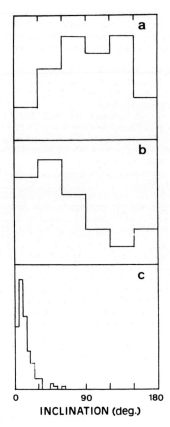

Figure C31 Inclination distributions of (a) long-period comets discovered after 1758. The 22 members of the Kreutz family of sungrazing comets have been considered as a single comet, (b) intermediate-period comets and (c) short-period comets, taken from Marsden's (1989) catalog.

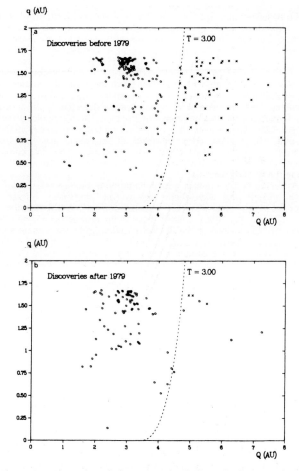

Figure C32 Plots of perihelion (q) and aphelion (Q) distances for asteroids (open circles) and comets (crosses) with $q < 1.67$ AU. The dashed curve marks the evolutionary track for $T = 3$ in the planar case (Hahn and Rickman, 1985).

low-velocity and long-lasting encounters of comets with Jupiter may lead to their capture as temporary satellites (Carusi and Valsecchi, 1981). On the other hand, most asteroids move in orbits of long-term dynamical stability. Most of them avoid close encounters with Jupiter, except a few objects like (944) Hidalgo, (3552) Don Quixote, 1982YA, 1983XF, 1984BC and 1991DA that move in 'cometary' orbits (Hahn and Rickman, 1985).

The Tisserand invariant derived from Jacobi's integral of the restricted three-body problem provides another useful criterion for distinction between comets and asteroids (Kresák, 1979). The Tisserand invariant T for a body moving on an orbit with perihelion distance q, semimajor axis a and inclination i is given by

$$T = 1/a + 2[2q(1 - q/2a)]^{1/2} \cos i \qquad (C11)$$

which is valid under the assumptions that the perturbing planet (Jupiter) has a circular orbit with unit radius and that other planets do not perturb the body. The encounter velocity U of the body with respect to the circular motion of Jupiter can be expressed in terms of T as

$$U = (3 - T)^{1/2} \qquad (C12)$$

which shows that encounters with Jupiter are possible only in the case $T < 3$. The great majority of SP comets indeed have $T < 3$, while the opposite holds for the asteroids, in accord with the previous discussion on orbital stability (Figure C32).

Main perturbing forces acting on comets

Planets

Comets entering the planetary region are subject to planetary perturbations. The three angular orbital elements: the inclination (i), argument of perihelion (ω) latitude of the ascending node (Ω) and the perihelion distance (q) experience only minor changes. By contrast, the orbital binding energy E can change drastically because the typical energy change is usually of the order of or larger than E. Bearing in mind that the binding energy is proportional to the reciprocal of the semimajor axis $1/a$, we usually take the latter variable as equivalent to the comet's binding energy. Typical energy changes of comet orbits per orbital revolution, $\epsilon = \delta(1/a)$, are shown in Figure C33. Comets in retrograde orbits will meet the planets at larger relative velocities than those in direct orbits, so that the former ones will on the average be less perturbed.

Through successive perihelion passages comets will undergo a random walk in energy space; they will evolve from initial near-parabolic orbits (binding energies ≈ 0) to periodic orbits (binding energies $\gtrsim 0.03$ AU^{-1}), after several hundred or thousand revolutions (Fernández, 1981). The basic assumption is that two successive perturbations ϵ of a LP comet are independent of each other, which is justified on the basis that the comet will meet during each passage a planetary configuration completely different from the previous one, due to its long orbital period. Let us now define $\nu(1/a,t)\,d(1/a)$ as the number of LP comets passing perihelion per year, with reciprocal semimajor axes in the range $(1/a, 1/a + d(1/a))$ at a certain time t. The total number of comets in this range of $1/a$ will thus be given by

$$N(1/a,t) = a^{3/2} \nu(1/a,t)\,d(1/a) \qquad (C13)$$

where $a^{3/2}$ is the comet's orbital period in years and a is expressed in AU. The increase in the number of comets of a certain $1/a$ per year will be given by $\partial N(1/a,t)/\partial t$. Let us also define $\Psi(\epsilon)$ as the probability that a comet experiences an energy change ϵ after a perihelion passage caused by planetary perturbations. Therefore the diffusion equation will be expressed as

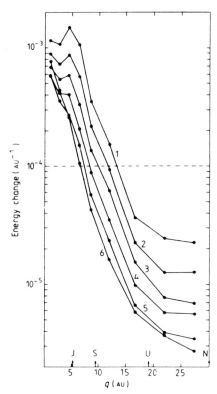

Figure C33 Typical energy change per perihelion passage as a function of the cometary perihelion distance and for inclination ranges of 30°. Thus curve 1 is for $0 < i < 30°$, ..., curve 6 for $150° < i < 180°$ (Fernández, 1981).

$$\partial N(1/a,t)\partial t = \int_{-\infty}^{+\infty} \nu(1/a-\epsilon,t)\,\Psi(\epsilon)\,d\epsilon - \nu(1/a,t) \quad (C14)$$

where the first term on the right hand side represents the number of comets acquiring a reciprocal semimajor axis $(1/a)$ per year and the second term those changing from $(1/a)$ to other values.

Using the previous analytical approach, van Woerkom (1948) showed that comets diffusing inwards from a field of parabolic orbits $(1/a = 0)$ would reach a uniform $(1/a)$ distribution after a large number of passages $(t \to \infty)$. Yet, as shown by Oort (1950), there is a clear excess of comets with original near-parabolic orbits $(0 < (1/a)_{\text{orig}} < 10^{-4}\text{ AU}^{-1})$, which he called 'new' comets since he supposed that they were entering the planetary region for the first time. The term 'original' orbit refers to the orbit the comet has before entering the planetary region and being perturbed by the planets. The observed concentration of comets in near-parabolic orbits led Oort to the conclusion that a huge swarm of $\approx 10^{11}$ comets surrounds the solar system at distances of a few tens of thousands of AU. This structure, called the Oort cloud, is generally supposed to be the source of LP comets.

From numerical experiments, Kerr (1961) approximated the probability distribution function of energy changes $\Psi(\epsilon)$ by a Gaussian or a double-exponential distribution. Everhart (1968) pointed out that $\Psi(\epsilon)$ deviates from a Gaussian distribution due to the presence of long tails of large values of ϵ that correspond to strong planetary perturbations in close encounters. Indeed close planetary encounters can greatly influence the dynamical evolution of LP comets (Stagg and Bailey, 1989; Gallardo and Fernández, 1992).

Most LP comets will be ejected to interstellar space during their random-walk in energy space (i.e. they acquire negative binding energies). In fact, from an initial population of N_0 comets on near-parabolic orbits, the number $N(n)$ still gravitationally bound to the solar system after n perihelion passages is (Everhart, 1976)

$$N(n) \approx \tfrac{1}{2}\,N_0\,n^{-1/2} \quad (C15)$$

Non-gravitational forces

These arise from the 'jet reaction' produced by the non-isotropic, sublimating gases of the comet nucleus, as discussed by Whipple (1950). Non-gravitational forces for a non-rotating nucleus would only give a radial component J_r. Yet in a more realistic situation of a rotating nucleus, thermal inertia will cause the region of maximum outgassing to shift towards the nucleus 'afternoon' by a certain lag angle λ, giving rise to additional transverse and normal components J_t and J_n.

The main non-gravitational effect that can be detected in a periodic comet observed at previous apparitions is a delay (or advance) in the time of the perihelion passage with respect to that derived from purely gravitational theory. In other words, non-gravitational perturbations will introduce a change ΔP in its orbital period P. For instance, for the last few apparitions P/Halley has arrived at its perihelion with an average delay of $\Delta P = 4.1$ days. By using Gauss' equations, Rickman (1986) derives for ΔP the expression

$$\Delta P = \frac{6\pi}{\eta^2 a}\int_0^P \left[\frac{e\sin f}{(1-e^2)^{1/2}}J_r + \frac{a(1-e^2)^{1/2}}{r}J_t\right]dt \quad (C16)$$

where η is the mean motion, e the eccentricity and f the true anomaly.

The standard view (Whipple, 1950) was that non-gravitational forces arise from the transverse component J_t due to a rotationally induced thermal lag. This would be the case for a comet with a symmetric lightcurve (with respect to perihelion), with the result that the first term of equation (C16) integrated over the whole orbital period P would vanish. Thus the only term remaining in this case will be the one containing J_t. Marsden, Sekanina and Yeomans (1973) carried out extensive numerical integrations using the standard model (see also Weissman, 1979). Yet most comet lightcurves are moderately or highly asymmetric, suggesting an asymmetric outgassing, so the integral of the term containing J_r will not vanish. Indeed it can be much greater than the other one. The probable dominant influence of asymmetries in comet lightcurves on nongravitational effects was noted by Rickman (1986) and developed afterwards by Yeomans and Chodas (1989).

Steady perturbers of the Oort cloud: background stars and galactic tides

Oort cloud comets are subject to the action of external perturbers such as passing stars, molecular clouds and galactic tides. Let us define as background stars those that pass by the solar system at distances greater than about 2×10^4 AU. Background stars and galactic tides exert a quasi-steady perturbing action on Oort cloud comets. By contrast, very close stellar passages (that mainly affect comets along the star's path) and penetrating encounters with molecular clouds occur sporadically over time scales of the order of several tens of millions of years (Fernández and Ip, 1991). The steady supply of new comets is thus due to the action of background stars and galactic tides. Very close stellar passages and penetrating encounters with molecular clouds will cause sudden enhancements in the flux of new comets that Hills (1981) called 'comet showers' (see below).

During an orbital revolution of period $P = a^{3/2}$ years a comet will be perturbed by many stars. Let $s(D_\odot)dD_\odot = 2n_*D_\odot dD_\odot$ be the rate of stellar passages with impact parameters in the range $(D_\odot, D_\odot + dD_\odot)$. n_* is the stellar flux in the Sun's neighborhood, which is about 7 stars per million years passing through a circle of radius 1 pc at an average encounter velocity with the Sun of $V = 30$ km s^{-1}. The cumulative change in the orbital velocity of the comet during P, Δv_*, will be expressed as

$$\Delta v_*^2 = P\int_{D_M}^{D_m}\overrightarrow{\Delta v}^{\,2}\,s(D_\odot)\,dD_\odot \quad (C17)$$

where $\overrightarrow{\Delta v}$ is the impulsive change in the comet's velocity due to a single star passage which is given by

$$\overrightarrow{\Delta v} = 2GM/V\,(\overrightarrow{D}/D^2 - \overrightarrow{D_\odot}/D_\odot^2) \quad (C18)$$

where G is the gravitational constant, M the stellar mass, and $D_\odot$, D are the distances of closest approach of the star to the Sun and the comet, respectively. $D_m = (2n_*P)^{-1/2}$ is the minimum distance of closest approach of a star to the Sun expected during P. D is the maximum distance of a passing star that may have some dynamical influence; it can be taken as infinity without too much error.

The motions of Oort cloud comets are also influenced by the gravitational field of the galactic disk, which can be modeled approximately by a homogeneous disk in the midplane of the galaxy. The respective potential can thus be simply expressed as $U = U_0 + 2\pi G\rho z^2$, where z is the distance from the galactic midplane and ρ is the density of the galactic disk in the Sun's neighborhood. Bahcall (1984) derives a value of $\rho = 0.185\,M_\odot\,\text{pc}^{-3}$ from the comparison of different gravitational potential models of the Galaxy with velocity dispersions of tracer stars. Recent studies have shown that the galactic tidal force directed into the galactic disk has a dominant role in the dynamical evolution of Oort cloud comets as compared to stellar perturbations (e.g. Byl, 1983; Heisler and Tremaine, 1986). From the above potential of the galactic disk, the change in the transverse component of an Oort cloud comet of semimajor axis a and aphelion point at a galactic latitude ϕ, during P, can be easily derived, leading to

$$(\Delta v_{\text{tide}})_T = 3 \pi G \rho a P \cos \alpha \sin 2\phi \quad (C19)$$

where α is the angle between the orbital plane and the plane perpendicular to the galactic disk containing the radius Sun–comet.

Equation (C19) shows that the change will be largest for mid-galactic latitudes ($\phi = 45°$) and negligible near the galactic poles and the galactic equator. This is in agreement with observations showing a concentration of LP comets at mid-galactic latitudes (Delsemme, 1987).

Passing stars are the most efficient perturbers of the inner portions of the Oort cloud, say for distances $r \lesssim 10^4$ AU, because of their ability to penetrate it (Bailey, 1986). A strong change in the transverse velocity, Δv_T, will generally correspond to drastic changes in both q and i. Therefore comets will get random orientations of their orbital planes when the cumulative change in the transverse velocity after n stellar perturbations becomes of the order of v_T, i.e. when $(\Delta v_T)^2_{\text{cum}} = \sum_{i=1}^{n} (\Delta v_T)^2_i \approx v_T^2$. Even if comets started out their dynamical evolution in orbits close to the ecliptic plane, a full randomization of their orbital planes should have been reached for semimajor axes $a > 3000$–4000 AU as a result of very close stellar passages. Numerical simulations carried out by Duncan, Quinn and Tremaine (1987) that included planetary perturbations, random stars and galactic tides showed randomization of orbital planes for $a > 5000$ AU.

Random perturbers of the Oort cloud: very close stellar passages and molecular clouds

A star penetrating deeply into the core of the Oort cloud will strongly perturb comets along the star's path; they will receive impulses $\vec{\Delta v} \approx 2GM/VD \; \vec{D}/D$. Therefore comets injected into the planetary region as a result of this passage will have their aphelion directions clustered in a sky area along the stellar path, and in particular toward the point of closest approach of the star to the Sun. Indeed, some anomalous clusterings of aphelion points of dynamically young comets have been regarded as signatures of past close stellar passages (Biermann, Huebner and Lüst, 1983).

Penetrating encounters with interstellar molecular clouds may also play a fundamental role in the dynamical evolution and stability of Oort cloud comets. Their dynamical influence and possible catastrophic consequences were analyzed by Biermann (1978), Napier and Staniucha (1982) and Bailey (1983) among others. For the purpose of illustrating their effects, let us neglect any kind of irregularity or clumpiness in their internal structure and assume a spherical molecular cloud of uniform density, radius R_{cl} and mass M_{cl}. The impulsive change in the velocity of a comet at a distance r from the Sun can in this case be approximated by (Biermann, 1978)

$$\Delta v_{cl} = 2GM_{cl}/v_{cl} \, r/b^2 \, [1 - (1 - b^2/R_{cl}^2)^{3/2}] \sin \zeta \quad (C20)$$

where $v_{cl} \approx 20$ km s^{-1} is the typical encounter velocity with molecular clouds, b is the impact parameter and ζ is the angle between $\vec{r}$ and $\vec{v}_{cl}$.

The most devastating effects will occur when the solar system encounters a giant molecular cloud (GMC), typically of a mass $M_{cl} \approx 5 \times 10^5 \, M_\odot$ and radius $R_{cl} \approx 20$ pc. There might have been only a few such encounters during the solar system lifetime (Talbott and Newman, 1977). Penetrating encounters with intermediate-size molecular clouds, with masses of a few 10^3–$10^4 \, M_\odot$, may be more frequent, although less dramatic, with time scales of several tens of millions of years.

Fluctuations in the frequency of comet passages: comet showers

Oort cloud comets have been thermalized over the age of the solar system under the action of external perturbers, certainly for semi-major axes $a \gtrsim 10^4$ AU (e.g. Duncan, Quinn and Tremaine, 1987). But not all directions of the velocity vectors of thermalized comets are possible. Oort cloud comets entering the inner planetary region will be quickly removed by planetary perturbations owing to the fact that their typical energy changes are greater than their binding energies in the Oort cloud (Fernández, 1981). Since the velocity vectors of these comets fall very close to the solar direction, there will be an empty region in the velocity phase space known as the loss cone (Hills, 1981). For a thermalized cometary population, the fraction of comets with semimajor axis a having perihelion distances $q < q_L$ will be approximately given by $F \cong 2q_L/a$ provided that $q_L << a$. Thus the loss cone will have an angular radius of $2F^{1/2}$ radians and the solar direction as the axis.

Oort cloud comets will diffuse within the loss cone under the combined action of external perturbers. These comets will then be injected into the planetary region and removed afterwards from the Oort cloud by planetary perturbations. Results for the efficiency in filling the loss cone by different perturbers show a slight predominance for galactic tides. The combined effect of stellar perturbations and the vertical galactic tidal force determines a limiting semimajor axis $a = a_{\text{fill}}$ such that comets with $a > a_{\text{fill}}$ will have their loss cones permanently filled. Fernández (1992) obtains $a_{\text{fill}} \cong 3.16 \times 10^4$ AU, which turns out to be in good agreement with previous analytical derivations (Heisler and Tremaine, 1986) and Monte Carlo simulations (Heisler, 1990).

A random perturber – either a very close stellar passage or a molecular cloud – will affect the inner portions of the Oort cloud where loss cones are to a large extent empty. Thus its effect will be a sudden 'refilling' of the loss cones, leading to a comet shower. This will last until the refilled loss cones are emptied, typically of the order of the orbital period of the injected comets (about 10^6 years). The intensity of a comet shower will depend on the degree of central condensation of the Oort cloud. For a heavily concentrated Oort cloud, a close stellar passage at $D_\odot \approx 10^4$ AU can trigger a comet shower with a frequency of passages ≈ 100 times the background comet flux during its phase of highest intensity. The same effect can be reached with a penetrating encounter with a GMC. Closer stellar passages at say $D_\odot \approx 5000$ AU might trigger showers about 10^2–10^3 times as intense, depending again on the degree of central condensation of the Oort cloud (Fernández, 1992). Stellar encounters at ≈ 5000 AU should be expected at average time intervals on the order of 1–2×10^8 years. Monte Carlo simulations carried out by Heisler, Tremaine and Alcock (1987) and Heisler (1990), including both stellar and galactic tide perturbations, illustrate very well the production of comet showers at average intervals of several tens of millions of years.

There have been several works arguing that evidence of past comet showers can be found in the impact cratering record (e.g. Alvarez and Muller, 1984). Yet studies of impact crater ages are not conclusive in this respect. The major problem seems to be that the terrestrial cratering rate is dominated by Earth-crossing asteroids, where comets may only account for less than 10% of all the observed terrestrial craters (Bailey, 1990). Simple calculations show that a comet shower must be at least 300 times greater than the steady state comet flux to show up in crater statistics, and this might only occur at intervals of some 10^8 years.

The observed aphelion distribution of new and dynamically young comets follows a pattern reflecting the influence of the galactic disk potential, with the highest concentration of aphelion points at mid-galactic latitudes. This was already noted by Byl (1983); subsequently Delsemme (1987) analyzed its statistical significance. The evidence that most new comets seem to be deflected to the inner planetary region by galactic tides – a steady perturber – suggests that the frequency of comet passages is at present close to its background level (Fernández and Ip, 1991). Comets injected during a shower might greatly exceed the steady supply of Oort cloud comets, brought mainly by galactic tides, thus erasing or severely weakening the galactic signature. As mentioned above, there are some weak aphelion clusterings that might reflect past close stellar passages, though they seem to be a minor fraction of the overall aphelion sample.

Dynamical stability of the Oort cloud

Under the action of external perturbers Oort cloud, comets will on the average gain energy. Once the cumulative change of the comet's velocity reaches the escape velocity, it will be lost to interstellar space, i.e. when

$$\Delta v_{\text{cum}}^2 = v_{\text{esc}}^2 = 2GM_\odot/r \cong 4GM_\odot/3a \quad (C21)$$

where Δv_{cum}^2 is obtained by summing quadratically the Δv contributions of external perturbers. If we only consider stellar perturbations, the limit imposed by equation (C21) will be at $a \approx 1.3 \times 10^5$ AU. For penetrating encounters with GMCs the upper limit is however at $a \approx 10^4$ AU; here the uncertainty comes from considering a possible range between 1 and 10 penetrating encounters (Bailey, 1983). The perturbing action of the tidal force of the galactic disk in the outgoing leg of the comet's orbit roughly compensates for that in the incoming leg, so that its net effect is very small, typically of order (q/a) (Byl, 1983). Therefore, the overall effect of galactic tides on the cometary energy is negligible.

We then see that penetrating encounters with GMCs are of fundamental importance in defining the stability boundary of the Oort cloud at $a \approx 10^4$ AU. This turns out to be significantly smaller than the boundary originally found by Oort (1950). The previous value can be compared with the observed maximum separations between members of wide binary stars in the Galaxy, which are coincidentally of the order of 10^4 AU (Bahcall and Soneira, 1981).

The fact that we define a stability radius does not mean that beyond it the Oort cloud is empty, since the outer region will be continuously replenished with comets from the inner cloud. In this regard we note that nearly 40% of the original semimajor axes of new comets have $a > 3 \times 10^4$ AU, indicating that they come from the outer or classical Oort cloud (as originally defined by Oort). Nevertheless we should expect a significant drop in the number of Oort cloud comets for larger a as comets have ever decreasing dynamical lifetimes in the outer region.

The Kuiper belt

To a certain extent, the dynamical evolution of comets can not be separated from their origins. There are several possible sources of the Oort cloud comets, namely the condensation of icy planetesimals in the solar nebula and the ejection of small bodies from the Uranus–Neptune zone. As mentioned before, in addition to the insertion of comets into the Oort cloud at a radial distance of a few 10^4 AU from the Sun, the formation of an inner core (or Kuiper belt) within a few thousand kilometers is possible. The Kuiper belt might have a flat disk-like configuration, with a concentration near the ecliptic plane. The recent discoveries of objects 1992QB1 and 1993FW located at about 40 AU (Jewitt and Luu, 1993; *IAU Circular*, 5370), give strong observational support to the existence of a transneptunian comet belt. Together with the small bodies of a few kilometers in size, a number of large planetoids up to lunar or Martian size might also exist in this region (Fernández, 1980; Ip, 1989). Several cosmogonical reasons suggest the presence of several 10^3 km size planetoids during the formation of Uranus and Neptune (Stern, 1991). That the Kuiper belt, with its inner edge possibly located just outside the orbit of Neptune, might be the main supplier of the SP comets was investigated by Duncan, Quinn and Tremaine (1988) who showed that the predominantly low-inclination distribution of the SP comets can be most naturally explained in terms of the orbital capture of Neptune-crossing comets from such a cometary population. A significant population of the LP comets with isotropic distribution in inclination always ends up in retrograde orbits after orbital capture (Duncan, Quinn and Tremaine 1988; Ip and Fernández, 1991). This is contradictory to the observed inclination distribution of the SP comets (Figure C31).

Given that the Kuiper belt may be the source region of the SP comets, the next question is then what might be the mechanism(s) responsible for the scattering of Kuiper belt comets to Neptune-crossing orbits. As suggested by Fernández (1980), the existence of a number of lunar-sized objects outside the orbit of Neptune might be able to provide the required gravitational perturbation. The gravitational scattering process of a few Earth-sized objects, which are remnants of the Uranus–Neptune accretion zone, could also be important in this respect (Ip and Fernández, 1991). One remaining problem to be resolved, however, concerns the orbital stability of the bodies stored in the Kuiper belt. That is, if the dynamical lifetimes against planetary secular perturbations are much shorter than the age of the solar system, no Moon- or Earth-sized planetoids might remain at present. Gladman and Duncan (1990) and Torbett and Smoluchowski (1990) examined the dynamical evolution of small, gravitationally noninteracting objects in the outer solar system by numerical integration methods. Torbett and Smoluchowski found that the orbits of bodies initially located at radial distances less than 50 AU will become chaotic and be quickly perturbed to Neptune-crossing orbits within 10^7 years. Gladman and Duncan also concluded that orbits of small bodies that start on circular orbits within 34 AU will become Neptune-crossing in 20 million years. This means that the inner edge of the Kuiper belt should be outside 34 AU. Using the statistical method of Markov chains to investigate the long-term dynamical behavior, Levison (1991) showed that comets with orbits with $q > 50$ AU would have a lifetime of more than 5×10^9 years before they leave the Kuiper belt and become Neptune-crossers. More recently, Holman and Wisdom (1993) and Levison and Duncan (1993) performed extensive numerical integrations over periods up to 10^9 years. These studies suggest that the Kuiper belt has been heavily depleted up to ≈ 45 AU.

The studies described above are important in the following senses. Firstly, the Kuiper belt could feed SP comets into the inner solar system over the age of the solar system simply by means of such stochastic orbital diffusion. Secondly, Moon- and Mars-sized objects could still contribute to the orbital diffusion of the SP comets via gravitational scattering, even though they might not play the major role as previously described. The possible presence of large planetoids and the size of the Kuiper belt population must await clarification by future observations. It is yet too soon to tell from the current survey work (Luu and Jewitt 1988; Kowal 1989).

The angular momentum of the Oort cloud

The spacecraft missions to comet P/Halley in 1986 provided the first opportunity to estimate the size and average density of a comet nucleus. The information that the surface albedo of P/Halley is very low ($A \approx 0.04$), and hence that its size is considerably larger than previously thought has prompted several renewed efforts in the calculations of the total mass and angular momentum of the Oort cloud. For example, Marochnik, Mukhin and Sagdeev (1989) estimated a total angular momentum of $J_i \approx 7.5 \times 10^{50} - 2 \times 10^{52}$ g cm^2 s^{-1} for the inner Oort cloud and $J_0 \approx 1.5 \times 10^{50} - 3 \times 10^{51}$ g cm^2 s^{-1} for the outer Oort cloud. The lower limits of these values are in fact consistent with the numbers given by Weissman (1991) by using different values of the number of comets in the Oort cloud and the average comet mass. According to the previous results, the total angular momentum of the inner and outer Oort clouds may exceed that of the planets (3.2×10^{50} g cm^2 s^{-1}) by a factor ranging between 3 and 70.

Although all these estimates are somewhat uncertain because of the many parameters entering into the calculations (i.e. orbital distribution, size distribution, etc.), it is important to note the implication of the large angular momentum of the outer and inner Oort clouds on the dynamical evolution of the solar nebula and protoplanets. As noted by Fernández and Ip (1984), the injection of a large number of planetesimals from the Uranus and Neptune accretion zone into the Oort cloud would unavoidably lead to the modification of the original orbital configurations of the outer planets including Jupiter, Saturn, Uranus and Neptune.

A related question concerns the perturbing effects on the orbital motion of the comets in the Oort cloud by passing stars, galactic tides and GMCs. Due to these external perturbers, we expect that the angular momentum of the Oort cloud has been appreciably altered from its original value. The time evolution of the angular momentum distribution and mass budget are a topic of active research.

Summary

Comets are continuously evolving dynamically under the action of several perturbers, such as planets, nongravitational forces, passing stars, molecular clouds and galactic tides. The observed concentration of original orbital energies at very small values (near-parabolic orbits) led to the concept of a huge cometary reservoir surrounding the solar system, known as the Oort cloud. It is possible that the Oort cloud stretches with no discontinuities from the vicinity of Neptune's orbit up to several 10^4 AU, passing from a flattened structure to a spherical one.

From the above discussion we can define three regions in the Oort cloud.

1. An inner core, for semimajor axes $a \lesssim 3000\text{--}4000$ AU, that might be concentrated towards the ecliptic plane (if comets formed in the protoplanetary disk). The flattened inner portions of the core closer to the planetary region may be identified with the Kuiper belt. The inner core constitutes a reservoir of the outer portions of the Oort cloud and, perhaps, of short-period comets.
2. A spherical, stable Oort cloud, for $4000 \lesssim a \lesssim 10\,000$ AU, containing comets dynamically stable over periods comparable to the age of the solar system.
3. An outer Oort cloud, for $a \gtrsim 10\,000$ AU, whose population has dynamical lifetimes shorter than the solar system age (about 4.6×10^9 years). This region is continuously replenished with comets coming from the inner core of the Oort cloud.

Comets can be transferred to the Earth's vicinity from both the Oort cloud and the Kuiper belt through different dynamical mechanisms.

J.A. Fernández and W.-H. Ip

Bibliography

Alvarez, W. and Muller, R.A. (1984) Evidence from crater ages for periodic impacts on the Earth. *Nature*, **308**, 718–20.

Bahcall, J.N. (1984) Self-consistent determination of the total amount of matter near the Sun. *Astrophys. J.*, **276**, 169–81.

Bahcall, J.N. and Soneira, R.M. (1981) The distributions of stars to $V = 16$th magnitude near the north galactic pole: normalization, clustering properties, and counts in various bands. *Astrophys. J.*, **246**, 122–35.

Bailey, M.E. (1983) The structure and evolution of the solar system comet cloud. *Mon. Not. Roy. Astron. Soc.*, **204**, 603–33.

Bailey, M.E. (1986) The near-parabolic flux and the origin of short period comets. *Nature*, **324**, 350–2.

Bailey, M.E. (1990) Comet craters versus asteroid craters. *Proc. 28th COSPAR meeting*.

Biermann, L. (1978) Dense interstellar clouds and comets, in *Astronomical papers dedicated to Bengt Stromgren* (eds A. Reiz and T. Anderson). Copenhagen Observatory, p. 327.

Biermann, L., Huebner, W.F. and Lüst, R. (1983) Aphelion clustering of 'new' comets: star tracks through Oort's cloud. *Proc. Natl. Acad. Sci. USA*, **80**, 5151–55.

Byl, J. (1983) Galactic perturbations on near-parabolic cometary orbits. *Moon and Planets*, **29**, 121–37.

Carusi, A. and Valsecchi, G.B. (1981) Temporary satellite captures of comets by Jupiter. *Astron. Astrophys.*, **94**, 226–8.

Delsemme, A.H. (1987) Galactic tides affect the Oort cloud: an observational confirmation. *Astron. Astrophys.*, **187**, 913–8.

Duncun, M., Quinn, T. and Tremaine, S. (1987) The formation and extent of the solar system comet cloud. *Astron. J.*, **94**, 1330–8.

Duncan, M., Quinn, T. and Tremaine, S. (1988) The origin of short-period comets. *Astrophys. J. Lett.*, **328**, L69–L73.

Everhart, E. (1968) Change in total energy of comets passing through the solar system. *Astron. J.*, **73**, 1039–52.

Everhart, E. (1976) The evolution of comet orbits, in *The Study of Comets* (eds B. Donn, M. Mumma, W. Jackson *et al.*). IAU Coll. 25, NASA SP-393 pp. 445–64.

Fernández, J.A. (1980) On the existence of a comet belt beyond Neptune. *Mon. Not. Roy. Astron. Soc.*, **192**, 481–91.

Fernández, J.A. (1981) New and evolved comets in the solar system. *Astron. Astrophys.*, **96**, 26–35.

Fernández, J.A. (1992) Comet showers, in *Chaos, Resonance and Collective Dynamical Phenomena in the Solar System*, (ed. S. Ferraz-Mello). *Proc. IAU Symp.*, 152, Kluwer. pp. 239–54.

Fernández, J.A. and Ip, W. (1984) Some dynamical aspects of the accretion of Uranus and Neptune: the exchange of angular momentum with planetesimals. *Icarus*, **58**, 109–20.

Fernández, J.A. and Ip, W.-H. (1991) Statistical and evolutionary aspects of cometary orbits, in *Comets in the Post-Halley Era* (eds. R.L. Newburn, M. Neugebauer and J. Rahe). Kluwer, pp. 487–535.

Gallardo, T. and Fernández, J.A. (1992) The dynamical path from the Oort cloud to periodic orbits: numerical studies, in *Periodic Comets* (eds J.A. Fernández and H. Rickman). Montevideo: Universidad de la República, pp. 35–43.

Gladman, B. and Duncan, M. (1990) On the fates of minor bodies in the outer solar system. *Astron. J.*, **100**, 1680–96.

Hahn, G. and Rickman, H. (1985) Asteroids in cometary orbits. *Icarus*, **61**, 417–42.

Heisler, J. (1990) Monte Carlo simulations of the Oort comet cloud. *Icarus*, **88**, 104–21.

Heisler, J. and Tremaine, S. (1986) The influence of the galactic tidal field on the Oort comet cloud. *Icarus*, **65**, 13–26.

Heisler, J., Tremaines, S. and Alcock, C. (1987) The frequency and intensity of comet showers from the Oort cloud. *Icarus*, **70**, 269–88.

Hills, J.G. (1981) Comet showers and the steady-state infall of comets from the Oort cloud. *Astron. J.*, **86**, 1730–40.

Holman, M.J. and Wisdom, J. (1993) Dynamical stability in the outer solar system and the delivery of short period comets. *Astron. J.*, **105**, 1987–1999.

Ip, W.-H. (1989) Dynamical processes of macro-accretion of Uranus and Neptune: a first look. *Icarus*, **80**, 167–78.

Ip, W.-H. and Fernández, J.A. (1991) Steady-state injection of short-period comets from the trans-Neptunian cometary belt. *Icarus*, **92**, 185–93.

Jewitt, D. and Luu, J. (1993) Discovery of the candidate Kuiper belt object 1992QB1. *Nature*, **362**, 730–2.

Kerr, R.H. (1961) Perturbations of cometary orbits. *Proc. 4th Berkeley Symp. of Mathematical Statistics and Probability*, vol. 3, Berkeley: University of California Press, pp. 149–64.

Kowal, C.T. (1989) A solar system survey. *Icarus*, **77**, 118–23.

Kresák, L. (1979) Dynamical interrelations among comets and asteroids, in *Asteroids* (ed. T. Gehrels). Tucson: University of Arizona Press, pp. 289–309.

Levison, H.F. (1991) The long-term dynamical behavior of small bodies in the Kuiper belt. *Astron. J.*, **102**, 787–94.

Levison, H.F. and Duncan, M.J. (1993) The gravitational sculpting of the Kuiper belt. *Astrophys. J.*, **406**, L35–8.

Luu, J.X. and Jewitt, D. (1988) A two-part search for slow-moving objects. *Astron. J.*, **95**, 1256–62.

Marochnik, L.S., Mukhin, L.M. and Sagdeev. R.Z. (1989) The distribution of mass and angular momentum in the solar system. *Astrophys. Space Phys. Rev.*, **8**, 1–55.

Marsden, B.G. (1989) *Catalogue of Cometary Orbits*, 6th edn. Cambridge, Mass: Central Bureau for Astron. Telegrams.

Marsden, B.G., Sekanina, Z. and Yeomans, D.K. (1973) Comets and nongravitational forces. *Astron. J.*, **78**, 211–25.

Napier, W.M. and Staniucha, M. (1982) Interstellar planetesimals – dissipation of a primordial cloud of comets by tidal encounters with massive nebulae. *Mon. Not. Roy. Astron. Soc.*, **198**, 723–735.

Oort, J.H. (1950) The structure of the cloud of comets surrounding the solar system and a hypothesis concerning its origin. *Bull. Astron. Inst. Neth.*, **11**, 91–110.

Rickman, H (1986). Masses and densities of comets Halley and Kopff, in *Comet Nucleus Sample Return Mission*, (ed. O. Melita). ESA SP-249, Noordwijk: ESTEC, pp. 195–205.

Stagg, C.P. and Bailey, M.E. (1989) Stochastic capture of short-period comets. *Mon. Not. Roy. Astron. Soc.*, **241**, 507–541.

Stern, S.A. (1991) On the number of planets in the outer solar system: evidence of a substantial population of 1000-km bodies. *Icarus*, **90**, 271–281.

Talbott, R.J. and Newman, M.J. (1977) Encounters between stars and dense interstellar clouds. *Astrophys. J. Suppl. Ser.*, **34**, 295–308.

Tancredi, G. and Lindgren, M. (1992) The vicinity of Jupiter: a region to look for comets, in *Asteroids, Comets, Meteors 1991* (eds. A. Harris, and E. Bowell). Houston: Lunar Planetary Inst., pp. 601–4.

Torbett, M.V. and Smoluchowski, R. (1990) Chaotic motion in a primordial comet disk beyond Neptune and comet influx to the solar system. *Nature*, **345**, 49–50.

van Woerkom, A.J.J. (1948) On the origin of comets. *Bull. Astron. Inst. Neth.*, **10**, 445–72.

Weissman, P.R. (1979) Nongravitational perturbations of long-period comets. *Astron. J.*, **84**, 580–4.

Weissman, P.R. (1991) The angular momentum of the Oort cloud. *Icarus*, **89**, 190–3.

Whipple, F.L. (1950) A comet model. I. The acceleration of comet Encke. *Astrophys. J.*, **111**, 375–94.

Yeomans, D.K. and Chodas, P.W. (1989) An asymmetric outgassing model for cometary nongravitational accelerations. *Astron. J.*, **98**, 1083–93.

Cross references

Galactic cycle
Kuiper belt
Oort, Erast Julias and Oort cloud

COMET: HISTORICAL PERSPECTIVE

When beggars die, there are no comets seen;
The heavens themselves blaze forth the death of princes.

So wrote Shakespeare, in *Julius Cæsar*; and this more or less summed up the general feeling about comets current at the time. And Milton, in *Paradise Lost*, wrote:

Unterrified, and like a comet burn'd,
That fires the length of Ophiuchus huge
In th' Artick sky, and from its horrid hair
Shakes pestilence and war.

It is true that a brilliant comet, with a gleaming head and a tail stretching right across the sky, must be an imposing and even a terrifying sight; it is a pity that so few people of the present day have seen them. They were fairly frequent in the 19th century, and the comets of 1811, 1843 and 1882, to name but three, were brilliant enough to cast shadows, but the last really 'great' comet dates back to 1910. (Comet Ikeya-Seki of 1965 enjoyed only a very brief glory.)

Cometary records go back a long way, and the Chinese, in particular, were meticulous in observing them; they were called 'hairy stars', and were generally regarded as indicating divine displeasure. The Chinese contented themselves with descriptions and follows their movements, but the early Greeks made the first efforts to explain them. Anaxagoras (500–428 BC) was inclined to regard them as due to clusters of faint stars, or possibly several planets close together, while the great philosopher Aristotle (384–322 BC) believed them to be atmospheric phenomena, due to exhalations from the Earth itself. (Warm exhalations were supposed to rise to great heights where they were ignited by friction, after which the comet was carried round the world by the circular motion of the celestial sphere.) Ptolemy of Alexandria, the last great astronomer of classical times, did not mention comets at all in his works dealing with celestial bodies, so that clearly he too regarded them as atmospheric.

There were two reasons why comets caused such alarm. One was purely astrological and superstitious, and it is worth quoting a comment by the Roman emperor Vespasian in AD 79, about a comet seen in that year: 'This hairy star does not concern me; it menaces rather the King of the Partians, for he is hairy, while I am bald' – though it must be added that Vespasian died soon after the comet had faded from view. There is also the famous description given by the French doctor Ambroise Paré in 1528:

This comet was so horrible, so frightful, and it produced such great terror that some died of fright and others fell sick. It appeared to be of extreme length, and was the colour of blood. At the summit of it was seen the figure of a bent arm, holding in its hand a great sword as though about to strike. At the end of the point there were three stars. On both sides of the rays of this comet there were seen a great number of axes, knives and blood-coloured swords, among which were a large number of hideous human faces, with beards and bristling hair.

Whether Dr Paré was actually describing a comet is not certain; he may have seen a brilliant display of aurora borealis, and it is true that there are no other records of bright visitor in 1528, but the account does demonstrate the fear which comets inspired.

The other reason for alarm was the possibility that a direct collision with a comet might destroy the world. This is not so unreasonable as it might sound in the light of modern knowledge, because a brilliant comet can look really striking, and there was no means of telling that it is really very insubstantial – the only solid part is the nucleus, which is icy in composition and is no more than a few kilometres in diameter. Neither was it possible to separate superstition from what was taken to be genuine science; this is shown by the views of William Whiston, who became Lucasian Professor at Cambridge in 1703 and was known to be an excellent mathematician.

Whiston was a curious mixture. He had carried out some very valuable scientific work (particularly in connection with the exact determination of longitude), but he was also convinced that events related in the Bible were linked with comets, and these views were set out in his book *New Theory of the Earth*, published in 1696. Like Immanuel Velikovsky in our own time, he made the classic mistake of confusing planets with comets, and went so far as to claim that the Earth used to be a comet moving in a path which took it close to the Sun. When God decided that the time had come for humanity to appear, the comet's path became less eccentric, the air was purified, oceans appeared and the Earth became a planet. (In fact, it would take around 60 000 million bodies the mass of Halley's comet to equal the mass of the Earth; and by cometary standards, Halley is large.)

This was not all. Whiston believed that a bright comet seen in 1680 was a sign that humanity was about to be punished for its sins, so that the comet would return in 1736 and signal the end of the world. His reputation ensured that he would be taken seriously, and on the appointed day there was something of a panic in London – it is even on record that the Archbishop of Canterbury felt bound to issue a public disclaimer.

Yet even in Whiston's time there had been some purely scientific observations of comets. In 1577 the great Danish astronomer Tycho Brahe (q.v.) proved, by parallax measurements, that they were much more remote than the Moon, thereby finally disposing of the idea that they were moving in the upper atmosphere. Then, in 1682, came the comet which led to a complete change in outlook. This particular comet was observed by Edmond Halley (q.v.), friend of Isaac Newton and later to become England's second Astronomer Royal. Halley realized that the 1682 comet moved in a path strikingly similar to those of comets previously seen in 1607 and in 1531, and concluded that the three bodies were one and the same. He predicted a return for 1758, and added modestly that if this should happen 'posterity will not refuse to acknowledge that this was first discovered by an Englishman'. The comet did indeed return; it was picked up on Christmas Night 1758 by the Saxon astronomer Palitzsch, and came to perihelion in the following year. Since then it has returned in 1835, 1910 and 1986; we know it, fittingly, as Halley's comet.

Other periodical comets were subsequently identified, and many dozens are now known, though only Halley's can ever become brilliant. (It failed to do so at the 1986 return only because it was so badly placed; when it should have been at its best, it and the Earth were on opposite sides of the Sun, so that the comet did not rival its splendour of 1835 or 1910 – or for that matter AD 837, when according to the *Anglo-Saxon Chronicle* it cast strong shadows, and presented a tail extending for well over 90° in length.) The periods of periodical comets which have been seen at more than one return range from 3.3 years (Encke's comet) to 156 years (comet Herschel-Rigollet), but very few of them ever attain naked-eye visibility. The really brilliant comets, such as those of 1811 and 1843, have periods which are so long that they are measured in thousands or even millions of years – and in some cases they may move in open paths, so that they may never return at all.

This was established during the 19th century. It was also found that some meteors are cometary debris. As a comet moves along it leaves a 'dusty' trail behind it; when the Earth passes through such a trail it collects meteors, which plunge into the atmosphere and burn away in the streaks of luminosity which we call shooting stars. Most annual meteor showers have known parent comets. For example the Perseids, which never fail to produce good displays in early August each year, are linked with the periodical comet Swift-Tuttle, which has a revolution period of 130 years and was last at perihelion in 1992.

The origin of comets was long shrouded in mystery, and it was widely believed that they came from interstellar space rather than being bona fide members of the solar system. That idea still has its supporters, but on the whole it seems unlikely; it would mean that comets would plunge sunward with velocities greater than they actually show. Nowadays the favored theory is that developed in 1948 by the late Jan Oort, one of Holland's most eminent astronomers.

According to Oort, there is a 'cloud' of comets orbiting the Sun at a distance of around 1 light year (9.46×10^{12} km). Basically they are icy, and thus they are completely unobservable. If one of them is perturbed for any reason – for example, by the pull of a passing star or even a remote solar planet – it may start to fall inward toward the Sun. Eventually it will invade the inner part of the solar system, and come within our range; nearing perihelion the ices in the nucleus will start to evaporate, and the comet will produce a head or coma, together with a tail (or even several tails). One of several things may then happen. The comet may plunge into the Sun and be destroyed;

in recent years several comets have been seen to self-destruct in this way. It may simply swing round the Sun and start on its journey back to the Oort cloud, not to return for an immensely long time. It may be perturbed by the pull of a planet (usually Jupiter) and ejected from the solar system altogether. Or the pull of the planet may force it into a short-period orbit which will bring it back to perihelion regularly (see Comet: origin and evolution).

We have no positive evidence of the existence of the Oort cloud (q.v.), but it does seem to be plausible, and there may even be a closer cloud (the Kuiper belt or cloud; q.v.) not too far beyond the orbits of the outermost known planets, Pluto and Neptune. At least the theory explains why all the periodical comets are relatively faint. At every perihelion passage a comet must lose material to form the head and tail; Halley's comet loses about 300 000 000 tons at each return. Obviously the supply of volatiles is not inexhaustible, and a short-period comet cannot last for long on the cosmic timescale.

There is direct observational confirmation of this conclusion, because several short-period comets which used to return regularly have now disappeared. The classic case is that of Biela's comet, which had a period of 6¾ years. It was seen in 1832; missed in 1839, because it was badly placed, and recovered in 1845, at which return it split in two. The twins came back in 1852, but that was their last appearance. They were missed in 1858–9, because of bad positioning but they should have been recovered in 1866 – and were conspicuous only by their absence. In 1872 a bright shower of meteors was seen coming from that part of the sky where the comet should have been, and the shower was repeated at intervals in after years, though by now they have virtually ceased. There can be no doubt that these meteors marked the funeral pyre of Biela's comet. The case is not unique; Brorsen's comet is another one, and Westphal's comet faded out as it neared the Sun in 1913.

It is thought that comets are very ancient, and represent material 'left over', so to speak, when the main planets were formed around 4 500 million years ago. Halley's comet, which is still quite bright, cannot have been moving in its current orbit for anything like that length of time – it would long since have lost all its volatiles – and so it must be a relative newcomer from the Oort cloud, and must have paid its first visit to the inner solar system only a few thousands, or at most tens of thousands, of years ago.

There may also be a link between periodical comets and what are termed close-approach (or near-Earth) asteroids. Most of the asteroids, or minor planets, keep strictly to that part of the solar system between the orbits of Mars and Jupiter, but there are some which do not; these tiny bodies – usually no more than a kilometer or two in diameter – come into the inner regions, and some of them may even make close approaches to the Earth. One dwarf, Phæthon, actually approaches the Sun to within the orbit of Mercury, so that at perihelion it must be red-hot; it moves in the same orbit as the Geminid meteors seen every December, and could well be their 'parent', in which case it is an ex-comet. Support for this view was gained when asteroid 4015, detected in 1979, was found to be identical with comet Wilson-Harrington of 1949. There can be no doubt that it had a tail in 1949, and was a genuine comet; today it looks exactly like an asteroid.

Until very recently there were two main theories about the constitution of comets. One, championed by R.A. Lyttleton, regarded them as 'flying gravel-banks', with nuclei made up of swarms of small particles. The other, proposed by F.L. Whipple, regarded a comet as a 'dirty ice-ball' which became active only when sufficiently close to the Sun. Whipple's theory was finally confirmed in 1986, when a whole armada of space craft flew past Halley's comet and studied it from close range. The nucleus did indeed prove to be icy, measuring $14 \times 8 \times 8$ km; there was a dark coating, probably of organic material, and superb pictures were taken by the European probe Giotto, which penetrated the head and imaged the nucleus from close range.

Despite the extent of our knowledge, it cannot be said that the fear of comets has entirely disappeared in the popular mind. In 1910 there was considerable alarm when some astronomers – who should certainly have known better – drew attention to the fact that the Earth would pass through the end of the tail of Halley's comet, which contains poisonous gases such as cyanogen; they omitted to add that the density of the tail is so low that even if it had been made of pure cyanogen it could have had absolutely no effect. (Some people even sealed up their windows and doors to protect themselves from the fumes!) Even in 1973 there was a mild scare caused by Kohoutek's comet, which at one time promised to become really brilliant though in the event it failed to do so. At the last return of Halley's comet there were the usual dire predictions from the prophets of doom. Human nature is slow to change.

Yet it must be admitted that there are scientific grounds for some concern, even if very slender ones. Few people will agree with Sir Fred Hoyle that life on Earth was brought here originally via a comet, and that even today comets may deposit bacteria in our air, producing epidemics; but it is quite definite that now and then the Earth must be struck by a wandering body – whether you call it a meteorite, a small asteroid, or a dead comet. The impact of an object the size of, say, the nucleus of Halley's comet would cause immense damage, and even climate change; there is persuasive evidence that an impact of this sort did occur around 65 000 000 years ago, and that as a result of the new conditions the once-dominant dinosaurs, together with many other species, were unable to adapt and died out. What can happen once can happen again. Fortunately the chances of a major impact in the foreseeable future are very slight. The only danger to the Earth as a habitable world seems to come from mankind's own activities!

We cannot tell when the next great comet will appear. It may be tomorrow, it may be next year, or it may not be for many decades. But at least we now have a good idea of what comets are really like, and what they signify. In the coming century we will certainly be able to obtain further close-range information from space craft, and perhaps even collect samples of cometary material. These wanderers in the sky have much to tell us, and no longer do we regard them as messengers of doom.

Patrick Moore

Bibliography

Hasegawa, I. (1980) Catalogue of ancient and naked-eye comets. *Vistas Astron.* **24**, 59.

Hellman, C.D. (1971) *The Comet of 1577. Its Place in the History of Astronomy.* New York: AMS Press. (Reprint of the 1944 edition with addenda, errata and a supplement to the appendix.)

Hughes, D.W. (1987) The criteria for cometary remarkability. *Vistas Astron*, **30** 145.

Marsden, B.G. (1989) *Catalogue of Cometary Orbits*, 6th edn. Cambridge: Smithsonian Astrophysical Observatory.

Yoke, H.P. (1964) Ancient and medieval observations of comets and novae in Chinese sources. *Vistas Astron*, **5**, 127.

Cross references

Halley, Edmond, and Halley's comet
Meteor, meteoroid
Meteor shower Meteor stream
Oort, Johann Hendrik, and Oort cloud

COMET: IMPACTS ON EARTH

Over 20 short-period comets are known to have orbits that cross that of the Earth. (Short-period (SP) comets refers here to those with orbital periods less than 200 years; long-period comets have periods greater than this.) The total number of SP comets currently in Earth-crossing orbits may be extrapolated from the observed population, taking into account selection effects. The actual number probably lies in the hundreds. Occasionally, one of these comets will collide with Earth. Combining these bodies' estimated number, sizes and orbital velocities with dynamical calculations of terrestrial collision probabilities suggests that an SP comet should impact Earth and excavate a crater 10 km in diameter or larger once every few million years.

Similarly, some ten long-period (LP) comets are estimated to cross Earth's orbit every year. LP comets are typically moving at much higher velocities relative to Earth than are SP comets (median terrestrial impact velocities are about 50 km s^{-1} and 25 km s^{-1} respectively), so that a smaller LP comet than an SP one is required to excavate a crater of a given size. LP comets should excavate a crater greater than 10 km in diameter on Earth every million years or so. A review of cometary impact frequencies with Earth has been given by Weissman (1990).

Cometary impacts (particularly the rarest, most energetic events) may be important for the evolution of life on contemporary Earth. It is likely that cometary impacts were much more frequent early in Earth's history. These early impacts may have played an important role in setting the stage for the terrestrial origins of life (Thomas, McKay and Chyba, 1996).

Earth-crossing orbits

For an object to impact the Earth, it is necessary, but not sufficient, for that object to be in an Earth-crossing orbit. Rigorously defined, any body whose orbit can be made to intersect that of the Earth, solely as a consequence of secular perturbations, is said to lie in an Earth-crossing orbit (Shoemaker et al., 1979). (Secular perturbations are the averaged perturbations experienced when effects that depend on the actual positions of the objects in their orbits are eliminated. However, the fundamental dynamics leading to terrestrial collisions with Earth-crossing objects may be illustrated by using a more restricted definition (Wetherill and Shoemaker, 1982): an Earth-crossing orbit is here taken to be one whose perihelion is less than the aphelion of the Earth's orbit (1.017 AU), and whose aphelion is greater than Earth's perihelion distance (0.983 AU). If this Earth-crossing orbit were coplanar with that of the Earth, it would necessarily also be an Earth-intersecting one. However, in general the orbit of an Earth-crossing object is inclined with respect to that of the Earth, so that the orbits do not intersect.

The orbit of an Earth-crossing object, of mass much smaller than the terrestrial mass, may be said to intersect that of the Earth if that object's orbit passes within one Earth's gravitational capture radius of Earth's orbit. The Earth's gravitational capture radius R_g is:

$$R_g = R_\oplus [1 + (v_{esc}/v_\infty)^2]^{1/2} \quad (C22)$$

where $R_\oplus$ is Earth's geometrical radius, 6371 km; v_∞ is the hyperbolic excess velocity of the object relative to Earth (the 'velocity at infinity') and $v_{esc} = 11.2$ km s^{-1} is the terrestrial escape velocity, given by

$$v_{esc} = (2GM_\oplus/R_\oplus)^{1/2} \quad (C23)$$

Here $G = 6.672 \times 10^{-11}$ N m^2 kg^{-2} is the gravitational constant and $M_\oplus = 5.974 \times 10^{24}$ kg is the terrestrial mass. The median value of v_∞ for SP comets is ~ 22 km s^{-1} (Chyba, Owen and Ip, 1994), and R_g is only $\sim 10\%$ larger than $R_\oplus$.

In general, a body on an Earth-intersecting orbit will pass through the orbit of the Earth while the latter is elsewhere in its orbit. Sooner or later, however, a collision will take place. The impact velocity v_i of the object with the Earth is then given by energy conservation as

$$v_i = (v_{esc}^2 + v_\infty^2)^{1/2} \quad (C24)$$

v_i for Earth ranges from 11.2 km s^{-1}, for $v_\infty = 0$, to 73.4 km s^{-1}, for a body falling in from infinity and colliding head-on with the Earth. Comets in highly retrograde orbits, such as comets Halley and Temple-Tuttle, would in fact attain impact velocities near the theoretical maximum in a terrestrial collision.

Of course, orbits are typically not coplanar, so Earth-crossing orbits are rarely Earth-intersecting. However, Öpik recognized (Öpik, 1951) that an Earth-crossing orbit would, as a result of precession of its major axis, become Earth-intersecting a discrete number of times (typically four) during each precessional period. The relevant geometry is illustrated in Figure C34. Here i is the inclination of the orbit of the Earth-crosser with respect to that of the Earth, and ω is the argument of pericenter for the Earth-crossing orbit, which is measured from the line of nodes (the line of intersection of the two orbit planes). When $\omega = \pi/2$ the major axis of the Earth-crossing orbit is exactly orthogonal to the line of nodes.

Gravitational perturbations from the Earth and other planets will cause precession (rotation) of this major axis, commonly called apsidal precession. Typical precessional periods for ω for Earth-crossing objects are $\sim 10^4$ years. This period may be compared to mean lifetimes of SP comets in Jupiter-crossing orbits of $\sim 10^4$ to 10^6 years (Weissman, 1990). It should be obvious from the geometry illustrated in Figure C34 that, as ω varies from 0 to 2π, intersection with Earth's orbit will occur at four values of ω. Sooner or later, the Earth and the Earth-crosser will be at one of these intersection points nearly simultaneously, and a collision will occur.

Another way of thinking about these intersections is in terms of the topology of the two overlapping orbits (Shoemaker et al., 1979). In the situation depicted in Figure C34, with ω near $\pi/2$, the two orbits are unlinked. However, a second topological relationship between the two orbits, a linked configuration, is also possible. When ω nears

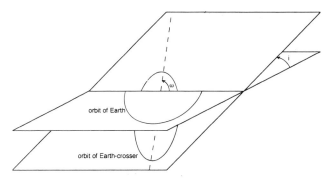

Figure C34 Geometry for collisions due to apsidal precession. The orbit plane of the Earth-crossing object is inclined to that of the Earth by some angle i. The argument of pericenter of the orbit of the Earth-crossing object, ω, is measured from the line of nodes. As ω precesses through 2π the orbit of the Earth-crosser intersects the orbit of the Earth four times. Collisions may occur at these intersections.

π the two orbits will be looped together like links in a chain. In a complete precessional period of ω through 2π, the transition between linked and unlinked configurations occurs four times, and at each of these transitions the orbits must intersect.

Öpik (1951) first calculated the probabilities and velocities of collision that result from this precession of Earth-crossing orbits, and the resulting formulae are commonly referred to as Öpik's equations. Öpik's original work (1951) allowed for the departure of the target planet from an exactly circular orbit. However, under certain circumstances, intersection may occur at as few as two or as many as eight values of ω, and for a correct calculation of impact probabilities, Öpik's original equations must be modified to take this into account (Öpik, 1951, 1976; Shoemaker et al., 1979; Shoemaker and Wolfe, 1982). The Öpik equations also break down for Earth-crossing objects with extremely low inclinations. Impact probability varies as $(\sin i)^{-1}$, and the derivation of this probability assumes the ratio of Earth's capture radius (in units of the semimajor axis a of Earth's orbit) to $\sin i$ is much less than 1. For nearly coplanar orbits, therefore, a different approach must be used.

Mean impact probabilities for the known Earth-crossing SP comets, calculated using Öpik's equations, are $\sim 7 \times 10^{-9}$ per perihelion passage, or $\sim 8 \times 10^{-10}$ per comet per year (Weissman, 1990). These low values should come as no surprise. A rough estimate of the probability of a terrestrial collision during any given intersection with Earth's orbit is just $\sim R_g/2\pi a$, or $\sim 8 \times 10^{-6}$. With four crossings typically occurring every 10^4 years, this naive calculation gives 3×10^{-10} collisions per year per Earth crosser.

Impact environment of early Earth

Because of the youth of Earth's surface, little is known directly about the terrestrial environment prior to 3.8 billion years (Ga) ago. (3.8 Ga is the age of the oldest, albeit metamorphosed, sedimentary rocks, those of Greenland's Isua formation.) The earliest terrestrial fossils indicate life originated on Earth before about 3.5 Ga ago. Thus there is little or no extant record of terrestrial conditions at the time of the origins of life. However, we know from radioactive dating of cratered surfaces on the Moon (from lunar samples returned to Earth by the US Apollo and Soviet Luna missions) that the early inner solar system was subject to a heavy bombardment of impactors, during which those planetesimals 'left over' from planetary formation were largely swept up or scattered.

The lunar cratering data are shown in Figure C35. Several somewhat disparate versions of the lunar cratering history have been published. The data shown here are from the compilation of the Basaltic Volcanism Study Project (1981). In Figure C35 the abscissa shows the age of a given surface, determined by the radioactive dating of returned samples. For each surface, the total number of craters larger than a certain diameter (in this case, larger than 4 km) is then found by crater counting, and the result expressed in terms of a 'crater density' (number of craters per 10^6 km^2). As Figure C35 shows, the early bombardment declined in intensity through orders of magnitude, dropping to its present comparatively low level by about

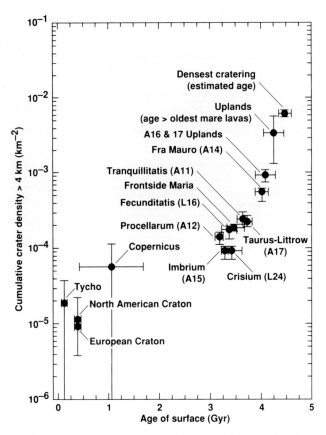

Figure C35 Cumulative lunar crater density as a function of surface age. The first billion years of inner solar system history shows a cratering flux decreasing roughly exponentially, which then levels off to its present comparatively low value. Data from the Basaltic Volcanism Study Project (1981).

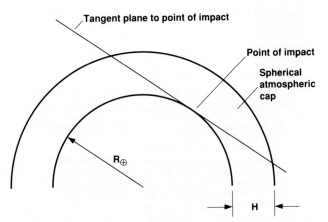

Figure C36 Geometry for atmospheric impact erosion. A sufficiently large and fast impactor will erode the entire spherical atmospheric cap above the tangent plane to the point of impact. In this simple model, due to Melosh and Vickery (1989), the thickness of the cap is taken to be $H = 8.4$ km, the scale height of the atmosphere.

3.5 Ga ago. These lunar results are readily extrapolated to Earth. It seems that life on Earth evolved not in a calm, quiescent environment, but rather in a violent, impact-ridden one.

What were the effects of these impacts on the environment of early Earth and, more specifically, on the origins of life? Broadly, there are two conflicting effects. On the one hand, sufficiently energetic impacts could have eroded the terrestrial atmosphere or even sterilized the Earth's surface. On the other, some fraction of the heavy bombardment population (see Comet; Carbonaceous chondrite) was rich in volatile elements and compounds such as carbon, nitrogen and water, and also in organic molecules. This fraction could therefore have played a role both in providing the Earth's surface with key 'biogenic' elements and compounds, and in helping to stock the terrestrial prebiotic inventory of organic molecules thought to have been necessary for the origins of life. (In modern usage, the word, 'organic' refers simply to molecules in which carbon is covalently bonded to hydrogen and perhaps to other atoms such as nitrogen, oxygen or sulfur. By this definition, hydrogen cyanide (HCN) is a simple organic, but carbon dioxide (CO_2) is not. All known life is based on organic molecules. However, the word no longer carries any implication of biological origin; organics are readily made abiotically.)

Impact erosion of the atmosphere

Sufficiently large and fast impacts can erode planetary atmospheres. Melosh and Vickery (1989) have proposed a simple model for this process. They find that large impacts will cause atmospheric erosion when two criteria are met. The first is that the impactor must strike the planet at a velocity high enough for a vapor plume to form and expand at a speed greater than the planet's escape velocity. The second is that the mass of the plume must exceed the air mass above the plane tangent to the impact. In effect, these criteria are a momentum balance requirement. Their treatment indicates that an impactor with mass m_{min} and velocity v_{min} above some threshold will expel all the air above the horizon, that is, will expel the atmospheric cap above the plane tangent to the point of impact. (Figure C36 shows the relevant geometry.) The mass of the atmospheric cap expelled by such an impact is a bit less than 0.1% of the total contemporary atmospheric mass of 5.1×10^{18} kg. For Earth, $m_{min} \approx 4 \times 10^{15}$ kg, corresponding to impactors in the 10 km radius size range, and $v_{min} \approx 25$ km^{-1}. This model may imply substantial erosion (~ 1 bar) of the early Earth's atmosphere, but this conclusion is strongly sensitive to uncertainties in the early bombarding flux, on the velocity distribution of that flux, and on the density of the early atmosphere (Chyba, Owen and Ip, 1994). In any case, it appears that condensed volatiles (e.g. Earth's oceans) would have been largely immune to impact erosion (Chyba, 1990). Volatiles that remained in the gas phase were preferentially eroded from the early Earth.

Impact frustration or annihilation of early life

Soon after returned Apollo samples established that lunar cratering had been much more intense ~ 4 Ga ago, speculations began to appear that early terrestrial impacts may have rendered Earth's surface inhospitable for the origin of life. Since fossil evidence for life in 3.5 Ga old rocks was known by this time, Sagan (1974) suggested that only a few hundred million years, or perhaps much less, may have been available for the origin of life on Earth. This conclusion, if correct, suggests that the origin of life can happen quickly, and therefore carries implications for the abundance of life elsewhere in the universe.

However, it was not until after the suggestion by Alvarez *et al.* (1980) that a large impact was responsible for the terrestrial mass extinction 65 Ma ago that investigators began to quantify the possibility that early impacts may have delayed the origin of terrestrial life. Maher and Stevenson (1988) coined the term 'impact frustration' for that time in Earth history during which the interval between devastating environmental traumas from impacts was shorter than the timescale needed to establish life. Sleep *et al.* (1989) modeled the results of the largest impacts in detail, and found that collisions by objects about 400 km in diameter (about the size of the asteroids Vesta and Pallas) would be sufficient to create a globe-encircling rock vapor atmosphere, leading to the evaporation of the entire terrestrial ocean. Such impacts may have globally sterilized the Earth, exterminating any life that may have previously evolved. Estimates of the frequency of such impacts on early Earth require small-number statistical extrapolations from the Moon, and are correspondingly uncertain. The poor statistics that are available suggest early Earth sustained several such impacts, implying that the period of time between the last such sterilizing event and the oldest known terrestrial fossils was probably about 100 million years.

Impact delivery of volatiles and prebiotic organics

The role of impacts in the delivery of volatile elements and prebiotic organic molecules has a speculative history extending back at least to the beginning of this century (Chamberlin and Chamberlin, 1908). Such speculation has, over the last two decades, received support from inhomogeneous terrestrial accretion models (see, for example, Anders and Owen, 1977), as well as dynamical models for outer planet formation in which Earth receives the bulk of its surface volatiles as a late-accreting impactor veneer (Fernandez and Ip, 1983).

Delivery of cometary volatiles

Comets may have been important sources for Earth's surface volatiles, and in particular for its oceans. Comets, the most volatile-rich small bodies known, appear to be composed of nearly 50% water and as much as 25% organic matter by mass. At the same time, chemical equilibrium models for terrestrial formation in the solar nebula predict that Earth should have accreted with orders of magnitude less water (and certain volatile elements) than are in fact present, indicating that some terrestrial accretion of volatile-rich material from greater heliocentric distances must have occurred (e.g. Delsemme 1991).

However, independently of solar nebula chemistry or planetary formation models, we may ask what the observed lunar cratering record, as well as lunar and terrestrial geochemical evidence, tells us about terrestrial accretion of volatiles and organics during the heavy bombardment. Such an approach has fundamental limitations, as the fraction of the heavy bombardment population that comets comprised remains uncertain. Thus any conclusions must be phrased in a contingent manner, e.g. 'If comets comprised x% of the heavy bombardment, they would have contributed y% of the contemporary oceans to Earth.' Moreover, the role of large impacts – especially high-velocity (generally cometary) ones – in eroding terrestrial volatiles must be incorporated. Taking all this into account leads to the conclusion that if comets comprised a significant mass fraction (some tens of percent) of the population responsible for early cratering, they could have delivered a terrestrial ocean of water (1.4×10^{21} kg), as well as the bulk of the terrestrial inventory of so-called 'biogenic' elements, those elements (such as carbon and nitrogen) essential for life. Impacts by carbonaceous asteroids would also have contributed. A scenario in which the bulk of Earth's surface volatiles was provided by cometary and asteroidal impacts subsequent to terrestrial core formation (probably around 4.4 Ga ago) is consistent with terrestrial mantle abundances of highly siderophile elements (Chyba, Owen and Ip, 1994).

Delivery of cometary organics

Even before the spacecraft missions to comet Halley in 1986, spectroscopic observations of carbon- and nitrogen-containing radicals in cometary comae led Oró (1961) to suggest that comets may have provided an important source of prebiotic organic material for early Earth. The long-standing objection to such proposals has been that fragile organic compounds would be destroyed by the heat of cometary atmospheric passage and the ensuing impact. (For example, an object 10 km in diameter – about the size of comet Halley – would hit the Earth with a kinetic energy equivalent to 10^8 megatons of high explosive.) In fact, shock heating of the impacting projectile is not uniform, so that some parts of the comet are less heated than others. Figure C37 shows results from a numerical simulation of an oblique cometary impact into an ocean (Chyba et al., 1990). Such simulations, coupled with laboratory data for the destruction of organic molecules in shock-tube experiments, indicate that most organic molecules will survive cometary impacts only for impact velocities less than ~ 10 km s^{-1}. Yet even objects as small as 100 m in radius cannot be decelerated by Earth's present atmosphere (barring a catastrophic disruption, or 'airburst', of the object in the atmosphere) down to this velocity.

Comets may nevertheless have played an important role in providing early Earth with prebiotic organics. Cometary volatiles sublime as comets approach the Sun, and this sublimation entrains dust particles rich in organics. Earth currently collects some 3×10^6 kg year^{-1} of interplanetary dust particles that are small enough to be sufficiently gently decelerated by the upper atmosphere so that at least some of their organics survive to reach Earth's surface intact. It remains unclear what fraction of these particles has a cometary and which an asteroidal origin, but direct inspection indicates that they average 10% organic carbon by mass. In an early solar system where many more comets were likely to be traversing the orbits of the inner planets, such cometary (and asteroidal) dust may have been a significant source of organics for the early Earth (Anders, 1989).

Impact shock synthesis of organics

Besides delivering intact exogenous organic molecules to early Earth, impactors could also have shock synthesized organics in the terrestrial atmosphere. This was suggested as early as 1963 by Gilvarry and Hochstim (1963). In 1970 Bar-Nun et al. demonstrated experimentally that shock waves in reducing (methane- and ammonia-rich) atmospheres led to the production of amino acids. However, production efficiencies of organics are much lower in intermediate oxidation state (carbon dioxide-rich) atmospheres, now generally regarded as more likely models for the atmosphere of early Earth. However, organics may nevertheless have been synthesized by recombination from reducing mixtures of gases resulting from the shock vaporization of comets or asteroids on impact, mixtures perhaps largely independent of the background atmosphere (Oró, 1961; Chyba and Sagan, 1992; Oberbeck and Aggarwal, 1992).

Balance sheet of prebiotic organics on early Earth

Sources of organic molecules on early Earth may be broadly divided into endogenous and exogenous categories. Straddling this division would be organics produced by impact shocks; although these would rely on the kinetic energy of heavy bombardment projectiles for their synthesis, some of the material synthesized would be of terrestrial origin. Direct delivery of exogenous organics, as well as shock synthesis, were considered above. Examples of endogenous sources include organic synthesis from the 'classical' energy sources of ultraviolet light and electrical discharges.

Which sources were quantitatively dominant depends strongly upon the composition of the early terrestrial atmosphere. In the event of a strongly reducing early atmosphere, production by impact-driven atmospheric shocks and by ultraviolet light appears to have dominated other sources. In the recently more popular case of an early terrestrial atmosphere of intermediate oxidation state, atmospheric shocks were likely of little importance for organic production. However, delivery of intact exogenous organics in interplanetary dust particles or synthesized in post-impact recombination may have been important sources. For either oxidation state endmember, impactors appear to have made a significant contribution to the terrestrial prebiotic organic inventory (Chyba and Sagan, 1992).

Mass extinctions driven by comets?

Alvarez et al. (1980) discovered high levels of iridium (Ir) in the clay layer marking the boundary of the Cretaceous and Tertiary periods (The 'K–T boundary'), and in 1980 suggested that this constitutes evidence for the impact of a ~ 10-km diameter comet or asteroid 65 Ma ago. They suggested the Ir had settled out of a global dust cloud raised by the impact (see Iridium anomaly). They also proposed that this cloud would have blocked the Sun and caused the collapse of the food chain, accounting for the extinction of ~ 75% of Earth's species (including the dinosaurs), as observed in the fossil record.

There is now substantial evidence for a giant impact 65 Ma ago. This evidence includes, in addition to the global distribution of Ir at the boundary layer, the discovery in this layer of shocked quartz, possible tektites, and soot believed to have resulted from global wildfires (a review of this evidence has been given by Alvarez and Asaro 1990). Moreover, the 200-km diameter Chicxulub crater on the Yucatán Peninsula in Mexico, suggested as the 'smoking gun' for the K–T impact (Hildebrand et al., 1991), has now been dated to have exactly the required age for this event (Sharpton et al., 1992).

While the evidence for a giant impact at the K–T boundary is firm, the hypothesis that this impact was the primary trigger for the mass extinction at that boundary remains somewhat controversial. More controversial is the claim for a periodicity in the terrestrial record of mass extinctions, and the suggestion that periodic comet showers are the cause. (As there is no known dynamical mechanism for the production of periodic showers of asteroids, claims for periodicity must focus on comets by default.) There appear to be serious objections to each of the mechanisms proposed for the origin of

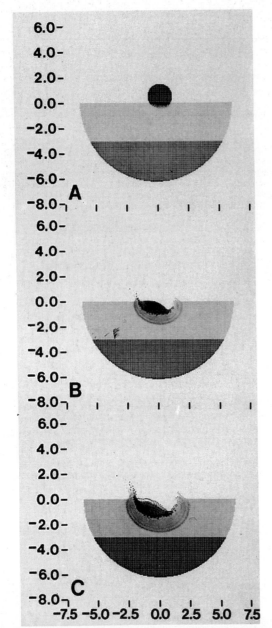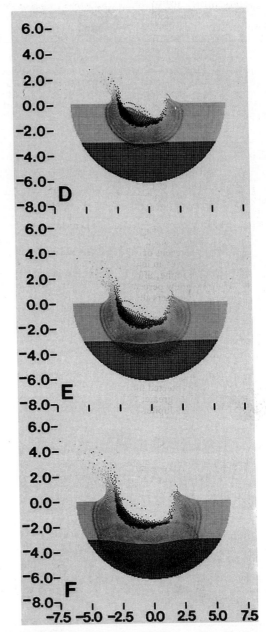

Figure C37 Step-by-step evolution of an oblique cometary impact onto an ocean 3 km deep underlain by a basaltic sea floor. The images identify particles from comet, ocean and ocean floor and are spaced at intervals of 0.2 s. Only the hardiest organic molecules are expected to survive shock heating to the resulting temperatures. (From Chyba et al., 1990.)

periodic comet showers. The statistics of alleged periodicities in terrestrial cratering have also been challenged, and in any case most of those craters for which the nature of the impactor can be identified seem to have been excavated by asteroidal, not cometary, bodies. These objections have been reviewed by Weissman (1990).

Zhao and Bada (1989) have reported a high abundance of two evidently extraterrestrial amino acids, α-amino-isobutyric acid (AIB) and isovaline, in sediments at the Stevns Klint, Denmark, K–T boundary site. (AIB and isovaline are extremely rare in the biosphere, but common in meteorites.) The AIB/Ir ratio at Stevns Klint is substantially higher than for the Murchison carbonaceous meteorite. This is remarkable, as impact models (Chyba et al. 1990), as well as the absence of meteoritic xenon in the K–T boundary sediments (Anders, 1989), both suggest amino acids would have been incinerated in the K–T impact. This discrepancy, in addition to the distribution of the AIB about the boundary, led Zahnle and Grinspoon (1990) to suggest that the amino acids may have reached Earth by the gentle collection of organic-rich dust evolved from an evaporating comet. (A fragment of this comet would eventually have been responsible for the giant impact itself.) Amino acid data from additional K–T sites are needed to test these models. Alternatively, the non-biological amino acids found at the K–T boundary may represent material that was synthesized in the post-impact fireball (Oberbeck and Aggarwal, 1992).

Contemporary impact hazard

On the morning of 30 June 1908 a small (~ 60 m diameter) bolide exploded over the Podkamennaya Tunguska River basin in Central Siberia. Trees were flattened over an area of some 2000 km² below the location (61°N, 102°E) of the explosion. The explosion is estimated to have occurred at an altitude of about 9 km, with an explosive energy in the 10–20 megaton range. The Tunguska object has been interpreted as a fragment of comet Encke, based on the former's temporal coincidence with the β Taurid meteor shower (Krésak, 1978), though this correlation has been questioned (Sekanina, 1983). Simulations of bolide deceleration, ablation and catastrophic fragmentation indicate that comets of appropriate energy explode far too high in the atmosphere fit the explosion's characteristics. Rather, the Tunguska event appears to represent a common fate for asteroids tens of meters in radius entering the Earth's atmosphere at typical hypersonic velocities (Chyba, Thomas and Zahnle, 1993; Hills and Goda, 1993). For a contrary view favoring a cometary explanation, see Bronshten and Zotkin (1995).

It is estimated that Tunguska-magnitude impacts with Earth occur once every several centuries. Impacts with energies of tens of kilotons may be annual occurrences (Morrison, Chapman and Slovik, 1994), though such explosions (such as that of the Revelstoke object in 1965), occur so high in the atmosphere they are scarcely noticed at the ground (Chyba, 1993). These explosions are observed by Earth-orbiting defense satellites (Tagliaferri *et al.*, 1994).

Impacts by Tunguska-sized objects could have devastating local effects on Earth. The results of collisions with larger objects would be considerably more grave: an impact by a kilometer-sized body could cause a global catastrophe, affecting all of human civilization. It is estimated that only a few percent of the present population of near-Earth objects with diameters above 0.5 km have so far been observed (Steel, 1991).

Finally, the large (~ 200 km diameter) trans-saturnian comet Chiron has an orbit known to be chaotic; numerical simulations suggest that it is likely to evolve into a short-period comet on a timescale shorter than that for its ejection from the solar system (Hahn and Bailey, 1990). Thus, even giant impacts rivaling those that occurred on Earth 4 Ga ago are not entirely ruled out. It is possible that Earth has been protected throughout its history from a cometary impact flux ~ 10^3 times greater than has in fact been the case due to shielding by the planet Jupiter (Wetherill, 1994).

Summary

Terrestrial impacts by comets and asteroids appear to have played an important role throughout Earth's history. Early Earth may have received much of its surface volatile inventory in this way. Large impacts prior to 4 Ga ago may have frustrated the origin of terrestrial life, or annihilated whatever life had previously arisen, 'resetting the clock' for life's origins. At the same time, interplanetary dust from comets and asteroids may have been a significant source of prebiotic organics. Subsequent to the origin of terrestrial life, large impacts may have been responsible for at least certain mass extinctions, thereby playing a critical role in the history of Earth's biosphere (as well as the eventual evolution of human beings). Terrestrial impacts continue to occur; the larger of these could still have devastating local, or even global, effects.

Christopher F. Chyba

Bibliography

Alvarez, L.W., Alvarez, W.A., Asaro F. and Michel H.V. (1980) Extraterrestrial cause for the Cretaceous–Tertiary extinction. *Science*, 208, 1095–1108.

Alvarez, W. and Asaro, F. (1990) An extraterrestrial impact. *Sci. Am.*, 263(4), 78–84.

Anders, E. (1989) Pre-biotic organic matter from comets and asteroids. *Nature*, 342, 255–7.

Anders, E. and Owen, T. (1977) Mars and Earth: origin and abundance of volatiles. *Science*, 198, 453–65.

Bar-Nun, A., Bar-Nun, N., Bauer, S.H. and Sagan, C. (1970) Shock synthesis of amino acids in simulated primitive environments. *Science*, 168, 470–3.

Basaltic Volcanism Study Project (1981) *Basaltic Volcanism on the Terrestrial Planets*. New York: Pergamon.

Bronshten, V.A. and Zotkin, J.T. (1995) Tunguska meteorite: fragment of a comet or an asteroid? *Solar System Res.*, 29, 241–5.

Chamberlin, T.C. and Chamberlin, R.T. (1908) Early terrestrial conditions that may have favored organic synthesis. *Science*, 28, 897–911.

Chyba, C.F. (1990) Impact delivery and erosion of planetary oceans in the early inner solar system. *Nature*, 343, 129–33.

Chyba, C.F. (1993) Explosion of small Spacewatch asteroids in the Earth's atmosphere. *Nature*, 363, 701–3.

Chyba, C. and Sagan, C., (1992) Endogenous, production, exogenous delivery and impact-shock synthesis of organic molecules: An inventory for the origins of life. *Nature*, 355, 125–32.

Chyba, C.F., Owen, T.C. and Ip, W.-H. (1994) Impact delivery of volatiles and organic molecules to Earth, in *Hazards Due to Comets and Asteroids* (ed. T. Gehrels). Tucson: University of Arizon Press, pp. 9–58.

Chyba, C.F., Thomas, P.J., Brookshaw, L. and Sagan, C. (1990) Cometary delivery of organic molecules to the early Earth. *Science*, 249, 366–73.

Chyba, C.F., Thomas, P.J. and Zahnle, K.J. 1993, The 1908 Tunguska explosion: atmospheric disruption of a stony asteroid. *Nature*, 361, 40–44.

Delsemme, A.H. (1991) Nature and history of the organic compounds in comets: an astrophysical view, in *Comets in the Post-Halley Era*, (eds R.L. Newburn, M. Neugebauer and J. Rahe). Dordrecht: Kluwer Academic Publishers, pp. 377–428.

Fernández, J.A. and Ip, W.-H. (1983) On the time evolution of the cometary influx in the region of the terrestrial planets. *Icarus*, 54, 377–87.

Gilvarry, J.J., and Hochstim, A.R. (1963) Possible role of meteorites in the origin of life. *Nature*, 197, 624–5.

Hahn, G. and M.E. Bailey (1990) Rapid dynamical evolution of giant comet Chiron. *Nature*, 348, 132–136.

Hildebrand, A.R., Penfield, G.T., Krieg, D.A. *et al.* (1991) Chicxulub crater: a possible Cretaceous/Tertiary boundary impact crater on the Yucatán Peninssula, Mexico. *Geology*, 19, 867–71.

Hills, J.G. and Goda. M.P. (1993) The fragmentation of small asteroids in the atmosphere. *Astron. J.*, 105, 1114–44.

Krésak, L. (1978) The Tunguska object: a fragment of comet Encke? *Bull. Astron. Inst. Czechosl*, 29, 129–34.

Maher, K.A. and Stevenson, D.J. (1988) Impact frustration of the origin of life. *Nature*, 331, 612–4.

Melosh, H.J. and Vickery, A.M. (1989) Impact erosion of the primordial atmosphere of Mars, *Nature*, 338, 487–9.

Morrison, D., Chapman, C.P. and Slovic, P. (1994) The impact hazard, in *Hazards Due to Comets and Asteroids* (ed. T. Gehrels). Tucson: University of Arizona Press, pp. 59–91.

Oberbeck, V.R. and Aggerwal, H. (1992) Comet impacts and chemical evolution on the bombarded Earth. *Origins of Life*, 21, 317–38.

Öpik, E.J. (1951) Collision probabilities with the planets and the distribution of interplanetary matter, *Proc. Roy. Irish Acad. Sect. A* 54, 165–99.

Öpik, E.J. (1976) *Interplanetary Encounters*. New York: Elsevier.

Oró, J. (1961) Comets and the formation of biochemical compounds on the primitive Earth. *Nature*, 190, 389–90.

Sagan, C. (1974) The origin of life in a cosmic context. *Origins of Life*, 5, 497–505.

Sekanina, Z. (1983) The Tunguska event: no cometary signature in evidence. *Astron. J.*, 88, 1382–414.

Sharpton, V.L., Dalrymple, G.B. Marin, *et al.* (1992) New links between Chicxulub impact structure and the Cretaceous/Tertiary boundary. *Nature*, 359, 819–821.

Shoemaker, E.M., Wolfe, R.F. (1982) Cratering timescales for the Galilean satellites, in *Satellites of Jupiter* (ed. D. Morrison). Tucson: University of Arizona, pp. 227–339.

Shoemaker, E.M. Williams, J.G., Helin, E.F. and Wolfe, R.F. (1979) Earth-crossing asteroids: orbital classes, collision rates with Earth, and origin, in *Asteroids* (ed. T. Gehrels). Tuscon: University of Arizona, pp. 253–82.

Shoemaker, E.M. Wolfe, R.E. and Shoemaker, C.S. (1990) Asteroid and comet flux in the neighborhood of Earth, in *Global Catastrophes in Earth History* (eds V.L. Sharpton and P.D. Ward. Geological Society of America Special Paper 247, pp. 155–70.

Sleep, N.H. Zahnle, K.F., Kasting, J.F. and Morowitz, H.J. (1989) Annihilation of ecosystems by large asteroid impacts on the early Earth. *Nature*, 342, 139–142.

Steel, D. (1991) Our asteroid-pelted planet. *Nature*, 354, 265–7.

Tagliaferri, E., Spalding, R., Jacobs, C. *et al.* (1994) Detection of meteoroid impacts by optical sensors in Earth orbit, in *Hazards Due to Comets and Asteroids* (ed. T. Gehrels). Tucson: University of Arizona Press, pp. 199–220.

Thomas, P.J., Mckay, C.P. and Chyba, C.F. (eds) (1996) *Comets and Evolution of Life.* New York: Springer-Verlag.

Weissman, P.R. (1990) The cometary impactor flux at the Earth, in *Global Catastrophes in Earth History* (eds. V.L. Sharpton and P.D. Ward. Geological Society of America Special Paper 247, pp. 171–80.

Wetherill, G.W. (1994) Possible consequences of absence of 'Jupiters' in planetary systems *Astrophys. Space Sci.*, **212**, 23–32.

Wetherill, G.W.J and Shoemaker, E.M. (1982) Collision of astronomically observable bodies with the Earth, in *Geological implications of Impacts of Large Asteroids and Comets on the Earth* (eds. L.T. Silver and T.H. Schultz. Geological Society of America Special Paper 190. pp. 1–13.

Zahnle, K. and Grinspoon, D. (1990) Comet dust as a source of amino acids at the Cretaceous/Tertiary boundary. *Nature*, **348**, 157–60.

Zhao, M. and Bada, J.L. (1989) Extraterrestrial amino acids in Cretaceous/Tertiary boundary sediments at Stevns Klint, Denmark. *Nature* **339**, 463–5.

Cross references

Comet: dynamics
Comet: origin and evolution
Comet: structure and composition
Cosmochemistry
Cretaceous–Tertiary boundary
Earth: atmosphere
Impact cratering Iridium anomaly
Life: origins
Mass extinction
Moon: geology

COMET: IMPACTS ON JUPITER

During the 1-week period from July 16 to July 22 1994, more than 20 fragments of comet P/Shoemaker-Levy 9 (SL9) slammed into the planet Jupiter at velocities of approximately 60 km s^{-1}. These impacts produced a variety of effects. Analysis of the observations will continue for some years; as this is written (1994) only the most preliminary results are available.

The impacts of comet SL9 are of historic significance. This was the first instance in human history where a major impact on a planetary body was predicted in advance of the event. SL9 is the first comet that has been detected orbiting the planet Jupiter. It is also the first comet that has been observed to break up into more than a few individual fragments (Figure C38) Significant cometary impacts on Jupiter are expected to occur about once in every thousand years or so, and thus the event was described as 'the astronomical event of the century.' The impacts did not disappoint the most hopeful of observers.

The study of cometary impacts has direct relevance for residents of Earth (see Comet: impacts on Earth). The giant impact that apparently wiped out the dinosaurs on Earth some 65 million years ago resulted from an object that was considerably larger than comet SL9; however, some thousands of small asteroidal bodies have orbits that cross that of the Earth (see Near-Earth object), and a significant future impact on the Earth is virtually certain to occur over the long term.

Discovery

Observers Gene and Carolyn Shoemaker and David Levy obtained the first image of comet SL9 on the evening of 23 March 1993; the image was examined and interpreted on 25 March. The observations were made with the 18-inch (45 cm) Schmidt telescope on Mt Palomar, California. The initial image showed a 'squashed comet' subtending an angle of about 50 arcsec (equivalent to a spread of 180 000 km). Multiple fragments were observed in later images,

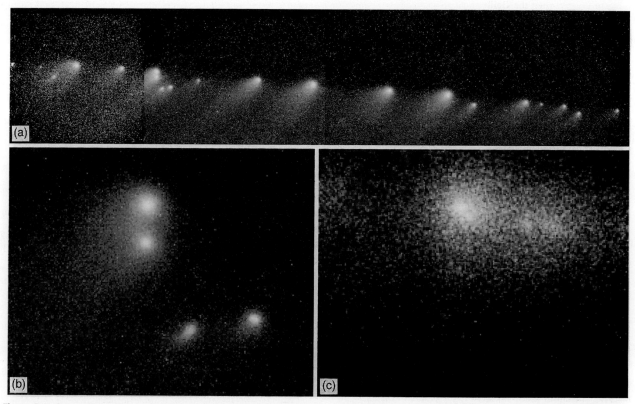

Figure C38 Comet P/Shoemaker-Levy 9, a 'string of pearls.' (a) Image was obtained by H.A. Weaver and T.E. Smith using the Hubble Space Telescope Wide Field Planetary Camera 2. Twenty distinct fragments are visible in this mosaic. Then two images of the region near the brightest nucleus obtained (b) January 1994, after the Hubble telescope servicing mission, and (c) July 1993, before the servicing mission.

along with a dust train of material, giving the impression of a 'string of pearls' (Figure C38; Plate 21).

The first determinations of the orbit of the comet were made before the end of March 1993. These indicated that the comet had approached within 1.3 Jupiter radii of the center of Jupiter on about 7 July 1992, in a close approach that had almost certainly resulted in the disruption of the original parent comet due to tidal forces (see Roche limit). The comet was in a highly elliptical, loosely bound orbit about Jupiter with a period of about 2 years. It is thought that the comet most probably orbited Jupiter for two decades or more before the close approach and disruption occurred. Most significantly, orbit determinations indicated that comet SL9 was on a collision course with Jupiter.

Pre-impact observations

Observations by the Hubble Space Telescope and the Keck Telescope of the University of Hawaii showed that the comet consisted of more than 20 individual nucleii (Weaver et al. 1994; Figure C38). The comae and dust of the comet fragments made it impossible to accurately determine the sizes of the fragments. This led to uncertainty concerning the energy and probable effects of the impacts on Jupiter. During the period from January to June 1994 some of the fragments were observed to split, and others became visible. The diameters of the larger fragments were estimated to be from 0.5 to perhaps 3.0 km. The diameter of Jupiter is about 140 000 km, and thus it was evident that the impacts could not cause any major disruption of that body. Simulations of impact processes (Sekanina, 1993; Boslough et al., 1994) suggested that fragment diameters of about 1 km or more would be required to cause major explosions with detectable fireballs and plumes. Both large and small fragments were expected to produce bright flashes in the upper atmosphere of Jupiter analogous to meteors on Earth.

The fragmentation of SL9 suggested an explanation for the presence of strikingly linear crater chains observed on the Galilean satellites (q.v.). Both Ganymede (q.v.) and Callisto (q.v.) have crater chains that might have been caused by impacts of objects like the comet fragments of SL9. Our knowledge of these bodies will increase greatly when the Galileo spacecraft (q.v.) begins to return data on the Jovian system in late 1995.

Orbital solutions for the comet fragments were successively improved in the period prior to the impact. The fragments were labeled with letters from A to W, and the impact times of each fragment were calculated by P. Choudas and D.K. Yeomans of the Jet Propulsion Laboratory. (As an aside, it is of interest to note that the time required for light to travel from Jupiter to observers on Earth was more than 42 min during the period of the impacts.) The impact site prediction was likewise refined in the period prior to the impacts. The fragments were expected to impact the planet at a latitude of about 45°S, just beyond the visible limb of the planet as viewed from Earth. The rapid rotation of Jupiter would bring the impact sites into view within about 30 min after the events.

As the predicted impact times grew closer preparations for observing the events were made at virtually every astronomical observatory on Earth. The Galileo (q.v.) and Ulysses (q.v.) spacecraft prepared to observe the impact effects. Galileo had the most favorable viewing geometry, with a direct view of the impact site, but was at a distance of 240 million km.

Impacts

The impact of fragment A was registered as predicted at about 20 h UTC on Saturday 16 July 1994 (Figure C39). It produced a fireball and plume rising some 1200 km above the surface cloud layers of the planet. Successive impacts produced greater or lesser fireballs, along with visible dark 'scars' that remained visible at the impact sites as the planetary rotation brought those sites into view. Observations at infrared wavelengths revealed the heat signatures of the impacts, while observations at ultraviolet wavelengths (Figure C40) revealed dark blotches of ejected materials. Both the dark blotches and the hot spots were observed to persist for several rotations of Jupiter.

Perhaps the largest single explosion was produced on 18 July by the impact of fragment G. This impact was estimated to pack the equivalent of the explosive power of 6 million megatons of TNT, a number dwarfing the total yield of humanity's nuclear arsenals. The explosion was brighter than the planet itself, and the fireball (Figure C41) extended thousands of kilometres into space.

The analysis of chemical compounds erupted in the explosions has only begun. Preliminary work shows the presence of sulfur compounds, carbon monoxide, methane, ammonia and various metals (Orton, 1994; Noll et al., 1994). Water was observed in the plumes but was not seen in the quantities expected. This was somewhat surprising as most comets are considered to contain appreciable quantities of ice (see Comet: structure and composition), and because models of the Jupiter interior suggest that a watery layer may be present relatively near the surface layers. It has been suggested that

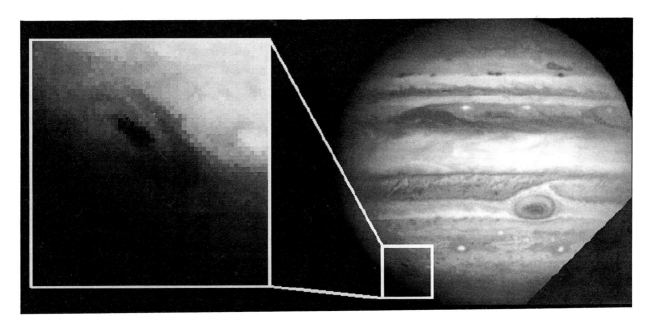

Figure C39 Hubble Space Telescope images of the impact site of fragment A, obtained on 16 July 1994. The large dark scar of the impact was unexpectedly large; fragment A was one of the smaller pieces of the comet. Jupiter's diameter is approximately 11 times greater than Earth's, so this feature was nearly the size of the Earth at the time this image was obtained. Inset shows site after impact, enlarged and enhanced. Image P-44434A, courtesy of NASA.

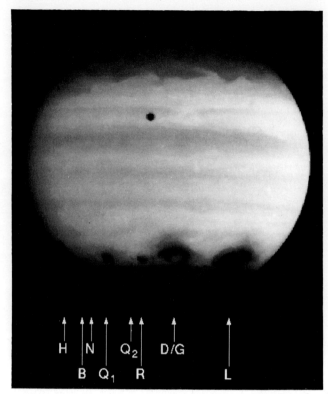

Figure C40 Jupiter imaged in ultraviolet wavelengths. This image was obtained after most of the fragments had impacted. The dark scars persisted throughout the 6 days of impacts. The dark spot in the northern hemisphere is one of the Galilean satellites of Jupiter. Image P-44431A, courtesy of NASA.

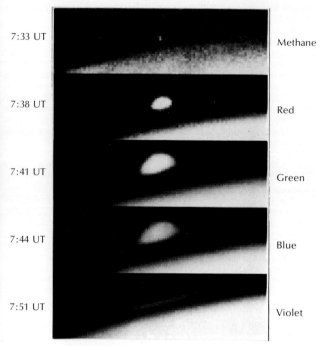

Figure C41 Fireball from the impact of fragment G of comet Shoemaker-Levy 9 on 18 July 1994. This series of time-lapse images was obtained over a period of 18 min using several filters. The initial expansion, rise and flattening of the impact debris cloud is well documented in this sequence. This was perhaps the most energetic impact, and it generated spectacular images at infrared wavelengths, where the fireball was brighter than the planet itself. Hubble Space Telescope image (P44438B), courtesy of NASA.

the original object may have come instead from the population of asteroids, but orbital characteristics are not favorable to this interpretation (Weissman, 1994). Pre-impact observations (Weaver et al., 1994) failed to detect fluorescence from the OH molecule, which is a key indicator of cometary activity. Failure to detect this signature suggested that the object might be deficient in water.

Preliminary observations showed a response to the cometary impacts in the form of enhanced Jovian auroral activity (see Aurora, planetary). Changes in the planet's synchrotron radiation were noted, along with unusual x-ray emissions (Waite et al., 1994). More complete discussions of the effects of the impacts of comet P/Shoemaker-Levy 9 will appear in the scientific literature in the near future.

James H. Shirley

Bibliography

Boslough, M.B, Crawford, D.A., Robinson, A.C. and T.G. Trucano, (1994) Watching for Fireballs on Jupiter. *EOS*, **75** (27), 305–7.
Noll, K.S., McGrath M., Weaver, H.A., et al. (1994) The Changed composition of Jupiter's atmosphere following the impact of comet SL9. *EOS*, **75**, 402.
Orton, G. (1994). Ground-based observations of the Shoemaker-Levy impact event. *EOS*, **75**, 401.
Sekanina, Z. (1993) Disintegration phenomena expected during collision of comet Shoemaker-Levy 9 with Jupiter. *Science*, **262** 382–7.
Waite, H.J., Jr, Gladstone, G.R., Na, C. et al. (1994) X-ray emissions produced as a result of the impact of comet Shoemaker-Levy 9 with Jupiter. *EOS*, **75**, 404.
Weaver, H.A., Feldman, P.D., A'Hearn, M.F., et al. (1994) Hubble Space Telescope observations of comet P/Shoemaker-Levy 9 (1993). *Science*, **263**, 787–91.
Weissman, P. (1994) Events After the events. *Nature*, **372**, 404–405.

Cross references

Aurora, planetary
Callisto
Collisions
Comet: structure and composition
Galilean satellites
Galileo mission
Ganymede
Impact cratering
Jupiter
Near-Earth object
Roche limit
Tide-raising force
Ulysses mission

COMET: OBSERVATION

Records of cometary observations with the unaided eye can be traced to the beginning of historical times (Yeomans, 1991). Cometary observations today benefit from all modern astronomical techniques, including space exploration (Plates 20, 21).

Remote sensing of comets

Discovery of comets

The average number of comets observed each year increased dramatically from about three in the 19th century to about 25 at the present time. Among these, 60% are short-period comets of period less than 200 years. A catalogue of comets and their orbits is maintained by Marsden (1992).

About one-fourth of all comet discoveries today are visual discoveries made with binoculars or small telescopes; most are made by trained amateur astronomers (Larson, Edberg and Levy 1991). Most of these discoveries are bright, unexpected comets. In contrast, photographic discoveries by professional astronomers are mainly chance detections of weak comets, or recoveries of periodic comets near their expected positions. Systematic surveys are currently made in connection with asteroid searches: for instance, the Palomar Asteroid and Comet Survey uses a 0.46-m Schmidt telescope; the Spacewatch programme (q.v) at Kitt Peak Observatory is performed with a 0.9-m telescope and CCD (charge-coupled device) detector (q.v). Such surveys, however, cover only a limited area of the sky and reveal only a small fraction of the faint comets population.

Distant comets and comets with low activity are difficult to distinguish from asteroids. Misidentifications may happen. A famous case is (2060) Chiron, a distant body first classified as an asteroid, which subsequently showed cometary activity. It may be anticipated that such ambiguous objects will be more and more frequently observed as the sensitivity limits of the surveys are improved.

Comets can (with difficulty) be observed from the Earth when they are very close to the Sun. Space solar coronagraphs are occasionally able to manage such observations: six comets were discovered by the Solwind coronagraph in 1981–84, and 10 by the Solar Maximum Mission coronagraph in 1987–89. These sun-grazing comets are very small objects coming within two solar radii from the Sun (Marsden, 1989). They apparently did not survive this passage.

Nomenclature of comets

Comets are usually named after their discoverers. In case of simultaneous discoveries, up to three names of discoverers may be grouped. No new name is given for the multiple returns of periodic comets; their names are preceded by P. In addition, each comet is provisionally designated by the year of its discovery (or recovery) followed by a small letter, in alphabetical order according to the order of the discovery in that year. Some time after, a definitive designation is given by a year and a Roman numeral, according to the order of the passage of the comets at perihelion. Thus, at its last passage, comet Halley was provisionally designated P/Halley (1982i) after its recovery in October 1982, and finally named P/Halley (1986 III) since it was the third known comet to pass perihelion in 1986, on 9 February.

Since 1995, a new system is used for the nomenclature of comets. There is no longer a provisional designation and a definitive designation is given as soon as a comet is discovered, consisting of the year, followed by a letter representing the fortnight in the year and a running number. Thus C/1995 A1 was the first comet discovered in 1995.

'Classical' ground-based observations

Astrometry

The most basic information obtained from cometary observations is their position. Precise astrometrical observations (classical photography is, more and more, being replaced by CCD electronic cameras for this) are a requisite to the determination of cometary orbits. They are necessary to understand and forecast orbital evolution due to planetary perturbations and to recover periodic comets at their successive passages. They allow one to evaluate the nongravitational forces affecting the orbits, which are caused by the 'rocket effect' due to anisotropic outgassing of the nucleus.

Imaging

Images of comets, at large and small scales, represent the most spectacular aspect of the cometary phenomenon. While drawings from visual observations of skilled observers may still be used in some cases (Larson, Edberg, and Levy, 1991), this is the place for wide-field photography with Schmidt telescopes and, more and more, smaller-field observations with CCDs. Imaging allows one to study the development of ion and dust tails, and their interactions with the solar wind and the interplanetary magnetic field. On short distance scales, one can investigate near-nucleus phenomena, such as the development of jets and the study of inhomogeneous outgassing of the nucleus.

It is important to study distant comets, which occasionally present outbursts of still unknown origin. One has to correct for the presence of background stars and galaxies, and then to discriminate between the stellar image of a bare nucleus and the fuzziness of a coma. There are many examples of cometary activity at distances larger than 10 AU; for instance, by comets Bowell (1982 I) and Cernis (1983 XII), and P/Halley for which an outburst was unexpectedly observed at 14 AU from the Sun (West, Hainaut and Smette, 1991).

Photometry

Long-term photometric observation of comets is the easiest way to study the evolution of cometary activity (Jewitt, 1991). A simplistic way to describe the magnitude evolution of a comet is through the law:

$$m_v = H + 5 \log \Delta + 2.5\, n \log r_h$$

where the visual magnitude m_v is 2.5 times the log of the brightness, plus an additive constant; H is the intrinsic magnitude of the comet; and Δ and r_h are the distances (in astronomical units) of the comet to the Earth and the Sun respectively. The index n depends upon the comet and its activity. For inactive bodies, such as asteroids or cometary nuclei far from the Sun, which are just scattering sunlight on their surface, $n = 2$. It is then possible to relate the brightness of such an object to the size of the nucleus and to its geometric albedo. Since cometary nuclei are too small to be resolved from the Earth, this is the only way to obtain information on their size, apart from space exploration. Active comets with a developed coma usually have n between 1.5 and 4. Departures from this simple law are frequently observed, such as different behaviours before and after perihelion, or stalling of activity around perihelion (Festou, Rickman and West,1993). They may be explained by seasonal effects or by the evolution of a dust crust over the nucleus surface.

The lightcurve of a comet on short time scales can help to characterize the rotation of the nucleus, in the same way as it is done currently for asteroids (Belton, 1991) (see Asteroid: lightcurve). The problem is that the presence of a dust and gas coma may obscure any periodicity due to rotation. As a consequence, rotation periods have been reliably determined for only a very small number of comets; They range from 6 to 15 h. The case for comet Halley's rotation period has been actively debated, since ground-based observations and space exploration lead to contradictory values (~ 2 days and ~ 7 days). It is possible that in this case the nucleus is precessing, or even shows a nearly chaotic motion (Belton, 1991).

The systematic monitoring of the total magnitude of comets is another significant contribution of amateurs. These observations are made visually, with binoculars or small telescopes, or even the unaided eye for the brightest comets. In this way, the lightcurves of many comets are available. These observations are continuously incorporated in a worldwide data base (Green, 1991). It is a basis for many statistical investigations and studies of cometary evolution (Figure C42).

Spectrophotometry

Spectrophotometry with broadband filters is a convenient way to monitor the gas and dust content of cometary atmospheres, even for faint comets and with modest telescopes. Standard filters (known as 'IHW filters', since this standard was first introduced for the 'International Halley Watch' observations of P/Halley), have been designed to cover the most important molecular bands as well as the continuum emission due to dust (Figure C43). This allows one to determine, via models, the production rates of the CN, C_2, C_3, and OH radicals, as well as the total content of dust. Such sets of observations are now available for many comets for a large range of heliocentric distances, allowing statistical studies.

Spectroscopy

The spectroscopy of comets has always been a challenge for spectroscopists and physicists. The bands of several radicals and molecular ions were discovered in cometary spectra well before they could be observed or identified in the laboratory. The Swan bands of the C_2 radical were first observed visually in comets 1864 II and 1868 II. The bands of the CN radical were first photographed in comet 1881 IV, together with other weaker bands which were linked with the C_3 radical only after 1950. One can today observe in cometary spectra forbidden lines and metastable energy levels that can only with difficulty be observed in the laboratory. High-resolution spectra enable us to resolve the rotational structure of the molecular electronic bands. In some cases (the OH, CN and C_2 bands) it is

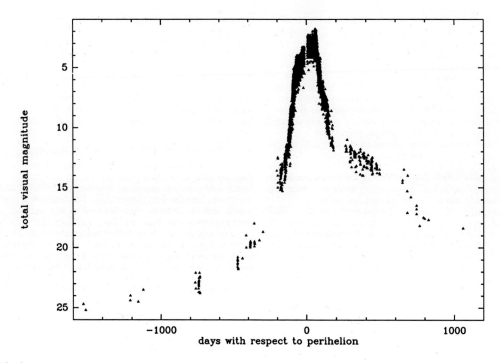

Figure C42 The lightcurve of comet P/Halley extracted from the data base of Green (1991). Most of the data come from a compilation of several thousand amateur observations.

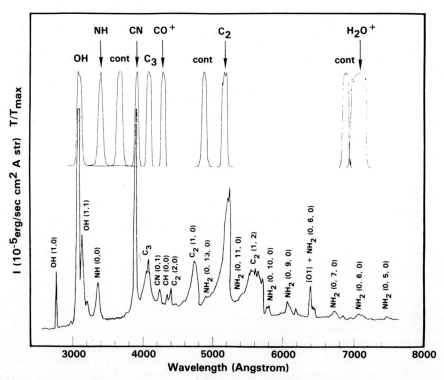

Figure C43 A typical cometary spectrum in the visible and near ultraviolet at medium resolution: comet P/Tuttle (1980 XIII) (bottom) with the wavelengths of standard cometary broadband filters (top). (From Osborn et al., 1990, Icarus, 88, p. 228.)

possible to reproduce this structure in almost every detail by models (Figure C44). It is also possible to distinguish between isotopic species such as ^{12}C and ^{13}C.

Table C6 gives a list of molecular and atomic species identified in atomic spectra. More details can be found in Wyckoff (1982) and in the compendium of cometary spectra compiled by Arpigny et al. (1993). It is important to note that in the visible domain, one only observes secondary products coming from the photodissociation or ionization of the molecules directly sublimed from the nuclei. There primary species have to be searched for in other spectral domains

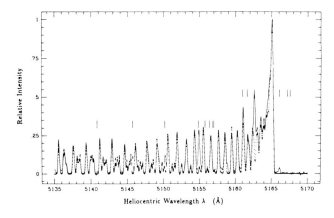

Figure C44 The C (0,0) Swan band observed at high resolution in comet P/Halley with the MMT (Multi Mirror Telescope, Arizona). The dotted line is a theoretical spectrum. From Gredel, van Dishoeck and Black 1989, *Astrophys. J.*, **338**, p. 1047.

Table C6 Molecular and atomic species identified in cometary atmospheres

Stable molecules
H_2O (IR), HCN (radio), CO (UV), H_2CO (radio, IR), CH_3OH (radio, IR), H_2S (radio), CO_2 (IR), S_2 (UV)

Radicals
CH (visible), NH (visible), NH_2 (visible), OH (radio, IR, UV), C_2 (visible, UV), CN (visible), C_3 (visible), CS (UV)

Molecular ions
CH^+ (visible), OH^+ (visible), H_2O^+ (visible), CN^+ (visible, UV), CO^+ (visible, UV), N_2^+ (visible), CO_2^+ (visible, UV)

Atoms
H (visible, UV), C (UV), O (visible, UV), S (UV)

Atomic ions
C^+ (UV)

Metals and refractories (in Sun-grazing comets)
Na, K, Ca^+, Cr, Mn, Fe, Ni, Cu (all in visible)

(see below). The metals listed in the last line of Table C6 were only observed in the Sun-grazing comet Ikeya-Seki (1965 VIII) when it passed at only 1.2×10^6 km from the Sun. Besides the signals of the species listed in Table C6, there is a wealth of weak bands still awaiting identification.

Polarimetry

The polarization of cometary light was first measured by F. Arago at Paris Observatory in the Great Comet of 1819. This polarization is present in both the continuum and the molecular bands. It indicates that cometary light is sunlight scattered on cometary dust or molecular fluorescence excited by the Sun's radiation. Polarization (and its variation with the comet phase angle) is related to the physical properties of dust.

New techniques

Radio

Reviews of the recent developments of cometary radioastronomy are presented by Crovisier and Schloerb (1991) and Crovisier (1992). The first successful radio observation of a comet was that of the lines of the OH radical at 18-cm wavelength in comet Kohoutek 1973 XII. Since that time cometary activity has been steadily monitored through OH radio observations at the Nancay radio telescope in

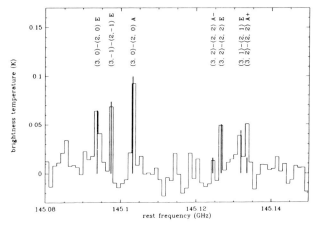

Figure C45 Rotational transitions of methanol in the 145-GHz region of methanol observed at the IRAM (Institut de Radioastronomie Millimétrique) 30-m radio telescope in comet Levy (1991 XX). (From Bockelée-Morvan *et al.*, 1990, ESA SP-315, p. 143.)

France. In some cases it is possible to measure the Zeeman splitting of the OH lines, which is the only remote-sensing way to measure the magnetic field within the coma.

With the recent availability of powerful millimeter radio telescopes, several cometary parent molecules were detected through their rotational lines: hydrogen cyanide, formaldehyde, hydrogen sulfide and methanol. Radio spectroscopy of comets is a powerful technique which permits the detection and unambiguous identification of species which have mixing ratios as small as 1/1000 (like HCN or H_2S), and which do not have vibrational or electronic transitions strong enough to be observed in the infrared, visible or ultraviolet. When it is possible to observe several rotational lines simultaneously (Figure C45), this allows one to determine the excitation conditions of the molecule and the physical conditions of the atmosphere. The high spectral resolution of radio observations permits resolution of the line shapes and permits one to probe (via the Doppler effect) the coma kinematics: besides *in situ* measurements, this is the most direct way to study the coma expansion. Radio continuum observations at millimeter and submillimeter wavelengths complement photometric studies at other wavelengths and are sensitive to the largest dust particles.

Radar

Due to the small sizes of cometary nuclei, only comets coming by chance close to the Earth are accessible to this technique. Comet P/Halley could hardly be detected. The most successful radar observation was that of comet IRAS-Araki-Alcock (1983 VII) at the Arecibo Observatory (Harmon *et al.*, 1989), when it was at only 0.033 AU from the Earth: besides a strong narrow-band echo from the comet nucleus, a broadband echo from large particles was observed.

Infrared

Broadband infrared photometry gives access to the thermal emission of cometary dust, whereas the continuum emission of comets in the visible and near infrared is only reflected Sun radiation (Hanner and Tokunaga, 1991). During its sky survey the Infrared Astronomical Satellite (IRAS) observed 11 comets, among which six were new discoveries. It also discovered 'cometary trails', which are dust trails associated with short-period comets, extending from several degrees to several tens of degrees and following cometary orbits.

Infrared spectroscopy is a powerful way to investigate the cometary chemical composition. A broad emission band attributed to silicate grains is observed at 10–12 μm. Cometary parent molecules emit through fluorescence of their fundamental bands of vibration (Weaver, Mumma and Larson 1991). Water could thus be directly observed for the first time in P/Halley from stratospheric airborne observations of its 2.7-μm band (Mumma *et al.* 1986). Carbon dioxide, methanol and possibly formaldehyde were also detected. Sensitive searches for methane were not conclusive. A strong

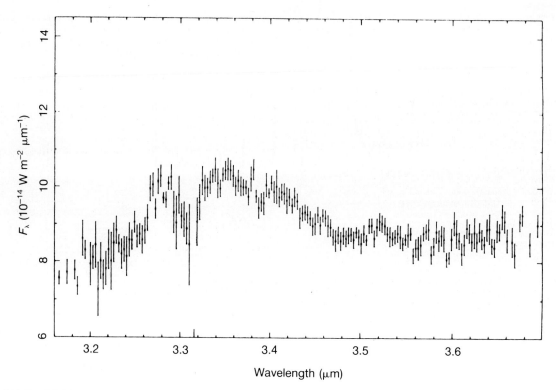

Figure C46 The 3.2–3.7 μm spectrum of comet Levy (1990 XX), observed at the UKIRT (United Kingdom Infrared Telescope). One can see the 3.28-μm feature attributed to aromatics, the 3.52-μm feature attributed to methanol, and the 3.33–3.45 μm emission which is yet unidentified. (From Davies et al., 1991, Mon. Not. Roy. Astron. Soc. **251**, p. 148.)

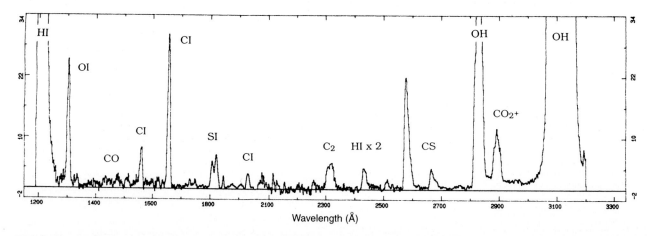

Figure C47 A composite UV spectrum of comet Bradfield (1979 X) observed by the *International Ultraviolet Explorer*. The abscissa scale is the wavelength in angstroms. From Festou, 1990, ESA SP-1134.

emission at 3.3–3.5 μm, first detected in P/Halley, was observed in several subsequent comets (Figure C46). It is attributed to the vibrational stretching mode of carbonaceous compounds, but the exact nature of the emitters and of the emission mechanisms is still debated. Aromatic species are suspected by the presence of a feature at 3.28 μm. Simple CHO species such as methanol are probable contributors, but large organic molecules and small carbonaceous grains may also be invoked.

Ultraviolet

Ultraviolet observations of comets (Feldman, 1991) were made from several rocket flights, satellite observations, and recently, from the Astro-1 payload of the Shuttle and from the Hubble Space Telescope (HST). The strong Lyman α line of atomic hydrogen at 115 nm was observed as early as 1970 in comet Tago-Sato-Kosaka (1969 IX) with the Orbiting Astronomical Observatory (OAO-2): it traces an hydrogen envelope which extends over 10^7 km. The ultraviolet spectrum of comets is dominated by the Lyman α line and the strong OH band at 305 nm. Two primary molecules have been identified from UV observations: carbon monoxide, first observed in comet West (1976 VI) during a rocket flight, has since been found to be an important constituent in several comets; diatomic sulfur was only identified in comet IRAS-Araki-Alcock (1983 VII) when it passed close to the Earth. In addition, the lines of several atomic species (oxygen, carbon, sulfur), as well as of several molecular radicals and ions (C_2, CS, CN^+, CO^+, CO_2^+) are also present (Figure C47). A comprehensive set of data has been collected since 1978 on more than 40 comets by the International Ultraviolet Explorer (IUE) satellite. The monitoring of the OH band at 305 nm is a convenient way to follow the water vapor production.

Coordination of observations

Each cometary apparition is a unique event. With a few exceptions like P/Halley, the known short-period comets are weak. Therefore the brightest comets, which are the more interesting for physical observations, are unpredictable and their observation must be organized on short notice. A rapid diffusion of cometary discoveries and crucial observations is thus essential. This is the objective of the Central Bureau of Astronomical Telegrams of the International Astronomical Union. Several hundreds of bulletins concerning timely astronomical events are circulated each year. Cometary discoveries and orbital predictions are sent to observers by telegram or telex. A much easier and more efficient distribution is now possible with the use of electronic mailing through computer networks.

The first observing campaign of worldwide importance was made in 1972–73 for comet Kohoutek 1973 XII, in support of observations from the NASA Skylab space station. But the most important effort of this kind was undoubtedly the organization of the International Halley Watch (IHW), to support the space exploration of comets P/Giacobini-Zinner and P/Halley (see below). The role of the IHW was to encourage and support scientific studies of comet Halley, to coordinate activities among the ground-based disciplines and the flight projects, to set up useful standards for the observations, and finally to ensure proper archiving of the observational data.

In situ measurements

A summary of *in situ* explorations of comets is given in Table C7. To date, only three comets have been now visited by space probes. These visits were flybys, i.e. the space probes encountered the comets with large relative velocities (about 70 km s^{-1} for P/Halley Plate 20). Therefore, the really useful observations spanned only a few hours. Due to these large relative velocities, the cometary dust particles were a serious hazard for the spacecraft, which could not go too close to their targets, and which had to be equipped with appropriate shielding.

The International Cometary Explorer (ICE) was a clever reuse of a probe, formerly designed to study the solar wind, to encounter comet P/Giacobini-Zinner. Although devoid of instruments specifically dedicated to cometary measurements (and especially without a camera), ICE was nevertheless able to explore the plasma of the cometary tail.

Suisei and Sakigake were two small twin probes designed by the ISAS (Institute of Space and Astronautical Science, Japan). They were equipped with plasma experiments and (for Suisei) an ultraviolet camera to get images of the cometary coma.

The two identical spacecraft Vega 1 and 2, en route to P/Halley, first visited Venus where they left a descent module. Giotto ultimately used the data collected by the Vega probes a few days before to correct its trajectory and to encounter the comet at only 600 km. Giotto and the two Vega probes were equipped with sets of instruments which encompassed almost all possible fields of investigation.

After its encounter with P/Halley, Giotto was retargeted to P/Grigg-Skjellerup, a moderately active comet with a production rate of about 1/100 that of P/Halley. Unfortunately, since half of its instruments (and especially the camera) were damaged at the P/Halley flyby, the scientific results were limited.

P/Halley was thus the only comet to be thoroughly investigated by space probes. A detailed description of the space exploration of P/Halley and of its scientific results may be found in Grewing, Praderie and Reinhardt (1988) and in Mason (1990).

Interaction with the interplanetary medium and solar wind

The plasma was investigated by wave plasma analyzers and energetic particle analyzers. Magnetometers measured the local magnetic field along the probe trajectories. This enabled investigators to study the relations between the comet and the interplanetary medium and the solar wind. Similar experiments were made successfully on P/Giacobini-Zinner by ICE.

Nuclear region imaging

The most impressive results for the layman were those of the cameras which made the first images ever obtained of a cometary nucleus and of its environment. The nucleus was found to be elongated, much larger ($15 \times 8 \times 8$ km) and darker than expected (albedo = 0.04). With its irregular shape and the presence of craters, P/Halley's nucleus is similar to small asteroids that have been imaged. The cometary activity was found to come from a few spots (about 10% of the total surface area), from which bright dust jets were originating, the remainder of the surface being covered by a non-volatile crust.

Cometary dust

Detectors measured the dust particle fluxes and mass spectra. A variety of sensors were used in order to cover about 15 decades of mass. Most of the mass of dust was found to be concentrated in large particles ($> 10^{-6}$ kg); however, an unexpectedly large population of very small ($< 10^{-19}$ kg) grains was revealed. The elemental compositions were determined by mass spectroscopy. In addition to particles similar to meteoritic carbonaceous chondrites, rich in iron, carbon and silicon, a population of grains rich in CHON elements was discovered: presumably grains with organic mantles, which could be responsible for the release of molecules and radicals in the cometary atmosphere.

Composition of cometary volatiles

The Giotto and Vega probes were equipped with mass spectrometers which could analyze the composition of neutral and ionic species. Due to limited spectral resolution, these instruments cannot separate species with the same mass number (e.g. CO and N_2), and some identifications may therefore be ambiguous. The determination of neutral abundances from the observation of ions requires delicate modeling. Despite these drawbacks, mass spectroscopy established that water was the dominant cometary neutral. CO_2, CO, H_2CO and several other minor constituents were identified. It was found that CO and H_2CO do not all come from the nucleus, but that additional extended sources are required. The Vega probes were also equipped with a 2.5–12 μm spectrometer, and a near-IR, visible and near-UV spectrometer, for remote sensing analysis of the near-nucleus region. Direct identification (in the infrared) of water, CO_2, and of the CH signature around 3.3 μm were achieved.

Future of cometary observations

Cometary observations have not yet revealed all the properties of these objects. For instance, the nucleus density, the detailed composition of cometary volatiles and refractories, the mechanism of

Table C7 Summary of space explorations of comets

Date	Comet	Spacecraft	Closest approach (km)
11 September 1985	P/Giacobini-Zinner	ICE (ESA–NASA)	7 800
6 March 1986	P/Halley	Vega 1 (URSS)	8 890
8 March 1986	P/Halley	Suisei (Japan)	150 000
9 March 1986	P/Halley	Vega 2 (URSS)	8 030
11 March 1986	P/Halley	Sakigake (Japan)	7 000 000
14 March 1986	P/Halley	Giotto (ESA)	596
10 July 1992	P/Grigg-Skjellerup	Giotto (ESA)	200

cometary activity at large distances and the relations between comets and asteroids are still poorly known. Future progresses on these topics will result from both long-term observing programs and from the use of sophisticated new techniques.

Systematic surveys with automatic image analysis will lead to the discoveries of many more comets and will especially expand the number of known Chiron-type objects. It may be anticipated that future telescopes such as the Very Large Telescope presently under construction at the European Southern Observatory will be able to follow short-period comets such as P/Halley all along their orbit.

Investigations of cometary composition require observations of the electromagnetic spectrum from microwaves to the ultraviolet. More efficient ground-based facilities are being developed at radio and infrared wavelengths, but some spectral domains can only be observed from space. The use of existing or future space observatories will be crucial: the Hubble Space Telescope in the ulraviolet, the Infrared Space Observatory (ISO) and the Shuttle IR Telescope Facility (SIRTF) in the infrared, the Far IR Space Telescope (FIRST) and Large Deployable Reflector (LDR) in the submilimeter domain.

Clearly, ultimate results on comets will come from space exploration and especially from rendezvous missions which could monitor the development of cometary activity along part of a cometary orbit, and from landing missions which could directly analyze the nucleus structure and composition. The Comet Rendezvous Asteroid Flyby (CRAF) project of NASA, unfortunately canceled, was to follow the activity of a short-period comet from aphelion to perihelion. The much more ambitious Rosetta project (or Comet Nucleus Sample Return) by ESA/NASA, was in its original form to return to the Earth for laboratory analysis an unaltered sample of cometary material. These projects are currently being redefined. In any case, only a very small number of objects could be explored in the near future. Space exploration and Earth-based remote sensing of comets are two complementary approaches.

Jacques Crovisier

Bibliography

Arpigny, C., Dossin, F., Woszczyk, A. *et al.* (1993) *Atlas of Cometary Spectra*. Dordrecht: D. Reidel Publishing Company.
Belton, M.J.S. (1991). Characterization of the rotation of cometary nuclei, in *Comets in the Post-Halley Era*, (eds R.L. Newburn, M. Neugebauer and J. Rahe). Dordrecht: Kluwer Academic Publishers, pp. 691–721.
Bockelée-Morvan, D., Crovisier, J., Colom, P. *et al.* (1991) Observations of parent molecules in comets at millimetre wavelengths: HCN, H_2S, H_2CO and CH_3OH, in *Formation of Stars and Planets, and the Evolution of the Solar System*. ESA SP-315, 143–8.
Crovisier, J. (1992). Radio spectroscopy of comets, recent results and future prospects, in *Asteroids, Comets, Meteors 1991* (eds A.W. Harris and E. Bowell). Houston: Lunar and Planetary Institute, pp. 137–40.
Crovisier, J. and Schloerb, F.P. (1991) The study of comets at radio wavelengths, in *Comets in the Post-Halley Era* (eds R.L. Newburn, M. Neugebauer and J. Rahe). Dordrecht: Kluwer Academic Publishers, pp. 149–73.
Davies, J.K., Green, S.F. and Geballe, T.R. (1991) The detection of a strong $3.28 \times \mu m$ emission feature in comet Levy. *Mon. Not. Roy. Astron. Soc.*, **251**, 148–51.
Feldman, P.D. (1991) Ultraviolet spectroscopy of cometary comae, in *Comets in the Post-Halley Era* (eds R.L. Newburn, M. Neugebauer and J. Rahe). Dordrecht: Kluwer Academic Publishers, pp. 139–148.
Festou, M.C. (1990) *IUE-ULDA Access Guide No. 2: Comets*. ESA SP-1134.
Festou, M.C., Rickman, H. and West, R.M. (1993) Comets. I. Concepts and observations. *Astron. Astrophys. Rev.*, **4**, 363–447.
Gredel, R., van Dishoeck, E.F. and Black, J.H. (1989) Fluorescent vibration-rotation excitation of cometary C_2. *Astrophys. J.*, **338**, 1047 70.
Green, D.W.E. (1991) *Int. Comet Q. Archive of Cometary Photometric Data*, magnetic tape, 4 ed. Cambridge, MA: Smithsonian Astrophysical Observatory.
Grewing, M., Praderie, F. and Reinhard, R. (eds) (1988) *Exploration of Halley's Comet*. Berlin: Springer Verlag (also 1987, *Astron. Astrophys.*, **187**).
Hanner, M.S., and Tokunaga, A.T. (1991) Infrared techniques for comet observations in *Comets in the Post-Halley Era* (eds R.L. Newburn, M. Neugebauer and J. Rahe). Dordrecht: Kluwer Academic Publishers, pp. 67–91.
Harmon, J.K., Campbell, D.B., Hine, A.A. *et al.* (1989) Radar observations of comet IRAS-Araki-Alcock 1983d. *Astrophys. J.*, **338**, 1071–93.
Jewitt, D. (1991) Cometary photometry. In *Comets in the Post-Halley Era* (eds R.L. Newburn, M. Neugebauer, and J. Rahe). Dordrecht: Kluwer Academic Publishers, pp. 19–65.
Larson, S.M., Edberg, S.J. and Levy, D.H. (1991) The modern role of visual observations of comets in *Comets in the Post-Halley Era* (eds R.L. Newburn, M. Neugebauer and J. Rahe). Dordrecht: Kluwer Academic Publishers, pp. 209–23.
Marsden, B.G. (1989) The sungrazing comet group. *Astron. J.* **98**, 2306–21.
Marsden, B.G. (1992) *Catalogue of Cometary Orbits*, 7th ed. Cambridge, MA: Smithsonian Astrophysical Observatory.
Mason, J. (Ed.) (1990) *Comet Halley, Investigations, Results, Interpretations*. Ellis Horwood.
Mumma, M.J., Weaver, H.A., Larson, H.P., *et al.* (1986) Detection of water vapor in Halley's comet. *Science*, **232**, 1523–8.
Osborn, W.H., A'Hearn, M.F., Carsenty, U. *et al.* (1990) Standard stars for photometry of comets. *Icarus*, **88**, 228–45.
Weaver, H.A., Mumma, M.J. and Larson, H.P. (1991) Infrared spectroscopy of cometary parent molecules, in *Comets in the Post-Halley Era* (eds R.L. Newburn, M. Neugebauer and J. Rahe). Dordrecht: Kluwer Academic Publishers, pp. 93–106.
West, R.M., Hainaut, O. and Smette, A. (1991). Post-perihelion observations of P/Halley. III. An outburst at $r = 14.3$ AU. *Astron. Astrophys.*, **246**, L77–L80.
Wyckoff, S., (1982) Overview of observations, in *Comets* (ed. L.L. Wilkening). Tucson: University of Arizona Press, pp. 3–55.
Yeomans, D.K. (1991) *Comets. A Chronological History of Observations, Science, Myth, and Folklore*. Wiley Science Editions.

Cross references

Giotto mission
Halley, Edmond, and Halley's comet
Sakigake and Suisei missions
Vega missions

COMET: ORIGIN AND EVOLUTION

Cometary orbits are classified as either long- or short-period, depending on whether their periods are greater than or less than 200 years respectively. The long-period (LP) orbits are randomly oriented on the celestial sphere, whereas the short-period (SP) comets are generally confined to direct orbits with inclinations less than $\sim 35°$. Most of the known SP orbits have periods between 5 and 20 years, while the LP orbits range up to $\sim 10^7$ years.

It is only in the last several decades that comets have been recognized to be true primordial members of the solar system. Approximately one-third of all LP comets observed passing through the planetary system are on weakly hyperbolic orbits. However, integration of the orbits backward in time to points outside the planetary region, and conversion from a heliocentric to a barycentric coordinate system (Bilo and van de Hulst, 1960), showed that those comets in fact had highly eccentric but still gravitationally bound orbits. Planetary perturbations, primarily by Jupiter, scatter the LP comets in orbital energy, either ejecting them on hyperbolic orbits or capturing them to more tightly bound ellipses.

The distribution of orbital energies for the observed LP comets is shown in Figure C48 (data from Marsden, 1990). The numbers of comets are plotted versus $1/a_o$, the inverse original semimajor axis of the orbit (prior to entry into the planetary region), which is proportional to orbital energy. Positive $1/a_o$ values indicate bound orbits whereas negative values of $1/a_o$ denote hyperbolic orbits. The distribution is characterized by a sharp spike of comets at near-zero but bound energies, and a low continuous distribution of comets in less eccentric orbits.

Oort (1950) showed that this unique distribution could be explained by a vast cloud of comets surrounding the planetary system

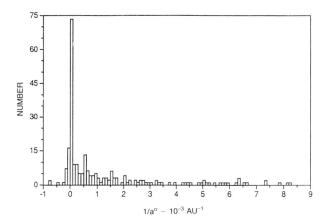

Figure C48 The distribution of original inverse semimajor axes for 190 long-period comets as found by Marsden (1990). The large spike at near-zero energy represents the dynamically new comets from the Oort cloud. The low continuous distribution is composed of returning comets which have been scattered in $1/a_o$ by planetary perturbations, primarily by Jupiter.

and extending half-way to the nearest stars. Comets in the Oort cloud are repeatedly scattered by perturbations from random passing stars. The orbits diffuse in velocity phase space and occasionally are perturbed to perihelia within the planetary region. However, once the comets enter the planetary region they are randomly scattered in $1/a_o$ by Jupiter, with a typical $1/a_o$ of $\pm 630 \times 10^{-6}$ AU^{-1} (van Woerkom, 1948; Everhart, 1968) on each perihelion passage. This is more than six times the width of the Oort cloud spike in Figure C48. Thus the comets rapidly diffuse in $1/a_o$. Only about 5% of 'dynamically new' LP comets are returned to Oort cloud distances after their first perihelion passage. Computer-based dynamical simulations show that a typical LP comet from the Oort cloud makes an average of five returns with a mean time of 6×10^5 years between the first and last perihelion passage. Approximately 65% of LP comets are hyperbolically ejected to interstellar space, 27% are randomly disrupted, and the rest are lost to a variety of mechanisms such as perturbation to a Sun impacting orbit or exhaustion of all volatiles (Weissman, 1979).

The few hyperbolic comets in Figure C48 are believed to be the result of small errors in their orbit determinations and/or nongravitational forces (i.e. jetting of volatiles from the nucleus surface), which cause the orbits to appear more eccentric than they actually are. True interstellar comets would have expected $1/a_o$ values on the order of -0.50 AU^{-1} (for hyperbolically encounter velocities of 20 km s^{-1}), more than 600 times greater than the largest value observed.

The dynamical evolution of LP comets is controlled by the combined action of planetary and stellar perturbations, by rare encounters between the solar system and giant molecular clouds (GMCs), by the tidal field of the galactic disk and (to a lesser extent) the galactic nucleus, and by nongravitational forces from jetting during perihelion passage. Computer-based, Monte Carlo simulations of the comets' dynamical evolution in the Oort cloud (Weissman, 1982, 1985a; Fernandez, 1982; Heisler and Tremaine, 1986; Heisler, 1990) are used to estimate the cloud population by comparing model results to the observed flux of LP comets through the planetary region, after correction for observational selection effects (Everhart, 1967). The current best estimate is $\sim 10^{12}$ comets (Weissman, 1990).

Studies of the total perturbations on comets in the Oort cloud have shown that their typical dynamical lifetime is only about half the age of the solar system (Hut and Tremaine, 1985). Thus it has been suggested that the cloud may need to be replenished. The two possible sources are an unseen inner Oort cloud which is not dynamically sampled except by the largest perturbations (those due to penetrating stellar passages or encounters with GMCs), or by capture of comets from interstellar space. Capture has been shown to be highly improbable (Valtonen and Innanen, 1982; Valtonen, 1983) at typical stellar encounter velocities. On the other hand, dynamical models (Duncan, Quinn and Tremaine, 1987; Shoemaker and Wolfe, 1984) have shown that an inner Oort cloud with a population of ~ 5 to 10 times that of the outer cloud is highly plausible. If this inner cloud exists, which is highly likely, the total population of the Oort cloud (inner and outer) is then between 6×10^{12} and 1.1×10^{13} comets. Assuming an average nucleus mass of 3.8×10^{16} g (Weissman, 1991a), the total mass of comets in the Oort cloud is between 38 and 70 Earth masses. Estimates of cometary masses are highly uncertain and some Oort cloud mass estimates are much larger, up to ~ 500 Earth masses (Bailey, 1990). A more complete review of Oort cloud dynamics is given in Weissman (1990).

Cometary origin

Hypotheses on the origin of the Oort cloud can generally be divided into two groups: (1) a primordial origin, coincident with the origin of the Sun and planetary system, or (2) episodic formation and/or capture, occurring either once or many times over the history of the solar system. The many different proposals for cometary origin are reviewed in Weissman (1985b). The major primordial origin hypotheses include (1) origin as icy planetesimals in and/or just beyond the outer planets zone, which were subsequently dynamically scattered to distant orbits in the Oort cloud by the growing proto-planets (Oort, 1950; Kuiper, 1951; Safronov, 1972; Cameron, 1978); (2) in situ accretion in the solar nebula at large solar distances (Biermann and Michel, 1978; Hills, 1982; Bailey, 1987); (3) formation in satellite nebula of the primordial solar nebula (Cameron, 1973); or (4) in some models, the Sun formed in a closely spaced cluster of stars and consequently comets formed around each star were dynamically mixed, populating each of their respective Oort clouds (Donn, 1976). The episodic hypotheses include (5) gravitational focusing of material in the solar wake after solar system passage through interstellar clouds (Lyttleton, 1948); (6) eruption from the giant planets and/or their satellites (Vsekhsvyatskii, 1967); (7) formation in compressed interstellar clouds at galactic spiral-arm shocks (McCrea, 1975); or (8) formation in GMCs and capture during encounters with the solar system (Clube and Napier, 1982).

In general, the episodic hypotheses have not found much support and will not be discussed further here (see discussion of problems and weaknesses in Weissman, 1985b). Of the primordial origin hypotheses, 1 and 2 are the two leading candidates. A cometary origin among the outer planets provides for formation in a relatively dense but still cold region of the solar nebula, with the problem that scattering to Oort cloud distances is dynamically inefficient. Conversely, formation at Oort cloud distances is dynamically efficient but is difficult to understand given the very low nebula densities expected at 10^3 to 10^4 AU from the protosun.

Oort (1950) suggested that comets were dynamically ejected from the asteroid belt by Jupiter and the other giant planets. Kuiper (1951) pointed out that the proposed icy composition of comets required that they be formed farther from the Sun, at the orbit of Jupiter or beyond. Safronov (1972) subsequently showed that gravitational scattering by Jupiter and Saturn tended to eject most planetesimals in their zones on hyperbolic orbits, rather than to bound orbits in the Oort cloud. However, Safronov showed that the lower masses and larger heliocentric distances of Uranus and Neptune would allow them to provide considerable dynamical scattering but with far fewer ejections. This work has served as the basis for the most widely accepted and widely studied theory of cometary origin.

Icy planetesimals formed in the outer solar system are believed to be the building blocks of the cores of the giant planets, as well as a substantial fraction of the total mass of Uranus and Neptune, which apparently formed after much of the solar nebula gas had dispersed. Safronov (1972) showed that if the current masses of Uranus and Neptune were dispersed into small planetesimals in their zones, their accretion times would be $\sim 10^{11}$ years, longer than the age of the solar system. To surmount this difficulty, Safronov suggested that the original mass of planetesimals in the Uranus–Neptune zone was ten times the current masses of the two planets combined. According to Safronov, most of this material was ejected to hyperbolic orbits but about 1 to 2% of it reached distant elliptical orbits to form the Oort cloud.

A dynamical simulation model of this scattering by Fernández and Ip (1981, 1983) confirmed many of Safronov's ideas, but also found some important differences. Fernández and Ip showed that Neptune was the most efficient at placing material in the Oort cloud, with 72% of all ejecta ending up there, the rest escaping to interstellar space or

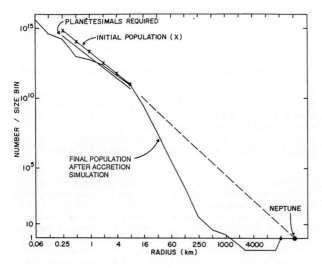

Figure C49 Size distribution of planetesimals in the Uranus–Neptune zone after 1.4×10^5 years, as modeled by Greenberg et al. (1984). The initial planetesimals' size distribution was chosen to match the number and size distribution of observed comets.

being sent to short-period orbits among the inner planets. The corresponding figures are 57%, 14%, and 3% for Uranus, Saturn and Jupiter respectively. These efficiencies are considerably more than the values of 2%, 1.2%, 0.5% and 0.2% for Neptune, Uranus, Saturn and Jupiter found by Safronov (1972). However, a portion of the disagreement can be attributed to Safronov using a much narrower capture range in orbital energy for the Oort cloud. Scaling Safronov's work to the same capture zone would lead to probabilities of 5.4%, 3.2%, 1.3% and 0.5%, still a factor of 6 to 10 less than Fernández and Ip. Other differences may be attributed to differences in the models. For example, Fernández and Ip considered capture to SP orbit as an end-state, whereas in reality the comets continue to circulate and will most likely be dynamically ejected.

Another accretion simulation model by Greenberg et al. (1984) studied the problem of simultaneously growing Neptune and populating the Oort cloud. They found that the planet and comet cloud had to be built from a primordial population of small bodies, ~ 4 to 8 km in radius; larger initial starting sizes of ~ 100 km radius did not lead to a sufficient number of comets in the Oort cloud. The resulting size distribution of comets is shown in Figure C49 from Greenberg et al. (1984). Note that a number of larger comets are accreted, some up to 10^3 km in diameter, or more. This compares favorably with recent suggestions by Stern (1991) that some icy bodies similar in size to Triton and Pluto may have accreted and may be resident in the Oort cloud or may have been ejected to interstellar space. The accretional and collisional evolution of this population of larger comets in the Uranus–Neptune zone, prior to their ejection to the Oort cloud, may provide for some degree of physical processing of the proto comets.

It is interesting to compare the size distribution of icy planetesimals in Figure C49 with the mass distribution for cometary nuclei found by Weissman (1991a), shown in Figure C50, based on Everhart's intrinsic distribution of cometary magnitudes, corrected for observational selection effects. That distribution has a break in slope at ~ 7 km radius, very similar to the break at 8 km in the Greenberg et al. (1984) work, and has similar slopes.

Ejection of comets to large aphelion distances is not, by itself, sufficient to create the Oort cloud, since the comets will return to small perihelion distances where they can again be perturbed by the planets. However, Duncan, Quinn and Tremaine (1987) showed that comets with aphelia of ~ 5 to 6×10^3 AU would be sufficiently perturbed by galactic tides during a single orbit to raise their perihelia out of the planetary region. In this manner, a large inner Oort cloud could be populated. Comets would continue to evolve in the cloud under the combined influence of stellar, GMC and galactic tidal perturbations. The net effect of those perturbations is to pump angular momentum into the cloud, randomizing the inclinations of the comet orbits and further raising their perihelia. A simulation of the evolution of cometary orbits in the Oort cloud over time is shown

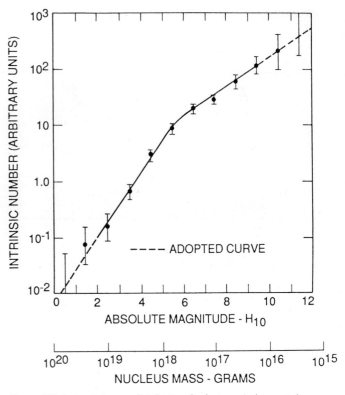

Figure C50 Estimated mass distribution for long-period comets by Weissman (1991a), based on Everhart's (1967) intrinsic distribution of cometary magnitudes, corrected for observational selection effects. Note the similarity to the accretion distribution found by Greenberg et al. (1984) in Figure C49.

in Figure C51. Duncan, Quinn and Tremaine showed that the Oort cloud would be essentially randomized at semimajor axes greater than about 8×10^3 AU. At semimajor axes $\geq 2 \times 10^4$ AU, typical perturbations by random passing stars and the galactic tide are sufficient to throw some comets back into the planetary region where they are observed as 'dynamically new' comets from the Oort cloud.

The dynamical problems associated with populating the Oort cloud from the outer planets region are eliminated if comets formed at Oort cloud distances, ~ 10^3 to 10^4 AU. A reasonable description of this process is provided by Hills (1982) and Hills and Sandford (1983a,b). They suggested that infalling, low-opacity dust clumps in the outer protosolar nebula experienced a net external radiation pressure which forced them to collapse and come together as protocomets. An important assumption in their study is the belief that the Sun was a member of a forming star cluster and that the dust clump experienced a roughly isotropic radiation field, forcing material in toward the center of the clump. Hills (1982) estimated that a clump with an initial radius of ~ 0.02 AU, a dust density of 4×10^{-20} g cm^{-3} and a gas density of 2×10^{-18} g cm^{-3} would collapse in about 400 years, considerably less than the 5×10^3 years required for the clump to fall into the proto-Sun. The resulting comet nucleus would have a radius of 1 km and a mass of 4×10^{15} g. Hills and Sandford (1983a, 1983b) showed that the radiation field did not need to be perfectly isotropic for collapse to proceed, but also that the radiation field from only two protostars would not focus material towards the center of the clump, and that no protocomet would form.

There are a variety of problems with this scenario which still need to be explored. What is the effect of turbulence in the nebula on the formation of irregular dust clumps? How efficient is the radiation–grain coupling in this irregular field of objects; i.e. do clumps shadow one another? What is the total number of comets that could be formed by this mechanism and what is their subsequent interaction with the inner primordial nebula? Late infalling clumps may have been delayed by their own angular momentum and this may also prevent them from falling directly into the proto-Sun. But is it

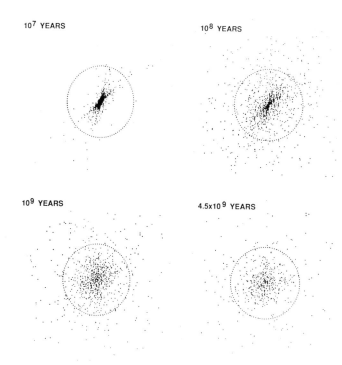

Figure C51 Dynamical evolution of a hypothetical cloud of comets ejected out of the Uranus–Neptune zone at several times during the history of the solar system, under a combination of galactic, stellar and planetary perturbations (projected onto a plane perpendicular to the galactic plane). The dotted circle is at a radius of 2×10^4 AU, the boundary between the inner and outer Oort clouds. (From Duncan et al., 1987.)

sufficient to prevent them from interacting gravitationally with the growing protoplanets, in which case they could be scattered to interstellar space just like the icy planetesimals formed in the inner nebula disk.

A somewhat different mechanism was proposed by Whipple and Lecar (1976) and developed by Bailey (1987). Whipple and Lecar suggested that a strong protosolar wind, akin to the observed T Tauri winds, would produce a turbulent boundary layer between infalling and outflowing material, and that the material would be driven into a thin circumstellar shell. Densities may be sufficient to allow rapid accretion of dust grains which are then sufficiently massive to decouple from the gas motion. These in turn may then achieve sufficient density to collapse gravitationally into comet sized bodies. In this manner it may be possible to produce a large number of comets, either bound to the proto-Sun or circulating in nearby interstellar space. The problems with this scenario are similar to those for the previous one; the effects of turbulence, heating at shock boundaries and subsequent dynamical evolution are not well understood or quantified.

If comets are formed in low eccentricity orbits in the protosolar nebula, the region between roughly 10^2 and 2×10^3 AU provides no dynamical mechanism for perturbing the comets either into the planetary region or out far enough to fall under the influence of stellar and galactic perturbations. Thus, this population of comets would circulate indefinitely without ever producing visible comets. One possible means for increasing the semimajor axes of such orbits would be rapid mass loss by the central protostar and nebula (Cameron, 1978). In particular, if the mass loss is slightly less than 50% of the total central mass, near-circular orbits are transformed into near-parabolic ones, leading to population of the Oort cloud. However, the range of values for the mass loss that would produce this transformation is very narrow, and thus the scenario is relatively unlikely. Dermott and Gold (1978) solved this problem by suggesting that the central mass loss occurred in a number of smaller, repeated steps. Then orbits could be slowly pumped up a bit at a time to Oort cloud distances. Also, the mass loss must be accomplished on a timescale long compared to the orbital periods of the planets, but short compared to the cometary periods. Scenarios for this to occur may be too constraining to be physically real.

Marochnik, Mukhin and Sagdeev (1988) estimated that the angular momentum of the Oort cloud is between 5×10^{52} and 2×10^{53} g cm^2 s^{-1}, two to three orders of magnitude greater than the total angular momentum of the current planetary system. They argued that this was too much to be ejected out of the planetary system without causing the giant planets to spiral in toward the Sun. However, Weissman (1991b) showed that most of the angular momentum in the present-day Oort cloud is the result of the action of external perturbers over the history of the solar system. In addition, some Oort cloud parameters used by Marochnik, Mukhin and Sagdeev are higher than current best estimates. Weissman showed that the total angular momentum of the current Oort cloud is likely between 6.0×10^{50} and 1.1×10^{51} g cm^2 s^{-1}, and the original angular momentum was likely a factor of five less than that. These more modest estimates are consistent with comets having been ejected from the Uranus–Neptune zone.

This point was also examined by Fernández and Ip (1984) who modeled the exchange of angular momentum between the giant planets and the evolving protocomets. They showed that the giant planets scatter material both inward and outward, resulting in little net movement in the semimajor axes of the protoplanets. Jupiter does the bulk of the ejecting because it has no larger planet to pass the evolving comets to, but because of its great mass, Jupiter only moves inward a few tenths of an AU.

Future physical and dynamical studies will continue to add detail to the various hypotheses with regard to cometary formation. At present the weight of evidence and physical plausibility favors cometary formation among the outer planets. However, detailed examination of cometary materials may shed the best light on the true formation site for the comets, and may provide sufficient impetus to quantify better the details of possible cometary formation at larger heliocentric distances.

Extrasolar Oort clouds and interstellar comets

Presumably, the same processes that led to the formation of our solar system's Oort cloud can also occur around other forming stars. Because the expected dimensions of Oort clouds are so large, it may be possible to actually detect and resolve such comet clouds. One method would be to look for thermal radiation from dust created by collisions and sputtering in the cloud. Stern, Stock and Weissman (1991) analyzed IRAS 60- and 100-μm sky flux images for 17 nearby stars (e.g. β Pictoris, ε Eridani), looking for infrared excess in summed circular annuli around the primaries. However, no detections were made; the expected optical depth of Oort clouds is below the IRAS detection limits. Dust at Oort cloud distances will be very cold, and radiation pressure and collisions with interstellar dust and gas should rapidly sweep fine dust from the comet clouds (Stern, 1990).

Stern, Shull and Brandt (1990) proposed searching for Oort clouds around red giant stars. These stars are sufficiently luminous that the comets in their Kuiper belts (see short-period comet discussion, below), if they had them, would be actively sublimating at rates comparable to the gas production rates of comets at 1 AU in our own solar system. Stern, Shull and Brandt suggested that observed OH–IR stars may be an example of this phenomenon.

It is interesting to speculate on the fate of the many comets ejected to interstellar space in forming the Oort cloud and over its history. Dynamical ejection is the most common loss mechanism for comets in the cloud, either due to close stellar and GMC perturbations, or as a result of Jupiter perturbations during passes through the planetary system.

No comet on a clearly interstellar trajectory has been observed passing through the planetary system. Sekanina (1976) showed that this fact sets an upper limit on the space density of interstellar comets of 6×10^{-4} $M_\odot$ pc^{-3} ≃ 4×10^{12} comets pc^{-3}, using Sekanina's mean nucleus mass of 3×10^{17} g. For comparison, this is ∼ 300 times less than the density of material in the solar neighborhood, ∼ 0.185 $M_\odot$ pc^{-3} (Bahcall, 1984), so interstellar comets cannot contribute significantly to the 'missing mass' problem in the galaxy. It is about half the space density of comets in the outer Oort cloud, assuming a population of 10^{12} comets in a sphere of radius 10^5 AU centered on the Sun. Thus the limit is not very strict.

The same problem was studied by McGlynn and Chapman (1989) who suggested that at least six interstellar comets should have been observed passing within 2 AU of the Sun in the past 150 years, even after accounting for the fact that only 7% of all LP comets passing within 2 AU are expected to be discovered. Their estimate was based on an average ejected population of 10^{14} comets per star. This is twice Sekanina's upper limit of 4×10^{12} comets pc^{-3} given above. The reason for this difference is not clear, though it may not be significant.

The above limits can be compared with the estimated space density of interstellar comets, if it is assumed that all stars produce cometary clouds. Dynamical models estimate that between three (Fernández and Ip, 1981) and 50 (Safronov, 1972) times as many comets are ejected by the protoplanets as are placed in the Oort cloud (though, as discussed above, Safronov's estimate is extreme as it assumes a very narrow range of semimajor axes for the Oort cloud). Another factor of two to three comes from the comets lost from the Oort cloud over the history of the solar system. Thus, taking a nominal current Oort cloud population of 7.0×10^{12} comets, the solar system has ejected $\sim 4 \times 10^{13}$ to 10^{15} comets to interstellar space. Taking a mean volume per star in the solar neighborhood of ~ 12 pc^3 (Allen, 1973), and assuming that all stars produce comet clouds, the predicted space density is 3×10^{12} to 9×10^{13} comets pc^{-3}. This is 0.8 to 23 times the upper limit determined by Sekanina.

Since half of all stars form in multiple systems, and that process may prevent the formation of a protoplanetary disk leading to cometesimals (though that conjecture has not actually been demonstrated), the factor of 0.8 excess is certainly not a problem. However, a factor of 23 excess is not consistent with the Oort cloud models and population estimates presented here, and thus represents a difference that clearly needs to be resolved. As noted above, a reduction factor of at least 2.7 (and probably twice that) can be obtained by assuming a wider range of Oort cloud semimajor axes than in Safronov's work. The remaining discrepancy clearly merits further study.

Origin of short-period comets

It has generally been thought that the SP comets are LP comets which have random-walked to small semimajor axis orbits as a result of perturbations by Jupiter and the other planets. Capture of comets to short-period in a single planetary encounter is highly unlikely (Newton, 1893) and cannot explain the observed number of SP comets. Planetary perturbations are largest for direct, low inclination orbits and it was suggested that this acted as a selection mechanism for producing the direct, low inclination SP comet population. However, evolution from an Oort cloud orbit to a short-period one is expected to take approximately 400 returns (Weissman, 1979), and it is most likely that a comet would be ejected over that time, or destroyed by one of several poorly understood physical mechanisms (e.g. disruption, sublimation). A variation which results in improved efficiency was proposed by Everhart (1972) who suggested that comets with perihelia among the outer planets random walk in $1/a_0$ without ever coming close to the Sun, and then are perturbed into small perihelia orbits late in their dynamical evolution. In this manner, the SP comets can be spared much of the physical degradation associated with solar heating or passage through the more densely populated regions of the planetary system. This also increases the planetary system's cross-section for capturing LP comets to SP orbits, thus better supplying the observed number of SP comets.

Estimates of the number of short-period comets produced by planetary perturbation from the Oort cloud have varied considerably (Joss, 1973; Delsemme, 1973), with some suggestion that the above mechanisms cannot produce the observed number of SP comets. An alternative suggested source of SP comets is a ring of comets beyond the orbit of Neptune (Kuiper, 1951; Whipple, 1964; Fernández, 1980), which may be up to 300 times more dynamically efficient than repeated perturbation of LP comets from the Oort cloud. Fernández (1980) pointed out that some massive comets $\sim 10^3$ km in diameter would be required in the distant comet ring, to slowly perturb other comets back into the planetary region, providing the SP comet flux.

Duncan, Quinn and Tremaine (1988) compared the two possible sources and found that LP comets tended to preserve their inclinations as they evolved inward to SP orbits. Thus, if SP comets were initially LP comets from the Oort cloud, far more high-inclination SP comet orbits would be expected. Duncan, Quinn and Tremaine (1988) showed that the observed inclination distribution was more consistent with a low-inclination trans-Neptunian population of comets as the source of the SP comets. They called this population the Kuiper belt in honor of Gerard Kuiper who first suggested its existence in 1951. Subsequently, Torbett (1989) showed that planetary perturbations would lead to chaotic motion in a disk of comets beyond Neptune, throwing those with perihelia near 30 AU into Neptune crossing orbits in only 10^7 years.

Stagg and Bailey (1989) and Bailey and Stagg (1990) noted two possible difficulties with the Kuiper belt scenario. First, they suggested that physical loss mechanisms might remove high-inclination LP comets during their longer evolution inward to SP comet orbits (longer because of the smaller average perturbations for high-inclination orbits), thus leading to a predominantly low-inclination population of observed SP comets. This would remove the problem found by Duncan, Quinn and Tremaine (1988) for this source for the SP comets. Second, Bailey suggested that the combined mechanisms: LP comet evolution plus Kuiper belt, might produce too many SP comets, and that this possibly implied a lower than estimated population for the Oort cloud or the Kuiper belt, or both.

Duncan, Quinn and Tremaine (1988) estimated that 0.02 $M_\oplus$ of comets are required in the Kuiper belt to maintain the current population of short-period comets. They used an average comet nucleus mass of 3.2×10^{17} g, implying a population of 3.8×10^8 objects. However, using the average nucleus mass of 3.8×10^{16} g noted above (Weissman, 1991a), the same number of Kuiper belt comets would have a total mass of ~ 0.0025 $M_\oplus$.

Initial observational searches for Kuiper belt comets were negative. Levison and Duncan (1990) set an upper limit of less than one object brighter than magnitude $V = 22.5$ (corresponding to a 50-km radius comet nucleus at 50 AU from the Sun) per square degree based on a search of 4.9 square degrees of sky near the ecliptic. This translates to $N \leq 10^{10}$ objects if the Kuiper belt comets are in orbits between 35 and 60 AU from the Sun. Yeomans (1986) set an upper limit on the unknown mass just beyond the orbit of Neptune of ~ 1 $M_\oplus$, based on a failure to observe gravitational perturbations on the orbit of comet Halley. An earlier limit of 5 $M_\oplus$ was found by Anderson and Standish (1986) based on tracking of the Pioneer 10 spacecraft. Future tracking of the Voyager and Pioneer spacecraft may further refine these limits.

More recently, Jewitt and Luu (1992) reported the detection of a $V = 23.5$ object at ~ 42 AU from the Sun. Designated 1992QB$_1$, this object may be the first detection of a Kuiper belt comet. A second, similar object, 1993FW, was discovered at a similar distance by Luu and Jewitt (1993). Initial orbit solutions for both objects suggest that they are in low eccentricity, low inclination orbits which do not cross or even approach the orbit of Neptune (Marsden, 1993). Given the small area of the sky searched to make these two discoveries, the total number of similar bodies in the Kuiper belt may be on the order of 10^4, with an even larger number of smaller remnant planetesimals. Assuming a cometary surface albedo, 1992QB$_1$ and 1993FW are each ~ 200 km in diameter. Assuming a density of 1 g cm^{-3}, 10^4 such objects would have a total mass of ~ 0.01 $M_\oplus$.

Paul R. Weissman

Bibliography

Allen, C.W. (1973) *Astrophysical Quantities*, London: Athlone Press, 310 pp.

Anderson, J.D. and Standish, E.M., Jr (1986) Dynamical evidence for Planet X, in *The Galaxy and the Solar System* (eds R. Smoluchowski, J.N. Bahcall, and M.S. Matthews). Tucson: University of Arizona Press, pp. 286–96.

Bahcall, J.N. (1984) Self-consistent determination of the total amount of matter near the Sun. *Astrophys. J.*, **276**, 169–81.

Bailey, M.E. (1987) The formation of comets in wind-driven shells around protostars. *Mon. Not. Roy. Astron. Soc.*, **69**, 70–82.

Bailey, M.E. (1990) Cometary masses, in *Baryonic Dark Matter*, (eds D. Lynden-Bell and G. Gilmore). Dordrecht; Kluwer, pp. 7–35.

Bailey, M.E. and Stagg, C.R. (1990) The origin of short-period comets. *Icarus*, **86**, 2–8.

Biermann, L. and Michel, K.W. (1978) On the origin of cometary nuclei in the presolar nebula. *Moon and Planets*, **18**, 447–64.

Bilo, E.H. and van de Hulst, H.C. (1960) Methods for computing the original orbits of comets. *Bull. Astron. Inst. Neth.*, **15**, 119–27.

Cameron, A.G.W. (1973) Accumulation processes in the primitive solar nebula. *Icarus*, **18**, 407–450.

Cameron, A.G.W. (1978) The primitive solar accretion disk and the formation of the planets, in *The Origin of the Solar System* (ed. S.F. Dermott) New York; John Wiley and Sons, pp. 49–75.
Clube, S.V.M. and Napier, W.M. (1982) Spiral arms, comets, and terrestrial catastrophism. *Quart. J. Roy. Astron. Soc.*, **23**, 45–66.
Delsemme, A.H. (1973) Origin of the short-period comets. *Astron. Astrophys.*, **29**, 377–81.
Dermott, S.F. and Gold, T. (1978) On the origin of the Oort cloud. *Astron. J.*, **83**, 449–50.
Donn, B. (1976) Comets, interstellar clouds, and star clusters, in *The Study of Comets*, NASA SP-393, pp. 663–72.
Duncan, M., Quinn, T. and Tremaine, S. (1987) The formation and extent of the solar system comet cloud. *Astron. J.*, **94**, 1330–8.
Duncan, M., Quinn, T. and Tremaine, S. (1988) The origin of short-period comets. *Astrophys. J.*, **328**, L69–73.
Everhart, E. (1967) Intrinsic distributions of cometary perihelia and magnitudes. *Astron. J.*, **72**, 1002–11.
Everhart, E. (1968) Changes in total energy of comets passing through the solar system. *Astron. J.* **73**, 1039–52.
Everhart, E. (1972) The origin of short-period comets. *Astrophys. Lett.*, **10**, 131–5.
Fernández, J.A. (1980) On the existence of a comet belt beyond Neptune. *Mon. Not. Roy. Astron. Soc.*, **192**, 481–91.
Fernández, J.A. (1982) Dynamical aspects of the origin of comets. *Astron. J.*, **87**, 1318–32.
Fernández, J.A. and Ip, W.-H. (1981) Dynamical evolution of a cometary swarm in the outer planetary region. *Icarus*, **47**, 470–9.
Fernández, J.A. and Ip, W.-H. (1983) On the time evolution of the cometary influx in the region of the terrestrial planets. *Icarus*, **54**, 377–87.
Fernández, J.A. and Ip, W.-H. (1984) Some dynamical aspects of the accretion of Uranus and Neptune: The exchange of angular momentum with planetesimals. *Icarus*, **58**, 109–20.
Greenberg, R., Weidenschilling, S.J., Chapman, C.R. and Davis, D.R. (1984) From icy planetesimals to outer planets and comets. *Icarus*, **59**, 87–113.
Heisler, J. and Tremaine, S. (1986) The influence of the galactic tidal field on the Oort comet cloud. *Icarus*, **65**, 13–26.
Heisler, J. (1990) Monte Carlo simulations of the Oort cometary cloud. *Icarus*, **88**, 104–21.
Hills, J.G. (1982) The formation of comets by radiation pressure in the outer protosun. *Astron. J.*, **87**, 906–10.
Hills, J.G. and Sandford, M.T., II (1983a) The formation of comets by radiation pressure in the outer protosun. II. Dependence on the radiation–grain coupling. *Astron. J.*, **88**, 1519–21.
Hills, J.G. and Sandford, M.T., II (1983b) The formation of comets by radiation pressure in the outer protosun. III. Dependence on the anisotropy of the radiation field. *Astron. J.*, **88**, 1522–30.
Hut, P. and Tremaine, S. (1985) Have interstellar clouds disrupted the Oort comet cloud? *Astron. J.*, **90**, 1548–57.
Jewitt, D.C. and Luu, J.X. (1992) *IAU Circular* 5611.
Joss, P.C. (1973) On the origin of short-period comets. *Astron. Astrophys.*, **25**, 271–3.
Kuiper, G.P. (1951) On the origin of the solar system in *Astrophysics*, (eds J.A. Hynek). New York: McGraw Hill, pp. 357–24.
Levison, H.F. and Duncan, M.J. (1990) A search for proto-comets in the outer regions of the solar system. *Astron. J.*, **100**, 1669–75.
Luu, J. and Jewitt, D. (1993) 1993 FW. *IAU Circular* 5730.
Lyttleton, R.A. (1948) On the origin of comets. *Mon. Not. Roy. Astron. Soc.*, **108**, 465–75.
Marochnik, L.S., Mukhin, L.M. and Sagdeev, R.Z. (1988) Estimates of mass and angular momentum in the Oort cloud. *Science*, **242**, 547–50.
Marsden, B.G. (1990) *Catalogue of Cometary Orbits*. Cambridge: Smithsonian Astrophysical Observatory, 108 pp.
Marsden, B.G. (1993) 1992 QB$_1$. *IAU Circular* 5855; 1993 FW. *IAU Circular* 5856.
McCrea, W.H. (1975) Solar system as space probe. *Observatory*, **95**, 239–55.
McGlynn, T.A. and Chapman, R.D. (1989) On the nondetection of extrasolar comets. *Astrophys. J.*, **346**, L105–8.
Newton, H.A. (1893). On the capture of comets by planets, especially their capture by Jupiter. *Mem. Natl. Acad. Sci.*, **6**, 7–23.
Oort, J.H. (1950) The structure of the cloud of comets surrounding the solar system and a hypothesis concerning its origin. *Bull. Astron. Inst. Neth.*, **11**, 91–110.
Safronov, V.S. (1972) *Evolution of the Protoplanetary Cloud and Formation of the Earth and Planets*, NASA TT-F-677, 206 pp. (Nauka Press, Moscow, 1969).
Sekanina, Z. (1976) A probability of encounter with interstellar comets and the likelihood of their existence. *Icarus*, **27**, 123–33.
Shoemaker, E.M., and Wolfe, R.F. (1984) Evolution of the Uranus–Neptune planetesimal swarm. *Lunar Planet Sci. Conf.* **15**, 780–1 (abstract).
Stagg, C.R., and Bailey, M.E. (1989) Stochastic capture of short-period comets. *Mon. Not. Roy. Astron. Soc.*, **241**, 507–41.
Stern, S.A. (1990) ISM induced erosion and gas dynamical drag in the Oort cloud. *Icarus*, **84**, 447–66.
Stern, S.A. (1991) On the number of planets in the outer solar system: Evidence of a substantial population of 1000-km bodies. *Icarus*, **90**, 271–81.
Stern, S.A., Shull, M.J. and Brandt, J.C. (1990) The evolution and detectability of comet clouds during post main sequence stellar evolution. *Nature* **345**, 305–8.
Stern, S.A., Stocke, J. and Weissman, P.R. (1991) An IRAS search for extra-solar Oort clouds. *Icarus*, **91**, 65–75.
Torbett, M.V. (1989) Chaotic motion in a comet disk beyond Neptune: the delivery of short-period comets. *Astron. J.*, **98**, 1477–81.
Valtonen, M.J. (1983) On the capture of comets into the inner solar system. *Observatory*, **103**, 1–4.
Valtonen, M.J. and Innanen, K.A. (1982) The capture of interstellar comets. *Astrophys. J.*, **255**, 307–15.
van Woerkom, A.F.F. (1948) On the origin of comets. *Bull. Astron. Inst. Neth.*, **10**, 445–72.
Vsekhsvyatskii, S.K. (1967) *The Nature and Origin of Comets and Meteors*. Moscow: Prosveschcheniye Press.
Weissman, P.R. (1979) Physical and dynamical evolution of long-period comets, in *Dynamics of the Solar System* (ed. R.L. Duncombe). Dordrecht: D. Reidel, pp. 277–82.
Weissman, P.R. (1982) Dynamical history of the Oort cloud, in *Comets* (ed. L.L. Wilkening). Tucson: University of Arizona Press, pp. 637–58.
Weissman, P.R. (1985a) Dynamical evolution of the Oort cloud, in *Dynamics of Comets: Their Origin and Evolution*, (eds A. Carusi and G.B. Valsecchi). Dordrecht: D. Reidel, pp. 87–96.
Weissman, P.R. (1985b) The origin of comets: implications for planetary formation, in *Protostars and Planets II* (ed. D.C. Black). Tucson: University of Arizona Press, pp. 895–919.
Weissman, P.R. (1990) The Oort cloud. *Nature*, **344**, 825–30.
Weissman, P.R. (1991a) The cometary impactor flux at the Earth, in *Global Catastrophes in Earth History* (eds V. Sharpton and P. Ward). Geol. Soc. Am. Special Paper, 247, pp. 171–80.
Weissman, P.R. (1991b) The angular momentum of the Oort cloud. *Icarus*, **89**, 190–3.
Whipple, F.L. (1964) Evidence for a comet belt beyond Neptune. *Proc. Natl Acad. Sci. US*, **51**, 711–18.
Whipple, F.L., and Lecar, M. (1976) Comet formation induced by solar wind, in *The Study of Comets*, NASA SP-393, pp. 600–62.
Yeomans, D.K. (1986) Physical interpretations from the motions of comets Halley and Giacobini-Zinner. In *20th ESLAB Symposium on the Exploration of Halley's Comet* (eds B. Battrick, E. J. Rolfe and R. Reinhard). ESA SP-250, **2**, 419–425.

Cross references

Kuiper belt
Oort, Jan Hendrik, and Oort cloud
Planetesimal

COMET: STRUCTURE AND COMPOSITION

Knowledge of the physical structure and chemical composition of comets is crucial to our understanding of the origin of comets and the solar system. Our current perception of comets has been derived from images of the nucleus of comet P/Halley, *in situ* measurements of the near-nucleus environment of three comets (P/Giacobini-Zinner, P/Halley, and P/Griggs-Skjellerup), and interpretation of comet coma (atmosphere) and tail observations obtained from remote sensing.

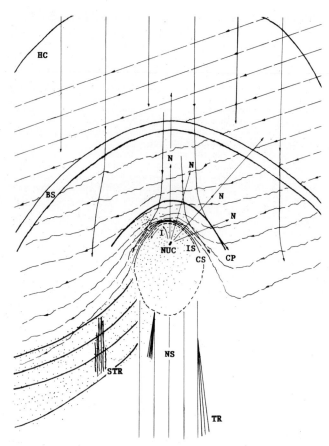

Figure C52 A schematic illustration of the components of a comet (not to scale). Solid lines extending from the top of the figure are solar wind particles that are diverted to flow around the comet. Long-dashed lines running from right to left represent interplanetary magnetic field lines that drape around the comet. NUC – nucleus, IS – inner shock, I – cometary ion, N – cometary neutral, CS – contact surface, CP – cometopause, BS – bow shock/transition region, NS – neutral sheet, TR – plasma tail ray, STR – dust tail striae, HC – hydrogen halo. The sun is located above the top of the figure, the plasma tail points to the bottom, and the dust tail extends to the lower left with exaggerated curvature. Between the bow shock and cometopause is an interaction region called the cometosheath. In the vicinity of the cometopause and in front of the contact surface, the mass-loaded solar wind piles up.

Physical structure

A comet consists of five distinct parts: a nucleus, a coma (constantly escaping atmosphere), an extended hydrogen halo, a plasma (type I) tail, and a dust (type II) tail. Associated with the dust is a trail of large dust particles distributed in the comet's orbit. Each component is discussed separately below (Figure C52).

Nucleus

The comet nucleus is a solid body, the source of activity and all the other features that constitute the phenomenon that is called a comet. The nucleus is composed of volatile ices (up to 90% water ice) and organic- and silicate-based dust. Nuclei are typically irregular in shape, and are best described dimensionally by a triaxial ellipsoid. Their sizes range from on the order of 10 km up to perhaps a few hundred km in the case of 2060 Chiron (q.v.), originally thought to be an asteroid upon discovery but which later displayed signs of cometary activity. In the case of the only directly measured nucleus (imaged by the Halley multicolour camera on board the Giotto spacecraft and the cameras on the Vega 1 and 2 spacecraft), the solid body of comet P/Halley (P/indicates a short-period comet with period less than 200 years) was found to be approximately $16 \times 8 \times 7$ km.

Other determinations of size are done indirectly, for instance by measuring the amount of sunlight reflected from the nucleus at large heliocentric distances where the light from the coma does not blanket that reflected from the nucleus, in conjunction with reflecting properties (albedo) of the nucleus that are assumed to be known. For comet P/Halley the albedo of the surface was directly measured and found to be only 3 to 4%. Its surface is among the blackest in the solar system. This low albedo is probably related to the composition of the surface but is also indicative of a highly fluffy surface structure, in which incident light can easily become trapped in pores. The fluffy nature of the surface suggests a similar state of the interior. The density of comet nuclei is estimated to lie between 100 and 1000 kg m^{-3}, therefore much of the solid body may actually consist of voids. This, in turn, implies a low packing fraction of material and the relatively fragile nature of the nucleus. Indeed, many observations of nuclei splitting or fragmenting have been recorded historically, particularly when the comet is near perihelion where solar tidal forces are greatest. All of these observations confirm the notion of the fluffy, icy conglomerate nucleus, an idea first proposed in 1950 by the American astronomer Fred L. Whipple.

The images of the nucleus of P/Halley clearly show that only about 10 to 20% of its surface has active, dust-producing areas. The remainder is covered by inactive, insulating regolith or crust. Two possible causes of regolith formation on the comet's surface are cosmic ray-induced polymerization of organic dust while the comet resides in the Oort cloud or in the Kuiper belt, and accumulation and sintering of coarse-grained dust particles during passages through the inner solar system. Since the dust must be entrained by gas to be transported into the coma, it is evident that the dust-producing areas must also produce gas. The combined gas production from these active areas is high enough to produce the observed gaseous coma of the comet (Keller et al., 1987). The localized and restricted activity on the surface of P/Halley's nucleus could generate the impression that the comet is about to become extinct. Once only a small fraction of activity is left, the relative decline from one orbit to the next might make itself clearly visible. The persistency and constancy of its recorded apparitions contradict this apparently obvious conclusion. As a consequence of non-uniform outgassing of the rotating nucleus, nongravitational forces arise that can slowly increase or decrease the comet's orbital period. However, from the constancy or the very small decline of the nongravitational forces, one can infer that a restricted level of activity is the normal mode and that it has persisted for a long time in the life of P/Halley. The level of activity of P/Halley is very high compared to other short-period comets. Therefore, it can be assumed that this restricted activity is typical for comets. The level of gas and dust release of most short-period comets is less than 10% of that of P/Halley and often in the range of about 1%. The sizes of these nuclei, such as those determined for comets IRAS-Araki-Alcock, Neujmin 1 and Arend-Rigaux, do not seem smaller than that of P/Halley. Therefore, their active areas must be even smaller in proportion. One may conclude that for all short-period comets only minor parts of the surface show activity. Comet P/Halley is not an exception but a typical example in this respect. The active regions provide most of our knowledge of the interior structure and composition of comet nuclei. They are the windows through which we gain access to the physical and chemical state of the interior.

The complete disintegration of a comet, observed several times for long-period comets, can be considered an extreme case of splitting. For example, comet Biela vanished after two apparitions with a dual nucleus, and only spectacular meteor showers were observed during the following anticipated returns in 1872 and 1885. Such a dissipation of a nucleus, and particularly of two nuclei simultaneously, suggests an irregular, loosely bound, low-density body of low tensile strength. Whether enhanced activity is the initial cause for the disruption is not known but tidal forces can play an important role. Comet West was observed to break apart during a close approach with the Sun in 1976 and comet P/Shoemaker-Levy 9 fragmented into at least 20 pieces as it passed close to Jupiter in 1992. These fragments impacted Jupiter in mid-1994. The large range in deduced cohesive strength, the active and inactive regions of the nucleus surface and the temporal variability of dust and gas release, as well as the diversity of comet appearance and behavior, indicates a very heterogeneous nucleus.

From coma observations of continuum flux and molecular bands, the mass ratio χ of dust to gas released from a comet nucleus can be estimated. Since the largest and heaviest particles cannot be entrained by the escaping gas, these particles remain on the nucleus. Thus χ is a lower limit of the dust-to-gas mass ratio in the nucleus and has been observed to lie in the range from nearly 0 (dust poor) to about 2 (dust rich). The distinction between dust-rich and dust-poor comets may not be clear cut. Dust-poor comets may not be truly poor in dust, but rather may have a large abundance of very fine (submicrometer) dust that is not observed because of the inefficiency of small particles in scattering visible light. This shifts the question from dust-rich versus dust-poor comets to large versus small particle size abundances in comets. Futhermore, large particles may consist of small particles sintered together that fragment upon release from the nucleus.

Comets are observed to vary in brightness at distances far from the Sun where light obstruction of the nucleus by the coma is a minimum. The observed variation of the light intensity is interpreted as evidence for the rotation of an aspherical cometary nucleus and was confirmed by analysis of spacecraft data from the P/Halley encounters. The rotation of the asymmetrical nucleus of comet P/Halley is complex and not completely understood. The most plausible explanation is based on a combination of the images returned from the Vega and Giotto spacecraft and remote observations of the coma 'lightcurve'. It supposes that the long axis of the nucleus is inclined to the angular momentum vector by approximately 66°. In this description the nucleus wobbles in a manner that causes the long axis to rotate around the angular momentum vector with a period of 3.7 days and with an oscillation period about the long axis of 7.1 days.

Coma

Upon heating, the frozen volatiles in the nucleus undergo a phase transition from solid to gas, bypassing the liquid state (sublimation), and the resulting gases flow away from the nucleus primarily on the sunlit side from discrete active regions. Because of pressure equilibration, the gases rapidly disperse into a nearly spherical flow to form the coma. The combination of the coma and nucleus is referred to as the head of the comet. The coma is roughly spherical in shape and can extend for several hundred thousand kilometers, where solar radiation pressure effects become significant and alter the spherical symmetry. One source of light from the coma arises from fluorescence of the gaseous molecules, C_2 fluorescence being the dominant source of visible light (Swan bands) followed by CN and C_3 fluorescence, even though they are all relatively minor constituents. The other component is reflected sunlight from the dust particles in the coma. In the visible continuum it displays the spectral features of the Sun. Additional emission features in the infrared spectrum indicate the temperature and composition of the dust.

Departures from spherical symmetry are caused by nucleus activity, distributed sources of gas molecules from dust 'jets' in the coma, solar wind effects on the ions, and radiation pressure from sunlight. Dust is a distributed source for some coma species (CO, H_2CO and probably CN, C_2 and others) and may provide clues to the organic inventory of the nucleus. At large distances from the nucleus, gaseous species progressively separate into individual coma components, as lighter molecules collisionally decouple from the neutral flow. As a result, the thermal evolution of the neutral species, the production and energetics of the ions and free electrons, and the dynamics of the nonthermal (fast) H and H_2 become increasingly important. The fast atomic hydrogen component leads to the extended hydrogen halo predicted by Ludwig Biermann (1963).

Jet-like dust features have been detected in many comets, but jet-like structures in the coma gas were first identified in P/Halley (Millis and Schleicher 1986; A'Hearn et al. 1986). The origin of these structures is not fully understood. They do not appear to be correlated with the observed dust features. However, very small dust grains had been detected by in situ measurements in P/Halley in large abundances. It is possible that the gas structures are related to dust 'jets' of organic (CHON) particles that are so small that they do not scatter sunlight but large enough that surface winds do not significantly accelerate them in a transverse direction. These CHON particles may disintegrate and release species that are the source of the observed 'jets' of CN, C_2 and others.

Closely associated with the dust are the extended sources of gas production in the coma. It was found that only about one-third of the CO that has been measured by ground-based instruments is released directly from the nucleus. Two-thirds are released slowly over a distance of about 10^4 km from a distributed source in the coma. This conclusion comes from the work of Eberhardt et al. (1987). It is suspected that the CO comes from the disintegration of the dust; however, the parent species is not known at present.

Hydrogen coma

Surrounding the comet at its largest scale is a halo of atomic hydrogen. It was first discovered from observations using the ultraviolet camera on board the OAO-2 satellite in 1970 by Arthur Code and collaborators. However, the hydrogen coma was first predicted theoretically 2 years prior to this by Ludwig Biermann, a German astrophysicist. The hydrogen coma is much larger than the normal coma surrounding the comet. It can be larger than 10^7 km (or about ten times the size of the Sun). Further observations have shown the existence of a somewhat smaller cloud made up of OH (hydroxyl) molecules. The two clouds originate from the dissociation of water, the primary volatile component of the nucleus. Ultraviolet photons from the Sun break apart the cometary water molecules on timescales of about 10^5 s (for quiet Sun conditions at 1 AU), yielding the major dissociation fragments, H and OH. These fragments carry off the excess energy of the photodissociation reaction (apportioned according to momentum conservation), which propels them to the observed large distances before they undergo further photolytic reactions or charge exchange reactions with the solar wind.

An H_2 coma, also produced by the photodissociation of water, should lie between the region of the conventional, heavier coma molecules such as OH and the atomic hydrogen coma. However, since H_2 is difficult to detect, it has not been identified as yet.

Radiation pressure acts on atomic and molecular species and dust particles differently, depending on their spectral properties in the visible. It produces accelerations of the order of one to 200 times that of solar gravity on light atoms or on small dust particles. In general, the force due to radiation pressure depends on the mass of the particle, the photon flux from the Sun, the momentum of a photon, the cross-section of the particle and the heliocentric distance of the comet. Molecules that have strong absorption lines, i.e. large oscillator strengths in the visible spectrum, are accelerated greatest in the antisolar distance. The acceleration is often expressed in terms of the fluorescence rate or g-factor.

For atomic hydrogen the only important transition is the resonance transition at Lyman α (1216 Å). In this case the acceleration is about 0.3 cm s^{-2}. With this acceleration acting over 10^6 s, the lifetime for photoionization and charge exchange of hydrogen at 1 AU, the velocity due to radiation pressure is about 3 km s^{-1} resulting in a tailward displacement of about 1.5×10^6 km.

These effects are particularly important for understanding the hydrogen Lyman α emission. The hydrogen atoms liberated in the reaction, $H_2O + h\nu \rightarrow H + OH$, have an average excess energy of about 3.4 eV, or a velocity of about 25 km s^{-1} if all the energy is transformed into kinetic energy. However, in the dissociation process the OH radical is left in excited states of rotation and vibration as was shown by Crovisier (1989). Since the photolytic rate coefficient for this process is about 10^{-5} s^{-1} for the quiet Sun at 1 AU heliocentric distance, some fraction of the hydrogen atoms are liberated inside the collisional sphere. However, only a small fraction of these are thermalized because of the small mass of the hydrogen atoms relative to the average mass of the molecules making up the bulk gas. Most of the hydrogen atoms are released outside the collision-dominated zone and appear as a population of fast atoms. In this region where collisions are rare (exosphere), the hydrogen atoms are ionized both by solar photons and by charge exchange reactions with solar wind protons. The expansion velocity and the typical lifetime of 10^6 s explains why the hydrogen coma should be of the order of 10^7 km in size, as originally pointed out by Biermann (1968).

There is a separate population of hydrogen atoms released in the reaction, $OH + h\nu \rightarrow O + H$, and these have an excess energy of about 1.3 eV, corresponding to an average velocity of 15 km s^{-1}. Since much of the OH is produced outside the collisional radius and since its mean lifetime at 1 AU is 1.5×10^4 s (but varying with solar activity and with the heliocentric radial velocity of the radical; van Dishoeck and Dalgarno, 1984), only a very small fraction of these atoms will be thermalized and both populations of hydrogen atoms will be influenced by radiation pressure.

Plasma tail and interaction with solar wind

As a comet approaches the Sun the molecules sublimate (evaporate) from the nucleus, primarily due to the energy carried by the visible sunlight, and dissociate and ionize, primarily by ultraviolet (UV) photoabsorption. The resulting flow of the cometary ions interacts with the solar wind. The explanation for the formation of the plasma tail was first given by Biermann in 1951 as a hypothesis for the existence of the solar wind. The plasma tail steadily grows as the comet approaches the Sun and shrinks as the comet recedes from the Sun, always pointing in the antisolar direction (like a wind sock indicating the wind direction), undergoing temporal changes such as condensations (enhancements in the plasma density) and disconnection events. The cometary plasma and neutral gas are coupled to the solar wind via collisions and the interplanetary magnetic field (IMF), for example by charge exchange, photoionization, dissociative recombination and ion–neutral friction. This interaction dramatically affects the plasma flow, resulting in the formation of the plasma tail and discontinuities such as the bow shock and the contact surface. The Lorentz force, caused by a charged particle moving in the interplanetary magnetic field, acts on the plasma to introduce additional asymmetries. 'Frictional' forces caused by the counter- and cross-streaming of the ions with respect to the neutrals further influence the plasma flow, especially inside the contact surface.

The plasma tail can be understood in terms of a model proposed by Hannes Alfvén (q.v.), a Swedish astrophysicist, in 1957. Molecules ionized in the cometary gas by solar irradiation and by charge exchange with the solar wind form an obstacle to the solar wind flow. This results in the quasi-stable capture of the interplanetary magnetic field. The motion of the solar wind drapes the magnetic field lines around the comet to form an elongated plasma structure, resulting in a magnetic field with two lobes of opposite polarity on opposing sides of a plane perpendicular to the original magnetic field through the tail. In the region where the two lobes approach one another along the plane, an abrupt change in magnetic polarity occurs. To maintain this structure a current sheet forms between the two magnetic lobes. As its name implies, the current sheet is a flat, planar distribution of plasma flowing across the tail axis of the comet.

The plasma tail is made visible by emissions from cometary ions, primarily CO^+, N_2^+ and H_2O^+, that are restricted to cyclotron motion around the draped magnetic field lines. Occasionally, a portion of the plasma tail appears to separate. The cause of this so-called disconnection event is controversial, but it is possible that it occurs when a rapid change in the direction of the IMF, such as a sector boundary, is encountered by the comet, severing and reconnecting the magnetic field lines that are draped around the comet. During such an event the portion of the tail that has disconnected is swept away from the comet and a new tail may form. Other forms of bulk motion are knots or condensations carrying cometary material down the plasma tail. Such plasma density fluctuations may be caused by outburst activity of the comet. Also suggested are magnetohydrodynamic (MHD) waves giving rise to density enhancements that move downstream in the plasma tail. Future observations may make it possible to distinguish between these wave and bulk motions by searching for Doppler shifts in the spectra of the cometary features.

The *in situ* measurements obtained by the spacecraft encounters with comet P/Halley confirmed Alfvén's theory and other hypotheses concerning large-scale plasma structures and phenomena. These observations showed evidence for a bow shock, the draped IMF configuration, a magnetic pile-up region where the field strength is enhanced over the normal solar wind value, a region of plasma stagnation, and a central magnetic field-free cavity. The bow shock was found to be at about 4.5×10^5 km from the nucleus, essentially the same distance as expected from theory. At times, the evidence of the shock nature was not clear and it appeared more like a bow wave similar to that detected at the much weaker comet P/Giacobini-Zinner. The magnetic pile-up region was measured at distances within 1.35×10^5 km from the nucleus on the inbound trajectory of the Giotto spacecraft (at a heliocentric distance of 0.89 AU) and at distances within 2.63×10^5 km on the outbound journey. The difference is due to the asymmetric orientation of the IMF in the solar wind and the flyby geometry. The field strength reached values of 57 and 65 nT in the respective regions. Normal solar wind values are 5–10 nT. The field-free cavity had a diameter of about 8500 km at the time of the Giotto encounter; this was the only spacecraft that came close enough to penetrate this feature. The cavity is formed by the pressure balance between the outflowing cometary gas on the inside and the mass-loaded solar wind with embedded magnetic field on the outside, taking into account the ion–neutral drag. The boundary between these two regions is called the contact surface, separating the pure cometary plasma near the nucleus from the mass-loaded solar wind plasma at greater distances. Mass loading occurs when the solar wind plasma, consisting mostly of electrons, protons and helium ions, picks up slow moving, heavier cometary ions consisting mostly of oxygen, carbon, nitrogen, water, OH and CO.

The proton population in the vicinity of the bow shock contains a significant fraction of H^+ ions from ionization of the coma gas. Their presence can be recognized by the phase space density distributions; the ion mass spectrometer (IMS) experiment on Giotto, for example, detected the shell-like structure of H^+ ions of cometary origin at about 8×10^6 km upstream of the bow shock.

The distribution of the neutral species in the coma is described well by global models out to cometocentric distances of a few hundred thousand kilometers, where the effects of radiation pressure need to be taken into consideration in multidimensional models. The effects of energetic ions and neutrals on the chemistry in the inner coma and sputtering on dust and the nucleus also need to be considered. Energetic ions were detected in the inner coma of P/Halley with energies up to 10 keV. This unexpected result may be explained as follows (Eviatar, Goldstein and Young, 1989). Neutral molecules with speeds of about 1 km s^{-1} move radially outward in the coma. Some of them are ionized by charge exchange with the solar wind. The neutralized solar wind ions (mostly hydrogen atoms) continue toward the inner coma, while at the same time the new, heavier ions are picked up by the solar wind, mass load it and slow it down. The energetic neutrals can now penetrate the contact surface without any major interaction by the ions or the piled-up magnetic field. Once they are inside the contact surface, the density of the coma gas increases, enhancing the probability of yet another charge exchange with comet ions that were produced by photoionization. Thus the energetic neutrals are converted to energetic ions that have been detected inside of the contact surface. The importance of this is twofold: there are not only energetic ions as measured by the Johnstone plasma analyser (JPA) instrument on board Giotto, but there must also be energetic neutrals. These energetic particles can undergo chemical reactions and can sputter on surfaces.

The radial expansion of cometary ionospheric plasma (inside the contact surface) has some similarities to the interaction of the solar wind with the interstellar medium in which the solar wind flow is thought to be terminated by the formation of a heliopause. In other words a contact discontinuity, a surface dividing plasma from the mass-loaded solar wind on the outside and pure cometary ions on the inside, should be formed between a pair of shocks (an inner and an outer shock) such that the supersonic flow of the ionospheric plasma can be diverted into the lateral direction near the boundary or slowed to such a degree that dissociative recombination neutralizes much of it. The best signature of the contact discontinuity comes from magnetometer measurements in which the magnetic field strength was observed to drop from a value of about 20 nT to zero over a distance of 25 km during the Giotto encounter with P/Halley. The concept of a magnetic field-free cavity was not a new idea in studies of solar wind–neutral atmosphere interaction, but the main issue having to do with the stability of the interface remains unresolved.

Dust tail

The mass of ejected dust can exceed several tons per second for an active dusty comet such as P/Halley. The dust, which is entrained by the outflowing gas in the inner regions of the coma, is subjected to repulsive, radiation pressure forces directed away from the Sun in the outer regions of the coma. In addition the dust is influenced by the gravitational attraction of the Sun. The result is the formation of a dust tail. According to Friedrich Bessel, a 19th-century German mathematician, the shape of the tail depends on the strength of the repulsion and the nature of the dust particles, such as particle size, shape and composition. Generally, large grains are less affected by the repulsive force than small ones. Unknown to Bessel, the repulsive force was found to be due to the pressure of solar photons (the momentum transfer of photons as they are scattered by a dust particle), although electrostatic forces can also influence the spatial distribution of charged dust particles through the interaction with the solar wind. Visible dust tails can stretch to distances exceeding 10^7 km, generally exhibiting some degree of curvature relative to the plasma tail. Many times structure can be seen within the dust tail, including streamers, sunward spikes and striae; these are caused by anisotropic dust emission from the nucleus (e.g. outbursts or the splitting of the nucleus), viewing geometry, electromagnetic forces and fragmentation of the dust particles.

The most readily available information on comet dust stems from ground-based optical observations of the dust coma and tail. A large number of comets have been observed in this way. Since it is light scattered by the dust particles that is observed, these data relate to particles with maximum scattering cross sections, i.e. 1 to 10 μm particles for visible light. For infrared (IR) observations the situation is somewhat improved with respect to larger particles. In addition, thermal emission depends not on detailed knowledge of the optical properties of the materials involved but only on the gross albedo. However, the mass distribution of cometary dust particles was best determined by the *in situ* detectors on board of the spacecraft. The statistically most significant data are the mass fluence measurements because they include all measurements along the trajectory. This averaging of the particle flux along the trajectory smooths out dynamical effects on the mass distribution in the distant parts of the trajectory as well as the dispersive effect of nucleus rotation near closest approach; however, information on particle fragmentation is lost. The spacecraft measurements are fairly consistent in slope, but they vary in absolute intensity by a factor of ten. This is caused by observation times that differ by several days (Vega 1, 6 March; Vega 2, 9 March; and Giotto, 14 March). During this time the total comet brightness varied proportionally. This demonstrates the strong variability of the dust emission by P/Halley.

Cometary dust grains of masses from about 10^{-20} kg to 10^{-5} kg were observed by the *in situ* detectors. The number of particles larger than the detector limiting mass steadily increased with decreasing limiting mass of various detectors. Remote, ground-based observations, on which the pre-encounter models were based, did not allow for the detection of particles smaller than about 10^{-16} kg (or a few tenths of a micrometer in size). Tiny particles with mass 3×10^{-20} kg, or just larger, were detected by the *in situ* detectors in large quantities. These particles are comparable to interstellar dust in size.

Clustering in the frequency of spacecraft detections of small dust grains in the outer coma cannot be explained by purely statistical fluctuations of the impact rates. It is interpreted as a result of fragmentation of larger particles during their passage through the coma. Observed changes in the size spectrum of dust particles are also explained in terms of fragmentation effects. Fragmentation of larger particles or grain clusters is a natural consequence if differential sublimation of grain material occurs. Another possible cause for fragmentation can be the charging of grains and their consequent electrostatic disruption.

A feature related to the dust release from the nucleus was discovered using observations from the Infrared Astronomical Satellite (IRAS) in the mid-1980s. Infrared images of the sky at 25 and 60 μm revealed narrow trails of dust that were found to coincide with several comet orbits. About 100 dust trails have been found. A few are not associated with any known comet. The narrowness of the trails indicates that the particles are large and were ejected from the nucleus at relatively low speeds (a few meters per second). Analysis of these images indicates further that they are the product of many emission events spread over many orbits. The age of the trails has been estimated to be on the order of hundreds of years. Dust trails are believed to be the source of meteor streams, reinforcing the long-known connection between comets and meteor streams. In all, over a dozen comets are associated with annual meteor streams. This relationship is important to the study of comets for several reasons. The nature of the meteoritic material yields information about the composition of cometary dust. Also, the orbital evolution of the meteor stream, which can be studied accurately, aids in finding the age of the associated comet in its present orbit. Lastly, a lower limit for the mass of a comet can be estimated from the mass contained in the associated meteor stream.

Chemical composition

Our knowledge of the chemical composition of comets is based on Earth-based observations and *in situ* measurements. Ground-based, airborne, rocketborne and Earth-orbiting satellite observations yield global averages of coma properties, while spacecraft measurements yield detailed data on local, instantaneous conditions. Physical and chemical processes in the coma are intimately related to each other and to the physical structure and chemical composition of the nucleus.

History

The suggestion in the 1930s by Karl Wurm (1943) that comets contain chemical compounds such as H_2O, NH_3, C_2N_2, CH_4, N_2 and CO_2 opened the possibility that chemical reactions occur in cometary comas. An advance in interpreting comet spectra was made in 1941 when Polydor Swings accounted for the Doppler shift of solar absorption lines caused by the comet's orbital motion relative to the Sun that determines the strength of cometary emission features. This phenomenon is now called the Swings effect. (In a similar manner in 1958, Jesse Greenstein corrected for the gas flow relative to the nucleus, now known as the Greenstein effect.) To explain the presence of CO and radicals such as CH, NH and CN in the coma, Delsemme and Swings in 1952 proposed that the nucleus contains solid hydrates of CH_4, CO_2 and others. The vapor pressure of pure CH_4 ice is several orders of magnitude larger than that of H_2O, so CH_4 would be released at much larger heliocentric distances than H_2O. However, these hydrates (clathrates) have approximately equal vapor pressures and so would be released into the coma at about the same rate. In the late 1950s Donn and Urey (1956, 1957) proposed that exothermic chemical reactions could explain observed brightness outbursts in comets. To obtain explosive energy release, they suggested a modification in the chemical composition: a mixture with stable molecules as well as reactive species such as free radicals and unstable molecules.

The great impetus for theoretical consideration of chemical reactions in the coma did nor occur until the number densities in the coma were established to be sufficiently high. In 1964 Biermann and Trefftz were the first to indicate this in their analysis of the observed emission lines from $O(^1D)$. When it became clear in the 1960s that the observed species, CN, C_2, C_3, NH, NH_2 and CO^+, were not the only constituents in the coma, but that water, which is very difficult to detect, must be present, the gas density could be estimated. These densities were confirmed by Huebner in 1965 and later by Malaise in 1970 by comparing a model of the CN spectrum in which not only pure fluorescence was considered but also the effects of collisions in exciting various rotational lines. The estimates indicated gas densities sufficiently high that the inner coma was dominated by collisions. This meant that chemical reactions could be expected between reactive species such as the above radicals and ions.

With the discovery that detected species in interstellar clouds can be explained by chemical reactions involving ions, it became clear that analogous reactions could occur in the comas of comets. In 1974 Aikin first applied ion gas-phase reactions to a water-dominated comet. He predicted a significant production of H_3O^+, making it the most abundant ion to distances of about 5000 km in the coma. He also pointed out the importance of the proton transfer reaction of H_3O^+ with ammonia (a special case of positive ion–atom interchange reactions, see Table C8). Abundant amounts of ammonia could shift the dominant ion in the innermost coma from H_3O^+ to NH_4^+.

The work of Oppenheimer in the mid-1970s stimulated large-scale investigations using chemical reaction networks of many species. He assumed complete steady state for the chemistry in the coma and chemical reactions fast enough that changes of species concentration (dilution of the gas as it streams outward in the coma) have no effect. The first assumption implies a negligible change in intercepted solar flux by the radial motion of the comet with respect to the Sun while gas flows from the nucleus to the boundary of the collision zone, and is justifiable as long as $r/v_r > 3 \times 10^4$ s, where r is the heliocentric distance of the comet and v_r is radial velocity component. But the second assumption is not satisfactory for all molecular species in the coma; the concentration of the reactants may be reduced faster than local chemical steady state can be established. Therefore, modern investigations treat the chemistry in a time-dependent manner.

Once a volume of gas has expanded and moved beyond the collision-dominated zone, free molecular flow applies. Free molecular flow may be also important on the nightside of comets, especially if the surface wind from the dayside is very weak. At large heliocentric distances a region devoid of particles will develop in the coma coincident with the light shadow.

Gas-phase chemistry in the coma

Coma chemistry involves the region from the bow shock to the nucleus–coma interface and the formation of the plasma tail. Surrounding the nucleus is a collision-dominated inner region with approximate spherical symmetry. Parent molecules leaving the surface are quickly transformed into a plethora of highly reactive radicals and ions by the attenuated solar UV flux and a variety of gas-phase reactions. At greater distances from the nucleus, species progressively separate into a multifluid ensemble. As a result the thermal evolution of the neutral species, the production and energetics of the ions and electrons, and the dynamics of non-thermal

Table C8 Gas-phase chemical reactions with examples

Photodissociation	$h\nu + H_2O \rightarrow H + OH$
Photoionization	$h\nu + CO \rightarrow CO^+ + e$
Photodissociative ionization	$h\nu + CO_2 \rightarrow O + CO^+ + e$
Electron impact dissociation	$e + N_2 \rightarrow N + N + e$
Electron impact ionization	$e + CO \rightarrow CO^+ + e + e$
Electron impact dissociative ionization	$e + CO_2 \rightarrow O + CO^+ + e + e$
Positive ion–atom interchange	$CO^+ + H_2O \rightarrow HCO^+ + OH$
Positive ion charge transfer	$CO^+ + H_2O \rightarrow H_2O^+ + CO$
Electron dissociative recombination	$C_2h^+ + e \rightarrow c_2 + H$
Three-body positive ion–neutral association	$C_2H_2^+ + M \rightarrow C_2H_4^+ + M$
Neutral rearrangement	$N + CH \rightarrow CN + H$
Three-body neutral recombination	$C_2H_2 + H_2 + M \rightarrow C_2H_3 + M$
Radiative electronic state de-excitation	$O(^1D) \rightarrow O(^3P) + h\nu$
Radiative recombination	$e + H^+ \rightarrow + h\nu$
Radiation stabilized positive ion–neutral association	$C^+ + H \rightarrow CH^+ + h\nu$
Radiation stabilized neutral recombination	$C + C \rightarrow C_2 + + h\nu$
Neutral–neutral associative ionization	$CH + O \rightarrow HCO^+ + e$
Neutral impact electronic state quenching	$O(^1D) + CO_2 \rightarrow O(^3P) + CO_2 + h\nu$
Electron impact electronic state excitation	$CO(^1\Sigma) + e \rightarrow CO(^1\Pi) + e$

atomic and molecular hydrogen must be considered. Dust is a distributed source for some coma species and may provide additional clues to the organic inventory of the nucleus.

Table C8 shows the chemical reactions for the inner coma. The reactions are listed approximately in decreasing order of importance. The coma is optically thin in the visible range of the spectrum so that solar radiation can freely penetrate to the nucleus, warming it and sublimating the icy component. However, in the UV range of the spectrum, molecular continuum cross-sections for dissociation and ionization make the coma optically thick for active comets at heliocentric distances of less than about 1.5 AU. For this reason, dissociation and ionization cannot occur near the surface of the nucleus. The gas must expand to about 100 km before radicals and ions are formed in sufficient quantities to initiate chemical reactions. At larger heliocentric distances the gas production is sufficiently low to decrease the attenuation of solar UV so that dissociation and ionization occur throughout the coma to the nucleus.

Cometary neutral molecules stream approximately radially outward from the nucleus. Deviation from exact radial motion is caused by asymmetric outgassing of the nucleus, collisions with ions (the ion–neutral drag), and radiation pressure with increasing cometocentric distance. The neutral species are progressively more ionized. Inside the contact surface, photoionization, photodissociative ionization and collisional ionization by energetic photoelectrons are the dominant ionization processes. Dissociative recombination is the major destruction mechanism of ions; other recombination processes are not very important. Ions can charge exchange or undergo chemical reactions, but this does not change the degree of ionization. Removal of ions into the plasma tail is also not very efficient in this region. Outside of the contact surface, photoionization is the most important ionization process until charge exchange and electron collisional ionization with the solar wind protons and electrons begin to compete effectively in the outermost regions of the coma.

Just as dissociation increases with distance from the nucleus, so does ionization. In the inner coma, ions are never more abundant than a few percent relative to the neutrals. However, even there, ions play an important role since they cannot cross the contact surface. Ions born inside the contact surface – primarily by photoionization and secondary photoelectron collisional ionization – respond to the solar wind pressure from the outside, are slowed down and increasingly recombine dissociatively with electrons. Since atomic ions can only recombine radiatively, by three-body recombination or by dielectronic recombination (all of these cross-sections are smaller than those of dissociative recombination of molecular ions), they tend to be deflected, together with a few molecular ions, into the tail. Outside the contact surface, ions are created by photoionization and charge exchange with the solar wind protons. Since these ions (primarily H_2O^+) are much heavier than the solar wind ions (mostly H^+), they mass load the solar wind and decelerate it. As they approach the contact surface from the outside, they are slowed down, pile up and recombine or are deflected into the plasma tail. The tail plasma consists primarily of ions that originate from outside the contact surface.

Ions are easily detected by mass spectrometers and they play a dominant role in the chemistry through ion–neutral reactions. Figure C53 illustrates creation and destruction of some of the most important ions. It should be noted that H_3O^+, NH_4^+ and CH_2OH^+ cannot be created by photoionization of mother molecules in the nucleus; they are protonated species of H_2O, NH_3 and H_2CO. Heavy arrows indicate the major production and destruction paths. Solid lines indicate paths of intermediate strength, with rate coefficients about a factor of ten lower than the major reactions. Dashed lines indicate reactions with a further reduction by a factor of ten. These reaction paths are typical for comets at a heliocentric distance of 1 AU and a cometocentric distance of approximately 1600 km in the coma. The relative importance of these reactions changes with heliocentric and with cometocentric distances (Huebner et al., 1991). A detailed reaction network can be found in Schmidt et al. (1988).

Interesting cometary ions deduced from the Giotto ion mass spectrometer (IMS) data obtained during the encounter with P/Halley have been found in mass channels 39, 45 and 47, corresponding to $C_3H^+_3$, HCS^+ and HCO^+_2, and H_3CS^+ and HNS^+, respectively. The cyclic ion of cyclopropenyl, $C_3H_3^+$, at $M/q = 39$ amu e^{-1} was identified by Korth et al. (1989) and they argued that cyclopropenyl must be released by the dust as a distributed source in the coma. Alternatively, $C_3H_3^+$ can be created in sufficient quantities from a small amount (0.1%) of the parent molecule allene, $H_2C_3H_2$, via the radiative association reaction $C_3H^+ + H_2 \rightarrow C_3H_3^+$ (Herbst, 1987).

Ion gas-phase reactions, including positive ion–atom interchange and positive ion charge transfer, are responsible for the creation of a variety of species in the inner coma. One must also consider radiative processes, such as electronic state de-excitation, including metastable states which contribute to the observed coma spectrum of forbidden emission lines. Protonation reactions and photolytic reactions for polymer condensates play an important role. Evidence from the Giotto spacecraft encounter with P/Halley supports the notion that organic dust grains are a distributed source of CO, H_2CO and probably other species. Other macromolecules, as distributed sources of coma species and new gas-phase species, such as negative ions, additional species in excited states, multiply ionized species, energetic particles and grain surface reactions, need to be considered in future studies.

Another chemical component of the coma is the electron plasma. In the inner coma the energy equipartition time for electron–electron and electron–molecule collisions is small compared to the dynamic time scale of expansion. In this region the fluid approximation for the electron gas is appropriate. At some point within the inner coma, electrons thermally decouple from the coma gas and must be treated as a separate component with unique temperature. Exterior to this region the energetic electrons interact weakly with the surrounding thermal electrons.

Table C9 lists chemical species that have been identified by their spectra at UV, visible, IR and radio wavelengths using ground-based

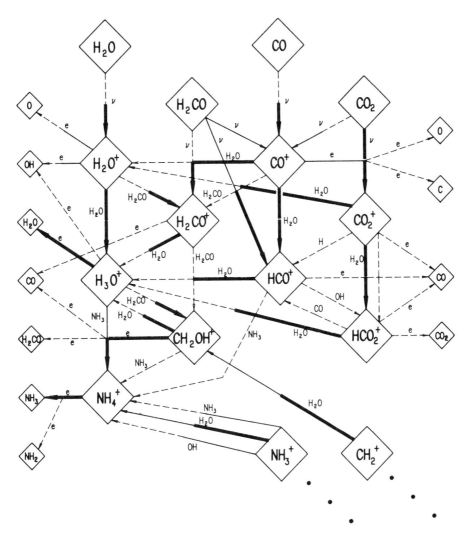

Figure C53 Major photolytic and gas-phase reactions involving the parent molecules, H_2O, H_2CO, CO and CO_2. Heavy arrows indicate the major production and destruction paths, solid lines represent paths of intermediate strength and dashed lines represent minor reaction paths. Symbols next to the arrows indicate the reactants, ν stands for photodissociation or photoionization, and e indicates electron dissociative recombination. The species H_3O^+, CH_2OH^+, HCO^+, HCO_2^+ and NH_4^+ are formed by protonation reactions. This reaction network is typical for comets at a heliocentric distance of about 1 AU and at cometocentric distances of approximately 2000 km in the coma.

Table C9 Chemical species identified in comets

Identification by microwave, infrared, visual, and ultraviolet spectra
H, C, O, S, C_2, $^{12}C^{13}C$, CH CN, ^{13}CN, CO, CS, NH, OH, S_2?, SH?, SO?, NO?, C_3, NH_2, H_2O, HCN, HCO, NH_3, H_2CO, NH_4, CH_3CN?, C^+, CO^+, CH^+, CN^+, N_2^+, CO_2^+, H_2O^+, H_2S^+

Identification by *in situ* **mass spectra**
H_2O, CO, N_2, C_2H_4, H_2CO, CO_2, H^+, C^+, CH^+, CH_2^+, N^+, CH_3^+, NH^+, O^+, CH_4^+, NH_2^+, OH^+, NH_3^+, CH_5^+, H_2O^+, NH_4^+, H_3O^+, H_3CO^+, C_3H^+, $C_3H_3^+$, H_3S^+

---

and remote observations, and in a few cases by *in situ* mass spectrometry during spacecraft flybys of P/Halley. Question marks in the table label those species that have been identified only once. In addition there are the metals, Na, K, Ca, V, Mn, Cr, Fe, Co, Ni and Cu, that are associated with sublimation (vaporization) of the dust detected mostly in Sun-grazing comets. Mass spectrometer identifications are somewhat problematic because of the overlap of masses from different species. A peak in mass channel 16 can be caused by O, CH_4 and NH_2, with lesser contributions from some isotopes. Similarly, mass channel 28 contains CO, N_2 and C_2H_4 as possible main constituents. There are similar problems with other mass channels. Although H_2O is so abundant that most of it must come from the nucleus, it is not known what fraction of it and the other species come from spatially distributed sources in the coma. These sources are most likely associated with dust. Protonated species, such as CH_5^+, NH_4^+ and H_3O^+, can only be produced by ion–molecule reactions.

Comets and the interstellar medium

Comparing the molecules in Table C9 with those identified in the interstellar medium, it is noted that all of the neutral cometary molecules (except CO_2, S_2 and the neutral cometary radicals NH_2 and NH) are also identified interstellar molecules. CO_2 and S_2 are symmetric species that have no dipole moment and therefore no pure rotational spectra by which most interstellar molecules are identified. On the other hand, S_2 does have a fine-structure transition, but a search for it in the interstellar medium turned out to be negative. From the presence of CO_2H^+ (protonated CO_2) in the interstellar medium, one can conclude that CO_2 is present. Furthermore, solid interstellar CO_2 has been identified by d'Hendecourt and Jourdain de Muizon (1989). Interstellar NH has been identified in the UV in diffuse clouds with background stars (Meyer and Roth, 1991). Because NH and NH_2 are light molecules, their rotational transitions are in the submillimeter region of the spectrum for which observational capabilities are being developed only now. This has led to the recent detection of NH_2 in the interstellar medium. Thus all molecules in comets, with the exception of S_2 that may be related to comet dust, appear to overlap with interstellar molecules. The molecule S_2 was detected by A'Hearn, Feldman and Schleicher (1983) in the coma of comet IRAS-Araki-Alcock. It has not been found in any other comet since then and its presence in comets is poorly understood.

In contrast to the neutral molecular species, there are many cometary radicals and ions that have not been identified in the interstellar medium. This is not surprising. Photolytic processes are the primary mechanisms for dissociation and ionization and the solar radiation field is very different from the interstellar radiation field.

The molecular similarities seem to indicate a close affinity between comets and interstellar clouds, but we must be cautious with such a conclusion as the relative abundances of species must be considered also. It is difficult to understand how comets might have formed at very low densities in the interstellar medium. On the other hand, interstellar molecules may have survived in the outer parts of the solar nebula where densities may have been high enough to form comets. Nevertheless, the interstellar molecules can be taken as a guide to identify new species from observed, but still unassigned, cometary spectral lines.

Composition of the nucleus

Assimilating the available comet observations, model results and the arguments presented above, Table C10 lists our best estimate for the icy composition of comet nuclei, given in percent number abundance. The data come from many sources and many, allegedly normal comets, although the *in situ* P/Halley measurements strongly influenced the relative abundances. As was suspected for many years, it is now clear that water is the dominant ice.

The development of infrared astronomy provided new insights into the composition of cometary dust. Spectral features around 10 and 18 μm led to the identification of silicates in the dust and emission around 3 μm is attributed to hyrocarbons. Additional clues concerning the dust composition were obtained from the spacecraft encounters with P/Halley. Particles were broadly classified into two types: CHON particles, rich in the elements C, H, O and N; and silicate particles, rich in Si, Mg and Fe. Most dust particles tended to be variable mixtures of these two types. The CHON-to-silicate ratio changed considerably during the flyby, possibly caused by dust ejected from different active areas on the nucleus. We know almost nothing about the ratio of organic dust (CHON particles) to silicate dust in comets other than P/Halley.

It must be also kept in mind that comet nuclei are inhomogeneous and that there may be compositional differences between comets. To be specific, all variations are associated with detections in the coma. Since the coma is the result of gas emission from an active area, which is only a small part of the surface of a comet nucleus, we must be very careful in the interpretation of these variations. Some claim that the variations are due to compositional differences of the nuclei, possibly related to the place and time of comet formation. On the other hand the differences may be only local, due to the inhomogeneous nature of a comet nucleus. The best-known variation is the mass ratio of dust-to-gas production. Some comets appear to be very dusty ($\chi > 1$), while others appear to be almost dust free ($\chi < 0.5$). Again, we must view such a simple interpretation with caution. A dust-free comet may actually contain a very large number of submicrometer-sized dust particles that are inefficient light scatterers, making them very hard to detect. However, such an argument only shifts the emphasis from dusty versus dust-free comets to comets with large versus small dust grains.

The ratio of CO to H_2O is also variable, based on UV observations of CO and observations of CO^+, the main plasma tail ion. Some comets have a very weak plasma tail or none at all, while others have a very dominant plasma tail. Finally, there is the variation of relative abundances of some minor species like CN and C_2. In most comets the ratio of production rates for C_2/CN varies from 1.2 to 1.5, but in some comets, like P/Giacobini-Zinner, this ratio is only about 0.3. This could be linked to the dust-to-gas ratio, which was also low for P/Giacobini-Zinner. Observations of the unusual comet Yanaka (1988r) at 0.91 AU heliocentric distance showed strong emissions of oxygen and NH_2, but no trace of either C_2 or CN, the two species that are among the strongest in most comets.

We should also note the coexistence of oxidized and reduced species in comets. Interstellar clouds are rich in CO and N_2. It is only natural to assume that the solar nebula was also rich in these species. On the other hand the regions of the solar nebula where the giant planets formed were probably rich in CH_4 and NH_3 because the temperatures and densities were high enough to convert CO to CH_4 and N_2 to NH_3. It is has been argued that the abundances of CH_4 and NH_3 may be very low in comets, but this conclusion is based primarily on data from only one comet, P/Halley. If the low CH_4 and NH_3 abundances can be confirmed in other comets, then comets must have formed outside of the subnebulae of the giant planets, probably in the trans-Neptunian region of the solar nebula.

It is quite possible that the ratios of CH_4/CO and NH_3/N_2 vary from comet to comet. The ratio of reduced to oxidized species may be a measure of metamorphosis toward chemical equilibrium, related to the place of origin of comets. Oort cloud comets may be more similar to the interstellar composition (i.e. richer in oxides) while short-period comets may be more similar to other solar system bodies (i.e. containing more reduced species). This would leave us with an apparent inconsistency, because P/Halley is a short-period comet while being rich in oxides. However, P/Halley has another characteristic which is unusual for short-period comets; its orbit is retrograde and highly inclined with the ecliptic. One may therefore speculate that it is a captured Oort cloud comet.

Various researchers have analyzed the elemental abundances of the gas-phase species and the CHON particles of P/Halley. Their conclusions are that the relative abundances of most elements in P/Halley are solar with two notable exceptions: hydrogen and nitrogen are depleted. To understand the hydrogen and nitrogen discrepancies, we must remember that all elements in the Sun are gravitationally bound. Thus the Sun reflects the elemental abundances of the solar nebula. Except for the very small fraction escaping in the solar wind (10^{-4} of the solar mass if the solar wind was constant over the age of the solar system), even the most volatile elements at the high temperatures of the Sun are bound by its gravitational field. The solar abundance of hydrogen is larger than the abundance of all other elements combined. The high abundance of hydrogen in the Sun must have existed also in the solar nebula, but the excess formed H_2 at the low temperature of the outer solar nebula. Apparently comets formed at a temperature high enough to prevent H_2 condensation. Any H_2 that was trapped in the ice has escaped since it is a very small and mobile molecule. Thus comets show a very large depletion of hydrogen relative to the other elements when compared to solar abundances.

The reasons for the nitrogen discrepancy are somewhat different. Nitrogen is less reactive than carbon or oxygen and tends to form N_2. At low temperatures N_2 is too volatile to condense if comets formed at temperatures between 25 K and 50 K. Carbon, on the other hand, is underabundant in the gas phase but condenses as hydrocarbon grains. The conversion of CO and N_2 in the solar nebula to form CH_4, which is also very volatile, and NH_3 was incomplete (except in the regions of the giant planet subnebulae). Additionally, oxygen is bound in water and silicates. Thus the temperature in the nebula region where comets formed was high enough to prevent condensation of CO and N_2, except for small amounts which were trapped in the formation of ice.

In the gas phase, carbon is also underabundant in the coma. However, when the carbon contained in the CHON particles is added, assuming $\chi \approx 1$ or 2, the relative carbon abundance in P/Halley is restored to the solar value. Oxygen is sufficiently abundant in the water ice so that the CHON contribution to the oxygen

Table C10 Estimated composition (%) of the ices in comet nuclei

H_2O	86
CO	4
CO_2	3
H_2CO	2
CH_3OH	1
N_2	1
Others (HCN, NH_3, C_2H_2, H_2CO_2, H_2S, CS_2, CH_3CN, $H_2C_3H_2$, CH_4, NH_2CH_3, etc.)	3

reservoir is negligibly small. Nitrogen is also underabundant in the CHON particles so that its relative abundance in comets remains below the solar value. It should be mentioned that some sulfur is contained in the CHON particles and its relative abundance in comets is very close to the solar value. Phosphorus, on the other hand, has not been detected in comets, but its relative solar abundance is so low that it may have eluded detection.

Daniel C. Boice and Walter F. Huebner

Bibliography

Aikin, A.C. (1974) Cometary coma ions. *Astrophys. J.*, **193**, 263–4.
Alfvén, H. (1957) On the theory of comet tails. *Tellus*, **9**, 92–6.
A'Hearn, M.F., Feldman, P.D. and Schleicher, D.G. (1983) The discovery of S_2 in comet IRAS-Araki-Alcock 1983d. *Astrophys. J.*, **274**, L99–103.
A'Hearn, M.F., Hoban, S., Birch, P.V. et al. (1986) Cyanogen jets in comet Halley. *Nature*, **324**, 649–51.
Biermann, L. (1951) Kometenschweife und solare Korpuskularstrahlung. *Z. Astrophys.*, **29**, 274–86.
Biermann, L. (1968) On the emission of atomic hydrogen in comets. JILA Report 93.
Biermann, L. and Trefftz, E. (1964) Über die Mechanismen der Ionisation und der Anregung in Kometenatmosphären. *Z. Astrophys*, **59**, 1–28.
Crovisier, J. (1989) On the photodissociation of water in cometary atmospheres. *Astron. Astrophys.*, **213**, 459–64.
Delsemme, A.H. and Swings, P. (1952) Hydrates de gaz dans les noyaux cométaires et les grains interstellaires. *Ann. d'Astrophys*, **15**, 1–6.
d'Hendecourt, L.B. and Jourdain de Muizon, M. (1989) The discovery of interstellar carbon dioxide. *Astron. Astrophys.*, **223**, L5–8.
Donn, B. and Urey, C.H. (1956) On the mechanism of comet outbursts and the chemical composition of comets. *Astrophys. J.*, **123**, 339–42.
Donn, B. and Urey, C.H. (1957) Chemical heating processes in astronomical objects. *Mem. Soc. Roy. Sci. Liege, Ser. 4*, **18**, 124–32.
Eberhardt, P., Krankowsky, D., Schulte, W. et al. (1987) The CO and N_2 abundance in comet P/Halley. *Astron. Astrophys.*, **187**, 481–4.
Eviatar, A., Goldstein, R., Young, D.T. et al. (1989) Energetic ion fluxes in the inner coma of comet P/Halley. *Astrophys. J.*, **339**, 545–57.
Greenstein, J.L. (1958) High-resolution spectra of comet Mrkos (1957d). *Astrophys. J.*, **128**, 106–13.
Herbst, E. (1987) Gas phase chemical processes in molecular clouds, in *Interstellar Processes* (eds D.J. Hollenbach and A. Thronson, Jr). Dordrecht: D. Reidel, pp. 611–29.
Huebner, W.F. (1965) Über die Gasproduktion der Kometen. *Z. Astrophys.*, **63**, 22–34.
Huebner, W.F. (ed.) (1990) *The Physics and Chemistry of Comets*. Berlin: Springer-Verlag.
Huebner, W.F., Boice, D.C., Schmidt, H.U. and Wegmann, R. (1991) Structure of the coma: chemistry and solar wind interaction, in *Comets in the Post-Halley Era*, Vol. 2 (eds R.L. Newburn, Jr, M. Neugebauer and J. Rahe). Dordrecht: Kluwer Academic, pp. 907–36.
Keller, H.U., Delamere, W.A., Huebner, W.F. et al. (1987) Comet P/Halley's nucleus and its activity. *Astron. Astrophys.*, **187**, 807–23.
Korth, A., Marconi, M.L., Mendis, D.A. et al. (1989) Probable detection of organic-dust-borne aromatic $C_3H_3^+$ ions in the coma of comet Halley. *Nature*, **337**, 53–55.
Malaise, D.J. (1970) Collisional effects in cometary atmospheres. I. Model atmospheres and synthetic spectra. *Astron. Astrophys*, **5**, 209–27.
Meyer, D.M. and Roth, K.C. (1991) Discovery of interstellar NH. *Astrophys. J.*, **376**, L49–52.
Millis, R.L. and Schleicher, D.G. (1986) Rotational period of comet Halley. *Nature*, **324**, 646–9.
Newburn, R.L., Jr, Neugebauer, M. and Rahe, J. (eds) (1991) *Comets in the Post-Halley Era.*, vols 1 and 2. Dordrecht: Kluwer Academic.
Oort, J.H. (1950) The structure of the cloud of comets surrounding the solar system, and a hypothesis concerning its origin. *Bull. Astron. Inst. Neth.*, **11**, 91–110.
Oppenheimer, M. (1975) Gas phase chemistry in comets. *Astrophys. J.*, **196**, 251–9.
Oppenheimer, M. (1976) Gas phase chemistry in comets, in *The Study of Comets* (eds) B. Donn, M. Mumma, W. Jackson, et al.). NASA SP-393, pp. 753–6.
Schmidt, H.U., Wegmann, R., Huebner, W.F. and Boice, D.C. (1988) Cometary gas and plasma flow with detailed chemistry.' *Comput. Phys. Commun.*, **49**, 17–59.
Swings, P. (1941) Complex structure of cometary bands tentatively ascribed to the contour of the solar spectrum. *Lick Obs. Bull.*, **19**, 131–6.
van Dishoeck, E.F. and Dalgarno, A. (1984) The dissociation of OH and OD in comets by solar radiation. *Icarus*, **59**, 305–13.
Whipple, F.L. (1950) A comet model I. The acceleration of comet Encke. *Astrophys. J.*, **111**, 375–94.
Whipple, F.L. (1951) A comet model II. Physical relations for comets and meteors. *Astrophys. J.*, **113**, 464–74.
Wurm, K. (1943) – Die Natur der Kometen. *Mitt. Hamburger Sternwarte*, **8**, No. 51.

Cross references

Chiron
Cosmochemistry
Giotto mission
Halley, Edmond, and Halley's comet
Kuiper Belt
Oort, Jan Hendrik, and Oort cloud
Sakigake and Suisei missions
Vega missions

COMMENSURABILITY

The orbital periods of two bodies are said to be commensurable when the ratio of their periods is close to that of two small integers. For instance, the orbital periods of Jupiter (11.86 years) and Saturn (29.46 years) are commensurate in that five orbits of Jupiter take very nearly the same amount of time as two orbits of Saturn.

There are many examples of commensurability of periods in the solar system. Roy and Ovendon (1954) showed that many more commensurabilities are present than would be expected on the assumption of random distribution of orbit periods. It now appears that processes such as tidal friction (q.v.) help systems to evolve toward a state characterized by commensurable mean motions (Peale, 1986).

Wilkins and Sinclair (1974) list many examples. For instance, 3 periods of Saturn = 1.052 periods of Uranus, 2 periods of Uranus = 1.020 periods of Neptune, and 3 periods of Neptune = 1.990 periods of Pluto.

Satellite systems include important examples. In the Jupiter system, for example, 2 periods of Io = 1 period of Europa, and 2 periods of Europa = 1 period of Ganymede. These three are linked in a triple commensurability, such that 1 period of Io minus 3 periods of Europa plus 2 periods of Ganymede = 0. This is termed the 'Laplace relation,' after the astronomer Pierre Simon de Laplace (q.v.), who was the first to explore its dynamic consequences. It appears that tidal dissipation in Io (q.v.), which is accompanied by spectacular volcanism, permitted the system to evolve to this state; orbital interactions are now maintaining the commensurability in a stable resonance (q.v.).

The dynamics of the Saturn–Jupiter resonance were summed up succinctly by Wilkins and Sinclair (1974):

> ... if two planets have a close commensurability of periods, then the amplitudes and periods of their mutual perturbations will be increased by a resonance effect. For an example we will consider the case of Jupiter and Saturn, where the periods are almost in the ratio 2 : 5 ... During 5 orbits of Jupiter (taking about 59 years), Saturn will make approximately 2 orbits, so the two planets will return to approximately their initial configuration. During this time Jupiter will have overtaken Saturn 3 times, at positions equally spaced around their orbits. In the next 59 years Jupiter will again pass Saturn at almost the same positions in their orbits ... Now the gravitational attraction of the two planets is greatest when they are in conjunction, and the result of this perturbation

repeatedly occurring near the same parts of their orbits is to greatly increase its effect. Thus we find that there are oscillations in the longitudes of Jupiter and Saturn, with period 900 years, and amplitudes 20′ and 48′ respectively.

Commensurabilities play important roles in the distribution of particles in ring systems, and in the distribution of asteroids in the main asteroid belt. The Kirkwood gaps in the distribution of asteroid orbits (see Asteroids) occur at distances corresponding to the 2 : 1, 7 : 3, 5 : 2 and 3 : 1 commensurabilities with Jupiter.

James H. Shirley

Bibliography

Goldreich, P. (1965) An explanation of the frequent occurrence of commensurable mean motions of the solar system. *Mon. Not. Roy. Astron. Soc.*, **130**, 159–81.
Peale, S.J. (1986) Orbital resonances, unusual configurations; and exotic rotation states among the planetary satellites, in *Satellites* (ed. J.A. Burns and M.S. Matthews) Tucson: University of Arizona, pp. 159–24.
Roy, A.E. and Ovendon, M.W. (1954) On the occurrence of commensurable mean motions in the solar system. *Mon. Not. Roy. Astr. Soc.* **114**, 232–41.
Wilkins, G.A. and Sinclair A.T. (1974) The dynamics of planets and their satellites. *Proc.Roy.Soc.London A.*, **336**, 85–104.

Cross references

Asteroid; resonance
Resonance in Saturn's rings
Resonance in the solar system
Tidal friction
Tidal heating

COORDINATE SYSTEMS

In this article we deal first with those geocentric and heliocentric coordinate systems which are used to describe the positions of stars and of bodies within the solar system, and then with planetographic coordinate systems which are used to describe the positions of points on the surfaces of planets.

Celestial coordinates

Altazimuth coordinates

To an observer on the surface of the Earth, the stars appear to be situated on the surface of a vast hemisphere, with the observer at the center of the base. The complete sphere (half of which is below the observer's horizon) is the celestial sphere.

Figure C54 shows the celestial sphere, with the observer at O. The points NESW are the north, east, south and west points of the horizon. Z is the zenith, Z′ the nadir. The direction to a star X is specified by two angles. One of these is the angle MX, which is the altitude or elevation of the star above the horizon. (The word 'altitude' is perhaps more commonly used; the word 'elevation' is perhaps preferable.) The complementary angle ZX is the zenith distance. The second angle is NM, the azimuth, A. Some writers observe a convention that azimuth is measured from the north point of the horizon eastwards, from 0° to 360°. However, not all writers observe this. Further, some observers in the southern hemisphere prefer to measure azimuth from the south point of the horizon. Therefore, to avoid confusion the convention used should always be explicitly stated. 'An azimuth of 285°' is open to misinterpretation; 'an azimuth of 75° measured from the north point of the horizon westwards' is unambiguous.

Great circles of fixed azimuth passing through the zenith and nadir are vertical circles. The vertical circle passing through S and N is the meridian. The vertical circle passing through E and W is the prime vertical.

Small circles of fixed altitude are parallels of altitude, or almucantars. The system of coordinates of vertical circles and almucantars,

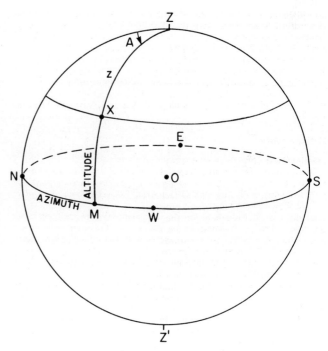

Figure C54 Altazimuth coordinate system.

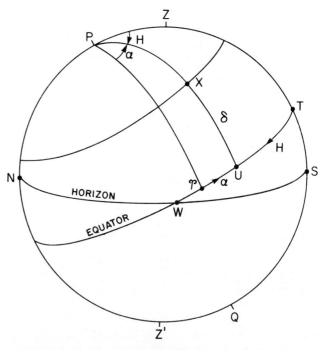

Figure C55 Equatorial coordinate system.

defining azimuth and altitude (or elevation) is called the altazimuth, or az-el, system.

Equatorial coordinates

During the course of a night the celestial sphere appears to rotate from East to West about an axis PQ. (Fig. C55) (The stars are, of course, fixed, and the apparent rotation of the celestial sphere is the reflection of the rotation of the Earth.) In Figure C55, P is the north celestial pole and Q is the south celestial pole. The figure is drawn for a northern hemisphere observer. For an observer in the southern hemisphere Q would be above, and P below, the horizon. The great circle 90° from P and from Q is the celestial equator.

There happens to be a fairly bright (second magnitude) star, called the Pole Star, or Polaris, or Alpha Ursae Minoris, less than a degree from the north celestial pole. There is no star as bright as this within 20° of the south celestial pole. The star σ Octantis, very close to the south celestial pole, is just visible to the unaided eye under good conditions.

The celestial sphere rotates through 360° in 23 h 56 min of solar time, or exactly 24 sidereal hours. During this period each star crosses the meridian twice. It has a lower meridian transit, and it culminates at upper meridian transit.

The position of a star in equatorial coordinates may be specified by two angles. One of these is the angle δ, measured from the equator to a celestial object. It is called the declination, and is in the range $-90°$ to $+90°$, positive declinations corresponding to positions north of the equator. The complement of the declination is the north polar distance. The second angle is the angle TU. It is called the hour angle, H, and is measured westwards from the meridian, usually from 0 hours to 24 hours, rather than from 0° to 360°. This convention is convenient because 1 sidereal hour after a star culminates at its upper meridian transit, its hour angle is 1 hour (or 15°).

The altitude of the north celestial pole is equal to the observer's geographic latitude, and stars whose north polar distance is less than that never rise or set for an observer in the northern hemisphere; such stars are circumpolar stars. For an observer in the southern hemisphere, a star is circumpolar if its south polar distance is less than the observer's southern latitude.

During the course of a night, a star's declination is unchanged, but its hour angle continuously increases (at a rate of 1 hour per sidereal hour). We have need of a second angular coordinate (in addition to declination) whose value does not change during the night. This second angle, α, is the right ascension, and is measured along the equator eastwards from a point on the celestial equator called the First Point of Aries, or the vernal equinox, or the March equinox, to be defined under the section on ecliptic coordinates. The First Point of Aries is traditionally denoted by the astrological symbol ♈ for the Ram.

Thus in star catalogs the positions of the stars are given by two coordinates, the right ascension α and the declination δ.

Great circles of fixed right ascension passing through P and Q are called hour circles, or declination circles, or colures. The colure passing through ♈ is the equinoctial colure. The colure 12 hours ahead of and behind this is the solstitial colure. Small circles of fixed declination are parallels of declination.

The right ascension of the First Point of Aries is, by definition, zero. The local sidereal time for an observer is defined as the hour angle of the First Point of Aries, and therefore it must also equal the right ascension of any star plus the hour angle of that star.

Because of precession (q.v.), the positions of the equinox and the equator are not fixed with respect to the stars, and therefore right ascensions and declinations must always be referred to the equinox and equator at a specified epoch. For catalogs and atlases, a standard epoch (at present usually chosen to be 2000.0) is chosen. In order to set the circles of a telescope, however, an observer needs to know the right ascension and declination referred to the equinox and equator of date (i.e. of the date when the telescope is being used).

If the equatorial coordinates (H, δ) of a body are known, its altazimuth coordinates (A, z) for an observe at latitude ϕ can be found from

$$\cos z = \sin \phi \sin \delta + \cos \phi \cos \delta \cos H \qquad (C22)$$

and

$$\tan A = \frac{\sin H}{\cos \phi \tan \delta - \sin \phi \cos H} \qquad (C23)$$

A simpler formula for A, once z is known, is

$$\sin A = \sin H \cos \delta / \sin z \qquad (C24)$$

However, since the arcsin function leads to an ambiguity of quadrant, this formula is recommended only as a numerical check on equation (C23).

If the altazimuth coordinates (A, z) are known, the equatorial coordinates (H, δ) can be found from

$$\sin \delta = \sin \phi \cos z + \cos \phi \sin z \cos A \qquad (C25)$$

and

$$\tan H = \frac{\sin A}{\cos \phi \cos z - \sin \phi \cos A} \qquad (C26)$$

A simpler formula for H, once δ is known, is

$$\sin H = \sin z \sin A / \cos \delta \qquad (C27)$$

However, since the arcsin function leads to an ambiguity of quadrant, this formula is recommended only as a numerical check on equation (C26).

Note that these formulas are valid for observers in the northern or the southern hemisphere, provided that the latitude ϕ is counted as negative for a southern observer, and the azimuth A is measured from the north point of the horizon for a northern or for a southern observer. For pre-meridian observations, H and A are measured eastwards from the meridian and the north point respectively; for postmeridian observations, H and A are measured westwards from the meridian and north point respectively.

Ecliptic coordinates

During the course of a year, the Sun moves eastwards relative to the stars along a great circle, the ecliptic, inclined at 23½° to the equator. This angle, ϵ, is the obliquity of the ecliptic. The point where the ecliptic crosses the equator where the Sun moves from north to south is the First Point of Aries, ♈, and is the zero point from which right ascension is measured. In Figure C56, K (at present in the constellation Draco) is the north pole of the ecliptic and P is the north celestial pole. The angles (λ, β) are called, respectively, the celestial longitude and latitude of a body or, better, the ecliptic longitude and ecliptic latitude.

Note that, because of precession (q.v.) the point P moves around K in a small circle in a period of 26 000 years, and the point ♈ moves westwards along the ecliptic. Consequently both the ecliptic and equatorial systems of coordinates are moving with respect to the stars, which is why, as mentioned in the previous section, it is essential to specify, when giving α and δ (or λ and β), the epoch to which these coordinates are referred.

When the ecliptic and equator were first described, in about 2300 BC, the ecliptic and equator intersected at the eastern edge of the constellation Aries. Precession has now carried the equinox well over into the constellation Pisces, although we retain the name First Point of Aries.

To convert from equatorial coordinates (α, δ) to ecliptic coordinates (λ, β), we use

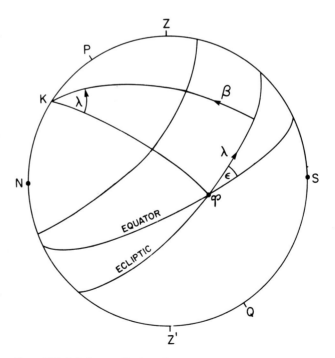

Figure C56 Ecliptic coordinate system

$$\sin \beta = \cos \epsilon \sin \delta + \sin \epsilon \cos \delta \sin \alpha \qquad (C28)$$

and

$$\cos \lambda = \cos \delta \cos \alpha / \cos \beta \qquad (C29)$$

To convert from ecliptic coordinates (λ, β) to equatorial coordinates (α, δ) we use

$$\sin \delta = \cos \epsilon \sin \beta + \sin \epsilon \cos \beta \sin \lambda \qquad (C30)$$

and

$$\cos \alpha = \cos \beta \cos \lambda / \cos \delta \qquad (C31)$$

Distance

The coordinates (α, δ) or (λ, β) give us the direction to a star or other body. We may also need to describe the distance to a body. The *geocentric distance* to a body is usually denoted by the letter Δ (sometimes ρ). Thus the position of a body in three-dimensional space relative to the Earth can be given by three geocentric equatorial coordinates (Δ, α, δ) or by three geocentric ecliptic coordinates (Δ, λ, β).

The coordinates (Δ, α, δ) are spherical coordinates in the sense of elementary mathematics. We can also express geocentric equatorial coordinates as rectangular coordinates (X_1, Y_1, Z_1), in which the X_1-axis is directed from the Earth towards Υ, the Y_1-axis towards a point on the equator 90° east of Υ, and the Z_1-axis towards the north celestial pole. (Fig. C56).

The rectangular and spherical geocentric equatorial coordinates are related by

$$X_1 = \Delta \cos \delta \cos \alpha \qquad (C32)$$

$$Y_1 = \Delta \cos \delta \sin \alpha \qquad (C33)$$

$$Z_1 = \Delta \sin \delta \qquad (C34)$$

$$\Delta^2 = X_1^2 + Y_1^2 + Z_1^2 \qquad (C35)$$

$$\cos \alpha + X_1 / (X_1^2 + Y_1^2)^{1/2} \qquad (C36)$$

$$\sin \delta = Z_1 / (X_1^2 + Y_1^2 + Z_1^2 \qquad (C37)$$

In a similar fashion a set of rectangular geocentric ecliptic coordinates (X_2, Y_2, Z_2) can be set up.

For precise solar system work it is necessary to distinguish between geocentric coordinates (centered at the center of the Earth) and topocentric coordinates (centered at the observer on the Earth's surface). This distinction is not trivial, for the Earth's diameter subtends an angle of 16" at an distance of 1 AU.

We also need heliocentric rectangular equatorial or ecliptic coordinates, (x_1, y_1, z_1) and (x_2, y_2, z_2). The heliocentric distance r, or the length of the radius vector from the Sun to a planet, is given by

$$r^2 = x_1^2 + y_1^2 + z_1^2 = x_2^2 + y_2^2 + z_2^2 \qquad (C38)$$

Heliocentric and geocentric rectangular coordinates are simply related by translational equations of the form

$$X_1 = x_1 - x_1 \text{ (Earth)} \qquad (C39)$$

$$x_1 = X_1 - X_1 \text{ (Sun)} \qquad (C40)$$

and so on. The geocentric rectangular equatorial coordinates of the Sun are published each year in *The Astronomical Almanac*, and are exceedingly important in the calculation of ephemerides. During computation of an ephemeris, the first stage is to calculate the position of a planet in the plane of its orbit. One then successively translates or rotates this to heliocentric ecliptic coordinates, heliocentric equatorial coordinates, geocentric (rectangular) coordinates and (finally) geocentric (spherical) coordinates, this last being the familiar right ascension and declination. For very fine work a further conversion to topocentric coordinates may be necessary.

Planetary coordinates

Earth

Positions on Earth's surface are specified by latitude, longitude and elevation. Elevation differences are very small compared with the planet's radius and are often omitted.

Earth rotates about an axis passing through its center of mass and intersecting its surface at two points called poles. The end of the rotation axis which now points in the general direction of Polaris is

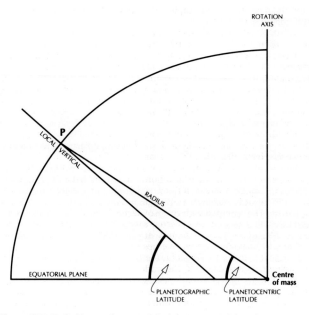

Figure C57 Definitions of geocentric (planetocentric) and geographic (planetographic) latitude.

traditionally called the north pole, and the other end is the south pole. A circle on the planet's surface whose center is at the center of mass is called a great circle, while any other circle on the globe is a small circle. A plane perpendicular to the axis at the center of mass intersects the surface along a great circle called the equator. Half a great circle running from one pole to the other is a meridian. East and west are directions perpendicular to a meridian, east being to the right for a north-facing observer, west to the left.

The terms latitude and longitude originally referred to the width and length respectively of the region known to classical geographers. Latitude is a measure of angular distance from the equator (0°) towards the poles ($\pm$ 90°), designated positive north of the equator and negative to the south. A point P on Earth's surface has a geocentric latitude ϕ equal to the angle between the equatorial plane and a radius connecting P to the center of mass (Figure C57). The same point has a geographic (sometimes called geodetic) latitude ϕ' equal to the angle between the equatorial plane and the local vertical at P (extended to the equatorial plane). On a spherical world geocentric and geographic latitudes would exactly coincide. Because Earth is slightly flattened at the poles its shape is better approximated as an oblate spheroid or ellipsoid of revolution (defined by rotating an ellipse about its minor axis), so ϕ and ϕ' differ slightly at P. Latitudes on most maps are geographic. ϕ and ϕ' are related by the equation:

$$\tan \phi = \left(\frac{C}{A}\right)^2 \tan \phi' = (1 - f)^2 \tan \phi' \qquad (C41)$$

where f is the polar flattening. If A is the equatorial radius and C is the polar radius,

$$f = \frac{A - C}{A} \qquad (C42)$$

The geocentric latitude may be measured from the radius at P to the rotation axis rather than the equatorial plane. This is the colatitude. The equator and small circles of constant latitude (however defined) are called parallels because they define a set of parallel planes. The term 'geocentric' is used for latitudes as defined here as well as for those celestial coordinates described in the previous section because both are measured from the center of the planet.

Longitude is a measure of angular distance around a parallel of latitude. A meridian is a line of constant longitude. For historical but otherwise arbitrary reasons the meridian passing through a transit circle at the Royal Observatory, Greenwich, in London, England, is designated 0° longitude, the prime meridian. Longitudes are measured from 0° to 180° both east and west from the prime meridian, east longitudes being designated positive. On some antique maps and in some modern scientific applications longitudes are measured up to 360°, increasing eastwards.

Latitude and longitude locate a point on the surface of a spherical (or ellipsoidal) idealized Earth, but the real planet has an uneven surface. To specify position fully we need a measure of distance from the center of mass or from a reference surface or datum of some kind. Elevations on Earth are traditionally measured from sea level. Cyclical changes in sea level caused by tides are avoided by defining a long-term average, and on most topographic maps a statement such as 'mean sea level' or 'mean high water mark' indicates which datum was adopted for that map.

Elevations in continental interiors are thought of as if measured from 'sea level' in hypothetical frictionless canals cut across the continent from the adjoining oceans. The shape of this idealized mean sea surface is estimated by measuring the strength of the gravitational field at many locations, a task best performed from orbit. The equipotential surface so defined, a reference surface having equal gravitational potential everywhere, is called a geoid (q.v.), an 'earth-shaped surface' (Kovalevsky, Mueller and Kolaczek, 1989; Jekeli, 1991). It differs from an ellipsoid of revolution by no more than a few tens of meters around the globe and is usually portrayed as deviations from an ellipsoidal surface. Since topography is measured relative to the geoid, the Earth's radius at any location is approximately the sum of the radius of the ellipsoid, the local deviation of the geoid from the ellipsoid, and local topography. This is still slightly incorrect because the latter two values are measured perpendicular to the ellipsoid rather than along an extended radius.

Other bodies

Forty worlds other than Earth had been mapped by December 1992 (Greeley and Batson, 1990; Stooke, 1991). Planetary cartographic history and procedures are described by Greeley and Batson (1990) (see also Cartography). The status of planetary geodesy is summarized by Thomas (1991). Planetary coordinates are based upon the same concepts as terrestrial coordinates with modifications to reflect special circumstances.

Two 'guiding principles' were established by the Working Group on Cartographic Coordinates and Rotational Elements of the Planets and Satellites of the International Astronomical Union (IAU, 1970, p. 128) to avoid a proliferation of inconsistent coordinate systems:

1. The rotational pole of a planet or satellite which lies on the north side of the invariable plane (see below) shall be called north, and northern latitudes shall be designated as positive.
2. The planetographic longitude of the central meridian, as observed from a direction fixed with respect to an inertial coordinate system, shall increase with time. The range of longitudes shall extend from 0° to 360°.

The invariable plane is perpendicular to the total angular momentum vector of the solar system. Latitudes are defined as for Earth (Figure C57), but termed planetocentric and planetographic respectively. Principle 1 is intended to resolve a long-standing confusion. Astronomers traditionally use the word north for the pole which appears to a distant observer to rotate counterclockwise. This makes the north pole of Venus point in almost the same direction as the south pole of Mercury and causes a jet of material emitted from the north pole of a retrograde-rotating comet nucleus to point south on the sky. Most planetary scientists have adopted the IAU principles but astronomers tend to retain their traditional usage even when drawing maps of objects like the nucleus of comet Halley. Regrettably, the literature is now confused in places, so the term 'north' should be properly explained in any description or map of a solar system body. This is especially important for objects like Uranus, Pluto and their satellites, whose rotation axes are near the planes of their orbits. If precession carried a rotation axis across the invariable plane, 'north' might need to be fixed where it was first defined since it would be hopelessly confusing to redefine directions every time that happened.

Longitudes may be defined in two ways. Planetocentric longitudes are measured positively towards the east (as on Earth) from a prime meridian defined relative to a fixed surface feature. Planetographic longitudes are measured positively in the direction opposite to rotation, in accordance with principle 2 above, but from a prime meridian assumed to rotate at the same rate as the spinning planet. If the rotation rate of the planet is well known the two types of prime meridian should coincide exactly, but a slight error would cause the planetocentric and planetographic coordinates to drift slowly apart.

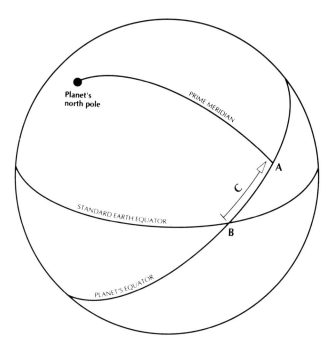

Figure C58 Location of the prime meridian of a planet.

Cartographic coordinates, those actually used on maps, are usually planetographic. Latitudes are equivalent to geographic latitudes on near-spherical bodies with significant polar flattening, while longitudes are planetographic in the sense of measurement but tied where possible to a small surface feature.

The full definition of a coordinate system on a planet begins with the right ascension and declination of the north pole of the mean axis of rotation at the standard epoch. Many satellites rotate synchronously (with periods equal to their orbital periods) about axes perpendicular to their orbital planes. The prime meridian of a satellite in this state is traditionally defined such that it faces the planet it is orbiting.

The prime meridian for any world with a known rotation rate may be defined with reference to Figure C58. A is the point where the prime meridian crosses the planet's equator. The standard Earth equator is a great circle parallel with Earth's equator at the standard epoch. The ascending node B is where a point carried around the equatorial plane by rotation passes from south to north of the standard Earth equator. The position of the prime meridian is given by the angle C, the argument of the prime meridian, measured eastwards along the equator from B to A. Angle C is assigned a specific value at the standard epoch and varies linearly with time, increasing for direct (prograde) rotation and decreasing for retrograde rotation. The prime meridian at any time is found by adding the total rotation since the standard epoch to the value of C at that epoch. Currently accepted values for the pole position, rotation rate and argument of the prime meridian are given in the *Astronomical Almanac* and in the most recent of the triennial reports of the Working Group on Cartographic Coordinates and Rotational Elements of the Planets and Satellites (Davies *et al.*, 1980; Davies *et al.*, 1986; Davies *et al.*, 1988).

For cartographic consistency it is desirable that coordinates remain tied to surface features, so control points (small circular craters) are used to define longitudes wherever possible. For instance, the prime meridian of Mars passes through the center of a small crater named Airy-0, while the 182° meridian of the Jovian satellite Europa passes through the center of a small crater named Cilix. The region of the 0° meridian of Europa was not observed in detail by the Voyager spacecraft. Using Cilix to define the 182° meridian ties the coordinate grid to the surface of Europa in case its rotation is not exactly synchronous or the icy crust shifts relative to the rocky interior, while still positioning the prime (0°) meridian very close to the sub-Jupiter point (at the time of the Voyager encounters) in accordance with tradition. Planetary maps produced by the US Geological Survey carry explanations of their coordinate systems.

In practice it is rarely necessary to calculate the position of the prime meridian for the major bodies of the solar system from first principles. Tables giving the longitudes of central meridians and other orientation data at various dates are published in the *Astronomical Almanac*.

Special cases

Moon
Early lunar charts did not contain a latitude–longitude grid. The rotation axis of the Moon was first located by Tobias Mayer in 1749. He applied a coordinate system to his lunar chart which was identical to that in use for Earth, with longitudes measured to 180° east and west from a prime meridian, positive in the direction of the Moon's rotation, and latitudes designated positive to the north. Mayer's prime meridian was the mean sub-Earth meridian. However, the terms east and west became confused on being applied to the Moon. The side of the Moon which faces in the direction of its apparent motion westwards across the sky was labeled west by astronomers. When spacecraft exploration was beginning in 1960 the IAU reversed this to bring lunar directions into agreement with terrestrial usage. Placenames were not affected by this change so *Mare Orientale*, the 'Eastern Sea', is on the western edge of the Moon as seen from Earth. The Moon's reference surface, a sphere of radius 1738.0 km, is used to derive positions on the lunar surface (Kopal and Carder, 1974). In order to minimize negative elevations, heights on most maps are measured from an arbitrary spherical datum of radius 1730.0 km. Experimental maps have been based on an equipotential reference surface derived from gravity data obtained by tracking orbiting spacecraft. Longitudes are defined relative to the small crater Mösting A, which has a longitude of 5° 9′ 53 ± 5″. Since longitudes increase in the direction of rotation the longitudes of the terminator and subsolar point decrease with time. It is sometimes convenient to reverse this by using the colongitude (90° or 450° minus the longitude).

Mars
The shape adopted for defining latitude and longitude on Mars is an oblate spheroid with a radius of 3393.4 km at the equator and 3375.7 km at the poles. An equipotential reference surface has been derived from gravity data for use as a topographic datum. The zero level is set at the elevation where the mean atmospheric pressure is 6.1 mbar, the triple point of water, an arbitrary elevation but one with potential geological or biological significance: below this elevation water could remain liquid for brief periods.

Venus
Venus rotates backwards relative to the other inner planets. Longitudes on modern maps increase in the direction opposite to rotation, towards the east rather than the west. The terms east and west are defined as for Earth so east is to the right on any map or image with north at the top. In other words, the terms east and west have nothing to do with the direction of rotation. Elevations on the most recent maps are given as radii or measured relative to a sphere of radius 6051.9 km.

Io
This large satellite of Jupiter was photographed in detail by Voyager 1 and found to be volcanically active. At the time it appeared that rapid resurfacing might make features seen by Voyager unrecognizable by the time the Galileo spacecraft arrived. It now seems likely that changes are slower and more superficial than previously thought, but it is still possible that any specific topographic feature used to define longitude might be hard to find in future images. Thus longitudes are defined relative to the mean sub-Jupiter point on Io. Longitudes on all other large satellites of the outer solar system are defined relative to small craters.

Irregularly shaped bodies
Conventional mapping methods and coordinate systems were devised for spherical or slightly flattened objects, but many small objects in the solar system have highly elongated or irregular shapes. Objects with radii greater than about 200 km are forced by gravity into roughly equipotential shapes, but the shapes of smaller objects are determined by impact, fragmentation and (in the case of comet nuclei) mass loss. The local vertical direction varies in a complex manner across the surface of an irregular body, so planetographic

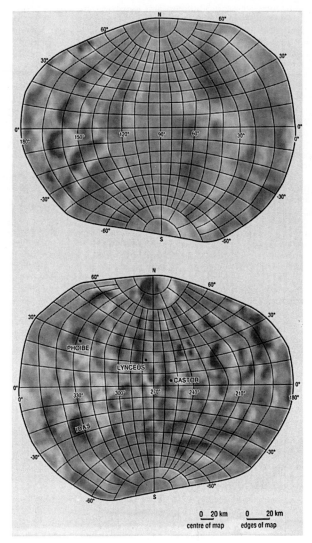

Figure C59 Map of Janus (maximum diameter about 200 km) on a morphographic conformal projection, the non-spherical equivalent of the familiar stereographic. The grid represents planetocentric latitude and longitude.

latitudes and elevations measured perpendicular to a reference surface are difficult to use. Most shape analysis and cartography of these bodies uses planetocentric coordinates with topography defined by radii (Stooke, 1992). Figure C59 shows a planetocentric coordinate system applied to the surface of the saturnian satellite Janus. It is based on a shape model derived by P. Stooke and M. Lumsdon.

Bodies having complex or unstable rotation
Hyperion, a small satellite of Saturn, appears to rotate chaotically, and the nucleus of comet Halley rotates in a complex manner which is still not understood in 1992 more than 6 years after its most recent perihelion. Other small bodies may be found with complex, unstable or indeterminate rotation states. Cartographic coordinates fixed to a rotation axis may be impossible to define and as exploration of small bodies proceeds new approaches will probably be required.

Jupiter
Jupiter has no solid surface but maps of ephemeral cloud configurations are frequently prepared. Since it exhibits a marked polar flattening, planetographic latitudes are usually used. Clouds are carried across the disk at rates which vary with latitude, traditionally

accommodated by assigning different rotation periods to equatorial and mid-latitude regions. Equatorial longitudes are designated system I and apply within about ± 9° latitude in the equatorial zone. The apparent system I rotation period is 9 h 50.5 min. Elsewhere system II longitudes with an average period of 9 h 55.7 min are used. A third rotation period of 9h 55.5 min, based on modulation of radio emissions, defines system III longitudes. System III is assumed to be tied to the deep interior of the planet where the magnetic field originates, and for most purposes can be taken as the planet's rotation period. Since the Voyager encounters with Jupiter, the concept of differential rotation has been largely replaced with a more physically plausible description of the atmosphere's motion: Jupiter rotates with the system III period and clouds are driven by winds which vary in speed and direction at different latitudes (Smith et al., 1979). Systems I and II are still used, but variations with latitude make them approximate at best (the periods given above are approximate mean values).

Saturn, Uranus and Neptune

Very few discrete clouds had been seen on these planets before the Voyager encounters, so longitude definitions were rarely needed. Maps of cloud patterns have now been prepared for Saturn and Neptune (e.g. Godfrey and Moore, 1986; Beebe et al., 1992). Planetographic latitudes are usually used since these planets are considerably flattened at the poles. Voyager observations of radio emissions are now used to define rotation periods, giving the equivalents of Jupiter's system III longitudes. System I is sometimes used for Saturn, but otherwise systems I and II are not used with these planets.

The Sun

Heliographic coordinates are used to locate features on the surface of the Sun. Since the photosphere, the visible 'surface', is not solid, features on or near it move relative to one another like cloud features on Jupiter. The apparent rotation rate varies with latitude and is fastest near the equator. The photosphere is spherical with a radius of 1 392 530 km, so heliographic and heliocentric latitudes are identical. Heliographic longitude is measured from a prime meridian which was at the ascending node of the solar equator on the ecliptic at 12:00 UT on 1 January 1854, and rotates with a sidereal period of 25.380 days (Green, 1985).

Philip J. Stooke and Jeremy B. Tatum

Bibliography

Astronomical Almanac (annual). Washington, DC: US Government Printing Office. London: Her Majesty's Stationery Office.
Beebe, R.F. *et al.* (1992) The onset and growth of the 1990 equatorial disturbance on Saturn. *Icarus*, **95**, 163–72.
Davies, M.E. *et al.* (1980) Report of the IAU Working Group on Cartographic Coordinates and Rotational Elements of the Planets and Satellites. *Celest. Mech.*, **22**; 205–30.
Davies, M.E. *et al.* (1986) Report of the IAU/IAG/COSPAR Working Group on Cartographic Coordinates and Rotational Elements of the Planets and Satellites: 1985. *Celest. Mech.* **39**, 103–13.
Davies, M.E. *et al.* (1988) Report of the IAU/COSPAR Working Group on Cartographic Coordinates and Rotational Elements of the Planets and Satellites: 1988. *Celest. Mechan Dynam. Astron*, **46**, 187–204.
Godfrey, D.A. and Moore, V. (1986) The Saturnian ribbon feature – a baroclinically unstable model. *Icarus*, **68**, 313–43.
Greeley, R. and Batson, R.M. (1990) *Planetary Mapping*. Cambridge: Cambridge University Press.
Green, R.M. (1985) *Spherical Astronomy*. Cambridge: Cambridge University Press.
International Astronomical Union (IAU) (1970) *Proceedings of the 14th General Assembly, Brighton, 1970*, (eds C. de Jager and A. Jappel). *Trans. IAU*, **4**. Dordrecht: D. Reidel Publishing Co.
Jekeli, C. (1991) Gravimetry and gravity field models. *Rev. Geophys.*, **29** (Suppl)., 123–31.
Kopal, Z. and Carder, R.W. (1974) *Mapping of the Moon. Astrophysics and Space Science Library*, Vol. 50. Dordrecht: D. Reidel Publishing Co.
Kovalevsky, J., Mueller, I.I. and Kolaczek, B. (1989) *Reference Frames in Astronomy and Geophysics. Astrophysics and Space Science Library*, Vol. 154. Dordrecht: Kluwer Academic Publishers.
Smith, B.A. *et al.* (1979) The Galilean satellites and Jupiter: Voyager 2 imaging science results. *Science*, **206**, 927–50.
Stooke, P.J. (1991) Lunar and planetary cartographic research at the University of Western Ontario. *CISM J.*, **45**(1), 23–31.
Stooke, P.J. (1992) A model and map of Amalthea. *Earth, Moon and Planets* **56** 123–39.
Thomas, P.C. (1991) Planetary geodesy. *Rev. Geophys.*, **29** (suppl.), 182–7.

Cross references

Astrometric observation
Cartography
Celestial mechanics
Map projections
Orbit

COPERNICUS, NICOLAUS (1473–1543)

The great Polish astronomer, born in Toruń (Thorn), Poland, Copernicus (born Mikolaj Kopernik) was destined to become, through the introduction of his heliocentric astronomy, one of the seminal figures in the history of scientific thought. He began his higher studies at the University of Cracow (Poland), at that time famous for its mathematics, astronomy and philosophy curricula, continued at Bologna (Italy) with canon and civil law (1496–1501), and with medicine for two years (1501–1503) at Padua (Italy), finally receiving his doctorate in canon law at the University of Ferrara in 1503. Then he returned to Poland where he resided in his uncle's palace in Lidzbark for 3 years and then in 1510 moved to the cathedral town of Frombork (Frauenberg) in Ermland with extended stays in Olsztyn, and then in 1521 settled permanently in Frombork. Copernicus performed his ecclesiastical duties, practiced medicine and prepared an in-depth treatise on monetary reform but devoted a large part of his time to research in astronomy. Copernicus became involved in defending the royal Prussian estates from the knights of the Teutonic Order, and in 1520–1521 he was responsible for securing the defense of Olsztyn against the Teutonic knights.

In Italy he lived for a while in the home of Domenico Maria da Novara (1454–1504) who maintained that the latitudes of the Mediterranean cities were 1° 10′ greater than recorded by Ptolemy. From this 'correction' Novara inferred a slow alteration in the direction of the Earth's axis, which implied that the Earth does move, contrary to Ptolemy. Nowhere does Copernicus inform us, however, of exactly when, where or how he arrived at his heliocentric–geokinetic hypotheses.

By 1514 Copernicus circulated his preliminary ideas privately in a manuscript titled *Commentariolus*, which later was greatly expanded and substantiated in *De Revolutionibus Orbium Coelestium* (1543). A detailed version (available perhaps as early as 1530) was privately circulated, but Copernicus continued to work on his masterpiece. Although the authentic parts of the entire work as printed were finally completed in 1542, Copernicus was only to see the printed volume on his deathbed. This seminal work challenged the geocentric cosmology that had been taken for granted since the time of Aristotle (384–322 BC) and Ptolemy (c. 90–168 AD). Copernicus revived the long-dormant theory of Aristarchus (c. 320–250 BC) of heliocentricity. The geometric Ptolemaic theory, based in part on the observation of falling bodies on Earth, introduced the need for ingenious mechanisms to explain the celestial observations. By contrast, Copernicus proposed that a rotating Earth revolving along with the other planets about a fixed central Sun could account for the observed phenomena of the apparent daily rotation of the heavens, the perceived annual movement of the Sun through the ecliptic, the observation of the maximum or bounded elongations of Mercury and Venus, the periodic seemingly retrograde motion of the planets, and above all the arrangement of the planets in a definite and determinate order.

In consequence of this new approach the old astronomical data had to be completely recalculated. Despite his totally new approach to the structure of the universe, Copernicus still adhered to the ancient Aristotelian doctrines of celestial spheres, and the perfect circular motion of heavenly bodies. To countless generations of astronomers, from Eudoxus (370 BC) to Copernicus and Tycho Brahe (1546–1601), the basic celestial orbit remained a circle. The solution of those classical issues had to wait for Kepler (1571–1630) to determine the ellipticity of planetary orbits, for Galileo (1564–1642) to invent a new concept of motion, and for Newton (1642–1727) to discover the theory of universal gravitation. It is a matter of fact that the *De Revolutionibus* of Copernicus was not based on new observations, nor was it written to explain observations which previous theory had failed to implement. He made his decision against accepting the planetary theory of Ptolemy and most other astronomers, although consistent with the numerical data, for the reason that he found their planetary theory not sufficiently pleasing to his mind and not in accord with the harmonious and commensurable universe that he believed the consummate artist of the universe had created: it was really a matter of intellectual elegance.

The heliocentric cosmology of Copernicus was criticized by fellow astronomers at that time, notably Tycho Brahe, who argued that when the Earth was moving, the fixed stars should also show an apparent movement by parallax. The argument of Copernicus that the stars were too far away for parallax to be apparent, was rejected for the reason that it left an empty space between Saturn and the stars, and was also inconsistent with the then accepted size of the universe. The implication of the heliocentric cosmology involving a moving Earth led the Church of Rome in 1616 to officially suspend *De Revolutionibus* until corrected. The Catholic Church did not remove it from its index of forbidden books until 1835.

One man, trained in Naples as a Dominican monk, Giodano Bruno (1548–1600) had the courage to read and respond to *De Revolutionibus* and went even further than Copernicus, seeing the fixed stars as innumerable suns with solar systems in an infinite universe. He abandoned the priesthood, travelled widely and in England was praised by Queen Elizabeth I. In 1591 he was lured back to Italy, imprisoned and tortured by the Inquisition, and finally burned at the stake in 1600.

The Copernican model with the Sun at the center of the solar system nevertheless gained ground from Galileo's observations of Jupiter's moons in 1609, and further from Bradley's discovery of the aberration of light (in 1729). The parallax of a fixed star was measured for the first time in 1838 by F.W. Bessel (1784–1846). The circular orbits for planets were replaced by Kepler's elliptical orbits by 1609. Nevertheless, the heliocentric theory of Copernicus marked the beginning of the scientific revolution that reached its climax with Newton about 150 years later.

Rhodes W. Fairbridge and Hans J. Haubold

Bibliography

Beer, A. (ed.) (1975) *Copernicus Yesterday and Today. Vistas in Astron.*, **17**.
Berry, A. (1898) *A Short History of Astronomy.* London: J. Murray (Dover Publications, Inc., 1961).
Biskup, M. (1977) Biography and social background of Copernicus. *Nicholas Copernicus, Studia Copernicana XVII*. Warsaw: Polish Academy of Sciences, pp. 137–52.
Dobson, J.F., and Brodetsky, S. (1947) *Nicolaus Copernicus' De Revolutionibus.* London: Royal Astronomical Society.
Gingerich, O. (ed.) (1975) *The Nature of Scientific Discovery: Copernicus Symposium.* Washington: Smithsonian.
Kuhn, T.S. (1957) *The Copernican Revolution.* Cambridge: Harvard University Press.
McMullin, E. (1987) Bruno and Copernicus. *Isis*, **78**, 55–74.
Rosen, E. (1959) *Three Copernican Treatises*. New York: Dover Publications, Inc.
Rosen, E. (1971) Copernicus, Nicholas. *Dict. Sci. Biogr.*, vol. 3, pp. 401–11.
Rosen, E. (1984) *Copernicus and the Scientific Revolution.* Malabar, Florida: Krieger.
Swerdlow, N., and Neugebauer, O. (1984) *Mathematical Astronomy in Copernicus's De Revolutionibus.* Two Parts. New York and Berlin: Springer-Verlag.
Taub, L. (1933) *Ptolemy's Universe.* Chicago: Open Court.
Weaver, J.H. (ed.) (1987) *The World of Physics*. New York: Simon and Schuster (Vol. I to III).
Westman, R.S. (ed.) (1975) *The Copernican Achievement*. Berkeley and Los Angeles: University of California Press.

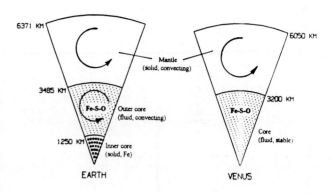

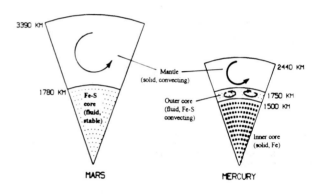

Figure C60 Schematic representation of probable present-day states for Earth, Venus, Mars and Mercury (after Jacobs, 1987; originally from Stevenson, 1983).

CORE, TERRESTRIAL PLANETARY

All the terrestrial planets, i.e. Earth (q.v.), Mercury (q.v.), Venus (q.v.), and Mars (q.v.), are believed to have a central metallic core in a liquid and/or solid state, surrounded by a mantle, (q.v.) and a crust (q.v.). Figure C60 schematically shows the profiles of probable present-day states for the four planets, including size, major chemical constituents and possible motions for their cores.

Since we have no way to observe directly the deep interior of the planets, we can only derive the properties of the interiors from observations made from space and at the surfaces, and from theories derived from a number of different disciplines such as physics, chemistry, astronomy, geology, geophysics, planetary physics and cosmology. In other words, constructing a core model is a theoretical 'inverse' problem, in which the available information is used to infer a possible core and its various properties (mass, density, size, elastic and viscous moduli, states of phase, temperature, pressure, electrical and magnetic properties and chemical constituents). Unfortunately, there is no unique solution for such an inverse problem, and we can only build a core model that is consistent with all reasonable constraints from our updated knowledge.

The Earth's interior has been the subject of extensive and detailed studies (Brush, 1982), from which we have developed various methods and models that may be used to study the interiors of the other planets. Conversely, new knowledge obtained from the other planets may also be applied to the Earth's interior. Until the advent of seismology, we could only make some guesses about the interior of the Earth (Bolt, 1982). Based on modern seismological observations and theory, Gutenberg in 1914 first accurately located the boundary of the central core of the Earth. In 1936 Lehmann produced the first evidence of the existence of the Earth's inner (solid) core. By 1940

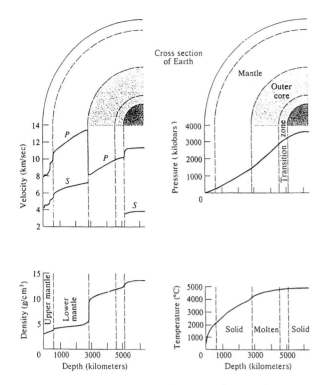

Figure C61 The average variation of seismic velocities, density, pressure and temperature inside the Earth (from Bolt, B., *Inside the Earth*. Copyright (c) 1982 by W.H. Freeman and Company. Used with permission).

Bullen was able to construct a fairly detailed picture of the Earth's core. Figure C61 summarizes our current knowledge of the variation of seismic velocities, density, pressure and temperature within the interior of the Earth as functions of radius.

The landings of Viking 1 and Viking 2 on Mars in 1976 placed the first seismographs on Mars but only one recorded event is likely to be a Marsquake (see Seismicity). Seismographs were also placed at six sites on the Moon during Apollo missions 12, 14, 15, 16, 11 and 17 (q.v.). From 1969 to 1977 the instruments detected between 600 and 3000 moonquakes every year. These seismic data have been used to resolve the Moon's interior structure (Bullen and Bolt, 1985; see also Moon: Seismicity). Various other spacecraft have also been launched to observe these planets (see Mariner missions; Apollo missions; and Venera missions). Figure C62 shows the likely density and temperature profiles of the cores (and mantles) for Mars, Venus, Mercury and Moon from our current knowledge.

Table C11 lists possible major chemical constituents and their masses for the cores (and mantles) of the four terrestrial planets (Surkov, 1990). Although the numbers in Table C11 do not always correspond with observations, the gross picture is generally accepted. It follows from the table that the farther from the Sun, the relatively smaller is the planetary core (e.g. Mercury's core takes 65% of the planet's mass but Mars' core only 12%). The cores of Mercury and Venus, the planets closest to the Sun, are composed of Fe – Ni melt while those of more distant planets (the Earth and Mars) are though to include sulfur in the form of troilite (FeS). The surface magnetic fields for Mercury, Venus, Earth, Moon and Mars are about 2×10^{-3}, 2×10^{-5}, 0.3, 2×10^{-6}, 10^{-4} Gauss respectively. Since the Earth's magnetic field is at least 3.5×10^{9} years old, it is believed that there has been a self-sustained magnetic dynamo in the metallic outer liquid core of the Earth. There is possibly a thin-shell dynamo within the outer liquid core of the Mercury; but it is unlikely that there are dynamos within the cores of Venus, the Moon and Mars (Jacobs, 1987).

The formation of the terrestrial cores is thought to be a part of the story of the formation and evolution of the whole solar system and its planetary system. When the solar system was formed, probably by the gravitational collapse of nebular clouds composed of gases and dusts, about 4.6 billion years ago, the clouds surrounding the new-born Sun had too much spin to be drawn into the Sun. Instead, they spread out into a circumstellar disk, the materials of which then presumably aggregated into planets (q.v.) or planetesimals (q.v.). Because the chemical composition and physical conditions (e.g., temperature and magnetic field) were grossly dependent on the distance from the Sun, the four inner planets are Earth-like, with some indications of order (for instance, the abundance of Fe decreases from Mercury to Mars). In contrast, the outer planets (the planets beyond the orbit of Mars, i.e. Jupiter, Saturn, Uranus, Neptune and Pluto) are colder and rich in volatiles.

The further evolution of the terrestrial planets involved internal differentiation, which resulted in a variety of layered internal structures ('spheres') and eventually formed central cores, mantles, crusts and atmospheres. The central cores could be either liquid or solid, depending on whether the temperature exceeds the melting point of the material under high pressure. The actual core formation processes for the four inner planets were complicated and are not yet fully understood. Various hypotheses are discussed in Tonks and Melosh (1992), Hamblin and Christiansen (1990), Cook (1980), Surkov (1990), Barsukov *et al.* (1992), Vilas, Chapman and Matthews (1988) and Runcorn (1988).

Yuan-Chong Zhang

Bibliography

Anderson, O.L. (1982) Are anharmonicity corrections needed for temperature-profile calculations of interiors of terrestrial planets? *Phys. Earth Planet. Int.* **29**, 91.
Barsukov, V.L. Basileusky, A.T., Volkov, V.P. and Zharkov, V.N. (1992) *Venus Geology. Geochemistry and Geophysics*. Tucson: University of Arizona Press.
Bolt, B.A. (1982) *Inside the Earth: Evidence from Earthquakes*. San Francisco: W.H. Freeman.
Brush, S.G. 1982. Chemical history of the Earth's core. *Eos*, **63** (47), 1185–8.
Bullen, K.E. and Bolt, B.A. (1985) *An Introduction to the Theory of Seismology*, 4th edn. Cambridge: Cambridge University Press.
Cook, A.H. 1980. *Interiors of the Planets*. Cambridge: Cambridge University Press.
Hamblin W.K. and Christiansen E.H. (1990) *Exploring the Planets*. New York: Macmillan Publishing Company.
Jacobs, J.A. (1987) *The Earth's Core*, 2nd ed. New York: Academic Press.
Runcorn, S.K. (1988) *The Physics of the Planets*. New York: John Wiley and Sons.
Stevenson, D.J. (1983) Planetary magnetic fields. *Rep. Prog. Phys.*, **46**, 555.
Surkov, Yu.A. (1990) *Exploration of Terrestrial Planets from Spacecraft: Instrumentation, Investigation, Interpretation*. New York: Ellis Horwood.
Tonks, W.B. and Melosh H.J. 1992. Core formation by giant impacts. *Icarus*, **100**, 326–46.
Vilas, F., Chapman, C.R. and Matthews M.S. (1988) *Mercury*. Tucson: University of Arizona Press.

Cross references

Apollo missions
Crust
Earth
Mantle
Mariner missions
Mars
Mercury
Moon: siesmicity
Planet
Planetesimal
Siesmicity
Terrestrial planets
Venera missions
Venus

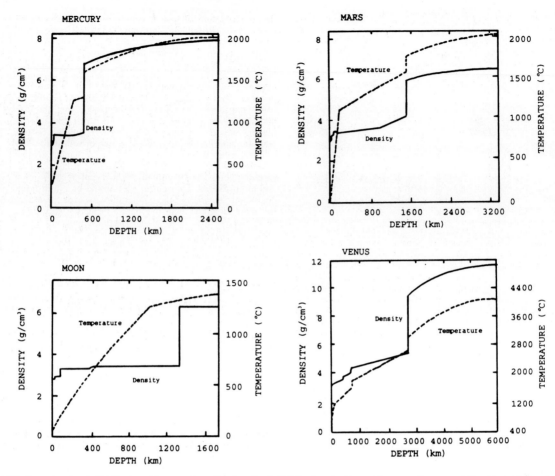

Figure C62 Core density (solid line) and temperature (broken line) distribution in Mars, Mercury, Venus and the Moon (after Jacobs, 1987; originally from Anderson, 1982).

Table C11 Composition of terrestrial planets (percentage by mass)

Component	Mercury		Venus		Earth		Mars	
	Planet in general	Mantle/core	Planet in general	Mantle/core	Planet in general	Mantle/core	Planet in general	Mantle/core
Mantle								
MgO	12.83	36.89	25.74	37.82	24.25	34.79	23.07	26.20
SiO_2	11.67	33.53	36.16	53.13	34.06	48.88	32.41	36.80
Al_2O_3	5.32	15.30	2.61	3.83	2.46	3.53	2.34	2.65
CaO	4.97	14.28	2.43	3.58	2.29	3.29	2.18	2.48
Na_2O	0.00	0.00	1.12	1.64	1.05	1.51	1.0	1.14
FeO	0.00	0.00	0.00	0.00	5.58	8.00	27.06	30.72
Total	34.79	100.00	68.86	100.00	69.69	100.00	88.07	99.99
Core								
Fe	61.75	94.69	30.25	94.69	24.16	79.73	6.09	51.00
Ni	3.46	5.31	1.70	5.31	1.60	5.27	1.52	12.74
S	0.00	0.00	0.00	0.00	4.54	15.00	4.32	36.25
Total	65.21	100.00	31.94	100.00	30.31	100.00	11.93	100.00

CORIOLIS FORCE, GEOSTROPHIC MOTION

Newton's second law states that acceleration is proportional to the net external force acting on a particle. This applies, however, only to a reference frame that is at rest or in uniform motion. Measurements made on rotating planets occur in an accelerating reference frame instead. Therefore, motion in a straight line as viewed from space appears as curved motion on the surface of the planet. This creates apparent, or inertial, forces seen by an observer on the surface of the planet. The most well-known apparent force is the Coriolis force, named after the 19th-century French mathematician Gaspard G. de Coriolis.

The Coriolis force is most easily illustrated by considering the path traced by a piece of chalk on a rotating turntable. When the turntable is at rest, a line drawn outward from the center traces a straight path.

When the turntable rotates, a line drawn straight from the center will trace a curved path as the turntable rotates underneath it. In the meteorological case, air moving equatorward from the pole has less angular momentum than air near the equator, since the latter is farther from the axis of rotation. Thus, since the air conserves its angular momentum it acquires a tangential motion slower than that of the underlying surface as it drifts equatorward. The argument can be generalized to motion in any direction. The general rule is that air deflects to the right of its original direction of motion in the hemisphere containing the positive rotation axis, and to the left in the opposite hemisphere. The Coriolis force changes only the direction of the wind, not its speed.

The mathematical form of the vector Coriolis force is $F_{Co} = -2\rho\Omega \times v$, where ρ is the density of air, Ω the planetary rotation vector and v the wind velocity vector. It can be seen from this equation that the Coriolis force has a vertical as well as horizontal component. The vertical Coriolis force is much weaker, however, than gravity and the vertical pressure gradient force and is usually neglected. Likewise, vertical motions can cause a horizontal Coriolis force, but vertical motions are typically too weak for this to be of significance.

In the absence of any real horizontal forces, a parcel of fluid on a rotating planet would undergo a circular motion, called inertial motion, due to the Coriolis force. The time period required to complete a full inertial circle is one half the rotation period divided by the sine of latitude. This is equivalent to the time required for a Foucault pendulum to rotate through 180°. Inertial motion can at times be observed in Earth's oceans.

The Coriolis force exerts a controlling influence on most large-scale aspects of the general circulations of all planetary atmospheres. On rapidly rotating planets it is the dominant force balancing the pressure gradient force. It therefore determines the general sense of wind flow around the planet, creating for example, the Earth's jet stream. The sense of flow around synoptic-scale low and high pressure centers is also governed by the Coriolis force. Even on slowly rotating planets such as Venus, where the Coriolis force is small, it is still believed to determine ultimately the direction of the superrotating winds, although it is not a major component of the force balance. The same is true for terrestrial tornadoes and, at least in part, for hurricanes, where centrifugal forces dominate instead.

Geostrophic motion

In atmospheres, winds blow predominantly in a direction determined by a balance of major forces acting in the zonal (east–west) and meridional (north–south) directions. Every atmosphere is subjected to spatial variations in heating. For example, more sunlight falls on the equator than on the poles on most planets, and this differential heating creates horizontal pressure gradients. The resulting force, if acting alone, would cause air to blow from high to low pressure. But planets rotate, and the Coriolis force is applied; it acts perpendicular to the direction of motion. On planets with prograde rotation, air is deflected to the right in the northern hemisphere and to the left in the southern hemisphere. On a planet like Venus with retrograde rotation, the opposite is true.

If the Coriolis force is large enough to balance the pressure gradient force, the result is called geostrophic motion (from the Greek geo-, earth, plus -strephen, to turn). This is the case if winds are significantly slower than the speed with which a point on the surface of the planet moves due to the planet's rotation. Earth, Mars, Jupiter, Saturn, Uranus and Neptune all rotate rapidly enough to satisfy the criterion for geostrophic motion, while the slowly rotating planet Venus and the moon Titan do not. Specifically, geostrophic motion prevails if the dimensionless Rossby number $Ro = U/fL$ $<< 1$, where U is a typical wind speed, $f = 2\Omega \sin \theta$ is called the Coriolis parameter, Ω is the angular velocity of the planet's rotation, θ is latitude and L is a typical length scale associated with a given meteorological phenomenon. Thus, geostrophic motion is more common in middle and high latitudes than in the tropics, where $\sin \theta \rightarrow 0$, and it is more likely to be associated with large-scale than with small-scale phenomena.

Geostrophy manifests itself in several obvious ways in Earth's atmosphere. Latitudinal differential solar heating creates a pressure gradient force directed from equator to pole. This is balanced by a Coriolis force directed from pole to equator. This can only be achieved if air moves from west to east on planetary scales in both hemispheres. The resulting wind is called the jet stream, and it accounts for the general eastward movement of weather patterns characteristic of mid-latitudes. Instability of the jet stream leads to the creation of individual low and high pressure centers on synoptic space scales (~ 1000 km). If air moves counterclockwise around the low and clockwise around the high (in the northern hemisphere), the regional balance is also geostrophic. Thus air tends to blow from the south in advance of a low pressure center, and shifts to blowing from the north as the low passes and high pressure approaches.

One consequence of geostrophic motion is that air moves along lines of constant pressure (isobars) rather than perpendicular to them. Thus the approximate wind direction is easily inferred from a standard weather map which contains isobars of the pressure field (or lines of constant geopotential height, which act in similar fashion). However, since geostrophic motion does not cross the isobars, it does no work and transports no heat. Instead, the poleward heat transport accomplished by weather systems is performed by the small (10–15%) deviations from geostrophic motion. These deviations are responsible for vertical motions, warm and cold fronts, and ultimately the clouds and precipitation that accompany mid-latitude weather patterns.

Anthony D. Del Genio

Bibliography

Panofsky, H.A. (1981) Atmospheric hydrodynamics, in *Dynamic Meteorology – An Introductory Selection* (ed. B.W. Atkinson). New York: Methuen, pp. 8–20.
Wallace, J.M., and Hobbs, P.V. (1977) *Atmospheric Science: An Introductory Survey*. New York: Academic Press.
Warsh, K.L., Echternacht, K.L. and Garstang, M. (1971) Structure of near-surface currents east of Barbados. *J. Phys. Oceanogr.*, **1**, 123–9.

Cross references

Atmosphere
Cyclone, anticyclone
Earth rotation

CORONA (VENUS)

A corona (pl. coronae) is a volcanotectonic landform that is apparently unique to Venus in the solar system. Named for the Latin word for garland or crown, coronae are large circular to oval features distinguished by a subconcentric ring of ridges and grooves that is elevated a few hundred meters above surrounding topography. Most coronae have numerous volcanic structures, particularly small domes, inside the ring and a topographic moat a few hundred meters deep just outside the ring. Coronae generally range from 100 to 600 km in diameter with a median of about 200 km (Ivanov and Basilevsky, 1990), although a few coronae over 1000 km have been identified. The ring of ridges and grooves and the surrounding moat are usually tens of kilometers wide.

Although recognizable in earlier Earth-based radar imagery, coronae were not identified as unique structures until after the Soviet Venera 15 and 16 missions, and were originally called ovoids for their oval shape (Barsukov et al., 1986; Basilevsky et al., 1986). The subsequent US Magellan mission showed that there are more than 200 coronae on Venus. Coronae are usually found in clusters and are often interconnected. Most coronae are at elevations near the planetary mean and are associated with large rift systems.

Several theories have been put forth to explain the origin of coronae, ranging from modified impact structures to rising or sinking mantle diapirs. The most popular theory involves doming of the surface followed by topographic collapse (Stofan and Head, 1990). The source of the doming is a subject of intense debate; the source may be thermal or thermochemical diapirs of hot material rising from as deep as the core-mantle boundary or as shallow as the lithosphere. Understanding these enigmatic features is central to understanding how mantle convection relates to surface deformation on Venus.

R.R. Herrick

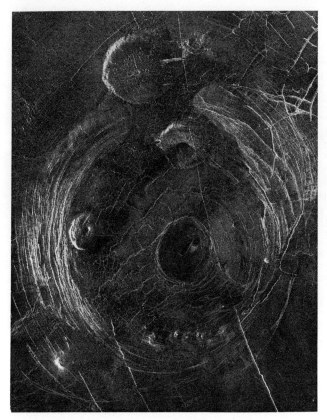

Figure C63 This Magellan radar image shows a region approximately 300 km across, centered on 59° S latitude, 164° E longitude and located in a vast plain to the south of Aphrodite Terra. The data for this image was obtained in January 1991. The large circular structure near the center of the image is a corona, approximately 200 km in diameter and provisionally named Aine Corona. JPL P-38340.

Bibliography

Barsukov, V.L. Basilersky, A.T., Burba, G.A. *et al.* The geology and geomorphology of the Venus surface as revealed by the radar images obtained by Veneras 15 and 16. *J. Geophys. Res.*, **91**, D378–98.

Basilevsky, A.T. Pronin, A.A., Ronca, L.B. *et al.* Styles of tectonic deformations on Venus: analysis of Venera 15 and 16 data. *J. Geophys. Res.*, **91**, D399–411.

Ivanov, M.A. and A.T. Basilevsky (1990) Coronae and major shields on Venus: comparisons of their areas, basal altitudes and areal distribution. *Earth, Moon, and Planets*, **50–51**, 409–20.

Stofan, E.R. and J.W. Head (1990) Coronae of Mnemosyne Regio: morphology and origin. *Icarus*, **83**, 216–43.

Cross references

Hot spot tectonics
Venus
Venus: geology and geophysics

CORONAL MASS EJECTIONS

Coronal mass ejections, CMEs, are spectacular manifestations of solar activity in which 10^{15}–10^{16} g of material from the solar atmosphere are expelled into interplanetary space (Gosling *et al.*, 1974; Hundhausen, 1988; Kahler, 1988). They were first detected optically in the early 1970s by telescopes flown on the OSO 7 satellite and on Skylab. CMEs have outward speeds ranging from less than 50 km s^{-1} in some events to greater than 1200 km s^{-1} in others. During their expansion outward from the Sun, they quickly attain dimensions far larger than the Sun itself. CMEs originate in regions in the solar corona that are magnetically closed and not previously participating in the solar wind expansion. Typically these closed field regions are found in the coronal streamer belt that both encircles the Sun and underlies the heliospheric current sheet. CMEs are often observed in close association with other forms of solar activity such as eruptive prominences and solar flares; however, careful timing measurements indicate that eruptive prominences and flares are not the causes of CMEs (Harrison *et al.*, 1990). Like other forms of solar activity, CMEs occur with a frequency that varies in a cycle of ~ 11 years. It has been estimated that on average the Sun emits about 3.5 CMEs per day near solar activity maximum, but only about 0.1 CMEs per day near solar activity minimum (Webb, 1991). Only a small fraction of all CMEs are directed earthward.

In the solar wind near the orbit of Earth, CMEs usually have distinct plasma and field signatures that distinguish them from the ordinary solar wind (Gosling, 1990). The most reliable of these signatures are those that are consequences of the closed magnetic field topology of CMEs. Virtually all transient shock waves in the solar wind are driven by CMEs; however, only about one out of three CMEs has a sufficiently high speed relative to the ambient solar wind ahead to produce a shock wave disturbance in the solar wind. The remainder of the CMEs simply ride along with the wind. Typically an interplanetary disturbance driven by a fast CME is 2–4 times broader in extent than the CME itself. Near solar activity maximum CMEs account for ~ 15% of all solar wind measurements in the ecliptic plane, while near solar activity minimum they account for less than 1% of all the measurements.

The prime importance of CMEs is related to their role in (1) the long-term evolution of the structure of the solar corona and the solar wind, (2) the production of large, transient disturbances in the solar wind, and (3) the stimulation of major disturbances in the geomagnetic field. For example, it is now clear that CMEs are the prime link between solar activity and geomagnetic activity (Gosling *et al.*, 1991). Virtually all of the largest geomagnetic storms observed near solar activity maximum are associated with Earth-passage of interplanetary disturbances driven by fast CMEs. The geomagnetic effectiveness of Earth-directed CMEs is directly related to the speed of the events at 1 AU and the strength of the southward component of the interplanetary magnetic field associated with them.

Research on CMEs, both observational and theoretical, is intensely pursued. The fundamental cause of CMEs and the factors that determine the size and speed of the ejections are still not well understood. One good possibility is that CMEs are the result of instabilities associated with the gradual, global evolution of the coronal magnetic field. The overall magnetic field topology of CMEs in interplanetary space is not yet firmly established. For example, it is not clear whether the field topology is that of simple magnetic loops or twisted flux ropes attached to the Sun at both ends, or of plasmoids magnetically detached from it. Each CME drags new magnetic flux out into the solar wind. In order to avoid a long-term build up of magnetic flux in interplanetary space, this addition of new flux must ultimately be balanced by the removal of magnetic flux elsewhere in interplanetary space. It appears likely that this removal is accomplished by the process of magnetic merging, either within the CMEs themselves or elsewhere in the solar wind or solar corona (McComas and Phillips, 1992).

Finally, because geomagnetic storms affect mankind's utilization of space in a variety of ways, there is a considerable interest in predicting the geomagnetic effects of solar activity and, in particular, CMEs. This problem is not yet solved. Although virtually all large geomagnetic storms near solar activity maximum are associated with CMEs, only about one in six Earthward-directed CMEs is effective in stimulating a large geomagnetic disturbance. Present evidence indicates that the initial speed of a CME close to the Sun is probably the most important factor in determining if an Earthward-directed event will be geomagnetically effective. Unfortunately, with present solar instrumentation it is extremely difficult even to detect Earthward directed CMEs, much less determine their speeds. Improvements in predicting the geomagnetic effects of solar activity will depend upon (1) progress in theoretical understanding of the physical processes that produce CMEs, (2) the development of techniques to predict the orientation of the interplanetary magnetic field, and (3) the develop-

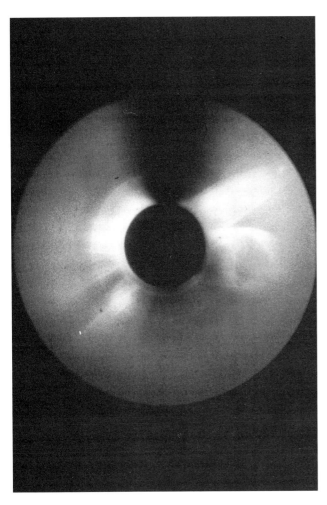

Figure C64 A coronal mass ejection above the east limb of the Sun (to the left) as observed with the coronagraph on Skylab. The Sun is obscured by the occulting disk, whose effective diameter is 1.5 times that of the Sun.

ment of techniques to detect Earthward-directed CMEs and their speeds close to the Sun.

J.T. Gosling

Bibliography

Gosling, J.T. (1990) Coronal mass ejections and magnetic flux ropes in interplanetary space; In *Physics of Magnetic Flux Ropes*. (ed. C.T. Russell, E.R. Priest and L.C. Lee). Washington, DC: American Geophysical Union, Geophysical Monograph Series, Vol. 58, pp. 343–364.

Gosling, J.T., *et al.* (1974) Mass ejections from the Sun: a view from Skylab. *J. Geophys. Res.*, **79** (31), 4581–4587.

Gosling, J.T., *et al.* (1991). Geomagnetic activity associated with Earth passage of interplanetary shock disturbances and coronal mass ejections. *J. Geophys. Res.* **96** (A5), 7831–7839.

Harrison, R.A., *et al.* (1990) The launch of solar coronal mass ejections: results from the coronal mass ejection onset program. *J. Geophys. Res.*, **95** (A2), 917–937.

Hundhausen, A.J. (1988) The origin and propagation of coronal mass ejections, in *Proc. 6th International Solar Wind Conference*, Technical Note 306 + Proc. (eds V. Pizzo, T.E. Holzer and D.G. Sime). Boulder: National Center for Atmospheric Research, pp. 181–214.

Kahler, S. (1988) Observations of coronal mass ejections near the Sun, in *Proc. 6th International Solar Wind Conference*, Technical Note 306 + Proc. (eds V. Pizzo, T.E. Holzer and D.G. Sime). Boulder: National Center for Atmospheric Research, pp. 215–231.

McComas, D.J. and J.L. Phillips (1992) The extension of solar magnetic fields into interplanetary space, in *Proc. 1st SOLTIP Symposium*, Vol. 1 (eds S. Fischer and M. Vandas). Prague: Astronomical Institute of the Czechoslovak Academy of Sciences, pp. 180–191.

Webb, D.F. (1991) The solar cycle variation of the rates of CMEs and related activity. *Adv. Space Res.* **11** (1), 37–40.

Cross references

Interplanetary magnetic field
Shock waves
Solar corona
Solar wind

CORPUSCULAR RADIATION

Of utmost importance for understanding the physics of the universe is the realization that the stars, the galaxies and the other celestial objects (including the interstellar matter) emit particles (corpuscular radiation) as well as photons (light radiation). Thus corpuscular radiation may be defined as the signature of nuclear processes taking place at many astrophysical sites. Two of these are of importance: the Sun as a star, and the rest of the universe. Gas atoms, ions, electrons and molecules fill space, and by measuring their flux and energy, we may obtain information about the emitting regions.

Particles streaming from outer space

Cosmic rays, before the space age, were studied by elementary particle physicists, who discovered in them positrons, several kinds of mesons and many other strange particles. Today the definition of cosmic rays has been broadened, to emphasize their role in astrophysics: they can be considered as 'the birth cry of atoms' in cosmic space. Protons dominate the nuclear component, contributing almost 90% of the flux at the top of the atmosphere. Helium nuclei contribute another 9% and the remaining 1% are heavier nuclei (up to $Z = 95$). The electron component is about 1% as abundant as nuclei in the same energy interval, and positrons form about one-tenth of this component. Cosmic rays originate in localized regions: those with an energy greater than 10^{16} to 10^{20} eV are produced in outer galaxies (and possibly from pulsars), whilst those of lower energy (less than 10^{11} eV) originate in supernovae sources or explosions in our galaxy. Perturbations in intensity are introduced by the Earth's magnetic field, and cosmic rays are modulated by the solar cycle, with a clear inverse correlation (Forbush effect). The fundamental question of how these cosmic ray nuclei and electrons are accelerated to their fantastic energies is still wide open.

Particles streaming from the Sun

Protons emitted by the Sun are often referred to as 'solar cosmic rays', but although protons constitute the principal component, a number of other particles travel from the Sun to the Earth. Except for a thin un-ionized layer around the photosphere, the Sun is a gaseous plasma, made up of protons, electrons, neutral particles and heavy ions, all coming from its constituents, which are hydrogen (76%), helium (21%) and the balance an admixture of heavier elements, chiefly carbon, nitrogen, oxygen, neon, magnesium, silicon and iron.

Solar flares

Along with enhanced bursts of UV and x-ray emission constituting major solar flares, copious quantities of nucleons are emitted from major active spot groups. The particles arrive in the vicinity of the Earth about 1 to 2 h after the flare eruption. The corpuscular emission may last 1 to 6 d. The energy is such that the particles penetrate to altitudes of 30 km or less in our atmosphere. Not all major flares are accompanied by corpuscular emission, but those that are (often referred to as proton flares), are also accompanied by type IV solar radio bursts (solar emissions occurring over a broad band of radio frequencies). These solar proton events are solar cycle modulated; their entry into the atmosphere is principally a high-latitude phenomenon. Low-energy particles cause geomagnetic storms and auroras.

Coronal mass ejection

The possibility that particles might leave the Sun was raised by observations of great eruptions, for which the speed of the particles exceeds the escape velocity (618 km s^{-1}). Such majestic eruptions are surprisingly common. The mass ejected is around 10^{15} to 10^{16} g d^{-1}, which is about the same order of magnitude as for the steady solar wind.

Solar wind

The average 'quiet' solar wind as observed at 1 AU has a speed of approximately 350 km s^{-1} with a density of about 1 to 10 cm^{-3}. This implies a particle flux of about 10^{36} particles s^{-1}. In coronal holes, where the high speed streams originate, the observed speeds are about 700–800 km s^{-1}, but such streams have lower densities of about 5 cm^{-3}. Hence the particle flux is about the same, although one major difficulty is to explain theoretically these observed speeds. The Ulysses spacecraft has already confirmed these two regimes through observations of the fast wind flowing from the polar coronal holes.

Correlation of corpuscular radiation with Earth's climate

Both cosmic rays and solar particles ultimately couple into meteorological processes through the alteration of the chemical or electrical properties of the atmosphere. Protons gradually lose energy through ionizing collisions with the atmospheric constituents. Specific air ionization rates can be computed, and a typical profile with altitude is deduced. Consideration of the energetics involved demonstrates that the energy, primarily in the form of corpuscular radiation supplied to the Earth's atmosphere by solar or galactic events, does not compete with the total solar irradiance. The disparity between the corpuscular radiation and solar insolation becomes less severe when we recognize that shielding by the geomagnetic field forces the major deposition of corpuscular radiation to focus at high latitudes where solar insolation effects become minimal, particularly under local wintertime conditions.

J.P. Rozelot

Bibliography

Herman, J.R. and Goldberg, R.A. (1978) *Sun, Weather and Climate*. Washington, DC: NASA. 360 pp.
Marsden, R.G., *et al*. (1996) The Ulysses mission. *Astron. and Astrophys*., special issue.

Cross references

Aurora, planetary
Cosmic rays
Forbush effect
Solar particle event
Solar wind
Sun

COSMIC RAYS

Cosmic rays were discovered in 1912 by Victor Hess during a balloon ascent to a height of 17 500 ft (~ 5500 m) – the objective being to determine the source of the 'radiation' that caused terrestrial ionization chambers to show a reading no matter how carefully they were shielded. Hess found a dramatic increase in signal as he proceeded above about 6000 ft (2000 m). and he correctly diagnosed the cause as a radiation coming from outside the atmosphere. Hess, understandably, thought that some form of ultra-penetrating gamma rays were responsible – hence the term 'radiation' – but later work showed that it was mostly particles (largely protons and heavier nuclei) that were causing the results. Most cosmic rays entering the atmosphere have velocities very close to the speed of light (*in vacuo*) and in consequence strong interactions occur with the atomic nuclei of the atmosphere. These interactions were the subject of intense study in

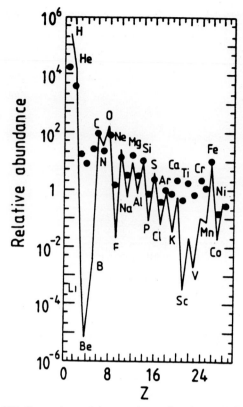

Figure C65 Comparison of the cosmic ray abundances (full circles) and universal abundances (indicated by full lines). The data are normalized at carbon. The CR results refer to the range 170–280 MeV per nucleon.

the 1930s, 1940s and 1950s, and some of the more important so-called 'elementary particles' were discovered; specifically, the positron, the muon, the pion and the strange particles. After this period artificial accelerators became increasingly useful, however, and interactions were, and indeed are, studied in this way. Although there is still some interest amongst cosmic ray physicists in interactions – at very high energies – most now work in areas related to astrophysics: where do the cosmic rays come from and how do they propagate through the galaxy and beyond?

Mass composition of cosmic rays

Although cosmic rays extend in energy as high as 10^{20} eV, and perhaps beyond, the bulk of our knowledge about their masses is confined to energies below about 10^{12} eV. The region below 10^9 eV has been studied rather comprehensively and Figure C65 shows the mass composition (of nuclei); the 'universal abundances' are also indicated. Comparison of the two shows some similarities and some remarkable differences. The differences can be understood in terms of the filling in of holes in the universal abundances by the fragments of heavier nuclei, the fragmentation being due to the break-up of heavy nuclei, accelerated elsewhere, on the atomic nuclei of the interstellar medium. At higher energies the information about the mass composition becomes increasingly imprecise and above 10^{15} eV there is great argument. The author's view is that the composition of Figure C64 probably extends to above 10^{17} eV, above which there is an increase in the relative fraction of heavy nuclei, with iron nuclei predominating. Above about 10^{19} eV there is a further change, as we shall see.

Energy spectrum

The energy content of the cosmic radiation is considerable, on the astrophysical scale. Table C12 gives a summary. Also included are the electron and gamma ray components. It will be noted that there is near equality between the energy densities of starlight, gas motions

Table C12 Energy densities of the cosmic ray components near the Earth

Component	Energy (eV)	Energy density (eV cm^{-3})
Protons and heavier nuclei	Above 10^9	$\sim 5 \times 10^{-1}$
	10^{12}	2×10^{-2}
	10^{15}	10^{-4}
	10^{18}	10^{-5}
Electrons and positrons	Above 10^9	$\sim 6 \times 10^{-3}$
	10^{10}	$\sim 5 \times 10^{-3}$
	10^{11}	$\sim 5 \times 10^{-4}$
γ-rays: diffuse background	Above 10^7	$\sim 1 \times 10^{-5}$
	10^8	2×10^{-6}

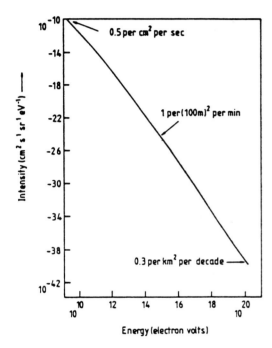

Figure C66 Energy spectrum of primary cosmic rays. The particles are thought to be mainly protons at all energies up to a few times 10^{17} eV, above which heavy nuclei appear to dominate. At lower energies, the fluxes of heavier nuclei ($\cong$ 10%) and electrons ($\cong$ 3% at $\sim 10^9$ eV) are small but, nevertheless, their significance is great. The rates shown relate to fluxes above the energies indicated.

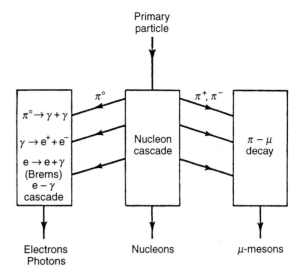

Figure C67 Schematic diagram of the propagation of an extensive air shower.

The important arrival directions of the primary particles are determined by timing the arrival of the shower particles at the various detectors. Uncertainties of several degrees are common.

Origin of the particles

From the time of their discovery it was realised that the Sun was responsible for only a few, if any, of the cosmic rays. However, it is now known that the Sun does produce some from time to time and their analysis is very instructive. At the very lowest energies the 'solar wind' can be considered. This is generated in the solar coronae; particles, protons and electrons and other ions stream out with velocities of several hundred km s^{-1}. On encountering the Earth's magnetic field this plasma causes a 'bow shock', a phenomenon of considerable interest in its own right. It seems that some of the ions are accelerated in the vicinity of the shock – and elsewhere – and some find their way into the Earth's atmosphere, particularly near the magnetic poles where they form the aurorae (in both the Arctic and Antarctic). Also contributing are particles which have leaked out of the Van Allen radiation belts where embryonic cosmic rays are kept, trapped by the geomagnetic field. Although the auroral particles are too low in energy to be classed as bona fide cosmic rays, the mechanism by which many are accelerated – the shock process – almost certainly plays a role at higher energies for much bigger systems, as we shall see.

Occasionally, great outbursts erupt on the surface of the Sun and from these giant solar flares particles of quite respectable energies appear (up to about 10^{10} eV). The manner in which the particles achieve their energy on the Sun is not completely clear but reconnection of magnetic fields almost certainly plays a role. Although stars like the Sun contribute far too few cosmic rays (by about a factor 100 000) to be responsible for the detected total flux – and their individual energies are too low – many of us believe that stars of different types which flare more frequently inject cosmic rays into space and these are then accelerated to higher energies by other processes.

The difficulty in determining cosmic ray origin is that the galaxy is threaded by a tangled magnetic field which deflects the charged cosmic ray particles hither and thither so that their arrival directions are very nearly completely random. The other side of the coin is that the very deflection of the very low mass electrons by the magnetic field causes radio waves to be emitted (the so-called synchrotron radiation), and this process is in fact responsible for the whole field of radio astronomy.

The giant radio telescopes tell us of the ubiquity of the cosmic ray electrons (and the magnetic field) in our own galaxy and in others. It is found that the cosmic ray electron intensity in some galaxies is far higher than in ours.

and magnetic fields in the local region of the galaxy; equality with the last two can be understood but that with starlight cannot.

Figure C66 shows the energy spectrum of the nuclear component and an indication of the rates of arrival at three energy thresholds. It will be noted that the rate falls off extremely rapidly with increasing energy. Particles of energy above about 10^{15} eV can only be detected at all by virtue of the great 'extensive air showers' that they cause in the atmosphere. The primary particle's energy is transferred through a cascade to secondary particles, principally electrons, 10% muons and a few percent of protons, neutrons and pions. Figure C67 shows the cascade. The total number of particles at ground level is very roughly the initial energy in electron volts divided by 10^{10}. The principle of the detection method at very high energies is to distribute as many detectors as can be afforded over a wide area (several km^2) and to sample the shower. The lateral distribution of the particles can be determined and integration over the whole shower together with an adopted model leads to an estimate of the primary energy. The estimate of mass is bound up with the adopted model – hence the argument.

It is poetic that cosmic gamma rays, first thought to comprise the cosmic rays, then written off, should make a comeback in the search for cosmic ray origin. It is true that they represent only a very tiny fraction of the cosmic radiation but their flux is finite ($\gamma/p \sim 10^{-6}$ at 10^9 eV). Gamma rays, unlike charged particles, travel in straight lines through magnetic fields and one can thus trace them back to their sources.

The gamma ray sky at energies above 70 MeV from the satellite-born detectors shows a form not too dissimilar to the synchrotron sky, near the galactic plane at least; there are a few 'discrete' sources together with a rather diffuse component distributed rather widely but again mainly confined to the plane.

The discrete sources are very interesting, although their contribution to the total gamma ray flux is only some 20% or so. Four have definitely been identified as being due to pulsars by virtue of the gamma ray signal pulsing at the same rate as that of the radio radiation. Electrons appear to be accelerated close to the pulsar surface and they are deflected to produce radio waves, the deflection taking place in the pulsar's magnetic field. The magnetic field is aligned at an angle to the rotation axis, thus giving a lighthouse effect. The gamma rays from the pulsars are also produced by electrons, probably accelerated further away from the pulsar surface, but there is some evidence at much higher energies for protons also being accelerated in pulsars, but the case for this is not yet watertight. The evidence about proton sources comes more particularly from the diffuse gamma ray flux, much of which we believe to come from the interactions of cosmic ray protons with the gas in the interstellar medium. Until recently there was argument as to whether the cosmic ray protons seen were coming mainly from sources within our own galaxy or were coming from outside (i.e. were 'extragalactic'). Analysis of the diffuse gamma rays seems to have sorted this problem out, for most protons (energies below 10^{10} eV or so), at least. The method is to work out on a large scale the distribution of cosmic ray proton intensity in the galaxy – assuming a simple dependence on distance to the center of the galaxy – which would give the observed distribution of gamma ray flux. A necessary ingredient is the distribution of interstellar gas in the galaxy and, although there are still some arguments about the gas, the result is a falling proton intensity as a function of distance from the galactic center.

Pulsars seem unlikely as the sources at these low energies (perhaps their contribution is important at much higher energies?). Instead, we have evidence for supernova remnant (SNR) shocks, rather than the supernova explosions themselves providing the energy. Models for cosmic ray acceleration in SNR give predictions quite close to what we appear to find and the result is that SNRs are probably the prime candidates for 'cosmic ray accelerators' at low energies where most cosmic rays are to be found.

The SNR mechanism probably works up to about 10^{14} eV above which, according to the modelers, SNRs become inefficient. At higher energies pulsars may be important sources, but again there is an energy (10^{17} to 10^{18} eV?) beyond which their efficiency is in doubt. It is here that very recent observations suggest that heavy nuclei may take over and Galactic sources may be efficient producers of these nuclei to 10^{19} eV or so. At even higher energies, specifically above about 10^{19} eV, various factors suggest strongly that the particles (mainly protons, we think) are not produced in our galaxy but are here extragalactic. Lack of knowledge of the conditions in other galaxies gives the theorists a field day to put forward a variety of options. Galaxies with very active nuclei may contribute and galaxies in clusters moving with great speed may cause 'bow shocks', or, at least cause considerable turbulence in the low-density ionized gas between the galaxies and in turn generate extremely energetic particles.

Cosmic rays and cosmology

The interaction of cosmic rays with cosmology is twofold – both bad and good. The 'bad' refers to the role of synchrotron radiation from high galactic latitudes in providing a 'foreground' against which fluctuations in the cosmic microwave background are sought. The result has been to require measurements at frequencies above 20 GHz in order to minimize the cosmic ray effects. The recent COBE claims to have detected 'ripples in the early universe' are indeed at higher frequencies, but another problems appear here – dust, heated by starlight and cosmic rays emitting radiation in the same high frequency region. The more acceptable face of cosmic rays and cosmology relates largely to the interactions of extragalactic

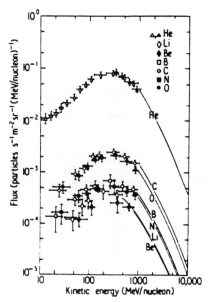

Figure C68 Primary spectrum in the low-energy region near sunspot maximum.

particles with the cosmic microwave background (CMB). The effects are twofold: the extragalactic proton spectrum will be reduced (starting at $\sim 2 \times 10^{18}$ eV with a modest reduction followed by a much bigger fall above 6×10^{19} eV) and the interaction products cascade through the universe, leading to significant fluxes of gamma rays at lower energies. The flux of extragalactic gamma rays in the 100-MeV region will be a sensitive indicator of the extragalactic intensity of protons above 10^{18} eV or so. Already, it has proved possible to rule out an otherwise promising model in which the ultra-energetic particles come from collapsing cosmic strings (Chi et al., 1992).

Cosmic rays and planetary sciences

Although the bulk of cosmic rays originate from beyond the planetary system, the interaction of the subjects is not negligible. Mention has been made already of the production of some cosmic rays by the Sun and processes in the solar corona whereby particles of different charge are accelerated; such phenomena represent examples of what happens elsewhere on much larger scales.

The effect of the solar wind on the galactic particle flux, thereby modifying the spectrum measured at Earth, is considerable at energies below about 10^{10} eV. Figure C68 gives an example. The modulation – a topic of considerable interest in its own right – is even of importance at an energy of 10^{11} to 10^{12} eV in the sense that it thwarts efforts to use the measured anisotropies of these abundant low-energy particles to give indications as to their origin or origins.

Turning to the planets, the role of Jupiter in accelerating some of the low-energy electrons (10^7 eV or so) seen at Earth has been realized. The advent of space craft encountering the planets and thereby able to study the acceleration mechanisms in their magnetospheres is leading to important advances.

The solar wind also has a role in generating the 'anomalous component' of cosmic rays: neutral atoms ionized by solar radiation, which are then accelerated in the solar wind and appear from otherwise forbidden directions.

Finally, there is the study of cosmic rays 'tracks' left in solar system bodies by the passage of cosmic rays particles. Studies of such tracks in lunar and meteoritic materials lead to the conclusion that the average cosmic ray intensity has not changed much over the past eons; specifically, variations of probably less than a few tens of percent over 10^7 years, and less than a factor of a few over 10^9 years.

Arnold W. Wolfendale

Bibliography

Chi, X., Dahanayake, C., Wdowczyk, J. and Wolfendale, A.W. (1992) *Astroparticle Phys.*, **1**, 129.
Dorman, L.I. (1974) *Cosmic Rays: Variations and Space Explorations*. Amsterdam: North-Holland Publ. Co., 675 pp.
Forbush, S.E. (1954) World-wide cosmic-ray variations, 1937–52. *J. Geophys. Res.*, **59**, 525–42.
Forbush, S.E. (1966) Time variation of cosmic rays in *Handbuch der Geophysik* (ed. J. Bartels) New York: Springer-Verlag, **3**, 159–247.
Rossi, B. (1964) *Cosmic Rays*. New York: McGraw-Hill, 268 pp.

Cross references

Forbush decrease
Forbush effect

COSMIC RAY EXPOSURE AGES

Isotopes can be used to obtain ages for solids in the solar system. The decay of very long-lived radioisotopes like ^{40}K to radiogenic isotopes like ^{40}Ar can give formation ages and often dates of major metamorphic or shock events. The interactions of the energetic particles in the cosmic rays produce certain isotopes that can be identified as cosmogenic and used to determine the time when a sample was exposed to cosmic rays. These nuclides are not normally abundant in these samples; examples are the minor stable isotopes of the noble gases (such as ^{21}Ne) and radionuclides with half-lives less than $\sim 10^8$ years (such as 7×10^5 years for ^{26}Al).

Most samples are shielded from cosmic ray particles by enough matter that production rates of cosmogenic isotopes are negligible. Exposure may occur as a result of a major collisional event. Only when a sample is within a few meters of the surface is the cosmic ray particle flux intense enough to produce measurable amounts of cosmogenic isotopes. The cosmic ray exposure age is determined from the measured concentration of a cosmogenic nuclide using a production rate, or from the ratio of a stable and radioactive nuclide pair (Marti and Graf, 1992; Vogt, Herzog and Reedy, 1990). Measurements for other cosmogenic nuclides are often needed to (1) determine the production rate (or ratio), which can vary due to an object's geometry and a sample's location in that object, and (2) establish that the sample had a simple exposure history. Some samples have been exposed to cosmic rays in more than one geometry; this is called a complex history. Meteorites found on the Earth's surface often fell long ago; short-lived radionuclides like 5730-year ^{14}C with concentrations below their expected production rates can be used to infer terrestrial ages.

Exposure ages for chondritic meteorites show several patterns (Marti and Graf, 1992). The H chondrites have a big peak in their distribution of exposure ages near 7×10^6 years (7 Ma) that appears to represent a major fragmentation event in a parent body. L chondrites and LL chondrites have smaller clusters in their exposure ages at about 40 and 15 Ma respectively; however, most samples do not fall in these peaks, suggesting many formation events. Very few chondrites or other stony meteorites have exposure ages greater than 100 Ma, and good records of complex histories are rare. Iron meteorites have exposure ages ranging back more than 1000 Ma; exposure ages of certain chemical groups have major clusters, as at 600–700 Ma for group IIIAB (Voshage, Feldman and Braun, 1983). Many iron meteorites have complex histories. Thus most meteorites appear to have been buried deep in parent bodies until fairly recently.

Particle exposure data have yielded exposure ages and histories for lunar samples (Burnett and Woolum, 1977). Most lunar rocks have exposure ages in the range 1–200 Ma and erosion rates of ~ 1 mm Ma^{-1}, and many have complex histories. Several lunar features have been dated with cosmogenic nuclides, such as the South Ray and North Ray craters at the Apollo 16 landing site with formation ages of 2 and 50 Ma respectively. Some cosmogenic nuclides produced *in situ* in terrestrial rocks, like ^{26}Al and 1.5-Ma ^{10}Be, can now be measured by accelerator mass spectrometry and used to get surface exposure ages and erosion rates (Lal, 1988).

Robert C. Reedy

Bibliography

Burnett, D.S. and Woolum, D.S. (1977) Exposure ages and erosion rates for lunar rocks. *Phys. Chem. Earth*, **10**, 63–101.
Lal, D. (1988) *In situ*-produced cosmogenic isotopes in terrestrial rocks. *Ann. Rev. Earth Planet. Sci.*, **16**, 355–88.
Marti, K., and Graf, T. (1992) Cosmic-ray exposure history of ordinary chondrites. *Ann. Rev. Earth Planet. Sci.*, **20**, 221–43.
Vogt, S., Herzog, G.F. and Reedy, R.C. (1990) Cosmogenic nuclides in extraterrestrial materials. *Rev. Geophys.*, **28**, 253–75.
Voshage, H., Feldman, H. and Braun, O. (1983) Investigations of cosmic-ray-produced nuclides in iron meteorites: 5. More data on the nuclides of potassium and noble gases, on exposure ages and meteoroid size. *Z. Naturforsch.*, **38a**, 273–80.

Acknowledgement

This work supported by NASA and performed under the auspices of the US Department of Energy.

Cross references

Chronology: meteorite
Dating methods
Meteorite
Moon: geology

COSMOCHEMISTRY

Cosmochemistry is mainly concerned with the study of the chemical, isotopic and mineralogical composition of extraterrestrial materials such as meteorites, interplanetary dust particles (IDPs), cometary dust particles, lunar samples and, when they become available, samples from other bodies in the solar system. It is generally acknowledged that the emergence of cosmochemistry as a separate discipline stems from the seminal work by Urey, Suess and Brown in the late 1940s and early 1950s on the chemical processes involved in the origin of the solar system and on the abundances of the elements (e.g. Urey, 1952; Suess, 1947a,b; Brown, 1949).

Historically, some of the major objectives of cosmochemistry were to determine the solar system abundances of the elements, the shape of the elemental abundance curve as a function of mass number, the chemical behavior of the elements and their compounds in solar composition material, and the mechanisms (nebular and planetary) responsible for producing the different chemical fractionations and mineral phase assemblages seen in meteorites. At present some of the major research topics in cosmochemistry involve searches for preserved interstellar grains (e.g. SiC, graphite) in meteorites (Virag *et al.*, 1992), the origin of isotopic anomalies in oxygen and other elements in meteorites (Thiemens, 1988; Lee, 1988), and the origin of oxidizing conditions at high temperatures in the solar nebula (Palme and Fegley 1990).

This article summarizes two major results of cosmochemical studies, namely the determination of the solar system abundances of the elements and their chemical behavior in a solar composition (i.e. H_2-rich) system. These two topics are related because the observed elemental depletions in the different classes of primitive meteorites (the chondrites) are generally correlated with the volatility of the elements and/or their compounds in a solar composition system. A significant fraction of the research work in cosmochemistry during the past four decades has been concerned either directly or indirectly with these topics.

Another important area of cosmochemistry, namely studies of the isotopic composition of meteorites, their components (chondrules, matrix and inclusions) and IDPs is not covered here because of a lack of space. However, good introductions to cosmochronology, isotopic anomalies in meteorites and extinct radionuclides are found in several books such as Faure (1986) and in Kerridge and Matthews (1988). (See also Meteorite; Dating methods; Dust.)

Finally, several terms (i.e. nomenclature) used in the following sections will be briefly defined here. Chondrites (q.v.) are stony meteorites which generally contain small, glassy objects known as chondrules. The chondrites are primitive in the sense that they are composed of constituents (metal, silicate and sulfide grains) formed

Table C13 Abundances of the elements in CI chondrites and the solar photosphere

Atomic Number	Element name and chemical symbol	CI chondrites (Si = 10^6)	Solar photosphere (H = 10^{12})	Abundance in Orgueil CI chondrite
1	Hydrogen (H)	2.79×10^{10}	1.00×10^{12}	2.02%
2	Helium (He)[a]	2.79×10^9	1.00×10^{11}	56 nL g^{-1}
3	Lithium (Li)	57.1	14.45	1.49 μg g^{-1}
4	Beryllium (Be)	0.73	14.13	24.9 ng g^{-1}
5	Boron (B)	21.2	398	870 ng g^{-1}
6	Carbon (C)[a]	9.90×10^6	3.55×10^8	3.45%
7	Nitrogen (N)[a]	2.60×10^6	9.33×10^7	3180 μg g^{-1}
8	Oxygen (O)[a]	2.07×10^7	7.41×10^8	46.4%
9	Fluorine (F)	843	3.63×10^4	58.2 μg g^{-1}
10	Neon (Ne)[a]	3.28×10^6	1.17×10^8	203 pL g^{-1}
11	Sodium (Na)	5.74×10^4	2.14×10^6	4900 μg g^{-1}
12	Magnesium (Mg)	1.074×10^6	3.80×10^7	9.53%
13	Aluminium (Al)	8.49×10^4	2.95×10^6	8690 μg g^{-1}
14	Silicon (Si)	1.00×10^6	3.55×10^7	10.67%
15	Phosphorus (P)	1.04×10^4	2.82×10^5	1180 μg g^{-1}
16	Sulfur (S)	5.15×10^5	1.62×10^7	5.25%
17	Chlorine (Cl)	5240	3.16×10^5	698 μg g^{-1}
18	Argon (Ar)[a]	1.11×10^5	3.98×10^5	751 pL g^{-1}
19	Potassium (K)	3770	1.32×10^5	566 μg g^{-1}
20	Calcium (Ca)	6.11×10^4	2.29×10^6	9020 μg g^{-1}
21	Scandium (Sc)	34.2	1259	5.83 μg g^{-1}
22	Titanium (Ti)	2400	9.77×10^4	436 μg g^{-1}
23	Vanadium (V)	293	1.0×10^4	56.2 μg g^{-1}
24	Chromium (Cr)	1.35×10^4	4.68×10^5	2660 μg g^{-1}
25	Manganese (Mn)	9550	2.45×10^5	1980 μg g^{-1}
26	Iron (Fe)[a]	9.00×10^5	3.24×10^7	18.51%
27	Cobalt (Co)	2250	8.32×10^4	507 μg g^{-1}
28	Nickel (Ni)	4.93×10^4	1.78×10^6	1.10%
29	Copper (Cu)	522	1.62×10^4	119 μg g^{-1}
30	Zinc (Zn)	1260	3.98×10^4	311 μg g^{-1}
31	Gallium (Ga)	37.8	759	10.1 μg g^{-1}
32	Germanium (Ge)	119	2570	32.6 μg g^{-1}
33	Arsenic (As)	6.56	–	1.85 μg g^{-1}
34	Selenium (Se)	62.1	–	18.2 μg g^{-1}
35	Bromine (Br)	11.8	–	3.56 μg g^{-1}
36	Krypton (Kr)	45	–	8.7 pL g^{-1}
37	Rubidium (Rb)	7.09	398	2.30 μg g^{-1}
38	Strontium (Sr)	23.5	794	7.80 μg g^{-1}
39	Yttrium (Y)	4.64	174	1.53 μg g^{-1}
40	Zirconium (Zr)	11.4	398	3.95 μg g^{-1}
41	Niobium (Nb)	0.698	26.3	246 ng g^{-1}
42	Molybdenum (Mo)	2.55	83.2	928 ng g^{-1}
44	Ruthenium (Ru)	1.86	69.2	714 ng g^{-1}
45	Rhodium (Rh)	0.344	13.2	134 ng g^{-1}
46	Palladium (Pd)	1.39	49	556 ng g^{-1}
47	Silver (Ag)	0.486	8.7	197 ng g^{-1}
48	Cadmium (Cd)[a]	1.61	58.9	680 ng g^{-1}
49	Indium (In)	0.184	45.7	77.8 ng g^{-1}
50	Tin (Sn)	3.82	100	1680 ng g^{-1}
51	Antimony (Sb)	0.309	10	133 ng g^{-1}
52	Tellerium (Te)	4.81	–	2270 ng g^{-1}
53	Iodine (I)	0.9	–	433 ng g^{-1}
54	Xenon (Xe)	4.7	–	8.6 pL g^{-1}
55	Cesium (Cs)	0.372	–	186 ng g^{-1}
56	Barium (Ba)	4.49	135	2340 ng g^{-1}
57	Lanthanum (La)	0.446	16.6	236 ng g^{-1}
58	Cerium (Ce)	1.136	35.5	619 ng g^{-1}
59	Praseodymium (Pr)	0.1669	5.1	90 ng g^{-1}
60	Neodymium (Nd)	0.8279	31.6	463 ng g^{-1}
62	Samarium (Sm)	0.2582	10	144 ng g^{-1}
63	Europium (Eu)	0.0973	3.2	54.7 ng g^{-1}
64	Gadolinium (Gd)	0.33	13.2	199 ng g^{-1}
65	Terbium (Tb)	0.0603	0.8	35.3 ng g^{-1}
66	Dysprosium (Dy)	0.3942	12.6	246 ng g^{-1}
67	Holmium (Ho)	0.0889	1.8	55.2 ng g^{-1}
68	Erbium (Er)	0.2508	8.5	162 ng g^{-1}
69	Thulium (Tm)	0.0378	1	22 ng g^{-1}

Table C13 Continued

Atomic Number	Element name and chemical symbol	CI chondrites (Si = 10^6)	Solar photosphere (H = 10^{12})	Abundance in Orgueil CI chondrite
70	Ytterbium (Yb)	0.2479	12	166 ng g^{-1}
71	Lutetium (Lu)	0.0367	5.8	24.5 ng g^{-1}
72	Hafnium (Hf)	0.154	7.6	108 ng g^{-1}
73	Tantalum (Ta)	0.0207	–	14.0 ng g^{-1}
74	Tungsten (W)	0.133	12.9	92.3 ng g^{-1}
75	Rhenium (Re)	0.0517	–	37.1 ng g^{-1}
76	Osmium (Os)	0.675	28.2	483 ng g^{-1}
77	Iridium (Ir)	0.661	22.4	474 ng g^{-1}
78	Platinum (Pt)	1.34	63.1	973 ng g^{-1}
79	Gold (Au)	0.187	10.2	145 ng g^{-1}
80	Mercury (Hg)	0.34	–	258 ng g^{-1}
81	Thallium (Tl)	0.184	7.9	143 ng g^{-1}
82	Lead (Pb)[a]	3.15	89.1	2430 ng g^{-1}
83	Bismuth (Bi)	0.144	–	111 ng g^{-1}
90	Thorium (Th)[a]	0.0335	1.9	28.6 ng g^{-1}
92	Uranium (U)	0.009	< 0.34	8.1 ng g^{-1}

Elemental abundance compilations do not list the following radioactive elements which have no stable isotopes and are not found in meteorites: Technetium (43), Promethium (61), Polonium (84), Astatine (85), Radon (86), Francium (87), Radium (88), Actinium (89).
The abbreviations used for abundances in the Orgueil CI chondrite have the following meanings: % = mass %, nL g^{-1} = 10^{-9} liters per gram, μg g^{-1} = 10^{-6} grams per gram, pL g^{-1} = 10^{-12} liters per gram, ng g^{-1} = 10^{-9} grams per gram.
The abundances in Table C13 are generally from Anders and Grevesse (1989) with the following values also being included:
[a] The photospheric abundance is from Grevesse and Noels (1993).

in the solar nebula and little altered since that time by planetary processes. In particular, the chondrites have not been subjected to igneous differentiation on a planetary body. They are subdivided into several classes such as the carbonaceous, ordinary and enstatite chondrites on the basis of their major element compositions and mineralogy. In turn, these classes are further subdivided into different petrographic types. The most primitive chondrites, i.e. those which best reproduce the elemental abundances in the Sun, are the C1 (or CI) chondrites. This classification scheme and the properties of meteorites are described in more detail in the articles on meteorites.

Solar system abundances of the elements

One of the major triumphs of cosmochemistry is the determination of the abundances of the elements and of the shape of the elemental abundance curve as a function of mass number (Table C13; Figure C69). These first-order results are important for at least four reasons. First, they show that to a very good first approximation the abundances of the elements in the Sun, which constitutes over 99% of the mass of the solar system, are the same as those in the primitive meteorites known as the C1 (or CI) chondrites. The few exceptions to this generalization are (1) light elements and isotopes, such as Li and D, which are destroyed by thermonuclear reactions in the Sun; (2) the highly volatile elements such as H, O, C, N and the noble gases, which are incompletely condensed in meteorites; and (3) some rare elements such as Hg, Ge, Pb and W which are either difficult to analyze in the Sun, or in meteorites, or in both. To a somewhat lesser degree there is also a correspondence between the elemental abundances in the Sun and the elemental abundances in all chondrites. This similarity is an important factor which has led cosmochemists to believe that the chondrites are relatively unaltered samples of material formed in the solar nebula. As such, the chondrites contain a record which, if properly interpreted, will provide information about the chemical and physical conditions existing in the solar nebula.

Second, a good knowledge of the abundances of the elements is necessary for modeling the chemical fractionations which occurred in the solar nebula and which determined to a large extent the compositions of the planets, their satellites and the minor bodies (asteroids, comets and meteorites) in the solar system. Third, determination of the solar system abundances of the elements is important for comparisons to elemental abundances in other stars and the interstellar medium in our galaxy. This is useful for modeling galactic chemical evolution over time. Fourth, a firm knowledge of the solar system elemental abundances and of the shape of the

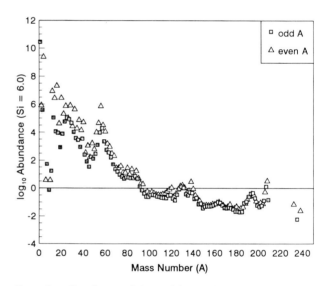

Figure C69 Abundances of the nuclides as a function of mass number (A). The even and odd mass number nuclides are indicated by separate symbols. The abundances, which are from Anders and Grevesse (1989), are plotted on a logarithmic scale normalized to Si = 6.00.

abundance curve as a function of mass number has been and continues to be an important constraint for models of nucleosynthesis in stars.

Attempts to determine the abundances of the elements date back at least to Clarke (1889), who attempted to find periodicities in the relative abundances of the elements in the Earth's crust. However, as we now know, the relative elemental abundances in the terrestrial crust have been modified by planetary differentiation and weathering processes and generally are not representative of the solar system abundances of the elements. Instead, it is necessary to analyze samples of primitive meteorites, such as the chondrites, which have been relatively unaltered by planetary processes, to obtain elemental abundances representative of the average composition of the solar system.

Chemical analyses of chondritic meteorites in the modern era date back to the work by Goldschmidt, the Noddacks and their colleagues in the 1920s and 1930s. This work was critically assessed by Goldschmidt (1937, 1954) who compiled the first table of elemental abundances in meteorites. This tabulation showed that to first approximation, the abundances of the non-volatile elements in meteorites and in the Sun were similar. It also served as a stimulus for the seminal work of Suess (1947a,b) who postulated that the abundances of the nuclides, and especially the odd mass number nuclides, are a smooth function of mass number.

Suess (1947a,b) used this postulate to adjust the elemental abundances to produce a smooth variation of abundance with mass number. In some cases, such as Re, adjustments of up to a factor of 100 were made and shown to be correct by subsequent chemical analyses of meteorites. A modern version of the elemental abundance curve as a function of mass number is shown in Figure C69. Many of the details of this curve are reviewed by Woolun (1988).

A later paper by Suess and Urey (1956) carried this approach even further and produced an influential table of solar system elemental abundances. One outcome of their table was the pioneering studies of stellar nucleosynthesis mechanisms by Burbidge *et al.* (1957). Another outcome was an increasing number of high-quality analytical studies of elemental abundances in chondritic meteorites. In the intervening 40 years the improvements in chemical analyses of meteorites, in the understanding of stellar nucleosynthesis mechanisms and in astronomical observations of elemental abundances in the Sun and other stars have led to vast improvements in our knowledge of the solar system abundances of the elements.

The result of all these efforts is displayed in Table C13, based on Anders and Grevesse (1989), which summarizes present knowledge of the solar system abundances of the elements. It shows the atomic abundances of the elements in CI chondrites (normalized to 10^6 atoms of Si) in the solar photosphere (normalized to 10^{12} atoms of H), and elemental abundances by mass in the Orgueil CI chondrite. The abundance compilations by Cameron (1982), Anders and Ebihara (1982) and Anders and Grevesse (1989) describe the details involved in assessing the solar system abundance table. In addition, two compilations by Mason (1971, 1979) provide comprehensive references to the elemental analyses of the different types of meteorites. More recent papers on elemental abundances in meteorites can be found in the journals *Earth and Planetary Science Letters*, *Geochimica et Cosmochimica Acta*, and *Meteoritics*, while more recent papers on astronomical observations of elemental abundances in the Sun and other stars can be found in *Astronomy and Astrophysics* and the *Astrophysical Journal*.

Cosmochemical behavior of the elements

Our knowledge of the cosmochemical behavior of the elements is based upon analytical studies of meteorites, IDPs and lunar samples; remote sensing and spacecraft observations of the composition of evolved planetary bodies (e.g. Mercury, Venus and Mars) and primitive bodies such as comets, asteroids and icy satellites; *in situ* spacecraft analyses of volatiles emitted from comet P/Halley; and theoretical calculations of the equilibrium chemistry of solar composition material as a function of pressure and temperature.

The latter calculations are reviewed in detail elsewhere (Grossman and Larimer 1974; Larimer, 1988) and will not be described here except for the following. Most condensation calculations have been done at pressures of either 10^{-3} or 10^{-4} bars, and thus the results in Table C14 are given at 10^{-4} bars pressure. The early physical models of the solar nebula by Cameron gave pressures of this magnitude for the formation region of the terrestrial planets and asteroids. Current models of the P,T structure of the solar nebula also give similar pressures at high temperatures (Wood and Morfill, 1988). An exemplary P,T model of the solar nebula, based on the work by Lewis and colleagues, is given in Figures C70 and C71.

This large body of information provides us with a fairly detailed understanding of the chemical behavior of the elements (and/or their compounds) in a solar composition (i.e. H_2-rich) system. In particular, it is now well established that volatility-controlled fractionations of the elements, or their stable compounds, in a H_2-rich system was an important process which to a large extent determined the distribution of the elements on the different bodies (i.e. the planets, their satellites, the asteroids, comets, etc.) in the solar system (e.g. Larimer, 1988). Thus we will now review the cosmochemical classification of the elements, which is based upon their volatility in a H_2-rich system. The important points of this classification are summarized in Table C14 and in Figure C70, which form the basis for much of the subsequent discussion.

Overview

Elements which are either the first to condense from a hot gas with solar composition or are the last to vaporize from a solid with CI chondritic abundances are classified as refractory elements. Both siderophile elements (which geochemically prefer to be in a metallic phase) and lithophile elements (which geochemically prefer to be in an oxide or silicate phase) fall into this category. The condensation of iron metal and the magnesian silicates, which together constitute most of the rocky material in the solar system, is a dividing line which separates the refractory elements from the moderately volatile elements. In turn, the moderately volatile elements are separated from the highly volatile elements by the condensation of troilite, FeS, which occurs at the pressure-independent temperature of ~ 690 K (Figure C69). Finally, the highly volatile elements are separated from the atmophile elements (H, C, N and the noble gases) by water ice condensation. Each of these categories will now be discussed in more detail.

Refractory elements

Refractory lithophiles

The lanthanides (rare earth elements), the actinides, the alkaline earths, Al and elements in groups 3b (Sc, Y), 4b (Ti, Zr, Hf), and 5b (V, Nb, Ta) of the periodic table are included in this category. The refractory lithophiles are indicated by RL in Table C14. These elements constitute about 5% by mass of the total rocky material in a solar composition system. Extensive studies of the chemical composition of stony meteorites show that these elements behave as a group in most meteorites, that is, their abundances in different types of meteorites are either enriched or depleted by about the same factor.

The condensation chemistry of these elements is generally well known, with a few exceptions such as Be, Sr, Ba and Eu. The data in Table C14 for Be are based on the calculations by Lauretta and Lodders (1996), which are the latest ones available for this element. These calculations predict $Ca_2BeSi_2O_7$ condensation in melilite and $BeAl_2O_4$ condensation in spinel. However, the principal Be host phases (if any) in chondrites are unknown. Condensation temperatures in Table C14 for Sr, Ba and Eu assume ideal solid solution in perovskite $CaTiO_3$, but condensation calculations indicate that perovskite has already reacted away when condensation of these trace elements is calculated to occur. However, perovskite is observed to be a major host phase for these elements.

Large enrichments (on average 20 times solar elemental abundances) of the refractory lithophile elements are found in certain inclusions in the Allende meteorite and other carbonaceous chondrites. These inclusions are known as Ca, Al-rich inclusions or CAIs because they have a mineralogy dominated by Ca-, Al- and Ti-rich minerals such as hibonite $CaAl_{12}O_{19}$, melilite, a solid solution of gehlenite $Ca_2Al_2SiO_7$ and åkermanite $Ca_2MgSi_2O_7$, spinel, $MgAl_2O_4$ and perovskite $CaTiO_3$ (MacPherson, Wark and Armstrong, 1988). Condensation curves for two of these minerals (corundum Al_2O_3 and hibonite) are illustrated in Figure C70. The condensation curves for the other refractory phases fall in the same region at slightly lower temperatures. Details of their condensation chemistry are given by Kornacki and Fegley (1984).

In contrast to bulk meteorites, where they behave as a coherent group, the refractory lithophiles in CAIs sometimes display large chemical fractionations from one another. These fractionations are consistently explained by the relative volatilities of the different refractory lithophiles (Kornacki and Fegley 1986); fractionations among the lanthanides form a classification scheme for the CAIs (Martin and Mason, 1974; Fegley and Ireland, 1991).

Refractory siderophiles

The refractory siderophiles are the Pt-group metals (except Pd), Mo, W and Re. They are indicated by RS in Table C14. Like the refractory lithophiles, they are enriched to about 20 times solar elemental abundances (on average) in CAIs. Refractory metal nuggets, which are rich in Pt-group metals, and complex multiphase assemblages of metal, oxide and sulfide, known either as Fremdlinge or as opaque assemblages, are the two principal host phases for the refractory siderophiles in Ca, Al-rich inclusions in Allende and other

Table C14 Cosmochemistry of the elements in the solar nebula

Atomic number	Element name and chemical symbol	Condensation temperature (K) ($P = 10^{-4}$ bar)	Initial condensate in the solar nebula	Major gases in the solar nebula	Notes and sources
1	Hydrogen (H)	180	$H_2O(s)$	H_2	A, 1
2	Helium (He)[a]	< 5	He(s)	He	A, 1
3	Lithium (Li)	1225	Li_2SiO_3 in $MgSiO_3$	LiCl, LiF	MV, 2
4	Beryllium (Be)	1490 (50%)	$Ca_2BeSi_2O_7$	Be, $Be(OH)_2$, BeOH	MV, 3
5	Boron (B)	964 (50%)	$CaB_2SiO_2O_8$	$NaBO_2$, KBO_2, HBO_2, HBO	MV, 3
6	Carbon (C)[c]	78	$CH_4 \cdot 6H_2O(s)$	CO, CH_4	A, 1
7	Nitrogen (N)[d]	120	$NH_3 \cdot H_2O(s)$	N_2, NH_3	A, 1
8	Oxygen (O)[e]	–	–	CO, H_2O	A
9	Fluorine (F)	736	$Ca_5(PO_4)_3F$	HF	MV, 4
10	Neon (Ne)[a]	~ 5	Ne(s)	Ne	A, 1
11	Sodium (Na)	970 (50%)	$NaAlSi_3O_8$ in feldspar	Na, NaCl	MV, 4
12	Magnesium (Mg)	1340 (50%)	$Mg_2SiO_4(s)$	Mg	ME, 5, 13
13	Aluminium (Al)	1670	$Al_2O_s(s)$	Al, AlOH, Al_2O, AlS, AlH, AlO, AlF	RL, 6
14	Silicon (Si)[f]	1529	$Ca_2Al_2SiO_7(s)$	SiO, SiS	ME, 6
15	Phosphorus (P)	1151 (50%)	$Fe_3P(s)$	PO, P, PN, PS	MV, 4, 7
16	Sulfur (S)	713	FeS(s)	H_2S, HS	MV, 4, 12
17	Chlorine (Cl)	863 (50%)	$Na_4[AlSiO_4]_3Cl(s)$	HCl, NaCl, KCl	MV, 4
18	Argon (Ar)	50	$Ar \cdot 6H_2O(s)$	Ar	A, 14
19	Potassium (K)	1000 (50%)	$KAlSi_3O_8(s)$ in feldspar	K, KCl, KOH	MV, 4
20	Calcium (Ca)	1634	$CaAl_{12}O_{19}(s)$	Ca	RL, 6
21	Scandium (Sc)	1652 (50%)	$Sc_2O_3(s)$	ScO	RL, 8
22	Titanium (Ti)	1600	$CaTiO_3(s)$	TiO, TiO_2	RL, 6
23	Vanadium (V)	1455 (50%)	Diss. in $CaTiO_3(s)$	VO_2, VO	RL, 8
24	Chromium (Cr)	1301 (50%)	Diss. in Fe alloy	Cr	MV, 9
25	Manganese (Mn)	1190 (50%)	Mn_2SiO_4 in olivine	Mn	MV, 2
26	Iron (Fe)	1337 (50%)	Fe alloy	Fe	ME, 7,9
27	Cobalt (Co)	1356 (50%)	Diss. in Fe alloy	Co	RS, 9
28	Nickel (Ni)	1354 (50%)	Diss. in Fe alloy	Ni	RS, 9
29	Copper (Cu)	1170 (50%)	Diss. in Fe alloy	Cu	MV, 2
30	Zinc (Zn)	684 (50%)	ZnS diss. in Fe	Zn	MV, 2
31	Gallium (Ga)	918 (50%)	Diss. in Fe alloy	GaOH, GaCl, GaBr	MV, 10
32	Germanium (Ge)	825 (50%)	Diss. in Fe alloy	GeS, GeSe	MV, 10
33	Arsenic (As)	1012 (50%)	Diss. in Fe alloy	As	MV, 10
34	Selenium (Se)	684 (50%)	FeSe diss. in FeS	H_2Se, GeSe	MV, 2
35	Bromine (Br)[b]	~ 350	$Ca_5(PO_4)_3Br(s)$	HBr, NaBr	HV, 4
36	Krypton (Kr)	54	$Kr \cdot 6H_2O(s)$	Kr	A, 14
37	Rubidium (Rb)[b]	~ 1080	Diss. in feldspar	Rb, RbCl	MV, 5, 13
38	Strontium (Sr)	1217 (50%)	Diss. in $CaTiO_3(s)$	Sr, $SrCl_2$, $Sr(OH)_2$, SrOH	RL, 8
39	Yttrium (Y)	1622 (50%)	$Y_2O_3(s)$	YO	RL, 8
40	Zirconium (Zr)	1717 (50%)	$ZrO_2(s)$	ZrO_2, ZrO	RL, 8
41	Niobium (Nb)	1517 (50%)	Diss. in $CaTiO_3(s)$	NbO_2, NbO	RL, 8
42	Molybdenum (Mo)	1595 (50%)	Refractory metal alloy	MoO, Mo, MoO_2	RS, 9
44	Ruthenium (Ru)	1565 (50%)	Refractory metal alloy	Ru	RS, 9
45	Rhodium (Rh)	1392 (50%)	Refractory metal alloy	Rh	RS, 9
46	Palladium (Pd)	1320 (50%)	Diss. in Fe alloy	Pd	MV, 9
47	Silver (Ag)	993 (50%)	Diss. in Fe alloy	Ag	MV, 2
48	Cadmium (Cd)[b]	430 (10^{-5} bars)	CdS in FeS	Cd	HV, 11
49	Indium (In)[b]	470 (50%)	InS in FeS	In, InCl, InOH	HV, 11
50	Tin (Sn)	720 (50%)	Diss. in Fe alloy	SnS, SnSe	MV, 2
51	Antimony (Sb)	912 (50%)	Diss. in Fe alloy	SbS, Sb	MV, 10
52	Tellurium (Te)	680 (50%)	FeTe diss. in FeS	Te, H_2Te	MV, 2
53	Iodine (I)	?	?	I, HI	MV/HV?
54	Xenon (Xe)	74	$Xe \cdot 6H_2O(s)$	Xe	A, 14
55	Cesium (Cs)	?	?	CsCl, Cs, CsOH	MV/HV?
56	Barium (Ba)	1162 (50%)	Diss. in $CaTiO_3(s)$	$Ba(OH)_2$, BaOH, BaS, BaO	RL, 8
57	Lanthanum (La)	1544 (50%)	Diss. in $CaTiO_3(s)$	LaO	RL, 8
58	Cerium (Ce)	1440 (50%)	Diss. in $CaTiO_3(s)$	CeO_2, CeO	RL, 8
59	Praseodymium (Pr)	1557 (50%)	Diss. in $CaTiO_3(s)$	PrO	RL, 8
60	Neodymium (Nd)	1563 (50%)	Diss. in $CaTiO_3(s)$	NdO	RL, 8
62	Samarium (Sm)	1560 (50%)	Diss. in $CaTiO_3(s)$	SmO, Sm	RL, 8
63	Europium (Eu)	1338 (50%)	Diss. in $CaTiO_3(s)$	Eu	RL, 8
64	Gadolinium (Gd)	1597 (50%)	Diss. in $CaTiO_3(s)$	GdO	RL, 8
65	Terbium (Tb)	1598 (50%)	Diss. in $CaTiO_3(s)$	TbO	RL, 8
66	Dyprosium (Dy)	1598 (50%)	Diss. in $CaTiO_3$	DyO, Dy	RL, 8
67	Holmium (Ho)	1598 (50%)	Diss. in $CaTiO_3(s)$	HoO, Ho	RL, 8

Table C14 Continued

Atomic number	Element name and chemical symbol	Condensation temperature (K) ($P = 10^{-4}$ bar)	Initial condensate in the solar nebula	Major gases in the solar nebula	Notes and sources
68	Erbium (Er)	1598 (50%)	Diss. in $CaTiO_3(s)$	ErO, Er	RL, 8
69	Thulium (Tm)	1598 (50%)	Diss. in $CaTiO_3(s)$	Tm, TmO	RL, 8
70	Ytterbium (Yb)	1493 (50%)	Diss. in $CaTiO_3(s)$	Yb	RL, 8
71	Lutetium (Lu)	1598 (50%)	Diss. in $CaTiO_3(s)$	LuO	RL, 8
72	Hafnium (Hf)	1690 (50%)	$HfO_2(s)$	HfO	RL, 8
73	Tantalum (Ta)	1543 (50%)	Diss. in $CaTiO_3(s)$	TaO_2, TaO	RL, 8
74	Tungsten (W)	1794 (50%)	Refractory metal alloy	WO, WO_2, WO_3	RS, 9
75	Rhenium (Re)	1818 (50%)	Refractory metal alloy	Re	RS, 9
76	Osmium (Os)	1812 (50%)	Refractory metal alloy	Os	RS, 9
77	Iridium (Ir)	1603 (50%)	Refractory metal alloy	Ir	RS, 9
78	Platinum (Pt)	1411 (50%)	Refractory metal alloy	Pt	RS, 9
79	Gold (Au)	1284 (50%)	Fe alloy	Au	MV, 2
80	Mercury (Hg)	?	?	Hg	MV/HV?
81	Thallium (Tl)[b]	448 (50%)	Diss. in Fe alloy	Tl	
82	Lead (Pb)[b]	520 (50%)	Diss. in Fe alloy	Pb, PbS	HV, 11
83	Bismuth (Bi)	472 (50%)	Diss. in Fe alloy	Bi	HV, 11
90	Thorium (Th)	1598 (50%)	Diss. in $CaTiO_3(s)$	ThO_2	RL, 8
92	Uranium (U)	1580 (50%)	Diss. in $CaTiO_3(s)$	UO_2	RL, 8

The condensation temperatures either indicate where the condensate first becomes stable or where 50% of the element is condensed and 50% is in the gas. The 50% condensation temperature is generally used when solid solutions are formed.
The major gases vary as a function of temperature and total pressure. The gas chemistry in Table C14 is generally valid at the condensation temprature of each element, and was either taken from the original references below or calculated as part of this work.
Sources cited in Table C14: (1) Lewis 1972; (2) Wai and Wasson, 1977; (3) Lauretta and Lodders, 1996; (4) Fegley and Lewis, 1980; (5) Grossman and Larimer, 1974; (6) Kornacki and Fegley, 1984; (7) Sears, 1978; (8) Kornacki and Fegley, 1986; (9) Fegley and Palme, 1985; (10) Wai and Wasson, 1979; (11) Larimer, 1973; (12) Lauretta et al., 1996; (13) Wasson, 1985; (14) Sill and Wilkening, 1978.
[a] This temperature is below cosmic background and condensation will not occur.
[b] The condensation temperature and initial condensate is uncertain and needs to be re-evaluated.
[c] Kinetic inhibition of the CO to CH_4 conversion yields either $CO \cdot 6H_2O(s)$ or $CO(s)$ as the initial condensate.
[d] Kinetic inhibition of the N_2 to NH_3 conversion yields either $N_2 \cdot 6H_2O(s)$ or $N_2(s)$ as the initial condensate.
[e] Oxygen is the most abundant element in rocky material and a separate condensation temperature is meaningless. The bulk of oxygen condenses as water ice; the remainder is present as CO or in rocky material.
[f] Most Si condenses when the silicates $MgSiO_3$ and Mg_2SiO_4 form (e.g. 1340 K at 10^{-4} bar). See the $MgSiO_3$ condensation curve in Figure C70.
Key to abbreviations used for cosmochemical classification of the elements: A = atmophile, HV = highly volatile, ME = major element, MV = moderately volatile, RL = refractory lithophile, RS = refractory siderophile.

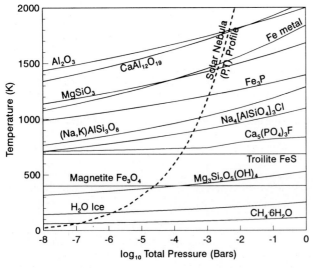

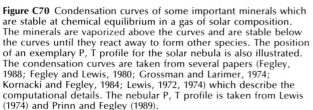

Figure C70 Condensation curves of some important minerals which are stable at chemical equilibrium in a gas of solar composition. The minerals are vaporized above the curves and are stable below the curves until they react away to form other species. The position of an exemplary P, T profile for the solar nebula is also illustrated. The condensation curves are taken from several papers (Fegley, 1988; Fegley and Lewis, 1980; Grossman and Larimer, 1974; Kornacki and Fegley, 1984; Lewis, 1972, 1974) which describe the computational details. The nebular P, T profile is taken from Lewis (1974) and Prinn and Fegley (1989).

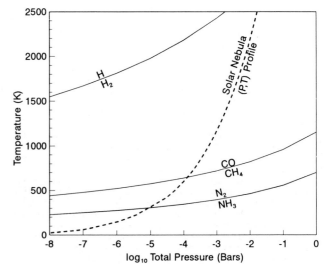

Figure C71 Important phase boundaries for H, C and N chemistry in a solar gas are compared with an exemplary P, T profile in the solar nebula. The different gases have equal abundances along the phase boundaries and are more abundant on their side of the boundary. Thus CO is the major carbon gas at high temperatures and low pressures, while CH_4 is the major carbon gas at low temperatures and high pressures. The nebular P, T profile is the same one shown in Figure C70.

carbonaceous chondrites (Palme and Wlotzka, 1976; Fuchs and Blander, 1980; El Goresy, Nagel and Ramdohr, 1978). The refractory metal nuggets are generally believed to be condensates from the solar nebula; their compositions are reproduced by equilibrium condensation calculations (e.g. Palme and Wlotzka, 1976; Fegley and Palme, 1985). The origin of the complex multiphase assemblages (Fremdlinge or opaque assemblages) is more controversial, and a variety of models ranging from formation in the meteorite parent bodies (e.g. Blum et al., 1988), to condensation in the solar nebula (e.g. Bischoff and Palme, 1987), to formation by the partial evaporation and melting of interstellar dust aggregates (e.g. Fegley and Kornacki, 1984) have been proposed to explain their origin. While each of these models has some attractive features, none can easily explain all of the complexities observed in the Fremdlinge/opaque assemblages, and their origin is still somewhat enigmatic.

Iron alloy and magnesian silicates

The condensation of the major elements Fe, Mg and Si as metallic iron and the magnesian silicates is an important phase boundary because it is the point where most of the rocky materials in solar composition material condense. These elements are indicated by ME in Table C14. As illustrated in Figure C70, at pressures above about 10^{-5} bar iron metal condenses before magnesian silicates (represented by $MgSiO_3$ in Figure C70, while at lower pressures the magnesian silicates condense first. In both cases the separation between the metal and silicate condensation temperatures increases as the pressure is either increased or decreased from the crossover point.

The magnesian silicates which condense at high temperatures are essentially pure $MgSiO_3$ and Mg_2SiO_4 because the large excess of H_2 in solar gas leads to extremely low oxygen fugacities. The temperature dependent oxygen fugacity (f_{O_2}) of solar gas is given by the expression $\log_{10} f_{O_2} = 2 \log_{10} (H_2O/H_2) + 5.59 - 25\,598/T$ which is valid from 300 to 2500 K (Rubin, Fegley and Brett, 1988). As a result the FeO content of the magnesian silicates is predicted to be insignificant until low temperatures of about 400–600 K are attained where the reactions

$2MgSiO_3$(enstatite) + $2Fe$(metal) + $2H_2O$(g) = Mg_2SiO_4(forsterite) + Fe_2SiO_4(fayalite) + $2H_2$(g)

Fe_2SiO_4(fayalite) + $2MgSiO_3$(enstatite) = Mg_2SiO_4(forsterite) + $2FeSiO_3$(ferrosilite)

$3Fe$(metal) + $4H_2O$(g) = Fe_3O_4(magnetite) + $4H_2$(g)

become thermodynamically favorable (Larimer, 1967; Grossman and Larimer, 1974). At these temperatures olivine and pyroxene solid solutions containing several tens of mole percent of fayalite and ferrosilite are predicted to form, and any remaining Fe metal is predicted to form magnetite at a pressure independent temperature of about 400 K (see Figure C70).

However, recent work by Fegley (1988) and Palme and Fegley (1990) has questioned this picture because the slow solid-state diffusion at the required low temperatures will inhibit gas–solid and solid–solid reactions such as the formation of magnetite and FeO-rich silicates over the estimated 10^5–10^6 year lifetime of the solar nebula. Instead, they find that the observed textural features and chemistry of FeO-rich olivines are plausibly explained by high-temperature condensation under oxidizing conditions in the solar nebula. These conditions are proposed to result in dust-rich regions, such as the nebular midplane, where heating of the dust releases the oxygen in rock into the gas and increases the local O/H elemental ratio above the solar value. However, although many workers have now discarded the notion that FeO-rich silicates formed at low temperatures in the solar nebula, an origin by metamorphic reactions on the meteorite parent bodies is still being debated as an alternative to an origin under oxidizing conditions in the solar nebula.

Moderately volatile elements

By convention, the moderately volatile elements are defined as having condensation temperatures intermediate between those of the major elements Fe, Mg and Si and of troilite FeS. The elements in this group are geochemically diverse and include Na, K, Rb, Cr, Mn, Cu, Ag, Au, Zn, B, Ga, P, As, Sb, S, Se, Te, F and Cl. These elements are indicated by MV in Table C14. The condensation chemistry of many of these elements is not well known because of uncertainties about the thermodynamic properties of the condensate minerals and their solid solutions in major host phases such as Fe metal, FeS and apatite minerals. For example, Ga is predicted to condense in solid solution in Fe metal, but Ga condensation in silicates such as feldspar cannot be modeled because the thermodynamic properties of the relevant minerals are unknown. Likewise, the thermodynamic properties of Rb aluminosilicates and their solid solutions with anorthite are also unknown. In some cases such as Cs, Br and I, periodic behavior suggests that an element should be classified as moderately volatile, but either no data are available on any plausible condensates (as is the case for Cs and I), or the available data suggest that the element is highly volatile (Br). However, despite these uncertainties, the calculated condensation temperatures are generally in accord with expectations based on observed elemental abundance trends in chondritic meteorites (e.g. Wai and Wasson, 1977).

Highly volatile elements

Only a few elements fall into this category. Mercury is probably a highly volatile element but it is calculated to condense as $Hg(s)$ at temperatures below 200 K (Larimer, 1967). It is unlikely that Hg is as volatile as water ice, and the low condensation temperature is an artifact due to incomplete thermodynamic data on plausible condensates. However, apparently all of the analytical data for Hg in chondrites are subject to contamination problems, so neither its abundance or host phases (if any) are known. The other highly volatile elements are Br, Cd, In, Tl, Pb and Bi. All the highly volatile elements are indicated by HV in Table C14. The condensation chemistry of these elements is not well known. Br is calculated to condense as bromapatite $Ca_5(PO_4)_3Br$ at about 350 K, but its abundance in chondrites suggests a higher condensation temperature. The elements Tl, Pb and Bi are calculated to condense in solid solution in Fe metal while Cd and In are calculated to condense in solid solution in FeS. However, it is difficult to model the chemistry of these elements because of incomplete and uncertain thermodynamic data, and the poor knowledge of their host phases in meteorites.

Chemically reactive atmophile elements

Hydrogen

As Table C13 shows, hydrogen is the most abundant element and therefore H_2 is the most abundant gas in solar composition material. At sufficiently high temperatures, dissociation to atomic H occurs. However, as illustrated in Figure C71, the phase boundary where equal abundances of H_2 and H occurs is at lower pressures and higher temperatures than those expected in the solar nebula. Conversely, H_2 remains in the gas until temperatures of about 5 K, where it will condense out as solid hydrogen (Lewis, 1972). It is unlikely that temperatures this low were ever reached in the solar nebula.

About 0.1% of all hydrogen condenses out as water ice at temperatures of 150–250 K, depending on the total pressure (Figure C70). Hydrated silicates such as serpentine and talc are also predicted to form by reactions such as

$2Mg_2SiO_4$(forsterite) + $3H_2O$(g) = $Mg_3Si_2O_5(OH)_4$(serpentine) + $Mg(OH)_2$(brucite)

$4MgSiO_3$(enstatite) + $2H_2O$(g) = $Mg_3Si_4O_{10}(OH)_2$(talc) + $Mg(OH)_2$(brucite)

at temperatures below 300 K at 10^{-4} bar (Figure C70). However, although they are thermodynamically favorable, these reactions probably did not occur in the solar nebula because the vapor phase hydration of rock in a near-vacuum is a very slow process (e.g. Fegley, 1988; Fegley and Prinn, 1989; Prinn and Fegley, 1989). Instead, the theoretical studies of hydration kinetics in the solar nebula and petrographic studies of water-bearing chondrites both suggest that the production of hydrated minerals occurred on the meteorite parent bodies. Thus it is very likely that water ice is the first H-bearing condensate to form.

Carbon

Carbon chemistry is significantly more complex. To a good first approximation, CO is the dominant carbon gas at high temperatures and low pressures and CH_4 is the dominant carbon gas at low temperatures and high pressures in solar composition material (Urey, 1953; Lewis, Barshay and Noyes 1979). The phase boundary where the abundances of CO and CH_4 are equal is shown in Figure C71. The two gases are converted by the net thermochemical reaction $CO(g) + 3H_2(g) = CH_4(g) + H_2O(g)$. Increasing the H_2 pressure (essentially the total pressure in solar material) and/or decreasing the temperature drives this reaction to the right and yields more CH_4. As shown in Figure C71, the nebular P,T profile crosses the CO/CH_4 boundary at about 600 K. CO is more abundant at higher temperatures, and CH_4 is more abundant at lower temperatures.

Detailed modeling by Lewis, Barshay and Noyes (1979) also shows that at high temperatures, similar to those for the H/H_2 phase boundary, CO dissociates to the elements and C(g) becomes the dominant C-bearing gas. Conversely, at sufficiently low pressures, both CO and CH_4 decompose to form graphite (gr) via the reaction $CO(g) + CH_4(g) = H_2O(g) + H_2(g) + 2C(gr)$. CO also disproportionates via the reaction $2CO(g) = CO_2(g) + C(gr)$ to form CO_2 plus graphite at low temperatures and pressures. However, the graphite and CO_2 stability fields are at much lower pressures than those along the nebular P,T profile in Figure C71.

As first noted by Urey (1953), and later quantified by Lewis and Prinn (1980), the kinetics of the CO → CH_4 conversion may be so slow under the (P,T) conditions expected in the solar nebula that CO cannot be converted to CH_4 within the lifetime of the solar nebula. An exception to this occurs in the giant protoplanetary subnebulae, which are higher density environments predicted around Jupiter and the other gas giant planets during their formation. The CO → CH_4 conversion is predicted to take place in these environments (Prinn and Fegley, 1981, 1989; Fegley and Prinn, 1989).

At low temperatures in the outer solar nebula and the giant protoplanetary subnebulae, CO and CH_4 may react with water ice to form the clathrate hydrates $CO \cdot 6H_2O(s)$ and $CH_4 \cdot 6H_2O(s)$. The condensation curve for $CH_4 \cdot 6H_2O$ is illustrated in Figure C70, and the curve for $CO_4 \cdot 6H_2O(s)$ is only slightly lower. The formation of these clathrate hydrates requires sufficiently rapid diffusion of CO or CH_4 through the water ice crystal lattice. Theoretical models, which use experimentally determined activation energies for clathrate formation, predict that CH_4 clathrate hydrate can form in the giant protoplanetary subnebulae but that CO clathrate hydrate cannot form in the much lower density environment of the outer solar nebula (Fegley, 1988; Fegley and Prinn, 1989). However, other workers have argued that clathrate hydrate formation can still occur under special circumstances (Lunine, 1989).

Nitrogen

The most important features of nitrogen chemistry are shown in Figure C71, which displays the phase boundary where N_2 and NH_3 have equal abundances. At any pressure this boundary is at a lower temperature than the analogous phase boundary for CO and CH_4. N_2 is the major nitrogen gas at high temperatures and low pressures while NH_3 is the major nitrogen gas at low temperatures and high pressures.

The two species are converted via the reaction $N_2 + 3H_2 = 2NH_3$, which is analogous to the reaction which converts CO and CH_4. Reduction of N_2 to NH_3 is also predicted to be kinetically inhibited in the solar nebula and to be both thermodynamically favored and kinetically facile in the giant protoplanetary subnebulae (Lewis and Prinn, 1980; Prinn and Fegley, 1981, 1989; Fegley and Prinn, 1989). This is true even when the possible catalytic effects of Fe metal grains are taken into account. Thus N_2 is predicted to be the dominant nitrogen gas throughout the solar nebula and NH_3 is predicted to be the dominant nitrogen gas throughout the giant protoplanetary subnebulae.

At low temperatures in the outer solar nebula, $N_2 \cdot 6H_2O(s)$ becomes thermodynamically stable, but its formation is probably inhibited by two factors. One is the limited availability of water ice, which may already be totally consumed by reactions to form other hydrates and clathrates. The other is the expected kinetic inhibition of clathrate hydrate formation in the outer solar nebula (Fegley and Prinn, 1989). In this case N_2, like CO, will not condense until temperatures of about 20 K (at 10^{-4} bar pressure), where the solid ices form. On the other hand, $NH_3 \cdot H_2O$ formation is predicted in the giant protoplanetary subnebulae, because it is both thermodynamically favorable and kinetically facile.

Noble gases

The noble gases He, Ne, Ar, Kr and Xe display fairly simple chemistry in solar composition material. All are present in the gas as the monatomic elements and Ar, Kr and Xe undergo condensation to either ices or clathrate hydrates at sufficiently low temperatures. Condensation of the pure ices will occur at slightly lower temperatures than condensation of the clathrate hydrates. The condensation temperatures for the noble gas clathrate hydrates in Table C14 are from Sill and Wilkening (1978). However, the formation of these species, like the clathrates of CO and N_2, may be kinetically inhibited. Temperatures of about 20 K (at 10^{-4} bar pressure) are required for quantitative condensation of Ar, Kr and Xe as pure ices. Neither He nor Ne will condense out of the gas because temperatures of 5 K or below are required for this to happen (e.g. Lewis, 1972).

Bruce Fegley

Acknowledgement

This work was supported by NASA Grant NAGW-2861.

Bibliography

Anders, E. and Ebihara, M. (1982) Solar system abundances of the elements. *Geochim. Cosmochim. Acta*, **46**, 2363–80.

Anders, E. and Grevesse, N. (1989) Abundances of the elements: meteoritic and solar. *Geochim. Cosmochim. Acta*, **53**, 197–214.

Bischoff, A. and Palme, H. (1987) Composition and mineralogy of refractory–metal–rich assemblages from a Ca,Al-rich inclusion in the Allende meteorite. *Geochim. Cosmochim. Acta*, **51**, 2733–48.

Blum, J.D., Wasserburg, G.J., Hutcheon, I.D., *et al.* (1988) Origin of opaque assemblages in C3V meteorites: implications for nebular and planetary processes. *Geochim. Cosmochim. Acta*, **53**, 543–56.

Brown, H. (1949) Rare gases and the formation of the Earth's atmosphere, in *The Atmospheres of the Earth and Planets* (ed. G.P. Kuiper) Chicago: University of Chicago Press, pp. 260–268.

Burbidge, E.M., Burbidge, G.R., Fowler, W.A., and Hoyle, F. (1957) Synthesis of the elements in stars. *Rev. Mod. Phys*, **29**, 547–650.

Cameron, A.G.W. (1982) Elemental and nuclidic abundances in the solar system, in *Essays in Nuclear Astrophysics* (eds C.A. Barnes, D.D. Clayton and D.N. Schramm). Cambridge: Cambridge University Press, pp. 23–43.

Clarke, F.W. (1889) The relative abundances of the chemical elements. *Bull. Phil. Soc. Washington*, **11**, 131.

El Goresy, A., Nagel, K. and Ramdohr, P. (1978) Fremdlinge and their noble relatives. *Proc. Lunar Planet. Sci. Conf.* **9**, 1249–66.

Faure, G. (1986) *Principles of Isotope Geology*. New York: John Wiley & Sons.

Fegley, B., Jr (1988) Cosmochemical trends of volatile elements in the solar system, in *Workshop on the Origins of Solar Systems* (eds J.A. Nuth and P. Sylvester). Houston: LPI Technical Report No. 88–04, pp. 51–60.

Fegley, B., Jr and Ireland, T.R. (1991) Chemistry of the rare earth elements in the solar nebula. *Eur. J. Solid State Inorg. Chem.*, **28**, 335–46.

Fegley, B., Jr and Kornacki A.S. (1984) The geochemical behavior of refractory noble metals and lithophile trace elements in refractory inclusions in carbonaceous chondrites. *Earth Planet. Sci. Lett.*, **68**, 181–97.

Fegley, B., Jr, and Lewis, J.S. (1980) Volatile element chemistry in the solar nebula: Na, K, F, Cl, Br, and P. *Icarus*, **41**, 439–55.

Fegley, B., Jr, and Plame, H. (1985) Evidence for oxidizing conditions in the solar nebula from Mo and W depletions in refractory inclusions in carbonaceous chondrites. *Earth Planet Sci. Lett.*, **72**, 311–26.

Fegley, B., Jr and Prinn, R.G. (1989) Solar nebula chemistry: Implications for volatiles in the solar system in *The Formation and Evolution of Planetary Systems* (eds H.A. Weaver and L. Danly). Cambridge: Cambridge University Press, pp. 171–211.

Fuchs, L. and Blander, M. (1980) Refractory metal particles in refractory inclusions in the Allende meteorite. *Proc. Lunar Planet. Sci. Conf.* **11**, 929–44.

Goldschmidt, V.M. (1937) Geochemische Verteilungsgesetze der Elemente IX. *Skrifter Norske Videnscaps-Akademiend*, Oslo I. mat. Natur. Kl. No. 4.

Goldschmidt, V.M. (1954) *Geochemistry*. Oxford: Clarendon Press.

Grevesse, N. and Noels, A. (1993) Cosmic abundances of the elements, in *Origin and Evolution of the Elements* (eds N.

Prantzos, E. Vangioni-Flam, and M. Cassé). Cambridge: Cambridge University Press, pp. 14–25.
Grossman, L. and Larimer, J.W. (1974) Early chemical history of the solar system. *Rev. Geophys. Space Phys.*, **12**, 71–101.
Kerridge, J.F. and Matthews, M.S. (eds) (1988) *Meteorites and the Early Solar System*. Tucson: University of Arizona Press.
Kornacki, A.S. and Fegley, B., Jr (1984) Origin of spinel-rich chondrules and inclusions in carbonaceous and ordinary chondrites. *Proc. Lunar Planet. Sci. Conf. 14, J. Geophys. Res.*, **89**, B588–96.
Kornacki, A.S. and Fegley B. Jr (1986) The abundance and relative volatility of refractory trace elements in Allende Ca, Al-rich inclusions: implications for chemical and physical processes in the solar nebula. *Earth Planet. Sci. Lett*, **75**, 297–310.
Larimer, J.W. (1967) Chemical fractionations in meteorites – I. Condensation of the elements. *Geochim. Cosmochim. Acta*, **31**, 1215–38.
Larimer, J.W. (1973) Chemical fractionations in meteorites – VII. Cosmothermometry and cosmobarometry. *Geochim. Cosmochim. Acta*, **37**, 1603–23.
Larimer, J.W. (1988) The cosmochemical classification of the elements, in *Meteorites and the Early Solar System* (eds J.F. Kerridge and M.S. Matthews). Tucson: University of Arizona Press, pp. 375–89.
Lauretta, D.S., Kremser, D.T. and Fegley, B., Jr (1996) The rate of iron sulfide formation in the solar nebula. *Icarus*, in press.
Lauretta, D.S. and Lodders, K. (1996) The cosmochemical behavior of beryllium and boron. *Earth Planet. Sci. Lett.*, in press.
Lee, T. (1988) Implications of isotopic anomalies for nucleosynthesis, in *Meteorites and the Early Solar System* (eds J.F. Kerridge and M.S. Matthews) Tucson: University of Arizona Press, pp. 1063–89.
Lewis, J.S. (1972) Low temperature condensation from the solar nebula. *Icarus*, **16**, 241–52.
Lewis, J.S. (1974) The temperature gradient in the solar nebula. *Science*, **186**, 440–3.
Lewis, J.S. and Prinn, R.G. (1980) Kinetic inhibition of CO and N_2 reduction in the solar nebula. *Astrophys. J.*, **238**, 357–64.
Lewis, J.S., Barshay, S.S. and Noyes, B. (1979) Primordial retention of carbon by the terrestrial planets. *Icarus*, **37**, 190–206.
Lunine, J.L. (1989) Primitive bodies: molecular abundances in comet Halley as probes of cometary formation environments, in *The Formation and Evolution of Planetary Systems* (eds H.A. Weaver and L. Danly). Cambridge: Cambridge University Press, pp. 213–42.
MacPherson, G.J., Wark, D.A. and Armstrong, J.T. (1988) Primitive material surviving in chondrites: refractory inclusions, in *Meteorites and the Early Solar System* (eds J.F. Kerridge and M.S. Matthews). Tucson: University of Arizona Press, pp. 746–807.
Martin, P.M. and Mason, B. (1974) Major and trace elements in the Allende meteorite. *Nature*, **249**, 333–4.
Mason, B. (ed.) (1971) *Handbook of Elemental Abundances in Meteorites*. New York: Gordon & Breach.
Mason, B. (1979) Cosmochemistry. Part 1. Meteorites, in *Data of Geochemistry*, 6th edn (ed. M. Fleischer), Geol. Surv. Prof. Paper 440-B-1, Washington, DC: US Government Printing Office.
Palme, H. and Fegley, B. Jr (1990) High-temperature condensation of iron-rich olivine in the solar nebula. *Earth Planet. Sci. Lett.*, **101**, 180–95.
Palme, H. and Wlotzka, F. (1976) A metal particle from a Ca, Al-rich inclusion from the meteorite Allende, and the condensation of refractory siderophile elements. *Earth Planet. Sci. Lett.*, **33**, 45–60.
Prinn, R.G. and Fegley, B., Jr (1981) Kinetic inhibition of CO and N_2 reduction in circumplanetary nebulae: implications for satellite composition. *Astrophys. J.*, **249**, 308–17.
Prinn, R.G. and Fegley, B., Jr (1989) Solar nebula chemistry: origin of planetary, satellite, and cometary volatiles, in *The Origin and Evolution of Planetary and Satellite Atmospheres*, (eds S.K. Atreya, J.B. Pollack and M.S. Matthews), Tucson: University of Arizona Press, pp. 78–136.
Rubin, A.E., Fegley, B., Jr and Brett, R. (1988) Oxidation state in chondrites, in *Meteorites and the Early Solar System* (eds J.F. Kerridge and M.S. Matthews). Tucson: University of Arizona Press, pp. 488–511.
Sears, D.W. (1978) Condensation and the composition of iron meteorites. *Earth Planet. Sci. Lett.*, **41**, 128–38.
Sill, G.T. and Wilkening, L.L. (1978) Ice clathrate as a possible source of the atmospheres of the terrestrial planets. *Icarus*, **33**, 13–22.
Suess, H.E. (1947a) Über kosmische Kernhäufigkeiten. I. Mitteilung: Einige Häufigkeitsregeln und ihre Anwendung bei der Abschätzung der Häufigkeitswerte für die mittelschweren und schweren Elemente. *Z. Naturforsch.*, **2a**, 311–21.
Suess, H.E. (1947b) Über kosmische Kernhaufigkeiten. II. Mitteilung: Einzelheiten in der Häufigkeitsverteilung der mittelschweren und schweren Kerne. *Z. Naturforsch*, **2a**, 604–8.
Suess, H.E. and Urey, H.C. (1956) Abundances of the elements. *Rev. Mod. Phys.*, **28**, 53–74.
Thiemens, M.H. (1988) Heterogeneity in the nebula: evidence from stable isotopes, in *Meteorites and the Early Solar System* (eds J.F. Kerridge and M.S. Matthews). Tucson: University of Arizona Press. pp. 899–923.
Urey, H.C. (1952) *The Planets*. New Haven: Yale University Press.
Urey, H.C. (1953) Chemical evidence regarding the Earth's origin. in *XIIIth International Congress Pure and Applied Chemistry and Plenary Lectures*. Stockholm: Almqvist & Wiksells, pp. 188–217.
Virag, A., Wopenka, B., Amari, S., *et al.* (1992) Isotopic, optical, and trace element properties of large single SiC grains from the Murchison meteorite. *Geochim. Cosmochim. Acta*, **56**, 1715–33.
Wai, C.M. and Wasson, J.T. (1977) Nebular condensation of moderately volatile elements and their abundances in ordinary chondrites. *Earth Planet. Sci. Lett.*, **36**, 1–13.
Wai, C.M. and Wasson, J.T. (1979) Nebular condensation of Ga, Ge and Sb and the chemical classification of iron meteorites. *Nature*, **282**, 790–3.
Wasson, J.T. (1985) *Meteorites*. New York: W.H. Freeman and Co.
Wood, J.A. and Morfill, G.E. (1988) A review of solar nebula models, in *Meteorites and the Early Solar System* (eds J.F. Kerridge and M.S. Matthews). Tucson: University of Arizona Press, pp. 329–47.
Woolum, D.S. (1988) Solar-system abundances and processes of nucleosynthesis, in *Meteorites and the Early Solar System* (eds J.F. Kerridge and M.S. Matthews). Tucson: University of Arizona Press, pp. 995–1020.

Cross references

Carbon, carbon dioxide
Chemical element
Comet: structure and composition
Interstellar grain
Iron
Silica, silicate
Solar system: origin

CRATER

A circular to slightly polygonal depression formed on the surface of a solid body by the impact of an interplanetary body. Craters are ubiquitous throughout the solar system. They range in size from microns through kilometers to over 1000 km in diameter, with the corresponding impacting bodies ranging from interplanetary dust through meteorites, asteroids, comets to planetesimals. Craters result from the transfer of the kinetic energy contained in the impacting body to the target rocks. This energy can be considerable, as the impactor is generally travelling at velocities of kilometers to tens of kilometers per second. Energy transfer to the target is by means of a hemispherically propagating shock wave. This shock wave increases the internal energy of the target rocks (leading to so-called shock metamorphism) and also sets the target rocks in motion. This motion leads to the excavation and displacement of the target rocks, forming a cavity. Subsequent modification and collapse of this cavity results in the final observed crater.

Craters have two basic forms; so-called simple and complex craters. Simple craters consist of a bowl-shaped depression with an uplifted rim area. Larger, complex craters are a more modified and collapsed form with a flatter cross-sectional shape. They have uplifted central structures, in the form of a central peak(s) and/or rings and a structurally faulted rim area. The transition from simple to complex forms occurs at a specific diameter range, which varies with planetary gravity and to a lesser extent target rock type. Depending on the nature of the uplifted central structures, complex craters are further subdivided with increasing diameter into central peak craters (interior peak), central peak basins (interior peak and rings), peak ring basins (one interior ring) and multiring basins (two or more interior rings). (See also Impact cratering.)

Craters are surrounded by a zone of continuous ejecta (q.v.), consisting of material excavated from the crater. Beyond the continuous ejecta is a zone of discontinuous ejecta, with evidence of secondary craters formed by the impact of large blocks excavated from the primary crater. Ejecta deposits exhibit a reverse stratigraphy, with the deepest excavated material occurring on top and closest to the rim. Original near-surface materials occur on the bottom of the ejecta deposit and furthest from the rim.

Richard A.F. Grieve

Bibliography

French, B.M. and Short, N.M. (1968) *Shock Metamorphism of Natural Materials*. Mono Book Corp., 433 pp.
Roddy, R.J., Pepin, R.O. and Merrill, R.B. (eds) (1977) *Impact and Explosion Cratering*. Pergamon, 1301 pp.
Grieve, R.A.F. (1990) Impact Cratering on the Earth. *Sci. Am.*, **262**, 66–73.

Cross references

Comet: impacts on Earth
Impact cratering
Mars: impact cratering

CRETACEOUS–TERTIARY (K–T) BOUNDARY

An instant in time approximately 65 million years ago corresponding to a stratigraphic break in the sedimentary rocks of the Earth. The boundary between the Cretaceous (K) and Tertiary (T) geological time periods was chosen on the basis of the occurrence of a mass extinction, the disappearance of many forms of terrestrial life as recorded in the fossil record, followed by the rapid emergence of new forms to reoccupy the vacant ecological niches. The stratigraphic level at the boundary is often marked by the occurrence of a thin (~ 3 mm thick) clay layer.

The cause of the mass extinction had been much debated but the debate took a new turn in 1980, when it was discovered that the boundary clay layer was rich in iridium, an element rare in the Earth's crust but relatively enriched in asteroids and comets. It was proposed that a large asteroid or comet had impacted the Earth, throwing sufficient quantities of dust into the stratosphere to block sunlight resulting in darkness and cold on the ground. The lack of light was postulated to halt photosynthesis, leading to mass extinction of life because of collapse of the food chain. Subsequently additional evidence that the boundary layer is an ejecta layer from a very large impact has been found, including that the iridium-rich layer occurs globally, that the layer contains the other platinum group elements in the same proportions as found in chondritic meteorites, and that shocked mineral grains, tektites and the thicker proximal ejecta blanket of the crater also occur at the K–T boundary. Finally, the impact crater itself was found buried on the Yucatán Peninsula of Mexico in 1990. The ~ 180 km diameter Chicxulub crater, the largest known impact crater to have formed in recent Earth history, was excavated in rocks of the correct composition to produce the boundary tektites, has a K–T boundary age and is correctly located to produce the observed ejecta blanket. It was probably produced by the impact of a comet based on the amount of iridium distributed globally and the size of the crater.

The Chicxulub impact may have caused the global extinction of species by a variety of processes: by cold and darkness from ejected dust, from the thermal pulse of re-entering ejecta, from the acid rain produced from burning atmospheric nitrogen or releasing sulfur dioxide from the sulfate-bearing target rocks, or from a 10 000–100 000 year long greenhouse warming caused by carbon dioxide released from the carbonate rocks in the target. Although the effects of the impact were much greater than those of any known terrestrial process a debate continues among Earth scientists as to whether or not the impact caused the observed K–T boundary extinctions.

Alan Hildebrand

Bibliography

Alvarez, W. and Asaro, F. (1990) An extraterrestrial impact. *Sci. Am.*, **263** (4), 78–84.
Hildebrand, A.R., Penfield, G.T., Kring, D.A., *et al.* (1991) Chicxulub crater: A possible Cretaceous–Tertiary boundary impact crater on the Yucatán Peninsula, Mexico. *Geology*, **19**, 867–71.
Hildebrand, A.R. (1993) The Cretaceous/Tertiary boundary impact (or the dinosaurs didn't have a chance). *J. Roy. Astron. Soc. Can.* **87**, 77–118.
Sharpton, V.L. and Ward, P.D (eds) (1990) *Global Catastrophes in Earth History; An Interdisciplinary Conference on Impacts, Volcanism, and Mass Mortality*. Geological Society of America Special Paper **247**, 631 pp.
Silver, L.T. and Schultz, P.H (eds) (1982) Geological implications of impacts of large asteroids and comets on the Earth: Geological Society of America Special Paper 190, 528 pp.
Swisher, C.C., Grajales-Nishimura, J.M., Montanari, A., *et al.* (1992) Coeval 40Ar/39Ar ages of 65.0 million years ago from Chicxulub crater melt rock and Cretaceous – Tertiary boundary tektites. *Science*, **257**, 954–958.

Cross references

Comet: impacts on earth
Iridium anomaly
Mass extinction

CROLL, JAMES (1821–1890)

A Scotsman, born in Perthshire, the son of a stone mason, and forced to drop out of school at the age of 13 and work as a millwright, Croll was self-educated. Nevertheless he eventually became a shopkeeper, which gave him the leisure to read physics, astronomy and geology. At the age of 38 he got a job at the Andersonian Museum in Glasgow (1859), and from 1867 to 1880 worked at the Geological Survey of Scotland, eventually taking charge of their Edinburgh office. His readings in astronomy and geology convinced him that long-term climatic changes on planet Earth were cyclic and externally forced, notably by small, changing parameters in orbital characteristics. His *Climate and Time* was published in 1875 and his *Climate and Cosmology* appeared in 1885.

In 1868 Croll presented powerful evidence in favor of the 'fluvial doctrine' which claimed denudation of land surface, in contrast to the popular marine planation model of the day (Davies, 1968). He even converted Charles Darwin to this view. So too, he convinced Sir Archibald Geikie, his superior at the Survey in Edinburgh.

He became intensely interested in the origin of ice ages, in glacier mechanics, paleoclimates and in oceanic circulation, past and present. Thanks to his belief in cyclic forcing he claimed there was evidence of seven glacial cycles in the ice age deposits of northern Europe; at that time it was customary to accept no more than four or as little as one. For the pre-Pleistocene, he reviewed evidence for no less than nine glaciations. He also made a valuable contribution by calculating that a major deglaciation involving Antarctica would raise sea level by 60–120 m (1875, p. 388).

The concept of cosmic causes for terrestrial phenomena such as ice ages goes back to Adhémar (1842), with a model that was based merely on the precession of the equinoxes, an oversimplification. In any case, extraterrestrial forcing was emphatically anathema to Sir Charles Lyell and the geological and meteorological 'establishment' of the day; it remained so until marine geological field evidence of overwhelming clarity emerged in the late 20th century (Imbrie and Imbrie, 1979). The idea was developed further by M. Milankovitch (q.v.) in the early 20th century, but that model was rejected with equal vehemence.

Croll's ice age theory rested on the astronomical evidence of a cyclic variation in the Earth's orbital eccentricity (90 000–100 000 years), combined with the precession of the equinoxes, whereas the effect of the variation in obliquity was underestimated and treated separately. This model enjoyed a period of popularity, but it suffered from a fatal flaw; it resulted in an alternation of glaciation in northern and southern hemispheres. Furthermore, although it tried to explain cyclic climate change, it made no attempt to tackle the problem of recurrent ice ages (on $> 10^8$ year time scale). Although he was a prodigious reader, Croll was no mathematician, and the whole problem had to await its exhaustive analysis by Milankovitch in the 20th century.

Rhodes W. Fairbridge

Bibliography

Burstyn, H.L. (1971) Croll, James. Dict. Sci. Biogr., Vol. 3, pp. 470–1.
Croll, J. (1875) *Climate and Time, in their Geological Relations; a Theory of Secular Changes of the Earth's Climate*. London and New York, 577 pp. (2nd ed, 1885.)
Croll, J., (1885) *Climate and Cosmology*.
Croll, J. (1889) *Discussions on Stellar Evolution and its Relations to Geological Time*. London.
Davies, G.L. (1968) *The Earth in Decay*. London: Macdonald Techn. & Sci., 390 pp.
Horne, J. (1891) Obituary notice of James Croll. *Trans. Edinburgh Geol. Soc.*, **6**, 171–87.
Imbrie, J. and Imbrie, K.P. (1979) *Ice Ages: Solving the Mystery*. Short Hills, N.J: Enslow Publ., 224 pp.
Irons, J.C. (ed.) (1896). *Autobiographical Sketch of James Croll with Memoir of his Life and Work*. London: Stanford, 53 pp. (Review: Geol. Mag., **34**, 71–7, 1897.)

CRUST

The upper part of the lithosphere, which is the outermost solid layer or concentric shell of the planet Earth, as used in geology (Plates 9, 10; Woollard, 1960). It constitutes $< 0.1\%$ of the Earth's volume. The term is also used in geomorphology and soil science for a hard superficial layer usually 10–100 cm thick that owes its origin to chemical concentration by capillarity and dehydration ($CaCO_3$, SiO_2, Fe_2O_3, etc.).

A genetic explanation for the Earth's crust was presented by Dana (1875, p. 147): 'cooling went forward until . . . a *crust* formed outside . . . The crust has since, through time, continued cooling and increasing in thickness.' However, this simplistic theory of a cooling Earth is not now accepted and the crust is now known to be a complex mixture of rocks of all ages.

In plate-tectonic terminology, crust has sometimes been used as a synonym for lithosphere, a term first used by Suess (1875, p. 158; in Dennis *et al.*, 1979, p. 50), which clearly carries the implication of rigidity, in contrast to the asthenosphere, a lower concentric shell of 'no strength' over which the crustal plates are believed to be transported by convection currents in the mantle (q.v.). A seismically defined lower boundary of crust is the Mohorovičić discontinuity (q.v.).

In the case of the Earth's continental crust, which is up to 65 km thick, the majority of dated rocks are < 2.6 billion years old, and since mantle differentiation began somewhat before 3.9 Ga, this age represents an upper bound to the age of the oldest crust. Examples of > 3 Ga dates are known in all of the major cratons (the long-stabilized parts of the continental crust), but they are mostly limited to restricted areas or to inclusions in younger rocks. Accordingly, as a generalization, it can be stated that most of the continental crust consists of an inhomogeous association of sedimentary and metamorphosed rocks, or plutonic igneous intrusions of granitic suites, most of which represent recycled sedimentary materials. Continental crust, from a geochemical point of view, has been identified in its upper part as 'sial' (dominated by the elements silicon and aluminum in complex mineral associations). This is in contrast to the lower part and upper mantle which was designated as 'sima' (i.e. a petrology dominated by silicon and magnesium complexes). The mean density of sial is < 2.7, as against sima which is mostly > 3.0. Thus the sialic rocks 'float' above the denser simatic material. In his theory of continental drift Wegener proposed that the continental sial floated 'like icebergs' on a 'sea' of sima. Unfortunately this concept contained an inherent fallacy because the sialic rocks melt at lower temperatures than the sima. The lower plate boundary is now believed to be a pressure-related discontinuity and not a geochemical one.

Beneath ocean basins the crust is very different from its continental equivalent. It varies in thickness from 0 to 10 km and in age from < 1 Ma near spreading centers to < 200 Ma except where slivers of it have been caught up and incorporated in the continental crust of orogenic belts. 'Lithosphere' beneath ocean crust acquires a distinctive role, because in plate tectonics it is still the lower boundary of a plate that may be up to 100 km deeper than the crust (McClain and Orcutt, 1989). Petrologically, freshly formed ocean crust is basaltic in character and geochemically similar to the classical sima. As the oceanic crustal terrain ages it becomes progressively capped by an upper layer of accumulated deep-sea sediments.

Rhodes W. Fairbridge

Bibliography

Dana, J.D. (1875). *The Geological Story, Briefly Told* . . . New York, 263 pp. [2nd edn, 1895, 302 pp.]
Dennis, J.D. *et al.* (eds) (1979). *International Tectonic Lexicon*. Stuttgart: Schweizerbart, 153 pp.
James, D.E. (ed.) (1987) *The Encyclopedia of Solid Earth Geophysics*. New York: Van Nostrand Reinhold, 1328 pp.
McClain, J.S. and Orcutt, J.A. (1989). Lithosphere, oceanic – formation and evolution, in *The Encyclopedia of Solid Earth Geophysics*. (ed. James) New York: Van Nostrand Reinhold, pp. 660–8.
Sanders, J.E. (1981). *Principles of Physical Geology*. New York: John Wiley, 624 pp.
Woollard, G.P. (1960). The crust of the Earth. *Science Teacher*, **27**(5), 6–11.

Cross references

Core, terrestrial planetary
Lithosphere
Mantle

CYCLONE, ANTICYCLONE

On rapidly rotating planets, wind blows in a direction controlled by the competition between two forces. The pressure gradient force, if acting alone, would cause air to blow from high to low pressure. The acceleration of the rotating reference frame induces an apparent force, the Coriolis force, which deflects air to the right of its direction of motion in the hemisphere containing the positive rotation axis, and to the left in the opposite hemisphere. If a circular region of low pressure exists, these forces can balance only if wind blows in a circular fashion around the low. The sense of rotation of the flow is the same as the sense of the planet's rotation in either hemisphere. In Earth's northern hemisphere, for example, the wind blows counterclockwise as viewed from above. (Plate 9). The resulting flow is called a cyclone (from the Greek *kyklon*, moving in a circle).

If a circular region of high pressure exists, the forces can likewise balance only if wind blows in a circular fashion around the high. The sense of rotation of the flow is opposite the sense of the planet's rotation in either hemisphere. The resulting flow is called an anticyclone (from the Greek *anti-* opposite, against, plus *kyklon*).

Although anticyclones appear to be simply the opposite of cyclones, there are several important differences. For example, the centrifugal force associated with circular motion always acts radially outward, regardless of the direction of rotation of the flow. This reinforces the Coriolis force in cyclonic flow, but mitigates it in anticyclonic flow. Thus the pressure gradients and winds associated with an anticylone are typically weaker than those accompanying a cyclone. Furthermore, anticyclonic flow around a high can only be balanced by Coriolis forces, since centrifugal forces always act outward; there are no high-pressure tornadoes.

On Earth the forces in cyclones and anticyclones balance to within 10–16%. Near the surface, where friction slows the wind and reduces the Coriolis force, the deviation is even larger. The imbalance is such as to cause air to spiral in toward the low. The converging air leads to rising motion, cloud formation and precipitation. Thus cyclones are the locations of stormy weather in mid-latitudes. The cloudiness and precipitation is often concentrated in narrow regions called fronts which emanate from the low. The fronts are regions where warm air moving poleward ahead of the low overtakes cold air, or where cold air moving equatorward behind the low overtakes warm air. These cyclones grow in strength as a result of a baroclinic instability of the jet stream. The spatial extent of mid-latitude cyclones typically decreases with increasing planetary rotation rate and with decreasing vertical stability to convection.

For anticyclones the imbalance is such as to cause air to spiral outward away from the high. The diverging air leads to sinking motion, suppressing clouds and precipitation. Thus anticyclones are typically locations of fair weather. The spatial extent of anticyclones tends to decrease with planetary rotation rate and with decreasing vertical stability to convection.

The pressure, temperature and wind fluctuations measured at the Viking lander sites on Mars are probably the result of passing cyclone–anticyclone pairs. On the Jovian plantets, long-lived anticyclones are a common occurrence, outnumbering cyclonic features by about 10 to 1. The most obvious examples are Jupiter's Great Red Spot (Plate 22) and White Ovals, and Neptune's Great Dark Spot and Oval D2. On the Jovian planets, several long-lived cyclones exist, most notably the 'brown barges' seen in Voyager images of Jupiter. Slowly rotating Venus has no regional cyclones because of its weak Coriolis force.

Anthony D. Del Genio

Bibliography

Charney, J.G. (1973) Planetary fluid dynamics, in *Dynamic Meteorolog* (ed. P. Morel). Dordrecht, Holland: D. Reidel Publshing Co., pp. 97–351.
Emanuel, K.A. (1988) Toward a general theory of hurricanes. *Am. Sci.*, **76**, 371–79.
Hatzes, A., Wenkert, D.D., Ingersoll, A.P. and Danielson, G.E. (1981) Oscillations and velocity structure of a long-lived cyclonic spot. *J. Geophys. Res.*, **86**, 8745–9.
Haines, K. (1989) Modons as prototypes for atmospheric blocking. *J. Atmos. Sci.*, **46**, 3202–18.
Smith, B.A., Soderblom, L.A., Johnson, T.V. *et al.* (1979) The Galilean satellites and Jupiter: Voyager 2 imaging science results. *Science*, 206, 927–50.
Smith, B.A., Soderblom, L.A., Banfield, D. *et al.* (1989) Voyager 2 at Neptune: imaging science results. *Science*, 246, 1422–1449.
Williams, G.P. and Wilson, R.J. (1988) The stability and genesis of Rossby vortices. *J. Atmos. Sci.*, **45**, 207–241.

Cross references

Atmosphere
Coriolis force, geostrophic motion
Earth: atmosphere
Hadley circulation
Jupiter: atmosphere
Mars: atmosphere
Neptune: atmosphere
Venus: atmosphere

D

DARWIN, GEORGE HOWARD (1845–1912)

George Darwin, fifth child of the famous naturalist Charles Darwin, was born at Down House in Kent (England) on 9 July 1845. He went to Trinity College, Cambridge, where he had a distinguished career. He first studied law and was called to the Bar in 1874, but returned to Cambridge in the following year, and then devoted the rest of his life to mathematics and astronomy. He became Plumian Professor of Astronomy and Experimental Philosophy at Cambridge in 1883.

Darwin's first paper, *on the influence of geological changes on the earth's axis of rotation*, appeared in 1876; this was the beginning of a long series of research papers dealing with tidal forces, the stability of rotating bodies and associated phenomena. In particular, he was concerned with the origin of the Moon, and proposed a theory which was widely accepted for many years. From similarities in composition and structure he concluded that the Moon was produced from the same materials as those in the Earth's mantle, in which case the two bodies were logically considered to have once been combined. Darwin postulated a sequence of events in which the spin rate of the original Earth/Moon body was so rapid that it became unstable; it became first pear-shaped and then dumbbell-shaped. Eventually the neck of the dumbbell broke, and the Moon moved away as an independent body, its distance from the Earth continually increasing. Darwin calculated that some 54 million years ago the Moon was close to the Earth, while the rotation period of the Earth itself was no more than 5.5 h.

Unfortunately, on such geological grounds as the evidence of paleotides very similar to those of today, the fission hypothesis was quite unacceptable, although many geophysicists assume that at an early stage in the evolution of the system the Moon's distance from the Earth was much less than it is today (see Moon: origins). In spite of numerous and heroic attempts to revive it, the tidal theory in the form originally proposed by Darwin is no longer accepted.

There is, in fact, still disagreement about the origin of the Moon. For some time during the present century the idea that it and the Earth were once combined fell out of favor, but it has since been revived in modified form; one recent theory is that the original body was struck by a large impactor, so that the end product was the Earth–Moon system of today. Tidal friction (q.v.) is still driving the Moon gradually outward, and at the same time braking the Earth's axial rotation.

Darwin also paid particular attention to the planets, again from a purely dynamical point of view (he was never a practical telescopic observer). For instance, he maintained that Saturn's ring system was produced by the tidal disruption of a former satellite which moved within the gravitational danger zone and was literally torn apart. This remains a possibility, though opinions differ, and it is also widely believed that the rings are made up from debris which never condensed into a satellite.

Darwin received many honors. He was president of the Royal Astronomical Society in 1899 and its Copley Medalist in 1911; he was president of the British Association in 1905, and was knighted in the same year. He was undoubtedly one of the leading mathematicians of his time, and his pioneer investigations have been of the utmost value to later theorists. He died at Cambridge on 7 December 1912.

Patrick Moore

Bibliography

Darwin, G.H. (1876) On the influence of geological changes on the Earth's axis of rotation. *Phil. Trans. Roy. Soc. London*, **A167**, 271–312.
Darwin, G.H. (1878) The nebular hypothesis. *Observatory*, **1**, 13.
Darwin, G.H. (1879) On the precession of a viscous spheroid and on the remote history of the Earth. *Phil. Trans. Roy. Soc. London*, **A170**, 447–538.
Darwin, G.H. (1881) On the tidal friction of a planet attended by several satellites, and on the evolution of the solar system. *Phil. Trans. Roy. Soc. London*, p. 491 ff.
Darwin, G.H., 1898. *The Tides and Kindred Phenomena in the Solar System*. London; Boston.
Darwin, G.H. (1902) On a pear-shaped figure of equilibrium. *Phil. Trans. Roy. Soc. London*, **A200**, 251–314.
Darwin, G.H. (1908) *Scientific Papers*, Vol. II (*Tidal Friction and Cosmogeny*). Cambridge, 516 pp.
Kopal, Z. (1971) Darwin, George. *Dict. Sci. Biogr.*, Vol. 3, 582–4.
Ringwood, A.E. (1979) *Origin of the Earth and Moon*. New York: Springer-Verlag, 229–31.
Stratton, F.J.M. 1913. Sir George Darwin. *Mon. Not. Roy. Astron. Soc.*, **73**, 204.

DATING METHODS

There are many approaches to estimating time in a geological or cosmological context on the basis of the accumulated effect of some natural process. The most widely used remote sensing technique for dating planetary surfaces, for example, is based on impact crater density and an inferred flux of impacting projectiles along with consideration of cratering mechanics, crater degradation, etc. (e.g. Chapman and Jones, 1977). This is our principal source of chronological information for all observable planetary surfaces outside the Earth–Moon system.

The approaches usually granted the greatest degree of confidence for quantitative determination of absolute time are those based on radioactive decay, since the underlying process – radioactive decay – proceeds at a known rate independent of environmental conditions (in essentially all relevant cases). The various specific methodologies based on radioactive decay generally require measurement of the abundance of specific isotopes or of isotope ratios, and so characteristically require laboratory (rather than remote sensing) observation. Even for remote sensing crater density ages, 'ground truth' calibration is achieved by comparing crater densities and radiometric ages for the one case where such a comparison is possible, i.e. the Earth's Moon.

One major class of radiometric approaches is based on accumulation of the radiogenic daughter. For example, decay of radioactive ^{87}Rb (the parent) to stable ^{87}Sr (the daughter) over a time t leads to the relation

$$(^{87}Sr/^{86}Sr)_m = (^{87}Sr/^{86}Sr)_I + (e^{\lambda t} - 1)(^{87}Rb/^{86}Sr)_m \quad (D1)$$

where subscript m designates a measured present ratio and I designates the 'initial' ratio at time t in the past, and λ is the decay rate of ^{87}Rb. The use of ^{86}Sr as normalization for all the terms in equation (D1) reflects the circumstance that these ratios are more easily and precisely measured than are absolute abundances, and especially that the initial ^{87}Sr is much more readily characterized by its ratio to ^{86}Sr than by its absolute abundance. The exponential term follows from the initial (at time t) abundance of the parent being a factor of $e^{\lambda t}$ greater than the present measured abundance, so that subtracting unity gives the amount which has decayed in time t.

Equation (D1) is readily generalized to other parent–daughter pairs; in some cases the second term on the right must be multiplied by an appropriate constant if the decay is branched (e.g. ^{40}K) or if multiple daughters are produced (^{4}He). If the initial daughter abundance is presumed known or negligible, one pair of measurements permits solution of equation (D1) for t (usually termed a 'model age'). More generally, measurement of multiple samples presumed to have the same age and initial daughter abundance leads to an overdetermined system which permits solution for both terms (t is then an 'isochron age') as well as a test for how well the system satisfies the assumptions. Typically the data are plotted on an 'isochron diagram' whose axes are the two measured quantities in equation (D1) or its analogs for other parent–daughter systems. If the assumptions are satisfied the data will define a linear array whose slope $e^{\lambda t} - 1$ determines the age and whose intercept is the initial daughter abundance ratio.

Although the daughter-accumulation methods outlined above are conceptually simple, the practical utility of any particular scheme is governed by several other factors, such as the abundance of parent and daughter elements in natural samples, how much variation there is in parent/daughter ratios, and whether the state of the art of analytical instrumentation permits the necessary measurements with useful precision. As a first approximation, such factors have comparable effect for both terrestrial and extraterrestrial materials. In the present state of the art, extensive profitable application has been made for methods based on the decays of ^{238}U to ^{206}Pb, ^{235}U to ^{207}Pb, ^{232}Th to ^{208}Pb, ^{238}U, ^{235}U and ^{232}Th to ^{4}He, ^{40}K to ^{40}Ar, ^{87}Rb to ^{87}Sr, ^{147}Sm to ^{143}Nd and (more recently) ^{187}Re to ^{187}Os. The particular features of such schemes are described in standard texts (e.g. Faure, 1986).

Special note should be made for the U–Pb system, the only case in which two extant isotopes of the same element (^{238}U, ^{235}U) decay to two daughter isotopes of the same element (^{206}Pb, ^{207}Pb). This permits the writing of two versions of equation (D1), one for each parent–daughter pair, which can be compared to see if they give the same ('concordant') age, or combined to eliminate the parent–daughter ratio and determine an age (a Pb–Pb or ^{207}Pb/^{206}Pb age) based solely on Pb isotopic ratios. A Pb–Pb age is also rather insensitive to most recent isotopic disturbances. Moreover, the decay constants for both ^{238}U and ^{235}U are better determined experimentally than those of other parents used in geochronology. Precise U–Pb dating generally requires quite high U/Pb ratios and so is not always applicable to a given problem, but when it is applicable this system can produce ages more precise and, if concordant, more robust than absolute ages determined by other techniques based on long-lived radionuclides.

Age determination from such isotopic data generally requires the assumptions that at some time in the past – the age to be determined – the samples under study had the same initial daughter abundance ratio and that subsequently they have remained closed systems to both parent and daughter elements. Much of the art of isotopic geochronology lies in evaluating the degree to which these assumptions are valid. The event being dated, by definition, is 'isotopic closure,' the transition from an open system (more specifically one in which the daughter isotope is able to mix with other isotopes of its element) to an isotopically closed system in which neither the parent nor daughter element migrates into or out of the system and the daughter isotope remains physically associated with the parent element. Another aspect of the geochronologist's art lies in inferring what an isotopic closure event corresponds to in the history of the samples. In different circumstances this event could be formation of solid from vapor, quenching or crystallization of a liquid, or cooling through a threshold temperature (the 'blocking' temperature) below which subsequent diffusion is inconsequential on the scale of the relevant sample size (cf. Dodson, 1973). Different parent–daughter systems need not necessarily record the same event, since it is possible for one isotopic system to be open in the same circumstances in which another is closed. It is even possible for the same parent–daughter system to record different events for different sample sizes of the same material, since isotopic redistribution may proceed on a mineral scale (≤ 1 mm) even in materials which remain closed on the hand-specimen or outcrop scale (≥ 10 cm).

In other approaches based on radioactive decay the chronological information comes not from the ratio of accumulated daughter to parent, as outlined above, but from the abundance of the radioactive parent itself. This class of methods uses relatively short-lived (up to a few Ma) species which would not be present at all except by virtue of their recent production by nuclear reactions, generally either induced by energetic cosmic rays or arising in the decay chains of long-lived U and Th. The event dated is a change from one environment to another and the age is determined by the degree to which the radioisotope's abundance has approached equilibrium in its new environment. In terrestrial geochemistry the best known and most extensively applied technique is 'radiocarbon' (^{14}C) dating; other such radionuclides profitably exploited include ^{230}Th, ^{10}Be, ^{26}Al, ^{53}Mn and ^{36}Cl (also Faure, 1986). For meteorites this same basic approach is sometimes used to determine the time of exposure to cosmic rays or the time of fall on Earth (see Cosmic ray exposure age). The range of such methods, basically governed by the relevant half-life, is from historical times to a few Ma in the past, a range in which the daughter accumulation methods based on long-lived parents are usually not very sensitive.

Another class of methods which is particularly applicable to meteorites is also based on short-lived radionuclides. In this case, however, the relevant radionuclides are not produced locally and recently but rather are products of stellar nucleosynthesis and were present in the interstellar medium from which the solar system formed. The relevant isotopes, usually collectively termed 'extinct radionuclides,' have lifetimes long enough to survive the interval between nucleosynthesis and the formation of meteoritic solids, but short enough that they are truly extinct or unmeasurably rare today. The prior presence of a now-extinct radionuclide is inferred from radiogenic excess of its daughter; since their abundances are generally quite low in comparison with nearby stable isotopes, they are usually detectable only in chemically fractionated samples with high parent/daughter elemental ratios. A handful of these extinct radionuclides have been identified in meteorites (Podosek and Swindle, 1988; Shukolyukov and Lugmair, 1993), with half-lives ranging from 0.72 Ma (^{26}Al) to 103 Ma (^{146}Sm).

Chronology by now-extinct radionuclides is essentially a limiting case of the daughter-accumulation approach outlined above, and the same interpretations of the dated events in terms of isotopic closure are applicable. For equation (D1) in the limiting case $\lambda t \gg 1$, the exponential term becomes very large (and the unity term negligible in comparison) and the present abundance of the parent becomes very (unmeasurably) small, but their product remains well defined: it is the parent abundance at time t. In the limit $\lambda t \to \infty$ equation (D1), rewritten for the example of ^{26}Al, becomes

$$(^{26}Mg/^{24}Mg)_m = (^{26}Mg/^{24}Mg)_I + (^{26}Al/^{27}Al)_t (^{27}Al/^{24}Mg)_m \quad (D2)$$

where the added normalization to (stable) ^{27}Al is introduced to permit characterization as the isotope ratio $(^{26}Al/^{27}Al)_t$, its value at time t. When the I term is important, as it usually is, an isochron approach is generally used. Extinct radionuclide methods thus determine the abundance of the radionuclide at the time of isotopic closure from the present excess of the daughter; on this basis it is not possible to determine the absolute age t. If, however, samples of age t_1 and t_2 are compared, then (for ^{26}Al)

$$(^{26}Al/^{27}Al)_{t_1}/(^{26}Al/^{27}Al)_{t_2} = e^{\lambda(t_1 - t_2)} \quad (D3)$$

from which age differences $t_1 - t_2$, i.e. relative chronologies, can be determined. Because of the high value of λ, i.e. the short lifetime, the relative chronologies can be quite precise, well within 1 Ma in favorable cases, especially impressive for events which occurred some 4.5 Ga ago and generally much better than precisions attainable with the long-lived radionuclides.

The most extensively studied extinct radionuclides are ^{129}I (half-life 16 Ma) and ^{26}Al. Direct chronological interpretations for those isotopes are uncertain and controversial, however, because in some cases they fail to correlate with or conflict with expectations based on other criteria (Swindle and Podosek, 1988). For ^{129}I interpretational difficulties may arise because the distribution of the parent element I in meteorites is largely unknown and also because the daughter, ^{129}Xe, is a noble gas and perhaps thereby more susceptible to open system behavior. Chronological interpretations based on ^{26}Al have also been challenged by hypothesizing that the distribution of ^{26}Al in the early solar system was strongly heterogeneous, so that the assumption of isotopic uniformity inherent in equation (D3) is violated. The hypothesis of strongly heterogeneous distribution has been extended to ^{129}I and also to other extinct radionuclides. Interpretation of extinct radionuclide data, and unraveling chronological and distributional effects, remains a major challenge in study of the early solar system.

While the direct radiometric methods considered above have provided most of the chronological information available for meteorites, several other methodological approaches, frequently requiring somewhat more complicated interpretational models, have been used to good effect in providing a variety of types of chronological constraints. Evolution of 'initial' ^{87}Sr/^{86}Sr (equation (D1)), for example, has been used to infer reservoir extraction ages in conjunction with models for reservoir evolution (e.g. Wasserburg *et al.*, 1977; Podosek *et al.*, 1991). Spontaneous fission of now-extinct ^{244}Pu can be detected more sensitively by fossil fission tracks than by daughter isotopes (heavy isotopes of Xe); the fission track method has been used to infer thermal histories of meteorites by constraining cooling rates (Pellas and Störzer, 1981). Cooling rates and thermal histories have also been studied via arrested diffusion profiles of Ni attendant on exsolution of Fe–Ni metal alloys (e.g. Scott and Rajan, 1981). Accumulation of stable cosmic ray-induced spallation products such as ^{21}Ne and ^{3}He, noble gases which are detectable because their background concentrations are so low, is the principal means for measuring the time of exposure of meteorites to cosmic rays (see Cosmic ray exposure age).

In summary, many different methods have been applied to meteorites to produce several different kinds of chronological constraints. Most but not all of the methods are based on radioactive decay; most but not all involve isotopic analysis. Some of the methods are thoroughly familiar tools to the terrestrially oriented geochronologist; others are essentially uniquely applicable to meteorites. As in many disciplines, some of the results appear to be formidably robust, others range to the edge of speculation, and still others are thoroughly enmeshed with other major unanswered questions about the nature of the solar system. As it has been for decades, this discipline is a very active research area and some generalizations seem quite subject to change in response to both intellectual and technological innovation.

Frank A. Podosek

Bibliography

Chapman, C.R. and Jones, K.L. (1977) Cratering and obliteration history of Mars. *Ann. Rev. Earth Planet. Sci.*, **5**, 515–40.
Dodson, M.H. (1973) Closure temperatures in cooling geochronological and petrological systems. *Contrib. Mineral. Petrol.*, **40**, 259–74.
Faure, G. (1986) *Principles of Isotope Geology*. New York: Wiley.
Pellas, P. and Störzer, D. (1981) ^{244}Pu fission track thermometry and its application to stony meteorites. *Proc. R. Soc. London*, **A374**, 253–70.
Podosek, F.A. and Swindle, T.D. (1988) Extinct radionuclides, in *Meteorites and the Early Solar System*, (eds J.F. Kerridge and M.S. Matthews). Tucson: Universiy of Arizona Press, pp. 1093–113.
Podosek, F.A., Zinner, E.K., MacPherson, G.J., *et al.* (1991) Correlated study of initial ^{87}Sr/^{86}Sr and Al–Mg isotopic systematics and petrologic properties in a suite of refractory inclusions from the Allende meteorite. *Geochim. Cosmochim. Acta*, **55**, 1083–110.
Scott, E.R.D. and Rajan R.S. (1981) Metallic minerals, thermal histories and parent bodies of some xenolithic, ordinary chondrite meteorites. *Geochim. Cosmochim. Acta*, **45**, 53–67.
Shukolyukov, A. and Lugmair, G.W. (1993) Live iron-60 in the early solar system. *Science*, **259**, 1138–42.
Swindle, T.D. and Podosek, F.A. (1988) Iodine–xenon dating, in *Meteorites and the Early Solar System* (eds J.F. Kerridge and M.S. Matthews). Tucson: University of Arizona Press. pp. 1127–46.
Wasserburg, G.J., Tera, F., Papanastassiou, D.A. and Huneke, J.C. (1977) Isotopic and chemical investigations on Angra dos Reis. *Earth Planet. Sci. Lett.*, **35**, 294–316.

Cross references

Chronology: meteorite
Cosmic ray exposure age
Meteorite

DAWES, WILLIAM RUTTER (1799–1868)

The Rev. W.R. Dawes was one of the finest 19th-century observers. He was renowned for his keen sight and was nicknamed 'the eagle-eyed'; owing to his wide contacts, he was able to make observations with most of the largest telescopes in England.

He was born in London on 19 March 1799, and trained for the ministry. He did indeed become minister of the Congregational Church at Ormskirk in Lancashire, and he established a small observatory there. He resigned his charge in 1839, and later he took a degree in medicine; he practiced at Haddenham in Buckinghamshire, but his main interest was always in astronomy.

He made excellent measurements of the separations and position angles of double stars, and also carried out regular studies of the Sun, drawing fine details of sunspots with the help of a solar eyepiece which he had himself designed; but his best work was in connection with the planets. He was an independent discoverer of the crêpe ring of Saturn (ring C) in 1850, and made many observations of Saturn and Jupiter; the first reliable map of Mars was published in 1869 by R.A. Proctor on the basis of Dawes' drawings. Dawes died on 15 February 1868.

Patrick Moore

Bibliography

Alexander, A.F.O'D. (1962) *The Planet Saturn*, London: Faber.
Dawes, W.R. (1848) *Month. Not. Roy. Astron. Soc.*, v.11, p. 21.
Flammarion, C. (1891) *La Planète Mars*. Paris: Gauthier-Villars, pp. 184–9, 205.
Hoskin, M. (1971) Dawes, William Rutter. *Dict. Sci. Biogr.*, Vol. 3, pp. 605–6.
Proctor, R.A. (1869) *Half-Hours with the Telescope*. London, Plate VI.

DEEP SPACE NETWORK

To enable the study of the planets of our solar system by sending spacecraft equipped with science instruments to the targets under study, it is necessary to have continuous, two-way communication capability between the Earth and the spacecraft throughout the mission. In view of this requirement, NASA initiated the design and development of the Deep Space Network in early 1958. The first three stations were located around the Earth 120° apart in longitude and were implemented by 1961, in time for the original launch date of 1 July 1961 for Ranger, the first spacecraft to take photographs of the moon.

Most of the objects of our solar system are within 30° latitude of the equator; therefore these stations are located just beyond the 30° latitude to minimize the movement of the antennas in declination. The antennas were polar mounted and were 26 m in diameter. Since that first set of stations was implemented, there have been several others implemented at each longitude with diameters of 34 m and 64 m; the 64 m stations were later converted to 70 m. The communication bands are all in the microwave part of the spectrum, between 1 and 10 GHz, because it provides the minimum interference from natural radio emissions. The Deep Space Network has provided

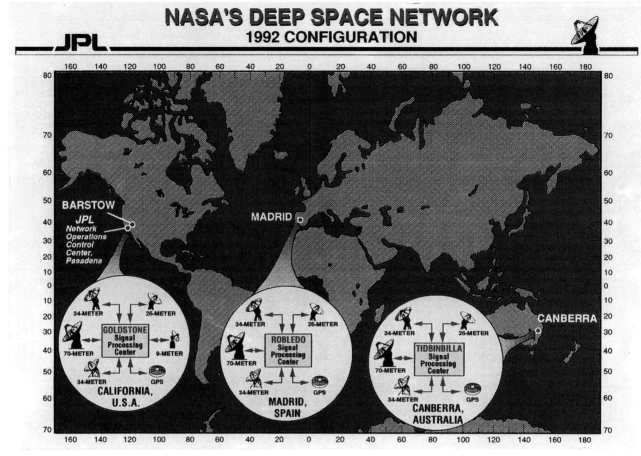

Figure D1 The Deep Space Network.

communications for practically all of the world's missions to the planets, returning all of the scientific data in the field of planetary science.

In addition to the primary communications function, each station of the Deep Space Network has instrumentation which generates the observable data used for navigating a spacecraft to its target. Early in the history of space research of the planets, it was discovered that the radio link between the spacecraft and the Earth can be used for direct science observations (see Radio science). The effects of the atmosphere and the solid bodies on the radio link provide data upon which one can make original contributions to understanding the nature of those bodies. The instrumentation also made it possible to use the radio link for tests of general relativity to accuracies better than those existing by any other technique. In fact, it has been said that the Deep Space Network has made 'enormous contributions to our confidence in Einstein's theory of general relativity.' Other science experiments with this same instrumentation have contributed to our knowledge of the nature of the space between the Earth and the spacecraft; the nature of the solar corona by its effect on the radio communication link; and the masses and inertial properties of the planets and their satellites, asteroids and, to a limited extent, comets.

In 1967, shortly after their completion, the 64-m diameter antennas were in great demand by radio astronomers of the United States, Europe and Australia because of their sensitivity. Of the five most sensitive radio telescopes used for study of natural radio sources throughout the world, three are in the Deep Space Network and are used regularly for astrophysical research of compact radio sources.

At the Goldstone Deep Space Communications Complex, the antennas and some of their mounted electronics are part of a planetary radar capability that has characterized the surfaces of Mercury, Venus, Mars, the Galilean satellites, the Saturnian satellite Titan, and main belt and Earth-approaching asteroids.

Nicholas A. Renzetti

Bibliography

Waff, C.B. (1993) The road to the Deep Space Network. *IEEE Spectrum*, April 1993, 51–6.

Posner, E.; Rauch, L. and Madsen, B. (1990) Voyager mission telecommunication firsts. *IEEE Commun. Mag.*, **28** (9), 22–7.

DEIMOS

Deimos is the more distant and smaller of the Martian moons and is the smallest of the three natural satellites of the terrestrial planets. Although it has not received the attention given to Phobos, it is a distinctive object whose properties have raised many questions about what is a typical asteroid or small rocky satellite.

Investigating Deimos

Deimos was discovered in 1877 by Asaph Hall. Earth-based observations of Deimos suffer less from scattered light from Mars than those of Phobos. Spectra of Deimos are decidedly superior to those of Phobos. Most data on Deimos come from the Viking Orbiter spacecraft, one of which passed within 50 km of the satellite in 1977 and returned images with resolutions of about 3 m, as well as infrared thermal measurements.

Basic characteristics

The physical and orbital properties of Deimos are given in Table D1.

Deimos's chief characteristics are an irregular overall shape, a smooth surface with filled-in craters, and prominent albedo features. Deimos has a mean radius of 6.2 km, yet it has a concavity 11 km across (Figure D2). It is not known whether this is a true crater, a

Table D1 Characteristics of Deimos and its orbit

a (km)	23 459
a (Mars radii)	6.92
Period	30.3
Eccentricity	0.000196 ± 0.000034
Inclination	1.789 ± 0.003
Mass (g)	$180 \pm 0.15 \times 10^{18}$
Density (g cm^{-3})	1.8 ± 0.2
Approximate axes (km)	7.5, 6.1, 5.2
Mean radius	6.2

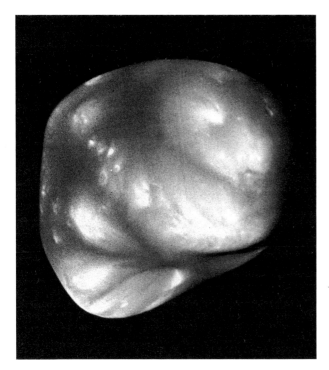

Figure D2 A low phase angle view of Deimos; the visible disk is about 12 km across, north is approximately at the top. Small craters are the source areas of most of the bright streamers visible here. Bright is only relative, however, as the dark areas have albedos of about 6%, the other areas about 8%. Note the very smooth appearance due to global redistribution of regolith. The south polar area (bottom) is in an 11 km wide concavity.

spallation feature, or evidence of accumulation of Deimos from two or more objects. The density of impact craters on Deimos is somewhat less than that on Phobos. The very highest resolution images show that nearly all the craters are partly to totally filled by sediment. The difference in crater density from Phobos may be largely a function of burial of craters rather than of overall age. Depths of filling are a few meters in craters with prominent rims, and perhaps up to 100 m in large craters with subdued rims. There appears to be substantially greater fill in those craters which have minimal obstructions to downslope creep of regolith into the crater. Albedo markings show that material creeps downslope to fill the craters and other low areas, remarkable in that gravity is effective even when only about 1/4000 that on Earth. From analogy with observations of lunar soil creep, impact shaking and thermal creep are the most likely drivers of the downslope motion. The debris is undoubtedly crater ejecta. Some of it probably reaccreted from Mars' orbit, while some appears to have traveled only a few hundred meters from source craters.

Unlike Phobos, Deimos has prominent albedo markings (Figure D2). They appear as brighter patches and streamers up to 4 km in length, and evidently mark the downslope movement of material that is slightly different from the majority of Deimos' surface. The simplest explanation of the 30% difference in albedo (8% versus 6%) is a difference in particle size, the brighter areas being made of finer material. Because of the compositional uncertainty, however, this cause of the albedo markings cannot be proved.

Deimos' orbit

Deimos' orbit is nearly circular at 6.9 Mars radii. It should evolve tidally only very slowly outward from Mars because of its distance from Mars and its small mass. Attempts to measure secular acceleration of its position have yielded small and conflicting solutions, but it would seem a safe assumption that it has not evolved far from its present orbit.

Composition

Deimos has a visual albedo of about 6%. Its spectrum is flat at wavelengths longer than 0.35 μm, but falls off sharply at shorter wavelengths. Like Phobos, it has been compared to carbonaceous meteorites, but the comparison may be misleading. Its density is 1.8 ± 0.2 g cm^{-3}, which is perhaps too low for meteorite analog materials. Porosities of 30–50% may be necessary to match carbonaceous or chondritic materials. Recent ground-based infrared spectra suggest that the surface is anhydrous. If this characteristic is typical of the bulk of the satellite, perhaps water is not a viable low-density component of Deimos.

Origin

The origin of the Martian satellites remains controversial. A carbonaceous composition may suggest capture of asteroidal material, but capture mechanisms have severe problems. Dynamically it is difficult to explain by capture the two Martian satellites in regular orbits, one inside, the other outside the synchronous point. Deimos' orbit is a substantial barrier to possible capture scenarios for Phobos, and if Phobos were captured first it is difficult to understand how Deimos could have been captured much farther out a significant time later. A carbonaceous composition would imply origin farther from the Sun and subsequent perturbation to Mars' orbital distance, but the uncertainties of the composition of both Martian satellites means that capture is not absolutely required.

Summary

Deimos is the smallest solar system satellite studied in detail. Its surface gravity of only 0.2 cm s^{-2} is still sufficient to guide downslope movement of debris that fills craters and that gives the whole satellite a smooth appearance. Its composition and origin remain controversial, as are the reasons it is so different from Phobos despite similar environments and likely similar compositions.

Peter C. Thomas

Bibliography

Burns, J.A. (1992) Contradictory clues as to the origin of the Martian moons. In *Mars* (eds H.H. Kieffer, C.W. Snyder and M.S. Mathews. Tucson: University of Arizona Press, 1283–1301.
Murchie, S.L., Britt D.T., Head, J.W. *et al.* (1991) Color heterogeneity of the surface of Phobos: relationship to geologic features and comparison to meteorite analogs. *J. Geophys. Res.*, **96**, 5925–45.
Thomas, P. (1979). Surface features of Phobos and Deimos. *Icarus*, **40**; 394–405.
Veverka, J. and J. Burns, (1980) The Moons of Mars. *Ann. Rev. Earth. Planet. Sci.*, **8**; 527–8.

Cross references

Mars
Phobos
Satellites, natural

DENNING, WILLIAM FREDERICK (1848–1931)

Denning was born on 25 November 1848 in Radstock, Somerset (west England), but spent most of his life in Bristol, where his father was an accountant. In his youth he was an excellent athlete (he was invited to play county cricket, though he never did so), but after 1906 he suffered so badly from ill health that he became something of a recluse.

His interest in astronomy dated from an early age; he acquired a small telescope, and later carried out his planetary work with a 12.6-inch (32 cm) Calver reflector. He made valuable and systematic observations of Jupiter, Saturn, Mars and Venus; his historical investigations into early observations of Jupiter's Great Red Spot were particularly noteworthy, and in 1903 he made a very accurate determination of the rotation period of Saturn. He discovered one nova (Nova Cygni 1920) and four comets (1881 V, 1890 VI, 1892 II and 1899 I), of which the first was recovered in 1978 by S. Fujikawa; it has a period of 9 years. However, Denning's main work was in meteor astronomy. He made many thousands of observations, and was a pioneer in determining true paths and radiants. For these investigations he was awarded the Valz Prize of the Paris Académie des Sciences in 1895 and the Gold Medal of the Royal Astronomical Society in 1898; in 1927 the University of Bristol awarded him an honorary MSc degree. He wrote one book, *Telescopic Work for Starlight-Evenings*. He died at Bristol on 9 June 1931.

Patrick Moore

Bibliography

Prentice, J.P.M. and Phillips, T.E.R. (1931) *W.F. Denning. Br. Astron. Assoc.*, **42**, 36–40.

DETERMINISM

Determinism is both a theory of nature and a theory of the way we know it. It holds that the present has already been decided (determined) by the past and therefore, given knowledge of the boundary conditions of all action (sometimes called the 'laws of nature') and the current state of affairs, the future can be known in all its details. Since, in fact, the current state of affairs is never known in its entirety and the boundary conditions are less than certain, the theory is often stated in hypothetical terms; however, doubt about achieving perfect knowledge does not, for determinists, alter their belief that the future has been determined. The optimism implicit in the theory is based in the belief that causal relations are linear, natural and open to investigation.

Determinism underlay the mechanics that emerged out of the theorizing of Copernicus (1473–1543), Kepler (1571–1630), Galileo (1564–1642) and Newton (1642–1727) (q.v.). In stressing the search for natural, as opposed to supernatural, causes, it played a, and perhaps the, primary, role in displacing miraculous and superstitious explanations of natural events. The great value of determinism has always lain in the impetus it has given to a thoroughgoing and exclusive search for natural causes.

The limitation and ultimate failure of determinism stems from its reliance upon a theory of causation borrowed from the ancient Greeks. In the Platonic version of that theory, which influenced Newton, this held that causes are Forms (reason) acting upon inert matter. For the western tradition, which has followed Plato, nature is rational and what happens must happen, i.e. natural events are determined by their causes just as a conclusion in a syllogism is necessitated by its premises. Science has had its great influence in the modern western world because its determinism incorporated beliefs central to the western tradition since ancient times.

The ideal of the perfect predictability of natural events contained in determinism is based upon two fallacies. The first confuses two quite different enterprises. To find causes is to discover 'when–then' relationships; to find explanations is to discover 'if–then' relationships (Grene, 1974). The former are sequences and occur in nature; the latter are implications and exist in logic. Plato confused the two when he said that ideas cause events, and it was this false identification that Newton accepted when he claimed that natural laws govern events. Newton's theory of gravity was based upon accurate accounts of sequences in nature, but he erred in thinking that the 'laws of nature' were its operative elements. Newton referred to the 'general Laws of Nature, by which the Things themselves are form'd' (Newton, undated). That natural laws 'govern' events remains even today the 'received view' of nature (Hempel, 1964), or even of scientific theorizing (Suppe, 1977).

The second fallacy follows closely upon the first. Just as no conclusion in a syllogism can contain anything not already in the premises, so effects can be no greater than their causes. As Aristotle put it, an effect is already potentially in its cause, as the oak is in the acorn. Causal relationships, in other words, are continuous, linear and proportional, as well as implicative.

These relics of ancient science are apparent in the principles implicit in Newtonian mechanics:

1. nature consists of particles that are simple, indivisible and indestructible;
2. all changes consist of continuous motions of these particles in space and time;
3. all motions of matter are caused by other particles acting upon it;
4. all effects are proportional to their causes;
5. all motions observe laws that can be stated mathematically.

When Newton, who has been called 'a seventeenth-century Pythagorean' (Fauvel *et al.*, 1988; cf. Ekeland, 1988), and Leibnitz (1646–1716) invented differential calculus to deal with rates of change, they thought it a perfect tool to deal with a world in which all changes are linear, continuous, rational and externally induced. In such a world it was credible to believe that the human understanding could, if it were only able to grasp the facts, make predictions and retrodictions with certainty.

The most forthright expression of strict determinism was made by Laplace (1749–1827; q.v.), whose great work was *Mécanique céleste* (1799–1825) and who formulated the principle of the invariability of planetary mean motions. Laplace asserted that an intelligence which 'could have a knowledge of all forces controlling nature together with the momentary conditions of all the entities of which nature consists . . . would be able to embrace in a single formula the movements of the largest bodies in the universe and those of the lightest atoms; for it nothing would be uncertain; the future and the past would be equally present to its eyes' (transl., Nathaniel Bowditch).

The first evidence taken to support determinism was provided by the solar system (q.v.), the cycles of which are highly regular and which is largely free from external influences. That system is close to equilibrium, and predictions can be made with uncanny accuracy, within one part in a hundred million for the motion of the Earth around the Sun. The presumed equilibrium of the solar system was taken to be a model for the universe and all systems in it. 'Classical physics was born as a dependent of astronomy, and . . . remained tied to it throughout its whole history' (Koyré, 1978). The importance of the model lay in its influence in sustaining a picture of the physical universe as perfectly balanced – an equilibrium of causes and effects, a balance of forces, a system that was rational because all its forces were proportional and additive.

Because it considered the solar system typical, classical physics treated other systems as isolated or closed. It treated matter as a set of mathematical points on a Euclidean space grid and disregarded features such as friction, which disturb the 'free' functioning of objects. These abstractions from experience led to results that conformed to the postulated 'laws' but did so at the expense of the detailed accuracy promised by determinism.

In 17th- and 18th-century physics, the search for linear relationships was exceedingly fruitful, and its results were taken as spectacular evidence that nature is a closed system in equilibrium. For classical physics determinism seemed to be both an assumption needed in scientific investigation and a conclusion implicit in its findings. Ultimately, scientists thought, it would be possible, at least in principle, to predict the future in detail and with certainty.

Determinism expressed the optimism of the Enlightenment, and scientists became exuberantly hopeful. Thus Michelson (1852–1931), in an often quoted and much derided statement, claimed in 1903 'The more important and fundamental laws and facts of physical science have all been discovered, and these are now so firmly established that the possibility of their ever being supplemented in consequence of new discoveries is exceedingly remote.' Similarly Max Born (1882–1970) concluded in 1928 that 'physics as we know it will be over in six months.' Even today the position is well entrenched. Stephen

Hawking (1942–) stated in 1988 that 'there are grounds for cautious optimism that we may now be near the end of the search for the ultimate laws of nature.'

However, determinism 'has lived very largely on the credit of classical mechanics,' and its hold has been loosened as the limits of classical physics have been discovered. Such apocalyptic views as the above have now been largely discounted. Sir James Lighthill, a successor to the Lucasian chair once held by Newton, has pointed out that 'we may have generally tended to believe [such generalizations about predictability] before 1960, but . . . we now recognize [they] were false.' (For sources of the above quotations and others like them see Coveney and Highfield, 1990, and Cassirer, 1939.) Natural events occur in systems that are open rather than closed and that maintain a dynamical stability rather than equilibrium. Discontinuities and disproportionalities are common; linear equations are inadequate to describe nature's complexity.

The mechanistic model was inadequate from the beginning. Newton himself was rightly accused of deserting it when he postulated his force of gravity; Leibnitz derided it as 'a scholastic occult quality or the effect of a miracle.' Newton was in effect describing a field of force at work, thus intimating that force is internal to the events being described rather than external to them, as theory required (Westfall, 1971). Such a field concept was later used by Faraday (1791–1867), Maxwell (1831–1879) and others to describe electrostatic, magnetic, electrical and light phenomena. A field of force, we now know, is a system rather than a sequence of energies; it is characterized by 'circular causality' (von Foerster, 1984) rather than causes that are discrete, prior and external.

If Newton himself had to depart from mechanistic and deterministic principles to state his position, his successors were under even greater pressure. In 1879 Henri Poincaré (1854–1912) showed that even in closed systems the introduction of a third body creates a condition in which Newton's equations cannot be solved precisely. He was discovering that any physical system experiences, not only perturbations caused by external events, but also fluctuations resulting from the interactions of the bodies in the system itself. Thus, for example, the planets 'wobble' in their orbits (q.v.) from time to time, asteroids are thrown out of orbit, and magnetic poles of planets are reversed at irregular intervals. The solar system itself is not at equilibrium, although it is close to it.

Other natural systems are farther from equilibrium, and even more marked deviations from mechanistic principles were required by those who investigated them. Darwin (1809–1882) and Wallace (1823–1913), in their account of biological evolution, described organisms and an ecosystem that gain in complexity over time. The facts of evolution were so contrary to classical physical theory that early in this century living organisms were called 'cheats in the game of entropy.' (See Needham, 1943, for reactions to this anomalous situation.) The 'co-adaptations' necessary to the survival of organisms, origin of species and stability of ecosystems represent systemic interadjustments that are nonlinear and unpredictable.

Boltzmann (1844–1906) found that descriptions can be no more than statistical and probable in thermodynamics. Einstein (1879–1955) (q.v.) discarded the concept of rigid bodies unaffected by their motion in his theory of relativity, and Rutherford (1871–1937) and Bohr (1885–1962) discarded the notion of simple, indivisible atoms. And quantum theory, resisting efforts to reduce it to classical terms, seems to lead to the conclusion that probabilism is a characteristic of all knowledge at the level of elementary particles, and perhaps of reality itself (Pais, 1986; Gigerenzer et al., 1989).

As a result of these advances, it becomes clear that reality in a number of fields cannot be explained in classical terms. Ernst Cassirer (1874–1945), reacting in 1939 to the unsettling effect of quantum mechanics upon scientific theory, referred to 'the lost paradise of classical concepts.' The 'traditional equilibrium-based view of the world' 'had to be given up' (Coveney and Highfield, 1990), and with it went the justification for strict determinism.

Nevertheless, determinism has remained as resistant to contrary evidence in scientific circles as *laissez-faire* economics has in the modern welfare state. The seeming impregnability of determinism rests upon the fear that chance is the only alternative to necessity, and this in turn depends upon the fallacies borrowed from Greek science, the identification of when–then with if–then relationships and the belief that all causal relationships are linear. Given these misconceptions, to deny necessary connections is to deny causal relationships and, since science is the search for causes, to give up the scientific enterprise itself.

Einstein remained a determinist even as his theories undermined its assumptions. Others, many of whom think science requires determinism, try to redefine the latter or, like Einstein, continue to use its language even as they rely upon concepts inconsistent with it. While the search for causes has revealed complexity, nonlinearity, circular causes, systemic unity and webbed interlockings in nature, the hope remains that all these phenomena can be 'reduced' to classical formulas. In other disciplines there is widespread hope, sometimes called 'physics envy,' that the facts can be explained by a few simple laws. The most common concession scientists make to the inadequacies of determinism is the claim that natural events are the products of 'chance and necessity' operating together (e.g. Monod, 1971).

The contradictory of necessity, however, is not chance but contingency. Contingent events abound in nature and are subject to scientific investigation, although not to linear equations. This discovery is indebted to research of the last two generations – in cybernetics (Wiener, 1949), systems theory (von Bertalanffy, 1968), information theory (Shannon and Weaver, 1949), nonequilibrium thermodynamics (Nicolis and Prigogine, 1977), chaos theory (Gleick, 1987), catastrophe theory (Thom, 1975) and complexity theory (von Neumann, 1966).

As a result of developments in these fields, as well as biology (Varela 1979), geophysiology (Lovelock, 1988) and physics (Davies, 1989), we find nature abounds with self-organizing (autocatalytic) and self-maintaining (autopoietic) systems. These contrast with engineered control systems, which are linear and predictable (deterministic) because they use instructions to manipulate materials (Rosen, 1987; Varela, 1979).

In autopoietic systems, by contrast, fluctuations create feedback such that minute forces sometimes cause gross, and unpredictable, effects. Poincaré, faced with such feedback in the three-body problem, commented 'These things are so bizarre that I cannot bear to contemplate them.' They were strange because they involved discontinuities and disproportionalities and reflected internally created fluctuations rather than externally induced perturbations, the only causes reigning theory recognized.

Natural systems are open and dissipative, i.e. they exchange energy and matter with their surroundings. Those far from equilibrium are exceedingly sensitive to conditions affecting them. An increase in energy intake may cause such systems to create a new form of order (such as a tornado, a 'chemical clock' or life). The energy provided has exceeded the capacity of existing systems to use it, and so a new system self-organizes and converts energy at a higher level through the use of feedback and feed-forward mechanisms (Nicolis and Prigogine, 1989; Prigogine and Stengers, 1984). Neither the amount by which the energy will be multiplied nor the form which the new system will take can be predicted in detail on the basis of physical laws and earlier boundary conditions (Cassirer, 1939; Born, 1949). Once formed, such systems operate at a steady state which is never at equilibrium; resilience and stability, fluctuations and perturbations are their characteristics (Jantsch, 1980).

Such autopoietic systems may be quite temporary (like a tornado) or relatively long lasting (like an organism, Jupiter's red spot or the Earth as an ecosystem). In their self-organization and collapse they create new boundary conditions, which cannot be inferred even from full knowledge of initial conditions and governing laws. Thus boundary conditions are not fixed and unchanging, as classical determinism required for detailed predictions (Conrad, 1983). Through the autocatalyzing of autopoietic systems, a common event in nature, new boundary conditions are being created all the time. Natural events all have causes but, even given perfect knowledge, an observer would find some of them to be emergents, that is to say, unpredictable. The scientific discovery that contingency is the rule in nature leads to the conclusion that science does not need determinism in its search for causes.

Sheldon Ackley

Bibliography

Born, M. (1949) *Natural Philosophy of Cause and Chance*. Clarendon Press.

Cassirer, E. (1939) *Determinism and Indeterminism in Modern Physics*. Yale.

Conrad, M. (1983) *Adaptability: The Significance of Variability from Molecule to Ecosystem*. Plenum.

Coveney, P. and Highfield, R. (1990) *The Arrow of Time*. Ballantine.
Davies, P. (1989) *The New Physics*. Cambridge.
Ekeland, I (1988) *Mathematics and the Unexpected*. University of Chicago.
Fauvel, J.K., Flood, R. Shortland, M. and Wilson R. (eds) (1988) *Let Newton Be!* Oxford.
Gigerenzer, G., Swijtink, Z. Porter, T., et al. (1989) *The Empire of Chance*. Cambridge University Press.
Gleick, J. (1987) *Chaos*. Viking.
Grene, M. (1974) *The Understanding of Nature*. Reidel.
Hempel, C. (1964) *Aspects of Scientific Explanation*. Free Press.
Hiley, B.T. and Peat, F.D. (eds) (1987) *Quantum Implications: Essays in Honor of David Bohm*. Routledge & Kegan Paul.
Jantsch, E. (1980) *The Self-Organizing Universe*. Oxford.
Koyré, A. (1939) *Galileo Studies*. Humanities Press.
Lovelock, J. (1988) *The Ages of Gaia*. Commonwealth Fund.
Monod, J. (1971) *Chance and Necessity*. Knopf.
Needham, J. (1943) *Time: the Refreshing River*. Macmillan.
Newton, I. (undated) Newton ms. University Library Cambridge, Add Mss. 3970.3 f. 296, p. 401.
Nicolis, G. and Prigogine I. (1977) *Self-Organization in Nonequilibrium Systems*. Wiley-Interscience, 1977.
Nicolis, G. and Prigogine, I (1989) *Exploring Complexity*. W.H. Freeman.
Pais, A. (1986) *Inward Bound: Of Matter and Forces in the Physical World*. Oxford.
Prigogine, I. and Stengers, I. (1984) *Order Out of Chaos*. Bantam.
Rosen, R. (1987) Some epistemological issues in physics and biology, in *Quantum Implications: Essays in Honor of David Bohm* (eds B.T. Hiley and F.D. Peat) Routledge & Kegan Paul, pp. 314–27.
Shannon, C. and Weaver W. (1949) *The Mathematical Theory of Communication*. University of Illinois.
Suppe, F. (1977) *The Structure of Scientific Theories*. University of Illinois.
Thom, R. (1975). *Structural Stability and Morphogenesis*. Addison-Wesley.
Varela, F. (1979) *Principles of Biological Autonomy*. Elsevier North Holland.
von Bertalanffy, L. (1968) *General Systems Theory*. Braziller.
von Foerster, H. (1984) *Observing Systems*. Intersystems.
von Neumann, J. (1966) *Theory of Self-Reproducing Automata* (ed. A.W. Burks). University of Illinois.
Westfall, R. (1971) *Force in Newton's Physics: The Science of Dynamics in the Seventeenth Century*. American Elsevier.
Wiener, N. (1948) *Cybernetics, or Control and Communication in the Animal and the Machine*. Wiley.

DIFFERENTIATION

Differentiation is the process of segregation of zones of different chemical or mineralogical properties within a celestial body. In planetary (or satellite) scale differentiation the zones are concentric, with denser materials at greater depth. Differentiated rocky bodies, such as the terrestrial planets, the Earth's Moon and some asteroids, typically have cores of metal (or metal plus sulfide) beneath silicate outer layers. In many cases the silicates are further differentiated into a crust (q.v.) of material of lower density and/or melting temperature overlying a mantle (q.v.). Differentiated icy bodies typically consist of a rocky core (silicates plus metal) beneath an exterior of ice. Although the term 'differentiation' is usually applied to solid bodies, analogous processes occur in the interiors and deep atmospheres of the Jovian planets. All of the planets, with the possible exception of Pluto, are believed to be differentiated. At least some satellites (e.g. the Moon) and asteroids (Vesta) certainly are, but the extent of differentiation in many others is unknown (Schubert, Spohn and Reynolds, 1985).

Magmatic differentiation of igneous rocks, as seen on planet Earth, usually referred to only as differentiation, is the production of rock types of contrasted composition from one originally homogeneous parent magma. Many different mechanisms have been suggested, e.g. crystal fractionation, elutriation, filter pressing, flowage differentiation, gravity separation or liquid immiscibility (Daly, 1933, p. 319 et seq.).

For materials to differentiate, they must be able to separate from one another, hence temperatures high enough to melt at least some of the phases are required. Possible heat sources for differentiation include the kinetic and gravitational potential energy released during accretion, radioactive heating and (particularly for satellites) tidal heating (see Tidal friction). The amount of melting required depends on the density difference between the phases and the strength of the gravitational field. In small asteroids it is apparently as high as 50% (Taylor, 1992). At the other extreme, differentiation may be retarded or inhibited even in a planet-sized, convecting, molten 'magma ocean' by the transport of crystals suspended in the liquid (Tonks and Melosh, 1990).

Several techniques can be used to study differentiation. Seismic data (available so far for only the Earth and Moon) can identify the boundaries between different layers, although the detailed composition of those layers is not determined. If samples are available (Earth, Moon and probably Mars and Vesta), chemical analyses of trace elements can, in principle, be used to determine the extent and timing of differentiation. The siderophile and chalcophile elements (see Chemical element) which would tend to concentrate in the core, are particularly useful for studying the primary metal/silicate differentiation, while the rare earth elements are the most useful tracers of differentiation of the silicate portions. Even for Earth, where the present internal structure is reasonably well known, the timing and mechanisms of planetary-scale differentiation are still not established with any certainty (Newsom and Jones, 1990). For unsampled bodies the primary constraint comes from the bulk density, although the surface morphology and composition and the detailed gravity field can also be used to deduce parts of the differentiation history.

Timothy D. Swindle

Bibliography

Daly, R.A. (1933) *Igneous Rocks and the Depths of the Earth*. New York and London: McGraw-Hill Book Company.
Newsom, H.E. and Jones, J.H. (eds) (1990). *The Origin of the Earth*. New York: Oxford University Press.
Schubert, G., Spohn, T. and Reynolds, R.T. (1986) Thermal histories, compositions and internal structures of the moons of the solar system, In *Satellites* (eds J.A. Burns and M.S. Matthews), Tucson: University of Arizona Press. pp. 224–92.
Taylor, G.J. (1992) Core formation in asteroids. *J. Geophys. Res.* 97, 14717–26.
Tonks, W.B. and Melosh, H.J. (1990) The physics of crystal settling and suspension in a turbulent magma ocean in *The Origin of the Earth* (eds H.E. Newsom and J.H. Jones). New York: Oxford University Press; pp. 151–74.

Cross references

Accretion
Chemical element
Earth–Moon system: origin
Igneous rock
Mantle convection
Thermal evolution of planets and satellites
Tidal friction
Vesta

DIURNAL VARIATION

As a planet rotates about its axis, a point on the surface experiences alternating heating by the Sun when it is on the dayside and cooling by radiation to space when it is on the nightside. This systematic variation over the length of a day is called diurnal variation or the diurnal cycle (from the Latin *diurnus*, of a day).

The length of the day (P_{day}) depends on a planet's rotation period about its axis (P_{rot}) and its revolution period about the Sun (P_{rev}) according to $P_{day} = (P_{rot}^{-1} + P_{rev}^{-1})^{-1}$. On most planets $P_{rot} \ll P_{rev}$, and the length of the day is therefore approximately equal to the rotation period. A major exception is Venus, whose rotation and revolution periods are comparable, yielding a day that is roughly half as long as either.

The fraction of the day that a point on a planet spends in sunlight depends on the planet's obliquity, or axial tilt, and the time of year.

At the equinoxes all points spend exactly half the day in sunlight. At the solstices, latitudes poleward of 90° minus the obliquity are in sunlight (summer) or darkness (winter) for the entire day.

The response of a planet's surface and atmosphere to diurnal variations in heating depends on its mass, composition and temperature. On Earth, for example, diurnal temperature variations at the ground penetrate about 10 cm below the surface. Below this depth temperatures vary little from day to night and change only with the seasons. In the part of the atmosphere adjacent to the surface the diurnal cycle produces large variations in temperature from day to night. During the day, near-surface air becomes turbulent up to an altitude of several kilometers, like a pot of boiling water; at night the turbulence dies down to a shallow layer or disappears completely. Weather phenomena which depend on surface solar heating, such as thunderstorms, also exhibit a pronounced diurnal cycle in their occurrence. Land heats more quickly than ocean during the day, leading to low pressure over land and a sea breeze from ocean onto land in the afternoon. In the free atmosphere well away from the surface, on the other hand, radiative heating/cooling takes about a month to occur, and so diurnal variations there are negligible.

Other planets experience a range of diurnal cycle strengths depending on their distance from the Sun and the thickness of their atmospheres. Mars' thin atmosphere, for example, undergoes extreme day–night variations in temperature, giving rise to planetary-scale waves known as thermal tides which may help lift dust off the surface. Venus' thick atmosphere, on the other hand, takes decades to respond to variations in sunlight. Therefore, day and night are indistinguishable at depth. Near the cloud tops, however, the atmosphere is thinner, and the second harmonic of the diurnal heating distribution produces a thermal tide which contributes to Venus' superrotation. The Jovian planets are so cold that responses to varying radiation are miniscule, and diurnal variations are therefore negligible on these planets.

Anthony D. Del Genio

Bibliography

Cooper, H.J., Garstang, M. and Simpson, J. (1982) The diurnal interaction between convection and peninsular-scale forcing over south Florida. *Mon. Weather Rev.*, **110**, 486–503.
Seiff, A., Kirk, D.B., Young, R.E., *et al.* (1980) Measurements of thermal structure and thermal contrasts in the atmosphere of Venus and related dynamical observations: results from the four Pioneer Venus probes. *J. Geophys. Res.*, **85**, 7903–33.
Sellers, W.D. (1965) *Physical Climatology*. Chicago: University of Chicago Press.
Zurek, R.W. and Leovy, C.B. (1981) Thermal tides in the dusty Martian atmosphere: a verification of theory. *Science*, **213**, 437–9.

Cross references

Atmosphere
Earth: atmosphere

DOME

In geomorphology, dome is a term for any landform determined by more or less ovoid rising contours and gently arched cross-section. Scale is not essential to the definition. In geology it is a structure defined by outward dipping layers giving a more or less circular outcrop pattern. Geological domes may or may not have a geomorphologic expression, being sometimes buried or truncated by erosion.

Domes can be produced in a number of ways.

1. Volcanic eruption of viscous (silicic or felsic) lavas may produce a landform with a high aspect ratio, with or without a central crater, or occurring within a larger crater, e.g. the 'tholoids' of the Lipari Islands and New Zealand, trachyte domes of Hawaii or the dacite dome in the crater of Mount St Helens.
2. Inflation by subvolcanic intrusion, which may be transient prior to an eruption, e.g. the pre-eruption doming at Kilauea, or may lead to the formation of a permanent dome with or without a central crater, e.g. the lava domes on the mare areas of the Moon.
3. Surface extrusion or subsurface intrusion of a low viscosity material, e.g. salt domes ('halokinesis'), gypsum domes, and ice/permafrost in a pingo or palsa.
4. Residual landforms above a peneplain produced by exfoliation of granitic-type intrusions, e.g. certain types of inselberg.
5. Small to meso-scale tectonic features (10^1 km) defined by outward dipping layers (bedding or foliation) in a double-plunging antiform, e.g. the Black Hills of South Dakota.
6. Large-scale tectonic features (10^2–10^3 km) defined by positive deviations from a hypsometric curve in topography, e.g. the Hawaiian bulge in the Pacific Ocean, Tharsis Ridge on Mars, or by the differential elevation of ancient peneplains, e.g. the domes of East Africa. This type of dome marks the site of a sub-lithospheric hot-spot upwelling.

Domes may occur singly or in groups, and several forms may occur together, e.g. the trachyte domes of Hawaii on a dome-like shield volcano, itself sited on top of the Hawaiian bulge. The 'domed highland' areas identified on Venus (Figure D3) are of unknown origin.

Adrian F. Park

Bibliography

Fairbridge, R.W. (ed.) (1968) *The Encyclopedia of Geomorphology*. New York: Van Nostrand Reinhold, 1295 pp.
Greeley, R. (1987) *Planetary Landscapes*. New York: Allen and Unwin.
Phillips, R.J., Grimm, R.E. and Malin, M.C. (1991) Hot-spot evolution and the global tectonics of Venus. *Science*, **252**, 651–4.
Sleep, N.H. and Phillips, R.J. (1985) Gravity and lithospheric stress on the terrestrial planets with reference to the Tharsis region of Mars. *J. Geophys. Res.*, **90**, 4469–89.

Cross references

Igneous rock
Volcanism in the solar system

DUST

Multitudes of small particles populate the inner solar system and asteroid belt. Comets shed particles along their orbits, particularly near the Sun. Collisions of solid objects in the asteroid belt yield small particles with high relative velocities, which may again collide with one another to form yet smaller particles. Particles with diameters less than about 1 mm are characterized as dust, or interplanetary dust particles (IDPs). Particles much smaller than about 1 μm tend to be rapidly expelled by solar radiation pressure. Small interplanetary objects with diameters larger than about 1 μm are most often termed meteoroids, or in the case of cometary trails, debris (Sykes, 1991).

Interplanetary dust particles are the cause of the zodiacal light (q.v.), which is visible as a faint brightening of the sky along the ecliptic plane near the horizon after sunset and before sunrise. In the outer solar system, dust-sized particles form the rings of Jupiter and Neptune, and thin bands of material in the ring system of Uranus. Interplanetary dust is captured and accreted by the Earth at a rate of about 4×10^7 kg per year (Love and Brownlee, 1993).

Studies of interplanetary dust shed light on a number of important questions, including the formation of the planetary system, the compositions of comets and asteroids and the dynamics and ages of comets. Analysis of interplanetary dust provides information which often cannot be obtained from any other source. This article touches on the collection, composition, origins and orbital dynamics of interplanetary dust.

Collection

Pristine interplanetary dust particles have been collected by high-altitude aircraft flights in Earth's stratosphere since the 1970s. Collectors coated with inert silicone oil have been exposed to the atmosphere at altitudes of about 20 km by U2 aircraft. Laboratory

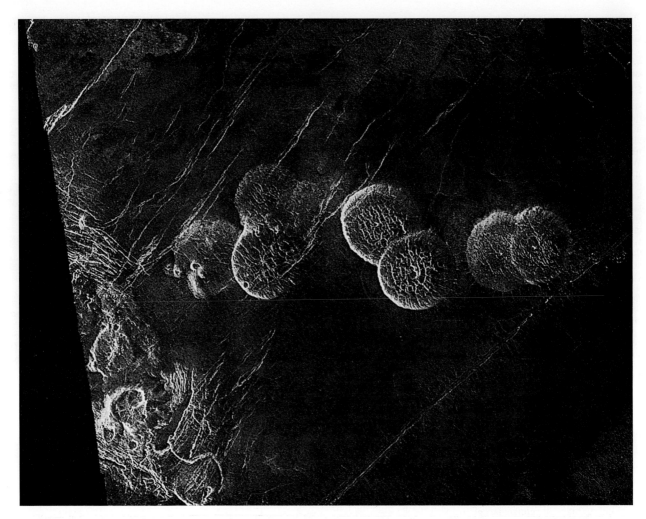

Figure D3 Domes: the eastern edge of Alpha Regio on Venus is shown in this image centered at 30° S latitude and 11.8° E longitude (longitude on Venus is measured from 0° to 360° east). Seven circular, dome-like hills, averaging 25 km in diameter with maximum heights of 750 m, dominate the scene. These features are interpreted as very thick lava flows that came from an opening on the relatively level ground, which allowed the lava to flow in an even pattern outward from the opening. The complex fractures on top of the domes suggest that if the domes were created by lava flows, a cooled outer layer formed and then further lava flowing in the interior stretched the surface. The domes may be similar to volcanic domes on Earth. Another interpretation is that the domes are the result of molten rock or magma in the interior that pushed the surface layer upward. The near-surface magma then withdrew to deeper levels, causing the collapse and fracturing of the dome surface. The bright margins possibly indicate the presence of rock debris on the slopes of the domes. Some of the fractures on the plains cut through the domes, while others appear to be covered by the domes. This indicates that active processes predate and postdate the dome-like hills. The prominent black area in the northeast corner of the image is a data gap. North is at the top of the image. (Image P-37125.)

techniques are then employed to isolate the captured particles (Brownlee, 1991). The particles obtained generally range in size from 1 to 50 μm. Modern laboratory equipment can successfully determine the compositions of these tiny samples.

IDPs are also obtained from seafloor sediments and from some polar ice environments. Larger particles (up to 1 mm) are usually melted during passage through the atmosphere, forming spherules. A small minority of these particles survive atmospheric entry intact; low incidence angles and low entry velocities are required for this to occur.

Spacecraft experiments may also be designed to collect interplanetary dust. The disadvantage of these is a lower density of particles, and the high impact velocities of many of the particles; the latter will degrade the particles unless special collectors are employed. The Long Duration Exposure Facility (LDEF) (McDonnell et al., 1990) orbited the Earth for 5.77 years, using 1.6 mm thick aluminum panel collectors. In this case particle fluxes and characteristics were studied on the basis of the craters formed in the surface by the impacts of the particles. The advantage of space-based collectors is that the directions and velocities of impacting particles may be determined accurately by electronic techniques.

Composition

The majority of IDPs have compositions similar to chondritic meteorites (Brownlee, 1987; see Carbonaceous chondrite; Chondrites, ordinary). The most common components are olivine, pyroxene, hydrated silicates, iron sulfide, glass, iron carbide and amorphous carbon. Organic carbon averages about 10% by mass.

Although there are compositional similarities between typical meteorites and typical dust particles, it is important to note that the two types of objects represent different samples of the interplanetary matter. Meteorites represent a more restricted (and necessarily biased) sample, for two reasons. First, only relatively durable rocks are capable of surviving atmospheric entry; less coherent samples are disrupted and vaporized. Second, meteoritic material is delivered to the Earth from the asteroid belt by selective, non-random gravitational mechanisms; certain regions of the belt (Kirkwood gaps) have been depleted of material due to orbital resonances with Jupiter (see Asteroid: resonance; Resonance in the solar system). This material may find its way to Earth, while objects that are not close to the gravitational resonance positions are much less likely to arrive as meteorites at the Earth's surface. There are important compositional differences present in the population of asteroids as a function of radial distance from the Sun (see Asteroid: compositional structure and taxonomy).

Interplanetary dust, on the other hand, represents a less biased sample of interplanetary material since the dynamical mechanisms for transport to the Earth (discussed below) are not selective in terms of particle origins. In addition, it is possible to collect relatively fragile particles of dust, and to obtain compositional information for these, while larger objects of similar composition cannot survive atmospheric entry.

The chondritic dust particles may be grouped in two classes. Hydrous minerals such as serpentine dominate one class, while anhydrous minerals such as olivine and pyroxene are most common in the other class. The anhydrous samples tend to be porous, while the hydrous type are compact, without pore spaces. The differences between the two are attributed to differences in particle origins: the compact hydrous minerals appear to have been processed by heat and pressure, conditions linked with an asteroidal origin, while the porous anhydrous samples have not been processed in this manner. They are thought to be of cometary origin, with the porosity developing due to sublimation of volatiles.

It was believed for many years that short-period comets were the primary source for dust in the inner solar system. After all, the streaming tail of a comet and the occasional disconnection events observed are visible indications of the loss of materials from cometary nuclei. In the last century a link between cometary orbits and terrestrial meteor showers was established, strengthening the hypothesis of a cometary origin for most IDPs (see Meteor shower, Meteoroid stream). However, recent results show that collisions in the asteroid belt produce considerable quantities of dust, and that this material makes its way to the Earth as well.

As previously noted organic carbon is present in IDPs. Some relatively complex polycyclic aromatic hydrocarbons (Clemett et al., 1993) have been reported. The delivery of organic matter to the Earth by comets, meteorites and dust particles may have played an important role in the origination of life on Earth (see Life: origins; Comet: impacts on the Earth).

Comets are believed to have formed from primordial materials in the outer reaches of the solar system. It is likely that they contain 'presolar' grains, or interstellar grains (q.v.). Determination of compositions of submicron-sized grains is difficult, but some isotope studies of tiny inclusions in anhydrous, cometary dust particles yield deuterium to hydrogen ratios that are considerably enhanced over the terrestrial value (Brownlee, 1991), suggesting a presolar or interstellar origin.

The dust mass spectrometer instrument aboard the Giotto spacecraft (q.v.) analyzed the chemical compositions of several thousand particles. Two populations (organic grains and mineral grains) were distinguished. The former were dominated by the elements H, C, N and O, while the mineral grains were composed of Na, Mg, Si, Ca, Fe and other elements. In aggregate the composition of comet Halley's dust is close to that of carbonaceous chondrite meteorites (q.v.).

Dynamics

A number of forces act on dust-sized particles in the solar system environment. The most important of these are the gravitational force, radiation pressure, Poynting–Robertson drag (q.v.) and corpuscular drag (due to the solar wind; Sykes, 1991). Radiation pressure results from the absorption and scattering of solar photons by the particle, and acts against the force of solar gravity, imparting a component of momentum directed away from the Sun. Both the radiation pressure and the solar gravity depend on the inverse second power of distance, but as the size of particles decreases, the surface to mass ratio drops, and radiation pressure becomes more important. For meteoroids at diameters of 1 cm or more, gravitation dominates. For dust particles, gravitation and radiation pressure effects are comparable. At sizes below 0.1 μm, however, as particle sizes become smaller than the wavelength of the radiation, radiation pressure becomes less important. At this point corpuscular drag, due to the outflowing solar wind, takes over to help expel the particles from the inner solar system.

Poynting–Robertson drag is possibly the most significant force for dust-sized particles. This is another effect linked with the solar radiation pressure. As previously noted, the dust particles orbit the Sun in prograde orbits. A component of the momentum transferred from the solar photons acts against this motion, and the orbit loses energy. This causes the particle to slowly spiral in toward the Sun, over a period of some 10^4–10^5 years. The eccentricity of the particle orbit is also decreased.

The life cycle of a dust particle might progress as follows. Collisions of objects in the asteroid belt grind down the materials until they become small enough for Poynting–Roberston drag to become effective. The particles spiral in toward the inner solar system, while still experiencing collisions and being heated by increased solar radiation. Particle sizes and masses may be reduced by sublimation and by collisions, until at some point they become small enough to be entrained along with the solar wind and pass to the outer regions of the solar system. In an alternative scenario, dust particles may be expelled from comets or accelerated by collisions with other particles, achieving hyperbolic orbits that permit them to escape the solar system (Sykes, 1991).

Thus the final result of these dynamical effects is the expulsion of dust particles from the inner solar system, over time periods much shorter than the age of the solar system. A source of new particles is required to explain the zodiacal dust cloud. As previously noted, cometary dust is one source. As hinted previously, collisions of objects in the asteroid belt represent another. This process has received greater attention in the last decade, following the discovery of dust bands in the asteroid belt during the sky survey at infrared wavelengths performed by the Infrared Astronomical Satellite (IRAS) in the 1980s (Dermott et al., 1984). The infrared observations at a wavelength of 25 μm showed two pairs of bands, centered on the ecliptic plane. (The bands appear bright at this wavelength due to thermal emission by the dust. A single band appears as a pair due to oscillations of the particles above and below the mean orbital plane of the particles; as in a simple pendulum, the particles spend more time at the extremes of their oscillation than at the center of the range.) The dust bands correspond to the orbits of the Koronis and Themis families of asteroids (see Asteroid: families). Thus these families appear to be a significant contemporary dust source. Recently the Koronis family asteroid Ida (q.v.) was discovered to have a small satellite, which may represent a collisional fragment of the larger body.

The estimated lifetimes of the dusty rings of the outer planets are much less than the age of the solar system, and so the question of the source(s) of the dust is important there also. The small satellite Galatea has been suggested as a possible source for particles in the Neptune ring system, and the satellites Adrastea and Metis appear to represent a source of dust for the Jupiter rings (see Planetary ring).

The future

Studies of interplanetary dust particles will continue to yield important results for many years to come. As laboratory techniques improve it will become possible to address many significant questions bearing on the formation and early evolution of the solar system. New results (as this is written) include the discovery of interstellar dust entering the solar system by the Ulysses spacecraft (q.v.), and hints of the important role of charged dust particles (or dusty plasmas) in the dynamics of ring systems and in cometary particle dynamics (Glanz, 1994).

James H. Shirley

Bibliography

Brownlee, D.E. (1987) Morphological, chemical, and mineralogical studies of cosmic dust. *Phil. Trans. Roy. Soc. London*, **A323**, 305.

Brownlee, D.E. (1991) Interplanetary dust, collection and analysis, in *The Astronomy and Astrophysics Encyclopedia* (ed. S.P. Maran). New York: Van Nostrand Reinhold, pp. 321–3.

Clemett, S.J., Maechling, C.R., Zare, R.N., *et al.* (1993) Identification of complex aromatic molecules in individual interplanetary dust particles. *Science*, **262**, 721–5.

Dermott, S.F., Nicholson, P.D., Burns J.A. and Houck, J.A. (1984) The origin of the solar system dust bands discovered by IRAS. *Nature*, **312**, 505–9.

Glanz, J. (1994) A dusty road for space physics. *Science*, **264**, 28–30.
Grun, E. *et al.* (1993) Discovery of Jovian dust streams and interstellar grains by the Ulysses spacecraft. *Nature*, 362, pp. 428–30.
Love, S.G. and Brownlee, D.E. (1993). A direct measurement of the terrestrial mass accretion rate of cosmic dust. *Science*, **262**, 550–3.
McDonnell, J.A.M., Sullivan, K., Stevenson, T.J. and Niblell, D.H. (1990) Particulate detection in the near-Earth space environment aboard the LDEF, in *Origin and Evolution of Interplanetary Dust* (eds A.-C. Levasseur-Regourd and H. Hasegawa). Dordrecht: Kluwer Academic Publishers, pp. 3–10.
Sykes, M.V. (1991) Interplanetary dust, dynamics, in *The Astronomy and Astrophysics Encyclopedia* (ed. S.P. Maran). New York: Van Nostrand Reinhold, pp. 323–6.
Sykes, M.V., Greenberg, R., Dermott, S.F., *et al.* (1989) Dust bands in the asteroid belt, in *Asteroids II* (eds R. Binzel, T. Gehrels and M. Matthews). Tucson: University of Arizona, pp. 336–67.

Cross references

Asteroid: compositional structure and taxonomy
Asteroid: families
Asteroid: resonance
Carbonaceous chondrite
Chondrites, ordinary
Comet: structure and composition
Comet observation
Giotto mission
Ida
Interstellar grain
Life: origin
Meteor, meteoroid
Meteor shower, meteoroid stream
Micrometeorite
Planetary ring
Resonance in the solar system
Zodiacal light

DYNAMO THEORY

Stars and many planets have magnetic fields. Dynamo theory is the branch of astrophysics (and planetary physics and geophysics) that deals with the generation and maintainance of those magnetic fields. The nature of the problem is easily illustrated with reference to the Earth. If the Earth's field were not continuously regenerated, it would disappear in a period of about 15 000 years (the magnetic resistive decay time). Yet we know that the Earth has had a magnetic field of about the present intensity throughout geologic time. This requires that the field must be continually regenerated in some way within the Earth. In other words, the energy source must be adequate to overcome the natural decay of electric currents in a conductor of finite conductivity. The magnetic Reynolds number must exceed a critical value. The motion must have a helical asymmetry just as in a self-exciting dynamo. This motion is attributed to the Coriolis force in a rotating body.

Larmor (1919) was the first to suggest that the magnetic field was generated by fluid motions in the electrically conducting fluid core of the Earth. Cowling (1934) stated the essentials of the dynamo problem and made considerable contributions to the field in the following decades. Mathematical models were presented by Parker (1955) and Babcock (1961) which accounted for observed features of solar activity on the Sun. Fundamental aspects of the dynamo problem are discussed in more detail in the article on Magnetism.

Many fine review papers on dynamo theory have been presented through the years (see for instance Weiss, 1971; Gubbins, 1974; Cowling, 1981; Parker, 1987; Busse, 1989); these review the development of the theory. Recent results from helioseismology (see Sun) constrain the internal rotation of the Sun; these constraints will help to refine new solar dynamo models. Although great progress has been made, questions remain and dynamo theory continues to evolve. Among the open questions are the problem of the reversals of the Earth's magnetic field (see Geomagnetic polarity reversals and the geological record). The possible association in time of these events and faunal mass extinctions, and cometary impacts, is under examination (Jacobs, 1992).

James H. Shirley

Bibliography

Babcock, H.W. (1961) The topology of the Sun's magnetic field and the 22-year cycle. *Astrophys. J.*, **133**, 572–87.
Busse, F. (1989) Geomagnetic field, main: theory, in *The Encyclopedia of Solid Earth Geophysics*, (ed. D.E. James) New York: Van Nostrand Reinhold, pp. 511–7.
Cowling, T.G. (1934) The magnetic field of sunspots. *Mon. Not. Roy. Astron. Soc.*, **94**, 39–48.
Cowling, T.G. (1981) The present status of dynamo theory. *Ann. Rev. Astron. Astrophys.*, **19**, 115–35.
Gubbins, D. (1974) Theories of the geomagnetic and solar dynamos. *Rev. Geophys. Space Phys.*, **12**, 137–154.
Jacobs, J.A. (1992) Causes of changes in reversals of the Earth's magnetic field: inside or outside the core? *Eos*, **73**, 89–91.
Larmor, J. (1919) How could a rotating body such as the Sun become a magnet? *Br. Assoc. Adv. Sci. Rep.*, pp. 159–60.
Parker, E.N. (1955) Hydromagnetic dynamo models. *Astrophys. J.*, **122**, 293–314.
Parker, E.N. (1987) The dynamo dilemma. *Solar Phys.*, **110**, 11–21.
Weiss, N.O. (1971) The dynamo problem. *Quart. J. Roy. Astron. Soc.*, **12**, 432–446.

Cross references

Eccentric dipole
Geomagnetic polarity reversals and the geological record
Magnetism
Paleomagnetism
Polarity reversals

E

EARLY BOMBARDMENT

The rate at which space debris collides with the Earth and other planets is currently quite low; the risk of being struck by a meteorite within one's lifetime is small. However, the heavily cratered terrains of other planets and moons in the solar system indicate that impact cratering rates were much higher in the past. Collisions between objects in space not only have produced the impact crater scars remaining today on most solar system bodies but also have been responsible for the actual formation and major modification of the planets, moons, asteroids and comets in our solar system.

Early heavy bombardment

The solar system is believed to have formed from a rotating cloud of gas and dust which collapsed into a thin disk with the protosun at the center. As the cloud began to cool, solid particles formed and began to collide with one another. At relatively low velocities, collisions between particles can cause those particles to stick together, a process known as accretion. The accreting bodies grew into larger bodies, and the larger gravity associated with these increasingly more massive objects attracted yet more particles to them. Much of the mass of the original cloud of gas and dust went into forming the planets and other solar system bodies in this way. Since the rate of collision between bodies was high during this time, many scientists refer to this period as either the *early bombardment period* or the *early heavy bombardment period*. This period of early heavy bombardment is believed to have taken about the first 0.5 billion years of solar system history.

Late heavy bombardment

After the initial period of accretion, the planets and most of the larger moons underwent a process of differentiation, where heavier elements such as iron sunk to their cores and lighter elements such as silica rose to the surface to form a crust. Although the actual chemical compositions of the planets and moons vary throughout the solar system, evidence from spacecraft encounters with these bodies suggest that most have undergone the process of differentiation. Any impact scars from the early heavy bombardment period (i.e. the accretion period) were destroyed by the differentiation process (although small bodies such as some moons, the asteroids and possibly comets may still retain their accretion scars since they do not appear to be differentiated). Although much of the debris left over from solar system formation was swept up by the growing planets and moons during accretion, some material still remained to impact the surfaces of the newly differentiated bodies. Some planets and moons still display the surfaces scarred during this early period of time, a period often referred to as the late heavy bombardment period. Study of these surfaces, such as the heavily cratered highlands of the Earth's Moon, the heavily cratered plains of Mercury and Mars, and the heavily cratered icy plains of many outer solar system satellites, indicates that impact rates during this late heavy bombardment period were at least 100 times higher than the current cratering rate. The size–frequency distribution of craters dating from this time period also differs from that associated with more recent craters, suggesting to some scientists that the size distribution of the impacting objects was quite different from that seen for the asteroids and comets which constitute the current impact population (Chapman and McKinnon, 1986). For example, the largest craters (called basins) on most planets and satellites appear to have formed during the late heavy bombardment period rather than more recently, suggesting that a greater number of large objects existed in planet-crossing orbits in the past than do today. Scientists are currently debating whether these large basins formed randomly throughout the late heavy bombardment period or if they all formed during a great cataclysm just before the end of this period (Ryder, 1990).

Duration of early bombardment

Six Apollo missions between 1969 and 1972 returned soil and rock samples from the Moon which have been dated using radioisotope dating techniques. Samples from heavily cratered and lightly cratered terrains (late heavy bombardment aged and post heavy bombardment aged units respectively) indicate that the end of the heavy bombardment period occurred about 3.8 billion years ago in the Earth–Moon system (Taylor, 1982). The cessation of the heavy bombardment period throughout the rest of the solar system is believed to have occurred at about the same time, based on numerous computer simulations. Thus the late heavy bombardment period is believed to have lasted for about 700 million years. Combining the time periods of the early and late heavy bombardment periods indicates that the duration of the early bombardment period in the solar system was just over 1 billion years.

Nadine G. Barlow

Bibliography

Chapman, C.R. and McKinnon, W.B. (1986) Cratering of planetary satellites, in *Satellites* (eds J.A. Burns, and M.S. Matthews). Tucson: University of Arizona Press, pp. 492–580.

Ryder, G. (1990) Lunar samples, lunar accretion and the early bombardment of the Moon. *EOS*, **71**, 313, 322–43.

Taylor, S.R. (1982) *Planetary Science: A Lunar Perspective*. Lunar and Planetary Institute, 481 pp.

Cross references

Earth–Moon system: origin
Impact cratering
Moon: geology
Moon: origin

EARTH

Since 1959 spacecraft have visited every planet but Pluto, and it is now possible to describe the Earth in its solar system context much as the planetologist on an interstellar expedition might do. The purpose of this article is to summarize the ways in which the Earth differs from the other silicate planets: Mercury, Mars and Venus. Where appropriate, the Earth will also be compared with natural satellites, in particular with the Moon. A convenient matrix for comparison is the sequence Moon, Mercury, Mars, Venus and Earth. As shown in Figure E1, this is a sequence of increasing mass, size, internal activity and degree of crustal evolution. The major characteristics of the planets are listed in Table E1.

The Earth's position in the solar system is an extremely important characteristic. Its distance from the Sun, in combination with its strong gravity and its heat-retaining atmosphere, permits liquid water to exist in large quantities on its surface (Plates 9,10). This liquid water is not only unique (at the present time) in the solar system, but has enormous importance for terrestrial geology. The retention of abundant water, both on the surface and internally, has the following consequences as a minimum.

Effects of water

First and most obvious, the existence of surface water as far back in the geologic record as there are datable rocks, about 4 billion years, has promoted a wide range of geologic processes including weathering, erosion and sedimentation (fluvial, lacustrine and marine). However, igneous processes, in particular the generation and differentiation of magmas, have also been influenced by water. Most regional metamorphism, prograde and retrograde, is also water catalyzed if not water dependent. Much continental tectonism, such as overthrust faulting, is greatly aided by interstitial fluids, chiefly water. It has been shown that hydrothermal circulation through oceanic crust contributes strongly to cooling of the oceanic lithosphere and thus to sea-floor spreading (Wolery and Sleep, 1976). The general absence of recognizable spreading centers on Venus, which shows no evidence of water, further suggests the necessity of abundant water for plate tectonics.

In addition, the availability of surface water has also dominated terrestrial geology indirectly by permitting life to arise. One theory (Oro, Miller and Lazcano, 1990) on how this happened involves the prebiotic synthesis of the components of nucleic acids, leading to formation of primeval ribonucleic and deoxyribonucleic acids (RNA and DNA). These in turn gave rise to the first proteins and then to living organisms. Life in turn may have helped to regulate the surface environment of the Earth, in particular its temperature, through a complex series of biological feedback mechanisms. This is the controversial Gaia hypothesis, first formulated by James Lovelock (1979), and which has been compared (Nisbet, 1987) to Darwin's theory of evolution in scope and importance.

Gaia

The fundamental mechanism of the Gaia concept is self-regulation, in which life has for at least 4 billion years controlled the atmospheric composition of the Earth, and thus its surface temperature, by a wide variety of feedback mechanisms. This process can be termed homeostasis, a word borrowed from physiology to describe the ways in which organisms maintain themselves in an approximately constant state despite external changes. The Earth is, in the Gaia hypothesis, an extremely complex self-regulating system, in which the surface environment (and probably much else) is controlled by life itself, in contrast to the older view in which life simply adapts to the surface environment.

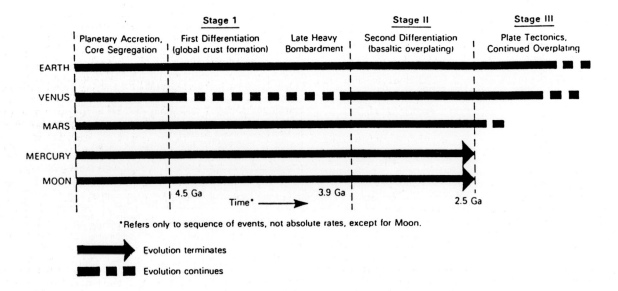

Figure E1 Diagram of inferred crustal evolution stages in the Moon, Mercury, Mars, Venus and Earth. Continental crust is considered to be formed in the stage 1 'first differentiation,' and to be analogous to the old cratered highlands of the Moon, Mercury and Mars. Stage III is expressed as 'plate tectonics' only on the Earth, and as a wide range of folding, faulting and igneous activity on Venus and Mars.

Table E1 Major characteristics of the plants and largest satellites (after Davies et al., 1989)

Name	Axis (AU)	Revolution Period	Diameter (km)	Rotation (days)	Mass (10^{20} kg)	Density (g cm^{-3})	Escape Velocity (km s^{-1})	Surface	Atmosphere
Mercury	0.39	0.24 years	4 878	58.65	3.3	5.4	4	Silicates	Trace Na
Venus	0.72	0.62 years	12 102	243.0 R[a]	48.7	5.3	10	Basalt, granite?	90 bar: 97% CO_2
Earth	1.00	1.00 year	12 756	1.00	59.8	5.5	11	Basalt, granite, water	1 bar: 78% N_2, 21% O_2
Moon	–	27.32 d	3 476	27.3	0.7	3.3	2	Basalt, anorthosites	None
Mars	1.52	1.88 years	6 787	1.03	6.4	3.9	5	Basalt, clays, ice	0.07 bar: 95% CO_2
Jupiter	5.20	11.86 years	142 984	0.41	18 991	1.3	60	None	H_2, He, CH_4, NH_3, etc.
Callisto	–	16.69 d	4 800	16.69	1.1	1.9	2	Dirty ice	None
Ganymede	–	7.15 d	5 262	7.15	1.5	1.9	3	Dirty ice	None
Europa	–	3.55 d	3 138	3.55	0.5	3.0	2	Ice	None
Io	–	1.77 d	3 650	1.77	0.9	3.6	3	Sulfur, SO_2	Trace SO_2
Saturn	9.54	29.46 years	120 536	0.44	5 686	0.7	36	None	H_2, He, CH_4, NH_3, etc.
Titan	–	15.65 d	5 150	15.95	1.3	1.9	3	Ice? hydrocarbons?	1.5 bar: N_2, trace CH_4
Uranus	19.18	84.07 years	51 118	0.72 R	866	1.2	21	?	H_2, He, CH_4, NH_3, etc.
Neptune	30.06	164.82 years	49 660	0.67	1 030	1.6	23	?	H_2, He, CH_4, NH_3, etc.
Triton	–	5.88 d	3 000 ?	5.88 ?	?	?	?	CH_4, ice, liquid N_2 (?)	Trace CH_4
Pluto	39.44	248.6 years	2 400	6.39 R	0.01	1.7	1	CH_4, ice	Trace CH_4

[a] R denotes retrograde rotation

The most important aspect of this supposed self-regulating temperature system is the composition of the atmosphere. It is well known that carbon dioxide (and other multi-atom gases such as methane) are infrared absorbers, largely responsible for the 'greenhouse' effect in the Earth's atmosphere. The extremely high temperatures on Venus are understood to result from the dense CO_2 atmosphere. The much milder terrestrial temperatures are due partly to the fact that the Earth's atmosphere is dominantly diatomic oxygen and nitrogen, which are not strong infrared heat absorbers. It has been realized for several decades that the Earth's oxygen is almost entirely of biological origin, from photosynthesis. However, the Gaia hypothesis carries this interpretation further, ascribing the maintenance of atmospheric composition (and hence temperature) to life. This state of affairs may have arisen in the following way, according to Lovelock (1990).

The earliest terrestrial atmosphere was probably chiefly carbon dioxide, as are those of Venus and Mars today. Early Archean cyanobacteria, or photosynthesizers, consumed this carbon dioxide. If the process had continued unchecked, this greenhouse gas would have eventually been removed from the atmosphere and the Earth's temperature would have dropped to the point that all water would have frozen, presumably ending life. However, the methanogens consumed organic matter produced by the cyanobacteria, generating atmospheric methane, which is also a greenhouse gas and which in combination with other processes would become abundant enough to keep the planet's temperature up. The planetary temperature would, because of the nature of terrestrial life, be kept in the range of liquid water. The abundance of water on the Earth is thus (in this model) due to terrestrial life.

The importance of life in controlling the surface conditions of the Earth is illustrated by comparison with Venus and Mars, both of which have, almost certainly, no life at all. The atmosphere of Venus is extremely dense (Table E1) and hot, and there is no evidence of fluvial processes in the present physiography. The mass of Venus has permitted it to retain this thick atmosphere, but the absence of life has made it drastically different from that of the Earth. The atmosphere of Mars is at present equally different; although also largely carbon dioxide, it is extremely cold and thin, with surface pressures of only a few millibars. Fluvial landforms demonstrate that Mars at one time had liquid water, but the low mass of Mars apparently made it impossible to retain enough of the atmosphere for life, if it arose, to survive and flourish as it did on the Earth.

The Gaia hypothesis views the Earth's collective life as a single organism, analogous to the billions of diverse cells making up a living human being. Although stimulating, it is not accepted by all geologists, many of whom (e.g. Holland, 1984; Gregor, 1988) consider inorganic recycling processes to be equally important or more so in regulating the Earth's environment. These will now be discussed briefly.

Chemical recycling

Another way in which the Earth differs from almost all other solid bodies in the solar system is its pervasive chemical recycling, due to a combination of tectonic, petrologic and biological processes. This can best be illustrated by the most obvious example, water, which is recycled by evaporation and precipitation, and less obviously by return to the mantle in subduction zones (Wolery and Sleep, 1988), as well as by other processes. Carbon dioxide is a major participant in the carbon cycle, in which life is agreed to play a dominant role at this time. Similar cycles have been recognized for nitrogen, sulfur and other elements. Most general of all is the rock cycle, which is discussed below under 'crustal reworking and recycling.' This pervasive recycling is an expression of the Earth's greater internal energy and consequently greater tectonic activity.

A conspicuous exception to the Earth's uniqueness in recycling is the Jovian satellite Io, found in 1979 to be covered with young overplated volcanic deposits and enormous active volcanos that are visibly renewing this volcanic cover (Lowman, 1989). This process obviously must involve some form of recycling, presumably remelting of the Ionian crust at depth with heat from the tidal reworking caused by Io's elliptical orbit around Jupiter. An analogous but much lower temperature process may have occurred on Europa, whose icy surface is clearly the result of some sort of resurfacing and conceivably a form of water ice recycling.

Effects of planetary mass

The mass of the Earth is greater than that of the other terrestrial (silicate) planets. This characteristic has had several major effects on terrestrial geologic evolution. First, the energy of accretion of the Earth was much greater than that of smaller bodies such as the Moon. The Earth thus began with a greater heat supply, augmented by short-lived radioactivity. Furthermore, the surface to volume ratio of the Earth, equal to the ratio of the diameter squared to the diameter cubed, is significantly less than that of smaller bodies. The Earth thus retained more of its original heat. It is widely recognized that tectonic processes are largely thermally driven, though regulated by other phenomena. The Earth's greater heat supply has promoted greater geologic activity and crustal evolution than experienced by Mars and the Moon. Venus appears to have a comparable degree of tectonic activity (Solomon et al. 1992), though in the absence of water it has taken the form of basaltic volcanism and non-plate tectonic crustal deformation.

The Moon

Our hypothetical interstellar planetologist would be interested by the fact that the Earth, alone among the inner planets, has a single large satellite. The origin of the Moon has been a difficult problem for many decades. Each of the formerly popular theories – intact capture, fission and coaccretion – has encountered fundamental difficulties (Melosh, 1992). This situation led several theorists in the mid-1980s to propose that the Moon accreted from debris ejected from the primordial Earth by the impact of a Mars-sized planetesimal. Despite its seemingly ad hoc nature, the collision hypothesis appears to meet constraints such as angular momentum, bulk chemistry and volatile depletion, and accordingly has been widely accepted at least as a working hypothesis.

Assuming its fundamental validity, we must ask if the supposed collision origin of the Moon affected the later evolution of the Earth. One consequence of the impact of a Mars-sized body would certainly be the catastrophic heating of the Earth, with much of this heat being retained after ejection of the Moon-forming debris. This heat presumably would have promoted early global differentiation, although the ensuing vigorous convection might have destroyed the initial crust thus formed, accounting for the apparent scarcity of rock older than about 4 billion years. It may be at least speculated in addition that the gross crustal dichotomy of the Earth – one-third sialic, two-thirds simatic crust – might in some way result from the supposed primordial impact. Perhaps an initial global crust was partly destroyed, or the converse – nucleation of a primordial Pangea – resulted. In any event the Earth's unique Moon must be kept in mind when considering the Earth's early crustal evolution. Some of the ways in which it differs from Venus may reflect the fact that Venus has no Moon.

Effects of impact cratering on the Earth

Impact craters are the most common landform in the solar system, occurring on every solid body except Io. The Earth clearly underwent such a bombardment in common with the other planets, over 120 terrestrial impact craters now being known (Millman, 1992). When allowance is made for different gravity and other conditions, terrestrial craters and basins are seen to be essentially similar to those of the Moon and other planets (Spudis, 1993). However, at least two consequences distinguish impacts on the Earth from impacts elsewhere. First, the impact of large but subplanetary sized bodies may have produced multiring basins and disrupted the primitive crust of the Earth, initiating mantle upwelling and sea-floor spreading and forming the Earth's first ocean basins. In this concept impacts would be responsible for the crustal dichotomy mentioned above. The converse process may have happened instead, with major impacts initiating formation of continental nuclei, with the same end result. It is now recognized that craters several kilometers or larger in diameter have substantial volumes of impact melt, i.e. local crust melted by the impact. From simple comparison with the primitive cratered crusts of the Moon, Mercury and Mars, and from the growing number of identified Precambrian craters, it seems clear that the early Earth must have been similarly impacted. The corresponding early impact melts may survive in the lowermost continental crust, though probably metamorphosed beyond easy recognition. Differentiation of these impact melts which, once formed, can be considered magmas, may have contributed to generation of primordial sialic crust, a process that seems to have occurred as recently as the early Proterozoic in the Sudbury Igneous Complex of Ontario (Lowman, 1992).

The evolution of life on the Earth may have been greatly influenced by impacts (McLaren and Goodfellow, 1990). A major impact would destroy many life forms. The now-confirmed Chicxulub impact on the Yucatan peninsula (Sharpton et al., 1992) evidently wiped out 90% of all living species at the end of the Cretaceous Period (Spudis 1993). However, similar impacts throughout the Phanerozoic would have had the effect of clearing the ecological stage, so to speak, for the next era of organic evolution, just as the extinction of the dinosaurs made room, eventually, for encyclopedia compilers.

Bimodal crust

One of the most obvious characteristics of the Earth is its bimodal topography, contrasting sharply with the unimodal topography of its near-twin Venus. The bimodal terrestrial topography corresponds to the two types of crust, oceanic (mafic) and continental (sialic). The general nature of these crusts is reasonably well understood, but their origin – in particular that of continental crust – is not as clear as commonly believed. The fundamental problem is often posed as the origin of continents, but this wording tends to prejudge the answer. A more flexible term would be the origin of continental crust, which is focused on the processes by which such crust is extracted from the mantle (Lowman, 1989). Still more flexible is the general question: why does the Earth have two types of crust?

Putting the problem in the last form suggests the possibility that the dominant global process has been not the growth of continents, but the growth of ocean basins, or 'oceanization' to use the term favored by V.V. Beloussov (1992). It is here that major impacts may have played a role by breaking up a global primordial crust and initiating sea-floor spreading, thus forming the first ocean basins, remotely ancestral to those of the present. Younger ocean basins, such as the North Sea, may have been produced by upwelling of the asthenosphere (Artyushkov and Baer, 1989).

The most popular view is that the continents have grown over geologic time by some form of lateral accretion around primitive continental nuclei (Condie, 1982). Island arcs such as Japan and the Aleutian Islands would in this concept represent additions to the continents of Asia and North America respectively. The Appalachians and the North American Cordillera have been interpreted as resulting from accretion of exotic terranes, carried to and docked with the continent by sea-floor spreading. Overplating and underplating of basaltic rocks have probably contributed to the vertical growth, or thickening, of continental crust.

An alternative hypothesis for the origin of continental crust holds that the present continents are the greatly altered remnants of a global primordial crust. Comparative planetology shows clearly that silicate planets (including the Moon) undergo global differentiation (the 'first differentiation' of Figure E1) very early in their history, and it has been argued (Lowman, 1989) that the Earth also developed a global crust early in its history. However, the composition of crusts formed by this 'first differentiation' is unknown except for that of the Moon, arguing against simple extrapolation to the Earth. The first differentiation on Earth may have produced sialic continental nuclei, later enlarged by lateral accretion as described above.

The fundamental origin of the bimodal terrestrial crust is, in summary, not yet understood, even though plausible explanations for most individual rock types and associations have been developed. Further terrestrial research, aided by comparative planetology, will be required to explain this apparently unique characteristic of the Earth.

Crustal reworking and recycling

One final characteristic of terrestrial geology that appears to set it apart from other planets, with the conspicuous exception of Venus, is the continual reworking, both petrologically and structurally, of the terrestrial continental crust. The oceanic crust appears to be relatively rigid and inactive, away from plate boundaries, and ephemeral in geologic terms, being generated, subducted and recycled through the mantle within a few hundred million years. The continental crust, in contrast, is fundamentally several billion years old, and has in most areas undergone repeated episodes of magmatism, metamorphism and deformation, often lasting hundreds of millions of years. These processes make up a large part of the rock cycle illustrated in introductory geology texts.

The analogous crusts of the Moon, Mars and probably Mercury have been largely undisturbed except for volcanic overplating, impact cratering and minor deformation since they were formed. Lunar highland rocks, for example, tend to have radiometric ages centered on 4.4 billion years, close to the age of the Moon itself. But the ages for most Precambrian shields are like a palimpsest, reflecting many periods of crustal reworking, a characteristic now known (Solomon et al., 1992) to prevail on Venus as well. A petrologic consequence, apparently unique to the Earth, has been the formation of abundant granites, now generally agreed to be largely formed by remelting of earlier crust rather than direct extraction from the mantle. (This is one reason that quartz is so common on the Earth but rare on the

Moon; the lunar crust has not been repeatedly remelted, producing rocks oversaturated in silica such as granites.)

Crustal evolution, Gaia and water

The phenomenon of crustal reworking brings us, perhaps surprisingly, back to the Gaia concept. To the extent that retention of water may be the indirect result of life on Earth, it can be argued that the continental crust in its post-Archean state is fundamentally biogenic. This generalization applies not only to organic rocks such as coal or limestone, but to all rocks in whose formation water has played a dominant role. This role is obvious for fluvial, glacial and marine sedimentary rocks, but it is almost as important for crystalline rocks, in particular the granitic rocks that make up most of the Precambrian shields. These rocks appear to have formed largely by anatexis, or partial melting, of pre-existing crustal rocks such as amphibolites, metabasalts or older granitoids, rather than by direct extraction from the mantle. It has long been known that water is extremely effective in lowering the eutectic melting point of rocks; a totally anhydrous crust would be unlikely to generate the granitic melts. Granites are a very minor fraction of the lunar crust, the Moon's lack of water and of crustal reworking being major factors.

Regional metamorphism, responsible for formation of the schists and paragneisses characteristic of shields and core complexes, is equally dependent on the presence of water, as first shown by Yoder (1955). The older facies concept of metamorphism was concerned primarily with the effects of temperature and pressure. Yoder showed that water content and water pressure were comparably important. Retrograde metamorphism similarly involves reintroduction of water into previously metamorphosed rock (Best, 1982). Hydrothermal alteration, as around igneous intrusions, is by definition water dependent.

The tectonic effects of water stem from the role of internal fluid pressure, now realized to be important in promoting rock deformation on all scales (Seyfert, 1987). High fluid pressure in soils, for example, decreases normal stress on shear planes, thus promoting failure. Overthrust faulting, far more widespread in crustal rocks than formerly appreciated, was shown by Hubbert and Rubey (1959) to depend strongly on fluid pressure although fold-and-thrust belts have now been recognized on Venus (Suppe and Connors, 1992). Water is thus nearly as important in tectonism as in magma generation and regional metamorphism, and to this extent the Earth's tectonic style is strongly influenced by the presence of life.

The overall result of crustal reworking, coupled with the dynamic effects of plate tectonic processes, has been to make the Earth the most highly evolved, in a geologic sense, of the silicate planets. This advanced and still continuing crustal evolution may be fundamentally the result of terrestrial life.

Summary

To summarize this description of Earth, our imaginary interstellar planetologist – whom we will assume to see the same colors we do, and to have made a few surreptitious rock-collecting landings – might report on the third planet this way.

This spectacular blue and white planet is unique in the solar system in several ways, largely stemming from its distance from the Sun, its greater mass compared to other silicate planets and the presence of life for several billion years. The most striking characteristic of the Earth is its abundant water: colloidally suspended in the atmosphere; covering two thirds of its surface; coating, falling on and flowing over the remaining third; and infiltrating the crust and mantle. It retains this water partly because of the surface temperature of the planet, but also because the Earth behaves like a living organism that maintains this temperature by a wide variety of feedback mechanisms, many of which are caused by life itself. The Earth is the only silicate planet with abundant life, probably the only one with any life at all. Life on the Earth, every variety of which contains DNA, apparently arose because the presence of abundant water permitted formation, probably over 4 billion years ago, of RNA and DNA, which in turn promoted the formation of complex carbon compounds that comprise terrestrial life. The Earth thus provides an example of self-regulation on a planetary scale: water permitted life to arise, and this life in turn regulates the surface environment to preserve conditions under which this water survives as liquid. Terrestrial geology is dominated by water, which promotes magmatic, metamorphic and tectonic processes, in addition to sculpturing the surface of the Earth into a greater variety of landforms than displayed by any other body in the solar system. Most rocks, structures and landforms of the continental crust are fundamentally biogenic in that water has been essential for their origin. Although the Earth has followed a crustal evolution scheme whose initial stages are common to other silicate planets, it has evolved and continues to evolve geologically much farther. It is unique, a uniqueness that may be due primarily to its life.

Paul D. Lowman Jr

Bibliography

Artyushkov, E.V. and Baer, M.A. (1989) Mechanism of formation of deep basins on continental crust, in *Origin and Evolution of Sedimentary Basins and Their Energy and Mineral Resources* (ed. R.A. Price). American Geophysical Union, Washington: Geophysical Monograph 48, pp. 175–185.
Best, M.G. (1982) *Igneous and Metamorphic Petrology*. San Francisco: W.H. Freeman, 630.
Beloussov, V.V. (1992) Endogenic regimes and the evolution of the tectonosphere, in *New Concepts in Global Tectonics*. (eds S. Chatterjee and N. Hotton III). Lubbock, T.X., Texas Tech University Press, pp. 411–20. Condie, K.C. (1982) *Plate Tectonics and Crustal Evolution*. New York: Pergamon Press: 310 pp.
Davies, M.E., Seidelman, P.K., Standish, E.M. and Tholen, D.J. (1989) Planets and large satellites, in *AGI Data Sheets*, (eds J.T. Dutro, Jr, R.V. Dietrich and R.M. Foose.) American Geological Institute AGI Data Sheet 80.1.
Gregor, C.B. (1988) Prologue: cyclic processes in geology, a historical sketch, in *Chemical Cycles in the Evolution of the Earth*, (eds. C.B. Gregor, R.M. Garrels, F.T. Mackenzie and J.B. Maynard. New York: John Wiley, pp. 5–16.
Holland, H.D. (1984) *The Chemical Evolution of the Atmosphere and Oceans*. Princeton, N.J. Princeton University Press. 582 pp.
Hubbert, M.K. and Rubey, W.W. (1959) Role of fluid pressure in mechanics of overthrust faulting and of landsliding, *Geol. Soc. Am. Bull.* **70**, 115–66.
Lovelock, J.E. (1979) Gaia: a New look at Life on Earth Oxford: Oxford University Press, 157 pp.
Lovelock, J. (1990) *The Ages of Gaia*. New York; Bantam Books, 252 pp. Lowman, P.D., Jr (1976) Crustal evolution in silicate planets: Implications for the origin of continents. *J. Geol.*, **84**, 1–26.
Lowman, P.D., Jr (1989) Comparative planetology and the origin of continental crust, *Precambrian Res.*, **44**, 171–95.
Lowman, P.D., Jr (1992) The Sudbury Structure as a terrestrial mare basin, *Rev. Geophys.*, **30**, 227–43.
McLaren, D.J. and Goodfellow, W.D. (1990) Geological and biological consequences of giant impacts, *Ann. Rev. Earth Planet. Sci.*, **18**, 123–71.
Melosh, H.J. (1992) Moon, origin and evolution, in *The Astronomy and Astrophysics Encyclopedia*, (ed. S.P. Maran). New York: Van Nostrand Reinhold: pp. 456–9.
Millman, P.M. (1992) Earth, impact craters, in *The Astronomy and Astrophysics Encyclopedia*, (ed. S.P. Maran.) New York: Van Nostrand Reihnold, pp. 187–9.
Morrison, D. and Owens, T. (1989) in *Planets and Large Satellites*, in *AGI Data Sheets*, (complied by J.T., Dutro, Jr, R.V., Dietrich and R.M. Fosse) Alexandria, V.A: American Geological Institute AGI Data Sheet 80.
Nisbet, E.G. (1987) *The Young Earth*, Boston; Allen and Unwin, 402 pp.
Oro, J., Miller, S.J., and Lazcano, A. (1990) The origin and early evolution of life on Earth, in *Ann. Rev. Earth Planet. Sci.*, (eds. G.W. Wetherill, A.L. Albee and F.G. Stehli) **18**., 317–56.
Seyfert, C.K. (1987) Fluid pressure and the formation of overthrusts and gravity slides in *The Encyclopedia of Structural Geology and Plate Tectonics* (ed. C.K. Seyfert) New York: Van Nostrand Reinhold, pp. 239–49.
Sharpton, V.L., (1992) New links between the Chicxlub impact structure and the Cretaceous/Tertiary boundary. *Nature,* **359**, 819–21.
Solomon, S.C., (1992) Venus tectonics: an overview of Magellan observations. *J. Geophys. Res.*, **97** (E8) 13199–255.
Spudis, P.D. (1993) *The Geology of Mult-Ring Impact Basins*, Cambridge: Cambridge University Press, 263 pp.
Suppe, J. and Connors, C. (1992) Critical taper wedge mechanics of fold-and-thrust belts on Venus: initial results from Magellan. *J. Geophys. Res.*, **97** (E8) 13545–61.

Veizer, J. (1988) The evolving exogenic cycle, in *Chemical Cycles in the Evolution of the Earth* (eds C.B. Gregor, R.M. Garrels, F.T. Mackenzie, and J.B. Maynard). New York: John Wiley, pp. 175–220.

Wolery, T.J. and Sleep, N.H. (1976) Hydrothermal circulation and geochemical flux at mid-ocean ridges. *J. Geol.*, **84**, 249–75.

Yoder, H.S., Jr, (1955) Role of water in metamorphism, in *Crust of the Earth* (ed. A. Polderwaart). Boulder, CO: Special Paper 62. Geological Society of America, pp. 505–23.

Cross references

Greenhouse effect
Impact cratering
Life: origin
Mass extinctions
Moon: origin
Plate tectonics
Surface processes
Venus: geology and geophysics
Water
Weathering

EARTH: ATMOSPHERE

The Earth is, in a great many respects, a most unusual planet (Plate 9). The atmosphere of the Earth is unique in the solar system in both the composition of its major and minor constituents, and in the role it plays in the complex system of chemical, radiative, geological and biological feedbacks known as 'the Earth system'. In discussing the Earth's atmosphere it is useful to call upon comparative planetology to provide a guide as to which features of this complex system are most notable and deserve coverage in a brief review article. For this comparison we have two other examples of atmospheres around Earth-like worlds: Venus and Mars. When looking for specific insight into the nature and evolution of thin atmospheres around small, rocky worlds, one is restricted at present to the three planets within 0.5 AU of our present location.

Table E2 shows the gaseous composition of Earth, Venus and Mars. The Earth's atmosphere is composed primarily of nitrogen and oxygen, with minor amounts of carbon dioxide, water vapor, methane, ozone, argon and a host of other species. This is in stark contrast to the composition of both the Martian and Venusian atmospheres, which are both dominated by carbon dioxide, with minor amounts of nitrogen, carbon monoxide, water vapor, argon and oxygen in the case of Mars, and sulfur compounds in the case of Venus. Although, as with all present day comparative planetological exercises, we are restricted to an uncomfortably small statistical sample from which to generalize, it is apparent that Earth stands out from the CO_2-dominated 'norm' established by both our neighboring planets. Venus and Mars have strikingly similar atmospheric compositions despite surface pressures which differ from that of Earth by two orders of magnitude. This, and the obviously very different thermal, interior, surficial and atmospheric evolutionary histories of these two planets, strengthens the significance of this compositional trend, and suggests that a meaningful way to examine this trio is to look for explanations of the Earth's divergence. It is irresistible to ask which of Earth's other apparently unique features are related by cause and effect to this oddball atmosphere. Among these features one might include an active hydrosphere, plate tectonics, one giant moon, a strong intrinsic magnetic field and, of course, life.

Atmospheric circulation

The Earth's complex weather patterns can be seen, on a planetary scale, as the atmosphere's dynamic response to solar heating. The Sun's energy input is localized on the dayside and concentrated towards equatorial latitudes. This inequity creates temperature differences which result in air motions as the atmosphere attempts to smooth out these differences. How successfully it does so, and the forms these motions take, depend largely on two planetary properties: the thickness of the atmosphere and the planet's rotation rate. Venus is an example of a planet with a thick atmosphere which very efficiently redistributes solar heat so that the whole surface is at nearly the same temperature, day and night, equator to pole. The relatively wispy atmosphere of Mars is much less effective at carrying heat away from the subsolar point and thus daily, seasonal and latitudinal temperature extremes are large on this planet. Earth's atmosphere is intermediate in extent, as seen in the surface pressures given in Table E2, and consequently its daily weather patterns and latitudinal variations are more interesting than those on Venus, but not so extreme as on Mars.

In the simplest form of global circulation, solar energy input causes air to heat up and rise in equatorial regions from where it is transported poleward at high altitudes. Cooler air sinks near the poles and a low-altitude flow returns air towards the equator. This is known as a Hadley circulation pattern. The Coriolis forces resulting from a planet's rotation, and to some extent the surface topography, can lead to considerably more complex motions. On Venus, which rotates very slowly (Table E2), Coriolis forces are weak and Hadley circulation dominates the redistribution of heat.

Earth's rotation is considerably more rapid than that of Venus and thus north–south moving air masses experience considerable Coriolis forces which deflect them in an east–west direction. Thus on Earth a single Hadley cell would be unstable, and each hemisphere is broken up into three cells. Condensation of water vapor rising from the ocean with the equatorial Hadley cell results in a belt of clouds girdling the planet. At higher latitudes the Coriolis forces have a larger effect and winds tend to blow in an east–west direction, and 'eddy' circulation becomes important. These circular motions – cyclones driven by pressure lows and anticyclones driven by pressure highs – are caused by the instabilities that result from large temperature differences across adjacent latitudes, known as baroclinic instabilities.

Temperature and composition

Table E2 also includes the average surface temperatures for each planet. At first glance there are no surprises to the general trend. As we would expect, surface temperature is inversely correlated with distance from the Sun. However, when we look at the planetary equilibrium temperatures in this table, the plot thickens. This calculated temperature indicates the portion of incident sunlight actually absorbed and reradiated by a planet. Earth's equilibrium temperature is actually the highest of all three planets. Venus has a low equilibrium temperature for the same reason that it shines so

Table E2 Atmospheric characteristics of Earth, Venus and Mars

	Earth	Venus	Mars
Major gases (%)	N_2 (78), O_2 (21), H_2O (1), Ar (0.93)	CO_2 (96.5), N_2 (3.5)	CO_2 (95.3), N_2 (2.7), Ar (1.6)
Minor gases (ppm)	CO_2 (330), Ne (18), He (5.2), Kr (1.1), Xe (0.087), CH_4 (1.5), H_2 (0.5), N_2O (0.3), CO (0.12), NH_3 (0.01), NO_2 (0.001), O_3 (0.4), SO_2 (0.0002), H_2S (0.0002)	H_2O (30), SO_2 (180), COS (4.4), Ar (70), CO (23), Ne (7), HCl (0.4), HF (0.01)	O_2 (1,300), CO (700), H_2O (100), Ne (2.5), Kr (0.3), Xe (0.08), O_3 (0.1)
Surface pressure (bars)	1.013	92	0.006
Surface temperature	288	735	218
Equilibrium temperature	250	235	211
Rotation period (relative to the Sun)	24 h, 4 min	117 (Earth) days	24 h, 36 min

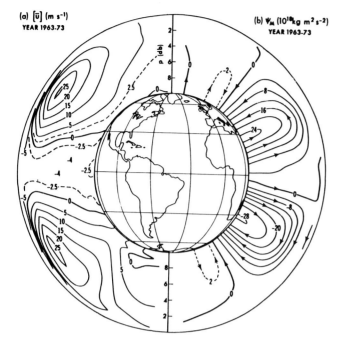

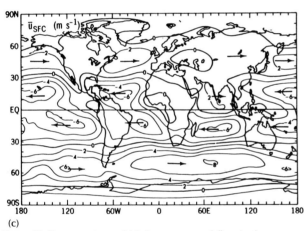

Figure E2 Cross-sections of (a) the mean zonal flow in the atmosphere (m s^{-1}), and (b) the streamlines of relative angular momentum in Hadley units, both for annual mean conditions 1963–73. Note that the vertical scale of the atmosphere is grossly exaggerated. (After Oort and Peixoto, 1983.) (c) Global distribution of zonal wind component u at the surface for annual mean conditions (m s^{-1}).

brightly in our evening and morning skies: the global cloud cover gives it an albedo of 0.75, meaning that 75% of the light is reflected into space. By contrast, Earth's albedo is 0.33.

It turns out that it is the radiative properties of the atmospheres and clouds of these worlds, not simple proximity to the Sun, which largely control their climates. Each planet has some degree of 'greenhouse' warming. This term describes the process in which infrared radiation emitted by a planet's surface and lower atmosphere is absorbed by the gases and aerosols in the atmosphere, heating up the surface and lower atmosphere until radiative balance is achieved at a temperature higher than the equilibrium temperature. The magnitude of the greenhouse effect on these three worlds can be gauged by comparing the equilibrium and surface temperatures on each. Here it can be seen that this warming is approximately 500, 38, and 7 K for Venus, Earth and Mars respectively. It is not a coincidence that the relative magnitudes of these temperature differences correlate with the surface pressures (also seen in Table E2), which are a rough gauge of total atmospheric abundance (although

surface gravity affects surface pressure). However, the composition of a planet's atmosphere is also a very important factor in determining how efficiently infrared radiation is absorbed, and thus in determining the climate.

Earth's dominant gases, N_2 and O_2, are both ineffective infrared absorbers. In general diatomic molecules do not absorb much infrared radiation as they lack complex vibrational modes. It is the trace gases H_2O, CO_2, CH_4 and some others that are responsible for most of the greenhouse warming on Earth. By contrast, the atmospheres of Venus and Mars are dominated by CO_2, a triatomic molecule which absorbs strongly in the infrared. Of the three, Earth is the only planet with a climate in the appropriate range for maintaining an active hydrosphere and for life, or at least life as we know it, which depends upon aqueous solutions of organic compounds. If Earth did not have its strange atmosphere, things would be uncomfortably hot for us. When discussing the evolution of the differing compositions of these planets, it is good to keep in mind the strong control of climate by atmospheric composition.

Atmospheric evolution

How did Earth arrive at its unique atmospheric composition? Most studies of comparative evolution start with the assumption that Venus and Earth, neighboring planets of approximately the same size, began with comparable 'volatile inventories', or total amounts of chemical elements capable of forming atmospheric gases under the right conditions. Whether or not this was strictly true depends on the details of terrestrial planet formation (Lewis and Prinn, 1984; Hunten, 1993). If we assume that this was the case, then three obvious questions arise: what happened to Venus' water, what happened to Earth's CO_2 and where did Earth's oxygen come from?

The answers to all of these questions may be closely related. Most of the Earth's carbon reserves are at present locked up in the crust in the form of carbonate rocks and organic sediments. However, estimates of the amount of CO_2 that would be produced if all of this carbon were converted to this form come fairly close to the total CO_2 content of Venus' atmosphere (Kasting, 1993). This observation, along with similar nitrogen abundances, helps justify the assumption of similar initial volatile inventories. So perhaps Earth's CO_2 has slowly been converted into carbonates and other crustal reservoirs. But why didn't this happen on Venus?

That brings us to the question of water. If Venus started out with an Earth ocean's worth of water, evolutionary models suggest that it could have been lost early in that planet's history by a 'runaway greenhouse' in which the greenhouse warming of water vapor feeds back positively with the tendency for more water to evaporate into the atmosphere at higher temperatures. This could have resulted in large amounts of water vapor high in the atmosphere where solar ultraviolet radiation could break up these molecules, allowing the hydrogen to escape into space. If Venus did have an ocean, but lost it early, this could help to explain the dramatic difference in atmospheric CO_2 abundances.

On Earth water is important in the weathering reactions which draw CO_2 out of the atmosphere and into carbonate rocks. If the water disappeared, CO_2 would probably build up in the atmosphere. Earth, due to its greater distance from the Sun, avoided this early runaway greenhouse. So it seems plausible that two similar planets at the distances from the Sun of Venus and Earth could end up with a CO_2-dominated atmosphere, and an ocean respectively. But, as suggested above, the Earth's current climate depends on the small amount of CO_2 which is in the atmosphere at present. Is it just a lucky coincidence that the Earth's climate evolved in such a way as to maintain abundant liquid water and a comfortable environment for life? It has been suggested that this is not at all a coincidence. Both abiological and biological feedbacks have been proposed as controlling factors in this evolution.

The story is complicated by the fact that the solar constant, the amount of radiation received from the Sun, has not been constant at all, but has increased by about 30% over the lifetime of the solar system, as the Sun slowly heats up, according to widely accepted astrophysical theory. If the Earth's atmospheric composition has remained constant over its whole history, then for the first 2 billion years of Earth's existence the magnitude of greenhouse warming indicated in Table E2 would have been insufficient to keep the surface above the melting point of water ice, and it would have been constantly frozen over (Ringwood, 1961; Kasting and Grinspoon, 1991). Yet there is abundant geological evidence suggesting that this was not the case. In addition, microfossils have been found dating back to 3.5 billion years ago, and it is generally believed that liquid water was necessary for the origin of life on Earth.

The easiest solution to this potential paradox is to postulate that the composition of the Earth's atmosphere has evolved and that there was a higher abundance of greenhouse gases in the past. Given the atmospheric composition of both our neighboring planets and the similar carbon reserves of Venus and Earth, it is natural to guess that Earth's earliest atmosphere was also dominated by carbon dioxide. Climate calculations indicate that the faintness of the young Sun could have been offset by an atmosphere of between 0.1 and 10 bars of CO_2 (Kasting, 1993). As solar luminosity increased, the surface of the Earth would have heated up. Yet paleoclimatic evidence suggests that, within a range of interesting and complex variations, the overall climate of the Earth has remained remarkably constant. This is where the above-mentioned feedbacks come into play.

It has been suggested that the CO_2 content of the Earth's atmosphere is regulated, in such a way as to maintain a relatively constant climate by the combined effects of silicate weathering and volcanism (Walker, Hays and Kasting, 1981). According to this model volcanic gases have provided a continual source of CO_2 to the atmosphere over Earth history, as they do today. If the Sun was fainter, the surface would have been colder, which would have decreased the rate of silicate weathering. Weathering draws CO_2 out of the system and makes carbonates, so under these circumstances volcanically supplied CO_2 would have built up in the atmosphere. This would have led to an enhanced greenhouse, which would heat up the surface, increasing the rate of weathering, and so on. This model suggests that as the Sun's luminosity has increased, the CO_2 content of Earth's atmosphere would have gradually decreased by this negative feedback mechanism, maintaining the average surface temperature close to, but comfortably above, the freezing point of water. If this mechanism has operated over the history of the Earth, it would help to explain both the relative constancy of climate over geological time, and the current lack of atmospheric CO_2 in comparison to our planetary neighbors. This model also predicts that this mechanism can continue to operate for approximately 1 billion years into the future before the continued warming of the Sun overwhelms the capacity of this system.

It has also been suggested that Life has played a controlling role in CO_2 regulation on Earth by facilitating carbonate deposition in the oceans and by increasing the CO_2 concentration in soils and thus accelerating the weathering process (Lovelock and Whitfeld, 1982). This is part of a larger set of hypotheses known collectively as the Gaia Hypothesis, in which the combined activities of life on Earth serve to regulate actively the biosphere in such a way as to maintain an optimum environment for life (Lovelock, 1988; Schneider and Boston, 1991). Although hampered by the lack of a consistent definition of the hypothesis, and by difficulties defining such terms as 'optimum environment for life', the Gaia hypothesis has stimulated a great deal of worthwhile scientific and philosophical study and debate over the role of life in the evolution of the Earth and its atmosphere, and an increased appreciation of the subtlety and complexity of many of these interactions. It is certainly true that the biota have had a major influence on the evolution of the Earth's atmosphere. This brings us to the third major evolutionary question mentioned above: the question of oxygen.

Taking another look at Table E2, we see that another way of describing the uniqueness of the Earth's atmosphere is that, compared with that of Venus or Mars, or indeed that of any other planet in the solar system, the mixture of gases is dramatically far from thermodynamic equilibrium. In particular, the large abundance of oxygen is clearly unstable without a dynamic process, in this case photosynthetic life, maintaining it. Proponents of the Gaia hypothesis maintain that this huge departure from equilibrium is a universal property of inhabited planets. This proposition, while not unreasonable, is difficult to test at present. At any rate, it is clear that biological activity is responsible for the high O_2 abundance of the Earth's atmosphere. The geologic record suggests that O_2 in the Earth's atmosphere first rose to significant levels about 2 billion years ago. The source of most of the oxygen is photosynthesis, followed by burial of organic carbon. Oxygen is a waste product of photosynthesis. Plants, using solar energy, convert carbon dioxide and water into food, liberating oxygen. Oxygen is actually a poison for simple organic life, as it reacts with and destroys complex organic molecules. The build-up of significant amounts of free oxygen in the atmosphere, largely due to the success of photosynthetic life, thus constituted Earth's first global environmental crisis. Life responded creatively, by evolving the ability to harness this disequibilibrium with a controlled burn known as oxygen respiration. The great energy reserves thus liberated in the service of further evolution allowed the development of eukaryotes, multicellular life, animals and eventually creatures smart enough to ponder the global environment and write about it, but not necessarily smart enough to avoid causing self-exterminating changes in it.

The build-up of oxygen in the atmosphere allowed another significant development: the formation of an ozone (O_3) layer as a by-product of ultraviolet destruction of O_2 molecules. As is well known the ozone layer absorbs solar ultraviolet radiation. The build-up of ozone effectively shielded the surface from this biologically harmful radiation, allowing life to colonize the Earth's land areas.

The existence of an ozone layer causes another unique feature of Earth's atmosphere, the stratosphere. This is a layer of the atmosphere 12 to 47 km above the surface where temperature rises with altitude due to heating resulting from solar ultraviolet radiation being absorbed by ozone. This 'inversion' type of temperature structure inhibits convection and results in long residence times for particles and gases which find their way into the stratosphere. The effects of this include relatively long-lived climate perturbations from volcanically generated aerosols, and dangerous accumulations of anthropogenic chemicals, such as chlorofluorocarbons, in the stratosphere. The latter are threatening the existence of the very ozone which causes this stable thermal structure (Solomon, 1990). The lack of significant quantities of ozone in the atmospheres of Mars or Venus results in the lack of any structure in these atmospheres analogous to Earth's stratosphere.

Conclusions

The presence and unusual nature of the Earth's atmosphere would be obvious to the casual extraterrestrial observer when observing over a range of timescales and electromagnetic wavelengths. Seen in the visible, the atmosphere reveals itself by the lovely hazy blue limb of the Earth against the blackness of space. Its lively dynamics and life-sustaining temperature range are revealed in the rapid and complex daily movements of condensed water clouds across the globe. Observations in the infrared reveal extreme spatial and temporal variations in the water vapor content on daily and seasonal timescales. Patient aliens observing over the Earth's history with infrared eyes might have noticed a long-term overall decrease in CO_2 content, and turning their attention to the ultraviolet, a dramatic increase in ozone and other oxygen compounds over the last 2 billion years. The latter, with its attendant extreme departure from chemical equilibrium, might alert them to the fact that Earth is inhabited, or perhaps even alive according to their definitions. Yet a puzzling set of rapid changes over the last several decades might cause them some concern over the nature of this life: a rapid decline in stratospheric ozone, depleted by half at certain latitudes and seasons, correlated with the rise in such unlikely compounds as CF_2Cl_2 and $CFCl_3$ (Solomon 1990), would reveal inhabitants who are certainly clever but may lack wisdom. Similarly, a recent rapid increase in atmospheric CO_2 would be apparent. This change, comparable in magnitude to other other historical excursions, but apparently lacking the orbital forcing usually responsible for such changes (Raynaud et al. 1993), and with a very different profile of oceanic and terrestrial sources and sinks than that which usually accompanies such episodes, would also be cause for alarm for concerned observers.

How is Earth's unique, biologically altered atmosphere related to the other distinctive planetary qualities listed above? The existence of Earth's unusually large moon is thought to have helped maintain the climate within 'healthy' limits for life. This is because the Moon's gravitational torque has protected Earth from the extreme quasi-periodic changes in orbital obliquity which have probably produced climate oscillations on Mars (Ward 1992). Also, the giant impact in the late stages of the Earth's accretion which formed the Moon (Hartmann, 1986) probably had important effects on the Earth's structure and composition. It may have partially depleted Earth in some volatile elements, although these effects have not been carefully modeled as yet. It is clear, for reasons described above, that the maintenance of Earth's climate in a range suitable for the existence of an active hydrosphere, and thus life as we know it, is intimately linked with the compositional evolution of the atmosphere. Whether the atmosphere and biosphere have had important effects on the evolution of Earth's apparently unique (at least in this solar system) style of plate tectonics is less certain. However, two reasons that have often been proposed to explain the divergence in tectonic styles between Venus and Earth are the likely lack of water in Venus' crust compared to Earth, which affects the material properties of rocks, and the large difference in surface temperature between these worlds. So, to the extent that the biota has affected, or possibly controlled, atmospheric evolution, plate tectonics has conceivably been biologically modulated to some extent. If we accept this statement then we must accept that the Earth's overall thermal evolution, and thus even such remote quarters as the molten outer core, which produces Earth's singular magnetic field, may not have been immune to the modifying effects of this unique gaseous envelope.

David H. Grinspoon

Bibliography

Hartmann, W.K. (1986) Moon origin: The impact trigger hypothesis; in *Origin of the Moon* (eds W.K. Hartman, R.J. Phillips and G.J. Taylor). Houston: Lunar and Planetary Institute.
Hunten, D.M. (1993) Atmospheric evolution of the terrestrial planets. *Science*, **259**, 915–9
Kasting, J.F. (1993) Earth's early atmosphere. *Science*, **259**, 920–5.
Kasting, J.F. and Grinspoon, D.H. (1993) The faint young Sun problem, in *The Sun in Time* (eds C.P. Sonnett, M.S. Giampapa and M.S. Matthews). Tucson: University of Arizona Press, pp. 447–62.
Lewis, J.S. and Prinn, R.G. (1984) *Planets and Their Atmospheres: Origin ad Evolution*. New York: Academic Press.
Lovelock, J. (1988) *The Ages of Gaia*. New York: Norton.
Lovelock, J. and Whitfield, M. (1982) Life span of the biosphere. *Nature*, **296**, 561–3.
Oort, A.H. and Peixoto, J.P. (1983) Global angular momentum and energy balance requirements from observations. *Adv. Geophys.*, **25**, 355–490.
Pollack, J.B. (1991) Kuiper Prize Lecture: present and past climates of the terrestrial planets. *Icarus*, **91**, 173–98.
Pollack, B. (1994) Atmospheres of the terrestrial planets, in *The New Solar System* (eds J.K. Beatty and A. Chaikin). Cambridge University Press, pp. 91–106.
Raynaud, D., Jouzel, J, Barnola, J.M, et al. (1993) The ice record of greenhouse gases. *Science*, **259**, 926–33.
Ringwood, A.E. (1961) Changes in solar luminosity and some possible terrestrial consequences. *Geochim. Cosmochim. Acta*, **21**, 295–96.
Schneider, S.H. and Boston P.J. (eds) (1991) *Scientists on Gaia*. Cambridge: MIT Press.
Solomon, S. (1990) Progress towards a quantitative understanding of Antartic ozone depletion. *Nature*, **347**, 347–54.
Sundquist, E.T. (1993) The global carbon dioxide budget. *Science*, **259**, 934–41.
Walker, J.C.G., Hays P.B. and Kasting J.F. (1981) A negative feedback mechanism for the long-term stabilization of Earth's surface temperature. *J. Geophys. Res*, **86**, 9776–82.
Ward, W.R. (1992) Long-term orbital and spin dynamics of Mars, in *Mars* (eds H.H. Kieffer, B.M. Jakosky, C.W. Snyder and M.S. Matthews). Tucson: University of Arizona Press.

Cross references

Angular momentum cycle in planet Earth
Atmosphere
Atmospheric thermal structure
Carbon, carbon dioxide
Coriolis force, geostrophic motion
Cyclone, anticyclone
Greenhouse effect
Hadley circulation
Insolation
Life: origin
Mars: atmosphere
Solar constant
Spectroscopy: atmospheres
Venus: atmosphere
Water
Weathering

EARTH: GEOLOGY, TECTONICS AND SEISMICITY

For a geologic description of the Earth, a convenient starting point will be a map showing its major crustal divisions, tectonically active features, and volcanic areas (Figure E3; Plate 10). Major tectonic boundaries were drawn primarily from the distribution of seismic activity (Figure E4). This map shows features active within roughly the last 1 million years, a period short enough to be considered the geologic present but long enough to give a representative picture of geologic activity. In addition, this is roughly the length of time that newly formed volcanic features withstand erosion well enough to be recognized. Volcanic areas of the central eruptive type were compiled on the basis of historic activity, geologic maps and orbital photographs. Oceanic ridges and continental rifts are also volcanically active, often producing fissure eruptions. This map unavoidably underrepresents the extent of young volcanism, for radar altimetry of the oceans has revealed thousands of previously unknown seamounts (Craig and Sandwell, 1988), some of them probably active within the last million years.

The map shows the two primary crustal divisions of the Earth, oceanic and continental, and its major tectonic components, the plates. 'Plate' in this context refers to horizontally extensive segments of the Earth's lithosphere, which may include both oceanic and continental crust, that are relatively rigid and inactive at present. Plates are bounded by ridges, rifts, trenches, transform faults or broad zones of diffuse seismicity and tectonism. Oceanic ridges and continental rifts are zones of crustal divergence and volcanic activity and are termed spreading centers. The spreading rates, expressed as horizontal crustal motions, have been directly measured for plates in the Pacific Basin by space geodesy (satellite laser ranging, very long baseline interferometry and the Global Positioning System) (Smith and Turcotte, 1993). Trenches are the corresponding zones of crustal convergence, or subduction zones, in which oceanic crust is thrust (or gravitationally pulled) downward into the mantle. For areas of converging continental crust, notably the Himalayas, the low density of continental crust prevents deep subduction and it is believed that shallow underthrusting occurs. Transform faults are fracture zones, dominated by horizontal movement, that connect segments of ridges or other tectonic elements.

Continents

Roughly one-third of the Earth's surface is underlain by continental crust, often termed 'sialic' (from its relatively high silicon and aluminum content). Continental crust can be described, at risk of oversimplification, as long lasting and granitic, in contrast to the geologically ephemeral and basaltic oceanic crust. Until modern deep sea exploration began, in the 1960s, 'geology' meant essentially continental geology, a subject filling entire libraries. To summarize it, we will describe the continents in a generic sense, by covering the major types of continental crust, with only enough examples to clarify the descriptions. Since most continents are largely underlain by Precambrian rock (Goodwin, 1991), it is clear that continental crust was largely formed by the end of the Precambrian, which has two major subdivisions, Archean and Proterozoic. It now appears that the Archean–Proterozoic boundary, about 2.5 Ga ago, marked a major change in tectonic style on the continents (Best, 1982). The continental crust can thus be conveniently described in chronologic order.

Archean crust

Where well exposed, the Archean crust appears to be 'vast formless seas of gneissic granites in which swim islands and rafts of metamorphic rocks . . .', in the evocative description of Turner and Verhoogen (1960). The term 'granite–greenstone terrain' also describes Archean crust concisely if somewhat incompletely. The Canadian Shield's Superior Province, probably the best-mapped large Archean terrain, is shown in Figure E5, with a generalized cross-section (Figure E6). The province was first defined (Stockwell *et al.*, 1976) partly by its dominant age, 2.5 Ga, as measured chiefly by the potassium–argon method, which for large areas gives essentially the time at which the

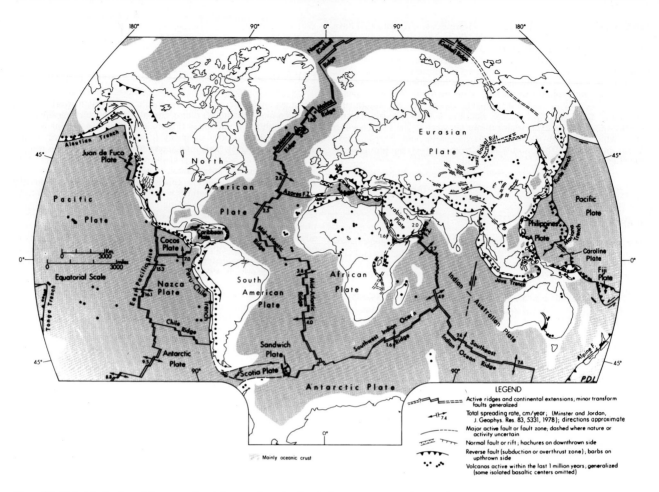

Figure E3 Global tectonic activity map, showing tectonic and volcanic features active within the last 1 million years. Shaded regions are predominantly oceanic crust. (From Lowman, 1982.)

crust cooled enough to retain radiogenic argon. This age is not necessarily, and in fact is generally not, the age at which the rocks of the province were extracted from the mantle. It is now known that the province includes rocks at least as old as 4.0 Ga. Regional provinces of the Shield, and of North America in general, are shown in Figure E7 (Hoffman, 1988). These provinces are based on more than potassium–argon dates, and accordingly give a better picture of the chronology of crustal evolution.

Areally, Archean crust is dominated by granitoid rocks of the tonalite–trondhjemite–granodiorite association. 'Granites' in the strict petrographic sense, essentially potassium-rich rocks with purely magmatic textures, are widespread, but any given outcrop in the Superior Province will probably be found to be gray ortho- or paragneisses of intermediate composition. Where lower continental crust is exposed, as in the Kapuskasing Zone (Percival and Card, 1983), the granitoid rocks grade downward into highly deformed granulite-grade gneisses. Seismic reflection profiling and other evidence suggests that such gneisses typically underlie Archean crust in most areas (Lowman, 1984). An unusually good exposure of these rocks occurs in the Scourian Complex of northwest Scotland (Figure E8), illustrating the layered structure and generally low dips of the lower crust. The fact that, where good exposures are available, these high-grade gneisses grade upward into the granite greenstone terrains, indicates strongly that they are essentially the highly metamorphosed equivalents of these rocks (Card, 1990).

The 'greenstones' of Archean areas are intensely deformed belts of volcanic and sedimentary rock, generally metamorphosed to chlorite grade or higher and characterized by greenish schists and gneisses. The more inclusive term 'supracrustal' is becoming widely used to describe this association, as in the authoritative compilation by Ayres *et al.* (1985). Because many Precambrian shield ore deposits occur in greenstone belts, these rocks have been extensively described although they are volumetrically a subsidiary component of Precambrian crust. The greenstones are often cut by Archean granite intrusions restricted to the belts (Trowell and Johns, 1986).

Age relationships between the 'granites' and the 'greenstones' remain one of the main unsolved problems of Precambrian geology. In some areas the greenstones are clearly cut by the surrounding granites (as distinguished from those inside the greenstone belts). In others the contacts are depositional, implying that the volcanics and sediments were deposited on an older crust. These relationships can probably be reconciled if the greenstone belts were deposited on the granitic crust, which was later remobilized by anatexis or diapiric intrusion (Schwerdtner and Lumbers, 1980) to produce the local cross-cutting relationships. Such a sequence has been proposed for various parts of the Superior Province by Verpaelst, Brooks and Franconi (1980) and Ayres and Thurston (1985).

The Archean crust is the fundamental rock of the continents, most other Precambrian rocks being either reworked Archean crust or additions to it. The origin of continental crust is thus essentially the origin of Archean rocks. The two major schools of thought on continent formation can be simplistically termed lateral accretion and early global differentiation. The most widely accepted view (e.g. Condie, 1980) is that continents were formed over geologic time by lateral accretion of island arcs and other terranes around the first-formed continental nuclei. Many have interpreted the granite–greenstone terrains in this framework (e.g. Card, 1990). The major subdivisions of the Canadian Shield (Figure E7) have been interpreted by Hoffman as terranes whose tectonic assembly formed the 'United Plates of America.' A contrasting view (Hargraves 1976; Shaw 1976; Armstrong 1981; Lowman 1989) is that the continental crust formed very early as a global shell, analogous to the lunar

EARTH: GEOLOGY, TECTONICS AND SEISMICITY

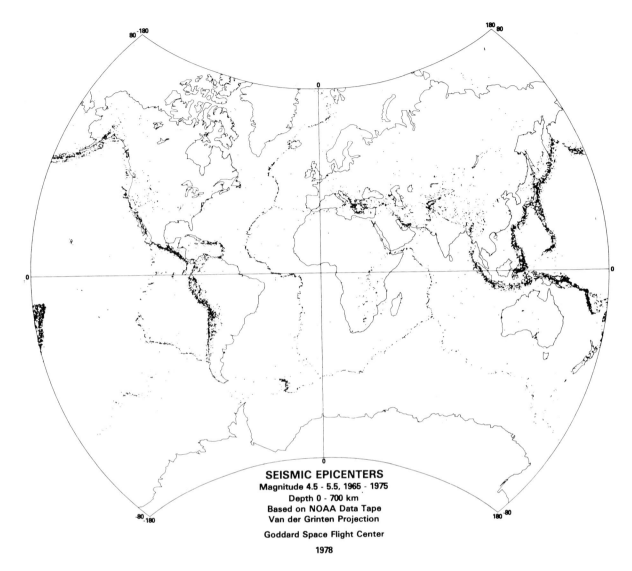

Figure E4 Seismicity map, used as base for Figure E3. (From Lowman, 1982.)

highland crust, later disrupted by impacts or other processes. Still another concept has been outlined by Cooper (1990), who has shown that in southern Africa the continental crust has undergone 11 'megacycles' of rifting, tectonism and magmatism, with a periodicity of roughly 320 Ma. Crustal evolution has in this view involved reworking of older crust in concentric belts, not lateral accretion. There is, in summary, an extremely wide range of views on how the Precambrian crust was formed.

Proterozoic crust

Beginning very roughly 2.5 billion years ago, terrestrial tectonic style underwent a fundamental change (Windley, 1984). The continental crust became more stable, and the Earth began to develop fold and thrust belts similar in many (but by no means all) ways to those of the Phanerozoic, such as the Appalachians, Caledonides and Alps. These Proterozoic fold belts, now deeply eroded, have been intensely deformed and metamorphosed, but their main structural features can be matched by younger fold belts. Whether these belts represent major new crust formation or reworking of previous (Archean) crust is debated (Best, 1982); a useful discussion of the problem based on isotope geochemistry has been presented by McCulloch (1987). Proterozoic crust can be conveniently discussed using as an example the area of Georgian Bay, Ontario, shown on a Landsat scene (Figures E9 and E10).

This area shows parts of three crustal provinces of the Canadian Shield: Superior, Southern and Grenville. (Manitoulin Island is underlain by gently dipping Paleozoic sediments making up the north rim of the Michigan Basin.) The Superior Province is the oldest crust, on which all other rocks were deposited after a prolonged erosional period, the Lipalean Interval. The Southern province is represented here by the Proterozoic Penokean Fold Belt (Card, 1978). This belt was formed, over several hundred million years, by northwest-directed folding and thrusting of the geosynclinal Huronian Series, several kilometers of sedimentary and volcanic rocks, cut by later granitic intrusions. Whether this fold belt occupies the former continental margin, or was on the interior of the contemporary continent, is a matter of interpretation, but it appears to have been ensialic (Card *et al.*, 1972).

The area shown on the Landsat scene is partly occupied by rocks of the Grenville Province, which joins the Southern (elsewhere the Superior) province along the Grenville Front. Only a very small part of the Grenville Province is shown here; rocks of similar age, structure and lithology outcrop or occur in subsurface all the way from Labrador to Mexico. The Grenville Province as a whole was formed from a thick accumulation of sediments and volcanic rocks,

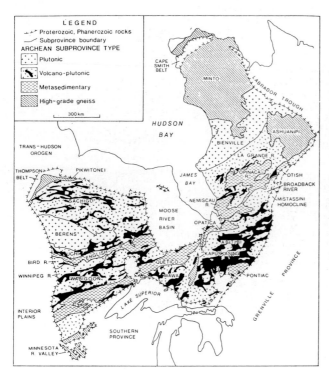

Figure E5 Map of Superior Province, from Card (1990). East boundary between Superior and Grenville Provinces is the Grenville Front, shown in Figures E9 and E10.

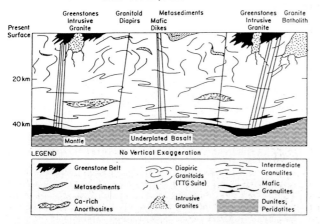

Figure E6 Cross-section of Archean crust, from Lowman (1989). Based on published maps and geophysical data covering the Superior Province in Ontario and Manitoba.

later intruded by granites. The Grenvillian rocks, 'imbricated crust' on Figure E7, were intensely metamorphosed, folded and overthrust toward the northwest (Davidson, 1986), with these events culminating about 1 billion years ago. The Grenville Front, a mylonitic reverse fault zone, is the northernmost of several such zones. It has been interpreted as a suture along which an exotic terrane was accreted to the contemporary continent, but field relations along the Grenville Front show clearly that the suture, if any, must lie well to the southeast. A problem still under study is how much of the Grenvillian rocks in this area are reworked Archean rocks and how much represent new additions to the crust. The persistence of pre-Grenville rocks and structures in highly modified form well southeast of the Grenville Front shows that much reworking has occurred.

The Penokean and Grenvillian belts in this area represent roughly 1 billion years of intermittent tectonic and magmatic activity. They thus illustrate a characteristic of many Proterozoic and Phanerozoic crustal areas, first brought out by Arduino in 1759 (Adams, 1938) and two centuries later by Wynne-Edwards (1969), namely the extremely prolonged and repeated nature of orogenic processes. These provinces, as well as the Superior Province, have recently been interpreted in terms of terrane accretion and continental collisions. Whether such events can account for, in this area, a billion years of tectonic activity is at least debatable.

A local Proterozoic feature, interesting both by itself and as an illustration of Proterozoic tectonism, is the Sudbury Igneous Complex. This large igneous body occupies a depression now thought to have been formed 1.85 billion years ago by the impact of a large meteorite (Dietz, 1964; Lowman, 1992a). It has been proposed that the Sudbury Structure is the terrestrial equivalent of a small lunar mare basin. The original shape of the crater is unknown, but it is clear that the Penokean Orogeny was responsible for at least some of the structure's present shape in that folding and thrusting to the northwest have displaced much of the southern rim.

The Penokean fold belt, only part of which is shown here, is similar to several other Proterozoic fold belts, such as the Trans-Hudson Orogen (Fig. E7), which occupies a sinuous path several thousand kilometers long from Quebec to Saskatchewan and perhaps farther. Similar fold belts are known on other continents (Windley, 1984). Collectively they have been interpreted as the beginning of tectonic processes essentially similar to those of the present.

Mafic dikes are expressed in the Landsat picture as some of the lineaments cutting rock of all three provinces. These dikes are a volumetrically minor but important component of Precambrian crust, often occurring as swarms occupying thousands of square kilometers (Fahrig and West, 1986). Similar to those underlying the great Phanerozoic basalt plateaus, the dikes are thought to have been feeders for enormous volcanic eruptions whose rocks are now largely eroded away. They represent a stage of recurring basaltic volcanism analogous to that on the Moon, Mercury, Mars and Venus, which has been termed (Lowman, 1976) the 'second differentiation' (see Earth). They are, like the greenstone belt metabasalts, presumably the terrestrial counterparts of the basaltic eruptions that formed the oceanic crust. Their subcrustal equivalents have probably underplated the continental crust, contributing directly and indirectly to the growth of continents (Lowman, 1989).

Phanerozoic crust

The beginning of the Paleozoic Era 570 million years ago marks the start of Phanerozoic time, for which the geologic record is both well exposed and relatively easy to read, compared to that of the Precambrian. The Paleozoic also saw the well-known explosive expansion and diversification of life on Earth, which may well have had direct influence on terrestrial geology as discussed elsewhere (see Earth).

Continental Phanerozoic crust is thus relatively familiar. Over most of the continental areas, Phanerozoic rocks form relatively thin, undeformed sedimentary blankets, as in the cratonic cover visible in Figure E9 as the rim of the Michigan Basin. The classic exposures of the Grand Canyon, shown in every geology text, reveal a cross-section of the Colorado Plateau that is itself an isolated segment of the North American craton. They illustrate typical cratonic structure, showing that such crust underwent relatively little deformation for several hundred million years.

Phanerozoic fold belts, whose currently active examples are shown on Figure E3, have been the inspiration for most major theories of the Earth's crustal evolution. The Paleozoic Appalachian chain (Figure E11), extending from Newfoundland to west Texas, has long been considered the type example of fold and thrust belts. The Appalachians were first interpreted in the 19th century as the folded traces of a geosyncline, from the great thickness of sedimentary rocks encountered as one goes from west to east. This geosyncline, a thick, laterally extensive sequence of shallow- to deep-water marine sediments, appears to have had two major subdivisions. The western part, the miogeosyncline, was characterized by relatively thin, stable shelf sediments such as limestone and quartz sandstones, with little or no volcanic component. The eastern part, the eugeosyncline, had a much thicker sedimentary section, including poorly sorted sediment-

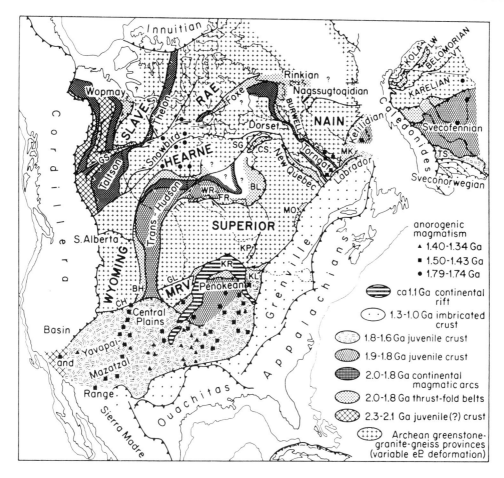

Figure E7 Terrane map of North America, from Hoffman (1988).

Figure E8 Cliff at Scourie, northwest Scotland, showing granulites of the Scourian Complex, interpreted as similar to rocks of the lower continental crust, from Lowman (1984). Layering is partly foliation resulting from deformation, with compositional variation as well.

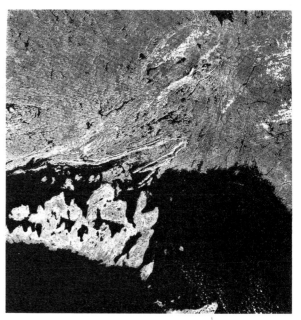

Figure E9 Landsat view of Sudbury area, from Lowman (1992a). Landsat 2 scene 999–15070–6, acquired 17 October 1977. Smoke plume from Inco smelter at Sudbury visible at top center. Copyright American Geophysical Union.

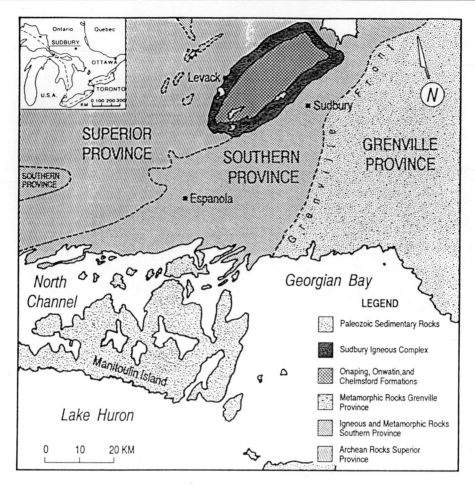

Figure E10 Map of Sudbury area shown in Figure E9. Copyright American Geophysical Union.

ary rocks such as graywackes, and abundant volcanics. (The term 'eugeosyncline,' though now rarely used in western literature, is nevertheless a useful shorthand for lithologic associations of the type described.) This twofold nature is seen in other Phanerozoic fold belts around the world. Structurally, the Appalachians are dominated by overthrusting and folding, directed from the Atlantic Ocean inward to the continent. The extent of this overthrusting has only recently become fully appreciated, with the discovery from reflection profiling that the entire Appalachian chain is allocthonous, i.e. an immense series of overthrusts that include Precambrian crystalline rocks such as the Blue Ridge.

The North American Cordillera, particularly in Canada, appears to be structurally and lithologically analogous to the Appalachians. Both have belts of batholiths, eugeosynclinal sediments and folded and overthrust miogeosynclinal sectors. Cordilleran geology has been explained in recent decades as the result of crustal convergence between various Pacific Ocean plates and the North American continent, resulting in accretion of many exotic terranes (Hillhouse, 1989).

The general acceptance of sea-floor spreading and subduction has led to a reinterpretation of folded mountains (and much else). As applied to the Appalachians, this belt is now considered by many geologists (Williams and Hatcher, 1982) to have been formed by the accretion and docking of several exotic terranes by means of sea-floor spreading. Similar interpretations have been made of the Mesozoic–Cenozoic North American Cordillera, where many exotic terranes have been mapped. Collectively, terrane accretion is thought to have led to lateral growth of the continent. Whether this is the dominant means by which sialic crust has been separated from the mantle is however debatable. Alternative views have been put forth: 'ensialic plate tectonics' (Kroner, 1981) and 'surge tectonics' by Meyerhoff et al. (1992). The geosyncline concept (Seyfert, 1987), once criticized as a theory for the origin of mountains with the mountains left out, remains a useful descriptive tool even in the light of new discoveries in tectonics.

Ocean basins

Roughly two-thirds of the Earth is occupied by the deep ocean basins, which should be distinguished from shallow epicontinental seas such as Hudson Bay. The ocean basins were until recently little understood geologically, not only because of their inaccessibility but because there are few deep sea sediments preserved in the continental geologic record. However, as a result of investigations carried out since the late 1950s, notably deep sea drilling and geophysical surveys, our general understanding of ocean basin geology and tectonics is now apparently better in many respects than our understanding of continental geology.

Oceanic crust

With a few important exceptions, such as the Seychelles and the Lord Howe Rise, the oceanic crust can be described briefly as basalt with a thin layer of sediment. In more detail, there appear to be three main layers (Christensen and Salisbury, 1975) (Figure E12). Layer 1, the topmost, consists of unconsolidated sediments, ranging in thickness from near zero at the mid-ocean ridges to several kilometers near the continents. These sediments are largely terrigenous near the continents, but the deep ocean basins are covered by oozes of siliceous or carbonate composition. Clays and manganese nodules are also important constituents. Layer 2, generally 1 to 3 km thick, consists largely of tholeiitic basalts of various types. The upper sublayers are dominantly flows, often pillowed. Layer 3, generally about 6 km thick, can be considered the plutonic foundation of the ocean crust, consisting of sheeted dike complexes composed of gabbros, meta-gabbros, serpentinites and related types.

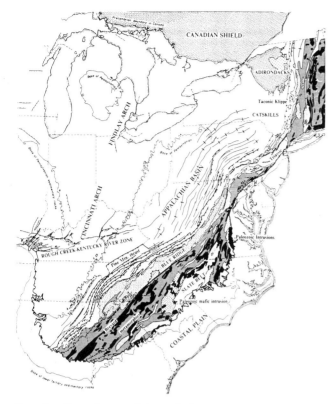

Figure E11 Tectonic map of the Applachians. (From Spencer, 1977.)

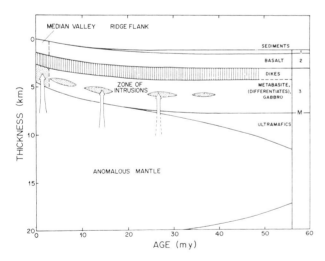

Figure E12 Cross-section of oceanic crust, showing main rock types and layers, with inferred age as a function of distance from spreading center (median valley, left). (From Christensen and Salisbury 1975, Copyright American Geophysical Union.)

Under the Mohorovicic discontinuity, generally a sharp boundary in the ocean basins, the mantle consists of ultramafic rocks and their serpentinized equivalents.

Oceanic tectonics

It could be reasonably argued as late as the early 1960s that the deep ocean floor was a primordial surface analogous to the face of the Moon. However, in that year Harry Hess' celebrated 'essay in geopoetry' was published, reconciling several lines of apparently contradictory evidence and forming the basis for modern tectonic theory. Hess showed, in brief, that the oceanic crust was mobile and relatively young, being formed at the top of mantle convection cells at the mid-ocean ridges and moving outward by sea-floor spreading (a term coined by R.S. Dietz). During the next few years this speculative concept evolved into 'the new global tectonics' and shortly thereafter 'plate tectonics' (q.v.). This subject is covered by another encyclopedia in this series (Seyfert, 1987) and will not be further discussed here.

With regard to the nature of oceanic crust, plate tectonic theory interprets most of the Earth's crust as mobile, actively moving and relatively ephemeral. The movement of oceanic crust, i.e. sea-floor spreading (Figure E3), has now been directly measured by two independent techniques, satellite laser ranging and very long baseline interferometry (Lowman, 1992a). The rates agree surprisingly well with those inferred from dated marine magnetic anomalies (Stein 1993). These measurements appear to provide direct support of the central elements of modern tectonic theory: sea-floor spreading, subduction and transform faulting. However, it has been pointed out (Lowman, Allenby and Frey, 1979) that to demonstrate continental drift by space geodesy it must be shown that the plates in question are in fact rigid and that the baseline changes thus represent plate movements. The plate rigidity requirement has been met for the Pacific Ocean crust, but so far not for the Eurasian, African and American plates.

There remain numerous anomalies not explained by the 'new global tectonics.' For example, Meyerhoff et al. (1992) have argued that since many deep sea drill holes that have encountered basalt have penetrated only intrusive contacts, the true age of the bulk oceanic crust at those points cannot be determined. Choi, Vasil'yev and Bhat (1992) present evidence for foundered continental crust in the northwest Pacific Ocean, in an area that by plate theory should consist of Mesozoic oceanic crust. Grant (1980) has shown that Precambrian continental structures can be traced by geophysical techniques past the continental margin into the supposedly young Labrador Sea, and proposed (Grant, 1987) 'inversion tectonics' as an alternative to sea-floor spreading. Lowman (1985) has shown that there are strong obstacles to continental drift as a corollary of sea-floor spreading, proposing instead 'plate tectonics with fixed continents.' The cause of earthquakes at supposed passive margins, such as those around the Atlantic Ocean, is unexplained. It is clear, therefore, that despite the enormous progress in ocean basin geology, much remains to be learned.

Summary

We now appear to have a clear concept of the Earth in its solar system context as a highly evolved and continually active planet, one with similarities and analogies to other planets but which is nevertheless unique. Exploration of the Earth's land surface has been essentially completed, and the focus of research in terrestrial geology has shifted to the ocean basins and to the deep continental crust. However, new analytical methods and new concepts continue to arise, making restudy of previously mapped areas and old data still profitable. Space exploration also produces a new context in which to view terrestrial geology (see Remote sensing).

Paul D. Lowman Jr

Bibliography

Adams, F.D. (1938) (Dover edition, 1954) *The Birth and Development of the Geological Sciences*. New York: Dover Publications, 506 pp.

Armstrong, R.L. (1981) Radiogenic isotopes: the case for crustal recycling on a near-steady-state no-continental-growth Earth. *Phil. Trans. Roy. Soc. London*, **301**, 443–72.

Ayres, L.D., Thurston, P.C., Card, K.D. and Weber, W. (1985) Evolution of Archean supracrustal sequences. Geological Association of Canada Special Paper 28, 380 pp.

Ayres, L.D. and Thurston, P.C. (1985) Archean supracrustal sequences in the Canadian Shield: an overview, in *Evolution of Archean Supracrustal Sequences* (eds L.D. Ayres, P.C. Thurston, K.D. Card and W. Weber). Geological Association of Canada Special Paper **28**, pp. 343–80.

Best, M.G. (1982) *Igneous and Metamorphic Petrology.* New York: W.H. Freeman, 630 pp.
Bird, J.M. (1980) *Plate Tectonics,* 2nd edn American Geophysical Union, 986 pp.
Card, K.D. (1978) Geology of the Sudbury-Manitoulin Area. Ontario Geological Survey, Report 166, 238 pp.
Card, K.D. (1990) A review of Superior Province of the Canadian Shield, a product of Archean accretion. *Precambrian Res.*, **48**, 99–156.
Card, K.D., Church, W.R., Franklin, J.M., *et al.* (1972) The Southern Province, in (eds R.A. Price and R.J.W. Douglas). Geological Association of Canada Special Paper 11, *Variations in Tectonic Styles in Canada*, pp. 335–80.
Choi, D.R., Vasil'yev, B.I. and Bhat, M.I. (1992) Paleoland, crustal structure and composition under the northwestern Pacific Ocean, in *New Concepts in Global Tectonics* (eds S. Chatterjee, and N. Hotton, III) Lubbock: Texas Tech University Press, pp. 179–92.
Christensen, N.I. and Salisbury, M.H. (1975) Structure and constitution of the lower oceanic crust. Rev. Geophys. Space Phys, **13**, 483–512 (In Bird, J.M., 1980.)
Condie, K.C. (1980) Origin and early development of the Earth's crust. Precambrian Res., **11**, 183–97.
Cooper, M.R. (1990) Tectonic cycles in southern Africa. *Earth-Sci. Rev.*, **28**, 321–64.
Cox, A. (ed.) (1973) *Plate Tectonics and Geomagnetic Reversals,* San Francisco: W.H. Freeman, 702 pp.
Craig, C.H. and Sandwell, D.T. (1988) Global distribution of seamounts from Seasat profiles. *J. Geophys. Res.*, **93**(B9), 10408–20.
Davidson, A. (1986) New interpretations in the southewestern Grenville Province, in: *The Grenville Province* (eds J.M. Moore, A. Davidson, and A.J. Baer). The Geological Association of Canada Special Paper 31, pp. 61–74.
Dietz, R.S. (1964) Sudbury Structure as an astrobleme. *J Geol.*, **72**, 412–34.
Fahrig, W.F. and West, T.D. (1986) Diabase dyke swarms of the Canadian Shield. Ottawa: Geological Survey of Canada, Map 1627A,
Goodwin, A.M. (1991) *Precambrian Geology.* London: Academic Press, 666 pp.
Grant, A.C. (1980) Problems with plate tectonics: the Labrador Sea. *Bull. Can. Petrol. Geol.*, **29**, 252–78.
Grant, A.C. (1987) Inversion tectonics on the continental margin east of Newfoundland. *Geology*, **15**, 845–8.
Hargraves, R.B. (1976) Precambrian geologic history. *Science*, **193**, 363–71.
Hillhouse, J.W. (1989) Deep Structure and Past Kinematics of Accreted Terranes. American Geophysical Union, Geophysical Monograph 50, 283 pp.
Hoffman, P. (1988) United Plates of America, the birth of a craton – Early Proterozoic assembly and growth of Laurentia. *Ann. Rev. Earth Planet. Sci.*, **16**, 543–603.
Kroner, A. (1981) Precambrian plate tectonics, in *Precambrian Plate Tectonics* (ed. A. Kroner). Amsterdam: Elsevier, pp. 57–90.
Lowman, P.D., Jr (1976) Crustal evolution in silicate planets: implications for the origin of continents. *J. Geol.*, **84**, 1–26.
Lowman, P.D. (1982) A more realistic view of global tectonism. *J. Geol. Ed.*, **30**, 97–107.
Lowman, P.D., Jr (1984) Formation of the earliest continental crust: inferences from the Scourian Complex of northwest Scotland and geophysical models of the lower continental crust. *Precambrian Res.*, **24**, 199–215.
Lowman, P.D., Jr, (1985) Mechanical obstacles to the movement of continent-bearing plates, *Geophys. Res. Lett.*, **12**, 223–5.
Lowman, P.D., Jr (1989) Comparative planetology and the origin of continental crust. *Precambrian Res.*, **44**, 171–195.
Lowman, P.D., Jr (1992a) The Sudbury Structure as a terrestrial mare basin. *Rev. Geophys.*, **30**, 227–243.
Lowman, P.D., Jr (1992b) Geophysics from orbit: the unexpected surprise. *Endeavour*, **16**, 50–8.
Lowman, P.D., Jr, Allenby, R.J. and Frey, H.V. (1979) Proposed Satellite Laser Ranging and Very Long Baseline Interferometry Sites for Crustal Dynamics Investigations. NASA Technical Memorandum 80563, 64 pp.
McCulloch, M.T. (1987) Sm–Nd isotopic constraints on the evolution of Precambrian crust in the Australian continent, in *Proterozoic Lithospheric Evolution*, (ed. A. Kroner). American Geophysical Union, Geodynamics Series **17**, pp. 115–30.
Meyerhoff, A.A., Taner, I., Morris, A.E.L., *et al.* (1992) Surge tectonics: a new hypothesis of Earth dynamics, in *New Concepts in Global Tectonics* (eds S. Chatterjee and N. Hotton, III). Lubbock: Texas Tech University Press, pp. 309–409.
Milkereit, B., Forsyth, D.A., Green, A.G. *et al.* (1992) Seismic images of a Grenvillian terrane boundary. *Geology*, **20**, 1027–30.
Nisbet, E.G. (1987) *The Young Earth.* Boston: Allen and Unwin, 402 pp.
Percival, J.A. and Card, K.D., 1983. Archean crust as revealed in the Kapuskasing uplift, Superior Province, Canada. *Geology*, **11**, 323–6.
Schwerdtner, W.M. and Lumbers, S.B. (1980) Major diapiric structures in the Superior and Grenville Provinces of the Canadian Shield, in The Continental Crust and Its Mineral Deposits. (ed. D.W. Strangway). Geological Association of Canada Special Paper 20, pp. 149–80.
Seyfert, C.K. (1987) *The Encyclopedia of Structural Geology and Plate Tectonics*, New York: Van Nostrand Reinhold, p. 876.
Shaw, D.M. (1976) Development of the early crust, Part 2: Pre-Archean, Proto-Archean, and later eras, in *The Early History of the Earth* (ed. B.F. Windley). New York: John Wiley, pp. 33–53.
Smith, D.E. and Turcotte, D.L. (eds) (1993) *Contribution of Space Geodesy to Geodynamics: Crustal Dynamics,* Washington: American Geophysical Union, Geodynamics Series, **23** 429 pp.
Spencer, E.W. (1977) *Introduction to the Structure of the Earth,* 2nd edn. New York: McGraw-Hill, 640 pp.
Stein, S. (1993) Space geodesy and plate motions, in *Contributions of Space Geodesy to Geodynamics: Crustal Dynamics,* (eds D.E. Smith and D.L. Turcotte). Washington: American Geophysical Union, Geodynamics Series, Vol. 23, pp. 5–20.
Stockwell, C.H., McGlynn, J.C., Emslie, R.F. *et al.* (1976) *Geology of the Canadian Shield,* in *Geology and Economic Minerals of Canada, Part A* (ed. R.J.W. Douglas). Dept. of Energy, Mines, and Resources, pp. 43–150.
Trowell, N.F. and Johns, G.W. (1986) Stratigraphical correlation of the western Wabigoon Subprovince, northwestern Ontario, in *Volcanology and Mineral Deposits,* (eds J. Wool and H. Wallace, H.) Ontario Geological Survey Miscellaneous Paper 129, pp. 50–61.
Turner, F.J. and Verhoogen, J. (1960) *Igneous and Metamorphic Petrology* 2nd edn, New York: McGraw-Hill 694 pp.
Verpaelst, P., Brooks, C. and Franconi, A., (1980) The 2.5 Ga Duxbury massif, Quebec: a remobilized piece of pre-3.0 Ga sialic basement. *Can. J. Earth Sci.*, **17**, 1–18.
Williams, H. and Hatcher, R.D., Jr (1982) Suspect terranes and accretionary history of the Appalachian orogen. *Geology*, **10**, 540–46.
Windley, B.F. (1984) The Evolving Continents, 2nd edn. New York: John Wiley 399 pp.
Wynne-Edwards, H.R. (1969) Tectonic overprinting in the Grenville Province, southwestern Quebec, in *Age Relations in High-grade Metamorphic Terrains* (ed. H.R. Wynne-Edwards). Geologic Association of Canada Special Paper 5, pp. 163–82.

Cross references

Mantle convection
Plate tectonics
Remote sensing
Seismicity
Tectonics

EARTH: MAGNETIC FIELD AND MAGNETOSPHERE

Earth is the most important planet for the study of the generation of planetary magnetic fields and the physics of planetary magnetospheres, not because of any special property of its magnetic field or its magnetosphere but simply because we have been able to study the magnetic field over a very long period of time. Understanding the temporal behavior of both interior and exterior sources of magnetic fields is the key to understanding the physics of those source regions.

Continuous measurements of the Earth's magnetic field have now been taken for over a century and a half from observatories on the surface of the Earth and for about three decades from spacecraft above the surface of the Earth. These data have been supplemented with less precise data based on inferred values of the magnetic field deduced from remanent magnetic fields in rocks, pottery, and similar objects that can become magnetized as they cool.

Planet and interior

The Earth is the most rapid rotator of the planets of the inner solar system. It rotates once on its axis relative to the stars every 23 h 56 min (and every 24 h relative to the Sun). It has a crust and a mantle surrounding a liquid metallic core which in turn encloses a solid metal core. The average radius of the Earth is 6371 km. Seismic evidence shows that the mantle is about 2886 km thick and the metallic core about 3485 km. By volume, 16% of the planet consists of core and 84% of mantle. The inner solid core is created by the solidification of the outer core as the Earth cools. Thus the inner core is believed to be still growing in size. At present it is about 1221 km in radius and occupies about 4% of the volume of the core. Thus it is still quite small. The energy released by the freezing of the inner core (both latent heat of fusion and gravitational settling) is thought to provide the source of energy for driving the Earth's magnetic dynamo. The density of the inner core, deduced from long-period gravimeters which can detect the pendulum-like motion of the inner core, rocking back and forth within the fluid core, is 13 g cm^{-3}. The outer liquid core density is thought to be about 0.5 g cm^{-3} less. Thus while the liquid core occupies only about 16% of the volume of the planet, it contains about 36% of the mass.

The heat released in the interior of the planet by the solidification of the core, by radiative decay and by the general cooling of the Earth, is transmitted to the surface of the planet by a combination of convection and conduction. Convection requires the motion of the material. In a fluid this results in the familiar cells of hot liquid rising, cooling at the top of the cell and sinking back down to the source of the heat, only to be heated and repeat the cycle. In a rotating fluid this motion becomes much more complex. The addition of a magnetic field whose strength is expected to be large enough to be dynamically significant will further complicate the motion. As a result of this complexity, progress in understanding the generation of the Earth's magnetic field has been slow (see Dynamo theory).

Convection occurs even in the solid mantle above the core. In turn this convection drives plate motions which drive tectonic activity on the surface of the Earth. However, it is difficult to deduce the interior processes from their surface manifestations in tectonic activity. In order to probe the interior, seismic waves are being used to create tomographs of the horizontal as well as radial structure. This field is in its infancy but it promises to allow us to measure the inhomogeneity of the mantle and thereby gain greater insight into how convection in the mantle is taking place.

Magnetic field

It has been known at least since W. Gilbert's *De Magnete*, written in the reign of Elizabeth I at the end of the 16th century, that the Earth was a giant magnet with a dipolar magnetic field. It was realized not long afterward that the internal magnetic field varied with time. While at present the principal variation is a westward drift, the field undergoes notable other variations including reversals. Today the magnetic pole that is in the northern hemisphere is a south pole. It attracts the north pole of a magnet. In earlier times the pole in the northern hemisphere has been a north pole, and compasses, had they been available at that time, would have pointed in the opposite direction. The frequency of these reversals is itself a function of time. For example, 60 million years ago reversals occurred about once every 500 000 years, whereas 10 million years ago reversals occurred three times as often, about every 150 000 years (Barton, 1989).

The magnetic field is generally represented by a spherical harmonic expansion which is a solution of Laplace's equation. Contributions to the field on the surface of the Earth arise from both internal and external sources. In the following discussion we will concern ourselves with only the interior sources. The spherical harmonic expansion expresses a scalar potential, V, whose gradients are the three vector components of the magnetic field.

Table E3 Spherical harmonic coefficient (in nT) of terrestrial magnetic field (IAGA Division I Working Group 1, 1987)

Coefficient	Degree (m)	Order (n)			
		1	2	3	4
	4				169
g_n^m	3			835	−426
	2		1 691	1 244	363
	1	−1 903	2 045	−2 208	708
g_n^0	0	−29 877	−2 073	1 300	937
	1	5 497	−2 191	−312	233
	2		−309	284	−250
h_n^m	3			−296	68
	4				−298

Table E4 Location and strength of the centered tilted magnetic dipole

Year	Dipole moment ($\times 10^{15}$ T m^3)	Colatitude (degrees)	East longitude (degrees)
1990	7.84	10.8	289.0
1980	7.91	11.2	289.2
1970	7.97	11.4	289.8
1960	8.02	11.5	290.5
1950	8.07	11.5	291.1
1900	8.27	11.5	292.0
1850	8.47	11.5	295.6
1800	8.61	10.8	301.0
1750	8.84	10.1	305.4
1700	9.00	8.3	314.6
1650	9.18	7.0	322.3
1600	9.36	5.4	330.3
1550	9.54	3.1	334.1

$$V = a \sum_{n=0}^{\infty} \sum_{m=0}^{n} \left(\frac{a}{r}\right)^{n+1} (g_n^m \cos m\phi + h_n^m \sin m\phi) P_n^m (\cos \theta)$$

where θ and ϕ are the Earth's longitude and latitude, a is the radius of the Earth and r is the distance from the center of the Earth. The index n is the degree of the term and the index m is the order. The function $P_n^m(\cos \theta)$ is the associated Legendre function with Schmidt normalization. Three normalizations are encountered in practice: Schmidt, abbreviated P_n^m; Gauss, abbreviated $P^{n,m}$; and Neumann, abbreviated $P_{n,m}$. The Schmidt normalization is standard for published coefficients. The coefficients g_n^m and h_n^m are called the Gauss coefficients. Table E3 gives the Gauss coefficients for the Earth's magnetic field to degree and order 4 (IAGA, Division I Working Group 1, 1987).

The dipole term $g^0{}_1$ is the dominant term of the spherical harmonic expansion. This coefficient gives the size of the effective dipole moment along the Earth's rotation axis. We can combine the three degree 1 terms to create a tilted dipole moment, M, whose magnetic moment is given by

$$M = a^3 (g_1^{0^2} + g_1^{1^2} + h_1^{1^2})^{1/2}$$

This moment would have a pole at a latitude

$$\lambda = \tan^{-1} [g_1^{0}/(g_1^{1^2} + h_1^{1^2})^{1/2}]$$

and at a longitude given by

$$\phi = \tan^{-1} (h_1^1/g_1^1)$$

Table E4 gives the colatitude, east longitude and magnitude of the magnetic moment of the Earth's dipole at selected times over the last four centuries (Barton, 1989). This table clearly shows that the strength of the dipole moment has been monotonically decreasing over the last 400 years, and has diminished almost 20% in that time. The pole has drifted 45° westward or about 0.1° per year and the colatitude has varied from a low of 3.1° to 11.5° and now is decreasing

Table E5 Characteristic times (in years) for spherical harmonic coefficients of terrestrial magnetic field (IAGA Division I Working Group 1, 1987)

Coefficient	Degree (m)	Order (n)			
		1	2	3	4
g_n^m	4				25
	3			8 350	304
	2		242	2 073	47
	1	190	601	480	1 300
g_n^o	0	1 288	151	255	9 370
h_n^m	1	244	191	59	61
	2		15	123	114
	3			27	27
	4				331
Median characteristic time		224	191	255	114

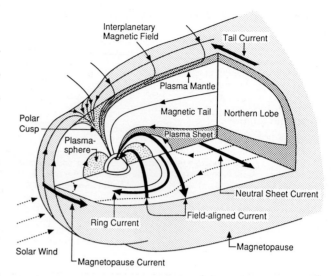

Figure E13 Cut-away drawing of the Earth's magnetosphere showing the major plasma regimes, current systems and flows.

rapidly again at a rate of about 0.02° per year, a rate similar to the maximum rate of change in colatitude seen in the historical record. Thus it is no longer appropriate to quote 11.5° as the tilt of the dipole axis, as is widely done.

Another means of obtaining insight into how rapidly the Earth's magnetic field is evolving is to examine the secular variation, the time rate of change of each of the terms in Table E3. We can obtain a charactertistic time by dividing the secular variation into the value of the coefficient. These characteristic times are given in Table E5. The minimum times of about 20 years are overestimates of the rapidity of the change because these represent coefficients passing near zero, but the median values of close to 200 years are meaningful. The interior field of the Earth is varying sufficiently rapidly that it must be remeasured on decadal timescales to be accurate enough for most research purposes. The most accurate way to do this is with low-altitude polar satellites, but no such program is planned by any of the major space agencies.

Secular variation is not just important for determining the most accurate magnetic field model at all times. The secular variation also provides a means of determining the behavior of the dynamo, because the magnetic field is to a high degree of approximation frozen into the fluid motions of the core. Thus, to understand the interior of the Earth, we need a vigorous program of space-borne measurements, in addition to the seismic networks. Finally, while the seismic studies depend on event-oriented studies, the magnetic surveys require long baselines of the highest accuracy measurements. Thus there are basically two quite different styles of observation in these two disciplines, despite quite similar objectives.

Magnetosphere

The magnetic field of the Earth acts as an obstacle to the magnetized flowing plasma from the Sun, called the solar wind (q.v.). The typical dynamic pressure exerted by the solar wind on the Earth's magnetic cavity, or magnetosphere as it is now called, is about 1.7 nPa. The point where the solar wind dynamic pressure and the pressure exerted by the Earth's magnetic field are in balance is at about ten Earth radii in the sunward direction along the Earth–Sun line. As shown in Figure E13, the Earth's magnetic field forms a cavity in the shape of a blunt body with its 'nose' at this position and an extended 'tail'. The cavity flares to about 15 Earth radii on either side of the Earth above the dawn–dusk terminators, and the solar wind stretches the terrestrial magnetic field for perhaps 1000 Earth radii in the antisolar direction. Since the solar wind is traveling relative to the Earth at speeds greater than the speed of compressional waves that are needed to divert the solar wind around the Earth's magnetosphere, a bow shock is formed in the solar wind upstream which slows the flow to 'subsonic' velocities, diverts it around the obstacle and heats the downstream solar wind plasma. The nose of the bow shock is about 13 Earth radii in front of the Earth. The properties of the bow shock determine the nature of the flow that interacts with the magnetosphere; these properties are thus important in the transfer of energy from the solar wind to the magnetosphere.

The magnetosphere in Figure E13 is cut away to reveal the different regions of charged particles in the magnetosphere and the electrical currents flowing therein. The region of least energetic plasma is the plasmasphere in the innermost region of the magnetosphere, which is the high-altitude extension of the ionosphere. This cold, dense plasma is caused by the ionization of the neutral atmosphere by solar ultraviolet and extreme ultraviolet radiation. The tenuous plasma mantle at high altitudes is formed by the entry of solar wind along the boundary of the magnetosphere. These charged particles reach the center of the tail, where processes can accelerate them to even higher energies. The dashed lines illustrate the flow of the average low-energy particles outside and inside the magnetosphere. The flow on the outside is away from the Sun and on the inside toward the Sun. The electric fields associated with these flows can accelerate the charged particles to very high energies and help populate the radiation belts, also known as the Van Allen belts, both inside and outside the plasmasphere.

The thick, dark lines in Figure E3 show the major electrical current systems. The magnetopause current is that which flows due to the pressure gradient in the shocked solar wind plasma when it reaches the magnetic field of the Earth. The tail current is the part of this current which connects to the current flowing across the center of the tail, dividing it into two lobes with oppositely directed magnetic fields. This current is often called the neutral sheet current. The current associated with the energetic particles in the radiation belts which encircle the planet is called the ring current. Finally, there are currents that move along magnetic field lines rather than across them. These field-aligned currents transmit stresses from the outer magnetosphere to the ionosphere. Thus if one were to push on the plasmas in the outer reaches of the magnetosphere, a current would flow along the magnetic field to the ionosphere. As the current flowed across magnetic fields in the ionosphere, it would exert a force on the ionosphere. Thus the force on the outer magnetosphere ultimately is transmitted to the ionosphere and thence to the atmosphere.

This energy transfer also depends greatly on the direction of the interplanetary magnetic field (q.v.). If the interplanetary magnetic field points northward, parallel to the Earth's magnetic field at the stand-off point in the equatorial plane, there is minimal energy transfer. The interaction is not completely non-viscous but under normal conditions the magnetosphere remains in a quiescent state. When the interplanetary magnetic field turns southward, the process known as reconnection begins near the subsolar point as illustrated in Figure E14. The magnetic field lines of the magnetosphere and the solar wind become linked and maximal energy transfer ensues. If the rate of energy transfer is sufficiently small, this energy can be stored for some time through a build-up of magnetic flux in the tail. Eventually, this storage region becomes unstable and releases the magnetic flux and its associated energy. In this example, M measures the rate at which the interplanetary magnetic field is connected to the

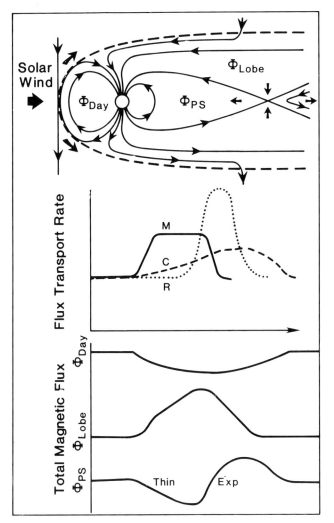

Figure E14 The near-Earth neutral point model of substorms. In the top panel is sketched the magnetic topology of the magnetosphere shortly after the interplanetary magnetic field turns southward. The newly connected magnetic flux is carried into the tail as it is merged on the dayside at rate M. The magnetic flux on the dayside Φ_{Day} decreases and the magnetic flux in the tail Φ_{Lobe} increases until tail reconnection begins at rate R. This removes flux from the lobe that is convected to the dayside at rate C (R.L. McPherron 1974, copyright by the American Geophysical Union).

magnetosphere's magnetic field and the rate at which magnetic flux is carried into the magnetotail. At this time the reconnection rate R in the tail increases and flux is returned from the lobes to the closed field line region. This process is known as a substorm; it provides the energy for auroral displays and ionospheric disturbances observed on the surface of the Earth. If the southward interplanetary magnetic field is strong and remains southward for a prolonged period (hours), energetic plasma moves deep into the magnetosphere, circling the Earth in a ring current. When the ring current causes a depression in the surface magnetic field of more than about 50 nT, a geomagnetic storm (q.v.) is said to have occurred.

The classical geomagnetic storm begins with a sudden jump in the Earth's magnetic field followed within hours by a large depression of the magnetic field as the ring current builds up. Then after several days the ring current returns to quiet values. We now know that this period of storminess is initiated when a large bubble of magnetized plasma is ejected from the Sun in a process known as a coronal mass ejection (q.v.). This plasma bubble plows through the solar wind causing a shock wave in front of it. The bubble itself has large magnetic components parallel and anti-parallel to the Earth's field.

When the leading shock wave hits the Earth, the magnetosphere is compressed and the surface field jumps. When the region of anti-parallel magnetic field encounters the Earth, strong reconnection begins and much energy flows into the magnetosphere from the solar wind, powering the geomagnetic storm and building up the ring current.

The arrival of the interplanetary shock can cause disruption of terrestrial power and communication systems. Power distribution systems and communication systems employ long (continent and ocean-wide) electrical conductors. When the shock wave strikes, the rapid rate of change of the magnetic field induces voltages in these conductors which may cause parts of them to fail. Historically, major power blackouts have often been the result of solar disturbances. Thus the solar–terrestrial environment has a very direct effect on our increasingly technological society.

Christopher T. Russell and Janet G. Luhmann

Bibliography

Barton, C.E. (1989) Geomagnetic secular variation, in *The Encyclopedia of Solid Earth Geophysics* (ed. D.E. James). New York: Van Nostrand Reinhold, 560–77.
Chapman, S. and Bartels, J. (1940) *Geomagnetism*. Oxford University Press.
IAGA Division I Working Group 1 (1987) The International Geomagnetic Reference Field revision, 1985. *J. Geomag. Geoelect.*, **39**, 773–9.
Jacobs, J.A. (1975) *The Earth's Core*. London: Academic Press.
McPherron, R.L. (1974) Current status of the growth phase controversy. *EOS*, **55**, 994–5.
Merrill, R.T. and McElhinny, M.W. (1983) *The Earth's Magnetic Field: It's History, Origin and Planetary Perspective*. London; Academic Press.
Russell, C.T. and McPherron, R.L. (1973) The magnetotail and substorms, *Space Sci. Rev.*, **15**, 205–66.
Stacey, R.D. (1977) *Physics of the Earth*. New York: Wiley and Sons.

Acknowledgement

The preparation of this review was supported by the National Sciences Foundation under grant ATM 90–16900.

Cross references

Geomagnetic polarity reversals and the geological record
Geomagnetic storm
Magnetism
Polarity reversals

EARTH OBSERVING SYSTEM

The need for the Earth Observing System

EOS, the Earth Observing System, is a complex program including satellite measurements, data analysis and modeling '. . . designed to provide comprehensive, long-term observations from space of changes that are occurring on the Earth from natural and human causes so that we can have a sound scientific basis for policy decisions to protect our future' (NASA, 1992). It is part of our response to concerns about climate change on Earth.

The idea that the climate of Earth has changed is not new. Much evidence has been collected from the sedimentary record indicating periodic as well as secular changes in climate parameters such as surface temperature, precipitation amount and ice extent and abundance, on many time scales. Human activity is now having an impact on climatic conditions, on a global scale. (Solomon 1990; Houghton, Jenkins and Ephraums, 1990. See Greenhouse effect).

Based upon the rapid advances in remote sensing instrumentation for aircraft, balloons and especially for Earth-orbiting satellites, we are at last capable of measuring globally, over periods of 10 to 20 years, key factors affecting conditions at the surface of Earth, well enough to interpret these observations in terms of changes in the climate of the planet. Given recent advances in computing hardware, numerical modeling and techniques for 'assimilating' environmental measurements into climate models, it may be possible over the next 10 to 20 years to develop mathematical models that can predict future climatic conditions on the 10- to 100-year time scale.

An international effort is being made to accomplish these goals. It is taking place under the aegis of the World Climate Research Program (WCRP) and the International Geosphere–Biosphere Program (IGBP). The United States is participating through its multi-agency Global Change Research Program (GCRP). The National Aeronautics and Space Administration's (NASA's) contribution to the GCRP is the Mission to Planet Earth, of which the largest element is the Earth Observing System (EOS).

Overview of the EOS program

Traditionally, the Earth has been studied by separate scientific communities, each interested in one of five 'systems': the atmosphere, the hydrosphere, the lithosphere, the cryosphere and the biosphere (Earth Systems Sciences Committee, 1988). These discipline-oriented research communities developed measurement and modeling techniques to explore the systems further, and by consensus determined which questions became the focus of subsequent research.

This view of Earth science research broadened in the 1970s and 1980s. The Landsat missions, designed to observe land surface processes, have also provided information about ocean surface structure and atmospheric cloud behavior. NASA's Seasat mission, which focused on oceanography and air–sea interactions, generated data such as near-surface wind velocity over the ocean, which were used in models of both atmospheric and oceanic circulation. These experiences pointed to possible advantages of 'multidisciplinary' instrument packages. Improvements in numerical modeling methods and computer hardware in the same time period made possible the first steps toward coupling atmosphere and ocean climate models, again lowering the barriers between disciplines.

With EOS, NASA gives additional emphasis to interdisciplinary scientific issues, especially those relating to exchanges of energy, momentum and material across the interfaces of the systems, and the consequences of these exchanges. 'From pattern to process' emerged as a slogan that captures the underlying intent of using global-scale data to understand the mechanisms of climate change. New concepts and areas of research have arisen, such as the study of biogeochemical cycles (which involve the budgets of chemicals, and their elemental constituents, as they pass through biological and non-biological planetary-scale systems). Over a decade, these ideas have been refined, a process that continues with the EOS mission itself. The early history of this effort is documented in a series of reports generated by committees of experts from multiple disciplines (e.g. National Research Council, 1979 and references therein, 1986; NASA, 1982; 1983a; Earth Systems Sciences Committee, 1988; for the related effort involving instrumentation, see Earth Observing System, 1995 and references therein).

Seven general areas were selected for their importance to understanding global climate change on Earth. The selection was based upon scientific, technical and programmatic considerations: trade-offs between (1) uncertainties in models and (2) what it is possible to measure, on what time and space scales, and at what cost (NASA, 1992). The intention is to treat each of the general subjects in a way that does not stop at the interfaces of the five traditional systems. The areas are

1. the role of clouds, radiation, water vapor and precipitation in global climate;
2. the biological productivity of the oceans, ocean circulation and air–sea interaction;
3. the sources and sinks of greenhouse gases, and their atmospheric transformations;
4. changes in land use, land cover, primary productivity and the water cycle;
5. the role of the polar ice sheets and sea level;
6. the coupling of ozone chemistry with climate and the biosphere;
7. the role of volcanoes in climate change.

There are three main components of EOS: (1) the EOS observatories, which are spaceborne platforms carrying instruments that will measure key parameters globally, (2) the EOS Scientific Research Program, which supports the interdisciplinary studies involving satellite data, in situ data, and numerical models to understand planetary-scale processes, and (3) the EOS Data and Information System, which must process, validate, document, archive, and distribute the vast amounts of data to be generated by EOS and other missions and related scientific research efforts.

The EOS observatories

The EOS observatories are designed to create a database of 'comprehensive global observations of Earth.' Parameters describing natural forcing of the climate (i.e. solar input and surface albedo), natural responses (e.g. temperature, biomass and cloud cover) and human activity (such as land-use patterns, gas emissions and fluid pollutants) need to be monitored, so that the cycles of energy, material and momentum can be analyzed. The choice of temporal and spatial scales for EOS measurements is a critical issue because of the complexity of the Earth environment and the nonlinear feedback processes involved (e.g. small 'causes' may have large effects). However, measuring every variable of potential interest to Earth scientists for which an instrument could be built, at the highest resolutions of possible interest, would require an unrealistically large budget and would result in unmanageably large amounts of data. Thus the scope of EOS was pared down in several ways, and the mission goals were assessed in the context of the long-term record of Earth observations provided by the satellites of many nations.

The time scale for climate changes of interest to EOS is set at 15 years, with 1- to 16-day temporal resolution for variabilities. Regular sampling of key global parameters is to be obtained at spatial scales of 1 to about 100 km, supplemented by local and regional coverage at tens to hundreds of meters for some quantities. The instruments ultimately included on the EOS observatories were restricted to those that could measure variables believed to address 'critical' global change issues within these guidelines. As part of the selection process, instruments which measure slowly changing variables (e.g. surface chemistry and mineralogy), those which observe parts of the Earth system less likely to impact the biosphere directly (e.g. the magnetosphere), and ones which depend upon the achievement of major technological advances (e.g. Doppler lidar for measuring winds), were eliminated or deferred. The availability of data from non-EOS satellites and from field studies, including the current and planned programs of non-US space agencies, has also been critical in EOS instrument selection. In spite of these limitations, the EOS observatories will provide systematically better space–time and spectral coverage, and a greater coincidence of different types of measurements, than any previous monitoring of Earth. (For detailed descriptions of the instruments selected for EOS see Earth Observing System, 1995.)

From an earlier strategy for deploying two giant platforms carrying about 15 instruments each, NASA has evolved a plan that now involves launching several series of smaller satellites. Each satellite will operate for about 5 years, after which time it will be replaced by a similar satellite. Taken together, the EOS observatories will provide 15-year coverage of the Earth. To optimize our ability to compare observations, certain groups of instruments were selected to fly together on each payload.

The first EOS Satellites: EOS-AM and EOS-PM

The first two series of EOS satellites, EOS-AM and EOS-PM, will carry about 5 instruments each, and will observe the Earth from near-polar orbits at altitudes of roughly 700 km above sea level. EOS-AM 1 and EOS-PM 1 are scheduled for launch in 1998 and 2000 respectively. The orbits are Sun synchronous; the local time on the surface of the Earth below the EOS-AM satellite will always be about 10.30 a.m. on the dayside (and 10.30 p.m. on the nightside). Cloud cover over land is generally lowest in the morning, making this a good orbit from which to make land surface measurements. For EOS-PM, the corresponding times are 1.30 p.m. and 1.30 a.m., to observe daily heating and cooling cycles and provide coverage complementary to EOS-AM

NASA is planning to fly nine instruments on the first EOS-AM and EOS-PM satellites (Table E6). The MODIS and CERES instruments will fly on both AM and PM satellites; they will contribute to the understanding of atmospheric radiation and clouds. CERES will measure broadband reflected and emitted radiation from the top of the atmosphere at about 20 km spatial resolution, at a variety of angles and several times of day. MODIS will produce broad-swath, multispectral images in the visible and near infrared, at 1-km and 250-m spatial resolutions. MISR, on EOS-AM, will add multi-angle

Table E6 The instrument payloads for the first NASA EOS-AM and EOS-PM satellites

MODIS	Moderate-resolution imaging spectroradiometer (EOS-AM and EOS-PM)
CERES	Clouds and Earth's radiant energy system (EOS-AM and EOS-PM)
MISR	Multi-angle imaging spectro-radiometer (EOS-AM)
MOPITT	Measurements of pollution in the troposphere (EOS-AM)
ASTER	Advanced spaceborne thermal emission and reflection radiometer (EOS-AM)
AIRS	Atmospheric infrared sounder (EOS-PM)
AMSU	Advanced microwave sounding unit (EOS-PM)
MHS	Microwave humidity sounder (EOS-PM)
AMSR	Advanced microwave scanning radiometer (EOS-PM)

visible imaging of the Earth's surface and aerosols, at 275-m resolution. Together these instruments should help elucidate the complicated relationships among surface properties, clouds, other aerosols and radiation.

The EOS-PM spacecraft will carry a suite of sensors for sounding the lower atmosphere and measuring its interaction with the surface. AIRS, AMSU and MHS represent major improvements in atmospheric sounding, particularly in the spectral domain. Taken together, these instruments will produce vertical profiles of temperature and humidity. MIMR, provided by the European Space Agency, will contribute global measurements of rain, snow cover, sea ice, soil moisture, and sea surface temperature, even in the presence of clouds. This collection of measurements, along with cloud, snow and radiation fields from MODIS and CERES, should inspire new insights into the Earth's energy budget and hydrological cycle.

A key to understanding the Earth's carbon cycle is measuring how the biomass of land vegetation and oceanic phytoplankton changes. Once again the MODIS instrument takes center stage, with its large number of spectral bands useful for deriving biological information from sunlight scattered by the land surface and upper ocean. The Canadian Space Agency is contributing MOPITT, which will measure vertical profiles of CO abundance and column abundances of methane. With the help of surface properties derived from MODIS, outstanding questions about the sources of these two important tropospheric gasses should be resolved.

Through multispectral visible and infrared imaging at very high spatial resolution (15 m to 90 m), ASTER will measure many of the quantities needed for understanding the transfer of heat and moisture between the land surface and the atmospheric boundary layer, cloud properties at high spatial resolution and the characteristics of volcanoes and their emissions. When combined with MODIS observations, ASTER's regional measurements could be useful for studying the entire globe.

Other EOS instruments

NASA plans to fly other instruments, not included on the first EOS-AM and EOS-PM satellites, as part of the EOS program (Table E7).

Table E7 Instruments not included on the first EOS-AM and EOS-PM satellites, that NASA plans to fly as part of the EOS program

LIS	Lightning imaging sensor
SAGE-3	Stratospheric aerosol and gas experiment
EOSP	Earth observing scanning polarimeter
ACRIM	Advanced cavity radiometer irradiance monitor
HIRDLS	High-resolution dynamics limb sounder
MLS	Microwave limb sounder
SOLSTICE-2	Solar stellar irradiance comparison experiment
TES	Tropospheric emission spectrometer
SeaWinds	Scatterometer
SSALT	Solid state altimeter
GLAS	Geosciences laser altimeter system

Several of these instruments are designed to extend our record of conditions in Earth's stratosphere, while others focus on solar radiation, lightning, aerosols, tropospheric chemistry, air–sea interactions, the polar ice sheets and ocean circulation.

The EOS scientific research program

Twenty-nine Interdisciplinary science (IDS) teams were selected by NASA as part of the EOS program. Their function is to begin coordinated research using multiple existing satellite data sets and models, and to help guide the development of EOS instruments and measurement strategies so that the EOS data can be applied to key interdisciplinary issues. A total of 366 researchers were selected as principal or co-investigators; the median number of members per interdisciplinary science team is 13, with a range from 3 to 25. (For a brief summary of each investigation, see Earth Observing System, 1995.)

The scope, content and structure of the EOS IDS investigations represent the scientific community's interpretation of earlier efforts to identify 'interdisciplinary' issues for global climate change. Most of the teams cover a wide range of expertise, reflecting the requirements of interdisciplinary work. Each selected investigation encompasses the use of both instrument data (remote sensing and *in situ*) and theoretical models; each is committed to exchanging information with a wider scientific community by contributing specific data products to the Earth Sciences Data and Information System.

The IDS investigations can be loosely divided into 'process' studies, which concentrate on details of the interactions among Earth systems in particular regions, and global studies. Of the selected US investigations, about 80% have a global focus. Of the international efforts selected by the NASA program, more than 70% are process studies for which the investigators' locations have particular relevance, such as the effects of deforestation and other changes on the Amazon ecosystem (Brazil), the role of terrestrial vegetation in mid- and high-latitude ecosystems (Canada), and the variability in southern ocean productivity due to the El Niño phenomenon (Australia).

There are two teams refining coupled chemical, dynamical and radiation models with the aid of assimilated global data, one for each of the middle and upper atmospheres; six such efforts deal with the global troposphere, with a variety of emphases. Several include work on improving biosphere parameterizations, and one concentrates on the derivation of atmospheric parameters from multiple instruments. There is a study dedicated to the global carbon cycle, one for the global budget of angular momentum and two on the global budget of water.

Many investigations examine elements of the global cycles in detail. Three studies look at surface–atmosphere exchanges of energy, momentum and water over oceans, and one each examines specifically the southern and polar ocean regions. Most of these efforts include biological productivity as a parameter of study. There is an investigation of exchanges between the land biosphere and the atmosphere, one focused on the water cycle in semi-arid regions, one that will produce estimates of volcanic contributions to the atmosphere, and at least three that estimate runoff (land–hydrosphere exchange), two in mountainous regions. All of these studies involve the collection and analysis of multiple data sets; several include sophisticated models of boundary layer physics and chemistry.

The high priority of the effort to constrain the global hydrologic cycle is also reflected in the IDS investigation dedicated to global cloud physical properties, and several of the global climate efforts place special emphasis on estimating precipitation from observations.

The issue of detecting changes in global climate is addressed in several ways by the IDS program. Two studies have components dealing with long-term climate change (10 000 year timescales), one making use of high-latitude ice core data, and the other looking at the climatic effects of the Andes mountains. Many studies involve searches for climate change during the 15-year period of EOS data collection, including focused efforts by the global carbon cycle and stratosphere groups, several of the global troposphere groups, and at least one group studying mid- and high-latitude oceans.

The principal aim of these efforts is to create better models for predicting the future climate of Earth, to develop what is called 'predictive understanding' (NASA, 1992). As such, emphasis is placed upon using data to improve models. This involves refining the details of already-sophisticated atmosphere and ocean models and making basic improvements in models of marine and terrestrial ecosystems, air–sea interactions and air–land interactions. The third component of EOS, the Earth Sciences Data and Information System, will make these tasks feasible (NASA 1986).

The Earth Sciences Data and Information System (ESDIS)

One implication of the requirement to obtain 'comprehensive global observations of Earth'is that huge volumes of data will be involved. A simple calculation illustrates this point. The surface area of the Earth is about 5×10^8 km^2. Relevant temporal scales range from a fraction of a day upward. One might sample on the order of 500 information channels (i.e. measurements used to infer geophysical parameters) at 16-bit encoding, four times per day (to resolve diurnal time scales). To obtain global coverage at 1-km spatial resolution would require more than 10^{13} bits per day. At this rate the archive would accumulate as much data as is currently stored in all the books in the Library of Congress every few weeks. However, the overall data volume can be reduced significantly by resorting to selective spatial and/or temporal sampling.The sheer volume of data poses serious challenges for the systems that must handle them. Even 'routine' operations that are performed on the data – reformatting of raw spacecraft data, radiometric correction, geographical registration, conversion to geophysically useful quantities, storage on archival media, cataloging, documenting and distributing to users, and possibly gridding to create uniform data products – place unprecedented demands on the storage and throughput capabilities of the data system. The plan is to build upon existing capabilities and experience, creating an ESDIS that will continuously evolve to meet the needs at the time, making maximum possible use of improved technology as it becomes available.

Structure of ESDIS

Several underlying principles govern the structure of ESDIS (NASA,1986):

- Since technology is expected to advance considerably over the 15-year nominal lifetime of EOS, the data system is designed in a modular fashion, making it possible to swap individual elements with minimal disruption of the overall system.
- The ESDIS must serve a geographically widely distributed community; the system is designed so that users can perform as many operations as possible at their home institutions, both for data analysis and for instrument support.
- Since EOS will produce a relentless stream of data for many years, any pile-up at a 'slow step' in the ESDIS could be very difficult to correct; as much of the data processing as possible is supposed to be automatic. It is planned that algorithms to produce the initial data products will be running at central sites, so that processed data can be delivered to users within a few days of acquisition. This raises an issue about operations that are not quite routine, and will require human intervention rather than being completely automated (see validation, below).
- To assure that the data will be handled thoughtfully, an effort is made to set up the main archives at places where there is institutional interest and expertise in studying the types of data to be stored there.
- Since data will be archived in many places, by communities familiar with different formats and content, ESDIS is designed to foster 'interoperability.' This means that a user will be able to access data from multiple sources from a single interface.

The ESDIS will have many components (Earth Observing System, 1995), only some of which are described here. The EOS Operations Center (EOC), located at the NASA Goddard Space Flight Center in Greenbelt, Maryland, is the site of mission control for NASA's EOS spacecraft and their instruments. The EOS Data and Operations System (EDOS) will gather data from the EOS observations, separate it from the telemetry stream, and deliver it via the NASA Communications network to the EOC and the Distributed Active Archive Centers (DAACs). The DAACs perform a major part of the initial data processing, archiving, and distribution to the user communities. About 100 data products containing geophysical parameters at the resolution of the instruments (called level 2 data) are to be generated routinely, starting at the beginning of the mission, as well as about 100 gridded and other, more highly processed data sets. Over the years prior to launch, large existing data sets from earlier NASA and National Oceanic and Atmospheric Administration (NOAA) missions are being archived and cataloged in these DAACs. Eight existing data repositories are being adapted to the

Table E8 The eight data repositories that are being adapted to the role of DAAC

1	Upper atmosphere, atmospheric dynamics, global biosphere and geophysical	NASA Goddard Space Flight Center, Greenbelt, Maryland.
2	Land processes imagery	US Geological Survey's EROS Data Center, Sioux Falls, South Dakota.
3	Ocean circulation, air–sea interactions	Jet Propulsion Laboratory, Pasadena, California.
4	Radiation budget, clouds and aerosols and tropospheric chemistry	NASA Langley Research Center, Hampton, Virginia.
5	Hydrology	NASA Marshall Space Flight Center, Huntsville, Alabama.
6	Cryosphere	University of Colorado, Boulder, Colorado.
7	Sea ice, polar processes imagery	University of Alaska, Fairbanks, Alaska.
8	Biogeochemical dynamics	Oak Ridge National Laboratory, Oak Ridge, Tennessee.

role of DAAC, each covering a unique area of scientific interest (Table E8). In addition to the data products that will be generated routinely, some data will be acquired only upon request, or in response to special events such as the eruption of a volcano. It is the policy of EOS that data will be available to user communities as soon as they are generated, at no more than the incremental cost of filling a user request.

The issue of validation

One of the great challenges of EOS is to take data that is traditionally used by one small community of researchers and make it scientifically meaningful to other communities of researchers who work in different disciplines. How will the meaning of the measured parameters – the precise definitions, the caveats and the various types of uncertainties – be characterized and communicated in a way that is useful to a broad swath of the scientific community? For example, different remote sensing measurements of sea surface temperature (SST) sample different horizontal and vertical scales, depending on technique and in some cases on atmospheric conditions. *In situ* measurements of 'the same' parameter have very different sampling characteristics from any of the remotely sensed determinations. To the extent that such differences affect scientific conclusions, any research which uses sea surface temperature derived by one method must take into account the detailed characteristics of the data set used. The need for an accurate description of the scientific meaning of the data is particularly important for climate change studies, which are concerned with long-term records of environmental conditions, and which usually require the simultaneous analysis of more than one measured quantity. 'Validation' is the process by which these descriptions are generated. Validation involves(1) identifying the assumptions made in deriving an environmental parameter from measured radiances, (2) testing the input data and derived parameter for statistical error, sensitivity and internal consistency, over the range of conditions under which the parameter is retrieved, and (3) comparing with similar parameters obtained from other sources using other techniques. A general methodology for performing these steps has yet to be developed. Because the handling and use of scientific data for EOS is both quantitatively and qualitatively different from previous experience, NASA initiated the Pathfinder program in 1991. The goal of this effort is to prepare several existing large, geophysical data sets for use by an interdisciplinary community of scientists. If this effort succeeds, it will presage the transformation that the larger Earth sciences community must undergo to meet the challenges of understanding and predicting global change.

Ralph Kahn and Daniel Wenkert

Bibliography

Earth Observing System. (1995). *1995 MTPE/EOS Reference Handbook*. Greenbelt, Maryland: NASA Goddard Space Flight Center.

Earth Systems Sciences Committee (1988) *Earth System Science. A Closer View*. Washington, DC: NASA Advisory Council.
Houghton, J.T., Jenkins, G.J. and Ephraums, J.J. (eds) (1990) *Climate Change: The IPCC Assessment*. Cambridge: Cambridge University Press.
Moore, B. and Dozier, J. (eds) (1992) Adapting the Earth Observing System to the projected $8 billion budget: recommendations from the EOS investigators. *The Earth Observer*, September/October.
National Aeronautics and Space Administration (1982) *Global Change: Impacts on Habitability – A Scientific Basis for Assessment*. Washington, D.C.: NASA, JPL D-95.
National Aeronautics and Space Administration (1983a) *Global Change: A Biogeochemical Perspective* Washington, DC: NASA, JPL 83–51.
National Aeronautics and Space Administration (1983b) *Land-Related Global Habitability Science Issues*. Washington, DC: NASA, TM-85841;
National Aeronautics and Space Administration (1986) *Report of the EOS Data Panel* Washington, DC: NASA, TM-87777.
National Aeronautics and Space Administration (9 March, 1992) *Report to Congress on the restructuring of the Earth Observing System*. Washington, DC: NASA.
National Research Council (1979) *Toward a US Climate Program Plan. National Academy of Sciences*. Washington, DC: National Academy of Sciences.
National Research Council (1986) *Global Change in the Geosphere–Biosphere*. Washington, DC: National Academy of Sciences.
Solomon, S. (1990) Progress towards a quantitative understanding of Antarctic ozone depletion. *Nature* **347**, 347–54.

Acknowledgements

We thank M. Schier, J. Shirley and an anonymous reviewer for thoughtful comments on an early version of this article. The work of R. Kahn and D. Wenkert was performed at the Jet Propulsion Laboratory/California Institute of Technology under contract to the National Aeronautics and Space Administration.

Cross references

Biosphere
Earth: atmosphere
Greenhouse effect
Remote sensing

EARTH ROTATION

Kinematics

The motion of the whole Earth in space consists of a translation of the center of mass plus a rotation around the center of mass. To an extremely good approximation to one part in 10^8, the Earth rotates in space every 23 h 56 m 4.091 s at a uniform angular velocity. One second was originally defined as 1/86 400 of the mean solar day. In 1964 the second was redefined in terms of the atomic time based on the transition of two hyperfine energy levels of the cesium 133 atom.

It is the minute, rather complex deviation from the uniform rotation that makes the Earth's rotation scientifically interesting and of practical concern in space navigation and military endeavors. This temporal variation, by definition a three-dimensional vectorial quantity, can be conveniently separated into two components: (1) the one-dimensional variation in the rotational speed; and (2) the two-dimensional variation in the rotational axis orientation. Variation (1) is often expressed in terms of the Universal Time (UT) or its time derivative, the length of day (LOD). Variation (2) with respect to inertial space is generically called the nutation (including precession, see below), while that seen by an observer on the surface of the Earth is referred to as the polar motion.

Measurement

The orientation, as a function of time, of the terrestrial reference frame relative to the celestial reference frame determines the three-dimensional Earth rotation parameters (ERP), from which UT and nutation/polar motion can be derived. A convenient unit for measuring the slight angular deviations of Earth orientation from a uniform rotation is the milliarcsecond (mas). One mas equals 4.85×10^{-9} radian, and is equivalent to 1/15 millisecond (ms) of UT, and about 3 cm distance on the Earth's surface. For a long time prior to the space age, conventional astrometry using optical instruments (e.g. the astrolabe, optical and photographic zenith telescopes) was the only means of measuring ERP. Since the 1970s the advent of several space geodetic techniques has quickly revolutionized the observation. These include satellite Doppler tracking, lunar laser ranging, satellite laser ranging (SLR), very long baseline interferometry (VLBI) and the Global Positioning System (GPS). SLR and VLBI have been utilized during the last two decades, whereas in recent years the use of GPS, with its ease and low cost, has proved a great addition in contributing to the measurement of ERP. These measurements are now made under the auspices of the International Earth Rotation Service headquartered in Paris.

In the technique of SLR, a geodetic satellite equipped with optical retroreflectors bounces back laser pulses shot from ground laser stations. The travel times of the laser pulses are then converted into ranges. Knowing the satellite orbit to a great accuracy (which requires sophisticated modeling of the Earth's gravitational field and other dynamic effects) and with multistation ranging, these range measurements can yield the determination of ERP, among other geodynamic parameters. In VLBI two radio antennas separated by an intercontinental distance receive simultaneously radio waves from distant quasars, and the difference in the arrival times of the waves is measured. These time differences, from multiple antenna pairs, also contain information to derive the ERP. The GPS system has found wide commercial, military, and scientific applications by providing instantaneous geographical location for ground users. In precise geodetic applications, it also yields ERP through locating fiducial stations relative to the GPS satellite constellation whose orbits are precisely determined. All these techniques have now achieved internal precision of well within 0.1 mas for daily ERP determinations; and between them they agree at the accuracy level of within 1 mas.

Dynamics

The Earth's rotational variations are of two distinct dynamic types. Astronomical variations refer to those caused by the direct tidal torque exerted on the Earth by the Moon and the Sun. The torque τ alters the Earth's angular momentum, H, governed by

$$d/dt\, H = \tau$$

The Earth's rotation ω varies accordingly because $H = I\,\omega$, where I is the inertia tensor of the Earth.

The best-known astronomical variation is probably the astronomical precession and nutation of the Earth's rotational axis in space. The precession was noticed by ancient civilizations as the slow but large 'secular' drift of the celestial constellations relative to the calendar year. It owes its origin to the Earth's permanent equatorial bulge (itself due to rotation) and obliquity (the 23.45° difference between the Earth's equatorial plane and the ecliptic plane), which result in a luni-solar tidal torque exerted on the Earth but perpendicular to the Earth's angular momentum. The latter is the classical condition giving rise to precessional motion, such is the case for a spinning top – hence the familiar statement that 'the Earth spins like a top'. The rotation axis traces out in space a complete cone of radius equal to the obliquity every 25 800 years. The Moon contributes more than twice as much as the Sun because of its proximity, which compensates for its smaller mass. Nutations are in a sense shorter-period perturbations in the precession. They are the result of periodic variations in the orbital elements, and hence in the torque. The largest nutation term has a period of 18.6 years due to the precession of the Moon's orbit. Other relatively prominent nutation terms have periods that are half of the main tidal periods, in particular semi-annual and fortnightly. Precession/nutation involves no mechanical work and does not affect LOD. Traditional celestial mechanics has provided precise precession/nutation predictions for a rigid Earth. Modern geodetic advances have made it necessary to consider small modifications with more refined and realistic Earth models. For example, the elasticity of the Earth has been incorporated into standard models, and a free nutation of the fluid core has been demonstrated to influence certain near-resonance tidal amplitudes.

Another important astronomical variation is the tidal braking. The rotation, coupled with the Earth's non-instantaneous response to the tidal forcing (due to inelasticity, e.g. fluidity of the ocean and the core, viscosity of the mantle), 'carries' the tidal bulge away from the Earth–Moon (or Earth–Sun) line by a small angle. This manifests itself as a tidal phase lag to an observer on Earth. Rotational kinetic energy is dissipated via friction in the process, and the Earth's rotation slows down as a result. Definitive evidence for this tidal braking has been disclosed through analysis of historical accounts (as well as more modern records) of astronomical events such as solar eclipses and lunar occultations. The rate of the secular lengthening of LOD thus obtained is approximately 2 ms per century. This rate is consistent with, and hence indirectly confirmed by, the lunar laser ranging measurement of the recession of the Moon: under the conservation principle, the angular momentum lost by the Earth is gained by the Moon, causing the Moon to recede from the Earth at about 3.7 cm per year. On a much longer timescale, the rate of tidal braking can be estimated by counting the number of daily ridges in a yearly growth ring on certain types of marine organisms, such as bivalves, corals and stromatolites, which lived in the geological past. For example, it has been established that 1 year some 400 million years ago consisted of about 400 days, meaning that a day at that time was about 10% shorter than today.

Other planets in the solar system also slightly perturb, in a quasi-regular way, the Earth's rotational and orbital parameters, particularly the obliquity, orbital eccentricity and precession of perihelion, typically with periodicities from tens to hundreds of thousands of years. Their impact on the insolation pattern on the Earth and hence global climate are called 'Milankovitch cycles', first proposed by M. Milankovitch in the 1920s. The periodicities have in recent years been identified in deep sea sedimentary cores, polar and glacial ice cores, loess magnetism and other kinds of geological records. They are believed to be the 'pacemaker' of ice ages.

Geophysical variations, on the other hand, arise from geophysical processes endogenic to the Earth. The net angular momentum (of the Earth as a whole) is conserved in the process, but the rotation of the solid Earth (which one measures) varies as a result. The governing equation is

$$H = I\omega + h = \text{constant}$$

where h is the 'relative' angular momentum of moving masses with respect to the solid Earth. Thus changes in I (due to mass redistribution, such as air mass or snow loading) or in h (due to mass movement such as wind, ocean currents) give rise to changes of ω in both LOD and polar motion. It is fair to say that everything that moves on or in the Earth will cause the Earth's rotation to change. Loosely speaking, LOD is more sensitive to mass movement whereas mass redistribution is the more efficient means to excite the polar motion.

Figure E15 shows a time series of daily averaged ΔLOD solutions derived by combining all independent space geodetic measurements. The signal content is rich. The conspicuous 'spiky' oscillations at monthly and fortnightly periods are clearly of tidal origin. The cause is not tidal torque as in the astronomical variations but rather the tidal deformations in the solid Earth and oceans, and the accompanying tidal currents in the oceans. Short-period (i.e. diurnal and semidiurnal) tides are absent in Figure E15 owing to the daily sampling and averaging in those data. Recent studies with high time-resolution data have demonstrated the dominant role of these tides in causing shorter period variations in both UT and polar motion.

Other than the tidal influences, the bulk of the ΔLOD on a timescale shorter than a few years is caused by the variation in the atmospheric angular momentum (AAM). Under the conservation of total angular momentum, the AAM increases at the expense of the solid Earth and vice versa. By comparing observed ΔLOD with the computed AAM using global meteorological data, this relationship has been established beyond any doubt. The El Nino/Southern Oscillation phenomenon in the ocean–atmosphere system of the tropical Pacific–Indian Ocean region, and the quasi-biennial oscillation in the tropical stratospheric wind field have been identified as the main causes of interannual ΔLOD. The oceanic angular momentum may play a progressively more significant role on longer timescales. On even longer timescale, the large decadal ΔLOD is believed to be caused primarily by activities in the fluid core (which is responsible for the generation of the Earth's magnetic field), again mechanically coupled to the mantle subject to the conservation of angular momentum. In fact, the decadal ΔLOD is the reason for leap seconds that are needed to keep our standard time as close to UT as possible. The

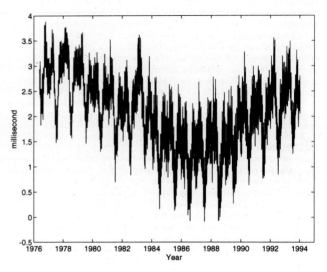

Figure E15 The excess length-of-day series at 1-day interval in units of milliseconds for the period 1976–1994 (Courtesy of NASA).

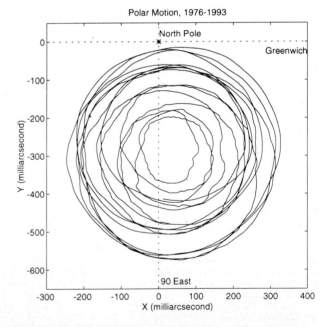

Figure E16 The locus of the polar motion during 1976–1994. The unit is milliarcsecond, the motion is counterclockwise. The origin indicates the conventional North Pole. The x axis is along the Greenwich Meridian, the y axis along the 90°E longitude. (Courtesy of NASA).

secular lengthening of LOD due to the tidal braking mentioned above is so small that it is completely swamped in short, modern records such as Figure E15.

The polar motion is a more complex phenomenon, in which the Earth acts as a linear harmonic resonant system with a (two-dimensional) free oscillation mode referred to as the Chandler wobble. The observed polar motion is the response of such a system to geophysical 'excitations'. To an observer on Earth, the Chandler wobble is a circling motion of the rotation axis around the North Pole (see below) in the counterclockwise (or 'prograde') direction at a period of about 14 months. Figure E16 shows the pole path in the terrestrial coordinates derived from the same data sources and spanning the same period as Figure E15. Several things are worth noting.

The polar motion is not around the North Pole, but rather around some 'mean' pole position about 10 m away in the general direction of 80° W. In a sense it can be said that today's 'North Pole' is no longer where it was originally defined as the mean pole during the years 1900–1905. This phenomenon is called the polar drift, believed to be the response of the Earth's rotation axis to the mass void left behind by the rapid disappearance of the ice sheets that covered much of the North America and Northern Europe during the last ice age ending some 8000 years ago. The polar drift should continue until the postglacial rebound of the mantle material gradually 'refills' the void. Water mass redistribution due to present-day melting of mountain glaciers or possibly polar ice sheets, sea-level rise and even water impoundment in artificial reservoirs may have also contributed to the polar drift. On a much larger spatial scale and longer temporal scale, geological evidence discloses an apparent polar wander with respect to the drifting continents and spreading sea floor associated with the global tectonic movement of lithospheric plates.

The smooth, prograde, circling motion of the pole (around the mean pole) is actually the combination of two wobbles: the Chandler wobble and an annual wobble. It is the beating between them that gives rise to the in-and-out spiraling evident in Figure E16. The annual wobble has long been recognized as being primarily caused by the seasonal redistribution of atmospheric mass. Seasonal redistribution of hydrospheric mass (via oceanic circulations and seasonal changes in continental storage of water and snow) presumably also contributes significantly although present data sets are incomplete. The prominent presence of the annual wobble in the polar motion is due to its nearness in period with the Chandler resonance.

A free oscillation but inevitably dissipative, the Chandler wobble is continually being excited; the excitations manifest themselves as the 'jumps' and 'bumps' in the pole path in Figure E16. The Chandler wobble has frustrated geophysicists for a century now by not revealing its excitation source(s). Short-period as well as interannual fluctuations in the ocean-atmosphere mass distribution have long been suspected to be an important culprit; and indeed significant evidences have been, and will continue to be found as global meteorological observations improve. Another culprit is earthquakes: mass redistribution associated with earthquake faulting, when integrated globally, is found to be potentially important in delivering 'kicks' to the Chandler wobble. Unfortunately these kicks due to earthquakes that have occurred during the past three decades, when accurate modern pole position data were available, were far too weak to explain the observed excitations. Nevertheless, truly great earthquakes (comparable to the 1960 Chile event or the 1964 Alaska event) will certainly leave unmistakable signatures in modern polar motion record.

Summary

Rapid advances in space geodetic techniques have made it possible to monitor the minute variations in Earth's rotation to a great accuracy. These variations contain a wealth of information on all timescales. Their investigation, together with other geophysical observations, can provide a great deal of knowledge for the understanding of our planet as a whole, the geophysical processes that occur on the surface and internally, and the dynamic coupling of the solid Earth to other parts of the Earth.

Benjamin Fong Chao

Bibliography

Hide, R (ed.) (1984) *Rotation in the Solar System.* Proc. Royal Soc. London: Royal Society.
Hide, R. and Dickey, J.O. (1991) Earth's variable rotation. *Science*, **253**, 629–37.
Lambeck, K. (1980) *The Earth's Variable Rotation.* New York: Cambridge University Press.
Munk, W.H. and MacDonald G.J.F. (1960) *The Rotation of the Earth.* New York: Cambridge University Press.
Wahr, J.M. (1988) The Earth's rotation. *Ann. Rev. Earth Planet. Sci*, **16**, 231–49.

Cross references

Chandler wobble
Global positioning system
Length of day
Milankovitch, Milutin, and Milankovitch theory
Obliquity
Plate tectonics
Precession and nutation
Tidal friction

EARTH: ROTATIONAL HISTORY

The problem of integrating the present lunar orbit with the history of the Earth's deceleration and the Moon's recession under the influence of tidal friction, (q.v.), whereby angular momentum is transferred from the Earth's rotation to the Moon's orbital motion, was first addressed by George Darwin in the 19th century. The consequences of backward extrapolation of the present rate of lunar recession are well known; the Moon would have been within the Roche limit, where tidal forces would disrupt the Moon, at ~ 1500 Ma. No geologic evidence exists for such a catastrophe, however, and stromatolite (algal) growth patterns indicate the presence of lunar tides at ~ 3000 Ma (Pannella, 1976).

Paleontological 'clocks'

Wells (1963) provided the first direct estimate of the number of days per year in the geologic past from his study of postulated annual and daily growth increments in Paleozoic corals. Wells' work raised the hope that at least the latter history of the Earth's rotation and the Moon's orbit could be accurately traced from the study of skeletal growth increments in fossils. In general, such studies of coral and mollusc fossils have indicated an increasing number of lunar (synodic) months per year (Figure E17), days per year (Figure E18) and days per lunar month going back to the early Paleozoic (Rosenberg and Runcorn, 1975; Brosche and Sündermann, 1978, 1990; Lambeck, 1980). These findings are consistent with the hypothesis of lunar tidal friction.

The great expectations held for paleontological 'clocks' have not, however, been entirely fulfilled. Reviewing the topic, Scrutton (1978, p. 182) concluded that none of the published data is wholly satisfactory. The common assumption that the data reflect the solar day and lunar month may not be justified; even data from bivalves, the most promising source, 'are likely to be inaccurate because of earlier misunderstanding of the interplay between lunar and solar day stimuli and the effects of tide type.' Scrutton stated that all data to that time 'should be treated as approximations rather than as precise quantities for mathematical analysis.' Furthermore, the reliability of Precambrian data obtained from growth patterns of stromatolites is unclear (Hofmann, 1973; Pannella, 1976). Consequently, work on paleontological 'clocks' reached an impasse, with little progress being made since the 1970s.

Tidal rhythmites

The study of vertically accreted, rhythmically laminated tidal deposits (tidal rhythmites) has rejuvenated historical analysis of the Earth – Moon system and extended such analysis into the Precambrian. Tidal rhythmites are now known from sedimentary rocks up to ~ 650 Ma and from modern deposits (Williams, 1991). The periods recorded by tidal rhythmites usually can be ascribed to tidal pattern and type, thus avoiding some of the uncertainties associated with paleontological data. Furthermore, tidal rhythmite sequences may span many years – the longest continuous sequence so far measured is 60 years (Williams 1989a,b, 1990, 1991) – thus providing more reliable paleotidal data and revealing long-term periods unobtainable from fossils.

Tidal rhythmites in the late Proterozoic (~ 650 Ma) Elatina Formation and coeval Reynella Siltstone, South Australia, provide an unsurpassed paleotidal record. These rhythmites comprise graded (upward-fining) laminae < 0.2 mm to 2 cm thick of very fine-grained sandstone, siltstone and mudstone. The laminae typically are grouped in 'lamina cycles' (Figure E19) that range from ~ 1 mm to $\geqslant \sim 6$ cm in thickness and contain 8–26 or more laminae. Lamina thickness is maximal near the center of lamina cycles and minimal

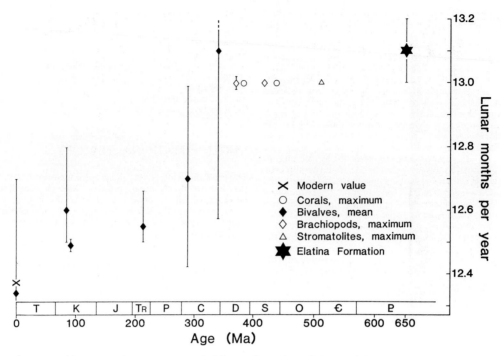

Figure E17 Plot of presumed lunar months per year compiled from paleontological data, and the value indicated by tidal rhythmites of the late Proterozoic Elatina Formation, South Australia. Abbreviations: P, Proterozoic; €, Cambrian; O, Ordovician; S, Silurian; D, Devonian; C, Carboniferous; P, Permian; Tr, Triassic; J, Jurassic; K, Cretaceous; T, Tertiary. (Modified from Williams, 1989b.)

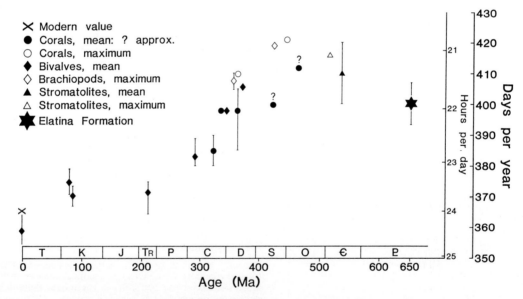

Figure E18 Plot of presumed solar days per year compiled from paleontological data, and the value indicated by tidal rhythmites of the late Proterozoic Elatina Formation and Reynella Siltstone, South Australia. Abbreviations as for Figure E17. (Modified from Williams 1989b.)

near cycle boundaries where thin, muddy laminae may crowd together to form conspicuous, darker bands.

The laminae represent diurnal increments, some of which comprise semidiurnal sublaminae (Figure E19b), and a lamina cycle is a grouping of such increments that records the lunar fortnightly cycle. Thicker, sandy laminae near the center of lamina cycles reflect the spring phase of the tidal cycle when strong tidal currents prevailed, and the muddy bands bounding cycles are mud drapes deposited in quieter waters at neap tides. Periodic changes in lamina thickness reflect variations in the amount of fine clastic material entrained and deposited by tidal currents in estuaries and on tidal deltas in response to periodic changes in the height, velocity and range of paleotides.

Longer tidal cycles are recorded by systematic vertical change in lamina cycle thickness.

Late Proterozoic tidal and rotational values

A continuous, 60-year-long sequence of lamina-cycle thickness measurements obtained from drill cores of the Elatina rhythmites contains strong periods that are evident visually (Figure E20) and are revealed by fast Fourier spectral analysis (Figure E21). These data, and measurements from the Reynella rhythmites, provide paleotidal and paleorotational values of unequalled accuracy (Table E9).

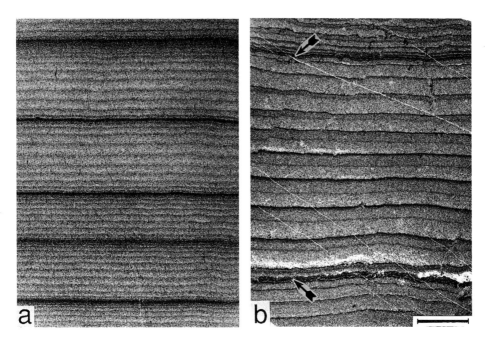

Figure E19 Late Proterozoic tidal rhythmites, South Australia. Muddy material appears darker than sandy to silty layers. Scale bar 1 cm. (a) Elatina Formation. Four fortnightly lamina cycles each comprising about 10 to 14 diurnal (lunar day) laminae are bounded by thin muddy bands deposited at neap tide. (b) Reynella Siltstone, showing one thick, fortnightly lamina cycle that contains 14 diurnal (lunar day) laminae of fine-grained sandstone, each with a muddy top. Most diurnal laminae comprise two semidiurnal sublaminae. Arrows indicate thinner, muddy laminae deposited at neap tide. (From Williams, 1993.)

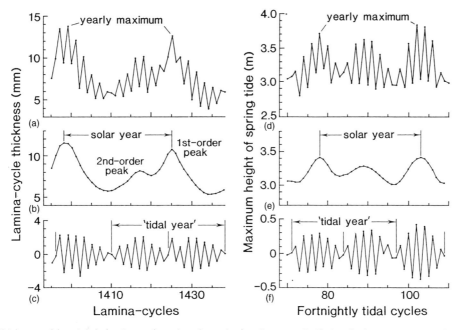

Figure E20 (a–c) Thickness of fortnightly lamina cycles taken from the late Proterozoic Elatina rhythmite sequence, South Australia (lamina cycle number increases up-sequence). (a) Unsmoothed curve, showing yearly maxima in thickness. (b) Smoothed curve, showing first-order peaks that define the solar year. (c) Residual curve (a minus b), showing the 'tidal year'; the vertical lines mark 180° phase reversals in the sawtooth pattern. (d–f) Tidal patterns for Townsville, Queensland. (d) Maximum height of the fortnightly tidal cycle from 19 October 1968 to 3 June 1970. (e) Smoothed curve, showing the solar year. (f) Residual curve (d minus e), showing the 'tidal year'; the vertical lines mark 180° phase reversals in the sawtooth pattern. Modern tidal data for Figures E20 and E21 supplied by the National Tidal Facility, Flinders University. (From Williams, 1990.)

Tidal patterns for the Elatina rhythmites and for Townsville, Queensland, are compared in Figure E20. A yearly signal largely reflecting annual variation of sea level is evident in each data set. Alternation of high- and low-amplitude fortnightly cycles (Figure E20d,f) gives a 'sawtooth' pattern (monthly inequality) resulting from the eccentricity of the lunar orbit; the same pattern is shown by the alternation of thick and thin lamina cycles (Figure E20a,c). The mean period for a 360° change of phase of the sawtooth pattern ('tidal

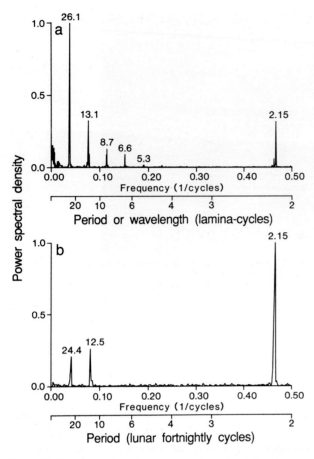

Figure E21 Fast Fourier transform smoothed spectra, with power spectral densities normalized to unity for the strongest peak in each spectrum and with linear frequency scales. (a) Spectrum for the Elatina continuous sequence of 1580 fortnightly lamina cycle thickness measurements. The strong peak at 26.1 lamina cycles represents an annual signal, with additional harmonics at 13.1 (semi annual), 8.7, 6.6 and 5.3 lamina cycles. The peak near 2 lamina cycles (the Nyquist frequency) reflects the monthly inequality of alternate thick and thin fortnightly lamina cycles. (b) Spectrum for the maximum heights of 495 spring tides between 1 January 1966 and 31 December 1985 for Townsville, Queensland. The periods of 24.4 and 12.5 fortnightly cycles represent annual and semi annual signals. The peak near 2 fortnightly cycles (the Nyquist frequency) reflects the monthly inequality of alternate high and low spring tides. (From Williams, 1993.)

year') is longer than the solar year because of the prograde rotation of the lunar perigee. The mean durations for the solar and tidal years at ~ 650 Ma (13.1 and 14.6 lunar months respectively) indicate a lunar apsides cycle of 9.7 ± 0.1 years. The period of the paleolunar nodal cycle also is revealed as an amplitude modulation of the semiannual cycle (represented by the second-order peak in Figure E20b).

Late Proterozoic values for lunar months/year and solar days/year are plotted in Figures E17 and E18 respectively, with values obtained from fossil data. The value of 13.1 ± 0.1 lunar months per year at ~ 650 Ma accords with the small decrease in the number of presumed lunar months per year since the early Paleozoic. The value of 400 ± 7 solar days per year at ~ 650 Ma conflicts, however, with some paleontological data for the early to middle Paleozoic (Cambrian–Devonian). As noted below, the internal consistency of the late Proterozoic paleotidal data strongly supports the validity of the values in Table E9, whereas some Paleozoic values for days per year are suspect (Williams, 1989b).

The values for solar days per sidereal month, sidereal months per year and sidereal days per year at ~ 650 Ma (Table E9) strongly suggest that the Earth's moment of inertia and, by inference, the

Table E9 Late Proterozoic (~ 650 Ma) and modern tidal and rotational values

Parameter	Late Proterozoic[a]	Modern
Lunar days/lunar month	29.5 ± 0.5	28.53
Solar days/lunar month	30.5 ± 0.5	29.53
Solar days/sidereal month[b]	28.4 ± 0.5	27.32
Lunar months/year	13.1 ± 0.1	12.37
Sidereal months/year[b]	14.1 ± 0.1	13.37
Lunar apsides cycle (years)	9.7 ± 0.1	8.85
Lunar nodal cycle (years)	19.5 ± 0.5	18.61
Solar days/year	400 ± 7	365.24
Sidereal days/year[b]	401 ± 7	366.24
Length of solar day (hours)	21.9 ± 0.4	24.00
Earth–Moon distance (R_E)	58.28 ± 0.30	60.27

[a] Values indicated by tidal rhythmites of the late Proterozoic Elatina Formation and Reynella Siltstone, South Australia (Williams, 1989a, b, 1990, 1991). The rhythmite data also display strong semiannual and annual periods (Figures E20 and E21).
[b] Units in invariable time required for dynamical calculations.

Table E10 Mean Earth–Moon distance at ~ 650 Ma and mean rate of lunar retreat since ~ 650 Ma indicated by tidal rhythmites of the Elatina Formation and Reynella Siltstone (from Williams, 1990)

Tidal/rotational value	Relative mean Earth–Moon distance (a/a_0)	Mean Earth–Moon distance (R_E)	Mean rate of lunar retreat (cm/year)[a]
19.5 ± 0.5 years (lunar nodal period)	0.969 ± 0.017	58.40 ± 1.02	1.83 ± 1.00
13.1 ± 0.1 lunar months/year	0.967 ± 0.005	58.28 ± 0.30	1.95 ± 0.29
400 ± 7 solar days/year	0.968 ± 0.007[b]	58.34 ± 0.42	1.89 ± 0.41

[a] Calculated mean rates of lunar retreat since late Proterozoic time are based on an estimated age of 650 Ma for the Elatina Formation and Reynella Siltstone. Because the widely accepted age of the Precambrian–Cambrian boundary has recently been reduced from 570 Ma to about 540 Ma, a better estimated age of these formations may be 620 Ma. Using this latter age gives mean rates of lunar retreat of 1.92–2.04 cm/year since late Proterozoic time.
[b] Makes allowance for the solar tides' contribution to the loss of angular momentum of the Earth's rotation (Deubner, 1990).

Earth's radius at ~ 650 Ma were similar to those of today (Williams, 1990).

History of the lunar orbit

Three independent determinations of mean Earth–Moon distance (semimajor axis of the lunar orbit) at ~ 650 Ma employing different tidal and rotational values show excellent agreement (Table E10), demonstrating the internal consistency and accuracy of the late Proterozoic values given in Table E9. The mean Earth–Moon distance of 58.28 ± 0.30 Earth radii (R_E) at ~ 650 Ma, indicated by the number of lunar months per year (the best constrained value in Table E10), gives a mean rate of lunar retreat of 1.95 ± 0.29 cm per year since ~ 650 Ma. This mean rate is only about half the present rate of lunar retreat of 3.7 ± 0.2 cm per year obtained by lunar laser ranging (Dickey, Williams and Newhall, 1990). The indicated increasing rate of lunar retreat since ~ 650 Ma accords with increasing oceanic tidal dissipation as the Earth's rotation slows (Webb, 1982). The present relatively high rate of lunar retreat may reflect a near-resonance of the oceans.

A mean rate of lunar retreat of 1.95 cm per year, if extrapolated backward nonlinearly, gives a close approach of the Moon at ~ 3000 Ma; as no geologic evidence exists for such an event, an even lower rate of lunar retreat likely prevailed prior to ~ 650 Ma (Williams, 1990). Early Proterozoic paleotidal values of high quality are required to illuminate the early history of the Earth's rotation and the lunar orbit.

George E. Williams

Bibliography

Brosche, P., and Sündermann, J. (ed.) (1978) *Tidal Friction and the Earth's Rotation.* Berlin: Springer-Verlag, 243 pp.
Brosche, P., and Sündermann, J. (ed.) (1990) *Earth's Rotation from Eons to Days.* Berlin: Springer-Verlag, 255 pp.
Deubner, F.L. (1990) Discussion on late Precambrian tidal rhythmites in South Australia and the history of the Earth's rotation. *J. Geol. Soc. London,* **147**, 1083–1084.
Dickey, J.O., Williams, J.G. and Newhall, X.X. (1990) The impact of lunar laser ranging on geodynamics. *Eos,* **71**, 475.
Hofmann, H.J. (1973) Stromatolites: characteristics and utility. *Earth-Sci. Rev.* **9**: 339–73.
Lambeck, K. (1980) *The Earth's Variable Rotation: Geophysical Causes and Consequences.* Cambridge: Cambridge University Press, 449 pp.
Pannella, G. (1976) Geophysical inferences from stromatolite lamination, in *Stromatolites* (ed. M.R. Walter). Amsterdam: Elsevier, Developments in Sedimentology Vol. 20, pp. 673–85.
Rosenberg, G.D. and Runcorn, S.K. (ed.) (1975) *Growth Rhythms and the History of the Earth's Rotation.* London: Wiley, 559 pp.
Scrutton, C.T. (1978) Periodic growth features in fossil organisms and the length of the day and month, in *Tidal Friction and the Earth's Rotation* (ed. P. Brosche and J. Sündermann). Berlin: Springer-Verlag, pp. 154–96.
Webb, D.J. (1982) On the reduction in tidal dissipation produced by increases in the Earth's rotation rate and its effect on the long-term history of the Moon's orbit, in *Tidal Friction and the Earth's Rotation II* (ed. P. Brosche and J. Sündermann). Berlin: Springer-Verlag, pp. 210–221.
Wells, J.W. (1963) Coral growth and geochronometry. *Nature* **197**, 948–50.
Williams, G.E. (1989a) Late Precambrian tidal rhythmites in South Australia and the history of the Earth's rotation. *J. Geol. Soc. London,* **146**. 97–111.
Williams, G.E. (1989b) Tidal rhythmites: geochronometers for the ancient Earth–Moon system. *Episodes* **12**, 162–71.
Williams, G.E. (1990) Tidal rhythmites: key to the history of the Earth's rotation and the lunar orbit. *J. Phys. Earth* **38**, 475–91.
Williams, G.E. (1991) Upper Proterozoic tidal rhythmites, South Australia: sedimentary features, deposition, and implications for the Earth's paleorotation. *Can. Soc. Petrol. Geol. Mem.* **16**: 161–78.
Williams, G.E. (1993) History of the Earth's obliquity. *Earth Sci. Rev.,* **34**, 1–45.

Cross references

Obliquity: terrestrial record
Tidal friction

EARTH–MOON SYSTEM: DYNAMICS

Geometry

In a first approximation the motion of the Moon around the Earth can be described by Kepler's laws of planetary motion, the Moon moving in an elliptic orbit having the Earth at one focus. But the departures from this description, called perturbations, are much greater for the Moon than for the planets. The principal departures have been known for a considerable time (although their description in comparison with Keplerian motion is historically recent).

The orbit of a planet is described by a set of geometrical elements that remain almost constant over many revolutions. At any instant the position and velocity of a planet specify a Keplerian orbit; the elements of that orbit are called osculating elements. Perturbed motion can be described by following the variation of these elements.

The elements relevant to this discussion are the semimajor axis a and eccentricity e, giving the size and shape of the ellipse, and three

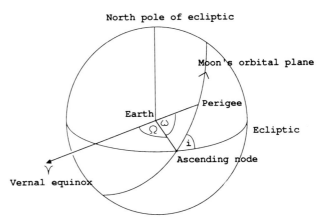

Figure E22 The orbit of the Moon in space.

angles that specify its orientation in space. These are shown in Figure E22.

The reference plane is the ecliptic; angles in this plane are measured eastward from the vernal equinox. The Moon's orbit crosses the ecliptic, going north, at the ascending node. The angle, Ω, specifying this direction is the longitude of the ascending node. The angle between the ecliptic and the orbital plane is the inclination, i. Finally, the angle measured from the ascending node to the direction of perigee is ω, the argument of perigee. (Perigee is that point in the orbit closest to the Earth; the most distant point is apogee.) We shall next itemize the principal perturbations in these elements.

Nodes

The line of nodes regresses, making a complete revolution relative to the vernal equinox in approximately 18⅔ years. There are periodic fluctuations in this rate of regression; when the Sun is at a node or 90° from a node, the rate has its mean value; but when the Sun is 45° from a node, the rate has its greatest or least value; the greatest difference is ± 1° 40′.

The time elapsing between successive passages of the Sun through the ascending node is called an eclipse year; it is approximately 346.62 days. The time elapsing between successive passages of the Moon through its ascending node is the nodical or draconitic month, its length being 27.212 days. The time elapsing between one new Moon and the next is the synodic month, of length 29.531 days. For a solar eclipse to occur, it is necessary that the Moon be new, and both the Sun and Moon be close to a node. It is a coincidence that

19 eclipse years   = 6585.78 days
223 synodic months = 6585.32 days
242 nodical months = 6585.35 days

The period 6585⅓ days, or about 18 years and 11⅓ days (depending on leap years), is called the Saros (q.v.); it was known to ancient astronomers and, reputedly, used in the prediction of eclipses.

Inclination

The inclination has a mean value of 5° 9′. It has a small fluctuation with total amplitude 18′ and period of one-half an eclipse year.

Argument of perigee and the eccentricity

The lunar perigee advances with period 8.85 years. This rate fluctuates with a principal period equal to half the time between two successive passages of the Sun through perigee, i.e. 206 days. The greatest difference is ± 12° 20′.

The eccentricity has a mean value of 0.0549. This fluctuates between 0.0549 ± 0.0117 with the period just mentioned.

As measured from the Earth, the angle between the Moon and the direction of perigee is the true anomaly of the Moon. The angle through which the Moon will have rotated since the preceding perigee passage is equal to the mean anomaly of the Moon. The difference between these angles produces the optical libration in longitude of the Moon; this difference is called the equation of the center. If e and ω were constant, this would fluctuate between ± 6°

17'. Because of the perturbations, the maxima and minima can vary between 5° 3' and 7° 31'. This effect is called the evection and has been recognized since the time of Hipparchus.

Semimajor axis

The semimajor axis of the Moon's orbit is 3.844×10^8 m, or 0.00257 AU. Its principal perturbation has period of one-half a synodic month. Observationally, this is apparent through perturbations in the longitude of the Moon, since the average rate of increase of the longitude depends on a (through Kepler's third law). Because of change in a, this angular rate changes; if U is the angle between the Sun and the Moon, the adjustment necessary is $39' \sin 2U$. This is called the variation.

The solar perturbations over one month will depend, in part, on the distance of the Earth–Moon system from the Sun. This distance fluctuates because of the eccentricity of the orbit of the Earth–Moon system about the Sun; it is least in early January and greatest 6 months later. This affects the rate of increase of longitude; the Moon is apparently running ahead of time for the first 6 months of the anomalistic year, and behind for the final 6 months. If V is the Sun's longitude, measured from perihelion, then the adjustment to the mean longitude is $-(11' \ 16'') \sin V$. This is the annual equation.

These are the principal pertubations; many hundreds more must be taken into account in any adequate lunar ephemeris.

Secular acceleration of the moon

Over long enough periods of time, the effects just described could be expected to average out, leaving a truly 'average' rate of increase in the longitude. This would lead to a formula for the increase of this average longitude of the form $L = L_0 + at$. But this has been found to be inadequate. If the time t is measured in Julian years, the required formula is

$$L = L_0 + at + b\left(\frac{t}{100}\right)^2$$

where b is the secular acceleration of the Moon, and has the approximate value 11''. This effect was first discovered by Halley, based on discrepancies in the timing of ancient eclipses. It was apparently explained by Laplace using the current diminution in the eccentricity of the Earth orbit. But a subsequent, more accurate investigation by Adams showed that only one-half of the value of b was accounted for in this way. It is now generally agreed that the remainder is caused by tidal friction in shallow seas between the sea beds and the water.

We have mentioned the nodical and synodic months. To those should be added three others. The sidereal month, 27.3217 days, measures a complete revolution relative to inertial space (the stars). The anomalistic month 27.5546 days, is the time between successive perigee passages. The tropical month is the time taken for an increase of 360° in longitude (measured with respect to the vernal equinox).

Lunar theories

Newton was the first mathematician to attempt to solve the dynamical problem of the motion of the Moon, with the publication of the *Principia* in 1687. At the time the British Admiralty was interested in acquiring accurate formulas for the Moon's position in the hope that they could be used in the determination of longitude at sea. But it is probable that Newton would anyway have attacked what was to prove to be one of the major challenges in mathematics for the next 200 years.

Any application of pertubation, methods must proceed by successive approximations, and will depend on some quantity or quantities being small. For instance, the perturbation of a minor planet by Jupiter will involve the ratio of the mass of Jupiter to the mass of the Sun: 10^{-3}. The small quantity, in the case of the lunar theory is the ratio of the length of the month to the length of the year: $m = 0.0748$. A solution will consist of formulas involving powers of this small quantity. Here another difficulty arises in lunar theory; the coefficients in such series can become large. As an example, the rate of advance of perigee is given by the formula

$$\frac{3}{4}m^2 + \frac{225}{32}m^3 + \ldots$$

Since the labor of applying a method increases geometrically (at least) with the highest power of the small coefficient it can be appreciated why the problem is so difficult.

Newton showed that the observed inequalities in the motion of the Moon were due to pertubations by the Sun, and he predicted some inequalities that had not been observed up to that time. He had difficulty over the motion of perigee. His analysis accounted for the first term in the formula just quoted. From examination of his papers, it seems that he did eventually discover the following term, accounting for the apsidal motion to within 8% of the observed motion. (Newton is reputed to have told Halley that the lunar theory 'made his head ache and kept him awake so often that he would think of it no more.) His arguments were cast in abstruse geometrical form, but were almost certainly first worked out analytically, using the theory of fluxions.

Sixty years later Clairaut applied analytical methods to the problem. He derived the differential equations and transformed them so that the true longitude was the independent variable. He succeeded in explaining the motion of perigee by using second-order approximations. In 1752 he published his *Théorie de la Lune* and in 1754 he published a set of numerical tables for the computation of the position of the Moon.

Euler (q.v.) published some tables in 1746. In 1753 he published a lunar theory that included (essentially) an exposition of the method of variation of elements. (Credit for this method is usually given to Laplace, who developed it systematically. It may be noted that it was, in fact, applied by Newton in his geometrical discussions.) Euler's most significant contribution came in 1772 when he published his second lunar theory. In this his equations of motion were referred to axes rotating with the mean motion of the Moon.

Laplace's theory was published in 1802. He used the true longitude as the independent variable. His reference orbit was a Keplerian ellipse modified to avoid terms proportional to the time in the coefficients. He gave the explanation for the secular acceleration of the Moon noted above. Laplace's methods were carried to a higher degree of accuracy by several mathematicians, including Damoiseau, whose theory was published in 1827, and who published tables that were used until they were superseded by those of Hansen.

Hansen's work extended over 40 years from 1829. His tables were published in 1857 and were used generally for over 50 years. The reference system that he invented for his method of calculating pertubations and his notation are effective though complex.

In 1860 Delaunay published the results of a lunar theory involving the removal, by analytical means, of the terms of the disturbing function, one by one, and the gradual building up of the solution. His expression for the disturbing function included 320 terms (including all terms of the seventh order of small quantities and some of the eighth). The analytical work took 20 years. The solution is not numerical but analytical, and can be applied not only to the motion of the Moon but also to any satellite motion (of the three-body type). It was the most perfect solution yet found for this type of motion. This theory was later made the basis of lunar tables. Delaunay's work has since been continued to a much higher order by the use of symbolic algebraic manipulation on the computer.

One difficulty in the methods so far mentioned is that they pose the problem in the notation of Keplerian motion. G.W. Hill abandoned this reference. He not only laid the foundations of modern lunar theory, but made substantial contributions to mathematics. In 1877 Hill published the first paper introducing a new method using rectangular coordinates based on axes rotating with the mean motion of the Sun. Hill first found a closed solution, symmetrical about the axes; this is an oval, with the longer axis perpendicular to the direction of the Sun, and is known as Hill's variational curve. This curve was used as a reference orbit in describing the actual motion of the Moon. To calculate the motion of the line of apsides Hill set up an equation, of the type now known as 'Hill's equation,' and invented infinite determinants in order to solve it. The work was further developed by Brown, who published tables in 1919 which have been used since 1923 in the preparation of lunar ephemerides. Refinements of these continue in use.

The formulas used today are adequate for the most precise observations. Approximate formulas exist that are good enough for many computations. Possible sources are the references by Meeus (1991) or Van Flandern and Pultkinen (1979).

Long-term evolution of the Moon's orbit

Because the Earth's rotational angular motion is more rapid than the orbital angular motion of the Moon, the tidal bulges produced by the Moon on the Earth are not in the line joining the Earth and Moon.

Instead they are displaced forward in the direction of the Earth's rotation. The tidal friction (q.v.) and the torque exerted by the Moon on the bulges slow down the rate of rotation of the Earth, and dissipate the total energy of the system. Angular momentum is conserved and this is transferred to the orbital motion of the Moon. Consequently, the Moon is currently receding from the Earth. The rate of increase in the semimajor axis has been put at 3.2 cm per year.

George Darwin calculated these effects backward in time to what he assumed to be the fission of the proto-Earth to form the two separate bodies. More recently, Gerstenkorn and MacDonald (MacDonald, 1966) among others, have calculated a past history where, going backward in time, the Moon approached the Earth to about a distance of three Earth radii; just outside the Roche limit of 2.89 Earth radii. This took place around 1.3×10^9 years ago. At a distance r the tidal heights vary with r^{-3} and the torques with r^{-6}. So very rapid perturbations took place during this close approach, resulting in an increase in the inclination, until the Moon's orbit was circumpolar and then retrograde; following that, its eccentricity increased. It is speculated that the orbit then became parabolic. One dynamical scenario of the history of the Earth–Moon system involves the tidal capture of the Moon by a rapidly rotating Earth. At the time of closest approach most of the kinetic energy of the Earth's rotation was converted to heat. Then the Moon receded to its present orbit. Needless to say, this scenario is not unchallenged (see Earth–Moon system: origin)

J.M. Anthony Danby

Bibliography

Brown, E.W. (1960) *An Introductory Treatise on the Lunar Theory*. New York: Dover Publications.
Danby, J.M.A. (1988) *Fundamentals of Celestial Mechanics*, 2nd edn. Richmond, VA: Willman-Bell.
MacDonald, G.J.F. (1966) The origin of the Moon: Dynamical considerations. In *The Earth–Moon System* (ed. B.G. Marsden and A.G.W. Cameron) New York: Plenum Press, pp. 165–209.
Meeus, J. (1991) *Astronomical Algorithms*. Richmond, VA: Willman-Bell.
Van Flandern, T.C. and Pulkinen, K.F. (1979) Low-precision formulas for Planetary positions. *Ap. J. Suppl.*, Nov 1979, 3941–411
Turncotte, D.T., Cisne, J.L. and Nordmann, J.C. (1977). On the evolution of the lunar orbit. *Icarus*, 30, 254–66.

Cross references

Celestial mechanics;
Earth–Moon system: origin
Moon: origin
Orbit
Planetary dynamical astronomy
Tidal friction

EARTH–MOON SYSTEM: ORIGIN

Because of the strong similarity in orbits, physical structure, and chemical composition of Mercury, Venus, Earth and Moon, and Mars, the origin of the Earth–Moon system is most naturally studied within the context of the formation of the terrestrial planets as a whole, which requires in turn an understanding of the events that led to the formation of the entire solar system. While we certainly cannot claim to have a fully developed theory of solar system formation at this time, we do know enough to outline at least one chain of events through which the terrestrial planets may have formed, namely, through the collisional accumulation of a population of smaller sized bodies, termed planetesimals.

The starting point for modern theories of planet formation is the solar nebula, a flattened, rotating disk of gas and dust surrounding the early Sun. The solar nebula (q.v.) was formed from the collapse of an interstellar cloud of gas and dust, a process that took on the order of 100 000 years (see Solar system: origin). The planets formed in the solar nebula, out of residual dust and gas that did not become a part of our Sun. This hypothesis of planetary formation in the solar nebula predicts that the resulting planets acquired orbital characteristics similar to those actually characteristic of our solar system, namely confinement to a nearly planar region and revolution in the same sense as the rotation of the Sun.

Elements destined to form the terrestrial planets (e.g. Si, O and Fe) were condensed into dust grains at the temperatures and pressures characteristic of stellar envelopes and then injected into interstellar clouds (see Cosmochemistry). The interstellar dust grains, with characteristic sizes of about 10^{-5} cm, began to undergo significant growth once they entered the solar nebula, where their greatly enhanced spatial (number) density led to more frequent collisions and hence to growth because of stickiness caused by intermolecular forces. Within about 1000 years the dust grains sedimented to the mid-plane of the nebula, growing to sizes of perhaps 1 cm in the process. The dust grains formed a thin disk, which may have become gravitationally unstable if the dust disk was sufficiently dense. The dust disk might then have broken up into a large number of self-gravitating planetesimals, on the order of 1 to 10 km in radius. Alternatively, growth may have continued by grain collisions and coagulation, particularly if these particles had a fluffy, porous structure. However growth proceeded in these early phases, it is generally thought that at some time there existed a swarm of about 10^{12} planetesimals of kilometer size, revolving about the Sun in nearly circular orbits, in the mid-plane of the solar nebula. According to the accumulation hypothesis, these planetesimals are the basic building blocks for forming the protoplanets (or planetary embryos) that grew into the final terrestrial planets.

Closely packed phase of terrestrial planet accumulation

The planetesimal accumulation process can be divided naturally into two phases. In the earliest phase (closely packed phase), the planetesimals are relatively small (perhaps 1–10 km in radius) and numerous enough that collisions can occur between planetesimals that are confined within a narrow ring of orbits about the Sun (≈ 0.0001–0.01 AU wide, where 1 AU = 1.5×10^{13} cm is the astronomical unit, defined to be the distance between Earth and the Sun). In this phase the planetesimal masses are too small to be able to gravitationally perturb the planetesimals out of their initially circular orbits or to high relative velocities. Collisions then lead to growth, because kilometer-sized bodies are massive enough ($\approx 10^{16}$ g) to trap gravitationally the debris of gentle collisions, i.e. those with low relative velocities, between the colliding planetesimals. In the later phase (loosely packed phase) the planetesimals are considerably larger in mass, but smaller in number and hence more widely spaced, so that in order for collisions to occur, planetesimals from different rings must collide, which requires eccentric orbits. One important reason for distinguishing between these two phases is that the study of each phase requires a distinctly different theoretical approach; in the closely packed phase the planetesimals can be treated in a statistical sense, as a collection of particles in a box (similar to the kinetic theory of gases). The loosely packed phase, however, requires that the three-dimensional orbits of the planetesimals about the Sun be accurately computed, and hence is best treated by the methods of celestial mechanics.

In either phase, perhaps the most fundamental question is whether runaway accretion occurs. Runaway accretion occurs when a single body in a swarm of planetesimals located within a given annulus of the nebula begins growing at a faster rate than the other planetesimals and becomes much more massive than the rest. This is possible because growth is dependent on the cross-section for collisions, and the self-gravity of planetesimals greatly increases their effective cross-sections. In the two-body gravitational scattering approximation and in the limit of small relative velocities, this cross-section is proportional to the fourth power of the planetesimal radius, rather than the second power appropriate for a purely geometrical cross-section. Thus the more massive a body becomes, the greater its cross-section for accumulating other planetesimals, compared to that for smaller planetesimals, and hence the more rapid its growth. In the absence of outside forces, runaway growth of the most massive body will only stop when the reservoir of accretable bodies is depleted. Recent work on the closely packed phase implies that runaway accretion can be expected to occur in general, largely because of the effects of the equipartition of kinetic energy. The tendency toward equipartition in systems composed of interacting bodies results in lower relative velocities for the larger mass bodies, an effect that further raises the effective cross-section for collisions for the massive bodies. The massive bodies can thus sweep up the much more abundant smaller bodies on a time scale of about 10^5 years in the terrestrial planet zone. This phase of the accumulation process leads to the formation of 10^{26} to 10^{27} g bodies (planetary embryos), spaced every 0.01 to 0.02 AU in the inner nebula. Because of the need for substantial orbital changes in order for further accumulation to proceed in the terrestrial region, it is unlikely that runaway accretion could occur all the way from approximately kilometer-sized bodies to the terrestrial planets.

Loosely packed phase of terrestrial planet accumulation

The initial phases of accumulation may well have occurred in the presence of the gaseous portion of the solar nebula, because evolution in the closely packed phase requires no more than about 10^5 years. This is about the same time interval as the minimum time during which young stars are thought to remove their gaseous envelopes and disks through strong stellar winds and accretion onto the central protostars. Thus the later, loosely packed phase for terrestrial planet accumulation may very well have occurred largely in the absence of significant gas; while gas drag is negligible for very large bodies, the effects of gravitational interactions between protoplanets and a significant gaseous nebula (i.e. the generation of spiral density waves and the opening of gaps in the nebula) could lead to substantial orbital evolution. In the remainder of this section we will assume that the loosely packed phase occurred in the absence of such gas.

The orbital evolution of a loosely packed swarm of planetary embryos is thought to occur as follows. The initially small orbital eccentricities are gradually increased following close encounters (near misses) between embryos. If the eccentricities become too large, subsequent collisions will occur at relative velocities such that collisional fragmentation dominates over collisional accumulation, thereby halting the growth process. At the same time that close encounters are pumping up the eccentricity, collisions damp the eccentricity of the accumulating embryos. If the eccentricity is damped too much, no further collisions will occur, again halting the growth process. It turns out that the combined effect of these two processes is to regulate eccentricities toward values that allow continued growth of the protoplanets.

Runaway accretion does not appear to be significant for the loosely packed phase, in part because the remaining mass range (e.g. from $\sim 10^{26}$ g to $\sim 10^{28}$ g ≈ Earth's mass) is considerably smaller than the range of masses encountered in the closely packed phase ($\sim 10^{16}$ g to $\sim 10^{26}$ g). Instead, a swarm of planetary embryos evolves in a more uniform fashion, with collisions often occurring between nearly equal mass protoplanets. In order for accumulation to proceed in the loosely packed phase, fairly large eccentricities must be maintained, leading to relative velocities upon impact on the order of 10 km^{-1}. This means, however, that the accumulation process involves potentially catastrophic collisions. In particular, the proto-Earth is quite likely to have been struck at about 10 km s^{-1} by a protoplanet as large as one-third the mass of the Earth. Other collisions that built up the Earth (say of Moon sized bodies) would have been somewhat less spectacular, but qualitatively similarly violent events. Considering that these giant impacts can be expected to result in considerable melting, vaporization and fragmentation, depending on the details of the impact, the planetary accumulation process may well involve occasional episodes of major mass loss as well as mass gain. Collisions of this magnitude also present intriguing possibilities for making the Moon. Accumulation of the Earth (and the other terrestrial planets) in this fashion requires about 10^7 years to accumulate 90% of the Earth's mass, and another 10^8 years to gather the remaining 10% of mass. Thus compared to the earlier phases of protostellar collapse, planetesimal formation and the closely packed phase, the loosely packed phase requires by the far the largest interval of time.

Because of the stochastic nature of the final phase of accumulation of the terrestrial planets, one can never hope to produce a theoretical model that is exactly equivalent to our solar system. Nevertheless, computer simulations of the accumulation of swarms of planetary embryos containing the total mass, angular momentum and energy of the present terrestrial planets have shown that it is quite probable for such a system to evolve into four planets with masses and orbits similar to those of the terrestrial planets. Indeed there is even a nearly unit probability for an Earth-mass planet to form around 1 AU from the Sun, a result with possibly profound implications for the existence of life on other planetary systems in our galaxy.

The chemical and isotopic composition of planets formed by collisional accumulation is determined primarily by the compositions of the planetesimals and planetary embryos that built up the planet, modified by whatever losses or fractionations occur during these possibly highly energetic collisions. The primordial composition of the solar nebula itself is generally assumed to be similar to that of chondritic meteorites, enhanced by sufficient gaseous species such as hydrogen and helium to reach solar abundance. Variations in the composition of any given planetesimal could result from condensation of different elements or minerals at different temperatures (i.e different distances from the Sun), providing that the primordial dust grains encountered temperatures high enough to produce vaporization. Such thermal processing might have occurred in the accretion shock surrounding the nebula or in the nebula itself. Condensation sequences in a solar nebula with temperature decreasing outward from the Sun have long been proposed as an explanation for the gross compositional gradient with heliocentric distance of the planets – the terrestrial planets are composed of iron and silicates, while the giant planets have ice and rock cores, implying lower temperatures as the distance from the Sun increases. Some solar nebula models predict mid-plane temperatures in the terrestrial planet region high enough to experience at least portions of the condensation sequence. However, orbital evolution during the loosely packed phase can be expected to redistribute in a stochastic sense the planetary embryos throughout the terrestrial planet zone at least, eroding any explanation of a compositional gradient within the terrestrial planet zone (i.e Mercury being composed largely of iron and the Earth largely of silicates). Well-defined feeding zones may not exist for the terrestrial planets, except in the general sense that the feeding zone for each terrestrial planet was the entire terrestrial zone. Mercury's largely iron composition is perhaps better explained as being possibly caused by a giant impact that removed the silicate mantle of proto-Mercury.

Earth's earliest thermal history

The concentration of radioactive isotopes can be used to date the time when a rock was last crystallized, and samples of the Earth, Moon and meteorites have been studied in order to find the oldest possible rocks. The age of the oldest rock gives a lower bound on the age of the solar system. Many primitive meteorites give ages clustering around 4.5 billion years strongly implying that this value is the approximate age of the solar system. Silicate inclusions in the Allende meteorite currently give the oldest accepted dates of crystallization, about 4.56 billion years ago.

Contrary to the ideas of a few decades ago, it is no longer necessarily thought that Earth's accumulation proceeded primarily through the impact of bodies very much smaller in size. This removes one fundamental problem: forming the Earth's molten outer core quickly enough to be able to provide a terrestrial dynamo capable of explaining the remanant magnetism found in the earliest terrestrial rocks. If accumulation by the impact of very small bodies had been the case, the impacts would bury little energy deep within the protoplanet, instead depositing the dissipated kinetic energy at the surface, where it could be readily radiated to space and lost from the planet's thermal budget. In such a case, the Earth would have formed relatively cold, i.e. with temperatures everywhere below the solidus, unless the formation time was very short (e.g. 10^6 years). The fact that such short formation times are hard to obtain in the accumulation hypothesis meant that it was difficult to explain how the iron core formed early in Earth's evolution. It now appears that the late phases of Earth formation involved energetic collisions between rather massive planetary embryos and hence formation of the initial iron core during the accumulation phase.

Formation of the Earth through collisional impacts involving very large protoplanets ($> 10^{26}$ g) must have been a violent process; the gravitational energy liberated is potentially sufficient to melt the entire planet. The degree of melting obtained in the proto-Earth as a result is a fundamental question, because melting can result in geochemical differentiation, greatly enhanced heat transport and qualitatively different outcomes with respect to dynamical questions such as the possibility of rotational fission. Rigorous models of impacts on this gigantic scale are being investigated, and the details of these solutions are certain to have profound importance for the early thermal history of the Earth in the context of the accumulation hypothesis of origin.

The most important differentiation event in Earth's evolution was the separation between iron and silicates and the migration of the

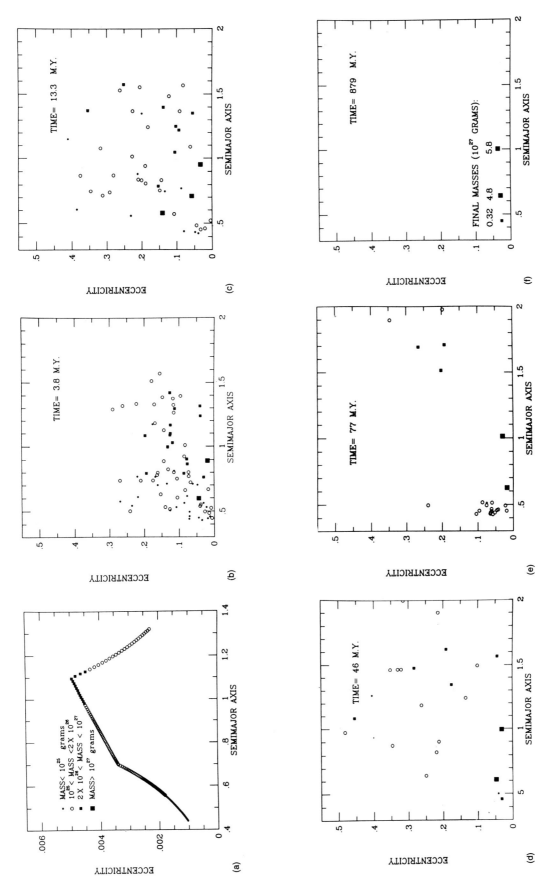

Figure E23 Simulation of the final phase of terrestrial planet accumulation, starting from runaway planetary embryos. (a) Arbitrary initial conditions (0 Ma) for a swarm of embryos with the total mass, energy and angular momentum of the terrestrial planets. (b)–(e) The evolution of the swarm with time. Eccentricities increase considerably, and radial mixing occurs throughout the region; a few bodies are ejected altogether from the solar system. (f) A system bearing a strong similarity to the present terrestrial planets (879 Ma). (Reprinted from Wetherill, 1990.)

iron to form the core. Once the proto-Earth grew to about the size of Mars, the energy released by the impacting planetesimals was sufficient to lead to pools of liquid iron, which then sank because of their high density compared to the silicates. The energy released by this sinking process, coupled with the ongoing accretional energy, was transported downward by the descending molten iron, and led to a partially molten, largely silicate mantle. The molten iron produced by subsequent impacts continued to sink to the core through the partially molten silicate mantle.

Formation of the iron core occurred at the same time that the Earth was accreting, and so the two processes must be considered together. However, it not so certain whether differentiation of the silicate mantle occurred to a significant extent during the formational period. The answer to this question depends primarily on the question of whether the silicate mantle, once it became partially molten, was able to resolidify (perhaps only locally) before becoming molten again as a consequence of the next large impact. If not, then the mantle may have remained largely molten throughout the formational period, and differentiation would only have been possible once accretion ended and the planet began to cool and resolidify. Differentiation in a partially molten mantle may also have been retarded by the resistance to motion inherent in a medium composed of a solid matrix threaded by molten veins.

Lunar formation: classical hypotheses

George H. Darwin (son of Charles Darwin) analyzed the orbit of the Moon over 100 years ago and found that at some time in the past the Moon must have been considerably closer to the Earth (the net effect of the tides raised on the solid and fluid portions of the Earth by the Moon is to transfer angular momentum from the Earth's spin to the Moon's orbit, thereby moving the Moon further away from the Earth as time goes on; see Tidal friction). Darwin therefore hypothesized that the Moon arose from the mantle of the Earth, and proposed a mechanism for removal of the Moon: rotational fission. Rotational fission is the first of three hypotheses that have come to be known as the classical hypotheses of lunar origin.

Rotational fission involves the splitting into two of a body which is rotating too fast to be stable. The concept was based upon studies of the possible equilibrium configurations of self-gravitating, incompressible (uniform density) fluid bodies in solid-body rotation. Clearly when such a body is spun fast enough, the centrifugal acceleration needed to hold the equatorial regions onto the body can no longer be provided by the finite gravitational acceleration inward. What happens thereafter could not be discerned on the basis of the equilibrium models available 100 years ago but, in part because of the existence of equilibrium models for tidally distorted bodies in orbit about one another, it was hypothesized that the natural outcome of rotational instability would be separation into a binary system. Recently numerical calculations have been able to follow the growth of the fission instability in rapidly rotating, inviscid bodies with compressibilities similar to that of the Earth. Contrary to the classical fission hypothesis, however, the outcome of at least one type of fission instability is not simple bifurcation into two bodies; instead, the embryo central binary loses its orbital angular momentum to growing spiral arms located further outward. The binary effectively decays into a single body before it can become well defined, while the spiral arms remove the excess angular momentum from the inner body, allowing it to reach a stable equilibrium. The spiral arms degenerate into a ring with a small fraction of the total mass, but a large fraction of the total angular momentum. While this unforeseen outcome appears to eliminate the fission hypothesis as a means of producing nearly equal mass binary systems (e.g. close binary stars), it is still a potentially attractive means for forming the Moon (providing the Moon can be assembled out of the ring), because the mass of the Moon is only about 1/80 that of the Earth. An alternate outcome of rotational instability is mass shedding, where the excess angular momentum is shed simply by ejecting a low-mass ring from the equator of the body – a similar outcome in either case. Mass shedding is more likely to occur if the body was sufficiently viscous to rotate as a solid body throughout the formation phase.

Rotational fission has several strong arguments in its favor. The first is that forming the Moon out of the Earth's mantle would produce a Moon with a low mean density (about 3 g cm^{-3}), similar to that of the Earth's mantle, as is observed. Furthermore, analysis of the lunar samples returned by the Apollo missions has shown that the lunar surface rocks are strongly depleted in iron, as is the Earth's mantle, consistent with the low mean density of both. Siderophilic elements have abundances in the Apollo samples that are very close to those of the Earth's mantle. In several ways then, the Moon's chemistry appears to be very similar to the Earth's mantle, which argues in favor of a fission origin. However, the Apollo samples also showed that the Moon is greatly depleted in volatiles such as water, compared to Earth's mantle, but it has been argued that devolatilization could have been caused by the energetic processes associated with the fission instability and ring formation.

While fission appears to be basically consistent with the history of the Moon's orbit, there are several dynamical problems which in sum appear to rule out the possibility of forming the Moon in this manner. One early objection to the fission hypothesis was that fission requires about four times more angular momentum in the initial proto-Earth than is present in the Earth–Moon system. It is hard to explain the amount of angular momentum in the Earth–Moon system, much less four times as much. For example, if the angular momentum is derived from the tangential impact at 10 km s^{-1} of Mars-sized bodies with the proto-Earth, then four successive impacts, each in the same spot on the limb of the proto-Earth, would be required to initiate mass shedding; this prospect is quite unlikely compared to the single giant impact hypothesis (to be discussed shortly). Second, assuming that a mechanism could be found for providing the requisite angular momentum, the excess angular momentum must be disposed of eventually. In the numerical calculations of fission, it is conceivable that inefficient formation of the Moon out of the ring of matter orbiting the Earth could account for the loss of the excess angular momentum (the ring is likely to be considerably more massive than the Moon), but this has not been investigated.

The second classical hypothesis of lunar origin is capture. This hypothesis proposes that the Moon formed elsewhere in the solar system, and then was captured in orbit about the Earth. The former assumption can be challenged on the basis of the strong chemical similarities between the Moon and the Earth's mantle; if the Moon was formed out of a different sample of planetesimals than the Earth, it is improbable that it would have acquired certain elemental and isotopic abundances so similar to those of Earth. The latter assumption (i.e. a mechanism allowing capture) is an equally crucial failing of this hypothesis.

Intact capture of a proto-Moon about the Earth requires the dissipation of the relative energy of motion between the proto-Moon and Earth. The most promising (and most often invoked) mechanism is tidal dissipation, which is certain to occur in a close encounter between partially molten or solid protoplanets. However, even the most optimistic estimates of the efficiency of tidal dissipation imply that in order to be captured, the proto-Moon must have approached the Earth from an orbit with a relative velocity of no more than about 0.1 km s^{-1}. Considering that the orbital velocity of the Earth is about 30 km s^{-1}, this means that the proto-Moon must have started from an orbit very nearly the same as that of the Earth. But then it is hard to see how the proto-Moon could have accumulated to its present size without having been subjected to a close encounter with proto-Earth earlier in their evolution, an encounter that would have kicked the Moon to a much higher relative velocity. We have seen that encounters between planetesimals and protoplanets in the loosely packed phase occur at higher relative velocities (up to 10 km s^{-1}), where capture by tidal dissipation can be ruled out.

The intact capture hypothesis was modified by the added hypothesis of tidal disruption of the proto-Moon into a shower of debris, some of which might remain in orbit about the Earth and later reaccumulate into the Moon. This hypothesis is based upon the absence of equilibrium configurations for a fluid satellite on a circular orbit within a critical distance from the primary body. The critical distance, named the Roche limit after the scientist who first derived it, is about three times the radius of the Earth for the Earth–Moon system. It was thus hypothesized that if a planetesimal passed within the Roche limit of a growing protoplanet, it would be torn asunder by strong tidal forces. One appeal of disruptive capture is that the probability of passing within the Roche limit is considerably larger than that of a direct collision, because of the roughly nine times larger cross-section. However, the Roche criterion applies to an inviscid satellite on a circular orbit, where there is an infinite amount of time to disrupt the satellite tidally. In terrestrial planet accumulation, a planetesimal or embryo passing close by the Earth will only spend on the order of an hour within the Roche limit, and furthermore is unlikely to be totally molten and hence an inviscid fluid. A highly viscous body will successfully resist tidal disruption during such a limited duration encounter. On the other hand an inviscid fluid body can be disrupted if its mass is small (roughly Moon-sized) and if it approaches with a small relative velocity. In the latter case the fraction of tidally disrupted mass that is trapped in Earth orbit is only a small fraction of the incoming body's mass, so tidal disruption appears to be unable to emplace a lunar mass in orbit in a single event. Multiple events would lead to orbiting debris in both prograde and retrograde senses, so that the net outcome would be a debris disk with insufficient angular momentum to form the Moon (see next hypothesis).

The third classical hypothesis of lunar origin is binary accretion, where the Moon forms in orbit about the proto-Earth as both grow to their final size. This hypothesis has the advantage of explaining the similar chemistries of the Moon and Earth's mantle as being caused by their accumulation from essentially the same sample of planetesimals. In order to account for the lack of iron in the Moon, however, significant compositional filtering must occur. It has been argued that a swarm of planetesimals orbiting the proto-Earth could preferentially trap silicate planetesimals, rather than iron planetesimals, because of the former's lower mean density. This mechanism assumes that a prior generation of protoplanets have already differentiated into mantles and cores and later had been disrupted into much smaller pieces, and that a swarm remains in place about the proto-Earth to accomplish the processing. The former requirement appears to be consistent with a scenario where runaway accretion to protoplanetary sizes occurs, with accumulation proceeding from a distribution dominated by many very small bodies, but this is at odds with the scenario developed so far in this section, where the terminal phase of accumulation does not involve runaway accretion. The latter requirement is even harder to achieve dynamically. Accumulation in Earth orbit is likely to proceed on a much shorter timescale than that for the introduction of new planetesimals from heliocentric orbit, because of the much shorter orbital period in geocentric versus heliocentric orbit, so that the compositional filter may be lost. A third objection, apparently fatal, is that it is impossible to achieve the angular momentum necessary for maintaining the orbiting swarm, unless a special population of orbits provides the bulk of the planetesimals interacting with the binary planet system. This is because incoming planetesimals will have roughly equal chances of colliding with the swarm and ending up in orbit in either possible sense of revolution about the proto-Earth; thus, on average, the angular momentum is nearly zero. Obtaining appreciable angular momentum in this way requires accreting planetesimals from very nearly circular orbits or from a population depleted at the Earth's heliocentric distance. Both of these possibilities appear to be unlikely, because of the relatively highly eccentric orbits of planetesimals late in planetary accumulation, and because of the tendency for stochastic evolution to fill in any depleted regions with more planetesimals. Thus all of the classical hypotheses for forming the Moon appear to be severely flawed.

Lunar formation: giant impact hypothesis

The frequent occurrence of giant impacts on the growing terrestrial planets provides a natural means for forming a major satellite like the Moon, and for providing the large amount of angular momentum necessary for its orbit. If a giant impact of a Mars-sized body at 10 km s^{-1} occurs close to the limb of the Earth, the angular momentum of the incoming body with respect to the center of the Earth is approximately equal to that of the Earth–Moon system. Providing that the colossal explosive impact that follows ejects substantial debris into stable Earth orbit, it may be possible for the Moon to later reaccumulate in geocentric orbit. The implications of this hypothesis for explaining the chemistry of the Moon are extremely attractive: the impact would presumably result in the injection of portions of the silicate mantles of the Mars-sized interloper and the Earth into Earth orbit, accounting for the low mean density and similarity to terrestrial elemental abundances of the Apollo samples. The thermal processing implied by a giant impact makes devolatilization of the prelunar matter quite understandable. Provided that the impact places a fair amount of proto-Earth mantle in orbit, it appears possible that the giant impact hypothesis can explain all of the chemistry of the Moon.

Dynamically speaking, the fact that the giant impact hypothesis involves collisions that are predicted to occur with high probability,

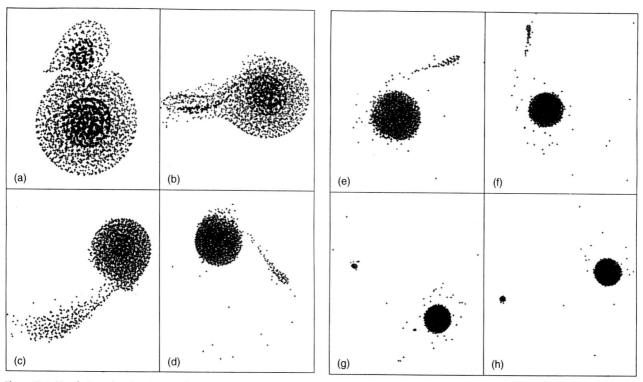

Figure E24 Simulation of a giant impact between a 0.16 Earth-mass body and the proto-Earth. A sequence in time is shown, covering a period of about a day. The smaller body approaches from zero velocity at infinity with an impact parameter such that the system has 1.3 times the angular momentum of the Earth–Moon system. The silicate mantles and iron cores of both bodies are evident. Following the giant impact, the iron core of the smaller body enters the Earth's core, while a silicate body with approximately the mass of the Moon remains on a stable orbit about the Earth. (Reprinted from Cameron and Benz, 1991.)

and that can account for the angular momentum of the Earth–Moon system, means that this hypothesis has already overcome two of the most formidable obstacles to a consistent theory of lunar formation. It is also consistent with the observation that of the four terrestrial planets, only the Earth has a major satellite. Venus may not have a Moon-like satellite for several good reasons; its largest impact simply may not have occurred at the right orientation to the Venusian limb to create a satellite system, or even if it did once have a large satellite, Venus may have lost the satellite through tidal decay of the satellite's orbit onto the slowly rotating Venusian surface.

Alan P. Boss

Bibliography

Boss, A.P., Cameron, A.G.W. and Benz, W. (1991) Tidal disruption of inviscid planetesimals. *Icarus*, 92, 165–78.
Brush, S.G. (1988) A history of modern selenogony: theoretical origins of the Moon, from capture to crash 1955–1984. *Space Sci. Rev.*, 47, 211–73.
Cameron, A.G.W. and Benz, W. (1991) The origin of the Moon and the single impact hypothesis IV. *Icarus*, 92, 204–16.
Hartmann, W.K., Phillips, R.J. and Taylor, G.J. (eds) (1986) *Origin of the Moon*, Houston, Texas: Lunar and Planetary Institute, 781 pp.
Newsom, H.E. and Jones, J.H. (eds) (1990) *Origin of the Earth*. New York: Oxford University Press, 378 pp.
Newsom, H.E. and Sims, K.W.W. (1991) Core formation during early accretion of the Earth. *Science*, 252, 926–33
Newsom, H.E. and Taylor, S.R. (1989) Geochemical implications of the formation of the Moon by a single giant impact. *Nature*, 338, 29–34.
Ringwood, A.E. (1989) Flaws in the giant impact hypothesis of lunar origin. *Earth Planet. Sci. Lett.*, 95, 208–14.
Stevenson, D.J. (1987) Origin of the Moon – the collision hypothesis. *Ann. Rev. Earth Planet. Sci.*, 15, 271–315.
Wetherill, G.W. (1990) Formation of the Earth. *Ann. Rev. Earth Planet. Sci.*, 18, 205–256.
Wetherill, G.W. (1991) Occurrence of Earth-like bodies in planetary systems. *Science*, 253, 535–8.

Cross references

Earths: rotational history
Moon: origin
Obliquity: terrestrial record
Solar system: origin
Tidal friction

ECCENTRIC DIPOLE

The external magnetic fields of planets are very complicated because they are caused by complex motions of their electrically conducting liquid cores. In the description of the field by expansion in spherical harmonics, the first term represents the dipole approximation to this field. It is usually found that the expansion that most accurately represents the observed external field is that of a dipole that is displaced from the geometric center of the planet.

It has been known since the 1830s that the geomagnetic dipole is eccentric. Its displacement from the Earth's center was 342 km in 1920, and it has been increasing almost linearly with time. In 1980 it was 488 km, and by now it may be approximately 520 km, or about 0.08 Earth radii.

Magnetic field measurements on spacecraft have determined the magnitudes and locations of the dipoles of the outer planets. Jupiter's field is very large and the displacement of its dipole is 0.12 ± 0.01 radii. The dipole of Saturn, on the other hand, is offset from the polar axis by only 0.04 ± 0.02 times the planet's radius, and its equatorial offset is also small. The dipole of Neptune is displaced by 0.48 radii but lies very close to the equatorial plane.

Uranus, which has its rotational axis very close to its orbital plane, also has a very strange magnetic field. Its dipole is centered almost on the rotation axis but oriented 60° away from it and displaced 0.33 radii from the center (see Uranus: magnetic field and magnetosphere).

The magnetic fields of the inner planets have not yet been measured with precision sufficient to determine the properties of their dipoles.

Conway W. Snyder

Bibliography

Chapman, S. and Bartels, J. (1940) *Geomagnetism*. London: Oxford University Press.
Connerney, E.P. and Acuña, M.H. (1991) The magnetic field of Neptune. *J. Geophys. Res.*, 96, 19023–42.
Ness, N.F., Acuña, M.H., Behannon, K.W., et al. (1986) Magnetic fields at Uranus. *Science*, 233, 85–9.
Wallis, D.D., Burrows, J.R., Hughes, T.J. and Wilson, M.D. (1982) Eccentric dipole coordinates for magsat data presentation and analysis of external current effects. *Geophys. Res. Lett.*, 9, 353–6.

Cross references

Dynamo theory
Earth: magnetic field and magnetosphere
Magnetism

ECCENTRICITY

In modern astronomy the eccentricity of a planetary or cometary orbit is the ratio of the focal distance of the orbit's center to the orbit's semimajor axis. This usage is the end-stage of a development that began with the practice in Greek astronomy of using eccentric circles (from εκ, out of + κεντρον, center) to account for apparent non-uniform motion of celestial bodies. Thus to explain why the four seasons of the solar year are unequal, Ptolemy (2nd century AD), following Hipparchus (2nd century BC), imagined the Sun to be moving uniformly on a circle (ASP in Figure E25), with the terrestrial observer (at E) displaced from the circle's center, so that the Sun appeared to be moving more slowly in the region of the apogee (at A) than in the region of the perigee (at P).

The term *eccentricitas* first appears in late Latin; it occurs in Copernicus's *De Revolutionibus* (1543), and is used as a standard term of art in Tycho Brahe's *Astronomiae Instauratae Progymnasmata* (1602). It meant the displacement of the center of a planet's circle from a point of reference, as compared with the radius of the circle. For Tycho, a geocentrist, the point of reference was the Earth's center; for Copernicus it was the mean Sun, or center of the Earth's circle. Kepler was the first to measure the eccentricity of a planet

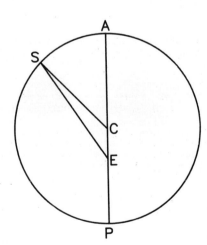

Figure E25 The eccentric circle of the Sun in the astronomy of Hipparchus and Ptolemy: ASP is the Sun's path, on which it moves uniformly; C is the circle's center; E is the place of the Earth; A is the apogee and P is the perigee.

from the true and apparent Sun, which was thus placed at E in Figure E25.

Ptolemy in his theories of Mars, Jupiter, Saturn and Venus had bisected the eccentricity; that is, he had placed the center about which the planet's motion was angularly uniform (the equant point), not at the center of the planet's circle, but twice as far from the Earth. Kepler in his early investigations of the motions of Mars discovered that the eccentricity of the Earth's (circular) orbit also had to be bisected, the center of the circle being placed not where Ptolemy had put it, at C, but halfway between E and C, and the remainder of the Sun's inequality being accounted for by the non-uniformity of the Earth's motion on its circle. Once Kepler learned that the planetary paths were ellipses, the points C and E became foci of the ellipse.

Until the 18th century, eccentricities were given in units of which the orbital radius contained a multiple number – 60 for Ptolemy, 10 000 for Copernicus, and 100 000 for Kepler and his 17th-century successors. In the late 18th century the practice emerged of giving orbital eccentricities as fractions with the semimajor axis as the unit.

Among the nine major planets, Pluto has the highest orbital eccentricity, about 0.25, Mercury the next highest, about 0.2, and all the rest have eccentricities less than 0.1. The departure of the planetary ellipses from circularity is given very nearly by $e^2/2$, where e is the eccentricity; in Pluto's case this amounts to 3% of the semimajor axis, in Mercury's case to 2%, and in the other cases to less than 0.5%. Halley's comet, the first periodic comet to be discovered, proved to have a much higher eccentricity (about 0.967), hence a much more elongated orbit.

The celestial mechanists of the 18th century discovered that the planetary eccentricities were subject to slow oscillatory changes, with periods measured in tens of thousands of years, due to the mutual perturbations of the planets. Lagrange and Laplace gave arguments purporting to prove that these changes would be confined to narrow ranges, but these arguments are now known to be inconclusive. The orbits of Pluto and the inner planets Mercury, Venus, Earth and Mars are sensitive to initial conditions in such a way that predictability of orbital motion is limited to about 100 million years; beyond such a time period, celestial mechanics can set no bound to the possible range of variation of the eccentricities (Sussman and Wisdom, 1988; Laskar, 1990; see Chaotic dynamics in the solar system).

In 1875 James Croll (1821–1890) (q.v.) proposed that small changes in the Earth's eccentricity, combined with the precession of the equinoxes, could explain the c. 100 000-year cyclicity of ice age glacial phases. The theory was expanded to include the factor of orbital obliquity, and quantified in detail in the early 20th century by M. Milankovitch. It is now the accepted model for these long-term changes in climate on planet Earth (Imbrie and Imbrie, 1979). Although there is a well-documented history of recurrent ice ages, there is no evidence of any major departure in the very small variations of the Earth's eccentricity in the last 4 billion years.

Curtis Wilson

Bibliography

Fairbridge, R.W. (1987) Climatic variation, geological record, in *The Encyclopedia of Climatology*, (eds. J.E. Oliver and R.W. Fairbridge). New York: Van Nostrand-Reinhold, pp. 293–305.
Imbrie, J. and Imbrie, K.P. (1979) *Ice Ages*. New York: Macmillan.
Kepler, J. (1609) *Astronomia Nova*, Chapters 22–28.
Laskar, J. (1990) The chaotic behavior of the solar system: a numerical estimate of the size of the chaotic zones. *Icarus*, **88**, 266–91.
Moulton, F.R. (1914) *An Introduction to Celestial Mechanics*, New York: Macmillan pp. 277–365. (Dover edition, 1970.)
Sussman, G.J. and Wisdom, J. (1988) Numerical evidence that the motion of Pluto is chaotic, *Science*, **241**, 433–7.
Toomer, G.J. (1975) Ptolemy. *Dict. Sci. Biogr.*, Vol. 11, pp. 186–206.
Toomer, G.J. (tr.) (1984) *Ptolemy's Almagest*, Book III. New York, Berlin: Springer-Verlag.

Cross references

Chaotic dynamics in the solar system
Ice age
Milankovitch, Milutin, and Milankovitch theory
Orbit

ECLIPSE

For any solar system body, this is the total or partial cut-off of sunlight from that body by the interposition of another body of similar or larger dimensions. Thus our Moon is eclipsed when it passes through the Earth's shadow; the satellites of the superior planets are eclipsed when they pass through the shadows of their parent planets or of other satellites. Strictly speaking, solar eclipses are actually occultations, since the solid body of the Moon partially or totally blocks our Earth-based view of the Sun's disk. However, they will be discussed here because of long usage.

Lunar eclipses

If the Moon's orbit were in the plane of the ecliptic, the Moon would pass through the Earth's shadow at each successive full phase, resulting in an eclipse every month. However, the orbit is inclined to the ecliptic by just over 5°, so that the full Moon usually passes above or below the shadow, although on occasion it will pass through it (see Figure E26). The points where the lunar orbit crosses the ecliptic are called the nodes. If the full Moon is at or near a node when these are aligned with the Sun, a lunar eclipse can occur. Because the Sun is not a point source and is larger than the Earth, the Earth's shadow is a long, fuzzy-edged cone stretching over 1.25 million km into space in the antisolar direction. If Earth had no atmosphere, this shadow, if cast on a huge screen at the Moon's distance, would appear somewhat as in Figure E26. The solid shadow, in which no sunlight would fall, is called the umbra. The annular area surrounding this is the penumbra (Latin *paene* – almost), which grades smoothly from the total shadow of the umbra to full sunlight at the outer edge.

The Moon's path can cross this shadow at any position. If the umbra never falls on the Moon, it is a penumbral eclipse; if the umbra covers part but not all of the Moon at the middle of the eclipse, it is a partial eclipse. When the Moon passes completely into the umbra, it becomes a total eclipse. The magnitude of the eclipse is reckoned as the fraction of the lunar diameter covered by the umbra at mid-eclipse. Thus magnitude 1.0 means that the Moon is just completely covered; magnitude 1.5 signifies that the edge of the umbra lies half a lunar diameter beyond the edge of the Moon at mid-eclipse. Lunar eclipses are visible, weather permitting, from anywhere that the Moon is above the horizon.

If the umbra were perfectly dark, the Moon would completely vanish during each total eclipse. A few such eclipses are so dark that the Moon is barely discernible, but much more frequently it remains visible, glowing with a reddish or coppery colored light. The illumination is usually uneven, with a lighter, whiter segment most often in the direction of the nearest part of the penumbra. This is caused by sunlight passing horizontally through the Earth's atmosphere and being refracted into the umbra. This light consists largely of the reddish sunset hues, with some whiter light refracted by the higher levels of our atmosphere. Cloudy or dusty conditions along all or parts of the sunrise–sunset line on Earth influence the general darkness of the eclipse, and of the locations of any brighter areas. Penumbral eclipses are much less eye-catching events, and are likely to pass unnoticed by the casual observer.

Lunar eclipses have rather limited scientific value, especially those that are penumbral or partial. Total lunar eclipses can yield valuable data on the nature of the lunar surface when observed in the infrared portion of the spectrum. Early observations indicated a very rapid drop in surface temperature as the light and heat from the Sun was gradually cut off by the Earth's shadow; this pointed to a highly insulating layer with low heat-retention capacity, conditions that were amply confirmed by subsequent research and particularly by the Apollo missions (q.v.). Later higher-resolution infrared observations showed that numerous areas and spots cooled more slowly than their surroundings; this was correctly interpreted as being due to areas strewn with rocky debris and larger blocks ejected from fresher impact craters.

Solar eclipses

Just as lunar eclipses can occur only when the Moon is near a node of its orbit and at the full phase, solar eclipses can occur only when the Moon is near a node and at the new phase. The Sun's diameter is about 400 times that of the Moon, but by an unusual circumstance it is about 400 times farther away than the Moon, so that from Earth these two bodies appear to be very similar in apparent diameter.

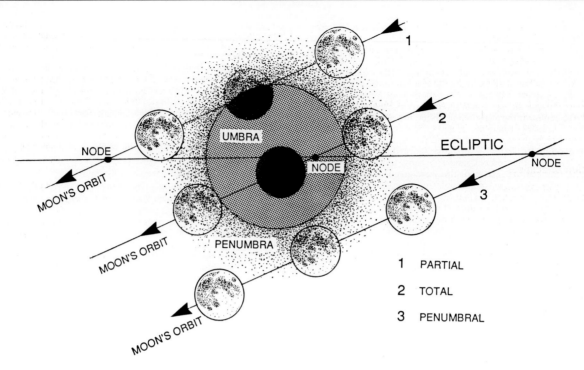

Figure E26 The three types of lunar eclipse. The umbra is shown in a light gray tone for clarity, and the inclination of the Moon's orbit to the ecliptic is exaggerated.

Because of the eccentric positions of the Sun and Earth within the Earth's and Moon's orbits respectively, the Moon can appear to be either slightly (1) larger or (2) smaller than the Sun.

In case 1, the tip of the Moon's shadow cone sweeps across the Earth's surface as the Moon travels in its orbit, the maximum diameter of the shadow being about 250 km. From locations within this path of totality, the bright body of the Sun can be completely covered anywhere from an instant up to a maximum of 7 min 40 s in the most favorable case. Locations outside the path of totality will experience a partial solar eclipse provided that they lie within the Moon's penumbra, the magnitude of the eclipse depending upon the distance of the location from the path of totality (Figure E27).

In case 2 above, the tip of the Moon's shadow cone does not extend as far as the Earth's surface, resulting in an annular (Latin *annulus* – a ring) eclipse for locations on the path of annularity, and a partial eclipse for other locations within the Moon's penumbra.

Scientifically, total solar eclipses are extremely important, as may be gathered from the armies of scientists who travel to remote locations with bulky equipment to amass data of all kinds during the few minutes of totality. Although instruments such as the coronagraph can create artificial eclipses, they cannot eliminate unwanted sunlight scattered by our atmosphere, which tends to swamp the relatively faint light emitted by the Sun's corona. Nothing can beat having a perfectly size-matched, slowly moving occulting disk outside our atmosphere, and our Moon fits the bill ideally! This allows study of the Sun's outer and inner corona, and also its chromosphere, almost down to the level of the photosphere.

As a spectacular astronomical phenomenon, a total solar eclipse viewed in a clear sky is unsurpassable. The gradual dimming of the light and heat radiation with no clouds intervening is already an odd sensation; immediately before totality begins, the disk of the Moon's shadow approaches at phenomenal speed, plunging viewers into twilight conditions. The horizon is not dark, however, being illuminated by a thin crescent of the solar disk. The brighter planets and stars can be seen, but the showpiece is the solar corona, along with any prominences that may be present, and the chromosphere.

Recurrence of eclipses

The line of nodes of the Moon's orbit on the ecliptic is not a fixed direction in space, but makes a complete revolution in 18⅔ years, so that it might be expected that the sequence of solar and lunar eclipses

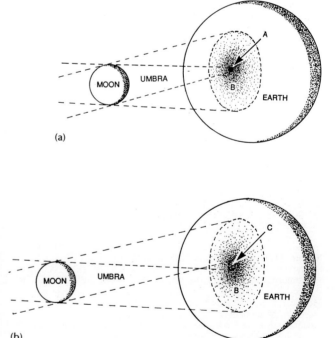

Figure E27 The three types of solar eclipse. Total eclipse seen at A, partial eclipse at B, and an annular eclipse at C.

would repeat on fairly similar dates after a period of about this duration. This is indeed the case, the actual period being 18 years 11 days. This is known as the Saros (q.v.), and was recognized by the Chaldeans some 40 centuries ago.

If the sequence of lunar and solar eclipses repeats after this period, may we assume that the type of eclipse, e.g. annular, total, penumbral, repeats after this interval? The answer is 'no', the reason being that the Moon's orbit around the Earth is slightly elliptical. Thus according to Kepler's laws, the Earth occupies an eccentric position within that orbit, which means that the apparent diameter of the Moon, as well as its angular velocity, varies during each revolution. As with the line of nodes, the line of apsides (q.v.), the line joining the points of nearest and farthest distance from the Earth (i.e. perigee and apogee), is not a fixed direction in space, but makes a complete revolution in 8.85 years. This is sufficiently less than half the Saros period that what was an annular solar eclipse on a given date could, 18 years and 11 days later, be a total eclipse.

Prediction of eclipses

The discovery of the Saros period obviously represented a great step forward in the art of eclipse prediction, although the perturbations just noted, together with imprecise knowledge of the Moon's motions, led to some incorrect predictions. By the second half of the 19th century, these and the Earth's motions were known with sufficient accuracy that the circumstances of lunar and solar eclipses could be calculated with considerable precision, and Oppolzer's famous 'Canon der Finsternisse' (1887) tabulates the elements and circumstances of all eclipses from 1206 BC to 2162 AD. This has proved to be an invaluable reference work not only for predictions of future eclipses, but also for dating ancient records of eclipses where no date was provided. Even more precise tables are now available, with extended past and future limits.

Eclipses of other planetary satellites

All the outer planets have satellites, which may be eclipsed by their parent planet on occasion. For Mars' minuscule moons, observation of such eclipses is virtually impossible because of the overwhelming light from the planet. For the major planets except Jupiter, the comparative proximity of the Earth to the Sun means that the eclipses take place almost entirely behind the planet from our viewpoint; in addition, the angle of tilt of those satellites' orbits means that such eclipses are very rare anyway. For Jupiter the angle of tilt of the orbits of the four Galilean moons is small enough that the inner three are eclipsed every revolution (the fourth less frequently), and the whole system is close enough that the phenomena are readily observable, and have yielded valuable scientific data. At Jupiter's equinoxes, i.e. when the Sun lies in Jupiter's equatorial plane, these four satellites can undergo mutual eclipses because their orbits lie very close to that plane.

Ewen Whitaker

Bibliography

Jones, H.S. (1934) *General Astronomy*. London: E. Arnold & Co. (Chapter 8, excellent for theory of eclipses.)
Pasachoff, J.M. (1989) *Contemporary Astronomy* (4th edn). Philadelphia: Saunders College Publishing. (Good general treatment of solar eclipses.)

Cross references

Ecliptic
History of planetary science I
Orbit
Synergetic tidal force

ECLIPTIC

Based on the Greek word *ekleiptikos* and the Latin *ecliptica*, the ecliptic is the great circle formed by the intersection of the plane of the Earth's orbit with the celestial sphere. It is thus a great circle representing the apparent annual path of the Sun upon the celestial sphere (sometimes called the 'solar circle'). This apparent annual motion of the Sun across the sky is caused by the actual revolution of the Earth around the Sun. The ecliptic plane is equal to the mean orbital plane of the Earth; it forms a convenient reference plane in astronomy and celestial mechanics, as the orbits of the planets all (with the exception of Pluto) have very small inclinations ($\leq 7°$) relative to this plane. (See Celestial mechanics; Coordinate systems; Orbit.)

To the ancient astronomers (of Greece and the Middle East) the equator of a spinning sphere was the most fundamental circle on that sphere. The ecliptic was then described as making a slanting or oblique angle to it. They also identified it as the great circle through the middle of the signs (i.e. of the zodiac). The Greek astronomers (notably Ptolemy, q.v.) gave angular coordinates for the perceived motion of the Sun with reference to the ecliptic ('ascensions'), not to the equator as one would today; the equatorial coordinates only came into general use in the 18th century.

Fergus J. Wood

Bibliography

Woolard, E.W. and Clemence, G.M. (1966) *Spherical Astronomy*. New York: Academic Press.

Cross references

Celestial mechanics
Coordinate systems
Orbit

EJECTA

Material thrown out of a crater. The mechanism of ejection may be expanding gases, as in a volcanic crater, or particle motions induced by a shock wave, as in an impact crater. The ejection process in impact is remarkably orderly, with early-time, near-surface ejecta being launched with high velocity and travelling great distances. At progressively later times in crater formation, ejecta come from deeper sources, are launched with relatively lower velocity, and are deposited nearer the crater rim. The final materials ejected have the deepest original source. Thus, the ejecta blanket displays a reverse stratigraphy. There is also a progression in grain size in the material ejected, with the early ejecta being finest and the late ejecta being the coarsest grained. Ejecta deposits at impact craters can be divided into near-rim continuous ejecta and further field discontinuous ejecta. Rays of ejecta also occur and can range out to considerable distances from the crater rim. For example, the ejecta rays from the crater Copernicus (diameter = 96 km) on the Moon extend for several hundred kilometers.

When the ejecta impacts the exterior ground surface, it is moving with considerable velocity and may erode and incorporate local material, thus forming a ground-surge deposit. The percentage of secondary local material incorporated in the primary ejecta is also a function of distance from the crater, with more secondary material at greater ranges. This is because the primary ejecta at these ranges are relatively high-speed ejecta and have more erosive power.

The presence of an atmosphere (Earth, Venus) can complicate ejection dynamics, as can volatiles in the target rocks (Earth, Mars). Volatile release and atmospheric lofting can lead to extensive dispersion of very fine-grained ejecta and, in major impact and volcanic events, distribution of fine ejecta may be continental or even global in extent (e.g. the K–T boundary layer). Volatiles and atmospheric entrainment can also lead to fluidized ejecta and produce unusual ejecta patterns such as around the so-called 'rampart impact craters' on Mars.

Richard A.F. Grieve

Bibliography

Oberbeck, V.R. (1975) The role of ballistic erosion and sedimentation in lunar stratigraphy. *Rev. Geophys. Space Phys.*, **13**, 337–362.
Hörz, F., Ostertag, R. and Rainey, D.A. (1983) Bunte Breccia of the Ries: continuous deposits of large impact craters. *Rev. Geophys. Space Phys.*, **21**, 1667–725.

Shoemaker, E.M. (1962) Interpretation of lunar craters, in *Physics and Astronomy of the Moon* (ed. Z. Kopal), Academic Press, pp. 283–359.

Cross references

Breccia
Crater
Cretaceous–Tertiary boundary
Impact cratering
Iridium anomaly
Moon: geology

ELECTROMAGNETIC RADIATION

X-rays, visible light, radiant heat and broadcast television signals are different forms of electromagnetic radiation. Each can be described as an electric vector *E* (or magnetic vector *H*; Figure E28) which propagates in free space with a velocity c close to 3×10^8 ms^{-1}. Monochromatic radiation has a characteristic frequency f and wavelength $\lambda = c/f$; in some cases orthogonal components (for example, E_x and E_y in Figure E28) are synchronized so that a polarization can be defined.

Most natural radiation is a composite of many frequencies and is (at most) only weakly polarized. In planetary sciences electromagnetic radiation is commonly associated with remote sensing, typically a passive and distant activity. But naturally occurring radiation is often a fundamental component of the physical processes taking place. Its distribution with frequency (wavelength) can be diagnostic of source characteristics (see Airglow) or properties of a medium with which it has interacted (see Reflectance spectroscopy). Proper characterization and interpretation of the radiation can be immensely important to understanding these processes.

Anthropogenic radiation typically is well defined in both frequency and polarization; it is often modulated for transmission of information (see Global Positioning System) or for environmental probing (see Radar astronomy). To maximize detectability such signals are generated over narrow ranges of frequencies; the desired signal strength compared with natural 'interference' can then be made quite high. Use of directional antennas can improve signal detectability along certain paths.

Radiation and matter interact, sometimes strongly. At very low frequencies energy can be exchanged between slowly propagating electromagnetic waves and charged particles (see Whistler; Plasma waves). Blackbody radiation is the broadband (multifrequency) emission of objects that are physically hot.

Richard A. Simpson

Bibliography

Ulaby, F.T., Moore, R.K. and Fung, A.K. (1981) *Microwave Remote Sensing: Volume 1 – Fundamentals and Radiometry*. Reading, MA: Addison-Wesley.
Elachi, C. (1987) *Introduction to the Physics and Techniques of Remote Sensing*. New York: John Wiley and Sons.
Kraus, J.D. (1966) *Radio Astronomy*. New York: McGraw-Hill.
Jackson, J.D. (1962) *Classical Electrodynamics*. New York: John Wiley and Sons.

Cross references

Airglow
Global Positioning System
Plasma wave
Radar astronomy
Reflectance spectroscopy
Whistler

ELLIPSOID

Some planetary bodies more closely approximate the shape of an ellipsoid than of a sphere. An ellipsoid is a second-degree geometric solid of which all plane cross-sections are ellipses. In its general triaxial form, with each of the three axes different in length, the analytic equation is

$$x^2/a^2 + y^2/b^2 + z^2/c^2 = 1$$

where a, b and c are its semiaxes in the directions of the x, y and z coordinate axes respectively. A sphere is a limiting form for which a, b and c are equal. Generally, major non-spherical planetary bodies take the form of an oblate ellipsoid of revolution. In this case, if the z axis is the polar axis, a and b are equal to each other and greater than c, and the surface may be generated by rotating an ellipse of semimajor axis a and semiminor axis c about its minor axis. This surface of revolution is also often called a spheroid since the shape is usually only slightly flattened from that of a sphere.

More commonly, the dimensions of the ellipsoid of revolution are given for the generating ellipse, and the above equation is changed to

$$x^2/a^2 + y^2/b^2 = 1$$

where the y axis extends along the polar axis and b is the semiminor axis. The flattening f is then $(1 - b/a)$, and the eccentricity e of the ellipse (and ellipsoid) is $(2f - f^2)^{1/2}$.

The most measured ellipsoidal shape in the solar system has of course been that of the Earth, beginning with the Cassinis in 18th-century France. After measuring the curvature of the meridian through Paris, but only in France, the Cassinis concluded that the Earth is a prolate ellipsoid, elongated rather than flattened at the poles. This contradicted Isaac Newton's conclusions from the law of gravitation, so the French Academy of Sciences sent survey teams to Lapland and Peru (now Ecuador). By 1738 their measurements had confirmed Newton's prediction of oblateness. After numerous ground-based surveys along various meridians, leading to many regionally adopted 'figures of the Earth' with varying dimensions for the reference ellipsoid, satellite geodesy began to be used in the 1960s for these measurements. This led to the current GRS80 figure, adopted by the IUGG at Canberra in 1979. For this ellipsoid a is taken as exactly 6 378 137 m, and the flattening f as approximately 1/298.257. The WGS84 figure frequently referenced may be considered identical to the GRS80.

Mars is the only other planet with a surface rigid enough for topographic mapping and a shape taken to be non-spherical. Its

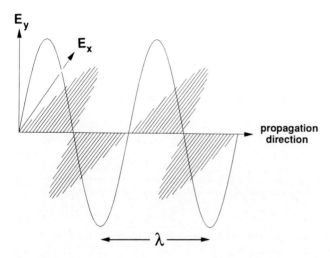

Figure E28 A representative plane electromagnetic wave with electric vector components in both the x and y directions, propagating in the $+z$ direction. The frequency f of the wave is the number of wave crests an observer fixed in space would detect during one second. The wave velocity c depends on properties of the medium; for vacuum $c = 3 \times 10^8$ m s^{-1}. The wavelength λ – the distance between adjacent crests or troughs – is then $\lambda = c/f$. Although only electric components are shown here, the wave could be described equally well by its magnetic components; the E_x electric field is paired with an H_y magnetic field, while the E_y field has a magnetic counterpart in the $-x$ direction.

adopted dimensions are $a = 3393.4$ m and $f = 1/192.8$. Saturn's satellites Mimas and Enceladus are also treated as ellipsoids of revolution for mapping, while Mars' satellites Phobos and Deimos are among those officially treated as triaxial ellipsoids, although they deviate substantially from the latter simplification.

Triaxial ellipsoids have also been considered for the Earth, Mars and even the more spherical Moon, but the locations of axes have varied widely with the investigator; such figures are considered more academic than usable.

John P. Snyder

Bibliography

Greeley, R. and Batson, R.M. (1990) *Planetary Mapping*. Cambridge: Cambridge University Press, 296 pp., pp. 141–76.

Cross references

Cartography
Coordinate systems
Earth rotation
Geoid
Planetary rotation

EMISSIVITY

With the possible exception of black holes, all celestial objects exchange radiation with other objects in their environments through two principal processes – emission and absorption. An ideal black body absorbs 100% of the energy incident upon its surface; its emission is dependent strictly on its temperature (see Blackbody radiation). No object behaves as a perfect black body; the degree to which an object approaches the ideal varies with a number of conditions such as the texture of its surface, details of the composition and the observing frequency or wavelength.

Emissivity is the ratio of the actual emitted energy to that expected from a black body with the same size, shape and temperature when viewed under the same conditions; it is often expressed as a function of frequency (or wavelength). For objects with smooth surfaces the emission may vary with an observer's view angle, and the emitted radiation may be stronger in one linear polarization than the other. For smooth surfaces the emissivity is the complement of the reflectivity under certain conditions (see reflectivity).

Most solar system bodies are too cold to radiate measurably at visible wavelengths, so emissivity is generally reserved to discussions of planetary behavior at infrared and radio wavelengths. Rocks and soils often exhibit emissivities of 90–95% over these wavelength ranges, but certain ices and high dielectric materials can have anomalously low values.

Richard A. Simpson

Bibliography

Elachi, C. (1987) *Introduction to the Physics and Techniques of Remote Sensing*. New York: John Wiley and Sons.
Pettengill, G.H., Ford, P.G. and Wilt, R.J. (1992) Venus surface radiothermal emission as observed by Magellan. *J. Geophys. Res.*, **97**(E8), 13091–102.
Ulaby, F.T., Moore, R.K. and Fung, A.K. (1981) *Microwave Remote Sensing: Volume 1 – Fundamentals and Radiometry*. Reading, MA: Addison-Wesley.
Ulaby, F.T., Moore, R.K. and Fung, A.K. (1986) *Microwave Remote Sensing: Volume 3 – From Theory to Applications*. Dedham, MA: Artech House.

Cross references

Absorption, absorption spectrum
Blackbody radiation
Reflectivity

ENCELADUS

Enceladus is the seventh of 17 moons of Saturn, measuring outward from the innermost (Plate 29). It is one of the smaller of the classical satellites, having a radius of 197 km but is the brightest moon in the Saturn system, with a bond albedo (total reflectivity) of 0.89 (89%). Enceladus' orbit is nearly circular (eccentricity 0.004) and lies four Saturn radii from the planet almost directly in the orbital plane (Morrison, Owen and Soderblom, 1986). Enceladus is an enigma, being not much larger than tectonically inactive Mimas but with regions which appear to have been very extensively resurfaced by mobile fluids. Some smooth plains and ridged plains areas have few or no craters at all (Rothery 1992). Equally intriguing is Enceladus' brightness: it must have been very recently coated with fresh, icy material. Its orbit is also the locus of the E ring of Saturn, which must be resupplied frequently over geologic time against loss of material.

Together with the presence of uncratered and hence youthful surfaces, the conclusion is that Enceladus very probably possesses volcanic activity, where the volcanic fluids are water or more volatile materials. A likely volcanic fluid for Enceladus is an ammonia–water mixture, which melts from ice and solid ammonia hydrate at 176 K or lower. If such volcanism continues to the present, or nearly so, a heat source for melting the ammonia–water must be identified. Enceladus, like Mimas, is too small for heating by radiogenic elements to keep the interior sufficiently warm. Enceladus is distinct from Mimas by being in a position to have been tidally heated in recent times. Its current eccentricity is not enough to cause melting, but if the eccentricity had been appreciably greater at some earlier stage it is possible that small amounts of tidal heating could maintain a thin liquid shell to the present. While a slim possibility, there is little else to go on in understanding this shiny and enigmatic object.

Jonathan I. Lunine

Bibliography

Morrison, D., Owen, T. and Soderblom, L.A. (1986) The satellites of Saturn, in *Satellites* (eds J.A. Burns and M.S. Matthews). Tucson: University of Arizona Press, pp. 764–801.
Rothery, D.A. (1992) *Satellites of the Outer Planets: Worlds in their Own Right*. Oxford: Clarendon Press.

Cross references

Satellite, natural
Saturn: ring system
Saturn: satellite system
Volcanism in the solar system

ENCKE, JOHANN FRANZ (1791–1865)

Johann Encke was the son of a Hamburg pastor. During the Napoleonic Wars he served in the horse artillery, and then went as an assistant at the Seeberg Observatory. In 1825 he became Director of the Berlin Observatory, and remained in this post for 40 years, finally retiring to Spandau in 1864.

Encke accomplished much useful work, and was an excellent observer. However, his fame rests mainly upon his cometary work and upon his association with the discovery of the planet Neptune. In 1818 J.L. Pons discovered a comet; Encke computed the orbit and concluded that the comet was identical with those previously seen in 1786, 1795 and 1805. The period was given as 3.3 years and Encke successfully predicted the next return, that of 1822. Fittingly, the comet now bears his name. It has since been seen at every return except that of 1944; in fact it can now be followed all round its orbit. The identification of Neptune, in 1846, was due to the calculations of the French astronomer U.J.J. Leverrier, who contacted the Berlin Observatory and requested astronomers there to search in the expected position. Encke granted permission, and the planet was identified, by J. Galle and H. D'Arrest, on the first night of the search.

Patrick Moore

Bibliography

Cronk, G.W. (1984) Comets. New York: Enslow Publishers, pp. 243–6.
Encke, J.F. (1846) Account of the Discovery of the Planet of Le Verrier at Berlin. *Mon. Not. Roy. Astron. Soc.*, **7**, 153.
Freiesleben, H.C. (1971) Encke, Johann Franz. *Dict. Sci. Biogr.*, Vol. 4, pp. 369–70.
Moore, P. (1996) *The Planet Neptune*, New York: John Wiley, pp. 22–4.

ENSTATITE METEORITES

The term 'enstatite meteorites' is used to describe collectively several small groups of meteorites (see Meteorite) which differ in many respects but which have in common a high state of reduction. Thus unlike other meteorites, the iron in the enstatite meteorites is entirely in the metallic and sulfide forms, while their silicates are remarkably Fe-free (see Chondrites, ordinary and Carbonaceous chondrite). The metallic phase of enstatite meteorites even contains Si (despite the strength of the Si–O bond), and there are a great many sulfide minerals in these meteorites which also involve elements normally bound to oxygen. These are absent from other meteorites and extremely rare or unknown on Earth. The enstatite meteorites even contain nitride minerals.

Another distinctive feature of these meteorites is the low Mg/Si ratio of the silicate fraction (approximately 0.85) compared to that of other early solar system materials (the solar ratio is 1.05). As a result of this, and the low oxidation state, the dominant silicate mineral is enstatite ($MgSiO_3$).

A recent list of the known enstatite meteorite stones was published by Keil (1989). Other enstatite meteorites are the Horse Creek iron meteorite and the Mount Egerton stony iron meteorite.

Classification

While sharing these unusual compositional properties, the enstatite meteorites show great diversity in texture and composition and include many very different classes. Horse Creek and Mount Egerton were mentioned above. Of the stones, some are chondrites, having essentially solar composition and often containing chondrules (see below), while others are igneous rocks. These igneous rocks are termed 'enstatite achondrites', the term 'achondrite' arising from their conspicuous lack of chondrules, or they are termed aubrites, after an important member of the class, Aubres. Enstatite meteorites thus span the full range of stone (achondrite and chondrite), iron, and stony iron meteorites.

Enstatite chondrites

The texture of the chondrites, as well as their compositions, suggests agglomeration of their components (silicates, metal and sulfides) in the early solar system without subsequent reprocessing. However, the relative proportions of the metal, sulfide and silicates vary widely. They have been subdivided into two groups, the EH (high iron) and EL (low iron) chondrites, depending on the amount of iron and other siderophile elements in the bulk analysis. The plot of the Fe/Si ratio against the Mg/Si ratio (Figure E29) is thus a convenient way of summarizing the enstatite meteorite classes in terms of their bulk compositions (Sears *et al.*, 1982). A recent count included 21 EH chondrites, 19 EL chondrites and 13 enstatite achondrites, but this number changes frequently as new discoveries are reported.

Like the other chondrite classes, (see Carbonaceous chondrite; Chondrites, ordinary), many enstatite chondrites contain chondrules. These are approximately spherical aggregates of silicate grains encased in glass which may subsequently have crystallized.

The space between the grains is filled with silicate, sulfide and metal grains. Very little of the fine-grained matrix characteristic of the other chondrite groups is present.

Aubrites

The aubrites are igneous rocks consisting of coarse-grained enstatite with small amounts of other silicates and trace amounts of metal and iron sulfide. This mineralogy is reflected in their bulk compositions;

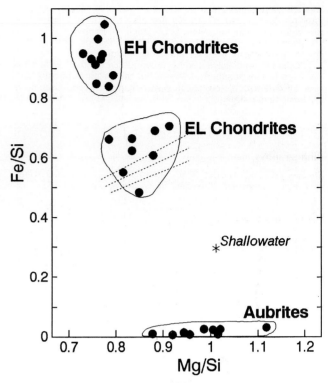

Figure E29 Plot of Fe/Si against Mg/Si for the bulk compositions of enstatite meteorites. These data enable the EH, EL and aubrite classes of enstatite meteorites to be resolved. Plotting in the field between the EL chondrites and the aubrites is the unusual metal-rich Shallowater aubrite. The broken lines refer to meteorites whose Si content is unknown, so their exact placement on the plot is unclear.

not only are their Mg/Si ratios lower than that of the Sun, but they are also highly depleted in siderophile and chalcophile elements (e.g. Fe/Si; Figure E29).

Except for Shallowater, the aubrites are all breccias. Some are fragmental breccias, composed of angular pieces, while a few are regolith breccias (see Breccia). Regolith breccias consist of rock fragments embedded in a soil-like matrix which is rich in inert gases. The elemental and isotopic patterns of the gases show that they were deposited by the solar wind. The process occurred while the meteorites were part of a regolith similar to that on the lunar surface (Keil, 1989). This suggests a near-surface location on the parent body for the aubrites.

Mineralogy

In addition to enstatite, metal and troilite (FeS), the enstatite chondrites contain minor amounts of graphite, schreibersite (($Fe, Ni)_3P$), sphalerite (ZnS), oldhamite (CaS), daubreelite ($FeCr_2S_4$), various crystallographic forms of SiO_2 and the silicates, plagioclase and olivine. There are several mineralogical differences between EL and EH chondrites, some reflecting differences in bulk composition while the others reflect differences in metamorphism on the parent body. Thus in general, only the EH chondrites contain djerfisherite (a complex potassium sulfide containing Na, Cu, Fe and Ni) and niningerite (MgS, with some dissolved FeS), while the EL chondrites contain sinoite (Si_2N_2O) and alabandite (MnS with some FeS). The EH chondrites also contain greater quantities of metal and sulfide than the EL chondrites, and the plagioclase in EH chondrites is albite ($NaAlSi_3O_8$) while in EL chondrites it is oligoclase (a solid solution of albite and 15 mol% anorthite, $CaAl_2Si_2O_8$; Keil, 1968). The composition of the unusual metal found in enstatite meteorites and of the so-called 'cubic sulfides' (FeS–MgS–MnS) is summarized in graphical form in Figure E30.

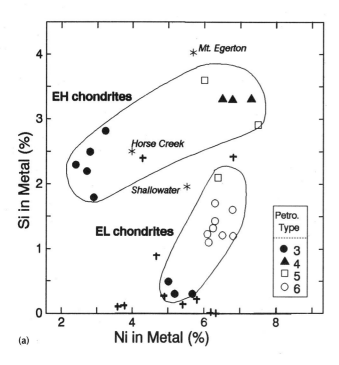

Aubrites consist predominantly of enstatite with lesser amounts of sulfide and metal, small amounts of plagioclase, olivine and diopside, and trace amounts of the sulfides discussed above and phosphides (Watters and Prinz, 1979). They resemble the EL chondrites more closely than the EH chondrites in several respects, for example the composition of their metal grains and cubic sulfides (Figure E30). Osbornite (TiN) has been observed in several aubrites.

Metamorphism

Petrographic types

Like most chondrites, the EH and EL chondrites have experienced elevated temperatures which caused various solid state changes ('metamorphism') on their parent bodies. The petrographic types of Van Schmus and Wood (1967) divide the chondrites into types 1 to 6 according to metamorphism, where the higher value refers to greater levels of metamorphic alteration. Types 1 and 2 are reserved for aqueously altered carbonaceous chondrites, so that enstatite chondrites cover types 3 to 6. The type is usually written alongside the class, e.g. EH5 or EL3. During metamorphism, mineral grains became compositionally homogeneous; the chondrule glass characteristic of the low types is absent. The enstatite has been converted from the monoclinic to the orthorhombic crystal structure; and textures were obliterated to the extent that it is very difficult to detect chondrules in type 6 enstatite chondrites. At the present time EH chondrites of types 3–5 are known while EL chondrites of type 3 and 6 are known. This is changing almost daily as new meteorites are being recovered, especially in the Antarctic.

Metamorphic temperatures

The temperatures experienced by the enstatite chondrites can be estimated from equilibria between the various sulfide minerals, and between the phosphides and metal (Zhang, Benoit and Sears, 1992). An example of the method, and an important feature of the results, is illustrated in Figure E30b. Inserted on Figure E30b are five isotherms showing the equilibrium compositions at these temperatures. Similar isotherms could have been drawn on Figure E30a. Type 4 and two type 5 EH chondrites appear to have equilibrated at around 600°C, one of the EH5 and the EH3 chondrites seem to have equilibrated well below 600°C, while for Shallowater the value is 900°C. In contrast, all the EL chondrites plot well below the 600°C value, even though some are type 6 and the aubrites are igneous. The simplest interpretation is that the meteorites lying well below the 600°C isotherm have either experienced (1) very little metamorphism (which is apparently the case for the type 3 chondrites), or (2) they cooled very slowly following metamorphism and were able to continue to react down to temperatures << 600°C. This suggests a very different physical history for the EH and EL chondrites (Skinner and Luce 1971; Zhang, Benoit and Sears, 1992). In contrast, equilibration temperatures involving the silicates and oldhamite are around 1000°C, suggesting that these systems were unaffected by metamorphism at low temperatures and may preserve a record of pre-accretionary (nebular) processes.

Type 3 enstatite chondrites

Enstatite chondrites of type 3 have been recognized only recently. Several EH3 chondrites are known (Prinz et al., 1984) but only three fragments of a single EL3 chondrite are known (Lin et al., 1991; Chang, Benoit and Sears, 1992) although there are clasts of EL3 material in the unusual Kaidun meteorite (Ivanov and Ivanov 1989). Type 3 enstatite chondrites contain heterogeneous minerals, including enstatite, which is sometimes unusually FeO-rich for the class, and abundant well-defined, glassy chondrules. Unlike the higher petrographic types, they also contain significant amounts of olivine (Mg_2SiO_4 containing small amounts of Fe_2SiO_4). There are also detailed differences in the composition of the metal and sulfides between type 3 and the higher types (Figure 30a,b), reflecting their different equilibration temperatures. Type 3 enstatite chondrites are enriched in volatile elements relative to the higher types and there is experimental evidence that this is because metamorphic heating caused volatile element loss (Biswas et al., 1980).

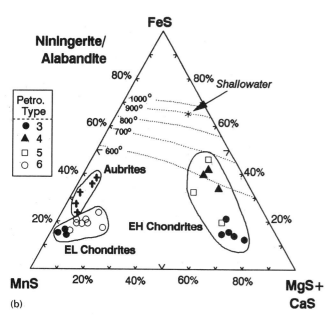

Figure E30 The compositions of metal and sulfide in enstatite meteorites are among their most characteristic properties. (a) The metal is noted for its Si content which is here compared with Ni. The EH and EL chondrites plot in discrete fields within which the Ni and Si in the metal increase with petrographic type. The aubrites (plus symbols) plot near or in the EL field, Mount Egerton and Horse Creek plot with the EH chondrites while Shallowater is intermediate. (b) The curved lines on this FeS–MnS–MgS/CaS triangular plot refer to experimentally determined values observed at the indicated temperatures (Skinner and Luce 1971). The EH chondrites and the Shallowater aubrite contain niningerite (Mg-rich solid solution) and the equilibration temperatures decrease along the series Shallowater, EH4, 5 and EH3. The EL chondrites and the other aubrites contain an Fe-rich alabandite (MnS-rich solid solution) with very low equilibration temperatures. The equilibrium curves below 600°C are unknown, but equilibration temperatures for EL3 appear to be lower than those for EL6 chondrites.

Unusual enstatite meteorites

Happy Canyon is an EL6 chondrite with an igneous texture (Olsen *et al.*, 1977). It is probably an impact melt (Keil, 1989). Shallowater is an unusual aubrite, rich in metal (3.3 vol%) and is unbrecciated. Keil *et al.* (1989) recently suggested that it has experienced a multistage cooling history on a separate parent body from the other enstatite meteorites.

The Horse Creek iron meteorite contains 2.5% Si, similar to the metal in EH chondrites, and there are lamellae of the Fe–Ni–Si phosphide, perryite (Buchwald, 1975). The metallic phase in the Mount Egerton stony iron meteorite resembles that of Horse Creek while its silicate portion resembles the aubrites (McCall, 1965). The silicates in the Tucson iron meteorite contain Fe-free olivine and pyroxene, and the metal phase contains Si, so an affinity towards the enstatite chondrites can be inferred (Nehru *et al.*, 1982). However, the abundance of olivine and different oxygen isotope patterns (see below) suggest that the link is weak.

There have been numerous reports of clasts resembling enstatite meteorite materials in other groups of meteorites, such as in the unusual Allan Hills A85085 and Kaidun carbonaceous chondrites and the Bencubbin stony iron meteorite. Allan Hills, Kaidun and Bencubbin are sometimes regarded as CR chondrites, so there may be a link between the enstatite meteorites and the CR chondrites.

History and origin

It is widely assumed that, like the other chondrite classes, the enstatite chondrites are aggregations of nebular materials including chondrules, metal and dust. However, their unusual compositions and their highly reduced state are a major challenge to the simple chemical models which work so well for the other chondrite classes. The observed mineral phases suggest a non-solar C/O ratio, but it is unclear how this was produced.

The enstatite meteorites have formation ages of about 4.5 Ga (Minster, Ricard and Allegre, 1979), comparable with the Earth, Moon and Sun, and they formed over a small interval of time, ~ 50 Ma, soon after the end of nucleosynthesis. There is some evidence that the EL chondrites formed about 4 Ma after the EH chondrites (Kennedy *et al.*, 1988) (see Chronology: meteorite).

Since the aubrites have igneous textures it has often been suggested that they were formed by a melting process acting on material resembling the EL6 chondrites (e.g. Fogel, Hess and Rutherford, 1988). This has been disputed on several grounds, most notably the Ti content of the FeS (which is much higher than can readily be explained by conversion of enstatite chondrites to aubrites) and the phase composition of the aubrites (Brett and Keil, 1986), but also the lack of instances where aubrite and enstatite chondrite material coexist in the same breccia, although the matter is still unresolved. Several authors have also discussed the possibility that the bulk composition of the aubrites was determined by primary (i.e. nebular) processes prior to the igneous phase of their history (Sears, 1980).

Most authors accept that the enstatite meteorites spent most of their solar system history in the asteroid belt, and several small rare asteroids have spectral reflectivity similar to that of the enstatite meteorites (Zellner *et al.*, 1977) (see Asteroid). However, others have discussed an origin for the enstatite meteorites much closer to the sun (Wasson and Wetherill, 1979; Sears, 1980). Oxygen isotope data for the aubrites, EH chondrites and EL chondrites distinguish them from the other meteorite classes and suggest not only a genetic link between the three classes of enstatite meteorite but also a possible link with the Earth and Moon (Clayton *et al.*, 1984).

The abundance of cosmic ray-produced isotopes indicates that the enstatite chondrites were fragmented to meter-sized objects 0.5–200 Ma ago, aubrites having been exposed to cosmic rays for longer than enstatite chondrites (Wasson and Wai, 1970) (see Cosmic ray exposure age).

Derek W.G. Sears

Bibliography

Biswas S., Walsh, T., Bart, G. and Lipschutz, M.E. (1980) Thermal metamorphism of primitive meteorites – XI. The enstatite meteorites: origin and evolution of a parent body. *Geochim. Cosmochim. Acta*, **44**, 2097–110.

Brett R. and Keil, K. (1986) Enstatite chondrites and enstatite achondrites (aubrites) did not come from the same parent body. *Earth Planet. Sci. Lett.*, **81**, 1–6.

Buchwald, V.F. (1975) *Handbook of Iron Meteorites*. Berkeley, California: University of Berkeley Press.

Clayton, R.N., Mayeda, T.K. and Rubin, A.E. (1984) Oxygen isotopic compositions of enstatite chondrites and aubrites, in *Proc. Lunar Planet. Sci. Conf.* **15**, *J. Geophys. Res., Suppl.*, **89**, pp. C245–9.

Chang, Y., Benoit, P.H. and Sears, D.W.G. (1992) Bulk compositional confirmation of the first EL3 chondrite and some implications. *Lunar Planet. Sci.*, **23**, pp. 217–8.

Fogel, R.A., Hess, P.C. and Rutherford, M.J. (1988) The enstatite chondrite–achondrite link, in *Lunar Planet. Sci. XIX*, pp. 342–3.

Ivanov, A.V. and Ivanov, B.A. (1989) The Kaidun meteorite: estimation of the impact velocity of the meteorite parent bodies, in *Lunar Planet. Sci. XX*, pp. 444–5.

Keil, K. (1968) Mineralogical and chemical relationships among enstatite chondrites. *J. Geophys. Res.*, **73**, 6945–76.

Keil, K. (1989) Enstatite meteorites and their parent bodies. *Meteoritics*, **24** 195–208.

Keil, K., Ntaflos, Th., Taylor, G.J. *et al.* (1989) The Shallowater aubrite: evidence for origin by planetesimal impacts. *Geochim. Cosmochim. Acta*, **53**, 3291–307.

Kennedy, B.M., Hudson, B. Hohenberg, C.M. and Podosek, F.A. (1988) 129I 127I variations among enstatite chondrites. *Geochim. Cosmochim. Acta*, **52**, 101–11.

Lin, Y.T., Nagel, H.-J., Lundberg, L.L. and El Goresy, A. (1991) MAC88136 – the first EL3 chondrite, in *Lunar Planet. Sci. XXII*, 811–2.

McCall, G.J.H. (1965) A meteorite of a unique type: Mount Egerton stony-iron. *Mineral. Mag.*, **35**, 241–9.

Minster, J.-F., Ricard, L.-P. and Allegre, C.J. (1979) ^{87}Rb–^{87}Sr chronology of enstatite chondrites. *Earth Planet. Sci. Lett.*, **44**, 420–40.

Nehru, C.E., Prinz, M. Weisberg, M.K. and Delaney, J.S. (1982) The Tucson iron meteorite and its relationship to enstatite meteorites, in *Proc. Lunar Planet. Sci. Conf.*, **13**, pp. A365–73.

Olsen, E.E., Bunch, T.E., Jarosewich, E. (1977) Happy Canyon: a new type of enstatite chondrite. *Meteoritics*, 12, 109–23.

Prinz, M., Nehru, C.E., Weisberg, M.K. and Delaney, J.S. (1984). Type 3 enstatite chondrites: a newly recognized group of unequilibrated enstatite chondrites (UEC's), in *Lunar Planet. Sci. XV*, 653–4.

Sears, D.W. (1980) The formation of E chondrites – a thermodynamic model. *Icarus*, 43, 184–202.

Sears, D.W., Kallemeyn, G.W. and Wasson, J.T. (1982) The compositional classification of chondrites – II. The enstatite chondrite groups. *Geochim. Cosmochim. Acta*, **46**, 597–608.

Skinner, B.J. and Luce, F.D. (1971) Solid solutions of the type (Ca, Mg, Mn, and Fe) S and their use as geothermometers for the enstatite chondrites. *Am. Mineral.*, **56**, 1269–96.

Wasson, J.T. and Wai, C.M. (1970) Composition of the metal, schreibersite and perryite of enstatite achondrites and the origin of enstatite chondrites and achondrites. *Geochim. Cosmochim. Acta*, **34** 169–84.

Wasson, J.T. and Wetherill, G.W. (1979) Dynamical, chemical and isotopic evidence regarding the formation locations of asteroids and meteorites, in *Asteroids* (eds T. Gehrels and M.S. Matthews). Tucson: University of Arizona Press, pp. 926–74.

Watters, T.R. and Prinz, M. (1979) Aubrites: their origin and relationship to chondrites, in *Proc. Lunar Planet. Sci. Conf.*, **10**, pp. 1073–93.

Zellner, B., Leake, M., Morrison, D. and Williams, J.G. (1977). The E asteroids and the origin of enstatite achondrites. *Geochim. Cosmochim. Acta*, **41**, 1759–67.

Zhang, Y., Benoit, P.H. and Sears, D.W.G. (1992). The thermal history of enstatite chondrites. *Meteoritics*, **27**, 310–1.

Cross references

Achondrite meteorites
Carbonaceous chondrite
Chondrites, ordinary
Cosmic ray exposure ages
Iron
Meteorite

EOLIAN TRANSPORT

Eolian (wind) processes are one mechanism by which a planetary surface can be altered, provided that planet has an atmosphere, particles on its surface and winds of sufficient strength to move them. Eolian processes are capable of producing distinctive features over a wide range of scales. Their importance in a terrestrial context has long been recognized; the fundamental starting reference for physics of eolian studies is still the famous study by Ralph Bagnold (1941), and eolian processes and products are well covered in geomorphology texts (e.g. Cooke and Warren, 1973; Bloom, 1991). The damage caused by the wind during the Dust Bowl days of the 1930s has prompted much study and major revisions in agricultural practices (Chepil and Woodruff, 1963). Spacecraft observations of eolian landforms on Mars and Venus have led to the consideration of eolian activity in a comparative planetology context (Plate 16; Greeley and Iversen, 1985). Eolian features have also been found on Neptune's moon Triton, and may also occur on Titan, the largest moon of Saturn.

Particles move in a fluid or gas in three different modes: suspension, where the particle is buoyed by the upward component of flow turbulence; saltation, where the particle is bounced along the surface; and traction, where the particle is rolled along the surface. The latter is not very significant in the eolian case, particularly on Earth. The boundary between each mode of transport is a function of particle size and weight, atmospheric density, and wind speed. 'Duststorm' is the proper term for a large quantity of material moving in suspension; in a true sandstorm the particles saltate and typically stay within a few meters of the surface, with clear air above.

Suspension and saltation produce distinctive surface morphologies, either depositional, transportational or erosional, or a combination of the three. In the terrestrial environment only clay and silt-sized particles can travel long distances in suspension. Suspended particles do not interact with the surface while in transport, nor do dust-laden winds cause significant erosion. Airborne particles usually settle out of the atmosphere without forming an identifiable deposit, with two important exceptions. Large-scale deposits, termed loess, can occur close to the source region of the suspended material. A famous example is found on the east side (immediately downwind) of the Mississippi River; the source was the large quantity of silt being produced by glacial grinding during the last ice age being transported by the river. The other exception, wind streaks, are more common on Mars (and Venus) than they are on Earth, where their presence is masked by the effects of running water. Martian and Venusian wind streaks are thought to be due to the addition or removal of a thin layer of dust (Thomas et al., 1981; Greeley and Iversen, 1985; Greeley et al., 1992). Martian duststorms are common, ranging in size up to storms that involve the entire planet.

Saltation of sand-sized particles ($\sim$ 50–2000 μm) produces a wide variety of surface features. Ergs, or sand seas, occur in places where a large quantity of sand accumulates over time. In regions of net sand transport, two sizes of landforms result: ripples and dunes. Ripples range in wavelength (separation) from 0.5 cm to 25 m and in height from 0.1 cm to 25 cm or more, depending on wind speed, average particle size and sand size distribution (Greeley and Iversen, 1985). Several models have been proposed for their formation mechanism. Bagnold (1941) believed ripple wavelength was directly related to the average distance traveled by the sand in each saltation bounce, Sharp (1963) proposed ripple height as a controlling factor on ripple wavelength, and Anderson (1987) separated the set of saltating particles into those that bounce repeatedly and those that are splashed from the surface by the nearby impact of a saltating particle and then cease moving, with the ripple wavelength being directly related to the latter. Ripples have been tentatively identified at the Viking Lander 2 site on Mars (Greeley and Iversen, 1985).

Dunes come in a variety of shapes and sizes. Longitudinal dunes are symmetric sand ridges aligned parallel with the prevailing wind direction. They are common on Earth but are found nowhere on Mars. Sand saltates along their length, but the dune itself does not migrate. Their formation mechanism has been a subject of controversy (Tsoar, 1982). Transverse dunes are parallel sand ridges that are aligned perpendicular to the prevailing wind and have asymmetric cross-sections. Sand is transported from dune to dune and the set of transverse dunes migrates downwind. Numerous examples can be found on Mars (Tsoar, Greeley and Peterfreund, 1979) and there appear to be two fields of transverse dunes on Venus (Greeley et al.,

Figure E31 Transverse barchanoid ridges in the Cadiz Valley, Mojave Desert, California, about 60 km ENE of the town of Twentynine Palms. The dark area above the dunes is part of the playa of Cadiz Dry Lake. The numerous creosote bushes (Larrea Tridentata) are 1–2 m across.

1992). Barchan dunes form in areas of net sand transport. They are crescent shaped with the horns of the crescent pointing downwind, partially wrapping around the slipface, a portion of the dune that lies at the angle of repose (longitudinal and transverse dunes often have small slipfaces along their tops). Sand is deposited on the stoss (upwind) side of the barchan, then it saltates over the crest, or highest part, of the dune to the brink (top of the slipface), where there is a separation of the airflow. Sand periodically avalanches down the slipface, and the dune slowly advances downwind. Barchan dunes have been found on Mars, and changes in them observed over the duration of the Viking mission indicate that they are presently active (Tsoar, Greeley and Peterfreund, 1979). Transverse dunes that have some barchan-like features are called transverse barchanoid ridges (Figure E31). A parabolic dune, also called a blowout dune, is similar to a barchan in shape, but not in orientation, as its horns point upwind and the slipface is on the outside of the dune's curve. Parabolic dunes form when the ends of a transverse dune are stabilized, usually by vegetation, but the middle portion keeps migrating downwind. Another common type of dune is the star dune, a large (up to 300 m high, 2 km in width) peaked mound with radiating arms. Star dunes are probably caused by prevailing winds from multiple directions. Other types of dunes and variations on those described above can be found in Cooke and Warren (1973), McKee (1979), Greeley and Iversen (1985) and Lancaster (1988).

One type of wind erosion is deflation, the removal of material from a flat surface. Deflation can uniformly lower a surface several meters or more and can affect very large areas. Where the material being deflated contains particles too large for the wind to move, deflation can remove the fine material and act to concentrate the coarse material on the surface (Cooke and Warren, 1973). Another type of wind erosion is eolian abrasion, where sand in saltation strikes and erodes rocks and less well-consolidated materials. Rocks so abraded are termed ventifacts, and the streamlining of their shapes can be quite spectacular. The susceptibility of rock materials to eolian abrasion is a function of target and impactor properties, impact angle and is proportional to the cube of the speed of the impacting particle (Greeley et al., 1982, 1985).

Wind erosion can produce large, elongated ridges, termed yardangs. Their streamlined shape has been likened to inverted boat hulls. They range in height from 1 m to over 100 m and in length from a few metres to over 30 km; yardang fields can cover large areas (Greeley and Iversen, 1985). Similar fields of yardangs have been found in several locations on Mars (Ward, 1979) and there appears to be at least one similar field of yardangs on Venus (Greeley et al., 1992).

Steven H. Williams

Bibliography

Anderson, R.S. (1987) A theoretical model for eolian impact ripples. *Sedimentology*, **34**, 943–56.

Bagnold, R.A. (1941) *The Physics of Blown Sand and Desert Dunes*. London: Meuthen.
Bloom, A.L. (1991) *Geomorphology: A Systematic Analysis of Late Cenozoic Landforms*, 2nd edn. Englewood Cliffs: Prentice Hall.
Chepil, W.S. and Woodruff, N.P. (1963) The physics of wind erosion and its control. *Adv. Agronomy*, **15**, 211–302.
Cooke, R.U. and Warren, A. (1973) *Geomorphology in Deserts*. Berkeley: University of California Press.
Greeley, R. and Iversen, J.D. (1985) *Wind as a Geologic Process*. Cambridge: Cambridge University Press.
Greeley. R., Leach, R.N., Williams, S.H. *et al.* (1982) Rate of wind abrasion on Mars. *J. Geophys. Res.*, **87** (B2), 10009–24.
Greeley, R., Williams, S.H., White, B.R. *et al.* (1985) Wind abrasion on Earth and Mars, in *Models in Geomorphology* (ed. M.J. Woldenberg). Boston: Allen & Unwin, pp. 373–422.
Greeley, R., Arvidson, R.E., Elachi, C. *et al.* (1992) Aeolian features on Venus: Preliminary Magellan results. *J. Geophys. Res.*, **97**(E8), 13319–45.
Lancaster, N. (1988) *A Bibliography of Dunes: Earth, Mars, and Venus*. NASA Contractor Report 4149.
McKee, E.D. (ed.) (1979) *A Study of Global Sand Seas*. United States Geological Survey Professional Paper **1052**.
Sharp, R.P. (1963) Wind ripples. *J. Geol.*, **71**, 617–36.
Thomas, P., Veverka, J., Lee, S. and Bloom, A. (1981) Classification of wind streaks on Mars. *Icarus*, **45**, 124–53.
Tsoar, H. (1982) Internal structure and surface geometry of longitudinal (seif) dunes. *J. Sediment. Petrol.*, **52**, 823–31.
Tsoar, H., Greeley, R. and Peterfreund, A.R. (1979) Mars: the north polar sand sea and related wind patterns. *J. Geophys. Res.*, **84**(B14), 8167–80.
Ward, A.W. (1979) Yardangs on Mars: evidence of recent wind erosion. *J. Geophys. Res.*, **84**(B14), 8147–66.

Cross references

Aerosol
Dust
Erosion
Mars: geology
Venus: geology and geophysics

EPHEMERIS

An ephemeris (plural: ephemerides, pronounced Eff-uh-MERR-i-Deez) is defined to be a tabular listing of the position of a celestial body at regular intervals. Throughout history scientifically observant cultures have sought to understand and predict celestial phenomena, most notably the motions of the Sun, Moon and planets.

Any effort to establish an ephemeris involves the development of some sort of mathematical model which reproduces past and current observations. This model is then evaluated at regular extrapolated intervals to obtain the desired predicted positions. The models devised by ancient astronomers consisted of elaborate systems of what were termed epicycles, wherein the Sun and planets were attached to a hierarchy of circles. The center of each of these circles rode on another circle, and the entire system ultimately was assumed to have the Earth as its center. Despite this unwieldy complexity the system described celestial motions with fair success over short periods, though it was unworkable over long time spans.

In the Renaissance the invention of the telescope, the rise of mathematics and Isaac Newton's theory of universal gravitation established celestial mechanics as a respectable scientific endeavor. Kepler deduced that planets follow elliptical orbits with the Sun at one focus; in actuality the orbits differ slightly from ellipses due to perturbations by other celestial bodies.

Any representation of the motion of a celestial body as a mathematical function to be evaluated is termed a theory. Because the Moon and planets follow orbits that are in fact nearly circular, the natural ingredients for any theory representing their motions are series of trigonometric functions, similar to Fourier series but more elaborate. The coefficients of the trigonometric terms are usually polynomials in time and functions of other parameters; the trigonometric arguments are themselves functions of time and of parameters describing the orbits of the perturbing bodies.

From the 17th century through to the present, lunar, planetary and satellite theories of increasing refinement have been developed. Until the advent of computers, theories were the only means of ephemeris representation. While theories are accurate over comparatively long time spans, all suffer from the limitations of the large amount of time required to evaluate them.

Numerical integration

The 1960s were a watershed for celestial mechanics and ephemerides: electronic computers became widely available, navigation of spacecraft to the Moon and planets required knowledge of celestial positions with unprecedented accuracy, and the precision and technology of observations increased substantially. The approximations given by analytic lunar and planetary theories usually cannot meet the accuracy requirements of space missions, necessitating the use of numerical integration.

All current high-precision lunar and planetary ephemerides are produced by numerical integration of the differential equations of motion. Even for complex models, the equations can be formulated concisely. The results of the integration are converted and stored as a set of interpolating polynomials. These polynomials can be evaluated at periodic intervals for purposes of printed tabulation (as published annually in many astronomical almanacs throughout the world), or made available to subsequent computer programs for obtaining the positions of celestial bodies at arbitrary times.

Most of the numerically integrated ephemerides in use throughout the world are produced by the Jet Propulsion Laboratory (JPL) of the California Institute of Technology in Pasadena, California. In computer form, JPL ephemerides typically cover a few decades to perhaps a century; DE102, a considerably longer high-precision integrated ephemeris, spans more than 44 centuries, from 1410 BC to 3002 AD (Newhall, Standish and Williams, 1983).

Mathematical model

Planetary ephemerides

The JPL planetary and lunar ephemerides are computed simultaneously by numerical integration. The planetary force model includes contributions from (1) relativistic point-mass interactions of the nine planets, the Sun and the Moon in the isotropic, parametrized post-Newtonian n-body metric (see Newhall, Standish and Williams, 1983, for the equations of motion); (2) the Newtonian effects of nearly 300 asteroids on those same bodies; (3) the interaction of the point-mass Sun on the figures of the Earth and Moon; and (4) the effects of the principal non-spherical gravitational term (J_2) of the Sun on the Moon and planets.

Lunar ephemeris

Treatment of the lunar ephemeris employs a more extensive model to accommodate the accuracy of laser ranging data. Besides the relativistic point-mass effects, the model of the Earth–Moon system includes (1) interaction of Moon and Sun on the Earth figure; (2) interaction of Earth and Sun on the Moon figure; and (3) the effects on the Moon on the Earth's solid and ocean tides raised by the Sun and Moon.

In addition to the position of the Moon in its geocentric orbit, the physical librations (the rotational position of the Moon about its center of mass) must be integrated numerically. The librations are represented as a set of three Euler angles defining the orientation of the Moon with respect to the Earth's mean equator and equinox of the epoch J2000: ϕ, the angle along the Earth's equator from the equinox to the ascending node on the Moon's true equator; θ, the inclination of the Moon's true equator to the Earth's equator; and ψ, the angle along the Moon's equator from the node on the Earth's equator to the selenographic prime meridian.

The mathematical model for librations represents the response of the Moon to torques arising from action of the point-mass Sun, the point-mass Earth, and the Earth figure on the lunar figure. The lunar moment of inertia is modeled as having a rigid-body component and a time-varying component arising from a delayed dissipative elastic response to the deformation caused by Earth-induced solid tides and by the lunar angular velocity.

Satellite ephemerides

The third area of development of ephemerides involves those of the satellites of Mars and of the giant planets Jupiter through Neptune. (Satellite ephemerides are not considered in the modeling of planetary and lunar ephemerides. Each of the outer planets and its family of satellites are treated as if combined into a point mass – a satisfactory approximation, as the perturbations on other planets due to the actual satellites are small.)

The ephemerides of Phobos and Deimos, the two satellites of Mars, are represented as theories. The most recent developments are by Sinclair (1989), based on Earth-based observations and on spacecraft imaging during the Mariner 9 mission in 1971 and the two Viking missions in 1976.

In 1977 two Voyager spacecraft were sent from Earth toward the outer planets. Accurate satellite ephemerides were essential to the success of the missions. Both Voyagers visited Jupiter and Saturn; Voyager 2 continued on past Uranus and Neptune. Up to that time the ephemerides of outer-planet satellites were determined solely from Earth-based telescopic observations. The two missions provided considerable improvement in ephemerides (and estimated masses) of the satellites.

Unlike the case of planetary and lunar ephemerides, numerical integration of satellite orbits has met with success only for the Uranus and Neptune systems; it has not been used for the Galilean satellites of Jupiter or for the eight principal satellites of Saturn. The numerical integration and parameter estimation procedure fails because of nonlinearities in the satellite systems, arising from (1) the large number of revolutions that occur for the satellites over a given time span, and (2) strong resonances between various satellite pairs. For the planetary and lunar ephemerides, a reasonably linear system, 25 years comprises about 330 revolutions of the Moon and 104 revolutions of Mercury. By contrast, Io, the innermost major satellite of Jupiter, completes nearly 5200 revolutions in 25 years; Mimas, the closest large satellite to Saturn, about 9700 revolutions.

Nonlinearity obstructs integration and estimation as follows: the integration process consists of specifying the position and velocity of each satellite at some initial epoch and then using the equations of motion to propagate the states forward with the integration. In general, any error in the initial states will grow over the span of the integration, yielding a progressively degraded ephemeris. In usual estimation techniques, corrections to the initial states are made by comparing observations of the satellites with computed values based on the integration and then adjusting the initial conditions accordingly. This technique works only if corrections to the initial states are proportional to measured errors in the orbit (i.e. if the system is linear), which is not the case for the Jupiter or Saturn systems.

The original theory for the Galilean satellites of Jupiter was developed in the early part of this century by Sampson (1921). Lieske (1976) at JPL produced an improved theory of the Galilean satellites using a computer for algebraic manipulation. The theory for the satellites of Saturn used for the Voyager missions was taken from the summary and references found in *Explanatory Supplement to the Ephemeris* (1961) and improved by fits to Earth-based and spacecraft observations; a more complete theory is that given by Duriez and Vienne (1991) and continued in Vienne and Duriez (1991).

Construction of ephemerides

The JPL ephemerides have been instrumental in the success of NASA planetary and lunar missions. In the early days of unmanned space exploration the ephemerides were derived from planetary and lunar theories. With time, significant advancements in spacecraft tracking systems and data accuracy required analysis beyond the scope of approximations offered by theories and necessitated the use of numerical integration.

Data types

The orbits and other parameters related to the planets are determined by a least squares fit to various types of observations. The observations span most of the 20th century (those from earlier times have large uncertainties). The following types of observations are used.

- Optical meridian transits. The disk of the Sun or a planet is observed through a telescope to cross the meridian. Between 1911 and 1982 the US Naval Observatory in Washington DC made several thousand transit observations of the Sun and Mercury through Neptune. In 1984 the photoelectric meridian transit was introduced, where a photocell replaced the human observer. Transits of Pluto have been obtained since 1988.
- Astrolabe. The disk of a planet is observed through a telescope to cross a fixed altitude, both ascending and descending. Observations of Mars through Uranus were begun in 1969. Accuracies are typically from 0.3″ to 1.6″.
- Photographic Astrometry. Before 1988 the observations available for Pluto were from photographic plates. From plate measurements, Pluto's coordinates are established relative to selected reference stars. This data type is also used for planetary satellites other than the Moon.
- Satellite Eclipses. Times of disappearance and reappearance of a planetary satellite due to passage through the shadow of its primary are recorded.
- Occultation Timings. Uranus occasionally passes between the Earth and a star. In 1977 the planet was found to have rings; subsequent measurements of the time and duration of the blocking or occultation of the star's image by the rings give the celestial coordinates of Uranus to less than 0.2″. Similar occulations by the disk of Neptune beginning in 1968 establish Neptune's coordinates to about 0.3″.
- Radio Astrometry. Beginning in 1983 the Very Large Array in New Mexico made radio observations of the thermal emission of Jupiter, Saturn, Uranus and Neptune; the observed celestial coordinates are accurate to about 0.03″.
- Ranges. The above observation types are measurements of angular position of the designated body as seen on the celestial sphere. Enter the ranging era, in which the measurement is the distance between an observing station and a celestial object. Beginning in 1970, Mercury and Venus were ranged by radar, with an uncertainty of about 1.5 km. (By comparison, a typical 1″ angular measurement error at Mercury is about 750 km in transverse linear distance.) Ranges to spacecraft in the vicinity of Mercury, Mars and Jupiter have uncertainties as small as 500 m. The Viking landers on Mars were ranged between 1976 and 1982 with uncertainties of 7 m.

  Between 1969 and 1972 the Apollo astronauts placed three arrays of laser retroflectors on the lunar surface; a fourth reflector array was landed by the Russian–French Lunakhod 2 spacecraft. Over the following years more than 8000 laser ranges between terrestrial observatories and the four lunar reflectors were acquired. Lunar ranges since 1990 have uncertainties as small as 2 cm or less; both the Viking ranges and the best lunar ranges have relative accuracies of better than five parts in 10^{11}.
- Optical Navigation. Ephemerides of the outer planets and their satellites had errors of 1000 km or more before the Voyager missions. During approach to each planetary system, cameras on board the spacecraft obtained images of the planet and its satellites against a background of cataloged stars. These images were then used to modify the spacecraft trajectory and to refine the satellite ephemerides. Optical-navigation data and radio observations have reduced the uncertainty in the Jupiter ephemeris to about 100 km and in the Saturn ephemeris to about 500 km.

Representation of ephemerides

In contrast to printed tabulations, ephemerides used by computers are stored as coefficients of an interpolating polynomial, allowing a user to obtain ephemeris values at arbitrary times. The polynomials of choice for the JPL ephemerides are Chebyshev polynomials (Newhall, 1989). They are stable during interpolation and provide a reliable estimate of the errors introduced due to truncating at a selected polynomial degree. The standard for interpolation error in the JPL ephemerides is 0.5 mm. (This figure denotes the precision to which interpolated values match the original numerical integration, not the actual dynamical state of the solar system.)

Ephemeris parameter estimation

Many physical parameters in the solar system can be determined to considerable accuracy from ephemeris data. The creation of the lunar and planetary ephemerides typically entails estimating about 150 parameters, including the initial positions and velocities of the planets and Moon, initial values of the lunar librations, station and reflector locations, lunar gravity model, lunar elasticity coefficients,

Earth and lunar tidal dissipation, polar motion, the angular position of the Earth (UT1), precession and nutation of the Earth's pole, the astronomical unit (average distance between the Earth and Sun), and the masses of the Moon, planets and major asteroids. (In practice the actual mass of a celestial body is difficult to determine; the quantity that is well determined and is instead estimated is GM, the product of the gravitational constant G and the mass M of the body.) A few results from parameter estimation follow.

Relativity

For the Earth–Moon system, laser ranging has provided an accurate confirmation of the Strong Principle of Equivalence (that gravitational mass M_G is equal to inertial mass M_I in the General Theory of Relativity. The formulation used expresses the ratio of the two mass quantities as

$$\frac{M_G}{M_I} = 1 + \eta \frac{U_G}{Mc^2}$$

where η is a dimensionless parameter, U_G is the gravitational self-energy of the Earth, M is the mass of the Earth and c is the velocity of light. (For the Earth, $U_G/Mc^2 \approx 5 \times 10^{-10}$). By exploiting the fact that a disparity between the two would cause a sunward displacement of the lunar orbit, estimation of the lunar ephemeris has found that, for the Earth, $\eta = 0.003 \pm 0.004$

Some theories of gravity suggest that the gravitation 'constant' G is not constant at all but may be time varying in some fashion. The usual form in which a variable G is expressed is the ratio $\dot{G}/G$. Laser ranging has found that $\dot{G}/G$ is zero to at least as small as a few parts in 10^{11} per year. Other relativistic quantities are the post-Newtonian parameters measuring curvature and superposition of gravitational fields and geodetic precession of the lunar orbit. All determinations are consistent with the predictions of General Relativity, both from lunar laser ranging and from planetary ranging.

Earth-related quantities

Another quantity improved by laser ranging is the effect of the Earth tides on the lunar orbit. The Moon and Sun raise both solid and ocean tides on the Earth, which in turn affect the lunar orbit. The most pronounced effect is the secular acceleration of the geocentric lunar longitude, whereby the tidal bulge on the Earth lies ahead of the Earth–Moon line and tends to add energy to the lunar orbit. This energy addition causes the average Earth–Moon separation to increase, with an attendant slowing of the Moon in its orbit. The measured acceleration of longitude is -25.9 arc sec/century^2, corresponding to the Moon receding at about 3.8 cm per year.

Planetary gravity

A fruitful use of the ephemerides is the navigation of robotic spacecraft and the determination of the gravity fields of the Moon and planets. The fine structure of those fields is irregular due to aspherical mass distribution. The fields are customarily represented by coefficients of spherical harmonics. Precise determination of the trajectories of planet-orbiting spacecraft provides estimations of a large set of gravitational coefficients.

Planetary masses

Any non-orbiting spacecraft passing near a planet or satellite provides a good determination of GM for the body. Over the course of the past two decades values of GM for all planets and major satellites (except for the Pluto–Charon system) have been reliably determined from spacecraft data.

Other ephemeris applications

Spacecraft mission planning

Planned spacecraft missions, particularly to the outer planets and their satellites, face severe navigation constraints. Payload-imposed fuel limitations often necessitate gravitational assists from intermediate planets in order to supply sufficient energy for the spacecraft to reach the target. Sensitivity of the final arrival trajectory to errors at intermediate planets and satellites can be extreme; accurate ephemerides and GM determinations are essential to devising reliable fuel and payload budgets and navigation strategies.

Observations and data reduction

Ephemeris accuracy is necessary for acquisition of data, as in planetary microwave observations, where angular position is important, and radar ranging, where round-trip travel times must be known *a priori* to within a few microseconds. The reduction of data also demands accurate ephemerides: for the Earth and target body in the case of radar ranging, and for the Earth alone when analyzing the timing of received signals from millisecond pulsars.

Planet X

For several years some astronomers stated that the best-fit predicted orbit of Uranus differed from the actual orbit. They attributed this discrepancy to a possible massive tenth planet, designated 'Planet X.' Several attempts to find it telescopically were made, but without success. Then in 1989 the close approach of Voyager 2 to Neptune provided a significant improvement in that planet's GM. When this new value was included in the numerical integration the Uranus orbit differences largely disappeared, casting substantial doubt on the existence of Planet X.

Summary

Accurate planetary, lunar and satellite ephemerides are essential for successful space missions. Ephemerides originally were produced from mathematical theories but, with the exception of the satellites of Jupiter and Saturn, are today generally computed by numerical integration. Accurate computation of ephemerides has provided estimates of numerous quantities relating to the planets and satellites and to the theory of relativity.

X.X. Newhall

Bibliography

Duriez, L. and Vienne, A. (1991). A general theory of motion for the eight major satellites of Saturn. I. Equations and method of resolution. *Astron. Astrophys.*, **243**, 263–75.

Explanatory Supplement to the Ephemeris (1961). Washington, DC: Nautical Almanac Office, US. Naval Observatory, pp. 362–86.

Lieske, J. H. (1976). Theory of motion of Jupiter's Galilean satellites, *Astron. Astrophys.*, **56**, 333–52.

Newhall, X.X. (1989) Numerical representation of planetary ephemerides. *Celest. Mech.*, **45**, 305–10.

Newhall, X.X., Standish, E.M. and Williams, J.G. (1983). DE102: a numerically integrated ephemeris of the Moon and planets spanning forty-four centuries, *Astron. Astrophys.*, **125**, 150–67.

Sampson, R.A. (1921) Theory of the four great satellites of Jupiter, *Mem. Roy. Astron. Soc.*, **63**, 1–270.

Sinclair, A.T. (1989). The orbits of the satellites of Mars determined from Earth-based and spacecraft observations. *Astron. Astrophys.*, **220**, 321–8.

Vienne, A. and Duriez, L. (1991) A general theory of motion for the eight major satellites of Saturn. II. Short-period perturbations. *Astron. Astrophys.*, **246**, 619–33.

Cross references

Celestial mechanics
Gravity-assist navigation
Hohman transfer orbit
Orbit
Planetary dynamical astronomy

ERATOSTHENES (276–195 BC)

A member of the so-called Alexandrian school, Eratosthenes was born in Cyrene (now Shahhat, North Africa), and studied in Athens under Callimachus. There he earned a reputation as a poet, playwright (of comedies) and as a literary man, but truly versatile, he also took an interest in astronomy and mathematics. In about 240 BC he was appointed chief of the great library at Alexandria in Egypt, which had become essentially Greek, following its conquest by Alexander the Great (in 332 BC). The burning of that library by a rebellious crowd in 391 AD destroyed much of Eratosthenes' original material, but fortunately some copies had reached Baghdad and have been passed down through Arabic translations.

Eratosthenes will always be remembered as the mathematician-astronomer who first computed the circumference of planet Earth. He was author of a book on mathematically based Earth studies called *Geographica* (Smith, 1986). He also worked out a system of chronology, fixing classical dates back to the conquest of Troy.

An account of Eratosthenes' classic measurement reached us through the Greek writer, Cleomedes (Taton, 1963, p. 324). Using a portable gnomon, Eratosthenes established that at Aswan (then known as Syene) on midsummer's day at noon, the summer solstice, the Sun was vertically overhead. This was confirmed by having a deep well dug there; at this critical hour the sides of the well showed no shadows (according to Pliny; Dicks, 1971). At Alexandria Eratosthenes used a scaphe to measure the zenith angle of the Sun, which he found to be $1/25$ of the diameter of the diameter of the scaphe's hemisphere, or $1/50$ of the circumference of a sphere (7.2°). Alexandria lies almost due south of Aswan, a horizontal distance reported to be approximately 5000 stadia. If a degree of arc is equal to 700 stadia, then the Earth's circumference is 250 000 stadia.

So how does the calculation of Eratosthenes match up with modern measurements? A meridional (great circle) circumference is 40 008 km, which would require the stadia value to be about 0.1587 km. The Egyptian stadia of that time was probably 157.5 m, which makes an extraordinarily close approximation. Some critics raise a question about its exact value. Values varied at different times and different places. It was generally based on the 'pace' ($\times 100$), the marching pace (one left + one right) of foot soldiers. However, the marching time from Aswan to Alexandria, partly along the Nile, partly across desert, would be an approximation at best. One standard for the stadia is the length of the race track at the Stadium of Athens and some other famous athletic arenas which can be measured today as 185 m (or 600 Greek feet), but other stadiums in the ancient world are different. Thus the problem remains.

Another astronomic measurement by Eratosthenes was the tilt of the Earth's axis of spin, or the obliquity of the ecliptic (q.v.). This was given as $1/83$ of a meridian circle, which is 23°51'20" (23·8550°), a value only 7' greater than the value calculated for the epoch of 240 BC, which is 23°43' (23·7240°). For comparison, the present-day value (epoch of 1994 AD) is 23.44°. As a result of precession and the motion associated with the Earth's node, the obliquity in these 2240 years has decreased by 0.42°. The yearly shift of precession is 17.3 m; that of the Earth's node, 0.976 m per year. Acting in the same direction, these two factors would shift the latitude of the Tropic of Cancer southward by 41 km, a small correction. It is interesting in this connection that the precession of the equinoxes (see Precession and nutation) was not known at that time and was discovered by Hipparchus (q.v.) only in about 125 BC.

Again, using the same measurement of angles and geometrical calculations customary in those days, Eratosthenes was able to calculate the mean distances to the Moon and to the Sun. He also prepared a unique map of the known world at that time but the work was severely criticized by Hipparchus, although only fragments remain (for details and methods see Taton, 1963).

The name Eratosthenes is also remembered in the label for a 61 km wide crater on the Moon.

Rhodes W. Fairbridge

Bibliography

Bunbury, E.H. (1879) *History of Ancient Geography*. London, Chapter 16 (Dover reprint of 1883 edn: New York, 1959, 2 vols).
Dicks, D.R. (1971) Eratosthenes. *Dict. Sci. Biogr.*, vol. 4, pp. 388–93.
Sanders, J.E. (1995) Astronomical forcing functions: from Hutton to Milankovitch and beyond. *Northeastern Geology and Environmental Sciences*, vol: 17(3), pp. 306–47.
Smith, J.R. (1986) *From Plane to Spheroid*. Rancho Cordova, California: Landmark Enterprises, 219 pp.
Taton, R. (1963) *Ancient and Medieval Science, from the Beginnings to 1450*. New York: Basic Books (transl. A.J. Pomerans), 551 pp.
Thalamas, A. (1921) *La Geographie d'Eratosthene*. Versailles, France.

EROSION

Based on the Latin word *erodere* to gnaw away and the French *érosion* for that process in nature, the term is used in geology for processes that involve the general wearing away of a land surface or the sea floor (submarine erosion); in planetary sciences it affects also the surfaces of any hard celestial bodies and those of artificial satellites on space vehicles (as by particle impacts).

In geology, erosion is part of a complex system with discrete steps: (1) weathering (the breakdown of rocks *in situ* by chemical or mechanical means), (2) Erosion (the acquisition and removal of those weathering products), (3) transport (the shifting of them to a different site), and (d) sedimentation (the settling and deposition of those products in a new locality). While the overall system is a unity (and is sometimes called the 'geocycle'), its intermediate steps involve up to six different agencies: (1) hydraulic, (2) chemical, (3) wind (eolian), (4) glacial, (5) gravitational, and (6) biological.

Extraterrestrial erosion

On the Earth-like planets and natural satellites of the solar system, photographic evidence discloses universal indications of erosion by asteroid and meteoroid impacts. Hydraulic or water-like erosion is evident on Mars, but may be limited only to certain long-period climatic cycles when liquid water was liberated from subsurface permafrost layers.

Rhodes W. Fairbridge

Bibliography

Derbyshire, E., Gregory, K.J. and Haik, J.R. (1979) *Geomorphological Processes*. Folkstone (UK): Dawson Press, 312, pp.
Fairbridge, R.W. (1968). *Encyclopedia of Geomorphology*. New York: Van Nostrand Reinhold, 1295 pp.
Scheidegger, A.E. (1991) *Theoretical Geomorphology* 3rd edn Berlin: Springer-Verlag, 434 pp.
Yatsu, E. (1988) *The Nature of Weathering: an Introduction*. Tokyo: Sozo-Sha Publ. (Maruzen Co.), 624 pp.

Cross references

Eolian transport
Mars: geology
Surface processes
Weathering

EUCRITE METEORITES

Eucrites are igneous meteorites that are characterized by subequal proportions of low-calcium pyroxene (pigeonite) and anorthitic plagioclase. There are two general textural varieties: basaltic eucrites which are fine to medium-grained and formed either in volcanic eruptions or hypabyssal intrusions; and cumulate eucrites, which are coarse-grained and formed in slowly cooled plutons. All eucrites show some indication of shock, the most obvious being pervasive maskelynite (shock-melted isotropic plagioclase). Among the basaltic eucrites, some retain the original Fe/Mg zoning in pyroxene, whereas others with similar textures show homogeneous Fe/Mg in pyroxene as well as exsolution lamellae of augite, indicating a prolonged episode of thermal metamorphism. Some eucrites may constitute an entire meteorite, while others form polymict breccias made up solely of eucrite clasts, and still others form breccias called howardites that are mixtures of eucrite and diogenite clasts (diogenites are a separate class of meteorites that consist primarily of coarse crystals of magnesian orthopyroxene with minor olivine).

Mineralogically eucrites are very similar to low-titanium lunar mare basalts. Chemical analyses show that eucrites are similar to mare basalts in terms of a low oxidation state and low abundances of water and volatile elements (Basaltic Volcanism Study Project, 1981). Eucrites also have siderophile element depletions relative to primitive chondrites that are similar but not identical to those in mare basalts; these depletions indicate that the eucrite parent body underwent separation of metal during core formation (Hewins and Newsom, 1988). Isotopic analyses show that the eucrite parent body formed from the same oxygen reservoir as the Earth and the Moon (Clayton, 1977), suggesting that all three bodies initially formed in the same part of the solar system. However, incompatible lithophile trace elements (e.g. the rare earths) in eucrites are nearly unfractionated from their chondritic ratios, indicating that the source region of the eucrites had not undergone a crust-forming event similar to that of the Earth or Moon.

The relation between eucrites and diogenites is a matter of considerable debate. The classic theory is that eucrites are the derivatives of more primitive magmas that first crystallized the diogenites (Mason, 1963). The more recent theory is that eucrites and diogenites formed as the result of two episodes of melting (Stolper, 1977): the first produced eucritic magmas by direct melting of an undifferentiated source, the second produced higher temperature diogenitic magmas by remelting of the eucrite-depleted source. Nearly continuous variation in pyroxene composition between diogenites and eucrites (Ikeda and Takeda, 1985) is the strongest evidence for the classic theory. The clustering of the compositions of most of the volcanic eucrites near a reaction point (olivine reacts with magmatic liquid to form pyroxene and plagioclase) is the major evidence for Stolper's theory because only melting and not crystallization is likely to produce a clustering of magma compositions at such a reaction point. Another objection to the classic theory is the absence of volcanic equivalents of diogenites, despite the fact that diogenite magmas are calculated to be less dense than eucrite magmas and hence easier to erupt (Longhi and Pan, 1989).

Eucrites have crystallization ages of approximately 4.5×10^9 years (Basaltic Volcanism Study Project, 1981). These very old ages and the lack of evidence of igneous fractionations in the eucrite source region suggest a small parent body that lost its heat rapidly and stopped its internal evolution soon after formation. An asteroid is thus a likely parent body. The rocky asteroid Vesta (q.v.), which has a diameter of 538 km and a reflectance spectrum similar to that of eucrites (Gaffey, Bell and Cruikshank, 1989), is often cited as a potential parent body or at least a large fragment of the parent. Calculations predict that collisions of $\sim$ 100-m sized asteroids with smaller debris can eject material from the asteroid belt and that subsequent perturbations of this material by the gravitational field of Jupiter are sufficient to deflect a small amount of it into Earth-crossing orbits (Zimmerman and Wetherill, 1973).

John Longhi

Bibliography

Basaltic Volcanism Study Project (1981) *Basaltic Volcanism on the Terrestrial Planets*. New York: Pergamon Press, 1286 pp.
Clayton, R.N. (1977) Genetic relationships among meteorites and planets, in *Comets, Asteroids and Meteorites* (ed. A.H. Delsemme), University of Toledo, Ohio, pp. 545–50.
Gaffey, M.J., Bell, J.F. and Cruikshank, D.P. (1989) Reflectance spectroscopy and asteroid surface mineralogy, in *Asteroids II* (eds R.P. Binzel, T. Gehrels and M.S. Matthews), Tuscon: The University of Arizona Press, pp. 98–127.
Ikeda Y. and Takeda, H. (1985) A model for the origin of basaltic achondrites based on the Yamato 7308 howardite. *Proc. Lunar Planet. Sci. Conf.* **15**, in *J. Geophys. Res.*, **90**, C649–63.
Hewins, R.H. and Newsom, H.E. (1988) Igneous activity in the early solar system, in *Meteorites and the Early Solar System* (eds J.F. Kerridge and M.S. Matthews). Tucson: The University of Arizona Press, pp. 73–101.
Longhi, J. and Pan V. (1989) The parent magmas of the SNC meteorites. *Proc. Lunar Planet Sci. Conf.* **19**, pp. 451–64.
Mason, B. (1963) Meteorites, *Am. Sci.*, **51**, 429–55.
Stolper, E.M. (1977) Experimental petrology of eucritic meteorites. *Geochim. Cosmochim. Acta*, 587–611.
Zimmerman, P.D. and Wetherill G.W. (1973) Asteroidal source of meteorites. *Science*, **182**, 51–3.

Cross references

Basaltic achondrite meteorites
Meteorite

Figure E32 Leonhard Euler.

EULER, LEONHARD (1707–1783), AND EULERIAN MOTION

The Swiss mathematician Leonhard Euler was born as the son of a Calvinist clergyman in Basel (Switzerland), who provided him much of his early education in mathematics, physics, astronomy, medicine, oriental languages and theology. He was sent to the University of Basel to prepare for the ministry. There he became a close friend of David and Nicolaus Bernoulli and their family of eminent mathematicians. Bernoulli was eventually able to convince Euler's father that Euler should do his major in mathematics. After graduation in 1727, Euler joined Bernoulli at Catherine I's Academy of Sciences at St Petersburg. After serving in the Russian navy and working in the medical section of the Academy, he became professor of physics in 1730 and professor of mathematics in 1733. At this time he also married and left the house of Bernoulli's family.

In 1741 Euler was invited by Frederick the Great to join the Berlin Academy of Science, where he remained for the next quarter of century. He became the director of its mathematics section in 1744 and was elected a foreign member of the Paris Academy of Science in 1755. In 1766 Euler accepted Catherine the Great's offer to become the director of the St Petersburg Academy of Sciences and this offer included the generous accommodation of the members of Euler's large family at St Petersburg. He left Prussia and remained in Russia for the rest of his life.

Euler may be the most prolific mathematician in history; he contributed to almost all areas of pure and applied mathematics established at that time. In pure mathematics he integrated Gottfried Wilhelm Leibniz's (1646–1716) differential calculus and Isaac Newton's (1642–1727) method of fluxions into mathematical analysis (*Introductio in Analysin Infinitorum*, 1748). He laid the foundation for the theory of special functions by introducing the beta and gamma transcendental functions. Euler developed the use of series solutions and elaborated on their convergence properties; he solved linear differential equations and developed partial differential calculus. He made number theory into a science, stating the prime number theorem and the law of biquadratic reciprocity. To the calculus of variation and to topology Euler made a great number of original contributions. Many mathematical notations and their logic base are due to Euler, including the number whose hyperbolic logarithm = 1 'e', i for $\sqrt{-1}$, the Greek letter 'π' for the ratio of circumference to diameter in a circle, and the sigma symbol.

Euler applied the mathematical techniques he invented for pure mathematical reasons to contemporary problems in mechanics and celestial mechanics and also introduced the principle of virtual work. In his book *Mechanica* (1736) he presented Newtonian dynamics for the first time in analytic form. In his theory of the motion of rigid bodies (1765) he laid the foundation of analytical mechanics. The three-body problem of the Earth, Sun and Moon system was solved approximately by Euler in 1753 and he elaborated on this approximations in 1772. With Alexis Claude Clairaut (1713–1765) he studied lunar theory.

Eulerian motion

The rotational motion of a rigid body was first studied by Leonard Euler through a set of angular momentum equations now known as Euler's equations. In particular, the theory describes a wobbling motion for an axially symmetric, rotating body about its center of mass, often referred to as the Eulerian motion, or the '(free) Eulerian nutation' or '(free) Eulerian precession', although it is neither a nutation nor precession in the classical sense.

This motion is a free motion, requiring no external torque. Planetary bodies can presumably undergo Eulerian motions as long as there are proper excitation sources, although only that of the Earth has been observed. The Earth's Eulerian motion is known as the Chandler wobble after its discoverer Seth Chandler (in 1891). However, the behavior of the motion can be quite different from that predicted by the rigid-body Euler theory. For example, the Chandler period, instead of being the characteristic period of 300 days as predicted, is approximately 434 days owing to the non-rigidity of the Earth.

Hans J. Haubold and Benjamin Fong Chao

Bibliography

Berry, A. (1961) *A Short History of Astronomy*. New York: Dover Publications.
Boyer, C.B. (1991) *A History of Mathematics*. New York: John Wiley and Sons.
Munk, W.H and MacDonald G.J.F. (1960) *The Rotation of the Earth*, New York: Cambridge University Press.
Segrè, E. (1984) *From Falling Bodies to Radio Waves: Classical Physicists and Their Discoveries*. New York: W.H. Freedman and Company.
Whittaker, E. (1989) *A History of the Theories of Aether and Electricity*. New York: Dover Publications. (Two Volumes Bound as One.)
Youschkevich, A.P. (1971) Euler, Leonhard. *Dict. Sci. Biogr.*, Vol. 4, pp. 467–84.

Cross references

Chandler wobble
Earth rotation

EUROPA

Europa is the smallest of the four Galilean satellites of Jupiter (Plate 24). Discovered by Galileo in 1610 and named by Simon Marius, these large moons were crucial testimony against the geocentric view of the Solar System. Prior to the *Voyager* encounters in 1979, understanding of Europa was limited to information derived from rotational lightcurves, spectrophotometry and deductions about bulk properties. Subsequently, knowledge of the satellite's basic characteristics greatly improved, and Europa has emerged as one of the most intriguing satellites in the solar system. Europa orbits Jupiter at a distance of 670 900 km (9.397 Jupiter radii), completing one revolution in 3.551 days; it apparently rotates synchronously with its orbital revolution, keeping the same face toward Jupiter. It also participates with Io and Ganymede in the geophysically important Laplace orbital resonance (see Commensurability).

With a radius of 1569 km and a bulk density of 2970 kg m^{-3}, Europa is smaller and somewhat less dense than the Earth's moon. In contrast with the Moon, its surface is nearly pure water ice, as determined by infrared spectroscopy, and structural calculations indicate that Europa is a largely silicate world whose surface ice may extend to a depth of approximately 100 km. Europa is currently geologically active and has very subdued topography. Its surface is covered with unusual markings (Figure E33). Unfortunately, Europa was the Galilean satellite imaged at the lowest resolutions by *Voyager*, so the geologic make-up of much of the satellite's surface is unknown. The orbital resonance leads to significant tidal heating within Europa, though, and it is thought that beneath the surface ice there is an ocean of liquid water. It is the nature of this possible ocean, and the potential for life within it, that most intrigues scientists.

Surface composition

Europa has the highest surface albedo of any of the Galilean satellites, close to 65% on average. At Europa's distance from the Sun, this implies maximum daytime temperatures of $\sim$ 140 K at the equator; diurnally averaged, subsurface temperatures are $\sim$ 105 K at the equator, declining to $\sim$ 50 K the poles (Ojakangas and Stevenson, 1989). The latter is nearly as cold as the surface of Pluto or Triton.

The visible appearance of the satellite is generally bright and white, but there are variations with position on the surface (McEwen, 1986a). Europa's trailing hemisphere (referring to the direction of travel in its orbit) is darker and redder than the leading hemisphere and is especially darker in the ultraviolet. The spatial dependence of these changes is consistent with a combination of causes. One is sputtering of the trailing side by Jovian magnetospheric ions. These ions are dominantly sulfur and oxygen, originating at Io, and as ions they must move with the rotating Jovian magnetosphere, preferentially impacting the trailing hemisphere at high speeds ($\sim$ 100 km s^{-1}). Sulfur atoms become permanently implanted in the surface ice

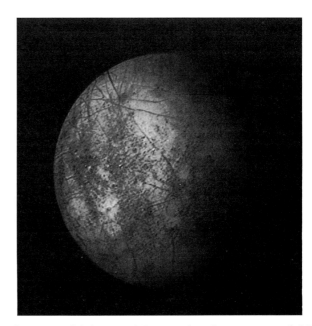

Figure E33 Global view of Europa taken from a range of 1.25 million km. Europa appears to be a smooth icy sphere. Two terrain types, bright plains and somewhat darker, mottled terrains can be seen, as well as abundant lineations. (Voyager 2 image FDS 20625.16; north is at top.)

by this process and form compounds such as SO_2. Sputtered ice is somewhat reddened in general, and sulfur compounds account for the strong ultraviolet absorption; they may also contribute to reddening and darkening in the visible (Johnson et al. 1988).

Added to this is preferential meteoritic bombardment of Europa's leading side (an effect somewhat similar to running through a rainstorm), which keeps the surface of that hemisphere churned up, preventing sputtering and implantation effects from accumulating. In contrast, smaller micrometeorites bombard the satellite from all directions, because their trajectories are strongly affected by charging and electromagnetic forces in the Jovian magnetosphere. These micrometeorites are probably mostly dark, reddish cometary dust particles, and may be primarily responsible for the reddening of the satellite's surface in the visible. All these physical processes are also consistent with hemispheric variations in Europa's water-ice infrared spectra. The water-ice absorption bands are, however, very strong in general, implying that whatever contaminants exist on the surface of the satellite are minor (Clark, Fanale and Gaffey, 1986).

Europa's subsurface has been probed by radar waves transmitted from and returned to terrestrial radio telescopes. Europa's albedo at radar wavelengths is extraordinarily high, well in excess of unity (compared with a perfectly diffusive reflecting sphere), and represents an endmember in the radar behavior of icy surfaces, given that all the icy Galilean satellites are unusually reflective at radar wavelengths. Cold ice is extraordinarily transparent to radar, and the relatively low temperatures and nearly pure ice composition of the Europan surface render it ideal for multiple scattering of radio waves, and in particular, a phenomenon known as coherent backscatter. In this, incoming photons multiply scatter and some reflect and retrace the scattering path, adding coherently to cause a strong signal return in the direction of illumination. The polarization behavior of radar echoes from Europa is also unusual, but matches the polarization properties of coherently backscattered electromagnetic radiation (Hapke 1990; Ostro et al., 1992).

The fine-scale structure of Europa's outermost surface layer also appears unusual for an icy satellite, and may be related to the satellite's active geology, discussed below. Voyager observations as well as those obtained from Earth orbit by the International Ultraviolet Explorer imply that Europa has a remarkably compact surface texture (e.g. Buratti, Nelson and Lane, 1988), while ground-based telescopic measurements show a pronounced increase in reflectivity when Europa is within $\sim 1°$ of opposition to the Sun. This latter has been interpreted as evidence for relatively high porosity, and reanalysis of Voyager photometry determined that Europa's small-scale surface roughness is low compared with that usually generated by impacts (Domingue et al., 1991). These interpretations are not yet reconciled and it may be that other phenomena are at work, such as unusually shaped particles or coherent backscatter at optical wavelengths.

Geology

The geology of Europa is fundamentally different from that of any other icy satellite (Malin and Pieri 1986). Europa's surface contains abundant darker mottlings and linear, arcuate and sinuous features, or lineations. Voyager image resolutions are in general insufficient to determine the nature of the lineations, but the best images (covering about 25% of the surface) indicate that most may be fractures and faults (Figure E34). The overall topographic scale of Europa's surface is quite subdued, and no topographic features appear to exceed 200 m in height above the mean. The most prominent among these are linear and arcuate ridges, seen at very low Sun elevation angles near the terminator (Figure E34). Indeed, if there is any morphological analog for Europa's surface, it is the extensive pack ice of the Earth's polar oceans.

No large craters are seen on Europa's surface, and only a few smaller craters (near or less than 15 km diameter) have been identified at all. The lack of craters and topography may imply that the crust of Europa is rheologically weak and cannot support any substantial geologic structures. There are a few possible sites that mark the sites of former craters (called palimpsests on Ganymede and Callisto), but generally there are no obvious circular gaps in the lineation patterns. This suggests that the explanation for Europa's low crater density is simply that the surface is relatively young. Estimates of how young are necessarily uncertain, but over the last decade a more complete census has been taken of so-called Jupiter-family comets (active and extinct) by E.M. Shoemaker and co-

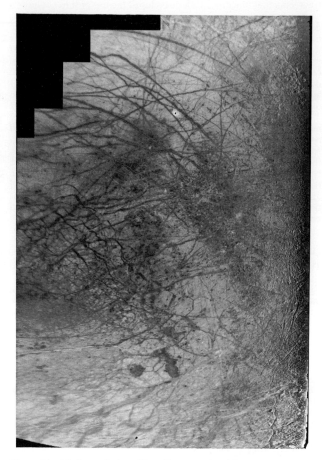

Figure E34 High-resolution, blue filter mosaic of Europa, covering the region from latitude −70° to +50° N and longitude 135° to 215°; projection is simple cylindrical, so that latitude and longitude lines are equally spaced. The image is corrected to normal albedo, and displays mainly plains units, with mottled terrains appearing at right. Global-scale lineations are seen in the upper portion, while the rift zone of wedge-shaped and arcuate dark bands, approximately 1500 km long and 500 km wide, extends from −40° S, 160° longitude to 0°, 210° longitude. Linear and arcuate ridges are seen at the terminator (far right). (Mosaic courtesy of A.E. McEwen, USGS, Flagstaff.)

workers. These comets are the principal impactors of the Galilean satellites at the present time, and if the present population can be taken as representative of recent epochs, no more than 10 to 100 million years are necessary to accumulate either Europa's observed crater or relict crater populations.

The conspicuous lineation pattern on Europa is not arranged in a random or incoherent fashion, but appears to follow an organized geometrical scheme. This is especially true for the so-called global lineations that arc for hundreds and in some cases thousands of kilometers (Figures E33 and E34). Viewed as a coherent fracture system, much effort has been made to deduce the source of the stresses that created the fractures. As an example, an early hypothesis had that Europa once possessed a deep liquid-water ocean that froze from the top down. As this freezing progressed to deeper levels, the accompanying volume increase caused the surface ice to fracture. Unfortunately, this does not result in any sort of oriented stress or fracture pattern. Consequently, a broad variety of other stress sources have been tested, such as shape changes caused by spindown of the satellite due to Jovian tides, orbital retreat of Europa from Jupiter, and present-day flexing as Europa travels in its ($\sim 1\%$) eccentric orbit.

Only one stress source has been shown to provide a satisfactory explanation for the lineation pattern: non-synchronous rotation of Europa's ice shell (McEwen 1986b). While it is most likely that the

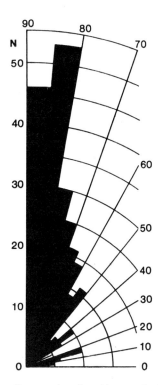

Figure E35 Rose diagram of angles of intersection between lineations on Europa and tensile stress trajectories calculated for a faster than synchronously rotating ice shell. N is the number of lineation segments at a given orientation. The generally high angles are consistent with extension fracturing. This particularly good fit shifts the tidal axis 25° to the east, consistent with fracturing occurring over the last roughly 50° of non-synchronous rotation. (Based on McEwen, 1986b.)

tion by silicates, salts, and even organic materials have been suggested.

Close to the terminator, many dark lineations become visible as low ridges, suggesting that either infilling water freezes and expands, infilling warm ice squeezes upward as a sort of viscous volcanic dyke, or that the region is later subjected to compression and the warmer, rheologically weaker infilling material is preferentially squeezed upward. Other lineations appear to be shallow troughs at the terminator, more directly consistent with extension. The arcuate to cycloidal ridges (Figure E34) are simply enigmatic, but are thought to indicate compression of the crust.

Europa's lineation pattern becomes less recognizable and finally disappears moving from the plains to the mottled terrains (Figure E33). The mottled terrains have a generally lower albedo than the plains, and near the terminator they appear topographically rough (Malin and Pieri, 1986; Figure E34). They may have been formed by multiple intrusion and disruption of plains material and thus be younger than the plains (Lucchitta and Soderblom, 1982), or the plains may have been created by more recent resurfacing and fracturing, and thus be the younger. This and other interpretations of Europa's geologic history must remain tentative, however, given the limited spatial and resolution coverage obtained by Voyager. Advances in understanding based on the planned multiple Galileo flybys of the satellite should be impressive.

Origin

Europa's make-up, predominantly silicate with a surface ice (or ice + water) layer, is unique in the solar system and deserves some comment. Outer solar system satellites, if condensed from material of solar composition, should be very icy. Ganymede and Callisto, circling Jupiter outside Europa's orbit, are roughly 50/50 rock/ice by mass. In contrast, Europa and innermost Io contain relatively little and no ice, respectively. This compositional gradient in some ways resembles that of the planets, whereby metal-rich, rock-rich and finally ice- and gas-rich compositions are encountered as one moves away from the Sun. It has been suggested that Europa would have formed with more ice were it not for the high infrared flux from the young Jupiter, which prevented ice condensation close to the planet (Pollack and Fanale, 1982). An alternative hypothesis suggests that Europa was once very ice rich, like Ganymede, and had differentiated into a rock core and ice mantle, but early in its history it suffered an intense cometary bombardment, focused by the powerful gravitational field of Jupiter, and literally had the bulk of its icy mantle stripped away. Ganymede suffered much less because of its greater distance from Jupiter. Possibly both processes operated, but until there are better theories and models of how the Galilean satellites formed and evolved, we will not know.

Ice shell and ocean

No silicate mountains, volcanoes or impact crater rims stick up through Europa's ice shell, so the overall H_2O layer thickness must exceed a few kilometers. The fracturing and rotation of crustal 'plates' of ice 50–100 km across (Figure E34) imply that the surface layer of brittle, fracturable ice is itself at least a few kilometers thick. The rift zone means that the surface ice is decoupled from the silicate bulk of Europa by warm ice or water in particular, and the fracture pattern evidence for non-synchronous rotation implies that the decoupling is accomplished by liquid water. Mechanical decoupling requires an ice layer at least ~ 10–20 km thick, and the theoretical model of the tidally heated ice shell (Ojakangas and Stevenson, 1989) predicts this thickness. The parameters of the water ocean below (depth, chemistry, etc.) are naturally very poorly constrained, but the very existence of another ocean in the solar system is profound in itself, for where there is water there may be life.

Early post-Voyager speculations focused on intermittent illumination of the ocean through cracks in the ice shell and the possibilities for photosynthesis. Given the plausible thickness of the ice layer and cracking mechanics, this does not appear to be a very likely scenario. It is, however, interesting to note that recent work on the origin of life indicates that life on Earth may have begun in submarine hydrothermal systems (Holm, 1992). The most ancient forms of life on the Earth currently known are, in fact, so-called hyperthermophilic bacteria that live in these environments, utilizing, in complete darkness, the chemical and thermal energy available there. Such life

large silicate core of the satellite is precisely tidally locked to face Jupiter (as is the Moon to the Earth), theoretical calculations predict that an ice shell floating on liquid water should creep forward slightly faster than synchronously, rotating completely once every 10 million years or so (Ojakangas and Stevenson, 1989). This occurs because of a slight mismatch in the position of the thickest part of the ice shell, which is determined by tidal flexing and heating at the base of the shell as Europa moves in its eccentric orbit, and that which is required for it to remain tidally locked. The ice shell itself is predicted to be ~ 10–20 km thick. The orientations of the lineations relative to the stresses caused by non-synchronous rotation are most consistent with tensional or extensional fracturing (Figure E35). This agrees with the limited structural evidence, because the parallel edges of most lineations and lack of observable offsets are most compatible with extension, as opposed to strike-slip motion or compression. Unfortunately, the amount of non-synchronous rotation predicted to occur between the Voyager encounters and the upcoming Galileo encounters in 1996–97 is only about 20 m, well below the resolution of Voyager images and thus undetectable.

The highest resolution images (Figure E34) also illustrate the break-up and separation of part of the Europan shell: fracturing with shear offsets and openings in the region close to the point directly opposite Jupiter have combined to create wedge-shaped and arcuate bands (Schenk and McKinnon, 1989). The sense of extension is in the NE–SW direction, but the source of stress is unknown. The rift openings have apparently been filled with slightly darker and redder (but still quite icy) material. This material has not overtopped the edges of the separated fractures, and so is consistent with water or warm ice filling in from below. In other areas on Europa, slightly darker material has apparently spilled outward from the edges of lineations. The causes of one behavior versus another are not known, nor are the causes of the differing albedo, spectral and photometric properties of the infilling material compared with those of the surrounding plains, although grain-size differences and contamina-

forms may have evolved in the past on Europa when its interior was substantially warmer, and would be stable there today if Europa's silicate core remains geologically active. The latter depends on how much tidal energy goes into heating Europa's core over time, which is related to the overall state and evolution of the three-way Laplace resonance between Io, Europa and Ganymede, a topic of current research.

Whatever the answer, the example of Europa points to the possibility of multiple habitable zones in other solar systems: a radiatively warmed zone within a suitable distance range from the parent star, and a tidally heated zone for satellites at suitable distances from the giant planets they orbit (Reynolds, McKay and Kasting, 1987). The anticipated arrival of the Galileo spacecraft in Jovian orbit in late 1995 may shed some light on whether Europa is within such a zone.

William B. McKinnon

Bibliography

Buratti, B.J., Nelson, R.M. and Lane, A.L. (1988) Surficial textures of the Galilean satellites. *Nature*, **333**, 148–51.
Clark, R.N., Fanale, F.P. and Gaffey, M.J. (1986) Surface composition of natural satellites, in *Satellites* (eds J.A. Burns and M.S. Matthews). Tucson: University of Arizona Press, pp. 437–91.
Domingue, D.L., Hapke, B.W., Lockwood, G.W. and Thompson, D.T. (1991) Europa's phase curve: implications for surface structure. *Icarus*, **90**, 30–42.
Hapke, B. (1990) Coherent backscatter and the radar characteristics of outer planet satellites. *Icarus*, **88**, 407–17.
Holm, N.G. (ed.) (1992) Special issue: marine hydrothermal systems and the origin of life. *Origins Life Evolution Biosphere*, **22**, 1–242.
Johnson, R.E., Nelson, M.L., McCord, T.B. and Gradie, J.C. (1988) Analysis of Voyager images of Europa: Plasma bombardment. *Icarus*, **75**, 423–36.
Lucchitta, B.K., and Soderblom, L.A. (1982) Geology of Europa, in *Satellites of Jupiter* (ed. D. Morrison) Tucson: University of Arizona Press, pp. 521–55.
Malin, M.C. and Pieri D.C. (1986) Europa, in *Satellites* (eds J.A. Burns and M.S. Matthews). Tucson: University of Arizona Press, pp. 689–716.
McEwen, A.S. (1986a) Exogenic and endogenic albedo and color patterns on Europa. *J. Geophys. Res.*, **91**, 8077–97.
McEwen, A.S. (1986b) Tidal reorientation and the fracturing of Jupiter's moon Europa. *Nature*, **321**, 49–51.
Ojakangas, G.W. and Stevenson, D.J. (1989) Thermal state of an ice shell on Europa. *Icarus*, **81**, 220–41.
Ostro, S.J., Campbell, D.B., Simpson, R.A. *et al.* (1992) Europa, Ganymede, and Callisto: new radar results from Arecibo and Goldstone. *J. Geophys. Res.*, **97**, 18 227–44.
Pollack, J.B. and Fanale, F. (1982) Origin and evolution of the Jupiter satellite system, in *Satellites of Jupiter* (ed. D. Morrison). Tucson: University of Arizona Press, pp. 872–910.
Reynolds, R.T., McKay, C.P. and Kasting, J.F. (1987) Europa, tidally heated oceans, and habitable zones around giant planets. *Adv. Space Res.*, **7**, 125–32.
Schenk, P.M. and McKinnon, W.B. (1989) Fault offsets and lateral crustal movement on Europa: evidence for a mobile ice shell. *Icarus*, **79**, 75–100.

Cross references

Callisto
Galilean satellite
Ganymede
Impact cratering
Io
Jupiter: satellite system
Spectrophotometry
Tidal heating

EUROPEAN SPACE AGENCY

The European Space Agency (ESA) was established on 31 May 1975 by merging the European Launcher Development Organisation (ELDO) and the European Space Research Organisation (ESRO), which had both existed since 1964. ESA presently has 13 European member states: Austria, Belgium, Denmark, France, Germany, Ireland, Italy, the Netherlands, Norway, Spain, Sweden, Switzerland and the United Kingdom. Finland is presently an 'associate member'; it will join the Agency on 1 January 1995 as the 14th member. Canada is a 'cooperating state'. The member states gave ESA the task 'to provide for and to promote, for exclusively peaceful purposes, cooperation among European States in space research and technology and their space applications, with a view to their being used for scientific purposes and operational space applications systems'.

ESA has five main centers in Europe: ESA Headquarters in Paris, the European Space Research and Technology Centre (ESTEC) in Noordwijk, the Netherlands, the European Space Operations Centre (ESOC) in Darmstadt, Germany, the European Space Research Institute (ESRIN) in Frascati (near Rome), and the European Astronaut Centre (EAC) in Cologne, Germany. ESTEC is the largest of the ESA centers. It operates several large satellite testing facilities and manages an extensive research and development program. ESA has a total staff of about 2000 and an annual budget of 2.8 billion Accounting Units (1 Accounting Unit = US$ 1.3 in 1993).

To carry out its programs, ESA spends the bulk of its budget on contracts awarded to industry in its member countries. Its policy on this allows each member state to be sure of receiving, in return for its contributions to ESA, both a financial return and a share in the technology spin-off. ESA does not itself develop and manufacture its own spacecraft. Whilst the definition stages of a program are the responsibility of ESA's engineers and scientists working closely with experts from the member states, development work itself is carried out under the watchful eye of ESA staff.

ESA's activities fall into two categories: the programs carried out under the general budget and the science program budget are 'mandatory', all other programs are 'optional'. All the member states contribute to the mandatory program according to their gross national product (GNP) and each has a vote. The optional programs are of interest only to some of the member states who are free to decide on their level of involvement.

Scientific programme

ESA's Directorate of Scientific Programmes and its predecessor ESRO have so far successfully launched 17 satellites for scientific purposes. Recent missions include the following (launch years are in brackets).

Solar system exploration

- Geos 1 (1977) and Geos 2 (1978), to study the Earth's magnetosphere;
- ISEE 2 (1977), as part of the NASA/ESA International Sun/Earth Explorer program, to study the Earth's magnetosphere;
- Giotto (1985) (q.v.), to encounter Halley's comet on 13/14 March 1986 (this was ESA's first interplanetary mission);
- Ulysses (1990) (q.v.), an ESA/NASA collaborative project, to study the interplanetary space out of the ecliptic and over the solar poles.

Astronomy and astrophysics

- COS-B (1975), to study in detail the sources of extraterrestrial γ-rays;
- Exosat (1983), to observe in detail cosmic x-ray sources, such as normal nearby stars, white dwarfs, neutron stars, black holes and supernova remnants;
- IUE (1978), the International Ultraviolet Explorer observatory, a joint project with NASA and the UK, to observe celestial objects in the ultraviolet;
- Hipparcos (1989), to measure very precisely the position, parallax and proper motion of more than 100 000 stars in our Galaxy;
- HST (1990), the cooperative NASA/ESA Hubble Space Telescope to which ESA provided the Faint Object Camera and the solar arrays in return for which ESA has a 15% share of the observing time;
- ISO (1995), the cryogenically cooled Infrared Space Observatory for photometric, spectroscopic and polarimetric observations of celestial objects.

ESA's science program also makes use of multidisciplinary platforms in space:

- FSLP (1983), the First Spacelab Payload, carried eight space science experiments
- Eureca (1992), the European Retrievable Carrier, carried five space science experiments.

The future of ESA's scientific program is laid down in ESA's Long Term Plan 'Horizon 2000' which includes four large 'cornerstone' projects and four less costly, medium-size projects to be launched in the next 10–15 years. The four cornerstones were selected at the outset, the four medium-size missions are selected competitively, following a Call for Mission Proposals which is issued about every three years. Horizon 2000 includes the following already approved missions.

Future cornerstone projects

- STSP, the joint ESA/NASA Solar–Terrestrial Science Programme, consisting of SOHO (1995), the Solar and Heliospheric Observatory, and four Cluster (1995) spacecraft to study the Earth's magnetosphere;
- XMM (2000), a high-throughput x-ray spectroscopy mission;
- Rosetta (2003), a rendezvous with comet Schwassmann-Wachmann 3 in 2010/2011 including one or two landers;
- FIRST (2006), the Far Infrared and Sub-millimetre Space Telescope;

Future medium-size projects

- Huygens (1997) (q.v.), a probe to explore the atmosphere and the invisible surface of Saturn's moon Titan in 2004, as part of NASA's Cassini mission to the Saturn system;
- INTEGRAL (2001), the ESA/NASA/Russian International Gamma-Ray Astrophysics Laboratory to observe the galactic plane and center and to search for extragalactic γ-ray sources;

The third medium-size project will be selected in 1996 and launched in 2004.

Earth observations

ESA's Directorate of the Observation of the Earth and its Environment has developed and launched a series of global weather satellites and, more recently, the ERS-1 (European Remote Sensing) satellite.

During the development phase of the Meteosat programme, ESA launched the weather satellites Meteosat 1, 2 and 3 in 1977, 1981 and 1988, respectively, into geostationary orbits. Since the creation of Eumetsat, the European Meteorological Satellite organization, in 1987 ownership and operations of the Meteosat satellites has been passed to that organization. The Meteosat Operational Programme (MOP) began in 1989 with the launch of Meteosat 4. Meteosat 5 was launched in 1991, Meteosat 6 should go in orbit in 1993 and Meteosat 7 in 1995/96. Working together with their counterparts, the American GOES and the Japanese GMS 'Himawari', the Meteosats have for more than 15 years been involved in surveillance of the planet's atmosphere as part of the World Weather Watch.

The ERS-1 satellite, launched in 1991 into a near-polar orbit at 800 km altitude, is ESA's first climatology satellite. With its advanced instrumentation it watches the changes in shorelines, the ocean currents and sea surface, monitors the growth of crops and vegetation, surveys and maps the polar icecaps, icefields and drifting ice, and watches over the exchanges between the oceans and the atmosphere. A follow-on satellite, ERS-2, will be launched in 1995.

For the future, two series of missions are under development: an environmental series of platforms (Envisat) and a meteorological climate series (Metop), oriented to more routine operational observations. The first Envisat will be launched in 1998, the first Metop in 2000, both into near-polar orbits at 800 km altitude.

Telecommunications

The Directorate of Telecommunications has up to now launched eight telecommunications satellites, all into geostationary orbits, beginning with the launch of OTS (Orbital Test Satellite) in 1978. The series of ECS (European Communication Satellites) consisting of four satellites, launched in 1983, 1984, 1987 and 1988, make up the first generation of European communication satellites using the experience gained with OTS. They carry ten Ku band transponders with a capacity of 12 000 telephone circuits or ten TV programs. The ECSs, renamed Eutelsat-1, etc. after they came into operational service are now run by Eutelsat, the European Telecommunications Satellite Organisation.

To provide communications between mobile stations, especially ships at sea, ESA developed the MARECS satellites (Maritime ECS). Two MARECS were launched, the first in 1981, the second in 1984. During their operational periods they were leased to Inmarsat, the International Organization for Maritime Telecommunications by Satellite.

Launched in 1989, the Olympus experimental telecommunications satellite was at that time the biggest civilian telecommunications satellite in the world. It provides a variety of services, including Ka band (20–30 GHz) telecommunications.

For the communications between satellites ESA is developing the Artemis (Advanced Relay Technology mission) geostationary pre-operational satellite to be launched in 1996, and the two DRS (Data Relay Satellites) in geostationary orbits to ensure European independence in passing data between satellites or platforms in low-Earth orbit and stations on the ground. The first DRS will be launched in 1998/99.

Space station, manned space flight and microgravity

From the outset, ESA took on a share in the US-crewed Space Shuttle program, first by building and supplying Spacelab and then, since 1985, in the Space Station Freedom program, with the 'Columbus Programme'.

ESA built a total of three Spacelabs which were flown on the Space Shuttle on a variety of missions, starting with Spacelab-1 in 1983. For the Spacelab program, ESA selected three European astronauts: Ulf Merbold from Germany who flew on Spacelab-1 in 1983 and on the IML-1 (International Microgravity Laboratory) in 1992, Wubbo Ockels from the Netherlands who flew on Spacelab D1 in 1985, and Claude Nicollier from Switzerland who deployed the Eureca platform from the Shuttle Atlantis in 1992. Six more astronauts, including one woman, were selected in 1992; they will take part in 'precursor flights' on the Space Shuttle, on Spacelab and on the Russian Mir station in preparation for the future operation of ESA's Columbus program. More astronauts will be selected in 1994.

At present, the main element of the Columbus program is a pressurized laboratory 12 m long and 4.5 m wide that will be permanently docked with the international Space Station Freedom. As a result of recent revisions, the Columbus program now takes in a number of flights designed to boost European experience in the fields of crewed spaceflight and in-flight operations until the Space Station Freedom is in place and fully operational.

Spacelab, the attached pressurized laboratory, and Eureca, provide the possibility of space research in the life and materials sciences under microgravity conditions. The free-flying Eureca is particularly suitable as it has a reduced level of parasitic accelerations (10^{-5} g as compared to 10^{-3} g in an orbiting space station or shuttle). Eureca was launched in 1992 into an orbit at 500 km altitude and was retrieved by the Space Shuttle Endeavour 11 months later. Its 1-ton payload consisted mainly of experiments in the life sciences and on the growth of crystals under microgravity.

Ariane program

Based on proven technologies and a European know-how gained in various national launcher development programs, the Ariane program has given ESA a reliable launcher of its own. The Ariane 1 rocket which first flew on 24 December 1979 has become a commercial launcher, taking bookings from many countries for launching their satellites. As the mass of satellites was getting ever larger Ariane 1 gave way, from 1984, to the more powerful Ariane 2 and Ariane 3, and these were in turn superseded when the Ariane 4s arrived in 1988.

As Europe's 'space workhorse', Ariane 4 is available in six versions, one 'bare', the others fitted (depending on the mass to be put in orbit) with two or four solid or liquid strap-on boosters. Arianespace, the company that markets the Ariane launchers, today holds more than half of the world's launch market. With its 55th launch in December 1992 Ariane had put a total of 92 satellites into orbit. The Ariane 5, a new generation launch vehicle (shorter and much wider), will be available from 1995 in several different versions to launch one, two or three satellites at a time.

All Ariane rockets are launched from the Centre Spatial Guyanais (CSG) in Kourou, French Guiana, South America. Being near the equator allows a given rocket to launch larger masses.

Rüdeger Reinhard

Bibliography

ESA's report to COSPAR, about 200 pages, describing the scientific projects, published bi-annually.
ESA Annual Report, about 100 pages, published annually.
ESA Bulletin, about 150 pages, published quarterly.
ESA's quarterly journals in the fields:

Earth Observations
Microgravity
Space Transportation Systems
Technology Programme

These publications are available free of charge from
ESA Publications Division

PO Box 299
2200 AG Noordwijk
The Netherlands

EXOSPHERE

The exosphere can be thought of as the uppermost region of a planetary atmosphere; its upper limit is the height where the density of atmospheric particles approximates that of the interplanetary medium (about 10^2 particles cm^{-3} at the orbit of the Earth). In the case of the Earth the exosphere encloses in succession the thermosphere (q.v.), the mesosphere (q.v.), the stratosphere (q.v.) and the troposphere (q.v.). The lower boundary of the Earth's exosphere is at about 500 km elevation; this is the upper boundary of the thermosphere, with a temperature of approximately 1500 K. At this elevation the density of particles (about 10^7 particles cm^{-3}) is such that collisions become less frequent, and particles with upward velocities have a chance to leave the atmosphere. Thus the exosphere is also known as the 'region of escape.' Hydrogen in particular has a short lifetime in the exosphere. Earth's exosphere extends to perhaps 5000 km elevation; it is enclosed by the planetary magnetosphere.

Earth's Moon has a tenuous atmosphere (see Moon: atmosphere). It is technically an exosphere since particle paths are largely collisionless, and the interplanetary medium (solar wind) interacts directly with the lunar surface. This is also the case with many smaller bodies in the solar system. In such cases the 'atmospheres' are termed 'surface boundary exospheres.'

James H. Shirley

Bibliography

Rasool, S.I. (1967) Planetary atmospheres, in *The Encyclopedia of Atmospheric Sciences and Astrogeology*, (ed. R.W. Fairbridge). New York: Van Nostrand Reinhold, pp. 730–41.

Cross references

Atmosphere
Mesosphere
Moon: atmosphere
Stratosphere
Thermosphere
Troposphere

F

FLOOD BASALT

Flood basalts (formerly 'plateau basalts') are voluminous sequences of fissure-erupted tholeiitic lava flows with associated dike swarms and shallow intrusives. Although generally monotonous in composition, most flood basalt provinces on planet Earth do contain minor amounts of other types of lavas (Table F1) such as nephelinites and picrites (Karoo) and rhyolites (Paraná). The classic flood basalt provinces – the Deccan Traps, Columbia River, Karoo – are situated on continents, hence 'continental' commonly modifies flood basalts in the literature. However, much of the Tertiary North Atlantic volcanic province lies on the ocean floor and there is some evidence that some ocean floor swells may be properly called flood basalt accumulations as well (Richards, Duncan and Courtillot, 1989).

Flood basalt provinces (Table F1) are typically several hundred thousand square kilometers in area, but when allowances are made for erosion and submersion some of the larger provinces exceed 10^6 km^2 in area: e.g. Deccan at 1.5×10^6 km^2 (Courtillot et al., 1986) and Karoo at 3×10^6 km^2 (Bristow and Saggerson, 1983). Some provinces consist of dozens of 10 to 50 m flows layered one on top of the next with little intervening sediment or even erosion, suggesting very rapid extrusion of the succession of flows. Radiometric dating is consistent with this interpretation: ^{40}Ar-^{39}Ar dating shows that a 2000 m thick section of the Deccan Traps erupted within 2 million years (Duncan and Pyle, 1988); similarly, much of the Karoo erupted

Table F1 Major flood basalt provinces

Province	Age[a]	Maximum thickness (km)	Estimated original area (km^2)	Approximate volume (km^3)	Tectonic setting	Associated hotspot
Columbia River Plateau (US)	Lower to upper Miocene 17–14 Ma	⩾ 4	2×10^5	2×10^5	Convergent plate margin	Yellowstone
Ethiopian	Oligocene/Miocene 32–21 Ma	3	7.5×10^5	7×10^5	Prior to continental break-up	Afar
North Atlantic (E greenland)	Lower Paleocene ⩽ 63 Ma	3.5	⩾ 10^5		Prior to continental break-up	Iceland
Deccan (Indian)	Cretaceous/Tertiary 65 Ma	2	2×10^6	1.5×10^6	Continental craton	Reunion
Ontong Java Plateau (western Pacific)	Lower Cretaceous 120 Ma		~ 0.75×10^6		Ocean floor	Louisville Ridge?
Paraná (Brazil)/ Entedka (Namibia)	Lower Cretaceous 133 Ma	⩾ 0.65	2×10^6	1.5×10^6	Prior to continental break-up	Tristan de Cuhna
Hartford/Newark Basin (US)	Lower/upper Jurassic 186 Ma	0.45	~ 2.4×10^4	~ 1×10^4	Continental rift	
Karoo (S. Africa)	Lower Jurassic 190–193 Ma	1.5	3×10^6	2×10^6	Continental craton	Marion
Siberian Platform	Permian/Triassic 240–250 Ma	3.5	> 1.5×10^6	9×10^5	Continental craton	Jan Mayen?
Keweenawan	Late Precambrian ~ 1100 Ma	12	> 1.25×10^5	> 3×10^5 km^3	Continental rift	

[a] Age of peak volcanism, if known
Data Sources: BVSP (1981), MacDougall (1988), Puffer et al. (1981), Renne et al. (1992), Richards, Duncan and Courtillot (1989).

Table F2 Compositions of flood basalts and picrites (oxides in wt%, trace elements in ppm)

	1	2	3	4	5	6	7
SiO_2	48.7	48.4	48.6	45.0	50.7	48.8	49.8
TiO_2	1.68	1.57	2.58	2.46	1.25	1.86	3.18
Al_2O_3	15.2	15.5	13.5	11.8	16.0	14.0	10.5
FeO^T	10.4	11.0	13.5	10.9	10.6	13.0	11.2
MgO	9.04	7.03	6.58	10.4	6.57	6.25	10.6
MnO	0.20	0.17	0.24	0.16	0.15	0.20	0.15
CaO	11.1	9.92	10.7	12.1	10.3	11.0	8.39
K_2O	0.09	0.51	0.10	1.11	0.76	0.17	1.10
Na_2O	2.46	2.76	2.76	2.08	2.17	3.09	2.27
P_2O_5	0.15	0.24	0.21	0.35	0.11	0.15	0.45
Ba	2.3	280	72	545	190	105	588
Th	0.22	0.57	1.17			1.77	
Nb	<1		10.5	50	5	7.5	16.5
La	1.12	7.03	17.9	25		10.1	
Sr	95	274	217	400	184	229	842
Nd	4.58		19.7		18.6	17.9	59
Sm	1.84	3.72	5.65		4.15	4.86	12.6
Zr	51		153	185	101	124	368
Yb	3.13	2.60	3.03			2.73	
$^{87}Sr/^{86}Sr$	0.7028[a]	0.7035[a]	0.7042[b]	0.7154[b]	0.7050[b]		0.7053[c]
$^{143}Nd/^{144}Nd$		0.51302[a]	0.51283[b]		0.51195[b]	0.51265[b]	0.51233[c]
ϵ_{Nd}		+7.4[d]			−13.0[b]	+0.4[b]	−4.4[c]

1: normal MORB, Sun, Nesbitt and Sharastein (1979); 2: Picture Gorge, Columbia River, major and trace elements from BVSP (1981), isotopes from Carlson *et al*.; 3: Deccan (Bas 224), Lightfoot and Hawkesworth (1988); 4: Deccan (picrite 259), Krisnamurthy and Cox (1977); 5: Deccan (ABC38), Cox and Hawkesworth (1985); 6: Deccan (ABC33), Lightfoot and Hawkesworth (1988); 7: Karoo (picrite KS3), major and trace elements from Bristow *et al.* (1984), isotopes from Hawkesworth *et al.* (1984).
[a] Uncorrected; [b] corrected to 65 million years; [c] corrected to 190 million years; [d] corrected to 15 million years.

within 3 million years (Fitch and Miller, 1984), and most of the Paraná erupted within 2 million years (Renne *et al.*, 1992). Morgan (1972), observing that many flood basalts are situated at or near continental margins, proposed that flood basalts marked the first arrival of a hot spot plume beneath crust which subsequently rifted and spread apart. Laboratory simulations of diapirism show that thermal instabilities in viscous fluids develop into spherical diapirs fed by thin trailing conduits (Whitehead and Luther, 1975). Given that the volume flux in the rising diapir is many times that in the trailing conduit, it is reasonable to expect that the volcanic flux from an incipient hot spot would be much larger than the flux from a mature hotspot tens of millions of years old. Richards, Duncan and Cortillo (1989) have calculated that average eruption rates of ~ 10^6 km^3 per year for several flood basalt provinces exceed estimates of volcanic production at associated hotspots by approximately 20 times, thus supporting the plume initiation hypothesis. Alternatively, Mutter, Buck and Zehnder (1988) and White and McKenzie (1989) have argued that flood basalts are produced when abnormally hot mantle is passively drawn upward in response to continental stretching and rifting. In rebuttal Richards, Duncan and Cortillo (1989) point out (1)) that some flood basalt provinces (Columbia River, Siberian Traps, Ontoong-Java Plateau) do not appear to be associated with rifting, (2) that in others (Karoo) feeder dikes appear to be randomly oriented rather than tectonically controlled, and (3) that when rifting is present (Deccan), it may postdate flood volcanism by tens of millions of years. However, as compelling as the arguments for the plume initial model appear, it seems possible that the tectonic model may be more appropriate for some flood basalt provinces. For example, in the Newark and Hartford Triassic/Jurassic rift basins, which are part of a much larger system of rifts, there are basalt flows that are chemically similar to other flood basalts and that appear to have covered tens of thousands of square kilometers (Puffer *et al.*, 1981); thus these are properly 'flood' basalts. On the other hand, there are only three major flows which are separated by hundreds of meters of continental sediment, indicating a lower intensity and longer intervals between eruptions than several of the other provinces. Furthermore, these basalts clearly postdate the rifting and their three feeder dikes are parallel to one another (Philpotts and Martello, 1986), suggesting tectonic control. Thus both plume initiation and tectonic processes may be responsible for flood volcanism in different cases, although the geological evidence seems to favor plume initiation in the case of the larger provinces.

In terms of major elements and mineralogy flood basalts are generally similar to mid-ocean ridge basalts (MORB), but typically have higher concentrations of K_2O and P_2O_5 at comparable levels of MgO (Table F2). They also typically have higher concentrations of the incompatible trace elements such as Ba, Th, Nb, Zr and the light REE, and in this respect flood basalts are more similar to transitional or 'T-type' MORB than to normal MORB (Lightfoot and Hawkesworth, 1988). Like transitional MORB the incompatible element patterns of most flood basalts are weakly to moderately fractionated with respect to chondrites. The most fractionated incompatible element patterns seem to belong to the most primitive lavas, the picrites (Krisnamurthy and Cox, 1977; Cox, 1988). This feature precludes these picrites (Table F2, cols 4 and 7) from being parental to the basalts and also rules out assimilation–crystallization mechanisms for explaining the elemental fractionations. In terms of isotopic compositions, there seems to be evidence of at least three components. Flood basalts display an overall negative correlation of ϵ_{Nd} versus $^{87}Sr/^{86}Sr$ with most basalts (Table F2, cols 2, 3 and 6) having positive ϵ_{Nd} (depleted source) and lying on the ocean island basalt (OIB) portion of the so-called 'mantle array' (Carlson, Lugmair and MacDougall, 1981; Hawkesworth *et al.*, 1985; Lightfoot and Hawkesworth, 1988; Cox, 1988). Some samples (e.g. col 5) form a clear trend toward very high $^{87}Sr/^{86}Sr$ (≥ 0.720) and negative ϵ_{Nd} (≤ -16). These samples also have highly radiogenic Pb, and it appears that assimilation of Precambrian continental crust was responsible for this trend. A second trend is formed by samples (e.g. col. 7) extending to negative ϵ_{Nd} (≤ -7), but with $^{87}Sr/^{86}Sr$ values (≤ 0.705) much closer to bulk Earth; these samples also have unradiogenic Pb. In this case a mixing of sources – depleted OIB asthenosphere and enriched subcontinental lithosphere – appears to be responsible.

Using laboratory simulations, Campbell and Griffiths (1990) have argued that circulation within the head of a rising plume of (OIB) asthenosphere would entrain surrounding (subcontinental) lithosphere, thus producing a mixture of mantle sources. Specifically, they suggested that picrites with strong incompatible element fractionations would be produced near the center of the plume where the head was fed by the trailing conduit of OIB material and was relatively free of admixed lithosphere, whereas the more voluminous basalts would be produced by melting of the mixed regions. However, the few isotopic data on Karoo picrites cited by Cox (1988) indicate that the distinction between picrite and basalt sources is not correct: picrites have ϵ_{Nd} in the range of -4 to -10 and appear to lie on the trend attributed above to subcontinental lithosphere, whereas most basalts have compositions near the OIB endmember. On the other hand, Fram and Lesher (1993), using pressure-dependent partition coefficients, have modeled polybaric melting processes and shown that the major contribution to the major and trace element compositions of the basalts of the North Atlantic Tertiary Province was the depth at which melting stopped, which is in turn related to lithospheric thickness (namely, the thicker the lithosphere, the deeper the level at which ascending mantle is likely to stop melting). In their model the most MgO-rich (picritic) magmas are restricted to melting below ~ 60 km in the garnet-stability region by a thick section of lithosphere through which the magma must rise, whereas the least MgO-rich magmas are produced by melting that extends to relatively shallow 5–20 km levels under thinned lithosphere. The MgO-rich magmas have the greatest thickness of lithosphere to traverse and thus the most opportunity for lithospheric contamination. The Fram and Lesher (1993) model explains the differences in composition between basalts and picrites very well, but does not the address what caused the initiation of lithospheric thinning. Despite the problems with the Campbell and Griffths (1990) interpretation, the general plume initiation model remains viable inasmuch as the entrainment in the plume head predicted by the laboratory models might be restricted to the asthenosphere and the plume head might spread out along the base of the lithosphere, producing uplift and melting rather than entrainment.

Among the oldest known flood basalts are the mid-Proterozoic Keweenawan volcanics of the Lake Superior district (Table F1), however, there may well have been many more earlier flood events, now eroded away, as evidenced by Precambrian dike swarms in the Canadian shield (BVSP, 1981). Perhaps of greatest interest to planetary scientists is the extraordinary coincidence of many of the flood basalt eruptions with mass extinctions. Rampino and Strothers (1988) have correlated 11 flood basalt eruptions with mass extinctions in the last 250 million years, including the event at the Cretaceous–Tertiary boundary which is synchronous with the Deccan Traps at 65 million years. Even if several of the correlations made by Rampino and Strothers (1988) prove incorrect, the remaining coincidences are still remarkable given that flood basalts are unlikely to have had the sort of catastrophic effects on global climate that could have caused the extinctions: the largest provinces make up only about 1% of the Earth's surface and it is not likely that a single flow ever reached this dimension; furthermore, the likeliest cause of catastrophic climate change is the sudden introduction of large amounts of dust into the upper atmosphere, yet by their very nature flood basalts must have formed from relatively quiescent eruptions with little explosive activity – the early stages of the Siberian traps being the only major exception (BVSP, 1981). At the same time evidence for a bolide impact at the Cretaceous–Tertiary boundary (global iridium layer in sediments, tektite fields, turbidite layers, etc.) seems overwhelming (Sharpton et al., 1993). Thus it is reasonable to conclude that bolide impacts were the ultimate cause of some flood basalt eruptions, most likely by triggering gravitational instabilities in the mantle.

John Longhi

Bibliography

Bristow, J.W., Allsopp, H.L., Erlank, A.J. et al. (1984) Strontium isotope characterization of Karoo volcanic rocks. *Spec. Publ. Geol. Soc. S. Afr.*, **13**, 295–329.
Bristow, J.W. and Saggerson, E.P. (1983) A general account of Karoo volcanicity in southern Africa, *Geol. Rundsch.*, **72**, 1015–60.
BVSP [Basaltic Volcanism Study Project] (1981) *Basaltic Volcanism on the Terrestrial Planets*, New York; Pergamon Press, 1286 pp.
Campbell, I.H. and Griffiths, R.W. (1990) Implications of mantle plume structure for the evolution of flood basalts. *Earth Planet. Sci. Lett.*, **99**, 79–93.
Carlson, R.W., Lugmair, G.W. and MacDougall, J.D. (1981) Columbia River volcanism: the question of mantle heterogeneity of crustal contamination. *Geochim. Cosmochim. Acta*, **45**, 2483–99.
Courtillot, V., Besse, J., Vandamme, D. et al. (1986) Deccan flood basalts at the Cretaceous/Tertiary boundary. *Earth Planet. Sci. Lett.*, **80**, 361–76.
Cox, K.G. (1988) The Karoo Province, in *Continental Flood Basalts* (ed. J.D. MacDougall), London: Kluwer Academic Publishers, pp. 239–71.
Duncan, R.A. and Pyle, G. (1988) Catastrophic eruption of the Deccan flood basalts, western India. *Nature*, **333**, 841–3.
Fitch, F.J. and Miller, J.A. (1984) Dating Karoo igneous rocks by the conventional K–Ar and $^{40}Ar/^{39}Ar$ age spectrum methods. *Spec. Publ. Geol. Soc. S. Afr.*, **13**, 149–69.
Fram, M.S. and Lesher, C.E. (1993) Geochemical constraints on mantle melting during creation of the North Atlantic Basin. *Nature*, **363**, 712–4.
Hawkesworth, C.J., Marsh, J.S., Duncan, A.R. et al. (1984) The role of continental lithosphere in the generation of the Karoo volcanic rocks: evidence from combined Nd- and Sr-isotope studies. *Spec. Publ. Geol. Soc. S. Afr.*, **13**, 341–54.
Hawkesworth, C.J., Mantovani, M.S.M., Taylor, P.N. and Palacz, Z. (1985) Evidence from the Parana of south Brazil for a continental contribution to DUPAL basalts. *Nature*, **322**, 356–9.
Krisnamurthy, P. and Cox, K.G. (1977) Picritic basalts and related lavas from the Deccan Traps of Western India. *Contrib. Mineral. Petrol.*, **62**, 53–75.
Lightfoot, P. and Hawkesworth, C.J. (1988) Origin of Deccan Traps lavas: evidence from combined trace element and Sr-, Nd-, and Pb-isotope studies. *Earth Planet. Sci. Lett.*, **91**, 89–104.
MacDougall, J.D. (1988) *Continental Flood Basalts*. London: Kluwer Academic Publishers.
Morgan, W.J. (1972) Plate motions and deep mantle convection, *Mem. Geol. Soc. Am.*, **132**, 7–22.
Mutter, J.C., Buck, W.R. and Zehnder, C.M. (1988) Convective partial melting 1. A model for the formation of thick basaltic sequences during the initiation of spreading. *J. Geophys. Res.*, **93**, 1031–48.
Philpotts, A.R. and Martello, A. (1986) Diabase feeder dikes for the Mesozoic basalts in southern New England. *Am. J. Sci.*, **286**, 105–26.
Puffer, J.H., Hurtubise, D.O., Geiger, F.J. and Lechler, P. (1981) Chemical composition and stratigraphic correlation of the Mesozoic basalt units of the Newark Basin, New Jersey, and the Hartford Basin, Connecticut. *Geol. Soc. Am. Bull.*, **92**, 515–53.
Rampino, M.R. and Strothers, R.B. (1988) Flood basalt volcanism during the past 250 million years. *Science*, **241**, 663–8.
Renne, P.R., Ernesto, M., Pacca, I.G. et al. (1992) Age and duration of Paraná flood volcanism in Brazil. *EOS*, **73**, 531–2.
Richards, M.A., Duncan, R.A. and Courtillot, V.E. (1989) Flood basalts and hot-spot tracks: plume heads and tails. *Science*, **246**, 103–7.
Sharpton, V.L., Dalrymple, G.B., Marin, L.E. (1992) New links between the Chicxulub impact structure and the Cretaceous/Tertiary boundary. *Nature*, **359**, 819–21.
Sun, S.-S., Nesbitt, R.W. and Sharaskin, A.Y. (1979) Geochemical characteristics of mid-ocean ridge basalts. *Earth Planet. Sci. Lett.*, **44**, 119–38.
White, R. and McKenzie, D. (1989) Magmatism at rift zones: the generation of volcanic continental margins and flood basalts. *J. Geophys. Res.*, **94**, 7685–729.
Whitehead, J.A. and Luther, D.S. (1975) Dynamics of laboratory diapir and plume models. *J. Geophys. Res.*, **80**, 705–17.

Cross references

Basalt
Igneous rock
Impact cratering
Moon: geology
Volcanism in the solar system

FORBUSH DECREASE

This phenomenon is a rapid decrease, followed by slow recovery in the magnetic plasma cloud of the solar wind. It is recorded by a sudden fluctuation in the cosmic ray (q.v.) flux received by planet

Earth. It was first identified during a geomagnetic storm (q.v.) in April 1937 (Forbush, 1939). The decreased is observed to follow the eruption of high-energy solar flares (q.v.) by one to several days. The high-energy plasma travels two to three times faster than the normal solar wind, producing a shock wave which, on reaching the Earth's magnetosphere (see Earth: magnetic field and magnetosphere), causes a 'sudden commencement', i.e. a temporary increase in strength of the magnetic field. This whole event is known as a 'geomagnetic storm' (q.v.), leading to considerable disruption of the usual lines of magnetic force. (Most importantly, the sudden increase in plasma pressure pushes the lines of force apart, thus decreasing the field strength.) When the shock wave is passing through the already magnetized solar wind plasma it carries with it a still stronger magnetic field, which screens off the particles of weaker galactic cosmic rays (q.v.): the Forbush decrease.

Incoming cosmic rays, now known to emanate from sources close to the center of the Milky Way Galaxy, are always partly screened by the Earth's magnetosphere, the potential of which is modulated by two factors: variations in the Earth's endogenetic magnetic field (in time frames of 10^2 to 10^7 years) and variations in the character and components of the solar wind. Forbush (1954, 1966) also confirmed the 1939 evidence of Monk suggesting that the Sun's 27-day spin rate is reflected in the cosmic ray flux. This is almost certainly related to the Sun's spin axis, having an approximately 7° tilt from the vertical with respect to the ecliptic, creating the so-called 'lighthouse effect' in the radiative distribution of the solar wind whereby for 13.5 days the Earth is rather more influenced by the Sun's northern hemisphere and in the next 13.5 days by its southern hemisphere.

The long-term solar wind flux intensity may also reflect some relationship to solar activity cycles and nonlinear events such as high-energy solar flares, which have a 'general' correlation with the 11.12 (± 6.0) year solar cycle and its magnetic twin the 22.24-year Hale cycle, together with near-random solar flare events. The latter are claimed to occur over a long-term period at about 150-day intervals, an average rate of 0.417 years (Bai and Sturrock, 1987). It may just be coincidental, but this 0.417-year mean is equivalent to 1/26 of the high-activity mode of the sunspot cycle (10.832 years) and 1/200 of the long-term (83.4-year) solar torque period (of Landscheidt, 1989). As pointed out by Dorman (1974), variations of the solar cycle do not cause any displacement of the cosmic ray equator with respect to the geomagnetic equator.

As anticipated by Rossi (1964), Forbush decreases should occur throughout interplanetary space because the Sun radiates in all directions, reaching out to 100 AU and more. Thus the space probe Pioneer 5 encountered a Forbush decrease some 5 million km from the Earth, marked by a sudden rise in the magnetic field by about one order of magnitude. Further observations were made by Pioneer 10 and 11, and by Voyager 1 and 2.

On planet Earth the solar flare–Forbush events, leading to decrease in cosmic rays, appear to correlate with a climatic effect in high latitudes. Both temperature and atmospheric pressure at certain stations have a negative correlation with cosmic rays. The higher the planetary latitude the larger is the climatic fluctuation. In a series of tree rings from the Mackenzie delta (northern Canada), Fan *et al.* (1986) found ring widths showing that favorable growth conditions were anticorrelated with sunspot number; likewise with the rings sampled for ^{14}C, which is generated by cosmic rays. The year 1942, specially noted by Forbush for its high flare activity, was followed in 1943 by rings with abnormally high ^{14}C.

Rhodes W. Fairbridge

Bibliography

Bai, T. and Sturrock, P.A. (1987) The 152-day periodicity of the solar flare occurrence rate. *Nature*, **327**, 601–3.
Dorman, L.I. (1974) *Cosmic Rays: Variations and Space Explorations*. Amsterdam: North-Holland Publ. Co., 675 pp.
Fan, C.Y., Chen, T.M., Yung, S.X. and Dai, K.M. (1986) Radiocarbon activity variation in dated tree rings grown in Mackenzie delta. *Radiocarbon*, **28**(2-A), 300–5.
Forbush, S.E. (1939) *Rev. Mod. Phys.*, **11**, 168.
Forbush, S.E. (1954) World-wide cosmic-ray variations, 1937–52. *J. Geophys. Res.*, **59**, 525–42.
Forbush, S.E. (1966) Time variation of cosmic rays, in *Handbuch der Geophysik*, (ed. J. Bartels). New York: Springer-Verlag, Vol. 3, 159–247.
Landscheidt, T. (1989) *Sun–Earth–Man: a Mesh of Cosmic Oscillations*. London: Urania Trust, 132 pp.
Rossi, B. (1964) *Cosmic Rays*. New York: McGraw-Hill, 268 pp.

Cross reference

Cosmic rays

FORBUSH EFFECT

Cosmic rays were discovered in 1912 but it was not until the 1930s that it was generally agreed that the 'radiation' comprised particles rather than gamma rays. Inevitably, the particles are deflected by the magnetic fields in space: those in the general interstellar medium, in the interplanetary medium and by the Earth's magnetic field itself. Results taken in the 1930s, 1940s and 1950s led Scott Forbush to claim a remarkable correlation of cosmic ray intensity with sunspot number, the 'intensity' relating to ionization chamber measurements at a variety of places on the Earth's surface.

Figure F1 shows the results given in Forbush's important 1954 paper. Since then many other observers have confirmed the effect, for detectors on the surface of the Earth, detectors flown in balloons and detectors beneath the Earth's surface. It will be noted that the correlation is a negative one, i.e. the rate is high when the sunspot number is low and vice versa.

The observations show that the magnitude of the change in cosmic ray intensity (in relative terms) diminishes as the energy of the primary particle giving rise to the detected secondary particle increases. The greatest variation occurs in the flux of neutrons – these particles being caused by primary protons of very low energy, and the neutrons (unlike charged particles) not losing energy by ionization in the atmosphere during their path from production at high levels in the atmosphere to the point of observation.

The explanation of the correlation (the so-called 'Forbush effect') is straightforward in principle: the interplanetary magnetic field associated with the solar wind is positively correlated with sunspot number, and this magnetic field acts to shield the Earth from the low-energy primary cosmic rays. The details are still the subject of considerable study; not surprisingly the topography of the interplanetary field and its variation with time is a matter of great complexity.

Although the average cosmic ray intensity is lowest at solar maximum, this is also the time when solar flares are most common. Thus spikes in intensity are seen from time to time associated with giant solar flares. Again, there is much contemporary interest in the problem.

Forbush has also given his name to the 'Forbush decreases' (q.v.), rapid intensity decreases followed by slow recoveries. The reduction takes typically an hour and the recovery several days. Again the solar wind is responsible, in this case the emission by the Sun of a magnetic plasma cloud which sweeps particles away as its expands; the particles can then leak back in more slowly. The size of the cloud can be as large as 1 AU.

Arnold W. Wolfendale

Bibliography

Forbush, S.E. (1954) World-wide cosmic-ray variations, 1937–52. *J. Geophys. Res.*, **59**, 525–42.
Forbush, S.E. (1966) Time variation of cosmic rays, in *Handbuch der Geophysik*, Vol. 3 (ed J. Bartels). New York: Springer-Verlag, pp. 159–247.
Rossi, B. (1964) *Cosmic Rays*. New York: McGraw-Hill, 268 pp.

Cross reference

Cosmic rays

FRACTURE, FAULT

In structural geology a fracture is a general term for any break in a rock due to mechanical failure by stress. It includes cracks, joints and faults (American Geological Institute: *Glossary of Geology*). A fault is any fracture involving displacement of one side relative to the other. The terms fracturing and faulting designate the processes involved. Both terms have numerous synonyms in the early literature; for fractures: 'break', 'fissure', 'rent', 'crack', 'cleft', etc. (Dennis *et al.*, 1979); for faults (an ancient mining term), there is 'failure', 'disruption', etc. 'Cleavage' is the tendency for brittle rocks to fracture along systematic secondary planes. Most rocks under confining pressures respond to stress in various degrees of ductility or 'flow' before they rupture, so that transitional behavior is often involved.

The sense of displacement in faulting is susceptible to geometric description. Thus, a simple, horizontal extension ('pull apart') is termed 'normal faulting', or 'dip-slip' (the direction of slippage being coincident with the dip, or tilt of the fault surface), and in miners' terminology its 'hanging wall' is depressed relative to its 'foot wall'.

'Reverse faulting' is also dip-slip, but in a horizontal, compressional sense, so that the hanging wall rides over the foot wall.

'Thrust' and 'Overthrust faulting' is similar to reverse faulting, but applied to displacements with a low dip angle; 'overthrust' is used for large-scale, long-distance displacements (on a scale of 1–100 km).

'Detachment' or 'décollement faulting' is a development of thrust faulting, where the displacement surface follows conformably one or more bedding planes, usually lubricated by a schist, shale, salt or gypsum layer.

'Listric faulting' is a further development of thrust faulting, but with the fault planes generally curving upward (concave-up), and in multiples comparable to fish scales.

'Synsedimentary' or 'growth faulting' is a combination of normal (dip-slip) faulting and detachment faulting that takes place during sedimentation; hence it tends to die out upward.

'Strike-slip faulting' is where the displacement is primarily in a horizontal sense, parallel to strike of the fault (i.e. normal to the dip). A horizontal torsion is involved. On a small or medium scale (< 100 km), these displacements are sometimes called 'wrench faults', and commonly found in échelon series; on a large scale (traversing mountain systems or continental blocks) they are better known as 'transcurrent faults.'

'Transform faulting' displaces sea-floor spreading centers in major fracture zones that may extend long distances (> 1000 km). Strike-slip motions occur but in opposite directions on different sides of the spreading axis. Transforms often extend into transcurrent systems; in major global fracture systems a zig-zag plan often links transforms, transcurrents and normal-fault (pull-apart) sectors.

Lineament is a term reserved for global or continental-sized fracture zones that may incorporate a variety of fractures and faults, while maintaining a more or less uniform trend over distances of hundreds or thousands of kilometers. In many cases some continent-sized lineaments have been active for extremely long periods (often exceeding 0.5 to 1.0 Ga), even though the fracture zone is without obvious vertical or horizontal displacement. Such lineaments are sometimes attributed to major changes in the form or orientation of the Earth's spheroid that are the result of changes in planetary spin rate or pole shift (caused by mass displacements resulting in shifts in the symmetry of the planet's moment of inertia). Lineaments are often linked over large regions by systematic joint systems (fractures of limited dimensions), the joints being picked out by lines of trees, springs and stream patterns.

On other planets (with brittle crusts) and on some major satellites both fracturing and faulting systems are observed, suggesting that these bodies are also subject to changes of spin rate and tilt (obliquity). The Moon is characterized by prominent north–south lineaments, which suggests brittle fracture under stress of small but constantly changing orbital parameters.

Rhodes W. Fairbridge

Bibliography

Billings, M.P. (1972) *Structural Geology*, 3rd edn. Englewood Cliffs, N.J.: Prentice-Hall (3rd edn.), 606 pp.

Dennis, J.G. *et al.* (eds) (1979) *International Tectonic Lexicon*. Stuttgart: E. Scheweizerbart., 153 pp.

Hills. E.S. (1963) *Elements of Structural Geology*. New York: J. Wiley and Sons, 483 pp.

Hobbs, W.H. (1911) Repeating patterns in the relief and in the structure of the land. *Geol. Soc. Am. Bull.*, **22**, 123–76.

Sonder, R.A. (1938) Die Lineamenttektonik und ihre Probleme. *Eclog. Geol. Helv.*, **31**, 199–238.

Stauffer, M.R. (ed.) (1983) *Fabric of Ductile Strain*. Stroudsburg, PA: Hutchinson Ross, Benchmark Vol. 75, 397 pp.

Voight, B. (ed.) (1975) *Mechanics of Thrust Faults and Décollement*. Stroudsburg, PA: Dowden, Hutchinson & Ross, Benchmark Vol. 32.

Wilson, J.T. (1965) A new class of faults and their bearing on continental drift. *Nature*, **207**, 343–7.

Cross references

Crust
Earth: geology, tectonics, and seismicity
Plate tectonics

FRAUNHOFER LINE

In 1817 Joseph von Fraunhofer (1787–1826), who had been experimenting with ways of defining the colors which are used to determine the refractive index of glasses, published the results of his observations of the solar spectrum made by an improved spectroscope. He noted that the spectrum was crossed by many dark lines. By repeating his observations with a range of optical elements, he proved that the lines were a real feature of the spectrum and were contained in the sunlight. He made a map of several hundred of the lines and designated the prominent ones with the letters A, B, C, etc., by which some of them are still known: for example the D lines at 589.6 and 589.0 nm or the H line at 396.8 nm and the K line at 393.3 nm. Some 44 years later, Kirchhoff and Bunsen identified a number of these dark lines to be caused by the absorption by metals: sodium for the D lines, calcium for the H and K lines, etc. Fraunhofer continued by inventing and manufacturing many diffraction gratings with which he succeeded in measuring accurately the wavelengths of the spectral features he observed, thus providing a basis whereby the results of different observers could be compared directly. He was the instigator of observational spectrometry.

Fraunhofer was also the first pioneer in stellar spectroscopy. He performed visual observations of the spectra of the planets and of the brightest stars, including Sirius, by using an objective prism. He noted that all the spectra exhibited dark lines, those of the planets being similar to the Sun, but those of the stars being different. Over the following years, intercomparison between Fraunhofer spectra of a large number of stars formed the basis of a system for the classification of stars by spectral types (Jaschek and Jaschek, 1990).

About 25 000 Fraunhofer lines are now recognized in the solar spectrum, seen in absorption against a bright continuum, and corresponding to 70 different elements; some of them are, in fact, due to oxygen or water in the Earth's atmosphere, or even to interstellar material (Moore, Minnaert and Houtgast, 1966). Observations of their intensities and widths may be compared with theoretical expectations in order to determine the excitation temperature, the turbulent velocity, the rotational broadening, the electron and gas pressures, the surface gravity, the chemical composition, and eventually the magnetic field through the Zeeman effect, in the stellar atmosphere. There are two classical model atmospheres (Mihalas, 1970): one in which the continuum spectrum is assumed to be formed in the photosphere and the line spectrum entirely in an overlying 'reversing layer'; in the other approximation it is assumed that both the line and continuum spectrum are formed in the same layers in such a way that the ratio of the line and continuum absorption coefficients is a constant. If it is also assumed that line formation takes place in an isothermal layer in local thermodynamic equilibrium, both approximations lead to the classical relation (Menzel, 1936) from which theoretical curves of growth are computed. These can be compared to empirical curves of growth relating the equivalent widths of the Fraunhofer lines with their oscillator strengths and wavelengths, in order to infer the different stellar parameters, in particular the relative abundances of the elements. In the case of the Sun, for instance the best approximation of the data is found for an excitation temperature of 5143 K and a turbulent velocity of 1.4 ± 0.2 km s^{-1} (Cowley and Cowley, 1964).

Roger Ferlet

Bibliography

Cowley, C.R. and Cowley, A.P. (1964) A new solar curve of growth. *Ap. J.*, **140**, 713.
Fraunhofer, J. (1817) *Denkschriften der Königlichen Akademie der Wissenschafter zu München*, **V**, 193–226.
Jaschek, C. and Jaschek, M. (1990) *The Classification of Stars*. Cambridge: Cambridge University Press.
Kirchhoff, G. and Bunsen, R. (1861) Chemical analysis by spectrum-observations. *Phil. Mag.*, **22**, 329.
Menzel, D.H. (1936) The theoretical interpretation of equivalent breadths of absorption lines. *Ap. J.*, **84**, 462.
Mihalas, D. (1970) *Stellar Atmospheres*. San Francisco: Freeman.
Moore, C.E., Minnaert, M.G.J. and Houtgast, J. (1966) The solar spectrum 2935 Å to 8770 Å. *Natl. Bur. Stands Mon.*, **61**.

Cross references

Sun
Zeeman effect

G

GALACTIC CYCLE

The solar system (including the Sun, planets, minor bodies and the surrounding Oort comet cloud) undergoes long-term cyclic motions as it moves through the Milky Way Galaxy. A long cycle involves the revolution of the solar system around the center of the disk-shaped Galaxy. This revolution is estimated to take about 200 to 250 million years (Rampino and Stothers, 1986). During its revolution, the solar system undergoes an in-plane epicyclic motion from its perigalactic position to its apogalactic position (semimajor axes = 400 pc and 600 pc (parsec) in the radial and transverse directions, respectively) in about 170–180 million years (Bailey, Clube and Napier, 1990). The solar system is now at a distance of ~ 8 kpc (kiloparsec) from the galactic center.

A shorter cycle involves the simple harmonic oscillation of the solar system perpendicular to the mid-plane of the Galaxy. This vertical oscillation of the solar system with respect to the Galactic plane, the so-called Z oscillation, can be estimated from the mass distribution in the galactic plane, and from observations of the motions of stars and clouds in the galactic disk. Astronomical models (which include missing mass) suggest a full period of about 52 to 66 million years, and therefore a half period of about 26 to 33 million years from one plane crossing to the next (Bahcall and Bahcall, 1985; Matese et al., 1995). The current amplitude of the solar Z oscillation is on the order of 80–90 pc (Bailey, Clube and Napier, 1990). At present the solar system lies relatively close to the galactic plane (≤ 10 pc above the plane).

Possible effects on the solar system

As the solar system moves through the galaxy in these basic cycles, it should experience variations in gravitational and magnetic forces, radiation and in the numbers and types of stars, clouds of gas and dust encountered. Many authors have suggested that these cyclic variations might be detected in cyclic changes in the geologic record. In recent years the discovery of significant periodicities in geologic events has led to a renewed interest in galactic cycles. A history of these proposals may be found in Williams (1981), Rampino and Stothers (1986) and Napier (1989) – what follows is a brief outline.

Holmes (1927) was probably the first author to propose a connection between long-term geologic cycles and the Sun's motion through the galaxy. Forbes (1931) suggested that an obscuring cloud, orbiting around the galactic center inside the solar galactic circle, periodically blocks out the radiation coming from the center. The radiation received from the galactic center would in any case vary because of the solar system's changing galactocentric distance (Gidon, 1970). However, the intensity of radiation reaching the Earth is apparently too small to produce a noticeable climatic or biological effect. Some have speculated that the spatially varying gravitational potential of the galaxy somehow induces a significant change of the strength of gravity in the solar system, and thereby might affect tectonism and biological evolution on Earth (Tamrazyan, 1967; Steiner, 1967, 1973, 1979; Kropotkin, 1970; Steiner and Grillmair, 1973).

Periodic interactions of the Earth with the galactic magnetic field (Steiner, 1967; Crain, Crain and Plaut, 1969; Crain and Crain, 1970) or with magnetically trapped galactic cosmic rays (Hatfield and Camp, 1970; Meyerhoff, 1973) have been proposed as causing measurable geomagnetic and biological disturbances. However, the magnetic field strength of the Galaxy is only 10^{-5} times that of the Earth.

Tamrazyan (1957) and Steiner and Grillmair (1973) suggested that large spatial variations of the galactic gravitational potential during galactovertical oscillation directly affected the strength of gravity in the solar system. A proposed ~ 80-Ma geomagnetic cycle was correlated with the full oscillation of the solar system through the galaxy's magnetic field, which lies moderately concentrated to the galactic plane (Crain, Crain and Plaut, 1969; Crain and Crain, 1970). Magnetically trapped galactic cosmic rays might possibly be related to mass extinctions (Hatfield and Camp, 1970; Meyerhoff, 1973). A possible ~ 70-Ma regularity in the geologic periods has also been correlated with the full period of galactovertical oscillation (Innanen, Patrick and Duley, 1978; Steiner, 1979). Schwartz and James (1984; also Hatfield and Camp, 1970), discussed the half-period of the Sun's galactovertical oscillation in connection with a model of mass extinctions at approximately 26 million year intervals that relies on spatial variations in galactic cosmic ray and soft x-ray intensity. Their model requires that mass extinctions occur at times when the solar system is farthest from the galactic plane.

Galactic cycles and comet showers

The search for long-term periodicities in the geologic record (e.g. mass extinctions, flood basalt eruptions, global tectonism, climatic change, sea level and possibly geomagnetism) has recently resulted in the discovery of two periods that may be significant – 30 ± 3 Ma and 260 ± 25 Ma (Negi and Tiwari, 1983; Raup and Sepkoski, 1984, 1986; Rampino and Stothers, 1984a,b, 1986, 1988; Raup, 1985; Stothers, 1986; Rampino and Caldeira, 1992). The ~ 30 million year cycle shows up in records of mass extinctions (Raup and Sepkoski, 1984, 1986) and in terrestrial impact craters (Rampino and Stothers, 1984a,b, 1986; Alvarez and Muller, 1984; Yabushita, 1992), suggesting that periodic or quasi-periodic showers of comets from the solar system's Oort cloud led to the extinctions. The most recent epoch of the mean cycle seems to lie close to the present time (Rampino and Caldeira, 1992) – the Earth may be in a comet shower at present (Stothers, 1985; Yabushita, 1992), and the solar system lies near the galactic plane.

Heylmun (1969) proposed a model in which interstellar comets, or other kinds of interstellar debris, are distributed within the plane of the galaxy. As the Sun moves through the plane, these objects enter the solar system where some hit the Earth. However, comets with hyperbolic orbits are unknown, and are unlikely to exist exclusively in the galactic plane. A model where most of the missing mass in the galactic disk (which has been inferred from stellar dynamical studies, e.g. Bahcall, Flynn and Gould, 1992) lies within the galactic plane and gravitationally perturbs the Oort comet halo during solar system passages (Goldsmith, 1985) assumes that the missing mass is much more concentrated in the plane than is allowed by the dynamical studies (Bahcall and Bahcall, 1985).

The compressional force exerted by the galactic disk induces a periodically varying gravitational perturbation of the Oort comet halo as the solar system oscillates about the galactic plane, and this might cause comet showers (Bailey, Clube and Napier, 1990). The Oort cloud may be made observable predominantly by the quasi-steady state tidal torque (Matese and Whitman, 1992). Recently, Matese et al. (1995) calculated that the changes in galactic tidal force as the solar system undergoes Z-oscillation could lead to significant periodic showers at ~ 30-Ma intervals as the galactic plane is crossed.

Gravitational interaction between the solar system and massive interstellar clouds of gas and dust during the solar system's vertical oscillation might also cause comet showers. Hills (1981) has shown that if comets in the solar system's inner comet reservoir, which is believed to feed the surrounding Oort comet halo, are gravitationally disturbed by a massive outside body, then some of these comets will lose energy and fall into small perihelion distances, where a few will collide with the Earth. Such disturbances might arise by encounters with passing stars (Hills, 1981) or with passing interstellar clouds (Rampino and Stothers, 1984a). Actual penetration of an interstellar cloud could also lead to flooding of the inner solar system with particles of interstellar gas and dust, which might disturb the Earth's climate.

Rampino and Stothers (1984a) found that intermediate-sized interstellar molecular clouds having mean density 10^2 to 10^3 particles cm^{-3}, radius 3 to 6 pc, and mass 10^3 to 10 $M_\oplus$ are the most frequently encountered objects that could give rise to significant perturbations of the inner comet reservoir or of Oort's comet halo.

For a galactic modulation to exist, the maximum amplitude of the Sun's vertical oscillation must be significantly greater than measures of the vertical density distribution of the perturbing clouds. If the Sun's Z_{max} amplitude is too small, and the clouds too dispersed above and below the plane, then galactic modulation would be unlikely. On the assumption that such molecular clouds are the dominant perturbers, the a priori probability of detecting the galactic signal in a record of length 600 Ma is calculated as ~ 50% (Stothers, 1985).

However, the Sun's present rms Z-velocity averaged over a vertical orbit is 6 km s^{-1}, which lies well below the ensemble mean of other solar-type stars (20 km s^{-1}), and hence Z_{max} is only ~ 80 to 90 pc. The occasional close encounter of the Sun with a giant molecular cloud is capable of deflecting the Sun's ballistic trajectory in space. As a result of the total vector change of the Sun's total velocity, the Z-component of velocity may change by more than 50% in a single close encounter. This may imply that the Sun's present low velocity and low Z_{max} are anomalous.

It is thus possible that the Sun's rms Z-velocity in the past was close to the ensemble mean observed for solar-type stars. If so, Z_{max} would have been typically 300 pc. This larger vertical trajectory would not have significantly changed the half-period of oscillation, but it would have led to a stronger modulation of the rate of random encounters with interstellar clouds of all sizes, which are more concentrated towards the galactic plane (Stothers, 1985).

Clube and Napier (see reviews in Bailey, Clube and Napier, 1990, and Clube, 1992) have argued for some time that cometary impacts on the Earth are responsible for mass extinctions and geophysical phenomena. They proposed that collisions between the solar system and giant molecular clouds led first to a stripping away of the Oort comet halo and then to the capture and infall of new comets manufactured in the giant molecular cloud itself; these episodes would take place when the solar system encountered galactic spiral arms on a quasi-periodic timescale of 50 to 150 Ma. It is not clear, however, whether a giant molecular cloud could either destroy (Hut and Tremaine, 1985) or supply (Valtonen, Zheng and Mikkola, 1992) the Oort comet halo. The shower comets might come instead from Hills' inner reservoir of comets which, if it exists, is probably mildly perturbed by close encounters with giant molecular clouds.

Napier (1989) proposed that a 15 million year cycle of comet showers can arise from the 30 million year Z-oscillation as the solar system passes through spiral arms. In this view, interpulses arise because there are two opposing tidal components: the tide due to the disk material is compressive, and that due to the spiral arm is tensile. These tend to cancel close to the arm, but as the Sun arises toward the peak of its orbit the tensile force of the arm may decrease, whereas the compressive force of the disk begins to dominate, increasing the overall stress. This secondary effect may be significant for arm/interarm contrasts in the range of 2 : 1 to 4 : 1. Napier suggests that such a 15 million year cycle can be isolated from the geologic record.

Cloud complexes will be encountered preferentially when the solar system crosses a spiral arm and other dusty sectors of the galaxy (McCrea, 1975, 1981; Innanen, Patrick and Duley, 1978), including the portion of the galactic disk upwarped by gravitational effects of the nearby Magellanic Clouds (Williams, 1975). The solar system's comet halo might then be lost to the dust clouds, while a new perturbed comet halo is subsequently picked up from them (Bailey, Clube and Napier, 1990). The gravitational attraction exerted by the dust clouds might merely perturb, without destroying, the comet halo and/or inner cometary reservoir (Rampino and Stothers, 1984a). The changing radial gravitational potential of the galaxy should lead to significant changes in the binding of distant comets to the Sun. These changes might possibly periodically alter the overall size of the Oort comet halo.

These perturbations might effect impact cratering on the Earth. Passages through the galactic spiral arms could serve as the pacemaker, and the cycle could be quasi-periodic. For example, if the spiral pattern speed happens to be approximately half the Sun's galactic rotational speed, and the Sun crosses two spiral arms in a circuit, the Sun could overtake a spiral arm every ~ 125 Ma.

Impact cratering on the Earth seems to show a long periodicity of about 250 Ma over Phanerozoic time (Rampino and Stothers, 1984b). A similar period of 250 ± 25 Ma (Hatfield and Camp, 1970) or possibly 300 Myr (Fischer 1979) has been inferred from data on mass extinctions. The existence of a long tectonic period is also suspected (Williams 1981; Bailey, Clube and Napier, 1990). Time-series analyses of geomagnetic reversal frequencies have yielded long-period estimates from 250 Ma to 350 Ma (Crain and Crain, 1970; Ulrych, 1972; Irving and Pullaiah, 1976; Negi and Tiwari, 1983).

A long period of about 180 Ma (close to the estimated period of the Sun's epicyclic motion in the plane) may be present in the most severe mass extinctions (Rampino and Caldeira, 1992). Four of these cycles at ~ 800 Ma, ~ 600 Ma, ~ 430 Ma and ~ 245 Ma seem to be correlated with evidence of inferred glaciation – tillites. However, some of these tillites may in actuality be ejecta of large impacts (Marshall and Oberbeck, 1992; Rampino, 1992), suggesting periodic or quasi-periodic major bombardment episodes.

Michael R. Rampino

Bibliography

Alvarez, W. and Muller, R.A. (1984) Evidence from crater ages for periodic impacts on the Earth. *Nature*, **308**, 718–20.

Bahcall, J.N. and Bahcall, S. (1985) The Sun's motion perpendicular to the galactic plane. *Nature*, **316**, 706–8.

Bahcall, J.N., Flynn, C. and Gould, A. (1992) Local dark matter from a carefully selected sample. *Astrophys. J.*, **389**, 234–50.

Bailey, M.E., Clube, S.V.M. and Napier, W.M. (1990) *The Origin of Comets.* Oxford: Pergamon, 577 pp.

Clube, S.V.M. (1992) The fundamental role of giant comets in earth history. *Celest. Mech. Dynam. Astron.*, **54**, 179–93.

Crain, I.K., and Crain, P.L. (1970) New stochastic model for geomagnetic reversals. *Nature*, **228**, 39–41.

Crain, I.K., Crain, P.L. and Plaut, M.G. (1969) Long period Fourier spectrum of geomagnetic reversals. *Nature*, **223**, 283.

Fischer, A.G. (1979) Rhythmic changes in the outer Earth. *Geol. Soc. Lond. Newsl.*, **8**(6), 2–3.

Forbes, W.T.M. (1931) The great glacial cycle. *Science*, **74**, 294–5.

Gidon, P. (1970) Glaciations majeures et revolution galactique du systeme solaire. *Comptes Rendus Acad. Sci. Paris Serie D*, **271**, 385–7.

Goldsmith, D. (1985) *Nemesis the Death Star, and other Theories of Mass Extinction.* New York: Walker.

Hatfield, C.B., and Camp, M.J. (1970) Mass extinctions correlated with periodic galactic events. *Geol. Soc. Amer. Bull.*, **81**, 911–4.

Heylmun, E.B. (1969) Geologic significance of the passage of the Earth through the galactic plane. *Geol. Soc. Am. Abstr. with Prog.*, **1**, 36.

Hills, J.G. (1981) Comet showers and the steady-state infall of comets from the Oort cloud. *Astron. J.*, **86**, 1730–40.
Holmes, A. (1927) *The Age of the Earth: An Introduction to Geological Ideas*. London: Benn.
Hut, P. and Treimaine, S. (1985) Have interstellar clouds disrupted the Oort cloud? *Astron. J.*, **90**, 1548–57.
Innanen, K.A., Patrick, A.T. and Duley, W.W. (1978) The interaction of the spiral density wave and the Sun's galactic orbit. *Astrophys. Space. Sci.*, **57**, 511–15.
Irving, E. and Pullaiah, G. (1976) Reversals of the geomagnetic field, magnetostratigraphy, and relative magnitude of paleosecular variation in the Phanerozoic. *Earth-Sci. Rev.*, **12**, 35–64.
Kropotkin, P.N. (1970) The possible role of cosmic factors in geotectonics. *Geotectonics*, **2**, 80–8.
Marshall, J.R. and Oberbeck, V.R. (1992) Textures of impact deposits and the origin of tillites. *EOS. Trans. Am. Geophys. Union*, **73**, 324.
Matese, J.J., and Whitman, P.G. (1992) A model of the galactic tidal interaction with the Oort comet cloud. *Celestial. Mech. Dynam. Astron.*, **54**, 13–35.
Matese, J.J., Whitman, P.G., Innanen, K.A. and Valtonen, M.J. (1995) Periodic modulation of the Oort cloud comet flux by the adiabatically changing galactic tide. *Icarus*, **116**, 255–68.
McCrea, W.H. (1975) Ice ages and the galaxy. *Nature*, **255**, 607–9.
McCrea, W.H. (1981) Long time-scale fluctuations in the evolution of the Earth. *Proc. Roy. Soc. London*, **A375**, 1–41.
Meyerhoff, A.A. (1973) Mass biotal extinctions, world climate changes, and galactic motions. *Can. Soc. Petrol. Geol. Mem.*, **2**, 745–58.
Napier, W.M. (1989) Galactic cycles, in *Catastrophes and Evolution: Astronomical Foundations* (ed. S.V.M. Clube). Cambridge: Cambridge University Press, pp. 133–67.
Negi, J.G. and Tiwari, R.K. (1983) Matching long-term periodicities of geomagnetic reversals and galactic motions of the solar system. *Geophys. Res. Lett.*, **10**, 713–6.
Rampino, M.R. (1992) Ancient 'glacial' deposits are ejecta of large impacts: the ice age paradox explained. *EOS, Trans. Am. Geophys. Union*, **73**, 99.
Rampino, M.R. and Caldeira, K. (1992) Episodes of terrestrial geologic activity during the past 260 million years: a quantitative assessment. *Celest. Mech. Dynam. Astron.*, **54**, 143–59.
Rampino, M.R. and Stothers, R.B. (1984a) Terrestrial mass extinctions, cometary impacts, and the Sur's motion perpendicular to the galactic plane. *Nature*, **308**, 709–12.
Rampino, M.R. and Stothers, R.B. (1984b) Geological rhythms and cometary impacts. *Science*, **226**, 1427–31.
Rampino, M.R. and Stothers, R.B. (1986) Geological periodicities and the galaxy, in *The Galaxy and the Solar System* (eds R. Smoluchowski, J.N. Bahcall and M.S. Matthews). Tucson: University of Arizona Press, pp. 241–59.
Rampino, M.R. and Stothers, R.B. 1988, Flood basalt volcanism during the past 260 million years. *Science*, **241**, 663–8.
Raup, D.M. (1985) Magnetic reversals and mass extinctions. *Nature*, **314**, 341–3.
Raup, D.M. and Sepkoski, J.J., Jr (1984) Periodicity of extinctions in the geologic past. *Proc. Natl Acad. Sci. USA*, **81**, 801–5.
Raup, D.M. and Sepkoski, J.J., Jr (1986) Periodic extinctions of families and genera. *Science*, **231**, 833–6.
Schwartz, R.D. and James, P.B. (1984) Periodic mass extinctions and the Sun's oscillation about the galactic plane. *Nature*, **308**, 712–3.
Steiner, J. (1967) The sequence of geological events and the dynamics of the Milky Way Galaxy. *J. Geol. Soc. Austr.*, **14**, 99–131.
Steiner, J. (1973) Possible galactic causes for synchronous sedimentation sequences of the North American and Eastern European cratons. *Geology*, **1**, 89–92.
Steiner, J. (1979) Regularities of the revised Phanerozoic time scale and the Precambrian time scale. *Geol. Rundschau*, **68**, 825–31.
Steiner, J. and Grillmair, E. (1973) Possible galactic causes for periodic and episodic glaciations. *Geol. Soc. Am. Bull.*, **84**, 1003–18.
Stothers, R.B. (1985) Terrestrial record of the Solar System's oscillation about the galactic plane. *Nature*, **317**, 338–41.
Stothers, R.B. (1986) Periodicity of the Earth's magnetic reversals. *Nature*, **322**, 444–6.
Tamrazyan, G.P. (1957) Geotectonic hypothesis. *Izvest. Acad. Nauk Azerbaid. SSR*, **12**, 85–115.
Tamrazyan, G.P. (1967) The global historical and geological regularities of the Earth's development as a reflection of its cosmic origin (as a sequence of interaction in the course of galactic movements of the Solar System). *Ostrav. Vysoke Skoly Banske Sbornik Ved. Praci Rada Horn Geol.*, **13**, 5–24.
Ulrych, T. (1972) Maximum entropy power spectrum of long period geomagnetic reversals. *Nature*, **235**, 218–9.
Valtonen, M., Zheng, J.-Q. and Mikkola, S. (1992) Origin of Oort Cloud comets in the intersteller space. *Celest. Mech. Dynam. Astron.*, **54**, 37–48.
Williams, G.E. (1975) Possible relation between periodic glaciation and the flexure of the galaxy. *Earth Planet. Sci. Lett.*, **26**, 361–9.
Williams, G.E. (1981) *Megacycles*. Stroudsburg: Hutchinson Ross.
Yabushita, S. (1992) Periodicity in the crater formation rate and implications for astronomical modeling. *Celest. Mech. Dynam. Astron.*, **54**, 161–78.

Cross references

Comet: dynamics
Comet: origin and evolution
Mass extinction

GALILEAN SATELLITES

The Galilean satellites are the four largest satellites of Jupiter. They were discovered in 1610 by Italian astronomer and physicist Galileo Galilei, after whom they are named. In order of increasing distance from Jupiter they are Io (Plates 23, 24), Europa (Plate 24), Ganymede and Callisto (Plate 25). The inner two satellites, Io and Europa, are 3630 km and 3138 km in diameter, respectively, values similar to the diameter of the Moon (3476 km). The outer two satellites, Ganymede and Callisto, are 5262 km and 4800 km in diameter, respectively, which is closer in size to the planet Mercury (4878 km). As does the Earth's Moon, each of these satellites rotates synchronously; they always show the same side to Jupiter. The geometric albedos for Io, Europa, Ganymede and Callisto are 0.6, 0.6, 0.4 and 0.2 respectively; these values are all considerably higher than the Moon's geometric albedo of only 0.12 (Burns, 1986).

The density and compositional and geologic complexity of these satellites is related to position with respect to Jupiter. Io has a density of 3.57 g cm^{-3}, whereas Europa, Ganymede and Callisto have decreasing densities of 3.04 g cm^{-3}, 1.94 g cm^{-3} and 1.86 g cm^{-3} respectively (Burns, 1986). The outer three satellites have densities intermediate between those of the Moon (3.34 g cm^{-3}) and water (1.0 g cm^{-3}). With increasing distance from Jupiter the bulk content of rock decreases and the content of water ice increases.

Analysis of telescopic spectra indicates that the dominant components of Io's surface are sulfur and sulfur dioxide (Nash *et al.*, 1986), whereas water ice or frost is a major constituent of the surfaces of Europa, Ganymede and Callisto (Pollack *et al.*, 1978; Clark and McCord, 1980). Spectra indicate that the fraction of water ice or frost on the surfaces of these three satellites decreases with increasing distance from Jupiter. The aerial coverage of water ice/frost is thought to be 90–100% for Europa, approximately 65% for the leading hemisphere of Ganymede and 20–30% for the leading side of Callisto (Clark and McCord, 1980). Although bulk water ice content increases with distance from Jupiter, the fraction of optical surface that is water ice or frost decreases with distance from Jupiter.

Images taken by the two Voyager spacecraft that flew by Jupiter in 1979 show that the surfaces of these satellites are geologically diverse. The discovery of active volcanoes on Io shows that the surface of this satellite is young and is currently undergoing resurfacing. The almost complete lack of craters on the surface of Europa indicates that the surface of this satellite may also be young. Numerous lineaments on Europa indicate recent tectonic activity, possibly driven by tidal deformation. The wider variety of terrain types and densities of craters on Ganymede indicate that this satellite has an older surface than either Io or Europa, and has been shaped by a greater variety of geologic processes. The high crater density on the surface of Callisto indicates that it may have the oldest surface of all the Galilean satellites. For more in-depth discussions of the surfaces of these satellites, see the articles on Io, Europa, Ganymede and Callisto.

Deborah L. Domingue

Bibliography

Burns, J.A. (1986) Some background about satellites, in *Satellites* (ed. J.A. Burns and M.S. Matthews). Tucson: University of Arizona Press., pp. 1–38.
Clark, R.N., McCord, T.B. (1980) The Galilean satellites: new near-infrared spectral reflectance measurements (0.65–2.5 μm) and a 0.325–5 μm summary. *Icarus*, **41**, 323–39.
Nash, D.B., Carr, M.H., Gradie, J. *et al.* (1986) Io, in *Satellites* (ed. J.A. Burns and M.S. Matthews). Tucson: University of Arizona Press, pp. 629–88.
Pollack, J.B., Witteborn, F.C., Erickson, E.F. (1978) Near-infrared spectra of the Galilean satellites: observations and compositional implications. *Icarus*, **36**, 271–303.

Cross references

Callisto
Europa
Ganymede
Io
Jupiter: satellite system

GALILEO GALILEI (1564–1642)

The Italian astronomer and physicist, born near Pisa, Galileo was one of the leading figures in the great revolution in the theories of physics and astronomy whereby the medieval, scholastic concept of nature came to be replaced by the conception of nature prevalent today. In 1581 he entered the University of Pisa to study medicine, but soon his interest moved to mathematics and physics. In 1589 Galileo became a professor of mathematics at Pisa; he moved to Padua in 1592, and later to Florence. Among his non-astronomical findings were the discovery of the parabolic trajectories of projectiles, the initiation of thermoscopy, and his reflections on the breaking strength of beams. His work on mechanics is laid down in the *Mathematical Discourses and Demonstrations Concerning Two New Sciences, Relating to Mechanics and to Local Motion* (1638), where the two new sciences are now known as 'strength of materials' and 'kinematics'.

Of immense importance for physics was Galileo's formulation of laws of falling bodies. He found that the speeds at which bodies fall are independent of their weights. It is widely believed that he tested this rule by dropping weights from the leaning tower of Pisa. Actually, the laws were a result of experiments with rolling balls and inclined planes, and with pendulums.

Galileo's fame rests particularly on his achievements employing the telescope, an optical instrument which he learned about from the Dutch. By the end of 1609 he used a telescope constructed by himself to discover the lunar mountains. In 1610 he discovered the four largest satellites orbiting Jupiter, and the starry nature of the Milky Way, which are all too faint to be seen with the naked eye. These telescopic discoveries are described in Galileo's book *Sidereus Nuncius* (1610, 'The Sidereal Messenger').

Towards the end of 1610 Galileo observed dark spots on the surface of the Sun, which he reported as a matter of curiosity to various friends. At the same period Thomas Harriot (1560–1621) in England was independently making telescopic discoveries similar to Galileo's. Sunspots were simultaneously observed by Johannes Fabricius (1587–1615) in Holland and by Christopher Scheiner (1575–1650) in Germany.

Continued observations of sunspots showed Galileo that the Sun itself was actually rotating, with a period of about a month. Scheiner, a powerful Jesuit, interpreted the sunspots as stars circling the Sun. Galileo's realization that he was observing change in the heavens, and seeing heavenly bodies circle bodies other than the planet Earth, along with his open support for the heliocentric cosmology of Copernicus, brought him into conflict with the Catholic Church. His great astronomical treatise *Dialogue on the Two Chief Systems of the World, the Ptolemaic and Copernican*, published in Florence in 1632, collected all the then available evidence in favor of the Copernican model, while it ignored the work of Tycho Brahe. The Church's problem with the *Dialogue* was that it was not only very clear and unambiguous, but it was also written in Italian, so that everyone could read it. Accordingly, Galileo's *Dialogue*, as well as Kepler's *Epitome* (published 13 years before the *Dialogue*), was banned by the Church; printers were forbidden to publish anything further by Galileo or even to reprint his previous works. Tried by the Inquisition, and under threat of torture, Galileo, frail and at the age of 70, was forced to recant, although he is said to have muttered 'E pur si muove' (and yet it *does* move!). Under house arrest, and often in ill health, Galileo never abandoned his opposition to the reigning Aristotelian orthodoxy.

Hans J. Haubold

Bibliography

Berry, A. (1898) *A Short History of Astronomy*. London: J. Murray (Dover Publications, Inc., 1961).
Drake, S. (1972) Galilei, Galileo. *Dict. Sci. Biogr.*, Vol. 5, 237–50.
Gamow, G. (1961) *The Great Physicists from Galileo to Einstein*. New York: Arper and Brothers, Publishers (Dover Publications, Inc., 1988).
Holton, G. (1988) *Thematic Origins of Scientific Thought – Kepler to Einstein*. Cambridge: Harvard University Press.
Naylor, R.H. (1990) Galileo's method of analysis and synthesis. *Isis*, **81**, 695–707.
Tauber, G.E. (1979) *Man and the Cosmos (Man's View of the Universe)*. New York: Greenwich House.
Weaver, J.H. (ed.) (1987) *The World of Physics*. New York: Simon and Schuster (Vol. I to III).
Winkler, M.G. and Van Helden, A. (1992) Representing the heavens: Galileo and visual astronomy. *Isis* **83**, 195–217.

GALILEO MISSION

The Galileo spacecraft is in orbit about the planet Jupiter, to conduct an intensive and comprehensive investigation of the Jovian system using an atmospheric entry probe and a Jupiter orbiter (Figure G1). Galileo is instrumented to achieve much more than was possible with the Voyager (q.v.) flyby missions. The Galileo Probe has obtained the first *in situ* measurements of the Jupiter atmosphere, and the Galileo Orbiter is providing the first long-term close observations of the Jupiter system. The Orbiter encounters the Galilean satellites repeatedly and typically 100 times closer than Voyager did.

Jupiter, of all the planets, asteroids, comets and satellites, holds a prime key to understanding the evolution of the solar system. Essential to exploring the history of the solar system are measurements of elemental abundances – the chemical make-up of a variety of objects. Because it retains its primordial composition, Jupiter is a better cosmological 'laboratory' than any of the other planets.

The Jovian system is also a solar system in miniature, with massive gaseous Jupiter, circled by a retinue of satellites, all enveloped in an intense magnetosphere. Each of the four major satellites – Io, Europa, Ganymede and Callisto – shows a different level of geological activity, and each has followed a different evolutionary path.

The Galileo mission has three major and coequal general scientific objectives, namely to investigate (1) the chemical composition and physical state of Jupiter's atmosphere, (2) the chemical composition and physical state of the Jovian satellites, and (3) the structure and physical dynamics of the Jovian magnetosphere.

The Galileo Orbiter carried the probe to Jupiter, releasing it on its ballistic trajectory about 150 days before arrival at Jupiter. Probe release took place on 13 July 1995, over 80 million km from Jupiter. The Probe had no control system and was entirely passive during the 5 months from release to Jupiter arrival. The orbiter's configuration at the time of release resulted in the Probe being spin-stabilized at 10 rpm to achieve entry at zero angle of attack within a tolerance of ±6.0°.

After Probe release, the orbiter performed trajectory corrections in order to (1) overfly the probe during entry in order to record probe data for later relay to the Earth and (2) be in position for Jupiter orbit insertion (JOI). This Orbiter deflection maneuver used the orbiter's main engine for the first time in flight.

The Orbiter's arrival at Jupiter on 7 December 1995 satisfied the many competing mission objectives for arrival day. On 7 December the Orbiter was to be in position over the entry site to receive the uplink signal from the probe. The arrival geometry for both the Orbiter and the Probe is shown in Figure G2.

GALILEO MISSION

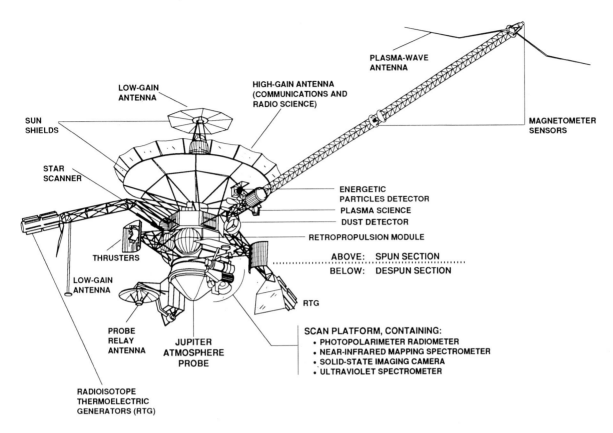

Figure G1 Galileo spacecraft configuration.

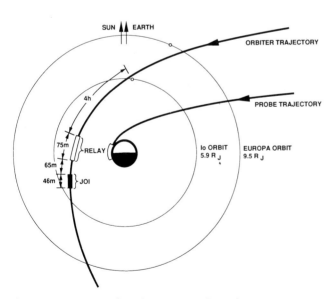

Figure G2 Jupiter arrival (Io flyby, probe relay and Jupiter orbit insertion): 7 December 1995.

The Probe entered Jupiter's atmosphere with a velocity of about 47 km s^{-1}, at a location just above the Jovian equator, at latitude approximately 7° N at 22:04 UT on 7 December. The entry and descent events are depicted in altitude/time space on Figure G3 Following the Probe's descent mission of 61.4 min duration, the Orbiter began a 7-month orbit about Jupiter. For the subsequent 23 months, near each periapsis passage, the Orbiter executes a close encounter with one of Jupiter's Galilean satellites, using a gravity assist from the encounter to change its orbit in order to achieve the desired satellite encounter near the next periapsis. This satellite gravity-assist tour is the foundation of the Orbiter mission design.

The Probe investigations and instruments are identified in Table G1. A comprehensive description of these investigations is given in Yeates *et al.* (1985) and Russell (1992). About 40 min after entry, the probe penetrated to a depth corresponding to a pressure of greater than 10 bars, below the lowest cloud layers, in the well-mixed region of the Jupiter atmosphere where the bulk composition was measured. A 46-min engine burn subsequently placed the Orbiter in a 7-month orbit about Jupiter. During the first phase of this orbit the Orbiter returned the probe data back to Earth.

The Neutral Mass Spectrometer detected smaller than expected amounts of several constituents, including oxygen, water and neon. Some elements are more abundant than was expected. These include carbon, nitrogen and sulfur. Since the amounts of these elements are larger than found for the Sun, these results may contain clues concerning the process of planetary formation. The Probe entered an unusually hot and dry area of the Jupiter atmosphere, and as a result only a few misty clouds were detected. The Probe determined that the high winds present in the upper atmosphere extend to considerable depths. This finding eliminated the theory that the winds were driven by energy from the Sun; instead these > 500 km h^{-1} winds must be driven by heat from within Jupiter.

The Io encounter was originally introduced as a mission element to provide the gravity assist to reduce JOI propellant cost. Galileo's Io encounter at 1000 km altitude was 20 times closer than Voyager's and presented a unique scientific opportunity. However, the observations planned for this Io encounter were cancelled due to uncertainties in the performance of the onboard tape recorder.

On each successive orbit of Jupiter, the Orbiter is precisely navigated to a close encounter of one of the three outer Galilean satellites – Europa, Ganymede or Callisto. The prime mission tour design precludes a return to Io because of the radiation hazard (see Jupiter: magnetic field and magnetosphere). For each encounter the satellite flyby aiming point is selected to result in the satellite gravity assist that will change Galileo's orbit to the next desired one.

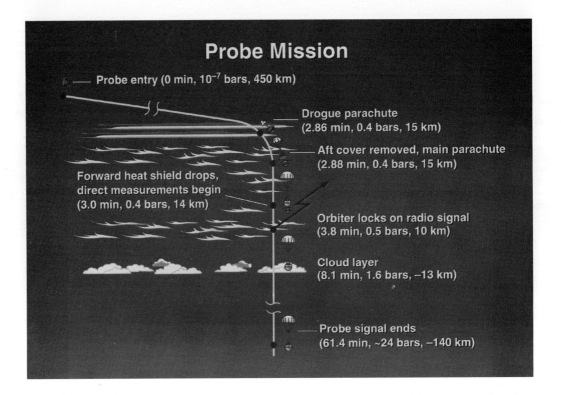

Figure G3 Predicted probe entry/descent events.

Table G1 Probe scientific payload

Experiment	Mass (kg)	Range	Objectives
Atmospheric Structure Instrument (ASI)	4	Temperature: 0–540 K Pressure: 0–28 bar	Determine temperature, pressure, density and molecular weight as a function of altitude
Neutral Mass Spectrometer (NMS)	11	Covers 1–150 AMU	Determine chemical composition of atmosphere
Helium Abundance Detector (HAD)	1	Accuracy: 0.1%	Determine relative abundance of helium
Nephelometer (NEP)	5	0.2–20 μm particles, as few as 3 per cm^3	Detect clouds and infer states of particles (liquid versus solid)
Net-Flux Radiometer (NFR)	3	6 infrared filters from 0.3 to > 100 μm	Determine ambient thermal and solar energy as a function of altitude
Lightning and Energetic Particles (LRD/EPI)	2	Fisheye lens sensors; 1 Hz–100 kHz	Verify the existence of lightning and measure energetic particles in inner magnetosphere

While the Orbiter's remote sensing (or telescopic) instruments observe primarily at the satellite encounters, and image Jupiter when inside 50 R_J, the fields and particles instruments gather data continuously during much of the Orbiter mission. The Orbiter scientific investigations and instruments are identified in Table G2. A comprehensive discussion of these is given in Russell (1992).

The satellite tour shown in Figure G4 was chosen after a long tour design and selection process. It represents an intricately constructed compromise between the competing scientific objectives of the Orbiter instruments (O'Neil et al., 1992).

Historical Development

Launch

Galileo was launched on 18 October 1989 aboard the shuttle Atlantis (STS-34, OV-104). The inertial upper stage (IUS-19) placed Galileo on its Earth-to-Venus trajectory. Following the IUS burns Galileo configured itself for solo flight and separated from the IUS on 19 October. All aspects of the launch were essentially perfect.

The Galileo Venus–Earth–Earth gravity assist (VEEGA) trajectory is illustrated in Figure G5 and is discussed in detail in D'Amario et al. (1989). The Venus gravity assist in February of 1990 was the first of three gravity assists designed to put Galileo into its transfer orbit to Jupiter. The second two gravity assists were with the Earth, both on 8 December in 1990 and 1992.

Venus gravity assist

The 10 February 1990 Venus gravity assist provided Galileo with its first target of opportunity. All of the orbiter's science instruments gathered data during the flyby. With its magnetospheric instruments, energetic particles were detected, bowshock crossings were indicated, and the plasma wave instrument saw evidence of lightning discharges. With the remote sensing instruments, Venus was observed over many days, providing new scientific insight as well as early indications of

Table G2 Orbiter scientific payload

Experiment	Mass (kg)	Range	Objectives
Solid-state imaging (SSI)	30	1500 mm, f/8.5 800×800 CCD, 8 filters $0.47°$ field of view	Map Galilean satellites at roughly 1 km resolution and monitor atmospheric circulation over 23 months while in orbit around planet
Near-Infrared mapping spectrometer (NIMS)	20	0.7–5.2 μm range, 0.03 μm resolution	Observe Jupiter and its satellites in the infrared to study satellite surface composition, Jovian atmospheric composition and temperature
Ultraviolet spectrometer (UVS)	5	1150–4300 Å	Measure gases and aerosols in Jovian atmosphere
Extreme ultraviolet spectrometer (EUV)	12	0.05–0.14 μm	Measure Io plasma torus temperature, scale height and composition; monitor aurora
Photopolarimeter–Radiometer (PPR)	5	Discrete visible and near-infrared bands, radiometry to > 42 μm	Determine distribution and character of atmospheric particles; compare flux of thermal radiation to incoming solar levels
Magnetometer (MAG)	5	32–16 384 gammas	Monitor magnetic field fore strength and changes
Heavy ion counter (HIC)	8	10 MeV–200 MeV	Monitor highly ionizing energetic particles
Energetic particle detector (EPD)	11	Ions: 0.020–55 MeV Electrons: 0.015–11 MeV	Measure high-energy electrons, protons and heavy ions in and around Jovian magnetosphere
Plasma detector (PLS)	13	1 eV–50 keV in 64 bands	Assess composition, energy and three-dimensional distribution of low-energy electrons and ions
Plasma wave (PWS)	7	6–31 Hz, 50 Hz–200 kHz, 0.1–5.65 MHz	Detect electromagnetic waves and analyze wave–particle interactions
Dust detector (DDS)	4	10^{-16}–10^{-6} g, 2–50 km s^{-1}	Measure particle mass, velocity and charge
Radio science (RS); Celestial mechanics	N/A	S-band signals	Determine mass of Jupiter and its satellites (uses radio system and low-gain antenna)
Radio science (RS): Propagation	N/A	S-band signals	Measure Jovian atmospheric structure and body radii (uses radio system and low-gain antenna)

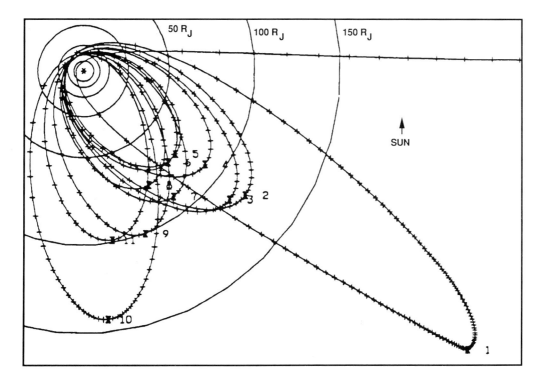

Figure G4 Galileo tour of Jovian satellites (R_J = 71 398 km).

the health and performance of Galileo's scan platform instruments. Because the high-gain antenna had to remain furled under its sunshade until Galileo traveled beyond Earth's orbit, all the Venus science had to be recorded. The vast majority of the Venus data were not received on Earth until mid-November 1990, when the Galileo-to-Earth communication range became short enough that science data rates could be achieved over one of Galileo's low-gain antennas.

Results from the Galileo Venus encounter were presented in a special issue of the journal *Science* (1991, **253**, 1515–50).

Earth gravity assist 1

The first Earth gravity assist required that Galileo fly virtually up the Earth's magnetotail (see Earth: magnetic field and magnetosphere).

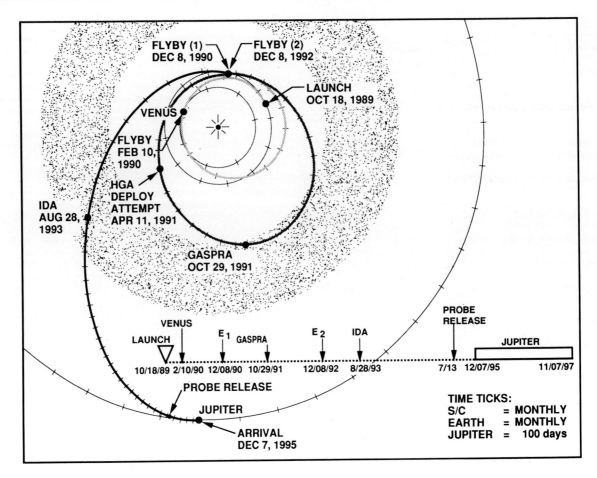

Figure G5 The Galileo VEEGA trajectory to Jupiter.

This provided unique measurements of the magnetotail. The fields and particles instruments provided nearly continuous measurements from 30 days before Earth encounter to 8 days after. From the day of closest approach and continuing for 7 days, periodic Earth and Moon observations were made. Beginning 2½ days after the flyby, Galileo photographed the Earth every minute for 25 h. These images were processed at the Jet Propulsion Laboratory to produce a color movie of the Earth making one full rotation on its axis. Excellent images were also obtained revealing part of the lunar farside not visible from the Earth. Multispectral global maps of lunar compositional units were also obtained. Nearly 3000 imaging frames were taken during the Earth–Moon encounter.

High-gain antenna deployment anomaly

Because of the VEEGA trajectory, which took Galileo into the inner part of the solar system, it was necessary to keep Galileo's high-gain antenna (HGA) furled under its tip shade for thermal protection. HGA unfurling (deployment) was scheduled for the earliest thermally acceptable time, 11 April 1991.

As seen in Figure G5, the antenna is analogous to an inverted umbrella. In a normal deployment the ribs unfurl symmetrically, in about 3 min. Analysis indicates that during Galileo's HGA deployment attempt, three adjacent HGA ribs stuck to the central tower, causing the deployment mechanism to stall and leaving the HGA in a partially and asymmetrically deployed configuration. This conclusion is based on extensive analysis of the flight telemetry, computer modeling and testing of the flight spare HGA at the Jet Propulsion Laboratory. The configuration of the antenna subsequent to the initial deployment attempt as estimated by the anomaly investigation team is illustrated in Figure G6 (O'Neil, 1991; O'Neil, 1993).

From mid-1991 until early 1993 thermal cycling as well as dynamic excitation of the antenna system were carried out in an attempt to free the antenna. None of the activities freed any of the stuck ribs. In

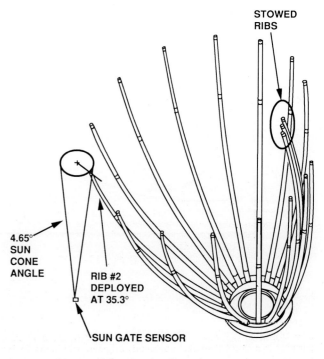

Figure G6 Sun gate obscuration by rib no. 2.

1993 work began to modify ground and flight software and to adapt operating plans to fly the Jupiter mission with the low-gain antenna 1 for telemetry, tracking and commands.

The full Galileo Atmospheric Entry Probe mission and the Orbiter's insertion into Jupiter orbit were accomplished without the HGA. Tracking, telemetry and command will continue over the low-gain antenna (LGA-1), albeit at low telemetry rates. The orbiter playback of the probe data took place in the first few months of the 7-month initial orbit at Jupiter. The orbiter's computer systems were loaded with new flight software which enables the satellite tour to be conducted using the LGA-1 and the on-board tape recorder for collection of high-priority science data (O'Neil et al., 1992).

Gaspra encounter

On 29 October 1991 (at 22:37:01 UTC) Galileo became the first spacecraft to visit an asteroid. Galileo flew by Gaspra (q.v.) at a distance of 1600 km. Gaspra is a main-belt, S-type (silicate: olivine-rich) asteroid. The greatest challenge of the Gaspra encounter arose from the limited accuracy of its ground-based ephemeris. Its position was uncertain by hundreds of kilometers, while Galileo's camera field of view for the highest resolution images is only tens of kilometers. Originally, the HGA was to be used to transmit the optical navigation pictures to Earth in real time. Due to the HGA anomaly, only four frames were taken, recorded on the on-board tape recorder, and then played back to Earth over the LGA-1. As planned, all of the orbiter science instruments, except the Heavy Ion Counter, collected data. In November 1991 Galileo returned the first-ever picture of an asteroid. All the Gaspra data were returned to Earth over the low-gain antenna on approach to the second Earth gravity assist (8 December 1992; see Gaspra).

Earth gravity assist 2

The second Earth flyby was particularly interesting because it provided Galileo with unprecedented observations of the north polar region of the Moon. Infrared spectroscopy with the Galileo Near Infrared Mapping Spectrometer (NIMS) was very successful. A magnetospheric survey was also made at the second flyby, providing a view of the Earth's magnetosphere complementary to that which other spacecraft have been able to provide. In response to the possibility that the HGA might never be successfully deployed, significant observing time was also spent exploiting the high data rates available over the LGA-1 to obtain comprehensive calibrations of the science instruments before returning to the lower data rates characteristic of the LGA-1 when Galileo is not near the Earth.

Earth–Jupiter cruise

On the direct Earth-to-Jupiter leg, Galileo encountered a second asteroid. Galileo flew by Ida on 28 August 1993. The observation strategy was based on the results of the Gaspra encounter. Unlike the Venus and Gaspra encounters, there was no opportunity to achieve high telemetry rates over LGA-1 any time after the Ida encounter in 1993. Ida data were returned during 1993 and 1994 when telecommunication capability enabled playback of the onboard tape recorder. Since the enhanced flight software was not available until after Jupiter arrival, then-current spacecraft capabilities were used. Therefore, only the very highest priority data from the Ida encounter were returned to Earth. Galileo successfully imaged Ida in 1993 and made the first discovery of a Moon orbiting an asteroid (see Ida).

During Earth–Jupiter cruise, Galileo continued to return periodic data from the Magnetometer, the Dust Detector and the Extreme Ultraviolet Spectrometer. Radio science investigations also continued, including a multi-spacecraft search for gravity waves.

In July 1994, one and a half years before Jupiter arrival, the Galileo instruments were trained on Jupiter to capture data during the impacts of the 'string of pearls' comet Shoemaker-Levy 9 (see Comet: impacts on Jupiter). Unlike Earth-based telescopes, Galileo could see the impact sites directly. Galileo observations helped determine the precise times of the impacts. Galileo observations also helped determine the sizes and temperatures of the impact fireballs, providing uniquely valuable information on these spectacular events.

In summer of 1995, Galileo flew through a series of interplanetary dust storms on its way to Jupiter. The dust particles, finer than particles in a cloud of smoke, apparently originate within the Jovian system. They may be a product of volcanic eruptions on the moon Io, or they may originate within the planet's rings. They must first be electrically charged, and then they may be accelerated by Jupiter's powerful magnetic field. The velocities of the particles may be as high as 200 km s^{-1}.

The Galileo mission is remarkable for the resilience and ingenuity of the mission team. The failure of the HGA and problems with other mechanical subsystems of the spacecraft have been mitigated, if not totally overcome, by remarkable efforts by the mission team. Galileo is the only spacecraft in history to receive a completely new version of the flight software used by its central computing system while in flight to the primary mission objective.

Galileo's rendezvous with Jupiter is providing detailed new information about the Jovian system and the evolution of our solar system.

Karen Buxbaum

Bibliography

D'Amario, L.A., Bright, L.E., Byrnes, D.V. et al. (1989) Galileo 1989 VEEGA Mission Description. San Deigo, CA: AAS Publication Office.

O'Neil, W.J. (1991) Project Galileo Mission Status. Paris, France: International Astronautical Federation.

O'Neil, W.J. (1993) Galileo: Challenges Enroute To Jupiter. San Diego, CA: AAS Publications Office.

O'Neil, W.J., Ausman, N.E., Johnson, T.V. and Landano, M.R. (1992) Galileo Completing VEEGA – A Mid-Term Report. Paris, France, International Astronautical Federation.

Russell, C.T. (ed.) (1992) *Space Sci. Rev.* (Galileo Issue), **60**, Nos. 1–4. Dordrecht, Belgium: Kluwer Academic Publishers.

Science (1991) The Galileo Venus encounter. Special issue, **253**, 1516–50.

Yeates, C.M., Johnson, T.V., Colin, L. et al. (1985) Galileo: Exploration of Jupiter's System. Washington, DC: NASA Scientific and Technical Information Branch.

Cross references

Gaspra
Gravity-assist navigation
Ida

GANYMEDE

One of the Galilean moons of Jupiter, Ganymede is the largest natural satellite in the solar system (Plate 25). With a diameter of 5262 km, it is larger than the planets Mercury and Pluto. Ganymede orbits Jupiter at a distance of 1.07 million km, completing one cycle in 7.155 days. Ganymede is in synchronous rotation, so that it always presents the same 'face' to Jupiter. Virtually everything that is known about the geology and geophysics of Ganymede is derived from observations made by the two Voyager spacecraft in 1979.

Ganymede has a bulk density of 1.94 g cm^{-3}, which is about twice that of liquid water. This density, in combination with cosmochemical considerations, implies that the satellite is composed of H_2O and silicate rock in roughly equal portions, ~ 55/45 rock/ice by mass. If the material within Ganymede is completely differentiated (with the least dense materials forming the outer layer and the most dense at the center), the boundary between a rocky core and a H_2O outer shell lies two-thirds of the distance from the center to the surface. The actual internal structure of Ganymede is unknown.

Ganymede's surface brightness and color hint that the distribution of material within the body may be complex. If the outer portion of Ganymede was composed of pure water ice, it should appear bright in reflected sunlight. Instead Ganymede is a dingy brown, rather like a mixture of ice and mud. Although it is possible that this dirty appearance is due to minor dark non-ice contamination (at the several percent level) of an otherwise clean ice layer, it is alternatively possible that the outermost crust of the satellite has a substantial component of silicate (rocky) material. This might mean that the crust was never warm enough to cause density segregation (while the rest of the interior was), or that the surface was plastered with meteoritic dirt and debris after density segregation took place, or that the ice at the surface largely sublimated away leaving behind a lag

Figure G7 Global view of Ganymede showing polygons of dark terrain separated by bands of bright terrain. Note the polar brightening (especially in the north where it stands out against dark terrain), and the bright rays of impact craters. (Voyager 2 image FDS 20608.11; north is at top.)

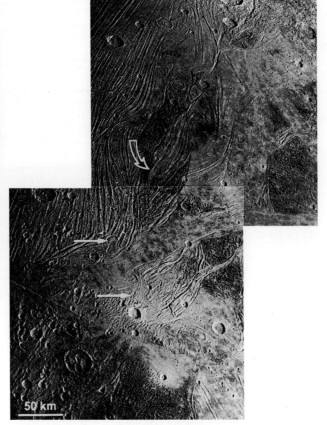

Figure G8 Grooved and smooth terrain on Ganymede. The smooth terrains exhibit a distinctive patchy or mottled appearance; circular bright patches may be barely flooded craters. Also seen are grooved dark terrain (large arrow) and two flooded crater complexes (small white arrows). The upper complex appears flooded by material significantly more viscous than water or slush. Taken near the south pole by Voyager 2, these are among the highest resolution images available. (Mosiac of frames FDS 20640.25 and FDS 20640.27; centered near 75°S, 170°W; north is at bottom.)

of rocky material. We simply do not know which, if any, of these hypotheses is correct.

Closer inspection of Ganymede's surface reveals that it is (to first order) dominated by two more or less equally extensive terrain types that are distinguished by their relative brightness (or albedo) in reflected sunlight and crater size–frequency distribution (per unit area). They are referred to as bright terrain and dark terrain. Dark terrain is inferred to be the older of the two because it has a greater number of large craters (per unit area) and is seen to be buried or destroyed where bright terrain has formed. Globally, dark terrain forms polygons separated by broad bands of bright terrain (Figure G7).

Bright terrain appears to be composed of material that erupted onto the surface, filling fault-bounded troughs (known as rifts or graben). The extensional tectonic regime implied by the global geometry of the graben was possibly caused by the internal expansion of Ganymede, associated with a youthful episode of heating and internal chemical or phase alteration. Global networks of graben bounding and separating large polygonally shaped blocks of old crust are observed on other icy outer planet satellites (such as the Uranian satellites Titania and Ariel), and are attributed to some combination of the internal expansion of these satellites and the cooling and thermal contraction of their outermost layers. The material of the bright terrain nearly completely fills the rifts on Ganymede, although a few Voyager images show blocks of dark terrain clearly standing above the surrounding bright terrain. The appearance of dark material in the ejecta excavated from depth by large impacts on bright terrain indicates that the bright terrain is not thicker than a few kilometers.

The physical state of the bright terrain material at the time of its eruption onto the surface is unknown. It is alternatively speculated that it came to the surface as a liquid, a slush of solid particles mixed with a liquid, or as relatively warm ice moving sluggishly (like glacier ice on Earth) across a much colder and more rigid surface. In a few places, where bright material spills over the graben and inundates the surrounding older dark surface, the bright terrain has a distinctive dark mottling, suggestive of a thin layer of bright material overlying dark terrain. This observation supports the hypothesis that bright material erupted as a fluid or slush. Elsewhere on Ganymede bright material forms thick rounded termini (flow fronts?), suggesting a glacier-like emplacement mechanism. It is possible that bright material erupted in several different forms, varying from place to place and time to time (Figure G8). It has been inferred that bright material is nearly pure water ice (Clark, Fanale and Gaffey, 1986), perhaps mixed with some ammonia or salts, and transported to the surface through dikes and conduits from the upper mantle of the satellite. As such, it is direct evidence for at least partial differentiation (density and chemical segregation) of Ganymede's interior.

Much of the bright terrain is overprinted with a pervasive topographic fabric commonly referred to as grooved terrain (Shoemaker et al., 1982). Grooved terrain is composed of parallel to subparallel alternating troughs and ridges. Ridge to trough relief averages ~ 500 m and can be as great as ~ 1 m. Ridge to ridge spacing varies between 4 and 15 km. The ridges and troughs form 'bundles' or sets whose widths vary from a few kilometers to ~ 100 km and can be many hundreds of kilometers long (Figure G8). Often groove sets interact with one another to form complex cross-cutting relationships. The relation of these groove sets with their neighbors can be used to derive the local sequence of groove set formation. In some cases the groove sets, running in different directions, are superimposed on each other to form a criss-cross pattern of hummocky hills called reticulate terrain. Reticulate terrain is often found along transition zones between older ungrooved dark terrain and younger grooved bright terrain. Groove sets are occasionally seen crossing dark terrain.

The relationship between the grooves and the bright terrain is poorly understood. Grooves on the bright terrain, except in rare

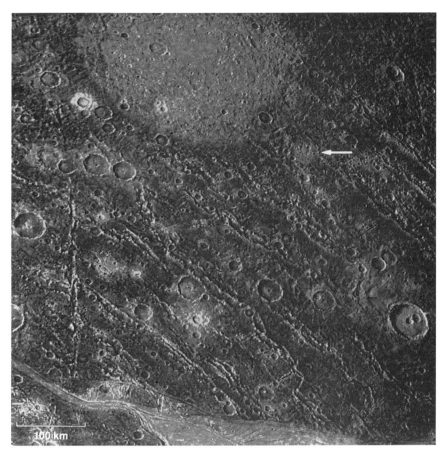

Figure G9 A high-resolution view of ancient dark terrain (note the bright grooved terrain entering along the bottom). Most of the craters in this view are highly flattened. Many craters have pits. One very flattened crater has a central dome (arrow). A large palimpsest is at the top of the frame. The rimmed furrows crossing the image diagonally are part of a multiring basin. Sections of this feature's ring system, as well as any original central structure, have been destroyed by the subsequent formation of bright terrain. (Voyager 2 image FDS 20638.33; centered near 10°N, 132°W; north is at top.)

cases, do not break up impact craters, which implies that grooves formed very soon after the emplacement of the bright terrain. Also craters on bright grooved terrain are still circular. This rules out the possibility that large amounts of distributed regional-scale strain are involved in groove formation, for example, that the ridges are compressional fold belts (like the Appalachian Mountains of the eastern US) formed by slow long-term cooling and contraction of the interior during the later part of Ganymede's history. Definitive evidence for large-scale strike-slip motion associated with bright terrain formation is generally lacking as well (although offsets of approximately 10 to 100 km can be inferred in a few locations). Grooves may be the expression of deep faulting and fracturing that extends upwards into the deposits of bright material from below. Alternatively they may be essentially confined to the bright material and are an expression of the poorly understood emplacement and cooling history of the bright terrain. The observation that grooves are concentrated on bright terrain (as opposed to being evenly distributed across the surface without regard for terrain type) may indicate that something other than youthful global expansion of the satellite is responsible for their formation. A few conclusions regarding groove formation can be made. The most significant are that extensional (pull-apart) faulting probably dominates groove formation, and that the concentration and timing of formation of grooves on bright terrain strongly indicates that the same internal mechanism is responsible for both bright material 'volcanism' and its subsequent faulting (McKinnon and Parmentier, 1986).

Like most solid-surface objects in the solar system, Ganymede exhibits numerous impact craters (Figure G9). Craters on Ganymede exhibit surprising diversity in their appearance. Although small morphologically fresh craters (< 6 km in diameter) are bowl-shaped and generally similar to fresh craters in this size range seen on the Earth's moon and other rocky worlds (albeit somewhat shallower), the depth-to-diameter ratios of larger fresh craters on Ganymede decline with increasing crater diameter and are shallower in comparison with fresh craters of similar size on rocky planets (Schenk, 1991). Craters on Ganymede's neighbor Callisto show the same trend. Many older craters on Ganymede and Callisto are anomalously flat even in comparison with these trends: some have floors that are bowed upward. Such flattening has been ascribed to slow viscous relaxation of the crater's topographic relief over geologic time or possibly to an inability of the early crust to support appreciable crater topography in the first place (Shoemaker et al., 1982). It is thought that the combination of internal heat and relative proximity to the sun resulted in sufficient crustal warmth to allow the viscous relaxation (or prompt collapse) of topography to take place. The craters on the surfaces of many of the other icy satellites of the outer planets do not show such pronounced flattening, because their crusts have been much colder and therefore more rigid. Also, for a given size, younger craters show less flattening than do older craters, reflecting the cooling and stiffening of Ganymede's crust with age (Passey and Shoemaker, 1982).

Craters greater than ~ 6 km across, in addition to their relative shallowness, exhibit a number of unusual landforms. Central peaks predominate between ~ 6 and 35 km while many above ~ 20 km and most above ~ 35 km have central pits rather than central peaks. (Central peaks form almost exclusively in the large craters on rocky worlds.) Furthermore, the floors of pits in craters on Ganymede larger than ~ 50 km are commonly occupied by smooth bright domes. The formation of central pits in lieu of central peaks is speculated to be due to the uplift of fluid or slushy material from depth at the time of crater formation, or alternatively, the fluidization of material at the center of the crater floor by the energy of the impact (Chapman and McKinnon, 1986; Schenk, 1991). The former mechanism may be better supported by the present evidence; there is a correlation between increasing pit diameter and crater diameter, very young craters above ~ 20 km diameter are more likely to have central peaks instead of pits, and craters in this size range on icy satellites with cooler histories do not exhibit central pits. Bright domes within central pits have been hypothesized to be either the tips of clean-ice diapirs rising up from the mantle (where they are nucleated by crustal movement associated with viscous relaxation), or alternatively, the domes may be created during the impact (Moore and Malin, 1988; Schenk, 1993). Thick tongue-like lobes along the flanks of some domes, suggestive of glacier-like warm ice flows, support the former hypothesis. The alternative explanation invokes prompt (or 'fast') diapirism associated with the general process of central pit formation.

A class of impact features that is virtually unique to Ganymede and Callisto is characterized by low relief (or flat), roughly circular high albedo patches or spots with diameters greater than ~ 30 km. Most are seen on old dark terrain and these spots (or faculae) are usually brighter than the surrounding terrain (Figure G9). Their centers are usually smooth but may become texturally rough around their peripheries. Most show no evidence of a rim or where a rim might have once existed. Beyond the bright faculae, fields of secondary craters can often be seen. This is the most compelling evidence that these features were indeed caused by comet or asteroid impacts. They have been given the name 'palimpsests', which is a term borrowed from archeology, where it is used to describe erased and overwritten but still faintly visible original writing on parchment.

The formation of palimpsests is a matter of some controversy. Suggestions for their origin are similar to those offered above for very flattened craters, that is, they may represent extreme cases of viscous relaxation, or perhaps owe their appearance to the inability of a warm, thin crust to retain any significant topographic relief. Any broad topography that might have formed would have been subsequently 'relaxed' regardless of which hypothesis is chosen. Palimpsests have also been attributed to the diapiric eruption and lateral spreading of clean ice from below (Thomas and Squyres, 1990), and thus may be an extreme form of the central dome formation mechanisms discussed above. Palimpsests are at any rate the oldest impact features on Ganymede and were formed at a time when the satellite was young and relatively warm.

The largest impact features on Ganymede are multiring basins. The oldest and largest of these features are manifested by a series or system of concentric furrows or troughs which extend over 1000 km from the center (Figure G9). One such multiring system appears on several different pieces of dark terrain, separated by broad bands of bright terrain. The center of this system lies within an area that is today occupied by bright terrain and therefore its surface expression is buried or destroyed. The continuity of the ring pattern's circularity across several different pieces of dark terrain has been used to argue that there has been very little lateral motion of dark terrain crust pieces relative to one another. This and other observations indicate that nothing akin to plate tectonics has taken place on Ganymede. The concentric rings of furrows are interpreted to be graben formed by radially symmetric extensional stresses at the base of a thin, relatively brittle crust as the underlying mantle viscously flowed inward to infill the central cavity formed by the impact (Chapman and McKinnon, 1986; McKinnon and Parmentier, 1986). The youngest of these multiring basins have fewer and less perfectly formed rings and show greater topographic relief. This time evolution of multiring basins on Ganymede is consistent with the above mechanism operating with an increasingly cool, hence thicker, lithosphere.

The uppermost surface or regolith ('soil') of Ganymede is composed of a layer of unconsolidated, impact generated and stirred debris. The regolith is thought to be no more than a few tens of meters thick. Below this depth, heating during Ganymede's youth is thought to have thermally annealed (welded together) the ice-rich debris, transforming it into a solid substrate. Across the surface of the regolith, ejecta from the youngest craters have created bright rays. The brightness of the rays is proportional to the brightness of the terrains they cross, demonstrating that rays are mostly locally derived material stirred up by secondary projectiles (which themselves represent only a small component of the ray material) (Shoemaker et al., 1982).

Impact-vaporized H_2O migrates to the polar regions of Ganymede resulting in the brightening of the polar regions (Figure G7), referred to as polar 'shrouds.' These shrouds have indistinct boundaries. Ganymede's surface is also subjected to high-energy atomic particles from Jupiter's intense radiation belts. This irradiation, along with solar heating, should operate to devolatilize and erode the surface. The most obvious effects are a relative darkness and redness of the trailing hemisphere of the satellite as it orbits Jupiter (Clark et al., 1986; Schenk and McKinnon, 1991). The trailing sides of all the Galilean satellites are subjected to continuous bombardment by charged particles, because the particles are forced to corotate with Jupiter's magnetosphere (see Jupiter). These particles slam into Ganymede at speeds of up to 190 km s^{-1}. The redness is hypothesized to be caused by the implantation of sulfur atoms originally injected into the magnetosphere by volcanoes on Io.

Overall, Ganymede offers a distinct lesson in the limits of our present-day understanding of planetary evolution, when compared to its neighboring satellite Callisto (q.v.). Although of similar size, density (and presumably composition), and sharing adjacent orbits, Ganymede and Callisto appear very different. Ganymede has had an active geological past, while Callisto apparently has not. Planetary scientists have wrestled with this dichotomy but have come to no clear explanation, as all credible theories have these two worlds evolving in a similar fashion, unless conditions are finely tuned (McKinnon and Parmentier, 1986; Kirk and Stevensen, 1987). Perhaps Callisto was geologically active, but 'died' so early that its surface became completely covered with impact craters. Or perhaps Ganymede was subjected to some unique source of heating that Callisto did not share. A prime candidate is tidal heating associated with Ganymede's orbital resonance with Io and Europa (Malhotra, 1991). While not important today, it could have been significant in the past, and not have affected Callisto at all. Scientists also do not have good evolutionary models for large ice – rock satellites. Consider that due to the high pressures inside Ganymede, H_2O may exist in up to seven different phases, with different densities, viscosities and so forth. The Galileo spacecraft, presently en route to Jupiter, will provide new information that may help unravel the mystery of the divergent evolution of Ganymede and Callisto.

Some of the unresolved issues concerning the geology and geophysics of Ganymede that the forthcoming Galileo mission (q.v.) may help answer are (1) what is the composition of the uppermost surface of the different terrains; (2) how was the bright terrain emplaced (water, slush or warm ice); (3) where are the source vents of bright material eruptions; (4) what is the geologic process responsible for the formation of grooved terrain (grabens, tilt blocks, crevasses or folds); (5) what is the composition and structure of the regolith and upper crust; and (6) how differentiated is the interior (does it have a dense core)?

Jeffrey M. Moore
William B. McKinnon

Bibliography

Chapman, C.R. and McKinnon, W.B. (1986) Cratering of planetary satellites, in *Satellites* (ed. J.A. Burns and M.S. Matthews). Tucson: University of Arizona Press, pp. 492–580.

Clark, R.N., Fanale, F.P. and Gaffey, M.J. (1986) Surface composition of natural satellites, in *Satellites* (ed. J.A. Burns and M.S. Matthews). Tucson: University of Arizona Press, pp. 437–91.

Kirk, R.L. and Stevensen, D.J. (1987) Thermal evolution of a differentiated Ganymede and implications for surface features. *Icarus*, **69**, 91–134.

Malhotra, R. (1991) Tidal origin of the Laplace resonance and the resurfacing of Ganymede. *Icarus*, **94**, 399–412.

McKinnon, W.B. and Parmentier, E.M. (1986) Ganymede and Callisto, in *Satellites* (ed. J.A. Burns and M.S. Matthews). Tucson: University of Arizona Press. pp. 718–63.

Moore, J.M. and Malin, M.C. (1988) Dome craters on Ganymede. *Geophys. Res. Lett.*, **15**, 225–8.

Passey, Q.R. and Shoemaker, E.M. (1982) Craters and basins on Ganymede and Callisto: morphological indicators of crustal evolution, in *The Satellites of Jupiter* (ed. D. Morrison). Tucson: University of Arizona Press, pp. 379–434.

Schenk, P.M. (1991) Ganymede and Callisto: complex crater formation and planetary crusts. *J. Geophys. Res.*, **96**, 15635–64.

Schenk, P.M. (1993) Central pit and dome craters: exposing the interiors of Ganymede and Callisto. *J. Geophys. Res.*, **98**, 7475–98.

Schenk, P.M. and McKinnon, W.B. (1991) Dark-ray and dark-floor craters on Ganymede, and the provenance of large impactors in the Jovian system. *Icarus*, **89**, 318–46.

Shoemaker, E.M., Lucchitta, B.K., Wilhems, D.E. *et al.* (1982) The geology of Ganymede, in *The Satellites of Jupiter* (ed. D. Morrison). Tucson: University of Arizona Press, pp. 435–520.

Thomas, P.J. and Squyres, S.W. (1990) Formation of crater palimpsests on Ganymede. *J. Geophys. Res.*, **95**, 19161–74.

Cross references

Callisto
Europa
Galilean satellites
Io
Jupiter: satellite system

GASPRA

951 Gaspra is a small (7 km mean radius) asteroid located near the inner edge of the main asteroid belt (orbital semimajor axis 2.21 AU) in the Flora region. It was discovered in 1916 by G. Neujmin and was named for a resort in the Crimea.

Ground-based reflectance spectra show Gaspra to be an unusually red, olivine-rich S-type asteroid, possibly a fragment of a differentiated parent body disrupted by catastrophic collisions. Earth-based light curves show uneven and variable minima, and a large amplitude of up to 1.1 magnitude, implying an irregularly shaped, elongated object with a rotation period of 7.04 h.

Gaspra became the object of the first ever close-up reconnaissance of an asteroid on 29 October 1991, when the Galileo spacecraft passed within 1600 km (Belton *et al.*, 1992). An extensive series of observations were performed by the spacecraft's science instruments during the flyby. High-resolution (up to 54 m per pixel) images show Gaspra to be an irregularly shaped object (about 18 × 11 × 9 km)

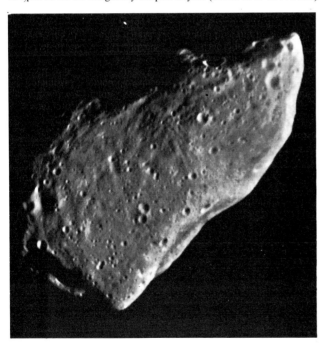

Figure G10 Highest-resolution (54 m per pixel) image of Gaspra obtained by Galileo spacecraft. Sunlight is incident from right; Gaspra rotates counterclockwise about the north pole, located near the terminator at upper left. Large concavity on lower right limb is about 6 km across; prominent crater near terminator, center left, is about 1.5 km across. (NASA/JPL.)

that appears to have been created by collisional fragmentation of a parent body. The overall shape is defined by smooth, slightly concave surfaces that meet along ridges. The aspect viewed is consistent with predictions of spin period and spin axis orientation derived from ground-based data. Images taken over a full Gaspra rotation reveal an overall peanut-like shape, suggestive of the binary shapes observed in asteroids imaged with Earth-based radar.

The surface is conspicuously cratered, and there is an abundance of fresh craters, but there are fewer craters per unit area than on most planetary satellites; there is a deficiency of craters larger than about 1.5 km in diameter. The low density of visible craters implies that the surface is relatively young, on the order of 200 million years. The steep slope of the crater size distribution supports the view that cratering by small objects is important in this region of the asteroid belt. The large shallow concavities may be impact features or sites of spallation or chipping away of fragments following large impacts. Other features include linear depressions 100 to 400 m wide and up to 2.5 kilometers long. If these are grooves like those on Phobos, they are likely evidence of nearly catastrophic impacts.

Although the surface morphology appears subdued and smoothed, consistent with many tens of meters of regolith on the surface, impact models for a low-gravity object like Gaspra predict that much less regolith would be retained, at most a few meters or perhaps much less. The smoothed appearance may be the result of erosion by many small impacts below the image resolution limit.

Multispectral images reveal subtle variations in albedo and color which appear to correlate with topographic features. Brighter deposits, found around some craters and on the ridges, have a 10% higher albedo and a stronger 1-μm absorption (attributed to the presence of the minerals olivine and perhaps pyroxene) than the global mean. Darker regions, covering some of the 'facets' between the ridges, are about 20% lower than average in albedo and have a significantly weaker 1-μm absorption. The origin of these differences is not yet clear.

H. Herbert Breneman

Bibliography

Belton, M.J.S. *et al* (1992) Galileo encounter with 951 Gaspra: first pictures of an asteroid. *Science*, **257**, 1647–52.

Cross references

Asteroid
Galileo mission

GAUSS, CARL FRIEDRICH (1777–1855)

Born in Brunswick (Germany), Gauss was raised in humble circumstances, but showed such precocity that he was granted a ducal stipend and by the age of 25 he was a famous mathematician and astronomer. In 1802 he became director of the observatory in Göttingen and stayed there for 47 years.

While his primary contributions were to mathematics, numbers theory and statistics, Gauss made valuable contributions to geodesy, geomagnetism and planetary science. He independently discovered the Titius–Bode law of planetary distances. According to May (1972), 'he thought numerically and algebraically, after the manner of Euler . . .' When Piazzi (q.v.) described (and 'lost') the new 'planet' Ceres (q.v.) seen in January 1801, Gauss applied elliptical orbit theory and the new least squares method to predict, successfully, its reappearance in January 1802. His full theory of orbital motion appeared in 1809. When his stipend was raised in 1801 he remarked: 'But I have not earned it. I have not done anything . . . .'

At Göttingen he was able to study the perturbation of Pallas when passing Jupiter (*Disquisitiones generales. . .*, 1813). This was followed (in 1818) by a work showing that the perturbation caused by a planet is the same as that of an equal mass distributed along its orbit in proportion to the time spent on an arc.

After 1817 Gauss was much concerned with geodesy, competing with the French for the measurement of one degree of arc on the meridian. He did much of the fieldwork for the triangulation of the state of Hannover (completed in 1847). He became interested in the measurement of sea level. In 1828 he was invited to Berlin for 'a little celebration' by Alexander von Humboldt which stimulated a renewed interest in geomagnetism. Gauss then set up a magnetic observatory, part of a network planned by von Humboldt. By 1834 there were 23 operating observatories and the first magnetic storms were demonstrated. He designed a bifilar magnetometer. Jointly with Weber, their *Atlas des Erdmagnetismus* was published in 1840. With declining health in later life he was forced to abandon most of his scientific work, but by using his statistical methods, he began trading in securities, eventually building his net worth to about 200 times that of his annual salary: mathematician to the end.

Rhodes W. Fairbridge

Bibliography

Dunnington, G.W. (1955) *Carl Friedrich Gauss: Titan of Science*. New York.
Gauss, K.F. (1809) Theory of the motion of the heavenly bodies moving about the Sun in conic sections (*Theoria motus* . . . transl. by C.H. Davis). New York: Dover Publ., 1963, 326 pp.
May, K.O. (1972) Gauss, Carl Friedrich. *Dict. Sci. Biogr.*, Vol. 5, pp. 298–315.
Stern, D.P. (1977) Carl Friedrich Gauss – 200th anniversary. *EOS*, 58, 186 [repr. *History of Geophysics*, Vol. 2, Washington, DC: Amer. Am. Geophys. Union, pp. 89–90.]
Von Waltershausen, S. (1856) *Gauss zum Gedächtnis* (transl. by Gauss, Helen W., 1966. 'Gauss, a Memorial', Colorado Springs).

GEOID

The gravitational potential of a planet is determined by the mass distribution within it (see Gravity fields of the terrestrial planets). Any non-radially symmetric density distribution will produce irregularly shaped equipotential or level surfaces.

In the absence of winds, currents and other oceanographic perturbations, the Earth's ocean surface is one such surface and, as such, it provides a natural definition for the shape of the Earth. This is the geoid; defined as that equipotential surface corresponding to the mean sea surface over the oceans and, on the continents, as the surface that would be defined by 'a series of criss-crossing free-flowing canals in open connection to the sea'.

Because of its rotation, the Earth is permanently deformed and the geoid can, in a first approximation, be represented by an ellipsoid of revolution, with its short axis along the rotation axis, and with a geometric flattening f of

$$f = (R_e - R_p)/R_p, \sim 1/300$$

where R_e and R_p are the mean equatorial and polar radii of the planet respectively. This ellipsoid provides a convenient reference with respect to which the geoid can be defined in terms of geoid height; the separation N between the two surfaces (Figure G11). For most geodetic work the parameters f and R_e are chosen so as to minimize these geoid heights over the globe. Best fitting values are $f = 1/(298.275 \pm 0.001)$ and $R_e = 6378.136 \pm 0.001$ km.

Departures of the geoid from this surface are then of the order of 100 m. For geophysical work a more appropriate reference figure is the hydrostatic equilibrium configuration of a rotating body whose size, radial density distribution and rotational velocity corresponds to the actual Earth, for now the geoid anomalies reflect directly the departures of the planet from hydrostatic equilibrium (see Gravity fields of the terrestrial planets). If the density distributions are specified, then these reference figures also serve as a model for the external gravity field.

While the geoid is defined in terms of the shape of the ocean, free from any oceanographic perturbations, this surface will lie partly within the crust of the continental areas and its mathematical extension to the continents does not have a unique definition, being a function of the lateral and radial density distribution within the crust. For most practical purposes, the geoid over continents is defined by geodetic leveling, which determines the height differences between surfaces of different potentials, and by simple mathematical models for the variation of gravity within the crust. Outside of the Earth the

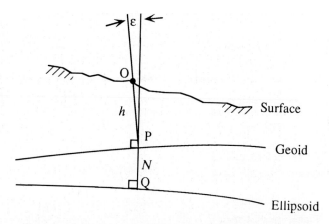

Figure G11 Definition of geoid height N. Gravity is measured at O on the Earth's surface at height h above the geoid and is projected along the vertical to P (g_P). The geoid lies at a distance N from the reference ellipsoid. The theoretical gravity at the projection of P onto this ellipsoid at Q is γ_Q. The angle ϵ is the deflection of the vertical.

gravitational potential is everywhere defined. Lines orthogonal to the equipotential surfaces specify the direction of the gravity vector, the vertical or direction of the plumb line.

The measurement of the shape of the geoid has traditionally been carried out from gravity measurements taken on the Earth's surface or from astronomical measurements of the deflection of the vertical (Figure G11), which is the angle between the local vertical and the normal to the reference ellipsoid. The global mapping of the external equipotential surfaces has also become possible from analyzing the gravitational perturbations in the motion of close Earth satellites and from a direct mapping of the shape of the ocean surface using satellite-mounted radar altimeters (see Gravity fields of the terrestrial planets). Figure G12 illustrates the long-wavelength characteristics of the geoid.

In gravity surveys and gravity field interpretations the concept of the geoid is important. Gravity is measured on the Earth's surface at a height h, measured by spirit leveling, above or below the geoid. Thus, measurements g_o can be reduced to values g_p on the geoid using the free-air and Bouguer corrections (see Gravity fields of the terrestrial planets). The geoid at P (Figure G11) lies at a distance N from the reference ellipsoid, but this will generally be unknown until computed from a regional geoid model, which itself requires gravity measurements throughout the region. The gravity anomaly Δg_p is therefore defined as

$$\Delta g_p = g_p - \gamma_Q$$

where γ_Q is the theoretical gravity corresponding to the adopted reference ellipsoid at the projection Q of the point P on to the ellipsoid. These gravity anomalies are typically of the order of 10^{-3} m s^{-2} or about 100 mgal (gravity anomalies are conveniently expressed in milligals where 1 mgal = 10^{-5} m s^{-2}).

The geoid over the oceans is also an important concept in oceanography. The shape of the sea surface will exhibit many irregularities because of anomalous density structures in the crust and mantle, but smaller irregularities, primarily of a time-dependent nature, are caused by oceanographic processes including water movements and density changes. To measure the longer-term changes in the shape of the oceans, the geoid must be defined with a high precision and high spatial resolution; this is now achieved with satellite-borne radar altimeters combined with other space methods for mapping the gravity fields (see Gravity fields of the terrestrial planets).

The concept of the geoid is also extended to the other planets, where the shape of their equipotential surfaces can be mapped from the analysis of the accelerations experienced by planetary orbiters, although the convenient analogy of one of these surfaces with a physical surface such as the ocean does not exist. For Mars the reference ellipsoid of revolution adopted usually corresponds to the expected hydrostatic equilibrium state for the planet, and the equipotential reference surface is mathematically defined as one with zero mean departure from the reference surface. The treatment for

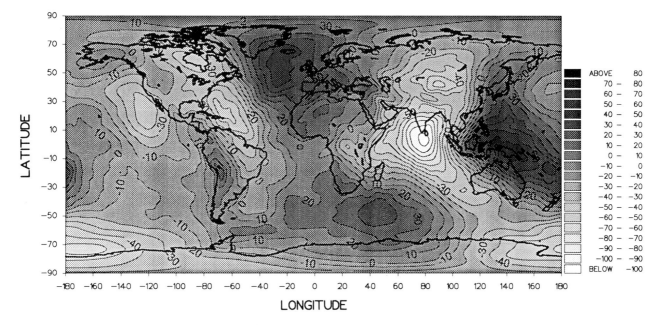

Figure G12 Long-wavelength variations in the shape of the Earth, the geoid, expressed relative to a best-fitting ellipsoid according to a recent solution by G. Balmino and C. Reigber. Contour interval in meters.

slowly rotating Venus is simpler: since it does not possess a rotational bulge, the adopted reference surface is a sphere. For the Moon, permanently deformed by the Earth's gravitational attraction, the reference figure is a triaxial ellipsoid with the major axis directed towards the Earth and the minor axis corresponding to the lunar rotation axis.

Kurt Lambeck

Bibliography

Bomford, G. (1988) *Geodesy*, 4th ed. Oxford.
Lambeck, K. (1980) *The Earth's Variable Rotation*. New York: Cambridge University Press.
Lambeck, K. (1988) *Geophysical Geodesy*. Oxford: Oxford University Press.
Torge, W. (1999) *Geodesy*, 2nd edn. Berlin: de Gruyter.

Cross references

Gravimetry
Gravity fields of the terrestrial planets
Mars: gravity
Moon: gravity
Venus: gravity

GEOMAGNETIC POLARITY REVERSALS AND THE GEOLOGICAL RECORD

Geologists commonly use two methods of dating rocks. The biostratigraphical method depends on the identification of fossils in a sedimentary rock sequence. It utilizes the evolutionary changes in species to provide a detailed *relative* age zonation in sedimentary rocks. Igneous rocks may be dated by their stratigraphic relationship to a dated sedimentary sequence, but are most usually dated radiometrically. The radiometric method uses the decay of commonly occurring radioactive isotopes, assuming known decay rates, to determine the absolute age of a rock specimen. Provided suitable stratigraphic correlations are available, it is possible to estimate the ages of major boundaries or events in the geological record. However, until recently the accuracy of the method was limited to about ± 1–2%, which is inadequate for fine-scale correlation.

The science of paleomagnetism is based on the well-established phenomenon that rocks can acquire a stable remanent magnetization parallel to the ambient magnetic field during their formation or subsequent alteration. The recording process and paleomagnetic sampling methodology result in time averaging of the ancient geomagnetic field so that the axial dipole component is preserved. The high stability of many rock magnetizations ensures that the record of the ancient geomagnetic field can endure throughout geological eons. In this way reversals of polarity of the Earth's magnetic field have been well documented in the remanent magnetization of rocks. In sedimentary and igneous sequences magnetozones of alternating uniform polarity are found that form a magnetostratigraphy. When used in conjunction with paleontological dating, magnetostratigraphy can provide increased resolution of biostratigraphical zonation and permits the association of fairly reliable absolute dates with fossil events.

The cause of a polarity reversal is not understood, and there is disagreement about the behavior of the field during the polarity transition. Observations show that the intensity of the field decreases to a low value during a transition. The tracks of the virtual geomagnetic pole during transitions appear to favor two broad bands of longitude, one over the Americas and the other antipodal. This evidence for axial symmetry of the transitional field may indicate that the dipole is still dominant during the reversal (Laj *et al.*, 1991). Other investigators have questioned the statistical significance of the preferred longitudinal paths; they postulate that the transitional field is dominated by higher order non-dipolar components with a non-axially symmetric geometry (Valet *et al.*, 1992). The divergent interpretations arise because the data base relating to the confinement of transitional paths is still too meager. Whether the transitional field configuration is dipolar or non-dipolar remains a matter of conjecture.

The transition of polarity is fast in geological terms, probably lasting only 4000–5000 years, so that it can be regarded as a virtually instantaneous event in the polarity record (Figure G13). Between reversals the geomagnetic field maintains constant polarity for irregular lengths of time, which are designated polarity intervals. In the past 10 Ma the mean length of a polarity interval has been about 200 000 years. The longest intervals last for 10^5–10^6 years and are called polarity chrons (Cox, 1982); they are often interrupted by shorter intervals of opposite polarity, called polarity subchrons, which last typically for 10^4–10^5 years. Occasionally the magnetic poles appear to roam for a few thousand years into intermediate latitudes; this displacement is called an excursion.

The polarity of the geomagnetic dipole field is a global characteristic; at a given time it is the same everywhere on Earth. Reversals are globally synchronous events that appear to occur randomly. The

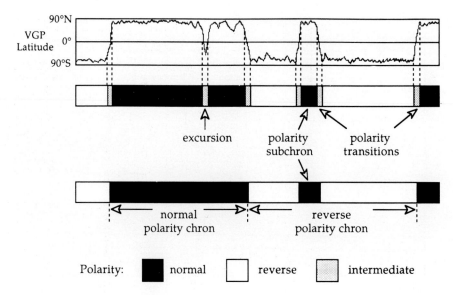

Figure G13 Schematic diagram to illustrate the definition of polarity transitions, excursions, chrons and subchrons (modified after Cox, 1982).

lengths of intervals ('chrons') in a given polarity sequence therefore form a diagnostic 'fingerprint' pattern that can be used to date the sequence by comparison with a standard global polarity timescale. These characteristics make geomagnetic reversal history an important tool for stratigraphic correlation and dating. The detailed reversal record for about the last 160 Ma, corresponding to the time during which the present oceans have formed, has been progressively improved, dated and expressed in the form of a standard global polarity timescale. There is abundant paleomagnetic evidence of polarity reversals before this time, but we do not yet have a continuous record of their history.

Geomagnetic polarity record

The best source of information about geomagnetic polarity during Cenozoic and late Mesozoic time comes from the interpretation of oceanic magnetic anomalies (e.g. Cande and Kent, 1992) formed by the quasi-continuous process of sea-floor spreading that accompanies plate tectonic motions. This record (Figure G14) shows two sequences of mixed geomagnetic polarity; the younger extends back from the present until the Late Cretaceous (0–83 Ma) and the older covers the Early Cretaceous and Late Jurassic (118–160 Ma). The two intervals of mixed polarity are separated in the Cretaceous by a long interval of constant normal polarity. The broad features and most of the details of this Cenozoic and late Mesozoic polarity record have been verified by magnetostratigraphic studies on sequences of sedimentary rocks that are now exposed on land or that have been sampled by drilling cores from the ocean floor. The resolutions of geomagnetic polarity histories interpreted from marine magnetic anomalies and from magnetostratigraphy are comparable, about 20 000–40 000 years. Several short subchrons reported from magnetostratigraphic studies are absent in the oceanic record and cannot yet be verified by independent confirmation. Moreover, it is difficult to distinguish the marine magnetic anomalies of very short chrons from geomagnetic intensity fluctuations. It is possible that the reversal sequence may contain many short polarity subchrons that have not been detected or confirmed.

The most valuable applications of magnetostratigraphy have been in conjunction with biostratigraphy. The two disciplines reinforce each other and yield a larger number of tie levels for stratigraphic correlation than either method alone. The correlation of major paleontological stage boundaries of known radiometric ages with the reversal record yields tie points for the development of a magnetic polarity time scale. In turn this enables the interpolation of absolute ages for important paleontological events, such as the first appearance datum (FAD) or last appearance datum (LAD) levels of species, which define paleontological zones.

An example of this use of geomagnetic polarity in the geological record is given by data for the Late Cretaceous (roughly 65–90 Ma ago) obtained from a pelagic limestone section in the Bottaccione gorge, near Gubbio, Italy (Lowrie and Alvarez, 1977). From paleomagnetic measurements of the directions of magnetization of rock samples taken at 1-m intervals, the latitude of the virtual geomagnetic pole (VGP) corresponding to each direction was computed as a function of stratigraphic position (Figure G15). After compensation for tectonic displacements of the sampling site the VGP latitude should fluctuate between 90°N (normal polarity) and 90°S (reverse polarity). The Gubbio record showed a sequence of normal and reversely magnetized magnetozones in a distinctive pattern that could be correlated unambiguously to the dated magnetic polarity sequence inferred from the marine anomaly record for the Late Cretaceous (Figure G14). Determination of the FAD and LAD levels of planktonic foraminifera identified key paleontological zones and tied the levels of the Late Cretaceous stage boundaries to the polarity sequence. The Cretaceous–Tertiary boundary was found to lie near (but not coincident with) the younger end of negative chron C29R. The ages of other magnetozone and foraminiferal zone boundaries may be estimated by interpolation.

Geomagnetic reversal timescales

The dated marine magnetic polarity sequence in Figure G15 is part of a geomagnetic reversal timescale (GRTS). Since the first complete marine anomaly record was obtained, there have been several attempts to improve the GRTS or to construct new versions. The polarity sequence itself is a composite that has been refined to allow for different sea-floor spreading rates in different ocean basins. Key stage boundaries with known absolute ages have been tied by magnetostratigraphic correlation to the polarity sequence. By stretching or squeezing the polarity record between the tie points and interpolating linearly, the ages of intervening reversals are calculated (Figure G14). Recent versions of the GRTS constructed in this way differ in the choice and number of calibration points and the absolute ages associated with them (e.g Lowrie and Alvarez, 1981; Cox, 1982; Cande and Kent, 1992). However, they agree quite closely for the Cenozoic and Late Cretaceous; ages associated with the same reversal differ by only a few percent from one recent timescale to another. The Late Jurassic–Early Cretaceous reversal sequence is less securely dated.

Prior to the well-dated, continuous late Mesozoic–Cenozoic reversal sequence the record of geomagnetic polarity is much poorer. Polarity reversals are known to have occurred but repeatable patterns of polarity have not been obtained. No precise geomagnetic polarity timescale is available. Magnetostratigraphic research, mainly in continental sediments, has identified long intervals ('superchrons') in

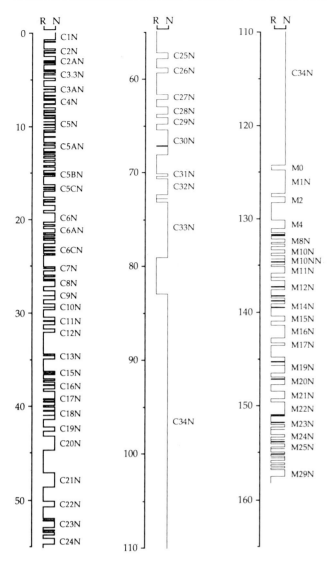

Figure G14 A recent magnetic polarity timescale (modified after Harland et al., 1989) for the time since mid-Jurassic (about 160 Ma ago), showing the ages of polarity transitions. The numbering system for polarity chrons is derived from the identification of oceanic magnetic anomalies. The chron number increases with increasing age. Normal polarity chron C1N is the youngest; it is preceded by reverse polarity chron C1R, etc.

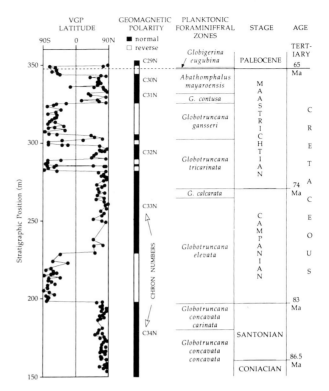

Figure G15 Results of a combined magnetostratigraphic and biostratigraphic study in Late Cretaceous limestones near Gubbio, Italy (modified after Lowrie and Alvarez, 1977). The magnetostratigraphy can be correlated to the dated reversals in Figure G14, which allows magnetozones to be identified by the corresponding chron numbers. Biostratigraphy gives the positions of paleontological zone and stage boundaries, which can also be dated by correlation to the reversal sequence.

which the geomagnetic field is biased to normal or reverse polarity, or both polarities are fairly equally represented (Figure G16).

Reversal frequency

A reliable timescale makes possible the statistical analysis of reversal rates, and allows models of reversal behavior to be tested. Although no definitive model exists for the physical process, the reversal of geomagnetic field polarity has been modeled as a statistical process involving interaction between the dipole and non-dipole fractions of the field (Cox, 1968). This model predicts that the lengths of polarity chrons will be distributed exponentially, corresponding to a reversal mechanism governed by a Poisson process. In fact, however, the observed polarity chrons are not Poisson distributed. The observed polarity record contains too few short chrons. Analysis of the statistical properties of the reversal sequence is hindered by the non-stationary character of the reversal process. The ages of reversals in the past 83 Ma (Figure G17a) can be fitted by a smooth curve (in this case a cubic polynomial), the slope of which at any point gives the local value of mean chron length and its inverse, the reversal rate, which can then be plotted as a function of age (Figure G17b). Mean reversal rate appears to have peaked about 12 Ma ago, after increasing throughout the earlier Cenozoic. The distribution of polarity chron lengths in this record can be fitted well by a Poisson process with a non-stationary mean value.

William Lowrie

Bibliography

Cande, S.C. and Kent, D.V. (1992) A new geomagnetic polarity time scale for the Late Cretaceous and Cenozoic. *J. Geophys. Res.*, **97**, 13917–51.

Cox, A. (1968) Lengths of geomagnetic polarity intervals. *J. Geophys. Res.*, **73**, 3247–60.

Cox, A.V. (1982) Magnetostratigraphic time scale, in *A Geologic Time Scale* (eds W.B. Harland, A.V. Cox, P.G. Llewellyn et al.). Cambridge: Cambridge University Press, pp. 63–84.

Harland, W.B., Armstrong, R.L., Cox. A.V. et al. (1990) *A Geologic Time Scale 1989*. Cambridge: Cambridge University Press, 263 pp.

Laj, C., Mazaud, A., Fuller, M. and Herrero-Bervera, E. (1991) Geomagnetic reversal paths. *Nature*, **351**, 447.

Lowrie, W. and Alvarez, W. (1977) Late Cretaceous geomagnetic polarity sequence: detailed rock and palaeomagnetic studies of the Scaglia Rossa limestone at Gubbio, Italy. *Geophys. J. Roy. Astr. Soc.*, **51**, 561–82.

Lowrie, W. and Alvarez, W. (1981) One hundred million years of geomagnetic polarity history. *Geology*, **9**, 392–7.

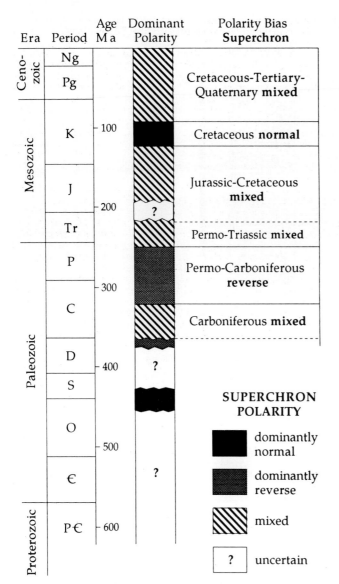

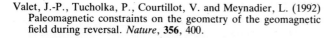

Figure G16 Polarity bias superchrons since the Late Precambrian (after Harland et al., 1989). The detailed reversal sequences in the Permo-Triassic and Carboniferous mixed polarity superchrons have not yet been established.

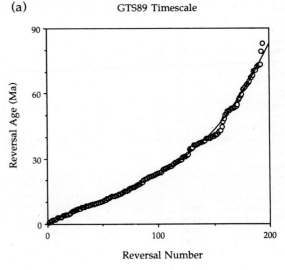

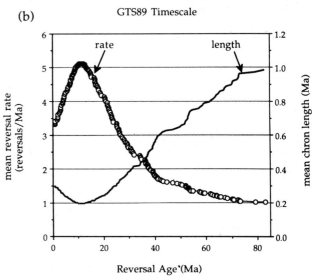

Figure G17 (a) Ages of 194 reversals in the Cenozoic and Late Cretaceous, according to the GRTS of Harland et al. (1989). (b) Mean chron length η and reversal rate in the past 83 Ma.

Valet, J.-P., Tucholka, P., Courtillot, V. and Meynadier, L. (1992) Paleomagnetic constraints on the geometry of the geomagnetic field during reversal. *Nature*, 356, 400.

Cross references

Earth: magnetic field and magnetosphere
Magnetism
Plate tectonics

GEOMAGNETIC STORM

Global, transient variations of the magnetic field observed at the surface of the Earth occur in response to solar activity and the passage of disturbances in the solar wind (q.v.). Changes in the solar wind dynamic pressure on the magnetosphere are communicated to the surface by hydromagnetic compression waves and transverse Alfvén waves. The nearly dipole large-scale geomagnetic field geometry controls the access of these waves to the surface to produce latitude dependence in storm-time geomagnetic variations. Additional latitude-dependent storm effects result from ionospheric and magnetospheric current systems which may be highly variable in strength and location.

At low latitudes an intense storm begins when the abrupt onset of elevated solar wind pressure pushes the magnetopause inwards to produce a general enhancement of magnetic flux density (see Earth: magnetic field). Surface fields remain slightly elevated for ~ 1 h during this onset phase, which is followed by a field decrease over several hours to an intensity $\sim 1\%$ below the pre-storm level. During this main phase, the depressed field intensity results from magnetospheric electric currents circulating westward, i.e. in the same sense as the dominant internal dynamo currents and opposite to planetary rotation. Field direction may vary by $\sim 1°$. After the storm main phase, the field intensity may recover to pre-storm levels over ~ 1 day if additional disturbances are not superposed.

Compared to low-latitude observations, high-latitude magnetic fluctuations are typically larger in both field intensity and direction and are more irregular. While global effects dominate low-latitude storm variability, polar fluctuations depend strongly upon local solar time. Auroral displays commonly accompany geomagnetic storms at high latitudes and extend to exceptionally low latitudes ($\sim 30°$) during the most intense storms. (see Aurora, planetary).

Transient features of solar activity identified as precursors to geomagnetic storms include chromospheric flares, disappearing filaments, evolving coronal holes and coronal mass ejections. Their common element is an impulsive reordering of the large-scale magnetic field of the solar corona (see Solar corona). The ensuing interplanetary disturbances reach Earth typically 2 days later and are detected in spacecraft observations of solar wind plasma and fields. An interplanetary magnetic field with a large vector component antiparallel to the geomagnetic field at the magnetopause is a principal cause of magnetic storms. Storms with sudden onsets are highly correlated with the incidence of interplanetary shocks on the magnetosphere.

Intense geomagnetic storms are associated with ionospheric disturbances affecting radio communications, abrupt decreases in the intensity of galactic cosmic rays reaching Earth, and upper atmospheric heating and expansion which increases the drag on orbital satellites. Pipelines and electric power lines can experience the effects of surface electric fields exceeding 10 V km^{-1} induced by time-varying magnetic fields. These electric fields can drive currents as great as 10 to 100 amperes between transformer grounds in transmission lines and produce power blackouts. In March 1989 6 million people in Quebec, Canada, lost power for 9 h during an extremely intense magnetic storm in which the sunward magnetopause was driven to less than half its normal distance from the Earth.

Gary D. Parker

Bibliography

Boteler, D.H. (1991) Predicting geomagnetic disturbances on power systems. *EOS, Trans. Am. Geophys. Union*, **72**, 159–60.
Hewish, A. (1988) The solar origin of geomagnetic storms. *Solar Phys.*, **116**, 195–8.
Sugiura, M. and Heppner, J.P. (1965) The Earth's magnetic field, in *Introduction to Space Science* (ed. W.N. Hess), New York: Gordon and Breach, pp. 5–92.
Tsurutani, B.T., Goldstein, B.E., Smith, E.J. et al. (1990) The interplanetary and solar causes of geomagnetic activity. *Planet. Space Sci.*, **38**, 109–26.

Cross references

Aurora, planetary
Cosmic rays
Forbush decrease
Ionosphere
Magnetosphere
Radiation belt
Solar corona
Solar flares
Solar wind

GILBERT, WILLIAM (1540–1603)

The most distinguished man of science in England during the reign of Queen Elizabeth I, Gilbert was born in Colchester of a long-established Suffolk family and studied medicine and mathematics at Cambridge. He set up practice in London and in 1601 was appointed personal physician to the Queen. He traveled extensively on the continent of Europe and brought home the Copernican idea of a heliocentric solar system. He may have met Kepler, who was born in 1571, and possibly also Galileo (born 1564). In fact, both Kepler and Galileo knew some of Gilbert's books and spoke with admiration of his scientific ideas concerning electric and magnetic phenomena.

Gilbert's best-known work was his *De Magnete* (1600), which was of immediate practical value to a seafaring nation. It was republished in several later editions in Germany, including Stettin (1628, 1633) and Frankfurt (1629, 1638). This work represented the culmination of many years' practical research which became the hallmark of the 'scientific method' and is often considered to be the first great scientific work written in England. It also marked a turning point in the cultural realm of science from the Mediterranean–Middle East centers to the northern European world of Britain, France, Germany and Scandinavia.

The planet Earth, as Gilbert saw it, was nothing but a giant bar magnet, as illustrated in his well-known diagram, albeit a static 'lodestone', showing the magnetic lines of force parallel to the Earth's surface near the equator, but gradually assuming steeper dips in the higher latitudes until a vertical angle is displayed at the north and south poles. He designed and built a vertical-plane magnetic dip needle specifically for determining this inclination, as distinct from declination, the azimuth angle between the geographic pole (spin axis) and the magnetic pole. The latter, in the 16th–17th century lay in the region of Spitsbergen (declination E of Greenwich); in the last three centuries it has migrated to the Canadian Arctic (that is, well to the W of Greenwich).

To further demonstrate the magnetic nature of a whole planet, Gilbert took a sample of lodestone (sometimes also spelled 'loadstone'), ground to a sphere (similar to that mentioned by Peregrinus in 1269); this was pivoted on a polar axis to show the orientation of the magnetic field. Unfortunately nature does not always fully cooperate, and regional deviations of the compass were recognized that Gilbert erroneously attributed to a supposedly greater magnetic intensity of the continents. It is said that this 'variance' (and its practical significance) was first discovered by Christopher Columbus in 1492 on his first trans-Atlantic voyage. It is true, however, that large deposits of the mineral magnetite (Fe_3O_4), the principal example of lodestone, are only found on continents. The phenomenon of ferromagnetism is widespread, in varying intensities, among the different iron-bearing minerals. Gilbert noted that a weakly magnetic object could be strengthened by contact with the lodestone, a fact already known to the makers of the marine compass.

In his view of the universe, Gilbert followed Giordano Bruno (1547–1600) and, in common with Galileo, believed it was of infinite dimensions, with no center and no boundary. In some way analogous to the magnetic field of force, according to Gilbert, there appeared to be a 'carrying virtue' that spread out from the Sun with or like the rays of light and heat, and which seemed to be related to the planetary motions around it, sharing the Sun's own motions. He remarked: 'There is therefore a conflict between the carrying power of the Sun and the impotence or material sluggishness (i.e. inertia) of the planet;' during the planetary motions he saw there was a constant exchange of this 'virtue' (i.e. angular momentum). This quotation is translated from Latin in Gilbert's *Epitome* (book IV, pt. 2) and the English is provided by Berry (1898, p. 196). In Gilbert's introduction to *Commentaries on the Motion of Mars* he postulated that there existed a more general 'gravity', which was expressed as 'a mutual bodily affection between allied bodies tending towards their union or junction'. Another example of this force was the lunar action on terrestrial tides, Gilbert clearly anticipating Newton.

On Gilbert's death, all his books and equipment were left to the Royal College of Physicians, but they were totally destroyed in the Great Fire of London (1666).

Rhodes W. Fairbridge

Bibliography

Berry, A. (1898) *A Short History of Astronomy*, London: Murray (also New York: Dover reprint, 1961), 440 pp.
Cajori, F. (1929) *A History of Physics*. New York.
Gilbert, W. (1600) *De Magnete, Magneticisque Corporibus, et de Magno Magnete Tellure*. London.
Gilbert, W. (1651) *De Mundo Mostro Sublunari Philosophia Nova*. Amsterdam (posthumous edn).
O'Reilly, W. (1984) *Rock and Mineral Magnetism*. Glasgow: Blackie (and New York: Chapman Hall), 220 pp.
Pumfrey, S. (1990) Neo-Aristotelianism and the magnetic philosophy, in *New Perspectives on Renaissance Thought: Essays in Memory of Charles B. Schmitt* (eds J. Henry and S. Hutton). London: Duckworth, pp. 177–89.
Singer, C. (1959) *A Short History of Scientific Ideas to 1900*. Oxford: Clarendon Press, 525 pp.

Acknowledgement

Manuscript kindly reviewed by H.H. Haubold.

GIOTTO MISSION

The Giotto spacecraft of the European Space Agency (ESA) encountered comet Halley on 14 March 1986, passing the comet nucleus at a distance of 596 km on the sunward side (Plate 20). Giotto carried a scientific payload of ten experiments, among them a high-resolution camera which provided the first clear images of the outline of the dark Halley nucleus against the brighter coma. The images also revealed some features on the nucleus surface.

During the Halley encounter the spacecraft and several experiments were damaged. It was, nevertheless, possible to change Giotto's orbit in such a way that it made an Earth gravity-assist swingby on 2 July 1990. Giotto then encountered comet Grigg-Skjellerup on 10 June 1992, passing the comet nucleus at a distance of about 200 km. Halley is a large, highly active comet, while Grigg-Skjellerup is an average sized, rather evolved, dust-poor comet (Table G3).

Halley was an ideal target for a first cometary mission: Halley is the most active of all comets with a well-known orbit, a mission to Halley required very little launch energy, and Halley could be observed from the Earth at the time of the encounter. It had only one disadvantage: because of its retrograde orbit the relative flyby velocity was extremely high.

The Giotto mission to comet Halley was ESA's first interplanetary mission. It was named after the Italian painter Giotto di Bondone who, in 1304, depicted Halley's comet as the 'Star of Bethlehem' in one of the frescoes in the Scrovegni chapel in Padua, Italy.

Table G3 Comparison of key parameters for comets Halley and Grigg-Skjellerup

		Halley	Grigg-Skjellerup
Orbit	Perihelion (AU)	0.587	0.99
	Aphelion (AU)	35	4.93
	Period (years)	76	5.09
	Inclination (deg.)	162.2 (retrograde)	21.1 (prograde)
Nucleus size		$16 \times 8 \times 7$ km irregular ellipsoid	A few km (estimate)
Gas production rate during the encounter (molecule s^{-1})		7×10^{29}	6×10^{27}
Time of closest approach (spacecraft onboard time)		00:03:02 UT 14 March 1986	15:18:43 UT 10 July 1992
Relative flyby velocity (km s^{-1})		68.37	13.99

The spacecraft

The Giotto spacecraft (Figure G18) was cylindrical in shape, with a diameter of 1.86 m and a height of 2.85 m, measured from the tip of the tripod to the adaptor ring at the bottom. In flight, Giotto was spin stabilized, nominally at 15 rpm. In its center Giotto carried a solid-propellant boost motor, used to inject the spacecraft from a geo-

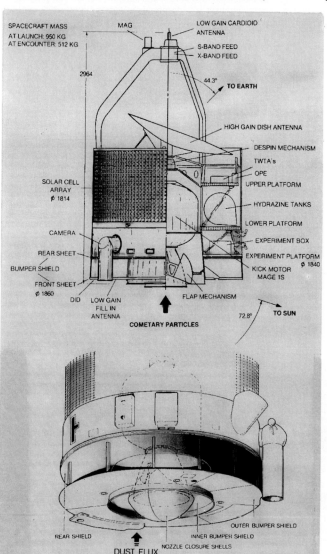

Figure G18 The Giotto spacecraft.

Table G4 Overview of the scientific experiments onboard Giotto

Experiment	Measurement
Halley Multicolor Camera (HMC)	Narrow angle camera for imaging the cometary nucleus and inner coma
Neutral Mass Spectrometer (NMS)	Energy and mass of neutrals M-analyzer: 1–37 amu e^{-1} E-analyzer, gas mode: 200–2150 eV e^{-1} (8–86 amu e^{-1}) ion mode: 20–1420 eV e^{-1} (1–56 amu e^{-1})
Ion Mass Spectrometer (IMS)	Energy and mass of ions HERS: 10–4500 eV e^{-1} and 1–35 amu e^{-1} HIS: 300–1400 eV e^{-1} and 12–57 amu e^{-1}
Particulate Impact Analyzer (PIA)	Mass (3×10^{-16}–5×10^{-10} g) and composition (1–110 amu) of individual dust particles
Dust Impact Detection System (DID)	Determination of mass spectrum of dust particles from 10^{-17} to 10^{-3} g with three different detectors
Johnstone Plasma Analyzer (JPA)	FIS: solar wind and cometary ions from 10 eV e^{-1} to 20 keV e^{-1} IIS: cometary ions from 100 eV e^{-1} to 86 keV e^{-1} and 1–45 amu
Rème Plasma Analyzer (RPA)	EESA: solar wind and cometary electrons from 10 eV to 30 keV PICCA: cometary singly charged ions from 10 to 213 amu
Energetic Particles Analyser (EPA)	Three small particle telescopes to measure the energy distribution of electrons, protons and heavier nuclei with $E \geq 20$ keV
Magnetometer (MAG)	Interplanetary and cometary magnetic field Outboard sensor: triaxial fluxgate (28.24 vectors s^{-1}) Inboard sensor: biaxial fluxgate
Optical Probe Experiment (OPE)	Coma brightness in four continuum (dust) bands and at four discrete wavelengths (gaseous emissions of OH, CN, CO^+, C_2)
Giotto Radio-science Experiment (GRE)	Dust column density in the coma

stationary transfer orbit into a heliocentric orbit. At launch Giotto weighed 960 kg. During the Halley flyby its mass was reduced to 573.7 kg, after the rocket motor had burned out and 9 kg of the available 69 kg of hydrazine had been used for three orbit correction maneuvers and a large number of attitude corrections.

Most of the spacecraft's outer cylindrical wall consisted of a solar cell array, providing 196 W of power at 0.9 AU heliocentric distance (as during the Halley encounter), which was not quite sufficient to power the spacecraft and all experiments. Giotto, therefore, carried four Si–Cd batteries in addition.

Giotto had no onboard data storage capability and all science data were transmitted to Earth in real time at a rate of 40 kilobits per second in the X band (8.4 GHz). The antenna beam was inclined 44.3° with respect to the spacecraft spin axis, and the antenna dish itself was despun so that the beam could point continuously at the Earth during the mission. The pointing requirements in the X band were rather stringent: if the antenna beam were to be misaligned by more than 1° the telemetry link to the ground receiving station would be lost. It was calculated before the mission that a 0.1 g dust particle impacting at 68 km s^{-1} on the rim of the bumper shield could change the spacecraft attitude and hence the antenna pointing direction by more than 1°.

Giotto was a fairly conventional spacecraft, following a low-cost approach and designed for maximum reliability, bearing in mind the uniqueness of the Halley encounter opportunity. However, it had three unusual features: a dual-sheet bumper shield (Figure G18, lower part) to protect it from being damaged by high-velocity dust impacts during the encounter, an inclined and despun high-gain dish antenna for high data rate telecommunications over a distance of 1 AU, and a specially designed scientific payload, consisting of ten experiments (Table G4).

All experiments were mounted on the 'experiment platform' (Figure G19), with three exceptions: the DID sensors were mounted on the outer face of the front bumper shield, the two MAG sensors were mounted on the carbon fiber tripod as far away from the disturbing spacecraft's magnetic field sources as possible, and OPE was mounted on the upper platform inside the spacecraft, looking rearward. All other experiments looked in the forward direction during the Halley flyby (downward in Figures G18 and G19), apart from some plasma experiments which looked to the side. The camera could be rotated by 180° which, together with the spacecraft spin, allowed viewing in all directions.

As the spacecraft traversed the coma it was decelerated by dust and gas impacts (atmospheric drag); a radio science experiment was designed to measure this deceleration by observing the Doppler frequency shift of the Giotto radio signal, and to derive from it the dust column density along the Giotto trajectory.

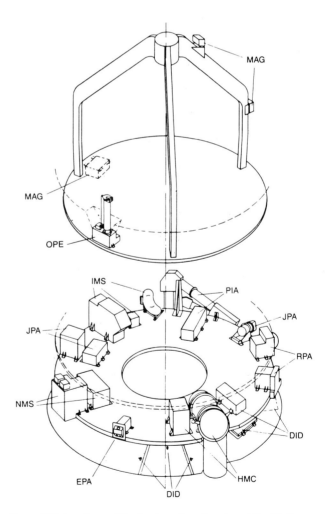

Figure G19 Locations of the ten experiments onboard the Giotto spacecraft.

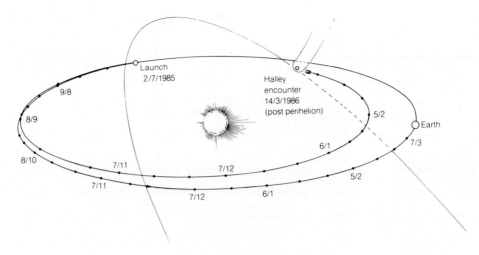

Figure G20 Giotto's interplanetary trajectory from launch to Halley encounter.

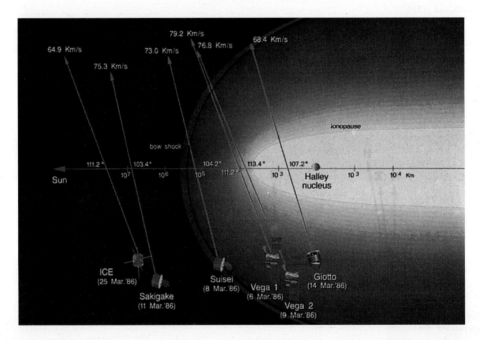

Figure G21 Flyby distances of the six Halley spacecraft. The Sun is to the left, the distance scale is logarithmic. For each mission the flyby date is given at the bottom, flyby distance at the center and flyby speed at the top.

Mission to comet Halley

Launch and cruise phase

Giotto was launched on 2 July 1985 by an Ariane 1 from Centre Spatial Guyanais (CSG) in Kourou, French Guiana, in South America, into a highly elliptical Earth orbit with 200 km perigee and 36 000 km apogee (geostationary transfer orbit). After a few revolutions in this orbit, Giotto's onboard rocket motor was fired at perigee such that Giotto left the Earth's gravitational field on course for a Halley encounter in mid-March 1986 (Figure G20).

The required launch energy for an encounter with comet Halley varies strongly with the heliocentric distance and the distance above or below the ecliptic, reaching a minimum for an encounter near the ecliptic. Halley's orbit crossed the ecliptic twice, the first time (ascending node) on 9 November 1985 at a heliocentric distance of 1.8 AU, and the second time (descending node) on 10 March 1986 at a heliocentric distance of 0.84 AU. The second encounter opportunity was chosen because it was closer to the Sun, where the comet is more active and, secondly, because the required launch energy was even lower than for a pre-perihelion encounter. It was actually one of the lowest energy of all possible cometary missions and it was, therefore, no surprise that not one but six spacecraft from four space agencies were sent to this point (in addition to Giotto: Vega 1 and 2 from the Russian-led Intercosmos, Suisei and Sakigake from the Japanese ISAS, and ICE from NASA). Figure G21 shows the flyby distances of this 'Halley armada'.

For the routine operations of the spacecraft during the cruise phase, ESA's 15-m ground station at Carnarvon, Australia was used, while the 64-m ground station at Parkes, Australia, was used for high data rate transmission of scientific data during the cruise and the encounter. The Parkes antenna is owned and operated by the Commonwealth Scientific and Industrial Research Organisation (CSIRO), and is normally used for radio astronomy. During the encounter, continuous data coverage for 70 h before and 30 h after closest approach in high data rate mode was provided by the Parkes ground station, together with the NASA 64-m Deep Space Network (ground stations at Goldstone, Madrid, Canberra).

Halley encounter

Giotto was targeted to fly by at a distance of 540 km on the sunward side. This distance was chosen as a compromise between the partially conflicting requirements of the various experimenters. Most experimenters wanted to approach the nucleus as closely as possible, even risking spacecraft and experiment survival, but at distances < 500 km the camera could not rotate fast enough to follow the apparent motion of the nucleus during the flyby. Therefore 500 km was selected, adding to this distance the estimated targeting uncertainty of 40 km.

Targeting Giotto to this preselected point was a difficult task, as the nucleus was too small to be resolved by ground-based telescopes and, moreover, at heliocentric distances < 2 AU it is masked by the coma of dust and gas. To make matters even more difficult, the dust and gas is mostly emitted into the sunward hemisphere and, consequently, the nucleus is accelerated away from the Sun. Compared to a purely Keplerian orbit, Halley's orbit is extended by 4 days. This outgassing effect is irregular and difficult to model. The asymmetric outgassing also introduces an offset between the center of light in the coma and the actual nucleus position which can exceed 1000 km. Based on a large number of astrometric observations from the ground the targeting uncertainty for the Giotto flyby was estimated to be several hundred kilometers.

Fortunately, the two Vega spacecraft encountered Halley a few days before Giotto did and in the good spirit of international scientific collaboration the Vega spacecraft position data and the Vega camera pointing angles were provided to the Giotto project immediately after the two Vega flybys. This by itself was not sufficient as the Soviet experts, using only conventional (6 GHz) ranging and Doppler techniques, estimated a Vega position uncertainty of about 400 km. Using the widely separated tracking stations of their DSN and very long baseline interferometry (VLBI) techniques, NASA was able to pinpoint the two Vega spacecraft with an accuracy of only 30 km, resulting in an overall targeting accuracy for Giotto of 40 km.

On its way towards the nucleus Giotto was repeatedly struck by sizeable dust particles causing small attitude perturbations. At 7.6 s before closest approach Giotto was hit by a 'large' (~ 0.1 g) dust particle, or by a shower of large dust particles, causing a shift of the spacecraft angular momentum vector by 1°. Consequently, the spacecraft performed a nutation around the new axis with an amplitude of 1°, which gave a maximum deviation from the Earth-pointing direction of 2° and, as had been expected in such an event, the telecommunications link to Earth was interrupted. For the next 32 min scientific data were received only intermittently. By that time the onboard nutation dampers had reduced the maximum deviation from the Earth-pointing direction to < 1° and continuous scientific data were received again.

About half the experiments worked flawlessly throughout the encounter while the other half suffered damage due to dust impacts. The spacecraft also suffered some damage during the encounter but it was nevertheless possible to redirect it to Earth before it was put in a hibernation configuration on 2 April 1986.

Scientific results

Halley nucleus

The Halley Multicolor Camera (HMC) onboard Giotto obtained a total of over 2000 images with ever increasing resolution during the flyby. The first image (after processing) on which the nucleus can already be identified was taken from a distance of 145 000 km, the last image (with a resolution of 50 m) that could be transmitted in full was taken at a distance of 2000 km from the nucleus. At 9.2 s before closest approach the camera stopped working.

The images taken by the camera onboard the Giotto spacecraft revealed a single solid nucleus of irregular, elongated shape (Figure G22). The length of the nucleus as seen in projection is 15 km, while its width is 8 km. On the sunward side the warmed nucleus releases several tons of gas and fine dust per second. The dust reflects the sunlight, and regions with a high dust concentration therefore appear bright on the images, masking the nucleus to a large extent on the sunward side. Only the nightside of the nucleus could be clearly seen against the somewhat brighter background of coma dust. An elongated bright area seen on the nightside is a mountain, about 500 m high, reflecting the sunlight. Several shallow craters about 1 km in diameter could be identified near the terminator (the line that marks the transition between the day- and the nightside). Several possibilities for their origin are being discussed, but no firm conclusion has yet been reached. It is quite possible that these craters or troughs are of crucial importance for our understanding of the physics of comets. They are the only sign of inhomogeneity on the surface and may be connected with the cometary activity, which is known to be inhomogeneous, or they may reflect the inner structure of the nucleus itself.

The images also show that almost all of the dust and gas emanates from just a few active regions on the nucleus. While the dust remains confined, forming 'dust jets' which are clearly visible on the sunward side, the gas diffuses very quickly and, in principle, there should be no 'gas jets'. CN (cyanogen) and C_2 gas jets were found, however, in images obtained by ground-based telescopes and the most likely explanation is that most of these molecules originate from the dust particles in the dust jets.

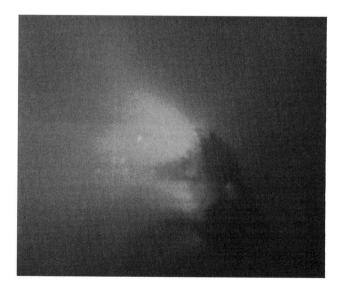 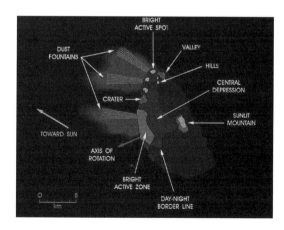

Figure G22 Composite image of the Halley nucleus. This image is composed of seven different images taken at different distances, ranging from 25 600 to 2700 km from the nucleus. The resolution varies from 570 m in the outer parts to 60 m in the central part of the image. The direction to the Sun is to the left and 17° above the horizontal.

The nucleus is covered by a layer of dust. This conclusion was inevitable after the infrared spectrometer onboard Vega 1 found that the surface of the nucleus was much warmer (~ 350 K) than expected. If the surface consisted mostly of water ice mixed with dust, the surface temperature would be controlled by the sublimation temperature of water ice (≤ 200 K). The thickness of the dust layer is still an open question. Estimates range from ≤ 1 cm to several tens of meters. In fact, the thickness of the layer may well vary over the surface, since the dust and gas are not blown off homogeneously, and it may also vary with time.

There is considerable evidence that the larger dust particles are aggregates of submicron-sized dust particles with ice in between or, after the ice has evaporated, as in the case of the dust layer, with voids in between. The dust crust would then be porous, allowing the gas to diffuse through it from underneath. Also, solar photons could be easily trapped in this lattice of small dust particles, which might explain why the nucleus appears so dark. It has an albedo of only 4%, which makes it one of the darkest objects in the solar system. Another consequence of the low packing density would be a low heat conductivity between the outer dust layer and the ice–dust mixture beneath, so that the layer would not have to be very thick to maintain a temperature difference of 150°.

The mass of the nucleus is inferred from nongravitational effects on its orbit to be a few times 10^{17} g which, if taken together with the nucleus volume of about 400 km^3, implies a mean density of about 1 g cm^{-3}, consistent with the idea that the nucleus consists of a mixture of solid water ice and dust.

The density of the particles in the dust layer is likely to be lower (~ 0.1 g cm^{-3}). The gravitational attraction of the comet nucleus is very small (less than one billionth that of the Earth) and it is thus conceivable that low-density structures on the surface of the cometary nucleus can easily be preserved.

Dust in the coma

During the encounter the spacecraft velocity decreased by 23 cm s^{-1} due to atmospheric drag, corresponding to a total mass of 0.32 g impacting on the spacecraft, almost all of it being dust.

The first dust particle impact was recorded by DID at a distance of 290 000 km from the nucleus, farther away than expected by dust models. In all, 12 000 dust particle impacts were recorded, ranging in mass from 10^{-17} to 1.4×10^{-4} g.

Neither the optical nor the infrared data on the ground provide any reliable information about particles smaller than 0.1 μm. Before the encounters it was unclear whether such small dust particles existed at all in the cometary coma. It was, therefore, a major surprise when the impact detectors onboard Giotto found a large number of these small particles. Moreover, they were found outside their bounding paraboloids defined by the classical theory based on radiation pressure deceleration. One possibility might be that these grains are electrically charged and are accelerated by the electric fields in the magnetized interplanetary plasma.

Determination of the total amount of dust that leaves the nucleus every second, while of fundamental importance, is not straightforward, because dust particles are observed over a limited range only, and it is not known up to which maximum mass the distribution must be integrated. If the maximum mass that can be lifted from the nucleus is 1 g, the total dust production rate at the time of the Giotto flyby was 3.3×10^6 g s^{-1}; if the maximum liftable mass is 1 kg, the total rate was 3.3×10^7 g s^{-1}.

The chemical composition of several thousand individual dust particles was analyzed by the dust mass spectrometer onboard the Giotto spacecraft. Some dust particles were found to be dominated by the light elements H, C, N and O ('organic grains'); others were rich in the mineral-forming elements Na, Mg, Si, Ca and Fe ('mineral grains'). Averaged over many grains, the composition is close to that of CI carbonaceous chondrites (q.v.), with a significant excess of C and N. (A CI carbonaceous chondrite is a special carbon-rich type of chondrite; a chondrite is a stony meteorite characterized by the presence of chondrules, small spherical grains predominantly composed of iron and magnesium silicates.) On average, the mineral grains are heavier than the organic grains and are seen closer to the nucleus.

H, C, N and O could be the building blocks of organic molecules. They may form tar-like substances, which could be another reason why the nucleus appears dark. Perhaps the dust particles are more complex than previously thought, with a silicate core and an organic mantle. The grains that are rich in H, C, N and O may be the 'parents' of the spiral CN and C_2 jets observed in the coma of Halley's comet from the ground.

Neutral gas and ions

The total gas production rate at the time of the Giotto flyby was about 7×10^{29} molecules s^{-1}, 80% of it being H_2O molecules; that is, the water production rate was about 16 tons s^{-1}. Water is obviously a parent molecule and this means that 80–90% of the nucleus consists of dust and water. The composition of the remaining 10–20% is not yet completely clear. The problem is that, unlike the dust grains, the composition of the gas changes significantly with distance from the nucleus in a chain of complex chemical reactions. Only in a very few cases can the parent molecules be observed directly; in most other cases they can be determined only by fitting the spectra observed *in situ* by the gas mass spectrometers to chemical models. This process has not yet been completed, but it is already clear that other parent molecules are CO (15% of H_2O at distances between 20 000 and 75 000 km, dropping to 7% at 1000 km from the nucleus), CO_2 (2.5%), NH_3 (1.5%) and at the 0.1% level HCN and various hydrocarbons (C_2H_n, C_3H_m). Fe and Na were also found, and S_2 and H_2S are strongly suspected parent molecules. From the radial increase of the CO abundance relative to H_2O it is concluded that CO evaporates not only from the nucleus but also from the dust grains in the coma.

Information on elemental and isotopic ratios can, in principle, be used to determine the origin or age of cometary material. Isotopic ratios of $^{32}S/^{34}S = 22 \pm 5$, $^{12}C/^{13}C = 65 \pm 9$, $^{14}N/^{15}N > 200 \pm 100$, $^{16}O/^{18}O = 435 \pm 100$ and $D/H = (0.6–4.8) \times 10^{-4}$ have been observed. Within their uncertainties, these values are consistent with both terrestrial and average solar system ratios, i.e. comet Halley is a member of the solar system and not of interstellar origin.

The cometary gas, streaming away from the nucleus at a velocity of ~ 1 km s^{-1}, is ionized by solar ultraviolet radiation, by electrons and by charge exchange with the solar wind plasma. Many different ionic species have been identified, including H_3O^+ (the dominant ion near the nucleus), H_2O^+, OH^+, C^+, CH^+, O^+, Na^+, C_2^+, S^+ and Fe^+. The spectra show a striking richness in C^+, which cannot be accounted for by photodissociation of CO, CH_4 and CO_2 to C, followed by photo-ionization of C. Perhaps carbon atoms are released directly at the surface, or the dust grains themselves may be the source.

Pick-up ions and plasma physical boundaries

Pick-up ions are cometary particles that travel out to large distances from the nucleus as neutral atoms or molecules before being ionized and 'picked-up' by the solar wind; they can be distinguished from solar wind ions by their energy. Water group pick-up ions were detected at distances up to several million kilometers from the nucleus.

As the supersonic solar wind is mass loaded by the heavy cometary ions, it is slowed down to subsonic speeds and a bow shock of hyperboloid shape forms. The bow shock was found by the Giotto plasma experiments at $\sim 1.1 \times 10^6$ km inbound and $\sim 0.7 \times 10^6$ km outbound.

Much closer to the nucleus there is another major plasma physical boundary, the ionopause, which separates the region of smoothly ('cold') outward-flowing cometary ions from a disturbed ('hot') transition region containing both cometary and solar wind ions. The ionopause was crossed by the Giotto spacecraft inbound at a distance of 4660 km and outbound at a distance of 3930 km from the nucleus. Outside the ionopause, a layer of stagnant cometary plasma was found. Inside, the cold cometary ions flowed smoothly outward at a velocity of ~ 1 km s^{-1}. The ion temperature dropped from 2000 K to as low as 300 K across the ionopause. Inside the ionopause the magnetic field strength was found to be essentially zero because the comet nucleus has no intrinsic magnetic field and the magnetic field that is carried by the solar wind plasma is kept outside the ionopause by the pressure of the cometary ionosphere. This leads to a compression of magnetic field lines outside the ionopause, and there the magnetic field strength reaches values exceeding 60 nT (typically ≤ 10 nT everywhere else in the coma).

Mission to comet Grigg-Skjellerup

From 2 April 1986 until 19 February 1990 Giotto was in a hibernation configuration, orbiting the Sun with a period of 10 months. In February 1990 Giotto was again close enough to the Earth to be reactivated; a detailed checkout of the scientific experiments was made in May 1990. It essentially confirmed the preliminary damage assessment made during the Halley flyby. Most unfortunately, the camera, while in principle fully operational, could not provide any images, presumably because the camera's aperture was blocked by a piece of the outer baffle which was destroyed during the Halley flyby by dust impacts.

Giotto made an Earth gravity-assist swingby on 2 July 1990, precisely 5 years after launch, passing at 22 700 km above the Earth's surface. During the passage through the magnetosphere two experiments, MAG and EPA, were switched on and provided a valuable 'snapshot' of the magnetosphere with clearly identified inbound and outbound bow shock and magnetopause crossings.

Giotto's new heliocentric orbit, after the Earth flyby, was also in the ecliptic but now outside of the Earth orbit with a period of 13 months. From 23 July 1990 until 4 May 1992 Giotto was again in hibernation.

The Giotto spacecraft and its scientific payload were designed for the specific geometry during the Halley encounter such that the dust particles impacted vertically on the dust shield while the high gain dish antenna pointed at the Earth. During the Grigg-Skjellerup encounter, the high-gain antenna again had to point at the Earth, but now the angle between the spacecraft spin axis and the relative velocity vector was 70° and the side of the spacecraft was fully exposed to dust particle impacts.

As during the Halley encounter, water group pick-up ions were the first signs of cometary activity, detected by the plasma instruments at a distance of 440 000 km from the nucleus. Smooth wave fields with a period of 70s, generated by the pick-up ions, were observed by the magnetometer. The bow shock was crossed at 20 000 km inbound and 25 400 km outbound from the nucleus. Close to the nucleus the magnetic field strength increased to 89 nT, higher than at Halley, but an ionopause with a magnetic cavity inside, as during the Halley encounter, was not detected. This was no surprise as the cavity, if it exists at all at Grigg-Skjellerup, was predicted to extend only to about 60 km around the nucleus.

From the intensity distribution of the brightness of the sunlight scattered by the dust particles in the coma a flyby distance of about 200 km was derived. Shortly after closest approach three distinct dust particle impacts were recorded. There were certainly many more dust impacts but one has to bear in mind the lower impact velocity compared to the Halley encounter and the unfavorable impact angle.

Giotto operations were terminated on 23 July 1992, after a final orbit adjustment and configuring the spacecraft for its third hibernation. On its present course Giotto will pass 200 000 km above the Earth's surface on 1 July 1999. Future operations of the spacecraft however are doubtful, partly because the remaining hydrazine (only a few kilograms) is insufficient for anything more than an Earth or Moon flyby, and partly because of the age of the spacecraft and the software in the ground systems used to support the Giotto mission.

Rüdeger Reinhard

Bibliography

(Each of the following contains a large number of papers on the Giotto mission and its scientific results.)

Battrick, B. and Guyenne, D. (eds) (1986) *Giotto Special Issue. ESA Bull.*, **46**, May 1986.
Encounters with comet Halley, the first results (1986). *Nature* (special issue), **321**, 259–366.
Halley's Comet (1987). *Astron. Astrophys.* (special issue), **187**(1/2), Nov (II).
Huebner, W.F. (ed.) (1990) *Physics and Chemistry of Comets.* Springer-Verlag.

Mason, J.W. (ed.) (1990) *Comet Halley: Investigations, Results, Interpretations.* Ellis Horwood.

Cross references

Carbonaceous chondrite
Chondrites, ordinary
Comet
Comet: observation
Dust
Halley, Edmond, and Halley's comet
Sakigake and Suisei Missions
Vega Missions

GLOBAL POSITIONING SYSTEM

In the early 1980s the US Department of Defense began deploying a satellite-based navigation system known as the Global Positioning System (GPS). This system enables an observer with the proper receiver to obtain instantaneous three-dimensional position information in virtually all parts of the globe with a precision of several meters. Under certain conditions this can be improved by up to three orders of magnitude. The scientific applications include precision ship and aircraft navigation for gravity measurements, precision aircraft and satellite positioning for altimetry, and high-precision geodesy for measurement of sea level, crustal strain and fault motion rates. This article briefly reviews these and other applications of high-precision GPS measurements. More comprehensive reviews can be found in Leick (1990), Dixon (1991) and Hager, King and Murray (1991).

The space segment of GPS is a constellation of satellites in high Earth orbit equipped with powerful radio frequency transmitters and highly stable atomic clocks. The constellation consists of 21 satellites plus three spares orbiting at an altitude of about 20 000 km in six orbital planes with 12-h periods. GPS satellites transmit two L band carrier frequencies, each modulated by several lower-frequency signals (Spilker, 1978). The carriers (L1 at 1.57542 GHz and L2 at 1.22760 GHz) are coherent multiples of a 10.23-MHz atomic clock, a stable oscillator that provides a frequency standard on each satellite (L1 = 154 × 1023 MHz, L2 = 120 × 1023 MHz). The L1 carrier has two components. The in-phase component is modulated by the precision (P) code. A lower-frequency coarse/acquisition (C/A) code is modulated in quadrature, i.e. on the same carrier frequency but phase shifted 90°. The L2 carrier is normally modulated only with the P code. All three carriers also are modulated with a low bit rate (50 Hz) data stream transmitting satellite health, ephemeris and other housekeeping information. These codes can be considered as square waves with values of ± 1, and are termed pseudo-random noise (PRN) codes because they have sufficiently long repeat times that they appear random to a user without knowledge of code structure.

The basic principle for point positioning with GPS is that an observer at or near the surface of the Earth can uniquely define his location in three dimensions by determining the distance between himself and three satellites whose orbital positions are known. Distance is inferred from the travel time of a unique coded signal that emanates from each satellite. Ignoring transmission media effects on the speed of light c, and any timing (clock) errors, the true range ρ between satellite and receiver is just $c(t_r - t_s)$, where t_r and t_s are the receive and transmit times. GPS receivers recover the signal propagation time by matching (correlating) the received code to an identical, receiver-generated code slaved to an internal clock. Errors in receiver or satellite clocks are present in this range estimate, which for this reason is referred to as pseudorange R:

$$R = \rho + c(\Delta t_r - \Delta t_s + \Delta t_p) \quad \text{(G1)}$$

where Δt_r is the receiver clock offset from 'true' (GPS system) time, Δt_s is the satellite clock offset and Δt_p is the delay associated with all other error sources, mainly due to atmospheric propagation effects. Information from a fourth satellite allows a first-order clock correction ($\Delta t_r - \Delta t_s$), and approaches discussed below can be applied to estimate and correct for Δt_p.

It is also possible to obtain distance information (strictly speaking, the change in distance) through phase measurements on the carrier signal itself, keeping track of the number of cycles after signal acquisition. Ignoring propagation effects and clock errors

$$\rho = n\lambda + \phi\lambda = (v_\phi/f)(n + \phi) \qquad (G2)$$

where n is the number of integer carrier wavelengths at signal acquisition (initially unknown), ϕ is the phase in cycles, λ is wavelength, f is frequency and v_ϕ is phase velocity. Since the wavelength of the carrier is considerably shorter than that of the lower frequency code modulations, the resulting 'distance' measurement, though ambiguous by the initial number of wavelengths, is considerably more precise than a pseudorange measurement. The carrier phase approach (sometimes called integrated Doppler) is one of the keys to high-precision geodesy with GPS. It should be noted that carrier phase is not measured directly by the receiver (the required sampling rate would be too high). Instead, the incoming signal is mixed with a signal generated by the receiver's internal clock and the lower-frequency 'carrier beat phase' is sampled.

Relative positioning

Relative (as opposed to point) positioning is the other key to high-precision geodetic measurements with GPS. Simultaneous observation of a group of satellites by a network of ground receivers defines a series of three-dimensional vectors (baselines) between all stations in the network, relative to one or more stations whose positions may already be known. The combination of simultaneous observations and the analytical techniques designed to accommodate such data enables us to eliminate or reduce greatly the common mode errors, including satellite and receiver clock uncertainties and atmospheric effects. Consider two receivers simultaneously observing the same satellite; a single difference can be formed from the two range equations that effectively eliminates satellite clock error. Similarly, if two or more satellites are observed, the receiver clock error is common mode and can be eliminated in the second or double difference. Several software systems designed for analysis of high-precision GPS data accomplish essentially the same thing (i.e. elimination of clock error) through the use of clock models without explicitly forming double differences.

For high-accuracy geodesy with GPS, we may also require precise knowledge of the Earth's orientation in inertial space during the observing session. This includes knowledge of changes in the Earth's spin axis direction with respect to inertial space (precession and nutation), spin axis changes with respect to the Earth's crust (polar motion), changes in rotation rate (UT1—UTC) and tidal effects. This information usually comes from models or from measurements by other geodetic techniques including very long baseline interferometry (VLBI) or satellite laser ranging (SLR). Recent experiments involving globally distributed GPS receivers suggest that the GPS data itself can be used to solve for Earth orientation parameters such as pole position and rotation rate with very high time resolution (Herring, Dong and King, 1991; Lindqwister, Freedman and Blewitt, 1992; Lichten, Marcus and Dickey, 1992; Figure G23).

Effect of the atmosphere

The GPS signal is affected by propagation through the atmosphere, through changes in velocity and by ray bending. There are two main regions to consider, the frequency-dispersive ionosphere ($\sim <1000$ km altitude), and the non-dispersive neutral atmosphere, especially the troposphere (~ 0–10 km altitude). Free electrons in the ionosphere interact strongly with the GPS signal. Ionospheric effects are proportional to the integrated electron content along the signal path, and thus depend on solar activity, the elevation angle of the observation, time of day and latitude. Fortunately they can largely be eliminated by exploiting the dual frequency nature of the GPS signal.

The tropospheric delay, the difference in travel time between actual signal propagation time in the troposphere and the theoretical transit time in vacuum, is usually expressed as equivalent path length by multiplying by the speed of light. All components of the atmosphere contribute to the delay, but it is convenient to consider separately the dry delay (~ 200–230 cm at zenith, or elevation angle, $\theta = 90°$, for altitudes near sea level) associated with molecular constituents of the atmosphere in hydrostatic equilibrium (including H_2O), and the wet delay, associated with water vapor not in hydrostatic equilibrium (~ 3–30 cm at zenith). The delays at other elevation angles are larger, increasing approximately as $1/\sin(\theta)$, but other effects, including the finite height of the atmosphere, the vertical distribution of components, Earth curvature and ray bending are usually incorporated in a 'mapping function' for greater accuracy (e.g. Davis et al., 1985). The total path delay can be written as

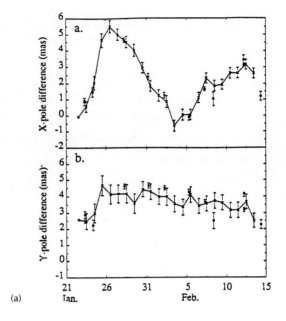

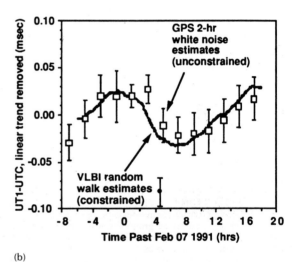

Figure G23 (a) Daily variations in x and y components of pole position measured by GPS (crosses and solid line) and VLBI (solid dots) from Herring, Dong and King (1991). (b) High-frequency variations in Earth rotation rate from GPS and VLBI. (from Lichten, Marcus and Dickey, 1992. Copyright American Geophysical Union.)

$$\rho(\theta) = \rho_d^0 M_d(\theta) + \rho_w^0 M_w(\theta) \qquad (G3)$$

where ρ^0 refers to the path delay at zenith, subscripts d and w refer to dry and wet components respectively, and $M(\theta)$ is the mapping function, assumed to be azimuthally symmetric.

The dry zenith delay is determined by measurement of surface pressure. The wet zenith delay can be calibrated in at least three ways: by measurement of surface temperature and relative humidity coupled with a simple atmospheric model; with a water vapor radiometer (WVR), an instrument that measures atmospheric blackbody radiation in the microwave region, which is affected by a rotational molecular transition of water vapor near 22.2 GHz; and by estimation techniques without prior calibration, exploiting the data strength of GPS and the known elevation angle dependence of the wet delay (Dixon and Kornreich Wolf, 1990; Dixon et al., 1991a). In

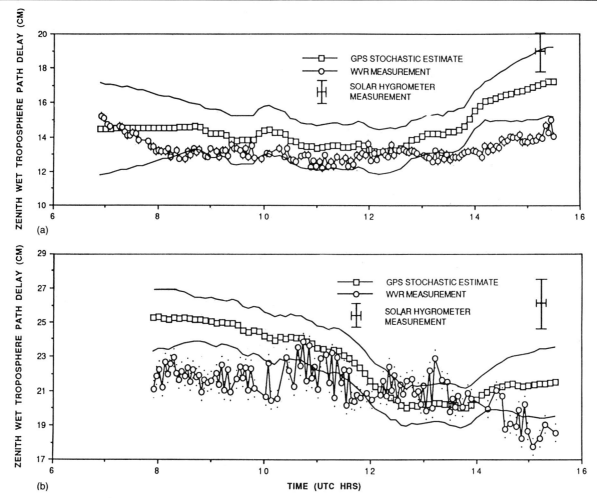

Figure G24 Comparison of zenith wet tropospheric path delay measured by a WVR, and estimated solely from GPS data using a stochastic model, for two successive days (25 and 26 January 1988) at a site (Liberia) in Costa Rica. (From Dixon and Kornreich Wolf, 1990. Copyright American Geophysical Union.)

fact, GPS data can be used to solve directly for the integrated water vapor content of the atmosphere (Figure G24). The sensitivity of the GPS signal to ionospheric electron content and tropospheric water vapor means that GPS-equipped ground stations can monitor temporal variations in these two atmospheric parameters with important applications in atmospheric sciences.

Orbits

For high-precision geodesy it is necessary to know precisely the positions of the satellites at the time of observation. This can be accomplished during data analysis by solving for orbits simultaneously with other geodetic parameters of interest. An estimate of the six orbital components (three positions and three velocities per satellite, or equivalently the six Keplerian elements) at the initial epoch is made, followed by numerical integration of the equations of motion, using accurate models for various perturbing forces (especially solar radiation pressure) and resulting accelerations, allowing prediction of locations and velocities of the satellites at later epochs. Ground tracking data from sites whose positions may be known independently can be used in a least squares adjustment to improve the initial position/velocity estimates, the model parameters and subsequent position/velocity estimates. In the near future global tracking networks will be operational to provide the necessary information.

Precision and accuracy

The precision and accuracy of high-precision GPS geodetic measurements can be estimated from the scatter of data taken over several days in a single experiment (short-term repeatability; Blewitt, 1989; Dong and Bock, 1989), the scatter of several experiments over several years (long-term repeatability; Davis et al., 1989; Larson, 1990), and by comparison to other techniques. Dual-frequency GPS measurements have at least three significant error sources, each with a different dependence on baseline length. A simple error model describing this is

$$\sigma = [a^2 + b^2(1 - e^{-L/\lambda})^2 + c^2L^2]^{1/2} \quad (G4)$$

where a is the receiver and set-up error, b is the asymptotic tropospheric contribution to baseline error for station separations exceeding several tropospheric correlation lengths, λ is the tropospheric correlation length (typically a few kilometers to a few of kilometers), c is orbit-related error and L is baseline length. Figure G25 shows this model compared to data on long-term repeatability for a series of GPS experiments in California. The vertical component is the most poorly determined, reflecting the geometric limitation that satellites are observed in the upper hemisphere only, and higher sensitivity to errors in tropospheric calibration.

Geological and geophysical results

One of the early scientific applications of GPS has been measurement of crustal deformation and plate motion. Significant compressional deformation along parts of the southern San Andreas fault system in California has been documented by comparison of GPS data to older triangulation or trilateration data (Feigl, King and Jordan, 1990) and by analyzing several years of GPS data (Donnellan, 1991). Measurement of the velocity vector describing relative motion of the Pacific and North American plates has also been accomplished through analysis of GPS data spanning several years (Figure G26) (Dixon et al., 1991b). The expansion of the GPS satellite constellation in the last few years has enabled high precision GPS geodetic measurements

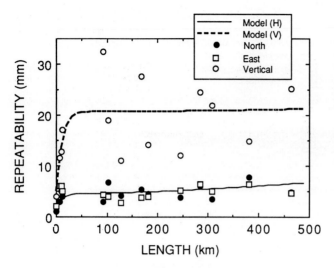

Figure G25 Total error from equation (G3) for the horizontal (solid line) and vertical (dashed line) components, respectively, of a hypothetical GPS experiment, with $a = 2$ and 5 mm, $b = 4$ and 20 mm, $\lambda = 10$ km, and $c = 1 \times 10^{-8}$. For comparison, the long-term repeatability (weighted rms scatter about a best-fit straight line) for GPS measurements in central and southern California between 1986 and 1988 are also shown as separate data points. (Modified from Dixon, 1991. Copyright American Geophysical Union.)

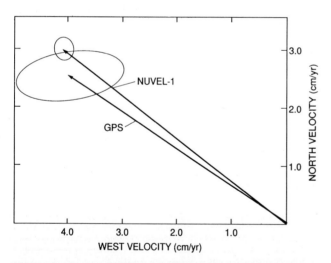

Figure G26 Vector diagram summarizing relative motion of Pacific plate (at Cabo San Lucas, Mexico) with respect to North America, based on GPS measurements spanning 4 years, and NUVEL-1 (DeMets et al. 1990), a global plate motion model which averages over several million years. Error ellipses are 1 standard error. (Modified from Dixon et al., 1991b. Copyright American Geophysical Union.)

in virtually all parts of the globe; long-term experiments are now under way or planned along virtually every major plate boundary for earthquake cycle studies, regional deformation and relative plate velocity measurements. The increasing coverage and data quality from GPS global tracking networks such as that provided by the National Geodetic Survey's CIGNET network and NASA's FLINN network also means that regional GPS experiments can be conducted with greater ease and enhanced accuracy, particularly those requiring long baseline measurements where orbit-related errors would otherwise be large. Global networks also provide a means for studying phenomena not previously amenable to the GPS approach, including long-wavelength tidal and atmospheric effects, polar motion and variations in the Earth's rotation rate.

Other applications

GPS is being used to measure post glacial rebound, to study tides and long-term sea level change, to measure the relative elevations of wave-cut terraces and ancient shorelines, to track drifting buoys for ocean current studies, and to monitor volcanoes, glaciers, floating icebergs and ground subsidence due to water and oil withdrawal. GPS is also used for accurate time transfer among widely separated stations. High-precision GPS positioning and navigation of ships and aircraft have enabled accurate calculation of non gravitational accelerations, allowing enhanced recovery of the gravity field from suitably equipped platforms. Tracking of low Earth orbiting satellites is also possible to extremely high precision. The Topex/Poseidon mission, a joint US–French radar altimeter satellite for studying ocean circulation, employs a GPS receiver to obtain ephemeris information accurate to about 10 cm. This enables study of basin-scale circulation from space for the first time, which was limited in earlier altimeter missions by long-wavelength orbit error. GPS can also be used to improve knowledge of spacecraft orientation for accurate instrument pointing.

The dual-frequency nature of GPS signals is exploited by NASA's Deep Space Network for monitoring ionospheric conditions, an important factor in accurate recovery of data telemetered from spacecraft. Given adequate station coverage, it is in principle possible to map ionospheric activity with GPS on a global basis. GPS receivers on low Earth-orbiting satellites can observe numerous ray paths through different levels of the atmosphere, and thus can be used for three-dimensional atmospheric tomography.

Timothy H. Dixon

Bibliography

Blewitt, G. (1989) Carrier-phase ambiguity resolution for the Global Positioning System applied to geodetic baselines up to 2000 km. *J. Geophys. Res.*, **94**, 10187–203.

Davis, J.L., Herring, T.A., Shapiro, I.I. (1985) Geodesy by radio interferometry: effects of atmospheric modelling errors on estimates of baseline length, *Radio Sci.*, **20**, 1593–607.

Davis, J.L., Prescott, W.H. Svarc, J.L. and Wendt, K. (1989) Assessment of Global Positioning System measurements for studies of crustal deformation. *J. Geophys. Res.*, **94**, 13635–50.

DeMets, C., Gordon, R.G., Argus, D.F. and Stein, S. (1990) Current plate motions. *Geophys. J. Int.*, **101**, 425–78.

Dixon, T.H. (1991) An introduction to the Global Positioning System and some geological applications. *Rev. Geophys.*, **29**, 249–76.

Dixon, T.H. and Kornreich Wolf, S. (1990) Some tests of wet tropospheric calibration for the CASA Uno Global Positioning System experiment. *Geophys. Res. Lett.*, **17**, 203–6.

Dixon, T.H., Gonzalez, G., Lichten, S. and Katsigris, E. (1991a) First epoch geodetic measurements with the Global Positioning System across the northern Caribbean plate boundary zone. *J. Geophys. Res.*, **96**, 2397–415.

Dixon, T.H., Gonzalez, G., Lichten, S.L. et al. (1991b) Preliminary determination of Pacific–North-America relative motion in the southern Gulf of California with the Global Positioning System. *Geophys. Res. Lett.*, **18**, 861–4.

Dong, D. and Bock, Y. (1989) GPS network analysis with phase ambiguity resolution applied to crustal deformation studies in California. *J. Geophys. Res.*, **94**, 3949–966.

Donnellan, A. (1991) A geodetic study of crustal deformation in the Ventura Basin region, southern California. PhD Thesis, California Institute of Technology, 220 pp.

Feigl, K.L., King, R.W. and Jordan, T.H. (1990) Geodetic measurement of tectonic deformation in the Santa Maria fold and thrust belt, California. *J. Geophys. Res.*, **95**, 2679–99.

Hager, B.H., King, R.W. and Murray, M.H. (1991) Measurement of crustal deformation using GPS. *Ann. Rev. Geophys.*, **19**, 351–82.

Herring, T.A., Dong, D. and King, R.W. (1991) Sub-milliarcsecond determination of pole position using Global Positioning System data. *Geophys. Res. Lett.*, **18**, 1893–6.

Larson, K. (1990) Evaluation of GPS estimates of relative positions from central California, 1986–1988. *Geophys. Res. Lett.*, **17**, 2433–6.
Lichten, S.M., Marcus, S. and Dickey, J.O. (1992) Sub-daily resolution of Earth rotation variations with Global Positioning System measurements. *Geophys. Res. Lett.*, **19**, 537–40.
Leick, A. (1990) *GPS Satellite Surveying*. Wiley-Interscience, 352 pp.
Lindqwister, U.J., Freedman, A. and Blewitt, G. (1992) Daily estimates of the Earth's pole position with the Global Positioning System. *Geophys. Res. Lett.*, **19**, 845–8.
Spilker, J.J. (1978) GPS signal structure and performance characteristics. *Navigation: J. Inst. Nav.*, **25**: 121–46.

Cross references

Earth rotation
Geoid
Plate tectonics

GODDARD, ROBERT HUTCHINGS (1882–1945)

Pioneer American experimenter in the field of rocketry. Goddard was born in Worcester, Massachusetts, and in 1909 began an interest in powder- or solid-fuel rockets, publishing a modest volume in 1919 entitled 'A method of reaching extremely high altitudes'. The rockets concerned were intended to carry scientific instruments for atmospheric exploration, but he also speculated that, if provided with sufficient fuel, a rocket could eventually be designed that could reach to the Moon.

The propellant first used was magnesium powder, but after considerable calculations and experiments, he achieved a successful launch on 16 March, 1926, using a liquid fuel, liquid oxygen and lamp oil (kerosene). Between then and 1937 Goddard experimented with various flight-control devices from steerable vanes and gyroscopes to a pendulum system. One of the latter achieved a velocity of 700 mph (1100 km h^{-1}), landing 9000 ft (2750 m) away, and during World War II he collaborated with the US Navy in guided missile development.

Goddard's fame was entirely posthumous. During his lifetime he was sustained by Clark University (his alma mater, in Worcester), becoming a full professor there in 1919. Early in the 1960s the US government awarded his estate the sum of $1 million for the rights of his more than 200 patents, and the National Aeronautics and Space Administration (NASA) named after him the Goddard Space Flight Center, which is the principal research center in Greenbelt, Maryland, as well as the Goddard Institute for Space Studies (near Columbia University in New York city).

Rhodes W. Fairbridge

Bibliography

Goddard, R.H. (1919) A method of reaching extreme altitudes. *Smithsonian Misc. Coll.*, **71**(2).
Goddard, R.H. (1946) *Rockets*. New York: American Rocket Society.
Goddard, R.H. (1959) *An Autobiography Astronautics*, **14**, 24–7, 106–9.
Swenson, L.S., Jr (1972) Goddard, Robert Hutchings. *Dict. Sci. Biogr.*, Vol. 5, pp. 433–4.
Von Braun, W. and Ordway, F.I., III (1969) *History of Rocketry and Space Travel*, 2nd edn New York.

GRAVIMETRY

Gravimetry is the measurement of gravity or the attraction on a proof mass at various locations around a planetary body. On Earth this is done with gravimeters located at many land stations as well as on board ships and submarines. However, for the planets and the Moon there is a technique that uses the motion of an orbiting satellite to acquire gravity field information. As the satellite orbits a body it experiences accelerations (gravity variations) due to mass variations within that body. Of course these gravity measures are at spacecraft altitude and not at the planet surface, as a gravimeter provides. These measures however can be mapped to the surface using analytic functions.

The measuring system is the same as that used to navigate the spacecraft. A radio signal at 2200 million cycles per second is transmitted from an Earth radio antenna and received at the spacecraft, where it is in turn relayed back to the Earth antenna. There is a frequency shift in the received signal due to the relative motion of the Earth-based radio antenna and the spacecraft. This differential shift in frequency is called the Doppler shift and it provides a speed measurement. When the speed changes the spacecraft has experienced an acceleration or a gravity effect. To obtain gravity field variations, however, large well-known motions must first be removed. These include such things as spacecraft orbital velocity, relative Earth–planet velocity, Earth spin velocity, gravitational effects of the Sun, Moon and other planetary bodies, and solar radiation pressure, as well as other smaller effects. What remains is then attributed to the unknown gravity field variations. These variations can be quite different for each planet, but usually variations of 100 mgal are observed, which are equivalent to accelerations of 1 mm s^{-2}.

William L. Sjogren

Bibliography

Ferrari, A.J., Sinclair, W.S., Sjogren, W.L. (1980) Geophysical parameters of the Earth Moon system. *J. Geophys. Res.*, **85**, (B7), 3939–51.
Sjogren, W.L. (1979) Mars gravity: high resolution results from Viking Orbiter 2. *Science*, **203**, 1006–10.
Kaula, W.M. (1966) *Theory of Satellite Geodesy*, Waltham, Massachusetts: Blaisdell.
Konopliv, A.S., Borderies, N.J., Chodas, P.W. *et al.* (1993) Venus gravity and topography: 60th degree and order model, *Geophys. Res. Lett.*, **20**, 2403–6.
Heiskanen, W.A. and Moritz, H. (1967) *Physical Geodesy*. San Francisco, California: W.H. Freeman and Company.
Muller, P.M. and Sjogren, W.L. (1968) Mascons: lunar mass concentrations. *Science*, **161**, 680–4.

Cross references

Geoid
Mars: gravity
Moon: gravity
Venus: gravity

GRAVITATION

Physicists identify four forces in nature: the gravitational force, the electromagnetic force and the strong and weak nuclear forces. Based on a reinterpretation of data obtained by Roland von Eötvös in 1889, Ephraim Fischbach suggested in 1986 that there might be experimental evidence for a gravity-like fifth force as well, but highly sophisticated experimentation over the next 6 years failed to find evidence for such a fifth force. Nevertheless, these experiments were fruitful in another sense; they improved the experimental limits for the Newtonian inverse square law (Fischbach and Talmadge, 1992).

Two of the four known forces, gravitation and electromagnetism, are similar in two ways. They are both long range, extending in principle to infinity; and, in the appropriate static limit, they obey an inverse square law – Coulomb's law for electrostatics, Newton's law (q.v.) for gravitation. Largely motivated by this similarity, Einstein (q.v.) and others between 1920 and 1950 tried to unite gravitation and electromagnetism into a single theory, but this goal proved elusive. Instead, quantum mechanical considerations led to the unification of electromagnetism and the weak nuclear force, a force governing radioactive decay in some atomic nuclei. In 1979 Sheldon Lee Glashow, Abdus Salam and Steven Weinberg shared a physics Nobel prize for explaining this 'electro-weak' particle interaction (see Langacker and Mann, 1989 for a review). Currently the strong nuclear force is described by a separate theory of quantum chromodynamics (Moriyasu, 1983). Physicists now combine the electro-weak theory and quantum chromodynamics into a paradigm known as the standard model, but they exclude gravitation. Yet theorists anticipate that eventually all four forces will be unified into a single quantum theory. The current line of research is superstring theory (Schwarz, 1987).

Gravitational theory and experiments

Our heritage for an empirical understanding of physical processes dates from the ancient Greek civilization, notably the ideas and methods of two early scientists, Aristotle (384–322 BC; q.v.) and Claudius Ptolemy (q.v.). Although other ancient civilizations made significant contributions, it was during the long era of Greek intellectualism that the framework for experimental science was defined, including experimental gravitation. These inquiries involved two categories of phenomena: (1) terrestrial manifestations of gravitation (for example, falling bodies) and (2) the motion of heavenly bodies, or celestial mechanics (q.v.) The Greek scientists seem to have kept these two realms separate, and because their ideas were so influential worldwide, their concept of separate terrestrial and celestial phenomena survived well into the scientific revolution spawned by the European Renaissance. Even Johann Kepler (1571–1630), who carried on the Greek tradition of celestial mechanics, seemed content to describe the motion of the planets in geometrical terms. His goal was to predict the positions of the planets to the same accuracy that Tycho Brahe (1546–1601) could measure them (about 1 minute of arc). On the other hand, Galileo Galilei (1564–1642; q.v.), perhaps best known as an observational astronomer and advocate of the heliocentric model of planetary motions, performed fundamental experiments in terrestrial dynamics. Using inclined planes, Galileo arrived at two cornerstones of modern gravitation: (1) the natural state of motion of an object is to move at constant velocity on a straight line, not to remain at rest as taught by Greek science, and (2) all bodies fall with the same acceleration $g = 9.798$ m s^{-2}, independent of their mass and internal composition. Galileo, by extrapolating his laboratory measurements to the ideal case of no frictional forces, was able to confine his results to gravitational forces only. By contrast, in not removing effects of atmospheric drag, Aristotle had concluded that more massive bodies fall faster than lighter bodies.

It was Isaac Newton (1642–1727; q.v.), who accomplished what some consider the first important unification in physics, that of terrestrial dynamics and celestial mechanics. While working at his mother's home at Woolsthorpe, Newton realized that both terrestrial objects, the legendary apple for example, and the Moon are falling in the gravitational field of the Earth. However the Moon is not falling with an acceleration g, but with an acceleration reduced by the inverse square of its distance r from the center of the Earth, or actually the square of the ratio (R/r), where R is the radius of the Earth. In arriving at this conclusion, Newton made the assumption – along lines suggested by Aristotle – that the Earth's gravitational attraction could be represented by concentrating all the Earth's mass at a single point at its center, an assumption he later proved valid for a spherical mass distribution. Later, after proving that all celestial bodies could be represented to a high degree of approximation by point masses, Newton went on to show that the inverse square law implied that all the planets orbited the Sun according to the laws that Kepler had derived from Brahe's observations. With these new empirical insights, plus the theoretical and experimental work of his predecessors, Newton arrived at his three laws of motion and the law of gravitation. He published his work in 1687 under the descriptive title *Philosophiae Naturalis Principia Mathematica*, or in English translation, *The Mathematical Principles of Natural Philosophy* (Physics).

As developed by mathematical physicists who followed Newton, his law of gravitation states that the acceleration of a test particle can be derived by taking partial derivatives of a scalar potential function V according to the following rule:

$$\vec{a} = \frac{d\vec{v}}{dt} = -\nabla V \tag{G5}$$

where the gradient operator is coordinate independent, but takes its simplest form for Cartesian coordinates. The potential function, defined here as potential energy, is obtained as a solution to Poisson's equation

$$\nabla \cdot \nabla V = 4\pi G \rho \tag{G6}$$

where G is the gravitational constant $(6.6728 \pm 0.0016 \times 10^{-11}$ m^3 kg^{-1} s^{-2}) and ρ is the density of matter in an infinitesimal region surrounding the test particle. For the external gravitational field of a planet or other body, the density is zero, and V is given by Laplace's equation, just Equation (G6) with the right-hand side set to zero.

Newtonian physics triumphed not only in terrestrial mechanics, but most spectacularly in sophisticated calculations in celestial mechanics. By the mid to late 19th century all astrometric observations of solar system bodies could be predicted to the level of their accuracy, about 1″. However, there was one exception. Both Urbain Jean Joseph Leverrier (1811–1877; q.v.) in France and Simon Newcomb (1835–1909; q.v.) in the United States found that they needed to introduce an unexplained excess precession into the location of Mercury's perihelion, the point in the orbit where Mercury is closest to the Sun. They accounted for the gravitational attractions by other planets – most importantly Venus, Jupiter and Earth – and found a theoretical precession of about 570″ per century, but it was not enough. According to Newcomb, an excess precession of about 43″ per century was needed. When astronomers heard of this problem, their first approach was to look for unmodeled sources of gravitation near the Sun, including possibilities of circumsolar disks of matter, an extreme solar oblateness, an undiscovered planet interior to Mercury's orbit, or even modifications to Newton's inverse square law. But ultimately no new sources of gravitation were needed, because in 1916 observation and theory were reconciled by recognizing that Einstein's theory of general relativity had observable consequences for Mercury's orbital motion. This reconciliation, taken together with the theory's prediction of an expanding universe, a prediction in agreement with Edwin Hubble's 1929 announcement of the linear relationship between galaxy redshifts and distances, and in addition some appealing theoretical arguments, led practically all astronomers by the 1940s to accept general relativity as the preferred theory of gravitation.

General relativity

Around the turn of the 20th century there was general agreement that Newtonian mechanics needed modification of some sort, although theoretical physicists were mostly unaware of the problems with Mercury's orbit or, if they were aware, they felt that the meticulous orbital computations and observations were so complicated that they shed little light on the validity of Newtonian mechanics. Rather, physicists were impressed with the success of James Clark Maxwell's 1864 theory of electrodynamics, both in that it predicted the right outcome of experiments, and that it predicted electromagnetic waves. Further, in the experimental arena several new results were available, most notably Albert Abraham Michelson and E.W. Morley's 1887 optical measurements showing the isotropy of the speed of light to an accuracy of 2×10^{-5}, the 1889 Eötvös experiment showing the equivalence of free fall – the weak equivalence principle – to an accuracy of 5×10^{-9}, and Heinrich Hertz's 1888 discovery, based on Maxwell's theory, of the generation and detection of radio waves. Taken together with Simon Newcomb's 1882 determination of Mercury's excess perihelion precession, these new results made the 1880s one of the most productive decades in experimental physics. It was Albert Einstein (1879–1955) who brought the classical mechanics pioneered by Newton into agreement with all these new experimental data.

The special theory of relativity was published in seminal form in 1905. It unified space and time into a new four-dimensional physics. It also elevated the coordinate transformations discovered by Hendrik Antoon Lorentz (1853–1928) to a new level of physical significance. The laws of mechanics and electrodynamics, though not unified, were made consistent in the sense they were all invariant under four-dimensional Lorentz transformations. Further, Einstein concluded that all these laws have the same form in any inertial coordinate system, a system or frame of reference that is fundamental to any dynamical discussion. Being so fundamental, these inertial frames of Newton and Einstein seemed to have a reality of their own, and they seemed closely tied to the background of distant stars and galaxies. Today we conclude that any system at rest or moving at constant velocity along a straight line with respect to the cosmic background is for all practical purposes an inertial system.

Astronauts orbiting the Earth are nearly inertial observers, even though they are in circular motion in the same gravitational field that binds us to the Earth's surface. However they are not quite inertial, because with sufficiently accurate instrumentation they could measure the Earth's gravitational gradient across the dimensions of their spacecraft, or even across the smaller dimensions of their instruments. So, in a manner reminiscent of Galileo's extrapolations, physicists imagine an ideal situation where the spacecraft dimensions are so small that the coordinate frame is inertial. We say that in a freely falling frame in a gravitational field, the laws of physics are locally Lorentz invariant. In extending special relativity to include gravitation, Einstein imposed this principle as a constraint. More specifically, it was imposed as an additional constraint to the universality of free fall, and also to a third constraint that in locally freely falling frames the laws of physics are independent of location and time.

Einstein developed field equations analogous to Equation (G5) and defined methods to derive equations of motion analogous to Equation (G6). Guided by the work of contemporaneous mathematicians such as H. Minkowski and Marcell Grossman, Einstein cast the field equations and the equations of motion in terms of a Riemannian four-dimensional geometry. In 1916 he published the final form of his theory in the German journal *Annalen der Physik* under the title as translated into English, 'The foundation of the general theory of relativity.' In recent times some have preferred to label Einstein's general relativity 'geometrodynamics,' because in a sense it has removed gravitation as a force of nature. Just as Galileo concluded that the natural state of motion is at constant velocity along a straight line, Einstein's theory says that the natural state of motion – undisturbed by outside forces – is the shortest path (geodesic) in a four-dimensional Riemannian geometry. The Einstein field equations determine the geometry in terms of ten symmetric components of a four-dimensional metric tensor $g_{\mu\nu}$.

A useful way to illustrate the practical differences between Newtonian mechanics and general relativity is to consider the external field generated by a spherical distribution of matter of total mass M. In Newtonian mechanics the field is given by the following solution to Laplace's equation, the vacuum case of equation (G6) ($\rho = 0$):

$$V = -\frac{GM}{r} \tag{G7}$$

Einstein's field equations, on the other hand, yield a solution obtained by K. Schwarzschild in 1916. In terms of the usual spherical polar coordinates and one time coordinate, Schwarzschild obtained the following expression for the four-dimensional line element ds:

$$ds^2 = \left(c^2 - \frac{2GM}{r}\right)dt^2 - \left(1 - \frac{2GM}{rc^2}\right)^{-1}dr^2 - r^2(\sin^2\theta\, d\phi^2 + d\theta^2) \tag{G8}$$

The constants G and c that appear in equation (G8) are so fundamental to general relativity that they are usually set to unity in theoretical discussions. In practice, the velocity of light has a fixed value 299 792 458 m s^{-1}, adopted by the International Astronomical Union (IAU) for its 1976 System of Astronomical Constants. In effect, the SI meter is defined in terms of the second of time, which is further defined in terms of a fundamental resonance in the cesium atom. Consequently, astronomers currently use atomic clocks to measure both length and time – an empirical recognition of the space–time unification adopted by Einstein.

The gravitational constant G is measured in the laboratory by using a torsion balance to measure small forces between masses. Henry Cavendish (1731–1810) pioneered this technique and measured G for the first time more than 100 years after the publication of Newton's *Principia*. Note that because highly accurate measurements of planetary attractions, whether on natural bodies or on spacecraft, yield GM for the planet, the planet's mass in metric units (kg) must be obtained by dividing GM by the laboratory-determined value of G.

One of the properties of general relativity, exemplified by equation (G8), is that the coordinates are not defined. However, once one adopts a convention for the line element ds, it is essential that all calculations be carried out in the adopted coordinates. This involves three basic calculations: (1) the derivation of the equations of motion (geodesic paths) and the calculation of the position of the body or bodies of interest with respect to time t in the adopted coordinates, (2) the calculation of the proper time ds/c kept by clocks – real or imaginary – attached to the bodies participating in the definition of the observations, and (3) the calculation of the paths ($ds = 0$, or null geodesics) for the electromagnetic waves that produce the observations. In modern celestial mechanics the radial coordinate of equation (G8) is transformed such that the Schwarzschild metric takes on the following isotropic form:

$$ds^2 = \left(\frac{1 - \frac{GM}{2c^2r}}{1 + \frac{GM}{2c^2r}}\right)^2 c^2 dt^2 - \left(1 + \frac{GM}{2c^2r}\right)^4 (dx^2 + dy^2 + dz^2) \tag{G9}$$

In 1961 H.P. Robertson pointed out that solar system tests of general relativity did not require the exact Schwarzschild solution of equation (G9), but only an expansion to the first term beyond the Newtonian approximation. Following an earlier suggestion by Arthur S. Eddington, Robertson expanded equation (G9) in powers of the Newtonian potential as follows:

$$ds^2 = \left(1 - 2\alpha\frac{GM}{c^2r} + 2\beta\frac{(GM)^2}{c^4r^2} + \cdots\right)c^2 dt^2 - \left(1 + 2\gamma\frac{GM}{c^2r} + \cdots\right)(dx^2 + dy^2 + dz^2) \tag{G10}$$

Robertson introduced the three dimensionless constants α, β, γ, all unity in general relativity, and he derived the perihelion precession, the bending of light by solar gravitation, and the gravitational redshift using the metric of equation (G10). This post-Newtonian expansion of metric gravitation, usually called the Eddington–Robertson–Schiff (ERS) formalism, has proved useful in testing general relativity in the solar system, as well as in some binary stellar systems. It has been extended to include n bodies by Frank B. Estabrook, thus making it applicable to solar system experiments, and it has been generalized, most notably by Clifford M. Will and Kenneth Nordtvedt, to include additional effects not predicted by general relativity. Also, in 1964 Irwin I. Shapiro suggested a fourth solar system test to supplement the three tests discussed by Robertson. Shapiro showed that a radar signal bounced off an inner planet, as well as an actively transponded ranging signal to spacecraft, would be delayed by solar gravitation by about 200 μs near the solar limb. This effect was soon verified by radar ranging to Venus and by transponded ranging to the Mariner 6 and 7 spacecraft at solar conjunction.

To date, general relativity has passed all tests mentioned in the preceding paragraph. In 1976 a hydrogen maser was flown on a Scout rocket to an altitude of about 10 000 km, and the gravitational redshift was verified to an accuracy of 2×10^{-4}. Radar ranging to planets, active spacecraft ranging, and laser ranging to corner reflectors on the Moon are all consistent with general relativity. These data show that Robertson's parameters β and γ are unity to an accuracy of about 0.2% (see Will, 1986, for a popular review; Will, 1993, for technical details).

However, the applications and testing of general relativity are not limited to solar system celestial mechanics. One of the predictions of the theory is the generation of gravitational radiation by accelerated masses, something completely absent from Newtonian gravitation. Although numerous attempts to detect gravitational radiation from astronomical sources have failed, there is indirect evidence for its existence. Because the binary pulsar PSR 1913+16 is losing orbital energy at the rate predicted by general relativity for gravitational radiation reaction, the agreement between gravitational radiation theory and observation has been established at the 1% level. Further, the failure to detect gravitational waves has been attributed to the limited sensitivity of the detectors built to date, hence theorists expect a future detection, perhaps using laser interferometric techniques (see Thorne, 1987, for a review).

Applied gravitation

Outside of precision celestial mechanics and astrometry, there is little or no application for general relativity in the solar system. In the future, if gravitational wave detectors can be made sensitive enough, a new field of observational astronomy may be opened up, but an unambiguous detection is needed first. Within the solar system, just the Newtonian gravitational parameters can tell us much about the physical properties of planets, satellites and small bodies such as comets and asteroids. At one level, a measurement of total mass GM/G, when combined with data on size, yields the mean density of a body. Over the past 30 years the microwave Doppler tracking of interplanetary spacecraft has yielded masses and densities for all the planets, except Pluto, as well as masses for all the larger satellites. The current values of planetary masses are listed in Table G5, along with the space missions responsible for their determinations. The masses of Pluto and its satellite Charon have recently been announced by George W. Null using astrometric data from the Hubble Space Telescope.

Table G5 Planetary masses as determined by microwave Doppler data

Planet	M_{Sun}/M_{planet} [a]	Spacecraft mission
Mercury	6 023 600 ± 250	Mariner 10
Venus	408 524 ± 1	Mariner 5, Mariner 10, PVO[b]
Mars	3 098 714 ± 9	Mariner 4
Jupiter	1 047.3486 ± 0.0008	Pioneer 10/11, Voyager 1/2
Saturn	3 497 90 ± 0.02	Pioneer 11, Voyager 1/2
Uranus	22 902.94 ± 0.04	Voyager 2
Neptune	19 412 25 ± 0.06	Voyager 2

[a] Ratio of mass of Sun to mass of planetary system including satellites.
[b] PVO is Pioneer 12, the Pioneer Venus Orbiter. Other missions are the Mariner series to the inner planets, Pioneer 10 to Jupiter, Pioneer 11 to Jupiter and Saturn, Voyager 1 to Jupiter and Saturn, and Voyager 2 to all four outer planets.

Planetary gravity fields

If a planet or satellite were perfectly spherical, its total mass would be sufficient to describe its external gravitational field (equation (G7)). However all natural bodies deviate from spheres at some level, so what is needed is a general potential function that satisfies Laplace's equation for an arbitrary distribution of mass. The intensity of the resulting gravitational field, as measured by nearby spacecraft or gravimeters on the surface, is referred to as the body's gravity, meaning the property of having weight.

Using the spherical distribution as a first approximation, we seek a general solution to Laplace's equation in spherical coordinates. With geocentric coordinates r for radius, φ for latitude, and λ for longitude, instead of the conventional spherical polar coordinates, Laplace's equation is written as follows:

$$\frac{\partial^2 V}{\partial r^2} + \frac{2}{r}\frac{\partial V}{\partial r} + \frac{1}{r^2 \cos\phi}\frac{\partial}{\partial \phi}\left(\cos\phi \frac{\partial V}{\partial \phi}\right) + \frac{1}{r^2 \cos^2\phi}\frac{\partial V}{\partial \lambda^2} = 0 \quad (G11)$$

This partial differential equation can be separated by writing V as a product of three functions:

$$V = R(r)\Phi(\phi)\Lambda(\lambda) \quad (G12)$$

The general solution, written as follows for the planetary field, involves Legendre's functions P_{nm}:

$$V(r,\phi,\lambda) = \quad (G13)$$
$$\frac{GM}{r}\left[1 + \sum_{n=2}^{\infty}\sum_{m=0}^{n}\left(\frac{R}{r}\right)^n (C_{nm}\cos m\lambda + S_{nm}\sin m\lambda) P_{nm}(\sin\phi)\right]$$

where the external field is referred to a reference radius R – usually the planet's mean equatorial radius, but sometimes the semimajor axis of a reference ellipsoid. The gravity coefficients C_{nm} and S_{nm} are determined from data by the method of least squares. The Legendre functions are often normalized by integrating over a unit sphere, in which case the normalizing factors are absorbed into the gravity coefficients. A particular coefficient with indices nm is referred to as a gravity harmonic of order m and degree n. Coefficients with $m = 0$ are called zonal harmonics, coefficients with $n = m$ are called sectorial harmonics, and the rest are called tesseral harmonics. Zonal harmonics divide the surface of the sphere into $m + 1$ zones of latitude, sectorial harmonics into sectors of longitude or 'orange slices,' and tesseral harmonics into a checker-board pattern.

Detailed gravity models are available for the Earth from a number of data sources, including satellite Doppler tracking, laser ranging to the LAGEOS satellite, satellite altimetry and terrestrial gravity measurements. Doppler tracking of spacecraft orbiting the Moon, Venus and Mars has yielded gravity models for those bodies as well, though not in as much detail as Earth. In the case of the terrestrial planets, their gravity fields are compared with surface topographies, a comparison that yields information on likely dynamical processes within the planets and on the compensation of their topographic features with gravity. However, because the terrestrial planets deviate significantly from equilibrium figures, in the sense that they support large stresses, their gravity fields yield little information about their internal structure. By contrast the outer planets, and their larger satellites, are effectively in hydrostatic equilibrium. As a result, measured gravity harmonics provide important boundary conditions on their interior structure.

Figures of equilibrium

Isaac Newton in his *Principia* computed an equilibrium figure for a rotating Earth. Newton imagined that in principle a hole of unit cross-section could be drilled from the equator to the Earth's center, and that a similar hole could be drilled from the pole. If the holes were filled with a fluid, a state of static equilibrium would eventually occur. After reaching equilibrium, the fluid in the equatorial hole would have the same weight as the fluid in the polar hole. However, along the equator the inertial centrifugal force would effectively reduce the acceleration of gravity on the fluid, and in order to make up for this deficit in acceleration the height of the fluid in the polar hole would be less than the height in the equatorial hole. Newton therefore concluded that the Earth was slightly oblate, and he went on to compute the numerical flattening for a homogeneous Earth.

Today it is almost obvious that a spinning body should be flattened at the poles, but in Newton's time this was controversial. Dynamicists of the Cassini school argued that the Earth was prolate. It was not until the French sent a geodetic expedition into Lapland in 1738 that the matter was settled in favor of Newton.

If the giant Jovian planets did not rotate, their gravity fields would yield no empirical information on internal structure. However, because they rotate rapidly, their shapes and gravity fields yield information on the distribution of density with depth. The two parameters for shape and rotation are respectively the flattening f and the rotation parameter q defined by,

$$f = \frac{a-b}{a}; \quad q = \frac{\omega^2 a^3}{GM} \quad (G14)$$

where a is the planet's equatorial radius, b its polar radius, ω its rotational angular velocity, M its mass and G is the gravitational constant. Even without gravity measurements, the ratio f/q would yield some limited information on interior structure because it has a minimum value of ½ for a planet with an extreme mass concentration at its center, and a maximum value of $\frac{5}{4}$ for a homogeneous planet. For a spinning planet in hydrostatic equilibrium only the even zonal gravity harmonics $J_n = -C_{n0}$ ($n = 2,4,6,...$) are stimulated. The importance of the gravity coefficients J_n is that they are related to the internal density distribution ρ by the following volume integral over the planet's interior:

$$J_n = -\frac{1}{MR^n}\int_V \rho(r,\phi,\lambda)\, r^n P_n(\sin\phi)\, dV \quad n = 2,4,6,... \quad (G15)$$

The J_n coefficients represent boundary conditions that must be satisfied for any viable interior model. Planetary scientists call this technique of matching theoretical interior models to the planet's gravity field 'gravity sounding.' The deep interior is sounded by the second zonal harmonic J_2, while the envelope is sounded by higher harmonics to a depth of about 3100 km for Jupiter and 3600 km for Saturn. For example, in a simple polytrope of index 1, not a bad assumption for Jupiter and Saturn at least, the pressure p and density ρ are related by

$$p = K\rho^2 \quad (G16)$$

and the constant K is determined from the measured gravity coefficients J_2 and J_4 by the expression

$$K = -\frac{2\pi G b^2}{35 J_4}\left(J_2 + \frac{q}{3}\right)^2 \quad (G17)$$

For nonpolytropic models as well, a general density distribution with depth can be derived from measured gravity coefficients J_2 and J_4. The coefficient J_6 may also be useful, although differential rotation and deep atmospheric winds may complicate its interpretation. With a given density distribution, the pressure can be computed under the assumption of hydrostatic equilibrium, and the temperature follows from the equation of state for the assumed material in the envelope.

Current data on geodetic parameters for the four Jovian planets are summarized in Table G6. Most of the tabulated values are from Pioneer and Voyager spacecraft flybys, except for the Uranian gravity harmonics determined by ring dynamics (French *et al.*, 1988). All gravity harmonics are referenced to the equatorial radius given in the table.

John D. Anderson

Bibliography

Anderson, A.J. and Cazenave, A. (eds) (1986) *Space Geodesy and Geodynamics*. New York: Academic Press.

Table G6 Geodetic parameters for four giant planets

Planet	Mass ($M_\oplus$)	a^a (km)	Rotation period (h)	J_2^b (10^{-6})	J_4^b
Jupiter	317.82840	71 492	9.92492	14 697	−584
Saturn	95.16114	60 330	10.6650	16 298	−915
Uranus	14.53571	25 559	17.24	3 513.2	−31.9
Neptune	17.14774	25 225	16.11	3 411	−35

^{a} Equatorial radius at an atmospheric pressure of 1 bar (10^5 N m^{-2}).
^{b} Coefficients of the first two zonal gravity harmonics stimulated by planetary rotation. Other harmonics, complete through degree and order four, have never been detected for the outer planets. A value for Jupiter's zonal harmonic J_6 is available from the Pioneer 11 flyby ($J_6 \times 10^6 = 31 \pm 20$).

Anderson, J.D., Armstrong, J.W., Campbell, J.K. et al. (1992) Gravitation and celestial mechanics investigations with Galileo, in *The Galileo Mission* (ed. C.T. Russell). Dordrecht: Kluwer Academic Publishers, pp. 591–610.

Anderson, J.D. and Hubbard, W.B. (1977) Gravitational fields and interior structure of the giant planets, in *Exploration of the Outer Solar System* (eds E.M. Greenstadt, M. Dryer and D.S. Intriligator). New York: American Institute of Aeronautics and Astronautics, pp. 71–83

Anderson, J.D., Hubbard, W.B. and Slattery, W.L. (1974) Structure of the Jovian Envelope from Pioneer 10 gravity data. *Astrophys. J.*, **193**, L149–50.

Anderson, J.D., Colombo, G., Esposito, P.B. (1987) The mass, gravity field, and ephemeris of Mercury. *Icarus*, **71**, 337–49.

Ashby, N. and Bertotti, B. (1986) Relativistic effects in local inertial frames. *Phys. Rev. D*, **34**, 2246–59.

Bills, B.G. (1987) Venus gravity: a harmonic analysis. *J. Geophys. Res.*, **92**, 10335–51.

Campbell, J.K. and Anderson, J.D. (1989) Gravity field of the Saturnian system from Pioneer and Voyager tracking data. *Astron. J.*, **97**, 1485–95.

Campbell, J.K. and Synnott, S.P. (1985) Gravity field of the Jovian system from Pioneer and Voyager tracking data. *Astron. J.*, **90**, 364–72.

Cook, A.H. (1973) *Physics of the Earth and Planets*. New York: Halsted Press.

Davies, P.C.W. (1980) *The Search for Gravity Waves*. Cambridge: Cambridge University Press.

Fischbach, E. and Talmadge, C. (1992) Six years of the fifth force. *Nature*, **356**, 207–15.

French, R.G., Elliot, J.L., French, L.M. et al. (1988) Uranian ring orbits from Earth-based and Voyager occultation observations. *Icarus*, **73**, 349–78.

Harwit, M. (1988) *Astrophysical Concepts*, 2nd edn. New York: Springer-Verlag.

Hubbard, W.B. (1984) *Planetary Interiors*. New York: Van Nostrand Reinhold.

Kaula, W.M. (1966) *Theory of Satellite Geodesy*. Waltham, MA: Blaisdell Publishing Company.

Kaula, W.M. (1968) *An Introduction to Planetary Physics: The Terrestrial Planets*. New York: John Wiley and Sons.

Kovalevsky, J. and Brumberg, V.A. (eds) (1986) *Relativity in Celestial Mechanics and Astrometry*. Dordrecht: D. Reidel Publishing Company.

Krisher, T.P. (1990) New tests of the gravitational redshift effect. *Modern Phys. Lett. A*, **5**, 1809–1813.

Krisher, T.P., Anderson, J.D. and Taylor, A.H. (1991) Voyager 2 test of the radar time-delay effect. *Astrophys. J.*, **373**, 665–70.

Langacker, P. and Mann, A.K. (1989) The unification of electromagnetism with the weak force. *Physics Today*, December, 22–31.

Misner, C.W., Thorne, K.S. and Wheeler, J.A. (1973) *Gravitation*. San Francisco: W.H. Freeman and Company.

Moriyasu, K. (1983) *An Elementary Primer for Gauge Theory*. Singapore: World Scientific.

Moyer, T.D. (1981) Transformation from proper time on Earth to coordinate time in solar system barycentric space–time frame of reference, parts 1 and 2. *Celest. Mech.*, **23**, 33–68.

Newhall, X.X., Standish, E.M., Jr and Williams, J.G. (1983) DE 102: a numerically integrated ephemeris of the Moon and planets spanning forty-four centuries. *Astron. Astrophys.*, **125**, 150–67.

Podolak, M., Hubbard, W.B. and Stevenson, D.J. (1991) Models of Uranus' interior and magnetic field, in *Uranus* (eds J.T. Bergstralh, E.D. Miner and M.S. Matthews). Tucson: University of Arizona Press, pp. 29–61.

Podolak, M. and Reynolds, R.T. (1985) What have we learned from modeling giant planet interiors? in *Protostars and Planets II* (eds D.C. Black and M.S. Matthews). Tucson: University of Arizona Press, pp. 847–72.

Ryabov, Y. (1961) *An Elementary Survey of Celestial Mechanics* (transl. G. Yankovsky). New York: Dover Publications, Inc.

Shapiro, I.I. (1964) Fourth test of general relativity. *Phys. Rev. Lett.*, **13**, 789–91.

Shapiro, I.I. (1966) Testing general relativity with radar. *Phys. Rev.*, **141**, 1219–22.

Schwarz, J.H. (1987) Superstrings. *Physics Today*, November 33–40.

Thorne, K.S. (1987) in *300 Years of Gravitation* (ed. S.W. Hawking and W. Israel). Cambridge: Cambridge University Press.

Vessot, R.F.C. and twelve others. (1980) Test of relativistic gravitation with a space-borne hydrogen maser. *Phys. Rev. Lett.*, **45**, 2081–84.

Vincent, M.A. (1986) The relativistic equations of motion for a satellite in orbit about a finite-size, rotating Earth. *Celest. Mech.*, **39**. 15–21.

Weber, J. (1961) *General Relativity and Gravitational Waves*. New York: Interscience.

Weinberg, S. (1972) *Gravitation and Cosmology: Principles and Applications of the General Theory of Relativity*. New York: John Wiley and Sons.

Will, C.M. (1986) *Was Einstein Right?* New York: Basic Books.

Will, C.M. (1993) *Theory and Experiment in Gravitational Physics*. 2nd edn. Cambridge: Cambridge University Press.

Zharkov, V.N. and Trubitsyn, V.P. (1978) in *Physics of Planetary Interiors* (ed. W.B. Hubbard). Tucson: Pachart.

Cross references

Barycenter
Celestial mechanics
Earth–Moon system: dynamics
Ephemeris
Mars: gravity
Moon: gravity
Planetary dynamical astronomy
Planetary geodesy
Solar motion
Surface gravity
Tide-raising forces
Venus: gravity

GRAVITY-ASSIST NAVIGATION

By means of a close flyby of a planet, it is possible to increase a spacecraft's orbital velocity far beyond the capability of its propulsion

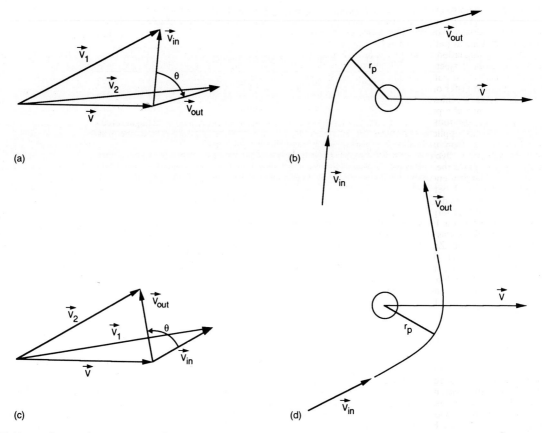

Figure G27 Vector diagram for gravity-assist flybys. The heliocentric velocity of the planet is V and the spacecraft velocities are V_1 and V_2 before and after the flyby. The effect on the planet-centered flyby is shown in (b) and (d), including the geometry of the bending angle θ. The energy-increasing flyby is illustrated (b) for a flyby behind the planet and (d) the energy-decreasing flyby for a flyby in front of the planet.

system. Although this may seem like getting something for nothing, in fact the spacecraft is taking some orbital energy from the planet – but only a tiny fraction. Except for minuscule dissipative forces on the spacecraft and planet, including effects of gravitational radiation, the total energy and angular momentum of the solar system are conserved during the gravity assist.

The earliest studies of the gravity-assist problem considered the orbital perturbations of comets making close approaches to Jupiter. In the 1890s F. Tisserand pointed out that the orbital elements of a comet could be radically different before and after the close approach. However, using an integral discovered by Carl Gustav Jacob Jacobi (1804–1851), he also showed that a combination of the semimajor axis a, the eccentricity e, and the inclination I to Jupiter's orbital plane are approximately conserved. Jacobi's integral says that neither the orbital energy nor the orbital angular momentum of the comet are conserved, but instead a combination of the two yields Tisserand's criterion,

$$\frac{1}{a} + 2[a(1-e^2)]^{1/2} \cos I = C \qquad (G18)$$

where C is a constant, and the semimajor axis is measured in units of the mean distance between the Sun and Jupiter. Equation (G18) allows different values of a, and therefore different values of orbital velocity, before and after the Jupiter flyby. Among several possibilities, a single flyby of Jupiter can alter the comet's trajectory from a parabola to an eccentric ellipse.

Some early attempts to analyze spaceflight flybys were based on the three-body problem studied in the last century. For example, with a close lunar flyby, Jacobi's integral showed that fuel expenditure could be minimized when escaping from the Earth–Moon system. Such an option was included in mission planning for the Pioneer 10 and 11 trajectories to Jupiter, but unavoidable launch delays prevented it. In 1963 two aeronautical engineers – P.A. Lagerstrom and J. Kevorkian at the California Institute of Technology – published a theory for uniformly valid asymptotic approximations to a spacecraft trajectory originating near a massive primary body and passing close to a less-massive secondary body. They assumed circular motion of the secondary mass about the primary, an assumption particularly valid for the Sun and Jupiter but also applicable in the Earth–Moon system. But by far the most powerful analysis tool for gravity-assist navigation was Walter Hohmann's 1925 method of patched conics. At JPL the gravity-assist problem was initially assigned to a 1961 summer student, Michael A. Minovitch, who later carried out extensive trajectory calculations on an IBM 7090 computer, both at JPL and at the University of California at Los Angeles (UCLA). He published the conclusions of his study in a JPL Technical Report dated 31 October 1963.

In the patched-conic technique the spacecraft arrives at the flyby planet with heliocentric velocity vector V_1. The planet has velocity V and the incoming flyby velocity is just the difference of the two heliocentric velocities, $V_{in} = V_1 - V$ (see Figure G27). The spacecraft's path will initially be along the incoming asymptote of the hyperbolic conic flyby, but because of the planetary mass, the flight path will bend through the angle θ. As a result, although the outgoing velocity vector V_{out} will have the same magnitude V_∞ as the incoming vector, it will differ in direction according to

$$\sin\frac{\theta}{2} = \left(1 + \frac{r_p V_\infty^2}{GM}\right)^{-1} \qquad (G19)$$

where r_p is the closest approach distance to the planet and M is the planetary mass. By Figure G27, the resulting heliocentric velocity after the flyby will be $V_2 = V + V_{out}$. In gravity-assist navigation, a mid-course maneuver can direct the spacecraft to pass either in front of the planet, with respect to V, or behind the planet. In the former case the heliocentric orbital energy is decreased, while in the latter, the energy is increased.

It is an instructive exercise to show that a ballistic transfer from the Earth's orbit to Jupiter can be used to cancel completely the spacecraft's heliocentric velocity. The resulting trajectory will fall to the Sun on an inertial rectilinear path, a useful trajectory for a solar flyby mission. Note that a utilization of the Jacobi integral [Equation (G18)] for this solar-probe trajectory yields less information on navigation requirements than the patched-conic technique illustrated by Figure G27. A ballistic transfer orbit can be found from equation (G18) with C = 2, but patched conics and equation (G19) are needed for the proof of the existence of a practical Jupiter gravity assist.

Gravity-assist navigation has found many applications for interplanetary trajectories. The first application occurred in 1973 when the Mariner Venus–Mercury mission used a flyby behind Venus to achieve a March 1974 Mercury flyby. Then the Mercury flyby was used as a gravity assist to place the spacecraft in orbital resonance with Mercury for more Mercury encounters. Three Mercury flybys were successfully accomplished before the spacecraft ran out of attitude-control fuel. Later in the decade the Voyager Mission relied on gravity assists to visit all four large outer planets within 12 years between 1977 and 1989. Between 1982 and 1985 the International Sun–Earth Explorer 3 (ISEE-3) was sent on a multiple gravity-assist trajectory to the comet Giacobini-Zinner. The renamed spacecraft – International Cometary Explorer (ICE) – made two close Earth flybys and five lunar flybys before embarking on its final trajectory to the comet. More recently the Galileo spacecraft, launched in October 1989, first made one Venus flyby and then two Earth flybys before entering its final transfer trajectory to Jupiter on 8 December 1992. Starting in December 1995, the Galileo mission is scheduled to use close flybys of Jupiter's large satellites to explore the entire Jupiter system, including the planet's atmosphere, the large satellites and the Jovian magnetosphere. Because of gravity assists, the entire mission at Jupiter can be flown with only 100 m s^{-1} of mid-course fuel capability, about 60 times less than required without gravity assists. Similarly, by using gravity assists from Titan, the Cassini mission is planning a tour of the Saturn system starting in 2002. Still in early planning stages, a Venus gravity assist during a return from Mars could shorten a manned mission to less than 2 years, while the stay at Mars could be shortened from the Hohmann transfer's 460 days to just a few weeks.

John D. Anderson

Bibliography

D'Amario, L.A., Bright, L.E. and Wolf, A.A. (1992) Galileo trajectory design, in *The Galileo Mission* (ed. C.T. Russell). Dordrecht: Kluwer Academic Publishers.

Wiesel, W.E. (1989) *Spaceflight Dynamics*. New York: McGraw-Hill.

Cross references

Cassini mission
Ephemeris
Galileo mission
Hohman transfer orbit

GRAVITY FIELDS OF THE TERRESTRIAL PLANETS

One of the important forces operating in the solar system is gravity, the force of mutual attraction between masses such as planets and satellites or the mutual attraction between small mass elements of a planet. Newton's law of gravitation, 'two particles attract each other with a central force in proportion to the product of their masses and inversely in proportion to the square of the distance between them,' has been found to be largely adequate to explain most gravitation phenomena in the solar system, whether it is orbital motions or the mass distributions within planets. The proportionality constant of Newton's law G, is 6.670×10^{-11} N m^2 kg^{-2}. An equivalent expression of Newton's law is in terms of the gravitational potential Φ, as the acceleration of gravity $\boldsymbol{a}$, imparted by gravity on a test particle according to

$$\boldsymbol{a} = -\nabla \Phi \qquad (G19)$$

where ∇ is the gradient operator. For a body of volume V the potential at a point P outside of V is

$$\Phi = G \int_v (\rho, \mathfrak{R}) \, dV \qquad (G20)$$

where $\mathfrak{R}$ is the distance of P from the volume element dV of density ρ. For a rapidly rotating planet such as Earth or Mars, gravity on its surface is the sum of this self-attraction and the centrifugal force f_c which acts in a direction perpendicular to the planet's rotation axis. This force is at a maximum at the equator and acts in a direction opposite to the direction of attraction from mass elements within the body. At the poles the centrifugal force vanishes. The sum of the two forces is usually referred to as the gravity vector although in many problems one is only concerned with its magnitude or simply gravity g. The corresponding potential will then include the potential Φ of the gravitational force and the potential of the centrifugal force f_c. On Earth, gravity at the equator is about 9.78 m s^{-2}.

If a rapidly rotating planet responded to the force of gravity as if it were a fluid, its form under the combination of self-attraction and the centrifugal force would be an ellipsoid of revolution with its major axis in the plane of the equator. The observed flattening of the Earth is close to that expected for such fluid behaviour, with the semimajor axis exceeding the semiminor axis by about 21 km. This ellipsoid of revolution provides a very satisfactory first-order description of the shape of the Earth and the best-fitting ellipsoid to a planet's shape constitutes a reference figure with respect to which departures in shape or figure are measured. A more precise description of the shape necessitates the introduction of the term geoid (q.v.).

Gravity on a planet's surface varies from place to place not only because of the centrifugal force but also because of the variation in distance of the surface from the planet's centre of mass and because of the asymmetrical density distribution within the planet. It is this last cause that is of importance for geological and geophysical studies and it is convenient to correct any observation of gravity made at its surface for the other factors. These corrections constitute the free-air and Bouguer correction, in which, in a first approximation, the gravity measurement on the surface is reduced to what it would be if projected onto the geoid with the intervening topographic mass removed: the free-air correction allows for the distance of the point of measurement above or below the geoid and the Bouguer correction allows for the attraction of the topographic mass between this point and the geoid. The difference between the 'corrected' gravity measurement and a theoretical value for the reference figure is referred to as the gravity anomaly (see Geoid). For the terrestrial planets this anomaly is typically of the order of 10^{-3} m s^{-2}.

On Earth, gravity is measured on the surface either as an absolute measurement using pendulum of 'free-fall' instruments, or it is measured as a relative quantity using spring-balance instruments or gravimeters that are calibrated against the absolute measurements. The relative measurements can be readily made with a precision of 10^{-8} m s^{-2} and are widely used in exploration surveys of the structure of the crust where the objective is to detect small density anomalies that may be associated with mineral or hydrocarbon deposits. Because gravity varies with distance from the centre of the mass of the Earth, by 3×10^{-6} m s^{-2} per meter change in elevation, the heights of the gravity measurement sites must be known with high precision, and this has often proven to be more difficult to achieve then the gravity measurement itself. Gravity measurements can be made at sea from moving ships or in the air from aircraft provided that any accelerations of the moving platforms can be independently measured. Large parts of the Earth's surface have been covered with gravity measurements of variable accuracy and a variable density of sites, but substantial areas remain inadequately mapped. Figure G28 illustrates gravity over the Australian continent as derived from surface measurements. Most of the spatial variability noted reflects the anomalous density structure of the continental crust and uppermost mantle and is related to tectonic structures such as ancient orogenic zones and sedimentary basins.

The other important method of establishing the gravity field is through the analysis of trajectories of satellites about a planet. If the planet is a radially symmetric body, an artificial satellite would move about it in an elliptical orbit according to Kepler's laws of motion. The effect of the oblateness of a rotating planet is that the equatorial bulge exerts a gravitational torque on the satellite and causes the orbital plane to precess about the equator. This is similar to the precession and nutation of the Earth as a result of the lunar and solar torques exerted on it (see Precession and nutation). Typically, the orientation in space of the orbit of a close satellite changes by a few

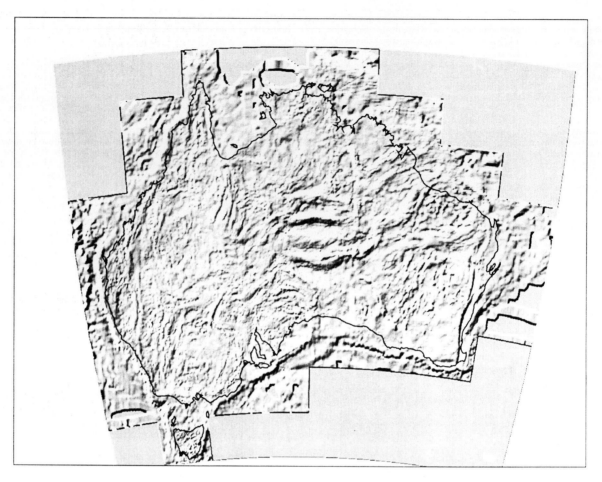

Figure G28 Gravity anomalies observed over the Australian continent and margins.

degrees per day and the precise tracking of the satellite then permits the oblateness to be measured. This oblateness can be defined as:

$$J_2 = (C - A)/MR_e^2$$

where C and A are the polar and equatorial moments of inertia of the planet of mass M and mean equatorial radius Re. Somewhat different definitions of the oblateness are used in the literature, and attention must be given to the choice of normalization functions adopted.

Precise tracking and orbit analysis of a satellite's motion about the Earth show that the orbit is constantly perturbed from this precessing state due to the irregular distribution of mass within the planet. As a satellite passes over a density anomaly it is accelerated in its motion, and the measurement of these accelerations allows the gravity field to be inferred. Because the force of gravity varies inversely with the square of distance from the density anomaly, the anomalous accelerations are small at the satellite altitudes and the resolution attainable by directly mapping these accelerations is limited. Instead, the cumulative accelerations of the satellite are measured over periods of hours to days and it is from these accumulated effects that the longer wavelengths in the planet's gravity field are deduced. The analysis of the orbits for the gravity field of the Earth requires a complete description of all the forces acting on the satellite, including the direct gravitational attraction of the satellite by the Moon and Sun, the gravitational attraction of the tidal deformation of the Earth itself under the gravitational attractions (see Tide-raising forces), and surface forces caused by the atmosphere and by solar radiation pressure.

The successful inference of the gravity field from satellite orbit perturbation analysis also requires the ability to precisely track the satellite. This is now done using either electronic or laser methods. In the latter case satellites carrying retro-reflectors are illuminated from the ground by short laser pulses and the return travel time is measured to determine the distance to the satellite at known instants of time. Tracking accuracies of 1 cm and better are now routinely obtained. Electronic tracking methods have been widely used in which either the distance is measured using radio frequencies rather than optical frequencies, or in which the line-of-sight velocity of the satellite relative to the observer is measured. Important recent examples include the tracking of the Global Positioning System (GPS) satellites developed in the USA and the DORIS tracking system developed in France.

The gravity field at the satellite's altitude is much attenuated when compared with that at the surface. Thus despite the high tracking precision, the resolution with which the field may be derived from the orbit perturbation analyses is limited. Another, less direct, measurement of the gravity field is from altimetric measurements of the satellite height over the ocean surface. The ocean surface, on average, approximates an equipotential surface (Φ = constant) known as the geoid, and if this surface can be mapped it contains, in principle, the same information as the gravity field. Because the precise tracking of the satellite enables the shape of equipotential surfaces to be determined at the satellite height, the altimeter measurement fixes the shape of the ocean surface and, ignoring or correcting for any oceanographic perturbations of this surface, the gravity field on the ocean surface. Several Earth-orbiting satellites have carried high-precision radar altimeters (e.g. SEASAT and GEOSAT) and the ocean surface has now been mapped in very considerable detail. Figure G29 illustrates results of the gravity fluctuations over the Tasman Sea to the east of Australia. Most of the observed anomalies reflect topographic features of the sea floor and their associated crustal structures, including subduction zones, ancient submerged continental fragments and ridges, and submerged volcanoes. Repeated mapping of the surface leads to measuring the time dependencies in the oceanographic perturbations and the altimeters are now an important instrument for studying the time dependence of oceanographic processes.

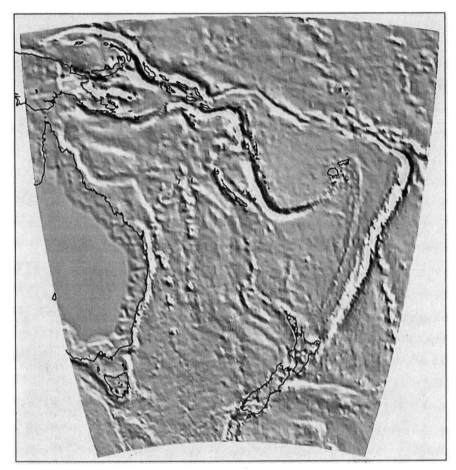

Figure G29 Gravity anomalies over the southwest Pacific as inferred from satellite altimeter data.

Optimum solutions for the Earth's gravity field combine the various data types; the results from the orbital perturbation analysis provide the long-wavelength structure of this field, surface gravity measurements provide the high-resolution variations primarily over the continents, and altimeter measurements over the oceans. Often the global results are presented not as gravity anomalies but as geoid anomalies (see Geoid), which is essentially a spatially integrated function of gravity.

Some of the satellite methods can also be used to measure the gravity fields of other bodies in the solar system, although there are important differences. One is that the planet-orbiting satellite is observed from Earth so that continuous tracking of the satellite along its orbit is possible except when it is eclipsed. Terrestrial satellites can often be tracked only along relatively short segments of their orbit unless a very high density of tracking stations is possible or, as in the case of GPS, the satellites are at great altitude, in which case the orbits contain little gravitational information. The ability to track long arcs continuously facilitates considerably the estimation of the gravity field for some of the planets. On the debit side, the orbital mechanics becomes more complex because the satellite motion is tracked from Earth and not from the planet's surface and high-precision planetary motion theories must be incorporated. One other difference is that the other terrestrial planets do not have a convenient ocean surface which approximates an equipotential surface, so that altimetric methods are not applicable for measuring gravity fields.

For the Moon, with its nearly constant orientation towards Earth, it has been possible to infer the gravity field of the near side with considerable precision and spatial resolution in large part because the absence of any atmosphere means that satellites can orbit this body at quite low altitudes without being subjected to the largely unpredicted and destructive drag forces that would occur for low orbiting Earth satellites. A successful method for mapping the lunar field has been by measuring the accelerations of the spacecraft as seen from Earth and transforming this information into estimates of the magnitude of gravity along the Moon's radial vectors. The analysis of accumulated orbital perturbations for measuring the lunar gravity field has been much less successful than for the Earth, in large part because the absence of a significant rotational flattening result in a poorly structured spectrum of perturbations, and the consequence is that the gravity field for the far side of the Moon remains poorly known.

The gravity field of Mars has been well mapped from orbital perturbation analyses of the Viking orbiters and the longer wavelength features of this field are well known (Plate 19). The long wavelengths of the gravity field of Venus (Plate 7) have also been mapped with considerable resolution from the tracking of the Pioneer Venus orbiters using both the direct mapping of the line-of-sight accelerations from Earth and orbital perturbation analysis.

All planetary bodies examined so far have revealed complex patterns of spatial variation in gravity. None corresponds to what may be expected solely from rotating fluid planets. Anomalous density structures occur in all instances and the planets are not in states of hydrostatic equilibrium. Because the inverse of the relation (G20), the estimation of density from the gravitational potential, is non-unique and any number of mass distributions within the planet can give the same external potential outside this surface, this is perhaps the only inference that can be drawn with absolute certainty from the observations for the gravity fields. The successful interpretation of gravity fields therefore depends on a knowledge of other geophysical and geological fields and on the ability to develop physically plausible models of the planet's interior. In the case of the Earth the supplementary constraints are considerable, particularly from the tomographic seismic modeling of the crust and mantle.

A simple example of the ambiguity of the gravity field interpretations is that the anomalies may be indicative of either an inert and cold planet or of a dynamic and warm planet. In the first instance the gravity anomalies reflect conditions of formation or of early evolution of the body and thus provide a measure of the strength and indirectly, of the temperature of the mantle. In the second instance the gravity anomalies reflect present or recent tectonic processes and contain little information on the planet's early history but much information on the dynamic processes and recent evolution. Before the satellite results of the Earth's gravity field became widely accepted it had often been argued that the Earth below the crust was in hydrostatic equilibrium, and that the gravity anomalies measured by gravity meters on the surface were largely the result of the topography and its isostatic compensation in which the surface loads were compensated at some depth by low-density crustal materials such that below this depth pressures are everywhere hydrostatic. The early satellite results yielded anomalies that exceeded these expectations and some scientists argued that this reflected an ancient past whereas others argued that the anomalies implied an ongoing evolutionary process. Few, if any, adherents of the first view would now be found insofar as the long-wavelength anomalies of the Earth is concerned. Nevertheless, many of the short-wavelength features found over the continents, such as seen in Figure G28, may reflect ancient conditions. This inability to distinguish between the two extreme hypotheses from gravity observations alone emphasises the non-uniqueness of the interpretation of the field and the dependence on other observations and hypotheses. For the Moon, the gravity field does appear to largely reflect long-past processes. The Martian gravity field may be indicative of both the distant and geologically more recent past, whereas the long wavelength gravity field of Venus, like that of Earth, reflects largely active dynamics within the planet.

Kurt Lambeck

Bibliography

Balmino, G.B., Moynot, B. and Vales, N. (1982) Gravity field model of Mars in spherical harmonics up to degree and order eighteen. *J. Geophys. Res.*, **87**, 9735–46.
Gegout, P. and Cazenave, A. (1993) Temporal variations of the Earth gravity field for 1985–1989 derived from Lageos. *Geophys. J. Int.*, **114**, 347–59.
Heiskanen, W.A. and Vening Meinesz, F.A. (1958) *The Earth and its Gravity Field*. New York: McGraw-Hill.
Jeffreys, H. (1970) *The Earth*, 5th edn. Cambridge: Cambridge University Press.
Lambeck, K. (1988) *Geophysical Geodesy*, Oxford University Press.
Marsh, J.G., Lerch, F.J., Putney, B.H. *et al.* (1990) The GEM-T2 gravitational model. *J. Geophys. Res.*, **95**, 22043–71.
Phillips, R.J. and Lambeck, K. (1980) Gravity fields of the terrestrial planets. Long-wavelength anomalies and tectonics. *Rev. Geophys. Space Phys.*, **18**, 27–76.
Reasenberg, R.D. and Goldberg, Z.M. (1992) High-resolution gravity model of Venus. *J. Geophys. Res.*, **97**, 14681–90.

Cross references

Geoid
Mars: gravity
Moon: gravity
Planetary geodesy
Venus: gravity

GREENHOUSE EFFECT

A greenhouse typically has a roof and sometimes walls of glass that permit sunlight to enter readily into its interior. There the sunlight is absorbed, raising the temperature of the absorbing surfaces. Infrared radiation is emitted from these surfaces and, in the absence of the glass enclosure, an equilibrium temperature for the surfaces is reached where the rate of solar heating is exactly compensated by the rate of infrared cooling. Because glass is rather opaque to infrared radiation, however, it tends to absorb and reradiate most of the infrared flux. This reradiated flux serves as an additional radiant heat source, and the temperature of the greenhouse interior becomes elevated to a higher equilibrium value. This is the greenhouse effect.

Planetary atmospheres are also heated at depth by the greenhouse effect. A semi-analytic treatment is contained in the article on atmospheric thermal structure (q.v.). The gaseous atmosphere is rather transparent to solar radiation, permitting much of it to penetrate to the surface (for the terrestrial planets) or to great depth (for the giant planets), where it is absorbed. On the other hand, any given atmosphere tends to be considerably more opaque to infrared radiation. Much of the radiation emitted from the surface is absorbed and re-emitted by the atmosphere, and some of the re-emitted radiation serves in turn to backwarm the surface. Eventually the lower atmosphere is heated to temperatures higher than the equilibrium temperature of a black body exposed to the same solar flux.

Cloudy atmospheres can also be heated at depth by the greenhouse effect. In this case the clouds act as frosted panes in a greenhouse. Sunlight can still penetrate, though diffusion by multiple scattering rather than direct transmission is the transport mechanism. Much of the radiation is lost to space by reflection, diminishing the effect. This decrease of the greenhouse effect by clouds is compensated to a degree by the blanketing that clouds provide as a result of being partially opaque to infrared radiation.

Most of the blanketing, however, is provided by a few gases that are very opaque to infrared radiation over wide spectral intervals. The ν_2 CO_2 band at 15 μm is especially important for the terrestrial planets Mars, Earth and Venus. Pressure broadening of CO_2 lines and CO_2–CO_2 collision-induced absorption are also very important for Venus because of its high surface pressure, while water vapor lines in the far infrared are significant in the Earth's atmosphere. Collision-induced absorption by H_2–H_2 and H_2–He pairs is responsible for much of the infrared opacity in the tropospheres of the major planets Jupiter, Saturn, Uranus and Neptune, while the tropospheric opacity of Titan is provided by collision induced absorption from different N_2–CH_4–H_2 combinations. In Titan's case the abundance of molecular hydrogen is low, and an atmospheric window at 18.5 μm reduces the greenhouse effect considerably.

Robert E. Samuelson

Bibliography

McKay, C.P., Pollack, J.B. and Courtin, R. (1991) The greenhouse and antigreenhouse effects on Titan. *Science*, **253**, 1118–21.
Samuelson, R.E. (1967) Greenhouse effect in semi-infinite scattering atmospheres: application to Venus. *Astrophys. J.*, **147**, 782–98.

Cross references

Atmosphere
Atmospheric thermal structure
Earth: atmosphere
Radiative transfer
Venus: atmosphere

H

HADLEY CIRCULATION

Because of the approximately spherical shape of planets, sunlight falls more directly on some locations than others. Except for Uranus, the axial tilt of planets is small enough so that, averaged over one revolution about the Sun, more sunlight falls on the equator than on the poles. This differential heating creates an energy imbalance between equator and pole that must be redressed by the atmospheric general circulation.

One means for doing so is the Hadley circulation. The latitudinally varying insolation induces pressure gradients which cause air to drift poleward in the upper troposphere. Warm air rises in low latitudes and cold air sinks at higher latitudes; the circulation is thermodynamically direct and acts as a heat engine. An equatorward return flow at lower levels completes the circulation cell. The net effect of the vertical motions is to transfer heat upward. Also, since the poleward branch is potentially warmer (i.e. after accounting for adiabatic compression) than the return branch in convectively stable atmospheres, the Hadley cell effects a net poleward heat transport.

Because planets rotate, the Hadley cell also redistributes angular momentum. The Coriolis force acting on the upper level poleward branch deflects air in the direction of planetary rotation. This helps to produce a region of strong zonal (east–west) winds known as the jet stream near the Hadley cell's high-latitude boundary, because the air leaving the equatorial region is farther from the axis of rotation, and is thus rotating faster than the planet's surface at higher latitudes. The opposite occurs in the return branch: air moving equatorward is deflected opposite the direction of planetary rotation because it is moving away from the rotation axis. On Earth, this produces the northeast to southwest surface wind pattern in the tropics of the northern hemisphere known as the trade winds. The southern hemisphere component of the trade winds blows southeast to northwest instead. This explanation of the trade winds was advanced by Hadley in 1735.

The Hadley cell shifts with latitude according to the seasons. On Earth it extends from surface to tropopause, and from equator to about 30° latitude. The return branch sweeps up moisture evaporated from the ocean and converges it near the equator. The rising branch is thus a preferred location for cloud formation and intense convective precipitation known as the intertropical convergence zone. The latent heat released in this region is actually the direct driver of the terrestrial Hadley cell, more so than sunlight itself. The descending branch of the terrestrial Hadley circulation is a region of drying and suppression of clouds and rain. This is the location of most of the world's deserts.

On Mars radiative heating by raised dust may play a role analogous to that of latent heating on Earth in driving that planet's Hadley circulation. The Martian Hadley cell may also be strengthened by the freezing out of the carbon dioxide atmosphere near the winter pole, which creates a pressure deficit at high latitudes. In the absence of the thermal inertia provided by an ocean, seasonal shifts in the Martian Hadley cell are thought to be much more extreme than on Earth.

In general, the latitudinal extent of the Hadley cell increases as a planet's rotation rate decreases. On slowly rotating Venus, the cell may extend almost to the pole. Venus, in fact, may have two Hadley circulations, one near the surface and another within its thick upper level cloud layer, where most of the sunlight is absorbed. The Hadley cell has been implicated as part of the mechanism which gives rise to Venus' remarkable superrotation.

Anthony D. Del Genio

Bibliography

Del Genio, A.D. and Suozzo, R.S. (1987) A comparative study of rapidly and slowly rotating dynamical regimes in a terrestrial general circulation model. *J. Atmos. Sci.*, **44**, 973–86.
Hadley, G. (1735) Concerning the cause of the general trade-winds. *Phil. Trans. Roy. Soc. London*, **39**, 58–62.
Held, I.M. and Hou, A.Y. (1980) Nonlinear axially symmetric circulations in a nearly-inviscid atmosphere. *J. Atmos. Sci.*, **37**, 515–33.
Pollack, J.B. (1981) Atmospheres of the terrestrial planets, in *The New Solar System* (J.K. Beatty, B. O'Leary and A. Chaikin), Cambridge, MA: Sky Publishing Corp., pp. 57–70.

Cross references

Atmosphere
Coriolis force, geostrophic motion
Earth: atmosphere
Venus: atmosphere

HALE, GEORGE ELLERY (1868–1938), AND HALE CYCLE

Pioneer among American astrophysicists, the name of Hale is permanently linked with his discoveries of the magnetic characteristics of the 'double sunspot cycle' (c. 22–23 years), now known as the 'Hale cycle'.

Hale was born and raised in Chicago, where he built his own little laboratory with a home-made telescope, and by 1884 began spectroscopic experiments. Experimental from the start, he was paving the way to the growing field of astrophysics. Majoring in physics at MIT, he got the idea of a spectroheliograph for photographing solar prominences.

In 1892 he became associate professor of astrophysics at the University of Chicago, and in 1895 he helped found the *Astrophysical Journal*. At a meeting at the Yerkes Observatory in Wisconsin in

1899 a new society was founded (with Hale as a vice-president) that in 1914 evolved into the American Astronomical Society.

In 1904 the Carnegie Institution of Washington provided Hale with funds to build the Mt Wilson Solar Observatory near Pasadena, where eventually he helped start the California Institute of Technology. Here the studies included the spectral lines of sunspots which showed that the spots are cool areas, becoming strengthened just as they are in an electric arc. This was in contrast to some intuitive assumptions based on observations that the high sunspot activity was reflected generally by warm climates on planet Earth. In the vicinity of the sunspots, in the hydrogen flocculi, vortices were observed which led to Hale's idea that the broadening of the spectral lines might be caused by intense magnetic fields in the spots. A splitting of the lines was attributed to the Zeeman effect.

This observation of extraterrestrial magnetic fields was greeted by R.S. Woodward (then president of the Carnegie Institution) as 'the greatest advance . . . since Galileo's discovery of those blemishes on the sun'. It led to Hale's discovery of the 22-year cycle of magnetic reversal in the spots, his so-called 'polarity law' (Hale, 1924; Hale and Nicholson, 1925).

Concerned also with the distant universe, Hale pushed for the building of big telescopes, eventually the 100-inch (254-cm) giant at Mt Palomar. After 1969 the Mt Wilson and Palomar centers were united as the 'Hale Observatories'. As a devoted believer also in the universality of humankind, he became the first president of the International Union for Cooperation in Solar Research (at Oxford, in 1905), later to become the International Astronomical Union, incorporating among others the International Union for Determination of Time and Latitude. When the National Research Council was formed in Washington, DC, in 1916 Hale became its first president.

Solar magnetic reversal

In 1889 F.H. Bigelow, observing the corona during an eclipse, had suggested that the Sun might have a magnetic field. Hale confirmed it, but it was not until 1952 that the two Babcocks (H.D. and H.W.) developed the solar magnetograph and measured a magnetic field of about 2 gauss. Its polarity was then opposite to that of the Earth, but it was reversed (i.e. to the north) during the next sunspot cycle (1957.9–1968.7, maximum to maximum). For this work Horace W. Babcock was awarded the George Ellery Hale Prize in 1992. In 1961 Babcock developed a hypothetical model to explain the changes through time of the lines of magnetic force in the Sun's photosphere. The dipolar field, which to start with is essentially symmetrical in the two hemispheres, has a flux potential of 8×10^{21} maxwell. Due to differential rotation within the photosphere, the lines develop into spirals that wrap around each hemisphere five times over the course of 3 years. These lines, on approaching the lower latitudes, appear to form 'ropes' of local concentration, where magnetic buoyancy brings bipolar magnetic regions (BMRs) to the surface.

This model essentially explains empirical rules associated with sunspot behavior, notably Spörer's law and Maunder's 'butterfly diagram', the fact that in any one cycle the spots grow in size and number as they migrate equatorward; and when plotted on a plane symmetrically bisected by the equator the spots mimic the shape of a pair of butterfly wings (Schove, 1983).

As a practical matter, it has been found by several specialists that if the sunspot cycle is plotted as positive or negative departures (Wolf numbers above or below zero), the resultant curve (the Hale cycle) is not only more convenient (Fairbridge, 1961), but also brings out further attributes (Figure H1).

The behavior of the Hale cycle can also be studied indirectly by analyses of the geomagnetic field (which is modulated by the solar wind, the auroras (historically documented for over two millennia), and of the Earth's long-term climate fluctuations (traced over nine millennia through tree rings and in ^{14}C fluctuations). The 22-year cycle in terrestrial magnetism is recognized by Currie (1973), Kane (1976) and Feynman (1982a,b); in auroras (Silverman and Shapiro, 1983); in ^{10}Be ice core data (Attolini et al., 1988); in temperature records (Folland, 1983) and other climatic data (Schove, 1983; Willett, 1987). An 88-year quadruple Hale cycle in the geomagnetic data has been reported by Feynman and Fougere (1984). A possible linkage between a 22.8-year period in the Stockholm air temperature (since 1756) with the North Atlantic thermohaline circulation is

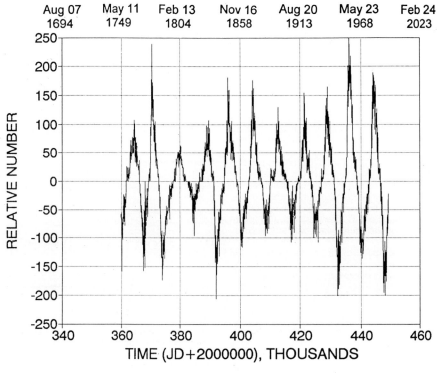

Figure H1 The Hale cycle, spanning the interval 1694 to 2023, but plotted sequentially in Julian days (JD) from monthly mean sunspot numbers (courtesy: John E. Sanders). Note that the two highest peaks in the positive series (1778 and 1957) are separated by 179 years, the sun's orbital precession (mean: 178.7338 years), but the negative series are less regular. For every 8 peaks of the 22-year cycle there are 9 epochs of the Saturn–Jupiter lap and 18 conjunction/oppositions of Jupiter. The timing of the Hale reversals is thus linked to planetary dynamics.

suggested by Moberg (1995), which itself may be a manifestation of the Hale cycle.

Rhodes W. Fairbridge

Bibliography

Attolini, M.R. et al. (1988) On the existence of the 11-year cycle in solar activity before the Maunder minimum. *J. Geophys. Res.*, **93**(A11), 12729–34.
Babcock, H.E. (1961) The topology of the Sun's magnetic field and 22-yr cycle. *Astrophys. J.*, **133**, 572–87.
Currie, R.G. (1973) Geomagnetic line spectra – 2 to 70yr. *Astrophys. Space Sci.*, **21**, 425–38.
Fairbridge, R.W. (1961) Solar variations, climatic change, and related geophysical problems. *New York Acad. Sci. Ann.*, **95** 1–740.
Feynman, J. (1982a) Geomagnetic and solar wind cycles. *J. Geophys. Res.*, **87**, 6153–62.
Feynman, J. (1982b) Solar cycle and long term changes in the solar wind. *Rev. Geophys. Space Phys.*, **21**, 338–48.
Feynman, J. and Fougere, P.F. (1984) Eighty-eight-year periodicity in solar–terrestrial phenomena confirmed. *J. Geophys. Res.*, **89**, 3023–7.
Folland, C.K. (1983) *Meteorol. Mag.*, **112**, 163–83.
Hale, E.G. (1924) Sunspots as magnets and the periodic reversal of their polarity. *Nature*, **113** (Suppl.), 105–12.
Hale, G.E. and Nicholson, S.B. (1925) Law of the sunspot polarity. *Astrophys. J.*, **62**, 270–300.
Hale, G.E. and Nicholson, S.G. (1938) *Magnetic Observations of Sunspots*. Washington: Carnegie Institute of Washington, 498 pp.
Kane, R.P. (1976) Geomagnetic field variations, *Space Sci. Rev.*, **18**, 413–540.
Kullmer, C.J. (1943) A remarkable reversal in the distribution of storm frequency in the United States in the double Hale solar cycles, of interest in long-range forecasting. *Smithsonian Misc. Collections*, **103**(60), 20 pp.
Moberg, A. (1995) Multi-taper spectral analysis of the Stockholm air temperature record: a significant 22.8-yr cyclic pulse observed. *J. Coastal Res.*, Special Issue 17, 39–44.
Schove, D.J. (ed.) (1983) *Sunspot Cycles*, (Benchmark Papers in Geology, Vol. 68). Stroudsburg: Hutchinson Ross, 393 pp.
Silverman, S.M. and Shapiro, R. (1983) Power spectrum analysis of auroral frequency, *J. Geophys. Res.*, **88**(A8), 6310–6.
Willett, H.C. (1951) Extrapolation of sunspot-climate relationships. *J. Meteorol.*, **8**(1), 1–6.
Willett, H.C. (1987) Climatic responses to variable solar activity – past, present, and predicted, in *Climate: History, Periodicity and Predictability* (eds M.R. Rampino et al.). New York: Van Nostrand Reinhold, pp. 404–414.
Wright, H. (1972) Hale, George Ellery. *Dict. Sci. Biogr.*, Vol. 6, pp. 26–34.
Wright, H., Warnow, J. and Weiner, C. (eds) (1972) *The Legacy of George Ellery Hale*. Cambridge, Mass.: MIT Press.

Acknowledgement

Manuscript kindly reviewed and improved by Horace W. Babcock.

Cross references

Solar activity
Sun

HALL, ASAPH (1829–1907)

Asaph Hall is remembered chiefly for his discovery of the two satellites of Mars, Phobos and Deimos, in 1877, but he made many other contributions to astronomy; for example, he made accurate observations of double stars, he discovered a bright white spot on Saturn in 1876, and gave a value of 8.81" for the solar parallax following studies completed in 1891.

He was born at Goshen, Connecticut, on 15 October 1829, and in 1856 accepted a junior post at the Harvard College Observatory at Cambridge, Mass. His ability was soon recognized, and in 1862 he became an assistant astronomer at the United States Naval Observatory in Washington; within a year of his arrival he was given a professorship. He was particularly interested in planetary work, and his observations of the white spot on Saturn allowed him to derive a rotation period of 10 h 14 min 24 s. His discovery of the Martian satellites was the result of a deliberate search with the 26-inch Washington refractor. He accepted the names suggested by an English schoolmaster named Madan. When the satellites were surveyed by space probes almost a century later the main crater on Phobos was named Stickney in honour of Hall's wife, née Chloe Stickney, who had persuaded him to carry on the search when he had been on the point of giving up.

Patrick Moore

Bibliography

Clerke, A.M. (1902) *History of Astronomy in the Nineteenth Century*. London: A.&C. Black, p. 419.
Gingerich, O. (1972) Hall, Asaph. *Dict. Sci. Biogr.*, Vol. 6, pp. 48–50.
Hall, A. (1876) The rotation of Saturn. *Mon. Not. Roy. Astron. Soc.*, **54**, 297.
Hall, A. (1877) The satellites of Mars. *Mon. Not. Roy. Astron. Soc.*, **28**, 206. See also Shapley, E. and Howarth, H.E. (1929) *A Source Book of Astronomy*. New York: McGraw-Hill, p. 320–3.
Hall, A. (1891) The Solar parallax. *Astron. J.*, Nos. 169, 170.

HALLEY, EDMOND (1656–1742), AND HALLEY'S COMET

Distinguished British scientist, probably second only to Newton in his breadth and achievements, Sir Edmond Halley began studying astronomy at St Paul's School and was later at Queen's College, Oxford. Halley was elected to the Royal Society of London at the age of 22. He was appointed Savilian Professor of Geometry at Oxford in 1702, and in 1720 he became Astronomer Royal. Halley had earlier been denied an Oxford professorship in astronomy on the grounds of alleged heresy. His first paper, on planetary orbits, was published at the age of 20; in the same year he made trip to observe the transit of Mercury from the island of St Helena, in the southern hemisphere. He was also interested in gravity and made many pendulum observations, recognizing the difference in gravity between St Helena and London, which was critically important for the pendulum clock, at that time an essential instrument in the determination of longitude.

Before he left school Halley was making determinations of the Earth's magnetic declination, and thus followed, apparently unknowingly, in the footsteps of the Chinese astronomer I-Hsing of around 720 AD [according to Needham's (1962) *Science and Civilization in China*, Vol. 4, Part 1]. The Earth's magnetic variations, regional and secular, were of great interest for navigators and widely recorded in ships' logs. Later, after 2 years of cruises (as a naval captain and chief scientist of HMS *Paramore*) to both the North and South Atlantic, Halley proposed the theory that the dipole field originated in the Earth's core (see Gilbert, William), believing that the secular westward drift could only be explained if that core were decoupled in some way from the outer shell of the Earth. He concluded 'the whole globe of the Earth is one great magnet,' but it was dynamic and quadrupolar; and thus not in the form of a permanently magnetized lodestone, symmetrical to the two poles, as visualized by Gilbert. For many years the isogonic lines (of equal magnetism) were known as 'Halleyan lines'.

Following his return to Britain, Halley became friendly with Isaac Newton, and eventually played a vital role in persuading Newton to finish his *Principia*, picking up the bill for its publication (1687). Using Newton's gravitational formulae he calculated the orbit of a comet observed in 1682, predicting its return in 1758, 75 years later. From its regular recurrence it has become known as Halley's comet.

Halley's trip to the South Atlantic, and other travels, stimulated an interest in the wind systems on planet Earth. He perceived that the general circulation of the atmosphere must be related to the distribution of the Sun's incoming radiation over the surface of the globe. In 1686 he published a description and chart of the trade winds and monsoons, observing the global wheel-round from westerlies to the trade-wind easterlies at around 30° latitude in both hemispheres. On the 1700 revision of his global chart he added also the lines of equal variation of the magnetic field.

Halley's interest in magnetism received a dramatic 'nudge' when on 6 March 1716 he observed a brilliant auroral display over London, and collating his impressions with reports from around the country, he was able to postulate a mechanism linking the optical display with the Earth's magnetic field through charged particles. Unfortunately unaware of the solar wind (q.v.), he had the particles emerge from the Earth.

He studied the Arabic language so that he could translate classical Middle Eastern astronomical texts into Latin (the only language for astronomy at that time). Halley also became interested in the 18.6-year nodal regression of the Moon, tracing its course through one complete cycle (1720–1738). A small secular acceleration of the Moon over a long period of time was suspected, but its origin was not identified. His final scientific paper (on a lunar eclipse) was in 1737, at the age of 81. Halley also made important contributions to the science of hydrology (Biswas, 1970).

In 1718 Halley discovered that the stars are not 'fixed' on the celestial sphere but move with their own proper motion. He did this by comparing his observations with those of the ancient Greeks. Halley noted that in the ~ 2000 years intervening, the positions of a few of the brightest stars had changed relative to others nearby, even after the motion due to the Earth's precession had been removed.

Halley further developed the idea of James Gregory (a Scottish mathematician who proposed using transits of Mercury and Venus to measure the astronomical unit) to show that Venus would provide a more accurate determination of the AU than Mercury. This led to an expedition of astronomers in 1761 (posthumously) to observe the Venus transit.

With his innumerable activities, Halley never did quite find time to finish his Oxford degree, but it was conferred nevertheless in 1678, by order of the King, Charles II, who was interested himself in science and gave the Royal Society its charter. Halley's last act, sitting in his arm chair, was to call for a glass of wine and, having drunk it, died peacefully.

Halley's comet

Comet Halley is the most famous of comets and is considered to be the touchstone for modern comet studies. It has been observed on 29 out of its last 30 returns to the inner solar system. With an orbital period slightly longer than 76 years, this record stretches back to 240 BC, when the Chinese made the first substantiated sighting. Throughout the intervening centuries, P/Halley (P/ indicates a short-period comet with period less than 200 years) has been seen by the Babylonians, the Romans, medieval Europeans, ancient Japanese and Koreans, Renaissance Europeans and many others up to its last apparition in 1985. It has been immortalized in the Bayeux Tapestry and by the Florentine painter, Giotto di Bondone, who included the 1301 apparition as the Star of Bethlehem in his fresco, *The Adoration of the Magi*. If Comet Halley kept a scrapbook, what magnificent snapshots of humanity it would contain!

With the coming of the space age, new vistas of cometary research were opened, overcoming the many limitations of ground-based observations. Investigations are no longer restricted to remotely sensed data; instruments can now be sent to comets directly, to gather *in situ* measurements. From the 6th to 25th March 1986, an international armada of six robotic spacecraft (in the order of flyby times: Vega 1, Suisei, Vega 2, Sakigake, Giotto and ICE) encountered and explored P/Halley on its return to the inner solar system. The spacecraft Giotto (named after Giotto di Bondone by the European Space Agency), passed closest to the nucleus at a distance of about 600 km. Traveling at 68 km s^{-1}, it sped through the coma to encounter only about 10^{14} molecules m^{-3}, less than 10^{-11} of the density of air at sea level! In spite of minor damages due to these 'shots through a void,' the investigations were spectacularly successful (Plate 20).

The Giotto and the two Vega (derived from Venus–Halley, using the Russian spelling of Halley) spacecraft investigated the nucleus (by remote sensing) and the gas and dust environment on the subsolar side of the coma. In an excellent international collaboration,

Figure H2 Comet Halley and meteor. This image of Halley's comet was obtained by observer Eleanor Helin at Palomar Observatory, California, on 7 January 1987. The comet was approaching perihelion when this image was obtained. The white streak is the path of a meteor entering the upper atmosphere of the Earth, probably at about 120 km above the surface. (JPL P-32348A.)

the two Vega spacecraft acted as pathfinders for the Giotto spacecraft. Most of the global physicochemical and plasma properties predicted by theory, such as ion pick-up by the solar wind (mass loading, deceleration, stagnation flow and ion pile-up) and the contact surface, were found to be in good agreement with observations. However, our knowledge of the dust composition and size distribution was poor compared to observations. In addition to silicate particles, it was found that a significant fraction of cometary dust is rich in organic molecules composed of hydrogen, carbon, nitrogen, oxygen and sulfur, and that there is an excess of particles with masses less than 10^{-17} kg. No detailed model existed for the shape of the nucleus. As expected, the images of the nucleus brought many surprises: the nucleus is much bigger than anticipated, very dark with a geometric albedo of about 3–4%, and deviates significantly from a spherical shape by being about twice as long as it is wide or thick, approximately $16 \times 8 \times 7$ km. When the much larger volume of the nucleus is combined with the observed nongravitational acceleration and the associated reaction force from the asymmetric vaporization of the frozen gases, it is found that the density of the nucleus is much smaller than estimates had indicated before the Halley encounters, in the range 100–1000 kg m^{-3}. The theory of the coma–nucleus interface, although not correct in the global details, was found to be in agreement for the active areas on the surface. About 90% of the coma gas and dust are emitted in jet-like features from a few compact, active source regions that comprised only about 20% of the illuminated surface area at the time of the Giotto encounter. In is now clear that the volatile composition is dominated by water (approximately 85% by number abundance). The measured dust-to-gas mass ratio of approximately two in the coma, when extrapolated back to the nucleus, strongly suggests that the nucleus is better described as a 'snowy dirtball' in contrast to the 'dirty snowball' originally proposed by Whipple. Several of the minor gaseous species have their origin, at least in part, in distributed sources linked to the evaporation of dust. Entirely unpredicted were the suprathermal ions with energies up to 10 keV and the heavy ions with masses up to about 120 amu that appear to be associated with the organic component of the dust.

Our understanding of comets has been dramatically enriched by the brief glimpses of comet Halley obtained from spacecraft flybys lasting only minutes to hours in the inner region of its coma. During the years following these encounters, our general concept of a comet has been substantially changed using comet Halley as the prototype. However, many interesting questions remain as justification for future missions to comets and, perhaps, to revisit P/Halley before its return in 2061!

Daniel C. Boice and Rhodes W. Fairbridge

Bibliography

Armitage, A.(1966) *Edmond Halley*, British Men of Science series. London: Nelson, 220 pp.
Biswas, A.K. (1970) Edmond Halley, F.R.S., hydrologist extraordinary. *Notes Rec. Roy. Soc. London*, **25**, 47.
Burstyn, H.L. (1966) Early explorations of the Earth's rotation in the circulation of the atmosphere and the ocean. *Isis*, 167–87.
Chapman, S. (1943) Edmond Halley and geomagnetism (Halley Lecture). *Nature*, **152**, 231–47. (Also publ. 1943, *Terr. Mag.*, **48**, 131–44).
Evans, M.E. (1988) Edmond Halley, geophysicist. *Physics Today*, 41, 41–5.
Grewing, M., Praderie, F. and Reinhard R. (eds) (1987) *Exploration of Halley's Comet*, Berlin Springer-Verlag.
Halley, E. (1683) A theory of the variation of the magnetical compass. *Phil. Trans. Roy. Soc. London*, **13**, 208–21.
Halley, E. (1692) An account of the cause of the change of the variation of the magnetical needle with a hypothesis of the structure of the internal parts of the Earth. *Phil. Trans. Roy. Soc. London*, **17**, 563–78.
Halley, E. (1714/16) A short account of the . . . saltiness of the ocean . . . with a proposal to discover the age of the world. *Phil. Trans. Roy. Soc. London*, **29**, 297–300.
Keller, H.U., Arpigny, C.B., Bonnet, C.M. *et al.* (1986) First Halley Multicolour Camera imaging results from Giotto, *Nature*, **321**, 320–6.
MacPike, E.F. (1932) *Correspondence and Papers of Edmond Halley*. Oxford: Oxford University Press.
Newburn R.L., Jr, Neugebauer, M. and Rahe, J. (eds) (1991) *Comets in the Post-Halley Era*, Dordretch: Kluwer Academic Publishing.
Rickman, H. (1989) The nucleus of comet Halley: surface structure, mean density, gas and dust production. *Adv. Space Sci.*, **9**, 59–71.
Ronan, C.A. (1970) *Edmond Halley, Genius in Eclipse*. London: Macdonald, 251 pp.
Ronan, C.A. (1972) Halley, Edmond. *Dict. Sci. Biogr.*, Vol. 6, 67–72.
Schaffer, S. (1977) Halley's atheism and the end of the world. *Notes Rec. Roy. Soc. London*, **32**, 17–40.
Singer, C. (1959) *A History of Scientific Ideas*. Oxford: Oxford University Press (and 1990, New York: Dorset Press).
Spencer Jones, H. (1957) Halley as an astronomer. *Notes Rec. Roy. Soc. London*, **12**, 175–192.
Williams, H.S. (1932) *The Great Astronomers*. New York: Newton Publishing Co., 618 pp. Chapter 15: 'Edmond Halley – unmasking the comet' (pp. 202–222).

Cross references

Comet
Comet: observations
Giotto mission
Sakigake and Suiset missions
Vega missions

HELIOSPHERE

The heliosphere can be broadly defined as the region of space in which the pressure of the solar wind exceeds the pressure of the streaming interstellar wind. The supersonic solar wind, consisting mainly of protons, electrons and alpha particles, radially expands in all directions, carrying with it the solar magnetic field. This wind creates a gigantic bubble (the heliosphere) in the interstellar medium. Processes within the heliosphere modulate the inward travel of cosmic rays originating outside the solar system, and thus investigations of cosmic rays may be employed to obtain information about the heliosphere. The size of the heliosphere is not presently known with certainty, but it extends to 100–150 AU, well beyond the orbit of Pluto.

Knowledge of the heliosphere has grown during the past three decades thanks in part to studies of cosmic rays over space and time. A major role has been played by four spacecraft, Pioneer 10 and 11 and Voyager 1 and 2. Thus far the heliosphere has been explored out to a radial distance of >45 AU in the ecliptic plane. Pioneer 10 is traveling along the (extended) heliomagnetic tail, almost parallel to the inferred direction of the interstellar wind and almost directly away from the other three spacecraft; the nuclear power of Pioneer 10 will probably be exhausted by the time it makes its exit from the heliosphere. Voyager 1 is proceeding almost directly into the interstellar wind at a latitude of 35°N, and is expected to encounter the heliospheric boundary (the 'heliopause') first. Pioneer 11, launched earlier than Voyager 1, is also traveling upstream, along a trajectory near the ecliptic plane. Voyager 2 has a longer path to reach the heliopause; its power supply will also cease to be operative before it reaches the boundary. It is traveling toward heliolatitude 48°S following its encounter with the planet Neptune in August 1989. The locations of the four spacecraft in the year 1990 and in the year 2000 are summarized in Table H1 (Parthasarathy and King, 1991).

Table H1 Spacecraft positions in the years 1990 and 2000

Year	1990		2000	
Spacecraft	Distance (AU)	Heliolatitude	Distance (AU)	Heliolatitude
Pioneer 10	48	3°N	74	3°N
Pioneer 11	30	17°N	55	17°N
Voyager 1	40	31°N	76	33°N
Voyager 2	31	2°N	60	21°N

It is expected that Voyager 1, if it survives until it crosses the heliopause, will be the first spacecraft to completely leave the solar system, pass into the interstellar medium and transmit back, for the first time, data describing conditions there.

Size and morphology of the heliosphere

Although the Earth and all the other planets orbit within the heliosphere, and many satellites have been launched around the Earth, we are as yet uncertain about the size and overall morphology of the heliosphere. It was at first thought that the heliosphere was spherical in shape. However, by study of the motion of nearby stars, we have found that the local interstellar medium flows past the Sun (from the general direction of the galactic center) with a speed of about 26 km s$^-$. This discovery changed the picture considerably. The density of the interstellar medium (q.v.) is rather low, but its speed relative to that of the heliosphere is sufficiently high to generate a bow shock on the upstream side. Also, the interstellar wind will create an elongated cavity or heliomagnetic tail in its flow direction. Thus the heliosphere has some similarities to the magnetosphere of the Earth.

Estimates of the size of the heliosphere have progressed upwards with time, especially since the 1960s and 1970s. The scale of the modulation region of cosmic rays around the Sun was initially considered to be about 2–5 AU in radius. At that time the interplanetary radial gradient of cosmic rays had been estimated to be about 50%, AU^{-1} (derived from spacecraft data of Mariner 2). In 1973 new data became available from Pioneer 10. Van Allen reported measurements consistent with a zero gradient per AU, but a few months later McDonald et al. (1974) came up with a gradient of only a few per cent per AU out to about 3 AU. This immediately presented a new perspective of a much larger heliosphere, stretching out to 10–20 AU. By 1977, with Pioneer 10 and 11, Voyager 1 and 2 and the Earth-orbiting satellites IMP7 and IMP8, a truly interplanetary network of monitors was in place. The new results from these probes raised estimates of the size of the heliosphere to a radial distance of 50–100 AU (Webber, 1987).

The question of the size of the heliosphere is intimately related to the question of the position and nature of the heliopause and an associated internal shock that arises to slow the solar wind flow before it collides with the interstellar flow. The existence of a standing shock prior to the heliopause was proposed three decades ago (Parker, 1961). On the upwind side after crossing the shock, the streamlines curve away from the original radial direction. This geometry is essentially caused by the nose-to-tail pressure gradient arising due to the motion of the heliosphere through the interstellar medium. The region between the terminal shock and the heliopause is called the heliosheath. In the direction opposite to the upwind direction the solar wind flows in the same direction as the interstellar flow, down what is known as the 'heliotail'. At the heliopause, which divides the solar wind plasma from the interstellar plasma, magnetic field reconnection may occur.

We may note here two estimates of the size of the heliosphere drawn from different sources. The first of these makes use of the known solar cycle variation of cosmic ray intensity. Forbush (1954) discovered a broadly negative correlation between solar activity and cosmic ray intensity (see Forbush effect), where solar activity was represented by sunspot numbers. The cosmic ray intensity minimum is delayed in time from the occurrence of solar activity maximum by about 9–15 months. Could this represent a time constant for the heliosphere? Since the solar wind takes about 4 days to traverse a distance of 1 AU, the distance corresponding to the interval noted above would correspond to 70–110 AU. The variation of several months may indicate that the position of the heliopause varies with time.

Such a phenomenon is included in a model presented by Van Allen. In this model the size of the heliosphere pulsates in a period that is negatively correlated with the sunspot cycle. At sunspot maximum the distance to the heliopause is about 48 AU, while at sunspot minimum the distance is estimated to be about 83 AU. Thus, somewhat unexpectedly, the heliosphere may decrease in size during increasing solar activity. During solar maximum it can be argued that there are a number of fast solar wind streams interacting with each other and with the slower streams, thus reducing the size of the interaction region. During the solar minimum, however, there are fewer streams. Thus the streams may extend over a larger distance before they interact with the background of slower solar wind. The modulation region may become more extended, and the region of cosmic ray modulation (heliopause) may become farther away.

More recently Kurth and Gurnett (1991) have discussed a plasma wave emission in the Voyager 1 and 2 data that they have interpreted as arising from the distant heliopause. In 1993 Gurnett et al. described radio emissions thought to emanate from the heliopause; these were registered by detectors on both Voyager 1 and 2. They estimated that the distance of the heliopause is between 116 and 177 AU from the Sun.

Present and future investigations

The recently launched Ulysses spacecraft (see Ulysses mission) enters a new region of the heliosphere as it leaves the neighborhood of the ecliptic plane and crosses over the poles of the Sun. It will provide new information about the morphology of the solar magnetic field, and will measure the composition, density and temperature of the ionized plasma, and various wave species.

In order to ensure that the heliopause is reached while spacecraft systems are still functioning, a spacecraft will have to travel much faster than our present probes. As presently conceived, such an interstellar probe would first travel to Jupiter, using that planet's gravity to assist its journey towards the Sun. The high-velocity trajectory would pass about 2×10^6 km (roughly three solar radii) above the solar surface. The probe would then be accelerated to an escape velocity sufficiently high to take it across Neptune's orbit within 2 years. Although such a flight would not be instantaneous, it would offer an excellent opportunity to cross the heliosphere in a small fraction of the solar cycle. If this probe were launched in the year 2000, it could reach a distance of 140 AU in 5 to 6 years, possibly encountering the heliopause in that time.

D. Venkatesan

Bibliography

Forbush, S.E. (1954) Worldwide cosmic ray variations 1937–52. *J. Geophys. Res.*, **59**, 525–542.

Gurnett, D.A., Kurth, W.S., Allendorf, S.C. and Poynter, R.L. (1993) Radio emission from the heliopause triggered by an interplanetary stock. *Science*, **262**, 199–203.

Kurth, W.S and Gurnett, D.A. (1991) Now observations of the low frequency interplantary radio emission. *Geophys. Res. Lett.*, **18**, 1801.

MacDonald, F.B., Teegarden, B.J., Trainor, J.H. and Webber, W.R. (1974) The anomalous abundance of cosmic ray nitrogen and oxygen nuclei at low energies, *Ap. J.*, **187**, L105.

Parker, E.N. (1961), *Astrophys. J.*, **134**, 20.

Parthasarathy, R. and King, J.H. (1991) Trajectories of Inner and Outer Heliospheric Spacecraft, NSSDC/WDC A.R. & S., 91–98.

Smith, E.J., Page D.E. and Wentzel, K.P. (1991) Ulysses: a journey above the Sun's poles. *EOS*, **72**, 241.

Venkatesan, D. (1985) Cosmic Ray Picture of the Heliosphere, Johns Hopkins University/Applied Physics Lab Technical Digest, Vol. 1.

Venkatesan, D. and Krimigis, S.M. (1990) Probing the helio-Magnetosphere, *EOS*, **71**, No. 44, 1755.

Webber, W.R. (1987) *Cosmic Rays in the Heliosphere from Essays in Space Science*. NASA Conference Publication 2464, p. 125.

Cross references

Interstellar medium
Pioneer 10 and 11 missions
Plasma wave
Solar wind
Voyager missions

HERSCHEL, WILLIAM (1738–1822)

One of England's greatest astronomers, Herschel was the discoverer of the planet Uranus in 1781, which earned him an FRS, and the highest award, the Copley Medal, of the Royal Society. He was awarded a personal stipend from King George III, who was greatly interested in astronomy. Herschel wanted to call the new planet *Georgium Sinus*, but Bode recommended the classical name and that became accepted. The minor planets also attracted his attention; he proposed the name *asteroid* for them, and with his new (40 ft) telescope measured some of their diameters. He may have been the first scientist to propose (1801) a correlation between the sunspot cycle and the climate of planet Earth.

Herschel had been born in Germany at Hanover, son of an oboist in a regimental band, and he too became an army musician. However, on a trip with the band to England he decided to emigrate, and in 1757 returned to London. As a musician he was very successful, also taking instruction in his spare time in mathematics and astronomy. He began to design and manufacture telescopes, eventually becoming quite wealthy, and able to devote himself entirely to astronomy. Sir William was greatly assisted by his sister, Caroline (1750–1848), who joined him from Hanover in 1772 and herself became quite a celebrated discoverer of comets. She also greatly expanded Flamsteed's catalog of stars. William's son, Sir John Herschel (1782–1871) was also a brilliant mathematician and distinguished astronomer, becoming especially interested in double stars and nebulae. In 1833, wanting to study the southern hemisphere sky, he moved to South Africa for 3 years and was able to study a return of Halley's comet.

After studying the evidence of solar activity, as suggested by the variations in sunspot numbers (observed from 1650–1800), Herschel (1801) proposed a climatic response which could be monitored by means of economic statistics. Thus a high-activity phase was matched by high crop abundance and low prices; during low or absent spot phases wheat was scarce and prices high. During the hemicycle 1798 to 1804, the mean sunspot number rose from 4 to 47 and the rise was paralleled by a dramatic improvement in British agriculture. Unfortunately, although followed up by the distinguished economist W. Stanley Jevons in 1875 (and subsequently many others), for modern stock market analysts the long-term results are not so consistent or simple. Nevertheless, Herschel's observations were 43 years ahead of Schwabe's observation of the sunspot cycle in 1844. The younger Herschel (John) proposed the electromagnetic nature of sunspots in 1833, the proof of which had to await Hale's work in 1908 (see Hale, George Ellery, and Hale Cycle).

Sir William played a key role in the foundation of the Royal Astronomical Society in 1820 and was its first president. Its seal depicts Herschel's 40-foot (12-m) telescope. The society's rooms and library are located in Burlington House, Piccadilly, a home it shares with the Geological Society.

Rhodes W. Fairbridge

Bibliography

Armitage, A. (1982) *William Herschel*. London.
Dreyer, J.L.E. (ed.) (1912) *The Scientific Papers of Sir William Herschel*. London (2 vols).
Herschel, W. (1801) Observations of the Sun. *Phil. Trans. Roy. Soc. London*.
Herschel, J.F.W. (1867) *Outlines of Astronomy*. London: Longmans Green, p. 261.
Hoskin, M.A. (1963) *William Herschel and the Construction of the Heavens*. London.
Hoskin, M.A. (1972) Herschel, William. *Dict. Sci. Biogr.*, Vol. 6, pp. 328–36.
Lubbock, C.A. (1933) *The Herschel Chronicle*. Cambridge.
Sidgwick, J.B. (1953) *William Herschel*. London.

HILDA ASTEROIDS

The Hildas are a group of approximately 30 asteroids located beyond the main asteroid belt, with a mean orbital semimajor axis of 4.0 AU. The asteroids' mean motions are in a 3 : 2 resonance with Jupiter. This resonance is stabilized by Jupiter's gravitation; neither close approaches nor collisions with Jupiter can occur (Nobili, 1989). Although chaotic motion could occur for asteroids near the 3 : 2 commensurability, the natural Hildas avoid that domain of phase space (Schubart, 1991).

153 Hilda, the namesake of the group and the first to be discovered, is also the largest with a diameter of 175 km. Other known members range in size from 50 to 170 km. Colorimetry, spectroscopy and radiometry of the Hildas indicate that they are composed of dark primitive material. No absorption feature due to hydrated silicates is seen in the spectra of Hildas, although such features are common in the spectra of C asteroids, found slightly closer to the Sun, and in the spectra of carbonaceous chondrite meteorites (q.v.). Jones *et al.* (1990) suggest that asteroids such as the Hildas are composed of substantial amounts of water ice mixed with rock. The surface layers would quickly lose their ice to sublimation or gentle solar heating, while the interiors have never achieved temperatures high enough to drive liquid or gaseous water to the surface. Bell (1986) proposes that the Hilda and Trojan asteroids are the true 'primitive' asteroid classes among the asteroids. The current model of distant asteroids as 'dirty snowballs' containing rock with substantial ice blurs the distinction between asteroids and comets. This is in keeping with the 1989 discovery of a coma around 'asteroid' 2060 Chiron (q.v.) as it currently approaches perihelion.

The variation of an asteroid's light with time is called the lightcurve, and it gives information on the rotational period and the shape of the asteroid. For a triaxial ellipsoid with axes $a : b : c$ $(a > b > c)$ rotating about the c-axis with the angular momentum vector perpendicular to the line of sight, the lightcurve amplitude gives a lower limit to the axis ratio a/b. Since the orientation of the rotation vector is not known a priori, however, further conclusions about an asteroid's shape cannot be drawn from a single lightcurve observation. The motivation for lightcurve studies of distant asteroids such as the Hildas comes from their dynamical and spatial isolation from most interactions with main belt objects. Differences between these groups of objects could reveal important clues about either their formation processes or their subsequent collisional evolution (French *et al.*, 1989). Preliminary studies of Hilda lightcurves suggested that Hildas might be more elongated, on average, than main belt asteroids. Lightcurves have been determined for 14 Hildas, summarized by Binzel and Sauter (1992). These authors conclude that the distribution of lightcurve amplitudes for Hildas is identical with that of main belt asteroids, consistent with a collisional history similar to that of the main belt.

Linda M. French

Bibliography

Bell, J.F. (1986) Mineralogical evolution of the asteroid belt, in *Proc. Lunar Sci. Conf.*, **17**, pp. 985–6.
Binzel, R.P. and Sauter, L. (1992) Trojan, Hilda, and Cybele asteroids: new lightcurve observations and analysis. *Icarus*, **95**, 222–38.
French, L.M., Vilas, F., Hartmann, W.K. and Tholen, D.J. (1989) Distant asteroids and Chiron, in *Asteroids II* (ed. R.P. Binzel, T. Gehrels and M.S. Matthews). Tucson: University of Arizona Press, pp. 468–86.
Jones, T.D., Lebofsky, L.A., Lewis, J.S. and Marley, M.S. (1990) The composition and origin of the C, P, and D asteroids: water as a tracer of evolution in the outer belt. *Icarus*, **88**, 172–92.
Nobili, A. (1989) The dynamics of the outer asteroid belt, in *Asteroids II* (ed. R.P. Binzel, T. Gehrels, and M.S. Matthews). Tucson: University of Arizona Press, pp. 862–879.
Schubart, J. (1991) Additional results on orbits of Hilda-type asteroids. *Astron. Astrophys.*, **241**, 297–302.

Cross references

Asteroid
Asteroid: lightcurve
Asteroid: resonance
Resonance in the solar system

HIPPARCHUS OF NICAEA (190–120 BC)

Often referred to as the founder of systematic observational astronomy and probably the greatest astronomer of Ancient Greece, Hipparchus is believed to have been born at Nicaea in Bithynia (NW Turkey); his development of the mathematics of trigonometry made it possible to quantify the simple geometry of the day and to apply this to determine the motions of celestial objects that moved too slowly to be resolved within an observer's lifetime. By collecting and analyzing the existing observational data he was able to establish the precession of the equinoxes (q.v.) He also confirmed the value of the obliquity of the ecliptic (q.v.), previously measured by Eratosthenes (q.v.). Thus he established two of the three parameters of the 'Milankovitch theory' that underlie the modern orbital theory of ice ages. His cosmology, however, was geocentric (see Berry, 1898, p. 41).

His observational work was mainly carried out at an observatory he built on the island of Rhodes. There he also completed a star catalog of 1080 stars, identified by celestial latitude, longitude and magnitude, and embracing also the names of the constellations assembled by Eudoxus, a catalog that was to remain a standard for the next 1½ millennia. His calculations of the length of the tropical year (365 d 5 h 55 min) and the sidereal year (365 d 6 h 10 min) were remarkably precise.

Rhodes W. Fairbridge

Bibliography

Berry, A. (1898) *A Short History of Astronomy*. London: Murray (Dover reprint, 1961), 440 pp.

Neugebauer, O. (1975) *A History of Ancient Mathematical Astronomy*. New York: 3 vols.

Toomer, G.J. (1978) Hipparchus. *Dict. Sci. Biogr.*, Vol. 15 (Suppl. 1), pp. 207–24.

HISTORY OF PLANETARY SCIENCE I: PRE-SPACE AGE

The story of planetary science can be divided into three fairly well-defined periods. Before 1610 all that could be done was to study the planetary movements; without optical aid it was impossible to learn anything about their physical characteristics. Between 1610 and 1962 equipment of all sorts could be used, and a great deal was found out, though it is true that some of our theories proposed as recently as the 1950s have proved to be very wide of the mark. Then, in 1962, came the first successful planetary probe – America's Mariner 2, to Venus – and since then we have depended very largely upon the results provided by the spacecraft. Only Pluto of the nine main planets has yet to be bypassed at close range (and there is, moreover, a serious doubt as to whether Pluto is worthy of true planetary status).

The difference between the 'fixed' stars and the 'wandering' stars, or planets, was recognized in very early times, and all the main civilizations of ancient times made careful studies of the motions of the naked-eye planets Mercury, Venus, Mars, Jupiter and Saturn. The Greeks were good observers, even though most of the philosophers continued to believe in a central Earth and that all orbits must be circular – simply because the circle is the 'perfect' form, and nothing short of perfection can be allowed in the heavens. The geocentric system was brought to its final form by Ptolemy (Claudius Ptolemæus), who lived from around 120 to 180 AD. In order to explain the somewhat complicated movements of the planets he was forced to adopt a cumbersome arrangement, in which a planet moved in a small circle or epicycle, the centre of which – the deferent – itself moved round the Earth in a perfect circle.

There was, of course, another reason for making careful studies of the planetary motions: astrology. It was only in the year 570 that Isidorus, Bishop of Seville, drew a definite distinction between astrology and astronomy, but astrologers continued to be powerful figures right up to the time of Newton. The revival of astronomy after the Dark Ages came with the Arabs, and in 813 Al-Ma'mun founded the Baghdad school, but it is fair to say that astrological considerations were paramount. Nonetheless, the work accomplished was very skilful indeed, and the Alphonsine Tables of planetary motion, published in 1270 by order of Alphonso X of Castile, represented a major advance. The stage was being set for a major revolution in outlook.

In 1543 Nicolas Copernicus published his book *De Revolutionibus Orbium Cælestium*, in which he revived an old theory according to which the Sun, not the Earth, was the center of the planetary system. The idea dated back to one of the most enlightened of the Greeks, Aristarchus, who lived around 270 BC, but it had been virtually forgotten. Copernicus made many mistakes – in particular he retained the concept of perfectly circular orbits, and he even brought back Ptolemy's epicycles – but he had taken the fundamental step. The story of the ensuing conflict has been told many times, but it came to a final conclusion in 1687 with the publication of Newton's immortal *Principia*. Earlier in the century, Johannes Kepler had used the detailed observations of Mars made by the Danish astronomer Tycho Brahe to prove that the orbits of the planets are not circular, but are elliptical.

One of the major protagonists in the story of the great revolution was Galileo Galilei, who was also one of the first to turn the newly invented telescope toward the skies. Bearing in mind the low power of even his largest telescope, which magnified a mere 30 times, he made some remarkable discoveries. He saw the mountains and craters of the Moon, and even made estimates of the heights of the lunar peaks which were of the right order. He observed the phases of Venus, and found that it showed a full range from crescent to full – which in the Ptolemaic system it could never do. He detected the phases of Mars, and found that there was something very unusual about the shape of Saturn, though he could not make out just what it was. Most important of all, he observed the four bright satellites of Jupiter. He was not the first to see them – they had been seen slightly earlier by Simon Marius, and for that matter it seems that the largest of them, Ganymede, had been recorded with the naked eye by Gan De, in China, as long ago as 364 BC – but he was the first to make systematic observations of them, which is why they are known collectively as the Galileans. It was not until our own century that astronomers started to use the four names given by Marius: Io, Europa, Ganymede and Callisto.

The important thing about the Galileans was that they moved round Jupiter, and this showed that there was more than one center of motion in the solar system. We now know that during the course of his observations Galileo also recorded the planet Neptune, which was close to Jupiter in the sky, though naturally he had no idea that it was anything but a star.

This may be the moment to say something about the transits of the two inferior planets, Mercury and Venus, because the first observations of the phenomena were made during this period. Mercury passed across the Sun's disk in 1631, and was observed by the French astronomer Gassendi. In 1639 the first observations of a transit of Venus were made, by two Englishmen, Jeremiah Horrocks and William Crabtree. Of course, only Mercury and Venus can pass in transit (the remaining planets are further away from the Sun than we are), and later on it was found that observations of transits of Venus could provide a reliable key to the distance of the Sun. The Earth–Sun distance, or astronomical unit, was not then known with any accuracy at all (the first reasonable estimate was made in 1682 by Cassini; he gave 86 000 000 miles) but if the distance of Venus could be found, the astronomical unit could be worked out by using Kepler's laws of planetary motion. Unfortunately, transits of Venus are rare. They occur in pairs, separated by 8 years, after which there are no more for over a century. The transits of 1761, 1769, 1874 and 1882 were carefully studied, though the results were not as good as had been hoped. The whole method is now obsolete, so that the next transits – those of 2004 and 2012 – will be regarded as of no more than academic interest.

Telescopic improvements gave rise to better maps of the Moon. In 1647 a reasonable chart was produced by Hevelius of Danzig (the modern Gdańsk), but this was superseded by a map drawn in 1651 by the Jesuit astronomer Riccioli, based upon observations by his pupil Grimaldi. The map was also notable because Riccioli introduced the system of naming lunar craters after personalities – usually, though not always, astronomers. The system has been retained and extended, though it is not without its drawbacks. Riccioli, who was no believer in the heliocentric system, 'flung Copernicus into the Ocean of Storms', and Galileo is represented by a very obscure, semi-ruined formation. On the other hand, the craters known in honour of Riccioli and Grimaldi are among the most prominent on the Moon!

The best planetary observer of the 17th century was without doubt Christiaan Huygens, of Holland, who is probably best remembered today as the inventor of the pendulum clock. Huygens equipped himself with the best telescopes available, and made intensive observations of all the known planets. Of course, his telescopes were anything but modern looking. A simple object-glass produces a great deal of false colour, and the only way of counteracting this, at the time, was to make the focal lengths very long; in some cases the object glass was fixed to a mast. Yet Huygens was able to discover Titan, the brightest satellite of Saturn, and to make the first drawing of Mars which showed surface detail. His sketch made in 1659 records the V-shaped marking known today as the Syrtis Major in recognizable form, and luckily he gave an accurate timing, so that we can show that the rotation period of Mars is constant; the modern value is 24 h 37 min 23 s.

Galileo had believed Saturn to be a triple planet, and he was very puzzled when the two 'companions' disappeared. What had happened was that the rings had become edgewise on to us, as they do every 15 to 17 years, and Galileo lost them. It was Huygens who established that, in his words, Saturn 'is surrounded by a thin, flat ring, inclined to the ecliptic, and which nowhere touches the body of the planet'. This announcement was made in 1655, the same year as the discovery of Titan. It met with sharp criticism and, surprisingly, the ring explanation was not finally accepted by all astronomers until 1665.

The next great planetary observer was G.D. Cassini, an Italian who became the first Director of the new Paris Observatory. With improved telescopes he determined the rotation periods of Mars and Jupiter with surprising accuracy; in the case of Mars his final figure was correct to within 3 min. He discovered the main gap in Saturn's ring system, which is still known as the Cassini division, and he also discovered four of Saturn's satellites: Iapetus (in 1671), Rhea (1672), and Tethys and Dione (1684). Iapetus is very variable. It is always brightest when west of Saturn, and much fainter when to the east, though it does not vanish completely, as Cassini claimed (it can always be followed with a modern 3-in (75-mm) refractor). We now know that its rotation period is the same as its orbital period (79 Earth days) and that one hemisphere is much brighter than the other, because it is coated with an icy deposit; when Iapetus is west of Saturn, it is always the more reflective hemisphere which is turned in our direction.

Venus, unlike Mars and Jupiter, showed no definite surface markings. During the transit of 1761 M.V. Lomonosov, the first great Russian astronomer, discovered that it has an extensive atmosphere, so that all we can see is the top of a cloudy layer. It was not until the space age that we found out what the surface of Venus is really like.

The Moon, of course, was under regular surveillance. Cassini, in 1693, established that it has synchronous rotation, so that there is a part of it which is always turned away from Earth, and in 1775 Tobias Mayer drew a small but accurate lunar map. However, the story of lunar and planetary observation around the turn of the 18th century is dominated by two men: William Herschel and Johann Hieronymus Schröter.

William Herschel, the Hanoverian-born organist who is often regarded as the best astronomical observer of all time, made his main contributions in the field of stellar astronomy, but he achieved initial fame in 1781 when he discovered the planet we now call Uranus. He was not searching for a planet, and did not even recognize it for what it was; his paper to the Royal Society was entitled *An Account of a Comet*, but before long it was found to be a planet orbiting the Sun at a distance much greater than that of Saturn. Six years later Herschel, using his great 40-foot (12-m) telescope, discovered two satellites of the planet, Titania and Oberon. He also studied Mars, making some reasonable sketches of the surface features, and establishing that the Martian atmosphere must be far thinner than ours; naturally, he believed that the polar caps were due to ice and snow.

He paid rather less attention to the Moon, but on several occasions he recorded bright glows which he ascribed to active volcanoes. In fact, lunar volcanism ended long ago, and there is no doubt that what Herschel saw was the brilliant lunar crater Aristarchus, illuminated by light reflected from the Earth.

Schröter was a man of very different type. He was Chief Magistrate of the town of Lilienthal, near Bremen, with enough funds and leisure to set up a major observatory; he equipped it with the best telescopes he could obtain, including one made by Herschel, and between 1778 and 1814 he carried out an energetic program of lunar and planetary research. He was the first great lunar observer, and made many drawings of the surface; he was the first to see the linear, crack-like features now called clefts or rills (German, *rilles*). His drawings of Mars were far better than any previously made, though it is true that he misinterpreted the nature of the dark markings, and regarded them as atmospheric. He made first-class sketches of Venus, though the markings he recorded were cloudy and vague; he found that there is a difference between the observed and the theoretical phase of the planet, due no doubt to the effects of the atmosphere; and he recorded the 'ashen light', or faint luminosity of the planet's dark side. A similar effect is often seen with the crescent Moon, but is no mystery; Leonardo da Vinci realized that it is due to reflected Earthlight. But Venus has no satellite, and it seems that the 'Ashen Light' is due to electrical effects in the planet's upper atmosphere, possibly at least superficially similar to our auroras.

Schröter's life came to a sad end. In 1814 the French invaded Germany; Lilienthal was captured, and Schröter's observatory was destroyed, with all his unpublished observations. Even the telescopes were plundered, because they had brass tubes, and the French soldiers assumed that they were made of gold.

The mantle of Schröter fell upon two of his countrymen, Wilhelm Beer and Johann Heinrich von Mädler. Beer, a Berlin banker, set up a private observatory near Berlin, and equipped it with a fine 3¾-inch (9-cm) refractor made by the great optical worker Fraunhofer. With his colleague Mädler, who actually did the main art of the observing, Beer worked untiringly between 1830 and 1838. The two observers made the first fairly accurate maps of Mars, but their main work was in connection with the Moon. In 1838 they published a first-class map, together with a large book containing a description of every named formation. It was a masterpiece of careful, accurate work, and it remained the standard for decades – so much so, indeed, that there was a general feeling that no more lunar work needed to be done, and for some time the Moon was shamefully neglected!

In the 1770s an otherwise obscure astronomer named Titius drew attention to a curious mathematical relationship linking the distances of the planets from the Sun. It was popularized by Johann Elert Bode, and is generally known as Bode's law. Today it is usually regarded as being mere coincidence, but formerly it was taken very seriously, particularly when Uranus was found to fit in with it. There was an apparent gap between the paths of Mars and Jupiter, and the suggestion was made that a planet ought to exist there. In 1800 a group of astronomers, headed by Schröter and the Baron Franz Xavier von Zach, set out to make a systematic hunt for it. Ironically, the first discovery was made by G. Piazzi, who was then drawing up a new star catalog from Palermo Observatory and was not a member of Schröter's 'celestial police'; on 1 January 1801 he discovered the first minor planet or asteroid, Ceres. Within the next few years three more were found: Pallas (1802), Juno (1804) and Vesta (1808). No more seemed to be forthcoming, and the 'police' disbanded; but in 1845 a fifth asteroid, Astraea, was found by a German amateur named Hencke, and other discoveries followed. Today thousands of asteroids are known, though only Ceres is over 800 km in diameter, and only Vesta is ever visible with the naked eye. Olbers, one of the 'Police', believed them to be the débris of a disrupted planet, but nowadays it is thought that no planet could ever form in that part of the solar system because of the powerful disruptive influence of Jupiter.

A much greater triumph occurred in 1846. Uranus was not moving as it had been expected to do; there was some perturbing influence. Two astronomers, J.C. Adams in England and Urbain Leverrier in France, independently concluded that the cause of the perturbations was a more distant planet. They worked out its proposed position, and in 1846 Galle and D'Arrest, from Berlin, found the new world very close to the position given by Leverrier; Adams was nearly as accurate. The chosen name was Neptune. Within a few weeks the British amateur William Lassell had discovered a satellite, Triton.

The second half of the 19th century showed striking advances in telescope construction, and photography began to play a major role; the first really good lunar photographic atlas was compiled in the 1890s by Loewy and Puiseux, at the Paris Observatory. Yet insofar as planetary observations were concerned, direct observation at the eye end of a telescope remained supreme.

Maps of Mars were improved, and name were given to the main features. In 1877, when the planet came to a particularly favourable opposition, G.V. Schiaparelli at Milan made a careful study, revising the nomenclature and also recording straight, artificial-looking lines which he called *canali*. He kept an open mind about their nature, but others were less cautious. Percival Lowell, a wealthy American, built a major observatory at Flagstaff in Arizona, and equipped it with a fine 24-inch (60-cm) refractor, mainly to observe Mars; he regarded the canals as artificial waterways, built by the inhabitants to preserve every scrap of water from the icy polar caps. The canal controversy raged for many years, and it was only with the arrival of the space craft that the linear features were finally found to be pure illusions.

However, Mars was still regarded as a planet with extensive tracts of vegetation, and the presence of more advanced life forms was certainly not ruled out. Venus remained a mystery; it might be a bone-dry dust desert, but it might also be a watery world, perhaps similar to the Earth in Carboniferous times. Mercury showed little even in the large refractors of the late 19th century; it was generally believed to be not unlike the Moon, with at best a very tenuous atmosphere.

Ideas of the four giant planets were very wide of the mark. It was usually thought that they were hot enough to warm their satellite systems; by 1900 Saturn was known to have nine attendants (the latest, Phoebe, was discovered in 1898 by W.H. Pickering – the first satellite to be discovered photographically) while Jupiter's fifth satellite was found visually by E.E. Barnard in 1892.

In 1882 R.A. Proctor, a noted astronomer and author, wrote *Saturn and its System*, in which he summed up the current view: 'Over a region hundreds of thousands of square miles in extent, the glowing surface of the planet must be torn by subplanetary forces. Vast masses of intensely not vapour must be poured forth from beneath, and rising to enormous heights, must either sweep away the enwrapping mantle of cloud which had concealed the disturbed surface, or must itself form into a mass of cloud, recognizable because of its enormous extent.' Nothing could be further removed from our views of today.

Various theories of the origin of the Moon and planets had been considered. G.H. Darwin's proposal that the Moon was thrown off the Earth by tidal forces was widely accepted; insofar as the planets were concerned, Laplace's nebular hypothesis had been discarded, to be replaced by the theory that the planets were ripped off the Sun by the action of a passing star. It is interesting to find that modern ideas have much more in common with Laplace than they do with the passing star picture.

All in all, it is fair to say that by the end of the 19th century a great deal of accurate information about the Moon and planets been obtained. It is not surprising that there were many errors; but for the development of space research methods, our knowledge would still be comparatively meager. We have learned more since 1962 than we had been able to do throughout the whole of human history, but we must always remember that the foundations of modern planetary science were laid by the dedicated observers and theorists of long ago.

Patrick Moore

Bibliography

Abbott, B.J. (1984) *Astronomers*. London: Bloud Educational.
Antoniadi, E.M. (1934) *L'Astronomie Egyptienne*. Paris: Gauthier-Villars.
Berry, A. (1898) *A Short History of Astronomy*. London: J. Murray (repr. Dover Publ., 1961), 440 pp.
Clerke, A.M. (1902) *History of Astronomy in the Nineteenth Century*. London: A. and C. Black.
Dreyer, J.L.E. (1906) *A History of Astronomy, from Thales to Kepler*. Cambridge University Press (reprint by Dover Books, New York, 1953).
Grant, R. (1852) *History of Physical Astronomy from the Earliest Ages to the Middle of the 19th Century*. London.
Heath, T.L. (1932) *Greek Astronomy*. London and Toronto: Dent.
Holton, G. (1988) *Thematic Origins of Scientific Thought – Kepler to Einstein*. Cambridge, Mass.: Harvard University Press.
Kuhn, T.S. (1957) *The Copernican Revolution*. Cambridge, Mass.: Harvard University Press. (And 1959, New York: Random House, 297 pp.)
Pannekoek, A. (1961) *A History of Astronomy*. New York: Interscience, 521 pp.
Sharpley, H. (1956). *Source Book in Astronomy*. Cambridge, Mass.
Singer, C. (1959) *A Short History of Scientific Ideas to 1900*. Oxford: Clarendon Press, 525 pp.
Taton, R. (1963) *Ancient and Medieval Science, from the Beginnings to 1450*. New York: Basic Books (transl. A.J. Pomerans), 551 pp.
ter Haar, D. and Cameron, A.G.W. (1967) Solar system: review of theories, in *The Encyclopedia of Atmospheric Sciences and Astrology* (ed. R.W. Fairbridge). New York: Reinhold Publ. Co., pp. 890–99.
Weaver, J.H. (ed.) (1987) *The World of Physics*. New York: Simon and Schuster (vol. 1–3).
Williams, H.S. (1932) *The Great Astronomers*. New York: Newton Publ., 618 p.

HISTORY OF PLANETARY SCIENCE II: SPACE AGE

Detailed investigation of the major planets at close range has been a classical aspiration of astronomers. Planetary satellites and rings, comets, minor planets and the interplanetary medium are readily included within this aspiration.

The challenge to the modern planetary sciences is to describe and understand the physical properties of each object, pressing continuously for greater detail, and to develop a coherent scenario for their origin and evolution.

Overview of planetary research by space techniques

Prior to the space age, scientific research in space, was conducted at great distances from the objects under study using progressively evolving techniques with optical and radio telescopes on the ground and carried by balloons and aircraft.

The development of high-performance rockets during World War II and thereafter, though primarily for military purposes, presaged important new possibilities. Observations of phenomena within the upper atmosphere of the Earth and beyond began in 1946, using German V-2 rockets, and soon thereafter, American Aerobee rockets. Such work during the period 1946–1957 laid the foundations, both technical and scientific, for much more ambitious undertakings with higher performance vehicles.

Geophysicists and astronomers alike were exhilarated by the successful launchings of the first artificial satellites of the Earth: Sputnik 1 (4 October 1957) and Sputnik 2 (3 November 1957) of the USSR; and Explorer 1 (31 January 1958) of the US. The latter carried the first instrument for geophysical observations, though the radio transmitters themselves on the two previous satellites provided valuable ionospheric, atmospheric and geodetic data. The observations by Explorer 1 and its prompt successor Explorer 3 (26 March 1958) yielded the discovery of the radiation belts of the Earth and led to the new subject of magnetospheric physics.

During 1958 24 space missions were attempted by the Soviet Union and agencies of the US Department of Defense. Of these, six were successful in achieving Earth orbit and yielding data of value. Two were aimed at the Moon. Despite failure to achieve their primary objective, these two space probes also yielded important observations of the Earth's radiation environment.

Legislation creating the National Aeronautics and Space Administration as an independent agency of the US government was developed by the Congress during the summer of 1958 and signed into law by President Eisenhower in October 1958. This National Aeronautics and Space Act of 1958 lists the following as the first of eight objectives of the aeronautical and space activities of the United States:

(1) The expansion of human knowledge of phenomena in the atmosphere and space.

It is within this thin two-line objective that the planetary program of the United States has been funded and conducted.

The frenetic activity in space during the early years was driven by competition between the US and the USSR for military superiority and for international prestige in the realm of high technology. Even the civil program of the US served a national security purpose. In short, the 'space race' between the US and the USSR formed the political rationale for the rapid expansion of space flight by both nations.

Systematic planning of the US scientific missions in space became the province of the Space Science Board of the National Academy of Sciences, created in the summer of 1958 in anticipation of such a need. The central emphasis of NASA itself was on the achievement of human space flight with the goal of placing one or more individuals briefly on the Moon and returning them safely to the Earth. The SSB and, later, NASA's own Lunar and Planetary Missions Board led the way to a program of exploring the planets. Comparable aspirations were developed in the USSR.

Venus and Mars were the earliest planetary targets because of their accessibility. (Exploration of the Moon has much in common with exploration of the planets but is not within the scope of this article.) Also, a special motivation for the exploration of Mars was based on

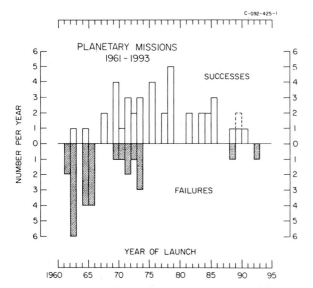

Figure H3 Graphical display of the year-by-year success/failure record of attempted planetary missions of all nations, by year of launch 1961–1993 (inclusive). The earliest attempts were in 1961. There were no attempts in 1963, 1966, 1968, 1974, 1976, 1979, 1980, 1982, 1986, 1987, 1991 and 1993. The dashed bar in 1989 represents a mission in progress to Jupiter with reasonable likelihood of success there (Galileo).

the judgment that the physical/chemical conditions there offered the best hope for the discovery of any form of extraterrestrial biological activity, however rudimentary.

In February 1961 the USSR attempted two flights to Venus. The first was a launch failure; the second achieved Earth-escape speed but telemetry was lost early in flight. During 1962 the US attempted two flights to Venus and the USSR attempted three flights to Venus and two to Mars. Of these, only the US Mariner 2 was successful. It flew by Venus in December 1962 and thus became the first successful mission, both ballistically and scientifically, to another planet. Its radial distance of closest approach (RCA) was 6.8 planetary radii.

The first mission to an outer planet was Pioneer 10's flyby of Jupiter on 4 December 1973.

During the 33-year period, 1961–1993, there have been 41 successful missions – out of 67 reported attempts – to Mercury, Venus, Mars, Jupiter, Saturn, Uranus and Neptune; and to three comets and two asteroids. One planetary mission is in progress as of early 1994. Figure H3 gives the success/failure record year-by-year. Most of the early failures occurred in the propulsion and guidance systems. There has been a vigorous effort to improve propulsion, guidance, spacecraft design and construction, scientific instrumentation, interplanetary navigation, telecommunications and ground control techniques.

Tables H2–H10 tabulate all planetary missions through 1993 which yielded significant scientific findings on the target object. Brief explanatory notes accompanying each entry give an indication of the nature of the mission. Note that some of the missions had multiple targets and are listed more than once in these tables whereas they were counted only once in preparing Figure H3. The dominant efforts of the US and the USSR have been joined recently by those of Japan and the European Space Agency.

Among the known major planets only Pluto and its satellite Charon remain unvisited by a spacecraft.

In every successive mission the formulation of scientific objectives has rested heavily on knowledge from Earth-based studies, from prior missions and from study of the characteristics of the Earth as a reference object.

Not evident in the tables is the accompanying and progressive increase in the quality, reliability and sophistication of the scientific instrumentation and the correspondingly enormous increase in the quality, quantity and significance of the observations. Many planetary missions have also yielded important interplanetary observations and several – Pioneer 10, Pioneer 11, Voyager 1, Voyager 2 and Ulysses – continue to do so.

Technical elements of planetary exploration by space techniques

Missions to the planets are enormously difficult and costly – far beyond the level of traditional, Earth-based astronomy. This section is intended to provide an appreciation of the technical achievements that are required to execute a successful planetary mission.

Ballistics and tracking

The physical principle of propelling a spacecraft from the Earth to another planet was understood by Isaac Newton over 300 years ago. The essential quantity, known as C_3 among ballisticians, is the square of the speed in km s^{-1} relative to the Earth that an object has after escape from its gravitational field. Neglecting the atmosphere of the Earth, the minimum speed for escape from its gravitational field is 11.2 km s^{-1}. Thus an initial speed of, say, 15.0 km s^{-1} relative to the Earth corresponds to a C_3 of $(15.0)^2 - (11.2)^2 = 100$ km^2 s^{-1} or an asymptotic speed of escape of 10 km s^{-1} relative to the Earth, a quantity also referred to as the hyperbolic excess speed. Relative to the Sun, the Earth's mean orbital speed is 29.8 km s^{-1}. In order to reach Venus, for example, the asymptotic velocity vector must be directed opposite to the Earth's velocity so that the apoapsis (q.v.) of the resulting ellipse relative to the Sun is at the Earth's orbit, whereas, in order to reach Mars, the asymptotic velocity vector must be in the same sense as that of the Earth's orbital velocity and the periapsis (q.v.) is at the Earth's orbit. The minimum value of C_3 that a spacecraft must have in order to reach each of the other planets on a direct flight is readily calculated and shown in Table H11 along with other relevant data. Any excess of C_3 over its minimum value reduces the flight time.

The second major consideration is the timing of the launch so that the spacecraft reaches a point on the orbit of the planet at the same time as does the planet itself. This is obviously a navigational matter of great difficulty.

Practical launch opportunities or windows occur at intervals of the synodic period of the target planet. The latter quantity is the time between successive identical (or nearly identical) configurations of the Sun–Earth–planet system, e.g. between successive inferior conjunctions or oppositions. For example, the mean synodic period of Mars is 780 days or 2.14 years. Values of the mean synodic periods of the major planets are included in Table H11.

During each day within the synodic window there is a diurnal window such that the asymptotic velocity vector has the desired direction and magnitude.

The foregoing considerations including Table H11 give a basic description of the ballistic requirements for flight to another planet but many important refinements are necessary in practice. For example, because the orbits of the planets are elliptical, not circular, the synodic period varies and the corresponding synodic opportunities vary substantially in attractiveness from the point of view of propulsion.

The durations of the synodic and diurnal windows are determined by how much the available C_3 exceeds the minimum necessary value and by the desired flight time and conditions of approach to the target planet. After the initial or primary period of propulsion, a spacecraft is essentially on a coasting Keplerian orbit but minor mid-course corrections of velocity are always necessary in practice.

The speed v of the spacecraft at any heliocentric distance r along a Keplerian ellipse is given by the *vis viva* equation

$$v^2 = GM_\odot (2/r - 1/a)$$

In this equation, derivable from Newton's law of gravitation and his three laws of motion, G is the universal gravitational constant, $M_\odot$ is the mass of the Sun, and a is the semimajor axis of the orbit. If r and a are measured in astronomical units (mean distance of the Earth from the Sun) and v in km s^{-1}, the numerical value of $GM_\odot$ is 887.

Minor corrections to a simple Keplerian orbit are necessitated by the gravitational forces of all of the planets and by solar light pressure; and, of course, major corrections are required during approach to the target planet (i.e. within its 'sphere of influence') because of its increasingly important gravitational force.

Accurate interplanetary navigation is, indeed, a profession in itself, employing massive computer facilities (see Ephemeris). An essential ingredient is tracking of a spacecraft using a coherent two-way radio link to combine Doppler data on the radial component of velocity with Newtonian theory. A spacecraft can now be guided to an absolute positional accuracy of the order of 100 km in its approach

Table H2 Successful missions to Venus (RCA = radial distance of closest approach)

Spacecraft	Launch date	Arrival date	Features
Mariner 2 (US)	27 August 1962	14 December 1962	Flyby RCA = 6.8 $R_♀$
Venera 4 (USSR)	12 June 1967	18 October 1967	Entry probe, hard lander, nightside
Mariner 5 (US)	14 June 1967	19 October 1967	Flyby RCA = 1.68 $R_♀$
Venera 5 (USSR)	5 January 1969	16 May 1969	Entry probe, hard lander, nightside
Venera 6 (USSR)	10 January 1969	17 May 1969	Entry probe, hard lander, nightside
Venera 7 (USSR)	17 August 1970	15 December 1970	Entry probe, soft lander, nightside
Venera 8 (USSR)	27 March 1972	22 July 1972	Entry probe, soft lander, dayside
Mariner 10 (US)	3 November 1973	5 February 1974	Flyby enroute to Mercury RCA = 1.94 $R_♀$
Venera 9 (USSR)	8 June 1975	22 October 1975	Orbiter, 1.26 × 19.5 $R_♀$ Entry probe, soft lander, dayside
Venera 10 (USSR)	14 June 1975	25 October 1975	Orbiter, 1.27 × 19.8 $R_♀$ Entry probe, soft lander, dayside
Pioneer 12 (US)	20 May 1978	4 December 1978	Orbiter, 1.03 × 11.9 $R_♀$ Continuous operation until atmospheric entry, 8 October 1992
Pioneer 13 (US)	8 August 1978	9 December 1978	Entry probes, 4 hard landers, dayside and nightside
Venera 11 (USSR)	9 September 1978	25 December 1978	Bus, flyby RCA = 5.13 $R_♀$ Entry probe, soft lander, dayside
Venera 12 (USSR)	14 September 1978	21 December 1978	Bus, flyby RCA = 5.13 $R_♀$ Entry probe, soft lander, dayside
Venera 13 (USSR)	30 October 1981	1 March 1982	Bus, flyby RCA = unknown Entry probe, soft lander, dayside
Venera 14 (USSR)	4 November 1981	5 March 1982	Bus, flyby RCA = unknown Entry probe, soft lander, dayside
Venera 15 (USSR)	2 June 1983	10 October 1983	Orbiter Radar mapping
Venera 16 (USSR)	7 June 1983	14 October 1983	Orbiter Radar mapping
Vega 1 (USSR)	15 December 1984	10 June 1985	Venus flyby enroute to comet P/Halley Lander, atmospheric balloon
Vega 2 (USSR)	21 December 1984	14 June 1985	Venus flyby enroute to comet P/Halley Lander, atmospheric balloon
Magellan (US)	4 May 1989	10 August 1990	High inclination orbiter Radar mapping, altimetry, radiometry Gravitational moments and anomalies
Galileo	18 October 1989	10 February 1990	Flyby enroute to Jupiter

Table H3 Successful missions to Mars

Spacecraft	Launch date	Arrival date	Features
Mariner 4 (US)	28 November 1964	15 June 1965	Flyby RCA = 3.91 $R_♂$ First surface imagery
Mariner 6 (US)	24 February 1969	31 July 1969	Flyby
Mariner 7 (US)	27 March 1969	5 August 1969	Flyby RCA = 2.04 $R_♂$
Mars 2 (USSR)	19 May 1971	27 November 1971	Orbiter Hard lander
Mars 3 (USSR)	28 May 1971	2 December 1971	Orbiter Soft lander
Mariner 9 (US)	30 May 1971	13 November 1971	Orbiter Entry probe
Mars 5 (USSR)	25 July 1973	(?)	Orbiter
Viking 1 (US)	20 August 1975	19 June 1976	Orbiter (19 June 1976–7 August 1980) Soft lander (20 July 1976–November 1982) Chemical/biological assay of surface material
Viking 2 (US)	8 September 1975	7 August 1976	Orbiter (7 August 1976–24 July 1980) Soft lander (3 September 1976–12 April 1978) Chemical/biological assay of surface material
Phobos 2 (USSR)	12 July 1988	29 January 1989	Orbiter Close approach to Phobos

Table H4 Successful mission to Mercury

Spacecraft	Launch date	Arrival date	Features
Mariner 10 (US)	3 November 1973	5 February 1974	Flyby of Venus
		29 March 1974	Flyby of Mercury (I) $RCA = 1.29\ R_{\mercury}$
		21 September 1974	Flyby of Mercury (II) $RCA = 21\ R_{\mercury}$
		16 March 1975	Flyby of Mercury (III) $RCA = 1.13\ R_{\mercury}$

Table H5 Successful missions to Jupiter

Spacecraft	Launch date	Arrival date	Features
Pioneer 10 (US)	3 March 1972	4 December 1973	First mission to an outer planet Flyby $RCA = 2.85\ R_{\jupiter}$
Pioneer 11 (US)	6 April 1973	3 December 1974	Flyby (retrograde) $RCA = 1.60\ R_{\jupiter}$ Redirected to Saturn
Voyager 1 (US)	5 September 1977	5 March 1979	Flyby $RCA = 4.9\ R_{\jupiter}$ Redirected to Saturn
Voyager 2 (US)	20 August 1977	9 July 1979	Flyby $RCA = 10.0\ R_{\jupiter}$ Redirected to Saturn
Ulysses (ESA/US)	6 October 1990	8 February 1992	Flyby $RCA = 6.3\ R_{\jupiter}$ Redirected into an orbit inclined at 79° to the ecliptic plane

Table H6 Successful missions to Saturn

Spacecraft	Launch date	Arrival date	Features
Pioneer 11 (US)	6 April 1973	1 September 1979	Flyby $RCA = 1.34\ R_{\saturn}$ Previous flyby of Jupiter
Voyager 1 (US)	5 September 1977	12 November 1980	Flyby $RCA = 3.08\ R_{\saturn}$ Previous flyby of Jupiter
Voyager 2 (US)	20 August 1977	26 August 1981	Flyby $RCA = 2.68\ R_{\saturn}$ Previous flyby of Jupiter Redirected to Uranus

Table H7 Successful mission to Uranus

Spacecraft	Launch date	Arrival date	Features
Voyager 2 (US)	20 August 1977	24 January 1986	Flyby $RCA = 4.19\ R_{\uranus}$ Previous flybys of Jupiter and Saturn Redirected to Neptune

Table H8 Successful mission to Neptune

Spacecraft	Launch date	Arrival date	Features
Voyager 2 (US)	20 August 1977	25 August 1989	Flyby $RCA = 1.18\ R_{\Psi}$ Previous flybys of Jupiter, Saturn and Uranus

Table H9 Successful missions to comets

Spacecraft	Launch date	Arrival date	Features
ICE (US)	12 August 1978		Originally called International Sun Earth Explorer 3 (ISEE 3) and stationed at L1 Lagrangian point of Sun–Earth system Later diverted to make first comet encounter and renamed International Comet Explorer (ICE)
		11 September 1985	Comet P/Giacobini-Zinner flyby
Vega 1 (USSR)	15 December 1984	10 June 1985	Venus flyby, lander and atmospheric balloon
		6 March 1986	Comet P/Halley flyby
Vega 2 (USSR)	21 December 1984	14 June 1985	Venus flyby, lander and atmospheric balloon
		9 March 1986	Comet P/Halley flyby
Sakigake (Japan)	7 January 1985	1 March 1986	Comet P/Halley flyby
Giotti (ESA)	2 July 1985	13 March 1986	Comet P/Halley flyby
		10 July 1992	Comet P/Grigg-Skjellerup flyby
Suisei (Japan)	18 August 1985	8 March 1986	Comet P/Halley flyby

Table H10 Successful flybys of two asteroids

Spacecraft	Launch date	Arrival date	Features
Galileo (US)	18 October 1989	29 October 1991	Flyby of asteroid Gaspra
		28 August 1993	Flyby of asteroid Ida
		(early December 1995)	En route to Jupiter via gravitational assists at Venus (10 February 1990), Earth 1 (8 December 1990) and Earth 2 (8 December 1992)

Table H11 Hohmann transfer ellipse from Earth orbit[a]

Target object	a^b (AU)	Mean synodic period (days)	a^b (AU)	$v_{heliocentric}$ at 1 AU (km s^{-1})	$v_{geocentric}$ (km s^{-1})	C_3 (km^2 s^{-2})	$P/2^c$ (years)
Sun	–	–	0.500	0.00	−29.78[d]	887.1	0.177
Mercury	0.387	115.9	0.694	22.25	−7.53	56.77	0.289
Venus	0.723	583.9	0.862	27.29	−2.50	6.25	0.400
Earth	1.000	–	–	29.78	–	–	–
Mars	1.524	779.9	1.262	32.73	2.95	8.68	0.709
Jupiter	5.203	398.9	3.102	38.58	8.79	77.31	2.731
Saturn	9.539	378.1	5.270	40.07	10.29	105.9	6.048
Uranus	19.182	369.7	10.091	41.07	11.28	127.2	16.03
Neptune	30.058	367.5	15.529	41.44	11.65	135.8	30.60
Pluto	39.44	366.7	20.22	41.60	11.81	139.6	45.46
Escape from solar system			∞	42.12	12.34	152.2	–

[a] The minimum energy (Hohmann) transfer ellipses in this table correspond to the assumption that the orbits of all planets are coplanar and circular with radii equal to their respective semimajor axes.
[b] The symbol *a* stands for the semimajor axis in astronomical units (AU) of the respective elliptical orbits in column 2 for the target and in column 4 for the transfer ellipse.
[c] $P/2$ is the one-way flight time from Earth to the target.
[d] Negative values of $v_{geocentric}$ are in the opposite sense to the Earth's orbital motion: positive values, in the same sense.

to a remote planet, using terminal guidance to correct for imperfect knowledge of the ephemeris of the planet itself.

The foregoing discussion has dealt with direct flights to a given planet. But in many cases indirect flights, those via another planet, require smaller values of C_3 and greatly reduce the flight time to the target planet. The ballistic principle involved is commonly, though perhaps misleadingly, called gravitational assist (see Gravity-assist navigation). This principle has formed the classical basis for discussing the conversion of, for example, a near-parabolic orbit of a comet to an elliptical one during a close encounter with Jupiter.

In a planetocentric coordinate system the near encounter trajectory of a spacecraft is an hyperbola. The asymptotic speeds of approach and recession are equal but the direction of the velocity vector is changed. Hence, because the planet is moving, the resulting velocity vector in the heliocentric coordinate system is changed in both direction and magnitude, and there is a net change, either an increase or a decrease, in the heliocentric kinetic energy of the spacecraft, this change being extracted from or added to the orbital energy of the planet.

The gravitational assist principle has had and will continue to have many important applications. For example, at a C_3 of 140 km^2 s^{-2} a direct flight to Pluto requires 45 years (Table H11) but a well-chosen intermediate encounter with Jupiter can reduce the flight time to less than 10 years.

Noteworthy applications of this principle were the Venus–Mercury tour of Mariner 10; the Jupiter–Saturn tour of Pioneer 11; the Jupiter–Saturn–Uranus–Neptune tour of Voyager 2; and the out-of-ecliptic deflection of Ulysses in a Jupiter encounter. It is planned to use gravitational assist in guiding the Jupiter-orbiting Galileo for an extensive study of the four Galilean satellites and the planet's magnetosphere.

The rocket equation

The achievement of interplanetary flight awaited the necessary value of C_3 in a practical rocket engine. The difficulties in developing such engines have been truly enormous. A brief examination of the basic rocket principle gives a sense of the propulsion problem of interplanetary flight.

Consider the vacuum, free space case. The linear momentum of the hot gas flowing out of the nozzle of a rocket engine is equal in magnitude and opposite in sense to that acquired by the surviving body of the rocket, thereby keeping the total linear momentum of all elements of the system constant. The resulting change in speed Δv of the surviving body is given by

$$\Delta v = U \ln (m_0/m_1)$$

In this equation U is the speed of the gas relative to the rocket body, m_0 is the initial total mass (propellant plus rocket body and payload), m_1 is the mass of the surviving rocket body and payload, and ln is the natural logarithm. The mass of the burned propellant is $(m_0 - m_1)$. U is essentially the mean speed of the hot gas, being proportional to the square root of the ratio of its temperature to its effective molecular weight. An alternative form of the foregoing equation is

$$\Delta v = g\, I_{sp} \ln (m_0/m_1)$$

I_{sp}, called the specific impulse, is the ratio of the propulsive force exerted on the rocket body per unit mass of propellant burned per unit time. Common practice is to state I_{sp} as pounds of thrust per pound of mass of propellant burned per second, but, of course, the numerical value is the same for grams of thrust per gram of mass per second, for kilograms of thrust per kilogram of mass per second, etc. – all of which ratios use 'improper' units for force. The factor $g = 9.80$ m s^{-2} or 9.80×10^{-3} km s^{-2} restores the proper relationship of force to mass and has no relevance to the circumstances of propulsion. In the SI system of units, thrust is measured in newtons (N), mass in kilograms (kg) and U in meters per second (m s^{-1}). In any system of units I_{sp} is measured in seconds (s).

The present state of rocket technology provides the following typical values of I_{sp}: 230–260 s for solid propellants and 270–450 s for liquid propellants. For the favorable values $I_{sp} = 450$ s and $m_0/m_1 = 5$, $\Delta v = 7.1$ km s^{-1} for the free space case. If the rocket were launched horizontally in a due easterly direction at the Earth's equator and if atmospheric drag were ignored (a very favorable assumption), there would be an additional increment of 0.46 km s^{-1} due to the eastward rotational motion of the Earth. Thus the resulting speed relative to a sidereal reference system centered on the Earth would be $v = 7.6$ km s^{-1}. Even this very optimistic value is less than the 7.86 km s^{-1} required for a low Earth orbit, and far less than the 11.2 km s^{-1} required for escape, i.e. for a C_3 = zero. In fact, for a one-stage rocket, the mass ratio m_0/m_1 must exceed 11 in order that escape speed be exceeded, even in this idealized case. No such rocket exists.

In the derivation of the basic rocket equation, it is evident that an idealized rocket should shed inert mass as the fuel burns. The use of multiple stages is the obvious approach to this ideal, because the mass of the rocket casing of the first stage is discarded before ignition of the second stage, etc.

Present rocket technology provides a value of C_3 of about 100 km^2 s^{-2} with a three-stage vehicle for a payload mass of about 0.5% of the gross lift-off mass. It is evident that an enormous amount of fuel and structure is required to launch a representative spacecraft to another planet.

Electronics and electrical power

In a miraculous coincidence of technological history, the discovery of the transistor and the development of highly reliable solid state electronics occurred contemporaneously with the advances in rocket propulsion. Modern solid state electronic circuitry requires far less electrical power for a given function than does equivalent vacuum tube circuitry. The factor is of the order of 10^{-3} for analog circuits and 10^{-6} for digital circuits. There are comparably favorable ratios of mass and volume.

Planetary missions would have been either impossible or, at best, rudimentary in producing scientific data without the enormous benefits of solid state electronics. Advances in propulsion were reasonably anticipated during World War II but the revolution in electronics was not. The discrete single element transistor was invented in 1947. The Explorer 1 satellite (1958) of the US was the first spacecraft to use transistor circuitry, followed soon thereafter by Vanguard 1.

Electrical power for early Soviet and early US satellites was derived from chemical batteries, whose energy storage was typically of the order of 20 watt hours per kilogram mass. Their active lifetimes in orbit were correspondingly limited to the order of 1 month.

Vanguard 1 was the first satellite of the Earth to derive its electrical power from arrays of solar cells, thin specially impregnated plates of silicon. The development of such cells was the second major technical revolution in making possible indefinitely extended lifetimes of satellites and planetary spacecraft. At a heliocentric distance of 1 AU, the power flux of sunlight is 1370 W m^{-2}. Some 10–15% of this can be converted to electrical power by solar cells. A typical figure of merit of practical solar panels is 50 W of electrical power per kilogram mass. All spacecraft to Mercury, Venus and Mars have derived their electrical power from solar cells.

Outer planet spacecraft are a different story. All such spacecraft – Pioneer 10, Pioneer 11, Voyager 1, Voyager 2, Galileo and Ulysses – derive their electrical power from devices called radioisotope thermal generators (RTGs). The isotope of choice has been plutonium 238, produced artificially in nuclear reactors. ^{238}Pu has a half-life of 86 years and is a (nearly) pure alpha particle emitter. The essential absence of beta and gamma rays in its decay scheme minimizes its adverse influence on nearby instruments, though the offensive isotope ^{239}Pu is usually present in a few parts per million. The kinetic energy of the alpha particles is absorbed by and therefore heats the multiple junction of a solid state thermopile, the 'cold' junction of which is thermally connected to fins on the exterior of the unit, thereby generating thermoelectricity. An exemplary RTG has a mass of 14 kg and generates 40 W at 24 V from some 20000 curies of ^{238}Pu.

The engineering break-even point between the mass of a practical solar cell array and the mass of an equivalent RTG occurs at about 3 AU. Beyond this heliocentric distance an RTG becomes increasingly more advantageous as the intensity of sunlight diminishes with increased distance. At 59 AU, the distance of Pioneer 10 in early 1994, the intensity of sunlight is only 3×10^{-4} of that at 1 AU. In fact, the operation of this RTG-powered spacecraft and the temperature of its components have been independent of the Sun for many years of flight.

No planetary mission has, thus far, used a nuclear reactor but such power sources are being considered.

Radio telemetry

The transmission of information from a planetary/interplanetary spacecraft is accomplished by beaming a modulated radio signal with a parabolic antenna pointed at the Earth. The signal is received by a large accurately pointed antenna at a ground station (see Deep Space Network). Typical frequencies of the carrier are in the S band (1550–5200 MHz) or X band (5200–10 900 MHz), chosen to lie in the radio spectrum so as to minimize interference by natural radio sources and the adverse influences of the interplanetary medium and the Earth's ionosphere and atmosphere. The output of a particular spacecraft sensor is converted to digital form and processed by an on board computer, then mixed with the output of other sensors to yield an encoded string of 0s and 1s – akin to the dots and dashes of a classical telegraph. The useful transmission rate is measured by the number of such bits per second that can be derived clearly from the signal that is received at the ground station.

A realistic current example (Pioneer 10) is that a radiated power of 8 W in the S band can carry 32 bits per second of digital data to a 70-m diameter antenna of the NASA/JPL Deep Space Network when the spacecraft is at a distance of 55 AU. Further improvements of the order of a factor of two in bit rate are in prospect and, of course, larger antennas on both the spacecraft and on the ground are possible. The maximum workable bit rate for a given space-craft and ground system is roughly inversely proportional to the square of the distance between the two.

Command and control

All planetary spacecraft, thus far and in the reasonably foreseeable future, carry no human crews as is obvious from their masses (Pioneer 10, 260 kg; Voyager 1, 800 kg). They are properly described as automated, commandable spacecraft. Many of their functions are autonomous but others are under remote control from ground stations.

Radio commands from ground stations cause reconfigurations of the electronics of the spacecraft, movement of mechanical parts, stopping and starting tape recorders, firing of gas jets for adjusting the velocity, the attitude and the spin rate of the spacecraft, etc.

Pioneer 10, Pioneer 11 and Ulysses are examples of spinning spacecraft whose transmitting antennas are kept pointed accurately at the Earth (to within about 1°) whereas the antennas of Voyager 1 and Voyager 2 are kept pointed at the Earth with similar accuracy but three-axis stabilization of the spacecraft is maintained by small gas jets actuated automatically by gyroscopic and stellar sensors. The Galileo spacecraft comprises both spinning and three-axis stabilized components.

In terms of scientific objectives, a spinning spacecraft is usually preferable for point-by-point measurements along the spacecraft's trajectory because it yields angular distributions. However, for imaging purposes three-axis stabilization is preferable. Some of the latter type spacecraft achieve part of the advantage of a spinning spacecraft by means of a platform, rotatable on command, usually by discrete steps.

During the course of a planetary mission the spacecraft is under essentially continuous monitoring and tens of thousands of commands are sent, acknowledged and executed. Certain sequences of commands are also stored electronically within the spacecraft and executed automatically at the desired time or actuated by a further command.

Categories of planetary missions

The simplest type of planetary mission is a flyby. Next in order of difficulty is the conversion of a flyby trajectory (an hyperbola relative to the planet) to a captive orbit (an ellipse) by firing a retarding gas jet at or near the periapsidal point.

A further type of encounter is represented by the 'soft landing' of equipment on the solid surface, if any, of a planet (or planetary satellite, asteroid or nucleus of a comet). A soft landing is one in which the landed equipment remains operable, perhaps for an extended period of time.

Hard, or destructive, landings are useful for observations during a brief, perhaps high speed, passage through an atmosphere. In the case of a gaseous planet (e.g. Jupiter) an entry probe can perform such functions on its one-way trip through the planet's outer atmosphere, having been slowed by atmospheric drag and then by parachute (see Galileo mission).

A next stage of difficulty is to soft land a powered roving vehicle for surveying surface, subsurface and atmospheric conditions during commandable traverses.

The Moon has been investigated by flybys, orbiters, hard landers, soft landers, rovers and landed human crews but neither of the two latter techniques has yet been applied to any planet.

The landing of human crews on planets, of which Mars is the most widely discussed example, and the establishment of habitable bases for scientific operations are visualized as possibilities for the remote future. It is evident from the earlier discussion that the technical difficulties of such enterprises are fundamental and truly enormous in magnitude.

Different types of approach to a planet make possible different classes of orbits or landing conditions. A type I approach is one that occurs during the first crossing of the planet's orbit by the spacecraft (i.e. outbound from the Sun); type II, during the second crossing (i.e. inbound); type III, during the third crossing (outbound); and type IV, during the fourth crossing (inbound). The spacecraft sweeps through a heliocentric longitude range of less than 180° for a type I approach, 180–360° for type II, 360–540° for type III and 540–720° for type IV. The longitude sweep is 180° for the minimum energy case (Hohmann transfer ellipse; q.v.) on the first encounter.

Current and recent status of planetary/interplanetary missions (as of early 1994)

Four spacecraft – Pioneer 10, Pioneer 11, Voyager 1 and Voyager 2 – have completed their programs of planetary flybys and continue to yield clean data as they coast outward through the interplanetary medium and approach the multifold boundary of the heliosphere, estimated to be at a heliocentric distance of about 150 AU. There is an optimistic, though dwindling, hope that one or more of these spacecraft will continue operating properly into the interstellar medium and make the first *in situ* measurements of its properties and of the cosmic radiation therein.

On 8 October 1992 Pioneer 12 (also known as Pioneer Venus Orbiter (q.v.)) completed nearly 14 years of continuous data gathering on the atmosphere and ionosphere of Venus, the planetary surface characteristics, the solar wind at Venus' orbit and the interaction of the solar wind with this essentially unmagnetized planet.

The Magellan spacecraft (q.v.) has completed the comprehensive high-resolution radar mapping of the topography of Venus and the survey of its detailed gravitational field. The arrival date at the planet was 10 August 1990.

The Galileo spacecraft (q.v.) is en route to Jupiter after gravitational assists at Venus, the Earth and the Earth for a second time in order to gain the necessary speed to reach Jupiter. It made the first close encounter with an asteroid (Gaspra; q.v.) on 29 October 1991 and an encounter with a second asteroid (Ida; q.v.) on 28 August 1993. During approach to Jupiter in late 1995, a deep entry probe will be released for the first attempt to penetrate and survey the structure and composition of Jupiter's atmosphere to a depth of about 10 bar. The mother spacecraft will be placed in orbit about the planet to make an extensive tour of the four Galilean satellites, to map their surfaces and to provide a comprehensive synoptic study of the planet's upper atmosphere and its enormous magnetosphere – all during a planned 2-year period in orbit. The intended data transmission rate of 134 000 bits per second has been severely degraded to about 10 bits per second because of the failure to deploy the 6 m diameter high-gain antenna. An intensive effort is being made to optimize the effective data return using the low-gain antenna.

The Ulysses spacecraft (q.v.) of the European Space Agency (q.v.) flew by Jupiter on 8 February 1992, utilizing the gravity-assist technique to change the plane of its orbit from one near the ecliptic to one inclined at 79° to the ecliptic. Important new data were obtained during this encounter though the primary objective is to survey, for the first time, the structure of the interplanetary magnetic field, the solar wind and the properties of cosmic rays and solar energetic particle bursts at high heliographic latitude at a distance of about 2.2 AU from the Sun.

Meanwhile, the ESA/Giotto spacecraft was reactivated and redirected to make a close flyby of comet P/Grigg-Skjellerup on 10 July 1992. The valuable observations there are being compared and contrasted to those obtained at comet P/Halley, its earlier, primary target.

The Mars Observer (q.v.), the first US spacecraft to be directed toward Mars since 1975, was launched on 22 September 1992 but experienced a catastrophic failure on 21 August 1993 as it approached the planet. Its primary objective was to have been a global survey of the detailed topography and mineralogical composition of the planet's surface, including a search for evidence of water therein.

Future of planetary exploration

The space program of studying the planets at close range and on their surfaces has also stimulated a much increased level of planetary observation by ground-based, balloon and aircraft instruments, utilizing great advances in infrared detectors, charge-coupled devices and adaptive optics. The synergistic effects of such work are noteworthy. A thriving scientific community of planetologists oversees the composite field of space and remote observations and interpretative work.

A US/ESA planetary mission, called Cassini (q.v.), is under development. Its purposes include an entry probe of the atmosphere of Saturn's satellite Titan and extensive observations of the ring system, magnetosphere, ionosphere and upper atmosphere of Saturn from the orbiting spacecraft; and a further search for small satellites. The launch of Cassini is scheduled for 1997.

A Russian/French mission to Mars is planned for launch in 1994 or 1996. Balloons, dragging long tails on the surface, will be used to study the wind system and the near-surface atmosphere, while a mother spacecraft orbits the planet.

Developmental work is being done on small, remotely controlled rovers to carry scientific instruments on traverses of the Martian surface.

It has been proposed to send a scientifically instrumented spacecraft on a high-speed direct flight to Pluto, the only major planet not yet visited. Although doubtless feasible, such a mission is only under preliminary consideration at present.

A NASA mission to rendezvous with and orbit the near-Earth asteroid Eros has been authorized.

Interpretation and consolidation of the great burst of new knowledge are today rounding out one of the most spectacular epochs of scientific achievement in history – the detailed investigation of planets, their satellites and rings; comets; and asteroids as individualistic physical bodies, not just as points of reflected sunlight.

The US has announced its long-range intention to send a human crew to Mars and return it to Earth, possibly as an international enterprise. Such a mission, because of its enormous costs and technical difficulties, lies in the remote future.

James A. Van Allen

Cross references

Deep Space Network
Ephemeris
European Space Agency
Galileo mission
Giotto mission
Gravity-assist navigation
Hohman transfer orbit
Magellan mission
Mariner mission
Mars Observer mission
NASA
Phobos mission
Pioneer Venus mission
Pioneer 10 and 11 missions
Sakigake and Suisei missions
Soviet Mars missions
Ulysses mission
Vega missions
Venera missions
Viking missions
Voyager missions

HOHMANN TRANSFER ORBIT

In 1925 Walter Hohmann considered the problem of finding optimum rocket trajectories for interplanetary flight. Earlier in the century, Konstantin Tsiolkovsky in Russia and Robert Goddard in the United States had realized that liquid-fueled rockets could be used to achieve extreme altitudes, including the possibility of space flight. Hohmann's contribution was to find a fuel-efficient trajectory that could be used to transfer a rocket and crew from the Earth to the vicinity of another planet and – after a predetermined wait of several months – return them to Earth. Hohmann showed that a manned Mars mission, limited by launch vehicle capability, requires 8.5 months to transfer from Earth to Mars, a waiting time at Mars of 14.9 months, and a return to Earth of 8.5 months, for a total mission duration of 2.7 years. As a demonstration of feasibility, Hohmann restricted the entire mission to a single plane, the ecliptic. His outgoing trajectory assumed two impulsive rocket burns, the first to escape the Earth's orbit and the second to enter the orbit of the target planet. Today we think of Hohmann's trajectory as a special case of the direct ballistic transfer (see Figure H4), which finds application not only to interplanetary trajectories, but also to the transfer of artificial satellites from one circular orbit at radius r_1 – a Shuttle altitude for example – to another circular orbit at radius r_2, perhaps at a much higher geosynchronous altitude.

Another remarkable Hohmann innovation was the approximation of the actual rocket trajectory by a finite sequence of Keplerian orbits. This 'patched-conic' approach is still used for the preliminary design of interplanetary trajectories, including non-ballistic and gravity-assist transfers. More specifically, the Hohmann maneuvers can be derived from the following two-body equation, which is just an expression of the conservation of total orbital energy per unit mass:

$$V^2 = \mu \left(\frac{2}{r} - \frac{1}{a} \right) \quad \text{(H1)}$$

The velocity of the orbiting body is V, its radial distance is r, the semimajor axis of its Kepler orbit is a, and μ is approximately equal to the gravitational constant G times the mass of the central body ($GM_{Sun} = 1.32715 \times 10^{20}$ m^3 s^{-2}). For the Hohmann ellipse, $a = (r_1 + r_2)/2$, and it follows by equation H1 that the two velocity increments of Figure H4 are given by

$$\frac{\Delta V_1}{V_1} = \left(\frac{2}{1+R} \right)^{1/2} - 1$$

$$\frac{\Delta V_2}{V_1} = R^{1/2} \left[1 - \left(\frac{2R}{1+R} \right)^{1/2} \right] \quad \text{(H2)}$$

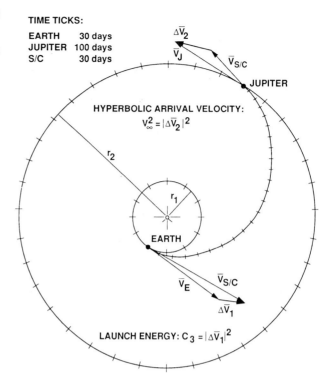

Figure H4 A realistic interplanetary trajectory closely approximated by the Hohmann ellipse. Time ticks are shown for a direct ballistic transfer of a spacecraft (S/C) to Jupiter. In the Hohmann transfer, the velocity increments ΔV_1 and ΔV_2 become collinear with the orbital velocities V_E and V_J respectively.

Table H12 Characteristics of Hohmann transfer from Earth to target planet

Planet	R^a	C_3 (km^2 s^{-2})	ΔV_2 (km s^{-1})	Flight time (years)
Mercury	2.583	56.7	9.6	0.29
Venus	1.382	6.2	2.7	0.40
Mars	0.656	8.7	2.6	0.71
Jupiter	0.192	77.3	5.6	2.73
Saturn	0.105	105.9	5.4	6.05
Uranus	0.052	127.3	4.7	16.03
Neptune	0.033	135.8	4.1	30.60
Pluto	0.025	139.6	3.7	45.46

^{a} Ratio of mean orbital radii of Earth and planet (r_1/r_2).

where the independent variable R is just the ratio of the Earth's orbital radius to the planet's orbital radius ($R = r_1/r_2$). For convenience, the required velocity changes of equation (H2) have been normalized to the Earth's orbital velocity ($V = 29785$ m s^{-1}). Note that the total velocity requirement $(\Delta V_1 + \Delta V_2)/V_1$ depends strongly on R, although it approaches the limiting case of $2^{1/2} - 1 = 0.414$ fairly rapidly for transfers to large orbital radii.

Another dependent parameter of interest is the time of flight T, which is just one-half the orbital period in the Hohmann ellipse. When normalized to the Earth's orbital period P_1 (1 year), T can be expressed as a function of R as follows:

$$\frac{T}{P_1} = \frac{1}{2} \left(\frac{1+R}{2R} \right)^{3/2} \quad \text{(H3)}$$

Although the special properties of the Hohmann transfer rule it out for most interplanetary missions, it provides useful reference values for launch energy and time of flight (see Table H12). Besides, by choosing launch times carefully, and by adding small mid-course maneuvers to change the plane, one can often find a direct transfer with orbital characteristics similar to Hohmann's.

In practice, the velocity change ΔV_1 is provided by the launch vehicle. Because the required launch energy is an important constraint, the parameter ΔV_1 is usually replaced by its square, $C_3 = \Delta V_1^2$, a measure of the excess energy needed at launch for the Earth–planet transfer (Table H12). According to Hohmann, the remaining velocity ΔV_2 is the amount of maneuver capability carried by the spacecraft itself for the targeting maneuver. Although ΔV_2 represents a realistic fuel requirement for transfers between satellite orbits, it does not in general apply to unmanned planetary missions. If the mission objective is to orbit the planet, the least fuel will be expended by first letting the spacecraft fall to the planet on a hyperbolic orbit with $V\infty = V_2$, and then by performing an orbital insertion maneuver when the spacecraft is as close to the planet as possible. For example, the Galileo mission's Jupiter orbital insertion (JOI), scheduled for 2.6 h after closest approach on 7 December 1995, will require only 630 m s^{-1} of retro-velocity increment.

Another consideration for interplanetary flight, not considered by Hohmann, is that the C_3 limitations of current launch vehicles rule out direct trajectories for all but the smallest of payloads. Missions such as Voyager, Galileo to Jupiter and Cassini to Saturn rely on one or more close planetary flybys between the launch and arrival. These gravity assists have the advantage of reducing C_3 below its Hohmann value, but they can also result in significantly increased flight times. For example, the Galileo mission, with one flyby of Venus and two flybys of Earth (VEEGA) reduces C_3 from its Hohmann value of 77 km^2 s^{-2} to about 13 km^2 s^{-2}, but the flight time is increased from 2.7 to 6.1 years. On the other hand, by using the full capabilities of the Titan III/Centaur rockets, plus a sequence of gravity assists by the planets Jupiter, Saturn and Uranus, Voyager 2 was able to reach Neptune within a relatively short flight time of 12 years. Even with the increased capability of the Titan IV or Proton rockets, and a relatively light payload of 200 kg, the fastest direct trajectory to Neptune takes about 15 years. Because Pluto is currently orbiting inside the orbit of Neptune, a large expendable rocket could send a 150 kg payload to that planet on a fast direct transfer around the turn of the century. An entire flyby mission could be completed 7–10 years after launch, far short of the 45.5-year flight time of Table H12.

John D. Anderson

Bibliography

d'Amario, L.A., Bright, L. and Wolf, A.A. (1992) Galileo trajectory design, in *The Galileo Mission* (ed. C.T. Russell). Dordrecht: Kluwer Academic Publishers, pp. 23–78.
Battin, R.H. (1964) *Astronautical Guidance*. New York: McGraw-Hill.
Roy, A.E. (1988) *Orbital Motion*, 3rd edn. Bristol: Adam Hilger.
Wiesel, W.E. (1989) *Spaceflight Dynamics*. New York: McGraw-Hill.

Cross references

Celestial mechanics
Ephemeris
Gravitation
Gravity-assist navigation

HOT SPOT TECTONICS

On Earth, a hot spot is persistent volcanic center (typically a 100 km wide) underlain by a broad swell (thousands of kilometers wide and hundreds of meters high). Movement of the lithosphere by plate tectonics over a hot spot produces discontinuous chains of volcanic structures, such as the Hawaiian islands, with active volcanism occurring only at the end of the chain currently over the hot spot. The term 'hot spot' was proposed by Wilson in 1963, and a mantle plume source was ascribed to hot spots by Morgan in 1972.

Because Venus and Earth have about the same mass and volume, their global heat fluxes should be similar, and hence both planets should be equally tectonically active; however, only Earth appears to have plate tectonics. Phillips and Malin (1983) proposed that the topography, heat flow and geology on Venus were instead dominated by the effects of large hot mantle plumes, or hot spots, and they termed this hot spot tectonics. The hot spot hypothesis requires that large-scale horizontal motion of the lithosphere is minor and that Venus loses most of its heat through lithospheric conduction, which is enhanced at the thin lithosphere associated with hot spots. As on Earth, more volcanism occurs over hot spots.

Several features of Venus support the idea of hot spot tectonics. Most of the large elevated regions are quasi-circular, as would be expected if uplift was caused by a cylindrical plume. Many of these regions have large free-air gravity anomalies that are suggestive of mantle depths of support for the topography. The areas that indicate the greatest depths of support are invariably topped by large shield volcanoes. However, the indicated depth of support varies considerably by area and some of the high regions on Venus are covered by tesserae (q.v.), not shield volcanoes.

To explain the varying depths of support, and the formation of tesserae, Herrick and Phillips (1990) proposed that mantle plumes were transient and that much of the tectonics on Venus is caused by the initiation of a plume by a large diapir of hot material, or blob tectonics. As the diapir rises there is initially rifting and shield volcanism, followed by massive flood volcanism when the diapir reaches the lithosphere, followed by topographic collapse. The blob tectonics hypothesis has not yet been adequately tested. Episodicity, also noted in terrestrial plume activity, may be forced by a thermal runaway mechanism (Rice and Fairbridge, 1975).

Robert R. Herrick

Bibliography

Herrick, R.R. and Phillips, R.J. (1990) Blob tectonics: a prediction for Western Aphrodite Terra, Venus. *Geophys. Res. Lett.*, **17**, 2129–32.
Morgan, P. and Phillips, R.J. (1983) Hot spot heat transfer: its application to Venus and implications to Venus and Earth. *J. Geophys. Res.*, **88**, 8305–17.
Morgan, W.J. (1972) Plate motions and deep mantle convection. *Geol. Soc. Am. Mem.*, **132**, 7–22.
Phillips, R.J. and Malin, M.C. (1983) The interior of Venus and tectonic implications, in *Venus*, (eds D.M. Hunten, L. Colin, T.M. Donahue, and V.I. Moroz). Tuscon: University of Arizona Press, pp. 159–214.
Phillips, R.J. and Malin, M.C. (1984) Tectonics of Venus. *Ann. Rev. Earth Planet. Sci.*, **12**, 411–13.
Phillips, R.J., Grimm, R.E. and Malin, M.C. (1991) Hot spot evolution and the global tectonics of Venus. *Science*, **252**, 651–8.
Rice, A. and Fairbridge, R.W. (1975) Thermal runaway in the mantle and neotectonics. *Tectonophysics*, **29**, 59–72.
Wilson, J.T. (1963) Continental drift. *Sci. Am.* **208**, 86–100.

Cross references

Mantle convection
Plate tectonics
Tesserae
Thermal evolution of planets and satellites
Venus: geology and geophysics
Volcanism in the solar system

HUYGENS, CHRISTIAN (1629–1695)

A Dutch physicist and mathematician, born in The Hague (The Netherlands), he is known for his wave theory of light, the invention of the pendulum clock and his discovery of the true nature of Saturn's rings.

The son of a well-to-do family, Huygens was educated at home, and later at the University of Leiden and the College of Orange at Breda, where he studied law before turning to natural philosophy and mathematics. Huygens' writings in the period 1650–1655 were devoted mainly to mathematical problems.

Figure H5 Christian Huygens. (Reproduced by permission of Rijksmuseum voor de Geschiedenis der Natuurwetenschappen, American Institute of Physics, Emilio Segre Visual Archives.)

Being a self-educated expert in lens grinding, Huygens discovered in 1655 the Orion Nebula and Titan, the largest moon of Saturn, using a home-made telescope. In the following year he discovered Saturn's ring system. In his *Systema Saturnium* (1659), Huygens described the phases and changes in the shape of the ring system.

In 1656 Huygens solved the problem of the dynamics of colliding elastic bodies based on Galileo's discovery of the constancy of a pendulum's period, refuting the laws of impact proposed by Descartes (1596–1650). Independently of Galileo, who in 1641 had conceived the idea of applying a pendulum to regulate clockwork, Huygens actually invented the first pendulum clock, thereby improving the accuracy of time measurement by several orders of magnitude. In his *Horologium Oscillatorium* (1673) he showed how to construct isochronous conical and cycloidal pendulums, gave the theory of compound pendulums and developed the mathematical theory of evolutes. It was Huygens (his work on pendulum motion) who made it possible for Newton to establish the Earth's figure as an oblate spheroid, and his recognition of the role of centrifugal force explained the poleward rise in the values of gravity.

Another of Huygens' important achievements was his wave theory of light (*Traité de la lumière*, 1678), describing light as a vibration spreading through an all-pervading medium called 'ether' which was imagined to consist of minute particles. Light did not actually consist of matter but was formed by the displacement of ether particles. He considered every point on a primary wave front to be the source of a series of secondary spherical waves, the envelope of which defined the wave front at the next instant. This insight into the propagation of waves has been named 'Huygens' construction' and made possible an account of the laws of refraction and reflection in terms of wave motion. The theory implied that light travels slower in denser media. In this way he was able to account for double refraction in Iceland spar (calcite), but not for the phenomenon of polarization or color (because in his theory the waves were neither periodic nor transverse). Huygens' wave theory of light had been put forward in sharp contrast to Newton's corpuscular theory of light. The two concepts of the nature of light came together only in the 20th century by revealing the particle–wave dualism accommodated by quantum theory.

Rhodes W. Fairbridge and Hans J. Haubold

Bibliography

Bell, A.E. (1947) *Christian Huygens and the Development of Science in the Seventeenth Century*. New York: Edward Arnold.
Bos, H.J.M. (1972) Huygens, Christiaan. *Dict. Sci. Biogr.*, Vol. 6, 597–613.
Newman, J.R. (ed.) (1988) *The World of Mathematics*. New York: Tempus Books (Vols I and II).
Weaver, J.H. (ed.) (1987) *The World of Physics*. New York: Simon and Schuster (Vols I to III).
Yoder, J.G. (1988) *Unrolling Time*. New York: Cambridge University Press.

HUYGENS TITAN ATMOSPHERIC PROBE

The Huygens atmospheric probe is an instrument package designed to explore the atmosphere of Saturn's moon Titan. It is an element of the Cassini/Huygens mission, a joint undertaking by NASA and the European Space Agency (ESA), whose goal is the detailed exploration of the Saturnian system. The Cassini spacecraft (q.v.) consists of the Huygens probe, provided by ESA, and the Saturn orbiter, provided by NASA.

The Huygens probe consists of two main parts: the descent module, which houses the scientific instruments and all the electrical subsystems, and the heat shield. During the cruise, when it is attached to the orbiter, and during the entry, the descent module is enclosed in the heat shield. After entry and ejection of the heat shield, the descent module will descend, under parachute, through the atmosphere down to the surface. The goals of Huygens are to measure the physical and chemical properties and to investigate the meteorology of the atmosphere of Titan, and to characterize (locally) the surface. Most of the scientific measurements will be made during the descent. Since it will hit the surface at a fairly low velocity (5–6 m s^{-1}) it is expected to survive and return data for a few minutes at least. Impact survival would permit direct surface composition measurements.

The scientific objectives of Huygens are to:

- determine abundances of atmospheric constituents (including any noble gases), establish isotope ratios for abundant elements and constrain scenarios of the formation and evolution of Titan and its atmosphere;
- observe vertical and horizontal distributions of trace gases, search for more complex organic molecules, investigate energy sources for atmospheric chemistry, model the photochemistry of the stratosphere and study the formation and composition of aerosols;
- measure winds and global temperatures; investigate cloud physics, general circulation and seasonal effects in Titan's atmosphere; search for natural lightning discharges;
- determine the physical state, topography and composition of the surface; infer the internal structure of the satellite; and
- investigate the upper atmosphere, its ionization and its role as a source of neutral and ionized material for the magnetosphere of Saturn.

Following the Probe mission, global remote sensing observations of Titan's atmosphere around the entry site will be made from the Saturn Orbiter. Global remote sensing observations from the orbiter and local *in situ* observations from the Probe will place the Huygens observations in their global context. Huygens data will provide the 'ground-truth' calibration of the orbiter data.

The payload

The scientific payload consists of six instruments, which all address in a complementary (and sometimes overlapping) manner one or

Table H13 The Huygens scientific payload

Instrument/principal investigator	Science objectives	Sensors/measurements	Mass (kg)	Power (c/o during cruise) (W)	Energy (during descent) (W h)	Data volume during descent 120–150 min (Mbits)	Data rate on surface (bps)
Huygens Atmospheric Structure Instrument (HASI) M. Fulchignoni, University of Paris 7 and Obs. de Paris-Meudon (France)	Atmospheric temperature and pressure profile winds and turbulence Atmospheric conductivity Search for lightning Surface permitivity, and radar reflectivity	T: 50–300 K, P: 0–2000 mbar, γ: 1 µg–20 mg AC E-field: 0–10 kHz, 80 dB at 2 $\mu V\, m^{-1}/Hz^{-1/2}$ DC E-field: 50 dB at 40 mV m^{-1} Conductivity: 10^{-15} ohm m^{-1} to ∞ Relative permitivity: 1 to ∞ Acoustic noise: 0–5/kHz, 90 dB at 5 mPa	6.7	20	41	4.1/5.1	625
Gas Chromatograph Mass Spectrometer (GCMS) H.B. Niemann, NASA/GSFC, Greenbelt (USA)	Atmospheric composition profile. Aerosol pyrolyzate analysis	Mass range: 2–146 amu Dynamic range: $> 10^8$ Sensitivity: 10^{-12} mixing ratio Mass resolution: $\sim 10^{-6}$ at 60 amu GC: 3 parallel columns, H$_2$ carrier gas Quadrupole mass filter 5 electron impact sources Enrichment cells ($\times 100$–$\times 1000$)	19.5	44.5	110	6.5/8.1	900
Aerosol Collector and Pyrolyzer (ACP) G.M. Israel, SA Verrieres-le-Buisson (France)	Aerosol sampling in 2 layers – pyrolysis and injection to GCMS	2 samples: 150–45 km, 30–15 km 3 step pyrolysis: 20°C, 250°C, 650°C	6.7	13.3 (71.5)	55	0.8/1.0	120
Descent Imager/Spectral Radiometer (DISR) M.G. Tomasko, University of Arizona, Tucson (USA)	Atmosphere composition and cloud structure. Aerosol properties Atmosphere energy budget. Surface imaging	Upward and downward visible (480–960 nm) and IR (0.87–1.7 µm) spectrometers, resolution 2.4/6.3 nm. Downward and side looking imagers (0.660–1 µm), resolution 0.06° to 0.20° Solar aureole measurements: 550 $\pm$ 5 nm, 939 $\pm$ 6 nm; Surface spectral reflectance with surface lamp	8.5	31	46	31.5/40.0	4250
Doppler Wind Experiment (DWE) M.K. Bird, University of Bonn (Germany)	Probe Doppler tracking from the Orbiter for zonal wind profile measurement	(Allan variance)$^{1/2}$: 10^{-11} (1 s); 5×10^{-12} (10 s); 10^{-12} (100 s) Wind measurement 2 m s^{-1} to 200 m s^{-1} Probe spin, signal attenuation	2.1	15	25	0.07/0.09 2.8/3.6 (*) (*) on Orbiter	10
Surface Science Package (SSP) J.C. Zarnecki University of Kent, Canterbury (UK)	Titan surface state and composition at landing site. Atmospheric measurements	γ: 0–100 g; tilt $\pm$ 60°; T: 65–110 K; T$_{th}$: 0–400 m W m^{-1} K^{-1} Sound speed: 150–2000 m s^{-1}, liquid density: 400–700 kg m^{-3} Refractive index: 1.25–1.45, liquid permittivity Surface acoustic sounding, sonar mode	4.2	15	26	3.5/4.1	660

several of the scientific objectives listed above. The principal characteristics of the Huygens instruments are listed in Table H13.

The mission

The Cassini spacecraft is scheduled to be launched in October 1997, arriving at Saturn about 7 years later. After insertion of Cassini/Huygens in orbit around Saturn in June 2004, Huygens will be dropped into the atmosphere of Titan by the Orbiter a few days prior to its first of many encounters with Titan. The entry date is 27 November 2004. After separation from the orbiter the probe will cruise on its own towards the largest satellite of Saturn at a speed of about 6 km s^{-1}. During the cruise the probe, except for a set of low-energy-consumption timers, is in a dormant state in order to conserve the energy stored in the batteries. When approaching Titan, the probe will enter the gravity field of the moon well before entering the upper layers of the atmosphere at about 1200 km above the surface. This will slightly accelerate the Probe until friction with the atmosphere overcomes the acceleration imparted by the Titan gravity field. Aerodynamic braking will decelerate the probe to a velocity of 400 m s^{-1} (Mach 1.5) in less than 2 min. The intense heat generated during the braking will reach a maximum of the order of 1 MW m^{-2}, and the peak deceleration will reach 150 m s^{-2}. The cruise timers will start up the flight computers just a few minutes before entry. The Mach 1.5 mark will initiate a complex series of activities inside the probe: parachute deployment, heat shield separation and experiment activation. During the entry the measurement of the deceleration profile will provide scientists with the data required to estimate the atmosphere density profile. All the scientific measurements will all be started within 1 min after heat shield separation at an altitude of about 170 km. The diameter of the main large parachute is sized to allow the heat shield to fall away underneath the descent module suspended under the parachute. If this parachute were retained for the entire descent to the surface, the descent would take about 7 to 8 h, which would require too many batteries. This large parachute will be jettisoned after about 15 min and a much smaller one will be

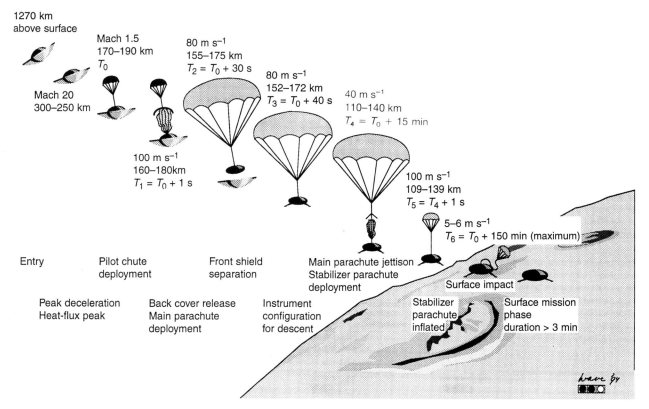

Figure H6 The entry and descent scenario of Huygens.

deployed to constrain the descent to 2.5 h maximum. This descent time reduces the required energy capacity of the batteries to about 1500–1800 Wh. The entry and descent scenario of Huygens is illustrated in Figure H6.

The nature of Titan's surface remains mysterious. Is it covered by lakes or large oceans? Is it an icy surface? Are rocks presents, and are there mountains? Is there any volcanic activity? It is not possible to design the probe so as to guarantee that it would survive surface impact. However, since the impact speed is rather low, there is a certain probability that the probe will survive and transmit data from the surface for at least a few minutes. The batteries will be sized to ensure that there will be enough energy left for a few minutes of operations after impact. The orbiter will be programmed to listen to the probe for up to 30 min after impact.

The probe will transmit its data through an antenna mounted on its top, facing up toward the sky. Since the radio receivers on Earth are not sensitive enough to capture and decode the probe telemetry signal, the orbiter will orient its high-gain antenna towards the probe to receive the probe signal. It will store the probe data on two solid state recorders and transmit them to Earth after the probe mission. Some speculate that the probe radio signal may be detectable on Earth by the most sensitive radio telescopes such as Arecibo, Goldstone and Bonn.

Conclusion

The *in situ* exploration of Titan by Huygens will be one of the major highlights of the Cassini mission. Although Voyager revealed some of Titan's mysteries, scientists today acknowledge that the opaque haze of Titan has kept most of the mysteries intact. When braking through the haze layers and the cloud cover of Titan, Huygens may reveal the unexpected. November 27 2004 will undoubtedly be an important date in the history of the exploration of the solar system. Huygens is expected to tell us a lot about the origin and evolution of the solar system and about the origin of life on Earth.

Jean-Pierre Lebreton

Bibliography

Hassan, H., McCarthy, C. and Wyn-Roberts, D. (1994) Huygens – a technical and programmatic overview. *ESA Bull.*, 77, 21–30.
Huygens: Science, payload and mission, ESA SP 1177 (1996); in preparation.
Lebreton, J.-P., Verdant, M. and Wills, R.D. (1994) Huygens – the science, payload and mission profile. *ESA Bull.*, 77, 31–41.

Cross references

Cassini mission
Saturn
Saturn: satellite system
Titan

HYDROSPHERE

Based on the Greek and Latin roots *hydro-*, water, and *sphaera*, globe or shell, the hydrosphere of planet Earth is unique by virtue of the fact that the water (q.v.) is mostly in its liquid phase. The Earth's hydrosphere (with a mean density a little over 1.00) underlies the atmosphere (q.v.) and overlies the lithosphere (q.v.). It constitutes over 0.02% of the Earth's mass. This liquid part is primarily sea water (96.5% pure water; 3.5% salts of sodium, magnesium, calcium, etc.), amounting to 1.4×10^{24} grams (or 10^{18} metric tons). A minor component of the liquid hydrosphere is fresh water that is distributed in rivers, lakes, and as subsurface waters (groundwater, etc., about 10^{23} g) distributed within the upper lithosphere (crust, q.v.). In the atmosphere, water is present in only small quantities (1.5×10^{17} g) as water vapor and droplets in clouds. Water at the surface of both land and ocean is involved in a constant recycling process, involving evaporation, transport as cloud and vapor, and reprecipitation as rain or snow: the hydrologic cycle. The hydrologic cycle is energized by the sun and gravitationally oriented processes at the Earth's surface.

In those land regions where the temperature is normally below freezing, a part of the hydrosphere ($2.5 - 10^{22}$ g) is represented by ice (termed Cryosphere). In most polar regions, sea ice floats on the surface of the ocean. Ice floats because it is about 10% less dense than liquid water, and at 4°C occupies about 10% more space. At this temperature liquid water reaches its maximum density, leading to a simple density overturn in lakes. In the ocean, when freezing of salt water takes place (at several degrees below zero), the sea salts are expelled into the water just below the ice surface, greatly increasing its salinity and density. This process, reaching a maximum in the autumnal and winter seasons, initiates a salinity–density-driven circulation that, in combination with other forcing agents, directs the circulation of the world ocean.

In regions of high precipitation, the rain contributes to the weathering (q.v.) of rocks and the leaching of soils. Here, the hydrosphere plays a vital role in the build-up of a biosphere (q.v.), the biotic environment and organic populations that make the Earth unique among planets. Although part of this biota is living in continuous relationship to the atmosphere, specific adaptations permit other parts of the biota to populate the water bodies themselves (oceans, lakes, rivers) or the moist soils (of the uppermost lithosphere).

The origin of the hydrosphere is generally attributed to the gradual 'degassing' of the primitive Earth that had accumulated from the cold accretion of planetesimals (q.v.). This 20th-century view contrasts with the old (18th-century) Kant–Laplace concept that called for condensation from a hot protoplanet. The degassing process can be visualized as volcanic exhalation and solfataras or hot springs, which provide modern 'actualistic' models, except that today most of the volatiles are recycled from subducted surface rocks and recent sediments. Modern degassing involves connate water, i.e. trapped during sedimentation, rather than juvenile water, i.e. coming from the mantle. If the mantle is in a state of nearly continuous convective overturn, however slow or interrupted, some recycling is inevitable.

On other planetary bodies (including the major satellites of Jupiter, Saturn and Uranus), water has been lost during the physical evolution of each body. The rocky planets Mercury, Venus and Mars do not now possess a hydrosphere, while some of the major (gas) planets' satellites have ice crusts. The case for an abundance of water on Mars at early stages of its history seems to be well established, although its present existence appears to be only as permafrost.

Rhodes W. Fairbridge

Bibliography

Carr, M.H. (1987) Water on Mars. *Nature*, **326**, 30–6.
Fairbridge, R.W. (ed.) (1967) *Encyclopedia of Atmospheric Sciences and Astrogeology*. New York: Reinhold, 1200 pp. [Hydrosphere, pp. 452–3.]
Goldschmidt, V.M. (1954) *Geochemistry*. Oxford: Clarendon Press, 730 pp.
Holland, H.D. (1984) *The Chemical Evolution of the Atmosphere and Oceans*. Princeton University Press, 582 pp. [see ms. Hydrosphere.]
Riley, J.P. and Skirrow, G. (eds.) (1975) *Chemical Oceanography*, 2nd edn. New York: Academic Press.
Rubey, W.W. (1951) Geologic history of sea water. An attempt to state the problem. *Geol. Soc. Am. Bull.*, **62**, 1111–48.
Squyres, S.W. (1989) Water on Mars (Urey Prize Lecture). *Icarus*, **79**, 229–88.
Wyllie, P.J. (1971) *The Dynamic Earth*. New York: J. Wiley and Sons, 416 pp.

Acknowledgement

Manuscript read and improved by Clare P. Marshall, Denver.

Cross references

Atmosphere
Biosphere
Crust
Ice age
Lithosphere
Mars: atmosphere
Planetesimal
Weathering

I

ICE AGE

The term 'ice age' was introduced (in German *Eiszeit*) in 1837 by the paleontologist Karl Schimper (1803–1867), although ideas about a former glacial epoch in Earth's history had begun to form at least 50 years earlier (North 1943; Carozzi 1966; Flint 1971). The full-fledged '*glacial theory*' evolved from the work of the Swiss geologist Louis Agassiz (1807–1873), whose vigorous expositions, supported with personal visits to Britain and North America, rapidly persuaded the scientific establishment of the day. It was one of the great intellectual revolutions of the 19th century, because it totally undermined the long-assumed uniformity and stability of the terrestrial environment with the present-day orientation of the 'uniformitarian' philosophy of Hutton and Lyell.

At first it was assumed that planet Earth had only suffered one 'ice age', but by the end of the 19th century traces were found not only of multiple glacial advances and retreats spanning the last 2 million years, but of former ice ages from the remote past, 200 to 600 million years ago and, more surprising still, from low-latitude regions of India, Africa, Australia and South America (Schwarzbach 1974; Frakes 1979; John 1979).

Key points of modern ice-age theory are as follows.

1. Most (about 90%) of geological time in the last ~ 4 billion years has been acryogenic, i.e. without significant glaciers except in high mountains. However, at no time in these 4 Ga was there a totally frozen planet; in fact ice rarely covered > 30% of the planet's land area.
2. The 'ice ages' have occupied periods of 10–30 Ma and mostly recur at intervals of 200–250 Ma. The principal ice ages so far established (Fairbridge, p. 504, in Oliver and Fairbridge, 1987) are: Quaternary (beginning about 2 Ma ago); Permo-Carboniferous (around 250 Ma ago); Late Ordovician (around 450 Ma ago); Eocambrian (around 650 Ma ago); Mid-Proterozoic (around 900 Ma ago); and Early Proterozoic (around 1250 Ma ago).
3. Continental ice sheets require the presence of large mid-latitude land masses, but do not approach the equator farther than 45° in the recent ice age. Changes in the Earth's obliquity may explain anomalous relations in earlier ice ages.
4. A paleogeographical 'ice-age preparation' (Schwarzbach 1974) is an essential prerequisite to achieve the necessary fall in global temperature. A plate-tectonic rearrangement of continents that would favor an ideal ice age is in the form of a meridional chain, blocking the warm circum-equatorial oceanic circulation and with mountains blocking the westerly ('zonal') jet streams. The orogenic building of such chains is strongly episodic.
5. The long-term mean temperature of planet Earth has been about 20 ± 5°C. In equatorial latitudes the mean SST (sea surface temperature) has been extremely stable, about 29 ± 2°C. However, the mean equator–pole temperature gradient varies greatly from a maximum glacial to a non-glacial state, with a temperature differential of > 50°C for the former, and a < 20°C range for the latter. Thus, for most of geologic time, most of the terrestrial environment has approached a subtropical state, although at glacial maxima up to one third of the globe would be covered by glaciers or sea ice (temperature < 0°C).
6. During any given ice age, there is a cyclic pattern of glacial advances and retreats, with a spectrum showing peaks at about 100 000 years, 42 000 years and 20 000 years, which correspond to minor orbital changes in the orbit of the Earth. These are the Milankovitch eccentricity, tilt and precession parameters (Imbrie and Imbrie 1979).
7. Besides orbital effects, the Sun's emissions may be variable on various timescales. The electromagnetic radiation (see Solar luminosity, Solar constant) varies over the period of the sunspot cycle, and likewise the particulate emission, carried by the solar wind (q.v.).

Rhodes W. Fairbridge

Bibliography

Berger, A., Imbrie, J., Hays, J.D. et al. (1984) *Milankovitch and Climate: Understanding the Response to Astronomical Forcing*, 2 vols. Dordrecht: D. Reidel, pp. 3–39.

Bucha, V. (1984) Mechanism for linking solar activity to weather-scale effects, climatic changes and glaciations in the Northern Hemisphere, in *Climatic Changes on a Yearly to Millennial Basis* (eds N.A. Mörner and W. Karlén). Dordrecht: D. Reidel, pp. 415–58.

Carozzi, A.V. (1973) Agassiz's influence on geological thinking in the Americas. *Arch. Sci. Genève*, **27**, 5–38.

Charvátová, I. and Střeštik, J. (1991) Solar variability as a manifestation of the Sun's motion. *J. Atmos. Terr. Phys.*, **53**, 1019–25.

Fairbridge, R.W. (1967) Carbonate rocks and paleoclimatology in *et al.* the biogeochemical history of the planet, in *Carbonate Rocks* (eds G.V. Chillinger, H.J. Bissell and R.W. Fairbridge). Amsterdam: Elsevier, pp. 399–432.

Fairbridge, R.W., and Sanders, J.E. (1987) The Sun's orbit, A.D. 750–2050: basis for new perspectives on planetary dynamics and Earth–Moon linkage, in *Climate–History, Periodicity, and Predictability*. New York: Van Nostrand Reinhold, pp. 446–71 (bibliography, pp. 475–541).

Flint, R.F. (1971) *Glacial and Quaternary Geology*. New York: Wiley.

Frakes, L.A. (1979) *Climates Throughout Geologic Time*. Amsterdam: Elsevier.

Imbrie, J. and Imbrie, K.P. (1979) *Ice Ages: Solving the Mystery*. Short Hills, NJ: Enslow Publishing.

John, B. (ed.) (1979) *The Winters of the World*. Newton Abbot, UK: David and Charles Publishing.

Kukla, G.J. (1972) Insolation and glacials. *Boreas*, **1**, 63–96.

North, F.J. (1943) Centenary of the glacial theory. *Proc. Geol. Assoc., London*, **54**, 1–28.

Oliver, J.E. and Fairbridge, R.W. (eds.) (1987) *The Encyclopedia of Climatology*. New York: Van Nostrand Reinhold, 986 pp.

Schwarzbach, M. (1974) *Das Klimate der Vorzeit*, 3rd eds Stuttgart: F. Enke Verlag, (English translation, 1963. New York: Van Nostrand).

Williams, G.E. (1981) *Megacycles: Long-term Episodicity in Earth and Planetary History*. Stroudsburg, P. (eds M.R. Rampino, J.E. Sanders, W.S. Newman and L.K. Königsson). Dowden, Hutchinson & Ross.

Cross references

Earth: atmosphere
Milankovitch, Milutin, and Milankovitch theory
Planetary ice
Polar cap

IDA

Minor planet 243 Ida was discovered by Johann Palisa in Vienna on 29 September 1884 and named by von Kuffner after a mythological character. Ida was a nymph from Crete who served as a wet nurse to Zeus.

Ida is a faint object viewed from Earth-based telescopes. In August 1993 the Galileo spacecraft (q.v.) encountered the asteroid at a relative velocity of 12.4 km s^{-1} and with a close approach distance of 2400 km. Ida is a member of the Koronis family which orbits in the outer main belt between 2.8 and 2.9 AU. Ida's orbit has a semimajor axis of 2.86 AU. It is an S-type asteroid; albedos of these objects imply a composition of olivine and pyroxene silicates. Ida has a mean diameter of 28 km, an irregular shape, and a rotation period of 4.63 h.

Galileo's science instruments collected data for approximately 6 h; this was stored on the spacecraft's tape recorder for later playback. The Ida encounter included one complete rotation of the asteroid. This data set was returned to Earth over a 10-month period following the observing phase. The first objective was to improve our knowledge of Ida: shape and size; morphology, composition and mineralogy of the surface. The detection of possible solar wind interaction effects was an additional goal. The collective data set will provide a means for comparative analysis with other asteroids and small bodies.

Images returned in September 1993 confirmed Ida's irregular shape. Ida was determined to be larger than the original estimate (53 × 23 × 18 km with a mean diameter of 28 km). The observed crater density on Ida implied that Ida is considerably older than originally estimated. A preliminary maximum surface temperature of 209 K was determined and the first Galileo Near Infrared Mapping Spectrometer spectrum of Ida was analyzed.

In mid-February 1994 it became apparent the data set had far surpassed the original expectations. A small body close to Ida, a satellite with a diameter of 1.5 km, was independently detected in data returned by two science instruments, the Near Infrared Mapping Spectrometer and the Solid State Imaging System. The detection of this small body, designated 1993(243)1 and later given the name Dactyl by the I.A.U. (q.v.), represents the first confirmed discovery of a satellite of an asteroid.

By April 1994 it was determined that Ida and 1993(243)1 do not have the same ratio of olivine to pyroxene. Only one side of Ida was compositionally imaged at high resolution, and the opposite side may contain materials more similar to those observed on the satellite. This fact nevertheless gives cause for some scientific speculation and debate as to the origin of the satellite. Is it a chip of Ida or the parent body? Was it captured in orbit around Ida, or was it just passing by Ida as Galileo flew by?

On 26 June 1994 the last of the data set was returned, some 10 months after the Ida encounter. At the time of writing, the Galileo Ida data analysis is ongoing.

Marcia Segura

Bibliography

Carlson, R.W., Weissman, P.R., Smythe, W.D., *et al.* (1994) Ida results from the Galileo Near Infrared Mapping Spectrometer. *EOS*, **75**, 224.

Kerr, R.A., (1993) Galileo reveals a badly battered Ida. *Science*, **262**, 33.

Klassen, K.P., Chapman, L.R. and Belton, M.J.S. (1994) Newly discovered satellite of 243 Ida. *EOS*, **75**, 224.

Cross references

Asteroid
Asteroid: families

Figure I1 Asteroid 243 Ida and its moonlet Dactyl, imaged on 28 August 1994 at a distance of 10 870 km. Resolution is about 100 m. Courtesy of the Jet Propulsion Lab (P-43731).

Galileo mission
Gaspra

IGNEOUS ROCK

Igneous rocks are those which have been formed by the solidification of molten or semi-molten material known as magma. This process is either in the Earth's crust (the plutonic or intrusive rocks) or at or near the surface (the volcanic or extrusive rocks). The name igneous is from the Latin *ignis*, fire; plutonic from the Greek *Pluto*, god of the infernal regions; and volcanic from the Roman god of fire, *Vulcan*. The principal plutonic rock is granite, a coarse-grained type that crystallized slowly in the crust in large masses known as batholiths. These rocks are now known on every continent where they have been exposed by denudation. However, they are never found in oceanic regions. Many occur in high mountain ranges, for example in the Himalayas (Le Fort, 1988; Shams, 1983). The main volcanic rock is basalt (q.v.), which is by far the commonest igneous rock on the surface of the Earth, where it forms large areas of lava, for example in Iceland, Hawaii, the Deccan Plateau, India, and extensively on the sea floor.

Composition and mineralogy

Most igneous rocks are composed of mixtures of separate crystalline minerals, or, where cooling has been very rapid, such as in some volcanic rocks, in the form of non-crystalline material ('glass'). Although about 4000 different minerals have been formally described, only about 20 rock-forming minerals make up the bulk of all igneous rocks (Deer, Howie and Zussman 1962–1986). These minerals are predominantly silicates, of which the feldspars are the most important, making up about 60% of all igneous rocks. There are two main groups, the alkali feldspars consisting of aluminosilicates of potassium, orthoclase and microcline, idealized formula $KAlSi_3O_8$, and the sodium feldspar, albite, $NaAlSi_3O_8$. The pure calcium feldspar anorthite (end-member, An, $CaAl_2Si_2O_8$) forms a series with the sodium member, albite (Ab), called the plagioclase feldspars. Specific names are used to denote compositional divisions between the endmembers, called oligoclase ($Ab_{90-70}An_{10-30}$), andesine ($Ab_{70-50}An_{30-50}$), labradorite ($Ab_{50-30}An_{50-70}$) and bytownite ($Ab_{30-10}An_{70-90}$). The feldspathoids have lower silica contents than the feldspars, occurring in the undersaturated rocks (i.e. undersaturated with respect to silica).

Quartz, SiO_2, is the single most important mineral in the Earth's crust, crystallizing typically in the oversaturated or 'acid' rocks. Among the ferromagnesian minerals are the olivines, pyroxenes, and amphiboles. Member of the olivine group $(Mg,Fe)_2SiO_4$ are major constituents of the basic and ultrabasic igneous rocks; their atomic structure consists of silica tetrahedra linked by the cations magnesium and iron. Olivine (Mg : Fe c. 10) is the main constituent of the upper mantle (Carr, 1984; Smith, 1979). The pyroxenes are generally dark, complex rock-forming silicate minerals having an atomic structure composed of chains of silica tetrahedra linked mainly by calcium, magnesium and iron. The varieties that have orthorhombic symmetry are the orthopyroxenes $(Mg,Fe)SiO_3$, notably enstatite and an intermediate member of the enstatite–orthoferrosilite series formerly known as hypersthene. Orthopyroxenes are common in the basic and ultrabasic plutonic rocks and in many volcanic rocks. They are probably important constituents of the upper mantle. The much commoner monoclinic pyroxenes have a very wide range of compositions, including the aluminous pyroxene augite which is especially characteristic of the basic igneous rocks. Minerals of similar complex compositions, but containing the hydroxyl ion, and having a double-chain silica structure, are the amphiboles. The commonest aluminous amphibole is hornblende which is particularly characteristic of the intermediate igneous rocks. The micas are sheet aluminosilicates with very complex formulae including hydroxyl ions. Muscovite is the colorless to pale yellow potassium mica; biotite the dark to black ferromagnesian mica. Both micas are common in granites and pegmatites, and biotite in the intermediate igneous rocks.

Apart from quartz, common oxide minerals in igneous rocks are magnetite, Fe_3O_4; hematite, Fe_2O_3; and ilmenite, $FeTiO_3$. The sulfide pyrite, or 'iron pyrites', FeS_2, and the calcium phosphate mineral, apatite, are of widespread occurrence in small concentrations, and occasionally in larger masses. Spinel, $MgAl_2O_4$, is only a minor mineral in igneous rocks, but gives its name to the extensive spinel group. These are isomorphous series of oxides containing a divalent and a trivalent ion; they form three series, spinel, magnetite and chromite. Minerals having spinel structure are important in the upper mantle.

Classification and nomenclature

Although many bases of classification have been suggested in the past (Johannsen, 1931–1938) there is now a fairly widely accepted agreement about naming rocks (Le Maitre *et al.*, 1989, 1990, 1991). Within the igneous rocks the primary division is into plutonic rock and volcanic rock. The plutonic rocks are relatively coarse-grained with individual crystals visible to the naked eye ('phanerocrystalline', 'phaneritic') and are presumed to have formed at considerable depth (10–20 km). Volcanic rocks are relatively fine-grained with most crystals not distinguishable to the naked eye. They often contain or consist of glass, and are seen or presumed to be associated with volcanic activity. Such activity is from eruptions (volcanoes or fissures) on the surface of the Earth, as lava flows, for example in Iceland, Hawaii and the Deccan Plateau, India, and on the sea bed. Volcanic rocks may also have been intruded at a high level as dikes (or dykes), which are tabular bodies, usually nearly vertical, cutting the bedding or foliation of the country rock; sills more or less parallel to the bedding planes of the country rock; and plugs, which are vertical feeder pipes of former volcanoes. If it is possible to determine the nature and quantity of the actual minerals, the so-called mode, this should be the basis of the classification. The primary modal classification is essentially based on the relative proportion of the following mineral groups (Streckeisen, 1967, 1973; Sabine, 1989; Le Maitre *et al.*, 1989, 1990, 1991):

Q = quartz;
A = alkali feldspar including albite, An_0 to An_5;
P = plagioclase feldspar including An_5 to An_{100};
F = feldspathoids or 'foids'.

These felsic minerals may be plotted on two triangles, Q–A–P or F–A–P, (since F and Q are mutually exclusive, Q + A + P is recalculated to 100 or F + A + P to 100). Modal mica, amphibole, pyroxene, olivine, opaque and accessory minerals constitute the mafic minerals, and rocks in which these amount to > 90%, the ultramafic rocks, are classified separately.

Among the more important plutonic rocks with mafic and related minerals M < 90%, are those shown in Table I1, based on the modal mineral content. This uses the quartz (Q) percentage, plagioclase (P) ratio = 100 × P/(P + A), and feldspathoid, or foid, (F), percentage. Granite contains quartz, potassium feldspar, oligoclase or andesine, with mica and/or hornblende. Hornblende and biotite are the typical dark minerals of granodiorite. Dolerite and diabase are used for a medium-grained gabbro. The feldspathoidal rocks (F 0–100) are usually named according to the particular feldspathoid present, e.g. nepheline syenite.

The ultramafic rocks, with M > 90, are usually classified according to the proportions of olivine, pyroxene and hornblende present. Dunite is a rock with olivine > 90%, peridotite has > 40% olivine with pyroxene or hornblende, while pyroxenite is a rock with pyroxene > 90%. Eclogite is a dense form of basalt crystallized at

Table I1 Classification and nomenclature of some plutonic rocks (M < 90%) based on their modal mineral content, using the Q and F percentages and P/(A + P) ratio (see text)

Name	Percentage of quartz (Q)	Ratio of plagioclase to alkali feldspar = 100 × P/(A + P)
Granite	20–60	10–65
Granodiorite	20–60	65–90
Syenite	0–5	10–35
Diorite	0–5	90–100[a]
Gabbro	0–5	90–100[b]

[a] Plagioclase more sodic than An_{50}.
[b] Plagioclase more calcic than An_{50}.

Table I2 Classification and nomenclature of some volcanic rocks (M < 90%) based on their modal mineral content

Name		Percentage of quartz (Q)	Ratio of plagioclase to alkali feldspar = $100 \times P/(A + P)$
Rhyolite		20–60	10–65
Dacite		20–60	65–100
Trachyte		0–5	10–35
Basalt	Q	0–20	65–100
	or F	0–10	
Andesite	Q	0–20	65–100
	or F	0–10	

high pressures. These rocks occur rarely on the Earth's surface and are thought to represent mantle material.

If the minerals of the volcanic rocks can be distinguished with the naked eye, an analogous classification to the plutonic rocks may be used (Table I2), but it presents more difficulties. An alternative chemical classification based on the total alkali versus silica plot (TAS) has accordingly been developed (Le Bas, Le Maitre and Woolley, 1992). Silica is used as a parameter to distinguish ultrabasic, basic, intermediate and acid rock groups at 45%, 52% and 63% by weight. The alkalis ($Na_2O + K_2O$) range from less than 3% to over 14%.

From Tables I1 and I2 it will be seen that rhyolite is the extrusive equivalent of granite, and trachyte the equivalent of syenite. Basalt and andesite, which form the great majority of volcanic rocks, are distinguished modally by their petrographical character, SiO_2 percentage and color index, but the TAS scheme may have to be used for them. Varieties of basalt, which dominate the igneous rocks of the Earth's crust, include olivine, alkali, subalkali, high-alumina, island arc, tholeiitic and mid-ocean ridge basalt (MORB).

Although the alkaline rocks constitute only a very small percentage by area of all igneous rocks, they show an extremely wide range of mineral combinations. As a result a bewildering variety of their rock types has been described. The rocks contain relatively high proportions of the alkalis, resulting in the presence of sodic pyroxenes or amphiboles, with or without feldspathoids (Sørensen 1974).

Occurrence

Granite

Granite and related rocks dominate the plutonic types in the Earth's crust. Mountain-building and subsequent denudation has exposed large areas, many in Precambrian terranes. Some granites ('A type', standing for anorogenic) are typical of rift zones and the interiors of stable continental plates. In one classification, granites are distinguished as I type, indicating that they may originate from igneous source rocks, or S type, of which the source rock is thought to have been sedimentary (pelitic) rocks. I-type granites are metaluminous calc-alkaline rocks, ranging through granodiorite to tonalite; S-type granites are peraluminous rocks characterized by the presence of muscovite (Chappell and White, 1992).

Granitic rocks occur throughout the world, in plate margin terranes from the Precambrian (> 560 Ma) to the Cenozoic (< 65 Ma), for example the extensive Mesozoic batholiths (c. 100 Ma) of North America, from Alaska to Lower California, and the large plutonic belts of the Himalaya, Karakoram and Southern Tibet, which are closely related to the Mesozoic–Cenozoic evolution of the region (Le Fort, 1988). In many places there is a relationship between emplacement of the granitic batholiths of plate-marginal orogenic belts and volcanism.

Basalt

Basalts represent the partial melting of planetary interiors (mantles). Examples from planetary bodies have been studied from the Earth, the Moon and from achondrite meteorites (Basalt Volcanism Study Project, 1981, p. 311), as well as by indirect evidence from other planets. Basalts span the history of the solar system from 4.5 billion years to the present. They are composed of calcic plagioclase and pyroxene, with or without olivine. Many different types have been distinguished. Pumice is a lava so full of gas that it floats in water. It is formed by the sudden release of pressure so that bubbles form which are trapped in the molten rock. Ash flows or pyroclastic flows are composed of fragments of magma, produced by the explosive expansion of gas. They are emitted from volcanic vents as airborne debris or flow from the vent at high speeds as glowing clouds (nuées ardentes). Fragments of volcanic rock blown out during volcanic explosions are called tephra. They are divided according to their size as volcanic ash (fragments less than 2 mm diameter), lapilli (2–64 mm diameter), and larger blocks or bombs. Tuff is a rock formed by compaction of volcanic fragments. If erupted in very hot form, these may result in a 'Welded tuff' or ignimbrite. Their submarine eruption creates a palagonite tuff.

Flood basalts (plateau basalts) are low-viscosity flows which have erupted along fissures in the continental crust. Examples of basalt plateaus are the Lake Superior Basin lavas (Late Precambrian), or the Tertiary basalts of the Deccan Traps, India, and the Columbia Plateau, USA. Ocean-floor basalts are easily the most abundant rocks on the Earth's surface and have been erupted from mid-ocean ridges (mid-ocean ridge basalts, MORB) and from transverse fracture zones. These rocks are largely tholeiitic, oversaturated with respect to silica (quartz-normative) and are orthopyroxene- (hypersthene-) normative. They contain low concentrations of K and Ti. Also tholeiitic in character but with a wider range of compositions are intraplate volcanoes, the shield volcanoes, e.g. Hawaii, the mountains of which are the highest individual mountains (from sea bed to summit) on Earth. The island-arc basaltic rocks are associated with convergent plate margins, the volcanism being mainly of intermediate (andesitic) composition (Basalt Volcanism Study Project, 1981).

The basaltic meteorites (q.v.) are achondrite types which originated in planetary bodies that had developed a basaltic crust. They include eucrites (q.v.) (about 4.6 billion years) and members of the shergotty–nakhlite–chassigny (SNC) meteorite class (q.v.) that are very much younger and came from a source by impact splash, perhaps from Mars, containing volatile chemical elements (Wasson 1985).

Lunar mare basalts cover about 17% of the Moon's surface, mainly on the near (Earth) side. They are flood basalts consisting of very gently sloping thin flows, reflecting a low magmatic viscosity. They are dark mafic rocks composed essentially of clinopyroxene and plagioclase with textures comparable to terrestrial basalts. However, they differ from terrestrial basalts in very low contents of alkalis, high TiO_2 and MgO, low Al_2O_3 and SiO_2, and no detectable H_2O. Iron is present as FeO or native iron; there is no Fe_2O_3 (Wilhelms, 1984). An account of the evolution of the Moon is given by Schmitt (1991).

By various remote sensing techniques basalts can be inferred to exist on other bodies in the solar system (Basaltic Volcanism Study Project, 1981, p. 439). These include photography, X-ray and γ-ray measurements, reflectance and spectral measurements. Many observations suggest that basalts occur on Mercury, Venus, Mars, Io (the innermost of the Galilean satellites of Jupiter) and Vesta (the brightest asteroid). Mercury has large dark plains similar to some lunar maria, and is believed to have a lunar-like soil. If basalts do occur, they are likely to be a low-Ti variety. Pyroxene and olivine are thought to occur in the dark areas of Mars, while the bright areas probably consist of ferric oxide or hydroxide and mafic silicate rock thought to be basalt. There is evidence of enormous shield volcanoes and extensive lava plains, notably in the Elysium and Tharsis regions; pyroclastic deposits may also occur (Basaltic Volcanism Study Project, 1981, p. 460; Strom, 1984; Carr, Kuzmin and Masson, 1993). Basalts, perhaps widely distributed, are likely to be present on the surface of Venus, as may granites, deductions which are based on measurements of U and K at two landing sites (Basaltic Volcanism Study Project, 1981, pp. 465, 788; Saunders and Carr, 1984). Vesta is the only asteroid to show a spectrum diagnostic of basalt, and may be the source of basaltic achondrite meteorites. The satellite of Jupiter, Io, very similar to the Moon in mass and size, is the most volcanically active body yet observed in the solar system (Basaltic Volcanism Study Project, 1981, p. 793).

Peter A. Sabine

Bibliography

Basaltic Volcanism Study Project (1981) *Basaltic Volcanism on the Terrestrial Planets*. New York: Pergamon Press, 1286 pp.

Bowes, D.R. (ed.) (1989) *The Encyclopedia of Igneous and Metamorphic Petrology*. New York: Van Nostrand Reinhold, 666 pp.
Carr, M.H. (1984) Earth in *The Geology of the Terrestrial Planets*, (ed. M.H. Carr). Washington, DC: NASA Scientific and Technical Information, pp. 79–105.
Carr, M.H., Kuzmin, R.A. and Masson, P.L. (1993) Geology of Mars. *Episodes*, **16** (1 and 2), 307–15.
Chappell, B.W. and White, A.J.R. (1992) I- and S-type granites in the Lachlan Fold Belt. *Trans. Roy. Soc. Edinburgh: Earth Sciences*, **83**, 11–26.
Deer, W.A., Howie, R.A. and Zussman, J. (1962a) *Rock-forming minerals*, Vol. 1, *Ortho- and Ring Silicates*. London: Longman, 333 pp.
Deer, W.A., Howie, R.A. and Zussman, J. (1962b) *Rock-forming minerals*, Vol. 3, *Sheet Silicates*. London: Longman, 270 pp.
Deer, W.A., Howie, R.A. and Zussman, J. (1962c) *Rock-forming minerals*. Vol. 5. *Non-Silicates*. London: Longman, 371 pp.
Deer, W.A., Howie, R.A. and Zussman, J. (1963a) *Rock-forming minerals*, Vol. 2, *Chain Silicates*. London: Longman, 379 pp.
Deer, W.A., Howie, R.A. and Zussman, J. (1963b) *Rock-forming minerals*. Vol. 4, *Framework Silicates*. London: Longman, 435 pp.
Deer, W.A., Howie, R.A. and Zussman, J. (1978) *Rock-forming minerals*, Vol. 2A, *Single-Chain Silicates*. London: Longman, 668 pp.
Deer, W.A., Howie, R.A. and Zussman, J. (1982) *Rock-forming minerals*, Vol. 1A, *Orthosilicates*, 2nd edn. London: Longman, 919 pp.
Deer, W.A., Howie, R.A. and Zussman, J. (1986) *Rock-forming minerals*, Vol. 1B, *Disilicates and Ring Silicates*, 2nd edn. London: Longman Sci. & Tech., 629 pp.
Johannsen, A. (1931–1938) *A Descriptive Petrography of the Igneous Rocks*, 4 Vols. Chicago: Chicago University Press.
Le Bas, M.J., Le Maitre, R.W. and Woolley, A.R. (1992) The construction of the total alkali–silica chemical classification of volcanic rocks. *Mineral. Petrol.* **46**, 1–22.
Le Fort, P. (1988) Granites in the tectonic evolution of the Himalaya, Karakoram and southern Tibet. *Phil. Trans. Roy. Soc. London*, **A326**, 281–299.
Le Maitre, R.W., Le Bas, M.J., Sabine, P.A. *et al.* (1989) *A Classification of Igneous Rocks and Glossary of Terms*. Oxford: Blackwell Scientific. 193 pp.
Le Maitre, R.W., Le Bas, M.J., Sabine, P.A. *et al.* (1990) *A Classification of Igneous Rocks* (wall chart). Beijing: Geological Publishing House (in Chinese).
Le Maitre, R.W., Le Bas, M.J., Sabine, P.A. *et al.* (1991) *A Classification of Igneous Rocks and Glossary of Terms*. Beijing: Geological Publishing House, 253 pp. (In Chinese).
Macdonald, G.A. (1972) *Volcanoes*. Englewood Cliffs, New Jersey: Prentice-Hall, 510 pp.
Sabine, P.A. (1989) Setting standards in petrology: the Commission on Systematics in Petrology. *Episodes*, **12**(2), 84–6.
Saunders, R.S. and Carr, M.H. 1984. Venus, in *The Geology of the Terrestrial Planets* (ed. M.H. Carr). Washington, DC: NASA Scientific and Technical Information, pp. 57–77.
Schmitt, H.H. (1991) Evolution of the Moon: Apollo model. *Am. Mineral.*, **76**(5&6), 773–84.
Sørensen, H. (ed.) (1974) *The Alkaline Rocks*. London: John Wiley and Sons, 622 pp.
Shams, F.A. (ed.) (1983) *Granites of Himalayas, Karakorum and Hindu Kush*. Lahore: Institute of Geology, Punjab University, 427 pp.
Smith, J.V. (1979) Mineralogy of the planets: a voyage in space and time. *Mineral. Mag.*, **43**, 1–89.
Streckeisen, A. (1967) Classification and nomenclature of igneous rocks (final report of an enquiry). *Neues Jahrb. Mineral. Abhandlungen*, **107**(2), 144–214, and **107**(3), 215–40.
Streckeisen, A. (1973) Plutonic rocks. Classification and nomenclature recommended by the IUGS Subcommission on the Systematics of Igneous Rocks. *Geotimes*, **18**(10), 26–30.
Strom, R.G. (1984) Mercury, in *The Geology of the Terrestrial Planets* (ed. M.H. Carr). Washington, DC: NASA Scientific and Technical Information, pp. 13–55.
Wasson, J.T. (1985) *Meteorites*. New York: W.H. Freeman & Co., 267 pp.
Wilhelms, D.E. (1984) Moon, in *The Geology of the Terrestrial Planets* (ed. M.H. Carr). Washington, DC: NASA Scientific and Technical Information, pp. 107–205.

Cross references

Achondrite meteorites
Basalt
Basaltic achondrite meteorites
Differentiation
Earth: geology, tectonics and seismicity
Flood basalt
Moon: geology
Silica, silicate
Volcanism in the solar system

IMAGING SCIENCE

Imaging science is the process of utilizing two-dimensional images to obtain or transmit information on the physical properties of an object or scene. As a presentation of information in an immediately identifiable form, an image will automatically call into play all the finely tuned, complex and autonomous capabilities of the human eye-brain system to perceive, recognize, synthesize and interpret its contents. Because we are exquisitely engineered to comprehend visual stimuli arrayed into two dimensions, images hold an eminent position in the vocabulary of human communication. Two people exchanging ideas on technical, spatial or even temporal relationships will often resort to drawing pictures.

It may be said that the first imaging scientists were Paleolithic people who attempted to convey the nature of their environment and activities through the creation of cave paintings. Other subsequent but early forms of imaging technology – such as the depiction on canvas of major historical events like battles, coronations, and the appearance of comets; the scientific drawings of da Vinci; the first images produced by the printing press – all relied heavily on the subjective point of view of the artist. It wasn't until the invention of photography that the recording of information could be objectively and faithfully rendered, and imaging data could be used in the pursuit of precise scientific information. The use of photographic glass plates placed at the focus of telescopes to record astronomical images is an example. However, even photography permitted very limited interaction with the image after its capture, and the results had to be physically and therefore slowly disseminated, as by mail.

Circumstances had changed considerably by the early 1920s when digitized photographs of world news events were being transmitted by submarine cable between New York and London. The earliest picture cable transmission systems encoded each position in the photograph into 5 distinct brightness levels; reconstruction at the receiving end was accomplished using a telegraph printer fitted with typefaces simulating a halftone pattern. By 1929 the reconstruction process and the encoding capability had both improved, the latter increasing to 15 levels. These were the first steps in image processing to enhance human interpretability. Improvements in the processing of image data continued over the next 35 years, but the application of image processing concepts did not become practical until the middle 1960s, when digital computers began to offer the speed and data storage capabilities required for the implementation of image processing techniques.

Like the large-scale digital computer, planetary exploration was also a child of the 1960s. Mariner 2, the first spacecraft dispatched from our planet to successfully reconnoiter another, was launched by the US from Cape Canaveral on 27 August 1962. During the next 7 years three Mariners (q.v.) flew past Mars and one past Venus, returning an impressive photographic diary of their travels. In the 1970s and 1980s there were equally dramatic events in the US space program. Pioneer 10 and 11 (q.v.) explored and photographed the Jupiter system; Pioneer 11 continued on to Saturn. Mariner 10 photographed Venus and Mercury. By 1977 the two Viking landers (q.v.) had taken detailed pictures of the surface of Mars, while the Viking orbiters continued to map the entire planet, one returning images for over 4 years. Two automated, reprogrammable spacecraft, Voyagers 1 and 2 (q.v.) spent the late 1970s and 1980s touring the outer solar system. Both spacecraft relayed back to Earth stunning pictures of the giant planets and examined their rings and moons at close quarters. During these three decades the Soviet Union was equally active, launching robotic craft to Venus and Mars. And both nations, through the year 1976, periodically visited the Moon, the later US missions piloted by humans.

The bounty of images returned by the space program provided fertile ground for the emerging fields of digital image processing and imaging science. The first fruitful application of computer image processing techniques began in 1964 at the Jet Propulsion Laboratory in Pasadena, California, when pictures of the Moon, transmitted by Ranger 7, were processed by a computer to correct various types of image distortion inherent in the onboard television camera. Since that beginning, both image processing and planetary imaging science have been inextricably intertwined and have grown together (though digital image restoration and enhancement is today being applied in many different scientific and practical areas).

In slightly under three decades, humankind successfully explored almost every planet in the solar system, and on almost every vehicle destined for another planet, the capability for two-dimensional digital imaging was carried aboard. Though there are many ways to gather data on the planets, imaging is by far the most widely applicable. One need only examine the annals of discoveries made by planetary probes to see the diversity of imaging science contributions. Images provide basic physical information on a variety of bodies – those with atmospheres and those without – over an almost limitless range of sizes, starting at a size equal to the wavelength of light. Because of the broad range of scientific objectives they can address, imaging data are often used in a corroborative fashion, providing the spatial context for the interpretation of data gathered by other scientific (e.g. spectral or radiometric) instruments not equipped to gather precise spatial relationship information. As a means of discovery, the science of imaging is unparalleled, replacing and removing our most sophisticated sense, human vision, to places too far and too formidable for humans to travel. Nowadays information beyond the range of human perception (in the infrared or extreme ultraviolet, for example) is often arrayed into two dimensions and displayed in visible colors to aid in human interpretation of these otherwise unseen spectral regions. Finally, the emotional impact of viewing a picture of a distant and alien world, in a way that the human eye would see it, cannot be overlooked. The sense of 'being there' provided by an image is an undeniable aspect of the appeal and power of imaging science.

In this article we review the general background and basic principles of image formation and digital image recording and processing, concentrating on those which are pertinent to the imaging of celestial bodies. A description of the processing techniques commonly used in planetary imaging science and a Voyager photograph collection illustrating some of these techniques are also presented.

Imaging devices: from photons to pictures

The nature of light

It was a Scot named James Clerk Maxwell who first recognized that light is a wave of sinusoidally fluctuating electric and magnetic fields. In formulating his electromagnetic theory in 1862, he succeeded in characterizing these fields and their relationships to charges and currents through a set of partial differential equations now known as Maxwell's equations. (q.v.) A great variety of macroscopic electromagnetic wave phenomena can be explained through Maxwell's equations: propagation, dispersion, reflection, refraction and interference. However, the interaction of electromagnetic disturbances with matter on an atomic and molecular level, and the fact that light is quantized into discrete packets known as 'photons', was not understood and described until the advent of modern quantum theory, built on contributions from Planck, Einstein, Bohr, Compton, de Broglie, Davisson and Germer, Heisenberg and Schrodinger. As a result of the evolution in understanding of the dual wave-particle nature of light, it is presently possible to furnish a consistent and unambiguous theoretical explanation for all optical phenomena using a combination of Maxwell's electromagnetic wave theory and modern quantum mechanics.

The electromagnetic spectrum is an ordering of radiation according to wavelength or energy: those waves with the shortest wavelengths (or highest frequency) have the largest energy. In vacuum, light of different wavelengths travels at the same speed. In matter, however, the speed of light is wavelength dependent: the interaction (which depends on the frequency of the wave) between light and the electrons bound to the atoms and molecules within a material can alter the propagation speed. Since the natural oscillation frequency of most bound electrons in matter is closest to the frequency of ultraviolet light, the ultraviolet waves will interact more strongly with matter than the blue, the blue more strongly than the red, etc. For example, the phenomenon of refraction – the bending of the direction of propagation of the wave in a material such as glass – will bend light of shorter wavelengths more than it will bend light of longer wavelengths, resulting in spectral dispersion. (It is the phenomenon of refraction, then, that explains how a prism breaks up white light into its component colors.) As a second example, the preferential scattering and removal from the line of sight of blue light, and the relatively undisturbed transmission of red light, by atmospheric molecules, turns the setting Sun red. The almost complete absorption of some colors and high transmittance of others is what gives stained glass its color: red glass is highly absorbing in the green and blue regions of the spectrum and transparent in the red.

Electromagnetic waves can also be 'bent' in vacuum through a process known as diffraction. Each point on any electromagnetic wavefront is a source of secondary spherical waves. Enunciated by Dutch astronomer and physicist Christian Huygens (q.v.) in the 17th century, this principle predicts the propagation of light around and behind a barrier: as a plane-parallel wave of light illuminates a sharp-edged opening in an opaque screen, all points along the wavefront passing through the opening serve as sources of secondary waves which propagate into all (forward) directions, including the region behind the screen. Each point in the shadow, especially those close to the opening, is illuminated by light rays having originated with the same phase but having traveled different distances and therefore arriving with different phases. This circumstance leads to the occurrence of constructive and destructive interference fringes in the shadow region. The angular width of the first circular fringe, referred to as the diffraction disk, is given by $\sin\theta = 1.22\lambda/a$, where a is the diameter of the opening and λ is the wavelength of light. Thus the larger the opening, the smaller the diffraction disk. The locations of the remaining interference fringes also depend on the size of the opening and the wavelength of the light. Because of these dependencies, the phenomena of diffraction and interference can be used to isolate certain spectral components from a white light source, as in the case of a diffraction grating (Rossi, 1957).

Optics

When a photon strikes an object, it can be absorbed, scattered or transmitted by the object. An image is simply a two-dimensional representation of the intensity (or brightness) of the electromagnetic radiation which has interacted with the object(s) in the imaged scene.

Refraction and reflection (a special form of scattering) can be used to focus and redirect these rays of light. Lenses which operate on refraction are of great importance in devices that collect and redirect light, such as telescopes, microscopes, film projectors and cameras. An object seen through a lens can appear to change in size because the rays emerging from the lens enter the eye forming a different angle than they would have formed without the lens. Mirrors, using reflection instead of refraction, can be shaped to redirect light rays and work in a fashion similar to lenses.

The resolution of an optical system is ultimately limited by the wave nature of light. Diffraction produces an unavoidable blurring of an object, even point sources, imaged through a lens or with a curved mirror: as with the opening in the diffraction screen, the smaller the diameter of a lens or mirror, the larger the diffraction disk. Aside from collecting as much light as possible from faint astronomical objects, the desire to increase the apertures of telescopes arises from the desire to improve resolving power. A somewhat arbitrary but commonly used criterion for the resolving capability of an optical element is called Rayleigh's criterion: two point sources are resolved if the centers of their diffraction disks are separated by a distance greater than the radius of the disks. Theoretically the minimum angle of resolution for two point sources, then, is given by the angular size of the diffraction disk, $1.22\lambda/a$, and is obviously dependent on wavelength. In practice the angular resolution is often taken to be twice the size of the blur circle of a point source, whether caused by diffraction or some other effect.

There are many other effects which can blur an image: spherical and chromatic aberrations, astigmatism, etc. They all affect the images produced by either lenses or mirrors. (Discussion of the basics of the optics of lenses and mirrors and their limitations can be found in any basic optics book; Rossi, 1957). However, none of these is as fundamental a limitation to the resolving power of an image-forming device as diffraction.

Telescopes and cameras

For most celestial bodies the only information available to us is the light they emit or scatter, and astronomical telescopes are constructed using very large lenses (refractors) or mirrors (reflectors) to collect as much light as possible from, and form an image of, these distant objects. Light rays emitted or reflected from a point on an object generally form a spherical wave. The portion of that wave intercepted by a very distant telescope is, to excellent approximation, a bundle of parallel rays. These are bent by the optical element in the telescope to form an image in the focal plane. The distance of the focal plane from the optical element is called the focal length. The larger the focal length, the greater the magnifying power of the optical system. Though telescopes obviously have very much larger focal lengths than hand-held cameras or the human eye, the image-forming principles are identical. One major difference is that a telescope is designed to intercept the parallel rays coming from very distant objects, rendering an adjustable focal length unnecessary. Cameras and eyes, which must intercept light from objects at a variety of distances, are designed with adjustable focal lengths. The prime focus of the 200-inch (5-m) telescope on Mt Palomar in southern California has a fixed focal length of 1000 in (25 m). A common focal length for hand-held cameras is 50 mm. Spaceborne telescopes, also referred to as 'cameras', generally have intermediate focal lengths, of 200–2000 mm, depending on the purpose for which they were designed: high magnification or large field of view.

For many scientific objectives it is necessary to separate the various spectral components comprising the white light coming from an imaged object. One can isolate radiation of a particular wavelength by utilizing the predictable interaction of light with certain materials and devices such as colored glass, prisms, gratings, etc. Spectral isolation can be most easily accomplished in a camera by placing a spectral filter in the optical train, generally immediately in front of the focal plane. The filter can be simply colored glass which will transmit a broad range of the electromagnetic spectrum, such as red or green light. Spectral images taken through these broadband filters are used to determine the overall color of a satellite, for example, which may be useful in diagnosing the mineralogical content of its surface. Or filters can be constructed of discrete and very thin multiple layers of different materials to make use of the principles of interference and can be designed to pass only a very narrow band of light, from a few angstroms up to tens of angstroms. These filters may be used to form an image in light emitted within certain spectral emission lines, such as the sodium (Na) D lines emanating from the Io plasma torus, or to form an image of a planet such as Saturn in light restricted to one of its relatively narrow and deeply absorbing atmospheric methane spectral bands. In the former case the bright areas in the image would be those containing a significant amount of Na; dark areas are relatively devoid of Na. In the latter case, atmospheric features appearing bright in the image would be those which are relatively high in the atmosphere and do not have a significant amount of (absorbing) methane above them. The stratigraphy of an atmosphere can be determined through such spectral imaging.

Detectors

At the focal plane of every image-recording device is a light sensitive material or detector. The function of the detector is to convert the radiation into a signal that can be stored or transferred to some other medium for storage and/or display. Different types of detectors are most effective at different wavelengths. There are two major technologies for image detection and recording: photochemical and photoelectronic (or optical-electronic). The former is exemplified by photographic film; the latter, by the television camera.

Image detection and recording by photographic film, the most common and presently readily accessible medium for recording and storing images, relies upon the properties of halide salts and silver. Silver halides are changed by exposure to light so that the action of mild reducing agents, called 'developers', results in the precipitation and deposition of grains of free silver (Mees, 1954). After development, the film is 'fixed' by chemical removal of the remaining silver halide grains. The more light that reaches a given area of the film, the more silver halide is rendered developable and the denser the silver deposit that is formed there. Because the silver grains are largely opaque to visible light, an image of gray tones is obtained in which the brightness levels are reversed, thus producing a film negative. The process is repeated to obtain a film positive.

Despite its common use and low cost, the photochemical process has severe limitations for quantitative analysis (Mees and James, 1966). Like the eye, most conventional black and white and color films are sensitive only to the visible portion of the electromagnetic spectrum from 0.4 to 0.7 μm. (However, film emulsions may be deliberately sensitized to a different spectral region, such as the ultraviolet or the near-infrared.) In addition, the mass of silver deposited is proportional to the logarithm of the total exposure time, resulting in a highly nonlinear mapping of the input intensity to the output brightness recorded by the film. Subsequent determination of the relative intensities of parts of a photographically recorded image becomes difficult. When all available silver is deposited due to very bright light, the film becomes saturated and increasing the brightness and/or exposure time makes no difference. On the other end, some small amount of silver is deposited even in the absence of any light exposure, causing fog. Another severe limitation of photography is that it allows very limited interaction with an image after it has been captured.

Photoelectric detectors make use of the photosensitive and electrical conducting properties of certain semiconductors to convert electromagnetic radiation to electrical signals. Such detectors have two great advantages over the photographic process. First, because the output is an electrical signal, it can be relayed or transmitted to a remote observer, a vital condition for spaceborne instrumentation. Second, the efficiency of photoelectric detectors in detecting quanta of light is generally higher, by a factor of 100 or more, than that of film. Subsidiary advantages include the possibility of signal amplification, linear response to light, a larger dynamic range (the difference between the minimum detectable light level and the level which produces saturation in the detector), and a generally larger range of wavelength sensitivity.

Whether or not the photoelectric detector is a single 'spot' detector, a linear array of small individual detectors, or a two-dimensional array of small detectors depends on the type of camera and the application. We focus our attention here on framing cameras using two-dimensional photoelectric detectors located in the focal plane of a telescope to record all picture elements or 'pixels' of an image simultaneously. These usually employ a shutter immediately in front of the focal plane to limit the exposure, and electronic circuitry to manipulate the output signals.

A two-dimensional photoelectric detector used in early television cameras and carried aboard planetary spacecraft in the 1960s and 1970s was the vidicon tube, consisting of a small (about 2.5 cm on a side) light-sensitive plate of semiconducting material, such as selenium–sulfur, whose conductivity is proportional to the intensity of light which strikes it. The electromagnetic radiation from the optics is focused on the plate, which is then raster scanned with an electron beam whose current changes depending on the conductivity of the pixel on which it falls. The changes in the analog output current, which is continuously varying as the electron beam sweeps across the detector, are amplified, sampled, quantized and digitized by electronic circuitry for later image reconstruction.

Despite the many advantages of vidicon tubes over photographic film, they are not without drawbacks. The range of light intensity over which their response is linear does not extend over the full dynamic range of the device, and the raster scanning process exhibits barrel distortion (discussed below). Both require computer data processing to correct.

A more complex two-dimensional detector which has greater sensitivity and is better photometrically and geometrically behaved than a vidicon tube is the charged-coupled device (CCD; q.v.). Commonly used in both Earth-based and spaceborne astronomical imaging devices today, CCDs are comprised of an array of individual metal oxide semiconducting diodes which change in electric charge when light strikes them. The charges within each diode (which form the basic pixels in a CCD image) are shifted in turn from one diode to the next and ultimately out of the device by periodically varying the voltages between diodes. As with the vidicon tube, the changes in the resulting analog current coming out of the device can be amplified, sampled, quantized and digitized.

Even for spaceborne cameras, which are not used to image through the Earth's atmosphere and which can be designed to minimize optical blurring effects like aberration and astigmatism, the inherent resolving power may be limited by effects other than diffraction. The size of the electron beam in a vidicon tube or the diode in a CCD was at one time the limiting factor in resolving power for such cameras. During the 30 years since the advent of CCDs, the size of the smallest manufacturable diode has shrunken: it is not uncommon now for CCDs to have diodes less than 10 μm on a side. Consequently, for those spaceborne cameras having relatively long focal lengths the diffraction limit can in practice be achieved by choosing the CCD so that the size of the diodes is matched to the size of the diffraction disk.

The physical dimension at the imaged object corresponding to a single diode in the CCD (or, equivalently, a pixel in the image) is determined by the inherent resolving power of the optical system and the distance to the imaged object. Obviously, the closer a spaceborne camera gets to its target, the smaller will be the smallest resolvable features. For example, the Voyager long focal length camera (with a focal length of 1500 mm) has a minimum angle of resolution, given by twice the width of the electron beam of its vidicon detector, of $\sim 18 \times 10^{-6}$ radians. An image taken by this camera from a distance of 1 million km has a resolution of 18 km: i.e. two features in the image closer than 18 km cannot be fully resolved as two separate features.

Digitization

The signals coming out of a detector must be quantized into discrete values for storage in digital form. This conversion between analog samples and discrete numbers is called quantization, and implies that the full dynamic range of possible signal values must be divided into discrete bins. The number of quantization levels must be sufficiently large to represent the smallest variations in brightness required of the imaging system, and is often dictated by a desire to keep the artificial effects of quantization unnoticeable to the human eye.

The number of bits required to represent the signal or brightness value of each pixel is the logarithm (to the base 2) of the number of quantization levels. If each pixel's brightness value is represented by an 8-bit number, the number of quantization or grey levels is 256. The difference between one level and the next is 0.4% of maximum brightness. When converted to visual response units, this quantization interval matches the contrast discrimination capability of the human eye only under the best and seldom-realized viewing conditions (Blackwell, 1946, Gonzalez and Wintz, 1987). Selective contrast enhancement in digital processing, which can bring these and smaller quantization differences within the range of detection of the eye, justifies digitization to even higher levels: 16 bits, for example, would result in 65 535 possible gray levels.

Spurious effects and imperfections

As mentioned above, imaging systems consisting of optics and detectors can be far from ideal, with various geometric and radiometric errors and distortions introduced into the output data. Likewise, methods used to manipulate the electronic signals may also distort and introduce defects into the image. Random noise may be introduced by the detector, periodic noise may be caused by interference from concurrently operating electronic equipment and when the image is acquired from a satellite, noise may be introduced during transmission back to Earth. When the noise level is higher than the quantization level, the noise becomes noticeable in the image. Image processing techniques are available to reduce or eliminate most types of noise.

Digital image processing

Interest in digital image processing techniques dates back to the early 1920s when digitized pictures of world news events were first transmitted by submarine cable between New York and London. Through continued developments, including the arrival of the digital computer which facilitated the widespread application of image processing concepts, digital image processing has become an important tool of research and interdisciplinary study in many scientific areas, from medicine to metallurgy.

Image processing is simply the manipulation of digital images by computer for more effective display, and its primary purpose is to aid the human analyst in the extraction and interpretation of information which may not be readily apparent in the original image. The various techniques employed in image processing are generally designed to enhance particular characteristics of the image so that they are well matched to the properties of the human visual system. The information of significance to a human observer may be definable in terms of the observable characteristics of contrast, texture (i.e. the spatial distribution of brightness), shape and color. The characteristics of the data and display medium, the properties of the human visual system and the final objectives of the image analysis determine the transformation from the original to the enhanced image.

Digital image processing typically proceeds in three steps: (1) image restoration, (2) image enhancement and (3) image analysis. Restoration techniques are oriented towards modeling the known degradations in the image quality and applying the inverse process(es) in order to recover the original scene as much as possible. Enhancement techniques are procedures which are designed to manipulate the characteristics of an image – contrast, texture, shape and color – to make it more easily interpretable by a human analyst. Image analysis is the final process of extracting the desired scientific information.

Image restoration techniques

Image restoration is the step in image processing which includes cosmetic improvement in image quality and correction for geometric distortions. Often it includes the rearrangement or mapping of the pixels into a specific geometry as well as the conversion of the raw brightness values (generally referred to as data numbers, DN) of each pixel to absolute radiometric units (radiometric correction). The cosmetic correction or 'clean-up' part of image restoration is extremely important because many enhancement operations will emphasize image imperfections to such an extent that useful information can be obscured (Andrews and Hunt, 1977). For this reason, image restoration precedes enhancement and analysis.

Initial image correction procedures vary from system to system and application to application. It is important to distinguish between systematic and scene-dependent distortions. Systematic distortions are caused by idiosyncracies of the spacecraft and camera system and are relatively easily identified and removed. Examples of scene-dependent distortions, which are more difficult to define and correct, are parallax (geometric) and local charge migration in vidicon cameras (radiometric).

We discuss here the three types of errors or distortions treated in the image restoration process: (1) geometric distortions, (2) radiometric errors and (3) cosmetic imperfections. (For more information on image restoration, see Andrews and Hunt, 1977)

Geometric corrections

The purpose of geometric correction or rectification is to rearrange the pixels in an image to conform to some standard cartographic projection (as in the creation of maps), to reflect a particular viewing geometry, or simply to remove distortions introduced by the imaging system and redefine internal image geometry. The geometric transformation used in this process can be an analytical or numerical mathematical function defining the relationship between the points in the raw image and those in the rectified image. When applied to the raw image, the transformation is analogous to printing the image on a rubber sheet, and stretching it until the relationship of the points in the rectified image matches a predefined and usually standard geometry.

A geometric transformation consists of two basic operations: (1) spatial transformation, which defines the rearrangement of pixels in the transformed image, and (2) resampling, which determines the correct assignment of brightness values to the transformed pixels. In a digital image the pixel locations are defined by two integer values, one specifying the row and one specifying the column in the two-dimensional image array. Under the first step in a geometric transformation, a pixel in the raw image will in general be mapped onto a different location in the rectified image plane specified by two non-integer numbers. This may apply to most or all transformed pixels in the rectified image. To turn this array of non-integer locations into a new digital image, data numbers must be evaluated at integer locations in the rectified image plane in a process known as resampling. Resampling techniques differ in the method by which the data numbers in the output image are derived from the data numbers in the input image. The principal resampling algorithms used are 'nearest neighbor', bilinear interpolation and convolution.

The initial geometric transformations applied to an image are generally those designed to correct gross internal distortions. Modern-day CCD detectors are virtually distortion free, but a common geometric distortion that afflicts vidicon tubes is barrel distortion. The electron beam which scans the image in the vidicon tube does not follow a perfectly square raster pattern, yet the output current is sampled, quantized and mapped onto a square grid, yielding an imperfect mapping from detector to digital image. The form of this distortion varies from system to system and can be scene dependent, varying from image to image as the electron beam is itself deflected by the image charge distributions. In the Voyager cameras, barrel distortion gave rise to scale changes of several percent near the perimeter of the image.

In order to measure and correct for this effect, a pattern of reseaux can be permanently burned onto the semiconducting surface of the detector. The position of each reseau on the detector is precisely measured. The final displacement between the apparent location of these marks, which appear as black dots in the raw image, and their actual premeasured location on the detector defines the geometric distortion and the spatial transformation to correct it.

Another smaller geometric error is often introduced in a vidicon tube near high-contrast features because the scanning electron beam is deflected by high charge accumulation on the vidicon plate. Thus the measured current does not accurately describe the image brightness where the beam was aimed, but rather at a point a short distance away. The size of such distortions is rarely more than one or two pixels.

The same techniques used in removing distortions introduced by an imaging system may be used to transform an image to a particular viewing geometry or cartographic projection. When imaging a body for which an absolute coordinate system has already been determined, a set of geologic features (of known latitude and longitude) may be used in the same way that reseaux on the detector's surface are used: to serve as a control grid by which the points in the final rectified image or map may be related to the points in the raw image. Detailed discussion of geometric rectification, spatial transformations and resampling algorithms and numerous examples of application of these techniques may be found in Castleman (1979), Rosenfeld and Kak (1982) and Green (1983).

Radiometric corrections
The goal of the radiometric correction process is to transform the raw data numbers to radiometrically defined units. Vidicon cameras exhibit a variety of problems which make the raw image data numbers unusable for accurate radiometric work such as photometry, photoclinometry or the generation of image mosaics comprised of normal albedos. In the case of a two-dimensional detector like a vidicon or CCD, the properties of a photosensitive detector – sensitivity to light, linearity of response, dark current, etc. – may vary spatially across the detector, some much more than others. The radiometric processing sequence addresses these effects for the entire array of pixels.

Because data numbers do not generally increase linearly with either radiant flux or with exposure time – the Voyager vidicon cameras displayed distinct nonlinearity of this nature – images must be corrected for nonlinear response. However, as nonlinearity is largely a function of the photosensitivity of the material from which the detector is made, the same correction for nonlinear response is generally applied to all pixels across the detector.

One of the principal difficulties in using spacecraft vidicon cameras to acquire photometrically useful data is the variability of the dark current (DC) or background level in the vidicon. Lack of knowledge of such DC levels results in large additive errors not only in the computed brightness of the scene but in color ratios which are effectively measures of slopes in different parts of the spectral reflectivity curve.

The DC as well as the absolute sensitivity of the detecting material to light generally vary across the field of view. The combined effect is termed 'shading'. As with nonlinearity, shading is a system-dependent effect and the approximations and corrections must therefore be computed for each imaging system and for each image acquired by that system. DC frames created by scanning the detector after an exposure of zero length provide a measure of the DC and its spatial variation. Such frames are deliberately taken during spaceflight for the express purpose of radiometric correction. To correct for the remaining contribution to shading, a uniformly illuminated field is imaged onto the detector: the remaining spatial variation in output DN values (after subtraction of the DC) is then only a function of the sensitivity of each pixel to light. DC and flat-field frames are used to normalize the absolute response of each pixel so that the final corrected image is one that would have been recorded by a perfect detector with no false background signal and with spatially uniform sensitivity to light.

After the relative radiometric errors have been removed, data numbers are on a linear scale proportional to the product of the exposure time and the radiant flux, and variation across the field of view has been removed. The image brightness values can now be scaled to absolute radiometric units.

Additional information on the subject of radiometric correction may be found in Andrews and Hunt (1977), Gonzales and Wintz (1987) and Rosenfeld and Kak (1982).

Cosmetic corrections
An image can be cosmetically imperfect for a variety of reasons: as described above, the detection and scanning process can introduce electronic noise, blemishes attributed to dust on the detector may appear, the loss of bits in the telemetry stream may occur, and reseau marks appear as a distracting grid of black dots all across the image. All of these can be either removed and/or modeled and accounted for. For example, the pixels contaminated by reseau marks or blemishes can be replaced by an average of the pixels surrounding the flaw. In general, cosmetic corrections attempt to make an image look better to the observer, even though the information content is not increased.

Image enhancement techniques

Enhancement techniques are designed to manipulate and display the information contained in an image to match the contrast, texture, shape, and/or color discrimination capabilities of the human visual system. However, the goals of image enhancement are not necessarily restricted to the analysis and interpretation of the image itself. Enhancement may be employed for a number of reasons directed at obtaining an idealized output: graphic overlay, the ultimate merging of different data sets, the production of eye-pleasing images, etc.

There are a great many image enhancement techniques – contrast stretching, spatial filtering, edge enhancement, multispectral processing, temporal analysis and the use of false color. The final objective will determine which is used. The discussion presented here is merely an overview of the most commonly used techniques. Detailed discussion of the techniques of digital image enhancement can be found in many standard texts, among them Pratt (1978), Castleman (1979), Moik (1980) and Gonzales and Wintz (1987).

Contrast stretching
An important consideration in the display and presentation of a processed image is the ability of the eye to discriminate different brightness levels in a complex scene.

The range of light intensity levels to which the human visual system can adapt is enormous – approximately 10^{10} between fully dark-adapted (scotopic) and fully daylight-adapted (photopic) vision. However, the eye is not capable of discriminating subtle brightness differences over this entire range simultaneously and, once adapted to a particular intensity level, can discern a range of brightness differences above and below that level that is far below its overall dynamic range. In examining a complex scene or image, the eye in fact does not adapt to any one brightness level but to an instantaneous and fluctuating brightness level as it roams across the image. Consequently, its ability to distinguish subtle differences in brightness (or contrast) between two objects set against a background depends not only on the average background brightness in the image but on the spatial variation in background brightness, the size of the objects and the nature of the boundary between objects and background (e.g. smooth versus sharp; Gonzalez and Wintz, 1987). The contrast threshold of the eye for well-illuminated scenes and fully resolved objects is $\sim 2\%$. Display devices, such as television monitors, are analog in nature and can display a continuum of different brightnesses. Consequently the ultimate limitation in the discrimination of contrasts in a typical image processing laboratory environment (in which a processed image is displayed on a television monitor) is the response of the eye. When the inherent contrast between the objects of interest in an image is greater than the quantization interval but below the eye's contrast threshold, it is beneficial to increase the contrast across the image. This process is called contrast stretching and is probably the most useful of all computer enhancement techniques.

Contrast stretching is accomplished by redistributing a range of input data numbers over a larger output scale. The redistribution can be linear or nonlinear. For example, in a linear stretch of an 8-bit image (which contains 256 different brightness levels), the 50 lowest brightness levels may be mapped to output brightnesses on a display monitor corresponding to the values between 0 and 250. The input data number 0 would remain at 0, the input value 1 would be assigned the output value 5, 2 assigned to 10, 3 to 15, and so on until the input value at 50 is assigned to 250. With this stretch, the 50 lowest data numbers now occupy nearly the full dynamic range of the original image, and the finest contrasts (0.4%) recorded within the low-brightness end of the image appear with contrast of ~ 2%, well within reach of the eye's discriminating abilities. Most displayed images have been stretched in one form or another to achieve optimum contrast discrimination.

Spatial filtering
Filtering an image in the spatial domain is performed to enhance particular spatial frequencies. An image of a natural scene typically contains a full spectrum of spatial frequencies. The low-frequency component arises from gradual changes in brightness over a relatively large area of the scene (or a large number of pixels in the image). The high-frequency component arises from rapid brightness changes over a short distance (or a small number of pixels) and defines the details of an image.

Digital spatial filtering is used primarily to either exaggerate or minimize abrupt changes in brightness. Algorithms that perform spatial frequency enhancement are called filters because they pass or emphasize certain spatial frequencies and suppress others. Spatial filters that pass high frequencies, emphasizing the details of an image, are called high-pass filters and are normally used for identifying and mapping these details. Conversely, low-pass filters produce image smoothing by suppressing the high spatial-frequency information. Spatial filtering is performed by convolution (or approximation to a convolution) with a box filter. A high-pass filter enhances features that are less than half the size of the filter box, and suppresses features that are more than half the box size. By varying the box size, one can vary the frequency range being enhanced. Typical box sizes for highlighting small-, intermediate- and large-scale structures in an image which is 800 pixels on a side would be approximately 7, 51 and 101 pixels square, respectively.

Edge enhancement
Edge enhancement algorithms are designed to exaggerate rapid changes in brightness levels from one pixel to the next. These very abrupt changes represent the highest spatial frequencies in the scene. Physically, an edge may be a sharp boundary between two different features or geological regions on a body. Edge enhancement has the effect of producing a sharper image and is accomplished with a two-step process. First, a high-pass filter image is generated using a small (i.e. 3×3 pixel to an 11×11 pixel) filter box. This image is referred to as the edge component. Second, the edge component is added back to the original image. The output image is edge enhanced.

Analogous to edge enhancement by high-pass filtering, the directional first difference, which approximates the first derivative, is also designed to highlight the edge information in an image. The first-difference algorithm enhances edges on a pixel-to-pixel scale, rather than an area-dependent neighborhood, by shifting the image by a pixel in either the horizontal, vertical or diagonal directions and subtracting the shifted image from the original image.

Color enhancement
Though the human visual system is limited in contrast discrimination, it is extremely efficient in distinguishing and comparing thousands of different color hues and intensities (Billingsley, Goetz and Lindsley, 1970). The use of color, then, provides a dramatic increase in the amount of information that can be displayed and immediately visually interpreted. Recognizing this, it is often practical to convert monochrome (black and white) image data to color, relating the various data numbers to discrete combinations of hue, saturation and brightness, and allowing the full power of the human visual system to be utilized in discriminating small brightness differences.

One type of conversion of a monochrome image into color is achieved with a pseudo-color transformation. One simple transformation is density slicing. Density slicing is the color equivalent of contrast stretching, where discrete density slices, or ranges in the brightness, are mapped to unique but arbitrary colors. Density slicing can be useful particularly where one wants to note and enhance subtle variations in brightness in a near-homogeneous surface.

Color composite enhancement is used to display the multispectral information inherent in the original scene. True color refers to adding (after spatial registration) identical images taken in different spectral bands, usually red, green and blue (RGB), from the visible part of the electromagnetic spectrum, and assigning these spectral images their 'true' color. False-color enhancement applies to identical images taken in spectral wavelength bands that are not necessarily restricted to the visible part of the spectrum, and assigning each an arbitrary color before combining them into one false-color image.

Color-compositing techniques can be useful tools when attempting to discriminate subtle differences in the scene and/or attempting to incorporate different kinds of information in one image. Typically, false-color composites may include multispectral ratio images, temporal difference images, output from transformation techniques such as principal components analysis, and so on. However, virtually any two or three images can be color composited for the purpose of enhancement.

Image analysis

The final step in image processing is image analysis, ie. extracting and utilizing the information contained in the image and made more readily discernible through the first two steps of image restoration and enhancement. The goals of image analysis can be as varied as nature itself. To illustrate just a few of these goals and some of the techniques for restoration and enhancement discussed above, we present three images taken with the Voyager vidicon-based cameras, and discuss the type of scientific analysis made possible by application of particular image processing techniques.

Saturn, with its magnificent system of rings, is arguably the most beautiful body in our solar system, and has been a source of admiration and intrigue for astronomers for hundreds of years. The Voyager flybys of Saturn in 1980 and 1981 brought the study of Saturn and its rings and satellites into the modern era. Plate 27a is a 1-s exposure of Saturn and its rings taken with the vidicon wide-angle camera carried aboard Voyager 2 from a distance of 1.5 million km. The image restoration procedure is demonstrated with a comparison of this unprocessed, raw image and a processed image, Plate 27b, in which the spatially variable dark current, shading and reseau pattern have been removed. Correction of the geometric distortion present in the vidicon camera, utilizing a mathematical mapping from the known locations of the reseaux to their observed locations in Plate 27a, is shown in Plate 27c. The effect of removing the inherent though barely noticeable barrel distortion is seen in the way the corners of the image have been extended outward. Once this stage is reached, an accurate Saturn-centered coordinate system may be mapped onto the globe of Saturn, or onto its rings, so that the locations of features in the image may be measured. When measurements of this type are made over many consecutive images, the motion of atmospheric clouds from one image to the other – i.e. wind speeds – or the orbital motion of features within the rings may be determined.

To accentuate atmospheric and ring features, different processing techniques may be applied. Saturn is a cold planet, situated 1.5 billion km away from the Sun, and its clouds lie deep in its atmosphere under an overlying layer of haze. The scattering of sunlight by the haze causes the clouds to appear with low contrast and little color. (They are significantly lower in contrast and less colorful than the clouds on Jupiter, which is closer to the Sun and warmer.) A nonlinear contrast stretch of Plate 27b (Plate 27d) maximizes the contrast of the scene for both planet and rings. No new information has been added: the information present in the raw image is simply more easily seen after the image restoration and enhancement stages. For example, the banded structure of Saturn's northern hemisphere becomes more readily apparent.

The amount of fine-scale structure present in the rings was one of the great surprises of Voyager's encounter with Saturn. High-pass spatial filtering applied to Plate 27b (Plate 27e) enhances the spatial detail in the rings and the differences in the distribution of these details across the entire ring system, while suppressing the differences in brightness apparent in Plate 27b. For example, the normally darker and more transparent C ring and Cassini division now appear

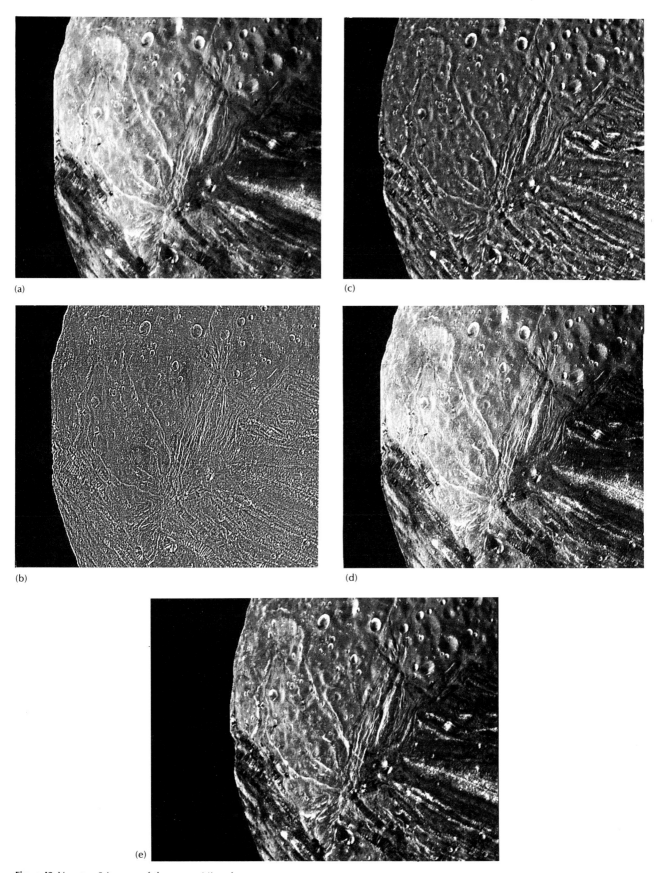

Figure 12 Voyager 2 images of the moon Miranda.

with the same overall brightness as the normally more brilliant A and B rings, making the measurement of ring feature locations all the easier.

To look for subtle variations in ring brightness, Plate 27b is processed using the density slicing technique, deliberately setting the globe of Saturn to black (Plate 27f). With a yellow–red–green–blue–purple color continuum, the lowest data numbers (darkest ring areas) are mapped to yellow and the highest data numbers (brightest ring areas) are mapped to purple. This pseudocolor picture helps distinguish many more levels of photometric detail than perceived in the initial monochrome image. Notice, for example, how easily visible now are the spokes in the B ring (the predominantly green/blue ring), as well as a curious variation in brightness with longitude in the outermost A ring, changing from mostly red in the upper part of the image to blue/green in the lower part. This effect, known as the A ring azimuthal asymmetry, is believed to be related to the manner in which particles in the A ring clump together as they orbit Saturn.

High-speed winds howl through Saturn's atmosphere. Concentrating on Saturn's northern hemisphere, Plate 27b has been more severely contrast enhanced than Plate 27d to produce Plate 28a. The dark areas are those in which the sunlight has been primarily absorbed and relatively little light is reflected; usually these regions are free of clouds. The bright areas are generally reflective clouds relatively high in the atmosphere. The stratigraphy of the atmosphere, as well as the structural definition within the clouds and eddies and the wavy jet stream, can be made more obvious by contrast enhancement.

The fine-scale details can be enhanced even further still (at the expense of brightness differences) by using the horizontal first-difference technique (Plate 28b). Here abrupt differences in brightness take on the appearance of 'edges' and are more readily apparent since the smoothly varying changes in brightness have been suppressed. A false-color composite of (Plate 28a and 28b (Plate 28c) clearly displays in a single image the information contained in both. One can see now the subtle differences in the clouds, haze and jet stream patterns, attainable with a first-difference image, together with the vertical stratigraphy of the atmosphere which is now encoded in color instead of brightness: blue is dark and low, pink is bright and high. Comparison of Plates 27a and 28c underscores the power of image processing in extracting scientific information.

Prior to the Voyager 2 encounter with Uranus in early 1986, little was known of the physical properties of the Uranian satellites aside from estimates of their sizes and reflectivities. Miranda (q.v.) certainly the most bizarre looking Uranian satellite and one of the most unusual in the solar system, has a dramatic and surprising degree of topography, with relief as large as 5% of its mean radius. This satellite, only 470 km in diameter and much smaller than the largest asteroids, is believed to have been internally differentiated, catastrophically disrupted and then reaccreted from its own fragments several times in the course of its history, explaining its extremely variegated appearance. In looking at the surface of Miranda, one is likely seeing, in places, material that was once at its core.

Figure I2a is a 2 s Voyager 2 exposure of Miranda taken from a distance of 31 500 km showing features as small as 300 m on a side. Figure I2a has been restored with the standard procedures for shading, geometric and cosmetic correction, and dark current subtraction. Applying a 5 × 5 high-pass filter to this image yields Figure I2b, showing clearly the high spatial frequency structures existing on Miranda's surface. Without the distracting variations in surface brightness, one can easily study Miranda's geomorphological record at the smallest size scale: the sizes, numbers and spatial distribution of the smallest craters are easily determined, the spacing of linear ridges and troughs can be measured, etc. Figure I2c is the result of filtering Figure I2a with a somewhat larger 31 × 31 box. It is obvious that medium to large-scale features, such as the ramparts of a somewhat obliterated large crater in the upper left part of the image, as well as some degree in brightness variation, may now be seen. Note that as the box size increases, fewer low-frequency components are removed from the original image, yet more high-frequency components are retained.

Figure I2d is an edge-enhanced image formed by adding Figures I2b and I2a and contains all the original information in the image with the fine-scale structure added in twice. Similarly, Figure I2e is the sum of Figure I2c and I2a. The reader can compare these two images to see how surface features of differing spatial scales may be exaggerated to differing degrees, depending on the size of the filter box. The scientific objective would determine which of these images would be preferable for analysis.

Of all the satellites observed and/or discovered by the Voyagers in their journeys across the outer solar system, none was so dramatic in its geological activity as Io (q.v.), the first Galilean satellite of Jupiter. Plate 23b is a nearly full-disk monochrome Voyager 1 image of Io, processed as were plates 27b and 27c to remove distortions and imperfections, showing its thoroughly mottled surface at an image scale of 4.5 km per pixel. The plume of Loki, one of the volcanoes active upon Voyager's arrival, can be seen extending above the limb of the satellite. Imaged in visible light, the plume's brightness is due to light reflected off particles ejected into Io's tenuous atmosphere. Plate 23c is a density-sliced image isolating the plume and showing clearly the distribution of brightness within the plume, a result of the variation in ejecta particle density. Scientifically valuable information like plume height and particle density variation with height can be readily determined from an image processed in this manner.

Plate 23a is a natural or 'true color' composite of three identical views of Io taken through the Voyager filters transmitting the blue, green and red regions of the spectrum, respectively. It displays Io approximately as the human eye, transported to Jupiter, would perceive it. Dull orange in overall color and covered with strange formations, the character of Io's surface is completely determined by its extraordinary volcanic activity and the yellow, orange, red, black and white sulfur and sulfur compounds that are its end products.

Carolyn C. Porco and Patricia T. Eliason

Bibliography

Andrews, H.C. and Hunt B.R. (1977) *Digital Image Restoration*. New Jersey: Prentice-Hall Inc.
Billingsley, F.C., Goetz A.F.H., and Lindsley J.N. (1970) Color differentiation by computer image processing. *Photo Sci. Eng*, **14**(1), 28–35.
Blackwell, H.R. (1946.) Contrast thresholds of the human eye. *J. Opt. Soc. Am.* **36**(11), 624–43.
Castleman, K.R. (1979) *Digital Image Processing*. New Jersey: Prentice-Hall.
Gonzalez, R.C. and Wintz P. (1987) *Digital Image Processing*. Reading, Mass: Addison-Wesley Publishing.
Green, W.B. (1983) *Digital Image Processing – A Systems Approach*. New York: Van Nostrand Reinhold.
Mees, C.E.K. (1954) *The Theory of the Photographic Process*. New York: Macmillan.
Mees, C.E.K. and James T.H., (1966) *The Theory of the Photographic Process*. New York: Macmillan.
Moik, J.G. (1980) *Digital Processing of Remotely Sensed Images*. Washington D.C: Goddard Space Flight Center, NASA Scientific and Technical Information Branch.
Pratt, W.K. (1978) *Digital Image Processing*. New York: John Wiley and Sons.
Rosenfeld, A. and Kak, A.C. (1982) *Digital Picture Processing*. New York: Academic Press.
Rossi, B. (1957) *Optics*. Reading, Mass.: Addison–Wesley.

Cross references

Charge-coupled device
Color
Miranda
Planetary ring
Saturn: atmosphere
Saturn: ring system

IMPACT CRATERING

The surfaces of most solid planets and moons in the solar system are scarred with circular craters produced by the impacts of smaller objects. More than 100 impact craters ranging in size from tens of meters to 140 km have also been recognized on Earth. Fresh impact craters are roughly circular rimmed depressions surrounded by hummocky blankets of debris. They form when an extraterrestrial body strikes the Earth or other planet at a velocity exceeding a few kilometers per second. Crater formation is an orderly, although rapid, process that begins when the impacting body first strikes the planet's surface and ends after the debris around and within the crater comes to rest. The crater is excavated by strong shock waves created as the impacting body plunges into the surface. These shock waves also cause diagnostic high-pressure mineralogical changes in the rocks surrounding the crater. The size of the final crater is a function of the speed and mass of the projectile that created it, as well as other factors such as the angle of impact and acceleration of gravity. Impact cratering is the dominant process sculpting the surfaces of asteroids and small satellites. Cratering also played a major role in the growth of all of the planets: it is now believed that impacts between planetary embryos and objects of perhaps 10% of their mass dominated every stage of the first 100 Ma of solar system history. A later era of heavy bombardment created the ancient cratered terranes on the Moon, Mercury and Mars between 4.5 and about 3.2 Ga ago. The Earth's early Archean history was probably strongly affected by large impacts. In the subsequent history of the Earth, a large impact caused at least one major biological extinction at the end of the Cretaceous era (see Mass extinction).

History of investigation

Craters were discovered in 1610 when Galileo pointed his first crude telescope at the Moon. Galileo recognized their raised rims and central peaks, but described them only as circular 'spots' on the moon. Although Galileo himself did not record an opinion on how they formed, astronomers argued about their origin for the next three centuries. The word 'crater' was first used in a non-genetic sense by the astronomer J.H. Schröter in 1791. Until the 1930s most astronomers believed that the Moon's craters were giant extinct volcanoes: the impact hypothesis, proposed sporadically over the centuries, did not gain a foothold until improving knowledge of impact physics showed that even a moderately oblique high-speed impact produces a circular crater rather than an elliptical crater, consistent with the observed circularity of nearly all of the Moon's craters. Even so, many astronomers clung to the volcanic theory until the high-resolution imagery and direct investigation of the Apollo program in the early 1970s firmly settled the issue in favor of an impact origin for nearly all lunar craters. In the current era spacecraft have initiated the remote study of impact craters on other planets, beginning with Mariner 4's unexpected discovery of craters on Mars on 15 July 1965. Since then craters have been found on almost every other solid body in the solar system.

The first terrestrial structure shown unambiguously to be created by a large impact was Meteor Crater, Arizona. This 1-km diameter crater and its associated meteoritic iron was investigated in detail by D.M. Barringer from 1906 until his death in 1929. After Barringer's work a large number of small impact structures resembling Meteor Crater have been found. Impact structures larger than about 5 km in diameter were first described as 'cryptovolcanic' because they showed signs of violent upheaval but were not associated with the eruption of volcanic materials. J.D. Boon and C.C. Albritton in 1937 proposed that these structures were really caused by impacts, although final proof had to wait until the 1960s when the presence of the shock-metamorphic minerals coesite and stishovite proved that the Ries Kessel in Germany was the result of a large meteor impact.

Theoretical and experimental work on the mechanics of cratering began during World War II and was extensively developed in later years. This work was spurred partly by the need to understand the craters produced by nuclear weapons and partly by the fear that the 'meteoroid hazard' to space vehicles would be a major barrier to space exploration. Computer studies of impact craters were begun in the early 1960s. A vigorous and highly successful experimental program to study the physics of impact was initiated by D.E. Gault at NASA's Ames facility in 1965.

These three traditional areas of astronomical crater studies, geological investigation of terrestrial craters and the physics of cratering have blended together in the post-Apollo era. Traditional boundaries have become blurred as extraterrestrial craters are subjected to direct geologic investigation, the Earth's surface is scanned for craters using satellite images, and increasingly powerful computers are used to simulate the formation of both terrestrial and planetary craters on all size scales. The recent proposals that the Moon was created by the impact of a Mars-size protoplanet with the proto-Earth 4.5 Ga ago and that the Cretaceous era was ended by the impact of a 10 km diameter asteroid or comet indicate that the subject of impact cratering is far from exhausted and that new results may be expected in the future.

Crater morphology

Fresh impact craters can be grossly characterized as 'circular rimmed depressions'. Although this description can be applied to all craters, independent of size, the detailed form of craters varies with size, substrate material, planet and age. Craters have been observed over a range of sizes varying from 0.1 μm (microcraters first observed on lunar rocks brought back by the Apollo astronauts) to the more than 2000 km diameter Hellas basin on Mars. Within this range a common progression of morphologic features with increasing size has been established, although exceptions and special cases are not uncommon.

Simple craters

The classic type of crater is the elegant bowl-shaped form known as a 'simple crater' (Figure I3a). This type of crater is common at sizes less than about 15 km diameter on the Moon and 3 to about 6 km on the Earth, depending on the substrate rock type. The interior of the crater has a smoothly sloping parabolic profile and its rim-to-floor depth is about one-fifth of its rim-to-rim diameter. The sharp-crested rim stands about 4% of the crater diameter above the surrounding plain, which is blanketed with a mixture of ejecta and debris scoured from the pre-existing surface for a distance of about one crater diameter from the rim. The thickness of the ejecta falls off as roughly the inverse cube of distance from the rim. The surface of the ejecta blanket is characteristically hummocky, with mounds and hollows alternating in no discernible pattern. Particularly fresh simple craters may be surrounded by fields of small secondary craters and bright rays of highly pulverized ejecta that extend many crater diameters away from the primary. Meteor Crater, Arizona, is a slightly eroded representative of this class of relatively small craters. The floor of simple craters is underlain by a lens of broken rock, 'breccia', which slid down the inner walls of the crater shortly following excavation. This breccia typically includes representatives from all of the formations intersected by the crater and may contain horizons of melted or highly shocked rock. The thickness of this breccia lens is typically one-half to one-third of the rim-to-floor depth.

Complex craters

Lunar craters larger than about 20 km diameter and terrestrial craters larger than about 3 km have terraced walls, central peaks and at larger sizes may have flat interior floors or internal rings instead of central peaks. These craters are believed to have formed by collapse of an initially bowl-shaped 'transient crater', and because of this more complicated structure are known as 'complex craters' (Figure I3b). The transition between simple and complex craters has now been observed on the Moon, Mars, Mercury and the Earth, as well as on some of the icy satellites in the outer solar system. In general the transition diameter scales as g^{-1}, where g is the acceleration of gravity at the planet's surface, although the constant in the scaling rule is not the same for icy and rocky bodies. This is consistent with the idea that complex craters form by collapse. with icy bodies having only about one-third the strength of rock ones. The floors of complex craters are covered by melted and highly shocked debris, and melt pools are sometimes seen in depressions in the surrounding ejecta blanket. The surfaces of the terrace blocks tilt outward into the crater walls, and melt pools are also common in the depressions thus formed. The most notable structural feature of complex craters is the uplift beneath their centers. The central peaks contain material that is pushed upward from the deepest levels excavated by the crater. Study of terrestrial craters has shown that the amount of structural uplift h_{su} is related to the final crater diameter D by

$$h_{su} = 0.06 D^{1.1}$$

where all distances are in kilometers. The diameter of the central peak complex is roughly 22% of the final rim-to-rim crater diameter in craters on all the terrestrial planets.

Complex craters are generally shallower than simple craters of equal size, and their depth increases slowly with increasing crater diameter. On the Moon the depth of complex craters increases from about 3 km to only 6 km while crater diameter ranges from 20 to 400 km. Rim height also increases rather slowly with increasing

Figure 13 Impact crater morphology as a function of increasing size. (a) Simple crater: 2.5 km diameter crater Linné in western Mare Serenitatis on the Moon (Apollo 15 Panametric Photo strip 9353). (b) Complex crater with central peak: 102 km diameter crater Theophilus on the Moon (Apollo 16 Hasselblad photo 0692). (c) Complex crater with internal ring: Mercurian craters Strindberg (165 km diameter) to the lower right and Ahmad Baba (115 km) to the upper left (Mariner 10 FDS 150, rectified). (d) Multiring basin: 620 km diameter (of most prominent ring) Orientale basin on the Moon (Lunar Orbiter IV medium-resolution frame 194).

diameter because much of the original rim slides into the crater bowl as the wall collapses. Complex craters are thus considerably larger than the transient crater from which they form: estimates suggest that the crater diameter may increase as much as 60% during collapse.

As crater size increases the central peaks characteristic of smaller complex craters give way to a ring of mountains (Figure 13). This transition takes place at about 140 km diameter on the Moon, 75 km on Mercury 45 km on Mars, and about 20 km on the Earth, again following a g^{-1} rule. The diameter of the central ring is generally about half of the rim-to-rim diameter of the crater on all the terrestrial planets.

The ejecta blankets of complex craters are generally similar to those of simple craters, although the 'hummocky' texture characteristic of simple craters is replaced by more radial troughs and ridges as size increases. Fresh complex craters also have well-developed fields of secondary craters, including frequent clusters and 'herringbone' chains of closely associated, irregular secondary craters. Martian and Venusian craters have flow-textured ejecta blankets that suggest fluidization by water or atmospheric gases. Very fresh craters, such as Copernicus and Tycho on the Moon, have far-flung bright ray systems whose nature is not well understood.

Multiring basins

The very largest impact structures are characterized by multiple concentric circular scarps, and are hence known as 'multiring basins'. The most famous such structure is the 930 km diameter Orientale basin on the Moon (Figure 13d), which has at least four nearly complete rings of inward-facing scarps. Although opinion on the origin of the rings still varies, most investigators feel that the scarps represent circular normal faults that slipped shortly after the crater was excavated. There is little doubt that multiring basins are caused by impacts: most of them have recognizable ejecta blankets characterized by a radial ridge-and-trough pattern. The ring diameter ratios are often tantalizingly close to multiples of √2, although no one has yet suggested a convincing reason for this relationship.

Unlike the simple/complex and central peak/internal ring transitions discussed above, the transition from complex craters to multiring basins is not a simple function of g^{-1}. Although multiring basins are common on the Moon, where the smallest has a diameter of 410 km, none at all have been recognized on Mercury, on which gravity is twice as large, even though the largest crater, Caloris basin, is 1300 km in diameter. The situation on Mars has been confused by erosion, but it is difficult to make a case that even the 1200 km diameter Argyre basin is a multiring structure. A very different type of multiring basin is found on Jupiter's satellite Callisto (q.v.), where the 4000 km diameter Valhalla basin has dozens of closely spaced rings that appear to face outward from the basin center. Another satellite of Jupiter, Ganymede (q.v.), has both Valhalla-type and Orientale-type multiring structures. Since gravity evidently does not play a simple role in the complex crater/multiring basin transition, some other factor, such as the internal structure of the planet, may have to be invoked to explain the occurrence of multiring basins.

Aberrant crater types

On any planetary surface a few craters can always be found that do not fit the simple size–morphology relation described above. These are generally believed to be the results of unusual conditions of formation in either the impacting body or the planet struck. Circular craters with asymmetric ejecta blankets or elliptical craters with 'butterfly wing' ejecta patterns are the result of very low impact angles. Although moderately oblique impacts yield circular craters, at impact angles less than about 6° from the horizontal the final crater becomes elongated in the direction of flight. Small, apparently concentric craters or craters with central dimples or mounds on their floors are the result of impact into a weak layer underlain by a stronger one. The ejecta blankets of some Martian craters show petal-like flow lobes that are believed either to indicate the presence of liquid water in the excavated material, or may be caused by ejecta interaction with atmospheric gases. Some Venusian craters show extensive flow units in their ejecta whose origin is not currently well understood. Incorporation of dense atmospheric gases or impact melt have been suggested as possibility. Craters on Ganymede and Callisto develop central pits at a diameter where internal rings would be expected on other bodies. The explanation for these pits is still unknown. Smaller craters on the Earth or Venus tend to form clusters of irregular craters, reflecting the effect of the atmosphere in breaking the original projectile into many smaller fragments before impact. In spite of these complications, however, the simple size–morphology relation described above provides a simple organizing principle into which most impact craters can be grouped.

Cratering mechanics

The impact of an object moving at many kilometers per second on the surface of a planet initiates an orderly sequence of events that eventually produces an impact crater. Although this is really a continuous process, it is convenient to break it up into distinct stages that are each dominated by different physical processes. This division clarifies the description of the overall cratering process, but it should not be forgotten that the different stages really grade into one another and that a perfectly clean separation is not possible. The most commonly used division of the impact cratering process is into contact and compression, excavation, and modification.

Contact and compression

Contact and compression is the briefest of the three stages, lasting only a few times longer than the time required for the impacting object (referred to hereafter as the 'projectile') to traverse its own diameter, $\tau = L/v_i$, where τ is the duration of contact and compression, L is the projectile diameter and v_i is the impact velocity. During this stage the projectile first contacts the planet's surface (hereafter, 'target') and transfers its energy and momentum to the underlying rocks (Figure I4). The specific kinetic energy (energy per unit mass,

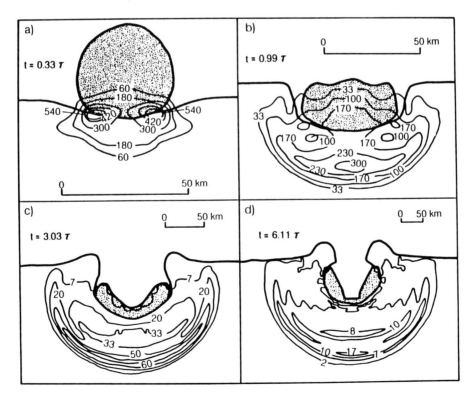

Figure I4 Three frames (a–c) showing the evolution of shock waves in the contact and compression stage of the vertical impact of a 46.4 km diameter iron projectile on a gabbroic anorthosite target at 15 km s^{-1}. The last frame (d) is a very early phase of the excavation stage. Pressure contours are given in GPa, and the times are given in units of τ, defined in the text. Note the change in lengths of the scale bar from frame to frame.

$\tfrac{1}{2} v_i^2$) possessed by a projectile traveling at even a few km s^{-1} is surprisingly large. A.C. Gifford, in 1924, first realized that the energy per unit mass of a body traveling at 3 km s^{-1} is comparable to that of TNT. Gifford proposed the 'impact–explosion analogy' which draws a close parallel between a high-speed impact and an explosion.

As the projectile plunges into the target, shock waves propagate both into the projectile, compressing and slowing it, and into the target, compressing and accelerating it downward and outward. At the interface between target and projectile the material of each moves at the same velocity. The shock wave in the projectile eventually reaches its back (or top) surface. At this time the pressure is released as the surface of the compressed projectile expands upward, and a wave of pressure relief propagates back downward toward the projectile–target interface. The contact and compression stage is considered to end when this relief wave reaches the projectile–target interface. At this time the projectile has been compressed to high pressure, often reaching hundreds of gigapascals, and upon decompression it may be in the liquid or gaseous state due to heat deposited in it during the irreversible compression process. The projectile generally carries off 50% or less of the total initial energy if the density and compressibility of the projectile and target material do not differ too much. The projectile–target interface at the end of contact and compression is generally less than a projectile diameter L below the original surface.

Contact and compression is accompanied by the formation of very high-velocity 'jets' of highly shocked material. These jets form where strongly compressed material is close to a free surface, for example near the circle where a spherical projectile contacts a planar target. The jet velocity depends on the angle between the converging surface of the projectile and target, but may exceed the impact velocity by factors as great as five. Jetting was initially regarded as a spectacular but not quantitatively important phenomenon in early impact experiments, where the incandescent streaks of jetted material amounted to only about 10% of the projectile's mass in vertical impacts. However, recent work on oblique impacts indicates that in this case jetting is much more important and that the entire projectile may participate in a downrange stream of debris that carries much of the original energy and momentum. Oblique impacts are still not well understood and more work needs to be done to clarify the role of jetting early in this process.

Excavation

During the excavation stage the shock wave created during contact and compression expands and eventually weakens into an elastic wave while the crater itself is opened by the much slower 'excavation flow'. The duration of this stage is roughly given by the period τ of a gravity wave (similar to a water wave in deep water in which the restoring force is gravity alone) with wavelength equal to the crater diameter D;

$$T = (D/g)^{1/2}$$

for craters whose excavation is dominated by gravity g (this includes craters larger than a few kilometers in diameter, even when excavated in hard rock). Thus, Meteor Crater ($D = 1$ km) was excavated in about 10 s, while the 1000 km diameter Imbrium Basin on the Moon took about 13 min to open. Shock wave expansion and crater excavation, while intimately linked, occur at rather different rates and may be considered separately.

The high pressures attained during contact and compression are almost uniform over a volume roughly comparable to the initial dimensions of the projectile. However, as the shock wave expands away from the impact site the shock pressure declines as the initial impact energy spreads over an increasingly large volume of rock. The pressure in the shock P as a function of distance r from the impact site is given roughly by

$$P(r) = P_0 \, (a/r)^n$$

where a ($= L/2$) is the radius of the projectile, P_0 is the pressure established during contact and compression, and the power n is between 2 and 4, depending on the strength of the shock wave (n is larger at higher pressures – a value $n = 3$ is a good general average).

The shock wave, with a release wave immediately following, quickly attains the shape of a hemisphere expanding through the target rocks (Figure 15). The high shock pressures are confined to the surface of the hemisphere: the interior has already decompressed. The shock wave moves very quickly, as fast or faster than the speed

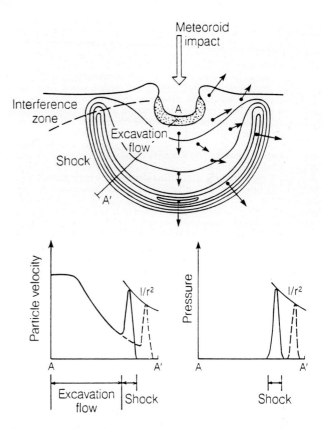

Figure 15 Schematic illustration of the expanding shock wave and excavation flow following a meteorite impact. The contours in the upper part of the figure represent pressure at some particular time after the impact. The region of high shock pressure is seen to be isolated or 'detached' on an expanding hemispherical shell. The insets show profiles of particle velocity and pressure along the section AA'. The dashed lines on these insets show particle velocity and pressure some time later than those shown by the solid lines, and the solid curves connecting the peaks are portions of the 'envelopes' of peak particle velocity and peak pressure.

of sound, between about 6 and 10 km s^{-1} in most rocks. As rocks in the target are overrun by the shock waves, then released to low pressures, mineralogical changes take place in the component minerals. At the highest pressures the rocks may melt or even vaporize upon release. As the shock wave weakens, high-pressure minerals such as coesite or stishovite arise from quartz in the target rocks, diamonds may be produced from graphite, or maskelynite from plagioclase. Somewhat lower pressures cause pervasive fracturing and 'planar elements' in individual crystals. Still lower pressures create a characteristic cone-in-cone fracture called 'shatter cones' (Figure 16) that are readily recognized in the vicinity of impact structures. Indeed, many terrestrial impact structures were first recognized from the occurrence of shatter cones.

The expanding shock wave encounters a special condition near the free surface. The pressure at the surface must be zero at all times. Nevertheless, a short distance below the surface the pressure is essentially equal to P, defined above. This situation results in a thin layer of surface rocks being thrown upward at very high velocity (the theoretical maximum velocity approaches the impact speed v_i). Since the surface rocks are not compressed to high pressure, this results in the ejection of a small quantity of unshocked or lightly shocked rocks at speeds that may exceed the target planet's escape velocity. Although the total quantity of material ejected by this 'spall' mechanism is probably only 1–3% of the total mass excavated from the crater, it is particularly important scientifically as this is probably the origin of the recently discovered meteorites from the Moon, and of the SNC (shergottite, nakhlite and chassignite) meteorites which are widely believed to have been ejected from Mars (see Lunar meteorite; SNC meteorite).

Figure 16 Shatter cones from the Spider Structure, W. Australia, formed in mid-Proterozoic orthoquartzite. This cone-in-cone fracture is characteristic of shattering by impact-generated shock waves. The scale bar on top is 15 cm long. (Courtesy of George Williams.)

The weakening shock wave eventually degrades into elastic waves. These elastic waves are similar in many respects to the seismic waves produced by an earthquake, although impact-generated waves contain less of the destructive shear-wave energy than earthquake waves. The seismic waves produced by a large impact may have significant effects on the target planet, creating jumbled terrains at the antipode of the impact site. This effect has been observed opposite Caloris basin on Mercury and opposite Imbrium and Orientale on the Moon. The equivalent Richter magnitude M caused by an impact of energy E ($= \frac{1}{2} m_p v_i^2$) is given approximately by $M = 0.67 \log_{10}E - 5.87$.

Target material engulfed by the shock wave is released a short time later. Upon release the material has a velocity that is only about one-fifth of the particle velocity in the shock wave. This 'residual velocity' is due to thermodynamic irreversibility in the shock compression. It is this velocity field that eventually excavates the crater. The excavation velocity field has a characteristic downward, outward then upward pattern (Figure 17) that moves target material out of the crater, ejecting it at angles close to 45° at the rim. The streamlines of this flow cut across the contours of maximum shock pressure, so that material ejected at any time may contain material with a wide range of shock levels. Nevertheless, the early, fast ejecta generally contain a higher proportion of highly shocked material than the later, slower ejecta. Throughout its growth the crater is lined with highly shocked, often melted, target material.

The growing crater is at first hemispherical in shape. Its depth $H(t)$ and diameter $D(t)$ both grow approximately as $t^{0.4}$, where t is time after the impact. Hemispherical growth ceases after a time of about $(2H_t/g)^{1/2}$, where H_t is the final depth of the transient crater. At this time the crater depth stops increasing (it may even begin to decrease as collapse begins), but its diameter continues to increase. The crater shape thus becomes a shallow bowl, finally attaining a diameter roughly three to four times its depth. At this stage, before collapse modifies it, the crater is known as a 'transient' crater. Even simple craters experience some collapse (which produces the breccia lens), so that the transient crater is always a brief intermediate stage in planetary crater formation. However, since most laboratory craters are 'frozen' transient craters, much of our knowledge about crater dimensions refers to the transient stage only, and must be modified for application to planetary craters.

Laboratory, field and computer studies of impact craters have all confirmed that only material lying above about one-third of the transient crater depth (or about one-tenth of the diameter) is thrown out of the crater. Material deeper than this is simply pushed downward into the target, where its volume is accommodated by deformation of the surrounding rocks. Thus, in sharp contrast to ejecta from volcanic craters, material in the ejecta blankets of impact craters does not sample the full depth of rock intersected by the crater.

The form of the transient crater produced during the excavation stage may be affected by such factors as obliquity of the impact

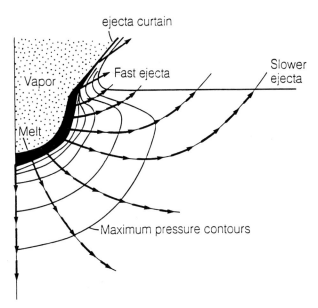

Figure 17 Geometry of the excavation flow field which develops behind the rapidly expanding shock front, which has moved beyond the boundaries of this illustration. The lines with arrows indicate 'stream tubes' along which material flows downward and outward from the crater. The stream tubes cut across the contours of maximum shock pressure, showing that material ejected at any given range from the impact site has been shocked to a variety of different maximum pressures. When material flowing through a stream tube crosses the initial surface it forms part of the ejecta curtain. Ejecta emerging near the impact site travel at high speed, whereas ejecta emerging at larger distances travel at slower velocities.

(although the impact angle must be less than about 6° for a noticeably elliptical crater to form at impact velocities in excess of about 4 km s^{-1}), the presence of a water table or layers of different strength, rock structure, joints or initial topography in the target. Each of these factors produces its own characteristic changes in the simple bowl-shaped transient crater form.

Modification

Shortly after the excavation flow opens the transient crater and the ejecta is launched onto ballistic trajectories, a major change takes place in the motion of debris within and beneath the crater. Instead of flowing upward and away from the crater center, the debris comes to a momentary halt, then begins to move downward and back toward the center whence it came. This collapse is generally attributed to gravity, although elastic rebound of the underlying, compressed rock layers may also play a role. The effects of collapse range from mere debris sliding and drainback in small craters to wholesale alteration of the form of larger craters in which the floors rise, central peaks appear and the rims sink down into wide zones of stepped terraces. Great mountain rings or wide central pits may appear in still larger craters.

These different forms of crater collapse begin almost immediately after formation of the transient crater. The timescale of collapse is similar to that of excavation, occupying an interval of a few times $(D/g)^{1/2}$. Crater collapse and modification thus take place on timescales very much shorter than most geologic processes. The crater resulting from this collapse is then subject to the normal geologic processes of gradation, isostatic adjustment, infilling by lavas, etc. on geologic timescales. Such processes may eventually result in the obscuration or even total obliteration of the crater, depending on the planetary surface on which it forms.

The effects of collapse depend on the size of the crater. For transient craters smaller than about 15 km diameter on the Moon, or about 3 km on the Earth, modification entails only collapse of the relatively steep rim of the crater onto its floor. The resulting 'simple crater' is a shallow bowl-shaped depression with a rim-to-rim dia-

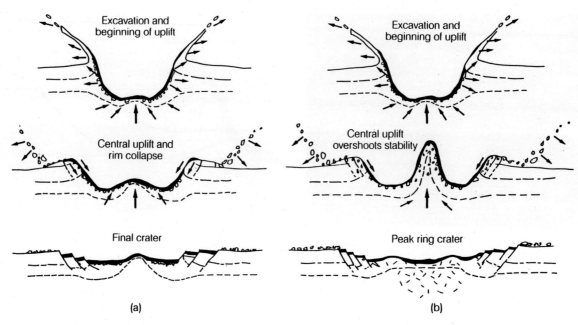

Figure 18 Schematic illustration of the formation of complex craters with either (a) central peaks or (b) peak rings. Uplift of the crater begins even before the rim is fully formed. As the floor rises further, rim collapse creates a wreath of terraces surrounding the crater. In smaller craters the central uplift 'freezes' to form a central peak. In larger craters the central peak collapses and creates a peak ring before motion ceases.

meter D about five times its depth below the rim H. In fresh craters the inner rim stands near the angle of repose, about 30°. Drilling in terrestrial craters shows that the crater floor is underlain by a lens of broken rock (mixed breccia) derived from all of the rock units intersected by the crater. The thickness of this breccia lens is typically half the depth of the crater H. Volume conservation suggests that this collapse increases the original diameter of the crater by about 15%. The breccia lens often includes layers and lenses of highly shocked material mixed with much less-shocked country rock. A small volume of shocked or melted rock is often found at the bottom of the breccia lens.

Complex craters (Figure I3b,c) collapse more spectacularly. Walls slump, the floor is stratigraphically uplifted, central peaks or peak rings rise in the center and the floor is overlain by a thick layer of highly shocked impact melt. The detailed mechanism of collapse is still not fully understood because straightforward use of standard rock mechanics models do not predict the type of collapse observed. The current best description of complex crater collapse utilizes a phenomenological strength model in which the material around the crater is approximated as a Bingham fluid, a material which responds elastically up to differential stresses of about 3 MPa, independent of overburden pressure, then flows as a viscous fluid with viscosity on the order of 1 GPa-sec at larger stresses. In a large collapsing crater the walls slump along discrete faults, forming terraces whose widths are controlled by the Bingham strength, and the floor rises, controlled by the viscosity, until the differential stresses fall below the 3-MPa strength limit. A central peak may rise, then collapse again in large craters, forming the observed internal ring (or rings). Figure I8 illustrates this process schematically. The rock in the vicinity of a large impact may display such an unusual flow law because of the locally strong shaking driven by the large amount of seismic energy deposited by the impact.

The mechanics of the collapse that produces multiring basins (Figure I3d) is even less well understood. Figure I9 illustrates the structure of the Orientale Basin on the Moon with a highly vertically exaggerated cross-section derived from both geological and geophysical data. Note that the ring scarps are interpreted as inward-dipping faults above a pronounced mantle uplift beneath the basin's center. One idea that is currently gaining ground is that the ring scarps are normal faults that develop as the crust surrounding a large crater is pulled inward by the flow of underlying viscous mantle material toward the crater cavity (Figure I10). An important aspect of this flow is that it must be confined in a low-viscosity channel by more

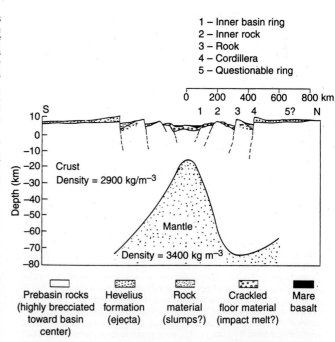

Figure 19 Geologic and geophysical structure of the Orientale Basin on the Moon, one of the freshest and best-studied multiring basins. A dense mantle plug underlies the center of the basin. The crustal thinning above the plug is due to the ejection of about 40 km of crustal material from the crater that formed the basin. The great ring scarps shown in cross-section formed during collapse of the crater. Note the 10× vertical exaggeration necessary to show the ring scarps.

viscous material below, otherwise the flow simply uplifts the crater floor and radial faults, not ring scraps are the result. Special structural conditions are thus needed in the planet for multiring basins to form on its surface, so that a g^{-1} dependence for the transition from

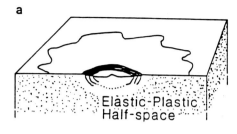
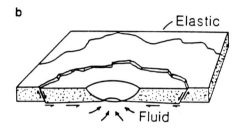
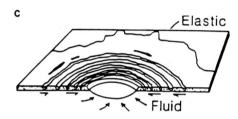

Figure I10 The ring tectonic theory of multiring basin formation: (a) shows the formation of a normal complex crater in a planet with uniform rheology; (b) shows the inward-directed flow in a more fluid asthenosphere underlying a lithosphere of thickness comparable to the crater depth and the resulting scarps; (c) shows a Valhalla-type basin developing around a crater formed in very thin lithosphere.

complex craters to multiring basins is not expected (or observed). This theory is capable of explaining both the lunar-type and Valhalla-type multiring basins as expressions of different lithosphere thickness.

Atmospheric interactions

As fast-moving meteoroids enter the atmosphere of a planet, they are slowed by friction with the atmospheric gases and compressed by the deceleration. Small meteoroids are often vaporized by frictional heating and never reach the surface of the planet. Larger meteoroids are decelerated to terminal velocity and fall relatively gently to the surface of the planet. The diameter of a meteorite that loses 90% of its initial velocity in the atmosphere is given by

$$L = 0.15 \, P_{surf}/(\rho_p \, g_{surf} \sin \theta)$$

where P_{surf} is the surface atmospheric pressure, ρ_p is the meteorite's density and g_{surf} is the planet's surface acceleration of gravity. L is typically about 1m for Earth and 60m for Venus. However, this equation assumes that the projectile reaches the surface intact, whereas in fact aerodynamic stresses may crush all but the strongest meteorites. Once fractured, the fragments of an incoming meteorite travel slightly separate paths and strike the surface some distance apart from one another. This phenomenon gives rise to the widely observed strewn fields of meteorites or craters on the Earth, which average about 1–2 km in diameter. On Venus, clusters of small craters attributed to atmospheric breakup are spread over areas roughly 20 km in diameter (Phillips et al., 1991).

The atmospheric blast wave and thermal radiation produced by an entering meteorite may also affect the surface: the 1908 explosion at Tunguska River, Siberia, was probably produced by the entry and dispersion of a 100 m diameter stony meteoroid that leveled and scorched about 2000 km² of meter-diameter trees (Chyba, Thomas and Zahnle, 1993). Radar-dark 'sploches' up to 50 km in diameter on the surface of Venus are attributed to pulverization of surface rocks by strong blast waves from meteorites that were fragmented and dispersed in the dense atmosphere (Zahnle, 1992).

Other atmospheric interactions dominate the post-impact evolution of craters produced by meteorites sufficiently large to strike the surface with a large fraction of their initial energy. Vaporized projectile and target expand rapidly out of the resulting crater, forming a vapor plume that, if massive enough, may blow aside the surrounding atmosphere and accelerate to high speed. In the impacts of sufficiently large and fast projectiles some of this vapor plume material may even reach escape velocity and leave the planet, incidentally also removing some of the planet's atmosphere. Such 'impact erosion' may have played a role in the early history of the Martian atmosphere (Melosh and Vickery, 1989). In smaller impacts the vapor plume may temporarily blow aside the atmosphere, opening the way for widespread ballistic dispersal of melt droplets (tektites) above the atmosphere and permitting the formation of lunar-type ejection blankets even on planets with dense atmospheres, as has been observed on Venus.

Scaling of crater dimensions

One of the most frequently asked questions about an impact crater is 'how big was the meteorite that made the crater?' Like many simple questions this has no simple answer. It should be obvious that the crater size depends upon the meteoroid's speed, size and angle of entry. It also depends on such factors as the meteoroid's composition, the material and composition of the target, surface gravity, the presence or absence of an atmosphere, etc. The question of the original size of the meteorite is usually unanswerable because the speed and angle of impact are seldom known. The inverse question, of how large a crater will be produced by a given size meteoroid with known speed and incidence angle, is in principle much simpler to answer. However, even this prediction is uncertain because there are no observational or experimental data on the formative conditions of impact craters larger than a few tens of meters in diameter, while the impact structures of geologic interest range up to 1000 km in diameter. The traditional escape from this difficulty is to extrapolate beyond experimental knowledge by means of scaling laws.

The first scaling laws were introduced in 1950 by C.W. Lampson, who studied the craters produced by TNT explosions of different sizes. Lampson found that the craters were similar to one another if all dimensions (depth, diameter, depth of charge placement) were divided by the cube root of the explosive energy W. Thus if the diameter D of a crater produced by an explosive energy W is required, it can be computed from the diameter D_0 of a crater produced by energy W_0 using the proportion:

$$D/D_0 = (W/W_0)^{1/3}$$

An exactly similar proportion may be written for the crater depth, H. This means that the ratio of depth to diameter, H/D, is independent of explosive energy W, a prediction that agrees reasonably well with observation. In more recent work on large explosions the exponent ⅓ in this equation has been modified to 1/3.4 to account for the effects of gravity on crater formation.

Although impacts and explosions have many similarities, a number of factors make them difficult to compare in detail. Thus explosion craters are very sensitive to the charge's depth of burial. Although this quantity is well defined for explosions, there is no simple analog for impact craters. Similarly, the angle of impact has no analog for explosions. Nevertheless, energy-based scaling laws were very popular in the older impact literature, perhaps partly because nothing better existed, and many empirical schemes were devised to adapt the well-established explosion scaling laws to impacts.

This situation has changed radically in the last decade, however, thanks to more impact cratering experiments specifically designed to establish scaling laws. It has been shown that the great expansion of the crater during excavation tends to decouple the parameters describing the final crater from the parameters describing the projectile. If these sets of parameters are related by a single, dimensional 'coupling parameter' (as seems to be the case), it can be shown that crater parameters and projectile parameters are related by power-law scaling expressions with constant coefficients and exponents. Although this is a somewhat complex and rapidly changing subject, the best current scaling relation for impact craters forming in competent rock (low porosity) targets whose growth is limited by gravity rather than target strength (i.e. all craters larger than a few kilometers diameter) is given by

$$D_{at} = 1.8 \, \rho_p^{0.11} \, \rho_t^{-1/3} g^{-0.22} L^{0.13} W^{0.22} (\sin \theta)^{1/3}$$

where D_{at} is the diameter of the transient crater at the level of the original ground surface, ρ_p and ρ_t are densities of the projectile and target respectively, g is surface gravity, L is projectile diameter, W is impact energy ($= \pi/12 \, L^3 \rho_p v_i^2$) and θ is the angle of impact from the vertical. All quantities are in SI units.

Crater depth H appears to be a constant times the diameter D_{at}. Although a few investigations have reported a weak velocity dependnce for this ratio, the experimental situation is not yet clear.

The amount of melt and vapor produced in an impact obeys a rather different scaling law, since it is determined solely by the physics of shock wave expansion and not by the gravity field of the planet (at the time of melt formation the shock pressures far exceed gravitational pressures for all but planet size projectiles). A widely used melt and vapor scaling law is:

$$\frac{\text{mass of melt}}{\text{mass of projectile}} = 0.14 \, \frac{v_i^2}{\epsilon_m} \text{ for } v_i \geq 12 \text{ km s}^{-1}$$

and

$$\frac{\text{mass of vapor}}{\text{mass of projectile}} = 0.4 \, \frac{v_i^2}{\epsilon_v} \text{ for } v_i \geq 35 \text{ km s}^{-1}$$

where $\epsilon_m = 3.4 \times 10^6$ J kg^{-1} is the specific energy of melting and $\epsilon_v = 5.7 \times 10^7$ J kg^{-1} is the specific energy of vaporization of silicate rocks.

Since the masses melted and vaporized are independent of surface gravity, while the transient crater diameter decreases as gravity increases, the relative volume of melt to material excavated increases with increasing crater diameter, as shown in Figure I11. Very large craters may differ substantially in morphology from small ones because of this difference in relative melt volume. Similarly, craters of similar size on different planets (e.g. lunar versus terrestrial craters) may have substantially different relative volumes of impact melt.

Impacts and planetary evolution

Modern theories of planetary origin suggest that the planets and the Sun formed simultaneously 4.6×10^9 years ago from a dusty, hydrogen-rich nebula. Nebular condensation and hydrodynamic interactions were probably only capable of producing c. 10 km diameter 'planetesimals' that accreted into planetary-scale objects by means of collisions. The timescale for accretion of the inner planets by mutual collisions is currently believed to be between a few tens and 100 million years. Initially rather gentle, these collisions became more violent as the random velocities of the smaller planetesimals increased during close approaches to the larger bodies. The mean random velocity of a swarm of planetesimals is comparable to the escape velocity of the largest object, so as the growing planetary embryos reached lunar size collisions began to occur at several kilometers per second. At such speeds impacts among the smaller objects were disruptive, whereas the larger objects had sufficient gravitational binding energy to accrete most of the material which struck them. Infalling planetesimals bring not only mass but also heat to the growing planets. In the past it was believed that the temperature inside a growing planet increased in a regular way from near zero at the center to large values at the outside, reflecting the increase in collision velocity as the planet became more massive. However, it now seems probable that the size distribution of the planetesimal population was more evenly graded between large and small objects, and that each growing planetary embryo was subjected to many collisions with bodies up to 10% of the mass of the embryo itself. Such catastrophic collisions deposit heat deep within the core of the impacted body, wiping out any regular law for temperature increase with increasing radius and making the thermal evolution of growing planets rather stochastic (Melosh, 1990).

The origin of the Moon is now attributed to a collision between the proto-Earth and a Mars-size protoplanet near the end of accretion 4.5×10^9 years ago. This theory has recently supplanted the three classic theories of lunar origin (capture, fission and co-accretion) because only the giant impact theory provides a simple explanation for the Moon's chemistry, as revealed in the lunar rocks returned by Apollo. One view of this process is that a grazing collision vaporized (by jetting) a large quantity of the proto-Earth's mantle along with a comparable quantity of the projectile. While most of the mass of the projectile merged with the Earth (incidentally strongly heating the Earth: if the Earth was not molten before this impact it almost certainly was afterward), one or two lunar masses of vapor condensed into dust in stable Keplerian orbits about the Earth and then later accumulated together to form the Moon.

Sometime after the Moon formed and before about 3.8×10^9 years ago the inner planets and their satellites were subjected to the 'late heavy bombardment', an era during which the cratering flux was orders of magnitude larger than at present. The crater scars of this period are preserved in the lunar highlands and the most ancient terrains of Mars and Mercury. A fit to the lunar crater densities using age data from Apollo samples (Melosh and Vickery, 1989) gives a cumulative crater density through geologic time of

$$N_{cum}(D > 4 \text{ km}) = 2.68 \times 10^{-5} [T + 4.57 \times 10^{-7} (e^{\lambda T} - 1)]$$

where $N_{cum}(D > 4 \text{ km})$ is the cumulative crater density (craters km^{-2}) of craters larger than 4 km diameter, T is the age of the surface in Ga ($T = 0$ is the present) and $\lambda = 4.53$ Ga^{-1}. Several other slightly different fits to the same data have been discussed in the literature. The current cratering rate on the moon is about 2.7×10^{-14} craters with $D > 4$ km km^{-2} year^{-1}. On the Earth the cratering rate has been estimated to be about 1.8×10^{-15} craters with $D > 22.6$ km km^{-2} year^{-1}, which is comparable to the lunar flux taking into account the different minimum sizes, since the cumulative number of craters $N_{cum}(D) \sim D^{-1.8}$. There is currently much debate about these cratering rates, which might be uncertain by as much as a factor of two.

The study of relative crater densities (number per unit area) on planetary surfaces provides an important tool for establishing relative ages of surface features. Since impacts are generally presumed to occur randomly on a surface (a minor exception is the leading–trailing edge cratering rate asymmetry on the tidally locked satellites of the giant planets), the spatial distribution of craters provides no information. However, older surfaces have been struck by more

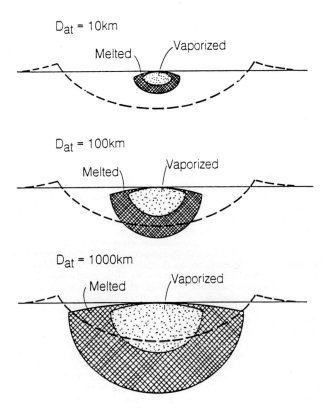

Figure I11 The different scaling laws for crater diameter and melt or vapor volume imply that as the crater diameter increases, the volume of melted or vaporized material may approach the volume of the crater itself. This figure is constructed for impacts at 35 km s^{-1} on the Earth.

impacts than younger surfaces, so that crater density is a useful indicator of relative age on a given planet or satellite. Absolute ages require calibration of the cratering rate as a function of time, which generally requires radiometric dating of surfaces of known crater density. Such information is currently available only for the Moon. Similarly, interplanetary correlations of crater density are difficult because the cratering rate may differ in different parts of the solar system. A limitation of the relative crater density technique is that once the crater density becomes high enough, the addition of a new crater of some size wipes out, on average, an old crater of the same size (not to mention many smaller ones). A surface reaching this condition is in equilibrium (sometimes the word 'saturation' is used interchangeably), and provides no age information subsequent to the time that equilibrium was achieved. The lunar highlands are widely believed to have reached this state.

The high cratering rates in the past indicate that the ancient Earth should have been heavily scarred by large impacts. Based on the lunar record it is estimated that more than 100 impact craters with diameters greater than 1000 km should have formed on the Earth. Although little evidence of these early craters has yet been found, it is gratifying to note the recent discovery of thick impact ejecta deposits in 3.2 to 3.5 Ga Archean greenstone belts in both South Africa and Western Australia. Since rocks have recently been found dating back to 4.2 Ga, well into the era of heavy bombardment, it is to be hoped that more evidence for early large craters will be eventually discovered. Heavy bombardment also seems to have overlapped the origin of life on Earth. It is possible that impacts may have had an influence on the origin of life, although whether they suppressed it by creating global climatic catastrophes (up to evaporation of part or all of the seas by large impacts), or facilitated it by bringing in needed organic precursor molecules, is unclear at present. The relation between impacts and the origin of life is currently an area of vigorous speculation. (See Comet impacts on Earth; Life: origin.)

The idea that large impacts can induce major volcanic eruptions is a recurring theme in the geologic literature. This idea probably derives from the observation that all of the large impact basins on the Moon's nearside are flooded with basalt. However, radiometric dates on Apollo samples made it clear that the lava infillings of the lunar basins are nearly 1 Ga younger than the basins themselves. Furthermore, the farside lunar basins generally lack any substantial lava fill. The nearside basins are apparently flooded merely because they were topographic lows in a region of thin crust at the time that mare basalts were produced in the Moon's upper mantle. Simple estimates of the pressure release caused by stratigraphic uplift beneath large impact craters make it clear that pressure release melting cannot be important in impacts unless the underlying mantle is near the melting point before the impact. Thus, to date, there is no firm evidence that impacts can induce volcanic activity. Impact craters may create fractures along which pre-existing magma may escape, but themselves are probably not capable of producing much melt. The massive igneous body associated with the Sudbury Ontario impact structure is now attributed to a thick impact melt sheet [Grieve, Stöffler and Dentoch 1991], rather than to impact-induced intrusive activity.

The most recent major impact event on Earth seems to have been a collision between the Earth and a 10 km diameter comet or asteroid 65 Ma ago that ended the Cretaceous era and caused the most massive biological extinction in recent geologic history (see Mass extinction). Evidence for this impact has been gathered from many sites over the last decade, and now seems nearly incontrovertible. First detected in an enrichment of the siderophile element iridium in the c. 3 mm thick K-T (Cretaceous–Tertiary) boundary layer in Gubbio, Italy, the iridium signature has now been found in more than 100 locations worldwide, in both marine and terrestrial deposits. Accompanying this iridium are other siderophile elements in chondritic ratios, shocked quartz grains, coesite, stishovite and small (100–500 μm) spherules resembling microtektites. All these point to the occurrence of a major impact at the K-T boundary. In addition, soot and charcoal have been found at a number of widely separated sites in abundances that suggest that the entire world's standing biomass burned within a short time of the impact. An impact of this magnitude should have produced a crater nearly 200 km in diameter. A very strong candidate for this crater has been found beneath the cover rocks of the Northern Yucatan Peninsula. Named Chicxulub Crater by its discoverer, Glen Penfield (Hildebrand et al., 1991), it is about 180 km in diameter. Although many details of the impact, and especially of the extinction mechanism, still have to be worked out, the evidence for a great impact at this time is becoming overwhelming, although a few geologists still adhere to some kind of volcanically induced extinction crisis. Future work should be able to resolve the mysteries surrounding this striking demonstration of the importance of impact craters in both solar system and Earth history.

H.J. Melosh

Bibliography

Nearly all of the material summarized in this article can be found explained at greater length, with original references in:
Melosh, H.J. (1989) *Impact Cratering: A Geologic Process*. Oxford University Press, 245 pp.

Chyba, C.F., Thomas, P.J. and Zahnle, K.J. (1993) Atmospheric disruption of small comets and asteroids and the 1908 Tunguska explosion. *Nature*, **361**; 40–4.
Grieve. R.A. Stöffler, D. and Deutsch, A. (1991) The Sudbury Structure: controversial or misunderstood? *J. Geophys. Res*, **96**, 753–64.
Hildebrand, A.R., Penfield, G.T., Kring, D.A. *et al.* (1991) Chicxulub Crater: A possible Cretaceous/Tertiary boundary impact crater on the Yucatán Peninsula, Mexico. *Geology*, **19**; 867–71.
Melosh, H.J., (1990) Giant impacts and the thermal state of the early, in *Origin of the Earth* (ed. J.H. Jones and H.E. Newsom) New York: Oxford University Press, pp. 69–83.
Melosh, H.J. and Vickery A.M. (1989) Impact erosion of the primordial atmosphere of Mars. *Nature*. **338**, 487–9.
Phillips, R.J., Arvidson, R.E., Boyce, J.M. *et al.* (1991) Impact craters on Venus: initial analysis from Magellan. *Science*, **252**, 288–97.
Zahnle, K.J., (1923) Airburst origin of dark shadows on Venus: *J. Geophys. Res.*, **97**, 10, 243–55.

Cross references

Accretion
Collision
Comet: impacts on the Earth
Cretaceous–Tertiary boundary
Iridium anomaly
Life: origin
Mars: impact cratering
Mass extinction
Mercury: geology
Moon: geology
SNC meteorites
Venus: geology and geophysics

INERTIA, INERTIAL FRAME

The word 'inertia', introduced by Kepler, is derived from Latin *iners* meaning idle, and implies that a body possesses some physical attribute which makes it resist any change (acceleration) from its existing state of rest or motion. Although Galileo, Kepler and Descartes (Herivel, 1965; Cohen, 1971) expressed similar concepts of inertia either vaguely or incompletely, it was Newton who first clearly defined inertia as a property of matter as stated in Newton's first law (the law of inertia): 'Every body perseveres in its state of rest or of uniform motion in a straight line, unless it is compelled to change that state by forces impressed thereon' (Motz and Weaver, 1988). Strictly speaking, this law cannot be derived directly from real experiments because we can never eliminate the influence of external forces on a body. However, we can find approximate examples from our daily lives: to stop a car, we have to apply the brakes but it must still travel some distance to stop. The heavier the car is, the greater is its inertia and the greater the braking force required. A satellite may maintain its speed and height for a few years because there is only very slight air friction (corresponding to a braking force) at high altitude in the atmosphere. Therefore, we accept Newton's first law since it becomes more and more valid as more external forces are reduced.

The quantitative measure of the inertia of a non-rotating body is its inertial mass m, which may be defined as $m = F/a$ from Newton's second law (see Newton, Sir Isaac, Newton's Laws), where F is the external force exerted on the body and a the acceleration of the body.

From this equation, we see the larger m is, the smaller is a produced by F, i.e. m does have the property of resisting the action of the force just as the inertia should have. Therefore the inertia may be taken to be proportional to the inertial mass. For the rotational motion of a body, the spatial distribution of mass with respect to the rotation axis also affects its inertia, as reflected in the inertia tensor.

However, under the action of gravitational force, all objects, 'heavy' or 'light', will have the same (gravitational) acceleration (neglecting air friction). This is because the gravitational force is proportional to the inertial mass, and the effect of mass is then canceled out.

A spacecraft may include a gyro-stabilized platform which is not influenced by the motions of the spacecraft. Based on such a platform, an inertial guidance system can be made to direct the vessel to a selected destination.

An important feature of the law of inertia is that it makes no distinction between rest and uniform velocity. For example, an astronaut in a spaceship can never know whether his spaceship is at rest or in a uniform motion if no external force acts on the ship and he does not look out of his window. Within his spaceship he has no way to determine ('feel') the speed of the ship, in the absence of acceleration.

Based on this fact, in classical mechanics an inertial frame is defined as a reference frame in which Newton's second law is valid. In other words, Newton's second law and other mechanical laws remain the same for all inertial frames, and all inertial reference frames are equivalent. Einstein extended mechanical laws to other physical laws. His principle of special relativity tells us that no experiment or observation (whether mechanical, optical or electromagnetic) that an observer can make within his own inertial frame of reference enables him to determine whether he is at rest in space or in a state of uniform translational motion. His principle of invariance states that the laws of nature must have the same algebraic form in all inertial frames, but can change their forms in non-inertial frames. In the general theory of relativity, Einstein removed all the restrictions on coordinate systems and further extended this principle in such a way that all frames of reference (inertial or non-inertial) are equivalent for laws of nature (Motz and Weaver, 1988).

In Newton's mechanics, inertia is viewed as an intrinsic property of matter and is independent of 'absolute space' as well as all other objects of the universe. Mach expressed his disagreement with Newton (Mach, 1893). In 'Mach's principle' as named by Einstein, it is thought that a body has inertia because it interacts in some way with all the matter in the universe (Sciama, 1959). In the letter to A.K. Schwarzschild in 1916, Einstein wrote: 'Inertia, in my theory, is, when all is said and done, just an interaction between masses, not an action in which, in addition to the masses in question, "space" as such participates' (Eisenstaedt, 1989). Discussions on physical, metaphysical or philosophical meanings of inertia may also be found in Wald (1992) and Sciama (1969).

Yuan-Chong Zhang

Bibliography

Cohen, I.B. (1971) *Introduction to Newton's 'Principia'*. Cambridge, Mass.: Harvard University Press.
Eisenstaedt, J. (1989) Cosmology: a space for thought on general relativity. In *Foundations of Big Bang Cosmology* (ed. F.W. Meyerstein). London: Word Scientific.
Herivel, J. (1965) *The Background to Newton's Principia*. Oxford: Clarendon Press.
Mach, E. (1893) *The Science of Mechanics: A Critical and Historical Account of its Development*. Lasalle: Open Court Publishing Company.
Motz, L. and Weaver, J.H. (1988) *The Concepts of Science, from Newton to Einstein*. New York: Plenum Press.
Sciama, D.W. (1959) *The Unity of the Universe*. New York: Doubleday & Co.
Sciama, D.W. (1969) *The Physical Foundations of General Relativity*. New York: Doubleday & Co.
Wald, R.M. (1992) *Space, Time and Gravity*. Chicago: University of Chicago Press.

Cross references

Angular momentum
Barycenter
Celestial mechanics
Newton, Sir Isaac, Newton's Laws
Torque

INFRARED RADIATION

Infrared radiation is invisible but is perceptible to humans as heat. The infrared portion of the electromagnetic spectrum begins at the lower limit of the visible light range, at wavelengths of about 1 μm (10^{-6} m). The infrared range extends to about 1 mm (10^{-3} m). This range is further subdivided into the near infrared (0.7–5.0 μm), the mid-infrared (5–100 μm) and the far infrared (100–1000 μm or 0.1 to 1 mm).

The radiation emitted by bodies with surface temperatures between about 5 K and 4000 K generally shows peaks in the infrared range. All of the known bodies in the solar system have temperatures within this range, except the Sun, which is hotter at about 6000 K. (Although solar radiation intensity peaks in the visible range, the Sun nevertheless radiates considerable energy in the infrared.) The planets emit most of their radiant energy in the mid-infrared.

Atoms, molecules and ions generally emit (and absorb) infrared radiation in characteristic wavelengths. Spectra of infrared emission can therefore provide information about the materials doing the emitting. Infrared absorption by H_2O and CO_2 in the Earth's atmosphere produces the greenhouse effect (q.v.), which raises the surface temperature of the Earth to a level permitting the existence of life as we know it.

Infrared observations provide considerable information about surface materials of solar system bodies. It is possible to obtain information about temperature, thermal inertia, composition and texture of solid and liquid surfaces from infrared observations. The following article (Infrared spectroscopy) discusses the applications (see also Spectroscopy: atmosphere).

The Earth's atmosphere acts as an almost opaque block to infrared radiation, raising problems for Earth-based infrared astronomers. However recent technological developments (Gillett and Houck, 1991) promise great improvements in infrared detection by orbiting astronomical spacecraft, and the development of a Space Infrared Telescope Facility is strongly favored by the Astronomy and Astrophysics Survey Committee (Bahcall, 1991). This instrument should be able to detect thermal radiation from a considerable portion of the mass in the universe, revealing features not yet seen and shedding light on processes such as the formation and evolution of planetary systems.

James H. Shirley

Bibliography

Bahcall, J.N. (1991) Prioritizing scientific initiatives. *Science*, **251**, 1412–3.
Gillett, F.C. and Houck J.R. (1991) The decade of infrared astronomy. *Physics Today*, **44**, 32–7.

INFRARED SPECTROSCOPY

The entire electromagnetic spectrum – radio, infrared, visible, ultraviolet and beyond – can be used to study planets. Each range of wavelength conveys a unique type of information about the object, and only by combining data from several regions of the spectrum can a complete understanding of a planet be gained. When observing a planetary atmosphere or solid surface, whether from a spacecraft, aircraft, or Earth-based observatory, several instruments are needed to cover all of the spectrum.

The infrared portion of the electromagnetic spectrum is usually defined as beginning just beyond the red end of the visible spectrum, at around 1 μm (10^{-3} mm) wavelength, and continuing to about 1 mm wavelength. The infrared is roughly divided into near infrared (0.7 to 5 μm), mid-infrared (5 to 100 μm), and far infrared (100 μm to 1 mm).

Planets reflect solar radiation in the visible and near infrared ranges, and emit thermal radiation in the mid- and far infrared. Infrared spectra contain information about the temperature, thermal inertia, composition and texture of solid and liquid surfaces. In the case of atmospheres, temperatures and composition can be mapped in horizontal and vertical distribution. From measured spectra it is possible to derive physical structure, dynamical activity, energy balance, chemical make-up and evolutionary history. A body of data is thereby developed for the Sun, planets, moons, asteroids and comets. By combining infrared spectroscopy with other disciplines, a comprehensive picture emerges of the past and present physical state of our solar system.

The field of infrared spectroscopy began with William Herschel, who discovered infrared radiation in the spectrum of sunlight. After passing light through a prism, Herschel placed a thermometer in the region just beyond the red end of the spectrum and found that it was heated. Infrared spectroscopy of the planets began about 1930 when photographic emulsions were developed which were sensitive to wavelengths longward of red to about 1 µm. With the advent of advanced infrared detectors in the 1940s, infrared astronomers were able to extend observations to longer wavelengths in the near infrared. By 1950 it had been shown (Kuiper, 1951) that the atmospheres of Venus and Mars contain carbon dioxide (CO_2), and that the atmospheres of Jupiter and Saturn contain methane (CH_4) and ammonia (NH_3). The atmospheres of Neptune and Titan (Saturn's largest moon) had also been shown to contain methane. During the next two decades, knowledge of planetary compositions increased steadily as ground-based telescopes and instrumentation improved.

Beginning in the late 1960s, infrared spectrometers were carried on planetary spacecraft (Hanel *et al.*, 1972a,b). Nimbus 3 and 4, Mariner 9 and Voyager 1 and 2 recorded detailed infrared spectra of Mars, Jupiter, Saturn, Uranus and Neptune, as well as many of their moons. The infrared spectrometers aboard Venera 15 and 16 (Moroz *et al.*, 1986) recorded spectra of Venus. Mars Global Surveyor, which will arrive at Mars in 1997, and Galileo, which arrived at Jupiter in December 1995, both carry infrared spectrometers.

Although standard optical telescopes work well for focusing light throughout the infrared, there are technical differences in the instrumentation and telescope facilities used in the various regions of the infrared. Types of detectors and radiation filters vary from one portion of the spectrum to the next. Cryogenic cooling of infrared instruments to liquid nitrogen or liquid helium temperatures is often required to operate detectors and to eliminate all sources of heat (which add noise to the data) in the instrument. Many regions of the infrared spectrum are blocked by absorption from molecules in the Earth's atmosphere, such as water and carbon dioxide. The observer can get above the Earth's atmosphere by placing his spectrometer on a spacecraft, on an aircraft such as the Kuiper Airborne Observatory, or at a high-altitude observatory like Mauna Kea on Hawaii.

A spectrometer separates electromagnetic radiation into its component wavelengths. The recorded spectrum is a plot of intensity versus wavelength. The four most common types of infrared spectrometer are the prism and grating spectrometers, the interferometer and the heterodyne spectrometer.

Simple prism spectrometers were used by Newton to study the color composition of light, and later by Herschel to demonstrate the existence of infrared radiation. A prism disperses radiation because the index of refraction of the prism material varies with wavelength. This causes the exit direction from the prism to be different for each wavelength. In a prism spectrometer radiation is passed through a slit, dispersed in the prism and focused to an image of the slit. A spectrum is formed by the separated images of the slit (each at a different wavelength), which are spread out from long to short wavelengths. Although not common today, prism spectrometers have often been used in ground-based planetary studies.

The grating spectrometer operates like the prism spectrometer, with a diffraction grating instead of a prism providing the spectral dispersion. A grating is usually a mirror like surface with evenly spaced, parallel grooves (between a few and several hundred per millimeter). The radiation is diffracted at each groove, and the interference among the diffraction patterns causes different wavelengths to leave the grating in different directions. A spectrum is formed when the slit is imaged, spreading the wavelengths out in the focal plane. Grating spectrometers are especially effective when an array of detectors is placed in the focal plane, because a large portion of the spectrum can be recorded simultaneously. Examples of grating spectrometers are the Near-Infrared Mapping Spectrometer (NIMS) on the Galileo spacecraft going to Jupiter, and the Visible and Infrared Mapping Spectrometer (VIMS) to be carried on the Cassini mission to Saturn. Grating spectrometers are also used with ground-based and airborne telescopes.

Interferometers also take advantage of the fact that different wavelengths can be distinguished by allowing a beam of radiation to interfere with itself. In an interferometer a fraction of the radiation is reflected at the surface of a partially transmitting material and, after traveling a different path length, is made to interfere with the original beam. Those wavelengths which are an even fraction of the path difference interfere constructively, and are transmitted more strongly. There are two kinds of interferometer used most commonly in planetary infrared spectroscopy: the Fourier transform spectrometer and the Fabry–Perot interferometer.

The Fourier transform spectrometer originated with the interferometer of A.A. Michelson (1881). The radiation beam is split at a beamsplitter into two equal parts and sent along two paths before being recombined. By changing the path difference continuously, the radiation at each wavelength in the recombined beam is modulated at a rate inversely proportional to the wavelength. The output signal, or interferogram, is the sum of these modulations. The spectrum is derived mathematically by taking the Fourier transform of the interferogram. A variation of this is the polarizing interferometer introduced by Martin and Puplett (1969), in which a linearly polarized beam is split into two polarization components and recombined after traveling different paths. Scanning then causes the polarization of the output beam to vary between two orthogonal states, which can be separated with another polarizer. The result is, again, an interferogram which is transformed to produce the spectrum.

The Infrared Interferometer Spectrometers (IRIS) carried on Nimbus 3 and 4, Mariner 9 and Voyager 1 and 2 were Fourier transform spectrometers, as is the Thermal Emission Spectrometer (TES) on Mars Observer. Cassini will carry the Composite Infrared Spectrometer (CIRS), combining a conventional and a polarizing interferometer in one instrument. At ground-based and airborne observatories Fourier transform spectrometers are generally used to attain high spectral resolution.

The Fabry–Perot interferometer (named after its inventors) consists of two parallel surfaces which are usually greater than 90% reflective, and transmit the remainder of the radiation. This causes a large number of reflections between the surfaces, with radiation traveling many different pathlengths recombining in the output beam. The different pathlengths are all multiples of a distance equal to twice the separation between the surfaces. Only radiation at wavelengths which divide evenly into this distance will constructively interfere and be transmitted. One of these transmitted wavelengths is selected (by using another, lower resolution spectrometer). A spectrum is created by changing the surface separation, which scans the output over a range of wavelengths. Fabry–Perot spectrometers are particularly suited to applications requiring two-dimensional imaging of a planet in a narrow spectral region. They have been used in ground-based and airborne planetary studies.

The heterodyne spectrometer differs from the types of spectrometer discussed above in that it directly measures the frequency of the radiation rather than its wavelength. Radiation from the source is combined with the monochromatic beam from a laser and focussed on a high-speed detector. The detector signal contains radio frequencies equal to the differences in frequency between the laser and source. Techniques borrowed from radio astronomy can then be used to record the infrared spectrum in the vicinity of the laser frequency. Heterodyne spectrometers are used in ground-based and airborne observations where very high spectral resolution is desired.

The infrared spectrum of a planet generally has the following characteristics. In the near infrared the albedo of the atmosphere or surface (whichever is observed) determines the fraction of solar radiation which is reflected. Reflected radiation dominates the spectrum from the visible out to about 5 µm, where the thermal radiation from the atmosphere or surface reaches a comparable intensity. The Planck thermal intensity distribution is determined by the temperature of the planet, and peaks in the mid- or far infrared. Compounds on the surface or molecules in the atmosphere absorb solar radiation in the near infrared and emit or absorb thermally in the mid- and far infrared. Surface compounds cause the appearance of broad features in the spectrum, and atmospheric molecules have band or narrow line signatures.

Figure I12 shows examples of spectra of Earth, Mars, Jupiter, Saturn and Titan (Saturn's moon) taken by the Infared Interferometer Spectrometer (IRIS) instruments on Nimbus 4, Mariner 9 and the Voyager spacecraft (Hanel, 1981). Comparison among the

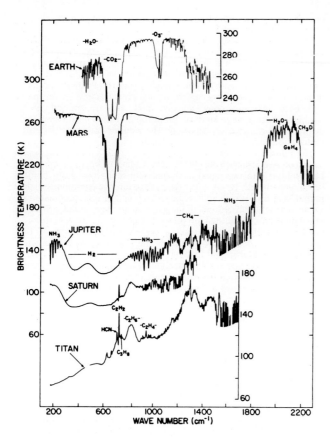

Figure I12 Infrared spectra of several planets with atmospheres. These spectra were recorded by the Infrared Interferometer Spectrometers (IRIS) on Nimbus 4, Mariner 9 and Voyager 1 and 2 (Hanel, 1981). Carbon dioxide (CO_2) absorption is common to the inner terrestrial planets Earth and Mars (and Venus, not shown). The outer giant planets Jupiter and Saturn (as well as Uranus and Neptune, not shown), exhibit prominent hydrogen (H_2) absorption and several features due to hydrocarbons and nitriles. Titan's atmosphere contains many complex molecules.

spectra shows some of the general characteristics of planetary atmospheres. The spectra of the inner planets are dominated by absorption from carbon dioxide and water. The main constituent in the atmospheres of Venus and Mars is carbon dioxide; the Earth's atmosphere is mostly molecular nitrogen and oxygen, but these have extremely weak spectral features. The atmospheres of the giant outer planets are predominantly molecular hydrogen, with some helium. Minor constituents such as methane, ammonia and acetylene are seen also. Titan (q.v.) is a special case, having an atmosphere of molecular nitrogen (no spectral features) and many complex minor constituents, including such exotic species as diacetylene, methylacetylene, propane, cyanoacetylene, cyanogen and hydrogen cyanide.

D.E. Jennings

Bibliography

Hanel, R.A., Conrath, B.J., Kunde, V.G. *et al.* (1972a) The Nimbus 4 infrared spectroscopy experiment, 1. Calibrated thermal emission spectra. *J. Geophys. Res.*, **77**, 2639–41.
Hanel, R.A., Conrath, B., Hovis, W. *et al.* (1972b) Investigation of the Martian environment by infrared spectroscopy on Mariner 9. *Icarus*, **17**, 423–42.
Hanel, R.A. (1981) Fourier spectroscopy on planetary missions including Voyager. *Proc. SPIE*, **289**, 331–4.
Kuiper, G.P. (ed.) (1951) *The Atmospheres of the Earth and Planets.* Chicago: University of Chicago Press.

Martin, D.H. and Puplett, E. (1969) Polarizing interferometric spectrometry for the millimeter and submillimeter spectrum. *Infrared Phys.*, **10**, 105–9.
Michelson, A.A. (1881) The relative motion of the Earth and the luminiferous ether. *Am. J. Sci. 3* **22**, 120–9.
Moroz, V.I., Spankuch, D., Linkin, V.M. *et al.* (1986) Venus spacecraft infrared radiance spectra and some aspects of their interpretation. *Appl. Optics.*, **25**, 1710–8.

Cross references

Galileo mission
Mariner missions
Reflectance spectroscopy
Spectroscopy: atmosphere
Venera missions
Voyager missions

INSOLATION

In meteorology and climatology the term insolation refers to the rate at which the spectrally integrated (total) solar irradiance is received by a unit horizontal area (UHA) at any point on or above the surface of the Earth, often at the surface of the Earth or at the top of the atmosphere (TOA). The same concept may also be applied to other planets of the solar system, e.g. the insolation on the surface of the Mars.

For a specified UHA at TOA, the insolation is a single-valued function of the solar constant (q.v.) S_o, the distance δ between it and the Sun, its solar zenith angle θ (the incidence angle of the Sun's rays) and its location (latitude and longitude). In the SI system insolation is expressed in W m^{-2}.

The solar constant S_o is the total radiative output of the Sun that, without any attenuation, reaches the UHA when θ is 0° and δ is 1 astronomical unit (AU; q.v.). The mean value of the solar constant measured during the period 1980–1983 was 1367.6 W m^{-2} (Willson, 1984). Although some features of solar activity (such as sunspots and faculae) affect solar irradiance (see Sun), the satellite observations of the Sun's total irradiance from 1978 to 1990 indicate that the variation of S_o is only a few tenths of a percent (Lean, 1991). Some empirical models (e.g. Foukal and Lean, 1990) indicate that the variation of S_o since the 1870s is also within a few tenths of a percent.

With a known S_o, the insolation of a UHA at TOA is equal to $S_o \cos\theta/\delta^2$ (δ in AU), which is the so-called astronomical insolation (at TOA). δ and θ can be uniquely determined by five astronomical elements describing the Earth's motion with respect to the Sun, namely the semimajor axis of the Earth's orbit, the orbital eccentricity (q.v.), the obliquity (q.v.), the longitude of the perihelion (see Periapse, perihelion, perigee, peribac) and the rotational rate of the Earth. These elements may be obtained by solving sets of differential equations in celestial mechanics. Therefore, with modern computers it is not difficult to determine quite accurately the insolation at TOA over the whole Earth as a function of time and latitude and longitude, provided that S_o is known. Since there are diurnal, seasonal, annual and secular changes for the astronomical elements, the insolation at TOA also has such characteristic variations. With a fixed S_o, the monthly astronomical insolation for different latitudes may have deviations as large as 13% of the long–term average in less than 10 000 years (Berger, 1988). Figure I13 shows the daily total and latitudinally averaged insolation at TOA as function of latitude and date. Figure I14 illustrates the daily and latitudinally averaged TOA insolation of 1990 plotted in a three-dimensional projection.

The insolation at the surface of the Earth, however, is highly unpredictable. When the solar radiation comes down through the atmosphere from TOA to the surface of the Earth, it is modified by scattering, absorption, reflection and emission of clouds, aerosols and gases. The radiation at the same time causes chemical and physical (including dynamic and radiative) changes of the atmosphere and the surface of the Earth. Figure I15 shows the monthly averaged insolation both at TOA and at the surface of the Earth for January and July of 1986 and illustrates how insolation is modified from TOA to the ground in a complicated way.

Insolation is of utmost importance to all the living things of the Earth. The visible radiation induces the photosynthesis reaction by plants that compound the carbohydrate molecules forming the cells in

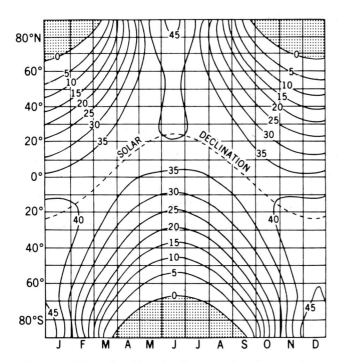

Figure 113 Daily total and latitudinally averaged insolation at the top of the atmosphere (TOA) as a function of latitude and date in 10^6 J m^{-2} (10^6 J m^{-2} is about 11.6 W m^{-2}). Shaded areas represent those regions not illuminated by the Sun. The Sun's declination is also shown (after Peixoto and Oort, 1992).

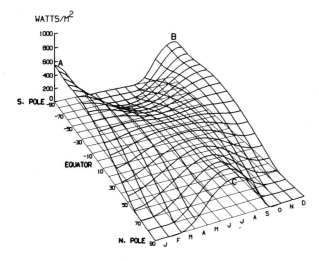

Figure 114 Three-dimensional daily averaged and latitudinally averaged insolation at the top of the atmosphere (TOA) as a function of latitude and date in W m^{-2} for 1990, produced by the author. The values of insolation at A, B and C are 552, 562 and 526 W m^{-2} for January 1, December 22 (the winter solstice) and June 21 (the summer solstice) respectively. A and B are located at the South Pole, and C at the North Pole.

plants. Photosynthesis also produces atmospheric oxygen (necessary for life in the animal kingdom) and the carbohydrates, which provide the basic food (energy) resource for herbivorous animals. More profoundly, insolation drives all physical, chemical, biological and geological processes in the whole dynamic climate system of the Earth. These processes involve complex interactions and feedback mechanisms among the three material layers of the outer Earth: atmosphere (q.v.), hydrosphere (q.v.) and lithosphere (q.v.), together with the thermally defined cryosphere and the ecologically constrained biosphere (q.v.).

Many mathematical models have been developed that simulate the global and regional climate processes in a highly simplified way and thus predict the responses of the climate system to the astronomical insolation as a primary forcing (e.g. Schneider and Dickinson, 1974; North et al., 1981; Schlesinger, 1988; Cess, 1989). Attempts have also been made to explain paleoclimates such as glacial–interglacial oscillations during ice ages, (see ice age) using variations of insolation. Milankovitch (q.v.) was the first to complete a full astronomical theory of Pleistocene Ice Ages. He computed the orbital elements and the subsequent changes in insolation using his simple climate model, and found some correlations between variations of insolation and Ice Age oscillations (Milankovitch, 1941). With more reliable paleoclimate data and better climate models, significant correlation and phase coherency between insolation and geological records have been confirmed (Berger 1988).

Yuan-Chong Zhang

Bibliography

Berger, A. (1988) Milankovitch theory and climate. *Rev. Geo.*, **26** (4), 624–658.
Cess, R.D. *et al.* (1989) Interpretation of cloud–climate feedback as produced by 14 atmospheric general circulation models. *Science*, **245**, 513–6.
Foukal, P. and Lean, J. (1990) An empirical model of total solar irradiance variation between 1874 and 1988. *Science*, **247**, 505–604.
Lean, J. (1991) Variations in the Sun's radiative output. *Rev. Geophysics*, **29**(4), 505–35.
Milankovitch, M. (1941) *Kanon der Erdbestrahlung und seine Anwendung auf das Eiszeitenproblem* (Canon of Insolation and the Ice-Age Problem, English translations by Israel Program for Scientific Translations, Jerusalem 1969). R. Serbian Acad. Spec. Publ. 132, Sect. Math. Nat. Sci., Vol. 33. Belgrade: Königlich Serbische Akademie.
North, G.R. *et al.* (1981) Energy-balance climate models. *Rev. Geophys. Space Phys*, **19**, 91–121.
Peixoto, J.P. and Oort, A.H. (1992) *Physics of Climate*. New York: American Institute of Physics.
Rossow, W.B., Zhang, Y.-C. and Lacis, A.A. (1990) Calculations of atmospheric radiative flux profiles. *Preprints of the Seventh Conference on Atmospheric Radiation*, July 23–27, 1990, San Francisco, California: American Meteorological Society.
Schlesinger, M.E. (1988) Physically based modeling and simulation of climate and climatic changes. *NATO ASI Series, Series C: Math. and Phys. Sciences*, Vol. 243. Dordrecht: Kluwer Academic.
Schneider, S.H. and Dickinson. R.E. (1974) Climate modeling. *Rev. Geophys. Space Phys.*, **12**, 447–93.
Willson, R. (1984) Measurements of solar total irradiance and its variability. *Space Sci. Rev.*, **38**, 203–42.

Cross references

Atmosphere
Earth: atmosphere
Greenhouse effect
Hadley circulation
Milankovitch, Milutin, and Milankovitch theory

INTERNATIONAL ASTRONOMICAL UNION

The International Astronomical Union (IAU), the world's organization of professional astronomers and astrophysicists, was established

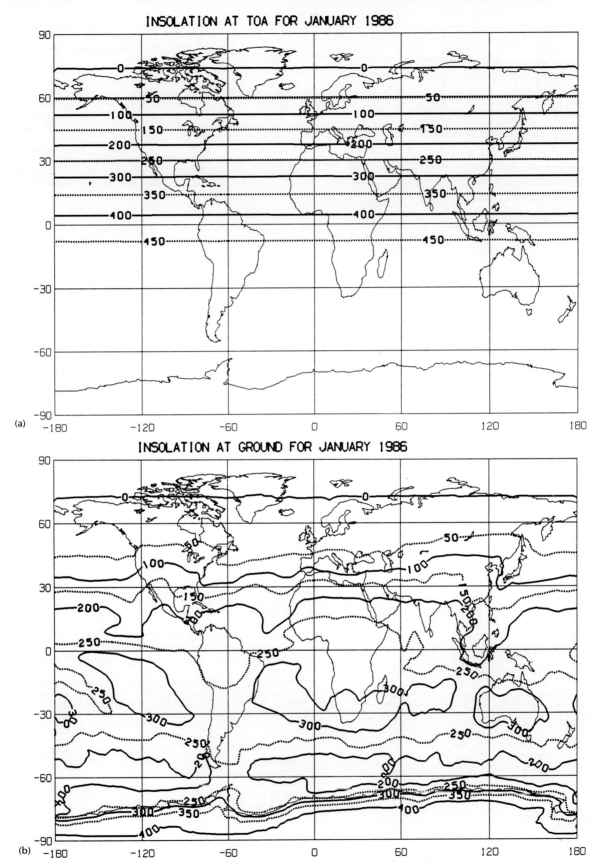

Figure I15 Four contour maps of monthly averaged insolation at TOA and the ground in W m^{-2} respectively for January (a,b) and July (c,d) 1986. The contour interval is 50 W m^{-2}. The insolation at TOA is calculated from the Sun's ephemeris formula by the author. The insolation

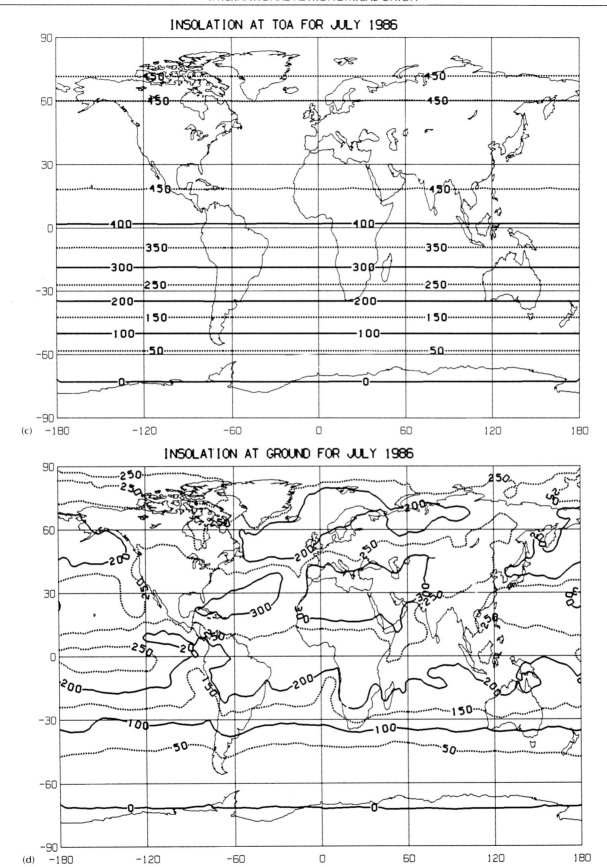

at ground level is calculated using the International Satellite Cloud Climatology Project (ISCCP) C1 data and the NASA Goddard Institute for Space Studies (GISS) radiation model (Rossow, Zhang and Lacis, 1990).

in 1919 and currently has some 7300 members in 65 countries. One of the 14 arms of the International Council of Scientific Unions, the IAU is unique in that it has permanent individual members, rather than national delegates with rotating terms. There is an Executive Committee (EC) consisting of the usual organizational officers and, in particular, six vice-presidents to ensure good international representation. The IAU's financial support comes largely through official adhering organizations – usually the national scientific academies – in 54 countries, and a secretariat is maintained in Paris. The scientific work of the IAU is conducted by its 40 commissions contained in 12 divisions and two union-wide working groups. The principal meetings are the triennial General Assemblies (GA, the most recent one having been held in The Hague in 1994), which represent a combination of policy decisions and purely scientific lectures and discussions; the IAU also sponsors (and cosponsors) scientific symposia and colloquia and supports the publication of the proceedings of the former (as well as of the GAs). Another triennial *Transactions* volume contains reports of the commissions and working groups, and in some cases these reports include comprehensive summaries of all relevant scientific research. Twice a year the IAU also publishes an *Information Bulletin*.

The principal scientific commissions whose interests concern the planetary system are Nos. 15 (Physical Study of Comets, Minor Planets and Meteorites), 16 (Physical Study of Planets and Satellites), 20 (Positions and Motions of Minor Planets, Comets and Satellites) and 22 (Meteors and Interplanetary Dust); current membership in these commissions ranges from 105 to 275. These commissions also form the backbone of I.A.U. Division III. Other relevant commissions include Nos. 4 (Ephemerides), 6 (Astronomical Telegrams), 7 (Celestial Mechanics) and 21 (Light of the Night Sky), and there is also the union-wide Working Group on Planetary System Nomenclature (WGPSN).

The basic purpose of the IAU is the promotion of international cooperation in astronomy and astrophysics. To facilitate such cooperation it is necessary for the IAU to set standards and to arrange for the appropriate collection and communication of relevant data. While all the commissions participate in this to some extent, in the area of planetary astronomy such activities are conducted on a routine basis mainly by Commissions 4, 6, 20, 22 and the WGPSN.

Commission 4 oversees the operation of the various national ephemeris offices. These include the US Nautical Almanac Office in Washington; HM Nautical Almanac Office in Cambridge, England; the Bureau des Longitudes (BdL) in Paris; the Institute for Theoretical Astronomy (ITA) in St Petersburg, Russia; and the Japanese Hydrographic Department in Tokyo. In their respective annual publications, such as the combined US–UK *Astronomical Almanac*, the offices ensure that reliable and accurate information is provided concerning planetary ephemerides – the predicted positions of the major planets (and also of their satellites) – as well as data relating to the observable disks and to the rotational axes of these bodies. Nowadays the information is also available on computer diskettes.

Commission 6 is responsible for the operation of the Central Bureau for Astronomical Telegrams (CBAT), which has since 1965 been located at the Smithsonian Astrophysical Observatory (SAO) in Cambridge, Mass. The CBAT receives, verifies and disseminates information about the discoveries of comets and the recoveries of periodic comets at their subsequent returns. It also announces new planetary satellites that might be found, e.g., by the Voyager missions. The discovery observations and preliminary orbital and ephemeris data are quickly disseminated by postcard and electronic circulars (known as the *IAU Circulars*). It is also possible for subscribers to log in to the CBAT's computers to obtain the information.

Commission 20 operates the Minor Planet Center (MPC), which since 1978 has also been located at SAO. The MPC works closely with the CBAT in connection with some of the newly discovered comets, but it collects and processes the astrometric and orbital data available for all comets and asteroids. The results are published in monthly batches of *Minor Planet Circulars/Minor Planets and Comets* and are also available by logging in to the CBAT/MPC computers (and in some cases also on magnetic tape). Urgent information is issued in *Minor Planet Electronic Circular*. Every year or so new editions (printed and electronic) are issued of the complete *Catalogue of Cometary Orbits* and *Catalogue of High Precision Orbits of Unnumbered Minor Planets*. Commission 20 also acknowledges the publication by ITA of the annual volume *Ephemerides of Minor Planets* (also now in printed and diskette form), which – in addition to giving ephemerides – updates the orbital elements of the numbered minor planets to a new standard epoch each year. In association with Commission 15, a working group monitors the determinations of absolute magnitudes for the numbered minor planets. An astrometric data center on satellites is in the process of being set up at the BdL, and it plans to provide data in printed and electronic form.

Commission 22 supports the Meteor Data Center at the Lund Observatory, Sweden. This center collects data on both the atmospheric trajectories and the heliocentric orbits of meteors and fireballs, on the basis of both photographic and radar observations. The data are disseminated by diskette.

The WGPSN is responsible principally for naming the surface features observed on planets and satellites. It was established in 1973 in response to the dramatic increase in the detection of such features from space probes. Until the space era such activity was restricted to Mars and the Moon and was organized by working groups in Commission 16 and the now-defunct Commission 17 (The Moon). The detailed work of the WGPSN is carried out in six task groups that deal with the Moon, Mercury, Venus, Mars (and its satellites), the outer planets (and their satellites and rings), and asteroids and comets. The WGPSN submits its recommendations of new names for approval by the EC, which meets annually, although names are not officially adopted until the next triennial GA. The adopted names are maintained in a *Gazetteer* published by the US Geological Survey in Flagstaff, Arizona.

The WGPSN is also responsible for naming newly discovered satellites themselves. This is done in association with Commission 20, which has a committee that decides if the orbits of the satellites have been determined sufficiently reliably to warrant a permanent designation and a name. Temporary designations for reported new satellites are provided by the CBAT, which also provides designations and names for comets, the latter in consultation with the Small Bodies Names Committee (SBNC) of Commission 20. Temporary designations and permanent numberings of minor planets are provided by the MPC, while the assignment of names for the newly numbered minor planets is also by the SBNC, generally, but not always, following the proposals of the discoverers.

Brian G. Marsden

Bibliography

Bergeron, J. (ed.) (1992) *Proceedings of the Twenty-First General Assembly. Transactions of the International Astronomical Union*, XXIB. Dordrecht: Kluwer.

Marsden, B.G. (1980) The Minor Planet Center. *Celest Mechan*, **22**(1), 63–71.

McNally, D. (ed.) (1991) *Reports on Astronomy. Trans of the International Astronomical Union*, XXIA. Dordrecht: Kluwer.

Oberbye, D. (1980) Life in the hot seat. *Sky and Telescope*, **60**(2), 92–6.

INTERPLANETARY MAGNETIC FIELD

Interplanetary space is permeated by a weak magnetic field – a remnant of the solar magnetic field 'frozen' into the flow of the solar wind (q.v.) as a result of the extremely high electrical conductivity of the solar wind plasma. This interplanetary magnetic field (IMF) plays a central role in many space processes. For example, the orientation of the IMF provides a preferred direction for solar wind thermal characteristics; enhancements in the strength of the IMF are effective in excluding low-energy cosmic rays from the solar system; energetic particles from the Sun are guided outward into interplanetary space along the IMF; and geomagnetic activity is excited when the IMF near Earth turns southward. The basic character of the IMF was first modeled by Parker (1958) prior to *in situ* observations by spacecraft. Close to the Sun the strength of the IMF in the ecliptic plane varies with heliocentric distance (R) approximately as $1/R^2$, while very far from the Sun it varies approximately as $1/R$. At the orbit of Earth the average strength of the IMF is about 6×10^{-5} gauss; however, the field strength is quite variable on a timescale of hours to days. This variability is caused principally by compressions and rarefactions of the solar wind plasma.

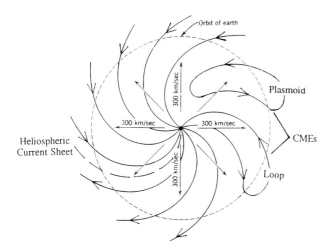

Figure I16 The interplanetary magnetic field configuration in the solar equatorial plane for a steady solar wind expansion (left) and for coronal mass ejections (adapted from Parker, 1963).

For the quiet solar wind expansion, field lines in interplanetary space are nearly identical to streamlines in the solar wind flow. Thus solar rotation causes interplanetary field lines to be drawn into Archimedean spirals in the solar equatorial plane (Figure I16). The pitch of these spirals varies with the speed of the solar wind. At the orbit of the Earth the average solar wind speed is ~ 400 km s^{-1} and the average field line spiral is inclined $\sim 45°$ to the radial direction from the Sun. At very large heliocentric distances the Archimedean spirals are inclined nearly perpendicular to the radial direction. In the Parker model the field lines out of the equatorial plane take the form of helixes wrapped about the rotation axis of the Sun. These helixes become ever more elongated at higher solar latitudes, finally approaching straight lines over the poles of the Sun. Measurements in the ecliptic plane show good agreement with Parker's model (Thomas and Smith, 1980); however, the model is currently being tested at high latitudes by the Ulysses mission. There is reason to believe that the model does not provide a correct description of the IMF at very large heliocentric distances at very high latitudes, and that interplanetary field lines there are actually probably mostly transverse to the radial direction (Jokipii and Kota, 1989).

The direction of the magnetic field in interplanetary space is determined by the magnetic polarity of the regions of the solar corona (q.v.) where the solar wind plasma originates. The average direction of the IMF in the ecliptic plane is pointed either away from or toward the Sun along the Archimedean spiral (Figure I16) and tends to remain constant for a number of days at a time (Wilcox and Ness, 1965). Major changes in field direction tend to be abrupt and are associated with crossings of the heliospheric current sheet (HCS) that both separates fields of opposite polarity and also encircles the Sun. The HCS has its origin in the overall dipole structure of the solar magnetic field and represents a rough mapping into interplanetary space of the location of the dipole equator (Hundhausen, 1977). Because the magnetic dipole is tilted relative to the Sun's rotation axis, the HCS is tilted relative to the solar equatorial plane. The HCS rotates with the Sun (with a period of ~ 27 days as viewed from Earth), and sweeps over the Earth at least twice per solar rotation. Solar wind characteristics generally vary in a regular fashion as a function of distance from the HCS. For example, the highest plasma densities and the lowest flow speeds, temperatures and helium abundances typically are observed near crossings of the HCS.

Even though the average IMF in the ecliptic plane lies nearly in that plane, the instantaneous field often points well out of the ecliptic. Substantial out-of-ecliptic perturbations in field direction lasting ~ 1 h or less are quite common; such perturbations often are associated with passages of Alfvén waves and other minor solar wind disturbances. Occasionally, a very strong out-of-ecliptic field component persists for many hours at a time; events of this nature often are found within solar wind disturbances driven by fast coronal mass ejections (CMEs; see Coronal mass ejection) and are of interest mainly because strong southward-directed fields in interplanetary space are the primary cause of major geomagnetic disturbances (Rostoker and Falthammar, 1967). The strong out-of-ecliptic fields often found within CME-driven interplanetary disturbances probably result from compression of the field within the CMEs themselves, from compression of the ambient IMF ahead of the CMEs, and from draping of the ambient IMF about the CMEs (Gosling and McComas, 1987).

Interplanetary magnetic field lines associated with the ambient solar wind expansion are considered 'open' in the sense that they connect to field lines of the opposite polarity only in the distant heliosphere very far from the Sun. By way of contrast, interplanetary magnetic field lines dragged into space by CMEs (Figure I16) typically are 'closed', either loops rooted at both ends in the solar atmosphere or partially or entirely disconnected from it as flux ropes or plasmoids (Gosling, 1990). The question of the overall topology of interplanetary magnetic field lines entwined within CMEs, and the related problem of magnetic flux balance in interplanetary space, are two of the major unsolved problems in solar wind physics today.

J.T. Gosling

Bibliography

Gosling, J.T. (1990) Coronal mass ejections and magnetic flux ropes in interplanetary space, in *Physics of Magnetic Flux Ropes* (ed. C.T. Russell, E.R. Priest and L.C. Lee). Washington DC: American Geophysical Union, Geophysical Monograph Series, Vol. 58, pp. 343–364.
Gosling, J.T. and McComas, D.J. (1987) Field line draping about fast coronal mass ejecta: a source of strong out-of-the-ecliptic magnetic fields. *Geophys. Res. Lett.*, **14**(4), 355–8.
Hundhausen, A.J. (1977) An interplanetary view of coronal holes. In *Coronal Holes and High Speed Wind Streams*, (ed. J.B. Zirker). Boulder: Colorado Associated University Press, pp. 225–329.
Jokipii, J.R. and Kota, J. (1989) The polar heliospheric magnetic field. *Geophys. Res. Lett.*, **16**(1), 1–4.
Parker, E.N. (1958) Dynamics of the interplanetary gas and magnetic fields. *Astrophys. J.*, **128**(3), 664–76.
Parker, E.N. (1963) *Interplanetary Dynamical Processes*. New York: Interscience.
Rostoker, G. and Falthammar, C.-G. (1967) Relationship between changes in the interplanetary magnetic field and variations in the magnetic field at the Earth's surface. *J. Geophys. Res.*, **72**(23), 5853–63.
Thomas, B.T. and Smith, E.J. (1980) The Parker spiral configuration of the interplanetary magnetic field between 1 and 8.5 AU. *J. Geophys. Res.*, **85**(A12), 6861–7.
Wilcox, J.M. and Ness, N.F. (1965) Quasi-stationary corotating structure in the interplanetary medium. *J. Geophys. Res.*, **70**(23), 5793–805.

Cross references

Coronal mass ejection
Geomagnetic storm
Heliosphere
Magnetospheres of the outer planets
Plasma wave
Shock waves
Solar wind
Ulysses mission

INTERSTELLAR GRAINS

Some meteorites consist of primordial materials from the early stages of formation of the solar system. Most solar system materials, for instance those that make up the Earth, have been atomically rearranged by melting and other violent processes. Laboratory analysis of meteoritic materials has led to the discovery of tiny grains that escaped this processing and retain the isotopic signatures of formation in stars other than our own Sun (Anders, 1989).

Silicon carbide, graphite and diamond grains isolated from objects such as the Murchison meteorite contain noble gas abundances and isotope ratios that were not characteristic to our own solar nebula (q.v.). Lewis, Amari and Anders (1990) and Gallino et al. (1990) determined that isotope ratios in the Murchison meteorite grains were diagnostic of formation in red-giant ('carbon') stars. It is estimated that as many as 1000 stars may have 'donated' material that formed the Sun and planets (Flam, 1991).

The study of interstellar grains is just beginning; it is likely to represent a fruitful area of investigation for many years to come. Comets are likely to include a higher proportion of unprocessed interstellar grains than do meteorites; this is one of the reasons for the European Space Agency's Rosetta mission, which is planned to visit a comet for sample return some time in the next century. Observations of Comet Halley by the Giotto spacecraft (q.v.) are consistent with the hypothesis that comets may consist largely of aggregates of interstellar grains (Greenberg and Hage, 1990).

Only a few years ago the materials of other stars were thought to be beyond the reach of humanity. Recently interstellar dust particles were detected entering the solar system by the Ulysses spacecraft (q.v.), which is currently on its way to investigate the polar regions of the Sun (Grun et al., 1993).

James H. Shirley

Bibliography

Anders, E. (1989) Prebiotic organic materials from comets and asteroids. *Nature*, **342**, 255–7.
Flam, F. (1991) Seeing stars in a handful of dust. *Science*, **253**, 380–1.
Gallino, R., Busso, M., Picchio G., and Raiteri C.M. (1990). On the astrophysical interpretation of isotope anomalies in meteoritic SiC grains. *Nature*, **348**, 298.
Greenberg, J.M. and Hage, J.I., (1990) From Interstellar dust to comets: a unifications of observational constraints. *Astrophys. J.*, **361**, 260–74.
Grun, E. et al., Zook, H.A., Baghul, M. (1993) Discovery of Jovian dust streams and interstellar grains by the Ulysses spacecraft. *Nature*, **362**, 428–30.
Lewis, R.S., Amari, S., and Anders E., (1990) Meteoritic silicon carbide: pristine material from carbon stars. *Nature*, **348**, 293–7.

Cross references

Comet: structure and composition
Cosmochemistry
Dust
Giotto mission
Meteorite
Solar nebula
Solar system: origin
Ulysses mission

INTERSTELLAR MEDIUM

The interstellar medium consists of the matter and associated fields (electromagnetic and gravitational) in the space between the stars in the Milky Way and other galaxies (Knapp, 1990). The elemental composition of the interstellar medium is very similar to that of the Sun: most of the atoms or ions are hydrogen, with about 10% helium by number. All of the elements heavier than helium together contribute only about 2% of the mass. Some of the elements, such as Si, Ca or Fe, are often 'depleted' from the gas phase, being instead located in small (submicron) dust grains which are mixed with the gas. A very small fraction of the nuclei and electrons – the 'cosmic rays' – have been accelerated to relativistic energies. The galactic magnetic field is an important component of the interstellar medium, and often plays a significant role in the gas dynamics. The total mass of the interstellar gas in the galaxy is $\sim 5 \times 10^9 M_\odot$, located in a disk of radius ~ 15 kpc and thickness of a few hundred parsecs. (Here $M_\odot = 1.99 \times 10^{30}$ kg is the mass of the Sun, and 1 pc = 3.09×10^{16} m.)

By convention, a distinction is made between the interplanetary medium within the solar system (the 'heliosphere') and the local interstellar medium. The boundary of the heliosphere is termed the 'heliopause'.

The Sun is moving with a velocity of ~ 20 km s^{-1} relative to the local interstellar medium. In the upstream direction both the solar wind and the interstellar medium are believed to undergo collisionless shocks near the heliopause. Neutral interstellar atoms are virtually unaffected by these collisionless shocks and thus stream into the interplanetary medium. Studies of backscattering of solar radiation in resonance lines of hydrogen and helium allow the density and temperature of the local interstellar medium to be determined. Lyman α observations indicate a streaming velocity 20 ± 1 km s^{-1}, $T = 8000 \pm 1000$ K, and $n(H^0) \approx 0.25$ cm^{-3}, while He 584 Å observations indicate $n(He^0) \approx 0.02$ cm^{-3} (Lallement, 1990; Lallement et al., 1991). It appears likely that the local hydrogen is predominantly neutral, with the properties of the 'warm neutral' phase (see below).

In addition to neutral atoms, relativistic cosmic ray particles from the interstellar medium penetrate the heliosphere and reach the inner solar system, where their energy spectrum and composition have been measured.

As we move away from the solar system, the interstellar medium is observed to be extremely inhomogeneous. Bright emission nebulae, although occupying only a very small fraction of the volume of the interstellar medium, are favored targets for astrophotographers. The brightest nebulae fall into a few standard types (Osterbrock, 1989):

- H II regions, such as the Orion Nebula, consist of gas which has been photoionized by radiation from one or more recently formed, hot, massive stars.
- Planetary nebulae, such as the Ring Nebula, consist of gas ejected by a star during an earlier stage of its evolution. The expanding gas, ejected $\sim 10^3$ years ago, is now photoionized by radiation from the star, which has shed its envelope and exposed its hot interior.
- Supernova remnants, such as the Crab Nebula or the Cygnus Loop, are produced when a supernova explosion drives a blast wave into the interstellar medium, creating a combination of relativistic particles and hot shocked gas.

Nearly all of the volume of interstellar space, however, is occupied by gas which is either too cold or too tenuous to be emitting strongly at optical wavelengths. This material may be divided into several distinct 'phases', which are in approximate pressure equilibrium with one another (McKee, 1990):

- Coronal gas – plasma at densities $n_H \approx 0.003$ cm^{-3} and temperatures $T \approx 10^6$ K, occupies perhaps $\sim 50\%$ of interstellar space (although the filling factor is controversial – Cox, 1990, argues for a much smaller value). This material has been heated by supernova blastwaves, and is observed through absorption lines of species like O VI, and through X-ray emission.
- Diffuse clouds – predominantly neutral gas with $n_H \approx 20$ cm^{-3}, $T \approx 30$–100 K, occupy $\sim 3\%$ of the volume. This material, studied primarily by 21-cm radio emission from atomic H, by far-infrared emission from dust grains mixed with the gas, and through absorption lines appearing in the spectra of bright background stars, is observed to be clumped into 'clouds'. Most of this gas is atomic, but in some of the denser diffuse clouds an appreciable fraction of the hydrogen is in H_2.
- Warm gas – partially ionized gas with $n_H \approx 0.3$ cm^{-3}, $T \approx 10^4$ K, occupies $\sim 50\%$ of the volume. Much of this gas is located on the boundaries of the clouds of neutral gas. The warm gas is observed using the same techniques as for the diffuse clouds; in addition, the free electrons and magnetic fields in these clouds are studied using dispersion and Faraday rotation toward pulsars.

In addition to the above 'phases', approximately 50% of the mass of the interstellar medium is found in a fourth state.

- Dense molecular clouds – cold ($T \approx 5$–30 K) regions where most of the hydrogen is in the form of H_2 (Scoville, 1990). Much of this gas is in self-gravitating clouds, with pressures which are somewhat higher than in the diffuse clouds. These molecular clouds have long been noted as 'dark clouds', and can be studied through the emission from molecules such as CO, as well as through infrared emission from the dust grains mixed with the gas. Dense, self-gravitating clumps within a molecular cloud gradually contract and ultimately are the sites of star formation.

Stars are the principal source of energy for the interstellar medium. Extreme ultraviolet photons from stars can dissociate molecules and ionize atoms. Starlight photons can also cause photoelectric emission from dust grains; these photoelectrons are an important source of heat in neutral regions. Energetic stellar winds deposit kinetic energy in the interstellar medium, but the dominant source of kinetic energy is probably the explosion of stars as supernovae; each supernova explosion injects $\sim 10^{51}$ ergs of kinetic energy of ejecta with velocities as large as 10 000 km s^{-1}. These supernova explosions drive shock waves into the surrounding interstellar medium. The complex, inhomogeneous structure of the interstellar medium is the result of this complex and intermittent injection of energy. The way in which these dynamic processes produce the observed interstellar medium is the subject of active theoretical study (McKee, 1990).

The interstellar medium of the Milky Way exhibits a spiral structure, with higher densities in the spiral arms. Star formation rates are enhanced in the spiral arms, and this is where most of the short-lived massive stars, and the H II regions they produce, are found.

The interstellar medium of the galaxy changes greatly as one moves away from the galactic plane, becoming more tenuous. There is also considerable radial structure in the galaxy, with a strong concentration of molecular clouds about 5 kpc from the galactic center (the Sun being located about 8.5 kpc from the center). The interstellar gas in the Milky Way has a gradient in the abundance of heavy elements, the gas in the central regions having been more extensively 'polluted' with the nucleosynthetic products of stars.

Other spiral galaxies of classification Sb or Sc have interstellar media roughly similar to that of the Milky Way. S0 and elliptical galaxies generally have less mass in the form of interstellar gas, but often harbor significant amounts of material in both atomic and molecular forms (Knapp, 1990).

The interstellar medium of the Milky Way (and other galaxies) is evolving. Stellar winds and supernova explosions gradually enrich the medium with heavy elements, although this may be partially compensated by infall of unenriched hydrogen and helium with near-primordial abundances. Continuing star formation, only partially compensated for by return of material in stellar ejecta and possible infall, is gradually consuming the interstellar gas. Understanding the evolution of the interstellar medium, particularly in order to interpret observations of galaxies at high redshift (seen as they existed in the distant past) is one of the current challenges to researchers in this field. Other topics of current research include the development of clumpy structure of molecular clouds; the process of star formation; energetic phenomena associated with star formation, such as high velocity jets and Herbig-Haro objects; the nature of interstellar dust grains, and their evolution; the origin of the interstellar magnetic field; and the acceleration of cosmic rays in interstellar space.

Bruce T. Draine

Bibliography

Cox, D.P. (1990) The diffuse interstellar medium, in *The Interstellar Medium in Galaxies* (eds H.A. Thronson, Jr and J.M. Shull). Dordrecht: Kluwer, pp. 181–200.
Knapp, G.R. (1990) The interstellar medium in galaxies – an overview, in *The Interstellar Medium in Galaxies* (eds H.A. Thronson, Jr and J.M. Shull). Dordrecht: Kluwer, pp. 3–37.
Lallement, R. (1990) Scattering of solar UV on local neutral Gases, in *Physics of the Outer Heliosphere* eds S. Grzedzielski and D.E. Page). Oxford: Pergamon, pp. 49–59.
Lallement, R., Bertaux, J.L., Chassefière, E. and Sandel, B. (1991) Interplanetary Lyman-α observations with UVS on Voyager: data, first analysis, implications for the ionization lifetime. *Astron. Astrophys.*, **252**, 385–401.
McKee, C.F. (1990) The three phase model of the interstellar medium: where does it stand now?, in *The Evolution of the Interstellar Medium* (ed. L. Blitz), San Francisco: Astronomical Society of the Pacific, pp. 3–29.
Osterbrock, D.E. (1989) Astrophysics of gaseous Nebulae and Active Galactic Nuclei. Mill Valley: University Science Books.
Scoville, N.Z. (1990) Dense gas in the galaxy, in *The Evolution of the Interstellar Medium* (ed. L. Blitz). San Francisco: Astronomical Society of the Pacific, pp. 49–61.

Cross references

Heliosphere
Solar wind

IO

Jupiter's innermost large satellite, Io, and the three other large satellites of Jupiter (Europa, Ganymede and Callisto), have had a special significance in the history of astronomical research ever since their discovery in 1610. In that year the Italian astronomer Galileo Galilei became the first human to undertake and record observations of celestial bodies with the assistance of a telescope. This technical achievement enabled him to study many objects that had not been previously seen with the unaided eye. After a few observations of Jupiter he concluded that the four newfound objects which he saw near Jupiter were not stars but instead are satellites which orbit about Jupiter as it orbits the Sun. They are now called the Galilean satellites in honor of their discoverer.

Galileo's study of the four large Jovian moons proved to have profound consequences for human civilization. His studies formed the cornerstone of a body of observational evidence which confirmed a Copernican view of the universe, that is, a solar system with the Sun rather than the Earth at its center. The Copernican revolution, which radically altered the self-perception of humans in the universe, is regarded as one of the great milestones in human philosophical development.

The early understandings of the Galilean satellites had important practical application as well. These satellites, like all objects in gravitationally bound orbits about a massive primary body, obey Kepler's laws of planetary motion as they move about Jupiter. Objects in Keplerian orbits have orbital periods which do not change. Therefore, shortly following their discovery it was realized that the very regular periods separating the times of eclipses and occultations of the Galilean satellites by Jupiter could be used to determine universal time at any point on the Earth's surface. Io, having the shortest period of the four satellites (1.769 days), was often used in this manner as a timepiece.

By 1675 the Danish astronomer Olaus Romer had observed that the times of the eclipses and occultations of the Galilean satellites by Jupiter exhibited a phase shift with about a 6.5-month periodicity. He correctly concluded that when Jupiter is at opposition (i.e Earth and Jupiter are closest, on the same side of the Sun), light from the Jovian system must travel a distance of approximately 4 astronomical units. (AU) to reach Earth (one astronomical unit is defined as the distance from the Earth to the Sun). However, a little more than 6 months later, when Jupiter was at conjunction (i.e Earth and Jupiter are farthest apart, on opposite sides of the Sun), the light reflected from the Jovian system traveled about 6 AU on its journey to the Earth. Romer interpreted this phase shift in the arrival time of Jovicentric events to mean that light has a finite velocity. Thus he was able to use the motions of the Galilean satellites to determine the speed of light, a fundamental physical constant.

The periodic orbits of the Galilean satellites played an important role in defining a standard for universal time. Knowledge of universal time was important because it could be used for determining the longitude of ships at sea. Navigators at sea were able to observe Galilean satellite events and then set a mechanical timepiece on board their ships. From the time setting and a star position, they could determine their latitude and longitude. The ability to precisely determine position on the Earth's surface was an important contribution to the advancement of intercontinental commerce.

During the next three centuries telescopic observations of many satellites in the outer solar system, including Io, indicated that their brightness varied regularly according to position in their orbit. This observation is consistent with the notion that while in orbit the object always keeps one face directed toward the primary body, just as does the Earth's Moon. Therefore, an Earth-based observer of the Jovian satellite system would see one complete rotation of each satellite in the time it takes to complete one orbit about the primary. However, little more could be inferred about the nature of the surfaces of these objects until a much higher level of technical capability could be achieved.

Even when viewed with the best Earth based telescopes, Io does not subtend enough area to permit an observer to resolve surface features directly. Prior the space age, knowledge about Io's surface was inferred from brightness variations in the sunlight reflected from Io as Io moved in its orbit about Jupiter. By calculating which part of Io was facing the Earth at the time of observation the properties of large surface regions could be inferred. In the 1920s telescopes equipped with photoelectric photometers confirmed and quantified

Io's orbital phase variation. Several decades later, more refined photometric observations at several distinct wavelengths showed that Io is covered with materials of vivid coloration and that it is one of the reddest objects in the solar system.

Spectroscopic studies have revealed that most planetary satellites in the outer solar system have surfaces which are covered with large quantities of water ice. Io, which has been found to be free of water in any form, is a notable exception. Io's coloration was found to be consistent with the color of sulfur or sulfur compounds. No other solar system object has this unusual coloration. From these observations it was already known at the dawn of the space age that Io was unlike any other planetary satellite.

One of the major research support tools that became available to planetary scientists in the last half of the 20th century was the development of rockets with the capability of launching deep space probes to the planets. These investigations led to major breakthroughs in efforts to understand the nature of the atmospheres and surfaces of planet-sized objects in the solar system. The Pioneer and Voyager missions to the outer solar system are responsible for returning much of what has been recently learned about the composition of Io's surface and the physical processes which modify it. One of the most significant discoveries of the Voyager mission was that Io has volcanoes which are kept active by a very intense internal heat source. The internal heating is caused by frictional dissipation of tidal energy generated by its orbital resonance with Europa. As a consequence of this intense heating, Io is believed to be the most volcanically active object in the solar system (Plate 23, 24).

Physical properties

Io's elliptical orbit about Jupiter has a semimajor axis of 422 000 km and the sidereal period of its orbit is 1 day 18 h 27 min 33.51 s. Its mass is 8.8899×10^{22} kg, making it about 1.2 times as massive as the Earth's Moon. It has a radius of 1818 ± 2.5 km, which is about 5% larger than the Moon. Io is the densest object in the outer solar system, having an average density of 3.53 g cm^{-3} Most other planetary satellites have much lower density.

Internal constitution

Most of the planetary satellites in the outer solar system have surfaces which are comprised principally of water ice. In addition, they have bulk densities close to 1 g cm^{-3} which is consistent with water ice being the predominant constituent throughout the interior of these objects. By comparison, Io's density is about 3.5 times as large as the density of most other planetary satellites in the outer solar system. This higher density is consistent with the hypothesis that Io's interior is composed principally of silicate material similar to inner solar system objects of comparable size such as the Moon.

It is generally believed that in the past Io was more massive than it is now and that Io's high density is due to the loss of low molecular weight volatile materials (including water) which have escaped Io's gravitational field. Early in Io's history this escape was probably due to heating of Io from nearby Jupiter as Jupiter radiated the energy it had acquired during gravitational condensation from the solar nebula. It is generally accepted that Jupiter was much hotter in the past than it is today and therefore the heat from Jupiter drove off most of the volatile materials from nearby Io. This scenario is very similar to what may have occurred in the early solar system as the young Sun drove volatiles off the inner planets.

At present a small number of molecules of low molecular weight species still escape via Io's volcanic processes. As a consequence of the early heating from Jupiter, combined with the additional long-term effect of tidally induced volcanism, all of the lower molecular weight species, such as nitrogen, oxygen and water, have probably been removed from Io. This evolutionary process has gone on for so long that Io may well be the most evolved object in the solar system.

There are several processes which provide heat to the interiors the planets and their satellites. The most important of these are gravitational energy acquired from the original accretion of the body, thermal heating from decay of radionuclides incorporated into the body as it condensed, and frictional heating from dissipation of tidal stresses induced within the body by a nearby large body. In all cases the heat generated by these processes is transferred to the surface of the object by conduction or convection by magma migration. From the surface the heat is radiated to space.

The amount of heat retained in the interior of a body depends on its size. The volume of the object, which determines the amount of heat generated, increases as the cube of the radius. The surface area of the object, which constrains the rate at which heat is radiated to space, increases as the square of the radius. Therefore, larger objects will retain heat for a longer time, have hotter interiors and be more likely to have a higher degree of differentiation by density between crust, mantle and core.

Io is small enough that if it had been heated only by gravitational accretion and radiogenic decay, it would probably not have a significant differentiation between crust and mantle. It would be essentially homogeneous and would have achieved thermal equilibrium with its environs eons ago. However, Io and its outer neighbor Europa occupy orbits about Jupiter such that they mutually perturb each other. These slight perturbations in their orbital motion cause Jupiter's gravitational field to exert a differential tidal force on Io. This tends to force Io's figure to assume different quasi-ellipsoidal shapes as it orbits Jupiter. The motion induced by such forces causes frictional heating within Io and, this tidal heating is the primary internal heat source. It is not definitely understood whether this tidal heating mechanism is continuous or episodic. However, it is operative now, and has been operating in the recent geologic past to the extent that Io has become extensively differentiated as a consequence of the tidal heating process.

There are differing opinions regarding how this extensive internal heating influences the nature of Io's interior. One view holds that Io has an entirely molten interior that is overlain by a thin solid crust. However, an alternative view suggests that Io has a partially liquid crust which feeds the volcanoes, and this crust overlies a solid mantle under which there is a partially molten core. In either view, the interior of Io is believed to be extensively stratified by density.

The global average heat flow from Io is 1–2 W m^2 but radiation of this heat to space does not occur uniformly across Io's surface. Observations of the thermal radiation emitted by Io indicate that most of the heat is emitted from a small number of hot spots which comprise about 2–3% of Io's total surface area. The hottest of these are located on the trailing hemisphere of Io as it orbits Jupiter and it is through the trailing side that Io radiates most of its heat. Half as much heat is radiated to space from Io's leading side as from its trailing side.

Geology of Io's surface

An examination of spacecraft images reveals that Io is completely devoid of impact craters. This is unusual because impact craters are the principal structures found on most solid surface objects in the outer solar system. There is no plausible mechanism that would immunize Io's surface from such impacts. Therefore, it is assumed that impacts are occurring on Io at a rate comparable to the other Galilean satellites but that the craters formed by such impacts are being rapidly covered over by material ejected from Io's many active volcanoes.

Io's surface can be divided into three distinct morphological regimes: vent regions, plains regions and mountains. The vent regions are evenly distributed across Io's surface and are the areas in which Io's volcanism is taking place. The plains regions are characterized by colors ranging from yellow to yellow-brown and they incorporate the flow-like patterns that may represent liquid sulfur moving down the slopes of the Io volcanoes (see Figure I17). In addition, the plains regions may consist of volcanic ejecta which has been deposited as a solid particulate accumulation. The mountain regions are quite steep and rugged. There are individual peaks as high as high as 9 km, and the existence of steep mountain slopes provides evidence that the mountains are not solid sulfur in composition. This is due to the fact that sulfur's bulk properties are such that massive sulfur accumulations could not support the steep mountainous slopes without extensive slumping. The mountains are most likely comprised of a silicate interior composition that is covered by sulfurous deposits emitted from the volcanoes.

The Voyager mission to the outer solar system identified nine active volcanoes on Io's surface. In addition, more than 100 features were identified as volcanic caldera which were not currently active but have been active in the recent past.

The volcanoes are believed to fall into two distinct classes, those which emit principally sulfur and those which emit sulfur dioxide. The sulfur volcanoes (Pele, 19°S, 257°W; Surt, 45°N, 338°W; and Aten, 48°S, 311°W) have vent velocities of ~ 1 km s^{-1} and their

Figure 117 Lava flows down the slopes of Ra Patera, a volcano on Io's trailing side. The feature is believed to be composed of liquid sulfur which has flowed from the volcano and frozen as a glass in the flow channel.

Figure 119 The white ring feature at the center of the image is the volcano Prometheus which is about 250 km across. The principal material ejected from this type of volcano is sulfur dioxide which condenses and is seen as a white frost in the ring around the vent.

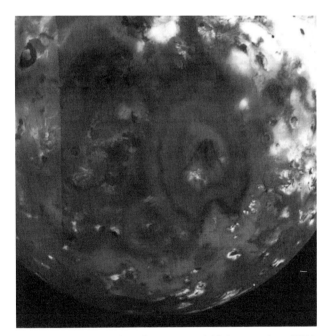

Figure 118 The heart-shaped structure, which is more than 1000 km across, is the volcano, Pele. The principal material ejected from Pele is elemental sulfur which covers the silicate material that most probably lies underneath the surface.

plumes extend 300 km above Io's surface (Figure I18). They are confined to Io's trailing side. The sulfur dioxide volcanoes (Prometheus, 3°S, 153°W; Volund, 21°N, 177°W; Amirani, 27°N, 118°W; Maui, 17°N, 122°W; Marduk, 28°S, 210°W and Masubi, 45°S, 52°W) have smaller vent velocities (~ 0.5 km s^{-1}) and are widely distributed on Io's leading hemisphere (Figure I19). It has been suggested that the Loki feature (19°N, 305°W) is a hybrid which exhibits intermediate characteristics between the low and high velocity volcanoes.

The area around each volcano is representative of the material being ejected. The spectral reflectance of the material around the low-velocity volcanoes is consistent with the material being condensed sulfur dioxide. Likewise, the material around the high-velocity plumes is spectrally consistent with the properties of elemental sulfur.

The material which returns to Io after being ejected from the volcanoes forms a very porous layer on Io's surface. This high-porosity layer, with perhaps as high as 90% void space, plays a very important role in determining the thermal properties of Io's regolith. On the sunlit side of Io, a thin layer of the upper surface will warm rapidly until it reaches equilibrium with the solar radiation, but the conduction of the heat downward into the regolith occurs slowly. At a small distance below the top of the regolith the temperature of Io's surface may be very close to Io's nightside temperature.

Composition of Io's surface

The principal means by which the composition of the surfaces of solar system objects is determined is through application of the technique of reflectance spectrophotometry. With this method the spectral features seen in telescopic observations of a solar system object can be compared to the spectral features seen in laboratory samples. If the features are similar, the presence of the material on the object is assumed to be confirmed. Using this technique, the presence of elemental sulfur on Io was suggested even before the arrival of the Voyager spacecraft at Jupiter. Io's other principal surface constituent, sulfur dioxide, was identified by Voyager observations as a gas in Io's very tenuous atmosphere, and ground-based observations identified SO_2 as a solid on Io's surface.

Figure I20 The Pele region, foreground, is Io's largest volcano. The surrounding material is believed to be elemental sulfur that has been ejected by the volcano. The white material on the surface is believed to be small amounts of condensed SO_2 frost. On the horizon Io's volcanic ejecta are seen above Io's surface after leaving the volcanic source.

Elemental sulfur occurs in many different forms called allotropes. Each allotrope is characterized by a different number of sulfur atoms per molecule. The most common allotrope of elemental sulfur found on Earth is cyclo-octal sulfur, S_8. However, in any bulk mass of elemental sulfur the dominant allotropic form will vary depending on the environment. The temperature and the radiation environment are the most significant factors in determining which allotropic form of sulfur will dominate a particular environmental circumstance. Therefore, it is quite possible that on Io's surface the principal allotropic form may not be cyclo-octal sulfur but another allotrope instead.

Sulfur melts at 119°C and has the unusual property that it increases in viscosity by five orders of magnitude at 160°C. The dominant allotrope at the melting point is S_8, which is yellow in color at room temperature. As the temperature increases, the number of short chain allotropes such as S_3, S_4 and S_5 increases. Above 160°C very long chain polymeric allotropes predominate. The color of the mixtures of long-chain polymers grows redder and darker and appears almost black at the boiling point of sulfur. It is possible that the black regions which are widely distributed on Io's surface are liquid sulfur pools or were once lakes which are now frozen. Laboratory studies report that the color of sulfur allotropes is preserved upon quenching the liquid to low temperature (i.e. some of the red color is retained in the low-temperature glass). Thus it is possible to infer that much of Io's surface is covered by mixtures of various allotropes of sulfur which may have been quenched.

It is known that the temperature of the high vent velocity volcanoes on Io exceeds ~ 300°C, well in excess of the melting point of elemental sulfur and well into the temperature regime where polymeric forms of sulfur would be expected. There have been isolated reports of volcano temperatures as high as 700°C. Therefore, it is quite probable that some sulfur is vaporized beneath Io's surface and that it is this sulfur vapor which drives the volcanoes. It is assumed that sulfur is heated in Io's subsurface layer and then explosively vents to the surface through the high-velocity volcanoes.

Even at the highest temperatures suggested for Io's volcanoes, the temperature is not high enough to impart a velocity to the ejected materials such that an appreciable fraction will escape Io's gravitational field. Nevertheless, the sulfur particles on ballistic trajectories may travel distances in excess of 1000 km from the source. At the other extreme, molten sulfur with little or no ballistic energy may flow down the side of the volcano as a lava and solidify on Io's surface. In either case the sulfur, once deposited, remains on Io's surface, where it is exposed to the ravages of the Io environment.

One environmental agent which acts on Io is a rain of atomic particles from the Jovian radiation belts. It is known that exposure of elemental sulfur to radiation causes the formation of metastable short-chain sulfur allotropes by sulfur bond sission. Thus, a steady state condition may exist wherein the sulfur on Io's surface is being converted to a variety of metastable allotropes, each with a slightly different red or red orange coloration. These short-chain allotropes eventually recombine to form S_8 on an indeterminate timescale. The materials lie exposed on the surface for only a short time before they are overlain by additional volcanic ejecta. The global average resurfacing rate of Io is estimated to the between 0.1 and 10 cm per year. Ultimately, the sulfur is recycled back into Io's crust where it is heated, melted and/or vaporized, and ejected once again through one of the volcanoes on Io's trailing side.

Io's other main surface constituent is sulfur dioxide. The infrared spectrometer onboard the Voyager spacecraft detected SO_2 as a gas over the Loki hot spot. Shortly after that discovery, spectral observations of Io from ground-based instruments were properly interpreted to identify SO_2 as a surface frost. The observations indicated that the spectral features due to SO_2 frost vary in strength depending on where Io is in its orbit about Jupiter. This orbital variation in strength is also observed in the spectral features of elemental sulfur. It is observed that whenever the SO_2 frost features are strong, the sulfur features are weak and vice-versa. This is because Io always keeps the same hemisphere facing toward Jupiter and therefore presents leading and trailing hemispheres to an Earth-based observer once each orbit. Thus when Io's trailing side is facing the Earth, the Earth-based observer is able to identify strong spectral features consistent with elemental sulfur, and when Io's leading hemisphere is facing the Earth-based observer the spectral properties of sulfur dioxide frost are more prevalent.

The principal product of the low-velocity, leading-hemisphere volcanoes is SO_2 gas. The temperature of the gas leaving the leading side volcanoes is about 100°C. The SO_2 molecules travel at best only a few hundred kilometer before they condense on Io's surface. From there it is believed they are covered over by subsequent deposits and ultimately recycled through the volcanoes again.

On Io's leading side the temperature of the volcanoes is too low for sulfur to participate significantly in the volcanism. On the trailing side it is possible that sulfur may be the only volcanic constituent; however, it is also conceivable that both sulfur and SO_2 participate in the trailing-side volcanism with sulfur playing the major role and rapidly covering over any SO_2 which may be collaterally deposited as a frost.

In addition to sulfur and SO_2 there is strong evidence that sodium and potassium bearing minerals are present on Io's surface. This is because Io's orbit is surrounded by a toroid-shaped distribution of sodium and potassium atoms. It is generally accepted that the atoms in the torus originate from the surface of Io. These alkali metal compounds are most likely to be found on Io's surface as sulfides of sodium and potassium. The spectral properties of these alkali sulfides have been investigated in the laboratory and are found to be consistent with Io's spectral geometric albedo. These materials need not be present as major constituents of Io's surface but they must be present at least as minor surface components if Io's surface is to be the origin of the sodium and potassium in the torus.

Hydrogen sulfide may be present as a minor surface constituent of Io. However its effect on the major physical and chemical interactions which occur in the Io environment is probably quite limited.

Io's atmosphere

There is strong evidence for an atmosphere on Io; however the density of the atmosphere and its long term stability have been the subject of a lively debate within the scientific community for more than a decade. A variety of Earth-based and space-based observations have produced evidence for an Io atmosphere, with the principal gas being SO_2. There are observations which indicate that Io's atmosphere occasionally precipitates. In addition, there is spectroscopic evidence which confirms the presence of atmospheric gases; and there is also evidence that Io has an ionosphere, which strongly implies that it has an atmosphere.

The evidence for precipitation is found in observations of Io's occasional post-eclipse brightening. In its orbit about Jupiter, Io spends about 5% of the time in the shadow of Jupiter. These instances, which last about 2.5 h each, represent a time in which the temperature of Io's dayside surface becomes as cold as the nightside. Telescopic observations of Io as it emerges from eclipse have reported that on occasion Io is about 15% brighter than usual immediately following emergence from eclipse by Jupiter. This effect is reported to be more pronounced toward ultraviolet wavelengths and it usually lasts about ~ 20 min. Credible observers have reported transitory brightening as large as 100%. Others have reported no measurable brightening during eclipse reappearance events. The phenomenon of post-eclipse brightening, if real, has been interpreted to mean that Io (at times) has a thick atmosphere which precipitates on Io during the eclipse and evaporates shortly thereafter as the sunlight warms the surface.

The evidence for an ionosphere is a product of the Pioneer mission to the outer solar system. During the Pioneer spacecraft encounter with the Jovian system, the radio signal from the spacecraft to Earth passed the edge of Io, creating an occultation of the radio signal. The attenuation of the radio signal as it passed near Io's limb was interpreted to mean that Io had an ionosphere, which implies an atmosphere.

The Voyager spacecraft had instruments which identified SO_2 gas over the Loki volcano area. If this sample of the region above Loki were taken to be typical for all of Io then Io would have a substantial atmosphere. However, because the Loki area is considerably hotter than Io's average surface, it is possible that this SO_2 constitutes a thick atmosphere only over the hot spots. There are independent ground- and space-based observations which indicate that the globally averaged SO_2 atmosphere is quite limited.

The thickness and extent of Io's SO_2 atmosphere is still under discussion and no proposed model has been able to explain all the information available. This may be because some of the observational data may be in error or that the atmosphere models are in need of improvement. Thick, intermediate and thin atmospheres have been proposed. Each proposal has its advantages and drawbacks.

If Io were to have a thick atmosphere which was in equilibrium with the surface frost, then the frost would ultimately concentrate at Io's north and south polar region where the temperature is probably the lowest. However, the images of Io from spacecraft indicate that most of the SO_2 is concentrated near Io's equator. In addition, an equilibrium atmosphere would distribute the SO_2 frost uniformly in longitude across Io, yet all observational evidence indicates that the SO_2 frost is concentrated preferentially on Io's leading hemisphere. It is possible that the ejecta from the sulfur volcanoes on Io's trailing side covers the SO_2 frost on short timescales; however, a very high volume of ejecta delivered almost continuously would be required in order to cover over a SO_2 frost layer deposited from an equilibrium atmosphere.

It has been suggested that Io's atmosphere may be in marginal equilibrium, with episodes of thick atmosphere equilibrium conditions followed by episodes of thin atmosphere conditions. This still does not explain the absence of SO_2 polar caps, which would build up during the episodes of thick atmosphere conditions unless another process were covering the polar deposits over again or a competing process were removing SO_2 from the poles and returning it back to the equatorial regions.

It has also been suggested that Io's very porous surface may play a role in controlling the surface distribution of SO_2 as a frost. SO_2 gas molecules ejected by the volcanoes and deposited on the regolith could easily pass through the upper regions and be deposited on the much colder subsurface layers. Once there, they would remain frozen until they were recycled into the volcanoes again. In this case Io's complete atmosphere would be quite thin. The SO_2 would not travel far from the source. Even in this situation, however, there is no mechanism which confines the SO_2 to the equatorial regions on Io's leading side.

The nature of Io's atmosphere remains poorly understood, and it promises to remain a topic of lively discussion among among planetary scientists for years to come.

Io's influence on the Jovian environment

Io influences its greater environment in several interesting and unusual ways, the two most notable being the removal of material from Io and its redistribution throughout the Jovian environment; and Io's influence on radio emissions from Jupiter.

The torus

Io's orbit about Jupiter traces a path which is the core of a torus-shaped distribution of neutral and ionized particles which travels about Jupiter at Io's orbital distance. From the torus, material is dissipated into the Jovian environment. Sodium was the first material identified in the Io torus. The current list of materials has now grown to include potassium, oxygen, sulfur and hydrogen. These elements are found in both neutral and ionized form.

The torus is supplied by material which is removed from Io's surface. It is estimated that 100–1000 kg of material is leaving Io each second and occupying the torus. Over the age of the solar system about 0.2% of Io's total present mass has been lost to the torus by this process. A steady state condition is assumed to exist and probably a comparable amount of material is leaving the confines of the torus and becoming part of the greater Jovian environment. It is believed that sulfur which has left the Io torus is the principal agent which colors the surfaces of the other Galilean satellites. The other satellites are composed principally of water ice, probably with the addition of small amounts of sulfur included as individual atoms within the ice.

The energy of the volcanoes is not believed to be of sufficient magnitude that volcanic activity can be the mechanism which supplies the material to the torus. The process by which the removal of this material occurs is not well understood. If it is assumed that Io has a very thin atmosphere, particles from the Jovian radiation belts can impact directly on Io's surface and material can be removed through the process of sputtering. The sputtering mechanism was widely accepted until the discovery of SO_2 gas over the Loki plume. If the density of the atmosphere over Loki is typical for all of Io then the particles from the radiation belts would be prevented from reaching Io's surface. If this were so, there would be no adequate mechanism to explain how material from Io's surface could be removed directly and supplied to the torus. It has been suggested that sputtering may remove material from the top of Io's atmosphere instead of its surface. However, this would require that the atmosphere contain sodium and potassium compounds in addition to the sulfur and oxygen compounds. The nature of a hypothesized sodium-bearing component in Io's atmospheric gas is not known.

Jovian decametric emission

Within two decades following the beginnings of the field of radio astronomy it was unexpectedly observed that Jupiter occasionally emits bursts of electromagnetic radiation at decametric wavelengths. These noise bursts are followed by periods of inactivity. This surprise was compounded by the realization that one component of these intermittent bursts was modulated by Io's position in its orbit. Observations from spacecraft above the Earth's atmosphere have extended the range of this emission to the hectometric wavelengths. This has been a subject of intense study and these bursts have been extensively classified. There is no mechanism yet proposed which provides a satisfactory explanation of this phenomenon (see Radio astronomy).

Summary

Jupiter's innermost Galilean satellite Io is one of the most unusual planet-sized objects in the solar system. Its close location to the giant planet Jupiter and the presence of its outer neighbor Europa, an object of comparable size to Io, combine to cause strong gravitationally induced tidal forces on Io. These forces dissipate their energy as frictional heat generated in Io's interior. The heat generated in Io's interior is carried to Io's surface, which has had hundreds of active volcanoes in the recent geologic past. Large active volcanoes abound on Io's surface today. The volcanoes emit two principal components, elemental sulfur and sulfur dioxide gas. After ejection by the volcanoes, these materials are recycled into Io's crust where they are reheated and ejected through the volcanoes again. Io lies within Jupiter's radiation belts and its surface is being exposed to bombardment by particles from Jupiter's magnetosphere. This causes material to be removed from Io's surface to supply a torus-shaped distribution which circles Jupiter at Io's orbit. The material ultimately escapes the torus region and is distributed about the Jovian system.

Robert M. Nelson

Bibliography

Ballester, G.E., Strobel, D.F., Moos, H.W. and Feldman, P.D. (1990) The atmospheric abundance of SO_2 on Io. *Icarus*, **88**, 1–23.

Binder, A.B. and Cruikshank, D.P. (1964) Evidence for an atmosphere on Io. *Icarus*, **3**, 299–305.

Brown, R.A. and Yung, Y.L. (1976) Io, its atmosphere and optical emissions, in *Jupiter*, (ed. T. Gehrels). Tucson: University of Arizona Press, pp. 1102–45.

Cheng, A.F. (1984) Escape of sulfur and oxygen from Io. *J. Geophys. Res.*, **87**, 5301–4.

Consolmagno, G. (1981) Io: thermal models and chemical evolution. *Icarus*, **47**, 36–45.

Durrance, S.T., Feldman, P. and Weaver, H. (1982) Rocket detection of ultraviolet emission from neutral oxygen and sulfur in the Io torus. *Astrophys. J. Lett.*, **267**, L125–9.

Fanale, F.P., Banerdt, W.B., Elson, L.S. et al. (1982) Io's surface, its phase composition and its influence on Io's atmosphere and Jupiter's magnetosphere, in *Satellites of Jupiter*, (ed. D. Morrison). Tucson: University of Arizona Press, pp. 756–81.

Franz, O.G. and Millis, R.L. (1974) A search for post eclipse brightening on Io in 1973. *Icarus*, **24**, 433–42.

Goertz, C.K. (1980) Io's interaction with the plasma torus. *J. Geophys. Res.*, **85**, 2949–56.

Goldberg, B.A., Garneau, G.W. and LaVoie, S.K. (1984) Io's sodium cloud. *Science*, **226**, 512–6.

Gradie, J. and Veverka, J. (1984) Photometric properties of powdered sulfur. *Icarus*, **58**, 227–245.

Haff, P.K., Watson, C.C. and Yung, Y.L. (1981) Ejection of matter from Io. *J. Geophys. Res.*, **86**, 6933–8.

Harris, D.L. (1961) Photometry and colorimetry of planets and satellites, in *Planets and Satellites* (eds G.P. Kuiper and B.M. Middlehurst). University of Chicago Press, pp. 272–342.

Ingersoll, A.P. (1989) Io meteorology: how atmospheric pressure is controlled locally by volcanoes and surface frosts. *Icarus*, **81**, 298–313.

Johnson, R.E. and Strobel, D. (1982) Charge exchange in the Io torus and exosphere. *J. Geophys Res.*, **87**, 10385–93.

Johnson, T.V. and Soderblom, L.A. (1982) Volcanic eruptions on Io: implications for surface evolution and mass loss, in *Satellites of Jupiter* (ed. D. Morrison). Tucson: University of Arizona Press, pp. 634–46.

Johnson, T.V., Lanzerotti, L.J., Brown, W.L. and Armstrong, T.P. (1981) Erosion of Galilean satellites surfaces by Jovian magnetospheric particles. *Science*, **21**, 1027–30.

Johnson, T.V., Morrison, D., Matson, D.L. et al. (1984) Volcanic hot spots on Io: stability and longitudinal distribution. *Science*, **226**, 134–37.

Johnson, T.V. (1990) The Galilean satellites, in *The New Solar System*, (eds K. Beatty and A. Chaikin). Cambridge, UK: Cambridge University Press and Cambridge, MA, USA: Sky Publishing Corp, pp. 171–88.

Kaula, W.M. (1968) *An Introduction to Planetary Physics: The Terrestrial Planets*. New York: Wiley.

Kieffer, S.W. (1982) Dynamics and thermodynamics of volcanic eruptions: implications for the plumes on Io, in *Satellites of Jupiter*, (ed. D. Morrison). Tulson: University of Arizona, pp. 647–723.

Kliore, A.J., Caine, D.L., Fjeldbo, G. et al. (1974) Preliminary results on the atmospheres of Jupiter and Io from the Pioneer 10 S-band occultation experiment. *Science*, **183**, 323.

Krimigis, S.M., Carbary, J.F., Keath, E.P. et al. (1981) Characteristics of the hot plasma in the Jovian magnetosphere: results from the Voyager spacecraft. *J. Geophys. Res.*, **86**, 8227–57.

Kupo, I., Mekler, Yu. and Eviatar, A. (1976) Detection of ionized sulfur in the Jovian magnetosphere. *Astrophys. J.*, **205**, L51–3.

Lane, A.L., Nelson, R.M. and Matson, D.L. (1981) Evidence for sulphur implantation in Eurpoa's UV absorption. *Nature*, **292**, 38–39.

Lanzerotti, L.J. and Brown, W.L. (1983) Supply of SO_2 for the atmosphere of Io. *J. Geophys. Res.*, **88**, 989–90.

Lanzerotti, L.J., Brown, W.L., Augustyniak, W.M. et al. (1982) Laboratory studies of charged particle erosion of SO_2 ice and applications to the frosts of Io. *Astrophys. J.*, **259**, 920–9.

Lewis, J.S. (1982) Io: Geochemistry of sulfur. *Icarus*, **50**, 103–14.

Macy, W. and Trafton, L. (1980) The distribution of sodium on Io's cloud: implications. *Icarus*, **41**, 131–41.

Matson, D.L. and Nash, D.B. (1983) Io's atmosphere: pressure control by regolith coldtrapping and surface venting. *J. Geophys. Res.*, **88**, 4771–83.

McCauley, J.F., Smith, B.A. and Soderblom, L.A. (1979) Erosional scarps on Io. *Nature*, **280**, 736–8.

McEwen, A.S. and Soderblom, L.A. (1983) Two classes of volcanic plumes on Io. *J. Geophys. Res.*, **86**, 736–8.

Moore, M.H. (1984) Studies of proton irradiated SO_2 at low temperature: implications for Io. *Icarus*, **59**, 114–28.

Moroz, V.I. (1968) *Physics of Planets*, NASA TT F-515. (Translation of *Fizika Planet*, Nauka Press, Moscow, 1957.)

Morrison, D. and Telesco, C.M. (1980) Io: observational constraints on the internal energy and thermodynamics of the surface. *Icarus*, **44**, 226–33.

Nash, D.B. and Nelson, R.M. (1979) Spectral evidence for sublimates and adsorbates on Io. *Nature*, **280**, 763–6.

Nash, D.B., Fanale, F.P. and Nelson, R.M. (1980) SO_2 frost: UV visible reflectivity and limits to Io surface coverage. *Geophys. Res. Lett.*, **7**, 665–8.

Nash, D.B., Carr, M.H., Gradie, J. et al. (1986) Io, in *Satellites*, (eds J. Burns and M.S. Mathews). Tucson: University of Arizona Press, pp. 629–88.

Nelson, R.M. and Hapke, B.W. (1978a) Spectral reflectivities of the Galilean satellites and Titan, 0.32–0.86 micrometers. *Icarus*, **36**, 304–29.

Nelson, R.M. and Hapke, B.W. (1978b) Possible correlation of Io's posteclipse brightening with major solar flares. *Icarus*, **33**, 203–9.

Nelson, R.M., Lane, A.L., Matson, D.L. et al. (1980) Io, longitudinal distribution of sulfur dioxide frost. *Science*, **210**, 784–6.

Nelson, R.M., Pieri, D.C., Baloga, S.M. et al. (1983) The reflection spectrum of liquid sulfur: implications for Io. *Icarus*, **56**, 409–13.

Nelson, R.M. and Smythe, W.D. (1986) Spectral reflectance of solid sulfur dioxide: implications of Jupiter's satellite Io. *Icarus*, **66**, 181–7.

Nelson, R.M. and Lane, A.L. (1987) Planetary satellites, in *Scientific Accomplishments of IUE* (ed. Y. Kondo). Dordrecht: Reidel, pp. 67–99.
Nelson, R.M., Smythe, W.D., Hapke, B.W. and Cohen, A.J. (1990) On the effect of x-rays on the color of elemental sulfur: implications for Jupiter's satellite Io. *Icarus*, **85**, 326–34.
Oliverson, R.J., Scherb, F. and Poesler, F.L. (1991) The Io plasma torus in 1981. *Icarus*, **93**, 53–63.
Peale, S.J., Cassen, P. and Reynolds, R.T. (1979) Melting of Io by tidal dissipation. *Science*, **203**, 892–4.
Pearl, J.C., Hanel, R. and Kunde, V. (1979) Identification of gaseous SO_2 and new upper limits for other gases on Io. *Nature*, **280**, 757–8.
Pearl, J.C. and Sinton, W.M. (1982) Hot spots of Io, in *Satellites of Jupiter* (ed. D. Morrison). Tucson: University of Arizona Press, pp. 724–55.
Pieri, D.C., Baloga, S.M., Nelson, R.M. and Sagan, C. (1984) The sulfur flows of Ra Patera, Io. *Icarus*, **60**, 685–700.
Pollack, J.B., Witteborn, F.C. and Erickson, E.F. (1978) Near infrared spectra of the Galilean satellites: observations and compositional implications. *Icarus*, **36**, 271–303.
Sagan, C. (1978) Sulfur flows on Io. *Nature*, **280**, 750–3.
Schubert, G., Spohn, T. and Reynolds, R. (1986) Thermal histories, compositions and internal structures of the moons of the solar system, in *Satellites* (eds J. Burns and M.S. Mathews). Tucson: University of Arizona Press, pp. 224–92.
Shemansky, D.E. and Sandel, B.R. (1982) The injection of energy into the Io plasma torus. *J. Geophys. Res.*, **87**, 219–29.
Sieveka, E.M. and Johnson, R.E. (1984) Ejection of atoms and molecules from Io by plasma ion impact. *Astrophys. J.*, **287**, 418–26.
Simonelli, D.P. and Veverka, J. (1986a) Disk resolved photometry of Io: I. Near opposition limb darkening. *Icarus*, **66**, 403–27.
Simonelli, D.P. and Veverka, J. (1986b) Disk resolved photometry of Io: II. Opposition surges and normal reflectances. *Icarus*, **66**, 428–54.
Sinton, W.M. (1981) The thermal emission spectrum of Io and a determination of the heat flux from its hot spots. *J. Geophys. Res.*, **86**, 3122–8.
Smith, R.A. (1976) Models of Jovian decametric radiation, in *Jupiter* (ed. T. Gehrels). Tucson: University of Arizona Press, 1146–89.
Smythe, W.D., Nelson, R.M. and Nash, D.B. (1979) Spectral evidence for SO_2 frost or adsorbate on Io's surface. *Nature*, **280**, 766.
Strom, R.G. and Schneider, N.M. (1982) Volcanic eruptions plumes on Io, in *Satellites of Jupiter* (ed. D. Morrison). Tucson: University of Arizona Press, pp. 598–633.
Trafton, L. (1975) High resolution spectra of Io's sodium emission. *Astrophys. J. Lett.*, **202**, L107–12.
Trafton, L. (1975) Detection of a potassium cloud near Io. *Nature*, **258**, 690–2.
Veverka, J.P., Simonelli, D.P., Thomas, P. *et al.* (1981) Voyager search for posteclipse brightening on Io. *Icarus*, **47**, 60–74.
Wamsteker, W., Kroes, R.L. and Fountain, J.A. (1974) On the surface composition of Io. *Icarus*, **10**, 1–7.
Yoder, C.F. and Peale, S.J. (1981) The tides of Io. *Icarus*, **47**, 1–35.

Acknowledgement

The author is a scientist at the Jet Propulsion Laboratory supported by NASA's Planetary Exploration Program. The author greatly appreciates the support of the management and staff of JPL and in particular the careful review given this manuscript by Linda J. Horn and Ellis D. Miner.

Cross references

Callisto
Europa
Galileo Galilei
Galileo mission
Ganymede
Ionosphere
Jupiter
Planetary torus
Thermal evolution of planets and satellites
Tidal heating

ION AND NEUTRAL MASS SPECTROMETRY

Mass spectrometry encompasses a broad range of techniques for separating ions according to their atomic mass per charge. In the laboratory, where mass spectrometers first appeared around 1920, the primary need is for high mass resolution and sensitivity with little restriction on instrument size, electrical power or sampling time. Satellite-borne mass spectrometers have a nearly diametric set of constraints.

Space age mass spectrometry has evolved rapidly, driven largely by the remarkable diversity of compositions and conditions found in the exploration of solar system atmospheres, ionospheres and magnetospheres. A wide variety of methods have been applied to obtain these important *in situ* measurements: roughly 80 mass spectrometers of some ten distinct types have already flown or will fly in the near future. Many draw heavily on laboratory techniques; others were developed in response to the peculiar requirements of the solar system's hot tenuous plasmas. Regardless of design heritage, however, all have had to conform to stringent mass, power, data rate and reliability requirements imposed by the rigors of space flight.

General considerations

It is useful to divide space mass spectrometry into thermal (a few eV or less) and energetic (a few eV to ≤ 100 keV) according to characteristic particle energies. Conversely, the distinction between spectrometers (in which mass is scanned) and spectrographs (in which all masses are displayed simultaneously) will be largely ignored except in the description of specific instruments.

Thermal mass spectrometry concerns ionized and neutral gases having relatively low energies typical of planetary atmospheres and ionospheres. Although momentum analysis is preferred, thermal particle mass/charge (M/Q) spectra can often be obtained using only energy/charge (E/Q) analysis if the relative spacecraft velocity is much larger than thermal velocities. This situation arises not only in the ionospheres of planets and comets, but also in the solar wind where electrostatic analyzers operating as 'pseudo' mass spectrometers have seen wide application (Vasyliunas, 1971).

Energetic mass spectrometry concerns ions sufficiently hot that both energy and momentum (or velocity) analysis are required to resolve composition. Energetic mass spectrometry targets very hot, tenuous and anisotropic plasma populations. These conditions create a demand for large apertures, good mass resolution at energies up to 100 keV and access to views of the entire sky. Traditional laboratory techniques have little applicability here and new ones have had to be developed over the past 20 years (Young 1989; Gloeckler, 1990).

Because of the chemical make-up of planetary atmospheres, the resolution of most spectrometers can be low relative to laboratory devices ($M/\Delta M < 60$), as can their range ($M/Q \leq 100$ amu e^{-1}). There are a few exceptions, particularly in situations where molecular, atomic or isotopic species have nearly the same mass (e.g. He^{++} versus H_2^+ or O^+ versus CH_4^+). However, recently developed energetic spectrometers capable of charge state and molecular fragment analysis alleviate the need for exceptional momentum resolution (Gloeckler, 1990).

Nearly all mass spectrometers employ one of two basic detection methods: either direct current sensing or particle counting. Instruments designed to admit large fluxes rely on current collection at the exit of the analyzer, followed by amplification and digitization of the amplified signal. Auto-ranging circuitry insures linear current response over a dynamic range of six or more decades. The practical sensitivity limit for current measurement is a few thousand ions cm^{-2} s^{-1}.

Particle counting offers greater sensitivity but more limited dynamic range at rates above a few million ions cm^{-2} s^{-1}. Individual particles may be detected by discrete dynode multipliers, continuous dynode channel electron multipliers or solid state detectors. Depending on details of detector shielding and coincidence circuitry, signal rates can be as low as a few particles per hour. Solid state detectors offer an additional benefit in that the amount of charge collected in each particle event is proportional to energy lost in the detector.

Nearly all planetary mass spectrometers can be assigned to one of three categories depending on whether mass separation is obtained by radio frequency, magnetic deflection or time-of-flight methods.

Gas chromatography has found application in a few instances but will not be discussed, nor will mass-discriminating cosmic ray instruments operating above 100 keV.

Radio frequency mass spectrometers

Quadrupole mass spectrometers belong to a class of instruments whose discrimination depends on the stability of selected ion paths in a radio frequency (RF) electric field (Blauth, 1966). As the name implies, ions travel along an axis centered among four rods (poles) to which RF and DC voltages are applied (Figure I21). Ion motion is either stable or unstable for a given frequency depending only on its M/Q ratio and the RF/DC voltage ratio. Mass resolution (ignoring ion charge, resolution is written $M/\Delta M$ and defined as an instrument's capability for separating a mass peak at M from one at $M + 1$ amu) depends on both the RF frequency and the length of the spectrometer, limiting spaceflight devices to resolutions < 100 (versus 10^4 for laboratory instruments).

Neutral mass spectrometry consists essentially of attaching an ionization source to the front of an ion mass spectrometer. While details of the neutral gas inlet and ionization source are critical (Figure I22), optical designs of both spectrometer types are much the same, taking advantage of high density and low thermal spread in the target populations (Nier, 1985). Some atmospheric gases are highly reactive (e.g. atomic oxygen), thus open source arrangements strive for minimum interactions between target gas and spectrometer, whereas closed sources bring the gas to equilibrium through collisions with the ion source walls.

Bennett (1950) devised a second widely used type of RF mass spectrometer. In its simplest form ions are initially accelerated into a set of three grids. An RF voltage is applied to the central grid such that ions with a selected M/Q value gain a small amount of energy (< 10%) due to resonance between the ion velocity and the RF field. Two, three or more grid sets (stages) are arranged in combinations designed to accelerate only the selected species at the expense of neighboring ones. At the end of the RF stages a retarding potential excludes all but the accelerated ions. Most Bennett spectrometers designed for space flight have three stages separated by distances that correspond to a small integral number of oscillations of the RF field. Resolution of the Bennett spectrometer is inherently modest (< 40) but both transmission and aperture are large and the device has good sensitivity for its size.

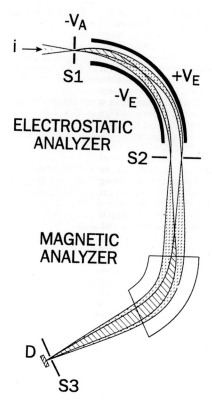

Figure I22 Double focusing deflection mass spectrometer. Ions (i) enter at top through the object slit S1 and are energy selected by applying voltages $\pm V_E$ to the electrostatic analyzer. Energy dispersion is limited by S2. Ions are momentum analyzed by the magnetic sector field and focused in both energy and angle at S3 where they are detected (D). Acceleration voltage $-V_A$ biases the entire analyzer and selects mass.

Magnetic deflection mass spectrometers

Magnetic deflection optics have found wide application in both thermal and energetic spectrometry. Deflection spectrometers can be categorized as either single or double focusing depending on whether they focus ions in angle only, or in both angle and energy (Duckworth, Barber and Venkatasubramanian, 1986).

Single focusing analyzers usually take the form of magnetic sectors of 60° or 90° following the early designs of Nier and coworkers (Nier, 1985) although any angle can be made to work. Ions enter normal to the magnetic field through a narrow slit. A small range of angles and energies are focused at a point on the far side of the magnet such that the object, image and vertex of the sector magnet all fall on a straight line (Figure I22 illustrates this principle for a double focusing spectrometer). A mass spectrum is obtained by accelerating the incoming ion beam over the range of voltages necessary to place each species at the object slit. Mass resolution is proportional to the ratio $R/(S1 + S3)$ where R is the radius of the ion path in the magnetic field and S1 and S3 are slit widths (Figure I22). The choice of magnet radius and slit width offer clear trade-offs of spectrometer size and sensitivity against resolution.

It should be emphasized that optical aberrations also limit spectrometer resolution. The image is broadened by angle and energy spreading of the beam and distorted by fringe fields. Special shaping of the magnetic field boundary can be used to reduce some blurring. A simpler approach obviously would be to reduce beam energy and angular spread to very small values; however, this limits spectrometer throughput, particularly for energetic plasmas.

Double focusing analyzers find wide application in planetary research because of their capability for energetic plasma spectroscopy. Double focusing is achieved by adding an electrostatic analyzer in tandem with the magnetic sector field (Figure I22). There are many possible combinations of sector angles and orientations that produce double focusing, but all operate on the principle that energy

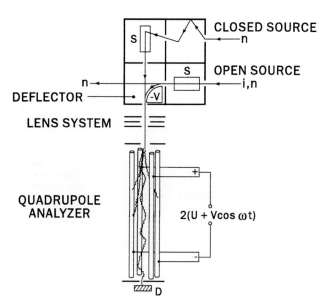

Figure I21 Quadrupole ion/neutral mass spectrometer. Ions (i) or neutrals (n) enter from right at spacecraft ram velocity. Either closed or open electron bombardment sources (S) can be used. Both DC and RF voltages $2(U + V \cos \omega t)$ are applied to the quadrupole. Transmitted ions are detected at D.

dispersion produced by the first analyzer is compensated for by an opposite dispersion in the second. The electrostatic analyzer causes higher energies to appear at larger radii and smaller energies at smaller radii. The magnetic field then reverses this dispersion, bringing ions of different energies and angles to focus at different points depending on the ion M/Q ratio. Mass spectra are obtained by scanning the acceleration and electrostatic analyzer voltages simultaneously. One particular arrangement that has found wide application is the Mattauch–Herzog design in which magnetic field curvature is reversed from that shown in Figure I22. This results in a true spectrograph with double focusing of all masses along a line near the magnet exit (Nier, 1985).

Wien filters are a special case of single focusing magnetic spectrometers in which an electric field is placed orthogonal to the magnetic field, and particles experience a Lorentz force proportional to $E \times B$. The selected trajectories are straight rather than curved, i.e. the magnet has parallel entrance and exit planes rather than being sector shaped. Wien filters can be made double focusing when combined with electrostatic analysis, an application found in some solar wind instruments (Ogilvie, Kittredge and Wilkerson 1968).

A number of double focusing spectrometers combine electrostatic fields with a second analyzer consisting of orthogonal magnetic and electric sector fields (Balsiger et al., 1976). These devices most resemble curved Wien filters, and although double focusing begins to break down at the upper end of their momentum range, they nonetheless resolve ions up to 30 keV with $M/\Delta M \approx 2$ (Young, 1989). A fixed pre-acceleration of a few kV reduces the range of momenta handled by the mass analyzer. Energy–mass selection proceeds by scanning simultaneously electric fields in both the electrostatic and mass analyzer sections. Although mass resolution is low, it has nonetheless proven adequate for investigation of magnetospheric plasmas where only major ion species are of interest (e.g. H^+, He^{++}, O^{++}, O^+). The most advanced device is a spectrograph based on toroidal geometry (versus cylindrical in Figure I22) with 30 keV energy range and 360° field of view (Shelley and Ghielmetti, 1990).

Time-of-flight mass spectrometers

The need for composition measurements at energies above a few tens of keV, and with progressively higher mass and charge resolution, has driven the design of a new generation of spectrometers based on time-of-flight principles. Ions are selected according to E/Q in an electrostatic analyzer (Figure I23) followed by post-acceleration by -10 to -30 kV. Angle-energy focusing is less of an issue due to the large post-acceleration voltage, although collisions in the foil inevitably add angle-energy scattering. Accelerated ions penetrate a thin (2.5 to 25 nm) carbon foil, ejecting one or more secondary electrons. Post-acceleration ensures that even the lowest energy ions typically lose $\lesssim 1$ keV in the carbon foil and scatter $\lesssim 5°$. Secondary electrons from the foil are collected onto a microchannel plate electron multiplier that transmits a 'start' signal to the fast (~ 1 ns resolution) timing electronics. Particles that exit the foil are usually charge neutral, with only a few percent to a few tens of percent still charged (some negatively), depending on foil thickness, particle species and velocity. In many time-of-flight designs the particle simply flies through a field-free region and strikes a second detector which provides a 'stop' pulse that ends the timing process. This combination of E/Q with time-of-flight measurement yields ion M/Q with relatively low mass resolution (< 10). In other designs the ion strikes a solid state detector where its total energy (as opposed to E/Q) is measured and a stop signal produced. The total energy measurement, together with E/Q and time-of-flight, uniquely determines M, E and Q separately. This triple coincidence technique also gives high immunity to background (Gloeckler, 1990) at the expense of sensitivity.

The most advanced space application of time of flight is found in spectrometers where ions exiting the foil are brought into an electric field whose properties cause them to execute an analog of simple harmonic motion (similar to the function of the DC field component in a quadrupole). Such a device can be made isochronous, i.e. temporally focusing. Existing designs are capable of producing resolutions approaching 100 even at energies of 20 keV and above (McComas and Nordholt, 1990; Hamilton et al., 1990).

Summary

Planetary mass spectrometry has evolved from classical thermal energy devices toward instruments with large apertures and excellent

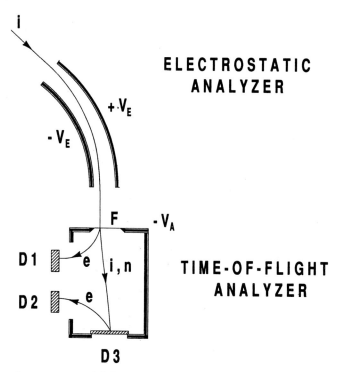

Figure I23 Time-of-flight mass spectrometer. Ions (i) enter and are energy/charge selected by the electrostatic analyzer biased at $\pm V_E$. Ions are then accelerated by voltage $- V_A$ into a thin (2.5 to 25 nm) carbon foil (F) which they penetrate, exiting as either ions (i) or neutrals (n). One or more secondary electrons (e) excite the 'start' timing detector (D1). The ion/neutral strikes the (solid state) detector (D3) after traveling across the time-of-flight analyzer. Secondary electrons excite the 'stop' detector (D2) to terminate the triple coincidence timing measurement.

resolution at energies thousands of times higher. Nonetheless, coverage of the full range of particle energies, and of neutral as well as ion species, still requires an assortment of spectrometer types for complete characterization of planetary atmospheres, ionospheres and magnetospheres. Nowhere is this broad and diverse range of spectrometric capabilities more apparent than in the Cassini/Huygens mission (q.v.) to Saturn and Titan. Scheduled for 1997 launch, the two spacecraft carry two time-of-flight mass spectrometers that together cover 1 eV to 300 keV for magnetosphere and ionosphere studies, and two quadrupole mass spectrometers devoted to atmospheric studies.

David T. Young

Bibliography

Balsiger, H., Eberhardt, P., Geiss, J. et al. (1976) A satellite-borne ion mass spectrometer for the energy range 0 to 16 keV. *Space Sci. Instrum.*, **2**, 499–521.

Bennett, W.H. (1950) Radiofrequency mass spectrometer. *J. Appl. Phys.* **21**(2), 143–9.

Blauth, E.W. (1966) *Dynamic Mass Spectrometers*. New York: Elsevier Publishing Co.

Duckworth, H.E., Barber, R.C. and Venkatasubramanian, V.S. (1986) *Mass Spectroscopy*, 2nd edn. New York: Cambridge University Press.

Gloeckler, G. (1990) Ion composition measurement technique for space plasmas. *Rev. Sci. Instrum.*, **61**(11), 3613–20.

Hamilton, D.C., Gloecker, G., Iparich, F.M. et al. (1990) New high-resolution electrostatic ion mass analyzer using time of flight. *Rev. Sci. Instrum.*, **61**(10), 3104–6.

McComas, D.J. and Nordholt, J.E. (1990) New approach to 3-D high sensitivity, high mass resolution space plasma composition measurements. *Rev. Sci. Instrum.*, **61**(10), 3095–7.

Ogilvie, K.W., Kittredge, R.I. and Wilkerson, T.D. (1968) Crossed field velocity selector. *Rev. Sci. Instrum.*, **39**(4), 459–65.
Nier, A.O. (1985) Mass spectrometry in planetary research. *Int J Mass Spectrom. Ion Proc.*, **66**: 55–73.
Shelley, E.G. and Ghielmetti, A.G. (1990) Angle, energy, and time-of-flight focusing with poloidal toroid electrostatic analyzers. *Nuclear Instrum. Meth. Phys. Res.*, **A298**, 181–8.
Vasyliunas, V.M. (1971) Deep space plasma measurements, in *Methods of Experimental Physics*, Vol. 9 Part B, (ed. R.H. Lovberg and H.R. Griem). New York: Academic Press, pp. 49–88.
Young, D.T. (1989) Space plasma mass spectroscopy below 60 keV, in *Proc. Solar System Plasma Phys.*, (ed. J.H. Waite, Jr. J.L. Burch and R.L. Moore). Washington, DC: American Geophysical Union, Geophysics Monograph Series **54**, pp. 143–57.

Cross references

Atmosphere
Spectrometry: atmosphere

IONOSPHERE

An ionosphere is the region of an atmosphere in which significant charged particle densities exist. The outermost regions of planetary atmospheres are not entirely neutral but contain partially ionized plasmas. Plasma is the name given to gases composed of charged particles (i.e. electrons and ions). The electron density is almost exactly equal to the positive ion density in an ionosphere and as a consequence the net electrical charge density is almost zero; this plasma property is known as quasi-neutrality. The plasma in the solar wind and in the magnetosphere are also quasi-neutral, as are almost all plasmas found in nature. However, unlike ionospheric plasma, the solar wind and magnetospheric plasmas are fully ionized and do not contain neutral atoms and molecules.

The terrestrial ionosphere starts at an altitude of about 90 km and extends up to about 1000 km. The ionosphere coexists with the region of the neutral atmosphere called the thermosphere. The density of electrons in the terrestrial ionosphere has a maximum value of the order of 10^6 cm^{-3}, which is typically attained near an altitude of 300 km. The Earth's ionosphere has been extensively studied, starting from the time that radio communication was first developed about a century ago; however, most of our understanding of the physics and chemistry of the ionosphere has come from experiments carried through the ionosphere by rockets and satellites over the past three decades.

All the planets in the solar system have ionospheres and almost all of these planetary ionospheres have been observed by instruments carried onboard space probes. Some planetary satellites are also known to possess reasonably dense ionospheres; these include Jupiter's satellite Io, Saturn's satellite Titan and Neptune's satellite Triton. Comets are also known to contain ionospheric plasma.

Ionospheric measurements

The ionosphere was first studied using radio waves at the turn of the 20th century when Marconi demonstrated that a radio signal transmitted from Cornwall, England, was received in Newfoundland. Kennelly and Heaviside independently suggested in 1902 that the radio waves in this demonstration must have reflected off an electrically conducting layer of ions located near a height of 80 km. The propagation of electromagnetic radiation through a medium can be described by means of the index of refraction, n, which for an unmagnetized plasma is given by:

$$n = \left(1 - \frac{\omega_p^2}{\omega^2}\right)^{1/2} \quad (I1)$$

where $\omega = 2\pi f$ and f is the wave frequency in Hertz. The plasma frequency ω_p is related to the electron density n_e by the following expression:

$$\omega_p^2 = \frac{n_e e^2}{\epsilon_0 m_e} \quad (I2)$$

where e and m_e are the electron charge and mass respectively, and where ϵ_0 is the permittivity of free space. This expression is in SI

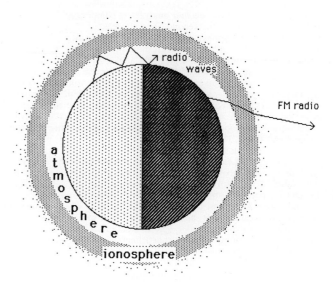

Figure I24 Schematic of radio wave propagation in the terrestrial ionosphere.

units. A useful expression for the plasma frequency is $f_p = 9 \times 10^{-3} n_e^{1/2}$ where the units of f_p are megahertz (MHz) and the units of n_e are cm^{-3}. For example, the maximum electron density in the Earth's ionosphere is approximately 10^6 cm^{-3} and thus the maximum plasma frequency is $f_p^{max} \approx 10$ MHz. Recall that the AM radio band is from 0.5 to 1.6 MHz (well below f_p^{max}) and the FM radio band is from 88 to 108 MHz (well above f_p^{max}). The index of refraction for a magnetized plasma, such as that in the Earth's ionosphere, is somewhat more complicated than the one given here and depends on the polarization of the waves as well as on the frequency, but the basic idea can be illustrated without this complication.

An examination of the index of refraction shows that it is always less than one and that for frequencies less than the plasma frequency ($f < f_p$) it is imaginary, which means that such electromagnetic waves cannot propagate. Radio waves directed at a plasma, such as the ionosphere, with $f < f_p$ are reflected! This radio wave property can be used to probe the ionosphere remotely. Radio waves over a range of frequencies are transmitted upwards and the delay time between their transmission and reception is measured. These delay times can be used to determine the altitude of reflection. Waves with low frequencies are reflected at lower altitudes than waves with higher frequencies because the electron density below the ionospheric peak (and thus the plasma frequency) increases with altitude. The electron density as a function of altitude can thus be determined for altitudes below the ionospheric maximum. Transmitters specifically designed to probe the ionosphere via radio wave reflection are called ionosondes and the data records they produce as the transmitter frequency is varied are called ionograms. Topside sounders carried onboard satellites and transmitting signals downward are used to measure the topside ionosphere (i.e., the part of the ionosphere above the peak). As indicated in Figure I24, AM radio waves are not only reflected from the ionosphere but also can be partially reflected by the ground, so that they are able to travel around the Earth. Radio waves with frequency greater than f_p^{max}, such as FM waves, are able to propagate through the ionosphere. For more information on this subject see Ratcliffe (1972) or Kelley (1989).

Radio waves have also been the chief means of observing planetary ionospheres. Space probes communicate with Earth by transmitting S and X band radio waves (f of 2.293 GHz and 8.6 GHz, respectively) that are received by the large ($\approx$ 30–60 m in diameter) receiving dishes of NASA's Deep Space Net (DSN). The radio wave path between the transmitter on a spacecraft and the Earth-based receiver dish will be occulted if the spacecraft moves behind the planet being encountered, as illustrated in Figure I25. Such radio occultations have been used to study both the neutral atmospheres and the ionospheres of almost all the planets in our solar system. The neutral atmosphere and the ionosphere both make contributions to the index of refraction. The radio waves are both refracted (i.e. bent) and have

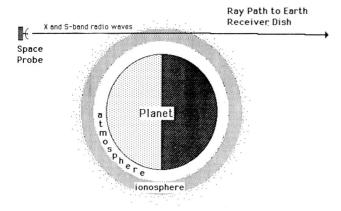

Figure I25 Schematic of radio occultation by a planetary ionosphere.

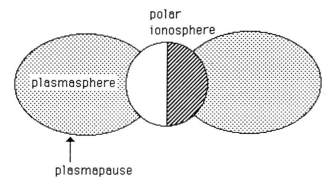

Figure I26 Schematic of the terrestrial plasmasphere.

their phase altered. The index of refraction is determined as a function of the impact parameter of the occultation by measuring the change in the wave phase. The electron density in an ionosphere is calculated by using equations (I1) and (I2) for the index of refraction.

Low-frequency (several kHz) electromagnetic waves, called whistlers, are also used to probe the electron density in an extended region containing cold ionospheric plasma, called the plasmasphere, that exists in the inner magnetosphere. The plasmasphere is the high-altitude extension of the topside ionosphere along closed magnetic field lines at low and mid-latitudes (see Figure I26). The boundary between the plasmasphere and the rest of the magnetosphere is marked by a sharp decrease of the electron density and is called the plasmapause. Whistlers are produced naturally by lightning and propagate along magnetic field lines from one hemisphere to the other. The index of refraction for whistlers is more complicated than expression (I1), and depends both on the electron density and on the magnetic field strength.

Over the last couple of decades a very important means of remote sensing the ionosphere has been incoherent radar backscatter, in which powerful high-frequency electromagnetic pulses are transmitted upwards and the weak time delayed signals scattered from very small-scale ionospheric irregularities are detected using very large receiver dishes. The ionospheric density profile both above and below the electron density maximum can be measured with this technique, as well as electron temperatures, ion temperature and plasma drifts.

In the 1950s *in situ* measurements of the ionosphere were first made by instruments carried onboard sounding rockets and, starting with Sputnik in 1957, *in situ* measurements have also been carried out in the ionosphere and magnetosphere by instruments onboard satellites. The Atmosphere Explorer C, D and E satellites and the Dynamics Explorer 1 and 2 satellites in particular have returned large quantities of useful thermospheric and ionospheric data. A large variety of physical quantities have been measured including ion composition, electron and ion temperatures, superthermal electron fluxes and electric and magnetic fields. *In situ* measurements have also been carried out in the thermosphere and ionosphere of Venus by several instruments onboard NASA's Pioneer Venus Orbiter, including a Langmuir probe (for electron densities and temperatures), a retarding potential analyzer (for ion composition and drifts, and superthermal electron fluxes), a magnetometer, a plasma wave detector and neutral and ion mass spectrometers (for neutral and ion composition respectively). The book *Venus Aeronomy* summarizes these measurements. *In situ* measurements were also made in the thermosphere and ionosphere of Mars by the neutral mass spectrometers and retarding potential analyzers on the Viking 1 and 2 landers (Nier and McElroy 1977; Hanson, Sanatani and Zuccaro, 1977).

Basic ionospheric processes

Ionospheres are formed by the ionization of thermospheric neutrals. Photoionization by solar extreme ultraviolet (EUV) photons is the most important ionization mechanism, although electron impact ionization is also important in the auroral regions of the Earth and other planets. These ionization processes can be represented by:

$$h\nu + M \rightarrow M^+ + e \tag{I3}$$

$$e_{fast} + M \rightarrow M^+ + e_{fast} + e \tag{I4}$$

where M represents a neutral species like N_2, O_2 and O for the Earth, CO_2, CO and O for Venus and Mars, H_2, H and CH_4 for the outer planets, H_2O, OH, O and CO for comets, and N_2 and CH_4 for Titan. Photons are represented by $h\nu$, where h is Planck's constant and ν is the photon frequency. The photon energy is given by $E_{photon} = h\nu$ and this energy must exceed the ionization potential of a neutral species (typically in the range 10 to 15 eV) in order for the photoionization process, (I3), to proceed. The wavelength of the ionizing EUV radiation is typically less than about 100 nm and is produced in the solar chromosphere and corona. Reaction (I3) produces a photoion, M^+, and a photoelectron. Fast electrons are represented by e_{fast} and must have energies in excess of the ionization potential of the neutral species M. The electrons produced by reaction (I4) are called secondary electrons and have energies of about 10 to 20 eV. The fast electrons can either be energetic auroral electrons, typically with energies of several keV, or can be those photoelectrons and secondary electrons which possess energies greater than the ionization potential.

The photoelectrons and secondary electrons produced by reactions (I3) and (I4) have energies of about 20 eV. These electrons quickly lose their energy by colliding with neutrals and with colder thermal electrons. The ionospheric electron gas is heated up in this manner, such that at higher altitudes the electron temperature is much higher than the neutral temperature. The high-altitude neutral thermospheric temperature (or exospheric temperature) for the Earth is typically in the range $T_n \approx 1000$–2000 K, whereas the electron temperature is $T_e \approx 5000$ K. The ion temperature above 300 km is typically $T_i \approx 2000$ K. For comparison, the exospheric temperature at Jupiter is about 1000 K. The exospheric temperature at Venus is only about 300 K, due to the high abundance of CO_2 which is an emitter of infrared radiation, and thus a good means of cooling the atmosphere. The electron temperature in the Venus ionosphere is much greater than the neutral temperature with $T_e \approx 5000$ K.

The cross-section for photoabsorption (and photoionization) is very large for EUV radiation, which is thus absorbed at high altitudes in the atmosphere. The unit optical depth (i.e. $\tau \approx 1$) for the absorption of solar EUV radiation impinging on the top of the atmosphere is reached near an altitude of 150 km for the Earth, 140 km for Venus and Mars, and an altitude of 600 km (above the cloudtops) for Jupiter. The production rate of ions from reaction (I3) is proportional to the product of the EUV photon flux and the neutral density and has a maximum, or a peak, near the altitude of unit optical depth, because the photon flux rapidly decreases below this altitude and because the neutral density decreases with altitude. Ion production from auroral electrons has a maximum value at an altitude which depends on the energy of the precipitating electrons, but an altitude of about 110 km is typical for the terrestrial aurora.

The composition of an ionosphere is not solely determined by which ions are most efficiently produced by reactions (I3) and (I4). An ion species, once created, generally participates in chemical reactions. Ion–neutral chemistry can quickly alter the ion composition of an ionosphere. Electron–ion recombination must also take place in order that ion production and loss can achieve equilibrium. At lower altitudes in an ionosphere, where the neutral densities are high, the chemical lifetime of an ion is generally short and local photochemical equilibrium (i.e. the local balance of ion production and chemical ion loss for each ion species) is a valid assumption. However, at higher altitudes the chemical lifetime of an ion is generally becomes larger than the time it takes the ion to move either up and down in the ionosphere (i.e. the vertical transport time) or to move horizontally a significant distance (i.e. the horizontal transport time). The motion, or transport, of plasma through the surrounding neutral gas is resisted by ion–neutral collisions in a process called ambipolar diffusion.

If the vertical transport time of the ionospheric plasma is much less than the chemical lifetime, as it is in the terrestrial topside ionosphere, an approximation called diffusive equilibrium becomes valid, and the electron density as a function of altitude is then described by the exponential function. For a simple isothermal ionosphere, the diffusive equilibrium electron density as a function of altitude z can be written as:

$$n_e(z) = n_{e0} \exp[-(z - z_0)/H_p] \tag{I5}$$

Where H_p is the plasma scale height, which is equal to

$$H_p = \frac{k_B(T_e + T_i)}{m_i g} \tag{I6}$$

k_B is the Boltzmann constant, T_e and T_i are the electron and ion temperatures, respectively, m_i is the mass of the dominant ion species and g is the gravitational acceleration. Diffusive equilibrium is the plasma equivalent of hydrostatic equilibrium for the neutral gas. Both the electron and ion temperatures appear in the plasma scale height formula, and this is a consequence of the ambipolar electric field which maintains quasi-neutrality between the electron and ion gases (i.e $n_e = n_i$ where n_i is the sum of the densities of all ion species).

Horizontal transport can also be important in ionospheres. For example, during geomagnetically active periods of time, fast $E \times B$ ion drifts ($v \approx 1$ km s^{-1}) are driven in the ionosphere over the terrestrial polar cap by electric fields which map down into the ionosphere from the magnetosphere along magnetic field lines. Also on Venus rapid dayside-to-nightside ion flow, with speeds of a few kilometers per second, is largely responsible for the maintenance of the nightside ionosphere of that planet.

Ionospheric plasmas are excellent electrical conductors and electrical currents are easily able to flow in them. Several current systems exist in the terrestrial ionosphere and produce magnetic field perturbations; in particular, there is the S_q current system at mid-latitudes and the equatorial electrojet at low latitudes. The auroral electrojet flows in the auroral zone, or auroral oval, and the strength of this current system depends on the level of solar activity. The current flow in all these systems is primarily horizontal, although at high latitudes some electrical current also flows along the magnetic field lines linking the ionosphere and magnetosphere. These field-aligned currents are called Birkeland currents. Electrical currents also flow in planetary ionospheres, such as in the ionosphere of Venus. In particular, they are present in the ionopause boundary layer that separates the solar wind and the Venusian ionospheric plasma. Induced electrical currents are also responsible for a characteristic layer of induced magnetic field observed by the magnetometer on Pioneer Venus Orbiter (PVO) near an altitude of 170 km during time periods when the solar wind is moving especially fast. Venus does not possess any significant intrinsic dynamo magnetic field, and ionospheric electrical currents and the resulting magnetic fields are induced by interaction of the solar wind with the ionosphere. Ionospheric processes are discussed in several of the references [Banks and Kockarts, 1973; Ratcliffe, 1972; Kelley, 1989; Rees, 1989; see also *Solar Terrestrial Physics, Principles and Theoretical Foundations* (Carovillano and Forbes, 1983)].

Terrestrial ionosphere

The terrestrial ionosphere has been divided into D, E and F regions, or layers, as illustrated in Figure I27. Often these 'layers' appear as distinctive peaked or ledge structures in the electron density profiles, but sometimes they do not. These regions and the physical and chemical processes that operate in these regions will now be briefly

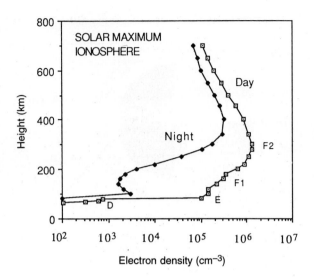

Figure I27 Typical electron density profiles versus height in the terrestrial ionosphere for both the daytime and nighttime. The different regions, or layers, are indicated.

discussed. (Also see Banks and Kockarts, 1973; Ratcliffe, 1972; Kelley, 1989; Rees, 1989.)

The D region lies in the altitude range 70–90 km. In this altitude range, N_2^+ and O_2^+ ions are produced by the x-ray part of the solar spectrum. NO^+ ions are also produced in the D region by photoionization of the minor neutral constituent, nitric oxide (NO), by solar Lyman alpha photons. Large ion production rates are also generated in the D region during solar proton events, in which energetic solar cosmic rays with MeV energies penetrate down to altitudes as low as 60 km.

N_2^+ ions produced in both the D and E regions rapidly react with O_2 to form O_2^+ ions or they react with atomic oxygen to produce NO^+ ions:

$$N_2^+ + O_2 \rightarrow N_2 + O_2^+ \tag{I7}$$

$$N_2^+ + O \rightarrow NO^+ + N \tag{I8}$$

The O_2^+ ions then react with NO molecules to create NO^+ ions:

$$O_2^+ + NO \rightarrow NO^+ + O_2 \tag{I9}$$

The most abundant ion species in both the D and E regions are O_2^+ and NO^+. Both these ions are removed from the ionosphere by very rapid electron–ion dissociative recombination, reactions. In fact, for NO^+ this is the only removal mechanism:

$$NO^+ + e \rightarrow N + O \tag{I10}$$

The D region is the only region in which negative ions (in particular, O_2^-) and cluster ions, such as $O_2^+ \ast H_2O$, are present in significant concentrations.

The E region is located in the altitude range of about 90–60 km. Reactions (I7)–(I10) are again important in this region, and O_2^+ and NO^+ are again the dominant ion species. Figure I28 is a schematic showing typical ion density profiles for the major ion species in the terrestrial ionosphere. The O_2^+ to NO^+ density ratio strongly depends on the nitric oxide abundance in the thermosphere, which is quite variable and sensitive to the levels of both solar and geomagnetic activity. O^+ ions are also produced in the E region but are removed by the reaction:

$$O^+ + N_2 \rightarrow NO^+ + N \tag{I11}$$

The E region ion and electron densities are controlled by chemistry rather than by transport. The chemical lifetimes of the major ion species in the E region are only a few minutes and thus the D and E-regions quickly disappear at dusk when the solar flux shuts off. However, a weak nighttime E region persists due to small night-time ionization sources such as photoionization by solar Lyman beta photons scattered from the hydrogen geocorona. Also, an E-type

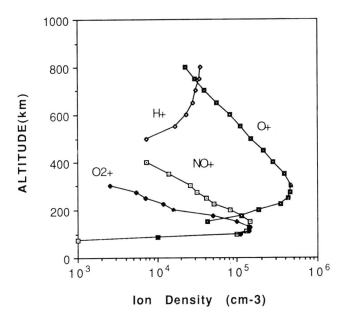

Figure I28 Schematic of typical terrestrial ion density altitude profiles (mid-latitude dayside ion composition) from information in Schunk (1983).

region can be present in the auroral zone at any time of day during which energetic electrons precipitate from the magnetosphere.

The major ion species in the F region is O^+ because the removal mechanism for this ion, (I11), weakens at higher altitudes due to the decreasing N_2 density. The electron density is still controlled chemically in the lower part of the F region (called the F_1 region), whereas above about 220 km, in the F_2 region, vertical transport becomes more important than chemistry. The peak ionospheric density occurs near the base of the F_2 region at an altitude where the O^+ chemical lifetime is comparable to the vertical transport time. Above the F_2 maximum, in the topside ionosphere, the electron density drops off exponentially as described by equations (I5) and (I6). Note in Figure I28 that the density in the topside terrestrial ionosphere is indeed exponentially decreasing with a scale height $H_p = 200$ km. The 'lifetime' of the F region plasma is several hours, so that this layer only slowly decays throughout the night. Ions 'stored' in the topside ionosphere slowly diffuse downward and resupply the F_1 region ionosphere during the night. The ions diffuse down to an altitude where these ions are able to be efficiently removed by reaction (I11).

The composition of the F_2 region gradually changes from O^+ to H^+ near an altitude of about 1000 km. The main neutral species at very high altitudes is atomic hydrogen and H^+ ions are produced by the charge transfer reaction.

$$O^+ + H \rightarrow H^+ + O \qquad (I12)$$

The high altitude extension of the F_2 region is often called the protonosphere (Figure I28). As mentioned earlier, at low and mid-latitudes, the high altitude extension of the ionosphere is called the plasmasphere (Figure I26). The plasmaspheric plasma is located on closed field lines and is able to flow between the northern and southern hemispheres. However, at high latitudes, and especially over the polar cap, ionospheric plasma is able to escape outward along open (or at least very extended) magnetic field lines, and the high-altitude plasma density is depleted in comparison with plasmaspheric densities. The flow speeds of this polar region plasma can reach supersonic speeds of several kilometers per second and has been called the polar wind. It is now thought that the polar wind supplies much of the plasma for the magnetosphere.

The ionosphere is strongly affected by the magnetosphere (i.e. ionosphere–magnetosphere coupling) in several ways: (1) auroral precipitation of energetic particles, especially electrons, that cause ionization and heating, particularly in the E region; (2) imposition of electric fields that can cause rapid horizontal motion of the F region plasma via $E \times B$ drifts (during magnetic storms the F region plasma is driven antisunward over the polar cap at speeds of the order of 1 km s^{-1}; (3) generation of both field-aligned and horizontal electrical currents.

Neutral winds can also affect the F region at all latitudes. The ionospheric plasma in the F region is 'tied to' the magnetic field lines and most easily flows along them. The frictional force exerted on the plasma by the moving neutral atmosphere can drive the plasma up or down an inclined magnetic field line, with the direction determined by both the neutral wind direction and the field orientation. The F_2 peak can be shifted either to a lower or a higher altitude in this manner. Neutral winds blowing at right angles to the geomagnetic field create dynamo electric fields that are responsible for the mid-latitude S_q current system and the equatorial electrojet.

A rich morphology and catalogue of phenomena exist for the terrestrial ionosphere. For example, a phenomenon called sporadic E occasionally occurs in which (it is thought) metallic ions, from meteoritic impact, are concentrated into narrow (1–2 km) layers in the lower E region by forces associated with neutral wind shears. Another phenomena called spread F has been observed in which small-scale density irregularities develop in the F region. In addition, traveling ionospheric disturbances (TIDs) are seen during magnetic storms when gravity waves in the neutral atmosphere are generated at high latitudes and propagate equatorward causing ionospheric effects via ion–neutral collisions.

Venus and Mars

Carbon dioxide is the major neutral species in the atmospheres of both Venus and Mars, although above an altitude of about 160 km atomic oxygen becomes increasingly abundant. The major ion produced at lower altitudes on both these planets (CO_2^+), was not the major ion measured at Venus by the ion mass spectrometer onboard the Pioneer Venus Orbiter or at Mars by the retarding potential analyzers on the Viking landers; instead O_2^+ was the major ion seen at both planets, although almost no neutral molecular oxygen is present in either planet's atmosphere. The O_2^+ ions in the ionospheres of Venus and Mars are mainly produced by the following ion–neutral chemical reaction:

$$(CO_2^+) + O \rightarrow O_2^+ + CO \qquad (I13)$$

At somewhat higher altitudes, where O^+ is produced, the following reaction is also important:

$$O^+ + CO_2 \rightarrow O_2^+ + CO \qquad (I14)$$

The effect of reactions (I13) and (I14) is to make O_2^+ the major ion below about 200 km on both planets. The CO_2^+ to O_2^+ density ratio is everywhere less than 10%.

The ionospheres of Venus and Mars below about 180 km are 'E region' ionospheres, by analogy with Earth and, as one might expect, the O_2^+ ions are removed from the ionosphere by dissociative recombination:

$$O_2^+ + e \rightarrow O + O \qquad (I15)$$

And, as in the terrestrial E region, NO^+ ions are also present. These ions result from reaction of O_2^+ with either NO or atomic nitrogen, and the NO^+ ions are removed by reaction (I12). However, the NO^+ to O_2^+ density ratio is much less at Venus and Mars than it is at Earth, due to the lower nitrogen abundance in the Martian and Venusian atmospheres. The peak electron density occurs near 140 km at Venus and 130 km at Mars; these are the altitudes where the ion production rates have their respective maxima for the two planets. The peak electron density in the dayside ionosphere of Venus is about 7×10^5 cm^{-3} and for Mars is about 2×10^5 cm^{-3}. Figure I29 shows typical ion density profiles for the dayside ionosphere of Venus.

The O^+ density increases with altitude in the lower ionosphere of Venus and Mars, just as it does in the terrestrial F region ionosphere. O^+ becomes the major ion near an altitude of 180 km on all three planets. Reaction (I11) of O^+ with N_2 acts as a sink of O^+ in the terrestrial F_1 region, whereas reaction (I14) of O^+ with CO_2 removes this ion species in the 'F_1 regions' of Venus and Mars. The O^+ density in the ionosphere of Venus reaches a maximum density of about 10^5 cm^{-3} near 200 km, at which altitude the vertical diffusion time becomes comparable to the chemical lifetime for this ion. The electron (and O^+) density follow a diffusive equilibrium type profile at higher altitudes [equation (I5)], except for disturbed conditions.

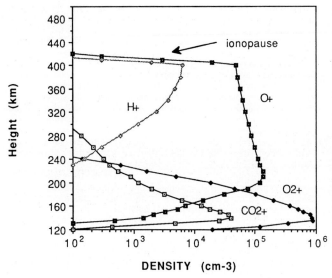

Figure I29 Vertical density profiles for the major ion species in the dayside ionosphere of Venus. The ionopause is indicated. Density is in units of cm^{-3}. Adapted from a theoretical model by Shinagawa and Cravens (1988), but the main features agree with the measurements made by the Pioneer Venus Orbiter.

This region is analogous to the terrestrial F_2 region. However, the 'F_2' electron density at Venus and Mars is less than the peak 'E region' density near 140 km; hence, unlike in the terrestrial ionosphere, the maximum electron density occurs in the 'E region' rather than in the 'F region'. The primary reason for this is that reaction (I14) is more efficient than reaction (I11).

An interesting feature in the Venus ionospheric profiles of Figure I30 is apparent. The density exponentially decreases between about 220 km and 400 km, but then at an altitude of 400 km the density suddenly decreases, becoming very small. The ionosphere essentially disappears at 400 km and this boundary is called the ionopause. The ionopause is a consequence of the solar wind interaction with Venus. The solar wind (q.v.) is a tenuous ($n_e \approx 10$ cm^{-3} in the inner solar system) plasma, consisting mainly of protons and electrons, that moves outward from the Sun at supersonic speeds of a few hundred kilometers per second. The ionosphere of Venus acts as an obstacle to the solar wind, which must flow around the planet (Russell and Vaisberg, 1983). The combined solar wind thermal and dynamic pressure are almost entirely converted to magnetic pressure in the magnetic barrier region that is located just outside the ionopause. This magnetic barrier contains 'compressed' interplanetary magnetic field and the field strength in it is high, whereas the field strength in the ionosphere is almost zero on average. The ionopause layer has a thickness of about 20 km and electrical currents flow in this layer. The external pressure outside the ionopause is balanced by the internal thermal pressure of the ionospheric plasma:

$$\text{external pressure} = n_e k_B (T_e + T_i) \quad (I16)$$

where the ionospheric density, electron temperature and ion temperature are evaluated at the altitude of the ionopause. The height of the ionopause is determined by this relation. As the solar wind pressure increases, a larger ionospheric pressure is required to maintain pressure balance, and the ionopause moves downward because a larger electron density is needed to achieve a larger pressure (T_e and T_i are approximately constant above 200 km). The ionopause typically moves up and down by hundreds of kilometers over the course of a few days as the solar wind conditions vary, although an ionopause height of about 500 km is typical.

Occasionally, the solar wind dynamic pressure is so large that it exceeds the maximum ionospheric thermal pressure. When this happens, the compressed interplanetary magnetic field lines associated with the magnetic barrier push deep into the ionosphere which then becomes magnetized. The ionopause moves down to an altitude of about 250 km and becomes very broad (50 km or more), losing its distinctive appearance. A layer of magnetic field near 170 km is also present when the solar wind dynamic pressure is high.

The solar wind interaction with Mars is not as well understood as the solar wind interaction with Venus. In particular it is still not known, even after several US and Soviet missions to that planet, whether or not Mars has a small intrinsic magnetic field! However, there are some indications that the solar wind interaction with Mars is analogous to that with Venus during time periods of high solar wind dynamic pressure.

Venus rotates in a retrograde direction with a period of 243 Earth days. The lifetime of the ionospheric plasma is orders of magnitude less than this time, yet a nightside ionosphere with peak densities of about 10^4 cm^{-3} has been measured on Venus by in situ experiments on PVO, as well as by radio occultation during several US and Soviet missions. The nightside ionosphere is primarily maintained, at least during solar maximum conditions, by the horizontal flow of ionospheric plasma from the dayside ionosphere to the nightside ionosphere with speeds of several kilometers per second. Some nightside ionization is also provided by a weak auroral precipitation of electrons coming from the induced magnetotail of Venus.

A rich morphology of ionospheric phenomena is present at Venus, just as for Earth. Our knowledge of these phenomena is mainly due to the PVO observations. A good reference for the ionosphere and upper atmosphere of Venus is the book *Venus Aeronomy* (Russell, 1991). Phenomena such as tail rays, ionospheric holes, disappearing ionospheres, magnetic flux ropes, and terminator waves have been observed.

Outer planets and satellites

Molecular hydrogen is the primary constituent of the thermospheres of the gas giant planets, Jupiter, Saturn, Uranus and Neptune. Atomic hydrogen is also present at higher altitudes in the exospheres of these planets, as a consequence of photodissociation or electron impact dissociation of H_2. Some methane (CH_4) is also present in the lower thermospheres of these planets.

The major ion produced is H_2^+, although some H^+ is also produced via dissociative ionization or by ionization of atomic hydrogen. At lower latitudes the main ion source is photoionization by solar EUV radiation:

$$h\nu + H_2 \rightarrow H_2^+ + e \quad 90\%$$
$$\rightarrow H + H^+ + e \quad 10\% \quad (I17)$$

The dissociative branch of (I17) accounts for only about 10% of the total ionization.

The electron impact or energetic ion impact equivalent of process (I17) is very important in the auroral regions of the outer planets. In the auroral regions of all the gas giants, auroral emissions in the Lyman and Werner bands of H_2 and in the Lyman alpha line of H were observed in the ultraviolet part of the spectrum by the Ultraviolet Spectrometers on the Voyager spacecraft and also by the International Ultraviolet Explorer (IUE) and Hubble Space Telescope (HST) satellites, which are in Earth orbit. These emissions were especially strong at Jupiter, somewhat weaker at Saturn and Uranus and quite weak at Neptune. These observations provided evidence for auroral precipitation of energetic particles from the magnetospheres of these planets into the high-latitude thermospheres.

Ionospheres were detected by the Voyager spacecraft on all the major planets by means of the radio occultation technique. The maximum electron densities observed were approximately 2×10^5 cm^{-3}, 2×10^4 cm^{-3}, 5000 cm^{-3} and 3000 cm^{-3}, for Jupiter, Saturn, Uranus and Neptune respectively (Atreya, 1986; Tyler et al., 1989). However, these densities are not necessarily the peak ionospheric densities because the observations did not necessarily extend deep enough into the respective ionospheres. Electron density profiles were also measured at Jupiter and Saturn during the Pioneer 10 and 11 missions.

The major ion species in the ionospheres of the outer planets is not H_2^+, the main ion produced. Just as in the E regions of the inner planets (Earth, Venus and Mars), ion-neutral chemistry alters the ion composition. In particular, the following reaction quickly removes H_2^+:

$$H_2^+ + H_2 \rightarrow H_3^+ + H \quad (I18)$$

The H_3^+ ions that result from this reaction are rapidly removed by dissociative recombination:

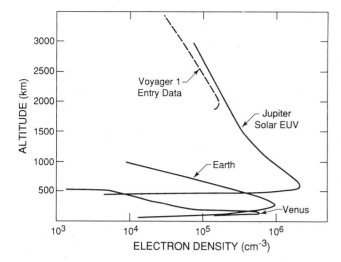

Figure I30 A comparison of typical electron density profiles for the dayside ionospheres of Venus, Earth and Jupiter are shown. 'Solar EUV' is solar extreme ultraviolet radiation, and the curve with this label is from a theoretical model of the ionosphere of Jupiter. Radio occultation data from the Voyager 1 mission is also shown for Jupiter. (Adapted from Cravens, 1983)

$$H_3^+ + e \rightarrow H_2 + H$$
$$\rightarrow H + H + H + \quad (I19)$$

H_3^+ ions can also be removed in the lower ionosphere by reaction with CH_4, thus forming CH_5^+, and subsequent photochemistry can also lead to the formation of heavier hydrocarbon ion species. The major ion species in the lower E region part of the ionosphere is either H_3^+ or hydrocarbon ion species. However, the ionospheric peak is located in a F_1 type region and the major ion species, for the outer planets, is H^+. The reason for this is that the chemical lifetime of ionospheric protons is quite long, whereas the other ion species have much shorter chemical lifetimes. The main chemical loss process for the H^+ ion is the slow radiative recombination reaction:

$$H^+ + e \rightarrow H + h\nu \quad (I20)$$

The H^+ density thus builds up to rather large values. In fact, theoretical models of the ionospheres of the outer planets have had to invoke additional loss processes such as removal of H^+ ions by reaction with water molecules or H^+ removal by reaction with vibrationally excited H_2, in order to bring calculated density values into line with the observed values. The peak regions of the outer planets are controlled by chemistry and are thus F_1 type ionospheres. Ambipolar diffusion operates only above the peak, that is, in the topside ionospheres of these planets.

Figure I30 shows a comparison of ionospheric density profiles for several planets including Jupiter. The ionosphere of Jupiter is quite extended because the thermosphere is quite extended with a large neutral scale height that results from a low molecular mass species (H_2) and rather hot thermospheric temperatures. The ionosphere of Saturn is even more extended than that of Jupiter.

The radio occultation technique was also used [during the Pioneer (q.v) or Voyager (q.v) missions; see, for example, Tyler *et al.*, 1989] to detect ionospheres at the satellites Io (Jupiter) and Triton (Neptune). The peak densities were 6×10^4 cm^{-3} and 4.6×10^3 cm^{-3} for Io and Triton respectively. The side of Io facing into the Io plasma torus flow (i.e. the ramside) had a lower ionospheric density than the wakeside of this satellite. The major ion species at Io is probably SO_2^+, because SO_2 is the major neutral in that satellite's tenuous atmosphere, coming from volcanic emissions (cf. Atreya, 1986). Triton's ionosphere is thought to be due mainly to ionization by energetic electrons coming from Neptune's magnetosphere. The identity of the major ion species is not known with any certainty but it might be N^+.

An upper limit for a peak ionospheric density of about 5000 cm^{-3} was set by the Voyager 1 radio occultation experiment for Saturn's satellite Titan (cf. Atreya, 1986). The major neutral species in the atmosphere of Titan is N_2, although CH_4 is also abundant. Theoretical models indicate that the ionosphere of Titan is an E region type ionosphere. The main ion produced is N_2^+, but these ions are converted to hydrocarbon ions via reaction with methane. The hydrocarbon ions are either removed by dissociative recombination or they react with HCN molecules and form H_2CN^+. The major ion species at Titan is probably H_2CN^+.

Thomas E. Cravens

Bibliography

Atreya, S.K. (1986) *Atmospheres and ionospheres of the outer planets and their satellites*. Berlin: Springer-Verlag.
Banks, P.M. and Kockarts, G. (1973) *Aeronomy*. New York: Academic Press.
Carovillano, R.L. and Forbes, J.M. (eds) (1983) *Solar–Terrestrial Physics, Principles and Theoretical Foundations*. Dordrecht: D. Reidel Publishing Company. (See the chapters by Schunk; Richmond and Cravens.)
Cravens, T.E. (1983) Comparative ionospheres, in *Solar-Terrestrial Physics, Principles and Theoretical Foundations* (eds R.L. Carovillano and J.M. Forbes). Dordrecht: D. Reidel Publishing Company, pp. 805–44.
Hanson, W.B., Sanatani, S. and Zuccaro, D.R. (1977) The Martian ionosphere as observed by the Viking retarding potential analyzers. *J. Geophys. Res.*, **82**, 4351–63.
Kelley, M.C. (1989) *The Earth's Ionosphere: Plasma Physics and Electrodynamics*. San Diego: Academic Press.
Mendis, D.A., Houpis, H.L.F. and Marconi, M.L. The physics of comets. *Fund. Cosmic Phys.*, **10**, 198–222.
Newburn, R.L., Jr, Neugebauer, M. and Raha, J. (eds) (1991) *Cosmets in the Post-Halley Era*. Dordrecht: Kluwer Academic Publishers.
Nier, A.J. and McElroy, M.B. (1977) Composition and structure of Mars' upper atmosphere: results from the neutral mass spectrometers on Viking 1 and 2. *J. Geophys. Res.*, **82**, 4341–9.
Ratcliffe, R.A. (1972) *An introduction to the ionosphere and magnetosphere*. Cambridge: Cambridge University Press.
Rees, M.H. (1989) *Physics and chemistry of the upper atmosphere*. Cambridge: Cambridge University Press.
Richmond, A.D. (1983) Thermospheric dynamics and electrodynamics, in *Solar-Terrestrial Physics, Principles and Theoretical Foundations* (eds R.L. Carovillano and J.M. Forbes). Dordrecht: D. Reidel Publishing Company, pp. 523–607.
Russell, C.T. (ed.) (1991) *Venus Aeronomy*. Dordrecht: Kluwer Academic Publishers.
Russell, C.T. and Vaisberg, O. (1983) The interaction of the solar wind with Venus, in *Venus* (ed. M. Hunten, L. Colin, T.M. Donahue and V.I. Moroz). Tucson: University of Arizona Press.
Schunk, R.W. (1983) The terrestrial ionosphere, in *Solar-Terrestrial Physics, Principles and Theoretical Foundations* (eds R.L. Carovillano and J.M. Forbes). Dordrecht: D. Reidel Publishing Company, pp. 609–76.
Shinagawa, H. and Cravens, T.E. (1988) A one-dimensional multi-species magnetohydrodynamic model of the dayside ionosphere of Venus. *J. Geophys. Res.*, **93**, 11263–77.
Tyler, G.L., Sweetman D.N., Anderson, J.D. *et al.* (1989) Voyager radio science observations of Neptune and Triton. *Science*, **246**, 1466–73.

Cross references

Magnetospheres of the outer planets
Plasma
Plasma wave
Radio science
Thermosphere
Whistler

IRIDIUM ANOMALY

An 'Iridium anomaly' generally refers to anomalously large concentrations of the element iridium (Ir) in one layer of a sequence of sedimentary rocks. Because Ir is highly siderophile (iron-loving), most of the Earth's inventory of Ir was scavenged from the silicate portion of the planet when the nickel–iron core formed. In consequence, over 99% of the terrestrial Ir inventory is in the core and typical crustal abundances of Ir are only 5 to 50 parts per trillion (ppt). Most asteroids and comets which impact the Earth contain much larger chondritic (meteoritic) abundances of Ir [~ 500 parts per billion (ppb)]. The sedimentary layers formed from the debris ejected from impact craters often contain large quantities of Ir reflecting the fraction of the ejecta that is composed of the impacting projectile. The global Cretaceous–Tertiary boundary clay layer is currently the most discussed layer with an associated Ir anomaly. Iridium concentrations of ~ 50 ppb, ~ 1000 times background levels, are commonly found in this layer. Because Ir anomalies are also caused by other geochemical processes, the occurrence of an Ir anomaly alone does not usually signal the impact of an extraterrestrial object. All the highly siderophile trace elements (ruthenium, rhodium, palladium, rhenium, osmium, iridium, platinum and gold) must be analyzed to determine whether they occur in chondritic proportions before this conclusion can be drawn. Because no known crustal geochemical processes produce large chondritic anomalies of the highly siderophile trace elements, such an anomaly usually signals the accretion of extraterrestrial material to the Earth.

Alan Hildebrand

Bibliography

Alvarez, W., Asaro, F. and Montanari, A. (1990) Iridium profile for 10 million years across the Cretaceous–Tertiary boundary at Gubbio (Italy). *Science*, **250**, 1700–2.
Bekov, G.I., Letokhov, V.S., Radaev, V.N. et al. (1988) Rhodium distribution at the Cretaceous/Tertiary boundary analysed by ultrasensitive laser photoionization. *Nature*, **332**, 146–8.
Evans, N.J., Gregoire, D.C., Goodfellow, W.D. et al. (1993) Ru/Ir ratios at the Cretaceous–Tertiary boundary: implications for PGE source and fractionation within the ejecta cloud. *Geochim. Cosmochimi. Acta*, **57**, 3149–58.
Ganapathy, R. (1980) A major meteorite impact on the Earth 65 million years ago: evidence from the Cretaceous–Tertiary boundary clay. *Science*, **209**, 921–3.
Kyte, F.T., Smit, J. and Wasson, J.T. (1985) Siderophile interelement variations in the Cretaceous–Tertiary boundary sediments from Caravaca, Spain.' *Earth Planet. Sci. Lett.*, **73**, 183–95.
Kyte, F.T. and Wasson, J.T. (1986) Accretion rate of extraterrestrial matter: iridium deposited 33 to 67 million years ago. *Science*, **232**, 1225–9.
Lerbekmo, J.F. and St Louis, R.M. (1986) The terminal Cretaceous iridium anomaly in the Red Deer Valley, Alberta, Canada. *Can. J. Earth Sci.*, **23**, 120–4.
Montanari, A., Asaro, F., Michel, H.V. and Kennett, J.P. (1993) Iridium anomalies of Late Eocene age, Messignano (Italy) and ODP Site 689B (Maud Rise, Antarctica). *Palaios*, **8**, 420–37.
Orth, C.J., Attrep, M. Jr and Quintana, L.R. (1990) Iridium abundance patterns across bio-event horizons in the fossil record: in *Global Catastrophes in Earth History; An Interdisciplinary Conference on Impacts, Volcanism, and Mass Mortality*, (eds V.L. Sharpton and P.D. Ward). Geological Society of America Special Paper 247, pp. 45–59.
Wallace, M.W., Gostin, V.A. and Keays, R.R. (1990) Acraman impact ejecta and host shales: evidence for low-temperature mobilization of iridium and other platinoids. *Geology*, **18**, 132–5.

Cross references

Chemical element
Cosmochemistry
Cretaceous–Tertiary boundary
Mass extinction

IRON

Iron is the most important metal in the universe and the fourth commonest element in the Earth's crust. The name is from the Old English *iren*, and the symbol Fe from the Latin *ferrum*. The atomic number is 26; the relative atomic mass 55.847; relative density at 20°C, 7.87; melting point 1535°C; boiling point 2800°C. In the periodic table, iron lies in Group VIII. There are three allotropic forms, known as alpha ('ferrite'), body-centered cubic; gamma, cubic close packed; delta, body-centered cubic, with transition points at 910°C ($\alpha \rightarrow \gamma$) and 1390°C ($\gamma \rightarrow \delta$). Four stable isotopes exist with the following natural abundances. ^{54}Fe, 5.8%; ^{56}Fe, 91.7%; ^{57}Fe, 2.2%; ^{58}Fe, 0.3%. In its pure form, iron is a lustrous silvery soft, malleable and ductile metal. Impurities, for example carbon, greatly affect the physical properties (Emsley, 1989).

Occurrence

Terrestrially, iron in its elemental form is a rarity. Although there are several oxidation states, only two, Fe^{II} (ferrous) and Fe^{III} (ferric) are common in the Earth's crust, where iron is widespread in a large number of minerals. The abundance in the Earth's crust is 6.2%, and it is generally agreed that the core of the Earth is composed essentially of an iron – nickel alloy containing 80% iron, making it the most abundant element in the Earth as a whole. The mantle of the Earth may contain about 15 wt% FeO, and the mantles of the bodies in the inner solar system typically contain 13–18 wt% FeO (Anderson, 1989). Iron is the dominant constituent of the iron meteorites (q.v.), whose occurrence suggests that the element is abundant in the solar system. They are differentiated objects that come from numerous sources and provide evidence for the early solar system (Taylor and Norman, 1990; Wasson, 1985). The planet Mercury, with a large iron core, contains proportionately twice as much iron as any other planet in the solar system (Strom, 1984). The Martian soil is iron-rich and aluminium-poor, perhaps consisting dominantly of Fe-rich smectites (Burt, 1989).

Mineralogy

The very small quantities of metallic iron in the crust of the Earth occur under very special conditions, but most iron is present in oxide and sulfide minerals, silicates and secondary ores and hydroxides. The main oxides and sulfides, all important ores of iron, are magnetite, Fe$_3$O$_4$; hematite, Fe$_2$O$_3$; and pyrite or iron pyrites, FeS$_2$. Magnetite or magnetic iron ore occurs in small accessory crystals in many different igneous rocks but also in large masses of deep-seated origin, probably magmatic segregations, e.g. the Kiruna ores of northern Sweden. Hematite is a blood-red earthy mineral including kidney iron ore (from its shape), but also in black lustrous foliated crystals called specular iron ore or micaceous hematite. Hematite also occurs in large deposits in Precambrian banded sediments and is very common as a replacement mineral in limestone. Pyrite or iron pyrite is a brassy yellow constituent of many sulfide ores and is a widespread accessory mineral in igneous rocks.

Iron occurs in the primary silicate minerals of igneous rocks, notably the olivines, pyroxenes, amphiboles and ferromagnesian micas. In the sedimentary rocks there are hydrous silicates and the chlorites, typically green minerals somewhat resembling the micas, are composed of hydrated magnesium, iron and aluminium silicates. The ferrous iron carbonate, siderite, was formerly an important source of low-grade iron ore. Various alteration products of iron minerals include the hydroxides goethite, limonite and lepidocrocite.

Large quantities of hot black fine-grained sulfides are emitted from hydrothermal vents, known as black smokers. They occur on the crests of oceanic ridges, and include sulfides and oxides of iron, manganese, zinc and other metals.

Peter A. Sabine

Bibliography

Anderson, D.L. (1989) Composition of the Earth. *Science*, **243**, 367–70.
Burt, D.M. (1989) Iron-rich clay minerals on Mars: potential sources or sinks for hydrogen and indicators of hydrogen loss over time, in *Proc. Lunar and Planet Sci. Conf.*, **19** (eds G. Ryder and V.L. Sharpton). Houston: Cambridge University Press; Lunar and Planetary Institute, pp. 423–32.
Emsley, J. (1989) The Elements. Oxford: Clarendon Press.

Strom, R.G. (1984) Mercury, in *The Geology of the Terrestrial Planets*, (ed. M.H. Carr). Washington: NASA, pp. 13–55.

Taylor, S.R. and Norman, M.D. (1990) Accretion of differentiated planetesimals to the Earth, in *Origin of the Earth* (eds H.E. Newsom and J.H. Jones). New York: Oxford University Press; Houston: Lunar and Planetary Institute, pp. 29–43.

Wasson, J.T. (1985) *Meteorites*. New York: W.H. Freeman and Company.

Cross references

Core, terrestrial planetary
Differentiation
Stony iron meteorites
Thermal evolution of planets and satellites

IRON METEORITES

Iron meteorites are perhaps the most spectacular samples of extraterrestrial materials, and may be the ones known to humanity for the longest time. Prehistoric humans used meteoritic iron as raw material for tools and artefacts. Some iron meteorites may have been venerated as sacred objects.

The abundance of iron meteorites in museum collections is not due to a high incidence of iron meteorite falls. Among the roughly 900 observed meteorite falls there are less than 50 irons. Most of the over 700 known iron meteorites owe their preservation to the fact that they weather more slowly and are more conspicuous to human collectors than other types of extraterrestrial rocks. A comprehensive overview of known iron meteorite samples has been given by Buchwald (1975).

Metallography and mineralogy

The major component of iron meteorites is nickel–iron (Fe, Ni), although practically all of them also contain various minor phases. The metal usually contains 5–15% Ni by weight, but samples with up to 60% Ni are known. No authenticated iron meteorites are known that contain less than 4% Ni. Aside from Co, which is usually around 0.5%, all other elements are usually present only in trace amounts (below 500 ppm) in meteoritic nickel–iron.

On polished and etched surfaces most (but not all) iron meteorites show an intricate texture known as Widmannstätten pattern (Figure I31). The formation of this texture can be understood by considering the Fe–Ni phase diagram (Swartzendruber, Itkin and Alcock, 1991). Above about 800°C meteoritic metal forms a single phase known as γ-Fe,Ni. During slow cooling a phase with lower Ni content, α-Fe,Ni, precipitates along the {111} planes of the γ-Fe,Ni parent. Depending on the bulk Ni content nucleation of α-Fe, Ni occurs between 500 and 800°C, lower temperatures corresponding to higher Ni contents. The body-centered cubic α-Fe,Ni, known as ferrite in metallurgy, bears the mineralogical name kamacite; γ-Ni,Fe is face-centered cubic, called austenite in metallurgy and taenite in mineralogy. A submicron intergrowth of Ni-poor and Ni-rich phase called plessite is present within the taenite regions of more Ni-rich samples. Near an Fe : Ni ratio of 1 : 1 taenite assumes an ordered tetragonal structure called tetrataenite.

When a polished section of a typical iron meteorite is etched, lamellae of kamacite, separated by varying amounts of taenite and plessite, become visible, forming the characteristic Widmannstätten pattern. Because kamacite lamellae are oriented along octahedral planes such meteorites are known as octahedrites. Some iron meteorites with Ni contents below 6% consist almost entirely of kamacite and show no Widmannstätten pattern; they are called hexahedrites (designated H). Samples with very high Ni content undergo the γ → α transformation at such low temperatures that only small kamacite spindles but no lamellae form, and the texture is only microscopically visible. These meteorites are known as ataxites (D).

Octahedrites are further subdivided according to the width of their kamacite lamellae into coarsest (Ogg, bandwidths > 3.3. mm), coarse (Og, 1.3–3.3 mm), medium (Om, 0.5–1.3 mm), fine (Of, 0.2–0.5 mm), and finest octahedrites (Off, < 0.2 mm). Plessitic octahedrites (Opl) are transitional to ataxites. Bandwidths tend to get finer with increasing Ni content.

A few iron meteorites do not fit this textural classification scheme, and are regarded as having an anomalous texture. Some of these are specimens in which the Widmannstätten pattern has been partly or wholly obliterated by reheating in space (e.g. Zerhamra).

The lamella width of the Widmannstätten pattern depends not only on the average Ni content of the metal, but also on the cooling rate. The higher the Ni content, the lower the temperature at which the α phase begins to precipitate; the faster the cooling, the less time the α lamellae have to grow. Therefore higher Ni content and faster cooling lead to finer patterns. By modeling the growth of α-phase, a metallographic cooling rate can be derived from Widmannstätten size. These model cooling rates generally range from a few degrees to a few hundred degrees per million years (Saikumar and Goldstein, 1988). Based on thermal models for typical parent bodies, these cooling rates would have been achieved in the cores of asteroids of a few tens to perhaps 250 km diameter.

Minor phases vary from sample to sample. The most common accessory minerals are troilite, hexagonal FeS; schreibersite, $(Fe,Ni)_3P$; cohenite, Fe_3C (cementite in metallurgy); and chromite, $FeCr_2O_4$. Several other sulfides and phosphates are also common. Silicates are frequent in certain groups, most notably in IAB and IIE (see below), but are rare or absent in most other iron meteorites. A large number of very rare minerals, some of them without terrestrial counterpart, have been found in iron meteorites, but typically they occur in only a few specimens.

Trace elements and chemical classification

Concentrations of minor and trace elements in almost all known iron meteorites have been published in a series of 11 papers (Wasson *et al.*, 1989, and references therein to earlier papers in the series). As expected, the more abundant trace elements are those with siderophile properties, i.e. a higher affinity to metallic rather than oxide phases (Co, Ge, Ga, noble metals, etc.). The abundance of elements such as P, which can occur in both oxidized and reduced state, varies between samples, perhaps reflecting differences in redox conditions during formation. The concentration of strongly lithophile elements (alkalis, Mg, Al, rare earth elements) is extremely low, although they may be enriched in some minor phases like phosphates.

On plots of log(E) versus log(Ni), where E denotes some well-determined trace element, about 80% of meteorites are seen to fall into one of 13 clusters or chemical groups. With two exceptions these groups fall within restricted areas on a log(Ge) versus log(Ni) plot (Figure I32a), whereas on log(Ir) versus log(Ni) plots (Figure I32b) the same samples form narrow, elongated fields spanning a considerable range in Ir contents. Samples that fall outside defined chemical

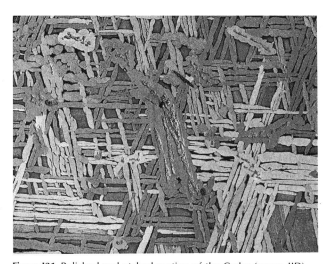

Figure I31 Polished and etched section of the Carbo (group IID) iron meteorite, a medium octahedrite. The texture known as Widmannstätten pattern, consists of kamacite plates aligned along octahedral surfaces and separated (in this case) by fields of mostly plessite. Irregular, deep-etched phase near center of picture is schreibersite, (Fe, $Ni)_3P$, around which 'swathing kamacite' has nucleated. The average width of kamacite lamellae in Carbo is 0.85 mm. (Credit: Smithsonian Institution.)

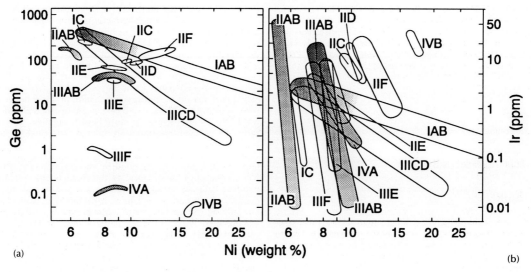

Figure 132 Plot of log(Ge) versus log(Ni) (a) and log(Ir) versus log(Ni) (b), showing the composition of the major iron meteorite groups as outlines. The most densely populated regions of the diagram are shaded. About 1 in 7 iron meteorites falls outside the fields and is classified as ungrouped. (After Wasson, 1985).

groups are called ungrouped (or sometimes anomalous) iron meteorites.

The largest chemical group, IIIAB, comprises about one-third of all known iron meteorites. This group, as well as a number of others (IIAB, IIC, IID, and IVA), show fields of similar shape on most log(E) versus log(Ni) plots, although they occupy different regions of the plot. As Scott (1972) has demonstrated, the behavior of these groups is consistent with fractional crystallization from a melt, and these groups are therefor called 'magmatic iron meteorite groups.' If the partition coefficient of element E, $k_E = C_S/C_L$ (C_S = concentration of E in the solid, C_L = concentration of E in the liquid) is constant throughout the crystallization process, successive fractions of crystallizing solid follow the Scheil (or Rayleigh) equation,

$$C = k\, C_0\, (1-g)^{k-1}$$

where C_0 is the concentration in the liquid before crystallization started, and g is the fraction that has already solidified. Mutually consistent partition coefficients can be derived from the fact that, given two elements A and B, the track of successive fractions of crystallizing solid on a plot of log(A) versus log(B) is a straight line with slope $(k_A-1)/(k_B-1)$.

Groups IAB and IIICD, which account for about 20% of all iron meteorites, exhibit a different behavior. For example, their Ni and Ge contents span a larger range, and their Ir contents a smaller range, than those of the other groups. These groups are also different in other ways and their origin is unclear. The remaining groups are too small to determine unambiguously whether they show igneous or non-igneous trends, but most of them (as well as the majority of ungrouped iron meteorites) seem to have formed by fractional crystallization of a parent melt. The most likely physical setting for fractional crystallization is the core of a small asteroid.

Isotopic composition and ages

The geological age of material formed by igneous processes can usually be determined from the decay of radioactive isotopes such as ^{40}K (half-life 1.3×10^9 years), ^{87}Rb (47×10^9 years), or ^{144}Nd (2.1×10^{15} years). Unfortunately there is only one radioactive isotope of a siderophile element suitable for such a determination, ^{187}Re, and its half-life (which is around 43×10^9 years) is not known with sufficient accuracy to compare Re/Os ages directly with the K/Ar or Rb/Sr ages of stony meteorites. In the case of silicate-bearing iron meteorites the silicate phase may be dated, with the caveat that one cannot be sure how this age relates to the origin of the metallic complement. The chemically primitive silicates in group IAB appear to be as old as other primitive meteorites, i.e. about 4.56×10^9 years, but the differentiated inclusions in group IIE give inconsistent ages, varying from 4.45 to 3.8×10^9 years.

Chronological information is also preserved in the daughter products of extinct radionuclides, although in this case no absolute age can be determined. I–Xe ages, based on the decay of ^{129}I (half-life 16×10^6 years), again place IAB silicate inclusions with the oldest and most primitive material in the solar system. There also appears to be a correlation between Ni content of the metal phase and I–Xe age of the silicate inclusions. This suggests a link between the two phases going back to the earliest history of the parent body, but the nature of the relationship is not yet understood.

Other extinct radionuclides of potential significance for iron meteorite genesis are ^{107}Pd (half-life 6.5×10^6 years) and ^{205}Pb (14×10^6 years). Evidence for live ^{107}Pd has been found in some iron meteorites (Chen and Wasserburg, 1990). However, the short formation ages implied by this would be difficult to reconcile with metallographic cooling rates, and the exact interpretation of extinct radionuclide systematics is open to debate.

The time that a meteorite spends as a meter-size or smaller object in space can be inferred from nuclides generated through spallation by cosmic rays. Noble gas isotopes are commonly used to determine this so-called exposure age, but the low abundance of lithophile elements in iron meteorites allows the use of the highly accurate $^{41}K/^{40}K$ method. Cosmic ray exposure ages of iron meteorites are typically in the range of 200 to 1000 million years, some 20–50 times longer than for stony meteorites. Exposure ages for IIIAB irons cluster tightly around 650 million years, indicating that a much larger body broke into meter-sized fragments at that time (Voshage, 1967).

Origin

Although the interpretation of the results is model dependent, the following scenario would seem to fit most of the data. Most (but not all) iron meteorites originated in cores of asteroids perhaps 50 to a few hundred kilometers diameter. Each magmatic group is thought to have formed by the fractional crystallization of one such core. If the majority of ungrouped iron meteorites also formed independently in separate cores, there had to be more than 50 asteroids which differentiated into mantle and core early in the history of the solar system. If most of these bodies were ≤ 250 km in diameter, their total mass would have been only a small fraction of the asteroid belt, in keeping with the relative scarcity of iron meteorites among observed meteorite falls. However, the energy source which heated such relatively small bodies to temperatures high enough to melt Fe,Ni remains unknown. Possibilities include heating by radioactive ^{26}Al (half-life 0.73×10^6 years) or some kind of inductive heating during a highly active stage of the early Sun. These and other hypotheses are reviewed by Wasson (1974).

Differentiation and crystallization was essentially complete 50 to 200 million years after the asteroids accreted. Some asteroids may have been heated to lower temperatures, leading to only partial differentiation. Some may also have had a complex history of impact mixing during or after cooling. Eventually the asteroids were broken up by collisions. Individual fragments spent several hundred million years in space, although the time of the original break-up cannot be established. A very small fraction of them suffered cosmic reheating, but most of them reached Earth preserving a record of their formation 4.6 billion years ago.

Alfred Kracher

Bibliography

Buchwald, V.F. (1975) *Handbook of Iron Meteorites*, 3 vols. Berkeley: University of California Press.
Chen, J.H. and Wasserburg, G.J. (1990) The isotopic composition of Ag in meteorites and the presence of ^{107}Pd in protoplanets. *Geochim. Cosmochim. Acta*, **54**, 1729–43.
Saikumar, V. and Goldstein, J.I. (1988) An evaluation of the methods to determine the cooling rates of iron meteorites. *Geochim. Cosmochim. Acta*, **52**, 715–26.
Scott, E.R.D. (1972) Chemical fractionation in iron meteorites and its interpretation. *Geochimica Cosmochim. Acta*, **36**, 1205–36.
Swartzendruber, L.J., Itkin, V.P. and Alcock, C.B. (1991) Fe–Ni (iron–nickel), in *Phase Diagrams of Binary Nickel Alloys* (ed. P. Nash). Material Park, OH: ASM International, pp. 110–132.
Voshage, H. (1967) Bestrahlungsalter und Herkunft der Eisenmeteorite. *Z. Naturforsch*, **22a**, 477–506.
Wasson, J.T. (1974) *Meteorites: Classification and Properties*. Berlin and New York: Springer-Verlag.
Wasson, J.T. (1985) *Meteorites: Their Record of Early Solar-System History*. New York: W.H. Freeman.
Wasson, J.T., Ouyang, X., Wang, J. and Jerde, E. (1989) Chemical classification of iron meteorites – XI. *Geochim. Cosmochim. Acta*, **53**, 735–44.

Cross references

Chronology: meteorite
Differentiation
Meteorite

ISOSTASY

Based on the Greek prefix *iso-* meaning 'even', or 'equal', and the suffix *stasia* for standing and *statikos* for stability, isostasy (coined by C.E. Dutton in 1889) is the process by which the Earth's crust adjusts to loading or unloading by vertical motions.

This condition of elevation compensation requires that the stress is hydrostatic below some depth within the Earth. When the deficiency of mass buried below the topography exactly equals the excess mass of the topography, the topographic feature is said to be isostatically compensated (McNutt, 1991). The compensation depth parameter plays an important role in studies of crustal characteristics (thickness, structure, strength, and rheology) of planets such as Mars and Venus (Mottinger, Schogren and Bills, 1985; Kaula, 1990; Solomon *et al.*, 1991; see also Gravity fields of the terrestrial planets; Thermal evolution of planets and satellites; Venus: geology and geophysics).

The ultimate isostatic state is rarely achieved in the Earth's crust because of constant changes in the loading conditions and the sluggish response of the upper mantle to changes in those loads. Crustal loading results from volcanic eruptions, sedimentary accumulation, orogeny, ice accumulation ('glacio-isostasy') and hydrologic transfer (as during glaciation and deglaciation, thus 'hydro-isostasy').

The isostatic state of any point on the Earth's crust is determined as a derivative from measurements of gravity. The average gravitational acceleration at the Earth's surface is approximately $g = 980$ cm s^{-2}.

From this it is possible to calculate the Earth's mean density as 5.5 g cm^{-3}. This value is much higher than for any rocks at the surface, and this has led to the model of a gravitationally layered or ordered planet: atmosphere, hydrosphere, lithosphere, mantle and core, with stepwise increases in density. It is deduced that the core is largely iron (density about 10), a deduction supported also by the existence of iron meteorites, believed to be the relics of the cores of differentiated parent bodies.

Table I3 Values for the international ellipsoid

Radius at the equator a (m)	6 378 388
Ellipticity v	1/297
Polar radius c (m)	6 356 911.946
¼ length of equator (m)	10 109 148.411
¼ length of meridian (m)	10 002 288.299
Average radius (m)	6 371 229.315
Radius of sphere of same surface (m)	6 371 227.709
Radius of sphere of same volume (m)	6 371 221.266
Surface (km^2)	510 100 933.5
Volume (km^3)	1 083 319 780 000

Measurements of gravity show that the acceleration varies over the Earth's surface. It is found to be subject to two variables: latitude and altitude. The latitude effect is a consequence of the Earth's rotation. The same principles will apply to any planet, natural satellite or any rotating body in any solar system. As a consequence of its rotation a degree of ellipticity develops. If it were not spinning, its figure might approach a perfect sphere and the latitude variable would disappear.

A theoretical equilibrium figure for the Earth can be calculated where a standardized value for gravity at sea level is determined as a function of latitude indicated by the Greek letter ρ. An internationally adopted value for 'normal' gravity is

$$g = 978.049 (1 + 0.0052884 \sin^2\rho - 0.0000059 \sin^2 2\rho) \text{ cm s}^{-2}$$

This equilibrium figure is theoretically a spheroid, but due to very small differences in density distribution it is an ellipsoid. An ideal ellipsoid of revolution would exist on an Earth that was in perfect hydrostatic equilibrium where the force of gravity (radially oriented to converge towards the center of the globe) is combined with the centrifugal force (radially oriented to diverge from the spin axis in the plane of the equator). Precisely at the equator these forces are opposite in direction, but at progressively higher latitudes the vector angle increases.

An 'international ellipsoid' is established for the Earth (by geodetic studies) and described by a set of agreed values (Table I3).

The second variable mentioned above is *altitude* or *elevation*. If one neglects the centrifugal force (in the plane of the equator), where M is a spherical mass, the gravitational attraction varies with a, the distance from the geocenter,

$$g = x \frac{M}{a^2}$$

where x is the gravitational constant. This formula applies to any change in altitude in the air, that is without consideration of rock density. The residual theoretical gravity value is therefore called a free-air anomaly.

If a uniform-density rock layer is assumed, then

$$\Delta g = 2 \pi x \rho h$$

where ρ = density of the rock, and h is the thickness (i.e. height above sea level). For an average density of crustal rock (2.67 g cm^{-3}) the 'induced' gravity change would be $\Delta g = 0.1119$ mgal m^{-1}. Combining the latitude and altitude formulae yields what is known as a Bouguer reduction. This is valid for a flat-Earth situation, but further corrections can be added to account for topographical variability. Furthermore, from place to place the crust and upper mantle contain density inhomogeneities that persist over appreciable lengths of time. Perturbations over an ideal hydrostatic Earth are also caused by the sluggishly fluid convection currents within the mantle.

A quasi-stable, equilibrium state is said to be isostatically compensated, and can be thought of in the sense of an iceberg floating in the ocean with its large, low-density 'roots' supporting its small projecting relief (a ratio of 1 : 8). Continents stand higher above sea level because continental crust has an average density (< 2.7) that is appreciably less than that of the oceans and their subjacent mantle (3.0–3.2). On a non-oceanic planet like Venus (e.g. Solomon, 1993) the compensation depths of several hundred kilometers are much greater than on Earth, and probably relate to deep convective circulation.

In the case of a mountain system a local compensation develops. Knowledge of this isostatic state was first obtained in the 1850s when geodetic surveyors in India observed when approaching the 8 km high Himalayas that although their plumb-boos (normally vertical) were visibly deflected by the giant mass of the mountains, the deflection was not as great as expected. There had to be a density deficiency within the mountain and its roots. Two models were offered (in 1855) to explain the anomaly:

1. the Airy model proposed a compensation by a thickening of the roots to balance the topographic uplift, thus

$$\omega = \frac{\rho_o h}{\rho_m - \rho_o}$$

where ω = crustal thickening, h = height of mountain; ρ_o = density of the topographic excess; and ρ_m = density of the subjacent mantle;

2. the Pratt model proposed that the mountain possessed on average a lower net density ($\Delta\rho$), enabling it to ride above an essentially flat mantle plane ($Z\rho$), thus

$$\Delta\rho = \rho_o \frac{h}{Z\rho + h}$$

It turns out that both models need to be adopted because a continuum exists between the two assumptions. The same conclusion has been reached with respect to the continent/ocean crust compensation, inasmuch as the continental crust is appreciably thicker but of lower density. Freshly erupted basaltic oceanic crust at a spreading center creates a 'mid-ocean ridge' because its heat lowers the density. Sluggish heat loss (over about 30 million years at about 500 km from the ridge axis) causes the ocean crust to reach an equilibrium surface elevation of about 5000 m depth.

Rhodes W. Fairbridge

Bibliography

Andrews, J.T. (ed.) (1974) *Glacial Isostasy*. Benchmark Vol. 10. Stroudsburg: Dowden, Hutchinson & Ross, 491 pp.

Barrell, J. (1914/1915) The strength of the Earth's crust. *J. Geol.*, **22** and **23** (11 parts).

Clark, J.A., Farrell, W.E. and Peltier, W.R. (1978) Global changes in postglacial sea level: a numerical calculation. *Quaternary Res.*, **9**, 265–87.

Fletcher, C.H., Fairbridge, R.W., Møller, J.J. and Long, A.J. (1993) Emergence of the Varanger Peninsula, arctic Norway, and climate changes since deglaciation. *Holocene*, **3**(2), 116–127.

James, D.E. (ed.) (1987) *The Encyclopedia of Solid Earth Geophysics*. New York: Van Nostrand Reinhold, 1328 pp.

Kaula, W.M. (1990) Venus: a Contrast in evolution to earth. *Science*, **247**, 1191–6.

Lambeck, K. (1990) Glacial rebound, sea-level change and mantle viscosity. *Quart. J. Roy. Astron. Soc.*, **31**(1), 1–30.

McNutt, M.K. (1983) Influence of plate subduction on isostatic compensation in northern California. *Tectonics*, **2**, 399–415.

Mörner, N.A. (1980) *Earth Rheology, Isostasy and Eustasy*. New York: Wiley, 599 pp.

Mottinger, N.A., Sjogren, W.L. and Bills, B.G. (1985) Venus gravity: a harmonic analysis and geophysical implications. *J. Geophys. Res.*, **90** Suppl. C739–56.

Scheidegger, A. (1982) *Principles of Geodynamics*. Berlin: Heidelberg, New York: Springer-Verlag.

Solomon, S.C. (1993) The geophysics of Venus. *Physics Today*, **46**(7), 49–55.

Solomon, S.C., Head, J.W., Kaula, W.M. *et al.* (1991) Venus tectonics: initial analysis from magellan. *Science*, **252**, 297–312.

Cross references

Geoid
Gravity fields of the terrestrial planets
Mars: gravity
Thermal evolution of planets and satellites
Venus: geology and geophysics
Venus: gravity

J

JONES, HAROLD SPENCER (1890–1960)

Spencer Jones was born in England on 29 March 1890. He was educated at Hammersmith Grammar School and Latymer Upper College; from there he obtained a scholarship to Jesus College, Cambridge, and rapidly showed his exceptional mathematical talent. In 1913 he was appointed Chief Assistant at the Royal Greenwich Observatory in succession to Arthur Eddington, who had gone to Cambridge as Plumian Professor.

During his first period at Greenwich he published important papers concerning the variation in latitude. He took part in an eclipse expedition to Russia, and also wrote the first of his celebrated popular books, *General Astronomy*. (Later on came two more: *Worlds Without End* and *Life on Other Worlds*.)

In 1923 he was appointed HM Astronomer at the Cape, following S.S. Hough, and he spent the next 10 years in South Africa, where he carried out many research projects. For example, he redetermined the elements of the Moon's orbit; he made parallax observations of Mars with the aim of redefining the length of the astronomical unit, or Earth–Sun distance; he made a series of spectroscopic observations of the bright nova RR Pictoris, and he published a catalog of the radial velocities of southern stars.

While at the Cape he collected over 1200 photographic observations connected with the solar parallax programme. In 1931 the tiny asteroid Eros made a close approach to the Earth, and Spencer Jones was entrusted with the task of using the observations to make a new estimate of the length of the astronomical unit. His painstaking work took many years, though unfortunately his final result is now known to be rather inaccurate; this was not Spencer Jones' fault – he had to manage without modern computers.

In 1933 he returned to England, and became Astronomer Royal in succession to F.W. Dyson. He continued with his theoretical work, and also became a noted popularizer of astronomy; in addition to his books he became known as an excellent speaker and broadcaster. He served as President of the British Astronomical Association (1934–35) and of the Royal Astronomical Society (1938–39); he was knighted in 1943, and was created KBE in 1955. After the end of the war he became President of the International Astronomical Union (1944–48).

Greenwich had gradually become unsatisfactory as an observing site in the 20th century, owing to the urban spread of London. Spencer Jones realized this as early as 1939, and was active in organizing a move away from the city. Herstmonceux Castle in Sussex was selected; the move took many years, and was not complete until 1958, 3 years after his retirement.

Spencer Jones investigated geophysical phenomena such as the rotation and oblateness of the Earth, and the variations in the magnetosphere. He proved that variations in the observed longitudes of the Sun, Moon and planets are due not to any real irregularities in their motion, but to changes in the speed of the Earth's rotation. His work on measuring the length of the astronomical unit was an outstanding contribution.

He received many honors, including the Royal Medal of the Royal Society and the Gold Medal of the Royal Astronomical Society. In 1947 he became the first President of the new Institute of Navigation, and served twice as Master of the Clockmakers' Company (an ancient guild of the City of London). After his official retirement in 1955 he served first as editor of the publications of the International Geophysical Year, and then as Secretary-General to the International Council of Scientific Unions. He died very suddenly on 3 November 1960.

Patrick Moore

Bibliography

Atkinson R.d'E. (1961), Harold Spencer Jones. *J. Br. Astron. Assoc.*, **71**, 84–9.

Kopal, Z. (1975) Spencer Jones, Harold. *Dict. Sci. Biogr.*, Vol. 12. pp. 573–4.

JULIAN CALENDAR, YEAR AND DAY

The Julian Calendar and Julian Year were initiated by the Alexandrian astronomer Sosigenes, in the reign of Julius Caesar in 46 BC, based on a year of 365.25 d. Every fourth year was a 'Leap Year' of 366 d. However, because of its cumulative error, the Julian calendar was gradually abandoned in the 16th–18th centuries and replaced (in 1582 in Catholic Europe) by the Gregorian system; Britain and the American colonies accepted the Gregorian calendar in 1752. Exceptionally, certain groups, e.g. the Eastern Orthodox Church, still use the 'old style' chronology; thus the Russian Christmas is on 7th January, 13 days later than the western holiday.

The Julian day is the basis of a system of astronomical dating that has nothing whatever to do with the Julian Year. It was invented by a French mathematician and amateur astronomer, Joseph Scaliger (1540–1609) and named in honor of his father Julius. The purpose was to provide scientists with a continuous dating system free from the confusion of irregular months, leap years and cumulative errors. It is now the standard system used in every ephemeris.

Julian day 1 began at noon GMT on 1 January 4713 BC, according to the Gregorian calendar. This would be -4712 in the astronomic timetable because there is no year 0 in the Christian calendar; thus the year -1 is 2 BC. In recent decades, 1970 began (1 January) with JD 244 0588; 1980 with 244 4240 . . . However, the Julian day is now taken to start at midnight, in Greenwich Mean Time or Universal Time. Tables of recent Julian day equivalents are provided annually in the *Astronomical Almanac*.

Rhodes W. Fairbridge

Bibliography

Harvey, O.L. (1983) *Calendar conversions by way of the Julian Day numbers*. Philadelphia: American Philosophical Society.

Ronan, C.A., *et al.* (1974) Calendar in: *Encyclopedia Brittanica*, 15th edn (Macropedia, Vol. 3, p. 595), Chicago.

JUPITER

Jupiter is the most massive of the planets of our solar system, with nearly two and a half times the mass of the rest of the planets combined (1.9×10^{27} kg). The mean density of Jupiter is 1.33 g cm^{-3} (where the density of ordinary water is 1 0), and (like the Sun) it is composed primarily of hydrogen and helium. Jupiter is the fifth planet from the Sun, with an orbital mean distance of 5.203 AU. Jupiter orbits the Sun in 11.86 Earth years, and rotates once on its axis in 9 h 55 min (one of the fastest rotation rates in the solar system). The rotation axis has a small inclination of 3.12° to the orbital plane. Jupiter emits from its interior roughly twice as much heat as it receives from the Sun. This heat flux is approximately 5.4 W m^{-2}, about 100 times as large as the heat flux from the Earth.

Jupiter is one of the brightest objects in the night sky, with a magnitude of −2.6. Telescopic investigation of Jupiter began in the years 1609–1610 with Simon Marius and Galileo Galilei, who discovered the four large moons Io, Europa, Ganymede and Callisto orbiting the planet. Although Galileo (q.v.) is often credited with the discovery of the Galilean satellites, Marius may have been the first to see them, and was the first to use the names employed today. (Galileo wanted to name them the Medician planets to honor his sponsor.) Telescopic observation of Jupiter revealed colorful cloud patterns and extraordinary features such as the Great Red Spot, which has persisted for many decades and perhaps centuries (Plates 22a, 24a). Later observers detected a number of smaller satellites orbiting the planet. Amalthea was discovered in 1892. Himalia and Elara were discovered by C.D. Perrine in 1904. Pasiphae (1908) and Sinope (1914) were next, followed by Lysithea (1938), Carme (1938), Ananke (1951) and Leda (1974). Jupiter's satellite system (q.v.) is remarkable in having a set of four moons in retrograde orbits; these are distinguished by names ending with an 'e.' They are presently thought to be captured objects (see Jupiter: satellite system).

Observations of the motions of the Galilean satellites led to the first estimates of the velocity of light. The Danish astronomer O. Roemer noted that occultations of the moons (by Jupiter) occurred systematically later as the distance separating the Earth and Jupiter increased; at the maximum Earth–Jupiter separation the eclipses were 16 min 40 s later than expected from observations made near minimum separation. (The difference in distance corresponds to about twice the radius of the Earth orbit, or 2 AU.) Roemer linked this delay with the travel time of light, and deduced that the velocity was about 300 000 km s^{-1}.

Earth-based observers were also able to determine that Jupiter has a massive magnetosphere which, if it were visible, would be comparable in size to Earth's Moon as viewed from Earth. Jupiter's magnetosphere (q.v.) is large enough to enclose the Sun, and the downstream 'tail' of the Jupiter magnetosphere is thought to extend to the region of the orbit of Saturn. In the 1950s astronomers discovered that Jupiter emits radiation at radio wavelengths (thermal, decimetric and decametric radiation). They deduced that the decimetric radiation originated in Jovian radiation belts, similar to Earth's Van Allen belts; this implied that a strong magnetic field must be present. The decimetric radiation is modulated by the motion of the satellite Io (q.v.), with a period of 21 h (half of Io's orbital period; see Jupiter: magnetic field and magnetosphere).

Exploration of the Jovian system began with the Pioneer 10 spacecraft, which passed within about 130 000 km of the planetary surface in 1973 (see Pioneer missions). Pioneer 11 (in 1974) came within 50 000 km of the planet. Both spacecraft carried multiple instruments to sample the interplanetary and Jovian environments. In 1979 the two Voyager spacecraft encountered Jupiter, returning the most detailed images yet obtained, along with a large volume of new data (see Voyager missions). The Galileo spacecraft (q.v.) will visit Jupiter in 1995–1996; this mission includes a probe designed to penetrate the atmosphere of the planet.

Spacecraft observations revealed a host of new and unexpected phenomena. Close views of the Galilean satellites were obtained, revealing a great diversity of surface features (see Galilean satellite; Io; Europa; Ganymede; Callisto). The moon Io (q.v.) has active volcanism; eruptions in process were imaged by Voyager scientists. Io also appears to be responsible for the presence of a large plasma torus that surrounds the equatorial regions of Jupiter (see Planetary torus). In addition an electric current of about 3×10^6 A was detected in association with a flux tube connecting Io and Jupiter.

Spacecraft observations also revealed Jovian auroras (q.v.) and the Jovian ring system (see Jupiter: ring system). The rings, which had not been detected from Earth, are comprised of dusty materials; this may be contrasted with the bright rings of Saturn, which are composed of icy materials.

Knowledge of the interior of Jupiter comes primarily from measurements of the planetary radius, shape, and gravitational moments (see Jupiter: interior). The region visible to us represents cloud tops of an atmosphere perhaps 1000 km in depth. Beneath this lies a thick fluid region principally composed of liquid hydrogen. At a depth of about 25 000 km, temperatures reach perhaps 11 000 K, and liquid hydrogen changes to a metallic form. Convection in this region gives rise to the strong Jovian magnetic field. Jupiter may have a small core of up to perhaps 15 Earth masses of iron and silicates.

The Jupiter atmosphere is characterized by alternating belts (dark bands) and zones (light bands) of remarkable structure and color (Plates 22, 24). In addition to belts and zones Jupiter possesses a varying number of persistent eddies. Best known of these is the Great Red Spot, which is an anticyclonic eddy with winds of up to 100 m s^{-1}. This feature may have been seen by Robert Hooke and J.D. Cassini in the early days of telescopic observations. It has undergone a number of changes during the past century or so. For instance, in the 1860s it became nearly invisible; a decade later it was very prominent, and then in 1883 it faded again. It was prominent during the Pioneer and Voyager flybys. Cyclonic storms are also present (the 'Brown Barges'), but they account for only about 10% of the eddies observed on Jupiter. (See Jupiter: atmospheric circulation).

Although we have obtained a great deal of information about Jupiter, the sum of our ignorance remains far greater. Flyby missions have been spectacularly successful, but the observations represent conditions prevailing during only a very brief interval of time. Unresolved questions, some of which may be addressed by information from the Galileo mission (q.v.), include the following.

- What is the detailed chemical composition of the atmosphere?
- What is the vertical structure of the deep atmosphere?
- What are the relationships of cloud layers?
- What is the nature of the cloud particles?
- What are the magnetospheric interactions of the satellites?
- What is the detailed structure of the magnetosphere?
- What are the distributions of energetic particles?

Also, We currently know very little about the structure, mineralogy, gravitational fields, magnetic fields and dynamics of the Jovian satellites. Clearly there is much more to be gained from investigations of the nature and properties of Jupiter and its satellites.

James H. Shirley

Bibliography

Belton, M.S., West, R.A., Rahe, J. and Pereyda M. (eds) (1989) *Time Variable Phenomena in the Jovian System*. Washington DC: NASA, 409 pp.

Gehrels, T. (ed.) (1976) *Jupiter*. Tucson: University of Arizona Press.

Ingersoll, A. (1981) Jupiter and Saturn, in *The New Solar System* (eds J.K. Beatty, B. O'Leary and A. Chaikin). New York: Cambridge University Press, pp. 117–28.

Morrison, D. (ed.) (1986) *Satellites of Jupiter*. Tucson: University of Arizona Press.

NASA (1982) *The Voyager Flights to Jupiter and Saturn*. Pasadena, California: JPL-CIT.

Science (Voyager I issue) (1979), **204**, pp. 945–1008.

Science (Voyager II issue) (1979), **206**, pp. 925–96.

Wolfe, J. (1975) Jupiter. Sci. Am., **233**, 119–26.

Yeates, C.M., Johnson, T.V., Colin, L. *et al.* (1985) *Galileo: Exploration of Jupiter's System*. Washington DC: NASA SP-479.

Cross references

Callisto
Europa
Galileo mission
Ganymede
Io
Pioneer 10 and 11 missions
Planetary ring
Satellite, natural
Voyager 10 and 11 missions

JUPITER: ATMOSPHERE

A discussion of Jupiter's atmospheric circulation divides naturally into two categories: topics dealing with the zonal winds and topics dealing with the eddies. The zonal (east–west) winds are made up of over a dozen alternating jet streams that vary in strength and direction as a function of latitude. The eddies are the vortices and waves that remain in the wind field after the zonal winds have been subtracted.

Zonal winds

Jupiter's atmosphere has the most color contrast of any atmosphere in the solar system, including Earth's Plates 22, 24. This makes the job of tracking cloud features on Jupiter a particularly rewarding one. Observations taken by the Voyager spacecraft, the Hubble Space Telescope and ground-based telescopes reveal 100 m s^{-1} winds swirling inside the prominent Great Red Spot, and equally strong alternating jet streams that correlate with the planet's banded appearance. Visible-light images show the motions of the cloud-top features, while infrared images reveal the three-dimensional temperature variations in the stratosphere above the cloud tops. Jupiter's weather is found to be simpler than Earth's weather in many respects. Storms on Jupiter remain unchanged for decades and even centuries. Instead of meandering unpredictably like terrestrial hurricanes, they simply drift at a constant rate in either the eastward or westward direction. One reason for the simpler situation on Jupiter is the lack of mountain ranges and air–sea interfaces that act on Earth to thwart the natural tendency for winds to settle into steady east–west patterns.

Belts and zones

The alternating bands of color on Jupiter are called belts and zones. The belts are the dark bands (dark like a belt worn around a waist) and the zones are the light bands. Starting in the north, the nomenclature is: North Polar Region (NPR), North North Temperate Zone (NNTZ), North North Temperate Belt (NNTB), north temperate Zone (NTZ), north temperate Belt (NTB), North Tropical Zone (NTrZ), north Equatorial Belt (NEB), north equatorial zone (EZn), equatorial Belt (EB), south equatorial Zone (EZs), South equatorial Belt (SEB), South tropical Zone (STrZ), south Temperate Belt (STB), South Temperate Zone (STZ), South South Temperate Belt (SSTB), South South Temperate Zone (SSTZ) and South Polar Region (SPR). The north Polar Region (NPR) and South Polar Region (SPR) are dusky hoods that do not have banded structure. The dark color of the belts fades in and out every few years, usually in a gradual manner. The exception is the South Equatorial Belt (SEB), which often displays dramatic outbursts of cloud activity. The South Tropical Zone (STrZ) can sometimes develop a darker section known as the South Tropical Disturbance (STD). Such was the case from 1901 until 1940. Peek (1981) gives a detailed history of the changes in the colors and positions of Jupiter's belts and zones.

Cloud-top winds

Because there is no solid surface on a giant planet except at its small rocky core, a problem arises when one wants to measure atmospheric winds relative to some meaningful reference frame. The problem is solved by measuring the winds relative to the rotation period of the planet's magnetic field, which is tied to the interior and indicates the bulk rotation. This magnetic field reference frame is historically called the system III reference frame, and corresponds to a rotation period of 9.92 h on Jupiter. The old System I and system II reference frames date back to the late 17th century, when observers like Cassini found it convenient to measure the rapid drifts of cloud features that were equatorward of 10° with a separate reference frame. The system I reference frame was chosen to match the fast equatorial drift rates, and the system II reference frame was chosen to match the slower drift rate of the Great Red Spot (Peek, 1981).

The zonal (east–west) wind profiles of the four giant planets and the Sun are shown in Figure J1. Jupiter's wind profile was found by digital pattern matching of successive Voyager images and is tabulated by Limaye (1986). Jupiter and Saturn both have eastward (also known as prograde or westerly) equatorial winds. Saturn's equatorial jet reaches velocities of over 400 m s^{-1} and its mid-latitude jets blow predominantly in the eastward direction, whereas Jupiter's blow in either the eastward or westward direction depending on the latitude. On the other hand, Uranus and Neptune have westward (also known as retrograde or easterly) equatorial winds, and only one broad

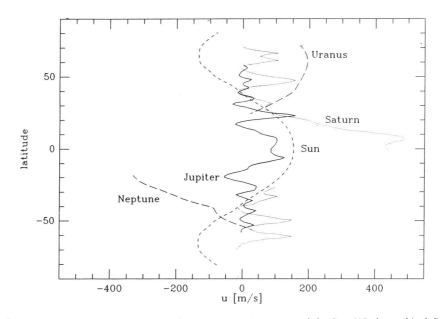

Figure J1 Observed surface zonal-wind profiles for Jupiter, Saturn, Uranus, Neptune and the Sun. Wind speed is defined relative to the rotation rate of the deep interior, which in the case of the giant planets is given by their respective magnetic field rotation rates, and in the case of the Sun is determined from helioseismicity.

eastward jet per hemisphere (assuming the large gaps in the data do not hide any important structure). A key goal in planetary atmospheric dynamics is to come up with a comprehensive theoretical model that reproduces all of the different curves in Figure J1.

Thermal winds

The heat budgets of the giant planets have been accurately determined by the Voyager infrared (IRIS) investigation (Hanel et al., 1981, Hanel et al., 1983, Pearl et al., 1990). Jupiter, Saturn and Neptune are like stars in the sense that they radiate more energy than they receive from the Sun, about twice as much in each case, except that stars radiate because of ongoing nuclear fusion and the giant planets radiate because the excess heat from the potential energy of the accreting matter that formed them continues to escape. In contrast to the other three giant planets, Uranus radiates at most only 6% more energy than it receives from the Sun (it can be speculated that the cataclysmic event that knocked Uranus on its side may have had a hand in extinguishing its internal heat by turning the planet inside out). Jupiter's internal heat is as important to its atmospheric circulation as the heat that it receives from the Sun.

Much of the weather on Earth is generated by the baroclinic (literally, 'pressure slope') instability mechanism, a process by which the excess solar insolation received at the equator is redistributed towards the poles by baroclinic eddies. On Jupiter the Voyager IRIS data indicate that the equator-to-pole temperature difference is nearly zero, which is insufficient to drive baroclinic eddies. Instead, the variations in incoming sunlight appear to be redistributed in the fluid interior.

Even though Jupiter's equator-to-pole temperature difference is small, the IRIS data reveal significant local variations of temperature in the stratosphere that correlate with the jet-stream patterns in the underlying troposphere. The significance of this correlation may be understood in the context of the thermal wind equation, which is derived by combining the equations describing hydrostatic balance, the balance between the gravitational force and the vertical pressure-gradient force, and geostrophic balance, the balance between the Coriolis force and the horizontal pressure-gradient force. The thermal wind equation relates horizontal temperature variations to changes in horizontal wind speed with height. The meridional (north–south) variations of temperature on Jupiter indicate that the zonal winds decay with height above the cloud tops. The cloud-top winds are located at a pressure level of 500–600 mbar (1000 mbar = 1 bar ≈ room pressure), and application of the thermal wind equation indicates that the level of no motion is near the 20 mbar pressure level, which is a height difference of 3 pressure scale heights or 60 km. The IRIS results for the other three giant planets indicate that their wind strengths also decay with height above the cloud tops. These observations imply that the momentum source for the zonal winds has a diminishing influence on the wind speeds with altitude.

Deep winds

The Voyager imaging and infrared data provide a three-dimensional picture of Jupiter's atmospheric circulation at and above the cloud tops. However, there is little information about the atmospheric circulation below the cloud tops. This has hampered progress in interpreting the cloud-top winds, because deep winds are associated, through the Coriolis force, with pressure ridges and troughs that act like variable bottom topography to the cloud-top winds.

The Galileo probe made the first descent into a giant planet atmosphere on 7 December 1995. This ambitious mission was a complete success, with preliminary results indicating that the zonal winds pick up speed by about 50% below the cloud tops, and that Jupiter's chemical composition resembles the solar composition. In addition to the probe findings, which are still being analyzed at the time of writing, there are other ongoing remote-sensing observations being pursued that promise to reveal the nature of Jupiter's deep circulation. Long wavelength radio observations yield useful constraints on the deep atmosphere that indicate that the surface banded structure extends into the interior. Unfortunately, the competing effects of temperature variations and cloud opacity variations are difficult to separate, and Jupiter's magnetic field contributes a large signal that must be subtracted. Surface expressions of atmospheric waves that extend below the cloud tops provide a complementary source of information. Analysis of one class of mesoscale waves on Jupiter suggests that the Richardson number, a measure of the stability of an atmosphere to vertical wind shear, is of the order of unity (Flasar and Gierasch, 1986).

Yet another approach is to observe vorticity (spin rate) variations while tracking clouds as they move across the planet. Any increase in vorticity following the motion indicates that there has been vertical stretching, since an increase in the spin rate implies that the vertical column of air has 'drawn in its arms' like an ice skater, and in so doing has stretched vertically in order to conserve its mass. This phenomenon is called vortex tube stretching, and the ratio of the vorticity to the vertical thickness is called the potential vorticity, which is usually denoted q:

$$q \equiv \frac{(\zeta + f)}{h} \tag{J1}$$

where $\zeta \equiv \hat{k} \cdot \nabla \times v$ is the relative vorticity, v is the velocity, $f \equiv \hat{k} \cdot 2\Omega = 2\Omega \sin$ (latitude) is the Coriolis parameter, Ω is the planet's angular velocity, and the effective thickness h is the mass per unit area between material surfaces. For inviscid (frictionless) and adiabatic (no heat sources or sinks) motions the potential vorticity is a conserved quantity:

$$\frac{dq}{dt} \equiv \left(\frac{\partial}{\partial t} + v \cdot \nabla\right) q = 0 \tag{J2}$$

The conservation of potential vorticity provides a means of estimating the pressure variations in Jupiter's deep atmosphere by observing the cloud-top winds (Dowling and Ingersoll, 1989). The indications are that Jupiter's winds are arranged in a special configuration that makes them neutrally stable. The next section contains further details about stability.

Stability

The theory of the stability of zonal-wind profiles is still under development, but there exists a set of theorems that provide a framework for analysis. These theorems provide constraints on the behavior of Rossby waves, which are large, slowly moving waves in an atmosphere that affect the meteorology. An instability occurs when an interaction between Rossby waves draws energy from the zonal winds and causes the Rossby waves to grow to large amplitudes. Stability theorems specify conditions that guarantee that there will be no such destabilizing interactions.

The restoring force for Rossby waves is proportional to changes in the background potential vorticity, so most stability theorems specify conditions on the potential vorticity gradient. The most commonly cited stability theorem is the Charney–Stern theorem, which states that if the meridional (north–south) gradient of the potential vorticity does not change sign, the flow is stable:

$$0 \leq q_y \text{ or } q_y \leq 0 \rightarrow \text{stability} \tag{J3}$$

where the subscript y denotes differentiation with respect to latitude. The barotropic stability theorem is a special case of (J3) that holds when the vertical thickness h in (J1) is a constant. It has been found that Jupiter's zonal winds have the property that the sign of q_y is always the same as the sign of the zonal winds. Since the zonal winds change sign at many different latitudes on Jupiter (Figure J1), the Charney–Stern stability criterion is violated. This would seem to suggest that the zonal winds are unstable. On the other hand, the zonal-wind profile determined from Voyager images taken in 1979 precisely matches the zonal-wind profile determined from Hubble Space Telescope images taken in 1992, and it is hard to understand how a complicated wind profile like Jupiter's could be unstable and yet not vary at all in 13 years.

In fact, there are at least two known ways for a zonal-wind profile to violate the Charney–Stern stability criterion and still remain stable. The more familiar of these is Fjørtoft's theorem, also known as Arnol'd's 1st theorem in the case of large amplitude perturbations. This theorem states that if in some reference frame the potential vorticity gradient and the zonal wind always have opposite signs, then the flow is stable:

$$-\infty < \frac{q_y}{u} < 0 \rightarrow \text{stability} \tag{J4}$$

where $u = u(y)$ denotes the zonal wind. Unfortunately, (J4) is opposite to the relationship found on Jupiter. However, there is a second, less well known theorem by Arnol'd that states that if in some reference frame q_y and u always have the same sign, and if in addition the ratio of the two is not too large, then the flow is stable:

$$0 < \frac{q_y}{u} < \frac{1}{L_d^2} \to \text{stability} \qquad (J5)$$

where L_d is an intrinsic length scale called the (first baroclinic) radius of deformation. For a detailed discussion of Arnol'd's theorems consult McIntyre and Shepherd (1987). Arnol'd's 2nd stability theorem (J5) precisely describes the situation found on Jupiter, where the ratio q_y/u equals the value $1/L_d^2$ needed for neutral stability (Dowling, 1993). This neutral stability configuration can explain the unchanging zonal-wind profile on Jupiter. At the same time, it allows for the generation of small vortices through a dynamical balance between the forces that destabilize the winds and the tendency for the instabilities to return the winds to a stable configuration.

The foregoing stability theorems all apply to thin atmospheres, and a different analysis is required for Jupiter's interior. One of the effects of Jupiter's rapid rotation on its interior dynamics is the suppression of motions in the direction of the rotation axis. This Taylor–Proudman effect tends to make the fluid move as rigid columns that are aligned with the rotation axis. The spherical shape of the planet causes these columns to stretch as they move towards the rotation axis, which by the vortex tube stretching mechanism gives rise to Rossby waves. Ingersoll and Pollard (1982) derived a barotropic stability criterion for this geometry that appears to be satisfied by a wide range of interior circulations that are possible for Jupiter.

Eddies

Anticyclones

Jupiter's most recognizable feature is the Great Red Spot, the large ruddy 'eye' that dominates its southern hemisphere. The Red Spot is an oval-shaped, high-pressure storm centered at 23° S that spans 20° in longitude by 10° in latitude (20 000 km by 10 000 km). Because of the Coriolis force, high-pressure storms circulate in the opposite direction to the planet's rotation direction and are called anticyclones. Interestingly, Neptune's Great Dark Spot is another giant anticyclone that is centered at the same southern latitude on Neptune and has the same angular size as Jupiter's Great Red Spot, although the Great Dark Spot is also known for its periodic changes in shape (see Neptune: atmosphere). The vertical extent of the Great Red Spot is surmised to be only about 200 km, based on application of the thermal wind equation to the 8 K cooler temperatures detected in its center relative to its surroundings by the Voyager infrared (IRIS) instrument, and on the assumption that cloud-top eddies in general do not extend significantly into the neutral interior. This means that the Red Spot is a pancake-shaped object with a breadth that is about 100 times larger than its height. For comparison, hurricanes have breadths about 10 to 20 times broader than their heights.

Robert Hooke first reported a large spot in Jupiter's southern hemisphere in 1664, and his discovery is recorded in the first issue of the *Philosophical Transactions of the Royal Society of London* (1666):

> A Spot in one of the Belts of Jupiter.
> The Ingenious Mr. *Hook* did, some months since, intimate to a friend of his, that he had, with an excellent twelve foot Telescope, observed, some days before, he than spoke of it, (*videl.* on the ninth of May, 1664, about 9 of the clock at night) a small Spot in the biggest of the 3 obscurer *Belts of Jupiter*, and that, observing it from time to time, he found, that within 2 hours after, the said Spot had moved from East to West, about half the length of the Diameter of *Jupiter*.

Hooke's spot was observed intermittently from 1664 to 1713, and in all likelihood was the Great Red Spot itself. According to Peek (1981), continuous observations of the Great Red Spot can be traced back to a sketch made by H. Schwabe on 5 September 1831. Computer simulations indicate that the Red Spot probably formed as a result of the merging of smaller vortices. Unlike three-dimensional turbulence where one expects a large eddy to break up into smaller and smaller eddies – the fate that befalls a smoke ring, for example – meteorology is governed by two-dimensional turbulence, which has the counter-intuitive property that smaller eddies merge together to create larger eddies, a process known as the backwards energy cascade. There is apparently no limit to the Red Spot's lifetime as long as frictional losses are continuously balanced by mergers with smaller eddies. Williams and Wilson (1988) ran one numerical simulation of the Red Spot for over 100 years and found virtually no change in the structure of their model vortex during the entire run.

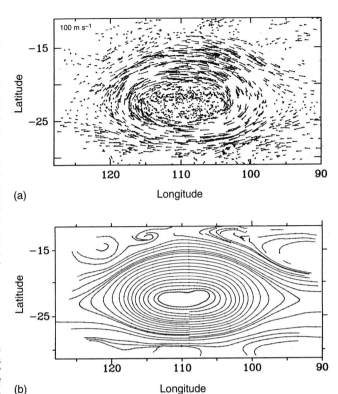

Figure J2 (a) Atmospheric circulation in and around Jupiter's Great Red Spot, as determined by tracking clouds in Voyager images. The initial positions of individual cloud features are indicated with the dots, and the lines point downwind. (b) Fluid trajectories computed from the wind data in (a).

Although the Great Red Spot was once thought to be an exotic, high-pressure storm that was special to Jupiter, it is now known that such long-lived vortices are the rule rather than the exception throughout the solar system. Compare Neptune's Great Dark Spot and Oval D2 (Smith *et al.*, 1989), Saturn's spots (Sromovsky *et al.*, 1983), and Earth's atmospheric blocking highs (Haines 1989), Gulf Stream rings (Flierl, 1987) and Mediterranean salt lens (Armi *et al.*, 1988). All these objects are large fluid vortices that take advantage of the inherent nonlinearities of fluid motion in order to behave almost as if they were solid. Comparative planetology has taught us that these large robust spots are not rare at all, but are a common feature of rapidly rotating atmospheres and oceans. There is even a class of rapidly rotating stars, the RS Canum Venaticorum stars, that have been surmised to have a giant, localized 'star spot' on one side – possibly a hydromagnetic analog of Jupiter's Great Red Spot.

As illustrated in Figure J2, peak winds of 110 m s^{-1} occur in an annulus around the outer edge of the Great Red Spot. Although the center is quiescent, there is no indication of the eye-wall structure associated with terrestrial hurricanes. The concentration of winds in an annulus is consistent with the fact that, with its 20 000 km major diameter, the Red Spot is much larger than the first baroclinic radius of deformation, which is estimated to be in the range 500–3000 km. The idea is that the smaller eddies that merge with the Red Spot contribute their momentum to only an outer annulus that is about the width of the radius of deformation.

Being a long-lived vortex, the Great Red Spot traps the atmospheric gases inside it, and this may help to sustain the chemical chromophores (coloring agents) that intensify its coloring. Interestingly, a short-lived 'Little Red Spot' occasionally appears in the northern hemisphere at 19° N (Beebe and Hockey, 1986). The chromophores responsible for the red and yellow colorations on Jupiter are not known, but three candidates have been suggested. From cosmochemical arguments one expects an ammonium hydrosulfide (NH_4SH) cloud to exist beneath the surface ammonia clouds. Perhaps sulfur-bearing chromophores are rising up to the surface from this underlying cloud. Alternatively, phosphorus-bearing chromophores may explain the colorations, since phosphine (PH_3) has been detected spectroscopically. A third possibility is that hydrocarbon material may be raining down on the surface clouds from the stratospheric haze layer.

The Red Spot is one of hundreds of different storms on Jupiter, 90% of which are anticyclones (MacLow and Ingersoll 1986). The next largest anticyclones after the Red Spot are the three White Ovals, which are located at 33° S. Their origins can be traced back to 1939, when a wavy disturbance developed in the South Temperate Belt (STB). The disturbance developed six pinched regions that were labeled A to F, and these eventually coalesced into the three White Ovals, which are named BC, DE and FA.

Cyclones

Cyclones account for only one in ten storms on Jupiter. This is unlike the case for Earth, where stormy weather is generally associated with low pressure. There are two differences between cyclones and anticyclones that may account for their small numbers on Jupiter. First, it has been shown mathematically that large anticyclones benefit from a balance between the nonlinear fluid dynamical effects of advection and dispersion – the balance operating in solitons – whereas cyclones do not (Flierl, 1987). Second cyclones are more susceptible to moist convection than anticyclones. The extra mass of an anticyclone depresses the atmospheric layers that it rides over, but cyclones have the opposite effect and can raise deep moist air past its lifting condensation level, thus triggering moist convection that (if it is strong enough) can completely disrupt the organized circulations of the cyclone. For example, a prominent chain of cyclones circles Jupiter at 40° S. These come in two varieties: oval-shaped cyclones and turbulent cyclones. One oval-shaped cyclone was observed by Voyager 2 to undergo a rapid alteration whereby its center brightened and the whole vortex then became turbulent (Smith et al., 1979). This cyclone life cycle may be governed by a moist convection instability that is triggered when a cyclone passes over a region of moist air. A related phenomenon is seen on Saturn, where convective activity at one particularly active region is triggered by cyclones that pass nearby (Sromovsky et al., 1983).

An important exception to the rule that cyclones are short-lived are the Brown Barges at 14° N. The strong cyclonic shear found at this latitude may help to stabilize the Barges. In infrared images the Barges are very bright, indicating that they are probably holes in the surface clouds that let out the warmer infrared radiation that is generated in the interior. The Barges exhibit an interesting 15-day oscillation in shape (Hatzes et al., 1981) that is similar to the one Neptune's Great Dark Spot undergoes.

Waves

Eddies include waves as well as vortices, and several different kinds of waves have been detected in Jupiter's atmosphere. Large thermal waves have been discovered that are notable for having about ten wavelengths around the planet's circumference and for having phase speeds that are much slower than the local wind speeds. The near-stationary nature of these waves suggests a dynamical tie with the deep interior. The first observations were reported by Magalhães et al. (1989) from an analysis of Voyager IRIS data. They found a mode-9 wave located in the lower stratosphere between 5 and 15° N, moving with a phase speed of $c = 10.5 \pm 12.4$ m s^{-1}. Soon thereafter, Deming et al. (1989) reported similar waves at deeper levels in the atmosphere from an analysis of ground-based infrared observations. Two mode-10 waves, one at the equator and one at 20° N, were discovered. The 20° N wave was estimated to be moving with a phase speed of $c = 25 \pm 33$ m s^{-1}.

Not all waves on Jupiter are slowly moving. At 8° N there are 11–13 evenly spaced clouds called equatorial plumes that encircle the planet, with an occasional one found at 8° S. The plumes drift with phase speeds that are similar to the fast eastward winds in the equatorial jet. They have been modeled in a self-consistent way as equatorially trapped Rossby waves (Allison, 1990).

Summary

Many of today's research topics in atmospheric dynamics have been motivated by the rich cloud-top dynamics revealed in Jupiter's atmosphere by the Voyager spacecraft. Progress in understanding the nature of long-lived anticyclones like the Great Red Spot and the atmospheric blocking highs on Earth has benefited from the high-quality planetary data. Where the study of giant-planet atmospheric dynamics lags farthest behind terrestrial meteorology and oceanography is in the area of vertical structure information, but in situ probes like the Galileo descent probe promise to illuminate the mysteries of Jupiter's deep atmosphere in the near future.

Present research efforts emphasize realistic numerical models as an indispensable tool in the study of Jupiter's weather. The lack of mountain ranges and air–sea interfaces means that modeling Jupiter's atmospheric circulations is more tractable than modeling Earth's, once more is known about Jupiter's interior circulations. General circulation models for Jupiter may be able to reveal important processes that would otherwise be obscured by the intricacies of terrestrial weather models, which are tuned specifically to the terrestrial problem. The basic questions for which no comprehensive answer yet exists – most notably the origin and maintenance of the zonal winds – continue to drive researchers towards the goal of a complete understanding of all atmospheric circulations.

Timothy E. Dowling

Bibliography

Allison, M. (1990) Planetary waves in Jupiter's equatorial atmosphere. *Icarus*, **83**, 282–307.

Armi, L., Herbert, D., Oakey, N. et al. (1988) The history and decay of a Mediterranean salt lens. *Nature*, **333**, 649–51.

Beebe, R.F. and Hockey, T.A. (1986). A comparison of Red Spots in the atmosphere of Jupiter, *Icarus*, **67**, 96–105.

Deming, D., Mumma, M.J., Espenak, F. et al. (1989) A search for pmode oscillations of Jupiter: serendipitous observations of non-acoustic thermal wave structure, *Ap. J.*, **343**, 456.

Dowling, T.E. (1993) A relationship between potential vorticity and zonal wind on Jupiter, *J. Atmos. Sci.*, **50**, 14–22.

Dowling, T.E. and Ingersoll, A.P. (1989) Jupiter's Great Red Spot as a shallow water system. *J. Atmos. Sci.*, **46**, 3256–78.

Flasar, F.M. and Gierasch, P.J. (1986) Mesoscale waves as a probe of Jupiter's deep atmosphere. *J. Atmos. Sci*, **43**, 2683–707.

Flierl, G. (1987) Isolated eddy models in geophysics. *Ann. Rev. Fluid Mech.*, **19**, 493–530.

Gill, A.E. (1982) *Atmosphere – Ocean Dynamics*. Academic Press.

Haines, K. (1989) Modons as prototypes for atmospheric blocking. *J. Atmos. Sci.*, **46**, 3202–18.

Hanel, R.A., Conrath, B.J., Herath, L.W. et al. (1981) Albedo, internal heat, and energy balance of Jupiter: preliminary results of the Voyager infrared investigation. *J. Geophys. Res.*, **86**, 8705–12.

Hanel, R.A., Conrath, B.J., Kunde, V.G. et al. (1983) Albedo, internal heat flux, and energy balance of Saturn. *Icarus*, **53**, 262–85.

Hatzes, A., Wenkert, D.D., Ingersoll, A.P. and Danielson, G.E. (1981) Oscillations and velocity structure of a long-lived cyclonic spot, *J. Geophys. Res.*, **86**, 8745–9.

Ingersoll, A.P. (1990) Atmospheric dynamics of the outer planets. *Science*, **248**, 308–15.

Ingersoll, A.P. and Pollard, D. (1982) Motion in the interiors and atmospheres of Jupiter and Saturn: scale analysis, anelastic equations, barotropic stability criterion. *Icarus*, **52**, 62–80.

Limaye, S.S. (1986) Jupiter: new estimates of the mean zonal flow at the cloud level. *Icarus*, **65**, 335–52.

McIntyre, M.E. and Shepherd, T.G. (1987) An exact local conservation theorem for finite-amplitude disturbances to non-parallel shear flows, with remarks on Hamiltonian structure and on Arnol'd stability theorems, *J. Fluid Mech.*, **181**, 527–65.

MacLow, M.M. and Ingersoll, A.P. (1986) Merging of vortices in the atmosphere of Jupiter: an analysis of Voyager images. *Icarus* **65** 353–69.

Magalhães, J.A., Weir, A.L., Conrath, B.J. et al. (1989) Slowly moving thermal features on Jupiter. *Nature*, **337**, 444.

Pearl, J.C., Conrath, B.J., Hanel, R.A. and Pirraglia, J.A. (1990) The albedo, effective temperature, and energy balance of Uranus, as determined from Voyager IRIS data. *Icarus*, **84**, 12–28.

Peek, B.M. (1981) *The Planet Jupiter*, 2nd eds. London: Faber and Faber.

Smith, B.A. Soderblom, L.A., Beebe, R. *et al.* (1979) The Galilean satellites and Jupiter: Voyager 2 imaging science results. *Science*, **206**, 927–50.

Smith, B.A., Soderblom, L.A., Banfield, D. *et al.* (1989) Voyager 2 at Neptune: imaging science results. *Science*, **246**, 1422–49.

Sromovsky, L.A., Revercomb, H.E., Krauss, R.J. and Suomi, V.E. (1983) Voyager 2 observations of Saturn's northern mid-latitude cloud features: morphology, motions, and evolution, *J. Geophys. Res.*, **88**, 8650–66.

Williams, G.P. and Wilson, R.J. (1988) The stability and genesis of Rossby vortices. *J. Atmos. Sci.*, **45**, 207–41.

Cross references

Atmosphere
Neptune: atmosphere
Saturn: atmosphere
Spectroscopy: atmosphere
Uranus: atmosphere

JUPITER: INTERIOR STRUCTURE

Jupiter is a rather starlike planet in many respects. Because of its exceptionally large mass and high surface gravity, it has retained a much higher fraction of the volatile gases hydrogen and helium than any other solar system body beside the Sun. To a rough first approximation, Jupiter is very similar in bulk composition to the Sun itself. If it were identical in composition to the Sun, Jupiter would contain (out of a total mass equal to 318 Earth masses) about 227 Earth masses of hydrogen, about 86 Earth masses of helium and about five Earth masses of all other elements. Careful modeling of the interior of Jupiter and comparison with observational constraints indicates that the actual amount of non-hydrogen – helium material in the planet is actually about three times larger, or about 15 Earth masses in all, and interestingly, similar to the total mass of Uranus or Neptune. Thus there is evidence that the formation of Jupiter involved some 'waste' of nebular gas, with about twice as much gas escaping as was incorporated in the planet.

Evidence that Jupiter's dominant interior constituents are hydrogen and helium comes primarily from measurements of its radius, shape and gravitational moments. Although Jupiter's mean density is similar to that of the Sun, Jovian interior temperatures are nearly 1000 times lower, and as a result the hydrogen behaves not as an ideal plasma of protons and electrons, but rather as an electron-degenerate metal of pressure-ionized protons and electrons, so-called metallic hydrogen. The compressibility of metallic hydrogen is little affected by its temperature (as long as it is low enough), and thus a low-temperature sphere of hydrogen with a mass similar to Jupiter's has a characteristic radius which is nearly independent of its mass or interior temperature; that is, metallic hydrogen is so compressible that a large increase in mass produces almost no change in radius. For pure hydrogen this characteristic radius is about 80 000 km, while for a solar-composition mixture of about 27% helium and 72% hydrogen by mass, the radius decreases slightly to about 70 000 km, very close to the observed mean radius of Jupiter. According to these calculations, even an object three times smaller in mass than Jupiter should have nearly the same radius, as is seen to be the case for Saturn. Theoretical calculations have shown that materials with higher mean atomic numbers than 1, i.e. anything heavier than hydrogen, have much smaller characteristic radii than Jupiter in the Jovian mass range (values are about 35 000 km for pure helium, and about 25 000 km for rock-forming minerals). Thus there can be little doubt that Jupiter (as well as Saturn) is composed predominantly of hydrogen.

The oblate shape of Jupiter, and the corresponding quadrupole component of its external gravitational potential, are produced by the response of the planet's interior structure to its rapid rotation. With a interior rotation period of 9.92492 h, the centrifugal acceleration at Jupiter's equator is 9% of the gravitational acceleration; correspondingly Jupiter's oblateness (relative difference between equatorial and polar radius) is 6.5%, large enough to be apparent in telescopic images of the planet. Jupiter's oblateness, axial moment of inertia (the moment of inertia is $0.26\ Ma^2$, where M is the mass and a is the equatorial radius), and its quadrupole and higher-degree gravity components are in excellent agreement with its density structure computed under the assumption of the predominance of metallic hydrogen.

Although the high abundance of hydrogen and helium in Jupiter has been known at least since the work of W.C. DeMarcus in 1959, an understanding of Jupiter's interior thermal state came only with measurements of intrinsic heat flow from the interior, which were first conducted in the 1960s by F.J. Low and associates via ground-based infrared observations. After allowance is made for thermalization of incident sunlight, the current best number for Jupiter's surface intrinsic heat flux is 5.4 W m^{-2}, nearly 100 times larger than the corresponding figure for the Earth. The observed heat flow is also 100 times too large to be explicable by radioactive heat production, even if Jupiter were composed entirely of terrestrial rocks rather than mostly hydrogen and helium.

The commonly accepted explanation of Jupiter's intrinsic heat flow is that it is derived from Kelvin–Helmholtz contraction of the planet from an initially distended and much hotter state. According to this model, Jupiter formed in much the same way as the Sun, with collapse of hydrogen gas onto a central nucleus. However, because of the importance of the primordial Sun's tidal forces, a fairly massive pre-existing central object, presumably composed of non-gaseous material such as rock and ice, was required to initiate this collapse. It is estimated that this non-gaseous nucleus may have been several times more massive than the Earth. Following the collapse of the hydrogen-rich envelope onto the nucleus, which must have occurred within the first few million years of the solar system, central temperatures on the order of 100 000 K may have been reached, and the primordial Jupiter would have had a significantly larger radius than at present.

The heat stored during the early collapse of Jupiter has been radiated into space at a continually decelerating rate during the remaining evolution of the planet. The effective temperature of Jupiter (i.e. the equivalent temperature of a black body which radiates heat at the same rate as the planet) is 124 K at present, while thermal evolution calculations indicate that the effective temperature would have been 300 K when the planet was approximately 0.5 billion years old, or 4 billion years before the present. The steady cooling of the planet results in a present central temperature of approximately 23 000 K, far above the melting point of metallic hydrogen. Thus Jupiter is entirely liquid, with the possible exception of a small rocky core. Interior temperatures have always remained well below the minimum temperature for thermonuclear reactions (about 500 000 K is required for deuterium reactions). According to calculations, the interior heat flux has remained sufficiently high during the evolution of Jupiter that the liquid metallic hydrogen interior is everywhere in a state of vigorous convection, with the result that most of the planet mixes thoroughly on a timescale of the order of a century.

Considerable uncertainty exists about the details of the hydrogen phase diagram pertaining to Jovian conditions. According to theory and data from diamond anvil experiments at high pressure, at low temperatures (300 K and lower), hydrogen undergoes a phase transition from an electrically non-conducting molecular solid to a new and possibly conducting solid molecular phase at a pressure of about 1.5 Mbar. True metallic hydrogen, a phase of pressure-ionized atoms with free electrons and protons, has been theoretically predicted to appear at a pressure of about 3 Mbar. Diamond cell experiments at pressures up to 2.5 Mbar have not yet shown clear evidence for such a phase. However, recent theoretical and experimental work indicates that hydrogen may become metallic in Jupiter at temperatures above 7000 K, and at pressures above 1 Mbar. The metallic zone in Jupiter may thus commence at depths no greater than 10 000 km below the observable surface.

Thus our current picture of conditions in the Jovian interior is as follows. Starting in the hydrogen-rich atmosphere, where the pressure is about 1 bar and the temperature is similar to the effective temperature, there is a monotonic increase in pressure, temperature and density with depth following the relation for adiabatic compression of a hydrogen–helium mixture. At a depth on the order of 700 km the pressure has risen to 2 kbar, the temperature has risen to about 1600 K and the molecular hydrogen gas has become sufficiently dense that it no longer acts like an ideal gas. At deeper levels the molecular hydrogen gradually behaves more like a fluid; there is no distinct transition from gas to fluid anywhere.

The chemical composition of Jupiter's atmosphere is profoundly affected by processes occurring at much deeper levels. Available measurements of atmospheric abundances are partially understood in terms of interior models, but much remains unclear. Both interior models and atmospheric measurements show that Jupiter deviates significantly from solar composition, in the sense that the planet has more of the elements heavier than hydrogen and helium. Part of these heavy elements may comprise a dense core of perhaps five Earth masses, but apparently another ten or so Earth masses is distributed throughout the hydrogen–helium envelope. Atmospheric measurements confirm that methane is overabundant (relative to solar composition) by about a factor of five. These results may be understood as a consequence of some degree of selective loss of hydrogen and helium during the initial collapse of Jupiter's gaseous envelope.

Atmospheric measurements of some species such as carbon monoxide, germane (GeH_4) and phosphine (PH_3), are in contradiction to strict thermodynamic equilibrium in a hydrogen-rich atmosphere, but may be understood if these species are produced in the observed abundances by equilibrium chemistry at much higher temperatures at the kilobar pressure level, and then carried into the observable atmosphere by rapid vertical convection and quenched in their disequilibrium state by the lower temperatures. Other atmospheric measurements, such as of water (H_2O), indicate a large depletion relative to solar composition, while all plausible formation models of Jupiter suggest that water should be a prime component of Jupiter's enriched heavy elements. However, atmospheric abundance measurements of species which condense in Jupiter's cold observable atmosphere are highly susceptible to the effects of such condensation, especially when it occurs at levels below the region probed. For the same reason, possibly abundant refractory oxides which may condense at kilobar pressures are completely unobservable.

The hypothesized phase transition between the molecular and metallic forms of hydrogen may play a major role in determining atmospheric abundances. No attempts have been made to calculate accurately the partitioning of elements between the two phases. Theoretical calculations have only been seriously attempted for the properties of helium in metallic hydrogen, where the results indicate that a solar abundance of helium cannot be dissolved in metallic hydrogen under Jovian temperature conditions, at the lowest pressure where metallic hydrogen is stable. The exact degree of solubility of the helium depends very sensitively on the assumptions of the model, however, and great uncertainty remains.

Direct observation of the composition of Jupiter's atmosphere does not show clear evidence for the predicted depletion of helium with respect to solar composition. The helium mass fraction in the observable atmosphere is 0.24, almost precisely the same as that in the present solar atmosphere. The primordial solar helium fraction is believed to be somewhat larger, possibly as large as 0.28, so there could still be helium depletion in Jupiter's outer layers if the Sun and Jupiter underwent similar processes of helium separation over geological time, with the Jovian global helium mass fraction corresponding to the primordial solar value. Atmospheric depletion of Jupiter's neon is observed, and according to some theories this would be a concomitant of helium phase separation.

The deep circulation of Jupiter's envelope is not fully understood. Metallic hydrogen is highly opaque to thermal photons and is not a particularly good thermal conductor for a metal; thus it is not able to transport enough heat to overcome the instability of the Jovian fluid to convection. However, it is suspected by some investigators that the photon opacity may be sufficiently small in the dense molecular hydrogen fluid at kilobar pressure levels to allow the presence of a thin radiative zone, which would serve as a barrier to convection. Likewise, the molecular–metallic interface, if it exists as a discrete phase boundary, may also serve as such a barrier.

Calculation of typical convective velocities deep in Jupiter's metallic hydrogen zone, coupled with the calculated electrical conductivity of liquid metallic hydrogen, shows that Jupiter is easily able to sustain a magnetohydrodynamic dynamo. The planet's powerful magnetic field is evidently produced in this region. From estimates of the convective velocities, one may estimate the timescale for appreciable changes in Jupiter's magnetic field geometry. This timescale may be on the order of a century, and thus accurate measurements of the evolution of the field may eventually provide clues to such deep currents.

William B. Hubbard

Bibliography

Hubbard, W.B. (1984) *Planetary Interiors*. New York: Van Nostrand Reinhold.
Gehrels, T. (ed.) (1976) *Jupiter*. Tucson: University of Arizona Press.
Hubbard, W.B. (1990) Interiors of the giant planets in *the New Solar System*, 3rd ed (ed. J.K. Beatty and A. Chaikin). New York, Cambridge University Press. pp. 131–138.
Atreya, S.K., Pollack, J.B. and Matthews, M.S. (1989) *Origin and Evolution of Planetary and Satellite Atmospheres*, part IV, *Outer Planets*. Tucson: University of Arizona Press.
Stevenson, D.J. (1982) Interiors of the giant planets, *Ann. Rev. Earth and Planet. Sci.*, **10**, 257–95.

Cross references

Neptune
Neptune: atmosphere
Solar system: origin
Uranus
Uranus: atmosphere

JUPITER: MAGNETIC FIELD AND MAGNETOSPHERE

Jupiter is the largest planet in the solar system with a radius of over 70 000 km. It rotates most rapidly of all the planets with a period of only 9 h 55 min 29.7 s. It also has the largest magnetic moment (computed as the product of the equatorial surface field and the cube of the planetary radius). Consequently it also has the largest magnetosphere in the solar system, large enough to encompass easily the Sun and the visible corona. If the Jovian magnetosphere were visible from Earth, it would be bigger than the Moon in the night sky. Jupiter is also a powerful emitter of radio waves. Its giant magnetosphere acts both as a trap and an accelerator of energetic charged particles. The most energetic of the trapped electrons radiate at radio frequencies, and it was the radio frequency radiation that led in 1955 to the discovery that Jupiter had a magnetic field (Burke and Franklin, 1955). Jupiter's magnetosphere differs importantly from the Earth's magnetosphere in that its energy is predominantly derived from sources internal to the magnetosphere rather than through its interaction with the solar wind.

Planet and interior

Jupiter's interior is very different from the interiors of the terrestrial planets and may even have important differences from the interior of Saturn because of its much greater mass. The planet consists mainly of hydrogen and helium. The enormous gravitational force exerted by the planet compresses the helium and hydrogen into the liquid state and converts the hydrogen to an electrically conducting metal at depths below about 0.75 Jovian radii (R_J). It is within this electrically conducting metallic hydrogen fluid that the Jovian dynamo is generated. The energy for the dynamo consists in part of primordial heat from the formation of the planet and in part the release of gravitational energy of denser material, drops of liquid helium, settling to the center of the planet. This process is analogous to the terrestrial dynamo power source, which is believed to be the solidification of the inner core.

Magnetic field

Even before the first probes to Jupiter much was known about the Jovian magnetic field from radio measurements. The moment was correctly estimated within a factor of two and the 10° tilt of the dipole moment correctly deduced. In 1973 and 1974 Jupiter was probed by Pioneer 10 and 11 (q.v.; which passed within 2.9 and 1.6 R_J), and again in 1979 when Voyager 1 and 2 flew within 5 and 10 R_J of the center of the planet. The most information about the planetary magnetic field came from Pioneer 11 which not only passed closest to Jupiter but did so in a retrograde sense that increased the range of planetary longitudes probed. These data revealed a magnetic field rich in multiple harmonics (in comparison to that of the Earth), presumably because the Jovian dynamo source region is closer to the

Table J1 Spherical harmonic coefficients (in μT) of Jovian field

Coefficient	Degree m	Order (n) 1	2	3
g_n^m	3			−24
	2		50	19
	1	−66	−68	−16
g_o^n	0	413	−15	−15
h_n^m	1	25	−48	−53
	2		11	45
	3			−27

surface of the planet. The dipole moment was found to be 1.55×10^{20} T m^3, almost 20 000 times that of the Earth, with a tilt of 10°. Table J1 gives the spherical harmonic coefficients of Schmidt normalized Legendre polynomials averaged over the solutions of the fits to the two data sets available from Pioneer 11 (Connerney, 1981; Smith and Gulkis, 1979).

Magnetosphere

The immense size of the Jovian magnetosphere is a result of the combination of three factors: (1) the strength of the planetary magnetic field, (2) the low density of the solar wind at 5.2 AU, and (3) the rapid rotation of the planet. If the magnetosphere were a vacuum, this latter effect would not be important. However, the moon Io's volcanically derived atmosphere is lost by sputtering to the magnetosphere. The trapped radiation belt particles collide with the atmospheric particles and knock them out of Io's gravitational sphere of influence and into orbit about Jupiter. There they are ionized by charge exchange, impact ionization and photoionization and spun up into corotation with the planet by the electric field associated with the rotating magnetized Jovian plasma (see Planetary Torus). The velocities associated with this process combined with the high mass loss rate from Io are sufficient to distort the magnetic field of Jupiter into a disk, or magnetodisk as sketched in Figure J3. The centrifugal force associated with this magnetodisk stretches the magnetosphere in all directions and increases the forward radius of the magnetosphere to close to 100 Jovian radii at times. Since a Jovian radius is more than ten Earth radii, the linear dimension of the Jovian magnetosphere is about 100 times that of the Earth and its volume a million times bigger.

The sputtering process leads to an interesting feedback process because the particles sputtered, lost to Jupiter and then accelerated, are eventually energized to high energies and return to sputter again and hence maintain the level of the radiation belts. The energy for all this acceleration is derived from the rotational energy of the planet. However, this energy reservoir is so large that no significant change in rotation has occurred due to magnetospheric processes over the age of the planet.

As with other magnetospheres, both intrinsic to the planet and induced by the solar wind interaction, Jupiter has a magnetic tail extending in the antisolar direction. In concert with vast size in the forward direction, the magnetotail is of enormous dimensions in the antisolar direction, stretching (at least) all the way to Saturn's orbit, over 5 AU downstream.

The Jovian magnetosphere is very dynamic. The magnetodisk configuration is much more sensitive to the variations in the solar wind pressure than other magnetospheres, and thus the magnetosphere is constantly in motion. Deep in the interior of the magnetosphere the mass and energy injections associated with the sputtering from the volcanic atmosphere of Io is sensitive to that volcanic activity. Thus the emission of radio waves is not constant but varies with time. Like the Earth, the radiation belts are not permanently trapped on the magnetic field lines but scatter and precipitate into the atmosphere, resulting in auroral emissions from the atmosphere.

The most recent mission to Jupiter is the Galileo mission (q.v.) launched in October 1989, with an arrival in December 1995, following two terrestrial gravity assists. Galileo carries a comprehensive payload that will follow up on the discoveries of the Pioneer and Voyager missions.

Christopher T. Russell and Janet G. Luhmann

Bibliography

Burke, B.F. and Franklin, K.L. (1955,) 'Observations of variable radio source associated with the planet Jupiter.' *J. Geophys. Res.*, **60**, 213–7.
Connerney, J.E.P. (1981) The magnetic field of Jupiter: a generalized inverse approach. *J. Geophys. Res.*, **86**, 7679–93.
Engle, I.M. (1991) Idealized Voyager Jovian magnetosphere shape and field. *J. Geophy. Res.*, **96**, 7793–802.
Dessler, A.J. (ed.) (1983) Physics of the Jovian magnetosphere. New York: Cambridge University Press.
Russell, C.T. (1987) Planetary magnetism, in *Geomagnetism*, Vol. 2 (ed. J.A. Jacob). London: Academic Press, pp. 457–523.
Smith, E.J. and Gulkis, S. (1979) The magnetic field of Jupiter: a comparison of radio astronomy and spacecraft observations. *Ann. Rev. Earth Planet Sci.*, **7**, 385–415.

Cross references

Aurora, planetary
Galileo mission
Io
Magnetospheres of the outer planets
Planetary torus

Acknowledgements

This work was supported in part by the National Aeronautics and Space Administration under research grant NAGW-2573.

JUPITER: RING SYSTEM

Jupiter's ring system was discovered on 4 March 1979, in a single image taken by Voyager 1 as it passed through the planet's equatorial plane. Voyager imaging scientists had realized that this edge-on viewing geometry would be ideal for detecting a faint ring that might have previously escaped detection by Earth-based astronomers, and their hunch was correct. This particular image is shown in Figure J4.

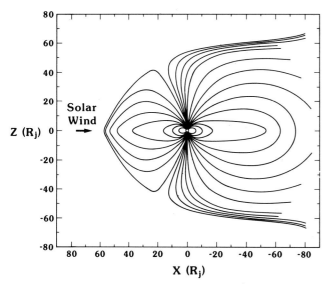

Figure J3 Cross-section of the magnetosphere of Jupiter in the noon–midnight meridian. Solid lines represent the direction of the magnetic field. The magnetic field lines point out of the planet in the northern hemisphere, opposite to the present day terrestrial field. (I.M. Engle, *J. Geophys. Res.*, **96**, p. 7793, 1991, copyright by the American Geophysical Union.)

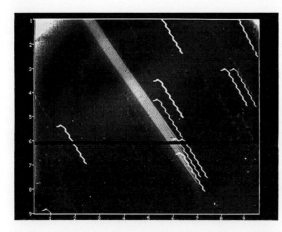

Figure J4 The Jovian ring discovery image returned by Voyager 1. The edge-on ring can be seen as a faint band of light extending from upper left to lower right. Erratic smear during this 11-min exposure accounts for the jagged star trails visible in the background.

In spite of the large and erratic smear (indicated by the jagged star trails in the background) during this 11-min exposure, a faint equatorial band of material is seen crossing the image from upper left to lower right. Subsequently, Voyager 2 was programmed to take 24 additional images of the ring with much greater clarity and from a variety of viewing geometries.

In retrospect, observations made by Pioneer 11 had also detected the ring several years earlier. A dip in the number of charged particles orbiting within the Jovian magnetosphere could be associated with absorption by ring material, and the meteoroid detector recorded several impacts during its passage through the ring plane. Nevertheless, other interpretations for these results were available at the time, so it took the Voyager 1 image to prove that the ring exists.

Compared to the Saturnian and Uranian rings known at the time, the Jovian ring was unusual in its extremely low opacity. In addition, the ring was found to brighten considerably in forward-scattered light, observed in some images taken looking back toward the planet as Voyager 2 passed through the Jovian shadow Plate J3. This brightening is the result of diffraction by particles comparable in size to the wavelength of light, and indicates that the ring contains a large population of micron-sized dust. The Jovian ring was therefore the first known example of a dusty ring, making it distinctly unlike the denser rings which contain predominantly centimeter- to meter-sized bodies. Today, we are well aware of many other dusty rings, including the D, E, F and G rings of Saturn, plus numerous dust belts interspersed among the major rings of Uranus and Neptune. Dusty rings are particularly interesting because they highlight a variety of physical processes and nongravitational forces that tend to be masked in the denser planetary rings.

Structure and particle properties

Jupiter's ring system is composed of three major components: the main ring, halo and 'gossamer' ring (Showalter *et al.*, 1985, 1987). The main ring is by far the most prominent component. However, it is only prominent in a relative sense; with an optical depth of $\sim 10^{-6}$–10^{-5}, it is nevertheless among the faintest rings known. The main ring is roughly 7000 km wide, with an abrupt outer boundary at radius 129, 130 ± 100 km from the center of Jupiter, and a far more gradual inner bound. It can be seen as the white band in Plate 22b. Two small satellites, Adrastea and Metis, are embedded within this ring. The former appears to orbit just at the ring's outer boundary, whereas the latter is embedded within a narrow, brighter belt in the ring.

The main ring's light scattering properties indicate that it contains a broad distribution of micron-sized dust grains, composed of silicates or carbonaceous material. However, the ring also seems to hold a separate population of larger bodies, perhaps meter-sized and larger, that comprise an optical depth comparable to that of the dust. These larger bodies are the ones originally detected by Voyager 1 in backscattered light, where they are fully illuminated by the Sun. They appear to be dark and red, much like the surface of Amalthea. Gradie, Thomas and Veverka (1980) have argued that sulfur contamination from Io, combined with impacts from magnetospheric ions and infalling micrometeoroids, act to darken and redden Amalthea's surface; it is perhaps not surprising that the ring, residing in the same environment, should appear to be the same color.

The halo arises at the bright ring's inner boundary and quickly expands to a full thickness of $\sim 20\,000$ km, but remains roughly symmetric about the ring plane (it is outlined in red and orange in Plate 22b). As such, it is the only ring aside from Saturn's broad E ring that has a significant vertical extent. It disappears from sight at an inner radius of $\sim 90\,000$ km, roughly halfway between the main ring and Jupiter's cloud tops.

Finally, the last Jovian ring component to be recognized is the 'gossamer' ring, so named because it is far fainter than the main ring and halo, typically by a factor of ~ 30 (Showalter *et al.*, 1985). This feature has only been seen in a single Voyager 2 image (see Plate 22b, where it is enhanced in blue and violet). This ring extends outward from the main ring and shows an almost uniform decay in brightness out to radius $\sim 210\,000$ km, well beyond the orbit of Amalthea and approaching the orbit of the tiny moon Thebe. Unlike the halo, this ring appears to be rather flat. Its only internal structure is a slight localized enhancement in brightness near 160 000 km, which may be related to the location of Jupiter's synchronous orbit. Both the halo and the gossamer ring have only been detected in forward-scattered light, suggesting that they are composed almost exclusively of micron-sized dust. However, a more detailed characterization of their particle sizes and properties is not currently possible.

Physical processes

Dust orbiting within the Jovian environment has a very limited lifetime due to a number of processes. Sputtering tends to destroy micron-sized particles on timescales of 10^2–10^4 years (Burns, Showalter and Morfill, 1984; Grün and Morfill, 1984). In the meantime, forces such as plasma drag and Poynting–Robertson drag cause orbital evolution on comparable timescales. Furthermore, the local plasma tends to charge the small grains electrically, so that perturbations from Jupiter's magnetic field can be quite important.

Because of their brief lifetimes, it seems clear that the dust grains we see in the Jovian ring must be continually replenished from some source. This source is most likely the larger bodies seen in backscattered light. In effect, the ring we see is probably in a steady state, in which new material ejected by micrometeoroid impacts into the larger parent bodies continuously replaces the dust removed or destroyed by other mechanisms. The ring system's overall morphology should then reflect, in some manner, the arrangement of the interior 'skeleton' of parent bodies, combined with the many processes defining the dust grains' subsequent evolution.

It is currently believed that most of the parent bodies in the Jovian ring reside in the outer section of the main ring; the moons Adrastea and Metis may simply be the largest of these. Dust released from their surfaces would tend to evolve inward due primarily to plasma drag, a force that arises from collisions between the ring dust and plasma ions. Since the plasma particles corotate with the magnetic field, they tend to drive dust particles away from the location of synchronous orbit, either inward or outward. At the synchronous orbit location ($\sim 160\,000$ km), dust and plasma ions orbit at the same rate so that this drag force vanishes. This model provides a natural explanation for this ring's relatively abrupt outer boundary and its more gradual inner limit.

Ring dust eventually evolves past the main ring's inner boundary and into the halo. Here the influence of Jupiter's inclined magnetic field disperses the material vertically, leading to the halo's observed breadth and thickness. However, it is perhaps surprising that the ring system's character should change so abruptly within such a limited radial domain. The key to understanding this transition seems to be Lorentz resonances. Burns *et al.* (1985; also Schaffer and Burns, 1992) were the first to note that, at the inner edge of the main band, dust grains complete exactly three orbits during every two rotations of the planet's magnetic field. The 3 : 2 resonant nature of the magnetic perturbations on a dust grain raises its orbital inclination rapidly, which likely explains the abrupt increase in the halo's vertical thickness. Furthermore, the 2 : 1 Lorentz resonance, where dust grains orbit twice per planetary rotation, falls very close to the halo's inner edge. It appears that the halo grains receive another large vertical perturbation as their orbits decay through this resonance, which disperses them so far that they enter the planetary atmosphere and are lost from the system. The concept of Lorentz resonances introduced here is very analogous to the gravitational resonances induced by external moons on the rings of Saturn, which can give rise to localized waves and gaps.

The gossamer ring remains the most puzzling of the ring system's components. Since the plasma drag force is inward in the main ring, it appears that the gossamer ring must arise from another source, perhaps an unseen band of parent bodies scattered outside the main ring. Since the ring straddles the location of synchronous orbit, it appears that the parent bodies must also straddle this location. The ring's one feature that may make some sense is the slight enhancement in brightness near the synchronous orbit; since the plasma drag force vanishes here, this would be the one location where dust could survive for a significant period of time. Hence, the brightness enhancement probably just reflects a local reduction in the strength of the key dust removal mechanism.

Future observations

In the last decade, technological innovations in Earth-based astronomy have made it substantially easier to detect the Jovian ring. Using a sophisticated infrared CCD, Nicholson and Matthews (1991) were able to detect the Jovian ring from the ground with a clarity beginning to approach that of the Voyager images. Nevertheless, the range of viewing and lighting geometries accessible by Earth-based observers is very limited compared to those achievable by a spacecraft *in situ*. The Galileo Orbiter will provide the next opportunity for a more detailed scrutiny of the Jovian ring.

Mark R. Showalter

Bibliography

Burns, J.A., Showalter, M.R. and Morfill, G.E. (1984) The ethereal rings of Jupiter and Saturn, in *Planetary Rings* (eds R. Greenberg and A. Brahic). Tucson: University of Arizona Press, pp. 200–72.
Burns, J.A., Schaffer, L.E., Greenberg, R.J. and Showalter, M.R. (1985) Lorentz resonances and the structure of the Jovian ring. *Nature*, **316**, 115–9.
Gradie, J., Thomas, P. and Veverka, J. (1980) The surface composition of Amalthea. *Icarus*, **44**, 373–87.

Grün, E. and Morfill, G.E. (1984) Dust–magnetosphere interactions, in *Planetary Rings* (eds R. Greenberg and A. Brahic). Tucson: University of Arizona Press, pp. 275–332.
Nicholson, P.D., and K. Matthews, (1991) Near-infrared observations of the Jovian ring and small satellites. *Icarus*, **93**, 331–46.
Schaffer, L. and Burns, J.A. (1992) Lorentz resonances and the vertical structure of dusty rings: analytical and numerical results. *Icarus*, **96**, 65–84.
Showalter, M.R., Burns, J.A., Cuzzi, J.N. and Pollack, J.B. (1985) Discovery of Jupiter's 'gossamer' ring. *Nature*, **316**, 526–8.
Showalter, M.R., Burns, J.A., Cuzzi, J.N. and Pollack, J.B. (1987) Jupiter's ring system: new results on structure and particle properties. *Icarus*, **69**, 458–98.

Cross references

Neptune: ring system
Planetary ring
Saturn: ring system
Uranus: ring system

JUPITER: SATELLITE SYSTEM

Jupiter has the richest satellite system of any planet but Saturn. Jupiter's 16 known moons conveniently divide into four quartets according to their orbits, listed in Table J2, and their physical characteristics, given in Table J3.

Orbits and physical properties

The inner satellites Metis (J XVI), Adrastea (J XV), Amalthea (J V) and Thebe (J XIV) orbit between $\sim$ 1.8 and 3.1 Jupiter radii from the center of the planet, with eccentricities less than 0.02 and inclinations less than 1° from Jupiter's equatorial plane. They are small and irregular in shape, with mean radii between $\sim$ 10 and 94 km. Metis and Adrastea are associated with Jupiter's ring system (q.v.) Like the Earth's Moon, Amalthea is in synchronous rotation, where its long axis always points toward Jupiter while its short axis stays perpendicular to its orbital plane. Presumably all four inner satellites are in the synchronous state, so that their spin periods equal their orbital periods. They are all quite dark, with albedos from 0.05 to 0.10. The colors of J XV and J XVI are unknown, but J V and J XIV are extremely red, possibly due to contamination by sulfur from J I (Io) (Thomas and Veverka, 1982; Pascu *et al.*, 1990).

The Galilean satellites Io (J I; Plates 23, 24), Europa (J II; Plate 24), Ganymede (J III; Plate 25), and Callisto (J IV; Plate 25) orbit between 5.9 and 26.7 Jupiter radii, with eccentricities of 0.01 or less, and inclinations of less than 0.5° relative to Jupiter's equatorial

Table J2 Orbits of Jupiter's satellites

Name	Number	Semimajor axis (km)	Semimajor axis (R_J)	Orbital period (d)	Eccentricity	Inclination (deg)
Metis	XVI	127 969	1.7922	0.2948	< 0.004	$\sim$ 0
Adrastea	XV	128 980	1.8065	0.2983	$\sim$ 0	$\sim$ 0
Amalthea	V	181 300	2.539	0.4981	0.003	0.40
Thebe	XIV	221 900	3.108	0.6745	$\sim$ 0.015	$\sim$ 0.8
Io	I	421 600	5.905	1.769	0.0041	0.040
Europa	II	670 900	9.397	3.551	0.0101	0.470
Ganymede	III	1 070 000	14.99	7.155	0.0015	0.195
Callisto	IV	1 883 000	26.37	16.69	0.007	0.281
Leda	XIII	11 094 000	155.4	238.7	0.15	27
Himalia	VI	11 480 000	160.8	250.6	0.16	28
Lysithea	X	11 720 000	164.2	259.2	0.11	29
Elara	VII	11 737 000	164.4	259.7	0.21	28
Ananke	XII	21 200 000	297	631	$\sim$ 0.17	$\sim$ 147
Carme	XI	22 600 000	317	692	$\sim$ 0.21	$\sim$ 163
Pasiphae	VIII	23 500 000	329	735	$\sim$ 0.4	$\sim$ 148
Sinope	IX	23 700 000	332	758	$\sim$ 0.3	$\sim$ 153

Table J3 Physical characteristics of Jupiter's satellites

Name	Number	Radius (km)	Albedo	Lightcurve amplitude (%)	Rotation period (h)
Metis	XVI	? × 20 × 20	0.05–0.10	?	7.075?
Adrastea	XV	12.5 × 10 × 7.5	0.05–0.10	~ 55	7.159?
Amalthea	V	135 × 82 × 75	~ 0.06	~ 64	11.95
Thebe	XIV	? × 55 × 45	0.05–0.10	> 22	16.19?
Io	I	1815	0.6	–	42.46
Europa	II	1569	0.6	–	85.22
Ganymede	III	2631	0.4	–	171.72
Callisto	IV	2400	0.2	–	400.54
Leda	XIII	8?	?	?	?
Himalia	VI	~ 90	~ 0.03	12	9.5
Lysithea	X	19?	?	28	12.8
Elara	VII	~ 40	~ 0.03	?	?
Ananke	XII	15?	?	27	8.3
Carme	XI	21?	?	22	10.4
Pasiphae	VIII	34?	?	?	?
Sinope	IX	20?	?	?	?

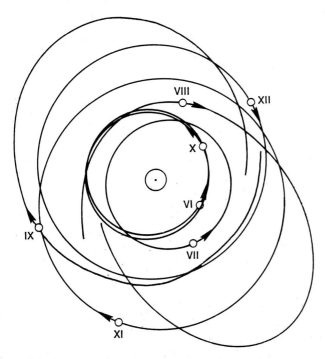

Figure J5 Orbits of Jupiter's outer satellites (except J XIII). For comparison, the dot represents Jupiter, while the small circle displays the orbit of Callisto (J IV), the outermost of the Galilean satellites. (Adapted from Wetterer, 1971.)

plane. They are all large and nearly spherical, with radii around 2000 km. They are also bright, with albedos from 0.2 to 0.6, and in synchronous rotation. J I, J II and J III are linked by the Laplace resonance, which maintains their orbital periods in a 1 : 2 : 4 commensurability. (See Io; Europa; Ganymede; Callisto.)

The prograde outer satellites Leda (J XIII), Himalia (J VI), Lysithea (J X) and Elara (J VII) orbit at mean distances between ~ 155 and ~ 164 Jupiter radii, while the retrograde moons Ananke (J XII), Carme (J XI), Pasiphae (J VIII) and Sinope (J IX) orbit at mean distances from ~ 297 up to ~ 332 radii (Figure J5). Both groups have relatively high eccentricities from 0.15 to 0.38, but the prograde swarm all have inclinations of 27° to 29° from the plane of Jupiter's orbit, while the retrograde swarm have inclinations of 147° to 163° (33° to 17° retrograde) from Jupiter's orbital plane. As Figure J5 indicates, the retrograde satellites are so strongly perturbed by the Sun that their orbits do not even close.

Himalia (J VI) and Elara (J VII) are again small and dark, with radii of ~ 90 km and ~ 40 km, respectively, and albedos of ~ 0.03. The sizes and albedos of the others are not known, but their estimated radii range from 8 km to 34 km. J VI, J VII, J XI and J XII vary in brightness with periods of ~ 10 h, suggesting that they are spinning rapidly and are irregular in shape (Degewij, Andersson and Zellner, 1980; Luu, 1991). Presumably the others are also variable, but the observations of J X and J XIII are inconclusive, while those of J VIII and J IX are inconsistent (Degewij and van Houten, 1979; Tholen and Zellner, 1984; Nakamura and Yasuda, 1989; Luu, 1991). The spectra of J XII and J XIII are unknown, but J VI, J VII, J VIII and J X have reddish spectra like C-type asteroids, while J IX is nearly as red as J I, J V and J XIV (Tholen and Zellner, 1984). In contrast, J XI appears to be bright in the ultraviolet, which Tholen and Zellner (1984) attribute to a subsurface layer of ice uncovered by a recent impact.

Origins

Most theories on the origin of satellite systems suggest that Jupiter's inner and Galilean satellites, like the regular satellites of most planets, accreted *in situ* from a protoplanetary disk surrounding the primary (Pollack and Fanale, 1982). This scenario has difficulty, however, in accounting for the orbits of irregular satellites, such as J VI–J XIII, S IX (Phoebe) and N II (Nereid). It may be plausible to explain Jupiter's prograde outer moons as remnants from the fringes of the proto-Jovian nebula, but this hypothesis is hard-pressed to accommodate the retrograde quartet. In fact, all of Jupiter's outer satellites have long been considered captured bodies (see Capture mechanism), possibly related to comets or the Trojan group of asteroids (q.v.) Although their colors and albedos do resemble those of distant asteroids (Cruikshank, Degewij and Zellner, 1986), their size distribution differs from that of Trojans and other asteroids (Gehrels, 1976, 1977).

The ages of Jupiter's outer satellites present another difficulty. Since their orbits cross (Figure J5), these objects can collide with one another at speeds of several kilometers per second. Kessler (1981) has calculated the mean time between destructive collisions among the retrograde quartet as 270×10^9 years, but more recent estimates of their sizes allow us to revise this lifetime to ~ 43×10^9 years. Even this reduced timescale is almost ten times the age of the solar system, and presents no problem. However, the collisional lifetime of the prograde outer swarm is a rather firm 0.9×10^9 years (Kessler, 1981), only a fifth of the age of the solar system. Furthermore, the above timescale is only an upper limit, because the figure implies that prograde and retrograde satellites can collide head-on at speeds around 10 km s^{-1}! This calculation suggests that Jupiter's outer satellite system is not primordial, but of relatively recent origin.

Anthony R. Dobrovolskis

Bibliography

Cruikshank, D.P., Degewij, J. and Zellner, B. (1986) The outer satellites of Jupiter, in *Satellites of Jupiter* (ed. D. Morrison). Tucson: University of Arizona Press. pp. 129–146.

Degewij, J. and van Houten, C.J. (1979) Distant asteroids and outer Jovian satellites, in *Asteroids* (ed. T. Gehrels). Tucson: University of Arizona Press, pp. 417–435.

Degewij, J., Andersson, L.E. and Zellner, B. (1980) Photometric properties of outer planetary satellites. *Icarus*, **44**, 520–40.

Gehrels, T. (1976) Discussion, in *Jupiter* (ed. T. Gehrels). Tucson: University of Arizona Press. p. 130.

Gehrels, T. (1977) Some interrelationships of asteroids, Trojans, and satellites, in *Comets Asteroids, Meteorites: Interrelations, Evolution and Origin* (ed A.H. Delsemme). Toledo, Ohio: University of Toledo Press, pp. 323–5.

Kessler, D.J. (1981) Derivation of the collision probability between orbiting objects: the lifetimes of Jupiter's outer moons. *Icarus*, **48**, 39–48.

Luu, J. (1991) CCD photometry and spectroscopy of the outer Jovian satellites. *Astron. J.*, **102**, 1213–25.

Nakamura, T. and Yasuda, N. (1989) CCD photometry of the outer Jovian satellites and their origin, in *Proc. 22nd ISAS Lunar and Planet. Symp.* (eds H. Mizutani, H. Oya and M. Shimizu). Tokyo: Institute of Space and Astronautical Science, pp. 57–61.

Pascu, D., Panossian, S.P. and Schmidt, R.E. (1990) B, V photometry of Thebe (JXIV). *Bull. Am. Astron. Soc.* **22**, 1117.

Pollack, J.B. and Fanale, F. (1982) Origin and evolution of the Jupiter satellite system, in *Satellites of Jupiter* (ed. D. Morrison). Tucson: University of Arizona Press. pp. 872–910.

Tholen, D.J. and Zellner, B. (1984) Multicolor photometry of outer Jovian satellites. *Icarus*, **58**, 246–53.

Thomas, P. and Veverka, J. (1982) Amalthea, in *Satellites of Jupiter* (ed. D. Morrison). Tucson: University of Arizona Press, pp. 147–73.

Wetterer, M.K. (1971) *The Moons of Jupiter*. New York: Simon and Schuster.

Cross references

Callisto
Europa
Galilean satellites
Ganymede
Io

K

KEPLER, JOHANNES (1571–1630)

A founder of modern astronomy, Kepler was born in Weil (Germany), and became a follower of Copernicus both in his vision of physical reality and his support of the heliocentric theory. Kepler studied theology, philosophy, and mathematics at the University of Tübingen and planned to pursue a position in religious life. Instead, however, he accepted the chair in mathematics and astronomy in the Protestant Seminary at Graz (Austria). In 1596 Kepler published his *Mysterium Cosmographicum*, in which he supported the Copernican cosmology and elaborated on his ideas on the structure of the solar system. Kepler viewed the 'Universe' (in his sense, i.e. our 'Solar System') as being governed by geometric relationships that conformed to spheres inscribed in and circumscribed around the five regular polygons, the tetrahedron, the cube, the octahedron, the dodecahedron and the icosahedron. These relationships determined the distances of the planets from the Sun. An astronomer, he was also a mystic and astrologer, believing in the 'music of the spheres' and the harmony of the Universe. In 1595 he issued a calendar – predicting a bitterly cold winter, peasant uprisings and a Turkish invasion – all of which came to pass, greatly enhancing his reputation. Nevertheless, he admitted that astrological forecasts were a matter of luck, although they made a valuable contribution to his income.

In 1600 Kepler was forced to leave his chair because the onset of the Counter-Reformation in Austria and the religious persecution (exiling) of all Protestants from Graz. He accepted an invitation from Tycho Brahe (1546–1601) to work with him in Prague. Brahe died the following year and Kepler was appointed his successor as Imperial Mathematician and continued working out the orbit of Mars using observations made by Brahe. Abandoning the ancient belief that the planets move in perfect circles, Kepler adopted the hypothesis that the planets were moved by the Sun, each with a 'force' inversely proportional to its solar distance, so that the velocity component at right angles to the radius also varied inversely as the solar distance. He combined this hypothesis with the assumption of a sinusoidal oscillation of the planet on a radius vector to arrive at what are now generally referred to as 'Kepler's first two laws of planetary motion'. He proved that the orbit of Mars is an ellipse, with the Sun occupying one of its two foci, thereby grasping the true implications of conics for celestial mechanics. This work superseded the circular orbits of traditional and Copernican cosmology. Kepler's first two laws of planetary motion are:

1. the orbits of the planets are ellipses, with the Sun at one of the foci of the ellipse; and
2. a line joining a planet and the Sun sweeps over equal areas during equal intervals of time.

Kepler announced these laws in his *Astronomia Nova* in 1609, with due credit given to Tycho Brahe's observations. Taken together, they yielded a more accurate representation of planetary motion than any earlier theory. This volume bears the subtitle *Physica Coelestis*, which might imply Kepler's search for one universal force law to explain the motions of Mars and other planets. Almost 10 years later he propounded a third law in his *Harmonice Mundi*, that is

3. the squares of the periods of revolution of any two planets are proportional to the cubes of their mean distances from the Sun.

Kepler's laws, revealing the harmonic motion of the planets around the Sun, laid the foundation for Newton's theory of universal gravitation. Whereas Copernicus had achieved a geometric rearrangement, Kepler was the first revolutionary astronomer, introducing physics into the celestial sphere.

Kepler published in 1604 a treatise on optics and astronomical refraction that was used as standard work at that time, and elabor-

Figure K1 Johannes Kepler. Reproduced by permission of American Institute of Physics, Emilio Segre Visual Archives.

ated ray optics as applied to telescope lenses in his *Dioptrice* (1611). His *De Stella Nova* (1606) presented information about and gave interpretation on the appearance of a (super)nova in 1604. His *Epitome Astronomiae Copernicanae* (1618–1621) was used as a standard textbook in astronomy over a long period.

After the publication of the three Keplerian planetary laws, Kepler proceeded to complete and publish, in 1627, the *Rudolphine Tables*, in which he provided tables in aid of calculating of the motions of the Sun, Moon and all the known planets, on the basis of his first two laws and Tycho Brahe's observation of planetary motion.

Hans J. Haubold

Bibliography

Beer, A. (1973) Kepler's astrology and mysticism. *Vistas in Astron.*, **16**, 399–426.
Caspar, M. (1959) *Johannes Kepler*. New York: Abelard-Schuman.
Dreyer, J.L.E. (1906) *A History of Astronomy from Thales to Kepler*. Cambridge: University Press (Dover Publications Inc., 1953).
Field, J. (1988) *Kepler's Geometrical Cosmology*. Chicago: The University of Chicago Press, and London: The Athlone Press.
Frisch, J. (ed.) (1858–1871) *Joannis Kepleri Astronomi Opera Omnia*. Frankfurt-Erlangen (8 vols); reprint: Hildesheim, 1971.
Gingerich, O. (1973) Kepler, Johannes. *Dict. Sci. Biogr.*, Vol. 7, pp. 289–312.
Holton, G. (1988) *Thematic Origins of Scientific Thought – Kepler to Einstein*. Cambridge: Harvard University Press.
Kuhn, T.S. (1957) *The Copernican Revolution*. Cambridge: Harvard University Press.
Stephenson, B. 1987. *Kepler's Physical Astronomy*. Berlin: Springer-Verlag.
Weaver, J.H. (ed.) (1987) *The World of Physics*. New York: Simon and Schuster (Vols I to III).
Wilson, C. (1972) How did Kepler discover his first two laws? *Sci. Am.*, **226**, 92–106.

KEPLER'S LAWS

Kepler's laws govern the motion of the planets in their orbits around the Sun. They are stated as follows.

- First law. The planets move in ellipses with the Sun at one focus.
- Second law. The line joining the Sun and a planet sweeps out equal areas in equal intervals of time (law of areas).
- Third law. The square of the period of revolution P of a planet is proportional to the cube of the semimajor axis of the orbit a. For all planets the ratio P^2/a^3 is equal to a constant (harmonic law).

Kepler's laws are satisfied precisely only when the mutual perturbations of the motions of the planets by the others are neglected.

Johannes Kepler (1571–1630; q.v.) announced the discovery of the first and second law in his *Astronomia Nova* (1609); the third law he propounded in his *Harmonices Mundi* in 1619.

Newton (q.v.) showed that the gravitational force between Earth and Moon and the gravitational attraction of the surface of the Earth were related by an inverse square law of force. If m_1 is the mass of the Earth, assumed to be spherically symmetrical with radius r_1, then the force exerted by the Earth on a small mass m_2 near the Earth is given by

$$F = G \frac{m_1 m_2}{r_1^2} \tag{K1}$$

where G is the gravitational constant, and for the acceleration of gravity on the Earth's surface, g, holds

$$g = G \frac{m_1}{r_1^2} \tag{K2}$$

Let m_2 be the mass of the Moon, a its mean distance from the Earth, and P the period of revolution of the Moon around Earth. Then Kepler's third law of motion applied to the Earth–Moon system, neglecting any perturbation by the Sun and other planets, leads to

$$\frac{a^3}{P^2} = G \frac{m_1 + m_2}{4\pi^2} \tag{K3}$$

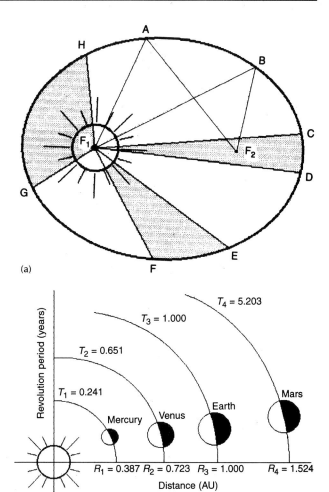

Figure K2 Schematic representation of the three Kepler laws of planetary motion. (a) Planets follow elliptical orbits with the Sun in one of the foci: $F_1 \times A \times t \times A \times F_2 = F_1 \times B \times t \times B \times F_2$. The radius vector connecting the Sun and the planet sweeps over equal areas of the planetary orbit in equal intervals of time: $C \times D \times F_1 = E \times F \times F_1$. (b) The squares of the periods of revolution of different planets around the Sun stand in the same ratio as the cubes of their mean distances from the Sun: $(T_i/T_j)^2 = (R_i/R_j)^3$.

Equation (K3) gives the constant of the harmonic law based on Newton's law of gravitation. Using equations (K2) and (K3) the acceleration of gravity is determined to be

$$g = 4\pi^2 \frac{m_1}{m_1 + m_2} \left(\frac{a}{r_1}\right)^2 \frac{a}{P^2} \tag{K4}$$

When Newton dealt with the problem of planetary motion applying his law of universal gravity, he obtained a fairly general equation for the motion of a planet around the Sun, describing the position of the planet given in terms of the distance r and the angle ϑ which the radius vector to the planet makes with the major axis,

$$r = \frac{p}{1 - e \cos\vartheta} \tag{K5}$$

where $p = a(1 - e^2)$, containing the semimajor axis a of the ellipse and its eccentricity e. Equation (K5) is the polar coordinate equation of any conic section depending on the appropriate value of e. For $e = 0$, equation (K5) describes a circle; for $e = 1$ it is the equation of a parabola; and for $e \geq 1$ it represents the equation of a hyperbola. In fact, Newton's law of universal gravitation gives a more general first law of motion than does Kepler's law, since it shows that any conic section is a permissible orbit for a celestial body. Equation (K4) has been used, taking into account the Earth's oblateness and perturbations due to the Sun, to compute the distance of the Moon from Earth. These results have been superseded only by radar measurements.

Observed perturbations in the motion of the planet Uranus, first seen in 1781 by W. Herschel (q.v.), enabled J.C. Adams in 1845 and U.J. Leverrier (q.v.) in 1846 independently to predict the existence of the until then unobserved planet Neptune. Applying Newton's theory, they both calculated the position of Neptune, confirming the agreement of theory and observation. In 1930 C. Tombaugh discovered the planet Pluto by observation. Afterwards its position and orbit could be determined by applying the same method used to predict Neptune's existence, based on Newton's theory.

F.W. Bessel (1838) observed non-uniform proper motions of the stars Sirius and Procyon and conjectured that both were gravitationally deflected by unseen bodies. Much later, observations confirmed that Sirius and Procyon have white dwarfs as companions.

Elaborating on the application of Newtonian gravitation to planetary motion, U.J. Leverrier (1859) discovered a discrepancy in the orbit of the planet Mercury. Because of the disturbance by other planets, the perihelion of Mercury's orbit advances. Taking into account all hitherto known gravitational effects still left an observed motion of about 43 seconds of arc per century not accounted for by Newton's theory. This discrepancy was unexplained before A. Einstein invented his general theory of relativity in 1916.

Hans J. Haubold

Bibliography

Misner, C.W., Thorne, K.S. and Wheeler, J.A. (1973) *Gravitation*. New York: W.H. Freeman and Company.
Motz, L. and Duveen, A. (1977) *Essentials of Astronomy*. New York: Columbia University Press.
Moulton, F.R. (1914) *An Introduction to Celestial Mechanics*. New York: Dover Publications, Inc.
Smart, W.M. (1953) *Celestial Mechanics*. London, New York, Toronto: Longmans, Green and Company.

KIRCHHOFF, GUSTAV ROBERT (1824–1887), AND KIRCHHOFF'S LAW

A professor in Breslau, Heidelberg and later, Berlin, Germany, Kirchoff is remembered for his many contributions to physics. At the age of twenty, he solved the problem of calculating the distribution of electrical currents in an arbitrary network, called today Kirchhoff's laws for electrical circuits. These laws constitute a generalization of Ohm's law and the law of conservation of charge.

In 1850, when Kirchhoff was an appointed professor at Breslau, Germany, he met the chemist Robert Wilhelm Bunsen (1811–1899) and both managed to obtain appointments at the University of Heidelberg in 1859. Their joint research work there revealed the relation between absorption and emission of light in general and opened the way for spectral analysis, leading to the detection of chemical elements on the Earth and in the atmospheres of the Sun and the stars. The latter accomplishment signaled the birth of observational astrophysics.

A luminous solid or liquid generates a continuous spectrum (as observed by Newton), but in a hot gas, such as found at the Sun's surface, bright lines are seen in the spectrum that correspond to the components present. Dark lines in that spectrum, such as first identified by the Munich optician Joseph Fraunhofer (1787–1826) and now known as Fraunhofer lines (q.v.), are explained by sunlight, emanating from the hot interior, which passes through cooler gases that absorb light in the wavelengths corresponding to the dark lines.

The physical problem of the relation between emission and absorption was explained by Kirchhoff employing thermodynamics. If the power emitted per unit surface in a unit frequency interval by any body is denoted by e and the absorbed fraction of power incident on a body in a unit frequency interval is a, Kirchhoff showed that the ratio of e to a is a function of the frequency v and the temperature T, independent of the nature of the body. For a body perfectly absorbing all radiations, $a = 1$ and $e(v, T)$ is a universal function independent of the nature of the body, that is a black body, which constitutes Kirchhoff's fundamental law of electromagnetic radiation.

In 1871, Helmholtz invited Kirchhoff to Berlin as a professor of theoretical physics, where he taught a famous course in this field which became a standard for many years.

Following Rosenfeld (1973) 'Kirchhoff's law was the key to the whole thermodynamics of radiation. In the hands of Planck, Kirchhoff's successor to the Berlin chair, the study of black body radiation proved to be the key to the new world of the quanta, well beyond Kirchhoff's conceptual horizon.'

Rhodes W. Fairbridge and Hans J. Haubold

Bibliography

Rosenfeld, L. (1973) Kirchhoff, Gustav Robert. *Dict. Sci. Biogr.*, Vol. 7, pp. 379–82.

KIRKWOOD, DANIEL (1814–1895)

An American astronomer of Scots–Irish descent, born on a farm in Maryland, Kirkwood was not interested in farming but, with an aptitude for mathematics, taught it at schools and colleges until he finally became professor of mathematics at Indiana University in 1867. He remained there until his retirement in 1886, when he moved to California and lectured in astronomy at Stanford.

Kirkwood's name has long been associated with his discovery of the 'gaps' in the distribution of the asteroids with respect to their distance from the Sun (in 1857, but it was not published until 1866). Asteroids in resonance with Mars were frequent but none were orbiting in resonant states with Jupiter. Further discontinuities were noted at distances corresponding to certain fractions of the period of Jupiter (1 : 3, 2 : 5, and 2 : 7). Later he found gaps at 1 : 2, 3 : 5, 4 : 7, 5 : 8, 3 : 7, 5 : 9, 7 : 11 and 4 : 9 (see Asteroid: resonance).

The explanation offered by Kirkwood was that an early (continuous) ring of asteroids or other particles around Jupiter would rapidly dissipate at resonance distances, the lowest order first. A crowding of particles ensued and resultant collisions would set many on eventual collision courses with the Sun. Nevertheless, he later discovered that some asteroids (at large distances) did in fact exist, notably at periods of 2 : 3 (the Hilda group) and at 3 : 4 of Jupiter's period. His name has subsequently been given to an asteroid in the Hilda group (no. 1578).

In connection with Saturn's rings, Kirkwood noted that 'planets and comets have not formed from rings, but rings from planets and comets'. He suspected that there was a relationship between asteroids and comets. He was puzzled, however, that the orbits of the major planets should be so widely separated if in fact they had originated in the Sun, according to the Laplace nebular hypothesis. Eventually he came to reject it.

Rhodes W. Fairbridge

Bibliography

Brouwer, D. (1963) The problem of the Kirkwood gaps in the asteroid belt. *Astron. J.*, **68**, 152–159.
Kirkwood, D. (1860) Instances of nearly commensurable periods in the Solar System. *Math. Monthly*, **2**, 126–32.
Kirkwood, D. (1869) On the nebular hypothesis, and the approximate commensurability of the planetary periods. *Mon. Not. Roy. Astron. Soc.*, **29**, 96–102.
Kirkwood, D. (1871) On the periodicity of the solar spots. *Proc. Am. Phil. Soc.*, **11**, 94–101.
Kirkwood, D. (1873) *Comets and Meteors*. Philadelphia.
Kirkwood, D. (1888) *The Asteroids*. Philadelphia.
Marsden, G.B. (1973) Kirkwood, Daniel. *Dict. Sci. Biogr.*, **7**, 384–7.
Swain, J. (1901) Daniel Kirkwood. *Publ. Astron. Soc. Pacific*, **13**, 140–7.

Cross references

Asteroid
Asteroid: resonance
Resonance in the solar system

KUIPER BELT

In 1951 the American astronomer, Gerard Kuiper (1905–1973; q.v.), pointed out the unlikelihood that the disk of material from which the solar system formed should abruptly end at the orbit of the last giant planet, Neptune. He suggested that there were a significant number of small bodies, the remnants of the primordial planetary disk, in orbit about the Sun outside the region of the planets. Kuiper claimed that these objects were cometary nuclei and suggested that gravitational scattering by Neptune might have ejected them from the ecliptic plane into the Oort cloud. This flat, disk-shaped region extending roughly from about 30 up to 1000 astronomical units (AU) is called the Kuiper belt. It is estimated to contain between 10^8 and 10^9 comets with a total mass less than one-tenth of that of Earth.

Fernández (1980) suggested that the Kuiper belt is a source of the observed short-period comets. Strong theoretical arguments have been presented that support this idea by Duncan, Quinn and Tremaine (1988), who performed numerical simulations of the dynamical evolution of Kuiper belt comets. Previously, short-period period comets were thought to originate as long-period comets from the Oort cloud that were captured by the gravitational scattering of the planets. However, modern studies have shown that it is not possible to reproduce the low-inclination distribution of short-period comets from the spherical distribution of the long-period comets and that the efficiency of capture is too low by several orders of magnitude. Furthermore, a significant fraction of objects in low-inclination, Neptune-crossing orbits near the plane of the ecliptic will eventually evolve into a population consistent with the observed short-period comets. Subsequent investigations have found stable regions where cometary orbits can exist over the age of the solar system in the Kuiper belt. However, the inner Oort cloud cannot be ruled out as another possible source of short-period comets (Bailey 1986).

Our picture of the Kuiper belt and its relation to comets is evolving rapidly. Strong observational support of the existence of the Kuiper belt came in late 1992 and early 1993 with the discovery of two small objects, designated 1992QB1 and 1993FW, found beyond the orbit of Pluto at a heliocentric distance of about 50 AU (Jewitt and Luu, 1993). As of March 1994, a total of eight Kuiper belt candidates have been detected between 30 and 50 AU. Their diameters are estimated to range between 100 to 200 km, assuming typical cometary surface properties. These exciting developments will undoubtedly lead to clarifications over the next few years in this active area of cometary research.

Daniel C. Boice

Bibliography

Bailey, M.E. (1986) The near-parabolic flux and the origin of short-period comets. *Nature*, **324**, 350–2.
Duncan, M., Quinn, T. and Tremaine, S. (1988) The origin of short-period comets. *Astrophys. J.*, **328**, L69–73.
Fernández, J.A. (1980) On the existence of a comet belt beyond Neptune. *Mon. Not. Roy. Astron. Soc.*, **192**, 481–91.
Jewitt, D. and Luu, J. (1993) Discovery of the candidate Kuiper belt object 1992 QB$_1$. *Nature*, **362**, 730–32.
Kuiper, G.P. (1951) On the origin of the solar system, in *Astrophysics*, J.A. (ed. J.A. Hynek). New York; McGraw-Hill, pp. 357–424.

Cross references

Comet
Comet: dynamics
Comet: origin and evolution
Solar system: origin

KUIPER, GERARD PETER (1905–1973)

One of the foremost astronomers of the 20th century, Kuiper, more than anyone else, was responsible for restoring solar system astronomy from an almost moribund science to one of respectable stature. His selection by NASA as Principal Investigator for the Ranger program (q.v.) of moonshots – the first successful missions by the USA to another solar system body – reflects the esteem in which he was held. As one recognition of this and many other achievements in planetary sciences, his name is the only one that designates a topographical feature on each of three separate solar system bodies – the Moon, Mercury and Mars.

Kuiper was born in the Netherlands, and completed his formal education at Leiden University. Arriving in the USA from the Netherlands in 1933, his primary objective was to research the nature and origin of the solar system, and as a prelude to this he did pioneer work in the field of binary stars and the mass–luminosity relation for main-sequence stars. In 1943, using infrared sensitive plates in conjunction with the 82-in (208 cm) McDonald reflecting telescope in SW Texas, he discovered that Titan, (q.v.) Saturn's largest satellite, possessed an atmosphere of methane gas.

Following an absence for military service, he turned his full attention to planetary observation and research, using the newly available Cashman infrared detectors to obtain improved IR spectra. This and other lines of investigation led to a whole string of new discoveries and interpretations ranging from the detection of carbon dioxide on Mars, the discovery of a new satellite for each of the planets Uranus and Neptune, and improved measures of the diameters of Neptune and Pluto, to papers on the origin of the asteroids, of the solar system in general, of planetary atmospheres and various other related topics. The existence of a zone of comets beyond Neptune but separate from the Oort cloud, postulated by Kuiper in 1951, has recently received powerful support.

Kuiper next turned his attention to the Moon, the only solar system object close enough to reveal its surface topography and other features in comparatively fine detail. In 1958 he inaugurated the production of various lunar atlases, maps and a comprehensive crater catalogue, which led to their recognition by the newly formed NASA as essential items for the various planned lunar missions. In 1960 he founded the Lunar and Planetary Laboratory at the University of Arizona, now the largest academic institution in the USA dedicated to planetary science.

Kuiper's enquiring nature led him to probe all aspects of solar system astronomy, from the evolution of planetary atmospheres to the problems of lunar nomenclature, from the origin and nature of comets to the optimum locations for large planetary telescopes. The major installations in Chile and especially on Mauna Kea in Hawaii stem from his pioneering tests of astronomical observation at these locations. In attempts to detect small amounts of water vapor in planetary atmospheres, he inaugurated IR spectral observations from high-flying aircraft. NASA's C-141 aircraft fitted with a 36-in (91-cm) reflecting telescope for just such researches is named the Gerard P. Kuiper Airborne Observatory in recognition of his pioneering work in this field.

Ewen A. Whitaker

Bibliography

Cruikshank, D.P. (1974) Twentieth century astronomer. *Sky and Telescope* **47**(2), 159–64.
Cruikshank, D.P. (1993) Gerard Peter Kuiper. *National Academy of Sciences, Biographical Memoirs*. (Includes bibliography of all of Kuiper's publications.)
Kuiper, G.P. (1962) Organization and programs of the Laboratory. *Commun. Lunar Planetary Laboratory*, **1**(1), 1–20.
Owen, T. and Sagan, C. (1974) Obituary: planetary astronomer, Gerard Peter Kuiper, 1905–1973. *Mercury*, **3**(1), 16–18.
Sagan, C. (1974) Obituary: Gerard Peter Kuiper, 1905–1973. *Icarus*, **22**, 117–8.
Tatarewicz, J.N. (1990) *Space Technology and Planetary Astronomy*. Bloomington: Indiana University Press.
Waldrop, M.M. (1981) Mauna Kea (I): halfway to space. *Science*, **214**, 1010–3.
Whitaker, E.A. (1986) *The University of Arizona's Lunar and Planetary Laboratory: Its Founding and Early Years*. (Unpublished 78-page booklet, obtainable from LPL, University of Arizona, Tucson, AZ 85721.)

L

LAGRANGE, JOSEPH LOUIS (1736–1813)

A French physicist, born in Turin (Italy), who was one of the most important mathematical and physical scientists of the 18th century. Lagrange became a professor of geometry at the Royal Artillery School in Turin (1755–1766) and was there one of the founders of the Royal Academy of Sciences in 1757. When Euler (1707–1783) left his post at the Berlin Academy of Sciences, Lagrange became the director of the mathematical section of the Academy (1766). In 1787 he moved to Paris to join the Paris Academy of Sciences and taught mathematics at the École Polytechnique, which he helped to found. Napoleon named him to the Legion of Honor, and he was honored as Count of the Empire in 1808.

Lagrange invented the calculus of variations and applied this new mathematical discipline to celestial mechanics, particularly for tackling the action of gravitational forces on non-spherical, extended bodies (first in 1764), and later tacking the three-body problem (1772). In his studies of planetary perturbations, planets were taken as point masses. He was the first to introduce the potential function into celestial mechanics (1774), and to apply the method of variation of orbital parameters to all six orbital elements (1784). In his *Mechanique Analytique* (1786) classical mechanics was transformed into a branch of mathematical analysis by applying to it the calculus of variations and the ideas of the space of generalized coordinates. The latter may be considered to have been an in-between creation in the gradual transition from the traditional space of experimental awareness, as met in Newton's thinking, to the purely mathematical space as exemplified in geometries of the 19th century. Lagrange contributed immensely to the evolution of the field of mathematical physics. In his work he summarized the main results of mechanics in terms of differential equations and stressed the importance of Taylor series and the concept of mathematical function. In this regard he provided groundwork for important discoveries to be made later by A.L. Cauchy (1789–1857), K.W.T. Weierstrass (1815–1897) and N.H. Abel (1802–1829).

In the second edition of the *Mecanique Analytique* (1788) he proved that the time derivatives of the orbital elements are given by time-independent functions of the partial derivatives of the disturbing function with respect to the orbital elements – an essential step toward the emergence of Hamiltonian dynamics.

Rhodes W. Fairbridge and Hans J. Haubold

Figure L1 Louis Lagrange.

Bibliography

Itard, J. (1973) Lagrange, Joseph Louis. *Dict. Sci. Biogr.*, Vol. 7, pp. 559–73.
Newman, J.R. (ed.) (1988) *The World of Mathematics*. New York: Tempus Books (Vols I, II and III).
Sarton, G. (1944) Lagrange's personality (1736–1813). *Proc. Am. Phil. Soc.*, **88**, 457–96.
Serret, J.A. (ed.) (1867–1892) *Oeuvres de Lagrange*, 14 vols, Paris.
Weaver, J.H. (ed.) (1987) *The World of Physics*. New York: Simon and Schuster (Vols II and III).

LAGRANGIAN POINT

While serving as director of the Berlin Academy of Sciences in 1772, the French mathematician Joseph Louis Lagrange developed a theory of equilibrium motion within a perturbed orbit, providing a special case of the three-body problem (Lagrange, 1772). This classic astronomical problem had occupied the attention of leading mathematicians of the period – notably Clairaut, Euler and Laplace, and later Lambert – without any real success. The immediate challenge was to solve the problems of disturbed motion in the solar system, and specifically the deviations of a planetary object from its predicted place – first noted in empirical observations by Kepler, Halley and others. Lagrange's observations, testing the newly established theory of Newton, had (1) revealed the effects of synergetic gravitational attraction by any of the planets of the solar system upon a smaller object moving around the Sun, but (2) failed satisfactorily to solve the observed mutual gravitational interactions between the two massive planets, Jupiter and Saturn.

Even in the first case, a difficulty of solution lay in the fact that the mathematical reduction involves three differential equations of the second degree, which are insolvable by any direct means, and must be accomplished by successive approximations. In this process the deviations induced by the disturbing planet are derived by integration; the accelerations and forces are then corrected by amounts that are small in comparison with the original values; the corrected values are integrated to yield new, second-order corrections still smaller than the first; and so on. To obviate the fact that the computed disturbing forces are altered in an irregular fashion as the solution proceeds and the corrected positions of the disturbing body are obtained, a series of periodic terms are introduced. This solution device results in an enormous number of individual calculations which were nearly insurmountable to the mathematicians of this early period before electronic computers.

Restricted three-body problem

Lagrange's theory was based on the principal tenet that one of the components of the three-body system (e.g. typically an asteroid) must be of near-negligible mass compared to either of the other two, assumed to be the Sun and the planet Jupiter.

L1, L2 and L3 radius-vector Lagrangian points

In the complete development of the theory, three other points of dynamic equilibrium exist – one at the current position of Jupiter (L_3) and the remaining two along the Sun–Jupiter axis, at locations respectively inside and outside Jupiter's orbit, and at distances from the Sun determined by relative mass factors (Blanco and McCuskey, 1961). Since L_1 and L_2 are only theoretical rather than physically occupied points in the Jupiter–Sun system, in contrast to L_4 and L_5, further discussion is not appropriate at this point. However, the L_4 and L_5 Lagrangian points are of interest in connection with the possible existence of subsatellite conglomerates of particulate matter in similar fixed positions surrounding the Earth's Moon.

Observational confirmation of the Lagrangian points L4 and L5

Lagrange's theory was largely ignored as implausible until the discovery of the asteroid Achilles (588) by the astronomer Max Wolf in 1906. At discovery its separation angle in celestial longitude from both the Sun and Jupiter was approximately 55½° (ahead of Jupiter), thus closely fulfilling Lagrange's theory. This was reinforced in the same year by discovery of a second asteroid, Patroclus (617) which occupied the L5 point (approximately 60° following Jupiter).

Thereafter, the discovery of additional asteroids of this type by the use of ever faster photographic plates came rapidly. At the present time, 15 named asteroids are known to occupy either the L4 or L5 Lagrangian points. They are identified as the Trojan asteroids and, in accordance with International Astronomical Union policy, they are given Hellenic names corresponding to participants in the Trojan Wars. Hundreds of Trojan asteroids remain unnamed.

Recent observations indicate that an asteroid's motion with respect to Jupiter is produced by a slow, relatively stable drift. The drift in longitude is associated with a libratory movement caused by forces imposed synergetically by the other planets (and their satellites). The position of the asteroid relative to both Jupiter and the Sun remains basically the same, at a common angular separation of approximately 60° in celestial longitude, as Lagrange's theory predicts.

Fergus J. Wood

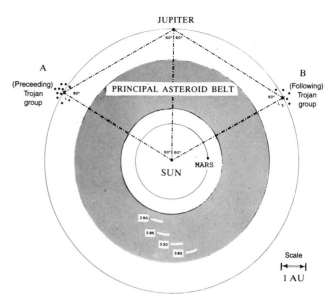

Figure L2 Stable distribution of the Trojan asteroids at the Lagrangian points L4 and L5, in accordance with the equilateral triangle solution of the three-body problem. Sun is shown (center) with Jupiter (top) and (A) the preceding Trojan group at 60° W with (B) the following Trojan group at 60°E.

The diagram also shows the locations of the principal examples of Kirkwood's gaps, at mean solar distances of 2.50, 2.95, 3.30 and 3.65 AU. (Note: the orbits are elliptical, but only depictable on this scale as circles.)

Bibliography

Blanco, V.M. and McCuskey, S.W. (1961) The restricted three-body problem, in *Basic Physics of the Solar System*. Reading, Mass: Addison-Wesley, 307 pp., pp. 179–85.
Brown, E.W. and Shook, C.A. (1964) The Trojan group of asteroids, in *Planetary Theory*. New York: Dover Publications, 302 pp., pp. 250–88.
Lagrange, J.L. (1772) Essai sur le probleme des trois corps (Essay on the problem of three bodies), Academie Royale des Sciences de Paris, **9**; in Serret, M.J.-A. (1873) *Oeuvres de Lagrange*, Vol. 6, Paris: Gauthier-Villars, pp. 229–331.
Moulton, F.R. (1920) Motion around the lagrangian Points, in *Periodic Orbits*. Washington, DC: Carnegie Institution of Washington, Publ. No 161 (Chapters 5–7).
Rabe, E. (1971) Trojans and comets of the Jupiter group, in *Physical Studies of Minor Planets* (ed. T. Gehrels). Washington, DC: NASA SP-267, 687 pp., pp. 407–12.
Schubert, J. and Bien, R. (1987) Trojan asteroids: relations between dynamical parameters. *Astron. Astrophys.*, **175**, 299–302.
Shoemaker, E.M., Shoemaker, C.S. and R.F. Wolfe (1989) Trojan asteroids: populations, dynamical structure and origin of the L_4 and L_5 swarms, in *Asteroids II* (eds R.P. Binzel, T. Gehrels and M.S. Mathews). Tucson: University of Arizona Press (rev. edn), 1258 pp., pp. 487–523.
Yoder, C.F. (1979) Notes on the origin of the Trojan asteroids. *Icarus*, **40**, 341–4.

Cross references

Celestial mechanics
Trojan asteroids

LAPLACE, PIERRE SIMON DE (1749–1827)

A French mathematician and astronomer, born on a farm in Beaumont-en-Auge (France), Laplace is well known for his seminal contributions to celestial mechanics and to probability theory, as well

Figure L3 Pierre Simon de Laplace.

as other work in mathematics, physics and astronomy that led to early ideas of the origin and formation of structure in the solar system. He is perhaps best known for his nebular hypothesis of the origin of the solar system printed in an addendum to his *Exposition du Système du Monde* (1796), a work sometimes regarded as one of the masterpieces of the French language. He postulated that the Sun and planets were originally formed from a rotating disk of gas, a hot cloud that eventually cooled into local condensations. He was not aware that Kant (1724–1804) had earlier suggested a similar scenario. Subsequently it has become known as the 'Kant–Laplace nebular hypothesis', although this attribution is not universally favored. The unifying criterium that appealed also to Kant, was the common counterclockwise motions in both planetary spin and orbital revolutions. Although replaced over the years by other theories, for example the Chamberlin–Moulton *'planetesimal hypothesis'* (q.v.), the Laplacian model has enjoyed some measure of revival in recent years (Prentice, 1978).

Laplace argued mathematically (1773) that gravitational perturbations of one planet by another would not lead to mechanical instabilities in their orbits, but his reasoning is not now appropriate. Using Newton's second law of motion together with Newton's law of gravity, he developed a comprehensive account of the gravitational interactions within the solar system. In particular, he was able to provide an explanation of three phenomena that had long puzzled astronomers and that appeared to threaten the stability of the solar system.

(1) The secular acceleration of the Moon. Certain perturbations which the moon experiences due to the Sun depend, inter alia, on the eccentricity of the Earth's orbit, which in turn is being altered by the action of the other planets, according to a 1787 paper. The perturbation which the Moon experiences is therefore slightly altered and as a consequence the Moon's average rate of motion is slowly decreasing and the length of the month increasing, by about 0.03 s per century. (An added factor, recognized a century later by Sir George Darwin was the secular deceleration of the Earth's daily rotation due to tidal friction.) Laplace explained the increase in lunar orbital rate by the contemporary decrease in the Earth's eccentricity now known as one of the Milankovitch parameters (q.v.) controlling the Earth's glacial cycles.

(2) The slow changes in the rates of the motion of Jupiter and Saturn ('the Great Inequality'). In the mid-1780s Laplace discovered that the apparent unidirectional secular changes in the mean motions of Jupiter, Saturn and the Moon were in fact oscillatory and periodic with very long periods, and so did not endanger stability. If the times of revolution of two planets are nearly proportional to two whole numbers (in this case approximately 5 : 2), then part of the periodic disturbing force produces an irregularity. This is the case for Jupiter and Saturn and, as first shown by Laplace, leads to an appreciable inequality with a period of about 850–900 years. The secular acceleration (actually only in part) of the Moon, Laplace claimed, was caused by a decrease in the eccentricity of the Earth's orbit, the latter being a phase of a periodic oscillation. In collaboration with Lagrange (q.v.), a new law was eventually established that defined the fundamental stability of the solar system, based on the collective sum of their orbital eccentricities. If the eccentricity of one planet increases, another decreases, the total angular momentum being conserved. Thus, for each planet he calculated a product:

$$\text{mass} = (\text{mean orbital axis})^{1/2} = (\text{eccentricity})^2$$

The sum of all the products is then a fundamental invariant within the constraints of Newtonian physics.

(3) The whole-number relations of the mean motions of the inner three Galilean moons of Jupiter, namely $n - 3n' + 2n'' = 0$. Laplace showed that this relation was a necessary consequence of the gravitational interactions of the three satellites.

Between 1799 and 1825 Laplace published the five volumes of his *Mécanique Céleste*, the first fundamental treatise on celestial mechanics and standard work for all 19th-century specialists. It incorporated not only the work of the preceding century (Euler, Clairaut, d'Alembert and Lagrange) which provided mathematical and physical explanations of the dynamics of solar system bodies based on Newton's gravitational theory, but also incorporated Laplace's own very important contributions. Only in the fifth volume, in a series of *Notices historiques*, did he acknowledge his indebtedness to predecessors.

In physics Laplace introduced the representation of the gravitational potential by the Laplace equation, thus preparing the way for a field theory of gravitational and other forces. In such a theory the force at any point in space is given by the 'Laplacian', or divergence of the gradient of a potential, a scalar function whose value at each point is the work required to bring a small particle from that point to a standard surface such as the surface at infinity.

In the 1780s Laplace assisted A.L. Lavoisier (1743–1794) in thermochemistry, conducting with him experiments on capillary action and on the determination of the specific heat of various substances. In physical optics he was opposed to T. Young's (1773–1829) wave theory of light. In other technical papers he reported on the figure of equilibrium of a rotating fluid mass, on the theory of the attraction of spheroids, and on the shape ('figure') of planets. In the area of spherical harmonics he introduced what are now known as 'Laplace's coefficients'. In 1812 he published his analytical theory of probabilities.

Over his lifetime Laplace made a successful career. J.L.R. d'Alembert (1717–1783) arranged for Laplace's appointment as professor of mathematics at the École Militaire in Paris (1768). He became an associate (1773) and then a pensioner (1785) of the Academy of Sciences in Paris. He contributed to the establishment of the metric system. Napoleon named him to the Legion of Honor and Count of the Empire in 1805 and 1806, respectively; Louis XVIII granted Laplace the title of Marquis.

Rhodes W. Fairbridge and Hans J. Haubold

Bibliography

Berry, A. (1961) *A Short History of Astronomy*. New York: Dover Publications, Inc.

Grant, R. (1852) *History of Physical Astronomy form the Earliest Ages to the Middle of the 19th Century*. London.

Grattan-Guinness, I. (1978) Laplace, Pierre-Simon, Marquis de. *Dict. Sci. Biogr.*, Vol. 15, pp. 273–403.
Prentice, A.J.R. (1978) Towards a modern Laplacian theory for the formation of the solar system, in *The Origin of the Solar System* (ed. S.F. Dermott). New York: John Wiley, pp. 111–61.
Singer, C. (1959) *A Short History of Scientific Ideas*. Oxford: Clarendon Press.
Taton, R. and Wilson, C. (eds) (1889) *Planetary Astronomy from the Renaissance to the Rise of Astrophysics* (Part A: *Tycho Brahe to Newton*). Cambridge: Cambridge University Press.
Weaver, J.H. (ed.) (1987) *The World of Physics*. New York: Simon and Schuster (Vols I to III).

LASER RANGING

Laser ranging is a technique that permits the determination of distances between an observing instrument and a target object that reflects laser light back to the instrument. The instrument may be fixed, as at a terrestrial astronomical observatory, or it may be carried aboard a satellite or planetary probe. The target may be a natural surface or a specially constructed reflector or reflector array. In this article we are concerned with three applications of laser ranging: (1) Lunar laser ranging (LLR) to corner-cube retroreflectors emplaced on the Moon; (2) satellite laser ranging (SLR) to Earth-orbiting satellites such as LAGEOS and STARLETTE; and (3) laser altimetry, as applied in the Mars Observer and Clementine planetary and lunar exploration missions.

The invariance of the velocity of light is the physical principal exploited in range measurements. The two-way travel time t of the light pulse is accurately measured, and a calculation of the form $r = ct/2$ yields the range r (here c is the velocity of light). The distance ranging technique may be employed with other forms of radiation; radar ranging from the Earth to the planets Mercury and Venus began in 1970, and the Mars Viking landers were ranged during their period of operation (1976–1982) with measurement uncertainties of only 7 m (see Viking missions).

The essential components of a laser ranging system are a laser transmitter, which generates a short pulse of laser light; a laser receiver, to collect the photons of the returning pulse; and an accurate time interval counter to measure the time between the emission of the pulse and the arrival of the returning pulse. Additional elements are introduced to refine the process; for instance, a spectral filter may be employed at the receiver to ensure that only photons of the proper color are counted, and a 'range gate' may be used to limit the time interval of operation of the receiver. In this case an 'expected' travel time is predetermined, and only photons arriving close to this expected time are counted.

Lunar laser ranging

The lunar laser ranging experiment was originally conceived as a method of testing for a possible time variation of the gravitation constant G (Bender *et al.*, 1973). By the early 1960s it was clear that deployment of laser retroreflectors on the Moon could yield considerable new scientific information about the Moon, its orbit, and its rotation and physical librations.

The astronauts of the Apollo 11 mission deployed the first laser reflector array on the lunar surface on 21 July 1969. This consisted of an array of 100 fused-quartz corner-cube reflectors, which return a pulse of incident light back along the path from which it came. Each retroreflector in the array had a diameter of 3.8 cm. On 1 August 1969 astronomers at Lick Observatory in California detected strong returned signals from this reflector for the first time. In subsequent years four more retroreflectors were deployed on the lunar surface; two were placed by the Soviet Luna 17 and Luna 21 spacecraft, and two more arrived with the astronauts of the Apollo 14 and 15 missions (see Apollo missions; Luna missions). The Apollo 15 array consisted of 300 retroreflectors. Bender *et al.* (1973) provide an overview of the Lunar Laser Ranging Experiment.

To record successfully laser photons returned from reflectors on the Moon is technically difficult. The laser beam enters an optical telescope where the eyepiece would be, and leaves with the diameter of the telescope's main mirror. This results in a spot on the Moon with a diameter of less than 2 km. Thus the aiming of the telescope is critical; this has been compared to aiming a rifle at a 1 cm diameter coin placed more than 3 km away. Since the area of the retroreflector is much smaller than the area illuminated by the beam on the Moon, only a small portion of the photons emitted are returned. In practice only about one photon of every 10^{19} transmitted is collected. To achieve this level of performance the power output of the laser must be extremely high, as much as 2×10^9 W.

More than 8000 lunar laser ranges have been obtained since 1969. The uncertainties in early ranges were measured in meters, but recent determinations have uncertainties of less than 2 cm. Since the average distance separating the centers of the Moon and Earth is about 385 000 000 m, the relative accuracy of the range values is better than one part in 10^{10}.

Mulholland (1980) and Dickey, Williams and Yoder (1982) summarized the scientific results obtained in the first decade of LLR measurements. Among these are confirmation of the Einstein general relativistic principle of equivalence, and the theory's prediction of the invariance of the gravitation constant G: the variation of G is constrained to be less than a few parts in 10^{11} per year. LLR has shown that the Moon is receding from the Earth at a rate of about 3.8 cm per year. LLR has also improved our knowledge of the mass of the Earth–Moon system, the lunar librations and moments of inertia, the lunar 'secular acceleration' (the observable result of tidal friction; q.v.), and the lunar gravity field. Improvements in our knowledge of the lunar orbit due to LLR now permit the accurate analysis of eclipses back as far as 1400 BC.

LLR has also provided improved knowledge of terrestrial parameters relating to observatory coordinates. The motions of the tectonic plates on which the observatories stand can be determined. LLR is also employed in sophisticated studies of the tides and the rotation of the Earth (see Earth rotation). A recent summary of past and present LLR investigations is presented by Dickey *et al.* (1994).

Satellite laser ranging

Satellite laser ranging is one of the three techniques making up the field of space geodesy (the others are GPS, applications of the Global Positioning System (q.v.), and VLBI, i.e. very long baseline interferometry). The objective of space geodesy is to make precise position measurements between points on Earth's surface. As before, a two-way travel time of a short laser light pulse is measured to determine a range. The satellites ranged must be equipped with laser retroreflectors. Although more than 15 satellites are so equipped, the LAGEOS (Laser Geodynamics Satellite) and STARLETTE satellites are most often employed for high-precision studies. Both of these are spherical and studded with retroreflectors; they have high-density cores to minimize the effects of drag forces, and lack propulsive or control systems. The LAGEOS satellite orbits at an altitude of about 6000 km, while the STARLETTE orbits at about 1000 km.

Multiple ranges to the reflecting satellite from multiple observing stations on the ground are employed to determine both the satellite orbit and the locations of the ground stations with high precision. Smith (1991) gives a recent review of the techniques employed and results obtained. A number of effects perturb the satellite orbit and must be included in solutions; these include Earth's gravity field, the gravity of the Sun and Moon, atmospheric drag, solar radiation pressure, additional attractions from Earth and ocean tides and planetary attractions due to Venus, Mars, Jupiter and Saturn. Nevertheless the uncertainties of LAGEOS ranges are near the 1-cm level, which is sufficient to resolve effects such as the slow motions of the tectonic plates of the Earth (a few centimeters per year). Comparisons between SLR solutions for plate motions with motions determined by geological and geophysical methods generally show good agreement. Thus SLR is a valuable technique for study of present-day tectonic plate motions (e.g. Gordon and Stein, 1992; see Plate tectonics).

Satellite laser ranging is also employed in the study of the rotation of the Earth (e.g. Yoder, *et al.* 1983; Christodoulidis *et al.*, 1988), and in studies of the temporal variations of the gravity field of the Earth (Chao and Yu, 1991; Nerem *et al.*, 1993).

The discipline of satellite laser ranging primarily relates to investigations of planet Earth, and is covered in more detail in other references, such as the *Encyclopedia of Solid Earth Geophysics* (James, 1991).

Laser altimetry

Laser ranging may be successfully accomplished without the benefit of retroreflectors at the target, so long as some of the transmitted photons are backscattered from a surface to the receiver. Two recent systems of this type were carried on the Mars Observer spacecraft and the Clementine probe.

The Mars Observer Laser Altimeter (MOLA) is described in detail by Zuber *et al.* (1992). Although contact with the spacecraft was regrettably lost in 1993 (see Mars Observer), a discussion of the nature of the instrument and the objectives of the experiment is appropriate here. The MOLA was to map the topographic relief of Mars at an accuracy of a few meters. This 26.2-kg instrument generated laser pulses at a wavelength of 1.064 µm, with laser power consumption of 13.7 W. The 'footprint' or illuminated area on the surface of Mars measured 160 m. The received pulse, returned from the surface, carries information in the pulse shape as well as in the travel time. Different topographic slopes generate different pulse widths, and the MOLA was equipped with low-pass filters designed to optimize detection of four values of surface slope (1.7°, 5°, 15° and 39°). The topography optimally detected by the four filters could be characterized as 'smooth,' 'moderate,' 'rough' and 'scarp' (i.e. precipitous).

The scientific objectives of the MOLA experiment were and are of critical importance in the understanding of the structure and dynamics of the planet. The present uncertainty of topographic data for the Mars surface is on the order of a kilometer in some regions. Better topographic information would help define and constrain models of the internal structure, bearing for instance on the questions of the thickness and strength of the lithosphere. This information in turn constrains models of heat flow and mantle convection within the planet. Topographic information aids in interpretation of crustal surface geologic features, such as volcanic landforms. In addition surface topography is an important constraint on atmospheric dynamics. The information returned from the MOLA would have led to great advances in our understanding of the red planet.

Preliminary analyses of laser altimeter data from the Clementine spacecraft Moon encounter are under way (Kerr, 1994). The Lidar Laser Transmitter on Clementine utilizes the same 1.064-µm wavelength as Mars Observer. The instrument was originally designed to detect 'dark, metallic objects' such as intercontinental ballistic missiles. The results from Clementine have already changed our perception of the Moon in important ways. For instance, the range of elevations known on the Moon was about 12 km based on data obtained during the Apollo missions. Clementine demonstrated that the actual range is nearly 25 km. One crater, the South Pole–Aitken impact basin, is 12 km in depth.

By combining the Clementine laser altimeter high-resolution topographic data with the best available gravity data, it is possible to determine the internal structure of the Moon and the strength of the crust in greater detail and with greater precision than before. There are suggestions that a portion of the core of the Moon is still molten. It is apparent that the Clementine laser altimeter data will yield new and perhaps startling insights for years to come.

James H. Shirley

Bibliography

Bender, P.L., Currie, D.G., Dicke, R.H. *et al.* (1973) The Lunar Laser Ranging Experiment. *Science*, **182**, 229–38.
Chao, B.F. and Yu, A.Y. (1991) Temporal variation of the Earth's low-degree zonal gravitational field caused by atmospheric mass redistribution: 1980–1988. *J. Geophys. Res.*, **96**, 6569.
Christodoulidis, D.C., Smith, D.E., Williamson, R.G. and Klosko, S.M. (1988) Observed tidal braking in the Earth/Moon/Sun system. *J. Geophys. Res.*, **93**, 6216–36.
Dickey, J.O., Williams, J.G. and Yoder, C.F. (1982) Results from lunar laser ranging data analysis, in *High Precision Earth Rotation and Earth-Moon Dynamics* (ed. O. Calame). Dordrecht: D. Reidel, pp. 209–16.
Dickey, J.O., Bender, P.L., Faller, J.E. *et al.* (1994) Lunar laser ranging: a continuing legacy of the Apollo Program. *Science*, **265**, 482–90.
Gordon, R.G. and Stein, S. (1992) Global tectonics and space geodesy. *Science*, **256**, 333–42.
James, D.E. (ed.) (1991) *The Encyclopedia of Solid Earth Geophysics*. New York: Van Nostrand Reinhold.
Kerr, R.A. (1994) Geophysicists celebrate two satellites, mourn a third. *Science*, **264**, 1666–7.
Mulholland, J.D. (1980) Scientific achievements from ten years of lunar laser ranging. *Rev. Geophys. Sp. Phys.*, **18**, 549–64.
Nerem, R.S., Chao, B.F., Chan, J.C. *et al.* (1993) Temporal variations of the Earth's gravitational field: measurement and geophysical modeling. *Ann. Geophys.*, **11**, C110.
Smith, D.E. (1991) Satellite laser positioning, in *The Encyclopedia of Solid Earth Geophysics* (ed. D.E. James). New York: Van Nostrand Reinhold, pp. 969–977.
Yoder, C.F., Williams, J.G., Dickey, J.O. *et al.* (1983) Secular variation of Earth's gravitational harmonic J_2 from LAGEOS and nontidal acceleration of Earth rotation. *Nature*, **303**, 757.
Zuber, M.T., Smith, D.E., Solomon, S.C. (1992) The Mars Observer Laser Altimeter investigation. *J. Geophys. Res.*, **97**, 7781–97.

Cross references

Apollo missions
Earth rotation
Global Positioning System
Luna missions
Mars observer
Plate tectonics
Tidal friction
Viking missions

LENGTH OF DAY

The rotation of the Earth was the best available timekeeping device throughout human history until fairly recently. The time it keeps is called the Universal Time (UT). Since the advent of the atomic clock in the late 1950s, the more uniform Atomic Time (AT) has replaced UT as the time standard. The slight difference between UT and AT, when differentiated with respect to time in some proper unit, gives the length-of-day (LOD) variations. The LOD variation reflects the temporal variation in the Earth's rotational speed.

LOD varies slightly on all temporal scales under tidal and geophysical influences. The typical amplitude is on the order of 1 millisecond, or about one part in 10^8, as has been well determined by modern space geodetic means. Subject to the conservation of angular momentum, the LOD variations at discrete tidal frequencies are caused by tidal deformation in the Earth and oceans, and the tidal currents in the oceans. Similarly, the LOD variations at seasonal and intraseasonal timescales are caused by variations of the atmospheric angular momentum, primarily in the form of changing winds. Interannual LOD variations also originate in the El Niño/Southern Oscillation phenomenon in the ocean–atmosphere system, and in the alternating winds of the stratospheric Quasi-biennial Oscillation. Based on indirect evidence, the decadal LOD variations are believed to be the results of activities in the fluid core (which is also responsible for the generation of the Earth's magnetic field) coupled to the solid Earth. On the longest timescale, the secular lengthening of LOD is caused by (external) frictional torque exerted by the lunisolar tides in the phenomenon known as the tidal braking of the Earth's rotation.

Benjamin Fong Chao

Cross references

Earth rotation
Tidal friction

LEVERRIER, URBAIN JEAN JOSEPH (1811–1877)

Leverrier was one of the leading French astronomers of the 19th century. He is always remembered chiefly for his greatest triumph, the tracking down of the new planet Neptune, but he accomplished much else besides.

He was born on 11 March 1811 at St Lô in Normandy, and went to various local schools before entering the Collège de St Louis in Paris. In 1831 he entered the École Polytechnique on a scholarship, and after spending some time in chemical research and teaching, turned to astronomy. His ability was soon recognized. His first paper, dealing with meteors, appeared in 1832, and he then began a long series of investigations into the stability and mechanics of the solar system. He suggested that slight irregularities in the motions of Mercury might indicate the presence of an inner planet, and later this hypothetical planet was even given a name – Vulcan; but it is now known that it does not exist. The problem of Mercury was solved by applications of Einstein's theory of relativity, long after Leverrier's death.

François Arago, the most famous astronomer in France at that time, suggested that Leverrier might tackle the problem of the movements of Uranus, which appeared to be straying from its predicted path. Leverrier accepted the challenge, and it was from his accurate calculations that Neptune was actually identified, in 1846, by Galle and D'Arrest in Berlin. Similar calculations had been made slightly earlier by John Couch Adams in England; Leverrier was unaware of these at the time, and Arago in particular was furious at what he regarded as a brazen English attempt to rob Leverrier of the honour. Fortunately, Leverrier himself took no part in the subsequent controversy.

The identification of Neptune made Leverrier's name widely known. A Chair of Astronomy was created for him in Paris in 1847, and in 1849 a Chair of Celestial Mechanics was created for him in the Sorbonne. In 1854 he became Director of the Paris Observatory, in succession to Arago. It cannot be said that he was popular with his staff; he was notoriously short-tempered, and one colleague commented that although he might not be the most detestable man in France, he was certainly the most detested! This may be rather extreme, but matters came to a head in 1870 and Leverrier was requested to resign. He did so, but was reinstated 2 years later when his successor, Charles Delaunay, was drowned in a boating accident. Despite all these personal problems, Leverrier achieved a great deal during his period as director; for example, he established a meteorological network covering much of mainland Europe, and he improved the instruments at the Observatory. He died there on 23 September 1877.

Patrick Moore

Bibliography

Baum, R.M. (1981) Le Verrier and the Lost Planet, in *1982 Yearbook of Astronomy*. London: Sidgwick and Jackson.
Grosser, M. (1962) *The Discovery of Neptune*. Harvard University Press.
Levy, J.R. (1973) Le Verrier, Urbain Jean Joseph. *Dict. Sci. Biogr.*, Vol. 8, pp. 276–9
Moore, P. (1996) *The Planet Neptune*. New York: John Wiley.

LIBRATION

The period of rotation of the Moon around its axis is equal to its sidereal period of revolution around the Earth. Consequently the same hemisphere of the Moon always faces the Earth. However, we can actually see about 59% of the surface of the Moon from the Earth, because of several phenomena collectively known as librations.

While the Moon rotates on its axis at an approximately constant angular speed, its angular speed of revolution around the Earth is faster at perigee than its rotational speed, and slower at apogee. Consequently, at different times during a lunation we can see a little further beyond the east or west limbs than a hemisphere. This is the libration in longitude.

The Moon's orbit is inclined at about 5° to the ecliptic, or between 18½° and 28½° to the equator. Consequently, during a lunation we can peek a little further beyond the north or south limbs than a hemisphere. This is the libration in latitude.

The radius of the Earth subtends a little less than 1° at the Moon's distance. Consequently we view the Moon at a slightly different angle when the Moon is on the meridian than at moonrise or moonset. This is the diurnal libration.

The rotation of the Moon is not absolutely constant, because the Moon is not perfectly spherical and it experiences varying torques from the Earth during a lunation, which cause a further small libration called the dynamical libration.

Jeremy B. Tatum

Bibliography

Moutsoulas, M.D. (1971) Librations of the Lunar Globe, in *Physics and Astronomy of the Moon* (ed. Z. Kopal). New York: Academic Press.

Cross references

Moon: origin
Planetary rotation
Tidal friction
Torque

LIFE: ORIGIN

Life is ubiquitous on the Earth and appears to have formed very early in Earth's history, suggesting that given the proper conditions life will readily evolve on any planetary surface. However, spacecraft exploration fails to indicate traces of any sort of life on the other planets in our solar system. The most essential requirement for life is liquid water. Evidence for it is found only on Mars, and then only in the distant past. Theories for the origin of life on planets must therefore rely heavily on our understanding of the nature and history of life on Earth.

Life on Earth

One of the profound results of modern science has been the unity of biochemistry. All life shares the same genetic code and is constructed from the same basic building blocks (Lehninger, 1975). The genetic code of life stored in nucleic acids, DNA and RNA, is largely made from the same five nucleotide bases (A, T, G, C and U). The common structure of life is built from proteins which are combinations of the same set of 20 amino acids. The unity of life is now understood in the evolutionary context implying that all life on Earth is related and therefore descended from a single common ancestor. Thus, there is really only one life form on Earth. We have only one data point toward a generalized understanding of life.

The life we find on Earth appears to have originated on the planet very early in its history. There is definite evidence for microbial life as old as 3.5 Ga. This evidence is in the form of stromatolites (see Figure L4) which are lithified microbial mat layers often many meters in size. In addition one finds microfossils in these old rocks. Stromatolites still form today (see Schopf, 1983, and the references cited therein). There is also chemical evidence suggestive of life in the carbon isotopes of sediments that are 3.8 Ga old (Schidlowski, 1988) – close to the period of the late impact bombardment toward the end of the accretion of the Earth. The data suggest that life emerged soon after conditions on the Earth's surface became habitable. In fact it may have been the case that life originated on Earth much earlier, only to be eradicated by giant impacts that sterilized the planet. The one essential criterion was the presence of liquid water, which would have been generated by condensation of volcanic steam, probably in an environment of global mean temperatures around 20°C. Certainly the emerging paradigm suggests that life originated on the early Earth quite rapidly.

Origin of life on Earth

There is no evidence in the geological record of how life appeared. The standard theory for the origin of life begins with the non-biological production of organic matter – the stuff of life (Horgan, 1991). Through a series of steps referred to as chemical evolution, this organic matter assembles into the first living cell. Because of the affinity of biochemistry for liquid water it is assumed that the chemical evolution phase occurred in an aquatic environment. For a collection of the key early papers in the origin of life field including the work of Urey, Miller and others, see Kvenvolden (1974). The possibility of a non-biological source for the primal organic material

Figure L4 A Precambrian stromatolite, 3.5 Ga old. Stromatolites are sedimentary structures formed as microorganisms move up through infalling sediments to reach the light. Since the environmental factors (sunlight and an atmospheric source of C, CO_2) and the biological response (phototaxis) that cause the formation of stromatolites can be expected to have been associated with any early Martian biota, these macroscopic fossils of early life may be found on Mars. Colored blocks on scale bar at bottom are 1 cm wide. (Photograph courtesy S.M. Awramik).

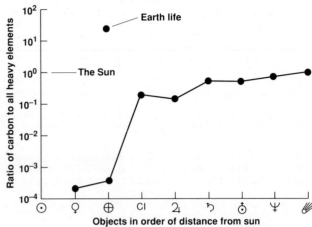

Figure L5 Ratio of carbon atoms to total heavy atoms (all atoms more massive than H and He) for various solar system objects illustrating the depletion of carbon in the inner solar system. The x axis is not a true distance scale, but the objects are ordered by increasing distance from the Sun. The planets are labeled by their astronomical symbol. CI refers to the carbon-rich asteroids and the final symbol stands for comets. Mars is not shown since the size of its carbon reservoir is unknown. (Figure adapted from McKay, 1991.)

was reinforced when Stanley Miller, in a series of experiments, passed an electrical discharge through an atmosphere rich in ammonia and methane, thought to be a simulation of an early reducing atmosphere on the Earth. He found that a number of organic compounds were produced including some of the amino acids needed for life. The ease by which organics could be produced under 'natural' conditions suggested that they should be widespread in the cosmos. This was shown by direct observation of non-biological organic material, first in meteorites and then in the interstellar medium, comets and the surfaces and atmospheres of the outer solar system – particularly Titan.

Since these initial experiments, the list of biochemical compounds essential to life that have been produced in non-biological reactors has increased (for a recent review, see Miller, 1992). However, no synthetic production of a living cell or organism has ever been achieved and no significant replica of chemical evolution has yet been reproduced in the laboratory.

Because of the preponderance of hydrogen in the Galaxy it was originally thought that the early atmosphere of the Earth was a reducing environment, implying that C was in the form of CH_4 and N was in the form of NH_3. However, it is now thought that for most of its early history Earth's atmosphere was dominated by CO_2 and N_2 – the oxidized forms of these elements. Under oxidized conditions the efficiency with which non-biological mechanisms produce organic material is greatly diminished. This poses a serious problem for the source of prebiological organic material on Earth.

While most of Earth's early history may have occurred under a CO_2-rich sky, during the accretional phase during which the Earth's core formed and meteoric material rained down on the Earth's surface, conditions may have been reducing for some period of time. However, given the rapidity with which life appears to have originated on the early Earth, a transient epoch of reducing conditions may have been adequate for life's origin.

An alternate, or possible additional, source of organic material to the prebiotic Earth could have been comets, interplanetary dust and meteors streaming in from the organic-rich outer solar system (Chyba and Sagan, 1992). Figure L5 shows a plot of the distribution of carbon in the solar system – a rough indicator of the distribution of abiotic organic material. From this can be seen the somewhat paradoxical situation that the regions around the Sun that are warm enough to support the environment needed for life (liquid water) are devoid of the stuff of life (organic material) while the regions rich in organics are too cold for liquid water. Comets may have bridged that gap carrying organics, as well as the components of the atmosphere and the ocean, to the early Earth. However, the efficacy of comet delivery of organics is still debatable since the high impact velocity with which they are expected to strike the Earth and the efficiency of the more gentle delivery by interplanetary dust may not be adequate to provide sufficient organics for life's origin.

Even with the source of prebiological organics uncertain, the usually accepted theory for the origin of life relies on organic material and chemical evolution occurring on the early Earth. There are, however, other theories for the origin of life. Most prominent of these is the suggestion by Cairns-Smith (1982) that the earliest life forms were not organic at all but were composed instead of clay minerals. The organic form we see today would then be the result of evolutionary progression. The fact that life appeared almost instantaneously on the early Earth has reinvigorated proponents of the theory that life was seeded on the Earth (Davies, 1988). This theory known as panspermia (proposed in 1901 by S.A. Arrhenius; q.v.), suggests that life originated elsewhere – perhaps taking billions of years – and that its rapid emergence on Earth was because spores or fragments of genetic material capable of self-replication were carried here, for example, by comets or passing interstellar dust.

In addition to the geological and paleontological records there is a record within life itself that helps us look backward to the common ancestor of life on Earth. This record is the genetic code and the evolutionary relationship between all life on the planet (Woese, 1987). Studies of the RNA sequences in all forms of life on Earth have led to the suggestion that the last common ancestor was a sulfur-metabolizing organism living in a hot-springs environment. It appears that the genetic and metabolic diversity we see today in micro-organisms appeared very early in Earth's history and were probably present in the 3.5 Ga old fossils (Schopf, 1983). It is interesting that for the next 2 Ga life on Earth was dominated by bacteria, and in fact in the modern world, replete as it is with higher life forms, the main agents in the biogeochemical cycles of N and S are still the microorganisms.

Definitions of life

Various definitions of life have been offered. Perhaps the one in use by most biologists would be that life is a material system that undergoes Darwinian evolution; this means that it is capable of self-reproduction, and mutation followed by selection based upon stored information. This definition does not lend itself to the search for life on other planets. In that case a more operational definition is required (McKay, 1991). In this regard the essential features of life are that it requires (1) a source of energy, usually sunlight; (2) carbon; (3) liquid water; and (4) some combination of other elements such as N, P and S.

That life requires energy is clear from thermodynamics. On Earth energy for life is provided largely by sunlight, which provides an

Table L1 Ten most abundant elements in the cosmos, in Earth's crust and in life (% by mass; Davies and Koch, 1991)

	Cosmos		Earth's crust		Humans		Bacteria	
1	H	70.7	O	46.6	O	64	O	68
2	He	27.4	Si	27.7	C	19	C	15
3	O	0.958	Al	8.13	H	9	H	10.2
4	C	0.304	Fe	5.00	N	5	N	4.2
5	Ne	0.174	Ca	3.63	Ca	1.5	P	0.83
6	Fe	0.126	Na	2.83	P	0.8	K	0.45
7	N	0.110	K	2.59	S	0.6	Na	0.40
8	Si	0.0706	Mg	2.09	K	0.3	S	0.30
9	Mg	0.0656	Ti	0.44	Na	0.15	Ca	0.25
10	S	0.0414	H	0.14	Cl	0.15	Cl	0.12

efficient (low entropy) energy source. Some specialized organisms can use chemical energy (e.g. $H_2 + CO_2 \rightarrow CH_4 + H_2O$), as in deep subsurface hydrothermal vents, but these systems are very limited in extent. Carbon is the most common atom in living systems and life on Earth is often described as carbon based. Carbon forms such an astonishingly complex number of molecules that an entire branch of science (organic chemistry) is devoted to this atom. However, most of these are irrelevant to life and only a tiny fraction form the basic biomolecules. Nonetheless, carbon is crucial to life and must be listed as an indispensible requirement. As mentioned above, the most significant requirement for life – and the one with the most implications in the planetary context – is the liquid-water criterion. Finally, life requires some other molecules besides C and H_2O. While many elements are required for life on Earth it is not clear that any particular element is an absolute requirement for life in general. The most important elements are certainly N, S and P. These are generally considered to be among the so-called biogenic elements (C, H, N, O, P, S). Life on Earth makes use of numerous other elements as well. Table L1 is a comparison of the elements in life with those in the cosmos (Davies and Koch, 1991).

Implicit in the above discussion is that life on another planet would be similar in broad respects to life on Earth. We have, however, no basis for this assumption and in radically different environments life, if it exists, could be radically different. With only one example of planetary life available for study, the best scientific approach to the problem is probably to consider environments similar to those on the Earth and then to look for life similar to the life we find here. The resulting systems, if we find them, could be similar enough to be recognizable but different enough to be enlightening.

There have been imaginative suggestions for life based on completely novel chemistry and energy sources. Speculative models of life based on silicon or ammonia and even cryogenic liquids have all been mooted. These notions cannot be proven false but have failed to make significant contributions to a general theory of biology or yield specific targets for a search strategy for life on other worlds. The conclusion from this may well be that our knowledge of life as a process is as yet too limited to allow for such a broad extrapolation from the single data point that is life on Earth (see Sagan, 1974). What are needed are more data points, not more imagination.

Life in the solar system

Currently Earth is the only planet in the solar system that has life or even the liquid water habitats necessary for life. Of the other planets only Mars shows evidence of liquid water habitats and thus the potential for life in the past.

Inner solar system

Mercury and the Moon lack a hydrosphere or an atmosphere; the prospects for life ever having been present on these airless worlds is remote. Under the thick CO_2 atmosphere of Venus conditions are much too hot for liquid water to exist on the surface. Milder conditions may be found in the upper atmosphere, possibly in the water clouds. However, clouds on Earth are not sites where life can grow: so it remains to be proven that the cloud environment on another planet could be conducive to life. We do not know what Venus' early history might have been but we have no evidence that suggests that is was ever a habitable planet. Its structural geology fails to reveal evidence of water-lain deposits.

Outer solar system

The gas giants Jupiter, Saturn, Uranus and Neptune lack surfaces that could collect liquid water. There have been suggestions that in water cloud layers in the atmosphere of these planets, particularly Jupiter, life could exist. However, no credible explanations exist of how life might flourish in clouds on Jupiter while in the more clement, energy-rich atmosphere of the Earth, life is not abundant in clouds.

Most of the moons of the outer solar system lack atmospheres and their surfaces are frozen, arguing against the possibility for life. The exception is Titan which has what is perhaps the most Earth-like atmosphere in the solar system. Its atmosphere is composed primarily of N_2 with a few percent CH_4 and trace amounts of other more complex hydrocarbons. The surface pressure on Titan is 50% greater than sea-level pressure on Earth. The non-biological production of organics on Titan naturally suggests a comparison to the standard theory for the origin of life on Earth (Raulin, Mourey and Toupance, 1982). However, the lack of liquid water due to the low temperatures is a serious deficiency on the road to the origin of life and seriously limits the analogy between the prebiological Earth and the present day Titan.

Europa is one of the Galilean moons of Jupiter and although it lacks any substantive atmosphere it has been suggested that underneath the icy surface is an ocean that may harbor life (Reynolds et al., 1983; Reynolds, McKay and Kasting, 1987). If there is liquid water on Europa, the ice shell which overlies it would be expected to crack occasionally allowing sunlight to reach the water column. This solar energy input or hydrothermal vents under the putative ocean could provide an energy source for biology. It must be emphasized however that there is no direct evidence yet of an ocean on Europa, much less of hydrothermal vents beneath such an ocean. Such evidence must await future spacecraft exploration.

Io, another of the Galilean moons, and possibly Europa, illustrate that tidal heating can be a significant source of energy for a small world orbiting a gas giant. With tidal heating providing an appropriate thermal regime and a central star to provide light for photosynthesis a habitable zone can exist around giant planets (Reynolds, McKay and Kasting, 1987).

In the outer reaches of the solar system, comets are a class of objects that also may have had liquid water at one time (Irvine, Leschine and Schloerb, 1980) due to radiogenic heating by radioactive decay early in their history. In order for such heating to have produced liquid water cores, the comets must be large and even then the liquid water epoch lasts only for a brief period after their formation. Other than this possible brief occurrence of a liquid state, water in comets is expected to be in the form of ice. The abiotic state of the large ice sheets on the Earth, such as the Polar Plateau and the Greenland ice sheet – salubrious by comparison to cometary surfaces – suggests that is is unlikely that life can grow on comet ice (Huebner and McKay, 1990).

The presence of ice, but not liquid water, is also the death knell for life on Triton, Chiron, Pluto and the other comet-like worlds of the outer solar systems. Even if there is, or was, liquid water on Europa and on comets, these environments are not similar to the early Earth and the argument by analogy fails.

Mars

Of all the other planets in our solar system, Mars is certainly the one that is the most interesting candidate for a search for life in the solar system. Since the advent of telescopes, astronomers had noticed that Mars exhibits changes in its polar caps and alterations in its surface coloration reminiscent of seasonal changes on Earth. The caps were incorrectly assumed to be water ice and the seasonal coloration due to vegetation. Thus Mars was destined to be the literary home to all manner of alien civilizations and life forms (see Lowell, Percival).

Viking results
However, spacecraft observations reveal that Mars also presents a poor prospect for life at the present time. The Viking missions which reached Mars in 1976 had as their primary objective the search for microbial life. Although it was known before Viking (based upon the results of Mariner 3, 4, 7 and 9) that Mars had a thin atmosphere and

was essentially a cold desert world, it was still felt that the Martian surface held the best hope of discovering extant life elsewhere in the solar system. Each Viking lander carried with it three biological experiments specifically designed to search for indications of life. The Gas Exchange Experiment (Oyama and Berdahl, 1977) was designed to determine if Martian life could metabolize and exchange gaseous products in the presence of water vapor and in a nutrient solution. The Labeled Release Experiment (Levin and Straat, 1977) sought to detect life by the release of radioactively labeled carbon initially incorporated into organic compounds in a nutrient solution. The Pyrolytic Release Experiment (Horowitz and Hobby, 1977) was based on the assumption that Martian life would have the capability to incorporate radioactively labeled carbon dioxide in the presence of sunlight (i.e. photosynthesis). In addition, the lander cameras looked for the presence of any obvious macroscopic life forms, the GCMS (Gas Chromatograph/Mass Spectrometer) searched for organics in the soil, while the X-ray Fluorescence experiment analyzed the elemental composition of the loose material at the Viking lander sites. Unfortunately, from the biological perspective, this later instrument could only detect elements with atomic number greater than Mg (Clark et al., 1982); thus there was no direct measurement of the important biogenic elements O, N, C or H in the soil material – although all are present in the Martian atmosphere. Nor was phosphorus directly measured because its signal was hidden by that of S and Si – both of which were present. Assuming the presence of P in the soil, all the biogenic elements (C, H, N, O, P and S) are present on the surface of Mars.

The most surprising result of the Viking soil analysis was that the GCMS failed to detect organics in any soil sample with a detectability reaching 1 part per billion (Biemann et al., 1977). The lack of organics in the Martian soil is one of the arguments against the presence of life at the Viking sampling sites.

The Viking Biology instruments all yielded evidence of positive activity but this has generally been interpreted as a result of non-biological processes (for a review see Klein, 1978, 1979; Horowitz, 1986; for an opposite view, see Levin and Straat, 1981; Levin, 1988). While the individual results of the biological experiments – particularly the labeled release experiment which showed uptake of ^{14}C-labeled CO_2 from a nutrient solution (Levin and Straat, 1981) – could be suggestive of life, when taken together the most likely explanation of the Viking results is the presence of one or more photochemically produced oxidants in the Martian soil such as H_2O_2. Furthermore, the present environment of Mars is not at all suitable for life. The key requirement that it appears to lack is liquid water (Horowitz, 1986; McKay, and Stoker, 1989). Mars is almost certainly sterile today, although we cannot at present rule out the unlikely case of subsurface hydrothermal liquid water habitats, or life frozen in the permafrost deposits.

Past water on Mars

Although the conditions on Mars are at present inimical to life, the photographs taken from orbit about Mars by the Viking and Mariner 9 orbiters show clear evidence of dry river channels and extinct volcanoes. Figure L6 shows an example of a riverbed feature on the heavily cratered region of Mars. These provide fairly convincing evidence that liquid water flowed on the surface of Mars sometime, and perhaps episodically, in its past (Carr, 1987; McKay and Stoker, 1989). In order for liquid water to have freely existed on the surface of Mars in the past, conditions must have been quite different from the present. Presumably Mars had a warmer climate but it is also possible that the mean temperature was below freezing and that only at certain places and during certain seasons did temperatures rise above freezing.

Water is the quintessence of life and the observation that there was liquid water on the surface of Mars in the past is the main motivation for considering the possibility of past life on Mars. The best argument for the origin of life on Mars is based on analogy to the origin of life on the early Earth. We do not know what was required for the origin of life or in what type of environment it occurred. Current speculation tends to favor sulfurous hydrothermal regions. However, active volcanism, small ephemeral ponds, large stable bodies of water, meteorite and cometary impact sites may also have been relevant. The argument for the origin of life on Mars by analogy with the Earth rests on the observation that all the major habitats and microenvironments that may have existed on the early Earth would be expected on early Mars as well. Even tidal pools would have existed on Mars, although the solar tide on Mars would be only about 10% of the lunar tide on the contemporary Earth. Similarly, the various possible sources of organic material and volatiles would have operated on both planets. Perhaps the major question, given these expected similarities, is the duration of such environments on Mars compared with the time required for the origin of life (McKay, 1986; McKay and Davis; 1991).

The events that led to the origin of life on the early Earth and which may have also occurred on early Mars may be better preserved in the sediments on Mars than on Earth. On Earth sediments that date back to the 3.5–4.0 Ga timeframe are rare and those that exist are usually severly altered. On Mars over half of the planet dates back to the end of late bombardment about 3.8 Ga ago and has been well preserved at low temperatures and pressures. Thus, while there may be no life there today, Mars may hold the best record of the events that led to the origin of life on Earth-like planets.

Evidence for life in a Martian meteorite

There are 12 meteorites on Earth that are thought to have come from Mars. That these meteorites came from a common parent body is evidenced by common ratios of the oxygen isotopes – values distinct from terrestrial, lunar, or asteroidal ratios. The Martian meteorites can be grouped into four classes. Three of these classes contain 11 of the 12 falls and are known by the name of the type specimen; the S (Shergotty), N (Nakhla) and C (Chassigny) class meteorites. The S, N and C meteorites are young, having formed on Mars between 200 and 1300 million years ago. Gas inclusions in two of the S type meteorites contain gases similar to the present Martian atmosphere as measured by the Viking Landers. The fourth class of Martian meteorite is represented by the single specimen known as ALH84001. In contrast to the SNC classes of meteorites, ALH84001 formed over 4 billion years ago on Mars and appears to have experienced aqueous alteration 3.6 billion years ago. This rock comes to us from the time period when Mars is thought to have had a warm, wet climate capable of supporting life.

McKay et al. (1996) suggest that ALH84001 contains evidence of life on Mars. They base their conclusion on four observations. Complex organic material (polycyclic aromatic hydrocarbons) consistent with a biogenic source is present inside ALH84001. Carbon isotope fractionation of carbonate globules, possibly of biological origin, is in the range which, on Earth, can result only from biogenic activity. Magnetite and iron-sulfide particles are present and the distribution and shape of these particles is most naturally explained by microbial activity. Microfossils are seen which could be actual traces of microbial life. This evidence of biotic activity in ALH84001 lends strong support to the suggestion that life was present on Mars about 3.5–4.0 billion years ago.

Importance of life on Mars

One of the direct questions that may be answered by the discovery of fossil life on Mars is the importance of sulphur hydrothermal environments to the origin and early evolution of life. On Earth we know that the common ancestor of all life lived in such an environment, which could imply either a critical role of such environments in the origin of life or a point when life in such environments was the sole surviving system on Earth. If we find that the earliest fossils on Mars are also associated with hydrothermally altered sediments and sulphur-bearing deposits, that could be an indication that such environments are critical to early life. On the other hand, if fossils on Mars show no association with sulphur deposits then chance may be the likely explanation.

There are other questions in biology that deal with the entirety of life on the planet and hence can only be answered from the perspective of data on other planets with life. An example is the Gaia hypothesis of James Lovelock which postulates that life on Earth acts as a self-regulating homeostatic system. If life existed on Mars we should be able to see the remnants of Gaian mechanisms that might have acted to prevent the inevitable extinction (McKay and Stoker, 1991).

Restoring life to Mars

Reasoning along these lines, if Mars did indeed once have a habitable climate it may be possible to restore those habitable conditions and restart life there (McKay, Toon and Kasting, 1991). The first step to making Mars habitable is warming the environment, possibly by producing on Mars small amounts of gases that are highly absorbing in the infrared, such as certain chlorofluorocarbons. Warming the planet will bring more carbon dioxide out of the soil and into the atmosphere. Since carbon dioxide is an effective greenhouse gas, this addition would result in a warmer planet. Warming would release even more carbon dioxide, and so on in a positive feedback loop. This effect contributes to making a thick warm carbon dioxide

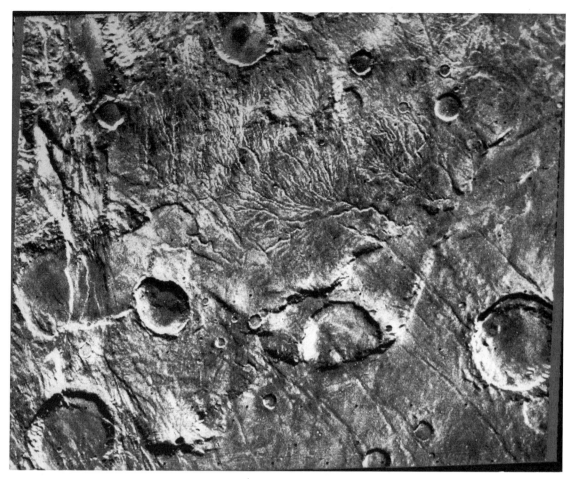

Figure L6 Liquid water on early Mars is indicated by well-developed dendritic channels in the ancient cratered terrain on Mars (48°S, 98°W). The scene is 250 km across. The craters overlying the channels indicate that these features formed about 3.8 Ga ago. This evidence for the existence of liquid water habitats on the Martian surface is the primary motivation for considering the possible origin of life on Mars. (NASA Photo; Viking frame 63A09.)

atmosphere. In order for Martian temperatures to reach those of Earth, this feedback process would have to continue for about 100 years. If this produced an atmosphere of carbon dioxide with a pressure twice that of Earth's atmosphere, the planet would be warm enough for life. At this point Mars would be habitable to many plants and microorganisms. This thick CO_2 atmosphere would be similar to the atmosphere that probably existed on Mars billions of years ago during the epoch when liquid water flowed on the Martian surface. To generate oxygen and make the atmosphere breathable by humans is much more difficult (McKay, Toon and Kasting, 1991).

Life beyond

The search for life in the solar system has been disappointing so far, with the remaining hope being possible fossils on Mars. But technology is advancing to the point where it will soon be possible to peer into other planetary systems and search for life there. Liquid water is the obvious thing to look for, but it may be impossible to detect remotely. It may appear that O_2 would be a good indicator of life, and certainly it is. However, we must keep in mind that for much of its early history Earth had life but did not have an O_2-rich atmosphere. An alternative suggestion has been to look for trace gases that are out of chemical equilibrium with the major gases. On Earth these are often associated with biogenic emissions (e.g. CH_4). However, caution is again warranted since we see photochemical production of non-equilibrium trace gases in many planetary atmospheres. The search for life outside the solar system may prove as challenging at the search for present and past life within it.

Christopher P. McKay

Bibliography

Biemann, K., Oro, J., Toulmin, P. III *et al.* (1977) The search for organic substances and inorganic volatile compounds in the surface of Mars. *J. Geophys. Res.*, **82**, 4641–58.

Cairns-Smith, A.G. (1982) *Genetic Takeover and the Mineral Origins of Life*, Cambridge: Cambridge University Press.

Carr, M.H. (1987) Water on Mars. *Nature*, **326**, 30–5.

Chyba, C. and Sagan, C. (1992) Endogenous production, exogenous delivery and impact-shock synthesis of organic molecules: an inventory for the origins of life. *Nature*, **355**, 125–32.

Clark, B.C., Baird, A.K., Weldon, R.J. *et al* (1982) Chemical composition of Martian fines. *J. Geophys. Res.*, **87**, 10059–67.

Davies, R.E. (1988) Panspermia: unlikely, unsupported, but just possible. *Acta Astronaut.*, **17**, 129–35.

Davies, R.E. and Koch, R.H. (1991) All the observed universe has contributed to life. *Phil. Trans. Roy. Soc. London*, **B334**, 391–403.

Horgan, J. (1991) In the beginning. *Sci. Am.*, **264**(2), 116–25.

Horowitz, N.H. (1986) *To Utopia and Back: The Search for Life in the Solar System*. New York: W.H. Freeman and Co.

Horowitz, N.H. and Hobby, G.L. (1977) Viking on Mars: the carbon assimilation experiments, *J. Geophys. Res.*, **82**, 4659–62.

Huebner, W.F. and McKay, C.P. (1990) Implications of comet research, in *Physics and Chemistry of Comets* (ed. W.F. Huebner). Berlin: Springer-Verlag, pp. 305–331.

Irvine, W.M., Leschine, S.B. and Schloerb, F.P. (1980) Thermal history, chemical composition and relationship of comets to the origin of life. *Nature*, **283**, 748–9.

Klein, H.P. (1978) The Viking biological experiments on Mars, *Icarus*, **34**, 666–74.

Klein, H.P. (1979) The Viking mission and the search for life on Mars. *Rev. Geophys. Space Phys.*, **17**, 1655–62.
Kvenvolden, K.A. (1974) *Geochemistry and the origin of life*. Stroudsburg, Pennsylvania: Dowden, Hutchinson and Ross, 422 pp.
Lehninger, A.L. (1975) *Biochemistry*. New York: Worth.
Levin, G.V. (1988) A reappraisal of life on Mars. Proceedings of the NASA Mars conference, *Am. Astron. Soc.*, **71**, 187–207.
Levin, G.V. and Straat, P.A. (1977) Recent results from the Viking labeled release experiment on Mars. *J. Geophys. Res.*, **82**, 4663–7.
Levin, G.V. and Straat, P.A. (1981) A search for a nonbiological explanation of the Viking labeled release life detection experiment. *Icarus*, **45**, 494–516.
McKay, C.P. (1986) Exobiology and future Mars missions: the search for Mars' earliest biosphere, *Adv. Space Res.*, **6**(12), 269–85.
McKay, C.P. (1991) Urey Prize Paper: planetary evolution and the origin of life. *Icarus*, **91**, 93–100.
McKay, C.P. and Stoker, C.R. (1989) The early environment and its evolution on Mars: implications for life. *Rev. Geophys.*, **27**, 189–214.
McKay, C.P. and Stoker, C.R. (1991) Gaia and life on Mars, in *Scientists on Gaia* (eds S. Schneider and P.J. Boston). Cambridge, Mass; MIT Press, pp. 375–81.
McKay, C.P. and Davis, W.L. (1991) Duration of liquid water habitats on early Mars, *Icarus*, **90**, 214–221.
McKay, C.P., Toon, O.B. and Kasting, J.F. (1991) Making Mars habitable. *Nature*, **352**, 489–96.
McKay, D.S., Gibson, E.K., Thomas-Keprta, K.L., Vail, H., Romanek, C.S., Clement, S.J., Chillier, X.D.F., Maechling, C.R. and Zare, R.N. (1996) Search for past life on Mars: possible relic biogenic activity in Martian meteorite ALH84001. *Science*, **273**, 924–30.
Miller, S.L. (1992) The prebiotic synthesis of organic compounds as a step toward the origin of life, in *Major Events in the History of Life*, (ed. J.W. Schopf). Boston: Jones and Bartlett, pp. 1–28.
Oyama, V.I. and Berdahl, B.J. (1977) The Viking gas exchange experiment results from Chryse and Utopia surface samples. *J. Geophys. Res.*, **82**, 4669–76.
Raulin, F., Mourey, D. and Toupance, G. (1982) Organic syntheses from CH_4–N_2 atmospheres: implications for Titan. *Origins of Life*, **12**, 267–79.
Reynolds, R.T., McKay, C.P. and Kasting, J.F. (1987) Europa, tidally heated oceans, and habitable zones around giant planets. *Adv. Space Res.*, **7**(5), 125–32.
Reynolds, R.T., Squyres, S.W., Colburn, D.S. and McKay, C.P. (1983) On the habitability of Europa. *Icarus*, **56**, 246–54.
Sagan, C. (1974) The origin of life in a cosmic context. *Origins of Life*, **5**, 497–505.
Schidlowski, M. (1988) A 3,800-million-year isotopic record of life from carbon in sedimentary rocks, *Nature*, **333**, 313–18.
Schopf, J.W. (ed.) (1983) *Earth's Earliest Biosphere; Its Origin and Evolution*. New Jersey: Princeton University Press.
Woese, C.R. (1987) Bacterial evolution. *Microbiol. Rev.*, **51**, 221–71.

Cross references

Carbon, carbon dioxide
Chemical element
Comet: impacts on the Earth
Cosmochemistry
Earth
Mass extinction
Miller–Urey experiment
Tidal heating
Viking mission

LITHOSPHERE

The hard, rigid outer shell of terrestrial planets. From the Greek *lithos* (rock) and *spheira* (globe or ball), the lithosphere is a structural component of planet Earth and Earth-like celestial bodies that is hard and rocky and thus distinct from the fluid hydrosphere (q.v.) and gaseous atmosphere (q.v.) that cover it, and the 'no-strength' asthenosphere (q.v.) that lies below it (Plate 10).

The thickness of Earth's lithosphere ranges from zero at an active sea-floor spreading center where the asthenosphere (in the form of basaltic lava eruptions) reaches the surface, to perhaps 100–150 km beneath the long-stable parts of the continents. Dennis *et al.* (1979, p. 25) review some of the historical literature on the lithosphere's thickness. Osmond Fisher (1889, other citations in Dennis *et al.*), author of the first treatise on geophysics, used 'crust' in a manner similar to lithosphere, and many others followed this lead. Fisher supposed it to rest on a 'vitreous substratum', which is what Barrell (1914, p. 659) later called 'asthenosphere'. The 'crust' was limited to that part of the lithosphere lying above the Mohorovičic discontinuity.

An isostatically responsive upper zone of lithosphere, generally about 20–30 km thick, is sometimes called an elastic or flexural lithosphere, and can be identified from its reaction to the application or removal of loads, such as glacial ice sheets; the response time is initially very rapid, a few centuries, but there is an exponential decay period of $> 10\,000$ years.

In contrast, below it, a thermal lithosphere is visualized in terms of a thermal gradient that is traceable (by calculation) to a depth of ~ 100 km.

Lateral inhomogeneity within the continental lithosphere is suspected from seismic data, and has long been deduced from structural and isostatic studies of mountain 'roots'. Both surface geology and geophysical studies of the upper mantle suggest that crust formation, through geologic history, has been a two-phase process: (1) an early period characterized by diapiric rise of cooling, lower-density crystal differentiates (traces of which are exposed today in the early Precambrian structures of the mature shields); and (2) a later period, since about 2.5 Ga, when inhomogeneities in the lithosphere began to develop, and thus led to the progressive growth of the continental crust during plate tectonics. In geohistory this growth is essentially irreversible, although continental crust is subject to stretching and foundering in basin–range situations and in the ocean in marginal basins (e.g. Japan Sea, South China Sea and Coral Sea), is subject to recycling with a limited degree of 'oceanization'.

Through time, continental lithosphere is marked by progressive growth, a physical evolution that is episodic, marked by relatively brief and violent accelerations (Umbgrove, 1947), interspersed by longer intervals of relative quiescence.

Rhodes W. Fairbridge

Bibliography

Anon. (1983) *The Lithosphere: Report on a Workshop*. Washington, DC: National Academy Press.
Barrell, J. (1914) The strength of the Earth's crust. *J. Geol.*, **22** (8 articles; also ibid., A15, 23, 3 articles).
Cloos, M. (1993) Lithospheric buoyancy and collisional orogenesis: subduction of oceanic plateaus, continental margins, island arcs, spreading ridges, and seamounts. *Geol. Soc. Am. Bull.*, **105**, 715–37.
Dennis, J.G. *et al.* (eds.) (1979) *International Tectonic Lexicon*. Stuttgart: E. Schweizerbart, 153 pp.
Dennis, J.G. and Atwater, T.M. (1974). Terminology of geodynamics. *Am. Assoc. Petrol. Geol. Bull.*, **58**, 1030–6.
Fisher, O. (1889) *Physics of the Earth's Crust*. London: Macmillan, 299 pp.
Gutenberg, B. (1959) *Physics of the Earth's Interior*. New York: Academic Press, 240 pp.
Maxwell, J.C. (1984) What is the lithosphere? EOS, *Trans. Am. Geophys. Union.*, **65**, 321–5.
Umbgrove, J.H.F. (1947) *The Pulse of the Earth*, 2nd edn, The Hague: Nyhoff, 258 pp. (1st edn 1942).

Acknowledgement

Manuscript kindly reviewed and improved by Clare P. Marshall, Denver.

Cross references

Asthenosphere
Atmosphere
Core, terrestrial planetary
Crust
Hydrosphere
Mantle
Plate tectonics

LOMONOSOV, MIKHAEL VASILYEVICH (1711–1765)

Particularly known for his achievement in being the first scientist to establish the existence of an atmosphere on the planet Venus, Lomonosov was also distinguished in many other branches of science. These fields included geology, oceanography, physics, chemistry and meteorology.

Lomonosov founded the University of Moscow and was one of the early members of the Russian Academy of Sciences. In geology (Tikhomirov, 1969) he is credited with theories on the origin of mountain building, volcanoes, the cause of earthquakes and the origin of petroleum and ore deposits. His name is honored by oceanographers in identifying the major submarine mountain belt in the Arctic, the Lomonosov Ridge. Of fundamental importance to planetary science was his insistence on the Earth's long, evolving history and its continual change, always in terms of an actualistic philosophy, which was constrained by the laws of physics and chemistry.

International attention of astronomers was focused on a transit of Venus predicted for 1761, on May 26 (June 6 of the new calendar). It was the first to be observed in many different places at the same time, and offered a unique opportunity to determine accurately the solar parallax, and thus its distance from the Earth. Russian observers were set up in Moscow (at the Observatory of the Science Academy), and also in two quite remote locations, Irkutsk and Selengisk. Lomonosov was able to observe during the transit.

In the words of Sharonov (1964, p. 7):

'As he watched Venus encroaching on the solar disk he noticed that the part of the planet's disk still seen against the background of the sky was edged by a narrow luminous rim nearly as bright as the photosphere; this was repeated when Venus moved off the solar disk. Lomonosov was perfectly right to attribute the rim to solar rays bending around Venus because of refraction in a gaseous atmosphere (he also drew a diagram of the ray paths).

Lomonosov concluded that 'the planet Venus is surrounded by an aerial atmosphere like (if not greater than) that which enfolds our terrestrial globe'. This refractive behavior is known as the 'Lomonosov phenomenon' (Sharonov, 1964, p. 298).

Rhodes W. Fairbridge

Bibliography

Khain, V.Y. (1963) Lomonosov and modern geology. *Internat. Geol. Rev.*, **5**, 706–15.
Menshutkin, B.V. (1952) *Russia's Lomonosov*. Princeton, NJ: Princeton University Press, 208 pp., transl. by J.E. Thal and E.J. Webster (2nd edn, 1970, Westport, Conn.: Greenwood Press, 208 pp).
Morosow, A.A. (1954). *Michail Wassiljewitsch Lomonosov 1711–1765*. German Translation (from the Russian) by W. Hoepp. With an introduction by S.I. Wawilow. Berlin: Rutten and Loening, 627 pp.
Parry, A. (1972) Mikhail Lomonosov. One-man university, in *The Russian scientist*. Russia Old and New Series. New York: Macmillan Co.; London: Collier-Macmillan, 196 pp., pp. 25–37.
Schutz, W. (1970) *Michail W. Lomonossow*. Biographien hervorragender Naturwissenschaftler und Techniker, Vol. 7. Leipzig, Germany: 104 pp. (2nd edn, 1976).
Sharonov, V.V. (1964) *The Nature of the Planets*. Moscow: Gosud. Izdat. Fiz. Mat. Lit. (1958: *Priroda Planet*), trans. by Israel Progr. Sci. Transl., 420 pp.
Tikhomirov, V.V. (1969) The development of the geological sciences in the U.S.S.R. from ancient times to the middle of the nineteenth century, in *Toward a History of Geology* (ed. C.J. Schneer). Cambridge, Mass.: MIT Press, pp. 357–385.
Vucinich, A. (1963) *Science in Russian culture. A History to 1860*. London: Peter Owen, 463 pp.

LOWELL, PERCIVAL (1855–1916)

An astronomer and writer, Lowell was born in Boston of a long line of distinguished forebears, many involved in the textile industry. The town of Lowell, Massachusetts, was named after one of them.

Lowell graduated from Harvard in 1876, with distinction in mathematics, and took the customary grand tour of Europe, even as far as Syria. He then settled down to work in the cotton industry, but after 6 years of shrewd investments he was free to travel again, including to the Far East. In Korea he was so well received that he was appointed their foreign secretary and counsellor for a Korean diplomatic mission to the United States in 1883. He wrote articles for the *Atlantic Monthly* and four books about the region, including *The Soul of the Far East* (1888).

Lowell was also interested in astronomy: his Harvard commencement speech had been on the nebular hypothesis, explaining the origin of the solar system, and he sometimes took a telescope on his travels. One unconfirmed story has Lowell learning in 1893 that the eyesight of the Italian astronomer Giovanni Schiaparelli, observer of 'canali' on Mars, was failing. Lowell, with his extremely keen eyesight, now decided that it was his manifest destiny to continue Schiaparelli's work.

In February 1894 Lowell announced that he would establish an observatory on a 7000-ft (2100 m) mountain near Flagstaff, in the Arizona Territory, and search for signs of intelligent life on Mars. In 1895 his controversial hypothesis of intelligent life on Mars was published, the subject of several articles in the *Atlantic Monthly* and in the book *Mars*. From visible changes, including changes in the polar cap and seasonal changes in the tint of dark areas, Lowell concluded that Mars had both an atmosphere and water, and together with mild temperatures thus could support life. Were there signs of actual inhabitants? Seemingly there was an irrigation system of straight canals (unfortunately visible only to Lowell and a few other enthusiasts, taken to be evidence for 'intelligent life'. It was postulated that following a progressive loss of atmosphere, the climate had become drier and a complex irrigation system had to be devised to bring water from the polar caps to the thirsty Martians living near the equator. For Lowell it was enough that the theory might be true, but professional scientists envisioned alternative explanations. Furthermore, scientists are normally expected to observe first and theorize secondly, if at all. W.W. Campbell at the Lick Observatory objected that 'Mr. Lowell went direct from the lecture hall to his observatory, and how well his observations established his pre-observational views is told in his book.'

Lowell's belief of canals on Mars is now thoroughly discredited. Perhaps it was the product of a psychological inclination to connect minute details. Perhaps it represents an instance of preconception influencing observation. In his obituary of Lowell, the Princeton astronomer Henry Norris Russell warned that 'if the observer knows in advance what to expect . . . his judgment of the facts before his eyes will be warped by this knowledge, no matter how faithfully he may try to clear his mind of all prejudice. The preconceived opinion unconsciously, whether he will or not, influences the very report of his senses . . . .

The planet Neptune was discovered in 1846 based on calculations attempting to resolve observed perturbations in the orbit of Uranus. Lowell believed that perturbations in the motions of Uranus, Neptune and comets suggested the existence of another planet, and in 1905 he began calculating the position of his 'Planet X'. Lowell and five mathematicians in Boston worked on the calculations while another seven observers at Lowell Observatory searched for the object. The search involved taking sets of photographs, each set of the same region of the sky but at different times, and then comparing photographs within each set, looking for objects that had changed position over time.

Lowell left more than a million dollars in trust for the observatory to ensure continuation of his work. When the search for Planet X resumed in 1929, a new astronomer, Clyde Tombaugh (q.v.), enthusiastically undertook the tedious task of photographing the same sections of the sky every few nights and comparing the images on two plates. Movement of an image could indicate a new planet. Tombaugh checked thousands of images before finally discovering the new planet in February 1930, on plates taken in January. The announcement was made on 13, March 1930, the 75th anniversary of Percival Lowell's birth. Continuing the tradition of naming planets for Roman gods, the director of the observatory chose the name Pluto, god of the regions of darkness, and PL (superposed) for the planetary symbol, Percival Lowell's initials. The name was first suggested by an 11 year old English girl named Venetia Burney.

Much controversy surrounded the discovery of Pluto and whether it really was Lowell's 'Planet X'. Pluto is less massive than the object predicted by Lowell and its very eccentric and inclined orbit does not agree well with that expected on the basis of Lowell's calculations.

Some of these controversies continue today (Hoyt, 1980), but everyone agrees that without Lowell's dedication to the search, the discovery of Pluto would have been delayed for quite some time.

Lowell left planetary scientists a rich heritage of observational data, and Lowell Observatory is today one of the foremost astronomical institutions in the world.

Nadine G. Barlow and Norriss S. Hetherington

Bibliography

Hetherington, N.S. (1971) Lowell's theory of life on Mars, *Astron. Soc. Pacific Leaflet*, pp. 1–8.
Hetherington, N.S. (1981) Percival Lowell: scientist or interloper? *J. Hist. Ideas*, **42**, 159–61.
Hoyt, W.G. (1976) *Lowell and Mars*. Tucson: University of Arizona Press.
Hoyt, W.G. (1980) *Planets X and Pluto*. Tucson: University of Arizona Press.
Lowell, A.L. (1935) *Biography of Percival Lowell*. New York: Macmillan.
Lowell, P. (1895) *Mars*. Boston and New York: Houghton, Mifflin and Co.
Marsden, B.G. (1973) Lowell, Percival. *Dict. Sci. Biogr.*, Vol. 8, pp. 520–3.
Sheehan, W. (1988) *Planets and Perception*. Tucson: University of Arizona Press.

Cross references

Chamberlin–Moulton planetesimal hypothesis
Mars
Nebular hypothesis (Kart–Laplace)
Planet X
Pluto
Tombaugh, Clyde William

LUNA MISSIONS

The Soviet Union flew 24 'Luna' missions from 1959 to 1976. These missions included three generations of launchers and a variety of payloads. The missions included flybys, orbiters, hard and soft landers, robotic rovers and even unmanned sample return missions (Table L2). Several are still of historical or scientific importance.

The first generation of Luna missions was launched in 1959 on a modified version of the booster that launched Sputnik. Luna 1, in January, flew past the Moon, becoming the first spacecraft to escape Earth's gravity. Luna 2, in September, crashed on the Moon to become the first terrestrial vehicle on another solar system body, but the most dramatic of the group was Luna 3, which returned the first photographs of the mare-poor far side of the Moon in October.

The second generation, in 1965–1968, started with several failures, but eventually included the first soft lander (Luna 9) and first orbiter (Luna 10), beating the American Surveyor and Lunar Orbiter by a few months.

The third generation, Luna 15–Luna 24, were launched from 1969 to 1976 on the larger D1e booster, and remain, in the mid-1990s, the most technologically sophisticated unmanned exploration missions to any extraterrestrial body. Two of them, Luna 17 and Luna 21, landed remote-controlled tank-like vehicles, the 'Lunokhods', which were driven on the lunar surface for a total of 16 months between them.

The most lasting scientific legacy of the Luna program comes from Luna 16, Luna 20 and Luna 24, which returned a total of about 300 g of material. The missions sampled three locations that are basically along a 400 km long north–south line at 60°E longitude. The locations are important, because they are about 30° east of the easternmost Apollo landing site (Apollo 17). Therefore, these samples greatly extend the surface area on the Moon for which samples are available on Earth. Some regional chemical trends can be discerned by comparing the Apollo and Luna samples.

An account of the Luna missions in the context of the US–USSR competition is given in Lewis and Lewis (1987). Results of analyses of many of the returned Luna samples are contained in collections of papers edited by Anders and Albee (1973) and Merrill (1978).

Timothy D. Swindle

Bibliography

Anders, E. and Albee, A.L (eds) (1973) Luna 20: a study of samples from the lunar highlands returned by the unmanned Luna 20 spacecraft. *Geochim. Cosmochim. Acta*, **37**, 719–1109.
Lewis, J.S and Lewis, R.A. (1987) *Space Resources: Breaking the Bonds of Earth*. New York: Columbia University Press.

Table L2 Luna missions[a]

Luna	Date	Result
1	2 Jan 1959	Flew within 6000 km – first interplanetary spacecraft
2	9 Sep 1959	Impacted – first spacecraft to reach another body
3	4 Oct 1959	Flyby – first photographs of lunar far side
4	2 Apr 1963	Missed Moon by 8500 km
5	9 May 1965	Crashed on Moon
6	8 Jun 1965	Crashed on Moon
7	4 Oct 1965	Crashed on Moon
8	3 Dec 1965	Crashed on Moon
9	31 Jan 1966	First successful soft landing on Moon
10	31 Mar 1966	First successful lunar orbiter
11	24 Aug 1966	Orbiter
12	22 Oct 1966	Orbiter – photographic
13	21 Dec 1966	Soft lander
14	7 Apr 1968	Orbiter – mapped gravity field
15	13 Jul 1969	Crashed on Moon
16	12 Sep 1970	First unmanned sample return (0°41'S, 56°18'E)
17	10 Nov 1970	Lunokhod 1 rover – first robotic rover
18	2 Sep 1971	Crashed on Moon
19	28 Sep 1971	Orbiter – photographic
20	14 Feb 1972	Sample return (3°32'N, 56°18'E)
21	8 Jan 1973	Lunokhod 2 rover
22	29 May 1974	Orbiter – photographic
23	28 Oct 1974	Failed sample return – drill arm damaged
24	9 Aug 1976	Drill core sample return (12°45'N, 60°12'E)

[a] Data from Lewis and Lewis (1987).

Merrill, R.B (ed.) (1978) *Mare Crisium: The View from Luna 24*. Elmsford, NY: Pergamon Press.

Cross references

Apollo missions
Moon: geology

LUNAR METEORITES

Lunar meteorites are lunar samples that arrived at the Earth by natural processes, not through the efforts of the space program. They were blasted off the Moon by impacts, traveled through space and survived the fiery entry through the Earth's atmosphere. They are the only meteorites for which the parent body is known with certainty, but this would not be the case if we had not already studied lunar samples returned by the Apollo and Luna missions.

The first lunar meteorite, Allan Hills 81005 (ALHA81005; Figure L7), was found in Antarctica in 1982. When the results of studies by 19 research groups were published (Bogard, 1983), it was clear that ALH81005 was a sample from the light-colored lunar highlands (see Moon: geology). The discovery of this first lunar meteorite was important both for meteoritics and for lunar science. It was the first meteorite from a known source and it provided new information about the lunar surface to complement that from returned lunar samples.

Between 1983 and 1989 new lunar meteorites were identified in the Antarctic collections (see Antarctic meteorites) at a rate of about one per year. All of these early lunar meteorites were thought to be highlands rocks. Then in late 1989 several investigators (Delaney, 1989; Warren and Kallemeyn, 1989) reported that EET87521 (Figure L8) was a basaltic meteorite from the dark lunar mare rather than a eucrite, as it had previously been classified (see Basaltic achondrite meteorite). During the next year three more basaltic lunar meteorites were identified: Y793169, which had also been classified as a eucrite; Y793274, which had been classified as a lunar highlands meteorite; and A881757 (formerly Asuka 31), a recent find. The most recently identified lunar meteorite is Calcalong Creek from Australia (Hill, Boynton and Haag, 1991), the first lunar meteorite found outside Antarctica. It is a KREEP-rich highland rock (see Moon: geology).

Table L3 lists all 12 lunar meteorites known as of mid-1993 (Lindstrom, 1991; Yanai and Kojima, 1991). The discovery date is the year the meteorite was identified as being of lunar origin, not the year of collection. Antarctic meteorite names include both a geographic location (which may be abbreviated to 1–4 letters) and a 5–7 digit sample number, the first two digits of which stand for the year the collection season began. Most of the lunar meteorites are

Figure L7 Allan Hills 81005, the first lunar meteorite. It is a highlands breccia consisting of white anorthositic clasts in a medium gray matrix. The cube is 1 cm.

Figure L8 Elephant Moraine 87521, the first basaltic lunar meteorite. It is a mare breccia with some light clasts in a dark gray basaltic matrix. The cube is 1 cm.

quite small (< 60 g), but three are over 400 g. The Antarctic lunar meteorites are curated by US and Japanese national programs (see Antarctic meteorite), while Calcalong Creek is privately owned.

The lunar highlands meteorites are all anorthositic breccias made up of a mixture of various rock types, but dominated by anorthositic rocks. They are similar to some rocks from the Apollo 16 site and have compositions very similar to the estimated 'average' highlands composition (Taylor, 1982). They vary somewhat in texture and composition: some are clearly regolith breccias (e.g. ALHA81005), while others (e.g. Y86032) are not. Some are ferroan, while others are magnesian. The four lunar mare meteorites also vary in texture and composition: two are coarse-grained igneous rocks (A881757 and Y793169) and two are breccias (EET87521 and Y793274) dominated by mare basalts. The mare meteorites are similar to some mare rocks collected at the Apollo 15, 17 and Luna 24 sites. Calcalong Creek is a breccia enriched in an incompatible element-rich KREEP component is similar to rocks from the Apollo 14 site.

Lunar meteorites are samples from unknown sites on the lunar surface and may include our first samples of the farside of the Moon. The significance of the lunar meteorites to our understanding of the nature and evolution of the Moon's crust depends in part upon how many distinct meteorites are collected and how many different sites they represent. If the 12 meteorites are all from different sites, they tell us much more about the lunar surface than if they are from just a few sites. The entire suite of Apollo and Luna samples are from just nine sites on the nearside of the Moon.

Meteorites often break apart as they fall. All meteorites collected in Antarctica are given separate names (see Antarctic meteorite), but some are later found to be samples of the same meteorite fall, or to be paired meteorites based on similarities in (1) location of find, (2) petrography and composition and (3) exposure to radiation. Table L4 shows that several of the lunar highlands meteorites are paired: Y82092, Y82093 and Y86032 are from a single fall, as are MAC88104 and MAC88105. Thus the seven highlands meteorites represent just four meteorite falls. The four mare meteorites and Calcalong Creek are all distinct specimens.

The proximity of the Moon to the Earth makes it possible for two unpaired meteorites to have been ejected from the Moon by the same impact but to have arrived at Earth as separate meteorites. These meteorites are called paired ejecta. It is more difficult to prove that two samples are paired ejecta than that they are paired meteorites. The meteorites must have the same time of ejection from the Moon but they need not have the same petrography, composition or crystallization age. The eleven well-studied lunar meteorites fall into four groups based on their times of lunar ejection. The first two are groups of paired meteorites: (1) Y82192, Y82193 and Y86032 at 11 Ma and (2) MAC88104 and MAC88105 at 230 ka. The remaining groups are (3) A881757 and Y793169 at 1 Ma and (4) ALHA81005, Y791197, EET87521 and Y793274 at < 0.1 Ma. These groups may represent paired ejecta or meteorites ejected at similar times from different sites. Our experience with samples collected during the Apollo missions is that there is some diversity in petrography and composition at each site, and that the amount of diversity increases with the geologic complexity of the site, but that we can generally distinguish samples from one site from those of another. Thus, the more similar two lunar meteorites are in petrography and composition,

Table L3 Lunar meteorites

Discovery	Location/number	Classification	Weight (g)
1982	Allan Hills (ALHA) 81005	Anorthositic breccia	31
1984	Yamato (Y) 791197	Anorthositic breccia	52
1985	Yamato (Y) 82192	Anorthositic breccia	37
1986	Yamato (Y) 82193	Anorthositic breccia	27
1987	Yamato (Y) 86032	Anorthositic breccia	648
1987[a]	Yamato (Y) 793274	Basalt-rich breccia	9
1989	MacAlpine Hills (MAC) 88104	Anorthositic breccia	61
1989	MacAlpine Hills (Mac) 88105	Anorthositic breccia	662
1989[b]	Elephant Moraine (EET) 87521	Basaltic breccia	31
1990	Asuka (A) 881757	Mare gabbro	442
1990[b]	Yamato (Y) 793169	Mare gabbro	6
1991	Calcalong Creek, Australia	KREEP-rich breccia	18

[a] Y793274 was classified as a lunar anorthositic breccia in 1987, and reclassified as a basalt-rich breccia in 1990.
[b] EET87521 and Y793169 were originally classified as eucrites.

Table L4 Lunar meteorite pairings

Proposed pairings	Based on	Lunar site type
Certainly paired meteorites		
Y82192–82193–86032	Same loc, min/pet, comp, c-age, exp, ejct	Ferroan highlands (rare mare)
MAC88104–88105	Same loc, min/pet, comp, c-age, exp, ejct	Ferroan highlands (rare mare)
Probably paired ejecta		
A881757–Y793169	Same c-age and ejct, similar but different min/pet, comp, exp, different loc	Low Ti mare
Possibly paired ejecta		
ALHA81005–Y791197	Same ejct, similar exp, different loc, min/pet, comp, c-age	Diverse highlands (rare mare)
EET87521–Y793274	Same ejct, similar exp, different loc, min/pet, comp, c-age	Diverse mare (some KREEP and highlands)
Unlikely, but possibly paired ejects		
ALHA81005–Y791197 and EET87521–Y793274	Same ejct, similar exp, very different loc, min/pet, comp, c-age	Very diverse mare–highlands interface

loc = location of find; min/pet = mineralogy/petrography; comp = bulk composition; c-age = crystallization age; exp = exposure to radiation; ejct = time of ejection from moon.

the more likely they are to have been ejected from the same site on the Moon. Table L4 lists group 3 samples A881757 and Y793169 as probably paired ejecta because they are similar basaltic meteorites (Yanai *et al.*, 1993). Group 4 meteorites are very diverse. ALHA81005 and Y791197 are both highland breccias, while EET87521 and Y793274 are both mare breccias. We consider both of these pairs of meteorites to be only possibly paired ejecta because they exhibit major differences in petrography and composition (especially Mg/Fe ratio) which require moderately diverse sites. It is unlikely, but remotely possible, that all four of these meteorites are from the same site, but if so it must have been a very diverse site.

Thus it is fairly certain that the 12 lunar meteorites represent nine distinct falls. It is possible that several of the unpaired lunar meteorites were ejected from the same site and they may represent from four to eight separate impacts on the Moon. Thus the lunar meteorites represent approximately as many locations on the Moon as the Apollo samples and their compositions shed important new light on lunar evolution (see Moon: geology).

Marilyn M. Lindstrom

Bibliography

Bogard, D.D. (ed.) (1983) A meteorite from the Moon. *Geophys. Res. Lett.*, **10**, 733–840.

Delaney, J.W. (1989) Lunar basalt breccia identified among Antarctic meteorites. *Nature*, **342**, 889–90.
Eugster O. (1989) History of meteorites from the Moon collected in Antarctica. *Science*, **245**, 1197–202.
Hill, D.H., Boynton, W.V. and Haag, R.A. (1991) A new lunar meteorite collected near Calcalong Creek Australia. *Nature*, **352**, 614–7.
Lindstrom, M.M., Schwarz, C., Score, R. and Mason, B. (1991) Mac Alpine Hills 88104 and 88105 lunar highland meteorites: general description and consortium overview. *Geochim. Cosmochim. Acta*, **55**, 2999–3007.
Nishiizumi, K., Arnold, J.R., Caffee, M.W. *et al.* (1992) Cosmic ray exposure histories of lunar meteorites Asuka 881757 and Yamato 793169, and Calcalong Creek. *Papers Sym. Ant. Met.*, **17**, NIPR, Tokyo.
Taylor, S.R. (1982) *Planetary Science – A Lunar Perspective*. Houston: Lunar and Planetary Institute.
Warren, P.H. and Kallemeyn, G.W. (1989) Elephant Moraine 87521: the first lunar meteorite composed of predominantly mare material. *Geochim. Cosmochim. Acta*, **53**, 3323–30.
Yanai, K. and Kojima H. (1991) Varieties of lunar meteorites recovered in Antarctica. *Proc. NIPR Sym. Ant. Met.*, **4**, pp. 70–90.
Yanai K., Takeda, H., Lindstrom, M.M. *et al.* (1993) Consortium reports of lunar meteorites Yamato 793169 and Asuka 881757, A new type of mare basalt. *Lunar Planet. Sci. XXIV*, 1555–6.

Cross references

Antartic meteorites
Basaltic achondrite meteorites
Impact cratering
Meteorite
Moon
Moon: geology

LUNAR ORBITER MISSIONS

The Lunar Orbiter program was one of NASA's two robotic spacecraft programs mounted in the late 1960s to provide the information necessary for planning the landings of men on the Moon. The other program comprised the Surveyor landers (q.v.).

The program was directed by the Langley Research Center in Hampton, Virginia. Its primary objective was to obtain detailed stereoscopic photographs of the equatorial area on the nearside of the Moon where Apollo landings were to occur. This objective was so capably accomplished by the first three orbiters, launched between August 1966 and February 1967, that the fourth and fifth missions were devoted to broader scientific objectives, including photographing virtually the entire lunar surface from polar orbit.

The photographs were recorded on 70-mm film, which was developed on board and passed through a photoelectric scanner. The resulting data were transmitted back to Earth. The orbiters also carried instrumentation to assess the probability of danger to the astronauts from radiation or micrometeorites. The precise orbital navigation data acquired by the Deep Space Network (q.v.) revealed the presence of gravitational variations caused by buried concentrations of mass ('mascons'; q.v.).

Conway W. Snyder

Bibliography

Hallion, R.P. and Crouch, T.D. (1979) *Apollo: Ten Years Since Tranquillity Base.* Washington: Smithsonian Institution Press.

Cross references

Apollo missions
Deep Space Network
Mascon
Ranger missions
Surveyor missions

LYMAN ALPHA

Lyman alpha transition of atomic hydrogen

The first transition, $1s^2S^0 - 2p^2P^0$, from the fundamental state of atomic hydrogen, a very simple atom, is perfectly known and could be calculated very precisely. The average value of this doublet corresponds to a spectral transition at 121.5664 nm, in the far ultraviolet. Theodore Lyman (1874–1954) discovered it as the first member of a series of lines ending at 91.2 nm and named it Lyman alpha. Because atomic hydrogen is the most abundant element in the universe this spectral line is expected to be observable in many astrophysical objects. However, since the Earth's atmosphere is entirely opaque at wavelengths shorter than 300 nm, the first astrophysical observations of the Lyman alpha line were made only when rocketborne instruments were flown above the Earth's atmosphere.

Solar Lyman alpha

The solar Lyman Alpha emission was observed for the first time by Rense (1953), and was easily identified due to its brightness; in effect there is as much energy in that single solar line as in the whole solar spectrum below 150 nm. Purcell and Tousey (1960) described all the line characteristics: 0.1 nm wide, this line presents a broad central reversal between two peaks separated by about 0.04 nm. Morton and Widing (1961) showed that this line is produced in the solar chromosphere and were able to reproduce through radiation transfer calculations all its main characteristics.

Being produced in a region of the solar atmosphere very sensitive to solar activity, the line intensity is variable with the solar 11-year cycle and 27-day rotation. Hinteregger (1965) observed these changes for the total line flux, while after the survey of the total and central line flux over almost a whole cycle, Vidal-Madjar (1975) was able to produce empirical relations linking these two fluxes to classical solar activity indices (the Zurich sunspot number and the solar flux at 10.7 cm). Variations of more than a factor of two were observed (Bonnet *et al.*, 1978). Observation of the flux at the center of the line is particularly important since it is related to the analysis and interpretation of most Lyman alpha observations made within the solar system.

Geocoronal Lyman alpha

An intense high-altitude night-time Lyman alpha emission was first discovered by Byram *et al.* (1957). Then Kupperian *et al.* (1959) showed that this emission had a minimum in the antisolar direction, and that surprisingly some emission was coming from the Earth, below the rocket itself. Johnson and Fish (1960) proposed the correct interpretation, that this emission is produced by hydrogen atoms present in the Earth's upper atmosphere (the geocorona), scattering back the solar Lyman alpha photons. With detailed radiative transfer calculations, Thomas (1963) and Donahue (1966) were able to confirm this hypothesis and showed that the hydrogen geocorona was very hot (more than 1000 K) and extends to very high altitudes (several thousand kilometers). More satellite observations (Bertaux and Blamont, 1971; Emerich and Cazes, 1977) showed the complex structure of the geocorona and even the existence of collision-induced Lyman alpha emissions.

Planetary Lyman alpha

Lyman alpha emissions have been observed for nearly all solar system planets, producing information on their outer layers through the scattering of the solar Lyman alpha photons by hydrogen atoms. For the terrestrial planets, more or less dense exospheres were observed. However, the surprises really came from the study of the giant planets.

The first rocketborne observation (Moos, Fastie and Bottema, 1969) opened the route for many observations from either observatories in Earth orbit (OAO, Copernicus, IUE) or probes travelling within the solar system (Pioneer, Voyager). The results revealed many unexpected structures, like the Jovian Lyman alpha bulge (Sandel, Broadfoot and Strobel, 1980; Clarke *et al.*, 1980) and very strong auroral activity in several planets, as well as an atomic hydrogen torus associated with the orbit of Titan around Saturn (Broadfoot *et al.*, 1981).

Lyman alpha in comets

The first observations of Lyman alpha emission from comets (Jenkins and Wingert, 1972; Bertaux, Blamont and Festou, 1973) confirmed for the first time the icy conglomerate model which produces through evaporation a huge coma of atomic hydrogen. These observations led to the evaluation of a production rate of 10^{29}–10^{30} water molecules per second in excellent agreement with the 'dirty snowball' model (Whipple, 1950).

Interplanetary Lyman alpha

The Lyman Alpha emission of an interstellar flow of hydrogen atoms entering the solar system was discovered with two experiments on board the satelite OGO-5 (Bertaux and Blamont 1971; Thomas and Krassa, 1971). It revealed that the local interstellar medium is relatively hot (8000 K), presents a density of about 0.05 atoms cm^{-3} and enters at a velocity of the order of 25 km s^{-1}. These observations may lead to a better understanding of the heliopause, the boundary between the solar wind and the local interstellar medium.

Deuterium Lyman alpha

Very interestingly, the heavy isotope of hydrogen, deuterium, is also observable using Lyman alpha. The isotope shift of the line, 0.033 nm, corresponds to a velocity shift of -81 km s^{-1}, large enough when compared to many astrophysical situations. The isotope ratio of the order of 10^{-5}–10^{-4} (Vidal-Madjar, 1991) however makes detection difficult in several sites. This ratio seems to be a tracer of the past formation and thermal history of solar system objects (Owen *et al.*, 1988).

Conclusion

The Lyman alpha line of atomic hydrogen is an extremely sensitive tracer of the most abundant element of the universe (10^6 times more sensitive than the famous 21-cm line). It reveals structures in the less dense regions of most astrophysical sites, and is particularly sensitive to phenomena in the energy range of several tens of electron volts (Vidal-Madjar *et al.*, 1987). It also produces the reference to abundance studies and the access to deuterium, a tracer of the past history of many astrophysical objects, including the universe itself.

Alfred Vidal-Madjar

Bibliography

Bertaux, J.L. and Blamont, J.E. (1971) Evidence for a source of an Extraterrestrial hydrogen Lyman-alpha emission: the interstellar wind. *Astron. Astrophys.*, **11**, 200–17.

Bertaux, J.L., Blamont, J.E. and Festou, M. (1973) Interpretation of hydrogen Lyman alpha observations of comets Bennett and Encke. *Astron. Astrophys.*, **25**, 415–30.

Bonnet, R.M., Lemaire, P., Vial, J.C. *et al.* (1978) The LPSP instrument on OSO8 II. In-flight performance and preliminary results, *Astrophys. J.*, **221**, 1032–61.

Broadfoot, A.L., Sandel, B.R., Shemansky, D.E. *et al.* (1981) Extreme ultraviolet observations from Voyager 1. Encounter with Saturn. *Science*, **212**, 206–11.

Byram, E.T., Chubb, T.A., Freidman, H. and Kupperian, J.E. (1957) in *The Threshold of Space*. Pergamon Press, pp. 203–17.

Clarke, J.T., Weaver, H.A., Feldman, P.D. *et al.* (1980) Spatial imaging of hydrogen Lyman α emission from Jupiter. *Astrophys. J.*, **240**, 696–701.

Donahue, T.M. (1966) The problem of atomic hydrogen. *Ann. Geophys.*, **22**, 175–88.

Emerich, C. and Cazes, S. (1977) Local perturbations of the atomic hydrogen density distribution near the exobase, inferred from D2A Lyman α airglow measurements, *Geophys. Res. Lett.*, **4**, 523–7.

Hinteregger, H.E. (1965) Absolute intensity measurements in the extreme UV spectrum of solar radiation. *Space Sci. Rev.*, **4**, 461–97.

Jenkins, E.B. and Wingert, D.W. (1972) The Lyman-alpha image of comet Tago-Sato-Kosaka (1969g). *Astrophys. J.*, **174**, 697–704.

Johnson, F.S. and Fish, R.A. (1960) The telluric hydrogen corona. *Astrophys. J.*, **131**, 502–15.

Kupperian, J.E., Byram, E.T., Chubb, T.A. and Freidman, H. (1959) Far Ultraviolet radiation in the night sky. *Planet. Space Sci.*, **1**, 3–6.

Moos, H.W., Fastie, W.G. and Bottema, M. (1969) Rocket measurement of Ultraviolet spectra of Venus and Jupiter between 1200 and 1800 Å. *Astrophys. J.*, **155**, 887–97.

Morton, D.C. and Widing, K.G. (1961) The solar Lyman-alpha emission line, *Astrophys. J.*, **133**, 596–605.

Owen, T., Maillard, J.P., de Bergh, C. and Lutz, B.L. (1988) Deuterium on Mars: the abundance of HDO and the value of D/H. *Science*, **240**, 1767–70.

Purcell, J.D. and Tousey, R. (1960) The profile of solar hydrogen Lyman-alpha. *J. Geophys. Res.*, **65**, 370–2.

Rense, W.A. (1953) Intensity of Lyman-alpha line in the solar spectrum. *Phys. Rev.*, **91**, 299–302.

Sandel, B.R., Broadfoot, A.L. and Strobel, D.F. (1980) Discovery of a longitudinal asymmetry in the H Lyman-alpha brightness of Jupiter. *Geophys. Res. Lett.*, **7**, 5–8.

Thomas, G.E. (1963) Lyman α scattering in the Earth's hydrogen geocorona, 1. *J. Geophys. Res.*, **68**, 2639–60.

Thomas, G.E. and Krassa, R.F. (1971) OGO 5 measurements of the Lyman alpha sky background. *Astron. Astrophys.*, **11**, 218–33.

Vidal-Madjar, A. (1975) Evolution of the solar Lyman alpha flux during four consecutive years. *Solar Phys.*, **40**, 69–86.

Vidal-Madjar, A. (1991) The cosmic D/H ratio. *Adv. Space Res.*, **11**, 97–103.

Vidal-Madjar, A., Encrenaz, T., Ferlet, R. *et al.* (1987) Galactic ultraviolet astronomy. *Rep. Progr. Physics*, **50**, 65–113.

Whipple, F.L. (1950) A comet model. I. The acceleration of comet Encke. *Astrophys. J.*, **111**, 375–94.

Cross references

Comet: structure and composition
Solar wind
Sun

M

MAGELLAN MISSION

The Magellan Venus mapping mission marked the end of the reconnaissance phase of NASA's exploration of the planets and the beginning of a new era of specifically focused missions to address scientific questions. Venus is the last planet to be mapped in detail. Magellan was approved in 1983; it was launched from Cape Canaveral aboard the shuttle Atlantis on 4 May 1989, and placed in Venus orbit 15 months later (Figure M1). The mission lasted four cycle periods (one cycle is 243 days, the rotation period of the planet beneath the spacecraft). Following the fourth cycle an aerobraking experiment circularized the orbit to enable the collection of high-resolution gravity data over the poles. The spacecraft was finally destroyed by atmospheric drag on 12 October 1994. The Magellan mission mapped more than 98% of the planet's surface.

Instruments

The Magellan spacecraft (Figure M2) has one instrument, a synthetic aperture radar capable of imaging the surface and mapping surface

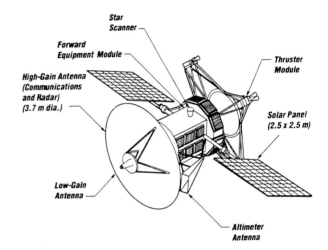

Figure M2 Magellan spacecraft. (Courtesy of JPL.)

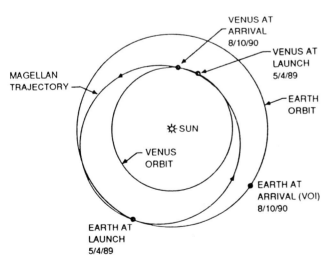

Figure M1 Magellan interplanetary trajectory. Magellan was launched on 4 May 1989 and was transferred to a trajectory that took the spacecraft around the Sun 1.5 times in 15 months. Magellan reached Venus on 10 August 1990, when the firing of a solid rocket slowed the spacecraft so it could be captured by Venus' gravity. (From Saunders *et al.*, 1992, *J. Geophys. Res.*, 97, p. 13067)

topography as an altimeter. The radar also measures radio emissions from the surface, revealing the electrical properties of the surface material. The 12.6 cm wavelength synthetic aperture radar (SAR), resolves features as small as 120 m, through the thick clouds that perpetually hide the planet.

The altimeter measures surface elevations accurate to about 30 m at a resolution of 10–20 km, and the main antenna, when used as a radiometer, records the natural thermal emissions from the surface to help determine the surface composition.

Magellan also measured the gravity field of Venus. Doppler shifts in the radio signal returned to Earth are measured and analyzed; these result from perturbations of the spacecraft orbit caused by gravity anomalies (see Gravimetry; Surface gravity; Venus: gravity). Mapping is not possible during the gravity experiment because the spacecraft's dish antenna must be pointed at the Earth in this phase.

Planetary mapping

The first cycle mission objective was to acquire imaging data from at least 70% of the planet. The mapping orbit had a periapsis (closest approach to the planet) altitude of 300 km and a period of 3 h 15 min (Figure M3). By the end of 1992 Magellan had mapped nearly 98% of the planet's surface (Plate 6; see Venus). During cycle 1 the radar was operated in left-looking mode (toward the east), while in cycle 2 about 50% of the planet was mapped in right-looking mode. Cycle 3 was again in left-looking mode. Data from the different cycles may be combined in stereo pairs to provide high-resolution topographic data.

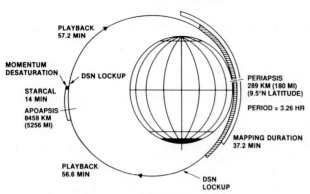

Figure M3 Magellan mapping orbit profile. Operations during a single orbit are described starting at the north pole. (1) Mapping data acquisition began near the north pole when the large antenna was pointed at the surface and the radar was turned on. (2) Mapping continued for 37.2 min to fill the onboard tape recorder. (3) The first point of tape recorder playback commenced near the south pole immediately after mapping. (4) Star calibration and momentum wheel unloading for attitude control were conducted at apoapsis. (5) The second portion of tape recorder playback was conducted between apoapsis and the start of mapping at the north pole. DSN stands for Deep Space Network (q.v.), the receiving end. The period for each orbit was 3.2 h. (From Saunders et al., 1992, J. Geophys. Res., **97**, p. 13067).

Scientific results

Magellan has revealed a unique global volcanic and tectonic style on Venus. Regional tectonism is evident in the widespread horizontal compression and stretching of the surface material (see Venus: geology and geophysics). Large deformed regions are the result of crustal motion in the planet's geologic past. Venus apparently has a dynamic mantle that drives ongoing crustal warping. There is little evidence of Earth-like plate tectonics, however. Volcanism on Venus is global. Magellan data and present knowledge of cratering rates in the inner solar system suggest that the age of the surface is on the order of 500 Ma. A global resurfacing event may have occurred in the geologically recent past.

The detailed topographic data from stereo (paired left-looking sets from cycles 1 and 3) radar data are valuable for unraveling stratigraphic relations in the volcanic terrains, and in understanding the relationships between craters (q.v.), tesserae (q.v.), and coronae (q.v.).

More generally, analysis of Magellan data will advance our understanding of the formation, structure and evolution of Venus. Comparisons of Venus and Earth may lead to new insights into the fundamental workings of our own planet.

R. Stephen Saunders

Bibliography

Arvidson, R.E., Phillips, R.J. and Izenberg, N. (1992) Global views of Venus from Magellan. *EOS*, April 14, 161–9.

Saunders, R.S. and Pettengill, G.H. (1991) Magellan: mission summary. *Science*, **252**, 247–9. (This issue contains numerous other papers describing Magellan results.)

Saunders, R.S., Spear, A.J., Allin, P.C. et al. (1992) Magellan mission summary. *J. Geophys. Res.*, **97**, 13067–90. (This volume contains numerous other papers describing Magellan results.)

Cross references

Gravimetry
Surface gravity
Venus
Venus: geology and geophysics
Venus: gravity

MAGNETISM

Magnetic fields are ubiquitous in the solar system. The Sun has a magnetic field, giving rise to an interplanetary field which permeates the solar system. The giant outer planets – Jupiter, Saturn, Uranus and Neptune – all have magnetic fields. Among the inner planets, the Earth and Mercury have internal magnetic fields, and Mars may have one. In the Sun, the giant outer planets and the Earth, the fields are maintained by magnetohydrodynamic processes taking place in regions of high electrical conductivity.

On Earth a paleomagnetic record of the geomagnetic field has been preserved by the natural remanent magnetization (NRM) of rocks. This has provided a history of the field, demonstrating both that it reverses polarity and that its morphology and strength fluctuate significantly. The geomagnetic field has played a role in the evolution of life on this planet – it partially shields life forms from the charged particles of the solar wind and cosmic rays, and it provides a possible means of navigation. Moon rocks also carry a paleomagnetic record of ancient fields, as do certain meteorites.

The Sun is not alone among the stars in having a magnetic field; far stronger fields are common. There is also an interstellar magnetic field which permeates the Galaxy. This article gives an introduction to these magnetic fields, with emphasis upon planetary fields, their origin, and their effect on magnetizable material.

Origin of magnetic fields

As a result of the work of a number of theoreticians in the mid-20th century, led by Walter Elsasser and Sir Edward Bullard, it became clear that the geomagnetic field is maintained by processes in the electrically conducting fluid of the outer core. The general principles of such field generation (or more correctly, regeneration) are now well established (Busse, 1989). Two processes are involved.

The first process is the creation of new magnetic fields from the geomagnetic field by motion of the fluid in the outer core. It takes place because magnetic field lines are trapped in electrical conductors, such as the fluid of the outer core. We can understand this trapping of field lines in electrical conductors with Michael Faraday's law of electromagnetic induction, which tells us that when a magnetic field changes, an electromotive force is generated, inducing a current to oppose that field change. Hence any change in field involved in the movement of magnetic field lines is inhibited by an induced current, so that any tendency for the field lines to move with respect to a good electrical conductor is opposed and the field line is trapped in the medium. This is the 'frozen field' effect. In the highly conducting fluid outer core of the Earth, the magnetic field is carried along with the fluid as it moves in response to the forces imposed upon it. In so doing, the field lines are deformed and new magnetic field is created. Two cases of special interest are illustrated in Figure M4.

The second process of interest taking place in the core is the diffusion of the magnetic fields. Just as a drop of colored dye in a swimming pool diffuses throughout the pool, so magnetic field lines diffuse throughout the outer core. Yet this diffusion must take place against the frozen field effect. The balance between these two competing processes determines the time dependence of the magnetic field – whether the field decays or is regenerated. On the large scale of astrophysical or planetary bodies, the field lines are caught up in the fluid motion, distorted and generate new magnetic field before they diffuse away. In the Earth's core the natural decay time of the field appears to be around 15 000 years.

It is clear that some form of dynamo operates in the fluid outer core of the Earth. One such scheme is illustrated in Figure M4. However, as is often the case in geology and geophysics, it is easier to develop models explaining nature than to demonstrate that a particular model has anything to do with what actually happens in nature. The details of the regenerative processes which maintain the geomagnetic field are obscure and unless some observation can be made that distinguishes between the possible models, the details may, as Walter Elsasser once noted, elude us for ever.

Observation of magnetic fields

The alignment of the compass needle by the geomagnetic field has been known and utilized for centuries. However, it was only during the 19th century that the intensity of the field was first measured by

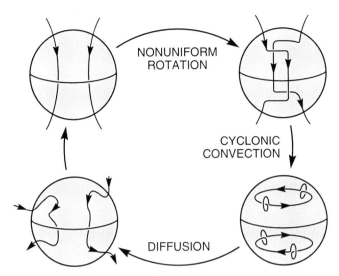

Figure M4 Field regeneration – simple regenerative scheme – $\alpha\omega$ dynamo. (1) Non-uniform rotation generates toroidal field from an initial poloidal field – α process. (2) Rising convective stream is twisted by the Coriolis force to generate small loops of poloidal field. (3) Diffusion increases size of loops of poloidal field to give initial large-scale dipole field.

Carl Frederich Gauss. Moreover, it is only relatively recently, in the decades since World War II, that the profusion of methods of field measurement now used have become available. Now, we can measure the geomagnetic field with fluxgate, proton precession, optical pumping and superconducting quantum interference device (SQUID) magnetometers, all of which are described in standard geophysics texts (see Magnetometry).

Jupiter's magnetic field was first recognized by observations (on Earth) of the polarization of Jovian radio emissions and from Jovian synchotron radiation. Space vehicles have now carried the instruments we use to measure the geomagnetic field to the planets. The fields of the Sun and certain stars are obtained from analysis of the shifts in spectral lines they cause (Zeeman splitting). The galactic field can be measured by its rotation of the plane of polarization of light (the Faraday effect).

All these observations are essentially instantaneous measurements, although in the case of the observation of stellar fields the light has travelled to us over the eons. For the Sun we have a very short historical sequence of measurements. However, for the Earth, the Moon and meteorites we can make use of the remarkable ability of rocks to record magnetic fields, so that a history of the fields can be built up for billions of years.

Magnetic fields in the solar system

The fields of the Sun and planets are reviewed elsewhere (Russell, 1989) and in this volume, so only a brief summary is given here. Without delving too deeply into the forbidding topic of units in electromagnetism, note that the Earth's field at the equator is about 30 μT, which is equivalent in the older units to 0.30 gauss, the tesla being equal to 10^4 gauss. The gamma, the unit used in the older system for weak fields, is 10^{-5} gauss, which is then equal to a nanotesla (10^{-9} T). It is also useful to speak of the dipole moment of the source of a field, which is usually expressed in units of A m^2 and sometimes, as T m^3 from the dipole field formula. The dipole moment provides a means of describing the magnetization of an object, such as the drill cores taken for paleomagnetic studies. This can then be expressed as dipole moment per unit volume or mass, to describe intensity of magnetization. The A m^2 kg^{-1} of the SI units is identical to the gauss cm^3 g^{-1} of the old units.

Sun

The Sun's magnetic field appears to be more complicated than the Earth's field because we observe it on the boundary of the generating region, whereas on Earth we are separated from the outer core by the mantle. However, the Sun does have a dipole moment. This dipole field reverses polarity every 11 years (the sunspot cycle). The complete 22-year magnetic cycle is known as the Hale cycle.

In the Babcock model of the magnetic solar cycle, the Sun's strong toroidal field is stretched by differential rotation. The tubes eventually become unstable and are lifted across the photospheric surface. This takes place at latitudes of about 30° to 40° N and S. The subsequent motion of the spots is illustrated by the well-known butterfly diagram, which reveals that the maximum activity is reached at latitudes of 10° to 20°. The dominant causes of the behaviour of the spots are convection and non-uniform rotation in the field generating region. Eventually, as the field evolves, the polarity of the polar regions changes in a way not understood in detail.

Magnetic fields of the planets

When considering spacecraft encounters with other planets, it is convenient to use the Earth's magnetosphere as a model for the fields in the space around the planets. The form of the magnetosphere is determined by the interaction between the charged particle plasma of the solar wind and the Earth's magnetic field and is discussed elsewhere in this volume (see Earth: magnetic field and magnetosphere). A space probe entering the Earth's vicinity from the Sun side would first cross the bow shock, to enter the magnetosheath, then pass through the magnetopause to reach the planetary field. Each event, or region, has a particular signature in terms of the fields and charged particles observed, so that by comparing what is seen at the various planets with the phenomena observed in Earth's magnetosphere, we can interpret the observations in terms of likely field structure.

Mercury

Mercury is the smallest of the terrestrial planets and rotates with a period of 59 days. Its heavily cratered crust, like that of the Moon, attests to its great age. It has a density of 5.4 g cm^{-3}, which is comparable with the Earth. It has been visited by a number of probes; Mariner 10 gave evidence of a magnetosphere similar to the Earth's. The range obtained for the magnetic dipole moment from the bow shock data and from sampling of field values was from 4×10^{18} to 1.2×10^{19} A m^2. This is about 10^4 times smaller than the geomagnetic dipole moment. The field has an internal origin, but whether it is a remanent field or is maintained by an active dynamo is not known.

Venus

Venus, whose radius (6051 km) is similar to the Earth's, rotates much more slowly, with a day the length of 243 Earth days. Mariner 2 carried a fluxgate magnetometer, a solar wind probe and an energetic particle collector, but detected no evidence of a magnetic field, setting an upper limit equivalent to 0.05 of the Earth's moment. Subsequent missions have reduced this limit to one part in 10^5 of the geomagnetic field moment. It is the atmosphere of Venus rather than the magnetosphere which interacts with the solar wind.

Earth

Earth, with a radius of nearly 6400 km, is the largest of the terrestrial planets; it has a core with a radius of almost 3500 km. The outer core forms a shell of highly conducting material roughly 1700 km thick, and it is here that the geomagnetic field is maintained by the geodynamo. The dipole moment is at present 8×10^{22} A m^2, with a tilt of the dipole axis of 11.4°. We will leave further discussion of the Earth's field for a later section.

Mars

Mars is intermediate in size between Earth and Mercury, with a radius of 3390 km. The Martian day is slightly longer than the Earth day. The mean density of Mars is 3.9 g cm^{-3} which is more like the Moon than Mercury or the Earth. Magnetic field observations have been made, but unfortunately a magnetometer was not included on key missions, and the interpretation of results remains controversial. However, with the future launch of a new Mars orbiter, we should soon have much better data on the field. The moment is at most comparable with the moment of Mercury, about 10^4 times smaller than Earth's dipole moment.

Jupiter

Jupiter, the largest of the planets ($R_J = 71\ 400$ km), also has the fastest rotation (roughly 10 h). It is composed predominantly of hydrogen, but the pressure at the center is sufficient to produce the metallic phase of hydrogen. As long ago as 1955 Burke and Franklin observed polarized Jovian radio emissions; later the discovery of synchotron radiation permitted recognition of the Jovian magnetic field before probes had visited the planet. Observations made with Pioneer 10 and 11 gave a dipole moment of $1.6 = 10^{27}$ A m^2. There is an enormous magnetosphere, so that the bow shock with the solar wind is 5 million km upstream. Downstream the magnetosphere stretches beyond the orbit of Saturn. Later Voyager 1 and 2 observations gave the ratio dipole: quadrupole: octupole as 1 : 0.24 : 0.21 with a 9.7° tilt of the magnetic axis.

Of the four large satellites of Jupiter, Io exhibited a magnetic signature consistent with an internal magnetic field during the Galileo spacecraft flyby, and speculations that Europa may have an internal magnetic field will be tested by Galileo's magnetometer.

Saturn

Saturn is the second largest planet with a radius of 60 000 km and a rapid rotation of about 10 h. Observations with Pioneer 11 indicated a field smaller than predicted – the dipole moment is nearly two orders smaller than that of Jupiter. The fall-off with distance is much closer to an inverse cube relationship than it is for the Jovian field, suggesting a purer dipole field. The tilt angle is about 1°.

The satellite Titan has a radius of 2575 km and has an atmosphere, but no magnetic field was detected by Voyager 1 which passed within 2.4 Titan radii. The interaction of Titan with the Saturn magnetosphere is similar to that of Venus with the solar wind.

Uranus

Uranus, discovered on 13 March 1781 by Herschel, is twice as far from the Sun as Saturn. It has an equatorial radius of 25 900 km and a density of 1.25. Its rotation period is 17 h. Observations by Voyager 2 (which passed within 4 radii) revealed a bow shock, magnetopause and magnetosphere. These observations provide the basis for our knowledge of its magnetic field. However, the solar system was to yield yet another surprise for magneticians – the dipole is inclined at 60° to the rotation axis. The field is about 50 times that of Earth, with strong quadrupole and octupole contributions.

Neptune

Neptune was discovered on 23 September 1846 by Galle and D'Arrest following Leverrier's analysis of perturbations of Uranus by Neptune. It is somewhat smaller than Uranus, but more massive. The rotation axis of Neptune is strongly inclined to the ecliptic and the magnetic dipole is off-center and inclined to the rotation axis.

Pluto

Pluto was discovered 18 February 1930 at 39.4 AU. Its radius of 1700 km and mass 0.2% of the Earth may be too small for magnetic field generation by dynamos, but it may carry a record of early fields of the solar system.

Magnetic field of the Earth

Present form and models of the field

As Gilbert pointed out, the Earth's magnetic field is similar to that of a large magnetized sphere. This is equivalent to the magnetic field of a current loop at the center of the Earth, or of a giant bar magnet. Such a field, when measured at the surface of the Earth, is dipolar, and so the field of a dipole can be used to describe the principal variation of the geomagnetic field over the surface of the Earth.

In describing the Earth's field vector at a particular place, a system of magnetic elements is used Figure M5. The angle between the meridian and the horizontal component of the field is the declination. The angle between the horizontal and the total vector is the inclination. The northerly, easterly and vertical components of the field are generally referred to as X, Y and Z, defining a right-handed coordinate system.

The properties of a dipole field predict the variation of the geomagnetic field over the surface of the Earth, so that if the field were a centered, axial dipole the following relationships would apply:

$$H = \frac{M \sin \theta}{r^3} = \frac{M \cos \lambda}{r^3}$$

$$Z = \frac{2M \cos \theta}{r^3} = \frac{M \sin \lambda}{r^3}$$

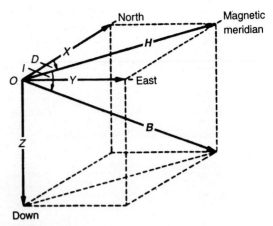

Figure M5 Geomagnetic field elements.

where M is the dipole moment, θ the polar angle and λ the latitude.
In terms of the latitude, the intensity of the field is

$$F = \left(\frac{M \cos^2 \lambda}{r^3} + \frac{2 M \sin^2 \lambda}{r^3} \right)^{1/2}$$

$$= \frac{M}{r^3} (1 + 3 \sin^2 \lambda)^{1/2}$$

The inclination (I) is the arctangent of Z/H and so its dependence on the latitude (λ) is as follows

$$\tan I = 2 \tan \lambda$$

For the field of an axial dipole, that is parallel to the rotation axis of the Earth and centered at Earth's center, the horizontal component H and lines of equal inclination will always be parallel to the meridian and the declination will be everywhere zero.

The most general approach to the description of the geomagnetic field is through spherical harmonics. The importance of spherical harmonics is twofold. First, they are solutions to Laplace's equation in spherical coordinates. It is in this role that they played a critical role in Gauss' analysis of the field. They are also convenient functions to describe the variation of a parameter on the surface of the Earth. The basis of Gauss' separation of the field into internal and external parts was first to show that Laplace's equation indeed applied on the Earth's surface, and then to show from the observed functional dependence of the field components that the source was internal.

The general expression of the radial field due to internal sources is

$$B_r = \sum_{n=1}^{\infty} \sum_{m=0}^{n} (n + 1)\left(\frac{a}{r}\right)^{(n+2)} P_n^m (\theta)(g_n^m \cos m\phi + h_n^m \sin m\phi)$$

where n is the degree and m the order of the harmonic, a the radius of the Earth, θ the polar angle or colatitude, and the longitude. Corrections must be made to compare the computed fields with observations on the Earth's surface because this surface departs from spherical shape.

The axial dipole is the field due to the first coefficient g_1^0. The inclined dipole can be described in terms of the first three coefficients of the spherical harmonic expansion:

$$F(\text{dipole}) = g_1^0 \cos \theta + g_1^1 \cos \phi \sin \theta + h_1^1 \sin \phi \sin \theta$$

The Gauss coefficients are most frequently replaced by the Schmidt quasi-normalized coefficients. The values of the first three coefficients, which describe the dipole contribution, were 29 989.6, 1958.6 and 5608.1 nT for 1979.85 (Langel, 1987). This defines an equatorial dipole whose longitude is

$$\tan^{-1} = \left(\frac{h_1^1}{g_1^1}\right)$$

and a tilt from the rotation axis given by

$$\tan^{-1}[(g_1^1)^2 + (h_1^1)^2]^{1/2} / (g_1^0)$$

With this simple model the lines of equal inclination depart from the lines of latitude and follow the curvature due to the inclined dipole.

The field can be separated into dipole and non-dipole fields, such that

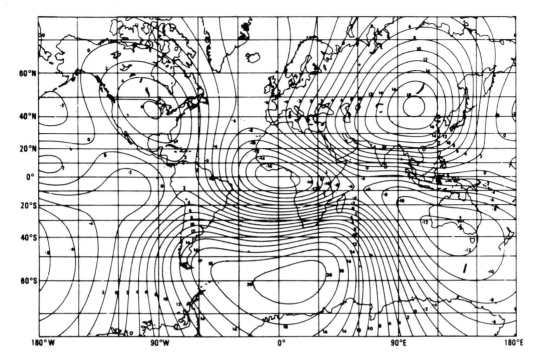

Figure M6 The vertical component of the non-dipole field in 1980. Contours are in units of 2000 nT. (Figure reproduced courtesy of D.R. Barraclough and Geomagnetism Unit, British Geological Survey, Edinburgh).

$$F(\text{obs}) = F(\text{dipole}) + F(\text{non-dipole})$$

The non-dipole field is illustrated in Figure M6, where features having wavelengths of a few thousand kilometers can be seen.

Surveys of the geomagnetic field are continuously being improved, and in the past decades a spectacular advance has been achieved by satellite surveys. The amount of the variation of the field accounted for by the various harmonics follows a distinctive pattern. The energy is dominantly in the first-degree terms which describe the dipole field. With increasing degree of the harmonic the amount of energy falls in a systematic manner. At about degree seven, the fall is arrested and the amount of energy remains roughly constant. The explanation of this phenomenon is that the low-degree harmonics are describing core fields with long-range coherence across the surface of the Earth, while the higher degree harmonics are representing fields due to the crustal magnetic anomalies which affect the surface of the Earth over much shorter distances.

The field can also be analysed in terms of virtual geomagnetic poles (VGP). This is principally used in paleomagnetic studies, but is also useful in describing aspects of the present non-dipole field and secular variation. The VGP is the pole of the dipole field which accounts for the values of declination and inclination observed at a particular site on the surface of the Earth. Like the other methods of describing the field the VGP can be used to focus attention upon the non-dipole components. For example, the scatter in the distribution of VGPs calculated from a region is a measure of the relative strength of the non-dipole versus the dipole field. The VGPs obtained for the Pacific show very little scatter, which tells us that the non-dipole field in this region is very subdued.

Recently models of the field on the core–mantle boundary have been generated. This 'downward continuation of the field' observed on the surface to obtain the field on the core–mantle boundary gives important insight into dynamo processes and motions at the surface of the core.

Secular variation

The highest frequency components of the geomagnetic spectrum are due to the external field, but at periods of tens of years this contribution becomes insignificant. Conversely, fluctuations due to the internal field are shielded by the 'skin effect' of the finite electrical conductivity of the mantle, if they have periods of less than about 10 years. The geomagnetic jerk, a very fast change in the field ascribed to internal sources, lies in a puzzling range, which will require a downward revision of the mantle conductivity, if it turns out to be a worldwide phenomenon of internal origin (Courtillot and Le Mouël, 1984).

At longer periods the internal sources account for the observed field variations. Fluctuations having periods of a few years to a few thousand years are called secular variation. We have detailed records of secular variation with good coverage over the Earth for the very recent past from observatories. However, for more remote times the record is much poorer.

The secular variation of the magnetic elements was discovered by Gellibrand in the 17th century, not long after the first field records became available. As early as 1692, Halley discussed the phenomenon of westward drift of the field. Bauer, at the beginning of this century, observed that the magnetic vector, when plotted in the form of declination versus inclination curves, exhibited clockwise loops. To visualize this, imagine that you are viewing the magnetic vector looking down along it in the northern hemisphere. The secular variation of the vector defines a cone which is traced in a clockwise sense. With the exception of the extreme fluctuations in declination near the poles, declination and inclination vary by about ± 10° of the mean values. Of this, a substantial part appears to be from localized core–mantle sources, but there are also longer-period dipole fluctuations. Secular variation is of interest to those who are working on the origin of the field. It also has practical importance for map makers and navigators.

An important component of secular variation is the westward drift of the non-dipole features of the field. However, westward drift cannot alone account for secular variation; there are standing anomalies which vary in strength. The studies of the field on the core–mantle boundary suggest that the westward drift is predominantly an expression of a strong feature in low to intermediate southern latitudes from approximately 90°E to 90°W (Bloxham and Gubbins, 1985).

Paleomagnetism

In addition to observations of the present fields of the planets, we also have the paleomagnetic record carried by rocks on Earth, by Moon rocks and by meteorites. The paleomagnetic record of the geomagnetic field carried by rocks has made fundamental contributions to the Earth sciences. Without it we might still be debating whether the continents drift. Although the record from the Moon and meteorites has not yet played a comparable role in planetary science, they may do one day.

The paleomagnetic record is somewhat fortuitous, depending as it does upon (1) the presence of magnetic mineral grains of just the right size to retain their magnetization over millions of years, and (2) processes of formation of the various rock types which generate the record of the geomagnetic field. The paleomagnetic record has been studied using sensitive magnetometers to measure the remanent magnetism of rock samples. Elaborate procedures have been established to measure and analyze this remanent magnetization.

In principle, there is the possibility of a paleomagnetic record of the direction and intensity of the field in which each rock or meteorite sample was magnetized. In reality the paleomagnetic record is difficult to establish accurately because the magnetization of a rock can be affected by the subsequent history of the rock. In particular, the determination of the intensity of the ancient fields has proved difficult. It is most successful in lavas, which can be reheated and cooled in the laboratory to give a magnetization similar to their natural remanent magnetization (NRM) in a known laboratory field. The magnetization process is known to be linear for weak fields and so the ratio of the ancient to laboratory fields can be obtained from the ratio of the magnetizations. Unfortunately, in heating the rock to give the laboratory magnetization the rock is frequently altered. In other rocks, such as sediments, it is not possible to simulate the magnetization process. The paleomagnetic record of geomagnetic field direction is therefore much better than the record of intensity. The lunar paleomagnetic record is enigmatic largely because the orientation of the lunar samples at the time of their magnetization is not known.

Magnetic anomalies caused by the magnetization of crustal rocks have also been studied on Earth and on the Moon. On Earth the magnetization has two sources: (1) the induced magnetism acquired by the rocks in the geomagnetic field, and (2) their natural remanent magnetization (NRM). Shipborne and airborne magnetic surveys have played a major role in establishing the structure of the crust. In particular, Vine and Matthews' recognition that the oceanic anomalies are due to the remanent magnetization of the ocean crust and record the reversals of the geomagnetic field was pivotal in the acceptance of the idea of sea-floor spreading. On the Moon the lunar field is so weak that the anomalies are caused by the NRM of crustal rocks. The magnetic anomalies measured on the lunar surface and with instrumentation on the Apollo subsatellites provides clear evidence of strong ancient fields on the Moon in its past.

Rock magnetism

Rock magnetism seeks to provide a physical understanding of paleomagnetism. The origin of magnetism lies in the orbital motion and spin of the electron. The orbital motion of the electrically charged electron about the nucleus constitutes an elementary current loop with a magnetic moment of magnitude 0.927×10^{-23} A m^2, if the electron is in the first Bohr orbit. The magnetic moment due to the spin of the electron is the same. This fundamental quantity is called the Bohr magneton μ_B. Atoms may either have spin and orbital contributions from their various electrons, which are self-compensating and sum to zero, or there may be only partial self-cancellation, so that the atom has a net magnetic moment.

Diamagnetic materials have zero net moment in the absence of a field. Quartz is an example of a diamagnet. When a magnetic field is applied the individual electron orbits respond to the change in magnetic field in accordance with the Faraday law and give rise to a moment opposed to the applied field. Hence, in the presence of the field (H), they exhibit a negative magnetization ($-M$)

$$M = kH$$

where k is the susceptibility, which is negative. It is also small, being of the order of 10^{-6} SI units. When the field is removed the diamagnetic material reverts to its zero net moment state.

Unlike diamagnetic materials, other materials have a net atomic moment because of imbalance of the various electron contributions to the magnetism. This imbalance arises due to the particular way in which the electron shells are filled to give the various atoms, which is discussed in physical chemistry texts. For a large number of elements there is such an imbalance and the individual atom has a net moment. For example, iron, nickel and cobalt are magnetic due to spin imbalance in the 3d shell. Amongst the various atoms which do have

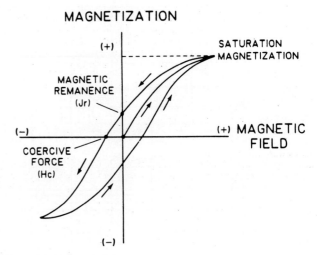

Figure M7 Hysteresis loop.

net moments, two fundamentally different types of magnetic behavior are found.

In paramagnetic materials the individual atoms do not interact, and their elementary moments are not preferentially aligned. When a magnetic field is applied the magnetization of such materials increases because spins align with the field to minimize the MH energy. However, when the field is removed they revert to random orientations. Hence these materials, like the diamagnets, are not of particular interest to us, because they do not record the fields in which they are magnetized after the field has been removed. Biotite and other ferromagnesian minerals are paramagnetic.

In contrast to diamagnetic and paramagnetic materials, ferromagnetic materials retain a memory of magnetic fields to which they have been exposed after the field has been removed. In ferromagnets the quantum mechanical exchange energy constrains the magnetic moments of neighboring atoms to be parallel. The behaviour of a ferromagnet exposed to magnetic field cycling is illustrated in Figure M7. As the magnetic field increases, the magnetization increases to a saturation value (M_s). On reduction and removal of the field the magnetization decreases to give the remanent magnetization (M_r). The application of a negative field reduces the magnetization to zero and the cycle repeats. To understand this, remember that in these materials the moments of individual atoms are strongly constrained to be parallel. In addition, there is a preferred direction of magnetization or, more correctly, an axis for magnetization of the individual grains. Together these two aspects of magnetism determine the sequence of events, which gives a hysteresis or a magnetic memory of the field for an assemblage of fine grains. At saturation all the atomic moments in the individual grains are parallel to the field. When the field is removed, the direction of magnetization of the various grains reverts to the nearest easy axis to the saturation field giving a non-zero net magnetization for the assemblage. At high temperatures, first the anisotropy energy and then the interaction energy are overcome by thermal energy. Thus the two effects giving remanence are lost at high temperature. The Curie point is the temperature at which the material loses its magnetic ordering due to the exchange force and reverts to paramagnetic behavior.

Terrestrial rocks contain iron oxides which are ferrimagnetic. Although ferromagnetic and ferrimagnetic materials have very different elementary order, their behavior in applied fields is similar. The important ferrimagnetic minerals which carry the bulk of the magnetic record in rocks are the various iron titanium oxides, of which the most important is magnetite. A lesser contribution comes from iron sulfides.

Magnetic material exhibits a great range of hysteresis loops. An important variable which determines magnetic behaviour is the grain size. Thus, whereas an assemblage of very fine grains of magnetite might exhibit a loop similar to that illustrated above, larger grains have magnetically softer characteristics, that is, they are magnetized and demagnetized by much smaller fields. This is because whereas small grains are magnetized homogeneously, large grains are subdivided by walls into domains of mutually opposed magnetization.

These domains arise because energy is expended in the external magnetic field of the grain. If the grain exceeds a critical size, so much energy must be expended in the external field that it becomes more favorable to expend excess energy in the configuration of the magnetization within the particle to reduce the energy in the external field. The presence of the domains reduces the dipole moment of the particle and hence its external field. The critical single-domain grain size for magnetite is of the order of tenths of a micron, but nucleation phenomena appear to preclude domain wall formation in this size range. Magnetization changes brought about by wall motion in multidomain material are lower-energy processes than those involving the rotation of the magnetization of single domain particles.

In the hysteresis loop illustrated, in Figure M7, three important states were noted. There is the state of saturation, at which all spins are aligned parallel with the applied field. There is the state of remanent magnetization, at which a memory of the field is retained, although the field has been reduced to zero. Finally, there is the coercive force field, the back field which reduces the magnetization to zero. For single-domain grains the remanent magnetization is a large fraction of saturation magnetization and the coercive force tends to be of the order of mT. In multidomain materials the coercive force is lower and remanent magnetization tends to be a small fraction of saturation magnetization. Clearly from the point of view of magnetic recorders, the single-domain small grains are more desirable than the large grains which are more easily reset.

The geomagnetic field is relatively weak, so that the magnetic particles which carry the paleomagnetic record have coercivities which are large compared with the geomagnetic field whose record they carry. Clearly some special mechanism of magnetization must be involved. The magnetization of many sediments, such as those laid down on the ocean floor, is easily explained by the orientation of the detrital magnetic particles in the geomagnetic field. This preferential alignment is locked in as the water-saturated sediment loses its water to become consolidated. Igneous rocks, whether they cool on the surface as lavas, or beneath the surface as intrusions, acquire their magnetization as they cool. In all but some rare examples (which are not useful for paleomagnetism) the rock solidifies before magnetization is acquired. As it continues to cool after solidification, the rock passes through the Curie points of the magnetic minerals it contains, and they become magnetic. The stable remanent magnetization carrying the record of the geomagnetic field is acquired a little below the Curie temperature in a manner first explained by the distinguished French theoretician Louis Néel (1955). He developed a model to calculate the magnetization of an assemblage of fine particles immediately below the Curie point and showed how this magnetization could be fixed, or blocked, as the temperature continues to fall.

In a similar way, the presence in lunar rocks and meteorites of native iron and other magnetic phases make possible in principle a record of lunar magnetic fields and the fields experienced by meteorites, and possibly of the fields of the parent bodies from which they originate. In the interpretation of lunar and meteoritic magnetism we are faced with problems not encountered in terrestrial paleomagnetism. The magnetic carriers of the record are metallic iron and its alloys with nickel, whose paleomagnetic behavior is not so well known as that of the iron oxides and sulfides carrying the record on Earth.

Paleomagnetic observations of the geomagnetic field

Paleosecular variation

One of the first tasks undertaken by paleomagnetists was to extend the record of secular variation, from the short timescale of observatory records to the geological time scale of millions of years. The NRM of archeological artefacts and lavas was one principal source of data, while the other was the NRM of rapidly deposited sediments, such as lake sediments. In general the former carries a reliable magnetization, but it consists of spot readings of the field, which are sometimes difficult to date with sufficient resolution. In contrast, the latter provide a continuous record of field changes but, because the time of acquisition of magnetization can be longer than some of the field changes of interest, there are complications in deconvolving the effects of the magnetization process from the true field changes. Ironically, this first task undertaken by paleomagnetists turned out to be particularly difficult, and is not yet satisfactorily resolved.

The main features of paleomagnetic secular variation for the past 10 000 years are similar to the observatory secular variation, although the fluctuation in intensity is greater. Lake sediments reveal a pattern of directional changes similar to those found in the observatory record. Archeomagnetic data and results from lavas also yield an intensity record with trends similar to the observatory records. An interesting recent development is the possibility of obtaining reliable relative intensity records, as well as field directions, from ocean sediments. This may eventually permit the compilation of detailed histories of the geomagnetic field for the past few million years. Preliminary results of this type show fluctuations in intensity of a larger magnitude than those seen in the paleosecular variation for the past few thousand years. There is also some indication from both lava and ocean sediment data that there may be repeated short events, during which the field may briefly reverse polarity in association with intensity lows. Paleosecular variation from more distant geological times is poorly known, but it does appear that there were systematic fluctuations in field strength to values substantially smaller than the present field.

Reversals

The most spectacular discovery to come from paleomagnetic studies of the geomagnetic field is the field reversal. This phenomenon is discussed in the articles Geomagnetic polarity reversals; and Polarity reversals in this volume.

Paleomagnetism of the Moon and meteorites

See Moon: magnetism and interior. Like lunar rocks, the various types of meteorites have distinctive magnetic characteristics permitting a magnetic classification of meteorites (Cisowski, 1987). Magnetization which antedates the arrival of the meteorites on the Earth has been demonstrated. In the primitive carbonaceous chondrites it is not clear how much of the NRM represents magnetization acquired billions of years ago. Similarly, in ordinary and enstatite chondrites the origin of the magnetization is unclear and intensity estimates are therefore hard to interpret. However, in the achondrites, stable magnetization is frequently found and some consistency of intensity estimates has emerged in the range 0.003–0.009 mT, roughly one order less than geomagnetic field intensities. These meteorites have crystallization ages of about 4 Ga. The achondrites include the shergottites, a group of meteorites with much lower crystallization ages of as little as 350 Ma. This suggests an origin on a planetary body which had volcanic activity in this time range. These meteorites exhibit low intensities of magnetization and may be the result of impact events in the low magnetic field of Mars. The paleomagnetism of meteorites is clearly a difficult topic which may however eventually provide key data for our understanding of the solar system.

Biomagnetism

One of the most surprising developments in the field of geomagnetism has been the emergence of the subject of biomagnetism (Kirschvink, Jones and MacFadden, 1989). The pioneering result was that of Lowenstam (1962) who demonstrated that magnetite is synthesized by chitons. Magnetotactic bacteria have strings of single domain particles of magnetite which have particular crystallographic habits. Northern hemisphere species are attracted to magnetic south poles, while southern hemisphere species are attracted to magnetic north poles. This may provide an evolutionary advantage in directing the bacteria to favourable depths in the water column. A number of species, including pigeons, appear to be able to sense the magnetic field and to use it as a navigational aid. Meanwhile magnetite has been found in numerous species. It remains to be seen whether the magnetite is playing a role in the nervous system, or whether it is simply a metabolic by-product having no function in the nervous system.

Conclusion

Magnetism exhibits a rich range of phenomena of interest to the planetary scientist. The study of the fields of the planets and their documentation and interpretation is still in its infancy. Among these, the fields of the Earth, Moon and Mars are open, or are likely to become open in the foreseeable future, to paleomagnetic study. The role of magnetic fields in the early solar system remains to be clarified, but access to these fields may eventually come through paleomagnetic studies of lunar rocks, meteorites and asteroids. Our understanding of the processes whereby rocks acquire and preserve their paleomagnetic records is improving, but much remains to be done. Paleomagnetism has enhanced geological studies, illuminating the movement of the tectonic plates and the deformation of the upper levels of the crust. Our models of field regeneration have become increasingly sophisticated, but face a possibly insurmountable problem of non-uniqueness. Finally, the interaction of the magnetic fields with life promises exciting discoveries. In a curious manner, magnetic fields are proving to be relevant to a wide range of studies.

Michael D. Fuller

Bibliography

Bloxham, J. and Gubbins, D. (1983) The secular variation of Earth's magnetic field. *Nature*, **317**, 777–81.
Busse, F.H. (1989) Geomagnetic field, main, theory, in *The Encyclopedia of Solid Earth Geophysics* (ed. D.E. James). Van Nostrand Rheinhold, pp. 511–17.
Cisowski, S.M. (1984) Magnetism of meteorites, in *Geomagnetism*, Vol. 2 (ed. J. Jacobs). Academic Press, pp. 527–60.
Courtillot, V. and Le Mouël, J-L. (1984) Geomagnetic secular variation impulses. *Nature*, **311**, 709–16.
Jacobs, J. (1984) *Geomagnetism*, Vols 1 and 2. Academic Press.
Kirschvink, J., Jones, D.S. and MacFadden., B.J. (1985) *Magnetite Biomineralization and Magnetoreception in Organisms*. Plenum Press.
Langel, R.A. (1987) The main field, in *Geomagnetism*, Vol. 1 (ed. J. Jacobs). Academic Press, pp. 249–492.
Lowenstam, H.A. (1962) Magnetite in denticle capping in recent chitons (polyplacophora). *Geol. Soc. Am. Bull.*, **73**, 435–8.
Merrill, R.T. and McElhinny, M.W. (1983) *The Earth's Magnetic Field, its History, Origin and Planetary Perspective*. New York: Academic Press.
Néel, L. (1995) Some theoretical aspects of rock magnetism, *Adv. Geophys.*, **5**, 99–136.
Russell, C.T. (1989) Planetary magnetic Fields, in *The Encyclopedia of Solid Earth Geophysics* (ed. D.E. James). Van Nostrand Rheinhold, 938–45.

Cross references

Dynamo theory
Earth: magnetic field and magnetosphere
Jupiter: magnetic field and magnetosphere
Mars: magnetic field and magnetosphere
Mercury: magnetic field and magnetosphere
Neptune: magnetic field and magnetosphere
Paleomagnetism
Saturn: magnetic field and magnetosphere
Uranus: magnetic field and magnetosphere
Venus: magnetic field and magnetosphere

MAGNETOMETRY

Using a splinter of magnetite for direction finding, the Chinese were studying the Earth's magnetic field in 2634 BC, over 4500 years ago. Though this information was brought to Europe by Marco Polo in the 13th century, it was Sir William Gilbert who published *De Magnete* some 300 years later, postulating that the Earth was 'a great magnet' and recording the first investigations of terrestrial magnetism.

Magnetometry represents the science of measuring the strength and direction of magnetic force fields. Sources of magnetic fields include electrical currents circulating in conductors and ionized gases, as well as 'permanent' magnets; those that occur naturally (magnetite or 'lodestone') and those constructed of materials like iron, iron oxides, cobalt and nickel alloys.

In the early years of the space program, measurements of Earth's magnetic field were collected using rockets and balloons. The instruments measured the strength of the equatorial electrojet, the auroral current system and other high-altitude magnetic phenomena. Using space flight measurement techniques adapted from instruments developed around the time of World War II, early space probes established the comet-like morphology of the distant Earth's magnetic field and discovered many of the features and boundaries of the magnetosphere (the bow shock, magnetopause and geomagnetic tail). Dolginov (1959, 1960), Heppner *et al.* (1960), Sonnett (1963), Cahill (1964) and Ness (1965) were among the first investigators to equip the early probes with magnetic field measuring instruments and carry out measurements in the Earth's magnetosphere and interplanetary medium.

Planetary magnetic fields like those of the Earth, Jupiter and Saturn are believed to be generated by currents circulating in the liquid metallic cores. Uranus and Neptune are not assumed to have formed metallic cores, and their magnetic fields are thought to be generated closer to the surface, where electrical currents can flow in the high-conductivity liquid crust (Ness *et al.*, 1986, 1989; Connerney Acuña and Ness, 1987, 1991). In the case of the terrestrial planets, Venus does not possess an intrinsic magnetic field, while Mercury is magnetized by the remains of an ancient dynamo which is decaying with time (Ness, 1979). Mars may have a weak magnetic field; it has not been explored in sufficient detail to establish the existence of a field (Dolginov, Zhuzgov and Shkolnikova, 1984; Luhmann, 1991; Riedler *et al.*, 1989). (See Mars: magnetic field and magnetosphere; Mercury: magnetic field and magnetosphere; Venus: magnetic field and magnetosphere).

Contemporary space missions such as Pioneer Venus, the Mariners, Voyager, Helios, Ulysses, Giotto and Galileo have carried out magnetic field measurements around most of the planets in the solar system, in the interplanetary medium, and near comets and asteroids. Many missions utilize magnetic field measurements for engineering applications. For instance, Earth-orbiting spacecraft apply magnetic field information to attitude determination and control, spacecraft momentum management and scientific instrument pointing. The Earth's magnetic field provides one of the basic 'natural' forces which modern systems and spacecraft utilize to establish their orientation with respect to a reference frame when inertial systems are too complex or costly to implement.

While numerous discoveries have been made in researching the sources and behavior of planetary magnetic fields, there is still much to be learned. Magnetic field measurements are essential to complement onboard energetic charged particle measurements in order to allow an understanding of the behavior of plasmas in the solar system and energetic trapped particles around magnetized planets.

Magnetic field units

The magnetic force between two magnetic poles is given by an expression identical to that of the gravitational force between two masses:

$$\boldsymbol{F} = (m_1 \, m_2 \, / \, \mu \, r^2) \, \boldsymbol{r} \tag{M1}$$

where $\boldsymbol{F}$ is the force in dynes (CGS) between the poles that are separated by r cm ($\boldsymbol{r}$ is the unit vector directed between m_1 and m_2). The permeability of the medium is denoted by μ. Magnetic poles cannot exist by themselves and always exist in pairs, while the force between them can be repulsive or attractive depending on the polarity of the poles involved.

The derived magnetic quantities are the magnetic field strength, $\boldsymbol{H}$, which is defined as the force $\boldsymbol{F}$ per unit pole and is measured in oersteds (cgs), and the magnetic field induction, which is the field $\boldsymbol{B}$ induced by the excitation $\boldsymbol{H}$ in a medium of permeability μ:

$$\boldsymbol{B} = \mu \, \boldsymbol{H} \tag{M2}$$

In the CGS system of units $\boldsymbol{B}$ is measured in gauss. In air or vacuum the permeability μ is unity so for all practical applications in space the Oersted and the Gauss can be used interchangeably. In the SI system of units the magnetic induction is measured in teslas (10^4 gauss), while the magnetic field strength is measured in ampere-turns per centimeter. To quantify weak magnetic fields the nanotesla (1 nT = 10^{-9} tesla) and the equivalent CGS unit the gamma (1 gamma = 10^{-5} gauss = 1 nT), are frequently used. The Earth's magnetic field at the surface and near the equator has an approximate strength of 31 000 nT or 0.31 gauss, while Jupiter's north magnetic pole field strength is more than 14 gauss. At the opposite end of the dynamic range, the interplanetary magnetic field (q.v.) at 1 AU is typically of the order of 5–10 nT, while at the orbit of Uranus it may as low as 0.05 nT or equivalently, 5×10^{-11} gauss. This very large dynamic range presents unique challenges when designing magnetic instrumentation for space missions to the planets.

Magnetic field measuring instrumentation

The instruments used to measure the strength of the magnetic field are called magnetometers. Since this is a *vector* quantity that has both magnitude and direction, we usually differentiate between two classes of instruments:

1. Scalar magnetometers which measure only the total strength or magnitude of the ambient magnetic field regardless of the orientation of the sensor; and
2. Vector magnetometers which produce an output proportional to the strength and direction of the magnetic field with respect to the sensing element. The polarity or sign of the output in general depends on the sense of the ambient field with respect to the magnetometer sensor orientation.

Both classes of instruments have been used aboard spacecraft to measure magnetic fields in space, but vector magnetometers are far more common due to their ability to provide directional information. We will discuss below some of the principal characteristics of these instruments, their advantages and disadvantages and the general problem of performing sensitive magnetic field measurements aboard an orbiting spacecraft or planetary probe. For a comprehensive review of early space research magnetometers the reader is directed to Ness (1970).

Scalar magnetometers

The most common scalar magnetic field measuring instrument in use is the proton precession magnetometer which is based on the phenomenon of nuclear magnetic resonance. Around 1945 it was discovered that many atoms possess a net magnetic moment and behave as small magnets. A sample of a liquid rich in protons (hydrogen nuclei) surrounded by a coil is magnetically polarized to align all of its magnetic moments in a given direction and is then allowed to 'relax' in the presence of an external magnetic field. The protons then precess like spinning tops around the ambient magnetic field and induce an AC signal in the polarizing coil whose frequency is proportional to the magnitude of the field. This frequency (hertz) is called the Larmor frequency and is given by

$$f = (\tau_p B / 2 \pi) \quad \text{(M3)}$$

where B is the magnitude of the external field and the proportionality constant τ_p is the gyromagnetic ratio of the proton. It is the ratio of the proton's magnetic moment to its spin angular momentum. Since this ratio is known very accurately from quantum mechanical principles, the proton precession magnetometer is very frequently used as a primary standard in the calibration of other magnetometers. The value of $(2\pi/\tau_p)$ is 23.4874 nT Hz^{-1}.

The polarize/count cycle of the typical proton precession magnetometer requires 1 s or more in traditional designs. In addition, the liquid sample volume is relatively large and massive, particularly when the polarizing coil mass is considered, and liquids that can operate over a wide temperature range are required. The polarizing power required to generate the 0.01 T or more polarizing field is significant and useful signals can only be obtained for ambient fields larger than approximately 20 000 nT. These limitations have restricted significantly the use of proton precession instruments aboard spacecraft. Recent developments like the Overhauser effect proton precession magnetometer, which uses an indirect technique to 'polarize' the sample and generate a continuous Larmor precession signal, promise further advances in this area but further work is required.

Optically pumped magnetometers are another class of magnetic field measuring instruments which have found considerable application aboard spacecraft, both as scalar as well as vector instruments (Slocum and Reilly, 1963; Farthing and Folz, 1967; Slocum, Cabiness and Blevins 1971; Slocum, 1972). In the scalar configuration they are capable of measuring magnetic fields over a wider range than the proton precession instruments and with much higher time resolution. These instruments use the energy required to transfer atomic electrons from one energy level to another as the mechanism for magnetic field detection. A cell containing a suitable gas is irradiated with light from a discharge lamp at the proper frequency to excite the atoms to a level which becomes overpopulated as a consequence of forbidden transitions in the system. Under these conditions the gas cell becomes transparent to the irradiating beam (the process used to achieve this is called optical pumping). If now the cell is subjected to an RF signal with the appropriate energy to cause a depopulation of this energy level, it will become opaque again, blocking the transmission of light at the discharge lamp frequency. The most commonly used elements for optically pumped magnetometers are helium and alkali metals like cesium, rubidium and sodium. Helium and its isotopes in particular are in wide use in high accuracy, high time resolution scalar magnetic field measurements for military applications. A modified version of these 'metastable helium' magnetometers was developed to perform wide dynamic range vector measurements aboard spacecraft and is also described below under vector instruments (Slocum and Reilly, 1963; Smith and Sonnett, 1976).

A generic optically pumped magnetometer is illustrated in Figure M8 and consists of a discharge lamp which irradiates one end of a gas cell containing the chosen element in gaseous form through a system of filters and polarizers. At the opposite end a solid state photodetector measures the intensity of the incident light. The electrons in the gas cell will precess about the axis of the external magnetic field at the Larmor frequency of the chosen element, modulating the intensity of the light incident upon the photodetector at the same rate. Thus the output from the photodetector is an AC signal at the Larmor frequency given by

$$f = g B / 2\pi \quad \text{(M4)}$$

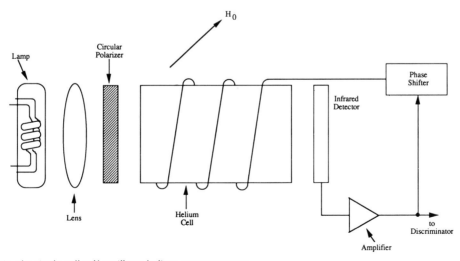

Figure M8 Schematic of a single-cell self-oscillator helium magnetometer.

where g is the electron gyromagnetic ratio for the chosen element. For helium $(g/2\pi)$ corresponds to 28 Hz nT^{-1}, considerably larger than the value of 0.04257602 Hz nT^{-1} obtained from proton precession instruments, making possible high time and amplitude resolution measurements. The corresponding figures for cesium and rubidium are 7 and 4.67 Hz nT^{-1} respectively.

In most instruments of this type the output of the photodetector is amplified and fed back to a coil wound on the cell, causing the system to oscillate continuously at the Larmor frequency given above. Because of the high frequencies involved, it is possible to measure very small magnetic field variations (≥ 0.01 nT) superimposed on large background fields.

Vector magnetometers

Vector instruments are, by a very large margin, the most widely used for magnetic field measurements aboard spacecraft, balloons and sounding rockets. In addition to providing information about the field strength they also indicate the direction and sense of the ambient field. Triaxial orthogonal arrangements of single axis sensors are used to measure the three components of the ambient field in a coordinate system aligned with the sensor magnetic axes. In contrast to proton precession and alkali vapor scalar magnetometers, whose accuracy is determined by quantum mechanical constants, vector magnetometers must be calibrated against known magnetic fields, both in strength and direction. Their output for zero field, scale factor and stability with temperature and time depend on electrical component values which may drift as the instruments age or are exposed to the effects of the space environment (Ness, 1970, Acuña, 1974). In spite of these shortcomings, vector instruments are capable of measuring magnetic fields over a very large dynamic range, 5×10^{-3} nT to over 2×10^6 nT, are lightweight and consume little power. In addition they are capable of operation over a wide temperature range, have proven to be extremely reliable and extremely resistant to the destructive effects of intense radiation from solar flares and energetic trapped particles in planetary magnetospheres. Ultraprecise instruments have been used to map the Earth's magnetic field from orbit with unprecedented accuracy, both in magnitude as well as direction – 5 nT and 3 arcsec respectively (Acuña et al., 1978).

The fluxgate magnetometer was first developed during World War II as a submarine detector (see Geyger, 1964, and included references). Today it is the most widely used vector magnetometer, not only aboard spacecraft, but also for innumerable industrial, military and scientific applications. The fluxgate sensor, as its name implies, is a device which is used to 'gate' the ambient magnetic field through a sensing coil and thus generate an induced voltage proportional to the strength and direction of the field. The gating of the latter is accomplished by winding a sense coil around a high-permeability material which is driven cyclically to saturation by an excitation signal. When the core material is saturated its permeability is very low and the total magnetic flux threaded by the sensing coil is not very different from that which would exist in the absence of a high permeability core. However, when the core is in its high permeability state, the flux threaded through it and the coil is greatly increased in proportion to the effective permeability of the core. Thus, by cyclically switching the core in and out of saturation at a frequency f, the ambient magnetic flux is modulated at twice this frequency and a corresponding voltage appears at the terminals of the sense coil. The amplitude of this signal is proportional to the magnitude and direction of the ambient field, and its phase with respect to the excitation signal (0° or 180°) depends on the sense of the field. In general balanced core arrangements are used to isolate the excitation signal that saturates the core from the sense coil to avoid generating a large signal at the excitation frequency. Many fluxgate sensor geometries have been used for space-based instruments, from commercial sensors using proprietary helical high-permeability cores to sensors using ultra-low noise ring cores developed by government laboratories. The very high performance of the latter has made them the sensors of choice for advanced space missions (Ness, 1970).

A block diagram of a typical fluxgate magnetometer is shown in Figure M9. A reference signal at frequency $2f_0$ is derived from a stable oscillator and applied to a divider, from which the excitation signal at a frequency f_0 is used to drive the sensor to saturation (in this case a ring core sensor with a toroidal excitation winding). A differential sense coil is wound around the outside of the ring core and tuned to the second harmonic of the drive frequency, $2f_0$. Thus, if both core halves are identical, no signal at the excitation frequency will appear at the sense winding terminals. The presence of an external magnetic field will cause the appearance of a signal at a frequency $2f_0$ at the terminals of the sense coil, due to the fact that one half of the ring core will be unbalanced with respect to the other. This signal will be amplified by the AC preamplifier and applied to the synchronous detector. The magnitude of the output of this detector is proportional to the amplitude of the signal present at the output of the preamplifier, while the polarity of the output depends on the phase of the signal. The output of the synchronous detector is applied to a high-gain integrating DC amplifier, whose output in turn is used to generate a current which is fed back to the sense winding in the fluxgate sensor, where it produces a magnetic field that opposes (and cancels almost completely) the original external field. This feedback arrangement produces an instrument with good linearity whose output is given by

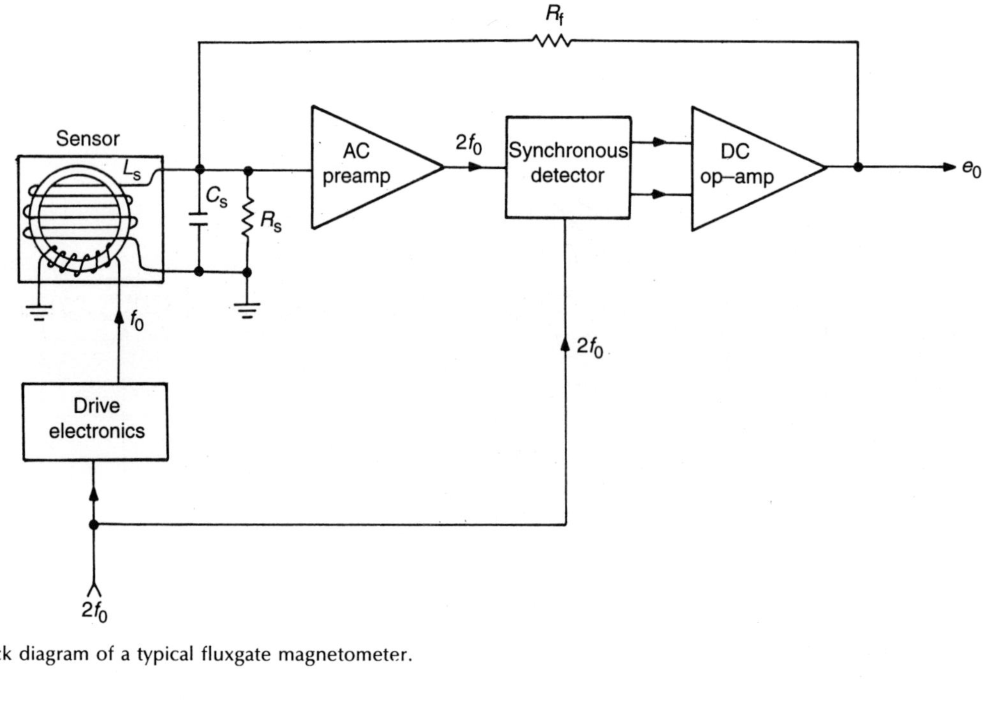

Figure M9 Block diagram of a typical fluxgate magnetometer.

$$e_o = k R_f |B| \cos(\theta) + V_z \tag{M5}$$

where e_o is the output voltage, k a constant related to the physical characteristics of the sense coil with dimension given in [ampere/gauss], R_f is the value of the feedback resistor, $|B|$ is the total magnitude of the applied field, and θ is the angle between the magnetic axis of the fluxgate sensor and the direction of the external field. V_z is a DC offset voltage which is present when the external field is zero and is produced by small asymmetries and imbalances in the system. Note that the scale factor of the magnetometer can be modified easily by changing the value of R_f. This is done in instruments that must cover a large dynamic range, like the Voyager planetary probes (Behannon et. al., 1977).

A second type of vector instrument that has been used in some planetary spacecraft is the vector helium magnetometer (Slocum and Reilly, 1963; Smith and Sonnett, 1976). Its operating principles are the same as the optically pumped magnetometers described previously under scalar instruments, except that a Helmholtz coil system has been added around the gas cell. This coil system is used to generate synchronous sweep fields in two planes that intersect along the optical axis by means of an auxiliary electronic system. These sweep fields produce a synchronous modulation in the photodetector light output which can be used to derive vector information about the external magnetic field. Typically, a frequency of 200 Hz is used for the sweep fields, and the resulting signal at the photodetector is amplified, synchronously rectified and fed back to the sensor assembly to cancel out the applied field. The operation is very similar to that implemented in the fluxgate magnetometers, except that the excitation and signal frequencies are identical and the vector information is derived not from individual sensors but from sweeping fields applied to a common sensor along two orthogonal planes. The output of the vector helium magnetometer is given by a relation identical to (M5).

Measurement of magnetic fields aboard spacecraft

The instruments described above are carried into space aboard spacecraft which include complex systems of mechanical, electrical and electronic components. These have the potential to generate strong magnetic fields of their own unless carefully controlled. Batteries, solar arrays, motors, wiring, materials, etc. must be especially designed and selected to minimize the generation of 'stray' magnetic fields that will affect the measurements. The design and implementation of a magnetically 'clean' spacecraft meeting the stringent requirements of an interplanetary mission is an extremely demanding task that has tested the fiber of many seasoned project managers and engineers. Since it is practically impossible to reduce the stray spacecraft magnetic field to the small levels required for sensitive measurements, the use of long booms to place the magnetic sensors away from the main body of the spacecraft is commonplace. This technique exploits the fact that magnetic fields decrease rapidly away from their source, proportionally to $1/r^3$ as a minimum, where r is the distance to the source. However, these booms must be rigid and preserve the required alignment between the magnetic sensors and the attitude determination sensors mounted in the main spacecraft body, and this limits their practical length. For a given spacecraft design, the trade-off between boom length and level of magnetic cleanliness required in the main body is a major decision that must take into account many conflicting requirements, including whether the spacecraft is spin- or three axis-stabilized. For details of magnetic interference, spacecraft testing and magnetic cleanliness programs, the reader is directed to Ness (1970).

The dual magnetometer technique was introduced in 1971 by Ness and coworkers to ease the problem of making sensitive magnetic field measurements in the presence of a significant spacecraft field (Ness et al., 1971). This method is based on the experimental observation that beyond a certain distance most spacecraft-generated magnetic fields decrease as expected for a simple dipole source located at the center of the spacecraft ($\approx 1/r^3$). Thus it can be shown that if two magnetometer sensors are used, mounted along a radial boom and located at distances r_1 and r_2 respectively, it is possible to separate uniquely the spacecraft-generated magnetic field from the external field being measured. If we denote by $\boldsymbol{B}_1$ and $\boldsymbol{B}_2$ the vector fields measured at radial locations 1 and 2 with $r_2 > r_1$, the ambient field and the spacecraft field at location r_1 are given by

$$\boldsymbol{B}_{amb} = (\boldsymbol{B}_2 - \alpha\boldsymbol{B}_1)/(1 - \alpha) \tag{M6}$$
$$\boldsymbol{B}_{s/c1} = (\boldsymbol{B}_1 - \boldsymbol{B}_2)/(1 - \alpha) \tag{M7}$$

where

$$\boldsymbol{B}_1 = \boldsymbol{B}_{amb} + \boldsymbol{B}_{s/c1} \tag{M8}$$
$$\boldsymbol{B}_2 = \boldsymbol{B}_{amb} + \boldsymbol{B}_{s/c2} \tag{M9}$$
$$\boldsymbol{B}_{s/c2} = \alpha \boldsymbol{B}_{s/c1} \tag{M10}$$
$$\alpha = (r_1/r_2)^3 \tag{M11}$$

Note that (M11) and (M12) imply that the spacecraft field can be correctly represented by that due to a dipole centered on the main body.

The dual magnetometer method is illustrated schematically in Figure M10. The spacecraft magnetic field decreases with distance

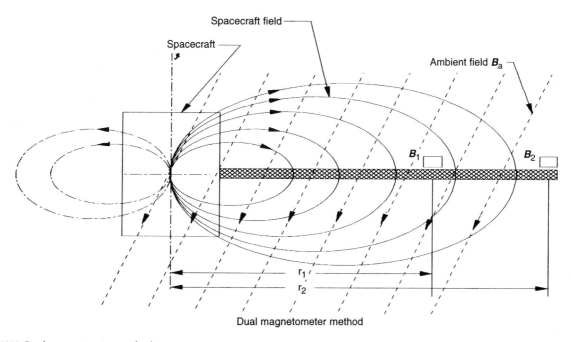

Figure M10 Dual magnetometer method.

due to the finite size of the sources, which are assumed to be located at the center of the spacecraft, leading to the existence of a gradient between the two magnetometer sensors. On the other hand, the ambient field being measured is identical at both sensors because its gradient is insignificant over the dimensions of the spacecraft and boom. Thus each sensor measures a different mixture of spacecraft and ambient field. For the special case of a dipolar spacecraft field, the fields due to the spacecraft at the two sensor locations are related by a simple proportionality constant. Thus the ambient and spacecraft field can be separated analytically as shown above.

A particular advantage of the dual magnetometer method is that it allows the unambiguous real-time identification and monitoring of changes in the spacecraft field. In addition, the use of two sensors provides a measure of redundancy which has proven extremely useful in long-duration missions like those to the outer planets.

Generally, to carry out the intended research the measurements must be expressed in a physical coordinate system which is different from that of the magnetic field sensors (Russell, 1971). Therefore, coordinate transformations are required which must take into account not only the orientation of the spacecraft in inertial space, but also the internal rotations associated with booms and other instruments and sensors aboard.

Finally, the study of the frequency spectrum of dynamic perturbations of the magnetic field is a powerful tool used to identify the types and characteristics of waves and other time-variable phenomena detected by spacecraft instruments.

Summary

The general concepts of magnetometry for space missions have been presented in a simplified form to illustrate the principles and some of the practical considerations involved. Several types of instruments used to measure magnetic fields aboard spacecraft and their capabilities and limitations are described according to whether they measure scalar or vector fields. The difficulties and challenges of performing sensitive measurements aboard spacecraft which may not be magnetically clean represent a fundamental trade-off which must be addressed right at the planning stages of any space mission which includes these measurements. The dual magnetometer technique, which allows the separation of fields of external and spacecraft origin, represents an important magnetometry tool which can result in significant savings in complex contemporary spacecraft built with minimum magnetic constraints.

Mario H. Acuña

Bibliography

Acuña, M.H. (1974) Fluxgate magnetometers for outer planets exploration. *IEEE Trans. Magnetics*, **10**, 519–23.

Acuña, M.H., Scearce, C.S., Seek, J.B. and Scheifele, J. (1978) The Magsat vector magnetometer – a precision fluxgate magnetometer for the measurement of the geomagnetic field. NASA TM-79656.

Behannon, K.W., Acuña, M.H., Burlaga, L.F. *et al.* (1977) Magnetic field experiment for Voyagers 1 and 2. *Space Sci. Rev.*, **21**, 235–45.

Cahill, L.J. (1964) The geomagnetic field, in *Space Physics* (eds D.P. LeGalley and A. Roser). New York: John Wiley, pp. 301–49.

Connerney, J.E.P., Acuña, M.H. and Ness, N.F. (1987) The magnetic field of Uranus. *J. Geophys. Res.*, **92**, 15234–48.

Connerney, J.E.P., Acuña, M.H. and Ness, N.F. (1991) The magnetic field of Neptune. *J. Geophys. Res.*, **96**, 19023–42.

Dolginov, Sh.Sh. and Pushkov, N.V. (1959) Results of measurements of magnetic field of the Earth with cosmic rockets. *Dokl. Akad. auk SSSR*, **129**, 77–80.

Dolginov, Sh.Sh., Zhuzgov, L.N. and Pushkov, N.V. (1959) Preliminary report on geomagnetic measurements carried out from the third Soviet artificial Earth satellite. *Artificial Earth Satellites*, **2**, 63–7.

Dolginov, Sh.Sh., Zhuzgov, L.N. and Selyutin, V.A. (1960) Magnetometers in the third Soviet Earth satellite. *Iskusstvennyye Sputniki Zemli*, **4**, 135–60 (*Artificial Earth Satellites*, 358–96).

Dolginov, Sh.Sh., Zhuzgov, L.N. and Shkolnikova, S.I. (1984) Magnetic field of Mars from data of simultaneous measurements in the magnetosphere of the planet and the solar wind. *Kosmich. Issled.*, **22**(5), 792–803.

Farthing, W.H. and Folz, W.C. (1967) Rubidium vapor magnetometer for near Earth orbiting spacecraft. *Rev. Sci. Instrum.*, **38**, 1023–30.

Geyger, W.A. (1964) *Non-Linear Magnetic Control Devices*. New York: McGraw-Hill.

Heppner, J.P., Stolarik, J.D., Shapiro, I.R. and Cain, J.C. (1960) Project Vanguard magnetic field Instrumentation and measurements. *Space Res.*, **1**, 982–99.

Luhmann, J.G. (1991) Space plasma physics research progress 1987–1990: Mars, Venus, and Mercury. *Rev. Geophys.*, suppl. **96**, 965–75.

Luhmann, J.G., and Brace, L.H. (1991) Near-Mars space. *Rev. Geophys.*, **29**, 121–40.

Ness, N.F. (1965) The Earth's magnetic tail. *J. Geophys. Res.*, **70**, 2989–3005.

Ness, N.F. (1970) Magnetometers for space research. *Space Sci. Rev.*, **11**, 459–554.

Ness, N.F. (1979) The magnetic fields of Mercury, Mars, and Moon. *Ann. Rev. Earth Planet. Sci.*, **7**, 249–88.

Ness, N.F., Behannon, K.W., Lepping, R.P. and Shatten, K.H. (1971) Use of two magnetometers for magnetic field measurements on a spacecraft. *J. Geophys. Res.*, **76**, 3564–73.

Ness, N.F., Acuña, M.H., Behannon, K.W. *et al.* (1986) Magnetic fields at Uranus. *Science*, **233**, 85–9.

Ness, N.F., Acuña, M.H., Behannon, K.W. *et al.* (1989) Magnetic fields at Neptune. *Science*, **246**, 1473–8.

Riedler, W., Möhlmann, D., Oraevsky, V.N., *et al.* (1989) Magnetic field near Mars: first results. *Nature*, **341**, 604.

Russell, C.T. (1971) Geophysical coordinate transformations. *Cosmic Electrodynamics*, **2**, 184–96.

Slocum, R.E. (1972) Zero-field level-crossing resonances in optically pumped $2^3S_1He^4$. *Phys. Rev. Lett.*, **29**, 1642–5.

Slocum, R.E. and Reilly, F.N. (1963) Low field helium magnetometer for space applications. *IEEE Trans. Nucl. Sci.*, **10**, 165–71.

Slocum, R.E., Cabiness, P.C. and Blevins, S.L. (1971) Self oscillating magnetometer utilizing optically pumped He. *Rev. Sci. Instrum.*, **42**, 763–6.

Smith, J.E. and Sonnett, C.P. (1976) Extraterrestrial magnetic fields: achievements and opportunities. *IEEE Trans. Geo. Electr.*, **14**, 154–71.

Sonnett, C.P. (1963) The distant geomagnetic field, 2, modulation of a spinning coil EMF magnetic signals. *J. Geophys. Res.*, **68**, 1229–32.

Cross references

Earth: magnetic field and magnetosphere
Mars: magnetic field and magnetosphere
Mercury: magnetic field and magnetosphere
Magnetospheres of the outer planets
Solar wind
Venus: magnetic field and magnetosphere

MAGNETOSPHERES OF THE OUTER PLANETS

Initial explorations of the giant planets and the outer solar system were carried out by four unmanned spacecraft, Pioneers 10 and 11 (q.v.) and Voyagers 1 and 2 (q.v.). Results from the Pioneer and Voyager spacecraft have completely transformed our knowledge and understanding of the interiors, atmospheres, magnetospheres, satellite systems and ring systems of the giant planets. This article will focus on the giant planet magnetospheres and their interactions with the atmospheres, satellites and ring systems as revealed by the Pioneer and Voyager missions. Table M1 summarizes these missions, showing the planetary closest approach dates and the heliocentric distances as of January 1995.

Mercury, Earth, Jupiter, Saturn, Uranus and Neptune are the six planets known to have 'intrinsic' magnetospheres, meaning that their planetary magnetic fields are strong enough to become an effective barrier to the solar wind. The solar wind is a tenuous ionized gas (a plasma) that consists mainly of hydrogen and helium, blowing hypersonically away from the Sun with an entrained solar magnetic field. The region of space dominated by the solar wind is called the

Table M1 Pioneer and Voyager spacecraft in the outer solar system

Spacecraft	Launch date	Planetary encounters	January 1995 distance from Sun (AU)
Pioneer 10	2 March 1972	Jupiter: 4 December 1973	61
Pioneer 11	5 April 1973	Jupiter: 3 December 1974	42
		Saturn: 1 September 1979	
Voyager 1	5 September 1977	Jupiter: 5 March 1979	58
		Saturn: 12 November 1980	
Voyager 2	20 August 1977	Jupiter: 9 July 1979	44
		Saturn: 26 August 1981	
		Uranus: 24 January 1986	
		Neptune: 25 August 1989	

'heliosphere' and is known to extend at least out to 54 AU, beyond the orbit of Pluto. Its full extent is currently unknown.

If the planetary magnetic field is strong enough, it deflects the solar wind and creates a cavity, called a magnetosphere, within which the planetary magnetic field controls the dynamics of the plasma and energetic particles. The solar wind is largely, but not entirely, excluded from entry into a magnetosphere. The magnetic fields of Venus and Mars are too weak to create intrinsic magnetospheres, so the solar wind blows directly onto their planetary ionospheres and forms 'induced' magnetospheres. The magnetic field of Pluto is unknown.

The magnetosphere of Jupiter, were it visible to the naked eye, would be one of the largest objects in the sky. As seen from Earth, it would have an angular extent measured in degrees, larger than the Sun and roughly the size of a thumb viewed at arm's length. The distance from Jupiter's center to the outer boundary of its magnetosphere, a surface called the magnetopause, is variable but is typically about 70 planetary radii at the subsolar point. Jupiter's magnetosphere is the largest in the solar system, both in absolute terms and relative to the size of the planet. The magnetosphere of Mercury, on the other hand, is the smallest in the solar system, again both in absolute terms and relative to the size of the planet. Mercury's magnetic field is barely strong enough to produce an intrinsic magnetosphere; the solar wind can reach the surface of Mercury when the wind pressure is unusually high.

A planetary magnetosphere contains large populations of charged particles trapped in the planetary magnetic field. Charged particles spiral tightly around a magnetic field line and tend to follow its curvature; moreover, they are repelled by strong field regions (diamagnetism). Charged particles can be trapped on field lines that extend from near one magnetic pole to near the opposite pole, because they bounce off the high field region near one pole, follow the field line, and then bounce again near the opposite pole. In addition there are slow drifts of charged particles, across the magnetic field, caused by electric fields and/or spatial inhomogeneities in the magnetic field. The latter effect is more important at higher particle energies and tends to produce drifts in longitude around the poles, for a nearly dipolar magnetic field. High-energy ions and electrons (above about 1 million eV), trapped on magnetic field lines in this way, comprise the Van Allen radiation belts within Earth's magnetosphere. Similar radiation belts have been discovered around all four of the giant outer planets by the Pioneer and Voyager spacecraft. Of all the intrinsic planetary magnetospheres, only that of Mercury may be too small and possibly too time variable to maintain large, durably trapped radiation belts. In addition to these high-energy particle populations, planetary magnetospheres contain low-energy plasmas with particle energies down to an electron volt or less. These particles are also magnetically trapped, but at low energies electric drifts are more important and typically lead to radial transport when driven by the solar wind. The total plasma density is typically dominated by particles with energy of a few to tens of electron volts, whereas the plasma pressure is typically dominated by particles above 10 000 eV.

From the above it is clear that a planetary magnetosphere is affected by the solar wind in which it resides; both the fields and the particle populations can vary in response to solar wind inputs. Likewise, it is clear that a planetary magnetosphere will be affected by interaction with the upper atmosphere of the planet, from which the controlling magnetic field originates. It is perhaps less obvious that planetary magnetospheres can be strongly affected by the presence of planetary satellites and rings. The clear demonstration of the importance of satellite–magnetosphere interations is one of the most significant results from the Pioneer and Voyager missions. Whenever a satellite has a significant bound atmosphere within a planetary magnetosphere, and to a lesser extent whenever a satellite or ring system has icy surfaces, satellite–magnetosphere interaction can be important.

Prior to the Pioneer and Voyager encounters, Jupiter was the only one of the giant planets definitely known to have an intrinsic magnetosphere. Jupiter's magnetosphere was first detected by radio astronomical observations, and analyses of Jupiter's decimetric radio emissions had produced correct predictions of the approximate magnitude and even the orientation of Jupiter's magnetic dipole moment. The existence and qualitative nature of Jupiter's radiation belts had been likewise inferred from the decimetric emissions prior to the first spacecraft encounter. The first observational evidence for satellite–magnetosphere interaction also came from radio astronomy, with the discovery (a decade prior to the Pioneer Jupiter encounters) that Jupiter's decametric emissions are modulated by the orbital phase of the satellite Io. Another aspect of satellite–magnetosphere interactions was predicted theoretically before the Pioneer Jupiter encounters, namely the existence of a new kind of ring around Saturn formed by gas escaping from the satellite Titan (such a ring would now be called a 'torus'; see Planetary torus). The first observational evidence for such a gas torus came around the time of the Pioneer Jupiter encounters, with the discovery of sodium D-line emissions from the vicinity of Io at Jupiter; this was the first indication of what would become known as the Io torus. However, the extent to which a planetary magnetosphere can be dominated by satellite–magnetosphere interactions was not at all appreciated prior to the Voyager 1 encounter with Jupiter.

Aside from establishing the existence and basic nature of magnetospheres at Saturn, Uranus and Neptune, the Pioneer and Voyager encounters highlighted a number of important issues, many of which remain unresolved. The first of these is the origin of plasmas in planetary magnetospheres, and specifically the relative importance of solar wind entry, planetary ionospheric sources and satellite–magnetosphere interactions. The second issue is energization of plasmas and how particles can be accelerated to energies far exceeding those found in any plasma source. The third is how and where particles are transported. The fourth is the existence and fundamental nature of global magnetospheric disturbances and specifically the question of whether magnetospheric substorms, as known at Earth, are a universal phenomenon.

In what follows, results from the Pioneer and Voyager encounters with the outer planets will be discussed in the light of these critical issues, highlighting our new understanding of the role of satellite–magnetosphere interactions. Table M2 lists some key characteristics of the known intrinsic magnetospheres and their satellite–magnetosphere interactions (if any). The subsolar magnetopause radius is in all cases variable, and Table M2 lists typical values for average solar wind conditions. The ion populations at Mercury are listed as unknown owing to the lack of spacecraft data, but protons and alpha particles from the solar wind are predicted to be present, as well as sodium ions from Mercury's atmosphere. Magnetospheric interactions with the icy Galilean satellites of Jupiter, namely Ganymede, Europa and Callisto, are predicted to produce water decomposition products (e.g. H and O), but this source is relatively minor and has not been specifically identified in spacecraft data. Titan is listed as the major satellite in Saturn's magnetosphere, but it spends part of the time

Table M2 Intrinsic planetary magnetospheres and satellite interactions

Planet	Subsolar magnetopause radius (planetary radii)	Principal satellite–magnetosphere interactions	Ion populations (source)
Mercury	1.4	None	Unknown
Earth	11	None	H, He (solar)
			H, O, etc. (iono.)
Jupiter	70	Io produces the Io plasma torus	S, O, SO_2, Na, etc. (Io torus)
		Icy Galilean satellites	(Minor)
			H, He, C, N, O etc. (solar)
			H, H_2, H_3 (ionospheric)
Saturn	21	Titan produces Titan torus	H, H_2, N (Titan torus)
		Icy satellites (Dione, Tethys, Rhea)	H, O, OH, etc. (water products)
		Icy main rings	Unknown
			H, He (solar)
			H, H_2, H_3 (ionospheric)
Uranus	18	Icy satellites	(Minor)
			H, H_2 (ionospheric)
Neptune	26	Triton produces Triton torus	H, H_2, N (Triton torus)
		Small satellites	(Minor)
			H, He (solar)

outside in the solar wind. Only three of Saturn's icy satellites are named in Table M2; these are the major plasma sources in the inner magnetosphere (the others are relatively minor). Saturn's main rings are also listed as an unknown plasma source, again because of the lack of spacecraft data, but they are predicted to produce water decomposition products. Likewise the numerous small icy satellites of Uranus are predicted to produce water decomposition products, but this source is relatively minor and has not been detected in the data. The composition of the small satellites of Neptune is unknown, but their contribution to the ion populations is predicted to be minor.

As is clear from Table M2, the intrinsic magnetospheres are populated by plasmas of exotic composition, meaning a composition very different from that of the solar wind (while no composition measurements have been made at Mercury, the same may be true there). If the primary plasma sources were the solar wind and the planetary atmosphere, the magnetospheric plasma populations should have compositions that reflect these sources. The solar wind is typically about 95% protons and 5% alpha particles by number, with all other species comprising much less than 1%. Jupiter's atmosphere is likewise about 90% molecular hydrogen and 10% helium by number. The helium abundance of Saturn's atmosphere is significantly lower, and that of Uranus and Neptune moderately greater (less than a factor of two), than that at Jupiter. Hydrogen- and helium-dominated compositions very similar to that of the solar wind characterize the solar interior, main sequence stars, interstellar medium and cosmic rays. These so-called 'cosmic abundances' of light elements are believed to be ubiquitous because they are a consequence of thermonuclear reactions following the big bang that created the present expanding universe.

Since both the solar wind and the planetary atmosphere are dominated by hydrogen and helium, it was natural to assume initially that these elements should also dominate the plasmas of planetary magnetospheres. For the outer planets, this idea was never seriously questioned until the initial results of the Voyager 1 Jupiter encounter, which revealed a sulfur- and oxygen-dominated magnetospheric composition. This surprising discovery led to a fundamental change in our understanding of planetary magnetospheres and specifically pointed to the importance of satellite–magnetosphere interaction. With the Voyager discoveries of active volcanoes and SO_2 on Io, it soon became clear that Io was the ultimate source of the sulfur and oxygen in Jupiter's magnetosphere. Subsequently, Voyager discovered different exotic plasma compositions at Saturn, Uranus and Neptune as well, but with the example of Jupiter these no longer came as a surprise.

Jupiter and Io

Prior to the first Pioneer spacecraft encounter with Jupiter, the existence of a magnetosphere and radiation belts at Jupiter had been inferred by interpreting Jupiter's decimetric radio emissions as being due to synchrotron radiation from relativistic electrons. As mentioned above, these predictions were verified by Pioneer measurements, which showed moreover that Jupiter's magnetic dipole moment is $4.2 \times 10^6 \, R_J^3$; the dipole is tilted 10 from its rotation axis, with a small offset from the planet center.

Given the existence of a Jovian magnetosphere, attempts were made to predict its nature using theories developed to explain Earth's magnetosphere. One basic issue was the competition between plasma corotation with the planet versus plasma flow driven by solar wind interaction ('convection'). Any rotating, magnetized conductor tends to impose corotation (rotation at the same angular velocity) upon a plasma in electrical contact with the rotating conductor, because the conductor acts as a homopolar generator and imposes electric fields that drive corotational drifts of particles. On the other hand, solar wind interaction also imposes electric fields in the magnetosphere, by mechanisms that are still poorly understood, and thereby drives convection of plasma generally towards the Sun (near the equatorial plane). Superposition of the corotational and convective electric fields, for the case of Earth, shows that corotation dominates close to the Earth and out to a radius of about six planetary radii, while convection dominates farther out. In the corotation-dominated region is found a relatively cold and dense plasma (the plasmasphere), while in the convection-dominated region is found a relatively hot and tenuous plasma (the plasma sheet). Application of these ideas to Jupiter lead to the prediction that the corotation-dominated region at Jupiter should extend out to the dayside magnetopause, because of the faster rotation (9.9 h), higher magnetic field and larger radius of Jupiter. A corollary of this prediction, that a convection-dominated region should exist on the nightside, is currently controversial.

The Pioneer Jupiter encounters supported the prediction of a corotation-dominated dayside magnetosphere, and the Voyager encounters provided definitive observations of approximate corotation in the dayside and dawn sector equatorial regions. The Voyager data furthermore showed that the magnetosphere lags significantly behind the rotation of the planet. This corotation lag is due to the mass loading and outward radial transport caused by the Jupiter–Io interaction (see below). The Pioneer Jupiter data also revealed a strong 'magnetodisk distortion' of the magnetosphere, meaning a centrifugally extended, flattened configuration with a thin equatorial current sheet. The initial suggestion following the Pioneer encounters was that the magnetodisk distortion was in fact due to centrifugal force on the rotating plasma, but subsequent analyses of Voyager data have shown that the magnetodisc distortion is actually maintained primarily by the pressure stresses (probably anisotropic) of particles at energies of several tens of keV.

Perhaps the most surprising result from the Voyager encounters was the extent to which sulfur and oxygen plasmas from Io interaction are dominant in Jupiter's magnetosphere. Sulfur and oxygen ions dominate the overall mass density in the magnetosphere, and they are comparable in number to protons. Within the Io plasma torus (which extends from about six to about eight planetary radii near the magnetic equator) sulfur and oxygen ions also dominate the total charge density. The source of the sulfur and oxygen is clearly Io, with its volcanoes, SO_2 atmosphere and condensed SO_2 on its surface. However, the volcanoes are not able to drive gas into space against Io's gravity.

It turns out that the magnetosphere of Jupiter is itself the agent that drives escape of sulfur- and oxygen-bearing species (SO_2 and various dissociation products) from Io. The magnetospheric plasma is also responsible for ionizing the sulfur and oxygen species to create the very ion population that dominates its own composition. The plasma is in this sense self-regenerating. The process that drives escape of sulfur and oxygen from Io is impact of magnetospheric ions onto Io's atmosphere which leads to numerous phenomena, including a cascade of elastic collisions that results in escape of atmospheric neutrals ('atmospheric sputtering'). Similar collision cascade sputtering occurs when ions impact solid surfaces at energies below several keV per nucleon. A total of about 2×10^{28} neutrals per second escape from Io.

After escaping from Io the sulfur- and oxygen-bearing neutrals remain gravitationally bound to Jupiter, so they orbit Jupiter near Io's orbit, forming neutral clouds in the shape of a torus extending completely around Jupiter with some enhancement near Io. The average density of atomic oxygen in this cloud is 30 cm^{-3}. The neutrals in this cloud are rapidly ionized and dissociated by electron impact and charge exchange reactions, yielding a source of plasma that maintains the Io plasma torus. The Io plasma torus has an average electron density of 2000 cm^{-3} and electron temperatures ranging from about 2 eV in its cold portion (inside Io's orbit) to about 7 eV in its warm outer portion.

The mass injection rate of plasma into the Io torus is about 10^6 g s^{-1}. This mass injection cannot be balanced by recombination of plasma and requires rapid radial transport out of the torus, on a time scale of 60 to 100 days. This transport may be driven by centrifugal interchange instability, a hydromagnetic instability driven by centrifugal force much like the Rayleigh–Taylor instability in ordinary fluids. Outward radial transport of 10^6 g s^{-1} causes the magnetosphere to extract angular momentum and energy from the rotation of Jupiter. This is because Jupiter maintains corotation of the plasma out to a distance determined by the degree of electrical contact with the plasma (the effective resistance of the ionosphere); this distance is about 20 planetary radii and is the characteristic distance beyond which there is significant corotation lag. In order to maintain corotation of the outward-moving plasma from the Io torus, an energy input of about 10^{14} W is required at the expense of Jupiter's rotational energy. This rotational energy loss does not cause a significant spin-down of Jupiter, even over geologic time.

The substantial ion and neutral densities within the Io torus cause electron energy losses to ionization and to excitation of extreme ultraviolet radiation (the radiation first observed by Voyager that revealed the nature of the Io torus). The total electron energy loss rate is a few times 10^{12} W and would lead to significant electron cooling within hours in the absence of any heating mechanism. One important energy source for the Io plasma torus is ion pick-up, which is another mechanism by which the plasma can tap the rotational energy of Jupiter. A neutral atom or molecule orbits Jupiter at a speed much lower than the corotation speed. Upon ionization, the new ion is accelerated by the corotational electric field to acquire a motion which is a superposition of corotation and gyromotion, with the gyrospeed given by the difference between the corotation and orbit velocities. The energy of ion gyromotion, corresponding to about 17 eV per nucleon for new ions, is available to heat the electrons via Coulomb collisions. This collisional energy transfer from hot ions contributes a major portion of the electron energy source in the Io plasma torus, but detailed modeling has shown that an additional energy source is needed. It has been suggested that collisional energy transfer from much higher energy ions, at ring current energies, may make up the deficiency.

Jupiter's magnetosphere also powers intense ultraviolet auroras from the polar regions of Jupiter's upper atmosphere. This aurora occurs at magnetic latitudes near to, or greater than, those connecting along magnetic field lines to the Io plasma torus. The aurora is excited by precipitating radiation belt heavy ions at ring current energies (oxygen or sulfur above a few hundred keV) or electrons of at least a few tens of keV. The relative contributions of ions and electrons are presently unclear. The total precipitating power input needed to drive the aurora is 5×10^{13} W.

The power requirement to excite the aurora is similar to the power extracted from Jupiter's rotation to maintain approximate corotation of outward transported magnetospheric plasma, as discussed above. These estimates are believed to set the overall scale of the energy budget for Jupiter's magnetosphere and have led to numerous studies of processes whereby plasmas or particles can extract energy from Jupiter's rotation. Quantitative understanding, however, is still lacking.

Pioneer and Voyager measurements in Jupiter's radiation belts have supported the idea that the aurora is powered by precipitation of energetic ions and electrons from the magnetosphere caused by scattering from plasma waves (e.g. ion cyclotron waves and whistlers). Charged particle measurements provided evidence for inward radial diffusion and drastic particle losses occuring within six to nine Jupiter radii, just outside the Io torus. If the radial diffusion occurs at the rate needed to balance the ion injection into the Io torus, then the energetic ion loss rate can be determined from the data. The ion losses occur at close to the maximum possible rate for charged particle precipitation into the atmosphere caused by pitch angle scattering from plasma waves (the so-called 'strong diffusion' limit). The precipitating power from energetic heavy ions may be adequate to power the ultraviolet aurora as well as the Jovian soft x-ray aurora discovered by the Einstein Observatory (an Earth-orbiting telescope).

While Jupiter's auroras can be powered by precipitation of radiation belt ions and/or electrons, the more fundamental question is how these particles were accelerated to the requisite high energies from the much lower energies found in the source region, the Io plasma torus. This question is still unresolved, although several mechanisms have been proposed to extract energy from Jupiter's rotation via various forms of radial transport across the magnetic field or via interactions with plasma waves. Even more challenging is the observation that Jupiter's magnetosphere accelerates electrons and ions to energies above 100 MeV; indeed, at energies of up to about 50 MeV, Jupiter's magnetosphere is the dominant source of relativistic electrons in the solar system.

Finally, it is clearly established that rapid radial transport of plasma occurs in Jupiter's magnetosphere, but the fundamental nature of this radial transport is controversial, particularly for the thermal plasma populations. The transport may be characterized by a continuum of scales, ranging from global-scale convection down to micro-scale diffusion. At higher radiation belt energies, radial diffusion generally dominates. Radial diffusion may be driven by Jovian upper atmospheric winds in the inner decimetric emission region but by centrifugal interchange-type instability outside the Io torus. It is currently controversial whether a global-scale solar wind-driven convection system exists at Jupiter and whether a global-scale wind outflow exists across the magnetotail.

Saturn and Titan

The existence of Saturn's magnetosphere was first established by Pioneer 11, although it was widely anticipated beforehand. After the discoveries of volcanoes on Io, neutral gas clouds in Jupiter's magnetosphere, and Io's plasma torus, it was also anticipated that similar interactions should occur between Saturn's magnetosphere and its major satellites and rings. Such is indeed the case, although the situation is more complicated at Saturn owing to the multiplicity of important satellites and rings. Moreover, planetary rotation is not such a dominant energy source at Saturn as it is at Jupiter, and energy input from the solar wind to the magnetosphere may be significant.

Saturn's magnetic dipole moment is 0.21×10^6 R_S^3, with no measurable tilt from the planetary spin axis ($< 1°$) and a small offset from the planet center. The near-perfect alignment of the dipole moment with the spin axis was surprising and posed difficulties both for theoretical models of the planetary dynamo generating the magnetic field and for understanding various spin-periodic magnetospheric phenomena. Most prominent of these is the Saturn kilometric radio emission, which is strongly modulated at the planetary rotation period and whose source is near local noon and in the auroral region. Very similar auroral kilometric emissions modulated at the planetary spin period are observed at Earth and all four of the giant planets; these latter emissions were discovered by Voyager. Precise source geometries and radio spectra vary from planet to planet. In all cases except Earth, the auroral kilometric emissions yield the best experimental determinations of the rotation period of the magnetic field, which is also that of the planetary interior. These measurements of the interior rotation rate allow determination of atmospheric wind speeds relative to the interior from optical tracking of cloud features.

Saturn's magnetosphere exhibits a significant magnetodisk distortion ('centrifugal' flattening) although this distortion is much less pronounced than at Jupiter. The magnetodisk distortion at Saturn, like at Jupiter, is not primarily supported by centrifugal force from corotating plasma, but the stresses of energetic ions at ring current energies play an important role. As at Jupiter, corotation dominates convection out to the dayside magnetopause of Saturn, but a significant corotation lag is observed.

Unlike the situation at Jupiter, where the single moon Io dominates the satellite – magnetosphere interactions, several moons have important interactions at Saturn. The largest of these is Titan (q.v.), which is larger than Mercury and which has a dense nitrogen-dominated atmosphere of its own. Methane and dissociation products are also present in Titan's atmosphere. Titan resides in Saturn's outer magnetosphere (at 20 R_S), but spends part of the time outside the magnetosphere (in the solar wind) during periods of high solar wind dynamic pressure. In addition several small icy moons reside in Saturn's inner magnetosphere (3–9 R_S); these have radii up to about half that of Earth's moon and have water ice on their surfaces.

Various processes lead to ejection of water molecules and various dissociation fragments from the surfaces of the small icy satellites into Saturn's inner magnetosphere. These processes include micrometeoroid impacts, charged particle impacts and ultraviolet photon-induced desorption, forming a water product neutral cloud in Saturn's inner magnetosphere. This cloud is ionized and dissociated to produce a heavy ion plasma with maximum ion density of about 50 cm^{-3} near the orbits of Tethys and Dione. This plasma torus was detected by both Pioneer and Voyager, which confirmed the presence of heavy ions but could not determine their composition experimentally. Moreover, no infrared, optical or ultraviolet emissions have been measured from this region. Only a Voyager measurement of energetic neutral atoms from charge exchanges provided tentative observational evidence for a water product neutral cloud, in addition to atomic hydrogen, in Saturn's magnetosphere.

As originally predicted before the first Pioneer Jupiter encounter, gas escaping from Titan maintains a vast neutral cloud called the Titan torus. This cloud contains atomic hydrogen, at an average density of 10 to 20 cm^{-3}, which has been observed in H Lyman α emission by Voyager and by Earth-orbital and suborbital telescopes. In addition the cloud is theoretically predicted to contain molecular hydrogen and atomic nitrogen. When Titan is inside Saturn's magnetosphere, magnetospheric plasma impinges onto Titan's ionosphere, producing a separate induced magnetosphere of Titan within Saturn's magnetosphere (analogous to the solar wind interaction producing an induced magnetosphere of Venus). A long tail of protons and heavy ions is extracted from Titan and carried downstream, causing a loss rate of 10^{24} heavy ions per second. In addition, about 3×10^{27} hydrogens per second (atoms and molecules, each molecule counting as two hydrogens) and about 3×10^{26} nitrogen atoms per second escape from Titan. Unlike the case at Io, however, escape of these neutrals from Titan is not driven by impact of magnetospheric ions, but is thermal escape of hydrogen and nonthermal escape of N from electron impact dissociation of N_2.

The neutrals escaping into the Titan torus are ionized to create an important, but possibly not dominant, plasma source for Saturn's magnetosphere. Satellite – magnetosphere interactions at Saturn are generally less well understood than at Jupiter. The total water product neutral source rate in the inner magnetosphere is at least about 10^{26} s^{-1} but may be greater, depending on poorly known micrometeoroid fluxes. The inner magnetospheric plasma source may be comparable to that in the outer magnetosphere. The identity of the heavy ions in the inner magnetosphere, that is, whether they are water products (O, OH, etc.) from the icy moons or N from Titan, has not been clearly established. The dominant source of atomic H in the Titan torus may actually be Saturn and not Titan; this is presently controversial. Also controversial is the density of atomic H near Tethys and Dione; chemical modeling suggests a maximum value of several tens per cubic centimeter in this region but Voyager Lyman α observations suggest much higher values. Finally the rate of radial transport in the inner magnetosphere, and the extent to which it balances the plasma injection, is controversial.

Saturn's ultraviolet aurora is also poorly understood. Unlike the case of Jupiter's aurora, which is magnetically connected to the vicinity of the Io plasma torus (at least in part), Saturn's ultraviolet aurora is magnetically connected to the outer magnetosphere or magnetic tail. The required auroral power input is about 2×10^{11} W, but it is unclear whether this power is derived primarily from planetary rotation or from solar wind interaction. Additional evidence of the significance of solar wind interaction at Saturn comes from Saturn's kilometric radio emissions, which are observed to be modulated by the solar wind.

If indeed solar wind interaction is important at Saturn, it might be expected that this interaction should drive a substorm cycle as it does in Earth's magnetosphere (and apparently at Mercury and Uranus, see below). Substorms at Earth involve sudden, localized intensifications of the nightside aurora, injections of energetic particles, global magnetic field disturbances such as dipolarization and disruption of a tail current sheet, and ejection of magnetic clouds or 'plasmoids' down the magnetotail (this last is controversial). No evidence for substorms was found at Jupiter or Saturn by the Pioneer and Voyager spacecraft, but none of these spacecraft remained within the magnetospheres long enough to make a definitive search for substorms. Whether substorms exist at Jupiter and Saturn is a fundamental unresolved question in magnetospheric physics.

Uranus and Neptune–Triton

The magnetospheres of Uranus and Neptune were discovered by Voyager 2. As these magnetospheres have some similarities, they will be discussed together. Both Uranus and Neptune have magnetic fields that are very different from those of Mercury, Earth, Jupiter and Saturn. For Uranus and Neptune the magnetic dipole moments are tilted by large angles, on the order of a radian, relative to the spin axis, and are offset by a large fraction of a planetary radius from the planet center. The dipole moment, tilt angle and offset magnitudes for Uranus are 0.23×10^6 R_U^3, 60° and 0.31 R_U respectively. The Neptune values are 0.13×10^6 R_N^3, 45° and 0.55 R_N respectively. For the other magnetized planets both the dipole tilts and the offsets are relatively much smaller. Moreover, at both Uranus and Neptune the degree of plasma loading in the magnetosphere is much less than at Earth, Jupiter or Saturn. There is no appreciable magnetodisk distortion or ring current distortion of the magnetic field at either Uranus or Neptune.

The competition between convection and corotation at Uranus and Neptune takes on qualitatively new aspects because of the peculiar orientations for the spin axes and magnetic moments. At Earth, Jupiter and Saturn the spin axis and magnetic moment are approximately aligned with each other, and both are approximately orthogonal to the direction of solar wind flow. The corotation and convection velocities are then roughly coplanar, leading to the competition described above. At Uranus the spin axis was roughly aligned with the solar wind flow at the time of the Voyager encounter, so the convection and corotation velocities were roughly orthogonal. Hence at Uranus the corotation velocity dominates (in magnitude) out to the dayside magnetopause but convection can penetrate deep within the magnetosphere.

At Neptune the situation is even more peculiar. The geometry of the spin axis, magnetic moment and solar wind velocity are such that a magnetic pole points approximately into the solar wind flow once in every 17-h planetary rotation; this produces a so-called 'pole-on' configuration. The magnetic pole for which this occurs is the one with the downward magnetic field, like Earth's north pole. The convection field, as viewed in a corotating reference frame, is inherently unsteady even with a steady solar wind. The nature of this convection and its implications for plasma transport are presently unclear. It is also unclear whether the magnetosphere can fully adjust to a pole-on configuration (e.g. forming a cylindrical tail current sheet) within the brief time that a magnetic pole points nearly into the solar wind.

The role of satellite–magnetosphere interactions is relatively minor at Uranus. Uranus has five small moons, at least four of which are generally within the magnetosphere, with diameters similar to those of the small icy moons of Saturn and with substantial water ice on their surfaces. Nevertheless, no heavy ions were observed in the Uranus magnetosphere, indicating that the satellite source of plasma is relatively minor. The large dipole tilt of Uranus is at least partly responsible for suppressing the effect of water product ion injection from the icy moon interaction; this is because the dipole tilt causes ions to be injected at high magnetic latitudes and therefore a large range of L values (the L value is the radial distance to a dipole field line at the magnetic equator), so the heavy ion plasma is diluted over an enormous volume whereas the neutral cloud is spatially confined near the satellite orbit plane. This situation at Uranus contrasts with that at Jupiter and Saturn, where the neutral clouds and associated plasma tori are roughly spatially coincident.

The plasma composition at Uranus is unique among known magnetospheres in that the only ion species so far detected are protons with a trace admixture of H_2 ions. No evidence for alpha particles has been detected in the Voyager data, indicating that the solar wind source of plasma must be relatively minor, like the satellite source of plasma. It is inferred that the upper atmosphere of Uranus must be the dominant plasma source; the specific process may be ionization of the extended atomic hydrogen corona of Uranus that was observed in H Lyman α emissions by Voyager.

In contrast with the situation at Uranus, satellite–magnetosphere interaction is again important at Neptune. The satellite Triton is a Pluto-sized body with a nitrogen-dominated atmosphere and active ejection of plumes containing some unknown dark material. Methane and dissociation products are again present in the atmosphere. Escape from Triton of hydrogen atoms and molecules plus nitrogen atoms maintains a gigantic neutral cloud in Neptune's magnetosphere, similar to the situation in Saturn's magnetosphere. The highly inclined, retrograde orbit of Triton around Neptune, together with the large magnetic dipole tilt and the inclination of Neptune's spin axis, combine to produce a complicated geometry for the Triton torus. Moreover, the Triton torus, like the Titan torus, turns out to be partially collisional.

The source of hydrogen (atoms and molecules, molecules counting as two hydrogens) in the Triton torus is about 7×10^{25} s^{-1}, yielding a proton source of about 10^{25} s^{-1}, the remaining hydrogens being ejected from the system, colliding with Neptune, or being ionized outside the magnetosphere. The source rate of nitrogen to the Triton torus and the resulting nitrogen ion source rate are comparable to or less than the corresponding hydrogen source rates. The Triton torus may be the dominant plasma source in Neptune's magnetosphere, as indicated by the composition of the cold plasma, which has a heavy ion density comparable to the proton density. Heavy ions were not observed at high energies, i.e. above several MeV.

Finally, both Uranus and Neptune display ultraviolet emissions that are conventionally referred to as auroras, although the processes responsible for these emissions are poorly understood. The required input powers are 4×10^{10} W and 10^9 W for Uranus and Neptune respectively. Uranus alone among the giant planet magnetospheres displayed evidence of magnetospheric activity during its Voyager encounter that was reminiscent of magnetospheric substorms at Earth. An energetic ion injection event with magnetic field dipolarization was observed; extremely strong whistler-mode waves were also measured that imply energetic electron precipitation lifetimes as short as an hour. Neptune, on the other hand, displayed no evidence for magnetospheric activity, despite the expectation that a global-scale reconfiguration might be required once in every rotation period when the magnetosphere goes pole-on. It is clear that much remains to be learned about the outer planet magnetospheres.

Andrew F. Cheng

Bibliography

Bigg, E.K. (1964) Influence of the satellite Io on Jupiter's decametric emission. *Nature*, **203**, 1008.
Cheng, A.F. and Johnson, R.E. (1989) Effects of magnetosphere interaction on origin and evolution of atmospheres, In *Origin and Evolution of Planetary and Satellite Atmosphere* (eds. S.K. Atreya, J.B. Pollack and M. Matthews) Tucson: University of Arizona Press, 682 pp.
Cheng, A.F., Haff., P.K., Johnson, R.E. and Lanzerotti, L.J. (1986) Interactions of planetary magnetospheres with icy satellite surfaces, in *Satellites* (eds J. Burns and M. Matthews). Tucson: University of Arizona Press, 403 pp.
Johnson, R.E., Pospieszalska, M., Sittler, E., *et al.* (1989) The neutral cloud and heavy ion inner torus at Saturn. *Icarus*, **77**, 311.
McDonough, T. and Brice, N. (1973) A Saturnian gas ring and the recycling of Titan's atmosphere. *Icarus*, **20**, 136.
Metzger, A.E., Gilman, D.A., Luthey, J. and Hurley, K.C. (1983) The detection of x-rays from Jupiter. *J. Geophys. Res.*, **88**, 7731.
Pang, K.D., Voge, C.C., Rhoads, J. and Ajello, J. (1984) The E-ring of Saturn and satellite Enceladus. *J. Geophys. Res.*, **89**, 9459.
Smith, R.A., Bagenal, F., Cheng, A.F. and Strobel, D. (1988) On the energy crisis in the Io plasma torus. *Geophys. Res. Lett.*, **15**, 545.

Vasyliunas, V.M. (1983) Plasma distribution and flow, in *Physics of the Jovian Magnetosphere* (ed. A.J. Dessler). New York: Cambridge University Press, 395 pp.

Cross references

Jupiter: magnetic field and magnetosphere
Neptune: magnetic field and magnetosphere
Planetary torus
Pioneer 10 and 11 missions
Saturn: magnetic field and magnetosphere
Uranus: magnetic field and magnetosphere
Voyager missions

MANTLE

The mantle of a terrestrial planet is that region of the planetary interior surrounding the core (q.v.), which is itself enclosed by the planetary lithosphere (q.v.) and crust (q.v.). This layering resulted from differentiation (q.v.) of primordial materials as a consequence of heating during planetary accretion (q.v.). (See Earth–Moon system: origin; Solar system: origin).

The internal structure of the Earth may be imaged by seismic waves, and modeled by gravitational and geochemical studies. It is well determined in comparison with the other planets. Earth has a metallic core of radius 3500 km, surrounded by a silicate mantle of thickness 2800 km. The mantle's present composition was reached following loss (degassing) of volatile elements that eventually went to build up the atmosphere and hydrosphere (Holland, 1984). Conditions of pressure and temperature within the mantle are sufficient to allow plastic flow of mantle rocks, and convective motions are thought to transfer heat from the core to the planetary surface regions (see Mantle convection). This is considered to be the fundamental driving force of Plate tectonics (q.v.; Garfunkel, 1985). The detailed structure and mineralogy of the mantle, and the pattern of convective flow within in the mantle are subjects of active research (e.g. James, 1989). A global network of mantle plumes appears to be the cause of observed volcanic surface hot spots (Sleep, 1992). The nature of processes occurring at the core–mantle boundary is also of considerable interest, as these may be involved in the generation of the Earth's magnetic field (q.v.) and may influence the rate of rotation of the Earth (see Magnetism; Earth rotation).

The interior structure of the Moon is known (though in much lesser detail) from seismological data returned by the Apollo program instruments (see Moon: seismicity). The data suggests the possible presence of a core of radius 170–360 km, surrounded by a mantle of radius 1300 km, which is divided into upper, middle and lower regions (see Figure M97 in Moon: seismicity). In the lunar case the rigid lithosphere is considered to extend to the base of the middle mantle, at a depth of 800–900 km.

Information concerning the internal structure of other planets may be derived from the known density of the body, from its gravity field and from the body's shape, i.e. departure from spherical symmetry (Hubbard, 1984). Mercury has a much higher mean density than the other terrestrial planets, and it is thought to have a relatively large metallic core; the mantle depth represents only about 25% of the radius of the planet.

The interior structure of Venus is not well known, but as its radius and density are similar to those of the Earth, a substantial silicate mantle is expected. Models of the interior of Mars suggest that the mantle of that body may have a depth representing between 50% and 67% of the radius of the planet (Cook, 1980).

The outer planets have deep fluid mantles, and may have solid cores (see Jupiter: interior; Saturn: interior; Uranus; Neptune).

James H. Shirley

Bibliography

Brandt, J.C. and Hodge, P.W. (1964) *Solar System Astrophysics*. New York: McGraw-Hill.
Cook, A.H. (1980) *Interiors of the Planets*. New York: Cambridge University Press.

Garfunkel, A. (1985) *Mantle Flow and Plate Theory* (Benchmark Papers in Geology Vol. 84). New York: Van Nostrand Reinhold.
Holland, H.D. (1984) *The Chemical Evolution of the Atmosphere and Oceans.* New Jersey: Princeton University Press.
Hubbard, W.B. (1984) *Planetary Interiors.* New York: Van Nostrand Reinhold.
James, D.E. (1991) *The Encyclopedia of Solid Earth Geophysics.* New York: Van Nostrand Reinhold, 1328 pp.
Romanowicz, B. (1991) Seismic tomography of the Earth's mantle. *Ann. Rev. Earth. Planet. Sci.*, **19**, 77–99.
Sleep, N.H. (1992) Hotspot volcanism and mantle plumes. *Ann. Rev. Earth Planet Sci.*, **20**, 19–43.
Wood, J.A. (1979) *The Solar System.* New Jersey: Prentice-Hall.

Cross references

Asthenosphere
Core, terrestrial planetary
Differentiation
Crust
Earth rotation
Earth–Moon system: origin
Jupiter: interior
Lithosphere
Magnetism
Moon: seismicity
Neptune
Plate tectonics
Saturn: interior
Solar system: origin
Ulysses mission
Uranus

MANTLE CONVECTION

Earth's mantle is the layer of solid rock between the underlying partially liquid core and the overlying solid rock crust. The thickness of the mantle is about 2885 km or about 45% of the Earth's radius; the mantle is about 84% of the Earth by volume and about 68% of the Earth by mass. The silicate mantle is compositionally distinct from the predominantly iron core and the silicate crust. Crustal rocks are intrinsically lighter than mantle rocks and certain elements such as uranium, potassium and thorium are concentrated in the crust. Mantle rocks are able to deform in a fluid-like manner over times in excess of about 1 million years and from a geological perspective the mantle is therefore a highly dynamic part of the Earth. Mantle convection is the process by which the excess heat in the Earth's deep interior is transferred to its surface through the fluid-like motions of the rocks in the mantle. It is generally believed that the other terrestrial planets and the Moon are similar in structure to the Earth, i.e. they have cores, mantles and crusts, and that subsolidus convection is also the dominant heat transfer process in the mantles of these bodies (Schubert, 1979). Mantle convection is therefore a fundamental mode of heat transfer in all the inner planets and the Moon. 'Mantle convection' is probably also an important process inside the larger rock–ice and rocky satellites of the outer solar system planets although the structures and compositions of these bodies may differ from those of the inner planets (Schubert, Spohn and Reynolds, 1986).

Convection occurs in the Earth's mantle because it is the most efficient way for the Earth to cool or dispose of its deep-seated heat content. In general, heat can be transferred by conduction, radiation or convection. Heat conduction is a diffusive process and it occurs at the molecular level through the collisions of more energetic (hotter) molecules with less energetic (colder) ones. The process can be represented at the macroscopic scale by the material property known as thermal conductivity. Rocks are poor conductors of heat; the thermal conductivity k of mantle rocks is typically 3 to 4 W m^{-1} K^{-1}, about two orders of magnitude smaller than the thermal conductivity of metals. Radiation is the transfer of energy by electromagnetic waves or photons; it is the means by which heat reaches us from the Sun. Radiative heat transfer also occurs through solids and the process can be described in terms similar to conduction, with the photons assuming the role of molecules. In the situation where photons travel relatively short distances before undergoing energy transfer by collisions, the case for the mantle, the phenomenon is essentially diffusive and can be represented macroscopically by a radiative thermal conductivity. Convection is the transfer of heat by the bulk motion or relative counterflow of hot and cold material. Convection requires fluid-like behavior of the material; it is commonplace in the atmosphere and oceans. Conduction and radiation require so much time to transfer heat across the mantle that even the slow, fluid-like motions of relatively hot and cold rocks at speeds of tens to hundreds of millimeters per year are more effective. Thermal convection, the buoyant upflow of hot material and the sinking of negatively buoyant cold material, is the principal mode of heat transfer in the mantle. Conduction heat transfer does occur in the Earth, particularly at places where physical circumstances preclude or discourage vertical motions of material. Such places include the near-surface region of the Earth and the lower boundary of the mantle at the interface with the core.

The excess heat in the Earth's interior derives from several sources. The main source of heating for the mantle is the energy released in the decay of the radioactive isotopes ^{235}U, ^{238}U, ^{40}K and ^{232}Th. The concentrations of these isotopes in mantle rocks are small in absolute terms, tens to about 100 parts per billion (ppb) by mass for ^{238}U, ^{232}Th and ^{40}K and about 0.11 ppb by mass for ^{235}U, but the aggregate rate of heat production of these isotopes, estimated at about 6.2×10^{-12} W kg^{-1} (Turcotte and Schubert, 1982), is sufficient to comprise the bulk, about 80%, of the heat loss at the Earth's surface, exclusive of surface heat loss due to radiogenic heating in the continental crust. The concentrations of the radiogenic elements in the rocks of the crust and mantle decrease with time according to the classical laws of radioactive decay. Thus radiogenic heat production was much higher, by factors as large as five, early in the Earth's evolution. Additionally, due to differences in the half-lives of the radioactive isotopes, radiogenic heating in the early Earth was mainly due to the decay of ^{235}U and ^{40}K, in contrast to the present wherein radiogenic heating is due mainly to the decay of ^{232}Th and ^{238}U.

The primary additional source of excess heat in the Earth's deep interior is the primordial heat left over from the formation of the Earth and from major events early in the Earth's evolution. The accretion of the Earth from initially dispersed mass releases gravitational potential energy, a percentage of which goes into heating the Earth (Kaula, 1979). This energy of accumulation was large enough to heat the Earth by thousands of kelvin and was apparently sufficient to trigger the melting and differentiation of the Earth into a mantle and core either contemporaneously with accretion or within a few hundred million years of the end of accretion (Stevenson, Spohn and Schubert, 1983; Wetherill, 1985; Stevenson, 1989). Separation of iron into the Earth's core in turn released additional gravitational potential energy equivalent to several thousand kelvin, some of which was retained as heat within the deep interior of the Earth (Birch, 1965; Tozer, 1965). Another cataclysmic event early in the Earth's evolution capable of substantially heating, even melting, the Earth could have been the collision of the Earth with a Mars-size body resulting in the formation of the Moon (Hartmann *et al.*, 1986; Stevenson, 1987).

The release of gravitational potential energy in the Earth's interior is occurring even today with the solidification of the inner core. This process is concentrating the light alloying element of the core into the outer liquid part, in effect concentrating heavier mass nearer the center of the Earth and releasing gravitational potential energy, some of which is manifest as heat. Solidification of the inner core also releases energy in the form of latent heat. These energy sources are responsible for driving the dynamo electric currents in the outer liquid metallic core that produce the geomagnetic field. They also contribute to the heat flux entering the mantle from below and to the heat flux at the top of the mantle and at the surface (Stevenson, Spohn and Schubert, 1983).

The radiogenic and primordial heat sources drive mantle convection. The Earth is a giant heat engine that converts a part of its internal thermal energy into the mechanical energy of mantle motions. As a result of mantle convection heat is able to efficiently escape from the Earth's deep interior to the surface, where it is deposited into the atmosphere–ocean system and eventually radiated to space. Mantle convection causes the surface to deform and it produces almost all the geologic structures at the surface. Mantle convection is the motive force behind plate tectonics. The cooling of the Earth is the driving force of geology.

The radiogenic and primordial heat sources that drive mantle convection on Earth are also believed to power convection in the interiors of other planets and satellites (Schubert, 1979; Schubert, Spohn and Reynolds, 1986). Generally, all terrestrial planets and satellites and the rock constituents of outer planet satellites are assumed to contain similar concentrations of radiogenic isotopes per unit mass of rock. There are significant differences among the planets and satellites in their primordial heat contents, with size being the main factor determining the amount of primordial heat in a planet. The larger the planet, the larger is the energy release upon accretion and core formation, and the larger is the planet's primordial heat content. There is one solar system body with an exceptional heat source in addition to radiogenic and primordial heat. Jupiter's satellite Io is strongly internally heated by tidal dissipation (Peale, Cassen, and Reynolds, 1979; Cassen, Peale and Reynolds, 1982; Schubert, Stevenson and Ellsworth, 1981; Schubert, Spohn and Reynolds, 1986; Ross and Schubert, 1985, 1986; Segatz et al., 1988; Ross et al., 1990).

The general acceptance of thermal convection as a process occurring in the Earth's mantle, and also in the interiors of other planets, has largely depended on the visibility of the process at the Earth's surface in the form of the motions of tectonic plates. Plate tectonics is the expression of mantle convection at the Earth's surface. The horizontal motions of the plates are the near-surface horizontal movements of the convecting mantle. Subduction zones mark the surface locations of mantle downflow with the descending slabs being the convective downflow elements. Mid-ocean ridges, however, are not the surface expressions of convective upflows from the deep mantle, but instead are features above shallow passive upflows where the forces on the plates from the sinking slabs have torn the plates apart (Lachenbruch, 1976; Schubert, 1992).

It is not required that mantle convection is manifest at the surface through plate tectonics or that it is in any way visible at the surface. Plate tectonics is a surface expression of mantle convection unique to the Earth. On some other planets, e.g. Mars and the Moon, mantle convection takes place beneath a thick, rigid, immobile outer spherical shell of rock known as the lithosphere (Schubert, Spohn and Reynolds, 1986; Schubert et al., 1992). The lithosphere is relatively cold, compared to underlying rock, and so viscous that it cannot undergo fluid-like deformation even on geologic times, hence its rigidity. The tectonic plates of the Earth comprise its lithosphere, which, instead of being an entire spherical shell, is broken into pieces that are quite rigid in their interiors but deformable at their boundaries. On Earth plates can move horizontally like rigid bodies over the surface, motions which are identical to rotations of spherical caps according to a theorem by Euler. The occurrence of mantle convection beneath a highly viscous lid that participates only minimally, if at all, in the convection is apparently the rule for all the terrestrial planets and satellites other than Earth. Earth is the exception in revealing the deeper-seated motions of its convecting mantle through the global movements of its plates. The acceptance of the idea that mantle convection occurs on other planets despite the lack of surface evidence for it, is based largely on the theoretical understanding of mantle convection as a fundamental response of mantle rocks to the heating they experience deep within planetary interiors.

The style and vigor of mantle convection depends on many factors in addition to the amount and mode of heating of the mantle (mode of heating refers to whether heat enters the mantle from below, i.e. from the core, or is generated directly within the mantle by radiogenic isotopes). One of the other important factors is the way in which mantle rocks deform when subjected to stresses. At the high temperatures and pressures of the deep mantle, and on geologic times, solid mantle rocks deform as a very viscous fluid (Weertman and Weertman, 1975; Turcotte and Schubert, 1982). The viscosity is typically about 10^{21} Pa s, though strong variations in viscosity do occur, partly because the viscosity of rocks depends on their temperature, pressure and state of stress. The temperature T dependence of rock viscosity μ is particularly strong, being of the Arrhenius type $\mu \propto \exp(A/T)$, where A is an activation enthalpy. The value of A is such that a difference in temperature of a few hundred kelvins can cause viscosity to vary by many orders of magnitude. Thus relatively hot convective upflow elements will be less viscous than their average-temperature surroundings and much less viscous than the relatively cold convective downflow elements, which in turn are more viscous than their average-temperature surroundings. Tectonic plates on Earth and the lithospheres of other planets are essentially rigid because of their relatively low temperatures and extremely high viscosities. Slabs descending through the Earth's mantle are more viscous than their surroundings because they tend to retain the relative coldness they possessed as plates at the Earth's surface.

The viscosity of mantle rocks can also vary with their mineralogical make-up. In the Earth's mantle the dominant mineral at shallow depth, olivine, undergoes a solid–solid phase transition to a more closely packed or denser structure known as spinel at about 410 km depth, and undergoes a further phase change to the still denser assemblage of perovskite and the simple oxides MgO and FeO (Ito and Takahashi, 1989; Ito et al, 1990; Boehler and Chopelas, 1991) at a depth of about 660 km (Shearer, 1991; Shearer and Masters, 1992). Viscosity changes may accompany these solid–solid phase changes (Karato and Li, 1992). Some evidence suggests that the lower mantle (the mantle below about 660 km depth) is more than an order of magnitude more viscous than the upper mantle (the mantle above about 660 km depth; Hager and Richards, 1989). The style of mantle convection would be affected by such an increase of viscosity with depth.

Yet one more factor influences rock viscosity with potentially important consequences for mantle convection. That factor is the water content of the mantle, particularly that of the shallow upper mantle. Wet rocks are less viscous than dry rocks under similar conditions of temperature, pressure and stress (Karato, Paterson and FitzGerald, 1986). Earth is known to have a low-viscosity layer, the asthenosphere, underlying the lithosphere, particularly beneath oceanic plates. The asthenosphere is at most a few hundred kilometers thick and may be as much as a factor of 100 less viscous than the typical mantle viscosity of 10^{21} Pa s (Cathles, 1975). The existence of a low-viscosity asthenosphere on Earth may be due in part to the water content of the shallow upper mantle, though the temperature and pressure dependence of mantle viscosity can also contribute to the occurrence of a shallow upper mantle low-viscosity zone. Subduction of sediments and hydrated minerals with the descending slabs on Earth offers a mechanism of recharging the upper mantle with water. The low-viscosity asthenosphere on Earth facilitates the sliding of tectonic plates and may in fact be responsible for plate tectonics. Venus is a dry planet with no water to speak of in its atmosphere or at its surface, it has no plate tectonics and there is evidence suggesting that it has no asthenosphere (Kaula, 1990).

Mantle convection is basically a viscous phenomenon since that is the primary mode of deformation of deep mantle rocks. However, at least on Earth, the style of mantle convection is strongly influenced by non-viscous modes of deformation, particularly near the surface. Plate tectonics, the near-surface form of mantle convection on Earth, is fundamentally non-viscous. The cold lithosphere of the Earth undergoes non-viscous deformation, e.g. by faulting, that makes plate tectonics possible. Subduction could not occur without the major thrust fault that separates the overriding and underthrusting plates. The cold lithosphere on Earth is so viscous that it succumbs to the build-up of large stresses by other available modes of deformation, e.g. it breaks or cracks at subduction zones and at rifts that sometimes are precursors to full-fledged spreading centers or mid-ocean ridges. The modification of deep mantle viscous convection by shallow mantle non-viscous behavior is a major complication in our attempt to understand convection in the Earth's mantle. Two fundamentally distinct modes of deformation must be combined in a self-consistent way in order to understand the nature of mantle convection on Earth. If mantle convection in other planets occurs beneath a largely intact, non-participatory lithosphere, then such forms of mantle convection may be entirely viscous in character.

Mantle convection occurs because relatively hot rocks are less dense and rise in a gravitational field while relatively cold rocks are more dense and sink. The rise of hot rocks advects heat upward while the fall of cold rocks advects cold downward; this counterflow is equivalent to an upward heat flux. Rocks change density with temperature because they possess a thermodynamic property known as the thermal expansivity or the coefficient of thermal expansion α. Thermal expansivity is the negative of the fractional change in the density per unit degree change in temperature at constant pressure. The thermal expansivity of mantle rocks is a positive quantity typically about 3×10^{-5} K^{-1}. Recent experimental and theoretical studies suggest that α decreases with increasing depth in the mantle, perhaps by a factor of about five over the entire thickness of the mantle (Anderson, Oda and Isaak, 1992; Chopelas and Boehler, 1992). The decrease of α with depth reduces the buoyancy of deep mantle rocks and so can influence the style and vigor of convection.

Mantle rocks also change density ρ under pressure, a property known as the isothermal compressibility χ, i.e. the fractional change in density per unit change in pressure at constant temperature. The increase in density from the Earth's shallow upper mantle to the bottom of the lower mantle is by about a factor of $\frac{5}{3}$, in part due to the olivine–spinel and spinel–perovskite + oxide phase changes and in part due to compressibility. Compressibility varies from about 8×10^{-12} Pa^{-1} in the Earth's shallow upper mantle to about 1.5×10^{-12} Pa^{-1} at the core-mantle boundary. The increase of density with depth is relatively unimportant for mantle convection except for the increases at the solid–solid phase changes at 410 km and 660 km depth. These density jumps can be important.

Increasing pressure with depth in the mantle not only compresses the rocks to higher density but also heats the rocks, a process known as adiabatic compression. The temperature of the mantle therefore increases with depth just due to adiabatic compression, independent of the increase of temperature with depth due to the heat sources discussed above. The adiabatic temperature gradient β_a, the rate of change of temperature T with depth, is given by $\alpha g T/c_p$, where g is the acceleration of gravity and c_p is the specific heat at constant pressure. The specific heat c_p is the amount of heat required to raise the temperature of a unit mass by 1 K. Mantle rocks have a value of c_p of about 1 kJ kg^{-1} K^{-1} and c_p varies little at the high temperature of the mantle. The magnitude of β_a in the Earth's mantle is about 0.5 K km^{-1}, giving a temperature increase with depth across the mantle due to adiabatic compression alone of about 1500 K. The adiabatic rise of temperature with depth due to compression does not contribute to the thermal drive of mantle convection. Only temperature increases with depth in excess of the adiabat or horizontal temperature variations drive mantle convection. Upflow elements in mantle convection will cool as they rise, in part due to adiabatic decompression, while downflow elements will heat in part due to adiabatic compression.

The influence of the above properties on mantle convection can, to a large extent, be quantified by combining these properties into a single dimensionless parameter known as the Rayleigh number Ra

$$Ra = \frac{\alpha g c_p \rho^2 d^3 \Delta T}{k \mu} \qquad (M12)$$

where d is the thickness of the mantle and ΔT is the superadiabatic temperature difference that drives convection. As discussed above, a number of the quantities contributing to the Rayleigh number vary within the mantle and in that case an average over the mantle can be used to define Ra.

If the main source of energy driving convection is radiogenic heating, it is appropriate to define a Rayleigh number based on the heat production rate per unit mass H. This internal heating Rayleigh number is given by

$$Ra = \frac{\alpha g c_p \rho^3 H d^5}{k^2 \mu} \qquad (M13)$$

Since convection in the Earth's mantle is driven both from below (by heat from the core) and from within (by radiogenic heating), the Rayleigh number for mantle convection should ideally combine both the above forms. The Rayleigh number for mantle convection (based on the overall thickness of the mantle and the values of parameters given above) is at least 10^7.

Seismic tomography (Woodhouse and Dziewonski, 1989; Romanowicz, 1991) has begun to give us a view of what mantle convection is like in the Earth's interior, but that view is presently a very fuzzy one. Numerical models of the process may provide insights into the nature of deep mantle convection in the Earth (Schubert, 1992). The models are simplistic in numerous ways and lack some of the real features of the Earth, but they are improving at a rapid rate and are becoming more realistic due to the continued development of increasingly powerful supercomputers. Figures M11 and M12 illustrate the morphology of convective elements in one numerical model of mantle convection (Glatzmaier, Schubert and Bercovici, 1990; see Plate 11). The model simulates fully three-dimensional thermal convection of a viscous fluid in a spherical shell for conditions that approximate the Earth's mantle. Heating is both from below and from within, but about 80% of the heat flowing through the upper surface is generated within the shell. The top and bottom boundaries of the shell are kept at constant temperature; no fluid can flow across these surfaces but flow parallel to the surfaces is uninhibited. The volume-averaged Rayleigh number for the case shown in Figures M11 and M12 is 1.6×10^6. Further details about the computation can be found in Glatzmaier, Schubert and Bercovici (1990).

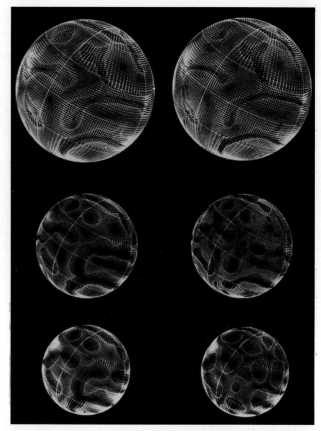

Figure M11 Plots (all at the same timestep) of convection velocities and temperature in three different constant-radius surfaces (5940, 4350 and 3770 km). These spherical surfaces are scaled according to their radii. The shades of gray in the three plots on the left represent the radial component of velocity with a contour increment of 1.5 mm year^{-1}. Lighter gray represents upflow (a maximum of 12.0 mm year^{-1}; darker gray represent downflow (a maximum of 19.5 mm year^{-1}). The shades in the three plots on the right represent the temperature relative to the spherically averaged value at each radius with a contour increment of 50 K. Lighter gray represent hot fluid (a maximum of +400 K); darker gray represent cold fluid (a minimum of −650 K). The arrows represent the direction and amplitude of the horizontal velocity in these surfaces. All are scaled in the same way, with a maximum of 20 mm year^{-1}. Velocities less than 2 mm year^{-1} are not plotted. (After Glatzmaier, Schubert and Bercovici, 1990.) See Plate 11.

Figure M11 shows the patterns of temperature variations and velocities in the fluid on three different spherical surfaces (near the top, the middle and the bottom of the shell). In the upper part of the shell, cold dense fluid tends to converge and sink in long narrow sheets surrounded by a weak background of warm buoyant diverging upflow. This thin sheet-like downflow tends to break up and broaden with increasing depth into cylindrical downflow plumes that squeeze the hot fluid between them as they impinge on the lower boundary, forcing hot cylindrical upflow plumes at the vertices of the connected network of warm fluid. These upflow plumes expand in the upper part of the spherical shell forming the weak background of upwelling there. The prominent morphological elements of convection in the model are cold, thin, arcuate downflow sheets and hot, cylindrical upflow plumes. These structures in the model may be similar to descending slabs and plumes beneath hot spots in the Earth (Sleep, 1992).

The morphology of convection in this model is further illustrated by the temperature variations and velocities shown in the meridional cross-sections of Figure M12 (Glatzmaier, Schubert and Bercovici, 1990). The three panels correspond to three different times separated by 200-Ma intervals; time increases from top to bottom. The shapes

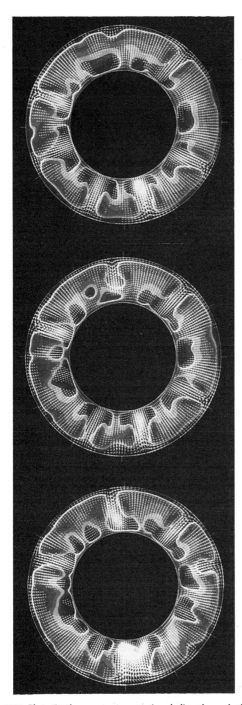

Figure M12 Plots (in the same cross-sectional slice through the three-dimensional shell) of convective velocities and temperatures at three different times separated by 200 timesteps (200 million years) beginning with the top plot. The shades of gray represent the temperature relative to the radially dependent adiabatic temperature profile with a contour increment of 50 K. Lighter gray represent hot fluid (a maximum of 300 K); darker gray represent cold fluid (a minimum of −750 K). The arrows represent the velocities in this cross-sectional surface and are scaled in the same way as those in Figure M11. (After Glatzmaier, Schubert and Bercovici, 1990.) See Plate 11.

of the downflow sheets and upflow plumes are seen in cross-section; such a cross-sectional view blurs the distinction between a cylindrical plume and a planar sheet. The upflow plumes and downflow sheets are seen to arise from instabilities of a thin, hot layer at the bottom boundary and a thin, cold layer at the top boundary, respectively. These layers are known as thermal boundary layers because of the large temperature changes that occur across them. The cold boundary layer at the top of the model may be analogous to the lithosphere or the plates of the Earth and the hot boundary layer at the bottom of the model may be analogous to a layer in the Earth's mantle just above the core–mantle boundary known as the D″ layer across which there are changes in the seismic velocities (Young and Lay, 1987).

As illustrated by Figure M12, convection in the model is strongly time dependent. The time dependence can in fact be rigorously described as chaotic. The patterns in Figure M11 change continuously with time, but the major types of morphological structures, e.g. downflows sheets, are persistent in character. The time dependence is driven by the evolving cold downflow sheets; the hot upflow passively reacts to the changing dominant downflow. The long downflow sheets in the upper part of the shell sometimes contract or break up, forming cylinder-like downwelling features and at other times the downflows sheets link up forming even longer sheet-like features. Downwellings form spontaneously over regions that are warm at depth. The chaotic nature of the convection in the model suggests that mantle convection may also be chaotic, especially since the Rayleigh number of the mantle is at least an order of magnitude higher than Ra of the model and convection is expected to be more vigorous with increasing Ra. That convection in the Earth's mantle is time dependent is not surprising given the changes that occur in the sizes and shapes of plates through geologic time and the associated drifts of ridges and subduction zones.

In assessing the relevance of models such as the one discussed above to the Earth's mantle it must be understood that the models often lack some elements of realism that may be important in the mantle. Such elements commonly include tectonic plates with their non-viscous modes of deformation, the strong dependence of mantle viscosity on the temperature, pressure and state of stress of the mantle, and the major solid–solid phase transitions in mineralogical structure that occur at depths of about 410 km (the olivine–spinel transition) and 660 km (the spinel–perovskite + magnesiowüstite transition) in the mantle and are marked by prominent discontinuities in seismic velocities.

The effects of the 660-km phase change have been taken into account in the numerical model of Tackley et al. (1993). Inclusion of this phase change in the model has a dramatic effect on the style of convection as illustrated in Figures M13 and M14. The phase change has a strong influence on convection because of its endothermic nature, i.e. to convert light material (located above 660 km depth) to heavy material (located below 660 km depth) requires the addition of heat (opposite to the condensation of water vapor to liquid water, for example, which is accompanied by a release of heat). The phase change also has a strong influence on convection because of the magnitude of the density change across the transition and the magnitude of the sensitivity of the pressure at which the phase change occurs to temperature. When cold sinking material encounters this phase change it tends to push the phase change to a greater depth (higher pressure) because of the lower temperature of the sinking material relative to its surroundings and the endothermic character of the phase transition, which implies that the transition will occur at greater pressures for lower temperatures. The downward displacement of the phase change below the cold sinking material provides an upward buoyancy force that inhibits the downflow because relatively lighter material in the downflow displaces relatively heavier material horizontally adjacent to the downflow. Whether the cold sinking material can actually penetrate the phase change and sink further depends on the relative magnitudes of the negative thermal buoyancy of the downflow (lower temperatures imply higher densities for the same phases) and the positive buoyancy of phase boundary displacement. The inhibitory effect of the endothermic phase change is proportional to the magnitude of the slope of the phase change equilibrium line on a pressure–temperature diagram (the Clapeyron slope) and the magnitude of the density jump across the phase boundary. The values of these phase change parameters in the model of Figure M13 and M14 are typical of the values believed to characterize the 660-km phase change in the Earth's mantle (Ito et al., 1990); other model parameters are also as close as possible to their values in the Earth and are essentially the same as in the model of Figures M11 and M12. The above arguments can be made to apply equally to hot upflows in the model.

The cold downflows in the snapshot of Figure M13a (a picture of the downwelling at one instant of time) form a network of interconnected sheets in the upper part of the spherical shell; these sheets do not penetrate the phase change. At the intersections of these sheets, large pools of cold material form above the phase change. This cold material is gravitationally unstable and after building up sufficiently it triggers a sudden avalanche or flushing into the lower

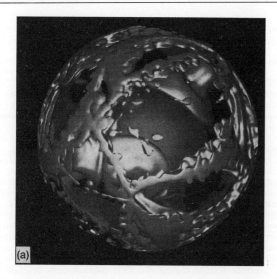
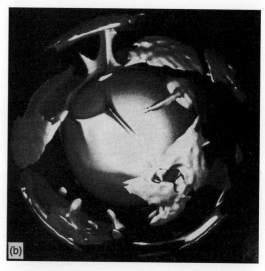

Figure M13 (a) A snapshot of the cold downwellings at one instant of time. The dark gray surface is an isocontour showing where the temperature is 110 K lower than the horizontally averaged value. The medium gray surface is the core. A network of interconnected linear downwellings is visible in the upper mantle, with three huge cylindrical downwellings in the lower mantle, spreading out into pools of cold material above the core–mantle boundary. (After Tackley et al., 1993.) (b) The hot upwellings at the same time as in (a). The medium gray surface is an isocontour of superadiabatic temperature, showing where the temperature is 110 K higher than the reference-state adiabat. A single plume from the core–mantle boundary feeds a hot region in the upper mantle. Most broad hot regions in the upper mantle are not directly linked to lower mantle structures. (After Tackley et al., 1993.) See Plate 11.

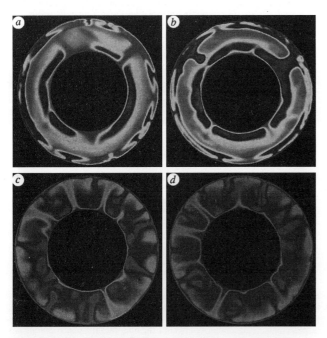

Figure M14 Superadiabatic temperature field on different cross-sectional slices. (a) and (b) The phase change simulations. Scale ranges from −1050 K to +350 K. (c) and (d) A typical case with no phase change. This is representative of mainly internally heated whole-mantle convection models with Rayleigh numbers for internal and basal heating of 1.4×10^7 and 5.5×10^5 respectively, a factor of 10 viscosity increase with depth, and a basal heat flow of ≈ 16% of the surface heat flow. Scale ranges from −780 K to +220 K. (After Tackley et al., 1993.)

part of the shell in the form of a large-diameter cylindrical downwelling plume. This downwelling acts as a conduit to the lower boundary, emptying material from the local region above the phase change to a large pool at the base of the shell. The downwelling then shuts off completely and does not recur in exactly the same place during the model simulation, although many such events may occur in the same general area. Convection in this model is highly time dependent and chaotic as in the model of Figures M11 and M12.

Figure M13b shows the corresponding hot upwelling regions at the same instant as in Figure M13. The most prominent features are the broad hot regions in the upper part of the shell which are generally not associated with any deep features in the lower shell. Ridges of hot material can be seen on the lower boundary. These are swept around by the enormous injections of cold material from the upper shell caused by the flushing or avalanche events. Occasionally a short-lived transient plume is formed at the intersection of these ridges, rising to the phase change and injecting hot material into the upper shell; one of these plumes is visible in Figure M13b. Thus, in both directions, cylindrical forms (plumes) are seen to penetrate the phase change whereas linear forms (sheets) do not. The upward flow in the region above the phase change and in the lower shell is generally a weak and distributed return flow (balancing the downflow) punctuated by occasional plumes.

The influence of the endothermic phase change on the convection model of Figures M13 and M14 may be similar to the effects of the 660-km spinel–perovskite transition on mantle convection. There is evidence from seismic tomography and global mapping of the 660-km seismic discontinuity that some descending slabs flatten along the discontinuity while others penetrate the discontinuity (van der Hilst et al., 1991; Fukao et al., 1992; Shearer and Masters, 1992). The flattening may be due in part to the inhibitory nature of the endothermic phase change as in the numerical model. Other effects in the mantle, such as an increase in viscosity across the 660-km discontinuity or some amount of density change across this boundary due to changes in chemical composition of the rocks (Jeanloz, 1991), may also contribute to the observed flattening of some slabs. The ability of some descending slabs or parts thereof to penetrate the 660-km discontinuity may be due in part to morphological similarities with the cylindrical avalanches of the model (there are plate intersections on Earth, e.g. the intersection of the Pacific, Philippine and Eurasia plates that are like the intersections of downwelling sheets in the model). Other characteristics of descending slabs, e.g. their ages at subduction (age determines the magnitude of the negative buoyancy of a slab) and their angles of descent into the mantle, undoubtedly also affect their ability to penetrate the 660-km discontinuity.

The issue of slab penetration into the lower mantle (the region of the mantle below the 660-km phase change) is related to the decades-old debate of whether the mantle convects as a single layer with exchange of material across the 660-km discontinuity or as two layers with little or no material exchange between the upper and lower parts of the mantle (Schubert, 1979). The possibility of two-layer mantle convection was originally attributed to a hypothetical change in the chemical composition of mantle rocks at the 660-km discontinuity. This no longer seems tenable in view of the acceptance of the 660-km discontinuity as a phase change, but the possibility of two-layer mantle convection due to the inhibitory nature of the 660-km phase transition cannot be ruled out. Though the seismic observations referred to above support the inhibitory behavior of the 660-km discontinuity, they also show the ability of at least some slabs to overcome the phase change effects and penetrate from the upper mantle into the lower mantle (as the avalanches or flushing events in the numerical model). It seems likely that mantle convection is of the whole mantle type but with a style strongly controlled by the 660-km transition.

Mantle convection is a rapidly changing field of study. With new advances in parallel computing it can be expected that numerical models will become increasingly able to simulate the real mantle. The realism of these numerical models will also be enhanced by the continuing acquisition of new laboratory data on the mechanical, thermal and structural behavior of mantle rocks at the high temperatures and pressures encountered in the mantle. Increasingly detailed seismic tomographic images of the mantle can also be anticipated to map the form of mantle flow structures and provide an actual snapshot of mantle convection. The debate about two-layer versus whole-layer mantle convection will soon be resolved.

Gerald Schubert

Bibliography

Anderson, O.L., Oda, H. and Isaak, D. (1992) A model for the computation of thermal expansivity at high compression and high temperature: M_gO as an example. *Geophys. Res. Lett.*, **19**, 1987–90.

Birch, F. (1965) Energetics of core formation. *J. Geophys. Res.*, **70**, 6217–21.

Boehler, R. and Chopelas, A. (1991). A new approach to laser heating in high pressure mineral physics. *Geophys. Res. Lett.*, **18**, 1147–50.

Cassen, P., S. Peale, J. and Reynolds R.T. (1982) Structure and thermal evolution of the Galilean satellites, in *Satellites of Jupiter* (ed. D. Morrison). Tucson, Arizona: University of Arizona Press, pp. 93–128.

Cathles, L.M., III (1975) *The Viscosity of the Earth's Mantle*. Princeton: Princeton University Press.

Chopelas, A. and Boehler, R. (1992) Thermal expansivity in the lower mantle. *Geophys. Res. Lett.*, **19**, 1983–6.

Fukao, Y., Obayashi, M., Inoue, H. and Nenbal, M. (1992) Subducting slabs stagnate in the mantle transition zone. *J. Geophys. Res.*, **97**, 4809–22.

Glatzmaier, G.A., Schubert G. and Bercovici, D. (1990) Chaotic, subduction-like downflows in a spherical model of convection in the Earth's mantle. *Nature*, **347**, 274–7.

Hager, B.H. and Richards, M.A. (1989) Long-wavelength variations in Earth's geoid: physical models and dynamical implications. *Phil. Trans. Roy. Soc. London*, **A328**, 309–27.

Hartmann, W.K., Phillips, R.J., Taylor, G.J. and Jeffrey, G. (1986) Moon origin; the impact-trigger hypothesis, in *Origin of the Moon* (eds W. Hartmann, R. Phillips, R.J. Taylor). Houston, Texas: Lunar Planetary Institute, pp. 579–608.

Ito, E. and Takahashi, E. (1989) Postspinel transformation in the system $Mg_2SiO_4-Fe_2SiO_4$ and some geophysical implications. *J. Geophys. Res.*, **81**, 10637–46.

Ito, E., Akaogi, M. Topor, L. and Navrotsky, A. (1990) Negative pressure–temperature slopes for reactions forming $MgSiO_3$ perovskite from calorimetry. *Science*, **249**, 1275–8.

Jeanloz, R. (1991) Effects of phase transitions and possible compositional changes on the seismological structure near 650 km depth. *Geophys. Res. Lett.*, **18**, 1743–6.

Karato, S. and Li, P. (1992) Diffusion creep in perovskite: implications for the rheology of the lower mantle. *Science*, **255**, 1238–40.

Karato, S.-I., Paterson, M.S. and FitzGerald, J.D. (1986) Rheology of synthetic olivine aggregates: influence of grain size and water. *J. Geophys. Res.*, **91**, 8151–76.

Kaula, W.M. (1979) Thermal evolution of Earth and Mars growing by planetesimal impacts. *J. Geophys. Res.*, **84**, 999–1008.

Kaula, W.M. (1990) Venus: a contrast in evolution to Earth. *Science*, **247**, 1191–6.

Lachenbruch, A.H. 1976. Dynamics of a passive spreading center. *J. Geophys. Res.*, **81**, 1883–902.

Peale, S.J., Cassen, P. and Reynolds, R.T. (1979) Melting of Io by tidal dissipation. *Science*, **203**, 892–4.

Romanowicz, B. (1991) Seismic tomography of the Earth's mantle. *Ann. Rev. Earth Planet. Sci.*, **19**, 77–99.

Ross, M.N. and Schubert, G. (1985) Tidally forced viscous heating in a partially molten Io. *Icarus*, **42**, 1982–90.

Ross, M. and Schubert, G. (1986) Tidal dissipation in a viscoelastic planet. *Proc. Lunar and Planet. Sci. Conf.* **16**, *J. Geophys. Res.* **91**, D447–52.

Ross M.N., Schubert, G., Gaskell, G.R.W. and Spohn, T. (1990) Internal structure of Io and the global distribution of its topography. *Icarus*, **85**, 309–25.

Schubert, G. (1979) Subsolidus convection in the mantles of terrestrial planets. *Ann. Rev. Earth Planet. Sci.*, **7**, 289–342.

Schubert, G., Stevenson, D.J. and Ellsworth, K. (1981) Internal structures of the Galilean satellites. *Icarus*, **47**, 46–59.

Schubert, G., Spohn, T. and Reynolds, R. (1986) Thermal histories, compositions and internal structures of the moons of the solar system, in *Satellites* (eds J.A. Burns and M.S. Matthews) Tucson, Arizona: University of Arizona Press, pp. 224–92.

Schubert, G. (1992) Numerical models of mantle convection. *Ann. Rev. Fluid Mech.*, **24**, 359–94.

Schubert, G., Solomon, S.C., Turcotte, D.L. *et al.* (1992) Origin and thermal evolution of Mars, in *Mars* (eds H.H. Kieffer, B.M. Jakosky, C.W. Snyder and M.S. Matthews). Tucson, Arizona: University of Arizona Press, pp 147–183.

Segatz M., Spohn, T. Ross, M.N. and Schubert, G. (1988) Tidal dissipation, surface heat flow, and figure of viscoelastic models of Io. *Icarus*, **75**, 187–206.

Shearer, P.M. (1991) Constraints on upper mantle discontinuities from observations of long-period reflected and converted phases. *J. Geophys. Res.*, **96**, 18147–82.

Shearer, P.M. and Masters, T.G. (1992) Global mapping of topography on the 660-km discontinuity. *Nature*, **355**, 791–6.

Sleep, N.H. (1992) Hotspot volcanism and mantle plumes. *Ann. Rev. Earth Planet. Sci.*, **20**, 19–43.

Stevenson, D.J., Spohn, T. and Schubert, G. (1983) Magnetism and thermal evolution of the terrestrial planets. *Icarus*, **54**, 466–489.

Stevenson, D.J. (1987) Origin of the Moon – The collision hypothesis. *Ann. Rev. Earth Planet. Sci.*, **15**, 271–315.

Stevenson, D.J. (1989.) Formation and early evolution of the Earth. In *Mantle Convection* (ed. W.R. Peltier). New York: Gordon and Breach, pp. 817–73.

Tackley, P.J., Stevenson, D.J., Glatzmaier, G.A. and Schubert, G. (1993) Effects of an endothermic phase transition at 670 km depth in a spherical model of convection in the Earth's mantle. *Nature*, **361**, 699–704.

Tozer, D.C. (1965) Thermal history of the Earth: 1. The formation of the core. *Geophys. J. Roy. Astron. Soc.*, **9**, 95–112.

Turcotte, D.L. and Schubert, G. (1982) *Geodynamics*. New York: John Wiley and Sons.

van der Hilst, R., Engdahl, R. Spakman, W. and Nolet, G. (1991) Tomographic imaging of subducted lithosphere below N.W. Pacific Island arcs. *Nature*, **353**, 37–43.

Weertman, J. and Weertman, J.R. (1975) High temperature creep of rock and mantle viscosity. *Ann. Rev. Earth Planet. Sci.*, **3**, 293–315.

Wetherill, G.W. (1985) Occurrence of giant impacts during the growth of the terrestrial planets. *Science*, **228**, 877–9.

Woodhouse, J.H. and Dziewonski; A.M. (1989) Seismic modelling of the Earth's large-scale three-dimensional structure. *Phil. Trans. Roy. Soc. London*, **A328**, 291–308.

Young, C.J. and Lay, T. (1987) The core–mantle boundary, *Ann. Rev. Earth Planet. Sci.*, **15**, 25–46.

Cross references

Asthenosphere
Core, terrestrial planetary
Differentiation
Plate tectonics
Thermal evolution of planets and satellites

MAP PROJECTIONS

Principles of map projection

A map projection is a systematic representation of all or part of the surface of a round body, especially the Earth, but also other planetary bodies, and the geocentric view of the heavens, on a flat surface. This representation usually includes lines delineating meridians of longitude and parallels of latitude, but they may be omitted, depending on the purpose of the map. Since this transformation cannot be achieved without distortion, the cartographer must choose the characteristic that is to be shown accurately at the expense of others, or a compromise of several characteristics. If the map portrays a continent or larger part of the Earth, distortion will be visually apparent. If the region is small, distortion may be barely measurable using any of the more appropriate projections. There is literally an infinite number of map projections that can be devised, and several hundred have been published, but most are rarely used novelties. Most projections can also be infinitely varied by choosing different points or lines on the body being mapped to define the projection.

No matter how carefully the projection is chosen, the inherent distortion is increased, not only by errors in placement of information on the map, but also because maps are commonly plotted and printed on paper, which is dimensionally unstable. Variations in atmospheric humidity can easily cause dimensional changes greater than those resulting from the choice between common projections for large-scale topographic quadrangle maps. Furthermore, the dimensional changes vary with direction on a given sheet. The use of stable plastic bases for maps is recommended for precision work. On large-scale maps, accurate determination of distances is facilitated by rectangular grid overprints, because the grid expands with the paper, and point positions are determined by reference to the nearest grid lines.

It cannot be said that there is one 'best' projection for mapping in general, or even for a given application, unless the parameters chosen are artificially constricting. The characteristics normally considered in choosing a map projection are either preservation of local shape or preservation of all areas.

1. Shape. Many of the common and most important projections, such as the Mercator and the Lambert conformal conic, are conformal (or orthomorphic), in that normally the relative local angles about every point on the map are shown correctly. (On a conformal map of an entire body there are usually one or more 'singular' points, for example the poles of the Mercator, at which local angles are still distorted.) Although a large region must still be shown distorted in shape, its small features are shaped essentially correctly. An important result of conformality is that the local scale in every direction around any one point is constant. Because local angles are correct, meridians intersect parallels at the same angles on a conformal projection that they do on the body represented. The linear and area scales are generally too great or too small throughout the map, but they are correct along certain lines, depending on the projection. Nearly all large-scale maps of the Earth prepared by the US Geological Survey and other mapping agencies throughout the world are now prepared on conformal projections.

2. Area. Many map projections, such as the Albers equal-area conic, are designed to be equal-area (or equivalent), so that a coin of any size on one part of the map covers exactly the same area of the actual body as the same coin on any other part of the map. Shapes, angles and scale must be distorted on most parts of such a map, but usually some parts of an equal-area map do retain these characteristics correctly, or very nearly so. Whereas conformality applies on a point or infinitesimal basis, an equal-area map projection shows areas correctly on a finite, in fact mapwide, basis. No projection can be both conformal and equal-area.

Some map projections, such as the azimuthal equidistant, are neither equal-area nor conformal, but linear scale is correct along all lines radiating from the center, along all meridians, or following other special patterns. No map projection can show scale correctly throughout the map. In addition, compromise projections, usually restricted to small-scale mapping, are used because they balance distortion in scale, area and shape.

Projections are often classified by the type of surface onto which the body may be mapped. The concept of cylindrical or conic projections, such as the Mercator or Lambert conformal conic, respectively, involves placing a cylinder or cone around a globe, projecting the map features onto the new surface, and then unrolling this map surface. If the axis of the cone or cylinder coincides with the polar axis of the globe, the projection has equally spaced straight meridians, parallel on the cylindrical projections and converging on the conics. The meridians are intersected at right angles by the parallels of latitude. The latter are straight on the cylindrical and concentric circular arcs on the conic projections. The spacing of the parallels is seldom actually based on geometric projection, but rather on mathematical formulations. There is normally no distortion along the line of tangency of the map surface to the globe.

A plane tangent to the globe at a pole leads to polar azimuthal projections, such as the polar stereographic, with the parallels mapped as arcs of concentric circles and meridians as equally spaced radii of the circles. The point of tangency is free of distortion. Scale remains constant along each parallel of latitude on a regular cylindrical, conic or polar azimuthal projection, but it changes from one latitude to another. Directions of all points are correct as seen from the center of an azimuthal projection.

If the cylinder or cone is secant instead of tangent to the globe, the projection conceptually has two lines instead of one that are free of distortion. A secant plane can provide a line rather than a point of no distortion only for a conformal projection (the stereographic). Wrapping the cylinder about a meridian leads to transverse projections. By placing a plane tangent to the equator instead of a pole, equatorial aspects of azimuthal projections result. Tilting the cylinder, cone or plane to relate to another point on the body leads to an oblique projection, and the meridians and parallels are not the straight lines or circular arcs they are in the normal aspect. The lines of constant scale are correspondingly rotated.

The shape of the body

Many planetary bodies, such as Mercury, are considered perfect spheres. For smaller-scale maps of large areas of less spherical bodies such as the Earth and Mars, the distortions resulting from mapping onto a flat surface are much greater at these scales than the slight additional corrections needed to compensate for the fact that the body is not a perfect sphere. For large-scale planimetric and topographic maps, the distortions resulting from projecting the round body onto a flat plane are less than the small shape distortion resulting from treating the body as a sphere instead of an ellipsoid.

For precision large-scale mapping, the Earth or Mars is taken to be an oblate ellipsoid of revolution, usually referred to as an ellipsoid or spheroid. This mathematical figure is formed by rotating an ellipse about its minor axis. The flattening of the ellipse in the case of the Earth is about 1 part in 298; for Mars it is taken as about 1 : 193. Other planetary bodies, such as some of the satellites and most asteroids, are irregular in shape and have been treated as triaxial ellipsoids or as bodies for which map projections have been specially designed. These projections will not be discussed here (Stooke and Keller, 1990)).

Coordinates on the sphere and ellipsoid

Longitude is measured from a chosen prime meridian, Greenwich for the Earth, counting in both directions on the Earth and Moon, from 0° to 180°, east longitude being plus and west minus. For other bodies, longitude is measured from 0° to 360° in the direction of rotation. All meridians on the sphere or ellipsoid are identical circles or ellipses, each in its own plane, and the longitude is the angle that the plane of the meridian passing through a given point makes with the prime meridian.

Regular or 'geographic' latitude is measured as the angle that a perpendicular to a given point on the surface of the ellipsoid makes with the plane of the equator, 0° to 90° north (+) or south (−) of the equator. This latitude is not the same as the 'geocentric' latitude, except at the equator, a pole or on the sphere. The length of a degree of latitude on the ellipsoid increases slightly away from the equator, and may be calculated as follows:

$$1° \text{ lat.} = \pi a (1 - e^2)/[180°(1 - e^2 \sin^2 \phi)^{3/2}]$$

where a is the semimajor axis of the ellipsoid, e is the eccentricity of the ellipsoid and ϕ is the latitude. To convert the flattening f to e,

$$e^2 = 2f - f^2$$

The length of a degree of longitude on the ellipsoid decreases to zero at the poles using the following formula:

$$1° \text{ long.} = \pi a \cos \phi/[180°(1 - e^2 \sin^2 \phi)^{1/2}]$$

For the sphere, e is zero and a is the radius.

The shortest distance between two points on the surface of a sphere is along a great circle, which is the intersection of the surface of the sphere with the surface of a plane passing through the two points and the center of the sphere. This curve is fairly simple to describe mathematically. On the ellipsoid, however, the shortest distance is a complicated curve unless the two points fall along the same meridian or along the equator. As a result, map projections that are azimuthal or that show great circles as straight lines, when applied to the sphere, often lose some of their properties when applied to the ellipsoid, except in an approximate form. The projections below are

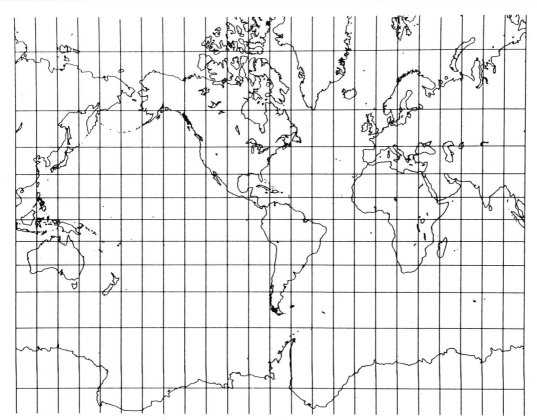

Figure M15 Mercator projection.

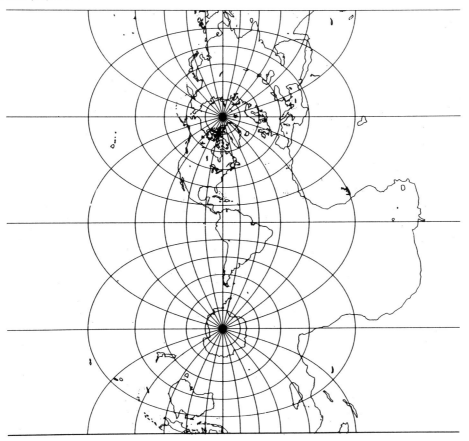

Figure M16 Transverse Mercator projection.

described as they relate to the ellipsoid. Generally these descriptions also apply to the sphere, but the spherical form may have additional characteristics as well. The formulas for the various projections discussed are not given here, since most are lengthy when applied to the ellipsoid. The forward (latitude/longitude to rectangular coordinates) and inverse forms, with worked examples and other references, are given in Snyder (1987).

Conformal map projections for large-scale maps

Mercator projection

Best known of all, the Mercator projection (Figure M15, figures will represent the Earth for convenience) was presented by Gerardus Mercator for navigational purposes in 1569. On this projection rhumb lines, or lines of constant bearing or direction, are plotted straight. This use of the projection for maps of regions away from the equator is justifiable, but the Mercator is generally unsuitable as a global map, although it is often so used.

It is a normal cylindrical projection, with the cylinder conceptually tangent to the equator. Lines of constant scale follow the parallels of latitude, all of which are straight and run parallel to the equator, except for the poles, which are at infinity. The scale on the Mercator increases away from the equator. The projection is conformal, and it is recommended for large-scale conformal mapping of regions bordering the equator. Even though the scale is commonly made correct along the equator, it may be shifted to two standard parallels that are symmetrically north and south of the equator. The Mercator projection has been used for planetary mapping extending in some cases to 57° from the equator.

Transverse Mercator projection

Rotating the cylinder of the Mercator so that it is conceptually tangent along a meridian of the ellipsoid leads to the very important conformal projection called the transverse Mercator (Figure M16). The central meridian, the equator, and each meridian 90° from the central meridian are straight lines. All other meridians and parallels are complex curves. The central meridian remains at a constant scale, usually reduced from the stated map scale to balance errors in measurement over the rest of the map. The lines of constant scale are nearly straight lines parallel to the central meridian, at which distortion is zero, except for the scale reduction.

The projection is recommended for conformal mapping of regions principally north to south in extent. It was developed for the sphere by Lambert in 1772, but Gauss and later Krüger developed the mathematics for the ellipsoidal form, which is sometimes called the Gauss–Krüger projection. It is used more than any other projection for large-scale topographic mapping on the Earth and Mars.

Oblique Mercator projection

A cylinder may be placed around a sphere so that it is tangent along a great circle that is neither a meridian nor the equator. Such a wrapping is not quite possible with the ellipsoid, but the oblique Mercator (Figure M17) may be thus conceptually projected for conformal mapping of a region chiefly extending along this oblique central line. Nearly all meridians and parallels are complex curves. Here the lines of constant scale run nearly parallel with the central line. There are several ways of adapting the oblique Mercator to the ellipsoid, although none is ideal: if there is perfect conformality, as is

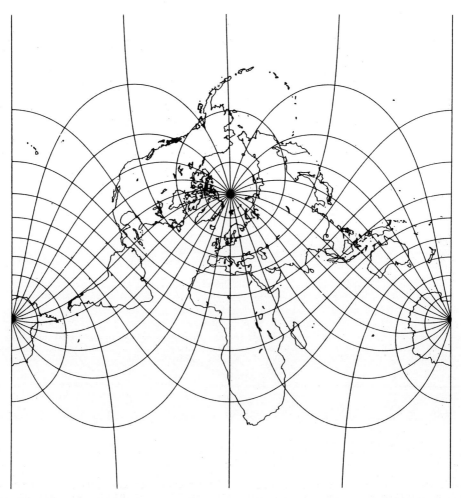

Figure M17 Oblique Mercator projection

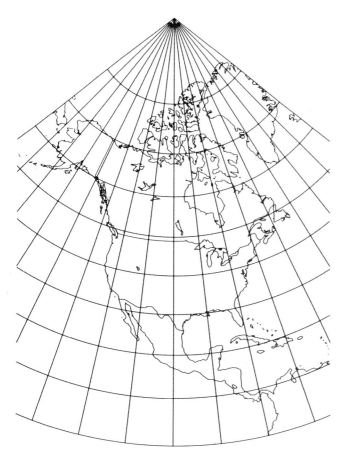

Figure M18 Lambert conformal conic projection, standard parallels 20° and 60°N latitude.

the case with Hotine's frequently used adaptation and most others, the central line does not remain at a precisely constant scale.

Lambert conformal conic projection.

Presented by Lambert in 1772, the Lambert conformal conic projection (Figure M18) shows parallels as concentric circular arcs and meridians as equally spaced radii of those circles. One pole is at the center of the circles, while the other pole is at infinity. The parallels are more closely spaced between the (normally) two standard parallels, which have no distortion of area or scale as well as of shape.

Because scale is constant along any given parallel of latitude, the projection is recommended for regions predominantly east to west in extent at latitudes other than the equator and poles. The Lambert is used second to the transverse Mercator for much of the large-scale mapping throughout the Earth, and it is regularly used for planetary mapping of intermediate ranges of latitude. The regular Mercator projection is its limiting form if the standard parallels are made symmetrical about the equator; the polar stereographic is the limiting form at the polar extremes.

Stereographic projection

For larger-scale maps of polar regions, the Stereographic projection (Figure M19) is commonly used. The oblique or equatorial aspect is also recommended for essentially circular or square regions centered away from either pole. The Stereographic is conformal in both spherical and ellipsoidal forms. The spherical form is a perspective and azimuthal projection of the sphere onto a tangent or secant plane. The point of perspective lies on the opposite surface of the globe. For the ellipsoid, the projection as normally used is not quite perspective and, if a pole is not the center, it is not quite azimuthal.

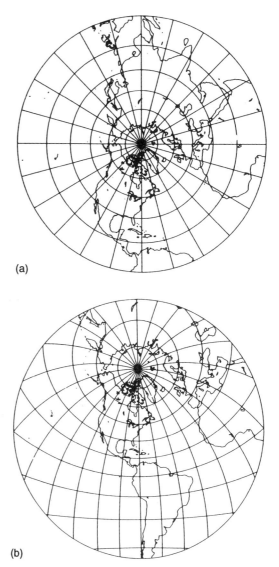

Figure M19 Stereographic projection: (a) polar aspect; (b) oblique aspect, centered at 40°N latitude.

All meridians and parallels on any aspect are straight lines or circular arcs, and the lines of constant scale are circles (nearly so for the ellipsoid) centered on the projection center. True scale may be at this center or along one of the concentric circles. The polar aspect, used for over 2000 years, displays straight equally spaced meridians radiating from the central pole, and the parallels of latitude are circular arcs centered on the pole, with their spacings gradually increasing away from the pole.

The stereographic projection has been used for polar maps of numerous planetary bodies and for maps of circular basin regions of the Moon, Mars and Mercury.

Equal-area map projections for large-scale maps

Cylindrical equal-area projection: normal, transverse and oblique aspects

The equal-area counterparts of the Mercator projection and its transverse and oblique aspects are the various forms of the cylindrical equal-area projection (Figure M20), first presented by Lambert in 1772. On the normal aspect, the parallels of the cylindrical equal-area become closer together with increased distance from the equator, but the meridians remain equidistant, just as they are on the

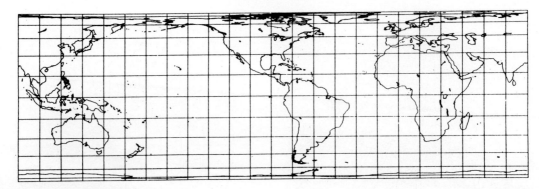

Figure M20 Cylindrical equal-area projection, normal aspect. (Transverse and oblique aspects bear the same relationship to this aspect that Figures M16 and M17 bear to Figure M15.)

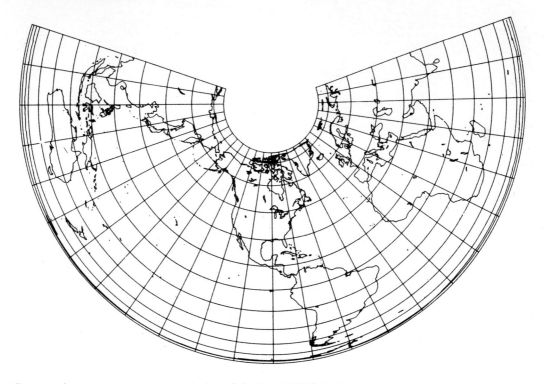

Figure M21 Albers equal-area conic projection, standard parallels 20° and 60°N latitude.

Mercator. The lines of constant scale are parallel to the equator or other central line, or nearly so. The projection is recommended for equal-area mapping of regions predominantly lying along the equator (normal aspect), a meridian (transverse) or oblique 'great circle' (oblique), but it has only recently been developed in the ellipsoidal form (Snyder, 1985, 1987) and has been almost unused for large-scale mapping.

Albers Equal-Area Conic projection

Albers developed the equal-area counterpart of the Lambert conformal conic (Figure M21) in 1805. The meridians are equally spaced radii of the concentric circular arcs representing the parallels, and there are normally two standard parallels, free of all distortion, just as on the Lambert. The parallels are, however, spaced farther apart between the standard parallels than they are beyond, and both poles are circular arcs. For a region of moderate latitude range, both projections appear similar; only careful measurements (or the label) indicate a difference.

Because meridians intersect parallels at right angles, it may at first be thought that there is no angular distortion. It exists, however, for any angle other than that between a meridian and parallel, except at the standard parallels. Scale is constant in a given direction along any given parallel; therefore, the Albers is recommended for equal-area mapping of regions predominantly E–W in extent. The normal cylindrical equal-area and the polar Lambert azimuthal equal-area are limiting forms.

Lambert Azimuthal Equal-Area projection

Although rarely used in the ellipsoidal form, the Lambert azimuthal equal-area projection (Figure M22), also presented in 1772, is commonly used in the spherical form and is available in the ellipsoidal, as the equal-area counterpart of the stereographic. The polar aspect shows meridians as radii of the concentric circles representing parallels, but the parallels are more closely spaced as the distance from the center increases, the opposite of the stereographic spacing. The meridians and parallels of the oblique and equatorial aspects are complex curves. The projection is recommended for equal-area mapping of a region circular or square in shape.

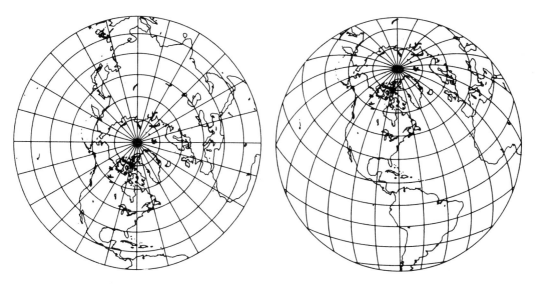

Figure M22 Lambert azimuthal equal-area projection: (a) polar aspect; (b) oblique aspect, centered at 40°N latitude.

There are numerous projections available with other characteristics and appearances, many for global maps. The Hammer is an equal-area elliptically bounded global map used for some celestial maps. Most are of less value for scientific mapping; a number are described in Snyder and Voxland (1989). Other ellipsoidal forms of projections have been developed, sometimes custom-made for a special application. Examples include a recent conformal map projection for Alaska designed to provide lines of constant scale that follow more closely its irregular shape. The distortion on such a projection for the region involved is substantially reduced by the use of a complex-algebra polynomial (Snyder, 1987). The availability of digital computers makes it possible to use projections such as these that would previously have been impractical.

John P. Snyder

Bibliography

Greeley, R. and Batson, R.M. (1990) *Planetary Mapping*. Cambridge: Cambridge University Press, 296 pp.
Maling, D.H. (1973) *Coordinate Systems and Map Projections*. London: George Philip & Son, 255 pp. 2nd edn, 1992, Oxford: Pergamon Press, 476 pp.
Richardus, P. and Adler, R.K. (1972) *Map Projections for Geodesists, Cartographers, and Geographers*. Amsterdam: North-Holland, 174 pp.
Snyder, J.P. (1985) The transverse and oblique cylindrical equal-area projection of the ellipsoid. *Ann. Assoc. Am. Geog.*, **75**(3), 431–42.
Snyder, J.P. (1987) *Map Projections: A Working Manual*. US Geological Survey Prof. Paper 1395, 383 pp.
Snyder, J.P. and Voxland, P.M. (1989) *An Album of Map Projections*. US Geological Survey Prof. Paper 1453. 249p.
Stooke, P.J. and Keller, C.P. (1990) Map projections for non-spherical worlds. The variable-radius map projections. *Cartographica*, **27** (2), 82–100.

Acknowledgement

Modified by the author from Van Nostrand Reinhold *Encyclopedia of Solid Earth Geophysics* by permission of the copyright holder.

Cross references

Cartography
Coordinate systems
Ellipsoid

MARALDI, GIACOMO FILIPPO (1665–1729)

Maraldi was born at Perinaldo, in Italy, on 21 August 1665. His uncle was G.D. Cassini (q.v.), the leading planetary observer of the time, who became the first Director of the Paris Observatory. After finishing his education, Maraldi joined him there, acting as assistant. He spent most of the rest of his life in France.

He was a careful observer of the planets, and made many drawings with the long-focus refractors then in use; the telescope used for most of his work was made by Campani, and had a focal length of 10.4 m (34 ft). Maraldi made observations of all the planets then known, but it is for his work in connection with Mars that he is best remembered.

There was an excellent opposition of Mars in 1704, and Maraldi made a long series of drawings, showing bright areas and dark patches. He confirmed the rotation period, given by Cassini as 24 h 40 min, and made the first observations of the polar caps, which he found to be variable. Further observations were made at the next favorable opposition, that of 1719, and the earlier results were fully confirmed; some of the sketches made in that year show markings in recognizable form, notably the triangular dark feature now known as the Syrtis Major. Maraldi died on 1 December 1729.

Patrick Moore

Bibliography

Flammarion, C. (1890) *La Planète Mars*, Vol. 1. Paris, pp. 35–45.
Maraldi, G. (1706) Observations des taches de Mars peut vérifier sa révolution autour de son axe. *Hist. et Mém. de l'Académie des Sciences*, Paris, p. 74.
Maraldi, G. (1720) Nouvelles observations de Mars. *Hist. et Mém. de l'Académie des Sciences*, p. 44.
Taton, R. (1974) Maraldi, Giacomo Filippo. *Dict. Sci. Biogr.*, Vol. 9, pp. 89–91.

MARINER MISSIONS

Mariner was the name given to the earliest set of American space missions to explore the planets and to the spacecraft developed to carry them out. The missions were planned and executed by the Jet Propulsion Laboratory (JPL) of the California Institute of Technology, which had been designated by the National Aeronautics and Space Administration (NASA) as its lead center for planetary missions.

Table M3 Mariner spacecraft

Spacecraft	Target	Mass (kg)	Science payload (kg)	Data rate (bits s^{-1})	Data storage (megabits)
Mariner 2	Venus	203	18	8.33	0
Mariner 4	Mars	261	16	33.3	5.24
Mariner 5	Venus	245	10	33.3	1
Mariner 6, 7	Mars	413	58	670	195
Mariner 9	Mars	998[a]	63	16 200	180
Mariner 10	Mercury	534	78	117 600	180

[a] This includes 466 kg of propellant to put the spacecraft into Mars orbit.

Birth pangs of the planetary program – 1960–1962

In January 1959, shortly after NASA had been established and JPL had become affiliated with it, the laboratory proposed a program of space missions that contemplated sending two spacecraft to Venus and two to Mars, starting in 1960. With some modifications, this program was adopted by NASA and titled Vega. In the next few months, as it became clear that launch vehicle development was falling short of expectations, Vega was canceled and replaced by less ambitious programs – Ranger for lunar and Mariner for planetary missions (see Ranger missions). The Mariner program, approved by NASA in July 1960, would be carried out by a series of increasingly capable spacecraft, beginning with Mariner A to fly by Venus in 1962 and Mariner B for missions to Venus and Mars (possibly with landers) beginning in 1964. The spacecraft were to be launched by the Atlas–Centaur launch system.

Considerable progress was made at JPL in developing the spacecraft and the missions for Mariner A and B, including the scientific payloads and the teams of scientists to implement them, but the development of the Centaur upper-stage rocket ran into continual problems; Mariner A was canceled in August 1961. To replace it, JPL proposed to use a simpler Ranger-type spacecraft on top of the Atlas–Agena launch system for the first observation of Venus. The scientific payload would have to be severely curtailed. In recognition of its Ranger heritage, the project was called Mariner R.

Mariner 1 was launched from Cape Canaveral on 22 July 1962. A small error in the launch guidance equations in the computer onboard the Atlas caused it to fly erratically, and it was destroyed by the range safety officer after 290 s.

Mariner 2 to Venus – 1961–1963

An identical Mariner 2 spacecraft was launched on 27 August 1962, and it carried out its mission by flying by Venus on 22 December at a distance of 34 762 km. It carried a payload of six scientific instruments weighing only 18 kg. Each of them operated essentially as intended. The microwave radiometer measured the temperature of the planet surface to be about 425°C, and the infrared radiometer registered the cloud-top temperature as −35°C.

The other experiments operated continuously during most of the flight, providing the first long-duration monitoring of conditions in interplanetary space. The micrometeorite detector, designed to register the impacts of dust particles, actually detected only two. The Charged-Particle Experiment monitored cosmic rays. The Magnetometer Experiment monitored the interplanetary field. The Solar-Wind Experiment determined for the first time the density, velocity, temperature and composition of the solar wind, i.e. plasma streaming out from the solar corona.

The spacecraft continued to operate nearly faultlessly for 21 days beyond Venus encounter (130 days total), reaching out to 87.4 million km from Earth. It thus demonstrated the efficiency of its three-axis stabilized design, its capability of effective two-way communication with Earth over vast distances with a transmitter power of only 3 W, and the ability of the Deep Space Network for precise tracking throughout the flight. It was the most successful space mission in history up to that time.

Mariner 4 to Mars – 1962–1965

The Mariner B mission was originally authorized in July 1960 for a landing on Venus or Mars in 1964. In February 1961 the Venus lander was dropped from consideration, but in April 1962 it was reinstated and the Mars landing was set aside. In May 1963 the Mars landing was postponed until 1966 and redesignated Mariners E and F, and these were effectively canceled in July 1964. This sad history resulted from the troubles with the Atlas–Centaur launch system.

Meanwhile, a much less ambitious Mars flyby mission, called Mariner C, with a spacecraft barely half the weight of Mariner B, was proposed in midsummer 1962, approved in March 1963 and was ready for launch by the Atlas–Agena in November 1964. The launch of Mariner 3 on 5 November appeared normal, but the signal for solar panel deployment was not received, and communication ceased as the battery ran down. It was soon determined that the nose cone (appropriately called the 'shroud') had failed to jettison. This information prompted a frantic effort to design and fabricate a new shroud in the 3 weeks before the launch period closed.

Mariner 4 was launched on 28 November 1964 and flew by Mars on 14 July 1965 at a range of 9844 km, taking a sequence of 21 photographs of the surface. The mission revealed two very important new facts: (1) that the surface is covered with impact craters; and (2) that the atmosphere is composed mainly carbon dioxide at a pressure of less than 7 mbar. The latter was determined by the Radio Occultation Experiment, which had not even been accepted as part of the mission until after launch. The other scientific experiments were three to detect charged particle radiation, a solar plasma probe, a magnetometer and a cosmic dust detector.

Mariner 5 to Venus – 1965–1967

A project to revisit Venus with a spare Mariner 4 spacecraft was authorized in December 1965, and Mariner 5 was launched on 14 June 1967. It flew past Venus at a distance of 10 151 km from the center of the planet on 19 October and successfully conducted seven experiments, 1 day after the Soviet Venera 4 had dropped on the surface a landing capsule that unfortunately failed to provide any data.

The plasma probe, the magnetometer and the energetic particle detectors were designed to investigate the interaction between the planet and the interplanetary plasma. They determined that the magnetic field of Venus is much weaker than the Earth's; it is too weak to hold off the solar wind and produce an Earth-like magnetosphere (see Venus: magnetic field). However, when the solar wind reaches the top of the atmosphere, it is deflected by the ionosphere, so that it flows around the planet without touching the surface. The ultraviolet photometer found large quantities of hydrogen in the upper atmosphere at a very low temperature, but no oxygen was detectable.

Three experiments utilized the radio links to and from the spacecraft. The S-band Occultation Experiment, by analyzing the downlink signal, obtained profiles of refractivity, temperature, density and pressure in the neutral atmosphere and of electron density in the ionosphere. These data demonstrated that the temperature was 475°C and the pressure was equivalent to 90 Earth atmospheres at the surface. The Dual-Frequency Occultation Experiment, using two radio transmissions from Earth that were analyzed on the spacecraft, obtained very detailed information on the ionosphere. Range and Doppler tracking of the spacecraft provided the data for the Celestial Mechanics Experiment, which determined the masses of Venus and the Moon with unprecedented accuracy.

Mariners 6 and 7 to Mars – 1965–1969

For the 1969 missions to Mars, the availability and effectiveness of the Atlas–Centaur launch system made possible major increases in the size and sophistication of the spacecraft and their scientific payloads. Mariner 6 was launched on 25 February and its twin, Mariner 7, on 27 March; they flew by the planet on 31 July and 5 August respectively. Each spacecraft carried two vidicon cameras – wide-angle and narrow-angle, and a much higher data transmission rate provided 200 times the amount of picture data of Mariner 4. The two spacecraft acquired 143 pictures before encounter that showed the entire visible disk (or a considerable fraction of it), and 59 pictures near encounter that covered small areas, about half of which had resolutions of 0.2 km or better. All the close-up pictures were in the southern hemisphere, and although they covered only about 10% of the surface, they did reveal some new surface features and laid to rest the myth of the Martian canals. Later missions, however, showed that most of the interesting features of Mars had been missed or misinterpreted.

The infrared radiometer measured a temperature of 150 K on the south polar cap, indicating it to be carbon dioxide ice and not water ice. The ultraviolet spectrometers on both spacecraft detected carbon monoxide in the atmosphere but, surprisingly, saw no trace of nitrogen. The Radio Occultation Experiment confirmed the Mariner 4 finding on the atmospheric pressure and detected an ionosphere. The Celestial Mechanics Experiment measured the planet's mass with unprecedented precision.

Mariner 9 to Mars – 1968–1972

For the next Mars opportunity in 1971, the plan was for a dual spacecraft orbiting mission that promised a large increase in information over the three earlier flybys. However, on 8 May the Centaur booster rocket failed to deliver Mariner 8 into orbit. Mariner 9, successfully launched on 30 May, arrived at Mars on 14 November to find the planet shrouded by the most intense global dust storm that had ever been observed. Settling into its intended orbit, the spacecraft simply waited out the storm, which had cleared by March, and in 349 days of operation it achieved virtually all of the objectives of the dual mission.

Mariner 9 revolutionized our understanding of Mars. The camera returned 7329 photographs and 54 billion data bits – 27 times as much as the three preceding missions. It discovered many volcanoes, including Olympus Mons, 27 km in elevation; an enormous system of deep canyons, now named Valles Marineris, that stretch about one-quarter of the way around the planet; a plethora of channels of five different types, many of them appearing to be ancient river beds; numerous evidences of eolian erosion and deposition; a variety of meteorological phenomena; and many other things.

The Infrared Radiometer Experiment measured the thermal inertia of the surface over much of the planet and found the surfaces of Phobos and Deimos to be very poor thermal conductors. The Infrared Spectrometer Experiment detected carbon dioxide, water, dust and ice crystals in the atmosphere, monitored the atmospheric temperature, and determined the atmospheric pressure at numerous points. The Ultraviolet Spectrometer Experiment measured the atmospheric pressure over much of the surface, thus determining the relative altitude in many places, and measured ozone and atomic hydrogen in the atmosphere.

The S-band Occultation Experiment, with more than 300 occultations to analyze, made many measurements of the atmospheric temperature profile, the surface pressure and the electron density in the ionosphere. The Celestial Mechanics Experiment, by analyzing the metric information in the tracking data, determined a more accurate description of the gravity field of the planet, improved the ephemeris of Mars by an order of magnitude, and obtained topographic data with a precision of 100 m.

Mariner 10 to Venus and Mercury – 1969–1975

A mission to Mercury was authorized by NASA in December 1969. For the first time in the Mariner program, the spacecraft was not to be built by JPL in-house, and the Boeing Corporation won the contract to build it. Mariner 10 was put on a course to Venus by an Atlas–Centaur rocket on 3 November 1973 and flew by the planet on 5 February 1974 at just the proper location to be deflected by the gravitational field into an orbit to Mercury. This was the first employment of the 'gravity-assist' technique of interplanetary navigation (see Gravity-assist navigation).

Mariner 10 passed Mercury on 29 March at a distance of 703 km and was deflected into a new orbit that took it around the Sun in 176 days and back to Mercury, at intervals of exactly 2 Mercurian years. The second encounter occurred on 21 September at about 50 000 km, the larger distance being chosen to extend the photographic coverage at good viewing angles. The third, on 16 March 1975, was targeted to the dark side of the planet at an altitude of 327 km in order to maximize the information about the magnetic field and the magnetosphere.

Mariner 10 exhausted its attitude-control gas supply on 24 March 1975, and the spacecraft transmitter was turned off for the last time, terminating the mission after 506 days and nearly 1.6 billion km of travel.

Mariner 10 was the most complex and sophisticated of all the Mariner spacecraft, and its mission achieved many space firsts. It was the first mission to visit two planets and to fly by the same planet three times. It was the first to return photographs in real time instead of storing them, as the tape recorder failed before the second Mercury encounter. Rotatable solar panels were used for the first time to maintain their temperature and power output through the large change in solar distance. 'Solar sailing' was attempted and used successfully for the first time, as the positions of the solar panels and the high-gain antenna were manipulated to maintain the orientation of the spacecraft by radiation pressure without the expenditure of attitude-control gas; this technique made the third encounter possible. Optical navigation was employed for the first time as the cameras provided the angles between Mercury and stars. The Celestial Mechanics Experiment had available, for the first time, radio signals in both X and S bands.

The TV camera was equipped with two Cassegrain telescopes to provide high- and low-resolution pictures and a filter wheel to permit photography in several wavelength bands, including the ultraviolet. It returned more than 1000 pictures of Earth and Moon, 3500 pictures of Venus and 3700 pictures of Mercury, a small number of which had resolution as small as 134 m. They provided very detailed coverage of the clouds of Venus (made possible by the ultraviolet sensitivity of the cameras) and the surface of Mercury. The motion of the clouds appeared to indicate that the primary circulation pattern on Venus is vertical upward motion at the subsolar point and subsidence over the poles (see Venus: atmosphere).

Mercury's surface was found to look generally like that of the Moon, but with some distinctly non-lunar features, including large scarps or cliffs nearly 3 km high and as much as 500 km long, probably indicative of crustal shrinkage. The density of small craters was surprisingly similar to that on Moon and Mars, implying that all these celestial bodies received similar intensities of early meteorite bombardment, contrary to some previous assumptions.

A major and unexpected discovery was an intrinsic magnetic field, which, although about 60 times weaker than Earth's, is sufficient to produce a bow shock and a magnetosphere that is a miniature copy of Earth's (see Mercury: magnetic field and magnetosphere).

The scientific payload also included a Radio Science Experiment, a Solar Plasma Experiment, a charged particle telescope, an infrared radiometer and two ultraviolet spectrometers. One of these was designed for detecting planetary airglow; the other, intended for sounding Mercury's atmosphere by looking at the Sun as it was occulted, also made observations on comet Kohoutek and the Gum Nebula. It detected a very sparse atmosphere of helium on Mercury (see: Mercury: atmosphere). Surface temperature extremes of 90 K and 460 K were measured by the radiometer. The Radio Science Experiment measured the diameter and mass of Mercury with unprecedented precision and detected no trace of an ionosphere.

The exploration of the inner solar system, ably initiated by the Mariner series, was continued by the spacecraft Viking and Voyager.

Conway W. Snyder

Bibliography

The scientific results of each successful mission were reported during and after the mission by papers in *Science* and *Journal of Geophysical Research*.

Ezell, E.C. and Ezell, L.N. (1984) *On Mars: Exploration of the Red Planet, 1958–1978*. Washington, DC: NASA Scientific Publication 4212.

Koppes, C.R. (1982) *JPL and the American Space Program*. New Haven and London: Yale University Press.

Snyder, C.W. and Moroz, V.I. (1992) Spacecraft exploration of Mars, in *Mars* (eds H.H. Kieffer, B.M. Jakosky, C.W. Snyder and M.S. Matthews). Tucson: University of Arizona Press, pp. 71–119.

Cross references

Mercury
Mercury: geology

Mars
Mars: geology
Venus

MARS

Mars has an equatorial diameter of 6794 km, making it the third smallest planet in the solar system. In addition, Mars is the fourth planet from the Sun, orbiting our star at an average distance of 228 million km. Because Mars lies further from the Sun than the Earth, it takes longer to complete one orbit: 687 days or almost 2 Earth years. However, the orbit of Mars is elliptical, not circular, resulting in the Mars–Sun distance ranging from 207 million km (perihelion) to 250 million km (aphelion). When Mars and Earth are lined up on the same side of the Sun, Mars is said to be in opposition. If Mars is at perihelion during this time, the opposition is said to be 'favorable' and many faint details can be seen on the planet's surface (Plates 16–19). If Mars is at aphelion during opposition, however, the configuration is said to be 'unfavorable' and little can be observed on the planet's surface.

Telescopic observations of albedo features on the planet's surface led to the determination that a day on Mars is 24 h 39 min 36 s, almost 45 min longer than a day on Earth. Astronomers also discovered that the Martian rotation axis was tilted about 25° from the perpendicular to the orbital plane, very close to the 23.5° tilt of the Earth. It is this tilt of the rotation axis which causes seasons, so Mars experiences four seasons just like Earth. However, since a year on Mars is almost twice as long as a year on Earth, the Martian seasons are correspondingly longer.

Atmosphere

Mars has an atmosphere composed primarily of carbon dioxide with traces of nitrogen, oxygen and water. This atmosphere is very thin, exerting a pressure of only 6 mbar on the Martian surface (less than 1% of the sea-level pressure on Earth). Because of this low atmospheric pressure, liquid water cannot currently exist on the Martian surface.

Even with its thin atmosphere, Mars experiences substantial winds caused by temperature differences in the atmosphere. These strong winds can raise dust from the surface, producing dust storms which are usually localized in extent, but which occasionally grow into global events. Conditions conductive to the formation of global dust storms usually occur during summer in the southern hemisphere.

Clouds have been reported on Mars for centuries and are of two general types: white and yellow. Yellow clouds are composed of dust and are associated with dust storms. White clouds are composed of water ice crystals and commonly form around topographic highs. In addition to the high clouds, early morning fog has been detected in the Valles Marineris canyon system.

Scientists believe that Mars may once have had a more extensive atmosphere (Pollack, Kasting and Poliakov, 1987), allowing liquid water to exist on the planet's surface. Because Mars is so small, lighter gases could easily escape and eventually the atmosphere was lost to space. This process may have been accelerated by erosion of the atmosphere by large impacts (Melosh and Vickery, 1989).

The question of life

Telescopic observations of Mars during the 18th and 19th centuries revealed that Mars had an atmosphere, exhibited clouds, had bright polar caps and dark surface markings, rotated in just over 24 h and went through four seasons, all properties similar to the Earth. Long-term observations revealed that the polar caps exhibited seasonal changes, with the maximum size being reached in winter and the minimum extent occurring in summer. In association with the polar cap changes, astronomers observed albedo changes on the surface. This was called the 'wave of darkening' since, as the cap shrunk, a dark band would move from the polar cap toward the equator. The process would reverse in the fall.

These observations led people to speculate on the possibility of Martian life. The wave of darkening, in particular, seems to suggest life since a plausible scenario was that water melting from the polar caps was used by plant life, causing a 'greening' of vegetation which progressed from the poles to the equator. The debate reached its peak around the end of the 1800s. In 1877 Giovanni Schiaparelli reported a network of thin lines crossing the Martian surface. Schiaparelli called these dark lines 'canali', which means channels in Italian, and initially insisted that they were natural surface features. 'Canali' was translated into English as 'canals', a term which implies an artificial origin. In 1893 Percival Lowell heard of Schiaparelli's discovery and began a program to observe and explain the Martian canals. Lowell rushed the completion of his observatory (Lowell Observatory) in Arizona so he could observe the favorable opposition of Mars in 1894. Over the next several years Lowell mapped hundreds of Martian canals and developed his theory of Martian life. According to Lowell, Mars once had a very warm, moist climate and life flourished. But as Mars began to lose its atmosphere, the climate turned cold and the Martians moved to the equator for warmth. However, the equator was very dry, so the Martians constructed a network of canals to bring water from the ice caps to their cities. The canals themselves were too narrow to be visible from Earth – instead, the dark lines observed by Lowell were the areas of vegetation on either side of the canals (Hoyt, 1976).

Many astronomers did not see Lowell's canals even under excellent seeing conditions and believed them to be dark fuzzy spots connected into lines by Lowell's imagination. The debate raged long after Lowell's death in 1916, and was not completely settled until spacecraft imagery of the planet in the 1970s revealed no evidence of canals. The consensus today is that the canals are optical illusions and the wave of darkening is simply caused by seasonal winds shifting bright dust from one region to another (Sheehan, 1988).

Even though spacecraft imagery revealed no evidence of Martian cities or canals, the question of life on Mars remained. Spacecraft imagery revealed many geomorphic indicators of water, such as large channels carved by flowing water. The discovery that water has existed on the surface of Mars led to speculation that perhaps microbial life exists on the planet. In 1976 the United States landed two robotic laboratories, Viking 1 and Viking 2 (q.v.), on the Martian surface. The primary objective of these two landers was to analyze the Martian soil for microbial activity. The landers contained three experiments to test for metabolic processes. The Pyrolytic Release Experiment measured carbon assimilation by the soil through the use of radioactive carbon. The transfer of large amounts of radioactive carbon from the air to the soil could indicate life. The Labeled Release Experiment moistened a sample of martian soil with radioactively labeled organic material. If life was present in the Martian soil, it should consume the organic material and eventually liberate gases with radioactive carbon. The final experiment, the Gas Exchange, moistened the soil with a solution enriched in a wide range of amino acids, salts, and vitamins, and tested for the evolved gases. Initial results from the Pyrolytic Release and Labeled Release Experiments appeared positive, leading to some initial excitement about the possibility that life had been detected. But further tests with those experiments and the results from the Gas-Exchange Experiment indicated that the results were not consistent with the behavior expected from life forms. The consensus of the Viking biologists was that the results from the three experiments could be explained by chemical reactions in the very oxidizing conditions present on the Martian surface (Klein, 1977). The issue of life on Mars was raised again in 1996 when scientists reported possible evidence of ancient life in a Martian meteorite. This evidence includes material produced by biologic activity and structures that look like microfossils. If these features indeed prove to be of biologic origin, this indicates that life began on Mars but did not survive to the present time.

Surface geology

The first close-up views of the martian surface were returned in 1965 when Mariner 4 flew by the planet. The 22 low-resolution images showed a surface covered with impact craters of varying sizes. The next two spacecraft, Mariners 6 and 7 (q.v.), flew by Mars in 1969, and impact craters again dominated the view. Mars was considered a geologically dead world, similar to the Moon. This view dramatically changed when Mariner 9 went into orbit around the planet in 1971. Mariner 9 revealed that Mars could be divided into two general hemispheres, roughly defined by a great circle tilted about 30° to the equator. South of this circle is the heavily cratered region imaged by the three previous Mariner missions. North of the circle are lightly cratered plains, two elevated volcanic regions, large channels formed by flowing water and a huge canyon system. Mars is obviously a geologically diverse world with many new questions to be answered (Mutch et al., 1976; Carr, 1980).

Impact craters are the most prevalent geologic feature on the planet; 60% of the planet displays heavily cratered surfaces formed during the earliest period of the planet's history. The remaining 40% of the planet's surface has been reworked by geologic processes and retains scars from only recent impact events. Craters on Mars range in size from 50 m (objects creating smaller craters burn up in the atmosphere before striking the ground) to over 2000 km in diameter. They look similar to craters on other planets and moons except for the appearance of the ejecta blanket around fresh craters. On bodies with no atmosphere (i.e. the Moon), material ejected during crater formation tends to clump near the crater rim and spread out into small secondary craters further away from the rim. On Mars most fresh craters are surrounded by a lobate ejecta pattern, apparently emplaced by flow processes. Either impact into an ice-rich substrate (Carr et al., 1977) or entrainment of the ejecta in atmospheric gases (Schultz and Gault, 1979) is believed to cause this fluidized ejecta pattern.

Volcanic features are common on Mars and range from volcanic plains to towering volcanoes (Greeley and Spudis, 1981). Three major centers of volcanic constructs exist on Mars. The largest of these centers is the Tharsis region, a huge bulge in the western hemisphere of the planet, which contains volcanoes of varying types, sizes and ages. The largest volcano in the solar system, the shield volcano Olympus Mons, dominates the Tharsis Bulge. Olympus Mons rises 27 km above the Mars datum (mean radius of the planet), approximately three times higher than Mauna Loa, Earth's tallest volcano. It is 600 km in diameter and is topped by a 65 km diameter caldera. Olympus Mons is one of the youngest features on the planet. Three other large shield volcanoes lie to the east of Olympus Mons, and the Tharsis region is dotted with smaller volcanic constructs called domes. The second major center of volcanism is the Elysium Region, an area of three volcanoes (shields and domes) in the eastern hemisphere of the planet. The third volcanic region is located in the southern hemisphere along the east side of the Hellas Basin. This area is characterized by very low-profile, highly eroded volcanoes called paterae which represent a very early stage of volcanism on Mars. There are many examples of other features formed by volcanic processes across Mars.

Stretching eastward from Tharsis is the Valles Marineris canyon system, a series of several canyons which together stretch over 4000 km in length and up to 700 km in width. In places the canyon is up to 7 km deep, over seven times deeper than Arizona's Grand Canyon. Valles Marineris is the largest manifestation of tectonic activity on Mars, but several smaller regions of extensional and compressional features are seen.

Several large channels debouch from the northern and eastern sides of Valles Marineris and flow into the northern plains. The sudden release of near-surface water created catastrophic floods which carved these large channels. Small dendritic channels, found almost exclusively in the heavily cratered regions, formed by surface collapse following removal of underground water (Baker, 1982). Many other Martian surface features are attributed to fluvial or glacial activity. Craters and other depressions with smooth floors may have once harbored lakes (Squyres, 1989; DeHon, 1992) and a number of geomorphic features have recently been attributed to the existence of oceans and glaciers (Baker et al., 1991).

The most ephemeral of the geologic features on the Martian surface are those caused by the wind. Among the eolian features seen by orbiting spacecraft and the Viking Landers are sand dunes, drifts, wind streaks and erosional landforms called yardangs. Wind streaks in the lee of craters and other obstacles were observed to change in appearance between the 1972 Mariner 9 and the 1976 Viking images, thus confirming that eolian activity is active on Mars today.

Surface composition

Scientists estimate the properties of the Martian soil by using information from the chemical analyses performed by the Viking Landers together with spectral data obtained from Earth-based telescopes and chemical analyses of a group of meteorites believed to be from Mars. The Viking Landers, designed primarily to test for metabolic processes indicative of life, performed only crude geochemical analyses of the Martian soil. Both found the soil to be highly oxidized with substantial amounts of iron and silicates. Reflectance spectroscopy using Earth-based telescopes reveals information about the mineral species present in the Martian soil. These spectra indicate that iron oxides and clays are the primary components of the Martian surface.

A third way to obtain information about Martian surface chemistry has become available recently with the discovery of a class of meteorites called the shergottites, nakhlites and chassignites (SNCs, pronounced 'snicks'). This group of meteorites is very different from other meteorites in that they have very young formation ages (about 1.3 billion years old, although one formed 4.5 billion years ago), have chemistries indicating they are from a large body, and contain gases which are isotopically identical to the Martian atmosphere as measured by Viking (McSween, 1985). The chemical evidence strongly suggests that these meteorites are pieces of Mars knocked off the planet by impacts. Chemical analysis of the SNC meteorites combined with geochemical modeling suggest that carbonates and hydrous minerals may be present in addition to the iron oxides and clays previously proposed (Karlsson et al., 1992).

Martian moons

Two small moons, called Phobos (q.v.) and Deimos (q.v.), orbit Mars. The two moons were discovered in 1877 by Asaph Hall who named them after the mythological attendants to Mars: Fear (Phobos) and Panic (Deimos). Both moons are very small, irregularly shaped objects: Phobos, the larger, measures $27 \times 22 \times 18$ km while Deimos is $15 \times 12 \times 11$ km. Because of their irregular shapes, small sizes and low densities, Phobos and Deimos are often suggested to be captured asteroids.

Deimos orbits Mars at a distance of 20 000 km above the Martian surface. Phobos, at a distance of 6 000 km above Mars, is so close that it rises and sets three times in a single Martian day. The orbit of Phobos is decaying due to tidal forces from Mars – in about 50 million years, Phobos will crash into the Martian surface, creating a new impact crater on the planet.

Because of their small sizes, neither Phobos nor Deimos show any signs of geologic activity. Both bodies are heavily cratered. A major feature on Phobos is the 10 km diameter impact crater called Stickney. Large grooves crossing Phobos probably resulted from seismic stresses produced during the Stickney impact. Features on Deimos are muted by a thick regolith produced by impacts, but features on Phobos are sharp, indicating little regolith cover. Scientists believe that the impact which created Stickney knocked loose material off Phobos, thus eliminating any regolith layer which previously existed.

Future missions to Mars

Future exploration of Mars is planned by the United States, Russia and the European Space Agency. The loss of the Mars Observer Mission (MOM) just prior to orbit insertion in August 1993, was a blow to US Mars exploration. However, two replacement missions will be launched in 1996 and 1998. These two missions are scheduled to study the surface chemistry and atmospheric circulation from orbit for an entire Martian year. The Mars Pathfinder mission will land on Mars in 1997, and it includes a small rover to explore its surroundings. In addition, Russia is launching the Mars 96 mission in 1996, which will include soft and hard landers to study surface properties and balloons to study the atmosphere.

Plans to explore Mars beyond MESUR Pathfinder are still being developed. NASA has committed itself to a ten-year program of detailed exploration of the planet. In 1989 former President George Bush announced his Space Exploration Initiative (SEI): establishment of a lunar outpost followed by a landing of humans on Mars. The general SEI scenario for the exploration of Mars includes (1) returning samples of Martian rocks and soil for chemical analysis, (2) sending a rover to the surface to determine the ease of movement, and (3) sending a human expedition for a short-term stay on the surface (Synthesis Group Report, 1991). However, Congressional funding of the SEI program is still to be realized and our future exploration of Mars remains uncertain.

Nadine G. Barlow

Bibliography

Baker, V.R. (1982) *The Channels of Mars*. Austin: University of Texas Press.

Baker, V.R., Strom, R.G., Gulick, V.C. et al. (1991) Ancient oceans, ice sheets and the hydrologic cycle on Mars. *Nature*, **352**, 589–94.

Carr, M.H. (1980) *The Surface of Mars*. New Haven: Yale University Press.

Carr, M.H., Crumpler, L.S., Cutts, J.A. et al. (1977) Martian impact craters and emplacement of ejecta by surface flow. *J. Geophys. Res.*, **82**; 4055–65.

DeHon, R.A. (1992) Martian lake basins and lacustrine plains. *Earth, Moon, and Planets*, **56**, 95–122.

Greeley, R. and Spudis, P.D. (1981) Volcanism on Mars. *Rev. Geophys. Space Phys.*, **19**, 13–41.

Hoyt, W.G. (1976) *Lowell and Mars*. Tucson: University of Arizona Press.

Karlsson, H.R., Clayton, R.N., Gibson, E.K. and Mayeda, T.K. (1992) Water in SNC Meteorites: evidence for a Martian hydrosphere. *Science*, **255**, 1409–11.

Klein, H.P. (1977) The Viking biological investigation: general aspects. *J. Geophys. Res.*, **82**, 4677–80.

McSween, H.Y. (1985) SNC meteorites: clues to Martian petrologic evolution; *Rev. Geophys.*, **23**, 391–416.

Melosh, H.J. and Vickery, A.M. (1989) Impact erosion of the primordial atmosphere of Mars. *Nature*, **338**, 487–9.

Mutch, T.A., Arvidson, R.E., Head, J.W., III et al. (1976) *The Geology of Mars*. Princeton: Princeton University Press.

Pollack, J.B., Kasting, J.F., and Poliakof, K. (1987) The case for a wet, warm climate on early Mars. *Icarus*, **71**, 203–24.

Schultz, P.H. and Gault, D.E., (1979) Atmospheric effects on Martiam ejecta emplacement. *J. Geophys. Res.*, **84**, 7669–87.

Sheehan, W. (1988) *Planets and Perception*. Tucson: University of Arizona Press.

Squyres, S.W. (1989) Urey Prize Lecture: water on Mars. *Icarus*, **79**, 229–88.

Synthesis Group Report (1991) *America at the Threshold*. Washington DC: US Government Printing Office.

Cross references

Deimos
Mariner missions
Lowell, Percival
Phobos
Soviet Mars missions
Viking missions

MARS: ATMOSPHERE

Nineteenth-century observers deduced that Mars must have an atmosphere. Observations of clouds moving across its disk provided convincing evidence, and seasonal changes in the size of its polar caps and brightness of various surface features further reinforced this notion. However, the mass and composition of the atmosphere remained controversial until the 1960s and the dawn of the spacecraft era. Prior to that time the surface pressure (the equivalent of atmospheric mass) was believed to be 85 mbar, and nitrogen (by analogy to Earth) was thought to be the principal constituent.

The 85-mbar estimate of surface pressure was based on the intensity and polarization of reflected sunlight. Scattering of sunlight by the surface and atmosphere contributes to the total brightness of the planet. Measuring that brightness at different wavelengths and viewing angles provides a means to estimate surface pressure if the composition is known. Early observers assumed that the atmosphere was gaseous and neglected the effects of suspended aerosols such as condensates and dust particles. The latter are now known to be almost always present. Consequently, the fraction of scattering attributed to the gaseous component, and hence its mass, was greatly overestimated.

Spectroscopic techniques gradually improved during the 20th century. Carbon dioxide was identified in 1947 and 16 years later water vapor was detected. Just prior to the first flyby in 1964, a better estimate of the surface pressure was obtained. Its low value (25 ± 15 mbar) implied that, contrary to early theories, the Martian atmosphere was thin and composed predominantly of CO_2. This was confirmed by Mariner 4's radio occultation measurements, which indicated a total surface pressure in the range of 4–6 mbar. Thus, by the mid-1960s the approximate mass and composition of the Martian atmosphere had been determined. Subsequent observations detected nitrogen (N_2), radiogenic argon (^{40}Ar), carbon monoxide (CO), ozone (O_3) and molecular oxygen (O_2), but all in amounts much less than carbon dioxide.

Table M4 Composition of the Martian lower atmosphere (< 120 km)

Constituent	Abundance
CO_2	95.32%
N_2	2.7%
^{40}Ar	1.6%
O_2	0.13%
CO	0.07%
H_2O	0.03% (variable)
Ne	2.5 ppm
Kr	0.3 ppm
Xe	0.08 ppm
O_3	0.04–0.2 ppm (variable)
Dust	0–≫ 5 (visible optical depth)

Mass and composition

The best quantitative information about the mass and composition of the Martian atmosphere comes from the Viking landers. Each lander carried a mass spectrometer which measured composition during the descent and while on the surface. The results of these measurements are given in Table M4. Carbon dioxide is indeed the principal constituent, followed by nitrogen, argon, oxygen, and carbon monoxide. Trace amounts of the noble gases are also present. Additional minor and highly variable constituents include water vapor, ozone, and dust particles. Together, these gases exert a mean surface pressure of about 7.5 mbar, which corresponds to an average column mass of 20.2 g cm^{-2}.

Water vapor and dust particles are important minor constituents of the Martian atmosphere. Their concentrations vary considerably with location and season. The average column abundance of water vapor is about 15 pr-μm and ranges from near zero in the the winter polar regions to as high as 100 pr-μm at the north pole during early summer. (A precipitable micron (pr-μm) is the depth (in 10^{-6} m) of water formed if the atmospheric water vapor were precipitated to the surface as liquid.) By comparison, the Earth's atmosphere normally contains several centimeters of water. The vertical distribution of Mars' water vapor is uncertain but is likely to be fairly well mixed.

Dust concentrations have not been measured directly. They have been inferred, however, from the measured optical depths at the Viking lander sites (Colburn, Pollack and Haberle, 1989; Figure M23). The inferred concentrations depend on the particle size distribution, which is believed to be dominated by submicron-size particles having a cross-section weighted mean particle radius of 2.5 μm. For a well-mixed atmosphere such a particle size distribution yields near-surface concentrations which range from 5 to 30 particles cm^{-3}.

The composition of Martian dust is uncertain, but spectroscopic studies indicate that the particles are dominated by igneous silicates, or weathering products like clays. Some basalt may also be present. Relatively high SiO_2 contents (> 60%) have been inferred from these studies. In contrast, lower silicate abundances (~ 40%) have been inferred from soil measurements at the Viking lander sites.

Orbital properties

Mars' orbital properties determine the amount of sunlight available to heat the atmosphere, which then determines the temperature structure, circulation and climate. They are listed in Table M5 together with those of Earth for comparison.

The most important points to note are: (1) Mars is further from the Sun than Earth and so receives about half as much annually averaged sunlight; (2) Mars' orbit is highly elliptical such that at perihelion it receives about 40% more insolation than at aphelion; and (3) Mars' rotation rate and obliquity (the angle between its spin axis and the

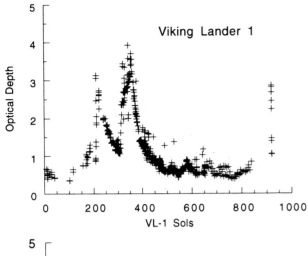

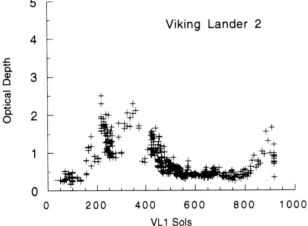

Figure M23 Seasonal variation of the measured visible optical depths at the two Viking lander sites. Viking Lander 1 was located at 22°N and 48°W, while Viking Lander 2 was at 48°N and 227°W. Time is measured in terms of a Martian solar day 'sol', which is equal to 88 775 s.

Table M5 Orbital parameters for Mars and Earth

Property	Mars	Earth
Mass (kg)	6.46×10^{23}	5.98×10^{24}
Radius (m)	3394	6369
Gravity at surface (m s^{-2})	3.72	9.81
Orbit eccentricity	0.093	0.017
Semimajor axis (AU)	1.52	1.0
Solar flux (W m^{-2})	590	1360
Length of year (Earth days)	687	365
Length of solar day (s)	88775	86400
Spin-axis inclination (deg)	25.2°	23.5°
Longitude of perihelion (deg)	250°	285°

orbit normal) are similar to Earth's, which means that daily and seasonal changes are also similar.

Temperatures

Except during dust storms, the atmosphere of Mars is nearly transparent to solar radiation. Consequently, its temperature structure is controlled by thermal emission from the surface. The globally averaged surface temperature on Mars is approximately 215 K.

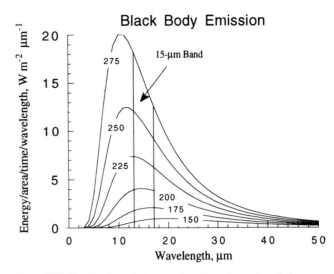

Figure M24 Blackbody emission as a function of wavelength for a range of temperatures representative of the Martian surface. CO_2 is the principal infrared absorbing gas in the Martian atmosphere and the approximate width of its 15-μm absorption feature is indicated.

However, because Mars lacks oceans its surface temperatures undergo considerable seasonal, diurnal and latitudinal variation. The lowest surface temperatures are ~ 150 K, occurring in polar regions during winter. These low temperatures are associated with the condensation of CO_2 onto the surface. The highest surface temperatures are ~ 275 K and occur in the southern subtropical regions during summer. At this time of year Mars is closest to the Sun. In these same regions, however, diurnal variations are also large as nighttime temperatures plummet to near 170 K. Thus daily temperature excursions at the surface can exceed 100 K.

As shown in Figure M24, a surface radiating between 150 K and 275 K emits most of its energy at infrared wavelengths between 5 and 50 μm. Since CO_2 is the principal constituent of the Martian atmosphere, and since it has a vibration–rotation band centered near 15 μm, some of the energy emitted by the surface will be absorbed in the atmosphere. The fraction absorbed depends on the width of the 15-μm band which, for average Martian surface pressures, is about 4 μm (13–17 μm). Thus, even though the peak in blackbody emission lies within the 15-μm band for typical Martian surface temperatures, the width of the band is so narrow that only about 10–20% of the radiation emitted by the surface actually gets absorbed in the atmosphere. The remainder is radiated directly to space.

Some of the radiation absorbed by the atmosphere is reradiated back to the surface. This additional radiation produces a modest greenhouse effect. A convenient measure of the greenhouse effect is the difference between the average surface temperature T_s, and the effective temperature T_e. The latter is the temperature at which a spherical black body would radiate away the amount of sunlight it absorbs and is readily calculated from

$$T_e = \left[\frac{S_o(1 - A_p)}{4\sigma} \right]^{1/4}$$

where $\sigma = 5.6696 \times 10^{-8}$ W m^{-2} K^{-4} is the Stefan–Boltzmann constant, $S_o = 590$ W m^{-2} is the average solar flux at Mars, and $A_p = 0.26$ is the planetary albedo and represents the total amount of sunlight reflected back to space by the surface and atmosphere. Given these values, $T_e = 210$ K. Thus the amount of greenhouse warming $(T_s - T_e)$ due to the Martian atmosphere is 5 K.

Compared to Earth's atmosphere, whose greenhouse effect is about 35 K, the Martian atmosphere produces a relatively weak greenhouse effect. This is due to the fact that even though it contains about 30 times more CO_2 per unit area than Earth's atmosphere, the Martian atmosphere absorbs radiation in a very narrow part of the spectrum (Figure M24). Earth's greenhouse effect is mostly due to water vapor, which has numerous absorption features throughout the infrared. Water vapor is too scarce on Mars to provide much infrared opacity. Furthermore, the higher surface pressure on Earth increases the frequency of collisions between molecules which 'pressure broadens' the absorption bands and further increases the total absorption.

Dust particles suspended in the atmosphere absorb in broad regions of the infrared and can therefore provide some greenhouse heating. However, they also counteract the greenhouse effect by reflecting sunlight back to space. The net temperature change depends on the surface albedo: bright surfaces will warm when dust is present, while dark surfaces will cool. The decline in mean temperatures at the Viking 1 Lander site during the passage of a dust storm suggests that at least for that location, dust particles have an 'antigreenhouse' effect.

Vertical structure

Because the Martian atmosphere is heated from below, temperatures decrease with height just as they do on Earth. As illustrated in Figure M25, the variation of temperature with height on Mars gives rise to a troposphere, a mesosphere and a thermosphere.

The troposphere is that region of the atmosphere extending upward from the surface where temperatures decrease with height. In the troposphere temperatures are controlled by radiative and convective processes. On Earth the troposphere is about 12 km deep and the temperature decreases with height at about 6.5 K km^{-1}. Based on the Viking lander entry measurements, the troposphere on Mars is much deeper ($\sim$ 40 km) and its average lapse rate is much smaller ($\sim$ 2.5K km^{-1}). (The rate at which temperature decreases with height is often referred to as the lapse rate.) On both planets the observed lapse rates are much less than the dry adiabatic lapse rate (9.8 K km^{-1} on Earth and 5.1 K km^{-1} on Mars).

On Earth the observed lapse rates are less than adiabatic because of latent heat release associated with the condensation of water vapor. For Mars, water vapor abundances are much too low for this to be a significant heat source. Instead the additional heating comes from the absorption of solar radiation by suspended dust particles. On both planets temperatures are further stabilized by vertical heat fluxes associated with large-scale circulation systems.

Theoretical studies (e.g. Gierasch and Goody, 1968) indicate that during the day on Mars the lapse rates produced by radiation greatly exceed the adiabatic value; intense convection is subsequently generated that could extend to very high altitudes ($\sim$ 15 km). In such regions the lapse rates are expected to be very close to the adiabatic value. Evidence for deep daytime convection on Mars was found in the Viking Lander 1 entry profile (Figure M25) which indicated a near-adiabatic lapse rate between the surface and 6 km. However, unlike Earth, convection on Mars does not appear to extend throughout the troposphere. Above 15 km temperatures continue to decrease with height, but are controlled almost entirely by radiation.

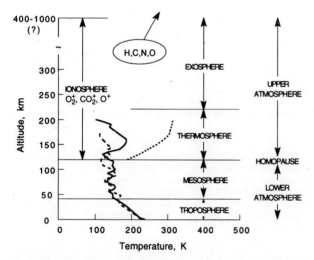

Figure M25 Vertical structure of the Martian atmosphere. Solid and dashed curves represent the temperature profiles measured by the Viking Lander 1 and 2 respectively, as they descended to the surface. Dotted curve above 120 km represents a theoretical profile consistent with Mariner 9 airglow measurements.

Above the troposphere on Mars is the mesosphere, where temperatures become nearly isothermal. This is different from the Earth where temperatures increase with height above the troposphere because of the absorption of solar ultraviolet radiation by ozone. This region of the Earth's atmosphere is known as the stratosphere. On Mars there is no analogous ozone layer and therefore no stratosphere.

Superimposed on the nearly isothermal structure of the Martian mesosphere are oscillations in temperature due to the adiabatic heating and cooling associated with vertically propagating planetary waves. These waves are associated with a global system of thermal tides. As the tides propagate upward their amplitude increases. Eventually, they produce superadiabatic lapse rates at which point the waves 'break' and generate local mixing. There are several locations in the Viking entry profiles where wave breaking is indicated.

Above the mesosphere, temperatures on both planets begin to increase because of heating by radiation in the far and extreme ultraviolet. This region is the thermosphere; it begins at about 80 km on Earth and about 100 km on Mars. The heating is due to the dissociation and ionization of various molecules and atoms. However, thermospheric heating was notably absent in the Viking profiles. Earlier measurements by the Mariner 9 spacecraft indicate a much warmer thermosphere than was seen by Viking. Temporal variation in thermospheric temperatures on Mars are not fully understood.

Photochemistry

Photochemical reactions occur throughout the Mars atmosphere as solar ultraviolet radiation penetrates all the way to the surface. Carbon dioxide, the main atmospheric constituent, is readily dissociated by this energetic radiation:

$$CO_2 + h\nu \rightarrow CO + O \ (\lambda < 2275 \text{ Å})$$

However, the three-body recombination reaction

$$CO + O + M \rightarrow CO_2 + M$$

(M is any non-reactive molecule) is very slow, such that the oxygen atoms tend to form O_2 and O_3 rather CO_2. The time required to convert the present CO_2 atmosphere into one composed predominantly of CO and O_2 is only several thousand years. Yet CO_2 is the dominant constituent while CO and O_2 are scarce. Thus some kind of recombination mechanism is required.

It is now believed that catalytic reactions involving the odd hydrogen species (H, OH, HO_2 and H_2O_2) are important for recombination (McElroy and Donahue, 1972; Parkinson and Hunten, 1972). Water vapor, though a minor constituent, is the source of this odd hydrogen. Like CO_2, water vapor is dissociated by ultraviolet radiation to produce atomic hydrogen (H) and the perhydroxyl radical (OH):

$$H_2O + h\nu \rightarrow H + OH$$

Excited oxygen [O(^{1}D)], formed by the dissociation of ozone (O_3), can also produce OH radicals:

$$H_2 + O(^1D) \rightarrow H + OH$$

Once produced, H and OH catalyze the recombination of CO and O in one of two ways: by reactions which directly utilize the oxygen atoms produced by CO_2 photodissociation,

$$H + O_2 + M \rightarrow HO_2 + M$$
$$O + HO_2 \rightarrow OH + O_2$$
$$CO + OH \rightarrow CO_2 + H$$

$$\text{Net: } CO + O \rightarrow CO_2$$

or by reactions in which the oxygen atoms are supplied by the photolysis of hydrogen peroxide (H_2O_2),

$$2 \times (H + O_2 + M \rightarrow HO_2 + M)$$
$$HO_2 + HO_2 \rightarrow H_2O_2 + O_2$$
$$H_2O_2 + h\nu \rightarrow 2OH$$
$$2 \times (CO + OH \rightarrow CO_2 + H)$$

$$\text{Net: } 2CO + O_2 \rightarrow 2CO_2$$

A schematic illustration of the odd hydrogen cycle is given in Figure M26. Each arrow is associated with a reaction partner and a

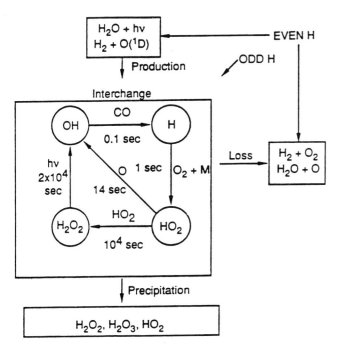

Figure M26 Flow diagram for odd hydrogen near the surface of Mars. Beside each arrow is the reaction time in seconds. From Chamberlain and Hunten (1987). Adapted from Hunten (1979).

reaction time. The inner loop, which corresponds to the first set of reactions, is the fastest with the rate-limiting reaction being $O+HO_2 \rightarrow OH+O_2$. The outside loop corresponds to the second set of reactions and is limited by the time it takes to photolyze hydrogen peroxide ($\sim 2 \times 10^4$ s).

Measurements of the hydrogen-containing species would help confirm the inferred importance of the water chemistry described above. However, such measurements have not yet been made. Support for this chemistry instead comes from the distribution and abundance of ozone.

Ozone abundances on Mars are much less than on Earth and range from below the threshold of detection in warm tropical regions to as high as 150×10^{15} molecules cm^{-2} in cold polar regions. Ozone is produced when O and O_2 combine, and is destroyed by ultraviolet dissociation. However, in the absence of additional atmospheric sinks, ozone would be much more abundant than observed. The H and OH produced by water photolysis provides this additional sink by using the same catalytic cycle that operates in Earth's stratosphere, namely,

$$H + O_3 \rightarrow OH + O_2$$
$$O + OH \rightarrow O_2 + H$$

Net: $O + O_3 \rightarrow 2O_2$

Since the source of odd hydrogen is water vapor, ozone will be depleted in regions where it is abundant, and plentiful in region where it is absent. This anticorrelation between ozone and water vapor has been observed and provides support for the importance of water as a key chemical ingredient of the Martian atmosphere (Barth 1985).

Escape processes

Escape occurs in the exosphere, which begins on Mars at about 230 km. In the exosphere the probability of collisions is so small that particles execute ballistic trajectories, some of which carry them away from the planet. The most important gases that can escape from Mars are hydrogen, oxygen and nitrogen.

When water vapor is photolyzed in the lower atmosphere one of the products is molecular hydrogen (H_2). Below the homopause, H_2 is well mixed and has a long lifetime (10^3 years). Above the homopause it is converted into atomic hydrogen, where it possesses enough kinetic energy in the form of thermal motion to escape to space. Ultraviolet spectrometers aboard the Mariner 9 spacecraft have detected atomic hydrogen escaping from Mars.

The escape of hydrogen implies that there must be a sink for O_2. Otherwise the amount of O_2 would double in about 2×10^5 years. Loss of oxygen can occur through the oxidation of surface materials and/or escape to space. Loss to the surface requires the continual exposure of surface materials and is not likely to be significant on the 10^5 year timescales. On the other hand, atomic oxygen is too heavy to escape on the basis of its thermal motion alone. However, it can escape when ionized oxygen molecules (O_2^+) in the ionosphere recombine with electrons. The recombination dissociates the molecule into its constituent atoms with enough kinetic energy to escape. This non-thermal escape mechanism – known as 'dissociative recombination' – yields an oxygen escape flux that adjusts itself until it balances the hydrogen loss, i.e. for every oxygen atom that escapes, two hydrogen atoms also escape. In effect, water is escaping the planet. If extrapolated over the 4.5 billion year age of the planet, the loss is equivalent to a layer of water covering the entire planet to a depth of about 2.5 m.

General circulation

Although the meteorological data base for Mars lacks the temporal and spatial coverage needed to characterize fully its general circulation, much can be inferred from it – particularly in connection with general circulation models (Haberle et al. 1993; Barnes et al. 1993). Both the data and the models indicate that the main components of the general circulation are a near zonally-symmetric mean meridional circulation, stationary and propagating planetary waves, thermal tides and a mass flow associated with the seasonal cycling of CO_2 into and out of the polar regions. The latter is a unique feature of Martian meteorology. The general circulation of Mars is schematically illustrated in Figure M27 and is discussed in detail by Zurek et al. (1992).

The mean meridional circulation dominates the lower latitudes and is characterized by Hadley cells which extend between 30°S and 30°N and which reach great heights (20–40 km). These Hadley cells undergo significant seasonal variations in structure and intensity. At the equinoxes, two roughly symmetric Hadley cells develop which share a common rising branch centered at or near the equator. At the solstices the two Hadley cells give way to a single cross-equatorial circulation. The intensity of the Hadley cell, as measured by the mass flux carried in its branches, is believed to vary significantly between equinox and solstice. Models indicate that the variation can be from 10^9 kg s^{-1} at equinox to 10^{10} kg s^{-1} at solstice.

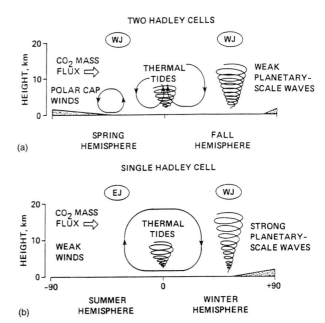

Figure M27 Schematic illustration of the general circulation of the lower Martian atmosphere at the equinoxes (a) and solstices (b).

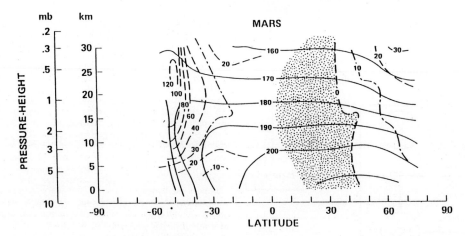

Figure M28 Isotachs of the zonal mean zonal winds (dashed lines) inferred from the thermal wind relationship using Mariner 9 temperature data (solid lines). Stippled region indicates easterly winds (winds blowing from east to west).

The zonal wind component of the mean meridional circulation has been inferred from temperature data through the thermal wind relationship. That relationship requires that the westerly component of the wind increase (decrease) with height at latitudes where zonally averaged temperatures decrease (increase) toward the pole. An illustration of winds derived in this manner is shown in Figure M28.

Application of the thermal wind relationship to Mars indicates that easterly winds prevail in the tropics at all seasons, and in the summer hemisphere at the solstices. Westerlies prevail in the winter hemisphere at the solstices, and at middle and high latitudes during the equinoxes. If zonal winds at the surface are relatively weak, as was indicated by the Viking lander measurements, then the thermal data indicate that the westerly jet stream in the winter hemisphere can reach speeds of 200 m s^{-1} at 40 km.

At the higher latitudes ($> 30°$), the general circulation is less symmetric with respect to longitude. During winter in the northern hemisphere, eastward-propagating disturbances of high and low pressure systems were detected by the Viking landers (Ryan et al., 1978). These traveling disturbances, called transient eddies, are very similar to terrestrial 'weather' systems in that southerly (northerly) winds are generally associated with falling (rising) pressures and warm (cold) air advection. Theory suggests that the transient eddies arise from baroclinic instability, an instability of the zonal jet stream. Both theory and observations indicate that the dominant zonal wavenumber of the transient eddies varies between 1 and 4, and that they propagate around latitude circles with phase speeds between 10 and 20 m s^{-1} (Barnes 1980).

The transient eddies detected by the Viking landers were observed to exhibit considerable seasonal variability. Eddy amplitudes were strongest during winter and spring, and occasionally exhibited a high degree of regularity. During summer, however, the eddies virtually disappeared. This disappearance is believed to be due to the small latitudinal variation in the large-scale temperature field that exists at this season. It is not known if transient eddies exist in the southern hemisphere.

Dust storms

The surface of Mars is mantled with a fine dusty material that is occasionally lifted into the atmosphere when surface winds become strong enough to initiate particle motion. Because of the low density of the Martian atmosphere dust-raising winds must be quite strong. Winds gusting to 30 m s^{-1} were measured at 1.6 m by the Viking 1 Lander during the passage of a dust storm. Apparently, this is the minimum surface wind speed required to initiate lifting. However, the dust-raising process is complicated and the threshold for lifting can vary in either direction depending on surface properties and atmospheric stability (Greeley and Iversen 1985).

Numerous dust storms occur each Martian year and they range in size from funnel-like dust devils at the small scale (~ 1 km^2) to truly global scale obscurations at the large scale ($\sim 10^8$ km^2). The smaller dust storms occur more frequently. About 100 dust devils have been detected in Viking images. Most of these developed over smooth terrain in the early afternoon during summer and extended to heights of 1–2 km. A similar number of larger ($\sim 10^6$ km^2), but still localized, storms were observed during the first Viking year. While these 'local' dust storms can occur at any location and season, they are most frequent when Mars is closest to the Sun (late northern fall).

The least frequently occurring of the Martian dust storms are those of regional scale (10^7 km^2) or larger. These truly spectacular events can literally envelop the entire planet. Such was the case when the Mariner 9 spacecraft arrived at Mars in November 1971. Only the tops of the Tharsis volcanoes were visible to the spacecraft's cameras. Dust particles were carried to heights in excess of 50 km during the storm which lasted for several months. Two planet encircling dust storms were also observed during the Viking mission. Since then the following characteristics have been deduced: (1) global dust storms occur during fall and winter in the north – none have been observed at other seasons; (2) they start as local storms in southern subtropical latitudes and expand first in the east–west direction, and then in the south–north direction; and (3) they occur in some years, but not in others. To date, only six well-documented planet-encircling dust storms have been observed.

While the mechanisms responsible for the life cycle of these storms are poorly understood, it is generally believed that the heating effects of suspended dust particles play an important role. This was clearly demonstrated in a theoretical simulation by Haberle Leovy and Poilack (1982) and is illustrated in Figure M29. Once in the atmosphere, dust particles strongly absorb sunlight. The subsequent heating of the atmosphere intensifies the circulation, which further lifts the dust and further intensifies the heating. This positive feedback continues until the dust is globally distributed. On the other hand, if the dust heating is ignored very little evolution occurs.

Clouds

Because water vapor in the Martian atmosphere is controlled by saturation, water-ice clouds are fairly common. However, their water content must be low (~ 1 pr-μm). During winter they can be seen in the subpolar regions as widespread diffuse hazes known as 'polar hoods'. Wave clouds forced by topographic obstacles (craters) are also seen at this season. The north polar hood appears to be more prominent than the south polar hood.

During summer, convective clouds have been observed over tropical uplands in the northern hemisphere. These clouds resemble cumulus clouds on Earth and indicate considerable vertical motions. In general, water-ice clouds are more common in the northern hemisphere than in the southern hemisphere.

Clouds of CO_2 ice may also form in the Martian atmosphere. During winter in both polar regions, atmospheric temperatures can reach the frost point of CO_2. The presence of CO_2 ice clouds has been inferred from thermal emission data.

Climate

The climate of Mars is characterized in terms of the seasonal cycles of CO_2, water, and dust (Haberle, 1986). Each of these cycles involves

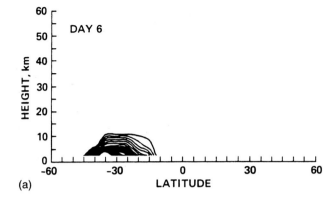

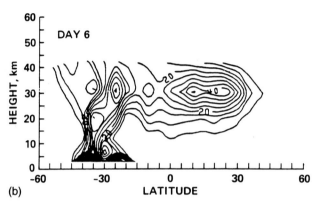

Figure M29 Computer simulations of a Mars global dust storm. Dust concentrations (g kg^{-1}) are contoured as a function of height and latitude for two different simulations after 6 simulated days of dust raising in the southern subtropical latitudes. (a) Dust was not allowed to effect atmospheric heating, Passive Tracer Experiment. (b) Dust was allowed to effect atmospheric heating, Radiatively Active Tracer Experiment. The results clearly indicate the importance of dust heating on the evolution of Mars dust storms. Adapted from Haberle, Leovy and Pollack (1982).

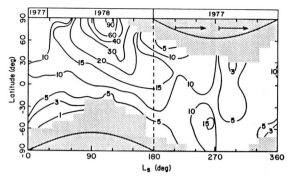

Figure M30 Seasonal and latitudinal variation of the column abundance of water vapor in the Martian atmosphere. Contours are in units of pr-μm (see text). Data cover 1 Mars year and are spliced at $L_s = 180°$ (dashed vertical line). Shaded regions indicate no observations. Smooth curves in the shaded regions indicate the latitude of the polar night. Horizontal arrows indicate the time span for the two global dust storms that occurred during the first year of the Viking mission.

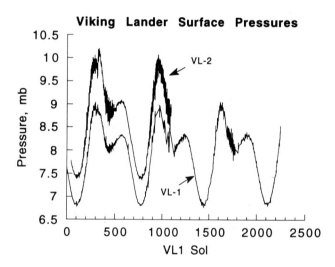

Figure M31 Seasonal variation of daily-averaged surface pressure at the two Viking landers. Time is given in terms of Mars solar days (sol) after Viking Lander 1 reached the surface.

the exchange of material between surface and atmospheric reservoirs. The exchange itself is driven by daily and seasonal variations in insolation. The atmosphere plays a major role in these cycles by serving as an agent of transport. Reviews of these cycles can be found in James, Kieffer and Page (1992), Jakosky and Haberle (1992) and Kahn *et al*. (1992).

During winter, temperatures in the polar regions fall to 150 K, the CO_2 frost point temperature of a 6-mbar atmosphere. Consequently, CO_2 condenses. The condensation of CO_2 during winter and its subsequent sublimation during spring is what gives rise to the familiar waxing and waning of the polar caps. Approximately 20% of the Martian atmosphere is cycled into and out of the polar regions each year by this process. The signature of this cycling can be seen in the measurements of surface pressure by the Viking landers (Figure M30).

At each site there is a pronounced semiannual variation. The variation is semiannual rather than annual because while one cap is growing the other cap is retreating. However, the variation is asymmetric with respect to season, being much more pronounced during southern winter and spring than during northern winter and spring. This asymmetry is a direct consequence of Mars' orbital eccentricity. Southern winters are much longer than northern winters such that much more CO_2 condenses out of the atmosphere. As a result, pressures are lowest during the middle of southern winter just before solar heating starts to vaporize the cap, and highest late in spring when the cap has nearly disappeared and returned all its CO_2 back to the atmosphere.

At both poles, however, the caps never completely disappear during summer. At the north pole CO_2 frost completely sublimes by summer, but in doing so it exposes an underlying water-ice cap. At the south pole CO_2 frost appeared to survive all summer long – at least for the year the Viking orbiters observed it. It is not known whether an underlying water-ice cap exists at the south pole. Thus, the summer caps appear to have different compositions: CO_2 frost in the south, and water ice in the north. The reason for this compositional asymmetry has been attributed to the south cap being brighter than the north cap (Paige and Ingersoll, 1985). Just why this occurs is uncertain, but may be related to the dust cycle in the sense that dust storms tend to occur most frequently during the time the north cap is forming (northern fall and winter). Thus the north cap should be 'dirtier' than the south cap and thus less reflective.

When water ice is exposed at the north pole, it sublimes into the atmosphere. The abundance of atmospheric water vapor was mapped by the Viking orbiters for over 1 Mars year (Jakosky and Farmer, 1982; Figure M31). Maximum observed abundances were about 100 pr-μm and occurred over the north pole during summer. The northern summer ice cap is therefore an important source of atmospheric water vapor. Minimum observed abundances were less than several pr-μm and occurred over both polar regions during winter. Because of their low temperature (~ 150 K), the seasonal CO_2 caps will act as a sink for any atmospheric water vapor that is brought in contact them. For the current epoch, this implies that on

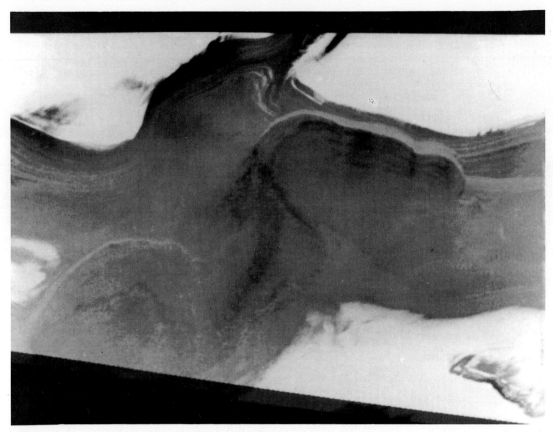

Figure M32 Layered terrain on Mars. A view of the north polar region of Mars, centered near 80° north latitude and 347° west longitude. The white areas are water ice, and the darker regions show a terraced structure. The layered structure apparently consists of strata about 30 m thick, stacked one upon another. The strata were most probably formed by atmospheric sedimentation processes that varied with time. The polar caps occupy the topographic highlands, above the layered terrains. The absence of fresh impact craters suggests that the surface is young (perhaps less than 10 million years in age). (Viking image courtesy of R. Haberle.)

an annual averaged basis there is a net transfer of water from the north cap to the south cap.

Quasi-periodic climate change

Both polar regions on Mars are characterized by extensive layered terrains which are among the youngest geological features on the planet. The layered structure consists of a series of plates approximately 30 m thick that are stacked one upon the other, and which decrease in diameter with altitude (Figure M32). The residual polar caps lie at the very top of these terrains. The fact that they are continuous and uniform suggests they were formed by atmospheric sedimentation processes that were somehow modulated in time. The lack of fresh impact craters indicates that they formed sometime during the past 10 million years.

The most widely accepted theory on the origin of the layered terrains is the astronomical theory (Toon et al. 1980). According to this theory, Mars' orbital parameters vary in a quasi-periodic fashion and this alters the seasonal and annual distribution of zonally averaged insolation which determines climate. The key orbital parameters are the obliquity, eccentricity and longitude of perihelion (see also Milankovitch, Milutin, and Milankovitch theory).

The obliquity is the most important parameter. Its current value is 25.2°, but it may have varied from between 0.2° to 51.4° during the past 10 million years (Bills, 1990). (See also Chaotic dynamics in the solar system.) As the obliquity increases, the annually averaged insolation increases significantly at the poles and decreases slightly at the equator. Consequently, the polar regions warm with respect to the equator and CO_2 stored within the high-latitude regolith is driven into the atmosphere. The subsequent increase in mean surface pressure may have been as much as 20 or 30 mbar, depending on the capacity of the regolith to store CO_2 and the presence or absence of permanent polar caps. Since dust lifting is easier at higher surface pressures, the atmosphere of Mars at high obliquity should also be dustier. This, in turn, would increase sedimentation rates in the polar regions. At low obliquities the opposite would occur. The polar regions would cool with respect to equatorial regions and permanent polar caps would form. The caps would act as a sink for both atmospheric and regolith CO_2, and the surface pressure would fall to as low as 0.1 mbar. Under these circumstances, dust storms would cease and sedimentation rates would be minimal. Thus by modulating surface pressure, the obliquity oscillations provide a mechanism to alter the transport and sedimentation of dust.

However, these oscillations would also affect the behavior of water. At low obliquities the permanent CO_2 caps would 'cold trap' any water brought in contact with them. Permanent water-ice caps would form whose thickness would be limited by the amount of water stored as permafrost, and its ability to diffuse through the soil and into the atmosphere. Then as the obliquity increases and dust storms become more frequent, the water-ice caps get covered with dust and are eventually sealed and insulated from the atmosphere. Thus, according to the astronomical theory, the layered terrains are ice sheets that were formed at low obliquity and buried by dust at high obliquity.

Origin and evolution

There are several lines of evidence that suggest Mars began with an atmosphere much different than it has today. The first is geochemical in nature and is based on measurements of the isotopic abundance of various gases in the atmosphere. The second is geological in nature and is based on the morphology of the surface and its implications for fluid flow. Both lines of evidence favor an initial atmosphere

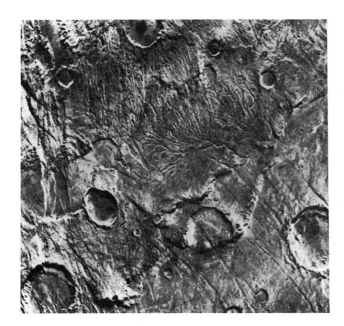

Figure M33 A valley network system in the southern hemisphere (48°S, 98°W). This is one of the densest drainage networks observed on the planet. The picture is approximately 250 km across.

containing much more CO_2, N_2 and water than it does today. However, they differ in the initial abundance of these gases; the geochemical evidence favors low initial abundances (hundreds of millibars), while the geological evidence favors high initial abundances (thousands of millibars).

An example of a geochemical argument can be found in McElroy, Kong and Yung (1977), who base their estimate of Mars' initial atmosphere on the observed enrichment of ^{15}N relative to ^{14}N. Like oxygen, nitrogen can escape from Mars non-thermally by dissociative recombination. Above the homopause (~ 120 km), the heavier isotope decreases in concentration more rapidly with height than the lighter one. Consequently, at the exobase (~ 200 km) there are fewer ^{15}N atoms available for escape. Over the lifetime of the planet, this leads to an enrichment of ^{15}N relative to ^{14}N. The observed value is $^{15}N/^{14}N=0.006$, or approximately 1.6 times the terrestrial value. McElroy et al. estimate that this implies an initial nitrogen partial pressure of 1–30 mbar, depending on the rate of nitrogen fixation in the soil. Assuming N_2, CO_2 and H_2O are present on Mars in the same relative abundances as they are on Earth, then 80–250 mbar of CO_2 and 6–185 m of H_2O were initially present.

An example of a geological argument can be found in Carr (1986), who based his estimate on the amount of water needed to carve out the outflow channels associated with the Valles Marineris canyon systems. These channels appear to have formed by the sudden catastrophic release of subsurface water. By assuming the water carried the maximum amount of debris possible, and that the channels were formed in a single event (no recycling), Carr estimated that the global equivalent of 35 m of water was required. By extrapolating this result to the rest of the planet, Carr inferred that Mars outgassed 500–1000 m of water, 10–20 bar of CO_2 and 0.1–0.3 bar of N_2.

There are other channel features that also support a massive early atmosphere. These are the so-called 'valley networks' which are found mostly in the heavily cratered terrain of the southern hemisphere (Figure M33). These terrains date back to the late heavy bombardment phase of the planet's evolution and suggest that the valley networks are at least 3.8 billion years old. Unlike the outflow channels, the valley networks resemble terrestrial drainage systems and appear to have formed by the slow but steady erosional action of recycled liquid water (Carr, 1981; Baker, 1982). Such erosion could have resulted from rainfall, but is more likely the result of groundwater sapping. In either case, a climate conducive to the presence of

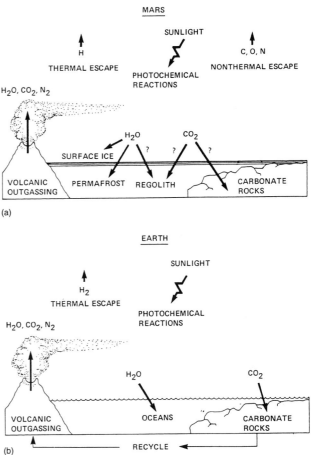

Figure M34 Schematic illustration of the CO_2 geochemical cycles on (a) Mars and (b) Earth. On Mars some of the outgassed H_2O, CO_2, and N_2 can escape to space as H, C, O and N. The remainder goes into various surface reservoirs. On Earth only hydrogen can escape. Most importantly, however, CO_2 can be recycled back into the Earth's atmosphere because of plate tectonics. Mars does not appear to have terrestrial-style plate tectonics so that the atmospheric CO_2 it loses to carbonates during the weathering process is irretrievably lost from the system.

liquid water at the surface and in equilibrium with the atmosphere appears necessary.

The most likely way to achieve warmer and wetter conditions at the surface of Mars is through a greenhouse effect, and the most likely greenhouse gases are CO_2 and water vapor. At present neither of these gases is present in sufficient quantities to produce much greenhouse warming. However, if the geological evidence for Mars' initial volatile inventory is correct, CO_2 and water vapor may have been present in ample quantities of the early Mars atmosphere. Initial estimates of the amount of CO_2 required to produce surface temperatures of 273 K early in Mars' history range from 1–5 bar (Pollack et al. 1987). The high value is that required for a globally averaged surface temperature of 273 K, and is based on the assumption that the early Sun was only 30% as luminous as it is today. However, it has been recently pointed out that such massive CO_2 atmospheres are greatly supersaturated, and that clouds of CO_2 would form which substantially inhibit the greenhouse effect (Kasting 1991). This implies that other greenhouse gases (CH_4, NH_3, dust, etc.) may be involved.

If Mars did start out with a more massive atmosphere than it has today, then how has it evolved to its present state? The most likely scenario involves the gradual conversion of CO_2 in the atmosphere into carbonate deposits on the surface (Figure M34). During the first billion years or so, the greenhouse effect was strong enough for liquid water to exist at the surface. In the presence of liquid water, CO_2 would have been removed from the atmosphere through the weathering of silicate rocks. However, the weathering lifetime of a massive CO_2 atmosphere is so short ($\sim 1.5 \times 10^7$ years), that some kind of recycling mechanism would have been required to keep it around long enough to form the channel features (Pollack et al., 1987).

On Earth carbonate minerals are recycled because of plate tectonics and sea-floor spreading. Carbonate sediments on the ocean floors are carried into the Earth's mantle where they are thermally decomposed back into CO_2 and water, and then returned to the atmosphere through volcanoes. On Mars volcanic activity was probably very intense during the first billion years, and this could have provided a means for recycling. Recycling by impact cratering might also have been important (Carr 1990). However with time, Mars' internal heat flow declined, volcanism and impact cratering subsided and the atmosphere was gradually removed. Even after the period when liquid water was stable on the surface, the atmosphere would continue to diminish through carbonate formation in transitory pockets of liquid water (Kahn 1985). Indeed, it may be no coincidence that the current mean surface pressure on Mars is so close to the triple point pressure of water (6.1 mbar).

The future

Our knowledge of the Martian atmosphere and climate would have been greatly advanced by the Mars Observer spacecraft which was to map the surface and atmosphere for 1 Mars year beginning in December 1993. Unfortunately, the spacecraft fell silent while preparing for orbit insertion and communication was never reestablished. Yet the science goals of that mission remain a high priority, and NASA is now planning to recover that science through a new 10-year program called Mars Surveyor.

The Survey program is to begin in 1996 with the launch of an orbiter and lander, and then to continue launching such spacecraft at every opportunity thereafter. (Favorable opportunities for launching to Mars occur about every 2 Earth years.) The lander for 1996 is called Pathfinder and is actually part of the Discovery program – a program of small, focused missions for solar system exploration. The orbiter is called Mars Global Surveyor, and it will map the surface and atmosphere for one Mars year. For atmospheric science, orbiting and landed platforms provide the opportunity to measure the structure and circulation of the atmosphere on both global and local scales.

Other countries are also planning Mars missions. In late 1996 Russia will send small stations and (possibly) penetrators to the surface of Mars as part of their Mars 96 mission. Each of the small stations will carry Finn-built meteorology packages to measure the near-surface environment. Finally, in 1998 the Japanese will launch Planet-B, an orbiting spacecraft that will focus on the upper atmosphere.

At present Pathfinder, Mars Global Surveyor, Mars 96 and Planet-B are the only approved missions. Yet the United States, Europe and Japan continue studying concepts for missions well into the 21st century, and it is very likely the some of these will be joint efforts. Ultimately, humans will go to Mars just as they have gone to the Moon. For all these missions the atmosphere and climate will be a major focus of scientific investigation.

Robert M. Haberle

Bibliography

Baker, V.R. (1982) *The Channels of Mars*. Austin: University of Texas Press.

Barnes, J.R. (1980) Time-spectral analysis of midlatitude disturbances in the Martian atmosphere. *J. Atmos. Sci.*, **37**, 2002–15.

Barnes, J.R., Pollack, J.B., Haberle R.M. *et al.* (1993) Mars atmospheric dynamics as simulated by the NASA/Ames general circulation model 2. Transient baroclinic eddies. *J. Geophys. Res.*, **98**, 3125–48.

Barth, C.A. (1985) Photochemistry of the atmosphere of Mars, in *the Photochemistry of Atmospheres* (ed. J.S. Levine). New York: Academic Press, pp. 337–92.

Bills, B.G. (1990) The rigid body obliquity history of Mars. *J. Geophys. Res.*, **95**, 137–44.

Carr, M.H. (1981) *The Surface of Mars*. New Haven: Yale University Press.

Carr, M.H. (1986) Mars: A water rich planet? *Icarus*, **68**, 187–216.

Carr, M.H. (1990) D/H on Mars: effects of floods, impacts, volcanism and polar processes. *Icarus*, **87**, 210–22.

Chamberlain, T.E. and Hunten D.M. (1987) *Theory of Planetary Atmospheres*, 2nd edn. Orlando: Academic Press.

Colburn, D.S., Pollack, J.B. and Haberle, R.M. (1989) Diurnal variations in optical depth at Mars. *Icarus*, **79**, 159–89.

Gierasch, P.J. and Goody R.M. (1968) A study of the thermal and dynamical structure of the Martian lower atmosphere. *Planet. Space Sci.*, **16**, 615–56.

Greeley, R. and Iversen, J.D. (1985) *Wind as a Geological Process*. Cambridge: Cambridge University Press.

Haberle, R.M. (1986) The climate of Mars, *Sci. Am.*, **254**, 54–62.

Haberle, R.M., Leovy, C.B. and Pollack, J.B. (1982) Some effects of global dust storms on the atmospheric circulation of Mars. *Icarus*, **50**, 322–67.

Haberle, R.M., Pollack, J.B., Barnes, J.R., *et al.* (1993) Mars atmospheric dynamics as simulated by the NASA/Ames general circulation model 1. The zonal mean circulation. *J. Geophys. Res.*, **98**, 3093–123.

Hunten, D.M. (1979) Possible oxidant sources in the atmosphere and surface of Mars. *J. Molec. Evol.*, **14**, 71–8.

Jakosky, B.M. and Farmer, C.B. (1982) The seasonal and global behavior of water vapor in the Mars atmosphere: complete global results of the Viking atmospheric water detector experiment. *J. Geophys. Res.*, **87**, 2999–3019.

Jakosky, B.M. and Haberle, R.M. (1992) The seasonal behavior of water on Mars, in *Mars* (eds H. Kieffer, B. Jakosky, C. Snyder and M. Matthews). Tuscon: University of Arizona Press, pp. 969–1016.

James, P.B., Kieffer, H.H. and Paige, D. (1992) The seasonal cycle of carbon dioxide on Mars, in *Mars* (eds H. Kieffer, B. Jakosky, C. Snyder and M. Matthews). Tucson: University of Arizona Press, pp. 934–968.

Kahn, R. (1985) The evolution of CO_2 on Mars. *Icarus*, **62**, 175–190.

Kahn, R., Lee, S.W., Martin, T.Z. and Zurek, R.W. (1992) The Martian dust cycle, in *Mars* (eds H. Kieffer, B. Jakosky, C. Synder and M. Matthews). Tucson: University of Arizona Press, pp. 1017–53.

Kasting, J.F. (1991) CO_2 condensation and the climate of early Mars. *Icarus*, **94**, 1–13.

Leovy, C.B. (1982) Martian meteorological variability. *Adv. Space Res.*, **2**, 19–44.

McElroy, M.B. and Donahue, T.M. (1972) Stability of the Martian atmosphere. *Science*, **177**, 986–8.

McElroy, M.B., Kong, T.Y. and Yung, Y.L. (1977) Photochemistry and evolution of Mars' atmosphere: a Viking perspective. *J. Geophys. Res.* **82**, 4379–88.

Paige, D.A. and Ingersoll, A.P. (1985) Annual heat budget of Martian polar caps: Viking observations. *Science*, **228**, 1160–8.

Parkinson, T.D. and Hunten, D.M. (1972) Spectroscopy and aeronomy of O_2 on Mars. *J. Atmos. Sci.*, **29**, 1380–90.

Pollack, J.B., Kasting, J.F., Richardson, S.M. and Poliakoff, K. (1987) The case for a warm, wet climate on early Mars. *Icarus*, **71**, 203–24.

Ryan, J.A., Hess, S.L., Henry, R.M. *et al.* (1978) Mars meteorology: three seasons at the surface. *Geophys. Res. Lett.*, **5**, 715–8.

Toon, O.B., Pollack, J.B., Ward, W., *et al.* (1980) The astronomical theory of climatic change on Mars. *Icarus*, **44**, 552–607.

Zurek, R.W., Barnes, J.R., Haberle, R.M. *et al.* (1992) Dynamics of the atmosphere of Mars, in *Mars* (eds H. Kieffer, B. Jakosky, C. Snyder and M. Matthews). Tuscon: University of Arizona Press, pp. 835–933.

Cross references

Atmosphere
Eolian transport
Erosion
Hadley cell
Polar cap
SNC meteorites
Surface process

MARS: GEOLOGY

Satellite imagery of the surface of Mars reveals features reminiscent of both the inner planets and the Earth. The southern highlands of Mars are very heavily cratered, like the highlands of the Moon and Mercury. These densely cratered terrains date back to an early period of intense bombardment of the inner solar system by planetesimals and other objects, which were swept up after the final stages of planetary accretion around 3.9 billion years ago (Shoemaker and Shoemaker, 1990). However, the martian surface has subsequently been considerably modified, in many areas, by impact cratering, volcanism, faulting, eolian and aqueous erosion, and mass wasting. Many landforms, such as the giant shield volcanoes, extensive lava flows, the equatorial canyon system, and outflow channels occur on a much larger scale than on Earth. Mars is divided into two sharply distinct physiographic provinces: a southern, older, densely cratered upland, occupying two-thirds of the surface area, and a northern, younger, sparsely cratered lowland, covering the remaining area (Plate 17). The time of formation of this fundamental crustal division and its exact cause remain controversial. Although Mars is presently very dry, and liquid water is unstable at the surface, considerable photogeologic evidence suggests the former existence of wetter conditions, under which a vast ocean occupied the northern plains episodically and much of the planet was glaciated. The geologic history of Mars, as reconstructed from crater densities and stratigraphic sequences, shows that, in contrast to the Moon or Mercury, volcanism, tectonism and surface modification continued throughout much of the last 4 billion years, although perhaps at declining rates (Table M6).

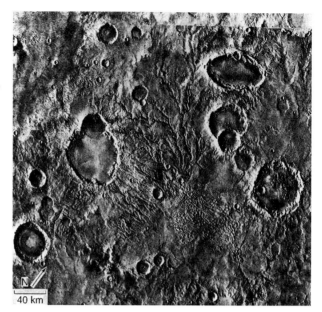

Figure M35 Dendritic valley networks in densely cratered terrain, 25°S, 10°W (NASA).

Highlands

The heavily cratered Martian highlands occupy most of the southern hemisphere, but extend up to 40°N, at around 320°W. The oldest exposed terrains comprise rugged, fractured, hilly ridges forming rims around large, ancient impact basins, such as Hellas, Argyre and Isidis Planitia (Table M6; Tanaka, 1986; Scott and Tanaka, 1986; Greeley and Guest, 1987). These represent uplifted blocks and ejecta surrounding the impact basins.

The most widespread terrain in the southern highlands is densely cratered, with large, often overlapping, low-rimmed, flat-floored depressions (see also Mars: impact cratering). Smaller craters (< 20 km), are generally less degraded, with sharper rims, clear-cut ejecta blankets and bowl-shaped interiors (Carr, 1981). The densely cratered terrain is frequently dissected by numerous narrow (< 1 km) dendritic valley networks or 'runoff channels' (Sharp and Malin, 1975; Figure M35), which vary widely in their degree of preservation. In places the cratered terrain is buried by lower relief intercrater plains (Tanaka, 1986).

Ridged plains are smoother, less heavily cratered than intercrater plains, and traversed by numerous wrinkle ridges (Figure M36). A wrinkle ridge (q.v.) is a sinuous broad arch (up to 10 km across), surmounted by a narrower, higher ridge (100 m), typically offset along en echelon segments. These ridges, common on the lunar

Table M6 Summary of Martian geologic history (adapted from Tanaka, 1986; Scott and Tanaka, 1986; Tanaka and Scott, 1987; Greeley and Guest, 1987; Tanaka et al., 1988; Strom et al., 1991)

Martian	Epoch	Major events
Amazonian	Late	Northern sand seas and mantling deposit
		Residual ice caps
		Polar layered deposits
		Latest lava flows around giant shield volcanoes
		Formation of outflow channels
	Middle	Glaciation of south polar regions
		Major lava flows, Arcadia Planitia and Tharsis
	Early	Oceanus Borealis
		Initial formation of Olympus Mons aureole, and lava flows around Elysium Mons, Alba Patera and Arcadia Planitia
Hesperian	Late	Deposition of northern plains (Vastitas Borealis Fm)
		Subsidence of chaotic terrain and catastrophic floods from outflow channels
		Deposition of layered sediments within Valles Marineris
		Further faulting, enlargement of Valles Marineris
		Beginning of major volcanism in Tharsis, Alba Patera, Elysium and Syrtis Major
	Early	Initial faulting of Valles Marineris; fracturing of Tharsis. Emplacement of ridged plains (lava flows). Declining cratering rate
Noachian	Late	Formation of intercrater plains (interbedded lava flows and sedimentary deposits – fluvial and eolian)
	Middle	Early highland volcanism and surface dissection by network channels
		Heavy cratering of highlands
	Early	Intense bombardment – large impact basins (Hellas, Argyre, Isidis Planitia) $\geq 3.9 \times 10^9$ years BP
		Planetary accretion – formation of solar system $\sim 4.6 \times 10^9$ years BP

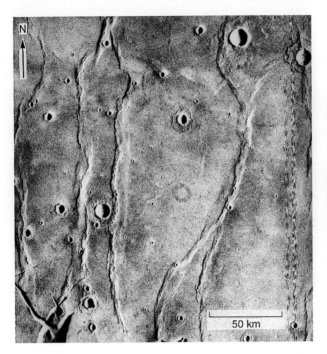

Figure M36 Wrinkle ridges in Lunae Planum region of Mars, 20°N, 66°W (NASA).

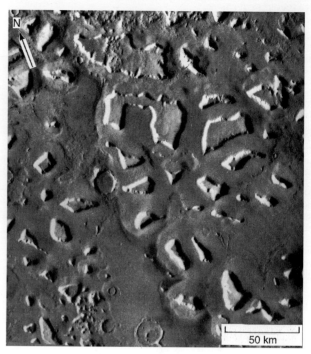

Figure M37 Flat-topped mesa outliers in 'fretted' terrain, 46°N, 332°W (NASA).

maria, may have formed by lava extrusion along fissures, or as compressional structures related to mare basin subsidence, or during regional deformation (Cattermole, 1989). Regardless of mode of origin, lunar wrinkle ridges are most commonly developed on the maria, which are basaltic lava flows. Thus by inference the ridged plains on Mars are probably also lava flows.

The highland–lowland boundary forms a sharp escarpment south of Elysium Planitia (220° to 250°W) and south of Amazonis Planitia (150° to 160°W). Over a wide area, for example north of Arabia and Syrtis Major (30°–45°N and 280°–350°W), the two terrains are separated by an irregular escarpment, 1–2 km high, with outlying, isolated flat-topped mesas and rounded hills. This transitional area, termed 'fretted terrain' (Sharp, 1973), represents remnants of the once-continuous heavily cratered plateau that have been eroded to varying degrees, largely by periglacial processes, as suggested by lobate debris aprons at the scarp bases (Figure M37; Squyres 1989). Elsewhere, the contact is buried under younger lava flows, or lacustrine sediments, presumably deposited by catastrophic floods (Parker, Saunders and Schneeberger, 1989).

The origin and time of formation of the Martian crustal dichotomy remain a major puzzle. Hypotheses of origin range from purely endogenic such as convective mantle overturn following early core differentiation (Wise, Golombek and McGill, 1979), or a much later thermal event, causing major crustal thinning (McGill and Dimitriou 1990), to exogenic associated with a single, enormous impact (Wilhelms and Squyres, 1984). A more likely explanation invokes an interaction of these processes, for example, early multiple overlapping impacts, modified by subsequent volcanism and crustal downwarping (Frey and Schultz, 1990, 1991).

Lowland plains

The equatorial plains are predominantly of volcanic origin (see below). The northern lowland plains are more complex and variable. The most extensive unit is the Vastitas Borealis Formation, which comprises four members (Tanaka and Scott, 1987). The knobby member, the oldest and most widespread, occurs in a circumpolar belt, around 60–70°N. It is characterized by kilometer-sized, dark, closely spaced rounded hills, which could be either volcanoes, highly degraded crater rims or older plains material. The mottled member also lies in a nearly circumpolar zone (around latitude 50–60°N). The mottling is produced by high-albedo crater deposits overlying low-albedo smooth plains of indeterminate origin (either lava flows, or alluvial or eolian deposits). The remaining two units include grooved and ridged members. The former is marked by polygonal grooves, 5 to 20 km across, which, except for a pronounced difference in scale, resemble terrestrial lava shrinkage or desiccation cracks, or periglacial patterned ground (Figure M38). The ridged member contains polygonal or arcuate protuberances, variously interpreted as volcanic dikes or other erosion-resistant material. Some of the erosional and depositional landforms may have been produced by large, transient bodies of standing water in the Borealis plains (Parker, Saunders and Schneeberger, 1989).

Volcanic features

The Tharsis region of Mars is a large topographic high (or bulge), which rises to 10 km above the Martian datum, at 5°S, 105°W. The Martian datum is defined by a fourth-order spherical harmonic gravity field and as the elevation at which the atmospheric pressure is 6.1 mbar, the triple-point pressure of water. This datum can be approximated by a triaxial ellipsoid with semimajor axes $A = 3394.6$ km, $B = 3393.3$ km and semiminor axis $C = 3376.3$ km (Topographic Map of Mars, 1: 15 000 000, US Geot. Surv., 1989). A hemisphere-wide system of fractures and grabens radiates from its center (see also Mars: structural geology and tectonics). Tharsis is also the largest volcanic province on Mars, with around a dozen major volcanoes and extensive lava fields. Olympus Mons is the highest and youngest Martian shield volcano, reaching a summit elevation of 25 km above the Mars datum. It is 600 km wide at its base, with a summit caldera 80 km across, and flank slopes of less than 4°. Its base is encircled by an escarpment which rises to 6 km, in places. (By way of comparison, the volume of Olympus Mons is over 50 times that of Mauna Loa, the Earth's largest volcanic shield). Lobes of ridged terrain (the 'aureole') surround the volcano, extending as far as 700 km to the northwest. The basal scarp and lobes may have been produced by massive gravity slides, forming near the base of the volcano (Cattermole, 1989), a possibility reinforced by the discovery of similar giant landslides on the flanks of shield volcanoes on the islands of Oahu and Hawaii (Moore et al., 1989).

Three giant shield volcanoes, Arsia Mons, Pavonis Mons and Ascraeus Mons, are aligned along a NE–SW trending line near the

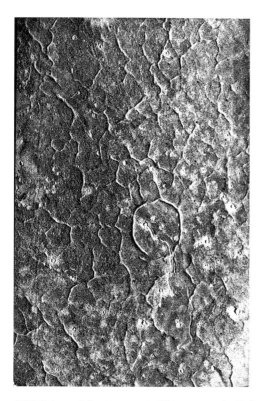

Figure M38 Polygonal fractures up to 20 km across, in Cydonia Region, 44°N, 18°W (NASA).

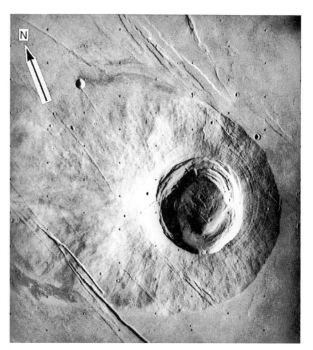

Figure M39 Biblis Patera – a smaller shield volcano in the Tharsis region. The summit caldera is 55 km across, and rises 4 km above the surrounding plains (NASA).

Tharsis crest. They range between ~ 350 and 400 km in diameter, with summit calderas 20–25 km above datum, and gentle flank slopes, averaging less than 5°. Several other smaller shield volcanoes, such as Biblis Patera, Ceraunius Tholus and Tharsis Tholus, also occur in the Tharsis region (Fig. M39).

Alba Patera, north of Tharsis, is another giant volcano, over 6 km high and ~ 600 km across, with very gentle slopes (average 0.5°). Lava flows have been traced for over 1300 km from the summit caldera. The huge dimensions of these flows indicate very high rates of extrusion of extremely fluid lava (Cattermole, 1989). The volcano is partially encircled by ring fractures, which belong to a set radial to the Tharsis bulge.

Martian volcanoes are also found in Elysium and around the Hellas basin. The latter are ancient, low-relief, central vent constructs, which typically exhibit a series of valleys and ridges radial to the central caldera. Some of these may be highly eroded ash flow deposits, as for example Hadriaca Patera (31°S, 267°W), northeast of Hellas, or Tyrrhena Patera (22°S, 253°W; Greeley and Crown, 1990).

In addition to the central eruptive vents, Mars is covered by extensive volcanic plains, which constitute 58% of the surface (Tanaka et al., 1988). Among the older probable lava flows are the ridged plains, described above. Younger volcanic plains are associated with major volcanic provinces in Tharsis and Elysium, and around Alba Patera. In general, extrusion of lava in massive sheets preceded formation of the giant shield volcanoes. In Arcadia Planitia and Amazonis Planitia, lava flows and small, scattered cones are abundant, but large constructs are generally absent. Possible pyroclastic deposits have been identified in southern Amazonis Planitia (Scott and Tanaka, 1986). Although lava flows may be widespread in the northern lowland plains, eolian, fluvial and lacustrine deposits may have obscured any diagnostic morphologies.

Canyons

Valles Marineris is the collective name given to a vast system of equatorial canyons, trending roughly WNW–ESW, between 30° and 110°W longitude, on the eastern flank of the Tharsis bulge (Plates 16–18). The canyon system is nearly 4000 km long, and 600 km wide at the center, encompassing three partially interconnected parallel canyons, with maximal depths of 7–9 km. The alignment of the canyons has been structurally controlled to a large extent. The prevailing trend follows a set of fractures radial to the Tharsis bulge. The overall linearity of the canyon walls, parallelism with plateau grabens, and occurrence of triangular faceted spurs at the bases of the canyons indicate that they were primarily formed by faulting (Figure M40a). However, the canyons have been substantially enlarged by mass wasting, sapping and karst-like subsidence of closed depressions (Figure M40b; Spencer and Fanale, 1990).

Chains of nearly circular collapse pits lie along many grabens. These coalesce into elongate depressions, and ultimately widen into canyons (Figure M40a). The canyons are capped by erosion-resistant lava flows (ridged plains) overlying thicker, more friable strata. Canyon walls retreat largely by mass wasting. The most common process creates a spur and gully topography. In many places canyon walls are intersected by arcuate alcoves, which are the scars of huge landslides (Figures M40 and M41). The blocky to hummocky texture, grading downslope into fine-striated lobes is suggestive of massive debris slides that were fluidized by groundwater or melted ground ice (Lucchitta 1979). The main canyons are locally dissected by dendritic, V-shaped tributaries, with rounded sources (for example, on the southern walls of Ius Chasma). Some degree of structural control is implied by the nearly orthogonal alignment (Figure M40b). These features may have formed by headward sapping along lines of weakness (Sharp and Malin, 1975; Baker, 1982).

Finely stratified deposits, forming flat-topped mesas or irregular hills, occur near the center of the canyon floors. That these deposits are not downdropped parts of the wall rocks is indicated by distinctly different erosional styles and, in places, onlap of the layered sediments against the canyon walls. The layered sediments could represent lacustrine (Nedell, Squyres and Andersen, 1987) or mafic volcanic ash (Geissler, Singer and Laccitta, 1990) deposits.

The fluidized debris flows, karst-like subsidence features and layered (possibly lacustrine) canyon floor deposits imply that the stratigraphic units underlying the canyon region contained substantial amounts of subsurface water, ice and/or carbonates (Carr, 1986; Spencer and Fanale, 1990). The significance of this potential reservoir is discussed further below.

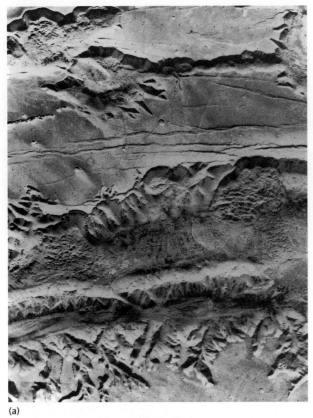

Figure M40 (a) The Ius Chasma section of Valles Marineris (3–8°S, 78–85°W). A series of grabens, collapse pits and closed depressions lie parallel to the main canyon. Note huge landslides along northern canyon rim (NASA). (b) Close-up of tributaries along the south rim of Ius Chasma, showing preferential erosion along lines of weakness. The frame is 110 km across (NASA).

Channels

Three types of channels have been recognized on Mars (Carr, 1981). These include:

1. valley networks (runoff channels, dendritic channels);
2. outflow channels;
3. fretted channels.

Valley network

Valley networks include numerous small, quasi-dendritic, often subparallel valleys, widely distributed throughout the heavily cratered southern highlands (Figure M35). The valleys represent an incipient drainage system, in that branching tributaries connect, and follow the local topography. However, the continuity of the valley networks has been disrupted in many places by subsequent erosion or burial. Thus the present valleys are probably relicts of a formerly more extensive network. While most of these valley networks are concentrated in densely cratered uplands, and are therefore believed to date back to an early period of high cratering rates, much younger channels occur locally, for example on the flanks of the volcano Alba Patera (Gulick and Baker, 1990). Although the widespread occurrence of the dendritic channels has been cited as evidence for a formerly wetter climate, with a denser atmosphere, permitting precipitation and surface runoff (Pollack *et al.*, 1987), these features could also be explained by groundwater or ice sapping under a climate similar to the present (Baker, 1982).

Outflow channels

The outflow channels originate from localized sources, generally from depressions or collapse zones within chaotic terrain at the eastern end of the canyons, where the crust has foundered and become highly fractured and dissected. Chaotic terrain grades directly into broad outflow channels, whose enormous dimensions and morphological features are difficult to explain by any other mechanism than by catastrophic flooding (Baker, 1982). The Kasei Valles System, northwest of Valles Marineris, for example, is over 3200 km long, up to 200 km across, and locally more than 3 km deep. Based on channel dimensions, maximum flood discharges exceeding 1 km^3 s^{-1} have been calculated, greater than any other known flood event on either Mars or Earth (Robinson and Tanaka, 1990). Geomorphological evidence for terrain scouring by vast floods includes braided channels, streamlined residual uplands (Figure M42), a low sinuosity and high width/depth ratio, pronounced widenings and constrictions of flow and stripping of surface regolith (Baker, 1982). These features, collectively, closely resemble the channeled scablands of eastern Washington State, USA, which formed by gargantuan floods that were released by the failure of ice-dammed Lake Missoula toward the end of the last glacial period (Baker, 1981).

The sudden release of huge volumes of water on Mars may have been triggered by rupturing a confined aquifer under high pore pressure by impacts, or by a major thermal event associated with Tharsis volcanism (Baker *et al.*, 1991). The anomalously high heat flux could have melted enormous volumes of ground-ice stored beneath the canyon region, activating hydrologic circulation and driving groundwater toward the Chryse basin lowlands, where subsidence and collapse of the chaotic terrain unleashed tremendous discharges toward the northern plains, causing large bodies of standing water to accumulate episodically (Parker, Saunders and Schneeberger, 1989; Baker *et al.*, 1991). These events occurred well after the end of the period of heavy bombardment (Table M6).

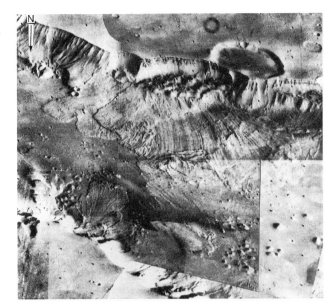

Figure M41 Massive landslide in Gangis Chasma section of Valles Marineris. Tilted slump blocks near the canyon wall give way downslope to finely striated flow lobes. The truncated crater on the canyon rim is around 17 km across (NASA).

Fretted channels

The last category of channels consists of irregularly shaped, flat-floored, steep-walled valleys, geographically concentrated within fretted terrain, mainly at the highland–lowland boundary in the northern hemisphere (Sharp, 1973; Sharp and Malin, 1975). The 'channels' or valleys represent more active zones of erosion within the fretted terrain. They are covered by debris blankets, probably produced by slow downslope movement of detritus mobilized by interstitial ice (Figure M43).

Glacial features

Landforms reminiscent of those produced by terrestrial glaciation appear widespread over much of the Martian southern highlands up to 33°S and also parts of the northern plains (Kargel and Strom, 1992). These landforms have been more thoroughly mapped in the regions surrounding the Argyre and Hellas basins, and include features interpreted as eskers, horns, cirques, aretes and also glacial outwash and glaciolacustrine plains. Although any individual feature may be explained by a non-glacial process, the regional association, ordered sequences of landforms and their relationships to topography are most consistent with a glacial hypothesis. Cratering studies suggest a middle Amazonian age (Johnson et al., 1991; Table M6). Ice may have covered between 4 and 18% of the Martian surface (Kargel and Strom, 1992).

Water reservoirs on Mars

Carr (1986, 1987) has estimated that Mars contains the equivalent of a 0.5 km thick layer of water, if spread uniformly over the surface. The major sink is near-surface ground ice at mid to high latitudes (greater than 30°N or S). Debris flows, fretted terrain and 'terrain softening' (i.e. smoothing of surface features, such as crater rims), suggest viscous creep of ice-rich regolith (Squyres, 1989). The Martian regolith consists primarily of a 1–2 km thick layer of relatively porous fragmental impact ejecta, overlying heavily fractured basement rock, that formed during the early intense bombardment (MacKinnon and Tanaka, 1989). Water, released to the surface by sapping or outflow at low latitudes, accumulated over the eons as ground ice in the regolith at higher latitudes (Carr, 1986; Fanale et al., 1986). Assuming average porosities of 10–20% and depths of 1 km, the high-latitude regolith could contain the equivalent of a 50–100-m planetwide water layer. At lower latitudes the evidence for

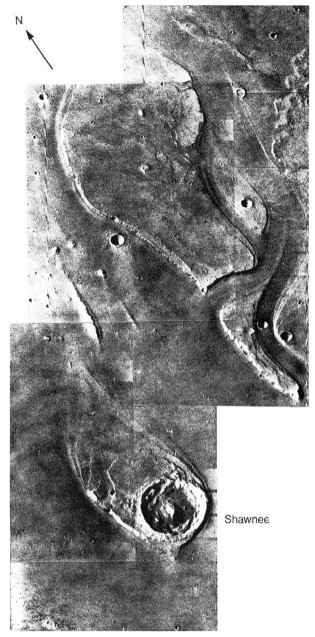

Figure M42 Streamlined islands in outflow channels, Ares Vallis, near the southern edge of Chryse Planitia. Shawnee crater is approximately 10 km across (NASA).

ground ice or other buried volatiles is indirect, but their presence is implied by the canyon enlargement mechanisms outlined above. Carr estimates another 25 m from the lower latitudes, bringing the upper regolith contribution to 75–125 m. The amount of erosion produced by the outflow channels suggests a water volume equivalent to another 350 m. Clay minerals formed by weathering or hydrothermal alteration of lava, volcanic ash or impact ejecta could chemically bind the equivalent of around 9 m of water. Finally, some water, exposed at the surface through slow seepage or channel formation, evaporated and recondensed into layered deposits at the poles. The planetwide water equivalent of the polar deposits is around 14 m. Greeley and Schneid (1991) estimate another 89 m, derived from 'juvenile' water released by outgassing from magmas. The overall total is thus between 537 and 587 m (Table M7).

Figure M43 Fretted channels, Nilosyrtis region (34°N, 200°W). The dark, NNW-trending valley is around 10 km across and 50 km long (NASA).

Table M7 Water inventory of Mars

Source	Amount (av. thickness of a planet-wide layer, m)	
High-latitude upper regolith (≤ 1 km)	50–100	Carr (1986)
Low-latitude upper regolith (≤ 1 km)	25	
Megaregolith (> 1 km)	350	↓
Clay minerals	9	
Polar layered deposits	14	
Subtotal	448–498	
Volcanism (juvenile H_2O)	89	Greeley and Schneid (1991)
Total	537–587	

Eolian features

Mars has a thin atmosphere (around 6–7 mbar), consisting of ∼ 95% CO_2, with lesser amounts of N_2 and Ar (see also Mars: atmosphere). Diurnal and seasonal temperature variations, as well as seasonal condensation and evaporation of CO_2 at the polar caps, set into motion the atmospheric circulation system. Winds, generated by warming in the southern hemisphere summer, close to perihelion, trigger dust storms that may rapidly (within days) expand to planetary scale.

Although Mars has clearly been subject to eolian activity for billions of years, the survival of the heavily cratered terrains attests to the relative inefficiency and geographical selectivity of wind as an agent of erosion. Average eolian erosion rates may not exceed 10^{-1} μm year^{-1} (Carr, 1981). Nevertheless, eolian processes have created numerous erosional and depositional features on Mars.

Yardangs provide unambiguous evidence for wind erosion. These are elongate, streamlined, wind-eroded ridges, which often occur in gently curved subparallel sets, aligned with the dominant wind direction, such as found in southern Amazonis, at 5–10°N, 135–185°W (Figure M44). 'Etched' and 'pitted' terrain, near the south pole, may have been stripped of friable or ice-rich surface layers by wind deflation. Alternatively, it may represent thermokarst (Tanaka and Scott, 1987). Pedestal craters, concentrated at high latitudes, may consist of more erosion-resistant ejecta blankets, which were selectively preserved, while intercrater material was removed by wind (Carr, 1981).

Depositional features are more widespread. Vast dark sand seas surrounding the north pole are the most prominent example (Figure M45). The north polar sand seas have been estimated to contain 1158 km³ of dune sediment (Lancaster and Greeley, 1990). Sand dunes also occur at low latitudes, in topographic traps such as craters, and in intercrater fields. Dune geometry (mean width, length and spacing) closely resembles terrestrial counterparts (Breed, Grolier

Figure M44 Yardangs, southern Amazonis (NASA).

and McCauley, 1979). The particle size most readily mobilized by the wind (∼ 0.1 mm), is also quite similar to that on Earth, although the threshold velocity to induce particle motion is significantly higher, because of the much lower Martian atmospheric pressure and gravity

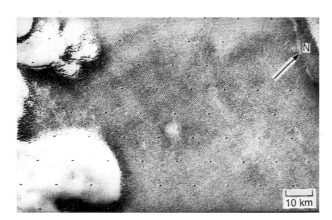

Figure M45 Sand dune fields, north polar region. Bright patches are residual ice. Note how sand dunes change orientation near ice patches, suggesting that the latter are older (NASA).

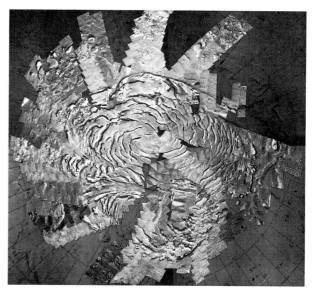

Figure M47 Overview of the north polar region, showing residual ice cap, dark frost-free slopes and valleys. Note the near-absence of craters in this region (NASA).

Polar regions

The seasonal changes on the Martian polar caps were first observed by telescopes. In the early 1970s Mariner space probes, confirmed that the seasonal ice caps are condensed CO_2. The maximum extent of the south polar cap is much greater than that of the north pole, because Mars is at aphelion during southern winter. Conversely, the relative warmth of southern summers, around perihelion, causes a more pronounced recession of the south polar cap. However, each pole has a residual cap that differs in size and composition. The north polar residual cap is composed chiefly of H_2O ice, with some admixed dust (Kieffer *et al.*, 1976). In contrast, the much smaller southern residual cap is mainly CO_2 ice, with very minor amounts of H_2O ice (Jakosky and Barker, 1984).

The permanent ice caps overlie a ~ 1–6 km thick sequence (Dzurisin and Blasius, 1975) of nearly crater-free, relatively flat, layered deposits which occupy most of the area poleward of 80° latitude, at both poles. These are among the youngest features on Mars (Table M6). The layered deposits are incised by numerous frost-free, stepped escarpments and valleys, spaced approximately 50 km apart, which spiral outward from the ice-covered poles [counterclockwise in the north (Figure M47), and clockwise in the south, the orientation being modulated by Coriolis forces; Carr, 1981; Tanaka and Scott, 1987]. Partially defrosted slopes display a fine, horizontal layering, 10–50 m thick, and continuous over hundreds of kilometers (Figure M48). The relative steepness of the exposed slopes (10–20°) suggests a fair degree of particle cohesiveness or cementation (Herkenhoff and Murray, 1990a).

It is generally agreed that the layered deposits represent eolian accumulation of dust and water ice over many cycles of climate change, caused by variations in the planet's orbital motion (Cutt, Blasius and Roberts, 1979; see also Mars: atmosphere). Deposition rates are estimated at ~ 10^{-1} mm year^{-1} (Plaut *et al.*, 1988). The presence of dark dunes, near scarps, in the north polar layered deposits suggests some admixture of sand (~ 1–10% by volume), together with the dust and ice. The color and albedo of the polar sand matches that of dune fields elsewhere on Mars (Thomas and Weitz, 1989; Herkenhoff and Murray, 1990b). The northern polar layered deposits overlie a smooth, plains-forming mantle deposit, probably dust, whereas in the south polar regions the layered deposits overlie densely cratered terrain, etched and pitted terrain and plains, which may have been glaciated (Kargel, Strom and Johnson, 1991).

The northern polar layered deposits are encircled by a circumpolar sand sea (or erg), north of 75°N latitude, which is visible, telescopically, as a dark 'collar' around the north polar cap, appearing as the seasonal ice retreats in spring. Transverse dunes account for almost

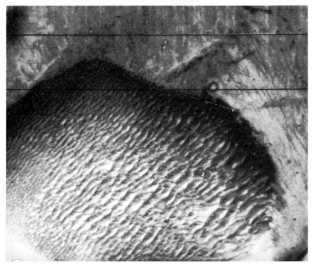

Figure M46 Dark dune field within 200 km diameter crater, Hellespontus region (NASA).

(Greeley and Iversen 1985). A smooth to hummocky deposit, up to several hundred meters thick, probably wind-blown dust, covers the northern lowland plains north of 66° N.

Bright and dark crater 'tails' and 'splotches' indicate ongoing eolian activity, in that changes have been observed over a period of several years. Bright streaks or tails from by deposition of a thin, light veneer of dust in the lee of topographic obstacles, such as crater rims (Greeley and Iversen 1985). Dark streaks result from removal of bright dust, due to turbulent motion downwind of a crater. Mixed-pattern streaks suggest multiple wind events. 'Splotches' are dark patches localized within craters, but occasionally extending beyond the crater rim. High-resolution imagery resolves the splotches into dune fields (Figure M46).

Comparison of streak patterns between the dust storms of 1971 and 1977 shows an overall constancy of the global circulation. Most bright streaks are produced by winds that maintain a fairly persistent orientation over a period of several years. On the other hand, dark streaks are more readily modified by dust fallout over a large area, followed by selective removal in the lee of craters (Thomas and Veverka, 1979).

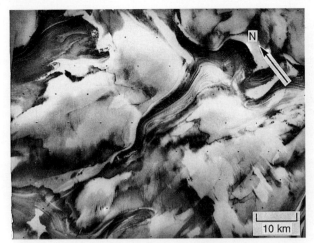

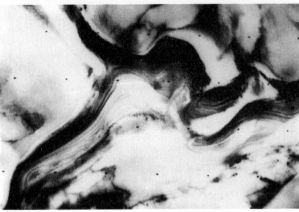

Figure M48 (a) Detailed view of layered deposits, north polar region. (b) Enlargement of upper right hand corner of Figure M48a. Note unconformity in finely layered sequence, which implies a period of erosion interrupting the depositional record, probably related to a change in the climate regime (NASA).

half the sand sea area of around 7×10^5 km^2 (Tsoar, Greeley and Peterfreund, 1979). Dune spacing and orientation remain nearly constant over hundreds of square kilometers. Barchan dunes occur on the southern periphery of the transverse dunes and in isolated fields. The dunes overlie smooth plains and polar layered deposits (Plaut et al., 1988). Dune materials may have been eroded from both units. Polar wind circulation patterns are complex, with seasonal changes in direction occurring both toward and away from the poles (Tsoar, Greeley and Peterfreund, 1979).

Surface composition

The view of the Martian surface from the Viking landers is one of a rock-strewn desert. While the Viking 1 landing site in Chryse Planitia (22.5°W, 47.8°N) reveals a gently rolling topography with encrusted, fine-grained drifts accumulating in the lee of large boulders (Figure M49), the Viking 2 site in Utopia Planitia (48°N, 225.6°W) is flatter, lacks drifts and is covered by angular, blocky, pitted rocks that are believed to be ejecta from a 100 km diameter crater 180 km to the east.

The chemical composition of the surface was analyzed by x-ray fluorescence spectrometers at the two Viking Lander sites (Clark et al, 1977). The remarkably similar results, in spite of the considerable separation of the sites, implies that there has been thorough mixing of the fine-grained component of the regolith, presumably by planetary dust storms.

The low SiO$_2$ (43–45% by weight) and high iron (18–20%) contents suggest derivation from mafic igneous parent rocks. On the other hand, the CaO (5.0–5.6%) and anomalously low Al$_2$O$_3$ (5–6%) contents are difficult to reconcile with any known terrestrial igneous rock. Yet the abundant SO$_3$ (6.5–9.5%) and Cl (0.6–0.9%) indicate a fairly substantial amount of volatiles. The instrument, however, cannot detect low atomic weight elements, such as sodium, nitrogen and carbon. Another disadvantage is that the mineralogy must be inferred indirectly. One plausible model is a mixture of iron-rich smectite (montmorillonitic) clay, with ferric oxides, carbonates and sulfates (Toulmin et al., 1977). The iron-rich clay may be a weathering or alteration product of palagonite, or basaltic volcanic glass, formed by interaction between lava and ice-rich regolith.

Additional information on surface composition comes from Earth-based telescopic and Viking orbiter spectral measurements. The pronounced red color of Mars has long suggested the presence of surficial ferric oxides. Furthermore, the brighter regions tend to be redder than the dark ones. Strong absorption band below 0.5 μm, and much weaker absorption features between 0.7 and 0.95 μm, detected in reflectance spectra (Singer, McCord and Clark, 1979), are consistent with the presence of ferric oxides. Faint bands near 1 μm suggest Fe^{+2} absorptions in mafic silicate minerals, such as pyroxene or olivine (Bibring et al., 1990). Many potential iron-bearing phases have been proposed, including goethite (αFeOOH), limonite (Fe$_2$O$_3 \cdot n$H$_2$O), Fe^{+3}-smectite clay, weathered palagonite, and maghamite (a magnetic form of iron oxide, γFe$_2$O$_3$). More precise spectral measurements clearly indicate several distinct absorption bands around 0.68 μm and 0.85 μm, attributed to crystalline hematite (αFe$_2$O$_3$; Bell, McCord and Owensby, 1990). However the match between the Martian spectra and terrestrial iron oxides or palagonite is not exact (Bell, McCord and Owensby, 1990). A close fit to the Martian data can also be obtained with a thin ferric oxide coating on weathered basalt (Figure 5 in Singer, McCord and Clark, 1979). The latter interpretation is consistent with the overall mafic surface composition, vesicular basalt-like appearance of the surface rocks, and photogeologic evidence for widespread mafic-type volcanism.

Spectral features around 3–4 μm have been ascribed to condensed water (Houck et al., 1973), or hydrated minerals (Bibring et al., 1990). Gas chromatograph–mass spectrometer analyses on the Viking Landers provided more direct indications for adsorbed and/or hydrated water on the Martian surface (Biemann et al., 1977). The presence of silicates in atmospheric dust was inferred from broad adsorption bands near 10 μ (Hanel et al., 1972). Other emission and absorption features in the 5.4–10.5 μm wavelength region have been tentatively identified with sulfates and carbonates (Pollack et al. 1990). If confirmed, these observations would constitute more direct evidence for the presence of minerals anticipated on theoretical grounds.

Summary of the geologic history of Mars

The geologic history of Mars can be divided into three major epochs – the Noachian, a period of intense impact bombardment, succeeded by the Hesperian, during which time the impact flux was rapidly declining, and lastly, the Amazonian, an interval of very low impact flux (Table M6).

The oldest Noachian basement, which dates back to the late stages of planetary accretion and differentiation (4.0–4.6 billion years ago), consists of highly fractured ridges and massifs that form the rims of large impact basins such as Hellas, Argyre and Isidis Planitia. Resurfacing by volcanic and fluvial processes began during the period of heavy cratering that produced most of the southern highlands. Lava flows and eolian deposits accumulated in smooth, intercrater plains toward the end of the Noachian Epoch.

Vast outpouring of lava flows from fissure eruptions, forming the ridged plains, characterized the early Hesperian. The onset of tectonism of the Tharsis bulge began during this period, which was succeeded by a major episode of volcanism in the Tharsis, Alba Patera, Syrtis Major and Elysium regions, as well as continued deformation of the Tharsis area and Valles Marineris. The thermal flux associated with this volcanotectonic activity may have activated groundwater circulation, leading to episodic break-outs of floodwaters, forming the chaotic terrain, outflow channels and sedimentary deposits of the northern plains (Parker, Saunders and Schneeberger, 1989; Baker et al., 1991).

As the cratering flux waned in the Amazonian Epoch, other processes predominated. Vast flood lavas were extruded in Tharsis

Figure M49 Viking 2 landing site, Utopia Planitia (NASA).

and Arcadia Planitia. A late episode of flooding and channel carving took place, and much of the southern highlands as far as 40°S were glaciated (Kargel et al., 1991; Johnson et al., 1991; Strom et al., 1991). Toward the end of the Amazonian, the style of volcanism on Tharsis shifted toward eruption of giant shield volcanoes (although smaller, older shields and domes occur there, and elsewhere on Mars). The late Amazonian is the only period when eolian processes were dominant (Tanaka et al., 1988). Climate-modulated dust and ice deposits accumulated in the polar layered deposits, covered by more recent residual ice caps, chiefly H_2O ice at the north pole and CO_2 ice at the south pole. Geologically recent and ongoing eolian activity is eroding the northern plains and polar layered deposits, forming a vast circumpolar sand sea.

Vivien Gornitz

Bibliography

Baker, V.R. (ed.) (1981) *Catastrophic Flooding: The Origin of the Channeled Scabland*. Stroudsburg, PA: Dowden, Hutchinson and Ross.

Baker, V.R. (1982) *The Channels of Mars*: Austin: University of Texas Press.

Baker, V.R., Strom, R.G., Gulick, V.C., et al., (1991) Ancient oceans, ice sheets and the hydrological cycle on Mars. *Nature*, **352**, 589–94.

Bell, J.F., III, McCord, T.B. and Owensby, P.D. (1990) Observational evidence of crystalline iron oxides on Mars. *J. Geophys. Res.*, **95**(B9), 14447–61.

Bibring, J.P., Combes, M., Langevin, Y. et al. (1990) ISM observations of Mars and Phobos: first results. Proc. *Lunar and Planetary Sci. Conf.* Houston: Lunar and Plantary Institute, **20**, pp. 461–71.

Biemann, K., Oro, J., Toulmin, P., et al. (1977) The search for organic substances and inorganic volatile compounds in the surface of Mars. *J. Geophys. Res.*, **82**, 4641–58.

Breed, C.S., Grolier, M.J. and McCauley, J.F. (1979) Morphology and distribution of common 'sand' dunes on Mars: comparison with the earth. *J. Geophys. Res.*, **84**(B14), 8183–204.

Carr, M.H. (1981) *The Surface of Mars*. New Haven: Yale University Press.

Carr, M.H. (1986) Mars: a water-rich planet? *Icarus*, **68**, 187–216.

Carr, M.H. (1987) Water on Mars. *Nature*, **326**, 30–5.

Cattermole, P. (1989) *Planetary Volcanism: A Study of Volcanic Activity in the Solar System*. Chichester, England: Ellis Horwood Ltd.

Clark, B.C., III, Baird, A.K., Rose, H.J., Jr et al. (1977) The Viking x-ray fluorescence experiment: analytical methods and early results. *J. Geophys. Res.*, **82**, 4577–94.

Cutts, J.A., Blasius, K.R. and Roberts, W.J. (1979) Evolution of Martian polar landscapes: interplay of long-term variations in perennial ice cover and dust storm intensity. *J. Geophys. Res.*, **84**, 2975–94.

Dzurisin, D. and Blasius, K.R. (1975) Topography of the polar layered deposits of Mars. *J. Geophys. Res.*, **80**, 3286–306.

Fanale, P.F., Salvail, J.R., Zent, A.P. and Postawko, S.E. (1986). Global distribution and migration of subsurface ice on Mars. *Icarus*, **67**, 1–18.

Frey, H.V. and Schultz, R.A. (1990) Speculations on the origin and evolution of the Utopia–Elysium lowlands of Mars. *J. Geophys. Res.*, **95**(B9), 14203–13.

Frey, H. and Schultz, R.A. (1991) Geologic and topographic constraints on the origin and development of the Martian crustal dichotomy: what they do and don't require, in *Lunar Planet. Sci. XXII*, Abstracts of papers submitted to the 22nd Lunar and Planet. Sci. Conf., Houston, TX, March 18–22, 1991, pp. 417–418.

Geissler, P.E., Singer, R.B. and Luccitta, B.K. (1990) Dark materials in Valles Marineris: indications of the style of volcanism and magmatism on Mars. *J. Geophys. Res.*, **95**(B9) 14399–413.

Greeley, R. and Crown, D.A. (1990) Volcanic geology of Tyrrhena Patera, Mars. *J. Geophys. Res.*, **95**(B5), 7133–49.

Greeley, R. and Guest, J.E. (1987) Geologic map of the equatorial region Mars, MAP I – 1802B, 1: 15 000 000, US Geological Survey.

Greeley, R. and Iversen, J.D. (1985) *Winds as a Geological Process on Earth, Mars, Venus and Titan*. Cambridge: Cambridge University Press.

Greeley, R. and Schneid, B. (1991) Magma generation on Mars: estimated volumes through time, in *Lunar Planet. Sci. XXII*. Abstracts of papers submitted to the 22nd Lunar and Planet. Sci. Conf., Houston, TX, March 18–22, 1991, pp. 489–490.

Gulick, V.C. and Baker, V.R. (1990) Origin and evolution of valleys on Martian volcanoes. *J. Geophys. Res.*, **95**(B9), 14325–44.

Hanel, R., Conrath, B., Hovis, W., et al. (1972) Investigation of the Martian environment by infrared spectroscopy on Mariner 9. *Icarus*, **17**, 428–31.

Herkenhoff, K.E. and Murray, B.C. (1990a) High-resolution topography and albedo of the south polar layered deposits on Mars. *J. Geophys. Res.*, **95**(B9), 14511–29.

Herkenhoff, K.E. and Murray, B.C. (1990b) Color and albedo of the south polar layered deposits on Mars. *J. Geophys. Res.*, **95**(B2), 1343–58.

Houck, J.R., Pollack J.B., Sagan, C., et al. (1973) High altitude infrared spectroscopic evidence for bound water on Mars. *Icarus*, **18**, 470–80.

Jakosky, B.M. and Barker, E.S. (1984) Comparison of ground-based and Viking orbiter measurements of Martian water vapor: variability of the seasonal cycle. *Icarus*, **57**(3), 322–34.

Johnson, N., Kargel, J.S., Strom, R.G. and Knight, C. (1991) Chronology of glaciation in the Hellas region of Mars, in *Lunar Planet. Sci. XXII*. Abstracts of papers submitted to the 22nd Lunar and Planet. Sci. Conf., Houston, TX, March 18–22, 1991, pp 651–652.

Kargel, J.S., and Strom, R.G. (1992) Ancient glaciation on Mars. *Geology*, **20**, 3–7.

Kargel, J.S., Strom, R.G. and Johnson, N. (1991) Glacial geology of the Hellas region on Mars, in *Lunar Planet. Sci. XXII*, pp. 678–88.

Kieffer, H.H., Chase, S.C. Jr, Martin, T.Z., *et al.* (1976) Martian north pole summer temperatures: dirty water ice. *Science*, **194**, 1341–44.

Lancaster, N. and Greeley, R. (1990) Sediment volume in the north polar sand seas of Mars. *J. Geophys. Res.*, **95**(B7), 10921–7.

Lucchitta, B.K. (1979) Landslides in Valles Marineris, Mars. *J. Geophys. Res.*, **84**(B14), 8097–113.

MacKinnon, D.J. and Tanaka, K.L. (1989) The impacted Martian crust: structure, hydrology, and some geologic implications. *J. Geophys. Res.*, **94**(B12), 17359–370.

McGill, G.E. and Dimitriou, A.M. (1990) Origin of the Martian global dichotomy by crustal thinning in the late Noachian or early Hesperian. *J. Geophys. Res.*, **95**(B8), 12595–605.

Moore, J.G., Clague, D.A., Holcomb, R.T. *et al.* (1989) Prodigious submarine landslides on the Hawaiian Ridge. *J. Geophys. Res.*, **94**(B12), 17465–84.

Nedell, S.S., Squyres, S.W. and Andersen, D.W. (1987) Origin and evolution of the layered deposits in the Valles Marineris, Mars. *Icarus*, **17**, 409–41.

Parker, T.J., Saunders, R.S. and Schneeberger, D.M. (1989) Transitional morphology in west Deuteronilus Mensae, Mars: implications for modification of the lowland/upland boundary. *Icarus*, **82**, 111–45.

Plaut, J.J., Kahn, R., Guinness, E.A. and Arvidson, R.E. (1988) Accumulation of sedimentary debris in the south polar region of Mars and implications for climate history. *Icarus*, **76**, 357–77.

Pollack, J.B., Kasting, J.F., Richardson, S.M. and Poliakoff, K. (1987) The case for a wet, warm climate on early Mars. *Icarus*, **71**, 203–24.

Pollack, J.B., Roush, T., Witteborn, F., *et al.* (1990) Thermal emission spectra of Mars (5.4–10.5 μm): evidence for sulfates, carbonates and hydrates. *J. Geophys. Res.*, **95**(B9), 14595–627.

Robinson, M.S. and Tanaka, K.L. (1990) Magnitude of a catastrophic flood event at Kasei Valles, Mars. *Geology*, **18**(9), 902–5.

Scott, D.H. and Tanaka, K.L. (1986) Geologic map of the western equatorial region of Mars, Map I-1802A, 1:15 000 000. US Geological Survey.

Sharp, R.P. (1973) Mars, fretted and chaotic terrain, *J. Geophys. Res.*, **78**, 4073–83.

Sharp, R.P. and Malin, M.C. (1975) Channels on Mars. *Geol. Soc. Am. Bull.*, **86**, 593–609.

Shoemaker, E.M. and Shoemaker, C. (1990) The collision of solid bodies, in *The New Solar System* (eds J.K. Beatty and A. Chaikin). Cambridge: Cambridge University Press, pp. 259–74.

Singer, R.B., McCord, T.B. and Clark, R.N. (1979) Mars surface compostion from reflectance spectra: a summary. *J. Geophys. Res.*, **84**(B14), 8415–26.

Spencer, J.R. and Fanale, F.P. (1990) New models for the origin of Valles Marineris closed depressions. *J. Geophys. Res.*, **95**(B9), 14301–13.

Squyres, S.W. (1989) Urey Prize Lecture: Water on Mars. *Icarus*, **79**, 229–88.

Strom, R.G., Kargel, J.S., Johnson, N. and Knight, C. (1991) Glacial and marine chronology on Mars, in *Lunar Planet. Sci. XXII*. Abstracts of papers submitted to the 22nd Lunar and Planet. Sci. Conf., Houston, TX, March 18–22, 1991, pp. 1351–1352.

Tanaka, K.L. (1986) The stratigraphy of Mars. *J. Geophys. Res.*, **91**(B13), E139–58.

Tanaka, K.L. and Scott, D.H. (1987) Geologic map of the polar regions of Mars, Map I-1802C, 1: 15 000 000, US Geological Survey.

Tanaka, K.L., Isbell, N.K., Scott, D.H. *et al.* (1988). The resurfacing history of Mars: a synthesis of digitized, Viking-based geology. In *Proc. Lunar Planet. Sci. Conf.* **18** (ed. G. Ryder). Cambridge University Press and Lunar Planetary Institute, pp. 665–78.

Thomas, P. and Veverka, J. (1979) Seasonal and secular variation of wind streaks on Mars: an analysis of Mariner 9 and Viking data. *J. Geophys. Res.*, **84**(B14), 8131–46.

Toulmin, P., III, Baird, A.K., Clark, B.C. *et al.* (1977) Geochemical and mineralogical interpretation of the Viking inorganic chemical results. *J. Geophys. Res.* **82**, 4625–34.

Tsoar, H., Greeley, R. and Peterfreund, A.R. (1979) Mars: the north polar sand sea and related wind patterns. *J. Geophys. Res.*, **84**(B14), 8167–80.

Wilhelms, D.E. and Squyres, S.W. (1984) The Martian hemisphere dichotomy may be due to a giant impact. *Nature*, 309, 138–40.

Wise, D.U., Golombek, M.P. and McGill, G.E. (1979) Tectonic evolution of Mars. *J. Geophys. Res.*, **84**, 7934–9.

Acknowledgements

The author wishes to thank Dr Michael H. Carr, Viking Team Leader. The images included here were provided by the National Space Science Data Center at NASA Goddard Space Flight Center.

Cross references

Eolian transport
Impact cratering
Polar cap
Volcanism in the solar system

MARS: GRAVITY

Mars has a mean surface gravity 2.63 times smaller than the Earth, at about 372 cm s^{-2}. However, it is the variation about this mean value that is of interest to the scientist, who uses this information to deduce the character of the internal structure of Mars (plate 19). Figure M50 displays the gravity variations in milligals (1000 mgal = 1 cm s^{-2}), where the large oblateness (J_2) due to the fast rotation rate of Mars has been removed. Several regional patterns are evident. Note the very densely packed contours at longitude −135° (this map has positive east longitudes, whereas other maps may have positive west longitudes). This feature is Olympus Mons, the largest volcanic structure on Mars, which generates the largest gravity anomaly amplitude (1500 mgal) of any feature on the Earth, Moon or Venus. In this same area, from −140° to −80° longitude, there is a large island of gravity highs associated with other volcanic features (Ascraeus Mons, Pavonis Mons, Arsia Mons and Alba Patera). This region, including Olympus Mons, is Tharsis, a high plateau that dominates Mars. To the east is a gravity low associated with Chryse Planitia and the outfall of Valles Marineris (the Mars Grand Canyon). At 85°E longitude and 10°N latitude there is a gravity high in the Isidis Basin, completely contrary to what would be expected for a topographic low (see Mascon). At 145°E and 20°N is another gravity high which is produced by Elysium Mons, a volcanic structure. The large amplitude of all these gravity variations have led scientists to conclude that Mars is a relatively cold body, having a thick and rigid crust capable of withstanding large loads.

These gravity data were obtained from the Doppler radio tracking of the United States Mariner 9 and Viking 1 and 2 orbiting spacecraft. Detailed explanations of the reduction and interpretation of these data are given in the references.

William L. Sjogren

Bibliography

Balmino, G., Maynot, B. and Vales, N. (1982) Gravity field model of Mars in spherical harmonics up to degree and order eighteen. *J. Geophys. Res.*, **87**, 9735–46.

Gapcynski, J.P., Tolson, R. and Michael, W.H., Jr (1977) Mars gravity field: combined Viking and Mariner 9 result, *J. Geophys. Res.*, **82**, 4325–7.

Jordan, J.F. and Lorell, J. (1975) Mariner 9P: an instrument of dynamical science. *Icarus*, **25**, 146–65.

Reasenberg, R.D. (1977) The moment of inertial and isostasy of Mars. *J. Geophys. Res.*, **82**, 369–75.

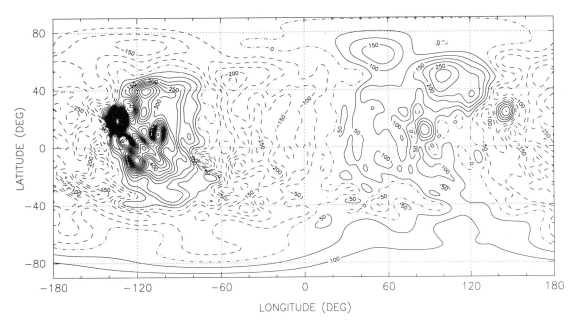

Figure M50 Mars gravity at the surface in milligals. (D.E. Smith *et al.*, *J. Geophys. Res.*, 1994. Copyright American Geophysical Union.)

Sjogren, W.L. (1979) Mars gravity: high resolution results from Viking Orbiter 2. *Science*, **203**.

Smith, D.E., Lerch, F.J., Nerem, R.S., *et al.* (1994) An improved gravity model for Mars: Goddard Mars model 1 (GMM-1). *J. Geophys. Res*, **98**, 20871–89.

Cross references

Gravimetry
Gravity fields of the terrestrial planets
Moon: gravity
Surface gravity
Venus: gravity

MARS: IMPACT CRATERING

Impact craters, created by collisions between two objects, were the major features identified in the 22 images sent to Earth by the first Martian spacecraft, Mariner 4, in 1966. In 1969 the Mariner 6 and 7 spacecraft returned over 200 additional pictures of the Martian surface and confirmed the Mariner 4 analysis that craters were the planet's dominant features. The view that Mars was a cratered, geologically dead world changed drastically in 1972 with images from the orbiting Mariner 9 spacecraft. Mariner 9 arrived at Mars in November 1971, only to be greeted by a global-scale dust storm which obscured the surface of the planet. As the dust began to settle in early 1972, Mariner 9 cameras revealed the presence of gigantic volcanoes, young plains, huge canyons, and channels formed by flowing water, in addition to impact craters. This new view of Mars rekindled interest in the planet and in 1976 the Viking 1 and 2 spacecraft, each composed of an orbiter and a lander, began an in-depth study of Mars which lasted for up to 6 years.

The imagery from these spacecraft revealed that, although Mars displays a variety of geologic features, impact craters are the dominant landform on the planet. More than 42 000 craters exist on the planet, but they are not evenly distributed. A great circle tilted approximately 30° to the equator in the western hemisphere divides the planet into a southern, heavily cratered region and a northern, lightly cratered plains region (Figure M51). The cause of this dichotomy is unknown: theories include an internal origin (Wise, Golombek and McGill, 1979), an ancient gigantic impact basin in the north (Wilhelms and Squyres, 1984), or a large number of overlapping basins in the north (Frey and Schultz, 1988).

Impact crater morphology and morphometry

The physics of impact crater formation is essentially the same throughout the solar system, with minor modifications occurring due to environmental effects (see Impact cratering). The general progression of crater morphology with increasing size is from simple, bowl-shaped craters to craters with more complex interiors to multiple ring basins. Simple craters have a depth–diameter ratio of approximately 0.2, while complex craters are shallower, approaching a value of 0.04 (Pike, 1988), although regional variations are seen (Hayashi-Smith and Mouginis-Mark, 1990; Barlow, 1992). The transition from complex craters to basins occurs around 50–70 km diameter. Unfortunately, all large Martian basins date from the early bombardment period and many are difficult to identify because of subsequent degradation caused by the active weathering environment on Mars (Schultz, Schultz and Rogers, 1982).

Although most Martian impact craters appear circular, a significant number are elliptical. Elliptical craters are rare on the Moon and Mercury – a study by Schultz and Lutz-Garihan (1982) suggests that Mars has a large number of elliptical craters because it was once surrounded by a large number of small moons. The orbits of these moons decayed with time, causing the objects to strike the Martian surface at an angle close to horizontal. Elliptical craters on Mars have received much attention in recent years as the potential originating sites of the SNC meteorites (Wood and Ashwal, 1981; Nyquist, 1983; Mouginis-Mark *et al.*, 1991).

Most morphologic features associated with Martian craters, such as central peaks and wall terraces, are similar to those features seen in lunar and Mercurian craters. Observed differences are attributable to unique characteristics of the Martian environment. Craters less than about 5 km in diameter display the simple, bowl-shaped morphologies characteristic of small craters (Pike, 1988). Martian craters larger than about 5 km in diameter display a range of interior features. Central peaks are common (Wood, Head and Cintala, 1978) and the peaks appear larger than peaks in lunar or Mercurian craters of identical size (Hale and Head, 1981). This result, together with the observation that central pits are more common in Martian craters (Wood, Head and Cintala, 1978), suggests that subsurface volatiles may be responsible. Concentrations of central pit craters have been reported in localized regions of the planet and may indicate regional reservoirs of subsurface volatiles (Barlow and Bradley, 1990).

Ejecta blanket morphologies of fresh Martian impact craters are distinctly different from ejecta morphologies of lunar and Mercurian craters, which are typically surrounded by a radial ejecta blanket,

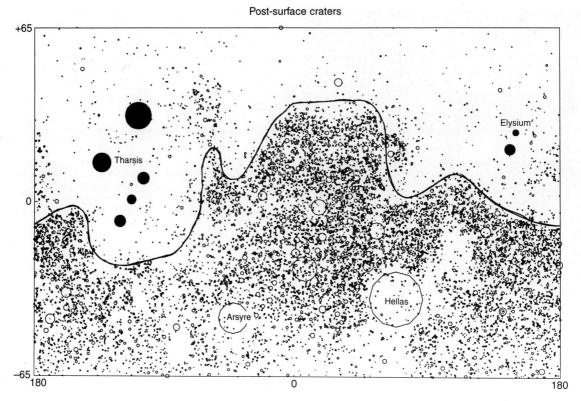

Figure M51 Map showing the distribution of all Martian impact craters greater than 15 km in diameter between ± 65° latitude. The solid line represents the approximate boundary between the heavily cratered southern hemisphere and the lightly cratered northern hemisphere. Large filled circles are major volcanoes.

emplaced by material ejected along ballistic trajectories from the primary crater. Most Martian craters, however, are surrounded by a lobate ejecta blanket emplaced by flow processes (Figure M52). Either heating of near-surface volatiles (Carr et al., 1977; Greeley et al., 1980) or the thin Martian atmosphere (Schultz and Gault, 1979) may provide the fluidizing medium in which the ejecta is entrained. Numerous studies have found correlations between ejecta morphology and crater size, location, terrain type and elevation (see Barlow and Bradley, 1990, for review). These studies provide strong support for the theory that differences in ejecta morphology result from impact into target material with varying concentrations of volatiles. However, the recent discovery of lobate ejecta morphologies surrounding impact craters on Venus has rekindled interest in the ejecta–atmosphere interaction theory, since Venus is very dry but does have an extensive atmosphere.

Impact craters on Mars display a range of degradation characteristics, from very fresh to almost completely destroyed (Figure M53). Craters are affected by a variety of geologic processes, including impact, volcanic, tectonic, fluvial, aeolian and glacial activity. Studies of crater size–frequency distributions in various regions of the planet suggest that crater obliteration by erosion and/or deposition was more common in the past that at present (Chapman and Jones, 1977). The greater incidence of highly degraded craters and the presence of dendritic channels in the heavily cratered southern hemisphere have led researchers to suggest that Mars had a thicker atmosphere during the period of heavy bombardment (Sagan, Toon and Gierasch, 1973; Pollack et al., 1987). The passage of large basin-forming objects through this atmosphere during heavy bombardment is thought to have eroded the atmosphere to its present tenuous state (Melosh and Vickery, 1989)

Although craters on younger units show less degradation than craters in older areas, low rates of obliteration have occurred across Mars since the end of heavy bombardment. Regional variations in degradation have been studied using both crater size–frequency distribution data (Tanaka et al. 1988; Grant and Schultz 1990) and morphometric measurements of craters (Craddock and Maxwell, 1990; Barlow, 1992). The results of these studies reveal the import-

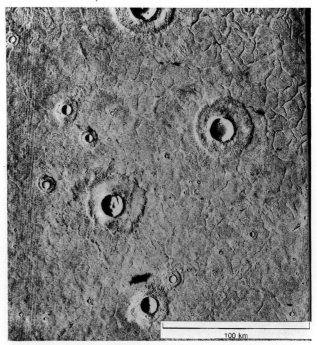

Figure M52 Examples of Martian impact craters. The fluidized appearance of the ejecta blankets surrounding the larger craters may result either from impact into subsurface volatiles or from atmosphere–ejecta interactions. (Viking Orbiter Frame 538A03.)

ance of volcanism, tectonism, fluvial, eolian and glacial processes as recent agents of degradation in localized areas of Mars.

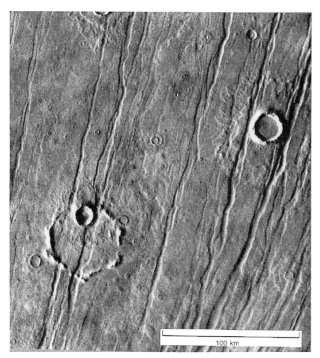

Figure M53 Examples of Martian impact craters exhibiting varying degrees of degradation. (Viking Orbiter Frame 701B82.)

Martian chronologies

Crater density, the number of craters per unit area, is greater for older regions than for younger areas. A lunar crater chronology has been established by comparing crater density with the radiometric ages derived for lunar samples returned by the Apollo astronauts. This lunar chronology indicates that the inner solar system has experienced two distinct episodes of impact cratering. During the first billion years of solar system history, planets and moons were exposed to an intense period of impact cratering (called the 'early bombardment or 'heavy bombardment period'), when impact rates were on the order of 100 to 500 times higher than at present. The impacting objects during this period resulted from material remaining after the formation of the solar system 4.5×10^9 years ago. After about 3.8×10^9 years ago the cratering rate decreased dramatically to its present low rate. Impact craters formed during this more recent time result from collisions with asteroids and comets (Shoemaker *et al.*, 1979).

Martian terrains display the same types of crater size–frequency distributions as do terrains on the Moon, indicating that both objects were exposed to the same impacting populations (Strom, Croft and Barlow, 1992). About 60% of the Martian surface appears to have formed during the heavy bombardment period (Tanaka, 1986; Barlow, 1988). The lightly cratered northern plains formed during the more recent post-heavy bombardment era, with the Tharsis volcanic region in western Mars being the youngest geologic unit on the planet.

Absolute ages for geologic units on Mars can be extrapolated from the lunar crater chronology provided three conditions are met (1) both objects were cratered by the same impact populations, (2) the end of heavy bombardment at both worlds is known and (3) the Martian impact rate is a known factor of the lunar impact rate. Crater statistical analysis indicates that condition 1 is met for Mars, and most dynamical models indicate that the heavy bombardment period ceased around 3.8×10^9 years ago throughout the inner solar system (Wetherill, 1975). Considerable controversy has existed regarding condition 3, however. Since Mars is closer than the Moon to the asteroid belt, we expect the cratering rate from asteroids to be higher at Mars. Estimates range from six times the lunar rate to the same as the lunar rate (Neukum and Wise, 1976; Neukum and Hiller, 1981), with twice the lunar rate being the most likely value (Strom, Croft and Barlow, 1992). Because of the range in estimates for the martian cratering rate, absolute ages derived for specific features on the planet can vary widely – Olympus Mons, the youngest volcano on Mars, is estimated to be between 3.3×10^9 and 0.1×10^9 years old. The current best model assumes that the Martian and lunar cratering rate were identical during heavy bombardment, but since the end of heavy bombardment the Martian cratering rate has been twice the lunar rate (Neukum and Hiller, 1981). Using this model, Olympus Mons is estimated to be 300 million years old.

Summary

Impact craters, ranging in freshness from pristine to very degraded, are the dominant geologic features on the Martian surface. Variations in crater morphology are attributed to environmental factors unique to Mars, such as the presence of subsurface volatiles and an atmosphere. Crater density has been used to determine relative and absolute chronologies for the various terrain units found across the planet. The results of these studies indicate that Mars has had an active geologic history, with volcanism occurring as recently as perhaps a few million years ago.

Nadine G. Barlow

Bibliography

Barlow, N.G. (1988) Crater size-frequency distributions and a revised Martian relative chronology. *Icarus*, **75**, 285–305.

Barlow, N.G. (1992) Quantifying crater degradation in Maja Valles and Memnonia, Mars, in *Lunar Planet. Sci. XXIII*, pp. 63–64.

Barlow, N.G. and Bradley, T.L. (1990) Martian impact craters: correlations of ejecta and interior morphologies with diameter, latitude, and terrain. *Icarus*, **87**, 156–179.

Carr, M.H., Crumpler, L.S., Cutts, J.A. *et al.* (1977) Martian impact craters and emplacement of ejecta by surface flow. *J. Geophys. Res.*, **82**, 4055–65.

Chapman, C.R. and Jones, K.L. (1977) Cratering and obliteration history of Mars. *Ann. Rev. Earth Planet. Sci.*, **5**, 515–40.

Craddock, R.A. and Maxwell, T.A. (1990) Resurfacing of the Martian highlands in the Amenthes and Tyrrhena regions. *J. Geophys. Res*, **95**, 14265–78.

Frey, H. and Schultz, R.A. (1988) Large impact basins and the mega-impact origin for the crustal dichotomy on Mars. *Geophys. Res. Lett.*, **15**, 229–32.

Grant, J.A. and Schultz, P.H. (1990) Gradational epochs on Mars: evidence from west-northwest of Isidis Basin and Electris. *Icarus*, **84**, 166–95.

Greeley, R., Fink, J., Gault, D.E. *et al.* (1980) Impact cratering in viscous targets: laboratory experiments, in *Proc. Lunar Planet. Sci. Conf.* **11**, pp. 2075–97.

Hale, W.S. and Head, J.W. (1981) Central peaks in martian craters: comparisons to the Moon and Mercury, in *Lunar Planet. Sci. XII*, pp. 386–8

Hayashi-Smith, J. and Mouginis-Mark, P.J. (1990) Morphometry of fresh impact craters in Hesperia Planum, Mars, in *Lunar Planet. Sci. XXI*, 475–576.

Melosh, H.J. and Vickery, A.M. (1989) Impact erosion of the primordial atmosphere of Mars. *Nature*, **338**, 487–9.

Mouginis-Mark, P.J., McCoy, T.J., Taylor, G.J. and Keil, K. (1991) Parent craters for the SNC meteorites, in *Lunar Planet. Sci. XXII*, 11. 931–2.

Neukum, G. and Hiller, K. (1981) Martian ages. *J. Geophys. Res*, **86**, 3097–121.

Neukum, G. and Wise, D.U. (1976) Mars: a standard crater curve and possible new time scale. *Science*, **194**, 1381–7.

Nyquist, L.E. (1983) Do oblique impacts produce Martian meteorites? *Proc. Lunar Planet. Sci. Conf.*, **13**, *J. Geophys. Res.*, **88**, A785–98.

Pike, R.J. (1988). Geomorphology of impact craters on Mercury, in *Mercury* (ed. F. Vilas, C.R. Chapman and M.S. Matthews). Tucson: University of Arizona Press. pp. 165–273.

Pollack, J.G., Kasting, J.F., Richardson, S.M. and Poliakoff, K. (1987) The case for a wet warm climate on early Mars. *Icarus*, **71**, 203–24.

Sagan, C., Toon, O.B. and Gierasch, P.J. (1973) Climatic change on Mars. *Science*, **181**, 1045–9.

Schultz, P.H. and Gault, D.E. (1979) Atmospheric effects on Martian ejecta emplacement. *J. Geophys. Res.*, **84**, 7669–87.

Schultz, P.H. and Lutz-Garihan, A.B. (1982) Grazing impacts on Mars: a record of lost satellites, in *Proc. Lunar Planet. Sci. Conf.*, **13**, *J. Geophys. Res.*, **87**, A84–96.

Schultz, P.H., Schultz, R.A. and Rogers, J. (1982) The structure and evolution of ancient impact basins on Mars. *J. Geophys. Res.*, **87**, 9803–20.

Shoemaker, E.M., Williams, J.G., Helin, E.F. and Wolfe, R.F. (1979) Earth-crossing asteroids: orbital classes, collision rates with Earth, and origin, in *Asteroids* (ed. T. Gehrels). Tucson: University of Arizona Press. pp. 253–82.

Strom, R.G., Croft, S.K. and Barlow, N.G. (1992) The Martian impact cratering record, in *Mars* (ed. H.H. Kieffer B.M. Jakosky, C.W. Snyder and M.S. Matthews). in press. Tucson: University of Arizona Press, pp. 383–423.

Tanaka, K.L. (1986) The stratigraphy of Mars. *Proc. Lunar Planet. Sci. Conf.* **17**, *J. Geophys. Res.*, **91**, E139–58.

Tanaka, K.L., Isbell, N.K., Scott, D.H. *et al.* (1988) The resurfacing history of Mars: a synthesis of digitized Viking-based geology, in *Proc. Lunar Planet. Sci. Conf.* **18** (ed. G. Ryder). New York: Cambridge University Press, pp. 665–78.

Wetherill, G.W. (1975) Late heavy bombardment of the Moon and terrestrial planets. *Proc. Lunar Sci. Conf.* **6**, pp. 1539–61.

Wilhelms, D.E. and Squyres, S.W. (1984) The Martian hemispheric dichotomy may be due to a giant impact. *Nature*, **309**, 138–40.

Wise, D.U., Golombek, M.P. and McGill, G.E. (1979) Tharsis Province, Mars: geologic sequence, geometry, and a deformation mechanism. *Icarus*, **38**, 456–72.

Wood, C.A. and Ashwal, L.D. (1981) SNC meteorites: igneous rocks from Mars? in *Proc. Lunar Planet. Sci. Conf.* **12**, pp. 1359–1375.

Wood, C.A., Head, J.W. and Cintala, M.J. (1978) Interior morphology of fresh martian craters: the effects of target characteristics, in *Proc. Lunar Planet. Sci. Conf.* **9**, pp. 3691–3709.

Cross references

Crater
Ejecta
Impact cratering
SNC meteorites

MARS: MAGNETIC FIELD AND MAGNETOSPHERE

Mars, at a heliocentric distance of approximately 1.5 AU, exemplifies a terrestrial body that is affected by both a slightly greater distance from the Sun than Earth and a lower gravitational field at the surface. The latter results from Mars' smaller size. The radius of Mars is about 3395 km on average (compared to the 6371 km radius of Earth). As a result, much of the inventory of the lighter gases of Mars has escaped to space over time. Some atmospheric constituents, such as the predominant CO_2 and a small contribution of H_2O, periodically freeze out into the polar caps in a seasonal cycle. Ground ice, like permafrost, is most probably present down to latitudes quite close to the equator.

A hypothesized early dense CO_2 atmosphere of Mars is considered to be today sequestered in carbonate rocks on the surface. These carbonates are thought to have formed in the presence of liquid water, when Mars had a greenhouse effect (q.v.) contributing to the warmth of its early climate. Mars is too small to have plate tectonics and the resulting volcanic activity that together would have recycled the CO_2 back into the atmosphere. It is likely that some of the early Martian atmosphere has been lost to space because Mars, like Venus, has no substantial intrinsic magnetic field to protect the atmosphere from solar wind scavenging.

Interior

The mean density of Mars deduced from spacecraft encounters is anomalously low at 3.9 g cm^{-3} (compared to $\sim$ 5.3 g cm^{-3} for Venus and Earth). However, estimates of the size of its iron-bearing core are more typical, at about half the planet's radius. Like the other planets, Mars probably had a molten core in the first billion or so years following accretion. The core of Mars may now be solid, if the small size of the planet led to more rapid evolution following the onset of solid core formation. Although the Viking landers carried a seismometer, which could have given some definitive information about the state of the Martian core, the device was not well coupled to the surface and suffered from vibrations caused by the winds. Surface features indicate that seismic activity related to plate tectonics is minimal, although volcanic activity was clearly present in the last billion years, as shown by the giant dormant volcano Olympus Mons and others.

Magnetic field

The first indication of the weakness of the magnetic field of Mars was obtained during the Mariner 4 spacecraft flyby in 1965. At a closest approach of 3.9 Mars radii, no indication of the Earth-like dipole magnetic field predicted by scaling arguments from theory was detected. Still, a shock-like disturbance in the solar wind signaled the presence of an obstacle approximately the size of Mars. Most subsequent magnetic field measurements in the vicinity of Mars were carried out on a series of five MARS spacecraft launched by the Soviet Union between 1971 and 1974 (see Soviet MARS missions). Several of these successfully operated in orbit for periods long enough to both confirm the Mariner 4 results and to measure the disturbance of the interplanetary magnetic field caused by the obstacle. However, none of these spacecraft approached Mars closer than $\sim$ 1300 km or $\sim$ 1.3 Mars radii from the center of the planet, and none probed the solar wind wake inside of the optical shadow, where the magnetotail of an intrinsic magnetosphere resembling a weak version of Earth's would be found. The Viking landers reached the surface of Mars in 1976, but did not carry magnetic field experiments as part of their scientific payloads, although they made ionospheric measurements of relevance to the magnetic field question. Because the available measurements could be interpreted from the viewpoint of either a small Earth-like magnetosphere, or a Venus-like ionospheric obstacle, different researchers have adopted both of these paradigms for over a decade. Their divergent views depended on the techniques and arguments used in analyzing the still ambiguous data (Luhmann and Brace, 1991).

These differences in opinion have to some extent been altered by the most recent magnetic field measurements on the Soviet Phobos-2 spacecraft in 1989 (e.g. *Nature*, **341**, 19 October 1989, describes the first results). The orbit of Phobos 2 went into the deep wake of Mars, for the first time providing magnetic field data in the optical shadow at distances as close as $\sim$ 2.7 Mars radii and as distant as $\sim$ 20 Mars radii. These data unambiguously showed that the magnetic fields in the wake of Mars are determined by the interplanetary field orientation, and are thus not Earth-like, at least in the near-equatorial spacecraft orbit plane. The current upper limit on the dipole moment remains at $\sim 10^{-4}$ times that of Earth, a value established on the basis of the previous observations. This moment is derived not from the wake data but from estimates of the subsolar altitude of the Martian obstacle to the solar wind of $\sim$ 400 km. Additional indirect information concerning the magnetic field of Mars derived from ionospheric observations and the understanding of solar wind interactions is described below.

Today, the only other 'direct' information that Martian magnetism is from a special class of meteorites known as the SNC meteorites (q.v.) which are thought to come from Mars. Magnetic field analyses of these possible samples of the Martian crust indicate that magnetic fields of $\sim$ 1000 nT may have been present on the surface of Mars at the time that these meteorites were ejected by a giant impact some 180 million years ago. (For comparison, the present field on Earth near the equator is about 3×10^4 nT. The present upper limit on the dipole moment implies surface fields of only a few tens of nanotesla.)

The dynamo theory of planetary magnetism indicates that Mars may have had a dipole moment of about one-tenth of Earth's when it was first formed (Schubert and Spohn, 1990). The rotation rate of Mars is approximately that of Earth and is thus sufficient for the operation of this initial dynamo. The other necessary ingredient of a convection driver in the core was supplied by heat left over from the accretion of the planet, which may have been effective for up to a few billion years. If such a field did indeed exist, evidence of it may still be present on the surface in the form of magnetized rocks and crustal regions like those observed on the Moon. No observations indicating the presence of such fields have been reported other than the aforementioned SNC meteorites' magnetization.

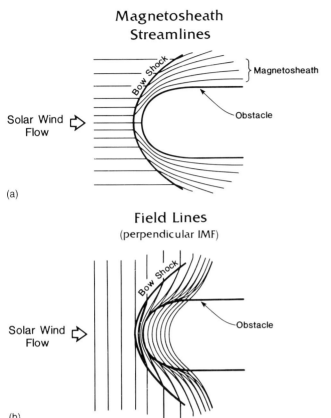

Figure M54 Illustration of the disturbance in the solar wind flow (a) and interplanetary magnetic field (b) produced by a planetary obstacle in the solar wind. Early spacecraft to Mars detected this disturbance, the size of which gave an upper limit to the strength of the Martian magnetic field. (J.G. Luhmann and L.H. Brace, *Rev. Geophys.*, **29**, 121, 1991, copyright American Geophysical Union).

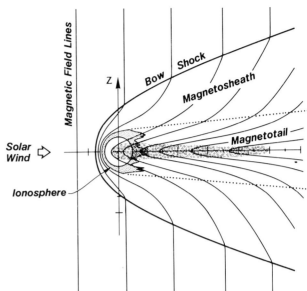

Figure M55 Illustration of the 'induced' magnetotail in the wake of Mars and its connection to the draped interplanetary magnetic field (J.G. Luhmann and L.H. Brace, *Rev. Geophys.*, **29**, 121, 1991, copyright American Geophysical Union).

Solar wind interaction

To date, the observations at Mars suggest that a great deal of similarity exists between near-Venus space and near-Mars space. Mars, like most of the other planetary obstacles, is preceded in the solar wind by a 'bow shock'-like structure that reflects the slightly greater than planet-size scale of the weakly magnetized Martian obstacle. The subsolar distance of the bow shock is $\sim 1.5\ R_M$, while its terminator position is $\sim 2.7\ R_M$. Within the bow shock, the solar wind plasma is diverted around the obstacle. Within it, the imbedded interplanetary magnetic field is compressed against the obstacle nose. The field distortion associated with the divergence of the flow gives the 'draped' configuration of field lines illustrated in Figure M54. This 'magnetosheath' is a common feature of all planetary obstacles.

Within the dayside obstacle boundary implied by the Mars bow shock position, the magnetic field geometry is unknown. However, both *in situ* measurements on the Viking Landers and radio occultation experiments on the Viking Orbiters and other spacecraft indicated the presence of a substantial dayside ionosphere below about 300 km. (The subsolar obstacle height inferred from the bow shock position is ~ 400 km.) The *in situ* measurements from the Viking Landers also provided information on the temperatures in the ionospheric plasma which were used to calculate the ionosphere's thermal pressure. This calculation resulted in the conclusion that Mars must have a planetary magnetic field of significance because these pressures were less than the incident solar wind pressure. Nevertheless, we know from the observations at Venus during disturbed solar wind conditions that an ionospheric obstacle can persist in the face of such levels of excess solar wind pressure. One possibility is that the Mars ionosphere, like that of the disturbed Venus counterpart, contains large-scale horizontal magnetic fields that are induced by the solar wind interaction. Models of these induced fields suggest that they should be several tens of nanotesla in strength. The electron temperatures measured in the ionosphere by the Viking Landers are also consistent with fields of this strength and orientation, but it is not clear from these whether the field is planetary or induced by the solar wind interaction.

As mentioned above, the Phobos 2 magnetic measurements in the near-equatorial wake of Mars showed that the fields in that region are controlled by the interplanetary field orientation. The relationship between the interplanetary field draped over the obstacle and the field in such an 'induced' magnetotail is illustrated by Figure M55. In spite of this finding, there are still advocates of an intrinsic field contribution to the Martian magnetotail because the magnetotail appears to be wider than that of Venus relative to the planet radius (~ 2.0 planetary radii in diameter compared to Venus' ~ 1.2). It is argued that this contribution has not been detected because the 'intrinsic' field tail features may be restricted to regions removed from the equator.

Phobos 2 did detect significant fluxes of planetary ions (mainly O^+, as at Venus) that had been scavenged from Mars by the passing solar wind (e.g. see the *Nature* special issue mentioned above). The details of the acceleration of these ions are not completely understood, but the electric field in the solar wind is expected to remove ions formed in the upper atmosphere that extends above the 'obstacle' boundary into the magnetosheath and undisturbed solar wind. The observed rates of escape for the oxygen suggest that the solar wind scavenging process has the potential to remove all of Mars' present inventory of atmospheric oxygen over the next 10^8 years. These observations also suggest that the solar wind interaction must have played some role in the Martian atmosphere's evolution over the past 4.5 billion years, or at least after the thermally driven planetary dynamo ceased to operate.

Janet G. Luhmann and Christopher T. Russell

Bibliography

Luhmann, J.G. and Brace, L.H. (1991) Near-Mars space. *Rev. Geophys.*, **29**, 121.

Russell, C.T., 1987, Planetary magnetism in *Geomagnetism*, Vol. 2 (ed. J.A. Jacobs) London: Academic Press, pp. 457–523.

Schubert, G. and Spohn, T. (1990) Thermal history of Mars and the sulfur content of its core, *J. Geophys. Res.*, **95**, 14095–104.

Acknowledgements

This work was supported by the National Aeronautics and Space Administration under research grants NAGW-1347 and NAG2-2573.

Cross references

Magnetism
Solar wind

MARS OBSERVER MISSION

(Editor's Note: At about 6 p.m. PDT on August 21, 1993, communication with the Mars Observer spacecraft was lost. The spacecraft was about to begin pressurizing its fuel tanks in preparation for orbit insertion. Although the cause of the loss of the spacecraft cannot be determined conclusively, it is thought that an explosion may have occurred due to gaseous diffusive leakage of a few milliliters of fuel through a faulty valve.

The scientific objectives of the mission were and are of considerable significance. Mars is after all the most desirable planet for future human exploration and colonization. Many of the objectives of the Mars Observer mission will be addressed by smaller missions of the Mars Surveyor program. The mission description presented here is of more than historic interest, as some of the experiments carried were unique. Instruments like Mars Observer's gamma-ray spectrometer and pressure modulator infrared radiometer will no doubt be flown on future missions in the next century.)

The Mars Observer mission was launched on 25 September 1992 (Albee, Arvidson and Palluconi, 1992). After the capture of the spacecraft by the planet and the adjustment into a low Sun-synchronous, polar-mapping orbit in late 1993 observations were planned to continue for a Mars year (687 days). The scientific mission included observations of the Mars atmosphere, surface and interior. The seven experiments carried by the spacecraft involve gamma-ray spectroscopy, magnetometry, surface and atmosphere imaging, atmospheric sounding, laser altimetry, gravity mapping and thermal emission spectroscopy.

Background

Mars is the most closely examined planet other than the Earth. Over 20 space vehicles have been sent to the vicinity of Mars in flyby, orbiter or lander modes, the most recent being the Soviet Phobos 2 Orbiter in 1989. A key objective in this exploration involves understanding the origin and subsequent evolution of Mars in the context of its nearest neighbors, Earth and Venus. These three terrestrial planets formed in the same region of the solar system at the same time, but they have subsequently followed quite different evolutionary paths. For example, the surface atmospheric pressure at Venus is 90 times that of Earth but on Mars the surface pressure is less than one-hundredth that of Earth. The atmospheres of both Venus and Mars are dominated by the greenhouse gas, carbon dioxide, and they are thus of considerable interest since this gas also figures prominently in concerns about climate change on Earth.

Although the Martian atmosphere is thin by terrestrial standards, it does contain measurable amounts of water vapor. At times, the Martian relative humidity reaches 100%, leading to formation of water clouds, fogs and frost in addition to carbon dioxide clouds and frost. However, it cannot rain on Mars at present because the atmospheric pressure is too low for liquid water to be stable. Apart from water vapor in the atmosphere, water is also known to exist on Mars in the extensive northern permanent polar cap. Surprisingly, the smaller southern permanent polar cap is covered, even during southern summer, by carbon dioxide frost and is not a source of water for the atmosphere at present. In a sense, we can think of Venus, Earth and Mars as immense laboratories where we may observe the results of natural atmospheric experiments.

Ample evidence exists indicates, that water in large amounts was present on the Martian surface in the past. What happened to it? Current thinking places some of this water beneath the surface in a frozen state, while some has escaped as vapor and has been carried off into interplanetary space by erosion of the Martian atmosphere in the solar wind. Unraveling the history of water on Mars is one of the underlying motivations for the Mars Observer mission. If we can understand the behavior of the atmosphere at present, we can more confidently extrapolate backward in time to understand conditions at earlier epochs.

The thinness of the Martian atmosphere is an advantage for a variety of remote sensing experiments because it permits a nearly unobstructed view of the surface from orbit. This thinness even permits some measurements, like gamma-ray spectroscopy, which are not possible from Earth orbit because of absorption of the emitted gamma rays by our thicker atmosphere. It is fortunate that measurements of surface properties can be made from orbit because, although Mars is a smaller planet than Earth, its 144×10^6 km² surface area is equal to the entire continental area of the Earth. For a long time to come, remotely sensed data will be the only type we will have from many regions of Mars.

The surface of Mars is especially important in understanding the evolution of the terrestrial planets because parts of its surface preserve direct evidence of processes going back all the way to the period of late bombardment following planetary formation (about 4 billion years ago). The first US mission to Mars, Mariner 4, returned images of this Moon-like, heavily-cratered region of Mars (see Mariner missions). On Earth, this early bombardment record has been either erased or heavily modified. Sea-floor creation and subsequent subduction has erased most traces of the Earth's early oceanic crust, and on the continents erosion has been nearly as effective. Subduction of the crust appears to be absent on Mars, and erosion has been considerably less effective. In this sense, the early history of Mars is more open to inspection than is the early history of Earth.

Key mission elements

The Mars Observer mission was organized around a set of spacecraft and mission choices that were applied for the first time to a planetary mission. Mars Observer's Sun-synchronous orbit is similar to that used by the Landsat, Spot and NOAA terrestrial polar orbiters. A low-altitude, near-circular, near-polar and near-Sun-synchronous orbit contributes to the measurement opportunities. The low altitude (400 ± 25 km) produces higher spatial resolution and improved signal to noise ratios for some experiments. The near-circular orbit (eccentricity < 0.01) allows nearly uniform spatial resolution at all latitudes and longitudes, facilitating intercomparison of measurements from different locations. The near-polar orbit (inclination of 93°) permits observations to be made at all latitudes and longitudes and is the key to a global mapping mission. The near-Sun-synchronous orbit (2 p.m. Sunward equator crossing time) makes possible repeated observations at the same time of day, thereby making it possible to separate diurnal and seasonal behavior. This orbit also readily accommodates continuous observation from experiments that must use radiators to cool detectors (i.e. Gamma-Ray Spectrometer and Pressure Modulator Infrared Radiometer). The orbit period is 118 min, which permits sampling two times of day for 13 longitudes each Martian day. The orbit design is such that the sub-spacecraft point on the surface of Mars (ground track) will trace out a non-repeating path allowing complete sampling of the surface. This ground track comes close to repeating at 7- and 26-day intervals, which allows near uniform coverage at these intervals. After one Mars year the average spacing of ground tracks along the equator will be about 3 km.

The Mars Observer spacecraft was designed to maintain a nadir-pointing orientation (the spacecraft will rotate once per orbit, keeping the instrument mounting face shown in Figure M56 pointed at Mars) for the entire mapping mission, so that each experiment will be able to view Mars continuously for an entire Martian year. The spacecraft was to support this continuous data collection with onboard tape recorders. A daily playback to Earth was planned for Mars' recorded data, along with supplemental real-time data transmissions for the high-data-rate experiments. Several experiments that need to look in more than one direction, e.g. atmospheric sensors that need to look from nadir to the limb, use internally driven electronic or mechanical mirrors.

The science instruments in total weigh 150 kg. The average total science instrument power consumption is about 121 W. The number of bits of recorded data returned each day depends on the Earth to Mars distance and would range from a low of 3.5×10^8 bits per day

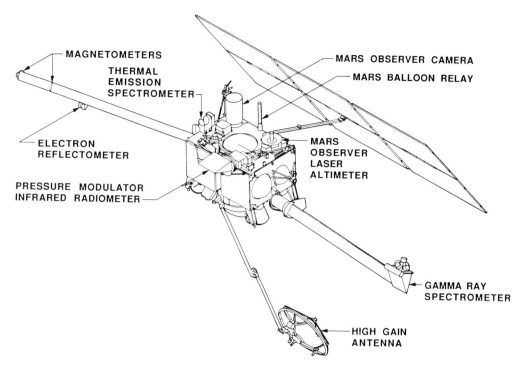

Figure M56 Schematic drawing of the Mars Orbiter spacecraft showing the location and configuration of the science instruments. The gamma-ray spectrometer and magnetometer booms are 6 m in length.

when Mars is farthest from the Earth to a high of 1.4×10^9 bits per day near opposition.

Science objectives

There were five scientific objectives for the Mars Observer mission. The first three encompass the geoscience objectives and involve measurements of the surface and interior (gravity and magnetics). The remaining two contain the climatology and atmospheric objectives and involve measurement of the atmosphere and surface. These objectives are:

1. determine the global elemental and mineralogical character of the surface material;
2. define globally the topography and gravitational field;
3. establish the nature of the magnetic field;
4. determine the time and space distribution, abundance, sources and sinks of volatile material and dust over a seasonal cycle; and
5. explore the structure and aspects of the circulation of the atmosphere.

Experiments and instrumentation

Each of the seven experiments selected for the mission contributes to meeting one or more of the scientific objectives.

Gamma-Ray Spectrometer

The Gamma-Ray Spectrometer (GRS) was designed to detect gamma rays emerging from within and near the Martian surface (Boynton et al., 1992). These high-energy photons are created by the natural decay of radioactive elements or are induced by cosmic rays that interact with atoms in the atmosphere and surface. The GRS measures the energy distribution of these photons, and the experiment team would use this information to establish the amounts of each element present in the surface material. Elements such as potassium, uranium, thorium, calcium, magnesium, aluminum, iron and others can be measured in the top m of the surface. Although the spatial resolution of this experiment is low (< 300 km for most elements), it is the only remote means of directly establishing elemental surface composition. The instrument also incorporated a neutron spectrometer for the measurement of the intensity of thermal and epithermal neutron flux. This measurement in conjunction with gamma-ray spectroscopy allows exploration of the stratigraphy of carbon and hydrogen in the upper meter of the surface. Cosmic gamma-ray spectra will also be recorded when the gamma-ray flux reaches a threshold level. Triangulation, using gamma-ray detectors in other parts of the solar system, permits location of these cosmic gamma-ray burst.

Magnetometer

The Magnetometer/Electron Reflectometer (MAG/ER), mounted on a 6-m boom, was designed to detect the presence of both global and local magnetic fields (Acuña et al., 1992). Mars is now the only planet from Mercury to Neptune whose magnetic field has not been measured. The magnetometer can detect the presence of a magnetic field directly, and the electron reflectometer, in conjunction with the magnetometer, can deduce the strength of the magnetic field in the region closer to the planet than the spacecraft by measuring the properties of electrons incident on the instrument. Previous measurements at Mars indicate that a global magnetic field, if present, is weak. Recent magnetic and particle measurements, made by the Soviet Phobos 2 mission in 1989, also supported the view that if a magnetic field is present it is very weak. The task of the magnetometer team was to sort out the many processes that can produce a magnetic field and to successfully identify the actual field generated by processes inside Mars.

Mars Observer Camera

The Mars Observer Camera (MOC) consisted of two wide-angle assemblies, which could photograph the planet from limb to limb, and one narrow-angle (1.4 m pixel^{-1}) system (Malin et al., 1992). The wide-angle cameras were to return low-resolution images of the entire planet every day to provide a record of the weather on Mars. A daily global image of Mars in the visible would provide for the first time an unbiased assessment of atmospheric phenomena. As an example, it would be possible to develop statistical information as a function of latitude, longitude, and season for local dust storms. This would permit an assessment of the role of local storms in the formation of global dust storms, should a global storm occur during the period of Mars Observer measurements. These cameras would also return moderate resolution images (of order 300 m pixel^{-1}) of the surface by returning only the central portion of the wide-angle images through editing done onboard the spacecraft. The high-resolution system could selectively return images from a number of areas where key questions can be better understood through detailed knowledge of surface morphology and albedo. Because of the volume of returned data involved in the high-resolution imaging, even with data compression, only a few tenths of 1% of the Martian surface were to have been examined with this mode. As a result the targeting of these high-resolution images would be very selective. These very high-resolution measurements could provide a critical test of ideas involving climate change. Suggestions that water or continental scale glaciers have shaped large areas of the surface can be tested by looking for the associated small-scale features, such as beaches and eskers, that accompany such processes.

Pressure Modulator Infrared Radiometer

The pressure modulator infrared radiometer (PMIRR) was designed to obtain data about atmospheric structure and dynamics by making measurements at the limb of Mars, in the thermal infrared wavelength region (McCleese et al. 1992). This instrument would concentrate its measurements where the path through the atmosphere from the location of the spacecraft is greatest. PMIRR would scan upward from the limb, sounding the atmosphere to produce altitude profiles of temperature, pressure, water vapor, dust opacity and cloud composition. These measurements could be used by the PMIRR experiment team to examine the structure and circulation of the atmosphere as a function of latitude, longitude, season and altitude. PMIRR could also do atmospheric sounding in a nadir-looking mode and make surface measurements in this mode. The band selection permitted full radiation budget measurements, including the solar reflected and Mars emitted components.

Mars Observer Laser Altimeter

The Mars Observer Laser Altimeter (MOLA) was designed to fire pulses of infrared light (wavelength = 1.06 μm) at the surface from a laser (Zuber et al., 1992). By measuring the travel time of the reflected pulse, it is possible to measure the distance from the spacecraft to the surface with a precision of several meters. By combining this measured distance with the distance from the center of the planet to the spacecraft, obtained from orbit reconstruction, the experiment team would gradually reconstruct the entire global topography of Mars. The laser pulses ten times per second and illuminates an area about 100 m in diameter at the surface. Although topography is basic to understanding the geophysics and geology of Mars, our absolute knowledge of this quantity is no better than a kilometer for much of the surface. The high precision of the Mars Observer altimeter, coupled with accurate orbits based on improved knowledge of the gravity field, could provide an improvement of many times in understanding topographic relationships.

Radio Science

In the Radio Science (RS) investigation, the experiment team would use an ultrastable oscillator, the spacecraft telecommunications system and ground station receiving equipment to probe the atmosphere and gravity field of Mars (Tyler et al., 1992). By carefully monitoring changes in the frequency of the radio signal from the spacecraft as it moves around Mars, the effect of the gravity field on the spacecraft velocity could be determined. The pole-to-pole coverage and low altitude of the Mars Observer orbit would permit a significant improvement in understanding of the gravity field of Mars. The radius of Mars could be accurately established at each of the occultation points, providing an independent measure of planetary shape. Changes in the radio signal as the spacecraft passed in and out of occultation by Mars, as viewed from Earth, could be used to construct very high vertical resolution temperature and pressure profiles of the atmosphere. In addition, on the way to Mars, a search for gravitational waves was conducted in conjunction with the Ulysses and Galileo missions.

Thermal Emission Spectrometer

The Thermal Emission Spectrometer (TES) operated primarily in the thermal infrared portion of the electromagnetic spectrum (Christensen et al., 1992). The nature of radiation from the Martian surface at these wavelengths depends on temperature, surface mineralogy, and other factors. The investigation team was to use spectrometer measurements to determine the thermal and mineralogical properties of the surface. The instrument could also provide data about Martian atmospheric properties, including cloud type (carbon dioxide or water ice) and dust opacity. TES is the third experiment (along with PMIRR and RS) designed to make atmospheric measurements. The differing atmospheric data sets obtained by these three instruments would permit intercomparison of the results of different measuring techniques, thus greatly strengthening confidence in the accuracy of the results obtained. TES used a 3×2 (three detectors crosstrack) detector array in each of the operating wavebands. The spatial resolution at the surface of each detector is 3 km. Like PMIRR, TES carried wavebands permitting full radiation budget measurements, which are of special interest in the polar areas.

Mars balloon relay

In addition to these seven experiments, the spacecraft carried an eighth device, supplied by the French Centre Nationale d'Etude Spatiales, which was to support penetrators and landers of the Russian Mars '94 mission. The equipment carried by the Mars Observer spacecraft consists of a receiver/transmitter combination operating continuously at frequencies near 400 MHz. A receiver attached to the Mars '94 surface stations will continuously monitor the transmitter frequency. When the signal strength reaches a threshold value indicating that Mars Observer is close enough to receive data, a transmitter on the station was to relay scientific and engineering information up from the surface to the receiver on the spacecraft. The data relayed up from the Mars '94 stations was to have been stored in the large solid state memory of the MOC, where it would then be encoded and processed for return to Earth. The primary data return path is through the Russian orbiter.

Organization and scientific personnel

The Mars Observer Project was managed for NASA by the Jet Propulsion Laboratory. NASA Lewis Research Center supplied the Titan III launch vehicle through a commercial launch services contract with Martin-Marietta Commercial Titan, Inc. NASA Marshall Space Flight Center supplied the upper stage, which was developed by the Orbital Sciences Corporation and built by the Martin-Marietta Astronautics Group. The spacecraft was developed through a system contract with General Electric Astro-Space Division (Potts, 1991). Integration of scientific instruments with the spacecraft began in July 1991 and was completed in May of 1992 prior to shipment of the spacecraft to the launch site in June of 1992.

The seven Mars Observer experiment teams were selected through a NASA announcement of opportunity in 1985 as were five Interdisciplinary Scientists. Eleven Russian Participating Scientists joined the Mars Observer scientific effort, and the Russian Mars '94 balloon experiment team includes several American Participating Scientists. In February 1992 NASA selected 32 additional Participating Scientists through an open NASA Research Announcement released in 1991. The Participating Scientist program is used to enlarge the scientific teams during the data collection and analysis period. The total number of scientists selected by NASA to participate directly in the mission was over 100.

Tracking and data acquisition for the mission was provided by the NASA Deep Space Network. Launch occurred from launch Complex 40 at the NASA Kennedy Space Center on the 10th day of a 28-day launch window which opened on 16 September 1992. The spacecraft arrived at Mars on 24 August 1993. Communication with the spacecraft was lost following the initial stage of preparation for Mars orbit insertion on 24 August 1993.

Summary

Although our knowledge of Mars is substantial, it is trivial when compared to our knowledge of Earth. By providing global measurements of the Mars atmosphere, surface and interior over a full Martian year, thus recording the full range of seasonal behavior, the Mars Observer mission would have consolidated the knowledge gained from both ground-based studies and previous spacecraft missions, added extensive new measurements, and provided a strong foundation for more intensive investigations of Mars in the future.

Arden L. Albee and Frank D. Palluconi

Bibliography

Acuña, M.H., Connerney, J.E.P., Wasilewski, P. et al. (1992) The Mars Observer magnetic fields investigation. *J. Geophys. Res., Planets*, **97**, 7799–814.
Albee, A.L., Arvidson, R.E. and Palluconi, F.D. (1992) The Mars Observer mission, *J. Geophys. Res., Planets*, **97**, 7665–80.
Boynton, W.V., Trombka, J.I., Feldman, W.C. et al. (1922) Science applications of the Mars Observer gamma-ray Spectrometer, *J. Geophys. Res., Planets*, **97**, 7681–98.
Christensen, P.R., Anderson, D.L., Case, S.C. et al. (1992) Thermal Emission spectrometer experiment: the Mars Observer mission, *J. Geophys. Res., Planets*, **97**, 7719–34.
Malin, M.C., Danielson, G.E., Ingersoll, A.P. et al. (1992) The Mars Observer Camera, *J. Geophys. Res., Planets*, **97**, 7699–718.
McCleese, D.J., Haskins, R.D., Schofield, J.T. et al. (1992) Atmosphere and climate studies of Mars using the Mars Observer Pressure Modulator Infrared Radiometer, *J. Geophys. Res., Planets*, **97**, 7735–57.
Potts, D.L. (1991) Mars Observer spacecraft description. *J. Spacecraft and Rockets*, **28**, 507–14.
Tyler, G.L., Balmino, G., Hinson, D.P. et al. 1992, Radio Science investigations with Mars Observer, *J. Geophys. Res., Planets*, **97**, 7759–79.

Acknowledgements

This research was conducted at the Jet Propulsion Laboratory, California Institute of Technology, Pasadena, CA, under contract with NASA. The authors wish to thank members of the Mars Observer Project Science Group and the Mars Observer Project Team for their work and support.

MARS: REMOTE SENSING

Spectroscopic remote sensing of Mars

Emission and absorption of energy by solids and gases provides an important mechanism for identifying the materials present on planetary surfaces, and in their atmospheres, using remote sensing techniques. Spectroscopic observations of Mars have been obtained in the ultraviolet (UV, 0.11–0.34 μm), visible (VIS, 0.4–0.7 μm), near infrared (NIR, 0.7–2.5 μm), mid-infrared (MIR, 2.5–5.0 μm) and far infrared (FIR, 5.0–200 μm). In the VIS and NIR, incident solar energy that is reflected by solids (both on the surface and in the atmosphere) is the dominant source of the energy measured from Mars. At wavelengths > 5 μm, thermal emission by the surface is the dominant source of the measured energy from Mars. This discussion will focus on the composition of the solids on the surface and in the atmosphere of Mars. Only a brief review is provided here and the interested reader wishing more details should see Soderblom (1992) and Roush, Blaney and Singer (1993).

The Martian surface can be divided into three broad categories based on VIS, NIR and MIR albedo of the region. These are called the 'classical' (i.e. known for many years from telescopic observations) bright regions (which are the principal contributor to Mars' red color), dark regions (Figure M57) and the polar caps, which are the brightest regions on Mars. This distinction between bright and dark regions exists on a wide range of spatial scales; from several thousand kilometers as viewed from Earth-based telescopes, to individual outcrops as seen from the Viking landers.

Iron oxides

Mars' reddish coloration has long been attributed to iron (Fe^{3+}) oxides. Yet the spectral properties of typical iron oxides, such as hematite or goethite, include several specific absorptions that are weakly present, if at all, in spectra of Mars' bright regions. This initially led to the suggestion that some terrestrial palagonites (alteration products of volcanic ash) provided the best spectral analogs to the Mars observations (Roush, Blaney and Singer, 1993). Careful laboratory studies have shown that nanophase iron oxides (NPIO), with crystal dimensions from a few to 10 nanometers, also do not exhibit the spectral behavior of typical bulk hematite or

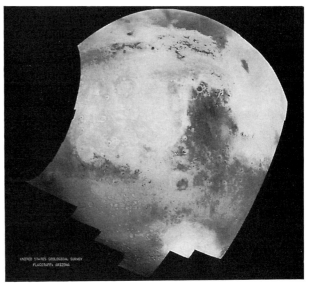

Figure M57 This digital mosaic is composed of about 100 individual Viking Orbiter images and has a spatial scale of 1 km pixel^{-1}. The images were acquired in 1980 during early northern summer on Mars. The center of the image is near latitude 0° and longitude 310°, and the limits of the mosaic are approximately −60° to 60° latitude and longitude 260° to 350°. The color variations have been enhanced by a factor of two, and the large-scale brightness variations (mostly due to Sun-angle variations) have been normalized by large-scale filtering. This mosaic illustrates the three general classes of albedo regions observed on Mars. The large circular bright area, located in the left central region of the image, is a high-albedo region and corresponds to the classical bright region known as Arabia. The low-albedo dark area to the right of Arabia is the classical dark region known as Syrtis Major. To the right of Syrtis Major is another classical bright region, Elysium. The bright white circular area to the south of Syrtis is the Hellas impact basin and it represents some of the brightest areas on Mars due to the presence of carbon dioxide frost or clouds. (Image produced by Tammy Becker at the United States Geological Survey in Flagstaff, Arizona.)

goethite (Morris et al., 1989); NPIO may therefore represent the coloring agent in some palagonites (Morris et al., 1990, Golden et al., 1993). This leads to the conclusion that much of the ferric iron on Mars exists in materials with poorly defined or extremely short range crystalline structure. Recent observations of Mars in the VIS and NIR have been interpreted as indicative of variations in the abundances and compositions of iron oxides present in the bright regions on Mars (Bell, Morris and Adams, 1992; Singer and Miller, 1992).

Primary crustal materials

The low-albedo dark regions on Mars are believed to represent areas which are less oxidized (or altered) when compared to the bright regions. These areas may represent regions where pristine igneous crustal materials occur. The spectra from these regions exhibit absorption features that are consistent with similar features observed in iron-bearing (Fe^{2+}) pyroxenes, a common mineral associated with basaltic lavas. Mineralogic interpretation of the spectra from low-albedo regions yield pyroxene compositions with high iron and calcium content that are consistent with the spectra of Shergotty, a meteorite believed to have come from Mars (McFadden, 1987; Singer and McSween, 1993, Mustard et al., 1993). Interpretation of some of the dark region spectra can also support the presence of olivine. Spectra of Mars obtained by the Mariner 9 spacecraft are consistent with the presence of plagioclase (Aronson and Emslie, 1975). Olivine, pyroxene and plagioclase are common constituents of terrestrial basaltic rocks, and thus the presence of these minerals suggests a basaltic crust for these areas of Mars.

Secondary crustal materials

Bright regions are believed to represent deposits of materials with spectral properties similar to the airborne dust observed during atmospheric storms. These regions also appear to have been chemically altered or oxidized. They appear to represent weathered material on Mars. Such weathering could occur due to gases released by volcanic activity. Meteorite impacts during the earliest history of Mars may have engendered a hydrologic cycle that included liquid water at the surface (see Mars: atmosphere). The gases are likely to be composed of water, carbon dioxide and sulfur dioxide; if they reacted with the primary igneous materials, chemically altered materials such as hydrates, hydroxylates, carbonates and sulfates could form. Thus the recognition of such materials provides an insight into the evolutionary history of both the surface materials and the atmospheric constituents. Initially, MIR Earth-based telescopic observations were used to identify the presence of hydrates associated with the surface materials on Mars (Sinton, 1967; Houck et al., 1973), and these observations have subsequently been supported by spacecraft observations (Bibring et al. 1989, Murchie et al., 1993) and airborne FIR observations (Pollack et al., 1990).

Carbonates (or bicarbonates) and sulfates (or bisulfates) have been identified in the NIR and FIR observations of Mars. Clark et al. (1990) suggested that bicarbonate and bisulfate located in the mineral scapolite were responsible for complex but weak features seen in the NIR spectra of Mars. Alternatively, Mg-bearing hydroxylates have also been suggested to explain these features (Singer, Owensby and Clark, 1985; Martin, 1985). Pollack et al. (1990) have attributed weak features observed in FIR spectra to an unidentified carbonate species (i.e. carbonate and/or bicarbonate) and another feature to an unspecified sulfate species (i.e. sulfate and/or bisulfate). Modeling of these data suggested that the carbonates and sulfates are associated with the aerosols suspended in the Martian atmosphere (Pollack et al., 1990).

There are two strong silicate absorption features present in Mariner 9 FIR spectra of Mars. Although these were originally interpreted as the result of the presence of the clay mineral montmorillonite suspended in the atmosphere (Hunt, Logar and Salisbury, 1973), the features seen in the Mars data do not exhibit the spectral complexity of terrestrial montmorillonites (Roush, 1989). Instead the features in the Martian spectra are more comparable to the FIR spectral properties of some palagonites (Roush, 1989, 1992; Roush, Blaney and Singer, 1993). It has recently been recognized that certain palagonites contain Mg-bearing clays (i.e. saponites) with very small crystal dimensions (Roush and Blake, 1991; Golden et al., 1993) and thus the spectral properties of amorphous or cryptocrystalline clays may also be consistent with the Martian data.

Polar caps

The seasonal polar caps on Mars are the brightest regions on the planet. The temperatures are cold enough that atmospheric carbon dioxide condenses at the surface and CO_2 frost or ice is formed. Spectral features due to CO_2 frost were first reported by Herr and Pimental (1969) based on features observed in the data returned by Mariner 7. The presence of an underlying H_2O ice deposit in the polar regions has also been suggested (Farmer, Davies and La Porte, 1976; Kieffer et al., 1976). Based on Mariner 9 FIR spectra it also appears that under certain atmospheric conditions CO_2, ice forms in the polar atmosphere and falls to the surface, resulting in the equivalent of a CO_2 snow storm (Paige, Crisp and Santee, 1990).

Summary

Spectroscopic observations, at a variety of wavelengths, provide direct information regarding the composition and mineralogy of both the surface solids and atmospheric aerosols of Mars. There is direct evidence for the presence of a variety of igneous minerals (pyroxenes, olivines and plagioclase) whose correspondence and composition would be consistent with the presence of basaltic materials on Mars. The secondary weathering products present on Mars include both ferric oxides and silicates having short-range or cryptocrystalline mineralogic structures, carbonates (and/or bicarbonates), and sulfates (and/or bisulfates) which appear to be associated with the atmospheric dust, hydrates associated with the surface materials, and possibly hydroxylates. There is direct spectral evidence for condensed carbon dioxide and water associated with the Martian polar caps.

Ted L. Roush

Bibliography

Aronson, J.R. and Emslie, A.G. (1975) Composition of the Martian dust as derived by infrared spectroscopy from Mariner 9. *J. Geophys. Res.*, **80**, 4925–31.

Bell, J.F., III, Morris, R.V. and Adams, J.B. (1992) Relative abundances of poorly-and well-crystalline ferric oxides in the Martian soil and dust from telescopic data and terrestrial spectral analog studies (abstract), in *Lunar Planet. Sci. XXIII*. Lunar and Planetary Institute, Houston: pp. 81–82.

Bibring J.P., Combes, M., Langevin, Y. et al. (1989) Results from the ISM experiment. *Nature*, **341**, 591–3.

Clark, R.N., Swayze, G.A., Singer, R.B. and Pollack, J.B. (1990) High-resolution reflectance spectra of Mars in the 2.3-μm region: evidence for the mineral scapolite. *J. Geophys. Res.*, **95**, 14463–80.

Farmer, C.B., Davies, D.W. and La Porte, D.D. (1976) Mars: northern summer ice cap – water vapor observations from the Viking orbiters. *Science*, **194**, 1339–41.

Golden, D.C., Morris, R.V., Ming, D.W., et al. (1993) Mineralogy of three palagonitic soils from the summit of Mauna Kea, Hawaii. *J. Geophys. Res.*, **98**, 3401–41.

Herr, K.C. and Pimental, G.C. (1969) Infrared absorptions near three microns recorded over the polar cap of Mars. *Science*, **166**, 496–9.

Houck, J.R., Pollack, J.B., Sagan, C. et al. (1973) High altitude infrared spectroscopic evidence for bound water on Mars. *Icarus*, **18**, 470–80.

Hunt, G.R., Logan, L.M. and Salisbury, J.W. (1973) Mars: components of infrared spectra and composition of the dust cloud. *Icarus*, **18**, 459–69.

Kieffer, H.H., Chase, S.C., Jr, Martin, T.Z., et al. (1976) Martian north pole summer temperatures: dirty water ice. *Science*, **194**, 1341–4.

Martin, T.Z. (1985) Improved access to Martian IR radiometry/spectroscopy data sets (abstract), in *Lunar Planet. Sci. XVI*, Houston: Lunar and Planetary Institute, p. 523.

McFadden, L.A. (1987) Spectral reflectance of SNC meteorites: relationships to martian surface composition, in *MEVTV Workshop on Nature and Composition of Surface Units on Mars* (eds J.R. Zimbelman, S.C. Solomon and V.L. Sharpton). Houston: Lunar and Planetary Institute, LPI Tech. Rpt. 88–05, pp. 88–90.

Morris, R.V., Gooding, J.L., Lauer, H.V., Jr and Singer, R.B. (1990) Origins of Marslike spectral and magnetic properties of a Hawaiian palagonitic soil. *J. Geophys., Res.*, **95**, 14427–34.

Morris, R.V., Agresti, D.G., Lauer, H.V., Jr et al. (1989) Evidence for pigmentary hematite on Mars based on optical magnetic, and Mössbauer studies of superparamagnetic (nanocrystalline) hematite. *J. Geophys. Res.*, **94**, 2760–78.

Murchie, S., Mustard, J., Bishop, J. et al. (1993) Spatial variations in the spectral properties of bright regions on Mars. *Icarus*, **105**, 454–68.

Mustard, J.F., Erard, S., Bibring, J.-P., et al. (1993) The surface of Syrtis Major: composition of the volcanic substrate and mixing with altered dust. *J. Geophys. Res.*, **98**, 3387–400.

Paige, D.A., Crisp, D. and Santee, M.L. (1990). It snows on Mars. *Bull. Am. Astron. Soc.*, **22**, 1075.

Pollack, J.B., Roush, T., Witteborn, F., et al. (1990) Thermal emission spectra of Mars (5.4–10.5 μm): evidence for sulfates, carbonates, and hydrates, *J. Geophys. Res.*, **95**, 14595–27.

Roush, T.L. (1989) Infrared transmission measurements of Martian soil analogs. in *MECA Workshop on Dust on Mars III* (ed. S. Lee), Houston: Lunar and Planetary Institute, LPI Tech Rpt. 89–01, pp. 52–4.

Roush, T.L. (1992) Infrared optical properties of Mars soil analog materials: palagonites. Houston: Lunar and Planetary Institute, LPI Tech Rpt. 92–40, pp. 32–3.

Roush, T.L., and Blake, D. (1991). Characterization of a Mauna Kea palagonite using transmission electron microscopy (abstract). Houston, Texas: Lunar & Planetary Institute, *Lunar Planet. Sci. XXII*, pp. 1139–40.

Roush, T.L., Blaney, D.L. and Singer, R.B. (1993) The surface composition of Mars as inferred from spectroscopic observations, in *Remote Geochemical Analysis* (eds C. Pieters and P.J. Englert). New York: Cambridge University Press, pp. 367–93.

Singer, R.B. and McSween, H.Y., Jr (1993) *The ingneous crust of Mars: compositional evidence from remote sensing and the SNC meteorites*, in *Resources of Near-Earth Space*, (eds J. Lewis and M. Matthews). Tucson: University of Arizona Press, pp. 708–36.

Singer, R.B. and Miller, J.S. (1992) Evidence for crystalline hematite as an accessory phase in Martian soils, in *MSATT Workshop on the Martian Surface and Atmosphere Through Time*, (eds R. Haberle and B. Jakosky). Houston: Lunar and Planetary Institute, LPI Tech Rpt. 92–02, pp. 134–5.

Singer, R.B., Owensby, P.D. and Clark, R.N. (1985) Observed upper limits for clay minerals on Mars (abstract), in *Lunar Planet. Sci. XVI*. Houston: Lunar and Planetary Institute, pp. 787–8.

Sinton, W.M. (1967) On the composition of Martian surface materials. *Icarus*, **6**, 222–8.

Soderblom, L.A. (1992) *The composition and mineralogy of the Martian surface from spectroscopic observations: 0.3 μm–50 μm*, in *Mars* (eds H. Kieffer, B. Jakosky, C. Snyder and M. Matthews). Tucson, Arizona: University of Arizona Press, pp. 557–93.

Cross references

Albedo
Reflectance spectroscopy
Remote sensing
Spectroscopy: atmosphere

MARS: STRUCTURAL GEOLOGY AND TECTONICS

Mars, like our Earth, has had an active and varied geologic history; however, its surface and topography have been much less affected by erosional processes because of its relatively thin atmosphere. Following the accretion and heavy impact cratering of the planet during its earliest geologic period (the Noachian), undetermined endogenic or exogenic processes divided its surface into a lowland and a highland region (Plate 18). The northern lowland plains, for much of their length, are separated from the rugged mountainous and plateau-forming highlands to the south by a highly dissected boundary scarp. This crustal dichotomy may have originated through mantle convection and crustal thinning beneath the plains (Wise, Golombek and McGill, 1979) or by a single giant impact (Wilhelms and Squyres, 1984) or multiple impacts (Frey and Schultz, 1989). Both of these hypotheses are consistent with gravity data (Phillips, 1988). Subsequently, tectonic forces have warped, fractured and faulted the crust of Mars.

The following overview of geologic structure and tectonics on Mars is primarily objective; it is based largely on features that can be recognized on Viking spacecraft images. Theoretical stress models and hypotheses relating to the origins of structural dislocations such as faults, fractures or ridges, though very important, are varied and change from time to time; detailed discussion of the deformation of the Martian lithosphere and various theoretical formulations used to model its behavior are provided by, among others, Banerdt, Golombek and Tanaka (1992).

Background

The structural geology of Mars, the study of its reshaped crust, deals with the geometrical shapes and positions of its deformed rocks and with forces that have displaced them. However, the tectonic activity of Mars is so closely related in space and time to the planet's volcanic evolution that it is difficult to assess the roles of individual events in terms of cause and effect.

The prime example of this volcanotectonic relationship is the Tharsis Montes, a huge volcanic mountain chain extending for more than 5000 km across the western equatorial region of Mars (Plate 18). Volcanoes forming the chain lie along the crest of a highly faulted northeast-trending structural arch that rises some 10 km above the average elevation. Deep-seated igneous intrusive bodies are also concentrated in the Tharsis region; their locations are inferred where fault swarms have been deflected around their central cores (Scott, 1982; Scott and Dohm, 1990a). The intrusives have no observable topographic or other surface expression and, with one exception, are not associated with lava flows. These features are too small to be shown on the small-scale maps (Plate 18), but their concentric fault patterns resemble those around some of the larger volcanic constructs such as Alba Patera. Large graben and normal faults produced by uplift and crustal extension form dense arrays both parallel and radial to the Tharsis arch. The most conspicuous of the radial structures are the huge rift systems of Valles Marineris (Plate 18), which extend several thousand kilometers from Tharsis, and the adjacent, densely fractured dome complex of Syria Planum (Lucchitta *et al.* 1992).

Although the origin and evolution of the Tharsis rise is poorly understood, gravity models indicate that Tharsis and its associated tectonic features can be explained by volcanic construction (Solomon and Head, 1982) or by isostatic uplift (Sleep and Phillips, 1985).

Faulting elsewhere on Mars is relatively minor and, with the exception of the Tharsis region and local areas around major impact basins, tectonism as expressed by faulting is almost nonexistent beyond the limits of volcanic fields. Although we cannot state categorically that all faults on Mars have normal separations produced by extensional forces, little direct evidence exists for reverse, thrust and strike-slip faults indicative of crustal compression. However, the occurrence of such faults has been postulated (Plescia and Golombek, 1986; Watters, 1988; Schultz, 1989; Golombek, Plescia and Franklin, 1991), especially in relation to the formation of wrinkle ridges. Relatively minor strike-slip faults may offset the wall rock of Valles Marineris in places (Witbeck, Tanaka and Scott, 1991).

Long, sinuous, somewhat uniformly spaced wrinkle ridges are ubiquitous on intercrater volcanic plains that cover large areas in the Martian highlands and almost encircle the entire Tharsis region. Although the ridges are recognized as structural features, their origin is controversial; like similar structures on the lunar maria and on the planet Mercury, they have been variously interpreted to be folds formed by compression (Howard and Muehlberger, 1973; Plescia and Golombek, 1986; Watters and Maxwell, 1986; Watters, 1988, Golombek, Plescia and Franklin, 1991) or volcanic extrusions and dikes intruded along extensional faults and grabens (Hartmann and Wood, 1971; Strom, 1971; Hodges, 1973; Scott, 1973; Young *et al.* 1973). Although most investigators believe that wrinkle ridges are formed by compressional stress, some evidence (Scott, 1989) suggests a causal relation between wrinkle ridges and extensional structures such as grabens and pit crater chains. Maxwell (1989) has also observed that not all ridges are of compressional origin and that the origin of many ridges is problematic.

Paleotectonic maps of the polar regions are not included here because relatively few faults occur at high latitudes. Ridges are common, however, in volcanic plains in the south polar region (Tanaka and Scott, 1987).

Methodology

The history of tectonic events on Mars is recorded by the global distribution of fault and ridge systems mapped from Viking Orbiter images. These structures were classified by stratigraphic age determined by mapping their occurrence in rock units emplaced during the three geologic periods of Mars (from oldest to youngest, these are the Noachian, Hesperian and Amazonian Periods), and the eight subdivisions of these periods (Scott and Carr, 1978; Tanaka, 1986). From these data a series of paleotectonic maps was made showing fault and ridge systems on Mars as they developed during the geologic past (Scott and Dohm, 1990a,b); the maps have been simplified for presentation here (Plate 18).

The boundaries of rock units in the three time-stratigraphic systems were extracted from the global geologic maps of Mars at 1 : 15 000 000 scale (Scott and Tanaka, 1986; Greeley and Guest, 1987; Tanaka and Scott, 1987). The chronological positioning of structures on the maps adheres to basic geologic principles: (1) faults and ridges of younger age (Amazonian) may extend across boundaries of older rock units (Hesperian, Noachian), (2) structures of intermediate age (Hesperian) may extend into older units (Noachian) but not into Amazonian rocks, and (3) Noachian structures occur only in Noachian rocks. However, some structures that are wholly confined to their Noachian (or Hesperian) host rocks and not in contact with younger units are indeterminate in age.

Volcanoes on Plate 18 have been mapped using Viking orbital images (Scott and Tanaka, 1986; Greeley and Guest, 1987). Most of the large volcanoes are surrounded by lava flows of relatively young age (Hesperian, Amazonian) that conceal any evidence of earlier events. The association of the Tharsis volcanoes with fault systems of Noachian age suggests, however, that volcanism in this region began early in Mars' history (Scott and Tanaka, 1981).

Tectonic history: fault and ridge development with time

Noachian Period

Rocks of the Noachian System are exposed in about 70% of the highland region; they are almost absent, however, north of the highland – lowland boundary where they presumably are buried by younger materials (Plate 18A). The boundary scarp separating these two regions is most prominent east of the prime meridian; farther west it is partly buried by young lava flows from the Tharsis Montes and Olympus Mons. Although no morphologic evidence of the origin and composition of Noachian rocks is recognizable on Viking images, the rocks are probably volcanic materials, impact breccias and eolian deposits that mantle and subdue parts of the heavily cratered terrain.

Faulting was concentrated along the northeast – southwest extensions of the Tharsis rise and formed a complex fan-shaped pattern south of Arsia Mons and Syria Planum; in these areas dense arrays of intersecting and cross-cutting fault systems dissect some of the oldest rocks on Mars (Scott and King, 1984; Scott and Tanaka, 1986; Tanaka and Davis, 1988; Rotto and Tanaka, 1990; Scott and Dohm, 1990b). Southwest of Tharsis, fault systems trend more westerly and appear to follow the highland – lowland boundary. Wherever patches of Noachian rocks are exposed around Tharsis volcanic centers, the rocks are highly faulted; this tectonic maze undoubtedly persists throughout the entire Tharsis region beneath the cover of young lava flows. Structures in this volcanic mountain chain were not produced by a single episode of deformation; they represent the cumulative effects of a long series of volcanotectonic events extending from the Early Noachian to the Amazonian. This protracted sequence of tectonic activity is shown not only by the stratigraphic position of faulted rocks but also by large departures of some fault systems and volcanoes from the dominant northeast structural trend of Tharsis. Large grabens of Middle and Late Noachian age partly encircle the ancient impact basin at Isidis Planitia, showing that structural deformation continued around the basin over an extended period; elsewhere in the eastern equatorial region very few faults have been mapped (Greeley and Guest, 1987). Fault density decreases in Upper Noachian rocks, but this observation may be more apparent than real because exposures of these rocks are sparse around tectonically active centers. We suspect that many more faults of Noachian age might be visible if the cover of young lava flows around the volcanoes in Elysium Planitia were stripped away. Unlike faulting in the Tharsis region, however, that in the eastern equatorial region was much less pervasive throughout the three periods of Martian geologic history.

Compared with the lunar-type wrinkle ridges that are profuse in Martian lava plains of Hesperian age, ridges in Noachian rocks generally are larger and farther apart, and intervening areas are more rugged and highly cratered. In places the Noachian ridges are asymmetric in profile with a steep front, and most appear to have been formed by normal faulting (Scott and Tanaka, 1986; Greeley and Guest, 1987). These older ridge systems also occur in lava plains, generally adjacent to or near the ridged plains of Hesperian age. In the western equatorial region, like their younger counterparts, the Noachian ridges are broadly concentric around the southwestern extension of the Tharsis arch. Most older ridge systems are concentrated, however, in the central equatorial region from Nochis Terra (−35° latitude, 15° longitude) to Arabia Terra (30° latitude, 330° longitude), where they commonly follow the trend of the highland–lowland boundary.

Hesperian Period

The Hesperian System consists mostly of lava flows of intermediate age in the Martian stratigraphic sequence (Plate 18b). The largest accumulations are the ridged plains flows that cover the Lunae Planum plateau, Hesperia Planum and many other large areas in both the highlands and the lowlands (Scott and Carr, 1978; Greeley and Spudis, 1981; Tanaka, Isbell and Scott, 1988). Characterizing the lava plains planetwide are wrinkle ridges of a type similar to those prevalent on the lunar maria and the plains of Mercury.

Many faults cut Hesperian rocks in the Tempe Terra and Terra Sirenum regions on the northeast and southwest extensions of the Tharsis rise. The fault patterns and the trends of wrinkle ridges are very similar to those in Noachian rocks, which suggests that either (1) structural growth of the Syria Planum and Tharsis rises continued over a long period of geologic time, or (2) deformation in Noachian rocks was largely initiated during the Early Hesperian. In the latter case, many faults occurring solely within Noachian rocks and thus attributed to this period may actually have been formed during the Hesperian or later. A more plausible explanation, however, is that tectonic activity waxed and waned within different areas throughout Noachian and Hesperian times.

Tectonic activity throughout the western equatorial region declined sharply in the Late Hesperian; Upper Hesperian materials largely resurfaced and obliterated fault and ridge structures in underlying Lower Hesperian rocks at Syria and Lunae Plana and on the Tharsis rise. Northeast of Alba Patera, however, radial faulting appears to have occurred in association with this volcano during its development in Late Hesperian or possibly earlier times (Rotto and Tanaka, 1990; Tanaka, 1990). These observations support the conclusions of a study of volcanic and tectonic events in the Tharsis region (Scott and Tanaka, 1980), which suggested that, although tectonic activity decreased with time in the Tharsis region, it was episodic and was concentrated in different localities at different times. In this respect, it is interesting to note that the floors of large outflow channels formed during the Late Hesperian (Scott and Tanaka, 1986; Tanaka, 1986) are mostly unaffected by faulting, but a few ridges appear to postdate the flooding events.

Amazonian Period

The decline in global tectonic activity and volcanism that began during the Late Hesperian continued through the Amazonian Period. The style of volcanic eruptions also began to shift from fissure vent flows that formed relatively level plains to central vent eruptions that produced large volcanic mountain constructs (Plate 18C). One of the most significant volcanic events postulated for this period was the extrusion of possible ash-flow tuffs in southern Elysium Planitia (Scott and Tanaka, 1981, 1986).

In the western hemisphere of Mars, Early Amazonian tectonism was largely restricted to the upper flanks of Alba Patera, where faults and large grabens are radial, tangential and concentric to the central core of this large volcano. Minor faulting occurred on the central constructs and lower flanks of the major Tharsis Montes volcanoes and on some of the older aureole deposits of Olympus Mons. Only a few ridges occur in Amazonian rocks, mostly in Arcadia Planitia. The decrease in the areal extent of lava flows and in the number of structures formed during the Middle to Late Amazonian reflects the interdependence of volcanic and tectonic activity on Mars.

In the eastern hemisphere, major faulting was concentrated in extensive Lower Amazonian lava flows that cover the plains around Elysium Mons and embay the older, smaller volcanoes of Albor and Hecates Tholi. Although some faults and grabens are concentric to Elysium Mons, most have northwesterly trends that closely parallel the highland–lowland boundary scarp. North–trending ridges are common in the Arcadia and Elysium Planitiae region, where they follow alignments of hills and knobby terrain of an older mountain range (Phlegra Montes) of possible tectonic origin (Elston, 1979). Tectonism and volcanism declined during the Middle and Late Amazonian Epochs in the eastern region as in the western region. A few ridges and faults cut younger Amazonian rocks in Elysium Planitia, and sparse ridges in Isidis Planitia are largely oriented normal to the highland–lowland boundary. The close association between faulting and volcanic activity is also demonstrated around Elysium Mons, the largest volcanic complex in the eastern equatorial region, during the Amazonian Period.

Summary and conclusions

1. Major fault systems on Mars are associated mostly with large volcanic centers and to a lesser degree with impact basins and the highland–lowland boundary.
2. Unlike Earth's diversity of fault types, only displacements indicative of normal faults and grabens have been directly observed on Mars; relatively small strike-slip faults, however, have been inferred in local areas of Valles Marineris.

3. The Tharsis region has undergone the most intense and the longest duration of faulting, extending from the Early Noachian to the Late Amazonian.
4. Tectonic adjustments around ancient impact basins, especially at Isidis Planitia, continued for some time – into the Late Noachian and possibly Early Hesperian.
5. The apparent parallel association of the highland–lowland boundary scarp with both faults and ridges in the highlands and lowlands supports a tectonic origin for the crustal dichotomy – possibly a giant impact or impacts or subcrustal erosion and crustal lowering.
6. Both faulting and volcanism have progressively decreased from the Noachian to the Amazonian Periods.
7. Martian ridge systems are of at least two types: wrinkle ridges similar to those on lunar maria and larger, older ridges with steep flanks resembling fault scarps in places.
8. Wrinkle ridges are everywhere associated with plains-forming lava flows mostly extruded during the Early Hesperian.
9. Both types of ridges partly encircle the Tharsis rise, supporting the concept of structural uplift that occurred largely during the Noachian and Hesperian Periods.

David H. Scott and James M. Dohm

Bibliography

Banerdt, W.B., Golombek, M.P. and Tanaka, K.L. (1992) Stress and tectonics on Mars in *Mars* (eds H.H. Kieffer, B.M. Jakosky and C.W. Snyder). Tucson: University of Arizona Press, pp. 249–97.
Elston, W.E. (1979) Geologic map of the Cebrenia quadrangle of Mars. *US Geol. Surv. Misc. Inv. Ser. Map I-1140*, scale 1: 5 000 000.
Frey, H.V. and Schultz, R.A. (1989) Overlapping impact basins and the origin of the Martian crustal dichotomy, Elysium, and Tharsis. Paper presented at *Fourth International Conference on Mars*, NASA, January 10–13, 1989, Tucson, Ariz.
Golombek M., Plescia, J.B. and Franklin, B.J. (1991) Faulting and folding in the formation of planetary wrinkle ridges, in *Proc. Lunar Planet. Sci. Conf.* **21**, pp. 679–93.
Greeley R., and Guest, J.E. (1987) Geologic map of the eastern equatorial region of Mars. US Geol. Surv. Misc. Inv. Ser. Map I-1028B, scale 1: 15 000 000.
Greeley R. and Spudis, P.D. (1981). Volcanism on Mars. *Rev. Geophys. Space Phys.*, **19**, 13–41.
Hartmann, W.K. and Wood, C.A. (1971) Moon: origin and evolution of multi-ring basins. *The Moon*, **3**(1), 3–78.
Hodges, C.A. (1973) Mare ridges and lava lakes, in *Apollo 17 Preliminary Science Report*. NASA SP-330, pp. 12–21.
Howard, K.A. and Muehlberger, W.R. (1973) Lunar thrust faults in the Taurus-Littrow region, in *Apollo 17 Preliminary Science Report*, NASA SP-330, pp. 22–5.
Lucchitta, B.K., McEwen, A.S., Clow, G.D., *et al.* (1992) The canyon systems on Mars, in *Mars* (eds H.H. Kieffer, B.M. Jakosky and C.W. Snyder). Tucson: University of Arizona Press, pp. 453–92.
Maxwell, T.A. (1989) Origin of planetary wrinkle-ridges – an overview (abstract), in *MEVTV Workshop on Tectonic Features on Mars* (eds T.R. Watters and M.P. Golombek). Houston: Lunar and Planetary Institute, LPI Tech. Rpt. 89–06, pp. 41–3.
Phillips, R.J. (1988) The geophysical signal of the Martian dichotomy. *EOS*, **69**, 389.
Plescia, J.B. and Golombek, M.P. (1986) Origin of planetary wrinkle ridges based on the study of terrestrial analogs. *Geol. Soc. Am. Bull.*, **97**, 1289–99.
Rotto, S.L. and Tanaka, K.L. (1990) Faulting history of the Alba Patera–Ceraunius Fossae region of Mars, in *Proc. Lunar and Planet. Sci. Conf.* **20**, pp. 515–23.
Schultz, R.A. (1989) Strike-slip faulting in the ridged plains of Mars (abstract), in *MEVTV Workshop on Tectonic Features on Mars* (eds T.R. Watters and M.P. Golombek). Houston: Lunar and Planetary Institute, LPI Tech. Rpt. 89–06, pp. 49–51.
Scott, D.H. (1973) Small structures of the Taurus-Littrow region, in *Apollo 17 Preliminary Science Report*, NASA SP-330, pp. 25–8.
Scott, D.H. (1982) Volcanoes and volcanic provinces: Martian western hemisphere. *J. Geophys. Res.*, **87**, 9839–51.
Scott, D.H. (1989) New evidence – old problems: wrinkle ridge origin (abstract), in *MEVTV Workshop on Tectonic Features on Mars* (eds T.R. Watters and M.P. Golombek). Houston: Lunar and Planetary Institute, LPI Tech. Rpt. 98–06, pp. 52–4.
Scott, D.H. and Carr, M.H. (1978) Geologic map of Mars. US Geol. Surv. Misc. Inv. Ser. Map I-1083, scale 1: 25 000 000.
Scott, D.H. and Dohm, J.M. (1990a) Faults and ridges: historical development in Tempe Terra and Ulysses Patera regions of Mars, in *Proc. Lunar Planet. Sci. Conf.* **20**, pp. 503–13.
Scott, D.H. and Dohm, J.M. (1990b) Chronology and global distribution of fault and ridge systems on Mars, in *Proc. Lunar Planet. Sci. Conf.* **20**, pp. 487–501.
Scott, D.H. and King, J.S. (1984) Ancient surfaces of Mars: the basement complex (abstract) in *Lunar and Planet Sci. XV* pp. 736–7.
Scott, D.H. and Tanaka, K.L. (1980) Mars Tharsis region: volcano-tectonic events in the stratigraphic record, in *Proc. Lunar Planet. Sci. Conf.* **11**, pp. 2403–21.
Scott, D.H. and Tanaka, K.L. (1981) Ignimbrites of Amazonis Planitia region of Mars. *J. Geophys. Res.*, **87**, 1179–90.
Scott, D.H. and Tanaka, K.L. (1986) Geologic map of the western equatorial region of Mars. US Geol. Surv. Misc. Inv. Ser. Map I-1802-A, scale 1:15 000 000.
Sleep, N.H. and Phillips, R.J. (1985) Gravity and lithospheric stress on the terrestrial planets with reference to the Tharsis region of Mars. *J. Geophys. Res.*, **90**, 4469–89.
Solomon, C.S. and Head, J.W. (1982) Evolution of the Tharsis province of Mars: the importance of heterogeneous lithospheric thickness and volcanic construction. *J. Geophys. Res.*, **89**, 9755–74.
Strom, R.G. 1971. Lunar mare ridges, rings and volcanic ring complexes. *Mod. Geol.*, **2**, 133–57.
Tanaka, K.L. (1986) The stratigraphy of Mars, in *Proc. Lunar Planet. Sci. Conf.* **17**, *J. Geophys. Res.*, **91**, E139–58.
Tanaka, K.L. (1990) Tectonic history of the Alba Patera–Ceraunius Fossae region of Mars, in *Proc. Lunar Planet. Sci. Conf.* **20**, pp. 515–23.
Tanaka, K.L. and Davis, P.A. (1988) Tectonic history of the Syria Planum province of Mars. *J. Geophys. Res.*, **93**, 14893–917.
Tanaka, K.L. and Scott, D.H. (1987) Geologic map of the polar regions of Mars. US Geol. Surv. Misc. Inv. Ser. Map I-1802-C, scale 1: 15 000 000.
Tanaka, K.L., Isbell, N.K. and Scott, D.H. (1988) The resurfacing history of Mars: a synthesis of digitized, Viking-based geology in *Proc. Lunar Planet. Sci. Conf.* **18**, 665–78.
Watters, T.R. (1988) Wrinkle ridge assemblages of the terrestrial planets. *J. Geophys. Res.*, 93, 10236–54.
Watters, T.R. and Maxwell, T.A. (1986) Orientation, relative age, and extent of the Tharsis plateau ridge system. *J. Geophys. Res.*, **91**, 113–25.
Wilhelms, D.E. and Squyres, S.W. (1984) The martian hemispheric dichotomy may be due to a giant impact. *Nature*, **309**, 7934–9.
Wise, D.U., Golombek, M.P. and McGill, G.E. (1979) Tharsis province of Mars: geologic sequence, geometry, and a deformation mechanism. *Icarus*, **38**, 4456–72.
Witbeck, N.E., Tanaka, K.L. and Scott, D.H. (1991) Geologic maps of the Valles Marineris region of Mars. *US Geol. Surv. Misc. Inv. Ser. Map I-2010*, scale 1: 2 000 000.
Young, R.A., Brennan, W.J., Wolfe, R.W. and Nichols, D.J. (1973) Volcanism in the lunar maria, in *Apollo 17 Preliminary Science Report*. NASA SP-330, pp. 31–1 to 31–11.

Cross references

Fracture, fault
Impact cratering
Seismicity
Tectonics
Volcanism in the solar system

MASCON

Mascon is a contraction of two words: 'mass' and 'concentration'. A mascon is a positive gravity anomaly located in a topographic lowland.

In 1968 P.M. Muller and W.L. Sjogren of the Jet Propulsion Laboratory/California Institute of Technology discovered unique gravitational features on the Moon. The gravitational effects of these

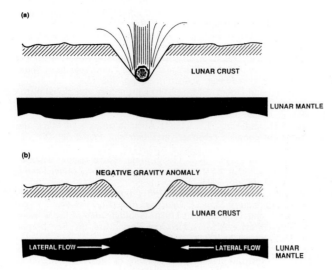

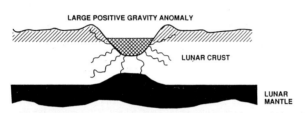

Figure M58 Mascon formation. (a) Impact of crater-forming body. (b) Rebound of mantle beneath crater. (c) Billions of years later: radioactive heating generated lavas, which filled the basins and crustal fractures. The mantle is cooler and more rigid.

features were revealed when the accelerations in the radio Doppler data from the unmanned Lunar Orbiter V were mapped across the lunar surface. Large positive gravity anomalies were centered over the ringed maria basins (Imbrium, Serentatis, Crisium, Nectaris and Humorum). When reporting their findings in *Science*, they named these unusual features mascons, for they could only explain their gravitational characteristics as a result of additional mass, or mass concentrations, in these basins.

Normally a basin has a deficiency of mass, but here just the opposite was observed. There have been many publications on the origin and internal structure of the mascons. For instance, Harold Urey had proposed that these large circular basins were caused by large dense impacting meteorites and that their remains now lay in the basins. Laboratory studies, however, have shown that this could not have occurred, for the objects would have vaporized and would have been dispersed in the gigantic explosions. The presently accepted explanation is that the initial basin (precursor to the mascon) was formed by a large impactor. This is thought to have happened some 3.5–4 billion years ago, when the Moon was much warmer. At that time the Moon is thought to have had a crust overlying a viscous inner layer. After the impact and the formation of a large crater, the Moon reacted to relax the large stresses at the surface by equalizing the gravitational pressure. Denser material was introduced beneath the crater by lateral transport (Figure M58). After this adjustment there was a small negative gravitational anomaly in and above the basin, as would normally be the case. During the next billion years or so, however, the Moon cooled considerably and became more rigid. During the next billion years the basins were flooded by lavas (from radioactive heating), which partially filled them with sufficient additional mass to cause the large positive gravitational anomalies observed today.

Analyses of gravitational data of the Earth, Venus and Mars has revealed only one similar feature. It is on Mars, in the Isidis basin. Mars is colder than the Earth or Venus, and much smaller in diameter. Its evolution could easily follow the lunar model.

William L. Sjogren

Bibliography

Gilvarry, J.J. (1970) The origin and nature of lunar mascons. *Radio Sci.*, **5**, 313.
Muller, P.M. and Sjorgren, W.L. (1968) Mascons: lunar mass concentrations. *Science*, **161**, 680.
Science (1968) Mascons interpreted, **162**, 1402 (Many authors and several articles).
Wise, D.U. and Yates, M.T. (1970) Mascons as structural relief on a lunar Moho. *J. Geophys. Res.*, **75**, 261.

Cross references

Gravimetry
Impact cratering
Moon: Geology
Moon: Gravity

MASS EXTINCTION

A mass extinction may be defined as a marked increase in rates of extinction of geographically widespread higher taxa during a relatively short interval. This results in a major decline in standing diversity (Sepkoski 1989). Mass extinctions during the last 600 Ma are estimated to have involved from ~ 30% to more than 95% of extant species. Mass extinctions of life are currently a subject of considerable debate. Some authors have concluded that mass extinction events are qualitatively similar to background extinctions, and hence reflect similar evolutionary processes (e.g. McKinney 1987). However, most studies of the fossil record suggest that large-scale evolutionary patterns have been shaped by two different regimes – the background extinction regime and the mass extinction regime. In this view, micro-evolutionary and macro-evolutionary processes of the background regime were occasionally disrupted, presumably by external forcing agents, and replaced by the mass extinction regime. During these intervals, many traits that had enhanced the survival of species became ineffective, and other traits that were not previously closely correlated with survivorship became important (Jablonski 1986).

Mass extinctions may be the result of environmental stresses so rare as to be beyond the 'experience' of the organisms, and extinction may be just a matter of the chance susceptibility of the organisms to these rare stresses, such as the effects of planetesimal impacts. The result might be a highly selective extinction, but one having little or no relationship to the general success of organisms in normal times (Raup 1986). Raup (1990) has recently suggested that perhaps all extinction events of various intensities may be the result of planetesimal impacts of various sizes (see also Rampino and Haggerty, 1994).

The mass extinction regime is apparently relatively short-lived – very often below the resolving power of the geologic record (< 10 000 years), and possibly instantaneous. By removing dominant taxa and enabling other, previously unimportant groups to undergo adaptive radiation, mass extinctions may play a larger role than is generally appreciated in creating opportunities for faunal change.

Sepkoski (1989) compiled a revised master list of the appearances and disappearances of taxa in the fossil record. The record of extinctions of marine invertebrates at the genus level over the last 600 M (for revisions, see Sepkoski, 1992) (the Phanerozoic) is shown in Figure M59, which shows 23 peaks in extinction rates (the average is one every 26 Ma). Five very severe reductions in global diversity have occurred during this interval, and these may be considered first-order mass extinctions: the Ordovician-Silurian boundary (peak 6), the Frasnian–Famennian boundary (Late Devonian; peak 8), the Permian–Triassic boundary (peak 13), the Late Triassic (peak 16, high proportional extinction rate) and the Cretaceous–Tertiary boundary (peak 20). These are all major upheavals in the history of life – the Cretaceous Tertiary boundary (65 Ma BP), for example, is marked by the extinction of some 75% of marine species, and the Permian–Triassic boundary (~ 250 Ma BP) seems to have involved

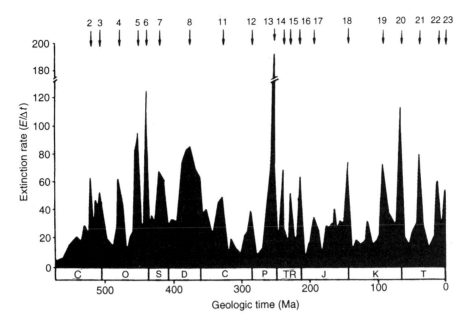

Figure M59 Absolute rate of extinction of marine genera per geologic stage (or substage) during the Phanerozoic (after McGhee, 1989, data from Sepkoski, 1989). Twenty-three significant peaks are recognizable, they are as follows: (1) Bottomian, (2) Dresbachian, (3) Trempealeauan, (4) Arenigian, (5) Caradocian, (6) Ashgillian, (7) Wenlockian, (8) Givetian, (9) Frasnian, (10) Famennian, (11) Serpukhovian, (12) Stephanian, (13) Guadalupian, (14) Olenekian, (15) Carnian, (16) Norian, (17) Pliensbachian, (18) Tithonian, (19) Cenomanian, (20) Maastrichtian, (21) Late Eocene, (22) Middle Miocene, (23) Pliocene. C = Cambrian, O = Ordovician, S = Silurian, D = Devonian, C = Carboniferous, P = Permian, TR = Triassic, J = Jurassic, K = Cretaceous, T = Tertiary.

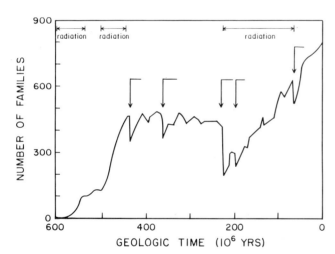

Figure M60 Phanerozoic record of diversity of families of marine invertebrates. Three principal radiations are indicated at the top; arrows mark the five major mass extinctions, and the bars delineate the recovery time for familial diversity. (After Erwin, Valentine and Sepkoski, 1987.)

the remarkable extinction of about 96% of marine species (Raup, 1979). Many geological boundaries are defined by these increases in extinction rates.

Long-term trends in diversity

Despite the occurrence of these mass extinctions, the record of marine familial diversity seems to show a general increase over the last 600 million years (Figure M60). On a family level at least, most of the marine mass extinctions seem to be minor setbacks in a climb toward greater diversity over the last 600 Ma. The background extinction rate may have been decreasing over this period, but sampling of the fossil record may be producing an artifact here.

Only the great Late Permian extinction shows up as a considerable readjustment in familial diversity, with a drop from more than 400 to less than 200 marine families. Numbers of marine families did not recover to the 400 level until the Late Jurassic, some 90 million years later. But family-level diversity may be somewhat misleading, since many families that survived the mass extinction events were greatly diminished at the genus and species levels.

Various workers have interpreted the life diversity curve in different ways. Valentine and Moores (1977) proposed that marine diversity was related to continental shelf area and sea level, hence times of continental accretion or drop in sea level marked a reduction of shelf area and decrease in diversity. By contrast, times of continental break-up afforded more shelf area, and the possibility of greater marine diversity.

Major changes in diversity over the last 600 million years have also been correlated with overturn of anoxic ocean waters, possibly brought about by major changes in climate and sea level (Rampino and Caldeira, 1992). In these interpretations, diversity may be dependent upon oceanographic and/or climatic conditions which are driven by tectonics and climatic change. However, the correlation of extinctions with inferred glaciations at ~ 435 Ma (Late Ordovician), ~ 365 Ma (Late Frasnian) and ~ 250 Ma (Late Permian) may be possibly explained in terms of impacts, as some of the evidence for glaciation – tillites – may really be impact ejecta (Rampino, 1992).

The diversity curve has also been interpreted in terms of equilibrium dynamics. Three distinctive major marine faunas occur during the Phanerozoic (Figure M60). Fauna I was dominant during the Cambrian, with a low equilibrium diversity. During the Ordovician, fauna II replaced it, and diversity reached a higher equilibrium level, forming a broad plateau through the rest of the Paleozoic. Fauna II lost its dominance at the Permian mass extinction, and was replaced by fauna III, which today seems to be climbing toward a higher equilibrium level. The plateau reached by fauna II suggests either some carrying capacity that can support only so much diversity, or a long-term inertia in the system which prevents the kind of biological innovation that leads to increased diversity. Erwin, Valentine and Sepkoski (1987) suggest that stasis during the Paleozoic may have been the result of (1) environments filled with life; (2) evolution of increasingly effective predators, making it difficult for new forms to evolve; or (3) existing organisms that were too specialized to give rise easily to new types.

Mass extinctions would open up adaptive space on a large scale, kill off predators and selectively remove specialized organisms. A prime example is the great radiation of the mammals following the dinosaur mass extinction at the Cretaceous Tertiary boundary (65 Ma). Had the dominant reptiles not disappeared, the mammals might have remained insignificant and marginally successful. In this way, mass extinctions may represent a major process that drives the evolution of life on Earth.

Evolution by catastrophe

Alvarez (1987), among others, offered the suggestion that survival during mass extinctions is largely the accident of the members of any species being so numerous and/or widespread that, even with very high 'kill rates', some of the population manages to escape destruction and lives through the crisis interval. It helps, of course, to have the luck of being 'preadapted to' or able to somehow avoid the heat blast and wildfires, cold, darkness, volcanic eruptions, acid rain and various other conditions caused by large impacts.

The idea of survival of the 'luckiest' in the case of catastrophic mass extinctions is in marked disagreement with conventional Darwinian concepts of speciation as a result of competition among organisms for food and living space. Darwin also stressed the gradual nature of evolution; more recently Eldredge and Gould (1972) proposed that speciation occurs rapidly at times of environmental stress – the intervals between speciation being marked by a general stasis, with little evolutionary change. In this view, any mutations during the equilibrium would be eliminated by the restoring force of the general ecological equilibrium. This 'punctuated equilibrium' model of speciation has generated much discussion, but with a growing consensus that rapid speciation events occur at times of rapid environmental changes and selective pressures.

By contrast, evolution by catastrophe means speciation through a different mechanism; every so often, many species are killed off in a global catastrophe, niches are vacated en masse, and the survivors are free to speciate to fill the many empty niches. The entire biosphere is affected, and the process of speciation following the catastrophe is intimately related to the recovery of the biosphere as a whole. Gould (1989) has called this new kind of evolution 'evolution by lottery', and suggests that it might be the most important form of evolution in terms of defining the major currents in the history of life on Earth.

Impact cratering

Impact craters, enhanced iridium content of stratigraphic layers, microtektites and shocked quartz provide evidence for the occurrence of large body impacts on the Earth (Rampino and Haggerty, 1994). The Alvarez group (1980) and other scientists present such physical evidence that the late Cretaceous extinctions (65 Ma) coincided with a large-body impact. Such indicators of impact have now been reported in rock layers at or close to the times of at least six other extinction events, at ~ 2.3 Ma (Pliocene), ~ 36 Ma (Late Eocene–Early Oligocene), ~ 92 Ma (Cenomanian–Turonian), ~ 20 Ma (Late Triassic), ~ 250 Ma (Permian–Triassic), and ~ 365 Ma (Frasnian–Fammenian) (Orth, 1989; Holser and Schonlaub 1990; Claeys, Casier and Margolis 1992), although the concentrations of iridium in some cases are in the range of hundreds of ppt as opposed to thousands of ppt Ir for the Cretaceous–Tertiary boundary. This is to be expected, however, as the iridium content of comets and different kinds of asteroids varies considerably.

The iridium anomalies at some of the geological boundaries are still considered equivocal. For example, five iridium anomalies were discovered associated with extinctions at the Cenomanian–Turonian boundary at ~ 92 Ma (Orth 1989). Other trace metals found along with the iridium peaks, however, suggest a possible volcanic source. A research group in China took samples at the most severe extinction event on record, the mass die-off that marks the boundary between the Permian and Triassic Periods, about 250 million years ago. Their analyses seemed to show a strong iridium anomaly in a clay layer at the boundary in China, but other groups have been unsuccessful thus far in replicating these measurements, although weaker anomalies have been found (Orth 1989). The same boundary in Austria shows two small iridium anomalies that coincide with significant changes in carbon and oxygen isotope ratios, indicating sudden decreases in oceanic biomass, and severe environmental perturbations (Holser and Schonlaub 1989).

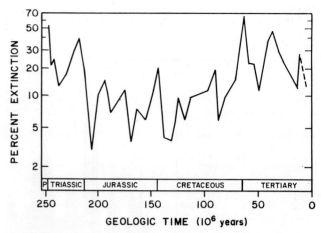

Figure M61 Mass extinction events (marine invertebrates) of the last 250 million years. Ordinate scale is percent extinction per geologic stage on a family level. (After Raup and Sepkoski, 1984.)

Another iridium peak was found associated with the Frasnian–Fammenian extinction at about 365 million years ago, but here the high levels of trace metals occurred within the fossilized remains of a calcareous algae, and biological concentration of the iridium and other elements was suspected (Orth 1989). The recent discovery of microtektites near that boundary, and the possibility that the algal mats are actually a disaster marker, however, supports the occurrence of an impact-induced extinction (Claeys, Casier and Margolis 1992)

Periodicity of extinctions

Over the last 250 Ma (the Mesozoic and Cenozoic Eras), where the geologic record is most clear, up to 12 marine extinction events of varying severity at the family and genus level have been noted (Raup and Sepkoski, 1984, 1986; Figure M61). The distribution of these times of increased extinction rate looks quite regular, with extinction events coming about once every 26 to 30 million years (Raup and Sepkoski, 1984, 1986). An earlier study by Fischer and Arthur (1977) had suggested that biological crises and associated changes in ocean circulation and climate came in an approximately 32 million year cycle.

Sepkoski's (1982 and revised 1989, 1992) data, however, are much more complete, and careful statistical studies of the marine mass extinction record of the last 250 million years have been carried out using these data. Raup and Sepkoski (1984, 1986) found that the die-offs fit very well into a periodicity of ~ 26 million years. Their statistical tests argued strongly that this period was not accidental, which suggested that the extinctions might have a single, cyclical underlying cause. Although the extinction periodicity has been criticized by some paleontologists and statisticians, the latest analyses, using more complete paleontological data, have strengthened the finding of a 26 to 30 million year cycle although the record seems to be a mixture of periodic and random events (Raup and Sepkoski, 1986; Sepkoski, 1989; Rampino and Caldeira, 1993). Extinctions of non-marine tetrapods (e.g. reptiles and mammals) (Benton, 1989) may follow approximately the same cycle (Rampino, 1988). A longer cycle of the most severe extinctions of about 180 million years may also be present (Rampino and Caldeira, 1992).

The discovery of a possible periodicity of 28 to 32 million years in terrestrial impact craters (Rampino and Stothers, 1984a,b; Alvarez and Muller, 1984; Yabushita, 1992) supports the idea that periodic impacts, possibly from comet showers related to galactic cycles, caused the periodic component of mass extinctions.

Michael R. Rampino

Bibliography

Alvarez, L.W. (1987) Mass extinctions caused by large bolide impacts. *Physics Today*, **40**, 24–33.

Alvarez, L.W., Alvarez, W., Asaro, F. and Michel, H. (1980) Extraterrestrial cause for the Cretaceous–Tertiary extinction. *Science*, **208**, 1095–108.
Alvarez, W. and Muller, R.A. (1984) Evidence from crater ages for periodic impacts on the Earth. *Nature*, **308**, 718–20.
Benton, M.J. (1989) Patterns of evolution and extinction in vertebrates, in *Evolution and the Fossil Record* (eds K.C. Allen and D.E.G. Briggs) Washington: Smithsonian. pp. 218–41.
Claeys, P., Casier, J.-G. and Margolis, S.V. (1992) Microtektites and mass extinctions: evidence for a Late Devonian asteroid impact. *Science*, **257**, 1102–4.
Eldredge, N. and Gould, S.J. (1972) Punctuated equilibria: an alternative to phyletic gradualism, in *Models in Paleobiology* (ed. T.J.M. Schopf). San Francisco: Freeman Cooper, pp. 82–115.
Erwin, D.H., Valentine, J.W. and Sepkoski, J.J., Jr (1987) A comparative study of diversification events: the early Paleozoic versus the Mesozoic. *Evolution*, **41**, 1177–86.
Fischer, A.G. and Arhur, M.A. (1977) Secular variations in the pelagic realm. *Soc. Econ. Paleont. and Mineral. Spec. Publ.* **25**, pp. 19–50.
Gould, S.J. (1989) The wheel of fortune and the wedge of progress. *Natural History*, No. 3, 14–21.
Holser, W.T., and Schönlaub, H.P. (eds) (1991) *The Permian–Triassic Boundary in the Carnic Alps of Austria (Gartnerkofel Region)*, Abhand. der Geologisch Bundesanstalt, **45**, 232 pp.
Jablonski, D., 1986. Causes and consequences of mass extinctions: A comparative approach, in *Dynamics of Extinction* (ed. D.K. Elliott). New York: Wiley, pp. 183–229.
McGhee, G.R., Jr, (1989) Catastrophes in the history of life, in *Evolution and the Fossil Record* (eds K.C. Allen and D.E.G. Briggs. Washington: Smithsonian, pp. 26–50.
McKinney, M.L. (1987) Taxonomic selectivity and continuous variation in mass and background extinctions of marine taxa. *Nature*, **325**, 143–5.
Orth, C.J. (1989) Geochemistry of the bio-event horizons, in *Mass Extinctions, Processes and Evidence* (ed. S.K. Donovan). New York: Columbia, pp. 37–72.
Rampino, M.R. (1988) Are marine and non-marine extinctions correlated. *EOS, Trans. Am. Geophys. Union*, **69**, 889–95.
Rampino, M.R. (1992) Ancient 'glacial' deposits are ejecta of large impacts: the ice age paradox explained. *EOS, Trans. Am. Geophys. Union*, **73**, 99.
Rampino, M.R. and Caldeira, K. (1993) Major episodes of geologic change: correlations, time structure and possible causes. *Earth Planet. Sci. Lett.*, **114**, 215–27.
Rampino, M.R. and Haggerty, B.M. (1994) Extraterrestrial impacts and mass extinctions of life, in *Hazards Due to Comets and Asteroids* (ed. T. Gehrels). Tucson, Arizona: University of Arizona Press, pp. 827–57.
Rampino, M.R. and Stothers, R.B. (1984a) Terrestrial mass extinctions, cometary impacts and the Sun's motion perpendicular to the galactic plane. *Nature*, **308**, 709–12.
Rampino, M.R. and Stothers, R.B. (1984b) Geological rhythms and cometary impacts. *Science*, **226**, 1427–31.
Raup, D.M. (1979) Size of the Permian–Triassic bottleneck and its evolutionary implications. *Science*, **206**, 217–8.
Raup, D.M. (1986) Biological extinction in earth history. *Science*, **231**, 1528–33.
Raup, D.M. (1990) Impact as a general cause of extinction: a feasibility test. *Geol. Soc. Am. Spec. Pap.* **247**, pp. 27–32.
Raup, D.M. and Sepkoski, J.J., Jr, (1984) Periodicity of extinctions in the geologic past. *Proc. Natl Acad. Sci. USA*, **81**, 801–5.
Raup, D.M. and Sepkoski, J.J., Jr, (1986) Periodic extinctions of families and genera. *Science*, **231**, 833–6.
Sepkoski, J.J., Jr, (1982) A compendium of fossil marine families. *Milwauk. Pub. Museum Contrib. Biol. Geol.*, **51**, 1–125.
Sepkoski, J.J., Jr, (1992) A compendium of fossil marine animal families, 2nd edn. *Milwauk. Pub. Museum Contrib Biol. Geol.*, **83**, 1–156.
Sepkoski, J.J. (1992) Periodicity in extinction and the problem of catastrophism in the history of life. *J. Geol. Soc. London*, **146**, 7–19.
Valentine, J.W. and Moores. E.M. (1970) Plate tectonic regulation of faunal diversity and sea level: a model. *Nature*, **228**, 657–59.
Yabushita, S., (1992) Periodicity in the crater formation rate and implications for astronomical modeling, in *Dynamics and Evolution of Minor Bodies with Galactic and Geological Implications* (eds S.V.M. Clube, S. Yabushita and J. Henrard). Dordrecht: Kluwer, pp. 61–78.

Cross references

Biosphere
Comet impacts on the Earth
Cretaceous–Tertiary boundary
Earth
Galactic cycle
Impact cratering
Iridium anomaly

MAUNDER, EDWARD WALTER (1851–1928), AND MAUNDER MINIMUM

Maunder was the son of a Wesleyan minister, best remembered for two things: his solar research, and as the founder of the British Astronomical Association.

He attended King's College in London, and later secured a post as photographic and spectroscopic assistant at the Royal Greenwich Observatory; he initiated a long series of daily photographic sunspot records there. Eventually he was able to prepare a famous diagram, known as Maunder's 'butterfly diagram', linking the latitudes of sunspot groups with the state of the 11-year solar cycle. He also carried out spectroscopic work. He publicized the apparent dearth of sunspots between the years 1645 and 1715, coinciding with a very cold spell in Europe; this is now formally known as the 'Maunder minimum'. His photographic records of sunspots were meticulous, and are still of great value today.

The Sun was by no means his only interest. He observed the planets, and in a famous experiment demonstrated that the 'canals' on Mars, claimed by Lowell and other observers, were optical illusions. He also made careful studies of the zodiacal light (q.v.).

He took part in various eclipse expeditions, notably the Royal Greenwich Observatory expedition to Mauritius in 1901, during which he obtained an excellent series of photographs. On other expeditions in 1896 and 1898 the weather was unkind.

Between 1877 and 1887 Maunder was coeditor of the *Observatory Magazine*; for many years he served on the Council of the Royal Astronomical Society, and was joint Honorary Secretary from 1892 to 1897. In 1890 he founded the British Astronomical Association, and subsequently served as President. He officially retired from Greenwich in 1913, but was recalled during World War I to carry on the sunspot record.

His first wife died in 1888. He then married Miss Annie Russell, herself a brilliant mathematician and an astronomer on the staff of the Royal Greenwich Observatory, who collaborated with him in much of his later work in connection with the Sun. He wrote several books, including a history of the Royal Observatory (1900), *Astronomy Without a Telescope* (1901) and *Astronomy of the Bible* (1908).

Maunder minimum

The Maunder minimum refers to a period from about 1645 to 1715 when sunspot activity virtually disappeared (Eddy, 1976; Schove, 1983).

The Maunder minimum is one of a number of protracted periods of low sunspot activity; among others identified are the Oort minimum (1010–1050), the Wolf minimum (1280–1340) and the Spörer minimum (1420–1530). It was following the work of Gustav Spörer that Maunder first called attention to the interruption of the normal course of sunspot activity during the second half of the 17th and first part of the 18 centuries. Maunder published his findings in 1894 but the research appeared to have created little academic interest. In 1922 Maunder published another paper with the same title as that produced earlier (*A prolonged sunspot minimum*), but it remained for more recent astronomers and astrophysicists to verify and discuss it.

Eddy (1976) noted that the sunspot dearth, 1645–1715, is supported by direct accounts in contemporary literature of the day. That the low sunspot activity was real, and not caused by a limitation in observing capability, is illustrated by drawings made of the Sun at the time. Additionally, a tree-ring anomaly spanning the same period shows evidence of a concurrent drought in the US Southwest

(Douglass, 1928) and when a ^{14}C lag time is assumed, there is a high degree of agreement in time between major excursions in world temperatures and excursions of solar behavior in the records of ^{14}C.

The coincidence of Maunder's prolonged sunspot minimum with the coldest excursion of the Little Ice Age has been noted by many researchers examining possible relations between solar activity and terrestrial climate (e.g. Schneider and Mass, 1975; Fairbridge, 1987; Grove, 1988). Given the time relationship between the Little Ice Age and the Maunder minimum, the role of solar variability and sunspot activity has also been examined in terms of recent and future climatic change (e.g. Kelly and Wigley, 1990) and as it may be related to possible changes in the Sun's brightness (Baliunas and Jastrow, 1990).

Patrick Moore and John E. Oliver

Bibliography

Anon. (1928) E.W. Maunder, *Mon. Not. Royal Astron. Soc.*, **89**, 318.
Anon. (1928) E.W. Maunder, *J. Br. Astron. Assoc.*, **38**, 231.
Anon. (1926/7) A.S.D. Russell (Mrs Maunder), *J. Br. Astron. Assoc.*, **57**, 238.
Baliunas, S. and Jastrow, R. (1990) Evidence for long-term brightness changes in solar-type stars. *Nature*, **348**, 520–3.
Douglass, A.E. (1928) *Climatic Cycles and Tree Growth*, Vol. 2. Washington, DC: Carnegie Institute.
Eddy, J.A. (1976) The Maunder minimum. *Science*, **192**, 1189–202.
Fairbridge, R.W. (1987) Little Ice Age, in (eds J.E. Oliver and R.W. Fairbridge). *The Encyclopedia of Climatology*, New York: Van Nostrand Reinhold, pp. 547–51.
Grove, J.M. (1988) *The Little Ice Age*. London: Methuen, 498 pp.
Kelly, P.M. and Wigley, T.L.M. (1990) The influence of solar forcing trends on global mean temperature since 1861. *Nature*, **347**, 460–2.
Maunder, E.W. (1894) A prolonged sunspot minimum. *Knowledge*, **17**, 173.
Maunder, E.W. (1922) A prolonged sunspot minimum. *J.Br. Astron. Assoc.*, **32**, 140.
Schneider, S.A. and Mass, C. (1975) Volcanic dust, sunspots and temperature trends. *Science*, **190**, 741–6.
Schove, D.J. (1983) *Sunspot Cycles* (Benchmark Papers in Geology, Vol. 68). Stroudsburg: Hutchinson Ross, 393 p.
Warner, D.J. (1974) Maunder, Edward Walter. *Dict. Sci. Biogr.*, Vol. 9, pp.183–5.

Cross references

Solar activity
Solar motion
Sun

MAXWELL THEORY

The solid, liquid and gaseous states of matter with which we are most familiar are accompanied by a fourth state in the space environment. The upper atmosphere of the planets, and the entire solar system, is filled with a rarefied plasma, in which the energies of the charged particles which make up this medium, mostly protons and electrons, are so great that the electrical forces that bind these particles together as atoms are overcome. Although these plasmas are generally so rarified that they do not interact through normal collisions, and hence are termed collisionless, they do interact with each other through the electromagnetic force. The laws which describe the behavior of the electromagnetic force were first compiled into a self-consistent theory in 1864 by the Scottish physicist James Clerk Maxwell (1831–1879), and since then the theory of electromagnetism has been called Maxwell theory, and the four key equations of electromagnetism called Maxwell's laws or Maxwell's equations.

The theory of electromagnetism divides its effects on electrically charged particles into two categories, electric and magnetic. The total force acting on a charged particle is given by the vector sum of the electric and magnetic forces. This is defined as the Lorentz force, given in System Internationale (SI) and Gaussian units, respectively, by

$$F = q(E + v \times B) \qquad [F = q(E + \frac{v}{c} \times B)] \qquad (M12)$$

where F is the electromagnetic force (in newtons), q is the magnitude of electric charge of a given particle (in coulombs), E is the electric field vector (in V m^{-1}), v is the velocity of the particle (in m s^{-1}), c is the speed of light (in m s^{-1}), and B is the magnetic field vector (in tesla). The equation on the left side of expression (M12) and quantities in parenthesis are in SI units. However, for historical reference, equations will also be given in Gaussian units, in which the electric fields and magnetic fields have equivalent units. In Gaussian units, electric fields are measured in statvolts, magnetic fields in gauss, and the electromagnetic force in dyne cm^{-2}. This system of units is no longer in widespread use, but has been utilized extensively in previous work and can be frequently found in the older literature.

In a plasma, especially, the electric and magnetic fields which act on charged particles are not separate entities, but are self-consistently related to one another. There are four basic laws which govern the behavior of the electric and magnetic fields, which are known collectively as Maxwell's equations (e.g. Jackson, 1962; Dendy, 1990). The first of these laws is Gauss' law, named after Carl Friedrich Gauss (1777–1855, Germany), which relates the electric field E to the electric charge density ρ_q of a volume

$$\nabla \cdot E = \frac{\rho_q}{\epsilon_0} \qquad [\nabla \cdot E = 4\pi \rho_q] \qquad (M13)$$

where ϵ_0 is the permittivity of free space. This expression shows that the amount of total electric flux through a given closed surface is proportional to the amount of electric charge in the volume contained by that surface. This also implies that a particle containing a given electric charge has an electric field associated with it.

The second of Maxwell's equations tells of the continuity of magnetic flux through a surface. Sometimes referred to as Gauss' law for magnetic fields, this expression states that the magnetic field B is divergenceless, and is given by

$$\nabla \cdot B = 0 \qquad (M14)$$

This law is similar to Gauss' law for electric fields, since it tells us about the amount of total magnetic flux through a given closed surface, which is zero. This states that all magnetic field lines which enter a particular closed surface must eventually leave the surface; thus there are no magnetic monopoles or sources of 'magnetic charge'.

Michael Faraday (1791–1867, England) discovered the law of electromagnetic induction, which describes how a magnetic field that changes in time can also act as a source for the electric field. Faraday's law is given by

$$\nabla \times E = -\frac{\partial B}{\partial t} \qquad [\nabla \times E = -\frac{1}{c}\frac{\partial B}{\partial t}] \qquad (M15)$$

André-Marie Ampere (1775–1836, France) discovered that current was a source of the magnetic field, thus the magnetic field is related to the current density j (in A m^{-2}) by

$$\nabla \times B = \mu_0 j \qquad [\nabla \times B = \frac{4\pi}{c} j] \qquad (M16)$$

where μ_0 is the permeability of free space and is related to ϵ_0 and the velocity of light c by the relation

$$\mu_0 \epsilon_0 = \frac{1}{c^2} \qquad (M17)$$

Ampere's law implies that electric charge is conserved since, if we take the divergence of both sides of (M16), we get

$$\nabla \cdot (\mu_0 j) = \mu_0 \nabla \cdot j = \nabla \cdot (\nabla \times B) \equiv 0 \qquad (M18)$$

However, Maxwell noticed that $\nabla \cdot j = 0$ is only valid for steady state situations and that the complete relation for the continuity of electric charge also includes the variation of the electric charge density ρ_q with time, which is given by

$$\nabla \cdot j + \frac{\partial \rho_q}{\partial t} = 0 \qquad (M19)$$

With this knowledge, Maxwell modified Ampere's law to relate the magnetic field to time-varying electric fields, as well as to the current density, obtaining

$$\nabla \times B = \mu_0 \left(j + \epsilon_0 \frac{\partial E}{\partial t} \right) \qquad \left[\nabla \times B = \frac{4\pi}{c} j + \frac{1}{c} \frac{\partial E}{\partial t} \right] \qquad (M20)$$

The second term of Ampere's law is called the Maxwell displacement current. Maxwell showed that it was needed in order to combine self-consistently the laws of electromagnetism. It was for this insight that equations (M13), (M14), (M15) and (M20) which explain the theory of electric and magnetic fields became known as Maxwell's equations.

These equations enable us to study electromagnetic phenomena ranging from waves, such as light waves in a vacuum, to the effect of the solar wind on a planetary magnetosphere. We note that particles having electric charge are subject to the forces which act upon them by the electric and magnetic fields. However, the electric charges contained by the particles are the source of the total electric field, and the motion of the charged particles, which creates electric current, is the source of the total magnetic field. Thus the charged particles in a plasma are constantly interacting with each other through the electric and magnetic fields which they help to create.

Michael H. Farris and Christopher T. Russell

Bibliography

Dendy, R.O. (1990) *Plasma Dynamics*. Oxford: Clarendon Press.
Jackson, J.D. (1962) *Classical Electrodynamics*. New York: John Wiley.

Cross references

Dynamo theory
Electromagnetic radiation
Magnetism
Magnetometry
Magnetosphere
Plasma
Plasma wave

MERCURY

History

Mercury, the closest planet to the Sun at a mean orbital semimajor axis of 0.3871 AU, has been observed from the times of earliest civilization. Egyptian, Greek, Roman and Mayan civilizations all included Mercury in their mythology and religions. The planet's close proximity to the Sun (from the perspective of an observer in Earth orbit, the planet never moves farther than 27.7° in angular separation from the Sun) hindered early astronomers, and hampers astronomers today, from observing Mercury visually. Detailed telescopic observations of the planet were first attempted by J.H. Shroeter and K.L. Harding in the early 1800s. The first serious attempt at mapping the surface features was conducted by G.V. Schiaparelli between 1881 and 1889. In 1934 E.M. Antoniadi published his book of observations of Mercury, producing another map of the planet. Both these and subsequent scientists, observing from the northern hemisphere, derived a rotational period equal to Mercury's orbital period of 88 Earth days.

Orbit and rotation

Radar observations in the early 1960s by Pettengill and Dyce showed that Mercury is locked into a 3 : 2 spin–orbit commensurability for which the spin angular velocity is 1.5 times the orbital mean motion, causing Mercury to have a rotational period of 58.6 Earth days, equal to two-thirds of its orbital period. Further study of the planet's orbital motion indicates that Mercury is locked into a quasi-commensurability with the Earth in which 54 Mercurian sidereal periods are equal to 13.00600 Earth years. As viewed from Earth orbit, the Mercurian physical ephemeris repeats itself once every 13 years. Earth-based observations are usually limited to the period around maximum elongation of Mercury from the Sun. Six elongations occur per Earth year, three when Mercury is located at northern declinations and three at southern declinations. Observations conducted in the northern hemisphere generally focus on Mercury when it is at northern declinations, and vice versa. During one 13-year interval, observations conducted when Mercury is at northern declinations favor Mercurian central meridian longitudes of 90 and 270°. Observations made when Mercury is at southern declinations, however, cover equally well all central meridian longitudes except 90 and

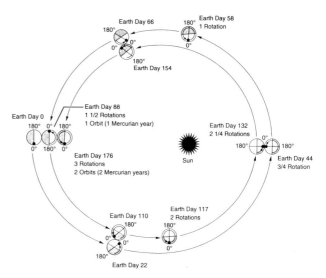

Figure M62 Mercury's 3 : 2 spin–orbit coupling is shown in this diagram, starting at aphelion with the 270° meridian facing the Sun. After half an orbit, Mercury has rotated three-fourths of a turn. After one complete orbit, Mercury has rotated 1.5 times. At the next perihelion passage Mercury has rotated 2.25 times. After two orbits, Mercury has rotated three times. (Figure adapted from Strom, 1987.)

270°. If the earlier telescopic efforts to map the Mercurian surface had been conducted in the southern hemisphere, the true rotational period might have been detected well before the radar observations occurred.

The spin–orbit commensurability coupled with Mercury's orbital eccentricity of 0.20563 produces another effect: at alternate perihelion passages, the same hemisphere faces the Sun. Likewise, at alternate aphelion passages, the same hemisphere faces the Sun. The Mercurian prime meridian (0° longitude on the planet's surface) was defined initially as the longitude facing the Sun when the first perihelion passage occurred in 1950. Based upon this definition, the central meridians of 0 and 180° always face the Sun at perihelion passage. At this close approach to the Sun, the surface temperature of Mercury is the most extreme (~ 740 K) and the subsolar longitudes are known as the 'hot longitudes' (also known in the literature as the 'hot poles'). Alternatively, the central meridians of 90 and 270° always face the Sun at aphelion passage. With a reduced temperature of ~ 525 K, the subsolar longitudes are called the 'cool longitudes' (also known as the 'cold poles' or 'warm poles'). On the night side of Mercury, temperatures drop to as low as 90 K.

Mariner 10

The Mariner 10 deep space probe flew past Mercury on 29 March 1974 at a closest approach distance of 705 km, 21 September 1974 at a distance of 50 000 km, and 16 March 1975 at a distance of 327 km. This mission effectively opened the planet to scientific study. Although Mariner 10 passed by Mercury three times, orbital dynamics limited the amount of the planet's surface visible from the spacecraft to ~ 45%. Mariner 10 conducted seven scientific experiments: television imaging, infrared radiometry, ultraviolet spectroscopy, magnetic fields, plasma science, charged particles and radio science (see also Mariner missions).

Physical properties and composition

With a radius of 2439 km, Mercury is the second smallest planet in the solar system. Mass estimations for Mercury were made initially by astronomers observing the interactions of Mercury with Venus and passing comets, and later refined by the interaction of Mariner 10 with the planet to be 3.302×10^{26} g. The mean bulk density of Mercury, 5.44 g cm^{-3}, is comparable to the value for the Earth (5.5 g cm^{-3}) and Venus (5.2 g cm^{-3}), but much larger than that for Mars (3.9 g cm^{-3}). The surface gravity of Mercury, 370 cm s^{-2}, is slightly greater than the value for Mars, due to the higher density and smaller radius. This high density of Mercury indicates that the interior of Mercury has a greater proportion of heavy elements than the other terrestrial planets. Iron is the most commonly found heavy element in the solar system, present in meteorites and rocks from the terrestrial planets. The high density suggests that the planet is composed of ~ 70% by weight of iron and 30% of rocky or silicate material. Most of the iron is probably concentrated in a core that extends to 75% of Mercury's entire radius. The process of forming this large iron core would cause extensive melting and some crustal deformation, affecting the formation of the surface features. Evidence for iron on Mercury's surface either in the form of iron oxides or metallic iron has been sought in spectral data of the planet; the results have been inconclusive.

Surface geology

Imagery from Mariner 10 shows that the dominant features on Mercury's surface are craters ranging in size from the lower limit of Mariner 10 resolution, 100 m, to Caloris Basin, 1300 km in diameter. The cratering record preserved on Mercury's surface has been interpreted to indicate that the same population of impacting objects affected all the terrestrial planets. These objects are considered to be responsible for the period of heavy bombardment in the inner planets, a period from 4.2 to 3.8 Ga ago when accretional remnants left over from the formation of the terrestrial planets collided with the planets, creating most of the basins and craters visible today.

Mercury's surface can be divided into two terrain types: highlands and lowland plains. The highlands contain heavily cratered areas mixed with rolling plains, called intercrater plains, which have a large number of craters $\leq$ 15 km diameter. These small craters often appear in chains or clusters, suggesting that they are impact craters formed by material ejected from a larger impact or impacts on Mercury's surface. The lowland plains, or smooth plains, occur within and around the Caloris Basin, in the north polar region, and on the floors of other large basins. The crater density on the smooth plains is substantially lower than that on the intercrater plains, suggesting that the smooth plains are younger in age. The extrusion of volcanic magma, which occurred when the Mercurian surface subsided following the impacts that created the basins on Mercury, is the favored origin for the smooth plains.

Lineament structures on Mercury's surface observed both on Mariner 10 imagery, and in radar observations of the unimaged hemisphere record the important aspects of lithospheric history, and changes on Mercury's surface. Early contraction of the lithosphere left a record in obscure lineaments found as crater rim segments, ridges, troughs and extrusive volcanic vents. Near the equator, this is represented by a grid of lineaments trending NW–SE and NE–SW. Closer to the poles, the lineaments trend more east–west, consistent with despinning of the planet from a more rapid rotational rate to the present slow rotation. Unique to the planet are the lobate scarps, which are thrust faults from 20 to 500 km in length, probably caused by a 1–2 km decrease in Mercury's radius during cooling of the planet's mantle and partial solidification of the core. Impact events, the most prominent being the impact creation of Caloris Basin, caused most of the other lineament features. Caloris is a large multiringed basin. Antipodal to Caloris is an area consisting of hills and depressions that break across other landforms. This hilly and lineated terrain could have formed by the antipodal focusing of seismic waves from the Caloris impact (see Mercury: geology).

Radar signatures at both the north and south poles of Mercury, have revealed bright features with radar characteristics similar to those seen of icy surfaces elsewhere in the solar system. Mercury may have two polar caps composed of dirty ice, apparently shielded sufficiently from the Sun to prevent melting in the intense heat.

Magnetic field

During the first and third encounters with Mercury, Mariner 10 passed through a small magnetosphere around the planet. The shape and characteristics of the magnetosphere appear to be similar to the Earth's magnetosphere, although smaller in overall dimension. A magnetopause and bow shock exist for Mercury, at 1.5 Mercurian radii (R_M) from the center of the planet. Solar wind interaction with the surface does not take place as it does on the Moon, due to the deflection of the solar wind by the magnetosphere. The magnetosphere data suggest that Mercury has an internal magnetic field with a dipole moment of ~ 300 nT R_M^3 and a surface strength of 350γ. One method of generating a magnetic field requires convection in an electrically conducting liquid core. The presence of a magnetic field in Mercury provides some constraints on the internal composition and structure of the planet. The partially molten large iron core suggested by Mercury's high density is sufficient to create the planet's magnetic field although, given Mercury's size and proximity to the Sun, the interior of the planet should have solidified over the lifetime of the solar system. An alternate scenario has Mercury's magnetic field created by remanent magnetism, a magnetic field frozen into the core. The large amount of iron expected in Mercury's core is much greater than that necessary to create the observed weak magnetic field.

Atmosphere

Mercury possesses a very tenuous atmosphere with five elements identified to date as components. The Mariner 10 ultraviolet spectroscopy experiment identified atomic oxygen, atomic hydrogen, and helium in the atmosphere from the presence of emission features at ultraviolet wavelengths. Ground-based telescopic observations at visible wavelengths added the sodium and potassium vapor components. The most abundant of these elements is oxygen, followed by sodium and hydrogen. The solar wind, entering via Mercury's magnetosphere, is the probable source for the hydrogen and helium. Infalling meteoroids also contribute both hydrogen and oxygen, since they contain up to 10% water that is photodissociated to atomic hydrogen and oxygen. Sputtering of surface material by ions from Mercury's magnetosphere is probably the dominant process creating the sodium and potassium components. Impact vaporization of surface and meteoroid materials, photon-stimulated desorption, and venting from subsurface sources have also been proposed as contributing mechanisms.

The dominant sink for all the atmospheric species is photoionization, followed by transport away from the planet on open magnetic field lines. At times when the radial velocity of Mercury is high, both sodium and potassium experience solar radiation pressure greater than half of Mercury's surface gravity. Some of these atoms may be swept completely off the planet by radiation pressure at these times.

Origin

With its position in the solar system as the planet closest to the Sun, Mercury is often studied as the planet representative of formation closest to the Sun. Mercury is believed to have formed from the condensation of material in the solar nebula and the subsequent accretion of material from the whole range of heliocentric distances encompassing the terrestrial planets, emphasizing the area near Mercury's orbit. Existing models of condensation and accretion as a function of distance from the Sun in the early solar system do not explain the high density of Mercury. If the planet formed in its present location, a mechanism such as the partial volatilization of Mercury by the early Sun, perhaps in a T Tauri-like phase, or removal of the crust and upper mantle of Mercury by large impacts must be invoked. Currently accepted theory holds that the Moon formed as the result of a major impact on the Earth, breaking off a portion of the planet; the creation of Mercury by a major impact should be considered. The origin of Mercury remains an enigma requiring future study.

Faith Vilas

Bibliography

Harmon, J.K. and Slade M.A. (1992) Radar mapping of Mercury: full-disk images and polar anomalies. *Science*, 258, 640–3.
Slade, M.A., Butler, B.J., and Muhleman, D.O. (1992) Mercury radar imaging: evidence for polar ice. *Science*, 258, 635–40.
Strom, R.G. (1987) *Mercury: The Elusive Planet*. Washington D.C.: Smithsonian Institution Press.
Vilas, F., Chapman C.R. and Matthews, M.S. (Eds) (1988) *Mercury* Tucson: University of Arizona Press.

Cross references

Mariner missions
Mercury: atmosphere
Mercury: geology
Mercury: magnetic field and magnetospheric

MERCURY: ATMOSPHERE

Mercury's atmosphere is very tenuous; the total weight of all known constituents is only 8 metric tons. By contrast, the Earth's atmosphere weighs nearly 5000 metric tons. Ground-based searches for atmospheric gasses such as CO_2 resulted only in upper limits. Actual detection with measurements of abundances awaited the results of two atmospheric experiments aboard the Mariner 10 spacecraft which made three encounters with the planet during 1974 and 1975. The two instruments designed for atmospheric measurements were an occultation experiment, which views a star as it passes behind the limb of the planet, searching for absorption from any gases close to the surface. This instrument, abbreviated (OE), measured four discreet channels in the ultra violet. The second instrument was a ten-detector ultra violet spectrometer (UVS). At the time of encounter, hydrogen (H) and helium (He) were detected and upper limits were placed on atomic oxygen (O), argon (Ar), neon (Ne) and xenon (Xe). Later analysis of the data revised values slightly and added upper limits for N_2, H_2O, CO_2 and H_2 (Broadfoot, Shemansky and Kumar, 1976). In 1985 the surprising discovery of significant abundances of neutral sodium in Mercury's atmosphere was made (Potter and Morgan, 1985), followed a year later by the discovery of neutral potassium (Potter and Morgan, 1986). Table M8 lists known atmospheric constituents at Mercury, and their abundances, as well as the upper limits on undetected species. Abundances are designated by giving the total number of atoms which are in a vertical column above one square centimeter of the surface (atoms cm^{-2}). This column of gas extends from the surface of the planet to the extent of the atmosphere, and is called the column abundance. Also shown in Table M8 are the instruments used for the detections and the emission line wavelengths of each atmospheric component.

The sources of atmospheric atoms are believed to be known, and vary from one constituent to another. Hydrogen and He originate from neutralized solar wind ions, with an additional radiogenic component of He outgassing from the interior. Sodium, K and O are likely to be gases released during regolith and meteoric vaporization following impact of micrometeorites on the surface, or released from the surface by impact of charged particles from the solar wind. Outgassing from the surface and subsurface materials is also probably a source for Na and K. Because Mercury's known atmosphere is similar to that of the Moon, it is likely that radiogenic Air and H_2 remain undetected. In addition, because micro meteoritic material is known to contain a wide range of silicates, there are probably atoms of magnesium (Mg), calcium (Ca) and silicon (Si) in the atmosphere but no detection of these has been made.

The principal loss of the atmosphere is a result of photo ionization by the solar ultraviolet flux. Once a neutral atom is in the atmosphere, it is subject to ionization. Ionization efficiency varies dramatically from species to species, with K being ionized most rapidly and H the slowest owing to differences in their electronic configurations. Because Mercury's eccentricity is high (0.2), the distance from the Sun varies over one orbital period (89 Earth days) from as close as 0.31 AU (astronomical unit: the average Earth–Sun distance) at perihelion to 0.47 AU at aphelion. This means the solar flux falling on Mercury's atmosphere varies by nearly an order of magnitude during the same period. One effect is a varying loss rate of ions which are swept out of the atmosphere into the interplanetary medium by electric fields in the solar wind. While ionization followed by loss to the interplanetary medium is the primary loss, it is also possible that H and other light elements are lost in Jeans escape. In addition, if there is a suprathermal (very hot) component to the atmosphere at high altitudes off the surface, these atoms could be subject to pressure from solar photons and be accelerated to escape trajectories and lost from the atmosphere. Migration of some neutrals to high latitudes, where temperatures are very low, may also result in some loss of atoms from the atmosphere because they stick permanently to the ground. Little is known about the storage of volatile and semi volatile elements at polar regions at Mercury; even less than is known about the same subject at the Moon. A number of publications discuss the sources and sinks of atmospheric constituents. Particulars can be found in Broadfoot, Shemansky and Kumar, 1976; Shemansky and Broadfoot, 1977; Goldstein, Seuss and Walker, 1981; McGrath, Johnson and Lanzerotti, 1986; Cheng et al., 1987; Potter and Morgan, 1985; Morgan, Zook and Potter, 1988, Tyler, Kozlowski and Lebofsky; 1988; Killen, Morgan and Potter, 1990; and Sprague, Kozlowski and Hunten, 1990, to name a few. Earlier papers discuss the distribution and losses of H and He, while papers after 1985 generally discuss Na and K.

Table M8 Mercury's atmospheric composition

Constituent	Average zenith column abundance (atoms cm^{-2})	Resonance line (Å)	Instrument[a]
Detections			
Oxygen (O)	$\sim 2 \times 10^{11}$	1304	M10: UVS
Sodium (Na)	$\sim 1.5 \times 10^{11}$	5890, 5896	Ground: HRS
Potassium (K)	$\sim 3 \times 10^{9}$	7664, 7699	Ground: HRS
Helium (He)	$\sim 6 \times 10^{10}$	584	M10: UVS
Hydrogen (H)	$\sim 2 \times 10^{8}$ day	1216	M10: UVS
	$\sim 2 \times 10^{10}$ night		M10: UVS
Upper limits	**LOS bright limb**		
Argon (Ar)	$<9 \times 10^{14}$	869	M10:OE
Neon (Ne)	$<3 \times 10^{13}$	740	M10:OE
Xenon (Xe)	$<1 \times 10^{12}$	1470	M10:OE
Carbon (C)	$<5 \times 10^{10}$	1657	M10:OE
CO_2[b]	$<4 \times 10^{14}$		M10:OE
N_2[b]	$<9 \times 10^{14}$		M10:OE
H_2[b]	$<9 \times 10^{14}$		M10:OE
H_2O[b]	$<8 \times 10^{14}$		M10:OE
	Ground-based		
Calcium (Ca)[c]	$<7 \times 10^{8}$	4227	HRS
Lithium (Li)	$<8 \times 10^{7}$	6708	HRS

[a] M10 = Mariner 10; HRS = high resolution spectrograph(s) used in ground-based observations; OE = occultation experiment; UVS = ultraviolet spectrometer
Upper limits from Broadfoot, Shemansky and Kumar (1976). [b] Upper limits were found by looking for absorptions in the continuum measured by the OE, [c] from Sprague et al. (1993).

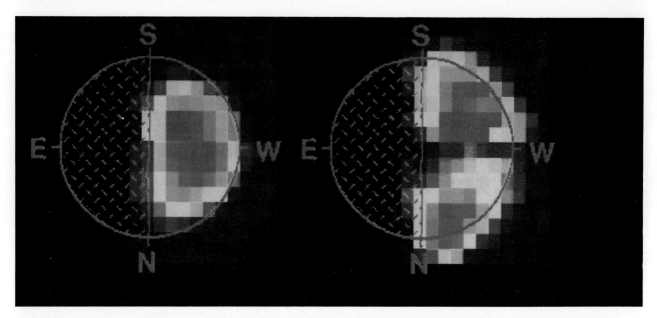

Figure M63 The two-dimensional images above are of Mercury on 8 December 1990, taken with an image slicer and stabilizer at the McMath telescope at the Kitt Peak National Observatory. The left-hand image is Mercury as seen in light reflected from its rocky surface. Mercury was at gibbous phase at the time of the exposure, which accounts for its non-circular shape. The right-hand image is Mercury in the light of emission from metallic sodium vapor in the thin atmosphere of Mercury. This remarkable image shows two bright spots at high north and south latitudes. These spots are attributed to enhanced emissions from atmospheric sodium vapor.

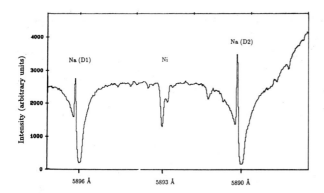

Figure M64 A spectrum showing resonantly scattered emission from neutral sodium in Mercury's atmosphere (courtesy of Sprague, Kozlowski and Hunten). Wavelength increases to the left in this spectrum. Solar Fraunhoffer lines are labeled. Notice how the emission spikes are shifted to wavelengths slightly longer than the center of the sodium solar Fraunhofer absorptions. This is because of the large radial velocity of Mercury relative to the Sun.

Because Na and K have particularly high probabilities for scattering solar photons, and because the wavelength re-emitted by both atoms is easy to observe from the ground these two constituents are the best studied. Figure M63 is a two-dimensional image of Mercury's atmosphere showing Na emission bright spots located on the dayside. Figure M64 shows typical Na emission spikes in a spectrum obtained by placing the HRS slit in a north–south direction across Mercury's disk. The neutral Na emission maxima can be seen to be offset from the center of the Na solar Fraunhofer absorption lines reflecting off the planet's surface. This is because there is a high relative radial velocity between Mercury and the Sun which enhances the ease with which these lines are viewed from Earth.

Observations of all constituents have shown non-homogeneities in abundance, both in time and location. It is clear that the only collisions atoms experience are with the surface; however, the atmosphere does not behave as an exosphere with the surface as the exobase. This was first noted for He by Smith *et al.* (1978), who found that there were significant departures in the distribution of He from that which would be expected from a purely Maxwellian distribution of atoms about the surface. Hydrogen also departs from the classical distribution and was found to be a two-component system consisting of a cold component whose scale height is characteristic of the night side temperatures, and a less abundant component with a distribution more characteristic of hotter, dayside surface temperatures (Shemansky and Broadfoot, 1977). Killen, Morgan and Potter (1990) Potter and Morgan (1990) and Sprague, Kozlowski and Hunten (1990) have measured enhanced Na and K concentrations at high latitudes and at varying longitudes (see Figure M63). Thus there must exist either significant additional source components or recycling mechanisms which concentrate Na and K in discreet locations to be released into the atmosphere and observed as bright emission enhancements. Several explanations have been suggested at to explain these observations. Ip (1990) suggested that different petrologic provinces on the surface of Mercury may account for the differences seen in Na abundance. Mercury's surface composition was suggested to be more feldspathic than that of the Moon at some locations by Tyler, Kozlowski and Lebofsky (1988), based upon the comparison of mid-infrared spectra obtained from both bodies.

A magnetospheric mechanism which recycles ions back into the neutral atmospheric population is another possibility for explaining the non-uniformities in abundances. Several authors have compared the magnetospheres of Mercury and Earth (Siscoe, Ness and Yeates 1975; Christon, 1987) and shown them to be similar in some respects. Like Earth's magnetic field, Mercury's is believed to be dipolar. There is a surface field strength of ~ 1/1000 that of Earth. Observations of sodium emission enhancements occurring in north–south pairs, roughly where the aurora cusp regions of a dipole field might be expected, have led to suggestions that magnetospheric recycling of heavy ions may increase sputtering (removal by knocking atoms from the surface) near the poles or provide an enhanced source of neutrals which adsorb (stick) on the surface and accumulate after ions impact the surface. Other studies of the interaction of sodium and potassium with the magnetic field have been made by Ip (1987). Recycling of ions by electric fields in the magnetosphere must occur. It is likely that a fraction of the Na^+ and K^+ atoms impact the nightside. Sprague (1992) has suggested that these ions may implant to depths of several hundred angstroms and accumulate during the long

Mercurian night (89 Earth days). When that region rotates beneath the Sun, subsequent outgassing of atoms to the atmosphere is likely. This could result in time-variable and spatially discrete enhancements. While it is not known if this is occurring, analogy to the Earth's magnetosphere leads one to believe it may.

Sprague, Kozlowski and Hunten (1990) suggested enhanced emissions of K may be attributed to an increased supply of K diffusing outward through impact-fractured regions in the subsurface and regolith material, in particular at Caloris basin and at the highly fractured terrain which lies antipodal to the huge Caloris basin. These locations nearly coincide with the hottest regions on the planet; this heat may contribute to a diffusion process. It is likely that the hottest, most extended component of the atmosphere is directly influenced by the effects of solar radiation pressure (Smyth, 1986; Potter and Morgan, 1987; Ip, 1990).

In summary, the complex radiation and magnetospheric environment known to exist at Mercury, combined with the lack of compositional information regarding the surface, makes it difficult to state the relative contributions of probable sources to the observed atmosphere. Theoretical and experimental physics of sputtering (removal of atoms from the surface of materials by energetic photons or charged particles) tells us a sputtered component must be present. By analogy to the Earth–Moon system, we believe a significant volatilized component from meteoric impact must exist. A source at depth, outgassing to the surface, cannot be ruled out and is likely in the highly fractured and hot Caloris basin and antipodal terrain, or from other radar-bright regions which indicate significant fracturing of the surface.

Ann L. Sprague

Bibliography

Broadfoot, A.L., Shemansky, D.E., and Kumar J.J., 1976. Mariner 10: Mercury atmosphere. *Geophys. Res. Lett.*, 3, 577–80.
Cheng, A.F., Johnson, R.E., Krimingis, S.M. and Lanzerotti, L.J. (1987) Magnetosphere, exosphere and surface of Mercury. *Icarus*, 71, 430–40.
Christon, S.P. (1987) A comparison of the Mercury and Earth magnetospheres: electron measurements and substorm time scales. *Icarus*, 71, 448–71.
Goldstein, B.D., Suess, S.T. and Walker, R.J. (1981) Mercury: magnetospheric processes and the atmospheric supply and loss rate. *J. Geophys. Res.*, 86(A7), 5485–99.
Ip, W.H. (1987) Dynamics of electron and heavy ions in Mercury's magnetosphere. *Icarus*, 71, 441–7.
Ip, W.H. (1990) On solar radiation-driven surface transport of sodium atoms at Mercury. *Astrophys. J.*, 356, 675–81.
Killen, R.K., Morgan, T.H. and Potter, A.E. (1990) Spatial distribution of sodium vapor in the atmosphere of Mercury. *Icarus*, 85, 145–67.
McGrath, M.A., Johnson, R.E. and Lanzerotti, L.J. (1986) Sputtering of sodium on the planet Mercury. *Nature*, 323, 694–6.
Morgan, T.H., Zook, H.A. and Potter, A.E. (1988) Impact-driven supply of sodium and potassium in the atmosphere of Mercury. *Icarus*, 74, 156–70.
Potter, A.E. and Morgan, T.H. (1985) Discovery of sodium in the atmosphere of Mercury. *Science*, 229, 651–3.
Potter, A.E. and Morgan, T.H. (1986) Potassium in the atmosphere of Mercury. *Icarus*, 67, 336–40.
Potter, A.E. and Morgan, T.H. (1987) Effects of radiation acceleration. *Icarus*, 71, 472–7.
Potter, A.E. and Morgan, T.H. (1990) Evidence for magnetospheric effects on the sodium atmosphere of Mercury. *Science*, 248, 835–8.
Shemansky, D.E. and Broadfoot, A.L. (1977) Interaction of the surface of the Moon and Mercury with their exospheric atmospheres. *Rev. Geophys. Space Phys.*, 85, 221–2.
Siscoe, G.L., Ness, N.F. and Yeates, C.M. (1975) Substorms on Mercury? *J. Geophys. Res.*, 80, 4359–63.
Smith, G.R., Shemansky, D.E., Broadfoot, A.L. and Wallace, L. (1978) Monte Carlo modeling of exospheric bodies: Mercury. *J. Geophys. Res.*, 83, 3783–90.
Smyth, W.H. (1986) Nature and variability of Mercury's sodium atmosphere. *Nature*, 323, 696–9.
Sprague, A.L. (1992) Mercury's atmospheric sodium bright spots and potassium variations: a possible cause. *J. Geophys. Res., Planets*, 97, 18257–64.
Sprague, A.L., Kozlowski, R.W.H. and Hunten, D.M. (1990) Caloris Basin: an enhanced source for potassium in Mercury's atmosphere. *Science*, 249, 1140–3.
Sprague, A.L., Kozlowski, R.W.H., Hunten, D.M. and Grosse, F.A. (1993) An upper limit on neutral calcium in Mercury's atmosphere. *Icarus*, 104, 33–7.
Tyler, A.L., Kozlowski, R.W.H. and Lebofsky, L.A. (1988) Determination of rock type on Mercury and the Moon through remote sensing in the thermal infrared. *Geophys. Res. Lett.*, 15(8), 808–11.

Cross references

Mariner missions
Moon: atmosphere
Spectroscopy: atmosphere

MERCURY: GEOLOGY

The Mariner 10 mission in 1974 mapped about half the surface of Mercury. On the basis of these data, we have a first-order understanding of the geology and history of the planet (Plate 5). Mercury's surface shows intercrater plains, basins, smooth plains, craters and tectonic features (Figure M65). The relations between these geological units allow us to reconstruct the history of the planet (Table M9).

Intercrater plains

Mercury's most densely cratered surfaces are not as heavily cratered as the lunar highlands. Mercury's surface has large exposures of intercrater plains (Figures M65 and M66), which are present (but much less extensive) on the Moon. At no place on Mercury do we find regions of overlapping large craters and basins like those that characterize the lunar highlands. The first-order implication of this observation is that Mercury may have experienced some type of resurfacing early in its history. This resurfacing was more intense than comparable activity on the Moon.

The intercrater plains are level to gently rolling terrain that occur between and around large craters (Figure M66). They show abundant elongate and shallow craters less than 10 km in diameter; clusters as well as chains are common, and in morphology and distribution they are similar to secondary impact craters. The only likely sources for these secondary craters are the large craters and basins of Mercury. Thus the intercrater plains predate the heavily cratered terrain (Table M9), and have obliterated many of the early craters and basins of Mercury.

The origin of the intercrater plains is unknown. They may be formed from old basin ejecta material, or they could be volcanic, although landforms of unambiguously volcanic origin have not been found. Some dome-like features in the intercrater plains may indicate volcanism, but many of these features are thought to be massifs associated with ancient multiring basins. Widespread volcanism early in Mercurian history is currently the best explanation for the intercrater plains.

Basins

A large population of multiring basins, all of which predate the intercrater plains, has been identified on Mercury. Ancient basins are delineated and mapped by a combination of criteria, including arcuate scarps and ridges, circular alignments of massifs, isolated massifs, and localized concentrations of smooth plains within otherwise heavily cratered terrain. At least 15 ancient basins have been mapped in the hemisphere of Mercury imaged by Mariner 10. They are apparently randomly distributed.

Several other basins postdate the intercrater plains. The oldest of these is Dostoevskij, a very degraded basin about 400 km in diameter. A radially textured ejecta blanket and long chains of radial secondary craters are evident in some regions. Tolstoj is a true multiring basin, displaying at least two, and possibly as many as four, concentric rings. Tolstoj has a well-preserved, radially lineated ejecta blanket extending outward as much as about one basin diameter (500 km). The interior of the basin is flooded with smooth plains that clearly postdate the basin deposits. Tolstoj ejecta are not radially

Figure M65 Mosaic by the Mariner 10 spacecraft of the planet Mercury. Although superficially similar to the Moon, Mercury displays some unique geological features, which testify to its distinct evolution and history.

Figure M66 Typical highlands region on Mercury. Rough intercrater plains cover the zones that occur between large craters. Younger smooth plains occur at top of frame. A variety of crater degradation types are evident in this view, ranging from relatively fresh (young; A) to nearly obliterated (very old; B). Irregularly shaped craters (arrows) are secondary impact craters from the Caloris basin impact.

symmetric with respect to the basin rim, as they appear to be absent in the northern and western sectors of the basin, possibly indicating basin formation by oblique impact.

Beethoven has only one ring, a subdued massif-like rim about 625 km in diameter, but displays an impressive, well-lineated ejecta blanket that extends as far as 500 km from its rim. As at Tolstoj, Beethoven ejecta are also asymmetric, but large expanses of younger, smooth plains occur in the areas where its ejecta appear to be absent. The subdued rim of Beethoven has suggested to some workers that it is an extremely old feature, but a surprisingly low crater density on the ejecta blanket suggests instead that Beethoven postdates Tolstoj and is slightly older than the Caloris basin.

The Caloris basin is the most striking basin yet seen on Mercury. It lies on the terminator as imaged by Mariner 10 and thus only half of its circumference is visible (Figure M67). The basin is defined by a ring of mountains 1300 km in diameter, consisting of smooth-surfaced blocks rising some 1 or 2 km above the surrounding terrain and having a width of between 100 and 150 km. Individual massifs are typically 30 to 50 km long; the inner edge of the unit is marked by basin-facing scarps. Much of this massif material probably consists of bedrock uplifted from deep within Mercury's crust. Small patches of plains occur in depressions between the mountains. This unit is distinguished from smooth plains by its more rugged surface morphology and its position within the Caloris mountains. It is probably a mixture of fallback material and impact melt.

Lineated terrain extends for about 1000 km out from the foot of a weak discontinuous scarp on the outer edge of the Caloris mountains. It consists of long hilly ridges and grooves subradial to the Caloris basin and is best expressed northeast of the basin (Figure M67). This lineated terrain is similar to the 'sculpture' surrounding the Imbrium basin on the Moon. Another unit of hummocky material forms a broad annulus about 800 km from the Caloris mountains. It consists of low, closely spaced to scattered hills about 0.3 to 1 km across and from tens of meters to a few hundred meters high. In some places the hills are aligned concentrically with the rim of Caloris, giving the plains a corrugated appearance. The outer boundary of this unit is gradational with the (younger) smooth plains materials that occur in the same region. Both the hummocky plains and the lineated terrain are facies of Caloris ejecta.

A hilly and furrowed terrain is found antipodal to the Caloris basin. This terrain was probably created by intense seismic waves that were generated by the basin-forming impact. These waves propagated through the planet and were focused by Mercury's core, causing intense surface disruption at the antipode of the basin impact.

Plains

The floor of the Caloris basin is a plains unit that shows intense secondary deformation in the form of sinuous ridges and fractures, giving the basin fill material a grossly polygonal pattern (Figure M67). These plains may be volcanic, formed by the release of magma as part of the impact event, or a thick sheet of impact melt produced during basin impact.

Table M9 The Mercurian chronostratigraphic scheme

System	Major units	Approximate age of base of system (Ga)	Lunar counterpart
Kuiperian	Crater material	1.0	Copernican
Mansurian	Crater material	3.0–3.5	Eratosthenian
Calorian	Caloris group; plains, crater, small-basin materials	3.9	Imbrian
Tolstojan	Goya Fm.; crater, small-basin, plains materials	3.9–4.0	Nectarian
Pre-Tolstojan	Intercrater plains, multiring basin, crater materials	>4.0	Pre-Nectarian

From Spudis and Guest (1988).

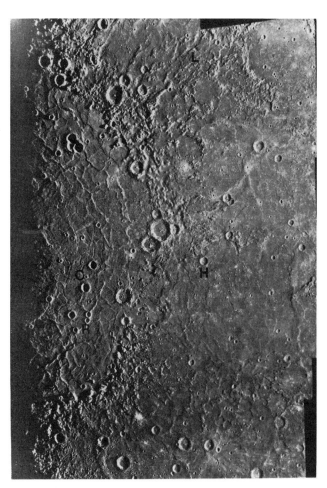

Figure M67 Mosaic of the Caloris basin. Main rim is defined by circular mountain range, 1300 km in diameter. Basin fill is smooth plains material, deformed by compressional ridges (R) and tensional cracks (C). Lineated terrain (L) is ejecta from the basin impact; this ejecta material is hilly in some areas (H). Swath of smooth plains concentrically surround the basin rim; because they have a much lower crater density than the basin rim materials, they are younger and are probably volcanic lavas erupted long after the impact which formed the basin.

Widespread areas of Mercury are covered by relatively flat, sparsely cratered plains materials. They fill depressions that range in size from regional troughs to crater floors. The smooth plains are similar to the maria of the Moon, an obvious difference being that the plains on Mercury differ little in albedo from the average cratered terrain and intercrater plains (Figure M66). Smooth plains are flat to gently rolling, possess abundant wrinkle ridges and are most strikingly exposed in a broad annulus around the Caloris basin (Figure M67). In high-resolution photographs, no unequivocal volcanic features, such as flow lobes, leveed channels, domes or cones are visible.

Careful study of crater densities on a variety of geological units indicates that the smooth plains are significantly younger than ejecta from the Caloris basin. Such an age difference strongly supports a volcanic origin for the Mercurian smooth plains, even in the absence of volcanic landforms. Some smooth plains appear to be transitional in age between those surrounding the Caloris basin and the older, intercrater plains. Those plains are not as extensive as the younger smooth plains, but their existence suggests that some mechanism of plains deposition operated throughout early Mercurian history. All of these observations are consistent with volcanism as the primary mode of origin for the smooth plains.

Craters

Mercurian craters have all the morphological elements of lunar craters: the smaller craters are bowl-shaped, and with increasing size they develop scalloped rims, central peaks and terraces on the inner walls (Figures M65 and M66). The ejecta sheets have a hilly, radial texture and swarms of secondary impact craters. Fresh craters of all sizes have dark or bright halos and well-developed ray systems. Although Mercurian and lunar craters are superficially similar, they also show subtle differences, especially in the effective range of ejecta. The continuous ejecta and fields of secondary craters on Mercury are far less extensive (by a factor of about 0.65) for a given rim diameter than those of comparable lunar craters. This difference results from the 2.5 times higher gravitational field on Mercury compared with the Moon.

As on the Moon, impact craters on Mercury are progressively degraded by subsequent impacts. There is thus a complete range of crater degradational morphologies (Figure M66). The freshest craters have ray systems and a crisp morphology. Secondary craters are well developed and the continuous ejecta show radial lineaments. Somewhat older craters have a similar form but, like their counterparts on the Moon, have lost their ray systems. With further degradation, the craters lose their crisp morphology, and features on the continuous ejecta become more blurred until only the raised rim near the crater remains recognizable, although it has many superposed smaller impact craters. Degradation continues until only a weak circular ring of hills is visible.

Because craters become progressively degraded with time, the degree of degradation gives a rough indication of the crater's relative age. On the assumption that craters of similar size and morphology are roughly the same age, it is possible to place constraints on the ages of other underlying or overlying units and thus to map globally the relative age of surface features. The stratigraphic significance of crater degradation is only approximate; a variety of evidence, including regional setting, proximity to other units, crater density and type of post-impact modification, must be used in concert to establish the relative ages of Mercurian features (Table M9).

Tectonic features

Lobate scarps are widely distributed over Mercury. They consist of sinuous to arcuate scarps that transect pre-existing plains and craters (Figure M68). They are most convincingly interpreted as thrust faults, indicating a period of global compression. All these scarps cut intercrater plains materials, which suggests that they began to form after the intercrater plains materials were emplaced. The lobate scarps typically transect smooth plains materials (early Calorian age; Table M9) on the floors of craters, but post-Caloris craters are

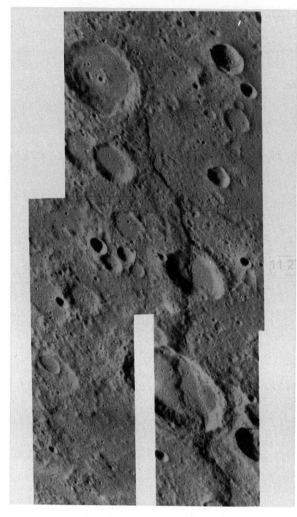

Figure M68 Discovery Rupes, one of the lobate scarps of Mercury. It is a compressional thrust fault, as shown by the deformed and shortened pre-scarp crater. Lobate scarps occur all over Mercury and indicate a period of global compression of the crust, a phenomenon not seen on any other terrestrial planet.

superposed on them. These observations suggest that lobate scarp formation was confined to a relatively narrow interval of time, beginning in the late pre-Tolstojan period and ending in the middle to late Calorian Period (Table M9).

In addition to lobate scarps, wrinkle ridges occur in the smooth plains materials (Figure M67). These ridges were probably formed by local to regional surface compression caused by lithospheric loading by dense stacks of volcanic lavas, as suggested for those of the lunar maria.

Geological history

The earliest decipherable event in Mercury's history was the formation of its crust. By analogy with the Moon, Mercury may have experienced early global melting, similar to the lunar 'magma ocean', whereby large-scale melting of at least the outer few hundred kilometers of the planet would concentrate low-density plagioclase into the uppermost part of the Mercurian crust. If this process operated during early Mecurian history, the crust of Mercury is probably composed largely of anorthosite and related rocks. Such a composition is consistent with full-disk spectra of Mercury, which suggest that the composition of Mercury's surface is similar to that of the Apollo 16 site in the highlands of the Moon.

The early cratering record of Mercury has been largely destroyed by the deposition of the intercrater plains. However, the largest impact features of this period (multiring basins) have been partly preserved and suggest that Mercury's surface may have originally resembled the cratered lunar highlands. Sometime during the heavy bombardment, the emplacement of massive quantities of intercrater plains materials largely obliterated the older crater population (Plate 5). The global distribution of the intercrater plains suggests that they may be at least partly volcanic in origin, although subsequent cratering has converted the original surface to breccia; primary surface morphologies were probably destroyed.

The impact that formed the Tolstoj basin marked the beginning of the Tolstojan Period (Plate 5; Table M9), still a time of high impact rates. Although some flood lavas may have erupted during this time, their preservation is sporadic and they may be largely covered by subsequent lavas. The basin Beethoven probably formed near the end of the Tolstojan Period.

The Caloris impact formed the largest well-preserved basin on Mercury's surface (Figure M67; Plate 5; Table M9) and provided an extensive stratigraphic datum on the planet. Catastrophic seismic vibrations from the Caloris impact probably formed the hilly and furrowed terrain on the opposite side of the planet. Some finite (but probably short) time after the Caloris impact came massive extrusions of flood lavas to form the Mercurian smooth plains. A rapidly declining cratering rate has produced minimal changes to Mercury's surface (Plate 5) since the final emplacement of the smooth plains. This low rate of cratering presently continues to produce regolith on all Mercurian surface units.

Paul D. Spudis

Bibliography

Gault, D.E., Guest, J.E., Murray, J.B., et al. (1975) Some comparisons of impact craters on Mercury and the Moon. *J. Geophys. Res.*, **80**, 2444–60.
Guest, J.E. and O'Donnell, W.P. (1977) Surface history of Mercury: a review. *Vistas in Astron.*, **20**, 273–300.
McCauley, J.F., Guest, J.E., Schaber, G.G., et al. (1981) Stratigraphy of the Caloris basin, Mercury. *Icarus*, **47**, 184–202.
McCauley, J.F. (1977) Orientale and Caloris. *Phys. Earth Planet. Interiors*, **15**, 220–50.
Melosh, H.J. and McKinnon, W.B. (1988) The tectonics of Mercury, in *Mercury* (eds F. Vilas, C.R. Chapman and M. Matthews). Tucson: University of Arizona Press, pp. 374–400.
Murray, B. and Burgess, E. (1977) *Flight to Mercury*. Columbia University Press, 162 pp.
Pike R.J. (1988) Geomorphology of impact craters on Mercury, in *Mercury* (eds F. Vilas, C.R. Chapman, and M. Matthews). Tucson: University of Arizona Press, pp. 165–273.
Schaber, G.G., Boyce, J.M. and Trask, N.J. (1977) Moon–Mercury: large impact structures, isostasy and average crustal viscosity. *Phys. Earth Planet Interiors*, **15**, 189–201.
Schultz, P.H. and Gault, D.E. (1975) Seismic effects from major basin formations on the Moon and Mercury. *The Moon*, **12**, 159–77.
Spudis, P.D. and Guest, J.E. (1988) Stratigraphy and geologic history of Mercury, in *Mercury* (eds F. Vilas, C.R. Chapman, and M. Matthews). Tucson: University of Arizona Press, pp 118–64.
Strom, R.G., Trask, N.J. and Guest, J.E. (1975) Tectonism and volcanism on Mercury. *J. Geophys. Res.*, **80**, 2478–507.
Strom R.G. (1987) *Mercury: The Elusive Planet*. Smithsonian Institution Press, 197 pp.

Cross references

Impact cratering
Mariner missions
Planetary ice
Radio astronomy

MERCURY: MAGNETIC FIELD AND MAGNETOSPHERE

Mercury is the smallest of the terrestrial planets. Its radius of 2440 km places it between the Earth's Moon and Mars in size. It is of great importance to those studying planetary magnetic dynamos and to those studying planetary magnetospheres. Its importance to the magnetic dynamo problem stems from its being the smallest and most slowly rotating planet with a presently active magnetic dynamo. Its importance to the physics of planetary magnetospheres stems from its lack of a dynamically important atmosphere or ionosphere. Currents generated by the solar wind interaction, which usually close in the ionosphere, cannot close in the same way at Mercury as they do in other planetary magnetospheres. It is thought therefore that the Mercury magnetosphere may be more strongly coupled to the solar wind than is the case for other planetary magnetospheres.

Planet and interior

Because of its small size and outward appearance, due largely to the absence of a significant atmosphere, it is most appropriate to compare Mercury with the Earth's Moon. Mercury rotates more slowly than the Moon, rotating with a period of 59 days compared with the Moon's 28-day period. Mercury also differs from the Moon in that its rotation is not synchronous with its orbital period. Mercury orbits the Sun every 88 days so that every 2 Mercurian years, the same side of the planet again faces the Sun. The slow rotation, close proximity to the Sun and lack of atmosphere causes a very high surface temperature (≥ 630 K) on the dayside of the planet and very cold temperatures on the nightside. The role of the atmosphere in controlling surface temperature can be appreciated by noting that Venus, with the most massive atmosphere of the terrestrial planets, has day and night temperatures that differ at most by a few degrees.

A unique characteristic of the rotation of Mercury is that its rotation axis is aligned along its orbital pole. Every other planet in the solar system has a rotation axis that is tilted with respect to its orbital pole, affecting, and in most cases dominating, seasonal changes. This oddity may influence both atmospheric and internal processes.

Mercury, with a density of 5.4 g cm^{-3}, is much denser than the Moon (whose density is 3.3 g cm^{-3}). In fact, it is much more dense than the Earth, when compared at constant internal pressure (5.3 g cm^{-3} versus 4.1 g cm^{-3} at 10 kbar). The high average density implies a metal-rich interior, perhaps 70% iron–nickel and 30% silicate. In the absence of measurements from the surface or from orbit, the interior properties of Mercury are constrained mainly by the total mass and size of the planet. Because of the relatively large density of Mercury, the core must occupy a larger fraction of the planet than is the case for the Earth. Moreover, since Mercury is smaller than the Earth, it should have cooled more rapidly and its solid inner core should be an even larger fraction of the radius of the liquid core than is the case for the Earth. Thus the remaining liquid core may be confined to a rather thin shell. As a result of these differences, it is possible that the dynamo that supports the magnetic field of Mercury differs substantially from the terrestrial dynamo. Only rudimentary constraints are presently available on the nature of the Mercury field.

As discussed elsewhere in this volume, a tenuous sodium and potassium atmosphere has been detected at Mercury. While various mechanisms have been proposed to explain this tenuous atmosphere, one possible source is outgassing of the planetary interior, suggesting in turn an internally active planet.

Magnetic field

Mercury has been visited by only one spacecraft, Mariner 10, which made three passes by the planet between March 1974 and March 1975. The first and third passes were suitable for studying the planetary field. On the first pass the spacecraft crossed the darkside of the planet within 723 km of the surface, at which point the field strength reached a maximum of close to 100 nT. The characteristics of the field resembled those of a mini-magnetosphere, in which the solar wind is deflected above the surface of the planet around a distorted dipole field. In contrast, the lunar magnetic field is so weak that the solar wind impinges on the surface and is absorbed. The third Mercury pass also traversed the darkside of the planet, approaching within 327 km of the surface and observing a maximum field of 400 nT. Again the characteristics of the observed field resembled those expected in a miniature version of the Earth's magnetosphere.

These two passes provided weak constraints on the magnitude of the intrinsic magnetic field, its orientation and its harmonic structure, in part because the coverage of the planetary field was poor and in part because of the lack of concurrent observations of the solar wind number density and velocity. The strength of a planetary magnetic field is measured in terms of its magnetic moment, the product of the equatorial surface field and the cube of the planetary radius. Estimates of the dipole moment of Mercury range from about 2 to 6×10^{12} T m^3, and the strength of the quadrupole moment and the tilt of the dipole moment are completely unconstrained. The dipole moment is known, however, to be pointed southward like the Earth's. Alternative sources to dynamo generation of the field have been proposed, such as remanent magnetization of an iron-rich crust, but it is difficult to obtain a strong enough field through these alternate mechanisms.

Magnetosphere

The weak magnetic moment of Mercury, about 4×10^{-4} of that of the Earth, combined with a solar wind pressure about seven times larger than the pressure at Earth, results in a very small planetary magnetosphere (in both absolute dimensions, and relative to the size of the planet). This magnetosphere is sketched in Figure M69, which shows the magnetic field lines in the plane containing the Sun and the magnetic dipole axis. The magnetic cavity deflects the solar wind at a distance of only 1.5 Mercury radii from the center of the planet. Since the solar wind moves faster (relative to Mercury) than the pressure wave needed to deflect the solar wind can propagate in the solar

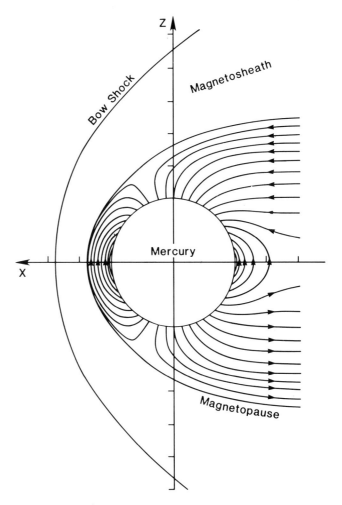

Figure M69 Cross-section of the magnetosphere of Mercury in the noon–midnight meridian. Solid lines anchored in the planet represent the direction of the magnetic field. The Mercury magnetic field lines point into the planet in the northern hemisphere as they do on Earth.

wind, a shock wave is formed in front of the magnetic cavity. This shock wave heats, slows and deflects the solar wind to allow it to flow around the magnetic cavity or magnetosphere. Similar bow shocks are found in front of all planetary magnetospheres. The energy dissipation that is required to heat the flow occurs through collisionless processes in which the electric and magnetic fields scatter the particles and the particles do not make direct collisions with each other. Thus these shock waves are often called 'collisionless' shocks.

An important question for all planetary magnetospheres is the coupling of the energy flux in the solar wind to the planetary magnetosphere. In the Earth's magnetosphere, stresses are communicated from the solar wind to the ionosphere and atmosphere and hence the solid body of the planet by electrical current systems which flow along magnetic field lines and then close across the magnetic field in the lower ionosphere. Mercury has no dynamically significant ionosphere or atmosphere, so the coupling must be quite different than in the terrestrial case. Based on dynamic events observed on the first Mariner 10 flyby, the Mercury magnetosphere is thought to be dynamic, varying markedly in the course of minutes. Clearly the energy transfer from the solar wind is much greater than on the Earth. As in the terrestrial magnetosphere, it is thought that the magnetotail is an important site for energizing the plasma of the magnetosphere. However, very little is known about the Mercury magnetosphere.

Christopher T. Russell and Janet G. Luhmann

Bibliography

Connerney, J.E.P. and Ness, N.F. (1988) Mercury's magnetic field and interior, in *Mercury* (eds F. Vilas, C.R. Chapman and M.S. Matthews). Tucson: University of Arizona Press, pp. 494–513.
Russell, C.T. (1987) Planetary magnetism, in *Geomagnetism*, Vol. 2 (ed. J.A. Jacobs). London: Academic Press, London pp. 457–523.
Russell, C.T., Baker, D.N. and Slavin, J.A. (1988) The magnetosphere of Mercury, in *Mercury* (eds by F. Vilas, C.R. Chapman and M.S. Matthews) Tucson: University of Arizona Press, pp. 514–61.

Acknowledgements

This work was supported in part by the National Aeronautics and Space Administration under research grant NAGW-2573.

Cross references

Earth: magnetic field and magnestosphere
Magnetometry
Mariner missions

MESOSPHERE

In Earth's atmosphere the temperature drops as altitude above the surface increases in the troposphere (q.v.). The stratosphere (q.v.) begins at the tropopause, where temperatures start to increase. Temperatures increase through the stratosphere; the mesosphere begins where this increase ceases, at an altitude of about 50 km. The mesosphere, like the troposphere, is characterized by temperatures that drop as altitude increases. At about 85–90 km altitude, temperatures begin to increase once more; this region is the thermosphere (q.v.). Not all of the planets possess a region of the atmosphere corresponding to Earth's mesosphere, but data indicates that mesospheres are present at Venus, Mars and the gas giants.

The physical behaviour of the mesosphere has been known only in the last few decades. A most striking feature of the mesosphere is its reverse meridional temperature gradient (Garcia, 1989). During summer the temperatures at 80–90 km are some 50 K colder than in the winter hemisphere. Figure M70 shows the temperature at various heights in the mesosphere. The annual cycle of temperature shows a steep decline during summer in the region of the mesopause. From data for 1962 a value of T_{min} = 140 K was obtained from observations of noctilucent clouds. Both summer and winter temperatures are far from radiative equilibrium. Recent satellite data shows high-latitude

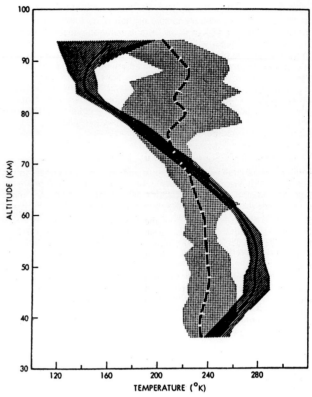

Figure M70 Seasonal average temperature profiles above Barrow, Alaska (71°N). The solid curve is the average of ten summer soundings, and the broken curve the average of 12 winter soundings conducted during 1965–67. The cross-hatched areas surrounding each curve represent the total range of temperatures included in the average. From Theon and Smith, *J. Atmos. Sci.*, **27**, 173–176, 1970; by permission of the American Meteorological Society.

summer mesopause temperatures near 150 K. On the other hand, colder temperatures are known to occur somewhat higher altitudes. Measurements by Theon, Nordberg and Smith (1967), Theon and Smith (1970) and others show a mesopause temperature of a little over 140 K at 85 km. Philbrick *et al.* reported temperatures in the interval of 120–140 K between 85 and 95 km. Other authors noted a temperature of 135 K near 86 km. Also, it has been reported that colder temperatures occurs occasionally. It is generally accepted that the mesospheric temperature profile varies smoothly with altitude if fluctuations due to gravity waves are filtered out (Garcia, 1989; McIntyre, 1989).

In contrast to summer, the mesosphere in winter is warmer, and the mesopause occurs at higher altitude. Various hypotheses have been discussed to explain the peculiar temperature of the mesosphere (Garcia and Solomon, 1985). It is now accepted that the cold summer mesopause is caused by adiabatic cooling produced by a strong summer-to-winter meridional cell (cf. Ebel, 1974; Memmesheimer and Blum, 1988). It follows from numerical studies that a strong meridional circulation can be driven by divergence of vertical momentum fluxes due to the breaking of vertically propagating gravity waves (cf. McIntyre, 1989).

There are in general two different atmospheric circulation systems in winter and in summer. The middle atmosphere region (20–30 km) changes its dynamical structure rather drastically twice a year. There are significant reversals of the circumpolar zonal wind in spring and autumn. The spring transition starts from an existing state of rather strong coupling of the atmospheric layers from below by the propagation of planetary waves in the westerly wind system (winds blowing from the west) of the stratosphere and the mesosphere. The spring reversal may start as a sudden breakdown of the polar vortex which rather abruptly terminates the upward propagation of planetary waves. The autumn transition arises from a slow and steady weaken-

Table M10 Thermal and dynamical structure of mesospheric variations

Winter time
Mesospheric temperature decrease stops at 82 km
Warm mesopause region
Strong prevailing westerlies from stratosphere to mesopause
Warm upper mesosphere
Low pressure in mesosphere

Spring conditions
Mesopause with temperature minimum at 85–90 km
Weak or no prevailing mesospheric winds
Mesospheric temperature decrease up to 85 km
Cold upper mesosphere
High pressure in mesosphere

ing of the stable middle atmosphere summertime westerly wind system (winds blowing from the east), mainly controlled by the heat balance of the upper stratosphere. The dynamical behavior of the mesosphere–mesopause region is important for the occurrence of noctilucent clouds.

Table M10 summarizes the main results concerning the structure and dynamics of the mesosphere. It follows from all the measurements that the mean motion of the mesosphere is zonal. Models of the mean circulation have been presented by Ebel (1974), Memmesheimer and Blum (1988) and Garcia (1989). The following general features are found. The summer circulation consists of an easterly current near 20–30° at 50–60 km height. The maximum velocity is more than 50 m s^{-1}. In the winter the circulation is westerly at about 40° at a height of 60–70 km. The maximum velocity is 80 m s^{-1}. It must be noted that there are important variations (e.g. diurnal, semidiurnal and shorter-period oscillations) in the mesosphere. Ebel (1974) noted, among the most dominant timescales, periods of 2, 5, 7 and 16 days which point to the existence of free planetary modes at mesopause heights. Quasi-biennial oscillations have been observed, for instance, using D-region data. Semiannual periodicities of tidal wave amplitudes and turbulence have been discussed by many authors. On the other hand, solar cycle variations must be considered. Many explanations of the momentum and energy balance of the mesopause are still tentative.

The water vapor concentration in the mesosphere is an important feature. It is a matter of historical interest that at the end of the 19th century and beginning of the 20th century Otto Jesse and Alfred Wegener suggested a relation between a dry upper atmosphere and volcanic eruptions. More recently, Hesstvedt (1961) and Chvostikov (1966) were among the first to present a hypothesis to explain the increase of water vapor concentration in the upper atmospheric region. Many photochemical models which considered the formation and destruction of such minor constituents as H_2O, CH_4, H_2, etc., under the control of solar ultraviolet radiation provide insight into the mechanism in the lower stratosphere.

From many observations during recent years it follows that the water vapor concentration in the mesosphere depends on the season and the latitude. Calculations show that the observed distribution of water vapor mixing ratio may be understood by taking into the account the combined effect of photochemical and transport processes.

Noctilucent clouds, which were observed first in 1885, are a remarkable phenomenon of the mesopause. Since 1885 regular observations has been made. The clouds occur only during summer, when the mesopause temperature is very cold. In southern latitudes they can be observed during the months December–February.

The height of the clouds is nearly constant at 82.3 km (82–84 km). They often show a well-defined wave structure, with wavelengths ranging from several kilometers to more than 100 km. The clouds are generally persistent and last for periods up to 5 h. On the other hand, individual parts (mostly billow structures) often form and decay within a few minutes or tens of minutes.

From the ground the noctilucent clouds can be observed in middle latitudes (65–50°N/S) when the solar depression angle is 6° to 16° below the observer's horizon. No generally accepted theory has been presented. However, it is clear that the clouds depend on the thermal-chemical structure and transition periods at mesopause altitude (Schröder, 1971; Gadsden and Schröder, 1989).

Recent satellite experiments have demonstrated the occurrence of polar mesospheric clouds (PMC). These PMCs are observed at latitudes above 70–75° and within 25–30 days about the peak, which itself is found to occur at 20–30 days after summer solstice. Thomas and Olivero reported variability on timescales from daily to annual. It is of interest that the northern hemisphere PMC are inherently brighter than the southern hemispheric clouds. The relationship between the noctilucent clouds and PMC is currently under study. Comparisons of visual noctilucent cloud data and satellite PMC shows some similarities.

In recent years many investigations of the atmospheres of the other planets have been organized. Measurements have been obtain by radio and star occultations and from *in situ* probes sensors. Information has also been obtained by visual and photographic telescope observations. From these data it was known that Venus and Mars have atmospheres which show various cloud phenomena. Color variations in the atmosphere of Mars have been noted. Astronomical observation also reveal the existence of dust storms and different layers of the atmospheres. The existence of blue (noctilucent?) clouds in the mesosphere of Mars has been suggested.

From more recent data by probes and other investigations the existence of a mesosphere and mesopause is known for Venus, Mars and the four Jovian planets. Nearly all the Jovian planets have a similar mesospheric structure and temperature profile, despite solar and planetary heat sources varying by a factor of 30 (cf. Chamberlain and Hunten, 1987). The height and the other physical features are probably comparable with the conditions of the Earth's mesosphere and structure. From this the existence of so-called mesospheric clouds (or blue clouds) in the atmosphere of Mars is not surprising.

Wilfried Schröder

Bibliography

Chamberlain, J.W. and Hunten, D.M. (1987) *Theory of Planetary Atmospheres*. London: Academic Press.
Chvostikov, I.A. (1966) Nature of noctilucent clouds. *Priroda*, **55**, 48–53.
Ebel, A. (1974). Heat and momentum sources of mean circulation of an altitude of 70 to 100 km. **26**, 325–33.
Gadsden, M. and Schröder, W. (1989) *Noctilucent Clouds*. New York: Springer.
Garcia, R.R. (1989) Dynamics, radiation and photochemistry in the mesosphere: implications for the formation of noctilucent clouds. *J. Geophys. Res.*, **94**, 14616.
Garcia, R.R. and Solomon, S. (1985) The effect of breaking gravity waves on the dynamics and chemical composition of the mesosphere and lower thermosphere. *J. Geophys. Res.*, **90**, 3850–68.
Hesstvedt, E. (1961) Note on the nature of noctilucent clouds. *J. Geophys. Res.*, **66**, 1985–7.
McIntyre, M.E. (1989) On dynamics and transport near the polar mesopause in summer. *J. Geophys. Res.*, **94**, 14617–28.
Memmesheimer, M. and Blum, P.W. (1988) Seasonal and latitudinal changes in atmospheric condition favoring the formation of polar mesospheric clouds. *Phys. Scripta*, **37**, 178–84.
Olivero, J.J. and Thomas, G.E. (1987) Clouds of the polar mesopause in summer. *Phys. Scripta*, **18**, 276–80.
Philbrick, C., Fiare, A.C. and Fryklund, D.H. (1984) The state experiment – mesospheric dynamics. *Adv. Space Res.*, **4**, 153–6.
Schröder, W. (1971) Transition periods in the mesosphere. *Gerlands Beitr. Geophysik*, **80**, 65–74.
Theon, J.S. and Smith, W.S. (1970) Seasonal transitions in the thermal structure of the mesosphere at high latitudes. *J. Atmos. Sci.*, **27**, 173–6.
Theon, J.S., Nordberg, W. and Smith, W.S. (1967) Temperature measurements in noctilucent clouds. *Science*, **157**, 419–21.

METEOR, METEOROID

A meteor is the bright streak or trail of light in the sky caused by the passage of a small chunk of extraterrestrial matter through the Earth's (or another planet's) atmosphere. While the object is in space it is termed a meteoroid, and should the object fall to Earth intact and be recovered, it is termed a meteorite (q.v.) or, if very small, a micrometeorite (q.v.). The upper size limit for meteoroids is not well defined; in general if the object is too small to be called an asteroid or comet, it is a meteoroid. The solar system contains large numbers of small objects, or meteoroids, moving in orbits around the Sun.

Meteors, also known as shooting stars, result from the atmospheric entry of objects ranging in mass from 0.1 g to many kilograms. The brightest meteors may be brighter than the full Moon (magnitude −15 to −20), and they are then termed fireballs. Occasionally the sound of their passage and break-up is audible. Meteors which explode are termed bolides.

Meteoroids encounter the Earth's atmosphere at high velocities (tens of kilometers per second). Friction heats the object, beginning at heights of 100 km or so, until (typically) most or all of the object is vaporized. At the same time some of the atmospheric molecules along the path are ionized. These trails of metallic ions (mostly sodium, calcium, silicon and iron) and ionized gases may be detected using radar techniques. Since the atmospheric ions and electrons are accelerated to near the speed of the meteoroid, a Doppler effect (q.v.) may be detected; this can in turn be used (in favorable cases) to determine the velocity and the orbit of the original meteoroid. Radar techniques can be employed to detect and study daytime meteors as well as those that occur at night. Meteor studies are also carried out with video recording observing systems. An important application of the study of meteor trails has been the description of the global wind patterns of the mesosphere (q.v.) of the Earth.

The presence of numbers of small, high-velocity objects in near-Earth space was and is a concern for spacecraft designers. Early meteoroid detection experiments suggested that there might be a significant 'cloud' of meteoroids orbiting Earth (Zook, 1991), but the first measurements turned out to be in error. More recently, the experiment of the Long Duration Exposure Facility (McDonnell et al., 1990) obtained the best ever sampling of the near-Earth meteoroid flux; these data are still being analyzed.

More meteors are observed after midnight than before, due to a geometric effect of the motion and rotation of the Earth; after midnight meteors meeting the Earth head-on with high relative velocities are seen, while before midnight we see objects which have caught up with the Earth from behind. A few sporadic meteors are to be seen practically every hour in good seeing conditions. Some meteors, however, appear in meteor showers (q.v.) which occur at predictable times during the year. Some of the showers have been shown to be associated with the orbits of known comets; thus it is apparent that some meteoroids (perhaps most) are icy fragments of comets. Study of meteors thus aids in characterizing comets and the materials that they shed in their passage through the inner solar system. The objects which are recovered on Earth (meteorites; q.v.) have widely varying compositions which provide considerable information about the planets, the asteroids and other small solar system bodies (e.g. Bhandari et al., 1989).

James H. Shirley

Bibliography

Bhandari, N. Bonino, G., Callegari, E., et al. (1989) The Torino, H6 meteorite shower. Meteoritics, 24, 29–34.

McDonnell, J.A.M., Sullivan, K., Stevenson, T.J. and Niblell, D.H. (1990) Particulate detection in the near-Earth space environment aboard the LDEF, in Origins and Evolution of Interplanetary Dust (eds A.C. Levasseur-Regourd and H. Hasegawa). Dordrecht: Kluwer Academic Publishers.

Mitton, S. (ed.) (1977) Cambridge Encyclopedia of Astronomy. New York: Crown Publishers.

Pasachoff, J.M. (1989) Contemporary Astronomy, 4th edn. New York: Saunders College Publishing.

Payne-Gaposhkin, C. (1954) Introduction to Astronomy. New Jersey: Prentice-Hall Inc.

Zook, H.A. (1991) Meteoroids, space investigations, in The Astronomy and Astrophysics Encyclopedia (ed. S.P. Maran). New York: Van Nostrand Reinhold.

Cross references

Comet
Dust
Halley, Edmond, and Halley's comet
Meteorite
Meteorite parent bodies
Micrometeorite

METEOR SHOWER, METEOROID STREAM

The rate of appearance of meteors varies; the average for sporadic meteors is about five to seven per hour under perfect conditions. At certain times many more meteors are seen, and these have the appearance of arriving from a particular point in the sky (the radiant). This occurrence is termed a meteor shower. Meteor showers are generally named for the stellar constellation that contains the radiant (for instance, the constellation Orion lends its name to the Orionid shower). Some showers are named after their parent comet (for instance, the Giacobinids are named after comet Giacobini-Zinner). The actual paths of the meteors in a shower are approximately parallel; they appear to originate at a point (the radiant) due to a perspective effect, analogous to the apparent convergence of a pair of (parallel) railroad tracks in the distance.

Table M11 lists the best-known meteor showers. Some of these occur dependably from year to year, while in others the number of meteors observed varies considerably. One of the most spectacular meteor showers in recent times occurred in 1966, when the rate of occurrence of meteors of the Leonid shower rose to approximately 150 000 per hour.

Records of meteor showers date back to 687 BC (Hawkes, 1991), but it was not until the 18th century that serious scientific investigations began. Giovanni Schiaparelli (q.v.) made an important contribution in 1866 when he demonstrated that the orbits of the Perseid meteors were nearly identical to that of comet 1862 III. This topic was under intensive investigation at the time. Burke (1986) provides an excellent historical perspective.

Many of the known meteor showers are associated with cometary orbits. Comet Halley, for instance, is linked with two showers, the eta Aquarids in May and the Orionids in October. The comet leaves a trail of material strewn along its path, and when the Earth's orbit intersects the path, large numbers of meteors may be seen. This trail of material is known as a meteoroid stream (the term 'meteor stream' is technically incorrect, since the term meteor refers to the visible phenomenon, whereas the objects are meteoroids). The stream linked with Halley's comet may contain five to ten times the total mass of the comet nucleus itself.

The distribution of material along the orbit is generally non-uniform, leading to variations in the rate of meteors seen in different years. Comets lose greater amounts of material near perihelion, when the solar radiation heating and pressure effects are greatest. And, as the meteoroid stream develops with time, it experiences gravitational perturbations and other effects that distribute the material more uniformly along the orbit. Hughes (1991) reviews the stages of dynamical evolution of meteoroid streams.

Meteoroid streams are short-lived phenomena in comparison to the age of the solar system. Comets that enter the inner solar system have generally been perturbed strongly by Jupiter; their dynamical lifetimes following this event are on the order of 10^4–10^5 years (see Comet: origin and evolution). A short-period comet may survive several thousand perihelion passages during its lifetime. The orbits may evolve such that streams that today intersect the Earth's orbit may no longer do so in the future; thus showers that are seen today may disappear in the relatively near future (Hughes, 1991). Over the long term, the fate of comets is to disappear entirely, or to become inactive, showing an appearance identical to that of an asteroid. In 1983 a connection was made linking the asteriod 3200 Phaethon with the Geminid meteor shower (Halliday, 1988). Phaethon is apparently a 'dead' or inactive comet nucleus, and the Geminids are the remnants of its passage. Another example is found in Biela's comet, which was observed to divide in two before 1846 and which disappeared by 1865. A portion survived for a time as a meteor shower, the Bielids, which has now disappeared.

On the other hand, some meteoroid streams have intersected the Earth's orbit for hundreds of years. The Lyrids were known in 687 BC, the Perseid stream was first recorded in 36 AD and the Leonids were first recorded in 902 AD.

Meteoroid streams are an active area of investigation. The trails may be seen on infrared images, for instance those produced by the Infrared Astronomical Satellite (IRAS). Meteor studies yield in-

Table M11 Important meteor showers

Name	Date (maximum)	Radiant RA	Declination	Zenith hourly rate
Nighttime showers				
Quadrantids	4 January	15 h 28 min	50	110
Corona Australids	16 March	16 h 20 min	−48	5
April Lyrids	22 April	18 h 08 min	32	12
Eta Aquarids	5 May	22 h 24 min	0	20
June Lyrids	16 June	18 h 32 min	35	8
Ophiochids	20 June	17 h 20 min	−20	6
Capricornids	25 July	21 h 0 min	−15	6
Delta Aquarids	27 July	22 h 36 min	0	35
		22 h 36 min	−17	
Pisces Australids	31 July	22 h 40 min	−30	8
Alpha Capricornids	2 August	20 h 36 min	−10	8
Tau Aquarids	6 August	22 h 32 min	−15	6
		22 h 04 min	−6	
Perseids	12 August	03 h 04 min	58	68
Kappa Cygnids	20 August	19 h 20 min	55	4
Orionids	21 October	06 h 24 min	15	30
Taurids	8 November	03 h 44 min	14	12
		03 h 44 min	22	
Cepheids	9 November	23 h 30 min	63	8
Leonids	17 November	10 h 08 min	22	10
Phoenicids	4 December	01 h 00 min	−55	5
Geminids	14 December	07 h 28 min	32	58
Ursids	22 December	14 h 28 min	78	5
Daytime showers				
Theta Cetids	15 May	02 h 00 min	−3	15
Psi Perseids	8 June	03 h 56 min	22	40
Arietids	8 June	02 h 56 min	23	60
Beta Taurids	29 June	05 h 40 min	17	24

After *Cambridge Encyclopedia of Astronomy*, S. Mitton (ed.), Crown Publishers, New York, 1977.

formation on the ages, compositions and other characteristics of comets. There are practical implications as well; the launch of the space shuttle Discovery was delayed in August 1993 due to the Perseid meteor shower. In that year the Earth made a close pass behind comet Swift-Tuttle, the source of the Perseids, and a replenished stream of debris produced record numbers of Perseid meteors.

James H. Shirley

Bibliography

Burke, J.G. (1986) *Cosmic Debris*. Berkeley: University of California Press.
Halliday, I. (1988) Geminid fireballs and the peculiar asteroid 3200 Phaethon. *Icarus*, 76, 29.
Hawkes, R.L. (1991) Meteors, shower and sporadic, in *The Astronomy and Astrophysics Encyclopedia* (ed. S.P. Maran). New York: Van Nostrand Reinhold.
Hughes, D.W. (1991) Meteoroid streams, dynamical evolution, in *The Astronomy and Astrophysics Encyclopedia* (ed. S.P. Maran). New York: Van Nostrand Reinhold.

Cross references

Comet
Comet: observation
Comet: origin and evolution
Dust
Ionosphere
Meteorite

METEORITE

Long before terrestrial spacecraft visited other bodies in the solar system, rocks from many of those bodies were landing on Earth. Ejected in violent impacts on their home bodies, and having survived a fiery passage through the Earth's atmosphere, these rocks from space – meteorites – provide us with samples of dozens of different solar system bodies.

The study of meteorites complements astronomical and spacecraft investigation of the solar system. With meteorites we can study detailed mineralogical structure, and determine isotopic and elemental compositions, on a level inaccessible to telescopes or spacecraft. On the other hand, the nature and characteristics of the world's meteorite collection are dictated by orbital dynamics, since we are limited to those rocks that happen to be on Earth-intersecting orbits. In fact, there are very few cases where we can match a meteorite with its parent body. Hence the study of meteorites involves both a study of the detailed history of meteorite parent bodies, and a search for the identities of those bodies whose history is being determined.

The interest in meteorites involves more than simply the fact that they are extraterrestrial material that we can literally get our hands on. Meteorites are the oldest material on Earth, with most of them dating from the time of the formation of the solar system. Some contain even older grains that have survived unaltered since their formation near other stars. Furthermore, most meteorites come from asteroids, small bodies that have not been investigated by spacecraft.

A body on an Earth-intersecting orbit is a meteoroid; once it enters the atmosphere, and is heated by friction to incandescence, it is a meteor. If it is large enough for some material to survive the passage, that material is a meteorite. Known meteorites range in mass from less than 1 g to more than 10 metric tons. Smaller particles can be decelerated high enough in the atmosphere that the heating is minimal. These micrometeorites, whose sizes are typically measured in microns, are compositionally distinct from 'macro'-meteorites, and are treated in a separate article (see Micrometeorite).

Recovery of meteorites

Each year, several hundred meteorites of greater than 1 kg in mass should fall on the Earth's land masses (Halliday, Blackwell and

Griffin, 1989). However, only about five to ten per year are recovered at the time of fall (these meteorites become known as 'falls'). A slightly larger number of meteorites of unknown fall date are also recovered each year ('finds'). Thus the world's collection of meteorites, which started in earnest in about 1800, contained slightly more than 2000 samples by the mid-1970s. Since then, targeted search programs have recovered more than 10000 meteorites (Plate 15).

The premiere meteorite-hunting locale is the Antarctic ice cap. This is partly because of the relative ease of finding an extraterrestrial rock on ice compared to finding it amid terrestrial rocks and soil. However, most Antarctic meteorites are between 10^4 and 10^6 years old, so there must be some concentration mechanism. This probably involves transport (within ice) to locations where the ice is ablated by wind. Indeed, the highest concentrations of meteorites are in areas where ice flow is stopped by mountain ranges. The United States and Japan have mounted expeditions nearly every Antarctic summer since the mid-1970s. European teams have also begun collecting meteorites in Antarctica.

Successful collection efforts have also been mounted in several of the world's other deserts. Hundreds of meteorites have been found in the Sahara in North Africa, in the Nullarbor Plain in southwestern Australia, and in the southwestern United States. Unlike the Antarctic meteorites, these meteorites are 'concentrated' only in that they are easier to find (and weather more slowly) than in areas with more vegetation, more rainfall or more rock outcrops.

The discovery of remote areas with copious quantities of meteorites has posed a nomenclature problem. Meteorites are traditionally named after the town nearest the discovery location. In the deserts and Antarctica there are far more meteorites than towns, so these meteorites are described by a geographic feature and a number (e.g. Allan Hills 81005 was the fifth meteorite from Allan Hills, Antarctica, found and classified in the 1981–82 field season, Roosevelt County 027 was the 27th meteorite to be classified from Roosevelt County, New Mexico, USA).

History of studies of meteorites

Many meteorite falls were surely witnessed prior to modern times. In fact, Pliny the Elder recounts stories of stones falling from the sky. However, he also tells of showers of milk, flesh and wool, so as the Middle Ages gave way to the Age of Enlightenment in Europe, the belief that stones fall from the sky was considered to be mere superstition (Burke, 1986).

The meteorite which has been preserved for the longest period of time is Nogata (Japan), which was placed in a Shinto shrine after its fall in 861 AD The second-oldest fall has been displayed in a church in Ensisheim, France, most of the time since its fall and recovery in 1492. With their passage from objects of mystery to objects of scientific scrutiny, the typical repository of meteorites moved from churches to museums.

The belief that meteorites existed, and furthermore that they were extraterrestrial, became widespread among scientists in about 1800. In 1794 E.F.F. Chladni analyzed reports of meteorite fireballs and fall phenomena, and concluded that the trajectories matched those of bodies entering the atmosphere. Then a series of chemical analyses by Howard, de Bournon and others showed that known meteorites were chemically similar to one another (in particular, the metal contained a significant fraction of nickel), but unlike other known terrestrial rocks. On an April afternoon in 1803, a shower of more than 2000 stones fell on L'Aigle, France, and specimens quickly reached Paris. This spectacular event, with hundreds of witnesses, and J.B. Biot's detailed investigation on behalf of the Académie Française, effectively ended the debate over the reality of meteorites. The ultimate source of the meteorites remained in doubt. Chladni thought they were remnants of disrupted planets or leftovers of planetary formation, but Biot and others argued that they were from lunar volcanoes. The lunar hypothesis was far more popular in the early 19th century, but Chladni's viewpoint is much closer to that of the late 20th century.

Advances in analytical techniques have often led to revisions in the understanding of meteorites (Burke 1986; Marvin 1986). The development (in the 1860s) of the polarizing microscope and of techniques to make thin sections of rocks were exploited by G. Tschermak and H.C. Sorby to explore the mineralogy of meteorites, and the development of microanalytical techniques starting after World War II (most prominently the electron microprobe and high-precision mass spectrometry) have led to a new level of discoveries. Also, just as the fall at L'Aigle was important, the availability of new material has led to leaps in understanding. For example, two large specimens of previously rare groups of meteorites, carbonaceous chondrites, fell in 1969. More than 100 kg of Murchison (Australia) and more than 1 metric ton of Allende (Mexico) was collected. The combination of microanalytical techniques and the availability of large quantities of material made it possible to search for components present at part-per-million levels, such as unprocessed interstellar grains. Similarly, the Antarctic collection has had the largest influence on the study of the rarest meteorite types. For example, the first true lunar meteorites (meteorites with chemical and isotopic properties matching those of the returned Apollo samples) were discovered in Antarctica.

Classification

Meteorites are traditionally described as stones, irons or stony irons, depending on the relative proportions of silicates and metal. A classification that says more about origins (Table M12) is to separate meteorites into undifferentiated meteorites and differentiated meteorites, depending on whether their formation required igneous processing (i.e. melting of some portion of a parent body). On a more detailed level, meteorites are separated into groups on the basis of their chemical compositions and mineralogy.

The best classification would be to separate meteorites by parent bodies. However, it is generally difficult to determine whether two meteorites from a group, much less two separate groups, came from the same parent body. One technique that can at least eliminate some potential relationships is based on the determination of the isotopic composition of oxygen, the most abundant element in rocks. Oxygen has three stable isotopes. Since oxygen is so abundant, nuclear reactions within the solar system have not affected its isotopic composition. Furthermore, the changes in isotopic composition as the result of chemical reactions, including those involved in differentation, are usually mass dependent – the ratio of ^{17}O to ^{16}O changes by about half as much as the ratio of ^{18}O to ^{16}O. Thus samples from a single body (e.g. the Earth) or from any bodies that have been in isotopic equilibrium with it (e.g. the Moon) tend to plot along a line of slope 0.5 on a plot such as Figure M71. Thus it came as something of a surprise when R.N. Clayton and coworkers (Clayton, Grossman and Mayeda, 1973) discovered that the oxygen isotopic composition of different groups of meteorites tend to plot along different lines of slope 0.5; some objects tend to fall along mixing lines with slopes other than 0.5. The oxygen isotope map of Figure M70 has become a standard in classifying meteorites. Although two meteorites that fall along the same slope 0.5 line do not necessarily come from the same parent body, two that do not fall on the same line cannot come from the same body unless it had never equilibrated isotopically.

In the rest of this section the major (and some of the minor) groups of meteorites will be briefly described.

Chondrites

The most common meteorites among those observed to fall are chondrites, undifferentiated meteorites consisting primarily of silicates, but also including Fe–Ni metal, sulfides and other phases. They

Table M12 Types of meteorites

Stones	Chondrites
	  Ordinary (H, L, LL)
	  Enstatite (EH, EL)
	  Carbonaceous (CI, CM, CO, CV)
		Undifferentiated

	Achondrites
	  Eucrites, Diogenites, Howardites
	  SNC (Martian?)
	  Lunar
	  Angrites
	  Enstatite achondrites (Aubrites)
		Differentiated

	Ureilites	??
Stony irons	Primitive achondrites (e.g., Acapulco)	??
	Pallasites Mesosiderites	Differentiated
Irons	(≥ 13 groups)	Differentiated

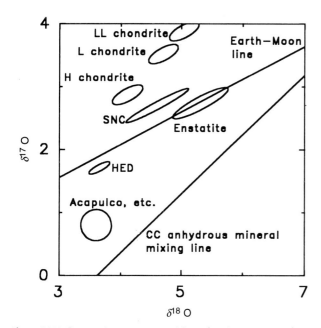

Figure M71 Oxygen isotope composition of various groups of meteorites. The quantities on the axes, δ^iO, are defined as the difference (in parts per thousand) of the $^iO/^{16}O$ ratio measured in the sample from that ratio in terrestrial ocean water. Compositions related by mass-dependent fractionation processes plot along lines parallel to the Earth–Moon line (where virtually all terrestrial and lunar samples plot). Most meteorite groups plot in clusters that spread out along such lines. Samples that plot along lines of different slope, such as the line shown for anhydrous minerals from some carbonaceous chondrites, were probably never part of a single isotopically homogenized reservoir. (Plot after Clayton, Onuma and Mayeda (1976), with more recent data from R.N. Clayton and coworkers included.)

are called 'chondrites' because they contain 'chondrules', small (typically 0.1 to 2 mm in diameter) spherules. In fact, many chondrites consist mostly of chondrules.

The most common of the chondrites are the ordinary chondrites (see Chondrites, ordinary), which comprise about 80% of all known meteorite falls. These are divided into three groups: H (high total Fe content, > 25 wt%), L (low Fe, 20–23 wt%) and LL (very low Fe, 17–19 wt%)), although there may be more than three ordinary chondrite parent bodies. The ordinary chondrites are progressively more oxidized in the sequence H–L–LL, such that the Fe content of the silicates increases even as the total amount of Fe (and the amount of metal) decreases.

Carbonaceous chondrites tend to be, as the name implies, richer in carbon (up to several weight percent) and other volatile elements such as sulfur and hydrogen than the ordinary chondrites. It was this volatile-rich nature that originally distinguished these meteorites. However, volatile contents are highly variable, even among closely related meteorites, and modern analytical techniques have shown that the original carbonaceous chondrites (as well as some carbon-poor meteorites) are closely related to one another. These are distinguished from other chondrites by being more oxidized and by being richer in refractory elements, such as aluminum and calcium. Thus, these four (or more) groups of meteorites are carbonaceous in name, though not always in chemistry (see Carbonaceous chondrite).

The last major type of chondrites is the enstatite chondrites, probably representing at least two parent bodies (Keil 1989). These meteorites are characterized by being extremely reduced. The name comes from the dominant mineral, enstatite ($MgSiO_3$), a virtually Fe-free pyroxene. Because they formed under such reducing conditions, the enstatite chondrites (and the possibly related enstatite achondrites) contain a host of bizarre minor minerals that are rare or unknown in any other rocks.

Achondrites

The most common achondrites are the closely related eucrites, diogenites and howardites (see Basaltic achondrites). These are magmatic rocks, representing the silicate portion of a differentiated asteroid. The eucrites are basalts (mostly made of the minerals pyroxene and plagioclase), somewhat similar to rocks formed by melting of the lunar or terrestrial mantles, while diogenites are pyroxene rich, somewhat analogous to rocks of the Earth's mantle. Howardites are 'regolith breccias', basically a mechanical mixture of eucrite and diogenite material. The indistinguishable oxygen isotopic composition of the three groups ('HED' on Figure M71), and the fact that howardites are mixtures of the other two groups, has led to the belief that all these meteorites may have come from the same body, perhaps the asteroid 4 Vesta (see Vesta).

Two groups of achondrites come from planet-sized bodies. Although the Moon has been suggested as a source of various meteorites since at least the days of Biot, when actual lunar samples were returned they were unlike any known meteorites. Since then about a dozen certifiably lunar meteorites have been found, mostly in Antarctica (see Lunar meteorite).

The SNC meteorites are three closely related groups of igneous rocks (typified by Shergotty, Nakhla and Chassigny). Crystallization ages range from 1.3 Ga to perhaps as young as 180 Ma, making these by far the youngest extraterrestrial samples known, and hence requiring a parent body with a long history of magmatic activity. Furthermore, there are glassy inclusions within one of the SNCs from Antarctica, Elephant Moraine 79001, which contain CO_2, nitrogen and noble gases with elemental and isotopic compositions that match the Martian atmosphere as measured by the Viking spacecraft. Although there are some problems with a Martian origin of the SNCs, most notably finding a crater young enough and big enough to eject the meteorites, these meteorites are generally thought to come from Mars (McSween, 1985; see SNC meteorites). The unique meteorite Allan Hills 84001 has also been identified as a piece of Mars, on the basis of its bulk and mineral chemistry and oxygen isotopic composition (Mittlefehldt, 1994), even though it is considerably older than the SNC meteorites.

There are two more significant groups of achondrites of unknown origins. The angrites, consisting of one large meteorite (Angra dos Reis) and two small (> 10 g total) Antarctic meteorites, have an importance disproportionate to their number. They are, like the eucrites, differentiated samples, but they are even older, with precisely determined formation ages within about 10 Ma of the oldest ages of chondrites (Lugmair and Galer, 1992). The ureilites are probably the least-understood achondrites (Goodrich, 1992; see Ureilite meteorites). They consist largely of silicates with compositions and textures indicative of differentiation. However, they also contain abundant carbon and noble gases, and their oxygen isotopes fall along the mixing line for anhydrous minerals from carbonaceous chondrites, rather than on a single mass fractionation line (Figure M71), suggesting a primitive origin. No theory of the origin of these enigmatic rocks has yet managed to match all their properties, although models involving impacts of carbon-rich projectiles into differentiating targets are currently being investigated.

Stony irons

Like chondrites, the meteorites classified as 'stony irons' are mixtures of metal and silicates, but with the metal much more abundant. Pallasites consist largely of metal, with interspersed olivine crystals. Mesosiderites are an intimate mixture of metal with eucrite-like silicates. Most mesosiderites and pallasites have oxygen isotopic compositions similar to those of eucrites (Figure M71), and so they may be from the same parent body. Another group of meteorites, including Acapulco (Mexico) and Lodran (Pakistan), consists of mixtures of metal with silicates of chondritic composition but achondritic textures. Some of these meteorites have previously been described as 'anomalous' or 'unique' chondrites, achondrites or stony irons. The recent discovery that several of them have oxygen isotopic compositions similar to one another but distinct from any other meteorites groups (Clayton, Mayeda and Nagehara, 1992; Figure M71) suggests the possibility that they are all closely related.

Irons

Iron meteorites are, as the name implies, metallic. However, the metal is not pure iron, but rather an iron–nickel alloy. The iron meteorites can, like the stones, be grouped on the basis of chemical composition (most commonly by using Ni, Ga and Ge). More than a dozen groups have been identified, and the large number of iron meteorites which do not fall into any of these groups suggests that many more parent bodies have been sampled. Most iron meteorites are apparently the cores (or fragments of the cores) of small asteroids, although some may represent localized melting events (e.g. from impacts; Scott, 1979).

Meteorite clues to the evolution of the solar system

Chondritic meteorites are both the oldest and the least differentiated rocks currently available, hence they can provide important insights into the processes which occurred very early in the history of the solar system. Achondrites, meanwhile, can be used to study the evolution of small differentiated bodies, while irons are our only samples of cores of solar system bodies.

Age and composition of the solar nebula

The ages of most chondritic meteorites, determined by radiometric dating techniques, are about 4560 Ma. This is generally taken to be the age of the solar system as a whole, since models show that solid materials should have condensed within the first few million years of the collapse of the protosolar cloud. Naturally, exceptions to the rule of identical ages for all meteorites are exceptionally important, and several of these will be discussed below.

Many achondritic meteorites, particularly the angrites, also formed at about the same time. This implies that some differentiation, at least, occurred within the first 10 Ma of the history of the solar system. The oldest terrestrial rocks are about 300 Ma younger, while the oldest well-dated lunar rocks are about 50 Ma younger. However, the Pb isotopic composition of the Earth is roughly consistent with the Pb isotopic composition of chondritic meteorites (Patterson, 1956), suggesting that the Earth (or the material that went into it) formed within about 50–100 Ma of the time the earliest meteorites formed.

Chondritic meteorites have relative elemental abundances that are very similar to those observed in the photosphere of the Sun for those elements whose photospheric abundances are known. The closest match to solar abundances is for the CI (carbonaceous chondrites like Ivuna) meteorites. Thus chondrites are probably nearly unfractionated samples of the rock-forming material in the solar system, and the relative abundances of the elements that cannot be determined accurately from the solar photosphere are usually assumed to have the CI or 'chondritic' values. Surprisingly, CI meteorites are far from the least altered meteorites. They have undergone extensive aqueous alteration – more extensive than in any other group – but it must have occurred in a closed system in which no overall loss or gain of any elements occurred.

Nebular processes

By about 1970 the consensus was that the contraction of the protosolar nebula had led to temperatures high enough to vaporize all solid material and thoroughly mix all isotopes of all elements. The calcium–aluminum-rich inclusions (CAIs) in carbonaceous chondrites have mineralogies and trace element patterns consistent with (1) condensation from a cooling high-temperature ($> 2000°$ C) gas, or (2) residual material resulting from high-temperature evaporation of solids. Similarly, the chondrules that are prevalent in chondrites require high formation temperatures (typically $1600°$ C or more), since their spherical shapes imply that they were once molten droplets. Furthermore, the isotopic composition of most elements in most meteorites is rather uniform.

More recently, detailed studies of the isotopic composition of selected phases have shown that the solar nebula cannot have been uniformly hot and uniformly mixed. The first observations were of noble gas (in particular, neon and xenon) isotopic compositions in some meteorites that could not be explained by any reasonable solar nebula process such as mass-dependent fractionation or nuclear decay. As analytical techniques have improved, such effects have been discovered in most multi-isotope elements. Perhaps the most significant effect is for oxygen, the most abundant element in rocks. As mentioned above, the variation in oxygen isotopic composition among different types of meteorites suggests that the bodies on which those meteorite formed acquired different amounts of oxygen from several distinct (unmixed) reservoirs.

Pursuit of the carrier phases of the isotopically anomalous noble gases ultimately led to the identification of some of the interstellar grains which survived nebular processing. The interstellar grains which have so far been identified (usually by their isotopic composition) are mostly carbonaceous, including graphite, diamond and SiC (Anders and Zinner, 1993). It has been more difficult to identify grains which might have formed in O-rich, rather than C-rich, stellar environments, because O is so much more common in meteorites. However, presolar aluminum oxide grains have been found (Nittler et al., 1994). Interstellar grains are usually isolated in the laboratory by a series of harsh chemical treatments. Of course, it is probably not surprising that the grains that resisted all processing in the solar nebula and on small asteroids are some of the most chemically stable and physically rugged materials known. The isotopic and elemental compositions of these grains are now being used to constrain models of the stellar processes which synthesize the elements.

The mechanism that produced chondrules is still unknown, but the ubiquity of chondrules means that it must have been a widespread process somewhere in the solar nebula. The variety of textures of chondrules have been reproduced by simulations in which solid material is heated to near the melting point, then cooled quickly, though not as quickly as radiative cooling to cold interplanetary space (Hewins, 1988). This suggests localized melting in a dusty environment. Such a process could also produce the variations in chemical composition that are seen. Transient events in the solar nebula, such as solar flares, lightning or magnetic field disturbances, seem to be the most promising chondrule-forming mechanisms, although none has been shown to work in detail.

Early processes on parent bodies

Although the chondrites are undifferentiated, they are not unprocessed. Most have undergone thermal metamorphism, aqueous alteration or, at the least, have been affected by impact-produced shocks. As well as the chemical classification discussed above, chondrites are also classified according to petrologic type on a scale of 1 to 6 (7 is sometimes used) designed to reflect the degree of equilibration and metamorphism (Van Schmus and Wood, 1967). Although the scale represents a continuous sequence, it is now generally believed that the most primitive chondrites are not type 1, but type 3, with types 2 and 1 (which are found only in carbonaceous chondrites) reflecting progressively greater degrees of aqueous alteration and types 4 through 6 (which are found mostly in ordinary and enstatite chondrites) reflecting progressively greater thermal metamorphism (McSween, 1979).

Most ordinary chondrites are types 5 and 6. The silicate minerals are in equilibrium with one another, the interchondrule matrix has been recrystallized and the outlines of chondrules are poorly defined, if present at all. These changes reflect thermal metamorphism. Cooling rates, inferred peak temperatures and ages are all consistent with an 'onion shell' structure, a stratified body in which the interior remained hot the longest, leading to the youngest ages and the highest degree of metamorphism (Pellas, 1992). Models suggest parent bodies of 100–300 km in diameter (i.e. the size of rather large asteroids).

The heat source for metamorphism of ordinary chondrites, and for early differentiation of irons and some achondrites, remains an enigma. The decay of ^{26}Al (half-life 0.75 Ma) has long been an attractive possibility, and evidence for freshly synthesized ^{26}Al has been found in many meteorites. But evidence for ^{26}Al has seldom been found in ordinary chondrites, and never in differentiated meteorites. The most popular alternative, heating by electromagnetic induction as a result of an enhanced T Tauri-like solar wind, has not been shown to be consistent with astrophysical models of the early Sun. Other possibilities, such as accretionary impacts or other short-lived radionuclides like ^{60}Fe (half-life 1.5 Ma) have also been suggested, but a completely satisfactory model has not yet been proposed (Wood and Pellas, 1991).

Most carbonaceous chondrites, as well as a few type 3 ordinary chondrites, have undergone aqueous alteration. Characterization of the resulting mineral assemblages is not complete, in part because many of the phases are extremely fine grained. Various lines of evidence suggest that aqueous alteration in carbonaceous chondrites involved low pressures, water/rock ratios of at least one, and temperatures of −20° C to 150° C. Again, the source of the heat necessary to mobilize the water is not known.

Some carbonaceous chondrites contain amino acids. Although terrestrial contamination at some level is always a potential problem, meteoritic amino acids do not show the favoritism for dextrorotary forms seen in terrestrial amino acids, and hence must be indigenous (to the meteorites). Their presence has obvious implications for the origin of life on Earth; either amino acids were present in the solar nebula, or they could be, and were, created in an abiological environment (perhaps as a by-product of the aqueous alteration) (see Life: origin).

With the recognition of meteorites from Mars and the continued study of the eucrites, it is now possible to use meteorites to make comparative chemical and chronological studies of terrestrial planets. The composition of these meteorites, coupled with knowledge of elemental partitioning behavior, can provide insights into the evolution of a differentiated body. Along with the Earth and Moon, the parent bodies of those two suites of meteorites are the only objects for which such detailed models can be made.

Collisional history of the solar system

Although they are primarily valued for their records of the early solar system, meteorites also carry a record of the ongoing collisional environment in the solar system.

One sense in which the collisional history can be determined is by studying meteorites which are regolith breccias (i.e. which have once been part of the 'soil' of an asteroid). For example, most meteorite regolith breccias tend to be mixtures of similar, or closely related, materials rather than random mixtures of many groups. That suggests a nebula where material was not thoroughly mixed after solids formed. Carbonaceous clasts are the most common 'exotic' component, in agreement with astronomical observations that show that carbonaceous asteroids have a much higher relative abundance in the asteroid belt than do carbonaceous meteorites in the meteorite collection. Many regolith breccias also contain records of exposure to the solar wind at some time in the past, but the timing of these exposures is very poorly constrained.

The timing of some impacts can be determined, in particular by potassium–argon ages, which are sensitive to the low-temperature thermal events produced by impacts. Many chondrites have ages of 4.5 Ga, although much younger ages are also found. On the other hand, most eucrites and mesosiderites contain evidence for impacts between 4.1 and 3.4 Ga ago, the same time that lunar basin-forming impacts were occurring (Bogard, 1995).

Finally, the recent collisional history is constrained by cosmic ray exposure ages. Material within about 1 m of the surface of a rock is bombarded by galactic cosmic rays, which produce many otherwise rare nuclides through spallation nuclear reactions. By measuring the build-up of these spallation products, it is possible to determine the length of time a meteorite has been a small object in space (some complications arise with material that has been in the regolith of a larger body). These exposure ages, in turn, carry information about the collisional lifetime of small bodies and, in some cases, the time of major collisional events on meteorite parent bodies. Stony meteorites typically have exposure ages of a few to a few tens of millions of years (Marti and Graf, 1992), while irons typically have exposure ages of more than 100 Ma. The decay of radioactive spallation products can be used to determine terrestrial ages. For non-Antarctic meteorites, these typically range from near zero to about 10^4 years, while many Antarctic meteorites have terrestrial ages greater than 10^5 years (see Collision; Cosmic ray exposure ages).

The traditional viewpoint is that most meteorite falls are essentially random events, but there may be meteorite 'streams', analogous to comet streams. Possible streams have been identified among the orbits of near-Earth asteroids (Drummond, 1991) and photographed fireballs (Halliday, Blackwell and Griffin, 1990). Furthermore, the Antarctic meteorites differ systematically from their non-Antarctic counterparts in several ways. Some of these differences (e.g. anomalous abundances of the elements Ce and I) are almost certainly the result of Antarctic weathering processes, but other differences (in the abundance of some other elements and in thermoluminescence properties) are more difficult to ascribe to weathering. These could represent differences in the available population of meteoroids 10^5 years ago (Lipschutz and Samuels, 1991), but dynamical studies to date have concluded that typical cosmic ray exposure ages suggest intervals which are too long for streams to remain intact. This issue has not been resolved.

Parent bodies of meteorites

One of the main differences between remote sensing of planets and studies of meteorites is that meteorites give much more detailed information on a much finer scale, but the context is often completely unknown. Hence, the search for the parent bodies of the meteorites is an ongoing and essential part of meteorite research.

Although models in which all meteorites or all chondrites came from different locations in a single parent body were once considered, it is now generally believed that most compositional groups represent distinct parent bodies. Thus we probably have meteorites from several ordinary chondrite parent bodies, at least two enstatite-rich parent bodies, several carbonaceous chondrite parent bodies and many differentiated (achondrite or iron) parent bodies. Two types of achondrites, the lunar meteorites and the SNC meteorites, probably come from the Moon and Mars respectively. None of the other 99.8% of the meteorites can be associated with any confidence with a particular parent body.

Most meteorites are generally believed to come from asteroids. Although comets are a potential source, E. Anders (1975) has pointed out that there are meteorites of virtually all compositional classes that show the effects of exposure to the solar wind at distances more typical of the main asteroid belt than of cometary orbits. Furthermore, camera networks designed to photograph fireballs caught three meteorites during atmospheric entry, and the orbits are quite similar, with perihelia near 1 AU and aphelia in the main belt (Figure M72).

More detailed associations have proved far more difficult (Wetherill and Chapman, 1988). Studies of orbital dynamics have provided plausible mechanisms for supplying meteorites from specific main belt locations, and optical reflectance spectroscopy of asteroids can provide some information about the mineralogy of the surface. Considerable effort has gone into trying to match various types of meteorites with specific asteroids. 4 Vesta seems to be the only large basaltic asteroid, and hence might be the parent of the eucrites, howardites and diogenites. Similarly, a main belt and a small near-

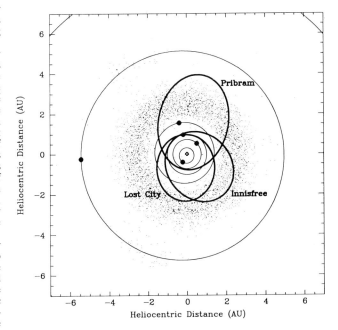

Figure M72 Orbits of three meteorites whose fireballs were photographed: Pribram (Czechoslovakia), Lost City (Oklahoma, USA) and Innisfree (Alberta, Canada). For comparison, the orbits of the terrestrial planets and Jupiter and the locations (as of 1 January 1993) of those planets and all numbered asteroids are also shown.

Earth asteroid have been suggested as the ultimate and recent source of the enstatite achondrites (Gaffey, Reed and Kelley, 1992). The most common type of meteorite, the ordinary chondrite, has the same basic mineralogy (olivine, pyroxene and metal) as the most common type of asteroid (S asteroids) in the inner part of the main belt. However, detailed spectroscopy has suggested that the relative abundances of the different minerals are not the same in the two classes of objects. A variety of explanations have been offered to either alter the spectra of chondritic asteroids to make them look like S asteroids, or to derive the ordinary chondrites from other, now rare or unseen, types of asteroids, but there is no consensus (e.g. Wetherill and Chapman, 1988). One of the early goals of spaceflight exploration of asteroids will be to determine the parent bodies of the ordinary chondrites, since those bodies are presently among the most intensely studied in the solar system, even though we do not know what or where they are.

Timothy D. Swindle

Bibliography

Anders, E. (1975) Do stony meteorites come from comets? *Icarus*, **24**, 363–71.
Anders, E. and Zinner, E. (1993) Interstellar grains in primitive meteorites: Diamond, silicon carbide, and graphite. *Meteoritics*, **28**, 490–514.
Bogard, D.D. (1995) Impact ages of meteorites: a synthesis. *Meteoritics*, **30**, 244–268.
Burke, J. (1986) *Cosmic debris, meteorites in history*. Berkeley: University of California Press.
Clayton, R.N., Grossman, L. and Mayeda, T.K. (1973) A component of primitive nuclear composition in carbonaceous meteorites. *Science*, **182**, 485–8.
Clayton, R.N., Mayeda, T.K. and Nagahara, H. (1992) Oxygen isotope relationships among primitive achondrites (abstract), in *Lunar Planet. Sci. XXIII*, Houston: Lunar and Planetary Institute, pp. 231–232.
Clayton, R.N., Onuma, N. and Mayeda, T.K. (1976) A classification of meteorites based on oxygen isotopes. *Earth Planet. Sci. Lett.*, **30**, 10–8.
Drummond, J.D. (1991) Earth-approaching asteroid streams. *Icarus*, **89**, 14–25.
Gaffey, M.J., Reed, K.L. and Kelley, M.S. (1992) Relationship of E-type Apollo asteroid 3103 (1982BB) to the enstatite achondrite meteorites and the Hungaria asteroids. *Icarus*, **100**, 95–100.
Goodrich, C.A. (1992) Ureilites: a critical review. *Meteoritics*, **27**, 327–52.
Halliday, I., Blackwell, A.T. and Griffin, A.A. (1989) The flux of meteorites on the Earth's surface. *Meteoritics*, **24**, 173–8.
Halliday, I., Blackwell, A.T. and Griffin, A.A. (1990) Evidence for the existence of groups of meteorite-producing asteroidal fragments. *Meteoritics*, **25**, 93–9.
Hewins, R.H. (1988) Experimental studies of chondrules, in *Meteorites and the Early Solar System* (eds J.F. Kerridge and M.S. Matthews). Tucson: University of Arizona Press, pp. 660–79.
Keil, K. (1989) Enstatite meteorites and their parent bodies. *Meteoritics*, **24**, 195–208.
Kerridge, J.F. and Matthews, M.S. (eds) (1988) *Meteorites and the Early Solar System*. Tucson: University of Arizona Press.
Lipschutz, M.E. and Samuels, S.M. (1991) Ordinary chondrites: multivariate statistical analysis of trace element contents. *Geochim. Cosmochim. Acta*, **55**, 19–34.
Lugmair, G.W. and Galer, S.J.G. (1992) Age and isotopic relationships among the angrites Lewis Cliff 86010 and Angra dos Reis. *Geochim. Cosmochim. Acta*, **56**, 1673–94.
Marti, K. and Graf, T. (1992) Cosmic-ray exposure history of ordinary chondrites. *Ann. Rev. Earth Planet. Sci.*, **20**, 221–43.
Marvin, U.B. (1986) Meteorites, the Moon and the history of geology. *J. Geol. Education*, **34**, 140–65.
McSween, H.Y., Jr (1979) Are carbonaceous chondrites primitive or processed? A review. *Rev. Geophys. Space Phys.*, **17**, 1059–78.
McSween, H.Y., Jr (1985) SNC meteorites: clues to Martian petrologic evolution? *Rev. Geophys.*, **23**, 391–416.
Mittlefehldt, D.W. (1994) ALH84001, a cumulate orthopyroxenite member of the martian meteorite clan. *Meteoritics*, **29**, 214–221.
Nittler, L.R., Alexander, C.M.O'D., Gao, X., Walker, R.M. and Zinner, E.K. (1994) Interstellar oxide grains from the Tieschitz ordinary chondrite. *Nature*, **370**, 443–446.
Patterson, C. (1956) Age of meteorites and the Earth. *Geochim. Cosmochim. Acta*, **10**, 230–7.
Pellas, P. (1992) Early chronology of the H chondrite asteroid (abstract). *Meteoritics*, **27**, 274.
Scott, E.R.D. (1979) Origin of iron meteorites, in *Asteroids* (ed. T. Gehrels). Tucson: University of Arizona Press. pp. 892–925.
Van Schmus, W.R., and Wood, J.A. (1967) A chemical-petrologic classification for the chondritic meteorites. *Geochim. Cosmochim. Acta*, **31**, 747–65.
Wetherill, G.W., and Chapman, C.R. (1988) Asteroids and meteorites, in *Meteorites and the Early Solar System* (eds J.F. Kerridge and M.S. Matthews). Tucson: University of Arizona Press. pp. 35–67.
Wood, J.A. and Pellas, P. (1991) What heated the parent meteorite planets? in *The Sun in Time* (eds C.P. Sonett, M.S. Giampapa and M.S. Matthews). Tucson: University of Arizona Press, pp. 740–60.

Cross references

Achondrite meteorites
Antarctic meteorites
Carbonaceous chondrite
Chondrites, ordinary
Enstatite meteorites
Eucrite meteorites
Iron meteorites
Lunar meteorites
SNC meteorites
Stony iron meteorites

METEORITE PARENT BODIES

There are several strong lines of evidence that most meteorites come from the asteroid belt. Meteoritic materials thus represent 'cheap' (compared to the cost of a space mission to sample an asteroid) samples of the highly diverse, numerous and effectively inaccessible asteroid population. The advantage of having these samples delivered to us is that they can be subjected to the level of sophisticated and highly detailed analysis that can only be done in a laboratory. This gives us access to priceless details on the earliest period of solar system formation, the mineralogy of the asteroid belt, and the processes that have shaped asteroids over the life of the solar system. However, the scientific price we pay for these cheap samples is that we lose what geologists call providence; information on the location, structural setting and context of the sample. That information is vital to understanding how it fits into the story of the origin and evolution of the solar system. Meteorites do not automatically provide the location and taxonomic class of their parent bodies. The very fact that a meteorite is literally 'in our hands' indicates some violent event occurred that may have fragmented and perhaps destroyed the parent body. The best that can be done is to link individual asteroid spectral classes with meteorite compositional groups. This is done by comparing laboratory visible and near-infrared reflectance spectra of meteorites to the telescopically obtained spectra and albedo of asteroids. This task is somewhat speculative, and all spectral matches between asteroids and meteorites, including the ones detailed here, need to be taken with healthy skepticism.

Delivery of meteorites

There are several factors that bias the population of meteorites arriving on Earth and therefore limit our sample of the asteroid belt. First, the dynamical processes that deliver meteorites from the asteroid belt to Earth are probably strongly biased toward sampling relatively narrow zones in the asteroid belt. Calculations suggest that the vast majority of meteorites and planet crossing asteroids originate from near the 3 : 1 Kirkwood gap and the $v6$ resonance. Both these zones are in the inner asteroid belt where the asteroid population is dominated by material that is differentiated or highly metamorphosed.

A second factor is the relative strength of the meteorites. Many meteorites begin the process of evolving into an Earth-crossing orbit by being ejected at high velocity from the parent body by a major impact. To survive the stress of impact and acceleration without being crushed into dust, the meteorite must have substantial cohesive strength. This would strongly select against the relatively weak carbonaceous chondrites. At the opposite extreme, the almost completely metallic iron meteorites and the mostly metallic stony irons have such great cohesive strength that it would be difficult to break off pieces for ejection. Cohesive strength is also an important selection factor for the meteorites that survive collisions while they are in near-Earth space, and for material that survives the deceleration and heating of atmospheric entry. Data from cosmic ray exposure ages of meteorites indicate that collisions in near-Earth space are common and that many meteorites are fragments of larger fragments that were broken up while in Earth-crossing orbits. Atmospheric entry usually involves a variety of thermal and dynamical stresses that typically break up most stony meteorites from one large individual into showers of much smaller stones. These selection effects introduce biases of unknown magnitude into the link between asteroids and meteorites that limit the usefulness of the meteorite collection as a representative sample of the asteroid belt. It is very likely that the meteorites available to us represent only a small fraction of the asteroids and that most asteroids either cannot or only rarely contribute to the meteorite collections. To put it another way, not all asteroid types are parent bodies of meteorites.

Meteorite parent bodies

Any link between asteroids and meteorites must be an extrapolation of our limited and biased sample of meteorites to a much larger and more complex population. With the caveats of the previous section in mind, mineralogical interpretations and possible meteoritic analogues for the Tholen (1984) asteroid classes are shown in Table M13.

The asteroid outer belt is dominated by the P and D classes (see Asteroid: compositional structure and taxonomy). The spectra of these types are dark, red to very red, generally anhydrous and relatively featureless. No direct analogs for these asteroid types exist in the meteorite collections. The analogs most commonly cited are cosmic dust or CI carbonaceous chondrites that are enriched in organics (Lebofsky et al., 1990). However, the spectral characteristics of these asteroids are difficult to duplicate with material that is delivered to the inner solar system. Probably P and D asteroids are composed of primitive materials that have had a different geochemical evolution than cosmic dust or CI chondrites. The Z class was proposed for the trans-Jovian asteroid 5145 Pholus which has the reddest visible and near-IR reflectance spectrum so far observed. Once again, this object does not have a meteorite analog but is thought to be dominated by primitive material such as complex organics. There is a general increase in 'redness' with heliocentric distance in the outer belt asteroids going from the moderately red P asteroids, through the redder D asteroids, to the much redder Z asteroids.

The dark inner belt asteroids of the B, C, F and G classes are characterized by relatively featureless flat spectra. The proposed analogues for most of these asteroids are the dark CI and CM carbonaceous chondrite meteorites. The spectral differences between the asteroid types are thought to represent varying histories of aqueous alteration or thermal metamorphism (Bell et al., 1989). The CI carbonaceous chondrites are rich in water, clay minerals, volatiles and carbon (see Meteorite). They probably represent primitive material that has been mildly heated and altered by the action of hydrothermal fluids. CM meteorites are less water rich, more heated and have less carbon. These two meteorite types are most analogous to the 'wet' C asteroids. The spectra of B, F and G asteroids also show hydration features similar to the CI and CM carbonaceous chondrites, but differences in their spectral slope and UV absorption suggest that these asteroids underwent varying degrees of aqueous alteration. About 40% of the C types are anhydrous, suggesting that the C class may be a collection of objects with roughly similar reflectance spectra but greatly varying mineralogy. Analogs for the 'dry' C asteroids include CM carbonaceous chondrites that have been devolatilized and/or ordinary chondrites that have been darkened by regolith processes.

Perhaps the best asteroid/meteorite spectral match involves the V-class asteroids and the basaltic achondrite meteorites. Spectrally, V types are interpreted to be a differentiated assemblage of (primarily) orthopyroxene, with varying amounts of plagioclase (Gaffey, Bell and Cruikshank 1989), which makes them very close analogs to the basaltic howardite/eucrite/diogenite (HED) association of meteorites. The petrology of these meteorites indicates that they are basaltic partial melts originating on asteroids that underwent extensive heating and differentiation. Thus, these meteorites probably represent the surface melts and upper crustal rocks of a differentiated asteroid. In addition, similarities in petrology, chemical trends and oxygen isotopes suggest that all the HED meteorites are closely related and probably come from a single parent body. A dozen small V-class asteroids have been identified in the main asteroid belt in orbits that stretch from near Vesta to near the 3 : 1 resonance that is thought to supply planet-crossing asteroids. These objects are almost certainly ejecta fragments from a major impact on Vesta. Three V-type asteroids have also been identified in Earth-crossing orbits. These objects have almost identical V-type spectra, have spectral features that suggest they are rocky fragments, and are small enough (less than 4 km in diameter) to be fragments of the crust of a disrupted differentiated asteroid. Interestingly, their orbits are very similar, suggesting that they were all part of a larger fragment that was collisionally disrupted while in an Earth-crossing orbit. In

Table M13 Meteorite parent bodies

Asteroid class	Inferred major surface minerals	Meteorite analogs
Z	Organics + anhydrous silicates? (+ ice?)	None (cosmic dust?)
D	Organics + anhydrous silicates? (+ ice?)	None (cosmic dust?)
P	Anhydrous silicates + organics? (+ ice?)	None (cosmic dust?)
C (dry)	Olivine, pyroxene, carbon (+ ice?)	'CM3' chondrites, gas-rich/black chondrites?
K	Olivine, orthopyroxene, opaques	CV3, CO3 chondrites
Q	Olivine, pyroxene, metal	H, L, LL chondrites
C (wet)	Clays, carbon, organics	CI1, CM2 chondrites
B	Clays, carbon, organics	None (highly altered CI1, CM2?)
G	Clays, carbon, organics	None (highly altered CI1, CM2?)
F	Clays, opaques, organics	None (altered CI1, CM2?)
M (wet)	Clays, salts?	None (opaque-poor CI1, CM2?)
V	Pyroxene, feldspar	Basaltic achondrites
R	Olivine, pyroxene	None (olivine-rich achondrites?)
A	Olivine	Brachinites, pallasites
M (dry)	Metal, enstatite	Irons (+ EH, EL chondrites?)
T	Troilite?	Troilite-rich irons (Mundrabilla)?
E	Mg-pyroxene	Enstatite achondrites
S	Olivine, pyroxene, metal	Stony irons, IAB irons, lodranites, winonites, siderophyres, ureilites, H, L, LL chondrites

addition, the orbits of these near-earth asteroids (NEAs) may be related to the peak fall times for HED meteorites. This may be a case where the study of meteorite chemistry, spectra and fall statistics has come together with remote sensing and orbital dynamics to describe the origin, evolution and current location of a major meteorite parent body.

The A-class asteroids are thought to represent the next lower zone of a differentiated asteroid. These asteroids are interpreted to be nearly pure olivine and may be derived from the mantles of extensively differentiated parent bodies. The best meteorite analog for this interesting asteroid type are the extremely rare brachinites. The 200-g meteorite Brachina is the only non-Antarctic member of this type. This illustrates what some workers (e.g. Bell et al., 1989) have termed 'the great dunite shortage'. If some asteroids are differentiated, there should be, along with the identified fragments of crust (V-types) and core (M-types) material, a substantial amount of olivine-rich mantle material similar to terrestrial dunite. However, A-type asteroids are relatively rare and dunite-like meteorites are very rare. Another suggested meteorite analog for the A-type asteroids are the somewhat more abundant pallasite meteorites (Gaffey, Bell and Cruikshank, 1989), but they are also a suggested analog for some S-type asteroids. The R class represents another possible mantle-derived asteroid (a single-member class made up of the asteroid 349 Dembowska). Analysis of its reflectance spectra suggests a mineralogy that contains both olivine and pyroxene and may be a partial melt residue of incomplete differentiation (Gaffey, Bell and Cruikshank, 1989). Unfortunately the meteorite collection contains no potential analogs for this mineralogy.

A more common asteroid class is the M class which has the spectral characteristics of almost pure iron–nickel metal, directly analogous to spectra of the metallic meteorites. This material may represent the cores of differentiated asteroids. However, there is a great deal of geochemical variety in the iron meteorite population. The 13 different classes of iron meteorites suggest origins from a number of different parent bodies and/or a variety of geochemical conditions. Several M-class asteroids have been identified in Earth-crossing orbits and their metallic composition has been strongly confirmed by radar. Recent observations have added new complications to the simple M = metal picture. Two M-class asteroids have been shown to have hydrated minerals on their surfaces. This raises the possibility that the 'wet' M asteroids are assemblages of clays, like the CI carbonaceous chondrites, but without the carbon-rich opaques that darken the CIs.

The T asteroids may be related to the metal-rich M asteroids. The T asteroids are inner asteroid belt objects which have featureless spectra, a strong red continuum slope, low albedo and no hydration features. Their spectra are similar to that of the mineral troilite (FeS) which is most common as an accessory mineral in iron meteorites. Although there are no pure troilite meteorites, some iron meteorites are as much as 35% by volume troilite. Theoretical work suggests that the metallic cores of asteroids would tend to concentrate troilite along radial dendrites during their cooling and crystallization. The troilite-rich zones would be much weaker than the bulk metal, and collisions would tend to split the core along these troilite-rich dendrites. This would create metallic core fragments with surfaces enriched in troilite.

The E-class asteroids are excellent analog for bright, red-sloped but spectrally featureless enstatite achondrites. These asteroids are probably composed of the same differentiated enstatite assemblages as the enstatite achondrites (Gaffey, Bell and Cruikshank, 1989).

Perhaps the most complex class of asteroids is the S class. Several authors have recently suggested breaking up this class into three (Howell, Merenyi and Lebofsky, 1994) or seven (Gaffey et al., 1993) subclasses. The intricacies of the subclasses are beyond the scope of this work, but it is enough to know that S-class spectra, on average, indicate subequal amounts of olivine and pyroxene with a substantial metallic component. A standard analogue for S asteroids is the pallasites, which show an interesting assemblage of large olivine grains set in a matrix of iron–nickel metal. This indicates that some S asteroids may represent the core–mantle boundary of a differentiated asteroid where the silicates (principally olivine) of the mantle are in direct contact with the metallic core. However, the S class is very large and includes a number of objects that do not conform to this standard interpretation. A number of S asteroids are rich in pyroxene and may represent a larger cross-section of the mantle and lower crust of an asteroid. Meteorite analogs for this 'mantle' material include ureilites, lordranites, brachinites, siderophyres and winonites. Some S asteroids have lower metal contents and may be the parent bodies of ordinary chondrite meteorites. Others (principally the Eos family) have already been split out to form the K class. The K-type asteroids have lower albedos and flatter spectral features than most S asteroids, and the CV and CO carbonaceous chondrites are interpreted as their analogs.

A long-running controversy in asteroid science has been the identification of the asteroidal source of the ordinary chondrite meteorites. Ordinary chondrites are by far the largest meteorite type, accounting for approximately 80% of observed meteorite falls, but so far only a few asteroids have been identified as Q class, the ordinary chondrite analog. A number of S-class asteroids have spectral absorption bands roughly similar to those of ordinary chondrites, but S asteroids typically have a strong red continuum slope that is not seen in ordinary chondrites. A number of explanations for the lack of ordinary chondrite parent bodies have been put forward.

1. At least some S types are ordinary chondrites, but regolith processes enrich the metal content of the regolith and increase their apparent spectral red slope (Bell et al., 1989).
2. Ordinary chondrite parent bodies are in the asteroid belt, but 4.6 billion years of collisions have ground them down to sizes that are too small to see with current telescopes.
3. Regolith processes can darken ordinary chondrites, so their parent asteroids actually have the spectra of the dark C-type asteroids.

Whatever the final answer, this subject promises to be a source of lively debate for some time in the future.

In general, the differentiated asteroids of the V, A, R, S and M classes probably represent a transect from the crust to the core of differentiated asteroids and as such they can tell us a great deal about the geochemical evolution of a differentiating body. The V-class asteroids would be the surface and crustal material. The A asteroids would be from a completely differentiated mantle while the R asteroids would represent a mantle that experienced only partial differentiation. S asteroids would be material from either some region in the mantle or at the core–mantle boundary. And finally, M class represents samples of the metallic cores of these asteroids.

Summary

There are several problems inherent in identifying any meteorite parent body. First, the fact that the meteorite is here on Earth points to some violent event on the parent body and the real possibility that the actual meteorite parent body no long exists. Second, because of limitations in orbital dynamics, collisional evolution and material strength most asteroids cannot supply meteorites to Earth and are not represented in the collections. Finally, meteorite parent bodies can only be inferred on the basis of similarities in their reflectance spectra. This provides a link between types of meteorites and classes of asteroids, but there is only one case where the evidence is strong enough to identify a particular asteroid as the parent of a particular meteorite (Vesta and the basaltic achondrites).

Daniel T. Britt and Larry A. Lebofsky

Bibliography

Bell, J.F., Davis, D.R., Hartmann, W.K. and Gaffey, M.J. (1989) Asteroids: the big picture, in *Asteroids II*, (eds R.P. Binzel, T. Gehrels and M.S. Matthews). Tucson: University of Arizona Press, pp. 921–45.
Gaffey, M.J., Bell, J.F. and Cruikshank, D.P. (1989). Reflectance spectroscopy and asteroid surface mineralogy, in *Asteroids II*, (eds R.P. Binzel, T. Gehrels and M.S. Matthews) Tucson: pp. 98–127, University of Arizona Press.
Gaffey, M.J., Bell, J.F., Brown, R.H. et al. (1993). Mineralogical variations within the S-type asteroid class. *Icarus*, **106**, 573–602.
Howell, E.S., Merenyi, E. and Lebofsky, L.A. (1994). Classification of asteroid spectra using a neural network. *J. Geophys. Res.*, **99**, 10847.
Lebofsky, L.A., Jones, T.D., Owensby, P.D., et al. (1990). The nature of low albedo asteroids from 3 mm spectrophotometry. *Icarus*, **83**, 12–26.
Tholen, D.J. (1984). Asteroid taxonomy from cluster analysis of photometry. Unpublished PhD Thesis, University of Arizona, Tucson.

Cross references

Asteroid
Asteroid: resonance
Asteroid: compositional structure and taxonomy
Vesta

MICROMETEORITE

A cone of light is visible on the horizon immediately after sunset and just preceding sunrise in favorable viewing conditions. This zodiacal light arises from sunlight scattered from interplanetary dust particles, lying within the orbit of Mars and concentrated on the ecliptic plane. Interest in this dust has been intense because the local cloud is a convenient place in which to examine such astrophysical processes as grain evolution and destruction, sputtering, magnetic fields and radiation pressure from the Sun. The latter process (called Poynting–Robertson drag) acts to expel the finest dust from the inner solar system and drag larger grains into the Sun, both with timescales on the order of 10^3 to 10^5 years. Thus, these particles must be continuously replenished by active comets, colliding asteroids, large impacts on the Moon and Mars and interstellar sources. About 10^8 kg of these particles enter the Earth's atmosphere annually, whereupon they are referred to as micrometeorites.

Interest in micrometeorites has also been spurred by the realization that their impacts onto spacecraft with typical velocities in excess of 7 km s^{-1} could be hazardous, resulting in the degradation of space-exposed surfaces, and even the breaching of spacecraft hulls. Consequently, there have been numerous attempts to measure the flux of micrometeorites in space, involving the HEOS, Helios and Pioneer spacecraft and, more recently, the Long Duration Exposure Facility (LDEF) which monitored the space environment in low-Earth orbit for 5.7 years (McDonnell, 1992). Micrometeorite collectors have also been deployed from the MIR space station, and are planned for Space Station Freedom. In spite of these efforts, the distribution and detailed sources of dust in the solar system are imperfectly understood.

Individually measuring less than 1 mm, micrometeorites are at the lower size end of the continuum of interplanetary materials. Unlike asteroids, which provide larger samples in the form of meteorites, comet samples are probably not present in the inner solar system in sizes larger than dust. Thus, when collected by spacecraft, in the Earth's stratosphere, from the oceans or polar ice caps, micrometeorites provide scientists with a unique sample of asteroids, our only samples of comets and, potentially, interstellar materials. The distinction between micrometeorites derived from asteroids versus comets is important, since comets are believed to preserve unaltered interstellar grains and nebular condensates while most asteroids experienced significant processing (from low-grade metamorphism to complete melting) early in solar system history (Brownlee, 1979).

The laboratory study of micrometeorites, though a new science only two decades old, is well advanced, involving highly precise analytical methods. The first micrometeorites were collected in the stratosphere in 1970 using a balloon; since 1974 they have been more efficiently collected by U-2, ER-2 and WB-57 NASA stratospheric aircraft. Subsequent studies have established that micrometeorites are widely variable in composition, mineralogy and structure, although only particles with refractory and chondritic compositions (the latter with approximately solar abundances of elements heavier than He) have received detailed attention (Zolensky, 1987). The chondritic micrometeorites, in particular, are the most interesting, and consist predominantly of ferromagnesian silicates (variously hydrous, anhydrous and amorphous), oxides, sulfides, carbonaceous materials and Fe–Ni metal (Mackinnon and Rietmeijer, 1987). These chondritic particles have a bimodal distribution of bulk densities, centered on 0.6 and 1.9 g cm^{-3} (Flynn and Sutton, 1990). Some of the lowest density particles, which are generally anhydrous and porous, are probably cometary grains, with the pores resulting from sublimation of ices. Chondritic micrometeorites contain more carbon than any other extraterrestrial samples now known, and a revolution in our understanding of carbon chemistry in the early solar nebula will occur when organic analyses are sufficiently developed to permit analysis of these nanogram-sized samples. Chondritic micrometeorites also contain micrometer-sized aggregates of much smaller (down to tens of angstroms) mineral grains, which appear to record the earliest stages of grain condensation and accretion in the solar nebula.

Several European and Soviet spacecraft performed analyses of dust shed from comet Halley in 1986, which revealed this presumably primitive material to be very similar to chondritic micrometeorites, although the results of these remotely performed analyses remain somewhat ambiguous (Brownlee and Kissel, 1989). Uncertainties concerning the composition and mineralogy of cometary material will be best put to rest when samples can be returned to Earth for detailed laboratory analysis. However, if dust grains can be collected in low-Earth orbit, with velocity and trajectory information preserved for each grain, it is theoretically possible to calculate whether individuals originated from either cometary or asteroidal sources.

Michael Zolensky

Figure M73 Collected in the stratosphere, this chondritic micrometeorite measures only 10 μm across, but is an aggregate of millions of far smaller grains. (Photograph Credit: Cosmic Dust Preliminary Examination Team, NASA Johnson Space Center.)

Bibliography

Brownlee, D. (1979) Interplanetary dust. *Rev. Geophys. Space Phys.*, **17**(7), 1735–43.
Brownlee, D. and Kissel, J. (1989) The composition of dust particles in the environment of comet Halley, in *Comet Halley Investigations, Results, Interpretations*. London: Horwood Ltd, pp. 89–98.
Flynn, G. and Sutton, S. (1990) Evidence for a bimodal distribution of cosmic dust densities, in *Lunar Planet. Sci. XXI*, Houston: Lunar and Planetary Institute, pp. 375–376.
Mackinnon, I. and Rietmeijer, F. (1987) Mineralogy of chondritic interplanetary dust particles. *Rev. of Geophys.*, **25**(7), 1527–53.
McDonnell, J.A.M. (ed.) (1992) *Origin and Evolution of Interplanetary Dust*. New York: Kluwer Academic Publishers.
Zolensky, M. (1987) Refractory interplanetary dust particles. *Science*, **237**, 1466–8.

Cross references

Comet
Dust
Interstellar grains
Meteorite
Poynting–Robertson drag

MICROWAVE SPECTROSCOPY

The microwave spectral range, which extends from wavelengths of fractions of millimeters to centimeters (corresponding to frequencies in the range 1–300 GHz), is an important window for investigating planetary and cometary atmospheres. Except for specific wavelengths associated with transitions of components of Earth's atmosphere, this spectral domain is accessible from the ground; large antennas can be used, and high-sensitivity measurements can be made. High spectral resolution (of the order of 1 part in 10^6) provided by commonly used heterodyne receivers makes microwave spectroscopy a very powerful tool for investigating thermal profiles and vertical distributions of constituents of planetary atmospheres, through the analysis of rotational molecular transitions. In the case of comets, microwave spectroscopy allows the detailed study of the OH radical, a major constituent in cometary atmospheres, and has led to the discovery of several minor parent molecules.

Heterodyne receivers mix the input signal at frequency f_S with a monochromatic local oscillator (LO) at a nearby frequency f_L and retain the difference of the two frequencies $f_{IF} = f_S - f_L$ (Harris, 1991). In the millimeter range, the mixer is typically a Shottky diode or a superconductor–insulating superconductor (SIS) tunnel junction. Values of f_{IF} range between 1 and 4 GHz, and the typical bandwidth is 0.5 GHz. Backend systems provide the spectral resolution: they consist of banks of individual filters, accousto-optical spectrometers (AOSs), or digital autocorrelators. A spectral resolution of about 100 KHz is typically achieved by all systems, which corresponds to a resolving power of 10^6 at a frequency of 100 GHz.

The first microwave detection of planets was made in 1956 (Mayer, McCullough and Sloanaker, 1958a,b), providing the first continuum measurements of Venus, Mars and Jupiter at 3.15 cm wavelength. With other continuum measurements at radio wavelengths, this led to important new results, such as the non-thermal (synchrotron) emission from Jupiter, and the discovery of a very high surface temperature on Venus. The first spectroscopic measurements were performed 20 years later, with the analysis of the CO (1–0) transition at 115 GHz on Mars and Venus (Kakar, Waters and Wilson, 1976, 1977). A few years later, microwave spectroscopic measurements allowed the analysis of CO and HCN (Muhleman, Berge and Clancy, 1984; Paubert et al. 1987) and, more recently, HC_3N and CH_3CN (Bézard, Marten and Paubert, 1992a,b) on Titan. It also led to the first detection of a stable SO_2 atmosphere on Io (Lellouch et al., 1990). The first cometary microwave measurements were performed on comet Kohoutek (1974 XII), both in the continuum (Hobbs et al., 1975) and in the OH transition at 18-cm wavelength (Biraud et al., 1974). Microwave spectroscopy later provided another very significant contribution to cometary research in the detection of several parent molecules on comet Halley and on subsequent bright comets (Crovisier and Schloerb, 1991).

Planetary and satellite atmospheres

In the microwave range the outgoing planetary emission is of thermal origin. Under the pressure and temperature conditions of planetary atmospheres (with the exception of Io), the local thermodynamical equilibrium (LTE) regime prevails for most of the probed atmospheric levels, so that the source function of each emitting atmospheric layer is the Planck function corresponding to the temperature of this layer. A useful unit of the outgoing radiance is the brightness temperature, which is the temperature of the black body which would emit the same radiance at the given wavelength. According to Eddington's approximation, in the case of emission coming from the center of a planetary disk, the outgoing brightness temperature is equal to the temperature of the atmospheric level for which the optical depth is equal to 1; in the case of emission integrated over the whole disk, the brightness temperature corresponds to the temperature of the atmospheric level where the optical depth is 2/3. As a consequence, the analysis of a line profile from the line center (where the optical depth is maximum) to the far wings (where the optical depth becomes negligible) allows one to probe a wide range of atmospheric pressure levels. If the temperature decreases as the altitude increases (as is the case in the convective part of a planetary atmosphere), the molecular line is observed in absorption. If a temperature inversion is present with a positive lapse rate in the stratosphere (as is the case for the giant planets and Titan), the

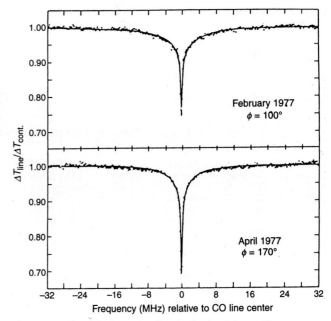

Figure M74 The CO absorption line in the $J = 1$–0 transition at 115 GHz in the atmosphere of Venus, observed at two different epochs corresponding to different planetocentric phase angles. The lines are formed in the altitude range 75–110 km. The retrieved CO profiles have mixing ratios in the range 10^{-3}–10^{-4} with respect to the major constituent CO_2 (Von Zahn et al., 1983). (From Wilson et al., 1981.)

molecular line profile shows an emission core, superimposed over broader absorption wings.

If the observed transition belongs to an atmospheric constituent of known constant mixing ratio throughout the atmosphere, the line profile can be used to retrieve the thermal profile. On the other hand, if the thermal profile is known from another means, the line profile can be used to retrieve the vertical distribution of the corresponding species. When several transitions of the same molecule are observed, both the temperature and density vertical profiles can be determined, or at least constrained, by the microwave observations. This analysis of the atmospheric composition and structure is especially powerful in the case of tenuous atmospheres (because the limited bandwidth of heterodyne spectroscopy is best suited for the study of narrow lines) and for atmospheres exhibiting a positive temperature lapse rate (because emission lines are easier to detect than absorption lines). The most favorable cases are the atmospheres of Titan and Io.

A promising technique which has recently been developed is the study of winds via the analysis of Doppler shifts over the planetary disks. For example, for the CO (1–0) transition at 115 GHz, a wind of 100 m s^{-1} corresponds to a Doppler shift of 38 kHz. Although less than the typical spectral resolution of heterodyne receivers, such shifts can be determined from a least mean square fitting of the shape of the absorption line, its absolute central frequency being accurately known from the frequency of the LO. This technique has been successfully applied to Venus (Shah, Muhleman and Berge, 1991) and Mars (Lellouch et al., 1991a).

Mars and Venus

In the case of Venus, observations of the CO (1–0) line have been performed by Kakar, Waters and Wilson (1976), Gulkis et al. (1977), Schloerb, Robinson and Irvine (1980), Wilson et al. (1981) and Clancy and Muhleman (1985). From these observations (Figure M74), the CO vertical distribution was retrieved above the cloud layers, in the altitude range 75–110 km, and showed significant diurnal variations between 75 and 85 km. Recently, CO interferometric mapping of the (1–0) transition has been performed by Shah, Muhleman and Berge (1991) and wind velocities have been derived in the Venusian mesosphere, showing evidence for westward horizontal winds in the order of 100 m s^{-1}. The other molecules observed on Venus in the microwave range are HDO, at 226 GHz (Encrenaz et al.

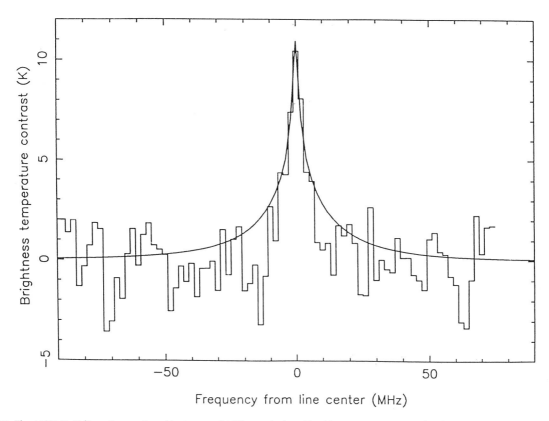

Figure M75 The HCN (3–2) line observed on Neptune at 2 MHz resolution. The histogram represents the data, and the solid line is the best fit model, corresponding to an HCN mixing ratio of 3×10^{-10} (above the condensation level), with respect to the major constituent H_2. (From Rosenqvist et al., 1992.)

1991a), and H_2O, at 183 GHz (Encrenaz et al. 1993). The spectral profiles, corresponding to the whole disk, have been used to retrieve the mean vertical distribution of water in the altitude range 75–95 km.

In the case of Mars, the CO (1–0) line has been observed by several authors including Kakar, Waters and Wilson (1977), Schloerb, Robinson and Irvine (1980) and Clancy, Muhleman and Berge (1990). The (2–1) and (3–2) transitions of CO were also observed (Lellouch, Paubert and Encrenaz, 1991; Lellouch et al., 1991b). All these observations were used to constrain both the temperature profile and the CO vertical distribution, which was found uniformly mixed up to about 60 km. No firm evidence was found for long-term temporal variations of CO. A CO mapping of the Martian disk has been made for the (1–0) and (2–1) transitions (Lellouch, Paubert and Encrenaz, 1991), showing no evidence for local CO variations larger than 40%. These data were also used to measure the winds in the southern atmosphere (at the time of southern summer solstice) and clearly indicate a global easterlies flow (Lellouch et al., 1991a).

Water vapor has been detected on Mars at 22 GHz (Clancy, Grossman and Muhleman, 1992); this observation, performed with the Very Large Array, has allowed a complete mapping of the water vapor distribution along the Martian limb. HDO and H_2O were also tentatively detected at 226 GHz and 183 GHz respectively (Encrenaz et al., 1991b; Cernicharo et al., 1993).

Giant planets

In the microwave spectrum of the giant planets, the continuum is due to the pressure-induced spectrum of hydrogen; in addition, in the case of Jupiter, the far wings of the ammonia rotation and inversion lines also contribute. As a consequence, the region probed in this spectral range is at a pressure of about 1 bar. The half-width of the pressure-broadened lines is thus several times larger than the bandwidth. For strong continuum sources, instrumental ripples affect the continuum level. As a result, molecular transitions can be detected only if the absorber is present in the stratosphere in sufficient amounts, so that an emission core can be observed. This is the case for CO and HCN on Neptune (Marten et al., 1993; Rosenqvist et al., 1992). These unexpected detections (Figure M75), combined with the lack of detection of these species on the other giant planets, provide important new constraints upon the formation scenarios of these planets.

Satellites

Microwave observations of Titan probe the stratosphere at pressures lower than 100 mbar; the temperature lapse rate is positive, leading to the formation of narrow emission cores. Both CO and HCN have been detected on Titan. The (1–0) CO line has been observed by Muhleman, Berge and Clancy (1984) and Marten et al. (1988). The latter results are consistent with a CO vertical distribution which is strongly depleted in the stratosphere by a mechanism which remains to be understood. The (1–0) line of HCN, at 88 GHz, has been observed on Titan by Paubert et al. (1987), and the HCN vertical distribution has been inferred by Tanguy et al. (1990). Recently, HC_3N has been detected at 146 and 218 GHz, and CH_3CN has been detected around 221 GHz (Bézard, Marten and Paubert 1992a,b). The lines are very narrow, indicating that they are formed at pressures less than 0.05 mbar. This result provides a new constraint on photochemical models of Titan.

The first evidence for a stable atmosphere of SO_2 on Io has been provided by the microwave detection of the 222-GHz transition of sulfur dioxide (Lellouch et al., 1990; Figure M76). From the intensity and the width of the transition, it was determined that SO_2 was present only over a small fraction (3–15%) of the disk, possibly near the subsolar point as a result of condensation effects, or possibly on discrete spots associated with volcanoes (Lellouch et al., 1992).

Cometary atmospheres

The principal excitation mechanism of cometary molecules is fluorescence by the solar radiation field; it is observed in the near-infrared in vibration–rotation bands. In the microwave range the emission from

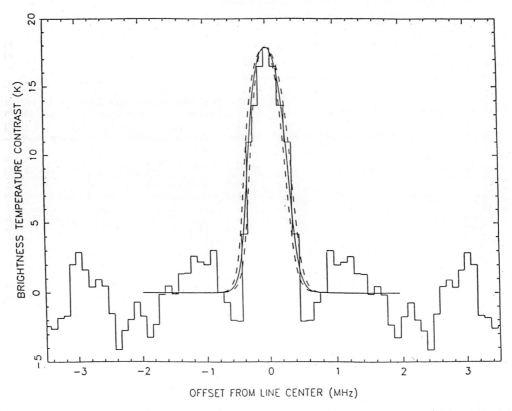

Figure M76 Detection of gaseous SO_2 on Io from its millimeter transition at 222 GHz. From the intensity and the width of the emission line, a pressure of 4–30 bar is inferred. SO_2 appears to be localized in discrete areas over Io's disk, possibly near the subsolar point or as a result of volcanic activity. (From Lellouch et al., 1990.)

rotational or hyperfine transitions are a natural consequence of the radiative decay following excitation at shorter wavelengths (Crovisier and Schloerb, 1991). Modeling of the microwave emission thus requires knowledge of the complete fluorescence scheme. In the case of the OH radical the observed transitions at 18 cm are produced by an hyperfine transition, governed by the decay of a fluorescence excitation produced in the near-UV. Because of the nature of the solar spectrum in this range, the OH transitions are observed in emission or in absorption depending upon the heliocentric radial velocity of the comet. These radio observations have thus played an important role in validating the OH fluorescence models. In the case of parent molecules, outgassed from the nucleus, the main excitation process is infrared excitation of their fundamental bands of vibration. Typically, radiative excitation is dominated by one or two vibrational bands, and the rotational population results from the competition between the infrared excitation rate and rotational spontaneous decay. After a few fluorescence cycles, a steady state situation ('fluorescence equilibrium') is established. A complete modeling is required, also including collisional effects in the inner coma; one may then derive the column density of the molecule in the beam and, through Haser's model, the molecular production rate (Crovisier and Schloerb, 1991).

Studies of the OH radical

The 18-cm wavelength transitions of OH arise from the $^2\Pi_{3/2}$ $J = \frac{3}{2}$ ground state, split into four levels (1612, 1665, 1667, 1721 MHz) by lambda doubling and hyperfine structure. The first cometary detection of OH was obtained on comet Kohoutek (1974 XII), showing a time variation of the lines, alternately in emission and absorption. This behavior was interpreted as the so-called 'Swings effect' – pumping of the OH radical by solar UV radiation (Biraud et al., 1974). The OH radical has been observed at 18 cm on all subsequent bright comets (Snyder, 1986) including comet Halley (Gérard et al., 1987), and these measurements have been used to monitor the water vapor production rate and kinematics through a vectorial model, OH being assumed to be the daughter product of the major gaseous component H_2O. OH imaging of comets Halley and Wilson at 18 cm has been performed with the Very Large Array (VLA) interferometer, showing some inhomogeneities in the OH spatial distribution (de Pater, Palmer and Snyder, 1991).

Observation of parent molecules

The first firm identifications of cometary parent molecules in the radio range were achieved at the time of the Halley apparition in 1985–86, with the detection of HCN (Despois et al., 1986; Schloerb et al., 1986) and H_2CO (Snyder, Palmer and de Pater, 1989). These two molecules were also detected in the microwave range in subsequent bright comets: Brorsen-Metcalf, Austin, Levy and – in the case of HCN – Wilson. In addition, CH_3OH and H_2S were also detected in comets Austin and Levy (Bockelée-Morvan et al., 1991; Crovisier et al., 1991).

Future of microwave spectroscopy

The application of microwave spectroscopy to planetology has led to important new discoveries over the past 10 years. This promising technique is expected to develop in the future in several directions.

The first expected improvement is the extended use of interferometry to the millimeter range. As mentioned above, interferometric images have been obtained with the VLA, in the radio range, on comets and on Mars. With the use of millimeter interferometers, the same studies are becoming possible at higher frequency. This should allow, in particular, mapping of minor constituents in planetary disks and cometary comas.

Another expected instrumental improvement is extension of the bandwidth available for heterodyne spectroscopy. As mentioned above, the present bandwidth (0.5 GHz) strongly limits the detectability of minor species in the tropospheres of the giant planets. Extending this bandwidth to 2 to 3 GHz should allow the search for tropospheric species on Jupiter, Saturn and Uranus.

A third major advance can be expected in the extension of the spectral range toward higher frequencies. This is becoming partly feasible from the ground, in atmospheric windows, but will be fully possible in Earth orbit with a submillimeter satellite, such as the FIRST mission, now under study at the European Space Agency. In the case of planets, the submillimeter range provides the advantage of a larger number of observable transitions, with stronger intensities. As a consequence, the expected detectability limits are, in many cases, at least an order of magnitude better than in the millimeter range.

Finally, an important improvement would be the inclusion of a heterodyne instrument aboard a planetary or cometary space probe. Such a device could be designed for monitoring species which are not observable from the ground, such as water vapor, which plays a key role in the photochemistry and dynamics of terrestrial planets, as well as in the kinematics of comets.

Therese Encrenaz

Bibliography

Bézard, B., Marten, A. and Paubert, G. (1992a) First ground-based detection of cyano-acetylene on Titan. *Bull. Am. Astron. Soc.*, **24**, 953–4.

Bézard, B., Marten, A. and Paubert, G. (1992b) International Astronomical Union Circular **5685**, December 30 1992.

Biraud, F., Bourgois, G. and Crovisier, J. (1974) OH observations of comet Kohoutek (1973 f) at 18 cm wavelength. *Astron. Astrophys.*, **34**, 163–6.

Bockelée-Morvan, D., Colom, P., Crovisier, J., *et al.* (1991) Microwave detection of hydrogen sulphide and methanol in comet Austin (1989c1). *Nature*, **350**, 318–20.

Cernicharo, J., Paubert, G., Encrenaz, T., *et al.* (1993) A tentative detection of the 183 GHz water vapor line in the Martian atmosphere. *Bull. Am. Astron. Soc.*, **23**, 1062.

Clancy, R.T. and Muhleman, D.O. (1985) Chemical–dynamical models of Venus mesosphere based upon diurnal microwave CO variations. *Icarus*, **64**, 183–204.

Clancy, R.T., Grossman, A. and Muhleman, D.O. (1992) Mapping the water vapor on Mars with the VLA. *Icarus*, **100**, 48–59.

Clancy, R.T., Muhleman, D.O. and Berge, G.L. (1990) Global changes in the 0–70 km thermal structure of the Mars atmosphere derived from 1975–1989 microwave CO spectra. *J. Geophys. Res.*, **95**(B9), 14543–54.

Crovisier, J. and Schloerb, F.P. (1991) The study of comets at radio wavelengths, in *Comets in the Post-Halley Era*, (eds R.L. Newburn, Jr, M. Nengebaner and J. Rahe). Dordrecht: Kluwer Academic Publishes, pp. 149–73.

Crovisier, J., Despois, D., Bockelée-Morvan, D., *et al.* (1991) Microwave observations of hydrogen sulfide and searches for other sulfur compounds in comets Austin (1989c1) and Levy (1990c). *Icarus*, **93**, 246–58.

de Pater, I., Palmer, P. and Snyder, L.E. (1991) A review of radio interferometric imaging of comets, in *Comets in the Post-Halley Era*, (eds R.L. Newburn, Jr, M. Nengebaner and J. Rahe). Dordrecht: Kluwer Academic Publishes pp. 175–207.

Despois, D., Crovisier, J., Bockelée-Morvan, D. *et al.* (1986) Observations of hydrogen cyanide in comet Halley. *Astron. Astrophys.*, **160**, L11–2.

Encrenaz, T., Lellouch, E., Paubert, G. and Gulkis, S. (1991a) First detection of HDO in the atmosphere of Venus at radio wavelengths: an estimate of the H_2O vertical distribution. *Astron. Astrophys.*, **246**, L63–6.

Encrenaz, T., Lellouch, E., Rosenqvist, J. *et al.* (1991b) 'The atmospheric composition of Mars: ISM and ground-based observational data.' *Ann. Geophys.*, **9**, 797–803.

Encrenaz, T., Lellouch, E., Cernicharo, J. *et al.* (1993) 'First detection of the 183 GHz water line in the atmosphere of Venus.' *Bull. Ame. Astron. Soc.*, **25**, 1097.

Gérard, E., Bockelée-Morvan, D., Bourgois, G., *et al.* (1987) 18-cm wavelength radio monitoring of the OH radical in comet P/Halley 1982 i. *Astron. Astrophys.*, **187**, 455–61.

Gulkis, S., Kakar, R.K., Klein, M.J., *et al.* (1977) Venus: detection of variations in stratospheric carbon monoxide, in *Proc. Symp. on Planetary Atmospheres*, (ed. A.V. Jones). Ottawa, Ontario: Royal Society of Canada, pp. 61–5.

Harris, A.I. (1991) Coherent and incoherent detection at submillimeter and far-infrared wavelengths, in *Coherent detection at Millimeter Wavelengths and their Applications* (eds. P. Encrenaz, C. Laurent, S. Gulkis *et al.*). Nova Science, pp. 7–34.

Hobbs, R.W., Maran, S.P., Brandt, J.C. *et al.* (1975) Microwave continuum radiation from comet Kohoutek (1974 XII): emission from the icy-grain halo? *Astrophys. J.*, **201**, 749–55.

Kakar, R.K., Waters, J.W. and Wilson, W.J. (1976) Venus: microwave detection of carbon monoxide. *Science*, **191**, 379–80.

Kakar, R.K., Waters, J.W. and Wilson, W.J. (1977) Mars: microwave detection of carbon monoxide. *Science*, **196**, 1020–1.

Lellouch, E., Paubert, G. and Encrenaz, T. (1991) Mapping of CO millimeter-wave lines in Mars' atmosphere: the spatial variability of carbon monoxide on Mars. *Planet. Space Sci.*, **39**, 219–24.

Lellouch, E., Belton, M.J.S., de Pater, I., *et al.* (1990) Io's atmosphere from microwave detection of SO_2. *Nature*, **346**, 639–41.

Lellouch, E., Goldstein, J.J., Bougher, S.W. *et al.* (1991a) First absolute wind measurements in the middle atmosphere of Mars. *Astrophys. J.*, **383**, 401–6.

Lellouch, E., Encrenaz, T., Phillips, T. *et al.* (1991b) Submillimeter observations of CO in Mars' atmosphere. *Planet. Space Sci.*, **39**, 209–12.

Lellouch, E., Belton, M., de Pater, I. *et al.* (1992) The structure, stability and global distribution of Io's atmosphere. *Icarus*, **98**, 271–95.

Marten, A., Gautier, D., Tanguy, L. *et al.* (1988) Abundance of carbon monoxide in the stratosphere of Titan from millimeter heterodyne observations. *Icarus*, **76**, 558–62.

Marten, A., Gautier, D., Owen, T. *et al.* (1993), First observations of CO and HCN on Neptune and Uranus at millimeter wavelengths and their implications for atmospheric chemistry. *Astrophys. J.*, **406**, 285–97.

Mayer, C.H., McCullough, T.P. and Sloanaker, R.M. (1958a) Observations of Venus at 3.15 cm wavelength. *Astrophys. J.*, **127**, 1–10.

Mayer, C.H., McCullough, T.P. and Sloanaker, R.M. (1958b), Observations of Mars and Jupiter at 3.15 cm wavelength *Astrophys. J.*, **127**, 11–6.

Muhleman, D.O., Berge, G.L. and Clancy, R.T. (1984), Microwave measurements of carbon monoxide on Titan. *Science*, **223**, 393–6.

Paubert, G., Marten, A., Rosolen, C., *et al.* (1987) CO abundance in the stratosphere of Titan from microwave observations. *Bull. Am. Astron. Soc.*, **19**, 873.

Rosenqvist, J., Lellouch, E., Romani, P.N., *et al.* (1992) Millimeter-wave observations of Saturn, Uranus and Neptune: CO and HCN on Neptune. *Astrophys. J.*, **392**, L99–102.

Schloerb, F.P., Robinson, S.E. and Irvine, W.M. (1980) Observation of CO in the stratosphere of Venus via its $J = 0–1$ rotation transition. *Icarus*, **43**, 121–7.

Schloerb, F.P., Kinzel, W.M., Swade, D.A. and Irvine, W.M. (1986) HCN production rates from comet Halley. *Astrophys. J.*, **310**, L55–60.

Shah, K.P., Muhleman, D.O. and Berge, G.L. (1991) Measurements of winds in Venus' upper atmosphere based on Doppler shifts of the 2.6 mm ^{12}CO line. *Icarus*, **93**, 96–121.

Snyder, L.E. (1986) The 18 cm OH lines in comets: preparation for Halley. *Astron. J.*, **91**, 163–70.

Snyder, L.E., Palmer, P. and de Pater, I. (1989) Radiodetection of formaldehyde emission from comet Halley. *Astron. J.*, **97**, 246–53.

Tanguy, L., Bézard, B., Marten, A., *et al.* (1990) Stratospheric profile of HCN on Titan from millimeter observations. *Icarus*, **85**, 43–57.

Von Zahn, U., Kumar, S., Niemann, H. and Prinn, R. (1983) Composition of the Venus atmosphere (eds D.M. Hunten, L. Colin, T.M. Donative and V.I. Moroz). Tucson: University of Arizona Press, pp. 229–430.

Wilson, W.J., Klein, M.J., Kakar, R.K., *et al.* (1981) Venus. I. Carbon monoxide distribution and molecular-line searches. *Icarus*, **45**, 624–37.

Cross references

Comet: observation
Mars: atmosphere
Radio astronomy
Spectroscopy: atmospheres

Titan
Venus: atmosphere

MILANKOVITCH, MILUTIN (1879–1958), AND MILANKOVITCH THEORY

Milutin Milankovitch developed the theory of orbital control of terrestrial insolation and climate change. He was of Serbian origin, born in Dalj (Slavonia) in what is now Croatia. He obtained his PhD in 1904 in Vienna and worked as a civil engineer for 5 years before going to the University of Belgrade as Professor of Applied Mathematics.

Milankovitch was destined to become one of those rare scientists who developed truly pivotal ideas. His inspirational conversion from engineering to orbital dynamics is elegantly told by Imbrie and Imbrie (1979), and may be followed with details and bibliography in an autobiographical work (1957).

An orbital control of climate was proposed in the 19th century (1842) by Adhémar, and developed by Croll (1821–1890; q.v.), but their concepts were not comprehensive or supported by the rigorous mathematical calculations that were provided by Milankovitch. The first version to reach the west was in French in 1920, with the full analysis in 1941 (and an English translation in 1969).

The 'Milankovitch parameters' are (1) the eccentricity (or ellipticity) of the orbit e, which measures the departure of the Earth's circumsolar orbit from a circle (with a period of about 90 000–100 000 years); (2) the obliquity of the ecliptic (or tilt) ϵ, which is the angle between the equator and the plane of the orbit (principal period about 41 000 years); and (3) the precessional parameter, related to the longitude of the perihelion, ω which is conveniently expressed as the angular distance of the spring equinox point from the perihelion. It has two principal terms, about 19 000 and 23 000 years. This is related to the 'general' precession that has recently been refined to 25 694 years (Berger, 1992). Stothers (1987) has shown that the principal terms are interrelated as a series of beat frequencies.

The tilt is 23°44′ at present. This angle (mean: 23°10′) defines the tropics, the highest latitudes where the Sun, at summer solstice, reaches the zenith. If the tilt were 0° there would be no seasons, and at the poles the Sun would never rise above the horizon. The equator at such times would receive the maximum and the poles the minimum annual insolation. The obliquity changes between approximately 22 and 24.5° in a cycle.

Since the Earth's circumsolar orbit is not circular but elliptical, and since the Sun is not in the geometric center of the ellipse, the Sun–Earth distance (and therefore the intensity of solar radiation) changes through the year. The incident solar energy is today strongest in early January, when the Earth is closest to the Sun (at perihelion), and weakest in early July when the Earth is at its greatest distance (aphelion). The value is close to the annual average, and therefore to the 'solar constant' (q.v.), in early April and early October. Although the variation in distance between perihelion and aphelion is only about 3%, the inverse-square dependence of the radiation intensity produces a 7% difference in solar energy between perihelion and aphelion. The Earth's orbit becomes more circular and less elliptical when the eccentricity is low. At such times the intensity of solar radiation at perihelion differs little from that at aphelion. Its last minimum, about 51 000 years ago, was during the last ice age.

Milankovitch used these orbital variables to calculate the incoming solar radiation ('insolation') at the top of the atmosphere as a function of latitude for the summer and winter half of each year (Figure M77). The two variables are the solar zenith angle and the Earth–Sun distance. He reached the conclusion that summer insolation at latitude 65°N was the critical value in the control of the glacial/interglacial fluctuation during ice ages. This northern hemisphere bias is due to its higher continentality – albedo characteristics.

Captured during World War I, but allowed to work at the Hungarian Academy of Sciences, Milankovitch completed the insolation theory for the Earth and had also worked out a climate history for Venus and Mars.

After the war Wladimir Köppen, the famous climatologist, was working on a book with his son-in-law Alfred Wegener (of continental drift fame), and what became a standard textbook on climate (1924) carried the Milankovitch message. Unfortunately it only reached a German-speaking audience. However, a German geologist and archaeologist from Breslau (now Vroclav), Frederick Zeuner,

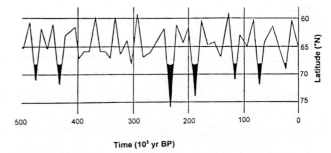

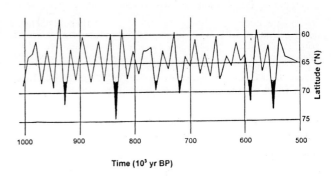

Figure M77 Variation of the total annual insolation on top of the atmosphere during the last 1 million years (Zeuner, 1959), as departures from a mean for latitude 65°N. Whereas the monthly insolation varies significantly throughout the world, the annual totals are effective in principle only in the high latitudes.

applied the insolation theory to the central European geological record, and after moving to London in the 1930s Zeuner explained to a world audience the Milankovitch ideas (Zeuner, 1959).

The 'establishment' of the day responded to the Milankovitch theory with severe criticism, an outrage almost paralleled by the hostile reception of Wegener's continental drift theory.

During World War II Milankovitch worked on a complete revision of his radiation theory, which was published (in German) by the Royal Serbian Academy of Sciences in 1941 as the *Kanon der Erdbestrahlung*. . . . (translated into English in 1969).

In the post-war decades, however, dramatic discoveries were being made at sea thanks to sediment coring and deep sea drilling (Imbrie and Imbrie 1979). Cesare Emiliani (1966) analyzed those sediment cores geochemically, discovering isotopic evidence confirming the warm–cold sequence. The postglacial eustatic curve of rising sea level was calibrated with radiocarbon dates (Fairbridge 1960, 1961). The sea level record paralleled the last phase of the Milankovitch radiation curve for 65°N. During the next two decades, dating systems were expanded, deep sea cores were obtained worldwide, and the results were conclusive: there was an exact match with the Milankovitch pattern (Fairbridge 1967), and it was recognized as 'the pacemaker of the ice ages' (Hays, Imbrie and Shackleton, 1976). An independent data set based on the ice-age wind-blown loess deposits and their interlayered fossil soils, worked out by Kukla (1970) in central Europe (and later, in China) brought the same results. The hypothesis has now become an accepted theory.

In 1979, to mark the 100th anniversary of the birth of Milankovitch, a symposium was organized in Belgrade by the Serbian Academy of Sciences, and a conference convened at the Lamont-Doherty Geological Observatory (Palisades, N.Y.). The evidence was overwhelming (see symposium volume, edited by Berger *et al.*, 1984).

It needs to be emphasized that although the chronology of the Milankovitch theory is almost universally accepted, the insolation component was rather weak. It appears that this aspect can be greatly strengthened by two additional factors, not considered by Milankovitch.

1. Feedbacks, notably albedo and atmospheric chemistry. The dimensions of the ice-and snow-covered areas, overwhelmingly in the northern hemisphere and therefore the principal albedo control, are now found to correlate rather persuasively with the orbital parameters (Liu, 1995). Gas bubbles trapped in polar ice

also show that major 'greenhouse gases' (CO_2 and CH_4) were greatly depleted during cold cycles, likewise effecting a positive feedback; their depletion was probably due to ice burial and aridification of former vegetated terrain, with the climate further modified by increased continentality by eustatic fall of sea level. A thermohaline feedback has also been proposed (Kukla and Gavin, 1992).

2. Solar activity variations. While Milankovitch assumed a constant solar emission rate, this is now known to fluctuate on both long- and short-term scales, which appears to correspond to orbital changes within the entire solar system. Berger's improved value for the general (axial) precession at 25 694 years (which is exactly 1425 × 18.0303 years, the 'Saros' and basic lunar tide cycle) is also exactly commensurable (× 2010) with the 12.7830-year Neptune–Jupiter lap (see Orbital commensurability and resonance).

The acceptance of the Milankovitch theory was far more than a quantification of dynamic change for the last ice age. Geologists are now applying it to the whole of Earth history. For climatologists, long-term modeling is now feasible (Kutzbach, 1985). For meteorologists it carries a crucial message: the terrestrial climate machine is neither chaotic nor unpredictable – it is forced by extraterrestrial agencies. The same message must apply also to other planets.

George Kukla and Rhodes W. Fairbridge

Bibliography

Berger, A.L. (ed.) (1981). *Climatic Variations and Variability: Facts and Theories*. Dordrecht: Reidel, 795 pp.
Berger, A.L. (1992) Astronomical theory of paleoclimates and the last glacial–interglacial cycle. *Quaternary Sci. Rev.*, **11**, 571–81.
Berger, A.L. and Loutre, M.F. (1991) Insolation values for the climate of, the last 10 million years. *Quaternary Sci. Rev.*, **10**, 297–317.
Berger, A., Imbrie, J. Hays, J., *et al.* (eds) (1984) *Milankovitch and Climate*. Dordrecht: Reidel, 895 pp.
Berger, A., Loutre, M.F. and Laskar, J. (1992) Stability of the astronomical frequencies over the Earth's history for paleoclimate studies. *Science*, **255**, 560–6.
Broecker, W.S. (1968) Milankovitch hypothesis supported by precise dating of coral reefs and deep sea sediments. *Science*, **159**, 297–300.
Emiliani, C. (1966) Isotope paleotemperatures. *Science*, **154**, 851–7.
Fairbridge, R.W. (1960) The changing level of the sea. *Sci. Am.*, **292**(5), 70–9.
Fairbridge, R.W. (1961) Eustatic changes in sea level in *Physics and Chemistry of the Earth*. Vol. 4 (eds L.H. Ahrens *et al.*). London: Pergamon Press, 99–185.
Fairbridge, R.W. (1967) Ice-age theory, in *The Encyclopedia of Atmospheric Sciences and Astrogeology* (ed. R.W. Fairbridge). New York: Reinhold Publ. Co., 1200 pp., pp. 462–7.
Hays, J.D., Imbrie, J. and Shackleton, N.J. (1976) Variations in the Earth's orbit: pacemaker of the ice ages. *Science*, **194**, 1221–332.
Imbrie, J. and Imbrie, K.P. (1979) *Ice Ages: Solving the Mystery*. Short Hills, NJ: Enslow Publ., 224 pp.
Köppen, W. and Wegener, A. (1924) *Die Klimate der Geologischen Vorzeit*. Berlin: Gebr. Borntraeger.
Kukla, G.J. (1970) Correlation between loesses and deep-sea sediments. *Geol. Fören. Förh.*, Stockholm, **92**, 148–80.
Kukla, G. and Gavin, J. (1992) Insolation regime of the warm to cold transition, in *Start of a Glacial*, (eds G. Kukla and E. We) NATO ASI Series I, **3**, 307–39.
Kutzbach, J.E. (1985) Modeling of paleoclimates. *Adv. Geophys.*, **28A**, 159–96.
Liu, H.-S. (1995) A new view on the driving mechanisms of Milankovitch glaciation cycles. *Earth Planet. Sci. Lett.*, **131**, 17–26.
Milankovitch, M. (1941) Kanon der Erdbestrahlung und seine Anwendung auf das Eiszeitenproblem. *Roy. Serb. Acad. Sp. Publ.* **133**, 633 pp. (English transl. 1969 as *Canon of Insolation and the Ice-Age Problem*, by Israel Progr. Sci. Transl., US Dept. Commerce, Washington, 484 pp.)
Milankovitch, M. (1957) Astronomische Theorie der Klimaschwankungen: ihr Werdegang und Widerhall. *Serbian Acad. Sci., Monogr.*, **280**, 1–58 (incl. comprehensive bibliography).
Sanders, J.E. (1995) Astronomical forcing functions: from Hutton to Milankovitch and beyond. *Northeastern Geology and Environmental Sciences*, **17**(3), 306–47.
Stothers, R.B. (1987) Beat relationships between orbital periodicities in insolation theory. *J. Atmos. Sky*, **44**(14), 1875–6.
Zeuner, F.E. (1959) *The Pleistocene Period*. London: Hutchinson.

Cross references

Ice age
Insolation

MILLER–UREY EXPERIMENT

In 1953 a paper was published in *Science* reporting on the non-biological production of some organic compounds in a flask containing a mixture of gases thought to be representative of the atmosphere of the primordial Earth. The experiments reported were conducted by Stanley L. Miller, then a graduate student working under the direction of Harold C. Urey at the University of Chicago chemistry department.

The mixture of gases used was methane, ammonia, water and hydrogen. The apparatus consisted of a large (5 L) and a small flask (0.5 L) connected with glass tubing so as to form a circuit. Water in the small flask was heated, circulating steam and gases around the apparatus. In the large flask an electric spark was created between a pair of tungsten electrodes through the action of a Tesla coil.

A variety of compounds were formed, notably biologically important amino acids which form the building blocks of proteins in all living things on Earth. Subsequent work with this apparatus showed that the most abundant products were formic acid, glycine, glycolic acid, alanine and lactic acid. One of the most intriguing results of the experiment was that the products were, in general, compounds of biological importance and not a random mixture of organics. Subsequent elaborations on the Miller–Urey experiment have demonstrated the production of 17 out of the 20 basic amino acids of life and the nucleotide bases adenine, guanine, cytosine and uracil that form RNA (for a review see Miller, 1992, and references therein).

The essential assumptions of the Miller–Urey experiment as applied to the question of the origin of life are (1) that the earliest life on Earth derived its energy by consuming organic matter and (2) that the organics that this life form required were made on the early Earth in a reducing atmosphere. (For a discussion of alternative sources of organics and the possibility that the atmosphere of the early Earth as not reducing, see Chyba and Sagan, 1992; see also Life: origin.)

Christopher P. McKay

Bibliography

Chyba, C. and Sagan, C. (1992) Endogenous production, exogenous delivery and impact-shock synthesis of organic molecules: an inventory for the origins of life. *Nature*, **355**, 125–32.
Miller, S.L. (1953) Production of amino acids under possible primitive earth conditions. *Science*, **117**, 528–9.
Miller, S.L. (1992) The prebiotic synthesis of organic compounds as a step toward the origin of life, in *Major Events in the History of Life* (ed. J.W. Schopf). Boston: Jones and Bartlett, pp. 1–28.

Cross references

Biosphere
Comet: impacts on Earth
Life: origin

MIRANDA

The smallest and innermost of the five 'classical' satellites of Uranus. Discovered telescopically in 1948 by Gerard Kuiper, little was known about Miranda until very recently due to its small size, its proximity to bright Uranus and its great distance from the Sun. The radius was believed to be 250 km but with an uncertainty of 110 km. Its mass

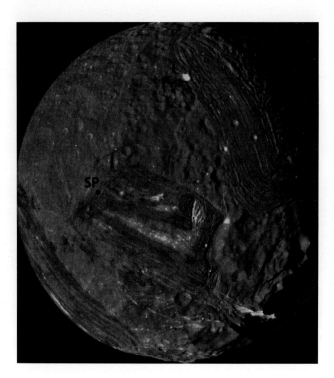

Figure M78 Global mosaic of southern hemisphere of Miranda, centered near the south pole (SP). Inverness corona is the polygonal ridged region at center, Elsinore and Arden coronae are located at upper right and lower left respectively.

and density were essentially unconstrained. Water ice was tentatively identified on the surface in 1984. Other ices such as methane or ammonia, although undetected, could also be present in quantity. Miranda's orbital period is 1.4 days. Although nearly circular, Miranda's orbit is inclined at ~ 4°, unusual for an inner satellite of a gas giant planet and unique among the Uranian satellites. This curiosity would later prove important in an unexpected way.

Exploration and general properties

This poor state of knowledge changed completely during the first spacecraft mission to Uranus by Voyager 2 in 1986. In order to visit Neptune in 1989, Voyager 2 was constrained to pass Uranus at a distance nearly coinciding with Miranda's orbit. Little excitement was expected from tiny Miranda although Voyager's approach distance was sufficient to result in some of the highest-resolution images of the entire mission (features as small as 350 m are resolved). Because the south poles of Uranus and its satellites were oriented almost directly toward the Sun in 1986 only the southern hemisphere of Miranda was visible. The evidence of extensive fracturing and volcanism came as a complete surprise. Although some progress has been made since 1986, the mystery of Miranda's complex geology continues to puzzle scientists.

The Voyager flyby resulted in the first characterization of Miranda's bulk properties. Miranda is a triaxial ellipsoid of mean radius 235.8 km; the difference between long and short axes is ~ 7.5 km, consistent with hydrostatic equilibrium in the Uranian gravitational potential. Voyager confirmed that Miranda's rotation is synchronous with its orbital period, keeping the same face toward the planet.

Miranda's relatively low density of ~ 1150 ± 150 kg m^{-3} and the abundance of water and other ices in the Uranian region implies a mostly icy composition (60 to 80 wt%), with some unidentified rocky component. Miranda, though, may be somewhat more ice rich than the other Uranian satellites which have ice compositions closer to 50 wt%. The global albedo of ~ 28% and the identification of water ice indicate the surface also is composed of water ice, with a minor but unidentified dark 'rocky' component. In visible light Miranda is a nearly uniform colorless gray. This implies that the dark rocky material is also spectrally gray, possibly carbonaceous.

Geology of Miranda

Miranda can be divided into two geologic terrains; lightly cratered rolling plains, and younger tectonically and volcanically modified coronae (q.v.). The visual albedo of the cratered plains is relatively uniform, averaging 30%. There is little evidence for geologic activity on these plains, except for a rectilinear fracture network partially associated with coronae formation and a mantling or softening of older craters and fractures. Crater density on these plains is variable but in general is as high or higher than on the lunar mare. The largest crater, Alonso, is ~ 25 km across and bowl shaped. (Place names on Miranda are derived from Shakespearean characters or locales.) Two irregular depressions 45 and 60 km across could be degraded impact craters. No viscously relaxed craters have been identified, indicating that Miranda's lithosphere has been essentially rigid over geologic time. Although the absolute age is very uncertain, this surface is very ancient (> 4 Ga).

Several linear troughs, or chasma, extend across the cratered plains. These troughs are 15 to 40 km wide, a few hundred kilometers long and several kilometers deep. Most troughs are fault bounded. Others are mantled but are also probably due to extensional tectonism. The most prominent trough, Verona Rupes, is on the order of 10 km deep. These extensional fractures imply that Miranda experienced a period of global areal expansion of up to 4%, assuming that the observed surface is representative of the whole. This tectonism appears to have taken place over an extended period of time, including that of corona formation.

Cross-sections of the plains regolith and crust are conveniently revealed in the faces of several of the large-scale fracture scarps. The upper kilometer of the Verona Rupes fracture scarp is relatively brighter than the material below. This could be regolith material that mantles older craters and fractures. Dark talus streaks can be seen trailing downhill from the crests of other scarps, as well as from the rims of several fresh craters. A few craters also have anomalously dark ejecta deposits. Most of these dark deposits are located near the margins of coronae, though, and are apparently exposures of otherwise frost-covered material that darkens on exposure to space. Their origins and composition are unknown.

Coronae

Large ovoid to polygon-shaped regions of the cratered plains have been extensively modified. These constitute the three large coronae. They have relatively low crater densities and are younger than the cratered plains. The relative ages of the three coronae are uncertain, however. The two largest coronae (each about 280 km across) are centered near the equator antipodal to each other. Named Elsinore and Arden coronae, they are located on the trailing and leading hemispheres, respectively. Each was viewed near the limb at the time of the Voyager encounter; thus we did not see their northern portions. The third corona, Inverness, is somewhat smaller (175 by 225 km), trapezoidal in shape and located near the south pole. Albedos within the coronae are variable, ranging from 25 to 32%.

Although distinct, each corona has a similar basic structure. Each is characterized by an inner zone of chaotic to parallel ridges and troughs, and an outer banded zone of concentric ridges and troughs. In Inverness corona this outer zone is only a few kilometers wide. Many ridges in the inner zone are truncated by ridges in the outer banded zones. Corona ridges can be up to 10 or 20 km wide and 500 m high. Many coronal ridges correspond to bright and dark albedo bands. The albedo contrasts could indicate compositional differences or differences in regolith or frost exposure ages.

The central chaotic regions of each corona are near the mean global elevation. The margins of Arden, eastern Elsinore and parts of Inverness, however, are bounded by wide peripheral troughs several kilometers deep coinciding with the outer banded zone. Parallel fault scarps in these bounding troughs, especially in Arden, result in a sawtooth topography, suggesting that extension and rotation of fault blocks has occurred. Several of these troughs extend into the cratered plains.

Along the southern margin of Elsinore and nearby margins of Inverness corona the bounding trough is absent. Many of the parallel ridges in the outer banded zone here have been interpreted as extrusions of either viscous icy 'lavas' or glacier-like flows extruded along parallel fractures. These ridges are a few percent darker than other terrains, have rough textures and partially embay small depressions. They appear to have formed on top of the cratered

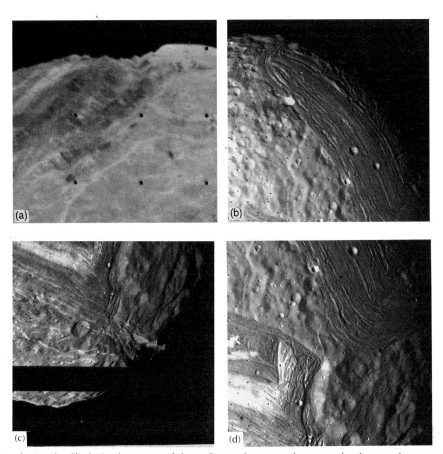

Figure M79 Four views of Miranda. Clockwise from upper left: (a) Sawtooth topography across the depressed western margin of Arden corona. This area has been extensively faulted. Width of depression is ~ 50 km. Dark downslope talus streaks are also visible. (b) Elsinore corona, showing outer banded zone of concentric ridges, which may be volcanic in origin. Largest impact crater is 17 km across. (c) Verona Rupes fault scarp, ~ 10 km deep. Upper surface of fault scarp is relatively bright. Horizontal dark bar is missing data. Width of view is ~ 250 km. (d) Parts of Inverness and Elsinore coronae. Rolling cratered plains separate the polygonal coronae. Width of view is ~ 250 km.

plains. Estimates of flow viscosities, based on flow shapes, indicate the flow ridges were unusually viscous during emplacement, and may have been as viscous as terrestrial andesite lavas or glacier ice (depending on uncertainties and assumptions inherent in the viscosity models). Estimated yield strengths of these flows are similar to those of terrestrial basaltic lavas. Apparently these flows were partially or nearly crystallized when extruded, and hence rather sticky.

The composition of the flows is unknown. It would be difficult to melt partially the interior of Miranda unless ice phases other than water ice were present. Also, liquid water would be too fluid to account for the thick topography of these flow ridges. Exotic ice phases such methane–clathrate or ammonia compounds could comprise 20% or more of Miranda. Relatively high viscosities similar to molten basalt have been measured experimentally for ammonia–water melts, for example, especially when mixed with minor amounts of low-order hydrocarbons, or 20 to 30% admixture of suspended crystals. These and similar experiments indicate that icy melt mixtures can have rheologies and flow behaviors as complex as silicate melts on Earth, leading to a variety of volcanic landforms depending on composition, flow history and eruption conditions. Ammonia or methane compounds on the surface can be photodissociated by solar or cosmic radiation into organic or other by-products over geologic time or may be masked by water frost. This may partly account for the failure to identify these species spectroscopically, if they are indeed present. Future high-resolution spectroscopy may identify these volcanic components.

Origin of coronae

There is as yet no clear perception of how coronae formed. The concentric ovoid pattern and parallel ridge and trough topography of the two large coronae suggest that they are related to an internal deformation process, most likely either upwelling from or downwelling into the interior due to density contrasts. Upwelling would produce a concentric zone of extensional features, downwelling a zone of compressional features. Because of its small size and proximity to Uranus, a very young differentiated proto-Miranda may have been broken apart in one (or more) large impact event. In this scenario coronae would form as a result of reaccretion of a few denser rocky core fragments into the upper crust. The sinking of this dense mass would induce regional downwelling into the interior. An alternative scenario is that partial melting of volatile ices such as ammonia–water due to heating of the interior gave rise to upwelling and associated surface volcanism locally controlled by pre-existing tectonic structures. Choosing between these alternatives depends on how Miranda's geologic record is interpreted.

The apparent extensional style of fracturing and associated volcanism in the outer banded zones favors upwelling from the interior, but this interpretation of the morphology is still controversial. Some of these extensional features may have formed by subsequent reactivation of older fault trends, complicating the interpretation. There is no independent evidence for the reaccretion of any large compositionally distinct proto-Miranda fragments on the observed surface, however (the controversial interpretation of Arden corona as a modified impact structure notwithstanding), although this may well have occurred without being preserved in the geologic record. The morphologic evidence would appear to favor upwelling from the interior and associated tectonism and volcanism in the overlying crust although future detailed analysis might reverse this conclusion.

Whatever the geologic mechanisms involved, the resurfacing of Miranda requires an unusual energy source to melt partially or soften the interior to permit fluidization and mass movement on a global scale. A plausible hypothesis is heating due to tidal dissipation in the interior, because accretional heating and the decay of early radioactive nuclides are insignificant in such a small body.

Tidal heating (q.v.) has been implicated as the source of energy for the extensive volcanism on Io. However, unlike Io and its orbital resonances with Europa and Ganymede, there are at present no orbital resonances amongst the Uranian satellites. The importance of resonances in the present context lies in the eccentricities they force on the orbits of the satellites. A high orbital eccentricity induces greater tidal forcing and consequently greater tidal heating. A resonance can also pump up the inclination of a satellite's orbit. Fortuitously, it is now established that Miranda's unusually high orbital inclination is positive evidence that it passed through a 3 : 1 orbital commensurability with Umbriel in the past. Numerical modeling shows that the same resonance would have pumped up Miranda's eccentricity, leading to an episode of chaotic orbital variations that eventually disrupted the resonance configuration. Thus, in addition to its very interesting geology, this small moon also has a very complex orbital history.

The 3 : 1 chaotic resonance between Miranda and Umbriel may have lasted several hundred million years, with a complicated feedback between the orbital dynamics and the geophysical evolution of the satellites. Miranda's internal temperature could have been raised by 100 K or so, although explicit thermal modeling has not yet been attempted. Tidal heating of this magnitude would be sufficient to melt partially small portions of the interior depending on initial composition. For example, water ice melts at 273 K, but ammonia–water melts at 175 K and methane is liquid at even lower temperatures. Small amounts of melts of these or other low-temperature ice phases could easily account for the small volume of volcanic materials on the surface ($\sim$ 1% of the total satellite volume). The density and compositional structure of Miranda's interior are unclear, however. The satellite may have been homogenous or composed of mixed lumps of rock and ice, depending on the formation and accretion history of Miranda. The presence of clathrates or other very low-temperature ices (such as methane) could reduce the temperatures necessary to mobilize material below that required for water or ammonia–water ice alone. The abundance or presence of these materials is speculative, although the stickiness of the extruded flows suggests that they are not simple compounds. The geology of the coronae, coupled with the complex orbital evolution, indicates that Miranda's history was complex, especially for such a small body.

Paul Schenk

Bibliography

Bergstralh, J.T., Mines, E.D. and Mathews, M.S. (eds) (1991) *Uranus*, Tucson: University of Arizona Press. (Especially, chapters by Brown *et al.*, Veverka *et al.*, Greenberg *et al.* and Croft and Soderblom).

Janes, D. and Melosh, H.J. (1989) Sinker tectonics: an approach to the surface of Miranda. *J. Geophys. Res.*, **93**, 3127.

Jankowski, D. and Squyres, S. (1988) Solid-state volcanism on the satellites of Uranus. *Science*, **241**, 1322.

Malhotra, R. and Dermott, S. (1990) The role of secondary resonances in the orbital history of Miranda. *Icarus*, **85**, 444.

Schenk, P. (1991) Fluid volcanism on Miranda and Ariel: flow morphology and composition. *J. Geophys. Res.*, **96**, 1887.

Smith, B., and 42 others, (1986) Voyager 2 in the Uranian system: Imaging science results. *Science*, **233**, 43.

Cross references

Imaging science
Satellite: natural
Uranus: satellite system
Voyager missions

MOON (EARTH'S MOON)

The Moon, the only natural satellite of the Earth and the second most impressive celestial body as seen from Earth, has long been an inspiration for scientists, artists, philosophers and romantics.

The Moon moves in a nearly circular orbit about the Earth, taking about 29½ days to go from one full Moon to the next. This cyclic motion of the Moon through the sky doubtless led to early definition of the month as an important unit of time. The Moon is quite large relative to the Earth when compared with other planet–satellite systems in the solar system; in fact, the Earth–Moon system is sometimes referred to as a double planet. Only the Pluto–Charon system has a larger ratio of satellite to primary mass ratio. The Moon's average distance from the center of the Earth is 284 000 km, or some 60 times the radius of the Earth. The Moon's radius (1738 km, about one-quarter the size of Earth) and its mass [$\frac{1}{81}$ (0.013) that of Earth], combine to produce a surface gravity that is only one-sixth that of Earth. The mean density of the Moon is 3.3 gm cm^{-3}, which is only 60% of the 5.5 gm cm^{-3} mean density of the Earth.

The Moon and Sun are the only celestial bodies that appear as disks to the naked eye, but only the Moon undergoes regular changes of phase in the course of its monthly cycle. These phase changes are due to the varying amount of the lunar surface that is in sunlight as seen from Earth – Figure M80 illustrates the different phases of the Moon as the Moon moves in its monthly orbit. Of course, half of the Moon's surface is in sunlight and half in darkness at any time. It is just that we see only part of the illuminated hemisphere except at full Moon. An interesting appearance of the Moon frequently occurs just after new Moon and is sometimes referred to as the 'old Moon in the arms of the new Moon', and is the faintly illuminated entire disk of the Moon nestled in the brightly lit crescent. This faint illumination is due to Earthshine, sunlight reflected from Earth to the Moon and then back to our eyes on Earth.

The Moon is in synchronous rotation, always presenting the same face or hemisphere toward Earth. The backside of the Moon was completely unknown before the space age when spacecraft with cameras opened up this unknown territory. The fact that we see only half of the Moon's surface is due to the fact that the orbital period of the Moon and its rotation period are the same. Most satellites in the solar system show synchronous rotation and, since it is unlikely that this occurs by chance, there must be some underlying reason why this condition occurs. The physical basis for synchronous rotation of satellites is to be found in the tides, a topic that we will discuss below. Spacecraft permit us to view the Moon from all angles (Figure M81).

The plane of the Moon's orbit is tilted slightly, about 5°, to the plane of the ecliptic, i.e. the plane of the Earth's orbit about the Sun. This means that the Moon's position in the sky follows the path of the planets as they appear to wander relative to the 'fixed' stars. The Earth's rotation axis, however, is tilted nearly 23.5° to the ecliptic plane, which produces a seasonal variation in the path of the Moon across the sky. Since the Moon is closer to the ecliptic than to the Earth's equatorial plane, the Sun and Moon more or less follow similar paths through the sky. But we tend to be more aware of the Moon at night when it is nearly full – when it is opposite the Sun, as viewed from the center of the Earth. Because the Sun is high in the sky during summer, the full Moon (being opposite) is low. In winter, when the Sun is low, we watch the full Moon follow a high course (similar to the Sun's path in summer).

Surface and interior of the moon

The application of the telescope to astronomical objects by Galileo in 1610 opened a new era of more detailed studies of the lunar surface. Studies of the lunar surface proceeded in an inverted fashion from studies of the Earth's surface: terrestrial studies went from the small-scale observations to ever larger scales culminating in studies from space, while lunar studies started from a global perspective and proceeded to ever more detailed observations, culminating (to date) with human exploration during the Apollo era.

The earliest telescopic observations revealed two types of terrain on the Moon: the heavily cratered, lighter colored regions called the lunar uplands, and the much less cratered, darker colored regions termed 'lunar maria', which took their name from the mistaken early impression that they were lunar seas (*mare* is Latin for sea). Refined studies led to the recognition that these areas are really vast basaltic plains on the Moon; however, the name mare (pronounced mar-e) has been retained (Plates 12–14). Early workers thought that the lunar craters were produced by lunar volcanoes; indeed, it wasn't until the middle of the 20th century that it was accepted by most planetary scientists that most lunar craters were produced by impacts of large projectiles (up to about 100 km in diameter). The fact that the maria

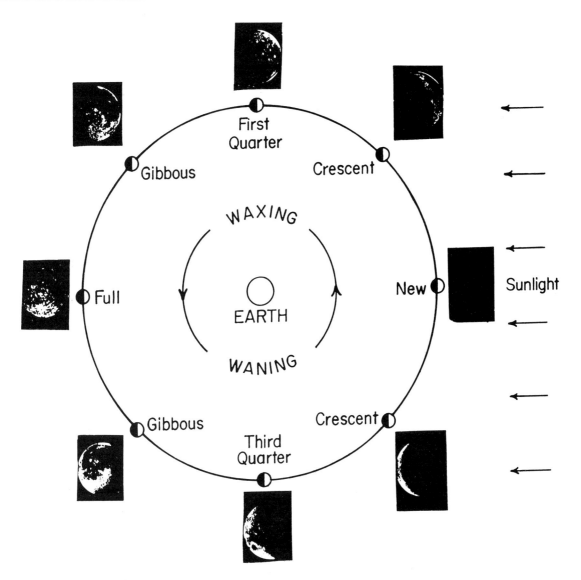

Figure M80 The position of the Moon in its orbit, corresponding to the different phases seen from Earth.

are lightly cratered means that the mare surfaces were created toward the end of the lunar formation period. With reliable ages of much of the lunar surface determined from rock samples returned by the Apollo astronauts, we now know that the lunar highlands were formed some 4.4 billion years ago, while the mare were formed up to 1.1 billion years later (see Moon: geology). During this time the frequency of projectiles hitting the Moon decreased dramatically, leaving the mare surfaces largely uncratered.

The surface of the Moon is composed of silicate rocks, basalts in the maria and anorthosite in the highlands, that are similar (but not identical, in some very important ways) to terrestrial rocks. The interior of the Moon, however, is quite different from the interior of the Earth. The lower overall density of the Moon relative to Earth means that it is made of lighter material. In particular, the Earth has a dense, iron-rich central core but the Moon is relatively low in iron and lacks a large central core. In fact, the density of the Moon is similar to the density of the upper mantle of Earth, an observation that led to 19th-century suggestions that the Moon was formed from the Earth's mantle when a large piece split off from the proto-Earth early in solar system history (see Earth–Moon system: origin).

The lack of an iron core and the slow rotation of the Moon would imply, based on our understanding of terrestrial magnetism, that the Moon should lack a magnetic field like that of Earth. However, measurements have shown that the Moon does have an appreciable magnetic field, although one that is considerably weaker than the Earth's. The origin of this field has been the object of considerable study, as described in the article on the lunar magnetic field.

Tides, the Moon's orbit and rotation rate

As noted above, the Moon and many other satellites of the solar system rotate with a period equal to the period of their orbit about their central planet. The reason for this synchronous rotation, as it is called, is tides. These tides are produced not only in oceans, as on Earth, but also in the solid body of the planet. These tides are generated by gravitational interactions between the satellite and planet (see Tide-raising forces) and exist in both the planet and the satellite. The action of tides produces changes in the rotation rate of both the planet and the satellite and acts to change the orbit of the satellite around the planet. In the case of the Earth–Moon system, tides acting over billions of years have resulted in the following.

1. Reduction of the Moon's rotation rate, so that it now equals the orbital period.
2. Reduction of the Earth's rotation rate (lengthening of our day). Current measurements show that the length of the day is increasing at the rate of 0.000007 s every year; so each day this year is 7×10^{-6} seconds longer than the corresponding day last year. Paleontological evidence from growth rings on certain corals

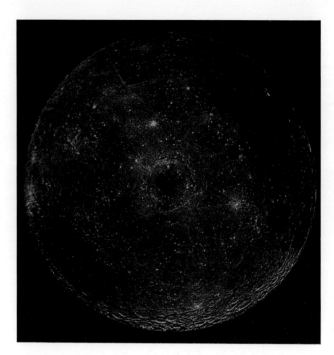

Figure M81 Earth's Moon, as imaged by the Galileo spacecraft, 9 December 1990. This view shows the western hemisphere of the Moon, with the Mare Orientale near the center of the image. The right half of the image is of the lunar nearside, which is visible from Earth, while the left side (the farside) is always turned away from the Earth. The large dark area to the upper right is Oceanus Procellarum. Mare Orientale, in the center of the image, was formed about 3.8 billion years ago by the impact of an asteroid-sized body. (NASA P-37327, courtesy of JPL.)

suggest that there were about 419 days in a year millions of years ago as compared with the current 365 days per year. Therefore, these corals lived in a world with a 21-h day.

3. Recession of the Moon from Earth. The current recession rate is about 3 cm year^{-1}, a surprisingly large value. Based on this value, models predict that the Moon would have been very close to the Earth only 1.5 billion years ago, yet we know from age dating of Apollo samples that the Moon is over 4 billion years old. Clearly the rate of recession over most of Earth's history was less than it is today. (See Roche limit; Tidal friction.)

All of these changes resulting from tides will continue into the future: the Moon is already locked into synchronous rotation but the Earth's day will continue to lengthen and the Moon will get further away. Finally, a state will be reached where the 'day' and 'month' are equal; the Moon will be nearly 50% further away from us by that time (billions of years into the future). At this point the Earth–Moon system, if left to its own devices, would be dynamically stable: the two bodies would orbit each other, always keeping the same hemisphere on each body pointed towards each other. However, the Sun and solar tides become dominant in such a system and these tides would act to reverse the recession of the Moon from Earth and cause the Moon to spiral inward toward Earth. This process would continue until the Moon came very close to Earth where the Earth–Moon tidal stress would rip the Moon apart, leading to our very own ring system, much more spectacular than the rings of Saturn. However, long before enough time has elapsed to allow this dramatic scenario to unfold, our Sun will have expanded into its red giant phase, consuming the terrestrial planets.

Exploration of the Moon

A mere 4 months after launching their Sputnik satellite, the Soviet Union became the first nation to send robotic spacecraft past the Moon with their Luna 1 mission, the first to send a spacecraft to another celestial body with their Luna 2 mission, and the first to return images of the far side of the Moon with their Luna 3 in 1959 (see Luna missions).

The US began lunar exploration with the Ranger program (q.v.). After a series of early embarrassing failures, Ranger yielded three missions that successfully televised thousands of pictures back to Earth up to the moment of each impact. The abrupt termination of these transmissions provided stark reminders that new craters had been added to the Moon's surface.

The US decision in 1961 to send astronauts to the Moon meant that a much more detailed reconnaissance of the Moon would be required by robotic spacecraft. The Lunar Orbiter program (q.v.) followed, designed to provide comprehensive photography of the lunar surface in order to identify suitable landing sites for the Apollo program and to provide data for planning the geologic exploration during these missions. Another robotic program was the Surveyor program (q.v.), which soft-landed spacecraft on the Moon and provided data on the lunar soil and other aspects of the lunar environment. Lunar Orbiter and Surveyor were both highly successful programs and resulted in ten successful missions to the Moon between May 1966 and January 1968.

The Soviet Union followed their early exploration missions with a series of lander missions with some unique features: their Lunakhod missions carried the first wheeled vehicle to explore another world. Looking like a bathtub on wheels, this robot traversed up to 30 km across the lunar surface, beaming pictures back to Earth and gathering interesting samples from the lunar surface. Lunar samples were placed into a capsule which was launched for the return trip to Earth. These are the only successful sample return missions by robotic spacecraft to date.

The US successfully met the goal set in 1961 by landing astronauts on the Moon and returning them safely to Earth in July 1969 (see Apollo missions). There were six lunar landing missions between 1969 and 1972. They returned a total of about 300 kg of lunar samples and many thousands of photographs; they left scientific stations on the Moon that returned data for many years after the astronauts left; and the command modules collected many other measurements from lunar orbit while the astronauts were working on the lunar surface.

With the termination of the Apollo program, lunar studies were again returned to the province of ground-based observations for the acquisition of new data, with the exception of occasional passing spacecraft bound for more distant targets, which turn their instruments toward the Moon in the fleeting moments as they pass by. The Galileo spacecraft made two flybys of the Moon on its laborious journey to Jupiter; the first was in December 1990 and the second in December 1992. On both occasions the spacecraft turned its suite of instruments toward the Moon, providing the first close-up view of the Moon in nearly 20 years.

Another spacecraft mission that has surveyed the Moon from lunar orbit is Clementine, a low-cost mission that was a joint effort between the Defense Department and NASA. In early 1994, this spacecraft orbited the moon for two months, returning an enormous number of pictures and other data of the lunar surface. Beyond Clementine there is an ambitious Japanese lunar mission, currently scheduled for 1997, which will deploy three penetrators into the lunar surface.

Data from the above projects, together with continued Earth-based studies, will help address many of the outstanding questions about the Moon. Are ices and other volatiles trapped near the lunar poles? Does the Moon have a core and, if so, how big is it? What is the nature of the projectile population that has hit the Moon (and the Earth) and how has it varied with time? What is the origin of the lunar paleomagnetism? How does crustal composition vary over the lunar surface? The answers to these and other questions that exist now will undoubtedly give rise to an entirely new set of scientific unknowns. The Moon will remain a challenging scientific object, a target of exploration and perhaps exploitation by humans, as well as a lovely and familiar companion in the night sky for as long as we can know today.

Donald R. Davis

Bibliography

Beatty, J.K. and Chaikin, A. (eds) (1990) *The Moon in the New Solar System*, 3rd edn. Cambridge, MA: Sky Publishing Corp.
Galileo Earth/Moon Encounter (1993) *J. Geophys. Res., Planets*, **98**(E9), 25 September (special section).

Hartmann, W.K., Phillips, R.J. and Taylor, G.J. (eds) (1984) *Origin of the Moon*. Proceedings of the Conference on the Origin of the Moon, Kona, Hawaii.

Wilhelms, D.E. (1993) *To a Rocky Moon: A Geologist's History of Lunar Exploration*. Tucson, AZ: University of Arizona Press.

Cross references

Apollo missions
Earth–Moon system: dynamics
Earth–Moon system: origin
Impact cratering
Luna missions
Lunar meteorites
Lunar Orbiter missions
Ranger missions
Surveyor missions

MOON: ATMOSPHERE

Attempts to detect a lunar atmosphere began with Galileo. In the 20th century, early searches set upper limits on the surface pressure of the lunar atmosphere ($< 10^{-4}$ Pa) from polarization effects near the lunar terminator. Sensitive occultation tests using the refraction of radio signals from lunar spacecraft and radio stars were used to set a surface pressure limit of 7×10^{-7} Pa. The discovery of specific atmospheric species resulted from observations made during the Apollo program.

The Apollo period

The first detection of a lunar atmosphere came during the Apollo program from instruments placed on the lunar surface. Contamination in the area of the Apollo surface detectors severely constrained the ability of landed instruments to make measurements during the lunar daytime. However, surface-based cold cathode gauges placed an upper bound of 1×10^{13} molecules m^{-3} for the surface number density of atoms in the daytime atmosphere and detected 2×10^{11} molecules m^{-3} at night. These results show that the lunar atmosphere is a collisionless planetary exosphere with its lower boundary directly at the lunar surface. Such an atmosphere is called a surface-boundary exosphere.

The Apollo surface instruments were able to identify two gases. During the lunar night and at sunrise and sunset, He and Ar were detected. The He is likely to be derived from the neutralization of solar wind alpha particles upon impact into the lunar surface; indeed, fluctuations in the measured. He concentration were strongly correlated with simultaneous fluctuations in the solar wind.

Both isotopes of Ar, ^{36}Ar and ^{40}Ar, were present at lunar sunrise and sunset, though not at night. Analysis of these data indicated a cold trapping of Ar on the lunar nightside, followed by a pre-sunrise increase presumably driven by a 'wind' emanating from the warming surface some kilometers east of the landing site. Typically, the ratio of ^{36}Ar : ^{40}Ar was about 1 : 10. Because the ratio of ^{36}Ar : ^{40}Ar is much greater than one in the solar wind, one can conclude that the contribution of the solar wind to the observed ^{40}Ar signal was small. The isotope ^{40}Ar is formed by the radiogenic decay of ^{40}K to ^{40}Ar, coupled with episodic release of ^{40}Ar from the lunar interior. Approximately 6% of the total lunar production of radioactive ^{40}K must be escaping the lunar interior into space.

An experiment onboard the Apollo Command Service Module detected alpha particles from the decay of ^{222}Rn and ^{210}Po. The radon isotope ^{222}Rn is a gas with a half-life of 3.8 days; the half-life of ^{210}Po is 21 years. Both species were positively detected. Rn was present throughout the lunar atmosphere but was concentrated over a small number of sites, most notably the Aristarchus/Marius Hills region, while the Po distribution also showed concentrations at mare boundaries.

Searches for other candidate species, including H, C, N, O Ne, N$_2$, CO, CO$_2$, CH$_4$ and NH$_3$ were undertaken using both surface instruments and sensitive spectrometers in orbit about the Moon, but only upper limits were obtained. The lunar ionosphere was also studied during Apollo. The principal detected species were in mass ranges consistent with Ne and Ar. The data indicate that lunar terminator ion number densities are in excess of 10^{11} m^{-3}, and that the terminator region exhibits a negative surface potential of 10–100 V, with a screening length near 1 km. The data indicate that ionospheric source and sink processes are controlled by the solar radiation and charged particle fluxes, as well as the position of the Moon in its orbit, particularly with respect to the Earth's magnetosheath.

Recent discoveries

For a period of some years following the last Apollo mission no new species were identified. However, in 1988 emissions due to both Na and K were detected above the bright limb of the Moon. These emissions are formed by resonant scattering of sunlight by the gaseous Na and K atoms in the lunar exosphere. The emissions due to these lines have now been observed as far as five lunar radii above the sunlit Moon, and a faint tail of Na has been reported. Thus the the outermost portions of the Na atmosphere look rather like the coma of a comet.

Relation of the lunar atmosphere to planetary science

The physical processes which produce the lunar atmosphere include sputtering by solar wind ions, vaporization of lunar surface materials following micrometeoroid impact, erosion of atoms from the surface by energetic solar photons, and outgassing from the surface. These processes all work elsewhere in the solar system, on surfaces as diverse as those of the planet Mercury, asteroids and the moons of the outer planets. However, only on the Moon do we have direct measures of the surface composition, and the fluxes of external agents. Thus the study of the lunar atmosphere is the study of a suite of processes which are present at any exposed surface in the solar system.

Limits of completeness

There is still much to learn about the atmosphere of the Moon. First, we do not know the full composition of the lunar atmosphere. The elements detected thus far and the models of source production rates indicate that other species, including both other metals (e.g. Mg, Al, Si, Fe and Ca) and molecular species such as CH$_4$ and CN, may also exist. In this regard, the Apollo-derived upper limit on the total gas density is orders of magnitude larger than the sum of all known constituents. Future studies of the lunar atmosphere will almost certainly detect new species and add to our understanding of the suite of processes which act on exposed surfaces in the solar system.

Thomas H. Morgan

Bibliography

Heiken, G., Vaniman, D. and French, B.M. (1991) *Lunar Sourcebook*. Cambridge: Cambridge University Press, pp. 736–83.

Mendillo, M., Baumgardner, J. and Flynn, B. (1991), Imaging observations of the extended sodium atmosphere of the Moon. *Icarus*, **99**, 115–9, 1992.

Morgan, T.H. and Shemansky, D.E. (1991) Limits to the lunar atmosphere. *J. Geophys. Res.*, **96**, 1351–67.

Potter, A.E. and Morgan, T.H. (1988) Discovery of sodium and potassium vapor in the atmosphere of the Moon. *Science*, **241**, 675–80.

Sprague, A.L., Kozlowski, R.W.H., Hunten, D.M. *et al.* (1992) The sodium and potassium atmosphere of the Moon and its iteraction with the surface. *Icarus*, **96**, 27–42.

Cross references

Apollo missions
Exosphere
Mercury: atmosphere

MOON: GEOLOGY

Even to the naked eye the Moon's face is not uniform: it is darker in some patches than others (Figure M82). Galileo, who was the first to

Figure M82 The lunar frontside as seen from the Earth. The light-colored, heavily-cratered regions are the ancient feldspathic highlands, and the darker smoother plains are low-lying areas covered with iron-rich lavas. Younger rayed craters such as Copernicus (left of center) and Tycho (bottom) are conspicuous. (Lick Observatory photograph.)

make telescopic observations of the Moon, distinguished the brighter areas as higher and more rugged, with mountains and valleys; the darker areas he distinguished as lower and smoother. These became known respectively, using Latin, the scientific language of the day, as *terrae* (lands) and *maria* (sing. *mare*) (seas); we now commonly refer to the lunar highlands rather than terrae, but maria persists. The ruggedness of the highlands records an ancient history of collisions that produced an abundance of overlapping craters at a variety of sizes. These collisions, or impacts, destroyed most of the landforms (such as lava flows) that might once have been present in the highlands. In contrast to the highlands, the mare plains are low-lying areas covered with lavas that spilled out at various times after the decline of impact cratering nearly 4 billion years ago. Most of the flows that compose Mare Imbrium were erupted several hundred million years after the formation of the Imbrium basin. The maria are not, nor ever were, water-filled basins or seas. Most of the lavas flowed out rapidly but gently from fissures in some ways similar to flood basalts on Earth (such as the Columbia River basalts of Washington State). They did not build up the conical edifices popularly associated with volcanoes on the Earth and Mars. The detailed recognition of different landforms, their time sequences and attempts to explain their origin, is the purview of lunar geology (Plates 12, 13).

Elements of lunar geology and its methodology

Geology in the space age is the scientific study of the origin, evolution, stratigraphy and structure of any planet or celestial object that has a solid surface. It focuses on observations of one kind or another made on rocks and landforms; and on scales ranging from smaller than a crystal to a planet wide view. These observations are combined to deduce a history of the outermost part of the planet in particular and, more inferentially, its interior.

The geology of the Moon is the description of its rock formations and physiography, and the attempt to comprehend the processes that produced them (Wilhelms, 1987). The geological record of the Moon is somewhat different from that of the Earth in three aspects. First, its initial chemical composition is different. Second, it is not today a dynamic evolving celestial object but a largely 'dead' one. Third its study has to be constrained by external observations and only a small number of sample analyses.

To the first of these points, although the Earth and Moon may have had a common beginning (see Moon: origin), and both are highly differentiated bodies (i.e. not uniform inside), their sizes and chemistry are different and so their internal evolutions diverged from the start. Hence the processes that produced lunar rocks and their landforms have a pattern different from terrestrial ones. Terrestrial geology has been strongly influenced by tectonics, including the horizontal movement of sections of the Earth's crust; the Moon currently has only minor evidence for structural movements having taken place, and then mainly in the vertical sense. Further, the lack of volatiles or any appreciable atmosphere on the Moon precluded the development of the water-lain and airborne sediments that form a ubiquitous cover to the Earth and constitute an abundant component of its crustal rock sequences. Instead, most lunar rocks are the product of igneous (magmatic) activity being either surface lavas or deeper-crystallized rocks, or the result of their redistribution, mixing and melting by asteroidal and meteoritic impacts (Figure M83). Both these processes are complex, and their relative importance has varied over lunar history. Lunar landforms reflect the balance of impact cratering (exogenic) and magmatic (endogenic) processes.

Second, the Moon is now to all intents and purposes a geologically dead body. There may be some activity in the deep interior, and there is a continued but sporadic drizzle of meteoritic dust and particles, with craters occasionally formed (the presence among meteorites collected in Antarctica of pieces thrown from the Moon demonstrates the geologically recent formation of some significant craters). But although some transient phenomena have been claimed, little significant geologic activity is taking place at the present time, so all processes have to be inferred from the landforms. In contrast, the Earth is an active planet, and the relationship between rock formations and the processes that produced them can be confidently deduced, not only in kind but often in detail and in intensity.

Third, the tools available for lunar study are different from those employed for Earth. Our sources of information about the Moon are dominated by photographs of the surface, with the most useful being those taken from lunar orbit or at the surface of the Moon itself. The Apollo landings (1969–1972; q.v.) allowed close-up human observation and the collection of samples for laboratory examination, but only for six locations; the scope and duration of the field work was strictly minimal and inflexible. Samples were also collected by three unmanned Soviet missions (1970–1975). In all cases the samples were not directly from specific geological formations, but instead were essentially grab-samples from the impact-mixed regolith (Figure M84). Some other orbital geochemical and surface geophysical and geochemical evidence is also available, as well as spectral reflectivity data recorded from Earth-based telescopes and from the Galileo spacecraft flyby. Geological maps of the Moon, such as those produced by the Astrogeology Branch of the US Geological Survey (in Flagstaff, Arizona) are mainly small scale (1 : 250 000 or smaller) (see plates 12 and 13). They depend mainly on variations of surface morphology, sequence relationships, and albedo (i.e. brightness) rather than on the detailed inspection of rock units, as is normally practiced on the Earth. There is much less availability (from orbit) of clues to the third dimension that direct close inspection allows. The photogeology mainly addresses the record that still has a preserved landform (however, for the Moon, some of this preservation goes back to at least 4 billion years ago). Most of the major geological differentiation took place before the surface morphology evolved. The major source of information about these older events is the collection of lunar rocks and regolith.

From Galileo to the space age

For three centuries the Moon remained dominantly an object limited to astronomical study. Data were collected about its shape, its size, its movements and its surface properties. The lack of an atmosphere and the absence of surface water were recognized; that it was

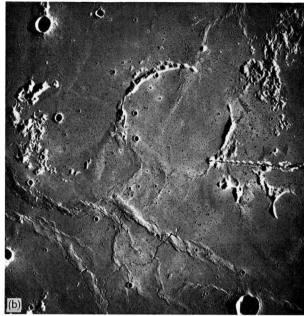

Figure M83 (a) Region of Apollo 15 landing site (*) in Palus Putredinis near Montes Apenninus (right). The heavily cratered highlands are embayed and superposed by the dark volcanic plains that include sinuous rilles. The mare plains are in turn penetrated by the younger craters Aristillus (topmost of pair, 55 km diameter) and Autolycus (39 km diameter). (Apollo 15 frame M-1537.) (b) Mare lavas in southern Oceanus Procellarum, flooding older highlands and craters, and showing complex mare ridges and rilles, as well as flow boundaries. Object to right is the Apollo 16 gamma-ray spectrometer boom. Remnant crater in center is about 60 km diameter. (Apollo 16 frame M-2837.)

Figure M84 Collection of a tube of regolith material in the ejecta material outside Van Serg crater on the Apollo 17 mission. The North Massif is in the background. (AS-17–143–21837.)

essentially an unchanging environment was also apparent. Geological aspects were barely considered by the early observers. Most writers of the last century assumed that the lunar craters were of volcanic origin. In 1893 the geologist G.K. Gilbert concluded on the basis of observations and experiments that the craters were unlike terrestrial volcanic craters in form. He inferred that they were instead impact produced, recognizing that the Imbrium basin was a giant impact scar (Gilbert, 1893). Few geologists pursued the idea. There was indeed little relevant information about the Moon, and impact craters on the Earth, which could have provided a stimulus for lunar studies, were not then recognized.

In the middle of the 20th century Baldwin made renewed observations and used a combination of natural and experimental analogs to make a strong case for the impact origin of most lunar craters and for the volcanic nature of the lowlands (Baldwin, 1949). (Nonetheless, some geologists disagreed with Baldwin's conclusions about craters, still preferring a volcanic interpretation.) In contrast, among others, Urey, who generally had a low opinion of the geological arguments, strongly and influentially advocated that the Moon not only was but always had been a cold body, incapable of significant volcanic activity (Urey, 1952). This pre-space age period was marked by observations of specific physiological features, and not by geological mapping.

The birth of lunar stratigraphy

The methods of geological mapping were first applied to the Moon in the early 1960s, pioneered by Shoemaker and Hackman (1962), but based on principles established by Gilbert (1893; see also Astrogeology). The essence was to establish the stratigraphy of rock units on the Moon, that is, study the spatial distribution, the time relationships and if possible infer the formative processes of morphological units. The information was telescope images of the lunar nearside, later greatly expanded by images from spacecraft, in particular the hugely successful Lunar Orbiter missions (q.v.) (1966–67). The fundamental method used was application of the law of superposition, which simply means that younger units overlie, cut or overlap older units (Figure M85). Using this method it became clear that older units had more craters than younger units, as expected on the assumption that craters were of impact origin: the longer a unit was exposed, the more impacts it would have suffered. This technique (of relative crater density as a method of assessing relative age of units that either did not touch or whose contact relationship was ambiguous) became standard, not only for the Moon but for Mars, Venus and Mercury (Figure M86). The new images, using stereoscopic pairs

Figure M85 Lobate mare basalt flows in Mare Imbrium, showing overlapping and superposition. The most prominent linear topographic features are wrinkle ridges (upper half of image). Older highland remnants cast shadows (center) and sparse younger craters disturb the flows. (Apollo 15 frame M-1556.)

Figure M86 Superposition relationships in the cratered nearside highlands east of crater Albategnius (right). Old degraded craters cut each other and are cut by radial Imbrium sculpture (grooving and elliptical craters trending approximately top to bottom). Planar fill such as in Albategnius appears to postdate the radial structures. Sharp-rimmed circular craters postdate all other events. (Apollo 16 frame M-392.)

and topographic information, also provided a wealth of detail on the landforms, such as flow fronts on lava flows, small volcanic domes, downslope movements, tiny craters and pits, crater walls and ejecta field morphology. The immense increase in research activity created by the availability of spacecraft images and by the stimulus of the anticipated Apollo exploration project resulted in the establishment of both a fairly detailed stratigraphic framework for lunar geology and reliable inferences about the magmatic and impact processes that had sculpted the landforms. The addition of samples and geophysical evidence obtained on the Apollo and Luna missions allowed confirmation, refutation, revision and greater insight into the nature of the processes that produced the lunar landforms. In particular the samples immediately confirmed the volcanic nature of the lowland plains, and the abundance of impact-produced features, particularly in highlands rocks. The samples also permitted a calibration of the relative stratigraphy against an absolute timescale from the application of radiometric dating techniques (Wilhelms, 1987).

Lunar stratigraphic column

The main divisions (systems) of the lunar stratigraphic column (Figure M87) consist of collections of units, and this framework has been modified only slightly since it was codified around 1963. The systems are intended to be time-stratigraphic, i.e. a system boundary is the same absolute age everywhere. That boundaries are defined does not imply that fundamental changes in geological processes took place at those times. Nonetheless the boundaries are reasoned, convenient markers that subdivide lunar history. The older boundaries are defined by the deposition of ejecta from specific impact events. The bases of the Eratosthenian and the Copernican are less precisely defined; although they depend on quantitative expressions of the relative degrees of crater degradation, the criteria are not entirely unambiguous in practical application.

The Copernican System is the youngest, and is named after the type crater Copernicus (which falls within the System, and is not a boundary for it). A crater formed today would be called Copernican. Only a small proportion of the Moon's face is Copernican. It is characterized by the presence of craters that retain prominent bright rays (Figure M88a), even when they are as small as about 5 km diameter. The rays are visible because there has not been time for them to have degraded. The landscape effects of the craters can be far reaching: the prominent crater Tycho in the southern highlands has rays that stretch almost around the Moon, and they appear to be responsible for a landslide and for a cluster of craters in the neighbourhood of the Apollo 17 landing site. The working definition for the base of the Copernican is actually founded on the degree of crater degradation, rather than ray degradation, and the base is not marked by a definable stratigraphic horizon. Copernican craters are scattered over the Moon, but there are only half as many in a given size range as in the preceding Eratosthenian era (44 larger than 30 km, compared with 88 in the Eratosthenian). Thus most workers would conclude that the base of the Copernican is about 1 to 1.5 billion years old, but is possibly older, perhaps as old as 2.0 billion years (Ryder, Bogard and Garrison, 1991). Very little except sporadic cratering took place in the Copernican, but some patches of mare basalt overlap some rayed craters and are thus of Copernican age. The crater Copernicus itself, on the basis of combined Apollo sample and geological considerations, is believed to be about 1 billion years old.

The Eratosthenian System is characterized by post-mare craters whose rays, even when large sets, have been lost over the course of time. This is the case with the type crater Eratosthenes (Figure M88b); rays from Copernicus cross Eratosthenes ejecta. The working definition of the base is made on crater degradation, as is that of the Copernican System. The Eratosthenian contains appreciable mare flows, mostly in the region of Oceanus Procellarum. Lava flows are about twice as extensive as crater deposits. The lava plains sampled on the Apollo 12 mission are in the Eratosthenian, but the older ones sampled on the Apollo 15 mission are not, hence its base is bracketed between 3.3 and 3.1 billion years old. Crater frequencies on Eratosthenian surfaces are appreciably greater than those of the Copernican.

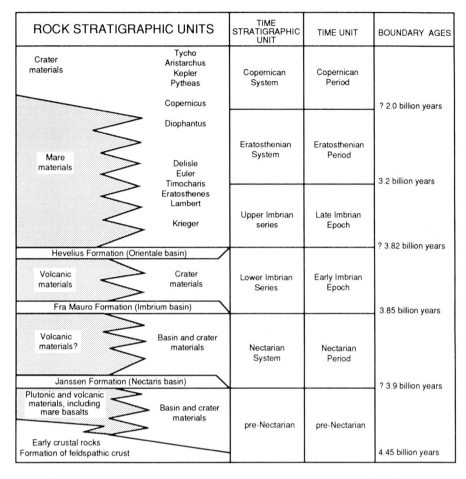

Figure M87 Lunar stratigraphic column illustrating relations among three classes of stratigraphic units, with the youngest at the top. Subdivisions do not even approximate equal time spans. Three named basin materials (Orientale, Imbrium and Nectaris) are the defining rock-stratigraphic units, each formed within a geologically brief period of time, of the oldest time-stratigraphic units. (Adapted from Wilhelms, 1987.)

The Imbrian System constitutes the rocks from the base of the deposits of the Imbrium basin until the start of the Eratosthenian. It is divided into an Upper Imbrian Series and a Lower Imbrian Series by the top of the deposits of the Orientale basin. The Upper Imbrian Series contains the most extensive mare plains on the Moon (two-thirds of the total mare), including all the older lava plains sampled on the Apollo and Luna missions. These lavas have a wide variety of compositions. The Upper Imbrian Series includes subordinate crater materials, but no basins at all (Orientale was the last). The mare units can be used to define the stratigraphy in more detail; the older units are the least exposed. These Upper Imbrian units were the most accessible for sample collections, and absolute ages based on actual samples are available.

The basin deposits that define both the base and top of the Lower Imbrian Series constitute the most extensive laterally continuous stratigraphic horizons on the Moon, and dominate a large part of the surface visible from the Earth (Figure M89). These horizons consist of ejecta from the basins but, especially towards their exteriors, local materials are mixed in. Thus the Fra Mauro Formation, the radially textured terrain sampled on the Apollo 14 mission, is a mixture (in uncertain proportions) of Imbrium ejecta and local materials added to it by secondary cratering. In between the basin deposits are the deposits of other smaller craters, and some light plains materials. None of the units identified as Lower Imbrian Series are mare plains, but some dark materials have been excavated by craters and may indicate buried mare materials. The light plains have varied ages and origins; most are probably related to the excavation of basins. The light plains sampled at the Apollo 16 site were broken, impact deposited materials, not volcanic lavas. However, at least the Apennine Bench Formation, near the Apollo 15 landing site (Figure M89), appears to be volcanic. Its lavas are unlike those of the mare plains, being rather more aluminous and hence paler in color.

The age of the Imbrium basin, and thus the base of the Lower Imbrian Series, is constrained by the age of the overlying Apennine Bench Formation and by materials within (hence older than) the Imbrium basin ring. It is believed to be close to 3.85 billion years old, according to sample analyses (Wilhelms, 1987; Dalrymple and Ryder, 1991). The Upper Imbrian Series represents in time most of the Imbrian System, inasmuch as some Apollo 11 and 17 basalts are almost as old as the Imbrium basin. Thus the Orientale basin is almost indistinguishable in age from Imbrium, and the Lower Imbrian Series probably occupies only a short period of time, perhaps less than 50 million years.

The Nectarian System is defined as those pre-Imbrium deposits that lie above the base of materials of the Nectaris basin. It includes the Nectaris, Serenitatis and Crisium basin deposits, plains materials and crater materials. It occupies tracts on the east and farside of the Moon, where there are no Imbrium and Orientale basin deposits. There is no direct evidence for volcanic deposits, although such have been proposed. Sample evidence from the Apollo 16 landing site, where Nectaris ejecta are possible, suggests that the Nectaris basin is about 3.9 billion years old; if so, the Nectarian, despite having endured an extensive impact history, did not last very long, perhaps only 50 million years.

The pre-Nectarian system (not a formally named system) consists of all landforms older than the Nectaris basin. It consists mainly of craters and their deposits, with most of it on the farside of the Moon. It has not been directly sampled, although undoubtedly remnants of pre-Nectarian rocks occur as fragments in later units; certainly small fragments of igneous plutonic and volcanic rocks and minerals have

Figure M88 (a) Copernicus, 93 km diameter crater, with rays and subconcentric chains of secondary craters. Rays overlie mare flows to upper right. (Lunar Orbiter 4 frame H-121.) (b) Eratosthenes, 58 km diameter crater, lacking prominent rays. Eratosthenian crater materials are partly overlain by mare flows. (Lunar Orbiter 4 frame H-114.)

been found and dated. Some are as old as 4.5 billion years. The age of craters in the pre-Nectarian cannot be directly calibrated, so the age of preserved pre-Nectarian landforms cannot be stated with any confidence. Igneous rocks are represented in samples older than 3.9 billion years, but undoubted impact melt rocks are not. In the earliest pre-Nectarian, soon after the birth of the Moon, the Moon was melted, cooled and differentiated to produce a crust and complementary mantle (Taylor, 1982). Fragments of crustal rocks from even this primordial era have been preserved.

Lunar geological processes

The presently visible landforms of the Moon were formed almost entirely from the extrusion of lavas from the interior and the impact of meteoritic objects from the exterior. Evidence for early plutonic igneous activity is preserved in rocks. Minor tectonic activity, such as small-scale faulting and downslope movement, is apparent for younger epochs. It results from interior action such as continued contraction and moonquakes during cooling, and from shock waves produced by meteorite impacts.

Extrusion of lavas

The smooth dark plains are the most Earth-like of lunar features, and were readily interpreted as lava flows prior to the space age. Their approximately level surfaces were taken as signs of fluid lava emplacement even in the 19th century, by G.K. Gilbert. Apollo sample collections confirmed that the dark plains did indeed consist of basaltic lava, although these lavas differed from terrestrial ones in being much richer in iron.

The dark plains cover about 30% of the nearside and 2% of the farside. They tend to flood large basins and depressions. Various morphological features, such as flow fronts, as well as telescopic properties such as color, allow the delineation of distinct units, interpreted as individual lava flow lobes. Mare Imbrium, for instance, consists of numerous individual flows of different compositions; it was not a single pool. The Imbrium lobes range from 10 to 35 m in height and extend for as much as 1200 km, indicating that each lobe erupted rapidly from a single location (probably fissures as well as some smaller vents) and flowed easily. The rock samples collected indeed show that lunar lavas, with their high iron content, would have been extremely fluid at magmatic temperatures. The rock samples also show that mare flows differ considerably in composition from each other. Not all mare flows are thick and extensive, and some may have ponded without traveling any great distance.

The smoothness of the dark plains is partly lost at higher resolution. Volcanic features such as domes and small cones (where lava partly accumulated *in situ* instead of running away), endogenic craters and sinuous rilles are present, if generally minor. The rilles are lava channels, or in some cases collapsed lava tubes. The lava plains are also interupted by tectonic features such as wrinkle ridges and grabens; such were probably induced some time after crystallization of the lava as the basin subsided under loading with basalt.

A distinct type of mare deposit is a dark mantling material, which partly covers both rugged highland terrain and mare surfaces. These layers are interpreted as deposits of volcanic glass produced in lava fountaining from vents. Such glasses have been identified at several Apollo sites, notably the famous orange glass at the Apollo 17 site and the green glass at the Apollo 15 site.

Detailed work on the Apollo samples has shown that the mare lavas were produced at depths up to 400 km by melting of a mantle that had already undergone a differentiation event. This mantle production took place at about 4.4 to 4.3 billion years ago; the lavas observed in the plains were extruded from about 3.8 Ga ago to perhaps as late as 2.0 Ga ago. However, a few samples show that

Figure M89 The Fra Mauro Formation, part of the Imbrium basin marker horizon that defines the base of the Lower Imbrian Series, forms the furrowed and ridged terrain that blankets and degrades older highlands mainly towards the left of the image. It represents a mixture of Imbrium ejecta and reworked local material with secondary cratering. The Apollo 14 mission (*) landed on the Fra Mauro Formation. Large crater in center is Fra Mauro, about 95 km diameter. (Apollo 16 frame M-1419.)

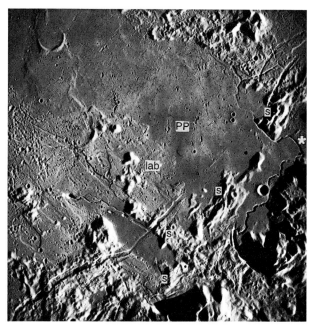

Figure M90 Region of the Apollo 15 landing site (*). The dark mare basalt plains overlap the Apennine Bench Formation (e.g. A) which is a post-Imbrium light plains unit that probably consists of aluminous volcanic rocks. Distance from the Apollo 15 landing site to A is about 150 km. (Apollo 15 frame M-416.)

mare lavas were extruded at even 4.3 Ga; any plains they may have formed have presumably been destroyed by impacting.

Impact processes

The lunar highlands are characterized by numerous overlapping impact craters, ranging from giant multiring basins down to microscopic sizes. Small craters form even today, and the Moon is still subject to a continuous rain of micrometeorites that contribute to the production of its regolith (q.v.). The lunar rock samples testify to this impacting, in that many of them are brecciated, mixed and melted materials. Indeed, there are virtually no landforms in the pre-Imbrian units that can be readily ascribed to processes other than impacts, and even in the Lower Imbrian, only the Apennine Bench unit (Figure M90) can at present be confidently ascribed to volcanism. Following the Apollo 16 mission and the recognition that the Cayley Plains consist of breccias, not volcanic rocks, most light plains and other units that were once ascribed to volcanic processes have now been reinterpreted as impact-related features.

Small craters (up to about 15 km diameter) are simple smooth bowl-shaped features, but larger ones have central peaks, flat floors and terraced walls. At even larger sizes the central peak is replaced by a ring, and basins larger than about 200 km diameter are multiringed features. Craters represent a combination of the immediate explosive excavation of the impact crater, the blanketing of the interior and surrounding area by ejected material, and subsequent modification (e.g. by rebound and slumping) that produces peaks and rings. Part of the target site is melted; in tiny craters this produces a glass lining, but in larger ones it produces ejected glassy material and sheets or pools of melt within the crater, and perhaps spilling out of it. The basement rocks beneath the excavated material are shattered and partly injected by melt and brecciated material.

The obvious geological effect of impacting is the production of a hole, associated with the deposition of varied ejecta units. Styles of ejecta deposition vary with size of the crater and distance from the rim. Ejecta consist of a mixture of fine-grained to coarse-grained material derived from the target area, and may include large blocks. An effect perhaps not quite so obvious is that of mixing of target materials in the ejecta blankets and melts. The ejected material is thus mixed to some extent with the materials in the surface it lands on to some extent. The Fra Mauro Formation, a prominent unit furrowed radially to the Imbrium basin, is an example of a unit that probably consists of both primary ejecta and local material.

Secondary cratering, that is, the cratering produced by the ejected material, adds to the mixing effect. Secondary cratering is particularly prominent surrounding multiringed basins, and its effect has to be taken into account when crater density measurements are being made to determine relative ages of units. The features of craters produced by slow-moving ejecta can be distinguished at least in part from those produced by primary fast-moving projectiles.

Over a period of time, continued impacting degrades the fresh form of a crater until eventually the crater or basin may be barely visible. Small craters are degraded faster than larger ones. Large young craters such as Tycho (about 110 Ma old) are extremely fresh looking, with prominent rays still visible.

Lunar regolith

The lunar surface is covered to the depth of a few meters by a layer of fine material produced by a rain of micrometeorites. Virtually all of our observations have actually been made on the regolith, not on exposed bedrock, and all the Apollo and Luna samples were collected from the regolith. The regolith consists of comminuted mineral, rock and glass fragments, and agglomerated mixtures of melted glass and mineral fragments (Taylor, 1982). Its average grain size is only a few tens of microns, although it contains rock fragments that are much larger. Many of the fragments show signs of shock deformation.

The regolith develops from fresh rock over a period of several hundred million years, and continually matures. It incorporates the remains of the micrometeorites, which constitute (in cryptic form, as micrometeorites vaporize on impact) about 2% of typical regoliths. It also is subjected to the solar wind and cosmic rays, hence the regolith builds up solar hydrogen, helium, other wind constituents and particulates, and displays radiation damage. The regolith might conceivably be mined one day for its accumulations of hydrogen and other solar wind constituents.

Future exploration and utilization of the Moon

Apollo and its robotic predecessors only started the exploration of the Moon, although the returned samples in particular remain a source of new information. While observations have continued from Earth-based spectroscopy and the Galileo spacecraft swingby, the next logical step is orbital observation to establish global chemistry, mineralogy, imaging and gravity. Such a robotic mission has been proposed for scientific purposes since Apollo ended.

It seems highly likely that sometime in the 21st century humans will return to the Moon and establish an outpost and then more permanently populated bases (Mendell, 1985). The purposes will be the further exploration of the Moon and expansion of human activities in the solar system, using resources available on the Moon. Precursor robotic missions, not only orbiters but also landers, will probably pave the way, providing scientific, engineering and resource data.

We have developed an outline of the origin of the Moon and its evolution far in advance of pre-Apollo ideas (Hartmann, Phillips and Taylor, 1986; Warren, 1985), but many questions remain unanswered. We do not know the bulk composition very well, nor much about the Moon's internal structure, even whether or not it has a small core. We need much greater understanding of its regional variation and its chronology. The Moon is a natural laboratory for studying impact and volcanic processes. Orbital and robotic landed missions, including instrumented rovers and even sample return, can provide much information. However, for some types of investigations the *in situ* cognitive and manipulative abilities of humans are ultimately needed, particularly those encompassing detailed geological fieldwork and the installation of complex instrumentation. The Moon forms an ideal stable platform for many types of astronomy, and humans would probably be necessary for installing and tending some of the instruments essential for such studies. Clearly global access, rather than constraint to a single site, is of great importance.

The Moon has resources, mainly in the regolith, that would be useful for activities both on the Moon itself and elsewhere in space. Thus mining and processing of lunar materials is a potential endeavour in the 21st century. Oxygen, available by decomposition from oxides or even silicate minerals, is a particularly important resource, especially as an oxidizer for propellants. Hydrogen is present in the regolith where it has been implanted from the solar wind, and is usable in rocket fuel, for water and for use in oxygen extraction schemes. Helium, also in the regolith from the solar wind, has potential in that ^{3}He has been proposed for the development of nuclear fusion energy; it is the only lunar resource proposed for potential use on the Earth itself, rather than for space activities. Iron and aluminum and other metals are available in rocks and soils for structural uses, although processing schemes will be unlike those on Earth where metals are rarely extracted from silicates, being available in forms easier to refine. While the raw regolith itself is a resource for construction and for shielding structures from radiation, much development work will be required to bring mining and processing schemes to fruition.

Graham Ryder

Bibliography

Baldwin, R.B. (1949) *The Face of the Moon*. Chicago: University of Chicago Press.
Dalrymple, G.B. and Ryder, G. (1991) ^{40}Ar/^{39}Ar ages of six Apollo 15 impact melt rocks by laser step heating. *Geophys. Res. Lett.*, **18**, 1163–6.
Gilbert, G.K. (1893) The Moon's face, a study of the origin of its features. *Phil. Soc. Washington Bull.* **12**, 241–92.
Mendell, W.W. (Éd.) (1985) *Lunar Bases and Space Activities of the 21st Century*. Houston: Lunar and Planetary Institute.
Ryder, G., Bogard, D. and Garrison, D. (1991) Probable age of Autolycus and calibration of lunar stratigraphy. *Geology*, **19**, 143–6.
Shoemaker, E.E. and Hackman, R.J. (1962) Stratigraphic basis for a lunar time scale, in *The Moon* (eds Z. Kopal, Zdenek, and Mikhailov). London: Academic Press, pp. 289–300.
Taylor, S.R. (1982) *Planetary Science: A Lunar Perspective*. Houston: Lunar and Planetary Institute.
Urey, H.C. (1952) *The Planets*. New Haven: Yale University Press.

Warren, P.H. (1985) The magma ocean concept and lunar evolution, in *Ann. Rev. Earth Planet. Sci.* (eds G.W. Wetherill, A.L. Albee, and F.G. Stehli). Palo Alto: Annual Reviews Inc., pp. 201–40.
Wilhelms, D.E. (1987) *The Geologic History of the Moon*. United States Geological Survey Professional Paper **1348**.

Cross references

Breccia
Apollo missions
Earth–Moon system: origin
Flood basalt
Impact cratering
Lunar meteorites
Mascon
Regolith

MOON: GRAVITY

The gravitational attraction that a planetary body has is directly proportional to its mass. The Moon has a relatively small mass as compared to the Earth (i.e. 81.30 lunar masses = 1 Earth mass). However, the gravitational attraction at the surface of the body is inversely proportional to the square of its radius. Since the Earth's radius is 3.67 times larger than the lunar radius, a factor of 13.47 (i.e. 3.67^2) is gained by the Moon. This makes the gravitational effect at the lunar surface ⅙ (i.e 13.47/81.30) of what it is on Earth, or 162 cm s^{-2}.

It is the variations about this mean gravitational value that interest the navigation engineer and the geophysicist, for the gravitational irregularity will greatly influence the lifetime of an orbiting spacecraft and the internal structure models that can be inferred. These variations are commonly called the lunar gravitational field. The contour lines in Figure M91 show how the gravity changes over different areas. The units of the contours are in milligals (1000 milligals = 1 cm s^{-2}). These changes are relatively small compared to the mean value, however they are large compared to variations on the Earth. Note the large positive values associated with the ringed basins; a most remarkable result, for one would expect negative values (see Mascon). The irregular maria basins do have negative values.

These results were obtained from Earth-based Doppler radio tracking of orbiting satellites (Lunar Orbiters 1–5 and Apollo modules and their two subsatellites). However, since the Moon always has its same side (near or visible side) toward the Earth, the spacecraft is not visible when it is on the farside. Since the spacecraft is not visible, the Earth-based radio communication link is lost and therefore no Doppler tracking data are obtained. As yet, no direct gravity observations have been made for the farside (i.e. longitudes −90 →180 →90). The results shown in Figure M89 are predictions made by the model using data after occultations. The farside field is very uncertain and results obtained by various investigators reveal very different results (see references). The nearside results are very consistent, however they do deteriorate in the higher latitudes (i.e. > 45° latitude).

The Moon's gravity field is similar to that of Mars (rather rough) and very unlike that of the Earth and Venus. This is primarily due to its size and thus its inability to retain internal heat. The fact that there is no lunar atmosphere also aids in its heat loss, making it a more rigid body.

William L. Sjogren

Bibliography

Konopliv, A.S., Sjogren, W.L., Wimberly, R.N., *et al.* (1993) A high resolution lunar gravity field and predicted orbit behavior, in AAS/AIAA Astrodynamics Specialist Conference, AAS 93–622, Victoria, BC, Canada. (AAS Publ. Office, PO Box 28130, San Diego, CA 92198.)
Bills, B.G. and Ferrari A.J. (1980) A Harmonic analysis of lunar gravity, *J.Geophys. Res.*, **85**, 1013–25.
Blackshear, W.T., Daniels, E.F. and Anderson, S.G. (1971) Lunar gravitational field: a thirteenth degree and order spherical harmonic estimate, *NASA, TMX-2260*, Langley Research Center.

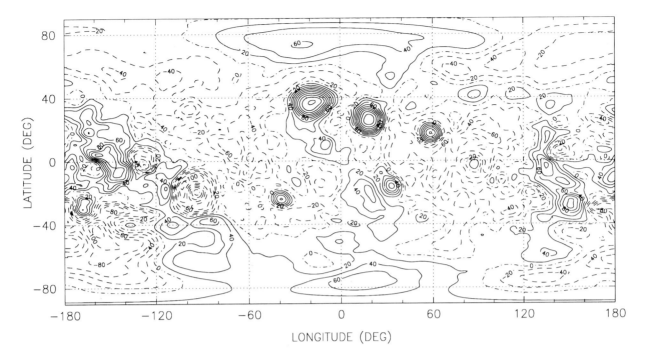

Figure M91 Moon Gravity at 100 km altitude in milligals (courtesy of A.S. Konopliv.)

Lorell, J. (1970) Lunar Orbiter gravity analysis. *Moon*, **1**, 190–231.
Liu, A.S. and Laing, P. (1971) Lunar gravity analysis from long-term effects, *Science*, **173**, 1017–20.
Muller, P. and Sjogren, W.L. (1968)Mascons: lunar mass concentrations, *Science*, **161**, 680.

Cross references

Clementine mission
Gravimetry
Gravity fields of the terrestrial planets
Mars: gravity
Mascon
Venus: gravity

MOON: MAGNETISM AND INTERIOR

The internal structure and magnetic field of the Moon are two basic properties of our nearest planetary body that may or may not be related. For the Earth, these properties are closely related because the geomagnetic field is generated by fluid motions in a metallic core. On the Moon the observed magnetic fields are due entirely to crustal magnetization; at present there is no core dynamo magnetic field and it is uncertain whether there ever was one. Some evidence suggests that the observed magnetization of lunar surface rocks could have been produced by a former core dynamo. However other evidence, as well as the small inferred size of the lunar core, suggest that most or all of the observed crustal magnetization can be explained by transient magnetic fields generated in large-scale meteoroid impacts on the lunar surface. In the following, the general constraints on the internal structure of the Moon are first briefly summarized and the implications of these constraints for lunar origin models are noted. Then the observed nature of the crustal magnetism is discussed, with emphasis on aspects that may have general implications for other airless silicate bodies in the inner solar system.

Internal structure

Crust and mantle

Analyses of returned samples combined with seismic data from the four station Apollo network showed that the Moon has a differentiated aluminous crust with a thickness of 60 ± 5 km beneath one of the Apollo mare sites. Inferences based on orbital gravity and topography data suggest that the crust is substantially thicker beneath the highlands. Since maria predominate on the nearside and highlands predominate on the farside, this would imply a larger mean crustal thickness on the farside than on the nearside (Figure M92).

The existence of such a thick aluminous crust implies that the Moon must have been melted to a depth of at least 500 km shortly after its formation (Taylor, 1982). Electromagnetic sounding of the lunar interior using surface and orbital magnetometers showed that the electrical conductivity increases rapidly with increasing depth in the upper mantle consistent with the rapid temperature increase predicted by thermal history models. The two successful Apollo surface heat flow experiments yielded somewhat high estimates of 21 and 16 m W m^{-2} (Langseth, Keihm and Peters, 1976). Initial interpretations required an excessive lunar abundance of heat-producing radioactive elements. However, later analyses point out that the two heat flow sites were near the edges of maria where heat flow is likely to be anomalously large relative to the global mean value. Inferred seismic velocities decrease with increasing depth in the upper mantle, again consistent with rapidly increasing temperatures. Although seismic velocity inferences at greater depths are less certain, available data indicate an increase in velocity below 500 km depth that could imply a composition or phase change (e.g. from spinel to garnet for aluminous silicates). The outermost 800 km of the Moon is characterized by very low seismic wave attenuation indicating a rigid, non-convecting lithosphere. At greater depths, wave attenuation increases rapidly and S waves were not observed to propagate suggesting a less rigid asthenosphere. Thus, the lunar interior may be considered as a stretched version of the outer 150 km of the Earth (Nakamura, Latham and Dorman, 1982). The zone where seismic wave velocities increase is usually referred to as the 'middle mantle' while the zone at depths below about 800 km where seismic waves are strongly attenuated is termed the 'lower mantle' (Figure M92).

Core

Evidence relating to the existence of a lunar metallic core is not yet definitive but several constraints point to the existence of a core with a radius of about 400 km (for a review see Hood, 1986). First, the lunar moment-of-inertia factor of 0.391 ± 0.002 derived from lunar laser ranging and orbital gravity data can be combined with observa-

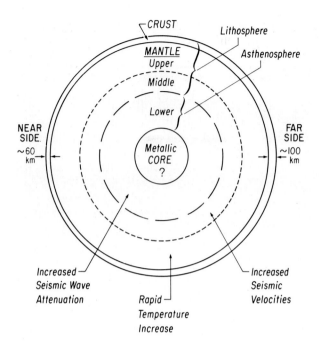

Figure M92 Schematic diagram of the lunar interior based on seismic, electromagnetic sounding and gravity/topography data (modified from Nakamura, Latham and Dorman, 1982). The crustal thickness is laterally variable but is believed to be less beneath the maria (dominating on the nearside) and greater beneath the highlands (dominating on the farside). The existence of a metallic core with radius of about 400 km is unconfirmed but is indicated by several lines of evidence.

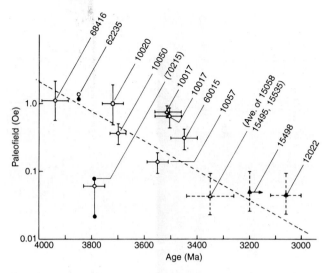

Figure M93 Lunar magnetic field intensities determined from laboratory experiments on Apollo samples. (From Runcorn, 1983.)

tionally constrained crust and mantle density profiles to show that a small dense core with radius > 300 km is probably necessary (Hood and Jones, 1987; Mueller, Taylor and Phillips, 1988). Second, electromagnetic sounding data provided an upper limit of 450–500 km for the radius of a highly electrically conducting core (Hood, Herbert and Sonett, 1982) while estimates of the lunar induced magnetic moment in the geomagnetic tail suggested a metallic core with radius of about 400 km (Russell, Coleman and Goldstein, 1981). Some analyses of lunar seismic data have also indicated the presence of a small, low-velocity core that would be consistent with a metallic iron composition. The relatively small size of the lunar core (consistent with its low mean density of 3.34 g cm^{-3}) suggests that the Moon may have experienced an exceptional origin that would not be predicted by solar nebula or coaccretion models. For example, it has been proposed that the Moon may have formed from material ejected during a giant impact on the Earth during the late stages of accretion. If this material came primarily from the impactor and the Earth's upper mantle, a Moon depleted in metallic iron would have resulted. The energy of the impact would also have been more than sufficient to account for melting of the lunar interior, which is needed to explain the thick aluminous crust (see also Earth–Moon system: origin; Moon: origin).

Magnetism

General properties

An unexpected outcome of Apollo manned and unmanned lunar exploration was the discovery of a pervasive magnetization of lunar surface materials with scale sizes up to about 100 km (Dyal, Parkin and Daily, 1974; Fuller and Cisowski, 1987). Like meteorites, the main ferromagnetic carriers in lunar samples are reduced Fe–Ni alloy grains whose magnetic properties are not fully understood. (This contrasts with the terrestrial case in which iron oxides such as magnetite are the ferromagnetic carriers.) These grains originated primarily by reduction of pre-existing iron silicates by shock and heat during impacts on the lunar surface (e.g. Strangway et al., 1973). Although the ferromagnetic properties of these grains are only partially known, available laboratory analyses of returned samples have been interpreted as implying that at least some lunar basalts acquired thermal remanent magnetization (requiring slow cooling in the presence of a steady magnetic field). There is also some evidence that the lunar paleofield amplitude may have varied with time (Figure M93). The highest lunar paleofields (about 0.1 to 1 gauss or 10^{-5} to 10^{-6} T) have been inferred by some analysts to have existed between about 3.8 and 3.6 Ga ago, suggesting a temporary core dynamo (Cisowski et al., 1983). Assuming that a former core dynamo existed, there have even been attempts to infer former changes in the position of the lunar rotational pole based on inferred directions of magnetization of the sources of lunar orbital magnetic anomalies (for a review see Runcorn, 1987).

However, a number of other observations have indicated an important role for impact processes in producing lunar crustal magnetization. For example, medium-amplitude orbital magnetic anomalies detected with the Apollo 16 subsatellite magnetometer were found to correlate with surface exposures of primary and secondary impact ejecta such as the Fra Mauro Formation peripheral to the Imbrium basin. Also, an Apollo 17 impact glass sample dated as < 200 Ma old yielded an estimated paleofield amplitude of 0.025 G or 2500 nT. (For comparison, the largest measured surface field was 327 nT near the Apollo 16 landing site.) The presence of such a large field in the relatively recent past suggests that a transient magnetic field event may have been produced by the impact that formed the glass sample. Finally, the largest concentrations of orbital magnetic anomalies detected by reflection of energetic electrons (and corroborated in many cases with direct magnetometer measurements) are located nearly antipodal (i.e. diametrically opposite) to four of the youngest lunar basins: Imbrium, Orientale, Serenitatis and Crisium (Lin, Anderson and Hood, 1988).

Impact field generation

The property of meteoroid impacts that leads to transient magnetic field generation is the production (at vertical velocities greater than about 10 km s^{-1}) of a cloud of partially ionized silicate vapor in addition to solid and molten ejecta. This impact plasma cloud can carry electrical currents in its interior (driven, for example, by thermal pressure gradients) and can interact with ambient plasmas and magnetic fields, leading to the generation of transient external magnetic fields (Srnka, 1977; Hood and Vickery, 1984). Laboratory-scale experiments have recently confirmed transient magnetic field generation both within and external to the impact plasma cloud (Crawford and Schultz, 1991). Transient field generation in meteoroid impacts combined with rapid resetting of magnetization in metallic iron remanence carriers by processes such as impact shock would provide a potentially efficient magnetization mechanism. As noted above, the largest lunar magnetization concentrations appear

Figure M94 Correlation of a relatively strong (22 nT) magnetic anomaly detected with the Apollo 16 subsatellite at a mean altitude of about 25 km with Reiner Gamma, an unusual, curvilinear albedo marking on western Oceanus Procellarum (section of Lunar Orbiter 4 frame 157). Model calculations suggest that surface fields of several thousand nanotesla (several hundredths of a gauss) are likely. For scale, the dimensions of the main albedo marking are approximately 30 by 60 km.

to occur antipodal to young basins. This observation has led to the construction of a model in which plasma expanding outward from a major impact converges at the antipode, compressing and amplifying a pre-existing (e.g. solar wind) magnetic field (Hood and Huang, 1991). At nearly the same time, compressional seismic waves from the impact converge in the antipodal zone producing shock pressures sufficient to allow magnetization acquisition, according to earlier laboratory experiments. If this model proves to be correct, large-scale impacts can potentially explain most of the magnetization mapped from lunar orbit by the Apollo subsatellites.

Lunar swirls

Even if impact processes and/or a possible former core dynamo ultimately account for most of the observed magnetization of the lunar crust, unresolved issues will remain. In particular, the strongest individual magnetic anomalies detected with the Apollo subsatellite magnetometers were found to correlate in locations, with peculiar swirl-like albedo markings whose origin is still debated. Numerous such albedo markings are found in basin antipode zones on the lunar farside (e.g. Hood and Williams, 1989). The strongest nearside anomaly was found to correlate with the Reiner Gamma swirl marking on western Oceanus Procellarum (Figure M94).

One model for the origin of the swirls suggests that they are surficial residues from one or more recent cometary impacts (Schultz and Srnka, 1980). In this model the associated magnetic anomalies were produced by transient magnetic fields generated during compression of the cometary ionized envelope against the Moon. However, ground-based spectral studies have not found evidence for unusual surface compositions at Reiner Gamma, as would be expected for a recent cometary impact site. Also, the tendency for swirls and strong magnetic anomalies to occur in basin antipode zones on the farside argues that they are not the result of randomly directed cometary impacts. The main alternate model for the origin of the swirls supposes that the albedo markings are a secondary consequence of the presence of very strong crustal magnetic fields (Hood and Schubert, 1980). In this model the magnetic fields deflect incident solar wind ions so that areas beneath strong magnetic field anomalies are not saturated with solar wind-implanted hydrogen. Solar wind-implanted hydrogen in turn is believed to contribute to the darkening with time (or 'optical maturation') of freshly exposed lunar surface materials (e.g. crater rays). This occurs via the production of microscopic metallic iron in glassy aggregate particles known as agglutinates that are formed during micrometeoroid bombardment of the lunar regolith (McKay et al., 1991). The generation of metal in agglutinates is in part a function of the amount of reducing solar wind hydrogen that is present in the uppermost part of the regolith. Thus one scenario for the formation of Reiner Gamma is as follows. First, a subsurface magnetization enhancement was produced by some process such as impact shock in the presence of a transient magnetic field. Second, volcanic lava flows covered the area with a thin veneer of mare basalt. One or more impacts in the adjacent highlands then produced secondary crater ejecta on the mare surface. Finally, the darkening with time of these freshly exposed ejecta occurred preferentially in areas unshielded by strong magnetic fields, leaving behind bright albedo markings with swirl-like shapes. The solar wind deflection model provides a natural explanation for the curvilinear shapes of the swirls, as has been shown using numerical simulations of solar wind ion trajectories in model magnetic fields.

Conclusions

Much remains to be done in understanding both the lunar interior and the origin(s) of lunar paleomagnetism. Additional lunar seismic data to complement that from the limited Apollo network combined with further electromagnetic sounding data should eventually confirm the existence and size of a lunar metallic core. The mass of such a core would in turn represent an important constraint on specific lunar origin models such as the giant impact model. During the last 10 years, appreciation has increased for the potential importance of impact processes in producing the pervasive but variable magnetization of the lunar crust. In part this is due to increasingly detailed experiments that have documented the production of impact plasmas and transient magnetic fields at the laboratory scale. It is also partly due also to the realization that the largest concentrations of orbital anomalies occur antipodal to the youngest large basins, leading to a physical model for the formation of these anomaly concentrations (sometimes called 'magcons'). Because impact cratering is a dominant geologic process in the solar system, these lunar results could have broad implications for the origin and nature of paleomagnetism on other airless silicate bodies. Among many unresolved issues is the origin of the lunar swirls, curvilinear albedo markings that correlate in location with the strongest lunar magnetic anomalies. Current evidence may be most consistent with a model involving a decreased implantation rate of solar wind hydrogen in areas shielded by strong anomalies. If this model is confirmed, it would represent macroscopic evidence for a role of the solar wind ion bombardment as well as micrometeoroid bombardment in the darkening with time of exposed silicate surfaces in the inner solar system.

Lon L. Hood

Bibliography

Cisowski, S.M., Collinson, D.W., Runcorn, S.K., *et al.* (1983) A review of lunar paleointensity data and implications for the origin of lunar magnetism, in *Proc. Lunar Planet. Sci. Conf.* **13**. *J. Geophys. Res.*, **88**, A691–704.

Crawford, D. and Schultz, P. (1991) Laboratory investigations of impact-generated plasma. *J. Geophys. Res.*, **96**, 18807–19.

Dyal, P., Parkin, C.W. and Daily, W.D. (1974) Magnetism and the interior of the Moon. *Rev. Geophys. Space Phys.*, **12**, 568–91.

Fuller, M. and Cisowski, S. (1987) Lunar paleomagnetism, in *Geomagnetism*, Vol. 2 (ed. J.A. Jacobs). pp. 307–456. London: Academic Press.

Hood, L.L. 1986. Geophysical constraints on the lunar interior. In *Origin of the Moon* (eds W.K. Hartmann, R.J. Phillips and G.J. Taylor). Houston: Lunar and Planetary Institute, pp. 361–410.

Hood, L.L. and Huang, Z. (1991). Formation of magnetic anomalies antipodal to lunar impact basins: two-dimensional model calculations. *J. Geophys. Res.*, **96**, 9837–46.

Hood, L.L. and Jones, J. (1987) Geophysical constraints on lunar bulk composition and structure: a reassessment, in *Proc. Lunar Planet. Sci. Conf.* **17**. *J. Geophys. Res.*, **92**, E396–410.

Hood, L.L. and Schubert, G. (1980) Lunar magnetic anomalies and surface optical properties. *Science*, **208**, 49–51.

Hood, L.L. and Vickery, A. (1984) Magnetic field amplification and generation in hyper velocity meteoroid impacts with application to lunar paleomagnetism, in *Proc. Lunar Planet. Sci. Conf.* **15**. *J. Geophys. Res.*, **89**, C211–23.

Hood, L.L. and Williams, C.R. (1989) The lunar swirls: distribution and possible origins, in *Proc. Lunar Planet. Sci. Conf.* **19**, (eds G. Ryder and V. Sharpton). Houston: Lunar and Planetary Institute, pp. 99–113.

Hood, L.L., Herbert, F. and Sonett, C.P. (1982) The deep lunar electrical conductivity profile: structural and thermal inferences. *J. Geophys. Res.*, **87**, 5311–26.

Langseth, M., Keihm, S. and Peters, K. (1976) Revised lunar heat flow values, in *Proc. Lunar Sci. Conf.* **7**, Houston: Lunar and Planetary Institute, pp. 3143–71.

Lin, R.P., Anderson, K.A. and Hood, L.L. (1988) Lunar surface magnetic field concentrations antipodal to young large impact basins. *Icarus*, 74, 529–41.

McKay, D.S., Heiken, G., Basu, A., *et al.* (1991) The lunar regolith, in *Lunar Sourcebook* (eds G.H. Heiken, D.T. Vaniman and B.M. French). New York: Cambridge University Press, pp. 285–356.

Mueller, S., Taylor, G.J. and Phillips, R.J. (1988) Lunar composition: a geophysical and petrological synthesis. *J. Geophys. Res.*, 93, 6338–52.

Nakamura, Y., Latham, G.V. and Dorman, H.J. (1982) Apollo lunar seismic experiment–final summary, in *Proc. Lunar Planet. Sci. Conf.* **13**. *J. Geophys. Res.*, **87**, A117–23.

Runcorn, S.K. (1983) Lunar magnetism, polar displacements and primeval satellites in the Earth–Moon system. *Nature*, 30, 589–96.

Runcorn, S.K. 1987. The Moon's ancient magnetism. *Sci. Am.*, 257, 60–8.

Russell, C.T., Coleman, P.J., Jr and Goldstein, B.E. (1981) Measurements of the lunar induced magnetic moment in the geomagnetic tail: evidence for a lunar core, in *Proc. Lunar Planet. Sci. Conf.* **12** (ed. R.T. Merrill). Houston: Lunar and Planetary Institute, pp. 831–6.

Schultz, P.H. and Srnka, L.J. (1980) Cometary collisions on the Moon and Mercury. *Nature*, 284, 22–6.

Srnka, L.J. (1977) Spontaneous magnetic field generation in hypervelocity impacts, in *Proc. Lunar Sci. Conf.* **8** (ed. R.T. Merrill). Houston: Lunar and Planetary Institute, pp. 785–92.

Strangway, D.W., Sharpe, H., Gose, W. and Pearch, G. (1973) Magnetism and the history of the Moon, in *Magnetism and Magnetic Materials – 1972*, (eds C.D. Graham, Jr and J.J. Rhyne). New York: American Institute of Physics, pp. 1178–87.

Taylor, S.R. (1982) *Planetary Science: A Lunar Perspective*. Houston: Lunar and Planetary Institute.

Cross references

Dynamo theory
Earth–Moon system: origin
Impact cratering
Magnetism
Moon: origin

MOON: ORIGIN

The origin of Earth's Moon has been a longstanding problem, partly because Earth's Moon seems unusual compared to satellites of other planets, and partly because most theories of lunar origin have been failures. A new theory, developed after the Apollo flights, suggests the Moon is the result of a very early collision between Earth and a giant interplanetary body.

The samples brought back by the Apollo expeditions, along with dynamical and chemical arguments, ruled out three pre-Apollo theories: the 'coaccretion' hypothesis, the capture theory and the fission theory. The idea that the Moon coaccreted alongside Earth failed to explain how the Earth acquired a large fraction of iron while the Moon gained almost none, presumably while being made from the same material. The hypothesis of a captured Moon might explain the different chemistry, if the Moon was formed in a different solar system zone, but this idea foundered in the face of the strong similarity of the Earth's and the Moon's oxygen isotope ratios, given that bodies in other parts of the solar system have different oxygen isotope ratios. Also, dynamical studies showed that capture of a large body is a very unlikely event. The proposal that the Moon spun off the Earth in a spontaneous fission event (perhaps when Earth's iron core formed) failed to explain the energy required for the Moon's ejection, or the angular momentum properties of the system.

The Moon has a chemical composition grossly similar to Earth's mantle, but with low iron and extremely low volatile content. The oxygen isotope equality implies that lunar material formed at about the same distance from the Sun as Earth material (or that the pre-Moon and pre-Earth materials were thoroughly mixed).

Inspired by Safronov's work on accretion (see Cold accretion theory), and cognizant of the above facts, Hartmann and Davis (1975) studied the size of the second largest bodies, third largest bodies, and so on, that might accrete near Earth's heliocentric orbit while Earth itself accreted. They concluded that the second largest such body might have reached a diameter of several thousand kilometers (the size of Moon or Mars) by the time Earth finished growing. They therefore suggested that the Moon may have originated in a collision of a giant planetesimal and the proto-Earth in its late stages of growth, after Earth's iron core formed. The impact blasted out mantle material, a fraction of which went into orbit and accreted to the Moon. In this hypothesis the hot mantle material would be iron deficient and the vaporized volatile component would be lost into space, explaining the gross properties of the Moon.

In the following year Cameron and Ward (1976) independently published an abstract with a similar result from a different approach. They considered the angular momentum of the Earth–Moon system, and concluded that a Mars-sized impactor might have hit proto-Earth in a glancing blow and ejected material to form the Moon.

This theory, that a huge planetesimal impacted Earth to eject the Moon's parent material, has come to be called the giant impact theory of lunar origin. Thompson and Stevenson (1983) concluded that material blasted off Earth in such fashion could rapidly form a disk and accrete to form the Moon. The accretion of the Moon in a circumterrestrial disk would have been very rapid, compared to planetary evolution, and the resulting heating would probably have melted most of the early Moon, in accord with models of the lunar magma ocean.

All this work was initially criticized as ad hoc, invoking a random impactor 'out of the blue.' Hartmann (1986) responded that the stochastic nature of the largest impact(s) was an advantage of the theory, because the stochastic effect of a few large impacts in the solar system, added to the statistically averaged effects of countless small impacts, helps to explain the range of obliquities, rotation states and satellites. For example, Safronov (1966) had already invoked impact of a large planetesimal to explain the obliquity and retrograde rotation of Uranus. In fact, it is now widely accepted that theories of solar system formation without impacts of some large planetesimals lead to difficulties in explaining the range of planetary properties.

Today, the giant impact theory is the leading hypothesis for lunar origin (Wood, 1986), and giant impact events have also been evoked to explain the large satellite of Pluto and odd properties of Neptune's satellite system. The theory has also led to a wider acceptance of catastrophic events in planetary history, superimposed in the long, slow march of more uniformitarian geologic evolution.

Current work on the theory centers on computer modeling of the impact (Melosh and Sonett, 1986; Benz, Slater and Cameron, 1986) and geochemical studies of Earth's mantle (in search of mineralogical evidence of such a high-temperature event). The computer simulations support the model and show that a giant impact could have blasted mantle material from both the Earth and the impactors into circumterrestrial orbit. These models predict extreme heating of Earth's mantle. Searches for mineralogical evidence of such heating have been ambiguous. Such work may help confirm or refute this model of the Moon's origin.

William K. Hartmann

Bibliography

Benz, W., Slattery, W. and Cameron, A. (1986) Short note: Snapshots from a 3-D modeling of a giant impact, in *Origin of the Moon* (eds W.K. Hartmann, R.J. Phillips and G.J. Taylor.) Houston: Lunar and Planetary Institute,

Cameron, A.G., and Ward, W.R. (1976) The origin of the Moon (abstract), in *Lunar Sci. VII*, Houston: Lunar Science Institute, pp. 120–2.

Hartmann, W.K. 1986. Moon origin: the Impact-trigger hypothesis, in *Origin of the Moon* (eds W.K. Hartmann, R.J. Phillips and G.J. Taylor). Houston: Lunar and Planetary Institute, pp. 579–60.

Hartmann, W.K. and Davis, D.R. (1975) Satellite-sized planetesimals and lunar origin. *Icarus*, **24**, 504–15.

Melosh, H.J. and Sonett, C.P. (1986) When worlds collide: jetted vapor plumes and the Moon's origin in *Origin of the Moon* (eds W.K. Hartmann, R.J. Phillips and G.J. Taylor), Houston: Lunar and Planetary Institute, pp. 621–42.

Safronov, V. (1966) Sizes of the largest bodies falling onto planets during their formation. *Sov. Astron. AJ*, **9**, 987.

Thompson, A.C. and Stevenson, D.J. (1983) Two-phase gravitational instabilities in thin disks with application to the origin of the moon (abstract), in *Lunar Planet. Sci. XIV*, Houston: Lunar and Planetary Institute pp. 787–8.

Wood, J.A. (1986) Moon over Maura Loa: a review of hypotheses of formation of Earth's Moon, in *Origin of the Moon* (eds W.K. Hartmann, R.J. Phillips and G.J. Taylor). Houston: Lunar and Planetary Institute, pp. 17–56.

Cross references

Earth–Moon system: origin
Impact cratering
Planetary rotation
Roche limit
Solar system: origin
Tidal friction

MOON: SEISMICITY

As a terrestrial planet evolves, its internal temperature changes. The initial accretional energy and the energy released by differentiation and by decay of radioactive materials dissipate via radiation to space. This causes strain to accumulate inside the planet. The strain may eventually be released as materials experience brittle failure; energy is released suddenly as stress waves, producing seismic events. In addition to these seismic events of internal origin, there are other types of seismic events induced by external causes, such as gravitational and radiational effects of the Sun and other planets, and impacts of small interplanetary objects. Quakes due to all of the above causes are observed on the Moon.

At present the only observational data concerning the seismicity of the Moon are from the network of seismic stations established on the lunar surface during the Apollo missions (q.v.; Plate 14). Using the data recorded between 1969 and 1977, researchers identified four types of seismic activities: deep moonquakes, shallow moonquakes, meteoroid impacts and thermal moonquakes. Figure M95 shows seismograms of the first three types.

Deep moonquakes

A large number of small moonquakes having nearly identical waveforms occur at great depths (700–1200 km) at highly regular monthly intervals (Lammlein *et al.*, 1974). The identical waveforms imply that they occur within a small hypocentral region. From a cross-correlation analysis of waveforms, Nakamura (1978) estimates that the hypocentral region is no greater than about 1 km in diameter. More than 3000 such events belonging to 109 separate hypocentral regions were identified (Nakamura, Latham and Dorman, 1982). The equivalent body-wave magnitude of individual events is estimated to be less than 2.

The focal depths of these moonquakes range from 700 km to 1200 km, about halfway to the center of the Moon and deeper than any known deep-focus earthquakes. Their epicenters appear to cluster within a few seismic belts, and they are nearly absent in the southeast quadrant of the Moon, which is mostly highland (Figure M96). All but one of the known epicentral regions are on the frontside of the Moon. This may indicate that there are few deep moonquakes on the mostly mountainous farside of the Moon, or may simply reflect an observational bias because all the Apollo seismic stations were on the frontside of the Moon. The deep interior of the Moon below the level of the deepest moonquakes hypocenters is thought to be partially molten (Figure M97), and this zone attenuates shear waves from moonquakes on the far side. The absence of shear-wave arrivals prevents us from positively identifying and locating farside moonquakes even if they exist.

The monthly periodicity of deep-moonquake activity suggests a strong relationship between deep moonquakes and tides raised on the Moon by the Earth and the Sun. At present, it is uncertain whether the tide merely triggers a release of accumulated tectonic strain, or if the tide is wholly responsible for these quakes. However, several lines of evidence favor the latter hypothesis. (1) There is a positive correlation between moonquake magnitudes and tidal amplitudes. (2) The orientation of the deep moonquake slip vector rotates with tidal

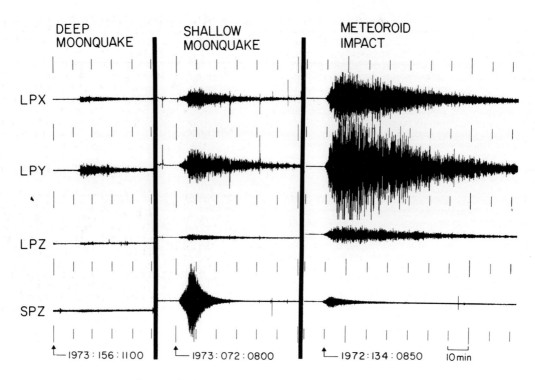

Figure M95 Seismograms in compressed timescale showing three major types of moonquakes: deep moonquakes (left), shallow moonquakes (center) and meteoroid impact (right). Tick marks are at 10 minute intervals. (From Nakamura et al., 1974).

phases (Nakamura, 1978). (3) Deep moonquake stress drops ($\sim$ 10 kPa; Goins, Dainty and Toksöz, 1981) are significantly lower than the maximum tidal stress inside the Moon ($\sim$ 100 kPa; Cheng and Toksöz, 1978). (4) The total energy release by all deep moonquakes is much less than the amount of energy tidally dissipated from the Earth–Moon system (see Tidal friction).

Shallow moonquakes

Rare but relatively strong moonquakes occur at shallow depths (Nakamura et al., 1979; Nakamura, 1980). These events are characterized by having unusually high frequencies for recorded seismic waves, compared with other lunar and earthquake events. Only 28 shallow moonquakes were identified during the 8 years of observation. Unlike deep moonquakes, their occurrence is not correlated with tides, and no multiple events have been detected from a single source region. Even though much fewer in number than deep moonquakes, their magnitude is much larger. The estimated body-wave magnitudes of the three largest ones observed in 8 years exceed 5.5.

The exact depths of shallow moonquakes are as yet unknown. This is because all those observed occurred outside the Apollo seismic array. Also, intense near-surface scattering of seismic waves prevented the identification of depth phases, the surface-reflected seismic arrivals which give a direct measure of focal depth for earthquakes. However, indirect evidence suggests that most of them occur in the upper mantle of the Moon, between 50 km and 200 km depths. Their epicenters (Figure M96) appear to occur near boundaries between dissimilar surface features, such as the edges of major impact basins. However, the paucity of the observed events precludes definitive conclusions.

The unusually high frequency content of seismic signals from shallow moonquakes suggests exceptionally high stress concentration at the source. Stress drops associated with large shallow moonquakes are estimated to exceed 100 MPa (Oberst, 1987). These events have seismic moment on the order of 10^{15} Nm. The high stress drop is consistent with a model in which thermoelastic strain within a thick lithosphere is released by occasional shallow moonquakes as the Moon cools.

Meteoroid impacts

Because the Moon has no atmosphere, meteoroids colliding with the Moon reach the surface without 'burning out'; the seismic waves generated by their impacts constitute yet another type of seismicity. Only a very small fraction of a meteoroid's kinetic energy is converted to seismic energy, because the majority is spent excavating a crater and heating the material. Nevertheless, many impacts have sufficient energy to be detectable at seismic stations. The long-period seismometers at Apollo stations detected more than 1700 such seismic events, representing meteoroids of masses of roughly 100 g to 100 kg, during the 8-year observation period. Those generated by impacts of smaller objects and detected only by short-period seismometers were too numerous to be counted.

Observations of impact-generated seismicity provide a unique census of the population of small objects in interplanetary space, and are complementary to Earth-based observations of meteors and meteorites. Two distinct types of objects crossing the Earth–Moon orbit have been identified from the seismic data: cometary and asteroidal (Oberst and Nakamura, 1991). The former objects are mostly lighter and less energetic, and many of them occur in clusters identifiable with known meteor showers. The latter include many massive objects, in mostly prograde orbits, and show a lesser degree of clustering. However, the asteroidal objects also include some prominent swarms, which may be concentrated debris from relatively recent disintegration of asteroids in Earth-crossing orbits (see Near-Earth object).

Thermal moonquakes

The Apollo seismometers detected numerous very small seismic events originating from many isolated locations within a few kilometers of each seismic station (Duennebier and Sutton, 1974). Similar to deep moonquakes, these events also possessed nearly identical waveforms repeating at regular monthly intervals. However, most of these micromoonquakes occurred during daylight hours, with high activity starting about 2 days after sunrise.

The clear diurnal periodicity suggests that these moonquakes are either induced or triggered by thermoelastic stresses due to temper-

PLATE 1 SUN

(a) The Sun White light image of the Sun, showing a number of sunspot groups. These rather large spots were recorded on 3 April 1969, a short time after the maximum phase of the 11-year sunspot cycle. Telescopic observers recorded sunspots in drawings, beginning in the early 17th century. More rigorous and systematic observations began following the recognition by Schwabe in 1843 of the 11-year cycle in sunspots. Today the Sun is imaged at many frequencies, at wavelengths both longer and shorter than those of visible light (about 0.4 to 0.7 µm); see *Electromagnetic radiation*. (Image courtesy of G.S. Chapman and C. Mach, San Fernando Solar Observatory and California State University, Northridge.) See *Maunder, Edward Walter, and Maunder minimum: Solar activity: Solar photosphere; Sun*.

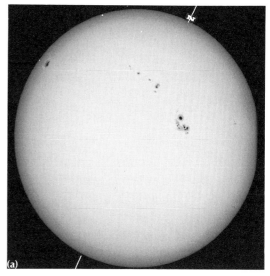

(b) X-ray image of the Sun X-rays are emitted at the extremely high temperature characteristic of the solar corona. Above the surface, in the corona, magnetic fields often trap the solar plasma in loop structures, seen clearly at many places in this high-resolution image. Various physical mechanisms, which scientists are studying with these images, heat the trapped plasma to temperatures of several million kelvin. This trapped plasma emits x-rays and appears bright in the image, showing the configuration of the magnetic field confining it. In areas near the solar poles, where the magnetic field extends into interplanetary space, the plasma is not trapped, and so does not become hot or dense enough to emit x-rays. Hence these parts of the image are dark. Much of the solar plasma that comes to the Earth (the solar wind) is believed to be emitted from these dark regions, called coronal holes. Some of the brightest regions in the image are near sunspots, where the solar magnetic field is strongest. However, many features, such as the tiny, transient, bright features called x-ray bright points, have no visible counterpart. This image was taken on 22 January 1992, near the peak of the 11-year sunspot activity cycle. At the limb of the Sun, many coronal features can be seen to extend out to a large distance. The top of the large loop at the lower right part of the image is 174 000 km above the solar surface (about 14 times the diameter of the Earth) and is expanding into interplanetary space. With x-ray images from satellites, such as this, the corona can be studied continuously. Ground-based telescopes can see the outer corona only during eclipses and cannot view it at all at x-ray wavelengths, which do not penetrate the Earth's atmosphere. This image was taken through a thin aluminum filter that admits a broad spectrum of 'soft' x-rays of about 1-keV energy. This image was acquired by the Soft X-ray Telescope on the Yohkoh solar research spacecraft of the Japanese Institute of Space and Astronautical Science. The Soft X-ray Telescope experiment is a Japan/US collaboration involving the National Astronomical Observatory of Japan, the University of Tokyo and the Lockheed Palo Alto Research Laboratory. The US work is supported by the National Aeronautics and Space Administration. See *Coronal mass ejections; Plasma; Solar corona; Solar wind; Sun*.

D.P. Cauffman

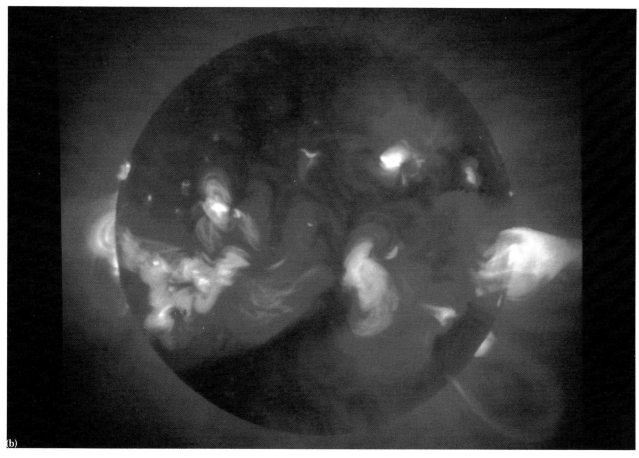

PLATE 2 RADIO IMAGES

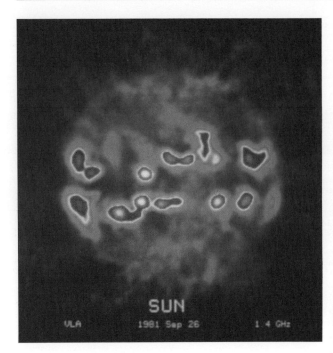

(a) Radio image of the Sun This image was obtained by G.A. Dulk and D.E. Gary on 26 September 1981, using the Very Large Array (VLA) radiotelescope. The wavelength was 20 cm for this observation. The solar active regions form two bands symmetric about the solar equator. Each active region corresponds to a volume in the corona, in the vicinity of but above sunspots, in which high-density hot material is confined by strong magnetic fields (hundreds of gauss). The brightness temperatures of the brightest regions are 2.4×10^6 K, about the same as the kinetic temperatures in the active regions; these sources are optically thick at 20 cm. Other active regions are less bright, probably because the density is lower and the optical thickness is less than unity. Away from active regions are quiet regions of moderate brightness (50 000 to 100 000 K), where the coronal kinetic temperature is about 10^6 K but the density is low enough that the corona is optically thin and most radiation is from the cooler transition region. Regions of very low brightness (less than about 50 000 K) are coronal holes and filament cavities, regions where the corona is of especially low density. Off of the limb of the Sun are wisps which are due to instrumental sidelobes.
See *Radio astronomy; Solar activity; Sun.* (NRAO CV81–SUN.VR16C, Courtesy NRAo/AuI.)

(b) Radio image of Jupiter This false-color radio photograph of Jupiter was obtained at a wavelength of 20 cm in May 1981. (de Pater, I. and Jaffe, W.J. (1984) VLA observations of Jupiter's non-thermal radiation. *Astrophys. J. Suppl.*, **54**, 405–19.) The resolution is 4.5″, or 0.25 Jovian radii. The planet Jupiter is shown by the disk in the center; this emission is thermal radio emission from the planetary atmosphere. The magnetic field of the planet traps and holds very fast-moving electrons; these radiate radio waves, producing the large extended features beyond the planet's disk. Most of the radiation is synchrotron radiation emitted by relativistic electrons. (CV 81-Jupiter. VR48, courtesy NRAO/AUI.) See *Jupiter magnetic field and magnetosphere; Planetary torus; Radio astronomy.*

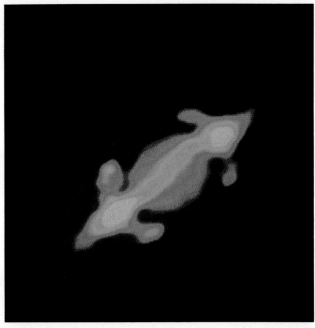

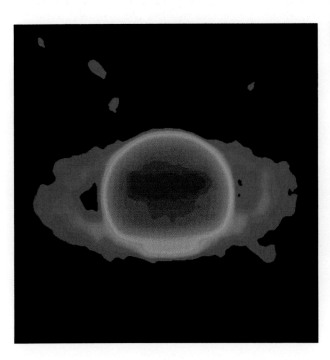

(c) Radio image of Saturn False-color radio photograph of Saturn at 6.2 cm wavelength. The data were taken with the VLA in June 1986, when the rings were wide open (ring inclination angle is 25°). (de Pater, I. and Dickel, J.R. (1991) Multi-frequency VLA observations of Saturn at ring inclination angles between 5° and 26°, *Icarus*, **94**, 474–92.) The resolution is 1.5″, and Saturn's disk is about 18″ across. Both the A and B rings are clearly visible, thanks to Saturn's radio emission reflected from the ring particles. Where the rings pass in front of the planet, they block off the planet's thermal emission. On the planetary disk itself, one can see a bright band across the disk at a latitude of ~ 30°. This implies that the ammonia abundance is slightly smaller at 30° latitude than at other latitudes, so warmer layers are probed in this region. (CV 82–SATURN. VR54. Courtesy NRAO/AUI.) See *Radio astronomy; Saturn: atmosphere.*

Imke de Pater

PLATE 3 PLANETARY AURORAS

(a) Terrestrial aurora The spatial distribution of auroral emissions on a global scale. This false-color image of the northern auroral oval at vacuum-ultraviolet wavelengths is overlaid with a coastline map to show how the nearly instantaneous auroral oval maps onto the polar regions. The image was obtained with imaging instrumentation on board the Earth-orbiting spacecraft Dynamics Explorer 1 during the 12-min period beginning at 0229 UT on 8 November 1981. This spacecraft was launched into a polar orbit with initial apogee altitude of about 23 000 km above the Northern Hemisphere in order to provide this splendid global view of the aurora. The sensitivity passband of the ultraviolet-wavelength photometer for this image extends from 123 to 155 nm. Auroral emissions at those wavelengths arise predominantly from the emission lines of atomic oxygen at about 180.4 and 185.6 nm and from the Lyman–Birgs–Hopfield bands of molecular nitrogen. The intensities of auroal emissions are similar to those from the sunlit hemisphere at large solar zenith angles as seen here in the upper-left portion of the image. (Courtesy of J.H. Waite.) See *Aurora, historical record; Aurora, planetary; Geomagnetic storm.*

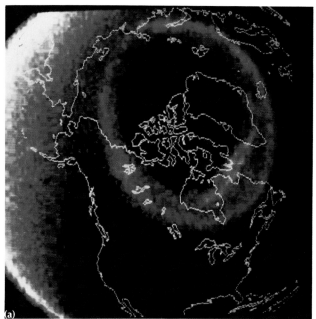

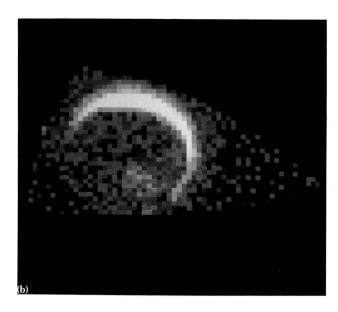

(b) Venus ultraviolet aurora This spin-scan image was acquired by the Pioneer Venus orbiter ultraviolet spectrometer on orbit 1067, on 7 November 1981. The spectrometer was commanded to accept light in a 13Å bandpass centered near 1304 Å, at the resonance triplet of atomic oxygen. The image is displayed in false color, with blue representing the faintest signal and white the brightest.

The bright crescent represents emission from the sunlit part of Venus' thermosphere. It is produced by resonance scattering of sunlight and by photoelectron impact on atomic oxygen. The brightest parts of the crescent are 3–5 kR.

The emission from the dark side of Venus represents the excitation of thermospheric atomic oxygen by soft electrons precipitating from Venus' 'magnetosphere' – a Venusian 'aurora'. This image contains the brightest 'aurora' seen by PV – about 100 R. The aurora is always patchy and is sometimes absent. Its brightness is very variable, but shows a positive correlation with solar activity. See *Aurora, planetary; Pioneer Venus mission; Venus: magnetic field and magnetosphere.*

(c) The Jovian Aurora A false-color image obtained by the Hubble Space Telescope of ultraviolet auroral emissions at Jupiter. A white grid shows Jovian latitude and longitude, while two partial circles near the pole indicate the mapping of planetary magnetic field lines onto the planet. Energetic particles precipitate downward along these field lines from the magnetosphere into Jupiter's upper atmosphere, where they excite auroral emissions at various wavelengths. The outer circle maps to the plasma torus created by the sulfur- and oxygen-emitting volcanoes of Io; the inner circle corresponds to magnetic field lines connected to the outer regions of the magnetosphere. The location of the auroral emissions thus furnishes clues about the origin of the auroral particles and the mechanisms that accelerate them, although uncertainties in our knowledge of the planet's surface magnetic field limit the accuracy of this approach. The auroral processes at Jupiter are in some respects similar to those at Earth. In contrast to Earth, however, the glowing atmospheric gas is H_2, not N_2 or O_2, and the energy driving the aurora is at least partially supplied by the rapid rotation of the planet and not solely by the interaction of the solar wind with the planetary magnetosphere. The auroral emissions in this example are extremely bright, with an estimated peak power input of 1 W m^{-2} over an area larger than the surface area of the Earth. For comparison, the peak power input for a bright Earth aurora is about 0.01 W m^{-2}. (Courtesy of J.H. Waite.) See *Aurora, planetary; Jupiter: magnetic field and magnetosphere; Magnetospheres of the outer planets.*

J.H. Waite

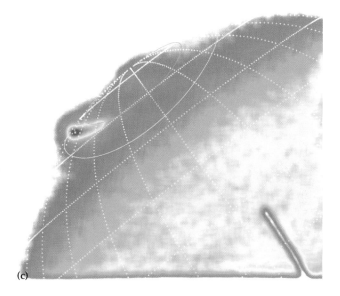

PLATE 4 PLANETARY CARTOGRAPHY

Construction of digital maps: Mars Diagram of a multidimensional digital image map. Maps can be combined in a variety of ways, and the resulting composite map can be reprojected in the computer to any desired scale and format. Here the bottom layer is an image map, similar to a photographic representation; the middle layer is a color-coded map of elevations, and the top layer is a geologic map. Any number of other kinds of maps can be included in the 'stack' below the bottom layer. (Image courtesy of R.M. Batson, US Geological Survey Branch of Astrogeology.) See *Cartography; Mars: geology*.

PLATE 5   MERCURY GEOLOGIC MAP

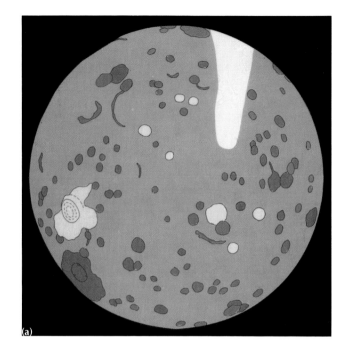

Geologic map of Mercury Geological maps of the planet Mercury at three different stages in its history. (a) Mercury as it was about 4 billion years ago, shortly after the impact that formed the basin Tolstoj (pale yellow multiringed feature at about 8 o'clock). The large expanse of brown–tan colors are the Mercurian intercrater plains, a global unit that resurfaced the planet early in its history.

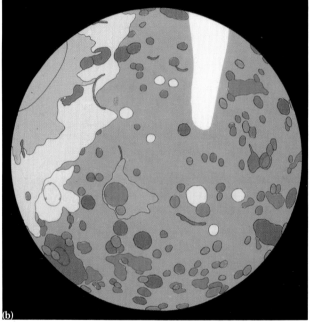

(b) A map of Mercury about 3.8 billion years ago, immediately after the impact that formed the Caloris basin (in blue, upper left side): craters (brown and blue) and the basin Beethoven (purple) have also formed. The red areas are smooth plains material beginning to be emplaced; these plains are probably volcanic flood lavas.

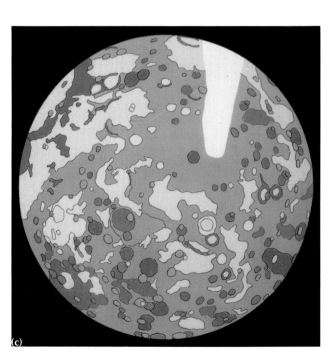

(c) The last image shows the geology of Mercury today. Regions of smooth plains (pink) have partly resurfaced the planet. Additional craters (green and yellow) are found on top of all units and continue to form today. See *Mercury: geology*.

P. Spudis

PLATE 6 VENUS, TWO GLOBAL VIEWS

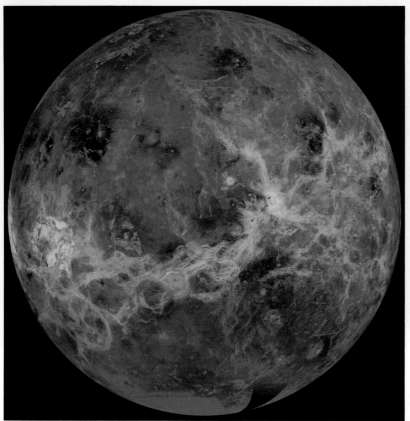

Venus These views of the surface of Venus are obtained from synthetic aperture radar data obtained by the Magellan spacecraft. The simulated color was chosen to agree with color images obtained by the Soviet Venera 13 and 14 landers. Bright areas represent areas of considerable surface roughness, such as impact craters and sites of geologically recent volcanism, while darker areas are thought to represent areas of more finely divided surface materials. A number of the dark areas are associated with impact craters. It is possible that the shock waves of large impacts helped to locally pulverize the surface materials in these areas. Volcanic deposits cover at least 85% of the planetary surface. Flat plains and small shields are the most common unit. The very bright area near the western (left) edge of the top image is Atla Regione, a part of the larger system of equatorial highlands known as Aphrodite Terra. These broad highland regions, 1000 to 3000 km in diameter, may be sites of mantle upwelling and associated volcanism. Magellan (P–39225 MGN 81, courtesy of NASA.) See *Corona (Venus); Impact cratering; Magellan mission; Tessera; Venus; Venus: geology and geophysics; Venera missions; Volcanism in the solar system*

Western Hemisphere

Eastern Hemisphere

PLATE 7   VENUS GEOID

Venus geoid The geoid is a theoretical equipotential surface. On a water-covered planet without atmospheric perturbations and oceanic current systems, the mean sea level would define the geoid. On real planets, mass is not distributed with perfect radial symmetry; inhomogeneities produce variations in the gravitational potential, which may be quantified by means of direct measurements (on Earth) or spacecraft measurements. The differences in potential relative to the theoretical equipotential ellipsoid (geoid) may thus be mapped as shown in this map of the planet Venus. Aphrodite Terra, the extensive equatorial highlands region of Venus, is near the center of the map. Ishtar Terra is another highlands region at high northern latitudes.

The map of departures from the geoid is a valuable tool for estimating the nature and extent of crustal structural differences. For the case of Venus such maps yield insight on questions such as the strength of the crust and the pattern of mantle convection within the planet. When the geoid is compared with topography, a strong correlation of topography and geoid height is evident. This is not the case for the Earth, where plate tectonics has produced a much more complex relationship in which geoid height and topography are not well correlated. (Map courtesy of M. Simons and S.C. Solomon. For more information, see Simons, M., Hager, B.H. and Solomon, S.C. (1994) Global variations in the geoid/topography admittance of Venus. *Science*, **264**, 798–803.) See *Venus: geology and geophysics*; *Venus*; *Gravity fields of the terrestrial planets*; *Geoid*; *Gravimetry*.

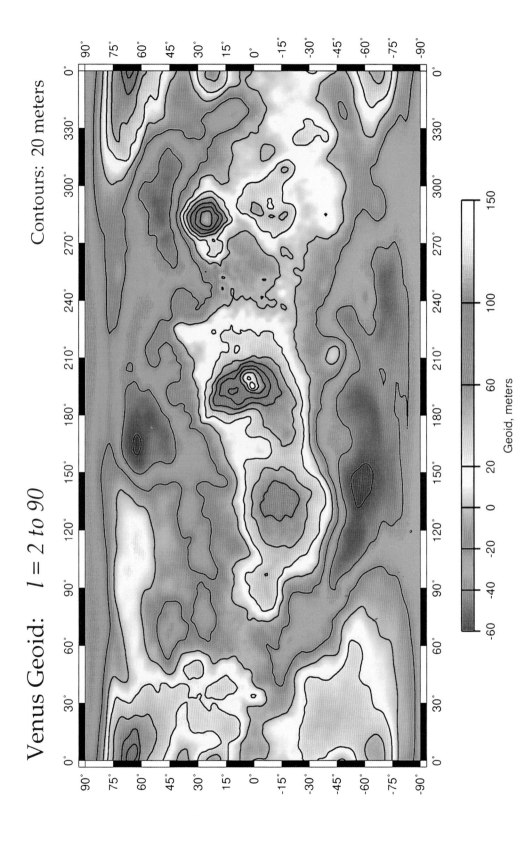

PLATE 8 VENUS ATMOSPHERE

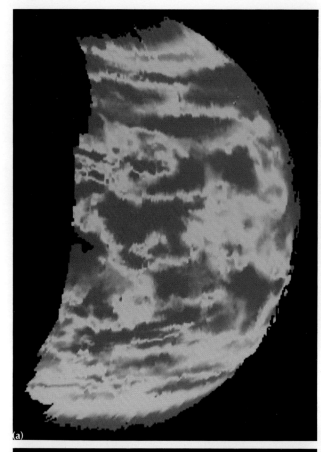

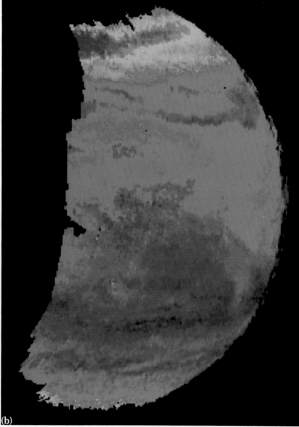

The Venus atmosphere The atmosphere of Venus is very dense. Pressures reach 92 bars at the surface, dropping to 1 bar at an altitude of 50 km and 0.01 bar at 75 km. Thick clouds, largely composed of sulfuric acid droplets, predominate at altitudes between 48 and 70 km. The upper portion of this cloud layer is spatially very homogeneous. Very little can be seen in the visible images taken by cameras in early flybys by Mariner, Pioneer Venus and the Venera spacecraft, which revealed the cloud tops only.

Recently it was discovered that the lower regions of the cloud layer could be sensed in infrared light by looking at the night side, away from the overwhelming solar illumination, and that these layers contained much more spatial structure than the upper layers (Allen, D.A. and Crawford, J.W. (1984) Cloud structure on the dark side of Venus. *Nature*, **307**, 222–4). The reasons why it is possible to see these deeper layers are complex. Firstly, the surface and lower atmosphere of Venus is hot enough to radiate strongly in the infrared, down to a wavelength of about 1 μm. Secondly, the sulfuric acid cloud particles, which absorb strongly at wavelengths longer than 3 μm, become transparent below that wavelength. Finally, the carbon dioxide molecules of the dense atmosphere, which blanket most of the infrared with their strong absorption bands, exhibit a few narrow 'windows' between these bands through which the radiation from the lower atmosphere can escape.

The data for Plate 8a were obtained by the near-infrared mapping spectrometer (NIMS) aboard the Galileo spacecraft when it flew by Venus on 10 February 1990 (Carlson, R.W., Baines, K.H., Encrenaz, Th. *et al.* (1991) Galileo infrared imaging spectroscopy measurements at Venus. *Science*, **253**, 1541–8). The observation covered almost one-third of the planet, from longitude 20° W to 80° E. This particular image was obtained at a wavelength of 2.30 μ meters, inside one of the windows between CO_2 bands. Radiation at this wavelength originates at an altitude of about 30 km, well below the cloud layers. Very large intensity variations are observed. These are thought to be due to dissolution of some of the lower cloud particles, most probably due to convective motions in the lower atmosphere, which bring hotter gas into the lower cloud layers. In the very brightest regions it is thought that the clouds have been dissipated up to an altitude of about 53 km, decreasing the total optical depth of the cloud layer by as much as one-third.

Other interesting features that can be seen in this image include large dark clouds in the equatorial region, which have been tracked at much coarser spatial resolution from the Earth (Crisp, D., McMuldroch, S., Stephens, S.K. *et al.* (1991) Ground-based near-infrared imaging observations of Venus during the Galileo encounter, *Science*, **253**, 1538–41). These have been shown to rotate across the planet with a period of about 5.5 days and to persist for periods longer than a month. The dark 'polar collar' which was previously observed by Pioneer Venus and Venera 15 may also be seen.

The image in Plate 8b is a result of processing of two images of the lower clouds: one is the image of Plate 8a, described above, while the other is an image taken simultaneously at a wavelength of 1.74 μm. The latter image is in light that is formed even deeper in the atmosphere (about 20 km in altitude). The cloud particles have slightly different absorbing properties at these two wavelengths, and thus their comparison can give information on the nature of these particles. Plate 8 is the result of evaluating a linear equation on these two wavelengths that measures the deviation from 'typical' behavior; it is interpreted as being a measure of mean cloud particle size. (Carlson, R.W., Kamp, L.W., Baines, K.H. *et al.* (1993) Variations in Venus cloud-particle properties, *Planet. Space Sci.*, **41**, 477–86). The false-color scheme used maps particles larger than normal as red, while the smallest ones are depicted as blue. (The actual size range of the particles likely to be responsible for this effect is about 1–3 μm.) The most striking result is that the regions of a given particle size are quite large and are roughly horizontally stratified, with a clear correlation with the cloud features. There is a marked enhancement of particle size at high northern latitudes, while a less pronounced diminution occurs in the corresponding southern region. It is likely that this is related to the global circulation patterns believed to exist in the Venus atmosphere. See *Venus; atmosphere; Spectroscopy; atmospheres; Infrared spectroscopy; Visible and near-infrared spectroscopy.*

Lucas W. Kamp

PLATE 9 EARTH

Earth The third planet from the Sun is unique; it possesses extensive oceans of liquid water and an oxygen-rich atmosphere that is far from chemical equilibrium. It is geologically active, continually recycling crustal materials, and it possesses life. The lower atmosphere and oceans, though only a few kilometers in depth, are the working fluids of a global heat engine that transfers absorbed solar energy from the equatorial regions to the poles. This image was obtained in April 1972 by the astronauts of the Apollo 16 mission. Extensive seasonal snow cover mantles much of the North American continent at the upper right; the western coastline from central America to Canada is visible near the center. A spinning low-pressure system, or cyclone, is producing inclement weather in the northeastern United States, while a circular patch of ocean in the Gulf of Alaska (top) marks the location of a strong atmospheric 'center of action'. This deep low-pressure center is spawning waves of cloudy storm systems, which are carried to the southeast across the continent by the prevailing westerly flow in the middle latitudes. The center of action in the Gulf of Alaska is driven by the transfer of energy from the relatively warm ocean to the much colder atmosphere. Hot air rising from the jungles of central America has produced extensive cloud cover over that region, in this mid-afternoon image. Thick cloud patches in the equatorial eastern Pacific are produced by strong convective motions, producing cloud towers, and driving the fundamental Hadley cell circulation of the terrestrial equatorial and middle latitudes. (NASA 72-HC-660.) See *Earth; Earth: atmosphere; Atmosphere; Hadley circulation; Cyclone, anticyclone; Polar cap; Angular momentum cycle in Planet Earth; Apollo missions.*

PLATE 10 EARTH TOPOGRAPHY

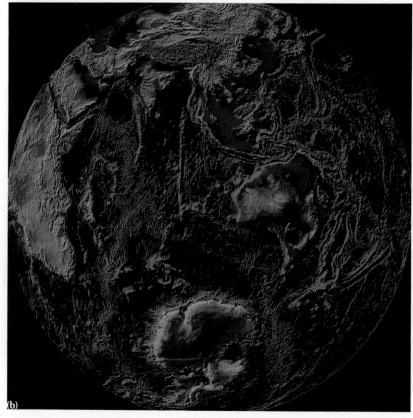

Earth: digital topography These images were generated from a computer file of topographic and bathymetric data at the US Geological Survey Branch of Astrogeology in Flagstaff, Arizona, USA. Color is used to indicate topographic elevation, and shading has been employed to help emphasize steep gradients, as at the boundaries of continental shelves. (a) Centered at longitude 80° W, latitude 30° N, showing North and South America and the north polar regions; (b) is centered on the opposite side of the planet, at 100° E, 30° S. Antarctica is shown at the bottom, with Asia at the top and Africa at the left.

These images make it easy to see the boundaries of tectonic plates, both above and below the ocean surface. The Atlantic mid-ocean ridge and a number of eastern Pacific fracture zones are prominent on (a), while deep trenches mark sites of subduction of oceanic crust in the western Pacific, on the second image. (Courtesy of K. Edwards and R.M. Batson, USGS.) See *Cartography*; *Earth*; *Earth: geology, tectonics and seismicity*; *Plate tectonics*; *Seismicity*; *Tectonics*.

PLATE 11 MANTLE CONVECTION

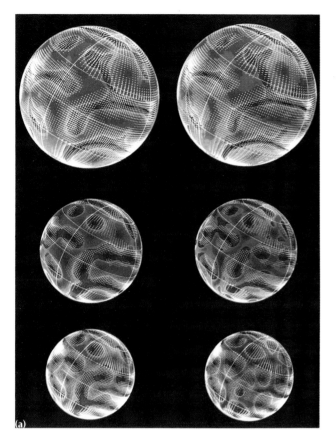
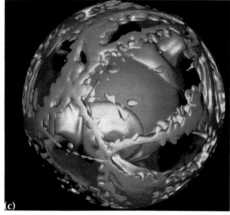
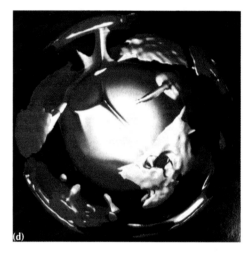
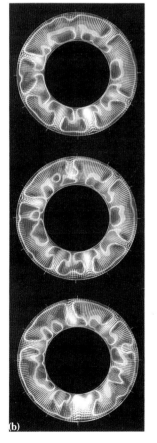

(a) **Mantle convection in the Earth** Plots (all at the same timestep) of convection velocities (left) and temperatures (right) in three different constant-radius surfaces (5940, 4350 and 3770 km). These spherical surfaces are scaled according to their radii. The colors in the three plots on the left represent the radial component of velocity with a contour increment of 1.5 mm year^{-1}. Reds and yellows represent upflow (a maximum of 12.0 mm year^{-1}); blues represent downflow (a maximum of 19.5 mm year^{-1}). The colors in the three plots on the right represent the temperature relative to the spherically averaged value at each radius with a contour increment of 50 K. Reds and yellows represent hot fluid (a maximum of +400 K); blues represent cold fluid (a minimum of −650 K). The arrows represent the direction and amplitude of the horizonatal velocity in these surfaces. All are scaled in the same way, with a maximum of 20 mm year^{-1}. Velocities less than 2 mm year^{-1} are not plotted. (After Glatzmaier, G.A., Schubert, G. and Bercovici, D. (1990) Chaotic, subduction-like downflows in a spherical model of convection in the Earth's mantle. *Nature*, **347**, 274–7).

(b) **Mantle convection for 600 million years** Plots (in the same cross-sectional slice through the three-dimensional shell) of convective velocities and temperatures at three different times separated by 200 million years beginning with the top plot. The colors represent the temperature relative to the radially dependent adiabatic temperature profile with a contour increment of 50 K. Reds and yellows represent hot fluid (a maximum of 300 K); blues represent cold fluid (a minimum of −750 K). The arrows represent the velocities of this cross-sectional surface and are scaled the same way as those in cold downflows and hot plumes in the Earth. (Also after Glatzmaier, G.A., Schubert, G. and Bercovici, D. (1990).

(c) A snapshot of the cold downwellings at one instant of time. The blue surface is an isocontour showing where the temperature is 110 K lower than the horizontally averaged value. The green surface is the core. A network of interconnected linear downwellings is visible in the upper mantle, with three huge cylindrical downwellings in the lower mantle, spreading out into pools of cold material above the core–mantle boundary. (After Tackley, P.J., Stevenson, D.J., Glatzmaier, G.A. and Schubert, G. (1993) Effects of an endothermic phase transition at 670 km depth in a spherical model of convection in the Earth's mantle. *Nature*, **361**, 699–704).

(d) The hot upwellings at the same time as (c). The red surface is an isocontour of superadiabatic temperature, showing where the temperature is 110 K higher than the reference-state adiabat. A single plume from the core–mantle boundary feeds a hot region in the upper mantle. Most broad hot regions in the upper mantle are not directly linked to lower-mantle structures. (Also after Tackley *et al.*, 1993). See *Mantle convection*.

PLATE 12 MOON GEOLOGY

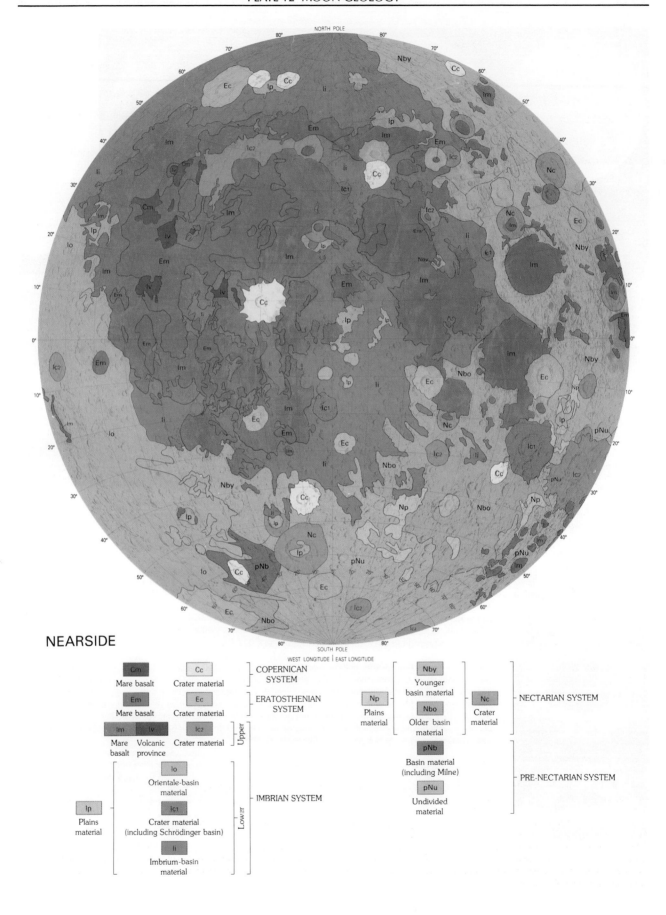

PLATE 13 MOON GEOLOGY

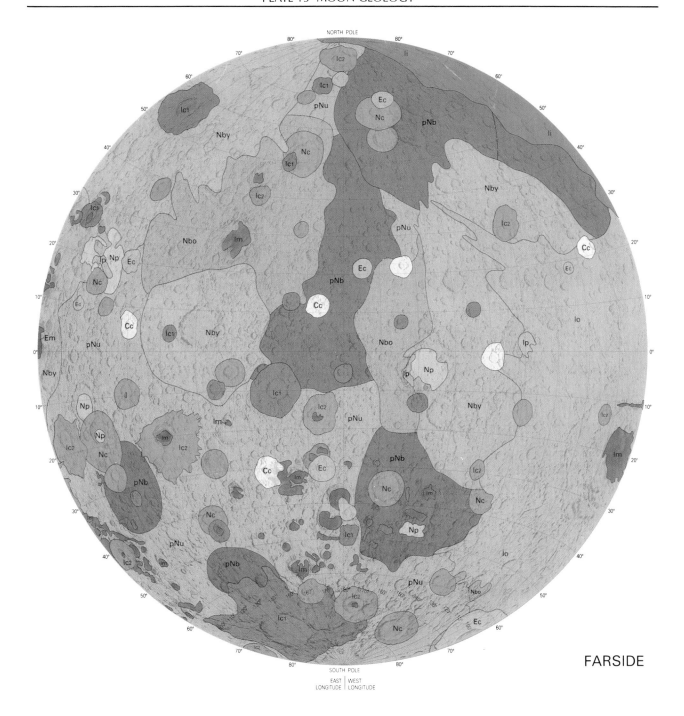

Geologic map of the Moon These maps detail the relative ages of surface materials on the nearside (Plate 12) and farside (Plate 13) of the Moon. As with all geologic maps, age relationships are based on the principle of superposition; younger units overlap, overlie, or cut older ones, and not the other way around. There are five main divisions of lunar geologic time. The Copernican System is youngest and extends to the present day. The Eratosthenian System is older, ranging from very roughly 1 to 3.2 billion years before the present. Both of these units are named for prominent craters; ejecta rays from the crater Copernicus lie atop the ejecta from the crater Eratosthenes.

The Imbrian System includes much of the mare basalt deposits of the lunar nearside. Its base is thought to have an age of about 3.85 Ga. Slightly older are the deposits that lie above the base of the Nectaris basin (Nectarian System). Finally, a few pre-Nectarian rocks are mapped, which were apparently formed more than about 3.9 billion years ago. The article Moon: geology includes a more detailed discussion of these divisions of the lunar stratigraphic column. (Map courtesy of D. Wilhelms.) See *Moon; Moon: geology; Moon: origin.*

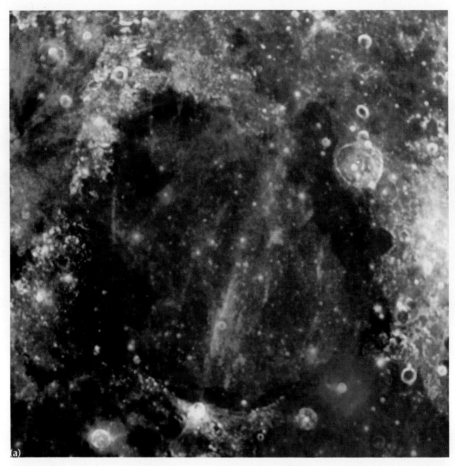

(a) **Mineralogy of Mare Serenitatis** This false-color image shows many of the major compositional units of the Moon's surface. White or pinkish areas indicate anorthositic highlands crust. Anorthosite is a relatively light mineral thought to have floated to the top of large magma chambers (or the 'magma ocean') early in lunar history. Blue regions are mare lavas rich in titanium, and the darkest blue areas are dark mantle deposits that resulted from explosive volcanism, also rich in titanium. Titanium-rich lunar soils preferentially trap solar-wind hydrogen, which can be extracted to produce water and fuel, so these locations are of interest for future lunar exploration and potential manned outposts. The Apollo 17 landing site is within the dark blue area near the southeast margin of the Serenitatis basin. A different type of dark mantling material produces the dark reddish areas near the southwest margin of Serenitatis. The greenish areas occur over mare lavas rich in iron and relatively poor in titanium. Fresh impact craters are bright due to immature soils, and provide compositional samples of subsurface layers. This image is constructed from an albedo (reflectivity) image merged with three color ratios: 756 nm/415 nm (displayed as red), 756 nm/990 nm (green) and 415 nm/756 nm (blue). The Apollo 15 landing site is near the left-hand margin of this picture. North is up. (Image courtesy of Alfred McEwen of the US Geological Survey, Flagstaff, Arizona, and the Galileo Solid-State Imaging Team.) See *Moon: geology; Galileo mission; Reflectance spectroscopy.*

(b) **Geophysics on the Moon** Deployment of the passive lunar seismic experiment package on the Moon by astronaut E. Aldrin on 20 July 1969. A network of four seismometers, placed by astronauts of the Apollo program, operated on the surface of the Moon for some 7 years. The cylinder shown in the center housed the seismometer, while the photovoltaic cell arrays on each side provided electric power for operation and transmission of data. A considerable number of geophysical experiments were carried out on the Moon and in lunar orbit; to date these represent scientific humanity's only in-person venture to another solar system body. (NASA 69-MC692.) See *Moon: seismicity; Apollo missions; Seismicity; Moon: geology.*

PLATE 15 METEORITES

(a) **A meteorite from Mars** A member of the class of SNC meteorites. This class (shergottites, nakhlites and chassignites) is named after type specimens that fell near Shergotty (India), Nakhla (Egypt) and Chassigny (France). All three types are relatively rare, with only a few specimens found. This photograph shows the saw-cut interior of the shergottite EETA 79001, showing black glassy inclusions in the volcanic host rock. The rough corner of the rock approximately marks the contact between a coarse-grained lava (left) and a finer-grained lava (right). The glass inclusions apparently formed through shock-induced melting during a meteoroid impact on Mars. As a result, Martian atmospheric gases were trapped in the glass, and were later discovered during laboratory analyses. (Courtesy J.L. Gooding, Johnson Space Center.) See *SNC meteorites; Impact cratering; Mars: impact cratering; Igneous rock*.

(b–e) **Ureilite meteorite photomicrographs** These four images illustrate important structural and mineralogical features of this class of meteorites.

(b) Transmitted light photomicrograph of the Kenna (New Mexico) ureilite showing pronounced mineral elongation lineation of olivine and pigeonite pyroxene (colored grains). Dark areas are graphite-rich 'carbonaceous matrix'. Scale bar equals 0.5 mm.

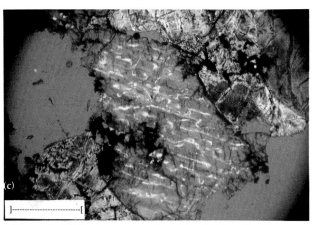

(c) Transmitted light photomicrograph of ALHA82130 ureilite. The central grain shows unmixing of ropy augite (light blue) in a pigeonite (darker blue) host. This ureilite is the only one to show any form of unmixing of pyroxene (produced during slow cooling). The lack of this feature in most ureilites shows that ureilites cooled very rapidly from high temperatures. Scale bar equals 1.0 mm.

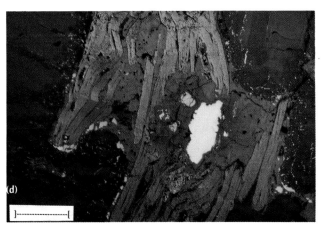

(d) High-powered reflected light photomicrograph of perfectly formed (euhedral) graphite crystals (light tan) in ALH78019 Antarctic ureilite. Graphite is surrounded by oxidized (terrestrial weathering) Ni–Fe metal (gray), an unaltered remnant of which lies right of center (white). Sulfides are bright yellowish grains. Tiny low-Ni metal grains embedded in Mg-rich silicate material (dark gray) result from silicate–graphite reduction reactions described in the article on *Ureilite meteorites*. Scale bar equals 100 µm.

(e) Transmitted light photomicrograph of 'mosaicized' (shattered) olivine grains in Goalpara, the first known ureilite. Such extreme deformation is evidence for high, impact-produced shock pressure late in the history of this meteorite. The large shattered grain (center-lower right) is about 2 mm long. Although originally a single, optically continuous grain, it is now composed of many multi-oriented grains. (Images courtesy of J. Berkley.) See *Ureilite meteorites; Meteorite; Igneous rock; Shock metamorphism*.

PLATE 16 MARS GLOBAL VIEW, MARS SURFACE

(a) **Mars** This view of the red planet is a mosaic of images from the Viking Orbiter spacecraft. The enormous rift near the equator is Valles Marineris, a system of canyons approximately 5000 km in length. Two large volcanic structures are visible at the left; these are Pavonis Mons and Ascraeus Mons, rising more than 20 km above the plains. The area around the craters, and in fact much of the northern hemisphere, has relatively few impact craters, suggesting that volcanic flows from the giant volcanoes may have erased many older features. The volcanoes are part of the Tharsis region, which also includes Olympus Mons, the largest volcano in the solar system. At the far right is a region of lower albedo. Patterns of bright and dark regions on Mars were detected by Earth-based astronomers (such as Schiaparelli) in the last century, leading to speculations regarding the presence of life, or even intelligent life, on Mars. (P-40222, courtesy of NASA.) See *Lowell, Percival; Mars: geology; Volcanism in the solar system; Schiaparelli, Giovanni Virginio.*

PLATE 17 MARS GEOLOGIC MAP

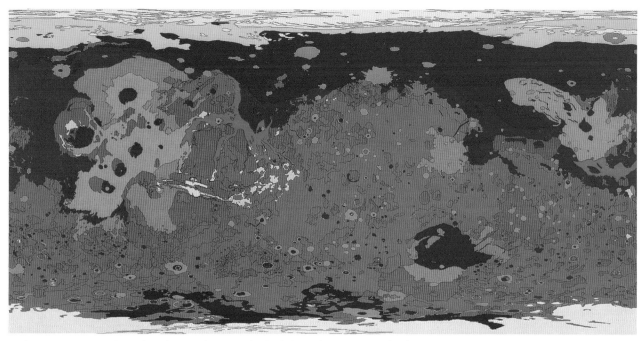

Geologic map of Mars The geologic history of Mars can be divided into three major epochs: the Noachian, a period of intense impact bombardment; the Hesperian, during which the impact flux was rapidly declining; and the Amazonian, an interval of much lower impact flux. See *Mars: geology* for a discussion of the events of these epochs. (Map courtesy of M. Kozak and R.M. Batson, US Geological Survey Branch of Astrogeology.) See *Cartography; Mars: geology.*

KEY TO MAP UNITS

CRATERS AND CRATERED TERRAINS

- Amazonian floor and rim units of Hellas basin
- Amazonian and Hesperian superposed impact-crater material
- Noachian and Hesperian cratered terrain materials, undivided
- Noachian crater and cratered terrain materials, undivided
- Noachian interbedded lavas and eolian deposits, mantled

SEDIMENTARY MATERIALS

- Amazonian channel and flood-plain deposits
- Hesperian channel and flood-plain deposits
- Hesperian chaotic terrain material
- Eolian, dune, and mantle materials
- Landslide materials
- Layered and floor deposits of Valles Marineris
- Polar layered
- Polar ice

VOLCANIC MATERIALS

HIGHLAND VOLCANIC MATERIALS
- Amazonian lava flows
- Hesperian lava flows
- Hesperian and Noachian lava flow materials, undivided
- Volcanic shields and domes of all ages, undivided

LOWLAND VOLCANIC MATERIALS
- Amazonian lava flows and other volcanic units, undivided
- Intercrater volcanic plains materials of Amazonian and Hesperian age, undivided
- Hesperian volcanic flow units, undivided

(b) **The surface of Mars** ◄This view from the landing site of the Viking 2 landing module shows a rocky surface. The image was obtained in winter, and traces of white condensate are visible on the surface. The condensate is either water or carbon dioxide ice (or both), and is only a few microns thick. Approximately 30% of the atmospheric carbon freezes out to form extensive polar caps in the winter season; the southern winter is longer and colder, since the planet is then farther from the Sun. The rocks may be ejecta from impact craters. Several small trenches dug by the lander's sampling arm can be seen in the lower center of the image, next to the discarded cover for the apparatus. A number of chemical experiments were performed in a search for evidence of life on Mars (see Viking missions). (NASA P-22458.) See *Life: origins; Mars: atmosphere; Planets; Polar cap; Viking mission.*

PLATE 18 MARS STRUCTURAL GEOLOGY

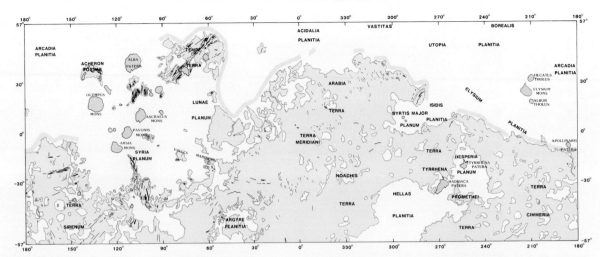

Mars: structural geology (a) The oldest exposed surface rocks are of the Noachian Period. Rocks of this age are shown in yellow on this map. The Martian surface has a pronounced structural boundary, shown in blue, which separates highland and lowland terrains (see *Mars: structural geology*). Shown in pink on the left are the volcanic centers of the Tharsis Rise. Faulting in this region (black lines) occurred in Noachian time, though the volcanoes are thought to be generally younger (Hesperian and Amazonian ages; see below). Ridges, shown in red, are another indication of tectonic activity during this early period.

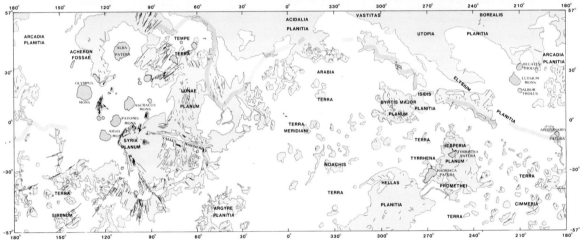

(b) Rocks and structures of the Hesperian Period. Once again, rocks of this age are shown in yellow, and other features (volcanoes, faults, ridges) are represented as in (a). The rocks are mostly lava flows. The flooding events that produced features like Valles Marineris occurred during this interval. Tectonic activity was waning, but a few features, notably ridges, clearly postdate the flooding events.

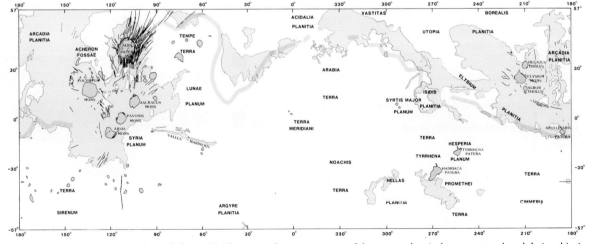

(c) Amazonian Period: structures are color coded as in (a). The large volcanic constructs of the western hemisphere were produced during this time. Although tectonism was decreasing, significant faulting did occur. (Maps by D.H. Scott and J.M. Dohm.) See *Mars: geology*; *Mars: structural geology*; *Tectonics*.

PLATE 19 MARS GRAVITY

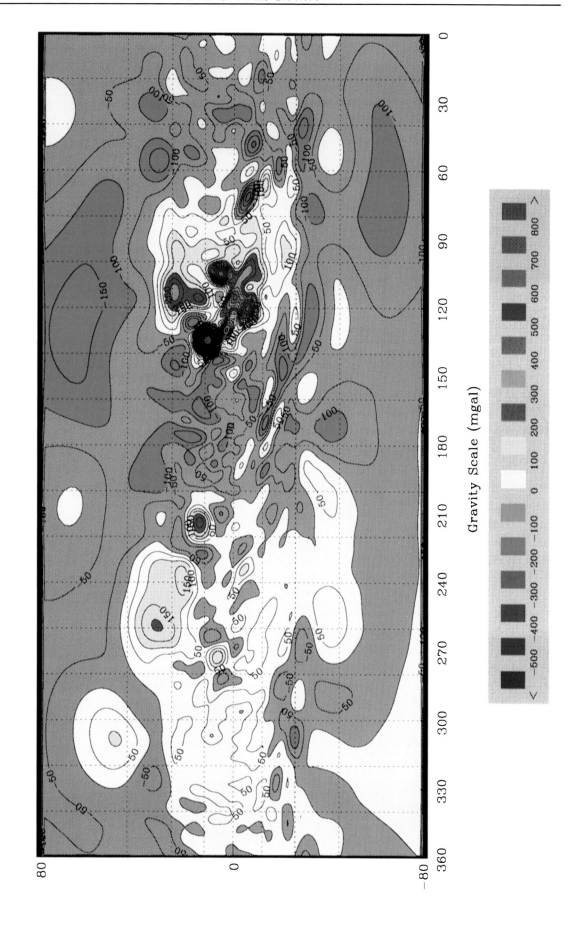

Mars: gravity This figure depicts Mars 'free-air' gravity anomalies derived from Viking Orbiter 1 and 2 and Mariner 9 tracking data. The gravity model is a spherical harmonic representation complete to degree and order 50, with horizontal resolution of about 300 km at the surface of Mars. The map shown is a Mercator projection. The regions of highest elevation on Mars, Tharsis and Olympus Mons volcanoes, show up as very strong positive gravity anomalies. Other weaker, but still prominent, positives lie over the Elysium volcanic region, the large Alba volcanic structure, and in the Isidis and Utopia impact basins. The strongest negative anomaly lies over the deep central Vallis Marineris canyons. (Courtesy D.E. Smith, Goddard Space Flight Center, NASA, 1992.) See *Gravity fields of the terrestrial planets; Mars: gravity; Radio science.*

PLATE 20 HALLEY'S COMET

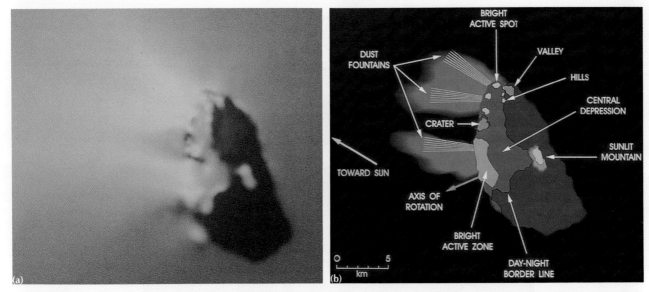

Nucleus of comet Halley The Giotto spacecraft of the European Space Agency flew by the nucleus of Halley's comet in March 1986. (a) A composite of 7 images taken at different distances, ranging from 25 600 to 2700 km. The resolution is about 570 m in the outer parts of the image, improving to about 60 m in the central part. The direction of the Sun is to the left, approximately 29° above the horizontal. (b) The features identified in this remarkable, first-ever close up view of a comet nucleus. (Image courtesy of R. Reinhard, ESA.) See *Comet; Comets: observation; Comets: structure and composition; Giotto mission; Halley, Edmond, and Halley's comet; Sakigake and Suisei missions; Vega missions.*

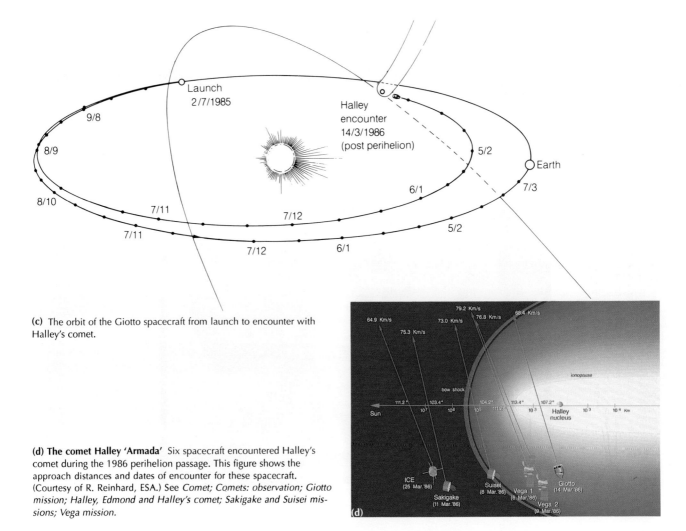

(c) The orbit of the Giotto spacecraft from launch to encounter with Halley's comet.

(d) **The comet Halley 'Armada'** Six spacecraft encountered Halley's comet during the 1986 perihelion passage. This figure shows the approach distances and dates of encounter for these spacecraft. (Courtesy of R. Reinhard, ESA.) See *Comet; Comets: observation; Giotto mission; Halley, Edmond and Halley's comet; Sakigake and Suisei missions; Vega mission.*

PLATE 21 COMET SHOEMAKER–LEVY 9

Comet Shoemaker–Levy 9 (a) In July 1992 an otherwise nondescript, unknown comet perhaps 10 km in diameter passed within about 50 000 km of the surface of the planet Jupiter. The comet was ripped apart by the strong tidal forces of the planet, and the fragments adopted a highly elliptical orbit about Jupiter. The comet was first observed in March 1993 by C. Shoemaker and D. Levy. It had been broken into perhaps 20 large fragments, giving the appearance of a 'string of pearls'. The comet collided with the planet Jupiter in July 1994.

This image, obtained in January 1994 by the repaired Hubble Space Telescope, shows the region near the brightest comet nucleus. The pressure of the solar wind is drawing out trails in the downstream direction. Although the disruption and even disappearance of comets has been observed previously, the impact of a cometary body on a planet has not. (Image based on observations with the NASA/ESA Hubble Space Telescope, courtesy of Space Telescope Science Institute, Association of Universities for Research in Astronomy Inc.) See *Comet*; *Comets: impacts on Earth*; *Comets: impacts on Jupiter*; *Comets: observations*; *Comets: origin and evolution*.

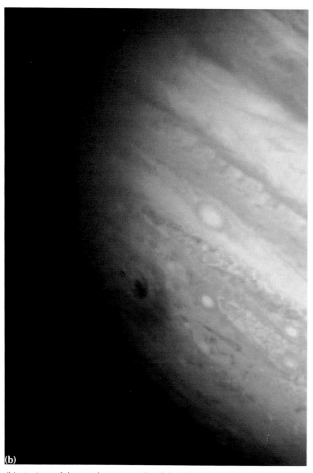

(b) A view of the southeastern side of the planet Jupiter following the impact of fragment G of comet P/Shoemaker–Levy 9 on 18 July 1994. The large dark impact scar is approximately the size of the Earth. To the left of the large impact site is a smaller dark spot that was produced by the impact of fragment D a day earlier. (Hubble Space Telescope image, NASA P-44430A).

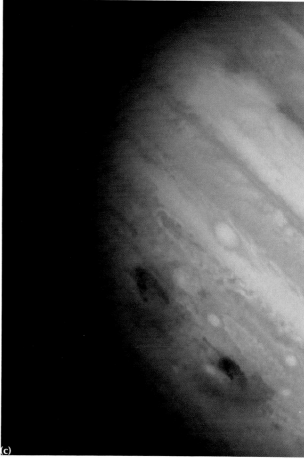

(c) Multiple cometary impact scars on Jupiter. The collision of the 20-plus fragments of comet P/Shoemaker–Levy 9 with the planet Jupiter in July 1994 produced multiple dark 'spots' at about 45° south latitude on the largest planet. The results of the impacts were observed by many astronomical observatories on Earth, and by the orbiting Hubble Space Telescope and the Galileo spacecraft. (NASA image P-44431B, obtained by the Hubble Telescope.)

PLATE 22 JUPITER GLOBAL VIEW, JUPITER'S RING

(a) **Jupiter** Images of Jupiter by the Voyager spacecraft reveal a striking atmosphere with circulating clouds that are organized into alternating dark and light stripes, called belts and zones respectively. The famous Great Red Spot was first described by Robert Hooke in 1664, and spans 20° of longitude by 10° of latitude. In it, small cloud features lap around every 6 days in the counterclockwise direction, signaling a high-pressure wind storm in the southern hemisphere.

Hundreds of previously unknown long-lived storms were discovered during the Voyager encounters, the majority of which are high-pressure storms or anticyclones. The ruddy hues in the otherwise white ammonia clouds that make up the surface are caused by an as yet undetermined trace contaminant, which may be upwelling sulfur, phosphorus or perhaps hydrocarbon material that has drifted downward from the overlying haze layer. (NASA P-37178). See *Atmosphere; Cyclone, anticyclone; Jupiter; Jupiter: atmosphere*.

Timothy E. Dowling

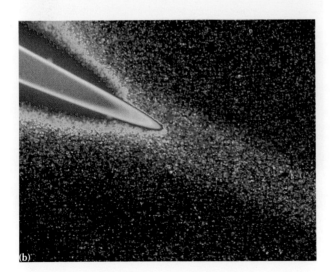

(b) **The Jovian ring system** This Voyager 2 image shows the Jovian ring very clearly in forward scattered light. The main ring (visible as a white belt) is especially bright in this viewing geometry due to the diffraction of light by micron-sized dust. Pseudocolor enhancement, in which the faintest material in the image is colored as a spectrum from violet to red, also shows the other two major components of the Jovian ring. The gossamer ring is the belt extending outward from the main ring in blue–violet. The interior halo can be seen arising at the main ring's inner edge; it is outlined in red–orange. (Courtesy of M. Showalter.) See *Dust; Jupiter: ring system; Planetary rings*.

PLATE 23 IO GLOBAL, IO VOLCANISM

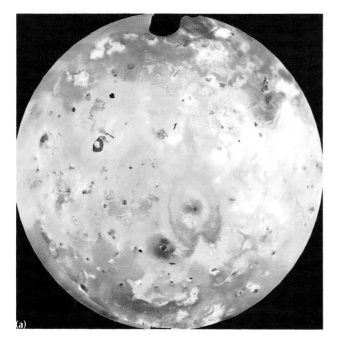

(a) **Io** Of all the satellites observed by the Voyager spacecraft in their journeys across the outer solar system, none was so dramatic in its geological activity as Io, the first Galilean satellite of Jupiter. This image is a natural or 'true color' composite of three identical views of Io taken through the Voyager filters transmitting the blue, green, and red regions of the spectrum, respectively. It displays Io approximately as the human eye, transported to Jupiter, would perceive it. Dull orange in overall color and covered with strange formations, the character of Io's surface is completely determined by its extraordinary volcanic activity and the yellow, orange, red, black and white sulfur and sulfur compounds that are its end products. (Image courtesy of C. Porco and P. Eliasen.) See *Galilean satellite; Io; Imaging science; Volcanism in the solar system.*

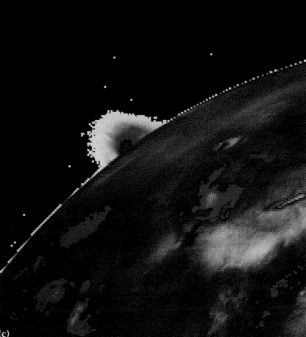

Volcanic eruption on Io These two views of an erupting volcano on Io serve to illustrate some of the techniques employed in image processing (see *Imaging science*).
(b) is a nearly full-disk monochrome Voyager image of Io showing its thoroughly mottled surface at an image scale of 4.5 km per pixel. The plume of Loki, one of the volcanoes active upon Voyager's arrival, can be seen extending above the limb of the satellite. Imaged in visible light, the plume's brightness is due to light reflected off particles ejected into Io's tenuous atmosphere.

(c) is a density-sliced image isolating the plume and showing clearly the distribution of brightness within the plume, a result of the variation in ejecta particle density. Scientifically valuable information like plume height and particle density variation with height can be readily determined from an image processed in this manner. (Courtesy of C. Porco and P. Eliasen.) See *Imaging science; Io; Volcanism in the solar system.*

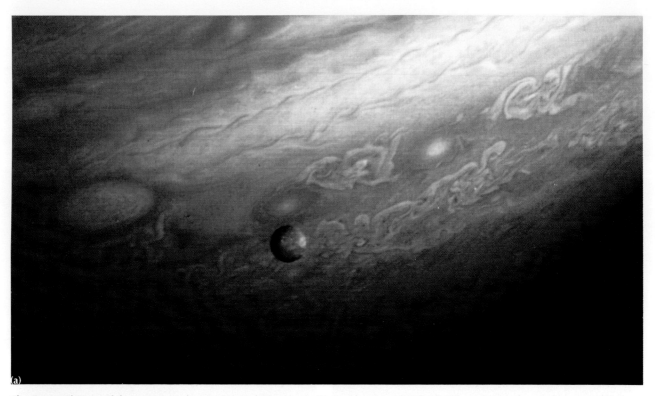

(a)

The Great Red Spot and the moon Io A dramatic image obtained by the Voyager 1 spacecraft in 1979. The volcanic moon Io is seen against the backdrop of Jupiter's turbulent atmosphere. Io is the innermost of the Galilean satellites, orbiting at a distance of 422 000 km. The Great Red Spot, an atmospheric disturbance that may have persisted for hundreds of years, is visible at left. (JPL P-21719, courtesy of NASA.) See *Io; Jupiter; Jupiter: atmosphere*.

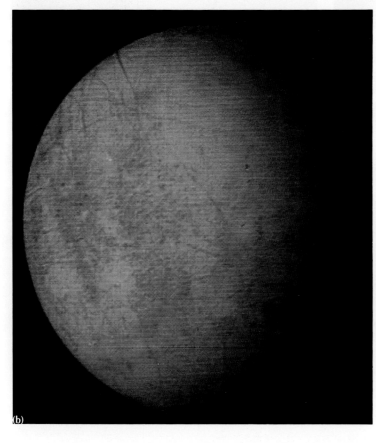

(b) **Europa** Europa is the brightest of the Galilean satellites of Jupiter. Its radius is 1569 km; with a density of 2970 kg m^{-3} Europa is slightly smaller and a little less dense than the Earth's Moon. Spectroscopic investigations suggest that the surface is nearly pure water ice. The lineations shown indicate that fracturing of the surface ice has occurred; darker and slightly redder subsurface material has welled up to fill the fractures. Europa is one of the most interesting of the moons of the solar system; it is possible that an ocean of liquid water may lie beneath its surface. (NASA P-21752, courtesy of JPL.) See *Europa; Galilean satellite; Galileo mission; Jupiter: satellite system*.

Callisto ▶ The outermost Galilean satellite of Jupiter, Callisto has a radius of about 2400 km, orbiting Jupiter at a distance of 1.88 million km. The density of Callisto suggests that it is composed of roughly equal proportions of water, ice and silicate rocks. The surface of Callisto is heavily cratered, reflecting a puzzling absence of geological activity since the period of early bombardment approximately 4.0–4.4 billion years ago. In contrast, the other Galilean satellites all show evidence of more recent geological activity that has erased many craters and modified others.

The bright features at the upper left center are near the center of the Asgard double multiring system, which is over 1600 km in diameter. The rings are barely visible in this global image. (NASA P-21748, courtesy of JPL.) See *Callisto; Galilean satellites*.

PLATE 25 GANYMEDE AND CALLISTO

Ganymede One of the Galilean moons of Jupiter, Ganymede is the largest natural satellite in the solar system. With a diameter of 5262 km, it is larger than the planets Mercury and Pluto. Ganymede orbits Jupiter at a distance of 1.07 million km, completing one cycle in 7.155 days. Ganymede is in synchronous rotation, so that it always presents the same 'face' to Jupiter.

Ganymede has a bulk density of 1.94 gcm^{-3}, which is about twice that of liquid water. This density, in combination with cosmochemical considerations, implies that the satellite is composed of H_2O and silicate rock in roughly equal portions. If the material within Ganymede is completely differentiated (with the least dense materials forming the outer layer and the most dense at the center) then the boundary between a rocky core and an H_2O outer shell should lie about two-thirds of the distance from the center to the surface.

The two principal terrain types, the bright and dark terrains, are evident in this image, along with impact craters, the brightest features, showing ejecta rays. The dark terrains are oldest, based on cratering frequencies, while the bright terrains appear to have erupted and filled surface grabens and depressions at a later time. Within the dark terrain are roughly circular patches of lighter albedo, termed 'palimpsests.' (Voyager P-21751.) See *Galilean satellites; Ganymede*.

Callisto

Saturn A true-color composite image of Saturn and four of its moons, obtained in August 1981 by the Voyager 2 spacecraft. The three bright moons, from left to right, are Tethys, Dione and Rhea. The shadow of Tethys appears on Saturn's southern hemisphere. A fourth moon, Mimas, appears as a bright spot near Saturn's lower limb, to the left of Tethys; Mimas also casts a shadow, just beneath the rings of Saturn. The banded structure of Saturn's atmosphere is well resolved. (NASA P-23887.) See *Saturn; Saturn: atmosphere; Saturn: satellite system.*

Imaging science: Saturn's rings▶ Saturn, with its magnificent system of rings, is arguably the most beautiful body in our solar system, and has been a source of admiration and intrigue for astronomers for hundreds of years. The Voyager flybys of Saturn in 1980 and 1981 brought the study of Saturn and its rings and satellites into the modern era.

(a) is a 1-s exposure of Saturn and its rings taken with the vidicon wide-angle camera carried aboard Voyager 2 from a distance of 1.5 million km. The image restoration procedure is demonstrated with a comparison of this unprocessed, raw image and a processed image (b), in which the spatially variable dark current, shading and reseau pattern have been removed. Correction of the geometric distortion present in the vidicon camera, utilizing a mathematical mapping from the known locations of the reseaux to their observed locations in (a), is shown in (c). The effect of removing the inherent, though barely noticeable, barrel distortion is seen in the way the corners of the image have been extended outward. Once this stage is reached, an accurate Saturn-centered coordinate system may be mapped onto the globe of Saturn, or onto its rings, so that the locations of features in the image may be measured. When measurements of this type are made over many consecutive images, the motion of atmospheric clouds from one image to the other – i.e. wind speeds – or the orbital motion of features within the rings may be determined.

To accentuate atmospheric and ring features, different processing techniques may be applied. Saturn is a cold planet, situated 1.5 billion km away from the Sun, and its clouds lie deep in its atmosphere under an overlying layer of haze. The scattering of sunlight by the haze causes the clouds to appear with low contrast and little color. (They are significantly lower in contrast and less colorful than the clouds on Jupiter, which is closer to the Sun and warmer.) A nonlinear contrast stretch of (b), shown in (d), maximizes the contrast of the scene for both planet and rings. No new information has been added: the information present in the raw image is simply more easily seen after the image restoration and enhancement stages. For example, the banded structure of Saturn's northern hemisphere becomes more readily apparent.

The amount of fine-scale structure present in the rings was one of the great surprises of Voyager's encounter with Saturn. High-pass spatial filtering applied to (b), shown in (e), enhances the spatial detail in the rings and the differences in the distribution of these details across the entire ring system, while suppressing the differences in brightness apparent in (b). For example, the normally darker and more transparent C ring and Cassini Division now appear with the same overall brightness as the normally more brilliant A and B rings, making the measurement of ring feature locations all the easier.

To look for subtle variations in ring brightness, (b) is processed using the density slicing technique, deliberately setting the globe of Saturn to black (f). With a yellow–red–green–blue–purple color continuum, the lowest data numbers (darkest ring areas) are mapped to yellow and the highest data numbers (brightest ring areas) are mapped to purple. This pseudo-color picture helps distinguish many more levels of photometric detail than are perceived in the initial monochrome image. Notice, for example, how easily visible the spokes in the B ring (the predominantly green/blue ring) have become. A curious variation in brightness with longitude is seen in the outermost A ring, changing from mostly red in the upper part of the image to blue/green in the lower part. This effect, known as the A ring azimuthal asymmetry, is believed to be related to the manner in which particles in the A ring clump together as they orbit Saturn. (Images courtesy of C. Porco and P. Eliasen.) See *Imaging science; Planetary rings; Saturn: ring system.*

PLATE 27 IMAGING SCIENCE: SATURN'S RINGS

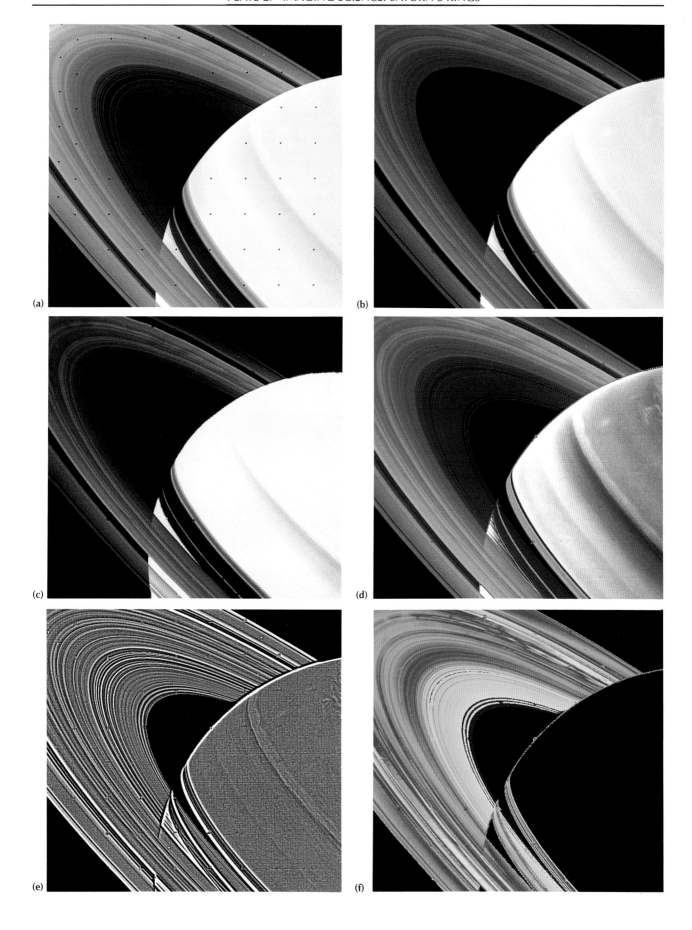

PLATE 28 IMAGING SCIENCE: SATURN ATMOSPHERE

Imaging science: Saturn atmosphere High-speed winds howl through Saturn's atmosphere. Concentrating on Saturn's northern hemisphere, illustration (b) of the previous plate has been more severely contrast-enhanced to produce (a). The dark areas are those in which the sunlight has been primarily absorbed and relatively little light is reflected; usually these regions are free of clouds. The bright areas are generally reflective clouds relatively high in the atmosphere. The stratigraphy of the atmosphere, as well as the structural definition within the clouds and eddies and the wavy jet stream, can be made more obvious by contrast enhancement.

The fine-scale details can be enhanced even further still (at the expense of brightness differences) by using the horizontal first-difference technique (b). Here, abrupt differences in brightness take on the appearance of 'edges' and are more readily apparent since the smoothly varying changes in brightness have been suppressed. A false-color composite of (a) and (b), shown in (c), clearly displays in a single image the information contained in both. One can see now the subtle differences in the clouds, haze and jet stream patterns, attainable with a first-difference image, together with the vertical stratigraphy of the atmosphere which is now encoded in color instead of brightness: blue is dark and low, pink is bright and high. Comparison of (a) from the previous plate and (c) underscores the power of image processing in extracting scientific information. (Images courtesy of C. Porco and P. Eliasen.) See *Imaging science; Saturn: atmosphere.*

(a)

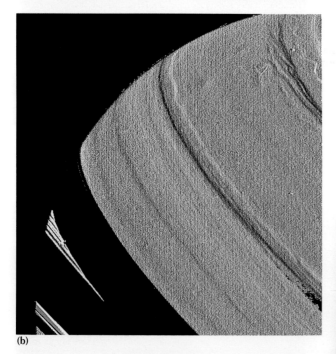

(b)

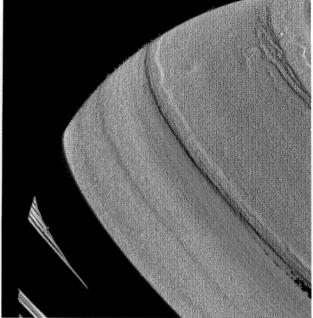

(c)

PLATE 29   ENCELADUS, TITAN

Enceladus The seventh of the 17 moons of Saturn, Enceladus is the brightest moon in the Saturn system, with an albedo of 0.89. It is relatively small, with a diameter of 394 km, and it orbits at a distance of about four Saturn radii. It shows both smooth surfaces and cratered areas, suggesting that some ice volcanism has occurred to resurface parts of the moon. The rims of several craters near the lower center of this image appear to have been flooded by smooth terrain materials. Enceladus has been suggested to be the major source of material for the E ring of Saturn (Hamilton, D.P. and Burns J.A., 1994, *Science*, **264**, 550–3). This Voyager 2 mosaic is the highest resolution image of Enceladus, obtained in November 1980. (NASA P-23955). See *Enceladus; Saturn: ring system; Saturn: satellite system.*

Titan Titan is the largest satellite in the Saturn system, with a radius of 2575 km, approximately 40% of that of the Earth. It is the only satellite with a dense atmosphere. This enhanced image shows thick haze layers obscuring the planet. It is possible that Titan has surface oceans, most probably composed of methane and ethane. Titan will be studied by the Huygens Probe, which will be deposited into the Titan atmosphere by the Cassini spacecraft early in the next century. Titan's reducing atmosphere is of scientific interest because of possible analogies with the atmosphere of the early Earth. This image was obtained on 12 November 1980 by the Voyager 1 spacecraft. (NASA P-23108.) See *Atmosphere; Cassini mission; Huygens mission; Saturn: satellite system; Titan.*

PLATE 30 URANUS

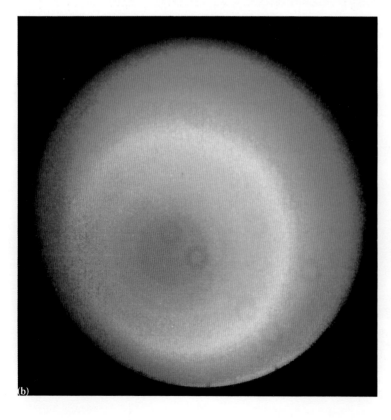

Uranus Natural and contrast-enhanced false color images of Uranus acquired by the Voyager 2 spacecraft. The globe appears featureless in natural color (a). A dark polar hood and zonal banding concentric about the pole of rotation appears in the contrast-enhanced version (b). Uranus's natural bluish color occurs as a result of the absorption of orange–red sunlight by methane gas and by clouds at the 2.7 bar level (see *Uranus: atmosphere*).

Uranus is the fourth most massive planet, orbiting at a distance of 19.18 AU from the Sun. The planet is tilted by 98° with respect to its orbital plane, so that at certain times in the orbit the poles of the planet point nearly at the Sun. The best hypothesis for this unusual orientation is that the impact of a planet-sized body into the primordial Uranus, during planetary accretion, caused the reorientation of the spin axis. The fact that the satellite system of the planet shares this unusual obliquity suggests that the event occurred in the early stages of solar system formation.

The extreme contrast enhancement has brought out donut-shaped features which are due to dust and flaws in the optics. The reddish rim is an artifact of the processing. (NASA P-29748.) See *Planet; Planetary rotation; Solar system: origin; Uranus; Uranus: atmosphere*.

PLATE 31 NEPTUNE

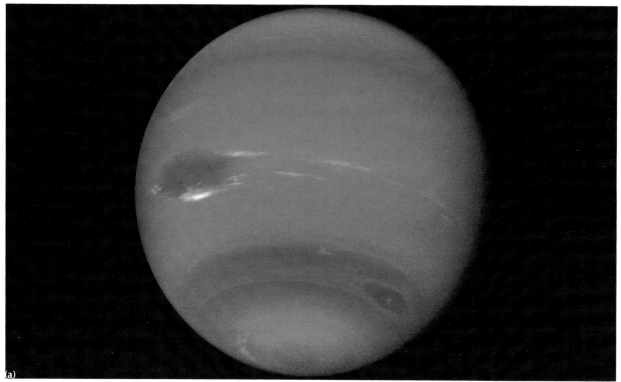

(a) **Neptune** Neptune is the most distant of the four giant planets, orbiting the Sun at a distance of 30.06 AU. This image was obtained by the Voyager 2 spacecraft during the Neptune flyby of August 1989. The 'Great Dark Spot' of Neptune is visible at the left-hand side of the image; the second dark spot, known as D2, is at the lower right. Voyager observations of the magnetic field of the planet determined a rotation period of 16.11 h. (NASA P-36611.) See *Neptune; Neptune: atmosphere; Neptune: magnetic field and magnetosphere.*

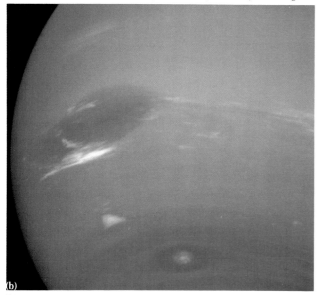

(b) **Neptune atmosphere** A higher-resolution view of the atmosphere of Neptune. This color image of Neptune was obtained by the Voyager 2 spacecraft in August 1989. Neptune's blue color is due to absorption of red light by methane gas and cloud material, predominantly within the troposphere. Prominent features include, progressing from center-left to center-bottom, (1) the Great Dark Spot (GDS), accompanied by (2) its bright, upper-level companion cloud feature, (3) the Scooter, and (4) the second dark spot, D2. The clouds of the Scooter lie considerably deeper within the atmosphere than do the bright companion clouds of the GDS. (NASA/JPL photograph P-34648.) See *Neptune: atmosphere.*

(c) **Clouds in Neptune's atmosphere** High spatial resolution (11 km per pixel) color image of Neptune near 27°N latitude showing cloud streaks and associated shadows stretched approximately along lines of constant latitude. This image, taken near the terminator in violet, green and orange filters 2 h before closest approach, provides evidence of vertical relief in Neptune's 50–200 km wide bright cloud streaks. The bright sides of the clouds which face the Sun are brighter than the surrounding cloud deck because they are more directly exposed to sunlight. Shadows can be seen on the opposite side. Rayleigh scattering causes shadows to be less distinct at short wavelengths (violet filter) and more distant at long wavelengths, thus giving the shadows a noticeable blue tint. Analysis of cloud shadow profiles extracted from the orange filter image indicate that the cloud streaks are 70–110 km above the underlying cloud deck. (NASA/JPL photograph P-34709.) See *Neptune: atmosphere.*

Triton Triton is the largest satellite of Neptune. This photomosaic of Triton was obtained during northern winter, during the Voyager 2 flyby in August 1989; the subsolar point is near latitude 55°S. In this image the equator runs more or less vertically near the contact between the bright frosted region on the left and the darker region on the right. The pinkish frost cap on the left is thought to represent sublimating deposits of nitrogen that contain a trace of methane and radiation-produced chromophores; the dark streaks in this region are thought to be wind-deposited dust or ice grains that have been lofted by geyser-like emissions. The more neutrally colored area on the right is probably covered by a thin veneer of annealed nitrogen ice. The large, diffuse, bluish patch in the middle of the image is probably fresh nitrogen frost. The fresh frost may be only a millimeter thick. Some scientists have suggested that this frost deposit was formed during the years just prior to the Voyager 2 flyby. The complex geologic province in the upper right is the 'canataloupe terrain' (properly named Bubembe Regio). (Photomosaic courtesy of Alfred McEwen, US Geological Survey Branch of Astrogeology.) See *Neptune: satellite system; Triton.*

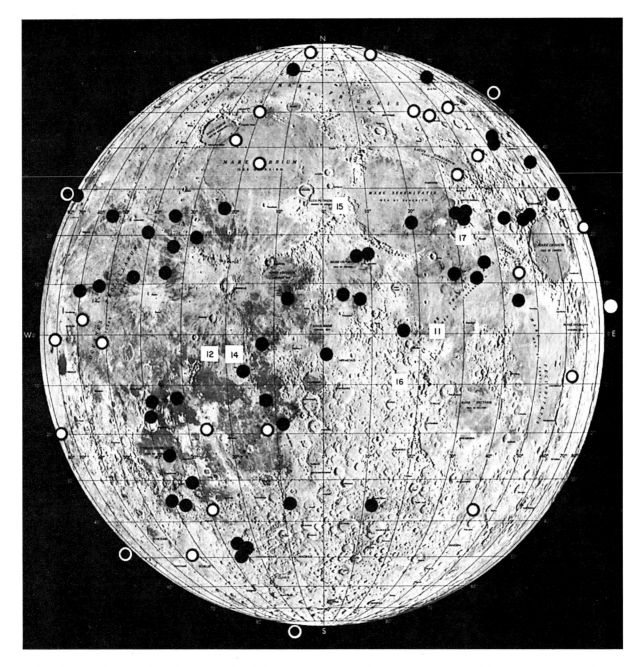

Figure M96 Moonquake epicenters. Each of the closed circles represents an epicentral region within which many deep moonquakes occur at monthly intervals. Open circles represent epicenters of observed shallow moonquakes. Those shown outside the rim of the Moon indicate epicenters located on the farside of the Moon.

ature variations on the lunar surface. It is possible that they represent slumping on lunar surface slopes triggered by thermal stresses.

Comparison with the Earth and other planets

The types of seismicity we observe on the Moon are different from those observed on the Earth. Both tidally induced earthquakes (the possible counterpart of the deep moonquakes) and thermal earthquakes may occur inside the Earth. However, even if they do, their magnitudes are too small to be detectable above the Earth's microseismic background noise. These two types of seismicity are observable on the Moon simply because the Moon's surface is so seismically quiet relative to the Earth. This is because the Moon has neither atmosphere nor ocean (the major sources of microseisms on the Earth). Practically all meteoroids encountering the Earth disintegrate as they enter the Earth's atmosphere, and those which reach the ground (meteorites) have decelerated sufficiently that they rarely generate globally detectable seismic waves.

Shallow moonquakes appear to be the only truly tectonic quakes in the Moon. However, even though the overwhelming majority of earthquakes are of tectonic origin, there are significant differences between shallow moonquakes and earthquakes. Most earthquakes occur along lithospheric plate boundaries, where large stresses caused by the relative motion of plates are concentrated. In contrast, there are no surface geologic features on the Moon that indicate the existence of moving lithospheric plates. Apparently, the extraordinary thickness of the Moon's lithosphere (Figure M97) precludes the occurrence of plate-like segmentations. Thus it is not surprising

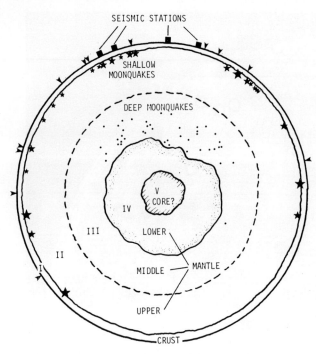

Figure M97 A schematic cross-section of the Moon showing structure of the Moon's interior, as deduced from observations of seismic waves from moonquakes and impacts, and locations of shallow (stars) and deep (dots) moonquake hypocenters as projected on the equatorial plane. The Earth is towards the top of the figure, and the small squares indicate locations of the Apollo seismic stations. The lithosphere extends from the surface to the boundary just below the level of deepest moonquakes. (From Nakamura et al., 1982).

that no moonquakes similar to typical interplate earthquakes have been observed. On the other hand, shallow moonquakes do have many similarities with intraplate earthquakes. These are the earthquakes that occur in the interiors of lithospheric plates, usually along some pre-existing weaknesses, and include some of the largest earthquakes ever observed. As the Moon has a single lithospheric plate, all moonquakes are 'intraplate'; thus such similarities are not surprising.

There is a significant difference in the overall level of seismicity between the Earth and the Moon. The total energy annually released by earthquakes is about 10^{18} J (10^{25} erg); in contrast, moonquakes annually release about 10^{12} J (10^{19} erg), mostly from shallow moonquakes. Because the Moon is smaller than the Earth, its interior has cooled faster and thus there is significantly less tectonic activity at the present time. However, the Moon is by no means a dead planet.

Comparisons with other planets are not straightforward because seismicity on other planets is yet to be observed. Limited observations on Mars during the Viking mission set certain limits on its seismicity, but these data only indicate that Mars is not as active seismically as the Earth. Whether it is more or less active than the Moon is not yet known. The present level of seismicity of various terrestrial planets may be estimated, based on analysis of surface geologic features, but the validity of such estimates needs to be verified with future direct observations.

Yosio Nakamura

Bibliography

Cheng, C.H. and Toksöz, M.N. (1978) Tidal stresses in the Moon. *J. Geophys. Res.*, **83**(B2), 845–53.
Duennebier, F. and Sutton, G.H. (1974) Thermal moonquakes. *J. Geophys. Res.*, **79**(29), 4351–63.
Goins, N.R., Dainty, A.M. and Toksöz, M.N. (1981) Seismic energy release of the Moon. *J. Geophys. Res.*, **86**(B1), 378–88.
Lammlein, D.R. et al. (1974) Lunar seismicity, structure and tectonics. *Rev. Geophys. Space Phys.*, **12**(1), 1–21.
Nakamura, Y. (1978) A1 moonquakes: source distribution and mechanism, in *Proc. Lunar Planet. Sci. Conf.*, **9**, pp. 3589–607.
Nakamura, Y. (1980) Shallow moonquakes: how they compare with earthquakes in *Proc. Lunar Planet. Sci. Conf.* **11**, pp. 1847–53.
Nakamura, Y. et al. (1974) High-frequency lunar teleseismic events, in *Proc. Lunar Sci. Conf.* **5.**, *Geochim. Cosmochim. Acta.*, **5**, (suppl), 2883–90.
Nakamura, Y. et al. (1979) Shallow moonquakes: depth, distribution and implications as to the present state of the lunar interior, in *Proc. Lunar Planet. Sci. Conf.* **10**, pp. 2299–309.
Nakamura, Y., Latham, G.V. and Dorman, H.J. (1982) Apollo lunar seismic experiment – final summary in *Proc. Lunar Planet. Sci. Conf.* **13**, Part 1. *J. Geophys. Res.*, **87** (suppl.), A117–23.
Oberst, J. (1987) Unusually high stress drops associated with shallow moonquakes. *J. Geophys. Res.*, **92**, 1397–405.
Oberst, J. and Nakamura, Y. (1991) A search for clustering among the meteoroid impacts detected by the Apollo lunar seismic network. *Icarus*, **91**, 315–25.

Cross references

Seismicity
Thermal evolution of planets and satellites
Tidal friction

N

NASA

The National Aeronautics and Space Administration (NASA) was established to direct the American space program in response to the launching of Sputnik in 1957. It was created by The National Aeronautics and Space Act of 1958, which was signed by President Eisenhower on 29 July 1958. The defined objectives of the agency were:

1. Expansion of knowledge of phenomena in the atmosphere and space.
2. Improvement of the usefulness, performance, speed, safety and efficiency of aeronautical and space vehicles.
3. Development and operation of vehicles capable of carrying instruments, equipment, supplies and living organisms through space.
4. Establishment of long-range studies of the potential benefits to be gained from the utilization of space activities for peaceful and scientific purposes.
5. Preservation of the role of the United States as a leader in aeronautical and space science and technology and in the application thereof to peaceful activities.
6. Making available to agencies concerned with national defense discoveries that have military value or significance, and the corresponding exchange from defense agencies to NASA.
7. Cooperation by the United States with other nations and groups of nations in space activities.
8. The most effective utilization of the scientific and engineering resources of the United States.

The act established the positions of Administrator and Deputy Administrator, to be appointed by the President, to direct the agency and a National Aeronautics and Space Council to advise the President with respect to the performance of the duties of NASA. The council included the President, the Secretaries of State and Defense, the Administrator of NASA, the Chairman of the Atomic Energy Commission and no more than four other members. NASA absorbed the functions, property and personnel of the National Advisory Committee for Aeronautics (NACA), which since 1915 had been charged with the coordination of research and development in aeronautics.

Dr T. Keith Glennan was appointed as the first administrator, and on 1 October 1958 NASA was proclaimed to be organized and prepared to discharge its duties. Later in the same month Project Mercury was organized to orbit men around Earth, and NASA's first space probe, Pioneer 1, was launched. In the three decades since, NASA has launched nearly 1000 vehicles into space.

Between 1961 and 1972 the human spaceflight program achieved 27 successful missions, including six one-person Mercury flights, 10 two-person Gemini flights and 11 three-person Apollo flights, which landed six crews on the Moon. In 1973 three crews spent a total of nearly 6 months in Skylab, and in July 1975 we witnessed the Apollo Soyuz Test Project, in which three American astronauts and two Soviet cosmonauts linked spacecraft and fraternized in space.

Since Apollo, all human missions have been performed by the Space Transportation System (STS), popularly known as the Space Shuttle, which had 50 successful launches from 1981 through 1992. Carrying crews as large as eight men and women, the Shuttle has been used to launch, retrieve and repair satellites, make observations of the Earth and the Sun, and conduct a variety of scientific and engineering experiments.

NASA has sent robotic spacecraft to investigate the Sun, the Moon and every planet in the solar system except Pluto in the Explorer, Pioneer Ranger, Mariner, Surveyor, Lunar Orbiter, Viking and Voyager Programs and the Magellan, Galileo, and Mars Observer missions. It has launched Earth satellites to provide instantaneous communication over the whole Earth and between satellites, to monitor the weather and the motion of the tectonic plates, to make possible precise navigation on land or sea, to measure the level of the oceans and to take photographs of the Earth that have myriads of applications for both military and civilian purposes.

NASA's expenditure in its first year (fiscal year 1959) was $331 million. With the initiation in 1961 of the Apollo program, the budgets began to rise dramatically to a peak above $5.1 billion in fiscal years 1964, 1965 and 1966. These sums are equivalent to more than $21 billion in 1991 dollars. After declining for several years, budgets began to climb again in FY 1975 to support the Shuttle, and the budget for FY 1993 is approximately $14.3 billion. The major items are approximately $5 billion for space flight, control and data acquisition (primarily for the Shuttle), $2 billion for the space station, $3 billion for space science and applications, and $1 billion for aeronautics research and technology.

Headquarters for NASA is in Washington, DC. Most of its work is done through its field centers. Most of the centers are engaged in a variety of research and development projects, but the principal responsibility of each is as follows:

- Lyndon B. Johnson Space Center in Houston, TX: human flight operations in space.
- George C. Marshall Space Flight Center in Huntsville, AL: development of launch vehicles for human and robotic spacecraft.
- Goddard Space Flight Center in Beltsville, MD: research with Earth satellites.
- Jet Propulsion Laboratory in Pasadena, CA (operated under contract with the California Institute of Technology): deep-space, lunar and interplanetary exploration.
- Ames Research Center at Moffett Field, CA: research in aeronautics and space environmental physics; robotic space missions.
- Langley Research Center in Hampton, VA: aeronautical and space structures, lunar and planetary missions.
- Lewis Research Center in Cleveland, OH: launch vehicles, power plant and propulsion.
- Dryden Flight Research Facility at Lancaster, CA: research in extremely high-performance aircraft and spacecraft.

- John C. Stennis Space Center at Bay St Louis, MS: rocket engine tests and Earth resources research.

In addition, NASA has two facilities for launching rockets – John F. Kennedy Space Center at Cape Canaveral, FL, for large rockets and Wallops Flight Facility Wallops Island, VA, for small rockets.

Besides conducting Earth-satellite and space missions, NASA sponsors a wide variety of research programs at its field centers and through grants to universities and industrial laboratories.

Conway W. Snyder

Bibliography

Alexander, K. (1989) *Countdown to Glory: NASA's Trials and Triumphs in Space.* Los Angeles: Price, Stern, Sloan.
Anon. (1991) *A Spacefaring Nation: Perspectives on American Space History and Policy.* Washington: Smithsonian Institution Press.
Benedict, H. (1989) *NASA: the Journey Continues.* Houston: Pioneer Publications.
De Waard, E.J. and DeWaard, N. (1984) *History of NASA, America's Voyage to the Stars.* New York: Exeter Books.
Glennan, T.K. (1993) *The Birth of NASA: the Diary of T. Keith Glennan.* Washington: NASA History Office, National Aeronautics and Space Administration.
Mirabito, M.M.H. (1983) *The Exploration of Outer Space with Cameras: a History of the NASA Unmanned Spacecraft Missions.* Jefferson, NC: McFarland.

Cross references

Apollo missions
Deep Space Network
History of planetary science II
Mariner missions
Mars Observer
Pioneer 10 and 11 missions
Pioneer Venus
Ranger missions
Ulysses mission
Viking mission
Voyager missions

NEAR-EARTH OBJECT

Our nearest neighbors

Asteroids as well as comets in near-Earth space collectively make up a population referred to as near-Earth objects (NEOs). Because these bodies are efficiently removed from the inner solar system by collisions or gravitational interactions with the terrestrial planets, the NEO population we see today must be continually resupplied. Any bodies that remained in the inner solar system immediately after the formation of the planets would have been depleted long ago. This process of depletion has had consequences for the geological evolution of the terrestrial planets, as evidenced by the existence of large impact basins and craters on the airless surfaces of the Moon and Mercury. The surfaces of Mars and Venus (even with its thick atmosphere) both show an abundance of craters. On our own planet, oceans plus relentless tectonic and weathering processes have mostly obscured the impact record. Depleting the NEO population through impacts may have also had profound consequences for biological evolution on Earth, such as the mass extinction that occurred at the end of the Cretaceous period 65 million years ago (see Mass extinction: Cretaceous–Tertiary boundary).

All that distinguishes the classification of a newly discovered NEO as either an asteroid or a comet is its telescopic appearance. Even objects discovered in orbits similar to those of known short-period comets are classified as asteroids if they do not display a coma or tail. Thus the population of objects colloquially called near-Earth asteroids (NEAs) most likely contains numerous low-activity or dormant comets. Extensive observations with increasing precision are providing evidence that cataloged asteroids such as 1566 Icarus, 2201 Oljato and 3200 Phaethon may have had a cometary genesis. One near-

Table N1 Types of near-Earth asteroidal objects

Amors – asteroidal objects which have semimajor axes greater than 1 AU and perihelion distances between 1.017 and 1.3 AU. Although Amor asteroids do not cross the Earth's orbit, their orbital characteristics and close approaches make them accessible spacecraft rendezvous targets

Apollos – asteroidal objects having semimajor axes greater than or equal to 1 AU and perihelion distances less than 1.017 AU. Apollo asteroids cross the orbit of the Earth

Atens – asteroidal objects which have orbital semimajor axes less than 1 AU and aphelion distances greater than 0.983 AU. Aten asteroids cross the orbit of the Earth

Earth object, cataloged as asteroid 4015, was subsequently realized to be the same object as a comet known as Wilson-Harrington.

Asteroidal bodies in near-Earth space are categorized as Atens, Apollos or Amors (Table N1). Names for each group come from their best-known members 1862 Apollo, 1221 Amor and 2062 Aten. At present about 200 NEAs have been discovered. There are probably 5000 to 10 000 of these objects larger than 0.5 km in diameter, suggesting that we have so far detected less than 1% of their total population. The largest is the 40-km 1036 Ganymed and the smallest are in the size range of 5–10 m across. One of these small NEAs, known as 1991 BA, passed within 0.0011 AU (one-half the distance to the Moon) in January 1991.

Objects displaying cometary properties are categorized according to their orbital period (see Comet). At present there are about 200 known short-period comets, representing a small fraction of their total population. Although most have orbits that carry them well above and below the plane of the ecliptic, approximately 30 short-period comets have inclinations low enough to make them accessible from Earth using spacecraft with modest propulsion systems. Objects in this category include Giacobini-Zinner, the first comet to be visited by a spacecraft, and comet Encke, perhaps the most carefully studied comet after Halley. Their nuclei, which are about 10 km across and may store virtually unaltered material from the early protoplanetary nebula, serve as reservoirs of volatile ices and dust. Heating by the Sun releases gas and dust from the body, thus giving comets their distinctive comas and tails.

The asteroidal and cometary near-Earth objects represent the upper size portion of a distribution that extends down to meteoroid-sized objects and dust. Bodies larger than ~ 10 m are able to survive atmospheric passage and deliver meteorites to the Earth's surface. Thus meteorites provide us with vital laboratory samples from objects in near-Earth space. High-altitude flights to collect interplanetary dust particles are yielding samples that are broadly expanding our knowledge of the chemistry of the solar system.

Applying the scientific method to achieve an understanding of the population of near-Earth objects leads us to frame a few key questions which must be answered before we can fully unravel their mysteries.

Origins

The question of the origin of NEOs is most properly considered in terms of the need to identify their source of resupply. One or more processes must be active to add new objects to the inner solar system at about the same rate with which they are removed by planetary encounters.

Cometary NEOs appear to be supplied from the Oort cloud, a hypothesized reservoir of primordial solar system material encircling the Sun more than 10 000 AU away. A second source for comets may be from a closer disk of material just a few times more distant than Pluto (See Kuiper belt). Although all such objects that randomly venture into the inner solar initially have long orbital periods, some are perturbed into short period orbits through interactions with Jupiter and the other planets (see Comet: origin and evolution).

One hypothesis for resupplying asteroidal NEOs is that they are derived from the main asteroid belt, the region between Mars and Jupiter where nearly 20 000 objects have known orbits. A process called chaotic dynamics (q.v.) may be responsible for removing objects from the asteroid belt and placing them on trajectories which bring them near the Earth. Suspect supply areas include the famous Kirkwood gaps, highly depleted regions within the main belt that were first noted in 1867 by the American astronomer Daniel Kirkwood. The locations of these gaps correspond exactly to positions of mean motion resonances with Jupiter. At these locations the orbital period of an asteroid is an exact integer ratio of Jupiter's, e.g. at the 3 : 1 resonance an asteroid completes exactly three revolutions in the time it takes Jupiter to complete one (see Asteroid: resonances).

It has been shown that objects orbiting in a 3 : 1 mean motion resonance with Jupiter (the location of one of the Kirkwood gaps at 2.5 AU), exhibit chaotic increases in their orbital eccentricity, thus allowing their orbits to cross that of Mars. Dynamical interactions with Mars can then further aid their delivery to the inner solar system. Apart from the dynamical calculations which support this hypothesis, observational evidence shows that many NEAs have aphelia within the asteroid belt. Also, most appear spectrally similar to main-belt asteroids.

Relationship between asteroids and comets

A second hypothesis for resupplying the Apollo, Amor and Aten asteroidal NEOs is from dormant or extinct comet nuclei. The end stages of a comet's life are poorly understood. One scenario is that as volatile ices in the upper layers are depleted, an inert mantle forms which effectively seals off and insulates volatiles within the interior. Without any outgassing to form a coma or tail, such a body would effectively display an asteroidal appearance. Apart from this model for cometary evolution, observational evidence which supports this hypothesis includes numerous asteroidal NEOs which have orbits similar to known short-period comets. At least one of these cataloged asteroids, 3200 Phaethon, is known to be associated with a meteor stream (the Geminids). Previously, meteor streams (q.v.) were known to be associated only with active comets. Further, the orbits of some asteroidal NEOs do not appear to follow strict gravitational dynamics. Intermittent outgassing could slightly modify their orbits in a manner analogous to the firing of small thruster rockets on a spacecraft. The action of such nongravitational forces has long been known to affect the orbits of active comets.

Relationship between NEOs and meteorites

The association of active comets and at least one asteroidal NEO with meteor streams is well established. It is also widely believed that the dynamical processes which can deliver asteroids from the main belt to the inner solar system are similarly effective over a continuum of sizes, down to that of meteorites. Analyses of the trajectories of meteorites have shown that many were traveling on orbits that extended out into the asteroid belt prior to their atmospheric entry.

Analyzing the spectral colors of NEOs allows us to investigate their mineralogy. Our mineralogical interpretations of NEO surfaces are intimately tied to our direct laboratory measurements of the compositions of meteorites. Both kinds of measurements are vital to our studies of solar system cosmochemistry. However, the pieces of the puzzle do not yet fit together neatly, as the observed samples of NEOs and meteorites appear to represent partially mismatched sets. The greatest apparent discrepancy is that the most common meteorite type, the ordinary chondrites, appear to be extremely rare among comet nuclei or asteroids in the main belt (see Meteorite parent bodies). There is a long-standing debate as to whether the most common (spectral class S) asteroids are the parent bodies of these most common meteorites. While they have the same qualitative mineralogy, the quantitative proportions of the constituents are difficult to determine. Based on their spectra, some researchers believe the S-type asteroids are more metal rich and are more closely related to the stony iron meteorites, objects that have undergone substantial heating. In contrast, ordinary chondrites are relatively primitive bodies that have undergone little heating. Comets do not appear to provide a match for the ordinary chondrites as they probably contain material that is of still more primordial composition. Perhaps this paradox results from some unknown form of 'space weathering' by solar radiation or by micrometeoroid impacts that alter the upper few microns of an NEO's surface sufficiently to affect its spectral characteristics. Another alternative is that the parent bodies for ordinary chondrite meteorites exist only among the smallest asteroids, which telescopic surveys are only beginning to investigate.

In summary, it is a fortunate circumstance that near-Earth objects, the bodies which may provide the missing links between meteorites, comets and main-belt asteroids, include some of the most easily accessible targets in the solar system. As the last major class of objects to be investigated by spacecraft, we can expect them to yield vital new details of the processes that have governed the formation and evolution of our solar system. Addressing these key scientific questions will be of fundamental importance for successfully making technological advances leading to the exploration, hazard assessment and exploitation of these bodies that share the inner solar system with us.

Richard P. Binzel

Bibliography

Beatty, J.K. and Chaikin, A. (1990) *The New Solar System.* Cambridge University Press.
Binzel, R.P., Gehrels, T. and Matthews, M.S. (eds) (1989) *Asteroids II.* Tucson: University of Arizona Press.
Binzel, R.P., Barucci, M.A. and Fulchignoni, M. (1991) The origins of the asteroids. *Sci. Am.* **265**, 88–94.
Cowen, R. (1994) Rocky relics. Getting the lowdown on near-Earth asteroids. *Sci. News*, **145**, 88–90.
Cunningham, C. (1988) *Introduction to Asteroids.* Willmann-Bell Inc.

Acknowledgement

This article first appeared in the *Planetary Report*, the magazine of the Planetary Society.

Cross references

Amor object
Apollo object
Asteroid
Aten object
Comet
Comet: impacts on the Earth
Cretaceous–Tertiary boundary
Impact cratering
Mass extinction

NEBULAR HYPOTHESIS: KANT–LAPLACE

Eighteenth-century belief in the orderliness of the universe made determination of that order an important theological, philosophical and scientific endeavor. William Whiston, Isaac Newton's successor at Cambridge University in 1703, argued that the system of the stars, the work of the Creator, had a beautiful proportion, even if frail man were ignorant of the order. Whiston was unable to propose an order for the Milky Way, a dense band of stars, but in 1750 the self-taught English astronomer Thomas Wright (1711–1786) did just that. Possibly misled by an incorrect summary of Wright's book, Immanuel Kant (1724–1804) later explained the Milky Way as a disk-shaped system viewed from the Earth located in the plane of the disk.

Thoroughly imbued with a belief in the order and beauty of God's work, Kant went on to suggest that nebulous patches of light in the Heavens are composed of stars and are other Milky Ways or so-called 'island universes'. The paradigm of the Newtonian solar system provided for Kant a model of the larger stellar system and a plausible physical explanation of its structure. The arrangement of the stars, he reasoned, might well be similar to the disk structure of the planets around the Sun, and the same cause that gave the planets their centrifugal force and directed their orbits into a plane could also have given the stars the power of rotation and have brought their orbits into a plane.

Kant was a pure philosopher, but he also indulged in some serious cosmological speculations. The cosmological implications of Newton's theory of gravitation were first discussed extensively by him. In Kant's cosmology, *Universal Natural History and Theory of the Heavens* (1755), the gravitational attraction of stars for each other was exactly balanced by orbital motions of the stars. Also, the inhabited portion of the universe began as a perturbation of initially static matter, distributed in a homogeneous and isotropic manner throughout infinite Euclidean space. This material perturbation was postulated as having initiated a condensation to form the planets and the stars. However, his manuscript of 1755 perished in the printer's bankruptcy, and only a condensed version appeared, hidden in the appendix of another of Kant's books, published in 1763. Johann Lambert (1728–1771), a self-taught German scientist, published a similar theory in 1761, and seems not to have learned of Kant's ideas until 1765. William Herschel (1738–1822; q.v.) published his own theories on the construction of the heavens, beginning in 1784, apparently with knowledge of neither Wright's, Kant's nor Lambert's work. Because the nebular concept was also adopted a half-century later by Laplace (q.v.), this model is usually referred to today as the Kant–Laplace nebular hypothesis.

In the late 18th century and during the French Revolution, reflecting the new atheistic approach to nature adopted by some of the scientists of the Enlightenment, the French astronomer and mathematician Pierre Simon de Laplace (1749–1827) attempted to replace the hypothesis of God's rule with a purely physical theory that could also explain the observed order of the universe, particularly the remarkable arrangement of the solar system. As he saw it, the universe was no more than an extension of the latter; the structure of the solar system was taken as a small-scale model of the structure of the universe as a whole. He was successful, at least in his own mind. According to legend, when Napoleon asked him whether he had left any place for the Creator, Laplace replied that he had no need of such a hypothesis (see Laplace, Pierre Simon de).

Instead, Laplace proposed his nebular hypothesis, which appeared in six versions between 1796 and 1835, the last after his death in 1827. Conceptually the idea had already been suggested a half-century earlier, but in no detail, by Kant, and so it is commonly known as the 'Kant–Laplace nebular hypothesis'. The theory postulated that the Sun had once been a hot fluid body extending beyond the present orbits of the planets. As the fluid cooled and condensed, gradually shrinking to the present size of the Sun, zones of material were left rotating around the Sun. These rings of gas shed by the contracting Sun condensed to form the planets. Thus Laplace was able to explain the structure and dynamics of the solar system as a direct consequence of physical laws.

From observations of nebulae, Laplace had reasoned downward to his theory of the solar system. Reasoning back up, he attributed rotation to stars. Furthermore, many stars were collected in groups, groups which could also be explained by the hypothesis of a contracting nebulous mass. Many of the observed nebulae were probably groups of stars that would look like the Milky Way if viewed from within their interiors. Not until around the middle of the 19th century, when observations of nebulae began independently to suggest rotation, however, would Laplace's hypothesis be seriously discussed.

Late in the 19th century problems with the nebula hypothesis were brought out by the American geologist T.C. Chamberlin, together with the astronomer Forest Moulton, which largely replaced Laplace's nebular hypothesis with a 'planetesimal' theory of their own that involved the creation of the nebulae scattered throughout space (see Chamberlin–Moulton planetesimal hypothesis).

Norriss S. Hetherington

Bibliography

Kant, I. (1968) *Universal Natural History and Theory of the Heavens*. New York: Greenwood Publishing Co. (Translation from German).
Munitz, M.K. (ed.) (1957) *Theories of the Universe*. New York: The Free Press.
North, J.D. (1990) *The Measure of the Universe: A History of Modern Cosmology*. New York: Dover Publications, Inc.

Acknowledgement

Reviewed by H.J. Haubold (UN, New York).

Cross reference

History of planetary science I: pre-space age

NEMESIS

The name Nemesis was suggested in 1984 by M. Davis, P. Hut and R.A. Muller for a hypothetical solar companion, possibly responsible for periodic comet showers in the inner solar system. Although in Greek mythology Nemesis was the goddess of retributive justice, it would be a mistake to carry her identification with a possible 'death star' too far. Her relationship to the ancient Greek society and religion, especially as both evolved over several centuries, was profoundly complex. Yet, recognizing the weakness of the metaphor, we may associate Nemesis with the mass extinction that occurred roughly 65 million years ago, an epoch when many species and whole families of plants and animals became extinct. These included ocean plankton, swimming mollusks, many large land animals weighing more than about 25 kg, and most notably all the large walking dinosaurs. Within the last decade many scientists have seriously suggested an astronomical origin for these extinctions. For example, the northern Yucatan peninsula provides evidence for at least one large impact at the Cretaceous–Tertiary (K–T) boundary. A large impact by a common asteroid about the size of 951 Gaspra – encountered by the Galileo spacecraft on 29 October 1991 – weighing 10^{14} kg, could release 10^{23} J of energy, roughly equivalent to 10^{14} metric tons of TNT, or roughly equivalent to about 2 million of the most powerful H-bombs exploded by the US and USSR between 1952 and 1961. It is possible that a sudden detonation of this magnitude could cause such extreme environmental trauma that a mass extinction would follow. However, the paleontological record is not clear on whether the K–T extinctions occurred over a few generations, or over hundreds of thousands of years or more. Consequently, not only is the environmental devastation caused by a major impact uncertain, but the subsequent response of flora and fauna is largely a matter of speculation.

Over the past 250 Ma there have been other mass extinctions of whole families of organisms, although the one at the K–T boundary is among the largest. In 1984 the paleontologists D.M. Raup and J.J. Sepkoski found a 26-Ma periodicity in the extinction record. Shortly after hearing of this result, geologists W. Alvarez and R.A. Muller found a similar 28.4-Ma periodicity in the dating of the Earth's major impact craters. The statistics of these findings were analyzed over the next few years, and although the periodicity was not established with a high degree of confidence, many geologists and paleontologists agreed that there probably are peaks in the impact record that correlate with peaks in the extinction record, most notably the mass extinction at the K–T boundary (see Mass extinction).

If there is a periodicity in the impact record, it is difficult to find a viable astronomical mechanism for it. The Nemesis hypothesis is a possibility. In this scenario the inner solar system suffers periodic bombardment as Nemesis, at its perihelion, passes through the Oort cloud of comets. Such a solar companion with a period of 28 Ma would have an orbital semi major axis of 92 000 AU. Statistical analysis of cometary orbits, supplemented by dynamical computer modeling, suggests an inner spherical cloud of comets and an outer cloud. Because the boundary between the inner and outer clouds is thought to lie around 20 000 AU, a reasonable guess for the perihelion distance of Nemesis is about 30 000 AU (Figure N1). Its orbit would be highly elliptical with an eccentricity of about 0.67. But there is difficulty with such an orbit. It would be so loosely bound to the Sun that perturbations by nearby stars could significantly change its orbital period. Such changes could occur on time scales much less than the 250 Ma of the impact and extinction records. It could also easily escape within a few hundred million years, which raises the intriguing possibility that shortly before it escaped, it produced the K–T mass extinction required for eventual human evolution some 60 Ma later. If so, we might view Nemesis as the bearer of good fortune, not ill.

But perhaps the impact record is not periodic, and the Nemesis hypothesis is unnecessary. Alternatively, significant peaks in the flux of new comets may be triggered by changing gravitational forces from the galactic tide, passing stars and giant molecular clouds (see Galactic cycle). On the other hand, random processes, supplying asteroids from the main belt between Mars and Jupiter, probably account for at least some major impacts. Although a comet from the Oort cloud is the most likely candidate for the K–T impact, an asteroid from the inner solar system cannot be ruled out.

John D. Anderson

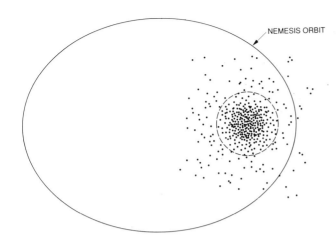

Figure N1 Schematic diagram showing a possible orbit for the hypothetical solar companion, Nemesis. The dashed circle at 20 000 AU represents the boundary between the inner and outer Oort cloud of comets. The dots schematically represent a rough distribution of comets gravitationally ejected from the Uranus–Neptune zone in the early history of the solar system.

Bibliography

Davis, M., Hut, P. and Muller, R.A. (1984) Extinction of species by periodic comet showers. *Nature*, **308**, 715–7.

Tremaine, S. (1990) Dark matter in the solar system, in *Baryonic Dark Matter* (eds D. Lynden-Bell and G. Gilmore). Dordrecht: Kluwer.

Weissman, P.R. (1990) The Oort cloud. *Nature*, **344**, 825–30.

Cross references

Cretaceous–Tertiary boundary
Mass extinction
Oort, Jan Hendrik, and Oort cloud

NEPTUNE

Neptune, one of the four giant planets, is the third most massive planet in the solar system (Plate 31). It orbits the sun with a period of 60 190 Earth days. With a semimajor axis of 4.5 billion kilometers, its orbit lies beyond the orbit of Uranus. Neptune is the most distant from the Sun of all the planets except Pluto. Neptune's orbit plane is inclined to the Earth's orbit plane by 1.774°, and thus lies close to the orbit planes of most other planets. Neptune's orbital motion is prograde; i.e. the planet moves counterclockwise in its orbit when viewed from above the north pole of the Sun, in the same sense as other planetary orbits. Neptune's orbital eccentricity $e = 0.0097$. Neptune's obliquity (angle of inclination between the spin axis and the normal to the orbital plane) is 29.6°, similar to the corresponding values for Earth, Mars and Saturn. Neptune spins with an interior rotation period of 16.11 h, which corresponds to the rotation period of its magnetic field as measured by the Voyager 2 spacecraft (q.v.) during the encounter with the planet in 1989.

Neptune cannot be seen by an Earth-based observer without the aid of a telescope. It is a faint bluish object with an apparent visual magnitude of 8 and an apparent diameter of about 2 arcsec. Since turbulence in the Earth's atmosphere typically blurs stellar images by about 1 arcsec, Neptune's extended disk can be discerned only with difficulty, and detailed features are rarely visible.

Neptune was discovered by a process of scientific analysis, and was the first planet discovered in this fashion. Following Herschel's chance discovery of Uranus in 1781, observations of that planet's motion showed that it did not follow the precise orbital path expected on the basis of all known perturbations. By 1840 the difference between the observed and predicted path had accumulated to about 100 arcsec. Two astronomers independently analyzed the differences and calculated the expected position of the undetected body responsible for the discrepancy. These were the British astronomer John Couch Adams (q.v.) and the French astronomer Urbain Jean Joseph Leverrier (q.v.). Both independently issued a prediction of the location where the new planet could be found, but only Leverrier's prediction was acted upon, by the German astronomer J.G. Galle, who first observed Neptune on 23 September 1846. Neptune was named after the Roman god of the sea, a name which proves to be most appropriate in the light of what we now know about the planet's probable water-rich composition.

Neptune's remoteness and faintness make it a difficult objective for ground-based investigations, although considerable progress has been made with visual-wavelength imagery and spectroscopy, infrared-wavelength bolometry and spectroscopy, and observations of occultations of stars by Neptune and its satellites at visual and infrared wavelengths. In August 1989 the spacecraft Voyager 2 encountered Neptune at close range and obtained important new data, concluding the spacecraft's 'grand tour' of the four giant planets. No further spacecraft will reach Neptune until at least well into the 21st century, and Earth-based investigations will continue to play a crucial role in studies of this planet. Table N2 presents basic data about Neptune.

Because Neptune has no discernible solid surface, its tabulated radius corresponds to an arbitrary pressure level (1 bar) in its atmosphere, a level which is accessible to direct observation. Neptune is non-spherical because of distortion induced by its rapid axial rotation rate, and its polar radius b at the 1-bar pressure level is about 450 km smaller than its equatorial radius a. This difference is expressed in terms of the dynamical oblateness $\epsilon = (a - b)/a$.

Table N2 also gives Neptune's dimensionless zonal harmonic J_2. This number is a measure of the relative distortion of Neptune's external gravitational potential V from the form for a point mass of mass M, (i.e. $V = GM/r$), and can be determined from its effect on the orbit of Neptune's satellite Triton, or from its effect on a spacecraft trajectory during a Neptune flyby. Knowing J_2 and the rotation period, it is possible to infer Neptune's moment of inertia about its rotation axis C. At still closer range, the influence of a higher-order component of Neptune's gravitational potential, J_4, can be measured. This number contains information about the density and rotation rate of Neptune's outermost layers.

For a body in hydrostatic equilibrium and rotating with a single rotation period P, the quantities e, J_2 and P are related, such that knowledge of any two implies a unique value for the third. Prior to the 1989 Voyager encounter, all three quantities had been measured for Neptune, but the results were inconsistent with the constraint imposed by hydrostatic equilibrium. The difficulty was caused by the result for P, which was determined to be about 18 h from Earth-based measurements of the rotation of cloud features in Neptune's atmosphere, while ground-based determinations of e and J_2 suggested that the rotation period was about 2 h shorter. High-resolution studies of cloud motions during the 1989 Voyager 2 encounter have shown that Neptune's atmosphere exhibits strong zonal (east–west) currents like the other giant planets, but that these currents vary more strongly with latitude than in any other planet. Atmospheric rotation periods vary from minimum values of about 12 h near the poles to about 18 h near the equator. Like Uranus, Neptune subrotates at low latitudes in a sense opposite to Saturn and Jupiter, which superrotate at low latitudes. The rotation period given in Table N2 gives the rotation period of Neptune's magnetic field, as measured during the Voyager 2 encounter. It is this period which is taken to correspond to rotation of the bulk of the planet, producing the lowest zonal harmonic J_2. The dynamical oblateness e given in Table N2, which is computed from J_2 and P, is consistent with independent ground-based measurements of the atmospheric oblateness.

Table N2 Neptune data

Mass M (Earth masses)	17.14
Equatorial radius a at 1 bar pressure (km)	24764
Interior rotation period P (h)	16.11
Mean mass density (g cm^{-3})	1.64
First even zonal harmonic J_2	0.003538
Second even zonal harmonic J_4	−0.000038
Dynamical oblateness ϵ	0.0182
Dimensionless axial moment of inertia C/Ma^2	0.24
Internal energy flux at surface (w m^{-2})	0.4

It is of interest to determine whether the substantial differential rotation observed in Neptune's atmosphere has deep roots, i.e. whether the zonal currents reflect fluid motions which penetrate deep into the interior and therefore involve significant mass. Calculations of the effect of these currents on the external gravitational potential V, as reflected in the values of J_2 and higher harmonics, give the following results. If the zonal currents are superficial, involving only negligible mass, such that virtually all the mass rotates with period P, then J_4 should be negative, and its value should be close to that given in Table N2. If the zonal currents corresponded to constant rotation rates on cylindrical surfaces lying at constant radii from the rotation axis, J_4 would be positive and would have an absolute value roughly twice that given in Table N2. Since the second model disagrees with the data on Neptune's gravitational harmonics, we conclude that the strong zonal currents involve only the outermost layers of Neptune.

Like the atmosphere of Uranus, the observable atmosphere of Neptune is composed mostly of hydrogen, but the atmosphere cannot extend to great depths nor comprise a major fraction of the planet's total mass. Under the pressures which would prevail in a hydrogen-rich Neptune, the hydrogen would be compressed to a mean density ~ 0.1 g cm^{-3}, far less than the observed mean density of Neptune (Table N2).

Knowledge of the mean density of Neptune and its zonal harmonics J_2 and J_4 does not unambiguously constrain its composition, other than ruling out hydrogen as a main component. Only a few remaining chemical elements are plausible major constituents. Helium is the second most abundant element, but it occurs in close association with primordial nebular hydrogen, and there is no plausible mechanism for forming a helium-rich Neptune from the primordial nebula. Thus the helium content of Neptune ought to be limited to the fraction relative to hydrogen that existed in the nebula, or that exists in the present solar atmosphere, i.e. a mass fraction of about 0.28, or about one helium atom for every 11 hydrogen atoms. Oxygen would have been abundant in the nebula and, under conditions prevailing in the outer parts of the nebula where Neptune formed, would be present either as H_2O ice (solid) or as CO (solid or gaseous, depending on conditions). Neglecting the formation of CO, one finds that under sufficiently low-temperature conditions, possibly those prevailing at Neptune's distance from the primordial Sun, virtually all solid material would have been in the form of H_2O ice ($\sim 40\%$), CH_4 ice ($\sim 25\%$) and magnesium silicates and iron ('rock', $\sim 25\%$). A remaining small fraction would include NH_3 ice. Thus the standard hypothesis for the bulk composition of Neptune, apart from its hydrogen-rich atmosphere, is that it is composed primarily of these solid components, which are conveniently denoted collectively as 'ice' and 'rock' respectively. Assuming that ice is made up of water, methane and ammonia in the stated proportions, its total mass would be about three times that of the rock component, and thus its properties would predominate in determining the interior structure of Neptune. (Atmospheric composition is discussed in detail in the following article.)

Although the predominance of ice in the interior of Neptune cannot be conclusively proven on the basis of present data, additional evidence exists to support this hypothesis. Figure N2 shows a plot of mass density ρ versus pressure in an interior model of Neptune (solid curve). This model has been adjusted to agree with the observational constraints given in Table N2. For comparison, Figure N2 also shows an interior model of Uranus calculated in the same manner (heavy dashed curve). The four curves shown with light dashes are theoretical compression curves corresponding to the following compositions. The lowest curve is the compression curve of a primordial nebular gas, primarily hydrogen and helium. The next highest curve is the compression curve of a mixture of nebular gas and 20% ice by mass, followed by a similar curve for a mixture of nebular gas and 70% ice by mass. The highest curve is the compression curve for ice. All these curves are computed under the assumption that the material is initially at a pressure of 1 bar (10^{-6} Mbar) and a temperature of about 70 K, and that the temperature increases with pressure from this point in accordance with the law of adiabatic compression.

Shock compression experiments have been conducted at Lawrence Livermore National Laboratory on a material called 'synthetic Uranus', a liquid solution of water, ammonia and isopropanol with molar abundances of H, O, C and N approximately equal to that of a solar-composition mixture of H_2O, CH_4 and NH_3. Results from these experiments are shown as triangles in Figure N2.

The main conclusions from the modeling studies and the shock

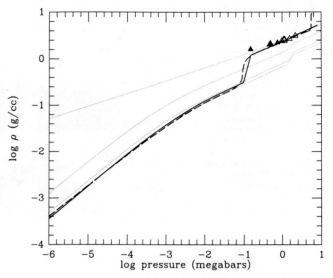

Figure N2 Mass density versus pressure on a log–log plot. Heavy solid curve shows a Neptune model; heavy dashed curve shows a Uranus model. Triangles show experimental shock compression points for 'synthetic Uranus'; solid triangles are single-shock data and open triangles are double-shock data.

experiments are: (1) Uranus and Neptune obey a very similar pressure–density relation and differ only slightly in gross interior structure; (2) the outer envelopes of Uranus and Neptune follow the pressure–density relation for nebular hydrogen and helium up to a pressure of about 100 kbar; (3) at pressures higher than 100 kbar both Uranus and Neptune closely follow the pressure–density relation for ice. The pressure–density relation for ice shown in Figure N2 includes effects of expected high temperatures (several thousand kelvin) at the highest pressures of several megabars. Such high temperatures are also observed in the shock compression experiments. Under these conditions, 'ice' is actually a hot fluid of the molecules H_2O, CH_4 and NH_3. Some molecular ionization and dissociation may be present as well.

Neptune's hydrogen-rich outer layer is fairly shallow, comprising no more than about 15% of the planet's radius, or a thickness of about 3500 km. The mass of this layer is quite small, comprising about 0.5 to 1 Earth mass. Although the deeper region of Neptune closely follows the ice compression curve, it is unlikely to be composed of pure ice since a significant portion of this material, perhaps 25% by mass, ought to consist of a much denser rock component. The material could still follow the ice compression curve if enough of a low-density component, such as nebular hydrogen and helium, is mixed with the ice to compensate for the rock. Such a low-density component could comprise another Earth mass or so of hydrogen and helium.

According to the results presented in Figure N2, Neptune is quite chemically homogeneous at pressures higher than 100 kbar. There is essentially no evidence that a dense rocky core has formed at the center of the planet. The accumulation of Neptune must have involved a protracted accretion sequence of ice–rock planetesimals, with relatively little chemical separation.

Like Uranus, Neptune shows evidence of enhancement of the deuterated methane molecule CH_3D in its atmosphere. From spectroscopic measurements, the deuterium-to-hydrogen number ratio (D/H) in Neptune is inferred to be $\sim 1.2 \times 10^{-4}$, much higher than the value for primordial solar system gas ($\sim 2 \times 10^{-5}$). This number should reflect the D/H value in the massive ice region below the hydrogen-rich atmosphere, and apparently shows that the icy planetesimals from which Neptune accumulated were similarly enriched in deuterium, possibly as a result of disequilibrium fractionation in interstellar ice particles.

Neptune has a substantial internal energy flux, about 0.3 W m^{-2}. This heat flow is far too large to be explained by radiogenic heating in the planet's presumed rocky component. It is also too large to be explained by heat dissipation from tides raised on Neptune by the large satellite Triton. The favored explanation is that the heat flow is derived from cooling of Neptune's high-temperature interior, which acquired its initial heat during the planet's initial accumulation from icy planetesimals. Ample energy is available from this source. However, it is interesting that the internal energy flux from Uranus is so much smaller than that from Neptune, despite the close similarity of the two planets' interior structure.

If the energy flux in Neptune's interior is accompanied by interior convective velocities ~ 0.1 to 1 cm s^{-1}, then there exists a plausible mechanism for generating Neptune's observed magnetic field. Shock-compression experiments on ice material have measured the electrical conductivities of the samples at their final compressions (at pressures on the order of megabars). The measured conductivities are on the order of 30 $(\Omega \text{ cm})^{-3}$, sufficiently large for magnetohydrodynamic dynamo action.

William B. Hubbard

Bibliography

Atreya, S.K., Pollack, J.B. and Matthews, M.S. (1989) *Origin and Evolution of Planetary and Satellite Atmospheres, Part IV, Outer Planets*. Tucson: University of Arizona Press.

Cruikshank, D.P. and Matthews, M.S. (1994) *Neptune and Triton* (eds D.P. Cruikshank and M.S. Matthews). Tucson: University of Arizona Press.

Geophysical Research Letters (1990), **17**, 1643–776 (Neptune system section).

Hubbard, W.B. (1984) *Planetary Interiors*. New York: Van Nostrand Reinhold.

Hubbard, W.B. (1990) Interiors of the giant planets, in *The New Solar System*, 3rd edn (eds J.K. Beatty and A. Chaikin). New York, Cambridge University press, pp. 131–8.

Journal of Geophysical Research (1991), **96**, 18903–9268 (Voyager 2 at Neptune issue).

Stevenson, D.J. (1982) Interiors of the giant planets. *Ann. Rev. Earth Planet. Sci.*, **10**, 257–95.

NEPTUNE: ATMOSPHERE

Neptune is the most remote of the distant group of large, hydrogen-rich worlds collectively known as the 'major planets', the others being Jupiter (the nearest and largest), Saturn and Uranus. Neptune and Uranus together comprise a distinct subclass. While the planets Jupiter and Saturn are composed predominantly of hydrogen and helium, these elements account for less than 80% of the masses of Uranus and Neptune. The outer pair have significantly higher percentages of heavier elements such as carbon, nitrogen, oxygen and iron. These major differences between the inner and outer pairs of the giant planets indicate distinct differences in their formation and evolution, as well as in their present-day interiors, atmospheres and dynamics.

Despite its remoteness, Neptune exhibits some of the most active meteorology and global variability in the outer solar system (Plate 31). Numerous haze and cloud layers distributed throughout the visible atmosphere have been observed to vary on timescales of minutes to years and on spatial scales from a few tens of kilometers (the Voyager resolution limit) to several thousand kilometers.

Sunlight is not the predominant source of power governing global temperature, circulation and meteorology in the visible troposphere. Instead, an unusually large source of internal heat source warms Neptune from the inside out (much as a hot kettle on a stove is warmed from the bottom up), causing convection, advection and associated storms and other meteorological activity. The feeble sunlight on distant Neptune amounts to just 0.1% of the Earth's solar insolation. On the other hand, the flux of radiant energy emitted by the internal power source is ~ 2.7 times greater than this solar component, warming the planet up to the temperature of Uranus, which is $\sim 35\%$ closer to the Sun. This factor in the ratio of internal to external power is nearly twice that found for Jupiter (~ 1.7) or Saturn (~ 1.8), the only other planets possessing substantial sources of internal heat.

To first order, the compositions and pressure levels of the major hazes and clouds on Neptune are quite similar to those in neighboring Uranus, an unsurprising result given their similar environments (e.g. comparable bulk planetary mass and gas composition, gravity and, as we have seen, thermal profile). Cloud layers of methane (CH_4), hydrogen sulfide (H_2S) and ammonium hydrosulfide (NH_4SH) are expected to reside in the troposphere at about 1, 7 and 37 bars respectively (if bar is the mean atmospheric pressure at sea level on the Earth). These clouds are condensates (in the case of CH_4 and H_2S) and chemical products (in the case of NH_4SH) of atmospheric trace gases. In the high stratosphere near 1 µbar, photochemistry initiated by shortwave solar ultraviolet radiation breaks up methane into reactive constituents which combine to form higher-order hydrocarbons, causing the formation of ethane (C_2H_6), acetylene (C_2H_2) and other species. These constituents then diffuse down to the lower stratosphere, where they freeze out as haze layers near the 10-mbar level.

A rather surprising aspect of Neptune's long-term variability is its near coherence with the solar cycle. During periods of high storm activity in Neptune's atmosphere, localized cloud outbursts dominate the planet's rotational lightcurve, as in the years 1989–1992 (Hammel et al., 1992a). Apart from such stormy periods, Neptune's mean global albedo varies by about 4% from solar minimum to maximum, becoming darker during periods of high solar activity (as characterized by increased numbers and strengths of solar storms and sunspots, increased solar wind flux and increased solar ultraviolet light flux), and brighter during more inactive periods (Lockwood and Thompson, 1979, 1986). Neptune is the only world which visibly responds to the solar variability.

The atmosphere of Neptune extends from space down to approximately 6000 km below the 1-bar level, where pressures exceed 1 Mbar (1 million times the mean atmospheric pressure at sea level on the Earth). There, due to this enormous pressure and concurrent high temperature (in excess of 2000 K), gases turn into superheated liquids, forming a deep, hot ocean comprised mainly of water, methane and ammonia. The upper 70 bar of the atmosphere has been remotely probed by ground-based and Earth-orbiting instruments, as well as by instruments onboard the Voyager 2 spacecraft. The atmosphere below that level is hidden from view, owing to the effects of an array of light-scattering and light-absorbing clouds and gases. We must then resort to thermodynamical and chemical theories to gain an understanding of the deep atmosphere.

Our direct knowledge is principally derived from measurements made over a large portion of the electromagnetic spectrum, from the far ultraviolet near 500 Å (0.000005 cm) wavelengths to radio wavelengths near 20 cm. In the ultraviolet, extinction by hydrocarbon molecules limits the view to the uppermost levels of the atmosphere. Shortward of 1350 Å, the uppermost atmosphere above 10 µbar is probed, limited by methane gas extinction. Between 1350 and 1500 Å ethane and methane absorption limits observations to the stratosphere near 0.1 mbar. Between 1500 and 2400 Å, solar radiation penetrates to between ~ 0.1 and ~ 80 mbar (depending monotonically on wavelength), thus reaching and perhaps, as noted above, photochemically altering the haze layers in the lower stratosphere. Since our own atmosphere readily absorbs UV radiation, the ultraviolet spectrum of Neptune can only be measured from space, as it has been from the Earth-orbiting International Ultraviolet Explorer (IUE) and the ultraviolet spectrometer (UVS) on the Voyager 2 spacecraft which flew by the planet on 25 August 1989.

In the visible, between approximately 3000 and 7000 Å, sunlight can penetrate through several thin haze layers down to an optically thick cloud deck at ~ 3.5 bar (Baines and Smith, 1990; Baines et al., 1995), well into the convective troposphere underlying the stratosphere. However, several methane-absorbing wavelength bands exist in the optical range which restrict the view to the lower stratosphere and upper troposphere, allowing haze characteristics in these regions to be determined from ground-based and spacecraft spectroscopy and imaging. In particular, special color filters onboard the Voyager 2 spacecraft have been used to image the planet's limb at close range to determine stratospheric aerosol properties (Smith et al., 1989). In the near infrared, from about 0.7 µm to ~ 3 µm, methane bands become progressively stronger, correspondingly limiting the view to successively higher stratospheric regions.

Beyond ~ 5 µm, the predominant radiation is not reflected sunlight but indigenous thermal radiation. From approximately 7 to 14 µm this radiation comes from thermally radiating molecules in the stratosphere (e.g. hydrocarbons such as methane, acetylene and ethane). Effective constraints on hydrocarbon abundances come from ground-based observations (Gillett and Rieke, 1977; Macy and Sinton, 1977; Orton *et al.*, 1992). Due largely to wavelength-dependent hydrogen gas opacity, various other atmospheric levels produce the predominant radiation seen at various other thermal wavelengths, extending out to the submillimeter and millimeter range, so that spectral observations yield the thermal profile from the high stratosphere down to 8 bar (Orton *et al.*, 1986; Conrath, Flasar and Gierasch, 1991). At radio wavelengths of about 3–20 cm the trace constituent ammonia is the predominant opacity source, allowing views to ~ 70 bar (de Pater, Romani and Atreya, 1991).

Additional information about the thermal structure and composition of the Neptunian ionosphere, thermosphere, stratosphere, and upper troposphere comes from the Voyager Radio Occultation Experiment (Tyler *et al.*, 1989; Lindal, 1992; see Radio science). Using the spacecraft communication system, this experiment pierced the atmosphere with Voyager's directional radio beam as the spacecraft, from the perspective of Earth-based receiving antennas, passed behind the planet. Variations in signal strength and phase were then used to determine atmospheric and ionospheric mass and ion number densities.

Composition

Neptune's composition yields important clues about the planet's formation and evolution, as well as present-day chemical and dynamical processes. For example, the methane-to-hydrogen ratio indicates how much of the planet was formed from the accretion of icy protoplanets rich in heavy elements (such as carbon and nitrogen) (Pollack *et al.*, 1986). The temperature at which those ices formed within the solar nebula can be determined from the relative amount of deuterium (a form of hydrogen which contains an extra neutron in the nucleus) compared to hydrogen (Hubbard and MacFarlane, 1980). The amount of helium is an indicator of the present state of the deep interior as well as its evolution (Fegley *et al.*, 1991). Spatial and temporal variability of photochemically generated species (e.g. ethane and acetylene), and the presence of species generated under non-equilibrium conditions (e.g. carbon monoxide), are indicative of a variety of chemical and dynamical processes occurring throughout the atmosphere.

Atmospheric composition data relevant to planetary formation and evolutionary processes are primarily obtained from examination of the deep troposphere, below the condensation level of the most prominent volatiles. In the stratosphere, photochemical as well as condensation processes result in significant depletions or enhancements of most minor gas constituents. In particular, photochemical processing of methane leads to a surprisingly rich set of higher-order hydrocarbon species.

Tropospheric abundances

The abundances of hydrogen, helium, methane and deuterium are particularly diagnostic of formation processes. Expressed in mole fractions, that is, by the fractional number of molecules of each species, the abundances are approximately $0.81^{+0.06}_{-0.03}$, $0.19^{+0.03}_{-0.06}$, 0.03 ± 0.01 and $7.2 \pm 1.0 \, 10^{-5}$ respectively. The hydrogen and helium mass fractions are much more uncertain here than on the other major planets due to the inference of a significant amount of molecular nitrogen (N_2) implied by the recent discovery of HCN (hydrogen cyanide) in the planet's stratosphere (Conrath *et al.*, 1993). HCN requires a source of elemental nitrogen for its photochemical formation, most likely from photolysis of previously unsuspected quantities of N_2. Since N_2 is a particularly heavy molecule compared to H_2 and He, relatively small abundances of N_2 (~ 0.001 molar fraction) could potentially contribute substantially to the mean molecular weight of the atmosphere, reducing the helium required to match the molecular weight deduced from thermal infrared and radio occultation observations.

Beyond hydrogen and helium, the three most abundant molecule-forming elements are oxygen, carbon and nitrogen, in that order. In the hydrogen-rich Neptunian environment, the predominant molecular forms of these elements are water, methane and, presumably, ammonia, all of which condense and/or participate in aerosol-forming chemical reactions within the planet's troposphere. The determination of molecular and elemental abundances requires observations beneath the relevant clouds for each species. The methane abundance is determined relatively easily from visible–near-IR methane band observations which probe methane gas residing below the ~ 1.4-bar methane condensation level (Baines and Smith, 1990; Baines *et al.*, 1995). Ammonia, which condenses at about 7 bar, is accessible only by microwave observations. Water condenses even deeper, and cannot be directly probed by remote sensing instruments. However, oxygen is cosmically abundant, so it is expected to be a major constituent of the Neptunian atmosphere. For a solar oxygen abundance, water clouds form near the 50-bar, 270-K level.

The tropospheric methane molar fraction of $2.3 \pm 0.6\%$ (Lindal, 1992; Baines *et al.*, 1995) represents a 25 to 44-fold carbon enrichment over the solar C/H value. This large inventory of carbon indicates that a major fraction of the planet was formed from the accretion of methane-, ammonia- and water-bearing ices and rocks (Podolak, Hubbard and Stevenson, 1991). Recent theories suggest that at least ~ 10% of the carbon in the solar nebulae at Neptune's distance was incorporated within icy planetesimals, of which at least 12% actually went into forming Neptune (Pollack *et al.*, 1986; Baines and Smith, 1990).

A small additional amount of carbon is present in the form of carbon monoxide. Recent millimeter wave observations are consistent with a uniformly mixed CO mole fraction of 0.3–1.2×10^{-6} (Marten *et al.*, 1992; Rosenqvist *et al.*, 1992), many orders of magnitude greater than expected under thermochemical equilibrium conditions in Neptune's cold upper troposphere and stratosphere. Its presence is thus direct evidence for the vertical transport of atmospheric gases via convection from the CO-forming region in Neptune's deep (several kilobars) atmosphere, powered by the planet's large internal heat source.

Ground-based microwave measurements indicate that ammonia (NH_3) is depleted by a factor of ~ 100 compared with the solar abundance ratio, when measured deep ($P > 100$ bar) within the atmosphere below its ~ 7-bar condensation level (de Pater, Romani and Atreya, 1991). This is somewhat surprising, given the large methane abundance noted above, since both ammonia and methane ices were presumably present in supersolar abundances within the progenitor protoplanetesimals, and since ammonia is present in near-solar abundances on less ice-rich Jupiter and Saturn. The result suggests chemical processing of NH_3 on Neptune. Specifically, sequestration of ammonia through reaction with H_2S to form deep (~ 37 bar pressure) NH_4SH clouds is indicated (requiring, in turn, an enhancement of sulfur over its solar value by at least a factor of five since sulfur is five times less abundant than nitrogen in the Sun).

Alternatively, the depletion of NH_3 may be due to the lack of elemental nitrogen within the planet (Lewis and Prinn, 1980). This could occur if, during the time of planetary formation, molecular nitrogen, N_2, was the dominant chemical form of nitrogen instead of ammonia, the dominant form expected for the outer part of the solar nebula. N_2, which condenses at extremely cold temperatures (~ 20 K), would have been in a gaseous state, preventing it from being efficiently entrained within the icy protoplanetesimals, and thus causing a depletion of nitrogen relative to carbon and oxygen within the planet.

A third hypothesis is that a substantial amount of elemental nitrogen is presently sequestered in molecular nitrogen (N_2) within Neptune rather than in ammonia (NH_3). This possibility has been raised recently due to the discovery of a significant amount of HCN (approximately 1 ppb) within the stratosphere, which is likely formed from the by-products of CH_4 and N_2 photolysis (Marten *et al.*, 1992). An N_2 mole fraction of ~ 0.0024 would account for the lack of ammonia while preserving the expected elemental nitrogen abundance, as well as allow for the formation of both H_2S and NH_4SH clouds without requiring an enhancement of sulfur above its solar value.

Hydrogen, the dominant constituent of the atmosphere, exists in two states, determined by the relative orientation of the molecule's two nuclear spin vectors (the *ortho* state denotes parallel nuclear spin vectors, and the *para* state antiparallel). The relative abundance of the *ortho* and *para* states of hydrogen is an indicator of vertical mixing. While the equilibrium ratio of *ortho* to *para* states is a strong function of temperature in the cold Neptunian atmosphere, equilibration takes several years in the absence of a catalyst. Thus the degree to which the observed *ortho/para* ratio is not in agreement with thermodynamic equilibrium indicates the timescales for vertical transport. The fraction of hydrogen observed in the equilibrium *ortho/para* state is 0.89–1.00 (Baines, *et al.*, 1985), indicating that a relatively small fraction of hydrogen is not in thermodynamic equilibrium. This implies time scales greater than 10 years for vertical mixing, if catalysts are unimportant. However, Neptune's vigorous circulation strongly suggests that vertical mixing occurs faster than this timescale, indicating that catalytic processes (e.g. Massie and Hunten, 1982) are important for the conversion of hydrogen within Neptune.

Stratospheric abundances

In the upper troposphere, above the 1.4-bar methane condensation level, the concentration of methane gas falls rapidly. At the minimum temperature of ~ 54 K (Conrath, Flasar and Gierasch, 1991), located at the tropopause near 100 mbar, a methane molar fraction of 2×10^{-4} just saturates the atmosphere.

Photochemical processing of this methane governs the composition of higher-order hydrocarbons within the stratosphere. Ethane (C_2H_6), acetylene (C_2H_2) and polyacetylenes ($C_{2n}H_2$, n = 2,3,4....), the most abundant stable methane photolysis products (Romani and Atreya, 1988), are produced at and above the 100-μbar level. Ultraviolet light photolysis of CH_4, produced near the 4-μbar level (Yelle et al., 1993), creates the radicals $1CH_2$, $3CH_2$ and CH, which quickly react with methane to create CH_3, which then reacts with itself to produce ethane, C_2H_6. These chemical reactions are expected to occur quickly, so that the production rate of ethane is nearly proportional to the destruction (photolysis) rate of methane. A number of ground-based and spacecraft observations, including thermal infrared spectroscopy (Orton et al., 1992), heterodyne measurements (Kostiuk et al., 1992), Voyager infrared spectrometry (Bézard et al., 1991) and Voyager UV spectrometry (Yelle et al., 1993) indicate a stratospheric molar fraction of ethane of about 10^{-6}, some two orders of magnitude greater than found on Uranus, consistent with the difference in stratospheric methane between the two planets. The observed acetylene mixing ratio is ~ 5×10^{-8} (Orton et al., 1992). Other detected species include ethene (C_2H_4; Maguire, 1992), hydrogen cyanide (HCN), and carbon monoxide (CO; Marten et al., 1992; Rosenqvist et al., 1992). The mixing ratio of hydrogen cyanide is ~ 3×10^{-10} (Romani et al., 1992), and is photochemically generated from methane and molecular nitrogen, the latter of which may be supplied either externally (primarily from the escape of N_2 from Triton's thin atmosphere; Romani et al., 1992) or internally via rapid vertical mixing from the deep, hot, relatively N_2-rich interior (Atreya et al., 1992). CO is also most likely conveyed from the interior, as its tropospheric mixing ratio, on the order of a part per million, exceeds by many orders of magnitude the concentration expected for thermochemical equilibrium.

Global-mean vertical cloud structure

Stratospheric hazes

As on Uranus, many of the trace gases found in the stratosphere and upper troposphere condense to form hazes and clouds. Theoretical phase equilibrium calculations (e.g. Romani and Atreya, 1988, 1989; Moses, Yung and Allen, 1992; Romani et al., 1993) indicate that ethane, acetylene and poly-acetylenes form discrete aerosol layers as these gases diffuse downward from their micro-bar formation regions within the temperature-inverted stratosphere. Ethane (C_2H_6) is the dominant component, comprising some 75% of the mass of the haze (Romani et al., 1993), with acetylene contributing ~ 20%. In addition to hydrocarbon condensates, recent evidence of HCN (cyanide) vapor in the high stratosphere at the part-per-billion level raises the possibility of an HCN haze layer near its saturation level at about 4 mbar. Figure N3 depicts the distribution of hazes as determined for the Equatorial Region (Baines and Hammel, 1994).

The vertically integrated global-mean stratospheric aerosol burden is ~ 1.6 μg cm^{-2}, corresponding to a 0.75-μm opacity of ~ 0.13 (Baines and Hammel, 1994), with an average particle radius of ~ 0.2 μm (Pryor et al., 1992). The stratospheric aerosol burden is diminished substantially at the equator, being only ~ 12% of the mean global value (Baines and Hammel, 1994). These aerosols absorb considerable amounts of ultraviolet radiation, and are characterized by an imaginary index of refraction of ~ 0.01–0.03 at 0.26 μm for spherical Mie-scattering particles. The particles settle out of the stratosphere on time scales of ~ 3 years. The deduced stratospheric haze precipitation rate of 10^{-17}–10^{-14} g cm^{-2} s^{-1} (Baines and Hammel, 1994) is consistent with the theoretical gas photolysis rate of ~ 4×10^{-15} g cm^{-2} (Romani and Atreya, 1989; Romani et al., 1993).

The formation and sedimentation of the hydrocarbon haze particles comprise an integral part of Neptune's carbon cycle. As depicted in Figure N4, the cycle begins in the troposphere. From there, carbon in the form of methane (CH_4) is conveyed into the stratosphere by (1) diffusive transport of the gas phase along with perhaps (2) convective transport of methane ice crystals, as a consequence of Neptune's large internal heat flux. There the carbon is converted into other molecules by photochemistry, which are then transported downward at a rate that, on average, balances the rate of carbon loss from methane gas photolysis. At their respective nucleation/condensation levels, the hydrocarbons condense out to form hazes which then precipitate, putting carbon back into the troposphere. Thus in steady state the aerosol formation rate, as denoted by the number of carbon atoms per second per square centimeter incorporated into hydrocarbon condensates, is equal to the methane gas photochemical destruction rate and is also equal to the precipitation loss rate back to the troposphere. As the hydrocarbon ices then fall through the troposphere they sublime and are converted back to methane in the hot planetary interior, to close the cycle.

Once nucleation is triggered, hydrocarbon hazes in the stratosphere form quickly (Romani et al., 1993). This means that stratospheric haze formation is probably an episodic process: (1) the haze particles nucleate, (2) the hydrocarbon supersaturations are reduced by loss of the vapor phase onto the ice crystals, (3) haze nucleation shuts down, (4) the haze particles grow and settle out of the lower stratosphere and (5) the supersaturations build up again to allow for nucleation of new particles.

Such episodic behavior may be one cause of the marked year-to-year variability in Neptune's albedo. As noted earlier, this variability seems to be correlated with solar activity. Modulation of the haze-inducing charged particle flux at Neptune caused by solar cycle variations in the solar wind may be one way to explain the correlation. An alternative hypothesis stems from consideration of the effect of solar ultraviolet irradiation on hydrocarbon particles. Studies of Uranus indicate that such particles may 'tan' significantly over several years, even under the relatively feeble solar UV illumination levels at the outer planets (Pollack et al., 1987). On Neptune it has been noted (Baines and Smith, 1990) that aerosol precipitation out of the UV-sensitive stratosphere can modify the albedo of the stratospheric haze on a timescale of a few months as fresh, unirradiated condensates continually replace older, irradiated ones. It is postulated that during their residence in the high stratosphere, the particulates are 'tanned' to a shade of grey proportional to the ultraviolet flux level, acquiring an imaginary index of refraction of ~ 0.02–0.04 (Pryor et al., 1992; Baines and Hammel, 1994). As the solar ultraviolet flux varies by a factor of two during the 11-year solar cycle, the stratospheric haze albedo varies accordingly, producing the observed correlation.

Tropospheric clouds and hazes

In the troposphere a number of volatile trace gases (CH_4, H_2S, NH_3, NH_4SH and H_2O) are expected to form discrete condensate cloud layers as the gases are convectively transported upward from the deep troposphere and adiabatically cooled. Figure N5 shows theoretical phase equilibrium estimates of the amounts of various cloud materials, disregarding sedimentation and coagulation effects. Cloud particles may be found at higher altitudes as well, driven upward by vigorous convective motions. Cloud-top altitudes depend on the strength of convective upwelling as well as on opposing sedimentation and coagulation rates; all of these are poorly understood.

Whether the clouds shown in Figure N5 actually form depends upon the details of atmospheric chemistry, thermal structure and the gaseous concentrations of the species, most of which are not understood well enough to make definite assessments. In particular there is significant uncertainty about the existence and relative masses of ammonia (NH_3), hydrogen sulfide (H_2S) and ammonium hydro-sulfide (NH_4SH) clouds, directly traceable to uncertainties in the amount of ammonia gas and the relative abundance of sulfur to nitrogen in the atmosphere. Microwave observations suggest that the H_2S cloud is most likely to occur (de Pater, Romani and Atreya, 1991), as implied by the absence of ammonia vapor absorption in 20-cm measurements. The implication is that NH_3 gas has been scavenged in the deep atmosphere by unexpected quantities of H_2S to form a massive NH_4SH cloud near 37 bar, thereby depleting the reservoir of ammonia gas available to form an ammonia ice cloud near 7 bar.

Above and below these possible clouds, substantial methane and water-ice clouds are expected near 1.5 and 45 bar respectively. Merging with the water cloud from below is an H_2O–NH_3–H_2S solution cloud (whose size again depends on the mole fractions of

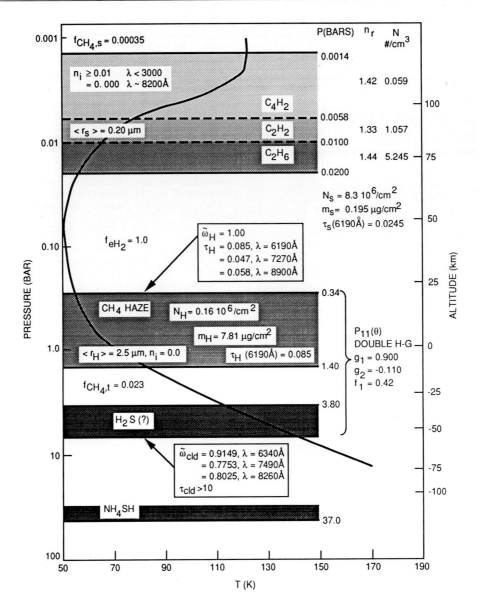

Figure N3 Nominal atmospheric structure for Neptune's Equatorial Region. Hydrocarbon condensates in the stratosphere are the end products of methane photochemistry occurring near the 1–10 μbar level. Diffusion transports hydrocarbon gases generated there downward to colder temperature levels. Condensation for individual species occurs at the specified pressure levels, forming optically thin haze layers. These hydrocarbon aerosols precipitate downward through the stratosphere and upper troposphere, evaporating near the 1-bar level. Methane condensation forms an optically thin haze near 1.4 bar. Parameter values for two types of CH_4 haze particles are shown: (1) 2.5-μm Mie scatterers and (2) particles adopting the nominal double Henyey–Greenstein (double H–G) phase function parameters empirically derived from Voyager multiple phase-angle observations (Pryor *et al.*, 1992). An optically thick cloud, most likely comprised of condensed hydrogen sulfide (H_2S), occurs near the 3.8-bar level. The thermal profile reported by the Voyager Radio Subsystem (RSS) occultation experiment (Tyler *et al.*, 1989) is shown with an adiabatic extrapolation into the deep atmosphere. (Figure adapted from Figure 12 of Baines and Hammel, 1994.)

ammonia and hydrogen sulfide in the lower troposphere). These clouds – perhaps loaded with as much as 0.1 to 10 kg cm^{-2} of material – may be 10–100 times more massive than the densest terrestrial clouds (Carlson, Rossow and Orton, 1988). Such clouds are heavily precipitating, with raindrops growing in tens to hundreds of seconds to mean sizes of about 10 and 200 μm in radius, respectively, for CH_4 and H_2O precipitates.

As determined by ground-based methane band spectroscopy and imaging, as well as by the Voyager spacecraft, the 1-bar methane cloud is surprisingly thin (Baines and Smith, 1990; Pryor *et al.*, 1992; Baines and Hammel, 1994), given that methane comprises ~ 15% of the mass of the atmosphere just below the condensation level. A visible opacity of less than unity is derived everywhere except within the ~ 2% of the planet covered by prominent discrete cloud features. Ground-based imagery and spectroscopy indicates disk-averaged opacities of about 0.05 (Baines and Hammel, 1994), more than two orders of magnitude less than the mean cloud opacity of the terrestrial atmosphere. If spherical Mie-scattering ice particles are assumed, with a mean radius of 2.5 μm, then the ground-based imagery indicates a mass column density of $7.8 \pm 1.1 \times 10^{-6}$ g cm^{-2}, corresponding to the rather miniscule amount of methane gas contained within a 0.1 cm thick column of air at the cloud base. Thus only a tiny fraction of the methane vapor available to form aerosol particles is actually incorporated into the haze at any one time.

An unexpected determination of relative altitudes for some cloud features came from unique images of bright wispy clouds observed by

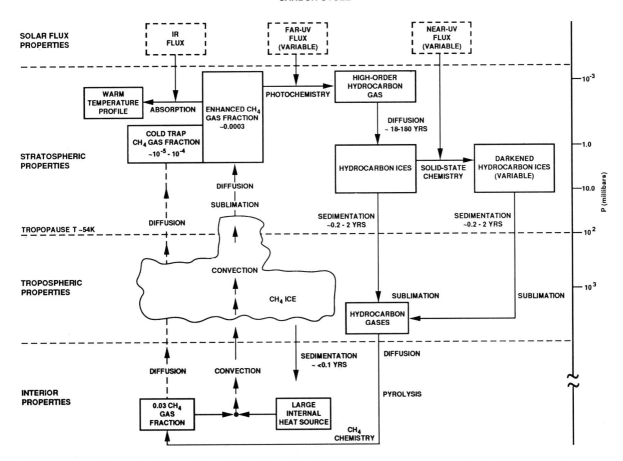

Figure N4 Methane carbon cycle. Neptune's large internal heat flux drives convection, augmenting molecular diffusion in delivering substantial quantities of methane into the stratosphere. There the methane absorbs sunlight, helping to warm the atmosphere. Photochemistry produces an array of hydrocarbons near the microbar level, which then diffuse down to the cooler condensation region near 10 mbar. Hydrocarbon ice particles formed there then precipitate out of the stratosphere into the troposphere, where they pyrolyze back into methane, completing the cycle. While in the UV-sensitive stratosphere, solar-induced solid state photochemistry on the surfaces of ice particles tans the aerosols to various shades of grey depending on the UV flux and on particle growth and sedimentation rates. Such darkened aerosols play an uncertain although undoubtedly key role in warming the stratosphere. (Figure modified from Baines and Smith, 1990.)

the Voyager 2 spacecraft near the terminator (Smith et al., 1989; Figure N6). Due to an unusual combination of targets and viewing and lighting conditions, distinct shadows of clouds were observed for the first time for any planet except Earth. The Voyager Science Imaging team utilized the geometry of one cloud's shadow cast onto another cloud to estimate the relative altitude of the clouds for two locations, 27°N and 71°S, deriving 100 ± 50 km and ~ 50 km for the northern and southern clouds respectively. It is suspected that the shadow-casting features are enhanced wisps of stratospheric ethane aerosols in the north and tropospheric methane aerosols in the south floating high above the optically thick H_2S cloud. In any event, their greatest significance is that they represent the first unambiguous resolution of cloud stratigraphy in an outer planet atmosphere.

The deepest detectable visible cloud is near 3.8 bar, as determined by the absorption observed within 4–0 S(0) and S(1) quadrupole lines of hydrogen near 0.64 μm and a methane feature near 0.68 m (Baines et al., 1995). As noted above, hydrogen sulfide (H_2S) is a thermochemically plausible candidate consistent with the lack of NH_3 vapor and the sluggish vertical mixing, but no positive identification has been made due principally to the lack of distinguishing spectral absorption features for H_2S or other candidate materials. Instead, this aerosol (similar to its counterpart in the Uranian atmosphere) has a bluish tint, being reflective in the blue–green and moderately absorbing at wavelengths longward of 0.6 μm (Baines and Smith, 1990; Baines and Hammel, 1994), which does not match any hypothesized condensate. This blue cloud, aided somewhat by the red-absorbing reservoir of methane above it, produces the bluish appearance of Neptune.

Circulation and dynamics

Observed temporal variability

Neptune is visually one of the most dynamic of the giant planets. Dramatic temporal variability has been witnessed over recent decades at many wavelengths and on many time scales, ranging from hours to years, despite the planet's remoteness from the Earth and Sun, which makes observations difficult and gives Neptune the least solar power (for driving dynamics) of any atmosphere-enshrouded world. This suggests that the observed spatial and temporal variability is ultimately powered by the planet's internal heat source, which dominates the solar energy input throughout the troposphere.

Diurnal variability is the major signature in the planet's rotational light curve. This led to early estimates of the planet's rotation rate (Slavsky and Smith, 1978; Cruikshank, 1978). Transient, short-term variability has been reported for more than a decade. Significant variability in full-disk observations at 2-μm region has been observed on hourly (Apt, Clark and Singer, 1980) and daily scales (de Bergh,

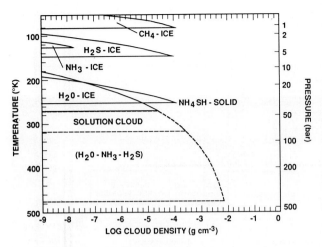

Figure N5 Predicted locations and densities of various condensate cloud layers in the troposphere of Neptune. Maximal values for densities are shown, as calculated from thermochemical equilibrium theory; cloud depletion mechanisms – such as coagulation and sedimentation – are not included. Deep-atmospheric composition used as inputs to the model include nitrogen (solar abundance ratio of [N]/[H]), sulfur (30 times solar abundance) and oxygen (solar, upper dashed curve; 30 times solar, lower dashed curve). In the absence of vertical transport (diffusion, upwelling and downwelling), up to four massive cloud layers could simultaneously exist, comprised principally of water (H_2O), ammonia hydrosulfide (NH_4SH), hydrogen sulfide (H_2S) and methane (CH_4). In addition, the NH_3 mixing ratio derived by the Voyager radio occultation experiment of 6×10^{-7} (Lindal et al., 1990) implies the formation of a small ammonia ice cloud at the ~ 4-bar pressure level. (From de Pater, Romani and Atreya, 1991.)

Lutz and Owen, 1987). Subsequent Voyager high spatial resolution imagery revealed variability on the scale of minutes to hours in localized, small-scale features (Smith et al., 1989; Limaye and Sromovsky, 1991). The longest timescale of variability appears to be linked to the solar cycle (Lockwood and Thompson, 1986; see discussion above), and thus is probably not directly a manifestation of planetary dynamics. However, on two occasions, unusually strong atmospheric disturbances significantly disrupted the typical solar cycle-correlated photometric variability in the mid-1970s (Joyce et al., 1977) and mid-1980s (Hammel et al. 1992a).

Observed spatial variability

For more than a decade, CCD images in the 0.89-μm methane band have revealed distinct bright features on 10 000-km scales (the sub-arcsecond seeing limit of the best terrestrial observatories). The distribution of such major features changed substantially during the 1980s. Imaging from 1981 to 1984 showed one or two features of equal brightness in both the northern and southern hemispheres (Terrile and Smith, 1983). Since 1986 a single bright feature in the south has dominated methane-band reflectivity (Hammel, 1989).

The Voyager 2 spacecraft provided unique close-up views of Neptune, with many images revealing features 1000 times smaller than observed from Earth, some just tens of kilometers across. As with the other major planets, a zonally symmetric banded structure predominates. For Neptune the equatorial region appears somewhat brighter in ultraviolet wavelengths and darker in near-infrared, methane band-sensitive wavelengths than the rest of the planet (Figure N7). Superimposed on the three major bands are several large (> 1000 km wide), localized features of widely varying spectral colors (Figure N7) and numerous smaller wisps of material confined to narrow ranges of latitude.

The most striking atmospheric feature observed by Voyager 2 was the Great Dark Spot (GDS; Figure N8), which had not been detected in ground-based images because of the lack of contrast at the red wavelengths used. The bright feature in simultaneous 1989 ground-based images – and presumably that seen since 1986 – was revealed to be the bright companion along the southern edge of the GDS. This

Figure N6 High spatial resolution (11 km per pixel) image of Neptune showing cloud streaks and associated shadows stretched approximately along lines of constant latitude near 27°N latitude. This image, taken near the terminator in violet, green and orange filters 2 h before closest approach, provides evidence of vertical relief in Neptune's 50–200 km wide bright cloud streaks. The bright sides of the clouds which face the Sun are brighter than the surrounding cloud deck because they are more directly exposed to sunlight. Shadows can be seen on the opposite side. Rayleigh scattering causes shadows to be less distinct at short wavelengths (violet filter) and more distinct at long wavelengths, thus giving the shadows a noticeable blue tint. Analysis of cloud shadow profiles extracted from the orange filter image indicate that the cloud streaks are 70–110 km above the underlying cloud deck. (NASA/JPL photograph P-34709.)

feature remained stationary with respect to the GDS throughout the several month-long encounter period.

Other conspicuous features included (1) the second dark spot, D2 (Figure N9), which developed a bright central core during the encounter, (2) Scooter, a small feature near 43°S which appeared to 'scoot' past the GDS when first discovered and which actually represented a collection of small streaks which varied with time, causing Scooter to change its shape, and (3) a bright south polar feature which moved at the speed of rotation of the interior. At all observable latitudes, many small wisps were seen which allowed a determination of the atmospheric wind profile (Limaye and Sromovsky, 1991; see discussion below). All of the features were superposed on a low-contrast banded pattern.

From the wavelength dependence (or color) of the various features (Figure N7), some idea of their relative altitudes can be derived (Smith et al., 1989). The bright clouds, with the exception of Scooter, are thought to be methane condensation clouds, probably located near the 1-bar level. The decreasing contrast of Scooter from blue to red wavelengths, and in particular its relative darkness in the Voyager 6190-Å methane band filter (methane-J, see the lower right-hand panel in Figure N7), suggest that it must lie lower in the atmosphere, possibly below the base of the methane cloud. The color variations of the GDS, especially the deep blue of the central core (Figure N9), are interpreted as either evidence for a generally clear atmosphere above this region, or alternatively as evidence that the 3.5-bar cloud is in fact somewhat deeper and/or bluer than elsewhere.

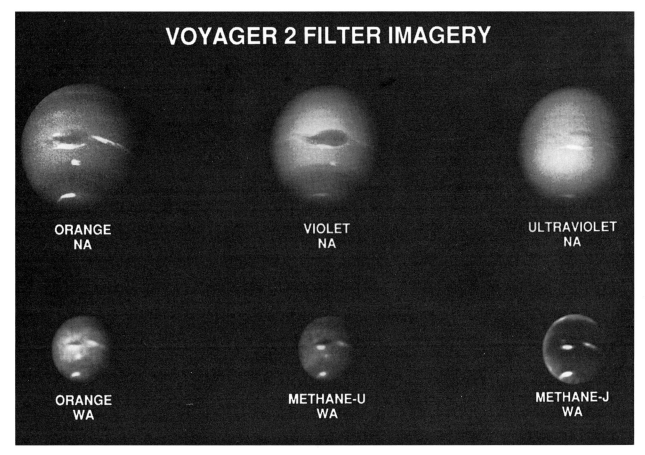

Figure N7 Multispectral images of Neptune acquired by the Voyager 2 spacecraft. Both narrow-angle (NA) and wide-angle (WA) images are shown, sequentially ordered by their susceptibility to molecular extinction. NA images in the violet and ultraviolet are progressively affected by molecular Rayleigh scattering, thus restricting aerosol sampling to higher altitudes. WA images in the methane-U (5460 Å methane band) and methane-J (6190 Å methane band) are progressively affected by molecular methane absorption, thus again restricting aerosol sampling to higher altitudes. The methane-J image shows remarkable limb brightening, revealing the presence of a high-altitude haze layer. A southern polar feature and the bright clouds accompanying the Great Dark Spot in the orange images remain the brightest features in the images. The Scooter, a bright spot at southern mid-latitudes, disappears in those images, indicating that it is at a relatively lower altitude.

Primary sources of dynamical power

On any atmosphere-enshrouded planet, sunlight and any internal heat primarily drive global circulation and dynamics by causing the atmospheric temperature to vary in latitude and longitude, thereby producing spatial gradients in atmospheric pressure forcing the air to move. Temperature variations may be caused by uneven solar heating, such as on Earth, and by uneven internal heating; the latter is likely to be the case for Neptune's troposphere.

In Neptune's stratosphere, atmospheric heating produced by the absorption of sunlight by aerosols is a significant source of dynamical power. On a temporal and spatial average, some 6–14% of the incident solar flux in the UV-to-visible spectral region is deposited in stratospheric aerosols between approximately 2 and 20 mbar (Baines and Smith, 1990). Such large localized heating increases mbar temperatures as much as 30 K over that predicted for an aerosol-free atmosphere (Appleby, 1986). Any temporal or spatial variability in the solar lighting or radiative properties of stratospheric hazes which in turn perturbs the temperature structure may thus produce dynamics. Seasonal variations in solar irradiance – due to Neptune's 29° obliquity – is one obvious example.

In the troposphere, both internal heat and solar fluxes influence the temperature. Hydrogen gas at the relatively high pressures of the troposphere (compared to the stratosphere) readily absorbs thermal infrared radiation in the 20–100 μm spectral range emitted from the interior, heating the atmosphere. Any horizontal variability in the 20–100 μm flux caused, for example, by variability in the uncertain internal processes releasing the heat, would presumably cause uneven temperatures and enhanced dynamics.

Horizontal circulation/winds

Atmospheric winds have been derived from the motions of clouds observed by cameras onboard the Voyager spacecraft (Hammel et al. 1989; Smith et al., 1989; Limaye and Sromovsky, 1991). However, Neptune's clouds exhibit unusual variability in their motions, much of which is attributable to the rapid formation and dissipation of the wispy features themselves rather than to actual atmospheric motions (Limaye and Sromovsky, 1991)

Neptune possesses the largest range of atmospheric rotation periods of any planet, with equatorial easterly winds exceeding 600 ms^{-1} and polar westerlies approaching 300 ms^{-1}. The smoothly varying equator-to-pole zonal wind structure lacks the alternating prograde and retrograde jets characteristic of Jupiter and Saturn. Moreover, at the equator the cloud level (1–4 bar) winds are retrograde rather than prograde as on Jupiter and Saturn. Such an uncomplicated wind profile structure fits a remarkably simple model wherein angular momentum is conserved within air parcels upwelling from the interior (Suomi, Limaye and Johnson, 1991). A notable aspect of this model is that the large-scale atmospheric circulation is independent of any energy input from the Sun, consistent with the internally heated thermal profile of the planet's upper troposphere noted earlier.

As does the Great Red Spot (GRS) in Jupiter's atmosphere, several of Neptune's major features oscillate, with periods ranging from 800 to several thousand hours (Sromovsky 1991). Latitudinal oscillations for the GDS, Scooter and D2 exhibit amplitudes of 2° to 4°, significantly greater than the GRS. Longitudinal oscillations in the GDS have not been determined; if present, the period is greater than

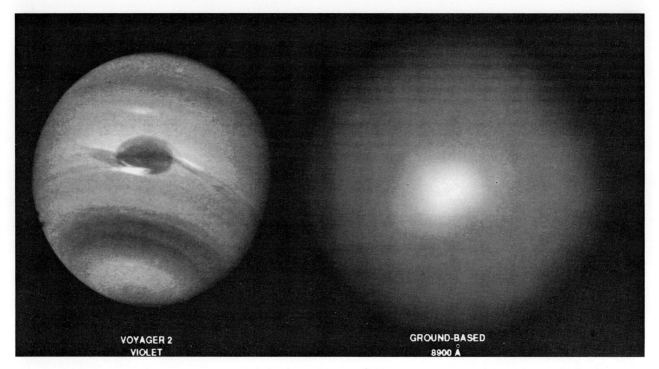

Figure N8 Simultaneous (left) Voyager 2 blue-filter and (right) ground-based 8890-Å CH$_4$ band imagery, acquired 24 August 1992. The companion cloud to the south of the Great Dark Spot in the Voyager 2 image (left) is the dominant feature in the ground-based imagery of Hammel (From Smith et al., 1989.)

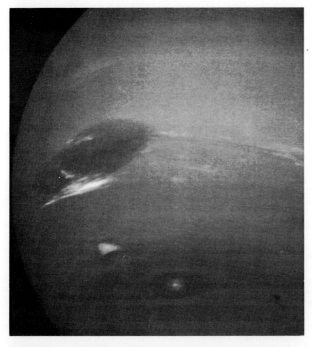

Figure N9 Color image of Neptune obtained by the Voyager 2 spacecraft in August 1989. Neptune's blue color is due to absorption of red light by methane gas and cloud material predominantly within the troposphere. Prominent features include, progressing from center-left to center-bottom, (1) the Great Dark Spot (GDS), accompanied by (2) its bright, upper-level companion cloud feature, (3) Scooter, and (4) the second dark spot, D2. Figure N7 shows that the clouds of Scooter lie considerably deeper within the atmosphere than the bright companion clouds (NASA/JPL photograph P-34648.)

several years. The smaller Scooter and D2 features oscillate in longitude by 8° and 98° (also larger than the GRS amplitude of ~ 1°), with periods of ~ 1200 and 880 respectively (compared to ~ 2160 h for the GRS). All these oscillations are currently unexplained.

Vertical dynamics

Generally, quantitative constraints on the vertical circulation within the Jovian planets are inferred from (1) the distribution of disequilibrium chemical species and (2) the vertical temperature structure. The presence of disequilibrium species – that is, species which are chemically-generated at high pressures and temperatures (typically, 1000 bar and 1000 K) but which cannot sustain themselves at higher and colder altitudes for long periods of time – is a direct indicator of vertical upwelling. On Neptune the part-per-million concentration of carbon monoxide (Marten et al., 1992; Rosenqvist et al., 1992) is an indicator of strong upwelling linking the deep, hot interior to the uppermost regions of the troposphere (e.g., Fegley et al., 1991). As noted earlier, Neptune's tropospheric temperature profile is adiabatic and close to that of Uranus, indicative of the warming from below it receives from its internal energy source. This temperature structure is further evidence of strong vertical circulation in the troposphere.

For Neptune, additional information about the nature of vertical updrafts can be derived from consideration of the carbon cycle discussed earlier (cf. Figure N4). The loss of stratospheric carbon due to the precipitation of hydrocarbon ice particulates implies that carbon is being resupplied to the stratosphere from the troposphere. For observed tropopause temperatures of ~ 54 K (Conrath, Flasar and Gierasch, 1991), upward diffusion of methane from the tropopause could maintain a stratospheric methane mixing ratio of ~ 2×10^{-4}, significantly less than the upper limits of recently derived values (Orton et al., 1992; Yelle et al., 1993; Baines and Hammel, 1994). Thus, as illustrated in Figure N4, convection may be an important transport mechanism in Neptune, conveying methane in the form of ice crystals upward more than 50 km from the troposphere (near 1.4 bar) to the lower stratosphere (near 60 mbar), where the crystals then sublimate into the observed gas.

Quantitative estimates of the resupply rate indicate that the required convection can be accounted for by one 2 km diameter updraft, upwelling at an areal mean velocity of 1 m s^{-1} at the 60-mbar sublimation level and entraining 1% of the methane vapor content

(Baines and Hammel, 1994). This would deliver the 1 metric ton per second of methane required to balance the carbon precipitation rate of 10^{-14} g cm^{-2} s^{-1} predicted from theory by Romani *et al.* (1993).

Kevin H. Baines

Bibliography

Appleby, J.F. (1986) Radiative–convective equilibrium models of Uranus and Neptune. *Icarus*, **65**, 383–405.

Apt, J., Clark, R. and Singer, R. (1980) Photometric spectra of the elusive Neptune haze. *Bull. Am. Astron. Soc.*, **12**, 705.

Atreya, S.K., Owen, T.C., Gautier, D. and Marten, A. (1992) HCN and CO on Neptune: an intrinsic origin. *Bull. Am. Astron. Soc.*, **24**, 972.

Baines, K.H. and Hammel, H.B. (1994) Clouds, hazes, and the stratospheric methane abundance in Neptune. *Icarus*, **109**, 20–39.

Baines, K.H. and Smith, W.H. (1990) The atmospheric structure and dynamical properties of Neptune derived from ground-based and IUE spectrophotometry. *Icarus*, **85**, 65–108.

Baines, K.H., Mickelson, M.E., Larson, L.E. and Ferguson, D.W. (1995) The abundances of methane and *ortho/para* hydrogen on Uranus and Neptune: implications of new laboratory 4–0 H$_2$ quadrupole line parameters. *Icarus*, **114**, 328–40.

Bézard, B., Romani, P.N., Conrath, B.J. and Maguire, W.C. (1991) Hydrocarbons in Neptune's stratosphere from Voyager infrared observations. *J. Geophys. Res.*, **96**, 18961–75.

Bishop, J., Atreya, S.K., Romani, P.N. *et al.* (1992) Voyager 2 UVS solar occultations at Neptune: constraints on the abundance of methane in the stratosphere. *J. Geophys. Res.*, **97**, 11681–94.

Carlson, B.E., Rossow, W.B. and Orton, G.S. (1988) Cloud microphysics of the giant planets. *J. Atmos. Sci.*, **45**, 2066–81.

Conrath, B.J., Flasar, F.M. and Gierasch, P.J. (1991) Thermal structure and dynamics of Neptune's atmosphere from Voyager measurements. *J. Geophys. Res.*, **96** Suppl., 18931–9.

Conrath, B.J., Gautier, D., Owen, T.C. and Samuelson, R.E. (1993) Constraints on N$_2$ in Neptune's atmosphere from Voyager measurements. *Icarus*, **101**, 168–71.

Cruikshank, D.P. (1978) On the rotation period of Neptune. *Astrophys. J.*, **220**, L57–9.

de Bergh, C., Lutz, B.L. and Owen, T. (1987) Neptune: short-term variations in the spectrum at 1.6 microns and the search for deuterated methane. *Bull. Am. Astron. Soc.*, **19**, 864.

de Bergh, C., Lutz, B.L., Owen, T. *et al.* (1986) Monodeuterated methane in the outer solar system. II. Its detection on Uranus at 1.6 microns. *Astrophys. J.*, **311**, 501–10.

de Pater, I., Romani, P.N. and Atreya, S. (1991) Possible microwave absorption by H$_2$S gas in Uranus' and Neptune's atmosphere. *Icarus*, **91**, 220–33.

Fegley, B., Jr Gautier, D., Owen, T. and Prinn, R.G. (1991) Spectroscopy and chemistry of the atmosphere of Uranus, in *Uranus* (ed. J.T. Bergstralh, E.D. Miner and M.S. Matthews). Tucson: University of Arizona Press, pp. 147–203.

Fink, U. and Larson, H.P. (1979) The infrared spectra of Uranus, Neptune, and Titan from 0.8 to 2.5 microns. *Astrophys. J.*, **233**, 1021–40.

Gillett, F.C. and Rieke, G.H. (1977) 5–20 μm observations of Uranus and Neptune. *Astrophys. J.*, **218**, L141–4.

Hammel, H.B. (1989) Discrete cloud structure on Neptune. *Icarus*, **80**, 14–22.

Hammel, H.B., Beebe, R.F., de Jong, E.M. *et al.* (1989) Neptune's wind speeds obtained by tracking clouds in Voyager images. *Science*, **245**, 1367–9.

Hammel, H.B., Lawson, S.L., Harrington, J. *et al.* (1992a) An atmospheric outburst on Neptune from 1986 through 1989. *Icarus*, **99**, 363–7.

Hammel, H.B., Young, L.A., Hackwell, J. *et al.* (1992b) Variability of Neptune's 12.2-micron ethane emission feature. *Icarus*, **99**, 347–52.

Hanel, R.A., Conrath, B.J., Herath, L., *et al.* (1981) Albedo, internal heat, and energy balance of Jupiter: preliminary results of the Voyager infrared investigation. *J. Geophys. Res.*, **86**, 8705–12.

Hanel, R.A., Conrath, B.J., Kunde, V.G. (1983) Albedo, internal heat, and energy balance of Saturn. *Icarus*, **53**, 262–85.

Hubbard, W.B. and MacFarlane, J.J. (1980) Theoretical predictions of deuterium abundances in the Jovian planets. *Icarus*, **44**, 676–81.

Joyce, R.R., Pilcher, C.B., Cruikshank, D.P. and Morrison, D. (1977) Evidence for weather on Neptune. I. *Astrophys. J.*, **214**, 657–62.

Kostiuk, T., Romani P.N., Espenak, F. and Bézard B. (1992) Stratospheric ethane on Neptune: comparison of groundbased and Voyager IRIS retrievals. *Icarus*, **99**, 353–62.

Lane, A.L., West, R.A., Hord, C.W. *et al.* (1989) Photometry from Voyager 2: initial results from the Neptunian atmosphere, satellites, and rings. *Science* **246**, 1450–4.

Lewis, J.S. and Prinn, R.G. (1980) Kinetic inhibition of CO and N$_2$ reduction in the solar nebula. *Astrophys. J.*, **238**, 357–64.

Limaye, S.S. and Sromovsky, L.A. (1991) Winds of Neptune: Voyager observations of cloud motions. *J. Geophys. Res*, **96**, 18941–60.

Lindal, G.F. (1992) The atmosphere of Neptune: an analysis of radio occultation data acquired with Voyager 2. *Astron. J.*, **103**, 967–82.

Lindal, G.F., Lyons, J.R., Sweetman, D.N. *et al.* (1990) The atmosphere of Neptune: Results of radio occultation measurements with the Voyager 2 spacecraft. *J. Geophys. Res. Lett.*, **17**, 1733–6.

Lockwood, G. and Thompson, D. (1979) The relationship between solar activity and planetary albedoes. *Nature*, **280**, 43–5.

Lockwood, G.W. and Thompson, D.T. (1986) Longterm brightness variations of. Neptune and the solar cycle modulation of its albedo, *Science*, **234**, 1543–5.

Lunine, J.I. and Hunten, D.M. (1989) Abundance of condensable species at at planetary cold traps: the role of moist convection. *Planet. Space Sci.*, **37**, 151–66.

Macy, W. and Sinton, W. (1977) Detection of methane and ethane emission on Neptune but not on Uranus. *Astrophys. J.*, **218**, L79–81.

Maquire, W. (1992) Identification and abundance of C$_2$H$_4$ on Neptune from Voyager 2 infrared observations. *Bull. Am. Astron. Soc.*, **24**, 974.

Marten, A. Gautier, D., Owen, T. *et al.* (1992) First observations of CO and HCN on Neptune and Uranus at millimeter wavelengths and their implications for atmospheric chemistry. *Astrophys. J.*, **406**, 285–97.

Massie, S.T. and Hunten, D.M. (1982) Conversion of para and ortho hydrogen in the Jovian planets. *Icarus*, **49**, 213–26.

Moses, J.I., Yung, Y.L. and Allen, M. (1992) Hydrocarbon nucleation and aerosol formation in Neptune's atmosphere. *Icarus*, **99**, 318–46.

Orton, G.S., Griffin, M.J., Ade, P.A.R. *et al.* (1986) Submillimeter and millimeter observations of Uranus and Neptune. *Icarus*, **67**, 289–304.

Orton, G.S., Lacy, J.H., Achtermann, J.M. *et al.* (1992) Thermal spectroscopy of Neptune: the stratospheric temperature, hydrocarbon abundances, and isotopic ratios. *Icarus*, **100**, 541–55.

Pearl, J.C. and Conrath, B.J. (1991) The albedo, effective temperature and energy balance of Neptune as determined from Voyager data. *J. Geophys. Res.*, **96**, 18921–30.

Podolak, M. Hubbard, W.B. and Stevenson D.J. (1991) Models of Uranus' interior and magnetic field, in *Uranus* (eds J.T. Bergstralh, E.D. Miner and M.S. Matthews). Tucson: University of Arizona Press, pp. 29–61.

Pollack, J.B., Podolak, M., Bodenheimer, P. and Christofferson, B. (1986) Planetesimal dissolution in the envelopes of the of the forming, giant planets. *Icarus*, **67**, 409–443.

Pollack, J.B., Rages, K., Pope, S.K. *et al.* (1987) Nature of the stratospheric haze on Uranus: evidence for condensed hydrocarbons. *J. Geophys. Res.*, **92**, 15037–65.

Price, M.J. and Franz, O.G. (1980) Neptune: limb-brightening within the 7300-angstrom methane band. *Icarus*, **41**, 430–8.

Pryor, W.R., West, R.A., Simmons, K.E. and Delitsky, M. (1992) High-phase-angle observations of Neptune at 2650 Å and 7500 Å: haze structure and particle properties. *Icarus*, **99**, 302–16.

Romani, P.N. and Atreya, S.K. (1988) Methane photochemistry and haze production on Neptune. *Icarus*, **74**, 424–45.

Romani, P.N. and Atreya, S.K. (1989) Stratospheric aerosols from CH$_4$ photochemistry on Neptune. *Geophys. Res. Lett.*, **16**, 941–4.

Romani, P.N., Bishop J., Bézard, B. and Atreya, S. (1993) Methane photochemistry on Neptune: ethane and acetylene mixing ratios and haze production. *Icarus*, **106**, 442–63.

Romani, P.N., Lellouch, E., Rosenqvist, J. *et al.* (1992) HCN on Neptune: origin and vertical distribution. *Bull. Am. Astron. Soc.*, **24**, 972.

Rosenqvist, J., Lellouch, E., Romani, P.N. *et al.* (1992) Millimeter-wave observations of Saturn, Uranus, and Neptune: CO and HCN on Neptune. *Astrophys. J. Lett.*, **392**, 99–102.

Savage, B.D. and Caldwell, J. (1974) Ultraviolet photometry from the Orbiting Astronomical Observatory. XIII. The albedoes of Jupiter, Uranus, and Neptune. *Astrophys. J.*, **187** 197–208.

Slavsky, D.B. and Smith, H.J. (1978). The rotation period of Neptune. *Astrophys. J.*, **226**, L49–59.

Smith, B.A., Soderblom, L.A., Banfield, D. *et al.* (1989) Voyager 2 at Neptune: imaging science results. *Science*, **246**, 1422–49.

Sromovsky, L.A. (1991) Latitudinal and longitudinal oscillations of cloud features on Neptune. *Science*, **254**, 684–6.

Stoker, C.R. and Toon, O.B. (1989) Moist convection on Neptune. *Geophys. Res. Lett.*, **16**, 929–32.

Suomi, V.E., Limaye, S.S. and Johnson, D. (1991) High winds of Neptune – a possible explanation. *Science*, **251**, 929–32.

Terrile, R.J., and Smith, B.A. (1983) The rotation rate of Neptune from ground-based CCD imaging. *Bull. Am. Astron. Soc.*, **15**, 858.

Tyler, G.L., Sweetman, D.N., Anderson, J.D. *et al.* (1989) Voyager radio science observations of Neptune and Triton. *Science*, **246**, 1466–73.

Yelle, R.V., Herbert, F., Sandel, B.R. *et al.* (1993) The distribution of hydrocarbons in Neptune's upper atmosphere. *Icarus*, **104**, 38–59.

Cross references

Atmosphere
Cosmochemistry
Radiative transfer
Radio science
Spectroscopy: atmospheres
Voyager missions

NEPTUNE: MAGNETIC FIELD AND MAGNETOSPHERE

Prior to the Voyager 2 encounter with Neptune in August 1989, the planet was thought to possess an ordinary magnetic field and magnetosphere. However, the intrinsic magnetic field of Neptune was equally as bizarre as that of Uranus (see Uranus: magnetic field and magnetosphere), with a symmetry axis inclined by 49° to the planetary rotation axis. Since the rotation axis itself was almost at right angles to the solar direction, the polar axis of the magnetosphere, called the polar cusp, was alternately nearly aligned with the flow and then nearly at right angles to the flow. At the time of the passage of the Voyager spacecraft, the polar cusp was so oriented relative to the encounter trajectory that Voyager (for the first time at any planet) entered the magnetosphere through the polar cusp (Belcher *et al.*, 1989).

Planet and interior

Neptune has an equatorial radius of 24 800 km, which is slightly less than the radius of Uranus, a mass of 17 Earth masses (which is slightly greater than the mass of Uranus), and a rotation period of 16.1 h, which is slightly shorter than that of Uranus. Thus we might expect that the internal structures of the two planets would be very similar, and from all accounts they are. Both are inferred to have a large rocky core, a middle layer of ice water and ammonia and an outer envelope of hydrogen and helium, accounting for the outer one-third of the radius of the planet. Unlike Uranus, the rotation axis of the planet is inclined by only 29° to the orbital plane, similar to the 23.5° obliquity of the Earth.

Magnetic field

The magnetic dipole axis of Neptune is tilted at an angle of 47° to the spin axis of the planet. The extrapolated near-equatorial surface field is 1.42 µT, corresponding to a magnetic moment (equatorial surface field times radius cubed) of 2.16×10^{17} Tm^3, close to 27 times greater than the terrestrial magnetic moment. The quadrupole moment of Neptune is quite large and makes a greater contribution to the surface magnetic field than at any other planet. The octupole and higher moments are essentially undetermined (Connerney, Acuña and Ness, 1991).

Magnetosphere

The Neptunian magnetosphere is perhaps the most quiescent magnetosphere in the solar system. The wave levels and energetic particle fluxes are both very low. No evidence for dynamical phenomena were seen in the Voyager flyby. Otherwise the magnetosphere is very similar to the terrestrial magnetosphere, with a bow shock, magnetopause and magnetotail of similar relative dimensions to those of the Earth. The most forward portion of the magnetopause is estimated to lie on average at about 26 Neptunian radii in front of the planet, and of the bow shock at about 34 Neptune radii.

Christopher T. Russell and Janet G. Luhmann

Bibliography

Belcher, J.W., Bridge, H.S., Bagenal, F. *et al.*, (1989) Plasma observations near Neptune: initial results from Voyager 2. *Science*, **246**, 1478–82.

Connerney, J.E.P., Acuña, M.H. and Ness N.F. (1991) The magnetic field of Neptune. *J. Geophys. Res.*, **96**, 19023–42.

Acknowledgement

This work was supported in part by the National Aeronautics and Space Administration under research grant NAGW-2573.

Cross references

Earth: magnetic field and magnetosphere
Magnetospheres of the outer planets
Uranus: magnetic field and magnetosphere

NEPTUNE: RING SYSTEM

Neptune's ring system, as seen by the Voyager 2 spacecraft in 1989, consists of six rings. The most conspicuous ones, the narrow Adams and Le Verrier rings, orbit at 62 930 and 53 200 km from the planet. A third narrow ring appears faintly near the Adams ring: it is co-orbital with the satellite Galatea, at 61 950 km from Neptune's center. A 4000 km wide plateau, 1989N4R, extends out of the Le Verrier ring and ends in a narrow and sharp-edged component, named 1989N5R. The innermost ring of the system, Galle, has a width of 1700 km and an orbital radius of about 42 000 km (Smith *et al.*, 1989). Three moonlets, Despina, Thalassa and Naiad, are embedded in the rings, between the Le Verrier and Galle rings at distances of 52 500, 50 000 and 48 000 km (Owen, Vaughan and Synnott, 1991). Their radii range between 30 and 80 km and their orbits are nearly circular with inclinations of 0.07°, 0.2° and 4.75°.

Just like Saturn's and Uranus' rings, this system contains both broad and narrow rings, together with irregular rings along which one can see brighter sectors, kinks or braids. The bright arcs of rings were detected in 1984 during a systematic campaign of observations of stellar occultations by Neptune (Brahic and Hubbard, 1989). The fact that a star disappeared behind some opaque material on one side of Neptune and not on the other side suggested that Neptunian rings are not continuous all around the planet, but that the material is gathered in a few 'arcs' on distinct orbits. As later seen by Voyager 2, four prominent arcs are indeed clustered in a 40° long sector of the Adams ring (Smith *et al.*, 1989). They have been named Courage, Liberté, Equalité and Fraternité (CLEF). Their length varies from 1000 to 10 000. The arcs are themselves irregular and consist of small clumps a few hundred kilometers long, which may be accumulations of dust particles around larger ones. The astrometry of these ground-based observations was corrected with the new value of Neptune's pole determined by Voyager 2 and these arcs were found to be clustered at the orbit of the Adams ring. From the comparison of ground-based

Figure N10 Neptune's rings. This image, taken from behind Neptune as Voyager 2 left the planet, shows the two main rings at about 53 000 km and 63 000 km from the planet. This geometry favors detection of particles that forward-scatter the solar radiation they receive; these ring images are thus brighter than those taken from the sunward side.

and space data sets, the arcs detected from Earth were identified with the CLEF arcs. They have been stable over 5 years in this 40° long sector. However they should be spread out over 360° in about 3 years because of the differential rotation due to the Keplerian motion. Without some confining force, such as gravitational effects from a nearby satellite or magnetic effects, these arcs would quickly diffuse around the full circumference of the ring.

The occultation data provided first estimates of the positions, widths and opacity of the Neptunian rings. The Voyager 2 spacecraft was thus successfully programmed to observe them extensively despite their faintness and the weakness of the incident solar flux. Voyager 2 mapped the distribution of material around the planet from different aspect angles. The amount of light scattered in different directions by the ring particles helps to constrain their physical properties (composition, size and thickness of ring particles layer). Our knowledge of this system was augmented during the flyby by UV observations of a star (σ Sagittarius), occultation of the spacecraft by the rings (PPS photometer and UVS spectrometer) and by probing the dust and plasma that surround the planet using the plasma instruments.

The chemical composition of these rings is unknown. Assuming compositions similar to bright silicates of Jupiter's rings or water ice slightly contaminated with impurities or dark 'coal' dust (as in Uranus' rings), the optical depths are as high as 0.1 for the arcs and 0.02 for Adams and Le Verrier rings (Smith *et al.*, 1989; Ferrari and Brahic, 1993, Showalter and Cuzzi, 1993). The Galle ring and the plateau are 100 times less opaque. The Adams and Le Verrier rings contain a large fraction of micron-sized particles (50 to 90%). The size distribution of particles inside the plateau is different but it is still difficult to determine whether it is much richer in submicron particles, or contains a larger fraction of macroscopic particles ($\gg 1$ μm; Showalter and Cuzzi, 1993). Plasma and Radio Astronomy experiments (Gurnett *et al.*, 1991, Pedersen *et al.*, 1991) have detected particle clouds with optical depths of 10^{-6} to 10^{-7} at distances of 85 000 and 103 000 km from Neptune in the equatorial plane but also at higher latitudes. As dust is very short-lived in planetary environments due to processes like Poynting–Robertson drag and gas and plasma drags, it must be constantly replenished to endure. Neptune's rings are about two orders of magnitude dustier than the other planetary ring systems where gas and plasma drags may be more efficient. Meteoroid bombardment is usually considered as the source of dust. The dust then falls rapidly towards the planet by the Poynting–Robertson effect. The bombardment of Proteus, orbiting at 117 650 km, and of the nearby satellites could feed the ring system with dust (Colwell and Esposito, 1990). The mutual collisions within the more opaque rings (Adams, Le Verrier) would also be an important source of dust.

Since 1984 most of the theoretical work on Neptune's rings has been devoted to the problem of the azimuthal confinement of the arcs. Many models have been proposed (Lissauer, 1985; Goldreich, Tremaine and Borderies, 1986; Lin, Papaloizou and Ruden, 1987) but none can readily explain the Voyager observations. Several explanations are plausible. Either some external forces due to gravitational or magnetic perturbations effectively confine the arcs at given longitudes, or the arcs are narrower than the 15 km wide ring and therefore less sensitive to the Keplerian shear. In this latter case their lifetime may exceed 5 years. A magnetic origin or confinement of such dusty arcs is efficient (Hamilton and Burns, 1992) but unlikely. On the other hand, the Adams ring is effectively gravitationally perturbed by the nearby satellite Galatea through a Lindblad resonance 42 : 43 (Porco, 1991). This resonance excites the eccentricities of the ring particles and produces a wavy radial distortion which travels in the ring following the orbital motion of the satellite. Moreover, the orbit of the arcs coincides with the location of a corotation resonance with this inclined satellite. If so, the arcs should be confined in 4° long sites regularly distributed along the ring, but most of these sites are empty and the Fraternité arc spreads over several consecutive sites. The actual strength of this type of confinement cannot be easily estimated, but it is probably weak because of the small mutual inclination of the ring and Galatea. Finally, the arcs could be confined at the Lagrangian points of a co-orbital moonlet (Lissauer, 1985; Sicardy and Lissauer, 1992) but such a moonlet has not been detected in the Voyager 2 images. With regard to the origin of arcs and narrow rings, Esposito and Colwell (1992) proposed that they result from the recent disruption of a small moon, the largest fragments of which would be trapped in the corotation sites. However the dynamical evolution and collective behavior of such a family of clumps and arcs is still unclear and we do not yet have the key to the question of the CLEF arcs.

Cecile Ferrari

Bibliography

Brahic, A. and Hubbard, W. (1989) The baffling ring arcs of Neptune. *Sky and Telescope*, June 1989.

Colwell, J.E. and Esposito, L.W. (1990) A model of dust production in the Neptune ring system. *Geophys. Res. Lett.*, **17**, 174–4.

Esposito, L.W. and Colwell, J.E. (1992) Neptune's rings and satellite system: collisional origin and evolution. Presented at AGU, Montreal, Canada.

Ferrari, C. and Brahic, A. (1993) Azimuthal brightness asymmetries in planetary rings: I – Neptune's arcs and narrow rings. *Icarus*, **111**, 193–210.

Goldreich, P., Tremaine, S. and Borderies, N. (1986) Towards a theory for Neptune's arc rings. *Astron. J.*, **92**, 490–4.

Gurnett, D.A. *et al.* (1991) Micron-sized particles detected near Neptune by the Voyager 2 plasma wave instrument. *J. Geophys. Res.*, **96** (suppl.), 19177–86.

Hamilton, D.P. and Burns, J.A. (1992) Orbital evolution of dust grains around Neptune. Presented at Neptune/Triton meeting, Tucson, January 1992.

Lissauer, J. (1985) Shepherding model for Neptune's arc ring. *Nature*, **318**, 544–5.

Lin, D.N.C, Papaloizou, J.C.B. and Ruden, S.P. (1987) On the confinement of planetary arcs. *Mon. Not. Roy. Astron. Soc.*, **227**, 75–95.

Owen, W.M., Jr, Vaughan, R.M. and Synnott, S.P. (1991) Orbits of the six new satellites of Neptune. *Astron. J.*, **101**(4), 1511–5.

Pedersen *et al.* (1991) Dust distribution around Neptune: grain impacts near the ring plane measured by the Voyager planetary radio astronomy experiment. *J. Geophys. Res.*, **96** (suppl.), 19187–96.

Porco, C. (1991). An explanation for Neptune's arcs. *Science*, **253**, 995–1001.

Sicardy, B. and Lissauer, J.J. (1992) Dynamical models of the arcs in Neptune's 63K ring (1989N1R). *Adv. Space Res.*, **12**(11), 85–95.

Showalter, M.R. and Cuzzi, J.N. (1993) Structure and particle properties of Neptune's ring system. *Icarus*, to be published.

Smith, B.A., Soderblom, L.A., Banfield, D.*et al.* (1989) Voyager 2 at Neptune: imaging science results. *Science*, **246**, 1422–49.

Cross references

Dust
Planetary ring
Shepherd satellite
Voyager missions

NEPTUNE: SATELLITE SYSTEM

Like the other gas giants, Neptune has a large family of satellites. Triton, Neptune's largest satellite, was discovered by Lassell in 1846, the year in which Neptune itself was discovered. Nereid was discovered by Kuiper in 1949. The inner six satellites were discovered in 1989 during the Voyager 2 flyby (Smith *et al.*, 1989). The orbital parameters and other characteristics of Neptune's eight satellites (Table N3; Figure N11) allow separate consideration in three groups: the six inner satellites, Triton and Neptune's outermost satellite (Nereid). Nereid's orbit is inclined and highly elliptical, Triton's orbit is circular, inclined and retrograde, and the inner six satellites have normal, almost circular orbits lying in Neptune's equatorial plane (except for Naiad's, which is inclined by nearly 5°). The unusual orbits of Nereid and Triton may reflect unusual origins and may relate to the vigor and youth of geological activity on Triton.

Inner satellites

The six inner satellites straddle the Roche limit (q.v.). These satellites increase in diameter and their orbits are increasingly widely spaced with increasing orbital distance (Thomas and Veverka, 1991). These progressions are similar to those of well-ordered satellite systems of the other gas giant planets and of the solar system itself. Neptune's rings (q.v.) are spread across the region interior to the Roche limit where the four innermost satellites orbit. Some inner satellites might help to control structure within the rings and may serve as the source of ring material.

Surface features and shapes are apparent only in the images of Proteus and Larissa. Both bodies are distinctly non-spherical and appear to be heavily cratered. One crater on Proteus (Figure N12) is 10 to 15 km deep and 255 km in diameter (greater than the radius of Proteus). Proteus is the largest satellite in the solar system to have a markedly non-spherical shape. Even so, many segments of its profile are rounded. Proteus apparently is a transitional object small enough that it never experienced substantial geologic activity, but large enough that slight internal heating might have given it a more compact shape (Croft, 1992). The inner satellites have low reflectivities (albedo ~ 0.06). Spectral identifications of the surface materials on these satellites and densities that would tell us something about the interior composition are lacking, but the low albedos of these objects suggest that they contain substantial amounts of carbonaceous compounds and rocky minerals in a matrix of water ice. Besides their low albedos, other indications that Proteus and Larissa (and probably the other small inner satellites) are probably composed of undifferentiated mixtures of primordial condensates include the presence of big impact craters and the non-spheroidal shapes of these satellites.

Table N3 Orbital and physical characteristics of Neptune's satellites

Satellite	Radius (km)[a]	Geometric albedo	Possible surface composition	Orbital distance (10^3 km)	Orbital eccentricity	Orbital inclination (deg)
Nereid	170 ± 25	0.155	Dirty ice	5513.4	0.75	27
Triton	1352.5 ± 5	0.70	Most N_2; also CH_4, CO, CO_2	354.8	0.000	159
N1 Proteus	208 ± 8 $218 \times 208 \times 201$	0.064	Carbonaceous	117.6	0.0004	0.04
N2 Larissa	96 ± 7 104×89	0.053	Carbonaceous	73.7	0.0014	0.20
N4 Galatea	79 ± 12	0.062	Carbonaceous	62.0	0.0001	0.05
N3 Despina	74 ± 10	0.061	Carbonaceous	52.5	0.0001	0.07
N5 Thalassa	40 ± 8	?	?	50.0	0.0002	0.21
N6 Naiad	29 ± 6	?	?	48.0	0.0003	4.74

[a] Proteus and Larissa are irregularly shaped; the tabulated radii are the mean values, followed by measurements of the *a*-, *b*-, and *c*-axes in the case of Proteus, and the *a*- and *c*-axes in the case of Larissa.

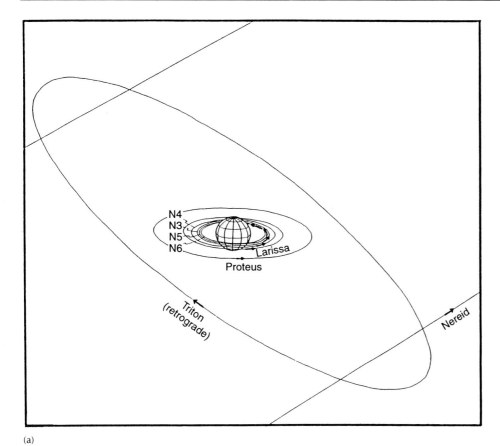

(a)

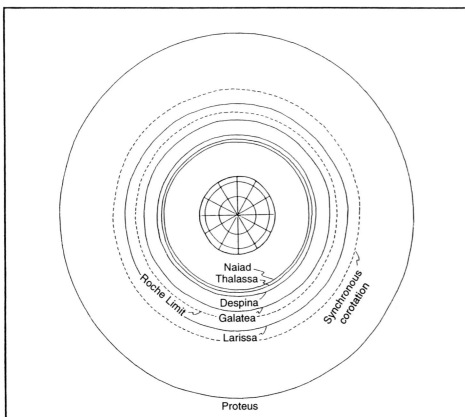

(b)

Figure N11 Orbits of Neptune's family of satellites. (a) Perspective view along a line of sight inclined 18° to Neptune's equatorial plane. All satellites have virtually circular orbits except for Nereid. The apparent crossing of Nereid's and Triton's orbits is an artifact of the projection. (b) Polar projection showing the orbits of the inner six satellites, the Roche limit (calculated for density 1.2 g cm^{-3}), and the corotational orbit.

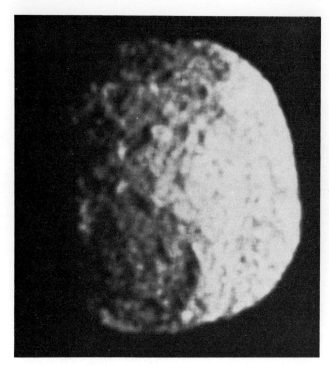

Figure N12 Voyager 2 image of Proteus showing a heavily cratered surface and a non-spherical shape.

Triton

Ground-based telescopic studies have added greatly to our knowledge of Triton, but most of what we know is a result of the Voyager 2 flyby (Smith *et al.*, 1989; Croft *et al.*, 1993). With a diameter of 2700 km, Triton is the seventh largest known satellite and is larger than Pluto (Plate 32). Triton has the second thickest atmosphere of all the satellites in the solar system; it is the third densest large satellite in the outer solar system. Triton's retrograde orbit distinguishes it from all other large satellites in the solar system.

Triton's atmosphere has a surface pressure of about 14 μbar, a surface temperature of 38 K and a surface density about 0.01% of Earth's. The major constituent of Triton's atmosphere is N_2, a composition shared by only two other atmospheres in the solar system – Earth's and Titan's. Methane is a trace gas amounting to just 0.001% of the atmosphere. These gas abundances are roughly consistent with the partial pressures at 38 K of pure nitrogen and methane ices exposed on the surface (Conrath *et al.*, 1989). Triton's atmosphere also contains smog-like hazes that are thought to be composed of tiny solid particles (~ 0.1 μm in diameter) that are produced by the action of solar ultraviolet rays on methane (Krasnopolsky, Sandel and Herbert, 1992).

Seasonality on Triton is thought to motivate cycles of global volatile transport, polar cap growth and sublimation, all of which are analogous to cycles of volatile transport on Mars. However, on Triton the important volatiles are nitrogen and methane. Voyager 2 images show ices apparently sublimating on the southern (summer) hemisphere (Figure N13). Ground-based spectroscopic observations have established that Triton's surface is covered by condensed solid nitrogen plus small amounts of methane, carbon monoxide and carbon dioxide (Cruikshank *et al.*, 1988; Spencer, Buie and Bjoraker, 1990; Brown *et al.*, 1996). Surprisingly, water ice has not been detected on Triton, but water ice is thought to exist at shallow depths beneath the surface.

Triton's mean density, 2.075 g cm^{-3}, requires a substantial rock component amounting to 65% to 83% of the body's mass. Its volcanic surface suggests that it is differentiated and that it probably includes a rocky core 2000 to 2200 km in diameter and an icy mantle and crust totaling 250 to 350 km in thickness. Several facts suggest that Triton contains substantial water ice beneath a veneer of frozen nitrogen and methane. Ice is widespread and abundant in the outer solar system, and it would be difficult to explain its absence in Triton. Furthermore, it has been suggested that the spectroscopically observed ices (other than CO_2) would be far too soft, even at 38 K, to maintain the observed relief even for short periods of geologic time. Water ice has the required strength. Finally, some of Triton's lava flows are very thick (approximately 200 m) and must have been quite viscous mixtures of once-molten ices (Croft *et al.*, 1993). Laboratory experiments show that the required viscosities could have been achieved by cryogenic aqueous mixtures of substances thought to be present in Triton, including water, ammonia and perhaps other substances such as methanol (Kargel *et al.*, 1991).

Prior to the Voyager 2 flyby, some spectroscopists interpreted evidence for surficial nitrogen to indicate that it might be in the liquid form, thus constituting a cryogenic ocean or lakes. Voyager 2 did not indicate such features and in fact proved that Triton is far too cold to have stable liquid nitrogen on its surface. Voyager 2's camera, however, did reveal an extraordinary diversity of geologic features on Triton. Some of the features most familiar to Earth-bound observers (at least superficially) include volcanic rifts; volcanic cones with craters on their summits and flanks; dome-like extrusions and smooth-floored, steep-walled depressions (planitiae). Planitiae are about 200 to 400 m deep and as much as 400 × 200 km in extent (Figure N14). Planitiae appear to have formed by extensive volcanism and collapse along circular or arcuate faults. Some aspects of Triton's planitiae suggest a kinship with volcanic calderas on Earth and other planetary bodies. However, the analogy is not perfect; Triton's planitiae are much larger than terrestrial volcanic calderas, and they do not sit on top of immediately obvious volcanic constructs. In the latter respect Triton's planitiae resemble the giant caldera on Mars' Alba Patera and many calderas on Io.

Enigmatic features on Triton include a global pattern of tectonic lineaments and the 'cantaloupe terrain.' So-called because of its resemblance to the melon's husk, cantaloupe terrain consists of ridges and fields of circular depressions that individually are about 30 to 40 km across (top right in Figure N13). The global pattern of lineaments and the cantaloupe terrain are thought to have formed by volcanic and tectonic processes, but their specific origins are unknown. One idea is that viscous diapiric flow of an unstable layered crust produced features analogous to salt domes on Earth (Schenk and Jackson, 1993). Another idea considers the cantaloupe terrain to have originated by folding and faulting caused by compression as Triton cooled (Boyce, 1993).

Among the most exciting of Voyager's findings was the discovery of active geyser-like plumes on Triton (Soderblom *et al.*, 1990). At least two plumes of particulate material formed columns about 8 km high. Several hypotheses attempt to account for the plumes. 'Greenhouse' models rely on solar heating of translucent nitrogen ice; subsurface ice warms by several degrees and its vapor pressure increases until gas suddenly jets through fractures and carries dust and ice particles into the atmosphere (Soderblom *et al.*, 1990). Alternatively, the plumes may be jets of relatively hot gases similar to terrestrial geysers or fumaroles (Croft *et al.*, 1993). Another idea is that the plumes are dust devils forming over warm, frost-free areas (Ingersoll and Tryka, 1990).

The southern hemisphere presents an unusual terrain composed of albedo patches, plateaus and depressions, which appear to be sublimating under the summer sun (Figure N13). Albedo streaks probably consist of wind-deposited ice grains and dark particles. Triton's tenuous winds might be strong enough to loft certain types of particles. More likely, geyser-like plumes lifted the material, which later was deposited by wind.

Even the most heavily cratered areas have much lower crater densities than most icy satellites (Croft *et al.*, 1993). Triton's low crater density and pristine volcanic surface give an impression of geologic youth. A crater production model based on comet impacts indicates an average age of less than 600 million years (less than one-seventh of the age of the solar system), but the absolute age is very poorly constrained (Strom and Croft, 1993). It is possible that Triton's youthful surface was formed mainly during a period of extensive melting and intense geological activity immediately following Triton's capture into orbit around Neptune.

It has been suggested that the wind streaks must be less than 1 Triton year old (less than 165 Earth years); otherwise they should be buried and then sublimated during Triton's annual global cycle of volatile transport. Geologically recent or even active volcanism on Triton is consistent with thermal models that indicate a molten

Figure N13 High-resolution mosaic of Triton centered near 20°N, 0°W. The sub-Neptune point at (0°, 0°) occurs slightly to the left of center. North is to the right. (Image processed by Alfred McEwen, US Geological Survey, Flagstaff.)

interior if its icy mantle contains ammonia (Stevenson and Gandhi, 1990). Alternatively, Triton might now be geologically dead except for climate-related atmospheric modifications of the surface and venting of gases from 'solar greenhouses'.

Nereid

Nereid's highly eccentric and inclined orbit is typical of distantly orbiting satellites in the outer solar system. Nereid's shape and surface details were not clearly revealed by Voyager 2, but its diameter and albedo were determined (Table N3). Pre-Voyager Earth-based observations of Nereid's rotational lightcurve had suggested that Nereid is either an unusually elongated object (almost pencil-like in shape) or that one side has more than ten times the reflectivity of the other side. However, Voyager 2 showed that neither idea was correct; instead, Nereid is slightly irregularly shaped and possesses a surface having a fairly uniform albedo. Although it is similar to Proteus in size, its orbit and higher albedo clearly distinguish Nereid from the inner satellites (Thomas, Veverka and Helfenstein, 1991). It is uncertain why Nereid is so distinguished, but in general terms these differences may be telling us that Nereid has a distinct origin.

Origin of Neptune's satellites

Theories of the origin of Neptune's satellites should account for similarities to and differences from the satellite systems of the other gas giants. All four of the gas giants have rings and many satellites. The total mass of Neptune's satellites (dominated by Triton) divided by the mass of Neptune is 0.00021. This value is almost the same as equivalent ratios for Jupiter (0.00021), Saturn (0.00024) and Uranus (0.00011; Pollack, Lunine, and Tittemore, 1991). But this is where most of the similarities end. The eight known Neptunian satellites are the fewest possessed by any of the gas giants, and Neptune lacks close equivalents of the large, regular satellites of the other planets.

The major satellites of Jupiter, Saturn and Uranus, with some important exceptions, have nearly circular orbits lying virtually within the equatorial plane of their respective planets. The diameters, orbital distances and other characteristics of these satellites tend to differ in smooth progressions. The closest cosmic analog to the regular satellite systems of Jupiter, Saturn and Uranus is the solar system itself. It is generally thought that these satellite systems and the solar system formed by very much the same processes. The irregular orbits of Neptune's two largest satellites (Triton and Nereid), especially the retrograde orbit of Triton, suggest that something radically different accounts for the origin of those objects.

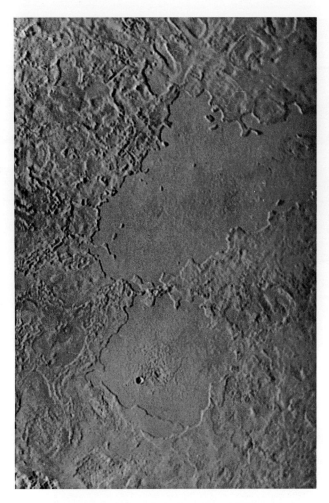

Figure N14 Mosaic of Monad Regio, a volcanic terrain dominated by Tuonela Planitia (center) and Ruach Planitia (below center). Dimensions of scene are about 940 × 490 km. Center coordinates are latitude 32°N, longitude 15°. North is toward the right.

Physical similarities between Triton and Pluto have spurred several hypotheses involving a common origin. They have nearly the same diameters and densities, and both contain nitrogen and methane frosts on their surfaces. Furthermore, Pluto has an unusual, Neptune-crossing orbit, while Triton has a peculiar orbit in the Neptune system. Capture hypotheses maintain that Triton was captured while ejection hypotheses hold that they once were sister satellites. Ejection hypotheses involve a near miss or an actual collision with an Earth-sized planet, causing Pluto to be ejected from orbit around Neptune and forcing Triton into its unusual orbit. Capture of Triton by Neptune could have occurred if some of Triton's orbital energy was dissipated as a result of (1) a gravitational encounter (near miss) or a physical collision with a regular satellite of Neptune or (2) gas drag as Triton swept through the fringes of the Neptunian nebula (Goldreich et al., 1989; Malcuit, Mehringer and Winters, 1992; McKinnon and Leith, 1992). Capture hypotheses now are favored because the dynamical problems are less severe than the problems in ejecting Pluto; ejection hypotheses are in disrepute also because they do not explain the existence of Pluto's large satellite, Charon.

If Triton was captured, its initial orbit would have been highly eccentric, but tidal interactions would have circularized it in a few hundred million years. Tidal dissipative heating during circularization would have melted and degassed Triton's interior, and it might have produced a dense, warm atmosphere and an ammonia–water ocean, or at least intense aqueous volcanism (Lunine and Nolan, 1992).

The capture hypothesis alone does not clearly explain the near coincidence of satellite/planet mass ratios of the four gas giants. One possibility is that the captured object initially was just a fraction of Triton's current mass but, during its eccentric travels through the Neptunian system, Triton cannibalized most of Neptune's original family of satellites (Goldreich et al., 1989). Hence the original ratio of satellite mass/Neptune mass would have been approximately the same as that observed today, although the mass would have been distributed very differently.

Nereid's highly eccentric and inclined orbit may indicate that it, too, was captured. Alternatively, Nereid's orbit may have been perturbed when Triton was captured.

The six inner satellites may be the intact remnants of Neptune's original family of satellites, or they may be objects reaccreted from a ring of debris produced during Triton's capture or Pluto's ejection. Four satellites orbit inside the Roche limit. Accretion inside the Roche limit is not straightforward, but it might have been possible if accreting particles adhered by nongravitational forces. It is more likely that these four satellites, plus Larissa, accreted outside the Roche limit and that tidal action later caused them to move to their current positions (Banfield and Murray, 1992).

Jeffrey S. Kargel

Bibliography

Banfield, D. and Murray, N. (1992) A dynamical history of the inner Neptunian satellites. *Icarus*, **99**, 390–401.

Boyce, J.M. (1993) A structural origin for the cantaloupe terrain of Triton (abstract), in *Lunar Planet. Sci. Conf.* **XXIV**, pp. 165–166.

Brown, R.H., Cruikshank, D.P., Veverka, J., Helfenstein, P. and Eluskiewicz, J. (1996) Surface composition and photometric properties of Triton, in *Neptune and Triton* (ed. D.P. Cruikshank). Tucson: University of Arizona Press, pp. 991–1030.

Conrath, B., Flasar, F.M., Hanel, R. et al. (1989) Infrared observations of the Neptunian system. *Science*, **246**, 1454–9.

Croft, S.K (1992) Proteus: geology, shape, and catastrophic destruction. *Icarus*, **99**, 402–19.

Croft, S.K., Kargel, J.S., Kink, R.L. et al. (1996) The geology of Triton, in *Neptune and Triton* (ed. D.P. Cruikshank). Tucson: University of Arizona Press, pp. 879–948.

Cruikshank, D.P. et al. (1988) Volatiles on Triton: the infrared spectral evidence from 2.0–2.5 microns. *Icarus*, **74**, 413–423.

Goldreich, P., Murray, N., Longaretti, P.Y. and Banfield, D. et al. (1989) Neptune's story. *Science*, **245**, 500–4.

Ingersoll, A.P. and Tryka, K.A. (1990) Triton's plumes: the dust devil hypothesis. *Science*, **250**, 435–7.

Kargel, J.S. Croft, S.K., Lunine, J.I. and Lewis, J.S. et al. (1991) Rheological properties of ammonia–water liquids and crystal–liquid slurries: planetological applications. *Icarus*, **89**, 93–112.

Krasnopolsky, V.A., Sandel, B.R. and Herbert, F. (1992) Properties of haze in the atmosphere of Triton. *J. Geophys. Res.*, **97**, 11695–700.

Lunine, J.I. and Nolan, M.C. (1992) A massive early atmosphere on Triton. *Icarus*, **100**, 221–34.

Malcuit, R.J., Mehringer, D.M. and Winters, D.R. (1992) Numerical simulation of retrograde tidal capture of a Triton-like planetoid by a Neptune-like planet: two-dimensional limits of a stable capture zone (abstract), in *Lunar Planet. Sci. Conf.*, **XXIII**, pp. 827–8.

McKinnon, W.B. and Leith, A.C. (1992) Gas drag and the evolution of a captured Triton. *Icarus*, 118, 392–413.

Pollack, J.B., Lunine, J.I. and Tittemore, W.C. (1991) Origin of the Uranian satellites, in *Uranus* (ed. J.T. Bergstralh, E.D. Miner and M.S. Matthews). Tucson: University of Arizona Press, pp. 469–512.

Schenk, P. and Jackson, M.P.A. (1993) Diapirism on Triton: a record of crustal layering and instability. *Geology*, **21**, 299–302.

Smith, B.A. et al. (1989) Voyager 2 at Neptune: imaging science results. *Science*, **246**, 1422–49.

Soderblom, L.A., Kieffer, S.W., Becker, T.L. et al. (1990) Triton's geyser-like plumes: discovery and basic characterization. *Science*, **250**, 410–5.

Spencer, J.R., Buie, M.W. and Bjoraker, G.L. (1990) Solid methane on Triton and Pluto: 3- to 4-micron spectrophotometry. *Icarus*, **88**, 491–6.

Stevenson, D.J. and Gandhi, A.S. (1990) Puzzles of Triton (abstract), in *Lunar Planet. Sci. Conf.*, **XXI**, pp. 1202–3.

Strom, R.G. and Croft, S.K. (1993) Triton's cratering record and its time of capture (abstract), in *Lunar Planet. Sci. Conf.*, **XXIV**, 1373–4.

Thomas, P. and Veverka, J. (1991) Neptune's small inner satellites. *J. Geophys. Res.*, **96** (suppl.), 19261–8.

Thomas, P., Veverka, J. and Helfenstein, P. (1991) Voyager observations of Nereid. *J. Geophys. Res.*, **96 (suppl.)**, 18903–9268.

Cross references

Nereid
Satellite, natural
Shepherd satellite
Triton

NEREID

Nereid (N II), is the second known satellite of Neptune, discovered in 1949 by G.P. Kuiper (Kuiper, 1949). It is the outermost of Neptune's moons and has the highest orbital eccentricity of any natural satellite in the solar system.

Orbit and origin

Nereid orbits Neptune at a mean distance (orbital semimajor axis) of 5 515 000 km ($\sim$ 219 Neptune radii) with a period of nearly 1 year (360 days). This orbit is prograde and inclined $\sim$ 10° to the plane of Neptune's orbit (Mignard, 1981; Veillet, 1982; note that this inclination is often misquoted). Nereid's most striking attribute is the high eccentricity of its orbit. Its value of 0.75 is by far the largest of any known natural satellite, and twice as great as the runner-up, Jupiter's outer moon Pasiphae (J VIII). Because of this eccentricity, Nereid's distance from the center of Neptune varies from $\sim$ 55 radii out to 383 radii. In contrast, Neptune's largest moon Triton (N I) remains at a practically fixed distance of 14 Neptune radii from the center (Figure N15).

Despite its high eccentricity and semimajor axis, Nereid is safe from escaping Neptune. A planet with an orbital period P can generally keep prograde satellites with periods up to $\sim P/9$ (Szebehely, 1978; Szebehely and McKenzie, 1978). This stability limit implies that Neptune, with a period of 165 years, can retain moons with periods up to $\sim$ 18 years, and semimajor axes up to $\sim$ 38 650 000 km $\approx$ 1560 Neptune radii, far beyond the orbit of Nereid. Analytic theory (Mignard, 1975) confirms that Nereid's orbit is quite stable with respect to solar perturbations. Furthermore, tidal interactions between Nereid and Neptune have negligible influence on Nereid's current orbit.

Because of its high eccentricity and semimajor axis, Nereid is usually considered a captured body rather than a primordial satellite of Neptune. Since its present orbit is so stable, however, it cannot be the result of simple gravitational capture from solar orbit (see Capture mechanism). Instead it may have been captured by gas drag in a primordial nebula surrounding Neptune, or by collision with a pre-existing satellite. It is often suggested that Neptune's massive retrograde moon Triton (q.v.) was itself captured into a very eccentric orbit, but that tidal interactions with Neptune have since reduced its eccentricity and semimajor axis to their current values. It is possible that Nereid was originally a regular satellite of Neptune, but that repeated close encounters with Triton during this evolution raised Nereid's orbital eccentricity, semimajor axis and inclination to their present values (Goldreich *et al.* 1989).

Physical properties

The shape, mass, composition and spin of Nereid are all unknown, but the Voyager 2 imaging team estimated its mean radius as 170 $\pm$ 25 km (Smith *et al.*, 1989). Together with observations of Nereid's brightness (mean opposition magnitude 19.4, absolute magnitude 4.7) and phase function, this size implies a geometric albedo of 0.18 $\pm$ 0.02 (Thomas, Veverka and Helfenstein, 1991). This albedo is similar to that of the icy moons of Uranus, but much brighter than the smaller satellites of both Uranus and Neptune. Nereid is relatively neutral (gray) in color, like Neptune's similar-sized satellite Proteus (1989N1) as well as the icy moons of Uranus (Thomas, Veverka and Helfenstein, 1991; see Uranus: satellite system). Altogether, Nereid's surface resembles rock or dirty ice much more than a dark carbonaceous body.

Shape and spin

Voyager 2 images did not reveal Nereid's shape, but some guidance may be obtained by comparison with other bodies. In general, moons with radii greater than $\sim$ 200 km are spherical, while smaller satellites are irregular (Thomas, Veverka and Dermott, 1986; Thomas, 1989). For example, Proteus appears irregular in outline, yet is nearly equidimensional, with principal semiaxes of 218 $\times$ 208 $\times$ 201 $\pm$ 8 km (Thomas and Veverka, 1991; Croft, 1992). Nereid's estimated radius of 170 $\pm$ 25 km falls between that of highly irregular Hyperion (S VII), with semiaxes of 175 $\times$ 120 $\times$ 100 $\pm$ 10 km, and the notably spherical Mimas (S I), with a radius of 197 $\times$ 3 km. Unfortunately this leaves Nereid in the transition zone between 'large' and 'small' bodies; we can conclude only that its longest axis may be up to twice as long as its shortest.

The rotation of Nereid is currently a topic of some controversy. From ground-based photometry spanning 8 days, Schaefer and Schaefer (1988a; also Veverka, 1988) reported brightness variations greater than a factor of four (1.5 magnitudes), with a period from 8 to 24 h, presumably due to rotation. This suggested either that Nereid is extremely non-spherical, or that its albedo varies drastically from one side the other, like Saturn's moon Iapetus (S VIII; q.v.) Follow-up observations (Schaefer and Schaefer, 1988b) seemed to confirm the large variability, but indicated long periods up to $\sim$ 200 h ($\approx$ 8.3 days). Bus and Larson (1989) also found periods longer than a day, but smaller variations of $\sim$ 60 % ($\approx$ 0.5 magnitude). In contrast, Voyager 2 images of Nereid spanning 12 days show 'no evidence for a lightcurve amplitude of more than 10%' ($\approx$ 0.1 magnitude; Smith *et al.*, 1989; also Thomas, Veverka and Helfenstein, 1991). The latter suggest either that Nereid is nearly spherical and uniform in albedo, or that it is rotating very slowly. Finally, the most recent ground-based observations (Williams, Jones and Taylor, 1991) again indicate large variations, by a factor of 3.3 ($\approx$ 1.3 magnitude) with a period of 13.6 h. At present the question of Nereid's shape and spin must be regarded as unresolved.

Theory provides some constraints on Nereid's spin state. Most bodies in the solar system seem to have originated with rotation periods of about half a day. However, tidal friction (q.v.) in satellites tends to slow their rotation. Tides have despun most satellites to the synchronous resonance, where the spin period equals the orbit period. This is why the same hemisphere of the Moon always faces the Earth. The only known exceptions are moons in distant and/or eccentric orbits, such as Jupiter's outer satellites or Saturn's Hyperion (S VIII) and Phoebe (S IX). If Nereid is made of solid ice, tides would take $\sim 10^{11}$ years to despin it from a period of 12 h to synchronous rotation (Dobrovolskis, 1995). This is still $\sim$ 20 times longer than the age of the solar system, but this timescale would be shorter if Nereid had once been partially melted, or closer to Neptune. Therefore tides may already have reduced Nereid's rotation rate.

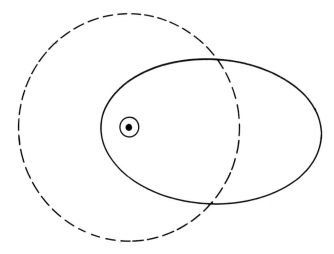

Figure N15 Relative orbits of Triton (N I) and Nereid (N II). The dot represents Neptune and the small circle shows Triton's orbit. For comparison, the ellipse displays Nereid's orbit, while the dashed circle indicates its mean distance (semimajor axis).

Even if Nereid has been despun, its rotation period is unlikely to be synchronous with its 360-day orbital period. For high eccentricities the synchronous resonance becomes quite weak. The same is true of other low-order commensurabilities, analogous to the 3:2 spin–orbit resonance of Mercury (q.v.), where the planet turns three times on its axis during two orbits of the Sun. In contrast, high-order resonances become quite strong; for $e = 0.75$ the 6:1, 13:2, 7:1 and 15:2 resonances are the strongest.

Even so, Nereid may not have been captured into any resonance. Because of librations (q.v.), the resonant periods are not perfectly precise, but have a finite 'bandwidth' depending on Nereid's shape. If its principal semiaxes differ by more than $\sim 1.66\%$ (≈ 3 km), as seems likely, the strongest spin–orbit resonances will overlap and give rise to a band of chaotic rotation (Dobrovolskis, 1995; see Chaotic dynamics in the solar system). Such chaos is characterized by irregular variations of the period and direction of spin, like those of Saturn's moon Hyperion (S VII) If Nereid has been despun, it may be tumbling chaotically with a period of a month or two. This might explain some of the discrepancies between ground-based and spacecraft observations of Nereid.

Anthony R. Dobrovolskis

Bibliography

Bus, E.S. and Larson, S. (1989) CCD photometry of Nereid. *Bull. Am. Astron. Soc.*, **21**, 982.
Croft, S.K. (1992) Proteus: geology, shape, and catastrophic destruction. *Icarus*, **99**, 402–19.
Dobrovolskis, A.R. (1995) Chaotic rotation of Nereid. *Icarus*, **118**, 181–98.
Goldreich, P., Murray, N., Longaretti, P.-Y. and Banfield, D. (1989) Neptune's story. *Science*, **245**, 500–4.
Kuiper, G.P. (1949) The second satellite of Neptune. *Publ. Astron. Soc. Pacific*, **61**, 175–6.
Mignard, F. (1975) Satellite à forte excentricité. Application à Néréide. *Astron. Astrophys.*, **43**, 359–79.
Mignard, F. (1981) The mean elements of Nereid. *Astron. J.*, **86**, 1728–9.
Schaefer, M.W. and Schaefer, B.E. (1988a) Large-amplitude photometric variations of Nereid. *Nature*, **333**, 436–8.
Schaefer, M.W. and Schaefer, B.E. (1988b) UVBRI photometry of Nereid. *Bull. Am. Astron. Soc*, **20**, 825.
Smith, BA. Soderblom, L.A., Banfield, O. *et al.* (1989) Voyager 2 at Neptune: imaging science results. *Science*, **246**, 1422–49.
Szebehely, V. (1978) Stability of artificial and natural satellites. *Celest. Mech.*, **18**, 383–9.
Szebehely, V. and McKenzie, R. (1978) Comparison between stability limits for satellite motion. *Celest. Mech.*, **18**, 391–4.
Thomas, P. and Veverka, J. (1991). Neptune's small inner satellites. *J. Geophys. Res.*, **96** (suppl), 19261–8.
Thomas, P. (1989) The shapes of small satellites. *Icarus*, **77**, 248–74.
Thomas, P., Veverka, J. and Dermott, S. (1986). Small satellites, in *Satellites* (eds J.A. Burns and M.S. Matthews). Tucson: University of Arizona Press, pp. 802–35.
Thomas, P., Veverka, J. and Helfenstein, P. (1991) Voyager observations of Nereid. *J. Geophys. Res.*, **96** (suppl) 19253–9.
Veillet, C. (1982) Orbital elements of Nereid from new observations. *Astron. Astrophys.*, **112**, 277–80.
Veverka, J. (1988) Taking a dim view of Nereid. *Nature*, **333**, 394.
Williams, I.P., Jones, H.P. and Taylor, D.B. (1991) The rotation period of Nereid. *Mon. Not. Roy. Astron. Soc.*, **250**, 1p–2p.

Cross references

Neptune: satellite system
Satellite, natural

NEWCOMB, SIMON (1835–1909)

Simon Newcomb was of Canadian origin, the son of a school teacher, and was born on 12 March 1835 at Wallace, Nova Scotia. He ran away from an apprenticeship, spent some time as a country schoolmaster, and eventually found work in the Nautical Almanac Office, then in Cambridge, Mass. He then went to Harvard, where he soon showed his exceptional mathematical ability, and graduated after 1 year. In 1861 he was appointed to the US Naval Observatory in Washington. In 1877 he was put in charge of the American Nautical Almanac Office (transferred to Washington), and in 1884 obtained the additional appointment of Professor of Mathematics and Astronomy at Johns Hopkins University in Baltimore, though he continued to live in Washington. When he retired from the Navy in 1897, he was given the rank of Rear Admiral.

His main work was in determining the positions of celestial bodies with meridian instruments, and for some time employed the Washington 26-inch (66-cm) refractor. In 1865 he used the parallax of Mars to make a new determination of the value of the astronomical constant, or Earth–Sun distance. When he took charge of the American Nautical Almanac Office, he began his greatest work – the calculation of the motions of solar system bodies; his planetary tables are still regarded as the official standards. Together with A. Downing, he also devised the universal standard system of astronomical constants.

In his biography, Marsden (1974) describes how Newcomb devised a procedure for determining the 'disturbing function' that is created by one planet's passage of another, the quantities being now known as the 'Newcomb operators'. Newcomb died on 11 July 1909 in Washington.

Patrick Moore

Bibliography

Archibald, R.C. (1924) Biographical memoirs, Washington. *Natl Acad. Sci.*, **17**, 19–69 (with complete bibliography).
Marsden, B.G. (1974) Newcomb, Simon. *Dict. Sci. Biogr.*, Vol. 10, pp. 33–6.
Newcomb, S. (1903) *The Reminiscences of an Astronomer.* Boston–New York.

NEWTON, SIR ISAAC (1642–1727), AND NEWTON'S LAWS

English physicist and mathematician Isaac Newton, born in Woolsthorpe, Lincolnshire (England), in the same year that Galileo died, was the culminating figure in the scientific revolution of the 17th century. In physical vision standing alongside Einstein, Newton made such fundamental contributions to physics and mathematics as to open completely new horizons in both sciences.

Newton entered Trinity College, Cambridge, in 1661. Beginning in 1664, Newton on his own undertook the study of the mechanical philosophy of René Descartes (1596–1650) and Robert Boyle (1627–1691) and the algebra and analytical geometry of F. Viète and J. Wallis. He must have been deeply impressed by Copernican cosmology as developed by Galileo. Newton's scientific genius emerged suddenly in the plague years 1665–1666 when he put into writing many of his revolutionary ideas in mathematics, optics, physics and astronomy within 2 years.

Newton laid the foundations for elementary differential and integral calculus, independently and almost simultaneously with the work of G.W. Leibniz, by inventing the method of fluxions. This incorporated the insight that the integration of a function (or finding the area under a curve) is the inverse procedure to differentiating it (or finding the slope of the curve at any point). From this emerged the differential and integral calculus of 19th-century mathematics. In 1669 he became Lucassian Professor of Mathematics, but in 1701 he resigned both it and his Fellowship at Trinity College. He was elected fellow of the Royal Society of London in 1672, serving as president from 1703 until his death.

During the period 1669–1676 Newton lectured extensively on optics, including his fundamental discovery that white light is not a single entity, as natural philosophers had believed since Aristotle (384–322 BC), but contains a spectrum of many different colored rays which are refracted at slightly different angles. Recognizing that differential refraction put an obstacle in the way of very accurate refracting telescopes, Newton constructed the first reflecting telescope.

Newton's greatest achievements were in physics and astronomy, leading to the formulation of the theory of universal gravitation. In

1666 he discovered the formula for a centrifugal force of a body moving uniformly in a circular path. From this law and Kepler's third law of planetary motion, Newton deduced that the centrifugal force of a celestial body must decrease as the inverse square of its distance from the center of its motion. In 1679 he began a correspondence with Robert Hooke (1635–1703) on the problem of planetary dynamics and the elliptical planetary motion already discovered by Kepler (1571–1630). Hooke conjectured that since the planets move in ellipses with the Sun at one focus (Kepler's first law), the centripetal force drawing them to the Sun varies as the inverse square of their distances from it. Newton showed mathematically that if a body obeys Kepler's second law, then the body is being acted upon by a centripetal force, demonstrating that a body moving in an ellipse and attracted to one focus must be drawn by a force that varies as the inverse square of the distance.

In 1684 Edmond Halley visited Newton and persuaded him to write a book on planetary dynamics and gravitation, which in July 1687 was to appear as his *Philosophiae Naturalis Principia Mathematica* (the '*Principia*'), one of the most important scientific books ever written. Incidentally, Halley even paid for the printing. In this book Newton united the work of Galileo, Copernicus and Kepler and provided a sound physical basis for Copernican cosmology. The historical background behind the naming of *Principia* is analyzed by Cunningham (1991).

In 1704 Newton published his *Opticks*, summarizing his experimental and theoretical work. It contains Newton's corpuscular theory of light, for a long time an intensely debated issue in the face of Huygens' wave theory of light, first expounded in 1678.

In Earth science Newton's first calculations were not an unqualified success. His estimate of terrestrial gravity was unfortunately based on incorrect data, specifically the Earth's radius that had been given as 13% too great. This was demonstrated later by the Abbé Picard's accurate triangulation of France, and Cassini's new map of France proved the domain of Louis IV to be almost one-third less than the 'Sun King' had imagined.

Newton was more successful, however, with the figure of the Earth. He used three measurable quantities: (1) Galileo's law of falling bodies; (2) Huygens' observations on pendulum motion; and (3) his own and Huygens' work on centrifugal force. His calculations suggested the Earth had the form of an oblate spheroid, i.e. a sphere with polar flattening and an equatorial bulge. From this it follows that the value of gravity rises towards the poles, which are closer to the Earth's center. However, more important in the latitudinal variation of gravity is the centrifugal factor (see Huygens, Christian).

Newton was also tempted to speculate on the gradual drying up of the world ocean following the traditional flood of Noah, which had been first estimated in Sweden by Celsius at 1.35 m per century. Alas, the land emergence there turned out (after water-level experiments by Celsius and Linneus) to be tilting, later shown to be postglacial isostatic uplift. Newton's contemporary, the naturalist–religious philosopher Emmanuel Swedenborg postulated that the emergence of northern lands was due to an acceleration of the Earth's rotation rate, which would lead to a change in the geoid height, falling in high latitudes and rising towards the equator. Newton, of course, could not have been aware at that time of the glacial theory (Wegmann, p. 392, in Schneer, 1969).

It is not surprising that the theologians of the day felt it necessary to criticize some aspects of the *Principia*, and Newton, 'the towering intellectual of the period' (Davies, 1968), felt it appropriate to reply in his *General Scholium*. The slowly evolving nature of Earth history was beginning to be suspected, and allowed no place for Bishop Ussher's calculations (made in 1650–1654) for the 'creation' in 4004 BC. Newton himself edited *Geographia Generalis* by Bernhard Varenius of 1650, required reading for his students at Cambridge. An English translation appeared in 1693. This and Steno's *Prodromus* (1669), while accepting Creation, set the scene for the eventual intellectual recognition of an evolving planet, although it had to await James Hutton (1726–1797), who did for Earth science what Newton did for planetary science.

Newton was knighted by Queen Anne in 1708, being the first scientist to be so honored for his achievements.

Newton's laws

Newton's laws of motion represent three fundamental principles which form the basis of classical or Newtonian mechanics. They are stated as follows:

Figure N16 Isaac Newton. (Reproduced by permission of Yerkes Observatory.)

- First law: a body at rest or in uniform rectilinear motion will continue in that state unless a force is applied.
- Second law: the applied force equals the rate of change of linear momentum of the body.
- Third law: if a body exerts a force on another body there is an equal but opposite force on the first body (law of action and reaction).

Newton, in his *Philosophiae Naturalis Principia Mathematics* (1686), stated these laws as axioms and showed them to be confirmed by an abundance of experiments. Newton's laws of motion have proved valid for all mechanical problems not involving velocities comparable to the velocity of light and not involving quantum mechanics. The more general methods of Hamilton and Lagrange are mathematical elaborations of the Newtonian principles.

Newton's second law of motion, stated analytically, becomes the differential equation,

$$\frac{d(m\boldsymbol{v})}{dt} = \boldsymbol{F} \qquad (N1)$$

where m denotes the particle's mass, v its velocity, t the time and $\boldsymbol{F}$ the force applied. The classical assumption of constancy of mass allows one to express equation N(1) as

$$m\boldsymbol{a} = \boldsymbol{F}$$

where $\boldsymbol{a}$ denotes the linear acceleration. Newton's law was restricted to the linear motion of an idealized body having negligible extension in space, commonly called a mass particle. As such a mass particle in physical space requires three cartesian coordinates, x, y and z, to fix its position. Its motion is described by time rates of coordinate change, called velocities, and time rates of velocity change, called accelerations, both being vectors. Newton's law of universal gravitation and his second law of motion thus provide the foundations of celestial mechanics.

Newton's third law is central to the dynamics of mechanical systems composed of two interacting particles or bodies. The equations of motion for systems of more than two interacting particles in space are mathematically intractable in the absence of geometrical restrictions,

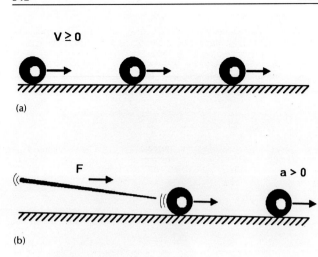

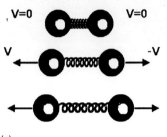

Figure N17 Schematic presentation of the three laws of Newton. (a) A ball on the horizontal plane continues in its state of rest, or of uniform motion in a straight line, unless it is compelled to change that state by forces impressed on it. (b) The change in motion of the ball is proportional to the motive force applied and is made in the direction of the straight line in which that force is applied. (c) Two balls are pushed with equal force by a compressed spring placed between them. If it is assumed that the balls have equal masses, they will move in opposite directions with equal velocities.

and for most n-body problems assumptions approximating the physical situation have to be made to arrive at solutions describing the problem.

Lagrange's equations expand on Newton's second law of motion and are partial differential equations of second order for quantities q as functions of time t. Thereby it is supposed that the configuration of the classical mechanical system under consideration is fully specified if, there being f independent values of q ($q_1, ..., q_f$), there are f equations of motion. These equations of motion relate the kinetic energy of the mechanical system to its generalized coordinates, generalized forces and the time. The equations of motion of a relativistic system may be written in Lagrangian form. The Lagrange equations can be derived from a function of generalized coordinates and velocities of a mechanical system, commonly called the Lagrange function L ($q_1, ..., q_f; \dot{q}_1, ..., \dot{q}_f; t$).

Hamilton's equations relate in turn to Lagrange's equations of motion and govern the motion of a classical mechanical system. Their advantage in comparison with Lagrange's equations is that they are first-order ordinary differential equations having a highly symmetrical form. They are referred to as the canonical equations for the general discussion of the motion of mechanical systems. Hamilton's equations can be derived from Lagrange's equations directly. The classical Hamilton function H (q, p, t), depending on the coordinates q, momenta p and time t, may be used to derive the quantum-mechanical Hamilton operator.

Hans J. Haubold and Rhodes W. Fairbridge

Bibliography

Cajori, A. (1962) *Sir Isaac Newton's Mathematical Principles of Natural Philosophy and His System of the World*. Berkeley and Los Angeles: University of California Press.
Cohen, I.B. (1958) *Isaac Newton's Papers and Letters on Natural Philosophy and Related Documents*. Cambridge: Harvard University Press.
Cohen, I.B. (1974) Newton, Isaac. *Dict. Sci. Biogr.*, Vol. 10, pp. 42–103.
Cunningham, A. (1991) How the *Principia* got its name; or, taking natural philosophy seriously. *Hist. Sci.*, **29**, 377–92.
Davis, G.L. (1968) *The Earth in Decay*. London: Macdonald Techn. & Sci.
Feingold, M. (1993) Newton, Leibnitz, and Barrow too. *Isis*, 84, 310–38.
Goldstein, H. (1959) *Classical Mechanics*. Reading, Massachusetts: Addison Wesley.
Holton, G. (1988) *Thematic Origins of Scientific Thought – Kepler to Einstein*. Cambridge: Harvard University Press.
Landau, L.D. and Lifschitz, E.M. (1959) *Mechanics*. Reading, Massachusetts: Addison-Wesley.
Misner, C.W., Thorne, K.S. and Wheeler, J.A. (1973) *Gravitation*. New York: W.H. Freeman.
Motz, L. and Duveen, A. (1977) *Essentials of Astronomy*. New York: Columbia University Press.
Newton, I. (1687) *Philosophiae Naturalis Principia Mathematica*. London (3rd edn, English translation by A. Koyré and I.B. Cohen, Cambridge, Massachusetts, 1972; and Cambridge, UK, 1972; Cambridge University Press, 2 vols).
Schneer, C.J. (ed.) (1969) *Toward a History of Geology*. Cambridge, Massachusetts: Massachusetts Institute of Technology Press.
Weaver, J.H. (ed.) (1987) *The World of Physics*. New York: Simon and Schuster (Vols I to III).

NOBLE GAS

The noble gases (Table N4) are the elements with filled outer shells of electrons. They are chemically unreactive, and hence condense only at low temperatures ($\ll 0°C$). This leads to their being severely depleted (relative to elements of similar mass) in most solid solar system matter. Since they are so depleted, nuclear processes which can produce noble gases from other elements can be detected easily, making the noble gases excellent tracers of many cosmochemical events. Their properties also lead to them being referred to as 'inert' or 'rare' gases.

Noble gas-producing nuclear decays can be used to date rocks, even if the noble gases produced are only a small portion of the decay products. The decay of ^{40}K to ^{40}Ar is the basis of a common dating technique for rocks (McDougall and Harrison, 1988; see Chronology), including extraterrestrial ones. Fissioning actinides (such as ^{238}U, ^{235}U and now-extinct ^{244}Pu) produce several isotopes of Xe, with each actinide producing a characteristic isotopic spectrum. Another now-extinct isotope, ^{129}I, decays to ^{129}Xe. Isotopic analysis of Xe in meteorites provided the first evidence for the presence in the early solar system of extinct nuclides, those with short half-lives that were incorporated into solids no more than a few half-lives after nucleosynthesis (Podosek and Swindle, 1988).

Table N4 Noble gases

Element	Symbol	Mass range (amu)	Boiling point[a] (K)
Helium	He	3–4	–
Neon	Ne	20–22	27.1
Argon	Ar	36–40	87.3
Krypton	Kr	78–86	119.8
Xenon	Xe	124–136	165.0
Radon	Rn	(223–228)[b]	211

[a] Normal boiling point (Ozima and Podosek, 1983).
[b] Radon does not have any stable isotopes; these are the longest-lived radioactive isotopes.

The decay of ^{40}K is also a valuable tracer of atmospheric history. Since ^{40}Ar was no more abundant than other Ar isotopes in the condensing solar nebula, but is the only one produced by decay, the ratio of ^{40}Ar to ^{36}Ar is essentially a measure of the relative importance of planetary outgassing compared to gases trapped either gravitationally or within rocks (the latter would have to have subsequently outgassed). Decay-produced ^{129}Xe can be used in a similar fashion, although the results apply only to the short timescales on which ^{129}I was still alive. The relative elemental abundances of the noble gases, which can vary significantly as a result of some fractionation processes, are also useful as tracers of atmospheric history (Ozima and Podosek, 1983; Pepin, 1989).

Nuclear spallation reactions caused by galactic cosmic rays or by solar flares also produce isotopes of all noble gases. The build-up of these isotopes can be used to determine the time of exposure of solid bodies to those energetic particles (see Cosmic ray exposure ages).

The low abundance of noble gases in materials condensed from the solar nebula also makes them prime tracers of unprocessed interstellar grains. In fact, noble gas isotopic anomalies were the first clues to several types of grains (including graphite, silicon carbide and diamond) later identified in meteorites.

In the outer solar system, temperatures do get low enough for some of the noble gases to condense or at least to form clathrates in ice. The abundances of noble gases in things like cometary comas and Titan's atmosphere, once they can be measured, can potentially be used to determine condensation temperatures of the parent bodies (Lunine, Atreya and Pollack, 1989; Green *et al.*, 1991).

Timothy D. Swindle

Bibliography

Green, J.C., Cash, W., Cook, T.A. and Stern, S.A. (1991) The spectrum of comet Austin from 910 to 1180 Å. *Science*, 251, 408–10.

Lunine, J.I. Atreya, S.K. and Pollack, J.B. (1989) Present state and chemical evolution of the atmospheres of Titan, Triton and Pluto, in *Origin and Evolution of Planetary and Satellite Atmospheres* (eds S.K. Atreya, J.B. Pollack and M.S. Matthews). Tucson: University of Arizona Press, pp. 605–65.

McDougall, I. and Harrison, T.M. (1988) *Geochronology and Thermochronology by the $^{40}Ar/^{39}Ar$ Method*. New York: Oxford University Press.

Ozima, M. and Podosek, F.A. (1983) *Noble Gas Geochemistry*. Cambridge: Cambridge University Press.

Pepin, R.O. (1989) Atmospheric compositions: key similarities and differences, in *Origin and Evolution of Planetary and Satellite Atmospheres* (eds S.K. Atreya, J.B. Pollack and M.S. Matthews). Tucson: University of Arizona Press, pp. 291–305.

Podosek, F.A. and Swindle, T.D. (1988) Extinct radionuclides, in *Meteorites and the Early Solar System* (eds J.F. Kerridge and M.S. Matthews). Tucson: University of Arizona Press, pp. 1093–113.

Cross references

Chronology: meteorite
Cosmic ray exposure ages
Interstellar grain

NOMENCLATURE

Communication of geographic concepts is not possible without place names, and humankind has been busily naming features for as many years as language has existed. We have not limited our naming of places to the mountains, valleys and rivers that surround us; we have named the stars and the planets and the features that we could see on their surfaces. As our ability to see these features has increased, so has the nomenclature; it facilitates description and discussion of planetary features. Nineteenth-century telescopic observations of the Moon and Mars began an explosion in planetary names that continues to this day.

History

The naming of features on the Moon and Mars became a common activity of astronomers once the telescope was invented. Several competing schemes were developed, and although the hopelessly chaotic ones fell by the wayside, many of the most conspicuous features on the nearside of the Moon were still known by at least three names in the early 1900s. When the International Astronomical Union (IAU) was formed in 1919, it charged a subcommittee with the task of standardizing the lunar nomenclature. The resulting report, 'Named Lunar Formations' (Blagg and Müller, 1935), was the first systematic listing of lunar names. It was later supplemented by 'The system of lunar craters, quadrants I, II, III, IV (Arthur *et al.*, 1963–1966). These documents were adopted by the IAU and became recognized as the authoritative source for lunar nomenclature. The lengthy list of names was greatly expanded as a result of the Apollo missions to the Moon; today there are more than 2200 named lunar features, nearly twice as many as on any other solar system body (Figure N18).

Telescopic observations of Mars in the late 19th century resulted in the publication of several nomenclature schemes, including those of Beer and Mädler, Secchi, Proctor, and Schiaparelli (Blunck, 1982). These schemes named albedo features only, because topographic features on Mars cannot be resolved with a telescope. Schiaparelli's system of names, based on ancient circumMediterranean geography and mythology, was also used by other observers, most notably E.M Antoniadi (Figure N19). In 1960 the IAU adopted 128 names for albedo features based on Antoniadi's (1929) publication '*La Planète Mars*,' thus establishing a basis for today's Martian nomenclature. The US Mariner 9 and Viking missions to Mars returned images of previously unknown landforms in addition to detailed representations of the albedo features previously observed by classical astronomers. The discovery of new landforms made it clear that a greatly expanded system of nomenclature would be required. For most features, Antoniadi's system was retained (Figure N20); new schemes were devised for craters and valleys.

In 1973 the nomenclature committees for the Moon and Mars were reorganized into the Working Group for Planetary System Nomenclature (WGPSN). Subsidiary task groups for the Moon, Mars, Mercury, Venus and the outer solar system were formed to conduct the preliminary work of choosing themes and proposing names for each body. A task group for comets and asteroids was added in 1984.

Venusian nomenclature began with Earth-based radar surveys of the cloud-shrouded planet in the 1960s. Radar-bright areas were designated Alpha, Beta, etc.; a few were named for scientists who were instrumental in the development of radar. As the American Pioneer Venus and Russian Venera programs began to return detailed images of the planet, a more comprehensive nomenclature scheme became necessary. The IAU chose a theme that would be in keeping with the age-old feminine mystique associated with Venus; all names would be those of females, both historical and mythological. (Names of females are not restricted to Venus; famous real and mythological women are recognized on other planets as well.) Only one male name remains on Venus: Maxwell Montes, named after James Clerk Maxwell, a pioneer in radio science. Prior to the Magellan mission, only a handful of features had been observed in sufficient detail to warrant naming; today Venusian nomenclature consists of over 800 names (e.g. Figure N21).

The spectacular successes of the Voyager program created the need for names for outer solar system bodies. The WGPSN devised naming themes for each of the bodies visited by the Voyager spacecraft, associating the theme with the name of the satellite or its properties. For example, features on Jupiter's satellite Europa are named after people and places associated with the myth of Europa (who was one of the illicit romantic conquests of Jupiter). And, when active volcanism was discovered on another of Jupiter's moons, Io (the original Io was also one of Jupiter's clandestine lovers), it seemed only appropriate to name certain Ionian features for volcanic and fire deities (Figure N22).

As the satellites of Jupiter are named after mythological characters associated with the king of the gods, so the moons of Saturn are named after characters connected with the ringed planet's namesake. Saturn's equivalent in Greece was Cronos, one of the mythic race known as Titans, and the father of Zeus. Discoverers of Saturn's satellites, therefore, named them after other Titans. When the IAU chose themes for features on Saturnian moons, however, epic stories from various cultures were selected to counterbalance the preponder-

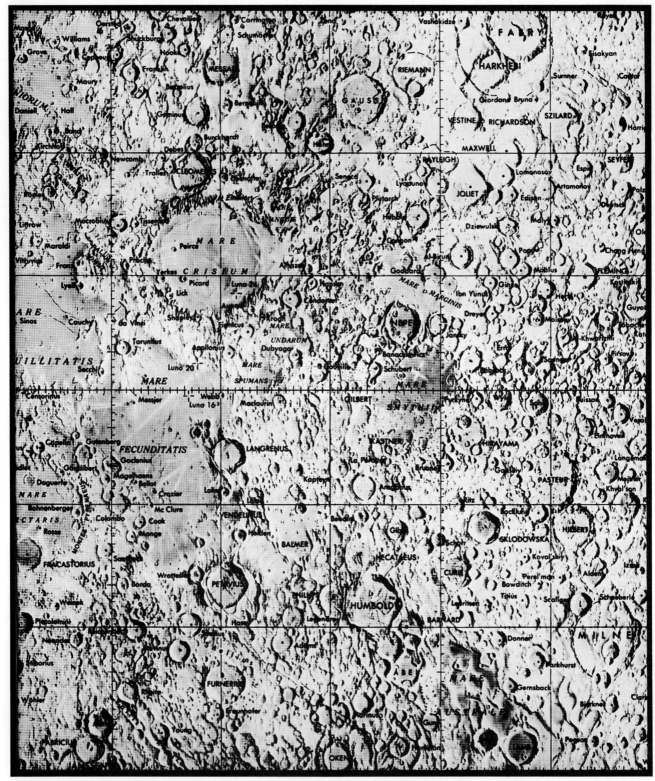

Figure N18 Part of NASA Map LPC-1 (1979), scale 1 : 10 000 000, showing the profusion of named features on the Moon.

ance of Greco-Roman names given to features elsewhere in the solar system.

The Neptunian system was the last place visited by Voyager. Because Neptune was the king of the seas, a 'watery' naming theme has been established for this part of the solar system. Four newly discovered satellites and 61 surface features on Triton (Figure N23) were named by the IAU in response to the Voyager 2 flyby of the Neptunian system.

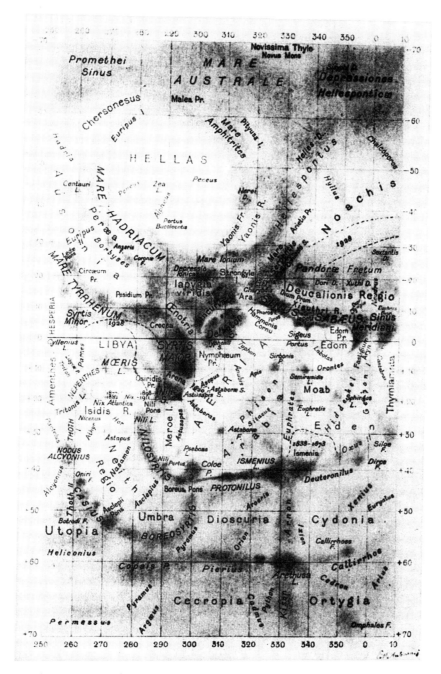

Figure N19 Plate II of E.M. Antoniadi's 1929 map of Mars. This map (with south at the top) was produced from telescopic observations of Martian albedo features.

Rules for planetary nomenclature

The IAU, through the WGPSN, has formulated rules and guidelines for naming features on every body in the solar system that has been visited by spacecraft (Masursky *et al.*, 1986). The IAU rules summarize nearly 300 years of experience in planetary nomenclature and are based on past, sometimes bitterly contentious, experience. Therefore, they are designed to make the naming process as fair and controveray-free as possible (Strobell and Masursky, 1990). By international agreement, all nations belonging to the IAU follow these rules when naming planetary features.

Specific rules govern the naming of certain feature types. Names of living persons are not given to planetary features. Craters on the Moon, Venus and Mars can be named after historical figures only after they have been dead for at least 3 years. Names of military or political leaders of the 19th and 20th centuries, or religious leaders of any era, are not acceptable. Solar system nomenclature must be kept international in scope; the IAU strongly supports equitable representation of a variety of national and/or ethnic groups in a planet's or satellite's nomenclature. Duplication of names (whether or not honoring the same person) is avoided. Spelling of a name must reflect the nation of origin, and diacritics (accent marks) are used where appropriate.

All planetary features, with the exception of craters, are given a feature-type designation (Table N5), which becomes part of its

Figure N20 Part of US Geological Survey Mars topographic map (1991), including the area of the 1929 Antoniadi map (Figure N19). Note that the names of many features reflect Antoniadi's nomenclature scheme.

formal name. These feature types are intended to be broad categories that identify form or shape but do not imply geologic origin.

The WGPSN has defined naming categories for all feature types on planets and satellites (Table N6). These categories help to keep the nomenclature uniform. Whenever possible they follow the traditional or historical aspects of the existing nomenclature of a planet or satellite. For example, the Uranian satellite Oberon was named (by John Herschel in the 19th century) after a character in Shakespeare's *A Midsummer Night's Dream*, features discerned by Voyager 2 were therefore named after Shakespearean characters (Figure N24).

The naming process

Most names are proposed by planetary investigators to facilitate discussions of important features. However, any person, scientist or layman, may suggest names or request that a specific feature be

Figure N21 Part of US Geological Survey topographic map (1989), showing features on Venus named after real and mythical women.

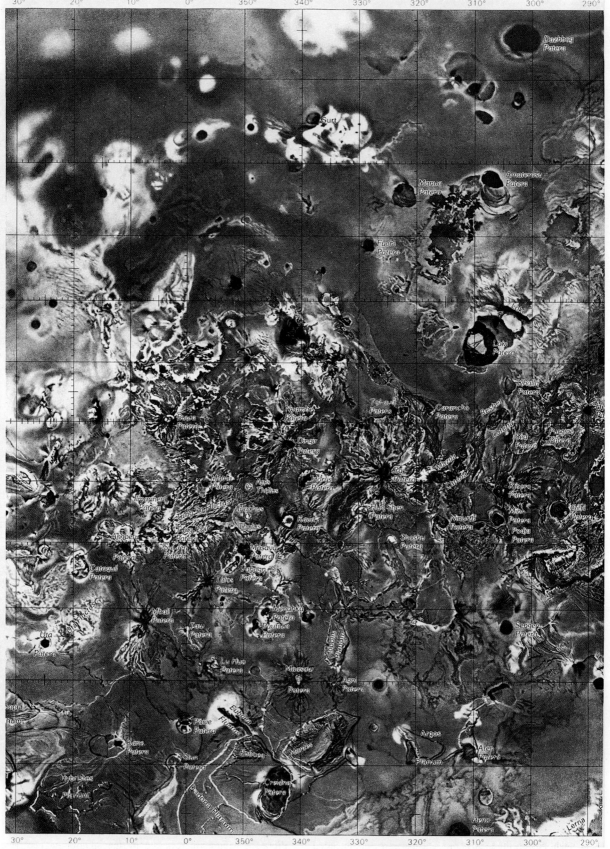

Figure N22 Part of US Geological Survey map (1987), showing several volcanic features on Io. Tung Yo Patera (latitude 19°S, longitude 2.5°) was named after the Chinese god of fire; Sengen Patera (latitude 33°S, longitude 304°) was named after the Japanese deity of Mt Fujiyama.

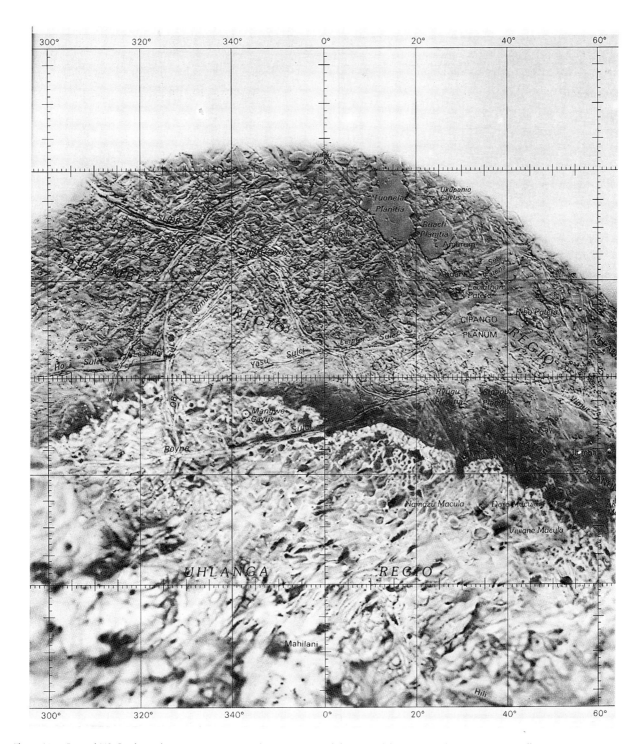

Figure N23 Part of US Geological Survey map (1992), showing some of the named features on the Neptunian satellite Triton.

named. Individuals wishing to submit names are encouraged to familiarize themselves with the rules and constraints. The use of names for identification takes precedence over their use for commemoration. Thus once a name has been used, the IAU avoids using it again even on another planet, although more than one worthy individual might possess the same name. The names are not intended to confer awards or honors. To do so would imply approval of the moral or ethical 'worthiness' of individuals, which is not the function of the IAU. While the adopted naming conventions attempt to eliminate personages offensive to some national or ethnic groups, unsavory characters remain. Certainly the moral and ethical values of many of the mythological personalities of planetary nomenclature would be considered unacceptable in most modern societies.

Proposed names are reviewed by the appropriate task group working under the WGPSN. When the task group has completed its review, the names are sent to members of the WGPSN. Any names

Table N5 Feature descriptor types

Feature(s)	Description	Feature(s)	Description
Catena, catenae	Chain of craters	Mons, montes	Mountain
Cavus, cavi	Hollow; irregular depression	Oceanus[a]	Ocean
Chaos	Distinctive area of broken, jumbled terrain	Palus, paludes[a]	Swamp
Chasma, chasmata	Canyon	Patera, paterae	Shallow crater with scalloped, complex edge
Colles	Small hills or knobs	Planitia, planitiae	Low plain
Corona, coronae	Ovoid feature	Planum, plana	Plateau or high plain
Crater, craters	Bowl-shaped depression; impact crater	Promontorium, -oria[a]	Cape
Dorsum, dorsa	Ridge	Regio, regiones	Region
Facula, faculae	Bright spot	Rima, rimae[a]	Fissure
Flexus, flexūs	Cuspate, linear feature	Rupes, rupēs	Scarp
Fluctus, fluctūs	Flow terrain	Scopulus, scopuli	Lobate or irregular scarp
Fossa, fossae	Long, narrow, shallow depression	Sinus, sinūs	Bay
Labes, labēs	Landslide	Sulcus, sulci	Subparallel furrows and ridges
Labyrinthus	Intersecting valley complex	Terra, terrae	Extensive land mass
Lacus[a]	'Lake'; small plain	Tessera, tesserae	'Tile'; polygonally grooved ground
Landing site name	Feature named on Apollo map or report	Tholus, tholi	Small domical mountain or hill
Linea, lineae	Elongate marking	Undae	Dunes
Macula, maculae	Dark spot	Vallis, valles	Valley
Mare, maria[a]	'Sea'; large plain	Vastitas, vastitates	Widespread lowlands
Mensa, mensae	Mesa, flat-topped elevation		

[a] Used only on the Moon.

Table N6 Naming categories for features on each solid surface planet and each satellite

The Moon

Craters, catenae, dorsa, rimae	Large craters: famous deceased scientists, scholars, artists. Small craters: common first names. Other features named from nearby craters
Lacūs, maria, paludes, sinūs	Latin terms describing weather and abstract concepts
Montes	Terrestrial mountain ranges or nearby craters
Rupes	Name of nearby mountain ranges (terrestrial names)
Valles	Name of nearby features

Mercury

Craters	Famous deceased artists, musicians, painters, authors, and one astronomer
Montes (one only)	Caloris, from Latin word for 'hot'
Planitiae	Names for Mercury in various languages
Rupēs	Ships of discovery or scientific expeditions
Valles	Radio telescope facilities

Venus

Chasmata	Goddesses of hunt; Moon goddess
Coronae	Fertility goddesses
Craters	Famous women; under 20 km in diameter, common female first names
Dorsa	Sky goddesses
Fluctūs	Goddess, miscellaneous
Lineae	Goddesses of war
Montes	Goddesses, miscellaneous (also one radar scientist)
Paterae	Famous women
Planitiae	Mythological heroines
Plana (1 only)	Lakshmi, goddess of prosperity
Regiones	Giantesses and Titanesses (also Greek alphabetic)
Rupēs	Goddesses of hearth and home
Tesserae	Goddesses of fate or fortune
Terrae	Goddesses of love
Valles	Name for planet Venus in various languages

Mars and its satellites

Mars

Large craters	Deceased scientists who have contributed to the study of Mars
Small craters	Villages of the world (less than 100 000 population, UN Yearbook)
Large valles	Name for Mars/star in various languages
Small valles	Classical or modern rivers
Other features	From nearest named albedo feature on Schiaparelli or Antoniadi maps
Deimos	Authors who wrote about satellites
Phobos	Scientists who helped discovery

Table N6 Continued

Satellites of Jupiter

Callisto	
Large ringed features	Homes of the gods and heroes
Craters	Heroes and heroines from northern myths
Catenae	Mythological places in high latitude
Europa	
Craters	Celtic gods and heroes
Flexūs	Places associated with the Europa myth
Lineae	People associated with the Europa myth
Maculae	Places associated with the Europa myth
Ganymede	
Craters	Gods and heroes of ancient (Fertile Crescent) people
Faculae	Places associated with Egyptian myths
Fossae	Gods (or principals) of ancient (Fertile Crescent) people
Regiones	Astronomers who discovered Jovian satellites
Sulci	Places associated with myths of ancient people
Io	
Active eruptive centers	Fire, sun, thunder gods and heroes
Catenae	Sun gods
Fluctūs	Name derived from nearby named feature
Mensae	People associated with Io myth
Montes	Places associated with Io myth
Paterae	Fire, sun, thunder, volcano gods, heroes, goddesses, mythical blacksmiths
Plana	Places associated with Io myth
Regiones	Places associated with Io myth
Tholi	Places associated with Io myth

Satellites of Saturn

Mimas	People and places from Malory's *Le Morte d'Arthur* legends
Enceladus	People and places from Burton's *Arabian Nights*
Tethys	People and places from Homer's *Odyssey*
Dione	People and places from Vergil's *Aeneid*
Rhea	People and places from creation myths
Hyperion	Sun and Moon deities
Iapetus	People and places from *Chanson de Roland*
Epimetheus	People from Castor and Pollux (twins) myth
Janus	People from Castor and Pollux (twins) myth

Satellites of Uranus

Miranda	Characters, places from Shakespeare's *The Tempest*
Ariel	Light spirits (individual and class)
Umbriel	Dark spirits (individual)
Oberon	Shakespearean tragic heroes and places
Titania	Female Shakespearean characters, places
Puck	Mischievous (Puck-like) spirits (class)
Small satellites	Heroines from Shakespeare and Pope

Satellites of Neptune

Proteus (1989 N1)	Water-related spirits, gods, goddesses (excluding Greek and Roman names)
Nereid	Individual nereids
Triton	Aquatic names (excluding Roman and Greek); worldwide aquatic spirits, famous terrestrial fountains or fountain locations, terrestrial aquatic features, famous terrestrial geysers or geyser locations, terrestrial islands
Small satellites	Gods and goddesses associated with Neptune/Poseidon mythology, or names of generic mythological aquatic beings

passing its review are voted upon at the annual meeting of the WGPSN. Provisional status is conferred upon names approved at this meeting; names are not considered fully approved until voted upon by the general membership at the triennial convention of the IAU. Provisional names may be used in publication, but their status must be indicated (as, for example, by the asterisk following 'Boadicea Patera' in Figure N21).

Summary

Planetary nomenclature is first and foremost a tool for planetary scientists, allowing them to discuss features of interest more easily. The function of the Working Group for Planetary System Nomenclature of the International Astronomical Union is to ensure that this nomenclature is useful, unambiguous and non-controversial.

Since the IAU began regulating planetary nomenclature in the early 20th century, the database of adopted planetary names has grown to over 5000. The data returned from the Magellan mission to Venus guarantees an increase in the number of named planetary features. Future missions to the Moon, Mars and the outer solar system will undoubtedly increase further the number of named features on other planets and satellites. A fair and well-regulated system of nomenclature is necessary to ensure that this increase does not recreate the chaos of the early 20th century.

Joel F. Russell and R.M. Batson

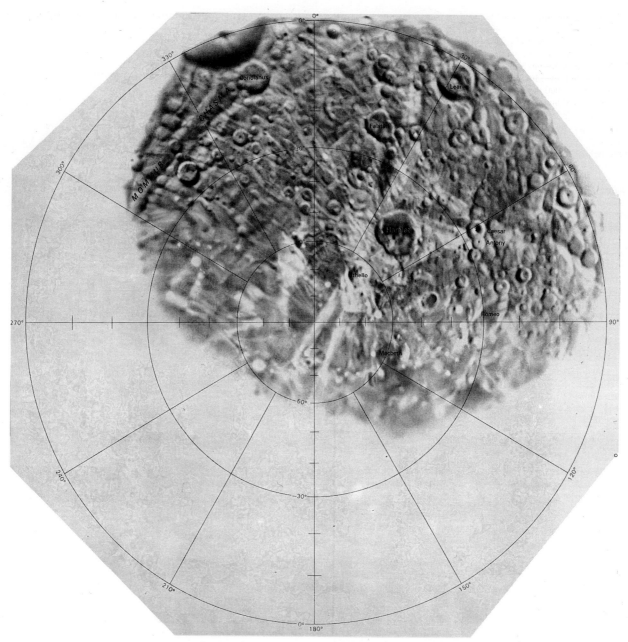

Figure N24 Part of US Geological Survey map (1988), showing features named after Shakespearean characters on the Uranian satellite Oberon.

Bibliography

Antoniadi, E.M. (1929) *La Planète Mars*. Paris: Libraire Scientifique Herman et Cie.

Arthur, D.G.W., Agnieray, A.P., Horvath, R.A. et al. (1963) The system of lunar craters, quadrant I. *Commun. Lunar and Planetary Laboratory*, **2**(30) 71–78.

Arthur, D.G.W., Agnieray, A.P., Horvath, R.A. et al. (1964) The system of lunar craters, quadrant III. *Commun, Lunar and Planetary Laboratory*, **3**(40), 1–59.

Arthur, D.G.W., Agnieray, A.P., Pellicori, R.H. et al. (1965) The system of lunar craters, quadrant III. *Commun. Lunar and Planetary Laboratory*, **3**(50), 61–2.

Arthur, D.G.W., Pellicori, R.H. and Wood, C.A. (1966) The System of lunar craters, quadrant IV. *Commun. Lunar and Planetary Laboratory*, **5**(70), 1.

Blagg, M.A., and Müller, K. (1935) *Named Lunar Formations*. London: Percy Lund, Humphries & Co. Ltd, 196 pp.

Blunck, J. (1982) *Mars and Its Satellites*. Smithtown, N.Y.: Exposition Press, 222 pp.

Masursky, H. et al. (1986) *Annual Gazetteer of Planetary Nomenclature*. US Geological Survey Open File Report 84–692.

Strobell, M.E. and Masursky, H. (1990) Nomenclature, in *Planetary Mapping* (eds Greeley R and Batson R.M.). New York: Cambridge University Press, pp. 96–140.

US Geological Survey (1987) Shaded relief and surface markings of Io. US Geological Survey Miscellaneous Investigations Series Map I–1713, scale 1 : 15 000 000.

US Geological Survey (1988) The southern hemispheres of Umbriel, Titania and Oberon. US Geological Survey Miscellaneous Investigations Series Map I–1920, scale 1 : 10 000 000.

US Geological Survey (1989) Topographic map of part of the northern hemisphere of Venus. US Geological Survey US Geological Survey Miscellaneous Investigations Series Map I–2041, scale 1 : 15 000 000.

US Geological Survey (1991) Topographic map of Mars. US Geological Survey Miscellaneous Investigations Series Map I–2179, scale 1 : 25 000 000.

US Geological Survey (1992) Pictorial map of Triton. US Geological Survey, Miscellaneous Investigations Series Map I–2154, scale 1 : 15 000 000.

Cross references

Antoniadi, Eugenios
Cartography
History of planetary science I: pre-space age
Map projections

O

OBERON

One of the moons of Uranus, Oberon was discovered telescopically by William Herschel in 1789, seven years after his discovery of Uranus. Oberon orbits Uranus in approximately 13.5 days at a distance of 583 400 km; it is the outermost of the Uranian moons. It has a diameter of 1523 km and a mass of 30.3×10^{20} km, making it the second largest moon in the Uranian system after Titania.

The surface of Oberon was observed by the Voyager 2 spacecraft in late January 1986 at the lowest resolution (12 km) of the major moons. Its surface is nearly saturated with large craters which most probably formed during the sweep-up of objects in heliocentric orbit early in the history of the Uranian system. The largest of these are the craters Othello and Hamlet, which display both dark basin floors reminiscent of the mare-filled lunar craters, and well-developed systems of bright rays emanating from them. Dark-floored craters may be due to impactors penetrating the surface to the depth of a dark viscous material, which was then extruded onto the surface to fill the basin. Crater counts by different investigators vary considerably and it is possible that Oberon's surface was also resurfaced at some point early in its evolution. A high, bright topographic peak was evident on the limb of Oberon as viewed by Voyager. It stands 11 km above the surface and is most likely the central peak of a large crater.

There is little evidence of extensive geologic activity on Oberon. Linear features with apparent vertical relief of as much as 1 to 2 km were observed; these may be faults. This interpretation is uncertain due to the relatively low resolution with which the moon was imaged.

Daniel M. Janes

Bibliography

Bergstralh, J., Miner, E. and Matthews, M. (1991) *Uranus*. Tucson: University of Arizona Press.
Miner, E. (1990) *Uranus: The Planet Rings and Satellites*. New York: Ellis Horwood.
Smith, B.A., Sederblom, B.A., Beebe, R. *et al.* 1986. Voyager 2 in the Uranian system: imaging science results. *Science*, **233**, 43–64.

Cross references

Satellite, natural
Uranus: satellite system

Figure O1 Highest resolution image of Oberon taken by the Voyager 2 spacecraft from a distance of 660 000 km on 24 January 1986. The smallest features visible on the surface are approximately 12 km across. The large dark-floored crater to the left of center is Hamlet, with Othello below it to the left. The south pole of the moon is near the midpoint of the two bright lines extending to the lower left from Othello. (NASA image P-29501.)

OBLIQUITY

The obliquity of the ecliptic is the angle between the plane of the Earth's orbit (i.e. the ecliptic) and the plane of the Earth's equator. The obliquity at present is about 23½°, though it is subject to a slow secular decrease of about 0.5 arcsec per year, superimposed upon which is a periodic variation of amplitude 9 arcsec and period 18.6 years. The periodic variation in the obliquity is called the nutation (see Precession and nutation) in the obliquity and is a result of the regression of the nodes of the Moon's orbit on the ecliptic with a period of 18.6 years. The regression of the Moon's nodes in turn is a result of planetary perturbations and the asphericity of the Earth.

The mean obliquity (i.e. not including the nutation) is

$$\epsilon = 23°27' \; 08''.26 + T \, [a + T(b + cT)]$$

where T is the time in Julian centuries from 1900 January 0.5; $a = -46.845$ arcsec century^{-1}, $b = -0.0059$ arcsec century^{-2} and $c = +0.00181$ arcsec century^{-3}.

The nutation in the obliquity is given by

$$\Delta\epsilon = (9''.210 + 0''.0009 \, T) \cos\Omega - 0''.090 \cos 2\Omega + 0''.551 \cos 2L + 0''.088 \cos 2M$$

and is tabulated in the *Astronomical Almanac*. Here Ω is the longitude of the ascending node of the Moon's orbit, L is the mean ecliptic longitude of the Sun (see Orbit), and M is the mean ecliptic longitude of the Moon.

The obliquity of the ecliptic has important effects on the seasons – i.e. on the seasonal insolation. Between latitudes $\pm 23\frac{1}{2}°$ (the tropics), the Sun can reach the observer's zenith. For latitudes greater than $90° - \epsilon$ or less than $-90° + \epsilon$ (the Arctic and Antarctic Circles) the Sun can be above the horizon all day or below the horizon all day. Between the polar circles and the tropics we have the familiar seasons of temperate latitudes.

The seasonal phenomena are quite different for planets whose obliquities are very different. For example, a planet moving in a circular orbit with zero obliquity would have no seasonal phenomena. At the poles, the Sun would be on the horizon all day and every day; at the equator, the Sun would rise vertically in the east, culminate at the zenith, and set due west every day. The angle between Mercury's equator and orbit is very close to zero, and it has been speculated that the bottoms of craters near the poles never see the Sun and are consequently very cold. However, the seasonal phenomena on Mercury are considerably complicated by the large eccentricity of its orbit and by the resonance between its spin and orbital periods.

For a planet moving in a circular orbit and having a 90° obliquity, the seasonal phenomena are different yet again, and over the course of a 'year' (or an orbital revolution) the poles receive more insolation than the equator. Certainly when a pole is facing the Sun it will be much warmer than the equator, and this is in fact the case for the planet Uranus, which has an obliquity of 98° (i.e. 82° and a retrograde rotation).

Jeremy B. Tatum

Bibliography

Smart, W.M. (1962) *A Text-Book on Spherical Astronomy*. Cambridge: Cambridge University Press.
Explanatory Supplement to the American Ephemeris and Nautical Almanac. United States Naval Observatory (1961).

Cross references

Coordinate systems
Insolation
Milankovitch, Milutin, and Milankovitch theory

OBLIQUITY: TERRESTRIAL RECORD

The obliquity of the ecliptic ϵ, which is the Earth's axial tilt (or the angle between the equatorial plane and the ecliptic), exerts a fundamental influence on terrestrial climate, controlling both the seasonal cycle and climatic zonation. The Earth's obliquity and that of the other terrestrial planets cannot be considered primordial; indeed, most of the planets evidently have undergone substantial change of obliquity or are subject to large obliquity variations (Laskar and Robutel, 1993; Touma and Wisdom, 1993; Williams, 1993). However, the Earth's mean obliquity of 23.3° and the presence of the Moon currently permit only small variations ($\pm 1.3°$).

As noted by Gold (1966), the Earth's mean obliquity may represent a balance between the effects of tidal friction, which slowly increase the axial tilt, and internal dissipation within the Earth, which tends to erect the axis. Dynamical calculations employing the modern, relatively large value of tidal dissipation have suggested that the mean obliquity is very slowly increasing under the action of lunisolar tidal friction and that early in Earth history the obliquity was only ~ 10–$15°$ (Goldreich, 1966; Mignard, 1982). Such models of obliquity evolution are incomplete, however, because they do not consider possible counter effects of dissipative core–mantle coupling or geologic evidence regarding paleoclimates unique to the Precambrian (prior to 540 Ma), and the history of the Earth's rotation.

Climatic effects of obliquity change

The global climate for a large obliquity ($\epsilon \gg 23°$) would be dramatically different from that of today (Williams, 1975, 1993).

1. Global seasonality would be greatly intensified, and large seasonal changes would extend into low latitudes. With an obliquity of 60°, for example, areas at 30–60° latitude would be within both the tropics and the polar circle, and the strongly seasonal global climate would be too stressful for all but small, primitive organisms. Furthermore, substantial atmospheric circulation across the equator would probably occur around solstices, when very cold air from the anticyclonic province in the winter hemisphere would flow toward the deep thermal depression in the summer hemisphere.
2. The ratio of solar radiation received annually at either pole to that received at the equator would be increased (Figure O2). For $\epsilon = 54°$, all latitudes would receive equal radiation annually and the climatic zones would disappear. Climatic zonation would reverse for $\epsilon > 54°$, with moderate to equatorial latitudes receiving less radiation annually than high latitudes. Under such a regime, low latitudes ($\leq 30°$), which would experience the additional cooling effect of frigid solstitial winds and no extreme summer temperatures, would be largely ice-covered during glacial epochs. In high latitudes, however, the cold, arid winter atmosphere would allow only limited snowfall which would melt entirely during the very hot summer.
3. The directions of zonal surface winds such as the tropical easterlies and mid-latitude westerlies would reverse for $\epsilon > 54°$ as circulation in 'Hadley cells' reversed direction (Hunt, 1982).

Conversely, decrease in obliquity ($\epsilon < 23°$) would cause an increase in mean annual insolation at low latitudes and a decrease in such insolation at high latitudes, thus steepening the equator-to-pole gradient of mean annual insolation (Hunt, 1982; Barron, 1984). Cooler polar temperatures and a reduction in amplitude of the seasonal cycle would be expected.

Paleoclimate record

Phanerozoic (since 540 Ma) paleoclimates are characterized by normal paleoclimatic zonation, normal zonal paleowind directions, circumpolar high-paleolatitude glaciation and little seasonality in low paleolatitudes, implying an obliquity like that of today (Williams, 1993). Furthermore, the broad agreement between paleoprecessional periodicity indicated by Ordovician–Silurian evaporites in Western Australia and the predicted precessional rate for that time based on tidal evolution of the Earth–Moon system and a constant mean obliquity implies that the mean obliquity at ~ 430 Ma was near the present value. The long-standing idea of a reduced obliquity during the Mesozoic – early Cenozoic (~ 200–50 Ma) to explain apparent light requirements of evergreen floras in high paleolatitudes is not supported by climate modeling (Barron, 1984), and other paleobotanical explanations that do not require annual equability of light are possible (Williams, 1993).

By contrast, the enigmatic late Proterozoic glacial environment (~ 800–600 Ma) presents a major paradox – frigid, strongly seasonal climates with permafrost and grounded ice sheets near sea level preferentially in low to equatorial paleolatitudes (mostly $\leq 12°$ paleolatitude, with no paleomagnetic evidence from glaciogenic rocks for coeval glaciation in high paleolatitudes). Such a paradoxical late Proterozoic glacial climate is indicated by paleoclimatic and paleomagnetic data for Australia, South Africa, West Africa and China and evidently also for Canada and the North Atlantic region (Williams, 1993, 1994). Particularly clear evidence comes from late Proterozoic (~ 650 Ma) glaciogenic strata in South Australia (Schmidt, Williams, and Embleton, 1991; Williams, 1993, 1994): detailed paleomagnetic studies confirm a low paleolatitude ($\leq 12°$) for glaciation; permafrost sand-wedge structures show that a strongly seasonal, frigid climate prevailed near sea level in low paleolatitudes; and the mean paleowind direction obtained for periglacial–eolian sandstone indicates paleo-westerlies in such paleolatitudes. The paradoxical late Proterozoic climate is best explained by glaciation with an obliquity $> 54°$ (assuming a geocentric axial dipolar magnetic field, which is supported by paleomagnetic data for 0–3500 Ma). Paleotidal patterns displayed by late Proterozoic tidal deposits are consistent with a substantial obliquity (Williams, 1993, 1994).

Early Proterozoic (2500–2300 Ma) paleoclimates also are enigmatic. For example, paleomagnetic studies indicate a moderate paleolatitude (26–35°) for early Proterozoic glaciation in Canada, and imply low paleolatitudes ($\sim 5°$) for broadly coeval glaciation and strongly seasonal paleoclimates in Australia (Williams, 1993).

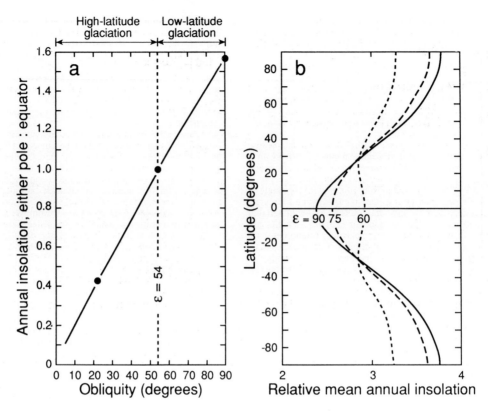

Figure O2 (a) Relation between the obliquity of the ecliptic ϵ and the ratio of annual insolation at either pole to that at the equator (solid line); the dashed line at $\epsilon = 54°$ separates the fields of potential low-latitude and high-latitude glaciation. (b) Latitudinal variation of relative mean annual insolation of a planet for various values of obliquity (ϵ, in degrees). The plots in (a) and (b) together show that for $\epsilon > 54°$, glaciation would occur preferentially in low latitudes. (From Williams, 1993)

Available data thus suggest reverse climatic zonation and strong seasonality in low paleolatitudes consistent with a large obliquity during the Proterozoic.

Origin and evolution of the Earth's obliquity

Present knowledge concerning the origin and early dynamics of the Earth–Moon system from ~ 4500–4000 Ma does not constrain the early Earth's obliquity. Indeed, the early Earth could have acquired a large obliquity by several envisaged mechanisms for the formation of the Earth–Moon system (see Moon: origin; Earth–Moon system: origin).

1. Laskar and Robutel (1993) and Laskar, Joutel and Robutel (1993) showed that, in the absence of the Moon, a proto-earth would have undergone chaotic variations of obliquity ranging from 0° to $\sim 85°$. If the Moon were then captured, the early mean obliquity could have been set at any value within that range; thereafter, the obliquity would have undergone reduced variation.
2. If the Moon formed close to a rapidly rotating proto-Earth with small obliquity, such as in the fission hypothesis, most of the angular momentum of the Earth–Moon system would then have resided in the Earth's spin, whereas about four-fifths now resides in the lunar orbit. During early, tidal transfer of much of the Earth's spin angular momentum to the lunar orbit when the Moon was driven rapidly from the Earth, the component of the Earth's angular momentum perpendicular to the lunar orbital plane would have been reduced and the component in the orbital plane would have been conserved. By such processes the early Earth could have acquired a large obliquity.
3. A single giant impact at ~ 4500 Ma, which may have formed the Moon (Hartmann, Phillips and Taylor, 1986), does not constrain the early Earth's obliquity. An impact-induced obliquity could have been any value from $\sim 0°$ to 180° or more, the most likely value depending on the impactor/target planet mass ratio, the velocity of approach, and the proto-Earth's pre-impact obliquity (Figure O3). Furthermore, Dones and Tremaine (1993) gave probabilities of 0.73–0.91 that a single giant impact would have induced an obliquity $> 24°$ for the early Earth, concluding that the Earth's obliquity may have evolved since formation.

Hence a large obliquity for the early Earth is possible, perhaps likely. Significantly, Venus, Uranus and Pluto have large obliquities (98°, 118° and 177° respectively) that have been attributed to several mechanisms including tidal torques and planetary impact (Williams 1993).

Subsequent evolution of the Earth's mean obliquity depends on the relative magnitudes of the rate of obliquity increase $\dot{\epsilon}_t$ caused by tidal friction, and the rate of obliquity decrease $\dot{\epsilon}_p$ due to dissipative core–mantle torques during luni-solar precession ($\epsilon < 90°$ is required for precessional torques to move ϵ toward 0°; Peale, 1976). Proterozoic paleotidal data imply $\dot{\epsilon}_t \leq 0.0006$ arcsec per century during most of Earth history (Williams, 1993), only about half the rate previously estimated using the modern value for tidal dissipation. The value of $\dot{\epsilon}_p$ resulting from the combined effects of viscous, electromagnetic and topographic core–mantle torques cannot be accurately determined because of uncertainties in estimating, at present and for the geologic past, the effective (eddy) viscosity of the outer core, the nature of magnetic fields at the core–mantle boundary and within the lower mantle, and the topography of the core–mantle boundary. Some geophysical estimates of $\dot{\epsilon}_p$ (e.g. Aoki and Kakuta, 1971) are, however, up to several orders of magnitude greater than the geologically indicated value of $\dot{\epsilon}_t$. If $\dot{\epsilon}_p$ did indeed exceed $\dot{\epsilon}_t$ in the past, the mean obliquity may have decreased during most of Earth history.

A proposed curve of mean obliquity versus time (Figure O4) integrates the geologic evidence for a moderate obliquity during the Phanerozoic and an apparent large obliquity ($\epsilon > 54°$) during the Proterozoic. Geologic data thus suggest that the early Earth acquired a large mean obliquity ($54° < \bar{\epsilon} < 90°$, arbitrarily taken here as $\sim 70°$), possibly by one of the mechanisms discussed above. Subsequent evolution of the Earth's mean obliquity evidently was influenced by dissipative core–mantle torques caused by luni-solar precession. The inflection in the curve between 650 and 430 Ma (Figure O4) suggests greatly increased core–mantle dissipation

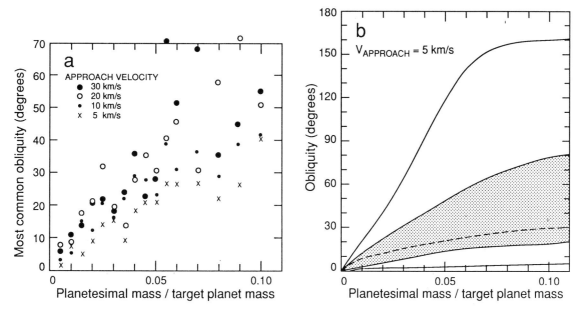

Figure O3 (a) Impact-induced obliquity of a target planet for a range of approach velocities and impactor/target-planet mass ratios. The target planet has physical parameters of the Earth, with pre-impact obliquity of 0° and pre-impact rotation period of 10 h. Each point plotted is the most frequent value in a run of 500 impacts at specified impactor mass. Results applicable to any planet, but assume the Earth's specific density and coefficient of angular momentum. (b) Variation in obliquity occurring when the impactor approaches the Earth at 5 km s^{-1}, with pre-impact obliquity of 0° and pre-impact rotation period of 10 h. The plot, which is smoothed, represents the results of 500-impact runs at intervals of 0.01 mass ratio. The shaded zone includes two-thirds of the values, centered on the most frequent values (dashed line); the outer envelope includes 99.8% of the values. (Adapted from Hartmann and Vail, 1986, by Williams, 1993)

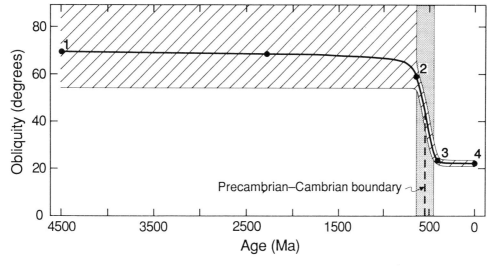

Figure O4 Curve of mean obliquity of the ecliptic $\bar{\epsilon}$ against time (unbroken heavy line), consistent with the early history of the Earth–Moon system and the geologic record. The main control points are (1) $\bar{\epsilon} \approx 70°$ at ~ 4500–4000 Ma (early dynamical processes); (2) $\bar{\epsilon} \approx 60°$ at ~ 650 Ma (late Proterozoic low-paleolatitude glaciation); (3) $\bar{\epsilon} \approx 24°$ at ~ 430 Ma (Ordovician–Silurian paleoprecessional period and high-paleolatitude glaciation); and (4) $\bar{\epsilon} = 23.3°$ at 0 Ma. A tentative point is shown for ~ 2300 Ma (early Proterozoic low- to moderate-paleolatitude glaciation). Striped area indicates the range of obliquity variation (Laskar, Joutel and Robutel, 1993). Stippled band shows the time of annual resonance of the fluid core's predicted 'nearly diurnal' free nutation (± 100 Ma, based on late Proterozoic paleorotational data) for a core–mantle boundary figure in hydrostatic equilibrium. The Precambrian–Cambrian boundary at ~ 540 Ma (dashed heavy line) coincides with the time of maximal core resonance and highest rates of obliquity decrease and reduction of global seasonality. (Modified from Williams, 1993)

during the late Proterozoic – early Paleozoic. Increase in such dissipation may have resulted from the precise annual resonance of the fluid core's predicted 'nearly diurnal' free nutation (Toomre, 1974; Hinderer and Legros, 1988) when the Earth's decelerating rotation rate at ~ 530 Ma (Williams, 1993) passed through the resonance value of ~ 22.2 hours per day for a core–mantle boundary figure in hydrostatic equilibrium (de Vries and Wahr, 1991). Consequent rapid decrease in mean obliquity and accompanying reduction in the range of obliquity variation (Figure O4) may be the first-order cause of the profound biotic changes that occurred across the Precambrian–Cambrian boundary at ~ 540 Ma; after more than 3 billion years of life on Earth, that time witnessed an unsurpassed evolutionary 'explosion' of multicelled Metazoa. Since then, the obliquity has remained near its present value and normal climatic zonation, moderate global seasonality and limited variation of obliquity have prevailed.

George E. Williams

Bibliography

Aoki, S. and Kakuta, C. (1971) The excess secular change in the obliquity of the ecliptic and its relation to the internal motion of the Earth. *Celest. Mech.*, **4**, 171–81.
Barron, E.J. (1984) Climatic implications of the variable obliquity explanation of Cretaceous–Paleogene high-latitude floras. *Geology*, **12**, 595–8.
de Vries, D. and Wahr, J.M. (1991) The effects of the solid inner core and nonhydrostatic structure on the Earth's forced nutations and earth tides. *J. Geophys. Res.*, **96**, 8275–93.
Dones, L. and Tremaine, S. (1993) Why does the Earth spin forward? *Science*, **259**, 350–4.
Gold, T. (1966) Long-term stability of the Earth–Moon system, in *The Earth–Moon System* (eds B.G. Marsden and A.G.W. Cameron). New York: Plenum, pp. 93–7.
Goldreich, P. (1966) History of the lunar orbit. *Rev. Geophys.*, **4**, 411–39.
Hartmann, W.K. and Vail, S.M. (1986) Giant impactors: plausible sizes and populations, in *Origin of the Moon* (eds W.K. Hartmann, R.J. Phillips and G.J. Taylor). Houston: Lunar and Planetary Institute, pp. 551–66.
Hartmann, W.K., Phillips, R.J. and Taylor, G.J. (eds) (1986) *Origin of the Moon*. Houston: Lunar and Planetary Institute, 781 pp.
Hinderer, J. and Legros, H. (1988) Tidal flow in the Earth's core: a search for the epoch and amplitude of exact resonance in the past. *Am. Geophys. Union Geophys. Mon.*, **46**, 79–82.
Hunt, B.G. (1982) The impact of large variations of the Earth's obliquity on the climate. *J. Meteorol. Soc. Jap.* **60**, 309–18.
Laskar, J. and Robutel, P. (1993) The chaotic obliquity of the planets. *Nature*, **361**, 608–12.
Laskar, J., Joutel, F. and Robutel, P. (1993) Stabilization of the Earth's obliquity by the Moon. *Nature*, **361**, 615–7.
Mignard, F. (1982) Long time integration of the Moon's orbit, in *Tidal Friction and the Earth's Rotation II* (eds P. Brosche and J. Sündermann). Berlin: Springer-Verlag, pp. 67–91.
Peale, S.J. (1976) Inferences from the dynamical history of Mercury's rotation. *Icarus*, **28**, 459–67.
Schmidt, P.W., Williams, G.E. and Embleton, B.J.J. (1991) Low palaeolatitude of late Proterozoic glaciation: early timing of remanence in haematite of the Elatina Formation, South Australia. *Earth Planet. Sci. Lett.*, **105**, 355–67.
Toomre, A. (1974) On the 'nearly diurnal wobble' of the Earth. *Geophys. J. Roy. Astron. Soc.*, **38**, 335–48.
Touma, J. and Wisdom, J. (1993) The chaotic obliquity of Mars. *Science*, **259**, 1294–7.
Williams, G.E. (1975) Late Precambrian glacial climate and the Earth's obliquity. *Geol. Mag.*, **112**, 441–65.
Williams, G.E. (1993) History of the Earth's obliquity. *Earth Sci. Rev.*, **34**, 1–45.
Williams, G.E. (1994) The enigmatic late Proterozoic glacial climate: an Australian perspective, in *Earth's Glacial Record* (eds M. Deynoux, J.M.G. Miller, E.W. Donack, *et al.*). Cambridge: Cambridge University Press, pp. 146–64.

Cross references

Chaotic dynamics in the solar system
Earth–Moon system: origin
Earth: rotational history
Ice age
Milankovitch, Milutin, and Milankovitch theory
Tidal friction

OCCULTATION

Occultations – the hiding of one celestial body by another – are simple, yet surprisingly informative, astronomical events. Stars – very compact, bright, and distant sources of light – can act as probes of

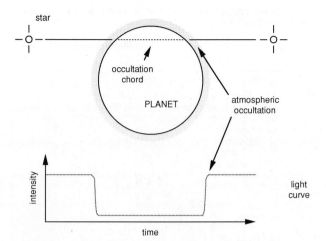

Figure O5 Occultation of a star by a planet includes apparent immersion of the star into the atmosphere (if any), blockage of all light by the opaque body of the planet, and then reappearance of the star through the atmosphere. Occultation chord length may be related to the planet radius. Lightcurve variations during immersion and emersion can reveal atmospheric structure and physical properties.

bodies which pass across their positions in the sky (Figure O5). The point-like backlighting from the star may be used to measure the dimensions of the closer body and to determine the structure of its atmosphere and ring system, if any.

The amount of time that light is extinguished allows calculation of the length of the corresponding chord across the disk of an occulting planet (Figure O5). An accurate chord length – plus precise measurement of the locations of the star and the (drifting) planet in the sky – permits an accurate determination of the planet's radius. More intriguing are occultations of asteroids, since astronomers cannot generally resolve their irregular shapes with Earth-based telescopes. The same occultation event seen from different points on the Earth's surface will yield a set of chord lengths. Compilation of results from many such observations can provide a surprisingly detailed silhouette, albeit only for the outline presented to Earth at the time of the occultation (Chapman, 1984).

When the occulting body is large enough to have an atmosphere, the star's intensity decreases slowly as it dips first into the atmosphere and then vanishes behind the opaque part of the body. The gradual decrease results from refraction by the gaseous atmospheric envelope and scattering or absorption by aerosols and molecules within it. Profiles of the measured stellar intensity can be used to infer structure and physical properties (such as temperature, pressure and density) of the atmosphere. Measurements during occultation egress – when the star reappears – are, of course, as valuable as those obtained during occultation ingress.

Occultations by planetary rings provide information on the spatial distribution of the ring material. When density of the ring material is high, the light will be absorbed or scattered most strongly; when the ring is tenuous, most of the signal will pass through unperturbed. The logarithm of the light curve from a ring occultation yields a profile of the ring optical depth (see Optical depth). The rings of Uranus and the arcs around Neptune were both discovered using stellar occultation techniques.

Occultation measurements involving spacecraft instruments can provide much higher spatial resolutions than Earth-based instruments because the spacecraft can be placed almost arbitrarily close to the body of interest. Very high quality observations of the atmospheres and rings of Jupiter, Saturn, Uranus and Neptune were made at ultraviolet and visible wavelengths during stellar occultations observed aboard the Voyager spacecraft (Allison *et al.*, 1991; French *et al.*, 1991). Occultations at radio frequencies provided some of the best data, however. In these cases the spacecraft became the signal source with observations being conducted from Earth. Since the radio signal transmitted from Voyager to Earth through the rings and atmospheres was coherent, the effects of diffraction as well as refraction and absorption could be included in the analysis. Ring profiles reconstructed from Voyager radio data have radial resolutions as small as 50 m.

Richard A. Simpson

Bibliography

Allison, M., Beebe, R.F., Conrath, B.J. et al. (1991) Uranus atmospheric dynamics and circulation, in *Uranus*, (eds J.T. Bergstrath, E.D. Miner, and M.S. Matthews). Tucson: University of Arizona Press, pp. 253–95.

Chapman, C.R. (1984) Pallas report. *Planet. Rep.*, **4** (4), 12.

French, R.G., Nicholson, P.D., Porco, C.C. and Marouf, E.A. (1991) Dynamics and structure of the Uranian rings, in *Uranus* (eds J.T. Bergstralh, E.D. Miner, and M.S. Matthews). Tucson: University of Arizona Press, pp. 327–409.

Cross references

Planetary ring
Radio science

OLBERS, HEINRICH WILHELM MATTHÄUS (1758–1840), AND OLBERS' PARADOX

An amateur astronomer and a gifted German theoretician, Olbers was professionally a medical doctor, a physician. Credited with the discovery of minor planets Pallas (in 1802) and Vesta (in 1807), he proposed a theory that Ceres (confirmed by him in 1802) and Pallas, which have the same period (4.61 years), and nearby asteroids were remnants of a primeval planet that exploded. This sort of celestial catastrophe later fell out of fashion for ideological reasons (probably a victim of the extremists in the Lamarck–Darwin debates).

Olbers had made his name by his development of a satisfactory method for calculating cometary orbits (1797). In 1815 Olbers discovered a comet that now bears his name, one with a period of 72.7 years. He also proposed that meteor streams could have long-period recurrence intervals, e.g. the Leonids with their 33-year cycle. The tails of comets were believed to be rarified particles lost from the comet and related in some way to electrical energy from the Sun, a quite remarkable anticipation of modern views about the solar wind (q.v.) and the illumination of cometary tails.

Olbers may also have anticipated the expanding universe hypothesis, assuming that the then-existing laws of physics were unchanging within a stable universe. If the sky is uniformly populated with stars to indefinite distances (as Newton proposed), the night sky should be as bright as the surface of an average star (the Sun), due to the geometry of overlapping star disks as seen from the Earth or other points. This is called Olbers' paradox (1926). His answer to this was that interstellar dust weakened the light of the stars. Olbers paradox was subsequently used as an argument against an infinite universe. The suggested solution to the paradox is to take into account the recession of distant stars and galaxies and the finite age of the universe. Because of the recession and redshift, the observer on Earth received virtually no light from beyond a certain distance and, because of the finite age, the radiation in the universe may not have come into equilibrium with the matter.

Rhodes W. Fairbridge

Bibliography

Jaki, S.L. (1969) *The Paradox of Olbers' Paradox*. New York.
Multauf, L.S. (1924) Olbers, Heinrich Wilhelm Matthias. *Dict. Sci. Biogr.*, Vol. 10, pp. 197–9.
Struve, O. (1963) Some thoughts on Olbers' paradox. *Sky and Telescope*, **25**, 140–2.

Acknowledgements

This entry was kindly reviewed and appreciably improved by Dr Hans J. Haubold (UN, Vienna).

OORT, JAN HENDRIK (1900–1992), AND OORT CLOUD

The 'father of modern Dutch astronomy', and one of the pioneers in 20th century radio astronomy, Jan Oort was born at Franeker, a little town in the northern part of the Netherlands. He attended the University of Groningen and graduated in 1926. After 2 years at Yale University, he was appointed professor of astronomy at the University of Leiden and remained there all his life. In later years he was widely recognized with honorary degrees from 10 universities, including Oxford, Cambridge and Harvard. He was a Fellow of the Royal Society of London. By his brilliance he attracted a large number of first-class minds to the study of astronomy and today Dutch-trained astronomers are to be found in every major country.

Oort's primary discovery was that the solar system rotates around the center of the Milky Way (about once every 250 million years, on a radius of some thousands of light years) and that the latter in turn rotates around its own axis. This epoch-making advance has sometimes been compared with the Copernican revolution of the 16th century. It may hold the key to understanding the cyclic return of great ice ages on planet Earth, because the solar system would pass through one of the galaxy's spiral arms (their dust dimming the Sun's effective radiation) twice in each 'galactic year'.

Emerging from his galactic studies came Oort's recognition that, with multitudes of galaxies, the total mass of the universe must be far greater than had previously been conceived. He believed that there was a 'hidden mass' or 'dark matter' that constitutes over 90% of the mass of the universe, which cannot be detected by ordinary optical telescopes. In the 1950s he came to the conclusion that as close as 1 light year from the solar system there is a giant cluster of rocky particles that appears to be the source of comets.

The Oort cloud

The Oort cloud is a nearly spherical distribution of comets surrounding the Sun at the fringe of the solar system. It serves as a reservoir for the new comets that are observed to visit our night skies. Oort presented the first observational evidence for its existence in 1950, substantiating the notion first suggested by Ernst Öpik in the early 1930s. Oort analyzed the distribution of semimajor axes of 19 comets and found an excess of long-period comets with aphelia beyond 20 000 AU and corresponding periods in excess of a million years.

The basic concept of the Oort cloud as the outer halo of the solar system has been confirmed by recent studies using improved samples of cometary orbits by Marsden and colleagues (1978). It is found that the mean aphelion distance of the classical Oort cloud is approximately 50 000 AU with some orbits reaching about one-third the distance to the nearest star. However, there is indirect evidence that the reservoir extends closer to the Sun than was previously thought, to a region referred to as the inner core, within about 20 000 AU of the Sun. Oort originally estimated that the cloud must contain about 2×10^{11} comets, with a total mass of less than one-half that of the Earth. Recent estimates extend these values to $3–5 \times 10^{12}$ comets with a total mass of about 30 times that of the Earth; about one-sixth of the mass is contained in the classical Oort cloud and the remainder is in the inner core.

The Oort cloud is the reservoir which supplies the observed long-period comets. The inclinations of the cometary orbits are distributed isotropically on the sky, aside from some clustering caused by stars and interstellar clouds passing through the solar neighborhood, and a depletion in a narrow band along the galactic equator caused by galactic tidal forces. These slight gravitational perturbations occasionally divert comets toward the inner solar system, where they are observed as dynamically new, long-period comets making their first visit near the Sun (in contrast to dynamically old comets that have returned to the inner solar system more than once). It appears that the new comets are dustier and brighten slower than the old comets as they approach the Sun. Galactic tidal perturbations appear to the the principle mechanism for providing this continuous supply of new comets from the classical Oort cloud (Fernández and Ip, 1991; see Comet: dynamics; Comet: origin and evolution).

Modern studies of the Oort cloud, including numerical simulations carried out on high-speed computers, have led to the realization that the classical Oort cloud would be completely dissipated over the age of the solar system from encounters with stars and interstellar clouds and by the galactic tides. Consequently, the idea that Oort cloud is the source of comets naturally requires a mechanism to replenishment it. This is the rationale for the existence of the inner core of the Oort cloud. Comets residing in the core are largely unaffected by passing stars; however, the passage of interstellar clouds can perturb comets in the inner core to larger aphelion distances, replenishing the classical Oort cloud and guaranteeing its survival. An additional reservoir of Oort cloud comets may reside in the ecliptic region beyond the orbit of Neptune in a disk-shaped distribution named the Kuiper belt (q.v.). This region may also be a source of short-period comets (Duncan and Quinn, 1993).

Daniel C. Boice and Rhodes W. Fairbridge

Bibliography

Bailey, M.E. (1986) The near-parabolic flux and the origin of short-period comets. *Nature*, **324**, 350–2.
Duncan, M.J. and Quinn, T. (1993) The long-term dynamical evolution of the solar system. *Ann. Rev. Astron. Astrophys.*, **31**, 265–95.
Fernandez, J.A. and Ip, W.-H. (1991) Statistical and evolutionary aspects of cometary orbits, in *Comets in the Post-Halley Era*, Vol. 1 (eds R.L. Newburn, Jr, M. Neugebauer and J. Rahe) Dordrecht: Kluwer Academic Publisher, pp. 487–535.
Huebner, W.F. (ed.) (1990) *The Physics and Chemistry of Comets*. Berlin, Heidelberg: Springer-Verlag.
Marsden, B.G., Sekanina, Z. and Everhart, E. (1978) New osculating orbits for 110 comets and analysis of original orbits for 200 comets. *Astron. J.*, **83**, 64–71.
Oort, J.H. (1950) The structure of the cloud of comets surrounding the solar system, and a hypothesis concerning its origin. *Bull. Astron. Inst. Neth.*, **11**, 91–110.
Öpik, E.J. (1973), Comets and the formation of planets. *Astrophys. Space Sci.*, **21**, 307–98.
Safronov, V.S. (1977), Oort's cometary cloud in the light of modern cosmogony, in *Comets, Asteroids, Meteorites*, (ed. A.H. Delsemme). Toledo: University of Toledo Press, pp. 483–4.
Weissman, P.R. (1990), The Oort cloud. *Nature*, **344**, 825–30.

Cross references

Comet
Comet: dynamics
Comet: origin and evolution
Kuiper belt

ÖPIK, ERNST JULIUS (1893–1985)

Ernst Julius Öpik was born at Port Kunda in Estonia, and became one of the most gifted and versatile astronomers of the 20th century, contributing original ideas to almost every branch of the science. His 1094 publications ranged from the dynamics and physics of the solar system to stellar evolution and the origin of the universe: truly a renaissance man. He also published one popular paperback: *The Oscillating Universe*.

He graduated in astronomy in 1916 from Moscow Imperial University. In the same year his first article used data on double stars with orbital motion to derive stellar densities in comparison with that of the Sun. He explained how this was possible assuming Stefan's law and with known spectral types for stars giving the surface temperatures. Most of the data gave good results, but one star, Omicron 2 Eridani, gave a density of 25 000 which Öpik regarded as 'impossible', indicating an unknown error: in fact he had discovered a white dwarf, whose existence was soon to be revealed by Walter S. Adams at Mount Wilson Observatory. Öpik's education spanned the days which led up to the October Revolution, and a brief career in the Red Army left him fanatically anti-Communist.

In 1919 he and several other academics left Moscow for the new Russian possessions of Central Asia, and here he helped to found the Turkestan University at Tashkent, chairing the astronomy department and restoring the observatory there.

In 1921 he returned to Estonia, now an independent state, and became head of the astronomy department at Tartu (Dorpat), where he received his PhD in 1923 with a thesis on the principles of his well-known 'double-count method' for the empirical determination of observational selection (completeness) in meteor observations.

About this time, when the true nature of the nebulae was still hotly debated, Öpik calculated the Andromeda system to possess 1.8×10^6 solar masses. He concluded: 'the coincidence of results obtained by several independent methods increases the probability that this Nebula is a stellar universe, comparable with our own Galaxy'.

In 1930–1934 Öpik was Research Associate at Harvard College Observatory and helped in planning the Harvard–Cornell expedition to Arizona for the study of meteors, using his ingenious 'rocking-mirror' method. Some 22 000 were recorded and their heights determined. He discussed the influence of stellar perturbations on nearly parabolic ellipses, which later was to be an important mechanism in Jan Oort's concept of the comet cloud surrounding the Sun. Fred Whipple always spoke of the 'Öpik–Oort comet cloud'.

In 1944, when the Baltic states fell to the Soviets. Öpik and his wife Alide escaped to Germany, where a 'Free Baltic University' was established at Pinneaberg for the benefit of the refugees, but it lacked funds, and in 1948 it was forced to close. Then Öpik had a stroke of luck, when Eric Lindsay invited Öpik to Armagh (Northern Ireland) as research assistant, where he remained until his retirement.

Generally, he spent a part of each year as Visiting Professor at the University of Maryland. In 1950 he became a vehicle for his prolific pen. He investigated the history of the Moon, convinced that the lunar craters were due to impact rather than volcanism; he predicted that Mars would be cratered too.

He was also deeply concerned with celestial mechanics, and he investigated the orbits of small bodies which made close approaches to Earth and Mars. He put forward theories to account for the past ice ages, based on the rotation of the galaxy and anomalous paleoclimatic evidence, possibly stimulated by discoveries made by his brother, Armin A. Öpik, a distinguished paleontologist, who had found refuge in Australia. In 1976 he followed up an earlier idea of Cameron (in 1973), suggesting changes in solar luminosity due to convective overturn.

Öpik was sometimes badly wrong. His model of Venus was faulty, and at one stage he was convinced that the dark areas (maria) on Mars were due to vegetation. In the field of astrophysics he maintained that the arms of spiral galaxies were leading rather than trailing, which is incorrect. Another of his theories was that sunspots were associated on Earth with political revolutions. It was eccentricities of this sort which sometimes led to his work being taken less seriously than it ought to have been. Nevertheless, he was undoubtedly one of astronomy's most original thinkers, and in many ways a true scientific pioneer.

Patrick Moore

Bibliography

Anon. (1972) *Irish Astronomical Journal*, Special Issue, 10, 1–92.
Anon. (1986) *Irish Astronomical Journal*, **17** (September), 411–442.
Moore, P. (1988) The turbulent life of E.J. Öpik. *Sky and Telescope*, February 149.
Muursepp, P. and Preem, R. (1968). Ernst Öpik and the Perseid Shower. *Sky and Telescope*, 376–7.
Öpik, E.J. (1965) Climatic change in cosmic perspective. *Icarus*, **4**, 289–307.
Öpik, E.J. (1967) Climatic changes. *Int. Dict. Geophys.*, Vol. 1, pp. 179–93.
Öpik, E.J. (1969) Stellar interiors: the source of life and death. *Irish Astron. J.*, **9**
Öpik, E.J. (1976) Solar structure, variability, and the ice ages. *Irish Astron. J.* **12**, 253–76.

OPPOSITION

A superior planet is in opposition to the Sun when its right ascension or ecliptic longitude is 180° from the Sun. The two cases are referred to as opposition in right ascension and opposition in longitude.

Jeremy B. Tatum

Bibliography

Green, R. M. (1985) *Spherical Astronomy*. New York: Cambridge University Press.
Woolard, E.W. and Clemence, G.M. (1966) *Spherical Astronomy*. New York: Academic Press.

OPPOSITION EFFECT

Solar system objects, unlike all other objects seen in the night sky, are only discernible by the sunlight reflected from their surfaces or from the top of the upper deck of the clouds in their atmospheres. Therefore, they exhibit a set of optical properties that are distinct from objects which have their own internal energy source, such as the Sun and stars.

The light reflected from a solar system body that is illuminated by the Sun and observed from the Earth undergoes a variety of geometrical and physical processes on its path from the Sun to the object and back to the Earth.

It has been known for more than a century that the intensity of this light when reflected from a solid surface varies as a function of the phase angle, the angle defined by the Sun–object–earth, at the time of observation. One source of this variation is the obvious geometrical effect caused by the fact that the object under observation is never fully illuminated except when the phase angle is 0°. However, even when a correction is made for this 'defect of illumination', the intensity of the reflected light is observed to increase nonlinearly as the phase angle decreases toward zero degrees. It is this nonlinear surge in reflected light seen when an object is observed at small phase angles that is called the opposition effect.

Research has revealed that the opposition effect is due to several independent geometric and physical processes acting in combination. The cumulative size of the opposition effect from all of these processes is a function of the particle size, albedo and the packing density of the regolith material. The principal geometric process that contributes to the opposition effect is called the shadow hiding hypothesis and the principal physical process is called the coherent backscattering hypothesis.

Shadow hiding hypothesis

As sunlight falls on a particulate planetary surface, the particles which comprise the regolith will cast shadows on one another. An Earth-based observer will see a combination of sunlight reflected from the particles on the surface, light reflected from a particle after being transmitted through one or more grains prior to reflection, light reflected from a grain after diffraction around the edge of one or more particles, and the portions of the surface that are lying in shadow and are only illuminated by multiply scattered light. The shadow hiding model posits that the size of the opposition effect will increase as the albedo of the particles gets smaller. This effect can be approximated by the following expression:

$$B(g) = \frac{B_0}{1 + (\frac{1}{h})\tan(\frac{g}{2})}$$

where B_0 is the amplitude of the opposition effect, the h is the width of the opposition effect and g is the phase angle of the observation. This expression has been used to model the opposition effect down to phase angles as small as a few degrees. The size of the opposition effect B_0 is a function of the single-scattering albedo and the angular scattering function of the regolith particles.

Coherent backscattering hypothesis

The shadow hiding model of the opposition effect does not predict a strong opposition surge for highly reflective particulate media. However, opposition effects are observed in such materials in the laboratory and these effects are most pronounced at phase angles between 0° and 1°. The principal cause of the opposition surge that is seen in highly reflective materials is the phenomenon of coherent constructive interference between light rays that are multiply scattered in the planetary regolith.

The coherent backscattering hypothesis argues that any two light rays that enter a medium at different points and travel the same path except in opposite directions will constructively interfere as the difference in total path length traveled by the two rays approaches the wavelength of the light. In this case, the rays will constructively interfere and a pronounced opposition effect will be seen. The angular width of this peak is related to the wavelength of the illuminating light, the index of refraction of the particles, and the packing density of the particles. Theoretical models estimate that the full-width half maximum of the opposition effect caused by coherent backscattering is $\lambda/2\pi D$, where λ is the wavelength of the light impinging on the surface and D is the photon mean free path in the medium.

Robert M. Nelson

Bibliography

Gehrels, T. (1956) The Opposition Effect. *Ap. J.*, **123**, 331.
Hapke, B.W. (1971) Inferences from optical properties concerning the surface texture and composition of asteroids, in *Physical Studies of the Minor Planets*. NASA, SP-267, pp. 67–77.
Hapke, B.W. (1986) Bidirectional reflectance spectroscopy, 4. The extinction coefficient and the opposition effect. *Icarus*, **67**, 264–280.
Hapke, B.W., Nelson, R.M. and Smythe, W.D. (1993) The opposition effects of the Moon: contribution of coherent backscattering. *Science*, **260**, 509–11.
Kaula, W.M. (1968) *Introduction to Planetary Physics*. New York: John Wiley and Sons.
Mishcenko, M.I. (1992). The angular width of the coherent backscatter opposition effect: and application to icy outer planet satellites. *Astrophys. Space Sci.* **194**, 327–33.
Nelson, R.M. Hapke, B.W. Smythe, W.D. *et al.* (1993) The coherent backscattering opposition effect in *Proc. Lunar Planet. Sci. Conf.* **24**, pp. 1061–1062.
Russell, H.N. (1916) On the albedo of the planets and their satellites. *Astrophys. J.*, **43**, 173–87.
Veverka, J. (1977) Photometry of satellite surfaces, in Planetary satellites (ed. J. Burns). Tucson: University of Arizona Press, pp. 171–209.

Acknowledgements

The author appreciates the review that his colleague Linda J. Horn gave to this manuscript.

Cross references

Albedo
Reflectance spectroscopy

OPTICAL DEPTH

When light or other radiation passes through a partially transparent medium, its intensity is diminished by scattering and/or absorption. The effect of the medium can be characterized by an attenuation coefficient α, defined as the fraction of intensity lost per unit distance traveled. The integration of this attenuation over the total path through the medium is called the optical depth:

$$\tau = \int_0^L \alpha(x) \, dx$$

where the attenuation in this expression is allowed to vary over the path $(0, L)$. The intensity decreases with increasing optical depth; a medium with unit optical depth ($\tau = 1$) reduces the intensity of the original signal by a factor e (to about 37% of its original strength). A completely opaque medium would have infinite optical depth.

Optical depth is commonly used in describing aerosol effects on atmospheric clarity; high concentrations of aerosols increase the optical depth and reduce visibility. Optical depth has also been used to describe the transmission properties of planetary rings – higher optical depths correspond to greater scattering and/or absorption by the ring particles.

Note that the definition of the attenuation coefficient α varies among fields. Those working primarily with coherent radiation prefer to define α in terms of the reduction of the electric or magnetic field strength. Those who work primarily with incoherent radiation prefer to measure the intensity of the radiation. Since the instantaneous intensity is proportional to the square of the field strength, the values of α will differ by a factor of two in the two cases. The definition for optical depth above is given for the incoherent (intensity) case.

Richard A. Simpson

Bibliography

Ulaby, F.T., Moore, R.K. and Fung, A.K. (1981) *Microwave Remote Sensing: Volume 1 – Fundamentals and Radiometry*. Reading, MA: Addison-Wesley.
Chandrasekhar, S. (1960) *Radiative Transfer*. New York: Dover.
Marouf, E.A., Tyler, G.L. and Rosen P.A. (1986) Profiling Saturn's rings by radio occultation. *Icarus*, 68, 120–66.
van de Hulst, H.C. (1981). *Light Scattering by Small Particles*. New York: Dover.

Cross references

Aerosol
Atmosphere
Planetary ring

ORBIT

The three laws of planetary motion were enunciated by Johannes Kepler in 1609 and 1619. Discovered empirically they are:

1. every planet moves around the Sun in an elliptical orbit with the Sun at one focus,
2. the radius vector from the Sun to the planet sweeps out equal areas in equal times;
3. the squares of the periods of the planets are proportional to the cubes of their semimajor axes.

The first and third of these laws were shown by Sir Isaac Newton to be a consequence of the inverse square law of gravitation. The second is a consequence of the fact that the gravitational pull of the Sun has no transverse component, and consequently the orbital angular momentum, which is proportional to the areal speed, is constant.

The size, shape and orientation of an elliptical orbit are described by five orbital elements (Figure O6):

a the semimajor axis of the ellipse;
e its eccentricity;
i the inclination of the plane of the orbit to the plane of the ecliptic. This can be in the range $0° < i < 180°$. Inclinations greater than 90° correspond to retrograde orbits – i.e. the motion is in the opposite direction to the motion of the Earth. Orbits with $i < 90°$ are direct or prograde.
Ω the longitude of the ascending node, Ω, measured eastwards from the First Point of Aries along the ecliptic (q.v.), from 0° to 360°.
ω the argument of perihelion, measured in the plane of the orbit, in the direction of planetary motion, from ☊, from 0° to 360°.

It is redundant (although of interest) to specify the period, *P*, since, from Kepler's second law of planetary motion, $P^2 = a^3$, where *P* is in sidereal years and *a* is in astronomical units (q.v.). Also of frequent use is the mean motion $n = 360°/P$, which is the mean angular speed in the orbit, expressed in degrees per solar day.

In addition to these five elements, a sixth element is necessary in order to calculate the position of a planet in its orbit. For a sensibly eccentric orbit, such as that of a comet, the sixth element is usually chosen to be *T*, the time of perihelion passage. For a nearly circular orbit *T* and ω are ill defined, and an alternative element, usually the mean anomaly at the epoch (see below), is chosen.

The angular elements *i*, Ω, ω, must always be referred to a specified equinox and equator (see Precession and nutation). The standard equinox and equator at present is that of J2000.0/FK5 – i.e. for the beginning of the Julian year 2000, and the fifth Fundamental Katalog. It is also necessary to specify, for all elements, an epoch of osculation, to be described below.

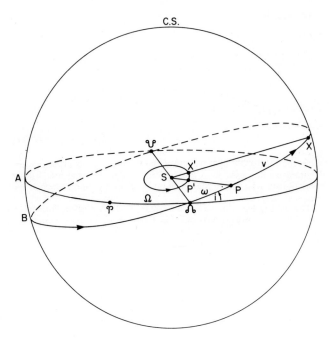

Figure O6 The circle CS represents the celestial sphere. The great circle A is the intersection of the plane of the Earth's orbit with the celestial sphere. The point marked ♈ is the First Point of Aries which the Earth passes on or near 22 September each year. Ecliptic longitudes are measured eastwards from ♈ – i.e. towards the right from ♈ in the figure. See the text, however, for some 'longitudes' which are measured in two different planes. The great circle B is the intersection of the plane of the planet's orbit with the celestial sphere. In this figure the planet is supposed to be traveling in the direction of the arrows. The angle *i* is the inclination of the orbit to the ecliptic.

The line ☊ S ☋ is the line of nodes, ☊ being the ascending node (where the planet passes from south to north of the ecliptic) and ☋ being the descending node. The angle Ω is the longitude of the ascending node. The small ellipse near the center of the figure, which is in the plane B, is the orbit of the planet. The orbit is an ellipse; in the figure we see the orbit projected at an angle rather than face-on. The projected orbit also appears as an ellipse, but the major axis of the projected ellipse is not the major axis of the true ellipse, nor is the Sun at a focus of the projected ellipse. Although the centers of the true and projected ellipses coincide, this is of relatively little importance. We suppose P' to be the position of perihelion, when the planet is nearest to the Sun, SP' being part of the major axis of the true ellipse. The point P is the projection of P' on the celestial sphere. The planet passes P at the instant *T*. At some time $t - T$ later, the planet is at X', whose projection on the celestial sphere is X. The several arguments, longitudes and anomalies may be understood by reference to the text and the figure.

There are several ways of describing the position of a planet in its orbit. The easiest to comprehend is the true anomaly *v*. Also used is the eccentric anomaly *E*. These are easily understood by reference to Figure O7. Note that the circle whose diameter is the major axis of the ellipse is the auxiliary circle. The angles *E* and *v* are related by

$$\cos v = \frac{\cos E - e}{1 - e \cos E} \tag{O1}$$

We can also imagine a fictitious body (the 'mean' planet, as opposed to the 'true' planet), coincident with the true planet at perihelion, and moving with constant angular speed *n*, the mean motion. At time $t - T$ after perihelion passage, the mean anomaly *M* is

$$M = n(t - T) \tag{O2}$$

The eccentric anomaly can be calculated from the mean anomaly by Kepler's (transcendental) equation

$$E = M - e \sin E \tag{O3}$$

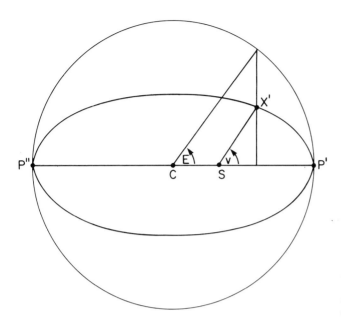

Figure O7 The true orbit seen face-on. The Sun S is at a focus. P' and P" are the perihelion and aphelion points. The ratio CS/CP' is the eccentricity e. The perihelion distance SP' is $q = a(1 - e)$. The aphelion distance SP" is $Q = a(1 + e)$. The circumscribing circle is the auxiliary circle. When the true planet is at X', its true anomaly is v and its eccentric anomaly is E.

and the true anomaly subsequently from equation (O1).

The angle $\theta = \omega + v$, measured from Ω, is called the argument of latitude of the planet. The quantity $l = \Omega + \theta = \Omega + \omega + v$, measured partly in the plane of the ecliptic and partly in the plane of the orbit, is the curiously named 'true longitude'. The word 'true' in this phrase refers to the true planet; the 'longitude', being the sum of two angles measured in different planes, is anything but 'true'. Likewise $L = \Omega + \omega + M$ is the 'mean longitude', which really means the 'longitude' of the mean planet. Further, the quantity $\tilde{\omega} = \Omega + \omega$ is called the longitude of perihelion, so that $l = \tilde{\omega} + v$ and $L = \tilde{\omega} + M$.

In the case of near-circular orbits of low inclination, for which T, Ω and ω are ill defined, an instant of time known as the epoch is specified, and, instead of T, either the mean anomaly at the epoch, M_o, or the true longitude at the epoch, l_o, or the mean longitude at the epoch, L_o, may be specified.

The inverse square law of gravitation requires that a particle will move around the Sun in an orbit that is a conic section – i.e. an ellipse, a parabola or a hyperbola. Any permanent member of the solar system moves in an ellipse. However, a long-period comet moves in a very large and elongated ellipse, of which we can observe only a very short arc near to perihelion. It is usually quite impossible to determine the semimajor axis or period with any pretense of accuracy, and the orbit is customarily represented by a parabola. In this case, rather than specify the major axis (which is infinite), the perihelion distance q is specified.

A comet approaching Jupiter may be temporarily perturbed into a hyperbolic orbit. In that case the element a is the 'semitransverse axis' and, in some conventions, is treated as negative.

In a two-body system (i.e. Sun and one planet), the planet does not move around the center of the Sun; rather the Sun and the planet move around their mutual center of mass. Since the mass of the Sun is only 1047 times that of Jupiter, this is not a negligible effect. In reality the solar system has more than one planet, so the situation is rather more complicated. When comets move a long way from the Sun, it is necessary to calculate their orbits with respect to the barycenter (q.v.) of the solar system.

Perturbations

There are many influences which will perturb a planet from a simple elliptical orbit. Perturbations may be periodic (for example in inclination or eccentricity) or secular (i.e. always in one direction).

For example, the advance of the line of apsides is a secular effect, so that ω is always increasing. The regression of the lunar nodes is a retrograde secular effect, so that Ω is always decreasing. The line of apsides is another name for the major axis, each end of which is an apsis. The word apse (plural apses), which is correctly an architectural term, is incorrectly used in this connection with distressing frequency.

There are several causes for perturbations. The most important is the influence of the other planets and, in the case of the Earth, its relatively massive Moon. For objects that closely approach the Sun, such as Mercury and some comets and asteroids, effects of general relativity must be taken into account; these measurably increase the rate of advance of Mercury's perihelion, for example, over that which would be predicted from planetary perturbations alone. The orbits of small dust particles expelled from comets are subject to the effects of solar radiation pressure. Since, like gravity, this effect falls off inversely as the square of the heliocentric distance, it is customary to deal with the effect by using a smaller 'effective' value of the gravitational constant G; the difference between the true gravitational constant and its effective value is greater for very small particles than for relatively large ones. Jets of matter are expelled from localized spots on the surface of a comet with sufficient momentum to impart both radial and transverse 'non gravitational forces' to the motion.

Other conceivable perturbations could arise from asphericity of the Sun, or from the effects of a retarding interplanetary medium; such effects are, however, believed to be wholly negligible.

The result of perturbations is that a true orbit is not a stationary ellipse. If we imagine a particle at some instant in the same position as a planet and having the same instantaneous velocity, but subsequently moving away in an unperturbed elliptical orbit, that orbit is called the 'osculating' ellipse; its elements are the 'osculating elements', and the instant is the 'epoch of osculation'. The word comes from the Latin *osculare*, to kiss; the osculating orbit is tangent to the true orbit, and there is presumably some fanciful likeness to the lips of a lover lightly touching the cheek of his or her loved one.

Ephemeris and orbit computation

Once the elements are specified for a modern epoch of osculation and referred to a specified equinox and equator, it is a fairly straightforward matter, though beyond the scope of this article, to compute an ephemeris (q.v.), though there are some refinements needed for accurate work. Chief among these are (1) a correction between geocentric positions (as seen from the center of the Earth) and topocentric positions (as seen from a particular location on the Earth's surface); (2) allowance for the light travel time between planet and Earth; and (3) a correction between the Universal Time used by observers and Terrestrial Dynamical Time used by orbit calculators.

Much more difficult than the computation of an ephemeris from the elements is the determination of the elements from the raw observations of the positions of a planet or comet at different times. One of the major difficulties is that, whereas we can make observations of the position of a planet or comet on the celestial sphere at different times, we have no idea of its geocentric distance. A powerful tool in attacking the problem, however, is provided by our knowledge that the heliocentric radius vector sweeps out equal areas in equal times.

We cannot treat the subject in detail here, but we can mention the principal key in the method outlined by C.F. Gauss (1777–1855) in his famous *Theoria Motus Orbium Coelestium*.

In Figure O8a we see the body moving in an elliptic orbit at positions P_1, P_2, P_3, around the Sun at focus S. The heliocentric distances are r_1, r_2, r_3. The time intervals τ_1 and τ_3 are as shown, expressed in dimensionless units in which $\tau_1 + \tau_3 = 1$. The sector areas b_1 and b_3 are as shown, expressed in dimensionless units in which $b_1 + b_3 = 1$. From Kepler's second law we know that $b_1/b_3 = \tau_1/\tau_3$, so that b_1/b_3 is a known quantity.

In Figure O8b we have joined P_1, P_2 and P_3 with straight lines. The triangle areas a_1 and a_3 are as shown, expressed in dimensionless units such that $a_1 + a_3 = 1$. It is a geometrical property of this figure that

$$r_2 = a_1 r_1 + a_3 r_3$$

We already know the ratio b_1/b_3. If only we knew the ratio a_1/a_3, we could solve this equation for the heliocentric distances. Much of Gauss' method is concerned with finding the sector-to-triangle ratios.

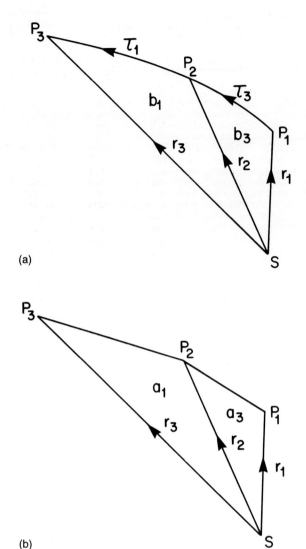

Figure O8 Illustrating (a) the sector areas b_1 and b_3 and (b) the triangle areas a_1 and a_3.

The closer the three observations are together, the closer these ratios are to unity.

In principle, just three observations (that is to say six angles – a right ascension and a declination for each observation) are necessary to establish an orbit, though there are some further constraints under particular circumstances of which orbit calculators are aware. In practice, however, many observations are available, and a question arises as to how to make the best use of them. An orbit based on the first few observations available is a preliminary orbit. When a large number of observations are finally available, the orbit must be differentially refined, and a definitive orbit must be computed. We cannot do better than to close this article with the words of Gauss as he described his celebrated method of least squares:

> ... an orbit based only on the six necessary data may be liable to considerable errors. In order to diminish these as much as possible, and thus to reach the greatest precision obtainable, no other method will be given except to accumulate the greatest number of the most perfect observations, and to adjust the elements, not so as to satisfy this or that set of observations with absolute exactness, but so as to agree with all in the best possible manner.

Jeremy B. Tatum

Bibliography

Dubyago, A.D. (1961) *The Determination of Orbits*. New York: MacMillan Co.
Herget, P. (1948) *The Computation of Orbits*. Cincinnati: Herget.
Roy, A.E. (1988) *Orbital Motion*. Bristol: Hilger.
Williams, K.P. (1934) *The Calculation of the Orbits of Asteroids and Comets*. Bloomington: Principia Press.

Cross references

Celestial mechanics
Chaotic dynamics in the solar system
Coordinate systems
Eccentricity
Obliquity
Planetary dynamical astronomy

ORBITAL COMMENSURABILITY AND RESONANCE

Commensurability is a mathematical condition that defines a simple relationship between whole numbers that have a common measure or divisor (from Latin roots: susceptibility to measurement in common). Thus, for example, the numbers 6 and 9 are said to be commensurable because they are both divisible by 3. The smallest whole integers that cannot be divided in this way are the prime numbers 3, 5, 7, 11, 13 . . . A second definition, more appropriate here, refers to a relationship of proportionality of integer values. In planetary terms the most important commensurabilities are in the ratios between the beat frequencies of planetary pairs. The best-known commensurabilities are the 5:2 ratio between the orbital periods of Jupiter and Saturn, and the roughly 2:1 ratio between the periods of Neptune and Uranus. Neither ratio is really precise (see Commensurability; Resonance in the solar system).

In a review of the Titius–Bode law (q.v.), Nieto (1972) was able to show that logarithmically the planets conformed rather closely but not perfectly to a straight line (mass against distance). There is a degree of order throughout the series which can hardly be shrugged off (King-Hele, 1975).

A new approach to planetary orbital relationships by Stacey (1963) proposed, instead of a distance relationship, one based on periods defined in terms of the synodic conjunctions between adjacent planetary pairs on a line oriented to the solar system's barycenter (q.v.).

Since the time of Newton it had been recognized that in approaching the conjunction of any planetary pair, the angular velocities of both bodies will be modified as they pass. For example, in the passage of a slower body by a faster one, the latter will be accelerated as it approaches the former, and vice versa as it departs. No lasting change in the orbital velocity of either body is involved. However, Stacey pointed out that the vector of gravitational attraction between a given planet and the Sun would be briefly augmented by that of the second planet, so that orbital radius as well as time would be involved. Stacey emphasized that the orbital period of any planet, defined in relationship to the 'fixed' stars, although an important observational quantity, is meaningless in dynamic terms, whereas the periods of the conjunctions document useful parameters in celestial mechanics. Inasmuch as the gravitational attraction between the adjacent planetary pair is enhanced during their passage, it may be assumed that the angular momentum of each is temporarily modified. Mass remains constant but both velocity and radius must change.

The periods of barycentric synodic conjunctions can easily be calculated as beat frequencies (BF), using the Helmholtz formula:

$$(P_o \times P_i)(P_o - P_i)^{-1} = BF$$

or, more simply:

$$P_{oi}^{-1} = P_i^{-1} - P_o^{-1}$$

where P_o is the orbital period of an 'outer' planet (i.e. a slower moving body) and P_i is that of an 'inner' or faster moving one. Each pair's BF may be designated as its 'lap period' by analogy with laps in an athletic arena.

Inasmuch as Saturn ($P = 29.4577$ years, tropical) and Jupiter ($P = 11.86223$ years) together possess 86% of the angular momen-

Table O1 Examples of prominent beat frequencies and related cycles of the planets and Sun

Years	1[a] Period Lap	2[b] 69.507	3 139.015	4 178.733	5 208.52	6 317.749	7 1668.18	8 6672.73	9 93,418.7
0.317386	EMeL	219	438	–	657	1 001	5 258	21 024	
0.396054	VMeL	–	–	–	–	–	4 212	16 848	
0.648909	JVL	–	–	–	–	–	–	10 283	
1.269545	4EMeL	–	–	–	–	–	1 314	5 256	
1.597876	EVL	–	–	–	–	–	1 044	4 176	
3.748725	In/OutBF	–	–	–	–	–	445	1 780	
5.560608	JEVL	–	–	–	–	–	300	1 200	
9.885526	Ja/p	–	–	–	–	–	–	675	
11.121217	SSC	–	–	–	–	–	150	600	8 400
12.783008	NJL	–	–	–	–	–	–	522	7 308
13.81518	UJL	–	–	–	–	23	–	483	6 762
17.376902	E/M:Ap/T	4	8	–	12	–	96	384	5 376
19.859317	SJL	–	7	9	–	16	84	336	4 704
27.80304	SSC spec	–	5	–	–	–	60	240	2 880
34.7538	EVMeRes	2	4	–	6	–	48	192	2 688
45.3920	USL	–	–	–	–	7	–	147	2 058
59.57795	SJTriad	—	—	3	—	—	28	112	1 568
69.5076	UJVL	1	2	–	3	–	24	96	1 344
139.0152	SJ/EVL		1	–	–	–	12	48	672
178.73387	SOP			1	–	–	–	(37⅓)	(522⅔)
208.5228	SE-1				1	–	8	32	448
230.0941	SE-2					–	–	29	406
278.0304	14SJ/EVL					–	6	24	336
297.8897	15SJ/EVL					–	–	–	(313.600)
317.749	16SJ/USL					1	–	21	294
417.045	21SJ/SE-1						4	16	224
476.6255	SJ/TLC						–	14	196
556.0609	SQ-1						3	12	168
893.6692	45SJ/Ecc						—	—	—
953.2472	48SJ/US						–	–	98
1 668.1825	SQ-3						1	4	56
3 336.365	SQ-6							2	24
4 627.220	OPRes							–	–
5 004.547	12×417							–	–
6 672.730	SQ-12							1	14
19 000.31	EMP-1								–
23 354.56	EMP-2								4
40 036.38	E:Ecl								

[a] Column 1: period, expressed in years, and identified by initials of principal lap (beat frequency) or related cycle, e.g. EMeL = Earth–Mercury Lap; In/Out BF = beat frequency of mean inner and mean outer planets; Ja/p = Jupiter aphelion/perihelion; SSC = mean sunspot cycle (±5 yr); E/M: ap/T = Earth/Moon beat frequency of apsides and 18.03-year tide (Saros) cycle; SSC spec = spectral peak in Schove's 2600-year sunspot proxies; SOP = Sun's barycentric inertial orbital progression; SE-1/2 = Solar emission cycle (based on ^{14}C flux in tree rings); SE-1 is also UJL–EVL resonance; SE-2 is also NJL–EVL; SJ:TLC = triad length cycle of SJL; SQ-1 through -12 = solar quadrature arrangements of outer planets, where each is a multiple of SJL and EVL; SQ-6 includes also 9 × 370.7, the NJL–inner planets (46.3384) resonance; SQ-12 includes also 147 USL and 144 × 46.3384; SJ/Ecc = SJ orbital eccentricity (=417 + 476 cycles); EMP-1/2 = Earth/Moon orbital precession terms; E:EcL = Earth's ecliptic tilt cycle (last three, with 93 418.22-year cycle, approximate the 'Milankovitch insolation parameters').

[b] Columns 2 through 9: prominent resonance intervals corresponding to the periods in column 1, given only as prime integers; other relationshps exist, such as exact halves, thirds, etc., but are omitted for simplicity; also omitted, for space reasons, is the almost perfect fundamental tone of 1 121 018.22 years, which would embrace also the all-planet Sun dynamic cycle of the SOP (178.73385 years × 6272).

Precision: note that the periods, in some cases indicated to six orders of magnitude, may exceed the limits of astronomic observations, but are calculated using the method of multiple resonance–commensurability construction. For reference convenience the periods are generally rounded off, but this action is avoided in calculations.

tum of the entire solar system, it is appropriate to use their beat frequency: 19.8593168 years (SJL) as a primary lap period. This unique rhythm has been labeled the 'pulse of the solar system' (Fairbridge and Sanders, 1987). It is found to occupy a key position in any idealized harmonic series constructed from planetary motions.

Beat frequencies for different planetary pairs were calculated and discussed by Mörth and Schlamminger (1979). They reported also a comprehensive inner: outer planet beat frequency of 3.75 years Neptune–Uranus–Saturn–Jupiter: Ceres–Mars–Earth–Venus, or (NUSJ: CMEV), now recalculated at 3.7487 years. A table for the periods, beat frequencies, angular motions and their principal commensurabilities was presented in Fairbridge and Sanders (1987, p. 452); small corrections have now been added (Table O1). Multiple resonances are identified, although their ultimate origin has long been controversial (see, e.g. Molchanov, 1968, 1969; Dermott, 1969; Gingerich, 1969; Garfinkel, 1970).

The SJL, because of its overwhelming angular momentum, cannot be temporally displaced by the action of perturbations, in any one synodic interval, more than ±0.9 years. Its observed length varies systematically over an 'SJL triad cycle' of 59.578 years, and over three triads (178.7338 years); and further, it is traced over alternating cycles of 417.0456 years (21 SJL) and 476.6236 years (24 SJL); a combination of these two (45 SJL) creates a 893.66925 years period, the 'great inequality' recognized already by Laplace.

The Sun's counterclockwise inertial motion around the system barycenter (Figure O9), as anticipated by Newton, follows a distinctive orbital geometry (Jose, 1965; Fairbridge and Sanders, 1987; Charvátová, 1995). This orbit reaches a maximum radius with respect

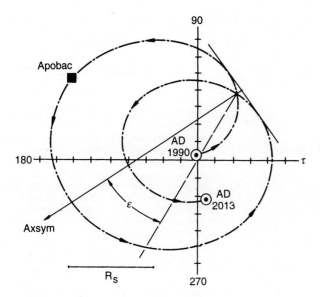

Figure O9 Example of a contemporary epitrochoid in the Sun's orbit of the barycenter (SOB), corresponding to a class E geometry, a pattern remarkable for its extreme asymmetry. The orbit is measured from a peribac in AD 1990 that comes close to the barycenter and briefly the Sun's motion is 'backward' with respect to that center of mass, the axis of spin being nearly coincident with orbital axis but in the opposite direction (Shirley, 1988). The last two such events, at 178-year intervals in the past, were marked on Earth by evidence of violent solar disturbances, such as extremes of warmth and cold. In the geometry of this orbit, note that the angle E (between axsym and barycentric projection) is very large; in category A orbits, it is always < 10°. Also the apobac is grossly displaced from its normal position close to the axsym, and the length of the cycle, AD 1990 to 2013, i.e. 23 years is more than 3 years longer than the usual SJL. Scale is indicated by R_s, the Sun's radius (c. 700 000 km).

to the barycenter at $> 1.5 \times 10^6$ km, equivalent to about 1% of an astronomical unit.

During the triad cycle the Sun's inertial orbit around the system barycenter executes a succession of three epitrochoids or cardioids in large and small loops that oscillate from a symmetric plan in the pattern of a three-leafed clover ('trefoil') to a chaotic arrangement (Charvátová and Střeštik, 1991; Charvátová, 1995). This geometry also undergoes a retrograde motion, a (clockwise) regression such that the axes of symmetry ('axsym') of the Sun's orbital loops shift systematically, due to Sun's tilt and the planetary torques set up by its oblate figure, to return to a close approximation (in celestial longitude) of their starting point in 178.73 years (the Sun's orbital progression, SOP); it is often called the 'Jose cycle' after P.D. Jose (1965), who first calculated its dynamics and plotted it graphically. Due to the Saturn–Jupiter 2:5 commensurability the 178.73-year SOP cycle corresponds approximately to six orbits of Saturn to 15 orbits of Jupiter, but precisely to 9 × SJL.

The small loops, noted above, develop a somewhat comparable periodicity, reflecting the aphelion/perihelion drift of the SJL, at 18 × 9.885526 years with a resonance between the Jupiter period (15 × 11.86 years) and the mean sunspot period (SSC, 16 × 11.12 years), and creating a 177.93476-year periodicity, sometimes known as the 'King-Hele cycle'.

When the SOP cycle is decomposed into three triads of nine SJLs, this arrangement leads to a phase jump in the normal sequence of alternating epitrochoids and cardioids; in most of the present millennium the SOP sequence has been e–c–e–e–c–e–c–e–c, but it drifts over longer periods. In the time domain, the defined epoch of the start of an SOP cycle (for practical purposes) can be set at the closest approach of the heliocenter to the barycenter ('peribac'). Again, for practical purposes, the Sun's orbit of the barycenter (SOB cycle), that averages about 39.859 years over the long term, can also be set into the same framework (peribac to peribac), but it should be understood that the successive loops are variable in dimensions, so that although the 'apobac' position (maximum heliocentric barycentric distance, which may exceed 1.5×10^6 km), may depart as much as 6 years from the mean epoch. In terms of planetary dynamics it has been recognized (Charvátová, 1995) that the SJL hemicycle (Saturn's conjunctions and oppositions with Jupiter) of 9.9296 years defines the mean periods of both large and small loops. A triple loop of 29.7889 years links the series and creates (× 7) a 208.52 years solar activity cycle that is important for terrestrial climate series.

Eight categories have been set up to identify the patterns of each orbit which ranges from highly symmetrical ones (class A to D) to various lop-sided shapes (classes E to H). Each orbit has been plotted from peribac, over a span of several millennia, an exercise that shows that no two successive orbits are ever identical (Fairbridge and Sanders, 1987). A given triad may be A–E–A, for example, an arrangement that occurred repeatedly in the same position within the 178.73-year SOP cycles of the Earth's 'Little Ice Age' (14th to 19th century). The sharp alternation in symmetry apparently triggers violent swings in solar emissions, the effect on the Earth climates fluctuating from 'unusual' cold to 'unusual' warmth (Figure O9).

Mörth and Schlamminger (1979) recognized that the SOP 178.73-year period would develop what they called a 'Jupiter–Saturn resonance drift' of 2681.008 years (15 SOP; 135 SJL) that is also commensurable with a number of other beat frequencies (e.g. 194 × 13.8196-year UJL; 59 × 45.392-year USL; 9 × 297.8897-year SJL/EVL; 3 × 893.6692-year 'Great inequality'). This is typical of the interlocking of commensurable planetary periodicities that can be expanded, as shown in Table O1. During recent millennia the epoch of maximum-length SJL (aphelion) was −1136.3 and minimum −4334.3 (−1 is 2 BC); the overall cycle is 6396 years, marking a return to the same celestial longitude. The next minimum (closest approach to the Sun) will be in AD 2061.7.

Approximately one-half of an SJ resonance drift is illustrated in Figure O10, which shows Sun–barycenter radii over seven SOP cycles, which are stacked to show the gradual evolution of the SOB cycle through time. The pentagonal stars, indicating apobac positions, correspond in number (9) to the SJL cycles, but only roughly to their midpoints (S–J oppositions). Another important resonance interval is the SJL perihelion/orbital precession (144 SJL = 2859.74 years) corresponding to 16 × 178.7 years SOP, and 6 × 476.624 years, the triad length cycle mentioned above.

The net axsym retrograde motion of the Sun's inertial orbit completes an approximately 360° realignment in 6672.73036 years (336 SJL). This value is commensurable with most planetary periods and thus represents a valuable 'fundamental tone' for studies of the last 10 000 years.

For longer-term studies, a 93 418.225-year period is useful. It corresponds closely to the calculation (by Milankovitch, 1941) for the Earth's eccentricity (ellipticity) cycle. In geological projections a fundamental tone of 1 121 018.7 years appears to be significant.

In astronomy the 93 418-year period corresponds to 8400 returns of the SSC (11.1212 years), 14 of the 6672.6-year alignments, and to 4 × 23 354.5 years, the E/M precession. It is also 47 × 178.73 years SOP.

Harmonic series

In music, harmony is an ear-pleasing sequence of notes as opposed to a repetitive rhythm ('aliquot' parts or 'beat'). In mathematics a harmonic series is one where the reciprocals are in arithmetic progression, e.g. 1, ½, ¼, ⅛, etc . . . A combination of such series constitutes a chord in music and its lowest frequency (longest wavelength) is the fundamental tone. This is the function $y = f(x)$ in a Fourier series, which constitutes the basis for harmonic analysis.

In the solar system all planetary periodicities may be represented as harmonics or fractions of resonance intervals within fundamental tones; the larger the number of planets involved, the lower the frequency, i.e. the longer the wavelength of that tone. All the giant planets' beat frequencies (plus most of the inner planets) divide evenly at 6672.73036 years (Table O1), except the NUL, while, as mentioned above, there is an ultimate fundamental tone for almost every combination at 1 121 018.7 years (SJL at 19.8593 × 56 448; USL at 45.392 × 24 696; NUL at 171.4096 years × 6540). In an expanded iteration of the Uranus–Jupiter terms, Milani and Nobili (1984) determined that the mutual perihelion return was approximately 1.1 million years. Although the very long-term resonances are very closely established by multiple resonance expansions, the short-term lap rates display remarkable variability; thus SJL = 19.859 ± 0.9 years, USL = 45.392 ± 3.75 years, and NUL =

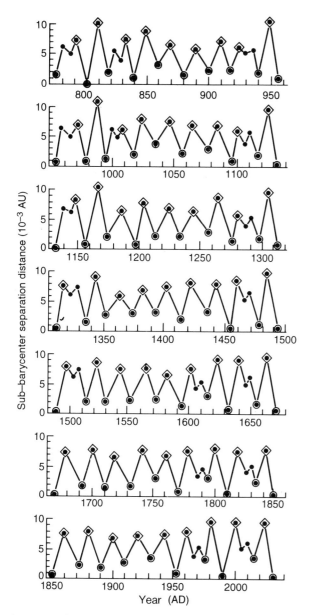

Figure O10 Plot of the Sun-barycenter separation distance through the time interval AD 770 to 2030 (Fairbridge and Sanders, 1987). This marks extremes of the Sun's inertial orbital radius, ranging from peribac, nearly zero, to apobac $> 10 \times 10^{-3}$ AU ($> 1.5 \times 10^6$ km). The mean period of this orbital circuit is 19.8593 years, corresponding to the synodic lap period of Saturn and Jupiter (SJL). The rows are arranged in 178-year intervals (the Solar Orbital Precession, SOP), corresponding to 9 SJL and 16 sunspot cycles (SSC). Stars mark apobacs and dots the peribacs. Intermediate fluctuations reflect phase interferences, which drift over centuries. (After Fairbridge and Sanders, 1987, p. 453.)

171.409 ± 1.6 years. Precise conjunctions are thus exceedingly rare, but in the case of slow-moving planets, there is only a small change in celestial longitude in the course of 1 year.

What is particularly interesting about the long harmonic series is that each 'step' in the 'ladder' represents dynamic reinforcement by various combinations of planetary laps. A major series is as follows: 17.3769 – 34.7538 – 69.5076 – 139.0152 – 278.0304 – 556.0608 – 1112.12 – 2224.24 – 4448.48 years ... The 17.3769-year period represents an E/M tidal beat frequency; the 139.01-year is an SJL/EVL resonance; the 556.06-year is an SJL/SSC resonance; the 4448.48-year period is the full quadrature cycle of the outer planets

(Stacey, 1963), and corresponds exactly to 400 SSC. Although the principal framework is provided by the major planets, there is synergism provided by the inner planets. The 278.03-year period corresponds precisely to 174 × EVL and to 702 × VMeL. Its subharmonics at 139.015, etc., represent reinforcements by various dynamic combinations (Figure O11).

Lunar cycles

The (Earth/Moon) pair constitutes the largest perturbing satellite relationship in the solar system, affecting both the Sun and the giant planets (Pettersson, 1912, 1930; Sanders, 1995). In Table O1 the Earth/Moon pair is treated as a single gravitational unit, and over extended periods the mean values are correct. However, when treated on the shorter terms (< 1000 years) there are very numerous lunar terms (Wood, 1985).

Most prominent of the lunar cycles is the 18.6134 (± 0.044) year nodal cycle that corresponds to a 360° clockwise rotation (regression) of the lunar nodes caused by the Sun's perturbation of the lunar orbit (that plane being inclined at a little over 5° to the ecliptic). It controls the inclination of the lunar orbit to the equator which varies from 18.3° to 28.6° (10.3°) during the hemicycle, 9.317 years. This inclination involves an approximately 1100 km latitude shift in the Moon's zenith positions over the Earth's surface which is particularly felt in the oceanic tides, sea level, atmospheric pressure, climate, etc. (Currie, 1987, 1995).

The axis of the Earth's rotation is also subject to several disturbing 'wobbles', such as the pole tide (1.1318 years), and Chandler wobble (1.1855 years), etc. An 18.604-year nutation period corresponds to a resonance of the Moon's synodic month and its anomalistic month. This period is present also in the spectra of the geomagnetic field which is believed to result from the stress on the core–mantle discontinuity. It is likewise seen in variations in the terrestrial aurora (q.v.) that also reflect the incoming energy from the Sun (Siscoe, 1980). Historical records of eight centuries of auroras in Scandinavia contain this 18.6-year periodicity (Ekholm and Arrhenius, 1898).

A second long-term period affecting the Moon is the prograde (counterclockwise) rotation of the line of apsides (see Apsis, apsides), an 8.849-year cycle. When this line is close to syzygy with the Sun and Moon, the ocean tidal range may be appreciably enhanced (up to 40%). The last maximum enhancement was in AD 1912.00 (Wood, 1985). This marks also a 186.04-year resonance cycle (21 × 8.849; 10 × 18.604).

The Moon's perigee-syzygy cycles are also of great significance in terrestrial oceanic tides (Wood 1985). The basic ('Saros') cycle is 18.03 years, with 31.008-year, 93.024-year (3 × 31; 5 × 18.6 resonance), 111.212-year, 222.4-year and 558.402-year resonance peaks.

Intersystemic resonances are found between the harmonic sequences of the E/M periodicities and of the solar system in general, for example 297.885 years (19.859 × 15; 18.61 × 16) and 893.6692 years (SJL × 45; NJL × 70; 18.61 × 48; 8.849 × 101). A 'Gabriel cycle' (discovered in the 1920s by a Jesuit priest, Père Gabriel) of 744.42 years, embraces all the major lunar periodicities.

Milankovitch orbital cycles

The long-term Earth–Moon orbital periods are well known on account of their role in the Milankovitch theory of insolation (q.v.) that is claimed to control the Earth's glacial/interglacial cycles (Milankovitch, 1941; Imbrie and Imbrie, 1979). Its history is reviewed in the biographical entries on Croll (q.v.) and Milankovitch (q.v.), and further in Berger et al. (1984) and Berger, Loutre and Laskar (1992).

The Earth's orbital eccentricity around the Sun, with a period of the order of 100 000 years, is small, but its seasonal effects and other factors can cause variations up to 3% in insolation for the most extreme ellipticity during several million years. Second, there is an obliquity variable with a period of about 40 000 years that intensifies summer–winter contrasts in both hemispheres. Third, is the precession of the equinoxes, a cycle that affects the longitude of the Earth's perihelion and thus the season that a given hemisphere is closest to

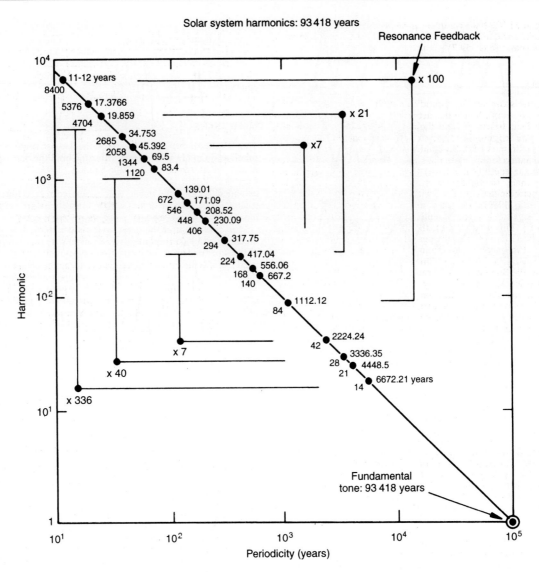

Figure O11 Graphical representation on triple log–log scales of the planetary–solar fundamental tone 93 418.7 years, its harmonics and resonances. Its 14th harmonic (6672.2 years) corresponds to an important quadrature symmetry of the outer planets around the Sun. Note the feedback potentials of the numerous resonance 'knots' in the network of synergetic relationships. Their dynamic affect on the Sun's radiation is suggested by the coincident harmonics of the mean sunspot cycle (11.1212 years). The 208.52-year period corresponds to the principal spectral peak of the ^{14}C flux on Earth (inverse measure of solar activity, from > 9000-year tree ring series). Higher frequencies, not shown on this illustration, relate mainly to the inner planets (Table O1).

the Sun; opposite hemispheres are out of phase, but the Earth's geographic asymmetry (continentality/oceanicity) causes the northern hemisphere to play the leading role. At the present epoch in time, perihelion is about January 3, so that the position of northern winter is anomalously closer to the Sun. This gyroscopic precession is clockwise and, perihelion to perihelion, possesses two peak terms, about 19 000 and 23 000 years. On Table O1 the latter are identified as EMP-1 and -2; the first, 19 000.3 years, is a simple fraction of the postulated 1 121 018.7-year fundamental (× 59 000) and the second, if taken as 23 354.5 years, would be ¼ × 93 418.2 and then is similar (× 48 000). The amplitude of the precession cycle varies greatly, sometimes in phase with the eccentricity period, with minimum to minimum creating an important 412 000 to 413 000-year cycle. Its last (intermediate) maximum was about 9000 years BP, and we are now approaching the next (intermediate) minimum.

For the gyroscopic precession, Berger offered six terms that are based on Earth–Moon orbital motions. These range from 18 976 years (nodes × 1020; apsides × 2144; also the 17.3769-year perigee–syzygy tide/apsides beat frequency × 1092; and 558.12-year PPSP tide × 34), to 23 716 years (approximately nodal period × 1274; apsides period × 2680). All appear to be locked into commensurabilities that appear to be ultimately linked into the entire planetary system.

The next obliquity minimum will be about 6000 years in the future, and the next eccentricity minimum will be in about 23 000 years. A long-term combination of such minima (about 56 500 years AP, 'after present') should herald the Earth's next major ice-age cycle. Reviewing all of Berger's orbital parameters, Stothers (1987) found that most of the precessional terms can be expressed in pairs that show beat frequencies in the 40 000 and 100 000-year groups and therefore they cannot be independent periods.

In contrast to the gyroscopic motions there is also a 'general or axial precession' period which is calculated by Berger to be 25 694 years. It is a counterclockwise motion of the orbit under the gravitational pull of the outer planets. Berger's value for the general precession is also exactly in resonance with important planetary and lunar periodicities (25 694 years = 2010 × Neptune–Jupiter Lap; 1425 × 18.03-year lunar perigee-syzygy tide period). Derived from it is a tilt cycle of 17 129 years (Sanders, 1995) and its fifth return resonating at 51 388 (two complete precessions). Finally there is the Earth's nodal cycle around the invariable pole which corresponds to 7 × 51 388 years, thus 359 716 years.

Solar cycles

The principal solar activity periodicity is the sunspot cycle (SSC: 11.1212 ± 6.0 years). While this activity is clearly forced by internal convective heat transport, and though its periodicity is commonly assumed to be triggered by 'a mysterious chronometer' deep within the Sun, its extreme variability combined with commensurability with external dynamic forcing potentials (i.e. planetary beat frequencies) makes an endogenetic (convective) triggering improbable. In support of an exogenetic timing device is the evidence of well-established planetary periodicities contained within the spectra of monthly mean spots (Verma, 1986), and they are also seen in the luminosity spectrum from satellite measurements (Shirley, Sperber, and Fairbridge, 1990). This latter record, though short, shows distinct peaks at EMeL and JVL, as well as the period of Mercury and Earth–Moon alignments. Prolonged sunspot minima relate in part to the 178.73 year SOP cycle (Fairbridge and Shirley, 1987).

Although the Sun's mean spin rate, seen from the Earth is 27.275 days, in a sidereal sense it is only 25.377 days. (0.06947855 year). Inasmuch as the spin rate and long-term mean SSC appear to be dynamically related, a small adjustment to 0.0695076 year (a difference of 0.000029 year) would make it commensurable with that of the mean sunspot cycle (×160 = 11.12 years), and likewise with an Earth/Moon beat frequency (1/250 × 17.3769) and with many planetary series (e.g. × 1000 = 69.507 years (UJVL) and 3000 = 208.52 years (the solar emission-controled ^{14}C flux rate maximum in tree rings.

An analysis of the last two solar orbit/barycenter cycles (c.357.5 years) in terms of sunspot cycle length shows statistical support for a c.178.73 year repeat period of an extremely short SSC mode of 9–10 years during high activity intervals (Fairbridge and Hameed, 1983), and a longer mode (avg. 12.266 years) during low activity. Extended to the roughly 2600-year length of the historical data series on sunspots and their auroral proxies (Schove, 1983), an overall consistency is shown with the 178 year period. This record has recently been confirmed by signal processing (Jelbring 1995), and the spectra disclose established planetary periodicities (e.g. 69.5 years: see Table O1). Charvátová and Střeštik (1991) indicate that there is a tendency for the activity peaks to match the crossover points in the solar epitrochoid orbits, points marked by rapid changes in the angular momentum (see Fig. O9). In any case there are nearly always 16 SSC for every 9 SJL cycles over at least the last two millennia. However, a true commensurability requires 25 SSC per 14 SJL, which generates a 278.03-year cycle, which as mentioned earlier is the SJL–E/M–Venus resonance period. Assuming these relationships are maintained makes it possible to estimate 100 800 SSC in 1.121018 Ma.

Closely associated with the approximately 11.12-year sunspot cycle is the ~ 22.24-year 'Hale cycle' (q.v.) or double-sunspot cycle that reflects the Sun's magnetic field reversal as well as that of the leading spots.

An interesting feature of the Sun's photosphere, sometimes confused with the Hale cycle, is the so-called 'solar jerk' acceleration (a sudden motion observed on the photosphere). The technical term employed for the rate of acceleration is the 'jerk'. The jerk for the dynamical motion of the Sun has been calculated. The jerks (Figure O12) appear to reflect abrupt changes in accelerations of the Sun with respect to the barycenter, which occur (positive and negative) in phase with the aphelion – perihelion cycle of Jupiter, 9.8855265 × 18, and the 177.9394-year King-Hele cycles. Note that although no two of the jerk patterns is identical, nevertheless the repeat period at about 178 years is remarkably similar. The Hale cycle and SSC are at times in phase with these accelerations of the heliocenter with respect to the center of mass (as shown by Sanders, 1995), e.g. around 1700–1850 and 1920 to 2000, but out of phase about 1850 to 1920.

Besides solar radiation in the electromagnetic spectrum, there are also emissions of subatomic particles or 'corpuscular radiation', associated particularly with solar flares (q.v.). These eruptions are highly episodic and nonlinear, but they have a mean periodicity, as calculated by Bai and Sturrock (1987), of 0.41587 years. A longer flare cycle (~ 13.34 years; 32 × 0.41704) appears to relate to accelerations of planetary torques (Landscheidt, 1988), as well as one about

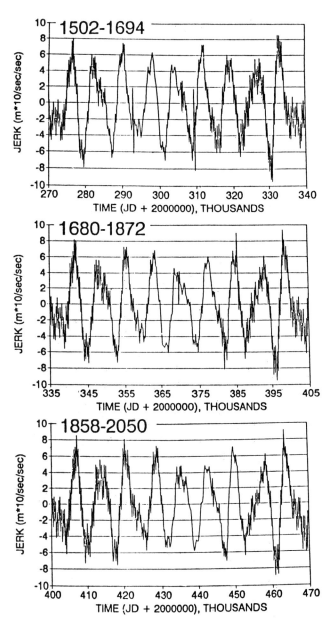

Figure O12 Examples of solar jerk patterns over three successive c. 178-year cycles (believed to be the 177.93948-year King-Hele cycles), AD 1502–2050, calculated in Julian days (courtesy J.E. Sanders). Each is provided with a 5000-d overlap.

83.4 years known as the 'Gleisberg cycle' (Schove, 1983), which may tie in with the 417-year UJV – E/M period (83.4 × 5).

Another period associated with solar radiation is the quasibiennial oscillation (QBO) of about 2.2 years which has been long-established in terrestrial weather series. It has also been measured in solar wind velocities, solar radio flux, UV emission and solar neutrino flux (Kulčar and Leftus, 1988). Linkage with planet Earth is suggested by the QBO period appearing in geomagnetic data, cosmic ray flux, ozone, tree rings, equatorial stratospheric wind reversals and velocity, El Niño (ENSO) incidence and other meteorological parameters. It appears to be related to the solar emission period identified in ^{14}C flux, the 208.52-year cycle ($\frac{1}{96}$ = 2.17211; also ⅛ of 17.3774 years). Thus, if correct, this key solar activity is systematically locked in to both outer and inner planet resonance 'ladders'. The interrelationships between the QBO and the sunspot cycle with the terrestrial climate have been brought out by Labitzke and Van Loon (1989).

Terrestrial proxy tests

More than 80 years ago, Sir Arthur Schuster (1911) analyzed planetary patterns in sunspot activity, while Otto Pettersson (1912) suggested that the Moon's long-term cycles might also be interacting with solar activity cycles, which in turn affected the climate of planet Earth. Independent tests of suggested solar – planetary links using terrestrial proxies now expand the time domain to the order of 10 000 years. Rigorous precision is provided by dendrochronology, which links it to solar activity provided by isotope geochemistry, principally through ^{14}C and ^{13}C analyses (Stuiver et al., 1991). Over a millennial time scale they can both be cross-checked with reference to the auroral record (Siscoe, 1980; Schove, 1983; Jelbring, 1995). The particulate and electromagnetic radiation associated with solar cycles, together with the solar flares (see Corpuscular radiation) is modulated by the intensity of the Earth's magnetic field, and partly controlled by the terrestrial nutation caused by the lunar cycles (notably 18.6 years), so that both solar and lunar cycles are present in long-term auroral records as well as climate series (Currie, 1987).

In the time frame of the last 10 000 years, long-term cycles of ^{14}C flux which provides a proxy for solar activity have been identified by several authorities, using spectral analyses based on a wide variety of statistical techniques (Stuiver and Braziunas, 1989; Thomson, 1990; Damon and Sonett, 1991; Stuiver et al., 1991). Clusterings of spectral peaks are interrelated in various harmonic series, e.g. 69.5, 72.47, 99.82, 208.52 and 230.09 years.

Conclusions

- Through series of beat frequencies, based on paired planetary periods, the entire solar system is linked by commensurable ratios of simple whole integers.
- Small variables that exist in each orbital period, e.g. SJL (19.8593 ± 0.88 years), help to create secondary periodicities, including some that are irregular (e.g. 4627.147 years = 233 SJL; also c. 102 USL and 27 NUL).
- In this way, simple harmonics, with resonance reinforcement from multiple planetary motions at commensurable intervals, appear to lock together the entire framework of celestial mechanics. Multiple resonance points represent knots or junctions in this network or 'scaffolding', which appears to become more rigid as its components increase.
- Fundamental 'tones' may be selected at progressively lower frequencies by an empirical testing of established beat frequencies (mainly of the giant planets), e.g. 93 418.22 years and 1 121 018.7 years.
- High frequency cycles are established using mainly the inner planet beat frequencies, but the Sun's behavior (sunspots, flares and spin rate) reflects a combination of all planets. The Sun's sidereal spin rate (approximately 0.0695076 years) reflects not only the giant planets (e.g. × 6000 = 417.045), but also the Earth–Moon periodicities, this natural satellite relationship being by far the greatest in the solar system.
- The mean sunspot cycle (11.1212 ± 6.0 years) likewise reflects the giant planets, but its secondary fluctuations show spectra that correspond to the inner planet beat frequencies.

Rhodes W. Fairbridge

Bibliography

Bai, T. and Sturrock, P.A. (1987) The 152-day periodicity of the solar flare occurrence rate. *Nature*, **327**, 601–3.
Berger, A.L. (1992) Astronomical theory of paleoclimates and the last glacial-interglacial cycle. *Quaternary Sci. Rev.*, **11**, 571–81.
Berger, A., Imbrie, J., Hays, J. et al. (eds) (1984) *Milankovitch and Climate*. Dordrecht: Reidel, 895 pp.
Berger, A., Loutre, M.F., and Laskar, J. (1992) Stability of the astronomical frequencies over the Earth's history for paleoclimate studies. *Science*, **255**, 560–6.
Charvátová, I., (1995) Solar–terrestrial and climatic variability during the last millennia. *J. Coastal Res.*, Sp. Issue, 343–354.
Charvátová, I. and Střeštík, J. (1991) Solar variability as a manifestation of the Sun's motion. *J. Atmos. Terrest. Phys*, **53**, 1019–25.
Currie, R.G. (1987) Examples and implications of 18.6- and 11-yr terms in world weather records, in *Climate: History, Periodicity, and Predictability* (eds M.R. Rampino, et al.). New York: Van Nostrand Reinhold, pp. 378–403.
Currie, R.G. (1995) Variance contribution of M_n and S_c signals to Nile River data over a 30–8 year bandwidth. *J. Coastal Res.*, sp. issue 17, 29–38.
Damon, P. and Sonett, C.P. (1991) Solar and terrestrial components of the ^{14}C variance spectrum, in *The Sun in Time* (eds C.P. Sonett, M.S. Giampapa and M.S. Matthews), Tucson: University of Arizona Press.
Dermott, S.F. (1969) The origin of commensurabilities in the solar system, III. *Mon. Not. Roy. Astron. Soc.*, **142**, 143–9.
Ekholm, N. and Arrhenius, S.A. (1898) Ueber den Einfluss des Mondes auf die Polarlichter und Gewitter. *Kongl. Svenska Vetenskaps Akademiens Handlingar* (Stockholm), **31**(2), 77–156.
Fairbridge, R.W., and Hameed, S. (1983) Phase coherence of solar cycle minima over two 178-year periods. *Astron. J.*, **88**(6), 867–9.
Fairbridge, R.W. and Sanders, J.E. (1987) The Sun's orbit, A.D. 750–2050: basis for new perspectives on planetary dynamics and Earth–Moon linkage, in *Climate: History, Periodicity, and Predictability* (eds M.R. Rampino et al.). New York: Van Nostrand Reinhold, pp. 446–71 (bibliography, 475–541).
Fairbridge, R.W. and Shirley, J.H. (1987) Prolonged minima and the 179-yr cycle of the solar inertial motion. *Solar Phys.*, **110**, 191–220.
Garfinkel, B. (1970) On the ideal resonance problem, in *Periodic Orbits, Stability and Resonances*. (ed. G.E.O. Giacaglia), Dordrecht: D. Reidel, pp. 474–81.
Gingerich, O. (1969) Kepler and the resonant structure of the solar system. *Icarus*, **11**, 111–3.
Imbrie, J. and Imbrie, K.P. (1979) *Ice Ages: Solving the Mystery*. Short Hills, NJ: Enslow Publ., 224 pp.
Jelbring, H. (1995) Analysis of sunspot cycle phase variations – based on D. Justin Schove's proxy data. *J. Coastal Res.*, sp. issue 17, 363–70.
Jose, PD, (1965) Sun's motion and sunspots. *Astron. J.*, **70**, 193–200.
King-Hele, D.G. (1966) Prediction of the dates and intensities of the next two sunspot maxima. *Nature*, **209**, 285–286.
Kulčar, L. and Leftus, V. (1988) Quasi-biennial oscillations of the solar wind. *Bull. Astron. Inst. Czechoslovakia*, **39**, 372–8.
Labitzke, K. and Van Loon, H. (1989) Association between the 11-year solar cycle, the QBO, and the atmosphere, Part III: Aspects of the association. *J. Climate*, **2**, 554–65.
Landscheidt, T. (1988) Solar rotation, impulses of the torque in the Sun's motion, and climatic variation. *Climatic Change*, **12**, 265–295.
Milani, A. and Nobili, A.M. (1984) Resonance locking between Jupiter and Uranus. *Nature*, **310**, 753–5.
Milankovitch, M. (1941) Kanon der Erdbestrahlung und seine Anwendung auf das Eiszeitenproblem. Roy. Serv. Acad., Sp. Publ. 133, 633 pp. (Engl. transl. 1969 as *Canon of Insolation and the Ice-Age Problem*, by Israel Progr. Sci. Transl., US Dept. Commerce, Washington, 484 pp.)
Molchanov, A.M. (1968) The resonant structure of the solar system: the law of planetary distances. *Icarus*, **8**, 203–15.
Molchanov, A.M. (1969) The reality of resonances in the solar system. *Icarus*, **11**, 104–10.
Mörth, H.T. and Schlamminger, L. (1979) Planetary motion, sunspots and climate, in eds B.M. McCormac and T.A. Selliga, *Solar – Terrestrial Influences on Weather and Climate*. Dordrecht: D. Reidel, 193–207.
Nieto, M.M. (1972) *The Titius–Bode Law of Planetary Distances: its History and Theory*. New York: Pergamon Press.
Pettersson, O. (1912) The connection between hydrographical and meteorological phenomena. *Roy. Meteorol. Soc., Quart.*, **38**, 173–91.
Pettersson, O. (1930) The tidal force: *Geogr. Annaler*, **12**, 261–322.
Sanders, J.E. (1995) Astronomical forcing functions: from Hutton to Milankovitch and beyond. Northeastern Geology and Environmental Sciences, **17**(3), 306–47.
Schove, D.J. (1993) *Sunspot Cycles*. Benchmark Papers in Geology, V. 68. Stroudsburg: Hutchinson Ross, 393 pp.
Schuster, A. (1911). The influence of planets on the formation of sunspots. *Roy. Soc. London Proc.*, **85A**, 309–23.

Shirley, J.H. (1988) When the Sun goes backward: solar motion, volcanic activity, and climate, 1990–2000. *Cycles*, **39**, 70–6.

Shirley, J.H., Sperber, K.R. and Fairbridge, R.W. (1990). Sun's inertial motion and luminosity: *Solar Phys.*, **127**, 379–92.

Siscoe, G.L. (1980) Evidence in the auroral record for secular solar variability. *Rev. Geophys. Space Phys.*, **18**, 647–58.

Stacey, C.M. (1963) Cyclical measures: some tidal aspects concerning equinoctial years. *Ann. New York Acad. Sci.*, **105**(7), 421–60.

Stothers, R. (1987) Beat relationships between orbital periodicities in insolation theory. *J. Atmos. Sci.*, **44**(14), 1875–6.

Stuiver, M. and Braziunas, T.F. (1989) Atmospheric ^{14}C and century-scale solar oscillations: *Nature*, **338**, 405–8.

Stuiver, M., Braziunas, T.F., Becker, B. and Kromer, B. (1991) Climatic solar, oceanic and geomagnetic influences on late-glacial and Holocene atmospheric $^{14}C/^{12}C$ change. *Quaternary Res.*, **35**, 1–24.

Thomson, D.J. (1990) Time series analysis of Holocene climate data. *Phil. Trans. Roy. Soc. London*, **A330**, 601–16.

Verma, S.D. (1986) Influence of planetary motion and radial alignment of planets on Sun, in *Space Dynamics and Celestial Mechanics* (ed. K.B. Bhatnagar). Dordrecht The Netherlands: D. Reidel, 143–54.

Wisdom, J. (1987) Urey Prize Lecture: chaotic dynamics in the solar system. *Icarus*, **72**, 241–75.

Wood, F.J. (1985) *Tidal Dynamics, Coastal Flooding and Cycles of Gravitational Force*. Dordrecht: D. Reidel, 712 pp.

Acknowledgements

This review is based on a multidecadal research that began with a symposium on solar–terrestrial relations that I organized at the New York Academy of Sciences in 1961, from which emerged friendships with a wide range of specialists, including the (late) amateur astronomer Clyde Stacey, the (late) antiquarian physicist Derek Justin Schove and the geologist John Sanders. Subsequent members of this dedicated circle were James Shirley (California), Hans Jelbring (Stockholm) and above all John Perras (New York), all skilled in various aspects of planetary–solar sciences and signal processing. James Hansen most kindly provided logistical support at the Goddard Institute for Space Studies (NASA), New York City.

Cross references

Apsides, apsis
Barycenter
Commensurability
Milankovitch, Milutin, and Milankovitch theory
Orbit
Precession and nutation
Resonance in the solar system
Saros cycle
Solar activity
Solar flare
Solar motion
Titius–Bode law

P

PALEOMAGNETISM

If the Earth's magnetism originated in a dipole at the center of the planet with its axis pointing due south, the magnetic inclination I would be related to latitude λ by the equation

$$\tan I = 2 \tan \lambda \tag{P1}$$

and the magnetic declination (or azimuth) D would be zero everywhere. In the northern hemisphere λ is positive, as would be the inclination I: by convention $I > 0$ when the magnetic field vector points below the horizontal. A measurement of D then gives the direction to the 'virtual geomagnetic pole' (VGP), and a measurement of I gives the angular distance θ from the VGP, where

$$\theta = 90° - \lambda = \tan^{-1}(2/\tan I) \tag{P2}$$

Many rocks are permanently magnetized by the Earth's field at the time they form, and this magnetism may still be stable and measurable hundreds of millions of years later. A VGP for a past time can be found if the rocks can be dated, and if strata horizontal when formed are still horizontal now, or if the tilt can be found and allowed for. McElhinny (1973) and Cox and Hart (1986) give good introductions to the subject. Both describe how to avoid errors of many possible kinds.

The Earth's field is not in fact a dipole aligned with the rotation axis, but over periods of about 10^4 to 10^5 years it may be considered so, to a good approximation. That period is long enough to average out the secular variation. Fortunately, it is also a minute fraction of the 10^8 year timescale for a continent to move to a significantly different position on the globe. For tectonic studies one thus estimates a mean VGP (called a paleomagnetic pole) for a given rock formation at a given locality, from numerous samples laid down over a time long enough for secular variation to be unimportant. By doing this for rocks of nearly the same age on different parts of the same continent one can check for tectonic stability (and show that the Earth's radius has changed very little in the last 4×10^8 years). By investigating rocks of different ages one can trace the movements of the paleomagnetic pole and obtain a polar wander path for that continent. The difference between the polar wander paths of Europe and North America provided the first geophysical evidence for continental drift (Runcorn, 1956), though its statistical significance and geological reliability took some time to confirm.

People finding a polar wander path for the purpose of studying continental movements ignore geomagnetic polarity reversals (q.v.), in order to obtain curves which are smooth on a million-year timescale. People studying the reversal process itself, of course, do not, nor do they average out secular variation, as periods of a few thousand years are crucial for both phenomena.

J. F. Harper

Bibliography

Cox, A., and Hart, R.B. (1986) *Plate Tectonics*. Palo Alto: Blackwell.
McElhinny, M.W. (1973) *Palaeomagnetism and Plate Tectonics*. Cambridge: Cambridge University Press.
Runcorn, S.K. (1956) Paleomagnetic comparisons between Europe and North America. *Proc. Geol. Assoc. Can.*, **8**, 77–85.

Cross references

Geomagnetic polarity reversals and the geological record
Plate tectonics
Polarity reversals

PERIAPSE, PERIHELION, PERIGEE, PERIBAC

The most general of these terms is 'periapse'; this signifies the point in an orbit nearest to the focus of gravitational attraction. (The line of apsides is the long axis of an elliptical orbit; see Apsides, apsis). The reverse of this condition is the point of greatest distance from the focus, the 'apoapse' (see also Aphelion).

The perihelion is the position in the elliptical orbit of a planet, asteroid, periodic comet, or spacecraft around the Sun, where the object reaches its closest approach to the Sun. Also, in a non-periodic parabolic orbit of a comet, it is the position at the instant when the comet passes closest to the Sun. From the Greek prefix *peri-* (about or near) plus *helios* (the Sun); in other cases, plus *gea* (Earth), and so on. Due to the motion of the apsides (q.v.) the date of Earth's perihelion drifts through time. It is now about 2–4 January; in the 19th century it was in late December.

The perigee is an analogous parameter referred to the Earth; it is the position in the orbit of the Moon on an artificial satellite where that body reaches its closest approach to the Earth in any single orbit. This distance varies with time. The lunar perigee occurs at mean intervals of 27.55455 days (the anomalistic month). At infrequent intervals (over several thousand years) the Moon achieves its maximum velocity/minimum radius orbit. At this stage the distance (Moon center to Earth center) is reduced to 355 880 km (with parallax at its maximum 61′53″), its mean distance being about 372 000 km. This phase requires a near coincidence with syzygy (direct alignment of Earth–Moon–Sun, approximation of which is called 'proxigee' (Wood, 1986). The closest example in recent centuries occurred at 1912.0 AD.

Peribac (suffix abbreviation: barycenter; q.v.) is applied to the nearest approach of the Sun to the center of mass of the solar system (barycenter) in its inertial motion around that center (Fairbridge and Sanders, 1987). The mean period, peribac to peribac, is 19.859 years (which corresponds to the mean synodic lap of Saturn and Jupiter), but with considerable variation (± 4 years). Only rarely does the heliocenter at peribac coincide with the barycenter; often they are up to about 100 000 km apart. A recent peribac was 1990.3 AD; the next will be 2013.8.

Fergus J. Wood and Rhodes W. Fairbridge

Bibliography

Brown, E.W. (1896) *Introductory Treatise on the Lunar Theory*. Cambridge University Press. Reprint, New York: Dover Publ. 292 pp.
Moulton, F.R. (1914) *An Introduction to Celestial Mechanics*. New York: Macmillan (see pp. 277–365).
Fairbridge, R.W. and Sanders, J.E. (1987). The Sun's orbit, A.D. 750–2050: basis for new perspectives on planetary dynamics and Earth–Moon linkage, in *Climate: History, Periodicity, and Predictability* (eds M.R. Rampino *et al.*). New York: Van Nostrand Reinhold, p. 446–71 (bibliography, pp. 475–541).
Wood, F.J. (1986). *Tidal Dynamics, Coastal Flooding and Cycles of Gravitational Force*. Dordrecht: Reidel, 712 pp.

PHOBOS

Phobos is the larger of the two Martian moons. As one of only three natural satellites of the terrestrial planets, it has been particularly well studied and provides a basis for many theories of processes on other small satellites and asteroids.

Investigating phobos

Phobos was discovered in 1877 by Asaph Hall at the US Naval Observatory. Because it is small and orbits close to Mars, Earth-based observations such as reflectance spectra are difficult to obtain. The first spacecraft data were images of Phobos against the disk of Mars returned by Mariner 7 in 1969. The first detailed images were returned from Mariner 9 in orbit about Mars, mostly during the period (late 1971) when Mars' surface was obscured by dust storms and the spacecraft cameras could be pointed elsewhere. Most data on Phobos come from the Viking orbiters which returned several hundred images, some of which showed details as small as 4 m across. In 1988 the Soviet Union launched two spacecraft to study Phobos. One returned low-resolution images and spectral data, but because of the loss of both spacecraft the hoped-for detailed data were not obtained.

Basic properties

The physical and orbital properties of Phobos are given in Table P1. The outstanding characteristics of Phobos are its irregular shape, heavily cratered surface and patterns of linear depressions. The irregular shape (Figure P1) is not surprising, because even very weak material could support large topography in the gravity field of about 0.5 cm s^{-2}, or 1/2000 that of the Earth. The heavily cratered surface was also expected because Phobos is subject to essentially the same population of impactors as Mars. The density of impact craters approaches that on parts of the Moon, and may be close to an equilibrium level. The morphology of the craters is also similar to those of lunar craters, and the size of the largest crater (10 km) is only slightly smaller than that thought likely to disrupt a satellite significantly. Morphologic evidence of ejecta blankets near the craters is lacking except for the largest crater, and may indicate that

Table P1 Characteristics of Phobos and its orbit

a (km)	9378.5
a (Mars radii)	2.76
Period (h)	7.65
Eccentricity	0.01515 (± 0.00004)
Inclination (deg)	1.068 (± 0.001)
Mass (g)	$1.08 \pm 0.01 \times 10^{19}$
Density (g cm^{-3})	1.90 ± 0.1
Approximate axes (km)	13.3, 11.1, 9.3
Mean radius (km)	11.1

the ejecta are spread very far in the low-gravity environment. The linear depressions or grooves (Figure P1) were not expected, and have so far have not been found on any other satellite. They form several patterns of roughly parallel members up to 10 km in length. They are usually less than 20 m in depth and 200 m in width, but the deepest have about 100 m of relief and are nearly 800 m wide. Their origin remains controversial, though most work suggests that they are linked to fracture displacement of portions of a loose regolith. While they are widest and deepest near the 10-km crater Stickney, their geometric pattern is closely tied to the shape of the satellite. The grooves appear to define planes that are roughly parallel to the intermediate axis of the satellite (which points essentially along the orbit). Thus if they are indeed traces of fractures, both the crater and the overall body shape or stress state may have influenced their formation. It has also been suggested that they are scars from impact ejecta, though this possibility has not been modeled in great detail.

The near surface of Phobos is inferred to be very loose fine-grained regolith, based on thermal inertias measured by the Viking orbiters. The inferred particle sizes and porosities are not greatly different from those of the lunar soil. The depth of some grooves and flat-bottomed craters suggests that the regolith may in places be over 100 m deep. This amount of debris is easily accounted for by ejecta from the visible population of impact craters on Phobos. Although some, perhaps most ejecta, would be initially lost because of the low escape velocity (about 8 m s^{-1}) nearly all should be reaccumulated quickly because they are trapped in orbits about Mars that are little different from Phobos'.

Composition

Among the surprising properties of Phobos is its relatively low density, about 1.9 g cm^{-3}. This density is difficult to associate with likely meteorite analogs based on albedo and spectrum. Phobos' albedo is low (about 6%). Its spectrum is similar to those of some carbonaceous meteorites and black chondrite meteorites. From the spectral comparison to carbonaceous objects one might expect a density of as low as 2.2–2.4 g cm^{-3}. It would require porosities of 20–25% if composed of carbonaceous material, and over 50% if made of the higher density chondritic material. These porosities might not be unreasonable, as the maximum internal pressure of Phobos is less than 1 bar. They would imply considerable fragmentation and minimal compaction by impacts. A low bulk density might also be fostered by the presence of ice. Although the presence of ice is difficult either to prove or to rule out, it appears that water ice could be preserved for geologic time below a few hundred meters of regolith. Indications of water loss from the Soviet Phobos 2 spacecraft data are still controversial.

Orbital characteristics

Phobos' orbit is noteworthy for its short period and rapid tidal evolution. The semimajor axis is decreasing due to tidal torques which extract energy from satellites in orbits with periods shorter than a synchronous one. The secular acceleration of Phobos' orbit has been measured and the results suggest that Phobos may impact Mars in less than 100 million years. It is not possible to extrapolate uniquely the orbit back because probable resonances and chaotic rotation would have drastically altered the rate of orbital evolution.

The close orbit also has interesting effects on the net acceleration at the surface of Phobos. The tidal and rotational effects at present are a substantial fraction of the acceleration due to the self-gravity of Phobos. Net surface gravity varies by nearly a factor of two over the satellite, and has changed substantially as Phobos has orbited closer to Mars.

Origin

The origin of the Martian satellites has not been resolved. The major difficulty is in reconciling the compositional suggestion of capture with the dynamical improbability of capture. If the satellites are carbonaceous material there is a strong presumption that they formed elsewhere and were subsequently captured into Mars orbit. Capture requires energy loss such as that by drag in a protoplanetary nebula. Such scenarios require capture to occur essentially as the nebula is collapsing, otherwise the orbit would decay to the planet's surface. The presence of Deimos, and the low inclinations of both orbits,

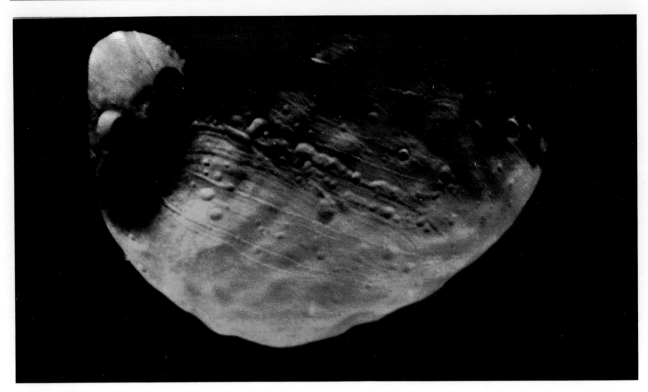

Figure P1 A view of Phobos from the Viking 1 spacecraft at a range of 1600 km. The large crater, Stickney, is about 10 km across; the sub-Mars point is near the bottom of the image. Note the series of grooves that range from thin straight lines to substantial troughs.

requires either two nearly simultaneous captures or complex fragmentation sequences. Because the compositions are so uncertain, there is not a compelling reason to require origin elsewhere as opposed to formation in orbit about Mars. This unresolved issue, so important for successful models of formation of satellites or of dispersal of asteroidal material, is a strong incentive for obtaining more compositional data on the Martian satellites.

Summary

Even after exploration by four spacecraft, many aspects of Phobos' properties and history remain unexplained. It is apparently composed of material very similar to Deimos, yet the two satellites have drastically different appearances. Because of the high resolution data, the Martian satellites have been used to predict properties of asteroids and even comet nuclei. With better data on other small solar system bodies, we may be able to tell which Martian satellite is more typical of small, irregularly shaped bodies. Better compositional information on the Martian satellites would allow much better testing of scenarios of satellite formation or capture.

Peter C. Thomas

Bibliography

Burns, J.A. (1992) Contradictory clues as to the origin of the Martian moons, in *Mars* (eds H.H. Kieffer, C.W. Snyder and M.S. Mathews). Tucson: University of Arizona Press.
Murchie, S.L., Britt, D.T., Head, J.W. *et al.* (1991) Color heterogeneity of the surface of Phobos: relationship to geologic features and comparison to meteorite analogs. *J. Geophys. Res.*, **96**, 5925–45.
Thomas, P. (1979) Surface features of Phobos and Deimos. *Icarus*, **40**, 394–405.
Veverka, J. and Burns, J. (1980) The moons of Mars. *Ann. Rev. Earth Planet. Sci.*, **8**, 527–8.

Cross references

Deimos
Mars
Satellite, natural

PHOBOS MISSION

The last planetary mission launched by the Soviet Union before its break-up was given the name Phobos (ФОБОС). It was the most ambitious unmanned space mission attempted by any nation up to that time, and was planned as the beginning of a concentrated long-term program of Mars exploration. The mission was under development for about 8 years, and its primary goal was the examination of the surface of the satellite Phobos from low orbit and on the ground.

The mission scenario (Sagdeef and Zakharov, 1989) was (1) to place a spacecraft into an elliptical, near-equatorial orbit around Mars, (2) to maneuver to an almost-circular orbit very close to the orbit of Phobos, and (3) to approach to within 50 to 100 m for several tens of minutes. During this time remote sensing observations of the surface would be made, and two small landing probes would descend to the surface to perform analyses of its physical properties.

In keeping with the new philosophy of *glasnost*, and in contrast to earlier Soviet practice, this mission was highly publicized in advance, and the onboard payload was prepared by scientific teams from 14 countries. The NASA Deep Space Network (DSN) participated in tracking and data acquisition. Two basically identical spacecraft carrying slightly different scientific payloads were launched – Phobos 1 on 7 July 1988 and Phobos 2 five days later. They were by far the largest craft ever used on a planetary mission, with a total mass of 6200 kg (of which 3600 kg was the propulsion system), and a scientific payload of about 500 kg. On 30 September an erroneous ground command was sent to Phobos 1, which switched off its orientation system and, as there was no program in the onboard computer to handle such an emergency, no further communication with the craft was possible.

Phobos 2 reached Mars on 29 January 1989 and was inserted into an orbit with period 79.2 h, periapsis altitude 865 km and eccentricity 0.90. A series of five orbit changes in the next 8 weeks brought the spacecraft into an orbit between 200 and 600 km from Phobos, but on 29 March, before the final orbit trims could be performed, communication was lost because of an onboard computer malfunction.

Between them, the two orbiters carried instrumentation for 25 experiments. Nearly all of these obtained some useful data but, because of the premature termination, much less than had been hoped for.

Eight instruments were designed to make measurements of radiation on the way to Mars, observing the Sun in visible, ultraviolet, x-ray, and gamma-ray wavelengths; monitoring solar oscillations and flares; and detecting cosmic rays, both solar and galactic. The relation between soft x-ray bursts and solar flares was investigated, and it was found that the latter are not always preceded by the former (Likin et al., 1991). In two experiments the temporal variations in intensity and energy spectrum of cosmic gamma-ray bursts were investigated at high and low energies. The duration of the bursts varied from less than 0.1 s to more than 100 s, and their time/energy structures were very complicated and variable (Mitrifanov et al., 1991, Barat et al., 1991). Photometric measurements of spectral irradiance from the whole solar disk were made in ultraviolet, visual and infrared wavelengths to investigate power spectra of the 5-min oscillations (Bruns and Shumoko, 1991). Measurements of electron and ion fluxes from 30 KeV to a few MeV elucidated the changes in the nature of solar wind patterns in a period of decreasing solar activity (McKenna-Lawlor et al., 1991). Recording of the flux, spectra and elemental composition of energetic particles from solar flares showed a large variability in the Fe/O ratio among events (Marsden et al. 1991).

A major effort was mounted to elucidate the nature of the interaction of the planet with the solar wind with two magnetometers and five plasma detectors. Because of the duration of the measurements and the relatively low altitude of periapsis, much new information was obtained. The altitude of the plasma bow shock at the subsolar point was found to be between 0.45 and 0.75 Mars radii, and the 'obstacle' that diverts the solar wind (which the Soviets call the planetopause) was placed at 0.12 radii from the data of one instrument and at 0.28 radii by another. In the region between these two boundaries (the planetosphere), unexpectedly large electron plasma densities were detected – up to 700 cm^{-3} (Riedler et al. 1991, Grard et al., 1991). A new process of atmospheric loss was discovered – a flux of planetary ions (mainly O$^+$) picked up by the solar wind and streaming down the magnetotail (Verigin et al., 1991). The detection of plasma and magnetic field effects along the orbits of Phobos and Deimos, even far from the satellites themselves, suggested the presence of rings of gas or dust emanating from them (Dubinin et al., 1991).

On the crucial question of the existence of an intrinsic planetary magnetic field, the data were still inconclusive. The low altitude of the planetopause proves that the solar wind interaction is mainly with the ionosphere, so that the magnetic moment (if any) is very small, and the discovery that the polarity of the magnetotail reverses with the reversals of interplanetary field suggests an induced character of the field at least beyond 2.8 Mars radii (Riedler et al., 1991). On the other hand, when the spacecraft was in circular orbit, it detected 12- and 24-hour periods in the power spectrum of the field – strong evidence for a corotating dipole (Möhlmann et al., 1991).

Three instruments obtained data on the Martian atmosphere. A spectrometer operating in the solar occultation mode obtained vertical profiles of ozone and aerosols in the upper atmosphere and detected clouds of water ice at about 50 km altitude (Blamont et al., 1991). An IR spectrometer with imaging capability observed the spatial variation of H$_2$O and CO on a 10-km scale (Combes et al., 1991), and another IR instrument studied the sizes of the particles in clouds and dust (Moroz et al., 1991)

The imaging spectrometer also measured the column density of CO$_2$, obtaining 36 000 spectra that determined the altitude of the surface over most of the geological formations except the polar caps (Bibring et al., 1991). Another radiometer obtained photographs of the surface in the visible and near IR that determine the temperature and thermal inertia (Murray et al., 1991). A gamma-ray spectrometer investigated the composition of the surface material, measuring the abundance of ten elements between O and U (Surkov et al., 1989).

Relatively little information was obtained about Phobos because the spacecraft failed when that phase of the mission had barely started. The TV camera obtained 37 images of Phobos, which had a greater range of wavelengths and (in some cases) higher resolution than those taken by the Viking orbiters. Significant spatial variations in color were detected, and the data provided an improved model of the shape and volume of the satellite (Avanesov et al., 1991; Hauber et al., 1991, Duxbury, 1991). An instrument that combined an UV-visible spectrometer and an IR radiometer detected three distinct types of surfaces (Ksanfomality et al., 1991). From the Doppler analysis of the radio signal, the gravitational perturbation of the orbit provided a new theory of the motion of Phobos and a more precise mass (1.08 × 10^{16} kg), which suggests that it is either somewhat porous or contains ice (Kolyuka et al., 1991, Avanesov et al., 1989).

The two Long-Term Small Landers (DAS) each carried a TV camera, a spectrometer of alpha backscattering and x-ray fluorescence to analyze the elemental composition of the surface layer, a Sun-angle position sensor to detect the libration of Phobos, a seismometer and a radio system to provide precise data on its orbit. The Movable Small Lander on Phobos 2 (called 'Hopper' because it was capable of changing its position on the surface several times) carried a gravimeter to measure its free-fall acceleration, a magnetometer and instruments to analyze the elemental composition, temperature, electrical conductivity and mechanical properties of the surface layer. These landers were, of course, never deployed.

Conway W. Snyder

Bibliography

Most of the published information about the mission in English is contained in two sources. *Phobos 2 at Mars: The First Results* (*Nature*, **341**, No. 6243, 19 October 1989), contains 15 papers. *Planetary and Space Science Special Issue: Phobos-Mars Mission.* (**39**, No 1/2, January/February 1991) contains 41 papers presented to the colloquium convened by CNES in Paris, October 1989.

Avanesov, G.A., Bonev, B.I., Kempe, F. et al. (1989) Television observations of Phobos. *Nature*, **341**, 585–7.
Avanesov, G., Zhukov, B., Ziman, Ya. et al. (1991) Results of TV imaging of Phobos (experiment VSK-Fregat). *Planet. Space Sci.*, **39**, 281–96.
Barat, C., Atteia, J.L., Jourdain, E. et al. (1991) On the absorption features in cosmic gamma-ray burst spectra recorded by the LILAS experiment. *Planet. Space Sci.*, **39**, 67–71.
Bibring, J.-P., Erard, S., Gondet, B. et al. (1991) Topography of the Martian tropical regions with ISM. *Planet. Space Sci.*, **39**, 225–36.
Blamont, J.E., Chassefiere, E., Goutail, J.P. et al. (1991) Vertical profiles of dust and ozone in the Martian atmosphere deduced from solar occultation measurements. *Planet. Space Sci.*, **39**, 175–188.
Bruns, A.V. and Shumoko, S.M. (1991) Short term variability of the power spectrum of 5-min oscillations of the sun. *Planet. Space Sci.*, **39**, 41–6.
Combes, M., Cara, C., Drossart, P. et al. (1991) Martian atmosphere studies from the ISM experiment. *Planet. Space Sci.*, **39**, 189–98.
Dubinin, E.M., Pissarenko, N.F., Barabash, S.V. et al. (1991) Plasma and magnetic field effects associated with Phobos and Deimos tori. *Planet. Space Sci.*, **39**, 113–22.
Duxbury, T.C. (1991) An analytic model for the Phobos surface. *Planet. Space Sci.*, **39**, 355–76.
Grard, R., Nairn, C., Pedersen, A. et al. (1991) Plasma and waves around Mars. *Planet. Space Sci.*, **39**, 89–98.
Hauber, E., Regner, P., Schmidt, K. et al. (1991) Color decorrelation for the Phobos mission camera experiment. *Planet. Space Sci.*, **39**, 297–310.
Kolyuka, Yu.F., Kudryavtsev, S.M., Tarasov, V.P. et al. (1991) International project Phobos. Experiment 'Celestial Mechanics'. *Planet. Space Sci.*, **39**, 349–54.
Ksanfomality, L., Murchie, S., Britt, D. et al. (1991) Phobos: spectrophotometry between 0.3 and 0.6 μm and IR-radiometry. *Planet. Space Sci.*, **39**, 311–26.
Likin, O.B., Golubkova, M.A., Dyachkov, A.V. et al. (1991) Phobos 2 mission: Observation of the solar soft X-ray radiation. *Planet. Space Sci.*, **39**, 15–22.
Marsden, R.G., Wenzel, K.-P., Afonin, V.V. et al. (1991) Energetic particle composition measurements from Phobos 2: results of the LET experiment. *Planet. Space Sci.*, **39**, 57–66.
McKenna-Lawlor, S.M.P., Afonin, V.V., Gringauz, K.I. et al. (1991) Interplanetary variability recorded by the SLED instrument aboard the Phobos spacecraft during that period of solar cycle 22 characterized by a transition from solar minimum- to solar maximum-dominated conditions. *Planet. Space Sci.*, **39**, 47–56.
Mitrofanov, I., Chernenko, A., Dolidze, V. et al. (1991) New insights on cosmic gamma-ray bursts from the APEX experiment. *Planet. Space Sci.*, **39**, 23–38.

Möhlmann, D., Riedler, W., Rustenbach, J. *et al.* (1991) The question of an internal Martian magnetic field. *Planet. Space Sci.*, **39**, 83–8.

Moroz, V.I., Petrova, E.V., Ksanfomality, L.V. *et al.* (1991) Characteristics of aerosol phenomena in Martian atmosphere from KRFM experiment data. *Planet. Space Sci.*, **39**, 199–208.

Murray, B., Naraeva, M.K., Selivanov, A.S. *et al.* (1991) Preliminary assessment of Termoskan observations of Mars. *Planet. Space Sci.*, **39**, 237–66.

Riedler, W., Schwingenschuh, K., Lichtenegger, H. *et al.* (1991) Interaction of the solar wind with the planet Mars: Phobos 2 magnetic field observations. *Planet. Space Sci.*, **39**, 75–82.

Sagdeev, R.Z. and Zakharov, A.V. (1989) Brief history of the Phobos mission. *Nature*, **342**, 581–5.

Surkov, Yu. A., Barsukov, V.L. Moskaleva, L.P. *et al.* (1989) Determination of the elemental composition of Martian rocks from Phobos 2. *Nature*, **341**, 595–8.

Verigin, M.I., Shutte, N.M., Galeev, A.A. *et al.* (1991) Ions of planetary origin in the Martian magnetosphere (Phobos 2/TAUS experiment) *Planet. Space Sci.*, **39**, 131–7.

Cross references

Mars
Soviet Mars missions

PHOTOCLINOMETRY

Photoclinometry is a technique used for the extraction of quantitative topography from spacecraft images. Unlike active ranging techniques (e.g. radar), which measure topographic heights, photoclinometric techniques estimate topographic slopes. A photoclinometrically derived topographic slope is determined by comparing the observed intensity scattered from a given location with the theoretical intensity which would be scattered by a horizontal surface. If the observed intensity is larger than the theoretical intensity the surface must be tilted towards the Sun (the illuminating source in spacecraft images). Conversely, if the observed intensity is smaller than the theoretical intensity the surface must be tilted away from the Sun. Qualitatively, this is the same process used by the human visual system. When we 'see' mountains, valleys or craters in a picture of the Moon's surface, we instinctively convert the different observed shades to topographic slopes.

The advantage of photoclinometry over our eyes is the ability to quantify the topographic results. In order to derive a quantitative slope value from an observed brightness, the manner in which the surface scatters incident solar radiation must be known. This relationship, called the 'photometric function' of the surface, relates the scattered intensity (units of $Wm^{-2} sr^{-1}$) from a surface to the incident solar flux (units of Wm^{-2}), the surface reflectance (sometimes called the 'albedo') and the lighting and viewing geometries (e.g. McEwen, 1991).

In practice, modern photoclinometric techniques use computer algorithms to analyze digital spacecraft images. Photoclinometry algorithms are either one-dimensional, deriving slope values for a line of points (and thus generating a topographic profile), or two-dimensional, deriving slope values simultaneously for a grid of points (generating a topographic map). Although one-dimensional techniques have historically seen much wider use, the popularity of two-dimensional techniques, which are quite computationally demanding, is likely to increase as computer capabilities improve. For an overview of two-dimensional techniques, see Horn and Brooks (1990).

Photoclinometric techniques suffer from three fundamental weaknesses (for a detailed discussion, see Jankowski and Squyres, 1991). First, although a single intensity value may determine a single slope magnitude, it does not uniquely determine a single slope orientation. For this reason, one dimensional techniques can only construct profiles for cases in which the topographic slope orientations are known a priori. Two-dimensional techniques attempt to solve for a 'best-fit' slope orientation along with the slope magnitude. Second, photoclinometry is generally incapable of distinguishing between a varying topographic slope and a varying surface albedo. As a result, human intervention is required to avoid clear albedo variations. Third, photoclinometry is generally insensitive to regional slopes.

These limitations must be considered when applying photoclinometry or interpreting its results.

David G. Jankowski

Bibliography

Horn, B.K.P. and Brooks, M.J. (eds) (1990) *Shape From Shading*. Cambridge, Mass.: MIT Press.

Jankowski, D.G. and Squyres, S.W. (1991) Sources of error in planetary photoclinometry. *J. Geophys. Res.*, **96**, 20907–22.

McEwen, A.S. (1991) Photometric functions for photoclinometry and other applications. *Icarus*, **92**, 298–311.

Cross reference

Cartography

PHOTOGRAMMETRY, RADARGRAMMETRY AND PLANETARY TOPOGRAPHIC MAPPING

Definitions

The word 'photogrammetry' is derived from three Greek roots – *photo*, *gram*, and *metron* – and means to measure graphically by means of light (photography). Conventional photogrammetry can be defined as the art, science, or technology of obtaining reliable measurements for making maps by means of photographs from frame cameras. Modern photogrammetry may be defined as the art, science or technology of obtaining geometric information for making maps and for other interpretation by means of imaging data from a variety of cameras such as frame, vidicon, or scanner cameras. At present, state-of-the-art photogrammetry involves the use of imaging data that are mostly in digital form acquired by cameras as well as by a broad spectrum of sensors, including optics, radar and interferometers. The data, which may be processed automatically, are used for making maps and for scientific and industrial applications.

Radargrammetry, according to Leberl (1990, p. 1), is the technology of extracting geometric information from radar images. Radargrammetry can be considered an extension or a branch of photogrammetry: photogrammetric measurements from radar imagery can be made with techniques and orientation procedures similar to those used for conventional photogrammetry. The mensuration principle, however, is different: conventional photogrammetry uses photographs taken by cameras, and radargrammetry uses radar imaging data collected by radar systems. As shown in Figure P2, a camera records angles (θ_1, or θ_2), and radar measures distance (R, or R_1). In other words, normal photographic imagery has a point-perspective geometry, whereas radar imagery has a range-and-time geometry (r_1, and Δt_1, or r_2 and Δt_2). The taking of photographs involves the collinearity of the center of the camera lens, image point, and ground point (c_1, p_1 and P, or c_2, P_2 and P). Recovery of three-dimensional information about a ground point (P) is made by locating conjugate image points on a pair of stereophotographs (p_1 on photo 1 and p_2 on photo 2). Radar, however, forms an image of a ground point through the use of range (distance – r_1 or r_2) and squint angle (azimuth – α_1 or α_2). The radar-projection lines are circles that are concentric to the flight paths. A ground point in object space can be located by the intersection of the two circular loci from images of the flight paths (Blackwell, 1981).

Radar collects terrain information by its own illumination and is therefore an invaluable all-weather, 24-hour mapping tool. It is perhaps the only practical method to map cloud-covered regions of Earth, the planet Venus and Saturn's satellite Titan (these last two bodies are entirely covered by clouds). On the other hand, due to problems of its unique geometry, layover and shadow, radargrammetry is more difficult than photogrammetry. For instance, to avoid shadow problems it is preferable to collect radar images from the same side; but same-side stereoradar reduces the length of base between two conjugate radar stations and weakens the base-to-range ratio.

The quality of an aerial camera and the base-to-height ratio of overlapped photographs are primary factors in stereophotogram-

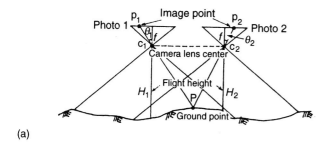

(a)

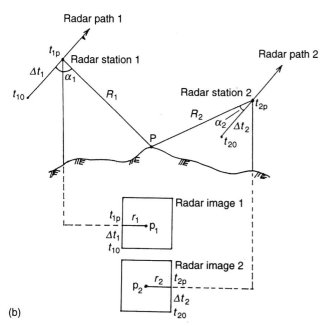

(b)

Figure P2 Diagrams comparing geometrics of photogrammetry and radargrammetry. (a) Photography with a point-perspective geometry combined with photogrammetry with a pair of overlapped photographs. (b) Radar imaging geometry (range and time) combined with radargrammetry.

metry, whereas radar wavelength and frequency, look angle and base-to-range ratio of overlapped radar imaging data are primary factors in stereoradargrammetry.

Planetary topographic mapping portrays surface topography and provides a quantitative representation of the landforms and relief of planetary bodies. Topographic data lead to a better understanding of the geologic processes and histories of planets and their satellites. Topographic information is also used for planning missions of planetary exploration. Photogrammetry and radargrammetry are the principal methods used for mapping the topography of planetary bodies. Other methods include radar and laser altimetry, both of which have high vertical resolution; however, because they have wide footprints, their horizontal resolutions are low, and a photogrammetric control network is often needed for locating their measurements.

Development and applications of photogrammetry

The history of photogrammetry began as early as the 15th century in Italy and Germany, when mechanical devices and optical lenses were constructed for producing perspective projections and stereoscopic drawings. Aerial photography was first used for military purposes by the French in the 18th century. In 1862 the American Union Army employed balloons for photography to obtain intelligence information. The US took its first aerial photographs in 1906. George

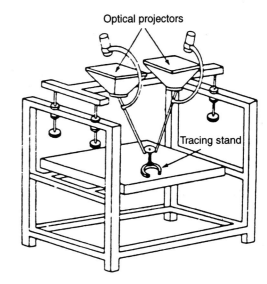

Figure P3 Structure of a Kelsh plotter with two projectors placed for a pair of conjugate photographic diapositives. The tracing stand is at the intersection of the rays from two conjugate points; it is used to draw contour lines on the map sheet.

Eastman and Sherman Fairchild were leaders in the early development of American photogrammetry: Eastman established a photographic factory and Fairchild developed aerial cameras (Thompson and Gruner, 1980).

Modern photogrammetry developed in three stages. From 1900 to 1960 maps were produced by conventional photogrammetry equipment such as the stereoplanigraph and the Multiplex, Bulplex and Kelsh (Figure P3) plotters. From 1960 to 1990, with analytical photogrammetry, maps were compiled on the AP/C, AP-II and the series of AS-11 analytical stereoplotters (Figure P4); secondary or auxiliary ground-control networks were established by analytical triangulation on computers. The analytical methodologies and equipment for aerial photographs make it possible to solve difficult photogrammetric problems. Spacecraft cameras, for instance, have a field of view that is narrower than that of cameras used in aerial photography, which results in weak geometry of the spacecraft images. Since 1990 systems for digital (soft copy) photogrammetry have been developed. Instead of producing photographs (hard copy), the systems produce imaging data in digital form, either by scanning photographs or by collecting data directly from a digital camera. In addition to analytical stereoplotters, photogrammetric instruments now include computers or digital workstations equipped with stereoscopic image display devices. With the development of automatic image-correlation algorithms, the first stage of the era of automatic digital mapping has been reached. The flexibility of soft-copy photogrammetry will be invaluable as scanning cameras become increasingly common in the near future. This type of camera was considered for the Mars Observer mission (q.v.), the Mars 94 mission and possibly on the lunar SCOUT mission. The inherent flexibility of digital techniques has also been vital to the development in recent years of soft-copy radargrammetry systems. Hard-copy radargrammetry is possible in principle, but no operational systems exist at present.

For planetary topographic mapping, photogrammetry can be used to derive such products as control networks, contour maps, orthophoto maps and digital terrain models. The topography of the entire Martian surface has been systematically mapped at various scales by using orbital images from the Viking mission (q.v.) (Wu and Doyle, 1990; Davies et al., 1991). Some topographic maps of other bodies, including the Moon and Uranus' satellite Miranda, have also been compiled by stereophotogrammetric methods. Also, with radargrammetric techniques and imaging radar data from the Magellan mission (q.v.), the topography of Venus is being mapped (See also Cartography; Photoclinometry).

Sherman S.C. Wu

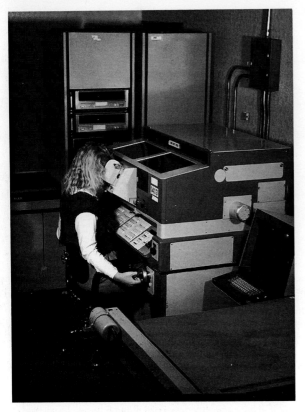

Figure P4 An AS-11 analytical stereoplotter. Main components are the viewer with which the operator views the stereomodel, a tracing table (right) for drawing contour lines, and computer peripherals (left).

Bibliography

Blackwell, B.H. (1981) Real-time math model for SAR imagery. Final Technical Report, RADC-TR-301, 70 pp.

Davies, M.E., Batson, R., and Wu, S.S.C., 1991, Geodesy and cartography, in *Mars* (eds H.H. Kieffer, B.M. Jakowsky, C.W. Snyder and M.S. Matthew (Space Science Series). Tucson and London: University of Arizona Press, pp. 321–42.

Leberl, F.W. (1990) *Radargrammetric image processing*: Norwood, Mass.: Artech House, 595 pp.

Thompson, M.M. and Gruner, H. (eds) (1980) Foundations of photogrammetry, in (eds C.C. Slama, C. Theurer and S.W. Henriksen) *Manual of Photogrammetry*, 4th edn. Falls Church, Va.: American Society of Photogrammetry, pp. 1–36.

US Geological Survey (1989) Topographic maps of the western, eastern equatorial, and polar regions of Mars: US Geological Survey Misc. Inv. Ser. Map I-2030, 3 sheets, scale 1:15 000 000.

Wu, S.S.C. and Doyle, F.J. (1990) Topographic mapping, in *Planetary Mapping* (eds R. Greeley and R. Batson) (Cambridge Planetary Science Series). Cambridge, New York: Cambridge University Press, pp. 169–207.

Cross references

Cartography
Magellan mission

PHOTOMETRY

Photometry is the measurement of light flux. It is one of the oldest and most fundamental techniques of astronomy. It has been used to compare the color and brightness of astronomical objects, and to search for temporal variations. Photometry measures radiation in the visible portion of the electromagnetic spectrum and is thus distinguished from spectroscopy and radiometry, which are more inclusive; spectroscopy is applied over a wide range of visible and non-visible wavelengths, while radiometry most often refers to studies in the infrared region. Bessell (1992) discusses technical aspects of astronomical photometric systems.

Photometric observations are typically compared with models of the target objects that incorporate physical parameters (Buratti, 1989). The agreement or lack of agreement between the observation and the model helps constrain the model parameters and leads to a more accurate characterization of the body. Some of the physical parameters include albedo (q.v.), the porosity of the reflecting surface and the degree of topographic roughness of the body.

The models must predict the brightness variations of the object as a function of solar phase angle (the angle between the Sun, the observer and the object). The brightness of objects varies with the phase angle, as with the phases of the Moon. As the phase angle nears zero (as at the full Moon), there is a nonlinear surge in brightness termed the opposition effect (q.v.), which is partly due to the disappearance of shadows between surface particles, and partly due to coherent backscatter. Observations taken at solar phase angles of about 12° or less are useful in characterizing the state of compaction of the surface materials.

A second effect related to the solar phase angle is due to the topography (or macroscopic roughness) of the object. At phase angles of greater than 40° the shadows of topographic highs may produce significant variations in the measured light flux, enabling a characterization of the degree of topographic relief present on the body.

Photometric techniques have made important contributions to the study of asteroids (see Asteroid: lightcurve; Asteroid: photometry). Lightcurves are graphic representations of the variations in brightness of an object over some period of time. Asteroid lightcurves have been used to determine the shapes, the rotation rates and the rotational axes and albedos of these bodies (Harris and Lupishko, 1989).

Prolonged series of photometric observations of planets, satellites or asteroids can be used to construct maps of albedo variations on the surfaces of these bodies. Over time the object will present essentially all aspects of its surface to the remote observer, and maps of brightness (albedo) may be derived as a function of the longitude of the body.

Charge-coupled devices (CCD; q.v.) have a very high quantum efficiency and are in the process of revolutionizing astronomical photometry. An important application of CCD photometry is the Spacewatch telescope at Kitt Peak, Arizona (see Spacewatch). This device is employed to scan the sky in search of small objects that might be missed by visual observers or by photographic surveys. It has already detected a considerable number of previously unknown near-Earth objects.

Spectrophotometry (q.v.) is a powerful tool for characterizing the composition and physical state of a planetary atmosphere or surface. As the name implies, a spectrum of the reflected light from the target object is obtained. Absorption features in the spectrum may be associated with particular atmospheric constituents or surface rocks (Hapke, 1981; see also Reflectance spectroscopy).

James H. Shirley

Bibliography

Bessell, M.S. (1992) Magnitude scales and photometric systems, in *The Astronomy and Astrophysics Encyclopedia* (ed. S. P. Maran) New York: Van Nostrand Reinhold.

Buratti, B.J. (1989) Planetary satellites, natural, in *The Encyclopedia of Physical Science and Technology*, 2nd edn (ed. R.A. Myers). San Diego: Academic Press.

Harris, A.W. and Lupishko, D.F. (1989) Photometric lightcurve observations and reduction techniques, in *Asteroids II*, Binzel, (eds R.P.T. Gehrels and M.S. Matthews). Tucson: University of Arizona Press.

Hapke, B.W. (1981) Bidirectional reflectance spectroscopy 1: theory. *J. Geophys. Res.*, **86**, 3039–54.

Cross references

Asteroid: lightcurve
Asteroid: photometry

Charge-coupled device
Opposition effect
Radiometry
Reflectance spectroscopy
Spacewatch
Spectrophotometry

PIAZZI, GIUSEPPE (1746–1826)

Piazzi, an Italian monk and astronomer, is known for the discovery of the first minor planet, Ceres (q.v.), in the Asteroid belt between the orbits of Mars and Jupiter, on the first day of the 19th century, 1 January 1801. He was engaged in star mapping and came across this object by accident, while he was director of the observatories at Palermo and Naples, and also professor of mathematics and astronomy at Palermo. Kopal (1979) described how, after the original sighting, Piazzi fell ill and could not confirm it. C.F. Gauss, who had recently invented the 'least squares' technique, applied it to Piazzi's data to extract the orbital elements and predict its future course. This permitted Olbers (q.v.) to rediscover Ceres on 1 January 1802, exactly 1 year later.

Piazzi's work in compiling a catalogue of stars (1814) and determining the proper motions for many stars played an important role in the recognition of the Titius–Bode law (q.v.) which predicted a planetary body in the position of the asteroid at 2.8 AU. Kepler had already noted that there seemed to be a gap at this spot.

Rhodes W. Fairbridge

Bibliography

Abetti G., 1974. Piazzi, Giuseppe. *Dict. Sci. Biogr.*, Vol. 10, pp. 591–3.
Kopal, Z. (1979). *The Realm of the Terrestrial Planets*. New York: John Wiley and Sons, 223 pp.
Royal Society *Catalogue of Scientific Papers*, **4**, p. 897.

Acknowledgements

Kindly reviewed by Dr H.J. Haubold (UN, Vienna)

PICKERING, WILLIAM HENRY (1858–1938)

W.H. Pickering, younger brother of the distinguished astrophysicist E.C. Pickering, was born in Boston, Massachusetts, on 15 February 1858. He was educated at the Massachusetts Institute of Technology, where his brother had been professor of physics. It was during this period that he went on an eclipse expedition to southern Colorado, and published his first research paper dealing with the polarization of sunlight.

From 1880 to 1887 he was an instructor in physics at the Institute of Technology, and was also associated with the Harvard College Observatory, where his brother was now Director. In 1891 he was concerned in setting up an observing station in Peru, and on Mount Chacocomam, at an altitude of 16 650 ft (5000 m), he set up what was then the highest meteorological station in the world. It was during this period that he made the photographic discovery of faint nebulosity covering much of the constellation of Orion.

In 1892 he began a long series of observations of Mars and drew features on the dark areas, proving that they were not seas; he believed them to be due to vegetation. He became associated with Percival Lowell and helped in the establishment of the Lowell Observatory at Flagstaff in Arizona, where he continued with his Martian studies. In 1900 he moved to Mandeville in Jamaica to set up a new observatory; the seeing conditions there were excellent and very suited to planetary work. At first the Jamaica station was connected with Harvard, but after Pickering's official retirement it became his private observatory.

In 1898 he discovered Phoebe, the eighth satellite of Saturn; this was the first satellite discovery to be made photographically. Later he suspected another faint satellite, moving round Saturn between the orbits of Titan and Hyperion; it was even given a name – Themis – but was never confirmed, and probably does not exist.

Like Lowell, Pickering was interested in the possibility of a planet moving beyond the orbit of Neptune. He made extensive calculations, changed several times, based on the motions of Neptune and Uranus. A search was instigated at Mount Wilson in 1919, by Milton Humason, for the position that Pickering had given. The result was negative, but after Pluto had been discovered in 1930, by Clyde Tombaugh, the planet's image was found on two of Humason's plates – once masked by a star and once by a flaw in the emulsion.

However, Pickering's main work was in connection with the Moon, and in 1903 he published a semi-popular book, as well as a photographic atlas which covered the whole lunar surface under several different conditions of illumination. Some of his ideas sound strange today; for example he believed in an appreciable lunar atmosphere, and thought that 'moving patches' inside some craters, such as Eratosthenes, were due to vegetation or even to swarms of insects or small animals. Yet his observational skill was never in doubt, and he made notable contributions to lunar and planetary studies. He remained in Mandeville for the rest of his life, and died there in 1938.

Patrick Moore

Bibliography

Hoyt, W. (1976) *Lowell and Mars*. Tucson: University of Arizona Press.
Hoyt, W. (1980) *Planets X and Pluto*. Tucson: University of Arizona Press.
Moore, P. (1988) *The Planet Neptune*. Chichester: Ellis Horwood.
Pickering, W.H. (1903a) *The Moon*. Harvard.
Pickering, W.H. (1903b) *A Photographic Atlas of the Moon*. Harvard.
Plotkin, H., 1974. Pickering, Edward Charles. *Dict. Sci. Biogr.*, Vol. 10, pp. 599–601.

PIONEER 10 AND 11 MISSIONS

The scientific descriptions of the outer solar system in the late 1960s and early 1970s were sparse and lacking in details, being limited to inferences made from ground-based observations made from Earth. There were many unknowns about deep interplanetary space beyond the orbit of Mars and about the very large and distant planets of the outer solar system. It was in the temper of these times that the National Aeronautics and Space Administration (NASA) decided to institute a low-cost exploratory mission to the outer planets. The Pioneer program that existed at the time included the Pioneer spacecraft (numbered 1 to 9) that made the first space flights toward the Moon and into interplanetary orbit about the Sun. As an extension of this program under the project management of the NASA Ames Research Center, two identical spacecraft, Pioneer 10 and 11, were designed and fabricated for the primary purpose of reaching Jupiter.

Mission objectives

The major objectives of the Pioneer 10 and 11 missions were:

- to investigate the interplanetary medium beyond the orbit of Mars;
- to study the asteroid belt and to assess possible hazards of the asteroid belt for missions to outer planets; and
- to measure the near environment of Jupiter *in situ*. An objective later added was:
- to measure the near environment of Saturn *in situ*.

Original mission

Pioneer 10 was launched at 8:49 p.m., EST on 2 March 1972 by an Atlas–Centaur launch vehicle from Cape Kennedy, Florida. The spacecraft weighed 258 kg and carried a payload of 11 instruments. The instruments weigh 33 km and consume 24 W of electric power. The magnetometer sensor is mounted on a long boom extending from

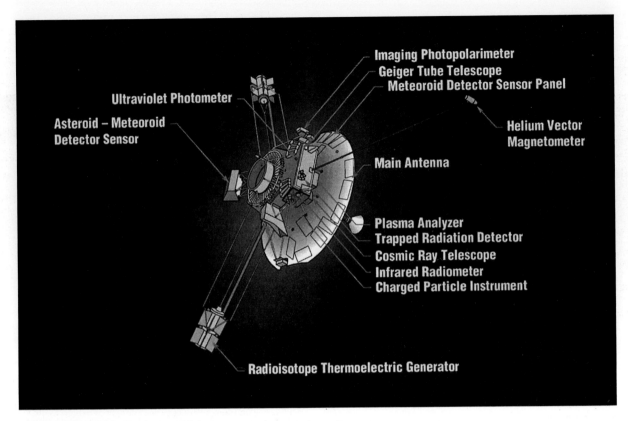

Figure P5 Diagram of Pioneer 10 spacecraft. (NASA/Ames Research Center.)

the scientific instrument compartment. The four radioisotope thermoelectric generators (RTGs) that furnish electric power are located at the end of two trusses extending from the body of the spacecraft. The diameter of the high-gain antenna reflector is 2.74 m. The spacecraft is spun about an axis parallel to the axis of the reflector at 5 revolutions per minute. Small thrusters (using hydrazine) located at the edge of the reflector control the orientation and rotation of the spacecraft providing in-flight velocity adjustments, alteration of the spin rate and changes in the direction of the reflector (so that it points toward the Earth and maintains maximum communication data rates). A diagram of Pioneer 10 is shown in Figure P5.

The 11 instruments and sponsoring organizations are:

- Helium Vector Magnetometer (Jet Propulsion Laboratory)
- Plasma Analyzer (Ames Research Center)
- Charged-Particle Instrument (University of Chicago)
- Geiger-Tube Telescope (University of Iowa)
- Cosmic-Ray Telescope (University of Maryland)
- Trapped Radiation Detector (University of California, San Diego)
- Ultraviolet Photometer (University of Southern California)
- Imaging Photopolarimeter (University of Arizona)
- Infrared Radiometer (California Institute of Technology)
- Asteroid–Meteoroid Detector (General Electric Company)
- Meteoroid Detector (Langley Research Center).

Pioneer 11 incorporated one additional instrument, a flux-gate magnetometer, in case the magnetic field encountered in a closer approach to Jupiter might exceed the range of the helium vector magnetometer. This magnetometer was supplied by the Goddard Space Flight Center.

Asteroid belt

In July 1972 Pioneer 10 became the first spacecraft to enter the asteroid belt. Since this belt is too thick to fly over without prohibitively expensive launch vehicles, all missions to the outer planets must fly through it. In February 1973 Pioneer 10 emerged having completed a 7-month passage through the belt without incident. The spacecraft carried two instruments to sense the particles encountered to determine the hazard posed by the asteroid belt. The Meteoroid Detector sensor panel directly registered the impact of particles that weighed as little as 10^{-9} g. The Asteroid–Meteoroid Detector consisted of an optical system to detect close passage of dust grains bigger than 0.1 mm by measuring the sunlight reflected from them. By the time Pioneer 10 reached Jupiter in early December 1973, although many impacts had been recorded, no apparent damage was incurred. The spatial density of small particles in the asteroid belt was found to be no greater than in the surrounding interplanetary space.

This passage by Pioneer 10 – later repeated by Pioneer 11 – demonstrated that future spacecraft (Voyager 1 and 2, Galileo and Ulysses) would be able to pass through the asteroid belt without special protection.

Jupiter

Pioneer 10 passed within 130 000 km of the Jovian cloud tops on 4 December 1973. Pioneer 11 was launched on a trajectory similar to that of Pioneer 10 on 5 April 1973 and passed by Jupiter within 42 000 km of the cloud tops on 3 December 1974. Its purpose was to serve as a back-up had Pioneer 10 failed for any reason and to permit an informed and optimized examination of Jupiter's environment based upon Pioneer 10's data. The redirection and acceleration of Pioneer 11 toward Saturn required a tightly curved passage from south to north around Jupiter. This path, which was highly complementary to Pioneer 10's passage, provided a deeper counterrotational penetration into Jupiter's radiation belt. Jupiter's magnetic field at the cloud tops was found to be more than 10 times stronger that Earth's field at its surface. The inner radiation belts of Jupiter measured by Pioneer had the highest natural radiation intensity that had ever been measured, 10 000 times greater than Earth's maximum.

Pictures of Jupiter were made by the imaging photopolarimeter onboard the Pioneer spacecraft. An example is shown in Figure P6 with the shadow of Jupiter's satellite Io present.

Figure P6 Picture of Jupiter with a shadow of its moon, Io. (NASA/Ames Research Center.)

Figure P7 Image of Saturn with its ring system illuminated from behind. (NASA/Ames Research Center.)

Saturn

The successful Jovian encounter of Pioneer 10 allowed Pioneer 11 to be retargeted halfway in its flight to Jupiter for an eventual encounter with Saturn. After Jupiter, Pioneer 11 looped high above the ecliptic plane and across the solar system on its voyage to Saturn. The spacecraft rose about 16° above the equatorial plane of the Sun, where it discovered a current sheet in space that separates the Sun's magnetic hemispheres. Pioneer 11 had its closest approach to Saturn (within 25 000 km of the cloud tops) on 1 September 1979. It penetrated the ring plane precisely where Voyager would later need safe passage by Saturn on its way to Uranus and Neptune. The Pioneer 11 spacecraft, after traveling more than 3.2 billion km on a journey of nearly 6.5 years, gathered a wealth of new information about Saturn and its ring system. The most fundamental discovery by Pioneer 11 was that Saturn has a magnetic field; there had previously been no conclusive evidence of the existence of a magnetic field. After penetrating the plane of the rings, Pioneer 11 passed very much closer to the planet, precisely mapping its magnetic field and finding a total lack of ion particles due to interaction of the rings with the magnetic field.

One of Pioneer 11's most spectacular pictures was an image of Saturn's ring system illuminated from behind (Figure P7). The rings that normally appear bright when viewed from Earth appear dark in the Pioneer picture, and the dark gaps in the rings (as seen from Earth) appear as bright rings.

Pioneer 10 became the first artificial object to 'leave the solar system' in June 1983 when it passed beyond the farthest known planet. Pioneer 11 left the solar system in February 1990. Further details of the scientific results from the Pioneers 10 and 11 mission can be found in *Science* (1974, 1975) and NASA (1980).

Extended mission objectives

An extended mission was declared after the successful completion of the planetary encounters. The objectives of the extended mission are:

- to continue the search for the heliospheric boundary with the interstellar medium;
- to search for previously undetected matter in the outer solar system; and
- to search for gravitational radiation (gravity waves).

Heliospheric boundary

The heliospheric boundary is the meeting surface of the solar wind and the medium outside the solar system. The solar wind is a 1 600 000 km/h^{-1} flow of gases expelled by the Sun, consisting of electrons, protons, helium nuclei and other heavier ions. The medium beyond the solar environment is of unknown composition, but includes energetic atomic nuclei whose flux is detected as far inward as Earth. Prior to Pioneer 10 and 11, the heliospheric boundary with interstellar space was thought to extend to the vicinity of Jupiter or perhaps somewhat farther. The Pioneer scientists now predict the distance from the Sun at which Pioneer 10 may encounter the heliopause is from 60 astronomical units (AU) to well over 100 AU. (The heliopause is defined as the boundary where the influence of the Sun ends and the region of interstellar space begins. This boundary may be sharp or extend over a region of many astronomical units. The boundary is also probably in motion, oscillating according to the solar cycle.)

Undetected matter

Unexplained perturbations in orbits of Uranus and Neptune were measured optically in the 19th century, and there are inconsistencies in planetary orbit fits between the 19th and 20th centuries. In addition, both the Pioneer spacecraft have experienced significant unmodeled forces since they passed the orbital radius of Saturn at about 10 AU. Previously undetected matter in the outer solar system as a possible cause for these deviations is being studied as part of the Celestial Mechanics Experiment (CM) conducted by the Jet Propulsion Laboratory. The experiment is based on Pioneer radio Doppler data generated by NASA's Deep Space Network at three complexes in California, Australia and Spain.

Possible sources of such matter include:

- a tenth planet (Planet X) about five times as massive as the Earth;
- a few small icy planets with diameters of about 1000 km; and
- a distribution of cometary material and debris left over from the formation of the solar system.

There may be other possibilities as well, but these three are given the most credence, based on other observations of the solar system phenomena (Wiedenschilling 1991).

Gravity waves

Gravitational waves are predicted by the theory of general relativity. The search for these waves is also part of the CM Experiment. During periods when Pioneer 10 is at solar opposition, the radio link between the Earth and spacecraft is most stable. This occurs during December for a period of about 10 days. At these times the Doppler data provide a clear signal for possible detection of gravitational radiation. Massive coalescing binary systems are candidate astronomical sources for detection by Pioneer 10. So far no gravitational waves from such sources have been detected, but drastically lower limits have been placed on a stochastic background of such sources in the millihertz band.

Star encounters and plaque

Both spacecraft will coast through interstellar space forever. By the year 34 602 AD, Pioneer 10 is projected to pass within 3.3 light years of the star Ross 248. During the next million years or so, it will travel past ten different stars.

Included on both spacecraft is a small gold-plated plaque on which the figures of a man and a woman are shown to scale next to a line silhouette of the spacecraft. The plaque shows a schematic representation of the path that Pioneers 10 and 11 took to escape the solar system. Also shown to help identify the origin of the spacecraft is a radial pattern etched on the plaque that represents the position of our Sun relative to 14 nearby pulsars and to the center of our galaxy. The plaque may be considered as the cosmic equivalent to a message in a bottle cast into the sea. Sometime, perhaps billions of years from now, it may pass through a planetary system of a remote stellar neighbor, one of whose planets may have evolved intelligent life. If that life possesses sufficient capability to detect the Pioneer spacecraft, it may also have the curiosity and the technical ability to pick up the spacecraft and inspect it. Then the plaque with its message from Earth may be found and deciphered.

Summary

Pioneer 10 and 11 paved the way for subsequent missions including Voyager, Galileo and Ulysses. The following list summarizes some of the Pioneers' firsts.

Pioneers' firsts

- First to travel through the asteroid belt;
- first to fly by Jupiter;
- first to fly by Saturn;
- first to leave the solar system;
- first to carry messages to extraterrestrials.

Lawrence E. Lasher

Bibliography

NASA (1980) '*Pioneer first to Jupiter, Saturn and beyond.*' NASA SP-466.
Science (1974) **183**, No. 4122, 25 January.
Science (1975) **188**, No. 4187, 2 May.
Weidenschilling, S.J. (1991) A plurality of worlds. *Nature*, **352**, 190–2.

Cross references

Asteroid
Heliosphere
Jupiter
NASA
Saturn
Voyager missions

PIONEER VENUS MISSIONS

The Pioneer Venus (PV) missions represent the United States' most extensive and intensive study of the planet Venus and its environment. The PV program consisted of two related spacecraft missions, an orbiter and a multiprobe. Both spacecraft were launched from Atlas–Centaur launch vehicles, on 20 May and 8 August 1978. They encountered Venus within 5 days of one another in early December 1978.

The spacecraft were built by the Hughes Aircraft Company for the National Aeronautics and Space Administration's Ames Research Center, which was the project manager. The orbiter and multiprobe

Table P2 Pioneer Venus scientific experiments

Multiprobe mission

Large probe
  Mass Spectrometer
  Gas Chromatograph
  Atmosphere Structure
  Nephelometer
  Cloud Particle Size Spectrometer
  Solar Flux Radiometer
  Infrared Radiometer

Small probes
  Atmosphere Structure
  Nephelometer
  Net Flux Radiometer

Probe carrier (bus)
  Neutral Mass Spectrometer
  Ion Mass Spectrometer

Orbiter mission

Neutral Mass Spectrometer
Ultraviolet Spectrometer
Cloud Photopolarimeter
Infrared Radiometer
Ion Mass Spectrometer
Electron Temperature Probe
Retarding Potential Analyzer
Magnetometer
Plasma Analyzer
Electric Field Detector
Radar Mapper
Gamma Burst Detector

missions shared a common spacecraft design – each had a basic cylindrical structure, about 8 ft (2.5 m) in diameter by 4 ft (1.2 m) high; and both were spinstabilized. (Table P2 lists the experiments integrated into the Multiprobe and Orbiter spacecraft.)

The purpose of the multiprobe mission was to sample the Venus atmosphere and to measure its properties, as a function of altitude, at four locations on the planet. The orbiter mission's objectives were to measure the properties of the upper atmosphere and clouds on a global basis (as a function of latitude and local Venus time). On 9 December 1978 data were collected simultaneously, with the multi-probe data serving as 'ground truth' for the orbiter data.

Both missions were designed to address key scientific questions about Venus: composition, structure, clouds, thermal balance, dynamics, ionosphere and solar wind, surface and interior. These questions were posed by mission scientists hoping to uncover critical information about Venus' origin 4.5 billion years ago and the evolutionary processes that led to its current state.

Multiprobe mission

The Multiprobe spacecraft consisted of a bus carrying four atmospheric entry probes – one large probe and three identical small probes. The bus and the four probes were instrumented with scientific experiments designed to make measurements, while descending to the surface, of the properties of Venus's upper and lower atmosphere and clouds. Survival on the surface for any of the probes was neither expected nor planned for because of the extremely high temperatures and pressures there.

The large probe contained seven experiments designed to obtain data on lower atmospheric chemical composition, temperature, pressure, density, winds, cloud structure, cloud chemistry and cloud particle size. Other instruments were designed to study the properties and distribution of solar and infrared radiation at various altitudes above the surface.

The three small probes contained three experiments each, designed to obtain data on temperature, pressure, density, winds, cloud structure and cloud particle size, and the sources and sinks of infrared radiation in the lower atmosphere within and below the cloud layers.

The probe carrier (or bus) carried two experiments to measure the chemical composition of the neutral upper atmosphere and charged-particle ionosphere well above the altitude of the highest clouds.

The Multiprobe was sent on a type I trajectory (in which the spacecraft changes longitude by less than 180° while in transit from Earth to Venus around the Sun). It encountered Venus on 9 December 1978 after traveling 330 million km at an average speed of 31 km s^{-1}. Twenty-four days before encounter, the spin-stabilized bus was oriented by firing thrusters, and the large probe was released toward its planned Venus entry location. About 4.5 days later the three small probes were released simultaneously, at a precise spin rate and precise time in the bus spin cycle, toward their planned Venus entry locations. All four probes were then silent until 22 min before entry on 9 December. The bus was given a final trim maneuver on 9 December to establish its own final upper-atmosphere entry trajectory.

The entry locations of the four probes across the hemisphere of Venus facing the Earth were selected to sample Venus' atmosphere at high and low latitudes, during both daylight and night-time conditions. This was done to enable scientists to estimate both vertical and horizontal global patterns of atmospheric and meteorological structures.

The four probes entered the Venus upper atmosphere and clouds staggered over a 10-min interval. Descent times to the surface from about 65 km altitude were about 1 h for each of the probes. The large probe descended on a parachute through the cloud layers to slow its descent. The parachute was released below the cloud deck at 47 km altitude and the large probe then descended in free fall to the surface, as did the three small probes beginning at entry. Data were collected from each probe during the descent interval. The information was telemetered in real time directly back to NASA's Deep Space Network on Earth. All the probes ceased transmitting upon impact with the Venusian surface, except for one small probe that continued to transmit for about 1 h.

Orbiter mission

The orbiter spacecraft carried 12 experiments designed to make measurements of the interaction of the solar wind with the ionosphere and upper atmosphere; the chemical composition, temperature and density structure of the upper atmosphere and ionosphere; the cloud structure and microphysical properties of the cloud tops; the surface topography and morphology; and the gravity field and magnetic field (if any) intrinsic to the planet.

The orbiter was sent on a type II trajectory (more than 180° around the Sun). It encountered Venus on 4 December 1978 (5 days before the multiprobe), after traveling 500 million km at an average speed of 29 km s^{-1}. The orbiter was inserted into orbit around Venus by firing the orbiter insertion motor near the point of closest approach to the planet. A series of small maneuvers established the final nominal highly elliptical orbit parameters: periapsis altitude 150–200 km; apoapsis altitude 66 900 km eccentricity 0.843; average period 24.03 h; inclination to the Venus rotation axis 15.6°; periapsis latitude 17°N.

These orbit parameters were chosen for several reasons. The low periapsis altitudes permitted several experiments to sample the ionosphere and upper atmosphere directly. Also, the radar experiment that mapped the surface of Venus required a low periapsis to perform well. High apoapsis permitted full-disk imaging by the remote sensing instruments. The 24-h period (one orbit per day) permitted regular and consistent data collection operations at the Deep Space Network (DSN) tracking stations on Earth.

The Pioneer Venus Orbiter (PVO) remained in this elliptical orbit around Venus; however, solar perturbations of the orbit are nontrivial. The major effect is to change the periapsis altitude as a function of time. During some periods the effect is to lower the periapsis altitude; at other times the opposite effect predominates. During the first 2 years following PVO orbit insertion, the effect of solar gravity was to lower periapsis altitude. It was necessary to fire thrusters on the spacecraft every week to raise periapsis altitude to avoid PVO descending into the atmosphere and being destroyed. This was done successfully through the middle of 1980; periapsis altitude was maintained in the range 150–200 km. These maneuvers were eliminated thereafter when solar perturbations began acting to increase the periapsis altitude. Periapsis altitude rose steadily from about 200 km in 1980 to 2200 km in 1986 and then decreased again. It reached 1978–1980 levels in early 1992. Maneuvers to increase periapsis altitude began again in September 1992. These maneuvers used up the remaining fuel and the spacecraft ceased transmitting in October 1992, indicating that the spacecraft entered the atmosphere and was destroyed, thus ending, after 14 years, perhaps the most successful missions to date undertaken by any spacefaring society to another planet.

Major scientific results

The Multiprobe mission experiments were designed to provide knowledge about the lower atmosphere from the surface to above the cloud tops, and the clouds themselves. The surface atmospheric pressure was confirmed to be 90 bar, and the surface temperature to be 480°C. The temperature on the surface was found to be within a few degrees of being the same on both the night and day sides of the planet, as well as at the equator and at the poles.

The chemical composition of the lower atmosphere was determined to consist of primarily carbon dioxide (96.5%) and nitrogen (3.5%). The relative abundance of many minor constituents was also determined in the parts per million range. Isotope ratios were measured, particularly those of the noble gases. The heavy hydrogen (deuterium) to normal hydrogen ratio was observed to be about 150 times higher than on Earth, suggesting that Venus may have had significant quantities of liquid water on its surface early in its history. It is believed that, if so, the water was lost by a catastrophic runaway greenhouse effect leading to the high surface temperature existing today.

The atmospheric circulation from the surface to about 100 km altitude is primarily retrograde, subsolar to antisolar flow. These zonal winds increase from a few meters per second at the surface to several hundred kilometers per second at the cloud tops. The wind speeds at the cloud tops are commensurate with 4 to 5 day circulation of ultraviolet features (see below). The equator-to-pole (or meridional) wind speeds are only a few kilometers per second throughout the lower atmosphere. There appear to be three stacked Hadley cells extending from the equator to the poles, the middle of which drives the meridional circulation in the cells above and below.

Clouds cover the entire planet in the altitude range 50–80 km. There is a three-layer vertical structure within this range. The major chemical constituent of the cloud layers is sulfuric acid contained in particles 2 to 3 mm in diameter.

About 2% of the solar energy incident on Venus reaches the ground. About 50% of the incident energy is absorbed in the clouds themselves and the rest by the thick carbon dioxide atmosphere below the clouds. Additional absorbers, yet unknown, beside sulfuric acid are required to explain certain other observations.

The Orbiter mission experiments were designed to provide knowledge about the surface and interior of Venus, the cloud tops, the ionosphere and thermosphere and the solar wind interaction with the planet.

Rolling hills dominate the topography of Venus' surface. Only two continental-scale features exist, both of which exhibit mountainous and volcanic terrain. Clouds are transparent to radio waves, which permitted radar mapping of the surface. About 90% of the surface was mapped at 50–100 km horizontal resolution; PVO was a precursor to higher-resolution mapping of the surface by the later USSR Venera 15 and 16 and the US Magellan missions.

It was found that subsurface gravitational anomalies compensate the major topographical features indicating, unlike at Earth, the absence of a global tectonic mechanism within the interior. No global, interior magnetic field was found. The slow rotation of the planet is believed to be the reason.

The ubiquitous high-altitude cloud cover precludes visual observations of the planet's surface from a vantage point above the cloud tops and in space or at the Earth's surface. Such visual observations indicate only featureless and yellowish cloud tops. Observations in the ultraviolet portion of the electromagnetic spectrum show large-scale features, due to an absorber believed to be dominated by sulfur, which superrotate with a 4 to 5 day period – some 50–60 times faster than the solid planet itself.

The spatial morphological structure of the solar wind–ionosphere interaction was mapped in detail on this unique magnetic field-free planet. Many details were studied and new features discovered: bow shock, magnetosheath, mantle and magnetotail regions; nightside ionospheric holes, tail rays, plasma clouds and streamers; magnetic flux ropes, nightward plasma flows, ion acceleration and escape and drift. Temporal variations of these features were studied over a 14-year period (more than one solar activity cycle of 11 years).

Evidence of subcloud lightning was suggested by the electric field detector low radio frequency observations. These results are controversial since the observations could also be explained by plasma instabilities in the ionosphere near the spacecraft.

The chemical composition of the lower thermosphere down to and including the turbopause was determined. Extremely low atmosphere temperatures were observed on the nightside of the planet. Highly variable abundances of sulfur dioxide in the atmosphere above the clouds were measured over long timescales. This result suggests the possibility of active volcanism on Venus, at the present epoch, ejecting fresh sources of that gas into the atmosphere.

The reader is referred to the encyclopedia articles on Venus for a more comprehensive discussion of our current knowledge of the planet.

Future missions to Venus

The Pioneer Venus missions were two of many robotic spacecraft missions to our sister planet, undertaken by the United States and the Soviet Union. From 1962 to 1978 there were five US missions (Mariners and Pioneers) and nine USSR missions (Venera) to Venus. Since 1978 there have been six Venera and Vega missions by the Soviet Union and one US mission (Magellan). This past emphasis on Venus derives from two facts. First there is the proximity of Venus to Earth, which leads to a relative simplicity of mission implementation. The second major impetus for Venus exploration is found in the fascinating features of Venus' environment that are so different from those of the Earth.

At present there are no further planned missions to Venus by any country. After the last successful Soviet mission to Venus, they announced that they would shift the focus of their planetary exploration program to Mars. Similarly, the US has shifted its scientific robotic-mission focus to other bodies in the Solar system: Saturn and Titan (Cassini), comets and asteroids (Comet Rendezvous and Asteroid Flyby or CRAF, since cancelled), the Moon (Lunar Orbiter) and Mars (Mars Observer).

Studies are under way to define future missions to Venus. Perhaps the highest priority would be a long-lived lander (with a lifetime of at least a few weeks). Designing the lander would be difficult because of the high surface temperature (750 K) and pressure (90 bar). Such a mission would permit detailed studies of surface–lower atmosphere chemical reactions, which are largely unknown at present. The mission would be very expensive and is beyond today's technology.

A relatively low-cost mission under study in the US is a single atmospheric entry probe, similar to the Pioneer Venus large probe, instrumented to make a detailed assessment of the chemical composition and structure and isotopic abundances of the gases in the lower atmosphere. Another relatively low-cost possibility presently under consideration is a series of very small probes to be deployed simultaneously around Venus, to assess global meteorological dynamics.

Lawrence Colin

Bibliography

Burgess, E. (1985) *Venus – An Errant Twin*. New York: Columbia University Press.

Colin, L. (1980) The Pioneer Venus program. *J. Geophys. Res.*, **85** (A13), 7575–98.
Colin, L. (1983) Basic facts about Venus, in *Venus* (eds. D.M. Hunten, L. Colin, T.M. Donahue and V.I. Moroz). Tucson, Arizona: University of Arizona Press, pp. 10–26.
Cruikshank, D.P. (1983) The development of studies of Venus, in *Venus* (eds. D.M. Hunten L. Colin, T.M. Donahue and V.I. Moroz). Tucson, Arizona: University of Arizona Press, pp. 1–9.
Fimmel, R.O., Colin, L. and Burgess, E. (1983). *Pioneer Venus*. NASA Special Publication NASA SP-461.
Hunten, D.M. Colin, L. Donahue, T.M. and V.I. Moroz (eds) (1983) *Venus*. Tucson, Arizona: University of Arizona Press.

Cross references

Galileo mission
Venus
Venus: atmosphere

PLANET

The recognition of differences between planets and stars is not a recent acquisition of mankind. Thousands of years ago it was noted that several 'stars' did not behave as all the others; they move across the stellar background along regular paths. In order to emphasize this difference, the 'erratic stars' were called οἱ πλάνητοι (the wanderers) by the Greeks.

Apart from the Moon and the Sun, whose nature was recognized thousands of years later, there were five bodies belonging to the category of planets: Mercury, Venus (Plate 6), Mars (Plate 16), Jupiter (Plate 22) and Saturn (Plate 26). Our own planet, the Earth (Plate 9), was not among them, being thought to stand in the centre of the universe. The five 'planets' were the only ones that could be seen without artificial aids.

A more detailed account of the discoveries of the other planets (Uranus (Plate 30), Neptune (Plate 31) and Pluto) is given elsewhere in this book: here we note only that, at least for some of them, it was necessary to wait for improved technology (telescopes) and mathematics (celestial mechanics) before they could be found.

In the last 50 years, especially after the discovery that Pluto – because of its small mass – can no longer be considered to be the cause of the perturbations of the orbit of Neptune, detailed searches for trans-Plutonian planets have been undertaken. They have not succeeded so far, but the debate about the existence of 'Planet X' (q.v.) is still going on. In fact, the meaning of the word 'planet' now seems a bit vague: there are satellites of planets (i.e. Ganymede) larger than the real planets (i.e. Pluto). Recently discovered bodies in the Edgeworth–Kuiper belt (100 km class objects orbiting the Sun beyond Neptune) are classified as 'minor' planets, like the asteroids. Table P3 provides some of the most important parameters of the nine planets, while Table P4 lists the characteristics of their satellites.

The controversy about the geocentric or heliocentric arrangement of planetary orbits (in fact, of the whole universe) was settled only after the foundation of modern physics by Galileo Galilei (q.v.), and the publication by Isaac Newton (q.v.) of his gravitational theory. Very sophisticated mathematical methods have been elaborated since

Table P3 Relevant parameters of planets

Planet	Distance from the Sun (AU)	Radius (km)	Density (g cm^{-3})	Period (years)	Rotational period	Discovered satellites	Rings
Mercury	0.39	2 420	5.13	0.24	58 d	–	No
Venus	0.72	6 200	4.97	0.62	243 d	–	No
Earth	1.00	6 378	5.52	1.00	24.0 h	1	No
Mars	1.52	3 400	3.94	1.88	24.5 h	2	No
Jupiter	5.20	71 370	1.33	11.86	9.8 h	16	Yes
Saturn	9.54	60 400	0.69	29.46	10.2 h	18	Yes
Uranus	19.18	23 500	1.56	84.01	17.2 h	15	Yes
Neptune	30.06	22 300	2.27	164.79	14.0 h	8	Yes
Pluto	39.44	1 140	2.00	247.69	6.4 h	1	No

Table P4 Satellites of planets

Planet	Satellite	Distance (10³ km)	Radius (km)	Period (days)	Year of discovery	Discoverer
Earth	Moon	384.4	1 738	27.3		
Mars	Phobos	9.4	11	0.32	1877	Hall
	Deimos	23.5	6	1.26	1877	Hall
Jupiter	Metis	128.0	20	0.20	1979	Synnott
	Adrastea	129.0	10	0.30	1979	Jewitt
	Amalthea	181.3	94	0.50	1892	Barnard
	Thebe	221.9	50	0.67	1979	Synnott
	Io	421.6	1 815	1.77	1610	Galileo
	Europa	670.9	1 569	3.55	1610	Galileo
	Ganymede	1 070	2 631	7.15	1610	Galileo
	Callisto	1 883	2 400	16.69	1610	Galileo
	Leda	11 094	8	238.7	1974	Kowal
	Himalia	11 480	93	250.6	1904	Perrine
	Lysithea	11 720	18	259.2	1938	Nicholson
	Elara	11 737	38	259.7	1904	Perrine
	Ananke	21 200	15	631[a]	1951	Nicholson
	Carme	22 600	20	692[a]	1938	Nicholson
	Pasiphae	23 500	25	735[a]	1908	Melotte
	Sinope	23 700	18	758[a]	1914	Nicholson
Saturn	1981 S 13	133.6	10	0.58	1981	Voyager 2
	Atlas	137.7	15	0.60	1980	Terrile
	Prometheus	139.4	50	0.61	1980	Collins
	Pandora	141.7	45	0.63	1980	Collins
	Epimetheus	151.4	60	0.70	1980	Fountain
	Janus	151.5	95	0.70	1966	Dollfus
	Mimas	185.5	196	0.94	1789	Herschel
	Enceladus	238.0	250	1.37	1789	Herschel
	Tethys	294.7	530	1.89	1684	Cassini
	Telesto	294.7	15	1.89	1980	Reitsema
	Calypso	294.7	13	1.89	1980	Pascu
	Dione	377.4	560	2.74	1684	Cassini
	Helene	377.4	16	2.74	1980	Lecacheux
	Rhea	527.0	765	4.52	1672	Cassini
	Titan	1 222	2 575	15.95	1655	Huygens
	Hyperion	1 481	145	21.28	1848	Bond
	Iapetus	3 561	730	79.33	1671	Cassini
	Phoebe	12 952	110	551[a]	1898	Pickering
Uranus	Cordelia	49.8	13	0.34	1986	Voyager 2
	Ophelia	53.8	16	0.38	1986	Voyager 2
	Bianca	59.2	22	0.43	1986	Voyager 2
	Cressida	61.8	33	0.46	1986	Voyager 2
	Desdemona	62.7	29	0.47	1986	Voyager 2
	Juliet	64.6	42	0.49	1986	Voyager 2
	Portia	66.1	55	0.51	1986	Voyager 2
	Rosalind	69.9	27	0.56	1986	Voyager 2
	Belinda	75.3	34	0.62	1986	Voyager 2
	Puck	86.0	77	0.76	1985	Voyager 2
	Miranda	129.8	236	1.41	1948	Kuiper
	Ariel	191.2	579	2.52	1851	Lassell
	Umbriel	266.0	586	4.14	1851	Lassell
	Titania	435.8	790	8.71	1787	Herschel
	Oberon	582.6	762	13.46	1787	Herschel
Neptune	Naiad	48.2	29	0.29	1989	Voyager 2
	Thalassa	50.1	40	0.31	1989	Voyager 2
	Despoina	52.5	79	0.33	1989	Voyager 2
	Galatea	62.0	74	0.43	1989	Voyager 2
	Larissa	73.6	96	0.55	1989	Voyager 2
	Proteus	117.6	208	1.12	1989	Voyager 2
	Triton	354.8	1 350	5.88[a]	1846	Lassell
	Nereid	5 509	170	359.6	1949	Kuiper
Pluto	Charon	20.0	600	6.39	1978	Christy

[a] Denotes retrograde orbit.

Figure P8 A picture of Mercury taken by Mariner 10 in 1974. In this area of the northern hemisphere some impact craters have probably been obliterated by subsequent volcanic activity.

then, leading to the construction of one of the most beautiful theories in mathematical physics: celestial mechanics (q.v.). The most brilliant scientists in the 18th and 19th centuries (like Gauss and Poincaré) have been active in this effort. As a result, modern mechanicians have at their disposal powerful tools to investigate the motion and mutual interaction of planets (see Planetary dynamical astronomy).

In recent decades, with the development of electronic computers, numerical analysis has become a major contributor to studies of planetary motion. It is now possible to track the dynamical history of the planetary system for a billion years, in order to investigate its behaviour in time.

Recent exploration

Another powerful means of investigation has become available in the last 30 years, due to developments in space science. With the exception of Pluto, all the planets of the solar system have been explored closely by automated space probes, leading to an enormous advancement of our knowledge of their properties. The color plates in this volume include many images of the planets and major moons obtained by spacecraft during this period.

At present the only active major mission to a planet is Galileo (q.v.), which, after having imaged for the first time two asteroids (Gaspra and Ida), and after releasing a probe into Jupiter's atmosphere, will study Jupiter and its moons in detail. In the next few years other probes will be launched towards Saturn and Mars.

The most successful space exploration undertaken so far has certainly been the Voyager mission (q.v.). The two Voyagers encountered Jupiter and Saturn between 1979 and 1981, and Voyager 2 visited Uranus in 1986 and Neptune in 1989, 12 years after launch. The amount of data obtained by the Voyagers is spectacular: besides imaging, almost all the relevant fields in planetary science have been addressed, and many long-standing problems clarified (if not solved).

The first reconnaissance missions to planets are over. The next steps include prolonged remote analyses, sample collection and return to Earth, emplacement of surface stations and geophysical observatories, and finally manned missions (to Mars). The aims of these subsequent enterprises will not be only scientific; the final goal is the establishment of permanent or semipermanent bases, as a prelude to possible colonization of other worlds. This will not happen during our life, however, and probably not within the next century.

The two groups of planets

For many years planets have been divided in two groups: the Terrestrial (or 'inner') and the Jovian (or 'outer') planets. The first (Mercury, Venus, Earth and Mars) are small, solid, rocky and dense; those of the second category (Jupiter, Saturn, Uranus and Neptune) are large, liquid (or more accurately, composed of an extended liquid mantle and an extended atmospheric envelope), and are composed of light elements and compounds (like hydrogen, helium or ice), with small densities (Table P3).

Pluto was formerly placed among the outer planets. Now, due to the small size of this planet, and to its peculiar double nature (the satellite Charon has a radius half that of the planet), Pluto is considered as a body of a distinct class, with a composition and possible evolution somewhat different from those of the Jovian planets.

There are other differences between the two groups of planets, as is evident from Table P3. The terrestrial planets have few or no satellites, while the Jovian planets are surrounded by many bodies, some of these being larger than the smallest planets. Moreover, the

Jovian planets exhibit ring systems, which are absent in the inner group. Other differences are present in the composition and dynamics of the atmospheres and the strength of magnetic fields (with the noticeable exception of Earth, whose magnetic field is comparable to that of most outer planets).

These differences may be due to the conditions in the places where the planets were formed, i.e. to the different composition of the solar nebula (q.v.) in those places. At the time of formation, in the innermost regions – inside about 3 AU – rocky material was predominant, and hence the planets which nucleated there are rocky.

There was little gas left, and hence these planets have atmospheres that are relatively insignificant in comparison with those of the Jovian planets. On the other hand, a large fraction of the original gas component was still present in the outer regions, where accreting, probably rocky, embryos were able to accumulate large volatile envelopes. It is important to note that the average composition of Jupiter is very similar to that of the Sun: it reflects the average composition of the solar nebula with little alteration.

Moreover, it has been proposed (although not proven) that a large fraction of the volatile component of Earth (including water) may not be primeval, but was carried onto the planet by impacting comets. If this is true, it would imply that the onset of life on our planet was triggered by processes external to it.

The present positions of planets in the solar system are by no means random. Even if strict relationships among distances (like the Titius–Bode law; q.v.) do not seem to be implicit in the formation theories, it is noteworthy that at least the orbits of the outer planets are coupled: the orbital periods of Jupiter and Saturn have a ratio very close to 2 : 5, and those of Uranus and Neptune close to 1 : 2. It is well known from celestial mechanics that these resonances may favour the stability of the system, but we still do not know if these properties are a natural consequence of the formation of planetary systems, nor if they originated at the beginning of the life of the system or during its evolution. The discovery of other systems around other stars may help to clarify this point.

Another striking difference between inner and outer planets is given by their rotational periods. As shown in Table P3, the inner planets have rather long rotational periods compared to those of the outer planets (again, with the exception of Pluto). The length of the 'day' of Mercury is almost exactly ⅔ of its 'year' (a good example of spin-orbit coupling). Venus' rotation is retrograde and very slow. Earth and Mars have almost the same rotational period, substantially longer than those of the outer planets. However, we know that the presence of the Moon is responsible for a slow deceleration of the rotation of our planet: this process, by transferring angular momentum from the Earth to the Moon, has caused the progressive enlargement of Moon's orbit.

Relationships with other bodies

The origin and dynamical properties of the individual planets are more thoroughly addressed elsewhere in this book, and the reader is referred to those sections for more details. Here we give only a short account of the characteristics of the system as a whole. These are indicative of the processes taking place at the time of formation, and in some cases help explain the present distribution of matter across the system.

Mass distribution

Almost all the mass of the solar system is contained in the Sun, all planets (and other bodies) together accounting only for about 0.2% of the total. On the other hand, the larger fraction of the total angular momentum is contained in the planets (mainly the outer group). This simple statement places strong constraints on formation theories, which have to explain why and when such a dichotomy should appear.

About half of the mass of the planetary system is contained in Jupiter. A tiny fraction of the remaining mass is in neither the planets nor their satellites. It is highly dispersed among many small bodies in four broad classes: asteroids, comets, meteoroids and dust.

Since the strength and radius of action of gravitational fields are dependent on the mass of the source, it is clear that the most powerful gravitational field in the solar system is that generated by the Sun (and this is why all the other bodies revolve around it). However, the gravitational fields generated by the four Jovian planets are by no means negligible: they control, for example, the input of cometary bodies to the inner regions of the system, giving rise to complicated evolutionary patterns which have been studied only in the last decades.

Late bombardment

At the time when Jupiter and its outer fellows were accreting (and the inner planets were probably already formed), a large number of 'boulders' were still orbiting the Sun in the outer regions of the solar system (beyond the orbit of Mars). It is generally agreed that the present asteroid belt (q.v.) contains the remnants of bodies disrupted and fragmented by catastrophic collisions. Although not yet defined, the role of the accreting Jupiter must have been relevant, with its increasing gravitational field 'pumping up' the relative velocities among the precursors of present asteroids. Once the relative velocities became larger than about 1 km s^{-1}, collisions started to fragment, instead of accrete, the parent bodies. This comminution has led to the enormous number of small fragments that we see today. This process is probably still active.

Between Jupiter and Neptune a great number of icy planetesimals were moved, either towards the inner regions of the system, or toward the edge of the gravitational sphere of action of the Sun (or outside it, to become interstellar messengers). Remnants of this early population are visible nowadays in the form of comets.

During this extended period of displacement of mass throughout the system quite large bodies impacted all the planets and satellites. We see the effects of these collisions as craters, whose ubiquity makes them one of the best presently available methods for dating planetary and satellite surfaces.

Relicts of this late bombardment are visible on almost all the terrestrial planets (maria on the Moon, basins on Mercury and Mars), and there are indications of similar events on the surfaces of satellites of outer planets.

The late bombardment has deeply modified the surfaces of planets and satellites (and maybe their atmospheres, where present), but has not changed their orbits very much. These are supposed to be the same as they were some billion years ago.

Different bodies, different surfaces and different interiors

Modifications of the surface of a planetary body (be it a planet or a satellite or any body with a solid surface) are due to several different processes. They may be triggered by internal movements, like thermal convection, or by external sources, like impacts. There are other subtle mechanisms by which the thermal profile of a body may be changed, and the thermal gradient across the surface is probably the most important cause of surface modifications.

When Voyager 1 imaged Io (the innermost Galilean satellite of Jupiter) in 1979, all planetary scientists were amazed by the strange structures visible, by its uncommon colours and by the almost total absence of impact craters. The answer to the many questions arising was already given, prior to the encounter of Voyager 1 with Io: the proximity to Jupiter must provoke a very strong fluctuating tidal distortion of the body. The heat originated by the tidal flexure is sufficient to melt most of the satellite, with the exception of a tiny crust. It is therefore not surprising that Io has no craters: its surface is continuously regenerated (and obliterated) by new crustal material coming from inside. The same mechanism is responsible for the flatness of Europa's surface. On the other hand, both Ganymede and Callisto exhibit extensive cratering, and there is the well-founded possibility that the surface of Callisto (the outermost of the Galilean satellites) is primeval, with modifications induced only by impacts.

At the time of the encounter of the Voyagers with Saturn and its satellite system, the small bodies surrounding the planet were mostly unknown. It was surprising to discover that some of them, notwithstanding the small dimensions, have surfaces with very complicated patterns. The surprise arose from the fact that small bodies cannot have very strong internal activity (there is not enough mass to justify that). It was therefore suggested that these satellites may have been disrupted and reaccreted several times in their lifetime. This conjecture is sustained by the irregular shape and peculiar orbital properties of Hyperion, the satellite immediately outside Titan (q.v.). Hyperion has an orbit whose period is in resonance with that of Titan, and is very irregular in shape. Apparently, Hyperion has been disrupted by collisions, and what we see now is one of the pieces of the original satellite, which has survived in the present location due to the orbital resonance. Other fragments may have been removed by the gravitational influence of Titan, and there is the possibility that these may have been responsible for heavy cratering and possible fragmentation of the innermost satellites like Mimas.

Planetary systems around other stars

In recent years searches for planets around nearby stars have demonstrated that our planetary system is not at all unique in our galaxy. No planets have been 'imaged' to date, due to the predominance of the light emitted by the nearby star. However, in many cases emissions in the infrared domain of the electromagnetic spectrum have been detected, which may be explained as indicative of the existence of 'dust rings'. These, in turn, may be interpreted either as protoplanetary disks, where planets are being accreted, or as cometary clouds, like the one surrounding the Sun at the edges of its gravitational sphere of action.

The existence of planets around other stars is almost certain on a statistical basis, and we think that we only need more sophisticated observational tools to have direct confirmation of. There are already strong indications of the presence of massive planets in some cases, where the apparent movements of stars are slightly perturbed by invisible companions.

The detection of other planetary systems is essential to provide answers to certain questions about the origin and evolution of our own system. It will also provide more insight into the possibility of development of life forms elsewhere in the universe.

Andrea Carusi

Bibliography

Beatty, J.K. and Chaikin, A. (eds) (1990) *The New Solar System*, Cambridge: Cambridge University Press, 326 pp.
Bergstralh, J.T., Miner, E.D. Matthews, M.S. (1991) *Uranus*, Tucson: University of Arizona Press.
Black, D.C. and Matthews, M.S. (eds) (1985) *Protostars and Planets II*, Tucson: University of Arizona Press, 1293 pp.
Carr, M.H. (edr) (1984) *The Geology of the Terrestrial Planets*, NASA SP–469, 317 pp.
Cruikshank, D.P. (1990), *Neptune and Triton*. Tucson: University of Arizona Press, Tucson.
Encrenaz, T., Bibring, J.-P. and Blanc, M. (1987) *The Solar System*, Astronomy & Astrophysics Library, Berlin: Springer-Verlag, 330 pp.
Gehrels, T. (ed.) (1976) *Jupiter*. Tucson: University of Arizona Press, pp. 1254.
Gehrels, T. and Matthews, M.S. (eds) (1984) *Saturn*. Tucson: University of Arizona Press, 968 pp.
Hunten, D.M., Colin, L., Donahue, T.M. and Moroz, V.I. (eds) (1983) *Venus*. Tucson: University of Arizona Press, 1143 pp.
Kieffer, H.H., Jakosky, B.M., Snyder, C.W. and Matthews M.S. (1992) *Mars*, Tucson: University of Arizona Press.
Levy, E.H. and Lunine, J.I. (1993) *Protostars and Planets III*, Tucson: University of Arizona Press.
Tholen, D. and Stern, A.S. (1996) *Pluto and Charon*, Tucson University of Arizona Press.
Vilas, F., Chapman, C.R. and Matthews, M.S. (eds) (1988) *Mercury*, Tucson: University of Arizona Press, 794 pp.

Cross references

Accretion
Comet: impacts on Earth
Earth
Planet: extrasolar
History of planetary science I
History of planetary science II
Jupiter
Life: origin
Mars
Mercury
Neptune
Planetary rotation
Pluto
Saturn
Solar system: origin
Uranus
Venus

PLANET: EXTRASOLAR

Planetary systems about other stars have been a subject of speculation for at least 2000 years. Black (1991) provides a popular review. Technology now makes possible the detection of such planets, and the more we learn about our own solar system, the more motivated we are to look for others. In our solar system no two planets are alike, even in gross parameters. The content of this encyclopedia testifies that the variety of known properties and phenomena grows richer the more the solar system is investigated. But together with this diversity there is a striking degree of order in the large-scale properties of the solar system that encourages us to describe the process by which the system formed. The planets all orbit the Sun in the same direction in nearly coplanar, nearly circular orbits. The Sun and most planets rotate about their axes in the same direction as the orbital motion, and the masses and compositions of the planets and satellites are correlated with distance from the Sun. Studies of dense interstellar clouds, young star-forming regions, young stars with their circumstellar material and this one known system of planets suggest strongly that planetary systems can be natural consequences of the evolution of self-gravitating, spinning clouds of interstellar matter (Black, 1980; Bodenheimer and Pollack, 1986; Levy *et al.* 1986, Shu, Adams, and Lizano 1987; Cameron, 1988; Boss, 1989; Wetherill 1990; Sargent and Beckwith 1993). Considerable variety in the masses and orbital radii are expected (Wetherill, 1991). Gas giant planets such as Jupiter may not necessarily form; instead the largest planets in some systems may be composed of ice with masses on the order of 10 to 15 times that of Earth. On the other hand, a gas giant planet could form as close as 3 AU from a star (Nakano, 1988). Uniqueness of our solar system has been suggested by Taylor (1992).

Current ground-based searches for planets

Measurements of stellar motions are capable of detecting planets indirectly. The basis of these measurements is the dynamical effect of planets on the stars they orbit. The orbital motion of a planet induces a reflex motion on the star about the common center of mass (barycenter) in inverse proportion to the relative masses of the two bodies. This means that the speed and displacement of the star about the barycenter are very small compared to those of the planet. For example, the Sun orbits its center of mass in common with Jupiter with a speed of 13 ms^{-1}. The component of this orbital motion projected onto the line of sight induces an oscillation of Doppler shift. Perpendicular to the line of sight (in the plane of the sky) and seen from a distance of 10 pc, our Sun's angular position oscillates 0.5 milliarcsec due to the gravitational effect of Jupiter.

These amplitudes of astrometric and Doppler shift effects are many times smaller than the errors of conventional observations of stars, so extraordinary measures are required in both the design and in the use of such instruments. Both radial velocity (McMillan *et al.*, 1985, 1986, 1990, 1992, 1993, 1994, Campbell, Walker and Yang, 1988; Latham *et al.*, 1989; Duquennoy and Mayor, 1991; Marcy and Butler, 1992; Cochran and Hatzes, 1994) and astrometric (Gatewood, 1987) techniques are being applied at ground-based observatories to search for planets, and these techniques complement each other. Astrometry is biased toward nearby (and coincidentally low mass) stars with planets in large orbits. Since planets in smaller orbits move faster, the radial velocity technique is biased in favor of finding those.

Observations of infrared and submillimeter radiation are sensitive to circumstellar dust and companion objects that are much cooler than the central stars, and optical imaging through coronagraphs also can detect circumstellar dust when the observing conditions are favorable. Substellar companions may have been discovered orbiting some stars (Becklin and Zuckerman, 1988; Latham *et al.*, 1989; Skrutskie, Forrest and Shure, 1989 and references therein, and references in Burrows and Liebert, 1993). Planetary systems are indirectly suggested by disks of material around some stars (Aumann *et al.*, 1984; Smith and Terrile, 1984; Skrutskie *et al.*, 1990; Sargent and Beckwith, 1991, 1993).

Wolszczan and Frail (1992) have reported a planetary system orbiting a pulsar; confirmation of this discovery is under way (Rasio et al., 1992). Since pulsars are the remnants of supernova explosions, the history of such planets must be very different from those in our own solar system.

Searches for planets from space

Ground-based observations cannot detect extrasolar planets directly, and even the indirect methods currently being used cannot easily detect planets with perturbation signatures smaller than those from 'Jupiters'. Since not all planetary systems are expected to have planets that massive, a conclusive survey for extrasolar planets must have more sensitivity. Possible spacecraft missions to detect planets from Earth orbit are described by Burke et al. (1993). Although these plans are too preliminary to describe in a permanent reference book, their common goal is to detect smaller planets than can the ground-based surveys.

Implications of completing a search for extrasolar planets

If surveys for extrasolar planets have adequate sensitivity and statistical completeness, and yet do not yield discoveries, then current theoretical expectations are incorrect. If planetary systems are found, the taxonomy or morphology of their planetary parameters should be pursued, so as to refine the theories of the formation of such systems.

Other solar systems could be the homes of extraterrestrial civilizations, and measurable characteristics of the planets in other solar systems are relevant to the formation and evolution of life. For example, Wetherill (1992) entertained the First International Conference on Planetary Systems in Pasadena, CA, with the implications of not having gas giant planets in a solar system. The absence of a 'Jupiter' exposes the terrestrial planets to a 1000 fold greater flux of destructive impacts by comets. This might make it extremely difficult for sophisticated forms of life to evolve. Thus, by looking for 'Jupiters' we are already searching for some of the circumstances favorable to a civilization like our own.

Update

Since the literature search for this article was made in late 1992 there have been exciting discoveries and technological developments. Adaptive optics and coronagraphy from groundbased observatories have made it possible to detect self-luminous substellar objects, commonly referred to as 'brown dwarfs'. Nakajima et al. (1995) used this technique to find such an object orbiting the star Gliese 229. Gliese 229B is the first object outside the solar system to show the absorption spectrum of methane, an indication of a temperature cooler than that of a star. Improvements also have been made in radial velocity technique. Mayor and Queloz's (1995) discovery of a Jupiter-mass companion orbiting close to the solar-type star 51 Pegasi has been confirmed by Marcy and Butler (1995). Guillot et al. (1996) show that even this close (0.05 AU) to the star, a gas giant planet could survive the lifetime of the star if it formed at a larger distance and somehow migrated inward to its present small orbit. Marcy and Butler (1996) also have discovered companions orbiting close to the solar-type stars 70 Virginis and 47 Ursa Majoris. The companions have masses of at least a few times that of Jupiter. Based on the number of stars surveyed by radial velocities by all investigators and the number of substellar-mass companions found with masses similar to Jupiter and larger, it is estimated that a few percent of solar-type stars may have such close companions (Mazeh, Latham and Stefanik, 1996).

Robert S. McMillan

Bibliography

Aumann, H.H., Gillett, F.C., Beichman, C.A. et al. (1984) Discovery of a shell around Alpha Lyrae. *Astrophys. J. Lett.*, **278**, L23–7.
Becklin, E.E. and Zuckerman, B. (1988) A low temperature companion to a white dwarf star. *Nature*, **336**, 656–8.
Black, D.C. (1980) In search of other planetary systems. *Space Sci. Rev.*, **25**, 35–81.
Black, D.C. (1991) Worlds around other stars. *Sci. Am.*, **264**, 76–82.
Bodenheimer, P. and Pollack, J.B. (1986) Calculations of the accretion and evolution of the giant planets: the effects of solid cores. *Icarus*, **67**, 391–408.
Boss, A.P. (1989) Low-mass star and planet formation. *Publ. Astron. Soc. Pacific*, **101**, 767–86.
Burke, B.F., Rahe, J.H., Beebe, R.F. et al. (1993) TOPS: toward Other Planetary Systems. Report by NASA's Solar System Exploration Division. US GPO Jacket No. 327–494. Alexandria, VA: DeLancey.
Burrows, A. and Liebert, J. (1993) The science of brown dwarfs. *Rev. Mod. Phys.*, **65**(2), 301–36.
Cameron, A.G.W. (1988) Origin of the solar system. *Ann. Rev. Astron. Astrophys.*, **26**, 441–72.
Campbell, B., Walker, G.A.H. and Yang, S. (1988) A search for substellar companions to solar type stars. *Astrophys. J.*, **331**, 902–21.
Cochran, W.D. and Hatzes, A.P. (1994) A high-precision radical-velocity survey for other planetary systems. *Astrophys. Space Sci.*, **212**, 281–91.
Duquennoy, A. and Mayor, M. (1991) Multiplicity among solar type stars in the solar neighborhood. *Astron. Astrophys.*, **248**, 485–524.
Gatewood, G.D. (1987) The multichannel astrometric photometer and atmospheric limitations in the measurement of relative positions. *Astron. J.*, **94**, 213–24.
Guillot, T., Burrows, A., Hubbard, W.B., Lunine, J.I. (1996) Giant planets at small orbital distances. *Astrophys. J. Lett.*, **459**, L35–L38.
Latham, D.W., Mazeh, T., Stefanik, R.P. et al. (1989) The unseen companion of HD 114762: a probable brown dwarf. *Nature*, **339**, 38–40.
Levy, E.H., Gatewood G.D., Stein, J.W. and McMillan, R.S. (1986) Astrometric telescope of 10 microarcsecond accuracy. *Proc. SPIE*, **628**, 181–7.
Levy, E.H., McMillan, R.S., Gatewood, G.D. et al. (1988) Discovery and study of planetary systems: in *Bioastronomy – The Next Steps* (ed. G. Marx), Proc. IAU Colloq, 99 Dordrecht: Kluwer, pp. 131–6.
Marcy, G.W., and Butler, R.P. (1992) Precision radial velocities with an iodine absorption cell. *Publ. Astron. Soc. Pacific*, **104**, 270–7.
Marcy, G.W. and Butler, R.P. (1995) *I.A.U. Circular* 6251.
Marcy, G.W. and Butler, R.P. (1996) Presentation to the 187th Meeting of the American Astronomical Society in San Antonio, Texas on January 17, 1996.
Mayor, M. and Queloz, D. (1995) A Jupiter-mass companion to a solar-type star. *Nature*, **378**, 355–9.
Mazeh, T., Latham, D.W. and Stefanik, R.P. (1996) Spectroscopic orbits for three binaries with low-mass companions and the distribution of secondary masses near the substellar limit. *Astrophys. J.*, in press.
McMillan, R.S., Smith, P.H., Frecker, J.E. et al. (1985) The LPL radial accelerometer, in *Stellar Radial Velocities*, (eds A.G. Davis Philip and D.W. Latham), Proc. IAU Colloq. 88, Schenectady: L. Davis Press, pp. 63–86.
McMillan, R.S., Smith, P.H., Frecker, J.E. et al. (1986) A Fabry–Perot interferometer for accurate measurement of temporal changes in stellar doppler shift. *Proc. SPIE*, **627**, 2–19.
McMillan, R.S., Smith, P.H., Perry, M.L. et al. (1990) Long-term stability of a Fabry–Perot interferometer used for measurement of stellar Doppler shift. *Proc. SPIE*, **1235**, 601–9.
McMillan, R.S., Smith, P.H., Moore, T.L. et al. (1992) Variation of the radial velocity of Epsilon Cygni A. *Publ. Astron. Soc. Pacific* **104**(682), 1173–6.
McMillan, R.S., Moore, T.L., Perry, M.L. et al. (1993) Radial velocity observations of the sun at night. *Astrophys. J.*, **403**(2), 801–9.
McMillan, R.S., Moore, T.L., Perry, M.L. et al. (1994) Long, accurate time series measurements of radial velocities of solar-type stars. *Astrophys. Space Sci.*, **212**, 271–80.
Nakajima, T., Oppenheimer, B.R. Kulkarni, S.R. et al. (1995) Discovery of a cool brown dwarf. *Nature*, **378**, 463–5.
Nakano, T. (1988) Formation of planets around stars of various masses – III. Massive and small-mass stars and the regions of planet formation. *Mon. Nat. Roy. Astron. Soc.*, **235**, 193–201.
Rasio, F.A., et al. (1992) An observational test for the existence of a planetary system orbiting PSR 1257+12. *Nature*, 355, 325–7.
Sargent, A.I. and Beckwith, S.V.W. (1991) The molecular structure around HL Tauri. *Astrophys. J. Lett.*, **382**, L31–5.

Sargent, A.I. and Beckwith, S.V.W. (1993) *Physics Today*, **46**(4), 22–9.
Shu, F.H., Adams, F.C. and Lizano, S. (1987) Star formation in molecular clouds: observations and theory. *Ann. Rev. Astron. Astrophys.*, **25**, 23–81.
Skrutskie, M.F., Forrest, W.J. and Shure, M. (1989) An infrared search for low mass companions of stars near the Sun. *Astron. J.*, **98**, 1409–17.
Skrutskie, M.F., Dutkevih, D., Strom, S.E. *et al.* (1990) A sensitive 10-micron search for emission arising from circumstellar dust associated with solar-type pre-main-sequence stars. *Astron. J.*, **99**, 1187–95.
Smith, B.A. and Terrile, T.J. (1984) A circumstellar disk around Beta Pictoris. *Science*, **226**, 1421–4.
Taylor, S.R. (1992) *Solar System Evolution: a new Perspective*. New York: Cambridge University Press)
Walker, G.A.H., Walker, A.R., Irwin, A.W. *et al.* (1995) A search for Jupiter-mass companions to nearby stars. *Icarus*, **116**, 359–75.
Wetherill, G.W. (1990) Formation of the Earth. *Ann. Rev. Earth Planet. Sci.*, **18**, 205–56.
Wetherill, G.W. (1991) Occurrence of Earthlike bodies in planetary systems. *Science*, **253**, 535–8.
Wetherill, G.W. (1992) Possible consequences of absence of a 'Jupiter' in planetary systems. Oral presentation to the 'First International Conference on Planetary Systems: Formation, Evolution, and Detection' held at the California Institute of Technology, December 8–10, 1992, Pasadena, CA.
Wolszczan, A. and Frail, D.A. (1992) A planetary system around the millisecond pulsar PSR 1257+12. *Nature* **355**, 145–7.

Cross references

Earth–Moon system: origin
Solar system: origin
Solar nebula

PLANET X

Sometime between 1905 and 1908, Percival Lowell coined the term 'Planet X' when referring to any undiscovered planet beyond the orbit of Neptune. Based on William Herschel's discovery of Uranus in 1781 and Johann Gottfried Galle's discovery of Neptune in 1846, Lowell conjectured that even more distant planets were awaiting discovery. Because Urbain Jean Joseph Leverrier (q.v.) had successfully predicted Neptune's position by computing its gravitational pull on Uranus, Lowell turned to similar orbital calculations to predict the location of Planet X. In 1930 Clyde Tombaugh, while working at the Lowell Observatory near Flagstaff, Arizona, found Pluto at celestial coordinates very near the location predicted by Lowell in 1914. At first some hailed this as another triumph for celestial mechanics, although Tombaugh himself was more circumspect. However, as it became increasingly clear that Pluto was far too small to induce any observable orbital perturbations on either Uranus or Neptune, astronomers agreed that the 1930 coincidence of Pluto's position with Lowell's prediction was fortuitous.

Although several observers insisted that there were unexplained motions in the orbits of both Uranus and Neptune, James W. Christy's 1978 discovery of Pluto's satellite Charon at the US Naval Observatory, and the subsequent mass estimate for the Pluto/Charon system by the Naval Observatory's Robert S. Harrington, ruled out Pluto as the cause. Recently a team at the Jet Propulsion Laboratory (JPL), led by George W. Null, used images transmitted from the Hubble Space Telescope to conclude that the individual masses of Pluto and Charon yielded mean densities of 2.1 and 1.4 g cm^{-3} respectively. Further, the mass of the system was only 1.46×10^{22} kg – about 0.0024 the Earth's mass – more in the category of icy satellites than planets. But although Pluto was ruled out, some astronomers insisted that the unexplained orbital motions of Uranus and Neptune had not gone away. Planet X was still a possible source of gravitational perturbation.

A few investigators, including Dennis Rawlins, Conley Powell, P. Kenneth Seidelmann and Robert S. Harrington of the Naval Observatory, and Adrian Brunini, R.S. Gomes and S. Ferraz-Mello in South America, concluded that the evidence for Planet X was suggestive, though far from conclusive. They agreed, some with more conviction than others, that unmodeled forces were acting on Uranus and Neptune, and most of them predicted the general location of Planet X if that were the cause, but they arrived at no consensus on either its location or its orbit. At JPL E. Myles Standish, Jr took exception to the conclusion that there were problems with the orbits of Uranus and Neptune. He insisted that any failure to fit the astrometric data over the past two centuries could be explained by systematic errors in the observations, and in some cases by faulty data reduction and interpretation.

The spinning spacecraft Pioneer 10 and 11 (q.v.) acted as sensitive probes of gravitational fields in the outer solar system as they receded from the solar system. Over an interval of 16 years, celestial mechanics investigators at JPL obtained good fits to the Pioneer orbital data without invoking Planet X. On the other hand, in 1986 A.S. Guliev in Azerbaidzhan used orbits of long-period comets to derive the orbital elements of a single Planet X at a distance of 36.2 AU, and more recently the orbital planes – with inclination of about 30 arc degrees to the ecliptic – for two Planets X at distances of 48.5–56.6 AU and 102–112 AU respectively. Unfortunately the Pioneer data could not rule out these planets, or for that matter other predictions based on the Uranus and Neptune data.

If the dynamical evidence points to at least one Planet X, why have optical surveys and the Infrared Astronomical Satellite (IRAS) failed to find it? Tombaugh conducted the most extensive optical search from 1929 to 1943. He became convinced that besides Pluto no planets existed to limiting magnitude 16, at least within a wide band surrounding the ecliptic. Between 1977 and 1984 Charles T. Kowal conducted an optical survey with the 48-inch (122-cm) Schmidt telescope at Palomar Observatory. He surveyed a region extending 15 arc degrees north and south of the ecliptic, and included significantly fainter objects than Tombaugh. He found no Planet X. Yet even these dedicated observers were unable to observe all the sky. A large region near the celestial south pole was inaccessible to Tombaugh. Other regions were at high ecliptic latitude where there was little motivation to survey the sky for planets. The possibility remains that Planet X was hiding in a region not photographed extensively, or that it was darker than magnitude 16 in one of Tombaugh's regions.

The problem with using the 1983 IRAS data is that there are many objects on the sky that emulate a planetary infrared spectrum, at least at the four IRAS wavelength bands centered at 12, 25, 60 and 100 μm. An unambiguous detection of Planet X requires a determination of its proper motion. Unless Planet X were moving fast enough to be identified in two confirmed IRAS scans of the same location (HCONs) separated by 1–2 weeks, it could easily go undetected. Otherwise, only 72% of the sky was covered with HCONs separated by 6 months or more. A group at the University of Toronto (Hogg, Quinlan and Tremaine, 1991) concluded that it was unlikely that any Planet X large enough to exert significant perturbations on the known planets would have escaped detection by IRAS; but they admitted that a smaller planet could have been missed for a variety of reasons. Considering all the evidence, we must conclude that the possible existence of Planet X is still an open question.

John D. Anderson

Bibliography

Anderson, J. (1988) Planet X: fact or fiction? *Planetary Rep.* **8**(4), 6–9.
Hogg, D.W., Quinlan, G.D. and Tremaine, S. (1991) Dynamical limits on dark mass in the outer solar system. *Astron. J.*, **101**(6), 2274–86.
Hoyt, W.G. (1980) *Planets X and Pluto*. Tucson: University of Arizona Press.
Littmann, M. (1988) *Planets Beyond*. New York: John Wiley and Sons.
Seidelmann, P.K. and Harrington, R.S. (1988) Planet X – the current status. *Celest. Mech.*, **43**, 55–68.

Cross references

Celestial mechanics
Leverrier, Urbain Jean Joseph
Lowell, Percival
Tombaugh, Clyde William

PLANETARY DATA SYSTEM

The Planetary Data System (PDS) is a distributed information system for digital planetary science data. PDS archives data from past and currently active spacecraft missions, ground-based astronomical observations and related laboratory measurements that can be applied to planetary studies. It is supported by the National Aeronautics and Space Administration (NASA) to facilitate and ensure long-term availability of planetary data and to stimulate new research and correlative analyses.

Structure

PDS comprises a Central Node and seven discipline nodes (Figure P9). The Central Node at the Jet Propulsion Laboratory (JPL) provides overall direction and coordination while the discipline nodes are responsible for developing and disseminating data archive standards, curating datasets specific to their expertise, maintaining catalogs and databases of planetary data and distributing those data to the planetary community. In order to cover all aspects of a discipline adequately, a consortium of subnodes is usually associated with each discipline node. One institution serves as the principal site for each node and one scientist at each site acts as the node manager. As the needs of the planetary science community change, the structure of PDS is expected to evolve slowly in response.

Five of the discipline nodes specialize in the science areas of atmospheres, geology and geophysics, fields and particles, small bodies and planetary rings. The Geosciences Node at Washington University, for example, handles Magellan radar and gravity data, spaceborne thermal and spectroscopy data, and data from studies on Earth that might be relevant to interpretation of planetary results. The Small Bodies Node at the University of Maryland concentrates on data about comets, asteroids, small satellites and dust. Imaging, the sixth discipline node, is split between the Jet Propulsion Laboratory and the US Geological Survey. It provides digital images, ancillary data, software tools and technical expertise necessary to utilize the vast collection of digital planetary images. With emphasis on imaging as a technique rather than the science content of the images, this node provides an important function without duplicating services provided by the five 'science' nodes. The seventh discipline node is the Navigation and Ancillary Information Facility (NAIF), which is responsible for archiving and distributing geometry information such as spacecraft trajectories and attitude information, planetary ephemerides and planetary constants.

Data flow

For spacecraft missions PDS seeks early consultation with mission managers to ensure that datasets and data systems will be compatible with previously adopted standards. Once data are acquired, PDS works with mission personnel to transfer the data into the PDS system, usually through one or more predesignated discipline nodes. After the active phase of a mission, PDS remains ready to ingest newly derived datasets, calibration data and relevant documentation. PDS also works with other scientists to restore old datasets, incorporate non-spacecraft observations of the planets and ingest laboratory data that could be used to support planetary studies.

Data presented to PDS must pass through several stages of scrutiny and adaptation before being approved for ingestion and distribution. First, datasets must adhere to standards intended to limit the proliferation of formats. The dataset must also be complete and be viable from a scientific perspective. Datasets that pass these reviews are entered into PDS on-line data catalogs. Long-term archive and distribution copies of the dataset are sent to the National Space Science Data Center (NSSDC).

Users in the planetary community may access the catalog information either by visiting a node or by making an electronic connection, usually via the World Wide Web. In searching the catalog the user will become familiar with the data available and, at the same time, will be able to narrow a future request (or 'order') for those data. Once the user is satisfied with the breadth and depth of the order, that request can be submitted. Some science data can be obtained immediately via the Web; all data can be obtained on physical media such as computer-compatible tape or CD-ROM.

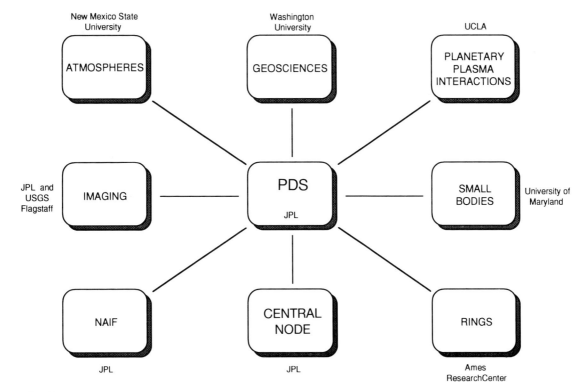

Figure P9 The Planetary Data System.

Functions and services

Catalogs

The PDS central dataset catalog is available at the Central Node. The catalog is similar in function and purpose to a card catalog in a library. Users can examine the general catalog to find datasets of interest and then navigate to discipline-specific catalogs for more detailed examination of the data holdings. The catalogs are generally accessible via electronic networks and modems, and a range of interfaces is available to support users with both simple and sophisticated computer equipment.

Data distribution

The on-line PDS catalogs make it possible for users in the planetary science community to locate data of interest for research projects. In some cases those data may be accessed and analyzed or manipulated immediately after being located, with results being delivered over the same lines as the initial queries were made. In other cases, where the user does not have the means to accept an immediate delivery or where the dataset may be too large for convenient transmission, PDS supplies a copy of the data on an appropriate medium – typically magnetic tape or CD-ROM. PDS does not have the resources to respond to requests for data from the general science community; those are handled through the NSSDC.

Data publishing

PDS has pioneered the use of CD-ROMs to distribute large datasets. It continues to lead in efforts toward standardization, readability and data compression within this medium. These efforts have meant that many planetary scientists now can access substantial archives of data with only a modest investment in computer equipment. PDS has archived several hundred CD-ROMs for various projects including Magellan, Galileo, Mars Observer, International Halley Watch and Phobos.

Standards

To ensure maximum utility of its archive collection PDS develops, promotes and maintains standards for documentation and preparation of datasets. These standards result in uniform presentation of general information about datasets and allow broad searches for data across missions, target bodies and disciplines. Adoption of standards for datasets means that a small set of software tools can be built to create, access and manipulate data. Standards also result in published products (CD-ROMs) that can be read by planetary scientists using a wide variety of computing equipment.

Software tools

PDS continues to develop software tools for creating, validating and manipulating data of interest. Initially these were needed to assist scientists in preparing datasets for submission to PDS, but they can also be used more widely to verify that products meet PDS standards. For some datasets, fully developed software packages for analysis of the data are available, occasionally included as part of the published dataset.

Susan McMahon and Sandy Dueck

Bibliography

Arvidson, R.E. (1986). *Issues and Recommendations Associated with Distributed Computation and Data Management Systems for the Space Sciences*. Committee on Data Management and Computation, Space Science Board, Assembly of Mathematical and Physical Sciences, National Research Council, Washington, DC: National Academy Press.

Cross references

NASA
Regional Planetary Image Facilities

PLANETARY DYNAMICAL ASTRONOMY

The existence and motions of the nearer planets have been known since classical times. The word 'planet' is derived from the Greek word to 'wander,' indicating that the planets were known to move with respect to the fixed stars. Knowledge of the motions of these bodies appears to have emerged independently in different parts of the world, such as the Middle East and China. Even the Mayan civilization was apparently very knowledgeable concerning the periods and motions of some of the planets. The motions of the Moon and the Earth (or equivalently, the Sun) were the bases for calendars, time keeping and the regulation of life patterns.

The advancement of science, in general, has been intimately tied to improvements in the ability to calculate the motions of the planets. Many powerful mathematical functions were first developed to solve problems of planetary motions. Problems in celestial mechanics have been a stimulus for development of computers, and improvements in computation (and observational technology) have in turn stimulated new investigations and improved models in celestial mechanics. The theory of relativity was developed in part to address a quantitative discrepancy between observations and celestial mechanics computations, e.g. an unexplained motion of the perihelion of Mercury. Our determination of the astronomical reference system and astronomical constants has been directly tied to the dynamics of the planets. Regular and accurate time scales are needed in this study. In short, a number of discoveries in both mathematics and fundamental physics have resulted from observations of the planets and from attempts to determine ephemerides for these bodies.

Ephemerides

The computation of ephemerides for solar system bodies requires a complete model for the solar system, the equations to represent that model and the computing capability to carry out the computation. If the ephemerides are to represent the true motions, they must be fit to accurate observational data covering an extended time period.

Accurate computations of the ephemerides became possible after Johannes Kepler (q.v.) developed the laws of planetary motion. Subsequently, Isaac Newton (q.v.) developed the universal law of gravitation, which was essentially consistent with Kepler's three laws. Einstein's theory of general relativity resolved a discrepancy with the motion of Mercury, and more accurately represented the true motions of the bodies of the solar system (Danby, 1988).

According to Newton's universal law of gravitation there is an attraction between every pair of bodies, so that while the planets each move in their orbit around the Sun they are also perturbed by each of the other planets in the solar system. Thus these perturbations must be included in the computations of the motion of each of the bodies. The computations can be made by the methods of general or special perturbations (Brouwer and Clemence, 1961).

Prior to the advent of electronic computing equipment in the 1940s, the computation of ephemerides was accomplished by means of general perturbations (Seidelmann, 1979). Thus a series of trigonometric terms called Fourier series was used to represent the motion of each one of the bodies. The principal sources of such theories were Newcomb (1898), Hill (1898) and Leverrier (1861a,b). With the advent of the electronic computing equipment, special perturbations or numerical integration could be performed accurately and rapidly by means of the computer (see Celestial mechanics). At the present time, accurate ephemerides are all calculated by numerical integration (Standish, 1982), since general theories have not been able to achieve the accuracies of numerical integration. The precision of the ephemerides is dependent on the number of digits available from the computer, the method of numerical integration and the truncation of the equations. On the other hand, the accuracy depends on the completeness of the model, the order of the integration and the accuracy and duration of the observational data.

Ephemerides were traditionally made available in the form of tables or almanacs. With the availability of computer equipment, the ephemerides could be made available on magnetic tapes, floppy disks, as computer subroutines and through computer networks. There is a continuing interaction between the accuracy of ephemerides and the accuracy of the observational data (Seidelmann, 1982).

Observations of planets and solar system bodies

Historically, observations were made in various ways. The best early observations were those of Tycho Brahe (q.v.) in the 16th century. His collection of observational data permitted Kepler to formulate his laws. It was fortunate that Kepler investigated the motion of Mars, because that was the only planet for which the observational data were sufficiently accurate, the eccentricity sufficiently large and the perturbations sufficiently small that Kepler could formulate his laws.

By 1830 the observational techniques and the reduction procedures improved significantly to produce more accurate observational data. Subsequent improvements in instrumentation, procedures and constants have permitted continuing improvements in the accuracy of the observations. Accurate positional observations have been made historically by means of telescopes which are restricted in their freedom of motion. For example, mural circles were mounted on a wall and meridian circles mounted so they only move along the meridian, etc. The techniques of making the observations and the precision of the instrumentation have improved over the years. The moving wire micrometer on a transit circle significantly improved the observational accuracy. The recognition of the personal equation (taking account of the fact that each individual observes slightly differently) led to additional improvements until finally electronic detectors eliminated the effect. Transit circle observations have provided most of the accurate optical angular observational data presently available for the Sun, Moon and planets. The observations were made along with star observations, so that the positions of the planets were determined with respect to the stars and the stars and planets were referred to the equator and equinox.

In 1930 Pluto was discovered as a 15th-magnitude body, which was too faint to be observed with transit circles. Therefore, observations of Pluto were made photographically, with respect to faint stars whose positions were not as well known as the brighter stars. Only since 1989 have observations of Pluto been possible by means of a transit circle instrument. This is possible because Pluto is near perihelion and transit circle detectors are more sensitive. The large uncertainty in the early observations, and the short observation period from prediscovery observations in 1914 to the present, result in a rather uncertain orbit for Pluto.

The observations of the Moon and planets are restricted in accuracy by magnitude differences, phase effects and the effective illumination, which means that the observed limb of the planet is different from that expected from a geometric determination of the phase effect. With accurate knowledge of the motions of the satellites around the planets and the stellar point-like images of the satellites, photographs of the satellites with respect to reference star fields can provide more accurate positions of the planets than direct observation of the planets themselves (Pascu and Schmidt, 1990).

In the 1960s radar ranging observations of the Moon, Mercury, Venus and Mars were initiated. For the first time accurate measurements of the distances from the Earth to the planets were possible, in addition to the angular optical observations. As with optical observations, the accuracy of these radar ranging measurements has significantly improved with time. In 1969 the Apollo landing on the Moon provided the means to place a corner-cube reflector on the Moon. This reflector could be used with lunar laser ranging equipment to bounce a light beam off the reflector and measure the return time. Thus very accurate measurements of the distance to the Moon were possible, and lunar laser ranging has continued, with the addition of other reflectors on the Moon, to provide the most precise distance measurements within the solar system. Current measurements of the distance to the Moon can be made at centimeter accuracies.

Spacecraft missions to the planets provide another source of observational data. With communication equipment on board the spacecraft, the speed of the spacecraft can be measured from the Doppler effect. The distance to the spacecraft, while it is in orbit around the planet or has landed on the planet, can also be determined. Also, as a spacecraft goes past, or into orbit around the planet, the gravitational effects of the planet on the spacecraft can be detected; these observations may be employed to determine the mass of the planet. Thus the spacecraft measurements provide very accurate observations, with the limitation that they only cover a short period of time (Standish, 1990).

Other sources of observations of solar system bodies include detection of the blackbody emissions of the minor planets by means of the Very Large Array; timings of eclipses, which provide unique measurements of the relative angular positions of the Sun and Moon; and occultation timings. Occultations provide the times of the Moon passing in front of a star, and thus measure the lunar limb effects and positions of the Moon with respect to the stars. The duplicity of binary stars can be detected, the difference between the star catalog equinox and the dynamical equinox can be determined, and the rotation of the Earth can be measured (Morrison, 1980). Occultations of stars by planets or minor planets permit the determination of the position of the planet with respect to the stars, measurement of the diameter of the planet and the detection of the presence or absence of an atmosphere.

Parameter solutions

The determination of an accurate ephemeris of a body requires the fitting of a preliminary ephemeris to accurate observational data, by means of the computation of partial derivatives and a solution, generally by means of the least squares method. From this fitting process accurate values of the osculating orbital elements, or the position and velocity, at a given epoch are determined. Also, the mass of another planet can be determined from its perturbations of the one being investigated, and the distance scale and other astronomical constants for the solar system can be determined.

Accurate computation of ephemerides requires the use of different types of observations, a fitting process and the appropriate weighting of the observations to achieve the most accurate ephemeris possible. The determination of the orbital motion of the Earth provides the means of defining the solar system reference planes. The mean orbital plane of the Earth is the ecliptic. The angle between the ecliptic plane and the mean equatorial plane is called the obliquity, and the intersection of the two planes, assuming a uniform motion of the ecliptic plane, is the equinox. Thus each ephemeris of the Earth determines these quantities.

Discoveries

The process of calculating ephemerides, making observations and fitting the ephemerides to the observations not only provides ephemerides and values of astronomical quantities, it also can lead to discoveries. Herschel was making observations when he detected a body that was moving with respect to the stars. This was the first discovery of a planet since classical times, and it was eventually named Uranus.

The inability to fit the observations of Uranus with an ephemeris, with residuals of over 100 arcsec between the observations and ephemerides, led both Leverrier and Adams to predict the existence of another planet. The prediction was sufficiently accurate that when observations were attempted, the body was discovered on the first night. This body was called Neptune (Grosser, 1979). By 1900 investigations of the orbits of Uranus and Neptune, in comparison with observations, led to predictions of another planet (Pickering, 1909; Lowell, 1915). After considerable searching, the planet Pluto was found relatively close to its predicted location (Tombaugh, 1946), but it was much fainter than expected, and has since been determined to be much less massive than necessary to justify its prediction (Hoyt, 1980). Thus Pluto's discovery close to its predicted position remains a mystery. Either the discovery was serendipitous and there are many such bodies out there, or there is another undiscovered body that caused the perturbations.

Observations of the planets have led to discoveries of satellites around the planets. The irregular image of Pluto on plates taken to determine its ephemeris led to the realization that it has a satellite, Charon (Christy and Harrington, 1978).

Investigation of the observations of Mercury, in comparison with its ephemeris, led to the recognition of an unexplained motion of the perihelion of Mercury. Attempts were made to explain this by means of a planet interior to Mercury, but eventually Einstein's general theory of relativity accounted for the discrepancy. The theory of relativity also required modifications of the equations of motion for all the bodies of the solar system, and thus provided the means for improving the accuracy of computation of the ephemerides (Will, 1985).

Timescales

Comparisons of the observations of the Moon with its ephemeris led to Newcomb's confirmation (1874, 1882) of Kent's hypothesis of the non-uniform rotation of the Earth (Felber, 1974). De Sitter (1927) determined that the discrepancies in the orbit of Mercury and the Moon were proportional to the mean motions of the bodies, thus strongly indicating that the Earth's rotation is variable. This variability in the rotation of the Earth meant that Universal Time, which was based on the rotation of the Earth with respect to the Sun, was not uniform.

In 1952 the concept of Ephemeris Time was introduced to provide a uniform time, or a time that could be used as the independent variable in the equations of motion for the solar system (Clemence, 1971). Unfortunately Ephemeris Time could not be predicted in the future; it could only be determined, after the event, from observations of the bodies of the solar system, particularly the Moon, which has the most rapid motion.

In 1956 atomic time became available to provide an accurate time, which was available in real time. Subsequently, for all intents and purposes, Ephemeris Time was effectively based on International Atomic Time (TAI). Prior to 1956 Ephemeris Time has to be determined from the observational data.

In 1976, Terrestrial Dynamical Time (TDT) was introduced as a timescale for geocentric ephemerides, and it was directly defined in terms of TAI. So, it is an idealized atomic time. Barycentric Dynamical Time (TDB) was defined as a timescale for the barycenter of the solar system for barycentric ephemerides and it was related to TDT; according to the relativistic metric being used for the equations of motion for the ephemerides. TDB was defined to have only periodic differences from TDT.

In 1991 the International Astronomical Union (IAU) adopted two new timescales. Geocentric Coordinate Time (TCG) is a coordinate time at the geocenter of the Earth and Barycentric Coordinate Time (TCB) is the coordinate time at the barycenter of the solar system. These timescales are to be determined with respect to TAI in strict accordance with the equations of general relativity. Thus TCB will have a secular difference with respect to TAI. The unit of time in all cases will be based on the System International (SI) second, such that the differences are consistent with time retardation as expected from the theory of relativity. The difference between the timescales for ephemerides, such as ET, TDT, or TCG, and Universal Time are given as Delta T. For times prior to 1956, Delta T must be determined from observational data as the difference between Ephemeris Time, determined from the orbital motion of the Sun, Moon and planets, and Universal Time, determined from the rotation of the Earth.

Computers in celestial mechanics

The problems of celestial mechanics are the determination of the ephemerides of the Sun, Moon and planets. They have presented a computational challenge from early times. Many developments in the field of mathematics have been directly due to the requirements of celestial mechanics (Szebehely, 1989). Likewise, the computational requirement has been a principal driver for computers (Seidelmann, 1976). Babbage constructed a model machine in 1822 to integrate second differences. Early computer equipment in the 1920s in England was used for computation of the lunar ephemeris. The early punch-card equipment was used for computation of the motions of the planets and preparation of almanacs. The numerical integration of the five outer planets by Eckert, Brouwer and Clemence (1951) was carried out on the Selective Sequence Electronic Computer by IBM. The numerical Orrery developed by Sussman (Applegate *et al.*, 1986) was a special-purpose computer designed and built to be able to numerically integrate the planets for a long period of time to investigate the stability of the solar system. Today, the speed of computers permits the solution of most problems of computation of the orbits. The limitations are in the accuracy that can be achieved. Modern mapping techniques provide improved means of satisfying both the speed and accuracy requirements (Wisdom and Holman 1991).

P.K. Seidelmann

Bibliography

Applegate, J.H., Douglas, M.R., Gursel, Y. *et al.* (1986) The outer solar system for 200 million years, *Astron. J.*, **92**, 176–94.
Brouwer, D. and Clemence, G.M. (1961) *Methods of Celestial Mechanics*. New York and London: Academic Press.
Christy, J.W. and Harrington, R.S. (1978) The satellite of Pluto, *Astron. J.*, **83**, 1005–8.
Clemence, G.M. (1971) The concept of ephemeris Time; a case of inadvertent plagiarism, *J. Hist. Astron.*, **2**, 73–9.
Danby, J.M.A. (1988) *Fundamentals of Celestial Mechanics*. Richmond, VA: Willman-Bell, Inc.
de Sitter, W. (1927) On the secular accelerations and the fluctuations of the longitudes of the Moon, the Sun, Mercury and Venus. *Bull. Astron. Inst. Neth.*, **4**, 21–38.
Eckert, W.J., Brouwer, D. and Clemence, G.M. (1951) Coordinates of the five outer planets. *Astron. Papers Am. Ephem.*, **12**. Washington, DC: US Government Printing Office.
Felber, H.-J. (1974) Kants Beitrag zur Frage der Verzorgerung der Erdrotation, *Die Sterne*, **50**(2), 82–90.
Grosser, M. (1979) *The Discovery of Neptune*. New York: Dover Publications Inc.
Hill, G.W. (1898) A New Theory of Jupiter and Saturn. *Astron Papers Am. Ephem.*, **4**. Washington.
Hoyt, W.G. (1980) *Planet X and Pluto*. Tucson, Arizona: University of Arizona Press.
Leverrier, V.J.J. (1858) *Ann. Obs.*, Paris, **4**, 1.
Leverrier, V.J.J. (1859) *Ann. Obs.*, Paris, **5**, 1.
Leverrier, V.J.J. (1861a) *Ann. Obs.*, Paris, **6**, 1.
Leverrier, V.J.J. (1861b) *Ann. Obs.*, Paris, **6**, 185.
Lowell, P. (1915) Memoirs on a trans-Neptunian planet, *Mem. Lowell Observatory*, **1**, 1.
Morrison, L.V. (1980) An analysis of total lunar occultations made in the years 1943–1974. *J. Br. Astron. Assoc.*, **91**, 14–24.
Newcomb, S. (1874) On the possible variability of the Earth's axial rotation. *Am. J. Sci. Arts, Sec. 3*, **8**, 45.
Newcomb, S. (1882) Transits of Mercury. *Astron. Papers Am. Ephem.*, **1**, 363–487. Washington.
Newcomb, S. (1898) Tables of the four inner planets, *Astron. Papers Am. Ephem.*, **6**, Washington.
Pascu, D, and Schmidt, R.E. (1990) Photographic positional observations of Saturn, *Astron. J.*, **99**, 1974.
Pickering, W.H. (1909) A search for a planet beyond Neptune, *Ann. Astron. Observ. Harvard College*, **41** (2). Cambridge: Mass.
Seidelmann, P.K. (1976) Celestial mechanics, in *Encyclopedia of Computer Science and Technology*, Vol. 4 (eds J. Belzer, A.G. Holzman and A. Kent). New York and Basel: Marcel Dekker Inc., pp. 243–67.
Seidelmann, P.K. (1979) The ephemerides: past, present, and future, in *Dynamics of the Solar System* (ed. R.L. Duncombe). Dordrecht, D. Reidel Publishing Co., pp. 99–114.
Seidelmann, P.K. (1982) Orbital motion of the planets, theoretical and observational, *Celest. Mech.*, **26**, 149–60.
Standish, E.M. (1982) The JPL planetary ephemerides, *Celest. Mech.*, **26**, 181–6.
Standish, E.M. (1990) The observational basis for JPL's DE200, the planetary ephemerides of the Astronomical Almanac, *Astron. Astrophys.*, **233**, 252–71.
Szebehely, V.G. (1989) *Adventures in Celestial Mechanics*. Austin: University of Texas Press.
Tombaugh, C.W. (1961) The trans-Neptunian planet search, in *Planets and Satellites* (eds G.P. Kuiper and B.M. Middlehurst). Chicago: University of Chicago Press.
Will, C.M. (1985) *Theory and Experiment in Gravitational Physics*. Cambridge: Cambridge University Press.
Wisdom, J. and Holman, M. (1991) Symplectic maps for the *N*-body problem, *Astron J.*, **102**, 1528–38.

Cross references

Astronomical constants
Celestial mechanics
Chaotic dynamics in the solar system
Ephemeris
Gravitation
Orbit
Solar system: stability

PLANETARY GEODESY

The objectives of planetary geodesy are to determine the gravity, topography and rotation of the planets. These measurements are put to two basic uses. In a somewhat pragmatic vein, the geodetic parameters are useful for navigational purposes: the motion of a

Table P5 Geodetic parameters of the planets

Planet	Mass (km³·s⁻²)	Radius (km)	Period (s)
Mercury	22 031.8 ± 1.0	2 440 ± 1	5 067 020 ± 430
Venus	324 858.15 ± 0.17	6 051.47 ± 0.03	20 996 800 ± 1 200
Earth	398 600.434 ± 0.002	6 371.01 ± 0.02	86 164 090 ± 0.004
Moon	4 902.799 ± 0.003	1 737.53 ± 0.03	2 360 591.554 ± 0.005
Mars	42 828.3 ± 0.1	3 389.92 ± 0.04	88 642.663 ± 0.002
Jupiter	126 712 767 ± 5	69 946 ± 6	35 729.70 ± 0.05
Saturn	37 929 100 ± 2 400	58 300 ± 6	38 364.12 ± 0.05
Uranus	5 783 900 ± 3 800	25 456 ± 60	61 200 ± 720
Neptune	6 809 000 ± 14 000	24 650 ± 100	64 200 ± 300
Pluto	908 ± 42	1 123 ± 4	551 850 ± 20

spacecraft in orbit around a planet, relative to features of interest on the planetary surface, depends on the gravity, topography, and rotation of the planet. On the other hand, these same geodetic parameters provide the only remotely accessible information on the internal mass distribution in the planet.

The methods used to obtain geodetic information on the planets are quite diverse. The earliest estimates of planetary masses came from observing the orbital motions of natural satellites. For a small satellite, the size of the orbit (semimajor axis a) and the mean orbital angular velocity (mean motion n) are related to the mass of the primary body M via $GM = a^3 n^2$, where G is the gravitational coupling constant. Through the gravitational constant $G = (6.6726 \pm 0.0005) \times 10^{-20}$ km³ s⁻² kg⁻¹ is only known to an accuracy of roughly one part in 10^4, the product GM is known for many of the planets to an accuracy of a few parts in 10^7.

Departures from spherical symmetry in the gravity field of a planet will change the shape and/or orientation of the orbits of satellites. Detailed observations of natural satellites (and rings) yield estimates of the orbital precession rates. These natural probes have to some extent revealed the low-degree gravity harmonics of Jupiter, Saturn, Uranus and Neptune. The limiting factor in most natural satellite observations is the $\sim$ 1-arcsec positional accuracy attainable with optical techniques.

Artificial satellites of other planets are much too small to be observed optically, but the Doppler shift in the radio communication links induced by the relative motions of the satellite and the ground tracking station can be accurately measured. Accuracies of a few millimeters per second are often available and have allowed fairly detailed mapping of the gravity fields of the Earth, Moon, Mars and Venus. For Earth gravity measurements these Doppler observations are supplemented by laser ranging to specially constructed geodetic satellites. Accuracies of a few centimeters in range allow very accurate determinations of the long- and intermediate-wavelength gravity anomalies.

The earliest determinations of the sizes and shapes of the planets relied on optical telescopic measurements of the dimensions of the projected disk. Much better accuracy is attainable by timing the duration of stellar occultations. Better still are the results of spacecraft occultations and radar altimeters. For nearby planets, Earth-based radar ranging to the planetary surface yields good results.

Estimates of the rotational states of the planets have also come from a variety of sources. Obviously, the rotational states of the Earth and Moon have been known fairly well for centuries, though recent technological improvements have helped reveal subtle variations. For the other planets, initial estimates of rotation rates and axial orientations came from optical telescopic observations of the motion of discrete features across the observed disk. Though the method worked well for Mars, Jupiter and Saturn, the initial results obtained this way for Mercury and Venus were wrong.

A small solid body can maintain whatever shape it acquired during formation. However, for sufficiently large bodies (or for sufficiently weak material), the mutual gravitational attraction of the constituent parcels of mass will cause the body to assume an essentially spherical form. Thus, in the absence of tidal or rotational distortions, the gravitational potential at an external point will depend only on the total mass of the planet and neither the gravity nor the topography will convey any information about the internal density distribution. However, a rotating hydrostatic body will be distorted by an amount that depends on the radial density variation $\rho(r)$ and the spin rate ω.

A dimensionless parameter that indicates the relative importance of rotational distortion versus the gravitational tendency toward sphericity is $q = \omega^2 R^3 / GM$. Table P5 gives current estimates for all of these parameters. It appears that, in terms of the importance of rotation, the planets can be divided into two basic categories. Saturn, Jupiter, Uranus and Neptune are rapid rotators ($q > 10^{-2}$), whereas Venus, Mercury and the Moon are slow rotators ($q < 10^{-4}$). The Earth, Mars and Pluto are in an intermediate regime where rotation is important but not completely dominant.

In many applications it is convenient to describe the topography and gravitational potential of a planet in terms of spherical harmonic functions:

$$H(\theta, \phi) = R \Sigma \Sigma H_{nm} \Lambda_{nm}(\theta, \phi) \quad (P3)$$

$$G(r, \theta, \phi) = \frac{GM}{R} \Sigma \Sigma G_{nm} \left(\frac{r}{R}\right)^{-(n+1)} \Lambda_{nm}(\theta, \phi) \quad (P4)$$

where (r, θ, ϕ) are radius, latitude and longitude, and $\Lambda_{nm}(\theta, \phi)$ is a complex surface harmonic of degree n and order m. Each such harmonic is a product of a latitude-dependent associated Legendre function $P_{nm}[\sin(\theta)]$ and a longitude-dependent complex exponential $e^{im\phi}$. The real and imaginary parts of the harmonic coefficients are customarily denoted C_{nm} and S_{nm} respectively, and the zonal harmonic ($m = 0$) coefficients are frequently reported in terms of $J_n = -C_{n,0}$. If R and M are chosen to be the mean radius and total mass respectively, the degree-zero harmonic coefficients are both equal to 1. If the center of mass is chosen as the coordinate origin, the first-degree gravity coefficients are identically zero.

For rotating bodies in hydrostatic equilibrium, the only nonzero coefficients are the even-degree zonals, $J_2, J_4, J_6, \ldots$, and these coefficients form a rapidly decreasing series with $J_{2n} = O(q^n)$. This high degree of symmetry (symmetry about the rotation axis and the equatorial plane) makes it possible to determine reliably the low-degree harmonics of topography and gravity for Jupiter, Saturn, Uranus and Neptune from relative few observations. No such symmetry applies to non-hydrostatic planets, and their topographic and gravitational variance spectra can only be determined from detailed observations over the entire surface.

For a spherical body with internal density variations described by

$$\rho(r, a, b) = \langle\rho\rangle \Sigma \Sigma A_{nm}(r) \Lambda_{nm}(a, b) \quad (P5)$$

where $\langle\rho\rangle = 3M/4\pi R^3$ is the mean density, the gravity harmonics are

$$G_{nm} = \frac{3}{2n+1} \int_0^1 A_{nm}(x) x^{n+2} \, dx \quad (P6)$$

with $x = r/R$. As a special case, the gravitational effect of topography in a surface layer of density $\rho_s = \tau \langle\rho\rangle$ can be approximated by $G_{nm} = 3\tau H_{nm}(2n + 1)$.

This formula conveys two basic ideas: if the surface layers have low density, the topography has relatively little gravitational impact and, for a given density, long-wavelength topography has the greatest effect.

For a fluid planet of uniform density, the relationship between rotation, gravity and topography is particularly simple. To lowest order in q, we have

$$-G_{2,0} = J_2(g) = \frac{q}{2}, \qquad -H_{2,0} = J_2(t) = \frac{5q}{6} \quad (P7)$$

These values are maximal in the sense that, for a given value of q, a hydrostatic body whose density increases with depth will have gravity and topography harmonics less than these limits. In fact, for a hydrostatic body the mean moment of inertia I is related to the gravitational oblateness via

$$\frac{I}{MR^2} = \frac{2J_2}{3J_2 + q}$$

It is primarily by just such comparisons of the gravity, topography and rotation that we can make inferences about the composition and internal structure of the Jovian planets. On the other hand, for a non-hydrostatic body the values of the low-degree harmonics will not bear any simple relationship to the rotation rate.

Additional information concerning individual planets is to be found in the corresponding articles in this volume (see cross references below). Bills (1989) provides a succinct review of other planetary geodetic quantities.

Bruce G. Bills

Bibliography

Anderson, J.D. (1975) Planetary geodesy. *Rev. Geophys.* **13**, 274–5.
Bills, B.G. (1989) Planetary geodesy, In D.E. James (Ed.), Encyclopedia of Solid Earth Geophysics, New York: Van Nostrand Reinhold, pp 931–8.
Bills, B.G. and Synnott, S.P. (1987) Planetary geodesy. *Rev. Geophys.* **25**, 833–9.
Bretagnon, P. (1982) Integration constants and mean elements for all the planets, *Astron. Astrophys.*, **108**, 69–75.
Davies, M.E., Bursa, M., Abalakin, V.K. *et al.* (1986) Cartographic coordinates and rotational elements of the planets and satellites. *Celest. Mech.*, **39**, 103–3.
De Vaucouleurs, G. (1964) Geometric and photometric parameters of the terrestrial planets. *Icarus* **3**, 187–235.
Ferrari, A.J., and Bills, B.G. (1979) Planetary geodesy. *Rev. Geophys.* **17**, 1663–77.
Hubbard, W.B., (1984) *Planetary Interiors*. New York: Van Nostrand Reinhold.
Sjogren, W.L., 1983, Planetary geodesy. *Rev. Geophys.*, **21**, 528–37.

Acknowledgement

Adapted with permission of the author and editor from Van Nostrand Reinhold *Encyclopedia of Solid Earth Geophysics*.

Cross references

Earth
Earth rotation
Gravity fields of the terrestrial planets
Jupiter
Mars
Mars: gravity
Mercury
Moon: gravity
Neptune
Planet·
Planetary rotation
Pluto
Saturn
Uranus
Venus
Venus: gravity

PLANETARY ICE

The high measured albedos of solar system bodies, together with calculations of surface temperatures, have long fostered the belief that ices must be present on the surfaces of many objects within the solar system. However, the present unambiguous observational evidence for the composition of these ices has only been available since the early 1970s, concurrent with rapid advances in detector technology for near-infrared astronomy.

Ices of condensed volatile materials have been detected on the surfaces of many of the objects in the solar system: from the mantles of distant comets to the surface of Pluto, and from the rings of Saturn to the poles of Earth. Ices may even exist on the poles of the innermost planet, Mercury.

The principal component of most ices found on the surfaces of solar system objects is ice formed from water. Evidence of ices composed of other materials is also found. CO_2 ice is present on Mars, SO_2 ice is found on Io and methane + nitrogen ice has been detected on Triton.

These ices are generally believed to occur in the form of frosts (i.e. ice having particle sizes less than a few millimeters). This interpretation arises principally from the high albedos that have been measured for ice-covered objects, which suggest significant scattering from the surface.

Historical development

Estimations of the nature of the surfaces of the outer planet satellites were made at least as early as the 1950s (Urey, 1952). Estimates of temperatures, calculations of the stability of volatiles on small satellites (Jeans, 1967), estimates of cosmochemical abundances (Rubey, 1952), considerations of the nebular hypothesis (q.v.) for solar system formation and the discovery of methane on Jupiter all contributed to the belief that ices would persist in the outer solar system and would consist of species which were reducing. Stabilities of minor species, in the form of clathrates, were estimated (Miller and Smythe, 1970). Broadband photometric observations (Johnson and McCord, 1971), together with sparse laboratory data, were consistent (but not diagnostic) with a hypothesis that water ice existed on the Galilean satellites Europa, Ganymede and Callisto. Telescopic observations of the persistent south polar cap on Mars, together with Mariner 7 observations in the infrared (Pimental, Forney and Herr, 1974) were consistent with the presence of a permanent cap of water ice. Radar observations revealed that both Europa and the rings of Saturn were extraordinarily bright.

Spectral evidence for water ice in the outer solar system was obtained for Jupiter's Galilean satellites over the spectral region 1–5 µm (Pilcher, Ridgeway and McCord, 1970; Lefobsky, 1976), and soon after for Saturn's rings. Wamestaker, Kroes and Fountain (1974) suggested that the presence of sulfur could account for the spectral properties of Io, which did not fit well with the water ice hypothesis. A deep absorption in Io's spectrum at 4.1 µm, observed by Pollack *et al.* (1978), was matched to a laboratory spectrum of SO_2 frost (Smythe, Nelson and Nash, 1979; Figure P9) – a surprising result inconsistent with the then-current notion that only reducing species should be prevalent in the outer solar system. The discovery of Io's volcanoes (Morabito *et al.*, 1979) suggested a reason for the unexpected composition. Observations of the Saturnian satellites (Clark *et al.*, 1984) and Uranian satellites (Brown, 1982) soon followed, showing the dominant species on the surfaces of these objects to be water ice. Observations of Pluto are consistent with presence of solid methane (Cruikshank, Morrison and Pilchar, 1976), although the observation is a difficult one due to Pluto's small size and great distance from the Sun. Perhaps the most intriguing discovery is that of an absorption attributed to nitrogen on Triton (Cruikshank *et al.*, 1988). The depth of this very weak absorption requires unusual pathlengths for light within the surface (on the order of tens of centimeters). This is inconsistent with the presence of normal surface frosts, which typically have very short mean pathlengths of light in the surface (less than 1 cm).

Measurements on comets explicitly show water ice. A variety of ionized species detected around comets suggest that the surface chemistry is much more complex. A recent radar observation of Mercury, showing a much stronger reflectivity at the poles, has been interpreted (Slade, Butler and Muhleman, 1992) as evidence for polar ice on the planet closest to the Sun.

The depth and composition of ice deposits has been estimated with models constrained by surface observations, density measurements, cosmochemical abundances and thermodynamics (Lunine and Stevenson, 1982). Ice depth estimates for the various bodies range from meters to hundreds of kilometers. Intriguing consequences of these models include the possibility of subsurface liquid oceans,

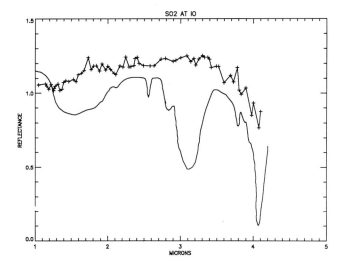

Figure P10 Spectra used to identify the presence of SO_2 frost on the surface of Io. The upper spectrum (Pollack et al., 1978; offset upward by 0.3 reflectance units, symbol is +) is an early astronomical observation of Io's leading side (the 'front' with respect to Io's orbital motion) which shows a strong absorption at about 4.1 μm. The lower curve is the laboratory spectrum of SO_2 frost used to identify the absorption. This particular sample of frost was slightly contaminated with H_2O frost, which is particularly evident in the 2.7–3.4 μm region of the spectrum.

clathrate-driven floods on Mars (Milton, 1974), greenhouse-driven ice volcanoes on Triton (Soderblom et al., 1990), and comets of dirty ice containing a wide variety of volatiles (e.g. Delsemme, 1992).

Observational limitations

Earth-based measurements have been employed for the discovery of the compositions of the icy surfaces of satellites. While these measurements have often been stimulated by observational programs associated with space probes, the instrumentation of the probes has not generally been designed to measure the spectral range most diagnostic for ices (about 1–5 μm). The ground-based observations are subject to limitations which include interference from atmospheric effects, low spatial resolution and observational geometries having small phase angles confined to the ecliptic plane.

There is a consensus that global transport of volatiles with polar trapping is an important process occurring on bodies with icy surfaces. Earth-based observations have very poor sampling of the poles of other bodies due both to the geometric constraints for Earth-based observations and to the lower illumination at the poles. Species more volatile than water may exist unobserved on the poles of the satellites.

The limited spectral range of atmospheric windows and noise introduced by the terrestrial atmosphere may prevent the observation of minor species. The low spatial resolution for Earth-based observations may also prevent the detection of minor species, since the weak absorptions are further disguised by areal averaging of the major and minor species.

Minor species have been detected for Europa and Triton. For Europa this was first seen as a difference in brightness between the leading and trailing hemispheres. A UV absorption (measured by the International Ultraviolet Explorer (IUE) satellite) suggesting sulfur implantation on the trailing side of Europa (Lane, Nelson and Matson, 1981) is also consistent with the presence of SO_2 in an ice matrix. Nelson et al. (1987) have reported a new absorption feature on Ganymede which is probably related to frozen volatiles. Triton (q.v.), whose major surface component appears to be solid methane with a substantial amount of solid nitrogen, has spectral features which strongly suggest the presence of carbon oxides. Minor species have yet to be identified in the other ice surfaces in the solar system, in part due to the observational selection effects discussed above. The intriguing albedo and color differences seen in the Voyager images of, for instance, Europa, imply a surface much more complex than is indicated by the spectra presently available.

Spectral considerations

Astronomical spectra of the planets and our understanding of laboratory studies of ice reflectance spectra have improved at about the same pace. In the early 1960s the astronomical spectra consisted of broadband filter photometry, and the laboratory spectra included few examples of reflectance spectra of ices. The effects of particle size and sample density on spectra were poorly understood, and the probable particle sizes for planetary surfaces had not been modeled. The cause of the opposition effect, a strong increase in brightness with decreasing phase angle (the Sun–body–observer angle), was unknown.

The study of planetary ice surfaces has advanced greatly in the last three decades. A nascent collection of laboratory reflectance spectra now exists (Fink and Sill, 1982). The opposition effect (q.v.) for bright materials has a good theoretical basis (coherent backscatter; Hapke, 1993). Scattering estimates have been made for the satellites (Buratti, 1991). Particle size evolution has been modeled (particle size increases with time; Eluskiewicz, 1991). Measurements of mixtures have been made (strong absorbers mask minor constituents: Kieffer, 1970). Ices have been identified on many of the bright (high-albedo) bodies in the solar system, and the spectral resolution for observations of these bodies have increased dramatically. The dominant constituent has been shown to be water ice – with notable additions of CO_2 on the Mars polar caps, SO_2 on Io, N_2 on Triton and CH_4 on Triton and Pluto, and probable minor occurrences of SO_2 on Europa and carbon oxides on Triton. Measurements of the ions near comets suggest minor quantities of more complex volatiles may exist on planetary bodies.

The spectral region from 1 to 5 mm is particularly rich in features for volatiles. This rotation–vibration region yields diagnostic spectra for the simple molecular condensates which have been identified on surfaces throughout the solar system. This wavelength region also has sufficient solar input to permit observations of reflectance. There are both broad and very narrow absorption features in this region, some of which have temperature-dependent shape and location (Kieffer and Smythe, 1974).

Band shapes are modified by observational geometry, since the scattering effects are different for high and low single-scattering albedos (Pleskot, 1981; Hapke, 1993). Particle size also has a strong effect on apparent band depth (Kieffer, 1970; Hapke, 1993), with small particle sizes giving bright, bland spectral curves and large particle sizes resulting in lower amplitudes. This arises principally from the accumulated pathlength within the material, with the larger grains providing larger effective pathlengths before light is scattered from the sample.

Some present challenges

Measurements of the darkening on the trailing side of Europa are consistent with the hypothesis of implantation of SO_2 in a water-ice matrix, but alternative explanations, such as CO_2 clusters, also closely match the observations. Measurements at higher spatial resolution across a broader spectral range may resolve the ambiguity. Similarly, such measurements may demonstrate the existence of surfaces having the complex condensate chemistry suggested by the ions surrounding comets, and by the present models of satellite interiors. These complex compositions may be associated with the striking patterns on satellites such as Europa (q.v.) and Ganymede (q.v.). Observations of the poles of icy satellites may show that the more volatile species have migrated to the colder parts of the planet.

There are elusive observations of post-eclipse brightening on Io (an increase in the brightness of Io as it exits Jupiter's shadow (O'Leary and Veverka, 1971). Suggested explanations for this brightening include deposition of SO_2, color shifts for sulfur (Franz, Missis and Pettauer, 1969), or observational uncertainties (Nelson, 1977). It is also uncertain to what extent volatiles drive the Ionian volcanoes.

Condensates of varying composition are present on the surfaces of many objects in the solar system. The compositions of these condensates have been identified, at least for molecules having relatively strong absorptions and dominant abundances, from Earth to Pluto. Considerable challenges remain to identify minor constituents, to map distributions, to understand the state of deposition (particle size, particle evolution and mixing) and to determine the relationship of the observations (which sample just the upper centimeter to meters) of these condensates to the composition and physical processes within the upper crusts and mantles of planetary bodies.

William D, Smythe

Bibliography

Brown, R.H. (1982) The satellites of Uranus: spectrophotometric and radiometric studies of their surface properties and diameters. Dissertation, Hawaii Institute of Geophysics, University of Hawaii.

Buratti, B.J. (1991) Ganymede and Callisto: surface textural dichotomies and photometric analysis. *Icarus* **90**, 312–23.

Clark, R.N, Brown, R.H., Owensby, P.D. and Steele, A. (1984) Saturn's satellites: near-infrared spectrophotometry (0.65–2.5 μm) of the leading and trailing sides and compositional implications. *Icarus*, **58**, 265–81.

Cruikshank, D., Morrison, D. and Pilcher, C. (1976). The coldest planet: methane ice found on Pluto. *Science*, **192**, 362.

Cruikshank, D.P., Brown, R.H., Tokunaga, A.T. *et al.* (1988) Volatiles on Triton: the infrared spectral evidence 2.0–2.5 microns. *Icarus* **74**, 413–23.

Delsemme, A.H. (1992). Cometary origin of carbon, nitrogen, and water on the Earth. *Origins of Life and Evolution of the Biosphere* **21**, 279–98.

Eluszkiewicz, J. (1991). On the microphysical state of the surface of Triton. *J. Geophys. Res.* **96**, 19217–29.

Fink, U. and Sill, G.T. (1982). The infrared spectral properties of frozen volatiles, in *Comets* (ed. L. Wilkening). Tucson: University of Arizona Press, pp. 164–202.

Franz, O.G., Missis, R.L. and Pettauer, T.V. (1969). A search for an anomalous brightening of Io after eclipse. *Bull. Am. Astron. Soc.*, **1**, 344.

Hapke, B. (1993). *Theory of Reflectance and Emittance Spectroscopy*. Cambridge: Cambridge University Press.

Jeans, J. (1967). *An introduction to the Kinetic Theory of Gas*. Cambridge: Cambridge University Press.

Johnson, T.V. and McCord, T.B. (1971) Spectral geometric albedo of the Galilean satellites, 0.3 to 2.5 microns. *Astrophys. J.*, **169**, 589–94.

Kleffer, H.H. (1970) Spectral reflectance of CO_2–H_2O frosts. *J. Geophys. Res.* **75**, 501–9.

Kieffer, H.H. and Smythe, W.D. (1974). Frost spectra: comparison with Jupiter's satellites. *Icarus*, **21**, 506–12.

Lane, A.L., Nelson, R.M. and Matson, D.L. (1981) Evidence for sulphur implantation in Europa's UV adsorption data. *Nature* **292**, 38–9.

Larson, H.P. and Fink, U. (1972) Identification of carbon dioxide frost on the Martian polar caps. *Astrophys. J. Lett.*, **171**, 91–5.

Lebofsky, L.A. (1976) Identification of water frost on the surface of Callisto. *Nature*, **269**, 785–7.

Lunine, J.I. and Stevenson, D.J. (1982) Formation of the Galilean satellites in a gaseous nebula. *Icarus*, **52**, 14–9.

Miller, S.L. and Smythe, W.D. (1970) Carbon dioxide clathrate in the Martian ice cap. *Science*, **170**, 531–3.

Morabito, L.A., Synnott, S.P., Kupferman, P.N. and Collins, S.A. (1979) Discovery of currently active extraterrestrial volcanism. *Science*, **204**, 972.

Milton, D.J. (1974) Carbon dioxide hydrate and floods on Mars. *Science*, **183**, 654.

Nelson, R.M. (1977) Search for color changes and brightening of Io upon eclipse reappearance. *Icarus*, **32**, 225–8.

Nelson, R.M., Lane, A.L., Matson, D.L. *et al.* (1987) Spectral geometric albedo of the Galilean satellites from 0.24 to 0.34 micrometers: observations with the International Ultraviolet Explorer. *Icarus*, **72**, 358–80.

O'Leary, B. and Ververka, J. (1971) On the anomalous brightening of Io after eclipse. *Icarus*, **14**, 265–8.

Pleskot, L.K. (1981) The opposition effect of particulate mineral surfaces and condensates: applications to Saturn's rings. PhD dissertation, Planetary and Space Physics, University of California, Los Angeles.

Pilcher, C.B., Ridgway, S.T. and McCord, T.B. (1972) Galilean satellites: identification of water frost. *Science*, **178**, 1087–9.

Pimentel, G.C., Forney, P.B. and Herr, K.C. (1974) Evidence about hydrate and solid water in the Martian surface from the 1969 Mariner infrared spectrometer. **79**, 1623–34.

Pollack, J.B., Witteborn, F.C., Erickson, E.F. *et al.* (1978) Near infrared spectra of the Galilean satellites: observations and compositional implications. *Icarus*, **36**, 271–303.

Rubey, W.W. (1952) Development of the hydrosphere and atmosphere, with special reference to the probable composition of the early atmosphere, in *Crust of the Earth*. Geol. Soc. America Special Paper 62, pp. 631–50.

Soderblom, L.A., Kieffer, S.W., Becker, T.L., (1990) Triton's geyser-like plumes: discovery and basic characterization. *Science*, **250**, 410–5.

Slade, M.A., Butler, B.J. and Muhleman, D.O. (1992) Mercury radar imaging: evidence for polar ice. *Science*, **258**, 635–40.

Smythe, W.D., Nelson, R.M. and Nash, D.B. (1979). Spectral evidence for SO_2 frost or adsorbate on Io's surface. *Nature*, **280**, 766.

Urey, H.C. (1952) *The Planets, their Origin and Development*. Newhaven: Yale University Press.

Wamsteker, W., Kroes, R.L. and Fountain, J.A. (1974). On the surface composition of Io. *Icarus*, **23**, 417–24.

Cross references

Callisto
Europa
Ganymede
Ice age
Io
Mars: atmosphere
Mercury
Milankovitch, Milutin, and Milankovitch theory
Pluto
Polar cap
Satellite, natural
Triton

PLANETARY LIGHTNING

Spacecraft observations have shown that lightning occurs not only on Earth, but on most of the planets in the solar system that have thick atmospheres. Lightning has not been discovered on Mercury, Mars or Pluto. The high temperatures and pressures produced in lightning discharges cause the formation of many trace gases in planetary atmospheres. In reducing atmospheres, and possibly in the Earth's early atmosphere, some of the gases produced by lightning are essential constituents needed for the evolution of life. On the present-day Earth, thunderstorms are responsible for maintaining the fair-weather electric field that exists between the upper atmosphere and the ground. On other planets the presence of lightning provides clues to the dynamics and cloud structures that occur below the upper cloud level of the atmosphere.

Because much more is known about terrestrial lightning than lightning on any other planet, it is best to start with a short review of what is known of terrestrial lightning and then consider how well this picture fits other planets in light of spacecraft observations.

Observed conditions for terrestrial lightning

Observed properties of lightning flashes

When a strong electric field develops in a dielectric medium such as air, the free electrons and ions that are normally present are accelerated by the electric field and collide with the air molecules to produce an avalanche of secondary electrons. A lightning discharge is the resultant high current discharge that flows through the atmosphere. Because the discharge channel is heated to very high temperatures, it expands in a shock wave that gives rise to the thunder we hear.

Most lightning discharges occur as either cloud-to-ground flashes or intracloud flashes. Occasionally flashes are observed between clouds and from clouds to the surrounding clear atmosphere. Recent images taken from the Space Shuttle have confirmed the existence of a rare form of lightning that moves upward from cloud tops.

The initial propagation of the conducting channel from the cloud base and the ground proceeds in a series of steps, each about 50 m in length until a connection is established with a short upward-propagating discharge from the ground. This process is termed a step leader. When the step leader has established a connection with the ground, the current rises from a value of about 100 A to a peak value of about 20 000 A; the temperature reaches 25 000 K and the discharge channel becomes intensely luminous. This bright, high-current discharge is labeled the return stroke. After the cessation of the high-current flow, the temperature, ionization level, conductivity and luminosity rapidly decrease. Often, before the integrity of the discharge column is lost, another charge center in the cloud becomes connected to the discharge column, and another leader (called a dart leader) and another return stroke occur. Typically there are five strokes during a flash, but the number has been observed to vary from one to 54. Although the duration of the return stroke is about 100 ms, the duration of a multiple-stroke flash can approach 2 s. Cloud-to-ground flashes often have a length of about 5 km and a luminous diameter of several centimeters.

The characteristics of the intracloud flash are somewhat different from those of cloud-to-ground flashes. An intracloud flash occurs when two or more charge centers within a cloud become connected through a discharge channel. The initiating process for the intracloud discharge is called a streamer. Most frequently, multiple streamers grow from a common origin. When the streamer taps large charge accumulations, the discharge channel brightens and currents rise to values of 1000 to 4000 A. This brightening of the discharge channel is called a K change and is analogous to the return stroke in a cloud-to-ground discharge. The most frequent number of K changes in a flash is six but a single flash can have as many as 20 K changes. The duration of the K change is about 1 to 3 ms and the electrical moment change is about 8 coulomb km.

During the first 50 to 100 μs of a stroke, the channel is sufficiently luminous that spectra can be obtained. Analysis of the spectra shows that most of the light is radiated by ionized and singly excited atoms of nitrogen and oxygen and by the continuum radiation from free free and free-bound processes. Faint radiation is also seen from atomic carbon and argon and from molecular nitrogen, hydroxyl and cyanogen radicals. Time-resolved spectra have been used to determine the time variation of the temperature and pressure in various types of lightning strokes. They show that during the first few microseconds of the discharge, temperatures near 30 000 K and pressures of the order of 8 atmospheres occur. Most of the energy dissipated by a stroke is used to heat the gas; only about 0.5% of the energy is radiated in the visible spectrum and about 1% of the energy is dissipated as sound waves.

The average flashing rate during a thunderstorm is 2 flashes min^{-1}; maximum values sometimes reach 100 flashes min^{-1}. On a global basis, there are approximately 100 lightning flashes per second with an average dissipation rate of 40 billion watts, which is equivalent to the power output from 400 100-MW nuclear reactors.

Observed properties of clouds that produce lightning

Most clouds that produce lightning contain both water drops and ice crystals and have updrafts strong enough to levitate large particles. Although many different processes have been proposed to explain how the particles become electrically charged, all that is known for certain is that the process is most effective at temperatures where supercooled water freezes. The clouds that produce the most violent storms often extend to the top of the troposphere, and consist of a number of individual cells that are a few kilometers in diameter. Although each cell lasts for about an hour, a storm along a weather front can last many hours as it moves into regions with new sources of warm moist air. Lightning is most frequently observed in cumulus clouds, rarely in stratus clouds and never in cirrus clouds. Lightning activity is observed at all latitudes between 60°N and 60°S, with the most frequent occurrence at low latitudes and over land. The reason for the high frequency over land is that convectively unstable conditions are frequent there. These conditions occur when air passes over ground strongly heated by the Sun and rises into the surrounding cooler atmosphere. At high latitudes the lightning frequency decreases because of the reduced convection resulting from cooler surfaces and reduced absolute humidities.

Methods of detection

To determine the characteristics of terrestrial lightning both *in situ* and remote techniques are employed. The *in situ* measurements include measurement of the extent and magnitude of electric fields from rocket- and balloon-borne sensors, the characteristics of the wind fields and rain droplets determined by radar beams, and the currents that flow through instrumented towers and rocket-lofted wires when they are struck by lightning. Remote measurements include optical imaging and spectroscopy, the measurement of electric field changes near thunderstorms and the analysis of the radio frequency emissions. At present all the data on planetary lightning has been obtained by remote techniques, i.e. from radio frequency emissions and optical imaging.

The most sensitive method of detecting the presence of lightning activity is to search for the radio waves emitted by the lightning strokes. Most of the radio frequency (RF) energy radiated by lightning is at frequencies below the ionospheric plasma frequency and is, therefore, trapped below the ionosphere. Consequently an excellent method of determining whether lightning exists on a planet is to place a low-frequency radio receiver on an entry probe that penetrates below the ionosphere. This is the method that was used by the Soviet Venera planetary probes that first detected the existence of lightning on Venus.

Some of the very low-frequency RF energy can escape through the ionosphere by propagating along magnetic field lines into space. These signals can then be detected by low-frequency receivers aboard a spacecraft when the spacecraft crosses the appropriate field lines. The mode of propagation of these radio waves modifies the characteristics of the signal to the extent that the source can be identified and the characteristics of the propagation medium can be determined. These signals are termed whistlers (q.v.) because that is what they sound like when they are recorded by audio-frequency equipment. The Pioneer Venus and Voyager spacecraft used this method to detect lightning on Venus, Jupiter and Neptune.

At frequencies of several MHz or greater, radio waves penetrate the ionosphere without needing to propagate along magnetic field lines. Although lightning radiates much less power at frequencies above 1 MHz than below, enough of the radiation gets through the ionosphere that spacecraft with sensitive receivers can often detect lightning at ranges in excess of 100 000 km.

Because the frequencies of whistler waves are very low, they can propagate many thousands of kilometers before they couple to magnetic field lines and penetrate the ionosphere. Therefore it is difficult to obtain an accurate lightning source location from whistler observations. This difficulty is also present when high-frequency receivers are used due to atmospheric refraction effects and the poor directivity of the spacecraft antennas. Only from optical images of lightning on other planets has it been possible to estimate the spatial dependence of lightning on latitude and longitude, the pressure level at which the activity is occurring, the total energy dissipated by lightning and the rate of trace gas production. Unfortunately, for all the planets that are found to emit lightning-generated radio waves, only on Earth and Jupiter are optical images of lightning activity available. It appears that the extensive cloud cover on most planets precludes the more detailed picture of lightning activity that optical imaging could provide.

Observations of extraterrestrial lightning

Venus

Both radio and optical searches have been conducted for lightning storms on Venus. All the radio searches have been successful, but most of the optical searches have failed. Venera 11, 12, 13 and 14 planetary probes descended through the ionosphere and down to the surface on the dayside near the subsolar point. During their descent they detected impulsive electrical signals in the 10-, 18-, 36- and 80-kHz bands of their low-frequency radio receivers. The measured discharge rate was often as high as 20 to 30 events per second.

The Pioneer Venus Orbiter (PVO) carried an electric field detector that could detect whistler signals in four frequency bands between 100 Hz and 30 kHz. This instrument detected strong impulsive signals whenever the PVO penetrated the night-time ionosphere. Although the results of this experiment have been controversial, recent analysis indicates that some of the signals have characteristics appropriate for lightning-generated whistlers. The magnetic field required for whistler penetration of the ionosphere must be supplied by the solar wind as it interacts with the atmosphere because Venus has no planetary magnetic field of its own.

A third experiment that confirmed the existence of lightning on Venus was conducted during the 1991 encounter with Venus by the Galileo spacecraft on its way to Jupiter. In this experiment a high-frequency receiver recorded the presence of noise bursts at frequencies of about 1 MHz. Nine events which had amplitude at least four standard deviations above the noise threshold were detected in the 100 kHz to 5.6 MHz frequency range. These events had amplitudes similar to that of terrestrial lightning at the same distance.

Searches for optical flashes from lightning were conducted with the spectrometers on the Venera 9 and 10 Orbiters, the Pioneer Venus Orbiter star sensor, and with the Galileo imager. Because the brilliant illumination on the dayside masks lightning signals, the instruments could search for lightning only on the nightside. The Venera 9 and 10 Orbiters searched an area of 3.5×10^7 km^2 between latitudes 32°N and 32°S and found only a single storm. The Venera 9 spectrophotometer recorded many short duration pulses for a 70-s period on 26 October 1975, when the instrument was searching near latitude 9°S. The pulses are interpreted as a lightning storm producing 2 flashes s^{-1} (1000)$^{-2}$ and which had an east–west extent of 450 km. If the average rate of lightning activity on Venus is based on this single observation, then the global flashing rate could be of the order of 400 flashes s^{-1}. However, the optical searches by both the Pioneer Venus Orbiter and the Galileo imager were not successful, even though one of the searches covered a much larger area than the Venera orbiters at very high sensitivity. Consequently, it is likely that the flash rate on the nightside is much lower than 400 flashes s^{-1}.

Added together, these results are consistent with a picture of lightning activity occurring frequently on the dayside, but rarely on the nightside. Although the radio experiments are extremely sensitive to the presence of radio emissions from lightning, the spacecraft antennas did not provide the spatial resolution needed to determine the location of the lightning activity. Low-frequency radio waves readily travel many thousands of kilometers within the ionosphere-surface waveguide. Hence lightning occurring on the dayside can travel to the nightside before escaping through the thin nightside ionosphere, and can then be detected by the PVO electric field detector. Since the optical searches were confined to the nightside, they generally reported no activity. These results are to be expected when we recall that (1) it is necessary to have processes that generate strong convection and condensation to produce lightning, (2) that there is extremely little water on Venus to drive convection and (3) that the daylight period of Venus is 58 days long. Hence the strong convection needed for the development of lightning activity is mostly likely to occur on the dayside after local noon. Although it is conceivable that lightning activity could be driven by blowing dust or by erupting volcanoes, the spectra of the lightning obtained by the Venera 9 Orbiter does not show the strong absorption in the blue portion of the spectrum that would be expected if the lightning occurred near the surface of Venus. It is therefore likely that the storms occur in the cloud layers 40 km above the planet's surface. This height is so great that cloud-to-ground lightning must be rare compared to intracloud lightning.

Mars

The strong dust storms on Mars suggest a possible source of electrical phenomena. Rather than the long spark-like discharges characteristic of lightning, glows might form in the low-pressure atmosphere. Charge separation by blowing dust results from collisions of airborne dust particles among themselves and with the ground (triboelectricity). On Mars the dust particles have been observed to reach an altitude of 50 km and to persist for many weeks. Because neither radio nor optical searches have been conducted for lightning, nothing is known about the frequency of such activity.

Jupiter

The discovery of lightning in two Voyager 1 images of the nightside of Jupiter represented the first optical detection of lightning on a major planet. The existence of lightning was confirmed by a Voyager plasma wave instrument that detected whistlers. Both Voyager 1 and 2 images show that the lightning activity is confined to two narrow latitude bands centered at 49°N and 13.5°N and to a single region near 60°N. No lightning activity is observed in the southern hemisphere even though a large area was observed with the same sensitivity as that used for northern hemisphere observations.

Because Jupiter has separate layers of clouds composed of ammonia, ammonia sulfhydride and water, it was initially unclear which cloud type was responsible for the lightning activity. That the water cloud was responsible was determined from a comparison of the measured brightness distribution within the spots produced by the lightning illumination at the top of the ammonia cloud with the brightness distributions predicted from model calculations. The analysis indicated that the lightning activity is occurring at the 5-bar pressure level, which is 40 km below the ammonia cloud tops and which coincides with the predicted level of the water cloud in the Jovian atmosphere.

The optical energy radiated per flash is 10^9 which is much greater than that of a typical terrestrial flash, but similar to that of the very brightest terrestrial flashes which are called 'superbolts'. Because only about 1/1000 of the energy dissipated in a lightning flash is radiated as visible energy, the total energy dissipated is 1000 times the measured optical energy. The planetary average energy dissipation rate per unit area due to lightning activity, as determined from the Voyager 2 images, is 3200 km^{-2} when the attenuation by the overlying clouds is considered. This value is approximately 40 times larger than that of the Earth. Because Jupiter is 10 times the size of the Earth, this means that Jupiter is producing 4000 times as much power in lightning activity as the Earth. It is clear that the meteorology on Jupiter is surprisingly efficient at converting thermal energy into organized lightning activity.

Saturn, Uranus and Neptune

Saturn, Uranus and Neptune are similar in that all have massive, cloud-shrouded atmospheres and that all show radio emissions from lightning, but do not show optical evidence of lightning.

Although the Voyager 1 plasma-wave detector, which detected whistlers at Jupiter, did not detect whistlers at Saturn, the radio receiver which scans from 20 kHz to 40 MHz did find impulsive, unpolarized electrostatic discharges believed to be radiated from lightning discharges on Saturn. The estimated power at the source was between 10^7 and 10^8 W. The signals had a periodicity of 10^h 10 min. Because this period is characteristic of the rotation rate of some locations of the rings of Saturn and because the radio waves were not at high enough frequency to penetrate the daytime ionosphere, it was originally believed that the lightning was occurring in the rings rather than in the atmosphere of Saturn. However later analysis showed that at some latitudes near the equator, cloud features carried by the winds also have a 10 h 10 min period. Because Saturn's rings cast a shadow on a portion of the dayside, the electron concentrations in this part of the ionosphere may be low enough to allow radio waves to penetrate the ionosphere and reach the spacecraft. Although the directivity of the radio detection system is too poor to prove which location is correct, many investigators believe that the location of the lightning activity is in the atmosphere because the frequency characteristics of the bursts can be explained by transmission through an ionosphere and because the meteorology of Saturn is similar to that of Jupiter.

When the Voyager 2 spacecraft encountered Uranus, strong bursts of high frequency emission with an instantaneous power of 10^8 W were detected. However only 140 bursts were observed compared to the 23 000 seen at Saturn. Again no whistlers were observed.

At Neptune only four possible RF bursts were seen and these were much less powerful than those observed at Saturn and Uranus. However 16 whistlers were found in addition to the RF bursts. The frequency dispersion of the whistler signals indicated that they had bounced back and forth from one hemisphere to another many times before being detected by the spacecraft plasma wave instrument.

Summary

There is strong evidence for lightning on all the solar system planets other than Mercury, Mars and Pluto. Pluto and Mercury are not expected to have lightning activity because their atmospheres are extremely tenuous. Mars has not been searched for lightning activity, but little is expected because of its thin atmosphere and the absence of deep clouds. Venus, Jupiter and Saturn appear to have a great deal of activity. Jupiter has much more lightning activity and much brighter lightning flashes than the Earth. On Jupiter the lightning occurs mostly in a narrow latitude band near 49°N latitude whereas on Earth most of the activity is at low latitudes.

The observation and understanding of lightning activity on other planets is just beginning. Although many of the characteristics, like the length, duration and current of the lightning discharges are

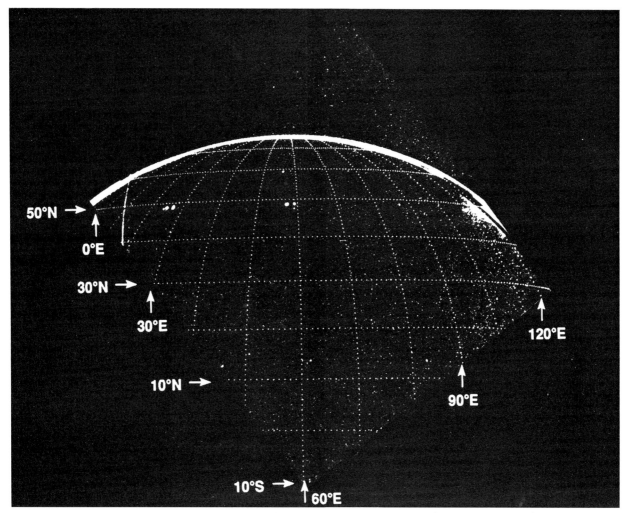

Figure P11 Lightning storms on Jupiter. Note the bright storms at 49°N at longitudes of 26°, 58°, 60°, 89° and 112°. Weaker storms can be seen at 60°N 57°E, 13°N 47°E and 14°N 62°E. Each storm is several thousand kilometers in diameter.

known for terrestrial lightning, these quantities are completely unknown for other planets. However, the Galileo and Cassini spacecraft should provide additional information on lightning activity on Jupiter and Saturn.

William J. Borucki

Bibliography

Borucki, W.J. and Chameides, W.L. (1984) Lightning: estimates of the rates of energy dissipation and nitrogen fixation. *Rev. Geophys. Space Phys.*, **22**, 363–72.
Borucki, W.J. and Magalhaes, J.A. (1992) Analysis of Voyager 2 images of Jovian lightning. *Icarus*, **96**, 1–14.
Burns, J.A., Showalter, M.R., Cuzzi, J.N. and Durisen, R.H. (1983) Saturn's electrostatic discharges: could lightning be the cause? *Icarus*, **54**, 280–95.
Golde, R.H. (ed.) *Lightning*, (1977) New York: Academic Press.
Ingersoll, A.P. (1990) Atmospheric dynamics of the outer planets. *Science*, **248**, 308–15.
Kaiser, M.L., Zarka, P., Desch, M.D. and Farrell, W.M. (1991) Restrictions on the characteristics of neptunian lightning. *J. Geophys. Res.*, **96**, 19043–7.
Krasnopolsky, V.A. (1983) Venus spectroscopy in the 3000–8000 Å region by Veneras 9 and 10, in *Venus* (eds D.M. Hunten, L. Colin, T.M. Donahue and V.I. Moroz). Tucson, Arizona: University of Arizona Press.
Ksanfomality, L.V., Scarf, F.L. and Taylor, W.W.L. (1983). The electrical activity of the atmosphere of Venus, in *Venus* (eds D.M. Hunten, L. Colin, T.M. Donahue and V.I. Moroz). Tucson, Arizona: University of Arizona Press, pp.565–603.
Levin, Z., Borucki, W.J. and Toon, O.B. (1983) Lightning generation in planetary atmospheres. *Icarus*, **56**, 80–115.
Park, C.G. (1982) Whistlers, in *CRC Handbook of Atmospherics*, Vol. II (Ed. R.H. Golde) Boca Raton, Florida: CRC Press, Inc., pp.21–97.
Rinnert, K. (1982) Lightning within planetary atmospheres, in *CRC Handbook of Atmospherics*, Vol. II (Ed. R.H. Golde) Boca Raton, Florida: CRC Press, Inc., pp.99–132.
Russell, C.T. (1991) Venus lightning. *Space Sci. Rev.*, **55**, 317–56.
Uman, M.A. (1987) *The Lightning Discharge*. New York: Academic Press.
Weidenschilling, S.J., and Lewis J.S. (1973) Atmospheric and cloud structures of the Jovian planets. *Icarus*, **20**, 465–476.
Williams, E.R. (1988) The electrification of thunderstorms. *Sci. Am.*, **259**, 88–99.
Williams, M.A., Krider, E.P. and Hunten, D.M. (1983) Planetary lightning: Earth, Jupiter, and Venus. *Rev. Geophys. Space Phys.*, **21**, 892–902.
Zarka, P. and Pedersen, B.M. (1986) Radio detection of Uranian lightning by Voyager 2. *Nature*, **323**, 605–8.

Cross references

Jupiter: atmosphere
Life: origin

Miller–Urey experiment
Venera missions
Whistler

PLANETARY RING

Planetary rings were first observed by Galileo Galilei in 1610 with his pioneering use of the telescope as a celestial research tool. He noted that Saturn appeared to have two large, nearby companion satellites and that these satellites, unlike the satellites he discovered circling Jupiter, appeared to remain stationary with respect to the planet. Two years after his discovery Galileo watched as these 'satellites' mysteriously disappeared (when the rings were positioned edge-on as seen from the Earth) then reappeared again the following year. For hundreds of years only Saturn was known to have a ring system. Then, in the decade beginning in the late 1970s, the number of known ring systems quadrupled. Ring systems were discovered with both ground-based and spacecraft observations, and a wealth of detailed information was garnered about the ring systems of Jupiter, Uranus and Neptune, as well as that of Saturn.

Ring systems were once thought to be composed of bland, featureless sheets or rings of material in orbit about the planet, but when more detailed observations became available they revealed ring structure at resolutions of less than 100 m. The next section is an historical overview of the rings, followed by a discussion on ring origin. The next four sections are devoted to the individual ring systems around Jupiter, Saturn, Uranus and Neptune. The final section discusses a future mission to the Saturn system. It is called Cassini in honor of one of the early astronomers who studied Saturn's ring system.

Historical overview

Forty years after Galileo's discovery, the Dutch astronomer Christiaan Huygens was the first individual to correctly interpret Saturn's rings as a discrete band of material separate from the planet itself. Huygens pictured the ring as a single, thick, solid band of material whose appearance varied as the tilt of the rings changed with respect to the Earth.

Pierre Laplace, a French mathematician, was able to show that a single solid ring would break apart under Saturn's gravitational forces, and by 1785 the rings were thought to consist of a number of solid, narrow bands of material. Astronomers wondered for many decades whether the rings were actually solid or composed of many individual particles. In 1859 a Scottish scientist, James Clerk Maxwell, mathematically demonstrated that the ongoing stability of Saturn's ring system (or any ring system) required separate orbiting particles. The particle orbital velocity, also called the Keplerian velocity, decreases with increasing radial distance from the planet. He hypothesized that a large number of orbiting particles, as seen from the Earth, would give the appearance of a continuous ring. The American astronomer James Keeler confirmed Maxwell's idea in 1895. He obtained spectroscopic observations of the rings which showed that the reflected sunlight was Doppler shifted by an amount equivalent to the orbital velocity of the ring particles as seen from the Earth. Keeler observed that the inner ring particles indeed orbit Saturn more quickly than the outer ring particles.

Today scientists believe that rings are composed of a collection of particles in a broad size distribution ranging from tiny dust grains to small embedded satellites, which orbit the planet in a thin disk. The particles themselves are composed of substances from water ice to rocky material, and range in brightness from 100% reflective to less than 5% reflective.

Planetary ring systems were thought to be a rare occurrence until about a decade ago, and in less than 10 years the number of known ring systems in our solar system increased to four. Before that time, only Saturn exhibited a detectable ring system, and for hundreds of years people wondered what made Saturn so unique. Finally, in 1977, while monitoring the variation in brightness of a star as it passed behind Uranus (called a stellar occultation) groundbased observers discovered that a second planet, Uranus, also had a ring system. It consists of narrow, opaque rings separated by large gaps, and it is very different from the broad disk of ring material which circles Saturn. Two years later the Voyager 1 spacecraft sent back pictures from Jupiter of a third planetary ring system. This system consists primarily of dust particles, distributed in an inner halo and outer ring. Ground-based stellar occultation measurements observed during the 1980s provided evidence for a fourth ring system, this time around Neptune. Neptune's ring system was very unusual and appeared to contain incomplete rings, or ring arcs. The Voyager 2 encounter with Neptune in August 1989 revealed that complete rings orbit Neptune. One ring contains several regions of relatively dense material or clumps which were the source of the ground-based detections. The remainder of this ring and the other Neptune rings were too tenuous to be detected from the Earth. Today, we know that all the giant outer planets, Jupiter, Saturn, Uranus and Neptune, have ring systems, each with its own distinct characteristics. Figure P12 displays each of these ring systems with the planets scaled to the same size. Figure P13 provides photographs of each of the ring systems. Comprehensive overviews of these ring systems are given in Cuzzi (1983), Burns (1990) and Nicholson and Dones (1991).

Origin of rings

Planetary rings form in a region close to a planet where gravitational tidal forces are stronger than the cohesive forces which hold particles together. Here large satellites are broken apart and new ones are prevented from forming. This concept was first proposed by Edouard Roche in the mid-19th century, and the radial boundary inside which this process begins is now called the Roche limit (q.v.).

Many ideas have been proposed to explain the initial source of ring particles. Two of these ideas are discussed in more detail below. The first is the 'circumplanetary disk' hypothesis, which speculates that the ring particles condensed directly from the same nebula which formed the central planet and that they were never part of a large, separate satellite. This same hypothesis is often applied to solar system formation with the Sun as the central body. The second idea is the 'parent body' hypothesis which involves the disruption of a large body to form the rings. This body may have been a captured asteroid or comet or a satellite whose orbit evolved inward, forcing it close enough to the planet to be tidally disrupted. The 'parent body' hypothesis may explain apparent compositional differences within the Saturn ring system if each ring is assumed to have had its own separate, unique parent body. A growing body of evidence on ring age seems to favor this hypothesis, because many aspects of the rings appear to be much younger than the age of the solar system. The parent bodies that formed the rings could have been disrupted long after the planets were formed.

Once the ring particles are created, the oblateness of the planet and the gravitational perturbations of nearby satellites cause them to evolve into a thin, flat disk centered near the planetary equatorial plane. If the particles are packed densely enough, differential Keplerian motion produces gentle interparticle collisions which result in spreading of the ring disk. The inner particles move toward the planet and the outer particles expand outward. The key result of collisions is first to circularize all of the orbits and, over a longer time period, to transport angular momentum from the inner ring particles to the outer ring particles. The ring disk will continue to spread in this fashion until the collisions become exceedingly infrequent unless some process, such as shepherding (confinement) by the nearby satellites, maintains the ring boundaries. A shepherd satellite can constrain a nearby ring through gravitational influence.

The gravitational interaction of ring particles with these small nearby satellites can both confine and contort ring edges, as well as create wave-like structure inside the rings. These satellites play an important role in ring structure and dynamics. Understanding rings involves a study of the interactions between the ring and nearby satellites either external to the ring system or within it.

The following sections provide a brief overview of the four ring systems. They share many similarities, but each ring system has its own unique characteristics and puzzles. These planetary ring systems are thought to be young, much younger than the age of the planets they circle. Since the beginning of our solar system, ring systems may have been created, evolved, dispersed and reformed over and over again. What we see now is simply a snapshot in time of very dynamic systems.

Jupiter ring system

Charged particle data from Pioneer 11 provided the first evidence for a ring around Jupiter (Plate 22). When the spacecraft flew close to

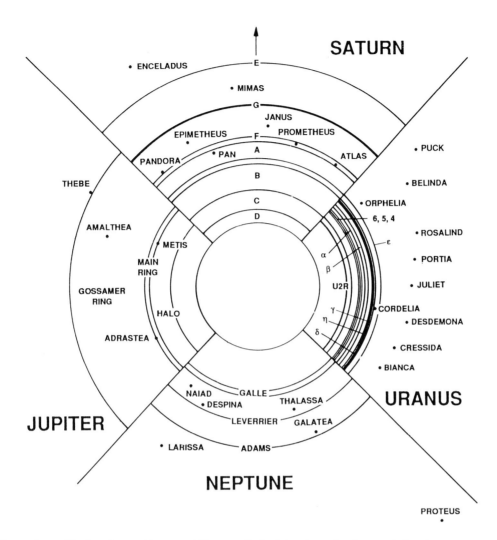

Figure P12 Ring systems of Jupiter, Saturn, Uranus and Neptune. Each of the planets has been scaled to the same size.

Jupiter in 1974, the high-energy charged particle density decreased as a result of either a hypothesized ring or a previously unseen satellite or both. The first pictures of the rings and new satellites were sent to Earth by the Voyager 1 and 2 spacecraft in 1979. The Jovian rings consist of three tenuous components (Figures P12 and P13). Comprehensive discussions of the Jovian rings can be found in Burns, Showalter and Morfill, (1984) and Showalter et al. (1986, 1987).

The central, main ring is approximately 7 000 km wide with an inner edge at about 1.72 Jupiter radii (R_J) and an outer edge at 1.81 R_J (1 R_J is equal to 71 400 km). The main ring outer edge is just inside the orbit of the satellite Adrastea. Another small satellite, Metis, resides in the central main ring, and both satellites are possible sources for the ring particles. Three brighter radial regions have been detected in the ring's outer 2000 km and may be influenced by the small satellites. Moving outward from Jupiter, Metis lies just outside the first bright region and Adrastea is outside the third. No detected body falls near the second region. The optical depth of this ring is very low, on the order of a few times 10^{-6} at visible wavelengths, which explains why the Jovian rings were not originally detected from Earth. The optical depth is a measure of the intensity of light passing through the ring. When the optical depth is close to zero, most of the incident light passes through the ring because only a small number of particles are present to absorb it.

The inner edge of the central, main ring gradually becomes a faint inner halo which reaches a maximum thickness of about 20 000 km. The halo is symmetric about the ring plane, and it disappears at about 1.3 R_J. The integrated optical depth of the halo is comparable to that of the central main ring. The density of particles in the halo decreases rapidly with increasing distance above the ring plane.

The third component is a fainter, outer 'gossamer' ring, which commences at the outer edge of the central main ring and fades from detectability at approximately 3 R_J, near the orbit of the satellite Thebe. The only detected feature in this ring is a slight brightening at the radial distance of synchronous rotation with Jupiter.

The central main ring is very bright in forward-scattered light, indicative of a significant population of small, micron-sized particles. In backscattered light the ring spectrum is red, much like the spectrum of Jupiter's inner satellites Amalthea, Metis and Adrastea. Larger ring particles in this ring also produce the backscattering. The red spectrum of the ring suggests that the particles are probably composed of silicates rather than ices. The ring appears to be ~ 30 km thick in backscatter and ~ 300 km thick in forward scatter.

Light from the halo is scattered somewhat differently than light from the main ring, indicating that the halo may have fewer large particles than the main ring. This is consistent with the magnetic resonance theory which suggests that the halo is comprised of small, charged dust grains in inclined eccentric orbits about Jupiter. The halo has not been detected in backscattered light.

The small Jovian ring particles (dust) have short lifetimes, approximately 100 to 10 000 years, because they experience a number of effects which cause them to leave the ring vicinity. These effects include Poynting–Robertson drag (loss of orbital energy as a result of absorption, and re-emission of sunlight), plasma drag (loss of orbital energy as a result of interaction with the Jovian magnetospheric ions and neutral atoms), erosion by energetic Jovian magnetic field

Figure P13 Photographs of the ring systems of (a) Jupiter, (b) Saturn, (c) Uranus and (d) Neptune, all taken by the Voyager cameras.

particles, and micrometeoroid impacts. Since the dust evolves away so quickly, some process must continually resupply it for the rings to continue to exist. Perhaps micrometeoroid bombardment of the small satellites is slowly eroding them away, producing the dust. Once these 'parent' bodies are gone the Jovian rings may no longer exist.

Electromagnetic resonances or Lorentz resonances are thought to influence most heavily the Jovian ring structure and evolution. Jupiter has the strongest planetary magnetic field in the solar system. Its field is tilted 11° with respect to its rotation axis (and ring plane) and it has significant non-dipolar terms which produce an asymmetric field. Dust particles are subject to charging, and once charged they experience changing Lorentz forces while moving through Jupiter's electromagnetic field. When electromagnetic forcing occurs at a frequency which is a rational multiple of the particle orbital frequency (e.g. $3/2$), the particles experience large radial and vertical displacements. The location where this forcing occurs is called a Lorentz resonance.

Some of these Lorentz resonances fall close to Jovian ring boundaries (e.g. the inner and outer boundaries of the halo). Dust particles drifting inward and, approaching the outer boundary, suddenly experience a Lorentz resonance which produces strong radial and vertical forcing, resulting in eccentric and inclined orbits which lift them out of the ring plane. Dust particles diffuse inward and form the halo. When the particles encounter a second Lorentz resonance at the inner boundary they experience an additional perturbation. Plasma and micrometeoroid interactions continue to erode the dust in this region. Eventually the dust particles become so small that the electromagnetic charging overcomes the force of gravity and the dust follows magnetic field lines into Jupiter's atmosphere.

The gossamer ring contains a smaller region of enhanced brightness near the radius of synchronous rotation. This increase may result from the absence of plasma drag as the particles orbit with the same frequency as the rotation of the planetary magnetic field and do not experience orbital decay. The Lorentz resonances are also very weak in this region so the particles remain in the equatorial plane. The outer boundary of the gossamer ring is near another Lorentz resonance which may vertically and radially displace the ring particles, spreading them apart to produce this visual boundary.

Saturn ring system

Saturn has the most extended and diverse planetary ring system in our solar system (Plates 26, 27). A schematic of the Saturn ring system is shown in Figure P12 and a photograph of the rings is shown in Figure 13. Its three main rings (A, B and C) span a radial distance of almost 70 000 km. Its large population of particles orbits Saturn in intricate patterns with recognizable structure on all scales down to the Voyager observational limit of less than 100 m. The total mass of the main rings is equivalent to the mass of the small Saturnian satellite Mimas. Four additional rings (D, E, F and G) have been discovered using ground-based or interplanetary spacecraft observations. Detailed discussions of Saturn's rings can be found in Cuzzi et al. (1984) and Esposito et al. (1984).

Spacecraft flybys by Pioneer 11, Voyager 1 and Voyager 2 provided detailed views of Saturn's complex and dynamic ring system. Most of the spacecraft observations were of the three main rings (A, B and C), all generally recognized from ground-based observations before the flybys. A small number of gaps are present within these rings, and some of them contain narrow opaque ringlets.

The outermost main ring, or A ring, spans a radial distance of roughly 14 500 km, and its optical depth ranges from about 0.4 to greater than 2 in some of the strong wave peaks. Irregular structure is present in the inner part of the A ring. This ring contains the largest number of gravitational waves driven by resonance interactions with external satellites. The A ring also exhibits a brightness asymmetry with azimuth that has been attributed to small-scale gravitational wake patterns created by the larger, embedded ring particles.

Two small gaps, the Encke gap and Keeler gap, are present in the A ring. The larger gap is the Encke gap with a width of 325 km. It contains both a small-embedded satellite named Pan (q.v.), which keeps the gap open, and one or more incomplete clumpy ringlets. Pan generates gravitational perturbations, called wakes, in the ring material on either side of the gap. These wakes propagate and change in wavelength as Pan moves in its orbit. Pan probably accreted outside the Roche limit and through some process evolved to its current location.

The Cassini division is located between the A ring and B ring. It is about 4000 m wide and has a much lower optical depth, typically 0.1, than the A and B rings which surround it. This region contains several narrower gaps, possibly created and maintained by embedded satellites, much like the Encke gap. The Cassini division is very similar in character to the C ring.

The central main ring, the B ring, is the widest ring and contains the bulk of the ring material. It is 25 500 km wide, with optical depths ranging from 0.4 to greater than 2. The radial structure in the B ring is irregular. Voyager 1 first observed narrow, primarily azimuthal brightness variations in Saturn's B ring which are called spokes. The greatest number of spokes were observed in the B ring region just emerging from Saturn's shadow. In backscatter the spokes are dark relative to the underlying ring material, while in forward scatter the spokes are much brighter than the ring material. This behavior is indicative of an abundance of micron-sized dust particles. Spokes may have first been noticed by ground-based observers as early as the late 19th century, when both the A and B rings sporadically exhibited dark radial bands.

The innermost main ring, or C ring, is 17 500 km wide and has an optical depth similar to the Cassini division. It has an average optical depth of about 0.2. Most of the gaps which contain narrow opaque ringlets are found in the C ring. Several broad regions 100 to 1000 km wide, of slightly greater optical depth are present in this ring also.

Earth-based observations suggested the possible existence of two other rings, the E ring and the D ring. The E ring, well outside the classical rings, was detected photographically when the Earth crossed through the ring plane in 1966. Voyager also photographed this ring in 1980 and 1981. It is a broad, diffuse ring which extends from about 3.5 Saturn radii (R_S, equivalent to 60 330 km) to 8 R_S; its greatest optical depth occurs near the orbit of Enceladus (3.95 R_S). The ring increases in vertical thickness away from the orbit of Enceladus.

The first report of a faint ring, or D ring, just inside the C ring was made in 1967. This ring was first clearly seen in forward-scattered light in the Voyager 1 and 2 images. The D ring is roughly 8000 km wide with an optical depth of much less than 0.05. Because of its low optical depth and close proximity to Saturn, positive detection of the D ring from Earth is probably impossible.

Pioneer 11 originally discovered a narrow ring, the F ring, orbiting several thousand kilometers outside the outer edge of the A ring. Its width varies from a few to several tens of kilometers, and its optical depth ranges from less than 0.1 to greater than 2 in its core. Two small satellites, Pandora and Prometheus, flank the F ring and confine or 'shepherd' it between them. These satellites, and possibly one or more embedded satellites as well, gravitationally interact with the ring material and distort the normal, circular ring shape by producing longitudinally and temporally varying kinks and twists in it. Voyager images first revealed the apparently unique, non-Keplerian behavior of the F ring. Depletions in the Pioneer 11 charged particle data suggest that a 1000 km wide satellite belt may surround the F ring in the intervening region between the two shepherding satellites. The F ring may be the product of the disruption of one or more of these satellites.

The G ring is tenuous and lacks the fine-scale structure of the main rings. Initially discovered by Pioneer 11 charged particle experiments, and verified by Voyager images, the G ring resides between the F ring and E ring. Voyager 2 flew extremely close to the outer edge of the G ring. It appears featureless, symmetric, optically thin and several thousand kilometers wide. It is the faintest ring observed in the Saturn system, with an optical depth of about 10^{-6}; it was first seen from Earth in near-infrared images taken in 1995 and was also observed in Hubble Space telescope images taken during the 1995 and 1996 ring plane crossings.

Visual and infrared spectroscopic analysis indicates that Saturn's rings are composed primarily of water-ice particles which range in size from large boulders to tiny dust grains. However, water ice alone cannot account for all of the spectral features observed in the rings, and small amounts of silicates or other non-icy absorbers must also be present.

The A and B rings strongly reflect ground-based radar, implying that the typical particle size is comparable to or larger than the radar wavelengths (around 10 cm). The results of ground-based radar, microwave and Voyager radio occultation measurements indicate that the main rings follow a broad particle-size distribution, with the bulk of the ring mass in the larger particles (radii > 10 m). The properties of this size distribution are compatible with the creation of ring particles from collisions or meteoroid bombardment.

Other rings are dominated by different particle sizes. For instance, Voyager images show that the F ring particles are strongly forward scattering. This behavior is indicative of a large population of small particles. The F ring appears to be composed of a dense core a few kilometers wide, consisting of larger particles similar in size to particles in the main rings, surrounded by a region dominated by micron-sized dust particles that may be produced by micrometeoroid bombardment of the larger core particles.

Radio occultation measurements indicate that some particles in the Cassini division are tens of meters in diameter. The E ring contains almost exclusively micron-sized particles, which indicates that this ring is not generated through collisions. The E ring particles must be continually replenished, because they have relatively short lifetimes (on the order of several thousand years) as a result of drag forces. The E ring particles may be continually renewed from possible geyser-like processes on Enceladus.

The G ring also appears to be composed primarily of micron-sized dust particles. Larger parent bodies must exist in this ring to provide a source for the short-lived dust particles and the charged particle absorptions.

Saturn's ring structure and evolution are thought to be dominated by gravitational resonance interactions (see Resonance in Saturn's rings). A gravitational resonance exists when the particle orbital period and satellite orbital period are a rational multiple (e.g. 2 : 1, 6 : 5, 5 : 3, etc.). The particles experience a repetitive, additive perturbation which creates a wave. The waves propagate radially through the rings for 10 to 1000 km, creating structure until they disperse as a result of interparticle collisions. Wave structure can take the form of spiral density (horizontal perturbations) waves, bending (vertical perturbations) waves and satellite wakes.

The theory of spiral density waves was first developed to explain the spiral structure observed in some galaxies. Planetary spiral density waves are like spiral galaxies, only on a much smaller scale. They are spiral arms of alternating compressed and rarefied ring material in orbit about a planet. Energy and angular momentum are transferred from the satellites (e.g. Mimas) to the rings at the resonances. This mechanism is probably not significant for all Saturn's rings given the low opacity (and hence infrequent ring particle collisions) in some of the rings. Wakes are also gravitational perturbations from moonlets within the rings. They produce a wavelike pattern that is different from a spiral density wave pattern.

Resonances can also confine ring edges. For example, the outer edge of the B ring is confined by a 2 : 1 horizontal resonance with the satellite Mimas while the outer edge of the A ring is confined by a 7 : 6 horizontal resonance with the co-orbital satellites Janus and Epimetheus.

Density waves result in a transfer of angular momentum between the rings and nearby satellites, and, as time passes, the satellites move further away from the rings. Moving backward in time, the small moons closest to the A ring could have been coincident with the A ring outer edge within the last 10^6 to 10^9 years.

Uranus ring system

The Uranian rings were discovered serendipitously in 1977 during an Earth-based stellar occultation observation designed for measurement of the Uranian atmosphere. Small, sudden decreases in star brightness both prior to and following the atmospheric occultation provided evidence for rings. The main rings consist of nine narrow, opaque rings named 6, 5, 4, α, β, η, γ, δ and ε (with increasing distance from the planet). Two rings, η and δ, have nearby, optically diffuse components. Voyager discovered two additional rings and ten small inner satellites during its flyby of Uranus in 1986. A small clumpy, dusty ring named λ falls midway between the δ and ε rings. The other ring, temporarily named 1986U2R, is a broad, optically diffuse band interior to the 6 ring. It is about 2500 km wide and has an optical depth of only 10^{-3}. A schematic of the Uranus ring system is shown in Figure P11. Comprehensive reviews of the Uranian ring system are given in Elliot and Nicholson (1984), French et al. (1991) and Esposito et al. (1991).

The Uranian ring system is very different from that of Saturn. The Uranian rings are widely separated and spread over a radial region of about 10 000 kilometers. Many of the key properties of the nine main rings were defined over the decade following their discovery by combining the results of numerous stellar occultation measurements. Most of these rings have small orbital eccentricities which are perhaps maintained by unseen satellites. Many of them also have small but non-zero inclinations to Uranus' equatorial plane. They range in width from 0.5 km to 100 km and in optical depth from 0.5 to more than 4.0. The internal fine-scale structure of the rings was revealed by Voyager stellar and radio occultations. Structure is present within each of the rings, down to the 100-m scale or finer.

Most of the intervening regions between the rings contain extensive dust bands of very low optical depth material. The source of the dust may be the rings themselves, or possibly a swarm of unseen satellites orbiting throughout this region.

The inner 6, 5 and 4 rings are similar in their main characteristics. Each ring is less than 10 km wide and exhibits a variation in width with orbital longitude of several kilometers. This width variation is not systematically correlated with orbital longitude. Even these narrow rings show considerable internal structure. The α and β rings are wider than the three inner rings and typically do not display such sharp inner and outer edges. The widest alpha ring profiles show a 'W' structure as well as a diffuse inner shoulder. The α ring edges tend to be very diffuse.

The η ring has two components, a narrow inner feature and a broad outer, low optical depth region which is about 55 km wide. In one of the Voyager stellar occultation profiles the η ring is so diffuse that it is indistinguishable in the data. The η ring has the smallest inclination and eccentricity of the nine major rings.

Both the γ and δ rings show self-excited oscillation modes superimposed on the usual Keplerian orbital ellipses. These 'normal' modes provide additional detectable distortions in the ring shape. The γ ring is one of the narrowest Uranian rings with a width of less than 1 km at some orbital longitudes. It has one of the highest optical depths when it is at its smallest width. The δ ring has a wavelike distortion near its inner edge which is probably the result of a density wave driven by an undetected satellite. The δ ring also has an inner diffuse companion roughly 10 km wide which appears to be unaffected by the δ ring oscillations.

The ε ring is the most eccentric, and among the most opaque, Uranian rings. It exhibits a systematic variation in radial width with orbital longitude which ranges from 20 to 96 km. The internal structure of the ε ring appears almost uniformly to expand and contract as the width varies. The inner and outer edges of the ε ring are extremely sharp and are controlled by gravitational resonances with the satellites Orphelia and Cordelia.

The Uranian ring particles are very dark, similar to the nearby satellites and primitive meteorites called carbonaceous chondrites. The material may be of primordial origin or may have been darkened by magnetospheric particle bombardment. Available data indicate that the bulk of the ring particles are probably large, with diameters of δ 1 m to several meters, with a very small population of micron-sized dust.

The dust bands were first detected in a long-exposure image taken by the Voyager cameras in forward-scattered light. The rings which are seen when illuminated in backscattered light are detectable in this image, but some of the brightest features show no correlation with any previously discovered rings. The η ring is the brightest feature in the image, in forward-scattered light, an indication that it has a significant component of micron-sized particles. Assuming an image brightness due primarily to micron-sized particles, its optical depth is estimated at about 10^{-5}.

The dust grains in the dust bands have short lifetimes and must continually be replenished, perhaps by unseen parent bodies orbiting in the dust bands. The dust grains experience Poynting–Robertson drag, plasma drag and erosion as a result of micrometeoroid impact. Poynting–Robertson drag along causes the dust to spiral into Uranus in about 10^7 years. Uranus' extended hydrogen atmosphere also acts as a source of drag for the ring particles, particularly for fine dust.

Gravitational resonances also play a role in the Uranian rings. Two of the ten small satellites discovered by Voyager, Orphelia and Cordelia, have resonances inside or near several of the rings. Orphelia and Cordelia shepherd (confine) the ε ring between them. These satellites also provide resonances which may confine the inner edge of the η ring and the outer edges of the δ and γ rings. However, no other resonances with the new satellites fall near the edges of the other Uranian rings. Still undiscovered satellites may be confining the other Uranian ring edges or perhaps these rings are being confined by some other unknown process. Some mechanism prevents these narrow rings from spreading.

Neptune ring system

Ground-based measurements provided the first evidence for rings around Neptune. These rings were poorly understood prior to the receipt of Voyager images and Voyager stellar occultation data in August 1989. During the 1980s over 20 stellar occultations by Neptune were observed and only a small fraction of these occultations produced conclusive ring detections. None of the Earth-based observations provided evidence for ring material on both sides of the planet. The favored explanation was that Neptune possessed incomplete rings or 'ring arcs'. These ring arcs were narrow and appeared to populate only a small fraction of a total ring circumference, otherwise more of them would have been observed.

Voyager 2 flew by Neptune in August 1989 and discovered that the Neptune system contains complete rings as well as ring arcs. Voyager images showed at least five rings in orbit around Neptune with tenuous material spread between some of them. Voyager also discovered six small satellites, four of which orbit near or within the ring system. These satellites range in radius from 30 to 150 km. These satellites most likely play a role in ring confinement and may be responsible for some of the unusual structure present within the rings. A schematic of the Neptune ring system and associated small satellites is shown in Figure P11. A comprehensive overview of Neptune's rings is given in Porco et al. (1994).

The outermost ring is the narrow Adams ring, and all the known ring arcs reside within it. The images show at least three arc segments, azimuthally confined regions of relatively high optical depth, embedded in the continuous, faint Adams ring. The ring arcs are about ten times less diffuse than the ring itself, and they have an optical depth of roughly 0.1. Voyager images showed that they are probably the sources that were detected in the ground-based observations; the more tenuous ring material residing between the arcs was not detected from Earth. The ring arcs are not evenly distributed around the ring but are grouped together over about one-tenth of the Adams ring circumference. One ring arc is about 11 000 km in azimuthal extent. The remainder are less than 5 000 km long. Small-scale azimuthal structure is present within the arcs. This structure may represent discrete clumps in the rings. A small satellite, Galatea, orbits near the ring and possibly controls the ring arcs.

The next ring inward is the narrow Leverrier ring; it is tenuous like the Adams ring. A small satellite, Despina, also orbits close to this ring but no arcs are seen. A region of diffuse material extends roughly 4 000 km outward from the Leverrier ring, terminating with a less diffuse boundary that may represent a separate ring. These two regions do not yet have official names and are temporarily designated 1989N4R and 1989N5R respectively. The optical thickness of these two rings is less than 0.001 times that of the ring arcs in the Adams ring.

An innermost ring, Galle, is roughly 1700 km wide, and much broader than the Adams and Leverrier rings. The Galle ring has an optical thickness similar to 1989N4R and 1989N5R. A sheet of material may also fill the inner part of the Neptune ring system.

Preliminary analysis of Voyager images indicates that the Neptune ring material is dark, similar to the Uranian rings. Unlike the Uranus rings, the Neptune rings appear brighter in forward-scattered light than in backscattered light. This is an indication that the Neptunian rings possess a larger fraction of dust particles than do the narrow rings of Uranus.

As at Uranus and Saturn, gravitational resonances play a role at Neptune, but with a new twist. The existence of ring arcs presents an intriguing puzzle. Unless some mechanism is confining the arc material, it should spread uniformly around the entire orbit in just a few years. A number of methods have been suggested for confining these arcs. Most involve some kind of gravitational interaction with one or more satellites. One hypothesis suggests that if the perturbing satellite's orbit were inclined, a single satellite could confine the ring arcs both in length and width. Galatea orbits within 1000 km of the arc-containing Adams ring, has a very small orbital inclination and may produce the hypothesized resonance effect. The long-term stability of the ring arcs is not known.

Dust in the Neptune system behaves much like the dust in the Uranus system. A source must be continuously resupplying the small dust grains and the dust will persist as long as this source is available.

Cassini – future mission to Saturn

The National Aeronautics and Space Administration (NASA) plans to send another spacecraft called Cassini (q.v.) for an extended visit to the Saturn system. Launched in 1997, Cassini will arrive at Saturn in 2004 and spend the next 4 years taking data while in orbit around Saturn. Cassini carries a sophisticated set of imaging, spectroscopic, photometric and field and particle instruments that are well suited for ring measurements. The highest-resolution observations of the rings will occur during the first Saturn flyby in 2004. The spacecraft will approach almost as close to Saturn as the inner C ring boundary and will never skim that close again. During this close pass of Saturn the spacecraft will fly above the rings and make direct field and particle measurements of the local ring environment for the first and only time. All the remote sensing instruments will obtain a single radial ring scan at unprecedented resolution.

Repeated flybys of Saturn's large satellite Titan will permit reorientation of the orbit. The first 3 years will be spent primarily in Saturn's equatorial plane rotating the spacecraft orbit about the Saturn–Sun line. The goal is to orient the orbit apoapsis near Saturn midnight by the end of the third year. In the fourth year of the tour the orbital inclination will be increased to as much as 85°.

The equatorial orbits do not provide extensive viewing of the rings because the rings are essentially positioned edge-on. During the first 3 years of the tour the orbital inclination will be increased twice for short periods of time in order to gather ring data. The inclination of these orbits will be optimized for radio science Earth occultations by the rings and planet. With each passing year of the orbital tour, the ring opening angle decreases, making it increasingly difficult for the radio beam to penetrate the optically thick regions of the A and B rings. Radio occultations early in the tour are advantageous for ring measurements in the centimeter to meter particle size regime.

During the inclined orbits, opportunities will also exist for numerous stellar occultations of the rings at a variety of ring opening angles. Two instruments aboard Cassini are capable of observing stellar occultations: an ultraviolet spectrometer and a visible–infrared mapping spectrometer. These two instruments will produce radial ring profiles at resolutions an order of magnitude better than those obtained by Voyager. Additional density waves, wakes and heretofore unimaginable ring structure remains to be uncovered by these occultations. Numerous stellar occultations will also provide temporal coverage of the rings, something unique to Cassini.

Both imaging and thermal infrared measurements of the rings will be made during the inclined orbits. The rings will be imaged at a variety of geometries through a number of ultraviolet, visible and infrared filters. Satellite searches will be performed to pinpoint new satellites in the near-ring environment. Detailed radial and circumferential structure will be imaged, and the B ring spokes will be carefully tracked. Thermal infrared measurements of the ring particles will provide information on ring particle sizes and texture. The Cassini mission will provide us with the most comprehensive, detailed information available on any planetary ring system.

Linda J. Horn

Bibliography

Alexander, A.F. O'D. (1962) *The Planet Saturn: A History of Observation, Theory, and Discovery*. London: Faber and Faber.
Borderies, N., Goldreich, P. and Tremaine, S. (1984) Unsolved problems in planetary ring dynamics, in *Planetary Rings*, (eds R. Greenberg and A. Brahic). Tucson: University of Arizona Press, pp. 713–34.
Borderies, N., Goldreich, P. and Tremaine, S. (1989) The formation of sharp edges in planetary rings by nearby satellites. *Icarus*, **80**, 344–60.
Burns, J.A. (1990) Planetary rings, in *The New Solar System*, 3rd edn (eds J.K. Beatty and A. Chaikin). Cambridge: Sky Publishing, pp. 153–170.
Burns, J.A., Showalter, M.R. and Morfill, G.E. (1984) The ethereal rings of Jupiter and Saturn, in *Planetary Rings* (eds R. Greenberg and A. Brahic). Tucson: University of Arizona Press, pp. 200–72.
Burns, J.A., Schaffer, L.E., Greenberg, R.J. and Showalter, M.R. (1985) Lorentz resonances and the structure of the Jovian ring. *Nature*, **316**, 115–9.
Colwell, J.E. and Esposito, L.W. (1990) A numerical model of the Uranian dust rings. *Icarus*, **86**, 530–60.
Colwell, J.E., *et al.* (1990) Voyager photopolarimeter observations of Uranian ring occultations. *Icarus*, **83**, 102–25.
Cuzzi, J.C. (1983) Planetary ring systems. *Rev. Phys. Space Phys.*, **21**(2), 173–86.
Cuzzi, J.N. and Burns, J.A. (1988) Charged particle depletion surrounding Saturn's F ring: evidence for a moonlet belt? *Icarus*, **74**, 284–324.
Cuzzi, J.N., Lissauer, J.J. and Esposito, L.W. (1984) Saturn's rings: properties and processes, in *Planetary Rings* (eds R. Greenberg and A. Brahic). Tucson: University of Arizona Press, pp. 73–199.
Dermott, S.F. (1984) Dynamics of narrow rings, in *Planetary Rings* (eds R. Greenberg and A. Brahic). Tucson: University of Arizona Press, pp. 589–637
Dobrovolskis, A.R. (1990) Tidal disruption of solid bodies. *Icarus*, **88**, 24–38.
Dones, L. (1992) A recent cometary origin for Saturn's rings? *Icarus*, **92**, 194–203.
Durisen, R.H. (1984) Transport effects due to particle erosion mechanisms, in *Planetary Rings* (eds R. Greenberg and A. Brahic). Tucson: University of Arizona Press, pp. 416–446.
Durisen, R.H. *et al.* (1989) Ballistic transport in planetary ring systems due to particle erosion mechanisms I. Theory, numerical methods and illustrative examples. *Icarus*, **80**, 136–66.
Elliot, J.L. and Nicholson, P.D. (1984) The rings of Uranus, in *Planetary Rings*, (eds R. Greenberg and A. Brahic). Tucson: University of Arizona Press, pp. 52–72.
Elliot, J.L., Dunham, E.W., and Mink, D.J. (1977) The rings of Uranus. *Nature*, **267**, 328–30.
Esposito, L.W. and Colwell, J.E. (1989) Creation of the Uranus rings and dust bands. *Nature*, **339**, 136–66.
Esposito, L.W., Harris, C.C. and Simmons, K.E. (1987) Features in Saturn's rings. *Astron. J. Suppl. Series*, **63**, 749–70.
Esposito, L.W., Brahic, A., Burns, J.A. and Marouf, E.A. (1991) Particle properties and processes in Uranus' rings, in *Uranus* (eds J. Bergstrahl, E. Miner and M.S. Matthews). Tucson: University of Arizona Press. pp. 410–68.
Esposito, L.W. *et al.* (1984) Saturn's rings: structure, dynamics, and particle properties, in *Saturn* (eds T. Gehrels and M.S. Matthews). Tucson: University of Arizona Press, pp. 463–545.
French, R.G., Nicholson, P.D., Porco, C.C. and Marouf, E.A. (1991) Dynamics and structure of the Uranian rings, in *Uranus* (ed J. Bergstrahl, E. Miner and M.S. Matthews). Tucson: University of Arizona Press, pp. 327–409.
French, R.G. *et al.* (1988) Uranian ring orbits from Earth-based and Voyager occultation observations. *Icarus*, **73**, 349–78.
Gehrels, T. *et al.* (1980) Imaging photopolarimeter on Pioneer Saturn. *Science*, **207**, 434–9.
Goertz, C.K. (1989) Dusty plasmas in the solar system. *Rev. Geophys. Space Phys.*, **27**, 271–92.
Goertz, C.K. *et al.* (1986) Electromagnetic angular momentum transport in Saturn's rings. *Nature*, **320**, 141–3.
Goldreich, P. and Tremaine, S. (1978) The formation of the Cassini division in Saturn's rings. *Icarus*, **34**, 240–53.
Goldreich, P., and Tremaine, S. (1982) The dynamics of planetary rings. *Ann. Rev. Astron. Astrophys.*, **20**, 249–83.
Goldreich, P., Tremaine, S. and Borderies, N. (1986) Towards a theory of Neptune's ring arcs. *Astron. J.*, **92**, 195–8.
Gresh, D.L., Rosen, P.A., Tyler, G.L. and Lissauer, J.J. (1986) An analysis of bending waves in Saturn's rings using Voyager radio occultation data. *Icarus*, **68**, 481–502.
Grün, E., Morfill, G.E., and Mendis, D.A. (1984) Dust-magnetosphere interactions, in *Planetary Rings* (eds R. Greenberg and A. Brahic). Tucson: University of Arizona Press, pp. 275–332.
Grün, E. Garneau, G.W. Terrile, R.J. *et al.* (1984) Kinematics of Saturn's spokes. *Adv. Space Res.* **4**(9), 143–8.
Harris, A.W. (1984) The origin and evolution of planetary rings, in *Planetary Rings* (eds R. Greenberg and A. Brahic). Tucson: University of Arizona Press, pp. 641–59.
Holberg, J.B., Forester, W. and Lissauer, J.J. (1982) Identification of resonance features within the rings of Saturn. *Nature*, **297**, 115–20.
Hubbard, W.B. *et al.* (1986) Occultation detection of a Neptunian ring-like arc. *Nature*, **319**, 636–40.
Lane, A.L. *et al.* (1982) Photopolarimetry from Voyager 2: preliminary results on Saturn, Titan, and the rings. *Science*, **215**, 537–43.
Lane, A.L., *et al.* (1986) Initial results from the Uranian atmosphere, satellites and rings, *Science*, **233**, 65–70.
Lane, A.L. *et al.* (1989) Photometry from Voyager 2: initial results from the Neptunian atmosphere, satellites and rings. *Science*, **246**, 1450–4.

Lin, C.C. and Shu, F.H. (1964) On the spiral structure of disk galaxies. *Astrophys. J.*, **140**, 646–55.
Lissauer, J.J. (1989) Spiral waves in Saturn's rings, in *Dynamics of Astrophysical Discs* (ed. J.A. Sellwood). Cambridge University Press, pp. 1–16.
Lissauer, J.J., Squyres, S.W. and Hartmann, W.K. (1988) Bombardment history of the Saturn system. *J. Geophys. Res.*, **93**, 13776–804.
Longaretti, P.-Y. (1989) Uranian ring dynamics: an analysis of multimode motions. *Icarus*, **82**, 281–7.
Marouf, E.A., Tyler, G.L. and Rosen, P.A. 1986. Profiling Saturn's rings by radio occultation. *Icarus*, **68**, 129–66.
Marouf, E.A., et al. (1983) Particle size distributions in Saturn's rings from Voyager 1 radio occultation. *Icarus*, **54**, 189–211.
Mendis, D.A., Hill, J.R., Ip, W.-H. et al. (1984) Electrodynamic processes in the ring system of Saturn, in *Saturn* (eds T. Gehrels and M.S. Matthews). Tucson: University of Arizona Press, pp. 546–89.
Nicholson, P.D. and Dones, L. (1991) Planetary rings. *Rev. Geophys., Suppl.*, April 1991, 313–27.
Nicholson, P.D., Matthews, K. and Goldreich, P. (1982) Radial widths, optical depths and eccentricities of the Uranian ring. *Astron. J.*, **87**, 433–47.
Nicholson, P.D. et al. Five stellar occultations by Neptune: further observations of ring arcs. *Icarus*, **87**, 1–39.
Nicholson et al. (1996) Observations of Saturn's ring-plane crossings in August and November 1995. *Science*, **272**, 509–18.
Porco, C.C. (1990) Narrow rings: observations and theory. *Adv. Space Res.*, **10**, 221–9.
Porco, C.C. (1991) An explanation for Neptune's ring arcs. *Science*, **253**, 995–1000.
Porco, C.C. et al. (1984a) The eccentric Saturnian ringlets at 1.29 R_S and 1.45 R_S. *Icarus*, **60**, 1–16.
Porco, C.C. et al. (1984b) Saturn's non-axisymetric ring edges at 1.95 R_S and 2.27 R_S. *Icarus*, **60**, 17–28.
Porco, C.C., Nicholson, P.D., Cuzzi, J.N. et al. (1994) Neptune's ring system, in *Neptune and Triton* (eds D. Cruikshank and M.S. Matthews). Tucson: University of Arizona Press.
Showalter, M.R. (1991) Visual detection of 1981S13, Saturn's eighteenth satellite, and its role in the Encke gap. *Nature*, **351**, 709–713.
Showalter, M.R., Cuzzi, J.N. and Larson, S.M. (1991) Structure and particle properties of Saturn's E ring. *Icarus*, **94**, 451–73.
Showalter, M.R., Burns, J.A., Cuzzi, J.N. and Pollack, J.B. (1987) Jupiter's ring system: new results on structure and particle properties. *Icarus*, **69**, 458–98.
Showalter, M.R., Cuzzi, J.N., Marouf, E.A. and Esposito, L.W. (1986) Satellite 'wakes' and the orbit of the Encke gap moonlet. *Icarus*, **66**, 297–323.
Shu, F.H. (1984) Waves in planetary rings, in *Planetary Rings*, (eds R. Greenberg and A. Brahic). Tucson: University of Arizona Press, pp. 513–61.
Shu, F.H., Cuzzi, J.N. and Lissauer, J.J. (1983) Bending waves in Saturn's rings. *Icarus*, **53**, 185–206.
Sicardy, B., Roques, F. and Brahic A. (1991) Neptune's rings, 1983–1989. Ground-based stellar occultation observations. I. Ring-like arc detections. *Icarus*, **89**, 220–43.
Smith, B.A. et al. (1979) The Jupiter system seen through the eyes of Voyager 1. *Science* **204**, 951–72.
Smith, B.A. et al. (1981) Voyager 1 imaging results. *Science*, **212**, 163–91.
Smith, B.A. et al. (1982) A new look at the Saturn system: Voyager 2 images. *Science* **215**, 504–37.
Smith, B.A. et al. (1986) Voyager 2 in the Uranian system: imaging science results. *Science*, **233**, 43–64.
Smith, B.A. et al. (1989) Voyager 2 at Neptune: imaging science results. *Science* **246**, 1422–49.
Stewart, G.R., Lin, D.N.C. and Bodenheimer, P. (1984). Collision-induced transport processes in planetary rings, in *Planetary Rings* (eds R. Greenberg and A. Brahic). Tucson: University of Arizona Press, pp. 447–512.
Tagger, M. and Henriksen, R.N. (1991) On the nature of the spokes in Saturn's rings. *Icarus*, **91**, 297–314.
Tyler, G.L. et al. (1981) Radio science investigations of the Saturn system with Voyager 1: preliminary results. *Science*, **212**, 201–6.
Tyler, G.L. et al. (1983) The microwave opacity of Saturn's rings at wavelengths of 3.6 and 13 cm from Voyager 1 radio occultation. *Icarus*, **54**, 160–88.
Tyler, G.L. et al. (1986) Voyager 2 radio science observations of the Uranian system: atmosphere, rings, and satellites. *Science*, **233**, 79–84.
Van Allen, J.A. (1984) Energetic particles in the inner magnetosphere of Saturn, in *Saturn*, (eds T. Gehrels and M.S. Matthews). Tucson: University of Arizona Press, pp. 281–317.
Van Allen, J.A. et al. (1980). Saturn's magnetosphere, rings and inner satellites. *Science*, **207**, 415–21.
Weidenschilling, S.J., Chapman, C.R., Davis, D.R. and Greenberg, R. (1984) in *Planetary Rings* (eds R. Greenberg and A. Brahic). Tucson: University of Arizona Press, pp. 367–415.
Wisdom, J. and Tremaine, S. (1988) Local simulations of planetary rings. *Astron. J.*, **95**, 925–40.
Zebker, H.A., Marouf, E.A. and Tyler, G.L. (1985) Saturn's rings: particle size distributions for thin layer models. *Icarus*, **64**, 531–48.

Cross references

Jupiter: ring system
Neptune: ring system
Radio astronomy
Saturn: ring system
Uranus: ring system

PLANETARY ROTATION

The Earth makes one complete revolution about its axis every 23 h 56 min 4 s; this is the length of the sidereal day. The Earth would have to spin about its axis once per year just to keep the same hemisphere pointed towards the Sun. Since the Earth rotates in basically the same direction as it orbits the Sun (this is referred to as prograde rotation), each year has one more sidereal day than solar day. The more familiar 24-h period of the mean solar day is thus (366.247/365.247) times as long as the Earth's sidereal rotation period. The Earth's rotation axis is tilted by slightly over 23° with respect to the normal to Earth's orbital plane; this is known as the obliquity of the ecliptic, or the tilt of Earth's spin axis, and is responsible for seasonal variations in the amount of sunlight received at any given latitude (see Obliquity; Earth rotation).

The rotational properties of the nine planets, as well as Ceres, the largest asteroid, are listed in Table P6. The rotational period and obliquity of Mars are very similar to those of the Earth. However, the rotation rates and directions of some of the planets differ significantly from those of Earth. Jupiter and Saturn are rapid prograde rotators, with spin periods of about 10 h. Mercury rotates very slowly in the prograde direction, taking 58.65 (mean solar Earth) days to complete one revolution with respect to the fixed stars. Venus spins in the opposite direction from which it orbits the Sun; this phenomenon is known as retrograde rotation. Venus is also a very slow rotator, but since it spins in the retrograde direction, a solar day on Venus is

Table P6 Planetary properties

Planet	Orbital period	Sidereal day[a]	Solar day[b]	Obliquity, ε (deg)
Mercury	88 d	58.65 d	176 d	~2
Venus	226 d	243 d	117 d	177.3
Earth	365.24 d	23.93 h	24 h	23.45
Mars	1.88 years	24.63 h	24.67 h	23.98
1 Ceres	4.61 years	9.08 h	9.08 h	~53
Jupiter	11.86 years	9.925 h	9.926 h	3.12
Saturn	29.46 years	10.675 h	10.675 h	26.73
Uranus	84.07 years	17.3 h	17.3 h	97.86
Neptune	164.8 years	16.11 h	16.11 h	~29.56
Pluto	248.6 years	6.387 d	6.387 d	~118.5

[a] The length of a sidereal day is the time it takes for a planet to spin 360° on its axis with respect to the positions of the fixed stars.
[b] A mean solar day is the average time between successive sunrises for a point on the planet's equator. A planet rotating in the prograde direction ($\epsilon < 90°$) has one fewer solar days per orbit than sidereal days, whereas a planet spinning retrograde ($\epsilon > 90°$) has one more solar day per orbit.

shorter than a sidereal day. Uranus' spin axis lies very close to the plane of its orbit; in more colloquial terms, Uranus spins on its side. As the angle between the spin and orbital angular momentum vectors of Uranus exceeds 90°, it is also considered to be a retrograde rotator.

It is apparent that there is some regularity to planetary rotation. The probability of at least six of the nine planets having obliquities less than 30° would be less than one part in 100 000 if planetary rotation axes were randomly distributed. The spin periods of six of the nine planets, including the five most massive ones, lie within a factor of 2.5 of each other. On the other hand, there are clearly exceptions to these patterns, with a few planets having very large obliquities or very long spin periods.

What process or processes can account for the observed rotational properties of the planets in our solar system? Were the planets formed with the spin periods and obliquities observed today, or has planetary rotation evolved significantly over the subsequent 4.5 billion years? Can a planet's rotation rate and direction change on time scales much shorter than the age of the solar system, and if so is there any evidence for such changes? These and other questions concerning the rotation of the planets have been active research topics for at least two centuries. Much progress has been made; however, some issues still remain poorly understood. In this article, I wish to present some of the more interesting theories of planetary rotation developed over the past 200 years and to summarize our current state of knowledge on the subject.

Historical overview

Questions concerning the origin of planetary rotation have historically played a major role in the development of theories of planetary formation. Until almost 1900, planetary cosmogonists (scientists who study the origins of planets) believed that planets which accumulated from material in Keplerian orbits would have retrograde rotation. It was reasoned that if two bodies in Keplerian orbits stuck together, the excess speed of the inner body (see Kepler's laws) would lead to retrograde rotation of the resultant assemblage (Figure P14). As the rotation directions of most planets were known to be prograde, various 'solutions' to this dilemma were proposed.

French mathematician and astronomer Pierre S. Laplace (1796) presented a model in which the protoplanetary disk split into several well-separated rings, each of which eventually accreted into a single planet. In Laplace's model each ring rotated with constant angular

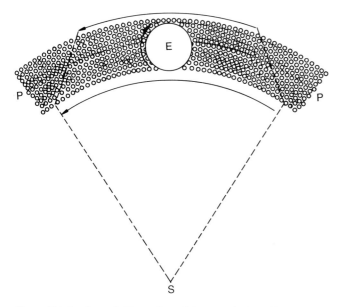

Figure P14 A schematic illustration of the canonical problem concerning the origin of planetary rotation: Keplerian shear in a disk of particles in orbit about the Sun seems to imply that planets should spin in the retrograde direction, whereas most planets are observed to rotate in the prograde direction (Table P5). (From Chamberlin, 1916.)

velocity (due to viscosity), just like a solid body. Thus the linear velocity of material within a given planet-forming ring increased outwards (since the outside of a ring has farther to travel around the Sun than does the inside), producing prograde rotation. However, even if the disk could have fragmented into distinct ringlets due to some unknown instability, the viscosity in either a gaseous or particulate ring is many orders of magnitude too small to maintain rigid rotation.

The American astronomer Daniel Kirkwood (1864) believed that planets formed with retrograde rotation, but were initially very large and distended. The substantial solar tidal forces acting on such bodies would produce synchronous prograde rotation, just as terrestrial tides locked the Moon into synchronous rotation, eons ago. As the planets shrank, tidal forces became less effective, and conservation of angular momentum resulted in an increase in each planet's rotation rate. Since tidal forces decrease rapidly with distance from the Sun, the outer planets did not have time to become tidally locked before shrinking, and thus remain in retrograde rotation. (In the 19th century Uranus and Neptune were both believed to be retrograde rotators due to the directions in which their largest satellites orbit.) Kirkwood's model of planetary rotation requires the planets to have passed through a stage during which they were distended gaseous objects. Such planets would presumably have had compositions similar to the surrounding solar nebula, i.e. predominantly hydrogen and helium. However, most planets are composed predominantly of elements and compounds which can condense at solar system temperatures, and thus are believed to have formed out of solid bodies (see Solar system: origin). Thus, Kirkwood's model has little relevance, except possibly in the cases of Jupiter and Saturn, both of which are more than half hydrogen and helium by mass.

The French astronomer Hervé Faye (1884) suggested that most of the planets formed within a nearly uniform density nebula. In such a configuration, orbital velocities grow with distance from the center, as the mass contained interior to an orbit increases rapidly with radius. Faye's theory also offered an explanation for the belief that the outer two planets were retrograde rotators: Uranus and Neptune were supposed to have formed in a lower-density outer region of the nebula, where the increase in mass interior to a given radius was insufficient to cause the radial gradient of the circular orbital velocity to be positive. However, the distribution of material described in Faye's theory is unstable and thus unrealistic.

The American geologist Thomas C. Chamberlin (1897) realized that the argument that planets which accreted from particles on Keplerian orbits would have retrograde spins depended on the assumption that the orbits were circular. He reasoned that particles were more likely to be on eccentric orbits than on circular ones. Thus the inner particle, being near aphelion (the point on its orbit farthest from the Sun), most probably orbited less rapidly than the particle on the outside, which was nearer perihelion (the point closest to the Sun). He surmised that the net result of a collision between the two particles would be prograde rotation. This was one of the major points in the Chamberlin (1905)–Moulton (1905) planetesimal hypothesis, which stated that the planets accreted from small solid bodies. This simple form of Chamberlin's argument is, however, also seriously flawed. Although in a collision between two bodies on eccentric orbits the particle with smaller semimajor axis moves faster, it is almost equally likely that either of the particles is farther from the Sun at the time of collision. Chamberlin presented an argument suggesting that collisions providing positive spin angular momentum were the more frequent, but was unable to conclusively demonstrate this point.

Prior to Chamberlin, the question had been 'why don't planets which accumulated from material in Keplerian orbits rotate in the retrograde direction?' Since Chamberlin, researchers assuming planetary growth via the accretion of solid planetesimals have faced the question 'given the high degree of cancellation inherent in the more or less random nature of the accretion of planetary spin angular momentum, why do the planets spin predominantly in the prograde direction and rapidly?'

The angular momentum of the particles in a given region of the protoplanetary disk around their collective center of mass is positive. Several 20th century researchers (e.g. Hoyle, 1946) have attributed the prograde rotations of most planets to this fact. However, when such planetesimals accumulate into a planet, the angular momentum is shared among spin and orbit. As planetary orbital angular momenta exceed their spin angular momenta by factors of the order of 10 000, and differences in the dynamics of the accretion process can affect the partitioning of angular momentum between spin and orbit, such arguments cannot be viewed as conclusive.

As is the case for most astronomical problems, questions concerning the origin of planetary rotation are more difficult to answer than those concerning its evolution. As various processes are believed to have significantly altered the rotation of some of the planets over geologic time, I shall next review such mechanisms for change with an eye towards estimating the initial conditions which any theory of the origin of planetary rotation must explain. I will then summarize recent theoretical results regarding the origin of planetary rotation.

Tidal forces

The rotation rates of four of the nine planets have been substantially slowed over geologic time due to tidal interactions with their moons and/or with the Sun. The gravitational pull of a moon in orbit about a planet is strongest on the portion of the planet located nearest the moon, and weakest on the portion farthest away. Gravitational tidal forces thus act to elongate the planet in the direction pointing towards and away from the moon (see Tide-raising forces). If planets were perfectly elastic bodies, these tidal bulges would point directly towards the moon. However, planets are not perfectly elastic; rather, they are dissipative, so the tidal bulges occur at locations on the planet which pointed towards the moon at an earlier time. Thus tidal bulges lag or lead depending on whether the planet rotates faster or slower than the apparent revolution of the moon. Provided the planet's rotation period is shorter than the moon's orbital period, this tidal lag causes the nearer bulge to lie in front of the moon, and the moon's greater gravity on the nearside bulge than on the farside bulge acts to slow the rotation of the planet (see Tidal friction). The reaction force upon the moon causes its orbit to expand. The above arguments remain valid if 'moon' is replaced by 'Sun', or if moon and planet are interchanged. Indeed, the stronger gravity of planets means that they have a much greater effect on the rotation of moons than vice versa. Most moons (including Earth's) have been slowed to a synchronous rotation state in which the same hemisphere of the moon always faces the planet; thus, no tidal lag occurs.

Evidence exists for the tidal slowing of Earth's rotation on a variety of timescales. Growth bands observed in fossil bivalve shells and corals imply that there were 400 days per year approximately 350 million years ago. Eclipse timing records imply that the day has lengthened slightly over the past two millennia. Precise measurements using atomic clocks also show variations in the Earth's rotation rate; however, care must be taken to separate secular tidal effects from the short-term periodic influences discussed below. Most of the secular decrease in Earth's rotation rate is caused by tides raised by the Moon, but at the present epoch $\sim 20\%$ is due to solar tides.

The Pluto–Charon system has evolved even farther. Charon is roughly one-seventh as massive as Pluto, by far the largest satellite-to-planet mass ratio observed in the solar system. The Pluto–Charon system has reached a stable equilibrium configuration, in which each of the bodies spins upon its axis in the same length of time they orbit about their mutual center of mass. Thus the same hemisphere of Pluto always faces Charon, and the same hemisphere of Charon always faces Pluto.

Solar tides have resulted in a stable spin–orbit lock for nearby Mercury, but one that is more complicated than the synchronous state which exists for most planetary satellites. Mercury makes three revolutions on its axis every two orbits about the Sun. The reason that equilibrium exists at exactly 1.5 rotations per orbit is that Mercury has a small permanent (non-tidal) deformation and a highly eccentric orbit. It is energetically most favorable for Mercury's long axis to point towards the Sun every time the planet passes perihelion, a configuration consistent with the observed 3:2 spin–orbit resonance (Colombo and Shapiro, 1966). Were it not for Mercury's permanent deformation, solar tides would slow Mercury's rotation further. However, because of the substantial eccentricity of Mercury's orbit, synchronous rotation would probably never be achieved. As required by Kepler's second law (which states that a line from the Sun to a planet sweeps out area at a constant rate; see Kepler's laws), a planet orbits the Sun much faster at perihelion than at aphelion. The variation is so large for Mercury that for a short time each orbit its angular velocity about the Sun is even faster than its rate of spin. During this brief interval the tidal bulge raised on Mercury by the Sun trails the Mercury–Sun line, so the Sun's gravity acts to speed up Mercury's rotation rate. Tidal effects on spin vary inversely with the sixth power of the distance between the two objects; thus the short period during which Mercury's rotation is rapidly accelerated is almost able to balance the much greater fraction of the time during which the Sun's tides act to slow the planet's spin. If Mercury's spin period were increased to roughly 70 days (significantly above its current value of 59 days, yet still well below the planet's 88-day orbital period), a balance between addition of spin angular momentum by solar tides near perihelion and removal during the remainder of the orbit would be achieved even in the absence of a permanent deformation.

Solar gravitational tides are probably the principal reason that Venus rotates very slowly, but they do not explain why that planet spins in the retrograde direction. A promising theory to explain the slow non-synchronous rotation of Venus was developed by Anthony Dobrovolskis and Andrew Ingersoll. Dobrovolskis (1980) argued that solar heating produces asymmetries in Venus' massive atmosphere, and the Sun's gravitational pull on such an asymmetric mass distribution prevents Venus' solar day from becoming excessively long.

Tidal forces slow the rotation rates of the other planets, but at rates too small to be significant, even over geologic time.

Short-term and periodic variations of planetary rotation

Tides are the principal cause of long-term changes in planetary rotation, but other processes can induce temporary (often periodic) variations in the spin rates and directions of the planets. The most important of these mechanisms involves gravitational perturbations by other planets; these can significantly alter planetary obliquities, but leave rotation rates essentially unchanged (Laskar and Robutel 1993). The largest such variations occur for Mars, whose obliquity varies chaotically from near 0° to as much as 60°, according to calculations by Laskar and Robutel (1993). Such oscillations, combined with similar changes in the orbital eccentricity and semimajor axis of Mars which also result from planetary perturbations, produce periodic variations in Mars' climate. The almost crater-free terrain observed near the Martian poles is believed to result from dust deposited along with the carbon dioxide ice which forms seasonal polar caps on Mars. The layering of Mars' polar terrain is evidence for climatic variations, which presumably result from the changes in Mars' obliquity, eccentricity and semimajor axis discussed above.

Ice ages on Earth may be due to analogous periodic variations in obliquity and orbital parameters, according to a model originally suggested by British astronomer John Herschel in 1830. This model was revived and developed by Serbian astronomer Milutin Milankovich early in the 20th century, and is referred to as the Milankovitch (or astronomical) theory of climate change (see Milankovitch, Milutin, and Milankovitch theory). Although planetary perturbations produce smaller changes in Earth's orbit and rotation than those produced for Mars, associated insolation changes and positive climatic feedback mechanisms (e.g. cooler temperatures increase the size of polar ice caps, and the high reflectivity of such ice caps leads to further cooling), could account for the difference between ice ages and the relatively warm 'interglacial' climate we now enjoy. Recent orbital calculations by Quinn, Tremaine and Duncan (1991) confirm the dynamical aspects of Milankovitch theory.

Analogous resonance-induced variations in the obliquities of the other planets are small at present, but may have been substantial during the planetary formation epoch (Harris and Ward, 1982).

Another well-known variation in the Earth's rotation is the precession of the equinoxes (q.v.). Lunar and solar torques on Earth's equatorial bulge cause the Earth's spin axis to precess about the normal to the ecliptic with a period of 26 000 years. Although the length of the day and the obliquity of the spin axis are unaffected, the celestial sphere appears to move due to this precession.

Non-periodic variations in planetary rotation can result from exchange of angular momentum with the atmosphere (e.g. the solid planet slows down when globally averaged wind speeds pick up in the direction of rotation), and from changes in the planet's moment of inertia (e.g. the planet's rotation speeds up when material becomes more concentrated near the polar axis, just as the spin rate of skaters increases when they bring their arms closer to their bodies). Although the magnitude of these effects are small (typically only one part in 10^9 for Earth), they are now routinely measured for the Earth. Larger variations may occur for the giant planets, which are mostly fluid. (Some ambiguity exists in defining the spin period of a planet which lacks an observable solid surface, since variations in wind speed result in latitude-dependent rotation rates for various cloud features. The most fundamental rotation period is usually taken to be that of the magnetic field, which is generated deep inside the planet's mantle or core.)

Origin of planetary rotation

We have seen that the rotation rates and obliquities of some of the planets have evolved significantly over the age of the solar system. Four billion years ago the length of the day on Earth was much shorter than at present, and Earth's spin rate continues to decrease due to a combination of lunar and solar tides. The Pluto–Charon system has reached a tidal equilibrium state which yields virtually no information about Pluto's initial rotation. Solar tides and dissipation at the core–mantle interface probably account for the slow rotation rates of Mercury and Venus, and the fact that the spin axes of these planets are nearly perpendicular to their orbital planes. However, tidal dissipation is not likely to have completely reversed the rotation directions of these planets. In other words, it is likely that Mercury initially rotated in the prograde direction (obliquity $< 90°$) and that Venus originally spun retrograde (obliquity $> 90°$). The obliquities of the planets change periodically due to mutual perturbations, but are restricted to fairly small variations in most cases.

A theory of the origin of planetary rotation must thus explain the following observed properties.

1. Planets rotate relatively rapidly. Those planets which have not been significantly slowed by tidal forces spin on their axes several hundred to several tens of thousand of times per orbit. Still, these rotation rates are only a third to a tenth as fast as would be required to cause the planets to be disrupted due to centrifugal force exceeding gravity at the equator.
2. The spin axes of the planets are not distributed randomly. Most planets rotate in the prograde direction with small obliquities. However, planetary obliquities are distinctly non-zero, and are quite large in some cases.

How did the planets obtain the rotational angular momentum required to produce their initial spins? As recognized by Immanuel Kant (1755) and Laplace (1796), the nearly coplanar and circular orbits of the planets imply planetary formation within a disk orbiting about the Sun. Recent theoretical and observational developments suggest that such disks are common by-products of star formation. Planetary accretion is now believed to have occurred via inelastic collisions between pairs of solid bodies referred to as planetesimals (see Planetesimal). Those planetary embryos which grew larger than 10–15 times the mass of the Earth were massive enough to accrete gravitationally large quantities of gas from the protoplanetary disk (Mizuno, 1980). The bulk of the rotational angular momenta of hydrogen/helium giants Jupiter and Saturn presumably arrived via hydrodynamic accretion of gas (Miki, 1982). The other planets, which are composed predominantly of condensable material, obtained their spin angular momenta from the relative motion of planetesimals which they accreted.

The leading Russian cosmogonist Victor Sergeivich Safronov proposed that the obliquities of the planets were produced by stochastic effects of large impacts. Safronov's (1966) analysis suggests that during the late stages of planetary growth, at least some of the planets accreted individual 'planetesimals' having up to 5–10% of the planet's total mass. Subsequent studies of the origin of planetary obliquities due to random impacts yield similar conclusions. Estimates of the largest impactors to strike each of the planets (other than Mercury and Venus, which have been slowed appreciably by solar tides and thus retain little evidence of their initial rotational properties) and Ceres are listed in Table P7 (from Lissauer and Safronov, 1991).

A.V. Artem'ev and V.V. Radzievskii (1965) realized that the 19th century argument that planets accumulating from bodies on circular orbits should rotate in the retrograde direction neglects the fact that a local analysis of the problem must include a term compensating for the planet's orbit about the Sun. By including this term in their analysis (which neglected the gravity of the planet), they showed that a planet which accretes from a uniform disk of planetesimals on circular orbits rotates very slowly in the prograde direction. They also argued that the focusing effects of the planet's gravity could increase its rotation rate substantially. However, when eccentricities are small, relative velocities between the planet and planetesimals are very small, and tidal effects of the Sun's gravity substantially alter the

Table P7 Estimates for largest planetary impactors

Planet	m_{max}/M_p	$m_{max}/M_\oplus$
Earth/Moon	0.07–0.1	0.07–0.1
Mars	0.006–0.02	0.0006–0.002
1 Ceres	0.05–0.1	0.00001–0.00002
Jupiter	0.002–0.02	0.6–6
Saturn	0.03–0.1	3–10
Uranus	0.01–0.06	0.15–1
Neptune	0.005–0.03	0.1–0.5
Pluto/Charon	~0.5	~0.001

planetesimal's trajectory relative to the planet. Such so-called three-body effects are extremely difficult to handle analytically, and are best studied using numerical integration.

R.T. Giuli (1968) studied the origin of planetary rotation by numerically integrating planetesimal trajectories in the restricted three-body problem. Giuli showed that a planet on a circular orbit which accretes from a uniform disk of planetesimals, also on circular orbits prior to their encounter with the planet, rotates rapidly in the retrograde direction. He also investigated the situation in which planetesimal orbits had a moderate initial eccentricity. He determined that such eccentricities would lead to rapid systematic prograde planetary rotation. Most randomly selected planetesimal trajectories do not impact the planet, and nearly the same number of impacts induce retrograde rotation as prograde. Thus the number of trajectories which need to be followed to give statistically significant results exceeded the computing power available in the 1960s. Giuli therefore relied on backwards integrations of trajectories from their points of impact on the planet. The resulting distribution of initial planetesimal orbits is difficult to characterize in a qualitative manner which can be compared to realistic expected disks; thus Giuli's eccentric planetesimal results must be viewed with some skepticism.

The origin of the rapid, predominantly prograde rotation of most of the planets is quite complex. Hydrodynamic models (Miki, 1982) imply rapid prograde rotation for very gas-rich planets, although detailed quantitative comparisons with the rotation rates of Jupiter and Saturn are still lacking.

Using a combination of analytic calculations and numerical experiments, Lissauer and Kary (1991) investigated the rotational properties of planets which accumulated from small planetesimals. The planet was always assumed to be on a circular orbit, and the planetesimals' perihelion angles (the positions of their closest approaches to the Sun) were uniformly distributed. For most simulations, the surface density (mass per unit area of the disk) of planetesimals was chosen to be uniform and all bodies orbited in the same plane.

The problem has proven to be deceptively simple. Even with the approximations we adopted, our solution involves dozens of equations, some of which involve integrals so long and complex that they require many pages of calculations to solve. The positive spin angular momentum provided by planetesimals impacting one side of the planet is almost exactly canceled out by negative spin angular momentum accumulated from planetesimals which hit the planet's other hemisphere. Since we must subtract two numbers of almost equal size in order to obtain a planet's net spin angular momentum, each of these numbers must be known quite precisely for our final results to be meaningful. Such stringent requirements forced us to restrict our analytic calculations to the case where we could ignore the planet's gravity. Fortunately, this corresponds to the limit of large planetesimal eccentricities (for which relative velocities are high and encounters are so fast that the planet's gravity has little effect on planetesimal trajectories), which is the most difficult case to handle numerically.

We found that a nongravitating planet formed in the manner described above rotates very slowly in the prograde direction, completing only slightly more than half a rotation every time it orbits the Sun. If the planetesimal orbits are inclined with respect to that of the planet, rotation rates are decreased by about one-third. We have also shown that rotation rates increase as the planet's gravity becomes more important, but the complexities of the three-body problem have prevented us from quantifying this result analytically.

We obtained our numerical results by integrating the orbits of tens of millions of test particles subject to the gravitational attraction of the Sun and the planet until the particles encountered the planet and either impacted or receded away. For those particles that hit, we computed the induced spin angular momentum as the product of the relative velocity and the moment arm (the degree to which the impact was off-center). We confirmed Giuli's result that when planetary gravity is important and all bodies start on circular orbits, then rapid retrograde rotation is produced. For large planetesimal eccentricities the planet rotates slowly in the prograde direction as predicted by our analytic results. When planetesimal eccentricities are diminished, more rapid prograde rotation results. There is a narrow range of planetesimal eccentricities for which prograde rotation at rates comparable to those observed in our solar system is produced. At still lower planetesimal eccentricities we find slow and/or retrograde rotation.

Do our results imply that most or all accreted planetesimals had eccentricities within the narrow ranges which led to rapid prograde rotation? We think not. Scatterings among planetesimals and by the planet should have produced a broad distribution of planetesimal eccentricities, and even if some unknown process favored a particular value, why should this value have exactly coincided with that necessary to produce rapid prograde spin? Using the shape of the distribution of eccentricities expected from these scattering processes, we have found that the observed rapid prograde rotation of the planets cannot be matched no matter what average eccentricity is simulated. This result has been confirmed by the independent calculations of Luke Dones and Scott Tremaine (1993).

Two alternative models have been proposed for the origin of the predominantly prograde rotations of those planets in the solar system which are believed to have accreted predominantly from planetesimals. Dones and Tremaine (1993) suggest that the rotations were produced entirely by stochastic processes, and the greater number of planets seen rotating in the prograde direction is a chance occurrence. Alternatively, Lissauer and Kary (1991) note that planetesimals from the outer extremities of the planet's accretion zone tend to provide positive spin angular momentum. As a planet grows, its accretion zone expands. Thus the planetesimal disk could have been depleted in bodies with semimajor axes similar to that of the planet, because most of the near planetesimals would already have been accreted. If the last phase of the planets' growth occurred within a planetesimal disk like the one depicted in Figure P15, the predominantly rapid prograde rotation of the planets could be understood.

Conclusions

Over the past two centuries the origin and evolution of planetary rotation has been an area of active research. Models of the origin of rotation have been integral parts of models of the origin of the solar system as a whole, and every theory of solar system formation has had to address this fundamental question. The nature of the problem has been confused by the subsequent evolution of planetary spins due to tidal effects of the Sun and moons and the short-term influences of other planets. In recent years the stochastic effects of a few giant impacts have been shown to be important factors in the formation of at least some of the planets. Nonetheless, strong prograde rotation may be a natural outcome of present models of planetary accretion, provided that the planet gets much of its material from the edges of its accretion zone, perhaps because of the growth of the zone as the planet grows.

Our understanding of the origin and evolution of planetary rotation has advanced greatly over the past two centuries, but there remain important problems in the field for future generations of astronomers to solve.

Jack J. Lissauer

Bibliography

Artem'ev, A.V. and Radzievskii, V.V. (1965). The origin of the axial rotation of the planets. *Sov. Phys. – Astron.*, **9**, 96–9.
Chamberlin, T.C. (1897) A group of hypotheses bearing on climate change. *J. Geol.*, **5**, 653–683.

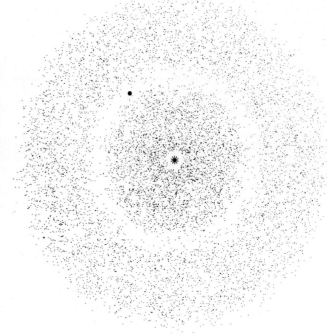

Figure P15 Schematic view of an orbital distribution of planetesimals which will produce rapid prograde planetary rotation according to the model of Lissauer and Kary (1991). The dots represent planetesimals, the circle lying within the nearly empty annulus is the planet and the Sun lies at the center of the disk. Orbital eccentricities cause some of the planetesimals to approach the planet on nearly tangential paths. (From Kary 1993.)

Chamberlin, T.C. (1905) Fundamental problems of geology, in *Carnegie Institution Year Book No. 3 for 1904*. Washington DC: Carnegie Institution, pp. 195–234.
Chamberlin, T.C. (1916) *Origin of the Earth*. Chicago: University of Chicago Press.
Colombo, G. and Shapiro, I.I. (1966) The rotation of the planet Mercury. *Astrophys. J.*, **145**, 296–307.
Dobrovolskis, A.R. (1980) Atmospheric tides and the rotation of Venus. II spin evolution. *Icarus*, **41**, 18–35.
Dones, L. and Tremaine, S. (1993) Why does the Earth spin forward? *Science*, **259**, 350–4.
Faye, H. (1884) La formation du système solar. *Astronomie*, **3**, 161–70.
Giuli, R.T. (1968) On the rotation of the Earth produced by gravitational accretion of particles. *Icarus*, **8**, 301–23.
Harris, A.W. and Ward, W.R. (1982) Dynamical constraints on the formation and evolution of planetary bodies. *Ann. Rev. Earth Planet Sci.*, **10**, 61–108.
Hoyle, F. (1946) On the condensation of the planets. *Mon. Not. Roy. Astron. Soc.*, **106**, 406–22.
Kant, I. (1755) *Allgemeine Naturgeschichte und Theorie des Himmels*. Königsburg und Leipzig: Johann Friederich Petersen. English translation, W. Hastie (1968). Universal natural history and theories of the heavens, in *Kant's Cosmology*. New York: Greenwood Publishing.
Kary, D.M. (1993) Ph D Dissertation, State University of New York at Stony Brook.
Kirkwood, D. (1864) On certain harmonies of the solar system. *Am. J. Sci.*, **38**, 1–18.
Laplace, P.S. (1796) *Exposition du Système du Monde*, Paris. English translation, H.H. Harte (1830) *The System of the World*. Dublin: University Press.
Laskar, J. and Robutel, P. (1993) The chaotic obliquities of the planets. *Nature*, **361**, 608–12.
Lissauer, J.J. and Kary, D.M. (1991) The origin of the systematic component of planetary rotation. *Icarus*, **94**, 126–59.

Lissauer, J.J. and Safronov V.S. (1991) The random component of planetary rotation. *Icarus*, **93**, 288–97.
Miki, S. (1982) The gaseous flow around a protoplanet in the primitive solar nebula. *Prog. Theor. Phys.*, **67**, 1053–67.
Mizuno, H. (1980) Formation of the giant planets. *Prog. Theor. Phys.*, **64**, 544–57.
Moulton, F.E. (1905) On the evolution of the solar system. *Astrophys. J.*, **22**, 165–81.
Quinn, T.R. Tremaine, S. and Duncan, M. (1991) A three million year integration of the Earth's orbit. *Astron. J.*, **101**, 2287–305.
Safronov, V.S. (1966) Sizes of the largest bodies falling onto the planets during their formation. *Sov. Astron.*, **9**, 987–91.

Cross references

Chaotic dynamics in the solar system
Earth rotation
Solar system: origin
Tidal friction
Torque

PLANETARY SAMPLING: *IN SITU* ANALYSIS

The chemical and physical nature of planetary surfaces can be only partially inferred from remote sensing measurements, whether the data are taken from spaceborne instruments orbiting the planet or from astronomical instruments on Earth. For many analytical techniques, *in situ* handling of the planetary surface material is required. Some also consume sample in the process. Without a detailed characterization of the chemical composition and physical characteristics of the mineral phases and atmospheric constituents, the data essential for understanding the origin and history of the planets and small bodies of the solar system will remain beyond reach. The ultimate scientific objective, of course, is to return samples for analyses in laboratories on Earth. However, this has been within the economical range of the space program only for the lunar surface, so far. Even when sample return becomes feasible for Mars, comets, asteroids and other planetary bodies, there will be a need for *in situ* analytical capabilities to help select samples so that not only will a representative suite of materials be acquired, but unusual and unique samples, often prerequisite to a full appreciation of the processes which have taken place, will be identified and sampled.

Constraints of space missions

Virtually all space missions to other planets have been and will continue to be highly constrained for scientific instrumentation. This stems from the high costs of launching payloads into space and the complexity of integrating and validating missions with a large number of science payloads of differing requirements.

Equipment which in the laboratory may weigh hundreds of kilograms must be reduced in mass by factors of 10 to 100. Volume is also at a premium. The instruments must be very robust to survive launch loads, exposure to interplanetary space and operation in difficult conditions on the target planet. Only a modest data rate is normally possible, often necessitating onboard data processing for compression. Finally, it is unlikely that sample preparation can be nearly as extensive as the standard protocols in analytical laboratories. For all these reasons, certain types of analyses seem beyond feasibility for *in situ* planetary analyses. The classical example of this has been isotopic age dating of igneous rocks because of the large number of sample processing steps and the analytical precision needed for success. This has not deterred some from trying, and a scheme for implementing an automated experiment for attempting Rb–Sr age dating at a comet or on a planet has been devised by Nyquist *et al.* (1980). As another example, it was often thought that electron microscopy would be out of the question for planetary surface exploration. However, a miniature, rugged electron microscope for space mission applications has now been brought to the brass-board stage of development and successfully demonstrated performance in both micro-imaging and x-ray fluorescence analysis (Conley *et al.*, 1983). Thus, although the challenges are great, there are many methods and advanced techniques which can be applied to the development of miniaturized, sophisticated instruments which can provide some (but not all) of the data that otherwise would require samples being brought back to Earth.

Past flight instruments

Moon

The first analytical instrument to be operated *in situ* on another planetary surface was the alpha backscatter spectrometer developed for the Surveyor spacecraft automated lunar landers which set the stage for the Apollo astronaut missions to the Moon (Turkevich, Patterson and Franzgrote, 1967). This investigation selected a nonstandard method for elemental analysis because of its ability to be miniaturized. A strong radioactive emitter of monoenergetic alpha rays, ^{242}Cm, was collimated to a narrow beam which would strike a soil or rock sample perpendicular to its surface. Detectors arrayed around the alpha sources measured the energies of alpha particles undergoing Rutherford scattering in the back (180°) direction. All elements in the periodic table, except hydrogen, produce a characteristic energy of backscatter. Because of progressive energy loss as the 6-MeV particles penetrate the samples, the backscatter actually results in a spectrum of energies. Each spectrum is flat up to a cut-off energy, determined by the atomic number of the target element (heavier elements and isotopes produce higher cut-off energies). By measuring the location and height of the steps in the spectra, various elements and their concentrations were inferred in the lunar soils. Elements with small differences in masses could not be readily distinguished. For this reason secondary protons from nuclear reactions were also analyzed, which improved the possibility of detecting certain elements. All detectors were Si solid state devices.

For three flights to the moon, Surveyors 5 through 7, the elements O, Mg, Al, Si, 'Ca', and 'Fe' were detected and measured. Quotemarks are used to indicate that the 'Ca' actually included the group S + K + Ca and the 'Fe' included Ti + Mn + Fe + Ni. Useful upper limits were also obtained for C and Na. The alpha instrument was notably successful in determining that the 'Fe' content was significantly lower for the lunar highlands rocks and soils compared to the mare soils of the earlier landers. The alpha instrument was deployed directly onto the ground and could even be moved to successive samples so that both soils and rocks could be analyzed.

The next analytical instrument to take measurements on another planetary surface also went to the Moon. It was an x-ray fluorescence spectrometer named RIFMA (Kocharov *et al.*, 1975), flown in 1970 and 1973 on two Soviet missions emplacing telerobotic rovers on the lunar surface. The first, Luna 17 carrying the Lunokhod I rover, included a hermetically sealed equipment bay for the instruments. RIFMA used energy-dispersive x-ray fluorescence, with a high specific activity tritium source (beta rays) to excite x-ray fluorescence and sealed gas proportional counters to obtain energy spectra. A somewhat improved version flew on the Lunokhod II. Elements reported included Al, Si, Ca and Fe, with a useful upper limit for K. These missions also determined the major dichotomy in Fe concentration between mare and highland surfaces on the moon.

Although not strictly qualifying as *in situ* science because they could also be conducted from orbit, passive x-ray and gamma ray spectrometers for detecting natural emissions from the lunar surface were included on Apollo flights 15 through 17. Fluorescent x-rays are stimulated by solar x-rays; gamma rays are emitted naturally by isotopes of K, U and Th, and several other elements are stimulated by neutrons from cosmic ray interactions with the surface materials. Using large-area, thin-window proportional counters, the x-ray experiment was successful in measuring the ratios of Mg/Si and Al/Si over those portions of the moon along the ground track of the orbiting Apollo command module. Similar results were obtained with the gamma instrument, with its 7 × 7 cm cylindrical NaI crystal detector and a 512-channel pulse height analyzer (PHA), for the naturally radioactive elements and also O, Mg, Si, Ti and Fe. Descriptions of the instruments and results are given in detail in the treatise by Adler (1986), which also includes thorough discussions of several other planetary analytical instruments up to the mid-1980s.

Mars

X-ray fluorescence spectrometers (XRFS) were flown on the two Viking lander spacecraft which touched down on the Martian surface in 1976. The design of the instrument (Clark *et al.*, 1977) was also based upon sealed proportional counters, but used detector optimization techniques, onboard calibration standards and more effective radioactive sources (^{55}Fe and ^{109}Cd) for excitation of fluorescent x-rays. As a result, not only were all of the elements of the RIFMA x-ray instrument and the alpha backscatter instrument detectable (except for oxygen), but also the elements S, Cl and Ti were analyzed. In addition, tighter upper limits were placed on K, useful values were obtained for a trace element group (Rb, Sr, Y and Zr), and the trace element Br was detected in several Martian samples. The spacecraft design included a thermally protected compartment at Martian ambient pressure but did not allow deployment of analytical instruments, with samples delivered to each experiment by a sampling scoop. The XRFS measurements revealed that Martian fines (soils) are extremely similar at the two landing sites, that all samples apparently contain relatively large quantities of salts mixed in with the expected basalt-like aluminosilicates and iron oxide minerals, and that crusty materials are enriched in the salt elements (S, Cl) compared to the non-indurated fines. It was also determined that igneous elements in Martian soil are in similar proportions as in the Shergotty meteorite, the type-name member of the SNC group of meteorites (q.v.) which are thought to have originated on Mars.

Analyses for organic compounds were accomplished on the Viking landers by soil pyrolysis (heating to temperatures as high as 500°C) and processing of released gases by gas chromatographic (GC) separation followed by mass spectrometric (MS) measurements. Hydrogen was used as the GC carrier gas. A Pd separator and a MS with electrostatic and permanent magnet sectors, scanned accelerating voltage and ion pump comprised the GCMS system. Two GC columns in series and an overload valving system were included to prevent atmospheric gases and H_2O from obscuring organic gases. No organics were detected, at levels of parts per billion for many compounds, with the exception of residues from an instrument cleaning fluid. Most importantly, H_2O at minimum levels of 0.1 to 1% was detected as vapor released from soils that were heated strongly (Biemann et al., 1977).

The MS portion of the GC instrument also included a direct inlet for measuring the composition of the Martian atmospheric gases. Sensitivity for minor constituents could be enhanced by a successive enrichment process where samples had their CO_2 and CO removed via uptake by LiOH. This experiment, an entry mass spectrometer, and the XRFS all found that the Ar content of the Martian atmosphere was much less than previously inferred. The GCMS also quantitatively measured CO_2, CO, N_2, and O_2 as well as the noble gases and a number of key isotopic ratios (Owen et al., 1977). It was found that Xe was unexpectedly high and the ratio of radiogenic ^{40}Ar to the intrinsic Ar (^{36}Ar) was much lower than on Earth, implying the possibility of less protracted volatiles outgassing on Mars. That the unique isotopic profiles in the Martian atmosphere coincides with that of gases that can be extracted from the SNC meteorites has now become generally accepted as conclusive proof that these meteorites are indeed from Mars.

The most important Viking experiments, comprising the most expensive and sophisticated instrument package yet flown to a planet, was the Life Detection System. A true wet chemistry system, it incubated various fluids and/or gases with soil samples and looked for the signs of biochemical metabolic transformations (e.g. Horowitz, 1977). As discussed by Klein (1979), life probably was not detected on Mars, but several unexpected results pointing to inorganically mediated oxidative capacity of the soils do indicate unusual chemical compounds on Mars.

The Soviet Phobos missions to Mars included alpha proton–x-ray experiments for deployment onto Phobos. They also included a laser-stimulation experiment for releasing ions from the surface to be analyzed by mass spectrometry. Unfortunately, these missions were not successful and no data was obtained from these advanced instruments.

The Mars Observer mission (q.v.) was intended to begin mapping the red planet from orbit in 1993. As an orbiter, it employed only remote sensing instrumentation. The GRS experiment utilizes a high-resolution Ge crystal with a 10 000-channel pulse-height analyzer for gamma-ray spectroscopy, which could also be flown on a lander for in situ measurements of element composition at a landing site. To be compatible with gamma-ray spectroscopy, a lander could not utilize a radioisotope thermoelectric generator (RTG) for power generation because such devices emit large amounts of gamma radiation. They also emit neutrons, but this might be advantageous in providing a source of higher-energy neutrons which would be thermalized by low atomic materials in the soil. The GRS on Mars Observer also includes a borated plastic scintillator sensor for neutron mapping from orbit.

Venus

X-ray fluorescence spectrometers have been flown by Soviet Venera missions (q.v.) through the atmosphere and also to the surface of Venus. Atmospheric monitoring by XRF was accomplished on Venera spacecraft. Yet more challenging was the XRF analysis of cytherean surface material after landings by Venera 13 and 14. Operating inside a pressure vessel protected from the extremely high ambient temperatures and pressures, samples were obtained by a drill and transported to an air lock by a screw device which then allowed presentation to x-ray fluorescence equipment. The instrumentation consisted once again of sealed proportional counter detectors, but utilized ^{55}Fe and ^{238}Pu as excitation sources (Surkhov et al., 1982). The elements reported included Mg, Al, Si, K, Ca, Ti and Fe, with significant differences between landing sites reported for three elements (Mg, K and Ca).

Gas composition analytical instruments have been flown on Soviet Venera 11 and 12 spacecraft, as well as the US Pioneer Venus large probe and orbiter. Together, they represent five MS and two GC gas analyzers which have performed in situ measurements in the upper and/or lower atmosphere of Venus. The most difficult challenges were for the lower, high-pressure regions of the atmosphere where cloud aerosols exist and condensation can occur as hot material enters a cooler instrument. The inlet for the Pioneer Venus gas analyzers was especially designed for this task, but nonetheless appeared to have been plugged by an H_2SO_4-rich aerosol droplet during a portion of the descent. Chemical getters and an ion pump were both employed in the MS instrument. The GC system, operating with He carrier gas, included pairs of long and short columns, with matched thermal conductivity detectors (Oyama, Carle and Berdahl, 1977). The design implementation benefited from the design and components utilized in the Viking GC experiment in the life detection package.

The use of two analytical techniques, i.e. both GC and MS instruments, stemmed from the recognition that each has strengths in avoiding certain ambiguities and that combining results from both should lead to an improved suite of measurements. In developing each instrument, the investigators improved on previous work and now have designs which can be flown on future missions. Together, analyses were made for CO_2, CO, N_2, O_2, Ar, Ne, SO_2, H_2S and C_2H_6, with useful upper limits on several other gases (Adler, 1986).

Comets

No attempts to land on the surface of a comet have yet been seriously planned. However, because the cometary coma consists of the gases and solid particles released from the nucleus, it is possible to sample materials from the cometary body without a landing. Three comets have been visited by spacecraft, but the most ambitious and scientifically rewarding were the flybys of comet Halley by five spacecraft in 1986. A particle-composition instrument designed especially for the European Space Agency's Giotto spacecraft (q.v.), the Particle Impact Analyzer (PIA), was also flown in versions known as PUMA on the two Soviet Vega spacecraft (q.v.). At the flythrough speed of 70 km s^{-1}, coma particles which struck a target plate were instantaneously vaporized and ionized. Via electrostatic fields, positive ions were withdrawn from the impact plasma, accelerated and traversed down a folded path time-of-flight mass spectrometer (TOFMS). The elements H, C, N, O, Mg, Al, Si, S, Ca and Fe, were detected in several thousand individual particle impacts. In addition, other elements were occasionally detected and some isotopic ratios could be examined. Particles were found to fall in several different classes, including silicate-rich grains, organic-rich grains (CHON particles) and, predominantly, mixtures of the two. Some exotic individual grains were also detected, with possible evidence for specific monominerallic grains of various types (Kissel et al., 1986).

Particle grain-size measurements of several instrument types were also flown on these three spacecraft (Reinhard and Battrick, 1986; Nature, 1986). In addition, several mass spectrometers were flown, but these operated on plasma – i.e. on charged particles – and therefore are of the class known as particles and fields instruments.

Missions in development

Mars

In the Russian space program an ambitious set of spacecraft are under development. Small stations were sent to the surface, containing a German-supplied version of the alpha proton–x-ray analyzer and a neutron detection device. The US is currently developing a soil oxidant detection experiment based upon chemical interaction of thin coatings on fiber optic cables, monitored for changes by reflection of light-emitting diode pulses. In 1996 a small instrumented rover will be sent, and the heavily instrumented French/Russian balloon will be launched.

Comets

For several years a group of composition analytical instruments have been under development for a comet rendezvous mission, specifically for the canceled 'CRAF' mission by NASA. A number of experiments qualify as *in situ* sampling devices. These include three instruments which individually collect dust grains in the coma for various types of analyses. The CIDEX instrument is a combination gas chromatograph (two pyrolysis ovens and four GC columns, each using metastable helium ionization detectors) and x-ray fluorescence spectrometer (x-ray tube and ^{244}Cm excitation sources, with a cooled high-resolution Si solid state detector). The CoMA instrument, supplied by German investigators, is a secondary ion mass spectrometer (SIMS) instrument, using a liquid metal ion source and a folded path TOFMS (Zscheeg *et al.*, 1992). The SEMPA is a miniaturized electron microscope for imaging individual microscopic dust particles, with secondary electron detector and HgI_2 x-ray fluorescence detector (Conley *et al.*, 1983). In addition, the CODEM instrument can monitor the flux, particle size distribution, electrostatic charge and velocity vector of coma dust. A neutral gas and ion mass spectrometer (NGIMS) is based upon quadrupole MS designs used in Earth-orbiting spacecraft.

A Penetrator device to have been rocketed into the nucleus from the CRAF spacecraft bus is designed to contain a gamma-ray spectrometer and a thermal/compositional analyzer for ices and low-temperature materials via differential scanning calorimetry (DSC) and evolved gas analysis (Gooding, 1989) using GC.

Titan

The Cassini mission (q.v.) to Saturn will include the Huygens probe (q.v.), which will descend through the atmosphere of Titan. Although the primary objective of the mission is to obtain atmospheric data, the probe may survive landing (or splashdown) and be able to transmit data for several minutes. Onboard is a Surface Science Package designed by British scientists (Zarnecki *et al.*, 1992). It includes an accelerometer, inclinometer, thermal and acoustic properties devices, a flotation level device, permittivity measurement device and refractometer. Through these measurements it will be possible to infer the characteristics and possibly the composition of an ocean. Originally part of this package was an x-ray fluorescence analyzer using Si(Li) and ^{55}Fe and ^{109}Cd isotope sources. Although selected on the basis of scientific merit, this portion of the experiment was beyond US budget constraints and is not being implemented.

Opportunities and options

Future *in situ* planetary experiments can include a great diversity of instruments. Based upon previous developments it is expected that x-ray, alpha, gamma ray, neutron, and mass spectrometers will be strong candidates for most future missions. Gas chromatographs and ion mass spectrometers are under advanced development, and GCMS combinations cannot be ruled out in spite of complexity. Exciting new developments in thermal analyzers, including differential scanning calorimeters and evolved gas analyzer systems, are under way for space application. Aqueous chemistry measurements capitalize on developments in solid state microchemical sensor technology. Mössbauer spectroscopy has been shown to have excellent potential for *in situ* mineralogical analyses on both the moon and Mars (Agresti *et al.*, 1992).

Because of the apparent geological diversity of Venus, a number of locations are candidates for future lander missions to determine the chemical composition of major geological units. These missions are technically very difficult because of unusually harsh environmental conditions and the nominally short lifetimes of landers.

X-ray and gamma-ray instruments are high priority for future missions where they are applicable, particularly polar orbiters. Both can also be operated on the ground, but then the x-ray would of course take advantage of artificial stimulation sources to increase sensitivity. Likewise, artificial neutron activation for gamma spectroscopy has been advocated and studied.

Advantages of sample return

For many scientific objectives, only returning samples to Earth will do, as noted above. This leads to more costly and extended missions, because of the multiplicity of events and usually the need for two separate spacecraft (outbound and return). Nonetheless, both the Soviet Union and United States have accomplished sample return from the Moon.

The array of complex, sophisticated measurements feasible only in laboratories on Earth certainly include isotopic ratio measurements and ultratrace element analyses (parts per billion or per trillion); transmission electron microscopy and atomic force imaging; Auger spectroscopy and two-step laser ion mass spectrometry; magnetic susceptibility and remnant magnetization; radiogenic nuclides assay and cosmic ray track etchings. These are, of course, only a few examples. As new techniques are developed, the value of pristine samples in reserve increases enormously. For example, the Apollo returned samples are only approximately 10% analyzed, yet still provide valuable research material for scientists and scientists in training all over the world and have been subjected to many analytical techniques that were not even available in the 1960s and 1970s when the samples were acquired.

However, the value of returned samples is much more than just the ability to accomplish intricate procedures using advanced, often finicky equipment. The quantity of data that can be generated and the number of scientists that can be involved in highly productive laboratory work can be increased by factors of tens to hundreds over that from spaceborne instruments, for the same science expenditures. In addition, new discoveries can be quickly verified by independent investigators, the foundation of the scientific method. Collaborative research combines a number of laboratories with their specialized techniques in studying the same or similar samples to magnify synergistically the understanding of the origin and history of the materials. Such collaborations have been responsible for innumerable recent scientific advances.

It must be remembered, of course, that there are many types of measurements that must be made *in situ*, even given the ability to return samples. Examples of this would be the detection and characterization of fragile components (e.g. highly reactive oxidants in Martian soil and atmosphere), samples too large to return (e.g. bedrock or sediment beds) and species that are transient (photochemical species) and would react with natural materials during the return trip. Also, samples will be subjected to a variety of unnatural (for them) environments during the rigors of the return flight, including higher temperatures, mechanical shocks, vibration and exposure to deep space radiation. In addition, the quantity of sample that can be practically returned from deep space on unmanned missions is typically of the order of 11 kg or less. The variety that can be achieved within this is significant if adequate sampling tools are included, such as chipping devices or coring drills. It is also strongly advisable to include compositional analysis instrumentation to screen samples in order to (1) determine if the samples chosen for return include a good representation of the major components and (2) assure the return of unusual materials. Sample return missions are intrinsically challenging; they are also the ultimate reward for planetary sampling science.

Benton C. Clark

Bibliography

Adler, I. (1986) *The Analysis of Extraterrestrial Materials*. New York: John Wiley and Sons.

Agresti, D.G., Morris, R.V., Wills, E.L. *et al.* (1992) Extraterrestrial Moessbauer spectrometry. *Hyperfine Int.*, **72**, 285–96.

Biemann, K., Oro, J., Toulmin, P. *et al.* (1977) The search for organic substances and inorganic volatile compounds in the surface of Mars. *J. Geophys. Res.*, **82**, 4641–58.

Clark, B.C., Baird, A.K., Rose, H.J. *et al.* (1977) The Viking x-ray fluorescence experiment: analytical methods and early results. *J. Geophys. Res.*, **82**, 4577–94.

Clark, B.C., Baird, A.K., Weldon, R.J., et al. (1982) Chemical composition of Martian fines. *J. Geophys. Res.*, **87**, 10059–67.
Conley, J.M., Bradley, J.G., Giffin, C.E. et al. (1983) Development of a miniature scanning electron microscope for in-flight analysis of comet dust, in *Microbeam Analysis – 1983* (ed. R. Gooley). pp. 177–81. San Francisco Press.
Gooding, J.L. 1989. Differential scanning calorimetry (DSC) and evolved-gas analysis (EGA) applied to planetary surface exploration, in *Proc. 18th North American Thermal Analysis Soc. Conf.* Vol. 1, (ed. I.R. Harrison) pp. 222–8.
Hoffman, J.H., Oyama, V.I. and von Zahn, L.L. (1980) Measurements of the Venus lower atmospheric composition. *J. Geophys. Res.*, **85**, 7871–81.
Horowitz, N.H. (1977) The search for life on Mars. *Sci. Am.* **237**, 52–61.
Kissel, J. Krenger, F., Clark, B.C. et al., (1986) Composition of comet Halley dust particles from Giotto observations. *Nature*, **321**, 336–337.
Klein, H.P. (1979) The Viking mission and the search for life on Mars. *Rev. Geophys. Space Phys.*, **17**, 1655–62.
Kocharov, G.E., Victorov, S.V., Kovalev, V.P. et al. (1975) *Space Research XV*, Berlin: Akademie-Verlag.
Nature (1986) Encounters with comet Halley – the first results. **321**, No. 6067.
Nyquist, L.E., Lugmair, G. Signer, P. et al. (1980) Mass spectrometry–isotope dilution (MSID) of cometary solids. NASA/Johnson Space Center.
Owen, T., Biemann, K. Rushneck, D. et al. (1977) The composition of the atmosphere at the surface of Mars. *J. Geophys. Res.*, **82**, 4635–40.
Oyama, V.I., Carle, G.C. and Berdahl, B.J. (1977) The Viking GEX results from Chryse and Utopia surface samples. *J. Geophys. Res.*, **82**, 4669–76.
Reinhard, R. and Battrick B. (1986) The Giotto mission – Its scientific investigations. ESA SP-1077.
Surkhov, Y.A., Schcheglov, O. Moskalyeva, L. et al. (1982) X-ray fluorescence spectrometry on the surface of Venus. *Anal. Chem.*, **54**, 957A–966A.
Turkevich, A.L. Patterson, J.H. and Franzgrote E.J. (1967) Chemical analysis of the Moon at the Surveyor V landing site. *Science*, **158**, 635–7.
Zarnecki, J.C. McDonnell, J.A.M., Hanner, M.F. et al. (1992) A surface science package for the Huygens Titan probe, in *Proc. Symp. Titan*, European Space Agency SP-338, pp. 407–9.
Zscheeg, H., Kissel, J. Natour, G.H. and Vollmer, E. (1992) CoMA – an advanced space experiment for *in situ* analysis of cometary matter. *Astrophys. Space Sci.*, **195**, 447–61.

Cross references

Apollo missions
Cassini mission
Giotto mission
Huygens mission
Luna missions
Phobos mission
SNC meteorites
Surveyor missions
Titan
Vega mission
Venera missions
Viking mission

PLANETARY TORUS

Since the first observation of neutral sodium around the Jovian satellite Io (Brown, 1974), the subsequent identification of a neutral cloud extending along the orbit of Io, which followed it in its rotation around Jupiter (Matson et al. 1978), and the discovery of ionized sulfur and oxygen tori surrounding Io's orbit (Kupo, Meckler and Eviatar, 1976; Pilcher and Morgan, 1979), a significant number of planetary tori have been discovered in the vicinity of the satellites of the giant planets Jupiter, Saturn and Neptune. These tori may consist of either neutral or ionized species. They may completely surround the planet or may only consist of partial arcs; they may follow the satellite's orbit or lie in a different plane. However, planetary tori are found only around magnetized planets, with satellites embedded in the magnetosphere. This indicates that, despite this wide variety of morphological and physical differences, the planetary tori are a single class of planetary objects. Reviewing the various tori observed so far, we will show that their differences are related to the compositional characteristics of the satellite, the dynamical characteristics of the planetary magnetosphere and the geometry of the satellite–magnetic field system.

Origin of planetary tori

The interaction between a satellite, its neutral environment, its plasma torus and the magnetosphere is based on feedback loops which insure the stability of the system. Suppose a satellite exists in a magnetosphere without an initial plasma population (this simplified scenario is in fact not realistic, since some plasma from the solar wind and from the planetary ionosphere is always present in a magnetosphere). Molecules and grains can be sputtered out from the surface of the satellite or from an atmosphere by micrometeoroid impact, by collisions with high-energy particles from the magnetosphere, by photosputtering, and by thermal escape of gas from an atmosphere. This type of source provides neutral gas in the vicinity of the satellite, roughly corotating with it. The fastest molecules can escape the satellite's gravitational field and flow into a toroidal volume around its orbit under the planet's gravity, forming a neutral torus (or a neutral torus segment depending on their lifetime). This neutral gas begins to be ionized, essentially by photoionization and interaction with the particles of the radiation belts. Although produced at a slow rate, these freshly created ions behave as a plasma 'frozen' into the planetary magnetic field and are dragged by the field lines into rotation. This is commonly called the 'pick-up' phenomenon. The magnetic field lines rotate roughly at the planet's velocity ('corotation'), i.e. significantly faster than the satellite on its Keplerian orbit, so that the newly picked up ions acquire an energy close to the corotation energy (in the range of several tens to a few hundred eV, depending essentially on the mass of the ion and the radial distance of the pick-up). Under the combined effect of the gravitational and centrifugal forces, the ions remain confined close to the centrifugal equatorial plane between the rotational and the magnetic equatorial planes (Richardson, Eviatar and Siscoe, 1986, and references therein). A fraction of these corotating ions impact their source satellite or its atmosphere and increase the population of neutrals in the corona and in the neutral cloud. Another fraction collide with electrons and energize them such that they become a new significant source of ionization of the neutrals, in addition to the initial photoionization. If this feedback loop (illustrated in Figure P16 for the case of Io) is efficient enough, the neutral cloud and the plasma can theoretically become self-supporting (Eviatar, Kennel and Neugebauer, 1978; Huang and Siscoe, 1987).

This oversimplified scenario does not take into account the necessary balance between plasma torus sources and losses required to attain a steady state equilibrium, nor the interactions of the tori with the surrounding magnetosphere (Figure P17). The hot ions can be lost to the torus by charge exchange with thermal neutrals, mainly from the neutral clouds. This results in the production of a thermal ion, immediately picked up by the magnetic field, and of a fast neutral which leaves the system at the corotation velocity. These neutrals are a significant source of plasma for the distant magnetosphere, where a fraction can be ionized again (Barbosa and Eviatar, 1986), and even for the interplanetary medium: a sodium nebula has recently been discovered around Jupiter extending beyond 400 R_J (Flynn, Mendillo and Baumgardner, 1992). The density of the torus is not affected, and the net effect concerns mainly the composition (the new ion can be different from the initial one) and the energy budget. Plasma can also be lost by recombination and radial transport into the magnetospheres. Centrifugally driven transport of heavy ion plasma controls the steady state Io plasma torus and is a major source of plasma in the magnetosphere of Jupiter (Siscoe and Summers, 1981). Radial transport also seems to play a major role in the control of the Triton torus–Neptune magnetosphere system (Richardson et al., 1991). At Uranus convective transport, externally driven by the solar wind, removes any plasma from the satellites, and prevents the formation of a plasma torus (Vasyliunas, 1986). By contrast, the density of the 'water group' plasma torus surrounding Dione and

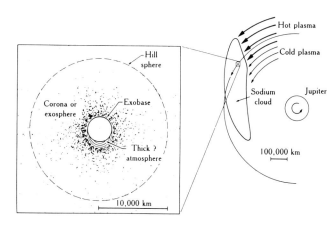

Figure P16 Schematic view of the atmosphere and sodium cloud of Io. To the left: Io's vicinity (the Hill sphere is the effective limit of Io's gravity). To the right, the neutral sodium cloud contour. The arrows represent the corotating plasma, traveling at 75 km s^{-1} past the neutrals that orbit Jupiter at 17 km s^{-1}. The cloud lies preferentially inside Io's orbit because the plasma is cooler there and less able to ionize the sodium atoms. Inside Io's orbit the cloud extends forward only, because atoms closer to Jupiter travel faster than Io. (From Schneider, Smyth and McGrath, 1989.)

Tethys is controlled by recombination (Richardson, Eviatar and Siscoe, 1986).

Finally, a plasma torus can also act as a sink for high-energy magnetospheric particles (among which some originally came from the torus plasma itself). The dense cold plasma of the torus favors resonant interaction of the particles with electromagnetic waves, resulting in pitch-angle diffusion, precipitation along field lines into the planetary high-latitude atmosphere and, ultimately, excitation of auroral emissions (Goertz, 1980; Thorne, 1983; Prangé, 1991; and references therein).

Observational techniques

Remote sensing by optical emissions

Although so far only applicable to the study of the Io torus (because of insufficient brightness of the other tori), remote sensing from ground-based telescopes is the most widely used tool for the study of planetary tori. It is through the detection of the D resonance doublet of sodium in the spectrum of Io that the Io sodium cloud was discovered in 1972 (Brown, 1974); all the information on the neutral and ionized Io tori was derived from observation of their optical emissions until the Voyager encounters in 1979. Remote sensing from space has also been widely used from 1979 on, with the observations of the Ultraviolet Spectrometer (UVS) during the Voyager 1 and 2 encounters with the Jupiter and Saturn systems, and of the Earth orbiting International Ultraviolet Explorer (IUE), still in operation, and since 1992, of the Hubble Space Telescope (HST) and of the Extreme Ultraviolet Explorer (EUVE). The neutral emission lines are due to resonance scattering of allowed transitions from the solar lines, whereas the emissions from the ionized species are excited by collisions with electrons. Remote imaging of the tori emission gives access to the morphology of the tori (the interpretation of the many 'asymmetries' of the Io tori and of their temporal variation has been fundamental in the understanding of the system). High-resolution spectroscopy provides information on the composition, density, temperature and thermodynamical equilibrium (by the study of the line intensity ratios), and on the velocity fields by the observation of Doppler shifts.

In addition, information on the integrated electron density in the Io torus has also been obtained by radio occultation techniques on the Voyager and Ulysses missions.

In situ measurements of plasma parameters

The deep space missions to the giant planets included instrumentation dedicated to the measurement of the charged particle distribution functions (density, composition, energy, flow direction, temperature) in various energy ranges, and in particular below a few keV. The *in situ* determination of the plasma frequency by the electromagnetic wave analyzers also provided the total plasma density. This has been the only means of identification and characterization of the distant or tenuous tori, whose emissions are too faint to be observed, such as the inner plasma torus in the magnetosphere of Saturn or the suspected Europa torus at Jupiter. Although not directly detected by the Voyager spacecraft, the Triton partial torus has been recognized by the analysis of the ions it has released in the 'nearby' magnetosphere.

A large body of information has been obtained on the Io plasma tori by comparing the *in situ* measurements of the Voyagers with the optical data. Similarly, the recent crossing of the torus by Ulysses has already begun to provide new and valuable data, in particular far from the equatorial plane, which can be correlated to remote observations (see Ulysses mission).

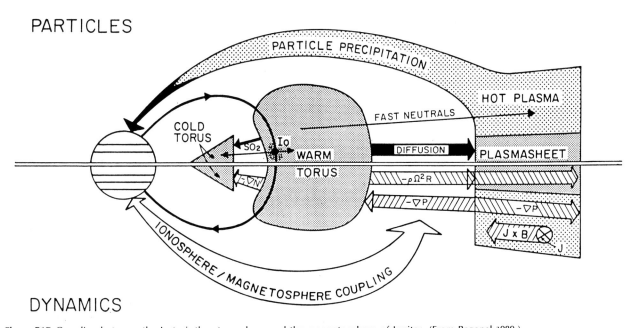

Figure P17 Coupling between the Io tori, the atmosphere and the magnetosphere of Jupiter. (From Bagenal 1989.)

Io tori: a model study of satellite–magnetosphere interactions

Io (q.v.), the innermost Galilean satellite of Jupiter, experiences strong time variable gravitational effects from Jupiter and Ganymede, resulting in tidal heating and volcanism. The volcanic activity was first discovered by imaging with Voyager in 1979, and provides to the tori mainly sulfur and oxygen, at a total rate of around 10^3 kg s^{-1} (a few 10^{28} atoms s^{-1}), with traces of sodium and potassium. This gives Io a well-developed atmosphere and ionosphere. The resulting tori are the densest and the brightest in the solar system. They are easily observable from Earth and they have been visited three times by deep space missions. Consequently a large body of data has been gathered on the Io tori, and an intense theoretical effort of modeling is taking place, making the Io tori a reference model for planetary tori.

Neutral clouds and tori

The observations and the physics of the neutral Io clouds have produced a large body of literature, including many review papers (Schneider, Smyth and McGrath, 1989; Thomas, 1992; also McNutt, 1991).

Io has a significant atmosphere of SO_2 (with a column density of 10^{15} to 10^{17} cm^{-2}), with some neutral sulfur and oxygen ($\geq 10^{12}$ cm^{-2}). Through the mechanisms detailed above, part of these neutrals escape and give rise to a neutral cloud extending along the orbit of Io, at roughly six Jovian radii (R_J) from Jupiter. However, these neutral clouds are extremely tenuous and their resonance lines very faint (a few rayleighs or less). Although sodium is a very minor constituent in the atmosphere of Io (10^{12} cm^{-2}), its resonance D lines at $\approx$ 5890–5896 Å are extremely bright (thousands of rayleighs in the vicinity of Io). The sodium cloud has been used as a model to study the neutral tori.

The 'banana-shaped' sodium cloud extends over several tens of degrees along Io's orbit in the rotational equatorial plane (Figure P15). Its size is limited by the lifetime of sodium atoms against ionization (a few hours in the equatorial plane). Models of its exact shape depend in particular on the distribution of the velocity of the escaping atoms and of the ionizing electrons (Smyth and Combi, 1988). By contrast, the lifetime of neutral oxygen and sulfur is much longer (tens of hours), giving rise to complete neutral tori.

An east–west asymmetry is present in the emission strength measured near Io, with an excess of ≈ 20% at western elongation. This has been attributed to solar radiation pressure effects, or to an asymmetric atmospheric shielding of the surface of Io to particle impact. It was also observed that the emission is strongest on the side of Io opposite to the Jovian magnetic equator (a north–south asymmetry modulated with a 13-h period). This has been explained by an increased ionization of the neutrals by the plasma torus particles which lie in the centrifugal equatorial plane, about 5° from the orbital plane of Io in the direction of the magnetic equator. Finally, the intensity varies also as a function of heliocentric orbital longitude, due to the variation of the solar flux when the resonance line is shifted along the gradient of the large solar Fraunhofer line, an effect not related to the physics of the system itself. Characteristic fast jets of neutral sodium (10 to 100 km s^{-1}) are regularly observed escaping from the cloud of slow sodium (Figure P18). They are now

Figure P18 Ground-based observations of the Io tori. Top: the S$^+$ (sulfur) plasma torus. The bright narrow feature extending on both sides of the equatorial plane is the 'ribbon' (inner edge of the warm torus). The cold torus, extending planetwards, is separated from the ribbon by an emission gap. Bottom: the Na cloud of slow sodium (elongated spot ahead of Io) and a torus of fresh Na atoms just charge exchanged from picked-up ions (faint bent trace). The bright spot in the left panel, connected to the sodium cloud, is the solar flux scattered by Io. Jupiter's light (center panel) has been attenuated by an absorbing film in both pictures. (N.M. Schneider, personal communication.)

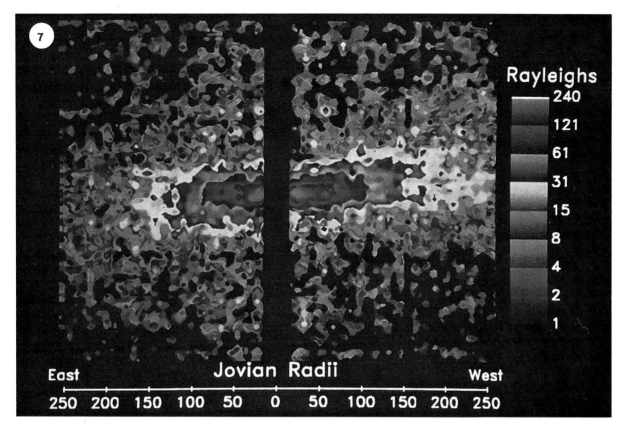

Figure P19 A ground-based composite image of the Jovian environment in the D1–D2 lines of neutral sodium from a series of drift scan spectra. The sodium nebula is evident in Jupiter's equatorial plane (tilted by 5° at the time of the observations), extending to several hundred Jovian radii in the interplanetary space. As in the case of the tori, an east–west asymmetry is visible. The dark region centered on Jupiter comes from the intensifier being turned off, to avoid damage due to light from Jupiter's bright disk. (From Flynn, Mendillo and Baumgardner, 1992.)

attributed to dissociative recombination of fresh sodium-bearing ions in the plasma torus, and are a source of Iogenic neutrals in the Jovian magnetosphere and the interplanetary medium (Barbosa and Eviatar 1986; Flynn, Mendillo and Baumgardner 1992), as shown on Figure P19.

Plasma tori

The remote optical observations and *in situ* plasma measurements have both given evidence of high plasma density in a toroidal region between 5 and 7.5–8 R_J from Jupiter, centered on the centrifugal equator, with a thickness scale height of $\approx 1\ R_J$. Both techniques have identified two main regions (Figures P17 and P20). The inner or 'cold' torus is dominated by singly ionized species (O^+, S^+), and the corresponding electron temperature is of the order of 10^4 K. It is located inside Io's orbit with a peak density of about 1000 cm^{-3} at the distance of $\approx 5.3\ R_J$, and is closely confined to the equatorial plane. In the outer or 'warm' torus multiple ionization dominates (S^{++}, O^{++}, S^{3+}), and the electron temperature is $\approx 6 \times 10^5$ K. The warm torus in fact extends inwards beyond Io's orbit, and one can define a third region, the 'warm inner torus' confined between 5.7 and 5.9 R_J, where the peak density, $\approx 3 \times 10^3$ cm^{-3}, exceeds by a factor of three the peak density in the outer torus (at 6 R_J). This narrow region had been independently observed as a very bright feature in the ground-based images and called the 'ribbon'. Efforts have been concentrated in the past decade on modeling of the Voyager data (Bagenal and Sullivan, 1981), of the optical images (e.g. Morgan, 1985) and of the thermodynamical equilibrium of the species (e.g. Smith and Strobel, 1985), in order to ascertain the characteristics of the Io tori from the observations. Recent reviews of such studies can be found in Strobel (1989), Bagenal (1989), Thomas (1992) and Spencer and Schneider (1996).

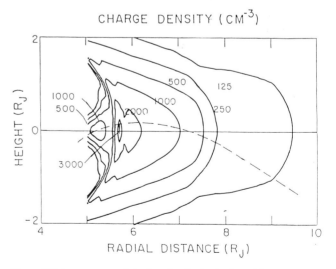

Figure P20 Isocontours of the ion density in the Io tori in a meridional plane derived from the plasma probe and UV spectrometer on Voyager. The large area on the right is the warm outer torus. The small secondary maximum on the left is the cold torus. The warm inner torus, or 'ribbon', observable from ground-based telescopes, is the region of sharp density gradients in between. The dashed line represents Voyager trajectory through the torus. (F. Bagenal, personal communication.)

The Io plasma torus radiates copious amounts of energy in the EUV, UV and visible lines through electron impact. At the epoch of the Voyager encounters, the total radiated power was of the order of $3-6 \times 10^{12}$ W. The traditional energy source (pick-up of fresh ions) is insufficient to fuel this power loss. To solve this 'energy crisis', it has been proposed that part of the high-energy magnetospheric ions diffuse inward into the hot torus and transfer their energy to electrons by collision (Smith et al., 1988). An alternative (or complementary) magnetospheric source of energy to the warm torus has been proposed in the form of a flow of hot ionospheric electrons secondary to the precipitation of magnetospheric particles in the outer torus (e.g. Thorne, 1983). These theories partly conflict with the early theories of a self-sustaining torus. Finally, the outer torus seems to be a sink for high-energy magnetospheric particles (Gehrels and Stone, 1983; Prangé, 1991), presumably through pitch-angle diffusion. Reviews of the complex coupling of the outer Io torus with the Jovian magnetosphere can be found in Thorne (1983) and Bagenal (1989).

Torus properties vary with longitude, latitude and local time. These variations, mostly poorly understood so far, stimulate theoretical studies on the interaction of the torus with the magnetosphere. A local time asymmetry in the EUV brightness of the torus, with maximum intensity at western elongation, has been attributed to a variation in electron temperature. The existence of a dawn-to-dusk electric field across the magnetosphere has been proposed to account for this increased electron temperature (Barbosa and Kivelson, 1983; Goertz and Ip, 1984).

Periodic longitudinal variations in the brightness of the short-lived species (such as S^{+1}) have been reported with maxima near 180–250°, not supported by the in situ Voyager measurements of total plasma density. These variations seemed to be better organized in a coordinate system rotating 3% slower than 'system III' (corotating with the magnetic field), giving rise to controversial interpretations. The very recent identification of denser plasma bubbles rotating in the inner and outer plasma tori by the radio experiment onboard the Ulysses spacecraft (Reiner et al., 1993) might provide an explanation of this phenomenon. Recently the radial distance of the torus from Jupiter has also been found to vary with longitude, roughly in phase with the variation of the surface magnetic field. A dawn-to-dusk asymmetry of this effect is also observed.

Finally, it is worth noting that the Io plasma tori were optically very faint at the epoch of the Pioneer encounters (1973–74). This considerable long-term variation of torus activity may be related to a similar variation of the Jovian UV auroral activity, which was hardly detectable by the Pioneer spectrometers, and reached tens to hundreds of kilorayleighs since the beginning of the 1980s.

Other planetary tori

Among the magnetized planets of the solar system, only Jupiter, Saturn and Neptune possess planetary tori. Mercury does not have a satellite. The Earth's Moon, in addition to releasing little matter through surface sputtering, orbits at a distance of 60 R_E from Earth, and spends less than 10% of its orbital period in the planet's magnetotail. By contrast, Uranus has five large satellites embedded in the magnetosphere, between 5 and 23 R_U. Ices seem to be present on all of them, and they must be sources of water and other volatiles. However, the motion of the plasma in the Uranian magnetosphere is controlled by solar wind-driven convection, rather than by corotation as in the case of Jupiter, and the odd geometry of the system (the axis of rotation is close to the ecliptic plane and, at the time of the Voyager flyby in 1986, sunward directed; the magnetic axis is tilted by $\approx 57°$ from the rotation axis, and rotating in ≈ 17 h) results in an efficient sweeping out of the satellite plasma within a few hours, preventing the formation of any plasma torus (Vasyliunas, 1986; Richardson et al., 1991), and hence of any self-sustaining neutral torus.

Jupiter

The existence of a plasma torus generated by the icy satellite Europa has long been suspected, but until recently there was no conclusive evidence of its existence. Very recent reanalysis of Voyager data now indicate the presence of hot, freshly picked-up ions in the vicinity of the orbit of Europa ($\approx 9.5 R_J$), an excess of O^{++} with respect to sulfur and the presence of H_3O^+, confirming the inference of a torus of 'water group' ions (Bagenal et al., 1992). This species can be expected from sputtering of the surface, but alternatively might confirm the role of tidal heating in exciting 'geyser-like' activity as suggested in Howell and Sinton (1989).

Saturn

One of the major differences between the magnetospheres of Jupiter and Saturn is the absence at Saturn of an Io-like satellite able to outgas large amounts of volatiles in the dense inner magnetosphere. By contrast, small icy satellites and rings can release smaller amounts of 'water-group' molecules in the inner magnetosphere (by sputtering and micrometeorid impact). The one satellite with an atmosphere and a significant gas output is Titan, which orbits at 20 R_s, well outside the main plasma bodies of the magnetosphere.

Essentially all observational information on the magnetosphere of Saturn comes from the Pioneer and Voyager encounters. This explains why, despite a large amount of data analysis and theoretical work, many unsolved questions remain still. The information comes from the in situ particle and wave data and, for the neutral Titan torus only, from remote sensing with the Voyager 1 and 2 UVS instrument. Figure P21 indicates schematically the location of the main neutral and plasma tori in the magnetosphere of Saturn. Recent work and reviews on the state of the art can be found in Richardson and Eviatar (1988), Eviatar, Podolak and Richardson (1990), McNutt (1991), Ip (1992) and Gombosi (1992).

Titan torus

As early as 1973 McDonough and Brice predicted the existence of a measurable neutral hydrogen torus in the vicinity of Titan's orbit. Although this was based on an erroneous composition of Titan's atmosphere, the Voyager 1 UVS instrument detected a 100 raleigh Lyman α emission from an extended region inside Titan's orbit (Broadfoot et al., 1981). This was interpreted as resonance scattering of solar Lyman α by an atomic hydrogen torus, of average density 10–20 cm^{-3}, $\approx \pm 3$–10 R_s thick, and extending radially from 8 to 25 R_s in the equatorial plane, consistent with a Titanogenic origin. From the discovery by Voyager of a thick atmosphere, composed mainly of molecular nitrogen and methane (with surface pressure ≈ 1 bar, and column densities $> 10^{16}$ cm^{-2} near the exobase), considerable efforts focused on the modeling of such a torus. Photodissociation and electron impact dissociation of CH_4 and N_2 provide mainly hot atomic hydrogen and nitrogen (H and N). A large fraction can escape Titan's gravity. Model estimates of the escape rates for each species have been derived in the range 2×10^{26} to 3×10^{27} atoms s^{-1} with escape velocities of a few kilometers per second (Strobel and Shemansky, 1981; Barbosa, 1987; Ip, 1992, and references therein). A colocated H_2 torus (undetectable so far) of similar density (a few tens per cubic centimeter) and a N torus, about five times fainter, are also predicted. The existence of local time asymmetries due to solar pressure effects are also discussed.

Ionization of the neutrals in these tori must provide keV picked up H^+ and N^+ ions (Eviatar, Podolak and Richardson, 1990; Barbosa, 1987; Gombosi, 1992). Warm, keV 'light' and 'heavy' ions in partial corotation, presumably H^+ and N^+, have been observed by the plasma probes on Pioneer 11, and Voyager 1 and 2, with low densities ≈ 0.1 to 1 cm^{-3} outward of 8 R_s (Lazarus and McNutt, 1983). Barbosa (1987) argues that this faint N^+ Titan torus, like the Io plasma torus at Jupiter, is a major plasma source for the magnetosphere, supplying energy for the observed auroral precipitation.

This 10 year old, well-documented picture of the Titan hydrogen torus (and other associated tori) might suffer major changes in the near future: Shemansky and Hall (1992) have reanalysed Voyager 1 and 2 UVS data. They state that Lyman α emission is distributed throughout the whole magnetosphere of Saturn inside the orbit of Titan; that the hydrogen atoms originate from Saturn's exosphere; and that any Titanogenic H torus would contribute at most 30 rayleighs of Lyman α, and fill the 18.5–20.5 R_s region only.

Inner plasma tori

In the inner magnetosphere the moons Enceladus (4 R_s), Tethys (4.9 R_s), Dione (6.2 R_s) and Rhea (8.7 R_s) have icy surfaces from which neutrals can be sputtered out. These neutrals can be dissociated or ionized. Removal of these ions from the tori as fast neutrals can occur from charge exchange or recombination at Tethys and Dione, or by transport processes at Rhea. The net result is the formation of tori containing H, H_2, O, OH, H_2O and O_2, and their ions. The Pioneer and Voyager plasma probes indeed measured a mixed proton and water-group ion plasma (O^+, OH^+ and H_2O^+)

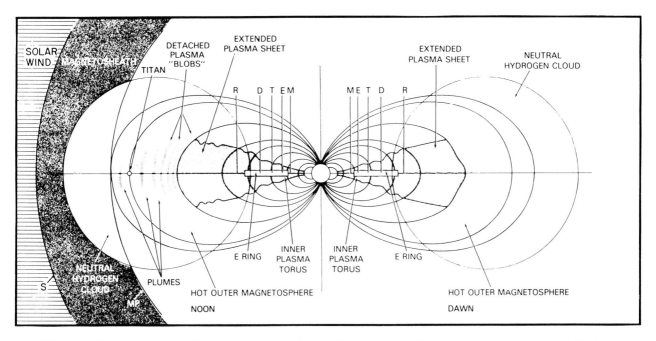

Figure P21 Schematic representation of the magnetosphere of Saturn with the location of the main neutral and plasma tori. The icy satellites are identified by their initial: Mimas (M), Enceladus (E), Tethys (T), Dione (D) and Rhea (R). (From Sittler, Ogilvie and Scudder, 1983.)

between 2.7 R_s (closest approach) and $\approx 7\ R_s$. The plasma is roughly corotating with the magnetic field (energies ≈ 100 eV), and is strongly confined to the equatorial plane with a half-width of the heavy ion tori $\leq 1\ R_s$ (with a somewhat larger thickness for the protons). Heavy ions densities of ≈ 20–40 cm^{-3} were observed near the equatorial plane at the orbits of Tethys and Dione, increasing to ≈ 150–200 cm^{-3} near the rings at 2.8 R_s (Lazarus and McNutt, 1983). The average heavy ion and proton density is roughly 12 and 2 cm^{-3}, respectively in the Tethys and Dione torus, and ≈ 2 cm^{-3} for both in the Rhea torus (Richardson, Eviatar and Siscoe, 1986). The presence of tenuous neutral tori of densities ≈ 2–8 cm^{-3} was theoretically predicted from these observations. As in the case of the Titan torus, part of this picture, and of the corresponding models, may be overthrown by the recent remote detection (with HST) of neutral OH in the vicinity of the orbit of Thetys with a density of 160 cm^{-3}, more than 20 times the previous estimates (Shemansky et al., 1993).

Ring hydrogen torus

An early indication of the presence of an atomic hydrogen torus surrounding the rings of Saturn has been obtained via the detection of Lyman α emission with a rocket flight and with Copernicus (Weiser, Vitz and Moos, 1977; Barker et al., 1980). The observation of 360 R of Lyman α by UVS during the Voyager 1 fly by confirmed the existence of a neutral hydrogen ring torus. The H average density should be 600 cm^{-3} if distributed in a torus of circular cross section of radius 1 R_s. This hydrogen was supposed to originate from sputtering of the ices in the ring material. However, the recent suggestion of a large exospheric source of atomic hydrogen filling the whole magnetosphere (see above), if confirmed, could probably also account for these observations without the need of an extra ring torus.

Neptune

Triton torus

The presence of nitrogen and methane frosts on Triton (at 14 R_N from Neptune), and the indication of a sizable magnetosphere, was inferred from IR spectroscopy and microwave observations a few years before the Voyager 2 encounter with Neptune. This analogy with Titan led Delitsky, Eviatar and Richardson (1989) to predict the existence of a measurable Triton plasma torus. The fly by of Neptune on 25 August 1989 found only indirect evidence of such a torus (whose density was below the detection threshold of the plasma detectors), in the form of the detection of H$^+$ and N$^+$ ions inside Triton's orbit, and of the observation of a faint, longitudinally asymmetric UV aurora. The discovery of the unexpected magnetic field geometry (dipole axis tilted by 47° from the planet's spin axis, and center offset by $\approx 0.5\ R_N$), combined with the inclination of the orbit of Triton by 22°, revealed a novel configuration, with Triton's orbit crossing a wide range of magnetic L shells from L = 14 in the magnetic equatorial plane, to more than 100 at high latitude (the McIlwain parameter L of a magnetic field line is the equatorial distance of the equivalent dipole field line, expressed in planetary radii; a magnetic L shell is the locus of the field lines with the same L

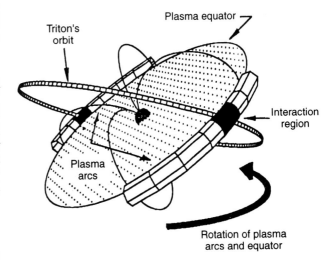

Figure P22 Formation of two torus segments (plasma arcs) in Neptune's plasma equator. Triton's orbit, tilted by $\approx 22°$ from Neptune's spin equator, intersects magnetic field lines from L ≈ 14 (near the plasma equator, tilted $\approx 40°$ from the spin equator) to L ≈ 100 (at high latitude). The subsequent ionized species are more confined and denser close to the minimum L flux tube, intersected in the longitude range 140–210° and 310–325°. (From Broadfoot et al. 1989, copyright 1993 by permission American Astronomical Society.)

parameter obtained by rotation around the planet). The subsequent formation of arcs of plasma in the centrifugal equator was inferred by Broadfoot et al. (1989). These arcs are longitudinally limited to the intersections of Triton's orbit with the minimum L shells, 14 to 15 (Figure P22), and could explain the morphology of the auroras. Voyager 2 also revealed the existence of a thick atmosphere (14 μbar at surface level), dominated by N_2, with traces of CH_4 and H_2 and of a dense ionosphere (with a peak density of $\approx 4 \times 10^4$ cm^{-3}, and composed primarily of N^+). Richardson et al. (1991) estimate that a neutral H cloud, similar to Titan's H torus, must form by thermal escape along Triton's orbit, extending inwards to L = 8, with average density of ≈ 500 cm^{-3} (other estimates range from 30 to 1000 cm^{-3}). The ionization of this torus could provide the source of the H^+ 'light ions' observed beyond L = 8, whereas the 'heavy' N^+ ions observed would be directly injected from the ionosphere at Triton. These estimates are consistent with the plasma measurements inside L = 13. By contrast, Richardson et al. (1991) could not find in the plasma data any confirmation of the longitudinal asymmetry of the plasma torus. However, the observations in the magnetosphere of Neptune have intrinsic limitations, and they are still very recent. Additional analysis and theoretical studies to come might change the present picture.

Renée Prangé

Bibliography

Bagenal, F. (1989) Torus–magnetosphere coupling, in *Time Variable Phenomena in the Jovian System* (eds M.J.S. Belton, R.A. West and J. Rahe) Washington: NASA SP-494, pp. 196–210.

Bagenal, F. and Sullivan J.D. (1981) Direct plasma measurements in the Io torus and inner magnetosphere of Jupiter. *J. Geophys. Res. A*, **86**, 8447–66.

Bagenal, F., Shemansky, D.E., McNutt, R.L., Jr et al. (1992) The abundance of O^{++} in the Jovian magnetosphere. *Geophys. Res. Lett*, **19**, 79–82.

Barbosa, D.D. (1987) Titan's atomic nitrogen torus: inferred properties and consequences for the Saturnian aurora. *Icarus*, **72**, 53–61.

Barbosa, D.D. and Eviatar, A. (1986) Planetary fast neutral emission and effects on the solar wind: a cometary exosphere analog. *Astrophys. J.*, **310**, 927–36.

Barbosa, D.D. and Kivelson, M.G. (1983) Dawn–dusk asymmetry of the Io plasma torus. *Geophys. Res. Lett.*, **10**, 210–3.

Barker, E.S., Cazes, S. Emerich, C. et al. (1980) Lyman alpha observations in the vicinity of Saturn with Copernicus. *Astrophys. J.*, **242**, 383–94.

Broadfoot, A.L., et al. (1981) Extreme ultraviolet observations from Voyager 1 encounter with Saturn. *Science*, **212**, 206–11.

Broadfoot, A.L., et al. (1989) Ultraviolet spectrometer observations of Neptune and Triton. *Science*, **246**, 1459–66.

Brown, R.A. (1974) Optical line emission from Io, in *Exploration of the Planetary System* (eds A. Woszczyk and C. Iwaniszewska) Boston: Reidel, pp. 527–31.

Delitsky, M.L., Eviatar, A. and Richardson J.D. (1989) A predicted Triton plasma torus in Neptune's magnetosphere. *Geophys. Res. Lett.*, **16**, 215–8.

Eviatar, A., Kennel, C.F. and Neugebauer, M. (1987) Possible origins of the variability in Jupiter's outer magnetosphere. *Geophys. Res. Lett.*, **5**, 287–9.

Eviatar, A., Podolak M. and Richardson, J.D. (1990) Atomic and molecular hydrogen from Titan in the Kronian magnetosphere. *J. Geophys. Res. A*, **95**, 21007–16.

Flynn, B., Mendillo, M. and Baumgardner, J. (1992) Observations and modeling of the Jovian remote neutral sodium. *Icarus*, **99**, 115–30.

Gehrels, N. and Stone, E.C. (1983) Energetic oxygen and sulfur ions in the Jovian magnetosphere and their contribution to the auroral excitation. *J. Geophys. Res. A*, **88**, 5537–50.

Goertz, C.K. (1980) Proton aurora on Jupiter's nightside. *Geophys. Res. Lett.*, **7**, 365–8.

Goertz, C.K. and Ip, W.H. (1984) A dawn–dusk electric field in the magnetosphere of Jupiter. *Planet Space Sci.* **32**, 179–85.

Gombosi, T.I. (1992) Mass loading at Titan and comets, in *Symposium on Titan*, (ed. B. Kaldeich) Paris: ESA SP-338, pp. 255–61.

Hill, T.W. and Dessler, A.J. (1990) Convection in Neptune's magnetosphere. *Geophys. Res. Lett.*, **17**, 1677–80.

Howell, R.R. and Sinton, W.M. (1989) Io and Europa: the observational evidence for variability, in *Time Variable Phenomena in the Jovian System* (eds M.J.S. Belton, R.A. West and J. Rahe). Washington: NASA SP-494, pp. 47–62.

Huang, T.S. and Siscoe, G.L. (1987) Types of planetary tori. *Icarus*, **70**, 366–78.

Ip, W.H. (1992) Plasma interaction of Titan with the Saturnian magnetosphere: a review of critical issues, in *Symposium on Titan* (ed. B. Kaldeich). Paris: ESA SP-338, pp. 243–53.

Kupo, I., Meckler, Y. and Eviatar, A. (1976) Detection of ionized sulfur in the Jovian magnetosphere. *Ap. J. Lett.*, **205**, L51–3.

Lazarus, A.J. and McNutt, R.L., Jr (1983) Low-energy plasma ion observation in Saturn's magnetosphere. *J. Geophys. Res. A*, **88**, 8831–46.

Matson, D.L., Goldberg, B.A., Johnson, T.V. and Carlson, R.W. (1978) Images of Io's sodium cloud. *Science*, **199**, 531–3.

McDonough, T.R. and Brice, N.M. (1973) A Saturnian gas ring and the recycling of Titan's atmosphere. *Icarus*, **20**, 136–45.

McNutt, R.L., Jr (1991) The magnetospheres of the outer planets. *Rev. Geophys. Suppl.* US National Report to IUGG, 985–97.

Morgan, J.S. (1985) Models of the Io torus. *Icarus*, **63**, 243–65.

Pilcher, C.B. and Morgan, J.S. (1979) Detection of singly ionized oxygen around Jupiter. *Science*, **205**, 297–8.

Prangé, R. (1991) Jovian UV aurorae, IR aurorae, and particle precipitation: a common origin? *Astron. Astrophys.*, **251**, L15–8.

Reiner, M.J., Fainberg, J., Stone, R.G. et al. (1993) Source characteristics of Jovian narrow-band kilometric radio emissions. *J. Geophys. Res. A*, **98**, 13163–76.

Richardson, J.D., and Eviatar, A. (1988) Observational and theoretical evidence for anisotropies in Saturn's magnetosphere. *J. Geophys. Res. A*, **93**, 7297–306.

Richardson, J.D., Eviatar, A. and Siscoe G.L. (1986) Satellite tori at Saturn. *J. Geophys. Res. A*, **91**, 8749–55.

Richardson, J.D., Belcher, J.W., Zhang, M. and McNutt R.L. Jr (1991) Low-energy ions near Neptune. *J. Geophys. Res. A*, **96**, 18993–9011.

Schneider, N.M., Smyth, W.H. and McGrath, M.A. (1989) Io's atmosphere and neutral clouds, in *Time Variable Phenomena in the Jovian System* (eds M.J.S. Belton, R.A. West and J. Rahe). Washington: NASA SP-494, pp. 75–99.

Shemansky, D.E., and Hall D.T. (1992) The distribution of atomic hydrogen in the magnetosphere of Saturn. *J. Geophys. Res. A*, **97**, 41143–61.

Shemansky, D.E., Matheson, P. Hall, D.T. et al. (1993) Detection of the hydroxyl radical in the Saturn magnetosphere. *Nature*, **363**, 329–31.

Siscoe, E.C. and Summers, D. (1981) Centrifugally driven diffusion of Iogenic plasma. *J. Geophys. Res. A*, **86**, 8480–4.

Sittler, E.C., Ogilvie, K.W. and Scudder, J.D. (1983) Survey of low-energy plasma electrons in Saturn's magnetosphere: Voyager 1 and 2. *J. Geophys. Res. A*, **88**, 8847–70.

Smith, R.A. and Strobel, D. (1985) Energy partitioning in the Io plasma torus. *J. Geophys. Res. A*, **90**, 9469–93.

Smith, R.A., Bagenal, F., Cheng, A.F. and Strobel, D.F. (1988) On the energy crisis in the Io plasma torus. *Geophys. Res. Lett*, **15**, 545–8.

Smyth, W.H. and Combi, M.R. (1988) A general model for Io's neutral clouds. II Appplication to the sodium cloud. *Astrophys. J.*, **328**, 888–981.

Spencer, J.F. and Schneider, N.M. (1996) Io on the eve of the Galileo mission. *Ann. Rev. Earth Planet Sci.*, **24**, 125–90.

Strobel, D.F. (1989) Energetics, luminosity, and spectroscopy of Io's torus, in *Time Variable Phenomena in the Jovian System* (eds M.J.S. Belton, R.A. West and J. Rahe). Washington: NASA SP-494, pp. 183–94.

Strobel, D.F. and Shemansky D.E. (1982) EUV emission from Titan's upper atmosphere: Voyager 1 encounter. *J. Geophys. Res.*, **87**, 1361–8.

Thomas, N. (1992) Optical observations of Io's neutral clouds and plasma torus. *Surv. Geophys.*, **13**, 91–164.

Thorne, R.M. (1983) Microscopic plasma processes in the Jovian magnetosphere, in *Physics of the Jovian Magnetosphere* (ed. A.J. Dessler). New York: Cambridge University Press, pp. 454–80.

Vasyliunas, V.M. (1986) The convection-dominated magnetosphere of Uranus. *Geophys. Res. Lett.*, **13**, 621–3.

Weiser, H., Vitz, R.C. and Moos, H.W. (1977). Detection of Lyman α emission from the Saturnian disk and from the ring system. *Science*, **127**, 755–7.

Acknowledgements

This article is dedicated to the memory of my friends Bob Smith and Chris Goertz who have contributed so much to our understanding of the physics of planetary tori, and whose lives and careers were brutally cut short a brief year ago.

Cross references

Io
Ionosphere
Magnetospheres of the outer planets
Plasma
Radio astronomy
Radio science
Ulysses mission
Voyager missions

PLANETESIMAL

According to the Chamberlin–Moulton planetesimal hypothesis (q.v.), the 'planetesimals' were identified as solid particles ranging from dust sized (hence 'dust-cloud hypothesis') to meteorite sized. Today, the term planetesimal is usually taken to include asteroids and even larger protoplanetary bodies (Lissauer, 1993). Chamberlin and Moulton (1900) developed the concept of a 'near-miss' encounter between two stars (an idea proposed earlier in 1745 by G.L.L. Buffon, 1707–1788), in which a large tidal protuberance, 'filament', or 'solar nebula' in a flattened disk, was generated from the surface of the major star that was destined to become our Sun. In Buffon's suggestion the collision was between a comet and the star. In the Chamberlin–Moulton model a series of violent eruptions created clouds of hot matter accelerated by the passing star. As the droplets cooled they became solid 'planetesimals' which possessed a high angular momentum. Interparticle collisions began and progressively built up into planet-sized bodies (see Planetary rotation). Stars have now been observed surrounded by what appear to be protoplanetary disks. These appear to consist of a mixture of gas and condensed matter, for which certain meteorites seem to offer the best geochemical model. While the rocky, inner (terrestrial-type) planets seemed to have accumulated in ever-larger pairs of planetesimal to protoplanet dimensions (Wetherill, 1989, 1990; Lissauer 1993), the Jovian planets seem to call for a two-stage growth, culminating in the gravitational capture of their gigantic gas envelopes.

What was the original nature of planetesimals? The Chamberlin–Moulton model of hot droplets is no longer acceptable (Safronov, 1969). The trail seems to lead to the interstellar dust that has been identified and spectroscopically analyzed (Greenberg, 1988). Two types of this dust are recognized: (1) highly diffuse clouds without volatiles like water; and (2) the protosolar nebulae which include dust particles mantled with volatiles of H_2O, CO_2, etc.

Three stages appear (Greenberg and Hage, 1991): (1) a silicate core with an organic refracting mantle (about 0.25 μm in diameter); (2) a 'precometary grain' on which there is a cover of ices, mainly H_2O; and (3) a cluster of the above forming a porous, low-density 'coma grain' (1 μm or more in diameter). In interstellar space the ices of simple chemical compounds are subjected to UV radiation, so-called 'photoprocessing', building up a variety of complex hydrocarbons (and depleting the free oxygen available).

It was postulated by Kuiper in 1951 that there should be a close-in disklike ring of planetesimals somewhere out from Neptune. In this Kuiper belt a very large, asteroid-sized body was identified in 1992 (reported in *Science*, **257**, 1865), some 200 km in diameter, and approximately 1.6×10^9 km beyond Neptune. Short-period comets could originate from this belt. Others appear to be low-volatile fragments of planetoid disruption.

In what is called the 'outer solar system' it seems inevitable that there should be diffusion of planetesimals that did not get caught up in the gravitational sweeping of the circumsolar disk during the accretion stage. Besides the Kuiper disk (at < 500 AU), there is also the Oort cloud (at $< 5 \times 10^4$ AU) which orbits the solar system and seems to be the source of the long-period comets. Stern (1990) postulates that in both of these vast clouds there is a 'substantial population' of asteroids and protoplanetary bodies of the order of 1000 km in diameter.

Slow accretion of planetesimals was necessary in the Earth's formative stages to accommodate the initial cooling required. The Earth's surface has always been cool (at least for the last 4 billion years, as shown by evidence of erosion by running water and sedimentation in bodies of standing, liquid water). All the geochemical components in the present Earth must have been present at the planetesimal stage; as Urey pointed out, volatile elements like bismuth, for example, would have been driven off by any overall melting. Volatile diffusion ('outgassing') has led to the progressive growth of the hydrosphere and atmosphere. Climatically, the Earth has, since about 4 billion years ago, always maintained the same mean surface temperature (about $20 \pm 5°C$), although the record shows cyclical recurrence of ice ages alternating with long intervals of more benign character. Occasional impacts by planetesimals or small asteroids have never been so catastrophic as to disturb the essential equilibrium required for terrestrial geological and biological evolution.

Rhodes W. Fairbridge

Bibliography

Chamberlin, T.C. (1900) An attempt to test the nebular hypothesis by the relations of masses and momenta. *J. Geol.*, **8**, 58–73. (See also *Astrophys. J.*, **14**, 17, 1901)

Chamberlin, T.C. (1927) *The Origin of the Earth*. Chicago: Chicago University Press.

Chamberlin, T.C. and Moulton, F.R. (1900) Certain recent attempts to test the nebular hypothesis. *Science*. (n.s.), **12**, 201–8.

Fairchild, H.L. (1904) Geology under the planetesimal hypothesis. *Geol. Soc. Am. Bull.*, **15**, 243–66.

Gold, T. (1963) Problems requiring solution, in: *Origin of the Solar System* (eds R. Jastrow and A.G.W. Cameron), New York: Academic Press, pp. 171–4.

Goldreich, P. and Ward, W.R. (1973) Formation of planetesimals. *Astrophys. J.*, **183**, 1051–61.

Greenberg, J.M. (1988) in *Dust in the Universe*. (eds M. Bailey and D.A. Williams. Cambridge University Press, pp. 121–00.

Greenberg, J.M., and Hage, J.I. (1990) *Astrophys. J.*, **361**, 260–74.

Jeffreys, H. (1924) *The Earth*. Cambridge University Press (and subsequent editions).

Lissauer, J.J. (1993) Planet formation. *Ann. Rev. Astron. Astrophys.*, **31**, 129–74.

Safronov, V.S. (1969) *Evolution of the Protoplanetary Cloud and Planets*. Moscow: Nauka Press (in Russian English translation by NASA, TTF-677, 1972). (See also *Icarus*, **94**, 260–71, 1991)

Stern, S.A. (1991) On the number of planets in the outer solar system: evidence of a substantial population of 1000-km bodies. *Icarus*, **90**, 271–81.

ter Haar, D. and Cameron, A.G.W. (1967) Solar system: review of theories, in *The Encyclopedia of Atmospheric Sciences and Astrogeology* (ed. R.W. Fairbridge). New York: Reinhold Publishing Co., 890–9.

Tonks, W.B. and Melosh, H.J. (1992) Core formation by giant impacts. *Icarus*, **100**, 326–46.

Urey, H.C. (1952) *The Planets, their Origin and Development*. New Haven: Yale University, Press.

Urey, H.C. (1959) Primary and secondary objects. *J. Geophys. Res.*, **64**, 1721–37.

Wetherill, G.W. (1990) *Ann. Rev. Earth Planet. Sci.*, **18**, 205–56.

Acknowledgements

This manuscript was kindly reviewed by J. Mayo Greenberg, Leiden; an early draft was kindly reviewed by Jack Lissauer.

Cross references

Chamberlin, Thomas Chrowder
Dust

Interstellar medium
Kuiper belt
Meteorite
Oort, Jan Hendrik, and Oort cloud
Planetary rotation
Solar nebula
Solar system: origin

PLASMA

A plasma is comprised of discrete electrically charged particles which are coupled through the long-range Coulomb potential so that they behave collectively. For a plasma system to exhibit this collective behavior (rather than each particle being unaffected by other charges) it must contain a sufficiently large number of particles. Formally, there must be a large number of particles within a sphere with radius equal to a Debye length, the scale length for the electrostatic screening potential,

$$\lambda_D = (kT_e\epsilon_o/n_e q_e^2)^{1/2}$$

where T_e, q_e are n_e are the electron temperature, charge and density; k is the Boltzmann constant; ϵ_o is the dielectric constant in free space. Thus the condition that particles behave collectively as a plasma is $N_D = \frac{4}{3}\pi\lambda_D^3 \gg 1$. On scale lengths larger than a Debye length, the plasma is electrically neutral when the number density of positive ions is on average equal to the electron number density.

The simplest plasma consists of electrons and one species of ions, protons (H^+), for example. More complex plasmas include several different ion species, neutral atoms and molecules. Solar system plasmas can be categorized by their sources: the solar wind, ionospheres, magnetospheres and comets. The solar wind is comprised of electrons and protons (the ionized part of the Sun's hydrogen atmosphere) which have sufficient energy to escape the Sun's gravity. Solar radiation photoionizes neutral atoms and molecules in planetary atmospheres to form, in most cases, an ionosphere with ion species reflecting the composition of the neutral atmosphere, mainly H^+ in the case of the giant planets and O^+, O_2^+ and NO^+ in the case of the Earth. An ionosphere can be simply described as a layer of ionization where the rate of photoionization balances the rate of recombination of the electrons and ions. Below the ionosphere layer, where the densities are higher, the recombination rate is higher and fewer ionizing photons penetrate. Above this layer the neutral densities and hence the source of particles for ionization, decrease. Magnetospheric plasmas, controlled by planetary magnetic fields, can have several sources which we discuss below. As a comet moves on its eccentric orbit towards the inner solar system, the escaping neutral gases form a very large cloud which is ionized by solar photons (and by solar wind impact) to form a very extended plasma tail of electrons and CO^+, H_2O^+ and OH^+ ions (see Comet).

In hot, tenuous plasmas collisions between particles are very rare, so that the plasma is primarily affected by electric and magnetic fields. In cold, dense plasmas collisions become important; one also has to consider ionization, charge exchange and recombination processes. While one expects cold, dense, collision-dominated plasmas to be in thermal equilibrium, even hot, tenuous plasmas in space are generally found not far from equilibrium, i.e. their particle distribution functions are observed to be approximately Maxwellian (though the ion and electron populations often have different temperatures). This fact is remarkable considering that the source mechanisms tend to produce particles with an initially very narrow range in energy, and timescales for equilibration by means of Coulomb collisions are usually much longer than transport timescales. At the same time, space plasmas support a variety of plasma waves which have various energy sources and cover a wide range of frequencies. Interactions between these waves and particle populations are thought to be responsible for bringing the bulk of the plasma towards thermal equilibrium as well as accelerating or scattering particles at higher energies. In addition to the thermal populations (which make up the bulk of the plasma by number density), all planetary magnetospheres contain populations of energetic particles which often dominate the energy density (Priest, 1985; Parks, 1991).

Plasma measurements

The most direct method of measuring the properties of plasmas in space is by placing a plasma detector on a spacecraft which flies through the plasma. By making the electric potential of a detector positive or negative, electrons or ions are correspondingly attracted or repelled. The electric potentials of a series of grids or metal surfaces are designed so that charged particles are guided through the instrument selecting only particles within a specific range of energy per unit charge (E/Q) (see Thermal plasma instrumentation). By varying the potentials, the spectrum of the particles' E/Q can be measured, ranging from a few volts to tens of kilovolts. If all the charged particles are of the same known species (e.g. electrons, protons), the E/Q spectrum is a direct measure of the velocity distribution of the particles. By pointing detectors in different directions, full three-dimensional velocity distributions can be measured in seconds. The velocity distributions of different ionic species can be measured with an electrostatic instrument if their spectral peaks are well separated in E/Q. Unfortunately, there are many space plasmas where the dominant species have the same or similar mass per charge ratios (e.g. O^+ and S^{++} in Jupiter's magnetosphere, or water dissociation products H_2O^+, OH^+, etc.) such that their E/Q spectra overlap. To separate these ion species it is necessary to add a deflecting magnet to the electrostatic analyzer, or to measure the time each particle takes to traverse a chamber in the instrument (Young, 1989; Young et al., 1989; Bame et al. 1989).

While plasma detectors provide detailed information about the particles' velocity distribution, from which bulk parameters such as density, temperature and flow velocity are derived, the plasma properties are only measured in the vicinity of the spacecraft. With the few existing spacecraft it is impossible to measure the changing properties of the many different plasmas in the solar system. Some of the most interesting space plasmas, however, can be remotely monitored by observing emissions of electromagnetic radiation. Dense plasmas, such as Jupiter's plasma torus, comet tails, Venus' ionosphere and the solar corona, have collisionally excited line emissions at optical or UV wavelengths. Similarly, when magnetospheric particles bombard the planets' polar atmospheres, various auroral emissions are generated from radio to x-ray wavelengths. Thus our knowledge of space plasmas is based on combining the remote sensing of plasma phenomena with spacecraft measurements which provide 'ground truth' details of the particles' velocity distribution as well as the local electric and magnetic fields that may be interacting with the plasma.

Obstacles in a flowing plasma

By the time plasma from the Sun reaches the planets its kinetic energy is largely bulk motion (i.e. the flow is supersonic) and, as the interplanetary magnetic field (IMF) is weak, the solar wind is super-Alfvénic (i.e. the solar wind speed is greater than the speed of Alfvén waves, magnetohydrodynamic waves that can propagate in a magnetized, collisionless medium). The fact that the solar wind is supersonic means that generally there is a bow shock upstream of the obstacle. The plasma is slowed down and heated as it passes through the shock and hence the flow around the obstacle is then sub- or trans-sonic. In contrast to the solar wind, plasma flows in planetary magnetospheres span wide ranges of sonic and Alfvénic Mach numbers, a consideration that must be kept in mind when comparing plasma interactions of different planetary satellites (e.g. the sonic/Alfvén Mach numbers for the flow near Io, Titan and Triton are 2/0.15, 0.6/2 and 1/0.2 respectively; Neubauer, Luttgen and Ness, 1991).

With regard to the characteristics of the obstacle, it is convenient to consider first the two extreme situations where the object is taken to be either a perfect insulator or a perfect conductor. It is then necessary to consider the effects of finite conductivity, since planetary bodies have effective conductivity in a range between these two extremes.

Non-conducting object

A magnetic field diffuses through an object with a time scale $\tau_d \sim \mu_o \sigma L^2$, where L is the size and σ the electrical conductivity of the body. If this diffusive timescale is much less than the timescale for changes in the ambient magnetic field, the field passes through the body largely unperturbed. For a magnetic field 'frozen' into a plasma flowing at a characteristic speed V the object sees the field change over the convective timescale $\tau_c \sim L/V$. Hence the magnetic interaction is weak for a non-conducting body with low magnetic Reynolds number ($R_m = \tau_d/\tau_c = \mu_o \sigma L V \ll 1$).

Although the magnetic field readily diffuses through the non-conducting body, the plasma particles obviously cannot penetrate the body and are therefore absorbed. Because the flowing plasma is absorbed on the upstream surface there is a cavity behind the object and a wake is formed downstream as the plasma expands into the low-pressure region (Figure P23a). The Earth's Moon and most of the icy satellites of the outer planets are non-conductors and hence simple absorbers of the flowing plasma in which they are embedded.

Perfectly conducting object

When there is a relative motion V between a magnetized plasma and a conducting body the Lorentz electric field $E = -V' \times B$ drives a current $J = \sigma E$ in the object (Figure 23b). The current in the body in turn produces a perturbation in the background magnetic field. Since the magnetized plasma is highly anisotropic (with parallel conductivity $\sigma_\parallel \gg \sigma_\perp$) the current is carried away from the flanks of the object along the magnetic field. In the magnetohydrodynamic (MHD) regime the plasma is coupled to the magnetic field and hence the plasma flow is also perturbed by the conducting body. In the case of a perfect conductor (i.e. $R_m \gg 1$) the resulting motion of the plasma in the tube of magnetic flux that intersects the body exactly matches the motion of the conductor. The surrounding plasma is then deflected around the body in a manner similar to incompressible hydrodynamic flow around a cylinder with essentially no wake downstream. The perturbations in the magnetic field (b) and the plasma flow (v) are an Alfvénic disturbance, which propagates along the ambient magnetic field with a characteristic speed of

$$V_A = B/(\mu_0 \rho)^{1/2}$$

and satisfies the Alfvén relation $v/V_A = \pm b/B$ (where the sign corresponds to propagation parallel or antiparallel to the ambient magnetic field). One can consider these Alfvén waves to be carrying field-aligned currents that complete the electrical circuit that flows through the object. If the object has a substantial ionosphere (with a conductance on the order of 1 mho), such as in the cases of Venus, Mars, Titan and Io, these currents flow through the ionosphere rather than the solid object. To first order, Jupiter's moon Io can be

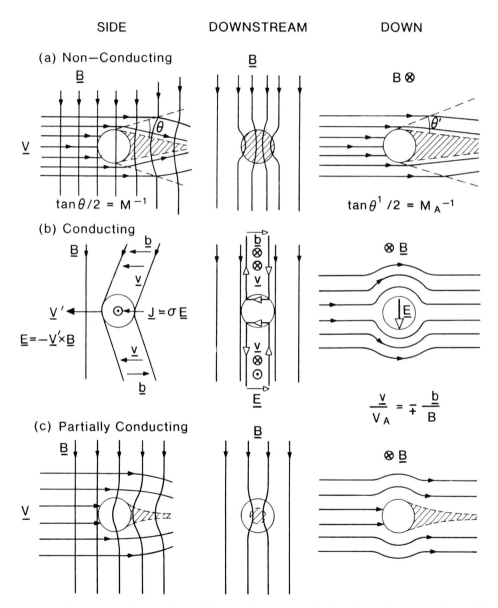

Figure P23 The interaction of magnetized plasma with an object that is (a) non-conducting; (b) conducting; and (c) partially conducting. The hatching indicates the region of low density in the object's wake. (The magnetic field B, the solar wind flow V, the motional electric field E and the induced current J are given by arrows. The symbols $\otimes$ and $\odot$ indicate arrows pointing into and out of the figure respectively. The perturbations in flow and magnetic field are shown by v and b)

described as a conducting object embedded in magnetospheric plasma flow (Belcher, 1987).

Partially conducting object

When the object has a finite conductivity the flow perturbation is insufficient to deflect the surrounding plasma to flow around the body and hence some is absorbed (Figure P23c). The plasma flow is reduced near the object and the magnetic field being 'hung up' in the stagnant flow. Since the magnetic field far from the object continues to be convected in the unperturbed flow the magnetic field lines become bent or 'draped' over the object to form a magnetic tail or wake downstream.

Object with an atmosphere

On the dayside of all objects having an atmosphere, the Sun's ultraviolet emissions ionize some of the neutral atoms. For objects that are embedded in a magnetospheric plasma (e.g. Titan, Triton and Io), ionization by particle impact may also be significant. In any case, the degree of ionization determines the conductivity of the upper atmosphere (ionosphere) and thus affects the nature of the interaction of the object with the plasma in which it is immersed. In the case of a planet with a dense ionosphere ($\sigma \to \infty$ or effectively $\sigma > \sim 1$ mho) the solar wind magnetic field is excluded and the flow is diverted around the flanks of the body (the case at Mars and Venus, to first order). Similar situations are found at Titan, Triton and Io, except that there is no bow shock for the sub- or trans-sonic magnetospheric flows. The boundary between the ionosphere and the surrounding plasma, the ionopause, is located where the combined magnetic and ram pressure of the external plasma is balanced by the particle pressure (P) in the ionosphere (Figure P24a). When the ionization is weak ($\sigma \to 0$) the magnetic field and plasma flow are dragged through the resistive ionosphere, causing a substantial downstream wake.

We must further consider the consequences of the ionization of any neutral material extending out into the streaming plasma. On ionization the particle 'sees' the Lorentz electric field and is accelerated up to the ambient flow. The momentum gained by the newly created ions comes from the surrounding plasma which correspondingly loses momentum. This effect is called 'mass loading' and contributes to the local deceleration of the flow and the draping of field lines over an object with a substantial atmosphere (e.g. comets, Venus and Titan; Luhmann, 1986; Luhmann et al., 1992; Cheng and Johnson, 1989.)

Magnetized object

Well before Biermann (1957) provided cometary evidence of a persistent solar wind, Chapman and Ferraro (1931) considered how a strongly magnetized body would deflect a flow of particles from the Sun and made an estimate of the location of the magnetopause stagnation point – the boundary between the magnetosphere in the direction of the Sun. They proposed that a dipolar magnetic field (of strength B_o at the planet's equatorial radius R_p) would 'stand off' the flow to a distance R_{CF}, where the external ram pressure of the solar wind balances the internal pressure of the planet's magnetic field (Figure 24b):

$$R_{CF} / R_p = \zeta \, (B_o^2 / 2 \, \mu_0 \, \rho_{sw} \, V_{sw}^2)^{1/6} \qquad (P8)$$

where ρ_{sw} and V_{sw} are the ion mass, density and flow speed of the solar wind. The dimensionless factor $\zeta \sim 1.3$ corrects for the various oversimplifications in the above description (Siscoe, 1979). The observed magnetopause stand-off distances, R_m, are found to be compatible with the values R_{CF} calculated using observed upstream conditions (see Table P8) except in the case of Jupiter where R_m is observed to be highly variable with an average value about twice the nominal value, owing to the presence of interior plasma with a pressure comparable to the magnetic field pressure.

The term magnetosphere was coined by Gold (1959) to describe the region of space wherein the principal forces on a plasma are electrodynamic in nature and are a result of the planet's magnetic field.

Plasma sources

The main source of plasma in the solar system is clearly the Sun. The solar corona, the upper atmosphere of the Sun (which has been heated by some, as yet undetermined, process to temperatures of 1–2 million K), streams away from the Sun at a more or less steady rate of 10^9 kg s^{-1} in equal numbers (8×10^{35} s^{-1}) of electrons and ions. Magnetospheres contain considerable amounts of plasma from various sources (Figure P25). Table P9 describes the primary and secondary sources of plasma for the six planetary magnetospheres.

First, the magnetopause is not entirely plasma-tight. Whenever the IMF has a component antiparallel to the planetary magnetic field, magnetic reconnection is likely to occur and solar wind plasma will leak into the magnetosphere across the magnetopause. Solar wind material is identified in the magnetosphere by its energy and characteristic composition of protons (H$^+$) with $\sim 4\%$ alpha particles (He^{++}) and trace heavy ions.

Second, although ionospheric plasma is generally cold and gravitationally bound to the planet, a small fraction has sufficient energy to escape up magnetic field lines and into the magnetosphere. Ionospheric plasma has a composition that reflects the composition of the planet's atmosphere (e.g. O$^+$ for the Earth, H$^+$ for the outer planets).

Third, the interaction of magnetospheric plasma with any natural satellites that are embedded in the magnetosphere can generate significant quantities of plasma. The ionization of the outermost layers of a satellite atmosphere by the impacting plasma flow can be a major source of plasma (nearly 1 metric ton per second in the case of Io). Energetic particle sputtering of the satellite surface or atmosphere produces less energetic ions directly, or an extensive cloud of neutral atoms which are eventually ionized far from the satellite. The distributed sources of water-product ions (totaling ~ 2 kg s^{-1}) in the magnetosphere of Saturn suggest that energetic particle sputtering of the icy satellites, particularly Rhea, Dione and Tethys, is an important process. Although the sputtering process, which removes at most a few microns of surface ice per thousand years, is probably insignificant in geological terms, sputtering has important consequences for the optical properties of the surface (Cheng and Johnson, 1989).

Plasma flows

Magnetospheric configuration is generally well described by magnetohydrodynamic theory (MHD) in which the magnetic field can be

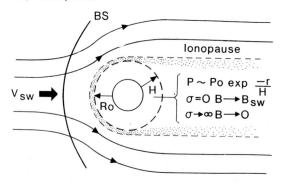

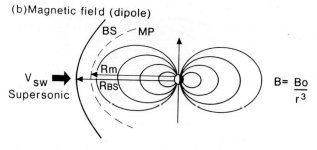

Figure P24 The interaction of a flowing magnetized plasma (V_{SW}) with an object possessing (a) an atmosphere of surface pressure P_0 and scale height H; (b) a magnetic field. The bow shock (BS) and magnetopause (MP) are shown.

Table P8 Properties of the solar wind and planetary magnetic fields

	Mercury	Earth	Jupiter[a]	Saturn[a]	Uranus[a]	Neptune[a]
Distance, a_{planet} (AU)	0.31–0.47	1[b]	5.2	9.5	19	30
Solar wind density (cm^{-3})	35–80	8	0.3	0.1	0.02	0.008
Radius, R_{Planet} (km)	2439	6373	71 398	60 330	25 559	24 764
Magnetic moment/M_{Earth}	4×10^{-4}	1[c]	20 000	600	50	25
Surface magnetic field B_0 (gauss)	3×10^{-3}	0.31	4.28	0.22	0.23	0.14
R_{CF}[d]	1.4–1.6 R_M	10 R_E	42 R_J	19 R_S	25 R_U	24 R_N
Observed size of magnetosphere	1.4 R_M	8–12 R_E	50–100 R_J	16–22 R_S	18 R_U	23–26 R_N

[a] Magnetic field characteristics from Acuña and Ness (1976), Connerney, (1987, 1991).
[b] 1 AU = 1.5×10^8 km.
[c] $M_{Earth} = 7.906 \times 10^{25}$ gauss cm^3 = 7.906×10^{15} T m^3.
[d] R_{CF} is calculated using equation (P8) for typical solar wind conditions of $\rho_{sw} \sim$ (8 amu cm^{-3})/a^2_{planet} and $V_{sw} \sim$ 400 km s^{-1}.

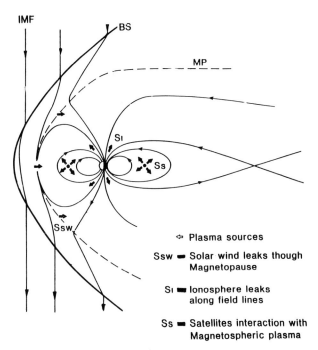

Figure P25 Sources of magnetospheric plasma.

considered to be frozen into the plasma flow. Thus we need to consider the processes controlling magnetospheric flows. The two largest sources of momentum in planetary magnetospheres are the planet's rotation and the solar wind. The nature of any large-scale circulation of material in the magnetosphere depends on which momentum source is tapped. For planetary magnetospheres, corotation of plasma with the planet is a useful first approximation, with any departures from strict corotation occurring when certain conditions break down. It may be helpful to think of plasma in the magnetosphere as mass that is coupled by means of magnetic field lines to a giant flywheel (the planet) with the ionosphere acting as the clutch.

Corotation

For magnetospheric plasma to rotate with the planet, the first two conditions are that the upper region of the neutral atmosphere must corotate with the planet and must be closely coupled to the ionosphere by collisions. The electrical conductivity of the ionosphere σ^i is large so that in a corotating ionosphere (with velocity V^i) any horizontal currents (perpendicular to the local magnetic field) are given by Ohm's law,

$$J_\perp^i = \sigma^i (E^i + V^i \times B)$$

Just above the ionosphere the conductivity perpendicular to the magnetic field in the (collision-free) magnetosphere, $\sigma_\perp^m$, is essentially zero and because the plasma particles are far more mobile in the direction of the local magnetic field, the parallel conductivity $\sigma_\parallel^m$ is large. Therefore, magnetic field lines can be considered to be equipotentials ($E \cdot B = 0$) in the magnetosphere and the electric field in the magnetosphere ($E^m = -V^m \times B$) can be mapped into the ionosphere (Figure P26a). Because the ionosphere is relatively thin, the electric field E^m just above the ionosphere is the same as E^i so that we can write

$$J_\perp^i = \sigma^i (V^i - V^m) \times B$$

The condition for corotation of the magnetospheric plasma is that the ratio J^i/σ^i is sufficiently small that

$$V^m = V^i = \Omega \times r$$

For a dipolar magnetic field that is aligned with the rotation axis, the corotational electric field (in the equatorial plane) is therefore radial with magnitude $E_{co} = \Omega B_o/r^2$.

It is clear that large ionospheric conductivities facilitate corotation. The large $\sigma_\parallel^m$ also means that any currents in the magnetosphere that result from mechanical stresses on the plasma are directly coupled by field-aligned currents to the ionosphere. Thus corotation breaks down when mechanical stresses on the magnetospheric plasma drive ionospheric currents which are sufficiently large that the ratio J^i/σ^i becomes significant. Such conditions might occur in regions of the magnetosphere where there are large increases in mass density due to local ionization of neutral material, where there are strong radial motions of the plasma or where there are sharp gradients in plasma pressure. When the magnetosphere imposes too large a load, the ionospheric clutch begins to slip.

Solar wind convection

Next let us consider how the momentum of the solar wind may be harnessed by processes occurring near the magnetopause where the external interplanetary magnetic field (B_{IMF}) interconnects with the planetary magnetic field. Figure P26b shows that at the poles the planetary magnetic field lines are open to the solar wind. The solar wind drives a plasma flow across the polar caps and the field lines from the polar region move in the direction of the solar wind flow, being pulled by the solar wind over the poles and back into the extended magnetotail. Conservation of flux requires that field lines are further cut and reconnected in the tail.

The MHD condition of the field being frozen to the flow can be written as $E + V \times B = 0$ which allows the convection electric field to be written

$$E_{cv} = -\eta \, V_{sw} \times B_{IMF}$$

where η is the efficiency of the reconnection process in harnessing the solar wind momentum, ~ 0.1 for the Earth. In simple magnetospheric models E_{cv} is assumed constant throughout the magnetosphere. The corresponding circulation is given by the $E \times B$ drift,

$$V_{cv} = \eta \, V_{sw} \, (B_{IMF}/B_o) \, (r/R_p)^3$$

After being carried tailward at high latitudes, the plasma then drifts towards the equatorial plane and eventually returns in a sunward flow to the dayside magnetopause.

Comparison of the corresponding electric fields indicates whether the magnetospheric circulation is driven primarily by the solar wind

Table P9 Plasma characteristics of planetary magnetospheres

	Mercury	Earth	Jupiter	Saturn	Uranus	Neptune
Maximum density (cm^{-3})	1	1–4000	>3000	~100	3	2
Primary sources	H^+	O^+, H^+	O^{n+}, S^{n+}	O^+, H_2O^+, H^+	H^+	N^+, H^+
	Solar wind	Ionosphere[a]	Io	Dione, Tethys	Ionosphere	Triton
Secondary sources	?	H^+	H	N^+, H^+	H^+	H^+
		Solar wind	Ionosphere	Titan	Solar wind	Solar wind
Source strength (ions s^{-1})	?	2×10^{26}	$>10^{28}$	10^{26}	10^{25}	10^{25}
(kg s^{-1})		5	700	2	0.02	0.2
Lifetime	Minutes	Days[a], hours[b]	10–100 days	30 days to years	1–30 days	~1 day
Plasma motion	Solar wind driven	Rotation[a], solar wind[b]	Rotation	Rotation	Solar wind + rotation	Rotation (+ solar wind?)

[a] Inside plasmasphere.
[b] Outside plasmasphere.

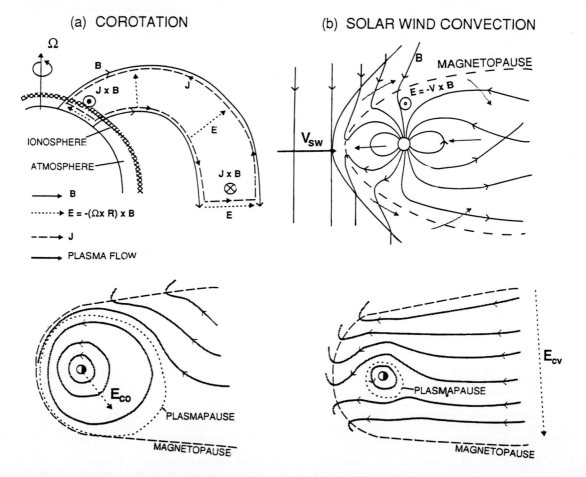

Figure P26 Large-scale magnetospheric circulation driven by (a) planetary rotation; (b) the solar wind. (The magnetic field **B**, the solar wind flow **V**, the motional electric field **E** and the induced current **J** are given by arrows. Ω labels the planet's spin axis. The symbols $\otimes$ and $\odot$ indicate arrows pointing into and out of the figure respectively).

or by the planet's rotation. Since E_{co} is proportional to R^{-2} and E_{cv} is constant, it seems reasonable to expect that corotation dominates close to the planet while solar wind driven convection dominates outside a critical distance.

$$R_c = (\Omega B_o/\eta \, V_{sw} \, B_{IMF})^{1/2}$$

This simply says that magnetospheres of rapidly rotating planets with strong magnetic fields are dominated by rotation while the solar wind controls the plasma flow in smaller magnetospheres of slowly rotating planets (Dessler, 1983; Hill and Dessler, 1991; Parks, 1991).

Plasma loss processes

For the planets with plasma transport driven by the solar wind, magnetospheric plasma is removed and carried away in the solar wind quite rapidly, on timescales of minutes (in the case of Mercury) to several days (in the case of Uranus). In rotation dominated magnetospheres radial transport of plasma is slow so that high plasma densities build up (Table P9) and other loss mechanisms can be significant. In particular, charge exchange occurs between corotating ions and neutral material (e.g. sputtered neutral clouds around

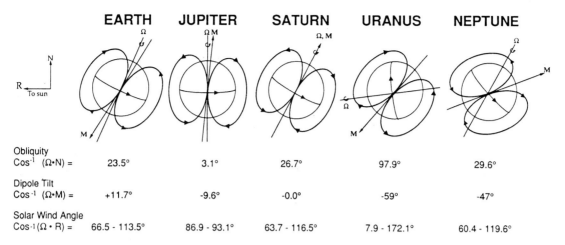

Figure P27 Orientations of the planets and their magnetic fields.

satellites and rings, or the extended neutral atmospheres of the giant planets). As a result of the charge-exchange process the corotating ion becomes neutralized so that it is no longer confined by the planet's magnetic field. The neutralized ion has maintained most of its corotational momentum and hence escapes from the planet. Such a wind of escaping fast neutral sodium atoms has been observed extending to several hundred Jovian radii (Mendillo et al., 1990). Charge exchange is also a dominant loss process in the magnetosphere of Saturn (Cheng and Johnson, 1989).

Plasma properties at the planets

Solar wind plasma

The solar wind is the outward expansion of the solar corona (Van Allen, 1990; Pizzo, Holzer and Sime, 1988; Marsch and Schwenn, 1992). At Earth's orbit and beyond, the solar wind has an average speed of about 400 km s^{-1}. The density of particles (mainly electrons and protons) is observed to decrease, from values of about 3–10 cm^{-3} at the Earth as the inverse square of the distance from the Sun, consistent with a steady radial expansion of the solar gas into a spherical volume, reaching values of 0.005 cm^{-3} at Neptune (Table P8).

The solar wind speed, while varying between about 300 and 700 km s^{-1}, always greatly exceeds the speed of waves characteristic of a low density, magnetized and completely ionized gas (MHD or Alfvén waves). Thus a shock is formed upstream of an obstacle (such as a planetary magnetosphere) that is imposed on the super-Alfvénic solar wind flow. A planetary bow shock can be described in fluid terms as a discontinuity in bulk parameters of the solar wind plasma in which mass, momentum and energy are conserved. Entropy, however, increases as the flow traverses the shock, with the solar wind plasma being decelerated and heated so that the flow can be deflected around the magnetosphere (see Shock wave). Thus a shock requires dissipative processes and the presence of a magnetic field allows dissipation to occur on a scale much smaller than a collisional scale length. Although planetary bow shocks do not play a significant role in magnetospheric processes, the crossings of spacecraft through planetary bow shocks have provided an opportunity to study the exotic plasma physics of high Mach number collisionless shocks that cannot be produced in a laboratory (Russell, Hoppe and Livesey, 1982; Slavin et al., 1985; Bagenal et al., 1987; Moses et al., 1990).

Magnetospheric plasmas

Spacecraft have explored the plasma environments of all the planets except Pluto. With the exception of Venus and Mars, all these planets have significant magnetic fields and magnetospheres. In the cases of Venus and Mars the solar wind interaction with ionospheres is confined to a small region close to the planet where the solar wind plasma is deflected around the planet, picking up O$^+$ ions from the upper layers of the planet's atmosphere (Luhmann, 1986; Luhmann et al., 1992). Given the roughly constant solar wind speed, the size of a planet's magnetosphere (R_{CF}) depends only on the ambient solar wind density and the planet's magnetic field [see equation (P8)]. Thus we expect a planet with a strong magnetic field to have a large magnetosphere, and even the weak fields of Uranus and Neptune produce moderate-sized magnetospheres in the tenuous solar wind of the outer solar system (see Magnetospheres of the outer planets). Table P8 shows that the measured sizes of planetary magnetospheres generally agree quite well with the theoretical R_{CF} values. Jupiter is the only notable exception, where the plasma pressure inside the magnetosphere is sufficient to 'inflate' further the magnetosphere. This makes the magnetosphere of Jupiter a huge object – about 1000 times the volume of the Sun, with a tail that extends at least 6 AU in the direction away from the Sun, beyond the orbit of Saturn. If the Jovian magnetosphere were visible from Earth, its angular size would be twice that of the Sun, even though it is at least four times farther away. The magnetospheres of the other giant planets are much more modest (while still dwarfing that of the Earth), having a similar scale of about 20 times the planetary radius, comparable to the size of the Sun. With only a weak magnetic field and being embedded in the denser solar wind close to the Sun, Mercury has a very small magnetosphere (Russell, Baker and Slavin, 1988; see Mercury: magnetic field and magnetosphere).

While the size of a planetary magnetosphere depends on the strength of a planet's magnetic field, the configuration and internal dynamics depend on the field orientation (illustrated in Figure P26) which is described by two angles: the tilt of the magnetic field with respect to the planet's spin axis and the angle between the planet's spin axis and the solar wind direction (which is generally within a few degrees of being radially outward from the Sun). Since the direction of the spin axis with respect to the solar wind direction only varies over a planetary year (many Earth years for the outer planets), and the planet's magnetic field is assumed to vary only on geological timescales, these two angles are constant for the purposes of describing the magnetospheric configuration at a particular epoch. Earth, Jupiter and Saturn have both small dipole tilts and small obliquities. This means that the orientation of the magnetic field with respect to the solar wind does not vary appreciably over a planetary rotation period and that seasonal effects are small. Thus Earth, Jupiter and Saturn have symmetric and quasi-stationary magnetospheres, with Earth and Jupiter each exhibiting only a small wobble due to their ~ 10° dipole tilts. In contrast, the large dipole tilt angles of Uranus and Neptune mean that the orientation of their magnetic fields with respect to the interplanetary medium varies considerably over a planetary rotation period, resulting in highly asymmetric and time-variable magnetospheres. Furthermore, Uranus' large obliquity means that the configuration of its magnetosphere will have strong seasonal changes over its 84-year orbit.

Table P9 summarizes the basic characteristics of plasmas measured in the magnetospheres of the planets which have detectable magnetic fields. The composition of the ionic species indicates the sources of magnetospheric plasma: satellites in the cases of Jupiter, Saturn and Neptune; the planet's ionosphere in the cases of Earth and Uranus. In the magnetospheres where plasma motions are driven by the solar wind, solar wind plasma enters the magnetosphere, becoming the primary source of plasma in the case of Mercury's small magnetosphere and secondary plasma sources at Earth, Uranus and Neptune. In the magnetospheres where plasma flows are dominated by the planet's rotation (Jupiter, Saturn and within the Earth's plasmasphere), the plasma is confined by the planet's strong magnetic field for many days so that substantial densities are accumulated.

Thus we can identify three categories of planetary magnetospheres:

1. the large, symmetric and rotation-dominated magnetospheres of Jupiter and Saturn;
2. the small magnetosphere of Mercury where the only source of plasma is the solar wind which drives rapid circulation of material through the magnetosphere; and
3. the moderate-sized and highly asymmetric magnetospheres of Uranus and Neptune, whose constantly changing configuration does not allow substantial densities of plasma to build up. The Earth's magnetosphere is an interesting hybrid of the first two types, with a dense corotating plasmasphere close to the planet and tenuous plasma, circulated by the solar wind driven convection, in the outer region (Bagenal, 1992; Russell, Baker and Slavin, 1988; Parks, 1990).

Fran Bagenal

Bibliography

Acuna, M.H., Ness, N.F. (1976) The main magnetic field of Jupiter. *J. Geophys. Res.*, **81**, 2917–22.

Bagenal, F. (1992) Giant planet magnetospheres. *Ann. Rev. Earth Planet. Sci.*, **20**, 289.

Bagenal, F., Belcher, J.W., Sittler, E.C. and Lepping, R.P. (1987) The Uranian bow shock: Voyager 2 inbound observations of a high Mach number shock. *J. Geophys. Res.*, **92**, 8603–12.

Bame, S.J., Martin, R.H., Comas D.J. *et al.* (1989) Three-dimensional plasma measurements from three-axis stabilized spacecraft, in *Solar System Plasma Physics* (eds J.H. Waite, J.L. Burch and R.L. Moore). American Geophysical Union Publications.

Belcher, J.W. (1987) The Jupiter–Io connection: an Alfven engine in space, *Science*, **238**, 170.

Biermann, L. (1957) Solar corpuscular radiation and the interplanetary gas. *Observatory*, **77**, 109.

Cheng, A.F., Johnson, R.E. (1989) Effects of magnetosphere interactions on origin and evolution of atmospheres, in *Origin and Evolution of Planetary and Satellite Atmospheres* (eds S.K. Atreya, J.B. Pollack and M.S. Matthews) Tucson: University of Arizona Press.

Chapman, S. and Ferraro, V.C.A. 1931. A new theory of magnetic storms. *Terr. Magn. Atmos. Elect.*, **36**, 77–97.

Connerney, J.E.P. (1987) The magnetospheres of Jupiter, Saturn, and Uranus. *J. Geophys. Res.*, **25**, 615–38.

Connerney, J.E.P., Acuña, M.H. and Ness, N.F. (1991) The magnetic field of Neptune. *J. Geophys. Res.*, **96**, 19023.

Dessler, A.J. (ed.) (1983) *Physics of the Jovian Magnetosphere*. Cambridge University Press.

Gold, T. (1959) *Symposium on the Exploration of Space* **64**, pp. 1665–1674

Hill, T.W. and Dessler, A.J. (1991) Plasma motions in planetary magnetospheres. *Science*, **252**, 410–415.

Luhmann, J.G. (1986) The solar wind interaction with Venus. *Space Sci. Rev.*, **44**, 241.

Luhmann, J.G., Russell, C.T., Brace L.H. and Vaisberg, O.L. (1992) The intrinsic magnetic field and solar wind interaction of Mars, in *Mars* (eds H.H. Kieffer, B.M. Jakosky, C.W. Snyder and M.S. Matthews). Tucson: University of Arizona Press.

Marsch, E. and Schwenn, R. (eds) (1992) *Solar Wind Seven*. Pergamon Press.

Mendillo, M., Baumgardner, J., Flynn, B. and Hughes, J. (1990) The extended sodium nebula of Jupiter. *Nature*, **348**, 312–4

Moses, S.L., Coroniti, F.V., Kennel, C.F. *et al.* (1990). Comparison of plasma wave measurements in the bow shocks at Earth, Jupiter, Saturn, Uranus, and Neptune. *Geophys. Res. Lett.*, **17**, 1653–6.

Neubauer, F.M., Luttgen, A. and Ness, N.F. (1991). On the lack of a magnetic signature of Triton's magnetospheric interaction on the Voyager 2 flyby trajectory. *J. Geophys. Res.*, **96**, 19171–5.

Pizzo, V.J., Holzer, T. and Sime, D.G. (eds) (1988) *Proc. Sixth Int. Solar Wind Conf.*, High Altitude Observatory, NCAR, Boulder.

Parks, G.K. (1991) *Physics of Space Plasmas*. Addison-Wesley.

Priest, E. (ed)(1985) *Solar System Magnetic Fields*, Dordrecht: D. Reidel, pp. 224–56.

Russell, C.T. Hoppe, M.M. and Livesey, W.A. (1982). Overshoots in planetary bow shocks. *Nature*, **296**, 45–8.

Russell, C.T., Baker D.N. and Slavin, J.A. (1988). The magnetosphere of Mercury, in *Mercury* (eds F. Vilas, C.R. Chapman and M.S. Mathews) Tucson: University of Arizona Press.

Siscoe, G.L. (1979). Towards a comparative view of planetary magnetospheres, in *Solar System Plasma Physics*, Vol II (eds C.F. Kennel, L.J. Lanzerotti and E.N. Parker). Amsterdam: North Holland, 402 pp.

Slavin, J.A., Smith, E.J., Spreiter, J.R. and Starhara, S.S. (1985) solar wind flow about the outer planets: gas dynamic modeling of the Jupiter and Saturn bow shocks. *J. Geophys. Res.*, **90**, 6275–86.

Van Allen, J.A. (1990) Magnetospheres, cosmic rays and the interplanetary medium, in *The New Solar System* (eds J.K. Beatty and A. Chaikin). Sky Publishing.

Young, D.T. (1989). Space plasma mass spectroscopy below 60 KeV, in *Solar System Plasma Physics* (eds J.H. Waite, J.L. Burch and R.L. Moore). American Geophysical Union Publications, pp. 143–58.

Young, D.T., Marshall J.A., Burch, J.L. *et al.* (1989). A 360° field of view toroidal ion composition analyzer using time of flight, in *Solar System Plasma Physics* (eds J.H. Waite, J.L. Burch and R.L. Moore). American Geophysical Union Publications, pp. 171–80.

Cross references

Alfvén wave
Aurora
Comet: dynamics
Ion and neutral mass spectrometry
Ionosphere
Magnetospheres of the outer planets
Planetary torus
Plasma wave
Radiation belt
Solar corona
Solar wind
Thermal plasma instrumentation

PLASMA WAVE

A plasma is an electrically neutral mixture of electrons and ions in which the kinetic energy greatly exceeds the interaction energy between the particles. Plasmas are produced (1) by collisions whenever a gas is heated to over a few thousand degrees, and (2) by photoionization, for example by ultraviolet radiation from the Sun. Plasmas are destroyed by recombination. Because of the very low densities that exist in interplanetary space and the correspondingly low recombination rates, almost all of the material that exists between the Sun and the planets is a plasma. This includes the solar corona, which is the hot ionized outer atmosphere of the Sun; the solar wind, which is an ionized gas streaming outward from the Sun at supersonic speeds; planetary magnetospheres, which are hot energetic plasmas surrounding planets with strong magnetic fields; and planetary ionospheres, which are layers of ionized gas in the upper regions of planetary atmospheres.

As in any fluid, waves can propagate through a plasma. Because of the electrical character of the plasma medium, plasma waves are very complex. Some of these waves have electric and magnetic fields, and are similar to the electromagnetic waves in free space. These are called electromagnetic waves. Others are more like sound waves and have no magnetic field. These are called electrostatic waves, since the electric field can be derived from the gradient of an electrostatic potential ($\vec{E} = -\vec{\nabla} \phi$). Usually, electromagnetic waves have propagation speeds near the speed of light, whereas electrostatic waves have propagation speeds near the speed of sound.

In most space plasmas the collision frequencies are very low. This type of plasma, with essentially zero collision frequency and infinite

Table P10 The most common plasma wave modes

Plasma wave mode	Frequency range	Electromagnetic/ electrostatic	Polarization	Free energy source
Free-spaced (L,O) mode	$\omega > \omega_{pe}$	Electromagnetic	L	Beam, loss cone
Free-space (R,X) mode	$\omega > \omega_{R=0}$	Electromagnetic	R	Beam, loss cone
Electron plasma oscillations (Langmuir waves)	$\omega \simeq \omega_{pe}$	Electrostatic	–	Beam
Z mode	$\omega_{UHR} > \omega > \omega_{L=0}$	Electromagnetic, electrostatic near ω_{UHR}	R for $\omega > \omega_{pe}$ L for $\omega > \omega_{pe}$	Beam
Electron cyclotron waves	Bands near $\omega \simeq (n + \tfrac{1}{2})\omega_{ce}$	Electrostatic	–	Ring distribution (electrons)
Whistler mode	$\omega < \mathrm{Min}\{\omega_{ce}, \omega_{pe}\}$	Electromagnetic, electrostatic near ω_{LHR}	R	Loss cone, beam above ω_{LHR}
Ion-acoustic mode	$\omega \lesssim \omega_{pi}$	Electrostatic	–	Drift between electrons and ions
Electrostatic ion cyclotron waves	Bands near $\omega \simeq (n + \tfrac{1}{2})\omega_{ci}$	Electrostatic	–	Ring distribution (ions) field-aligned currents
Electromagnetic ion cyclotron waves	$\omega \ll \omega_{ci}$	Electromagnetic	L	Pressure anisotropies

$\omega_{R=0} = \omega_{ce}/2 + ((\omega_{ce}/2)^2 + \omega_{pe}^2)^{1/2}$
$\omega_{UHR} = (\omega_{ce}^2 + \omega_{pe}^2)^{1/2}$
$\omega_{L=0} = -\omega_{ce}/2 + ((\omega_{ce}/2)^2 + \omega_{pe}^2)^{1/2}$
$\omega_{LHR} \simeq (\omega_{ce}\,\omega_{ci})^{1/2}$ if $\omega_{pe} \gg \omega_{ce}$

mean free path, is called a collisionless plasma. The absence of collisions effectively eliminates the basic mechanism of energy and momentum exchange that normally exists between particles in a fluid. Under this circumstance, waves provide the primary mechanism for energy and momentum exchange. Waves then play a role somewhat similar to collisions in an ordinary gas. Whenever a sufficiently large deviation from thermal equilibrium occurs, waves grow spontaneously in the plasma. The non-equilibrium feature that gives rise to the wave growth is called a free energy source. Examples of free energy sources are beams and anisotropies in the velocity distribution of the particles. Once generated, the waves are eventually reabsorbed via a process known as collisionless damping. The wave growth and damping lead to an energy and momentum exchange. From very general principles it can be shown that the energy and momentum exchange acts to drive the plasma toward thermal equilibrium, very similar to collisions in an ordinary fluid. Waves therefore play a crucial role in maintaining the equilibrium state of the plasma.

Many different plasma wave modes exist in a plasma, particularly if the plasma has a magnetic field. These wave modes are usually associated with certain characteristic frequencies. The two primary characteristic frequencies of a plasma are the plasma frequency ω_p and the cyclotron frequency ω_c. A plasma frequency and a cyclotron frequency can be defined for each species present in the plasma. The electron plasma frequency is given by

$$\omega_{pe} = \left(\frac{e^2 n_e}{\epsilon_0 m_e}\right)^{1/2} \tag{P9}$$

where e is the electronic charge, n_e is the electron number density, ϵ_0 is the permittivity of free space and m_e is the electron mass. The electron plasma frequency is the characteristic oscillation frequency that occurs whenever electrons are perturbed from their equilibrium position in the plasma. The electron cyclotron frequency is given by

$$\omega_{ce} = \frac{eB}{m_e} \tag{P10}$$

where B is the magnetic field. The electron cyclotron frequency is the characteristic rotation frequency that occurs whenever an electron has a component of velocity perpendicular to the magnetic field. Comparable equations for the ion plasma frequency and ion cyclotron frequency are obtained by changing (e) to (i) in equations (P9) and (P10). In addition to the electron and ion plasma frequencies and cyclotron frequencies, it is convenient to define four additional characteristic frequencies. These are: the upper hybrid resonance frequency,

$$\omega_{UHR} = (\omega_{pe}^2 + \omega_{ce}^2)^{1/2} \tag{P11}$$

the lower hybrid resonance frequency,

$$\omega_{LHR} = (\omega_{ce}\,\omega_{ci})^{1/2} \tag{P12}$$

the right-hand cut-off,

$$\omega_{R=0} = \omega_{ce}/2 + ((\omega_{ce}/2)^2 + \omega_{pe}^2)^{1/2} \tag{P13}$$

and the left-hand cut-off.

$$\omega_{L=0} = -\omega_{ce}/2 + ((\omega_{ce}/2)^2 + \omega_{pe}^2)^{1/2} \tag{P14}$$

The relationships that these characteristic frequencies have to the various wave modes that exist in a plasma are summarized in Table P10. This table lists the commonly accepted name of the mode, the frequency range over which the mode can propagate, the electromagnetic/electrostatic character of the mode, the polarization (R,L) with respect to the magnetic field (when applicable) and the free energy source that can cause wave growth. It should be noted that Table P10 only applies to a plasma consisting of electrons and one positive ion species. If more than one positive ion species is present, then additional modes appear between adjacent pairs of ion cyclotron frequencies. For a further detailed discussion of the wave modes that can exist in a plasma, see Stix (1962) or Krall and Trivelpiece (1973).

Instrumentation

Space plasma wave measurements have been carried out by spacecraft-borne instrumentation for over 30 years. The first instruments specifically designed to study naturally occurring space plasma waves were launched on the Earth-orbiting Alouette 1 and Injun III satellites in 1962 (Barrington and Belrose, 1963; Gurnett and O'Brien, 1964). Since then many different types of plasma wave instruments have flown on Earth-orbiting and interplanetary spacecraft. These instruments usually have several characteristics in common. In order to distinguish between electromagnetic waves and electrostatic waves, both electric and magnetic fields must be measured. (The absence of a wave magnetic field indicates the wave is electrostatic.) Electric fields are usually detected by an electric dipole antenna that extends in opposite directions from the center of the spacecraft, as illustrated in Figure P28. The quantity measured is the voltage difference, $\Delta V = V_2 - V_1$, between the two antenna elements. The electric field component along the axis of the antenna is then given by $E = \Delta V/l_{eff}$, where l_{eff} is a quantity called the effective length. For wavelengths λ longer than the tip-to-tip length L of the antenna, the effective length is given by $l_{eff} = L/2$. A wide range of electric antenna lengths can be used, ranging from a fraction of a meter to over 200 m. Because the measured voltage ΔV increases with the antenna length, longer antennas are generally preferred since they give better sensitivity. A variety of mechanisms are used to extend the antenna. One technique uses centrifugal force to pull a fine wire

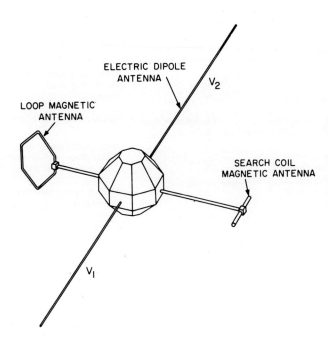

Figure P28 A typical antenna geometry for detecting space plasma waves. Electric fields are usually detected by an electric dipole antenna, and magnetic fields are detected by either a loop antenna or a search coil magnetometer.

radially outward from a fishing-reel type of dispenser in the spacecraft. This technique only works on spinning spacecraft. Another technique uses a motor-driven device to extrude a thin metal tape through a guide to form a rigid metal tube. This type of antenna works equally well on both spinning and non-spinning spacecraft. Sometimes small metal spheres with internal high impedance amplifiers are placed on the ends of the antenna to sense the potential in the plasma (Fahleson, 1967). In this case, the effective length is the center-to-center distance between the spheres.

Wave magnetic fields are usually detected using the magnetic induction principle, wherein a voltage is induced in a coil of wire by a time-varying magnetic field. The voltage induced is given by $V = Nd\Phi/dt$, where $\Phi = AB$ is the magnetic flux through the coil, N is the number of turns, A is the cross-sectional area and B is the magnetic field strength. Two types of magnetic sensors are used. The first type is a loop antenna. Usually a loop antenna consists of a single turn, which minimizes the inductance and gives the maximum bandwidth. A transformer is usually used to couple the antenna to the electronics. The second type is a search coil magnetometer, which consists of a high-permeability rod surrounded by a sensing coil. The high-permeability rod acts to concentrate the magnetic flux through the coil, thereby increasing the sensitivity. Generally, loop antennas provide better sensitivities at higher frequencies, particularly above a few tens of kilohertz, whereas search coils provide better sensitivities at lower frequencies, below a few hundred hertz. To reduce interference from electrical systems on the spacecraft, magnetic field antennas must be mounted on booms away from the spacecraft body, as illustrated in Figure P28. In some cases multiple axis antennas are also used. Full three-axis measurements give information on the direction of propagation of a wave.

The signals from the electric and magnetic antennas can be processed in a variety of ways. A typical block diagram of a plasma wave instrument is shown in Figure P29. Usually the antennas are connected to preamplifiers located close to the antennas. The preamplifiers are designed to provide low noise levels and to optimize the transmission of signals from the antennas to the main electronics package. The frequency range over which the antenna/preamplifier system must operate extends from the lowest characteristic frequencies of interest (usually f_{ci}) to the highest frequencies of interest (f_{pe} or f_{ce}). For planetary plasma wave investigations, this frequency range typically extends from a few hertz to a few megahertz.

Two different techniques are employed to process signals from electric and magnetic field sensors. In the first technique an onboard spectrum analyzer is used to generate spectrum amplitudes at a series of frequencies, $f_1, f_2, \ldots, f_n$. A spectrum analyzer of this type is shown in the top portion of the block diagram in Figure P29. The purpose of the onboard spectrum analysis is to provide continuous low-resolution survey spectrums using relatively modest telemetry rates, typically a few hundred bits per second. In the second technique a wideband receiver is used to transmit electric or magnetic field waveforms directly to the ground. The onboard signal processing is minimal, and the spectrum processing (Fourier analysis) is performed by ground-based computers. A wideband receiver is shown at the bottom of the block diagram in Figure P29. The main purpose of the wideband receiver is to limit the bandwidth of the signal and to control the amplitude of the signal by means of an automatic gain control. The waveform transmission can be either analog or digital. The advantage of the waveform measurements is the very high resolution. Since the entire waveform is transmitted, the resolution in frequency and time is limited only by the uncertainty principle ($\Delta\omega\Delta t \approx 1$). The disadvantage is that the telemetry rates are extremely high, often several hundred kbits per second or more. For this reason, wideband waveform transmissions are often of limited duration (60 s or less), thereby restricting the waveform measurements to a few specific samples, rather than continuous coverage. In this respect the onboard spectrum analysis and the wideband technique are complementary. The spectrum analyzer provides continuous low-resolution survey measurements, and the wideband receiver provides high-resolution spectrums for selected time intervals.

Observations

Spacecraft plasma wave observations have now been obtained at seven planets (Venus, Earth, Mars, Jupiter, Saturn, Uranus and Neptune). The most extensive measurements have been performed in the vicinity of Earth. Since the first such measurements in 1962 many spacecraft have provided plasma wave measurements in Earth orbit. These spacecraft have explored most of the near-Earth environment, with trajectories ranging from low-altitude orbits near the Earth, to highly eccentric orbits extending well beyond the orbit of the Moon. Plasma wave observations at the other planets are much more limited, and it is these measurements that will be emphasized here, since they are at the frontier of present day research. Of the various spacecraft that have flown to the other planets, the Voyager 1 and 2 mission to the giant planets, Jupiter, Saturn, Uranus and Neptune, has probably contributed the most to our expanding knowledge of space plasma waves. The giant planets, like the Earth, have strong magnetic fields and intense radiation belts, which make them a rich source of plasma waves. For the initial Voyager reports of plasma wave observations at the giant planets, see Scarf, Gurnett and Kurth (1979), Scarf et al. (1982), Gurnett, Kurth and Scarf (1979a, 1981) and Gurnett et al. (1986, 1989). The only other spacecraft that has provided plasma wave measurements at the giant planets is Ulysses, which flew by Jupiter in 1992. For the initial report of the Ulysses plasma wave observations, see Stone et al. (1992). The remaining two planets, Venus and Mars, have negligible internal magnetic fields and therefore fewer types of plasma wave phenomena. The first measurements of plasma waves in the vicinity of Venus were provided by the Pioneer Venus spacecraft, which was placed in orbit around Venus on 4 December 1979. The first report on the Pioneer Venus plasma wave observations was given by Scarf, Taylor and Green (1979). The only other spacecraft that has provided plasma wave observations in the vicinity of Venus is Galileo, which flew by Venus on 10 February 1990. For a report on the Galileo Venus plasma wave observations, see Gurnett et al. (1991). At Mars the first, and only, plasma wave measurements were obtained by the Phobos 2 spacecraft, which was placed in orbit around Mars on 29 January 1989. An initial report on the Phobos 2 plasma wave observations is given by Grard et al. (1991).

Since there are so many planets to review, no attempt will be made to describe the observations in detail at each planet. Instead, the observations will be organized according to the various types of plasma waves observed, ordered according to decreasing distance from the planet, starting from the Sunward side of the planet, and ending in the region near the closest approach. No discussion is given of electromagnetic radiation that can escape to great distances from the planet, since these waves are usually regarded as radio astronomi-

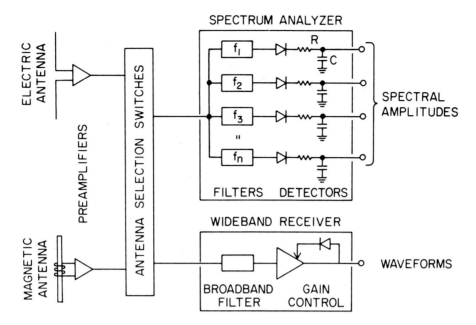

Figure P29 A block diagram of a typical plasma wave instrument. These instruments often consist of an onboard spectrum analyzer which gives low-resolution continuous spectrums, and a wideband waveform receiver which gives very high-resolution spectrums for selected intervals.

cal emissions. For a review of planetary radio emissions see Gurnett (1992).

Electron plasma oscillations and ion acoustic waves

The solar wind flows outward from the Sun at a nearly constant speed of about 400 km s^{-1}. At this speed the solar wind flow is supersonic. When the solar wind encounters a large object such as a planet, a shock wave is formed, very similar to the shock wave that forms upstream of an airplane in supersonic flight. This shock is called the bow shock. The approximate shape of the shock is shown in Figure P30. If the planet has no internal magnetic field, as in the case of Venus and Mars, the planet and its surrounding atmosphere and ionosphere act as the obstacle. The radial distance to the nose of the shock is then only slightly larger than the radius of the planet. If the planet has a strong internal magnetic field, as in the case of the Earth and the giant planets, the magnetic field acts as the obstacle. The position of the shock is then controlled by the strength of the planetary magnetic field. The interface between the solar wind and the planetary magnetic field is called the magnetopause (Figure P30). At Jupiter, for example, the nose of the shock is typically at 80 to 120 R_J (where R_J is the radius of the planet), and the magnetopause is at 50 to 70 R_J.

At the shock the plasma is strongly heated and some of the electrons and ions escape upstream into the solar wind. Because the escaping particles are guided along the magnetic field lines by magnetic forces, these particles are confined to a region upstream of the shock called the foreshock. Usually the escaping electrons have very high speeds, typically 10^4 to 10^5 km s^{-1}, which is much greater than the solar wind speed. At these very high velocities, the region accessible to the backstreaming electrons is essentially delineated by the magnetic field lines tangent to the shock (Figure P30). This region is called the electron foreshock. The escaping ions, because of their higher mass, have much lower velocities, more nearly comparable to the solar wind speed. The region accessible to the backstreaming ions is therefore angled backward substantially from the tangent field line (Figure P30). This region is called the ion foreshock.

Because the backstreaming electrons constitute a beam, these particles can excite electron plasma oscillations, also sometimes called Langmuir waves (Table P9). Electron plasma oscillations excited by electrons streaming into the solar wind were first discovered by Scarf et al. (1971) upstream of the Earth's bow shock. Since then similar electron plasma oscillations have been discovered at Venus and Mars and at all four of the giant planets. A multichannel plot illustrating the occurrence of electron plasma oscillations upstream of Jupiter's bow shock is shown in Figure P31. These data are from the low-rate spectrum analyzer onboard Voyager 1. The enhanced emissions in the 5.62-kHz channel from about 12 : 18 to 12 : 27 UT are electron plasma oscillations. The electron plasma frequency f_{pe} during this interval was about 5.5 to 6.0 kHz. The onset of the plasma oscillations at 12 : 18 UT corresponds to the crossing of the tangent field line, and the termination at 12 : 27 UT corresponds to the crossing of the bow shock.

The frequency of upstream electron plasma oscillations generally decreases with increasing distance from the Sun. As can be seen from equation (P9), the electron plasma frequency is proportional to the square root of the electron density. Since the solar wind density varies roughly as $1/R^2$, where R is the distance from the Sun, the

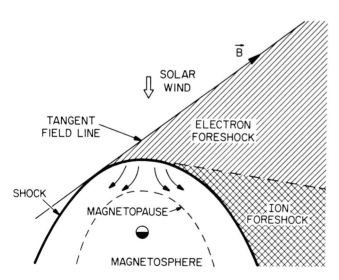

Figure P30 A sketch of the various boundaries and regions that occur in the solar wind upstream of a planet. Since the solar wind is supersonic, a shock wave forms upstream of the planet. Electrons and ions energized at the shock escape upstream into regions known as the electron foreshock and the ion foreshock.

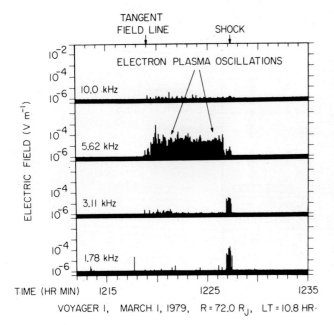

Figure P31 An example of electron plasma oscillations in the solar wind upstream of Jupiter. These waves occur in the electron foreshock and are produced by energetic (~ 1 to 10 keV) electron beams escaping into the solar wind from the bow shock.

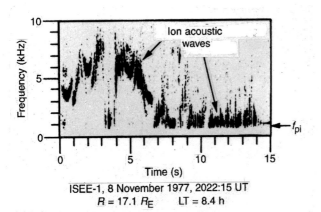

Figure P32 A high-resolution frequency–time spectrogram of ion acoustic waves observed in the solar wind upstream of the Earth's magnetosphere. These waves are produced by energetic (~ 10 keV) ions escaping into the solar wind from the bow shock.

terminations indicate that the mode is very close to marginal instability. The peak frequencies of the ion acoustic waves (~ 2 kHz) are well below the electron plasma frequency (f_{pe} ~ 5 kHz) but still above the ion plasma frequency (f_{pi} ~ 120 Hz). As can be seen from Table P9, the ion acoustic mode can only propagate at frequencies below the ion plasma frequency. This discrepancy is believed to occur because the waves have very short wavelengths, thereby introducing Doppler shifts due to the motion of the solar wind. For a wave of wavelength λ and frequency f in the plasma rest frame, the frequency f' detected in the spacecraft rest frame is given by

$$f' = f + \frac{V_{sw}}{\lambda} \cos \theta_{kv} \qquad (P15)$$

where V_{sw} is the solar wind speed and θ_{kv} is the angle between the propagation vector $\vec{k}$ and the solar wind velocity. The shortest wavelength that can exist in a plasma is $\lambda_{min} = 2\pi\lambda_D$, where λ_D is a characteristic length called the Debye length. For the plasma parameters that exist in the solar wind upstream of Jupiter, the shortest wavelength is about $\lambda_{min} \simeq 240$ m. The maximum Doppler shift, which is given by the second term on the right-hand side of equation (P11), is then about 1.7 kHz, which is comparable to the highest frequencies observed. Ion acoustic waves have only been reported upstream of the bow shocks at Earth, Mars and Jupiter. For unknown reasons, possibly due to instrumental limitation, ion acoustic waves have not been observed upstream of the bow shocks at Venus, Saturn, Uranus or Neptune.

The bow shock crossings at Venus, Earth, Mars, and all four of the giant planets can be easily identified in the plasma wave data by an intense broadband burst of electric field noise at the shock. This noise was first discovered in the Earth's bow shock by Fredricks *et al.* (1968). A wideband frequency–time spectrogram showing the shock-related electric field noise observed during the Voyager 1 crossing of Jupiter's bow shock is given in Figure P33. This is the same shock crossing shown in Figure P31. Note the electron plasma oscillations at ~ 6 kHz, increasing slowly in frequency as the spacecraft approaches the shock. The electric field noise at the shock extends up to a frequency of about 3 kHz and has a peak broadband intensity of about 1 mV m^{-1}. This noise is believed to be caused by solar wind ions that are magnetically reflected by the shock, thereby forming a gyrating ion beam that excites electrostatic waves via a two-stream instability. Currents flowing along the shock surface may also in some cases contribute to the generation of electrostatic waves. Earlier it was thought that these electrostatic waves played the dominant role in heating the plasma at collisionless shocks (Fredricks *et al.*, 1968). However, more recent studies by Scudder *et al.* (1986) and others suggest that these waves probably act only to thermalize the particle distribution, and that other processes, such as acceleration by quasi-static electric fields and magnetic reflection, are primarily responsible for converting the directed solar wind flow into a heated distribution at the shock.

Trapped continuum radiation

After the shock, the next boundary of interest is the magnetopause. This boundary forms the effective obstacle for the solar wind flow around the planet and is shown by a dashed line in Figure P30. Because the planetary magnetic field provides most of the pressure inside of the magnetosphere, an abrupt drop in the plasma density occurs at the magnetopause, thereby forming a low-density magnetospheric cavity. Since the electron plasma frequency is lower in the magnetosphere than in the solar wind, electromagnetic radiation can be trapped in the magnetospheric cavity. This trapped radiation was first discovered in the Earth's magnetosphere by Gurnett and Shaw (1973) and is called continuum radiation. Since then, trapped continuum radiation has been observed at three of the giant planets, Jupiter, Saturn and Uranus. No trapped continuum radiation was observed at Neptune, probably because the Voyager plasma wave instrument did not have sufficient sensitivity to detect this radiation at Neptune. The trapped continuum radiation at Jupiter is one of the most intense emissions observed at any of the planets. Since no magnetospheric cavity exists at Venus and Mars, trapped continuum radiation cannot occur at either of these planets. For a review of continuum radiation in planetary magnetospheres see Kurth (1991).

An example of trapped continuum radiation is shown in Figure P34. This spectrogram shows the Voyager 1 crossing of the magneto-

electron plasma frequency varies roughly as $1/R$. At Venus the electron plasma frequency is typically about 30 kHz, whereas at Neptune the electron plasma frequency is about 700 Hz. The electric field strength of the plasma oscillations also decreases with increasing distance from the Sun. At Venus and Earth the peak field strengths are about 1 mV m^{-1}, whereas at Neptune the peak field strengths are about 30 to 100 μV m^{-1}.

In addition to electron plasma oscillations, another type of wave also occurs upstream of planetary bow shocks. These waves were first detected upstream of the Earth's bow shock by Scarf *et al.* (1970) and are called ion acoustic waves (Gurnett and Frank, 1978). Ion acoustic waves are very similar to sound waves in an ordinary gas and are driven by ions escaping from the shock. Since these waves are driven by ions, they are confined to the ion foreshock. A wideband spectrogram of ion acoustic waves detected by the Voyager 1 spacecraft upstream of Jupiter's bow shock is shown in Figure P32. As can be seen, the ion acoustic waves have relatively narrow bandwidths and switch on and off abruptly. The abrupt onsets and

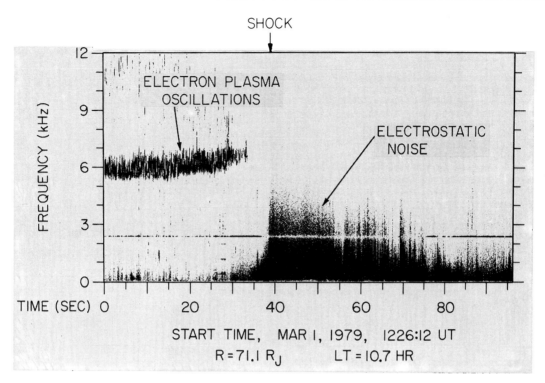

Figure P33 The Voyager 1 inbound crossing of Jupiter's bow shock. An abrupt burst of broadband electric field noise can be seen at the shock. This noise is believed to be caused by ion beams gyrating back into the solar wind from the shock. Electron plasma oscillations can also be seen upstream of the shock.

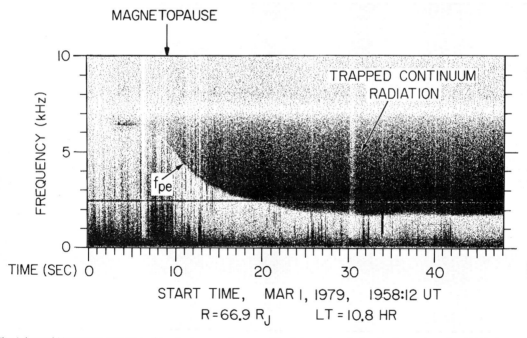

Figure P34 The inbound Voyager 1 crossing of Jupiter's magnetopause. The intense band of noise from about 2 to 7 kHz is continuum radiation trapped in the low-density magnetospheric cavity. The low-frequency cut-off of the continuum radiation is at the electron plasma frequency, f_{pe}.

pause at Jupiter. The continuum radiation consists of the dark band of noise extending upward from about 1 kHz, gradually fading into the receiver background noise above about 7 kHz. The sharp, low-frequency cut-off of the radiation at f_{pe} is believed to be caused by the reflection of free space (L,O) mode electromagnetic waves at the local electron plasma frequency. As can be seen from Table P10, the free space L,O mode can only propagate at frequencies $f > f_{pe}$. Free space (R,X) mode radiation is also most likely present. However, the low-frequency cut-off of the R–X mode is always above f_{pe}, so the L–O mode determines the low-frequency cut-off. The monotonic decrease in the low-frequency cut-off, from about 6.2 to 1.8 kHz over a period of about 20 s, is caused by the rapidly declining plasma

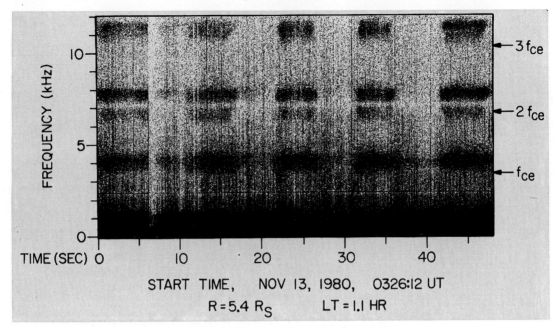

Figure P35 An example of electrostatic electron cyclotron harmonic (ECH) waves in Saturn's magnetosphere. These emissions occur in narrow bands slightly above harmonics of the electron cyclotron frequency, f_{ce}.

density as the spacecraft passes through the magnetopause. Note from equation (P9) that the electron plasma frequency is proportional to the square root of the electron density. The thickness of the magnetopause is controlled mainly by the cyclotron radius of magnetosheath ions as they gyrate into the region of strong field inside the magnetosphere. Continuum radiation comparable to Figure P33 is observed throughout the magnetospheric cavity of Jupiter. Once generated, the radiation is believed to undergo repeated reflections from the walls of the cavity, eventually building up to an equilibrium level throughout the cavity. Small random Doppler shifts caused by repeated reflections from the walls of the magnetospheric cavity, which are continuously in motion, are believed to spread the radiation into a nearly continuous spectrum, hence the term continuum.

Electron cyclotron and upper hybrid waves

For the magnetized planets, the magnetic field within the magnetosphere is generally much stronger than in the solar wind. The electron cyclotron frequency then plays an important role in controlling the types of waves that are generated. In the Earth's magnetosphere it has been known for many years that strong electrostatic emissions are generated near harmonics of the electron cyclotron frequency (Kennel et al., 1970; Shaw and Gurnett, 1975). These emissions are part of a band structure that is often referred to as electron cyclotron waves (Table P10). The free energy source of these waves consists of electrons with a loss cone or ring type of distribution function. Loss-cone velocity distributions are a characteristic feature of planetary radiation belts. Charged particles moving within a well-defined cone of angles around the magnetic field (the loss cone) strike the atmosphere and are lost from the system, thereby producing a hole in the particle velocity distribution.

Electron cyclotron waves are found in the magnetosphere of the Earth and all the giant planets. Typically these waves are most intense near half-integral harmonics $(n + \frac{1}{2})f_{ce}$ of the electron cyclotron frequency. Usually the $(n + \frac{1}{2})f_{ce}$ waves occur in two distinct frequency ranges, the first near low-order half-integral harmonics of the electron cyclotron frequency (i.e. $\frac{3}{2}f_{ce}$, $\frac{5}{2}f_{ce}$, etc.), and the second near the upper hybrid resonance frequency, when $(n + \frac{1}{2})f_e \simeq f_{UHR}$. The low-order harmonics are often called electron cyclotron harmonic (ECH) waves, and the emissions near the upper hybrid frequency are called upper hybrid resonance (UHR) waves. The emission frequencies depend in a complicated way on the densities and temperatures of the cold and hot components of the plasma, and are almost never exactly at $(n + \frac{1}{2})f_{ce}$. The half-integral notation, $\frac{3}{2}$, $\frac{5}{2}$, etc., is mainly just a convenient label to identify the emission band.

A spectrogram illustrating examples of low-order ($\frac{3}{2}$, $\frac{5}{2}$ and $\frac{7}{2}$) ECH emissions in the magnetosphere of Saturn is shown in Figure P35. The emission frequencies in this case are slightly above the electron cyclotron harmonics. Considerable fine structure can be seen within the emission bands. Electron cyclotron harmonic emissions of this type are typical of all the ECH observations at the giant planets. Usually, the emissions are strongest in a narrow band slightly above the electron cyclotron harmonics. A spectrogram illustrating an example of UHR emissions in the outer region of Jupiter's magnetosphere is shown in Figure P36. The UHR emissions in this case consist of very sharply defined bands near the lower edge of the trapped continuum radiation. The bands switch on and off as plasma density variations cause the upper hybrid resonance frequency to sweep past half-integral harmonics of the electron cyclotron frequency. Strong emissions occur whenever the condition $(n + \frac{1}{2})f_{ce} \simeq f_{UHR}$ is satisfied. The frequency spacing between the bands is roughly the electron cyclotron frequency.

A striking characteristic of both the ECH and UHR waves is their close confinement to the magnetic equator. Figure P37a shows a multichannel plot of electric field intensities from the Voyager 1 pass through the inner region of the Jovian magnetosphere. The ECH and UHR emissions are identified by circles. Figure P37b shows the magnetic latitude λ_m. The magnetic latitude oscillates up and down due to the rotation of Jupiter's magnetic dipole field, which is tilted at an angle of about 10° with respect to the rotational axis. As can be seen, the ECH and UHR waves occur in sharply localized regions centered almost exactly on the magnetic equator crossings. This narrow confinement to a region only 1 or 2 degrees from the magnetic equator is a characteristic feature of all the ECH and UHR observations at the giant planets. A similar effect also occurs in the Earth's magnetosphere, although not as dramatically as at the giant planets.

The reason that the ECH and UHR waves are confined to a narrow region near the magnetic equator is still a subject of investigation. Based on terrestrial studies it is believed that two factors are responsible. First, it is known that the electrons responsible for generating the waves have pitch angles near 90°. Due to the laws governing the motion of trapped radiation belt particles (conservation of the first and second adiabatic invariants), particles with pitch angles near 90° are closely confined to the vicinity of the magnetic equatorial plane. Since this type of highly anisotropic velocity distribution (with pitch angles near 90°) is required to generate the ECH and UHR waves, large wave growth can only occur near the magnetic equator. Second, ray tracing studies show

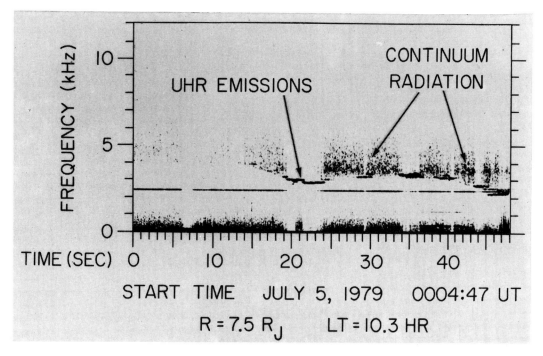

Figure P36 An example of upper hybrid resonance (UHR) emissions in Jupiter's magnetosphere. These emissions occur in narrow bands near the upper hybrid resonance frequency, f_{UHR}.

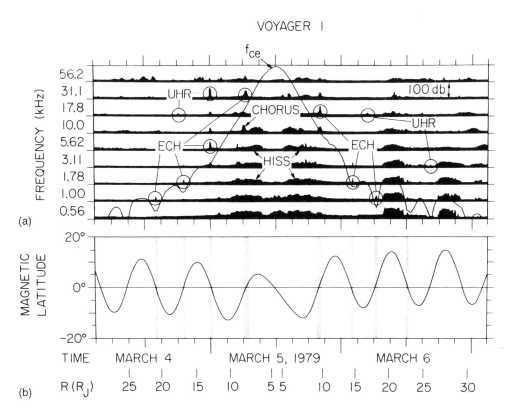

Figure P37 (a) Electric field intensities observed during the Voyager 1 pass through the inner region of the Jovian magnetosphere and (b) the magnetic latitude. Both ECH and UHR waves always occur very close to the magnetic equator.

that the electron cyclotron waves tend to be guided along the magnetic equator. This guiding effect is believed to confine the wave growth to a narrow region along the magnetic equatorial plane.

One may ask what role these electrostatic waves play in the magnetospheres of the giant planets. In the terrestrial magnetosphere, for many years it has been thought that electron cyclotron waves control the loss of trapped radiation belt electrons by scattering particles into the loss cone (Kennel et al., 1970). It seems likely that similar processes are operative at the giant planets. Unfortunately, adequate measurements are not available from the Voyager plasma data in the proper electron energy range (a few hundred eV to several tens of keV) to evaluate this loss mechanism. In the Earth's magnetosphere, UHR emissions are also believed to be a source of free-space electromagnetic radiation. The generation mechanism is believed to involve a mode conversion process by which UHR waves are converted to escaping electromagnetic (L, O mode) radiation. The mode conversion process can be either linear (Jones, 1980) or nonlinear (Melrose, 1981). Trapped continuum radiation is thought to be produced by this mode conversion process (Kurth, 1991).

Whistler-mode emissions

Whistlers are one of the oldest and best-known terrestrial plasma wave phenomena. Whistlers were first observed by ground-based radio receivers (Barkhausen, 1919). The modern theory of whistlers was first proposed by Storey (1953). According to Storey's theory, low-frequency electromagnetic radiation from a lightning discharge is guided along the magnetic field lines through the magnetospheric plasma. Because of the peculiar nature of electromagnetic wave propagation at frequencies below the electron cyclotron frequency, the higher frequencies propagate faster than the lower frequencies. Thus, the broadband impulsive signal produced by a lightning flash is converted into a whistling tone, hence the term 'whistler'. The plasma wave mode involved in the propagation of whistlers is called the whistler mode. Whistler-mode waves are right-hand polarized and propagate at frequencies below either f_{ce} or f_{pe}, whichever is smaller (see Table P10). The whistler mode is highly anisotropic and has a number of unusual characteristics, one of which is that the index of refraction goes to infinity along a cone of directions called the resonance cone (Stix, 1962). This highly anisotropic characteristic accounts for the fact that the wave energy is guided approximately along the magnetic field (see Whistler).

In addition to lightning-generated whistlers, whistler-mode waves can also be spontaneously generated in a magnetized plasmas. These waves are called whistler-mode emissions. Whistler-mode emissions are a common feature of the terrestrial magnetosphere and occur in the magnetospheres of all the giant planets. These emissions are mainly generated in the inner regions of the magnetosphere where the loss cone in the trapped energetic electron distribution provides an effective free energy source. From very general principles (Brice, 1964), it can be shown that the growth of whistler-mode waves leads to a decrease in the pitch angle of resonant electrons, thereby driving the particles toward the loss cone. The growth of whistler-mode waves is widely believed to be the dominant mechanism responsible for the loss of energetic electrons from planetary radiation belts. In a classic paper Kennel and Petschek (1966) showed that the growth of whistler-mode waves puts an upper limit on the energetic electron intensities that can exist in planetary radiation belts.

A representative spectrum of whistler-mode emissions in the inner region of Jupiter's magnetosphere is shown in Figure P38a. This spectrum was obtained in the Io plasma torus, which is a dense torus of plasma produced by gases escaping from Jupiter's moon Io (see Planetary Torus). The plasma in the Io torus is extremely energetic and produces very intense whistler-mode emissions, among the most intense ever observed in a planetary magnetosphere. Two types of emissions are observed, called 'hiss' and 'chorus'. The hiss is an essentially structureless emission. When the hiss signals are played through an audio speaker, they make a steady hissing sound, hence the term 'hiss.' According to current ideas, whistler-mode hiss is believed to represent a fully developed turbulent spectrum in which the wave growth and loss has achieved a steady state equilibrium. In contrast to the whistler-mode hiss, chorus emissions are highly structured. A wideband frequency–time spectrogram of chorus is shown in Figure P39. The term 'chorus' is an old term (Allcock, 1957), and has its origins in the term 'dawn chorus' which refers to the sounds made by a roosting flock of birds at daybreak. The reasons for the complex spectral structure, usually consisting of many discrete

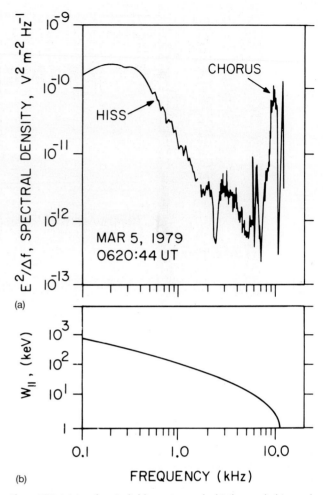

Figure P38 (a) An electric field spectrum of whistler-mode hiss and chorus emissions in Jupiter's Io torus. (b) The energy $W_\parallel$ of electrons that are in cyclotron resonance with these waves. The hiss tends to interact with very energetic electrons (~ 100 to 1000 keV), whereas the chorus interacts with much lower energies (~ 1 to 10 keV).

narrowband tones rising in frequency, is poorly understood. The current view is that the waves grow to such large amplitudes that local nonlinear processes play a dominant role in controlling the evolution of the wave. Computer simulations show that particles trapped in the wave field produce isolated wave packets, each of which evolves somewhat differently in time and space.

It is instructive to comment on the electron energies involved in the generation of hiss and chorus. Whistler-mode wave growth proceeds via a resonant process in which a constant force is experienced by a particle undergoing cyclotron motion along a magnetic field line, thereby leading to a deceleration (or acceleration) of the particle and a growth (or damping) of the wave. This process is called cyclotron resonance. The general condition for cyclotron resonance is

$$v_{\parallel \text{Res}} = \frac{\omega - n\,\omega_{ce}}{k_\parallel} \qquad (P16)$$

where $v_{\parallel \text{Res}}$ is the parallel resonance velocity (the symbol $\parallel$ refers to the component parallel to the magnetic field), ω is the wave frequency, $k_\parallel$ is the parallel component of the wave vector and n is an integer. For whistler-mode waves, the $n = 1$ term is usually most important. This resonance is called the first-order cyclotron resonance and occurs when both the wave and the particles (electrons) are rotating in the right-hand sense with respect to the magnetic field. From the propagation characteristics of the wave, $\omega(k)$, one can calculate the parallel energy $W_\parallel$ of the resonant electrons. The parallel resonance energy for whistler-mode emissions at Jupiter is

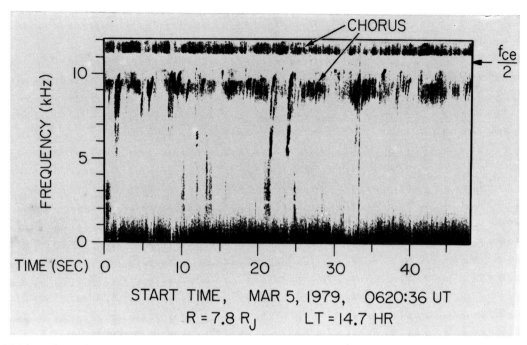

Figure P39 A high-resolution frequency–time spectrogram of chorus emissions. These emissions are highly structured and often consist of narrowband tones rising in frequency with increasing time. Chorus often has a sharp notch in the spectrum at one-half of the electron cyclotron frequency, $f_{ce}/2$.

shown in Figure P38b. As can be seen, the resonance energy decreases rapidly with increasing frequency. The energy of the electrons interacting with the hiss tends to be very high, 100 to 1000 keV, whereas the energy of the electrons interacting with the chorus tends to be much lower, 1–10 keV. This trend, for hiss to resonate with high energies and chorus to resonate with low energies, is typical of whistler-mode emissions at all of the giant planets.

In addition to Earth, three of the giant planets, Jupiter, Saturn and Uranus, have intense whistler-mode hiss and chorus emissions. These emissions occur in the inner regions of the magnetosphere where the trapped radiation belt electron intensities are the highest. The existence of whistler-mode emissions at Neptune is unclear. Some very weak emissions were observed in the spectrum analyzer data that are probably whistler-mode hiss. However, no chorus was observed in any of the wideband data. The absence of chorus at Neptune could be due to the low radiation belt intensities, which were the lowest of any of the giant planets. It can be shown that the growth rate of whistler-mode emissions increases in direct proportion to the intensity of the resonant electron. The extremely low whistler-mode emission intensities at Neptune could therefore be due to the low radiation belt electron intensities. On the other hand, the spacecraft did not pass through the equatorial region of the radiation belt where the highest wave amplitudes would be expected. Thus, it may be that strong whistler-mode emissions were present in the magnetosphere of Neptune, but the spacecraft did not pass through the proper region to observe these waves.

A third type of whistler-mode emission also occurs in planetary magnetospheres. This emission occurs in the auroral regions and is called auroral hiss. Auroral hiss is a nearly structureless emission and is believed to propagate at wave normal angles near the resonance cone. Near the resonance cone the whistler mode is very nearly electrostatic, with small magnetic fields, short wavelengths and low propagation speeds. These short-wavelength, quasi-electrostatic, whistler-mode waves are sometimes called lower hybrid waves, since they become completely electrostatic at the lower-hybrid resonance frequency, f_{LHR}. Because of the low propagation velocity, auroral hiss can be excited by beams, very similar to electron plasma oscillations. Auroral hiss has been extensively studied in the Earth's magnetosphere, where it has been shown that the emissions are produced by the same electron beams that produce the auroral light emission, hence the term 'auroral hiss.' Auroral hiss has also been observed at Jupiter by Voyager 1 (Gurnett et al., 1979) and by Ulysses (Stone et al., 1992). In both cases the identification was based on the similarity to terrestrial auroral hiss and not on a direct correlation with the aurora on Jupiter. No auroral hiss was observed at Saturn, Uranus or Neptune, most likely because the spacecraft did not pass through the proper region to observe such emissions.

Electrostatic ion cyclotron waves

Electrostatic ion cyclotron waves occur in discrete bands between harmonics of the ion cyclotron frequency (Table P10), very similar to electron cyclotron waves, which occur between harmonics of the electron cyclotron frequency. One of the unique features of the electrostatic ion cyclotron mode is that it is driven unstable by relatively weak field line currents. This feature led Kindel and Kennel (1971) to predict that electrostatic ion cyclotron waves would be produced by field-aligned currents over the Earth's auroral regions. The existence of such waves was subsequently confirmed by Kintner, Kelley and Mozer, (1978), using data from the polar orbiting S3–3 satellite. A representative spectrum of electrostatic ion cyclotron waves observed along the Earth's auroral field lines is show in Figure P40. Strong enhancements can be seen just above the lowest three harmonics of the proton (H^+) cyclotron frequency. Originally it was thought that these waves were driven by field-aligned currents. However, more recent studies suggest that these waves are produced by ion beams accelerated upward along the auroral field lines by the same quasi-static electric fields that produce the electron precipitation responsible for the aurora. Electrostatic ion cyclotron waves are also sometimes observed near the magnetic equatorial plane. These waves are believed to be driven by energetic ions trapped near the magnetic equator.

Observations of electrostatic ion cyclotron waves in other planetary magnetospheres are very limited. Since the Voyager spacecraft did not pass through the high-latitude auroral regions at the giant planets, with the possible exception of Neptune, no opportunity existed to search for electrostatic ion cyclotron waves driven by auroral processes. Barbosa and Kurth (1990) have interpreted a narrow band of low-frequency waves observed in the cold plasma torus at Jupiter as electrostatic ion cyclotron waves. They suggest that these waves are produced by a charge-exchange interaction between neutral gas emissions from volcanoes on Io and the rapidly rotating Io plasma torus, which is locked to the rotation of Jupiter. This charge-exchange process produces a ring-type ion distribution (sometimes called pick-up ions) and is expected to provide a very effective free energy source for generating electrostatic ion cyclotron waves. Barbosa and Kurth (1990) have also interpreted a band of low-

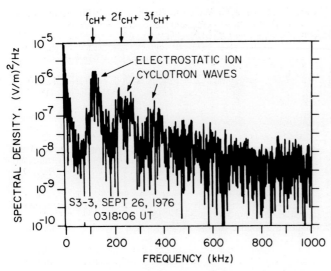

Figure P40 An electric field spectrum of electrostatic ion cyclotron waves observed in the Earth's magnetosphere by the S3–3 spacecraft. These waves occur between harmonics of the proton cyclotron frequency (f_{CH^+}) and are driven by currents flowing along the auroral field lines. (Kintner, Kelley and Mozer, 1978; copyright American Geophysical Union.)

frequency electric field noise in Neptune's magnetosphere as electrostatic ion cyclotron waves, also driven by the same charge-exchange process. Unfortunately, in neither case is it possible to confirm the electrostatic character of the waves, so the identification of the mode is not completely certain.

Electromagnetic ion cyclotron waves

Electromagnetic ion cyclotron waves are very similar to whistler-mode waves, except that they are left-hand polarized and propagate below the ion cyclotron frequency (Table P10). Since the ion cyclotron frequency is much lower than the electron cyclotron frequency [equation (P10)], ion cyclotron waves necessarily occur at extremely low frequencies, typically a few hundred hertz or less. Since the wave field of an electromagnetic ion cyclotron wave rotates in the same sense as positive ions (i.e. left-hand with respect to the magnetic field), these waves interact strongly with positively charged ions. Almost all of the ions observed in planetary magnetospheres are positively charged. The cyclotron resonance condition is identical to equation (P16), except that (e) is replaced by (i). Electromagnetic ion cyclotron waves are driven unstable by a loss cone in the energetic ion distribution. Since a loss cone is always present in a planetary radiation belt, the growth of these waves provides a mechanism for scattering energetic ions into the loss cone, thereby controlling the loss of radiation belt ions.

Despite the intense theoretical interest in the generation of electromagnetic ion cyclotron waves in planetary magnetospheres, relatively few observations are available. The first report of spontaneously generated electromagnetic ion cyclotron waves in the Earth's magnetosphere was by Taylor, Parady and Cahill (1975). These and other subsequent observations (Kintner, Kelley and Mozer, 1977; Roux et al., 1982) have shown that electromagnetic ion cyclotron waves are generated in the Earth's radiation belt during magnetic storms, when intense fluxes of energetic (10 to 100 keV) ions are injected deep into the inner regions of the magnetosphere. Electromagnetic ion cyclotron waves have also been observed at Jupiter by Thorne and Scarf (1984) using Voyager 1 measurements, and by Stone et al. (1992) using Ulysses measurements. In both cases intense waves were observed at frequencies below the proton cyclotron frequency. The Ulysses observations are particularly important because the magnetic field of the wave was measured, which confirms that the waves are electromagnetic and not electrostatic (Voyager had only an electric antenna). The ion precipitation produced by these waves is believed to be responsible for the extreme ultraviolet (EUV) aurora at Jupiter (Thorne and Moses, 1983). Using Voyager 2 Neptune data, Gurnett et al. (1989) reported observations of a strong band of electric field noise at Neptune at frequencies below the proton cyclotron frequency. This band of electric field noise was tentatively identified as electromagnetic ion cyclotron waves. However, since no wave magnetic field measurements were available, it was not possible to establish definitely the mode of propagation.

Conclusion

This review has described the primary types of plasma waves observed in the vicinity of the planets Venus, Mars, Earth, Jupiter, Saturn, Uranus and Neptune. These observations are summarized in Table P11. By necessity we have not attempted to describe the detailed nature of the observations at each planet. For a more detailed description, see the review by Kurth and Gurnett (1991). In making comparisons between these planets it must be recognized that the observations are in many cases incomplete, particularly at Uranus and Neptune, where the available data are limited to only one pass by the planet. At the giant planets almost no information is available at high magnetic latitudes, a region that we know from terrestrial observations has many complex aurora-related plasma wave emissions. No plasma wave observations have been obained at Mercury and Pluto. Thus there are very significant gaps in our knowledge. It is likely to be many years before these gaps are filled. The most promising missions for future plasma wave investigations are Galileo, which is to orbit Jupiter in late 1995, and Cassini, which is to orbit Saturn early in the 21st century; both spacecraft include plasma wave instruments.

Donald A. Gurnett

Bibliography

Allcock, G. McK. (1957) A study of the audio-frequency radio phenomena known as 'dawn chorus,' *Aust. J. Phys.*, **10**, 286–98.

Barbosa, D.D., and Kurth, W.S. (1990) Theory and observations of electrostatic ion waves in the cold ion torus. *J. Geophys. Res.*, **95**, 6443–50.

Table P11 A summary of the observations of various types of plasma waves at all the planets except Merccury and Pluto

Type of plasma wave	Venus	Earth	Mars	Jupiter	Saturn	Uranus	Neptune
Upstream electron plasma oscillations	×	×	×	×	×	×	×
Upstream ion acoustic waves		×	×	×			
Electrostatic noise at the bow shock	×	×	×	×	×	×	×
Electron cyclotron harmonic waves		×		×	×	×	×
Upper hybrid resonance waves		×		×	×	×	×
Whistler-mode hiss		×		×	×	×	×(?)
Whistler-mode chorus		×		×	×	×	
Whistler-mode auroral hiss		×		×			
Electrostatic ion cyclotron waves		×		×(?)			×(?)
Ion cyclotron whistlers (lightning)		×					
Electromagnetic ion cyclotron emissions		×		×			×(?)

Barkhausen, H. (1919) Zwei mit Hilfe der neuen Verstärker entdeckte Erscheinungen. *Phys. Z.*, **20**, 401.
Barrington, R.E. and Belrose, J.S. (1963) Preliminary results from the very-low frequency receiver aboard Canada's Alouette satellite. *Nature*, **198**, 651–6.
Brice, N. (1964) Fundamentals of very low frequency emission generation mechanisms. *J. Geophys. Res.*, **69**, 4515–22.
Fahleson, U.V. (1967) Theory of electric field measurements conducted in the magnetosphere with electric probes. *Space Sci. Rev.*, **7**, 238–62.
Fredricks, R.W., Kennel, C.F., Scarf, F.L. *et al.* (1968) Detection of electric-field turbulence in the Earth's bow shock. *Phys. Rev. Lett.*, **21**, 1761–4.
Grard, R., Nairn, C., Pedersen, A. *et al.* (1991) Plasma and waves around Mars. *Planet. Space Sci.*, **39**, 89–98.
Gurnett, D.A. (1992) Planetary radio emissions, in *Astronomy and Astrophysics Encyclopedia* (ed. S.P. Maran). New York: Van Nostrand Reinhold, p. 535–7.
Gurnett, D.A. and Frank, L.A. (1978) Ion acoustic waves in the solar wind. *J. Geophys. Res.*, **83**, 58–74.
Gurnett, D.A., Kurth, W.S., Poynter, R.L. *et al.* (1989) First plasma wave observations at Neptune. *Science*, **246**, 1494–8.
Gurnett, D.A., Kurth, W.S., Roux, A. *et al.* (1991) Lightning and plasma wave observations from the Galileo flyby of Venus. *Science*, **253**, 1522–5.
Gurnett, D.A., Kurth, W.S. and Scarf, F.L. (1979a) Plasma wave observations near Jupiter: initial results from Voyager 2. *Science*, **206**, 987–91.
Gurnett, D.A., Kurth, W.S. and Scarf, F.L. (1979b) Auroral hiss observed near the Io plasma torus. *Nature*, **280**, 767–70.
Gurnett, D.A., Kurth, W.S. and Scarf, F.L. (1981) Plasma waves near Saturn: initial results from Voyager 1. *Science*, **212**, 235–9.
Gurnett, D.A., Kurth, W.S., Scarf, F.L. and Poynter, R.L. (1986) First plasma wave observations at Uranus. *Science*, **233**, 106–9.
Gurnett, D.A. and O'Brien, B.J. (1964) High-latitude geophysical studies with satellite Injun 3, 5. Very-low-frequency electromagnetic radiation. *J. Geophys. Res.*, **69**, 65–89.
Gurnett, D.A., and Shaw, R.R. (1973) Electromagnetic radiation trapped in the magnetosphere above the plasma frequency. *J. Geophys. Res.*, **78**, 8136–49.
Gurnett, D.A., Shaw, R.R., Anderson, R.R. *et al.* (1979) Whistlers observed by Voyager 1: Detection of lightning on Jupiter. *Geophys. Res. Lett.*, **6**, 511.
Jones, D. (1980) Latitudinal beaming of planetary radio emissions. *Nature*, **288**, 225–9.
Kennel, C.F., and Petschek, H.E. (1966) Limit on stably trapped particle fluxes. *J. Geophys. Res.*, **71**, 1–28.
Kennel, C.F., Scarf, F.L., Fredricks, R.W. *et al.* (1970) VLF electric field observations in the magnetosphere. *J. Geophys. Res.*, **75**, 6136–52.
Kindel, J.M. and Kennel, C.F. (1971) Topside current instabilities. *J. Geophys. Res.*, **76**, 3055–78.
Kintner, P.M., and Gurnett, D.A. (1977) Observations of ion cyclotron waves within the plasmasphere by Hawkeye 1. *J. Geophys. Res.*, **82**, 2314–8.
Kintner, P.M., Kelley, M.C. and Mozer, F.S. (1978) Electrostatic hydrogen cyclotron waves near one Earth radius altitude in the polar magnetosphere. *Geophys. Res. Lett.*, **5**, 139–42.
Krall, N.A., and Trivelpiece, A.W. (1973) *Principles of Plasma Physics*. New York: McGraw-Hill.
Kurth, W.S. (1991) Continuum radiation in planetary magnetospheres, in *Planetary Radio Emissions III* (eds H.O. Rucker, S.J. Bauer and M.L. Kaiser). Vienna, Austria. Verlage der Osterreichischen Akademie der Wissenschaften, p. 329–50.
Kurth, W.S., and Gurnett, D.A. (1991) Plasma waves in planetary magnetospheres. *J. Geophys. Res.*, **96**, 18877–991.
Melrose, D.B. (1981) A theory for the nonthermal radio continuum in the terrestrial and Jovian magnetospheres. *J. Geophys. Res.*, **86**, 30–6.
Roux, A., Perraut, S. Rauch, J.L. *et al.* (1982) Wave–particle interactions near ω_{He+} observed on board Geos 1 and 2, 2. Generation of ion cyclotron waves and heating of He^+ ions. *J. Geophys. Res.* **87**, 8174–90.
Scarf, F.L., Gurnett, D.A. and Kurth, W.S. (1979) Jupiter plasma wave observations: an initial Voyager 1 overview. *Science*, **204**, 991–5.
Scarf, F.L., Taylor, W.W.L. and Green, I.M. (1979) Plasma waves near Venus: initial observations. *Science*, **203**, 748–50.
Scarf, F.L., Fredricks, R.W., Frank, L.A. and Neugebauer, M. (1971) Nonthermal electrons and high-frequency waves in the upstream solar wind, 1. Observations. *J. Geophys. Res.*, **76**, 5162–71.
Scarf, F.L., Gurnett, D.A., Kurth, W.S. and Poynter, R.L. (1982) Voyager 2 plasma wave observations at Saturn. *Science*, **215**, 587–94.
Scarf, F.L., Fredricks, R.W., Frank, L.A. *et al.* (1970) Direct correlations of large amplitude waves with suprathermal protons in the upstream solar wind. *J. Geophys. Res.*, **75**, 7316–72.
Scudder, J.D., Mangeney, A., Lacome, C. *et al.* (1986) The resolved layer of a collisionless, high β, supercritical, quasi-perpendicular shock wave, 3. Vlasov electrodynamics. *J. Geophys. Res.*, **91**, 11075–97.
Shaw, R.R. and Gurnett, D.A. (1975) Electrostatic noise bands associated with the electron gyrofrequency and plasma frequency in the outer magnetosphere. *J. Geophys. Res.*, **80**, 4259–71.
Stix, T. (1962) *The Theory of Plasma Waves*. New York: McGraw-Hill, 110 pp.
Stone, R.G., Pedersen, B.M., Harvey, C.C. *et al.* (1992) Ulysses radio and plasma wave observations in the Jupiter environment. *Science*, **257**, 1524–30.
Storey, L.R.O. (1953) An investigation of whistling atmospherics. *Phil. Trans. Roy. Soc. London*, A, **46**, 113–41.
Taylor, W.W.L., Parady, B.K. and Cahill, L.J. Jr, (1975) Explorer 45 observations of 1- to 30-Hz magnetic fields during magnetic storms. *J. Geophys. Res.*, **80**, 1271–86.
Thorne, R.M. and Moses, J. (1983) Electromagnetic ion-cyclotron instability in the multi-ion Jovian magnetosphere. *Geophys. Res. Lett.*, **10**, 631–4.
Thorne, R.M., and Scarf, F.L. (1984) Voyager 1 evidence for ion-cyclotron instability in the vicinity of the Io plasma torus, *Geophys. Res. Lett.*, **11**, 263–6.

Acknowledgement

This research was supported by NASA through contract 959193 with the Jet Propulsion Laboratory.

Cross references

Heliosphere
Magnetospheres of the outer planets
Planetary torus
Shock waves
Solar wind
Thermal plasma instrumentation
Voyager missions
Whistler

PLATE TECTONICS

The Earth's solid surface behaves in most places as if it were divided into a number of almost rigid 'plates'. Any horizontal motion of a rigid plate on a spherical Earth is necessarily a rotation about an axis through the center. This axis cuts the surface at the 'pole of rotation'. The plates move relative to one another over the asthenosphere (q.v.) at speeds of the order of 10–100 mm per year. Two good modern textbooks on the subject are Cox and Hart (1986) and Fowler (1990).

Figure P41 is a map of shallow earthquakes, which mark the plate boundaries well in the oceans but less so in continental areas. At mid-ocean ridges (Figure P42) the ocean is typically 2.5 km shallower than average; the plates are moving apart; hot, soft asthenosphere rises and turns into hard, cold sea floor; and there are only shallow earthquakes. The opposite sides of oceanic plates usually have 'subduction zones' where the plates bend and go down into the mantle. These zones are marked by trenches several kilometers deeper than normal ocean nearby; by earthquakes which are shallow near the trenches, and become deeper with distance away from them; and by lines of andesitic volcanoes above the earthquakes 100–200 km deep.

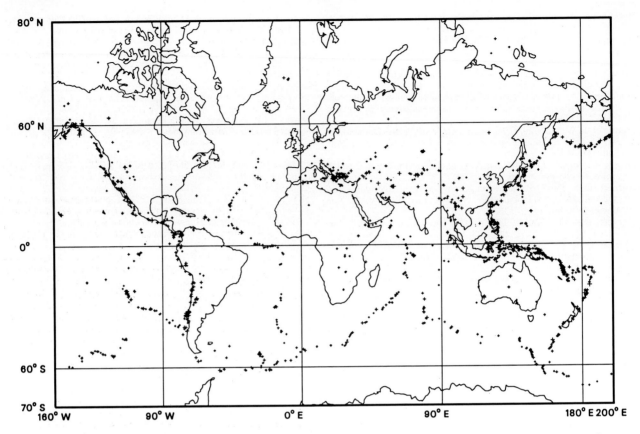

Figure P41 Earthquakes over magnitude 5.5 and shallower than 20 km, from the USGS Global Hypocenter Data Base. Larger + marks indicate larger earthquakes.

Continental plate collision zones are marked in the same way as subduction zones in Figure P42; the difference is that ocean plates subduct, to depths of at least several hundred kilometers, whereas continental plates do not. They normally pile up in a mountain range instead; the largest example is the Himalayas. At transform faults, the remaining type of plate boundary, one plate slides horizontally past the other, with little or no rising or sinking of crustal material.

All plate boundaries are seismically active: subduction zones the most, transform faults less so (there may even be continuously creeping segments with no earthquakes, as in parts of the San Andreas fault in California), and ridges least of all.

Figure P43 is a schematic, true-scale cross-section of a typical region of the Earth, showing a subduction zone, a ridge and parts of three plates. On this scale oceans and mountains are invisible, and only the most violent volcanic eruptions throw ash high enough to be seen.

No two experts in the field would be likely to show exactly the same map of the plates. For example, DeMets *et al.* (1990) separate the Indian and Australian plates along a nearly east–west line between the Carlsberg and Central Indian ridges, and do not show the Scotia or Sandwich plates as separate from the Antarctic. (The Sandwich plate is the small D-shaped area just east of the Scotia plate in Figure P42.) Other authors show more 'microplates', e.g., separating an Adriatic plate from the African and Eurasian plates (Anderson, 1987), or the Caroline basin from the Pacific plate (Weissel and Anderson, 1978), or inserting Easter and Juan Fernandez microplates on the Nazca–Pacific plate boundary (Hey *et al.*, 1985). The location of both dashed boundaries in Figure P42 is controversial.

In continental regions being compressed by plate motion, the deformation may be so widespread that it is doubtful whether the concept of plate behavior is useful at all. Molnar and Tapponnier (1975) suggested that the eastward motion of Tibet and nearby areas is better explained in terms of plastic flow throughout the region, as India continues to be pushed north into it. For people trying to understand the Earth as a whole, plate tectonics is a useful large-scale model. Even for those concentrating on a deforming region like Tibet or New Zealand, plate tectonics provides at least the boundary conditions at a large distance, and helps to explain why the deformation occurs.

History

Plate tectonics appeared during the 1960s as a synthesis of much previous work. The history is well described by Cox (1973) and Emiliani (1981). References for this section not listed below can be found in one or other of those books. Both show the subject as a fine example of how major changes occur in scientific understanding.

The general idea of currents inside the Earth leading to surface displacements dates back to the 19th century. In the early 20th century Wegener introduced the concept of drifting continents, which neatly explained a variety of topographic, sedimentological, paleontological, botanical and zoological observations. Wegener had no convincing driving mechanism, however, and so failed to convince many people (except in the southern hemisphere, where the field evidence for former connections of continents now separated was much stronger than in the northern hemisphere).

Hills (1934) drew the analogy between froth floating on boiling jam and moving continents floating on the mantle of the Earth, in each case the driving mechanism being thermal convection. Jeffreys (1934) admitted that this sort of driving mechanism would avoid his earlier objection (on mechanical grounds) to continents ploughing their way across basaltic ocean floors. Both authors thought that such movements had stopped early in the history of the Earth.

The seismological results of Gutenberg and Richter (1949) did not immediately lead to the emergence of plate tectonics, in spite of their maps which were very like Figures P41 and P42, mainly because oceanic geology was too little known at the time. Neither did the work of Runcorn in 1962 on paleomagnetism (q.v.). He and many others deduced polar wander paths from observations of the direction (in three dimensions) in which dated rocks in different parts of the world are found to be magnetized. From land-based paleomagnetic work the major conclusion was that about 200 million years ago the

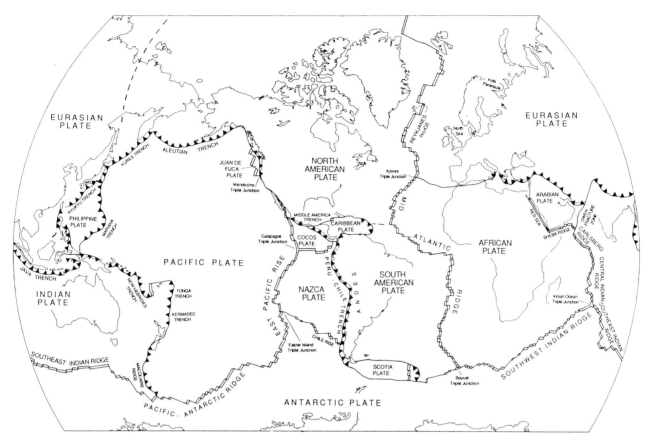

Figure P42 The world system of plates. Double lines: mid-ocean ridges; single lines: transform faults; dashed lines: uncertain plate boundaries; lines of triangles: subduction zones and continental collision regions, the triangles being on the upper plate. From Fowler (1990), with permission of the author and Cambridge University Press.

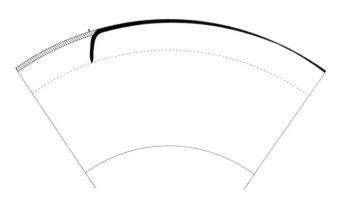

Figure P43 A schematic cross-section of the Earth at true scale. Solid regions: oceanic plates (whose thickness increases with square root of age); shaded region: a continental plate (thickness assumed constant at 100 km); dashed circle: the 700-km discontinuity, which may be the bottom of the asthenospheric convection cells, solid circle: the mantle–core boundary.

former supercontinent Pangaea began to split, first to Laurasia and Gondwana, then to the present collection of continents and fragments. Some, like India, have since reunited with larger blocks. It should be mentioned that Pangaea was not primordial: Laurasia was assembled from separate parts during the Paleozoic, Gondwana probably earlier.

The evidence that finally convinced Earth scientists of plate tectonics was the pattern of magnetic anomalies in the ocean floors, confirming the seismological conclusions. The Earth's magnetic field reverses itself at irregular intervals, and mid-ocean ridge rocks everywhere are magnetized in the direction of the field prevailing when they cool through the Curie point. As they move away from the ridges, at a speed which is nearly the same on each side, a pattern of normal and reversed magnetism develops which is its own mirror image across the ridge. The irregularities in timing of reversals become irregularities in spacing of the magnetic anomalies which, since the work of Hess (1962), Morley (first published by Emiliani, 1981) and Vine and Matthews (1963), have been identified and correlated over most of the oceanic crust.

The plate tectonic synthesis was expressed by Morgan (1968), Le Pichon (1968), and Isacks, Oliver and Sykes (1968). The theory was triumphantly confirmed by the first few Deep Sea Drilling Project cruises in the late 1960s. These found ages of fossils from the bottom of the ocean sedimentary layer steadily increasing with distance from mid-ocean ridges; the magnetic and paleontological time scales could then be calibrated against each other. The latter scale had already been calibrated radiometrically.

The driving mechanism remained in dispute much longer. Gilluly (1971) began a paper with 'So far as I know, no one has yet suggested a model for the generation of plate motion that is acceptable to anyone else', but by now there seems to be a consensus, largely based on work of Turcotte and Oxburgh in 1967 and McKenzie and Elsasser (independently) in 1969, that thermal convection in an Earth hotter inside than at the surface gives rise to plate motions, given that both the surface layers and the mantle below 700 km are much stiffer than the asthenosphere.

Modern developments

Kinematics

Work far too voluminous to list here (but see Fowler, 1990, Chapter 3; DeMets *et al.*, 1990; and DeMets, 1992) shows both how the plates are moving now and how the system behaved in the past. As ocean floors are eventually subducted, oceanic reconstructions become more conjectural as one goes backwards in time. One can be fairly sure of the pattern of plate motions in the Cenozoic, but information for periods before the Jurassic has disappeared. Continents, however, can be followed paleomagnetically back through the Paleozoic, except in regions with many subduction zones, such as Indonesia and the Philippines. The work is still unfinished, and the timescale is still being refined (Harland et al., 1989).

Dynamics

Many different physical plate-driving mechanisms have been proposed (Harper, 1989, 1990). The most plausible is thermal convection caused by radioactive heat generation in the interior of the Earth, but that statement does not explain how plates form and why no other planet seems to have plate tectonics like ours, even though they are all hot inside. The key considerations appear to be as follows:

1. The asthenosphere (q.v.), when pushed in the same direction for thousands of years, behaves like a viscous fluid (attested by studies of the rate of uplift of Canada and Fennoscandia after their Pleistocene ice sheets melted), and the lower mantle like a more viscous one. It is still controversial how much more. Peltier (1989) suggested a factor of two (upper mantle viscosity 10^{21} Pa s, lower mantle 2×10^{21} Pa s); James and Morgan (1990) suggested factors of 20 or 200 (upper mantle 10^{21} or 5×10^{20} Pa s, lower mantle 2×10^{22} or 10^{23} Pa s).
2. The lithosphere above is strong enough not to deform significantly under forces large enough to push the asthenosphere at typical plate speeds, but it is not perfectly rigid. Oceanic lithosphere for instance bends and then sinks at subduction zones.
3. Oceanic lithosphere is essentially cold asthenosphere rock. Being cold, it is dense, but it still floats on the asthenosphere, as ships made of steel float on water. The lithosphere thickness increases due to thermal diffusion as $\sqrt{T}$ where T is its age (i.e. the time since it appeared at a mid-ocean ridge), at least up to 70 Ma. The ocean depth also increases, being roughly given by $d = 2.5 + 0.35\sqrt{T}$ if d is measured in km and T in Ma. (This does not hold precisely above ridges, where there is often an axial valley due to resistance to upward flow, nor for $T > 70$ Ma, where d increases more slowly with age, for reasons which are not yet clear.)
4. In a system convecting under gravity, the major driving forces act where the major horizontal density differences occur, at the top and bottom thermal boundary layers. In the Earth these are the plates and the mantle–core boundary, but hydrodynamic image effects (Harper, 1990) imply that the latter will have a much smaller effect on plate motions. In oceanic plates the thickening with age implies a horizontal force per unit area proportional to ∇T away from the ridges (ridge push).
5. When a plate subducts it is pulled down by its own weight. This 'slab pull' is the strongest plate-driving force: the fastest-moving plates all have slabs attached.
6. New subduction zones are presumably generated if large amounts of rock break through oceanic lithosphere (like a plateau basalt, but at sea), weigh it down and cause it to founder (Turcotte, Haxby and Ockendon, 1977).
7. Other forces also help to move plates about, such as the reaction to slab pull, which must be applied by either the asthenosphere, the non-subducting plate or the mantle under the asthenosphere. Some authors have advocated friction at the bottom of slabs where they attempt to push into the higher-viscosity mantle; others have included the essentially hydrostatic push that each plate exerts on the other at a continental collision zone; others have suggested that the upflow from a hot spot, spreading out in all directions under the lithosphere, will push it outwards. If a plate boundary is nearby, the result is a force on each plate away from the other.

Hot spots

The Hawaiian islands have long been known to be a line of volcanoes becoming younger to the ESE, and there are several parallel lines of seamounts and islands in various parts of the Pacific plate. They appear to be due to 'hot spots' which remain nearly fixed in the mantle and which send up plumes of magma that punch through as a volcano. Hot spots also occur on other plates, and in each case their volcanic lines are nearly antiparallel to the local plate motion relative to the mantle; the hot spots themselves move relative to one another much more slowly than plates do. They can thus be used, along with magnetic anomalies, to elucidate past plate motions.

Back-arc spreading and marginal basins

In many places, especially the western Pacific, sea-floor spreading occurs above a subduction zone which is generating a new, young, marginal ocean basin behind the volcanic arc. Examples are the Lau–Havre Basin west of the Tonga–Kermadec arc and the Mariana Trough west of the Mariana Islands (not to be confused with the Mariana Trench east of them). There are also inactive marginal basins (e.g. the Sea of Japan) and other marginal basins which are generated without subduction nearby (e.g. the Cayman Trough in the Caribbean Sea, which is generated by E–W movement between the North American and Caribbean plates on a short N–S ridge segment on their boundary). Wherever back-arc spreading occurs above a subduction zone, the speed of subduction is faster than the speed of approach of the major plates. It can be much faster, as at the South Sandwich Islands (if the Sandwich–Scotia spreading is treated as back-arc spreading; Pelayo and Wiens, 1989). Several possible dynamical reasons have been advanced for such spreading (Figure P44), ranging from upwelling due to heat generation in the subducting slab (model 1) or mantle beneath (model 2), to the purely kinematic (model 3: the back-arc plate retreating for other reasons), to asthenosphere flow from elsewhere pushing the slab aside (model 4), or the slab sinking under its own weight and pushing the asthenosphere aside (model 5).

Other planets

The Earth is the only planet known to have plate tectonics. Mercury, Venus, the Moon and Mars seem to be one-plate planets (Solomon, 1978; Kiefer and Hager, 1991; Herrick and Phillips, 1992). Though Venus has active mantle convection, and even subduction zones if Sandwell and Schubert (1992) are right, it seems to lack a mid-ocean ridge system.

J. F. Harper

Bibliography

Anderson, H. (1987) Is the Adriatic an African promontory? *Geology*, **15**, 212–15.
Cox, A. (1973) *Plate Tectonics and Geomagnetic Reversals*. San Francisco: Freeman.
Cox, A. and Hart, R.B. (1986) *Plate Tectonics: How it Works*. Palo Alto: Blackwell.
DeMets, C. (1992) A test of present-day plate geometries for northeast Asia and Japan. *J. Geophys. Res.*, **97**, 17627–35.
DeMets, C., Gordon, R.G., Argus, D.F. and Stein, S. (1990) Current plate motions. *Geophys. J. Int.*, **101**, 425–78.
Emiliani, C. (1981) A new global geology, in *The Oceanic Lithosphere (The Sea, Vol. 7)* (ed. C. Emiliani). New York: Wiley-Interscience, pp. 1687–728.
Fowler, C.M.R. (1990) *The Solid Earth*. Cambridge: Cambridge University Press.
Gilluly, J. (1971) Plate tectonics and magmatic evolution. *Bull. Geol. Soc. Am*, **82**, 2382–96.
Gutenberg, B. and Richter C.F. (1949) *Seismicity of the Earth and Associated Phenomena*. Princeton: Princeton University Press.
Harland, W.B. Armstrong, R.L., Cox, A.V. et al. (1989) *A Geologic Time Scale 1989*. Cambridge: Cambridge University Press.
Harper, J.F. (1989) Forces driving plate tectonics: the use of simple dynamical models. *Rev. Aquat. Sci.*, **1**, 319–36.
Harper, J.F. (1990) Plate dynamics: Caribbean map corrections and hotspot push. *Geophys. J. Int.*, **100**, 423–31.
Herrick, R.R. and Phillips, R. (1992) Geological correlations with the interior density structure of Venus. *J. Geophys, Res.*, **97**, 16017–34.
Hess, H.H. (1962) History of ocean basins, in *Petrological Studies: a Volume in Honor of A.F. Buddington*, (eds A.E.J. Engel, H.L. James and B.F. Leonard), New York: Geological Society of America, pp. 599–620.
Hey, R.N., Naar, D.F., Kleinrock, M.C. et al. (1985) Microplate tectonics along a superfast seafloor spreading system near Easter Island. *Nature*, **317**; 320–5.

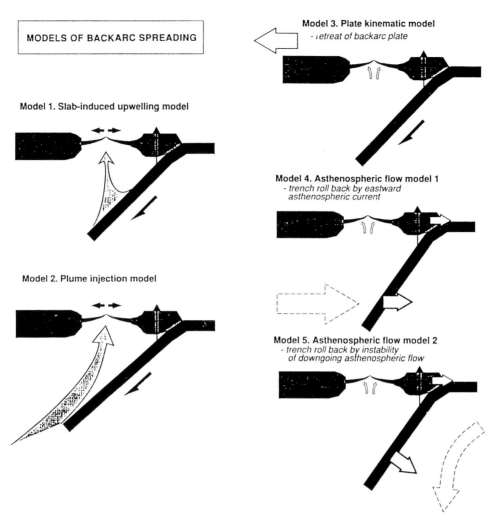

Figure P44 Five possible models of back-arc spreading. (From Tamaki and Honza (1991, with permission of the International Union of Geological Sciences.)

Hills, G.F.S. (1934) The granitic and basaltic areas of the Earth's surface. *Geol. Mag.*, **71**, 275–6.
Isacks, B., Oliver J. and Sykes, L.R. (1968) Seismology and the new global tectonics. *J. Geophys. Res.*, **73**, 5855–99.
James, T.S. and Morgan, W.J. (1990) Horizontal motion due to post-glacial rebound. *Geophys. Res. Lett.*, **17**, 957–60.
Jeffreys, H. (1934) Note on Mr. Hills's paper. *Geol. Mag.*, **71**, 276–80.
Kiefer, W.S. and Hager, B. (1991) A mantle plume model for the equatorial highlands of Venus. *J. Geophys. Res*, **96**, 20947–66.
Le Pichon, X. (1968) Sea floor spreading and continental drift. *J. Geophys. Res.*, **73**, 3661–97.
Molnar, P. and Tapponnier P. (1975) Cenozoic tectonics of Asia: effects of a continental collision. *Science* **189**, 419–26.
Morgan, W.J. (1968) Rises, trenches, great faults and crustal blocks. *J. Geophys. Res.*, **73**, 1959–82.
Pelayo, A.M. and Wiens, D.A. (1989) Seismotectonics and relative plate motions in the Scotia Sea region. *J. Geophys. Res.*, **94**, 7293–320.
Peltier, W.R. (1989) Mantle viscosity, in *Mantle Convection: Plate Tectonics and Global Dynamics* (ed. W.R. Peltier). New York: Gordon and Breach, pp. 389–478.
Sandwell, D.T. and Schubert G. (1992) Evidence for retrograde subduction on Venus. *Science*, **257**, 766–70.
Solomon, S.C. (1978) On volcanism and thermal tectonics on one-plate planets. *Geophys. Res. Lett.*, **5**, 461–4.
Tamaki, K. and Honza, E. (1991) Global tectonics and the formation of marginal basins: role of the Western Pacific. *Episodes*, **14**, 224–30.
Turcotte, D.L., Haxby, J.F. and Ockendon, J.R. (1977) Lithospheric instabilities, in *Island Arcs, Deep Sea Trenches and Back-Arc Basins* (Maurice Ewing Series 1) (eds M. Talwani and W.C. Pitman III). Washington, DC: American Geophysical Union, pp. 63–9.
Vine, F.J. and Matthews, D.H. (1963) Magnetic anomalies over oceanic ridges. *Nature*, **199**, 947–9.
Weissel, J.K. and Anderson R.N. (1978) Is there a Caroline plate? *Earth Planet. Sci. Lett.*, **41**, 143–58.

Cross references

Crust
Earth: geology, tectonics and seismicity
Isostasy
Lithosphere
Mantle convection
Seismicity
Tectonics

PLUTO

Perhaps the most enigmatic of the outer solar system planets, Pluto was discovered in 1930 by Clyde W. Tombaugh. Its discovery was no accident. Early in the 20th century a number of astronomers, most notably Percival Lowell, theorized that the gravitational influence of an undiscovered planet was responsible for perceived irregularities in the orbital motions of Uranus and Neptune. It now seems likely that these orbital discrepancies merely reflected uncertainties in measurements and calculations (Standish, 1993); nevertheless, the resulting search for a ninth planet led directly to the discovery of Pluto.

Early observations and theories

Although the physical nature of Pluto remains poorly understood, the orbital characteristics of the planet were quickly established. Compared to orbits of the other known planets, the orbit of Pluto is characterized by extreme values of eccentricity (0.25) and inclination (17°). (The former is a measure of orbital ellipticity, and the latter represents the tilt of the orbital plane with respect to the orbital plane of the Earth.) The inclination of Pluto is more than seven times the planetary average, and the orbit is significantly non-circular. Although the average Pluto–Sun distance is 39.44 AU, the perihelion and aphelion separations are 29.6 and 49.3 AU respectively. Despite popular perceptions of Pluto as the 'most distant planet,' Neptune holds this distinction for 20 years of each 247.7-year Pluto orbit. Pluto's most recent perihelion occurred on 5 September 1989, and it will be closer to the Sun than Neptune for the period from 21 January 1979 to 14 March 1999.

These peculiar orbital characteristics prompted several early attacks upon the planetary status of Pluto. A number of astronomers noted that Pluto's orbit would not be regarded as unusual if it were classified as an inactive comet rather than a planet. Others attributed genetic significance to the observation that near perihelion Pluto crosses within the orbit of Neptune, and proposed that Pluto represents an escaped satellite of its much larger neighbor. It was suggested that Pluto may have been pulled away by gravitational interactions with a massive passing object, or gravitationally slung outward during an ancient close encounter with the currently retrograde Neptunian satellite Triton. In either case, some astronomers advocated that Pluto did not qualify as a 'true' planet.

Issues of planetary origins cannot be resolved on the basis of orbital characteristics alone; knowledge of physical characteristics is also required. Of particular importance is the bulk density (mass/volume ratio). Initially, Pluto was expected to be quite massive because it was assumed to be responsible for the alleged gravitational perturbations of Uranus and Neptune. Two observations, however, suggested that the newly discovered planet was significantly smaller (and therefore less massive) than anticipated: it was very faint and, even when observed through the most powerful telescopes, it did not resolve into a well-defined disk (as did all other known planets), merely a starlike point of light. These observations, unfortunately, did not produce quantitative estimates of the size of Pluto; they simply demonstrated that it was not as large as expected.

In the 1950s planetary astronomers began to monitor variations in the amount of sunlight reflected by Pluto. Brightness variations on the order of 10% occurring with a period of 6.387 days were identified and reasonably assumed to correspond to the rotational period of the planet (Walker and Hardie, 1955). Curiously, expectations that Pluto would steadily increase in brightness as it approached perihelion (its closest approach to the Sun) were flatly contradicted by observations, which indicated a substantial decrease in brightness. Eventually, new discoveries would render this puzzling observation less mystifying.

Charon

During the summer of 1978, while examining high-resolution photographs of Pluto to refine determinations of its orbital trajectory, James W. Christy of the US Naval Observatory discovered that Pluto possesses a relatively large satellite. Christy had analyzed a number of photographic plates classified as defective because of apparent north–south-oriented elongations in the image of Pluto. These elongations had previously been attributed to distortion commonly associated with imprecise tracking of astronomical objects during photographic exposure. Christy noticed, however, that the stars in these images were undistorted, thereby eliminating the possibility of inaccurate telescopic tracking. Furthermore, the specific shape of the distortions suggested an object with a protruding, migrating bulge rather than blurred photographic images. Christy then examined earlier images of Pluto, and observed similar north–south-oriented elongations in photographic plates also labeled as defective. Systematic variations in the position of the bulge suggested the presence of a satellite orbiting at high inclination with a period of approximately 6.4 Earth days. Christy named this satellite Charon (pronounced 'care on' or 'share on').

That Charon inevitably appeared to the north or south of Pluto suggested an extreme orbital inclination for the satellite. Subsequent observations have determined that the orbital axis of Charon about Pluto is presently inclined approximately 122° with respect to the orbital axis of Pluto about the Sun, and it is generally assumed that the rotational axis of Pluto is similarly oriented (a condition that is, for Pluto and Charon, probably an inevitable end-state of tidal dynamics; see Tide-raising force; Torque). An inclination of this magnitude indicates that the Pluto–Charon system essentially 'lies on its side,' a rotational–orbital configuration shared only by Uranus and its satellite system (imagine the Earth–Moon system 'tipped on edge' and you will have a qualitative analog of the Pluto–Charon system). As Pluto orbits the Sun, conservation of angular momentum requires that the orbital plane of Charon maintains a fixed orientation in space (although slight oscillations occur on a long timescale). Twice within each 247.7-year orbital period of Pluto, Earth-based observers will, therefore, witness the Pluto–Charon system both 'head-on,' with the system resembling a bull's-eye target, and 'edge-on,' with Charon alternatively passing directly in front of and behind Pluto.

The orbital period of Charon is virtually identical to the rotational period of Pluto previously inferred from brightness variations (Walker and Hardie, 1955). (Because the contribution of Charon to the lightcurve amplitude is only $\simeq 15\%$, lightcurve variations are dominated by the rotation of Pluto rather than the orbital period of the system.) The coincidence of orbital and rotational periods indicates that Pluto and Charon are mutually tidally locked; each presents the same face toward the other and is suspended motionless in the other's sky. The locking of satellites into tidally maintained orbital–rotational synchronicity is common throughout the solar system (the Earth's Moon is an obvious example). The Pluto–Charon system is unique, however, because both bodies have mutually despun one another, an indication that the mass of Charon must be significant with respect to Pluto.

Physical characteristics

Mass

An immediate consequence of the discovery of Charon was that it constrained mass estimates of Pluto. On the basis of Newton's law of gravitation, Charon's orbital period implies a mass of close to 1.45×10^{22} kg for the Pluto–Charon system. This is less than 20% of the mass of the Earth's Moon, supporting earlier suspicions that Pluto could not have been responsible for orbital disturbances of Uranus or Neptune. Assumptions regarding the size and density of Charon allow individual mass estimates of Pluto.

Radius

Because telescopic observations of Pluto do not produce a well-defined disk, the radius of the planet cannot be accurately inferred by the usual method of measuring the angular diameter of an object and calculating size as a function of distance. Accurate determination of the radius of Pluto requires an alternative method. One such method is to record the amount of time an observable object requires to pass in front of or behind another object whose size is to be determined. Assuming that all relevant velocities are known, the size relatively easy to determine using transit and/or occultation times. An obvious scenario would be to measure the amount of time that Pluto required to pass in front of a star (see Occultation). In the absence of a resolvable planetary disk, a stellar occultation must be monitored from several locations on Earth; a single observation can only establish a lower limit on the radius because there is no way of determining whether the star passed directly behind the planet or merely 'grazed an edge.' Prior to the occurrence of such an occultation, the discovery of Charon presented an alternative method of determining the size of Pluto.

As noted above, the heliocentric motion of the Pluto–Charon system produces an edge-on alignment with respect to Earth-based observers twice per orbit. During such an alignment, Charon can be observed to pass directly in front of and behind Pluto. The former is referred to as a transit of Pluto by Charon, the latter as an occultation

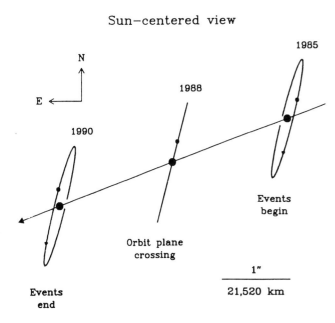

Figure P45 The geometry of mutual occultations and transits (mutual events) between Pluto and Charon that occurred in the 1980s. (From Stern 1992.)

of Charon by Pluto, and the combination as 'mutual events.' Shortly after the discovery of Charon, planetary scientists realized that a sequence of mutual events was imminent (Figure P45). Astronomers began monitoring Pluto in the early 1980s, anticipating the sudden, temporary reduction in brightness indicating that either Pluto or Charon was partially blocking the reflected light of the other. The first unambiguous confirmation that the images of the two bodies were beginning to graze one another was obtained in early 1985 (Binzel et al., 1985). Thus, 55 years after the discovery of the planet, but only 7 years after the discovery of its satellite, a series of measurements was initiated that would finally constrain the size of Pluto. Had Charon been discovered only a decade later, more than a century would pass before such an opportunity would present itself again. Charon, it seems, was discovered in a timely fashion.

Although the notion of measuring the time required for an object to pass in front of or behind another object whose size is to be determined is conceptually quite simple, observations involving the Pluto–Charon system were not straightforward. Because images of the two objects cannot be individually well resolved with ground-based telescopic observations, transit and occultation times had to be indirectly inferred on the basis of brightness variations. These measurements were converted into radius estimates for Pluto and Charon by computer simulations, and indicate a radius of about 1175 km for Pluto, and 625 km for Charon. In addition, ground-based speckle-interferometry observations and Hubble Space Telescope images imply that the semi major axis of the Pluto–Charon system is approximately 19 650 km.

Atmosphere

The earliest direct evidence concerning the composition of Pluto was acquired in 1976 when analysis of detailed, wavelength-dependent variations in the intensity of light reflected from Pluto (see Spectrophotometry) revealed the presence of methane (CH_4; Cruikshank, Pilcher and Morrison, 1976). Because this common, carbon-bearing compound often occurs in a gaseous state, a consequence of this discovery was considerable speculation concerning the existence of an atmosphere surrounding Pluto. At typical outer solar system temperatures, however, methane can exist as a gas, liquid or a solid (similar to water in an Earth-like environment). For this reason, the presence of methane did not conclusively demonstrate the existence of a significant atmosphere.

Confirmation of the presence of an atmosphere was provided by a completely different type of observation. Planetary scientists had long realized that a planet passing in front of a star offered the opportunity for probing the atmospheric properties of the planet (see occulatation). Accurate measurements of the manner in which starlight diminished in intensity as the source intersected the planetary limb offered vital clues toward inferring atmospheric structure and, indirectly, composition. Alternatively, an abrupt cessation of starlight, with no preceding diminution in intensity, would presumably indicate the complete absence of an atmosphere. The issue was settled on 9 June 1988 when Pluto passed directly between the Earth and a 12th-magnitude star in the constellation Virgo. All measurements revealed that the intercepted starlight did not terminate abruptly, but rather diminished gradually before disappearing behind the planet. Starlight emerging from the other side of Pluto attained its pre-occultation intensity gradually (Figure P46). These observations immediately confirmed the presence of an atmosphere (Elliot et al., 1989). Recently, the ices of molecular nitrogen (N_2) and carbon monoxide (CO) have been detected spectroscopically (Owen et al., 1993). Because these are materials which readily sublimate, these compounds probably constitute the bulk of the atmosphere.

The atmosphere of Pluto is apparently quite tenuous, with a basal pressure that is probably less than 0.01% of the atmospheric pressure at the surface of the Earth. Brightness variations observed during the mutual event occultations (when Charon was completely obscured by Pluto) suggest that the atmosphere of Pluto is incapable of obscuring surface features and is therefore likely to be relatively transparent. Analysis of the most detailed lightcurve obtained (Figure P46), however, indicates that the lowermost atmosphere of Pluto possesses a thin haze, rapid variations in temperature with altitude, or both. Either feature complicates lightcurve-based estimates of the solid surface radius of Pluto. At present the radius so estimated is somewhat larger than that determined from the mutual events, about 1200 km. Another stellar occultation will probably be required to pin down firmly the radius of Pluto (and by inference that of Charon).

Surface characteristics

Spectroscopic measurements recorded during the mutual events provide a clear indication that the surface compositions of Pluto and Charon are significantly different. The spectral signature of Pluto was isolated during the total occultations of Charon. Although Charon is incapable of completely obscuring Pluto, an uncontaminated spectral signature of Charon was obtained by simply subtracting the signature associated with Pluto from the combined signature. The individual spectra reveal that Pluto is spectrally dominated by methane frost and Charon is dominated by water frost (Buie et al., 1987; Marcialis, Rieke and Lebofsky, 1987; Fink and DiSanti, 1988). Modeling of the recent spectra of Pluto by Owen et al. (1993) shows that most of Pluto's surface is actually nitrogen ice. Methane and carbon monoxide ice are minor contaminants at the percent level. The simplest explanation for these observations is that Charon's smaller mass proved gravitationally insufficient to prevent the escape of extensive amounts of sublimating nitrogen, carbon monoxide and methane. The shape of the molecular nitrogen spectral adsorption band is also very sensitive to temperature, and indicates that much of Pluto's surface is at a temperature close to a frigid 40 K.

Mutual event analyses have provided a crude image of the surface features of Pluto. Even the most preliminary observations (i.e. when the shadows of the two objects were merely grazing the edges of one another) revealed that brightness reductions are enhanced when Charon casts a shadow onto Pluto as opposed to the times when Pluto casts a shadow of similar size onto Charon. This has been interpreted as an indication that the surface of Pluto is considerably brighter than that of Charon, a conclusion consistent with the spectrally inferred existence of nitrogen ice on the former and water ice on the latter. Combining lightcurve records acquired during the period 1954–1986 with selected mutual event observations, Buie, Tholen and Horne (1992) and others determined that Pluto possesses relatively bright polar regions and a darker equatorial band (Figure P47). The features have been confirmed in very recent HST images.

The inference of highly reflective polar regions or caps, combined with the extreme obliquity of the Pluto–Charon system, suggests an explanation for earlier observations of brightness reductions associated with the approach of perihelion. At the time that Pluto was discovered, its southern pole pointed toward the inner solar system (i.e. the Earth). Heliocentric motion, combined with the physical requirement that the equatorial plane of Pluto maintains a fixed orientation in space, gradually introduced a more equatorial profile with respect to Earth-based observers. It is reasonable to conclude,

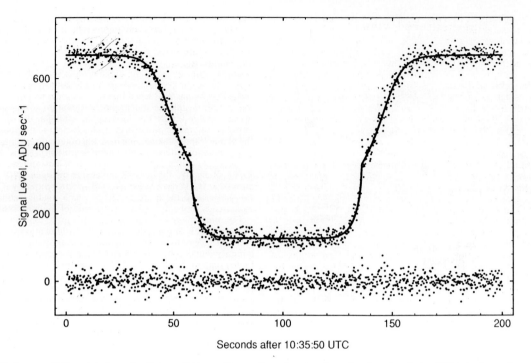

Figure P46 Stellar occultation data from the Kuiper Airborne Observatory showing the refractive signature of Pluto's atmosphere. The lightcurve is plotted as points and the best-fitting model as a line. The residuals from the model fit are the points at the bottom. The steepening of the lightcurve around the half-light level indicates either a haze layer or a steep temperature gradient in Pluto's atmosphere. (From Elliot and Young, 1992.)

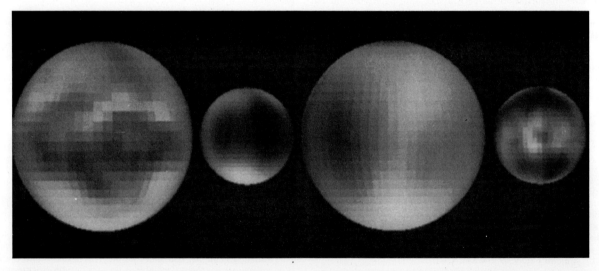

Figure P47 Reflectance images of Pluto and Charon, computed using maximum entropy reconstruction, from lightcurve and mutual event observations. The images show equatorial views of the system at a sub-Earth longitude of 0° (left) and 180° (right). Pluto's eclipsed hemisphere is at 0°, and north is up. Pluto's bright polar regions are evident (with the southern 'cap' being the brightest); the equator is darker with a pronounced 'spot' at about 100° longitude, and Charon is darker than Pluto. (Kindly provided by M. Buie.)

therefore, that secular brightness reductions are largely a consequence of the progressive presentation of relatively dark equatorial regions to Earth-based observers.

Origin and evolution

Radius constraints obtained during mutual event and stellar occultation observations imply a density for Pluto that is approximately 1750–2150 kg m^{-3} (McKinnon and Mueller, 1988; McKinnon et al., 1996). Although such a density range is consistent with values typical of large outer solar system satellites such as Ganymede, Callisto or Titan, Pluto must be considerably more rocky than these objects. A large ice–rock body, such as Ganymede, possesses sufficient mass to compress significantly the constituent ices and elevate substantially interior densities. Pluto, however, is much less massive and self-compressed. Thus, that Pluto's bulk density is comparable to that of the largest ice–rock satellites indicates that it must possess a greater silicate mass fraction. Calculations based on mutual event and stellar occultation measurements indicate that the bulk silicate mass fraction of Pluto is between 0.6 and 0.7, greater than Ganymede, Callisto or Titan (McKinnon and Mueller, 1988; Simonelli et al., 1989; McKinnon et al., 1996).

Any acceptable scenario encompassing the origin of Pluto must account for the relatively excessive silicate mass fraction in contrast to outer solar system ice–rock satellites. Suggestions that Pluto originated in the inner solar system, where relatively dense, rock–metal compositions are the norm, and evolved into its present-day orbit, are difficult to support. Numerical calculations indicate that, although the orbit of Pluto may be formally chaotic (e.g. small differences in past positions and trajectories can evolve into large differences in future behavior), it is highly unlikely that Pluto evolved into its present orbital configuration from a starting point within the inner solar system (Sussman and Wisdom, 1988).

More promising scenarios involve the accretionary process believed to be responsible for the formation of all large solar system bodies. A simple accretionary mechanism that might account for an elevated rock/ice ratio would be the preferential loss of water during energetic impacts associated with accretion. Volatile loss during an extremely energetic impact event, possibly associated with the creation of Charon, is plausible, particularly if the colliding bodies were differentiated prior to impact (McKinnon, 1989). Such an impact also presents an attractive explanation for the excessive angular momentum of the Pluto–Charon system.

Pluto's enhanced silicate component may also be a consequence of non-equilibrium chemical reactions involving the element carbon. Methane (CH_4) represents the stable form of this element in the outermost solar system. Extremely low ambient nebular gas densities and temperatures throughout this region, however, largely precluded the energetic molecular collisions required to promote chemical reactions between carbon monoxide (CO), the dominant carbon-bearing molecule inherited from presolar molecular clouds, and hydrogen, necessary to produce methane. In the neighborhood of Pluto, therefore, the metastable persistence of carbon monoxide may have dominated the array of carbon-bearing substances, primarily at the expense of methane. Carbon-based chemical reactions influenced the rock/ice ratio of solid objects because methane-producing reactions involving carbon monoxide released oxygen which immediately reacted with ubiquitous hydrogen to form H_2O. A direct by-product of methane production in the outer solar system, therefore, was water ice. Now, minor amounts of methane were inherited from presolar molecular clouds, which is consistent with observations of methane on Pluto. The important point, however, is that kinetic inhibition of methane production would have significantly elevated solid-body rock/ice ratios.

Why, then, are outer solar system satellites less rocky than Pluto? The governing factor seems to have been the pressure–temperature conditions prevailing during the formation of these bodies. Gravitational forces associated with giant protoplanets induced local perturbations in the solar nebula. Nebular conditions were, in fact, sufficiently perturbed that gas envelopes in the neighborhood of the forming giant protoplanets are termed protoplanetary nebulae. Of particular importance to our discussion, substantially elevated pressures and temperatures within these protoplanetary nebulae favored the prerequisite molecular collisions associated with the production of methane and water ice. It is for this reason that nearly all satellites of the giant outer solar system planets possess lower rock/ice mass ratios than Pluto. (There are three notable exceptions to this statement, the Galilean satellites Io and Europa, and Triton, the largest satellite of Neptune. In the case of the former, it is likely that H_2O was available during formation, but only in the vapor form. Excessive temperatures characteristic of the innermost regions of massive planetary nebulae prevented the condensation that is a prerequisite for accretion. The elevated silicate content of Triton is probably a consequence of origin within a heliocentric outer solar system orbit, similar to that of Pluto, and see below.)

It is instructive at this point to briefly reconsider arguments regarding the planetary status of Pluto. Recent discoveries concerning the composition of Pluto contradict suggestions that Pluto represents an errant satellite of Neptune. Pluto is not simply 'rockier' than outer solar system satellites, but in fact possesses a composition theoretically consistent with an object that condensed directly from the solar nebula in a heliocentric, outer solar system orbit.

The most reasonable interpretation is that Pluto represents a remnant planetesimal, a rare survivor from the earliest solar system. Planetesimals were kilometer-sized and larger objects that coalesced, through impact accretion, to form planets. The fate of a typical outer solar system planetesimal was determined by interactions with one or more of the giant protoplanets. Planetesimals not directly accreted onto these objects were routinely ejected from the solar system by gravitational close encounters with them (many eventually settling within the Oort cloud). Technically, Pluto probably existed at the edge of the 'feeding zone' of Neptune (according to the calculations of Holman and Wisdom, 1993) and, by either accretion or ejection, would almost certainly have been eliminated by the much more massive planet were it not for a peculiar orbital relationship that has existed between the two planets. Pluto orbits the Sun almost exactly twice during the period of time required by Neptune to complete three solar revolutions. This integer ratio of orbital periods, combined with the respective positions of the planets, guarantees that Pluto and Neptune do not approach one another too closely. When Pluto is in the neighborhood of perihelion or, in other words, within the orbit of Neptune, the larger planet is invariably far away. In fact Pluto routinely approaches Uranus more closely than it will ever approach Neptune, given current orbital configurations (Dobrovolskis, 1989). Although it is not known with certainty that the 'orbital resonance' between Neptune and Pluto has persisted since the beginning of the solar system (e.g. Sussman and Wisdom, 1988), the simple fact that Pluto has not shared the fate of the multitudes of planetesimals originally populating this region of the solar system attests to this probability.

Pluto may not be the sole surviving remnant planetesimal. It is likely that Neptune's relatively large, retrograde satellite Triton represents one of these rare early solar system relicts as well. The retrograde orbit of this satellite is regarded by many astronomers as direct evidence of capture by Neptune from heliocentric orbit, possibly during a period when the planet possessed an extended atmosphere capable of slowing and capturing any object that dared venture too closely, or alternatively, later by a collision of solar-orbiting Triton with an indigenous, regular satellite of Neptune. If Triton were indeed captured from a heliocentric, outer solar system orbit, Voyager 2 observations of this satellite may represent the next best thing to a similar mission to Pluto (a possibility currently under consideration by NASA). Measurements obtained during the Voyager 2 rendezvous with Neptune, in fact, confirmed predictions that Triton would prove to be a Pluto-like object. Comparisons cannot be too strict, however, because Triton almost certainly experienced a traumatic period of severe heating associated with post-capture tidal evolution. Finally, the recent discovery of numerous $\simeq$ 100–400 km diameter objects orbiting at $\simeq$ 30–50 AU may represent additional remnant planetesimals (Jewitt and Luu, 1995).

Steve Mueller and William B. McKinnon

Bibliography

Binzel, R.P., Tholen, D.J., Tedesco, E.F. *et al.* (1985) The detection of eclipses in the Pluto–Charon system. *Science*, **228**, 1193–5.

Buie, M., Tholen, D.J. and Horne, K. (1992) Albedo maps of Pluto and Charon: initial mutual events results. *Icarus*, **97**, 211–27.

Buie, M.W., Cruikshank, D.P., Lebofsky, L.A. and Tedesco, E.F. (1987) Water frost on Charon. *Nature*, **329**, 522–3.

Cruikshank, D.P., Pilcher, C.B. and Morrison, D. (1976) Pluto: evidence for methane frost. *Science*, **194**, 835–7.

Dobrovolskis, A.R. (1989) Dynamics of Pluto and Charon. *Geophys. Res. Lett.*, **16**, 1217–20.

Elliot, J.L., and Young, L.A. (1992) Analysis of stellar occultation data for planetary atmospheres. I. Model fitting with applications to Pluto. *Astron. J.*, **103**, 991–1015.

Elliot, J.L., Dunham, E.W., Bosh, A.S. *et al.* (1989) Pluto's atmosphere. *Icarus*, **77**, 148–70.

Fink, U., and DiSanti, M.A. (1988) The separate spectra of Pluto and its satellite Charon. *Astron. J.*, **95**, 229–36.

Holman, M.J. and Wisdom, J. (1993) Dynamical stability in the outer solar system and the delivery of short period comets. *Astron. J.*, **105**, 1987–99.

Jewitt, D. and Luu, J.X. (1995), The solar system beyond Neptune. *Astron. J.*, **109**, 1867–76.

Marcialis, R.L., Rieke, G.H., and Lebofsky, L.A. (1987) The surface composition of Charon: tentative identification of water ice. *Science*, **237**, 1349–51.

McKinnon, W.B. (1989) On the origin of the Pluto–Charon binary. *Astrophys. J.*, **344**, L41–4.

McKinnon, W.B. and Mueller, S. (1988) Pluto's structure and composition suggest origin in the solar, not a planetary, nebula. *Nature*, **335**, 240–3.

McKinnon, W.B., Simonelli, D.P. and Schubert, G. (1996) Composition, internal structure, and thermal evolution of Pluto and Charon. In *Pluto and Charon* (eds S.A. Stern and D.J. Tholen). Tucson: University of Arizona Press.

Owen, T.C., Roush, T.L. Cruikshank, D.P. *et al.* (1993) Surface ices and atmospheric composition of Pluto. *Science*, **261**, 745–8.

Simonelli, D.P., Pollack, J.B. McKay, C.P. *et al.* (1989) The carbon budget in the outer solar nebula. *Icarus*, **82**, 1–35.

Standish, E.M. (1993), Planet X: no dynamical evidence in the optical observations. *Astron. J.*, **105**, 2000–6.

Stern, S.A. (1992) The Pluto–Charon system. *Ann. Rev. Astron. Astrophys.*, **30**, 185–233.

Sussman, G.J. and Wisdom J. (1988) Numerical evidence that the motion of Pluto is chaotic. *Science*, **241**, 433–7.

Walker, M.F. and Hardie, R. (1955) A photometric determination of the rotational period of Pluto. *Publ. Astron. Soc. Pacific*, **67**, 224–31.

Cross references

Charon
Occultation
Planetesimal
Spectrophotometry
Tidal friction
Triton

POLAR CAP

A polar cap is a surface deposit of frost, ice or snow on higher (i.e. polar) latitudes of a planet or moon. The condensate forms from volatiles present in the atmosphere of the body in question. On Earth and Mars the polar caps wax and wane in seasonal cycles which are dependent upon the changes in solar heating of high-latitude regions on these bodies. On other solar system bodies the temporal variations of the polar caps are less well understood. Mercury may have icy deposits on the cold night side of the planet. The planet Pluto shows a 'polar brightening' (see Pluto), as does the Jovian satellite Ganymede. Neptune's satellite Triton (q.v.) has an icy polar cap, with dark streaks interpreted to be the result of geysers.

The chemistry of this condensate on Earth is essentially H_2O, water ice and snow; in the case of Mars it is mainly a CO_2 snow or 'dry ice' (with a small amount of water and ammonia ices). On Ganymede is could be impact-vaporized H_2O that migrates to the polar regions. The polar cap in each case is white, providing a high albedo, but whereas on Mars it may only be about 10 cm thick, on Earth it can be several meters thick. The Martian polar cap, then about 1000 km wide, was first imaged at close range by NASA's Mariner 9 in 1972. Condensation occurs in winter when the temperature falls to 150 K, but sublimates in spring; about 20% of the atmosphere is involved in this annual exchange. The Martian south pole is brighter than the north pole, which may be due to an underlying stratum of water ice.

The area of seasonally low or negligible radiation (darkness throughout the day) on any given planet is largely a question of the ecliptic angle and the latitude of the Arctic or Antarctic Circle (66° 31', in the case of Earth; about 65° for Mars). While the seasonal cycle on Earth is 12 months (365 d), on Mars it is nearly double that figure, 687 d. During the sunlit season on Mars the temperature range is from about 27°C at noon to −73°C at midnight.

The precise limit of the ice will be subject to variations in altitude and, on the Earth, the distribution of land and sea. The polar regions of the Earth are curiously asymmetric; most of the northern hemisphere sector being water, the Arctic Ocean, whereas most of the southern hemisphere sector is land, the Antarctic continent. The Earth's considerable topographic relief (over 8000 m above sea level to over 10 000 m below it) means that the cryosphere, or lower boundary of regular freezing, extends in both hemispheres, on isolated mountain peaks, all the way to the equator. Because of geographic variables (relief, insolation, ocean currents, etc.) the Earth's cryosphere boundary at sea level varies in winter from about 55 to 65° (exceptionally 70°, north of Norway), but in summer from 60 to 80°. Perennial ice, notably in Antarctica, Greenland, Iceland, Spitzbergen and the Canadian Arctic archipelago, may have been accumulating up to about 0.5 million years; ice-cap drilling discloses at least 200 000 years of ice layers, seasonally accumulated. Perennial ice is also found in mountain glaciers at high elevations, even to the equator (above about 500 m).

Present-day polar cap conditions on Earth and Mars do not necessarily apply to long-term changes which are known or predictable from the evidence of large orbital variations (eccentricity, obliquity and precession cycles, ranging on Earth from about 20 000 to 100 000 years (see Milankovitch, Milutin, and Milankovitch theory; Mars: atmosphere). Because of much larger long-term orbital variations in the case of Mars, it is speculated that at times there was liquid water at the surface (probably now buried as permafrost), and at such times there would be H_2O snow and ice in its polar caps.

Rhodes W. Fairbridge

Bibliography

James, P.B., Kieffer, H.H. and Paige, D. (1992). The seasonal cycle of carbon dioxide on Mars, in *Mars* (eds H. Kieffer, B. Jakosky, C. Snyder, and M. Matthews), Tucson: University of Arizona Press.

Oliver, J.E. and Fairbridge, R.W. (eds) (1987) *The Encyclopedia of Climatology*. New York: Van Nostrand Reinhold, 986 pp.

Paige, D.A. and Ingersoll, A.P. (1985). Annual heat budget of Martian polar caps: Viking observations. *Science*, **228** 1160–8.

Paterson, W.S.B. (1969). *The Physics of Glaciers*. Oxford: Pergamon Press, 250pp.

Slade, M.A., Butler, B.J., and Muhleman, D.O. (1992). Mercury radar imaging: evidence for polar ice. *Science*, **258**, 635–640.

Trafton, L. (1984). Large seasonal variations in Triton's atmosphere. *Icarus*, **58**, 312–24.

Cross references

Ice age
Mars: atmosphere
Mercury: atmosphere
Milankovitch, Milutin, and Milankovitch theory
Pluto
Triton

POLARIMETRY

Optical polarimetry of planetary bodies: principles

When unpolarized light is scattered by a rough surface or by a cloud of gas, particles, or crystals, it becomes partially linearly polarized. The plane of the polarization is usually found to be either normal to the plane containing the incident and observation rays, or it is parallel to this plane. If the scattered light has intensities I_1 and I_2, polarized in the planes normal to and parallel to the plane of the rays, the degree of polarization is defined as $P = (I_1 - I_2)/(I_1 + I_2)$. P is usually found to vary with the angle V between the incident and observation rays, usually known in astronomy as the phase angle.

Optical polarimetry has been extensively used in telescopic astronomy for investigations of planetary bodies (Lyot, 1929; Dollfus, 1957, 1961, 1985, 1990, 1992; Coffeen and Hansen, 1974; Gehrels, 1974; Geake and Dollfus, 1985; Steigmann, 1988). For objects without atmospheres – such as Mercury, the Moon, asteroids, the Galilean satellites, and the Saturnian rings and small satellites – optical polarimetry allows characterization of the solid surface. For planets with atmospheres – such as Venus, Jupiter and Saturn – polarimetry allows characterization of the turbidity of the gaseous atmosphere, the aerosols, the clouds and their behavior. Mars is an intermediate case in which clouds or hazes are usually partly superimposed over a solid surface. Investigations of comets center on the physics of the dust grains which are released by the nucleus.

Techniques

Polarimetry requires measurement accuracies which are ten times stricter than for conventional photometry. The visual fringe polarimeter (Lyot, 1929) produces a system of parallel interference fringes, which are seen at the surface of the planet where the light is polarized. They are matched visually by producing an opposite

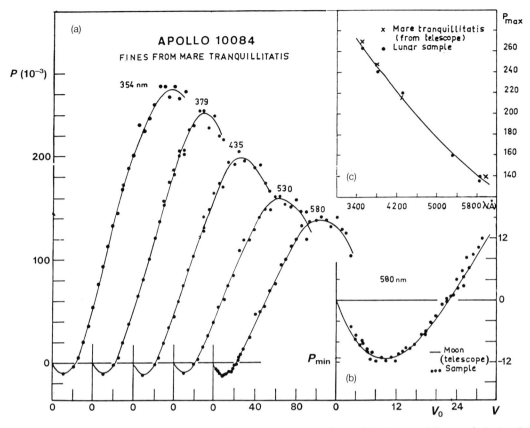

Figure P48 Polarization for the Mare Tranquillitatis region of the Moon. (a) Dependence of the degree of linear polarization P (expressed in units of 10^{-3}) on phase angle V for five wavelengths. (b) Enlargement of the negative branch of the 580-nm wavelength curve, illustrating the definition of parameters P_{min}, V_0 and h. (c) Wavelength dependence of the maximum degree of polarization P_{max}.

polarization with an optical compensator in order to cancel the polarization. The instrument enables local compensation on small areas over a planetary disk and is well adapted for the study of regional polarization.

Photoelectric polarimeters have been used in planetary studies since 1963. Because photomultipliers have a large range of spectral sensitivity, they allow measurement from the infrared to the ultraviolet. They integrate the light through a hole and are suitable for analyzing faint or star-like sources such as asteroids, planetary satellites and comets.

Since 1990 CCD video polarimetry has produced images which are formed exclusively from the part of the light which is polarized. Thus two-dimensional display of polarization parameters over planetary surfaces is now possible.

Some spacecraft have also been equipped with photopolarimeters. Venus has been studied by Pioneer Venus, Mars has been analyzed polarimetrically by Mars 5 and Jupiter's polarization has been measured during the Pioneer flyby missions.

Rough surface polarization

Light rays striking a solid surface may be reflected, refracted, diffracted, scattered or absorbed. Usually one ray is involved in several of these processes during an encounter with a surface. Because the processes are polarizing, the light which results from the interaction is polarized in a way which depends on the nature of the surface.

Polarization curves (Figure P48) express the dependence of degree of linear polarization on the phase angle. There is also a wavelength dependence. Note that the degree of polarization P is often negative – an inversion – for small phase angles V before the polarization increases to a maximum for larger angles. Such polarization curves can be reduced to a small number of characteristic parameters: the polarization minimum P_{min}, the inversion angle V_0, the slope at the inversion point h and the maximum polarization P_{max}. Together with the albedo A, these parameters have been found to be diagnostic of the texture and of some of the physical properties of a surface that can be studied remotely.

Laboratory measurements on a large variety of rocks, meteorites and lunar samples in both solid and pulverized form (Geake, Geake and Zellner, 1984; Dollfus 1985; Geake and Dollfus, 1986) have led to the establishment of empirical relationships among these polarimetric parameters. These relationships can be understood as follows:

P_{min} and V_0 are related to surface texture as shown in Figure P49. Bare chips, large fragments and rocks are confined to the specific region designated 'I' in Figure P48a. Very finely divided siliceous powders and lunar fines cluster in the distinct region 'II' of the same figure. Particles pulverized at sizes 60–300 microns fall between the other two domains (Figure P48b). The clear discrimination on the basis of P_{min} and V_0 allows remote inference of microtextures at the surface of planetary bodies.

The slope h of the polarization curve can be related directly to the albedo A, as shown in Figure P50, irrespective of the surface. This relationship allows accurate determination of the surface albedo of a planetary object even if it is observed as a point source. The method has been used extensively for asteroids.

The maximum polarization P_{max} may be related to the size and compaction of the grains forming the planetary surface, as has been described by Geake and Dollfus (1986).

Soil physics on planetary surfaces

Interpretation and simulation of polarization curves (Dollfus, 1962; Dollfus and Bowell, 1971; Geake and Dollfus, 1986; Shkuratov and Opanasenko, 1992), such as those of Figure P48, permitted prediction of the nature of the lunar surface 15 years before manned exploration (Dollfus, 1957). The soil was found to be covered by a layer of very small rock fragments, tens of mm in size, with a basaltic composition. The model was used for technical design of the Apollo lander and was then confirmed *in situ*. Now, telescopic CCD video

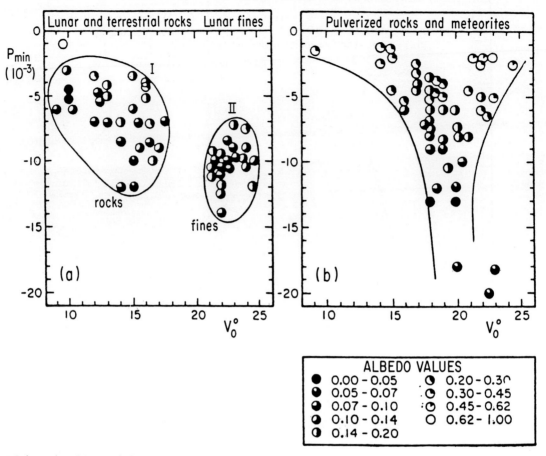

Figure P49 Relationships between polarization parameters P_{min} and V_0 for powdered surfaces with different grain sizes and albedos. (a) Relationships for the largest and smallest grain sizes. Domain I corresponds to blocks and fragments of at least 1 mm in diameter; domain II is for samples of lunar fines, with average grain size of about 40 μm (b) Relationship for intermediate particle sizes (between 60 and 300 μm).

polarimetry allows remote mapping of grain size on the lunar surface (Dollfus, 1990).

A similar finely comminuted surface has also been inferred on Mercury and Mars, but with a different composition. For asteroids (Bowell and Zellner, 1974; Dollfus and Zellner, 1979; Dollfus *et al.*, 1979, 1989), polarization indicates larger grain sizes. For Vesta, at 550 km in diameter the largest of the asteroids with a siliceous composition, polarimetry indicates a regolith made up of a broad mixture of particles larger than 50 mm, partially coated with smaller particles Lebertre and Zellner (1980).

The overall analysis of these telescopic polarimetry results has demonstrated that all planetary bodies in the solar system without atmospheres are covered by a layer of small grains and debris. This texture, a regolith of fines, is produced by meteoritic impacts which create a layer of ejected fragments that are periodically 'gardened' by new impacts. For objects of lunar size and average impact velocities of several tens of kilometers per second, the explosions produce an accumulation of small grains plus some liquid droplets of silicates which solidify and break to produce dark fragments of 'agglutinates' intermixed with the regolith grains. Solar proton bombardment also darkens the regolith generally.

On asteroids the depletion of small grains (compared with the Moon) is explained by the difference in escape velocities, which are at least ten times smaller for the asteroids. The average impact velocity, which is about 5 km s^{-1} in the asteroid belt, typically releases 15–20 times less energy at impact than for the Moon and is not sufficient to melt the ejected fragments; there are no agglutinates.

From all these polarimetry results, we see that the formation of a surface regolith by meteoric bombardment is a ubiquitous process when there is no atmosphere. All bodies larger than 1500 km retain the impact ejecta and build up a thick layer of very small fines over the whole surface. These fines, in turn, are subjected to subsequent impact and radiation from space. For the smaller objects (such as asteroids) the smallest grains are ejected into space and are lost; a coarser regolith remains.

Surface of Mars

In addition to comminution by meteoric bombardment, soil on Mars is subject to tectonic and volcanic activity. It is also influenced by permafrost under the surface and by interactions with the planet's atmosphere, including wind (Dollfus, *et al.*, 1983, 1992; Lee, *et al.*, 1990; Ebisawa and Dollfus, 1992).

Earth-based polarimetry observations are limited to phase angles of less than 45°; the corresponding segment of the polarization curve is not diagnostic of grain size variation. To extend measurements to smaller areas and to larger phase angles, Soviet–French photopolarimeters (VPM, or Visual Polarimeter Mars) were placed on board Mars 5, which orbited the planet in 1974. At a scale of a few tens of kilometers and despite the great variety of Martian terrains and geomorphologic units examined, the surface of Mars appeared almost everywhere as a comminuted soil. The planum-type terrains and dark-hued surfaces have non-identical textures. The more 'tectonized' areas – which include fossae, rupes or graben-types features – often have coarser soils with grain sizes in excess of 100 μm. The rims of craters sometimes produce an enhanced polarization signature, either because the soil has a larger grain size or because of the presence of rocks which are incompletely covered by dust.

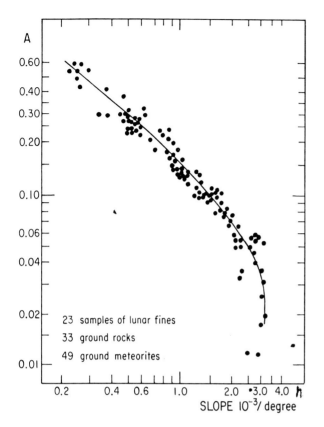

Figure P50 Relationship between the slope h of the polarization curve and the albedo. Data are from laboratory measurements on a variety of lunar samples, meteorites and terrestrial and pulverized terrestrial samples.

Cometary physics and dust grains

Polarimeters with filters which select radiation at specific emission wavelengths are used to understand the physics of gas released by cometary nuclei. Analysis of the continuum light, on the other hand, yields the properties of the dust ejected from the nucleus (Dobrolovsky et al., 1986).

Usually the coma which surrounds a cometary nucleus exhibits, in the continuum light, a variation of polarization with phase angle which is reminiscent of the polarization from rough surfaces. The interpretation is that this light comes from large fluffy flakes – an aggregation of micron-sized very dark grains. These large flakes have to be intermixed with a population of submicron-sized grains, such as those which have been sensed in situ by the spacecraft Vega and Giotto. The cataclysmic event which disrupted the nucleus of comet West in 1976 disturbed completely the polarization in favor of small grains.

Variations have been detected in polarization signatures with angular distance from the nucleus of comets. The variation has been interpreted in terms of evolution of the grains after release from the nucleus. Apparently there is first a mixing with ice (which vaporizes quickly), then a slower evolution over several hours which suggests vaporization of another less-volatile component.

After release from the nucleus, the smallest grains are subjected to radiation pressure from the Sun and are driven to the comet's tail. Characterization of these grains in the tail by ground-based instruments has been very difficult because of corrections needed for twilight and sky brightness effects. An Earth-orbiting telescope equipped with a CCD video polarimeter has been proposed as being particularly relevant to solving problems of planetary physics with polarimetry.

Mists, clouds and storms in planetary atmospheres

For Venus, which is cloud covered, the degree of linear polarization has been extensively studied as a function of phase angle and wavelength with ground-based telescopes and with the Venus-orbiting Pioneer spacecraft (Hansen and Hovenier, 1974; Kawabata et al., 1974; Santer and Dollfus, 1980a, b). Interpretations, based upon Mie theory, imply a high-altitude layer of transparent spherical particles of radius 1.05 μm with refraction index 1.44. Sulfuric acid has been assumed to be the main contributor. Some temporary departures from this model have been observed, lasting for several months. These have been related to the effect of condensates tentatively attributed to the effects of volcanic activity on Venus' surface.

Ground-based and Pioneer Venus polarization mapping over the planetary disk have allowed derivation of a vertical distribution for aerosols in the upper Venusian atmosphere. Variations occur at all time scales, reflecting an active atmospheric circulation and several meteorological processes at work in the atmosphere.

On Jupiter the upper atmosphere contains a dense layer of ammonia crystal clouds, of which the top layers can be directly observed by telescope. The degree of linear polarization, its distribution over the disk and its variation with wavelength have been analyzed with telescopes and onboard the Pioneer spacecraft which flew past Jupiter (Mischenko, 1990). The polarization is essentially concentrated around the two poles; its direction, or azimuth, is nearly perpendicular to the limb. The crystal clouds are less dense around the poles, and multiple scattering occurs between the molecules of the overlying atmosphere. The effect is variable with time over both poles.

Saturn's rings are subject to transient polarization anomalies producing deflections in the polarization azimuth. The anomalies occur unpredictably and persist for several days or weeks. They may result from the antagonistic and variable effects of alignment and randomization of small grains in the rings.

On Mars the clean gaseous atmosphere produces negligible polarization. However, aerosols, dust and crystals in the atmosphere do produce disturbances. Very faint hazes can be localized even when they escape detection with direct imaging. Crystal clouds increase the degree of polarization but dust veils produce the opposite effect; the nature of the haze may thus be immediately discerned from the degree of polarization observed.

An extensive survey of polarization by clouds in the Martian atmosphere has been conducted in France, Greece and Japan over several decades (Ebisawa and Dollfus, 1992). A description of the seasonal behavior has been derived for the crystal clouds, which are interpreted as small cirrus-like crystals made of water ice, except over the north pole in winter, where CO_2 ice is indicated (Lee et al., 1990). Dust clouds are occasionally raised in the atmosphere, which were sensed by polarimetry in terms of grain sizes, increasing from submicron to millimeter diameters, depending on the lifetime in the atmosphere. Recently, with the availability of CCD video polarimetry, the degree of polarization has been imaged over the Martian disk for both types of cloud.

Audouin Dollfus

Bibliography

Bowell, E. and Zellner, B. (1974) Polarization of asteroids and satellites, in *Planets, Stars and Nebulae Studies with Photopolarimetry* (ed. T. Gehrels) Tucson: University of Arizona Press, pp. 381–404.

Coffeen, D.L. and Hansen, J.E. (1974) Polarization studies of planetary atmospheres, in *Planets, stars and nebulae studied with photopolarimetry* (ed. T. Gehrels). Tucson: University Arizona Press, pp. 518–81.

Dobrolovsky, O.V., Kiselev, N.N. and Chernova, G.P. (1986) Polarimetry of comets, a review. *Earth, Moon, Planets*, **34**, 18.

Dollfus, A. (1957) Etude des planètes par la polarisation de leur lumière. Thèse, Paris and *Ann. Astrophys. Suppl.*, **4**. English Translation NASA-TTF **188**.

Dollfus, A. (1961) Polarization studies of planets, in *Planets and Satellites III* (eds G.P. Kuiper and B.B. Middlehurst). Chicago: University of Chicago Press, p. 343.

Dollfus, A. (1962) The polarization of moonlight, in *Physics and Astronomy of the Moon* (ed. Z. Kopal). Academic Press.

Dollfus, A. (1984), The Saturn's ring particles from optical reflectance polarimetry, in Anneaux de Planètes – Planetary Rings (ed. A. Brahic) France: Publ. CNRS, pp. 121–45.

Dollfus, A. (1985) Photopolarimetric sensing of planetary surfaces. *Adv. Space Res.*, **5**, 47–56.

Dollfus, A. (1988) Micro-texture of the surface of satellite Callisto: implications about the formation and evolution of icy satellites, in *The Physics of Planets* (ed. S.K. Runcorn). John Wiley and Sons, pp. 133–40.

Dollfus, A. (1990) Une nouvelle méthode d'analyse polarimétrique des surfaces planétaires. *C. R. Acad. Sci.* Paris 311, 119.

Dollfus, A. (1992) Planetary investigation by polarimeter. *Adv. Space Res.*, **12**, 167–75.

Dollfus, A. and Bowell, E. (1971) Polarimetric properties of the lunar surface and its interpretation. Part I – telescopic observations. *Astron. Astrophys.*, **10**, 29–53.

Dollfus, A. and Geake, J.E. (1977) Polarimetric and photometric studies of lunar samples. *Phil. Trans. Roy. Soc. London*, **A285**, 397–402.

Dollfus, A. and Zellner, B. (1979) Optical polarimetry of asteroids and laboratory samples, in *Asteroids* (ed. T. Gehrels). Tucson: University Arizona Press, pp. 170–83.

Dollfus, A., Deschamps, M. and Ksanfomaliti, L.V. (1983) The surface texture of the Martian soil from the Soviet spacecraft Mars-5 photopolarimeter. *Astron. Astrophys.*, **123**, 225–37.

Dollfus, A., Deschamps, M. and Zimbelman, J.R. (1992) Soil texture and granulometry at the surface of Mars. *J. Geoph. Res, Planets*, **98**, 3413–29.

Dollfus, A., Mandeville, J-C. and Duseaux, M. (1979) The nature of the M-type asteroids from optical polarimetry. *Icarus*, **37**, 124–32.

Dollfus, A., Bastien, P., Leborgne, J.-F. *et al.* (1988) Optical polarimetry of P/Halley: synthesis of the measurements in the continuum. *Astron. Astrophys*, **206**, 348–56.

Dollfus, A., Wolff, M., Geake, J.E. *et al.* (1989) Photopolarimetry of asteroids, in *Asteroids II* (eds P. Binzel and T. Gehrels). Tucson: University of Arizona Press, pp. 594–616.

Ebisawa, S. Dollfus, A. (1992) Dust in the Martian atmosphere: polarimetric sensing. *Astron. Astrophys.*, **272**, 671–87.

Geake, J.E. and Dollfus, A. (1986) Planetary surface texture and albedo from parameter plots of optical polarization data. *Mon. Not. Roy. Astron. Soc.*, **218**, 75–91.

Geake, J.E., Geake, M. and Zellner, B. (1984) Experiment to test theoretical models of the polarization of light by rough surfaces. *Mon. Not. Roy. Astron. Soc.*, **210**, 89–112.

Gehrels, T. (1974) Introduction and overview, in *Planets, Stars and Nebulae Studied with Photopolarimetry* (ed. T. Gehrels) Tucson: University of Arizona Press, pp. 3–44.

Hansen, J.E. and Hovenier, J.W. (1974) Interpretation of the polarization of Venus. *J. Atmos. Sci.*, **31**, 1137–60.

Kawabata, K., Coffeen, D.L., Hansen, J.E. *et al.* (1974) Cloud and haze properties from Pioneer Venus polarimetry. *J. Geophys. Res.*, **85**, 8129.

Lebertre, T. and Zellner, B. (1980) Surface texture of Vesta from optical polarimetry. *Icarus*, **43**, 172–80.

Lee, P., Ebisawa, S. and Dollfus, A. (1990) Crystal clouds in the Martian atmosphere. *Astron. Astrophys.*, **240**, 520–32.

Lyot, B. (1929) Recherches sur la polarisation de la lumière des planètes et de quelques substances terrestres. Thèse Paris, and *Ann. Observatoire de Meudon*, 8, fasc. 1. English Translation NASA-TTF **187**.

Mischenko, M.I. (1990) Physical properties of the upper tropospheric aerosols in the equatorial region of Jupiter. *Icarus*, **84**, 296–304.

Santer, R. and Dollfus, A. (1980a) Wavelength dependence of polarization: XXXIX – Venus UV cloud model from polarimetry, *Astron. J.*, **85**, 564.

Santer, R. and Dollfus, A. (1980b) Venus upper atmosphere aerosols from polarimetry. *Astron. J.*, **85**, 751.

Shkuratov, Yu. G. and Opanasenko, N.V. (1992) Polarimetric and photometric properties of the Moon: telescopic observations and laboratory simulations. 1 – 'The negative polarization'. *Icarus* **95**, 283–99; 2 – 'The positive polarization'. *Icarus* **99**, 468–84.

Steigmann, G.A. 1988, Optical polarimetry of the surface microstructure of airless planetary bodies. Part I: astronomical data, *J. Br. Astron. Assoc.*, **98**, 106–110; Part II: laboratory investigation, 140–145; Part III: theoretical models, 205–208.

Cross references

Comet
Jupiter: atmosphere
Moon: geology
Pioneer 10 and 11 missions
Pioneer Venus missions
Saturn: ring system

POLARITY REVERSALS

Early 20th-century researchers postulated that the polarity of Earth's magnetic field had reversed in the geological past. The evidence came from measurements of the remanent magnetization of igneous rocks. Skeptics were reluctant to accept the observations as a geomagnetic feature and attributed the observed anomalous polarities to a magnetomineralogical self-reversal process, which is indeed possible in some iron oxide minerals. However, by the late 1960s data from several different sources – radiometrically dated lavas, long continuous cores of deep sea sediment and the interpretation of lineated marine magnetic anomalies – established beyond doubt that geomagnetic polarity reversals have occurred. A detailed well-dated reversal history for the last 5 Ma was developed. The marine magnetic record was established and subsequently extended and refined. It has been confirmed and dated by magnetostratigraphy and now serves as the optimum record of geomagnetic polarity during the last 160 Ma (see Geomagnetic polarity reversals and the geological record). The long-term history of reversals in this time interval is well established, but the behavior of the geomagnetic field during a polarity reversal is a matter of controversy, and the mechanism that causes a reversal is unknown.

Geomagnetic field behavior during reversals

Many observations in igneous rocks and sediments have demonstrated that the total intensity of the field decreases during the polarity transition (Figure P51). Opinion is divided about whether the intensity decrease has the same duration as the time needed for the polarity transition, variously estimated to last about 4500 ± 1000 years (Prévot *et al.*, 1985). Some studies have suggested that the intensity decrease starts before the directional change begins and that it has a longer duration, but this behavior is not supported by all transitional records.

A matter of greater contention is the configuration of the geomagnetic field during a polarity reversal. The transitional field geometry is described in terms of the spatial geometries of individual terms in the spherical harmonic expansion of its potential. As recorded at the surface of the Earth, the paleomagnetically important axial dipole component of the field reverses sign, but it is uncertain what happens to the equatorial dipole component and to the non-dipole components while the reversal is taking place. The relative importance of these terms changes with distance from their sources. At the surface of the Earth the dipole terms are dominant, but closer to the source of the field in the Earth's liquid outer core higher-order terms increase in importance.

The paleomagnetic evidence about transitional behavior is derived in part from igneous rocks but most of the data are from sediments. The paleomagnetic record of a reversal (Fig. P50) recorded in a Miocene lava sequence (age about 15.5 Ma) shows that the field directions vary progressively from a dipolar direction before a reversal to the antipodal dipolar direction after the transition (Prévot *et al.*, 1985). A more common portrayal of directional behavior uses the location of the virtual geomagnetic pole (VGP). For transitional directions the VGP is calculated as if the field remains dipolar and the dipole axis changes orientation during the transition. Whether or not this is the case, the method allows data from different sites to be combined in a common reference framework. Attempts have been made to model transitional field behavior with axially symmetric (zonal) quadrupole and octupole fields that dominate while the axial dipole is diminished and, by developing at different rates in the northern and southern hemispheres, control the reversal process

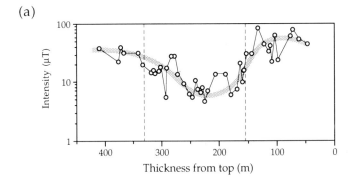

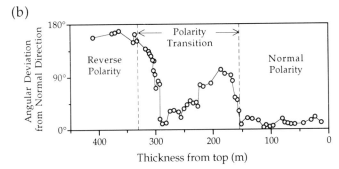

Figure P51 Geomagnetic field behaviour during a Miocene polarity transition recorded in a continuously extruded lava sequence at Steen's Mountain. (a) Intensity of the paleomagnetic field; the continuous gray curve is a visual fit to emphasize the intensity fluctuation. (b) Angular deviation of the transitional direction from the normal axial dipole field direction. (After Prévot et al., 1985.)

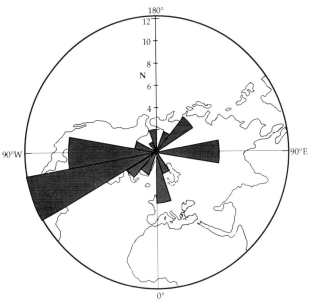

Figure P52 Rose diagram of the mean longitudes of transitional paths in 48 records of reversals and short events younger than about 12 Ma from several sampling sites. The radius of each segment corresponds to the number of transitions in a 20° longitudinal band. (After Tric et al., 1991.)

(Fuller, Williams and Hoffman, 1979). The accumulation of data from many reversals now indicates that the transitional field geometry has a non-zonal symmetry.

Geometry of transitional paths

The field configuration has been investigated by plotting the transitional paths of the VGP recorded for the same reversal at several sites. Consistent VGP transitional paths would argue in favor of a non-axial dipolar geometry during the field reversal. Analysis of the secular variation of the geomagnetic field during the past 5 Ma reveals that the time-averaged field has a non-zonal bias, and simulation of a reversal of the present non-axial dipole field would indeed give a transitional path over the Americas like that observed in the paleomagnetic data (Constable, 1992).

Unfortunately, this matter cannot yet be settled unambiguously. The most extensively studied transitions are younger than 12 Ma. As well as the transitions that bound well-established polarity chrons, this data base also includes transitions that bound very short events not found in all reversal records. Plots of the transitional paths show a striking preference for a broad band over the Americas and an antipodal band (Fig. P52). This evidence may indicate that the dipole is still dominant during the reversal (Laj et al., 1991). Other investigators have questioned the statistical significance of the preferred longitudinal paths and have postulated that the transitional field is dominated by higher-order non-dipolar components with a non-axially symmetric geometry (Valet et al., 1992). However, more rigorous analysis of the longitudes of equatorial crossings has indicated that the preference is statistically significant.

There remain several other objections to the hypothesis of preferred transitional paths. Some transitional paths are described by few intermediate directions or are quite ragged, so that the longitude of each equatorial crossing is not stringently confined. Moreover, the number of reversals yet studied is rather small for statistical analysis. Most seriously, the geographic distribution of the sampling sites is uneven, being biased to the northern hemisphere and unequally distributed in longitude. Most of the data come from sedimentary records. Consequently, the preferred paths have been interpreted skeptically as an artefact of smoothing of the paleomagnetic records during the slow acquisition of remanent magnetization in a sediment (Langereis, van Hoof and Rochette, 1992). The definite resolution of the geometry of the transitional field may require many detailed studies of individual reversals, each documented by data from a widespread distribution of sites.

The longitudinal bands preferred by transitional paths overlie regions of the core in which north–south fluid flow in the outer core is interpreted in models, assuming the frozen flux approximation. The preferred bands also overlie regions of anomalously high seismic velocities in the lower mantle that may reflect reduced temperatures. These coincidences led to the suggestion that core–mantle coupling may account for the apparent confinement of the transitional paths (Laj et al., 1991).

Cause of polarity reversals

The cause of geomagnetic polarity reversals is even less well understood than the transitional geometries. This is perhaps not surprising, since the origin of the main geomagnetic field is still inadequately understood. Although theorists agree that a magnetohydrodynamic dynamo process is necessary for generation of the main field, there is not unanimous agreement on the nature of the dynamo process. As a result, there are several views on what may cause polarity reversals.

An interesting speculation is that the high-speed impact of an extraterrestrial body, such as a meteorite or comet, might perturb the otherwise stable geodynamo, triggering a reversal (Muller and Morris, 1986). The correspondence of polarity reversals with the stratigraphic occurrence of microtektites produced in the impact might lend support to this idea, but Schneider and Kent (1990) showed that coincidence has not been clearly demonstrated. Moreover, there are comparatively few documented large meteoritic impacts, although there are numerous known reversals.

Another model that relies on disruption of the steady state dynamo has been proffered by McFadden and Merrill (1984). The main source of energy that drives the geodynamo is thought on energetic grounds to result from the solidification of a pure iron inner core from the fluid of the outer core; the depleted core fluid has a lower density and its buoyancy causes compositional convection (Loper, 1989). McFadden and Merrill (1984) have suggested that instabilities in core convection may be triggered by an anomalously hot thermal rising from the inner core boundary, or by a cold thermal sinking from the core–mantle boundary.

Events at the core–mantle boundary appear to be the most popular suspects for triggering reversals. Recent speculation has linked the emission of mantle plumes from the seismic D″ layer just above the core–mantle boundary with the thermal state and convective activity of the outer core, and thus with reversal rate (Larson and Olson, 1991). However, apart from exceptionally high crustal production during the Cretaceous long normal interval (in which no reversals occurred), the correlation of the volumetric oceanic crustal (and by inference mantle plume) production with reversal frequency is poor.

Analysis of reversal frequency suggests that the reversal-triggering mechanism is distinct from the source of the stable geomagnetic field (McFadden and Merrill, 1984). The non-dipole and dipole fields are thought to originate from separate mechanisms. Whereas rotational shear and convection in the fluid core are required for generation of the main field, it has long been suspected that the topography and temperature of the core–mantle boundary play a role in the generation of the non-dipole field and its secular variation. Cox (1968) hypothesized that the randomly varying non-dipole field might trigger a reversal at a time of low intensity of the dipole field. The plausibility of this model is supported by investigations of the physical processes controlling the recent field. Analysis of secular variation from historical records shows that the geomagnetic field intensity is currently decreasing. The radial geomagnetic field computed at the surface of the core by downward continuation of the surface field displays prominent non-dipolar features (Gubbins, 1987). Some large standing anomalies comparable in magnitude to the dipole field are growing in intensity and could conceivably grow large enough to overwhelm the dipole component.

Cox's phenomenological model relates polarity reversals to a random renewal process. Each reversal is considered to be an independent event that occurs without influence from preceding reversals. Statistical analysis of the lengths of polarity intervals since the Late Cretaceous (about 80 Ma ago) suggests that polarity reversals may indeed be controlled by a random but non-stationary (Poisson) process (McFadden and Merrill, 1984).

William Lowrie

Bibliography

Constable, C. (1992) Link between geomagnetic reversal paths and secular variation of the field over the past 5 Myr. *Nature*, **358**, 230–4.
Cox, A. (1968) Lengths of geomagnetic polarity intervals. *J. Geophys. Res.*, **73**, 3247–60.
Fuller, M., Williams, I. and Hoffman, K.A. (1979) Paleomagnetic records of geomagnetic field reversals and the morphology of transitional fields. *Rev. Geophys. Space Phys.*, **17**, 179–203.
Gubbins, D. (1987) Mechanism for geomagnetic polarity reversals. *Nature*, **326**, 167–9.
Laj, C., Mazaud, A. Fuller, M. and Herrero-Bervera, E. (1991) Geomagnetic reversal paths. *Nature*, **351**, 447.
Langereis, C.G., van Hoof, A.A.M. and Rochette, P. (1992) Longitudinal confinement of geomagnetic reversal paths as a possible sedimentary artefact. *Nature*, **358**, 226–30.
Larson, R.L. and Olson, O. (1991) Mantle plumes control magnetic reversal frequency. *Earth Planet. Sci. Lett.*, **107**, 437–47.
Loper, D.E. (1989) Earth's core, in *The Encyclopedia of Solid Earth Geophysics* (ed. D.E. James) New York: Van Nostrand Reinhold, pp. 315–8.
McFadden, P.L. and Merrill, R.T. (1984) Lower mantle convection and geomagnetism, *J. Geophys. Res.*, **89**, 3354–62.
Muller, R.A. and Morris, D.E. (1986) Geomagnetic reversals from impacts on the earth. *Geophys. Res. Lett.*, **13**, 1177–80.
Prévot, M., Mankinen, E.A., Grommé C.S. and Coe, R.S. (1985) How the geomagnetic field vector reverses polarity. *Nature*, **316**, 230–4.
Schneider, D.A., and Kent, D.V. (1990) Ivory Coast microtektites and geomagnetic reversals. *Geophys. Res. Lett.*, **17**, 163–6.
Tric, E., Laj, C., Jéhanno, C. et al. (1991) High-resolution record of the Upper Olduvai transition from Po Valley (Italy) sediments: support for dipolar transition geometry. *Phys. Earth. Planet. Int.*, **65**, 319–36.
Valet, J.-P., Tucholka, P. Courtillot, V. and Meynadier, L. (1992) Paleomagnetic constraints on the geometry of the geomagnetic field during reversal. *Nature*, **356**, 400–7.

Cross references

Dynamo theory
Earth: magnetic field and magnetosphere
Geomagnetic polarity reversals and the geological record
Magnetism

POYNTING–ROBINSON DRAG

Poynting–Robinson drag (or effect) is a drag force on orbiting bodies due to a non-radial component of solar radiation pressure. Poynting–Robinson drag was discovered by Poynting in 1903 and thoroughly examined in 1937 by Robertson. Thus Poynting–Robertson drag is occasionally referred to as the 'ROBERTSON EFFECT', or simply 'P–R drag'.

Poynting–Robinson drag is the result of a small non-radial component of solar radiation pressure on an orbiting body. This non-radial component is a result of the finite speed of light, and is a result of a process similar to 'aberration' in astronomy. The relative velocity between a solar photon and an orbiting body is the vector sum of the body's orbital velocity and the photon's velocity (ignoring the relativistic factor). Thus, in the frame of reference of the body, the photon appears to arrive somewhat forward (along the direction of motion) of the purely radial direction. The result is a net drag force, since the photons preferentially arrive from the front, but are re-emitted uniformly, transferring the photon's momentum in the direction opposite to the direction of motion. This drag force reduces the orbital kinetic energy of the body, decreasing its semimajor axis, and causing it to spiral inward at a rate

$$\frac{da}{dt} = 2ka^{-1}$$

For a black body in orbit around the sun, $k = 3.5 \times 10^{-7}/\rho r$, where r is the particle radius in meters, ρ is the density in kg m^{-3}, and a is the semimajor axis in astronomical units. The orbital eccentricity also decreases.

This effect transports millimeter-sized particles from the asteroid belt to the vicinity of the Earth in 10^7 years, and 0.1 m-sized particles in 10^9 years (Öpik, 1951). As a result, it may be responsible for many of the meteors observed on Earth. Since the time required for particles to spiral in to the Sun due to Poynting–Robertson drag is much less than the lifetime of the solar system, centimeter sized particles must be continuously produced, probably by collisions in the asteroid belt or by disruption of comets, to maintain the population currently observed as meteors and as the zodiacal light.

Michael C. Nolan

Bibliography

Öpik, E.J. (1951) Collision probabilities with the planets and the distribution of interplanetary matter. *Proc. Roy. Ir. Acad.*, **54A**; 165–99. Also as *Contributions from the Armagh Observatory*, **6**; 165–199.
Poynting, J.H. (1903) *Phil. Trans. Roy. Soc.*, London **A202**, 525.
Robertson, H.P. (1937) Dynamical effects of radiation in the solar system. *Mon. Not. Roy. Astron. Soc.*, **97**(6), 423–38.

Cross references

Dust
Meteor, meteoroid

PRECESSION AND NUTATION

The astronomical influence of precession was discovered by Hipparchus about 125 BC. The accompanying phenomenon known as the precession of the equinoxes consists of a slow, westward movement of the vernal and autumnal equinoxes produced by the combined gravitational attractions of the Sun and Moon upon the equatorial bulge of the rotating Earth. As a result of this joint influence, the Earth's axis of rotation is made to depart from a fixed orientation in space and, like the conical circling action of a spent toy top, the mean celestial pole (i.e. the average direction of the axis of rotation of the Earth) describes an approximately circular path around the pole of the ecliptic. This precessional motion (from the Latin *praecessus*, 'to go before') progressively shifts the positions of the equinoxes, and also alters the orientation of the Earth's rotational axis with respect to the familiar pole star. Its influence has changed the axially aligned pole star from Alpha Draconis (Thuban) of the period around 8000 BC to the star Alpha Ursae Minoris (Polaris) of the present time. It will bring the pointing of the Earth's polar axis fairly close to the bright star Alpha Lyrae (Vega) some 12 000 years from now.

Dynamical causes

The Earth is spinning on its axis with an angular velocity of 15° per hour, equivalent to a linear velocity of 1 669 km h^{-1} at the equator, and 0 km h^{-1} at its poles. The equatorial bulge is a dynamic circumstance which, in common with that of all the planets and the Sun, was probably created and has existed since the early history of the Solar System as coalescing matter was displaced toward the equator by the greater centrifugal force in this region. Thus the Earth is, rather than a perfect sphere, an oblate spheroid, which with its axis of spin tilted with respect to the ecliptic, presents an element of asymmetric mass attraction in this equatorial bulge.

The Earth's equator is inclined at an angle of approximately 23.44° to the plane of the Sun's mean apparent annual motion in the sky (the ecliptic). The plane of the Moon's orbit is inclined, in turn, to the Earth's equatorial plane at an angle varying from 18.28° to 28.58°, depending upon whether the Moon's ascending node coincides, respectively, with the position of the autumnal or the vernal equinox. Accordingly, except for twice a year at the vernal or the autumnal equinox when the Sun crosses the equator, or twice a month when the Moon crosses the equator, there is a tendency for the gravitational attractions of both the Sun and Moon to pull the Earth's equatorial bulge into alignment with their respective planes and establish dynamic equilibrium. Since the Moon is much closer to the Earth than is the Sun, the Moon's precessional influence accounts for about 80% of the total effect. However, because the Moon's orbital inclination from the Sun's path in the ecliptic is only 5° 9', the separate forces acting may, for most practical purposes, be regarded as one, and the combined result is known as luni-solar precession.

Through this combined angular pull of the Moon and Sun upon the Earth's equatorial bulge, the Earth's axis of spin is also subjected to a precessional motion. The effective angle of combined gravitational force application is the most influential in producing precessional motion the more nearly it is perpendicular to the Earth's equatorial plane. The precession-inducing force – in pure mechanics normally thought of as applied directly at right angles to the axis of spin of the rotating body – is, in the case of the Earth, made up only of those gravitational force components of the Moon and Sun which are exerted at right angles to the Earth's axis of spin. These are treatable as geometric complements of those applied to the Earth's equatorial plane. In accordance with the gyroscopic principle governing a spinning object whose axis of spin is subjected to an applied torque (q.v.), the anticipated rotation from the applied torque does not occur. Instead, a third motion results, around an axis mutually perpendicular to the axis of spin and the axis of torque.

Because the inclination angle between the path of the Sun in the ecliptic and the equatorial plane of the Earth is 23.44°, the line of torque action exerted by the Sun's gravitational force on point A in Figure P53 likewise makes an angle of 23.44° with the equator and the equatorial bulge. For the purpose of demonstration of a common situation, the Moon is assumed to be at its maximum Sun-reinforcing position – crossing the plane of the ecliptic, as it does twice a month.

Astronomical results

In space, i.e. with reference to the 'fixed stars,' a conical precessional motion of the Earth's axis of rotation is produced (Figure P53), the radius of the cone being equal in arc measurement to this same angle of 23.44° between the mean pole of the equator (i.e. the long-term or

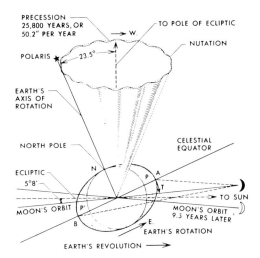

Figure P53 The Earth is represented by the circle in the lower portion of the figure, with the north pole at N. The Earth's equator is along the line from A to B, and the shaded regions represent the (greatly exaggerated) equatorial bulge of the Earth. Both the Sun and the Moon are positioned at the right of the figure, in the plane of the paper; the dashed line ('To Sun') represents the ecliptic, which is the plane of the Earth's motion around the Sun. The Moon is shown twice; the two positions represent the extremes in declination at one-half each nodal cycle. [In this cycle, the two points of intersection of the Moon's orbital plane with the ecliptic (the lunar nodes) slide, with a retrograde motion, along the ecliptic, taking 18.6134 years to complete a revolution.]

Both the Moon and the Sun tug on the mass in the equatorial bulge of the Earth. A component of the tangential force (T) attempts to align the bulge with the orbital planes of the Moon and Sun. However, as discussed in the text, the effect of this torque is instead to cause a revolution of the pole (N) about a line perpendicular to the ecliptic; this revolution (the precession) is depicted by the cone shown here. The small wave oscillations on the cone are the nutation, a nodding motion linked with the 18.6134-year revolution of the lunar nodes westward along the ecliptic.

secular average position of the Earth's axis of rotation) and the pole of the ecliptic. The Earth's polar axis tracing the sides of the cone similarly describes, subject to the additional effects of nutation (see below), a non-re-entrant quasi-ellipse of 23.44° mean 'radius' upon the celestial sphere. (In Figure P52 this elliptical figure is further foreshortened toward the viewer to show the wavelike deflections caused by nutation.) Matching conical figures are generated by the Earth's instantaneous (true) polar axis in both the northern and southern hemispheres.

The described motion of precession takes place around an axis which is mutually perpendicular to the plane of the ecliptic (and hence, approximately, the plane of the Moon's orbit) and the axis of torque – which passes through the center of the Earth and extends perpendicularly out toward the viewer in Figure P52. The Earth's axis of spin moves counterclockwise around the axis of precession as seen from the north pole of the ecliptic (i.e. in the same sense as the Earth's rotation around the pole of the equator).

The combined luni-solar gravitational components acting upon the Earth's equatorial bulge would seemingly tend to rotate this bulge around an axis joining the vernal and autumnal equinoxes. But, as seen above, the steady spinning of the Earth on its axis resists this applied rotation. The result of the substituted conical motion is a slow revolution of the mean poles of the equator around the poles of the ecliptic (here assumed to be fixed) in a period of about 25 694 years, the principal or luni-solar precession.

Several types of precession exist. Lunisolar precession shifts the two equinoxes steadily westward along the ecliptic, with an angular velocity of approximately 50.2" per year. In this motion the obliquity of the ecliptic remains unaltered. Planetary precession is a composite perturbing action, produced by ecliptic-orthogonal components of gravitational force of the major planets of the solar system, which strive to pull the Earth from its orbital plane into their individual planes of revolution around the Sun. The result is a clockwise rotation of the equinoxes (the points of intersection of the ecliptic with the celestial equator), and a shift eastward along the equator at an angular rate of about 0.11″ per year. Planetary precession also reduces the obliquity of the ecliptic by about 0.47″ a year which, in turn, causes a further variation in the luni-solar precession.

The vectorial sum – by no means constant – of the luni-solar and planetary precessions gives the general precession, having an average period of 25 870 years. All three quantities are thus known as secular or long-period variations. In addition, certain short-period variations exist within the precessional motion, inward and outward from the mean path (hence the term nutation or 'nodding').

Nutation

The progressive motions of the Earth's mean poles of rotation around the poles of the ecliptic have been seen to be responsible for the precession of the equinoxes. These motions also have, superimposed upon them, certain modulating, periodic variations caused by the continuously changing positions, alignments, declination angles and distances of the Moon and Sun. These variations result, in turn, in corresponding changes in the magnitudes of lunar and solar gravitational forces which are impressed as force moments (torques) upon the Earth's equatorial bulge.

The epochs of maximized nutational influence occur, for the most part, at recurring intervals of half a year for the Sun, and half a month for the Moon. These maximized nutation-producing influences correspond to (1) the semiannual positions of the Sun which establish a maximum torque angle with respect to the Earth's equator (the result of a maximum positive or negative solar declination, at the summer and winter solstices respectively); and (2) the Moon's semimonthly positions of maximum northern and southern declination – when the direction of gravitational force application upon the Earth's equatorial bulge similarly results in a maximum angle of torque.

The effects of such transiently maximized force vectors result in corresponding variations in the magnitudes of three ephemeris quantities known as nutation in longitude, nutation in the obliquity, and obliquity of the ecliptic. These variations are complex, being intermingled with all of the previously mentioned periodic influences as well as progressive changes. Still, the following are close generalities: both the second and third quantities, for example, attain detectable localized minima accompanying syzygy and perigee-syzygy alignments that occur at high declination angles; contrastingly, localized maxima occur in both quantities, under the same syzygy situations, when the Moon is close to the equator. Minimum values for the year in the obliquity of the ecliptic occur at the summer and winter solstices, and maximum values at or near the two equinoxes. The highest possible values of the obliquity of the ecliptic are reached at the times of the extreme maxima in lunar declination associated with two annual peaks in each 18.6134-year lunar nodical cycle (most recently on 5 April and 30 September 1987).

The variances introduced by nutation require a slightly altered concept over that outlined in the discussion of precession alone. The various perturbative force influences associated with nutation give rise to a modification of the precessional curve into a more nearly elliptical figure, which may be regarded as centered in a 'mean' pole, around which the 'true' pole revolves. The semimajor axis of this elliptical figure is designated as the nutation constant. The maximum value of this constant, which occurs at the time of one of the declination peaks in the lunar nodical cycle, is approximately 9.21″. An explanation of the cause of nutation was first given by the British astronomer James Bradley in 1748, and related theoretical, observational and statistical aspects have been intensively treated in subsequent studies (see references below).

Fergus J. Wood

Bibliography

Blanco, V.M. and McCuskey, S.W. (1961) Precession and nutation, in *Basic Physics of the Solar System*. Reading, Ma.: Addison-Wesley, 307 pp., pp. 206–10.

Kulikov, K.A. (1964) Precession (pp. 119–36); Nutation (pp. 178–83) in *Fundamental Constants of Astronomy*. (Translated from the Russian under the auspices of NASA and the National Science Foundation.) Washington, DC: US Dept. of Commerce, Office of Technical Services, 211 pp.

McCarthy, D.D., Seidelmann, P.K. and Van Flandern, T.C. (1980) On the adoption of empirical corrections to Woolard's nutation theory, 117–124, in *Nutation and the Earth's Rotation* (eds E.P. Fedorov, M.L. Smith and P.L. Bender). IAU Symposium No. 78, Kiev, USSR, 1977. Dordrecht and Boston: D. Reidel, 266 pp., pp. 8117–24.

Mueller, I.I. (1969) Precession (pp. 60–8); Nutation (pp. 68–71) in *Spherical and Practical Astronomy as Applied to Geodesy*. New York: Frederick Ungar, 615 pp.

Newcomb, S. (1906) Precession (pp. 225–46); Nutation (pp. 246–56), in *A Compendium of Spherical Astronomy*. New York: MacMillan, 444 pp.

Smart, W.M. (1936) Precession and nutation, in *Text-Book on Spherical Astronomy*. Cambridge: Cambridge University Press, 430 pp., pp. 226–48.

Stacey, F.D. (1977) Precession of the equinoxes, in *Physics of the Earth*, 2nd edn. New York: John Wiley, 414 pp., pp. 57–61.

Cross references

Celestial mechanics
Earth rotation
Ellipsoid
Hipparchus
Milankovitch, Milutin, and Milankovitch theory
Planetary rotation
Torque

PTOLEMY (CLAUDIUS PTOLEMAEUS, AD c. 100–c. 170)

Little is known of his birth (probably in Upper Egypt, of Greek ancestors) but according to Arabic sources he lived 78 years and flourished in Alexandria during the 2nd century AD. Known as the last of the great scientists of classical Greece, his fame rests chiefly on a great astronomical treatise the *Almagest*, which got its name from Arabic translators (a corruption of 'the Greatest'). It only came down to the western world by subsequent retranslation into Latin (by Gherard of Cremona, 1175; other versions followed). It was to remain the standard work on astronomy for more than 1400 years. It consisted of 13 books, dealing with the assembled facts about motions of the Earth, Moon, planets and stars; with the solution of mathematical problems; and with physical geography, for which he devised an interesting orthoconic projection (Toomer, 1975). Many maps were originally presented, but appear to have been lost, although they can be approximately reconstituted today using his coordinates (Singer, 1959, p. 95).

Just as the astronomy of Ptolemy became the accepted cosmology, so also did his *Geography*, with the leading map of the known world for many centuries. The book and map marked 'a turning point in the history of science' (Taton, 1957, p. 327); this mathematical treatment rested largely on the data assembled by earlier workers (e.g. Marinos of Tyre) and thus suffered from their imperfections.

Ptolemy used the Babylonian system of dividing the circle into 360°, with 60-minute and 60-second subdivisions. Much of the astronomical data he took over from earlier astronomers, with proper acknowledgment, mainly from Hipparchus (190–120 BC, q.v.). However, it was Ptolemy who was able to develop a mathematical theory to explain the motions of the five planets known at that time. Every scientist who attempts to collect, organize and synthesize the data of his specialty is forced to borrow extensively and make adjustments. Some of Ptolemy's actions have been questioned in a best-seller (Newton, 1978) but the case remains controversial.

Ptolemy was first and foremost a mathematician, and followed the simplicity principle in science, later to be known as 'Occam's razor'. It may, indeed, have led to his adherence to the current (Aristotle's) geocentric theory of the universe.

From this premise developed what became known as the 'Ptolemaic system', which in turn became a basic pattern for astrology (in which the planets became associated traditionally with color, metals, minerals and parts of the human body). Ptolemy's arguments on that 'certain power' of celestial bodies over 'earthly happenings' are quoted verbatim in Pannekoek (1961, p. 160).

The planet Earth, according to the Ptolemaic system, was the central and by far the largest celestial body. It was successively encircled by the orbits of the Moon, Mercury, Venus, the Sun (!), Mars, Jupiter, Saturn and the 'crystalline heavens and stars' (see Figure 41 in Singer, 1959). The same picture of the universe, but modified for theological purposes (with Paradise, Purgatory and Hell) was presented by Dante in about AD 1300 (e.g. Figure 72 in Singer, 1959). The actual orbits of these bodies were not concentric but could be constructed geometrically by an ingenious series of steps. First, the Earth's position was set eccentrically to the center of the cosmos. At an equal distance from that center in the opposite direction is a point called the 'equant', around which each of the other planets revolve. To accommodate what we now recognize as the Earth's heliocentric revolution, Ptolemy calculated a series of 'epicycles' superimposed on the general circular path of each planet. The motion was comparable to that of an eccentric valve gear on a steam locomotive. The important thing is that analytically it worked, with a reasonable degree of accuracy, because in mathematics it really makes no difference if one assumes the Earth or the Sun is in the fixed position. In modern celestial mechanics it is the system barycenter (q.v.) that is the 'fixed' point.

One of Ptolemy's most important discoveries is found in book IV of the *Almagest*, which deals with the Moon and the length of the month. Although the Moon's elliptical orbit and 8.849-year apsides cycle (q.v.) were well established, Ptolemy was able to discover an additional inequality in the Moon's motion, the evection cycle that also related the Moon–Sun position. For this he constructed a further epicycle with a 'deferant', the center revolving eccentrically around the Earth. Even so, small departures were observed at lunar intermediate phases (¼ and ¾ which were accommodated by an oscillation in the epicycle called a 'prosneusis'.

Other books of the *Almagest* included no. V, devoted to Ptolemy's principal astronomical instrument, the astrolabe, which for many centuries remained the basic instrument for celestial observation. Book VI explored the eclipse (q.v.) in depth, and with Ptolemy's tables lunar eclipses could be predicted with high precision. Solar eclipses were less accurate because the epicycle method failed to obtain a correct solar distance. He had observed the equinox of 132 AD, and was able to calculate the Sun's positions back to his datum ('epoch') year, which was Thoth 1 (first day of the Egyptian year), on 26 February 747 BC, and year 1 of the reign of the Babylonian king Nabonassar.

Books VII and VIII contain a catalog of the known stars and discussed the precession of the equinoxes. The last five volumes dealt with the theory of the motions of the planets, covering Ptolemy's original contributions to astronomy. Mars, Jupiter and Saturn were seen as oscillating to and fro on either side of a fictitious body. Similarly, Venus and Mercury oscillated about the Sun.

Berry (1898, p. 73) remarked: 'The history of Greek astronomy practically ceases with Ptolemy'. Centers of activity shifted gradually to Baghdad and elsewhere, reflecting the growth of Islam. The Alexandrian School declined, the library was burned and in 640 AD the city fell to the Arab invaders.

Rhodes W. Fairbridge

Bibliography

Berry, A. (1898) *A Short History of Astronomy*. London: Murray (also New York: Dover reprint, 1961), 440 pp.

Grasshoff, G. (1990) *The History of Ptolemy's Star Catalogues*. New York: Springer-Verlag, 347 pp.

Newton, R.R., 1978. *The Crime of Claudius Ptolemy*. Baltimore and London: Johns Hopkins University Press, 412 pp.

Pannekoek, A. (1961) *A History of Astronomy*. New York: Interscience, 521 pp.

Sarton, G., (1927/1931) *A History of Science*. Cambridge, MA: Harvard University Press, 2 vols, 615 pp, 528 pp.

Singer, C. (1959). *A Short History of Scientific Ideas to 1900*. Oxford: Clarendon Press, 525 pp.

Taton, R. (1957) *La Science Antique et Mediévale*. Paris: Presses Université de France (translated by A.J. Pomerans, 1963). *Ancient and Medieval Science*. New York: Basic Books, 551 pp.

Toomer, G.J. (1975). Ptolemy. *Dict. Sci. Biogr.*, Vol. 11, pp. 186–206.

Toomer, G.J. (1984) *Ptolemy's Almagest* (translation and annotation). London: Duckworth; and New York: Springer-Verlag.

Acknowledgement

Kindly read and improved by Hans J. Haubold, UN, Vienna.

R

RADAR ASTRONOMY

Planetary radar astronomy is the active, radio-frequency probing of solar system targets. Precisely formed signals are directed toward, scattered from and sometimes passed through surfaces, rings, atmospheres and plasmas. Changes in the signals resulting from the distant interactions can be used to determine the location, motion and intrinsic properties of the target. The advantages of radar in astronomy result from (1) the observer's control over all attributes of the signal used to illuminate the target – especially its time – frequency structure and polarization; (2) the ability of radar to resolve objects spatially via measurements of the distribution of echo power in time delay and Doppler frequency; and (3) the use of centimeter to meter wavelengths, which easily penetrate optically opaque clouds and cometary comas, permit investigation of near-surface macrostructure and bulk density and are sensitive to high concentrations of metal or ice. Radar astronomy may be conducted with Earth-based equipment, spacecraft instrumentation or both. It is limited only by the distances over which the echo signals can be detected and the precision with which they can be measured.

History

Radar technology was developed during World War II. In 1946 groups in the United States and Hungary obtained the first echoes from the Moon in experiments motivated by interest in propagation of radio waves through the ionosphere and by the possibility of using the Moon as a 'relay' for radio communication. Radar investigations of the Moon itself began in earnest when huge radio telescopes became available in the 1950s.

By the early 1960s radar had refined our knowledge of the astronomical unit through precise measurement of the distance to targets such as Venus, ensuring successful guidance of the first interplanetary space probes. By the end of that decade radar had also revealed that Venus' rotation is retrograde, that Mercury's rotation is in a 3:2 resonance with its orbital revolution and that Mars has considerable topographic diversity. Radar echoes from the Moon and Venus were first translated into images of their respective surfaces in the 1960s; that work has continued to the present, with maps by the Magellan spacecraft radar now revealing most of Venus at resolutions as fine as 100 m.

The detection of echoes from Saturn's rings in the early 1970s offered the first strong evidence that a large fraction of its ring particles are centimeter-sized or larger 'particles' of ice. In the mid-1970s echoes from Europa, Ganymede and Callisto revealed that those icy moons of Jupiter have radar properties totally outside the realm of previous experience. The permanent south polar cap on Mars (also presumably ice) and the polar regions of Mercury have since been shown to be unusually bright as well.

Radar first detected a near-Earth asteroid (1566 Icarus) in 1968, a main-belt asteroid (1 Ceres) in 1977 and a comet (Encke) in 1980. During the 1980s small-body radar astronomy expanded dramatically and today 37 main-belt asteroids, 39 near-Earth asteroids and six comets have been detected. Radar has discovered clouds of large particles around three comets, has revealed one Earth-approaching asteroid to have a contact-binary shape, and has provided the most persuasive identifications to date of metallic objects in the main-belt and near-Earth populations.

Telescopes

Large (60–70 m diameter) parabolic reflector antennas, transmitters with output of more than 100 kW and state-of-the art receivers are now required for Earth-based planetary radar studies. Only two such facilities are presently in regular use: the Arecibo Observatory in Puerto Rico and the Goldstone Tracking Station in southern California.

The Arecibo telescope is part of the National Astronomy and Ionosphere Center, operated by Cornell University for the National Science Foundation. The instrument is built around a 305 m diameter reflector, which is a section of a 530 m diameter sphere. Movable feeds suspended from a triangular platform some 130 m above the reflector correct for spherical aberration. The feeds can be aimed toward various positions on the reflector enabling the telescope, despite its fixed reflector, to follow celestial objects within 20° of zenith. For targets at declinations near + 18° (Arecibo's geographic latitude) this translates to about 2.5 h of data collection time each day. The Arecibo radar operates at wavelengths of 13 and 70 cm, with the shorter wavelength providing better sensitivity.

Upgrading of the Arecibo telescope began in 1992 to increase its sensitivity by more than an order of magnitude. The improvement was achieved by constructing a ground screen around the periphery of the dish, replacing the higher-frequency line feeds with a more efficient Gregorian subreflector configuration, doubling the transmitter power, and installing a fine-guidance pointing system.

The Goldstone radar is part of NASA's Deep Space Network (q.v.) which is operated by the Jet Propulsion Laboratory/California Institute of Technology. The Goldstone main antenna is a fully steerable, 70-m parabolic reflector with horn feeds. It operates at wavelengths of 3.5 and 13 cm, with the shorter wavelength also providing the better sensitivity. Although the peak Arecibo sensitivity is twice that which can be attained at Goldstone, the Goldstone antenna can track targets continuously for much longer periods and has access to the whole sky north of −40° declination. Recent experiments in 'aperture synthesis radar astronomy,' which employ 3.5-cm transmission from Goldstone and reception of echoes at the 27-antenna Very Large Array (VLA) (q.v.) in New Mexico, have been extremely successful. The Goldstone-VLA combination is about 50% more sensitive than the pre-upgrade Arecibo system.

Radar astronomy has also been conducted using spacecraft. Voyager spacecraft transmitting to Earth stations (including those of NASA's Deep Space Network, the Australian facility at Parkes and the Japanese antenna at Usuda) have allowed the study of targets at the edge of the solar system. Although the spacecraft transmitted signal power is typically 10–20 W, the forward propagation and sensitive Earth receiving equipment in these experiments have allowed probing of the atmospheres of the gas giants and two of their major satellites (Titan and Triton) and the ring systems of Saturn and Uranus (see Radio science). The self-contained radar mapper aboard the Magellan spacecraft orbiting Venus functional much as do Earth-based systems; being much closer to its target, however, it could afford to operate at lower transmitter power (350 W peak) and with less receiver sensitivity (see Magellan mission).

Measurements and interpretation

The simplest radar astronomical deductions are based on round-trip time of flight and Doppler shift of the signals, from which distance to and motion of a target body can be determined (Figure R1). For example, since the speed of light is known very accurately, a precise determination of the elapsed time between a transmission and reception of the corresponding echo yields the distance to the target. Round-trip time to the Moon from an Earth-based radar is about 2.5 s; round-trip time to Saturn is more than 2 h. It is possible to measure the distribution of echo power in time to a precision as fine as 10^{-7} s, corresponding to a range precision of about 15 m. Also, if echo signals are slightly higher in frequency than the signals transmitted, the target is approaching the radar. Measurement precisions of 0.01 Hz can be obtained, yielding velocities accurate to fractions of 1 mm s^{-1}. The most accurate planetary ephemerides make use of such distance and velocity derivations.

Measurements of echo amplitude, polarization and dispersion in both time and frequency are also useful. These are less readily interpreted, but they are believed to relate to the material properties of the target such as surface roughness, atmospheric homogeneity and the sizes and distribution of particles. When combined with other information they allow for improved understanding of the state of the target and the processes that have shaped its evolution.

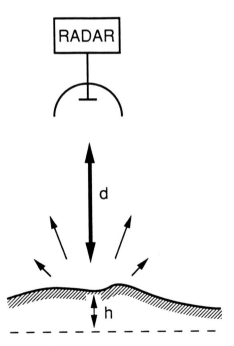

Figure R1 Nadir-viewing radar measuring height h relative to a reference surface (dashed line). Time delay between transmission and reception of a pulse is proportional to altitude d above the surface. Topography (h) is obtained by subtracting the measured altitude and the target radius (reference surface) from the estimated radar to target center distance, which itself is based on very long-term averages of such measurements. For quasi-specular (mirrorlike) surfaces, the strongest echo for a nadir-viewing radar is back toward the radar; less energy will be scattered in other directions, the amount depending on the centimeter- to meter-scale surface roughness.

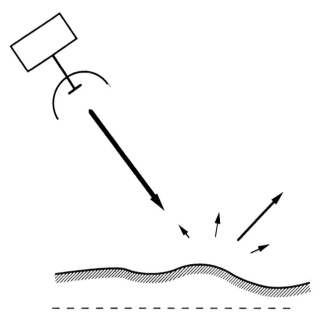

Figure R2 Oblique viewing radar. For quasi-specular scattering, the strongest echo is in a direction away from the radar but at a similar angle with respect to local vertical. A bistatic radar receiver in that direction would detect a strong echo. The amount of energy scattered back toward the transmitter decreases sharply with departure from nadir viewing (Figure R1). At highly oblique angles unusual surface tilts, highly irregular structure or an increase in the density of the reflecting material may increase the backscattered energy, brightening images derived from radar data such as those produced by Magellan.

Signal power P_R observed at a radar receiver is given by the bistatic 'radar equation'

$$P_R = \frac{P_T G_T \lambda^2 G_R}{(4\pi)^3 R_T^2 R_R^2} \sigma \qquad (R1)$$

where P_T is the transmitted power, G_T is the gain of the transmitter antenna, λ is the radar wavelength, G_R is the gain of the receiving antenna, R_T is the transmitter-to-target distance, R_R is the target-to-receiver distance, and σ is the target radar cross section. The bistatic form (R1) allows for the possibility that transmitter and receiver are not collocated (Figure R2). Terrestrial planets typically have radar cross-sections that are about 10% of their projected disk area when viewed monostatically – that is, with all equipment on Earth. With Mars at its closest approach to Earth ($R_T = R_R = 0.35$ AU) observations from the Arecibo telescope ($P_T = 450$ kW, $G_T = G_R = 3 \times 10^7$, and $\lambda = 0.13$ m) have yielded an echo of approximately 2×10^{-15} W spread over about 7400 Hz. The background noise in the same bandwidth is about three orders of magnitude less, which means that large signal-to-noise ratios are possible.

Spreading of the Mars echo in frequency arises from the planet's rotation. Near equatorial points on the approaching limb are moving toward Earth at about $v = 240$ m s^{-1}, while those near the equator on the receding limb are moving away at the same speed (both relative to Mars' center). Doppler shift (q.v) can be calculated from

$$f_D = \frac{2v}{\lambda} \qquad (R2)$$

If echoes from both limbs can be detected, their frequencies will be at ± 3700 Hz from the center frequency. The retrograde rotation of Venus and the resonant rotation of Mercury were both discovered by application of such methods.

Planetary radar echoes are not distributed uniformly in frequency. From geometrical considerations it is clear that the only points on Mars contributing to an echo at ± 3700 Hz must be both on the limb and at the equator. On the other hand, the plane formed by the Earth–Mars line and Mars' rotation axis defines a great circle on the red planet. Points on this great circle have velocities with no

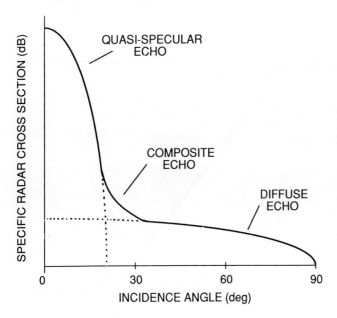

Figure R3 Specific radar cross-section σ_0 for a representative planetary surface, showing its variation with incidence angle. Integration of σ_0 over the target surface area yields the total radar cross-section σ. Quasi-specular (mirrorlike) scattering dominates σ_0 at small incidence angles; diffuse scattering occurs at all angles.

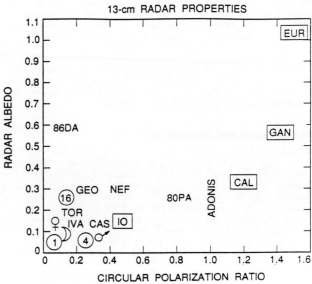

Figure R4 Radar properties for selected planetary targets at 13-cm wavelength. The albedo of targets in the expected sense of circular polarization (vertical axis) is plotted against the ratio of the echo powers in unexpected and expected senses. Relatively smooth, rocky surfaces should have albedos and polarization ratios that are both on the order of 0.1. Symbols are used for the Moon, Venus and Mars, which fall within this category. The circled numbers correspond to the main-belt asteroids 1 Ceres, 4 Vesta and 16 Psyche. Note the disparate properties of the Galilean satellites Europa, Ganymede and Callisto (rectangles). Also note the diversity of the near-Earth asteroids 1986 DA, 3199 Nefertiti, 1620 Geographos, 1980 PA, 1685 Toro, 1627 Ivar, 4769 Castalia and 2101 Adonis. (Courtesy NASA/JPL.)

component either toward or away from Earth ($f_D = 0$). Since there are many more points contributing near $f_D = 0$ than near $f_D = \pm 3700$ Hz, the echo spectrum will be broadly peaked as a result. Certain asteroids, as well as the icy Galilean satellites of Jupiter, have broad echo spectra indicating that their radar brightness depends primarily on the amount of surface area visible to the radar and not on the angle at which that surface is viewed.

Although there are variations, most terrestrial planets return echo spectra with much narrower peaks, suggesting that their surfaces are not uniformly bright; that is, the radar scattering is 'quasi-specular' or 'mirrorlike,' rather than being essentially uniform with angle as is the rule at optical wavelengths (Figure R3). A planetary surface viewed with radar eyes would resemble a slightly rippled water surface – the quasi-specular echo would appear as the glints of sunlight from the small waves. Spectra become broader (and the associated surface tilts presumably larger) as the radar wavelength decreases, in agreement with notions that the surface structure becomes rougher as the scale of examination becomes finer.

Monostatic radars operating at very oblique incidence angles respond to a combination of uncommon (but highly tilted) meter-scale reflecting facets and to irregular structure which may include cracks, blocks and inhomogeneities in the bulk surface material. Although weak from individual areas, this 'diffuse' component to planetary echoes can rival the quasi-specular contribution when summed over the whole body. The icy Galilean satellites of Jupiter represent a major exception to this rule of thumb – the diffuse echo is 10–20 times larger from these bodies than from the inner rocky planets, and no quasi-specular component has been identified in Earth-based experiments whatsoever. Current hypotheses attribute this unusual behavior to subsurface volume scattering within the icy surfaces; the process, called 'coherent backscatter' or 'weak localization,' was first discovered in laboratory studies of the scattering of electrons and of light. Recent experiments suggest that the residual south polar cap on Mars, and previously unsuspected icy polar deposits on Mercury behave in similar ways.

One of the key aspects of the unusual echoes from icy surfaces is that the echo polarization has a sense opposite to that expected from a specular target. It is as though the signal were gently 'turned around' and sent back toward the radar rather than simply being bounced off its surface. Circular polarization has been employed for most Earth-based work to avoid problems with rotation of linearly polarized signals during passage through the Earth's ionosphere. If the transmitted signal has a right rotational sense, the backscattered echo from flat surfaces is expected to have a left rotational sense, and indeed that was the experience for the Moon and other targets until the mid-1970s when echoes from icy surfaces were first encountered. Echo power in the same circular sense of polarization as transmitted (the unexpected sense) has been taken to be a measure of surface characteristics not accommodated in the quasi-specular model, presumably related to a diffuse scattering mechanism.

The ratio of echo power in the two polarizations can be a useful gauge of the target's near-surface, wavelength-scale complexity or roughness. Figure R4 shows values of the radar albedo, defined as the ratio of the expected sense (opposite circular) radar cross-section to the target's projected physical area, and the polarization ratio (unexpected to expected) for selected planetary objects. For targets with low polarization ratio the albedo is proportional to the surface reflection coefficient, which in many applications is a nearly linear function of bulk density. If the polarization exceeds about 0.1, physical interpretations are more complicated because models must consider not just the nature of the surface–space interface but also the regolith's structural and electrical properties, including the size distribution, spatial distribution and scattering properties of subsurface rocks. The icy Galilean satellites of Jupiter (EUR, GAN and CAL in Figure R4) are clearly in a category by themselves. Asteroids are diverse, while the terrestrial targets are 'conventional' in having both low albedo and low polarization ratio.

Future directions

Increasingly sophisticated signal processing techniques have expanded the horizons of planetary radar astronomy. Using receivers at the Very Large Array (New Mexico) in conjunction with the high-powered transmitter at the Goldstone Tracking Station, radar astronomers are now able to study targets from Mercury to Saturn. Signal processing specialists are developing modulation/demodulation techniques which allow detection of weak signals previously discarded as inseparable from background noise and clutter. The extraordinary success of the Magellan mission to Venus is opening other doors:

a mapping radar for the NASA Cassini mission will probe the surface of Saturn's cloud-covered satellite Titan. Meanwhile laboratory, theoretical and analog studies continue to improve the understanding of radar echoes and their interpretability for planetary science.

Richard A, Simpson and Steven J. Ostro

Bibliography

Campbell, D.B., Chandler, J.F., Ostro, S.J. *et al.* (1978) Galilean satellites: 1976 radar results. *Icarus*, **34**, 254–67.
Elachi, C., Im, E., Roth, L.E. and Werner, C.L. (1991) Cassini Titan radar mapper. *Proc. IEEE*, **79**, 867–80.
Evans, J.V. and Hagfors, T. (1968) *Radar Astronomy*. New York: McGraw-Hill.
Goldstein, R.M., Green, R.R., Pettengill, G.H. and Campbell, D.B. (1977) The rings of Saturn: two-frequency radar observations. *Icarus*, **30**, 104–110.
Hapke, B. and Blewett, D. (1991) Coherent backscatter model for the unusual radar reflectivity of icy satellites. *Nature*, **352**, 46–47.
Harmon, J.K. and Slade, M.A. (1991) An S-band radar anomaly at the north pole of Mercury. *Bull. Am. Astron. Soc.*, **23**, 1197.
Harmon, J.K., Campbell, D.B., Hine, A.A. *et al.* (1989) Radar observations of comet IRAS-Araki-Alcock 1983d. *Astrophys. J.*, **338**, 1071–93.
Harmon, J.K., Sulzer, M.P., Perillat, P.J. and Chandler, J.F. (1992) Mars radar mapping: strong backscatter from the Elysium basin and outflow channel. *Icarus*, **95**, 153–6.
Hudson, R.S. and Ostro, S.J. (1990) Doppler radar imaging of spherical planetary surfaces. *J. Geophys. Res.*, **95**, 10947–63.
Muhleman, D.O, Grossman, A.W., Butler, B.J. and Slade, M.A. (1990) Radar reflectivity of Titan. *Science*, **248**, 975–80.
Muhleman, D.O., Butler, B.J., Grossman, A.W. and Slade, M.A. (1991) Radar images of Mars. *Science*, **253**, 1508–13.
Ostro, S.J. (1992) Planetary radar astronomy. *Rev. Modern Physics*, **65**, 1235–79.
Ostro, S.J., Chandler, J.F., Hine, A.A. *et al.* (1990) Radar images of asteroid 1989 PB. *Science*, **248**, 1523–8.
Ostro, S.J., Campbell, D.B., Chandler, J.F. *et al.* (1991) Asteroid 1986 DA: radar evidence for a metallic composition. *Science*, **252**, 1399–404.
Peters, K.J. (1992) The coherent backscatter effect: a vector formulation accounting for polarization and absorption effects and small or large scatterers. *Phys. Rev. B*, **46**, 801–12.
Pettengill, G.H. (1978) Physical properties of the planets and satellites from radar observations. *Ann. Rev. Astron. Astrophy.*, **16**, 265–92.
Saunders, R.S. and Pettengill, G.H., (1991) Magellan: mission summary. *Science*, **252**, 247–9.
Saunders, R.S., Pettengill, G.H., Arvidson, R.E. *et al.* (1990) The Magellan Venus radar mapping mission. *J. Geophys. Res.*, **95**, 8339–55.
Shapiro, I.I., Chandler, J.F., Campbell, D.B. *et al.* (1990) The spin vector of Venus. *Astron. J.*, **100**, 1363–8.
Simpson, R.A. and Tyler, G.L. (1982) Radar scattering laws for the lunar surface. *IEEE Trans.*, **AP-30**, 438–49.
Simpson, R.A., Harmon, J.K., Zisk, S.H. *et al.* (1992) Radar determination of Mars' surface properties, in *Mars* (ed. H.H. Kieffer, B.A. Jakosky, C. Snyder and M.S. Matthews) Tucson: University of Arizona Press.
Slade, M., Butler, B., Muhleman, D. and Jurgens, R. (1991) Mercury Goldstone/VLA Radar; Part I. *Bull. Am. Astron. Soc.*, **23**, 1197.
Tyler, G.L. (1987) Radio propagation experiments in the outer solar system with Voyager. *Proc. IEEE*, **75**, 1404–31.
Yeomans, D.K., Chodas, P.W., Keesey, M.S. *et al.* (1992) Asteroid and comet orbits using radar data. *Astron. J.*, **103**, 303–17.

Cross references

Magellan mission
Near-Earth object
Planetary ring
Radio science

RADIATION BELTS

Radiation belts are regions containing trapped, energetic charged particles (electrons and ions) in a planetary magnetosphere. The radiation belts in the terrestrial magnetosphere are called the Van Allen belts after their discoverer. Van Allen used observations from a Geiger tube flown as part of the Explorer 1 (the United States' first orbiting spacecraft) scientific payload to determine that the region above about 400 km altitude is characterized by a vast population of very energetic electrons and ions. Some of the earliest measurements of primarily higher-energy particles showed variations in the counting rate as a function of altitude, and an inner and outer radiation belt were identified centered near 1.5 and 3.5 Earth radii, respectively, from the center of the Earth (Figure R5). One of the important distinctions between the two is that the inner zone is temporally much more stable than the outer one. Subsequent measurements showed that the energy of the trapped particles ranges from less than 1 eV to as much as several hundred MeV.

The fundamental physics governing the trapping of energetic charged particles in a dipolar magnetic field similar to that of the Earth was worked out by Störmer long before the discovery of the Van Allen belts. In a dipolar magnetic field, charged particles execute a number of cyclical motions with a progression of natural frequencies. The most basic of these is a cyclotron or circular motion about a magnetic line of force. Another cyclic motion of a charged particle trapped in a dipolar magnetic field is called the bounce motion, in which particles spiraling around the magnetic field lines also bounce between 'mirror points' in opposite hemispheres. The third fundamental cyclic motion of a particle trapped in a dipolar magnetic field is a drift in longitude around the planet. The cyclotron period can be as small as 10^{-5} s for electrons, whereas the bounce and azimuthal drift periods are of the order of 1 and 10^3 s respectively for electrons.

The primary source for the highest energy particles in the inner belt is a process known as cosmic ray albedo neutron decay (CRAND), in which galactic cosmic rays interact with the atmosphere, producing neutrons. As the neutrons escape, some of them decay into electrons and protons which are trapped by the planetary magnetic field. Intermediate energy constituents of the belts are the result of solar wind plasmas which enter the magnetosphere through the magnetopause (via a poorly understood mechanism) and gain energy as they are diffused inward toward the Earth. The lowest energy populations of the radiation belts probably have their origin in the upper atmosphere and ionosphere. High-altitude nuclear explosions in 1958 and 1962 induced artificial radiation belts which persisted for periods of many months. Variations in the solar wind during the 11-year solar activity cycle provide a major natural source of temporal variations in the particle population and character of the belts.

Radiation belt particles are lost through interactions with the Earth's exosphere, and through instabilities which increase the relative proportion of their kinetic energy parallel to the magnetic field so that they collide with atmospheric gas. This latter loss process can be thought of as one that moves the 'mirror points' downward into the atmosphere.

Earth is not the only planet with radiation belts. All of the planets with strong intrinsic magnetic fields (Earth, Jupiter, Saturn, Uranus and Neptune) have radiation belts. Jupiter has the most intense radiation belts of any of the planets in the solar system. At least for higher-energy particles, its inner magnetosphere (within about 15 planetary radii) is the locale of the Jovian radiation belts. Energetic particles are transported inward toward the planet and gain energy through the conservation of the first adiabatic invariant (associated with the cyclotron motion of the particles in the dipole-like magnetic field). The particle intensities are great enough and the energies high enough that significant amounts of synchrotron radiation are radiated. These synchrotron emissions form the characteristic Jovian decimetric radiation, observable at the Earth.

William S. Kurth

Bibliography

Schardt, A.W. and Goertz, C.K. (1983) High-energy particles in *Physics of the Jovian Magnetosphere* (ed. A.J. Dessler) New York: Cambridge University Press, pp.157–96.

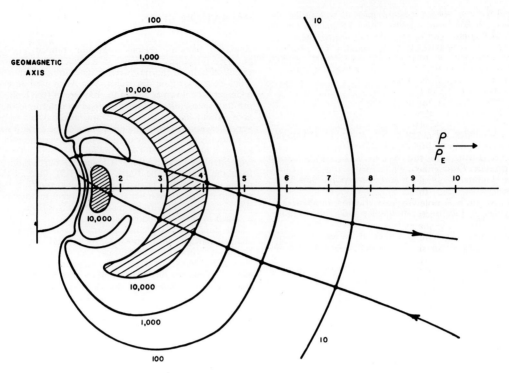

Figure R5 An early model of the distribution of particles in the Earth's radiation belts based on Pioneer 3 and 4 observations. (After Van Allen, 1959; Copyright American Geophysical Union.)

Störmer, C. (1907) Sur des trajectoires des corpuscles électrises dans l'space sons l'action du magnétisme terrestre, chapitre 4. *Archives des Sciences Physiques et Naturelles*, **24**, 317–64.

Van Allen, J.A. (1959) The geomagnetically-trapped corpuscular radiation. *J. Geophys. Res.*, **64**, 1683–9.

Van Allen, J.A. (1983) *Origins of Magnetospheric Physics*. Washington: Smithsonian Institution Press.

Acknowledgement

This research was supported by the National Aeronautics and Space Administration through Contracts 959193 and 958779 with the Jet Propulsion Laboratory.

Cross references

Earth: magnetic field and magnetosphere
Jupiter: magnetic field and magnetosphere
Magnetism
Magnetospheres of the outer planets
Whistler

RADIATIVE TRANSFER IN PLANETARY ATMOSPHERES

Radiative transfer: a physical process and a tool for atmospheric physicists

The transfer of electromagnetic radiation (mostly reflected sunlight and emitted thermal infrared) in planetary atmospheres is a physical process that plays a role in determining the dynamical state of an atmosphere; it is singularly important for the information it conveys. Almost everything we know about the atmospheres of the giant planets and most of what we know about other planetary atmospheres has been acquired from remote measurements of radiation scattered, emitted, absorbed or refracted by those atmospheres. Ground-based and spacecraft experiments are designed to exploit the information content of the radiation. From these measurements we are able to investigate the chemical composition of the atmosphere, its temperature and the properties (optical thickness, particle mean radius and altitude) of cloud and haze layers.

Physics of radiation

Wave and particle descriptions of radiation

Our understanding of electromagnetic radiation has its roots in James Clerk Maxwell's equations for electromagnetism, developed in the 19th century. The electric vector of an electromagnetic wave traveling in the x direction is always in the y–z plane and its direction oscillates with the frequency of the wave. The concept of light as a particle (a photon) came in the early part of the 20th century with the discovery and elucidation of the photoelectric effect (for which Albert Einstein received the Nobel prize) and the advent of quantum mechanics. Photons have energy (E) proportional to the frequency (v) of vibration of the electric vector: $E = hv$, where h is Planck's constant (6.625×10^{-27} J s). The frequency and wavelength of light are related: $\lambda = c/v$, where c is the speed of light.

A single photon has either right or left circular polarization, which means the electric vector rotates in a clockwise or counterclockwise direction in the y–z plane as it propagates along x. Photons with left and right circular polarizations can mutually interfere and form linearly polarized light whose electric vector oscillates along a line in the y–z plane. A large ensemble of photons (e.g. sunlight) can have polarizations with random phases, in which case the light is unpolarized. A convenient formalism to describe the polarized radiation field was derived by Stokes who proposed a four-element vector notation. The first element of the Stokes vector is the intensity, the second and third account for the degree and direction of linear polarization, and the fourth accounts for the degree of circular polarization.

Radiative properties of atoms and molecules

Photons are emitted, absorbed and scattered by atoms and molecules. The description of photon interactions with atoms and molecules is best approached from a quantum mechanical framework. Atoms consist of a nucleus and an electron cloud. The energy states of the electron cloud are quantized into a set of fixed energy values which depend mostly on how many electrons surround the nucleus and the charge of the nucleus. Molecules have additional energy states corresponding to vibrational and rotational motions of the individual atoms that make up the molecule. Typical electronic excitation states emit radiation at ultraviolet wavelengths (100–300 nm), while vibrational transitions usually occur in the infrared (700–10 000 nm) and rotational transitions occur at far-infrared and microwave wavelengths (see also Spectroscopy: atmosphere).

Observations of atomic and molecular spectral lines in planetary atmospheres are the most direct and informative means for determining atmospheric composition and temperature. However, quantitative interpretation of data requires thorough treatment of the physics of the radiation process. Among the most important processes which contribute to formation of spectral lines are (1) the intrinsic cross-section of the spectral line; (2) the temperature of the gas which determines the distribution of molecules among the available energy levels and the shape of the spectral line at low pressure; (3) the pressure, which determines the shape of the line at modest to high pressure and which allows some pressure-induced spectral lines to be observed; (4) the strength and sign of thermal gradients in the atmosphere for thermal-infrared transitions; (5) the optical thickness of the atmosphere as a function of wavelength; and (6) scattering by aerosol particles or gas molecules.

A spectral line is formed when a molecule absorbs or emits a photon and increases or decreases its internal energy by the difference between upper and lower energy levels. Isolated atoms and molecules not in their ground state have a natural lifetime for making a radiative transition to a lower-energy state. The radiative cross-section or lifetime depends on whether a change from one state to the other is accompanied by a change in the electric dipole moment of the molecule. If so, the cross-section is relatively high and the lifetime is relatively short. This is not the case for the hydrogen molecule at low pressure and so the next available radiative transitions correspond to changes in the electric quadrupole moment. As a result, the hydrogen quadrupole lines are weak in the atmospheres of the giant planets, even though hydrogen is the most abundant species. At modest pressure (about 0.1 bar for hydrogen on Jupiter) molecular collisions can induce a temporary dipole moment, and the molecule can radiate. These are called pressure-induced transitions and are important for molecular hydrogen in the infrared.

Local thermodynamic equilibrium

For an emitted photon the initial energy state of the molecule is higher than the final state. The population of energy states for a gas in local thermodynamic equilibrium (LTE) is established primarily by collisions among the molecules and is described by the Boltzman distribution. In order for high energy levels to be populated the gas must be sufficiently warm. At low temperatures only the ground stated is populated. Thus measurement of the relative strengths of lines corresponding to higher and higher initial energy states allows a determination of the temperature.

A key concept for radiative transfer for a gas in thermodynamic equilibrium is the idea of detailed balancing. Detailed balancing means that energy state transitions occur at the same rate in both the forward and reverse direction. Under steady state conditions (where the timescale for temperature change is slow compared to a typical radiative or collision timescale of $\sim 10^{-6}$ s) photons are absorbed at the same rate that they are emitted, and molecules lose energy at the same rate as they gain energy, either by collisions or by radiative transitions. As a consequence, one can calculate emitted thermal radiance from an atmosphere or surface by knowing its temperature and absorption properties. This principle is embodied in Kirchoff's law (in steady state an object emits radiation at the same rate as it absorbs it). An object that totally absorbs radiation incident on it is called a black body. Black bodies are, by Kirchoff's law, also perfect emitters and emit thermal radiation according to Planck's law:

$$W_\lambda = c_1 \lambda^{-5}/[\exp(c_2/\lambda T) - 1]$$

where c_1 and c_2 are known as the first and second radiation constants, and have the values 3.74×10^{-12} W cm^2 and 1.44 cm deg respectively, and W_λ is the radiant energy (W cm^{-2}) per cm wavelength. No object is a perfect black body and some absorb almost no light, so for real objects one must multiply the Planck function by an emissivity term ϵ which depends on the object (see also Emissivity).

Spectral lines

Much of the work in radiative transfer goes toward calculating the absorption coefficient of the gas as a function of frequency or wavelength. The absorption coefficient k_ν determines the rate that radiation is absorbed along the path. The ratio of intensity that has traveled along a path of length l to the incident intensity is $I/I_0 = \exp(-k_\nu ln)$, where n is the number density of gas molecules in the atmosphere (molecules cm^{-3}). The product ln is the column abundance of molecules along the path and is sometimes measured in meter amagat. At standard temperature and pressure, 1 cm^3 contains Loschmidt's number ($n_0 = 2.69 \times 10^{19}$) of molecules, and 1 cm amagat is n_0 molecules cm^{-2}. Other forms of k_ν (e.g. dimensions of inverse mass density or inverse length) are sometimes used.

The central frequency (ν_0) of a spectral line is determined by the difference in energy between the initial and final states of the atom or molecule which emitted the photon: $\nu_0 = \Delta E/h$. Absorption and emission occurs over a range of frequencies centered on ν_0. Lines are broadened both by the Doppler shift from thermal motions of the molecules and by collisions with other molecules.

The line-of-sight velocity distribution, $p(v)$ of a gas in thermal equilibrium is described by the Maxwell–Boltzmann distribution:

$$p(v) = (m/2\pi KT)^{1/2} \exp(-mv^2/2KT)$$

where K is the Boltzmann constant (1.38×10^{-16} erg deg^{-1}) and T is the temperature in K. Photons emitted from molecules traveling toward or away from an observer will be Doppler shifted to a new frequency $\nu' = \nu_0 (1 \pm v/c)$. The Maxwellian velocity distribution combined with the Doppler shift produce a Gaussian line shape whose width

$$\alpha_D = (\nu_0/c)(2KT/M)^{1/2}$$

is proportional to the square root of temperature. The absorption coefficient for a Doppler broadened line is given by

$$k_\nu = \frac{S}{\alpha_D(\pi)^{1/2}} \exp\left[-\left(\frac{\nu - \nu_0}{\alpha_D}\right)^2\right]$$

where S is the integrated line strength.

When molecular collisions are frequent (at pressures greater than about 10 µbar) collisional broadening of the line becomes important. Collisions perturb the structure of the electron cloud and thereby influence the radiative transitions in a number of ways. The simplest picture of the perturbation is to imagine the collision as truncating the radiating wave. The radiative lifetime is thereby shortened and, by the Heisenberg uncertainty principle, the energy of the radiated photon becomes less certain. The mathematical form for the line shape resulting from this process is a Lorentz shape:

$$k_\nu = \frac{S}{\pi} \frac{\alpha_L}{(\nu - \nu_0)^2 + \alpha_L^2}$$

The parameter α_L is called the pressure broadening half-width and depends on the upper and lower energy states of the transition and on the broadening gas. The broadening gas in the atmospheres of Venus and Mars is CO_2, while the broadening gases in the outer planet atmospheres are H_2 (80% or greater) and He. The broadening half-widths must be measured in the laboratory for each line.

The Lorentz expression is valid for most atmospheric gases at relatively low pressure (less than a few bar), and to frequencies not far ($\Delta\nu/\nu_0$ less than about 0.01) from the line center. Further from the line center the Lorentz expression overestimates the opacity. For some lines (microwave transitions of NH_3 and H_2O in the giant planet atmospheres) the Lorentz shape is not a good approximation and other expressions must be used. More experimental and theoretical work is needed in these areas.

Doppler and Lorentz broadening are both important over a range of pressure. The proper expression for the line shape in this regime is given by the convolution of the Doppler profile with the Lorentz profile. The German physicist Woldemar Voigt (1850–1919) was probably the first to make use of this function and it is named after him. The expression for the Voigt function is not as simple as the Gauss or Lorentz formulas, and a considerable body of literature has been devoted to finding efficient methods to compute it (typically several million calculations of the Voigt function are required to calculate a model atmospheric spectrum).

Non-LTE

At high altitudes collisions between molecules become sufficiently infrequent that absorption and re-emission of sunlight play an important role in determining relative energy level populations, and consequently the heating and cooling rates. Detailed balancing is used to determine state populations, but collisional transitions no longer dominate the rates. As collisions become infrequent the balance turns more toward a purely radiative one, with transitions to the upper state (proportional to the incident solar flux) balanced by spontaneous and stimulated emissions to lower levels. Populations can become inverted (higher numbers in the upper energy state, relative to the lower state) and can lead to masing (where emitted radiation at some wavelength is amplified by coherent stimulated emission) in the high Martian atmosphere.

To calculate the populations of each state and the rate of radiative transitions, one inverts a matrix describing the relative transition rates among states. Rotational transitions have lower energies than vibrational or electronic transitions, and collisional cross-sections are relatively high for rotational transitions. The rotational levels of the gas can therefore be in local thermal equilibrium while the vibrational and electronic levels are in equilibrium with solar pumping.

Radiative properties of small particles

Thus far the discussion has been concerned with the behavior of gas-phase atoms and molecules. When atoms and molecules are in the liquid or solid phases the near-neighbor interactions provide a means of transferring energy, and narrow spectral lines give way to broad absorption features. Particles that are comparable to or larger than the wavelength of light scatter and absorb in ways that isolated molecules cannot.

Very small solid or liquid particles (generically called aerosols if airborne) can reside in the atmosphere for long periods. Larger particles (cloud droplets or dust) have shorter but still significant residence times. The optical properties of cloud and aerosol particles can be calculated (at least for some shapes and sizes) from Maxwell's electromagnetic equations, coupled with theories of the interaction of electromagnetic waves with dense media. Dielectric materials like glass at visible wavelengths do not permit electrical currents to flow but do allow for a polarizability of the material in response to an electric field. Electrons are free to flow in metals. Both these types of behaviors can be described by common empirical physical laws which make use of the dielectric constants or, alternatively, refractive indices of the material.

The important optical properties for atmospheric radiation calculations are the absorption and scattering cross-sections (their sum is the extinction cross-section), and the scattering phase matrix which describes how polarized or unpolarized light scatter in various directions, as well as the polarization state of the scattered light. The phase matrix $P(\theta)$ is a 4×4 matrix whose elements are functions or the scattering angle (θ). The P_{11} term is called the phase function and describes the angular scattering of the intensity component of the Stokes vector.

If the particles are much smaller than the wavelength of light (as are air molecules at visible wavelengths) they radiate as a dipole oscillator driven at the frequency of the incident light wave. The dipole term is the first in a series of multipole terms, the higher-order multipoles become significant as the particle size increases. The intensity part of the Stokes vector is scattered symmetrically about $\theta = 90°$, with $P_{11}(\theta) = 0.75[1 + \cos^2(\theta)]$. Lord Rayleigh deduced this law in 1871 and scattering of this type is called Rayleigh scattering. Scattered radiation from a dipole oscillator is also highly linearly polarized at scattering angles near 90°.

Scattering from particles whose size is comparable to or larger than the wavelength is more difficult to calculate. For the special case of spherical particles, Maxwell's wave equations can be written in spherical coordinates. Gustave Mie was the first to solve this problem (in 1908), and the so-called Mie theory is now widely used for calculations of the optical properties of spherical cloud and haze droplets.

Many particles (dust and ice particles) are not spheres, and the scattering problem is at present not tractable for arbitrarily shaped particles larger than a few times the wavelength. Various methods have been devised to attack the problem. In one method (the discrete-dipole array approximation) the particle is subdivided into many small sub-units, each a small fraction of the wavelength. The response of each of these to the incident electromagnetic field, and to the fields of all the other elements, is calculated. This method can accommodate particles of arbitrary shape, but as the particles grow larger than a few times the wavelength the number of elements increases as the cube of the diameter, and the time required for a numerical solution becomes extremely large.

If a particle is much larger than the wavelength one can use a ray-tracing code to calculate the optical properties. A ray-tracing code follows the path of a light ray as it enters through a bounding surface and is refracted or reflected. The method is only an approximation to Maxwell's wave equations and is therefore not valid for particles smaller than about 30 times the wavelength of the incident light. There is presently no technique to fill the gap between the size of the largest particle that can be handled by the dipole array technique and the smallest one that can be calculated by ray tracing.

Methods to compute radiative transfer

With the fundamental principles of the radiative properties of molecules and particles mentioned above, we are now in a position to examine the process of radiative transfer in a planetary atmosphere. The equation of radiative transfer describes how radiation is absorbed, emitted or scattered from a beam propagating through an atmosphere. The rate of change of the intensity I along a path is given by

$$dI/ds = -Ik_e + S$$

where ds is a differential path length, k_e is the extinction coefficient (km^{-1}) and S is a term called the source function which accounts for emission and scattering into the beam. Light is removed from the beam either by absorption or scattering, denoted respectively by k_a and k_s, whose sum is k_e. The ratio of scattering to extinction coefficient is called the single-scattering albedo, ω. In the above equation I and S are four-element Stokes vectors and k_e is a 4×4 extinction tensor. In the majority of situations it is acceptable to ignore the polarization part of the equation and treat I, S and k_e as scalars.

Under most conditions the emitting part of the source function is from thermal radiation, given by the Planck function, $k_a B(\nu, T)$. The scattering part of the source term is an integral over solid angle of the intensity field times the scattering phase function times the scattering part of the extinction coefficient:

$$k_s \int_\Omega I(\Omega) P(\Omega) d\omega/(4\pi)$$

where Ω is a vector denoting the angle for scattering into the beam.

The equation of transfer is a simple first-order linear differential equation if there are no scattering particles. The solution depends on the atmospheric temperature profile. In the visible part of the spectrum where the source function is negligible (if there is no scattering), sunlight is attenuated in the atmosphere according to Beer's law:

$$I/I_0 = \exp(-\tau)$$

where $\tau(\nu)$ is the line-of-sight optical depth $= \int k_e ds$. It is usual to approximate the atmosphere as a stack of plane-parallel layers and to designate $\tau(\nu)$ as the optical depth in the vertical. The frequency dependence of τ and the phase function and source terms will be assumed implicitly in what follows to simplify the notation.

For a purely absorbing atmosphere the ratio of intensity at the ground to the incident intensity is $I/I_0 = \exp(-\tau/\mu_0)$, where μ_0 is the cosine of the solar zenith angle and τ is the vertical optical depth to the ground. For a sensor above the atmosphere, sunlight that is scattered from the surface will be further attenuated by the factor $\exp(-\tau/\mu)$, where μ is the cosine of the zenith angle of the upward propagating ray.

In the absence of scattering, thermal radiation can be computed readily. If an atmosphere is isothermal, the emitted thermal radiation from the top or bottom is given by $B(\nu, T)[(1 - \exp(-\tau/\mu)]$. This can be obtained either by integrating the differential equation above or by first solving for the transmitted light and applying Kirchoff's law. The latter method is especially useful when computing emitted thermal radiation from a scattering layer, as discussed below.

Single and multiple scattering

The scattering part of the source term in the equation for radiative transfer introduces an integral over the angle of the intensity field. The resulting integro-differential equation is more difficult and vastly complicates the problem of inverting observed radiances to infer atmospheric structure (temperature and/or aerosol distribution). If the aerosol optical depth is very small ($\leq 10^{-2}$) the radiance can be computed from the equations for single scattering. For a plane-parallel layer having extinction optical depth τ, the intensity reflected from the top is

$$I(\Omega) = \frac{F}{4\pi} \frac{\mu_0}{\mu + \mu_0} (1 - e^{-\tau/\mu} e^{-\tau/\mu_0}) P(\Omega)$$

for incident solar flux F.

A variety of techniques have been developed to calculate multiple scattering in the plane-parallel atmosphere approximation. The plane-parallel approximation allows the three-dimensional transfer problem to be reduced to a sum of two-dimensional problems by decomposing the azimuthal dependence of the scattering into a Fourier series. A vertically inhomogeneous atmosphere can be treated by subdividing it into a stack of layers, each of which is considered to be homogeneous. A conceptually simple method for calculating multiple scattering is to follow the photons as they travel through the atmosphere and interact with each of the layers. The solution after one scattering was shown above. The solution for twice-scattered light can then be calculated by starting with the solution for the internal radiation field which has been scattered once. Solutions for higher orders of scattering can be obtained by successive applications of the formula and then summed. This method is called successive orders of scattering. It is computationally fast provided the product of the optical depth and single-scattering albedo are small (≤ 1). It is slow to converge for large optical depth and high single-scattering albedo because the number of times a photon scatters is large in that case.

Another conceptually simple but computationally faster method for multiple scattering is the doubling and adding algorithm. Suppose we know the diffuse reflection (R_A) and transmission (T_A) functions (which are matrices for computations on a discrete grid in μ and μ_0) for layer A of optical depth τ_A. To calculate the reflection function for a layer of optical depth $2\tau_A$ (or two layers of optical depth τ_A) we first calculate the reflection matrix (D) for multiple bounces between the layers. The matrix D is an infinite series starting with R_A ($I + R_A \times R_A + (R_A \times R_A \times R_A \times R_A)$ + higher powers of $R_A \times R_A$) where I is the identity matrix and $\times$ implies an angle integration which is carried out by an intermediate multiplication by a weighting matrix. The sum can be taken to several terms and truncated by a geometric series approximation

$$1 + \Sigma_{n=1}^{\infty} x^n = 1/(1 - x); 0 \leq x < 1$$

for each matrix element. Alternatively the matrix $[I - R_A \times R_A]$ can be inverted, subtracted from I, and multiplied by R_A, making use of the matrix formula for the geometric series (in that case it is called the matrix operator method).

Having calculated the multiple bounces between layers, we sum the following terms to calculate the total reflection function from the two layers.

1. The reflection function R_A for light reflected from the top layer alone;
2. D exp$[-\tau_A(1/\mu + 1/\mu_0)]$ for light transmitted directly through the top layer, scattered at the interface and transmitted directly back through the top;
3. $T_A \times D \times T_A$ for light diffusely transmitted through the top layer, reflected at the interface, and diffusely transmitted back through the top layer;
4. $T_A \times D$ exp$(-\tau_A 1/\mu_0)$ for light directly transmitted through the top layer, scattered from the interface, and diffusely transmitted back through the top; and
5. exp$(-\tau_A 1/\mu)D \times T_A$ for light diffusely transmitted through the top layer, reflected at the interface, and directly transmitted through the top layer.

The doubling method starts with a layer of very thin optical depth where single and twice-scattered light (which is also easy to calculate) account for the total. Layers are added together, or doubled, until the desired optical thickness is reached. The doubling method can be generalized to add two layers of different optical thickness or different scattering properties. In this way the radiative properties of vertically inhomogeneous atmospheres with large optical depth can be computed. The doubling/adding method has also been used to calculate the transfer of multiply scattered polarized light.

A number of other methods are available for computing multiple scattering. Some of the are as efficient as the doubling/adding method, and some more efficient under certain circumstances. There is insufficient space to describe them here. More details about radiative transfer in planetary atmospheres can be found in the bibliography.

Robert A. West

Bibliography

Chandrasekhar, S. (1960) *Radiative Transfer*. New York: Dover.
Goody, R.M. and Yung, Y.L. (1989) *Atmospheric Radiation Theoretical Basis*, 2nd edn. New York: Oxford University Press.
Lenoble, J. (ed.) (1985) *Radiative Transfer in Scattering and Absorbing Atmospheres: Standard Computational Procedures*. Hampton, Virginia: A. Deepak Publishing.
Liou, K.-N. (1980) *An Introduction to Atmospheric Radiation*. New York: Academic Press.
McCartney, E.J. (1976) *Optics of the Atmosphere. Scattering by Molecules and Particles*. New York: John Wiley and Sons.
Mihalas, D. (1978) *Stellar Atmospheres*, 2nd edn. San Francisco: W.H. Freeman and Company.

Cross references

Jupiter: atmosphere
Neptune: atmosphere
Saturn: atmosphere
Spectroscopy: atmosphere
Uranus: atmosphere

RADIO ASTRONOMY

Ground-based radio astronomical observations of planets, satellites asteroids and comets provide information on these objects which is complementary to that obtained from data at visual and infrared wavelengths (Plate 2). One typically measures thermal emission from planetary bodies; in addition, Jupiter emits also non-thermal or synchrotron radiation at wavelengths longwards of 6 cm. The thermal emission provides information on a planet's atmosphere and crust, while the non-thermal radiation can be used to map out the planet's magnetic field and energetic particle distributions.

Any object with a temperature above absolute zero emits a continuous spectrum of electromagnetic radiation at all wavelengths. This emission is referred to as thermal or 'blackbody' radiation. A black body is defined as an object which absorbs all radiation which falls on it at all frequencies and all angles of incidence; none of the radiation is reflected. The radiation from such an object can be described by Planck's radiation law (see Blackbody radiation) which, at radio wavelengths, can usually be approximated by the Rayleigh–Jeans law:

$$B_\nu(T) = \frac{2\nu}{c^2} kT \qquad (R3)$$

where $B_\nu(T)$ is the brightness, ν is the frequency, T is the temperature, k is the Boltzmann constant and c is the velocity of light. Using radio telescopes, radio astronomers measure the power flux density emitted by the object. A common unit is the flux unit or Jansky, where 1 Jy = 10^{-26} W m^{-2} s^{-1} Hz^{-1}. If a planet emits blackbody radiation and has a typical size of $a \times b$ arcsec, the flux density can be related to the temperature of the object:

$$S = \frac{abT}{490\lambda^2} \qquad (R4)$$

with λ the observing wavelength in cm. Usually planets do not behave like a black body, and the temperature T in equation (R4) is called a brightness temperature, defined as the temperature of an equivalent black body of the same brightness.

The radio power emitted by an object is received by a radio telescope (Figure R6), which basically consists of an antenna and a receiver. The sensitivity of the system depends upon many factors, but the most important are the effective aperture and system temperature. The effective aperture depends upon the size of the dish and the aperture efficiency. The sensitivity increases when the effective aperture increases and/or the system temperature decreases. The resolution of the telescope depends upon the size of the dish: the width of the antenna pattern at half power is approximately equal to

Figure R6 Aerial view of the Very Large Array in New Mexico. (Courtesy NRAO/AUI.) The National Radio Astronomy Observatory (NRAO) is operated by Associated Universities, Inc. (AUI) under contract with the National Science Foundation (NSF).

λ/D in radians, with D the dish diameter in the same units as λ. At 6 cm wavelength, an antenna with a diameter of 25 m has a resolution of approximately 8 arcmin. A large planet has a diameter of 0.5–1 arcmin and would be 'unresolved'. By connecting two radio antennas into an interferometer, the single antenna output is modulated on a scale λ/L; thus the resolving power of the instrument, in the direction of the projected baseline, depends upon the separation L of the two dishes. The VLA (Very Large Array) in Socorro, New Mexico, consists of a Y-shaped track, with nine antennas along each of the arms (Figure R6). With this instrument one gathers data from 351 individual interferometer pairs, each with its own resolution and projected baseline. With such an array of antennas one can build up an image which shows both the large-and small-scale structure of a radio source. Typical resolutions for planetary observations are in the range 0.5–4″.

For additional information on radio astronomy and interferometry the reader is referred to Kraus (1986), Thompson, Moran and Swenson (1986) and Perley, Schwab and Bridle (1989).

Terrestrial planets

Radio observations of the terrestrial planets can be used to extract information on the (sub)surface layers of these planets and, for Venus and Mars, on their atmospheres (de Pater 1990, 1991; and references therein for additional information). The temperature structure of the (sub)surface layers of airless bodies depends upon a balance between solar insolation, heat transport within the crust and reradiation outward. During the day a planet's surface heats up, and reaches its peak temperature near noon, while at night the object cools off. Its lowest temperature is reached just before dawn. Since it takes time for the heat to be carried downwards, there will be a phase lag in the diurnal heating pattern of the subsurface layers with respect to that at the surface, and the amplitude of the variation will be suppressed. At night heat is carried upwards and radiated away from the surface. Hence, while during the day the surface is hotter than the subsurface layers, at night the opposite is true. The diurnal heating pattern of the subsurface layers is largely determined by a combination of the thermal conductivity, the heat capacity, the density of the material and the planet's rotation period. At radio wavelengths one typically probes a depth of ~ 10 wavelengths into the crust. Hence, by observing at different wavelengths, one can determine the diurnal heating pattern of the Sun in the subsurface layers. Such observations can be used to constrain the thermal and electrical properties of the crust, which can be related to the compactness (i.e. rock versus dust) and mineralogy (i.e. metallicity) of the crustal layers (Mitchell and de Pater, 1994).

Venus and Mars have atmospheres which consist for more than 95% carbon dioxide gas (CO_2). The surface pressure of Venus' atmosphere is approximately 90 bar (surface pressure on Earth is 1 bar), and the microwave opacity of this much CO_2 gas is substantial. The opacity decreases with wavelength, and Venus' surface can be probed at wavelengths longwards of ~ 4 cm. The surface pressure of Mars' atmosphere is approximately 7 mbar, and the atmosphere is transparent throughout most of the microwave region. CO_2 gas is photodissociated by sunlight into carbon monoxide (CO) and oxygen (O). Carbon monoxide gas has strong rotational transitions at millimeter wavelengths, which can be utilized to determine the atmospheric temperature profile and the CO abundance on Venus and Mars in the altitude regions probed.

Moon

At radio wavelengths one probes the surface and subsurface layers of the Moon. As mentioned above, the temperature structure of these layers depends upon a balance between solar insolation, heat transport within the crust and reradiation outwards. Measurements of the

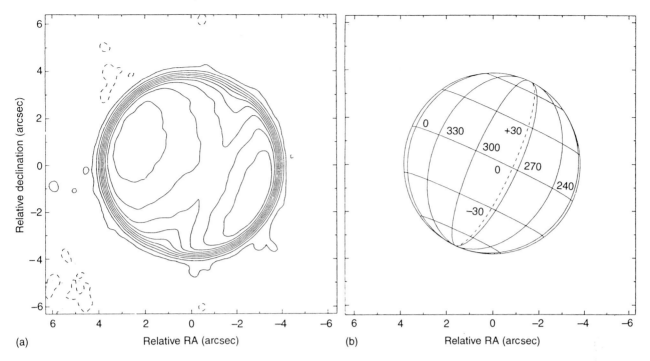

Figure R7 (a) Thermal emission from Mercury observed at λ = 3.6 cm with the VLA. Contours are at 42-K intervals (10% of maximum), except for the lowest contour, which is at 8 K (dashed contours are negative). The beam size is 0.4″. (b) The aspect of Mercury during the observation. The dashed line is the morning terminator, and the night hemisphere is to the east (left). (From de Pater and Mitchell, 1993.)

variations in brightness temperature with lunar phase, and the change in temperature over time when the Moon goes into an eclipse can be used to constrain the thermal and electrical properties of the upper 5–15 cm of the lunar regolith, also called 'soil' in analogy to terrestrial soil, although it does not contain organic matter.

In the mid-1970s lunar radio astronomy received new interest because radio receivers had improved dramatically and Apollo samples provided a ground truth for several sites on the Moon. For example, lunar core samples have been used to derive a density profile with depth near the Apollo landing sites: the regolith is very porous (density of ~ 1 g cm^{-3}) in the upper ~ 2 cm, after which it rises sharply to a density of 1.8 g cm^{-3}. This structure is likely caused by bombardment of small meteorites, which maintains the top layer at a low density while compacting deeper layers. Analysis of microwave and infrared measurements showed that the Apollo-based thermophysical properties are representative of a large portion of the lunar nearside hemisphere. However, radio images of the Moon reveal differences in brightness temperature between the maria and highlands. The maria are typically slightly (~ 5 K) warmer than the highlands at full Moon. This may result from a real difference in the subsurface temperature caused by the difference in albedo (reflectance) between maria (dark: low albedo) and highlands (light: high albedo). It may also be caused by a difference in radio emissivity between the maria and highlands or a difference in microwave opacity. There is evidence from lunar samples that the microwave opacity in the highlands is somewhat (by a factor of ~ 2) lower than in the maria. If true, deeper cooler layers will be probed in the lunar highlands during full Moon (as observed); at new Moon the temperature contrast should be reversed (no observations have yet been published), since the temperature increases with depth at night. It has further been found that the loss tangents of lunar fines increase with the abundance of iron and titanium bearing minerals, in particular ilmenite (FeTiO$_3$), which is the most common titanium-bearing mineral on the Moon. Because of the lower microwave opacity in the lunar highlands, ilmenite may be largely absent in these region compared to the lunar maria (e.g. Mitchell and de Pater, 1994, and references therein). Ilmenite is also opaque at optical wavelengths and is largely responsible for the dark appearance of maria compared to the highlands.

Mercury

As on the Moon, radio observations of Mercury probe the planet's surface and subsurface layers. Due to the 3 : 2 resonance between Mercury's rotational and orbital periods in combination with Mercury's large orbital eccentricity, the average diurnal insolation varies significantly. Regions along Mercury's equator near longitude ϕ = 0° and 180° (the subsolar longitudes when the planet is at perihelion) receive roughly 2.5 times more sunlight on average than longitudes 90° away. As a result of this non-uniform heating, the diurnal temperature variation depends upon the planet's longitude. The night-time surface temperature is approximately 100 K, independent of longitude, but the peak (noon) surface temperature varies between 700 K at ϕ = 0° and 180° to 570 K for ϕ = 90° and 270°.

Radio images of the planet show a brightness variation across the disk which displays the history of the solar insolation. At short wavelengths, where the surface layers are probed, the dayside temperature is usually highest. However, at longer wavelengths where deeper layers are probed, the diurnal heating pattern is less obvious, and one can distinguish two hot regions: one at a longitude of 0° and one at 180°. Figure R7 shows a radio image at 3.6 cm, where one probes ~70 cm into the crust. The viewing geometry is indicated in Figure R7b. The dashed line shows the morning terminator, and the nightside is to the east (left on the figure). The hot region at longitude 0° is clearly visible on the nightside, while the high temperature on the dayside is caused by solar insolation.

Radio spectra and images (like that shown in Figure R7), together with infrared data from the Mariner 10 spacecraft, have been used to derive Mercury's surface properties. The planet's surface appears to be quite similar to that of the Moon. The top few centimeters consist of a low-density (ρ ~ 1.0 g cm^{-3}) powder, which increases with depth in the crust, to ρ ~ 2 g cm^{-3} at a depth of about 2.5 ms. Like on the Moon, this structure is expected to result from small meteorite bombardment, which maintains the upper few centimeters at a very low density, while compacting deeper layers. The primary difference at radio wavelengths between Mercury and the Moon is in the microwave opacity: Mercury is much more transparent (by a factor of 2–3 in opacity) than the Moon. This suggests that the ilmenite abundance on Mercury is much smaller than on the Moon, which also explains why Mercury is much brighter than the Moon (Mitchell and de Pater, 1994).

Venus

Radio observations of Venus show a steep temperature increase between wavelengths of a few millimeters, where the brightness temperature is ~ 300 K, and ~ 6 cm, where the temperature is close

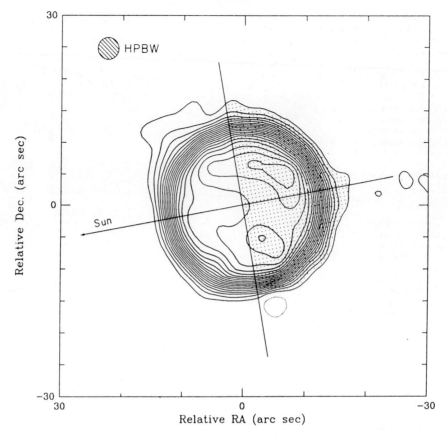

Figure R8 Radio image of Venus at 3 mm. The beam size is 3.5″. The direction to the Sun and the terminator are indicated. The contour levels are in 5% intervals, with a maximum of ~ 365 K. (From de Pater, Schloerb and Rudolph, 1991.)

to 700 K. This increase in temperature is due to the fact that Venus' atmosphere is opaque at millimeter and short centimeter wavelengths, but gets gradually more transparent at longer wavelengths. Longwards of ~ 4 cm the planet's surface can be probed. Due to the adiabatic temperature structure in Venus' atmosphere, the radio spectrum shows a steep temperature increase from millimeter to centimeter wavelengths. There are no diurnal temperature variations because of the atmosphere's large heat capacity.

Passive microwave as well as active radar observations have been used to study Venus' surface. Such observations allow determination of the planet's emissivity, which is related to the dielectric constant ϵ of the surface layers. The dielectric constant appears to be between 4 and 5 in the lowlands, and may range up to over 30 in the highlands. These values are well above those measured for the Moon, Mercury and Mars, which are typically in the range 2–3. A value of ϵ of ~ 2 implies porous surface materials; ϵ ~ 5–9 is typical for solid rocks (granite–basalt), while much higher dielectric constants can be caused by the inclusion of metallic and sulfide material. Hence Venus' surface is overlain, at most, by only a few centimeters of soil or dust, and probably consists of dry solid rock. The highlands may contain substantial amounts of minerals and sulfides close to the surface. On Earth such material is often formed in volcanic areas.

About half the microwave opacity in Venus' atmosphere is caused by CO_2 gas; the other half may be attributed to gaseous sulfuric acid (H_2SO_4) and sulfur dioxide (SO_2) gas (Steffes, 1986). The continuum emission at millimeter wavelengths originates approximately in the cloud layers. At 3 mm wavelength the nightside of the planet is approximately 10% brighter than the dayside (Figure R8). This is probably due to a lower opacity on this hemisphere, so one probes deeper warmer layers in the atmosphere. Spatial variations in opacity might be caused by local changes in cloud humidity (note that the clouds consist primarily of liquid sulfuric acid). In the upper part of the atmosphere CO_2 is photodissociated into CO and O. The ground and first excited rotational transitions of CO have been observed at wavelengths of 3 and 1 mm respectively. Since CO is formed in the

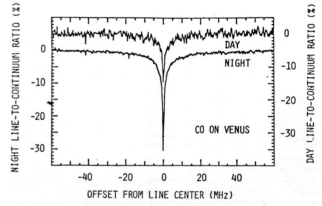

Figure R9 Spectra of Venus in the $J = 1$–0 line: the upper curve is for the dayside hemisphere (when Venus is near superior conjunction), the lower curve for the nightside hemisphere (when Venus is near inferior conjunction). (From Schloerb, 1985.)

upper part of the atmosphere, the line is seen in absorption against the warm continuum background. Typical altitudes probed in the CO lines are between 70 and 120 km, in the mesosphere, just above Venus' visible cloud layers. Examples of CO spectra on Venus' day and night side hemisphere are shown in Figure R9. It has been possible to determine the planet's temperature structure and altitude profile of the CO abundance from such observations. The spectra in Figure R9 show the nightside line to be approximately three times deeper and also narrower than the dayside line, which is suggestive of a larger concentration of CO at high altitudes on the nightside of the planet than the dayside. This is just opposite to what one would expect if CO is formed by photodissociation of CO_2.

Venus' cloud layers exhibit strong retrograde zonal winds, at approximately 100 m s^{-1} (see Venus: atmosphere). In contrast to the lower atmosphere, where the temperature does not show diurnal variations, at higher altitudes a large difference in temperature exists between the day and nightside. This temperature gradient causes winds to blow from local noon to midnight. Such winds have been suggested to blow the CO right after its formation from the dayside to the nightside, but this has not yet been proven. The mesosphere above the cloud layers is a transition region between the zonal winds in the cloud deck and the solar-to-antisolar winds in the upper atmosphere. This region is probed in the CO lines and, via measurements of the Doppler shift of these lines at different locations on Venus' disk, the mesospheric winds can be studied. This project is feasible in the near future, using the sensitive arrays of telescopes at millimeter wavelengths, which are currently being built.

Mars

Due to the eccentricity in Mars' orbit, the planet's surface brightness temperature varies approximately as $1/r^{1/2}$ where r is the heliocentric distance. The temperature of the crust may be slightly increased above that expected from solar illumination alone due to the atmospheric greenhouse effect. In addition, although atmospheric dust storms are transparent at radio wavelengths, these storms have a direct influence on the Martian flux: the dust particles increase the opacity at wavelengths shorter than ~ 40 μm, affecting the temperature gradient in the atmosphere, reducing the amount of solar radiation to the surface, and causing a decrease in the surface brightness temperature.

In addition to the diurnal heating cycle, the disk-averaged radio brightness temperature of Mars varies as a function of central meridian longitude of the planet, by up to 5–10 K. These variations are suggestive of a non-uniformity in the Martian surface properties, although no completely satisfactory explanation has yet been given. Radio images of the planet at millimeter wavelengths show that Mars is hottest in the afternoon, as expected from the the diurnal insolation pattern. Images at centimeter wavelengths show a reduced brightness temperature in the polar regions, by $\sim 20\%$ compared to the average brightness temperature. This can be explained by the presence of a thick layer of CO_2 frost.

Both the ground and first excited state of CO gas in the Martian atmosphere have been observed. An example of CO spectra is shown in Figure R10. Whether the line is seen in emission, absorption or a combination thereof, depends mainly on the temperature–pressure profile in the atmosphere and how this compares to the brightness temperature of the surface. The temperature profiles measured by the Viking spacecraft show a roughly constant temperature of 140 K down to an altitude of about 60 km, below which it increases to ~ 210–220 K at the surface. The CO lines are optically thick; thus the core of the line is formed high up in the atmosphere where it is cold, and hence is seen in absorption against the continuum background from the planet's surface. The wings of the line are formed in the lower atmosphere, just above the surface. As a consequence of the surface emissivity (< 1) the brightness temperature of the surface is somewhat less than the kinetic temperature in the atmosphere just above it. The wings of the line are therefore seen in emission against the continuum background. Since the CO abundance and the temperature structure in the Martian atmosphere both influence the line profiles, a unique solution of the CO abundance can only be found if measurements are made of both the optically thick ^{12}CO lines and of its isotope ^{13}CO, the lines of which are optically thin.

The CO abundance does not seem to vary over time, while the atmospheric temperature structure varies considerably. At present we do not have information on possible spatial variations in the CO abundance. After completion of new sensitive millimeter arrays this can be investigated. With such arrays one can also measure the wind velocity field in the Martian atmosphere.

Giant planets

At radio wavelengths in the millimeter–centimeter range one generally probes regions in the atmospheres of the giant planets Jupiter, Saturn, Uranus and Neptune which are inaccessible to optical or infrared wavelengths. One typically probes pressure levels of ~ 0.5–10 bar in Jupiter and Saturn's atmospheres, and down to 50–100 bar on Uranus and Neptune (de Pater, 1990, 1991; de Pater

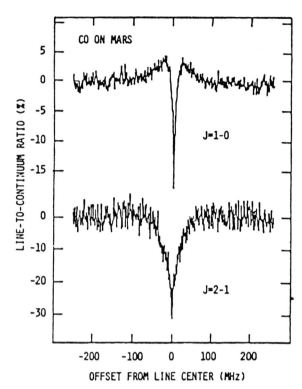

Figure R10 Spectra of Mars in the CO $J = 1$–0 and $J = 2$–1 transitions, during the 1984 opposition. (From Schloerb, 1985.)

and Mitchell, 1993). Much information is contained in the planet's radio spectrum: a graph of the disk-averaged brightness temperature of the planet as a function of wavelength. These spectra generally show an increase in brightness temperature with increasing wavelength beyond 1.3 cm, due to the combined effect of a decrease in opacity at longer wavelengths, and an increase in temperature at increasing depth in the planet. At millimeter–centimeter wavelengths the main source of opacity is ammonia gas, which has a broad absorption band at 1.3 cm. At longer wavelengths (typically > 10 cm) absorption by water vapor and droplets becomes important, while at short millimeter wavelengths the contribution of collision-induced absorption by hydrogen gas becomes noticeable. On Uranus and Neptune there may be additional absorption by hydrogen sulfide gas.

Radio spectra of the planets can be interpreted by comparing observed spectra with synthetic spectra, which are obtained by integrating the equation of radiative transfer through a model atmosphere. At first approximation the spectra of both Jupiter and Saturn resemble those expected for a solar composition atmosphere, while the spectra of Uranus and Neptune indicate a depletion of ammonia gas compared to the solar value by about two orders of magnitude. Resolved images of both Jupiter and Saturn show bands of enhanced brightness temperature on their disks implying latitudinal variations in the precise ammonia abundance. Both Uranus and Neptune show a brightening towards the visible (south) pole.

Jupiter

Radio signals from Jupiter were first detected in 1955 at a frequency of 22.2 MHz (Burke and Franklin, 1955). This emission was sporadic in character, and confined to frequencies of less than 40 MHz. It is commonly referred to as decametric radiation. In subsequent years the planet's thermal emission was detected at short centimeter wavelengths, and its synchrotron radiation at wavelengths between ~ 6 cm and roughly 3 m. The latter radiation is emitted by high-energy electrons in a Jovian Van Allen belt. Figure R11 shows a schematic representation of Jupiter's spectrum, showing the relative intensities and the frequency ranges where atmospheric, synchrotron and decametric emissions dominate.

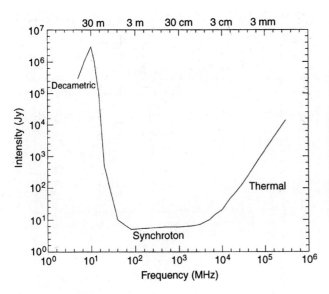

Figure R11 An approximate spectrum of Jupiter between 1 MHz and 300 GHz.

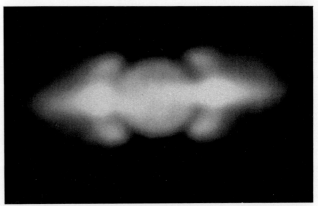

Figure R13 A radio photograph of Jupiter at a wavelength of 20 cm. The resolution is 3″ and the size of the disk is 41.6″. The data were obtained in 1981. The north pole is facing Earth (longitude λ_{III} = 200°). (From de Pater and Dickel, 1986.)

zonal regions, so deeper and warmer layers are probed in the belts. This is suggestive of gas rising up in the zonal regions; when the temperature drops below ~ 140 K ammonia gas condenses out. In the belts the air, now depleted in ammonia gas, descends. This general picture is in agreement with that suggested by other people based upon analysis of visible and infrared data (West, Strobel and Tomasko, 1986).

Synchrotron radiation

Synchrotron radiation is emitted by relativistic electrons gyrating around magnetic field lines (Legg and Westfold, 1968). The radiation is beamed in the forward direction within a cone $1/\gamma$, with $\gamma = 2E$, and E the energy in MeV. A typical energy of electrons in Jupiter's magnetic field emitting at 20 cm wavelength is about 25 MeV.

Figure R13 displays an image of Jupiter's synchrotron radiation at 20 cm. It shows that the radiation is confined to the magnetic equatorial plane out to a distance of approximately three Jovian radii, the orbital distance of the satellite Thebe. Several intriguing features are visible on this and other images. One of the main radiation peaks at each side of the planet usually appears to be brighter than the other peak. This can be visualized (see Figure R14) by imagining a torus of radiating electrons, which is denser at a certain longitude (~ 255° in system III, 1965.0 coordinates). The torus rotates with the planet, and the side at which the denser region is located will produce an enhanced emission. Thus the pattern varies with Jovian rotation. This enhancement in emission is probably caused by higher-order moments in Jupiter's field, although no detailed model to explain this enhancement yet exists. Other intriguing features are the secondary emission peaks just north and south of the main peaks. They must be produced by electrons at their mirror points, implying the presence of a rather large number of particles which cross the magnetic equator between 2.5 and 3 Jovian radii. These particles have not been seen by spacecraft.

The total flux density of the planet varies significantly in time. Variations on timescales of years seem to be correlated with solar wind parameters, in particular with the square root of the ram pressure, $(\rho v^2)^{1/2}$ where ρ is the density and v the velocity of the solar wind. This correlation suggests that the solar wind influences the supply and/or loss of electrons into Jupiter's inner magnetosphere (Botton et al., 1989; de Pater and Goertz, 1994).

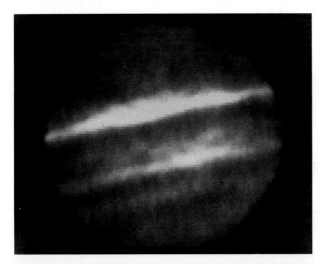

Figure R12 A radio photograph of Jupiter at a wavelength of 2 cm. The resolution is 1.2″ and the size of the disk is 32″. The image was obtained in the spring of 1983. (From de Pater and Dickel, 1986.)

Thermal emission

A detailed analysis of radio spectra of Jupiter's thermal emission suggests that the ammonia (NH_3) abundance in the planet's deep atmosphere is enhanced by a factor of approximately 1.2 compared to the solar nitrogen value. The abundance decreases at higher altitudes, due to condensation into an ammonium hydrosulfide (NH_4SH) cloud layer and an ammonia-ice cloud. The NH_4SH cloud is formed upon condensation of ammonia gas together with hydrogen sulfide (H_2S) gas. The decrease in the abundance of NH_3 gas together with the absence of H_2S gas at higher altitudes has been used to determine an abundance of 6–7 times the solar sulfur value for H_2S. The cloud layers seen at visible wavelengths are the ammonia-ice clouds.

A radio image of the planet at 2 cm wavelength is shown in Figure R12. At the time of the observations the resolution was 1.2″ and the disk diameter 32″. The image shows bright horizontal bands across the disk, which coincide with the brown belts seen at visible and infrared wavelengths. These bands have a higher brightness temperature, probably due to a depletion in ammonia gas relative to the

Decametric radiation

At frequencies below 40 MHz, or wavelengths longwards of 7.5 m, Jupiter is a strong emitter of sporadic non-thermal emission (Carr, Desch and Alexander, 1983). From the ground, emissions at decametric wavelengths can be observed. These emissions are characterized by a complex, highly organized structure in the frequency–time domain and depend upon the observer's position relative to Jupiter. In addition, the satellite Io strongly modulates the emission. This component of Jupiter's radiation is probably due to cyclotron emission from electrons which have their mirror points close to Jupiter's ionosphere. The details of the exact nature of the emission are not yet understood.

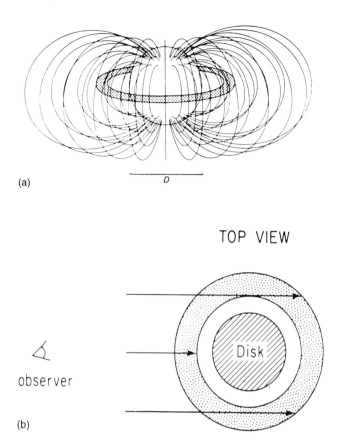

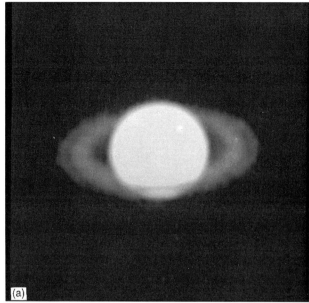

Figure R14 A schematic representation of the energetic electrons in Jupiter's magnetic field. (a) Front view, and (b) top view.

Saturn

Thermal radio emission can be detected both from Saturn's atmosphere and its rings. As with Jupiter, radio spectra of the atmospheric emission can be interpreted in terms of ammonia abundance and local variations with altitude and latitude. The ammonia and hydrogen sulfide abundances on Saturn are slightly higher than on Jupiter: NH_3 gas is enhanced by a factor of about three compared to the solar nitrogen value, while H_2S gas is enhanced by a factor of 10–15 compared to solar sulfur. Radio images of the planet are shown in Figure R10. At 2-cm wavelength the disk is extremely smooth, and no latitudinal structure can be distinguished. At 6-cm wavelength, however, a bright band is visible at mid-northern latitudes, indicating a depletion of NH_3 gas over the altitude region probed at this wavelength. The region at mid-latitudes is probably a region of subsiding gas, like the bright belts seen on Jupiter. A similar conclusion was reached from an analysis of infrared data.

As shown in Figure R15, in addition to atmospheric emission, we receive radiation from Saturn's main rings, the A, B and C rings (see Saturn: ring system). This emission consists primarily of Saturn's radiation scattered by the ring particles. The ring brightness temperature can be measured at either side of the planet; on average, at microwave frequencies the temperature is roughly 6 K. In front of the planet the rings block out part of Saturn's radio emission, resulting in an absorption feature. From this feature one can determine the opacity or optical depth of the rings. The optical depth is the integral along the line of sight of the particle density multiplied by the total absorption coefficient. One usually considers the normal optical depth, where the integral is taken along a path perpendicular to the ring plane. The optical depth is largest in the B ring, where it is approximately unity (1–1.5); it is about half that value in the A ring, and only 0.03 in the C ring. Variations in the optical depth as a function of wavelength and ring inclination angle, together with a radio spectrum of the ring brightness temperature have been used to

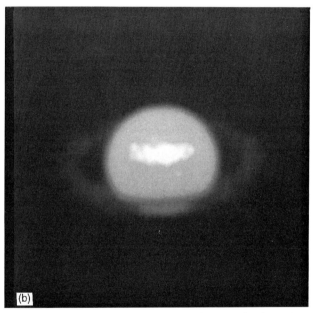

Figure R15 Radio photographs of Saturn at 2 cm (a) and 6 cm (b) wavelength. The resolution is 1.5″, and the sizes of the disks are ~15 and 18″ respectively. The observations were made in 1986. (From de Pater and Dickel, 1991.)

determine the scattering properties of the ring particles. Such properties are unique for certain size distributions and compositions of the particles. The rings consist primarily of icy particles, with sizes between a few millimeters up to ~5 m. The size distribution varies approximately as a^{-3}, where a is the particle radius (Cuzzi et al., 1980).

Uranus and Neptune

Uranus is unique among the giant planets in having its rotation axis closely aligned with the plane of the ecliptic. With its orbital period of 84 years, the seasons on Uranus last for 21 years. In the late 20th century, Uranus' south pole is on the Sunward side and facing Earth. In another 20 years we will see the opposite pole. This geometry must have a pronounced effect on the large-scale circulation of the Uranian atmosphere.

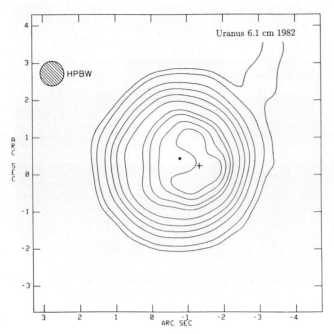

Figure R16 Contour map of Uranus at a wavelength of 6 cm. The cross indicates the position of the pole, the dot that of the subsolar point. The resolution of the image is 0.65", and contour values are 14, 28, 69, 110, 151, 193, 248, 261 and 270 K. (From de Pater and Gulkis, 1988.)

The opacity in the atmospheres of Uranus and Neptune is dominated by collision-induced absorption by hydrogen gas at millimeter wavelengths and by ammonia gas and hydrogen sulfide at centimeter wavelengths. The radio spectra of Uranus and Neptune suggest an overall depletion of ammonia gas in their atmospheres, by roughly two orders of magnitude compared to the solar nitrogen value. Considering models of planetary formation, it is highly unlikely that NH_3 gas is depleted throughout the outer planet atmospheres. A possible explanation for the apparent depletion is that NH_3 gas is completely condensed out in the NH_4SH cloud. This can happen if there is more H_2S gas present than NH_3 gas. Thus, in contrast to Jupiter and Saturn, we will find H_2S gas and no NH_3 gas above the NH_4SH cloud layer on both Uranus and Neptune. The ammonia abundance may be similar to the solar nitrogen value at deeper levels of Uranus and Neptune's atmospheres, while H_2S gas may be enhanced above the solar sulfur value by a factor of 10–30 on Uranus and ~ 30–60 on Neptune (de Pater, Atreya and Romani, 1991). Around the ~ 140-K level in the atmosphere, H_2S gas freezes out and forms an ice cloud, as NH_3 gas does on Jupiter and Saturn. The H_2S-ice cloud, however, is not the cloud seen at visible wavelengths; at visible wavelengths one sees the methane-ice cloud. Near the tropopause the temperature on Uranus and Neptune drops well below 80 K, the approximate level at which methane gas freezes out.

Uranus' brightness temperature has been monitored since 1966. A pronounced increase in brightness temperature was noticed when the south pole came into view. This was the first indication that the polar regions are considerably warmer than the equatorial region, a theory later confirmed by radio images obtained with the VLA. Figure R16 shows an image of Uranus at 6-cm wavelength. The brightest point on the disk is displaced towards the pole, and the temperature contrast between the pole and the equator is roughly 60 K. In analogy to the bright regions on Jupiter and Saturn, the pole is probably warmer than the equator due to a relative lack of absorbing gases. Such a scenario implies descending gas in the polar region and rising gas at other latitudes. A similar scenario has been postulated for Neptune, where the brightness temperature around the equator is depressed compared to that at the (visible) south pole (Gautier et al., 1994).

The vertical rising and subsiding motions in the deep atmospheres of the giant planets as deduced from radio data generally agree with the vertical motions derived from visible/infrared data at higher altitudes. Hence the vertical motions are true large-scale motions, generally extending from the cloud tops at levels of a few hundred millibars, down to at least the bottom of the NH_4SH clouds, if not deeper. On Jupiter and Saturn the NH_4SH cloud forms at approximate levels of 2.5 and 5 bar respectively, while on Uranus and Neptune it is closer to 40 bar.

Asteroids and satellites

Asteroids

In analogy to the terrestrial planets, radio spectra of asteroids and airless satellites provide information on the (sub)surface properties of the material, in particular the compactness (rock versus dust) and metallicity. A small number of asteroids have been observed between $\lambda = 350$ μm and 20 cm. Unfortunately, the measurement uncertainties are rather large. Most asteroids observed show a pronounced decrease in brightness temperature between ~ 10 μm and centimeter wavelengths. This is due to the fact that at longer wavelengths one probes deeper and colder layers in the body's crust (note that we can only observe the dayside of objects which orbit the Sun at distances larger than 1 AU). The precise shape of the spectrum can be interpreted in terms of the compactness of the material and variations therein with depth. Most asteroids are covered to a depth of many meters by a layer of regolith, overlain by a few centimeters of fluffy dust, similar to the structure seen on the Moon and Mercury (Webster and Johnston, 1989).

Galilean satellites

The surface temperature of an object depends upon its bond albedo, which is the ratio between total reflected and incident sunlight. Satellites with low albedos are warmer than bodies which reflect most of the incident sunlight. The surface temperature of a satellite can be determined from measurements at infrared wavelengths, while radio observations can be used to determine the temperature at depth in a solid object. Usually, the infrared and radio emissivities are slightly less than unity, ~ 0.9, which means that 90% of the thermal (blackbody) radiation is radiated into space. One can easily correct the measurements for this emissivity, which slightly raises the physical temperature of the object.

The situation for the Galilean satellites is quite different, however (de Pater, Brown and Dicket, 1984). The infrared emissivity is ~ 0.9, and the brightness temperature at infrared wavelengths can be converted directly into a surface temperature. Radio observations of the satellites, however, showed that the radio emissivity for Ganymede and Europa is far from unity at about 0.5. These low emissivities agree with radar measurements. In a radar experiment microwaves are transmitted to an object, and the reflected signal is measured. The sum of the radar reflectivity and radio emissivity must be approximately unity. Indeed for Ganymede and Europa the radar reflectivities are very high: 0.33 for Ganymede and 0.65 for Europa. The high radar reflectivities, and consequently low emissivities, are probably caused by a coherent backscattering mechanism (Ostro, 1993).

Titan and Io

Titan and Io each have an atmosphere which has been observed at radio wavelengths. Titan's atmosphere consists primarily of nitrogen gas, with traces of methane, argon, CO and a variety of hydrocarbons (Chamberlain and Hunten, 1987). Titan's surface can be probed at radio wavelengths. The surface temperature is 94 K and the pressure is 1.5 bar. At this temperature ethane will form a liquid, and it has been suggested that Titan might be covered by a 1 km thick ocean of liquid ethane. Radio measurements obtained since that time contradict a global ethane ocean, although lakes of liquid ethane may exist (Muhleman et al. 1990). Measurements of the CO line on Titan can be used to derive the temperature structure in Titan's atmosphere.

Io's atmosphere is probably dominated by SO_2 gas, with a surface pressure of approximately 10–40 nbar (Lellouch et al., 1992). The first global measurement of Io's atmosphere has been obtained in the rotational SO_2 line at a frequency of 222 GHz (a wavelength of

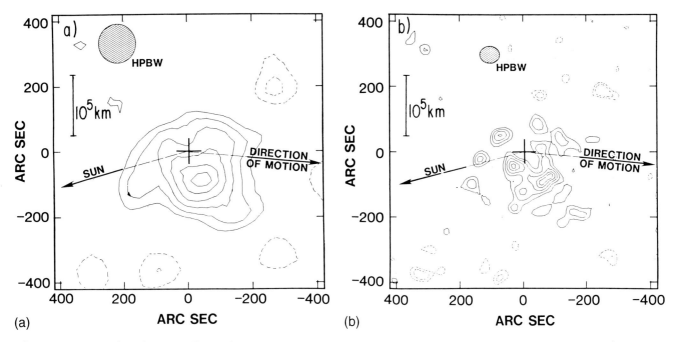

Figure R17 Contour plots of comet Halley, 13–16 November 1985. The image is taken at the peak flux density of the line (0.0 km s^{-1} in the reference frame of the comet). (a) A low-resolution image (3') and (b) a high-resolution image (1'). Contour levels for the low-resolution image are 4.9, 7.8, 10.8, 13.7, 16.7 and 18.6 mJy per beam. For the high-resolution image the contours are 4.4, 4.4, 6.0, 7.7, 9.3 and 10.4 mJy per beam. Dashed contours indicate negative values. The beam size, a linear scale, the direction of motion and the direction of the Sun are indicated in the figures. The cross indicates the position of the nucleus at the time of the observations. (From de Pater, Palmer and Snyder, 1986.)

1 mm). The radio data suggest that Io's atmosphere covers only a fraction (5–20%) of its surface, is relatively hot (500–600 K at 40 km altitude) and exhibits a global, temporal stability. It is not clear whether the atmosphere is of volcanic origin, or in vapor pressure equilibrium with SO_2 frost on its surface.

Comets

Radio observations of comets provide information which complements studies at other wavelengths. Continuum measurements are sensitive to the thermal emission from the cometary nucleus and any bulk material around the comet. Observations of a few comets suggest that the temperature gradient in the nucleus may be very steep, or that the radio brightness temperature at depth may be substantially depressed by subsurface scattering.

The most significant advances in cometary radio research have been obtained from spectroscopic studies. The cometary nucleus consists primarily of water ice, which sublimates off the surface when the comet approaches the Sun. It is difficult to detect H_2O directly from the ground; the first definitive detection of water was not obtained until the apparition of comet Halley in 1986. The detection was made at infrared wavelengths, using the Kuiper Airborne Observatory (Mumma *et al.*, 1993). Water dissociates into OH and H. It is relatively easy to observe the hydroxyl (OH) molecule at a wavelength of 18 cm. Since the early 1970s OH has been monitored for several comets. The line sometimes appears in emission (maser emission), and sometimes in absorption against the galactic background. Whether it appears in emission or absorption depends upon the comet's velocity with respect to the Sun (heliocentric velocity). The OH radical is excited by solar UV photons; however, when the heliocentric velocity of the comet is such that solar Fraunhofer lines are Doppler shifted into the excitation frequency, the radical is not excited. In that case it will absorb 18-cm photons from the galactic background and be seen in absorption against the galactic background. If the line is excited, 18-cm background radiation will trigger its de-excitation, and the line is seen in emission. A detailed study of the OH line as a function of heliocentric velocity has revealed many clues regarding the OH, and hence water production rate of comets. The production rate varies drastically on short (less than 1 day) timescales. Spectra of the OH line suggest that the emission is usually not isotropic. Images of the OH emission show that the spatial brightness distribution can be very irregular; it appears to change on short timescales in spatial as well as velocity coordinates. An image of comet Halley in the OH line is shown in the Figure R17.

Radio observations are carried out at wavelengths between a few tenths of a millimeter and 18 cm. Many potential parent molecules in comets have rotational transitions in this region. Hence one of the potential strengths of radio astronomy is the detection of parent molecules in a cometary coma. Such detections are crucial for our understanding of the cometary composition, and the formation of the entire solar system. To date, many molecular species have been detected at radio wavelengths, e.g. hydrogen cyanide (HCN), formaldehyde (H_2CO), hydrogen sulfide (H_2S) and methanol (CH_3OH). Whereas most molecules seem to originate on the cometary nucleus, the source of formaldehyde seems to be distributed over a larger space, and may originate on dust grains in the coma. This observation was confirmed by *in situ* measurements with the Giotto spacecraft of Comet Halley (Krankowsky, 1991). With the advent of new powerful millimeter arrays, the distribution of parent molecules in the cometary coma can be imaged directly. This will provide additional information to constrain cometary models and theories regarding the formation and evolution of our solar system.

Imke de Pater

Bibliography

Bolton, S.J., Gulkis, S., Klein, M.J., de Pater, I. and Thompson, T.J. (1989) Correlation studies between solar wind parameters and the decimetric radio emission from Jupiter. *J. Geophys. Res.*, **94**, 121.

Burke. B.F. and Franklin, K.L. (1955). Observations of a variable radio source associated with the planet Jupiter. *J. Geophys. Res.*, **60**, 213.

Carr, T.D., Desch, M.D. and Alexander, J.K. (1983) Phenomenology of magnetospheric radio emissions, in *Physics of the Jovian Magnetosphere* (ed. A.J. Dessler). Cambridge University Press, Cambridge: pp. 226–84.

Chamberlain, J.W. and Hunten, D.M. (1987) *Theory of Planetary Atmospheres*. New York: Academic Press.

Cuzzi, J.M., Pollack, J.B. and Summers, A.L. (1980) Saturn's rings: particle composition and size distribution as constrained by observations at microwave wavelengths, II. Radio interferometric observations. *Icarus*, **44**, 683.

de Pater, I. (1990) Radio images of planets, *Ann. Rev. Astron. Astrophys.*, **28**, 347–99.

de Pater, I. (1991) The significance of radio observations for planets, *Phys. Rep.*, **200:1**.

de Pater, I., Atreya, S.K. and Romani, P.M. (1991) Possible microwave absorption by H_2S gas in Uranus and Neptune atmospheres. *Icarus*, **91**, 220.

de Pater, I. and Dickel, J.R. (1986) Jupiter's zone–belt structure at radio wave-lengths: I. Observations. *Astrophys. J.*, **308**, 459–71.

de Pater, I. and Dickel, J.R. (1991) Multi-frequency VLA observations of Saturn at ring inclination angles between 5° and 26°. *Icarus*, **94**, 474–92.

de Pater, I. and Geortz, C.K. (1994) Radial diffusion of energetic electrons and Jupiter's synchrotron radiation: II. Time variability. *J. Geophys. Res.*, **99**, 2271–87.

de Pater, I. and Gulkis, S. (1988) VLA observations of Uranus at 1.3–20 cm. *Icarus*, **75**, 306–323.

de Pater, I. and Mitchell, D.L. (1993) Microwave observations of the planets: the importance of laboratory measurements. *J. Geophys. Res., Planets*, **98**, 5471–90

de Pater, I., Brown, R.A. and Dickel, J.R. (1984) VLA observations of the Galilean satellites. *Icarus*, **57**, 93–101.

de Pater, I., Palmer P. and Snyder, L.E. (1986) The brightness distribution of OH around comet Halley. *Astrophys. J. Lett.*, **304**, L33–6.

de Pater, I., Schloerb, F.P. and Rudolph, A. (1991) CO on Venus imaged with the Hat Creek Radio Interferometer. *Icarus*, **90**, 282–98.

Gautier, D., Conrath, B.I., Owan, T. *et al.* (1995) The troposphere of Neptune, in *Neptune and Triton* (eds D.P. Cruikshank and M.S. Matthews). Tucson: University of Arizona Press, pp. 547–612.

Krankowsky, D. (1991) The composition of comets. In *Comets in the post-Halley era* (ed. R.L. Newburn Jr., M. Negebauer and J. Rake), p. 855.

Kraus, J.D. (1986) *Radio Astronomy*. Powell, Ohio: Cygnus Quasar Books.

Legg, M.P., and Westfold, K.C. (1968) Elliptical polarization of synchrotron radiation. *Astrophys. J.*, **154**, 499–514.

Lellouch, E., Belton, M., de Pater, I. *et al.* (1992) The structure, stability, and global distribution of Io's atmosphere. *Icarus*, **98**, 271–95.

Mitchell, D.L. and de Pater, I. (1994) Microwave imaging of Mercury's thermal emission: observations and models. *Icarus*, **110**, 2–32.

Muhleman, D.O., Grossman, A.W., Butler, B.J. and Slade, M.A. (1990) Radar reflectivity of Titan. *Science*, **248**, 975–80.

Mumma, M.J., Weissman, P.R. and Stern, S.A. (1993) Comets and the origin of the solar system: reading the Rosetta Stone. In *Protostars and Planets III* (eds E.H. Levy and J.I. Lunine). Tucson: University of Arizona Press, p. 1177.

Ostro, S.J. (1993) Planetary radar astronomy. *Rev. Modern Phys.*, **65**, 1235–79.

Perley, R.A., Schwab, F.R. and Bridle, A.H. (1989) Synthesis imaging in radio astronomy NRAO Workshop No. 21, Astronomical Society of the Pacific.

Schloerb, F.P. (1985) Millimeter-wave spectroscopy of solar system objects: present and future. Proceedings of the ESO-IRAM-Onsala Workshop on (sub) Millimeter Astronomy, Aspenas, Sweden, 17–20 June 1985 (eds P.A. Shaver and K. Kjar). pp. 603–616.

Steffes, P.G. (1986) Evaluation of the microwave spectrum of Venus in the 1.2–22 centimeter wavelength range based on laboratory measurements of constituent gas opacities. *Astrophys. J.*, **310**, 482–9.

Thompson, A.R., Moran, J.M. and Swenson, G.W., Jr, 1986. *Interferometry and Synthesis in Radio Astronomy*. New York: John Wiley and Sons.

Webster, W.J. and Johnston, K.J. (1989) Passive microwave observations of asteroids, in *Asteroids II* (eds R.P. Binzel, T. Gehrels and M.S. Matthews). Tucson, Arizona: University of Arizona Press, pp. 213–269.

West, R.A., Strobel, D.F. and Tomasko, M.G. (1986) Clouds, aerosols and photochemistry in the Jovian atmosphere. *Icarus*, **65**, 161–217.

RADIO SCIENCE

Planetary radio science focuses on the use of radio signals traveling between spacecraft and an Earth terminal for scientific investigation of planets and their environs. These signals provide an extremely precise measurement of the radio path between the ground station and the spacecraft, and such measurements in turn are employed to infer important characteristics of planetary systems. The technique is applied to the study of planetary atmospheres (including ionospheres), rings, surfaces and gravity. Much of our fundamental knowledge of these subjects has been derived from radio science observations. Examples of recent and current radio science investigations are those conducted with Voyager (Eshleman *et al.* 1977; Tyler, 1987), Galileo (Howard *et al.*, 1992) and Mars Observer (Tyler *et al.*, 1992). Earlier missions which incorporated radio science investigations included the Mariners, Pioneers and Viking, as well as Soviet missions.

Radio science experiments fall into two broad categories of investigation. First, for the study of planetary environments, the orbit or trajectory of the spacecraft is arranged so that the spacecraft passes behind the planetary body as seen from the tracking station on the Earth. In this case the spacecraft is said to be occulted by the planet. In the Saturn occultation experiments with Voyager 1, for example, the spacecraft, when approaching Saturn, flew behind the satellite Titan and then, about 18 h later, passed behind the ball of Saturn itself and finally, after emerging from behind Saturn, passed behind Saturn's rings (Stone and Miner, 1981). In the occulted intervals radio signals were not blocked entirely, but passed through the atmospheres of Titan and Saturn, and through the rings before being received on Earth (Figure R18). During occultation events one 'senses' the media of interest – atmospheres or rings – by use of the radio signal as an active probe (Eshleman, 1973). The geometry and other experimental conditions must be controlled so that the only unknown factors are the properties of the medium along the radio path. Changes in the radio signals that are not associated with known factors are used to infer the properties of the unknown medium.

Second, when the radio path is well clear of occulting bodies, the spacecraft can be considered as a test 'particle' falling in the gravity field of the planetary system, and the component of its velocity along the line of sight to the tracking station can be measured by the Doppler effect. In contrast with occultation experiments, gravity experiments are based on determining the motion of the spacecraft in response to the gravity field of a planet and its satellites.

Both approaches have counterparts in classical astronomy. Stellar occultations are used to determine the presence or absence of an atmosphere on other bodies by observing the rate of extinction of a star as it passes into occultation by a planet (e.g. Elliot *et al.* 1989)

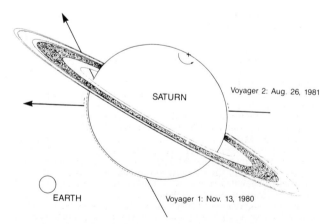

Figure R18 Voyager occultation geometry at Saturn. Figure shows the Earth view of Voyager trajectories at Saturn; stippled portions represent the track of the virtual image of spacecraft in the atmosphere of Saturn. Voyager 1 was occulted by Titan about 18 h before the start of the Saturn occultation (not shown); the spacecraft emerged from Saturn occultation near the equator, then passed behind the rings. Voyager 2 was occulted by the atmosphere only. The geometry is typical of spacecraft occultation experiments.

and to study planetary rings by observing the extinction of starlight (e.g. Elliot and Nicholson, 1984). Earth-based measurements of the motions of natural satellites are used to determine the gravity fields of planets. When available, radio science methods provide much greater accuracy and dynamic range than these classical approaches.

Short of *in situ* measurements by entry probes, radio occultation studies of atmospheres provide the most detailed information available on the vertical structure of the neutral atmosphere, the ionospheric structure and atmospheric waves. However, the information obtained applies only at the location of the occultation event. Radio occultation measurements strongly complement and extend the use other spacecraft and Earth-based remote sensing techniques, such as infrared spectroscopy, which provide detailed information on atmospheric constituents and low vertical resolution structure over wide regions by instrumental scanning. The best results are obtained when radio occultation and these other observations are combined.

Radio tracking studies of gravity are uniquely suited to the determination of interior structure of planets, for those cases for which the appropriate geometrical conditions can be obtained. To date, the intense radiation field of Jupiter's magnetosphere and the hazard of planetary rings at Saturn and Uranus have prevented close approach to these planets by spacecraft, thereby limiting the utility of this technique at the outer planets. In these cases the best information is obtained by classical methods. However, spacecraft methods appear to be superior in all instances for determination of total system mass and the masses of individual satellites.

History

Radio occultation experiments for the study of planets were first suggested by V.R. Eshleman of Stanford University and described in Booker *et al.* (1962). Initial implementations grew out of engineering studies of the effects of atmospheres on radio tracking data led by D.L. Cain at the Jet Propulsion Laboratory. The first radio occultation experiment was conducted at Mars using Mariner 4 on 14 July 1965 (Kliore *et al.*, 1965; Fjeldbo and Eshleman, 1968; see Mariner missions). Prior to Mariner 4, ground-based observers had concluded that the atmosphere of Mars consists predominantly of nitrogen, with a surface pressure in the range of perhaps 100 mbar, or about one-tenth that of Earth, which possibly could be capable of supporting terrestrial life provided also that oxygen was present. After the launch of Mariner 4, doubts regarding the interpretation of the ground based results led the NASA/JPL team controlling the mission to reprogram the spacecraft to pass behind Mars for the purpose of resolving this question. During the brief periods of occultation immersion and emersion the radio signals probed Mars' atmosphere. Careful analysis of the radio data showed that at the time the radio path grazed the surface of Mars the spacecraft appeared to be roughly 2 m farther from Earth than was calculated from the trajectory during the approach to Mars. This apparent increase in distance was caused by an extra delay of 0.007 μs as the radio signal passed through the thin Martian atmosphere. Further analysis of the data revealed an atmosphere comprised predominantly of carbon dioxide at a pressure of about 5 mbar and a temperature of about 180 K (or $-93°C$), conditions which were likely to be the result of an atmosphere in vapor pressure equilibrium with the polar caps. The polar caps themselves were inferred to be made up largely of dry ice (see Mars: atmosphere).

Subsequent development of the theory and experimental methods led to use of the radio occultations to explore further the atmosphere of Mars with Mariners and the Viking orbiters (q.v.; Lindal *et al.*, 1979), and the atmosphere of Venus with Mariners and with Pioneer Venus Orbiter (q.v.; Lipa and Tyler, 1979; Kliore and Patel, 1982). The most important contributions to the theory of occultation were made during this time and later by G.F. Lindal. Limited occultation measurements were conducted by the Soviets with Venera spacecraft (Ivanov-Kholodny *et al.*, 1979). The first occultations of Jupiter and Saturn were conducted by Pioneer 10 and 11 (q.v.; Fjeldbo *et al.*, 1976; Hunten and Veverka, 1976; Kliore and Woiceshyn, 1976).

Voyager 1 and 2 were the first spacecraft for which the detailed design was optimized for radio science observations. Substantial improvements in measurement capabilities over previous missions were achieved for both occultation and tracking data. These spacecraft explored the atmospheres and rings of the four gas giants, and the atmospheres of the satellites Titan (q.v.) and Triton (q.v.; Eshleman *et al.*, 1979a,b; Tyler *et al.*, 1981, 1982, 1986, 1989; Tyler, 1987). The greatest atmospheric depth sensed by Voyager was at Neptune, where the occultation probed to a depth of 6.3 bars (Lindal, 1992); the same experiment was also able to measure surface conditions at Neptune's moon, Triton, where the pressure was only about 14 mbar. The greatest depth probed to date is ~ 7 bar, at Venus by Mariner 10 and Magellan (q.v.).

Ulysses (q.v.) obtained significant occultation data from the Jupiter–Io torus (Bird *et al.*, 1992), but was not occulted by the ball of Jupiter. Galileo (q.v.) has planned occultation measurements of Jupiter commencing in December, 1995. The planned Cassini (q.v.) mission to Saturn includes an elaborate series of occultation studies of Saturn's atmosphere and rings, and of the satellite Titan, beginning about 2004.

Gravity investigations have been conducted as part of all the missions cited above. Only the Pluto–Charon system has not been visited by spacecraft and observed through use of radio science methods.

Equipment

Radio science instrumentation comprises the combination of equipment on the ground and onboard spacecraft needed to create and maintain a highly stable and precise radio link. Most commonly, radio signals generated on the ground and transmitted through the large antennas of the NASA Deep Space Network (q.v.) – dubbed 'uplink' signals – are received by planetary spacecraft, shifted in frequency, and then retransmitted to the Earth – as 'downlink' signals – where they are received either at the original site or at a second site, possibly located on another continent. The frequency shift at the spacecraft is introduced so that the downlink signals can be received simultaneously while transmitting without interference between the uplink and downlink systems. The spacecraft relay equipment is called a transponder. (The transponder employed for radio science is the same device used for communication with the spacecraft.) The frequency of the downlink signal retransmitted by the spacecraft is related to the received uplink frequency by an integer ratio which is known exactly. Because the downlink signal frequency is derived in a precise way from that of the uplink, it is possible to measure changes in the radio pathlength by comparison of the received, downlink signal with the ground oscillator that originally generated the uplink signal. An increase in the radio pathlength retards (decreases) the phase of the received downlink signal relative to the ground oscillator, while a decrease in pathlength has the opposite effect. Measurement of the downlink phase thereby provides an extremely precise method of determining changes in the round trip propagation time to the spacecraft. The rate of change of phase gives the rate of change of the radio path in units of wavelengths, and is usually expressed as a Doppler frequency. A 1 Hz difference between the frequencies of the uplink and downlink signals means that the total radio pathlength is changing at the rate (or line-of-sight velocity) of one wavelength per second; larger or smaller frequency differences correspond to proportionally larger or smaller rates of pathlength change. Overall, the accuracy of the measurement procedure depends on the signal-to-noise ratio achieved and, ultimately, on the stability of the ground station oscillator over the round trip flight time of the radio signals to the spacecraft and back (Eshleman and Tyler, 1975; Lipa and Tyler, 1979).

On the ground, US stations for communication over interplanetary distances are built around the large antennas of the NASA Deep Space Network. These stations are used primarily for uplink transmission of commands and downlink reception of spacecraft data (Yuen, 1983). Antennas up to 70 m in diameter, transmitting up to 400 kW, are available. On the spacecraft, typical antenna sizes are only a few meters, at most, and the transmitted downlink signal power ranges from about 1 to 20 W. At the present time hydrogen maser atomic clocks are used for the ground station frequency reference while microwave frequencies in the 2 and 8 Gz bands, corresponding to 12–13-cm and 3–4-cm wavelengths respectively, are used for the radio signals. Either band can be used separately or both used simultaneously. Future trends are to the use of even higher frequencies. Frequency changes as small as about 0.001 Hz can be measured, corresponding to a fractional accuracy in the range of a few parts in 10^{14} (Tyler *et al.*, 1992). In the absence of other effects, this would lead to an accuracy in the measurement of spacecraft velocity, for example, in the range of 30 μm s^{-1} when the 8-Gz band is used. Tracking accuracy in this range has been achieved with the Magellan and Mars Observer spacecraft. The use of a two-way, uplink/downlink radio path is suitable for study of gravity and for spacecraft navigation purposes. Because it relies on the reception by the spacecraft transponder of an uncorrupted uplink signal, the two-way technique can be used reliably only under free-space uplink propagation conditions, however (Tyler, 1987).

As an alternative, signals generated onboard a spacecraft can be transmitted directly to the ground, in which case the observations rely on comparison of signals derived from a spacecraft oscillator with those from ground-based oscillators at the tracking station. For this 'one-way' mode the relative stability of the two oscillators is important, rather than the stability of a single oscillator over time. 'One-way' observations are advantageous for occultation measurements during which the uplink signals undergo more rapid changes in signal amplitude and frequency than present spacecraft transponders can reproduce reliably; the one-way technique also greatly simplifies and strengthens the analysis and interpretation of occultation experiments since only a single path through the media of interest need be considered. Practical spacecraft oscillators currently are based on quartz crystal resonator technology that can rival or exceed the stability performance of atomic clocks on timescales of a few to a few hundred seconds. As these time intervals are well matched to the duration of many occultation events, reliable occultation measurements are obtained. Over time intervals greater than about 15 min, however, the performance of quartz oscillators deteriorates, so the one-way technique currently is not suitable for tracking or gravity experiments on longer timescales. Atomic clocks based on rubidium, under development in Europe for future planetary radio science applications, could overcome this limitation.

Applications

Gravity

The gravity field of a planet is the result of its mass and depends in detail on the manner in which the mass is distributed. Rapidly rotating planets bulge at the equator as a result of centrifugal forces, for example, and the associated concentration of mass in the equatorial plane results in a stronger gravitational acceleration on a spacecraft when it is near the pole than when it is at the same distance from the planet's center but located over the equator. Similarly, variations in the internal distribution of mass (or density) are expressed as departures of the external gravitational field from that of a uniform sphere or a radially stratified spherical distribution. In response to these irregularities in the gravity field, space probes in the vicinity of planetary bodies follow trajectories that differ significantly from the ideal orbits described by Kepler's laws.

Two-way radio tracking provides an accurate measurement of spacecraft velocity along the line of sight to the tracking station; measurements of line-of-sight velocity can be differentiated with respect to time to obtain the corresponding gravitational acceleration, thus sensing the mass of a planet or satellite. Such data, through the connection between mass and gravitational force, are used to study planetary interiors. The utility of the data for study of mass variations is dependent upon the sensitivity of the radio system (see above), the details of the geometry and the characteristics of the overall spacecraft trajectory. The method is capable of detecting surprisingly small effects.

As two examples, consider the Voyager mission to the outer planets (Campbell and Anderson, 1989; Jacobson et al., 1992) and the Mars Observer mission (Tyler et al., 1992). In the case of Voyager, 'flyby' trajectories were used to survey Jupiter, Saturn, Uranus and Neptune, each of which is at the center of a complex system of small bodies including ensembles of satellites and rings. Typically, Voyager flew within 10 000 to 100 000 km of the central planet and of one or more satellites. The total time spent within each system was approximately 1 Earth day. Yet these encounters permitted determination of the masses of the total system and of the central planet to an accuracy of about 0.001%. The degree of dynamical oblateness, or the equatorial mass bulge, was found, and in some cases the next order of distortion, the so-called quadrupole moment, could be estimated. It was also possible to determine the masses of all the major satellites at the level of about 1% or better, even though only a few of the satellites were approached closely. The latter determination follows from the effect the satellites have on the motion of the primary body, which in turn perturbed the motion of Voyager as it passed through the system. Since each satellite has a different orbital period, their effects can be separated by analyzing the time variations in the tracking data. These results substantially improved many values for satellite masses obtained from centuries of ground-based observations.

The ratio of density to size of a planetary body or satellite provides important clues as to its composition and internal structure because this parameter depends strongly on material properties. Analyses of density show Neptune's moon Triton to be very similar to Pluto, while Titan, the largest satellite of Saturn, is more like the Galilean satellites of Jupiter (Tyler et al., 1989).

Orbiting spacecraft can provide considerably more detail on internal structure than can flybys (e.g. Muller and Sjogren, 1968). Such an experiment was planned for Mars Observer (q.v.) Tyler et al., 1992), which was scheduled to arrive at Mars in late 1993 to build on earlier results for Mars (Reasenberg, 1977; Sjogren, 1979; Balmino, Moynot and Vales, 1982; Zuber et al., 1991). Contact with Mars Observer was lost during maneuvers to place the spacecraft in orbit about Mars. Mars Observer was designed to orbit Mars every 117 min at altitudes between about 360 and 380 km above the surface. At these heights the typical speed of the spacecraft would be a little more than 3 km s^{-1}. From such low altitudes, even relatively minor variations in the mass density of surface and near-surface features would be detectable. For example, the measurement accuracy of the two-way Doppler effect would correspond to a sensitivity to changes of the line-of-sight spacecraft velocity of 100 µm s^{-1} or less for measurements made about once per second, so that accelerations of the order 100 µm s^{-2} could be observed readily. This sensitivity could be improved by taking advantage of a sequence of measurements in the approximately 2 to 3 min required by Mars Observer to pass over any surface point. At this level of measurement accuracy the 'missing mass' associated with a 200 km diameter Martian crater only 400 m deep may be easily detected. Considered another way, a reduction in density of about 0.3 g cm^{-3} (or about a 10% change in the average Martian crustal density) of a cube of rock 50 km on a side and near the surface would be measurable and would have about the same effect as the crater example. Figure R19 presents a map of the gravity field of Mars as it is presently known from Viking orbiter and Mariner 9 radio tracking data (Plate 19).

The gravity field of Venus has been measured by Pioneer Venus (Ananda et al., 1980; Sjogren et al., 1980) and Venera (q.v.) spacecraft. These measurements are less accurate and complete than those planned for Mars due to the much more elliptical orbits employed and to use of an earlier generation of radio tracking equipment. In 1992 the Magellan spacecraft initiated an improved set of Venus measurements from an elliptical orbit and obtained further data from a low circular orbit in 1993 and 1994. Detailed gravity maps of the Earth are available from surface and satellite measurements for comparison with those of other planets (see Geoid).

Atmospheres

Radio occultation studies of atmospheres can be understood in terms of 'geometric' or 'ray' optics refraction of signals traveling between spacecraft and ground stations. In a spatially varying medium in which the wavelength is very short compared with the scale of variation in refractive index, the direction of propagation of an electromagnetic wave always curves in the direction of increasing refractivity. Consequently, in a spherically symmetric atmosphere with gas refractivity (which is proportional to density) constantly decreasing with height, the radio path remains in a plane and bends about the center of the system. The degree of bending depends on the strength of the refractivity variation. This simple model approximates a real atmosphere and is useful for understanding the basic phenomena of radio occultation.

The geometry is illustrated in Figure R20, where the atmosphere is represented by the refractivity as a function of radius from the center $\mu(\rho)$, and the bending can be described in terms of a bending angle α and a ray asymptote a. The variation of the bending angle, ray asymptote and refractivity are linked elegantly through an Abel transform relationship (Fjeldbo and Eshleman, 1971):

$$\alpha(a) = -2a \int_{\rho=\rho_0}^{\rho=\infty} \frac{1}{\mu} \frac{\partial \mu}{\partial \rho} \frac{\partial \rho}{[(\mu r)^2 - a^2]^{1/2}}$$

where $\rho_0 = a/\mu(\rho_0)$, from Bouguer's rule, is the ray periapse and

$$\mu(\rho_{01}) = \exp\left\{ -\frac{1}{\pi} \int_{a=a_1}^{a=\infty} \ln\left\{ \frac{a}{a_1} + \left[\left(\frac{a}{a_1}\right)^2 - 1 \right]^{1/2} \right\} \frac{d\alpha}{da} \, da \right\}$$

with $\rho_{01} = a_1/\mu(\rho_{01})$. In this last expression a_1 represents the asymptotic miss distance for a ray whose radius of closest approach is ρ_{01}.

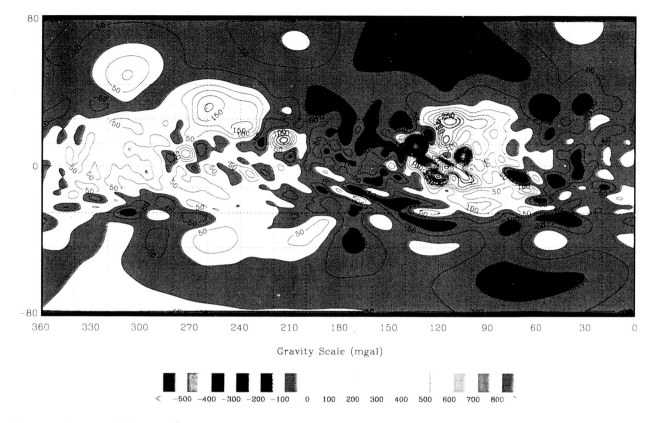

Figure R19 The gravity field of Mars. Figure depicts Mars 'free-air' gravity anomalies derived from Viking Orbiter 1 and 2 and Mariner 9 tracking data. The gravity model is a spherical harmonic representation complete to degree and order 50, with horizontal resolution of about 300 km at the surface of Mars. The map shown is a Mercator projection. The regions of highest elevation on Mars, Tharsis and Olympus Mons volcanoes, show up as very strong positive gravity anomalies. Other weaker, but still prominent, positives lie over the Elysium volcanic region, the large Alba volcanic structure and in the Isidis and Utopia impact basins. The strongest negative anomaly lies over the deep central Vallis Marineris canyons. (Courtesy D.E. Smith, Goddard Space Flight Center, NASA, 1992.)

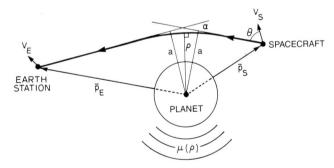

Figure R20 Occultation ray path geometry. Signals passing between spacecraft and ground are refracted by a planetary atmosphere. The refraction angle is α. The position and velocity of spacecraft and Earth station are known from tracking data. Angle θ can be found from the Doppler effect, leading to solution for α and a (see text).

Thus for spherical atmospheres, if $\alpha(a)$ is known, the corresponding refractivity profile can be found exactly. For non-spherical geometry alternative numerical solutions are available.

The bending angle and the ray asymptote can be determined accurately by radio occultation measurements to create an experimentally derived table of α versus a, or $\alpha(a)$. This is accomplished as follows. Referring again to Figure R20, the position and velocity of the spacecraft with respect to the Earth tracking station and the planet's center can be found from tracking data during periods when the radio path is well clear of the atmosphere. Given knowledge of the geometry, a measurement of the Doppler shift over the spacecraft-to-ground path is sufficient to find θ and a, from which α then can be determined. It was possible with Voyager to measure α to an accuracy of about 1×10^{-8} radians, and better accuracy is expected for planned future missions; a is known typically to about 1 km. These high levels of measurement accuracy in turn permit determination of an accurate refractivity profile $\mu(\rho)$, as outlined above.

In order to interpret the refractivity in terms of gas parameters, the pressure and temperature are calculated assuming hydrostatic equilibrium, for example, from

$$p(h) = \overline{m} \int_h^\infty g(h')n_t(h')dh'$$

and

$$T(h) = \frac{p(h)}{k_B n_t(h)}$$

where p and $T(h)$ are the pressure and temperature as a function of height h, respectively, g is the acceleration of gravity, k_B is the Boltzmann constant, $\overline{m}$ is the mean molecular mass, and n_t is the molecular number density. Formal use of these equations requires a priori knowledge of the atmospheric composition. On several occasions, however, independent estimates of atmospheric temperature, in combination with the equations above, have been used to infer the principle components by a process of elimination from likely constituents. The preponderance of carbon dioxide in the atmospheres of Venus and Mars, and the primarily nitrogen composition of the atmospheres of Titan and Triton were first revealed experimentally in this way. Alternatively, the absorption of radio energy in the atmosphere can also be measured, and the results used to determine the amounts of certain gases, such as ammonia and sulfuric acid vapor, which are microwave absorbers. Currently our best information on the composition of the outer planets has been obtained by combined use of infrared spectroscopy (q.v.) and occultation profiles.

Rings

Ring occultation measurements are a direct extension of the methods used to study atmospheres. The initial attempt at a ring occultation measurement using radio techniques was conducted with Voyager 1 during its 1979 flyby of Jupiter, where a geometric occultation by Jupiter's gossamer ring actually occurred, but was not detectable in the radio data (Tyler, Marouf and Wood 1981). At Saturn, on 13 November 1980 a long-planned occultation experiment by the same spacecraft was successful, and is responsible for much of our observational knowledge of the physical properties of the ring particles and dynamical processes within the rings (Marouf and Tyler, 1982; Marouf, Tyler and Eshleman, 1982; Tyler et al., 1983; Cuzzi et al., 1984; Esposito et al., 1984; Zebker, Marouf and Tyler, 1985; Rosen et al., 1991). Voyager 2 provided ring occultation measurements at Uranus (Gresh et al., 1989; French et al., 1991), where the ring material could also be detected easily in the radio signals. A ring occultation was attempted again at Neptune, where the chance timing of the occultation caused the occultation signals to pass between fragments of the beaded rings of that planet. These four attempts represent the total measurements of radio ring occultations to date. Additional ring occultation experiments at Saturn beginning in about 2004 are planned for the Cassini mission.

Ring occultation experiments are similar in geometry and execution to atmospheric occultations; the spacecraft trajectory is chosen so that the spacecraft passes behind the rings as viewed from Earth (Marouf and Eshleman, 1982; Figure R18) During the occultation interval radio signals transmitted from the spacecraft pass through the rings and are received by a tracking station on the ground. The analysis of ring observations is different from that of atmospheres, however, as the occulting ring medium is in the form of discrete particles orbiting the planet but held in fixed statistical relationships by dynamical processes. Consequently, the data are interpreted in terms of the propagation characteristics of electromagnetic waves through moving, randomly distributed particulate material. A radio wave passing through such media is attenuated by a combination of two main effects. Some of the energy intercepted by the ring particles is converted to heat and lost to the radio beam by that process, while most of the intercepted energy is reradiated in a variety of directions. The strength of the beam continuing to Earth is reduced by both mechanisms. The first process is called absorption, the second scattering.

Collectively, although made up of discrete particles, a ring can be represented as a thin slab of material with average electromagnetic properties that are determined by such parameters as the dielectric constant, number density, size distribution of the ring particle ensemble and the total thickness of the ring. In an occultation experiment one measures the average attenuation and electrical phase shift associated with this equivalent slab (Tyler, 1987).

It is also possible to measure that portion of the energy scattered in the near-forward direction which parallels the original beam direction to Earth. (In order to visualize this, picture a narrow beam of energy from the Voyager spacecraft spreading out to illuminate a small region in Saturn's rings. In the absence of scattering, only the energy initially directed towards Earth would be received on the ground. Scattering redirects some of the spreading energy – that would otherwise miss Earth – into the direction of Earth.) Because the scattered signals follow paths that are slightly different from that of the 'direct' ray which travels along the line of sight, and as a result of the relative motion of the spacecraft, Earth and ring particles, the scattered signals typically undergo a small Doppler shift relative to the direct signal. The small differences between the Doppler shift of the direct and the scattered signals provide the basis for separation and analysis of these two signal components. After removal of the direct signal by filtering, the frequency spectrum of the scattered signal can be converted to an angular scattering function which describes the strength of the scattering in various directions; for Voyager the scattering was dominated by the forward-scattering diffraction lobe of particles greater than about 1 m in diameter, and was limited to a few degrees about the Earth direction.

The theory of radiative transfer (q.v.), when applied to the combination of the direct signal extinction and the strength and angular distribution of the scattered signals, provides a framework for determining the characteristics of the rings. A ring comprising particles larger than about a wavelength in radius and sufficiently separated by the larger of roughly ten diameters or ten wavelengths (this condition assumes that the particles are only weakly coupled electromagnetically, and appears to apply in Saturn's rings) attenuates the direct signal only as a function of the total area of the ring particles projected along the direction of propagation, or the 'optical depth.' The strength of the signals scattered in the near-forward direction depends on a more complex, combined function of the particle size and the optical depth. The distinctly different dependence of the extinction and the near-forward scattering on particle size and total amount of ring material permits solutions for two fundamental ring properties, the thickness of the rings and the particle size distribution. Results from study of the Voyager 1 data for Saturn's rings A and C indicate that the rings are extremely thin, with the particle orbits typically confined to within a few meters of the average ring plane – this in a system more than 270 000 km in diameter about Saturn (Zebker and Tyler, 1984). The particle sizes are found to follow approximately a power law type distribution over the size range from microns to several meters, with the largest particles in the range of 10 m diameter. Both the particle size distribution and the maximum particle size vary somewhat with location in the ring, however (Zebker, Marouf and Tyler, 1985). Ring B could not be studied very extensively by Voyager due to the large value of its optical depth.

Precision dynamical studies of rings are based on stellar and radio occultation ring profiles, for which the radio results provide the highest radial resolution (at a specified signal-to-noise ratio). Extremely fine resolutions can be achieved with proper treatment of the radio signals (Rosen, Tyler and Marouf, 1991; Rosen et al., 1991).

The natural resolution of ring occultation extinction measurements is given approximately by the Fresnel zone size of the radio beam in the ring plane. For Voyager at Saturn the Fresnel zone was approximately 30 and 60 km in the radial coordinate of the rings at wavelengths of 3.6 and 13 cm respectively, characteristic of the usual diffraction limit to the smallest ring features that could be discerned. The coherent nature of the radio wave probe allows considerable improvement on these natural limits, however (Marouf, Tyler and Rosen, 1986). Consider for the moment only the geometry and the illumination of the rings with the radio signals. The effect of using coherent, i.e. highly monochromatic, waves for occultation experiments is that the diffraction pattern of the rings is formed on the Earth side of the ring plane. The relative motion of the rings, the spacecraft and the tracking station allows measurements on the ground of a one-dimensional sample of this diffraction pattern along the radial dimension of the ring system. To the degree that the rings are azimuthally symmetric, it is possible to obtain from such measurements a diffraction-corrected 'image' of the rings with resolution limited primarily by signal-to-noise ratio, the coherence properties of the oscillators employed, the degree of azimuthal symmetry and uncertainties in spacecraft position relative to the planet. In effect the occultation measurements provide a one-dimensional sampling of an interplanetary scale microwave hologram of the rings; this sampling is sufficient for reconstruction of the electromagnetic waves at the rings, given the azimuthal symmetry.

In practice, radio occultation resolutions as fine as 200 m were achieved in Saturn's ring C and in the Cassini division, and as fine as 1 km in ring A. Limited regions of ring B, primarily near the inner edge with ring C, were studied at 1-km resolution. Comparable resolutions were achieved with Voyager 2 at Uranus. For comparison, the best Voyager images of the rings have a resolution of about 10 km.

At the finest resolutions the Voyager radio occultation observations of the rings of Saturn and Uranus reveal the interior structure with considerable clarity. For Saturn 30 wave features, as well as all gaps and edges, have been identified in the approximately 60 000-km radial extent of the main ring system. There are three known types of waves in Saturn's Rings – density waves, bending waves and wakes. The first two are manifestations of resonant gravitational perturbations by satellites of Saturn, where a density wave occurs when the perturbing force lies in the plane of the ring and a bending wave occurs when the perturbing force has a significant component perpendicular to the ring plane. Density waves are similar in their effects to dilatational waves wherein the particles are alternatively more and less tightly bunched; the wave propagates radially outward. Overall it forms a spiral in the ring plane. Bending waves appear as warps in the ring plane wherein the sheet of ring material is distorted and can deviate by a few hundred meters from the mean plane. These waves propagate inward, toward the center of the system, also forming a spiral. Wakes are a form of density wave associated with the presence of a nearby moon orbiting inside a gap or division within the rings or next to a ring edge. All these wave types are detected in radio occultation data by the variation they impose on the extinction of the radio signals with radial position in the ring. The properties of these waves provide information on the total quantity of material, i.e. the mass density, the collision frequencies and other physical properties of the rings and ring particles. Accurate measurements of wave location provide tests of theories of ring formation and dynamics. Of the 30 waves observed in Saturn's rings, 23 have been identified as to their source and type (one tentatively). The sources of the remaining seven wave features remain unknown. Identification of the gaps in Saturn's rings allows the rings to be counted. On this basis there are 22 rings in the main ring system. Otherwise all parts of the rings are completely connected on the scale of 1 km. Observationally, Saturn's rings can be understood readily in terms of established concepts, although there are many important questions remaining to be addressed (see Planetary ring; Saturn ring system).

The rings of Uranus, in contrast with those of Saturn, comprise nine distinct, very narrow rings typically only several kilometers up to 300 km in extent, spread out over a distance of approximately 10 000 km in radius from the planet. They differ from Saturn's rings in other important ways. First, no wave structures are identified unambiguously in the radio occultation data. Second, the nature of the particle distributions and internal dynamics remains somewhat obscure, but is distinctly different from Saturn. Standard radiative transfer calculations for 'classical' ring models comprising well-separated, non-interacting particles are inconsistent with the data for the innermost five rings (rings 6, 5, 4, α and β). Similarly, for the outermost of the nine primary rings (ring ϵ), the radio occultation data cannot be reconciled readily with current ideas regarding ring stability and evolution. The three remaining rings (rings η, δ and γ) are 'optically' very thick, and again not well understood theoretically (Gresh et al., 1989).

Surfaces

Radio science techniques have been applied to the study of planetary surfaces in limited experiments. As an extension of the occultation technique, the scattering properties of planets and satellites can be studied by directing the transmitted signals from nearby spacecraft toward the surface and receiving the carom signal on Earth. Oblique scattering experiments complement the older radar astronomy technique which utilizes transmitters and receivers on the Earth (Simpson, 1993).

Prospects for the future

Although initially conceived as an exploratory tool, radio science techniques have provided considerable, originally unanticipated, detail regarding the atmospheres and gravity of the planets. Continued improvement in the stability of radio systems, particularly for oscillators, is expected to enable some further performance gains.

Radio science investigations can be improved radically through the use of powerful ground transmitters with signal reception onboard the spacecraft, reversing the direction of signal flow with respect to most previous planetary experiments (Eshleman et al., 1987). Gains of 1000 to 10 000 in signal-to-noise ratio are readily achievable. For occultation measurements, such improvements would make possible detailed probing of the densest portions of the Saturnian and Uranian rings, and precise global measurements of the dynamics and cloud structures in the atmospheres of the giant planets. Examples of other, future applications might include (1) the study of Mars' weather, including dust storms as analogs of the asteroid-collision hypotheses for the Cretaceous–Tertiary boundary; (2) investigation of the middle cloud region of Venus including possible volcanically driven changes, and wave dynamics of the Venus atmosphere; (3) exploratory measurements of the Pluto-Charon system; and (4) eventual application to monitoring of terrestrial global change.

G. Leonard Tyler

Bibliography

Ananda, M.P., Sjogren, W.L., Phillips, R.J. et al. (1980) A low-order global gravity field of Venus and dynamical implications. J. Geophys. Res., **85**(A13), 8303–18.

Balmino, G., Moynot, B. and Vales, N. (1982) Gravity field model of Mars in spherical harmonics up to degree and order eighteen. J. Geophys. Res. **87**, 9735–46.

Bird, M.K., Asmar, S.W. Brenkle, J.P. et al. (1992) Ulysses radio occultation observations of the Io plasma torus during the Jupiter encounter. Science, **257**(5076), 1532–5.

Booker, H.G., Dessler, A.J., Eshleman, V.R. et al. (1962) Bistatic radar astronomy, in A Review of Space Science. Space Studies Board of National Academy of Science.

Campbell, J.K. and Anderson, J.D. (1989) Gravity field of the Saturnian system from Pioneer and Voyager tracking data. Astron. J., **97**(5), 1485–95.

Cuzzi, J.N., Lissauer, J.J., Esposito, L.W. et al. (1984) Saturn's rings: properties and processes in Planetary Rings (eds R. Greenberg and A. Brahic). Tucson: University of Arizona Press.

Elliot, J.L. and Nicholson, P.D. (1984) The rings of Uranus in Planetary Rings (eds R. Greenberg and A. Brahic). Tucson: University of Arizona Press.

Elliot, J.L., Dunham, E.W., Bosh, A.S. et al. (1989) Pluto's atmosphere. Icarus, **77**(1); 148–170.

Eshleman, V.R. (1973) The radio occultation method for the study of Planetary atmospheres. Planet. Space Sci., **21**, 1521–31.

Eshleman, V.R. and Tyler, G.L. (1975). Radio Occultation: Problems and Potential Solutions. Stanford Electronics Laboratories, Stanford University. Report No. 3241–3, September.

Eshleman, V.R., Hinson, D.P., Lindal, G.F., and Tyler, G.L. (1987) Past and future of radio occultation studies of planetary atmospheres. Adv. Space Res., **7**(12), 29–32.

Eshleman, V.R., Tyler, G.L., Anderson, J.D. et al. (1977) Radio science investigations with Voyager. Space Sci. Rev., **21**(2), 207–32.

Eshleman, V.R., Tyler, G.L., Wood, G.E. et al. (1979). Radio science with Voyager 1 at Jupiter: preliminary profiles of the atmosphere and ionosphere. Science, **204**(4396), 976–8.

Eshleman, V.R., Tyler, G.L., Wood, G.E. et al. (1979b). Radio science with Voyager at Jupiter: initial V-2 results and a V-1 measure of the Io torus. Science, **206**(4421), 959–62.

Esposito, L.W., Cuzzi, J.N., Holberg, J.B. et al. (1984) Saturn's rings: structure, dynamics, and particle properties, in Saturn (eds T. Gehrels and M.S. Matthews). Tucson: University of Arizona Press.

Fjeldbo, G. and Eshleman, V.R. (1968) The atmosphere of Mars: radio occultation measurements and interpretations, in The Atmospheres of Venus and Mars (eds J.C. Brandt and M.B. McElroy). New York: Gordon and Breach.

Fjeldbo, G. and Eshleman, V.R. (1971) The neutral atmosphere of Venus as studied with the Mariner V radio occultation experiment. Astron. J., **76**(2), 123–140.

Fjeldbo, G., Kliore, A.J., Seidel, B. et al. (1976) The Pioneer 11 radio occultation measurements of the Jovian ionosphere, in Jupiter (ed. T. Gehrels). Tucson: University of Arizona Press.

French, R.G., Nicholson, P.D., Porco, C.C., and Marouf, E.A. (1991) Dynamics and structure of the Uranian rings, in Uranus (eds J.T. Bergstralh, E.D. Miner and M.S. Matthews). Tucson: University of Arizona Press.

Gresh, D.L., Marouf, E.A., Tyler, G.L. et al. (1989) Voyager radio occulation by Uranus' rings. 1. Observational results. Icarus, **78**(1), 131–68.

Howard, H.T., Eshleman, V.R., Hinson, D.P. et al. (1992) Galileo radio science investigations. Space Sci. Rev., **60**, 565–90.

Hunten, D.M. and Veverka, J. (1976) Stellar and spacecraft occultations by Jupiter: a critical review of derived temperature profiles in Jupiter (ed. T. Gehrels). Tucson: University of Arizona Press.

Ivanov-Kholodny, G.S. et al. (1979) Daytime ionosphere of Venus as studied with Veneras 9 and 10 radio occultation. Icarus, **39**(2), 209–213.

Jacobson, R.A., Campbell, J.K., Taylor, A.H., and Synnott, S.P. (1992) The masses of Uranus and Its major satellites from Voyager tracking data and Earth-based Uranian satellite data. Astron. J., **103**(6), 2077–8.

Kliore, A.J. and Patel, I.R. (1982) Thermal structure of the atmosphere of Venus from Pioneer Venus radio occultation. Icarus **52**(2), 320–34.

Kliore, A.J. and Woiceshyn, P.M. (1976) Structure of the atmosphere of Jupiter from Pioneer 10 and 11 radio occultation measurements, in *Jupiter* (ed. T. Gehrels). Tucson: University of Arizona Press.

Kliore, A.K., Cain, D.L., Levy, G.S. *et al.* (1965) Occultation experiment: results of the first direct measurement of Mars' atmosphere and ionosphere. *Science*, **149**, 1243–8.

Lindal, G.F. (1992) The atmosphere of Neptune: an analysis of radio occultation data acquired with Voyager 2. *Astron. J.*, **103**(3), 967–82.

Lindal, G.F., Hotz, H.B., Sweetnam, D.N. *et al.* (1979) Viking radio occultation measurements of the atmosphere and topography of Mars: data acquired during one Martian year of tracking. *J. Geophys. Res.*, **84**(B14), 8443–56.

Lipa, B.J. and Tyler, G.L. (1979) Statistical and computational uncertainties in atmospheric profiles from radio occultations: Mariner 10 at Venus. *Icarus*, **39**(2), 192–208.

Marouf, E.A. and Tyler, G.L. (1982) Microwave edge diffraction by features in Saturn's rings: observations with Voyager 1. *Science*, **217**(4556), 243–5.

Marouf, E.A., Tyler, G.L. and Eshleman, V.R. (1982) Theory of radio occultation by Saturn's rings. *Icarus*, **49**(2), 161–94.

Marouf, E.A., Tyler, G.L. and Rosen, P.A. (1986) Profiling Saturn's rings by radio occultation. *Icarus*, **66**(1), 120–66.

Muller, P.M. and Sjogren, W.L. (1968) Mascons: lunar mass concentrations. *Science*, **161**, 680–4.

Reasenberg, R.D. (1977) The moment of inertia and isostasy of Mars. *J. Geophys. Res.*, **82**, 369–75.

Rosen, P.A., Tyler, G.L. and Marouf, E.A. (1991) Resonance structures in Saturn's rings probed by radio occultation I: methods and examples. *Icarus*, **93**(1), 3–24.

Rosen, P.A., Tyler, G.L., Marouf, E.A. and Lissauer, J.J. (1991) Resonance structures in Saturn's rings probed by radio occultation II: results and interpretation. *Icarus*, **93**(1), 25–44.

Simpson, R.A. (1993) Spacecraft studies of planetary surfaces using bistatic radar. *IEEE Trans. Geosci. and Remote Sensing*, **31**(2), 465–82.

Sjogren, W.L. (1979) Mars gravity: high resolution results from Viking Orbiter 2. *Science*, **202**, 1006–10.

Sjogren, W.L., Phillips, R.J., Birkeland, P.W. and Wimberly, R.N. (1980) Gravity anomalies on Venus. *J. Geophys. Res.* **85**(A13), 8295–302.

Stone, E.C. and Miner, E.D. (1981) Voyager 1 encounter with the Saturnian system. *Science*, **212**(4491), 159–63.

Tyler, G.L. (1987) Radio propagation experiments in the outer solar system with Voyager. *Proc. IEEE*, **75**(10), 1404–31.

Tyler, G.L., Marouf, E.A. and Wood, G.E. (1981) Radio occultation of Jupiter's rings: bounds on optical depth and particle size. *J. Geophys. Res.* **86A**(10), 8699–703.

Tyler, G.L., Eshleman, V.R., Anderson, J.D. *et al.* (1981) Radio science investigation of the Saturn system with Voyager 1: preliminary results. *Science*, **212**(4491), 201–6.

Tyler, G.L., Eshleman, V.R., Anderson, J.D. *et al.* (1982) Radio science with Voyager 2 at Saturn: atmosphere and ionosphere and the masses of Mimas, Tethys, and Iapetus. *Science*, **215**(4532), 553–8.

Tyler, G.L., Marouf, E.A., Simpson, R.A. *et al.* (1983) The microwave opacity of Saturn's rings at wavelengths of 3.6 and 13 cm from Voyager 1 radio occultation. *Icarus*, **54**(2), 160–88.

Tyler, G.L., Sweetnam, D.N., Anderson, J.D. *et al.* (1986) Voyager 2 radio science observations of the Uranian system: atmosphere, rings and satellites. *Science*, **233**(4759), 79–84.

Tyler, G.L. *et al.* (1989) Voyager radio science observations of Neptune and Triton. *Science*, **246**(4936), 1466–73.

Tyler, G.L., Balmino, G., Hinson, D.P. *et al.* (1992) Radio science investigations with Mars Observer. *J. Geophys. Res., Planets*, **97**(E5), 7759–79.

Yuen, J.H. (ed.) (1983) *Deep Space Telecommunications Systems*. New York: Plenum.

Zebker, H.A. and Tyler, G.L. (1984) Thickness of Saturn's rings inferred from Voyager 1 observations of microwave scatter. *Science*, **223**(4634), 396–8.

Zebker, H.A., Marouf, E.A. and Tyler, G.L. (1985) Saturn's rings: particle size distributions for thin layer models. *Icarus*, **64**(3), 531–48.

Zuber, M.T., Smith, D.E., Lerch, F.J. *et al.* (1991) A 40th degree and order gravitational field model for Mars. *Lunar Planet. Sci. Conf.* XXII, Clear Lake, Texas. Lunar and Planetary Science Institute.

Cross references

Galileo mission
Ionosphere
Mars: gravity
Planetary ring
Radar astronomy
Saturn: ring system
Titan
Voyager missions

RADIOMETRY

Radiometry is the measurement of optical radiation. In planetary science the term most often refers to the measurement and characterization of radiation emitted by planetary, satellite and asteroid surfaces in the infrared portion of the electromagnetic spectrum. A closely related (and nearly synonymous) term is spectroscopy, in which the intensity of the incident radiation from some source is described as a function of wavelength, permitting the construction of a spectrum. Many articles in this volume describe spectroscopic investigations of solar system bodies (see Infrared spectroscopy; Microwave spectroscopy; Visible and near-infrared spectroscopy).

The quantitative characterization of radiation requires a measuring apparatus or instrument. Radiation typically passes from a source, through an aperture, to detectors. Optical filters and other devices may operate in the path of the radiation, and electronic circuitry is required to accept the detector output and convert this information to the desired format for storage or transmission. Considerable sophistication is required in the design of radiometric instruments. The objective is the characterization of the target in terms of size, shape, location, composition, temperature, radiant properties, energy levels, and so on. However, these target properties cannot be measured directly; they must be inferred from the instrument response to the flux incident on the sensor aperture (Wyatt, 1992). An ideal radiometer or spectrometer has yet to be constructed. Instead the instrument must be optimized for the particular application. The types of sensors employed, their linearity or non-linearity in terms of spectral response, the calibration of the instrument, the precision of the instrument and the possible sources of noise and errors in the system must all be characterized before interpretation of any observational results can begin. Wyatt (1992) provides an excellent overview of these topics.

To further complicate the problem, one must also realize that the radiation incident on the aperture is composed of emitted and reflected radiant energy, not only from the target body but also from the 'background' and from the intervening medium. After these effects are taken into account, one may derive considerable information about the objects under observation. In the following we will be primarily concerned with radiation that has been first absorbed, and then re-emitted, at thermal wavelengths (from about 5 μm to about 0.1 mm in wavelength).

In the simplest case, solar system bodies with solid surfaces may be expected to be in a state of thermal equilibrium, where a balance exists between the incoming radiant energy absorbed and the outgoing energy emitted. In such a case the important parameters are the solar flux at the distance of the body, and its emissivity (q.v.) and albedo (q.v.). The equilibrium temperature may be estimated; for most satellites in the solar system the measured temperatures correspond closely to the theoretical value (see Satellite, natural). This condition is not fulfilled, however, if the body is radiating significant internal heat from the interior, or if the object's atmosphere generates a significant greenhouse effect (q.v.). Saturn's satellite Titan is an example of the latter effect; its thick atmosphere, like that of the Earth, impedes the radiation of thermal energy to space, resulting in a warming of the surface and lower atmosphere.

Radiometry can provide information about the thermal response of planetary satellite surfaces. Observations of satellites during occultations by the primary characterize the rate of cooling (and thus thermal conductivity) of the surface; this can yield information about the composition and porosity of surface materials. The Galilean satellites of Jupiter have been studied in this way.

Radiometry is an important technique for obtaining information about asteroids (Lebofsky and Spencer, 1989) (see Asteroid: thermal infrared studies). The LANDSAT satellites have provided the longest continuous series of radiometric measurements of the surface of the Earth. More detailed discussions of radiometric techniques and the results of radiometric investigations are provided in the articles on Remote sensing and Infrared spectroscopy.

James H. Shirley

Bibliography

Lebofsky, L.A. and Spencer, J.R. (1989) Radiometry and thermal modeling of asteroids, in *Asteroids II* (eds R.P. Binzel, T. Gehrels and M.S. Matthews). Tucson: University of Arizona Press.
Wyatt, C.L. (1992) Radiometry, in *Encyclopedia of Physical Science and Technology* 2nd edn (ed. R. A. Myers). San Diego: Academic Press (Harcourt Brace Jovanovich).

Cross references

Asteroid: thermal infrared studies
Galilean satellite
Infrared spectroscopy
Microwave spectroscopy
Remote sensing
Spectroscopy: atmosphere
Titan
Visible and near-infrared spectroscopy

RANGER MISSIONS

Program initiation

Ranger was the name given to the earliest set of American space missions authorized by the National Aeronautics and Space Administration (NASA) to explore the Moon and to the spacecraft developed to carry them out. The missions were planned and executed by the Jet Propulsion Laboratory (JPL) of the California Institute of Technology, which NASA had designated as its lead center for lunar and planetary missions.

In January 1959, shortly after NASA had been established and JPL had become affiliated with it, the laboratory proposed a 5-year program of space missions to the Moon, Venus and Mars, starting in 1960. With some modifications, this program was adopted by NASA and titled Vega. In the next few months, as it became clear that launch vehicle development was falling short of expectations, Vega was canceled and replaced by less ambitious programs – Ranger for the lunar missions and Mariner for planetary missions.

As agreed upon in December 1959, five Ranger spacecraft were to be designed, fabricated and built by JPL and launched on lunar reconnaissance missions in 1961 and 1962. The program would begin with two engineering flights ('Block 1') to check out the new Atlas–Agena B booster system, and to test the spacecraft, in a highly elliptical Earth orbit that would take it out to about 1 million km from Earth – far beyond the Moon's orbit. Block 2 would include three missions with mid-course correction capability to impact on the Moon. Block 3 would come later with spacecraft and scientific missions of greater complexity. By the end of 1960, Ranger spacecraft were nearing the assembly stage.

The design of the spacecraft was significantly influenced by the fact that JPL was also involved with planetary missions. These required spacecraft with three-axis stabilization that could keep a large parabolic antenna pointed at the Earth and solar panels to provide electric power for months or years. Earlier spacecraft (Pioneers and Explorers) had been spin-stabilized and operated on batteries. These features would have sufficed for the brief Ranger missions, but JPL opted to use the same basic design concepts for Ranger and Mariner spacecraft in order to save time and effort.

Block 1

For Block 1, since there was no requirement to get near the Moon, the scientific instruments chosen were chiefly to sample conditions in interplanetary space at a large distance from Earth. A set of electrostatic charged-particle analyzers would measure the properties of the 'solar wind' – the plasma presumed to be streaming out from the Sun. A rubidium vapor magnetometer would monitor the magnetic field. A group of four detectors would analyze the trapped particles in the Van Allen belts, and cosmic rays would be monitored by a triple-coincidence cosmic ray telescope and an ionization chamber. A Lyman alpha telescope would observe the hydrogen geocorona. Dust particle detectors would register micrometeorites. There was also an engineering experiment to measure friction in the vacuum of space.

Just before the final design freeze on the first two spacecraft, the Atomic Energy Commission asked to add an experiment, code-named Vela Hotel, to detect above-ground nuclear explosions. Against the vigorous objections of the laboratory management, who were trying desperately to meet the launch schedule, NASA agreed to the request, and the instrument was added to the payload. It brought the total weight of scientific instruments to 51 kg; the spacecraft at launch would weigh 307 kg.

As launch preparations for Ranger 1 at Cape Canaveral approached completion, NASA officials estimated the probability of successful performance by the Atlas–Agena at only 0.5 because of difficulties that had been encountered in developing the second-stage Agena. The launch at dawn on 23 August 1961 appeared satisfactory, with the first burn of the Agena placing the rocket into its intended parking orbit. But in a few minutes it became clear that the Agena had malfunctioned during its second burn, leaving Ranger in a satellite orbit with an apogee altitude of 501 km. The spacecraft performed as well as it could in these anomalous conditions, but the data from most of the experiments was either partially or completely useless.

Ranger 2 was launched on 17 November 1961 and, to the consternation of the scientists with experiments on the spacecraft, it was decided again to test out the Agena's second burn. Its performance was even worse than before, and the spacecraft burned up in less than 24 h. Only the friction device and the Vela Hotel experiment obtained a significant amount of useful data. Block 1, which had the capability of achieving a number of important scientific discoveries, was a dismal failure.

Block 2 – three disappointments in 1962

The three spacecraft for Ranger Block 2 were designed for hard landing on the lunar surface. Each carried a vidicon camera to take close-up pictures during approach, a surface-sensing pulse radar to measure the distance to the lunar surface and trigger the camera at the proper time (and incidentally to collect data on the surface texture), and a gamma-ray detector to determine the concentration of various radioactive elements in the surface. These instruments totaled 14 kg. There was also a 148-kg landing capsule designed to survive the impact; it enclosed a seismometer to search for moonquakes.

In the launch of Ranger 3 on 26 January 1962, a failure in the guidance system caused the Atlas to place the Agena and its passenger in too high a parking orbit. The Agena then functioned as programmed, but an error in its computer program caused its second burn to have the wrong duration. Ranger successfully executed the first mid-course maneuver in history, but again a computer program error sent it away from the Moon instead of toward it, Ranger missed the Moon by 36 785 km (more than ten lunar diameters). It was hoped still to obtain some useful photographs of the Moon, but a failure in the attitude-control system allowed the spacecraft to roll, and the pictures showed only empty space.

The launch sequence of Ranger 4 on 23 April 1962 was so perfect that no mid-course maneuver was necessary, and it made its impact on the farside of the Moon after a 64-hour flight. However, the master clock failed during the launch sequence so that Ranger could neither execute its programmed commands nor accept new commands from Earth.

In August 1962 the planetary program was inaugurated with the launch of the very successful Mariner 2 to Venus, but disaster continued to be the fate of the lunar program. Ranger 5 was launched normally on 18 October 1962 but, after about an hour a malfunction in the switching and logic unit short-circuited the solar panels. In the few hours available before the batteries ran down, a trajectory maneuver was attempted to make the dead spacecraft hit the Moon. This also failed, and Ranger flew past the Moon at 725 km altitude.

Both JPL and NASA convened special failure review boards, which issued reports critical of various aspects of engineering design and testing and of laboratory management. Among the results of these reports were various changes in procedures and organization, abandonment of the practice of heat sterilization of the spacecraft (which had been implemented to assure that no live organisms would be deposited on the moon), the adoption of television as the sole scientific objective of Block 3, and its postponement for a year.

Block 3

For Block 3, the 1-year grace period and the availability of somewhat more spacecraft weight (365 kg) permitted numerous design changes, the incorporation of more redundancy and much more intensive spacecraft testing. The television subsystem, the only scientific instrument aboard, weighed 171 kg and comprised two cameras with wide-angle lenses and four with telephoto lenses.

The lift-off of Ranger 6 finally took place on 26 January 1964. Two minutes later the Atlas main-stage engines shut down as scheduled and the television system turned on by itself, operated for a minute, and shut itself off. No explanation for this anomaly could be imagined, but the telemetry revealed no further problems. The mid-course maneuver adjusted the trajectory to very near the chosen impact point. At impact minus 18 min the camera turn-on command was issued as planned, but there was no response!

This disaster was the ultimate nadir of JPL's fortunes. Criticisms flooded in from all directions. New NASA and JPL review boards concluded that arcing had destroyed a power supply, but they adopted conflicting theories on why it had happened and what to do about it. Congress also launched an investigation, which concocted a still different theory and was extremely critical of both JPL and NASA. Not until 6 months later was the cause determined. A cloud of plasma produced by the ignition of the propellants venting from the Atlas had penetrated the shroud that covered the spacecraft, making a conductive path between two terminals, thus turning on the television and burning out a power supply.

After nearly 6 months of the most thorough testing and inspection that any unmanned spacecraft had undergone, Ranger 7 lifted off only 8 s after its intended launch time of 12 : 50 EDT on 28 July 1964. It crashed onto the lunar surface at its planned impact point near the Crater Guericke at 06 : 25 PDT on 31 July. A total of 4316 pictures were taken during the 17-min sequence and transmitted to the Deep Space Network without loss of a single frame – the last one from an altitude of about 520 m. Their quality could only be described as 'superb', surpassing in resolution the best Earth-based photographs by three orders of magnitude. In September 1964 the International Astronomical Union officially designated the impact area as Mare Cognitum.

Ranger 8 was launched on 17 February and Ranger 9 on 21 March 1965. These spacecraft returned 7137 and 5815 pictures respectively. The former landed in the Sea of Tranquility (near where Apollo 11 would touch down 4 years later), and the latter in the Crater Alphonsus. Ranger 9 finished the program with a flourish by providing its pictures in real time to Earthbound viewers of domestic television.

Other lunar missions

Even before Block 1 was completed, NASA authorized Ranger Block 4 to include four missions with considerably augmented science payloads to provide information for Project Apollo, but it was canceled in July 1963. The exploration of the moon was continued by the Lunar Orbiter program, managed by the NASA Langley Research Laboratory, Hampton, VA, (five missions from August 1966 to August 1967), by JPL's Surveyor program of seven landers (May 1966 to January 1968), by the Apollo program of six manned landings (July 1969 to December 1972), and by 24 Soviet 'Luna' (or 'Lunik') spacecraft (1959 to 1976).

Conway W. Snyder

Bibliography

Hall, R.C. (1977) *Lunar Impact: A History of Project Ranger*. Washington: NASA Scientific and Technical Information Office (NASA SP-4210).

Koppes, C.R. (1982) *JPL and the American Space Program*. New Haven and London: Yale University Press.

Newell, H.E. (1980) *Beyond the Atmosphere: Early Years of Space Science*. Washington: NASA Scientific and Technical Information Branch.

Cross references

Apollo missions
Lunar Orbiter missions
Moon
NASA

REFLECTANCE SPECTROSCOPY

The amount of light scattered by substances in bulk mass form changes with wavelength. At wavelengths visible to the human eye, this phenomenon is the basis for the perception of color. These spectral variations are caused by wavelength-dependent absorption, and can be used to infer such properties of the material as composition and electronic or molecular structure.

As with gases, the absorption usually occurs in discrete bands, the exception being metals, which absorb over a broad range of wavelengths. However, solid state bands are much broader than gaseous absorption bands, with $\Delta\lambda/\lambda \sim 10\%$ being typical; here λ is the wavelength of the band center and $\Delta\lambda$ is its width. For this reason the bands usually overlap, complicating their recognition and producing a continuum of absorption.

The bands arise from a variety of causes (Hunt, 1980; several of the chapters in Pieters and Englert, 1993, e.g. Salisbury, 1993). In metals and semiconductors absorption is due to collisions of mobile electrons with the solid state lattice. In insulators, absorption of photons in the far UV occurs by valence-conduction band transitions. In the near-UV, visible and near-IR ranges, absorption occurs by electronic transitions in which the electron either remains localized on one ion or jumps from an ion to its neighbor. In the thermal infrared range absorption occurs by photon-induced transitions in the vibrational and rotational states of the lattice.

Spectral reflectance measurements are usually made in one of two geometries: bidirectional reflectance, in which the source radiance is highly collimated (e.g. sunlight) and the detector subtends a small solid angle seen from the surface of the medium (e.g. the eye or a telescope); or directional hemispherical reflectance, in which the source is collimated but the light scattered into the entire hemisphere above the surface is sampled. Other geometries are possible, such as hemispherical-directional reflectance, when the eye observes radiance scattered from a surface illuminated by light from an overcast sky. Virtually all remotely sensed measurements are bidirectional, but many laboratory measurements are directional-hemispherical.

When the surface being studied is a plane interface between two media of differing optical properties, the reflectance is specular or mirrorlike, and the fraction of light reflected is given by the Fresnel reflectivity equations. If the upper medium is air or vacuum, and the lower has a complex refractive index $m(\lambda) = n(\lambda) + ik(\lambda)$, the specular reflectivity at normal incidence is

$$r_s(\lambda) = \frac{(n-1)^2 + k^2}{(n+1)^2 + k^2} \quad (R5)$$

If $k \lesssim 0.1$, r_s is sensitive only to n. However, in strong absorption bands $k \gtrsim 1$ and r_s is strongly dependent on k. Thus specular reflectance can be used to measure $n(\lambda)$ and $k(\lambda)$ in the region of strong bands. The analysis of such measurements is greatly facilitated by a technique known as the Kramers–Kronig method, which allows n and k to be separated (Wooten, 1972).

In the process of diffuse reflectance from a particulate material, the spectral information is contained in the light-scattering properties of the particles making up the medium. The most important of these properties is the single-scattering albedo $w(\lambda)$, which is the fraction of incident light scattered by a particle. Diffraction of light around the particle is important if the particles are separated by many times their size, as in a cloud, but can be ignored if the particles are close together, as in a soil.

The single-scattering albedo is determined by a number of factors, but primarily by the complex refractive index $m(\lambda)$ and the particle size D. Ignoring diffraction, two processes are involved in the

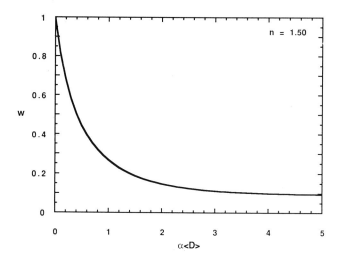

Figure R21 The single-scattering albedo of a particle versus the internal absorbance. This figure is calculated for a particle whose real part of the refractive index is $n = 1.50$.

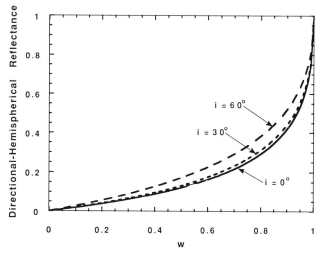

Figure R22 The directional-hemispherical reflectance plotted against single-scattering albedo, for three different values of the angle of incidence.

scattering of light by a single particle: surface scattering, which is Fresnel reflection from the surface of the particle, and volume scattering, the internal scattering of light that has been refracted and transmitted into the particle. If the particles are perfect spheres, $w(\lambda)$ can be calculated using an exact solution of Maxwell's equations known as Mie theory (Bohren and Huffman, 1983) For irregular particles an approximate expression for w is

$$w(\lambda) = S_e(\lambda) + [l - S_e(\lambda)] \frac{1 - S_i(\lambda)}{e^{-\alpha(\lambda)<D>} - S_i(\lambda)} e^{-\alpha(\lambda)<D>} \quad (R6)$$

where $S_e(\lambda)$ is the integral over a hemisphere of the Fresnel reflection coefficients for light externally incident on the surface, $S_i(\lambda)$ is the integral over a hemisphere of the Fresnel coefficients for light incident on the surface from the interior, $\alpha(\lambda) = 4\pi k(\lambda)/\lambda$ is the absorption coefficient, and $<D>$ is an effective particle diameter which depends on the particle shape but is typically of the order of $D/2$. Equation (R6) is plotted against the particle internal absorbance $\alpha<D>$ in Figure R21. When $\alpha = 0$, the particle absorbs no radiation and $w(\lambda) = 1$. As α increases, $w(\lambda)$ decreases exponentially and saturates at $w = S_e$.

For geometrically complex surfaces, such as particulate media, the reflectance and its variation with the positions of the source and detector is complicated and depends on a number of physical properties of the surface. The exact solution of Maxwell's equations describing the interaction of an electromagnetic wave with a non-uniform medium is not possible with today's computational techniques. Hence several simplifying assumptions are necessary.

If it is assumed that a particulate medium, such as a powder or planetary regolith, is uniform on scales larger than the particle separation, and that the particles scatter light isotropically, then the bidirectional reflectance is given by

$$r_{bd}(\lambda) = \frac{w(\lambda)}{4\pi} \frac{\cos i}{\cos i + \cos e} H(w, \cos i) H(w, \cos e) \quad (R7)$$

where i is the zenith angle of incident light, e is the zenith angle of the emerging light and $H(w,x)$ is a function given by the solution of the integral equation

$$H(w,x) = 1 + \frac{w}{2} x H(w, x) \int_0^1 \frac{H(w,x')}{x+x'} dx' \quad (R8)$$

This function is tabulated in Chandrasekhar (1960). A useful approximation is

$$H(w,x) \simeq \frac{1 + 2x}{1 + 2x(1-w)^{1/2}} \quad (R9)$$

Similarly, the directional-hemispherical reflectance is

$$r_{dh}(\lambda) = 1 - (1-w(\lambda))^{1/2} H(w(\lambda), \cos i) \simeq \frac{1 - (1-w)^{1/2}}{1 + 2(1-w)^{1/2}\cos i} \quad (R10)$$

Equation (R9) is plotted in Figure R22. Approximate expressions for other geometries are given in Hapke (1981, 1984, 1986, 1992, 1993). The dependence of $r(\lambda)$ on $w(\lambda)$ shown in Figure R22 is typical of all reflectances. At small values of w, r is low and only light scattered once by a particle is important. As w increases, multiply scattered light becomes important and r increases nonlinearly.

The reflectances of mixtures of different kinds of particles may be calculated from these equations by using an average single scattering albedo $\bar{w}(\lambda)$ given by the mixing equation

$$\bar{w}(\lambda) = \left[\Sigma_j \frac{M_j}{\rho_j D_j} w_j(\lambda)\right] / \left[\Sigma_j \frac{M_j}{\rho_j D_j}\right] \quad (R11)$$

where M_j is the mass fraction of particles of type j, ρ_j is the solid density of those particles, D_j is their size and $w_j(\lambda)$ their single-scattering albedo.

If m and D of a particulate medium are known, the reflectances and scattered radiances may be calculated using equations (R5)–(R11). Conversely, if the reflectance is measured, S_e, S_i and $<D>$ can often be estimated, and $\alpha(\lambda)$ can be calculated from the equations.

Reflectance spectroscopy is a major tool in the study of the surfaces of other bodies of the solar system. Often the major minerals making up a planet's surface can be identified from the wavelengths of the bands in the reflectance spectrum. For example, Figure R23 shows a spectrum of a lunar rock, in which diagnostic absorption bands caused by Fe and Ti in several minerals are identified. This technique has been exploited by many investigators to carry out remote compositional analyses of the surfaces of the Moon, asteroids and other bodies of the solar system. Recent comprehensive reviews of applications of reflectance spectroscopy to planetary remote sensing are given in Pieters and Englert (1992).

Bruce W. Hapke

Bibliography

Bohren, C. and Huffman, D. (1983). *Absorption and Scattering of Light by Small Particles*. New York: John Wiley and Sons.

Chandrasekhar, S. (1960) *Radiative Transfer*. New York: Dover Press.

Hapke, B. (1981) Bidirectional reflectance spectroscopy. 1. Theory. *J. Geophys. Res.*, **86**, 3039–54.

Hapke, B. (1984) Bidirectional reflectance spectroscopy. 3. Correction for macroscopic roughness. *Icarus*, **59**, 41–59.

Hapke, B. (1986) Bidirectional reflectance spectroscopy. 4. Extinction and the opposition effect. *Icarus*, **88**, 407–17.

Hapke, B. (1992) Combined theory of reflectance and thermal emission, in *Remote Geochemical Analysis* (eds C. Pieters and P. Englert). New York: Cambridge University Press.

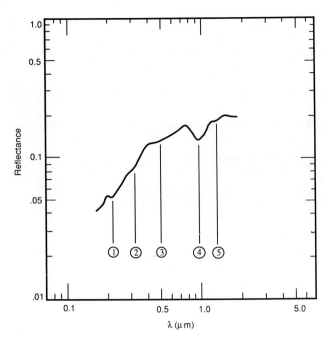

Figure R23 The spectral reflectance of a pulverized lunar rock (NASA sample number 70017). Several diagnostic absorption bands caused by Fe^{+2} in different minerals can be identified, and are labeled with the numbered lines: (1) Fe^{+2}–O^{-2} charge transfer band in augite and anorthite; (2) Fe^{+2}–Ti^{+4} charge transfer band in augite; (3) Ti^{+3} d–d band in augite; (4) Fe^{+2} d–d band in augite; (5) Fe^{+2} d–d band in anorthite.

Hapke, B. (1993) *Theory of Reflectance and Emittance Spectroscopy*. New York: Cambridge University Press.
Hunt, G. (1980) Electromagnetic radiation: the communication link in remote sensing, in *Remote Sensing in Geology* (eds B. Siegal and A. Gillespie). New York: John Wiley and Sons, pp. 5–46.
Pieters, C. and Englert, P. (eds) (1993) *Remote Geochemical Analysis*. New York: Cambridge University Press.
Salisbury, J. (1993) Mid-infrared spectroscopy: laboratory data, in *Remote Geochemical Analysis* (eds C. Pieters and P. Englert). New York: Cambridge University Press.
Wooten, F. (1972) *Optical Properties of Solids*. New York: Academic Press.

Cross references

Remote sensing
Spectrophotometry

REFLECTIVITY

When radiation is incident on an object it is either transmitted into the object (where it can be absorbed, scattered or re-emitted) or it is reflected. The relative proportions depend on the frequency (wavelength) of the radiation, the electrical and magnetic properties of the object, and the geometry. A hypothetical 'perfect reflector' would return 100% of the incident radiation to the original medium.

Reflectivity is a measure of the amount of energy actually reflected to that which would be expected from a perfect reflector. Smooth, perfectly conducting surfaces reflect like mirrors; not only is all the incident energy returned to the original medium, but the angle at which the reflection takes place (with respect to the local normal) is equal to the angle of incidence. As a perfectly reflecting surface becomes roughened, directions for the reflected energy become more widespread, but the total reflectivity remains 100%. If the material is then changed, so that it is no longer a perfect reflector, energy will be partially transmitted through the boundary into the second medium and the reflectivity will be less than 100%.

Reflectivity of polished metal surfaces is close to 100% for radio waves and optical waves, largely because free electrons within the metal can respond quickly to incident electric and magnetic fields. For insulators, electrons are bound tightly but a temporary offset between positive and negative charges can be induced within the material. Such dielectric phenomena yield reflectivities on the order of 5–10% for polished rock surfaces when incidence is near normal, increasing to nearly 100% at very oblique angles.

Richard A. Simpson

Bibliography

Evans, J.V. and Hagfors, T. (1968) *Radar Astronomy*. New York: McGraw-Hill.
Jackson, J.D. (1962) *Classical Electrodynamics*. New York: John Wiley and Sons.
Ulaby, F.T., Moore, R.K. and Fung, A.K. (1981) *Microwave Remote Sensing: Volume 1 – Fundamentals and Radiometry*. Reading, MA: Addison-Wesley.

Cross references

Albedo
Opposition effect

REGIONAL PLANETARY IMAGE FACILITIES

The Regional Planetary Image Facilities (RPIFs) are located throughout the United States and in Europe and Japan (Table R1). The facilities are jointly funded through the National Aeronautics and Space Administration (NASA) and the institution where the facility resides. The RPIFs were established through NASA's Planetary Geology Program in 1979, as a mechanism for archiving planetary imaging data sets. Today they continue to be funded jointly through NASA's Planetary Geology and Geophysics Program and their home institution to serve the ever-broadening research, educational and public communities.

The RPIF collections encompass images from the earliest Ranger mission photographs of the moon (dating from the 1960's) through the more recent data returned from the Voyager encounter with Neptune in 1989, the Magellan Venus mapping mission starting in 1990, and the current Galileo mission flybys of Venus and the Earth and Moon. The collection continues to grow at a tremendous rate; the Magellan Venus images alone increased the total by a factor of two.

Each facility contains image data of the planets and their moons taken from both Earth and space. RPIF users have access to over a million images and other related products. For each mission data set there are basically three types of image data: public release, systematic and special. The press release images are the highest quality, 'best' pictures, and usually include a detailed descriptive caption. The systematic (also called standard) images include every image a given spacecraft returned. This data set represent the most complete collection of images from a given mission. The special products, also known as 'second-order' products, are produced by applying specialized processing techniques which might for instance result in a reprojection of a collection of images into a mosaic or a stereo view. Many of these products are used in scientific literature summarizing the mission's accomplishments. In addition to the image data, other ancillary information exists such as targeting and navigational data, and information on the camera and spacecraft systems and their processing capabilities. A variety of media providing both digital and non-digital data is housed within the RPIF collection. Most of the facilities include black and white and color photographs and film, microfilm, hardcopy paper listings, slides, analog videodisks, videotapes, magnetic tapes and compact-disc read-only memories (CD-ROMs). There are also spacecraft models, globes, cartographic products, on-line catalogs and the continually growing collection of slides. Cartographic and research data are usually contained in maps and journals in a reference area of most facilities.

The RPIFs have computer systems that allow rapid searching and selection of available digital and analog images. Each RPIF has a workstation available that can display and provide some basic image

Table R1 Regional Planetary Image Facility locations

Arizona State University Space Photography Laboratory Department of Geology Tempe, AZ 85287-1404, USA +1 (602) 965-7029	Jet Propulsion Laboratory Regional Planetary Image Facility MS 202-101 4800 Oak Grove Drive Pasadena, CA 91109, USA +1 (818) 354-3343
Washington University Regional Planetary Image Facility Department of Earth and Planetary Sciences Campus Box 1169 One Brookings Drive St Louis, MO 63130-4899, USA +1 (314) 935-5609	Smithsonian Institution Regional Planetary Image Facility National Air and Space Museum, 3773 MRC 315 Washington, DC 20560, USA +1 (202) 357-1457
Lunar and Planetary Institute Regional Planetary Image Facility 3600 Bay Area Blvd. Houston, TX 77058-1113, USA +1 (713) 486-2180	CNR Istituto Astrofisica Spaziale Southern European Regional Planetary Image Facility Reparto di Planetologia Viale Dell'Universita, 11 00185 Roma, Italy +39 6-445-69.51
Brown University Regional Planetary Image Facility Department of Geological Sciences Box 1846 Providence, RI 02912, USA +1 (401) 863-3243	University of London Observatory Regional Planetary Image Facility 33/35 Daws Lane Observatory Annexe London, NW7 4SD, England +181 959-7367
University of Hawaii Pacific Regional Planetary Data Center Planetary Geosciences Division/ Hawaii Institute of Geophysics 2525 Correa Road Honolulu, HI 96822, USA +1 (808) 956-6488	Institute of Space and Astronomical Sciences 3-1-1 Yoshinodai Sagamihara-shi, Kanagawa 229, Japan +81 427-51-3911
US Geological Survey Regional Planetary Imaging Facility 2255 N. Gemini Drive Flagstaff, AZ 86001, USA +1 (602) 556-7262	Universite Paris-Sud Phototheque Planetaire d'Orsay Laboratoire de Geologie Dynamique de la Terre et des Planetes Department des Sciences de la Terre Batiment 509 F-91405 Orsay Cedex, France +33.1 60.19.14.46
Cornell University Spacecraft Planetary Image Facility Center for Radiophysics and Space Research 317 Space Sciences Building Ithaca, NY 14853-6801, USA +1 (607) 255-3833	University of Oulu Regional Planetary Image Facility Department of Astronomy 900570 Oulu, Finland +358-81-352106
University of Arizona Space Imagery Center Lunar and Planetary Laboratory Tucson, AZ 85721, USA +1 (602) 255-9002	Deutsche Forschungsanstalt fuer Luft- und Raumfahrt e.V (DLR) Regional Planetary Image Facility Institute for Planetary Exploration Rudower Chaussee 5 0-1199 Berlin, Germany +49-30-69545-300

processing or copying capability. An analog videodisk, which stores up to 100 000 images per disk, allows users to take a quick browse through many mission data sets, while other image browsing is possible through the CD-ROMs. The CD-ROMs have become the favorite among the image data producers in recent missions because of the ability to store large quantities of data (650 Mbytes) in a single small product (5¼″ disc). The RPIF has additional data access equipment that includes microfilm viewers, stereo viewers, videotape players and data analysis aids like light tables and map tables.

The RPIFs are public facilities and are commonly used by the science and education communities, by students of all ages and grade levels, and by planetarium directors and employees, as well as by the general public. Though the RPIFs are open facilities, prior arrangements for a visit may be necessary because the RPIF's host facility may be a government institution (such as the Jet Propulsion Laboratory). RPIF personnel are available to answer questions and provide assistance in identifying, locating and researching planetary imaging products. Since the RPIFs act as NASA's regional reference center for planetary image data, they can provide additional information on where to obtain copies of the planetary products. NASA's central source for the purchase of planetary image data is:

National Space Science Data Center (NSSDC)
Goddard Space Flight Center
Code 633.4
Greenbelt, Maryland 20771, USA
Telephone: +1 (301) 286-6695

Leslie J. Pieri and Marian E. Rudnyk

REGOLITH

A general term for any kind of superficial layer or blanket of loose particulate rock material found on planet Earth or any other hard celestial object, e.g. lunar regolith, Martian regolith, etc. (Arking, in Fairbridge 1967, p. 913; Hapke, ibid., p. 641). It is usually less than 10 m thick. Derived from Greek *rhegos*, blanket, plus *lithos*, stone; it is sometimes spelled 'rhegolith'. As defined by Merrill (1897), it is any natural material covering the bedrock. The term regolith is preferable to 'soil' and 'pedon' of soil science, which often carry biological and agricultural connotations; and also to 'mantle' (ambiguous with the Earth's internal layer beneath the lithosphere) and 'overburden' (a mining term for any formation covering the ore deposit).

It may be transported, for example, by gravity such as landslide debris, by glaciers such as till, erratics ('drift', boulder clay), by periglacial processes such as colluvium or solifluction debris, by eolian action (wind) such as loess or dune sand, by water in slope wash such as in a bajada, by water in a flood-plain deposit (alluvium), by settlement following volcanic pyroclastic eruption (ash fall), or by fallout of ejecta from a meteorite impact.

Alternatively, the regolith may be formed *in situ*, as a cumulative deposit, such as a peat accumulated by biological processes, a spring deposit (sinter, tufa, travertine) accumulated by *chemical processes*, precipitated from supersaturated subsurface waters (usually $CaCO_3$, SiO_2, $CaSO_4.2H_2O$). The regolith may also consist primarily of saprolite (Becker, 1895) which is literally 'rotted rock', i.e. bedrock that has been chemically weathered *in situ*; on some ancient landscapes (e.g. in Australia, Africa and Brazil) the thickness of saprolite may exceed 100 m. In the process of *weathering* (q.v.) the saprolite is progressively leached, as the weathering solutions selectively attack the more soluble minerals, leaving the 'resistate' components behind.

Finally, included in the regolith of geological cartography, is the soil, which is the product of the interaction between the agencies of the atmosphere, hydrosphere and biosphere upon the surface layer of regolith, be it transported material of saprolite. Soil types vary greatly according to climate, vegetation, relief and land utilization (Bridges 1970; Duchaufour 1978). An ancient, deactivated soil, buried beneath geological deposits (or posthumously re-exposed), is a paleosol.

Lunar regolith (also called 'lunar soil') was studied by polarimetry, radar and laboratory calibration experiments long before actual samples were obtained (Hapke, in Fairbridge, 1967, p. 641). Hapke wrote:

> the photometric properties of the moon are so grossly different from those of familiar terrestrial materials that it is clear that the lunar surface could not be solid rock . . . (because) the lunar surface materials have been exposed to a far different sort of 'weathering' than have terrestrial rocks. The moon possesses no atmosphere, and hence erosion by air and water are completely absent. However, the surface is continuously being bombarded by meteorites which impact with velocities of many miles per second and which range from tiny micron-sized grains to asteroids several miles in diameter . . . . The lack of an atmosphere allows the lunar surface to be exposed to solar radiations of all forms and wavelengths, including ultraviolet and X-rays, as well as corpuscular radiation in the form of hydrogen ions with energies of the order of a few thousand electron volts.

Lyot (1929) and Dollfus (1957, 1985) in France had found that only finely divided, dark powders similar to certain volcanic ashes or pulverized basalts match the lunar polarization curves. Temperature measurements showed that the thermal conductivity is similar to rock flour *in vacuo*. Radar studies showed that on a scale of centimeters to meters the lunar surface is characterized by very smooth, gentle slopes with no jagged outcrops. Further, it was seen that the entire lunar surface was like this; thus the lunar weathering had to be the result of an external homogeneous force, and could not be the result of local processes such as volcanism. Large numbers of high-velocity meteorite and asteroid impacts would have created a fine rock flour that would have settled over the entire topography. This blanketing effect on a landscape may be observed on planet Earth in such places as Guatemala and New Zealand, where it is caused by volcanic rather than meteoritic ejecta. On the Moon, however, the rock flour is even darker than volcanic ash. After several billion years of exposure to solar hydrogen ion irradiation and impact vitrification, the material is bleached to a light grey. This is exactly what the first lunar walkers encountered. The regolith had the consistency of a light snow and footprints sank into it by 1–2 cm. Regolith was thinner (a few centimeters) over the minor topographic features which proved to be major ejecta (q.v.) from impacts.

Regolith on Mars and other planets has not yet been investigated first-hand by human exploration, but it is probably intermediate between the regolith of Earth and that of the Moon. Paleoclimatic changes on Mars from time to time have made it possible to have liquid water present over limited periods, thus permitting chemical weathering and water transport of both meteoritic and volcanic ejecta. The principal regolith is an eolian blanket, of yellow-red colors (chemically mobilized iron oxides).

Rhodes W. Fairbridge

Bibliography

Becker, G.F. (1895) Reconnaissance of the gold fields of the southern Appalachians. US Geol. Surv., 16th Annual Report, part 3, pp. 251–331.
Bridges, E.M. (1970) *World Soils*. Cambridge: Cambridge University Press, 128 pp.
Dollfos, A. (1957) Étude des plaustes pour la polarisation de leur lumière. Thèse, Paris and *Am. Astrophys. Suppl.*, **4**; English translation NASA-TTF-188.
Dollfus, A. (1985) Photopolarimetric seasing of planetary surfaces. *Adv. Space Res.*, **5**, 47.
Duchaufour, P. (1978) *Ecological Atlas of Soils of the World*. New York: Masson (translated from French by G.R. Mehuys *et al.*), 178 pp.
Fairbridge, R.W. (ed.) (1967) *The Encyclopedia of Atmospheric Sciences and Astrogeology*. New York: Reinhold Publ. Co., 1200 pp.
Fairbridge, R.W. (ed.) (1968) *The Encyclopedia of Geomorphology*. New York: Reinhold Book Corp., 1295 pp.
Fairbridge, R.W. (ed.) (1972) *The Encyclopedia of Geochemistry and Environmental Sciences*. New York: Van Nostrand Reinhold Co., 1321 pp.
Lyot, B. (1929) Recherches sur la polarisation de la lumière des planètes et de quelques substance, terrestres. Thèse, Paris and *Am. Observatoire Meudon*, 3 fasc. 1; English translation NASA-TTF-189.
Merrill, G.P. (1897) *A Treatise on Rocks, Rock-Weathering and Soils*. New York, London: Macmillan, 411 pp.
Short, N.M. (1975) *Planetary Geology*. Englewood Cliffs: Prentice-Hall, 361 pp.

Cross references

Mars
Moon
Reflectance spectroscopy

RELATIVISTIC COSMOLOGY

The general theory of relativity provides the best available description of the fundamental nature of gravitation and the best available description of the origins and evolution of the universe of which the planetary system is a part. This article introduces a number of the most important concepts to provide a context for topics such as the origin and evolution of planetary systems.

The cosmological world models are solutions of Einstein's general relativistic gravitation equations

$$R_{ik} - \tfrac{1}{2} g_{ik} R + \lambda g_{ik} = \frac{8\pi G}{c^4} T_{ik} \; (i,k = 0,1,2,3), \; X^0 = ct \quad (R12)$$

This is the case of isotropic matter distributions in which the matter tensor T_{ik} is a function of the mean energy density $u(t)$ and the mean pressure $p(t)$. In the gravitation equations $\lambda \geq 0$ is Einstein's cosmological constant, G is the Newtonian gravitational constant and c the velocity of light. For such isotropic matter the Einstein equation (R12) become

$$\left(\frac{\dot r}{r}\right)^2 = \frac{\lambda c^2}{3} - \frac{\epsilon c^2}{r^2} + \frac{8\pi G}{3c^4} u$$

and

$$\dot u + \frac{3\dot r}{r}(u + p) = 0 \quad (R13)$$

together with an equation of state $p = p(u)$. (R13) is the Friedmann equation for the measure $r(t)$, the mean distance of the space points, and (R14) is Tolman's conservation law of (adibatic) expanding matter.

In the cosmos, only the rest-mass density $\rho(t)$ and density of radiation energy $\omega(t) = 3 p(t)$ are important. We have

$$u = \rho c^2 + \omega = \rho c^2 + 3p \qquad (R14)$$

with the conservation laws $\rho r^3 = A =$ constant and $\omega r^4 = B =$ constant.

With these the Friedmann equation (R11) gives the Friedmann–Lemaitre equation:

$$\dot{r}^2 = \frac{\lambda c^2}{3} r^2 + \frac{8\pi G}{3c^4}\left(\frac{Ac^2}{r} + \frac{B}{r^2}\right) - \epsilon c^2 \qquad (R15)$$

The parameter ϵ has the values $\epsilon = +1$, $\epsilon = 0$ or $\epsilon = -1$. In the case $\epsilon = +1$ the space is spherically closed and r means the finite radius of the universe with total mass $M = 2\pi^2 A$. For $\epsilon = 0$ the space becomes flat and for $\epsilon = -1$ we have an infinite space with hyperbolic curvature. The term $\frac{\lambda c^2}{6} r^2$ is the potential of a repulsive force $\sim \lambda r$ and the term

$$\frac{4\pi G}{3c^4}\left(\frac{Ac^2}{r} + \frac{B}{r^2}\right)$$

it the potential of gravitational attraction.

The most widely accepted world model, the Einstein–de Sitter cosmos, describes a flat space ($\epsilon = 0$) without the repulsive force ($\lambda = 0$). We have a deceleration of the (expanding) velocity r in consequence of the gravitational attraction. For $r = 0$ the densities $\rho \sim 1/r^3$ and $\omega r \sim 1/r^4$ are infinite and the leading term is the radiation energy. In later times the radiation energy goes to zero. According to Planck's radiation law the equation $\omega r^4 =$ constant gives von Laue's relations for the frequencies ν and the temperature ν of the radiation:

$$\nu r = \text{constant} \qquad (R16a)$$

and

$$\Theta = \text{constant} \qquad (R16b)$$

and the number of quanta is conserved. Equation (R16a) implies Hubble's cosmological red shift. (R16b) involves the background radiation as blackbody radiation if the number of quanta is 10^9 times larger than the number of particles with rest masses. Asymptotically, the Einstein–de Sitter cosmos becomes a matter cosmos

$$r = (18\pi\, G\, A)^{1/3}\, t^{2/3}$$

with the Hubble parameter

$$H = \frac{\dot{r}}{r} = \frac{2}{3t} \qquad (R17)$$

According to this scenario the most important processes in the evolution of the universe operate in the first moments after the singularity ('big bang') at $t = 0$. The main point is the interpretation of the isotropic background radiation as a relict of the primary state of the cosmos. However, this scenario has the problem that the duration of very large cosmic structures (galaxies, clouds and super clouds of galaxies) has relaxation times, $\gg H^{-1}1$.

Another scenario uses the concept of infinitely large time in the past. According to the Eddington–Lemaitre model the space is finite ($\epsilon = +1$) with the radius $r(t)$ and the total mass is

$$M = 2\pi^2 A = \frac{\pi}{2} \lambda^{-1/2}\, G^{-1}\, c^2 \qquad (R18)$$

without cosmologically important radiation. Then, the Friedmann–Lemaitre equation is

$$\dot{r}^2 = \frac{\lambda}{3} C^2 r^2 + \frac{8\pi GA}{3c^2\, r} - c^2 \qquad (R19)$$

Following these the radius r has a minimum $r = 0$ for $r = \lambda^{-1/2}$, which is realized at time $t \to \infty$. At this time the universe is an Einstein cosmos with a balance $\ddot{r} = 0$ of the cosmic repulsion and the gravitational attraction.

The primary matter in this Einstein cosmos is a hydrogen gas with very low temperature. The time of building of the astrophysical structure is infinitely large. But these processes destroy the balance between repulsion and attraction. The cosmos expands with increasing acceleration r and velocity $\dot{r}$. Asymptotically for $t \to +\infty$ the Eddington–Lemaitre cosmos becomes the infinite de Sitter cosmos with vanishing matter density, and

$$H^2 = \left(\frac{\dot{r}}{r}\right)^2 = \frac{\lambda c^2}{3} \qquad (R20)$$

The principal point of this model is the determination of the cosmic mass M. The number N of the particles becomes a universal constant, Eddington's number $N \approx 10^{80}$. A difficulty is found in the thermalization of the cosmic background radiation, which is a secondary effect according to this scenario.

The Friedmann–Lemaitre models – favored by radio astronomers – are a synthesis of the Emden–de Sitter and the Eddington–Lemaitre scenarios. A Friedmann–Lemaitre cosmos is a finite space ($\epsilon = +1$) with a mass

$$M = 2\pi^2\, \rho\, R^3 \qquad (R21)$$

This universe starts at the time $t = 0$ with a big bang and infinite $H \sim 1/t$, and later has a long phase with a minimal velocity of expansion $H^2 \approx \lambda\, c^2 - c^2/r^2$. For $t \to +\infty$ this cosmos becomes the de Sitter cosmos with $H^2 = \frac{\lambda c^2}{3}$.

Generalizations of the general relativistic world models are analog solutions of generalized Einstein equations with the Newtonian G as a new field variable according to Dirac's hypothesis. Such scalar–tensor theories of gravitation are involved in the Kaluza–Klein theories of relativity with higher dimensional manifolds.

The space–time metrics ds^2 of each relativistic cosmos are conformal to the Lorentzian metrics

$$ds_0^2 = c^2\, dt^2 - dx^2 - dy^2 - dz^2$$

of special relativity with $ds^2 = \phi^2(r, t)\, ds_0^2$ (Weyl's cosmological principle).

Wilfried Schröder and Hans-Jürgen Treder

Bibliography

Eddington, A.S. (1933) *The Expanding Universe*. Cambridge: Cambridge University Press.
Einstein, A. (1955) *The Meaning of Relativity*, 5th edn. Princeton, Princeton University Press.
McVittie, G.C. (1965) *General Relativity and Cosmology*, Urbana: University of Illinois, Press.
Zel-dovich, Y.B. and Novikov, I.D. (1974) *Relativistic Astrophysics*, Vol. 2: *The Universe and Relativity*. Chicago: Chicago University Press.

Cross references

Celestial mechanics
Gravitation

REMOTE SENSING

Remote sensing is a body of scientific techniques for investigating the properties of a surface or material without touching it. The eye is a remote sensing device, gathering information about objects it sees for processing by the brain. Similarly, remote sensing equipment and techniques are used to gather information about the color, brightness, composition and material properties of objects, or to discern the spatial arrangement of features of a surface. The surface or object to be studied may be as small as a single mineral grain or as large as a planet or galaxy. It may be studied in the field, in a laboratory setting or from aircraft or spacecraft. The results may take the form of photographs, digital images or other types of numeric data. Remote sensing is broadly applicable to a variety of scientific disciplines, including (but not limited to) geology, hydrology, forestry and agriculture, marine and atmospheric studies (including weather), cartographic and urban studies, environmental assessments and monitoring and intelligence; readers may wish to consult the *Manual of Remote Sensing* (Colwell, 1983) for information on specific applications. Astronomy and planetary geology are conducted primarily via remote sensing using both ground-based and orbiting telescopes and planetary probes. Remote sensing is primarily used as a tool in support of individual scientific discipline studies; however, as the science progresses, it is becoming more and more central to global environmental monitoring and assessment, and serves as the basis for modeling of global processes.

Types and history of remote sensing

Direct visual observations of material properties, locations and other phenomena can be thought of as the origins of remote sensing. The invention of the telescope and the microscope in the 17th century made it possible to view objects that were either too far away or too small for investigation by the unaided eye, while the invention and refinement of photography in the 19th century made it possible to record observations directly. Pigeons and balloons were both employed to carry cameras aloft in early remote sensing efforts, but photographic remote sensing took a major leap forward with the advent of controlled aerial flight. By the time of World War I aerial photography was in use for intelligence as well as for scientific purposes, and by World War II photo reconnaissance and the interpretation of aerial photography were well developed. The need to obtain precise measurements from aerial photographs for map-making gave rise to the science of photogrammetry (q.v.).

Space-based photography was acquired from Earth-orbiting spacecraft such as Apollo–Soyuz and Skylab, and both panoramic and metric photographs were an important component of the Apollo missions to the Moon. Subsequently, digital images gained prominence with the launch of the Landsat (formerly ERTS, the Earth Resources Technology Satellite) and other Earth-orbiting satellites, with planetary probes such as Mariner and Viking, and with Earth-orbiting astronomical platforms like the Hubble Space Telescope. Digital images record information by dividing the imaged area into discrete picture elements, called 'pixels,' whose brightness and color can be represented numerically; these offer the advantage of electronic transmission directly from the spacecraft, eliminating the need for film return. Digital data also lend themselves to computer enhancement and statistical analysis (collectively called image processing), and thus are of prime importance in scientific research. However, photography still offers advantages in spatial resolution, and thus film-return missions and instruments like the Large Format Camera and the Soviet KFA-1000 flown aboard the Cosmos satellites are of importance.

For military and strategic applications, reconnaissance aerial photography (from balloons and aircraft) has been in use since the US Civil War, and by the 1960s the high-altitude U-2 aircraft was heavily in use, for instance returning critical information in the Cuban Missile Crisis. Early surveillance satellites in the late 1960s used high-resolution film cameras, which necessitated film return from orbit, recovery and processing before images could be used. Current satellite systems such as the KH-11 employ cutting-edge technology to acquire and transmit near real-time images with extremely high resolution; however, the price paid for such fine-scale information is a very small field of view which has been likened to viewing the world through a soda straw.

Regardless of spatial resolution and field of application, visible and reflected infrared light form the basis for much photography and digital image data such as that returned by Landsat or SPOT; however, other regions of the electromagnetic spectrum have special utility, especially the thermal infrared and microwave or radar wavelengths. In multispectral remote sensing, a series of different, narrow parts of the spectrum (often called 'bands') are used in combination. The use of multispectral data can provide more information about the composition and properties of surfaces and materials than would be possible with either single spectral bands or panchromatic study.

Relevance of remote sensing

The advantages of remote sensing can include time- and cost effectiveness for studies of large or inaccessible regions of Earth, when compared with the logistics of direct field study of those regions, e.g. the study of ocean circulation patterns or continental-scale vegetation monitoring. In addition, both photographic and digital remote sensing can provide information that is not available over large areas using any other method (for example, chlorosis, a symptom of stress in conifer forests, can be detected through its spectral response in infrared wavelengths well before damage is visible to the human eye). Statistical techniques can be applied to multispectral digital data in particular, so that quantitative information can be extracted and compared to other data (for example, a time series of image data for a given location).

The primary disadvantages of remote sensing have been equipment and maintenance costs, data storage and resolution limitations; in addition, questions of security arise with the use of extremely high-resolution data. The computer hardware and software required to analyze digital data have historically been bulky, expensive, and have required specialized operating environments; however, continual advances in workstation-environment computing and the development of high-speed, high-storage personal computers are bringing on-site image processing within the reach of a much wider audience. The costs of data acquisition (e.g. aerial overflights or satellite data acquisitions) can be high, depending on the data source. Data that must be acquired especially for a given project usually cost the user more than data that already exist in a data archive. Until recently, data storage media have been both bulky and have required specialized storage (archival conditions for photographic prints and negatives, and clean environments for magnetic reel computer tapes for digital data). Many digital data are now becoming available on CD-ROMs (the data from recent planetary missions such as Magellan are routinely distributed this way) and photographic images are increasingly being made available on laser disks (for example, hand held Hasselblad and Linhof photographs of Earth taken by Space Shuttle astronauts).

The spatial and spectral resolution of digital image data have been constrained by cost, data transmission and storage limitations, and questions of security. While the technology exists to fly extremely high-resolution sensors, the handling and distribution of the data from such systems is one of the greatest challenges facing the remote sensing community. The US planetary science community has approached this question through the use of high-density storage media as the CD-ROM, and by developing a system of archival 'nodes' at university and government laboratories throughout the country, each of which has responsibility for curating and distributing a particular type of data. The US Earth Observing System (EOS) program scheduled for the late 1990s will fly a variety of high-resolution instruments, and the development of an efficient data information and distribution system is a significant and challenging part of EOS mission development.

Aerial photographs are routinely taken by private firms for specific purposes and specific clients, but remote sensing sensor development is often conducted by government laboratories, or through public–private partnerships. The Landsat sensor systems, for example, were developed in this manner and were administered as an experimental system by the National Aeronautics and Space Administration (NASA), with data distribution handled by the US Geological Survey. When the technology was proven and the system declared operational, administration was transferred first to the National Oceanic and Atmospheric Administration (NOAA), and then to a private corporation, EOSAT, by the Landsat Commercialization Act of 1984. The National Landsat Policy Act of 1992 amended the 1984 legislation and laid the groundwork for management of the Landsat program to be transferred jointly to NASA and the Department of Defense.

Planetary missions are deemed experimental and thus both the design and operation of planetary exploration probes are administered by NASA and its affiliated field centers and laboratories (e.g. Goddard Space Flight Center in Maryland or the Jet Propulsion Laboratory in Pasadena, California). The French SPOT (System Probatoire d'Observation de la Terre) satellite system is another example of a public–private partnership, with government support for satellite launches and data distribution handled by SPOTImage, Inc. Other satellite systems are managed entirely within government, such as the US weather satellite system or the Japanese Marine Observation Satellite (MOS). The end user of remotely sensed images may acquire them directly from these agencies, or he/she may contract with one of many private companies that offer 'value-added' images (images that have been custom-processed and analyzed to the user's specifications). Processed images are used for a variety of purposes, ranging from academic research to resource evaluation, and including environmental management, weather prediction, urban planning or strategic concerns. In most cases the remote sensing data analyses are supported by 'ground truth,' i.e. field investigations, laboratory measurements (for example, soil texture and moisture, or high-resolution spectroscopy), or experimental models that simulate the surfaces and materials being studied via remote sensing. On its own, remote sensing is a useful tool, capable of discriminating materials that are different from one another and determining spatial relationships (i.e. mapping). When underpinned by ground-based information (e.g. soil moisture measurements or high-resolution laboratory spectroscopic data), remote sensing can provide a means for extending and extrapolating ground-based data through time or to geographically large regions.

Historically, while sensors with extremely high spatial resolutions have been invaluable in intelligence applications, the potential availability of such high-resolution data in the public sector has raised considerable concern that national security would be compromised. For many years the accepted resolution limit on publicly-available satellite images was 10 m. In the mid-1980s the Soviets launched the Cosmos KFA-1000, a film-return mission with 5-m resolution. However, the trade-off between high resolution (in not only the spatial domain but in spectral resolution and time domains as well) and the volume of data so generated remains, and thus most current and planned public or commercial missions compromise on moderate resolution. For instance, 15-m resolution was chosen for the 'high-resolution' panchromatic band on Landsat 6, and the High-Resolution Imaging Spectrometer (HIRIS) planned for the Earth Observing System (EOS) will operate at 30 m. Even in the post-Cold War era, security issues will continue to surround both the launching of truly high-resolution public/commercial systems, and also the release for scientific/commercial use of archival data originally acquired by the intelligence community. Initiatives for the declassification and release of such data are in discussion at the time of writing; if the security issues can be addressed and overcome, the remote sensing users' community may eventually acquire access to a significant and substantial archive of unique scientific and historical information.

Electromagnetic waves

Most remote sensing applications involve the detection of radio waves, microwaves or infrared or visible light waves that have been emitted (or reflected) by the environment being studied. These types of radiation are referred to as electromagnetic waves, since they are all composed of time-varying electric and magnetic fields. The physics of electromagnetic waves is described by the electromagnetic wave equation, which is in turn derived from Maxwell's equations (Jackson, 1962).

The types of electromagnetic waves referred to above differ only in frequency (f), or alternately wavelength (λ). These two quantities are related by the equation

$$f = c/\lambda \qquad (R22)$$

where c is the speed of light (3×10^8 m s^{-1} in vacuum). One also sometimes uses the wavenumber k, defined as the reciprocal of the wavelength.

An electromagnetic wave consists of an electric field and a magnetic field which are normal to each other and to the direction of propagation. These fields undergo time variation that is usually taken to be sinusoidal, so that we have for the electric field strength E:

$$E = E_0 \cos(kz - \omega t) \qquad (R23)$$

where z is distance in the direction of propagation, $\omega = 2\pi f$, and E_0 is the field amplitude. The quantity $(kz - \omega t)$ is the phase; in general we can add a constant term to the phase to describe a wave that 'starts out' in a different point of its cycle. The magnetic field obeys the space-time variation given by equation (R23), but with a $\pi/2$ phase difference.

The waves used in remote sensing span a wide range of wavelengths as shown in Figure R24. Different regions of the spectrum are used for different applications, depending on the emittance (or reflectivity) of the substance to be detected, the transmittance of the atmosphere, the resolution needed, etc. These considerations will be explained in more detail in the sections to follow.

Wave phenomena – interference, diffraction and Doppler shift

Several phenomena used in remote sensing follow directly from the wave nature of light. For example, electromagnetic waves obey the superposition principle, so that if a point in space is simultaneously

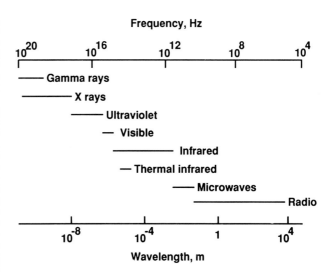

Figure R24 Regions of the electromagnetic spectrum.

occupied by two different waves, the resultant electric and magnetic fields can be obtained by vector addition of the fields of the two original waves. There may be regions in which the fields are in phase and augment each other, and other regions in which the fields partially cancel. Such interference effects give rise to the phenomena of fading and scintillation, which can complicate radar measurements (Dunn and Howard, 1970). A visible manifestation is the speckle effect seen with laser light.

Electromagnetic waves are also subject to diffraction, which causes a wave to 'bend' around an obstacle or to 'spread out' after passing through an aperture. This places an ultimate limit on the resolution attainable with a camera, telescope, radar, etc. An imaging system with an aperture of radius R cannot resolve two distant point sources if their angular separation is less than

$$\Delta\theta \approx 0.61 \, \lambda/R \qquad (R24)$$

Because of diffraction, waves cannot be used to detect objects that are much smaller than one wavelength. For example, a laser radar (which uses short-wavelength light waves) can detect small objects that could not be observed with a radio-frequency radar.

The Doppler effect is a shift in the observed frequency of a wave when there is relative motion between the source and the receiver. The frequency shift (for speeds well below the speed of light) is

$$\Delta f = \pm \, fv/c \qquad (R25)$$

where f is the original frequency and v is the velocity component of the source toward or away from the receiver. If the velocity is solely in the transverse direction, there is no Doppler shift (apart from relativistic effects). We take the plus sign (frequency increase) if the source is moving toward the receiver, and the negative sign if the source is moving away from the receiver.

A radar signal reflected from a moving object will display a Doppler shift. In this case, however, the reflected is moving relative to both the source and the receiver. The resulting Doppler shift will therefore be twice the value given by equation (R25) (for colocated source and receiver).

The wave phenomena discussed in this section are also observed with sound waves (i.e. sonar). Because sound travels much more slowly than light, noticeable Doppler shifts occur at relatively low velocities. For speeds much below the speed of sound, equation (R25) can be used, with the speed of sound substituted for c. At higher speeds the Doppler shift for sound is different for the cases of moving source and moving receiver (Halliday and Resnick, 1970).

Polarization

If an electromagnetic wave is propagating along the z-axis of a Cartesian coordinate system, the E vector can point along the x-axis, along the y-axis, or at any angle in between. The orientation of the E vector defines the polarization of the wave. For plane polarized light, the plane of the E vector does not change as the wave propagates,

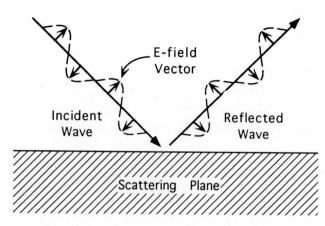

Figure R25 Polarization of an incident and reflected wave.

and the electric field undergoes sinusoidal variations in a single plane. In circular polarization, the direction of the E field continuously changes, and the tip of the E vector describes a helix in space.

The polarization of the wave is maintained during reflection from a smooth, flat surface. For example, in Figure R25 both the incident and reflected waves are vertically polarized. (If the E vectors were pointing into and out of the page, the polarization would be horizontal.) In some situations, e.g. multiple scattering from numerous small objects or reflection from a rough surface, the polarization of the incident wave may not be preserved.

Earth-sensing radars often use waves of different polarizations to obtain more information than could be obtained by using a single polarization. For example, a vegetation canopy gives a large signal when the waves are polarized in the direction of the plants' stems (i.e. vertically), while the surface under the canopy could be studied with horizontally polarized waves.

Natural sources of electromagnetic radiation

Passive remote sensing makes use of naturally emitted radiation. Natural radiation sources can also complicate active remote sensing by acting as background noise that can overwhelm the desired signal. Individual atoms produce radiation at discrete wavelengths, in agreement with the calculations of quantum mechanics. Such spectral lines are observed from heated gases since the density of the gaseous state is usually low enough to permit atoms or molecules to radiate without significant perturbation from collisional interactions. Transitions between the electronic energy levels of atoms and molecules generally produce visible or ultraviolet light. Rotational and vibrational transitions in molecules result in infrared or microwave emission.

In solids and liquids, interactions between neighboring atoms or molecules smear out the electron energy levels, resulting in continuous absorption and emission spectra. Many passive remote sensing applications involve detection of continuum radiation from thermal sources (including the Sun). For a surface that is a good infrared emitter (high emittance), and whose emittance is independent of wavelength, one can take the emission to be 'blackbody.' The energy radiated per unit wavelength per second per unit area of the emitter (in W m^{-2} Å^{-1}) is then

$$w_\lambda = \frac{2\pi hc^2}{\lambda^5}\left(\frac{1}{e^{hc/\lambda kT}-1}\right) \quad (R26)$$

where h is the Planck constant (6.62×10^{-34} Js), k is the Boltzmann constant (1.38×10^{-23} J deg^{-1}), and T is the absolute temperature. This blackbody emission spectrum does not depend on the composition of the emitter. By expanding the exponential term in equation (R24), we find that when λT is much greater than hc/k,

$$w_\lambda = 2\pi ckT/\lambda^4 \quad (R27)$$

the Rayleigh–Jeans law. Equation (R27) can generally be used for microwave emission. By integrating equation (R26) over λ, we can find the total power radiated per unit area, (in W m^{-2})

$$w = \sigma T^4 \quad (R28)$$

where σ, the Stefan–Boltzmann constant, has the value 5.67×10^{-8} W m^{-2} K^{-4}.

Equation (R26) indicates that thermal emission peaks at a wavelength that decreases as temperature increases. At 800 K the maximum is at 3.6 μm, with considerable emission in the red part of the spectrum; at this temperature a body is 'red hot.' At room temperature the maximum is about 10 μm. Hence the wavelength region from 7 to 15 μm is referred to as 'thermal infrared.'

The Sun radiates as a black body with a temperature of 6000 K. Its maximum emission occurs at about 4800 Å.

Interactions and radiative transfer

Reflection and refraction at solid and liquid surfaces

Much of remote sensing involves detection of solar radiation that is reflected from solid objects or water. The basic physics of the reflection process can be formulated in terms of just a few parameters such as the index of refraction, a dimensionless quantity defined as

$$n = c/v_p \quad (R29)$$

where v_p is the speed of light in the substance in question. When electromagnetic waves encounter an interface between two media with differing indices of refraction, some of the energy is transmitted through the interface and some is reflected back. The transmitted waves propagate at an angle given by Snell's law of refraction. The ratios between the electric fields are given by reflectance and transmittance coefficients. The power in an electromagnetic wave is proportional to the square of the electric field strength; hence the power ratios (termed reflectivity and transmissivity) are obtained by squaring the reflectance and transmittance coefficients. Formulas for these coefficients depend on the polarization of the incident waves and can be found in standard references (Klein, 1970).

Textbook examples, however, are usually not adequate to obtain useful predictions of reflectivity for remote sensing applications. Indeed, one is usually interested in deducing detailed quantities such as composition, surface texture, etc., that are not treated explicitly in the simple theory. However, the basic physical principles allow the construction of more elaborate models of reflection behavior.

Transmission through the atmosphere

Energy is lost from a beam of electromagnetic energy that travels through a gas. The fraction of energy that is transmitted a distance Δx is termed the transmittance

$$T = \exp[-k(\lambda)\Delta x] \quad (R30)$$

where $k(\lambda)$ is the extinction coefficient. The quantity $k(\lambda)\Delta x$ is the optical thickness. The extinction coefficient can be broken down into separate terms for molecular and aerosol scattering and absorption:

$$k(\lambda) = k_m + \sigma_m + k_a + \sigma_a \quad (R31)$$

where the absorption terms (k_m and k_a) represent energy that is converted into heat, and the scattering terms (σ_m and σ_a) represent energy that is removed from the beam by a change in propagation direction. Values for these quantities at various laser wavelengths are tabulated in McClatchey et al. (1978).

The Earth's atmosphere is strongly absorbing over most of the ultraviolet spectrum. It is transparent at visible wavelengths. Molecular constituents of the atmosphere cause strong absorption in various infrared and microwave bands. At other wavelengths 'windows' exist which may be exploited for spacecraft remote sensing (Table R2).

Scattering processes can be discussed in terms of a volume scattering coefficient σ, as in equation (R31). Scattering can result from interactions with molecules or with aerosols, small solid or liquid particles suspended in the air.

The mathematical treatment of scattering depends on the size of the scatterers. For small particles of (diameter less than approximately $\lambda/10$), Rayleigh scattering theory applies. The volume scattering coefficient for Rayleigh scattering is

$$\sigma = \frac{4\pi Nr^6}{\lambda^4}\left(\frac{n^2-1}{n^2+2}\right)^2 \quad (R32)$$

where N, r and n are the density (particles per volume), radius and refractive index of the scatterers respectively. For small particles the

Table R2 Infrared and microwave atmospheric windows (Fraser and Curran, 1976)

Near-infrared (μm)	1.0–1.12
	1.19–1.34
	1.55–1.75
	2.05–2.4
Mid-infrared (μm)	3.5–4.16
	4.5–5.0
Thermal infrared (μm)	8.0–9.2
	10.2–12.4
	17.0–22.0
Microwave (mm)	2.06–2.22
	3.0–3.75
	7.5–11.5
	20.0+

Table R3 Radiometric terminology

Quantity	Definition	Units
Radiant flux, power	Energy per time	W
Flux density	Power per area	W m^{-2}
Irradiance	Power per solid angle	W sr^{-1}
Spectral irradiance	Power per solid angle per wavelength interval	W sr^{-1} μm^{-1}
Radiance	Power per area per solid angle	W m^{-2} sr^{-1}
Spectral radiance	Power per area per solid angle per wavelength	W m^{-2} sr^{-1} μm^{-1}

Table R4 Photometric terminology

Photometric quantity	Photometric unit	Corresponding radiometric quantity
Luminous flux	lumen	Radiant flux
Illuminance	lux (= lumen meter^{-2})	Flux density
Luminous intensity	candela (= lumen steradian^{-1})	Irradiance
Luminance	candela meter^{-2}	Radiance

Table R5 Some commonly used space-based imaging sensor systems

Sensor	Total spectral range	Spatial resolution
Landsat MSS	0.4–1.1 μm	79 m
Landsat TM	0.45–2.35 μm	30 m
	10.4–12.5 μm	120 m
SPOT	0.5–0.89 μm	20 m
	Panchromatic	10 m
NOAA AVHRR	Visible, reflected IR	1 km
HCMM	0.5–1.1 μm	500 m
	10.5–12.5 μm	600 m
Seasat	23.5 cm	25 m
SIR-A,B,C	23.5 cm	40 m

scattering is generally isotropic; equation (R32) shows a strong dependence on the ratio between particle size and wavelength.

For larger particles (of radius up to several times λ), the more complicated Mie scattering is encountered. In the Mie regime we see a transition from Rayleigh scattering to a situation where the scattering is wavelength independent and the scattered light is concentrated in the forward direction (Born and Wolf, 1970).

For very large scatterers, such as raindrops, Mie theory can be replaced by simpler geometrical optic calculations (Kruse, McGlaughin and McQuistan, 1962). In this approach one calculates the scattered field by summing the effects of reflection, refraction and diffraction.

The propagation of electromagnetic waves through the atmosphere is often modeled by the use of sophisticated computer codes, such as LOWTRAN, developed by the Geophysics Directorate at Hanscom Air Force Base.

Radiation and reflectance terminology

Many specialized terms and concepts are used (not always consistently) in remote sensing. This section provides some of the more important definitions.

Table R3 lists some radiometric quantities. Units are based on the SI (mks) system. These quantities can also be expressed in photometric units, which are based on the response curve of the human eye. The conversion between photometric and radiometric quantities therefore depends on the wavelength. Photometric quantities and units are summarized, and related to analogous radiometric quantities, in Table R4.

Additional terminology is used to describe the reflectance properties of surfaces. The albedo is the fraction of electromagnetic radiation (of all wavelengths) reflected from a surface. The reflectance is the ratio of reflected to incident energy at a particular wavelength. The bidirectional reflectance distribution function (BRDF) is denoted as $f(\theta,\phi;\theta',\phi')$ and is the ratio between the reflected radiance (W m^{-2} sr^{-1}) in the direction θ, ϕ' and the incident irradiance (W m^{-2}) of parallel rays in the direction θ, ϕ. The BRDF has units of sr^{-1}.

The bidirectional reflectance $d\rho(\theta,\phi;\theta',\phi',)$ is the ratio between the reflected radiance in the direction θ',ϕ' and the incident irradiance in the direction θ,ϕ. The bidirectional reflectance is related to the BRDF by

$$d\rho(\theta,\phi;\theta',\phi') = f(\theta,\phi;\theta',\phi')\cos\theta' d\omega' \tag{R33}$$

where $d\omega'$ is a differential solid angle.

A Lambertian surface is one that is perfectly diffuse, i.e. one that reflects light equally in all directions. The bidirectional reflectance factor (BRF) is the ratio of flux reflected by the target of interest to the flux reflected by a Lambertian target. A specular surface is one that reflects light in a single direction, like a mirror. Few Earth materials and planetary surfaces exhibit specular behavior; for remote sensing in the visible and reflected infrared, most are considered to be Lambertian or nearly so. At radar wavelengths, however, some materials are more nearly specular, so that the radar beam is reflected with little scattering; such materials may look dark if the radar beam is reflected away from the antenna. If the geometry of the reflecting surface allows the beam to return to the antenna, producing an extremely bright return, it is termed a 'corner reflector.'

Platforms and sensor systems

Remote sensing instruments can be carried aboard a variety of air- or spacecraft, collectively termed platforms (Table R5). Astronomical remote sensing, for example, uses telescopes based on Earth, and instruments mounted on Earth-orbiting platforms like the Hubble Space Telescope and the Gamma Ray Observatory. For Earth observations, aerial photographs are routinely collected using cameras mounted on small aircraft. High-altitude aerial photography utilizes aircraft such as the U-2. Earth-looking sensors can be carried aboard manned spacecraft, or mounted on satellites in telecommunication with the ground. Satellite orbits for Earth-observation satellites like Landsat are generally near-polar and Sun-synchronous, so that data are always acquired at nearly the same time of day, in long north–south 'tracks' that completely and repetitively cover the planet. The orbit altitude for the Landsat craft has been in the range approximately 800–915 km. Geosynchronous satellites (mostly weather satellites and communications satellites) are placed in orbit at an altitude of approximately 35 800 km from Earth that is synchronous with the Earth's rotation, so that they continually 'look' at a particular part of the Earth. If the orbit is also equatorial, the satellite always appears in the same part of the sky and is referred to as a geostationary satellite. In principle, Earth-looking radar systems can be carried aboard any type of platform, but with the exception of the Shuttle Imaging Radars (SIR-A, SIR-B and SIR-C) and NASA's 1978 Seasat mission, most are mounted aboard aircraft.

Planetary remote sensing is conducted from Earth-based telescopes, or from planetary probes. These probes may operate in orbit around a planet, by landing on it, or they may simply fly by it, acquiring data as they pass. The Viking missions to Mars provide an example of the former two cases; each craft was composed of both an orbiter and a lander, which sent back data for several years after reaching Mars. The Voyager spacecraft, on the other hand, encountered and sent back images of Jupiter, Saturn, Uranus and Neptune in their long encounter and flyby mission.

Film return or direct data recording are usually possible from aerial and manned orbital missions. Satellites and probes, however, must preprocess and transmit their data to Earth receiving stations, perhaps having stored it on board until within reach of a receiving station. The early Landsats, for example, stored data using onboard tape recorders until within range of one of several ground receiving stations. Landsat 4 and 5 carry no tape recorders; instead, they rely on continuous transmission to the orbiting Tracking and Data Relay Satellites (TDRSS) to relay data to ground receiving stations.

The sensor systems carried aboard remote sensing platforms can be divided broadly into two categories. Active sensors, like radar, lidar or sonar systems, emit either electromagnetic radiation or sound waves and then measure the signal returned after it interacts with a target. Passive systems simply measure the behavior of ambient electromagnetic radiation, for example sunlight, as it interacts with the target medium. Passive systems most frequently are designed to look at the visible, reflected infrared or thermal infrared wavelengths, in order to detect and monitor land, ocean and cloud surface features and processes. Radar systems use microwave radiation and thus can penetrate some optically thick media such as water vapor or, for L-band radar, vegetation canopies, to interact with the surface beneath. Lidar systems typically use coherent (laser) light to measure, for example, the amount of dust suspended in air.

Digital image processing

Image processing equipment and techniques are used to transform raw digital image data into visual image format, and to extract statistical relationships from the data for analysis. The picture elements, or pixels, which make up raw satellite or airborne digital image data can be thought of as the tiles in a mosaic whose colors and patterns need to be assembled before the resulting picture is visible or useful. Each pixel has both a unique geometric location in the image data array that corresponds to a particular place on the Earth or planetary surface being studied, and a digital number or data number (often referred to as a DN) that corresponds to the brightness of that pixel relative to others imaged by the sensor system. In multiband or multispectral digital image data, several sensors that look at different colors of light may be used in parallel, resulting in a series of DNs that describe the brightness and color of a ground location over a range of wavelengths. Corrections to the raw data thus may be in either the spatial (geometric) or spectral (radiometric) domains. A raw rectangular array, or matrix, of pixels must be adjusted geometrically to fit commonly used map projections, and to remove distortions that can be introduced by some sensors (e.g. the 'smearing' of data that occurs near the edges of some airborne detector systems). Radiometric corrections involve adjustments of the brightness to account for flaws in the data (e.g. 'dropped' or missing pixels or 'striping' due to variations in detector response in a given sensor system), to correct for atmospheric contributions to brightness or to transform the integer DNs into physical quantities. For example, spectral radiances can be calculated from the raw DNs of Landsat Thematic Mapper data through the equation:

$$L_\lambda = \frac{L_{min\lambda} + (L_{max\lambda} - L_{min\lambda})Q_{cal}}{Q_{calmax}} \quad (R34)$$

where Q_{cal} is the calibrated and quantized scaled radiance in DNs, $L_{min\lambda}$ is the spectral radiance at $Q_{cal} = 0$, $L_{max\lambda}$ is the spectral radiance at $Q_{cal} = Q_{calmax}$, Q_{calmax} is the range of rescaled radiance in DNs, and L_λ is the spectral radiance (Markham and Barker, 1986).

Statistical manipulation can be useful to standardize and compare image data to other images acquired by different sensor systems or at different times, and to detect trends within the data. Techniques range from simple calculation and histogram display of the range, mean and standard deviation of brightnesses in an image data set to sophisticated factor analysis, fractal calculations or clustering techniques to determine spatial and spectral relationships among pixels within an image or set of images.

One of the most powerful approaches in image processing and remote sensing data analysis is to combine the data from several different sensor systems and/or data acquired at different times, using the unique information afforded by each sensor to build up a model of the reflectance of a given region over a wide spectral range or through time. Image data can also be combined with non-image information (e.g maps, field measurements or population statistics) via the use of a geographic information system (GIS). A GIS treats each information source as a 'layer' of data that can be stacked with other layers, and then analyzed statistically to extract relationships that would otherwise be too complex to resolve. Geographic information systems have shown particular utility in Earth resources monitoring and management and in urban planning; however, the power of the approach suggests that its use will continue to spread in the remote sensing community.

Remote sensing of the Earth's surface

Aerial and spaceborne photography

The major difference between photography and digital remote sensing is that photography involves the return to Earth of film from handheld or automated cameras. Photography is thus mostly limited to aerial flight or manned space missions. Film return from unmanned orbital craft is rare; the Soviet Cosmos KFA-1000 was of this type. The major advantages of aerial photography include its high spatial resolution and the opportunity for obtaining stereo coverage for topography and terrain analysis.

Aerial photography is a mainstay of many Earth resources and science applications. It is commonly used in urban planning, soil conservation, archeology, forestry and basic geographic mapping. Photographs acquired in the visible and infrared regions of the spectrum are most common; films and filters can be combined to yield results that are panchromatic, panchromatic with an extended red response, infrared, color, false-color or multiband (multispectral). Typical mapping cameras use film with an image area of 23×23 cm (9×9 in). The scale of the resulting photographic images is dependent on the focal length of the camera and the altitude of the aircraft; aerial photographs may be acquired at low altitude (less than 9 km), medium altitude (9–15 km) or high altitude (15–25 km).

Although much aerial photography is acquired specifically for local or state government or by private parties, an archive of US photographs, including those acquired by the National High Altitude Photography (NHAP) Program, is kept at the US Geological Survey's Earth Resources Observation Systems (EROS) Data Center in Sioux Falls, South Dakota. The goal of the NHAP Program is to provide complete aerial photographic coverage of the continental US on a repetitive basis. NHAP coverage includes black-and-white (panchromatic) photographs at 1 : 80 000 scale, and color-infrared photographs at 1 : 58 000 scale (Williams, 1983).

Spaceborne photography was an integral component of the Apollo missions; Apollo panoramic and metric photographs served as a primary basis for mapping the lunar surface. Spaceborne Earth photography was part of the Skylab experiment, with both the Earth Terrain Camera and the S190A Multiband camera. The Large Format Camera flown on the Space Shuttle carried 230 mm × 460 mm format film and thus acquired very large, high-resolution Earth images. Much recent Earth photography has been taken by Shuttle astronauts using both Hasselblad and Linhof cameras. These photographs include ocean and storm systems as well as land areas, and are archived at Johnson Space Center in Houston, Texas, at the Technology Applications Center in Albuquerque, New Mexico, and at the National Air and Space Museum in Washington, DC.

Satellite-based digital images

Although there are now several US and international civilian remote sensing systems in use, the most commonly used data are from the US Landsat program and the French SPOT system. The sensors on these two satellite systems record Earth images over a range of wavelengths in the visible and infrared portions of the spectrum, and utilize spatial resolutions ranging from 80 m (in the case of the Landsat Multispectral Scanner) to 10 m (for the SPOT panchromatic mode images). The multitemporal, multiscale, multispectral images of Earth acquired by these two satellite systems provide a primary tool for a wide variety of Earth observations, mapping and monitoring applications.

While Landsat and SPOT comprise the bulk of commercially available Earth land surface observations, several other missions and sensor systems are worth noting. The Advanced Very High Resolution Radiometer (AVHRR) carried aboard the National Oceanic and Atmospheric Administration (NOAA) satellites has four channels sensitive to visible, reflected infrared and infrared radiation (an additional IR channel exists on NOAA-7 and subsequent platforms). While primarily intended for meteorologic, hydrologic and oceanic measurements, AVHRR data are also used for continental-scale and global-scale mapping of some land processes (e.g. Tucker, Townshend and Goff, 1985). The Heat Capacity Mapping mission (HCMM) flew in 1978 and acquired day and night images in the visible and thermal infrared (0.5–1.1 and 10.5–12.5 μm respectively). These image pairs were used to calculate temperature differences and apparent thermal inertia of ground regions. The Coastal Zone Color Scanner operated in six visible, reflected IR and thermal IR channels and acquired images of many coastal regions.

Radar

Radar observations of the Earth's surface have several advantages: they do not rely on solar illumination, can reveal subsurface features and can penetrate clouds that block shorter wavelengths. Some radar systems simply function as altimeters and are used for topographic mapping or ocean wave measurements (Seasat). Other radars produce images that can be used for the same purposes as photographic images in such fields as geology, polar ice studies and vegetation monitoring.

The two basic types of radar systems are real aperture radar and synthetic aperture radar. In both types of radar systems, the ground resolution in the range direction is a function of pulse length (or equivalently, bandwidth). Short pulses provide good ground range resolution, but may provide inadequate energy for long-range measurements. Improved range resolution may be obtained by a variety of pulse coding or compression techniques. The azimuth resolution (parallel to the travel direction of the radar instrument) is limited by the antenna length for a real-aperture system due to diffraction effects as discussed above. Synthetic aperture radar (SAR) overcomes this limitation by using the relative motion of the antenna and target to make a short antenna act like a much longer antenna array. SAR systems utilize the Doppler shift of the returning signals to improve resolution in the along-track direction. SAR systems thus are the more widely used for mapping and other surface observations (Ulaby, Moore and Fung, 1982).

Most recently, airborne multipolarization radars have shown great promise in discriminating Earth materials and surface characteristics. Surface factors like roughness, moisture content, vegetation cover and composition will influence the intensity and polarization of the reflected radar signal, and different transmitted polarizations may produce very different returns depending on the properties of the surface from which the reflection comes. These differences allow surfaces that might appear similar when viewed at a single polarization to be distinguished from one another.

Much radar imagery is collected from airborne platforms. Examples of orbital imaging SARs include Seasat and the Shuttle Imaging Radars SIR-A, SIR-B and SIR-C. The Seasat imaging radar operated for 100 days in 1978 and carried both a radar altimeter and a scatterometer. Although designed to test radar utility in oceanographic studies, Seasat also obtained L-band (23.5 cm) radar images of land regions. The Shuttle Imaging Radars SIR-A, SIR-B and SIR-C also operate in L-band (SIR-C has two wavelengths). Such long wavelengths can penetrate to some depth in dry Earth materials such as desert sand, revealing subsurface structure. Applications for imaging radar data are numerous, but they are particularly useful in geologic and hydrologic mapping, polar and sea ice studies, oceanography and vegetation studies (Elachi, 1987).

Atmospheric remote sensing

Weather radar

The use of radar for monitoring weather patterns is familiar to almost everyone, thanks to television weather reports. Radar energy can be scattered by cloud droplets, and by solid and liquid forms of precipitation. To detect the small droplets that make up clouds, one needs to use a fairly short radar wavelength (1 cm = 30 000 MHz). Larger precipitation particles are detectable by longer waves (10 cm). Weather surveillance radars typically have operational ranges of about 200 km (Knight, 1983).

Weather radar images are transmitted electronically along communications land lines to facsimile receivers. By this means, users of weather data have access to most of the country's major weather surveillance radar stations.

Laser remote sensing of the atmosphere

Lidar ('light detection and ranging') is the use of laser pulses to detect reflecting or scattering targets, in a manner analogous to radar. Lidar technology is useful in the study of the atmosphere.

Laser light can be scattered from atmospheric constituents in a variety of ways (Measures, 1984). Rayleigh and Mie scattering processes have already been described in this article, and can be used to measure molecular densities and to detect aerosols, respectively. The velocity of the scattering particles can be determined by measuring the Doppler frequency shift of the scattered light. Such Doppler lidars are used to measure upper-atmosphere wind speeds (Hardesty, Post and Banta, 1991).

The monochromatic nature of laser light allows specific atomic or molecular transitions to be pumped. By tuning the frequency of the laser to match a particular atomic or molecular transition, one can obtain enhanced sensitivity to a particular atmospheric component; this process is termed resonance scattering. A number of atoms have resonance transitions at accessible laser wavelengths and can be detected by this technique. For example, a layer of atomic sodium, which exists at altitudes around 90 km, has been studied by a number of investigators using this technique.

The laser light is also absorbed by atmospheric constituents during its transit to the scattering region and back. In the technique of 'differential absorption lidar' (DIAL), the scattering is non-resonant (Rayleigh), but the absorption is used to infer path-integrated concentrations of species of interest. Near-simultaneous measurements are made at the resonant wavelength and at a wavelength slightly off-resonance, to allow the resonant contribution to the round-trip absorption to be inferred. A number of molecular constituents and pollutants can be detected by this means (Collis and Russell, 1976). The measurement of high-altitude ozone levels with the DIAL technique is of particular current interest (Browell, 1991).

Rayleigh scattering is also denoted 'elastic' scattering, because the photons exchange no energy with the scatterers. Inelastic scattering also occurs, in which the photons give up or receive energy from the scatterer molecules. The frequency of scattered photons is decreased or increased in such inelastic, or Raman, scattering. The frequency shift allows the scatterer molecules to be identified. Unfortunately the Raman scattering process results in a much smaller scattered signal than is observed in Rayleigh scattering. Nevertheless, with high-power Q-switched lasers, sufficient signal can be obtained to allow measurement of atmospheric constituents such as H_2O, as well as a number of pollutants (SO_2, NO, CO, H_2S, etc.).

Remote sensing of the ionosphere

Ground-level measurements of ionospheric parameters are made by reflection or scattering of radio waves. The ionosonde technique involves detection of reflected radio waves. The time delay allows the height of the reflection point to be determined. To a good approximation, the reflection occurs where the electron density n (per cubic centimeter) is such that plasma frequency, given by

$$f_p = 9000\, n^{1/2} \qquad (R35)$$

equals the radio wave frequency. The peak electron density is a few million per cubic centimeter, giving a maximum f_p of about 12 MHz. If the radio wave is launched at an angle other than vertical, the wave's propagation direction is gradually changed as the wave propagates upward, causing the reflection point to be different from the prediction of equation (R35). The presence of the Earth's magnetic field also complicates the physics by producing two types of electromagnetic wave (the ordinary and extraordinary wave) that reflect at somewhat different points.

More detailed information can be obtained from scattering measurements such as incoherent scatter radar (ISR; Kelley, 1989). This technique uses relatively high frequencies (above 50 MHz), to avoid reflection and allow probing of ionospheric layers above the level of maximum electron density. A small amount of energy is scattered by electrons and is detected by large radar antennas. The amount of scattered energy contains information on electron densities. A frequency spread results from Doppler shifting caused by electron and ion motions, allowing temperatures and flow velocities to be inferred. Density fluctuations caused by plasma instabilities in the ionosphere can also be measured by the coherent scatter radar (CSR).

Changes in amplitude and/or phase of a transmitted radio signal also can be used to study the ionosphere. Such scintillation techniques have been carried out with beacon satellites that radiate a number of different frequencies.

Future developments and emerging issues

The trend in space borne remote sensing, both Earth and planetary, is toward sensors with higher spectral and spatial resolution. The Magellan spacecraft, for example, returned some 480 billion bits of Venus image data between 16 August 1990 and 15 January 1992 (the first 17 months of its mission; Saunders et al., 1990). Planned missions such as EOS (see Earth Observing System) will carry numerous highly specialized instruments on multiple small- and intermediate-sized platforms; both data return rates and volumes will be high. High data volume missions will require reliable but increasingly sophisticated downlinks and ground systems for data handling and distribution. Multinational efforts are becoming common as a means of sharing costs and benefits of space exploration, but realistic binding agreements among parties are essential to the success of such missions. These issues, faced already in planetary missions such as the CRAF/Cassini comet rendezvous and in planning for the US Space station, will be critical to continued cooperation as nations rely on each other to share instrument development and flight costs and data distribution; the utility of one partner's instrument and data may be severely compromised by the failure or cancellation of another. Security questions also emerge as commercial sensors achieve higher spatial resolution.

A precedent was set in Earth observing satellite systems by the Landsat Commercialization Act of 1984, which transferred control of the operational Landsat satellite and responsibility for future satellite development from the US Government to a private corporation, EOSAT. With privatization came questions of data ownership, archival responsibility and profitability of the US civil remote sensing system. As future Earth observing missions are developed, flown and become operational, the same issues of control and profitability will emerge, as will the responsibility for long-term archiving of data. Secure curation of high-resolution remote sensing data on stable, accessible and reproducible media, with attendant issues of data access, distribution and cost, will be key elements in the development of future civil remote sensing policy.

Patricia A. Jacobberger and Gerald P. Jellison

Bibliography

Born, M., and Wolf E. (1970) *Principles of Optics*. New York: Pergamon, pp. 653–6.
Browell, E.V. (1991) Ozone and aerosol measurements with an airborne lidar. *Optics and Photonics*, 2(10), 8–11.
Collis, R.T.H. and Russell P.B. (1976) Lidar measurement of particles and gases by elastic backscattering and differential absorption, in *Laser Monitoring of the Atmosphere* (ed. E.D. Hinkley) Berlin: Springer-Verlag, p. 106.
Colwell, R.N. (ed) (1983) *Manual of Remote Sensing*. Falls Church, VA: American Society of Photogrammetry.
Dunn, J.H. and Howard, D.D. (1970) Target noise, in *Radar Handbook* (ed. M.I. Skolnik). New York: McGraw-Hill, p. 28–1.
Elachi, (1987) *Introduction to the Physics and Techniques of Remote Sensing*. New York: John Wiley.
Fraser, R.S. and Curran, R.J. (1976) Effects of the atmosphere on remote sensing, in *Remote Sensing of Environment* (eds J. Lintz and D.S. Simonett). Reading, MA: Addison-Wesley, p. 35.
Halliday, D. and Resnick R. (1970) *Fundamentals of Physics*, New York: Wiley, pp. 660–1.
Hardesty, R.M., Post, M.J. and Banta, R.M. (1991). Observing atmospheric winds with a Doppler lidar. *Optics and Photonics*, 2(10), 12–5.
Jackson, J.D. (1962) *Classical Electrodynamics*. New York: John Wiley.
Kelley, M.C. (1989) *The Earth's Ionosphere: Plasma Physics and Electrodynamics*, San Diego: Academic Press.
Klein, M.V. (1970) *Optics*. New York: John Wiley.
Knight, P.E. (1983) Weather observations and forecasting, in *Van Nostrand's Scientific Encyclopedia*, 6th edn (ed. D.M. Considine) New York: Van Nostrand Reinhold, p. 2994.
Kruse, P., McGlaughin L. and McQuistan, R. (1962) *Elements of Infrared Technology*. New York: John Wiley.
Landgrebe, D.A. (1978) The quantitative approach: concept and rationale, in *Remote Sensing: The Quantitative Approach* (eds P.H. Swain and S.M. Davis). New York: McGraw-Hill, p. 5.
Markham, B.L. and Barker, J.L. (1986) Landsat MSS and TM post-calibration dynamic ranges, exoatmospheric reflectances and at-satellite temperatures. *EOSAT Landsat Data Users Notes* 1, pp. 1–8.
McClatchey, R.A. Fenn, R.W. Selby, J.E.A. *et al.* (1978) Optical properties of the atmosphere, in *Handbook of Optics* (eds W. Driscoll and W. Vaughan.) New York: McGraw-Hill, pp. 14-14–14-21.
Measures, R.M. (1984) *Laser Remote Sensing: Fundamentals and Applications*. New York: John Wiley.
Saunders, R.S., Pettengill, G.H., Arvidson, R.E. *et al.* (1990) The Magellan venus radar mapping mission. *J. Geophys. Res.*, 95 (B6), 8339–55.
Tucker, C.J., Townshend, J.R.G. and Goff, T.E. (1985) African land-cover classification using satellite data. *Science*, 277, 369–75.
Ulaby, F.T., Moore, R.K. and Fung, A.K. (1982) *Microwave Remote Sensing: Active and Passive*. Reading, Massachusetts: Addison-Wesley Publishing Company.
Williams, S., Jr (ed.) (1983) Geological applications, in *Manual of Remote Sensing* (ed. R. Colwell), Falls Church, Virginia: American Society of Photogrammetry, pp. 1667–953.

Cross references

Comet: observation
Electromagnetic radiation
Imaging science
Infrared spectroscopy
Mars: remote sensing
Microwave spectroscopy
Photoclinometry
Photogrammetry, radargrammetry and planetary topographic mapping
Radar astronomy
Radio astronomy
Reflectance spectroscopy
Spectrophotometry
Spectroscopy: atmosphere
Ultraviolet spectroscopy
Visible and near-infrared spectroscopy

RESONANCE IN SATURN'S RINGS

The Saturn ring system (Alexander, 1962; Smith, *et al.*, 1982; Yoder *et al.*, 1983; Esposito *et al.*, 1984) consists physically of an aggregation of exceedingly diffuse, but gravitationally bound, small, solid particles, disposed mainly within several concentric bands surrounding Saturn (Plates 26, 27). Together, these rings form a nebulous, disklike configuration only some 16 km in thickness, located precisely in the plane of the planet's equator. Within this projection of the equatorial plane, the myriads of small particles revolve around the planet at velocities determined by their individual distances from its center.

A wide variety of physical mechanisms are cited by modern-day planetologists to explain the formation of the many features – including gaps, ring moons, ring shepherds, co-orbital satellites and ringlets – within the complex ring structure of Saturn. Comprehensive discussions of these various processes are included within the

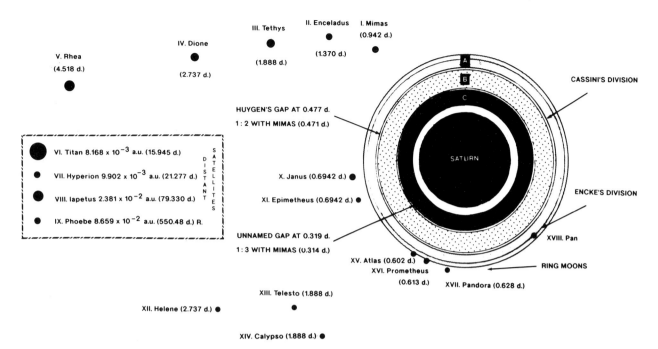

Figure R26 Saturn: satellites and ring system, illustrating the synergetic resonance origin of the Huygens gap in the Cassini division and other features. The plane of the paper corresponds to Saturn's equatorial plane. Due to the inclination of that equatorial plane to the usual line of sight from the Earth, there may appear to be some reduction in apparent ring widths in this orthogonal projection. The Roman numbers increase serially according to the satellite's distance from the planet (I through IX); thereafter, in order of discovery (with the tropical period of revolution given in days, in parentheses; see Saturn: satellite system). Except for the four outermost satellites (shown in box), all satellites are shown at the appropriate radius from the planet's center, with their relative sizes also indicated.

definitive works listed in the bibliography. A time-honored resonance theory which credibly accounts for certain confirmed gaps not previously known in the ring system of Saturn is that developed by Daniel Kirkwood (1869) to explain similar gaps in the orbits of the asteroids. It is discussed at greater length elsewhere (see Asteroid: resonance). This theory will be applied in the present article to the quantitative verification of a resonance gap detected by spacecraft in a radial profile plot of Saturn's rings (Greenberg and Brahic, 1984).

Within Saturn's complex ring system, by no means all the dark gaps which form divisions between the rings are of a simple resonance origin. Among the various annular zones where an absence of light-reflecting particles results in a visible separation between the rings is the so-called 'Cassini division,' named after its discoverer, Jean-Dominique Cassini (q.v.), who described it in 1675. Investigations analyzing many thousands of close-up photographs, instrument recordings and graphs from the Voyager 1 and Voyager 2 flybys of Saturn in 1980 and 1981 have revealed substantial additional details concerning the planet. While the action of resonance-driven gravitational waves has been suggested in connection with the Cassini division, the hypothesis of the origin of this major gap being due solely to a commensurability between the periods of ring particles and any satellite of Saturn is considered questionable (Cuzzi *et al.*, 1984).

However, a much smaller discontinuity known as the Huygens gap (also sometimes spelled Huyghens), located between narrow ringlets at the inner edge of the Cassini division (Figure R26) has been determined from these same observational data to be of resonance origin. Because of its more general familiarity, the Cassini ring feature may be used as a guide to the location of the Huygens gap. The Cassini division consists of a telescopically quite distinct gap lying between the relatively dark, outermost band (ring A) and a considerably brighter central band (ring B) in Saturn's main ring system. The distance from the center of the planet to the inner edge of the Cassini division is 117 583 km. The Cassini division occurs as a gap of some 246 km separating this particle boundary from the inner edge of ring A. The Huygens gap, located about 117 824 km from the planet's center, between narrow ringlets, is some 285–400 km wide. (Note: These figures represent order-of-magnitude averaged values relative to notably uncertain boundary surfaces, and are subject to successively more accurate re-evaluations of Voyager 1 and 2 spacecraft flyby data.)

Theory of origin of the Huygens gap

In the analytic substantiation of a resonance origin for the Huygens gap, it may be assumed that initially a random particle in the aggregating mass of particulate matter which composed the forming ring system of Saturn found itself revolving around the planet at a distance which produced a revolutionary period commensurable with that of one of the satellites of Saturn. (Since, with a few recently discovered exceptions known as the 'ring satellites,' the rings are interior to the satellites, the commensurable period of the original particle would logically be a fractional value of the revolutionary period of the satellite.) A particle coming into this position of synchronous orbital motion with one of Saturn's satellites would, through resonance, repeatedly receive bursts of energy forcing it to a higher energy level and greater speed than Kepler's third law permits. The resulting dynamic adjustment would cause the particle to vacate its previous position for one in which its speed of motion is balanced by the existing energy state. In a presumed multiplicity of similar actions, the results would leave an annular gap in Saturn's ring system. The calculated basis for the existence of the Huygens gap, known to be caused by resonance – as opposed to other possible gap-producing factors (Kirkwood, 1872; Franklin, *et al.* 1984; Greenberg, 1984) – and its commensurable relationships with the periods of certain of Saturn's satellites, are given below.

Quantitative corroboration of a resonance origin

According to the same procedure used in determining the periods of the asteroids around the Sun (see Asteroid: resonance), the revolutionary period of any small particle at a distance a from the center of Saturn may be established by using Kepler's laws (q.v.). Also, as before, selecting the intermediate of the three gravitationally involved bodies as a reference, we may use the well-established period of Saturn's satellite Mimas (0.942422 d). This satellite revolves around the planet at a distance of 185 489 km from its center. Saturn has a radius of 60 330 km.

Table R6 Relationships between Saturn's major satellites and the Huygens gap

Satellite	Comparative mass (Saturn = 1)	Tropical period of revolution (days)	Comparison with period of particle located at distance of Huygens gap:	
			Commensurable unit-fraction of satellite's period	Resulting value compared with particle's period (0.477 d)
Mimas	8.0×10^{-8}	0.942421813	½	0.471
Enceladus	1.3×10^{-7}	1.370217855	⅓	0.457
Tethys	1.3×10^{-6}	1.887802160	¼	0.472
Dione	1.8×10^{-6}	2.736914742	⅙	0.456

Then, from Kepler's law:

$$(P_0)^2 = (0.942422)^2 \times (117\,824)^3/(185\,489)^3$$

where P_0 is the period of a particle revolving around Saturn at a distance from the planet (117 824 km) equal to that of the Huygens gap. Solving the equation, $P_0 = 0.4771542$ d = 11.452 h.

Thus, on average, it may reasonably be assumed that the period of revolution for a point at the inner edge of the Huygens gap is very close to 11.452 h or 0.477 d. (A point at the middle of the 246 km gap will revolve with a slightly smaller period.)

This calculated value of the hypothetical particle's period of revolution may now be compared, and its mathematical commensurability established, with certain unit-fractional values of the revolutionary periods of Saturn's four earliest discovered satellites, as shown in Table R6.

As in the case of the Kirkwood gaps in the asteroid belt, the relationship of the formation of the Huygens gap in Saturn's ring system to a commensurability between (1) the period of revolution of the initial particulate bodies around their primary and (2) the revolutionary period of a second, more massive, nearby body, is clearly evident from these examples.

An unnamed gap lying just beyond the outer boundary of the B ring likewise has been determined to be caused by the mathematically commensurable relationship (1 : 3) between particles which would otherwise occupy this gap, and the revolutionary period of Saturn's satellite Mimas.

From the Voyager 2 flybys of the outer planets, it was evident that Uranus and Neptune, too, are surrounded by ring systems. However, to date, no such similar gaps due to resonance have been discovered among these formations.

Fergus J. Wood

Bibliography

Alexander, A.F.O'D., (1962) *The Planet Saturn: A History of Observation, Theory and Discovery*. London: Faber and Faber, 474 pp.

Cuzzi, J.N. *et al.*, (1984) Saturn's rings: gravitational resonances, in *Planetary Rings* (eds R. Greenberg and A. Brahic). Tucson: University of Arizona Press, 784 pp., pp. 119–20, 137–9.

Esposito, L.W. *et al.*, (1984). Saturn's rings: structure, dynamics, and particle properties (pp. 463–545); Resonance with rings (pp. 485–90, 498–500), in *Planetary Rings* (eds R. Greenberg and A. Brahic). Tucson: University of Arizona Press, 784 pp.

Franklin, F. *et al.* (1984). Ring particle dynamics in resonances, III. Development of gaps at resonance, in *Planetary Rings* (eds R. Greenberg and A. Brahic). Tucson: University of Arizona Press, 784 pp., pp. 578–88.

Gehrels, T. and Matthews, M. (eds) (1984) *Saturn*. Tucson: University of Arizona Press, 968 pp.

Greenberg, R. (1984) Orbital resonances among Saturn's satellites, in *Saturn* (eds T. Gehrels and M. Matthews). Tucson: University of Arizona Press, 968 pp.

Greenberg, R. and Brahic, A. (eds.) (1984) *Planetary Rings*. Tucson: University of Arizona Press, 784 pp.

Kirkwood, D. (1869) On the nebular hypothesis, and the approximate commensurability of the planetary periods. *Mon. Not. Roy. Astron. Soc.* **29**: 96–102. An analytical development of the original theory of Kirkwood's Gaps in the asteroid belt and in Saturn's rings.)

Kirkwood, D. (1872). On the formation and primitive structure of the solar system (resonance between Mimas and Saturn's rings). *Proc. Am. Phil. Soc.*, **12**, 163–7.

Morrison, D. (1982) *Voyages to Saturn*. Washington: National Aeronautics and Space Administration, SP-451, 227 pp.

Smith, B.A. *et al.* (1982) A new look at the Saturn system: The Voyager 2 images. *Science*, **215**, 504–37.

Smith, B.A., *et al.* (1986) Voyager 2 in the Uranian system: imaging science results. *Science*, **233**, 43–64.

Yoder, C.F., *et al.* (1983) Theory of motion of Saturn's co-orbiting satellites. *Icarus*, **53**, 431–43.

Cross references

Commensurability
Planetary ring
Saturn
Saturn: ring system
Shepherd satellite

RESONANCE IN THE SOLAR SYSTEM

Resonances in dynamical systems occur when two angular variables describing the system (for instance, the longitude of two planets orbiting around the Sun) have commensurable mean velocities (see Commensurability). In such a case, the geometry of the system repeats itself periodically, allowing a building up of the interactions between the various parts of the system. For conservative dynamical systems especially (for which no dissipative force counteracts this reinforcement), resonances are often the most prominent and important features of the system.

A very simplified model of a resonance is the pendulum (or the swing). Let us consider a problem in which two angular variables (let us say ϕ_1 and ϕ_2) have commensurable mean velocities (say in the ratio p/q). The 'resonant angle' $\phi = q\phi_1 - p\phi_2$, (i.e. the angle the mean velocity of which is zero by virtue of the commensurability) behaves approximately like the angle between the string of a pendulum and the vertical. The orbits of the pendulum are shown in Figure R27a. We see in the so-called 'phase diagram' (in which one plots the angular velocity $d\phi/dt$ with respect to the angle ϕ) a 'libration' region (the eye-shaped region inside the thick curve in Figure R27a) where the mean velocity of ϕ is exactly zero. This models the resonance region of the dynamical problem we started with. There are also of course 'circulation regions': positive circulation (in the top part of Figure R27a), negative circulation (in the bottom part), according to the sign of $d\phi/dt$. These regions are separated from the resonance region by the 'separatrices' (the thick curves of Figure R27a). These are trajectories which take an infinite time to swing from unstable equilibrium (pendulum exactly upside down and represented twice – as the points $\phi = \pm \pi$, $d\phi/dt = 0$ – in Figure R27a) back to it.

This is an overly simplified model of resonance. In most systems this model would be distorted in one (or both) of the following ways. A small dissipative force (such as tidal interactions between satellites and planets) can lead to a distortion, as shown in Figure R27b. The

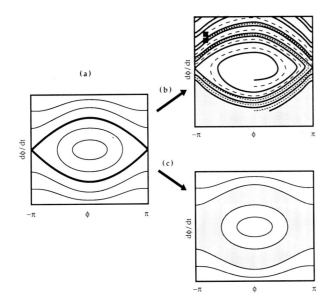

Figure R27 Typical distortions of the pendulum phase diagram. (a) The usual pendulum diagram with the 'libration' region (the eye-shaped region in the center) where the pendulum swings back and forth, the 'positive circulation' region (the top region) where the pendulum swings over the top, always in the same direction, with a positive angular velocity and the 'negative circulation' region (the bottom region) where the pendulum swings over the top in the reverse direction. These regions are separated by the 'separatrices' (thick curves) which represent the trajectories which take an infinite time to swing from the unstable equilibrium (at $\phi = \pm \pi$, $d\phi/dt = 0$) back to itself. (b) The separatrices open up and spiral. Two typical trajectories are shown as dashed curves: one (long dashes) ends up in the resonance region (not shaded) while the other 'jumps' over the resonance region and escapes to the negative circulation region (shaded). (c) The separatrices and the trajectories next to them have dissolved in a 'chaotic' layer (shaded) filled up randomly by any trajectory which starts within its confines.

separatrices have opened up and wind around each other in spirals. They guide the trajectories in such a way that some of them starting from the positive circulation region (at the top of the figure) end up in the resonance region (as the trajectory in the non-shaded strip, shown as a long dashed curve starting from the top square), while others escape from it (as the trajectory in the shaded strip, shown as a short dashed curve starting from the bottom square). When the dissipative forces are very small (as in the solar system) the strips are very thin and, in view of the uncertainties concerning the initial conditions and the parameters of any physical system, it is practically impossible to decide whether a particular object will be 'captured' by the resonance or will 'escape' from it. One measures a 'probability of capture' by comparing the width of the strip leading to capture (the non-shaded area) to the width of the strip avoiding capture (the shaded area).

Another typical distortion of the simplified pendulum model is the one shown in Figure R27c. Here one should view the diagram as a section of a three-dimensional torus (the third dimension representing another angular variable). Each trajectory cuts the diagram repeatedly and is thus represented by a succession of points. In some parts of the diagram (the non-shaded parts in Figure R27c) the points align themselves nicely on an orbit of the pendulum, but in a layer (the 'chaotic layer') around the positions of the separatrices in Figure R27a, the points randomly fill an area of finite size.

All these exotic phenomena of general dynamical systems are well represented in the solar system (see Peale, 1986, for an in-depth review).

Resonance in the solar system is as plain as the full Moon on a clear night. It is evident that the Moon always shows the same face to us; this is a 'spin–orbit' 1 : 1 resonance whereby the mean angular velocity of the Moon rotation is exactly equal to its mean angular velocity around the Earth. Almost all regular satellites do the same with their respective planets. Of course, this is not just by chance. The spin–orbit resonances were established by capture due to the dissipative tidal interactions between the satellites and their planet. One planet, Mercury, the closest to the Sun (and thus the only one for which the tidal interactions with the Sun are large enough), is also captured in a spin–orbit resonance, but of a different type: Mercury makes three turns on itself while circling twice around the Sun (3 : 2 resonance). The probability of capture into this 3 : 2 resonance is significant only for Mercury (about 10%) and not for the satellites because the eccentricity of Mercury ($e = 0.2$) is much larger than the eccentricities of the regular satellites.

Another type of resonance for which there are several well-established examples in the solar system are the 'orbit–orbit' resonances whereby the mean angular velocity of two satellites are commensurable: typically with a ratio of 2 : 1 (Figure R28). Again this is not sheer luck, but (at least for many of them) due to the dissipative tidal interactions between the satellites and the planets (Goldreich, 1965). The most interesting of these resonances is the double resonance between the first three Galilean satellites (q.v.) of Jupiter. This is the Laplace relation (see Commensurability). A likely scenario for the establishment of this resonance was proposed by Yoder (1979); see also Malhotra (1991) for a discussion on variants of this scenario. An interesting variation on the basic scenario of capture into resonance has been investigated recently by Tittemore and Wisdom (1990) and by Malhotra and Dermott (1990): according to their theories, satellites of Uranus may have been captured in the past into orbit–orbit resonances and escaped later from them because of the interplay of secondary resonances (which are not present in the cases of the satellites of Jupiter and Saturn). This could explain the anomalous high inclination of Miranda, and episodes of resurfacing on some satellites of Uranus.

Perhaps the most intriguing manifestations of resonances in the solar system are the peculiarities of the distribution of asteroids in the belt between the orbit of Mars and the orbit of Jupiter (see Figure R29). In the main belt we see well marked gaps (the Kirkwood gaps; see Asteroid: resonance) at all the active resonances (2 : 1, 3 : 1, 5 : 2, 7 : 3), while just outside the main belt, at the location of the resonances 3 : 2 and 4 : 3, we see on the contrary a nest of asteroids. It is believed that these gaps and nests and also more subtle peculiarities in the distribution of asteroids (for instance, differences in distribution for asteroids of different size or of different taxonomic type) may give clues to the origin and evolution of the solar system. Why, in the first place, did the belt remain a belt of small objects and not accumulate into a planet?

The fact is that we do not yet understand why there are gaps; we do not understand the difference between a resonance like the 2 : 1 resonance, characterized by a gap, and the 3 : 2 resonance, characterized by a group. Work by Scholl and Froeschlé (1974, 1975) has shown that these resonances are characterized by large chaotic layers (as in Figure R27c) and large excursions in eccentricity. Wisdom (1983) found that in the 3 : 1 resonance the increase in eccentricity is large enough to force the asteroid to cross the orbit of Mars, resulting eventually in its demise as a member of the belt. Indeed, even if such a 'Mars crosser' escapes direct collision, the close encounters with the red planet will completely change its orbit sooner or later. The other smaller gaps (5 : 2, 7 : 3) can be accounted for in the same way, but not the largest gap, the Hecuba gap at 2 : 1 resonance (Wisdom, 1987). The search is still going on. The dynamics of the 2 : 1 (gap) and the 3 : 2 (group) resonances are being investigated with more and more refined models to try to bring to light an effect capable of strongly increasing the eccentricity of most test particles located in the 2 : 1 resonance, but which does not have the same effect on test particles in the 3 : 2 resonance (e.g. Morbidelli and Moons, 1993).

Another possible explanation for the gaps involves the dissipation of the primitive solar nebula in the early stages of the solar system. The dissipation of this nebula introduces into the solar system a non-conservative effect which pushes asteroids out of the resonances (an effect very similar, but in reverse, to the capture into resonance described above for the satellites). This effect is proportional to the mass of the part of the nebula which is located between the orbit of the asteroid and the orbit of Jupiter and which disappears after the formation of both bodies (Henrard and Lemaitre, 1983). A mass of a few percent of the mass of the Sun is enough to explain the formation of the Kirkwood gaps. The 3 : 2 resonance, being closer to the orbit of Jupiter, may not have been cleaned up completely as the mass contained between it and the orbit of Jupiter is evidently smaller than

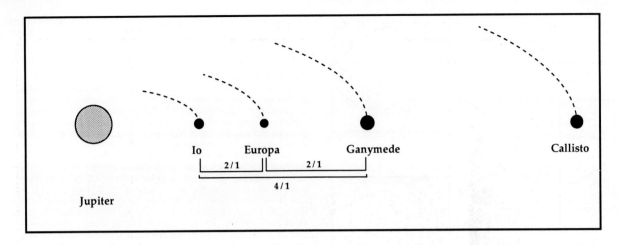

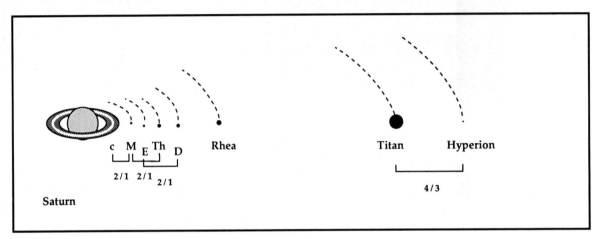

Figure R28 The orbit–orbit resonances among the two main systems of satellites. The planets and distances of the satellites to their planets are drawn to scale. The sizes of the satellites are exaggerated by a factor of ten. In the Saturn system, 'c' stands for the Cassini division between the A and B ring, 'M' for Miranda, 'E' for Encelades, 'Th' for Thetis and 'D' for Dione.

for the other gaps, especially if Jupiter, during its formation, made a 'dent' in the nebula by accreting part of it in its atmosphere, thus decreasing the density of the nebula in its vicinity.

This scenario is plausible but remains speculative. We do not know enough about the physics of the early solar system to really test it. Also, it does not explain why the gaps stay clean; i.e. why the debris of collisions between nearby asteroids does not slowly fill them up, at least partially. A conjunction of the two mechanisms, an early formation of the gaps by the 'cosmogonic' mechanism, followed by a continuing housekeeping of the boundaries of the gaps by the other mechanism, could be the answer.

The overall shape of the asteroid belt is also affected by another kind of resonance: the so-called 'secular resonances', whereby the angular quantities involved in the resonance are no longer the angular positions of the bodies but the angular positions of their perihelion (see Periapse, perihelion, perigee, peribac) or their nodes. These resonances are also responsible for the chaotic dynamics (q.v.) of the planets of the solar system. With respect to the asteroid belt, these secular resonances are responsible for large increases in the eccentricity of material entering them; they form the boundary of the main belt (as shown in Figure R29). The debris which is thrown into them by collisions between nearby asteroids can cross not only the orbit of Mars but the orbit of the Earth as well and end up as meteorites (q.v.) hitting the Earth.

The Hilda group (q.v.) at resonance 3 : 2 in the asteroid belt shows that resonances can be a protective mechanism and not necessarily a destabilizing mechanism. Tiny Pluto owes its continuing status as a planet to this mechanism. Indeed, Pluto can come closer to the Sun than Neptune; so it seems that the two orbits must cross each other and indeed they can; but a 5 : 2 resonance insures that when Neptune is at the crossing point, Pluto is at 90° from it, thus preventing any collision or even close encounter.

Last but not least in our list of the various ways by which resonances are shaping the solar system, we have to mention the planetary rings (q.v.). Voyager's pictures have revealed the unbelievable richness of detail in the structure of Saturn's rings; most of them are due to the delicate interplay between resonances (see Resonance in Saturn's rings) with outside satellites or embedded 'moonlets' (large particles inside the ring) and other phenomena shaping the rings (collision between particles responsible for a kind of viscosity, self-gravitation of the ring, etc.). Particles in Saturn's F ring and in the narrow rings of Uranus are believed to be herded by 'shepherd satellites' (q.v.). The delicate mission of these satellites is much helped by high-order resonances $p\lambda_{satellite} - q\lambda_{particle}$, with (p, q) of the order of several tens, between the angular position of the satellite and that of the particle.

Jacques Henrard

Bibliography

Goldreich, P. (1965) An explanation of the frequent occurrence of commensurable mean motions in the solar system. *Mon. Not. Roy. Astron. Soc.*, **130**, 159–81.

Henrard, J. and Lemaitre, A. (1983) A mechanism of formation for the Kirkwood gaps. *Icarus*, **55**, 482–94.

Malhotra, R. (1991) Tidal origin of the Laplace resonance and the resurfacing of Ganymede. *Icarus*, **94**, 399–412.

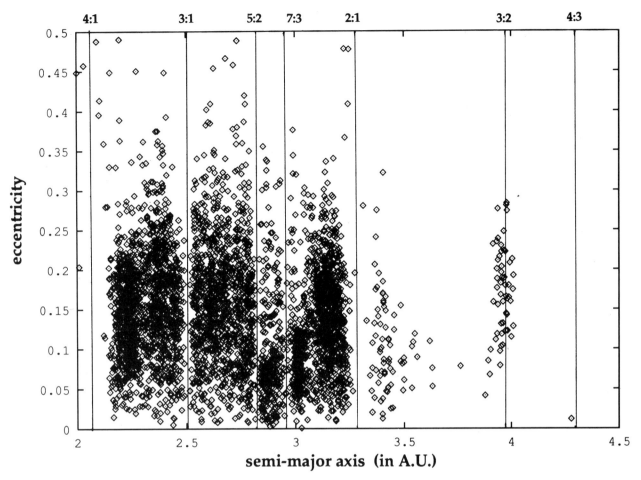

Figure R29 (a) Characteristics of the 5200 numbered asteroids in the asteroid belt. Eccentricity versus semimajor axis; the Kirkwood gaps (at resonances 1 : 3, 2 : 5, 3 : 7 and 1 : 2) stand out clearly. The Hilda group (at the resonance 2 : 3) is also seen.

Malhotra, R. and Dermott, S.F. (1990) The role of secondary resonances in the orbital history of Miranda. *Icarus*, **85**, 444–80.

Morbidelli, A. and Moons, M. (1993) Secular resonances in mean motion commensurabilities: the 2/1 and 3/2 cases. *Icarus*, **102**, 316–32.

Peale, S.J. (1986) Orbital resonances, unusual configurations and exotic rotation states among the planetary satellites, in *Satellites*, (eds J.A. Burns and M.S. Matthews). Tucson: University of Arizona Press, pp. 159–224.

Scholl, H. and Froeschlé, C. (1974) Asteroidal motion at the 3/1 commensurability, *Astron. Astrophys.*, **33**, 455–8.

Scholl, H. and Froeschlé, C. (1975) Asteroidal motion at the 5/2, 7/3 and 2/1 resonances. *Astron. Astrophys.*, **42**, 457–63.

Tittemore, W.C. and Wisdom, J. (1990) Tidal evolution through the Miranda–Umbriel 3 : 1 Miranda–Ariel 5 : 3 and Ariel–Umbriel 2 : 1 mean motion commensurabilities. *Icarus*, **85**, 394–443.

Wisdom, J. (1983) Chaotic behaviour and the origin of the 3/1 Kirkwood gap. *Icarus*, **56**, 51–74.

Wisdom, J. (1987) Urey Prize Lecture: chaotic dynamics in the solar system. *Icarus*, **72**, 241–75.

Yoder, C.F. (1979) How Tidal heating of Io drives the Galilean orbital resonance locks, *Nature*, **279**, 767–70.

Cross references

Asteroid
Asteroid: resonance
Chaotic dynamics in the solar system
Commensurability
Orbital commensurability and resonance
Planetary ring

ROCHE, ÉDOUARD ALBERT (1820–1883)

Édouard Albert Roche was born in Montpelier, France; he worked for a while at the Paris Observatory, but then returned to Montpelier, becoming a professor of mathematics at the university, and eventually died there.

His specialties were celestial mechanics, geophysics and meteorology. He was able to explain correctly the shape of comets, due to repelling force, long before the discovery of radiation pressure. Laplace's nebular hypothesis interested him considerably and he was able to provide a mathematical analysis for it. In 1848 he proposed a law of the differential variation of terrestrial density which is still valid, although his Earth core was solid.

His name is universally known for the Roche limit, which was established in 1849, determining the inner limit of stability for an approaching celestial body.

Rhodes W. Fairbridge

Bibliography

Anon (1883) Obituary Professor A. Roche. *Nature*, **28**, 11–2.
Lévy, J.R. (1975) Roche, Éduard Albert. *Dict. Sci. Biogr.*, vol. 11, p. 498.
Tisserand, F. (1889/1896) *Traité de Mécanique Céleste*. Paris, 4 vols.

ROCHE LIMIT

After considering the principles of gravitational stability that develop between orbiting celestial bodies, the French mathematician

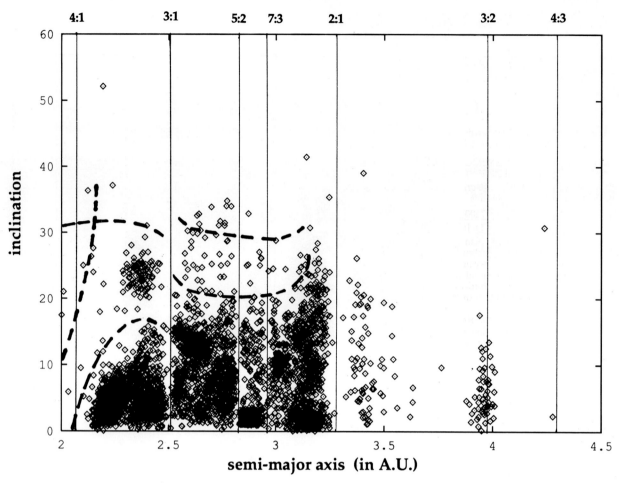

Figure R29 (continued) (b) Characteristics of the 5200 numbered asteroids in the asteroid belt. Inclination versus semimajor axis; the secular resonances, shown as dashed curves, cut deeply into the distribution.

Édouard Albert Roche (1820–1883) defined what has become known as the 'Roche limit', being the critical distance at which an approaching smaller body becomes structurally unstable and disintegrates. Depending upon the radius (R_m) and density of the major or massive body (ρ_m) and that of the smaller body (ρ_m),

$$R_{\text{Roche}} = 2.44\, R_m\, \rho_m^{1/3}$$

Every celestial body possesses a cohesive gravitational force which maintains its structural integrity. However, the major body exerts a tidal force or gravitational differential over the smaller one, which at this critical distance is disrupted (Kopal, 1959, 1966).

In the case of planet Earth, the Roche limit is about 2.89 Earth radii (c. 18 400 km). According to well-established astronomical observations, the Moon has been slowly receding from the Earth over the last three centuries. If the observed rate of recession is uncritically extrapolated back in time (c. 1.8×10^9 years) it would bring the Moon to the Roche limit. According to a highly imaginative and totally unsubstantiated hypothesis of Gerstenkorn (1955), at this moment there was a superficial disintegration of low-density crust postulated for the Moon which would have been gravitationally drawn to the Earth, to become the future continents. Gerstenkorn's idea was taken up and modified by the distinguished Swedish astronomer Hannes Alfvén (1908–1995). It was postulated that the Moon was originally a wayward planet captured, by chance, by the Earth. Alfvén (1963, 1965) argued the case without reference to the well-established geological history and composition of planet Earth, which would oppose any catastrophic model in the face of a quasi-equilibrium state that has survived > 4 billion years.

Even more dramatic and catastrophic than the Gerstenkorn–Alfvén model is the Pacific Moon-birth disruption concept (the Darwin–de Sitter–Wise model). In an earlier encyclopedia (Fairbridge, 1967, p. 617), it was stated that 'the geological data presently available do not offer a shred of evidence in favor of either.' A quarter of a century later, that situation does not seem to have improved. Calculations by Lyttleton (1967) suggest that Moon extraction from planet Earth under rotational instability is dynamically impossible. Although the Gerstenkorn–Alfvén model was consistent with the tidal friction theory of George Darwin and others, that hypothesis is subject to serious criticism (see Tide-raising forces), and furthermore is totally inconsistent with our knowledge of crust and mantle composition and with accepted models for the secular differentiation of both Earth and Moon (Lowman, in Fairbridge, 1967, p. 661). An added problem for the Gerstenkorn model (e.g. as developed by Kopal, 1966) is that its Roche limit approach was just enough to cause violent eruptions on the nearside lunar surface, but not on the farside. Any closer, and the Moon would have disintegrated into fragments. Thus the theory carries a serious ad hoc quality.

The above comments do not reflect on the Roche limit, in principle. Elsewhere, in the solar system it is interesting that the rings of Saturn (q.v.) lie within that planet's Roche limit. Were it not so, the innumerable particles that constitute the rings would probably have accreted and coalesced to form a satellite.

The Roche limit is also involved in considerations of the origin of planets in general during the formation of the solar system (e.g. Jastrow, in Fairbridge, 1967). Within the initial nebular material a certain critical mass is required to initiate the gravitational binding of gas and dust particles, which is subject to the temperature and density of the cloud. The temperature may be taken as being of the order of 100 K, while the density can be calculated in terms of the Roche limit. For the essential binding any two particles or 'elements of mass' (m_1 and m_2) separated by distance (r) must exert on one another a gravitational attraction (G) that is greater than the Sun's. This is achieved by taking the Sun's mass (M) at distance (R) from m_1 and at $R + r$ from m_2, thus

$$MG\left[\frac{m_1}{R^2} - \frac{m_2}{(R+r)^2}\right] < \frac{m_1 m_2}{r^2} G$$

If one takes $m_1 = m_2$ expressed simply as m, then the minimum density is

$$2\frac{M}{r^3} < \frac{m}{R^3}$$

which is also the Roche limit for density (ρ_{Roche}).

In the case of planet Earth, R is 1.5×10^{13} cm (= 1 AU) and thus

$$\rho_{Roche} = 2 \times \frac{2 \times 10^{33}}{(1.5 \times 10^{13})^3} = \frac{1.1 \times 10^{-6}}{(R/R_E)^3} \text{ g cm}^{-3}$$

From this value (at 1 AU) the critical mass for condensation (M_c) to begin from the nebular cloud is

$$M_c = 10^{23} \frac{(100)^{3/2}}{(1.1 \times 10^{-6})^{1/2}} \approx 10^{29} \text{ g}$$

However, the present mass of planet Earth is only 6×10^{27} g. This would need to be multiplied by about 50 to approximate the mass of the proto-Earth initial cloud, because of the overwhelming mass of hydrogen and helium which would have been lost from the protoplanet. Thus the mass of the proto-Earth would be about 3×10^{29} g and thus reasonably compatible with the (admittedly rough) model of Jastrow.

Outside of the solar system the Roche formula is also applicable in the problems of binary stars. An expanding star is likely to transfer matter to its smaller companion by means of its solar wind, but within that limit, via the so-called 'Roche lobe', matter is drawn the other way.

Rhodes W. Fairbridge

Bibliography

Alfvén, H. (1963) The early history of the Moon and the Earth. *Icarus*, 1, 357.

Alfvén, H. (1965) Origin of The Moon. *Science*, **148**, 476–7.

Fairbridge, R.W. (ed.) (1967). *The Encyclopedia of Atmospheric Sciences and Astrogeology*. New York: Reinhold Publ. Co., 1200 pp.

Gerstenkorn, H. (1955) Über Gezeitenreibung beim Zweikörperproblem. *Z. Astrophys.*, **36**, 245–74.

Jastrow, R. (1967) Solar system: origins, in *The Encyclopedia of Atmospheric Sciences and Astrogeology* (ed. R.W. Fairbridge). New York: Reinhold Publ. Co., pp. 885–90.

Kopal, Z. 1959. *Close Binary Systems*. New York: John Wiley and Sons.

Kopal, Z. (1966) On the possible origin of the lunar maria. *Nature*, **210**, 188.

Lévy, J.R. (1975) Roche, Édouard Albert. *Dict. Sci. Biogr.*, Vol. 4, 11, p. 498.

Lowman, P.D. (1967) Moon – theories of origin, in *The Encyclopedia of Atmospheric Sciences and Astrogeology* (ed. R.W. Fairbridge). New York: Reinhold Publ. Co., pp. 658–62.

Lyttleton, R.A. (1967) Moon-capture by the Earth. *Int. Dict. Geophys.*, Vol. 2, pp. 995–7.

MacDonald, G.J.F. (1964) Tidal friction, *Rev. Geophys.*, **2**, 467–541.

S

SABINE, EDWARD (1788–1883)

A British geophysicist, Major-General Sir Edward Sabine (FRS, KCB) was an artillery officer, who eventually became president of the Royal Society. He served in the Napoleonic wars in Canada and elsewhere. After the peace in 1815 he accompanied John Ross on his Northwest Passage expedition in 1818, and was again in the Arctic with William Edward Perry in 1819–1820. The Royal Society, interested in determining the true figure of the Earth, sent him on a pendulum expedition around the Atlantic in 1821–1822. This led him into studies of the Earth's magnetic field. Contemporaries called his multipole idea 'magnetic fever'.

He is perhaps best known for his extensive observations that related the sunspot cycle to the fluctuation of the magnetic field on planet Earth, as well as correlating individual events of only a few days' duration. This area was being studied at the same time in Switzerland by Rudolf Wolf (1816–1893) and Alfred Gautier (1793–1881); thus many of the discoveries were in fact shared. The observations, in the mid-19th century, that two apparently totally unrelated phenomena (on the Sun and on the Earth) could fluctuate in identical periods and almost precisely in phase was greeted at the time with some astonishment, and even today there are still skeptics, notably in the field of meteorology, who vigorously deny any solar linkage with the Earth's dynamic systems.

Sabine's name is also honored in connection with the 179-year cycle of low sunspot intervals that marked the Little Ice Age (13th–19th centuries), known as the Wolf, Spörer and Maunder minima. The low solar activity interval of the period about 1800–1830 has accordingly been identified as the 'Sabine minimum' (Fairbridge and Shirley, 1987), although some others refer to it as the 'Dalton minimum'.

Rhodes W. Fairbridge

Bibliography

Chapman, S. and Bartels, J. (1940) *Geomagnetism*. Oxford.
Charvátová, I. (1990) The relations between solar motion and solar variability. *Bull. Astron. Inst. Czechoslovakia*, **41**(1), 56–9.
Fairbridge, R.W. and Shirley, J.H. (1987) Prolonged minima and the 179-yr cycle of the solar inertial motion. *Solar Phys.* **110**, 191–220.
Georgi, J. (1959) Edward Sabine, ein grosser Geophysiker des 19 Jahrhunderts. *Deutsche Hydrographische Zeitung*, **11**, 225–39.
Reingold, N. (1975) Sabine, Edward, *Dict. Sci. Biogr.*, Vol. 12, pp. 49–53.

SAKIGAKE AND SUISEI MISSIONS

Two Japanese spacecraft, Sakigake and Suisei, were members of the international fleet of robotic spaceprobes that encountered and explored P/Halley on its return to the inner solar system in 1986 (Plate 20). The test spacecraft, Sakigake (Japanese for pioneer or forerunner), was Japan's first venture into interplanetary space. It was launched on 7 January 1985 and approached P/Halley within a distance of 7 million km on 11 March 1986. Suisei (Japanese for comet) was the dedicated spacecraft. It was launched on 18 August 1985 and approached the comet within 200 000 km on 8 March 1985. Such large encounter distances were planned to avoid the hazard of impacting cometary dust at comet–probe relative velocities exceeding 70 km s^{-1}. Due to the weight limitation of the launch vehicle, these spacecraft could not have any dust shielding. Despite its large encounter distance, Suisei collided with dust particles near closest approach. These were recorded as abrupt changes of the spacecraft's attitude; however, the impacts did not impair the operation of its instruments.

The spacecraft were almost identical in their lightweight design, except for the scientific instruments they carried. Sakigake was outfitted with three instruments: a plasma wave probe, a magnetometer and a solar wind ion detector. Its scientific objective was to measure the solar wind conditions ahead of the comet. Suisei carried an ultraviolet (UV) imager and an ion energy analyzer. Its objective was to study the hydrogen halo and the interaction between the solar wind and cometary plasma.

During the encounter both spacecraft performed flawlessly and returned valuable scientific information. The UV camera aboard Suisei detected a pulsation in the Lyman α images of the coma that has been linked through the atomic hydrogen production (and therefore the water vapor production from the nucleus) to the complex motion of the comet nucleus. At 4.5×10^5 km from the nucleus, Suisei detected a 10–30° deflection of the solar wind from the direction of the undisturbed flow, indicating the crossing of the bow shock by the spacecraft. It also detected shell structure in velocity space of cometary protons and water-group ions at 2.3×10^5 km from the nucleus. At 1.5×10^5 km its ion energy analyzer detected severe perturbations and a slowing of the solar wind flow from about 400 km s^{-1} to about 56 km s^{-1}, caused by the ionization of heavier cometary molecules and their assimilation into the solar wind (mass loading of the solar wind). The Sakigake spacecraft passed through the hydrogen halo at its closest approach. The fluxgate magnetometer on Sakigake detected electron cyclotron waves and Alfvén waves with frequencies close to the gyrofrequency of the water ions. This measurement suggests that water-group molecules from the comet were ionized at distances of about 10^7 km from the nucleus. Its instruments also gathered evidence that the spacecraft passed through regions of plasma turbulence associated with the pick-up of ions by the solar wind.

Daniel C. Boice

Bibliography

Hirao, K. and Itoh, T. (1987) The Sakigake/Suisei encounter with comet P/Halley. *Astron. Astrophys.*, **187**, 39–46.

Cross references

Comet
Giotto mission
Halley's comet
Vega mission

SAROS CYCLE

An important lunar periodicity that was probably first observed and named by Babylonian astronomers about two millennia BC, the Saros cycle came down to us through the compilations of Hipparchus and Ptolemy. It was successfully employed to predict a solar eclipse on 28 May 585 BC by Thales of Miletus (q.v.). According to Stephenson (1991) probably every eclipse from about 750 BC to 100 AD was anticipated and recorded on clay tablets, although many have been lost. The Saros length is 18.0303 years or 6585.32 days (223 'lunations' or synodic months), which represents the interval after which either solar or lunar eclipses of the same series are repeated. During this time the relative positions of the Earth, Moon and Sun have returned to almost the same relationships. Long-term changes, however, are still poorly understood (Kopal and Mikhaelov, 1962; Lustig, 1967).

The period is approximately 18 years 10⅔ d, and because of that ⅓ fraction the following eclipse shifts 120° west in longitude, to return to roughly the same path at 54.09-year intervals. According to the *Oxford English Dictionary*, the term 'Saros' in the original Chaldean (the elite astrologer group in Babylon) meant the number 3600, or 120 months of 30 days.

Rhodes W. Fairbridge

Bibliography

Kopal, Z. and Mikhaelov, Z.K. (eds) (1962) *The Moon* (Int. Astron. Symp. 14, Pulkovo, 1960). New York: Academic Press.
Lustig, L.K. (1967) Earth–Moon relations, in *The Encyclopedia of Atmospheric Sciences and Astrogeology* (ed. R.W. Fairbridge). New York: Reinhold Publ. Co., pp. 332–5.
Stephenson, F.R. (1991) The Earth's rotation as documented by historical data, *in New Approaches in Geomagnetism and the Earth's Rotation* (ed. S. Flodmark). Singapore: World Scientific Publ., pp. 87–113.

SATELLITE, NATURAL

A natural satellite is any one of the celestial bodies in orbit around one of the nine principal planets of the solar system. The central planet is sometimes called the primary and the satellite its secondary or moon. Together they are referred to as a system. Planetary satellites range from large, planetlike, geologically active worlds with significant atmospheres to tiny irregular objects tens of kilometers in diameter. The satellites of the inner solar system are composed primarily of rocky material, whereas the satellites of the outer solar system contain frozen volatiles as major components. These volatiles include water ice, methane, ammonia, nitrogen, carbon monoxide, carbon dioxide or sulfur dioxide existing alone or in combination with other volatiles. The planets have among them a total of 61 known satellites (Plates 22, 24, 25, 29, 32). There probably exist more undiscovered small satellites in the outer solar system. The relative sizes of the natural satellites are illustrated in Figure S1. Table S1 is a summary of their properties.

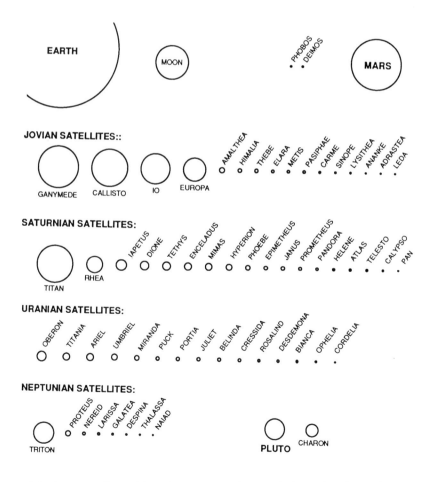

Figure S1 The relative sizes of the known satellites of the solar system are shown with the Earth and Mars for comparison.

Table S1 Summary of the properties of the planetary satellites

Satellite	Distance from primary (10^3 km)	Revolution period (days) (R = Retrograde)	Orbital eccentricity	Orbital inclination (degrees)	Radius (km)	Density (g cm^{-3})	Visual geometric albedo	Discoverer	Year of discovery
Jupiter									
J14 Adrastea	128	0.30	0.0	0.0	10		<0.1	Jewitt et al.	1979
J16 Metis	128	0.30	0.0	0.0	20		<0.1	Synott	1979/8
J5 Amalthea	181	0.49	0.003	0.4	131×86×73		0.05	Barnard	1892
J5 Thebe	221	0.68	0.0	0.0	50		<0.1	Synott	1979/8
J1 Io	422	1.77	0.004	0.0	1821	3.55	0.6	Galileo	1610
J2 Europa	671	3.55	0.000	0.5	1565	3.04	0.6	Galileo	1610
J3 Ganymede	1070	7.16	0.001	0.2	2634	1.93	0.4	Galileo	1610
J4 Callisto	1880	16.69	0.010	0.2	2403	1.83	0.2	Galileo	1610
J13 Leda	11110	240	0.416	26.7	5			Kowal	1974
J6 Himalia	11470	251	0.158	27.6	85		0.03	Perrine	1904/5
J10 Lysithea	11710	260	0.130	29.0	12			Nicholsons	1938
J7 Elara	11740	260	0.207	24.8	40		0.03	Perrine	1904/5
J12 Ananke	20700	617R	0.17	147	10			Nicholson	1951
J11 Carme	22350	692R	0.21	164	15			Nicholson	1938
J8 Pasiphjae	23300	735R	0.38	145	18			Melotte	1908
J9 Sinope	23700	758R	0.28	153	14			Nicholson	1914
Saturn									
S18 Pan	133	0.56	–	–	–	–	–	Showalter	1990
S15 Atlas	138	0.60	0.000	0.3	19×17×14		0.4	Voyager	1980
S16 Prometheus	139	0.61	0.002	0.0	74×50×34		0.6	Voyager	1980
S17 Pandora	142	0.63	0.004	0.1	55×44×31		0.6	Voyager	1980
S10 Janus	151	0.69	0.007	0.14	97×95×77	0.65	0.6	Dollfus	1966
S11 Epimethus	151	0.69	0.009	0.34	69×55×55	0.65	0.5	Fountain and Larson	1978
S1 Mimas	186	0.94	0.020	1.5	199	1.4	0.8	Herschel	1789
S2 Enceladus	238	1.37	0.004	0.0	249	1.2	1.0	Herschel	1789
S3 Tethys	295	1.89	0.000	1.1	523	1.2	0.8	Cassini	1684
S14 Calypso	295	1.89	0.0	1?	15×8×8		0.6	Space Telescope Tm.	1980
S13 Telesto	295	1.89	0.0	1?	15×13×8		0.9	Smith et al.	1980
S4 Dione	377	2.74	0.002	0.0	560	1.4	0.55	Cassini	1684
S12 Helene	377	2.74	0.005	0.2	17×16×15		0.5	Laques and Lecacheux	1980
S5 Rhea	527	4.52	0.001	0.4	764	1.3	0.65	Cassini	1672
S6 Titan	1220	15.94	0.029	0.3	2575	1.88	0.2	Huygens	1655
S7 Hyperion	1480	21.28	0.104	0.4	205×130×110		0.3	Bond and Lassell	1848
S8 Iapetus	3560	79.33	0.028	14.7	718	1.2	0.4–0.08	Cassini	1671
S9 Phoebe	12950	550.4R	0.163	150	110		0.06	Pickering	1898
Uranus									
U6 Cordelia	49.7	0.33	0.0005	0.14	13			Voyager 2	1986
U7 Ophelia	53.2	0.37	0.001	0.09	15			Voyager 2	1986
U8 Bianca	59.2	0.43	0.0009	0.2	21			Voyager 2	1986
U9 Cressida	61.8	0.46	0.0002	0.04	31		~0.04	Voyager 2	1986
U10 Desdemona	62.7	0.47	0.0002	0.2	27		~0.04	Voyager 2	1986
U11 Juliet	64.6	0.49	0.0006	0.06	42		~0.06	Voyager 2	1986
U12 Portia	66.1	0.51	0.0002	0.09	54		~0.09	Voyager 2	1986
U13 Rosalind	69.9	0.56	0.00009	0.3	27		~0.04	Voyager 2	1986
U14 Belinda	75.3	0.62	0.0001	0.03	33			Voyager 2	1986
U15 Puck	86.0	0.76	0.00005	0.3	77		0.07	Voyager 2	1985
U5 Miranda	130	1.41	0.017	3.4	236	1.2	0.35	Kuiper	1948
U1 Ariel	191	2.52	0.003	0.0	579	1.6	0.36	Lassell	1851
U2 Umbriel	266	4.14	0.003	0.0	585	1.5	0.20	Lassell	1851
U3 Titania	436	8.71	0.002	0.0	789	1.7	0.30	Herschel	1787
U4 Oberon	583	13.46	0.001	0.0	761	1.6	0.22	Herschel	1787
Neptune									
N8 Naiad	48	0.30	0.003	4.74	27			Voyager 2	1989
N7 Thalassa	50	0.31	0.0002	0.21	40			Voyager 2	1989
N5 Despina	52.5	0.33	0.0001	0.07	74		0.05	Voyager 2	1989
N6 Galatea	62	0.429	0.0001	0.05	79			Voyager 2	1989
N4 Larissa	73.6	0.554	0.001	0.20	96		0.06	Voyager 2	1989
N3 Proteus	117.6	1.12	0.0004	0.039	208		0.06	Voyager 2	1989
N1 Triton	354.8	5.875R	0.000015	157	1350	2.08	0.73	Lassell	1846
N2 Nereid	5509	360.1	0.753	6.7	170		0.14	Kuiper	1949

Discovery of satellites

Except for the Earth's Moon, none of the natural satellites were known before the invention of the telescope. Although the four large satellites of Jupiter are in principle visible to the naked eye, scattered light from Jupiter renders them invisible. When Galileo turned his telescope to Jupiter in 1610, he discovered these four bodies. His observations of their orbital motion around Jupiter in a manner similar to the motion of the planets around the Sun provided important evidence for the acceptance of the heliocentric (Sun-centered) model of the solar system. These four moons – Io (q.v.), Europa (q.v.), Ganymede (q.v.) and Callisto (q.v.) – are sometimes called the Galilean satellites.

In 1655 Christian Huygens (q.v.) discovered Titan (q.v.), the giant satellite of Saturn. Later in the 17th century, Giovanni Cassini (q.v.) discovered the four next largest satellites of Saturn. It was not until over 100 years later that the next satellite discoveries were made: the Uranian satellites Titania (q.v.) and Oberon (q.v.) and two smaller moons of Saturn. As telescopes acquired more resolving power in the 19th century, the family of satellites grew (Table S1). The smallest satellites of Jupiter and Saturn, and all the small satellites of Uranus and Neptune (except Nereid), were discovered during flybys of the Pioneer and Voyager spacecraft (Table S1). Charon (q.v.), the large moon of Pluto, was discovered photographically with an Earth-based telescope in 1978. Its existence was confirmed when it was observed to undergo a series of mutual eclipses with Pluto during the period 1985–1991.

The natural planetary satellites are generally named after figures in classical Greek and Roman mythology who were associated with the namesakes of their primaries. Another system of nomenclature is to designate them by the first letter of their primary and an Arabic numeral assigned in order of discovery: Io is J1, Europa J2, etc. When satellites are first discovered but not yet confirmed or officially named, they are known by the year in which they were discovered, the initial of the primary, and a number assigned consecutively for all solar system discoveries, e.g. 1980J27. Official names for all objects in the solar system, including satellites, are assigned by the International Astronomical Union (IAU). After planetary scientists were able to map geologic formations of the satellites from spacecraft images, they named many of the features after characters or locations from both Western and Eastern mythologies (see Nomenclature).

Dynamical properties

The motion of a satellite around the center of mass (q.v.) of itself and its primary defines an ellipse with the primary at one of the foci. The orbit is defined by three primary orbital elements: (1) the semimajor axis, (2) the eccentricity and (3) the angle made by the intersection of the plane of the orbit and the plane of the primary's spin equator (the angle of inclination). The orbits are said to be regular if they are in the same sense of direction (the prograde sense) as that determined by the rotation of the primary, and if their eccentricities and inclinations are low. The orbit of a satellite is irregular if its motion is in the opposite (or retrograde) sense of motion, if it is eccentric, or if it has a high angle of inclination. The majority of satellites move in regular, prograde orbits. Many of the satellites that move in irregular orbits are believed to be captured objects.

Most of the planetary satellites present the same hemisphere toward their primaries, a situation which is the result of tidal evolution. When two celestial bodies orbit each other, the gravitational force exerted on the nearside is greater than that exerted on the farside. The result is an elongation of each body to form tidal bulges, which can consist of either solid, liquid or gaseous (atmospheric) material. The primary tugs on the satellite's tidal bulge to lock its longest axis onto the primary–satellite line. The satellite, which is said to be in a state of synchronous rotation, keeps the same face toward the primary. Since this despun state occurs rapidly (usually within a few million years), most natural satellites are in synchronous rotation. Because tidal forces (q.v.) act as the inverse third power of the distance from the primary, those satellites which are not in synchronous rotation, such as Hyperion and Nereid (q.v.), are outer satellites. These bodies may be in captured orbits, not having had enough time to despin tidally.

Overview of physical properties

The satellites of the solar system are unique worlds, representing a vast panorama of physical processes. The small satellites of Jupiter and Saturn are irregular chunks of ice and rock, perhaps captured asteroids, which have been subjected to intensive meteoritic bombardment. Many of the satellites, including the two moons of Mars, the Saturnian satellite Phoebe, and areas of the Uranian satellites, are covered with C-type material, the dark, unprocessed, carbon-rich material found on the C class of asteroids. The Martian satellites, Deimos (q.v.) and Phobos (q.v.), may even be captured C asteroids. The surfaces of other satellites such as Hyperion and the dark side of Iapetus contain D-type primordial matter (named after the D class of asteroids), which is spectrally red and believed to be rich in organic compounds. Both D- and C-type materials are common in the outer solar system. Because these materials represent the material from which the solar system formed, understanding their composition, state, and origin will yield clues on the conditions and early evolution of the solar system. Iapetus presents a particular enigma: one hemisphere is ten times more reflective than the other. For the Pluto–Charon system, the secondary represents a significant fraction of the mass of its primary, about 17%. For the Earth–Moon system the fraction is about 2%. All other satellites are much less than 1% of the mass of their primaries.

Before the advent of spacecraft exploration, planetary scientists expected the icy satellites of the outer planets to be geologically dead worlds. They assumed that heat sources were not sufficient to have melted their mantles to provide a source of liquid or semiliquid ice or ice – silicate slurries. Reconnaissance of the icy satellite systems of the four giant planets by the two Voyager spacecraft uncovered a wide range of geologic processes, including currently active vulcanism on Io and Triton (q.v.). At least two additional satellites, Europa and Enceladus (q.v.), may have current activity. The medium-sized satellites of Saturn and Uranus are large enough to have undergone internal melting with subsequent differentiation and resurfacing. Among the Galilean satellites, only Callisto lacks evidence for periods of activity after formation. Charon's level of past or present activity is unknown; Pluto is the only one of the nine planets that has not been observed by spacecraft.

Recent work on the importance of tidal interactions and subsequent heating has provided the theoretical foundation to explain the existence of widespread activity in the outer solar system (see Tidal friction). Another factor is the presence of non-ice components, such as ammonia hydrate or methanol, which lower the melting point of near-surface materials. Partial melts of water ice and various contaminants – each with their own melting point and viscosity – provide material for a wide range of geologic activity. The realization that such partial melts are important to understanding the geological history of the satellites has spawned an interest in the rheology (viscous properties and resulting flow behavior) of various ice mixtures and exotic phases of ices that exist at extreme temperatures or pressures. Conversely, the types of features observed on the surfaces provide clues to the likely composition of the satellites' interiors.

Because the surfaces of so many outer planet satellites exhibit evidence of geologic activity, planetary scientists now think in terms of unified geologic processes that function throughout the solar system. For example, partial melts of water ice with various contaminants could provide flows of liquid or partially molten slurries which in many ways mimic terrestrial or lunar lava flows formed by the partial melting of mixtures of silicate rocks. The ridged and grooved terrains on satellites such as Ganymede, Enceladus, Tethys and Miranda (q.v.) may all have resulted from similar tectonic activities. Finally, explosive volcanic eruptions occurring on Io, Triton, Earth and possibly Enceladus may all result from the escape of volatiles released as the pressure in upward-moving liquids decreases.

Scientists believe the solar system formed 4.6 ± 0.1 billion years ago. This age is derived primarily from radiometric dating of meteorites, which are believed to consist of primordial, unaltered matter. The Sun and planets formed from a disk-shaped rotating cloud of gas and dust known as the protosolar nebula. When the temperature in the nebula cooled sufficiently, small grains began to condense. The difference in solidification temperatures of the constituents of the protosolar nebula accounts for the major compositional differences of the satellites. Since there was a temperature gradient as a function of distance from the center of the nebula, only those materials with high melting temperatures (e.g. silicates, iron, aluminum, titanium and calcium) solidified in the central (hotter) portion of the nebula. The Earth's Moon consists primarily of these materials. Beyond the orbit of Mars, carbon, in combination with silicates and organic molecules, condensed to form the carbonaceous material found on C-type asteroids. Beyond the outer region of the asteroid belt, formation temperatures were sufficiently cold to allow water ice to condense and remain stable. Thus the Jovian satellites are primarily ice – silicate admixtures (except for Io, which has apparently outgassed all its water). On Saturn and Uranus these materials are joined by methane and ammonia, and their hydrated forms. For the satellites of Neptune and Pluto, formation temperatures were low enough for other volatiles, such as nitrogen, carbon monoxide and carbon dioxide to exist in solid form. In general the satellites which formed in the inner regions of the solar system are denser than those of the outer planets, because they retained a lower fraction of volatile materials.

The formation of the regular satellite systems of Jupiter, Saturn and Uranus is sometimes considered to be a smaller-scale version of the formation of the solar system. A density gradient as a function of distance from the primary does exist for the regular system of small, inner Neptunian satellites and for the Galilean satellites (Table S1). This fact implies that more volatiles were included in the bulk primordial composition as the distance from the primary increases. However, this simple scenario cannot be applied to Saturn or Uranus because their regular satellites do not follow this pattern.

After the satellites accreted they started to heat up from the release of gravitational potential energy. An additional heat source was provided by the release of mechanical energy during the heavy bombardment of their surfaces by remaining debris. The satellites Phobos, Mimas and Tethys all have impact craters caused by bodies that were nearly large enough to break them apart; probably such catastrophes did occur. The decay of radioactive elements found in silicate minerals provided another major source of heat. The heat produced in the larger satellites was sufficient to cause melting and chemical fractionation; the dense material, such as silicates and iron, went to the center of the satellite to form a core, while ice and other volatiles remained in the crust. A fourth source of heat is provided by tidal interactions. When a satellite is being tidally despun, the resulting frictional energy is dissipated as heat. A different mechanism involving orbital resonances among satellites is believed to cause the heat production required for more recent resurfacing events (see Tidal heating)

Some satellites, such as the Earth's Moon, Ganymede and several of the Saturnian satellites, underwent periods of melting and active geology within a billion years of their formation and then became quiescent. Others, such as Io and Triton, are currently geologically active. For nearly a billion years after their formation, the satellites all underwent intense bombardment and cratering. The bombardment tapered off to a slower rate and presently continues. By counting the number of craters on a satellite's surface and making certain assumptions about the flux of impacting material, geologists are able to estimate when a specific portion of a satellite's surface was formed. Continual bombardment of satellites causes the pulverization of both rocky and icy surfaces to form a covering of fine material known as a regolith (see Impact cratering).

Many scientists expected that most of the craters formed on the outer planets' satellites would have disappeared due to viscous relaxation. The two Voyager spacecraft revealed surfaces covered with craters which in many cases had morphological similarities to those found in the inner solar system, including central pits, large ejecta blankets and well-formed outer walls. Scientists now believe that silicate mineral contaminants or other impurities in the ice provide the extra strength required to sustain impact structures for long periods in these materials.

Planetary scientists classify the erosional processes affecting satellites into two major categories: endogenic, which includes all internally produced geologic activity; and exogenic, which encompasses the changes brought by outside agents. The latter category includes the following processes: (1) meteoritic bombardment and resulting gardening and impact volatization, (2) magnetospheric interactions, including sputtering and implantation of energetic particles, (3) alteration by high-energy ultraviolet photons and (4) accretion of particles from sources such as planetary rings. These processes change the optical properties (color and albedo) of the satellites' surfaces.

Bonnie J. Buratti

Bibliography

Beaty, J.K., O'Leary, B. and Chaikin, A. (eds) (1990) *The New Solar System*, 3rd edn. Cambridge, MA: Sky Publishing Corp.

Bergstralh, J. and Miner, E. (eds) (1991) *Uranus*. Tucson: University of Arizona Press.

Binzel, R.P. (1990) Pluto. *Sci. Am.*, **262**, 50–8.

Buratti, B.J. (1987) Moon (astronomy), in *Encyclopedia of Physical Science and Technology* (ed. R.A. Meyers). San Diego: Academic Press, pp. 553–64.

Burns, J. and Matthews, M. (eds) (1986) *Satellites*. Tucson: University of Arizona Press.

Gehrels, T. (ed.) (1984) *Saturn*. Tucson: University of Arizona Press.

Hartmann, W.K. 1983. *Moons and Planets*, 2nd edn. Belmont, CA: Wadsworth.

Morrison, D. (ed.) (1982) *The Satellites of Jupiter*. Tucson: University of Arizona Press.

Johnson, T.V., Brown, R.H., and Soderblom, L.A. (1987) The moons of Uranus. *Sci. Am.*, **255**, 48–60.

Stone, E. and the Voyager Science Teams (1989) *Science*, **246**, 1417–501.

Acknowledgements

This work was performed at the Jet Propulsion Laboratory, California Institute of Technology, under contract with the National Aeronautics and Space Administration.

Cross references

Callisto
Charon
Deimos
Enceladus
Europa
Ganymede
Io
Miranda
Moon (Earth's Moon)
Nereid
Phobos
Titan
Titania
Triton
Umbriel
Voyager missions

SATURN

The Saturn system consists of the planet, its rings, 17 moons and a particle-filled magnetosphere (Plate 26). The intricacy of the rings and presence of a planet-sized moon (Titan) with a thick atmosphere distinguish Saturn from the other giant planets of the outer solar system. Saturn is presently the focus of a major international planetary mission (see Cassini mission).

Saturn orbits the Sun at a distance of 9.5 AU (1 AU = distance from the Earth to the Sun), with a period of 29.5 years, in a nearly circular orbit (eccentricity 0.06). The planet is the second-largest in the solar system, with a radius nine times that of the Earth, and mass of 95 Earth masses. Much has been made of the resulting low density of the planet, which is less than that of liquid water. As described below, while Saturn consists mostly of hydrogen and helium, it actually contains more heavy elements than does Jupiter. While the inclination of the orbit is only 2° relative to the Earth, the equatorial plane of Saturn is tilted by 27° to the orbit, making for significant seasons which are marked out from Earth by the opening and closing of the ring plane. Finally, the equatorial rotation period of the planet of 10.2 h is only 24 min slower than Jupiter's and leads to a significant flattening or oblateness of the planet of 10%.

The planet Saturn was known to the ancients as a naked-eye star which moved against the background of fixed constellations, as did the other four classical planets. Galileo (q.v.) interpreted his first telescopic observations of Saturn in 1610 as revealing a triple planet, a consequence of the telescope's small size and the narrow opening angle of the rings as Saturn approached opposition. He observed and recorded the disappearance of the strange companion bodies as the rings became edgewise, then described them later in the decade as two arms or handles as the ring plane opened up (Alexander, 1962). He never correctly deduced their true nature. Christiaan Huygens in 1659 was the first to explain the observations: a thin, flat ring. Four years earlier, the Dutch scientist had discovered the largest of Saturn's satellites, Titan. Over the next three centuries, increasingly detailed telescopic studies revealed four main rings (labeled A through D from outer to inner) and a diffuse band of material, called the E ring, which is centered on Enceladus' orbit.

Following Huygens' discovery of Titan, it was less than two decades before the Italian astronomer Jean-Dominique Cassini (q.v.) discovered four of Saturn's major moons, Iapetus, Rhea, Tethys and Dione, between 1671 and 1684. William Herschel in England discovered Mimas and Enceladus in 1789, and in the 19th century the last of the major moons, Hyperion and Phoebe, were identified by US and English astronomers (Burns, 1986).

The first spacecraft to fly through the Saturn system was Pioneer 11 (q.v.) in 1979, taking advantage of a serendipitous opportunity to use a gravity assist from Jupiter, the spacecraft's primary goal. Although spacecraft power output and instrument complement were not optimized for the Saturn system, Pioneer 11 discovered a thin ring, called the F ring, outside the A ring, and confirmed the existence of Epimetheus, a small satellite previously detected from Earth. Pioneer also made measurements of atmospheric temperatures at Saturn and Titan, and mapped the Saturn magnetosphere (Northrup, Opp and Wolfe, 1980).

The Voyager 1 and 2 flybys of the Saturn system in 1980 and 1981 constituted primary mission goals of this ambitious program. Much of what we now know about the Saturn system, summarized below and in the companion articles, has been derived from the sophisticated instrument complement aboard the two spacecraft. Most notable are the discovery of intricate structure in the rings, identification of a thick nitrogen atmosphere on Titan, discovery of diverse tectonics on several icy satellites, and exploration of a rich interaction between Titan's atmosphere and the Saturnian magnetosphere. Complementary to these spacecraft investigations have been ground-based and Earth-orbital observations which have added significantly to the knowledge base of the Saturn system.

Saturn's atmosphere

The atmosphere of Saturn generally presents a more subdued appearance than that of Jupiter, largely because the clouds, composed of ammonia, occur deeper in the colder atmosphere of this more distant planet. Nonetheless, long-lived oval spots, turbulence associated with shear zones, and alternating eastward and westward velocities associated with belts and zones are seen in the Voyager images (Smith et al., 1981). Although Saturn receives proportionately more thermal energy from internal sources than thermalized sunlight, when compared with Jupiter, the overall energy emission is only a third of that of Jupiter, and hence a somewhat more quiescent atmosphere is expected. It therefore was something of a surprise when ground-based observations in 1990 detected a major brightening on an equatorial portion of Saturn's disk, which then began to spread (Beebe et al., 1992). Hubble Space Telescope observations clearly showed an area of bright clouds being sheared along the direction of rotation. The simplest interpretation is of a disturbance deep below the visible surface of Saturn triggering vertical convection and resultant ammonia cloud condensation (with possible thunderstorm-style deep vertical convection); the ammonia clouds then track the direction and velocity of the high altitude winds above the convective part of the atmosphere. Other brightenings of Saturn's disk have been observed since 1793 (Beebe et al., 1992). As with the continually changing appearance of the atmosphere of Jupiter over decadal timescales, these Saturn brightenings indicate that the process of transport of thermal energy through the atmosphere is variable on short timescales.

The temperature profile of the atmosphere shows the typical pattern of decreasing with altitude up to a minimum, referred to as the tropopause (at which pressure is of order 0.1 bar and temperature 82 K), beyond which the temperature increases steeply as the altitude increases. The bulk constituents of the atmosphere, based on radio occultation and IRIS experiments on Voyager, are hydrogen and helium. Carbon, nitrogen and oxygen are present as the most abundant heavy elements; in the atmosphere they appear primarily as methane, ammonia and water respectively, enhanced several times above solar abundances. Methane finds its way into the stratosphere and is photolyzed into heavier hydrocarbons, some of which condense to form aerosols. Much of the work to detect atmospheric species has been through ground-based and airborne infrared telescopes (Gautier and Owen, 1989). A schematic illustration of atmospheric processes, as coupled to Saturn's interior, is shown in Figure S2.

Saturn's interior

The structure and composition of Saturn's interior is deduced from measurements of the planet's oblateness and its gravitational field, determined primarily by optical tracking of the satellite orbits and positions of the rings. This tracking has been accomplished both from Voyager observations and ground-based observations of stars passing behind the rings. Given these, an equation of state relating the pressure to the density at any given temperature and composition, and the atmospheric abundances of the two most abundant constituents, hydrogen and helium, one may solve for the distribution of the hydrogen, helium and heavy elements throughout the interior, as well as the run of pressure, temperature and density. In practice this procedure has driven experimental and theoretical studies of a profound physical problem: the nature of hydrogen and hydrogen–helium mixtures, at megabar pressures. The most recent such study (Chabrier et al., 1992) requires a core of roughly one Earth mass of elements heavier than hydrogen and helium, above which lies a region of hydrogen–helium with 22 Earth masses of heavier material, and an upper zone of hydrogen and helium with five Earth masses of heavy elements which grades imperceptibly into the atmosphere. The boundary between the middle and upper layer is distinguished by a transition in the state of hydrogen from a molecule of two hydrogen atoms to a higher-pressure phase in which electrons in conduction shells move among hydrogen nuclei (protons). This latter state is electrically conducting and is called metallic hydrogen; it has been predicted but not yet detected in laboratories due to the difficulties of achieving the megabar pressures required.

All Saturn models have this transition, while varying in the details of the layers (e.g. Zharkov and Gudkova, 1991). A complication with Saturn and also probably Jupiter is that helium, while fully soluble in hydrogen under molecular conditions, is predicted to form a separate immiscible phase in metallic hydrogen below a critical temperature. The model internal temperature profiles for both Jupiter and Saturn appear to cross this boundary, and so helium is separating out from hydrogen in the deep interiors of both bodies, but more extensively in Saturn (Stevenson and Salpeter, 1977; Stevenson, 1980). It is in fact the case that the internal energy output of Saturn is larger than would be expected based simply on the gravitational collapse of gas to form this body 4.5 billion years ago (the age of the solar system); for Jupiter the internal heat leaking outward is roughly consistent with virialized energy of collapse. The separation of helium from hydrogen, to form an enriched inner region of helium, releases gravitational potential energy which leaves the interior as heat. The extensive separation of helium predicted for Saturn is consistent with the excess heat. The deciding factor in the argument is the prediction that the helium abundance in Saturn's atmosphere should be depleted relative to that in the Sun; Jupiter's helium abundance should be solar or slightly below. The helium/hydrogen ratio $20 \pm 10\%$ (relative to solar) for Saturn measured by Voyager, confirm the soundness of these models of helium depletion (Gautier and Owen, 1989).

Figure S2 schematically illustrates processes expected to be ongoing in the interior and atmosphere of Jupiter and Saturn, along with observational techniques used to infer them. Internal energy and thermalized sunlight are transported outward by convection, followed by radiative processes near the stratosphere–troposphere boundary. The detection of significant amounts of carbon monoxide in the atmosphere, where it should be all but absent, is an indicator of rapid vertical upwelling of hot interior gas to the cold atmosphere (while molecular nitrogen is difficult to detect from Earth, it is also expected to be present as a tracer of vertical convection). Below the visible ammonia clouds there are expected to be ammonia–sulfur cloud compounds and water clouds. Much deeper in the molecular envelope, silicates and iron condense to form clouds; deeper still in

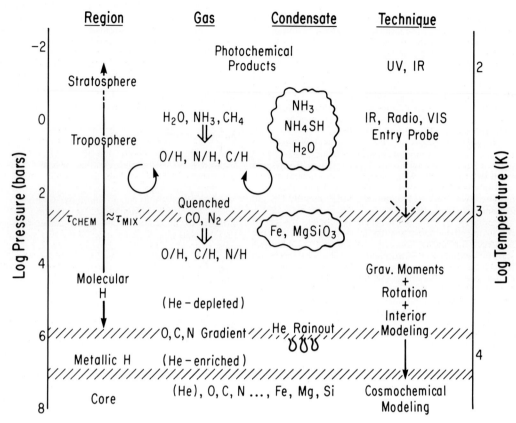

Figure S2 Schematic illustration of processes and species in the interiors of Jupiter and Saturn. The depth below the surface is expressed as the logarithm of the pressure, with an auxiliary rough scale of log temperature on the right. The diagram is divided from left to right into the region of the atmosphere or interior, secondary species present in the region, condensates at each level and the technique used to observe or infer the information. Only the core is solid, the other regions are fluid. Hatching divides the various regions except for the upper hatching, which is the 'quench' level above which the rate of mixing exceeds the rate of chemical reactions. Species which are out of equilibrium with the atmosphere are brought upward by convection from the quench zone. The actual level will vary from species to species.

the metallic regions droplets of helium form clouds at megabar pressures.

Saturn's magnetic field and magnetosphere

Based on Pioneer 11 and Voyager measurements, Saturn has a magnetic field whose strength at the equatorial surface is 0.21 gauss, roughly two-thirds that of the Earth's. Of course, the larger volume of Saturn implies that the corresponding magnetic moment is much greater than the Earth's, by the ratio of the volumes. Saturn's magnetic field more closely resembles that of a dipole than does that of any other giant planet, and is very closely centered on the axis of rotation and tilted from it by less than 1°. The Earth, by contrast, has a tilt in its dipolar field of 11°. The magnetic field of all the planets except Venus and Mars (for which no definitive detection has been made) must be continually regenerated through the action of a so-called magnetic dynamo; any imposed magnetic field would decay away on timescales that are short compared to the age of the solar system. Mechanisms for generating dynamos lie beyond the scope of this article, but they require net motions of electrically conducting fluids relative to the motion of planetary rotation (Garland, 1979). In Saturn convective motions in the metallic hydrogen layer are probably responsible. The wide range of magnetic field tilts and displacements in the magnetized planets is not fully understood, and may be due to structural differences, styles of convection or time variability, as is seen for the Earth. In the last regard one should perhaps not attach much significance to the simplicity of the Saturn field geometry.

The magnetic field of Saturn deflects charged particles in the solar wind streaming toward the planet, carving out a cavity called the magnetosphere of Saturn. Its bow is marked by a shock roughly 20 Saturn radii from the planet. This distance is variable, as the solar wind (q.v.) fluctuates with solar cycle. On the lee side the magnetosphere extends as a tail many astronomical units downstream of the solar wind direction. Particles from the solar wind occasionally become trapped in the Saturn magnetic field lines, and atoms stripped off the surfaces of satellites by bombardment of charged particles may themselves become ionized. The net result is that Saturn's magnetosphere is filled with ions of hydrogen, helium, oxygen and other species which are compelled to move along field lines, eventually being lost by diffusive processes out of the magnetosphere or collision with the atmosphere as they move to high-latitude regions where the field lines converge on the planet.

The implications of such a particle-filled magnetosphere are profound. Bombardment onto satellite surfaces ('sputtering') damages the microstructure of the outermost material, altering its spectroscopic appearance. Chemical reactions among some surface materials may alter the surface composition relative to that of the bulk interior. Sputtering of ring particles gradually reduces their mean size and results in net loss of ring material as various drag mechanisms act on the smallest particles. Ring bombardment might also play a role in generating the still-mysterious spokes of the ring system. Particle injection into the Saturn atmosphere causes auroras. Titan, which is sometimes in and sometimes out of the magnetosphere depending on the strength of the solar wind, has its atmosphere bombarded by charged particles in both environments, leading perhaps to interesting and time-dependent upper atmospheric chemistry. Nitrogen molecules are broken apart by magnetospheric particles, as well as by cosmic rays and solar short wavelength photons. The resulting atomic nitrogen escapes slowly over the age of the solar system.

The satellites of Saturn absorb charged particles which diffuse inward toward the planet. Because the inward drift rate depends on the particles' kinetic energy, a curious effect results, whereby those electrons that have drift rates in longitude which match the speed of the satellite in its orbit do not get absorbed, and hence inward of each satellite there is a population of electrons predominantly of one energy, selected in much the way that a set of color filters screen out all but a selected narrow set of photon energies (Van Allen, 1981).

Two main types of radio bursts, Saturn kilometric radiation and Saturn electrostatic discharges, were seen by the Voyager radio astronomy instrument. The origin of each of these sporadic bursts is unclear, but recent thinking puts the origin of Saturn electrostatic discharges in the atmosphere of Saturn. Both types of event are modulated by a number of processes within the Saturn magnetosphere and perhaps the solar wind.

Saturn's rings

The magnificent beauty of Saturn's rings as seen from Earth is a pale hint of the profoundly intricate structure revealed by Voyagers 1 and 2. Upwards of 1000 narrow rings, each of the order of 10 km in width, could be counted in the high-resolution Voyager images. These features are in turn probably composed of much narrower structures below the resolution limit of the images, and Voyager photopolarimeter studies of stars passing behind the rings revealed features as narrow as 1 km, limited again by the experiment resolution. The so-called gaps between the major rings identified from Earth also possess numerous narrow rings. The thickness of the rings, measured by imaging, radio and photopolarimetry experiments, is less than 1 km and for some ring edges less than 150 m (Cuzzi et al. 1984), extraordinary considering that the main ring structures (the classical A, B and C rings), have widths upwards of tens of thousands of kilometers. The optical depth, a measure of the penetration of light and hence amount of material in a cross-section of the ring, ranges from 0.1 to much greater than 1, so that in the thickest part of the rings essentially no light gets through. Voyager observations of the rings looking back at the Sun from beyond Saturn thus appear to be a negative image of the view we get from Earth, with small particles in the emptier ring gaps scattering sunlight toward the camera, and the main rings blocking off sunlight.

A wide range of techniques, from Voyager observations to ground-based radar, reveal that the bulk of Saturn's ring material is composed of particles ranging from centimeters to meters in size; however the spokes and forward scattering of light in the gaps indicate a population of much smaller particles as well (down to microns). Larger, house-sized bodies also form by temporary accretion of smaller boulders; these 'dynamic ephemeral bodies' are probably quickly broken apart by tidal effects. The compositions of the major classical rings appear to differ from each other, the C ring and the material in the classical Cassini division being less red than the A and B rings (Cuzzi et al. 1984). Ground-based optical and near-infrared spectra reveal features consistent with water ice. Supported by their high reflectivity or 'albedo', it appears that most ring particles are composed in large part of water ice. The water ice is volatile enough at Saturn to form a ring atmosphere, which has been detected through its ultraviolet emissions.

Individual ringlets exhibit a diverse range of phenomena, including orbital eccentricity and braiding. Propagation of waves through the rings can be seen in images, and some rings, such as the isolated F ring, are 'shepherded' by small satellites just inward and outward of the ring orbit. These features and other details of the ring structure are sculpted by the force of gravity which, through the massive gravitational field of the planet and presence of matter over a wide range of masses in orbit around Saturn, can be manifest in diverse ways (Burns, 1981). Radial positions in the rings which correspond to resonances with the major satellites such as Mimas (i.e. where the ratio of orbital periods of the ring material and the major satellite has small integers in the numerator and denominator) should be relatively devoid of ring material. Such resonances have thus been invoked to explain ring gaps, but the simplest mechanisms for clearing such gaps, associated with resonances, have not been completely successful in accounting for such structures. Alternatively, satellites can cause ring orbits to become elliptical, setting up a propagating wave through neighboring ring material. Such 'spiral density waves' can clear a wide gap just outside a resonance position with a satellite. Various kinds of propagating waves may be set up in rings by the major satellite system, perhaps accounting for much of the structure. Shepherding satellites may be responsible for the formation and maintenance of some narrow rings, such as the F ring. The presence of two satellites radially inward and outward from a ring can produce net gravitational repulsion as well as attraction on the ring material, forcing confinement into a narrow ring. The search for shepherds (q.v.) has been frustrating except in a few cases; such objects may be rare, or typically small enough that they were below the resolution of the Voyager cameras. Gravitational interactions between the ring material and moons implies significant orbital evolution of ring material over the age of the solar system; whether the rings must be relatively young by virtue of the substantial radial evolution continues to be controversial.

In addition to gravitational effects, electromagnetic processes may play a role in ring appearance. The discovery in Voyager images of ephemeral features resembling spokes in the B ring was a surprise. The comparative brightness relative to the main ring implicates small particles, which appear to be levitated above the ring plane, perhaps by electromagnetic forces. Other properties of the spokes, such as their prominence after emergence from Saturn shadow, remain important clues to the details of their formation, which continue to be debated (Cuzzi et al., 1984).

Finally, tenuous rings beyond the main ring system have been identified. The E ring is a band of very small particles which is roughly centered on the orbit of Enceladus. The short lifetime of material confined to that region suggests that the ring material originates in repeated and recent volcanic activity from that satellite. The source of material for the G ring, which is inward of the orbit of Mimas, remains a mystery.

Satellites of Saturn

Eight new satellites of Saturn were discovered by Voyager or complementary ground-based studies, the latter taking advantage of the rings being 'edge-on' around the time of the Voyager encounter (Morrison, Owen and Soderblom, 1986). Two of the new objects, Janus and Epimetheus, are in the same orbit. There are two 'coorbital' satellites of Tethys, and one of Dione. The presence of these objects is evidence for an early epoch of collisional processes within the regular satellite and ring system. All of the satellites are likely to have formed in place around Saturn with the exception of Phoebe, the outermost satellite at a distance of 215 Saturn radii. Phoebe's highly inclined and eccentric orbit suggests capture. Its extremely low albedo is more typical of primitive asteroids and carbonaceous meteorites than of an icy object, and the other Saturnian satellites for which albedos are determined are much brighter (except the darkside of Iapetus). Hyperion, situated between Iapetus and Titan, is something of an oddball, being irregular in shape, moderately dark, possessing a high eccentricity forced by nearby larger satellites and currently rotating (or tumbling) in chaotic fashion. Its origin and history are very poorly understood, in part because of the low resolution of images of its surface, but it is probably the remains of a larger body which was disrupted (Morrison, Owen and Soderblom, 1986).

The satellites of Saturn equal to or larger than Mimas are thought of as the 'regular' satellite system of Saturn. Each of these is bright (except for Titan, covered in photochemical smog), spherical and in orbits of low eccentricity and inclination. The densities suggest mostly ice with a lesser amount of rock and iron, with the ice being predominantly water (based on cosmochemical considerations), but with lesser amounts of ammonia, methane, carbon dioxide and perhaps carbon monoxide and nitrogen. Water ice has been identified on some satellites spectroscopically (Clark, Fanale and Gaffey, 1986).

Saturn's regular satellites have had a diverse range of geologic histories, which have been determined only in part by size, since satellites of similar size show differences in the amount of resurfacing of old, cratered terrains. The most straightforward division of these objects is that based on Rothery (1992). 'Dead' worlds, those which show a surface dominated with craters but essentially no tectonics, include Mimas, Rhea and Iapetus. This last object has been resurfaced on its leading hemisphere with a reddish material of very low (0.1) albedo. The pattern of the dark material along its edges, and appearance of nearby craters suggests an external source, perhaps Phoebe, for the material. The inclusion of both Rhea and Mimas in the dead category illustrates the failure of size as a guide to tectonic activity: Rhea is the largest of the regular satellites except for Titan at 764 km radius, while Mimas is the smallest (197 km).

Tethys and Dione, two similarly sized objects (~ 500 km radius), are satellites which were active in the early part of their history but show no evidence of current activity. Both show significant ridge and trench systems, with Dione's grading into bright wisps which suggest fresh ice exposed at the surface. Both objects are heavily cratered, but Dione possesses significant plains regions with smaller crater density.

Saturn's satellite Enceladus is an enigma, being not much larger than Mimas but with regions which appear to have been very extensively resurfaced by mobile fluids. There are some smooth plains and ridged plains areas which have either few or no craters at all. Equally intriguing is the fact that Enceladus is the brightest object in the Saturn system, reflecting 90% of the light it receives (Morrison, Owen and Soderblom, 1986). This object must have been very recently coated with fresh, icy material. Its orbit is also the locus of the E ring of Saturn, which must be resupplied frequently over geologic time against loss of material. Together with the presence of uncratered and hence youthful surfaces, the conclusion must be that Enceladus possesses volcanic activity, where the volcanic fluids are water or more volatile materials. A likely volcanic fluid for Enceladus, based on cosmochemistry, its low melting point, density and rheology, is an ammonia–water mixture (Stevenson, 1982), which melts from ice and solid ammonia hydrate at 176 K or lower. If such volcanism continues to the present, or nearly so, a heat source for melting the ammonia–water must be found. Enceladus, like Mimas, is too small for heating by radiogenic elements to keep the interior sufficiently warm. Enceladus is distinguished from Mimas by being in a position to have been tidally heated in recent times. Its current eccentricity is not enough to cause melting, but if its eccentricity had been pumped up earlier it is possible that its current interior configuration could focus tidal heating in a thin shell and maintain a small amount of liquid to the present. While a slim possibility, there is little else to go on in understanding this shiny and enigmatic object.

The cratering population on the older surfaces of the regular satellites is distinct in its size–frequency distribution from that of the Earth's Moon. One model which may explain the crater distribution invokes two populations. The first consists of left-over planetesimals from the formation of Saturn and other outer solar system objects; the second population later resulted from the disruption of a satellite by collision. The large craters on Mimas and Tethys, caused by impactors nearly large enough to disrupt these objects, supports the notion of such a later population.

Titan

We consider Titan separately because it is so much larger than the other satellites of Saturn, being intermediate in size and density between Ganymede and Callisto. Titan's density implies a world which is 50% ice by mass, the remainder rock and metals, with the ice again being dominated by water but with the other constituents cited above (Lunine, 1989). G. Kuiper discovered gaseous methane around Titan in 1944 by ground-based spectroscopy, and other hydrocarbons were discovered by other workers.

The Voyager 1 flyby of this planet-sized globe revealed a smog-covered globe, with no holes to allow the cameras to penetrate to the surface. However, the encounter with Titan enabled Voyager to use its remote sensing instruments to maximum extent, with a combination of ultraviolet observations of a solar occultation, infrared measurements and transmission of radio waves through the atmosphere during an Earth occultation revealing the nature of the atmosphere. As summarized in Figure S3, the surface pressure is 1.5 bar of molecular nitrogen, with a surface temperature of 95 K. Methane is the second most abundant constituent of the atmosphere with a fraction of 2–9%, while there may be up to 10%, argon and molecular hydrogen is present in the lower atmosphere with an abundance of a few tenths of a percent.

The chemistry and thermal properties of this atmosphere are complex (Lunine, Atreya and Pollack, 1989). Methane is photolyzed in the upper stratosphere to form acetylene, ethane, propane and heavier hydrocarbons; molecular nitrogen is photolyzed as well to create a series of nitriles. The heaviest hydrocarbons and nitriles form the global orange haze; the light hydrocarbons condense in the lower stratosphere, and all aerosols fall slowly to the surface. The photochemical destruction of methane is irreversible because the hydrogen produced escapes upward at a prodigious rate. Over the age of the solar system the equivalent of a layer of condensed methane 1 km thick on the surface has been photochemically destroyed in the stratosphere (Yung, Allen and Pinto, 1984). Because ethane and propane are liquid at the surface temperature of 95 K, and because

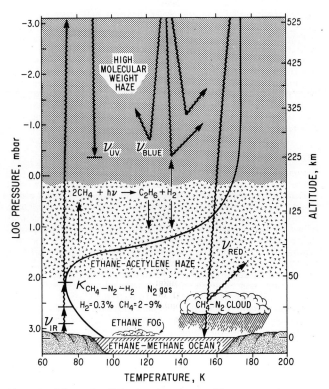

Figure S3 Schematic of Titan's atmosphere, with a temperature profile (curved line) plotted versus altitude and the logarithm of the pressure. Approximate regions of the high molecular weight haze and ethane–acetylene haze are indicated. Wavy lines indicate paths of photons in the blue, red and infrared parts of the spectrum, with infrared being heavily absorbed by gas-phase opacity (κ of methane, nitrogen and hydrogen), and blue light being absorbed and scattered by the haze. The surface may have oceans of light hydrocarbons and nitrogen, based on interpretation of Voyager data, though this remains controversial.

ethane is the primary photolysis product of the methane chemistry, the current surface should have a layer about 500 m thick of liquid hydrocarbons (Lunine, Stevenson and Yung, 1983). Beneath this ocean is the solid hydrocarbon and nitrile material. The hydrocarbon ocean, based on the atmospheric composition, also holds dissolved methane, which could be the dominant constituent; atmospheric nitrogen is also dissolved in the ocean.

Such a model of a globally widespread ocean is consistent with a suite of Voyager data, but runs afoul of observations of the surface by radar, which suggest a high reflectivity surface more typical of solid material. Possibilities are numerous (Lunine, 1992) and include an ocean which had solid particles mechanically stirred up within it (the solid hydrocarbons being denser than the liquid), significant regions of water-ice crust exposed above an ocean no more than a few hundred meters deep, or an ocean which has become trapped under the icy regolith (Stevenson, 1992). One final constraint on the presence of an ocean is Titan's significant orbital eccentricity, which is not maintained by resonances with other satellites. Too thin an ocean could be incompatible with the maintenance of such an eccentricity over the age of the solar system (Sagan and Dermott, 1982). The issue is likely to remain unresolved until the arrival of the combined Cassini/Huygens spacecraft (q.v.). In any event, Titan's surface–atmosphere system is likely to be so complex and rich in organics that it may be a good natural analogy to the prebiotic Earth (Thompson and Sagan, 1992).

Origin of the Saturn system

It is now generally accepted that giant planet formation is distinct from binary star formation. In contrast to the latter rocky and icy planetesimals accrete to form a large core, which attracts hydrogen and helium gas at an increasing rate. Thus an initial segregation between the heavy elements and the hydrogen–helium component gives way to the dissolution of the incoming icy and rocky planetesimals in the growing envelope, accounting for the inference of such material in present-day interior models (Pollack and Bodenheimer, 1989). Saturn and Jupiter formed in environments with enough gas to create massive envelopes; by either timing or position Uranus and Neptune were unable to capture as much material. The formation process gave rise to a disk of gas and dust through either infall of high angular momentum material during the accretion of Saturn, or by spin-off of high angular momentum material as the early envelope contracted. The disk was composed largely hydrogen and helium with, perhaps, a solar mix of rock- and ice-forming materials, though this is controversial (Pollack, Lunine and Tittemore, 1991; Stevenson, Harris and Lunine, 1986). Accretion of the satellites from the rock and ice would have been very rapid compared to the formation of the planets, with some of the satellites lost into the early envelope of Saturn by gas drag. The presence of giant Titan among a retinue of smaller regular satellites is not well explained by any formation model.

The subsequent history of the system is poorly understood. Clearly collisions and perhaps disruption of early satellites played a role in defining the nature of the present system we see, in particular with regard to the formation of the ring system. Perhaps because of the growth of smaller satellites in the regular satellite system of Saturn (compared to Jupiter), disruption and the establishment of a massive ring system was easier. Evolution continued to the present with occasional impacts from solar-orbiting comets altering the satellite surfaces and Titan's surface–atmosphere system, with tidal evolution changing the orbits of the rings and satellites (and perhaps heating Enceladus), and with the slow radiation of internal heat from Saturn leading to helium differentiation of the interior.

The future

Following the detailed exploration of the Jupiter system by Galileo in 1995–1998, the United States and the European Space Agency will undertake a most ambitious mission to explore the Saturn system, called Cassini. To be launched in 1997, the United States-built orbiter will carry the European probe Huygens into Saturn orbit in 2004, then release it for a 2-h-plus descent through Titan's mysterious atmosphere. Instruments built by US and European teams will analyze the chemistry, dynamics and physical properties of the atmosphere, and image the surface. Direct surface sampling may then take place if the descent probe survives its designed hard landing. Following the probe mission, the orbiter will conduct a 2-year investigation of Saturn's atmosphere, rings, satellites and magnetosphere, and continue the investigation of Titan by global remote sensing. Sophisticated US and European instruments will improve upon the brief but fruitful missions of Voyagers 1 and 2 at Saturn, and will include radar mapping of Titan's surface, sophisticated imaging and spectroscopy of satellite and ring surfaces, and direct imaging of the magnetosphere. Cassini's odyssey to the Saturn system, even before its launch, has not been an easy one because of the constrained fiscal climate in the United States. The payoff for completing development of and launching this ambitious endeavor is the opportunity for international exploration of the solar systems's most beautiful and complex planetary system.

Jonathan I. Lunine

Bibliography

Alexander, A.F.O'D. (1962) *The Planet Saturn*. New York: Macmillan Co.

Beebe, R.F., Barnet, C., Sada, P.V. and Murell, A.S. (1992) The onset and growth of the 1990 equatorial disturbance on Saturn. *Icarus*, **95**, 163–72.

Burns, J.A. (1981). Planetary rings, in *The New Solar System* (eds J.K. Beatty, B. O'Leary and A. Chaikin). Cambridge, Mass.: Sky Publishing Corp., pp. 129–42.

Burns, J.A. (1986) Some background about satellites, in *Satellites* (eds J.A. Burns and M.S. Matthews). Tucson: University of Arizona Press, pp. 1–38.

Chabrier, G., Saumon, D., Hubbard, W.B. and Lunine, J.I. (1992) The molecular–metallic transition of hydrogen and the structure of Jupiter and Saturn. *Astrophys. J.*, **391**, 817–26.

Clark, R.N., Fanale, F.P. and Gaffey, M.J. (1986) Surface composition of natural satellites, in *Satellites* (eds J.A. Burns and M.S. Matthews). Tucson: University of Arizona Press, pp. 437–91.

Cuzzi, J.N., Lissauer, J.J., Esposito, L.W. *et al*. (1984) Saturn's rings: properties and processes, in *Planetary Rings* (eds R. Greenberg and A. Brahic). Tucson: University of Arizona Press, pp. 73–199.

Garland, G.D. (1979) *Introduction to Geophysics: Mantle, Core and Crust*. Philadelphia: W.B. Saunders Co.

Gautier, D. and Owen, T. (1989) The composition of outer planet atmospheres, in *Origin and Evolution of Planetary and Satellite Atmospheres* (eds S.K. Atreya, J.B. Pollack and M.S. Matthews). Tucson: University of Arizona Press, pp. 487–512.

Lunine, J.I. (1989) Volatile processes in the outer solar system. *Icarus*, **81**, 1–13.

Lunine, J.I. (1992) Plausible surface models for Titan, in *Proc. Symposium on Titan*. Noordwijk, The Netherlands: European Space Agency SP-338, pp. 233–9.

Lunine, J.I., Stevenson, D.J. and Yung, Y.L. 1983. Ethane ocean on Titan. *Science*, **222**, 1229–30.

Lunine, J.I., Atreya, S.K. and Pollack, J.B. (1989) Evolution of the atmospheres of Titan, Triton and Pluto, in *Origin and Evolution of Planetary and Satellite Atmospheres* (eds S.K. Atreya, J.B. Pollack and M.S. Matthews). Tucson: University of Arizona Press, pp. 605–65.

Morrison, D., Owen, T. and Soderblom, L.A. (1986) The satellites of Saturn, in *Satellites*, (eds J.A. Burns and M.S. Matthews). Tucson: University of Arizona Press, pp. 764–801.

Northrup, T.G., Opp, A.G. and Wolfe, J.H. (1980) Pioneer 11 Saturn encounter. *J. Geophys. Res.*, **A11**, 5651–3.

Pollack, J.B. and Bodenheimer, P. (1986) Theories of the origin and evolution of the giant planets, in *Origin and Evolution of Planetary and Satellite Atmospheres* (eds S.K. Atreya, J.B. Pollack and M.S. Matthews). Tucson: University of Arizona Press, pp. 564–602.

Pollack, J.B., Lunine, J.I. and Tittemore, W.B. (1991) Origin of the Uranian satellites, in *Uranus* (eds J. Bergstralh and M.S. Matthews). Tucson: University of Arizona Press, pp. 469–512.

Rothery, D.A. (1992) *Satellites of the Outer Planets: Worlds in their Own Right*. Oxford: Clarendon Press.

Sagan, C. and Dermott, S.F. (1982) The tides in the seas of Titan. *Nature*, **300**, 731–3.

Smith, B.A., Soderblom, L., Beebe, R. *et al*. (1981) Encounter with Saturn: Voyager 1 imaging results. *Science*, **212**, 163–91.

Stevenson, D.J. (1980) Saturn's luminosity and magnetism. *Science*, **208**, 746–8.

Stevenson, D.J. (1982) Volcanism and igneous processes in icy satellites. *Nature*, **298**, 142–4.

Stevenson, D.J. (1992) Interior of Titan, in *Proc. Symposium on Titan*. Noordwijk, The Netherlands: European Space Agency SP-338, pp. 29–33.

Stevenson, D.J. and Salpeter, E.E. (1977) The dynamics and helium distribution properties for hydrogen–helium fluid planets. *Astrophys. J. Suppl. Ser.*, **35**, 221–237.

Stevenson, D.J., Harris, A.W. and Lunine, J.I. (1986) Origins of satellites, in *Satellites* (eds J.A. Burns and M.S. Matthews). Tucson: University of Arizona Press, pp. 39–88.

Thompson, R. and Sagan, C. (1992) Organic chemistry on Titan – surface interactions, in *Proc. Symposium on Titan*. Noordwijk, The Netherlands: European Space Agency SP-338, pp. 167–76.

Van Allen, J.A. (1981) Magnetospheres and the interplanetary medium, in *The New Solar System* (eds J.K. Beatty, B. O'Leary and A. Chaikin). Cambridge, Mass.: Sky Publishing Corp., pp. 23–32.

Yung, Y.L., Allen, M. and Pinto, J.P. (1984) Photochemistry of the atmosphere of Titan: Comparison between models and observations. *Astrophys. J. Suppl. Ser.*, **55**, 465–506.

Zharkov, V.N. and Gudkova, T.V. (1991) Models of giant planets with a variable ratio of ice to rock. *Ann. Geophysicae*, **9**, 357–66.

Cross references

Cassini mission
Enceladus
Huygens Titan atmospheric probe
Imaging science
Magnetospheres of the outer planets

Planetary ring
Titan
Voyager missions

SATURN: ATMOSPHERE

Physical conditions

Saturn appears to have formed from the same cloud mass as our Sun, thus it must be rich in the lightest elements, hydrogen and helium. The fact that the mass of Saturn is more than 95 times that of the Earth while the density is 0.69 g cm^{-3}, about one-eighth that of the Earth, substantiates this assumption. Because Saturn orbits the Sun at an average distance of about 9.54 times the distance of our Earth, the sunlight that would fall on a square centimeter of the Earth spreads out over an area of 9.54×9.54 cm^2, causing the sunlight to be almost 100 times more dilute than at Earth. This leads to temperatures between $-130°$ and $-200°C$ in the atmospheric region of the planet. In addition, there is no reason to believe that a planet of this size and composition has a well-defined solid surface. Instead the pressure increases with depth because the force of gravity acting on the overlying material compresses the hydrogen-helium-rich atmosphere, forcing the gas molecules closer together and, eventually, the material becomes a liquid. We consider the outer semi-transparent region and the gaseous underlying region that transports energy by convective mixing as the atmosphere. A study of Saturn's atmosphere involves the chemistry and energy balance of these regions.

A visual inspection of Saturn's atmosphere with an Earth-based telescope reveals a pale yellow deck of clouds with bands of slightly different intensities and shades of yellow which are oriented parallel to the plane of the rings, or Saturn's equatorial plane (Plates 26, 28). This pronounced banding, as in Jupiter's atmosphere, suggests there are strong east-west winds and that there is a global circulation where solar heating may play a significant role. In addition, Saturn is radiating 1.78 ± 0.09 times more energy in visible and infrared light (Hanel et al., 1983) than it absorbs from the incoming sunlight. This energy must be coming from the core of the planet and must also be transported through the lower atmosphere by convection or turbulent mixing.

Although the atmosphere is rich in hydrogen and helium, at these frigid temperatures small amounts of other elements should combine into molecules that condense and form ices and smog. The chemistry that is involved is complex and has been studied in detail (Prinn et al., 1984). Because the overlying layers of atmosphere absorb escaping infrared radiation and act as an insulating blanket, it is necessary to know how the temperature and pressure increase with depth in the atmosphere before our knowledge of chemistry can be applied to determine the composition of the ices that make up the cloud layers. Once this general structure is determined, the mechanism through which incident ultraviolet radiation generates smog in the upper atmosphere can be understood.

Temperature and pressure versus height

One method of determining how temperature and pressure vary in Saturn's atmosphere is to measure the light from a star or the well-known radio signal from a spacecraft as, relative to our line of sight, Saturn moves in front of the star or the spacecraft goes behind the planet. If the composition is assumed to be solar-like, the manner in which the signal fades can be interpreted to determine how the temperature of the atmosphere varies with depth. If, in addition, the atmosphere is assumed to be in pressure equilibrium, the relationship between temperature and pressure can be derived (Tyler et al., 1982). Because there is no solid surface from which to measure altitude, the temperature and pressure variation is frequently referenced to the depth in the atmosphere where the pressure is equal to sea-level pressure on Earth, approximately 1 bar, and the pressure is expressed in millibars (Atreya, 1986; Chamberlain and Hunten, 1987). Figure S4 compares the manner in which temperature varies with pressure on Saturn and on Earth. Assuming the atmosphere is composed of 94% H$_2$ and 6% He, Tyler et al. (1982) derived the temperature-pressure relationship for Saturn's atmosphere from the Voyager radio signal. This is compared to a standard Earth atmosphere (Allen, 1973). In both cases the temperature decreases with height at lower levels. This seems logical because, even though the surface of the Earth is heated by absorbed visible light and Saturn has an internal heat source, the atmosphere should lose heat to space. However, Figure S4 indicates that temperatures increase again at higher altitudes. This is due to absorption of ultraviolet light in the upper region before the UV is depleted from the incoming sunlight. The troposphere, the lower region where temperature decreases with altitude, is unstable to convective mixing and is the region within which vertical mixing

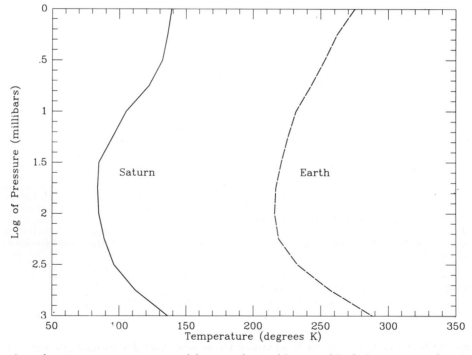

Figure S4 A comparison of pressure versus temperature of the atmospheres of Saturn and Earth. Saturn's atmosphere is indicated as a solid line while the Earth is denoted by the dashed line. Note that the pressures range from that at sea level at the bottom of the graph to 1 mbar at the top.

occurs, generating planetary weather. The temperature inversion at higher levels tends to suppress convection and thus, above this level, the stratosphere, there is little mixing and thin layers of clouds and hazes form. Although the temperature regimes of the atmosphere of Earth and Saturn do not overlap, these processes establish the same general trends within these two different environments.

Chemical composition

If the temperature and pressure are known as functions of height in the atmosphere, laboratory data concerning the rate of formation of specific molecules can be used to predict which molecules will form at each level in the atmosphere. This approach indicates that the visible cloud deck should be composed of ammonia ice (Prinn et al., 1984) and that the oxygen in the atmosphere would combine with the hydrogen, forming water which would freeze and the particles would snow out of the semitransparent region. These studies predict that the water-ice clouds form at a depth in the clouds where the pressures are greater than 20 bar and would not be visible from above the clouds. The gases above the clouds would be composed of about 94% molecular hydrogen (H_2) and 6% atomic helium (He) with traces of methane (CH_4) and other gases.

This model can be tested by using spectroscopic techniques. In this frigid environment the spectroscopic features that characterize these molecules occur in the infrared, requiring specialized observing techniques. The near approach and the design of the Voyager spectrographs were best suited for this study. Hanel et al. (1991 a, b) observed molecular hydrogen, methane and ethane features. By comparing their data with predictions of numerical models, they confirmed that the temperature–pressure relation deduced by Tyler et al. (1982, and others) is consistent with their results.

The presence of small amounts of ethane was established. This did not come as a surprise. When the effect of UV radiation on the upper atmosphere (Atreya, 1986) is considered, the ethane can be explained by the following set of equations:

$$CH_4 \xrightarrow{UV} CH_3 + H + \text{energy}$$
$$CH_3 + CH_3 \rightarrow C_2H_6$$

where any UV photon with energy greater than or equal to the energy necessary to break the hydrogen bond can trigger the first reaction. The energy balance is such that the low-mass particles, the hydrogen atoms, have higher velocities. They tend to move away from the reaction site instead of recombining with the CH_3 molecules. Some acquire velocities large enough to allow them to escape from the atmosphere.

These reactions will occur in Saturn's upper atmosphere and at the local temperatures the molecules will condense into solid particles. Additional complex reactions will occur on the grain surfaces and a smog will be generated. The highest concentration of this smog will be produced at the latitude where the direction of the incident sunlight is most nearly perpendicular to the surface of the cloud deck. Because Saturn is tilted 26.7° on its axis and the Saturnian year is 29.458 Earth years, at a given latitude, this concentration will vary with Saturn's seasons. The Voyager images (Smith et al., 1981, 1982) were obtained at the beginning of spring in the northern hemisphere. The southern hemisphere, which had not yet recovered from the effect of smog build-up during the summer, was more highly reflective. The decrease in details in the cloud structure and the color dependence of scattering of sunlight as the instruments looked back toward the Sun confirmed a smog build-up (West, 1983).

Global winds

Saturn suffers seasonal variations in solar heating and is losing internal heat to space. In addition, it rotates about its axis in 10.675 h. Because the radius of this large planet is 9.45 times that of the Earth, a parcel of atmospheric gas, moving with the rotation of the planet, would be swept around the planet more than 20 times faster than a similar particle on Earth. Our study of the circulation of the Earth's atmosphere has revealed that the global wind structure is strongly influenced by the rotation of the planet (Holton, 1972). In the case of Saturn the outward transfer of energy from the interior must respond to forces imposed by this rapid rotation (Ingersoll, 1990). Although few individual clouds are visible in the ground-based images, the Voyager imaging science data could be used to deduce horizontal motions. The longitude and latitude of selected individual features within the cloud deck could be determined on two or more images from a time sequence that was obtained as the planet rotated in front of the cameras. When the spacecraft was near the planet, the resulting spatial resolution was adequate to allow the determination of the east–west and north–south motions (Ingersoll et al., 1984). The result revealed strong east–west winds. Figure S5 shows the latitudi-

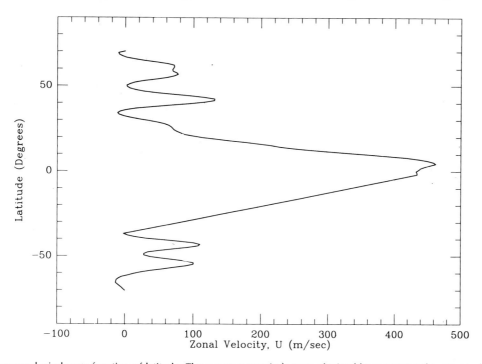

Figure S5 Average zonal winds as a function of latitude. These east–west winds were obtained by measuring the rate at which clouds moved longitudinally within time sequences of Voyager images. Latitudes are plotted in planetocentric coordinates, similar to the latitudinal system used on Earth.

nal dependence of the average east–west winds. Even though the southern hemisphere had just experienced summer and the northern hemisphere was emerging from winter, a time when accumulated effects due to differential solar heating should have been greatest, the extent of the symmetry of the wind pattern about the equator indicates that the global circulation is most strongly influenced by the rotational forces.

Occasional storms

In general, the structure of Saturn's atmosphere is such that the active mixing of the troposphere is shrouded by the overlying clouds. The features that were measured in the Voyager data appeared to be the tops of active convective systems that pushed their way up into the overlying semitransparent region. Occasionally a larger storm system appears to form and emerge through the cloud deck. Sanchez-Lavega (1981) noted that these storms seemed to have a seasonal dependence, and predicted that another was due. In September 1990 a bright spot formed near 4°N latitude. This brilliant white cloud interacted with the prevailing winds, disturbing the atmosphere. As in the Earth's atmosphere, this generated cloud patterns associated with waves that propagated around the planet. Ground-based (Sanchez-Lavega *et al.*, 1991; Beebe *et al.*, 1992) and Hubble Space Telescope observations (Barnet *et al.*, 1992) revealed that this disturbance appeared to be the atmospheric response to a single convective cell that expanded upward and outward in the atmosphere. Additional Hubble Space Telescope observations in June 1991 revealed that, although the equatorial region was brighter and whiter than before the onset of the storm, only a few cloud streaks remained.

To characterize the atmosphere in a disturbed state, horizontal wind speeds were derived from the Hubble Space Telescope data. Using near-infrared images (889 nm), where the light is absorbed high in the atmosphere so that reflected light must come from elevated clouds, and green images, where the sunlight penetrates to lower depth, Barnet *et al.* (1992) determined that the winds in the equatorial region decrease with height. These results agree with those derived from the Voyager infrared data for higher latitudes (Conrath and Pirraglia, 1983), indicating that the rate of decay of east–west winds is similar at low and mid-latitudes. This information contributes to our knowledge of the energy balance within Saturn's atmosphere.

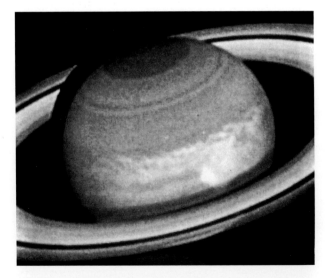

Figure S6 A Hubble Space Telescope image of the 1990 equatorial storm. This image was obtained in filtered green light (filter F547M) and shows the remains of the clouds associated with the original storm on 9 November 1990, about 48 days after the onset. This figure is based on observations with the NASA/ESA Hubble Space Telescope that were obtained at the Space Telescope Science Institute, which is operated by the Association of Universities for Research in Astronomy, Inc., under NAS5-26555.

In the past there have been three very similar equatorial storms that were recorded by ground-based observers in 1876, 1933 and 1990 and, although they were separated by approximately 2 Saturnian years (1 Saturnian year = 29.46 Earth years), no equatorial storms were reported in the intervening years. The combined effects of ellipticity of Saturn's orbit and the tilt of the axis of rotation cannot be used to explain why the northern hemisphere is preferred. Instead, the fact that the onset of the storm is arriving earlier and earlier suggests that the response time of Saturn's atmosphere to such disturbances is slightly less than 2 Saturnian years and that the increased solar heating may make the upper atmosphere more unstable and thus enhance the convection. If this is the case, the quasi-biannual pattern may not be repeated until about 2047.

Summary

Saturn's hydrogen–helium-rich atmosphere contains layered cloud decks. The visible cloud deck is composed of ammonia ices. Absorption of radiation by traces of methane gas in the upper region of the atmosphere plays an important role in the atmospheric heat balance. Photochemical reactions generate hazes in the stratosphere and active vertical mixing of the troposphere leads to the formation of storm systems.

This large, rapidly rotating atmosphere is dominated by strong east–west winds that decrease with height. Currently, there is no direct evidence to indicate the rate of vertical motions in the atmosphere. The response of the atmosphere to rapid rotation and transport of internal heat dominates the circulation of the atmosphere; however, absorption of sunlight by the upper atmosphere may trigger convective disturbances similar to the 1990 equatorial storm. The chemical reactions induced by ultraviolet radiation add greatly to the complexity of this frigid, hydrogen–helium-rich atmosphere.

Reta Beebe

Bibliography

Allen, C. (1973) *Astrophysical Quantities*. London: Athlone Press.
Atreya, S. (1986) *Atmospheres and Ionospheres of the Outer Planets and Their Satellites*. Berlin: Springer-Verlag.
Barnet, C.D., Westphal, J.A., Beebe, R.F. and Huber, L.F. (1992) Hubble Space Telescope observations of the 1990 equatorial disturbance of Saturn. *Icarus*, **100**, 499–510.
Beebe, R.F., Barnet, C.D., Sada, P.V. and Murrell, A.S. (1992) The onset and growth of the 1990 equatorial disturbance on Saturn. *Icarus*, **95**, 163–72.
Chamberlain, J. and Hunten, D. (1987) *Theory of Planetary Atmospheres*. San Diego: Academic Press.
Conrath, B.J. and Pirraglia, J.A. (1983) Thermal structure of Saturn from Voyager infrared measurements: implications for atmospheric dynamics. *Icarus*, **53**, 286–92.
Hanel, R.A., Conrath, B., Flaser, F.M. *et al.* (1981a) Infrared observations of the Saturnian system from Voyager 1. *Science*, **212**, 192–200.
Hanel, R.A., Conrath, B., Flaser, F.M. *et al.* (1981b) Infrared observation of the Saturnian system from Voyager 2. *Science*, **215**, 544–8.
Hanel, R.A., Conrath, B.J., Kande, V.G. *et al.* (1983) Albedo, internal heat flux, and energy balance of Saturn. *Icarus*, **53**, 262–85.
Holton, J. (1972) *An Introduction to Dynamic Meteorology*. London: Academic Press.
Ingersoll, A. (1990) Atmospheres of the giant planets, in *The New Solar System* (eds J.K. Beatty and A. Chaikin). Cambridge: Sky Publishing Corp., pp. 139–52.
Ingersoll, A.P., Beebe, R.F., Conrath, B.J. and Hunt, G.E. (1984) Structure and dynamics of Saturn's atmosphere, in *Saturn* (eds T. Gehrels and M.S. Matthews). Tucson: University of Arizona Press, pp. 195–238.
Prinn, R.G., Larson, H.P., Caldwell, J. and Gautier, D. (1984) Composition and chemistry of Saturn's atmosphere, in *Saturn* (eds T. Gehrels and M.S. Matthews). Tucson: University of Arizona Press, pp. 88–149.
Sanchez-Lavega, A. (1981) A white spot in Saturn's equatorial zone in 1978. *Publ. Astron. Soc. Pacific*, **93**, 134–8.

Sanchez-Lavega, A. (1989) Saturn's great white spot. *Sky and Telescope*, **78**, 141–3.
Sanchez-Lavega, A., Colas, F., Lecacheur, J. *et al.* (1991) The Great White Spot and disturbances in Saturn's equatorial atmosphere during 1990. *Nature*, **353**, 397–401.
Smith, B.A., Soderblom, L.A., Beebe, R. *et al.* (1981) Encounter with Saturn: Voyager 1 imaging science results. *Science*, **212**, 163–91.
Smith, B.A., Soderblom, L.A., Batson, R.F. (1982) A new look at the Saturnian system. *Science*, **215**, 504–37.
Tyler, G.L., Eshleman, V.R., Anderson, J.D. *et al.* (1982) Radio science with Voyager 2 at Saturn. *Science*, **215**, 553–7.
West, R.A. (1983) Spatially resolved methane band photometry of Saturn. II. Cloud structure models at four latitudes. *Icarus*, **53**, 301–309.

Cross references

Atmosphere
Jupiter: atmosphere
Neptune: atmosphere
Spectroscopy: atmosphere

SATURN: INTERIOR STRUCTURE

Saturn's mean density of 0.69 g cm^{-3} is the lowest of any object in the solar system. Like Jupiter and the Sun, Saturn is composed primarily of the light gases hydrogen and helium. If Saturn had the same chemical composition as the Sun, out of a total mass of 95 Earth masses, it would contain about 68 Earth masses of hydrogen, about 26 Earth masses of helium and about one Earth mass of all other elements (the so-called heavy elements). Models of the interior of Saturn indicate, however, that the planet actually contains about 29 Earth masses of heavy elements, with about 48 Earth masses of hydrogen and 18 Earth masses of helium. Saturn is, therefore, considerably more depleted in the light gases hydrogen and helium than Jupiter. If this is the case, why is Saturn's mean density so low, only half that of Jupiter's? The answer comes from consideration of the very high compressibility of hydrogen under the conditions in the interior of Saturn or Jupiter. Saturn's mass is less than one-third that of Jupiter, and pressures in its deep interior are about one-third that of Jupiter. But because hydrogen is so compressible, Saturn expands to fill nearly the same volume as Jupiter. Its mean radius, about 60 000 km, is similar to that of Jupiter (70 000 km). This result is in good agreement with the theory of hydrogen at high pressures, and thus shows that hydrogen, although it is depleted with respect to solar composition, is the dominant constituent of Saturn. If other, heavier, constituents were predominant, Saturn's radius would be about half the observed value, or about 30 000 km.

Saturn is the most oblate object in the solar system, a result of its exceptionally low mean density and relatively fast spin rate. The interior rotation period, as determined from measurements of the rotation of non-axisymmetric components of its magnetic field, is 10.65622 h. The response of the planet's figure to this rotation is pronounced and leads to an oblateness (relative difference between equatorial and polar radius) of 9.6%, which is apparent in images of the planet. This relatively large distortion also produces measurable second, fourth and sixth-degree distortions of the external gravitational potential from spherical symmetry, which are used to probe the planet's interior structure. Because Saturn's ring system includes narrow eccentric ringlets which precess in the planet's distorted gravitational potential, spacecraft and ground-based measurements of the motions of these ringlets can give detailed information about high-degree gravity components. From data available thus far, it is deduced that Saturn's moment of inertia about its spin axis is 0.22 Ma^2, where M is Saturn's mass and a is its equatorial radius. Since a sphere of uniform density would have a moment of inertia equal to 0.40 Ma^2, this result indicates that Saturn's interior density distribution is substantially concentrated toward its central regions. It also is incompatible with a uniform composition dominated by hydrogen; such a planet would have a larger moment of inertia. Thus there is evidence from Saturn's gravity field that the planet has a core rich in heavy elements.

Like Jupiter, Saturn has a measurable net flow of heat from its interior. This intrinsic heat flow was first detected in ground-based infrared measurements of Saturn made in the 1960s by F.J. Low and associates. After allowance is made for thermalization of incident sunlight, the current best estimate of Saturn's surface intrinsic heat flux is 2.0 W m^{-2}. While this is smaller than the corresponding figure for Jupiter, it is too large to be explained completely by the Kelvin–Helmholtz cooling mechanism that appears to work for Jupiter. Radioactive heat production from the approximately 29 Earth masses of heavy material in Saturn could account for no more than 1% of the observed heat flux.

Thermal evolution models for Saturn have assumed that, like Jupiter, it formed by the initial rapid collapse of hydrogen and helium gas onto a dense central nucleus composed of heavy material, and that the heat produced during this collapse has been slowly radiated into space over the rest of the planet's lifetime (the so-called Kelvin–Helmholtz mechanism). However, regardless of how high an initial temperature was produced during this initial collapse, detailed calculations show that virtually all the stored heat is radiated into space in about 2.6×10^9 years, or about 2.0×10^9 years less than the age of the solar system. Thus, unlike Jupiter, Saturn's present heat flow cannot be solely explained by the radiation of primordial stored heat; an additional source is required.

The most likely alternative mechanism for prolonging the lifetime Saturn's heat radiation is slow release of additional gravitational binding energy during Saturn's subsequent evolution after its initial collapse. The duration of release of this energy is critical because release which is too rapid, occurring only during the first few million years of Saturn's evolution, does not solve the energy shortfall problem. The planet is luminous during this early epoch and quickly disposes of any extra thermal energy which is made available at this time. A mechanism which postpones the release of gravitational energy until a stage when the interior is relatively cool is required instead. Naturally regulated chemical unmixing of an abundant dense constituent such as helium has been suggested as a possible mechanism. Helium is dense (compared to hydrogen) and abundant, and is predicted to become immiscible in the dense metallic phase of hydrogen that exists near Saturn's center. Calculations indicate that helium has substantially limited solubility in the metallic phase of hydrogen. While helium also has limited solubility in solid molecular hydrogen, which exists at pressures lower than about 3 Mbar, experiments show that the solubility is essentially unlimited in molecular hydrogen at the much higher temperatures which prevail in Saturn.

Gradual precipitation of dense helium droplets within Saturn is believed to be occurring at present, leading to unmixing of helium and maintenance of a high interior temperature by frictional heating as the droplets sink toward denser regions, where they are once again soluble. With 18 Earth masses of helium available for this process, it is possible to double Saturn's heat radiation lifetime without difficulty.

Saturn's atmosphere is observed to be strongly depleted in helium with respect to solar composition. The helium mass fraction in the observable atmosphere is about 0.06, while for solar composition it would be about 0.27. If the outer layers of Saturn have been depleted in helium because of helium–hydrogen immiscibility at deeper levels, the atmospheric helium mass fraction must represent the saturation value at these deeper levels.

According to models of Saturn's interior, the helium-depleted molecular hydrogen envelope extends to a depth of about 30 000 km, or about half the planet's radius. It also comprises almost half of Saturn's mass. At deeper levels, where the pressure exceeds 2 M bar, the hydrogen is instead in the metallic (pressure-ionized) phase, and it is in this region that helium precipitation, as well as magnetic field generation, is believed to occur. The partitioning of heavy elements across the molecular–metallic phase boundary is not well understood. There is some evidence from interior models that heavy elements are strongly enriched in both the molecular and metallic hydrogen phases. Of the 29 Earth masses of heavy elements, about 5 Earth masses is contained in the molecular mantle, while the remainder is situated in the metallic hydrogen layers and in a possible dense core. This strong partitioning of heavy elements toward the center of the planet is required to account for Saturn's small moment of inertia. But curiously, the mean density of Saturn's molecular hydrogen mantle does not appear to be low in comparison with a solar composition mixture of hydrogen and helium despite the presumed strong helium depletion. For this reason the mantle is thought to be strongly enriched in elements heavier than helium.

Apart from the substantial helium depletion, the observed composition of Saturn's atmosphere provides only weak constraints on the possible composition of deeper layers. Carbon (in the form of CH_4, methane) is enriched with respect to hydrogen by a factor of two to six over the value expected for solar composition. Nitrogen is enriched by a similar factor. As in Jupiter's atmosphere, the chemical disequilibrium species PH_3 and GeH_4 have been detected, and their presence is explained by assuming that they are produced by equilibrium chemistry in a region of high pressure and temperature (pressure of about 1 kbar, temperature of about 1000 K), and then quenched in a disequilibrium state by rapid vertical convection as they are carried into the cooler observable atmosphere.

Saturn's heavy-element component is expected to include substantial amounts of H_2O. However, water condenses deeper in Saturn's atmosphere than it does in Jupiter's, and this important molecule has not yet been reliably detected.

Fluid circulation in Saturn's deep interior may well be constrained by strong chemical gradients. Like Jupiter's interior, Saturn's interior must maintain a sufficient temperature gradient to transport the observed heat flow, and as a result prevailing temperatures are calculated to be about 6000 K at the molecular–metallic interface, and about 13 000 K at the base of the metallic hydrogen zone. These temperatures lie well above the melting curve of dense hydrogen, and thus Saturn must be entirely in the liquid state. Although Saturn's intrinsic heat flow is large enough to drive convection in most of its liquid interior, the boundary between the molecular and metallic phases of hydrogen, as well as the large change in composition which doubtless exists there, serves as a formidable barrier to convective motions. Transport of heat across this boundary may occur by conduction. Deeper in the metallic layers, convective motions are sufficiently rapid to excite a magnetohydrodynamic dynamo, producing Saturn's magnetic field. However, the stably stratified conducting layer at the top of the metallic zone may screen out all but axially symmetric components of Saturn's magnetic field external to the metallic core. Saturn's magnetic dipole axis is observed to align with the planet's spin axis to within 1^0. Because of this symmetry, it is difficult to determine Saturn's magnetic field rotation period.

Atmospheric winds on Saturn are almost entirely zonal, i.e. they flow in an east–west direction, and they are roughly north–south symmetric in the sense that a zonal flow velocity at a given northern latitude tends to have a counterpart at the corresponding southern latitude. The zonal flows are almost entirely prograde (toward the east) with a peak velocity in excess of 400 m s^{-1} at the equator relative to a frame in which the magnetic field is stationary. If Saturn's interior is a barotrope, i.e. if the temperature is everywhere constant on surfaces of constant pressure, then steady zonal flow velocities must be constant on cylindrical surfaces at a constant distance from Saturn's rotation axis. In this case, zonal flows at high latitudes would map onto cylinders extending deep into Saturn's interior. The observed approximate north–south symmetry is consistent with this model, as is the fact that large zonal flows do not persist on cylinders with radii smaller than about 25 000 km; cylinders with smaller radii would intersect a substantial fraction of the metallic and heavy element-enriched core.

William B. Hubbard

Bibliography

Atreya, S.K., Pollack, J.B. and Matthews, M.S. (1989) *Origin and Evolution of Planetary and Satellite Atmospheres*, Part IV, *Outer Planets*. Tucson: University of Arizona Press.

Gehrels, T. and Matthews, M.S. (eds) (1984). *Saturn*. Tucson: University of Arizona Press.

Hubbard, W.B. (1984) *Planetary Interiors*. New York: Van Nostrand Reinhold.

Hubbard, W.B. (1990) Interiors of the giant planets, in *The New Solar System*, 3rd edn (eds J.K. Beatty and A. Chaikin). New York, Cambridge University Press, pp. 131–8.

Morrison, D. (1982) *Voyages to Saturn*. Washington, DC: US Government Printing Office, NASA SP-451.

Stevenson, D.J. (1982) Interiors of the giant planets. *Ann. Rev. Earth Planet. Sci.* **10**, 257–95.

Cross references

Core, terrestrial planetary
Jupiter: interior structure
Planetary rotation
Solar system: origin

SATURN: MAGNETIC FIELD AND MAGNETOSPHERE

Saturn is the second largest planet in the solar system with a radius of over 60 000 km. It rotates almost as fast as Jupiter, having a rotational period of 10 h 39 min. Moreover the composition of Saturn is thought to be similar to that of Jupiter, consisting of mainly hydrogen and helium. Nevertheless, Saturn and Jupiter have many differences. First, Saturn's intrinsic magnetic field is much weaker than that of Jupiter. Second, the magnetic field, instead of being rich in harmonics, has a very simple structure, one that is axisymmetric to a very high degree of accuracy. Further, the magnetosphere of Saturn does not have an efficient energizing mechanism deep in the heart of the magnetosphere as does Jupiter. In short, Saturn has a relatively simple magnetic field and a relatively quiescent magnetosphere.

Planet and interior

The composition of Saturn is thought to be similar to that of Jupiter but its smaller radius and hence much smaller mass leads to an interior with a much smaller conducting core. As occurs at Jupiter, the cooling of the Saturn's interior causes condensation of helium within the hydrogen–helium fluid. This process then releases heat due to the change of phase and due to gravitational settling, and provides a heat source for powering convection in the interior of the planet and for powering the magnetic dynamo.

Magnetic field

While Saturn does generate radio waves, these waves are not strong enough to be detected on Earth. Thus until Pioneer 11 reached Saturn in 1979 it was not known whether Saturn had an intrinsic magnetic field. The passage of Pioneer 11 within 1.4 Saturn radii of the center of the planet was soon replicated by Voyager 1 in 1980, passing within 3.1 R_S, and Voyager 2 in 1981, passing within 2.7 R_S. These spacecraft found a magnetic field quite unlike that at any other planet. To the accuracy that could be obtained, there was no tilt to the rotation axis and the interior magnetic field was perfectly axisymmetric. The magnetic moment, the surface field strength at the equator times the cube of the radius, was also somewhat smaller than expected at 4.6×10^{18} T m^3. While this value is 580 times larger than that of the Earth, it is over 30 times smaller than that of Jupiter despite the small (15%) difference in radius. The contribution of the quadrupole moment at Saturn is also small compared to the quadrupole contribution at Jupiter and the Earth. For example, the ratio of the terrestrial quadrupole to dipole moment is 0.14 but at Saturn it is 0.07 (see Earth: magnetic field and magnetosphere).

Magnetosphere

Saturn also has an immense magnetosphere, whose linear dimension is about one-fifth that of the Jovian magnetosphere. This magnetosphere is more similar to the terrestrial magnetosphere than that of Jupiter. The magnetosphere traps radiation belt particles, and these particles reach levels similar to those of the terrestrial magnetosphere. On their inner edge the radiation belts are terminated by the main (A, B and C) rings of Saturn, which absorb any particles that encounter them. The radiation belt particles also are absorbed if they collide with one of the moons. Hence there are local minima in the energetic particle fluxes at each of the moons. Unlike Jupiter, but like the Earth, there is no internal energy and mass source deep in the Saturnian magnetosphere. However Titan, which orbits just inside the average location of the magnetopause, in the far reaches of the magnetosphere, has an interesting interaction.

Titan (q.v.) is the most gas-rich moon in the solar system, having an atmospheric mass per unit area much greater than even that of the Earth. At its upper levels this atmosphere becomes ionized through charge exchange, impact ionization and photoionization. This newly created plasma adds mass to the magnetospheric plasma, which attempts to circulate in the Saturnian magnetosphere at a velocity similar to that needed to remain stationary with respect to the rotating planet. Since this velocity is much faster than the orbital velocity of Titan, the added mass slows the 'corotating' magnetospheric plasma. The magnetic field of the planet that is effectively frozen to the magnetospheric plasma is then stretched and draped about the planet, forming a slingshot which accelerates the added mass up to corotational speeds. Thus the interaction between the Saturn magnetosphere and the Titan atmosphere resembles the interaction of the solar wind with comets and with Venus (Kivelson and Russell, 1983).

The Saturn magnetosphere, like the other planetary magnetospheres, is an efficient deflector of the solar wind. The solar wind at Saturn flows more rapidly with respect to the velocity of compressional waves than at Jupiter and the terrestrial planets. Thus the shock that forms at Saturn is very intense. Ironically this strength may weaken at least one form of coupling of the solar wind with the magnetosphere, that due to reconnection. However, some aspects of the interaction of the solar wind plasma should be much stronger than at Jupiter or at Earth because of the increased strength of the shock and the scale size of the interaction, which can accelerate charged particles to very high levels.

Saturn is also expected (like Jupiter) to have a very large tail, possibly one that could be dynamic like that of the Earth. However, observations of the tail are quite limited and we must wait until the Cassini mission (q.v.) in the early 21st century for further studies of the magnetic field, magnetosphere and magnetotail, and the answers to many of the questions that the Pioneer and Voyager data have generated.

Christopher T. Russell and Janet G. Luhmann

Bibliography

Connerney, J.E.P., Acuna, M.H. and Ness, N.F. (1984) The Z_3 model of Saturn's magnetic field and the Pioneer 11 vector helium magnetometer observations. *J. Geophys. Res.*, **89**, 7541–44.

Kivelson, M.G. and Russell, C.T. (1983) The interaction of flowing plasmas with planetary ionospheres: a Titan–Venus comparison. *J. Geophys. Res.*, **88**, 49–57.

Russell, C.T. (1987) Planetary magnetism, in *Geomagnetism*, Vol 2 (ed. J.A. Jacobs). London: Academic Press, pp. 457–23.

Acknowledgements

This work was supported in part by the National Aeronautics and Space Administration under research grant NAGW-2573.

Cross references

Cassini mission
Earth: magnetic field and magnetosphere
Jupiter: magnetic field and magnetosphere
Magnetospheres of the outer planets
Pioneer 10 and 11 missions
Planetary ring
Radiation belt

SATURN: RING SYSTEM

Saturn's rings, while no longer unique, remain by far the most massive and diverse ring system known (Plates 26, 27). Alexander (1962) discusses early studies. Pollack (1975) reviews modern work on the rings prior to the Voyager flybys. Cuzzi et al. (1984) and Esposito et al. (1984) provide comprehensive overviews of the Voyager results; Nicholson and Dones (1991) and Esposito (1993) treat more recent developments. Marouf, Tyler and Rosen (1986) discuss the many types of structure revealed by the Voyager radio occultation experiment. Showalter, Cuzzi and Larson (1991), Showalter and Cuzzi (1993), and Showalter et al. (1992) respectively investigate the tenuous outer E and G rings and the clumpy F ring. Goldreich and Tremaine (1982) and Araki (1991a) discuss the dynamics of planetary rings. Burns, Showalter and Morfill (1984) treat the dynamics of dusty rings.

Overview

The structure of Saturn's rings is summarized in Table S2. The rings fall into two main categories, depending on their mean optical depth $\bar{\tau}$, which measures the fraction of the ring's area covered with particles or, more precisely, the average number of particles that a ray perpendicular to the ring would traverse. The so-called main rings – the C ring, B ring, Cassini division and A ring – reside within or near Saturn's Roche limit (q.v.), have $0.1 < \bar{\tau} < 2$ and consist primarily of particles larger than about 1 cm. Particles in these rings follow near-circular orbits about Saturn; the orbits are governed by Saturn's gravitational force, and to a much lesser extent by gravity exerted by Saturn's satellites and ring particles, and by interparticle collisions. By contrast, the G and E rings, which lie external to the Roche limit, and Saturn's innermost ring, the D ring, are tenuous ($\bar{\tau} \leq 10^{-4}$) and consist primarily of microscopic 'dust'. For these rings, Saturn's gravity, solar radiation pressure and electromagnetic forces together determine the orbits (Burns, Showalter and Morfill, 1984; Horanyi, Burns and Hamilton, 1992). The F ring, which lies near the Roche limit, is intermediate in properties between the main rings and tenuous rings.

Size of ring particles

The size of particles in the main rings is derived from an experiment in which radio signals at wavelengths of 3.6 and 13 cm were transmitted by Voyager through the rings to Earth (Marouf et al., 1983). The particles have a broad range of sizes: $N(r)$, the number of particles larger than r, satisfies $N(r) \propto r^{-2}$. There are few particles larger than 5–10 m or smaller than about 1 cm. Most of the mass of the rings resides in particles larger than 1 m. Voyager photometry at small scattering angles reveals that the main rings contain very little microscopic dust (Doyle, Dones and Cuzzi, 1989; Dones, Cuzzi and Showalter, 1993). Longaretti (1989) provides the most detailed theory of the size distribution.

The E and G rings consist almost exclusively of microscopic particles (Showalter, Cuzzi and Larson, 1991; Showalter and Cuzzi, 1993). The G ring consists of very tiny particles whose size ranges down to 0.03 μm or smaller. Unique among all known rings, particles in the E ring have a narrow range of sizes, $r = 1 \pm 0.3$ μm. Hamilton and Burns (1993) propose that the E ring is self-sustained: solar radiation pressure pumps up the eccentricities of micron-sized particles in the ring to ≥ 0.7; the particles then collide with Saturn's inner satellites (Table S2), resulting in the creation of more ring particles. This model predicts that the E ring is broad, and densest near the orbit of Saturn's moon Enceladus, as is observed, and also predicts the optical depth of the ring and the characteristic size of the ring particles. Previous models invoking volcanic eruptions or geysers on Enceladus to produce the E ring appear less likely to be correct.

The F ring, too, consists primarily of dust (Showalter et al., 1992), but the Voyager Radio Science Experiment also detected a narrow core of centimeter-sized particles. Both the F and G rings must contain certain number of moonlets, roughly 1 km in size, to resupply small ring particles, which are otherwise rapidly sputtered away or lost by various drag forces (Burns, Showalter and Morfill, 1984). Analysis of Pioneer 11 magnetospheric data provides possible evidence for the absorption of charged particles by clumps of ejecta resulting from collisions between the hypothetical moonlets (Cuzzi and Burns, 1988).

Composition

Earth-based near-infrared spectra of the main rings detect absorption bands due to water ice, which remains the only compound definitely identified in the rings. However, at visual wavelengths the rings' relatively low albedo and reddish color must be due to impurities such as silicates, iron oxides or possibly organics like those present on the surface of the dark planetesimal Pholus (Estrada and Cuzzi, 1996). Earth-based studies indicate that the rings are very poor emitters of microwaves, implying, by Kirchoff's law, that the particles are poor absorbers as well. These results imply that less than 1% of the rings' mass consists of lossy material such as silicates or carbon. This finding is consistent with the low to moderate albedos found at visual wavelengths, as only small amounts of contaminants are needed to make ice or snow 'dirty'. The purity of the ring particles is surprising, given nominal estimates of the rate at which they are impacted by dark, carbonaceous meteoroids (Doyle, Dones and Cuzzi, 1989).

numerous density waves in the A ring should transfer angular momentum from the rings to the inner satellites, just as the Moon is inexorably moving outward due to tides exerted by the Earth. According to theory, the moons Atlas, Prometheus and Pandora should have been at the A ring's outer edge less than about 10^8 years ago. In the same way, the A ring should collapse to its inner edge in only 10^8 more years, far shorter than the age of the solar system, 4.5×10^9 years. Second, the bombardment of the rings by interplanetary dust should erode, darken and drag the ring particles inward, especially in the C ring, in about 10^8 years. If the rings are billions of years old, these short timescales are puzzling; advocates of recent origin suggest instead that the rings were created within the last 10^8 years.

Unfortunately, no one has found a plausible way to make Saturn's rings within the recent past. While the ring systems of Jupiter, Uranus and Neptune probably form by erosion or disruption of inner moons (Burns, Showalter and Morfill, 1984; Esposito, 1993), Saturn's ring system is more massive than all the satellites interior to Mimas combined (Table S2), and is at least 1000 times more massive than any other ring system. Even if a satellite as massive as the rings had existed near the planet, the chance that a cometary impact would have disrupted the satellite within the last 10^8 years is only about 1%, based on scaled crater counts (Lissauer, Squyres and Hartmann, 1988) and estimates of the population of comets at Saturn. In an alternative possibility (Dones, 1991), the rings might originate by the tidal disruption of a large short-period comet during a chance passage far within Saturn's Roche limit, followed by collisions between the fragments. However, in most such encounters, little debris remains bound to the planet, making this scenario equally unlikely. Further, the ice in Saturn's rings appears far purer than that in comets.

The upcoming Cassini mission to Saturn should be able to test at least one prediction of recent origin scenarios. Over the 4 year lifetime of the mission, the angular momentum transferred to Saturn's inner moons by spiral waves should make them gradually spiral out from Saturn, causing them to trail further and further behind the positions in their orbits they would occupy in the absence of interactions with the rings. For Prometheus and Atlas the lag will be 20–30 km, easily measurable in Cassini images. In the 1995 ring-plane crossing observations, Prometheus lagged its predicted position by 19°. This lag may be caused by interactions with an unseen, small co-orbital satellite or collisions with the F Ring. This surprising finding must be understood before the lag expected from ring torques can be measured reliably by Cassini. Only time will tell whether Saturn's rings are but a bright flame that flickers for an instant.

Luke Dones

Bibliography

Alexander, A.F.O'D. (1962) *The Planet Saturn: A History of Observation, Theory, and Discovery*. New York: Dover.
Araki, S. (1991a) Dynamics of planetary rings. *Am. Sci.*, **79**, 44–59.
Araki, S. (1991b) The dynamics of particle disks. III. Dense and spinning particle disks. *Icarus*, **90**, 139–71.
Borderies, N. (1989) Ring dynamics. *Celest. Mech. Dynam. Astron.*, **46**, 207–30.
Borderies, N., Goldreich, P. and Tremaine, S. (1984) Unsolved problems in planetary ring dynamics, in *Planetary Rings* (eds R. Greenberg and A. Brahic). Tucson: University of Arizona Press, pp. 713–34.
Brahic, A. (1977) Systems of colliding bodies in a gravitational field. I. – Numerical simulation of the standard model. *Astron. Astrophys.*, **54**, 895–907.
Burns, J.A. (1990) Planetary rings, in *The New Solar System*, 3rd edn (eds J.K. Beatty and A. Chaikin). Cambridge University Press, pp. 153–70.
Burns, J.A., Showalter, M.R. and Morfill, G.E. (1984) The ethereal rings of Jupiter and Saturn, in *Planetary Rings*, (eds R. Greenberg and A. Brahic). Tucson: University of Arizona Press, pp. 200–272.
Cuzzi, J.N. and Burns, J.A. (1988) Charged particle depletion surrounding Saturn's F ring: evidence for a moonlet belt? *Icarus*, **74**, 284–324.
Cuzzi, J.N. and Durisen, R.H. (1990) Bombardment of planetary rings by meteoroids: general formulation and effects of Oort cloud projectiles. *Icarus*, **84**, 467–501.
Cuzzi, J.N. and Scargle, J.D. (1985) Wavy edges suggest moonlet in Encke's gap. *Astrophys. J.*, **292**, 276–90.
Cuzzi, J.N., Lissauer, J.J., Esposito, L.W. *et al.* (1984) Saturn's rings: properties and processes, in *Planetary Rings* (eds R. Greenberg and A. Brahic). Tucson: University of Arizona Press, pp. 73–199.
Dones, L. (1991) A recent cometary origin for Saturn's rings? *Icarus*, **92**, 194–203.
Dones, L., Cuzzi, J.N. and Showalter, M.R. (1993) Voyager photometry of Saturn's A ring. *Icarus*, **105**, 184–215.
Doyle, L.R., Dones, L. and Cuzzi, J.N. (1989) Radiative transfer modeling of Saturn's outer B ring. *Icarus*, **80**, 104–35.
Esposito, L.W. (1986) Structure and evolution of Saturn's rings. *Icarus*, **67**, 345–57.
Esposito, L.W. (1993) Planetary rings. *Ann. Rev. Earth Planet. Sci.*, **21**, 487–523.
Esposito, L.W., Cuzzi, J.N., Holberg, J.B., *et al.* (1984) Saturn's rings: structure, dynamics, and particle properties, in *Saturn* (eds T. Gehrels and M.S. Matthews). Tucson: University of Arizona Press, pp. 463–45.
Estrada, P.R. and Cuzzi, J.N. (1996) Voyager observations of the color of Saturn's rings, *Icarus*, **122**, 251–72.
Goertz, C. and Morfill, G. (1983) A model for the formation of spokes in Saturn's ring. *Icarus*, **53**, 219–29.
Goldreich, P. and Tremaine, S. (1982) The dynamics of planetary rings. *Ann. Rev. Astron. Astrophys.*, **20**, 249–83.
Hamilton, D.P. and Burns, J.A. (1994) Saturn's E ring: self sustained, naturally. *Science*, **264**, 550–3.
Harris, A. (1984) The origin and evolution of planetary rings, in *Planetary Rings* (eds R. Greenberg and A. Brahic). Tucson: University of Arizona Press, pp. 641–59.
Horanyi, M., Burns, J.A. and Hamilton, D.P. (1992) The dynamics of Saturn's E ring particles. *Icarus*, **97**, 248–59.
Lissauer, J.J. (1989) Spiral waves in Saturn's rings, in *Dynamics of Astrophysical Discs* (ed. J.A. Sellwood). Cambridge University Press, pp. 1–16.
Lissauer, J.J., Shu, F.H. and Cuzzi, J.N. (1981) Moonlets in Saturn's rings? *Nature*, **292**, 707–11.
Lissauer, J.J., Squyres, S.W. and Hartmann, W.K. (1988) Bombardment history of the Saturn system. *J. Geophys. Res.*, **93**, 13776–804.
Longaretti, P.-Y. (1989) Saturn's main ring particle size distribution: an analytic approach. *Icarus*, **81**, 51–73.
Lynden-Bell, D. and Pringle, J.E. (1974) The evolution of viscous discs and the origin of the nebular variables. *Mon. Not. Roy. Astron. Soc.*, **168**, 603–37.
Marouf, E.A., Tyler, G.L., Zebker, H.A. *et al.* (1983) Particle size distributions in Saturn's rings from Voyager 1 radio occultation. *Icarus*, **54**, 189–211.
Marouf, E.A., Tyler, G.L. and Rosen, P.A. (1986) Profiling Saturn's rings by radio occultation. *Icarus*, **68**, 120–66.
Murray, C.D. and Giuliatti Winter, S.M. (1996) Periodic collisions between the Moon Prometheus and Saturn's F Ring, *Nature*, **380**, 139–71.
Nicholson, P.D. and Dones, L. (1991) Planetary rings. *Reviews of Geophysics Supplement*, US National Report to IUGG 1987–1990, pp. 313–27.
Nicholson, P.D., Hamilton, D.P., Matthews, K. and Yoder, C.F. (1992) New observations of Saturn's coorbital satellites. *Icarus*, **100**, 464–84.
Nicholson, P.D. *et al.* (1996) HST observations of Saturn's ring plane crossings in August and November 1995, *Science*, **272**, 509–15.
Ohtsuki, K. (1993) Capture probability of colliding planetesimals: dynamical constraints on accretion of planets, satellites, and ring particles. *Icarus*, **106**, 228–46.
Pollack, J.B. (1975) The rings of Saturn. *Space Sci. Rev.*, **18**, 3–93.
Porco, C.C. (1990) Narrow rings: observation and theory. *Adv. Space Res.*, **10**(1), 221–9.
Rosen, P.A., Tyler, G.L., Marouf, E.A. and Lissauer, J.J. (1991) Resonance structures in Saturn's rings probed by radio occultation II: results and interpretation. *Icarus*, **93**, 25–44.
Salo, H. (1992a) Gravitational wakes in Saturn's rings. *Nature*, **359**, 619–21.
Salo, H. (1992b) Numerical simulations of dense collisional systems. II. Extended distribution of particle sizes. *Icarus*, **96**, 85–106.
Showalter, M.R. (1991) Visual detection of 1981S13, Saturn's eighteenth satellite, and its role in the Encke gap. *Nature*, **351**, 709–13.
Showalter, M.R., and Cuzzi, J.N. (1993) Seeing ghosts: photometry of Saturn's G ring. *Icarus*, **103**, 124–43.
Showalter, M.R. and Nicholson, P.D. (1990) Saturn's rings through a

microscope: particle size constraints from the Voyager PPS scan. *Icarus*, **87**, 285–306.

Showalter, M.R., Cuzzi, J.N. and Larson, S.M. (1991) Structure and particle properties of Saturn's E ring. *Icarus*, **94**, 451–73.

Showalter, M.R., Cuzzi, J.N., Marouf, E.A. and Esposito, L.W. (1986) Satellite wakes and the orbit of the Encke gap moonlet. *Icarus*, **66**, 297–323.

Showalter, M.R., Pollack, J.B., Ockert, M.E. *et al.* (1992) A photometric study of Saturn's F ring. *Icarus*, **100**, 394–411.

Sridhar, S. and Tremaine, S. (1992) Tidal disruption of viscous bodies. *Icarus*, **95**, 86–99.

Tagger, M., Henriksen, R.N. and Pellat, R. (1991) On the nature of the spokes in Saturn's rings. *Icarus*, **91**, 297–314.

Weidenschilling, S.J., Chapman, C.R., Davis, D.R. and Greenberg, R. (1984) Ring particles: collisional interactions and physical nature, in *Planetary Rings* (eds R. Greenberg and A. Brahic). University of Arizona Press, Tucson, pp. 367–415.

Wisdom, J. and Tremaine, S. (1988) Local simulations of planetary rings. *Astron. J.*, **95**, 925–40.

Acknowledgements

I thank Doug Hamilton and Jack Lissauer for comments on the manuscript, and Linda Horn for providing her manuscript in advance of publication.

Cross references

Imaging science
Jupiter: ring system
Neptune: ring system
Planetary ring
Radio science
Shepherd satellite
Uranus: ring system
Voyager missions

SATURN: SATELLITE SYSTEM

In addition to having the most extensive ring system of any planet, Saturn has the most known satellites (18). These include the only moon with a significant atmosphere, one in retrograde orbit, satellites that share orbits, a pair that exchanges orbits, ring-associated moons, chaotic rotation, possible active water volcanoes and an enigmatic hemispherical albedo variation of a factor of five. In this summary the satellites are divided into four classes: Titan, intermediate-sized, small and retrograde.

Discovery of the Saturn satellites

Christian Huygens recognized Titan as a satellite of Saturn in 1655. The other satellites are considerably more difficult to detect, but Cassini found Iapetus, Rhea, Tethys and Dione in the early 1670s, and correctly inferred the drastic albedo differences of Iapetus' leading and trailing sides. William Herschel detected Enceladus and Mimas in 1789. Lassell and Bond independently discovered Hyperion in 1848. W.H. Pickering discovered Phoebe in 1899 using photographic plates. The existence of the much smaller satellites Janus, Epimetheus, Helene, Telesto and Calypso were confirmed by many observers in 1980. Voyager images revealed Prometheus, Pandora and Atlas in late 1980. Pan was detected in Voyager images several years after its existence had been predicted on the basis of ring perturbations.

The satellites are named after deities associated with the god Saturn; this practice began in the mid-19th century only after numerical designations of discovery order and distance from Saturn had caused much confusion.

Observing the satellites of Saturn

The orbits of most of the satellites can be determined well from Earth-based observations. Mutual gravitational perturbations of the orbits allow the calculation of the masses of some of the satellites. Surface features and precise sizes of the satellites are derived from Voyager images taken in 1980 and 1981. Tracking of the Voyager spacecraft helped refine the values of masses of some of the satellites. Earth-based spectroscopy is still the best source of data on surface compositions of the satellites. Radar has been used to characterize part of the surface of Titan.

Table S3 Satellites of Saturn

Name	Mean orbital radius (10^3 km)	Period (d)	Mean body radius (km)	Mean density (g cm^{-3})
Pan	133.58	0.575	10	
Atlas	137.64	0.602	15	
Prometheus	139.35	0.613	51	
Pandora	141.70	0.629	43	
Epimetheus	151.42	0.694	59.3	0.7 ± 0.2
Janus	151.47	0.694	89.5	0.7 ± 0.2
Mimas	185.52	0.942	198.8	1.14 ± 0.03
Enceladus	238.02	1.370	249.4	1.01 ± 0.02
Tethys	294.66	1.888	530	1.00 ± 0.02
Telesto	294.66	1.888	11	
Calypso	294.66	1.888	12	
Dione	377.40	2.737	560	1.44 ± 0.07
Helene	377.40	2.737	16	
Rhea	427.04	4.518	764	1.33 ± 0.10
Titan	1221.85	15.945	2575	1.88 ± 0.01
Hyperion	1481.1	21.277	141	
Iapetus	3561.3	79.331	718	1.21 ± 0.12
Phoebe	12952	550.48R[a]	110	

[a] R = retrograde orbit.

Titan

Because of its size and unique characteristic of having a dense atmosphere, Titan is dealt with separately (see Titan); here its relationship to the whole satellite system is very briefly reviewed. As the largest of the Saturn satellites, it contains over 95% of the mass of the satellite system. Its surficial geology is unknown because the aerosol hazes prevent optical imaging of the surface. The detection of nitrogen and carbon species in its atmosphere is an important chemical benchmark for materials that may have been included in other satellites of the system, but are more difficult to detect in solid surfaces.

Intermediate-sized icy satellites

This group includes Mimas, Enceladus, Tethys, Dione, Rhea and Iapetus; they range from 200 to 764 km in mean radius, and from a density of 1.0 to 1.4 g cm^3 (Table S3). The sizes increase outward to Titan, a pattern similar to that of Jupiter's and Uranus' satellites. The inner three, Mimas, Enceladus and Tethys, are less dense (Table S3) than the outer satellites, indicating a lower ratio of rock/ice in the inner satellites. With the exception of the leading side of Iapetus, all these satellites have albedos above 50%, and Enceladus' albedo is close to unity. These values indicate surfaces covered with fairly clean ice and, in the case of Enceladus, a distinctive texture as well.

Mimas

Mimas is heavily cratered and shows little indication of internal activity other than some linear fractures or fault traces. It is the smallest satellite observed that has viscously relaxed into an ellipsoidal form. Its relaxed shape suggests that it may have suffered some tidal heating because radiogenic heating of such small bodies with low fractions of rock has probably been inadequate to raise the internal temperature sufficiently. The difference in its long and short axes also suggests that Mimas has a modest central mass concentration, perhaps due to differentiation or to very deep porous surface layers.

Enceladus

Enceladus (q.v.) is distinguished by substantial resurfacing, extremely high albedo and association with the E ring (Plate 29). Parts of the surface are heavily cratered, but these areas are cut by regions of troughs and ridges, which are usually interpreted as showing both resurfacing and global expansion caused by freezing of the interior. Because Enceladus is small and has a small component of silicates, some form of tidal heating may have been necessary to allow melting and eruption of material. This process would be considerably aided by inclusion of NH_3 in the ice mantle, as it reduces the melting point, and would allow the formation of a fluid in the mantle of low enough density to be forced to the surface through a water-ice crust.

The very high albedo (geometric albedo of 1.04) for Enceladus indicates an extremely clean surface, and a texture that encourages significant backscatter of light, an unusual property for a high-albedo object. In combination with the observation that Enceladus is embedded in the E ring, which is apparently composed of submicron particles of water ice, there is the strong suggestion that Enceladus is still somewhat active (geysers?) and is the source of the E ring.

Tethys

Tethys has a density similar to that of Enceladus, and thus has a very small fraction of rock. The major surface feature is Ithaca Chasma, a trough 2000 km long, up to 5 km deep and 100 km wide, that may be associated with the formation of a 400-km impact basin, Odysseus. There are areas of distinctly different ages on the satellite, indicating periods of partial resurfacing. This object is large enough that internal heating could be effective. The resurfacing that is seen is very old, as estimated by the density of impact craters, though some postdates the major rifts.

Dione

Dione has the greatest density of the intermediate-sized satellites, a variety of surface morphologic features and the most prominent albedo markings of the satellites except for the hemispheric dichotomy on Iapetus. The surface exhibits regions of great variation in crater density, which indicate substantial periods of internal activity and regional resurfacing. Structural features such as chasma are associated largely with the younger plains units, and may result from water-ice volcanic processes that obliterated older features. Bright albedo features, radiating from an area on the trailing hemisphere, suggest some form of deposition of erupted or ejected water ice. The markings are of unknown thickness, and most cannot be directly associated with morphologic structures.

Rhea

The second-largest satellite of Saturn is heavily cratered but shows some provinces of different age and structure. The craters themselves appear to have suffered less viscous relaxation than many on other icy satellites. Although some structural features, such as troughs, grooves and pit chains, are present by far the dominant surface features are craters, including three multiring basins up to 700 km across their outer scarps.

Iapetus

Iapetus is most noted for its hemispherical albedo contrast: the trailing side has an albedo of about 0.5, much of the leading hemisphere has an albedo of about 0.05. Iapetus is heavily cratered, but the Voyager images of the dark areas are of such low resolution that crater densities on that hemisphere are unknown.

Debate on the origin of the dichotomy has concentrated on internal or external origin. Because of the high correlation of the low albedo with the apex of motion, there is considerable favor for theories that derive the marking by impact of material. It might be directly derived from dark infalling material, or indirectly by having the greater impact rate expose a dark component of Iapetus' crust. The satellite Phoebe, in a more distant and retrograde orbit, has been suggested as a possible source of infalling dark material. The dark area of Iapetus has spectral characteristics fairly similar to other dark, outer solar system materials, but differs in detail.

Small, irregularly shaped satellites

Ten Saturn satellites fall in the category of small and irregularly shaped. They occur in several different dynamical environments. Six are associated with the rings: Pan, Atlas, Pandora, Prometheus, Janus and Epimetheus. Pan disturbs particles in the A ring to form the Encke division. Pandora and Prometheus shepherd the F ring; Atlas shepherds the outer margin of the A ring. Janus and Epimetheus are unusual in that they effectively share an orbit. Periodically they pass one another and exchange leading and trailing positions, as well as slightly changing orbital periods. This gravitational interaction has allowed their masses to be inferred from ground-based observations, and thus their densities, which are both about 0.7 g cm^{-3}. This value suggests that they are composed of porous water ice with little included rock, which is consistent with the ring composition and that of the inner intermediate-sized satellites.

Two satellites orbit in Lagrangian positions ahead and behind of Tethys: Telesto and Calypso. Although, like the other small objects, they appear to be the result of collisional fragmentation, it has not been resolved as to how they came to be trapped in the Lagrange positions. Dione also has one Lagrange satellite, Helene.

Of the small satellites Hyperion is perhaps the most mysterious, largely because of its apparently chaotic rotation. The chaotic rotation results from torques applied by near resonance with Titan, the object's elongate shape and its orbital eccentricity. Although the rotation overall appears to change and be chaotic, it can have a slowly changing rotation rate for many observation periods, thus true detection of the chaotic rotation is difficult. However, it is definitely not in synchronous rotation. The satellite also appears to have suffered a massive spallation of 10–20 km of material from a significant fraction of its surface, as evidenced by a long series of linked arcuate scarps, and fractures extending many tens of kilometers in the exposed area near the scarps. It has been suggested that the resonance with Titan may prevent Hyperion from sweeping up debris ejected by collisions, and thus the surface may be fundamentally different, and the overall shape more rugged, than many of the other small satellites.

Phoebe

Phoebe is in a distant, retrograde, inclined orbit that strongly suggests it is a captured object. It has a mean radius of about 110 km, is roughly equidimensional, but has topography of about 10% of the diameter, and has an albedo of about 0.06, with variations up to 50% of this value. Its color is less red than that of Hyperion, which is in turn less red than the dark material on Iapetus. This color difference suggests that if material ejected from Phoebe is responsible for the albedo dichotomy on Iapetus, it is in some way altered during the impact process. Because of the low resolution of the Voyager images, no surface features other than the albedo markings (50 km or larger) are seen. It rotates every 9.4 h, as it is too far from Saturn for tidal locking into synchronous rotation even on geologic timescales.

Origins

The regular satellites of Saturn probably formed from a circumplanetary nebula, though it is not clear how much of the system may have migrated inward and broken up before final accretion of the planet. Much of the mass of these satellites is water ice, but the presence of nitrogen and methane on Titan implies that at least the outer parts of the system may have ammonia or methane ice or clathrates as well. Regardless of the kinds of ices, rock volume fractions in all the satellites are below 50%.

The satellites smaller than Mimas (200 km radius) have probably been subject to catastrophic or nearly catastrophic collisions. The collision rate increases rapidly toward Saturn because of the greater velocity of infalling materials and geometric focusing resulting in a higher number flux of particles. Thus the objects near the ring system are not thought to be of their original accreted sizes, though estimates of the reduction in size of the average population vary widely. The placement of the Lagrangian satellites in their positions remains controversial, although certainly related to large impacts on satellites.

Orbital history

The orbital history of the satellites of Saturn may not be fully determinable. Although a number of resonances exist among the satellites, it is likely that these developed only after expansion of the initial orbits under tidal torques. Resonances have probably been formed and disrupted, in particular the resurfacing of Enceladus suggests tidal heating in a resonance (with high eccentricity) not now present. The history of orbits is also important in restricting the age of the Saturnian rings and the processes (collisions and reaccretions) that may reform the rings, which appear to have lifetimes short compared to the age of the solar system.

Figure S7 This family portrait shows the smaller satellites of Saturn as viewed by Voyager 2 during its sweep through the Saturn System. The following chart corresponds to this composite photograph (distance from the planet increases to the right) and lists names and approximate mean radii (in km).

	Pandora	Janus	Calypso	
	43	90	12	
Atlas				Helene
15				16
	Prometheus	Epimetheus	Telesto	
	51	59	11	

These images have been scaled to show the satellites in true relative sizes. These small objects range in size from those of small asteroids to nearly that of Saturn's moon Mimas. They are probably fragments of somewhat larger bodies broken up during the bombardment period that followed accretion of the Saturnian system. They may be mostly icy bodies with an admixture of silicates or other materials (P-24061 BW).

Summary

The satellites of Saturn include the only satellite with a substantial atmosphere, a variety of surface geologic features and a full array of orbit types and interactions. The Cassini mission is scheduled to conduct a 4 year orbital survey of the system in the next decade, and should revolutionize our somewhat crude views of the geology and dynamics of this system.

Peter C. Thomas

Bibliography

Greenberg, R. (1984) Orbital resonances among Saturn's satellites, in *Saturn* (eds T. Gehrels and M. Matthews) Tucson: University, of Arizona Press, pp. 593–608.

Morrison, D., Johnson, T.V., Shoemaker, E.M., et al. (1984) Satellites of Saturn: geological perspectives, in *Saturn*, eds. T. Gehrels and M. Matthews). University of Arizona Press, pp. 609–39.

Morrison, D., Owen T., Soderblom L.A. (1986). The satellites of Saturn, in *Satellites*, eds. (J.A. Burns and M. Matthews). Tucson: University, of Arizona Press, pp. 764–801.

Soderblom, L. and Johnson T.V., (1982) The moons of Saturn. *Sci. Am.*, **246**, 100–116.

Van Helden, A., (1984). Saturn through the telescope: a brief historical survey, in *Saturn* (eds T. Gehrels and M. Matthews Tucson: University, of Arizona Press, pp. 23–46.

Wisdom, J. 1987. Chaotic dynamics in the solar system. *Icarus*, **72**, 241–75.

Cross references

Enceladus
Satellites, natural
Shepherd satellite
Small satellite
Titan

SCHIAPARELLI, GIOVANNI VIRGINIO (1835–1910)

Born at Savigliano, in Piedmont (Italy), Schiaparelli studied at Turin University and later in Berlin, being appointed to Milan (Brera Observatory) in 1860.

One of the first astronomers to study the surface features of planets, specifically Mars and Mercury, Schiaparelli believed he was able to determine the axial rotation of Mercury as approximately 88 days, the same as its orbital revolution. Although the surface of Venus is not visible at optical wavelengths, Schiaparelli observed shadows that suggested its rotation was 225 days and, like Mercury, the same as its orbital period. Thus both planets would have the same hemisphere facing the Sun, just as the Moon faces the Earth. Unfortunately neither was correct.

The concept that meteor swarms consisted of cometary debris was an outcome of Schiaparelli's observation, in 1866, that the comet 1862 III P/Swift-Tuttle possessed the same orbit as the Perseid meteors and that comet 1866 I corresponded to the Leonid meteors.

The detailed observation of physiographic features on planets was made possible by improved telescopes at the Brera Observatory, where Schiaparelli worked for most of his life. He served as a senator to the Kingdom of Italy and was awarded the gold medal of the Royal Astronomical Society in 1872. His description of 'canali' (which appeared then simply as black lines) on the surface of Mars stimulated some of the early science fiction in which they were interpreted as the irrigation canals of a mysterious extraterrestrial civilization. A map of the features and their names (mostly from Latin and Greek mythology), with latest updates, was published by the International Astronomical Union in 1955. After the NASA space mission Mariner 9, the question of the Martian canals was reviewed and examined by Sagan and Fox (1975). The general physiography of Mars is summarized in a reprint volume edited by Gornitz (1979).

Rhodes W. Fairbridge

Bibliography

Abetti, G. (1975) Schiaparelli, Giovanni Virginio. *Dict. Sci. Biogr.*, Vol. 12, 159–62.
Gornitz, V. (ed.) (1979) *Geology of the Planet Mars* (Benchmark Vol. 48). Stroudsburg; Dowden, Hutchinson and Ross, 414 pp.
Sagan, C. and Fox, P. (1975) The canals of Mars. *Icarus*, **25**, 602–12.
Schiaparelli, G. (1893) *The Planet Mars. Natura ed Arte*. Milano: Casa Edit. F. Vallarde (transl. E.C. Pickering, 1894).

SEISMICITY

The term seismicity, from the Greek *seismos* ('earthquake'), is employed to characterize the level of natural earthquake activity in a planetary body or satellite, or in a particular region of such a body, in terms of frequency and magnitude or other measures. It may refer to and include a wide range of effects and phenomena linked with seismic activity. When Gutenberg and Richter published their monumental catalog of worldwide earthquakes in 1954, including extensive descriptions, maps and statistical analyses, the volume was titled '*Seismicity of the Earth.*'

Although information concerning the seismicity of other solar system bodies is limited, it is nevertheless possible to make comparisons and draw certain inferences. The seismicity of the Moon was monitored for about 7 years by means of instruments deployed by the Apollo astronauts, and a seismometer was operated on the surface of Mars during the Viking missions (see Apollo missions; Moon: seismicity; Viking mission). Two Venera mission landers carried seismometers, which operated for a short time on the surface of Venus (see Venera missions). In subsequent sections we will review some of the results from these experiments.

In order to say that the seismicity of a planet or region is high or low one must first measure the size, frequency of occurrence and energy of seismic events. Thus a quantitative perspective is implied in the term. The distribution of seismic events in space and time is likewise essential in descriptions of the seismicity of a body; earthquakes are not uniformly distributed over the Earth, and moonquakes occur only in certain regions of the Moon. This article provides a brief introduction to what is known of the seismicity of the Earth, Moon, Venus and Mars. A later section reviews a few studies of the distribution of important seismic events in time.

Seismicity of the Earth

A global plot of earthquake epicenters reveals a number of important features of terrestrial seismicity (plots of this sort are included with the articles Plate tectonics, and Earth: geology, tectonics and seismicity). Earthquakes cluster along the boundaries of the tectonic plates of the crust of the Earth. This was one of the observations that led to the acceptance of the theory of plate tectonics, which has revolutionized our understanding of the geological evolution of the Earth (see Plate tectonics). Earthquakes accompany the subduction of great slabs of oceanic crust, for instance along the Aleutian Islands and beneath Kamchatka and Japan. More diffuse seismicity is seen in regions of continental collision, such as the Himalayan belt. Linear belts of epicenters mark the positions of mid-ocean ridges, where new oceanic crust is formed by eruption of molten rock.

Preparation of global maps of earthquakes requires the ability to resolve the locations and sizes of earthquakes, based on instrumental recordings of ground motions at distant sites. The science of seismology focuses on the description of the seismicity and internal structure of the Earth through the analysis of recorded earthquake waves. An overview of this discipline is beyond the scope of this article. Richter (1958) discusses the basic elements and the history of seismology, and the encyclopedia by James (1991) includes a number of articles that survey the field.

Sizes of earthquakes

Instrumental measures of the sizes of seismic events are important for evaluating seismicity. Charles F. Richter (1935) developed a magnitude scale for earthquakes in southern California. This measure of earthquake size is obtained by measuring the maximum amplitude of the trace generated by the earthquake on a standard seismograph, taking the logarithm of the amplitude and adding a correction for the distance of the event (Richter, 1958, p. 342).

The Richter scale was embraced by other investigators and by the news media. In subsequent decades a number of other scales were introduced, including the surface wave magnitude M_s, the body wave magnitude m_b and the moment magnitude M_w. The latter is much superior to the Richter scale because the wave measured for the Richter magnitude saturates (reaches a limit) for very large earthquakes (Kanamori, 1977, 1978). One great earthquake might be twice as large as another, but be assigned the same (or a lesser) Richter magnitude. Today the reported magnitudes of great earthquakes are usually M_w, though the press continues to use the term 'Richter magnitude.' The seismic moment, the basis of the moment magnitude M_w, is the product of three quantities: the area of the fault plane that experiences slippage, the amount of the slippage and the rigidity or strength of the material surrounding the fault (Kanamori, 1977).

The largest earthquake recorded on modern instruments occurred on 22 May 1960 in Chile. In this event a 1000 km long segment of ocean floor was subducted beneath the South American continent. The width of the fault zone that slipped was at least 60 km (in an east–west direction), and the amount of slip determined was a remarkable 20 m (Plafker and Savage, 1970). The moment magnitude of this earthquake is 9.5.

Frequency and magnitude

Small earthquakes occur much more frequently than great earthquakes. Gutenberg and Richter introduced a power law relation to describe the rate of occurrence of earthquakes of a particular magnitude. If N is the rate of occurrence of earthquakes and M is the magnitude of interest, then

$$\log_{10} N = a + b\,(8-M)$$

serves to describe the rate of occurrence of events for a particular region or for the globe (Gutenberg and Richter, 1954). Here a and b are constants determined by least-squares solutions; a describes the level of seismic activity of a region, and b is the slope of the graph. The value of b most often lies in the range from 0.8 to 1.2. Gutenberg and Richter found that for shallow earthquakes, a decrease of one unit in magnitude corresponds to an approximately eightfold increase in frequency. The nature of the frequency–magnitude relationship is under continuing study (e.g. Pacheco, Scholz and Sykes, 1992).

While there are many more small earthquakes than large ones, the global seismic energy release of the Earth is dominated by the largest events. A single great earthquake may release 10–100 times more energy than is represented in the sum of all smaller events of a region. Thus the hypothesis that many small events may release enough energy to reduce the hazard of larger events is not supported by the evidence. Figure S8 shows the level of worldwide seismic energy

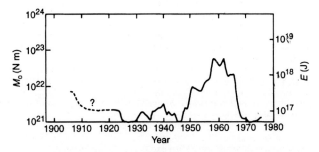

Figure S8 Worldwide seismic energy release 1906–1980. This curve (Shirley, 1986c, Gerlands Beitrage Zur Geophysik, **95**, p. 515; after Kanamori, 1977) gives the global annual average seismic energy release (unlagged 5-year running average) in terms of the seismic moment (left-hand scale) and the seismic energy (right-hand scale). The average value for the years 1950–1965 is nearly two orders of magnitude larger than that found for the years before or since.

release for the period from 1900 to 1980. The period from 1950 to 1965 was remarkable, with energy release nearly two orders of magnitude greater than the period before or since. This pulse of seismic energy was due to the occurrence of 13 great earthquakes during the period. Among these were seven of the eight largest earthquakes of this century (Kanamori, 1977).

Types of earthquakes

The earthquakes of planet Earth can be categorized in a number of ways. The depth of the event, the location of the event, the geometry of faulting in the event and the character of the seismogram of the event all serve to distinguish between varieties of earthquakes. Most earthquakes occur near plate boundaries (interplate earthquakes), but an important minority (intraplate earthquakes) occur within the interiors of the tectonic plates. These intraplate events may be disproportionately destructive, for two reasons. First, the relatively unfractured, coherent crust within plate interiors permits seismic waves to propagate large distances with little attenuation, thus producing large areas of strong shaking. Second, the rarity of these events may result in a lack of preparation for and experience with earthquakes on the part of the population affected; buildings of traditional construction may collapse. Important large intraplate earthquakes have occurred (1) recently, in Latur, India (Gupta, 1993), and (2) historically, in the central part of the United States, in 1811–1812 (Johnston and Kanter, 1990).

Earthquakes can be categorized on the basis of depth. Most earthquakes occur in the brittle upper crust of the Earth, at depths less than 50 km. Elevated temperatures below this level permit rock to flow in a ductile manner, without fracturing. Nevertheless a number of earthquakes are observed to occur at depths from 300 to more than 600 km. These deep-focus earthquakes are found in association with cool sinking slabs of ocean floor in subduction zones (see Plate tectonics). Fracture and frictional sliding cannot explain earthquakes at these depths, and mineral phase changes now represent the best model for their occurrence (Green, 1993; Kirby, Durham and Stern, 1991). On 9 June 1994 a M_w 8.3 event occurred at a depth of 630 km beneath Bolivia; this earthquake was felt as far away as Toronto, Canada (Kirby, 1994).

The geometry of crustal block motions in earthquakes also serves as a means of categorizing events. Areas of convergent plate motion, such as subduction zones and zones of continental collision, are characterized by thrust-type events, with one block or wedge of crust overriding another. This is an important mechanism of mountain building. Where the crust is pulled apart, so-called normal faulting occurs. Rifting and graben result from geometries involving extension (i.e tension); basins develop between bounding faults. A third geometric category is transform motion, where adjacent blocks of crust slide past one another with little vertical motion involved. Important examples are the great faults of New Zealand and California (Figure S9), which are characterized by predominantly strike-slip (horizontal) motion in earthquakes. Combinations of these pure types of block motions are also seen; they may be termed oblique, transtensional or transpressional in different cases. Modern seismological analysis can often distinguish between these types of earthquakes on the basis of remote observations alone.

At times many small earthquakes may occur in clusters in a limited region. 'Swarms' of earthquakes are often studied to gain an understanding of the stresses acting in the region. A particular type of clustering of seismic events is observed in association with active volcanism; this is termed 'volcanic tremor'. Seismology is an important tool in volcanology.

Tsunami earthquakes form yet another category of seismic events (Abe, 1979). These earthquakes occur with significant vertical displacements of the ocean bottom, and with relatively long event durations (up to 500 s). A set of events with still longer durations is termed 'slow earthquakes' (e.g. Beroza and Jordan, 1990). The existence of these events has been controversial, but it appears from very long-period seismic instrument records that on occasion the Earth moves in such a way as to excite significant long-period waves, without an accompanying short-period seismic pulse that may be recorded on conventional instruments.

One may also categorize some earthquakes on the basis of their spatial and temporal association with a larger nearby event, termed the mainshock. Smaller earthquakes that precede a mainshock are termed preshocks or foreshocks. Although it is not yet possible to recognize foreshocks as such before the mainshock, it is often

Figure S9 The San Andreas fault. This plate boundary transform fault experiences predominantly strike-slip (horizontal) motion. The location is in the Carrizo Plain, central California. (Photograph by R.E. Wallace (No. 194), courtesy of the US Geological Survey.)

valuable to investigate foreshocks as a means of studying the earthquake generation process. Following a large mainshock one typically observes a long series of shocks of lesser magnitude, presumably representing the process of crustal adjustment to the new state of stress created by the mainshock displacement. In some cases aftershock sequences can continue for decades.

Causes of earthquakes

Heat from within the Earth causes convective motions of the rocks of the mantle of the Earth (see Mantle convection). Coupling of the mantle flow to the lithosphere (q.v.) of the Earth may drive the motions of crustal plates (see Plate tectonics). Gravitational sinking of subducted slabs, 'ridge push' from extruded mid-ocean ridges, and other mechanisms play important roles in driving the motions of crustal plates. Earthquakes result from the relative motions of tectonic plates in this view.

Earthquakes result from stresses in the Earth. In the current view, stresses first accumulate and are then subsequently released as seismic energy in earthquakes when some threshold of stress is exceeded. Stresses within the Earth arise due to a number of causes in addition to the plate-driving forces. An important source of crustal stresses in some regions is due to the presence of elevated topography that is not compensated at depth, i.e. the crust is not in isostatic balance (see Isostasy). Although programs to identify and monitor stresses within the Earth are ongoing (Zoback, 1992), we still do not understand the root causes of earthquakes well enough to forecast the times of occurrence and probable sizes of future earthquakes.

Seismicity of the Moon

An array of sensitive seismometers was installed on the lunar surface by astronauts of the Apollo program in the 1970s. These operated for a period of about 7 years, recording both natural seismic events and meteoroid impacts. The article by Nakamura in this volume summarizes the results of the Apollo Lunar Seismic Experiment (see Moon: seismicity).

Two types of events were observed. Deep moonquakes are the most common form of lunar seismic activity; these occur at depths of about 900 km, roughly halfway to the center of the Moon. The deep moonquakes are small, having a maximum magnitude approximately equal to 1.3 on the Richter scale. They tend to recur at well-defined locations and their occurrence times show a correlation with tidal stresses (Lammlein, 1977).

The second category of moonquakes is the shallow, HFT (high-frequency teleseismic) events (Nakamura, 1977). Fewer than 30 of these events were recorded during the Lunar Seismic Experiment. However, the shallow moonquakes are much larger than the deep events, having maximum magnitudes equivalent to about 5.1 on the Richter scale. No clear tidal correlation is present in this sample of events.

The energy release in moonquakes is considerably less than is released by earthquakes. The annual average seismic energy release in the Moon was found to be 2×10^{10} J year^{-1} which may be compared with about 10^{18} J year^{-1} for the Earth (Goins, Dainty and Toksoz 1981). The total contribution from the deep moonquakes was less than 10^7 J year^{-1}; thus the small number of shallow moonquakes dominates the total seismic energy release for the Moon. The level of seismic activity observed was a surprise to investigators, since the Moon was considered for centuries to be a 'dead' world with a cold, thick, tectonically inactive shell.

The spatial distribution of moonquakes is not random. The vast majority are located on the side of the Moon facing the Earth. The deep moonquake foci show an association with the basalt-filled lunar maria; few were recorded in the southeastern quadrant of the visible portion of the Moon, which is largely composed of highlands (Figure 12 of Lammlein, 1977). The majority of shallow moonquake foci also seem to be associated with maria.

The Moon has a form elongated in the direction of the Earth; there is an Earth-side topographic bulge, and mass concentrations ('mascons,' q.v.) are present at relatively shallow depths beneath some of the maria. Thus it is plausible to associate lunar seismic activity with the slow subsidence of the floors of the maria, under the action of lunar gravity, as the Moon cools and contracts with time. However, it should be emphasized that the fundamental causes of lunar seismicity are not fully understood.

Seismicity of Mars

Mars is of a size intermediate between the Earth and the Moon, and should therefore have retained more internal heat than the Moon. Mars has been a geologically active world (see Mars: geology); in particular, gigantic shield volcanos such as Olympus Mons were erupted onto the crust of Mars during at least two episodes of tectonism (see Mars: structural geology). Loading by the elevated mass of large volcanoes stresses and deforms the lithosphere of the planet (Banerdt, Golombek and Tanaka, 1992).

The Viking missions to Mars included both orbiting spacecraft and landers. The latter were deployed on the surface of Mars in 1976. The two Viking landers included seismometers; however the seismometer on the first lander failed to deploy. Thus the available record of Mars seismicity comes from a few years of data recorded by a single instrument on the Viking 2 Lander (Anderson et al., 1977). A number of compromises were introduced in the design of the instrument; weight considerations precluded a design that could be installed on the Mars surface away from the lander. Instead the seismometer was mounted on the top of the lander equipment bay; following landing, it was found that one of the lander's feet was balanced on a rock.

The Viking seismometer failed to detect even minor seismic activity on Mars (Anderson et al., 1977). Due to the considerations noted above, and other factors, many investigators are not convinced that Mars is in fact a seismically inactive planet (see for instance Golombek et al., 1992). Calculations indicate that cooling of the Mars lithosphere should give rise to stresses sufficient to generate greater seismic activity than is seen on the Moon. Furthermore, measurements of fault offsets on the Mars surface, combined with reasonable estimates for the age of the faulting, suggest that faulting and associated marsquakes should be occurring today (Golombek et al., 1992). The extent and nature of faulting on Mars is discussed in the article Mars: structural geology.

The scientific importance of the question of the seismicity of Mars derives from the utility of seismology in describing the interior structure of a body. If Mars has moderate levels of natural seismic activity, seismometers placed on the planet will produce data enabling us to resolve the interior structure of the planet with an accuracy unobtainable by any other means.

Seismicity of Venus

The high surface pressure and temperature on Venus, along with the composition of the Venus atmosphere, results in an extremely hostile environment for seismometers or other instruments. However, the Russian Venera 13 and 14 Landers included vertical-axis seismometers designed to detect seismic activity on Venus (Moroz, 1983). The landers operated for only a short time (a few hours) on the surface of the planet. Two possible microseisms were recorded by the Venera 14 instrument (Ksanfomality et al., 1982).

The very high surface temperatures on Venus might be expected to restrict the possible locations and depths where brittle failure could occur. On the other hand, very little water is present in the surface rocks, a factor which in general elevates melting temperatures and reduces the ductility of crustal materials. We just do not know whether Venus is a highly seismic body or not. Magellan images reveal a silicate surface with a great variety of faults and other tectonic structures (see Venus: geology and geophysics; Magellan, mission). However, no faults with large strike-slip displacements (as on Earth) are seen on Venus (Solomon et al., 1991).

Comparative planetology

The review presented here provides a basis for comparisons between a few important solar system bodies. The Earth is the most seismic body for which data are available. We have seen that the seismicity is widespread; that there is a great diversity of types and forms of terrestrial seismic activity; and that a relatively small number of great earthquakes dominate the seismic energy release of the Earth. For the Moon, the largest events likewise dominate statistics of seismic energy release, but both the number and the diversity of seismic events is less, reflecting the lesser heat flow and thicker lithosphere.

From a planetary perspective, single earthquakes (or moonquakes) may appear relatively insignificant. The disruption of the surface (Figure S10) is generally fairly minor, affecting limited areas and involving small changes in elevations. Nonetheless it appears that single earthquakes may perturb the rotation of the Earth (Mansinha, Smylie and Chapman, 1979; Chao and Gross, 1987). There are indications that the occurrence times of moonquakes may bear a relationship to the rotation of the Moon; Lammlein (1977) notes that shallow moonquakes tend to occur near extremes of the lunar librations. Despite the many differences between moonquakes and earthquakes, a few investigators have presented comparative studies (Nakamura, 1980; Shirley, 1986a).

The cumulative effect of many earthquakes is overwhelmingly evident in the topographic relief and geographic distribution of the landmasses of the Earth, and is easily discerned from space. Plate tectonics however seems to be a uniquely terrestrial phenomenon; there is no evidence of past horizontal motions of crustal plates on the Moon or Mars, though a form of subduction may operate on Venus (Sandwell and Schubert, 1992).

Remote sensing of fault offsets is an important technique for estimating levels of seismicity on the Earth and on Mars. We are now recognizing that such estimates are best considered lower limits, because some important large earthquakes, such as the 1989 Loma Prieta (northern California) earthquake, may take place without generating detectable fault offsets at the surface. Thus the current estimates of the seismicity of Mars (Golombek et al., 1992) are probably conservative.

We presently have no direct evidence of endogenic seismic activity on solar system bodies other than the Earth, the Moon and Venus. However, this does not mean that seismic activity is not present elsewhere in the solar system. The conditions for generation of seismic energy include the presence of materials capable of brittle failure, and the presence of stresses within the body. It would not be too surprising to discover that natural seismic activity occurs on the moon Io, due to the intense geologic activity of that body. The planet Mercury has much in common with Earth's Moon, and seismic activity may be present.

'Impact seismology'

The phenomenon of impact cratering is not generally mentioned in the context of planetary seismicity, since the causes of the two

Figure S10 Surface fracturing due to earthquake-induced ground failure in the M_w 9.3 1964 Alaska earthquake. (Photograph copyright R. Samuels.)

phenomena are entirely distinct. Still, as an aside it is of interest to note the possible effects of impact-generated seismic waves within solar system bodies.

Impacts generate seismic waves within planetary-scale bodies; about 300 meteoroid impacts were detected per year during the Apollo Lunar Seismic Experiment. Although the frequency of large impacts has declined since the period of heavy bombardment, the ubiquity of craters on solar system bodies with solid surfaces indicates that all of these bodies have experienced fracturing and the generation and transmission of associated seismic energy.

A particularly remarkable effect of impact-generated seismic energy transmission has been suggested as an explanation for the formation of disrupted terrains antipodal to major impact sites (Schultz, 1972; Schultz and Gault, 1975). One likely example is found on Mercury, where the Caloris basin represents the site of a giant impact, and where the terrain on the opposite side of the planet shows a chaotic texture. Models suggest that a focusing of seismic energy may occur if the interior structure of the impacted object is favorable (Watts, Greeley and Melosh, 1991).

A necessary consequence of multiple large impacts on a planet or moon is the generation of systems of fractures in the outer regions of the body. If the body is later subjected to fluctuating stresses, some level of natural seismic activity might result even if the magnitudes of the stresses are insufficient to fracture the crystalline surface materials of the body.

Spatiotemporal patterns in lunar and terrestrial seismicity

Investigators have searched for evidence of order in the timing and spatial distribution of earthquakes since the first catalogs were published. The detection of non-random patterns in seismic events is important because the analysis and interpretation of the patterns may shed light on the fundamental physical mechanisms of seismic activity. Unfortunately most studies of global seismicity have failed to uncover evidence of significant periodicities or other non-random behavior (see references in Shirley, 1988).

On timescales of decades, the concepts of earthquake migration and the seismic gap provide valuable information regarding the most likely locations for future large earthquakes. Richter (1958) and Mogi (1968) described migrations of large earthquakes along fault zones, where later events in the sequence tended to occur progressively farther from the site of an initial event, and where the aftershock zones of the earthquakes tended not to overlap. Regions where no earthquake had occurred are now termed seismic gaps, and are considered to be likely locations for future events. Thatcher (1989) has recently discussed the relationships of earthquake occurrence and seismic gaps for the circum-Pacific region.

A number of studies reveal suggestions of pattern in the occurrence of very large events. Abe and Kanamori (1979) showed that the temporal variation of the number of large intermediate and deep-focus earthquakes closely follows that of large shallow shocks in the same years. This suggests that the two populations of very different events are somehow coupled on a global scale. The clustering of great earthquakes in the years 1950–1965, resulting in the curve of energy release of Figure S8, might be a result of a random statistical fluctuation or it might represent evidence of some planetary-scale physical mechanism influencing the distribution of great earthquakes in time.

Romanowicz (1993) has shown that the seismic energy release in great thrust and strike-slip earthquakes varies systematically in the period since 1920. Periods characterized by frequent large strike-slip earthquakes have few great subduction events, and vice versa. The periods during which one type of mechanism dominates over the other last about 20 years. This evidence once again indicates that the occurrence of the largest earthquakes is not entirely random in time.

A small number of studies report examples of non-random temporal ordering of larger earthquakes at periods of a year or less. Gutenberg and Richter (1954) looked for periodicities in their catalog

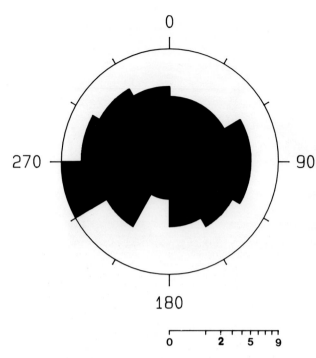

Figure S11 Distribution of solar hour angle (times of day) for 52 southern California earthquake mainshocks, $M \geq 6.0$, 1800–1994. In this plot local noon is at the top, 6 p.m. is at the right, and midnight is at the bottom (phase 180°). The area of each 30° sector is proportional to the number of earthquakes from the catalog that occurred during the corresponding 2-h period. For instance, the smallest count of events is for the sector following midnight (one event), while the largest is for the period preceding sunrise (240–270°, nine events). The majority of these earthquakes took place near 6 a.m. and 6 p.m. A random distribution of times would give a nearly circular pattern (Figure S12a).

of $M \geq 7.0$ earthquakes, and found a weak peak in the worldwide daily distribution corresponding to about 6 hours local time (i.e. about 6 a.m.). This dovetails with a similar indication for the southern California region; for many years it has been known that the largest earthquakes seem to take place near times of sunrise and sunset. Shirley (1988) performed a statistical evaluation of this tendency using the complete sample of $M \geq 6.0$ earthquakes for the region. The distribution was determined to be statistically non-random at the 5% significance level. Figure S11 shows an updated version of the distribution. Here the astronomical hour angle of the Sun is plotted, with noon at the top and midnight at the bottom of the diagram. The preference for times just before sunrise and around sunset is shown by the maxima seen at these times. No similar distribution is found for the distribution of lunar hour angles, as one might expect if the pattern is caused by tidal forcing.

As previously noted, the record of lunar seismic activity obtained during the Apollo Lunar Seismic Experiment includes a subset of shallow events that dominate the seismic energy release for the Moon. An analysis of the times of occurrence of these events revealed the presence of a temporal pattern with a very high statistical significance. Shirley (1986a) reported that the series of geocentric ecliptic longitude values for the times of the events shows clustering with a random probability of occurrence of 0.5%. (The geocentric ecliptic longitude is just the direction of the Moon, seen from the Earth, using the ecliptic coordinate system. Over short periods this system is approximately 'fixed in space,' over long periods precession must be considered.)

The investigation also considered the distribution of ecliptic longitudes for the times of occurrence of the largest shallow earthquakes of the same years. Remarkable similarities were noted between the two distributions, and the combined distribution of earthquakes and moonquakes shows a still higher level of statistical non-randomness. This study is the only evidence to date suggesting a direct relationship of large-magnitude lunar and terrestrial seismicity. Figure S12 provides a graphical synopsis of the results of that investigation.

Difficulties arise in the interpretation of these results. The tides are known to stress the bodies of the Earth and Moon, but the magnitude of the stresses is small, and further the parameter employed (the ecliptic longitude) is not correlated with tidal amplitude or phase in any consistent way. The cycle of the ecliptic longitude is the lunar sidereal month of 27.321 d, and seismic events do not occur like clockwork in every cycle.

In subsequent studies statistically non-random clustering of the ecliptic longitude was found for the catalog of large historic earthquakes in Alaska and the Aleutian Islands (Shirley, 1988), and in the series of great earthquakes making up the 1950–1965 pulse of worldwide seismic energy release (Shirley, 1986c; Figure S8). These studies show evidence of non-random temporal ordering of the largest earthquakes on time scales of a month or less.

Conclusion

Investigations of the seismicity of planetary-scale bodies can yield otherwise unobtainable information on the interior structure and characteristics of the body under investigation. Studies of the seismicity of the Earth, Moon and planets have only scratched the surface and much remains to be discovered.

James H. Shirley

Bibliography

Abe, K. (1979) Size of great earthquakes of 1837–1974 inferred from tsunami data. *J. Geophys. Res.*, **84**, 1561–8.

Abe, K. and Kanamori, H. (1979) Temporal variation of the activity of intermediate and deep-focus earthquakes. *J. Geophys. Res.*, **84**, 3589–95.

Anderson, D.L., Miller, W.F. Latham, G.V. *et al.* (1977) Seismology on Mars. *J. Geophys. Res.*, **82**, 4524–46.

Banerdt, W.B., Golombek, M.P. and Tanaka, K.L. (1992) Stress and tectonics on Mars, in *Mars* (eds H.H. Kieffer, B.M. Jakosky, C.W. Snyder and M.S. Matthews). Tucson: University of Arizona Press.

Beroza, G.C. and Jordan T.H. (1990) Searching for slow and silent earthquakes using free oscillations. *J. Geophys. Res.*, **95**, 2485–510.

Chao, B.F., and Gross, R.S. (1987) Changes in the Earth's rotation and low-degree gravitational field induced by earthquakes. *Geophys. J. Roy. Astron. Soc.*, **91**, 569–96.

Goins, N.R., Dainty, A.M. and Toksoz, M.N. (1981) Seismic energy release of the Moon. *J. Geophys. Res.*, **86**, 378–88.

Golombek, M.P., Banerdt, W.B. Tanaka, K.L. and Tralli, D.M. (1992) A prediction of Mars seismicity from surface faulting. *Science*, **258**, 979–1.

Green, H.W. (1993) The mechanism of deep earthquakes. *EOS*, (12; January), 23.

Gupta, H.K. (1993) The deadly Latur earthquake, *Science*, **262**, 1666–7.

Gutenberg, B. and Richter C.F. (1954) *Seismicity of the Earth*. Princeton: Princeton University Press.

James, D.E. (ed.) (1991) The Encyclopedia of Solid Earth Geophysics. New York: Van Nostrand Reinhold.

Johnston, A.C. and Kanter, L.R. (1990) Earthquakes in stable continental crust. *Sci. Am.*, **262**(3), 68–75.

Kanamori, H. (1977) The Energy release in great earthquakes. *J. Geophys. Res.*, **82**, 2981–87.

Kanamori, H. (1978) Quantification of earthquakes. *Nature*, **271**, 411–4.

Kirby, S. (1994) Bolivian earthquake ranks as largest deep quake on record. *EOS*, **75**, (26), 289.

Kirby, S.H., Durham, W.B. and Stern, L.A. (1991). Mantle phase changes and deep-earthquake faulting in subducting lithosphere. *Science*, **252**, 216–25.

Ksanfomality, L.V., Zubkova, V.M., Morozov, N.A. and Petrova, E.V. (1982) Microseisms in the landing sites of Venera 13 and Venera 14. *Pisma Astron. Zh.*, **8**, 444–7.

Lammlein, D.R. (1977) Lunar seismicity and tectonics. *Phys. Earth Planet. Int.*, **14**, 224–73.

Mansinha, L., Smylie, D.E. and Chapman C.H. (1979) Seismic excitation of the Chandler wobble revisited. *Geophys. J. Roy. Astron. Soc.*, **59**, 1–17.

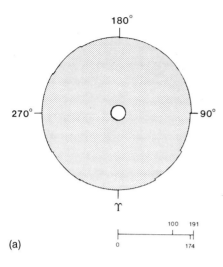

(a)

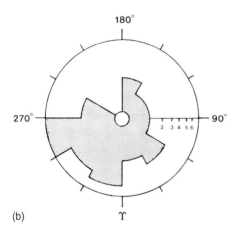

(b)

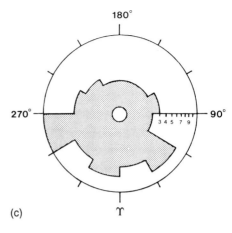

(c)

Figure S12 These three diagrams summarize the results of the investigation of large shallow moonquakes and shallow earthquakes by Shirley (1986a; *Earth Planet. Sci. Lett*, 76, p. 241). The parameter plotted is the ecliptic longitude of the Moon, which completes a counterclockwise revolution in the sideral lunar month of 27.321 days. The first, nearly uniform distribution (a) gives daily values of the Moon longitude for the 7 years of the study period. If the seismic events occur randomly in time, their distribution should be similar to that shown here. A strongly asymmetric distribution would indicate that the events may not be randomly distributed in time. The probability of occurrence for such distribution is estimated using Schuster's test (Shirley, 1986a, 1988). (b) Distribution of lunar ecliptic longitudes for the 27 shallow moonquakes recorded during the Apollo Lunar Seismic Experiment. The area of each sector is proportional to the number of events taking place while the Moon is passing through that part of the range of longitudes. A strong asymmetry is seen. Statistical testing shows that clustering of this sort will turn up about one time in every 200 trials with random data; thus the distribution is statistically significant at better than the 1% level. (c) Distribution of lunar ecliptic longitudes for the combined sample of 27 shallow moonquakes and the 27 largest shallow earthquakes of the same years (1971–1976). Addition of the earthquakes has strengthened the pattern shown for the moonquakes alone in (b). The random probability of occurrence for this distribution is 0.003. Pattern is thus evident in the distribution of the times of occurrence of the largest seismic events of the Earth–Moon system for the years 1971–1976.

Mogi, K. (1968) Migration of seismic activity. *Bull. Earthq. Res. Inst.*, **46**, 53–74.

Moroz, V.I. (1983) Summary of preliminary results of the Venera 13 and Venera 14 missions, in *Venus* (eds D.M., Hunten, L.Colin, T.M.Donahue and V.I. Moroz). Tucson: University of Arizona Press.

Nakamura, Y. (1977) HFT events: shallow moonquakes? *Phys. Earth. Planet. Int.*, **14**, 217–23.

Nakamura, Y. (1980) Shallow moonquakes: how they compare with earthquakes. *Proc. Lunar. Plan. Sci. Conf.* **11**, 1847–53.

Pacheco, J.F., Scholz, C.H. and Sykes, L.R. (1992) Changes in frequency–size relationship from small to large earthquakes. *Nature*, **355**, 71–3.

Plafker, G. and Savage J.C. (1970) Mechanism of the Chilean earthquakes of May 21 and 22, 1960. *Geol. Soc. Am. Bull.*, **81**, 1001–30.

Richter, C.F. (1935) An instrumental earthquake scale. *Bull. Seism. Soc. Am.*, **25**, 1–32.

Richter, C.F. (1958) *Elementary Seismology*. San Francisco: W.H.Freeman and Co.

Romanowicz, B. (1993) Spatiotemporal patterns in the energy release of great earthquakes. *Science*, **260**, 1923–6.

Sandwell, D.T. and Schubert, G. (1992) Evidence for retrograde subduction on Venus. *Science* **257**, 766–70.

Schultz, P.H. (1972) A preliminary morphological study of the lunar surface. PhD Thesis. Austin: University of Texas.

Schultz, P.H. and Gault, D.E. (1975) Seismic effects from major basin formation on the Moon and Mercury. *The Moon*, **12**, 159–77.

Shirley, J.H. (1986a) Shallow moonquakes and large shallow earthquakes: a temporal correlation. *Earth Plan. Sci. Lett*, **76**, 241–53.

Shirley, J.H. (1986b) Temporal patterns in historic major earthquakes in Chile, in *Investigating Natural Hazards in Latin American History* (ed. R.H.Claxton). *S. Ga. Coll. Stud. Soc. Sci.*, **25**. Carrollton, GA (USA): W. Georgia College.

Shirley, J.H. (1986c) Lunar periodicity in great earthquakes. *Gerl. Beitr. zur Geophys.*, **95**, 509–15.

Shirley, J.H. (1988) Lunar and solar periodicities of large earthquakes: southern California and the Alaska–Aleutian islands seismic region. *Geophys. J.*, **92**, 403–20.

Solomon, S.C., Head, J.W. Kaula, W.M. et al. (1991) Venus tectonics: initial analysis from Magellan. *Science*, **252**, 297–312.

Thatcher, W. (1989) Earthquake recurrence and risk assessment in circum-Pacific seismic gaps. *Nature*, **341**, 432–4.

Watts, A.W., Greeley, R. and Melosh, H.J. (1991) The formation of terrains antipodal to major impacts. *Icarus*, **93**; 159–68.

Zoback, M.L. (1992) First and second-order patterns of stress in the lithosphere: the world stress map project. *J.Geophys. Res.*, **97**, 11 703–28.

Cross references

Apollo missions
Earth: geology, tectonics and seismicity
Ecliptic
Fracture, fault
Impact cratering
Io
Isostasy
Libration
Lithosphere
Magellan missions

Mars: geology
Mars: structural geology and tectonics
Mercury: geology
Moon: seismicity
Plate tectonics
Subduction
Tide-raising forces
Venera missions
Venus: geology and geophysics

SHARONOV, VSEVOLOD VASILIEVICH (1901–1964)

Vsevolod Vasilievich Sharonov was born in St Petersburg, Russia, the son of an opera singer. Planning to study physics and mathematics at the university there, following the 1917 revolution, he was obliged to serve with the Red Army (1919–1924) but returned to graduate in 1926. He was particularly interested at first in solar research and sunspot behavior. He served for some time at the observatories in Leningrad, Pulkovo, Simeiz and Tashkent.

On the experimental side he was very skilled in developing instruments dealing with photometry, and studying both atmospheric optics and the colorimetry of both the Moon and planets. His first dissertation was on the photometric wedge (1929) and his senior doctoral thesis (1940) was on indexes of visibility. He determined the solar light constant as 135 000 lux. Observing seven successful total solar eclipses (1936–1963), he published work on the solar corona, and obtained photometric properties on more than 100 features on the Moon. With these observations he confirmed the 'meteor-slag theory' for the lunar soil (see Regolith) that had been proposed by his wife, N.N. Sytinskaya. He authored two important general works, on the Planets (1958) and Venus (1965).

Rhodes W. Fairbridge

Bibliography

Kulikovsky, P.G. (1975) Sharonov, V.V., *Dict. Sci. Biogr.*, Vol. 12 pp. 352–4.
Sharonov, V.V. (1964) *The Nature of the Planets*. Moscow: Gosud. Izdat. Fiz. Mat. Lit. (1958: *Priroda Planet*), transl. by Israel Progr. Sci. Trans., 420 pp.

SHEPHERD SATELLITE

The existence of shepherd satellites was first proposed in 1979 (Goldreich and Tremaine, 1979) to explain nine narrow, sharp-edged rings of Uranus that had been discovered from ground-based observations in 1977. According to the original theory of ring shepherding, a small satellite orbits on either side of a narrow ring and they act

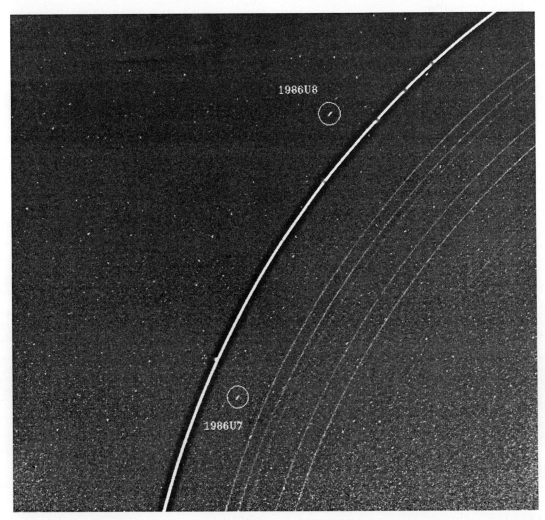

Figure S13 Shepherd satellites. Voyager 2 discovered two shepherd satellites associated with the rings of Uranus. The two moons, Cordelia (1986U7) and Ophelia (1986U8), orbit on either side of the bright ε ring. All nine of Uranus' rings are visible. (NASA P-29466, courtesy of JPL.)

together to prevent the ring from spreading in radius. The first apparent confirmation of the mechanism came in 1980 from Voyager 1 observations of two small (radius ~ 50 km) satellites, Pandora and Prometheus, orbiting on either side of the narrow F ring of Saturn. In 1986 Voyager 2 images of the rings of Uranus led to the discovery of two small (radius ~ 15 km) satellites, Ophelia and Cordelia, which appear to confine the ε ring of the planet (Figure S13).

Narrow rings had posed a significant problem for ring dynamicists since they should spread on timescales which are short ($< 10^7$ years) compared to the age of the solar system. Either the rings were relatively recent phenomena (which may still be a possibility for some rings) or some mechanism was acting to prevent the spreading caused by interparticle collisions and Poynting–Robertson light drag. A satellite orbiting close to a narrow, near-circular ring exerts a torque on the ring. These torques are particularly effective at resonant locations where the ratio of the orbital periods of the ring and the satellite is close to $(p + 1): p$ where p is an integer. As each ring particle encounters the satellite there is an exchange of angular momentum between the two objects which increases their radial separation. The net result is a process which appears to create a 'repulsion' between the ring and the satellite, although this is simply a consequence of Newtonian gravitation. If there is a satellite on each side of the ring then a confinement mechanism is produced; it is thought that the observed ring-shepherd systems have reached an equilibrium configuration. The clearest example of the shepherding mechanism in operation is the ε ring of Uranus where the period of the inner shepherd satellite Cordelia has a 24 : 25 resonance with the inner edge of the ring while the outer shepherd Ophelia has a 14 : 13 resonance with the outer edge (Porco and Goldreich, 1987).

Although the shepherding mechanism originally invoked two satellites, it is now believed that in some circumstances one satellite may be sufficient, at least for temporary confinement. This appears to be the case with the narrow Adams ring of Neptune, where the ring arcs imaged by the Voyager 2 spacecraft in 1989 are thought to be maintained by the shepherding action of a single satellite, Galatea, which orbits just interior to the ring (Porco, 1991). In this case the mechanism is more elaborate and two resonances between the ring and Galatea are involved, one of which prevents the arc from spreading along the ring's orbit.

Since a satellite effectively 'repels' nearby ring particles (with the possible exception of those which lie in the same orbit as the satellite), a small moon in a broad ring system would tend to create a gap and the satellite could be said to be shepherding the ring material on either side of its orbit. It has been suggested that some of the distinct gaps in Saturn's main ring system are due to the presence of small (radius < 10 km) satellites. One method of detecting such satellites is to look for evidence of waves at the gap edges or 'wakes' in the surrounding ring material. These are the signs of a recent passage of a satellite which has given each ring particle a small eccentricity e given by

$$e = 2.24 \left(\frac{m}{M}\right)\left(\frac{a}{\Delta a}\right)^2$$

where m and M are the masses of the satellite and planet respectively, a is the semimajor axis of the ring particle and Δa is the separation in semimajor axis between the particle and the satellite. Although each particle moves on its own independent Keplerian ellipse, the net effect is the formation of a leading or trailing wave of wavelength $3\pi\Delta a$ at the ring edge and a wake nearby. Particle collisions cause the eccentricity (and hence the wave) to dampen until the process begins again at the next encounter. The satellite Pan (named after the Greek god of shepherds) was discovered in 1990 orbiting in the Encke gap in Saturn's A ring (Showalter, 1991). The search for Pan was undertaken after an analysis of images near the gap showed the presence of waves and wakes thought to be associated with a nearby satellite.

While there is some indirect evidence for additional, undiscovered satellites in the ring system of Uranus (Murray and Thompson, 1990), shepherd moons have yet to be detected for most of the narrow rings of the outer planets. However, their existence cannot be ruled out on the basis of currently available images and occultation data.

Carl D. Murray

Bibliography

Goldreich, P. and Tremaine, S. (1979) Towards a theory for the Uranian rings. *Nature*, **277**, 97–9.

Murray, C.D. and Thompson, R.P. (1990) Orbits of shepherd satellites deduced from the structure of the rings of Uranus. *Nature*, **348**, 499–502.
Porco, C. (1991) An explanation for Neptune's ring arcs. *Science*, **253**, 995–1001.
Porco, C. and Goldreich, P. (1987) Shepherding of the Uranian rings. I. Kinematics. *Astron. J.*, **93**(3), 724–9.
Showalter, M. (1991) Visual detection of 1981S13, Saturn's eighteenth satellite, and its role in the Encke gap. *Nature*, **351**, 709–13.

Cross references

Neptune
Planetary ring
Small satellite
Saturn
Uranus

SHOCK METAMORPHISM

Permanent changes in the structure and nature of minerals and rocks due to the passage of a shock wave. As impact is the only known natural process resulting in a shock wave, shock metamorphic effects are considered diagnostic indicators of the occurrence of a hypervelocity impact event. A number of shock metamorphic effects have been duplicated in shock recovery experiments in the laboratory and in nuclear explosions. It is thus possible to derive an approximate calibration scheme of the shock pressures required to produce a particular feature. As the passage of shock wave is highly transient, over periods of microseconds to seconds, disequilibrium assemblages in shocked rocks are not uncommon. Shock metamorphic effects occur in allochthonous crater materials (breccias, melt rocks and ejecta) and in autochthonous crater materials, where they are limited to the central area of the crater floor.

Shock metamorphic effects are first manifested at shock pressures in the 2–5 GPa (20–50 kbar) range. At these relatively low shock pressures, shatter cones can form. Shatter cones are striated conical fractures and are the only known diagnostic shock feature that is megascopic in scale. Shatter cones can range in size from centimeters to meters and are best formed in fine-grained, structurally homogeneous rocks such as carbonates, quartzites, etc.

At shock pressures of 7.5–30 GPa (75–300 kbar) microscopic planar deformation features (PDF) are developed in quartz and feldspar. These closely spaced planes have specific crystallographic orientations, which are developed preferentially at different shock pressures. PDFs also develop in other minerals, e.g. olivine, pyroxene, apatite and zircon, at somewhat higher pressures but their orientations and pressure of occurrence are less well calibrated. PDFs represent the progressive increase in entropy and breakdown in mineral structure by the passage of the shock wave. By approximately 30 GPa (300 kbar), plagioclase begins to transform to glass. This solid state glass is called maskelynite. Quartz transforms to glass at approximately 40 GPa (400 kbar). These solid state glasses are known generically as diaplectic or thetomorphic glasses and represent the complete destruction of mineral order due to shock. Accompanying the progressive destruction of order, as observed optically, are corresponding changes in x-ray and other properties of shocked minerals.

Above 40 GPa (400 kbar) individual minerals begin to break down thermally or melt. By 45–50 GPa (400–500 kbar), rocks melt and form so-called impact melt rocks or impactite. On cooling they can form glasses, cryptocrystalline or crystalline igneous-textured rocks. They tend to be charged with mineral and lithic clasts and are generally distinguishable from endogenic igneous rocks by their composition, which corresponds to a mixture of the rocks in the target area. At higher shock pressures the target rocks are vaporized.

Richard A.F. Grieve

Bibliography

French, B.M. and Short, N.M. (eds) (1968) *Shock Metamorphism of Natural Materials*. Mono Book Corp., 433 pp.

Roddy, R.J., Pepin, R.O. and Merrill, R.B. (eds) (1977) *Impact and Explosion Cratering*. Pergamon, 1301 pp.

Stöffler, D. (1972) Deformation and transformation of rock-forming minerals by natural and experimental shock processes. I. Behavior of minerals under shock compression. *Fortschr. Mineral.*, **49**, 50–113.

Stöffler, D. (1974) Deformation and transformation of rock-forming minerals by natural and experimental shock processes. II. Physical properties of shocked minerals. *Fortschr. Mineral.*, **51**, 256–89.

Cross references

Impact cratering
Silica, silicate
SNC meteorites

SHOCK WAVES

Shock waves are ubiquitous phenomena in fluid dynamics, arising from velocity–amplitude nonlinearity, so that for finite amplitude waves where velocity increases with amplitude, the tendency exists for a wave to 'overrun' itself, leading to steepening. The local speed of sound is essential for defining the shock wave. For a body moving with speed in excess of the local sound speed with respect to the neighboring gas a leading edge shock will also arise. Shocks can also be generated by the interaction of gas streams with relative velocity in excess of the sound speed. One of the most impressive shocks in nature is that produced by the discharge of lightning. Other examples of shocks are the bow wave in front of a projectile or an aircraft in supersonic flight, and that produced by an explosion. More generally, analogs of shocks arise in fluid flow where the velocity differential between an obstacle and flow field is greater than the local characteristic wave speed.

Needless to say, shock waves are pervasive in the astrophysics of cosmic gas dynamics, though historically (prior to the space age) observational data was based upon optical and radio observation alone. This article is, with one exception, confined to the solar system which excludes shocks which exhibit optical radiation. Major interest in shock waves in the solar system is currently divided into two principal areas, the essentially fluid dynamical aspects and secondly their association with cosmic ray acceleration. The following is aimed primarily at a brief review of the former.

Basic fluid equations

The existence of shock waves has been known for more than a century. The basic equations governing fluid dynamical shocks were developed in the 19th century. They express the basic conservation laws for momentum and energy to which is usually added a caloric equation of state. But it was not until 1950 that de Hoffman and Teller first extended the theory to conducting fluids, possibly as an outgrowth of research on nuclear explosions in World War II. Such shock waves (generally in collisionally dominated flow) are called MHD or magnetohydrodynamic shocks. In the case of the MHD shock the magnetic field and its associated currents add complication. These Rankine–Hugoniot equations, e.g. Courant and Friedrichs (1976), are expressed in compact form and, as is customary, in coordinates at rest in the shock, by

$$[\rho v_n] = 0 \qquad (S2)$$

$$\left[\rho v_n + (p + \frac{B^2}{8\pi})n - \frac{1}{4\pi}B_n\boldsymbol{B}_t\right] = 0 \qquad (S3)$$

$$[v_n\boldsymbol{B}_t - B_n\boldsymbol{v}_t] = 0 \qquad (S4)$$

$$[B_n] = 0 \qquad (S5)$$

$$\rho v_n(h + v^2/2) + v_n\frac{B^2}{4\pi} - \frac{1}{B_n}\boldsymbol{v} \cdot \boldsymbol{B} = 0 \qquad (S6)$$

Vectors are bold faced and scalars italic; where not specified italics are scalar counterparts of defined vectors. ρ is density, p is pressure, n and t represent components respectively normal and parallel to the shock surface (the tangential component is bold faced since it can be viewed as a two-dimensional vector), $\boldsymbol{B}$ is magnetic induction, h is enthalpy, and $\boldsymbol{v}$ is velocity with respect to the frame at rest in the shock wave. The square brackets denote a jump, i.e. difference between initial pre- and final post-shock states. In such shocks there exist an electrical current density given by

$$\nabla \times \boldsymbol{B} = 4\pi\boldsymbol{J} \qquad (S7)$$

where $\boldsymbol{J}$ is the surface current density in the face of the shock. Although this formalism is complete for a simple shock, collision-free shocks are far more complex. Pressure is anisotropic, the plasma state is not necessarily transformed smoothly from initial to final state, and turbulence and cosmic ray acceleration may extract energy from the shock at the expense of heat. The simple caloric equation of state [equation (S6)] is too simple an expression of the internal shock kinetics.

Evolution of collision-free hydromagnetic shock

The next major development, shocks in rarefied gas, was triggered by increasing interest in the magnetized plasmas of space and astrophysics. The key difficulty was the enormous mean free path, which was the hindrance in conceptualizing the kinetics necessary to understand propagating waves, though as early as 1959 Gold discussed waves in rarefied cosmic gas and made the seminal observation that since the rise time for disturbances arriving at the Earth (signaled by the geomagnetic sudden commencement) was about 5 min, this was compatible not only with an extended interplanetary medium but also a steep-fronted wave.

The reality of shock waves in rarefied conducting gas was hotly debated in the 1960s and on into the beginning of the next decade. In a way, the general disbelief in such collision-free shock waves was the companion piece to the problems of a solar wind itself. For the shock (now commonly called the hydromagnetic shock) the matter was settled by observation of the Earth's bow shock wave and of traveling interplanetary shocks, with calculation disclosing that the conditions required by the de Hoffman–Teller extension of the Rankine–Hugoniot equations [equations (S2)–(S6)] were satisfied, though many details of the physics remained unexplored. Since magnetized plasmas are characterized by several wave speeds, the classification of hydromagnetic shocks is necessarily more complex than for the case of the counterpart fluid dynamical shock. The existence of what are now called collision-free shocks rests upon the secure observations from space, though observations were made concurrently in the laboratory at Culham (United Kingdom).

The solution to the interaction problem rests upon the imposition of dynamical constraints arising from the embedded magnetic field in electrified gas (plasma). But even here early work left unanswered just how an entropy jump would take place; such models often involved cold plasma and a steep pulse usually returned isentropically to its initial state (Adlam and Allen, 1958; Davis, Lüst and Schlüter, 1958; Baños and Vernon, 1960). The presence of particle orbits in the wave which contained multiple crossings suggests that eventually a kind of 'scrambling' would take place leading to irreversibility. With the hindsight of time, it is today not so remarkable that the hydromagnetic shock displays many of the basic large-scale features characterizing fluid dynamical shocks, save for the addition of the electromagnetic field. For example, in the convenient 'test bed' of the solar wind, observations (Sonett et al., 1964; Chao and Olbert, 1970; Hundhausen, 1972; Bavassano, Mariani and Ness, 1973) confirm the prediction that shock structures are present with the initial disturbance propagating outward. Following waves are predicted that are time reversed (in the sense that they propagate toward the source but are convected out at a higher speed (Sonett and Colburn, 1965; Colburn and Sonett, 1966; Simon and Axford, 1966). These shocks presumably originate as blast waves in the solar atmosphere, and correspond to the re-expansion phase of a spherical or cylindrical blast wave (Brode, 1955, 1959; Friedman, 1961).

Steep-fronted waves are also seen at the leading edge of long-lived corotating structure in the solar wind and embedded within the structure (Razdan, Colburn and Sonett, 1965). Here the forcing source probably lies in the interaction of streams of differing speed (Figure S14). In addition to magnetized shocks, observation in the laboratory of electrostatic shocks has been made by Honzawa (1973) and conjectured to be present in the Earth's magnetosphere (Goertz, 1979). Shock waves arise in the solar system via four distinct 'channels'. These are solar flare blast waves, corotating interaction regions

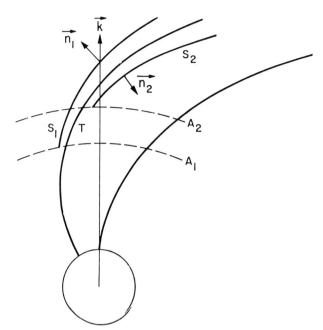

Figure S14 Formation of a recurrent shock pair (corotating interaction region or CIR) at the leading edge of the impingement of a fast stream upon a slower one. S_1 and S_2 are respectively the fast forward and reverse shocks. n_1 and n_2 are the unit normals to the two shock surfaces. The vector k is the propagation vector giving the shock direction. A_1 and A_2 indicate respectively the radial distance from the Sun of the onset of the forward and reverse shock waves. They begin where the radial component of flow in the shock frame is sufficient for a shock to develop. T denotes the tangential (contact) discontinuity. (After Figure 23 of D.S. Colburn and C.P. Sonett, 1966, *Space Sci. Rev.*, 5, 439–506, reprinted by permission of Kluwer Academic Publishers.)

(CIRs), standing bow shock waves upstream of planets and finally the heliospheric termination shock.

Planetary bow shock

The existence of a planetary magnetosphere has the consequence that the solar wind impact results in the generation of a bow shock wave on the upstream hemisphere with the solar wind slowed, heated and diverted to the flanks of the planet. The presence of the interplanetary magnetic field means that the shock is conditioned by the field and the shock properties vary over the face of the shock, depending upon the local angle between the shock normal and the field direction. Thus there is often a region locally where the magnetic field is normal to the surface; this shock is sometimes identified with the original switch-on shock.

Nearly all planets are endowed with global magnetic fields and therefore exhibit a bow shock wave. Venus is an interesting exception but still yields a bow wave as the conducting ionosphere provides a barrier to the incoming solar wind. Since the detectable lunar magnetic field consists of patchy small regions the manifestation of a steady state lunar shock wave is not present and lunar shock waves take the form of minor limb shocks near the terminator (Sonett and Mihalov, 1971; Mihalov et al., 1972).

Cometary shock

A relatively recent development has been the discovery of the cometary shock which, as for Venus, is the result of solar wind impact upon photoionized conducting gas (Figure S15). But, for the case of the comet, the gas is volatilized matter from the nucleus, which means that a characteristic time of the order of many minutes is involved in the ionization and the shock is formed from fresh gas actively undergoing ionization. This leads to the concept of 'mass loading'; the shock structure is highly modified and the general structure of the shocked region is more complex than in the case of magnetized bodies. The reader is referred to the ever increasing bibliography on this subject, e.g. Flammer (1991).

Multiple shocks and shock interactions

The forcing system from which shocks arise is more complex than that inferred from the simple observation of a single shock wave.

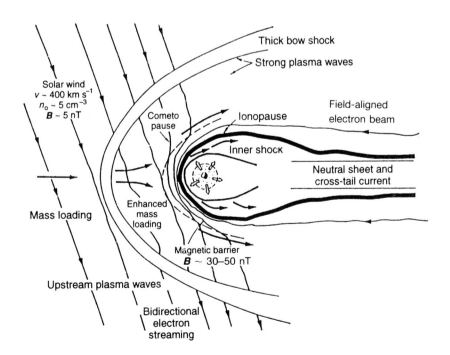

Figure S15 The global morphology of the interaction of the solar wind with a typical cometary atmosphere. (After Figure 1 of Flammer, K.R., 1991, in *Comets in the Post-Halley Era*, Vol. 2, eds R.L. Newburn, M. Neugebauer, and J. Rahe, reprinted by permission of Kluwer Academic Publishers.)

Spherical explosions (of which the solar flare is an inexact but useful analog) may yield a sequence of waves, a forward shock and a series of reflections leading to subsequent forward and reverse shocks and contact surfaces (tangential discontinuities in hydromagnetics). Examples of such systems are large, e.g. nuclear explosions on Earth and supernovae as well as solar flares. The latter inference arises from the prediction and observation of forward–reverse shock ensembles in the solar wind (Sonett and Colburn, 1965), which are characteristic of the spherical overexpansion involved in cylindrical and spherical geometry explosions (e.g. Brode, 1955, 1959; Friedman, 1961).

A possible example of the consequences of a spherically overexpanded system has been proposed by Sonett, Morfill and Jokipii (1987) and Sonett (in press), where steep short-term ^{10}Be and ^{14}C enhancements on Earth were attributed to multiple shock cosmic ray acceleration, in turn increasing terrestrial cosmogenic isotope production. The proposed source was a supernova at a distance of about 100 pc. In this case the forcing shocks are interstellar but still collision free. The reality of this proposal appears to rest currently upon whether a significant geomagnetic excursion or a field reversal took place approximately 35 000 years ago, a matter now under study, e.g. Tric et al. (1992).

Not all multiple shock systems necessarily arise in this way. Shock and contact surface interactions were intensively studied during World War II regarding fluid dynamical explosion-forced shock waves. Parker (1963) has summarized these interactions, which lead to a variety of wave systems.

Terminal shock

The largest-scale shock wave in the solar system is the putative heliospheric termination shock where the solar wind with a speed of ~ 400 km s^{-1} impinges upon the background interstellar plasma (Figure S16). Some form of interaction is the simple consequence of the supersonic velocity of the solar wind coming to rest (at the stagnation point). Assuming that the termination is at about 100 AU (a value that varies radically with small difference in solar wind momentum flux), the solar wind 'thermal' particle flux, which decreases as the inverse square of distance, is reduced from the 1-AU value of $\sim 2 \times 10^{-23}$ g cm^{-3} by a factor of 10^4, which infers a thermal energy density of about the same magnitude as that of the background cosmic rays. But the latter represents a relativistic gas of extreme temperature. Thus the Mach number relative to the cosmic ray background may be less than unity and one is faced with dual estimates of Mach number, only one perhaps being greater than unity. However, the existence of the anomalous component of the galactic cosmic radiation lends support to the hypothesis of the heliospheric shock, as this component is expected to be produced by shock wave acceleration.

Charles P. Sonett

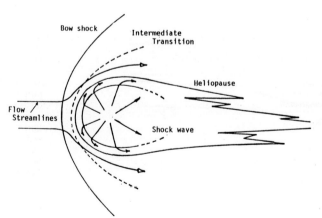

Figure S16 Conceptualization of the heliospheric termination shock. (From Perez de Tejeda, H. (1990) Signature of a Viscous Interaction at the Heliopause in *Physics of the Outer Heliosphere*, eds S. Grzedzielski and D.E. Page, COSPAR Colloquium Series, Vol. 1. Reproduced with permission of COSPAR.)

Bibliography

Adlam, J.H. and Allen, J.E. (1958) The structure of strong collision-free hydromagnetic waves. *Phil. Mag.*, **3**, 448–55.
Banõs, A. and Vernon, A.R. (1960). Large amplitude waves in a collision-free plasma, I. Single pulses with isotropic pressure. *Nuovo Cimento*, **15**, 1–20.
Bavassano, B., Mariani, F. and Ness, N.F. (1973) Pioneer 8 observations and interpretations of sixteen interplanetary shock waves observed in 1968. *J. Geophys. Res.*, **22**, 4535–46.
Brode, H.L. (1955) Numerical solutions of spherical blast waves. *J. App. Phys.* **26**, 766–75.
Brode, H.L. (1959) Blast wave from a spherical charge. *Phys. Fluids*, **2**, 217–29.
Chao, J.K. and Olbert, S. (1970) Observations of slow shocks in interplanetary space. *J. Geophys. Rev.*, **31**, 6394–7.
Colburn, D.S. and Sonett, C.P. (1966) Discontinuities in the solar wind, *Space Sci. Rev.*, **5**, 439–506.
Courant, R. and Friedrichs, K.O. (1976) *Supersonic Flow and Shock Waves*. New York: Springer-Verlag. (Reprinted unchanged from Interscience, 1948.)
Davis, L., Jr, Lüst, R. and Schlüter, A. (1958) The structure of hydromagnetic shock waves, I. Non-linear hydromagnetic waves in a cold plasma. *Z. Naturforsch.*, **13a**, 916–36.
de Hoffman, F. and Teller, E. (1950) Magneto-hydrodynamic shocks. *Phys. Rev.*, **80**, 692–703.
Flammer, K.R. (1991) The global interaction of comets with the solar wind, in *Comets in the Post-Halley Era*, Vol. 2 (eds. R.L. Newburn, M. Neugebauer and J. Rahe). Dordrecht: Kluwer, pp.1125–44.
Friedman, M.P. (1961) A simplified analysis of spherical and cylindrical blast waves. *J. Fluid Mech.*, **11**, 1–15.
Goertz, C.K. (1979) Double layers and electrostatic shocks in space. *J. Geophys. Res.* **17**, 418–26.
Gold. T. (1959) Plasma and magnetic fields in the solar system. *J. Geophys. Res.*, **64**, 1665–74.
Honzawa, T. (1973) Observations of stable, high Mach number collisionless electrostatic shocks. *Plasma Phys.*, **15**, 467–74.
Hundhausen, A.J. (1972) *Coronal expansion and the solar wind*. Springer.
Mihalov, J.D., C.P., Sonett, J.H. Binsack, and M.D. Mitsoulas, (1972) Possible fossil lunar magnetism inferred from satellite data. *Science*, **171**, 892.
Parker, E.N. (1963) *Interplanetary Dynamical Processes*. New York and London: Interscience-Wiley.
Perez de Tejeda, H. (1990) Signature of a viscous interaction at the heliopause, in *Physics of the Outer Heliosphere* (eds S. Grzedzielski and D.E. Page). Pergamon, pp. 301–9.
Razdan, H., Colburn, D.S. and Sonett, C.P. (1965) Recurrent SI+–SI− impulse pairs and shock structure in M region beams. *Planet. Space Sci.*, **13**, 1111–23.
Simon, M. and Axford, W.I. (1966) Shock waves in the interplanetary medium. *Planet. Space Sci.*, **14**, 901–8.
Sonett, C.P. (in press) A supernova shock ensemble model using Vostok ^{10}Be radioactivity. *Radiocarbon*.
Sonett, C.P. and Colburn D.S. (1965) The SI+–SI− pair and interplanetary forward–reverse shock ensembles. *Planet. Space Sci.*, **13**, 675–92.
Sonett, C.P. and Mihalov, J.D. (1971) Lunar fossil magnetism and perturbations of the solar wind. *J. Geophys. Res.*, **77**, 588.
Sonett, C.P., Morfill, G.E. and Jokipii, J.R. (1987) Interstellar shock waves and Be 10 from ice cores. *Nature*, **330**, 458–60.
Sonett, C.P., Colburn, D.S., Davis, L., Jr et al. (1964) Evidence for a collision-free magnetohydrodynamic shock in interplanetary space. *Phys. Rev. Lett.*, **13**, 153–6.
Tric, E., Valet, J.-P., Trucholka, P. et al. (1992) Paleointensity of the geomagnetic field during the last eighty thousand years. *J. Geophys. Res.*, **97**, 9337–51.
Zank, G.P. and Oughton, S., (1991) Properties of mass-loading shocks 1. Hydrodynamic considerations. *J. Geophys. Res.*, **96**, 9439–53.

Cross references

Comet: structure and composition
Earth: magnetic field and magnetosphere
Magnetospheres of the outer planets

Plasma wave
Solar wind

SIDEREAL PERIOD

The sidereal period of revolution of a planet around the Sun, or of a satellite around its planet, or the rotation period of a planet about its axis, is its revolution or rotation period with respect to the fixed stars (i.e. those with no measurable proper motion). A sidereal day is the interval between two consecutive upper transits of a fixed star past the meridian, and is 23 h 56 min 4 s. A sidereal year is 365.25636 solar days.

Jeremy B. Tatum

Bibliography

Green, R.M. (1985) *Spherical Astronomy*. New York: Cambridge University Press.
Woolard, E.W. and Clemence, G.M., (1966) *Spherical Astronomy*. New York: Academic Press.

SILICA, SILICATE

Silica is the oxide of the element silicon (SiO_2), and is one of the most common components of the Earth's crust. It is probably important in all 'hard' celestial objects, except those of water ices. Its most familiar form is the mineral quartz, which occurs in colorless, transparent hexagonal (trigonal) crystals (e.g. in granite), in fine-grained aggregates (as in the metamorphic rock quartzite) or crypto-crystalline masses (as in chalcedony, chert and flint). Fragmented or clastic silica, in one form or another, is the principal component of common sand (such as beach sand) and of sandstone, which is the same but usually with a silica cement. Commercial window glass is 99% silica.

Silica occurs naturally in eight crystalline polymorphs: besides quartz (by far the most usual) there is tridymite, cristobalite, coesite and stishovite. The last two pseudomorphs are of particular interest in connection with asteroid or large meteorite impact, because they are exclusively related to extremely high pressure conditions for which there is no known cause at the Earth's surface except as a result of these extraterrestrial collisions.

A silicate is any compound of silica in the form of tetrahedra of SiO_4, which may be isolated or in chains, sheets or three-dimensional compounds. Apart from the iron-rich core, most of the Earth's mantle and crust consists of silicates of various metals such as aluminum, magnesium and iron, either in crystallized or ionized form. Most igneous rocks are said to be silicic, that is, containing over 65% silica. Those rocks that are 'saturated' with respect to silica are much more widespread than those that are 'undersaturated'.

Silica in nature, especially in the form of quartz, is chemically most stable, so that during weathering in soils (especially under acid conditions) it becomes differentially concentrated as the other ions become gradually leached away. Cold-climate, high-latitude soils (e.g. 'podsols') are typically dominated by quartz, whereas low-latitude soils, especially in the subtropics, which are subject to evaporation often become alkaline, under which conditions silica may go into solution. On reprecipitation it is often as a hydrated gel which sometimes forms the gem mineral opal.

Rhodes W. Fairbridge

Bibliography

Fairbridge, R.W. (1972) *The Encyclopedia of Geochemistry and Environmental Sciences*. New York: Van Nostrand Reinhold Co., 1321 pp.
White, D.E. *et al*. (1963) *Chemical composition of subsurface waters*. US Geological Survey Prof. Paper 440-F.

Cross references

Crust
Differentiation
Igneous rock
Mantle
Shock metamorphism

SLIPHER, EARL CARL (1883–1964)

Earl Carl Slipher was born 25 March 1883 and distinguished himself as the foremost planetary photographer of the day in his 81 years. Slipher grew up in Mulberry, Indiana, and attended Indiana University where he learned astronomy under the tutelage of John A. Miller and Wilbur A. Cogshall. He graduated in 1906 and received a Lawrence Fellowship at Lowell Observatory in Flagstaff, Arizona.

Slipher's first assignment was to continue a project begun by John Duncan to photograph systematically the sky along the ecliptic in search of a trans-Neptunian planet. Neither effort revealed any indication of such a planet and the search was discontinued in 1907. The project was revived in 1928 by the Lowell Observatory staff and led to the 1930 discovery of Pluto by Clyde W. Tombaugh.

Slipher became a permanent member of the Lowell Observatory staff in 1907 and remained so until his death. His attentions were quickly diverted when in 1907 Percival Lowell, director of the observatory, sent him to Alianza, Chile, to photograph the planet Mars during its favorable opposition. Slipher was the observer-photographer of a five-member expedition led by David P. Todd of Amherst. Using an 18″ refractor at an elevation of 4000 (1200 m) ft, Slipher obtained over 13 600 images of Mars during June, July and August 1907. Features interpreted by Lowell and Slipher as canals appeared in most of the images, the first time that the Martian 'canals' had been photographed. These images, as well as others taken by Lowell and Carl. O. Lampland at Lowell Observatory, were used to support further Lowell's insistence of the existence of the canals and their implications for Martian life.

Slipher did not fanatically embrace Lowell's idea of intelligent life on Mars, but he did accept the framework of Lowell's theories as a working hypothesis. Throughout his life he strongly believed that vegetation did exist on Mars and was responsible for the seasonal changes in darkness across the planet.

Following the Chile expedition in 1907, Slipher continued a rigorous program of planetary photography. Besides over 200 000 images of Mars, Slipher's efforts included the discovery and micrometric measurements of several ring divisions in the A, B and C rings of Saturn. In his systematic photography of the brighter planets in the solar system he developed a technique which incorporated multiple images of a planet into one printing. This technique enhanced the quality of the planetary photographs to obtain fine detail. In 1954 Slipher (at age 72) led a joint Lowell Observatory–National Geographic Society expedition to Bloemfontein, South Africa, for another favorable opposition of Mars. This expedition produced 20 000 photographs of the planet and led to the discovery of a 'blue-green area' almost the size of Texas on the planet. His best images of Mars over 55 years of photography were collected into his 1962 book *The Photographic Story of Mars*, written for the US Air Force in support of the fledgling space program.

Slipher served on the International Astronomical Union and was chairman of the International Mars Commission. He also participated in numerous activities besides astronomy. He was elected as a city councilman and as mayor of Flagstaff and was an Arizona state senator. For many years he served on the board of education of Arizona State College (now Northern Arizona University) and he often participated in public lectures both in Flagstaff and around the world. Slipher remained active in his extracurricular activities as well as his astronomical pursuits up to the time of his death on 7 August 1964.

Nadine G. Barlow

Bibliography

Hoyt, W.G. (1976) *Lowell and Mars*. Tucson, AZ: University of Arizona Press.

Hoyt, W.G. (1980) *Planets X and Pluto*. Tucson, AZ: University of Arizona Press.

Richardson, R.S. (1964) Reminiscences of E.C. Slipher, *Sky and Telescope*, October.

SMALL SATELLITE

Irregularly shaped satellites with radii < 200 km have been discovered orbiting the planets Mars, Jupiter, Saturn, Uranus and Neptune. The small satellites which orbit close to the planet tend to have near-circular, low-inclination orbits and they are probably either fragments of larger bodies which have been broken up by impact, or residual material from the formation of the main satellite system. Spectral observations of the highly inclined outer satellites of Jupiter and the Martian satellites Phobos and Deimos suggest that they are captured asteroids.

Most of the small satellites of Saturn have albedos and reflectance spectra which are consistent with a composition of mainly water ice, although a notable exception is Phoebe, which is as dark as the small satellites of Uranus and Neptune discovered by the Voyager 2 probe. Since the densities of most of the small satellites have yet to be determined, it is not clear if these observations indicate different bulk compositions or surface coatings.

Impact craters have been detected on the surfaces of all the small satellites which have been imaged at sufficient resolution. Large satellites undergo relaxation of their topography after an impact because of the effects of gravity and hence their near-spherical shapes are preserved. Small satellites are not thought to be geologically active since it is material strength that dominates; large impacts tend to lead to fragmentation which results in an irregularly shaped satellite. Tidal forces acting on a satellite over the age of the solar system force it towards a configuration in which its spin period is equal to its orbital period, with the satellite's long axis pointing towards the planet. Although this is the case for most small satellites, a notable exception is the Saturnian moon Hyperion, which does not have a fixed spin period and undergoes chaotic rotation and tumbling (Wisdom, Peale and Mignard, 1984). This is due to a combination of Hyperion's irregular shape and the gravitational effect of the large moon Titan which orbits nearby.

Under certain conditions an orbiting body can maintain objects near stable positions which are located 60° ahead of and behind its own orbital position. These are referred to as the L4 and L5 equilibrium points respectively. In the Saturnian system a number of the large moons are involved in such dynamical relationships with small satellites. Helene orbits near the L4 point of the large satellite Dione, while Telesto and Calypso are close to the L4 and L5 points respectively of Tethys. Janus and Epimetheus, the coorbital satellites of Saturn, represent a special case of this phenomenon where the orbit of the smaller Epimetheus encompasses both the L4 and L5 points associated with the larger Janus. In this unusual configuration, referred to as a 'horseshoe orbit', the satellites experience a close approach every 4 years during which their orbits are exchanged (Dermott and Murray, 1981).

Some of the small satellites of the outer planets are associated with planetary rings. Adrastea lies close to the outer edge of Jupiter's dusty ring while another moon Metis lies within the ring. Such objects could also act as a source for ring material. The satellites Pandora and Prometheus lie on either side of the narrow F ring of Saturn and are thought to be confining the ring by a shepherding mechanism. The same is true for the satellites Cordelia and Ophelia which shepherd the outermost ϵ ring of Uranus. Galatea lies just interior to the orbit of the Adams ring of Neptune and it is believed to be responsible for the ring arcs as well as providing a radial confinement mechanism (Porco, 1991). In 1990 analysis of Voyager images led to the discovery of the satellite Pan orbiting in the Encke gap in Saturn's A ring (Showalter, 1991). The existence of Pan had been predicted in 1985 based on the detection of waves at the edge of the Encke gap. In a similar way it may be possible to use other ring observations to predict the location of small, undiscovered satellites of the outer planets.

Carl D. Murray

Bibliography

Dermott, S.F. and Murray, C.D. (1981) The dynamics of tadpole and horseshoe orbits. II. The coorbital satellites of Saturn. *Icarus*, **48**, 12–22.

Porco, C. (1991) An explanation for Neptune's ring arcs. *Science*, **253**, 995–1001.

Showalter, M. (1991) Visual detection of 1981S13, Saturn's eighteenth satellite, and its role in the Encke gap. *Nature*, **351**, 709–13.

Thomas, P., Veverka, J. and Dermott, S.F. (1986) Small satellites, in *Satellites* (eds J.A. Burns and M.S. Matthews). Tucson: University of Arizona Press, pp. 802–35.

Wisdom, J., Peale, S.J. and Mignard, F. (1984) The chaotic rotation of Hyperion. *Icarus*, **58**, 137–52.

Cross references

Jupiter
Mars
Neptune
Planetary ring

Figure S17 Small satellites of Saturn shown to the same scale. Starting at the top center and proceeding clockwise, the satellites are Janus (largest dimension 220 km), Telesto, Helene, Calypso, Epimetheus, Prometheus, Atlas, mid-Pandora (see Saturn: satellite system). NASA P-24061, courtesy of JPL.

Shepherd satellite
Saturn
Uranus

SNC METEORITES

Meteorites are extraterrestrial rocks that fall to Earth (or onto the surfaces of other planets) from space. Most meteorites probably come from asteroids that were broken into fragments through mutual collisions but which still represent primitive materials left over from the early history of the solar system (Plate 15). Since the early 1980s, however, it has become apparent that a few meteorites are probably derived by ejection of rocks from the surfaces of the terrestrial (Earth-like) planets through meteoroid impact. In particular, the so-called SNC meteorites are now widely believed to represent rocks so derived from the surface of Mars.

'SNC' has been adopted as a popular, although unofficial, abbreviation for the meteorites that are formally classified as shergottites (S), nakhlites (N), or chassignites (C). The group names were derived from type specimens that fell in Shergotty (India), Nakhla (Egypt) and Chassigny (France) respectively (Table S4). (In fact, Chassigny is the only known example of a chassignite, although there are five shergottites and three nakhlites.)

Evidence for Martian origin

The earliest argument for a Martian origin of the SNC meteorites (Wood and Ashwal, 1981) was based not on any piece of direct evidence but on a confluence of indirect evidence that seemed unexplainable in any plausible way. As igneous rocks formed by high-temperature melting and crystallization, the SNCs were first thought to be related to somewhat similar meteorites, the so-called HED group (howardites, eucrites and diogenites), that were inferred to come from asteroids such as Vesta. However, the SNCs all gave radiometric ages (about 200–1300 million years; Table S4) that were extraordinarily young compared with those of the HEDs, which are on the order of 4000–4500 million years. Even lunar rocks tend to be no younger than about 3100 million years. Consequently, it was apparent that the SNC rocks formed on a planetary body that was sufficiently large (i.e. larger than the Moon) to sustain volcanic activity until geologically recent times.

Oxygen isotopic analyses of the SNCs showed that they are distinct from all other meteorites, including the HEDs, but that they are geochemically related to each other (Clayton and Mayeda, 1983). Because the SNCs contain small portions of their indigenous iron in the ferric (Fe^{3+}) state (for example, as traces of the mineral titanomagnetite), it is clear that they evolved on a more oxidizing, Earth-like planet than did the HEDs or lunar rocks in which indigenous iron occurs exclusively in the ferrous (Fe^{2+}) or metallic (Fe^{o}) states.

The summation of the circumstantial evidence made Mars the most popular, although controversial, hypothetical source of the SNCs. In 1983, however, conjecture was replaced by firm, direct evidence when the shergottite EETA79001 yielded trapped noble gases that proved to be nearly a perfect match for those in the near-surface atmosphere of Mars as measured by the Viking landers in 1976 (Bogard and Johnson 1983). Detailed analyses of the trapped noble gases and nitrogen (Bogard, Nyquist and Johnson 1984; Becker and Pepin 1984) showed that the match with the Mars atmosphere was unmistakable. The trapped Mars-like gases reside principally in dark glass inclusions that are believed to have formed through shock-induced melting of the rock during an energetic meteoroid impact at the surface of Mars (Fig. S18). Therefore, because the SNC meteorites are closely linked with each other through independent geochemical evidence, the direct evidence for Martian origin of any one of them (in this case EETA79001) extends to the group as a whole.

Delivery to Earth

During the early 1980s, when the Martian origin of the SNCs was vigorously debated, planetary dynamicists maintained that the SNC rocks as we know them would be impossible to eject from Mars. It was argued that ejecta propelled to Mars escape velocity would be so highly shocked that they would be completely melted or vaporized; large volcanic rocks were held as not being able to survive in recognizable forms. Further skepticism from the dynamicists rested on the fact that there existed no meteorites from the Moon, even though the Moon's lower gravity and close proximity to Earth should make lunar meteorites more abundant than Martian meteorites. Coincidentally, the first certified lunar meteorite was recognized in 1982 and, in the following 10 years, a total of 12 pieces of lunar rocks were found as meteorites. With the conundrum of the missing lunar meteorites resolved, dynamicists revisited the Martian meteorite problem and discovered certain conditions that might permit the interplanetary launching of rock material during impact cratering on Mars. It has been pointed out that either oblique (low, grazing-angle) impacts (Nyquist, 1983) or more normal impacts that form 100 km sized craters (Vickery and Melosh, 1987) offer tenable hypotheses for the derivation of Martian meteorites. In particular, Vickery and Melosh called attention to a near-surface spall zone in which the crater-producing shock wave might actuate high-speed ejection of fragments with minimal melting.

The cosmic ray exposure ages of the SNCs indicate that they traveled through space, in the current sizes that we recognize, for about 10 million years in the case of the nakhlites/chassignites and for about 3 million years in the case of the shergottites. Taken at face value, those observations suggest two separate ejection events at 3 and 10 million years ago respectively. However, it is also possible that only one ejection event occurred at some unknown older date (but possibly much older than 10 million years) and that the observed cosmic ray exposure ages correspond to much later break-up of the ejecta into smaller pieces. Identity of the calendar days of fall for Chassigny and Zagami (on 3 October 1815 and 1962 respectively) has been argued as evidence for a single ejection event (Treiman, 1992).

The Antarctic members of the SNC clan apparently arrived on Earth less than 100 000 years ago, according to terrestrial residence ages determined from the decay of cosmogenic nuclides in the meteorites. The non-Antarctic SNCs have fallen within the last 200 years. Therefore, if a single ejection event applied, the ejecta became

Table S4 Identities and properties of SNC meteorites

Meteorite	Fall/find	Mass (kg)	Radiometric age (Ma)	Cosmic ray age (Ma)	Petrology
Shergotty	Fell 1865, India	5	165–205; 350	1.6–2.8	Shock-metamorphosed basaltic lava
Zagami	Fell 1962, Nigeria	23	116–230	1.9–3.1	Similar to Shergotty
ALHA77005	Found 1977, Antarctica	0.48	187	1.9–3.0	Resembles Shergotty; cumulate minerals
LEW88516	Found 1988, Antarctica	0.013	Not yet measured	3.0	Resembles ALHA77005
EETA79001	Found 1979, Antarctica	7.9	150–185	0.35–0.99; 2.9	Resembles, Shergotty; igneous contact, shock melt and cumulate minerals
Nakhla	Fell 1911, Egypt	40	1240–1370	10.7–11.9	Pyroxene–olivine cumulate or lava
Lafayette	Recognized 1931, Indiana, USA	0.60	1330	10.5–11.9	Similar to Nakhla
Governador Valadares	Found 1958, Brazil	0.16	1320	6.3–9.7	Similar to Nakhla
Chassigny	Fell 1815, France	4	1230–1390	9.6–11.8	Olivine cumulate

Figure S18 Saw-cut interior of the shergottite, EETA79001, showing black glassy inclusions in the volcanic host rock. The rough right side of the rock approximately marks the contact between a coarse-grained lava (right) and a finer-grained lava (left). The glass inclusions apparently formed through shock-induced melting during a meteoroid impact on Mars. As a result, Martian atmospheric gases were trapped in the glass and later discovered during laboratory analyses.

significantly dispersed before intersecting the orbit of Earth. It is not known how many SNCs remain to be discovered in Antarctica or elsewhere, or how many additional SNC rocks can be expected to fall as meteorites.

What SNC meteorites might tell us about Mars

If the SNC meteorites are accepted as genuine Martian samples, they have already revealed much about the geologic history of Mars. The SNCs are mafic igneous rocks produced by volcanism through processes of melting, differentiation and crystallization such as those operating on Earth (McSween, 1985). The nakhlites/chassignites clearly prove igneous activity as recently as 1300 million years. If the radiometric ages of the shergottites are igneous ages, rather than metamorphic ages defined by impact-related shock, then volcanism persisted on Mars until as recently as about 200 million years before present (Jones, 1989). Based on the SNC evidence, it is reasonable to expect volcanic activity on Mars even today.

Hydrous magmas are clearly indicated by igneous inclusions of primary amphibole and mica minerals but a separate origin is indicated for salt and clay minerals that postdate igneous crystallization (Gooding, 1992). The latter phases consist principally of calcium carbonate, calcium sulfate, iron-rich silicate clay minerals and closely associated iron oxides. Even though photogeology supports the interpretation of ancient wet conditions on Mars, the SNCs present the first direct evidence for low-temperature, water-based Martian chemistry. Oxygen isotopic analysis of water thermally extracted from the SNCs, which undoubtedly resides in the clay and salt precipitates, shows clear isotopic signatures for extraterrestrial origin (Karlsson et al., 1992). The ages of the parental water systems are unknown but are constrained to be within the 200–1300 million year range defined by the radiometric ages of the rocks. Considering that photogeologic evidence for water-carved features on the Mars surface is limited to apparent ages of 3000 million years or more, the SNCs offer compelling confirmation for continued, if less conspicuous, action by water in more recent geologic times. Such evidence opens the possibility that the SNCs might also support the first direct tests for chemical conditions that might have supported the synthesis of biogenic compounds on Mars.

James L. Gooding

Bibliography

Becker, R.H. and Pepin, R.O. (1984) The case for a Martian origin of the shergottites: nitrogen and noble gases in EETA79001. *Earth Planet. Sci. Lett.*, **69**, 225–42.

Bogard, D.D. and Johnson, P. (1983) Martian gases in an Antarctic meteorite? *Science*, **221**, 651–654.

Bogard, D.D., Nyquist, L.E. and Johnson, P. (1984) Noble gas contents of shergottites and implications for the Martian origin of SNC meteorites. *Geochim. Cosmochim. Acta*, **48**, 1723–39.

Clayton, R.N. and Mayeda, T.K. (1983) Oxygen isotopes in eucrites, shergottites, nakhlites, and chassignites. *Earth Planet. Sci. Lett.*, **62**, 1–6.

Gooding, J.L. (1992) Soil mineralogy and chemistry on Mars: possible clues from salts and clays in SNC meteorites. *Icarus*, **99**, 28–41.

Jones, J.H. (1989) Isotopic relationships among the shergottites, the nakhlites, and Chassigny, in *Proc. Lunar Planet. Sci. Conf.*, **19** pp. 465–74.

Karlsson, H.R., Clayton, R.N., Gibson, E.K., Jr and Mayeda, T.K. (1992) Water in SNC meteorites: evidence for a Martian hydrosphere. *Science*, **255**, 1409–11.

McSween, H.Y., Jr (1985) SNC meteorites: clues to Martian petrologic evolution? *Rev. Geophys.*, **23**, 391–416.

Nyquist L.E. (1983) Do oblique impacts produce Martian meteorites? in *Proc. Lunar Planet. Sci. Conf.* **13**, *J. Geophys. Res.*, **88**, A785–98.

Treiman, A.H. (1992) Fall days of the SNC meteorites: evidence for an SNC meteoroid stream, and a common site of origin. *Meteoritics*, **27**, 93–5.

Vickery, A.M. and Melosh, H.J. (1987) The large crater origin of SNC meteorites. *Science*, **237**, 738–43.

Wood, C.A. and Ashwal, L.D. (1981) SNC meteorites: igneous rocks from Mars? in *Proc. Lunar Planet. Sci. Conf.* **12**, pp. 1359–75.

Cross references

Chronology: meteorite
Igneous rock
Impact cratering
Mars: impact cratering
Meteorite
Viking mission

SOLAR ACTIVITY

The Sun is a variable star. A variety of transient phenomena are observed on and near its visible surface, including dark sunspots, bright faculae (Latin for 'little torches'), looping prominences and explosive flares (Plate 1). 'Solar activity' is a general, inclusive term employed to characterize these and other phenomena, along with their variations in time. Some effects of solar activity are propagated throughout the solar system by the streaming solar wind (q.v.). In the vicinity of the Earth, results of solar activity include auroras (q.v.), geomagnetic storms (q.v.) and on occasion regional power outages (Allen et al., 1989). Telecommunications may be disrupted and the orbits of satellites perturbed. In order to understand the nature of the near-Earth and interplanetary environment it is necessary to gain an understanding of the nature and effects of solar activity.

Basic considerations

The visible surface of the Sun is the photosphere. This is a layer only about 300 km thick, where the solar plasma becomes opaque to visible light. Below the photosphere is the convective zone, which is permeated by magnetic fields originating yet deeper in the Sun. Above the photosphere is the chromosphere, a tenuous, normally invisible layer about 2500 km in thickness, where temperatures soar to more than 10^4 K. Above the chromosphere is the still-hotter solar corona (q.v.) with temperatures of about 2×10^6 K, which may be seen as a halo of light surrounding the Sun during an eclipse. These three regions of the Sun (photosphere, chromosphere and corona) each have characteristic forms of solar activity that will be outlined below.

Solar magnetic fields cause the phenomena of solar activity. The Sun, like the Earth, has a bipolar magnetic field (like that of a bar magnet). The Sun's field is rapidly and continually changing, undergoing a cycle in which the polarity reverses every 11 years or so. The magnetic lines of force of the Earth's field run roughly north–south (a poloidal field), but in the Sun a combination of effects leads to fields in the equatorial and middle latitudes that trend more nearly east–west (a toroidal field). This is a consequence of the differential rotation of the Sun; our star rotates more rapidly in the equatorial regions than near the poles, and the effect on deeply buried, originally poloidal field lines is to wind them around the Sun's equatorial region, leading to the observed east–west pattern.

The vigorous convection in the outer portions of the Sun brings internal magnetic fields up to surface regions of the Sun, and leads to irregularities and disruptions of the fields. The phenomena of solar activity result from the evolution of these disturbed fields.

Sunspots and the photosphere

Sunspots are dark markings on the visible surface of the Sun (Figure S19). They may occasionally be seen with the naked eye under conditions when the Sun is partially obscured by haze or dust. Records of sunspot observations from the Orient date back many centuries (Stephenson and Wolfendale, 1988). Galileo (q.v.) was first to make telescopic observations of sunspots, and regular observations of high quality have been obtained since early in the 19th century. The images of Figure S19 show the progress of a large sunspot group across the face of the Sun over an interval of 4 days in 1968.

Sunspots generally consist of a dark region, the umbra, surrounded by a lighter but still dark region termed the penumbra. They tend to occur in pairs, though single spots and complex groups are also seen. The leading spot of each pair has the same polarity as the other leading spots on the same hemisphere of the Sun, while the trailing spots have polarities opposite to the leading spots. Thus spot pairs are thought to mark the emerging and returning locations of loops of magnetic field lines through the photosphere. The spots are dark due to their relatively low temperatures (about 2000 K lower than the surrounding photosphere). The fields are thought to block the convective motion of hot gases from below.

Bright areas in the vicinity of sunspot groups are termed faculae. These emit more radiation on average than the surrounding photosphere, approximately canceling the deficit associated with the sunspots (Chapman, Herzog and Lawrence, 1986; Foukal and Lean, 1986). Other aspects of photospheric activity, such as the granulation, are discussed in more detail in other entries (see Solar photosphere; Sun).

The sunspot number is one of the most ancient and fundamental measures of solar activity. Introduced by R. Wolf (q.v.), the sunspot number is proportional to $(10g + s)$, where g is the number of sunspot groups visible, and s is the number of individual spots counted on the surface of the Sun (see Wolf, Rudolf, and Wolf number). Following the recognition of the 11-year cycle in sunspot occurrence by Schwabe in 1843, Wolf reconstructed the cycle back to 1610. Figure S20 is a plot of daily sunspot numbers from 1818 to 1988; this shows the 11-year cycle. Daily, monthly and annual mean sunspot number curves have been analyzed in searches for the fundamental periodic components of solar activity (e.g. Kuklin, 1976; additional references in Fairbridge and Shirley, 1987, and Shirley, Sperber and Fairbridge, 1990). As previously noted, the polarities of leading sunspots and the polarity of the main dipolar magnetic field of the Sun are reversed in the subsequent cycle, indicating that the fundamental period of activity is in fact approximately 22 years.

Several techniques have been employed to construct a record of solar activity over longer times (Stephenson and Wolfendale, 1988). Observations of auroras and of visible sunspots have been used to reconstruct solar activity cycles before 1600 (e.g. Wittman, 1978; Stothers, 1979, Siscoe, 1980). The record of atmospheric ^{14}C preserved in tree rings yields a record of solar activity spanning more than 8000 years (Sonett, 1984; Damon and Linick, 1986). Most recently records of the isotope ^{10}Be preserved in polar ice cores have been employed to obtain a high-resolution record of solar activity in prehistoric times (Attolini et al., 1988; Beer et al., 1990). We will return to a discussion of characteristics and implications of the record of solar activity in a later section.

Chromospheric solar activity

The chromosphere may be imaged using light emitted at a wavelength of 6563 Å (0.6563 μm). This wavelength is a characteristic of a transition of the electron of hydrogen atoms from the second to the first discrete level of excitation; it is an advantageous wavelength for observation (Gibson, 1972), since hydrogen is the most abundant element in the solar atmosphere. The 6563-Å line is termed 'hydrogen alpha.' Fig. S21 shows chromospheric structures on the Sun on 17 August 1989, near the maximum phase of the solar activity cycle.

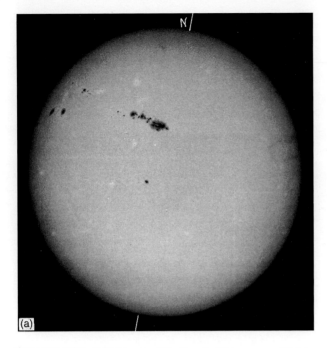 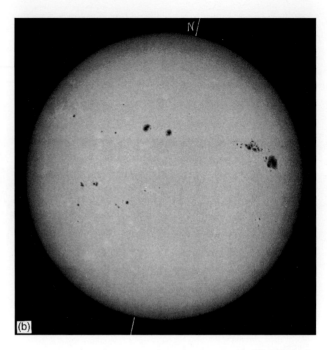

Figure S19 White light images of the Sun for 29 January and 2 February 1968, showing a number of large sunspots. The sequence illustrates the rotation of the Sun for a period of 4 days. Changes in the shape and structure of several of the sunspots are evident in these images. (a) 29 January 1968, 20:21 UT (b) 2 February 1968, 20:50 UT. (Both courtesy of G.A. Chapman, San Fernando Solar Observatory of California State University, Northridge.)

This image shows a number of long dark filaments, a few dark, nearly circular sunspots and bright plages (French for 'beaches'). The plages often occur in association with the sunspots; both help define the extent of solar active regions. The distribution of bright features defines two bands, north and south of the solar equator, which were the zones of solar activity when this image was obtained.

Solar flares (q.v.) are often evident in H_α images, indicating that they are an important component of chromospheric solar activity. Flares are explosive outbursts resulting from the release of magnetic energy in active regions; they emit radiation at many wavelengths (including energetic x-rays and gamma radiation) and may accelerate particles beyond solar escape velocity. Solar flares may cause nuclear reactions in the photosphere and may perturb the solar wind, triggering geomagnetic disturbances some time later at the Earth.

The corona

The solar corona (q.v.) is best imaged at x-ray wavelengths; here the cooler photosphere and chromosphere are dark, but hotter coronal structures show up clearly (Plate 1b). Such images show two different types of regions: those in which the coronal loop structures are closed, emerging from and returning to the Sun, and regions where the fields are open, stretching outward into interplanetary space. The former regions are identified with the same solar active regions seen in photospheric and chromospheric images, while the latter are broad regions of (predominantly) a single magnetic polarity ('unipolar regions'). These regions may also be coronal holes, which are thought to be the source of high-speed streams in the solar wind (q.v.).

Coronal mass ejections (q.v.) involve the expulsion of vast bubbles of low-density plasma from the Sun. They sometimes occur in association with solar flares, and are of undoubted importance in determining conditions in near-Earth space.

Solar activity cycle

Sunspot minimum is characterized by the presence of only a few active regions and sunspots, and by relatively large and well-developed coronal holes. The first sunspots of a cycle appear in relatively high latitudes (about 45° north and south), with leading spot polarities opposite to those of the preceding cycle. With time the sunspots and active regions appear in successively lower latitudes, until at sunspot maximum most spots appear within 10–15° of the solar equator. More complex active regions, with many sunspots, are found at higher levels of solar activity, and these complex regions generate most of the solar flares occurring in a given cycle. The large-scale solar magnetic fields are more complex during the high-activity phases of the cycle, and coronal holes may dominate the corona only near the poles.

The level of solar activity as measured by sunspot numbers increases relatively rapidly in the first few years of the cycle, with the subsequent decay to sunspot minimum requiring a longer period of time (Figure S20). During the decaying phase of the cycle, extensive regions of magnetic flux migrate from the sunspot zones to higher latitudes, first neutralizing the fields prevailing there and then establishing new polar caps of opposite polarity (Newkirk and Frazier, 1982; Wang, Nash and Sheeley, 1989).

A detailed discussion of the solar cycle would require much more space than is available here. Interested readers should consult the volumes by Zirin (1988) and Foukal (1990a), and the articles by Gibson (1972), Newkirk and Frazier (1982), Foukal (1990b) and Orrall (1991).

As we have seen, there are many different varieties of solar activity. It is important to remember that these diverse phenomena are related, and that they are all linked with the fundamental 11-year sunspot cycle, which itself represents a harmonic of the more fundamental 22-year Hale magnetic cycle (see Hale, George Ellery, and Hale cycle).

The solar activity cycle is clearly the expression of fundamental physical oscillatory processes within the Sun. Recent results underscore this point; solar activity is not merely a quasi-random, quasi-periodic expression of the bubbling of the solar cauldron. Significant changes in the rotation of the Sun occur in step with the solar cycle (Howard, 1981; Gilman and Howard, 1984; Woodard and Libbrecht, 1993). It has been known for many years that portions of the solar output at a number of non-visible wavelengths show solar cycle variations (White 1977; Holweger, Livingston and Steenbock, 1983; Hudson, 1987; Lean, 1989). Kuhn, Libbrecht and Dicke (1988) may have demonstrated that the solar surface temperature undergoes significant changes during the sunspot cycle. Perhaps most significantly for residents of Earth, it is now apparent that the Sun's total output varies by a small fraction of a percent over the sunspot cycle

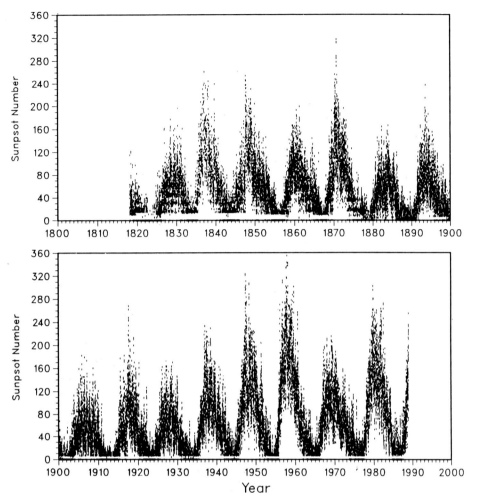

Figure S20 Daily sunspot numbers, 8 January 1818–31 December 1988. (From EOS, Transactions of the American Geophysical Union, 70 (32) Copyright American Geophysical Union.)

(Foukal and Lean, 1988; Willson and Hudson, 1991). Long-term changes of a fraction of a percent in the solar output could lead to important climatic changes on Earth (Stephenson and Wolfendale, 1988).

Periodicity and prediction

The source of the 11-year cycle of solar activity is an outstanding problem in solar physics. There appears to be nothing in the physical make-up of the Sun that would give a fundamental oscillation of this period.

The time series of sunspot numbers has been analyzed by numerous investigators, with a remarkable diversity of results. Rozelot (1994) provides a recent review, while Kuklin (1976) summarizes much earlier work. Most investigators might agree that the 11-year period is real (Dicke, 1978), though Wilson suggests that there are in fact two modes, one of 120 ± 4 months and another of 140 ± 5 months, which when averaged yield the 11-year cycle. There is less agreement on longer periods of about 60 and about 80–90 years (the 'Gleissberg cycle'). A number of investigators find a period of about 178 years (Jose, 1965; Cohen and Lintz, 1974; Fairbridge and Hameed, 1983). Still longer periods are seen in proxy data such as that obtained from radiocarbon fluctuations in tree rings (Sonett, 1984; Damon and Linick, 1986).

Fluctuations of solar activity with periods shorter than that of the basic sunspot cycle have also been described. A 155-day period in solar flare occurrence is well documented (Ichimoto et al., 1985; Bai and Sturrock, 1987; Silverman, 1990), and a 25.6-month ('quasi-biennial') period is noted by several authors (Westcott, 1964; Apos-

tolov, 1985). (The new science of helioseismology deals with oscillations of the solar surface with much shorter periods; see Sun.)

Advance prediction of the level of solar activity is a desirable goal, since astronauts, satellites, space probes and the Earth itself are not immune to the effects of solar phenomena such as large solar flares and coronal mass ejections. The periodicity of sunspot activity forms a basis for simple predictions, and methods such as that of McNish and Lincoln (1949) have been successively improved to yield forecasts of activity levels. Unfortunately the past is not a perfectly reliable guide to the future, and none of the available methods has been an unqualified success. Some recent models are described in Schatten and Sofia (1987), Wilson (1988) and Butcher (1990).

Thus the underlying causes of the observed periodic variations of sunspot activity are not known. A planetary influence on sunspot activity has been suggested many times; however, the planetary tides on the Sun are exceedingly small. Nevertheless some correlations have been found (Jose, 1965; Fairbridge and Shirley, 1987; Shirley, Sperber and Fairbridge, 1990; Charvátová 1990; Charvátová and Strestik, 1991). A possible role of nonlinear dynamics and chaos in the excitation of the solar variation is now under investigation (e.g. Feynman and Gabriel, 1990).

Another controversial subject is the possible relationship between solar activity and weather and climate on Earth. Literally hundreds of investigations on this topic have been performed, with a diversity of results. The bibliography by Fairbridge (1987) includes dozens of such studies. In recent years the hypothesis of a relationship linking solar activity and terrestrial weather has received considerable support in the form of strong positive statistical correlations reported by Labitzke and van Loon (1989) and Friis-Christensen and Lassen

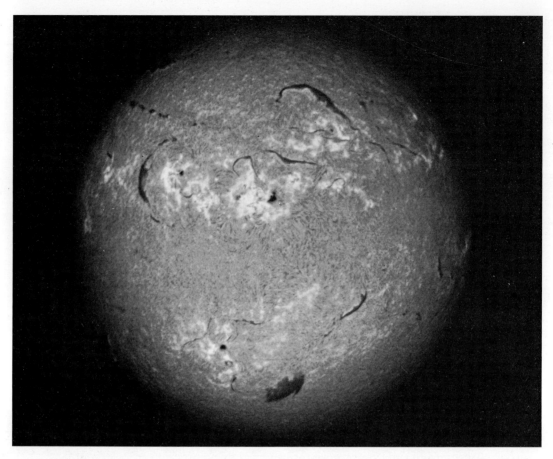

Figure S21 Hydrogen alpha image of the Sun, 17 August 1989. Dark filaments are prominences seen against the disk. They are areas of higher density of plasma within the chromosphere and corona, confined by magnetic fields. Two bands of activity, above and below the solar equator, are evident in this image. (Courtesy of C. Mach and G.A. Chapman, San Fernando Observatory of California State University Northridge.)

(1991). The question has considerable socioeconomic significance. As previously noted, a drop in the solar output of perhaps a large fraction of a percent, occurring over several decades, would lead to significant climatic cooling on the Earth. This is presently the best explanation for the occurrence of the Little Ice Age, a period of reduced temperatures and concomitant climatic and socioeconomic stresses in the 17th and 18th centuries. This episode was contemporaneous with a prolonged minimum of solar activity now known as the Maunder minimum (q.v.; Eddy, 1983).

James H. Shirley

Bibliography

Allen, J., Sauer, H., Frank, L. and Reiff, P. (1989) Effects of the March 1989 solar activity. *EOS*, **70** (46), 1479.

Apostolov, E.M. (1985) Quasi-biennial oscillation in sunspot activity. *Bull. Astron. Inst. Czech.*, **36**, 97–102.

Attolini, M.R., Cecchini, S., Gastagnoli, G.C. *et al.* (1988) On the existence of the 11-yr cycle in solar activity before the Maunder minimum. *J. Geophys. Res.*, **93**, 12729–34.

Bai, T. and Sturrock, P.A. (1987) The 152-day periodicity of the solar flare occurrence rate. *Nature*, **327**, 601–3.

Beer, J., Blinov, A., Bonani, G., *et al.* (1990) Use of ^{10}Be in polar ice to trace the 11-year cycle of solar activity. *Nature*, **347**, 164–6.

Butcher, E.C. (1990) The prediction of the magnitude of sunspot maxima for cycle 22 using abnormal quiet days in Sq(H). *Geophys. Res. Lett.*, **17**, 117–8.

Chapman, G.A., Herzog, A.D., and Lawrence, J.K. (1986) Time integrated energy budget of a solar activity complex. *Nature*, **319**, 654–5.

Charvátová, I. (1990) The relations between solar motion and solar variability. *Bull. Astron. Inst. Czech.*, **41**, 56–9.

Charvátová, I. and Strestik, J. (1991) Solar variability as a manifestation of the Sun's motion. *J. Atmos. Terrest. Phys.*, **53**, 1019–25.

Cohen, T.J. and Lintz, P.R. (1974) Long term periodicities in the sunspot cycle. *Nature*, **250**, 398–400.

Damon, P.E. and Linick, T.W. (1986) Geomagnetic–heliomagnetic modulation of atmospheric radiocarbon production. *Radiocarbon*, **28**, 266–78.

Dicke, R.H. (1978) Is there a chronometer hidden deep in the Sun? *Nature*, **276**, 676–80.

Eddy, J.A. (1983) The Maunder minimum: a reappraisal. *Solar Phys.*, **89**, 195–207.

Fairbridge, R.W. (1987) A comprehensive bibliography, *Climate – History, Periodicity, Predictability* (eds M.R. Rampino *et al.*). New York: Van Nostrand Reinhold.

Fairbridge, R.W. and Hameed, S. (1983) Phase coherence of solar cycle minima over two 178-year periods. *Astron. J.*, **88**, 867–9.

Fairbridge, R.W. and Sanders, J.E. (1987) The Sun's orbit, AD 750–2050: basis for new perspectives on planetary dynamics and Earth-Moon linkage, in *Climate – History, Periodicity, and Predictability* (eds M.R. Rampino *et al.*). New York: Van Nostrand Reinhold, pp. 446–71 (bibliography, p. 475–541).

Fairbridge, R.W. and Shirley, J.H. (1987) Prolonged minima and the 179-yr cycle of the Sun's inertial motion. *Solar Phys.*, **110**, 191–210.

Feynman, J. and Gabriel, S.B. (1990) Period and phase of the 88-year solar cycle and the Maunder minimum: evidence for a chaotic Sun. *Solar Phys*, **127**, 393–403.

Foukal, P. (1990a) *Solar Astrophysics*. New York: John Wiley and Sons.

Foukal, P. (1990b) The variable Sun. *Sci. Am.*, **262**(2), 34–41.
Foukal, P. and Lean, J. (1986) The influence of faculae on total solar irradiance and luminosity. *Astrophys. J.*, **302**, 826–35.
Foukal, P. and Lean, J. (1988) Magnetic modulation of solar luminosity by photospheric activity. *Astrophys. J.*, **328**, 347–57.
Friis-Christensen, E. and Lassen, K. (1991). Length of the solar cycle: an indicator of solar activity closely associated with climate. *Science*, **254**, 698–700.
Gibson, E.G. (1972) Description of solar structure and processes. *Rev. Geophys. Space Phys.*, **10**, 395–461.
Gilman, P.A. and Howard, R. (1984) Variations in solar rotation with the sunspot cycle. *Astrophys. J.*, **283**, 385–91.
Holweger, H., Livingston, W. and Steenbock, W. (1983) Sunspot cycle and associated variation of the solar spectral irradiance. *Nature*, **302**, 125–6.
Howard, R.W. (1981) Global velocity fields of the Sun and the activity cycle. *Am. Sci.*, **69**, 28–36.
Hudson, H.S. (1987) Solar variability and oscillations. *Rev. Geophys.*, **25**, 651–62.
Ichimoto, K., Kubota, J., Suzuki, M. *et al.* (1985) Periodic behavior of solar flare activity. *Nature*, **316**, 422–4.
Jose, P.D. (1965) Sun's motion and sunspots. *Astron. J.*, **70**, 193–200.
Kuhn, J.R., Libbrecht, K.G. and Dicke, R.H. (1988) The surface temperature of the Sun and changes in the solar constant. *Science*, **242**, 908–911.
Kuklin, G.V. (1976) Cyclical and secular variations of solar activity, in *Basic Mechanisms of Solar Activity* (eds Bumba and Klezek). Dordrecht: International Astron. Union, pp. 147–90.
Labitzke, K. and van Loon, H. (1989) Recent work correlating the 11-year solar cycle with atmospheric elements grouped according to the phase of the quasi-biennial oscillation. *Space Sci. Rev.*, **49**, 239–58.
Lean, J. (1989) Contribution of ultraviolet irradiance variations to changes in the Sun's total irradiance. *Science*, **244**, 197–200.
McNish, A.G., and Lincoln, J.V. (1949) Prediction of sunspot numbers. *Trans. Am. Geophys. Union*, **30**(5), 673–85.
Newkirk, G., Jr and Frazier, K. (1982) The solar cycle. *Phys. Today*, April 1982, 25–34.
Orrall, F.Q. (1991) Solar activity, in The Astronomy and Astrophysics Encyclopedia (ed. S.P. Maran) New York: Van Nostrand Reinhold, pp. 627–31.
Rozelot, J.P. (1994) On the stability of the 11-year solar cycle period (and a few others). *Solar Phys.*, **149**, 149–54.
Schatten, K.H. and Sofia, S. (1987) Forecast of an exceptionally large even-numbered solar cycle. *Geophys. Res. Lett.*, **14**, 632–5.
Shirley, J.H., Sperber, K.R. and Fairbridge, R.W. (1990) Sun's inertial motion and luminosity. *Solar Phys.*, **127**, 379–92.
Silverman, S.M. (1990) The 155-day solar period in the sixteenth century and later. *Nature*, **347**, 365–7.
Siscoe, G.L. (1980) Evidence in the auroral record for secular solar variability. *Rev. Geophys. Space Phys.*, **18**, 647.
Sonett, C.P. (1984) Very long solar periods and the radiocarbon record. *Rev. Geophys. Space Phys.*, **22**, 239–54.
Stephenson, F.R. and Wolfendale, A.W. (eds) (1988) *Secular Solar and Geomagnetic Variations in the Last 10,000 Years*. Dordrecht: Kluwer Academic Publishers.
Stothers, R. (1979) Solar activity cycle during classical antiquity. *Astron. Astrophys.*, **77**, 121–7.
Wang, Y.-M., Nash, A.G. and Sheeley, N.R. Jr (1989) Magnetic flux transport on the Sun. *Science*, **245**, 712–8.
Westcott, P. (1964) The 25- or 26-month periodic tendency in sunspots. *J. Atmos. Sci.*, **21**, 572–3.
White, O.R. (ed.) (1977) *The Solar Output and its Variation*. Boulder: Colorado Assoc. University Press.
Willson, R.C. and Hudson, H.S. (1991) The Sun's luminosity over a complete solar cycle. *Nature*, **351**, 42–4.
Wilson, R.M. (1987) On the distribution of sunspot cycle periods. *J. Geophys. Res.*, **92**, 10 101–4.
Wilson, R.M. (1988) A prediction for the size of sunspot cycle 22. *Geophys. Res. Lett.*, **15**, 125–8.
Wittman, A. (1978) The sunspot cycle before the Maunder minimum. *Astron. Astrophys.*, **66**, 93–7.
Woodard, M.F. and Libbrecht, K.G. (1993) Observations of time variation in the Sun's rotation. *Science*, **260**, 1778–81.
Zirin, H. (1988) *Astrophysics of the Sun*. Cambridge: Cambridge University Press.

Cross references

Carrington, Richard Christopher
Coronal mass ejection
Dynamo theory
Forbush effect
Forbush decrease
Fraunhofer line
Geomagnetic storm
Hale, George Ellery, and Hale cycle
Interplanetary magnetic field
Maunder, Edward Walter, and Maunder minimum
Sabine, Edward Albert
Solar constant
Solar flare
Solar luminosity
Solar motion
Solar particle event
Solar photosphere
Solar wind
Sun
Wolf, Rudolf, and Wolf number
Zeeman effect

SOLAR CONSTANT

The solar constant is defined as the total irradiance of the Sun at the mean orbital distance of the Earth. It has a value of about 1368 W m^{-2}. The solar constant is a fundamental quantity in atmospheric physics since it represents the amount of solar energy arriving at the top of the atmosphere.

The concept of the solar constant was introduced by A. Pouillet in 1837 (Pap, 1986). S.P. Langley devised a method for measuring the solar constant in the 1880s, and surface measurements of the solar constant were made routinely beginning in the early years of this century (Hoyt, 1979). Space-age measurements of the solar irradiance have been obtained by instruments such as the active cavity radiometer irradiance monitor (ACRIM) on the Solar Maximum mission and the Earth Radiation Budget (ERB) instrument on the Nimbus-7 satellite (Mechikunnel *et al.*, 1988).

It was recognized that the solar output might not be constant by Langley and others. However, in the first half of the 20th century the greatest number of investigators favored the notion of an unchanging Sun. Evidence suggesting variation of the solar output, obtained by C.G. Abbot and others, was thus very controversial. The difficulties in measuring the solar constant through the ever-changing atmosphere of the Earth precluded a resolution of this question prior to the space age.

Today there is irrefutable evidence of variations of the solar output of up to a few tenths of a percent, over time periods of days to weeks (Willson *et al.*, 1981). In addition it appears that the solar output varies in step with the 11-year cycle of sunspot activity (Willson and Hudson, 1991), with the highest irradiance and luminosity recorded near the maximum of the sunspot cycle (see Solar luminosity). The amplitude of this effect approaches 0.1%. This result was somewhat unexpected, since dark sunspots are cooler than the surrounding photosphere and are known to block a portion of the solar radiation. The larger number of sunspots at the maximum phase of the sunspot cycle would be expected to result in a decrease in the solar irradiance at that time. It now appears that the deficit is more than made up by radiation from bright faculae (Foukal and Lean, 1988).

If the solar output in addition experiences long-term variations, these are likely to have important implications for terrestrial climatic change (Schatten and Arking, 1990). Thus the subject of solar output variations on all timescales is an active area of research. In one study Shirley, Sperber and Fairbridge (1990) found evidence of a relationship linking solar luminosity with the solar inertial motion (see Solar motion).

By far the largest fraction of the solar output is found in the visible range of the electromagnetic spectrum. Some portions of the continuum of solar emissions at higher and lower wavelengths have been known to vary with the phase of the sunspot cycle (White, 1977). It has been suggested that variations in the solar ultraviolet flux (in particular) may play a disproportionate role in changes of the total irradiance of the Sun (Lean, 1989).

Observations of the planets Uranus and Neptune reveal systematic changes in brightness which appear to be correlated with the solar cycle (Lockwood, 1975; Lockwood and Thompson, 1986). These were originally interpreted as evidence of variable solar brightness, but that hypothesis is not widely supported. The intensity of sunlight at Neptune is only about 1/900 of the 'solar constant.' Alternatively the planetary albedos may change with the level of solar activity; the mechanism for such changes is unknown.

James H. Shirley

Bibliography

Foukal, P. and Lean, J. (1988) Magnetic modulation of solar luminosity by photospheric activity. *Astrophys. J.*, **328**, 347–57.
Hoyt, D.V. (1979) The Smithsonian Astrophysical Observatory solar constant program. *Rev. Geophys. Space Phys.*, **17**, 427–453.
Lean, J. (1989) Contribution of ultraviolet irradiance variations to changes in the Sun's total irradiance. *Science*, **244**, 197–200.
Lockwood, G.W. and Thompson, D.T. (1975) Planetary brightness changes: evidence for solar variability. *Science*, **190**, 560–2.
Lockwood, G.W. (1986) Long-term brightness variations of Neptune and the solar cycle modulation of its albedo. *Science*, **234**, 1543–5.
Mechikunnel, A.T., Lee, R.B. III, Kyle, H.L. and Major, E.R. (1988) Intercomparison of solar total irradiance data. *J. Geophys. Res.*, **93**, 9503–9.
Pap, J. (1986) Variation of the solar constant during the solar cycle. *Astrophys. Space Sci.*, **127**, 55–71.
Schatten, K.H. and Arking, A. (eds) (1990) Climate impact of solar variability. Washington DC: NASA, NASA CP-3086.
Shirley, J.H., Sperber, K.R. and Fairbridge, R.W. (1990) Sun's inertial motion and luminosity. *Solar Phys.*, **127**, 379–92.
White, O.R. (Éd.) (1977) *The Solar Output and its Variation.* Boulder: Colorado Assoc. University Press.
Willson, R.C., Gulkis, S., Janssen, M. *et al.* (1981). Observations of solar irradiance variability. *Science* **211**, 700–2.
Willson, R.C. and Hudson, H.S. (1991) The Sun's luminosity over a complete solar cycle. *Nature*, **351**, 42–4.

Cross references

Atmosphere
Insolation
Milankovitch, Milutin, and Milankovitch theory
Solar activity
Solar luminosity

SOLAR CORONA

The corona is a highly rarefied and hot ionized gas (a plasma), observed to extend from just above the solar chromosphere to a distance of several solar radii in the form of long streamers (Plate 1). Its outer extension supplies mass to the solar wind, the continual outflow of matter from the Sun. The corona is visible only during the times of total solar eclipse, but it can be observed at other times with special equipment called a coronagraph. The equatorial coronal extensions merge into the zodiacal light band. Physically, the hot solar corona shows more remarkably phenomena than the cooler chromosphere and photosphere below it. At a fine scale the coronal material presents great inhomogeneities, in densities, temperatures and magnetic field parameters. The general shape of the corona varies with the sunspot cycle of around 11 years. At sunspot maximum it is nearly circular, with streamers extending out in all directions. Polar plumes and equatorial streamers are visible as long coronal extensions, while arch structures appear in the low corona. At sunspot minimum the streamers near the poles of the Sun are short and often curved, like the lines of the magnetic field, while those near the equator are unusually long and pointed. At all periods bright and dense prominences are seen merging at the base of the corona from the chromosphere.

Coronal holes

Observations have shown that there are broad bands near the solar surface devoid of corona for certain time periods. There are regions where the local magnetic field is connected to very distant areas. This phenomenon is particularly visible at the solar poles where the magnetic field lines are opened towards interplanetary space. Solar wind data show that equatorial coronal holes are associated with high-velocity streams in the solar wind, and geomagnetic storms are associated with the passage of the high-speed streams.

Solar wind

The solar wind (q.v.) a constant stream of ionized particles (primarily hydrogen) from the Sun into interplanetary space. These particles are moving outward with velocities of 400 to 800 km s^{-1}. The stream occurs because high conductivity and high temperatures exist at a considerable distance from the Sun. At this distance solar gravity is no longer able to prevent particles from moving into outer space. Streams of these particles cause magnetic storms and auroras (q.v.) upon reaching the Earth.

Coronal mass ejection

At the present time there is no complete theory which explains the heating of the solar corona. Small-scale turbulent convection at the top of the photosphere and discontinuities in the lines of the magnetic field may produce Alfvén waves which heat the coronal material. Small-scale instabilities in the coronal arch structures may produce dissipation of the wave energy and part of the heating. There is some reason to believe that instabilities in the coronal arches are also responsible for the large-scale coronal mass ejections. These extremely important and often violent coronal mass ejections from the solar corona towards interplanetary space were first observed by the Skylab mission. These phenomena are invaded in the solar wind production and the amount material lost by the Sun. The phenomena observed in the solar corona are proof of the existence of closed solar magnetic field lines far away from the Sun's surface in interplanetary space. Instabilities in this system on a large scale are responsible for very energetic and violent phenomena in the lower parts of the solar atmosphere. But the inverse process is also possible: instabilities on a small scale, like disruptions of prominences very close to the Sun's chromosphere, may produce large-scale phenomena at a high altitude, like coronal mass ejections.

F, K and E components of the corona

Visible coronal light is photospheric radiation scattered from electrons in the ionized coronal medium, whose density ranges between 10^9 particles cm^{-3} near the limb to about 10^6 at a distance of two solar radii above it. The light seen from the corona originates in three distinct processes, the E (emission), K (continuous) and F (Fraunhofer) components. The emission line (E) corona is the true emission of the ions in the corona due to interaction of the ions with electrons, protons and photospheric radiation. Some of these lines (around 20) are emitted in the visible part of the spectrum and may be analyzed with ground-based equipment. The corona also emits ultraviolet light and x-rays, which are generated by ionized elements such as magnesium, silicon and iron, when collisions occur with fast-moving electrons in the coronal plasma. These short-wavelength lines must be observed from spacecraft. In x-ray images the active, dense and hot regions of the corona may be seen against the solar disk.

The continuous (K) spectrum of the corona is photospheric light scattered by the free electrons in the solar corona. The K corona is also called the inner portion of the solar corona. Although this light does not originate in the corona, its analysis leads to insight about the distribution of matter in the corona.

The Fraunhofer (F) corona is produced by scattering of sunlight by interplanetary dust between the Sun and the Earth. Sometimes the F corona is called the outer portion of the solar corona.

The great inhomogeneities in the corona represent a serious difficulty for the correct interpretation of observational data. Very precise measurements must be performed, assuming a knowledge of the morphology of the structures along the line of sight. The observation of the solar corona is among the most difficult astronomical observations. The brightness of the white corona is of the order of one-millionth of that of the solar disk. The contrast in the emission lines is better but the intensities do not exceed 100 times this value. Ground-based classical telescopes produce too much scattered light and do not permit the detection of the corona. The only optical instrument which is used to observe the corona without total eclipse conditions is the coronagraph. It was conceived by the astronomer B. Lyot and used at the Pic-du-Midi observatory (France) in 1930. In this way the structures and spectra of the inner part of the corona may be observed without a total eclipse. Many of the interesting phenomena in the corona are situated in this altitude range, which is the region of energy and mass exchange between the Sun and its atmosphere. The observation of the corona above an altitude of one solar radius is not possible with ground-based coronagraphs because the level of the light scattered by the Earth atmosphere is too high. Only instruments aboard spacecraft may be used to observe the coronal properties above this altitude without total eclipse conditions.

Jacques-Clair Noëns

Bibliography

Antonucci, E. and Somov, B.V. (eds) (1991) Solar corona and solar wind. *Adv. Space Res.*, **11** (1). Oxford: Pergamon Press, 1–416.
Billings, D. (1966) *A guide to the Solar Corona*. New York: Academic Press.
Durrant, C.J. (1988) *The Atmosphere of the Sun*. Bristol: IOP Publishing Ltd, Adam Hilger.
Hundhausen, A.J. (1972) *Coronal Expansion and Solar Wind*. Berlin: Springer-Verlag.
Pecker, J.C. (1984) *Sous l'étoile Soleil*. Paris: Fayard.
Priest, E.R. (1989) *Dynamics and Structure of Quiescent Solar Prominences*, (Astrophysics and Space Science Library). Dordrecht: Kluwer Academic Publishers.
Shklovskii, L.S. (1965) *Physics of the Solar Corona*, Oxford: Pergamon Press.
Tanberg-Hanssen, E. (1974) *Solar Prominences*, Dordrecht: D. Reidel Publishing Company.
Zirin, H. (1988) *Astrophysics of the Sun*. Cambridge: Cambridge University Press.

Cross references

Coronal mass ejection
Shock waves
Solar wind

SOLAR DAY AND TIME

The solar day is defined as the time interval between two successive transits of the Sun across the terrestrial meridian at any point on the Earth's surface. This could serve as an adequate measure of the passage of time in prehistoric times; however, the solar day defined in this way is not constant in duration. The apparent motion of the Sun relative to the fixed stars varies over the year, as the Earth moves more rapidly near perihelion than at aphelion. This causes the length of the solar day to vary by several seconds. In the past, astronomers or others needing precise time measurements defined a mean solar day and year, based on an averaged motion of the Sun. The mean solar day consists of 86 400 seconds, or 24 hours of 60 minutes each.

The mean solar day was the basis of timekeeping for several centuries, until precise astronomical observations forced a re-evaluation of the assumption that the Earth's rotation speed was constant. The Earth's rotational speed is not constant, and thus 86 400 (solar) seconds in one particular year will not be the same length as the 86 400 s in some other year. The time employed for general timekeeping purposes today is UTC or Coordinated Universal Time, which is based on the rotation of the Earth, but which employs a unit (the second) which is based on atomic oscillations and which does not vary with time. Thus from time to time UTC must be corrected by the addition of 'leap seconds.' More sophisticated definitions of time are required for precision applications. Sadler (1968) provides an entertaining discussion of the history of time measurement.

In astronomy and general usage the solar day has been traditionally taken as the 24 h from midnight to the next midnight. In defining the Julian day (q.v.), however, it is taken from midday to the next midday. In civil law, which involves, for example, lighting regulations, the solar day varies from locality to locality and time zone to time zone, and spans a constantly changing interval from local sunrise to local sunset. In the Moslem tradition the solar day begins with the zero hour at mean sunrise (6 a.m.), in order to coordinate calls to prayer.

The small variations in the rotation of the Earth mentioned earlier are of considerable geophysical interest. These are discussed in the articles on Earth rotation, Chandler wobble, and Tidal friction.

As a result of tidal friction, a concept developed by G. Darwin (1911) and others (Munk and MacDonald, 1960), there is a secular deceleration of the rotation of the Earth. This was confirmed over periods of several thousand years by analysis of Babylonian records of eclipses. The rate found for the recent period cannot be reliably extrapolated into the past. This would imply a very high spin rate in the Precambrian and a close approach of the Moon in Archean time. The geological evidence, however, rules out these events. Thus the slowing of the Earth's rotation is not constant with time (see Earth: rotational history).

From sedimentological evidence of paleotides about 650 Ma ago, Williams (1989) found that there were 30.5 ± 0.5 solar days in the lunar month (compared to 29.53 today), and 400 ± 7 solar days per year (365.25 today). The solar day was then 21.9 ± 0.4 h in duration.

Rhodes W. Fairbridge

Bibliography

Darwin, G.H. (1911) *The Tides and Kindred Phenomena in the Solar System*. London: John Murray.
Munk, W.H. and MacDonald, G.J.F. (1960) *The Rotation of the Earth*. London: Cambridge University Press.
Sadler, D.H. (1968) Astronomical measures of time. *Quart. J. Roy. Astron. Soc.*, **9**, 281–93.
Williams, G.E. (1989) Tidal rhythmites: geochronometers for the ancient Earth–Moon system. *Episodes*, **12**, 162–71.

Cross references

Chandler wobble
Earth rotation
Earth: rotational history
Julian calendar, year and day
Tidal friction

SOLAR FLARE

Solar flares are energetic eruptions from the surface of the Sun. They typically originate within active regions, which are areas where internal solar magnetic fields have emerged through the solar surface (the photosphere), forming bright faculae and dark sunspots. The release of energy in flares accelerates particles to high velocities (e.g. Chupp, 1990). In some cases these particles may later be detected at the surface of the Earth, even when the flare occurs on the opposite side of the Sun. Some geomagnetic storms (q.v.) and auroras (q.v.) may result from strong solar flares, while others appear to be caused by coronal mass ejections. Large flares release tremendous amounts of energy (up to about 10^{26} J in less than an hour).

The largest flares seem to pass through three stages. The preflare (or precursor) stage involves brightening of a portion of the solar surface within an active region. Loops of plasma confined by magnetic tubes emerging from the solar surface expand, sometimes over a period of several days. The impulsive phase (or flash) lasts from a few seconds to a few minutes. This phase includes the emission of bursts of x- and gamma rays and other radiation; a shock wave forms. The main, or extended, phase follows. The flare becomes visible in optical wavelengths, and soft x-rays and radio emissions continue. Flares are often observed in the hydrogen alpha spectrum line, which provides an image of the structure of the solar chromosphere (a thin layer immediately above the photosphere). (The H_α line is found at a wavelength of 6563 Å). (See Solar activity; Solar photosphere; Sun). Rust (1993) provides a recent review of observations of the development of flares.

The physics of solar flares is not well understood. They are linked with collisions and twisting of magnetic strands of force generated within the solar interior. The energy is thought to derive from a magnetic tension.

There are many varieties of flares. Some are confined to closed loops, while others disrupt loop structures; some are linked with coronal mass ejections (q.v.) that may change the structure of the corona (see Solar corona). Flares may be classified by the area of the chromosphere that brightens in the event; by the signal seen in solar radio wave observations (i.e. bursts); or by the enhancement in solar x-ray flux that accompanies the flare, or by other schemes (Orrall, 1991).

Flares, in common with many other solar phenomena, are more frequent at times of higher solar activity. In addition there is a well-developed cycle of flare occurrence, with a period of 155 days, whose origin is not well understood (Ichimoto et al., 1985; Bai and Sturrock, 1987).

A record of solar flare activity spanning about 10^7 years is preserved in near-surface rocks on the Moon. Protons emitted from the Sun in flares may produce radioactive isotopes such as ^{26}Al and ^{53}Mn in lunar rocks; radiometric dating then provides an estimate of the time of occurrence of the flare. The data suggest that the level of intensity of flare activity has not changed significantly in recent millenia.

James H. Shirley

Bibliography

Bai, T. and Sturrock, P.A. (1987) The 152-day periodicity of the solar flare occurrence rate. *Nature*, **327**, 601–13.
Chupp, E.L. (1990) Transient particle acceleration associated with solar flares. *Science*, **250**, 229–36.
Foukal, P.V. (1990) *Solar Astrophysics*. New York: John Wiley.
Ichimoto, K., Kubota, J. Suzuki, M. et al. (1985) Periodic behavior of solar flare activity. *Nature*, **316**, 422–4.
Orrall, F.Q. (1991) Solar activity, in *The Astronomy and Astrophysics Encyclopedia* (ed. S.P. Maras). New York: Van Nostrand Reinhold.
Rust, D. (1993) Solar flare prediction needed, *EOS*, **74** (47), 553–9.
Zirin, H. (1988) *Astrophysics of the Sun*. New York: Cambridge University Press.

Cross references

Coronal mass ejection
Solar photosphere
Shock waves
Solar activity
Solar corona
Solar particle event
Solar wind
Sun

SOLAR LUMINOSITY

The solar luminosity is the total power output of the Sun radiated to space. Solar irradiance, in contrast, is the total power per unit area at a distance of 1 AU (see Solar constant). Luminosity is determined empirically by measuring the total solar irradiance by spacecraft using sensors that absorb radiation over most of the electromagnetic spectrum, from x-rays to radio wavelengths; it has been accurately monitored in this way for less than two decades. Knowing the total solar irradiance and the mean Earth–Sun distance, the luminosity can be determined. The recently determined mean value of irradiance from the Solar Maximum mission satellite (1980–1989) is about 1368 W m^{-2}, although there are short-term fluctuations (Frölich et al. 1991). This translates into a solar luminosity of 3.85×10^{26} W (or 3.85×10^{33} erg s^{-1}). The largest uncertainty (about 0.3%) is due to disagreements between the measurements of different satellite detectors (Mechikunnel et al., 1988; see Solar constant).

The solar luminosity is thought to represent the total production of energy from nuclear fusion in the Sun's core. It is usually assumed that this production rate is constant and is equivalent to the luminosity. This assumption may not be true since some energy from the core could be stored in the convection zone or elsewhere. Further, if the production rate in the core were lower than is inferred from the luminosity, the Sun could be shrinking, converting gravitational energy into heat and making up for the deficiency in nuclear energy sources. Attempts to measure changes in the solar radius spanning the last three centuries have been equivocal and, in fact, theoretical interpretations of possible changes in radius are conflicting (Ribes et al., 1991; Spruit, 1991).

An attempted determination of the nuclear generation rate through detection of neutrinos from the solar core has shown a deficiency with respect to theoretical model predictions, but this is presently thought to be due to a lack of understanding of neutrino physics (Bahcall, 1989; Gough and Toomre, 1991). Theoretical models of solar evolution suggest that the solar luminosity has increased by about 30% since the solar system was formed (Kasting and Grinspoon, 1991). This change is expected to continue for another few billion years, after which the evolution of the Sun will become more rapid.

Use of the Earth's climate history as a proxy record of solar luminosity on a long-term basis has resulted in a confusing conclusion. The terrestrial insolation proxies do show long-term cycles (e.g. ice ages every $2-3 \times 10^8$ years), but there is no evidence of an important secular trend (Fairbridge, 1987). In contrast, the stellar evolution model would imply that the Earth should have been getting warmer over its 4.5-Ga history; this is referred to as the 'faint young Sun paradox' (Kasting and Grinspoon, 1991).

Gary A. Chapman

Bibliography

Bahcall, J.N. (1989) *Neutrino Astrophysics*. Cambridge: Cambridge University Press.
Fairbridge, R.W. (1987) Climatic variation, geological record. In Oliver, J.E. and Fairbridge, R.W., ed.). *The Encyclopedia of Climatology* (eds J.E. Oliver and R.W. Fairbridge). New York: Van Nostrand Reinhold, pp. 293–305.
Frölich, C. et al. (1991) Solar irradiance variability from modern measurements, in *The Sun in Time* (eds C.P. Sonett, M.S. Giampapa and M.S. Matthews). Tucson: University of Arizona Press, pp. 11–29.
Gough, D. and Toomre, J. (1991) Seismic observations of the solar interior. *Ann. Rev. Astron. Astrophys.*, **29**, 627–84.
Kasting, J.F. and Grinspoon, D.H. (1991) The faint young Sun problem, *The Sun in Time* (eds C.P. Sonett, M.S. Giampapa and M.S. Matthews). Tucson: University of Arizona Press, pp. 447–462.
Mechikunnel, A.T., Lee, R.B. III, Kyle, H.L. and Major, E.R. (1988) Intercomparison of solar total irradiance data from recent spacecraft measurements. *J. Geophys. Res.*, **93**, 9503–9.
Ribes, E. et al., 1991. The variability of the solar diameter, in *The Sun in Time* (eds C.P. Sonett, M.S. Giampapa and M.S. Matthews). Tucson: University of Arizona Press, pp 447–62.
Schatten, K.H. and Arking, A. (eds) (1990) Climate impact of solar variability. Washington D.C: NASA, NASA CP-3086.
Spruit, H.D., 1991. Theory of luminosity and radius variations, in *The Sun in Time* (eds. C.P. Sonett, M.S. Giampapa and M.S. Matthews). Tucson: University of Arizona Press, pp. 118–58.
Stix, M. (1989) *The Sun*. Berlin: Springer-Verlag.
Wolfsberg, K. and Kocharov, G.E. (1991) Solar neutrinos and the history of the Sun, in *The Sun in Time* (eds C.P. Sonett, M.S. Giampapa and M.S. Matthews). Tuscon: University of Arizona Press, pp. 288–313.

Cross references

Atmosphere
Earth: atmosphere and climate
Solar activity
Solar constant
Sun

SOLAR MOTION

The center of gravity of a thrown dumbbell describes a regular arc, while the weighted ends undergo complex gyrations. Similarly the

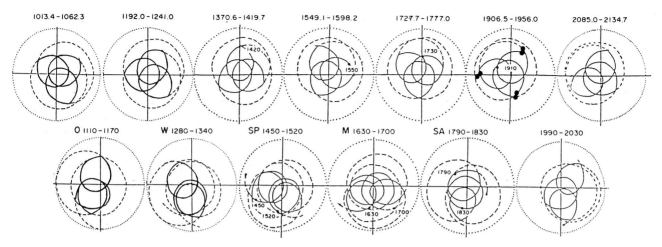

Figure S22 The ordered (top) and chaotic (bottom) motion of the Sun, alternately recurring every ~ 180 years from 1015 to 2135 AD (The Sun begins the more orderly 'trefoil' patterned orbits with a 178.7-year spacing on average – see the times at the top of Figure S22). This value corresponds to the basic period of solar motion (Jose, 1965; Fairbridge and Shirley, 1987). The more chaotic solar orbits during the Oort (O), Wolf (W), Sporer (SP), Maunder (M) and Sabine (SA) minima of solar activity are shown at bottom. The dotted line, a radius of 2.2 solar radii, limits the area in which the Sun moves. The black circles indicate the positions of sunspot maxima in the 20th century.

barycenter of our solar system follows a regular path as we orbit the galactic center, while solar system bodies, particularly the Sun, describe complex orbits about the barycenter. The center of the Sun has a maximum possible displacement of 2.2 solar radii from the barycenter of the solar system. The motion is principally determined by the relative motions of the giant planets Jupiter and Saturn, which together contain > 86% of total solar system angular momentum. The synodic period of this pair is 19.86 years, which has been characterized as a fundamental 'pulse' of the solar system. Each successive conjunction of the pair is approximately 120° removed from the position of the last.

In Figure S22 the solar motion is represented in a polar view for different time intervals during the past millenium. The solid lines represent the 'loop' portion of the orbit nearest the barycenter, while the connecting dashed lines represent the more distant 'arc' portions of the motion. The motion is prograde (clockwise), except for rare intervals of retrograde motion. The mean period of each of these ('loop' and 'arc') portions is half the Jupiter–Saturn synodic period, or 9.93 years. Other authors have used different definitions in their investigations of the solar motion (Fairbridge and Shirley, 1987).

The interest in the solar motion derives from the possible relationship linking solar motion and solar activity (Jose, 1965; Fairbridge and Shirley, 1987; Charvátová, 1988, 1989, 1990a, b; Charvátová and Střeštik, 1991a, b).

Varying geometry of the solar orbit

Figure S22 is arranged to show two types of solar orbit patterns. The upper set show a 'trefoil' pattern. The approximate 120° symmetry results from the fact that successive conjunctions of Jupiter and Saturn are separated by this angle. The lower set, here termed 'chaotic,' lack this symmetry; in general this is due to combined action of Uranus and Neptune, which, when the pair are near conjunction, are together able to distort the solar orbital path as shown.

The time over which the Sun returns to the trefoil orbit is practically constant and equal to 178.7 years (refer to the description of Figure S22 above). This offers an explanation of the basic period of solar motion, coincident also with the time interval after which the pattern of motion repeats (Jose, 1965, Fairbridge and Shirley, 1987).

The motion trefoil rotates through a full circle in approximately 2200 years (Figure S22; Charvátová, 1990b; Charvátová and Střeštik, 1991b).

Relations between solar motion and solar activity

Over long time periods solar activity shows extended periods of high and low solar activity. The latter are termed prolonged minima. The important epochs of prolonged minima are termed the Oort, Wolf, Sporer, Maunder and Sabine minima. Figure S22 illustrates the differences in solar orbital motion for high and low activity periods. The trefoil pattern appears in epochs of prolonged maxima; these are separated by epochs of chaotic motion corresponding to the prolonged minima. The two types of epochs follow roughly a 180-year cycle (Charvátová, 1988, 1989, 1990a, b). In antiquity and in the modern era the deviations of the Sun's motion from the Jupiter–Saturn (JS trefoil) pattern are smaller and the prolonged minima of solar activity are more moderate and shorter (e.g. the Sabine minimum, 1790–1830). We are now probably, as in antiquity, in a long-term maximum of solar activity. Stothers (1979), studying solar activity records from 373 BC to 333 AD, concludes that '. . . solar activity two millennia ago was not markedly different from what it is today.' The 'bottoms' of prolonged minima probably reoccur with a 171-year spacing reflecting the synodic orbit period of Uranus and Neptune. As all types of records (auroras, large sunspots, ^{14}C, etc.) indicate, the long-term trend of solar activity decreased towards the Middle Ages and increased from then on.

There are two intervals in which the Sun moved along the trefoil, in the time period corresponding to the series of the Wolf sunspot numbers recorded since 1700: from ~ 1730 to 1780 and from ~ 1910 to 1960 (Figure S22). Since the chaotic sections of orbit are different, these two intervals provide the best possibility of verifying the direct relation between the Sun's motion and the observed solar activity in detail. It was found that the motion of the Sun along nearly the same orbit did create almost the same series of five 10-year solar cycles in these two intervals (Charvátová, 1990b). Figure S23 shows the coefficients of correlation between the groups of five cycles corresponding to the trefoil orbits and successive groups of five cycles from 1700 to the present. The highest and only significant coefficient ($r = 0.81$) is the one between the series corresponding to the trefoil motion; the smallest correlations were found for the series of cycles of the Sabine minimum. The basic cycle of 179-year in both phenomena is thus evident.

The differences between both the series could be caused by the trefoil orbits of the Sun not being quite identical, by the inner planets not being considered or by the lack of our knowledge of the outer boundary of the solar system. But, above all, they could be ascribed to the substantially lower quality of observations of Wolf sunspot numbers in the 18th century.

Long-term prediction of solar activity

The above results provide a possible method of prediction of solar activity. The current cycle (No. 22) is probably the last from the prolonged maximum of the 20th century. It should be followed by an epoch of about 30 years in which the motion of the Sun will be chaotic (Figure S22, bottom) and solar activity, therefore, should be low. The cycles will probably be longer and irregular. No concrete

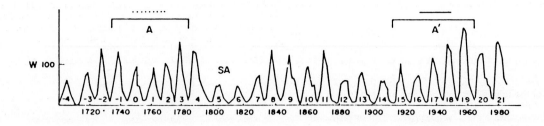

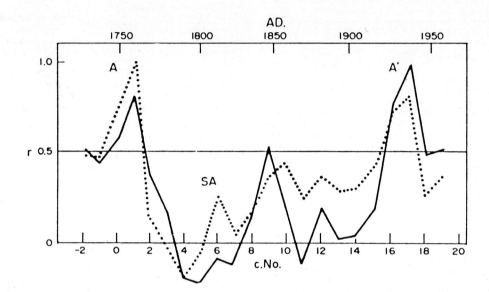

Figure S23 The series (A) of sunspot cycles −1 to 3 (1733–1784) and the series (A') of sunspot cycles 15 to 19 (1912–1963) in the most recent intervals of orderly motion of the Sun. The correlation coefficients between the series A and the successive groups of five cycles (in steps of one cycle) are always related to the central cycle of the group (dotted line) and the same for the series A' (solid line). The highest and only significant coefficient is that between series A and A' (r = 0.81). The smallest coefficients (+ or −) were found for the intervening epoch of chaotic motion of the Sun [the Sabine (SA) minimum].

prediction for these cycles can now be made. After about 2040, high-activity 10-year cycles will probably occur. The Sun will begin to move along nearly the same orbit as in the years 1730–80 and 1910–1960 after 2090. This probably means that in the years 2091–2142 a series very similar to the series of cycles 15–19 could occur with maxima close to the sixth years of the respective decades (Charvátová, 1988, 1990b).

Conclusion

The results show that the solar variability could be caused by inertial motion of the Sun. The 11-year solar cycle amplitude variation appears to be a successive 'excitation' and 'attenuation' due to the solar motion along the orbit determined by the planetary system as a whole. It seems that a key to a proper 'mechanism' required to explain solar variability may be hidden in the trefoil pattern (Charvátová, 1988). Besides the characteristics describing dynamics of the Sun's motion, the geometry of the solar orbit should be taken into account.

Ivanka Charvátová

Bibliography

Charvátová, I. (1988) The solar motion and the variability of solar activity. *Adv. Space Res.*, **8** (7), 147–50.
Charvátová, I. (1989) On the relation between solar motion and the long-term variability of solar activity. *Studia Geophys. Geod.*, **33**, 230–41.
Charvátová, I. (1990a) The relations between solar motion and solar variability. *Bull. Astron. Inst. Czech.* **41** (1), 56–9.
Charvátová, I. (1990b) On the relation between solar motion and solar activity in the years 1730–80 and 1910–60. *Bull. Astron. Inst. Czech.*, **41** (3), 200–4.
Charvátová, I. and Střeštík, J. (1991a) Long-term variations in duration of solar cycles. *Bull. Astron. Inst. Czech.*, **42** (2), 90–7.
Charvátová, I. and Střeštík, J. (1991b) Solar variability as a manifestation of the Sun's motion. *J. Atmos. Terrest. Phys.*, **53**, (6),1019–25.
Damon, P.D. and Linick, T. (1986) Geomagnetic–heliomagnetic modulation of atmospheric radiocarbon production. *Radiocarbon*, **28** (2A), 266–78.
Fairbridge, R.W. and Shirley, J.H. (1987) Prolonged minima and the 179-yr cycle of the solar inertial motion. *Solar Phys.*, **110**, 191–220.
Jakubcová, I. Pick M. (1987) Correlation between solar motion, earthquakes and other geophysical phenomena. *Annales Geophys.*, **58** (2), 135–41.
Jose, P.D. (1965) Sun's motion and sunspots. *Astron. J.*, **70**, 193–200.
Rabin, D., Wilson, M.R. and Moore R.L. (1986) Bimodality of the solar cycle. *Geophys. Res. Lett.*, **13**, 352–4.
Schove, D.J. (1983) *Sunspot Cycles*. Hutchinson-Ross.
Stothers, R. (1979) Solar activity cycle during classical Antiquity. *Astron. Astrophys.*, **77**, 121–7.

Cross references

Angular momentum
Barycenter

Celestial mechanics
Solar activity
Sun

SOLAR NEBULA

The flatness of the solar system implies that the planets were formed in a rather thin protoplanetary disk by coalescence of initially dispersed matter. The early solar composition of Jupiter, and the nearly solar relative abundance of non-volatile elements in the most primitive meteorites, carbonaceous chondrites, imply that the disk initially had the solar bulk composition and originated from the same distribution and volume of the interstellar matter as the Sun. This disk, composed of gas and dust and surrounding the Sun in an early stage of its evolution, is the solar nebula.

The solar nebula evolved with time (see Solar system: origin). Parameters characterizing the solar nebula at the time of planetary formation were reconstructed by Hoyle, Weidenschilling, Hayashi and others by adding sufficient amounts of H and He to the planetary masses to restore the initial solar composition. The reconstruction resulted in a solar rebula with a mass about 0.01 $M_\odot$ and angular momentum about 4×10^{51} g cm^2 s^{-1}. The mass should probably be increased by 0.02 to 0.05 $M_\odot$, and the angular momentum to one to a few times 10^{52} g cm^2 s^{-1}, to take into account the first that some unknown number of solid bodies (or planetesimals) were ejected from the solar system during the formation of the giant planets.

Based on the reconstruction, the surface density σ in the solar nebula may be approximated by a power law of the form $\sigma \propto R^{-3/2}$ to R^{-1}, where R is the radius with a wide and deep minimum in the region of Mars and the asteroid belt. The minimum most probably resulted from an interruption in (or significant reduction of) the rate of planetary accumulation caused by rapid growth of Jupiter.

The temperature distribution in the solar nebula at the stage of planetary accumulation was inferred by Lewis from the compositions of terrestrial planets, the asteroids Ceres and Vesta and the satellites Ganymede and Titan. The temperature was about 1400 K for the region of formation of Mercury, 1000 K for Venus, 700 K for Earth, 500 K for Mars, 400 to 150 K for the asteroid belt, 150 to 70 K for the region of Jupiter and 60 to 30 K for the region of formation of Titan.

During the evolution of the solar nebula the gas and solid material take separate pathways: solid particles grow due to collisions, possibly forming fluffy fractious aggregates; they then precipitate toward the central plane of the disk and migrate toward the Sun. The settling of solid particles is probably followed by the formation of a dust sublayer, the development of gravitational instability in the sublayer and the formation of dust condensations. The larger solid bodies or planetesimals, formed either by direct coalescence of particles or due to contraction of the dust condensations, accumulate to form the planetary system. The gas, on the other hand, is dissipated from the solar nebula, plausibly under the impact of a strong solar wind, in a timescale of the order of 10^7 years.

Kant (1755) was the first to guess that the structure of the solar system implies that the planets were formed due to conglomeration of a number of a much smaller bodies moving around the Sun in a disk flattened due to its rotation. The idea of a linked origin of the Sun and a surrounding solar nebula was first introduced by Laplace. It has evolved significantly since that time.

Disk-shaped, opaque gas–dust envelopes have been discovered around several young solar type stars, e.g. HL Tau and DG Tau which are similar to the Sun in term of mass. These stars are much younger, with ages of 10^5 to 10^7 years and are thought to be comparable to the young Sun. The radii of the non-transparent regions in these disks are estimated at ~ 10 to 10^2 AU, with masses estimated at 0.01 to 0.1 $M_\odot$. Elongated, plausibly disk-like envelopes are sometimes extended up to 10^3 AU. The distribution of the velocities in the disk around HL Tau is consistent with Keplerian rotation. A number of indirect data obtained by methods of infrared astronomy imply that up to 50% of young stars may be surrounded by gas–dust protoplanetary disks. Later stages of planetary formation about older stars have been observed. Extended transparent dust disks were discovered about some stars with ages of $\sim 10^8$ years, such as α Lyr, and β Pic. The central region with a radius of 30 AU in these disks is dust free. The empty region could not have been retained after the star's formation stage, since the dust grains from the surrounding envelope drift inward under the Poyting–Robertson effect (Aumann et al. 1984) and enter the volume near the star in a few hundred thousand years. The empty regions observed could be produced by large invisible planetesimals that sweep up small particles or shift them rapidly to the star.

Tamara V. Ruzmaikina

Bibliography

Aumann, H.H., Gillett, F.C. Beichman C.A. et al. (1984) *Astrophys. J.*, **278**, L23–7.
Kant, I. (1755) Allgemeine Naturgeschichte und Theorie des Himmels. (English Translation), in *Nebular hypothesis*. Edinburgh: Scottish Academic Press, 1981.
Lewis, J. (1974) The temperature gradient in the solar nebula. *Science*, **186**, 440–3.
Sargent, A. and Berkwith S. (1987) Kinematics of the circumstellar gas of H.L. Tauri and R. Monocerotis. *Astrophys. J.* **323**, 294–305.
Strom, S. Edwards S., and Skrutskie, M. (1993) Evolutionary time scales for circumstellar disks associated with intermediate and solar-type in *Protostars and Planets III* (eds E.H. Levy, and J.I. Lunine). Tucson: University of Arizona Press, pp. 837–86.

Cross references

Accretion
Cold accretion theory
Planetesimal
Planet: extrasolar
Poynting–Robertson drag
Solar system: origins

SOLAR NEUTRINO

An elementary particle, the neutrino is designated by the Greek symbol ν, with zero electric charge and zero rest mass. The latter has not yet been proved experimentally, however, in a conclusive manner. The neutrino is classified as a lepton governed by Fermi–Dirac statistics. Elementary particles interact with each other by four types of fundamental forces: strong, electromagnetic, weak and gravitational, in order of decreasing strength. The neutrino is the only known particle that has only weak and gravitational interactions.

The existence of the neutrino was postulated in 1930 by W. Pauli to explain the apparent non-conservation of energy in the beta-decay process $n \rightarrow p + e^- + \bar{\nu}$. The existence of the neutrino was experimentally verified by observing the reaction $\nu + n \rightarrow n + e^+$ (which is the inverse of the beta-decay process) by C.L. Cowan and F. Reines in 1957. Studying neutrino interactions, it was discovered that there are two distinct neutrinos – the electron neutrino ν_e, associated with beta decay, and the muon neutrino ν_μ, associated with pion decay $\pi^+ \rightarrow \mu^+ + \nu_\mu$. In 1975 a third charged lepton, the tauon ($\tau^\pm$) was discovered to exist in addition to the electron ($e^\pm$) and the muon ($\mu^\pm$). It is thus believed that six distinct neutrinos, associated with the respective leptons, exist: the electron-neutrino ν_e, the muon-neutrino ν_μ, the tau-neutrino ν_τ, and their antiparticles $\bar{\nu}_e$, $\bar{\nu}_\mu$, $\bar{\nu}_\tau$.

From the absence of neutrinoless double beta decay it follows that the electron neutrino ν_e is not identical to its antiparticle, the electron antineutrino $\bar{\nu}_e$. Considerations of angular momentum conservation in pion decay establish the spin of the neutrino to be ½ in units of $h/2\pi$, where h is the Planck constant. Direct mass measurements of the three known species of neutrinos have so far yielded only upper bounds, 9.5 eV for the ν_e, 170 keV for the ν_μ and 35 MeV for the ν_τ. The experimental upper bound on the charge of the neutrino is less than 10^{-15} times the charge of the electron. Taking into consideration these upper bounds it is generally assumed that the rest mass and electric charge of the neutrinos are identically zero. However, there has been experimental indication and theoretical speculation that the mass of at least some of the neutrino species is not identically zero but has some finite, although very small, value. If confirmed experimentally, this would lead to fundamental astrophysical and cosmological implications. Particularly, non-zero mass neutrinos may open the possibility that different types of neutrinos can transmute into each other (neutrino oscillations).

Table S5 Solar neutrino experiments in operation 1970–1992[a]

Name of collaboration	Location	Target vessel	Reaction for neutrino detection	Neutrino energy threshold	Theoretical neutrino capture rate of standard solar model (SSM)	Observational result	Time of operation	Half-life of target atom	Exposure time
Chlorine experiment (USA)	Homestake gold mine, lead, South Dakota, USA 1480 m underground	615 t C_2Cl_4 containing 133 t ^{37}Cl	Radiochemical experiment $\nu_e + {}^{37}Cl \rightarrow {}^{37}Ar + e^-$	0.814 MeV, able to detect neutrinos from $^7Be + e^- \rightarrow {}^7Li + \nu_e$, $^8B \rightarrow {}^8Be + e^+ + \nu_e$	7.9 ± 0.9 SNU[b] 1.8 Ar atoms per day	2.1 ± 0.3 SNU 0.472 ± 0.037 Ar atoms per day	Since 1970.3 starting with run 18...	35 d for ^{37}Ar	1–3 months
Kamiokande II experiment = Kamioka nucleon decay experiment in its second version (Japan)	Kamiokande metal mine, Gifu prefecture, Japan, 1000 m underground	2142 t water 680 t fiducial volume	Real-time experiment $\nu_e + e^- \rightarrow \nu_e + e^-$	9.3 MeV in the time period Jan. 1987–May 1988 7.5 MeV in the time period June 1988–Apr. 1990 able to detect ^8B neutrinos		0.46 ± 0.05 (stat.) ± 0.06 (syst.) of SSM's signal	Jan. 1987–May 1988 at $E_\nu \geq 9.3$ MeV, 450 days; June 1988–Apr. 1990 at $E_\nu \geq 7.5$ MeV, 590 days		
SAGE = Soviet-American Gallium Experiment (FSU + USA)	Mount Andyrchi, Baksan, Northern Caucasus, Russia, 4700 m w.e.	30 t liquid metallic ^{71}Ga	Radiochemical experiment $\nu_e + {}^{71}Ga \rightarrow {}^{71}Ge + e^-$	0.233 MeV, able to detect neutrinos from $p+p \rightarrow d + e^+ + \nu_e$, $p + e^- + p \rightarrow d + \nu_e$ ^7Be ^8B	132^{+20}_{-17} SNU (3σ) 1.2 Ge atoms per day	$20^{+15}_{-20} \pm 32$ SNU	Experiment began operation in May 1988; measurements in Jan. Feb. Mar. Apr. and July 1990 = 5 runs	11.4 d for ^{71}Ge	
GALLEX 1 = Gallium Experiment in its first series of runs (Europe + USA + Israel)	Gran Sasso d'Italia Apennines, northeast of Rome, Italy 3300 m w.e.	101 t $GaCl_3$ containing 30.3 t ^{71}Ga	Radiochemical experiment $\nu_e + {}^{71}Ga \rightarrow {}^{71}Ge + e^-$	0.233 MeV, able to detect neutrinos from – pp, ^7Be, ^8B	132^{+20}_{-17} SNU (3σ)	83 ± 19 (stat.) ± 8 (syst.) SNU	14 runs in the period 14 May 1991 to 31 Mar. 1992 = 295 days	11.4 d	^{71}Ge 3 weeks

[a] The question remains whether deficiencies in the solar model or new properties of the neutrino are the source of the solar neutrino problem.
[b] 1SNU = 10^{-36} interactions per target atom per second.

The main driving force of solar system evolution is the continuous loss of energy by the Sun into surrounding space. Photons, generated in the gravitationally stabilized solar fusion reactor, are the carriers of the escaping energy. Because of the electromagnetic interaction of solar photons with matter, being generated, scattered, absorbed and reemitted, the timescale for the transport of photons from the center to the solar surface is of the order of 10^7 years (the Helmholtz–Kelvin timescale). The Sun is a surface source of photons. As a by-product of thermonuclear reactions deep inside the Sun, neutrinos are generated by the proton–proton chain and the carbon–nitrogen–oxygen cycle. These elusive particles, whose interaction cross-section with matter is of the order of $\sim 10^{-44} x^2$ cm^2 (where x is the neutrino momentum in MeV/c), are able to leave the Sun almost instantaneously. The Sun is a volume source of neutrinos, although only 2% of the total amount of energy emitted by the Sun is due to neutrino emission. The total number of solar neutrinos reaching the Earth's surface per square centimeter per second can be estimated as

$$N_\nu = \frac{2L_\odot}{E} \approx 2 \times 10^{38}$$

where $L_\odot$ denotes the photon luminosity of the Sun and E is the energy generated by the overall net reaction of solar energy generation through the proton–proton chain (4p → α + e$^+$ + 2ν_e), including the energy of the two neutrinos. The solar neutrino flux on Earth is

$$\Phi_\nu = \frac{N_\nu}{4\pi(AU)^2} \approx 6.5 \times 10^{10} \text{ cm}^{-2}\text{ s}^{-1}$$

Painstaking detections on Earth show fewer of those elusive particles than the theoretical standard solar model predicts. This deficit is called the solar neutrino problem.

In the late 1930s two thermal fusion processes for converting hydrogen into helium were suggested as the energy source of the Sun: The proton–proton chain (PP-I segment), and the carbon–nitrogen cycle (CNO cycle). In the late 1940s it became clear that the Sun generates its energy by employing the proton–proton chain, and that less than 2% of the Sun's energy was produced by the carbon–nitrogen cycle. After the discovery of the fact that the proton–proton chain consists of three subchains (PP-I, II and III), all of them producing neutrinos at distinctive energies, it became urgent to test the hydrogen fusion theory for the Sun and to detect (at least) the high-energy neutrinos. In the 1950s and 1960s much experimental and theoretical work had been done to estimate the solar neutrino fluxes precisely enough to determine the physical characteristics of a neutrino detector needed to observe the solar neutrino emission. Since 1967, attempts have been made to establish a series of detectors for measuring the solar neutrino emission in different parts of the solar neutrino energy spectrum. To date four solar neutrino observatories are in operation, commonly known as the chlorine, Kamiokande II, SAGE and GALLEX 1 experiments (see Table S5). The classical solar neutrino result is that the chlorine experiment, in operation since 1970, has observed solar neutrino events at a rate two to three times smaller than that expected from the standard solar model (Davis et al., 1989). In addition, the data of the chlorine experiment taken in the past 20 years suggest a variation of the ^{7}Be and ^{8}Be solar neutrino flux over time. The time varying flux plotted together with the 11-year cycle of sunspot activity shows an apparent anticorrelation between the abundances of sunspots and solar neutrinos. Kamiokande II detected the Cerenkov light produced by recoil electrons when solar neutrinos are scattered elastically off electrons in water (Hirata et al., 1990). The scattering of neutrinos from electrons shows a strong directionality and hence the direction of the incoming neutrinos from the Sun was clearly identified. The Kamiokande II experiment confirmed the results of the chlorine experiment but has not demonstrated variations of the solar neutrino flux over its time period of operation. Its reaction rate, only 46 ± 8% of the rate predicted by the standard solar model, has been staying at a constant value to within the statistical error of 30%. SAGE and GALLEX 1 are primarily sensitive to the dominant component of the neutrino flux from the Sun that is directly linked to the solar luminosity through the proton–proton reaction. The capture rate reported by SAGE after five uncontaminated observing runs allows the interpretation that no solar neutrino signal has been seen (Abazov et al., 1991). The upper limit of SAGE's result of 79 SNU at the 90% confidence level is the theoretical lower limit on the gallium event rate, assuming only (1) that the Sun radiates energy as fast as it generates it (the Sun in thermal equilibrium) and (2) that the

Figure S24 The 37-Cl Neutrino Detector.
The photograph shows the massive tank employed in the chlorine neutrino detection experiment by Raymond Davis Jr at the Homestake Gold Mine in South Dakota, USA. This experiment has run since 1970, providing important constraints on solar neutrino flux models (Courtesy of Raymond Davis Jr.)

neutrinos emitted as a result of thermonuclear energy production travel unaffected to the Earth. After 295 days of operation, GALLEX 1 reported a neutrino capture rate 63 ± 16% of that predicted by the standard solar model (Anselmann et al., 1992). The GALLEX 1 experiment observed neutrinos from the proton–proton reaction. The neutrino capture rate of 83 SNU is to be compared with the value 79 SNU that is attributed to solar models in which all the solar luminosity is generated by the PP-I segment of the proton-proton chain. However, the GALLEX 1 result is not in agreement with those of the chlorine and Kamiokande II experiments. In summary, the solar neutrino problem is still with us.

Evolutionary sequences of the Sun as a main-sequence star show that the luminosity of the sun has increased by 40% over its lifetime of 4.6 Ga. Climatic fluctuations may be linked with possible variation of the solar energy output over intervals of a few million years. During the lifetime of the Sun, hydrogen has been converted into helium in the solar core, thereby increasing the opacity of matter, the central temperature and, as a consequence, the emission of photons and neutrinos. It is rather difficult to find records for the long-term variations of the photon luminosity of the Sun. However, there are possible ways of measuring the past neutrino luminosity of the Sun. Haxton (1990) proposed that this might achieved by measuring the conversion of tellurium into xenon isotopes, over time, in the Earth's crust under the irradiation of energetic solar neutrinos. In the standard solar model the increase of the Sun's luminosity is accompanied by an exponential growth of the boron 8 neutrino flux within the proton–proton chain (PP-III). A geochemical scheme for this purpose is based on the conversion of tellurium 126, into indium 126 by neutrino interaction, which in turn decays into xenon 126, the rarest of the xenon isotopes. Xenon 126 has a natural abundance of 0.089%. Measuring the excess abundance of xenon 126 in terrestrial rocks containing tellurium may permit the estimation of the flux of energetic solar neutrinos in the past. The xenon 126 concentration of tellurides thus measures the solar neutrino flux since the formation or recrystallization of the rocks concerned. The comparison of measurements from various rocks of different age may reveal the history of variations of the solar neutrino flux. Although an exciting project for the detection of the solar neutrino flux over geologic times, it has not yet been implemented due to experimental difficulties.

The Sun emits neutrinos through fusion, and the Earth emits antineutrinos through fission. Radioactive decay of potassium and uranium, the natural radioactivity of the still warm 'ashes'left over from the creation of the universe, make the Earth a rich source of antineutrinos. Detecting these neutrinos would provide information on the interior processes governing the Earth's evolution and dynamics. The flux of terrestrial antineutrinos is a direct reflection of the associated energy production of the Earth; similar questions may be asked for other planets of the solar system.

Hans J. Haubold

Bibliography

Abazov, A.I. et al. (1991) Search for neutrinos from the Sun using the reaction ^{71}Ga $(v_e, e^-)^{71}$Ge. *Phys. Rev. Lett.*, **67**, 3332–5.
Anselmann, P. et al. (1992) Solar neutrinos observed by GALLEX at Gran Sasso. *Phys. Lett.*, **B285**, 376–90.
Anselmann, P. et al. (1992) Implications of the GALLEX determination of the solar neutrino flux. *Phys. Lett.*, **B 285**; 390–7.
Bahcall, J.N (1989) *Neutrino Astrophysics*. Cambridge: Cambridge University Press.
Davis, R., Jr, Mann, A.K. and Wolfenstein, L.(1989) Solar neutrinos. *Ann. Rev. Nuclear Particle Sci.* **39**, 467–506.
Davis, R., Jr et al. (1989) In *Proceedings of the Thirteenth International Conference on Neutrino Physics and Astrophysics* (eds J. Schneps et al.). World Scientific: Singapore, pp. 518.
Haxton, W.C.(1990) Proposed neutrino monitor of long-term solar burning. *Phys. Rev. Lett.*, **65**, 809–12.
Hernandez, J.J., Stone, J., Porter, F.C. et al. (1990) Review of particle properties – Particle Data Group, *Physics Letters*, **239**, 1–516.
Hirata, K.D., Inoue, K., Kajita,T. et al. (1990) Results from one thousand days of real-time, directional solar-neutrino data. *Phys. Rev. Lett.*, **65**, 1297–1300.
Mathai, A.M. and Haubold, H.J (1988) *Modern Problems in Nuclear and Neutrino Astrophysics*. Berlin: Akademie-Verlag.

Cross reference

Sun

SOLAR PARTICLE EVENT

Solar activity is manifested in many ways, including the occasional emission of energetic particles by large solar events. The amount of energy involved in large solar events is huge (up to $\sim 10^{26}$ J in a fraction of an hour), and many types of radiation are emitted from the Sun by the larger events, including electromagnetic radiation over most of the spectrum from radio waves to gamma rays. The largest events also accelerate particles, usually electrons and occasionally atomic nuclei, to high energies (Chupp, 1987). Details of the accelerator mechanism(s) are not well known. A fraction of these solar energetic particles (SEPs) escape into interplanetary space and reach the vicinity of the Earth, where they are measured. These particles interact with objects in the solar system, including the Earth, and can be a serious radiation hazard in space.

Events emitting atomic nuclei with energies above a few MeV (1 MeV = 1.6022×10^{-6} ergs) occur fairly frequently at the Sun, on the order of several per month (except at times of very low solar activity, about 4 years around the time of minimum solar activity during a typical 11-year solar cycle) (Feynman et al., 1990). Events emitting higher fluxes of SEPs or having particles with higher energies occur less frequently. About 200 events since 1956 have had solar proton fluxes at the Earth of over 10 protons (cm^2 s sr)$^{-1}$ with energies above 10 MeV (Shea and Smart, 1990), and about 45 of these events had enough particles above 500 MeV to be detected by neutron detectors on the Earth's surface (Flückiger, 1991). The ability to predict solar particle events is poor, especially for the biggest events.

About 98% of the particles in a typical solar particle event are protons, with the rest being ^{4}He and some heavier nuclei. Larger events tend to normal composition (coronal abundances with corrections for the first ionization potential and ionic charge-to-mass ratio) (Breneman and Stone, 1985). These energetic particles travel out from the Sun, mainly along the field lines of the interplanetary magnetic field. The highest-energy particles begin to arrive at the Earth within about 15 min after being emitted from the Sun, while lower-energy particles take hours to start reaching the Earth. The flux of SEPs increases for a period of hours to days, depending on the event's location on the Sun, the particles' energies and conditions in interplanetary space. The fluxes then usually decrease roughly exponentially for several days. The propagation through interplanetary space is simple during 'quiet' times but can be complicated if strong shocks are present, with particles either being accelerated or decelerated, depending on the conditions.

Energetic particles from the Sun were discovered with the aid of ionization chambers in 1942, but systematic studies of SEPs did not start until about 1960 (Shea and Smart, 1990). Samples returned from the Moon contain nuclides like 7×10^5 year ^{26}Al produced by SEPs. These have provided us with a record of SEP fluxes for several time intervals up to $\sim 10^7$ years ago (Reedy and Marti, 1991). These nuclides are made by nuclear reactions induced by SEPs in the top few millimeters of lunar material with unique activity versus depth profiles. The long-term average SEP fluxes deduced from these nuclides in lunar samples are consistent with measurements of solar particle fluxes made since 1956.

Robert C. Reedy

Bibliography

Breneman, H.H., and E.C. Stone (1985) Solar coronal and photospheric abundances from solar energetic particle measurements. *Astrophys. J. Lett.*, **299**, L57–61.
Chupp, E.L. (1987) High-energy particle production in solar flares (SEP, gamma-ray, and neutron emissions). *Physica Scripta*, **T18**, 5–19.
Feynman, J., Armstrong, T.P. Dao-Gibner, L. and Silverman, S. (1990) Solar proton events during solar cycles 19, 20, and 21. *Solar Phys.*, **126**, 385–401.
Flückiger, E.O. (1991) Solar cosmic rays. *Nucl. Phys. B (Proc. Suppl.)*, **22B**, 1–20.
Reedy, R.C., and Marti, K. (1991) Solar-cosmic-ray fluxes during the last ten million years, in *The Sun in Time* (eds C.P. Sonett, M.S. Giampapa and M.S. Matthews). Tucson: University of Arizona Press, pp. 260–87.
Shea, M.S. and Smart, D.F. (1990) A summary of major solar proton events. *Solar Phys.*, **127**, 297–320.

Acknowledgement

This work supported by NASA and performed under the auspices of the US Department of Energy.

Cross references

Coronal mass ejection
Solar flare
Solar wind

SOLAR PHOTOSPHERE

The photosphere is the layer of a stellar atmosphere that emits the continuous radiation carrying most of the star's luminous energy (Plate 1). The effective temperatures of stellar photospheres range between about 50 000 K for massive young supergiants to below 2000 K for cool dwarf stars.

The photosphere of the Sun is the best studied, but with increased development of powerful techniques (interferometry, Doppler imaging), the photospheres of certain other particular stars are becoming known: for instance, spots have been detected on young stars such as T Tauri (Joncour, 1992). The outer limit of the solar photosphere is taken to be the boundary of the visible solar disk as seen in white light.

Until the beginning of the 20th century, the Sun's surface was assumed to be a hot liquid. From sunspot behavior C.G. Abbot (1900) postulated that it was gaseous, and now it is recognized as a plasma. The visible surface of the Sun is the outer limit of the photosphere, where the solar plasma becomes opaque to visible light. The thickness of this layer is about 300 km. Physically, the photosphere is limited below by the convective zone. The large scale mass motions in the convective zone produce many phenomena in the photosphere, including sunspots and solar activity. Above the photosphere lies the chromosphere. Through this region the energy surging from the interior of the Sun escapes to outer space at a rate of $E = 6.322$ kW cm^{-2}. By Stephan's law, this corresponds to an effective temperature of 5755 K. The photosphere is the coolest layer of the Sun, with the temperature decreasing outward from 8000 K to 4500 K at the upper boundary. The density is 3.4×10^{-7} g cm^{-3} (which must be compared to the mean density of the Sun itself, i.e. 1.4 g cm^{-3}). The radius of the Sun to the limit of the photosphere is 695 500 km (i.e. 959.63 arcsec, according to the *Astronomical Almanac*). The work of Dicke and Goldenberg (1967) suggested an equatorial diameter 84 milliarcsec larger than the polar diameter, confirmed by recent works (Rozelot and Rösch, 1996). Historical data (if correct) would imply that, through the solar cycle, the diameter varies by a few seconds of arc (Laclare, 1987). Consistent with this postulate is the observed fact that solar irradiance varies by up to about 0.3% through the 11-year cycle. Mean total irradiance $L = 4\pi R^2 E$ is 3.78×10^{23} kW (1368 W m^{-2} at 1 AU). If L varies through time, both the solar diameter and the Earth's climate should document such variations.

Granulation

Except for sunspots and accompanying solar activity, the so-called quiet photosphere shows a pattern of bright irregular polygonal structures, separated by thin dark lines, constantly changing. The granulation is formed by a great number of convective cells, the granules, which represent the smallest clearly defined pattern on the surface of the Sun. Figure S25 shows such a typical granulation or grainlike structure. Bray, Loughead and Tappere (1976) established that the bright regions contain an upward motion estimated at

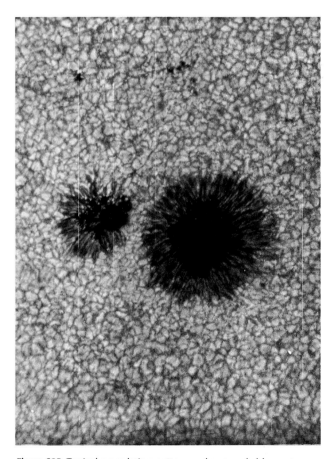

Figure S25 Typical granulation pattern and a remarkable spot near the center of the disk, taken at Pic du Midi Observatory by R. Müller and G. Cepptelli, with excellent seeing conditions (4 June 1980). Resolution is roughly 0.25 arcsec (180 km). Filter: $\lambda = 575$ nm, $\Delta\lambda = 10$ nm. (Courtesy R. Müller, OPMT.)

0.7 km s^{-1} and the dark lines downward motions estimated at 1.1 km s^{-1}. In terms of spatial scale, the granules have a typical horizontal size of 1000 km, the average distance between granules being about 1400 km at the center of the disk. The distribution of granulation is described in terms of its statistical characteristics; the rms temperature fluctuation in the granules (determined by the brightness fluctuation) is about 100 K, the lifetime is 9 min (larger granules last longer) and granules sometimes explode with a velocity of about 2 km s^{-1}. Destruction occurs by fragmentation (35%), decay (60%) and by merger (5%). The granular contrast is wavelength dependent (lower in the red), reflecting the combination of Planck function and height effects. Granulation theories depend on how much of the buoyancy work done by a rising warm element goes into vertical acceleration of that element, and how much is immediately converted into kinetic energy by turbulent cascades. One model was proposed by Nordlund (1980) but open questions remain.

Supergranulation, mesogranulation and filigree

Closer examination of the chromospheric pattern shows even larger convective cells, the supergranules, flowing horizontally with a velocity of about 300–500 m s^{-1} and with a typical horizontal distance between their centers of about 30 000 km. They live much longer than granules, from 20 h to several days, but little is known concerning the details of growth and decay. While granules appear to extend no farther than the upper photosphere, supergranule motions penetrate the chromosphere, which has the same spatial scale. Between granules and supergranules, a cellular motion intermediate in scale appears called mesogranulation. It has a spatial scale of about 5000 to 10 000 km, with a lifetime of about 2 h. Filigree is seen in the granulation as bright chains of very small features whose characteristic size is only 0.25 arcsec, with transverse velocities of the order of 1.5 km s^{-1}.

Sunspots and active regions

Sunspots appear at the surface of the photosphere as dark and irregular spots, isolated or in groups and closely associated with the solar magnetic field. They form depressions in the photosphere of typically 650 km, the largest covering an area comparable to supergranules. The variation in time of the total spotted area (well correlated with the so-called International Zurich Number) is known as the sunspot cycle, with a marked periodicity of 11 ± 6 years. During periods of minimum activity, no more than 1% of the hemisphere is covered by sunspots; during periods of maximum activity, a single, very large sunspot group may reach 3000 millionths in area (and be visible to the naked eye). The first spot of a cycle appears near 30° N and S latitudes and the last ones near the solar equator. There are few spots above 30° latitude. Galileo in 1610 was the first to discover spots in the photosphere, and George Ellery Hale (1913) found that leading spots in opposite hemispheres showed opposite magnetic polarity, and that these polarities reversed in each successive 11-year cycle. Thus it takes 22 years on average for the magnetic fields to come back to the same polarity.

Measurements of the magnetic field in the photosphere (quiet Sun) give intensities ranging from 5 to 10 gauss, to 50–300 gauss in the faculae and 1000–4000 gauss in the active regions. Thus spots appear dark because the strong field inhibits convective energy transport, thereby lowering the temperature by about 1000 K. Spots are surrounded by a dark umbra and a bright penumbra (still darker than the photosphere). The penumbra appears striated with roughly alternating bright and dark lines of filaments extending more or less radially from the umbra to the surrounding granulation. The granules disappear and are replaced by a fibrous structure, the fibrils, due to roll convection, showing motions of matter of 0.5 km s^{-1}. Spots have a lifetime of between a few hours and several solar rotational periods. Their growth and decay are generally related to those of active regions, in which they are embedded.

The 11-year activity cycle of the Sun is a magnetic phenomenon, caused by the interaction of convection and differential rotation. Convection is responsible for the transport of energy in the outer layers of the Sun; differential rotation refers to the fact that the Sun rotates more rapidly at the equator than near its poles. The interaction between the two causes the large-scale field of the Sun to vary from a predominantly polar field at sunspot minimum to a toroidal field at sunspot maximum (Walker, 1984).

Photospheric models

The photosphere appears to darken from the center to the limb over practically the whole of the visual spectrum. This limb-darkening is readily understood for the continuum in terms of the outward decrease of temperature in the photosphere. Thus the principal tool for building photospheric models is the analysis of this limb darkening as a function of wavelength. An example of such a model can be found in Vernazza, Avrett and Looser (1976), computed with detailed attention to deviations from local thermal equilibrium.

Fraunhofer lines

Spectral analysis of the photosphere reveals a continuum crossed by a great number of dark lines, discovered by Joseph von Fraunhofer in 1814. The continuum is seen as a juxtaposition of colours, from the violet to the red, but extending far into the x-ray and infrared range. Its spectral energy distribution in the visible mimics a 6000 K Planckian curve, but as one scans the spectrum at all wavelengths, one sees to different depths in an atmosphere having a steep temperature gradient. It arises from free–bound and free–free transitions of the negative hydrogen ion in the photosphere. Fraunhofer lines are caused by absorption by the chemical elements in the outer, cooler (i.e. lower excitation temperature) region of the Sun of the continuous spectrum emitted by its hot interior. More than 26 000 lines can be detected from 293.2 nm to 1349.5 nm. Identification of these lines is achieved by comparison of spectra of known chemical elements on Earth. The analytical chemistry at distance reveals the composition of the Sun: practically all the elements can be found plus some molecules, such as TiO. Solar-stellar abundances of chemical elements are summarized by Meyer (1985).

Solar waves and oscillations

After the discovery in the 1960s that the visible gas layers of the Sun oscillate with short periods (5 min), it was necessary to wait until 1975 before understanding that the surface phenomenon was only the visible part of free oscillations of the whole solar sphere. At that point helioseismology was born. These oscillations exist on a broad spectrum: at one end, the oscillations with very long horizontal wavelength penetrate to the solar interior, down to the core if they are radial; at the other end, those with very short wavelength are mainly surface phenomena. The Sun, as many other stars, vibrates under the effect of two forces: one corresponds to acoustic waves (called pressure modes or p-modes) and the other to gravity waves (called g-modes). Acoustic wave generation occurs within the convection zone, while gravity waves are produced at the upper boundary, where rising gas impacts the overlying convectively stable layers. The 'seismic' method for their study has arisen only recently, mainly because the amplitude of each wave at the visible surface is extremely small. The typical values are 10 to 20 m on the solar radius, or 1 to 2 mK in surface temperature, or 10 km s^{-1} in upward or downward surface velocity. However, experiments will be routinely operated until the year 2001 AD to probe both the solar surface and the solar interior. The same technique is also suitable for observing this phenomenon on other stars: δ Scuderi, ε Cyg, and α Boo are currently observed.

J.P. Rozelot

Bibliography

Abbot, C.G. (1900) Preliminary statement of the results of the Smithsonian Observatory Eclipse expedition, *Astrophys. J.* **12**, 69–73.
Bray, R.J., Loughead, R.E. and Tappere, E.J. (1976) Convective velocities derived from granules contrast profiles in Fe I. *Solar Phys.*, **49**, 3–18.
Dicke, R.H. and Goldenberg, H.M. (1967) Solar oblateness and general relativity. *Phys. Rev. Lett.*, **18**, 313–6.
Dickinson, R.E. (1986) Effects of solar electromagnetic radiation on the terrestrial environment, in *Physics of the Sun*. Vol. III (ed. Sturrock et al.). Reidel Pub. Comp., p. 155.
Hale, G.E. (1913) The Earth and Sun as magnets. *Smithsonian Report* for 1913, pp. 145–158. (Address delivered at the semicentennial of the National Academy of Sciences, at Washington, DC May 1913.)
Joncour, I. (1993) Imagerie Doppler d'une étoile T Tauri. PhD Thesis, Grenoble University.
Laclare, F. (1987) Mesure du diamètre du Soleil à l'astrolabe solaire, *C. R. Acad. Sci. Paris* Serie II, **305**, 451–4.
Meyer, J.P. (1985) Solar-stellar outer atmospheres and energetic particles and galactic cosmic rays, *Ap. J. Suppl.*, **57**, 173–204.
Nordlund, A. (1980) Numerical simulations of the solar granulation, *IAU Coll.*, **51**, 17.
Rozelot, J.P. and Rösch, J. (1996) Is the Sun changing shape? *C.R. Acad. Sci. Paris*, **322**, 637–44.
Vernazza, J.E., Avrett E.H. and Looser, R. (1976) Structure of the solar chromosphere, *Ap. J. Suppl.*, **30**, 1–60.
Walker, A.B.C. (1984) A golden age for solar physics, in *Astrophysics Today*, (ed. A.G.W. Cameron), American Institute of Physics, New York, pp. 74–81.
Zirin, H. (1988) *Astrophysics of the Sun*, Cambridge University Press, New York, 433 pp.

Cross references

Solar activity
Sun

SOLAR SYSTEM

The solar system may be broadly defined as consisting of all those objects that are ultimately governed by the gravitational field of the Sun (Encrenaz and Bibring, 1990). In addition to the planets, moons, minor planets and dust of the planetary system, the solar system includes the distant bodies of the Kuiper belt (q.v.) and Oort cloud (q.v.). The latter extends to distances of perhaps 50 000 AU (see Astronomical unit). The gravitational influence of the Sun extends perhaps three times as far (about 225×10^{12} km), a distance of about 2 light years, or about halfway to the nearest star.

The Sun comprises more than 99.8% of the total mass of the solar system, although on the other hand the largest part of the angular momentum of the solar system (98%) resides in the orbital motion of the planets.

The solar system is moving in a nearly circular orbit about the center of the Milky Way galaxy. The mean velocity of the galactic rotation is about 240 km s^{-1}. The solar system has a velocity of about 19 km s^{-1} relative to nearby stars in the galaxy. As explained by William Herschel (q.v.), the nearby stars appear to be shifting apart in the direction of the Sun's motion, its so-called standard apex; it is oriented at the present time in right ascension 19 h and declination +36°.

The question of the possible existence of other solar systems is explored in the article Planet: extrasolar. Theoretical considerations suggest that planetary systems may be relatively common in our universe, and observations may have already pinpointed the locations of forming planetary systems about some stars.

James H. Shirley

Bibliography

Brandt, J.C. and Hodge, P.W. (1964) *Solar System Astrophysics*. New York: McGraw-Hill.
Encrenaz, T. and Bibring, J.-P. (1990) *The Solar System*. New York: Springer-Verlag.
Wood, J.A. (1979) *The Solar System*. New Jersey: Prentice-Hall.

Cross references

Angular momentum
Astronomical unit
Kuiper belt
Oort, Jan Hendrik, and Oort cloud
Planet
Planet: extrasolar

SOLAR SYSTEM: ORIGIN

The solar system includes the Sun, nine major planets, with satellites around seven of them, a number of minor planets (asteroids and comets), other small solid bodies and interplanetary dust and plasma.

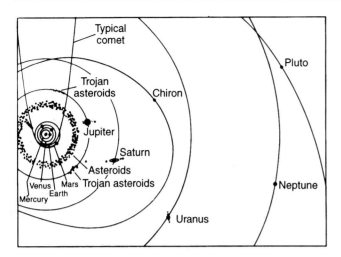

Figure S26 The arrangement of the nine major planets, the non-circular orbits of the small bodies Pluto and Charon, and representative comets and different types of asteroids. Orbits drawn to scale. (After W. Hartmann).

The planetary system is rather regular (Figure S26). Planetary orbits are nearly coplanar, nearly circular and spaced regularly. The direction of orbital motion is the same for all of the planets and is the same as the direction of rotation of the Sun. The bulk of the mass of the solar system is concentrated in the Sun, whose mass is 1 $M_\odot = 2 \times 10^{33}$ g. The total mass of the planets is about $1/743\ M_\odot$. However, more than 95% of the angular momentum in the solar system resides in the orbital motion of the planets. The properties of planets change regularly with their distance from the Sun. Near the Sun are small, stony terrestrial planets, which have slow rotation and few, if any, satellites. Farther from the Sun are large, light giant planets. They possess extensive satellite systems and rotate relatively rapidly. These satellite systems are similar in some features to the planetary system.

The flat and regular structure of the planetary system implies that the planets were formed in a flattened protoplanetary disk composed initially from gas and dust, which is often called the solar nebula (e.g. Wetherill, 1989).

Historical review of cosmogonical theories

The origin of the solar system is perhaps the oldest unsolved problem of scientific philosophy (Ter Haar and Cameron, 1963). Historically, there have been two different ways of approaching the problem of the origin of the solar system: theories considering a joint development of the Sun and circumsolar protoplanetary disk, and theories assuming that the interaction between the Sun and some external body has been the first step in the history of our planetary system.

Descartes was the first to attempt a physical approach to the problem of the origin of the solar system. In 1644, assuming that space is filled with an ether, he proceeded to show that the vortex shape of the solar system is the only stationary one. The friction between eddies would smooth out the rough shape of the primeval matter, which tended towards the center of the vortex, forming the Sun. The coarse bodies are captured in the vortex, forming the planets, and finer matter is swept out of the vortex by radiation pressure.

Significant progress in the development of cosmogonical ideas was initiated by Kant and Laplace. Kant (1755) suggested that both galaxies and the proto-Sun would be formed due to gravitational concentration of material in a region with a slightly larger density. Kant was the first to guess that the co-planarity and roundness of the planetary orbits imply that the planets were formed due to conglomeration of a number of much smaller bodies, moving in a flattened protoplanetary disk around the Sun, and that the disk was flattened due to its rotation (Figure S27). He did not assume, however, that the solar nebula was a part of the Sun. The idea of a linked origin for the Sun and a surrounding solar nebula was introduced by Laplace (1796). He assumed that during cooling and contraction of the proto-Sun, planetary material is left behind in the form of rings. This occurs when the centrifugal force at the equator of the contracting protosolar nebula becomes larger than the gravitational force. The ideas of Kant and Laplace, supplemented by the ideas of redistribution of the angular

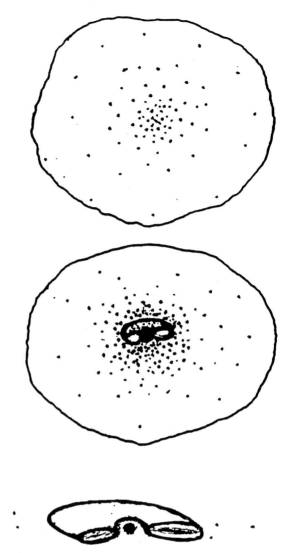

Figure S27 Schematic diagram showing the process of formation of the Sun and the solar nebula inside a collapsing protosolar cloud.

momentum in the solar nebula and braking of the solar rotation, laid the foundations of the modern cosmogony of the solar system.

It was understood by the 1920s that the dissipative processes in the solar nebula act to move its outer parts out and cause the inner parts to move in. In 1943 von Weizsäcker reached the conclusion that the effect of turbulent viscosity would be to separate the solar nebula into two parts – a central core containing most of the mass, and the rest in the form of an extended disk, containing most of the angular momentum. In 1962 Cameron argued that thermal convection could be the mechanism supporting the turbulence in the solar nebula; later he also discussed a mechanism associated with a shear flow between the accreting envelope and the solar nebula. Hoyle suggested in 1960 that the magnetic field linking the Sun to the solar nebula could brake the solar rotation by transferring angular momentum from the Sun to the inner part of the solar nebula. In 1962 Schatzman introduced the idea that the olar rotation has been slowed down by the magnetized solar wind. These findings revealed that the strong non-homogeneity of the angular momentum in the solar system could be natural result of its early evolution.

The non-homogeneity in distribution of the angular momentum gave birth to many tidal theories. The first tidal theory was put forward by Buffon in 1745, who suggested that a comet hitting the Sun might have been responsible for tearing from the Sun sufficient material to produce planets. In 1917 Jeans, and a year later Jeffreys, postulated that the passage of a star close to the Sun would cause one long filament of matter to be drawn out, and to break up into separate fragments condensing into planets. In tidal theories by

Lyttleton, Alfvén and Schmidt the capture of interstellar material by the Sun is suggested. Kaula and Kobric proposed that tidal interactions between collapsing protostellar clouds in the forming stellar cluster could accelerate the rotation of their outer envelopes; they also suggested that such an interaction could be responsible for the origin of the angular momentum of the solar system.

Two classes of hypothesis were emerging in studies of planetary formation: (1) Laplace, Jeans, Kuiper, Fesenkov and Cameron supposed that planets were formed initially as massive, gravitationally bound bodies with nearly solar composition; (2) Ligondes, Moulton, Chamberlain, Schmidt, and Alfvén and Arrhenius postulated formation of planets from the solid material. One of the main purposes of the models was explanation of the empirical ratio found between the number of a planet i (excluding Mercury) and the radius of its orbit R_i. For R_i measured in astronomical units (1 AU = 1.5×10^{13} cm is the average distance of the Earth from the Sun), the ratio could be written as

$$R_i = 0.4 + 0.3 \times 2^{i-2}$$

This ratio was first discovered by Titius and popularized by Bode (1777), and is called the Titius–Bode rule (q.v.).

In 1944 Schmidt and his colleagues began a systematic elaboration of the theory of planetary formation from non-volatile material of the solar nebula. First, Schmidt assumed that the Sun captured a swarm of small particles and bodies from the interstellar cloud; these then grew to form the planets by collisional accretion. The Titius–Bode law is thought to be a result of competition between the largest bodies accumulating dispersed material. Gurevich and Lebedinsky (1950) proposed that the particles from which the planets accumulated were formed in a gas cloud surrounding the Sun. In the region close to the Sun the heat from the Sun would allow only non-volatile materials to condense into particles which could accumulate into planets. Farther out, the volatile materials, which are much more abundant, may condense. Dust particles settle to form the thin disk in the equatorial plane of the solar nebula, and then the dust disk is fragmented, forming a number of dust condensations. Then asteroid-sized bodies are formed from the gravitational instability of the dust disk itself. Perturbations during close passages with other asteroid-sized bodies would cause an increase in the ellipticity of their orbits. Sometimes collisions result in accumulation and sometimes in the fragmentation of the bodies. The process would end with the formation of a few major planets (the largest of them, Jupiter and Saturn, would suck in both solids and gas), and some remaining asteroid-like bodies between the orbits of Mars and Jupiter. This scenario was developed quantitatively by Safronov (1991). A similar scenario of planetary formation is under study by many scientists throughout the world.

In 1952 Urey established a new direction in cosmogonical research. He deduced the conditions that could give rise to the observed properties of meteorites, suggesting that the chondrites were average samples of the non-volatile material in the solar system. At the present time, laboratory studies of meteorites have become the main method for studies of elemental and isotopic compositions, chronology and thermal evolution of the solar system (see also Cosmochemistry; Dating methods).

There are five specific problem areas in our current model of the manner in which the solary system formed:

1. the initial collapse of a protosolar cloud to form the Sun and solar nebula;
2. the growth of dust particles and formation of planetesimal-sized bodies;
3. accumulation of planets from planetesimals, which was interrupted in the asteroid belt;
4. the addition of planetary volatiles to the outer planets (including massive atmospheres of Jupiter and Saturn); and
5. the subsequent early evolution of the planets.

Figure S28 shows the main stages of planetary formation in the solar nebula, as they were drawn by Levin (1964). In general, this picture has been considered to be reasonable up to now.

Origin of the sun and solar nebula

The coincidence of isotopic compositions of the Sun, the Earth and meteorites for basic non-volatile elements, and the similarity of chemical compositions of the Sun and Jupiter, are evidence that the Sun and the solar nebula originated from the same concentration of interstellar matter. From the astrophysical point of view, the Sun is a

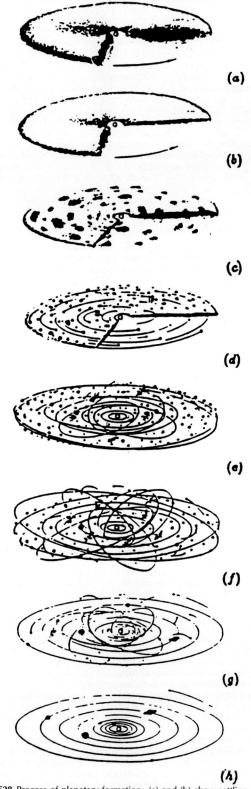

Figure S28 Process of planetary formation: (a) and (b) show settling of the dust to the central plane of the solar nebula and thin dust disk formation; (c) formation of dust condensations, followed by development of gravitational instability in the dust layer; (d) formation of solid bodies (planetesimals) from dust condensations; (e) beginning of accumulation and increase of random velocities; (f) appearance of embryos at separate orbits; (g) the final stage of planetary accumulation; (h) the contemporary solar system. (After B. Levin.)

typical main sequence star of spectral type G2 with mass $M_\odot = 2 \times 10^{33}$ g and age about 4.5×10^9 years. Some molecular clouds, recognized as regions of star formation, contain stars that are similar to the Sun in terms of mass and chemical composition, but are much younger, with ages of 10^5 to 10^7 years. Contemporary theory of stellar evolution holds that the fate of a star is determined by its mass and chemical composition. The similarity of these parameters justifies the identification of T Tauri stars with the young Sun, and the use of the observational data on these stars to construct a theory of the Sun and solar nebula formation.

T Tauri stars are formed in the molecular clouds: low-temperature (≈ 10 K) clumps of interstellar matter with relatively high density (10^2 to 10^4 cm^{-3}), in which hydrogen and other gases (beside the noble gases) are in a molecular state, and condensed matter is included in grains. These clouds appear on the sky as dark, obscuring patches of matter. (The Taurus molecular cloud is well-known example of a dark cloud.) The dark clouds contain many small ($\sim 10^{17}$ cm), and dense ($\geq 3 \times 10^4$ cm^{-3}) cores, whose masses range between $\sim 10^{-1}$ and 10 $M_\odot$. The detection of infrared objects as well as optically visible T Tauri stars suggested that solar-type stars are forming in these clouds. According to the theory developed by Jeans early in the 20th century, stars are formed as the result of the collapse of compacted regions of the interstellar medium under the influence of gravitational effects. The collapse occurs when the forces of self-gravity exceed the sum of forces restricting compression.

The maximum mass of a cloud with temperature T and density n that could be supported by thermal pressure against gravitational collapse is given by the Jeans mass:

$$M_J = 1 \times \left(\frac{T}{10 \text{ K}}\right)^{3/2} \left(\frac{10^5 \text{ cm}^{-3}}{n}\right)^{1/2} M_\odot$$

and the free-fall timescale for gravitational collapse is

$$\tau_{ff} = 1 \times 10^5 \left(\frac{10^5 \text{ cm}^{-3}}{n}\right)^{1/2} \text{ years}$$

Much theoretical work has been done on the collapse of protosolar clouds (molecular cloud cores with $M \sim 1 M_\odot \geq M_J$, where M_J is the mass of Jupiter) to understand the origin of the Sun and solar nebula, by Larson, Bodenheimer, Tscharnuter, Black, Boss and others. The entire process, from the beginning of the dynamical stage of contraction all the way to the formation of a star, has thus far only been examined for a non-rotating cloud whose collapse was first simulated by Larson (1969). The collapse of a cloud starts isothermally; contraction occurs close to free-fall velocities and is accompanied by an increase in the concentration of matter toward the center. When the central part of the cloud becomes non-transparent for thermal radiation, i.e. at a density of about 10^{-13} g cm^{-3}, the pressure increase triggers a temporary deceleration in the contraction of a central region of radius about 10^{14} cm. This is followed by one more stage of dynamical contraction that is initiated by dissociation of molecular hydrogen, which starts at a density of about 10^{-8} g cm^{-3}, and results in formation of a starlike core (with the initial mass of $M_c \sim 10^{-2} M_\odot$ and central density $\rho_c \sim 10^{-2}$ g cm^{-3}). This core was possibly born pulsating, subsequently coming to hydrostatic equilibrium. The core is surrounded by an envelope which initially contains about 99% of the mass, which falls into the core over a period about equal to the initial free-fall time (10^5 years). A single star is formed in this case.

The solar system possesses an angular momentum and hence was formed in a rotating cloud. The minimum angular momentum of the protosolar nebula is estimated as the sum of that angular momentum of the proto-sun (which could not be much in excess of 10^{51} g cm^2 s^{-1}) and that of the minimum-mass solar nebula. This nebula is reconstructed by adding sufficient H and He to the present planets to recover the solar chemical composition, and by the addition of a considerable mass to take into account the ejection of material by the gravitational effects of the giant planets. The reconstruction gives a solar nebula with a mass of 0.02 to 0.05 M and an angular momentum of about 10^{52} g cm^2 s^{-1}. Disks about some of the T Tauri stars have comparable angular momenta. It is not unusual for the angular momenta of solar mass cores in the molecular clouds to be about 100 times larger.

The collapse of a rotating cloud with an angular momentum larger than the maximal angular momentum that a single star could possess results in the formation of either a binary (or multiple system of stars) or a single star with a circumstellar disk (Bodenheimer, Ruzmaikina and Mathiew, 1993). The development of the Sun and solar nebula inside the collapsing protosolar cloud is shown schematically in Figure S25.

The strongly non-homogenous distribution of angular momentum in the solar system implies significant redistribution of the initial angular momentum. There are three stages of slow contraction when redistribution could be effective: before hydrodynamical collapse has started; when a central region becomes opaque but dissociation of the molecular hydrogen has not begun; and also when a hydrostatic starlike core starts to form. As was suggested by Ruzmaikina (1981), the redistribution of the angular momentum results in embryo disk formation about the core; further, the disk expands inside the accreting envelope (due to the redistribution of the angular momentum by the turbulent viscosity) up to a radius of the order of the present-day solar system. In the same year, viscous growth of the disk at the accretion stage of the collapse was also studied by Cassen and Moosman (1981).

Evolution of the disk at the accretional stage, and the ratio of disk and core masses, are sensitive to the ratio of the accretional rate to the angular momentum transport rate in the disk, and also to the total angular momentum of the protosolar cloud.

Turbulent viscosity ν is usually considered as the main cause of the redistribution of the angular momentum in the disk. If a viscous timescale is short compared with the accretional one, then the disk radius R_{SN} grows with time t as

$$R_{SN} \sim (\nu t)^{1/2}$$

In this case an extended low-mass disk, with parameters similar to those given by the reconstruction from the present structure of the solar system, could be formed at the end of the accretional stage. If redistribution of the angular momentum is not effective over the timescale of accretion, the radius of the forming solar nebula will be about the maximal centrifugal radius of accreting material R_{ce}; i.e. the radius at which centrifugal force in the equatorial plane of the infalling matter is equal to the protosolar gravitation, and the disk mass M_{SN} is equal to

$$M_{SN} \sim 0.2 \cdot \left(\frac{R_{ce}}{R_{SN}}\right)^{1/2} M_\odot$$

The distribution of the mass and angular momentum between the Sun and disk at the end of accretion is different from that in the solar system. The disk is much more compact, or more massive, than the solar nebula. However, transport of the angular momentum and mass over the viscous timescale would continue after the end of accretion, resulting in the reduction of the disk mass and an increase of its radius. Such a model of solar nebula formation was developed by Lin and Papoloizou (1980), Ruden and Pollack (1991), Shu, Adams and Lizano (1987) and others. It is not known which of these two scenarios was realized during solar nebula formation.

Dust and the formation of solid bodies

The mass content of elements heavier than helium is about 2% in gas of solar composition. In molecular clouds an appreciable fraction of these elements is tied up in solid dust grains. During the formation of the solar nebula, grains fell in together with gas. They evaporated in the vicinity of the Sun, and survived in the outer parts of the solar nebula. Also, new solid particles were formed in the solar nebula from condensible material. While the solar nebula was turbulent, the particles were mixed with gas and grew due to collisions, possibly forming fluffy fractal aggregates. As soon as the turbulence decayed, the particles would precipitate toward the central plane of the disk and migrate to the Sun, because of the difference in the density of the particles and the gas. During these processes the particles collide and grow. With efficient clumping, the models suggest that the particles will settle to the central plane over a time equivalent to about 10^3 revolutions around the Sun (if they grow as dense spherules). The process of settling could be up to 10^3 times slower if particles, do not stick each other, or if they form extremely fluffy clumps (with a ratio of the surface area to volume independent of the clump's mass). Studies of dust particle coagulation encountered a problem in the estimation of the sticking efficiency of particles with size larger than 10^{-3} cm. A fluffy aggregate can plausibly grow to centimeter size and larger; small particles are held together by the electrostatic van der Waals force. Furthermore, these fluffy clumps could become more compact and denser and evolve into solid bodies.

The settling of solid particles, followed by the formation of a thin dust sublayer, and the development of a gravitational instability in the sublayer, is considered to be an important stage in the progression from dispersed solid matter to the planets (Figure S29a–d).

A generalization of the classic theory of gravitational instability in homogeneous non-rotating media to the case of a rotating disk has been made by Toomre (1964) for an infinitely thin disk, and by Safronov (1960) for a disk of finite thickness. It was shown that an axisymmetric mode of gravitational instability, resulting in the appearance of ringlike structures in the dust disk, develops first at a critical density $\rho_{cr.} \simeq 2\rho_* = 3M_\odot/(2\pi R^3)$ for ring-like perturbations with widths about ten times larger than the disk thickness. The presence of the gas in dust layers decreases the critical density. However, the relative drift of particles with different sizes, due to gas friction, causes radial spreading and the appearance of ringlike perturbations. The spreading can be prevented, and the development of the gravitational instability allowed, only if a major fraction of the ring mass is contained in particles of a certain predominant size, e.g. if particles of a certain size stop growing because of erosion by impacts of smaller ones. During contraction and increases in density, rings become gravitationally unstable and fragment into numerous local rotating dust condensations with masses $m_o \sim \sigma_p^3/\rho_*^2$, where σ_p is the surface density of the solid material in the disk. Further coagulation of the condensations causes their contraction and transformation into ordinary solid bodies – planetesimals. It is not clear which mechanism (a direct growth of the protoplanetary bodies from the small particles, or their formation from the dust condensations) dominated in the early solar system. Independently of the manner in which planetesimals formed, their coalescence through collisions was the next principal stage in the planetary formation process.

Accumulation of the planets

Planetesimals were formed in the central plane of the solar nebula, having almost circular orbits. Interacting gravitationally as they approach each other, planetesimals would acquire randomly oriented eccentric orbits, and random relative radial velocities (Safronov, 1991). Some of the orbits will intersect, and the bodies will collide. The collisional cross-section for a body of mass m_1 and radius r with a much smaller body, moving with respect to it with a velocity v at infinity, is equal to

$$\Sigma_{coll} \simeq \pi r_1^2 \left(1 + \frac{2Gm_1}{v^2 r_1}\right)$$

The second term describes an increase of the collisional cross-section compared with a geometrical one under the gravitational influence of the body. Orbiting around the Sun, a large body can collide with bodies from a volume with width about $\Delta = 2(\Sigma_{coll}/\pi)^{1/2}$, which is called the feeding zone.

In early studies average values of random velocities and mass distributions of planetesimals were considered independently, and a relation between them was found (Safronov, 1969). This value of the velocity v is expressed in terms of the mass m_1 and radius r_1 of the largest body as $v = (Gm_1/\theta r_1)^{1/2}$, where θ is a dimensionless parameter, characterizing the ratio of the average random velocity of bodies to the escape velocity at the surface of the largest body. The average value of v is equal in the order of magnitude to the escape velocity from the surfaces of the bodies. For a system of bodies with a distribution of mass described by an inverse power law $n(m) \sim m^{-q}$ with $0 \leq q \leq 2$, the major portion of the total mass is concentrated in the large bodies, and the parameter θ becomes ~ 1 to 5.

The question of the distribution of mass between the largest bodies has a tendency to be different because of the sensitivity of the gravitational cross-section to the body radius, i.e. $\propto r_1^4$. With a few large bodies located in the same feeding zone for a long time, and growing under the same conditions, the 'embryo' of one typically grows faster than the others. A runaway growth of the embryo (with respect to other bodies), may be triggered which could continue until the ratio of its mass to that of the body is $< (2\theta)^3$.

In 1978 Levin supposed, and Greenberg, Wacker, Hartmann and Chapman found by numerical simulation, that starting from small relative velocities a system of planetesimals does not evolve to the power-law mass distribution, with the bulk of the mass concentrated in large bodies. Instead, several bodies grew rapidly, while most of the bodies remained small, with the bulk of mass remaining in the small bodies. The relative velocities were small, i.e. θ was large. In 1987 Lissauer emphasized the plausible importance of runaway accretion for the accumulation of giant planets, and stimulated activity in the investigation of conditions for the realization of runaway accretion. A plausible quantitative solution has been found by Wetherill and Stewart (1989), who came to the conclusion that runaway growth occurs when the accumulation starts from low random velocities, and when larger bodies have lower random velocities than smaller ones. This could be caused by the combined action of gravitational interactions and collisions, tending to the equipartition of the average kinetic energies of bodies of different masses; $v \propto m^{-1/2}$. This condition is never actually achieved because another process, 'viscous stirring' tends to give v independent of mass. A competition between viscous stirring and the equipartition process yields a value of v that decreases with the mass increasing more slowly than $m^{1/2}$. These lower velocities increase the gravitational cross-section of the larger bodies, causing their fast coalescence and reducing their number, resulting in a runaway growth of embryos. The finding that the average random velocity of bodies depends on their masses results in a change of the concept of the feeding zone. The feeding zone is now considered as a system (or continuum) of rings about the embryo's orbit, and the widths of the rings are different for bodies of different masses. In the case of the equipartitional distribution of the average kinetic energies, the width of the feeding zones is proportional to the embryo mass.

In current scenarios of planetary accumulation three main stages are considered. In the earliest stage, planetesimals are considered to be uniformly distributed spatially over a range of distances through which they can interact. Bodies are numerous, and their size distribution can be treated as continuous. Later the planetesimal swarm becomes spatially inhomogeneous. The largest bodies, 'tentative embryos' surrounded by many small bodies, begin to form in separate orbits. Embryos, separated by a distance larger than the width of their feeding zones, grow independently of each other and have a chance to accumulate comparable masses, forming a multi-embryo system. However, the width of the feeding zone of the embryos increases in time ($\propto m_1^{1/3}$). Hence the zones of neighboring embryos coalesce, and then the largest embryo begins to dominate, eventually sweeping up the less successful ones.

These processes could lead to two different scenarios (Safronov, 1969; Wetherill, 1990).

1. Formation of many large bodies of comparable size. The bodies contain most of the mass, and have a relatively large random velocities ($\theta < 10$). This results in a relatively low rate of accumulation, with a number of large bodies at the late stage of planetary formation. This scheme is shown in Figure S28e–h.
2. Fast runaway growth of a relatively small number of embryos. The embryos contain a small fraction of the total mass and have no influence on the velocity distribution, which is determined by much smaller bodies and corresponds to a large value of θ. This stage could continue until a significant fraction of the bodies in the embryo's feeding zone is swept up. It is followed by a stage of much slower embryo growth. It is governed by diffusion into the embryo's feeding zone of bodies or other embryos due to gravita-

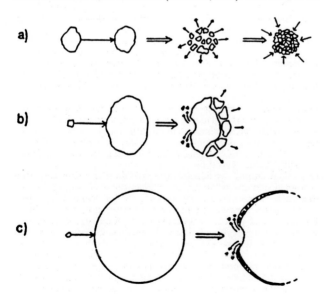

Figure S29 Effects of interplanetary hypervelocity collisions for variable size ratios of impactor and target: (a) disruption of both bodies and a possibility of reaccretion (b) cratering and spallation, and (c) cratering and reaccretion of debris, i.e. regolith formation. (After Stöffler *et al.* 1988).

tional interactions. Hence the runaway scenario is changed to a scenario of orderly growth.

In the region of terrestrial planets, runaway growth results in fast (up to 3×10^4 years) formation of embryos of 5×10^{25} to 5×10^{26} g, revolving about the Sun in orbits of low eccentricity and inclination, i.e. possessing small random velocities. Further, the increase of the random velocities due to gravitational interactions of the embryos slows down their growth, which approaches orderly growth, and lasts for a period of the order of 10^8 years. However, the memory of the initial stage is left in the increased probability of giant impacts (collision of two embryos of the comparable size) at the late stage of planetary accumulation. In the region of giant planets the role of a runaway accretion could be even more fundamental. Between 10^8 and 10^{10} years are necessary to form a core of a few Earth masses (needed for triggering the rapid hydrodynamical accretion of the solar nebula gas) by orderly growth. This timescale is in conflict with the astronomical evidence for removal of the envelopes around T Tauri stars within 10^6 to 10^7 years after star formation. The runaway accretion of a core of a few Earth masses can occur in the giant planet region on a timescale $\leq 10^6$ years. The further stage of hydrodynamical accretion could continue another few million years for Jupiter.

Runaway accretion continues until cores reach critical mass if the density in the solar nebula is somewhat higher than is estimated from the present-day masses of giant planets. However, a modest increase of the density in the region of the giant planets is not unreasonable (a significant fraction of this material could be later lost from the solar system due to perturbations by the giant planets). This contributes only about 10^{-1} $M_\odot$ to the nebula mass. (Runaway accretion would produce larger embryos if the process of planetary formation began in a more massive and more extended solar nebula, because the feeding zone is larger (volume $\propto R^2$) and because the surface density of solids is larger.) A local enhancement of H_2O ice (or both ice and gas) in Jupiter's and, plausibly, Saturn's region is possible because of the radial transport of the matter in the solar nebula.

The planets, except Venus and Uranus, rotate in a prograde direction (see Planetary rotation). The rotation, and a tilt of the rotational axis, may be acquired in giant impacts. The axial tilt of the Earth, $\sim 23.5°$, could be produced by impacts of bodies with a mass about 10^{-3} of the Earth mass ($M_\oplus = 6 \times 10^{27}$ g), and the steep axial tilt of Uranus (97.9°) could be produced by bodies of about one-tenth of the mass of Uranus. It follows from the value of the rotational angular momentum of planets that the bulk of the angular momentum was acquired either from large bodies with relatively small eccentricities, or from bodies captured first in a protosatellite swarm of planetesimals (or planetary nebula) surrounding the planets. Jupiter and Saturn acquired their angular momenta mainly from the accreted solar nebula gas.

The origin of the regularity in the spacing of planets described by the Titius–Bode rule is related to the planetary accumulation stage. Two possibilities are now considered possible: competition of embryos in the rate of growth, and the resonance capturing of planetesimals, greatly accelerating the early stages of planetary accretion.

Tamara V. Ruzmaikina

Bibliography

Bailey, M., Clube, S. and Napier, W.M. (1990) *Origin of Comets*. Oxford: Pergamon Press.
Binzel, R. (1989) An overview of the asteroids, in *Asteroids II* (eds R. Binzel, T. Gehrels and M. Matthews). Tucson: University of Arizona Press, pp. 3–17.
Bode, J. (1777) *Anleitung zur Kenntniss des gestirnten Himmels*, 3rd edn. Berlin: den Christian Friedrich Himburg.
Bodenheimer, P. Ruzmaikina, T. and Mathiew, R. (1993) Stellar multiple systems: constraints on the mechanism of origin, in *Protostars and Planets III* (eds E. Levy and J. Lunine). Tucson: University of Arizona Press, pp. 367–404.
Buffon, G.L. (1745) *De la Formation des Planets*. Paris.
Burns, J. (1986) Some background about satellites, in *Satellites* (eds J. Burns and M. Matthews). Tucson: University of Arizona Press, pp. 1–38.
Cameron, A. (1962) The formation of the Sun and planets. *Icarus*, 1, 13–69.
Cassen, P. and Moosman, A. (1981) On the formation of the protostellar disks. *Icarus*, 48, 353–76.
Drake, M. (1986) Is Lunar bulk material similar to Earth's mantle? in *Origin of the Moon* (ed. W. Hartmann, R. Phillips and J. Taylor). Houston: Lunar and Planetary Institute, pp. 105–24.
Descartes, R. (1644) *Principia Philosophiae*. Amsterdam. (English Translation: in *Studies in the History of Philosophy*. New York: E. Mellen Press, 1988.)
Greenberg, R., Wacker, J., Hartmann, W. and Chapman, C. (1978) Planetesimals to planets: numerical simulation of collisional evolution. *Icarus*, 35, 1–26.
Gurevich, Lev and Lebedinsky, A. (1950) On the planet formation. *Izvest. Acad. Nauk USSR: Seria Phys.*, 14, 765–99. (In Russian.)
Hartmann, W. (1983) *Moons and Planets*, Belmont: Wadworth Publ. Co.
Hoyle, F. (1960) On the origin of the solar system. *Quart. J. Roy. Astron. Soc.*, 1, 28–55.
Hubbard, W. (1989) Structure and composition of giant planets interiors, in *Origin and Evolution of Planetary and Satellite Atmospheres* (eds S. Atreya, J. Pollack and M. Matthews). Tucson: University of Arizona Press, pp. 539–63.
Hunten, D., Donahue, T., Walker, J. and Kasting, James (1989) Escape of atmospheres and loss of water, in *Origin and Evolution of Planetary and Satellite Atmospheres* (eds S. Atreya, J. Pollack and M. Matthews). Tucson: University of Arizona Press, pp. 386–422.
Jeans, J. (1917) The part played by rotation in cosmic evolution. *Mon. Not. Roy. Astron. Soc.*, 77, 186–99.
Kant, I. (1755) *Allgemeine Naturgeschichte und Theorie des Himmels*. (English translation: in *Nebular Hypothesis*. Edinburgh: Scottish Academic Press, 1981.)
Laplace, P.S. (1796) *Exposition du Système du Monde*. Paris: (English translation: in *Celestial Mechanics*. Bronx, NY: Chelsea Publ. Co., 1966.)
Larson, R. (1983) Numerical calculations of the dynamics of a collapsing protostar. *Mon. Not. Roy. Astron. Soc.*, 145, 271–95.
Laskar, J., Quinn, T. and Tremaine, S. (1992) Confirmation of resonant structure in the solar system. *Icarus*, 95, 148–152.
Levin, B. (1964) *Origin of the Earth and Planets*. Moscow: Nauka.
Levin, B. (1978) Relative velocities of planetesimals and the early accumulation of the planets. *Moon and Planets*, 19, 289–96.
Levine, J. (1987) The atmospheres of the Earth and other planets: origin, evolution, and composition, in *Workshop on the Origin of the Solar System* (eds J. Nuth and P. Sylvester). Houston: Lunar and Planetary Insitute, pp. 69–80.
Levy, E. and Sonnett, C. (1978) Meteorite magnetism and early solar system magnetic fields, in *Protostars and Planets* (ed. T. Gehrels). Tucson: University of Arizona Press, pp. 516–32.
Levy, E., Ruzmaikin, A. and Ruzmaikina, T. (1991) Magnetic history of the Sun, in *The Sun in Time* (eds C. Sonnett, M. Giampapa, and M. Matthews). Tucson: University of Arizona Press, pp. 589–632.
Lin, D. and Papaloizou, J. (1980) On the structure and evolution of the primordial solar nebula. *Mon. Not. Roy. Astron. Soc.*, 191, 37–48.
Lissauer, J. (1987) Time scales for planetary accretion and the structure of the protoplanetary disk. *Icarus*, 69, 249–65.
Melosh, H. (1989) *Impact Cratering: A Geologic Process*. Oxford: Oxford University Press.
Oort, J. (1950) The structure of the cloud of comets surrounding the solar system and hypothesis concerning its structure. *Bull. Astron. Inst. Neth.*, 11, 91–110.
Oparin, A. (1924) *Origin of life*. Moscow: Moscovsky Rabotchii. (English translation: in Bernal, J.D., *The Origin of Life*. Cleveland: World, 1967.)
Oro, J., Miller, S. and Lazcano, A. (1990) The origin and early evolution of life on Earth. *Ann. Rev. Earth Planet. Sci.*, 18, 317–56.
Pepin, R. (1991) On the origin and early evolution of terrestrial planet atmospheres and meteoritic volatiles. *Icarus*, 92, 2–79.
Proctor, R. (1898) *Other Worlds than Ours*. New York: D. Appleton.
Ruden, S. and Pollack, J. (1991) The dynamical evolution of the protosolar nebula. *Astrophys. J.*, 375, 740–60.
Ruzmaikina, T. (1981) On the role of magnetic field and turbulence in the evolution of the presolar nebula. *Adv. Space Res.*, 1, 49–53.
Safronov, V. (1960) On gravitational instability in flat rotating systems with an axial symmetry. *Problems of Cosmogony*, 8, 168–79.
Safronov, V. (1969) *Evolution of the Protoplanetary Cloud and Formation of the Earth and Planets*. Moscow: Nauka Press. (English translation: NASA-TT-F-677, 1972.)
Safronov, V. (1991) Kuiper Prize Lecture: some problems in the formation of the planets. *Icarus* 94, 260–71.

Schatzman, E. (1962) A theory of the role of magnetic activity during star formation. *Annales d'astrophysique*, **25**, 18–29.
Schmidt, O. (1944) Meteoritic theory of origin of the Earth and planets. *Doklady Acad. Nauk USSR*, **45**, 245–9. (In Russian.)
Shu, F., Adams, F. and Lizano, S. (1987) Star formation in molecular clouds: observation and theory. *Ann. Rev. Astron. Astophys.* **15**, 23–81.
Stöffler D., Bischoff, A., Buchwald, V. and Rubin, A.E. (1988) Shock effects in meteorites, in *Meteorites and the Early Solar System* (eds J. Kerridge, and M. Matthews). Tucson: University of Arizona Press, pp. 165–202.
Ter Haar, D., and Cameron, A. (1963) Historical review of the origin of the solar system, in *Origin of the Solar System*, New York: Academic Press.
Toomre, A. (1964) On the gravitational stability of a disk of stars. *Astrophys. J.*, **139**, 1217–38.
Von Weizsäcker, C. (1943) Über die Enstehung des Planetensystems. *Z. Astrophys.*, **22**, 319–55.
Wänke, H. (1981) Constitution of terrestrial planets. *Phil. Trans. Roy. Soc. London A*, **303**, 287–303.
Wetherill, G. (1989) The formation of the solar system: consensus, alternatives, and missing factors, in *The Formation and Evolution of Planetary Systems* (eds H. Weaver and L. Dandy). Cambridge: Cambridge University Press, pp. 1–30.
Wetherill, G. (1990) Formation of the Earth. *Ann. Rev. Earth Planet. Sci.*, **18**, 205–56.
Wetherill, G. and Stewart, G. (1989) Accumulation of a swarm of small planetesimals. *Icarus*, **77**, 290–303.

Cross references

Accretion
Asteroid
Asteroid: resonance
Chaotic dynamics in the solar system
Comet: origin and evolution
Cold accretion
Collision
Dust
Earth–Moon system: origin
Interstellar grains
Interstellar medium
Kuiper belt
Life: origin
Meteor, meteoroid
Moon: origin
Oort, Jan Hendrik, and Oort cloud
Planetesimal
Solar nebula
Titius–Bode law

SOLAR SYSTEM: STABILITY

The problem of the stability of the solar system has fascinated astronomers and mathematicians since antiquity, when it was observed that among the fixed stars there were 'wandering stars' – the planets. Efforts were first focused on finding a regularity in the motion of these wanderers, so their movement among the fixed stars could be predicted. For Hipparchus and Ptolemy the ideal model was a combination of uniform circular motions ('epicycles') which were continually adjusted over the centuries to conform to the observed course of the planets. Astronomy had become predictive, even if its models were in continual need of adjustment.

From 1609 to 1618 Kepler fixed the planets' trajectories: having assimilated the lessons of Copernicus, he placed the Sun at the center of the universe and, based on the observations of Tycho Brahe, showed that the planets described ellipses around the Sun. At the end of a revolution, each planet found itself back where it started and so retraced the same ellipse. Though seductive in its simplicity, this vision of a perfectly stable solar system in which all orbits were periodic would not remain unchallenged for long.

In 1687 Newton published his law of universal gravitation. By restricting this law to the interactions of planets with the Sun alone, one obtains Kepler's phenomenology. But Newton's law applies to all interactions: Jupiter is attracted by the Sun, as is Saturn, but Jupiter and Saturn also attract each other. There is no reason to assume that

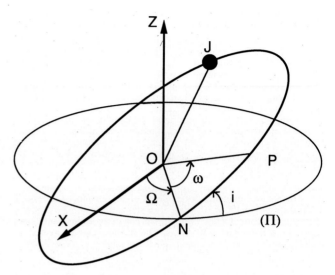

Figure S30 Elliptical elements. For any given time, a planet (J) can be considered to move on an elliptical orbit, with semimajor axis a and eccentricity e, with the Sun at one focus (O); N is the line of nodes. The orientation of this ellipse with respect to a fixed plane Π, and a direction of reference OX, is given by three angles: the inclination i; the longitude of the node Ω; and the longitude of perihelion $\varpi = \Omega + \omega$, where ω is the argument of perihelion (P). The position of the planet on this ellipse is given by the mean longitude $\lambda = M + \varpi$, where M (the mean anomaly) is an angle which is proportional to the area OPJ (Kepler's third law; see also Celestial mechanics; Coordinate systems; Orbit).

the planets' orbits are fixed invariant ellipses, and Kepler's beautiful regularity is destroyed.

In Newton's view, the perturbations among the planets were strong enough to destroy the stability of the solar system, and divine intervention was required from time to time to restore the planets' orbits to their place. Moreover, Newton's law did not yet enjoy its present status, and astronomers wondered if it was truly enough to account for the observed movements of bodies in the solar system.

The problem of solar system stability was a real one since, after Kepler, Halley was able to show, by analyzing the Chaldean observations transmitted by Ptolemy, that Saturn was moving away from the Sun while Jupiter was moving closer. By crudely extrapolating these observations, one finds that 6 million years ago Jupiter and Saturn were at the same distance from the Sun. In the 18th century Laplace took up one of these observations, which he dated 1 March 228 BC: 'At 4 : 23 am, mean Paris time, Saturn was observed "two fingers" under Gamma in Virgo.' Starting from contemporary observations, Laplace hoped to calculate backward in time using Newton's equations to replicate analytically and verify this 2000 year old observation.

The variations of planetary orbits were such that, in order to predict the planets' positions in the sky, de LaLande was required to introduce artificial 'secular' terms in his ephemeris tables. Could these terms be accounted for by Newton's law?

The problem remained open until the end of the 18th century, when Lagrange and Laplace correctly formulated the equations of motion. Lagrange started from the fact that the motion of a planet remains close, over a short duration, to a Keplerian ellipse, and so had the notion to use this ellipse as the basis for a coordinate system (Figure S30). Lagrange then wrote the differential equations that govern the variations in this elliptic motion under the effect of perturbations from other planets, thus inaugurating the methods of classical celestial mechanics. Laplace and Lagrange, whose work converged on this point, calculated secular variations: long-term variations in the planets' semi-major axes under the effects of perturbations by the other planets. Their calculations showed that, up to first order in the masses of the planets, these variations vanish (Poisson and Poincaré later showed that this result remains true through second order in the masses of the planets, but not through third order).

This result seemed to contradict Ptolemy's observations from antiquity, but by examining the periodic perturbations between Jupiter and Saturn, Laplace discovered a quasi-resonant term ($2\lambda_{\text{Jupiter}}$

$-5\lambda_{Saturn}$) in their longitudes. This term has an amplitude of 46′50″ in Saturn's longitude, and a period of about 900 years. This explains why observations taken in 228 BC and then in 1590 and 1650 could give the impression of a secular term.

Laplace then calculated many other periodic terms, and established a theory of motion for Jupiter and Saturn in very good agreement with 18th-century observations. Above all, using the same theory, he was able to account for Ptolemy's observations to within 1 arcmin without additional terms in his calculations. He thus showed that Newton's law was in itself sufficient to explain the movement of the planets throughout known history, and this exploit no doubt partly accounted for Laplace's determinism.

Laplace showed that the planets' semimajor axes undergo only small oscillations, and do not have secular terms. At the same time, the eccentricity and inclination of planetary trajectories are also very important for solar system stability. If a planet's eccentricity changes appreciably, its orbit might cut through another planet's orbit, increasing the chances of a close encounter which could eject it from the solar system.

Laplace revised his calculations, taking into account only terms of first order in the perturbation series, and showed that the system of equations describing the mean changes of eccentricity and inclination may be reduced to a system of linear differential equations with constant coefficients. He also showed that the solutions of this system are quasi-periodic (linear combinations of periodic terms), and that the variations in eccentricity reduce to a superposition of uniform circular motions (Figure S31). The inclinations and eccentricities of the orbits are therefore subject to only small variations about their mean values. But it must be stressed that Laplace's solutions are very different from Kepler's, because the orbits are no longer fixed. They are subject to a double precessional motion with periods ranging from 50 000 to several million years: precession of the perihelion, which is the slow rotation of the orbit in its plane, and precession of the nodes, which is the rotation of the plane of the orbit in space.

Later Le Verrier, famed for his discovery in 1846 of the planet Neptune through calculations based on observations of irregularities in the movement of Uranus, took up Laplace's calculations and considered the effects of higher-order terms in the series. He showed that these terms produced significant corrections and that Laplace's and Lagrange's calculations 'could not be used for an indefinite length of time.' He then challenged future mathematicians to find exact solutions, without approximations. The difficulty posed by 'small divisors' showed that the convergence of the series depended on initial conditions, and the stability of the solar system remained an open problem.

Between 1892 and 1899 Poincaré formulated a negative response to Le Verrier's question. In so doing he rethought the methods of celestial mechanics along the lines of Jacobi's and Hamilton's work. In his memoir *On the three-body problem and the equations of dynamics*, Poincaré showed that it is not possible to integrate the equations of motion of three bodies subject to mutual interaction, and not possible to find an analytic solution representing the movement of the planets valid over an infinite time interval, since the series used by astronomers to calculate the movement of the planets were not convergent.

In the 1950s and 1960s, the mathematicians Kolmogorov, Arnold and Moser took up Poincaré's work and showed that, for certain values of the initial conditions, it was nonetheless possible to obtain convergent series. If the masses, eccentricities and inclinations of the planets are small enough, many initial conditions lead to quasi-periodic planetary trajectories. But the actual masses of the planets are much too large for this result (known as the KAM theorem) to apply directly to the solar system and thereby prove its stability.

In recent years work on the problem of solar system stability has advanced considerably, due largely to computers which allow extensive analytic calculations and numerical integrations over model timescales approaching the age of the solar system.

One part of these efforts consists of direct numerical integration of the equations of motion (Newton's equations, sometimes with additional relativistic corrections or perturbations due to the Moon). Initial studies were limited to the motion of the outer planets, from Jupiter to Pluto. In fact, the more rapid the orbital movement of a planet, the more difficult it is to integrate its motion numerically. To integrate the orbit of Jupiter, a step size of 40 days will suffice, while a step size of 0.5 days is required to integrate the motion of the whole solar system. The project Longstop (Carpino, Milani and Nobili, 1987; Nobili, Milani and Carpino, 1989) must be counted among the most significant recent studies; this used a Cray to integrate the system of outer planets over a model time of 100 million years. At about the same time, calculations of the same system were carried out

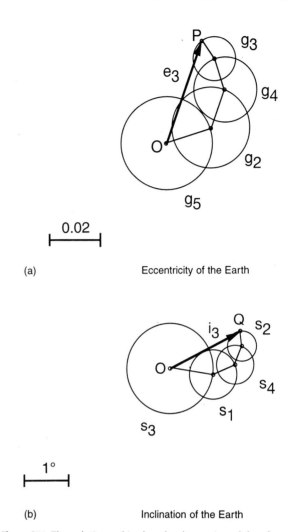

(a) Eccentricity of the Earth

(b) Inclination of the Earth

Figure S31 The solutions of Laplace for the motion of the planets are combinations of circular and uniform motions with frequencies corresponding to the precession frequencies g_i and s_i of the solar system, which have periods from about 50 000 years to several million years. The eccentricity e_3 of the Earth is given by OP in (a), while the inclination of the Earth with respect to the invariant plane of the solar system (i_3) is OQ in (b) (Laskar, 1992b).

at MIT over even longer periods, corresponding to times of 210 and 875 million years. These calculations were carried out on the Digital Orrery, a vectorized computer specially designed for the task (Applegate *et al.*, 1986; Sussman and Wisdom, 1988). This latter integration showed that the motion of Pluto is chaotic, with a Liapunov exponent of 1/20 million years. But since the mass of Pluto is very small, (1/130 000 000 the mass of the Sun), this does not induce macroscopic instabilities in the rest of the solar system, which appeared relatively stable in these numerical studies.

Laskar took a semi-analytical approach. Using perturbation methods first developed by Laplace and Le Verrier, he derived an extended averaged system for the whole solar system except Pluto, including all contributions up to second order with respect to the masses, and through degree five in eccentricity and inclination. For the outer planets, some estimated corrections of third order were also included. The system of equations thus obtained comprises some 150 000 terms, and does not model the motion of the planets, but rather the mean motion of their orbits. It may be integrated numerically on a computer using a very large step size, on the order of 500 years. An integration over 200 million years showed that the solar system, and more particularly the system of inner planets (Mercury, Venus, Earth and Mars), is chaotic, with a Liapunov exponent of 1/5 million years (Laskar, 1989). An error of 15 m in the Earth's initial position gives rise to an error of about 150 m after

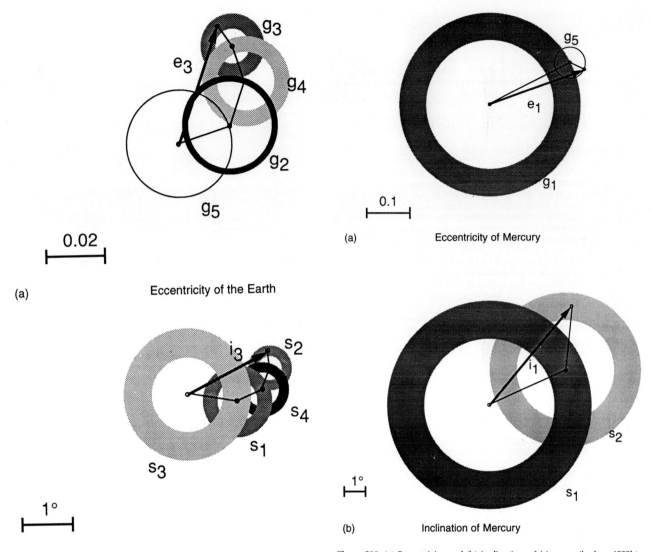

Figure S32 Solutions for the eccentricity and inclination of the Earth. The effect of the chaotic dynamics in the solar system can be shown by enlarging each circle of Figure S31, which corresponds to a fundamental frequency of the solar system, into a band corresponding to the chaotic zone. The position of the junction points inside this zone will be unpredictable after several million years (Laskar 1992b).

10 million years; but this same error grows to 150 million km after 100 million years. It is thus possible to construct ephemerides over a 10 million year period, but it becomes essentially impossible to predict the motion of the planets with any precision beyond 100 million years.

This chaotic behavior originates in two secular resonances among the planets: $\theta = 2(g_4 - g_3) - (s_4 - s_3)$, which is related to Mars and the Earth, and $\sigma = (g_1 - g_5) - (s_1 - s_2)$, related to Mercury, Venus and Jupiter (the g_i terms are the secular frequencies related to the perihelia of the planets, while the s_i terms are the secular frequencies of the nodes; Laskar, 1990). The two corresponding arguments change several times from libration to circulation over 200 million years, which is also a characteristic of chaotic behavior.

Quinn, Tremaine and Duncan (1991) published a numerical integration of the full solar system, including the effects of general relativity and the Moon, which spanned 3 million years in the future and in the past. Comparison with the secular solution (Laskar, 1990) shows very good quantitative agreement, and confirms the existence of secular resonances in the inner solar system (Laskar et al., 1992).

Figure S33 (a) Eccentricity and (b) inclination of Mercury (Laskar, 1992b).

Using mapping techniques, Sussman and Wisdom (1992) published an integration of the solar system over 100 million years which confirms the existence of the secular resonances as well as the value of the Liapunov exponent for the solar system.

The solar system, and more particularly the inner solar system, is strongly chaotic, but on a rather long timescale. This is because the fundamental periods of the motions involved are the precessional periods of the orbits (on the order of 100 000 years), and not the orbital periods (on the order of a year). Over periods on the order of 100 million years, the variations of planetary eccentricities and inclinations are dominated by the quasi-periodic components already present in the solutions of Laplace and Le Verrier. The effects of chaos over 400 million years may nevertheless be estimated to be about 0.01 for the Earth's eccentricity and 1° for its inclination (Figure S32). These are of course only lower estimates, since we do not yet have the means to bound these variations over more than several hundred million years, which would require a more global knowledge of the phase space of motion.

The most perturbed planet is Mercury; the effects of its chaotic dynamics are clearly visible over 400 million years (Figures S33 and S34). The chaotic component consists of the variations in the envelope bounding the eccentricity and inclination curves of Figures S34. These variations reach several degrees for the inclination, and this mechanism no doubt explains the current high values of Mercury's eccentricity and inclination.

Instabilities of another sort also manifest themselves in the motion of the planets. These motions are not present in the orbits, but rather in

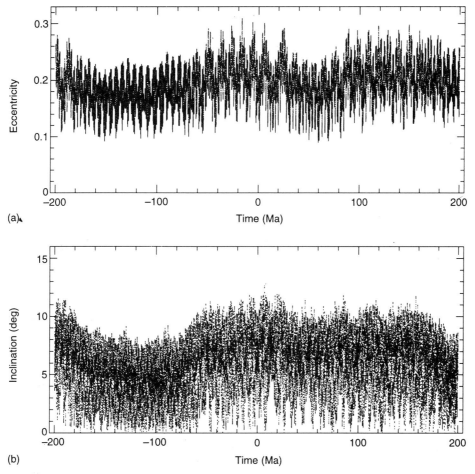

Figure S34 The chaotic motion of Mercury. Mercury is the planet for which the effect of chaotic dynamics is the most spectacular. The figures show the computed evolution of the eccentricity (a) and inclination (b) of Mercury with respect to time from −200 to +200 million years. On each of these curves, two kind of variations can be seen: a rapid variation, with a period of about 100 000 years, which corresponds basically to the regular part of the solution, as described by Laplace, and a slow variation, which shows the effect of the chaotic dynamics. This variation reaches 0.05 for the eccentricity of Mercury, and several degrees for its inclination. The regular variations (discovered by Laplace) are bounded for an infinite time, but we do not presently know the possible chaotic variations of the eccentricity and inclination of the planets, over time spans of 5 billion years, comparable to the age of the solar system. (Laskar, 1992a).

the orientation of the planets' axes of rotation. Because of their equatorial bulges the planets are subject to torques arising from the gravitational forces of their satellites and of the Sun. This causes a precessional motion, which in the Earth's case has a period of 26 000 years. Moreover, the obliquity of each planet – the angle between the equator and the orbital plane – is not fixed, but suffers a perturbation due to the secular motion of the planet's orbit. In particular, for Mars these perturbations translate into a chaotic oscillation of high amplitude ranging from 0 to 60° (Laskar and Robutel, 1993; Touma and Wisdom, 1993). For the Earth these oscillations, which appear to be the determining factor in the onset of the ice ages, are regular, varying only ± 1.3° about the mean obliquity value of 23.3°. But in the absence of the Moon, the Earth's obliquity would no doubt be chaotic, experiencing strong oscillations ranging from 0° to nearly 85°, which would profoundly modify the surface climate (Laskar, Joutel and Robutel, 1993). Such behavior was perhaps also experienced by Mercury and Venus in the course of their history, before their rotations were slowed through dissipative effects. In particular, for Venus, this possible instability provides a mechanism for inverting the planet without resorting to large impacts (Laskar and Robutel, 1993).

Jacques Laskar

Bibliography

Applegate, J.H., Douglas, M.R., Gursel, Y., Sussman, G.J. and Wisdom, J. (1986) The solar system for 200 million years, *Astron. J.*, **92**, 176–94.

Arnold, V.I. (1983) Small denominators and problems of stability of motion in classical celestial mechanics, *Russian Math. Surveys*, **18**(6), 85–193.

Carpino, M., Milani, A. and Nobili, A.M. (1987) Long-term numerical integrations and synthetic theories for the motion of the outer planets. *Astron. Astrophys.*, **181**, 182–94.

Laskar, J. (1980) A numerical experiment on the chaotic behaviour of the solar system. *Nature*, **338**, 237–8.

Laskar, J. (1990) The chaotic motion of the solar system. A numerical estimate of the size of the chaotic zones. *Icarus*, **88**, 266–91.

Laskar, J. (1992a) A few points on the stability of the solar system, in Symposium IAU 152 (ed., S. Ferraz-Mello). Dordrecht: Kluwer, pp. 1–16.

Laskar, J. (1992b) La stabilité du système solaire, in *Chaos et Déterminisme*, (eds A. Dahan et al.), Paris: Seuil.

Laskar, J., Quinn, T. and Tremaine, S. (1991) Confirmation of resonant structure in the solar system. *Icarus*, **95**, 148–52.

Laskar, J. and Robutel, P. (1993). The chaotic obliquity of the planets. *Nature*, **361**: 608–12.

Laskar, J., Joutel, F. and Robutel, P. (1993) Stabilization of the Earth's obliquity by the Moon. *Nature*, **361**, 615–7.

Nobili, A.M., Milani, A. and Carpino, M. (1989) Fundamental frequencies and small divisors in the orbits of the outer planets. *Astron. Astrophys.*, **210**, 313–36.

Quinn, T.R., Tremaine, S. and Duncan, M. (1991) A three million year integration of the Earth's orbit. *Astron. J.*, **101**, 2287–305.

Sussman, G.J. and Wisdom, J. (1988) Numerical evidence that the motion of Pluto is chaotic. *Science*, **241**, 433–7.
Sussman, G.J. and Wisdom, J. (1992). Chaotic evolution of the solar system. *Science*, **257**, 56–62.
Touma, J. and Wisdom. J. (1993) The chaotic obliquity of Mars. *Science*, **259**, 1294–7.

Acknowledgement

Translated by H.S. Dumas.

Cross references

Celestial mechanics
Chaotic dynamics in the solar system
Coordinate systems
Earth–Moon system: dynamics
Earth: rotational history
Ephemeris
Lagrange, Louis Joseph
Laplace, Pierre Simon de
Newton, Sir Isaac, and Newton's laws
Obliquity: terrestrial record
Planetary dynamical astronomy

SOLAR WIND

The solar wind is an ionized, magnetized gas which continuously emanates from the Sun in all directions (Plate 1). The flow accelerates to supersonic, super-Alfvénic speeds, filling the heliosphere. This flow of plasma was first predicted by E.N. Parker in 1958 as an inevitable consequence of a hot solar corona (with a temperature of several million degrees) and a very low fluid pressure in interstellar space. This prediction was very controversial at the time, particularly since it could not be tested directly until particle detectors capable of measuring such fluxes could be flown beyond the Earth's magnetic field. The existence of a steady, supersonic solar wind was ultimately confirmed by M. Neugebauer and C.W. Snyder in 1962 with measurements from the Mariner 2 spacecraft on its way to Venus. Observations since then have shown that the solar wind has speeds, densities and magnetic fields which vary on all timescales from the 22-year solar cycle, through the 25-day period of the solar rotation, to plasma oscillations with periods of less than a second. The flow of the solar wind past planets, comets and other solar system objects is of fundamental importance to the magnetospheric structure and energetic processes of these bodies.

Large-scale properties

The solar wind plasma is composed primarily of protons and electrons, with a variable fraction of alpha particles (typically a few percent) and trace quantities of heavier ions. It has been observed by *in situ* particle and magnetic field detectors over a range of heliocentric distances from the 0.29-AU perihelion of the Helios probes out to the current position of the Pioneer 10 spacecraft beyond 60 AU. The period of observation has exceeded one 22-year solar cycle. The latitudinal coverage has not been as extensive, being largely confined to the ecliptic plane. However, remote sensing techniques such as radio scintillation measurements and analysis of backscattered solar radiation have provided some information on the structure of the wind out of the ecliptic plane. Furthermore, the recently launched Ulysses spacecraft has been obtaining direct measurements of the high-latitude wind since 1993.

Figure S35 shows a typical example of solar wind plasma observations at 1 AU compiled by the National Space Science Data Center. The traces are hourly average values of the proton speed, density and temperature for a 27-day period. Over this time span the solar wind speed ranged from 270 km s^{-1} to 652 km s^{-1} while the proton density varied from 0.7 cm^{-3} to 42.0 cm^{-3} and the proton temperature from 1.2×10^4 K to 8.9×10^5 K.

Solar wind properties can be organized generally in terms of the flow speed of the plasma. As shown in Figure S35, low-speed wind tends to be cool and dense while the high-speed flow is hotter and more tenuous. It is thought that the two types of wind originate in different regions on the Sun, where they are subject to different degrees of acceleration and heating.

The magnetic field in the solar wind is also routinely measured. On average this field is the extension of the solar coronal field which is carried away by the flow. The electrical conductivity of the solar wind is so high that the diffusion of magnetic field through the plasma is negligible over the scales usually considered and one can treat the field as being 'frozen' in the plasma. It is then useful to think of field lines embedded in the fluid, such that the expanding flow pulls the solar magnetic field into heliospheric space with it. Since the solar wind flow is essentially radial, the magnetic field lines would also be radial if the Sun was not rotating. Because the Sun does rotate, the field lines are drawn into a spiral pattern. In heliocentric spherical coordinates, a steady radial solar wind with no azimuthal variation in speed would result in a global magnetic field, B, described by

$$B_r(r, \theta, \psi) = B(r_0, \theta, \psi + \Omega r)(r_0/r)^2$$
$$B_\psi(r, \theta, \psi) = -B(r_0, \theta, \psi + \Omega r) r_0^2 \, \Omega \sin\theta / rV \quad (S8)$$
$$B_\theta(r, \theta, \psi) = 0$$

where V is the flow speed, $B(r_0)$ is the coronal field at the base of the solar wind and $\Omega = 2.9 \times 10^{-6}$ rad s^{-1} is the angular rotation rate of the Sun. At the orbit of Earth, with a wind speed of $V = 400$ km s^{-1}, (S8) gives a field angle of 45° to the radial near the equatorial plane. It is clear that, for a given value of θ, the field is wound tighter as r increases. At high latitudes the effect of solar rotation is smaller and the field is expected to be essentially radial in the polar regions.

In principle, the coronal fields $B(r_0)$ can be arbitrary as long as the total magnetic flux leaving the Sun is zero. During solar minimum, these fields are organized into large unipolar regions covering the polar hemispheres, the field pointing outward in one hemisphere and inward in the other. These fields are essentially 'open' in that they extend out into the solar wind, rather than returning to the Sun. The unipolar regions are separated by a wavy band of closed field lines near the equator, across which the average radial field changes sign. At these times the large-scale field in the solar wind points either toward or away from the Sun along the spiral defined by (S8), depending on whether the observation point is above or below the thin, warped current sheet which separates the opposite polarities far from the Sun.

The open, unipolar structures are called coronal holes, due to their dark appearance in the soft x-ray pictures of the Sun where they were first seen. Coronal holes are now known to be the source of fast solar wind ($V > 500$ km s^{-1}). Thus the large polar coronal holes observed during solar minimum imply that the high-latitude solar wind is consistently fast, low-density flow. This inference has been borne out by the results of scintillation and solar backscatter investigations (Coles and Rickett, 1976; Lallement, Bertaux and Kurt, 1985) and by direct Ulysses observations (Phillips *et al.*, 1995). The source of the slow, dense solar wind is not as well determined, but it must come either from the edges of the coronal holes or from regions above the closed field lines, such as coronal streamers. This configuration of solar fields and flows is shown schematically in Figure S36.

As the Sun rotates, the plasma emitted in a given direction in the equatorial plane alternates between slow and fast flows. The boundaries between the two types of flow follow the same spiral pattern seen in the magnetic field. However, as the wind moves away from the Sun, the fast plasma along a radial line catches up the slower plasma ahead of it. The magnetic field embedded in the plasma prevents the streams from interpenetrating, so they form a compressed boundary region as illustrated in Figure S37. This boundary is very thin close to the Sun, but grows as the material continues to accumulate and compress. By the time the flow has traveled several AU, the internal pressure of the boundary region can increase to the point that shock waves form at the edges. These regions, when bounded by shocks, are called corotating interaction regions or CIRs (Smith and Wolfe, 1976), and these structures dominate the dynamics of the equatorial solar wind beyond 3 AU during solar minimum periods. As the shock waves propagate into the surrounding plasma, the CIRs grow in radial extent, eventually becoming so broad that they overlap. It has been estimated that the entire equatorial solar wind has passed through at least one of these shocks by 25 AU (Burlaga, Schwenn and Rosenbauer, 1983).

As the solar cycle progresses to solar maximum, the polar coronal holes shrink, leaving a jumbled mixture of closed-field active regions and isolated coronal holes at low latitudes on the Sun. The alternation of fast and slow wind becomes less periodic and the wind is further disrupted by the increased frequency of coronal mass ejections which send shock waves and large amounts of chromospheric and prominence material into heliospheric space.

As the solar wind expands in a spherical manner, the density falls

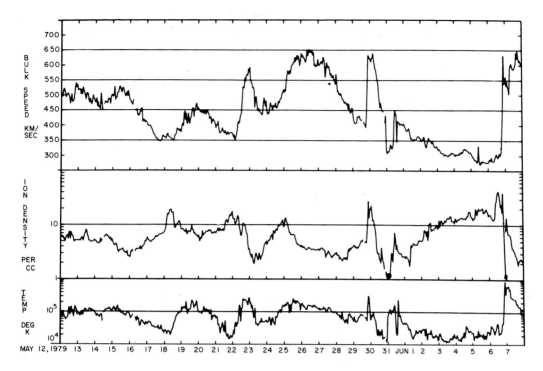

Figure S35 Solar wind properties at 1 AU over one solar rotation. Data are hourly averages from the National Space Science Data Center (Couzens and King, 1986).

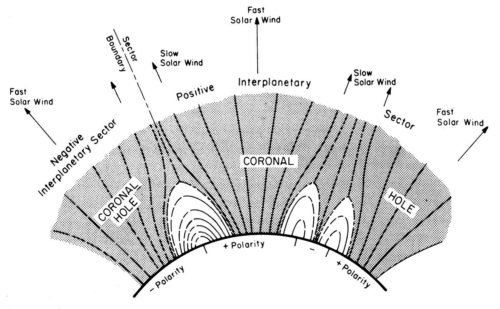

Figure S36 A schematic diagram of large-scale coronal magnetic structure in the solar equatorial plane, showing the associated solar wind flows (Hundhausen, 1977).

off and eventually the ram pressure of the flow becomes comparable with the pressure of the interstellar medium. At that point the flow is thought to decelerate at a termination shock and the subsequent slower and denser material is deflected into a tail, which trails the heliosphere as it moves through the interstellar medium. Predictions for the distance to the termination shock currently fall in the range of 70–100 AU from the Sun, but there are many complicating factors to include in the theory, such as the motion of the Sun through interstellar space, the unknown plasma parameters immediately beyond the heliosphere and the effects on the flow of penetrating particles like cosmic rays and interstellar neutral atoms. To date, all that is known for sure is that the Voyager and Pioneer spacecraft have not yet reached the shock, and the termination of the solar wind remains an active field of study (see Plasma wave).

Theory of coronal expansion

To explain why there is a solar wind at all, it is sufficient to discuss the ideal case of a steady state, one-fluid, spherically symmetric system. The fact that the solar corona has a temperature of several million degrees means that there must be an outward flow since no hydro-

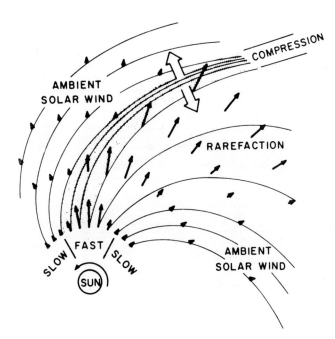

Figure S37 Illustration of the solar wind stream interaction in the equatorial plane. Light spiral lines represent the magnetic field, dark arrows denote the bulk flow speed, and the large open arrows indicate the non-radial flows driven by pressure gradients built up in the stream interaction. (Pizzo, 1985; copyright by the American Geophysical Union.)

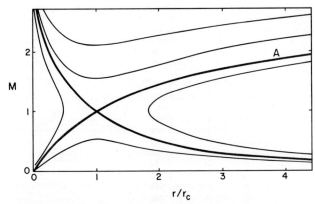

Figure S38 Topology of the solutions to equation (S13). The only solution which satisfies the requirements of low speed at the Sun and high speed at large radii is the one passing through the critical point and labeled A.

static equilibrium can exist in this system. This is because the conditions on a static atmosphere would yield a plasma pressure far from the Sun which is much larger than the pressure of the local interstellar medium.

If the corona is taken to consist of a fully ionized gas of protons and electrons, the steady one-fluid equations for conservation of particles and momentum can be written

$$\frac{d}{dr}(r^2 n V) = 0 \tag{S9}$$

$$V \frac{dV}{dr} + \frac{1}{mn} \frac{dP}{dr} + \frac{GM_S}{r^2} = 0 \tag{S10}$$

where $n = n_p = n_e$, $V = (m_p V_p + m_e V_e)/m$, $m = m_p + m_e$, $P = P_p + P_e$, the subscripts p and e refer to protons and electrons respectively, M_S is the mass of the Sun, and we have neglected the effects of viscosity, magnetic stresses or source terms in the momentum equation. We also require an equation of state. For this illustration, we will assume simple polytropic behavior

$$\frac{d}{dr}\left(\frac{P}{n^\gamma}\right) = 0 \tag{S11}$$

where γ is the polytropic index. For $\gamma = 5/3$, (S11) is an adiabatic equation of state, while $\gamma = 1$ represents isothermal flow. Any $\gamma < 5/3$ implies some unspecified form of heating within the system.

Equations (S10) and (S11) can be combined to form a complete differential, which gives the total energy E as a constant of the motion

$$E = \frac{1}{2} V^2 + \frac{\gamma}{\gamma - 1} \frac{P}{mn} - \frac{GM_S}{r} \tag{S12}$$

Equations (S9)–(S12) then yield an equation for the square of the sonic Mach number, $M^2 = mnV^2/\gamma P$,

$$\frac{M^2 - 1}{M^2} \frac{dM^2}{dr} = \frac{\left(1 + \frac{\gamma - 1}{2} M^2\right)\left(4E + \frac{3\gamma - 5}{\gamma - 1} \frac{GM_s}{r}\right)}{r\left(E + \frac{GM_s}{r}\right)} \tag{S13}$$

Although this equation must, in general, be solved numerically, the character of the solution can be illustrated by a topological diagram as in Figure S38. Consider the sign of the derivative dM^2/dr. The only way the sign of this quantity can change is when the right-hand side (RHS) of (S13) goes through zero or when M^2 goes through 1. It is clear that, for $1 < \gamma < 5/3$, the RHS equals zero at a single critical radius,

$$r_c = \frac{3\gamma - 5}{\gamma - 1} \frac{GM_s}{4 - E}$$

Let us assume that E is small enough that this critical radius is above the solar surface. Then the solution plane is divided into four regions defined by the sign of the slope of $M^2(r)$. Apart from a pair of singular solutions, (S13) requires that a solution curve crossing the $M^2 = 1$ line crosses with infinite slope, and that solution curves passing through $r = r_c$ must have zero slope there. The exceptions are the critical solutions, which pass through the critical point $(M^2, r) = (1, r_c)$. At this point, both sides of (S13) vanish simultaneously and the slope is determined by an expansion of the equation about that point. Equation (S13) has two of these solutions, drawn in bold in Figure S37.

The appropriate physical solution for the coronal expansion is then determined by the boundary conditions. At the inner boundary, low in the corona, the flow speed should be small. At large radii the pressure should decline to low values, matching the interstellar pressure. Furthermore, spacecraft observations have shown a highly supersonic flow. The only solution which satisfies these conditions is the one which passes through the critical point from low Mach number at small radii to $M > 1$ at large radii. Thus the flow speed is given by the critical solution labeled A in Figure S38.

There is a further condition on the solutions of (S13). A closer look at this equation shows that the flow speed goes as $r^{(3-2\gamma)/(\gamma-1)}$ as $r \to 0$, so we also require $\gamma < 3/2$ to yield small flow speeds at the inner boundary. This condition is equivalent to requiring a certain amount of heating in the low corona, maintaining the necessary temperature. Since the polytropic index has been estimated as $\gamma \sim 1.1$ in this region, this is not a problem, unless one inquires as to the physical mechanism which produces this heating. The heating of the corona in coronal hole regions is, to date, poorly understood in detail. A reader interested in the problem of coronal heating could consult Narain and Ulmschneider (1990) and references therein.

The illustrative model above assumes that the solar wind is driven solely by the thermal energy of the corona. However, such models have failed to produce realistic flows with speeds greater than about 300 km s^{-1}. The difficulty in producing faster flows is that, for the range of densities and temperatures appropriate for the corona, any energy addition to these flows below the critical point (by increasing the coronal temperature, for instance) results in more particles entering the flow rather than an increase in the flow speed (Leer and Holzer, 1980). Attempts to overcome this difficulty by incorporating more accurate expressions for the conductive heat flux or relaxing the assumption of spherical symmetry have not lead to significant improvements (Holzer and Leer, 1980 Leer and Holzer, 1980). Furthermore, two-fluid versions of this model (allowing the temperatures $T_p \neq T_e$) invariably produce hot electrons and much colder protons, since the electrons conduct the coronal heat much more efficiently. However, fast solar wind is always observed to have $T_p > T_e$. It is generally concluded that fast solar wind requires some form of energy input to the ions above the critical point. At that height, once the flow has become supersonic, no further mass can be added and further energization results directly in increased flow speed.

This energization is commonly thought to be provided to the fluid by hydromagnetic waves which have propagated out of the Sun and are damped near 5–6 solar radii, beyond the critical point. Most 'wave-driven' models assume that the primary energy is carried by Alfvén waves, which are non-compressive hydromagnetic fluctuations propagating at the Alfvén speed, $V_A = B (4\pi\rho)^{-1/2}$, where ρ is the total plasma density. Photospheric turbulent motions will generate hydromagnetic oscillations in all modes, but the Alfvén mode is the only one which can propagate through the transition region into the corona without being substantially reflected or damped. Furthermore, most of the wave power observed in the solar wind appears to be in low-frequency, non-compressive fluctuations such as would be produced by the Alfvén mode. In the corona, observations of solar spectral line broadening and Faraday rotation of spacecraft signals indicate the presence of substantial coronal fluctuations. If these motions are interpreted as being due to Alfvén waves, they can account for enough energy flux to heat the corona and accelerate the solar wind.

Alfvén waves propagating into the expanding solar wind without dissipation will decrease in intensity, and this gradient results in a wave-pressure force

$$F_w = -\frac{d}{dr}\left(\frac{\langle \delta B^2 \rangle}{8\pi}\right) \qquad (S14)$$

where $\langle \delta B^2 \rangle$ is the mean-square wave amplitude (Belcher, 1971). One-fluid models of coronal expansion including the wave force (S14) can produce a reasonable high-speed wind at 1 AU from reasonable values of the coronal parameters. However, as with the other one-fluid models, the equivalent two-fluid results show protons at 1 AU which are too cold. Thus heating of the protons is required as well as the momentum addition from (S14).

Wave heating of a plasma is a consequence of wave dissipation, so a rigorous model of high-speed solar wind driven by Alfvén waves requires a damping mechanism for these waves. Unfortunately, Alfvén waves in the solar wind damp only nonlinearly, and a suitable theory of Alfvén wave dissipation does not yet exist. Nonlinear dissipation and the related phenomena of magnetohydrodynamic turbulence are currently very active areas of study.

This lack of a dissipation mechanism has been circumvented by various ad hoc constructions. One technique, due to Hollweg (1978), is to allow the wave flux to evolve in radius without damping until the relative amplitude (which increases since the field intensity is also falling) reaches some 'saturation' value, say $\langle \delta B^2 \rangle / B^2 = 1/2$. At this point it is assumed that nonlinear processes act to limit the relative amplitude to this value and the excess energy of the waves is fed to the protons as heat. This procedure glosses over the physics of the microscopic damping processes, but it may simulate the overall effect of the wave heating on the solar wind. With reasonable plasma and wave parameters at the Sun, this model can easily obtain the densities, temperatures and velocities of the wind at 1 AU, and variations of the wind can simply be created by changing the wave flux at the Sun. However, this general success only indicates that wave interactions are a plausible explanation for the acceleration of the high-speed solar wind. The actual processes responsible for this acceleration are still not definitively known.

Microstructure of solar wind plasma

To this point, the solar wind plasma has been treated as a fluid, or a superposition of several fluids. As such, the plasma was described by specifying certain bulk properties like velocity, density and temperature without consideration of the detailed particle distribution function. However, above the corona the solar wind plasma is essentially collisionless and generally exhibits a large variety of kinetic features.

Electrons

Electron distributions in the solar wind are composed of two superimposed populations: a high-density, low-temperature core and a much hotter halo, as illustrated in Figure S39 from Pilipp et al. (1987). The core distribution is nearly isotropic with a typical temperature of 10 eV at 1 AU. The halo distribution consists of about 5% of the particles, and it is these particles which carry the bulk of the solar wind heat flux. The halo distribution is much more variable than the core and often contains a narrow beam, or strahl, pointing away from the Sun along the magnetic field. The strahl becomes narrower at the higher energies, with a width of 5° or less at the highest energy measured. The strahl is observed to be most intense in the smooth flows originating in the center of coronal holes. Toward the edges of the coronal hole flow, the strahl becomes progressively broader until it disappears in the vicinity of the current sheet where all the species are seen to be isotropic.

The kinetic theory of solar wind electrons is complicated by the fact that the thermal speed of the distribution is substantially larger than the solar wind speed. Thus, in the absence of collisions, the distribution at one point can in principle be affected by conditions at other points in the flow. A starting point for such a non-local theory is

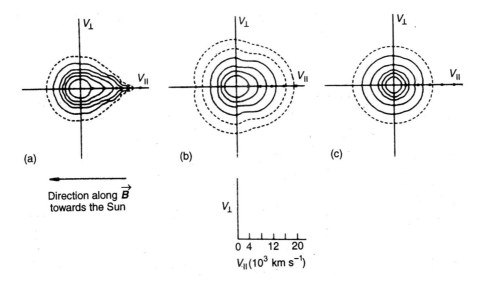

Figure S39 Contours describing three electron distribution functions measured on the Helios spacecraft. The figures show typical examples of distributions with (a) a narrow strahl, (b) a broad strahl and (c) a nearly isotropic structure (Pilipp et al., 1987; copyright by the American Geophysical Union.)

the exospheric picture which considers these collisionless electrons to move in single-particle orbits in the steady heliospheric fields (e.g. Lemaire and Scherer, 1973). In order to limit the free escape of negative charge, an electrostatic potential forms which inhibits the electrons from outrunning the ions. Most of the exospheric electrons are then trapped in the potential well created by the magnetic-moment mirror force close to the Sun and the electrostatic forces at large distances. A small number of energetic electrons will be able to overcome the potential barrier and escape, matching the charge flux of solar wind ions. In this picture the escaping electrons will stream freely along the magnetic field, their pitch angles decreasing adiabatically with distance from the Sun. If the escaping electrons correspond to the observed halo distribution and the trapped electrons correspond to the core, a natural explanation for the existence of two electron populations, and of the character of the high-energy strahl, is provided.

However, the exospheric picture as it stands implies that the peak of the trapped electron distribution sits at rest with respect to the Sun, and this large velocity difference between the proton and core electron populations is not observed. Furthermore, the relative isotropy of the non-strahl portions of the halo distribution cannot be explained in the absence of some mechanism to scatter the particles out of the beam.

One form of scattering is due to wave–particle interactions. The exospheric distribution is unstable to the generation of small-scale oscillations which tap the streaming energy of the distribution (Forslund, 1970). These waves, in principle, scatter the particles so as to reduce the relative drift of the core electrons and the beaming of the strahl. Much of the work to date on this process has been concerned with the linear problem of identifying the instabilities and modes most likely to be generated, but the investigation is hampered by the fact that we cannot observe the onset of the instability. Essentially, we only observe what would correspond to the final state of marginal stability, and the initial unstable distribution must be guessed. In addition, the subsequent problem of the detailed evolution of the particles and waves towards the observed state is a difficult one, and progress has been slow.

Another possible scattering mechanism is that of Coulomb collisions, which will certainly have some effect. Scudder and Olbert (1979a,b) have presented a model of the non-local effects of elastic Coulomb collisions in a solar wind which is neither exospheric or collision dominated. They point out that the velocity dependence of the Coulomb cross-section results in collisional mean free paths which are small compared to 1 AU for low-energy electrons and much larger than 1 AU for the higher-energy particles. This situation leads naturally to a bifurcation of the electron distribution similar to the observed core and halo populations. The transition energy between the two populations is predicted to be comparable to the electrostatic potential of exospheric theory, but while the exospheric picture predicts halo electrons only near the magnetic field direction, the Scudder and Olbert model calls for such hot electrons at all pitch angles, as is qualitatively observed. This model, though encouraging, is just a start on the difficult problem of non-local kinetic effects and needs to be pursued further. A realistic description of solar wind electrons is likely to include inelastic collisions as well as wave–particle effects.

Ions

Solar wind proton distributions display characteristics similar to those of electrons, in that they generally consist of a dense core and a more energetic halo carrying a field-aligned heat flux. As in the electron theory of Scudder and Olbert, the action of Coulomb collisions appears to govern the low-velocity regions of the proton distribution, particularly in the slow, dense wind. In the fast wind, and in the energetic portions of the distributions in the slow wind, the protons are essentially collisionless and are able to deviate significantly from Maxwellian shapes. However, the interpretation of these ion properties is considerably different from that of the electrons. Because the ion thermal speeds are much smaller than the flow speed (which is what is meant by calling the solar wind supersonic), ion distributions can truly be treated locally, determined by the conditions near the point of observation.

The variety of shapes displayed by the proton distribution is illustrated by the Helios data shown in Figure S40 (Marsch et al., 1982a). The proton core is basically isotropic in slow solar wind, but for faster flows it becomes progressively enhanced in the direction perpendicular to the magnetic field. This perpendicular anisotropy is the opposite of the collisionless expectation, where preservation of the magnetic moment of each particle leads to distributions aligned with the magnetic field. The energetic portion of the proton distribution usually displays a considerable skewing along the magnetic field in the direction away from the Sun. The skewing can consist of a high-energy tail or shoulder, but 20–30% of the data taken during the primary missions of the Helios spacecraft showed a well-defined second peak in the distribution (as seen in the distributions labeled (b), (k) and (l) in Figure S40). The relative speed, ΔV, between the main peak of the core and the secondary peak is correlated with the local Alfvén speed. While the relative speed in slow wind is comparable with the Alfvén speed, in fast wind ΔV typically exceeds the Alfvén speed with a mean value of $\Delta V/V_A \sim 1.8$.

The origin of these double-peak distributions is not clear, though several suggestions have been made. One hypothesis is that of interpenetrating streams due to a time-dependent flow speed at the Sun (Feldman et al., 1974). Another study (Livi and Marsch, 1987) has shown that double-peak distributions can result from the competition between magnetic moment conservation (which focuses particles into a parallel beam when moving into a decreasing magnetic field) and Coulomb collisions (which scatter the slower particles out of the beam). However, this process would only explain the small fraction of double-peak distributions observed in those regions of the solar wind where collisions are non-negligible.

The perpendicular anisotropy of the proton core has been taken as evidence for local wave heating of the distribution (Bame et al., 1975). This heating could result from cyclotron damping of the ambient waves of solar origin, and/or of locally generated waves due to streaming instabilities. This latter process would also naturally yield the observed correlation of beam speeds with the Alfvén speed, since super-Alfvénic beams will excite these instabilities and provide a self-regulation of the secondary population.

The distributions of heavier ion species in the solar wind also exhibit unusual properties. Helium is the most abundant of the heavy ions and therefore is the most readily observed. Practically all of the solar wind helium is fully ionized, and these alpha particles therefore have half the charge-to-mass ratio of protons. As such, it is expected that they would have greater difficulty flowing away from the Sun than the protons under the combined action of the heliospheric electrostatic field and the solar gravitational field. Coulomb collisions would limit this effect, allowing the protons to drag the alpha particles along to a certain extent, but a slower alpha particle flow would still result. This is not at all what is observed. In slow solar wind the alpha particle population flows at the same speed as the protons, and in the faster wind the alpha particles actually stream faster than the proton bulk speed (e.g. Neugebauer, 1981). Again, the drift of the alpha particle population relative to the protons is field aligned and correlated with the Alfvén speed. In fast wind this drift can equal the Alfvén speed, and has been observed to reach values of 170 km s^{-1} at the orbit of Mercury (Marsch et al., 1982b). At such times, given a 4% abundance of helium, the alpha particles can carry almost 20% of the total momentum of the solar wind. The temperature of the helium population is also anomalous in the high-speed wind, being typically three to five times hotter than the protons.

Other heavy ions exhibit the same behavior as the alpha particles (e.g. Schmidt et al., 1980). In slow wind they comove with the protons and in fast wind they essentially stream with the alpha population. The temperatures of these heavy ions tend to be proportional to the ion mass in the fast wind, implying that all ions are given comparable thermal speeds. In the case of iron, temperatures in excess of 80 million K have been observed (Mitchell et al., 1981), clearly indicating that these particles have been preferentially heated after leaving the solar corona.

This preferential acceleration and heating of heavy ions has not been explained. The most promising suggestion for the responsible mechanism has been a resonant interaction with the ambient ion cyclotron waves (Isenberg, 1984), but attempts to model this process have yet to match the observations successfully. In principle, this interaction can substantially heat and accelerate ions – this is the basis for most wave-driven models of the solar wind – and there is ample energy in the low-frequency waves to produce the extra energization, but it remains a puzzle as to how these waves can preferentially interact with the heavy ions to the extent observed. Perhaps the answer to this question will also lead to the mechanism responsible for the proton double-peak distributions.

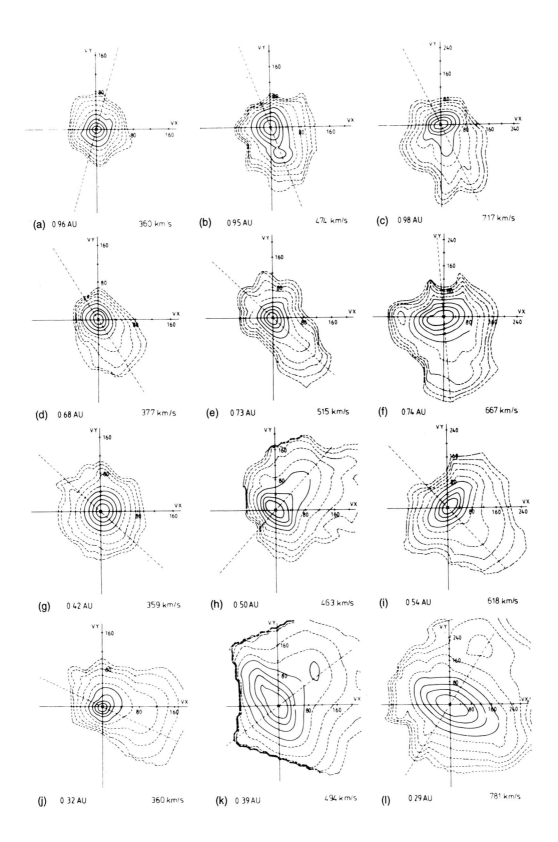

Figure S40 Contour plots of a variety of solar wind proton distributions as measured by the Helios 2 instruments. The cuts through the distributions are in the plane containing the magnetic field (dashed line) and the flow vector (vx-axis). (Marsch et al., 1982a; copyright by the American Geophysical Union.)

Summary

The solar wind is a supersonic flow of plasma which emanates from the Sun in all directions and fills the heliosphere. It is important as an energy source for most magnetospheric phenomena and as a demonstration laboratory for a huge variety of plasma interactions. The flow is understood as a consequence of a hot solar corona and the low pressure of the interstellar medium, but there are many aspects of the generation of the wind and of plasma processes observed in the flow which are not fully explained. Further information can be found in the monographs of Parker (1963) and Hundhausen (1972), the review of Isenberg (1991), the compilations of Kennel, Lanzerotti and Parker (1979) and Schwenn and Marsch (1991), and the proceedings of the Solar Wind Conferences (Mackin and Neugebauer, 1966; Sonett, Coleman and Wilcox, 1972; Russell, 1974; Rosenbauer, 1981; Neugebauer, 1983; Pizzo, Holzer and Sime, 1988; Marsch and Schwenn, 1992).

Philip A. Isenberg

Bibliography

Bame, S.J., Asbridge, J.R., Feldman, W.C. *et al.* (1975) Evidence for local ion heating in solar wind high speed streams. *Geophys. Res. Lett.*, **2**(9), 373–5.
Belcher, J.W. (1971) Alfvénic wave pressures and the solar wind. *Astrophys. J.*, **168**(3), 509–24.
Burlaga, L.F., Schwenn, R. and Rosenbauer, H. (1983) Dynamical evolution of interplanetary magnetic fields and flows between 0.3 AU and 8.5 AU: entrainment. *Geophys. Res. Lett.*, **10**(5), 413–6.
Coles, W.A. and Rickett, B.J. (1976) IPS observations of the solar wind out of the ecliptic. *J. Geophys. Res.*, **81**(25), 4797–9.
Couzens, D.A. and King, J.H. (1986) *Interplanetary Medium Data Book*, Supplement 3. Greenbelt, MD: Natl Space Sci. Data Center, NSSDC/WDC-A-R&S 86–04.
Feldman, W.C., Asbridge, J.R., Bame, S.J. and Montgomery, M.D. (1974) Interpenetrating solar wind streams. *Rev. Geophys. Space Phys.*, **12**(4), 715–23.
Forslund, D.W. (1970) Instabilities associated with heat conduction in the solar wind and their consequences. *J. Geophys. Res.*, **75**(1), 17–28.
Hollweg, J.V. (1978) Some physical processes in the solar wind. *Rev. Geophys. Space Sci.*, **16**(4), 689–720.
Holzer, T.E. and Leer, E. (1980) Conductive solar wind models in rapidly diverging flow geometries. *J. Geophys. Res.*, **85**(A9), 4665–79.
Hundhausen, A.J. (1972) *Coronal Expansion and Solar Wind*. New York: Springer-Verlag.
Hundhausen, A.J. (1977) An interplanetary view of coronal holes, in *Coronal Holes and High Speed Wind Streams* (ed. J. Zirker) Boulder, CO: Colorado, pp. 225–329.
Isenberg, P.A. (1984) Resonant acceleration and heating of solar wind ions: anisotropy and dispersion. *J. Geophys. Res.*, **89**(A8), 6613–22.
Isenberg, P.A. (1991) The solar wind, in *Geomagnetism*, Vol 4 (ed. J.A. Jacobs) New York: Academic Press, pp. 1–85.
Kennel, C.F., Lanzerotti, L.J. and Parker, E.N. (eds) (1979) *Solar System Plasma Physics*. New York: North-Holland.
Lallement, R., Bertaux, J.L. and Kurt, V.G. (1985) Solar wind decrease at high heliographic latitudes detected from Prognoz interplanetary Lyman alpha mapping. *J. Geophys. Res.*, **90**(A2), 1413–24.
Leer, E., and Holzer, T.E. (1980) Energy addition in the solar wind. *J. Geophys. Res.*, **85**(A9), 4681–8.
Lemaire, J. and Scherer, M. (1973) Kinetic models of the solar and polar wind. *Rev. Geophys. Space Phys.*, **11**(2), 427–68.
Livi, S. and Marsch, E. (1987) Generation of solar wind proton tails and double beams by Coulomb collisions. *J. Geophys. Res.*, **92**(A7), 7255–61.
Mackin, R.J. and Neugebauer, M. (eds) (1966) *The Solar Wind*. New York: Pergamon.
Marsch, E. and Schwenn, R. (eds) (1992) *Solar Wind Seven*. New York: Pergamon.
Marsch, E., Mühlhäuser, K.-H., Schwenn, R. *et al.* (1982a) Solar wind protons: three-dimensional velocity distributions and derived plasma parameters measured between 0.3 and 1 AU. *J. Geophys. Res.*, **87**(A1), 52–72.
Marsch, E., Mühlhäuser, K.-H., Rosenbauer, H. *et al.* (1982b) Solar wind helium ions: observations of the Helios solar probes between 0.3 and 1 AU. *J. Geophys. Res.*, **87**(A1), 35–51.
Mitchell, D.G., Roelof, E.C., Feldman, W.C. *et al.* (1981) Thermal iron ions in high-speed solar wind streams. 2. Temperatures and bulk velocities. *Geophys. Res. Lett.*, **8** (7), 827–30.
Narain, U. and Ulmschneider, P. (1990) Chromospheric and coronal heating mechanisms. *Space Sci. Rev.*, **54**(4), 377–445.
Neugebauer, M. (1981) Observation of solar-wind helium. *Fund. Cosmic Phys.*, **7**(2), 131–99.
Neugebauer, M. (ed) (1983) *Solar Wind Five*. NASA Conf. Proc 2280.
Parker, E.N. (1963) *Interplanetary Dynamical Processes*. New York: Interscience.
Pilipp, W.G., Miggenrieder, H., Montgomery, M.D. *et al.* (1987) Characteristics of electron velocity distribution functions in the solar wind derived from the Helios plasma experiment. *J. Geophys. Res.*, **92**(A2), 1075–92.
Phillips, J.L., Bame, S.J., Barnes, A., *et al.* (1995) Ulysses solar wind plasma observations from pole to pole. *Geophys. Res. Lett.*, **22**, 3301–4.
Pizzo, V.J. (1985) Interplanetary shocks on the large scale: a retrospective on the last decade's theoretical efforts, in *Collisionless Shocks in the Heliosphere: Reviews of Current Research*, (eds B.T. Tsurutani and R.G. Stone.) Washington, DC: American Geophysical Union, pp. 51–68.
Pizzo, V.J., Holzer, T.E. and Sime, D.G. (eds) (1988) *Solar Wind Six*. Boulder, Co.: NCAR Tech. Note TN-306 + Proc.
Rosenbauer, H. (ed.) (1981) *Solar Wind Four*. Lindau, FRG: Max-Planck-Inst. für Aeron. Rep. MPAE-W-100–81–31.
Russell, C.T. (ed.) (1974) *Solar Wind Three*. Los Angeles, Ca.: UCLA.
Schmidt, W.K.H., Rosenbauer, H., Shelley, E.G. and Geiss, J. (1980) On temperature and speed of He^{++} and O^{6+} ions in the solar wind. *Geophys. Res. Lett.*, **7**(9), 697–700.
Schwenn, R., and Marsch, E. (eds) (1991) *Physics of the Inner Heliosphere*. New York: Springer-Verlag.
Scudder, J.D. and Olbert, S. (1979a) A theory of local and global processes which affect solar wind electrons, 1. The origin of typical 1 AU velocity distribution functions – steady state theory. *J. Geophys. Res.*, **84**(A6), 2755–72.
Scudder, J.D. and Olbert, S. (1979b) A theory of local and global processes which affect solar wind electrons, 2. Experimental support. *J. Geophys. Res.*, **84**(A11), 6603–20.
Smith, E.J., and Wolfe, J.H. (1976) Observations of interaction regions and corotating shocks between one and five AU: Pioneers 10 and 11. *Geophys. Res. Lett.*, **3**(3), 137–40.
Sonett, C.P., Coleman, P.J. and Wilcox, J.M. (eds) (1972) *Solar Wind*. NASA Spec. Publ. 308.

Cross references

Coronal mass ejections
Geomagnetic storm
Heliosphere
Plasma wave
Shock waves
Ulysses mission

SOVIET MARS MISSIONS

Soviet Mars program – stage one

In the USSR, plans for an ambitious program of planetary exploration began immediately after the launching of the first Sputnik in October 1957. The program was under the direction of the military, and it contemplated sending at least two spacecraft to both Venus and Mars at each opportunity. Well ahead of the US in the development of large launching systems, as attested by the 1329-kg payload of Sputnik 3 in March 1958, the Soviets were able to mount their first attempt at a Mars mission in 1960.

Secrecy about the program was extreme. In 1962 the Administrator of NASA reported to Congress that unsuccessful attempts to launch Mars probes had occurred in 1960 on 10 October and 14 October.

However, Soviet scientists that were involved now say that only the latter launch was intended for Mars, and they have no knowledge of the former.

The launch system consisted of the 1–1/2-stage rocket that was launching Sputniks, topped by a second stage and an 'escape stage'. The spacecraft, which was called an 'Automatic Interplanetary Station' (AIS), weighed 640 kg, and its 10-kg scientific payload was similar to that of NASA's first planetary spacecraft, Mariner 2; it included a magnetometer, ion traps for measuring solar plasma, cosmic ray counters and micrometeoroid sensors.

In the next launch period in 1962, Mars probes were launched on 29 October, 1 November and 4 November. The first and third were stranded in Earth orbit, but the second became the first AIS to head for Mars, and it was therefore announced and given the name Mars 1 (in Russian, MAPC 1). It weighed 895 kg. Its primary objective was to photograph Mars in several colors, and its scientific payload also included 'magnetometers to detect the presence of a martian magnetic field and magnetic fields in space, gas-discharge and scintillation counters to detect martian radiation belts and study the spectrum of cosmic radiation, . . . special instruments to record micrometeorite impacts, . . . a spectroreflectometer for detecting vegetation, and a spectrograph for study of the ozone absorption bands' (Anonymous, 1963). In 37 telemetry sessions the AIS sent back significant data on the interplanetary medium, and it flew past Mars on 19 July 1963 at a distance of 195 000 km. However, communication with it had ceased on 21 March because of a problem with the orientation system.

The next attempt began on 30 November 1964 (2 days after the very successful Mariner 4) with the launch of ЗОНД-2 (Zond-2). The name (which translates as 'probe') was apparently a futile attempt to conceal the purpose of the mission. It carried the same scientific experiments as MAPC 1 and suffered the same fate, with communication being lost in early May 1965. It passed Mars on 6 August at 1500 km altitude. A second identical spacecraft had been prepared, but the launch period expired before it could be made ready, so it was launched a year later on 18 July 1966. ЗОНД-3 took some photographs of the far side of the Moon and then approached the Martian orbit even though the planet was very far away.

In the next opportunity, January 1967, the US was occupied with the Mariner 5 mission to Venus and preparations for Apollo, and the Soviets were also occupied with missions to Venus and Moon, so no Mars missions were attempted, looking forward to major efforts in 1969.

Stage two

Up to this time the Soviets had continued to employ the original Sputnik launch vehicle, but now they had the new Proton with nearly five times the payload capability (the spacecraft weighed 4860 kg); it had already launched three successful lunar missions. Another innovation was the involvement of the Institute for Space Research (IKI), which had been founded in 1967. It was responsible for the scientific payloads on the spacecraft, which were intended to be placed into Mars orbits. In launch attempts on 27 March and 14 April, both Protons failed and were never announced officially.

On 10 May 1971 the Soviets again attempted a Mars orbiter. The weight of the AIS was modest by Soviet standards in order to permit a quick trip and thus beat Mariner 8 into Mars orbit. The Proton first stage reached its parking orbit, but the second stage failed because of an erroneous ground command. The launch was identified as (KÓCMOC) 419 (COSMOS). The launch of Mariner 8 also failed. Launches on 19 and 28 May were successful and placed MAPC 2 and MAPC 3 into orbit around planet Mars a few days after Mariner 9. Each 'orbiting station' consisted of a carrier vehicle of 3440 kg with a 1210 kg descent module, enclosing a 350-kg landing capsule (Mariner 9 weighed 998 kg). In addition to two cameras ('phototelevision units'), the orbiters carried instruments for ten scientific investigations: three infrared photometers for measuring surface albedo, water vapor abundance and altitude profiles; an infrared radiometer for measuring surface temperature; a Lyman alpha sensor to detect hydrogen in the upper atmosphere; a magnetometer; two types of solar wind detectors; a detector of solar bursts at metric wavelengths; and a radio occultation experiment to measure altitude profiles of atmospheric and ionospheric density. Essentially the same complement had been on the 1969 spacecraft.

Landers (the Russian terminology is 'descent modules' – DM) were released 4.5 h before injection of the orbital stations into orbit. The one on MAPC 2 crashed and succeeded only in depositing on the surface 'a pennant depicting the coat of arms of the Soviet Union' for the first time.

The MAPC 3 DM executed all of its intended landing sequence, but it may have landed in a non-standard fashion because of some anomaly in the radar altimeter. The radio transmitter sent back 20 s of calibration data on the television camera and then ceased operation. The other experiments were never activated. They were intended to measure temperature, pressure, wind velocity, atmospheric composition and the chemical and physical properties of the surface material as well as to look for organic compounds that might indicate current or former living organisms.

While these three probes were en route, Mars became enshrouded in the most intense global dust storm that had yet been observed. The cameras on the Russian orbiting stations operated on preprogrammed sequences and saw only a featureless orange ball. Although both stations continued to operate for 4 months, their scientific output was disappointing, but for different reasons. The telemetry from MAPC 2 was of very low quality, and most of the data were useless except from the radio occultation experiment. The MAPC 3 propulsion system malfunctioned, leaving the orbiter in a 12.7-day orbit with an apapsis of 190 000 km, so that it had very few opportunities for quality observations. In addition to the phototelevision units and the radio occultation system, each orbiter carried seven instruments for remote sensing (four in the infrared and one each in ultraviolet, centimeter and meter wavelengths), two plasma sensors and a magnetometer.

The major scientific results included detection of a magnetic field (4000 times weaker than Earth's) and a plasma bow shock; detection of hydrogen, atomic oxygen and water vapor in the atmosphere; temperature measurements on the surface, the polar cap and the cloud tops; and altitude profiles from radio occultation of electron concentration in the ionosphere and temperature in the upper neutral atmosphere. They were surpassed by the results of Mariner 9 by many orders of magnitude.

1973 was to have been the biggest year yet for Mars exploration, with multiple orbiters and landers, but the American Viking mission was postponed, and the Soviet missions again fell short of expectations. Orbiters MAPC 4 and MAPC 5 were launched on 21 and 25 July 1973 and lander carriers MAPC 6 and MAPC 7 on 5 and 9 August, the Proton launch system performing faultlessly by now. All four arrived at Mars between 10 February and 12 March, but MAPC 4 failed to go into orbit, and MAPC 5, while orbiting successfully, suffered a transmitter failure after only 22 orbits. The flyby modules (FBM) on MAPC 6 and MAPC 7 were not intended to orbit. Each released its descent module (DM) 4 h before encounter, but the MAPC 7 DM missed the planet completely.

The DM on MAPC 6 apparently executed a perfect landing sequence, transmitting data continuously, during its descent on pressure, temperature, altitude, velocity and acceleration, but its transmitter failed at approximately the instant of touchdown. From some engineering measurements on the mass spectrometer during the descent, it was inferred that the atmosphere contained about one-third argon, but subsequent Viking measurements determined the argon abundance to be about 1.6%.

The orbital stations were very similar to MAPC 2 and MAPC 3 and carried essentially the same scientific payload with the addition of a line-scan imaging device with wide spectral sensitivity, a spectrometer and a polarimeter in the infrared band and an ultraviolet photometer. The information obtained surpassed that of all earlier Soviet Mars probes. About 60 photographs were returned, mostly in the same area. Scans of a limited area of the surface yielded the brightness in the visible and near infrared, polarization in visible and radio wavelengths, gamma radiation and relative altitudes. The abundances of water vapor and ozone in the lower atmosphere and atomic hydrogen and other trace elements in the upper atmosphere were measured. Among the most interesting new discoveries were that the ozone concentration peaked at 40 km altitude instead of very close to the surface as expected; the electron concentration in the night-time ionosphere was measured; and the most detailed data to date on the magnetic field and plasma supported the earlier conclusion that an intrinsic magnetic field holds off the solar mind.

Stage three

At this point that USSR decided to concentrate on Venus for the next decade, seeing little chance of competing favorably with the Viking missions.

Having completed the very successful series of mission called BEHÉPA (Venera 9 to 16) and BEΓA (Vega 1 and 2), the USSR elected to concentrate on Mars again for the next few years with a new spacecraft design that was considerably larger (6200 kg) and more sophisticated than before. Their first objective concentrated its attention on the satellite Phobos, as there was a widespread belief that the small bodies were the key to understanding the origin and evolution of the solar system. The spacecraft were given the name of the satellite.

Two spacecraft were launched on 7 July and 12 July 1988. On 30 September ФОБОС 1 responded to an erroneous command from Earth, switched off its attitude-control system, and was not heard from again. ФОБОС 2 entered Mars orbit on 29 January 1989 and operated successfully until 27 March. Further information on its results will be found in the entry on the Phobos mission.

Conway W. Snyder

Bibliography

Anonymous (1963) Enroute to the planet Mars, *Priroda*, No. 1, pp. 16–7.

Snyder, C.W. and Moroz, V. (1992) Spacecraft exploration of Mars, in *Mars* (eds H.H. Kieffer, B.M. Jakosky, C.W. Snyder and M.S. Matthews). Tucson: University of Arizona Press, pp. 71–119.

Cross references

Mariner missions
Phobos mission
Venera missions
Viking mission

SPACEWATCH

'Spacewatch' is the name of an astronomical observing program at the University of Arizona. The Spacewatch program has two goals: to discover new objects, and to develop new techniques for such discovery. The technique of scanning with charge-coupled devices (CCDs) has been refined significantly as a result of the Spacewatch program. The technique has been described by Gehrels (1991) and by Rabinowitz (1991). The telescope scans the sky, either by moving the telescope in any chosen direction and at any rate, or by simply turning off the drive of the telescope, whereby the Earth's rotation provides the scan. Mounted on the telescope is a CCD, wherein the charges generated by incident photons are transferred from pixel to pixel at exactly the rate at which the field moves in the focal plane (due to the scanning of the telescope). The pixel charges that are transferred from row to row end up in a register, and this information is subsequently transferred into a computer. The CCD has 2048 rows of 2048 pixels each. The pixel size is 24 × 24 mn, which translates at the 0.9-m Spacewatch Telescope to 1.076 × 1.076 arcsec. Spacewatch uses three consecutive scans, typically of 30-min length, for the same region.

The computer records an x–y pixel position for each object detected. The computer is programmed to compare these x–y positions and thereby to recognize a moving object; the three consecutive positions have to be in consistent places. In addition, the observer inspects each of the scans visually, searching particularly for fast and faint trails of objects, which may be close to the Earth.

The comets and asteroids that are close to the Earth constitute a primary interest of the Spacewatch programs, but more distant objects are also detected. In addition to comets and asteroids that come close to the Earth, Spacewatch also finds large numbers of main-belt asteroids between the orbits of Mars and Jupiter. Objects at yet greater distance are also found, such as 1992 AD. Table S6 shows a typical set of discoveries. The first column of the table gives the identification. Perihelion is the calculated distance for the object's closest approach to the Sun, and aphelion is the greatest distance attained in the elliptical orbit. The inclination is the angle between the plane of the orbit and the plane of the major planets (the ecliptic plane). The first line of the table has a periodic comet called Spacewatch. Lower down, the second 'Spacewatch' entry is also a comet, but it is a 'new' one, a first apparition, in a parabolic orbit coming in from the outer reaches of the solar system. 1992 h was its preliminary notation.

The other lines in Table S6 are for asteroids. The system of identification is most easily explained through an example. For instance, 1992 AD has the 'A' because it was discovered in the first half of January, and the 'D' because it was the fourth object discovered in these 2 weeks. 'Amor' (q.v.) is a type of asteroid that is not yet close enough to the Earth to be menacing. An 'Apollo' (q.v.) crosses the orbit of the Earth. Its orbit may change later. Asteroid (5145) has received that number as a permanently assigned designation in the Asteroid Catalog; it has also now been named, 'Pholus.'

Very small objects are also found by Spacewatch. The sizes are given in the fifth column of Table S6. One of them, 1991 VG, is a possible target for future spacecraft missions because its orbit is so similar to that of the Earth. It would therefore require a relatively small amount of energy to launch to it. Spacewatch also finds large objects, such as 1992 AD. This one is unique in that it is one of the reddest objects in the solar system, possibly due to the presence of organic material. Its orbit is chaotic: a chance proximity, to Uranus for instance, could throw it out of the solar system. Alternatively, it might become a near-Earth asteroid, with a chance of impacting the Earth.

An asteroid is not merely discovered. It must be observed again and again, in order to have its positions determined with precision. The orbit can then be computed. Only then are future positions of the object reliably predictable, enabling later detection and study. Follow-up of known objects is another of the Spacewatch programs. The Spacewatch Telescope is used during the bright half of the

Table S6 A sample of Spacewatch discoveries

Identification	Perihelion distance (AU)	Aphelion distance (AU)	Inclination (deg)	Diameter (km)	Date of discovery	Remarks
P/Spacewatch	1.54	4.8	10.0	–	10 September 1991	
1991 TT	1.00	1.4	14.8	0.03	6 October 1991	
1991 TU	0.94	1.9	7.7	0.009	7 October 1991	
1991 VA	0.93	1.9	6.5	0.02	1 November 1991	
1991 VG	0.97	1.1	0.2	0.01	6 November 1991	[a]
1991 XA	0.98	3.6	5.3	0.09	3 December 1991	
1992 AD	8.7	32.3	24.7	> 140	9 January 1992	(5145)
1992 AE	1.13	2.2	5.8	2.9	10 January 1992	Amor
1992 BA	1.25	1.4	10.5	0.4	27 January 1992	Peculiar orbit
1992 DU	0.96	1.4	25.1	0.05	26 February 1992	
1992 HF	0.61	2.2	13.2	0.6	24 April 1992	
Spacewatch	3.2	–	125.1	–	1 May 1992	1992 h
1992 JG	1.30	3.9	5.6	0.6	2 May 1992	Deep Mars crosser
1992 JD	1.00	1.1	13.6	0.05	3 May 1992	

[a] This may be an upper stage of a spacecraft or, if natural, a new type of asteroid with an orbit nearly the same as that of the Earth.

month, centered on full moon, for a search for planets of other stars by McMillan *et al.* (1993).

Tom Gehrels

Bibliography

Gehrels, T. (1991) Scanning with charge-coupled devices. *Space Sci. Rev.*, **58**, 347–75.
Gehrels, T. (1996) Collisions with comets and asteroids. *Sci. American*, **274**, 34–39.
McMillan, R.S., Moore, T.L., Perry, M.L. *et al.* (1994) Long, accurate time series measurements of radial velocities of solar-type stars. *Astrophys. Space Sci.*, **212**, 271–280.
Rabinowitz, D.L. (1991) Detection of Earth-approaching asteroids in near real time. *Astron. J.*, **101**, 1518–29.

Cross references

Amor object
Apollo object
Asteroid
Aten object
Charge-coupled device
Near-Earth object

SPECTROPHOTOMETRY

Reflectance spectrophotometry is a fundamental technique for determining the composition and physical state of the constituents of a planetary atmosphere or the mineral composition and texture of the regolith of a solar system object. The methodology which underlies the technique involves the acquisition and analysis of reflected electromagnetic radiation from a sunlit solar system object as a function of viewing geometry and wavelength. This provides important information about the reflectivity of the object under investigation. Spectral absorption features in the electromagnetic radiation reflected from a solar system object are generally due to either charge transfer, electronic absorptions or to the vibrational, rotational and translation motions of the scattering molecules. Therefore, absorption bands in the spectral geometric albedo of an object under investigation can be compared to reflection spectra of samples measured in the laboratory and estimates of the abundances of the spectrally active components can be made. In addition, the angular scattering properties of the reflected electromagnetic radiation from an object with no atmosphere can be related to the physical properties of the object's surface.

Bond albedo

The Bond albedo of a non-luminous object is defined as the ratio of the total radiative flux reflected in all directions to the total incident flux in a given wavelength range. Following the formalism of Russell (1916), if S_λ is the solar flux at the Earth's orbit at some wavelength λ, the amount of energy E_λ impinging on a solar system object of radius r at a distance R (in astronomical units) from the Sun is given by

$$E_\lambda = \frac{S_\lambda \pi r^2}{R^2}$$

If an observer were exactly on the Sun–object line (solar phase angle $\alpha = 0$) a fully illuminated disk would be seen when observing the object. The intensity measured in this case is defined to be $C_\lambda(0)$. In general, an observer is not exactly on the Sun–object line, and therefore a different intensity $C_\lambda(\alpha)$ will be observed, where α is the solar phase angle. The phase function, $B_\lambda(\alpha)$ is defined as the ratio of the intensity at zero phase angle to the intensity at phase angle α. In this more general observational circumstance, the rate at which the object reflects radiation in all directions is then:

$$I_\lambda = 2\pi d^2 C_\lambda(0) \int_0^\pi B_\lambda(\alpha)\sin(\alpha)d\alpha$$

where d is the distance between the observer and the object.
The Bond albedo A_λ is then:

$$A_\lambda = \frac{d^2 R^2 C_\lambda(0)}{r^2 S_\lambda} 2 \int_0^\pi B_\lambda(\alpha)\sin(\alpha)d\alpha$$

The term on the right in the above expression is often divided into two parts, the phase integral and the geometric albedo.

Phase integral and geometric albedo

The phase integral q_λ and the geometric albedo p_λ are defined as

$$q_\lambda = 2 \int_0^\pi B_\lambda(\alpha)\sin(\alpha)d\alpha$$

and

$$p_\lambda = \frac{d^2 R^2 C_\lambda(0)}{r^2 S_\lambda}$$

The geometric albedo is the ratio of the brightness of the object to the brightness of a perfectly diffusing disk (often called a Lambert disk) of the same radius at the same distance from the Sun. When the geometric albedo is presented as a function of wavelength it can be easily used to determine the composition of spectrally active absorbers in the atmosphere or on the surface a solar system object.

The bolometric Bond albedo is the Bond albedo integrated over all wavelengths. It is an important indicator of the energy balance on a planetary body. Knowledge of the bolometric Bond albedo provides a direct measurement of the amount of internal heating within a solar system object. Bolometric Bond albedo measurements have been used to demonstrate that certain large solar system objects, such as the giant planets, have internal heat sources.

Spectral geometric albedo

For the outer planets and their satellites, Earth-based observations can only be made over a narrow range of solar phase angles. Thus, from Earth-based observations alone the phase integral q cannot be determined and therefore, of course, neither can the Bond albedo. However, the geometric albedo as a function of wavelength [i.e. the spectral geometric albedo $p(\lambda)$] is obtainable and is of great use in making comparisons to laboratory spectra of hypothesized surface materials or atmospheric constituents (either gases or aerosols) in order to access compositional information. It is generally assumed that if the spectral absorption features in a laboratory reflection spectrum of a candidate mineral species match those in a spectral geometric albedo of a solar system object, then the presence of the mineral as a surface constituent of the object can be assumed with great confidence. The spectral geometric albedo is of greater use than the relative reflection spectrum normalized at an arbitrary wavelength because it is possible for laboratory spectra of particular sets of materials to have a spectral features which match in a relative spectral sense and yet the materials may have grossly different albedos. For such cases the likelihood of such a material being abundant on a planetary surface is quite small.

Most planetary satellites studied are in synchronous rotation (the same face always is oriented toward the primary body). Therefore, when studying bodies that do not have dense atmospheres, an observing program can be developed to determine the variation in the spectral geometric albedo as a function of orbital phase angle θ, and thus the variation in abundance of spectrally active absorbers on the surface of the object can be mapped (Figure S41). By ratioing the spectra of an object at one observational aspect (orbital phase angle) to those at other aspects, mineralogical mapping in one spatial dimension (longitude) can be undertaken if a spectral data base of sufficient size is obtained.

Prior to the space age, spectrophotometric studies drew their conclusions from observations of the spectral geometric albedo and the solar phase curve of an object as observed from the Earth. In the case of outer solar system objects, only a limited portion of the phase curve was observable. Full solar phase curves are only obtainable from interplanetary spacecraft. With the successful completion of the Voyager mission to the outer solar system, there is phase curve information for all the large solar system objects except the Pluto–Charon system.

Opposition effect

It has long been known that the reflected brightness of many solar system objects increases sharply as the phase angle between the source of illumination and the observer becomes small. This brightness surge at low solar phase angles is called the opposition effect. The effect is generally believed to be due to the elimination of mutual

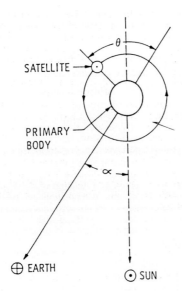

Figure S41 Typical observational configuration of an Earth-based observer when studying a typical outer solar system planetary satellite. The solar phase angle α is the Earth–object–Sun angle and θ is the rotational (or orbital) phase angle. For a satellite in a phase-locked orbit (synchronous rotation), the orbital phase angle at the time of observation designates the longitude of the sub-Earth point on the satellite. Thus it is possible to study the spectral variation as a function of longitude on a planetary satellite by undertaking observations at many orbital phase angles.

shadows between the grains of the regolith particles as the angular separation between the illumination source and the observer diminishes. The magnitude of the opposition and the angular width of the peak of the opposition surge are believed to be determined by textural properties of the regolith. Knowledge of the size and width of the peak is required for determining another important physical property of the regolith, the single-scattering albedo (see Opposition effect).

Single-scattering albedo

For bodies that have no atmosphere, the variation of the intensity of reflected light as a function of viewing geometry is, at small phase angles, an indicator of the microstructure of the regolith and at large phase angles is an indicator of large-scale surface topography. Spatially resolved data, such as the images returned from a planetary probe, can be corrected for viewing geometry to permit the intercomparisons to be made between each surface unit. Inferences can be made about the surface microstructure and macrostructure at scales much less than the spatial resolution of the images.

The reflectance of a regolith as a function of viewing geometry as seen in a typical spacecraft image of a planetary body can be expressed by:

$$r(\mu_0,\mu,g) = \frac{w}{4\pi} \frac{\mu_0}{(\mu_0+\mu)} \left\{ [1+B(g)]P(g) + H(\mu_0)H(\mu) - 1 \right\}$$

where μ_0 is the cosine angle of incidence, μ is the cosine angle of emission, w is the average single-scattering albedo and

$$B(g) = \frac{B_0}{1 + \frac{1}{h} \tan \frac{g}{2}}$$

where B_0 is the opposition effect amplitude, h is the opposition effect width, $P(g)$ is the average single particle phase function,

$$H(x) = \frac{(1+2x)}{1+2\gamma x}$$

and $\gamma = (1-w)^{1/2}$.

By measuring the reflected radiation from the individual compositional units at each position identified in a wavelength and solar phase specific spectral image, the value of the parameters which describe regolith properties in the above expression can be uniquely determined. Other images at other illumination geometries further constrain the parameters. Once a unique solution is determined for each compositional unit, the single scattering albedo as a function of wavelength can be determined. This term is particularly important because it permits the reflectance of a composite surface to be expressed easily in terms of the average single-scattering albedo of a multicomponent mixture. It is simply a linear sum of projected surface areas of the different particles.

$$w = \frac{\sum_{i=1}^{n} \frac{m_i Q_i}{\rho_i D_i}}{\sum_{i=1}^{n} \frac{m_i}{\rho_i D_i}}$$

where m_i is the bulk density of the ith component, ρ_i is the solid density of the ith component, D_i is the diameter of the ith component, Q_i is the scattering efficiency $/1+\alpha_i(\lambda)D_{pi}$, α_i is the absorption coefficient of the ith component and D_{pi} is the effective diameter.

The application of this technique to the spectral image data returned from spacecraft observations of planetary objects during a typical planetary flyby or while in orbit about a primary body leads directly to production of a map of the wavelength variation of the single-scattering albedo of the particles which comprise the surface. This can be compared to laboratory spectra. The resulting spectral identification converts the spectral albedo map to a mineral distribution map.

At present and in the near future, deep space exploration missions will carry imaging spectrophotometers which will provide the raw data necessary to complete the compositional mapping of the exposed planetary surfaces of the solar system.

Robert M. Nelson

Bibliography

Hapke, B.W. (1971) Inferences from optical properties concerning the surface texture and composition of asteroids, in *Physical Studies of the Minor Planets*. NASA SP-267, pp. 67–77.
Hapke, B.W. (1981) Bidirectional reflectance spectroscopy, 1, theory. *J. Geophys Res.*, **86**, 3039–54.
Hapke, B.W. (1984) Bidirectional reflectance spectroscopy, 3, correction for macroscopic roughness. *Icarus*, **59**, 41–59.
Hapke, B.W. (1986) Bidirectional reflectance spectroscopy, 4, the extinction coefficient and the opposition effect. *Icarus*, **67**, 264–80.
Kaula, W.M. (1968) *Introduction to Planetary Physics*. New York: John Wiley and Sons.
Russell, H.N. (1916) On the albedo of the planets and their satellites. *Astrophys. J.*, **43**, 173–87.
Veverka, J. (1977) Photometry of satellite surfaces, in *Planetary Satellites*, (ed. J.A. Burns). Tucson: University of Arizona Press

Acknowledgement

The author's research endeavors at the Jet Propulsion Laboratory are supported by NASA's Planetary Exploration Program. The author greatly appreciates the support of the management and staff at JPL and in particular the careful review of this manuscript by R.W. Carlson, Linda, J. Horn and Ellis D. Miner.

Cross references

Opposition effect
Photometry
Reflectance spectroscopy
Spectroscopy: atmosphere

SPECTROSCOPY: ATMOSPHERES

Spectroscopy, defined here as the study of characteristic interactions of electromagnetic radiation with atoms and molecules, is an important remote sensing tool used to investigate planetary and cometary atmospheres. In fact, most of our information concerning the composition and structure of these atmospheres is derived from observations of their spectra in the microwave, infrared, visible and ultraviolet regions. Indeed, without spectroscopic measurements we would not know even the major species of planetary atmospheres, let alone the distribution of trace components. Deducing such information is possible because individual atomic and molecular species possess unique and identifiable spectral signatures established by the quantum structure of atoms and molecules. In addition, the temperature profile of an atmosphere is established by the energy absorption and emission properties of the constituent molecules, and these properties are related to their spectral characteristics.

At the molecular level, matter exhibits wavelike, as well as particlelike, properties. Wave effects, described by Schrodinger's wave equation and resulting wave functions, produce a set of discrete states (eigenstates) in which the atom or molecule can exist, with each of these states possessing a distinct, quantized value of energy E. These energy levels are characterized by a set of distinct quantum numbers. Transitions between one state and another, with corresponding changes in one or several of the quantum numbers, result in emission (or absorption) of a photon of frequency v, given by Einstein's frequency condition

$$h v = E_2 - E_1$$

where h is the Planck constant. Individual transitions are called lines, a term historically derived from their appearance on the photographic plates used in spectrographs. For a given atom or molecule, the lowest energy state is called the ground state; higher energy states are called excited states.

The spectrum of a planet can exhibit both emission and absorption features, depending upon the formation mechanism. For example, sunlight incident on an atmosphere can be absorbed at discrete frequencies, with the absorbed energy subsequently transferred to the atmosphere through collisions, heating up the atmosphere and producing one or more spectral absorption features. However, if collisions are infrequent compared to the radiative lifetime, the radiation can be re-emitted and would appear as an emission feature, termed resonance scattering. If the final state is different from the original, then it is called fluorescence. Collisional excitation due to the thermal motions of atmospheric molecules (thermal excitation) can produce emission, with the shape of the feature determined by the temperature structure of the atmosphere. Chemical reactions and energetic particle excitation (e.g. aurora) also produce emission spectra.

The amplitudes of these spectral features depend not only on atmospheric conditions, but also more fundamentally upon (1) the quantum properties of the eigenstates and (2) the coupling efficiency of radiation to the electrical charge distribution of the atom or molecule. The strongest of these transitions are termed allowed transitions, while weaker transitions are often referred to as forbidden transitions. Forbidden transitions can be strictly forbidden due to impossible symmetry changes (for example), or simply weak, as when due to inefficient coupling to the radiation field. The most efficient radiation coupling occurs for species with a natural electric dipole moment – a net linear separation of electronic and ionic charges. Electric quadrupole and magnetic dipole moments also couple to the radiation field, but more weakly. In addition, collisions can induce charge asymmetries in otherwise spectrally inactive molecules, allowing pressure-induced or collision-induced transitions to occur.

Characteristic atomic and molecular spectral features occur over a large frequency (or wavelength) range and are due to a variety of transition types: rotational, vibrational, electronic, or inversion. The energy levels of a given molecule are the sum of the rotational, vibrational and electronic energies, with modifications due to their mutual interactions and also to nuclear spin and quantum-electrodynamical effects. In general, rotational energies and their spacings are much less than those for vibration, which in turn are much less than those for electronic transitions. In the microwave and far infrared regions transitions between molecular rotational levels are seen, as well as inversion transitions. Molecular vibrational transitions, with accompanying changes in the rotational state, occur in the infrared and, to a lesser extent, in the visible. Changes in electronic energy states, with possible changes in vibration and rotation quanta, produce spectral features mainly in the visible and ultraviolet regions.

There are many frequency, wavelength and energy units used in spectroscopy, along with a variety of notations. The frequency, s^{-1} or Hz (and multiples as kHz, MHz, GHz,) is used in the radio and microwave region, and often denoted by v or f. Related to the frequency is the wavenumber $\bar{v} = v/c = 1/\lambda$ where c is the velocity of light and λ is the wavelength. This represents the number of waves per unit length (cm^{-1}) which is occasionally called the Kayser. This measure tends to be used in the infrared and denoted by $v, \nu, \bar{v}, k$ and σ. Alternatively, one can discuss the radiation in terms of the wavelength $\lambda = c/v$. For the microwave region, centimeter or millimeter units are appropriate while in the near-and far-infrared regime the micrometer (μm) or simply micron (μ) is commonly used. Discussions of infrared, visible and ultraviolet radiation sometimes use units of nanometers (nm). More commonly, visible and ultraviolet wavelengths are described in Angstrom units Å = 10^{-10} cm. Energy levels are given in wavenumber units or in units of the electron volt (eV). The latter is useful because chemical and electron binding energies are in the few eV range. A 1-eV photon possesses 1.6022×10^{-12} ergs of energy, a wavelength of 1.2399 μm and a wavenumber of 8065.5 cm^{-1}.

Types of spectra

Pure molecular rotational transitions

For illustrative purposes, let us consider the case of a diatomic molecule and approximate its rotational properties as a rotating rigid dumbbell with moment of inertia I. In solving Schrodinger's equation for this rotational motion, the energy levels E_r and angular momentum $\mathbf{J}$ are found to be quantized and given by

$$E_r = h^2/(8\pi^2 I)\, J(J + 1) \qquad |\mathbf{J}| = h/2\pi\, (J(J + 1))^{\frac{1}{2}}$$

where J is the rotational quantum number $J = 0, 1, 2, \ldots$ If we refer to the energy of a level in wavenumbers, this is then called the *term value* $F(J)$, for which

$$F(J) = E_r/hc = B\, J\, (J + 1)$$

and B is the rotational constant.

An additional quantum number M describes the projection of rotational momentum along a spatial axis. There are $2J + 1$ possible values for M. In the absence of an externally applied field, the energy for a state of given J is independent of M (i.e. degenerate), and we ascribe a *statistical weight* of $2J + 1$ to that level.

In molecular spectroscopy it is usual to refer to upper (lower) energy levels with single (double) primes, and to write the upper level first. Thus the frequency (in wavenumbers) of a rotational transition is

$$\bar{v} = F(J') - F(J'') = BJ'(J' + 1) - BJ''(J'' + 1)$$

Lighter molecules, such as hydrides, have lower moments of inertia and therefore their rotational spectra occur at higher frequencies than heavier species.

There are only certain permissible or preferred changes in quantum states, which are described by *selection rules*. For an electric dipole transition there is the general selection rule

$$\Delta J = J' - J'' = \pm 1$$

giving rise to two *branches*, the P branch ($\Delta J = -1$) and the R branch ($\Delta J = +1$). For pure rotational electric dipole spectra, $J' - J'' = +1$ only, yielding a series of equally spaced R branch lines with wavenumber

$$\bar{v} = 2B\, (J'' + 1)$$

In homonuclear molecules consisting of the same isotopes, such as H_2, O_2 and N_2, there is no natural electric dipole moment, and therefore no strong rotational spectrum. Even if the nuclei consist of different isotopes, the transitions are still very weak.

For quadrupole transitions, the general quantum selection rules are $\Delta J = \pm 2$, giving rise to the more widely spaced O and S branches, corresponding to $\Delta J = -2$ and $+2$ respectively. For pure rotational quadrupole spectra, $J' - J'' = +2$ (S branch).

Of course, molecules do not behave as rigid dumbbells, and the above treatment is too simplistic. Furthermore, for molecules which are not diatomic or linear, additional quantum numbers are necessary, which describe for example the angular momentum components along molecular symmetry axes. Expressions for the term values and corresponding spectra are more complex in these cases. These considerations are beyond the scope of this article.

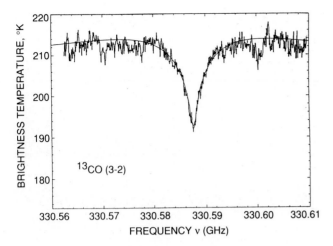

Figure S42 The 3–2 rotational line of the ^{13}CO molecule. This line was obtained from observations of Mars using the 10-m CSO antenna on Mauna Kea. The line profile is resolved and shows both emission and absorption-like behavior. At the center of the line one probes higher in the atmosphere where temperatures are colder, producing less emission than for the wings of the line which are formed in the hotter, lower regions. (Data from Encrenaź et al., *Ann. Geophysical*, **9**, 797, 1991.)

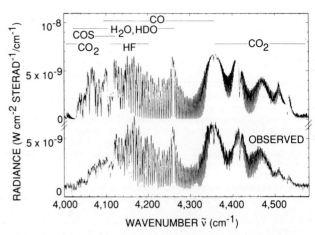

Figure S43 Vibration–rotation spectrum of Venus in the 2.3-μm window. Observed and calculated spectra are compared for this window, where the Venus atmospheric absorption is low, allowing one to probe the deep atmosphere. Radiation from the deep atmosphere is absorbed by molecules in the cooler upper regions of the atmosphere, forming many vibration-rotation lines shown in the figure. (Data and analysis are by Bézard et al., *Nature*, **345**, 508, 1990.)

An important rotational spectrum is that of H_2 in the atmospheres of the Jovian planets. These are collision-induced transitions that obey quadrupole selection rules. Because they are pressure induced, these lines are very broad (see below), dominating the infrared absorption spectra in the 300–800 cm^{-1} range. Another example of planetary rotational spectra is that of CO, whose transitions occur in the millimeter region and are useful for studying structure and wind velocities of the atmospheres of Venus and Mars. An example is shown in Figure S42. CO has also been detected as a minor constituent in the atmospheres of Neptune and Titan through rotational spectra. Water vapor rotational features are found in Mars spectra, and rotational lines due to HDO have been recently detected in spectra from both Venus and Mars. Rotational features of HCN have been seen from Neptune and Titan. Titan also shows rotational features from HC_3N, while Jupiter's satellite Io exhibits SO_2 features. Rotational spectra of cometary atmospheres include HCN, H_2S, CH_3OH and H_2CO features.

Vibration–rotation spectra

In order to introduce molecular vibrational spectra, consider the case of a diatomic molecule. In addition to rotation, the nuclei vibrate about their average separation; the energy of that motion can be approximated by an anharmonic oscillator with quantized energy levels given by

$$E_v/hc = G(v) = \omega(v+\tfrac{1}{2}) - x\,\omega(v+\tfrac{1}{2})^2 + \ldots$$

where $v = 0, 1, 2, \ldots$ is the vibrational quantum number, ω is the vibrational constant in wavenumber units and x is an anharmonicity constant. Due to anharmonicity, there are no stringent selection rules for changes in vibrational quanta, although a change of v by one unit tends to be strongest (this does not hold if there is also a change in electronic energy, see below). The transition from the ground state (the lowest energy state, $v'' = 0$) to the next highest vibrational state ($v' = 1$) is called the *fundamental* transition; changes in v of 2, 3, .. are referred to as *overtones*. For a given change in vibrational quanta, there will be many lines which correspond to differing rotational energy changes. This collection of lines is referred to as a *band*, and each band consists of various branches (O, P, Q, R, S with $\Delta J = -2, -1, 0, 1, 2$) whose presence depends upon selection rules.

For polyatomic molecules, additional modes of oscillation are possible. For example, for a linear triatomic molecule there are three modes of vibration possible, denoted (v_1, v_2, v_3), corresponding to symmetric stretching, bending and asymmetric stretching of the molecule. Vibrational spectra can involve changes in all of these vibrational quanta. Changes in only one vibrational quantum number are fundamental or overtone bands, whereas a simultaneous change in two or more vibrational modes is termed a combination band. Thus, a $2v_1 + v_2$ absorption band corresponds to the transition $(2,1,0) \leftarrow (0,0,0)$. More complicated molecules can have additional modes of oscillation, depending upon the number of nuclei and symmetry properties of their arrangement.

The absorption spectra of planets can be used to identify both major and minor atmospheric constituents. For example, numerous vibration-rotation bands of the dominant CO_2 molecule are seen in the infrared spectra of Venus and Mars, while the presence of trace species such as H_2O, CO, HCl, HF and COS is indicated by infrared vibration-rotation features in the spectrum of Venus. An example is given in Figure S43. Red and infrared spectra of the Jovian planets exhibit absorption in numerous CH_4 bands. H_2 quadrupole and induced-dipole vibration-rotation transitions are also seen. Jupiter and Saturn spectra also contain absorptions due to ammonia. Minor species observed spectroscopically from some of the outer planets include H_2O, GeH_4, CH_3D and PH_3.

Vibrational features can also be observed as emission features in planetary spectra. For example, if there is a temperature inversion, a thermal emission feature (rather than absorption) can be produced. Jupiter shows the CH_4 v_4 fundamental in emission at 7.8 μm which is produced in a warm stratosphere. Saturn and Titan also show this methane feature, along with other molecular emission bands. Titan has a rich thermal emission spectrum, showing the presence of C_2H_2, C_2H_4, C_2H_6, C_3H_8, CO_2 and HCN (Figure S44). Auroral excitation and heating produces Jovian H_2 and H_3^+ infrared vibrational spectra.

Vibration–rotation bands are observed from comets as emission features, formed by resonance scattering and fluorescence of sunlight. Observed cometary spectra include H_2O (the 2.7 μm v_3 band), CO_2 (4.3 μm v_3) and H_2CO (v_1 and v_5 bands at 3.5–3.6 μm).

Electronic spectra – atoms

Just as rotational and vibrational energies are quantized, so also are the energies of the electrons surrounding a nuclear core, and changes in these energy states can occur through emission or absorption of photons. In the following we briefly describe the quantum properties of electronic energy states and their spectroscopic nomenclature, and then present some examples of planetary atomic spectra.

Electronic energies are determined by the orbital distances of the electrons, their orbital angular momentum and the intrinsic angular momentum of the electrons themselves – the electron spin. The orbital radius is described by an orbital quantum number $n = 1, 2, 3, \ldots$ (K, L, M, . . . *shells*) with the ground state of an individual electron corresponding to $n = 1$ and the smallest orbital radius. In addition, two properties of the orbital angular momentum of an

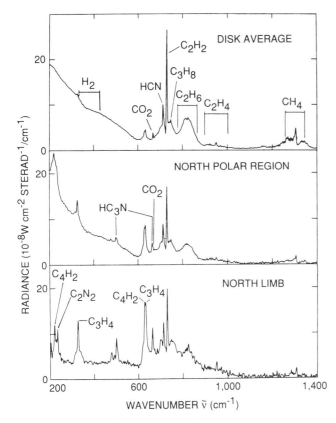

Figure S44 Vibrational band spectra of Titan in the thermal region. The warm stratosphere produces distinct thermal emission features allowing identification of molecular species present in the atmosphere. Each of these features is a distinct band and consists of many individual rotational lines not resolved here. (From Samuelson et al., J. Geophys. Res., **88**, 8709, 1983.)

the resonance lines, which are allowed transitions that couple to the ground state. For most atoms and atomic ions these lines occur in the ultraviolet; however for some metals such as sodium they can occur in the visible region of the spectrum. Another important class of transitions are the forbidden lines, where the selection rule that precludes the change in quantum numbers is only approximate. An example is a change in total spin S, which is forbidden only to first order; the resulting spectral lines are called *intercombination* transitions. The radiative lifetimes (see below) for producing forbidden lines can be quite long, and these long-lived states are called *metastable* states. These metastable states, and the corresponding radiation, are important in cometary atmospheres and the upper atmospheres of planets, where deactivating collisions are infrequent. Since the metastable states tend to have low excitation energy, the corresponding spectral lines often occur in the visible, accessible to ground-based observations and forming convenient monitors of atomic densities and processes.

In general, atoms are highly reactive and readily recombine into molecules. Consequently, free atoms (discounting the inert gases) are preferentially found in low-density upper atmospheres. They are produced from molecules by photodissociation, charged particle impact or photochemical reactions. Emission spectra can be produced directly in these processes, or by subsequent mechanisms such as resonance scattering and fluorescence of sunlight. Some planetary examples of atomic spectra are briefly discussed below.

The ubiquitous hydrogen atom is seen from comets and most planetary atmospheres. The atom is formed by dissociation of molecules such as water, molecular hydrogen and methane and is observed as emission features in the far ultraviolet: the Lyman series $1s^2S_{1/2}-np^2P^o_{1/2,3/2}$. The strongest of these is called the Lyman alpha line and corresponds to the transition $1s^2S-2p^2P^o$, occurring at 1216 Å. The excitation mechanisms for these lines include resonance scattering as well as auroral processes.

Another resonance scattering feature is the multiplet of atomic sodium, $3s^2S_{1/2}-3p^2P^o_{1/2,3/2}$ at 5896 and 5890 Å, which is seen from the mesosphere of the Earth, the tenuous atmospheres of Mercury and the Moon, comets and the Io torus around Jupiter. Potassium has also been seen in some cases. Other alkali metals such as Li, Mg and Ca$^+$ have been seen in the Earth's twilight spectrum.

The spectra of oxygen, and its ions, are well represented among planets and comets. For example, spectra of Venus, Earth and Mars show two ultraviolet features, one an allowed triplet $2p^4\ ^3P_{0,1,2}-2p^3 3s\ ^3S^o_1$, λ 1302 Å and another spin forbidden pair of lines $2p^4\ ^3P_{1,2}-2p^3 3s\ ^5S^o_2$, λ 1356 Å. Since the strength of the latter transition is too small to form a resonance scattering feature, it serves to indicate the amount of electron excitation and photodissociative excitation. Other useful oxygen emission features include the forbidden [OI] λ 6300 Å $^3P_2-^1D_2$ red line and the [OI] $^1D_2-^1S_0$ λ 5577 Å green line, both produced by a variety of processes. (The brackets indicate forbidden transitions.) In cometary atmospheres the red line is formed by the photodissociative excitation of H$_2$O and is therefore an indicator of water vapor abundances and fluxes.

Oxygen ion lines are present in the Earth's auroral and dayglow spectra, and in the Io torus in various stages of ionization (see Figure S45). These, and atomic sulfur ions, both of which are derived from SO$_2$ emanating from Io, were ionized and excited by electron impact by the hot Io plasma torus.

Carbon, nitrogen and helium are also seen in rocket and spacecraft ultraviolet spectra; carbon from Venus and Mars spectra, and nitrogen from spectra of Titan and the Earth, which also shows a forbidden line in the visible, [NI] λ 5200 Å. Helium has been observed in extreme ultraviolet spectra of Mercury, Venus, Earth and outer solar system planets.

As a final example, we mention transitions within a term. The ground term for atomic oxygen has three levels $^3P_{2,1,0}$ with $J=2$ being the lowest in energy. Collisional excitation to the upper J levels, only 20–30 meV higher, can result in magnetic dipole emission in the far infrared. This is one of the processes by which the Earth's thermosphere is cooled.

Electronic spectra – molecules

When atoms combine to form molecules, there are a large number of possible electronic energy levels, more so than for the separated atoms. In addition, transitions between electronic states can also include changes in vibrational and rotational quanta, so molecular electronic spectra can be quite complex. The nomenclature describ-

electron are quantized: its magnitude and projection along a spatial axis.

The angular momentum quantum number l can vary from $l = 0, 1, 2, 3, \ldots, n-1$ and is customarily referred to as s, p, d, f, . . . states. A given value of n and l is termed a *subshell*, and denoted nl = 1s, 2s, 2p, 3s, 3p, 3d, . . . For a multi-electron atom or ion, each subshell can contain only a fixed number of electrons owing to a property of electrons described by the Pauli exclusion principle.

The orbital angular momenta and spins of atomic electrons add vectorially to produce quantized values for the total orbital angular momentum L and spin S. A given L and S (and set of orbital numbers n) gives a *term*, denoted ^{2S+1}L. As in the single electron case, L is described in the notation S, P, D, F, . . . (L = 0, 1, 2, 3, . . .) The spin state of the atom $2S+1$ is called the *multiplicity* and referred to as singlets, doublets, . . . for $2S+1$ = 1, 2, . . . The angular momenta L and S combine to form quantized values of the total angular momentum J. For each term, there are several possible values for J, each with different energy due to spin–orbit interactions (fine structure splitting). An energy level of an atom is described in the notation $^{2S+1}L_J$, often with the occupancy of the subshells preceding. For example, the ground state of singly ionized oxygen ion OII (I = neutral, II = singly ionized, etc.) which contains two 1s electrons, two 2s electrons, and three 2p electrons is OII $1s^2 2s^2 2p^3\ ^4S^o_{3/2}$. (The superscript 'o' denotes the parity of the state, o = odd.)

If we ignore the small nuclear spin interactions (hyperfine structure), an atomic spectral line corresponds to the transition between two levels. In describing a transition it is customary in atomic spectroscopy to write the lower level first. The group consisting of all of the transitions between two terms is called a *multiplet*.

There are a variety of selection rules which determine the intensity of individual transitions, but a detailed discussion is beyond the scope of this article. In planetary spectra the strongest features tend to be

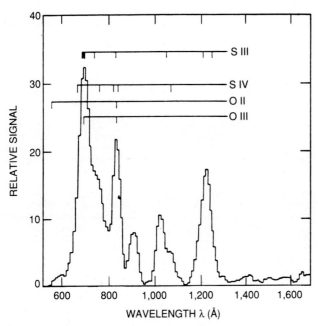

Figure S45 Atomic spectra of the Io torus. These Voyager I extreme ultraviolet spectra show the presence of sulfur and oxygen ions, which are excited by electron impact. Each of these features corresponds to a multiplet consisting of several lines not resolved here. Some of the features are also blends of multiplets from various species. The feature at 1216 Å is partly due to interstellar atomic hydrogen Lyman α emission. (From Sandel *et al.*, *Science*, **206**, 962, 1979.)

ing electronic energy states includes the total spin S and multiplicity $2S + 1$, similar to the atomic case, and an angular momentum value Λ which is the projection along a molecular symmetry axis. For $\Lambda = 0, 1, 2, \ldots$ we refer to these as $\Sigma, \Pi, \Delta \ldots$ states, and write as $^{2S+1}\Lambda$. In some cases symmetry properties are indicated (u, g and $+$, $-$). Further distinction is provided by a somewhat arbitrary alphabetical designation, with X reserved for the ground state, and A, B, C, ... generally denoting excited states with the same multiplicity as the ground state. Lowercase letters usually denote states of differing multiplicity. As an example, the ground state of molecular oxygen is $X\ ^3\Sigma_g^-$ and the lowest lying excited state is the $a^1\Delta_g$ state.

In general, transitions involve changes in the electronic state, accompanied by corresponding changes in the vibrational and rotational energies. Most transitions are from one bound state to another, and therefore involve discrete energy changes and produce discrete spectra. However, transitions involving *continuum* states are possible, as in photodissociation and ionization, and continuum spectra are produced.

For a given change in electronic state, there are selection rules and quantum effects related to changes in vibrational and rotational quanta. For changes in vibrational energy there are no strict selection rules. Instead, the intensity for a given change in vibrational quanta ($\Delta v = v' - v''$) is determined by the overlap of vibrational wave functions, the Franck–Condon factor. For a constant Δv, a series of bands called a *sequence* is found, while if the lower (or upper) vibrational state is fixed, then the resulting set of bands is called a *progression*; absorption spectra from the ground vibrational state produces the $v'' = 0$ progression. The totality of the bands between two electronic states is called a *band system*.

The strong electric dipole-allowed transitions often occur in the ultraviolet regions, while longer wavelength transitions (visible, infrared) often involve changes in multiplicity (intercombination bands). There are also radio transitions that involve coupling between rotation of the molecule and orbital motion of the electrons, termed Λ-doubling. Molecular electronic bands can appear in absorption or emission, with many possible mechanisms populating the upper emitting levels. A complete listing of observed planetary features is beyond the scope of this article; a sampling of such is given below and was chosen to illustrate the types of formation processes.

Resonance scattering and fluorescence is an important molecular electronic band emission mechanism. This is particularly true for comets (Figure S46), where most of the observed molecular bands arise from this process. This category also includes the radiowavelength emission features seen from the hydroxyl radical (OH) from comets due to a slight coupling between orbital and rotational angular momenta (Λ-doubling) and hyperfine (nuclear spin interaction) structure. The formation of these radio emission features is attributed to *optical pumping*, wherein solar ultraviolet photons are absorbed into an upper electronic state, followed by a two-step fluorescence process, the latter step producing the 18-cm radio emission. Planetary examples of resonance fluorescence include the N_2^+ first negative system ($B\ ^2\Sigma_u^+ \to X\ ^2\Sigma_g^+$), the NO γ bands ($A\ ^2\Sigma^+ \to X\ ^2\Pi$), and the molecular hydrogen Lyman ($B\ ^1\Sigma_u^+ \to X\ ^1\Sigma_g^+$) and Werner ($C\ ^1\Pi_u \to X\ ^1\Sigma_g^+$) bands. Auroral and electron excitation processes also contribute.

Solar photons can also produce atomic and molecular emission features through *photodissociative* and *photoionization* excitation. Examples include the photodissociation of ozone (O_3) which produces $O(^3P) + O_2(a\ ^1\Delta_g)$, the latter producing the Infrared Atmospheric band ($a\ ^1\Delta_g - X\ ^3\Sigma_g^-$) at 1.27 μm at Venus, Earth and Mars. Photoionization of CO_2 in the upper atmosphere of Mars can leave the CO_2^+ ion in an excited state, giving the Fox–Duffendack–Barker bands ($\tilde{A}\ ^2\Pi_u \to \tilde{X}\ ^2\Pi_g$) and the doublet bands ($\tilde{B}\ ^2\Sigma_u^+ \to \tilde{X}\ ^2\Pi_g$).

Auroral and electron excitation can also produce such emissions, as well as others since electron excitation can more easily change the spin state of the target molecule than photon excitation. The N_2 second positive system, ($C\ ^3\Pi_u \to B\ ^3\Pi_g$) is an example, where the excited triplet state is produced by electron impact from the singlet ground state.

Molecular electronic emission spectra as well as atomic spectra can also be produced through *photochemical reactions*. An example is the association of oxygen atoms to form molecular oxygen, often in an excited state. Emission from these states can produce airglow in a variety of bands. Dissociative recombination of molecular ions and electrons can also produce atoms and molecules in excited states. One example of many is the Cameron bands of CO ($a\ ^3\Pi \to X_1\Sigma^+$) observed in the atmosphere of Mars and due to the formation of excited CO by recombination of CO_2^+. Direct *radiative association* is also possible, but rare. One terrestrial example is the formation of NO_2 emissions from the recombination of NO and O.

Molecular bands are also seen as absorption features. Examples include the O_2 Infrared Atmospheric bands and the Atmospheric bands ($b\ ^1\Sigma_g^+ - X\ ^3\Sigma_g^-$) which are seen as $v'' = 0$ progressions in terrestrial solar absorption spectra, despite the fact that these are forbidden transitions (and therefore weak absorbers). The presence of these bands is simply due to the large abundance of terrestrial molecular oxygen. Allowed transitions, which would lead to strong absorption spectra, tend to occur in the ultraviolet and are therefore not easily observed. An example of molecular electronic absorption spectra is that of SO_2 from Venus, occurring in the near ultraviolet.

Another example of absorption spectra, which also illustrates a different molecular process, is that due to O_2 wherein the molecule is excited to the $B\ ^3\Sigma_u^-$ state, which is an unbound state. This results in photodissociation of the molecule; the broad absorption feature is called the Schuman–Runge continuum. Another planetary example of a photodissociation continuum is that of ozone in the ultraviolet, where the Hartley band (2000–3000 Å) occurs; this is seen from spacecraft as a broad absorption continuum in both Earth and Mars UV spectra.

Inversion spectra

An unusual class of transitions called inversion transitions occurs for the ammonia molecule NH_3, and is an important feature in the atmosphere of Jupiter. The ammonia molecule resembles a pyramid, with the three hydrogen nuclei (protons) forming the base and the N atom at the apex, where it executes small amplitude vibrational motions about the equilibrium apex position. Although classically the vibrational energy is too small to surmount and pass through the repulsive barrier of the three protons, quantum mechanically it can tunnel through and appear on the other side. Consequently, the wave functions give equal probability of finding the N atom on one side or the other of the plane of the protons. The wave functions can be either symmetric or antisymmetric with respect to this plane, and there are slight energy differences between them. Transitions between the lower (symmetric) state and the upper (asymmetric)

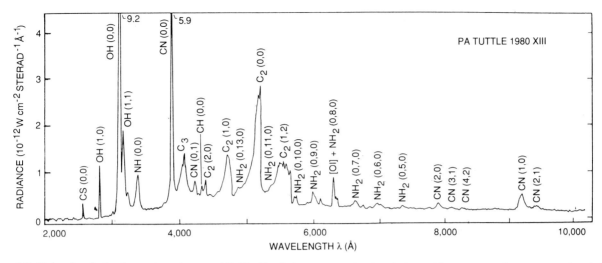

Figure S46 Molecular electronic spectrum for comet Tuttle. The features are predominately formed by resonance fluorescence of sunlight. Each feature is a molecular band and consists of many rotational lines which are unresolved here. The molecules represented here are photolysis products of parent cometary species. The NH_2 features correspond to the $\tilde{A}$–$\tilde{X}$ transition, the so-called α bands of ammonia (the parent molecule), and the vibrational states (v_1, v_2, v_3) of the upper electronic energy level are indicated in the figure. For the diatomic molecules the upper and lower vibrational states (v', v'') are indicated. The CN violet system (B $^2\Sigma^+$–X $^2\Sigma^+$) and the red bands (A $^2\pi$–X $^2\Sigma^+$) are evident, along with other molecular bands. This figure was adapted from A'Hearn and Festou, in *Physics and Chemistry of Comets* (ed. W.F. Huebner) Springer-Verlag, Berlin, 1990.

state give rise to inversion spectra. If there is no corresponding change in rotational or vibrational quanta, the pure inversion spectrum is produced and occurs in the microwave region. The energy differences depend upon rotational energy, resulting in a series of lines occurring in the 1.3-cm range which, in Jupiter's lower atmosphere, are greatly broadened by collisions, forming the dominant opacity source in the centimeter region.

Spectral strengths and shapes

Strengths

There are a variety of equivalent means to describe the emission and absorption properties of atomic and molecular transitions. Perhaps the most fundamental of these is the line strength S, sometimes called line intensity, which describes the coupling of two states through electromagnetic perturbation. The line strength can be regarded as the frequency-integrated absorption cross-section (see below), although it describes emission properties as well. A number of different units are used for line strengths, including cm^2-Hz and wavenumbers per unit column density. For molecules there are also band strengths (or band intensities), which are related to line strengths through rotational line strengths. For diatomic molecules these are called Hönl–London factors.

These strengths are related to the Einstein transition probabilities A_{21}, B_{12}, and B_{21}, which in turn, are uniquely related to each other. A_{21} is the spontaneous transition probability – the rate (s^{-1}) at which an upper level (2) decays to a lower level (1). The natural lifetime of a state is the inverse of the sum of the transition rates from that state. The Einstein absorption coefficients B_{12} and B_{21} respectively describe absorption from a lower level and stimulated emission (negative absorption) from an upper level. The latter forms the basis for masers and lasers, and is generally unimportant in planetary spectra.

In describing atomic absorption processes the oscillator strength or f-value is often used. Strong transitions have f-values of the order of unity. One occasionally encounters an oscillator strength for a molecular band system, but it is dependent upon the vibrational population of the lower state and therefore not unique.

Widths and shapes of spectral lines

In any experimental measurement the observed shape is the convolution of the instrumental bandpass and the profile of the feature. Except for the microwave region, the actual line profile is usually not resolved in planetary spectra. Nevertheless, in order to derive quantitative values of atmospheric properties, knowledge of the line shapes is necessary since this is an important factor in the formation of atmospheric spectral features. In this section we discuss the line profiles for single atoms and molecules. The absorption cross-section can be expressed as $\sigma(\nu) = Sf(\nu - \nu_0)$ where S is the line strength, ν_0 is the center frequency, and the line shape is described by the function f, whose integral is normalized to unity. The breadth and shape of these profiles are influenced by three factors – the intrinsic width of energy states, the Doppler effect and the influence of collisions by other particles.

In *natural broadening* the finite lifetime of an excited state corresponds to an intrinsic energy width of that state through Heisenberg's uncertainty principle ($\Delta E \, \Delta t \sim h/2\pi$) and this spread in energy leads to a finite frequency width for the lines. Classically, one can think of the line being produced by a damped harmonic oscillator, the damping related to the radiative energy loss. The resulting frequency distribution, identical to that obtained quantum mechanically, is the *Lorentz profile* $f(\nu) = (1/\pi) \, (\Gamma/4\pi)/[(\nu-\nu_0)^2 + (\Gamma/4\pi)^2]$ where Γ is the sum of the spontaneous transition rates from the upper and lower states. The Lorentz profile is illustrated in Figure S47a. The half-width of the line, at half-maximum, is $\Delta\nu_N = \Gamma/4\pi$ and is too narrow to be observed in planetary spectra. For example, in the near-ultraviolet region, $\nu_0 \sim 10^{15}$ Hz, and typical allowed transition rates are $A \sim 10^8$ s^{-1}, giving $\Delta\nu_N/\nu_0 \sim 10^{-7}$.

The translational motion of atoms and molecules leads to broadening through the Doppler effect. If the distribution of velocities is Maxwellian, corresponding to temperature T and most probable speed $v_D = (2kT/M)^{1/2}$, then the *Doppler profile* is Gaussian with a 1/e half-width $\Delta\nu_D = \nu_0(v_D/c)$. For typical atmospheric temperatures (100–1000 K) and molecular masses (1–50 AMU), the fractional half-width $\Delta\nu_D/\nu_0$ is in the approximate range 10^{-6}–10^{-5} and is again difficult to measure.

The combined contributions of natural and Doppler broadening are described by the convolution of these two profiles; the result is called the *Voigt profile*. At the central portions of the line it shows the gaussian shape of the Doppler profile, but far from line center the Lorentzian wings dominate.

Collisional broadening, or pressure broadening, can lead to much larger widths than those described above. Physically, one can imagine a collection of radiating oscillators which suffer collisions more frequently than the natural radiative lifetime. Loosely speaking, each optical collision results in immediate completion of the transition and termination of the radiated wave train. Random impacts give varying wave train lengths, resulting in a Lorentzian profile, for both emission and absorption. Although this simple physical picture is far from being accurate, Lorentzian profiles are found to be a fairly good approximation in many cases. The collision broadening width is

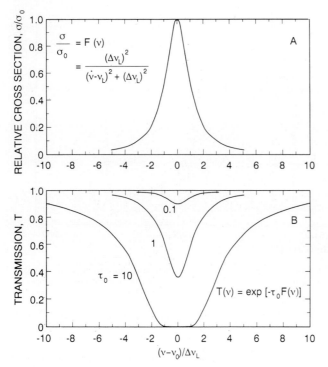

Figure S47 Absorption profile and transmission for a single Lorentzian line. (a) The absorption profile as a function of relative frequency difference $\nu - \nu_0/\Delta\nu_L$ where ν_0 is the center frequency. The transmission profiles for various amounts of absorbing gas are shown in (b). We use the optical depth at line center τ_0 to describe the gas abundance. For small amounts of absorbing gas, the integrated absorption area – the equivalent width – varies linearly with column density. At large column densities the core of the line becomes saturated, and the equivalent width varies more slowly, as the square root of amount of absorbing gas.

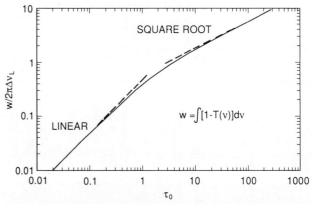

Figure S48 Curve of growth for a Lorentzian line. The variation of normalized equivalent width $W(\tau_0)$ with column density, or optical thickness at line center τ_0, shows two asymptotes, linear at low column abundance and as the square root at high densities. Other line profiles and assemblages of lines show different curves of growth.

proportional to the rate of impacts, and is therefore proportional to the pressure, as well as showing a dependance on temperature and the identity of the impacting particle. The rate of optical collisions at standard temperature and pressure is $\sim 10^{10}$ s^{-1}, giving a typical collisionally broadened width of $\Delta\tilde{\nu} \sim 0.05$ cm^{-1}, or $\Delta\nu \sim 1.5$ GHz (both at STP). Because of this breadth it is practical to observe such profiles in the microwave region, and they can be used to estimate the pressure at the level of line formation. In this spectral region there is a modification to the Lorentz profile that better describes the statistical properties of the collisions. This profile, the *van Vleck–Weisskopf profile*, reduces to the Lorentz profile for $(\nu-\nu_o << \nu_o)$.

Although the above line shapes enjoy wide use, it must be realized that the actual shapes depend upon the details of the interaction forces, the time of interaction and how the wave train is modified. These are especially important in the far wings and can result in sub- or super-Lorentzian behavior.

Finally we mention the unusual case of line narrowing, which can occur when the rate of kinetic collisions (i.e. velocity changing) is greater than for optical collisions, resulting in the *Dicke profile*. This effect is seen in the vibrational quadrupole bands of molecular hydrogen in the outer planets.

Atmospheric equivalent widths

In the previous sections we have considered the strength and profiles of individual atoms and molecules. In atmospheric spectra involving many molecules the resulting spectral profile and amount of absorption is modified, and depends upon the column density of absorbers, often given as the optical depth or opacity: the product of a cross-section and column density. (Multiple scattering is a complication that is not discussed here, but must be given due consideration.)

Consider the case of a single line for which the molecular absorption profile is Lorentzian and shown in Figure S47a. If continuum radiation passes through varying amounts of this gas, one finds transmission profiles such as those illustrated in Figure S47b. Although these profiles are hardly ever measured, the integrated absorption can be obtained, since it is invariant with the instrumental profile. This is called the *equivalent width*, and corresponds to the width of an opaque (zero transmission) rectangular profile with the same integrated area. At small optical depth the absorption occurs in the core of the line, and the equivalent width varies linearly with optical depth. At larger optical depth the core of the line is *saturated* – all of its energy being removed – and absorption in the weaker wings becomes important. For the Lorentz profile the equivalent width in this regime varies as the square root of the opacity, as shown in Figure S48. This variation of equivalent width with the amount of absorbers is called the *curve of growth* and is different for different line profiles and parameters. For a Lorentz profile it is sometimes called the *Landenberg–Reiche* function.

In many planetary spectra, not only are absorption profiles not resolved but also there can be many lines within the instrumental profile. In this case it is useful to discuss the mean absorption and the corresponding transmission over some spectral range, which can be crudely thought of as the sum of the equivalent widths within this range, relative to the width of the interval. These absorption and transmission values are generally complex functions of the absorber column density, and depend upon the frequency distribution of lines as well as their individual strengths. In interpreting planetary spectra, it would be desirable to have laboratory measurements taken at the same instrumental resolution and at the same pressure, temperature, and relative abundances as the planetary atmosphere. It is however often difficult to emulate planetary conditions in the laboratory, so an alternative approach would be line-by-line calculations of transmission based upon high-resolution laboratory spectroscopic data. Even this is often impractical as the number of lines can become excessive, even for modern high-speed computers. A remaining alternative, often very successful, is to use *band models*, which attempt to approximate the actual distribution of line centers and strengths with analytically tractable distributions.

Robert W. Carlson

Bibliography

Chamberlain, J.W. and Hunten, D.M. (1987) *Theory of Planetary Atmospheres*. Orlando: Academic Press.
Goody, R.M. and Yung, Y.L. (1989) *Atmospheric Radiation*. Oxford: Oxford University Press.
Herzberg, G.H. (1944) *Atomic Spectra and Atomic Structure*. New York: Dover.
Herzberg, G.H. (1945) *Infrared and Raman Spectra*. Princeton: Van Nostrand.
Herzberg, G.H. (1961) *Spectra of Diatomic Molecules*. Princeton: Van Nostrand.

Herzberg, G.H. (1961) *Electronic Spectra of Polyatomic Molecules*. Princeton: Van Nostrand.
Herzberg, G.H. (1988) *The Spectra and Structure of Simple Free Radicals*. New York: Dover.
Thorne, A.P. (1988) *Spectrophysics*. London: Chapman & Hall.
Townes, C.H. and Schalow, A.L. (1955) *Microwave Spectroscopy*. New York: McGraw-Hill.

Cross references

Absorption, absorption spectrum
Airglow
Aurora: planetary
Comet: structure and composition
Earth: atmosphere
Fraunhofer line
Infrared spectroscopy
Jupiter: atmosphere
Lyman alpha
Mars: atmosphere
Mercury: atmosphere
Microwave spectroscopy
Moon: atmosphere
Neptune: atmosphere
Planetary torus
Radiative transfer
Saturn: atmosphere
Titan
Ultraviolet spectroscopy
Uranus: atmosphere
Venus: atmosphere
Visible and near-infrared spectroscopy
Zeeman effect

STONY IRON METEORITES

Stony iron meteorites consist of approximately equal amounts by weight of silicates and iron–nickel metal plus troilite, although the ratio of silicate to metallic material can be quite variable. There are two major groups of stony iron meteorites, the mesosiderites and the pallasites. In addition, there are a few unique meteorites and minor groups of meteorites that are classified as stony irons, but these will not be discussed here.

Pallasites

The pallasites are composed of roughly equal amounts of coarse-grained, magnesian olivine and iron–nickel metal. They contain minor troilite and accessory amounts of orthopyroxene, clinopyroxene, chromite and various phosphates (Buseck, 1977). The texture of olivine varies considerably from highly angular, fragmental texture to rounded, equigranular texture (Scott, 1977a). Metal–silicate textures are variable as well. Some pallasites contain large masses of olivine, while in others olivine is approximately evenly dispersed in metal. One spectacular specimen of the Brenham pallasite contains a large central region devoid of olivine and smaller regions of typical pallasitic texture along parts of the edge of the specimen. These textures are believed to have been generated by the fragmentation and mixing of solid olivine with liquid metal, followed by variable degrees of textural equilibration (Scott, 1977a).

The oxygen isotopic composition of olivine in pallasites shows that two distinct groups are present (Clayton and Mayeda, 1978). The majority of pallasites have an O isotopic composition similar to that of mesosiderites and the basaltic achondrite groups the eucrites, howardites and diogenites (see Basaltic achondrite meteorites). These pallasites are referred to as the main-group pallasites. Three pallasites have distinct O isotopic compositions and must have come from a different parent body. These pallasites are called the Eagle Station trio after one of them.

Olivine compositions in main-group pallasites are remarkably uniform (Buseck and Goldstein, 1969). Most of them contain olivine with molar $100 \times (MgO/(MgO + FeO))$ (hereafter mg#) of between 89–87. A few anomalous members of the main-group pallasites have more ferroan olivines, with mg# in the range 84–82. The Eagle Station

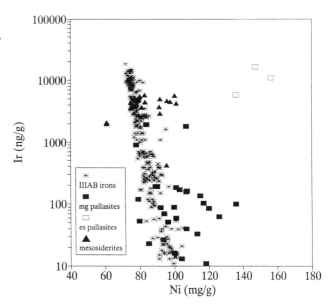

Figure S49 Ir versus Ni for IIIAB iron meteorites, pallasites and the metallic fraction of mesosiderites. Low-Ni, high-Ir IIIAB irons were among the first irons to crystallize from a molten Fe–Ni asteroid core. Fractional crystallization of Fe–Ni metal from the core caused the residual molten metal to become depleted in Ir and enriched in Ni. The metal of the main-group (MG) pallasites may represent residual liquid from the IIIAB iron core. Note that the Eagle Station trio (ES) pallasites have metal compositions distinct from main-group pallasite metal. (All data shown are from publications too numerous to list from J.T. Wasson's research group at UCLA.)

trio pallasites contain still more ferroan olivines with mg# of about 81–80 (Buseck and Goldstein 1969).

The silicates and metal of pallasites have maintained equilibrium, and redox reactions have produced some of the minor phases. Buseck and Goldstein (1969) calculated that the compositions of olivine and metal in pallasites indicate that these phases maintained equilibrium down to about 800 °C. This is also shown by intrinsic oxygen fugacity measurements of several pallasites, which show that the FeO content of olivines and the Fe/Ni ratio of the metal are correlated with the measured intrinsic oxygen fugacity (Righter et al., 1990). Davis and Olsen (1991) determined rare earth element (REE) contents of phosphates in eight pallasites and showed that the REE patterns in seven of them are consistent with formation of phosphates by oxidation of phosphorus dissolved in the metal and reaction with the silicates. Only in the main-group pallasite Springwater, which has ferroan olivine, do the REE contents of the phosphates suggest that they were formed by crystallization from a magma. However, Buseck and Holdsworth (1977) studied the textural relationships and major element compositions of phosphates from many pallasites and concluded that the phosphates were formed from residual melts.

The metal phase of main-group pallasites is compositionally closely related to that of the IIIAB iron meteorites (see Iron meteorites). The IIIAB irons exhibit a magmatic, fractional crystallization trend in major and trace element composition (Figure S49). Iridium is enriched in solid relative to liquid Fe, Ni, while Ni is slightly enriched in the liquid phase (Scott, 1972). Hence, iron meteorites from an asteroidal core that formed by fractional crystallization of molten Fe, Ni will vary from early, high-Ir, low-Ni irons to late, low-Ir, high-Ni irons as observed for the IIIAB irons (Figure S49). The metal phase of main-group pallasites is generally low in Ir and high in Ni (Figure S49) and may represent residual liquid left after crystallizing a molten IIIAB iron core (Scott, 1977b).

It is generally accepted that the pallasites, both the main group and the Eagle Station trio, were formed at the boundary between core and mantle of differentiated asteroids. However, Harold Urey argued that the number of pallasites is too great to support this model because the core–mantle boundary is a volumetrically insignificant fraction of an asteroid. He proposed that the pallasites formed around individual metal pods dispersed throughout an asteroid

(Urey, 1956). By having the metal dispersed in small pods in the parent body, a higher volume of metal–silicate boundary regions is obtained. If Urey's model is correct, there should be a range in cooling rates for the main-group pallasites because the individual metal pods would probably be formed at various depths in the parent body. Buseck and Goldstein (1969) determined metallographic cooling rates for most of the pallasites, and found that all of them cooled at the same rate of about 0.5–2°C Ma^{-1}. (These cooling rates may be a factor of five too low; Saikumar and Goldstein, 1988.) This includes both main-group and Eagle Station trio pallasites. Buseck and Goldstein (1969) concluded that Urey's model was incorrect.

Pallasites are somewhat implausible meteorites because the large density difference between olivine (~ 3.2 g cm^{-3}) and metal (~ 7.1 g cm^{-3}) should promote rapid separation of these phases even in the weak gravity field at the core–mantle boundary of an asteroid. Several models have been advanced to explain this curious mixture. Some pallasites contain large regions of olivine that appear to have been quenched in the process of breaking up. These regions are cut by metal veins and contain fragmented olivines (Scott, 1977a). This has led to the suggestion that during the waning stages of core crystallization, metal intruded into the lower portion of the solid olivine mantle, forming pallasites. An alternative model holds that the olivine mantle collapsed into the liquid outer portion of the core. Solid metal occupies about 2% less volume than an equal mass of liquid metal. Therefore, if an asteroid's core crystallizes from the center out, a vacancy would develop at the core–mantle boundary into which the mantle may collapse and mix with the residual molten metal. A third model assumes impacts that caused mixing of olivine and metal in pallasites. Many processes that affected meteorites are related to shock associated with meteoroid impacts on asteroids. Some researchers have suggested that a large impact event on the pallasite parent body caused mixing of metal and olivine deep in the asteroid's interior by shock processes.

Regardless of the precise mechanism of olivine–metal mixing, the pallasites are believed to represent samples of the core–mantle boundary region of igneously differentiated asteroids. They are therefore composed of materials that formed within a limited region of an asteroid. At least two such asteroids are required; one for the main-group and one for the Eagle Station trio pallasites.

Mesosiderites

The other major type of stony iron meteorite is the mesosiderite group. Unlike the relatively simple pallasites, the mesosiderites are complex breccias containing many different components. Also unlike the pallasites, the mesosiderites are breccias from the surface of their parent body, and it is likely that the metallic and silicate fractions originated on different parent bodies. The mesosiderites are composed of roughly equal amounts by weight of metal plus troilite and diverse mafic silicates.

Mesosiderite silicates are brecciated mixtures of igneous lithologies dominantly from the crust of an asteroid. The silicates are quite similar to the basaltic achondrite meteorites (see Basaltic achondrite meteorites), although there are significant, though subtle, differences. Mesosiderite silicates are composed of mineral and lithic clasts in a fine-grained matrix (Powell, 1971; Floran, 1978). The lithic clasts are primarily composed of basaltic and gabbroic rocks. Mineral clasts are dominantly coarse-grained orthopyroxene fragments up to about 10 cm in size. Coarse-grained olivine, up to about 5 cm in size, and coarse-grained plagioclase fragments a few millimeters in size are common in many mesosiderites. The silicate assemblage in mesosiderites is similar to that of the howardites; the most significant difference is in the common occurrence of coarse-grained olivine in the former, and its absence in the latter (Prior, 1918).

The mesosiderites can be divided into two major petrologic groups based on the amount of orthopyroxene present (Hewins, 1984). Type A mesosiderites contain more plagioclase and clinopyroxene, and therefore more basaltic and gabbroic components than type B mesosiderites (Figure S50). Type B mesosiderites contain more orthopyroxene than do type A mesosiderites, and are therefore richer in an orthopyroxenite component. One mesosiderite, RKPA79015 from Antarctica, contains almost exclusively orthopyroxene as its silicate phase (Prinz, Nehru and Delaney, 1982), and may be a third petrologic type, type C. These three petrologic types therefore reflect compositional variations in mesosiderites caused by varying ratios of basaltic to ultramafic material.

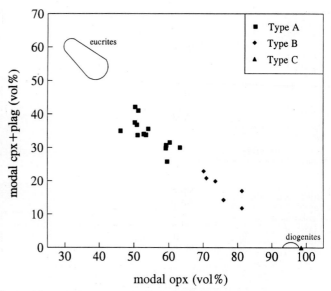

Figure S50 Modal clinopyroxene + plagioclase versus orthopyroxene for the three mesosiderite types compared with fields for basaltic eucrites and diogenites. (Data from Prinz *et al.*, 1980, and Prinz, Nehru and Delaney, 1982, except that the diogenite field was estimated from literature descriptions.)

In addition to petrologic distinctions based on the amount of basaltic–gabbroic material in the breccias, mesosiderites have been further subdivided on the basis of silicate textures. The silicate textures are thought to reflect increasing metamorphic equilibration of the silicates (Powell, 1971). The lowest grade (1) exhibits fine-grained fragmental matrix. Successively higher grades (2 and 3) have recrystallized matrices, and the highest grade examples (4) are melt-matrix breccias (Floran *et al.*, 1978). Textural grades 1 through 4 are found in both types A and B mesosiderites: the sole type C mesosiderite is of textural grade 2. Some mesosiderites have variable textures indicating more than one classification. For example, different samples of Estherville show textures indicating grade 3 and 4 (Hewins, 1984). The textures and mineralogies of the mesosiderites have been interpreted to indicate that these meteorites are polymict breccias formed by impact processes on the surface of a differentiated asteroid (Powell, 1971).

The composition of the metal phase of mesosiderites is generally similar to iron meteorites, particularly the IIIAB irons. However, mesosiderite metal does not exhibit the strong igneous fractionation trend evident in magmatic iron meteorite groups (Figure S49). This led Hassanzadeh, Rubin and Wasson (1990) to suggest that the metal was derived from a still-molten, and therefore homogeneous, core that was accreted onto the mesosiderite parent body and mixed into the silicates. Larger metallic nodules in mesosiderites have metallographic textures that allow cooling rates to be calculated. The calculated cooling rates for the temperature interval of $\sim 500–350°C$ is of the order of 1°C Ma^{-1} (Powell, 1969; recalculated to conform to recent diffusion data, see Bogard *et al.*, 1990). These are among the lowest cooling rates determined for meteorites, and are inconsistent with a near-surface origin as suggested by the silicate lithologies and textures. This is a long-standing conundrum and has driven numerous models for mesosiderite formation.

The composition of mesosiderite silicates is broadly similar to that of howardites, as would be expected for breccias composed of similar minerals. One major difference is a lower abundance of FeO relative to howardites. This difference, however, is due to redox reactions that occurred between the metal and silicates of mesosiderites (Mittlefehldt, Chou and Wasson, 1979). A reducing agent, probably P, in the metal caused reduction of FeO from the silicates to metallic Fe. There are few measurements of P_2O_5 for mesosiderite silicates, but the high modal abundance of merrillite compared to howardites (Prinz *et al.*, 1980) supports this model. In addition to differences caused by metamorphic equilibration between metal and silicate, mesosiderites are generally lower in the most incompatible trace elements, such as the light rare earth elements. This difference may

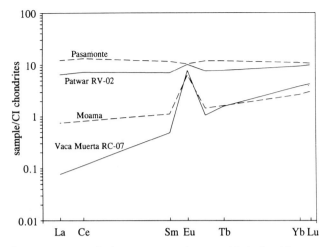

Figure S51 Rare earth element patterns for mesosiderite basaltic and gabbroic clasts (solid lines) compared to typical basaltic and gabbroic eucrites (dashed lines). The basaltic eucrite Pasamonte has a relatively flat REE pattern, while a basalt clast from Patwar shows a positive Eu anomaly and a depletion of the LREE. The mesosiderite cumulate gabbro clast from Vaca Muerta shows a more pronounced LREE/HREE fractionation than does the cumulate eucrite Moama. Moama has a more fractionated REE pattern than do most cumulate eucrites. (Data from Mittlefehldt, 1979, 1990.)

indicate that the mesosiderite breccias sampled more of the deeper crustal rocks, such as gabbroic cumulates, than did the howardite breccias (Mittlefehldt, 1988). The differences between types A, B and C mesosiderites may also have arisen from sampling different depths of the crust, with type A being shallow materials and type C deep. Alternatively, Hassanzadeh, Rubin and Wasson (1990) have suggested that olivine was added to the crustal silicates along with the metal during metal–silicate mixing, and that olivine was converted to orthopyroxene through reduction of FeO to metal and/or through reaction with tridymite in the matrix. They suggested that originally all mesosiderites were type A, and that type B mesosiderites were produced by these reactions.

Extensive investigation has been made of the petrology and composition of igneous lithic clasts in the mesosiderites. Clasts in mesosiderites are superficially similar to igneous materials in howardite, eucrite and diogenite (HED) meteorites (see Basaltic achondrite meteorites). Basaltic and gabbroic clasts are petrologically similar to basaltic and gabbroic eucrites and to mafic clasts in howardites, and the orthopyroxenite clasts are similar to diogenites and to orthopyroxenite clasts in howardites. However, detailed study of mesosiderite clasts has shown that most of the mafic clasts are distinguishable from similar HED meteorite materials. Most of the mesosiderite gabbroic clasts show extreme depletions in the most incompatible elements, such as the light rare earth elements (LREE), when compared to cumulate eucrites, and most of the basaltic clasts are also depleted in LREE when compared with basaltic eucrites (Figure S51; Rubin and Mittlefehldt, 1992). Pyroxene compositions in mesosiderite mafic clasts are also distinct in that they indicate that FeO was being reduced to metal in the magma during crystallization (Mittlefehldt, 1990). Because the reducing agent was brought in with the metallic phase during metal–silicate mixing, the mafic clasts are believed to be igneous products of the remelting of an earlier, differentiated crust (Mittlefehldt, 1990). A few mafic clasts have bulk and mineral compositions that indicate that they were not formed by crustal remelting, and these are believe to be primary crustal rocks from the mesosiderite parent body (Rubin and Mittlefehldt, 1992). Most of the orthopyroxene clasts also appear to be primary, in that they show no evidence for FeO reduction effects. Because the primary magmatic rocks are indistinguishable from eucrites and diogenites, the original crust of the mesosiderite parent body appears to have been essentially identical to that of the basaltic achondrites.

Age determinations on mesosiderites have shown that the original mesosiderite parent body crust was formed about 4.56 Ga ago, at essentially the same time as that of the basaltic achondrites, and shortly after the formation of chondritic meteorites (see Basaltic achondrite meteorites, and review by Rubin and Mittlefehldt, 1993). Ar–Ar ages have shown that mesosiderites were heated and thoroughly outgassed about 3.9 Ga ago, possibly in a single event (Bogard et al., 1990). Bogard et al. (1990) have suggested that this event was collisional disruption and reassembly of the mesosiderite parent body, in which mesosiderite breccias were deeply buried in the reassembled asteroid and allowed to cool slowly. This event then established the low metallographic cooling rates and reset the Ar–Ar ages, without affecting the silicate textures that preserve evidence for surficial processes.

Because mesosiderites are composed of crustal silicates and iron metal presumed to represent core material of an asteroid, the formation of these meteorites requires complex processes. This complexity is compounded by estimates for cooling rates of the metal which indicate slow cooling, and hence deep burial (Powell, 1969), and by the textures and compositional zoning of the silicates which give evidence for surficial processing (Delaney et al., 1981). Several models have been advanced to explain the metal–silicate mixing 'event' including internal and external mixing (Hewins, 1983). However, almost all models assume that the metallic and silicate fractions of mesosiderites originated in distinct regions, either the core and crust, respectively, of a single asteroid, or of two asteroids. Possibly the most plausible model is that of Wasson and Rubin (1985) and Hassanzadeh, Rubin and Wasson (1990), in which an asteroidal core, stripped of its silicates (except for some mantle olivine) by impacts, is accreted at low velocity onto the mesosiderite parent body and mixed into the regolith through impact gardening.

Regardless of the precise mechanism of metal–silicate mixing, the mesosiderites are believed to represent regolith samples of an igneously differentiated asteroid that were mixed with a metallic fraction from a distinct location. They are therefore composed of materials that formed in diverse regions of one or two asteroids.

David W. Mittlefehldt

Bibliography

Bogard, D.D., Garrison, D.H., Jordan, J.L. and Mittlefehldt, D.W. (1990) ^{39}Ar–^{40}Ar dating of mesosiderites: evidence for major parent body disruption < 4 Ga ago. Geochim. Cosmochim. Acta, 54, 2353–62.

Buseck, P.R. (1977) Pallasite meteorites – mineralogy, petrology and geochemistry. Geochim. Cosmochim. Acta, 41, 711–40.

Buseck, P.R. and Goldstein, J.I. (1969) Olivine compositions and cooling rates of pallasitic meteorites. Geol. Soc. Am. Bull., 80, 2141–58.

Buseck, P.R. and Holdsworth, E. (1977) Phosphate minerals in pallasite meteorites. Mineral. Mag., 41, 91–102.

Clayton, R.N. and Mayeda, T.K. (1978) Genetic relations between iron and stony meteorites. Earth Planet. Sci. Lett., 40, 168–74.

Davis, A.M. and Olsen, E.J. (1991) Phosphates in pallasite meteorites as probes of mantle processes in small planetary bodies. Nature, 353, 637–40.

Delaney, J.S., Nehru, C.E., Prinz, M. and Harlow, G.E. (1981) Metamorphism in mesosiderites, in Proc. Lunar Planet. Sci. Conf., 12, pp. 1315–42.

Floran, R.J. (1978) Silicate petrography, classification, and origin of the mesosiderites: review and new observations, in Proc. Lunar Planet. Sci. Conf., 9, pp. 1053–81.

Floran, R.J., Caulfield, J.B.D., Harlow, G.E. and Prinz, M. (1978) Impact-melt origin for the Simondium, Pinnaroo, and Hainholz mesosiderites: implications for impact processes beyond the Earth–Moon system, in Proc. Lunar Planet. Sci. Conf., 9, pp. 1083–114.

Hassanzadeh, J., Rubin, A.E. and Wasson, J.T. (1990) Compositions of large metal nodules in mesosiderites: links to iron meteorite group IIIAB and the origin of mesosiderite subgroups. Geochim. Cosmochim. Acta, 54, 3197–208.

Hewins, R.H. (1983) Impact versus internal origins for mesosiderites, in Proc. Lunar Planet. Sci. Conf. 14, Part 1. Geophys. Res., 88, Suppl., B257–66.

Hewins, R.H. (1984) The case for a melt matrix in plagioclase–POIK mesosiderites, in Proc. Lunar Planet. Sci. Conf. 15, Part 1. J. Geophy. Res., 89, Suppl. C289–97.

Mittlefehldt, D.W. (1979) Petrographic and chemical characterization of igneous lithic clasts from mesosiderites and howardites and

comparison with eucrites and diogenites. *Geochim. Cosmochim. Acta*, **43**, 1917–35.
Mittlefehldt, D.W. (1988) Petrogenesis of the mesosiderite regolith (abstract), in *Lunar and Planet. Sci. XIX*, pp. 788–9.
Mittlefehldt D.W. (1990) Petrogenesis of mesosiderites. I: origin of mafic lithologies and comparison with basaltic achondrites. *Geochim. Cosmochim. Acta*, **54**, 1165–73.
Mittlefehldt, D.W., Chou, C.-L. and Wasson, J.T. (1979) Mesosiderites and howardites; igneous formation and possible genetic relationships. *Geochim. Cosmochim. Acta*, **43**, 673–88.
Powell, B.N. (1969) Petrology and chemistry of mesosiderites – I. Textures and composition of nickel–iron. *Geochim. Cosmochim. Acta*, **33**, 789–810.
Powell, B.N. (1971) Petrology and chemistry of mesosiderites – II. Silicate textures and compositions and metal–silicate relationships. *Geochim. Cosmochim. Acta*, **35**, 5–34.
Prinz, M., Nehru, C.E. and Delaney, J.S. (1982) Reckling Peak A79015: an unusual mesosiderite (abstract), in *Lunar Planet. Sci. XIII*, p. 631.
Prinz, M., Nehru, C.E., Delaney J.S. *et al.* (1980) Modal studies of mesosiderites and related achondrites, including the new mesosiderate [sic] ALHA 77219, in *Proc. Lunar Planet. Sci. Conf.*, **11**, pp. 1055–71.
Prior G.T. (1918) On the mesosiderite–grahamite group of meteorites: with analyses of Vaca Muerta, Hainholz, Simondium, and Powder Mill Creek. *Mineral. Mag.*, **18**, 151–72.
Righter, K., Arculus, R.J., Delano, J.W. and Paslick, C. (1990) Electrochemical measurements and thermodynamic calculations of redox equilibria in pallasite meteorites: implications for the eucrite parent body. *Geochim. Cosmochimi. Acta*, **54**, 1803–15.
Rubin, A.E. and Mittlefehldt, D.W. (1992) Classification of mafic clasts from mesosiderites: implications for endogenous igneous processes. *Geochim. Cosmochim. Acta*, **56**, 827–42.
Rubin, A.E. and Mittlefehldt, D.W. (1993) Evolutionary history of the mesosiderite asteroid: a chronologic and petrologic synthesis. *Icarus*, **101**, 201–12.
Saikumar, V. and Goldstein, J.I. (1988) An evaluation of the methods to determine the cooling rates of iron meteorites. *Geochim. Cosmochim. Acta*, **52**, 715–26.
Scott, E.R.D. (1972) Chemical fractionation in iron meteorites and its interpretation. *Geochim. Cosmochim. Acta*, **36**, 1205–36.
Scott, E.R.D. (1977a) Formation of olivine–metal textures in pallasite meteorites. *Geochim. Cosmochim. Acta*, **41**, 693–710.
Scott, E.R.D. (1977b) Geochemical relationships between some pallasites and iron meteorites. *Mineral. Mag.*, **41**, 265–72.
Scott E.R.D. (1977c) Pallasites–metal composition, classification and relationships with iron meteorites. *Geochim. Cosmochim. Acta*, **41**, 349–360.
Urey H.C. (1956) Diamonds, meteorites, and the origin of the solar system. *Astrophys. J.*, **124**, 623–37.
Wasson, J.T. and Rubin, A.E. (1985) Formation of mesosiderites by low-velocity impacts as a natural consequence of planet formation. *Nature*, **318**, 168–70.

Cross references

Basaltic achondrite meteorites
Differentiation
Igneous rock
Iron
Iron meteorites
Meteorite parent bodies
Meteorite

STRATOSPHERE

The region of the atmosphere lying directly above the troposphere is called the stratosphere (from the Greek *strato-*, layered, plus *-sphaira*, globe). The stratosphere was discovered in 1902 by the French meteorologist Teisserenc de Bort and the German meteorologist Assmann. In the lower stratosphere temperature is almost uniform with height, while in the middle stratosphere the temperature begins to increase sharply with height until reaching a maximum at about 50 km altitude. This level, called the stratopause, is the upper boundary of the stratosphere.

The temperature maximum, and thus the character of the stratosphere below, is caused by absorption of ultraviolet radiation by ozone molecules. The stratospheric ozone layer is formed by ultraviolet photodissociation of oxygen molecules. Since atmospheric oxygen is the result of biological activity, the ozone layer is a uniquely terrestrial phenomenon. Consequently, not all planets have stratospheres in the conventional sense of increasing temperature with altitude. On Venus, for example, the troposphere simply gives way gradually to an extended layer of nearly isothermal temperature. On the other hand, absorption of sunlight by methane creates stratospheres on the Jovian planets, playing a role similar to that of terrestrial ozone.

The temperature profile of the stratosphere is highly stable with respect to vertical perturbations. This region of the atmosphere is therefore quite layered and quiescent relative to the highly turbulent troposphere below. Stratospheric air is quite rarefied (about 15% of the total mass of the atmosphere) and dry relative to the troposphere. Because density decreases sharply with height through the atmosphere, small disturbances originating in the troposphere can become large perturbations in the stratosphere, analogous to the cracking of a whip. Thus the circulation of the stratosphere is controlled as much by phenomena propagating upward from the troposphere as by absorption of sunlight within the stratosphere itself.

Volcanic eruptions can sometimes inject large amounts of sulfur-bearing gases into the stratosphere, leading to the formation of a fine haze of sulfuric acid droplets. Since the stratosphere lies above most clouds, precipitation and vertical mixing, this volcanic haze can remain in the stratosphere for years, blocking sunlight and cooling the Earth's climate. A thinner permanent sulfuric acid haze called the Junge layer also exists within the stratosphere. Although clouds are generally absent from the stratosphere, tenuous polar stratospheric clouds have been observed. These clouds are thought to act as catalysts for the destruction of ozone by anthropogenic chlorofluorocarbons mixed upward from the Earth's surface. The result is the formation of an ozone 'hole' over the Antarctic, and perhaps eventually over the Arctic, in the spring.

Anthony D. Del Genio

Bibliography

Goody, R.M. and Walker, J.C.G. (1972) *Atmospheres*. Englewood Cliffs: Prentice-Hall.
US Standard Atmosphere (1976) National Oceanic and Atmospheric Administration Report NOAA-S/T 76–1562. Washington, DC: US Govt Printing Office.

Cross references

Atmosphere
Atmospheric thermal structure
Mesosphere
Troposhere

SUN

The Sun is a main-sequence star, one of over 100 billion stars in the Milky Way galaxy (Plates 1,2). It takes the Sun over 200 million years to complete one orbit of the galaxy. It is located at present close to the Sagittarius–Carina spiral arm, in what is called the Orion spur.

The orbital trajectory is not planar, but undulating, the Sun passing through the principal plane of the galaxy once every 30 million years or so. The planets of the solar system revolve about the Sun in a plane normal to the Sun's circum-galactic trajectory. Thermonuclear reactions in the Sun's core convert hydrogen into helium, so that the helium fraction is increasing through time. The reactions produce the energy to make the Sun a yellow-orange main-sequence star of spectral type dG2 (d indicates dwarf). The central temperature is about 15 million K; the surface temperature is about 6000 K.

The Sun is a nearly perfect sphere of gas with a diameter of 1 400 000 km, held together by its own gravity. Its dimensions are at present not static and recent findings have shown that the Sun's diameter is at the present time decreasing about 1 m h^{-1}. This decrease may be a long-term oscillation which may be one effect of

Table S7 Physical characteristics of the Sun

Radius	$R_\odot = 6.960 \times 10^5$ km
Mass	$M_\odot = 1.991 \times 10^{33}$ g
Mean density	$\bar{\rho}_\odot = 1.410$ g cm^{-3}
Gravity at surface	$g_\odot = 2.738 \times 10^4$ cm s^{-2}
Total energy output (luminosity)	$L_\odot = 3.860 \times 10^{33}$ erg s^{-1}
Effective surface temperature	$T_{\text{eff}} = 5780$ K
Solar age	$t_\odot = 4.5 \times 10^9$ years

the long-term stabilization of the Sun's output of energy. It has also been found that every 2 h 40 min the Sun's surface pulses at a speed of 6 km h^{-1}. Thus its surface moves in and out to change the diameter by nearly 10 km.

The Sun contains 99.86% of all matter in the solar system. It is the source of light and heat for all planets and the support of life on Earth. A number of properties of the Sun are listed in Table S7 (Allen, 1973; Lang, 1980).

Solar structure

Internal structure

The structure of the Sun is determined by the conditions of mass conservation, momentum conservation, energy conservation and the mode of energy transport. The Sun is an oblate spheroid, like all the major bodies in the solar system, but in a first simplifying approach to describe the solar structure, the effects of rotation and magnetic fields will be neglected here so that the Sun is taken to be spherically symmetrical. Calculating a solar model involves the determination of pressure, temperature and chemical composition as a function of the mass or radius of the Sun (Chandrasekhar, 1967; Kourganoff, 1973).

Two forces keep the Sun in hydrostatic equilibrium in its current stage of evolution: the gravitational force directed inward and the total pressure force directed outward. The equation of hydrostatic equilibrium is

$$\frac{dP}{dr} = -\rho \frac{GM_r}{r^2} \tag{S15}$$

where P is the pressure, r the radial distance from the center, M_r the mass within a sphere of radius r, ρ the matter density and G the gravitational constant. This equation is consistent with radius changes, but requires the kinetic energy involved in expansion or contraction of the solar body to be small compared to the gravitational potential of the Sun. For an order-of-magnitude estimate, equation (S15) can be written

$$\frac{dP}{dr} \approx \frac{P_c - P_o}{R_\odot} \approx \frac{P_c}{R_\odot} \approx \frac{GM_\odot \bar{\rho}}{R_\odot^2} \tag{S16}$$

where $R_\odot$ is the solar radius, $M_\odot$ the solar mass, $\bar{\rho}$ the mean matter density of the solar gas sphere, P_c is the central and P_o the surface pressure respectively, where the latter can be neglected.

The equation of mass conservation

$$\frac{dM_r}{dr} = 4\pi r^2 \rho \tag{S17}$$

constrains the integral of the density over the volume to be equal to the mass and leads to the estimation of the mean matter density for the Sun

$$\frac{M_\odot}{R_\odot} \approx R_\odot^2 \bar{\rho} \rightarrow \bar{\rho}_\odot \propto \frac{M_\odot}{R_\odot^3} \tag{S18}$$

where the symbol $\propto$ means 'varies as'. In the general case, $\rho \propto M/R^3$, the constant of proportionality depends on the radial mass distribution and the radial distance R (Schwarzschild, 1958; Haubold and Mathai, 1987, 1992). Using equations (S18) and (S16), the central pressure of the Sun can be estimated to be

$$P_c \propto G \frac{M_\odot^2}{R_\odot^4} \tag{S19}$$

In the general case, $P \propto GM^2/R^4$, the constant of proportionality is determined by the radial distribution of mass in the Sun, and the particular radial distance R at which P is measured (Schwarzschild, 1958; Haubold and Mathai, 1987, 1992).

The interior of the Sun is entirely gaseous and the great majority of atoms are stripped of their electrons. The solar gas behaves under these physical conditions nearly like a perfect gas, governed by the 'equation of state'

$$P = \frac{k}{\mu m_p} \rho T \tag{S20}$$

where m_p is the mass of the proton, k is the Boltzmann constant and μ is the mean molecular weight. This equation of state relates the pressure, temperature, density and chemical composition, and is related to other thermodynamic quantities. Then the central temperature of the Sun can be estimated from the perfect gas law in equation (S20), that is

$$T_c \approx \frac{\mu m_p}{k} \frac{P_c}{\rho} \rightarrow T_c \propto \frac{m_p G}{k} \mu \frac{M_\odot}{R_\odot} \tag{S21}$$

This formula determines the temperature at the centre of the Sun according to its mass, radius and the mean molecular weight of the solar matter. In the general case, $T \propto \mu M/R$, the constant of proportionality depends on the mass distribution and the radial distance R (Schwarzschild, 1958; Haubold and Mathai, 1987, 1992).

When X, Y and Z are the mass fractions of hydrogen, helium and heavy elements respectively, it holds by definition $X + Y + Z = 1$. The mean molecular weight μ in equation (S20) can be calculated when the degree of ionization of each chemical element of solar matter has been specified. For solar gas composed of fully ionized hydrogen, there are two particles for every proton and $\mu = \frac{1}{2}$. For a gas composed of fully ionized helium $\mu = \frac{4}{3}$. For all elements heavier than helium, usually referred to by astronomers as metals, it holds that their atomic weights are twice their charge and accordingly $\mu = 2$. Thus the mean atomic weight for fully ionized gas is

$$\mu = \frac{1}{2X + (\frac{3}{4})Y + (\frac{1}{2})Z} \tag{S22}$$

The solar matter is at present approximately 75% hydrogen, 23% helium and 2% metals by mass fraction. Throughout the solar interior, $\mu_\odot$ is approximately 0.59, except at the surface, where hydrogen and helium are not fully ionized, and in the core, where the chemical composition is altering due to nuclear burning (Table S8; Kavanagh, 1972; Bahcall, 1989).

An equation of continuity must also be satisfied by the radiation

$$\frac{dE}{dt} + \frac{dL_r}{dr} = 0 \tag{S23}$$

where dE/dt is the rate of energy production per unit thickness of the shell of radius r. The equation of energy conservation is

$$\frac{dL_r}{dr} = 4\pi r^2 \rho \epsilon = -\frac{dE}{dt} \tag{S24}$$

where L_r denotes the total net energy flux through a spherical shell of radius r and ϵ is the net release of energy per gram per second by thermonuclear reactions occurring in the gravitationally stabilized solar fusion reactor. It is assumed in equation (S23) that the energy produced by nuclear reactions equals the photon luminosity of the Sun, thus neglecting gravitational contraction and subtracting energy loss through neutrino emission. The mean energy generation rate for the Sun can be inferred from equation (S24), that is

$$\bar{\epsilon} \approx \frac{L_\odot}{M_\odot} \tag{S25}$$

Finally the thermonuclear energy produced in the solar core is transported by radiation through the solar body to the surface. The force due to the gradient of the radiation pressure equals the momentum absorbed from the radiation streaming through the gas

$$\frac{dP}{dr} = -\frac{\kappa \rho}{c} \frac{L_r}{4\pi r^2} \tag{S26}$$

where $P = aT^4/3$ is the radiation pressure, κ is the opacity of the solar matter, $1/\kappa\rho$ is the mean free path of photons and c is the velocity of light. The coefficient of the radiation density a is related to the Stefan–Boltzmann constant σ since $\sigma = ac/4$. Equation (S26) is the energy transport equation taking into account the fact that energy transport in the deep interior of the Sun is exclusively managed by radiation. From equation (S25) follows the temperature gradient driving the radiation flux, that is

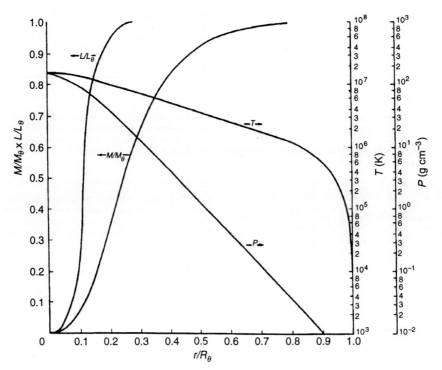

Figure S52 A standard solar model of the present solar interior: $X = 0.708$, $Y = 0.272$, $Z = 0.0020$, $\rho_c = 158$ g cm^{-3}, $T_c = 1.57 \times 10^7$ K. (Courtesy of R.L. Sears of University of Michigan; Sears, 1964.)

Table S8 Internal structure of the Sun ($R_\odot = 6.96 \times 10^5$ km)

Internal region	Extension in terms of solar radius	Chemical composition
Core	0.20 $R_\odot$	Center only: He, 0.63; H, 0.35; metals, 0.02 (almost entirely ionized matter)
Radiative zone	0.50 $R_\odot$	He, 0.23; H, 0.75; metals, 0.02 (highly ionized)
Convective zone	0.30 $R_\odot$	Same (less ionized)
Photosphere	0.002 $R_\odot$	Same (less ionized)
Solar surface	1.000 $R_\odot$	
Chromosphere	0.02	Same (less ionized)
Corona	≈ 5	Same (highly ionized)

$$\frac{dT}{dr} = -\frac{3}{4ac}\frac{\kappa\rho}{T^3}\frac{L_r}{4\pi r^2} \quad (S27)$$

allowing an estimate of the solar luminosity

$$\frac{T_c}{R_\odot} \approx \frac{1}{ac}\frac{\kappa\bar{\rho}}{T_c^3}\frac{L_\odot}{R_\odot^2} \rightarrow L_\odot \propto ac\left(\frac{Gm_p}{k}\right)^4 \frac{\mu^4}{\kappa} M_\odot^3 \quad (S28)$$

taking into account equations (S18) and (S21). The luminosity is independent of the radius; it depends on the opacity and increases with mass. Equation (S28) is an important result of the theory of the internal structure of solar-type stars, called the theoretical mass-luminosity relationship. The fundamental result as given by equation (S28) is that the luminosity of the star is simply determined by its mass, since this rule is based on the fact that the transfer of energy from the stellar interior towards the surface is managed by radiation. The luminosity of a solar-type star is determined largely by photon opacity and not by the energy source.

Gamma-ray photons produced in thermonuclear reactions in the core of the Sun are being scattered, absorbed or re-emitted by free electrons, ions and atoms on their way to the surface of the Sun. The opacity κ in equation (S26) is the measure of the solar material's efficiency at inhibiting the passage of the photons through the solar interior. The actual value of the opacity depends on various processes which may operate simultaneously: bound–bound transitions, bound–free transitions, free–free transitions and scattering of photons by free electrons, ions and atoms. Scattering of photons by free electrons is the most important process for the solar core. Approaching the solar surface, bound–free transitions take over to determine the opacity of solar matter. The structure of the Sun depends in a sensitive way on the opacity, for if κ changes, the Sun must readjust all its parameters to allow the energy generated in the core to stream to the surface, not being blocked at any point in the solar interior.

Boundary conditions for the system of nonlinear differential equations [equations (S15), (S17), (S24) and (S26)] have to be specified to arrive at specific solutions: at the solar centre it is $r = 0$, $M_r = 0$ and $L_r = 0$, and at the assumed solar surface (this is actually the photosphere) it holds that $M_r = M_\odot$, and for an age of $t_\odot = 4.5 \times 10^9$ years, $r = R_\odot$ and $L = L_\odot$. Mass, radius, surface temperature, surface chemical composition and luminosity of the Sun are known by observation. Using the conservation laws and known properties of gases (equation of state, opacity and energy generation rates), the internal structure of the Sun can be calculated in matching the observed properties at the solar surface. However, because the equations of solar structure form a system of first-order nonlinear simultaneous differential equations, they have to be integrated numerically to obtain a very detailed picture of the range of physical variables throughout the Sun. Order-of-magnitude estimates provided in equations (S18), (S19), (S21), (S25) and (S28) can be considered only to be a first approach to the problem (Mathai and Haubold, 1988). Figure S53 shows the numerical results of a standard solar model based on the system of differential equations as described above (Sears, 1964; Sackmann, Boothrayd and Fowler, 1990; Guenther et al., 1992).

Chemical composition changes with time [equation (10)] due to thermonuclear reactions in the solar core that result in a continuously evolving structure, the calculation of which adds another system of differential equations (kinetic equations) to the set of differential equations described above (Schwarzschild, 1958; Kourganoff, 1973).

Core

The core of the Sun is a gravitationally stabilized fusion reactor. There, energy is produced by conversion of hydrogen into helium.

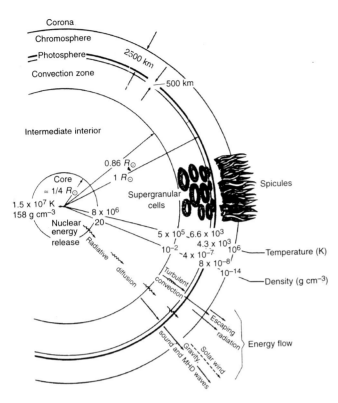

Figure S53 Schematic view of the structure of the Sun and modes of outward flow of energy. (Courtesy NASA Goddard Space Flight Center.)

Each hydrogen atom weighs 1.0078 atomic units and each helium atom is made from four hydrogen atoms, thus weighing 4.003 atomic units. Accordingly, the difference of 0.0282 atomic units, or 0.7% of the mass m, is converted into energy E according to Einstein's formula $E = mc^2$, where c is the velocity of light. Most atoms in the core of the Sun are entirely stripped of their electrons by the high temperature, and opacity is governed by scattering of photons by free electrons, by inverse bremsstrahlung on ionized hydrogen and helium, and by bound–free scattering by elements heavier than helium.

Radiative zone

The radiative zone is a region of highly ionized gas. There the energy transport is primarily by photon diffusion and is described in terms of the Rosseland mean opacity (this is a weighted inverse mean of the opacity over all frequencies, which can be used when the optical depth is very large and radiative transport reduces to a diffusion process).

Convective zone

In the outer regions, atoms may keep their electrons because of the low temperature, and ions and even neutral hydrogen exist. Here many atomic absorption processes occur, mainly bound–free transitions. The high opacity makes it difficult for photon radiation to continue outward and steep temperature gradients are established which lead to convective currents. The outer envelope of the Sun is in convective equilibrium. It is the location where sunspots and other solar activity phenomena are generated. Observationally, the outer solar atmosphere following the convective zone has been divided into three spherically symmetric layers – the photosphere, chromosphere and corona – lying successively above one another (Zirin, 1988).

Photosphere

The outer limit of the photosphere is the boundary of the visible solar disk as seen in white light. Most of the radiation emitted by the Sun originates in the photosphere, which is only about 500 km thick. This radiation is in equilibrium and the Stefan–Boltzmann law can be applied to calculate the effective temperature of the solar photosphere, which is $T_e = 5780$ K. According to the Stefan–Boltzmann law each square centimeter of the solar surface having temperature T emits, in all directions, light of σT^4 ergs per second. Subsequently, the total emission of the Sun in 1 s, i.e. the luminosity, equals

$$L_\odot = 4\pi R_\odot^2 \sigma T_e^4 \tag{S29}$$

This fundamental relation also determines the radius of the Sun when its luminosity and surface temperature are known. The spectrum of the photosphere consists of absorption lines superimposed on an approximately blackbody continuum (see also Fraunhofer line; Blackbody radiation).

Chromosphere

A thin transition region extending 5000 km above the photosphere is called the chromosphere. Considerably hotter than the photosphere, the chromosphere is heated by hydromagnetic waves and compression waves originated by spicules and granules. The temperature of the chromosphere is about 10 000 K and it has an emission spectrum.

Corona

During a total solar eclipse the outermost atmosphere of the Sun can be seen. Called the solar corona (q.v.), this is a hot gas merging gradually into the transparent interplanetary medium, and flowing outward from it is the solar wind (q.v.). Current theories indicate that the corona is heated by the dissipation of mechanical energy stemming from the convection zone, or by dissipation of magnetic energy by field-line reconnection. The kinetic temperature of the solar corona is about 2×10^6 K and its gas has a density of about 10^{-15} g cm^{-3}. Solar x-ray radiation originates in the corona.

Solar activity

The Sun emits radiation in a wide range of the energy spectrum from long radio waves (300 m) to x-rays (0.1 nm), including high-energy particles (cosmic rays, q.v.). Almost 95% of the radiated energy is concentrated in a relatively narrow band between 250 nm and 2500 nm. The total radiation received from the Sun is called the solar constant; it was formerly regarded as a fixed value, 1.36×10^6 erg cm^{-2} s^{-1} (although difficult to measure), but from satellite observations it is now confirmed as variable (by up to about 0.5%) (Herman and Goldberg, 1978; Sofia, 1981; Schatten and Arking, 1990). The transient phenomena occurring in the solar atmosphere can be grouped together under the term solar activity: sunspots and faculae occur in the photosphere; flares and plages belong to the chromosphere; and prominences and coronal structures develop in the corona. All solar activity phenomena are connected in one way or another with the 11- and 22-year sunspot cycles.

Granules

Granules are huge convective cells of hot gases, 400–1000 km in diameter, spread in a cellular pattern over the entire photosphere except at sunspots. Granulation supports the transfer of energy from the convective zone outward into space. Granules behave as short-lived bubbles, lasting only 3 to 10 min, that rise and fall at a velocity of about 0.5 km s^{-1}, thereby moving vertically a distance of the order of 200 km.

Spicules

Spicules look like hairs of gas rising and falling at the upper chromosphere, reaching into the corona. They last as long as 10 min, attaining vertical speeds of up to 20 km s^{-1}, rising as high as 15 000 km. Spicules array themselves into chromospheric networks, establishing giant supergranulated cells, with gases rising in the center and descending at their outer boundaries.

Sunspots

Sunspots are relatively cool and dark markings on the Sun's photosphere, and exhibit distinct cycles. They are concentrations of strong magnetic fields (200–3000 gauss), with diameters less than about 50 000 km and lifetimes of a few days to weeks. A sunspot generally develops a very dark central region, called the umbra, which is surrounded by the penumbra. The 11-year sunspot cycle consists of variations in the size, number and position of the sunspots (Fairbridge, 1987a). It is extremely variable in length (actually 7 to 17 years); the high-activity cycles (to ≥ 200 spots) are generally short (9–10 years), and the low-activity cycles (with maxima sometimes ≤ 50 spots) are long (12–13 years). In the sunspot cycle the number of sunspots usually peaks 2–3 years after the beginning of each cycle and decays gradually, but low-activity cycles may have a reversed symmetry. The first spots of the cycle appear at higher latitudes, mostly between 20° and 35°, and as the spots increase in size and number they occur closer to the equator. Very few spots are observed outside the latitude range 5–35°. The magnetic polarity of the sunspot groups reverses in each successive cycle so that the complete cycle lasts 22 years, the so-called 'Hale cycle' (q.v.). Recent observations have indicated that the magnetic solar cycle is a coherent phenomenon throughout the solar surface. For each pair of 11-year cycles, the one with north magnetic orientation is usually stronger than that south orientation. Between 1645 and 1715 very few sunspots were seen, a time period called Maunder minimum (q.v.). This period was associated with a long cold spell in Europe, known as Little Ice Age. Carbon 14 measurements from tree rings and Beryllium 10 measurements from arctic ice cores confirm the low solar activity level at that time (Sonett, 1984; Beer, 1987; Fairbridge, 1987b). The solar 11-year cycle has been recorded on a regular basis since the beginning of the 18th century, classified by Wolf's quantity N of the number of sunspots N_s plus ten times the number G of sunspot groups: $N = W(N_s + 10G)$, where W is a weighting factor assigned to an individual observer to account for variation in equipment, atmosphere conditions and observer enthusiasm (Gibson, 1973). The quantity N is widely used as an indicator of sunspot activity and is commonly called the Zurich sunspot number. More recently Bracewell (1989) was able to show that the quantity $N = \pm (N_s + 10G)^{3/2}$, with ± denoting the dipole orientation, is nearly a sinusoidal function of time with a period of 22.2 ± 2 years. Using proxy data for ancient sunspot periods (such as auroral frequency), the average of the 11-year cycle is 11.12 years, and is thus correlated with the magnetic period.

Prominences

These are regions of cool (10^4 K), high-density gas embedded in the hot (10^6 K), low-density corona. Prominences can be observed as flamelike tongues of gas that appear above the limb of the Sun when observed in the light of the Hα line. They occur in regions of horizontal magnetic fields, because these fields support prominences against the solar gravity and indicate the transition from one magnetic polarity to the opposite.

Flares

Sunspots are accompanied by large eruptions called solar flares, emitting high-energy particles and radiation in a very broad spectrum of energy. A solar flare is actually the result of an intensely hot electromagnetic explosion in the corona; The flare produces vast quantities of x-rays which brighten the chromospheric gases. Typical lifetimes of solar flares are 1–2 h and the temperature in flares can reach several million degrees. Flare particles ejected into outer space reach the Earth in a few hours or days and are the cause of disruptions in radio transmission. Aurora and magnetic storms are due to strong solar flare eruptions. The peak of solar flare activity is lagged by the sunspot cycle, usually by 1–2 years, but some high-energy eruptions may occur at any time; the mean cycle of flare frequency is 0.417 years (see Coronal mass ejection).

Solar rotation

Solar rotation was first accurately measured in the 19th century (in 1863) by R.C. Carrington (q.v.), who used the position of prominent spots as marker points to determine a synodic period of about 27 days. Beginning with the first year of observation the solar rotations are identified by 'Carrington numbers'. The solar surface, however, exhibits differential rotation, as well as a coherent pattern of activity related to magnetic fields and globally coherent oscillation modes. All three phenomena can be employed to shed light on the structure and dynamics of the Sun. Helioseismology, the study of solar oscillations, has made it possible to measure the depth of the solar convection zone, the internal rotation profile, the speed of sound throughout the Sun and the solar helium abundance (Deubner and Gough, 1984; Hill and Kroll, 1992). Employing a standard model for the internal structure of the Sun, it has been shown (with linear adiabatic perturbation theory) that small-amplitude oscillations of

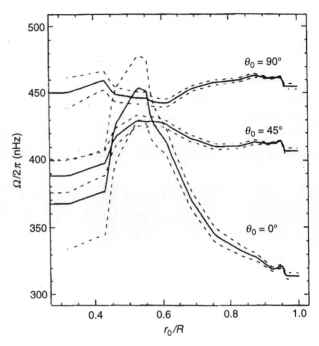

Figure S54 Solar interior rotation profile, as inferred from an inversion of p-mode splitting data, displayed at these latitudes: $\theta_0 = 0°$ (polar), $\theta_0 = 45°$, and $\theta_0 = 90°$ (equatorial). Dashed lines indicate 1σ error bars based on the observer's estimates of the uncertainties in the measured a_j coefficients (Courtesy J. Christensen-Dalsgaard; Schou, Christensen-Dalsgaard and Thompson, 1992.)

the solar body about its equilibrium state can be classified into three types: (1) pressure modes (p-modes), where the pressure is the dominant restoring force; (2) gravity-modes (g-modes), where gravity or buoyancy is the dominant restoring force; and (3) a class of surface or interface modes (f-modes), which are nearly compressionless surface waves. The existence of all three modes has been confirmed by solar observations. The solar rotation rate through a large part of the solar interior has been estimated, utilizing for the most part observations of the p-mode frequency splittings. Each mode is characterized by an eigenfunction with frequency eigenvalue ν_{nlm}, where n, l and m are integer 'quantum' numbers; n is the number of radial nodes in the wavefunction, and l and m describe the nodes in colatitude and longitude respectively. Rotation breaks the spherical symmetry of the Sun. Because of that the p-mode frequencies are not completely degenerate in m, and the frequencies ν_{nlm} in an nl-multiplet are said to be split, in analogy with the Zeeman splitting of degenerate atomic energy levels. Because of observational limits it is not yet possible to observe values of splittings for individual m to be used for inversion. However, results of observations are available in terms of coefficients a_j ($j \leq 5$) of least squares fits of the splittings:

$$\nu_{nlm} - \nu_{nl0} = L \sum_j a_j(n,l) P_j^{(l)}\left(\frac{m}{L}\right) \quad \text{(S30)}$$

where $P_j^{(l)}$ is a polynomial of degree j and $L = (l[l+1])^{1/2}$. The coefficients $a_j(n,l)$ of odd j reveal information about the internal rotation of the Sun. The analysis of observational data reveals that the latitude-dependent solar rotation profile as observed at the solar surface extends down through the convective envelope. In the radiative zone the rotation seems to have a solid-body profile. (Hill, Oglesby and Gu, 1992; Schou, Christensen-Dalsgaard and Thompson, 1992). To date there exists no obvious theoretical explanation for this helioseismologically inferred solar rotation profile.

Measurements of the individual frequencies of normal modes of the oscillating Sun may reveal the internal rotation profile. The ultimate goal of helioseismology is, however, to use all available pulsation data, including growth rates, phases and different modes –

and not just observed frequencies – to search for the internal structure and evolution of the Sun. Those data will definitely contribute to improvement of the inadequate treatment of convective transport of energy in the envelope of the Sun by the mixing length theory, as well as to solving the solar neutrino problem for the gravitationally stabilized solar fusion reactor. Eigenmodes of pulsations of different degree carry information about physical conditions in quite different parts of the Sun. High-degree modes are restricted to solar subsurface layers, where solar activity phenomena have their origin. By contrast low-degree modes propagate all the way through the solar body to the regions where the solar neutrino flux is generated.

According to observation and the theory of stellar evolution, young stars rotate rapidly. If the central part of the Sun still rotates rapidly, this should lead to a small oblateness in the Sun's disk, about one part in 10^5. The extreme observational values reported for the solar oblateness lie between $5.0 \pm 0.7 \times 10^{-5}$ (Dicke and Goldenberg, 1967) and $9.6 \pm 6.5 \times 10^{-6}$ (Hill and Stebbins, 1975), with a proposal that this quantity varies with the solar cycle (Dicke, Kuhn and Libbrecht, 1987). The oblateness of the Sun is still a hotly debated issue in observational and theoretical solar physics.

Solar magnetic field

All transient phenomena occurring in the solar atmosphere are connected with magnetic fields leading to a 22-year Hale cycle. To date, all observed phenomena due to subsurface solar magnetic fields are inferred from the laws of magnetohydrodynamics. In sunspots the magnetic field lines are bundled and magnetic fields reach values of 2000 to 3000 gauss. The mean magnetic-field intensity measurable at the solar surface is only approximately 1 gauss. The small-scale features of magnetic activity on the solar surface are continuously changing with a degree of randomness as a result of complicated turbulent and ordered convective motions in the envelope of the Sun. The large-scale sunspot cycle, however, shows a well-defined behavior as a result of convection and generation of poloidal and toroidal magnetic fields within the differentially rotating Sun. Near the base of the convection zone the magnetic field may reach an amplitude of 10^5 gauss.

The existence and generation of magnetic fields in the deep interior of the solar body is still a very controversial issue. The generally accepted view is that the convective envelope of the Sun is a converter of turbulence and differential rotation into an oscillating magnetic toroid and dipole. The magnetic field is confined to the convective envelope and is generated there by a dynamo mechanism, thereby consuming energy liberated by thermonuclear reactions in the gravitationally stabilized fusion reactor of the Sun. Energy generated in the core of the Sun is used to drive convection and differential rotation in the envelope of the Sun. Dynamo models successfully explain the periodic amplification of the solar magnetic field and the observed butterfly diagram of sunspots, respectively. Almost all these models rely on assumptions that employ stochastic mechanisms for the explanation of the 22-year solar activity cycle (Stix, 1989).

Contrary to the stochastic approach to the generation of the solar magnetic field, it is possible in principle to explain the magnetic field as the result of the collapse of the primitive solar nebula. The radiative core of the Sun may have conserved its primordial magnetic field, locked into matter. It can be supposed that the radiative core of the Sun has a high electric conductivity conserving its low-order magnetic multipoles. Because a magnetic dipole existing in a fluid conductor is unstable towards a splitting along its symmetry planes and rotation about 180°, the dominant magnetic field in the core has a quadrupole configuration. This quadrupole model for the solar magnetic field could explain many of the observed solar magnetic activity phenomena, but has not yet been confirmed by observations (Kundt, 1992).

Solar thermonuclear energy generation

The Sun shines because of the process of fusion where four protons fuse to form an alpha particle α, two positrons (e$^+$ and two neutrinos (ν_e), that is, $4p \rightarrow \alpha + 2e^+ + 2\nu_e$. In this fusion process of hydrogen nuclei into helium nuclei, the latter also known as alpha particles, the fusion can be accomplished through two different series of principal reactions: 98.5% of the energy generation in the present-day Sun comes from the proton–proton chain (p–p chain); 1.5% of the solar

Table S9 Principal reactions of the proton–proton chain and the CNO cycle in the Sun

Number	Reaction	Termination (%)	Neutrino energy (MeV)
p–p chain			
1	$p + p \rightarrow {}^2H + e^+ + \nu_e$	99.75	0.420 (spectrum)
or			
2	$p + e^- + p \rightarrow {}^2H + \nu_e$	0.25	1.44 (line)
3	${}^2H + p \rightarrow {}^3He + \gamma$	100	
4	${}^3He + {}^3He \rightarrow {}^4He + 2p$	88	
or			
5	${}^3He + {}^4He \rightarrow {}^7Be + \gamma$	12	
6	${}^7Be + e^- \rightarrow {}^7Li + \nu_e$	99.98	0.861 (90%) 0.383 (10%) (both lines)
7	${}^7Li + p \rightarrow 2{}^4He$		
or			
8	${}^7Be + p \rightarrow {}^8B + \gamma$	0.02	
9	${}^8B \rightarrow {}^8Be^* + e^+ + \nu_e$		14.06 (spectrum)
10	${}^8Be^* \rightarrow 2{}^4He$		
CNO cycle			
1	${}^{12}C + {}^1H \rightarrow {}^{13}N + \gamma$		
2	${}^{13}N \rightarrow {}^{13}C + e^+ + \nu_e$		1.2 (spectrum)
3	${}^{13}C + {}^1H \rightarrow {}^{14}N + \gamma$		
4	${}^{14}N + {}^1H \rightarrow {}^{15}O + \gamma$		
5	${}^{15}O \rightarrow {}^{15}N + e^+ + \nu_e$		1.7 (spectrum)
6	${}^{15}N + {}^1H \rightarrow {}^{12}C + {}^4He$		

energy output is due to the carbon–nitrogen–oxygen cycle (CNO cycle). The p–p chain and the CNO cycle are shown in Table S9; there the third column indicates the percentage of the solar terminations of the p–p chain in each reaction. Since the dependence of the energy generation rate ϵ [compare equation (S22)] on the temperature is quite different between p–p chain and CNO cycle, the p–p chain dominates at low temperature ($T \leq 18 \times 10^6$ K) and the CNO cycle does not become important until a high temperature is reached. At the present stage in the evolution of the Sun, the CNO cycle is believed to play only a rather small role in the energy and neutrino production budget (Bahcall, 1989).

In the first reaction of the p–p chain, a proton decays into a neutron in the immediate vicinity of another proton. The two particles form a heavy variety of hydrogen known as deuterium, along with a positron and an electronneutrino. There is a second reaction in the p–p chain producing deuterium and a neutrino by involving two protons and an electron. This reaction (pep reaction) is 230 times less likely to occur in the solar core than the first reaction between two protons (pp reaction). The deuterium nucleus produced in the pp or pep reaction fuses with another proton to form helium 3 and a gamma ray. About 88% of the time, the p–p chain is completed when two helium 3 nuclei react to form a helium 4 nucleus and two protons, which may return to the beginning of the p–p chain. However, for 12% of the time a helium 3 nucleus fuses with a helium 4 nucleus to produce beryllium 7 and a gamma ray. In turn the beryllium 7 nucleus absorbs an electron and transmutes into lithium 7 and an electronneutrino. Only once for every 5000 completions of the p–p chain, beryllium 7 reacts with a proton to produce boron 8 which immediately decays into two helium 4 nuclei, a positron and an electronneutrino.

The net result of either the p–p chain or CNO cycle is the production of helium nuclei and minor abundances of heavier elements as ${}^7Be$, ${}^7Li$, ${}^8Be$ and ${}^8B$ (in the case of the p–p chain) or ${}^{13}N$, ${}^{14}N$ and ${}^{15}N$ (in the case of the CNO cycle). The energy generated by thermonuclear reactions in the form of gamma rays is streaming (actually, diffusing) toward the solar surface, thereby getting scattered, absorbed and re-emitted by nuclei and electrons. On their way outward, the high-energy gamma rays are progressively changed to x-rays, to extreme ultraviolet rays, to ultraviolet rays and finally emerge mainly as

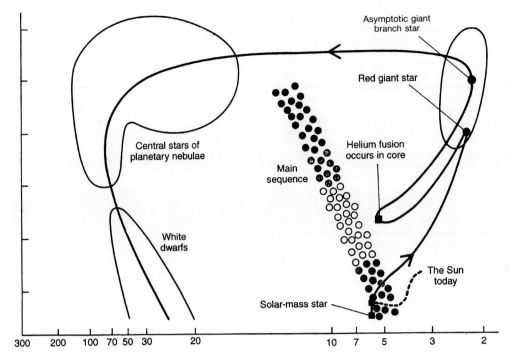

Figure S55 The path of a $1M_\odot$ star in the Hertzsprung–Russell diagram.

visible light from the solar surface and radiates into outer space. Only the weakly interacting neutrinos can leave the solar core with almost no interaction with solar matter. However, the chlorine experiment of Davis and collaborators, the Japanese Kamiokande experiment, and the GALLEX experiment at Gran Sasso to detect solar neutrinos show that the Sun emits fewer of these elusive particles than the standard solar model predicts (Iben, 1969; Lande, 1989; Hirata et al., 1991; Anselmann et al., 1992a,b). Since the beginning of the 1970s this deficit has challenged current understanding of solar and neutrino physics and of the process by which the Sun shines. The mystery of the missing solar neutrinos is commonly referred to as the 'solar neutrino problem' (Bahcall and Davis, 1982).

Solar evolution

The general evolution scheme of the Sun postulates a progressive contraction of gas by self-gravitation which is periodically interrupted by thermonuclear burning. After particular types of nuclear fuel (hydrogen and helium) are exhausted, the contraction-burning cycle will be repeated, but at higher temperatures. The stages of the Sun's evolution from primitive solar nebula contraction to the black dwarf stage can be followed in the Hertzsprung–Russell (H–R) diagram (Figure S55). There is a rapid movement of the Sun toward the main sequence, where the Sun spends the major part of its life, and then an eventual movement toward the black dwarf evolution stage, which is the final stage in its evolution (Schwarzschild, 1958; Gibson, 1973).

Presolar evolution stages

Cloud
Over 4.5 billion years ago, the gas cloud which would become the Sun had a diameter of over 480 trillion km, which is approximately 50 light years. This cloud was not dense, containing only a few thousand atoms per cubic centimeter of space. The total mass of the cloud would have been sufficient for building up several solar systems. Its temperature was that of interstellar space, of the order of 3 K, not radiating any light into the surrounding space. The fragile equilibrium state of the cloud, having only the choice of dissipating further into outer space or contracting into a denser configuration, eventually became disturbed either by an impact from outside or by random condensation of a large number of cloud particles, and finally began to condense.

Globule
After a period of the order of several thousand years, random concentrations of matter called globules formed at various places in the giant condensing matter cloud. The cloud collapsed almost in free fall; however, due to the influence of pressure, the motion was non-homogenous. The free-fall time of the cloud

$$t_{\text{ff cloud}} = \left(\frac{3\pi}{32G\rho_0}\right)^{1/2} \quad (S31)$$

where ρ_0 denotes the initial mean matter density of the cloud. The temperature in the cloud was rising very slowly, and was still not able to radiate light. Later, one of those globules, now having a dimension of several hundred solar systems, would become the Sun. The globule continued condensing, with the effect of increasing its temperature.

Protostar
Within 400 000 years the globule had condensed to a millionth of its original size, still over four times the size of the present solar system. At the center of the globule a core had developed, heated by the concentration of its matter, already able to radiate a substantial amount of energy into the less dense outer regions of the former globule. The emission of radiation by the core began to slow the condensation of its matter. The matter became opaque and free fall was stopped by the pressure. This core had now developed into a stable and well-defined configuration, called a protostar or proto-Sun. With the birth of the proto-Sun the evolution of this matter configuration advanced more rapidly. After the formation of a core, its free fall time is

$$t_{\text{ff core}} = \left(\frac{\pi^2}{8G}\frac{R^3}{M}\right)^{1/2} \quad (S32)$$

where M is the mass and R the radius of the core, respectively. Within a few thousand years it collapsed to a size of the diameter of the orbit of planet Mars. The interior temperature reached values of 56 000 K, leading to an ionization of atoms. The red light emitted at the surface of the proto-Sun was not produced by fusion of atomic nuclei but by gravitational contraction of matter. Gravitation released the potential energy of the globule, 7×10^{48} erg, during the condensation of the proto-Sun. According to the Virial theorem ($2T_k + \Omega = 0$) one half of the released gravitational energy Ω of the system was radiated from the proto-Sun while the other half was transformed into heat of the central core; T_k denotes the total kinetic energy of the particles.

Sun

Finally the proto-Sun contracted further until its temperature was high enough for the burning of deuterium to form helium 3. The Sun was fully convective in the contraction phase and the chemical composition was always uniform. Through deuterium burning the contraction was momentarily slowed down. As the Sun continued to contract, the central temperature increased and the radiative temperature gradient decreased relative to the convective gradient. Convection ceased and a radiative core grew outward. With the ignition of hydrogen the proto-Sun became a star, characterized by the gravitationally stabilized fusion reactor located at its center. Its binding gravitational energy $|\Omega| \approx GM_\odot^2/R_\odot$ was initially stored in the extended globule, called the primitive solar nebula. If the Sun were to shine by its store of thermal energy $|T_k| = |\Omega|/2$ (Virial theorem), then its lifetime would be given by the Kelvin–Helmholtz time scale, that is

$$t_{kw} = \frac{T_k}{L_\odot} \approx G\frac{M_\odot^2}{R_\odot L_\odot} \qquad (S33)$$

As the nuclear reactions began to release vast amounts of subatomic energy, the Sun was a quite variable star, varying in luminosity and surface activity as the result of the development of a radiative core and convective currents in its outer layers of gas. After a period of some 30 million years, its structure stabilized into the structure of a main-sequence star of one solar mass. The newly born Sun possessed enough fuel in the form of hydrogen to keep shining steadily for a time period of the order of

$$t_{nuc} = \frac{E_{nuc}}{L_\odot} \approx 10^{-3} \times \frac{M_\odot c^2}{L_\odot} \qquad (S34)$$

where the factor 10^{-3} is due to the product of the proportion of mass of the Sun available for hydrogen burning (0.1) and the fraction of mass converted into energy in hydrogen burning $\Delta m/4m_p \approx 0.01$, where Δm is the mass difference in the net reaction $4p \rightarrow \alpha + 2e^+ + 2\nu_e$.

That means also that the present Sun is right in the middle of its life span as a main-sequence star (4.5 billion years).

Postsolar evolution stages

Red giant

As the Sun ages, helium collects in its center. After a lifetime of 9 billion years as main-sequence star, approximately 10% of the hydrogen in the Sun's core will have been converted into helium and nuclear fusion reactions will cease producing energy. The equilibrium between the total pressure force directed outwards and the gravitational force directed towards the center of the Sun will be disturbed. The core of the Sun will start to collapse slowly under its own gravitational attraction. Fusion moves outward to a shell surrounding the core, where hydrogen-rich material is still present. The gravitational energy from the collapse will be converted into heat, causing the shell to burn vigorously and so the Sun's outer layers will swell immensely. The surface will then be far removed from the central energy source; it will cool and appear to glow red. The Sun will now evolve into the stage of a red giant. For a few hundred million years the expansion of the outer solar layers will continue, and the Sun will engulf the planet Mercury. The temperature on Venus and Earth will rise tremendously. Hydrogen fusion in the shell will continue to deposit helium 'ash' onto the core, which becomes even hotter and more massive.

In the Sun's core nuclear fusion of helium into carbon and oxygen will start to trigger even further the expansion of its outer layers. The helium-rich core is unable to lose heat fast enough and will become unstable. In a very short time of a few hours the core gets too hot and is forced to expand explosively. Outer layers of the Sun will absorb the core explosion but the core will no longer be able to produce energy by thermonuclear burning. Helium fusion then continues in a shell, and the structure of the Sun will look like an onion: an outer, hydrogen-fusion layer and an inner, helium-fusion layer which surrounds an inert core of carbon and oxygen.

The old Sun may repeat the cycle of shrinking and swelling several times. In this stage of evolution the Sun is called an asymptotic giant branch star. Finally enough carbon will accumulate in the core to prevent the core explosion. Helium-shell burning will add heat to the outer layers of the Sun, mainly containing hydrogen and helium. The asymptotic giant Sun will eventually generate an intense wind that begins to carry off its outer envelope. The precise mechanism behind this phenomenon is not yet well understood. The Sun will expand a final time and after about 30 million years it will swallow Venus and Earth; the outer layers will keep expanding outward and as much as half of the Sun's mass will be lost into space.

White dwarf

The solar core will keep shrinking and, because it is no longer able to produce radiation by fusion, the further evolution of this configuration is governed by gravitation. All matter will collapse into a small body about the size of the Earth. Thus the Sun will have become a white dwarf; this is a dense-matter configuration, having radiated away the energy of its collapse. Then the white dwarf will begin to cool rapidly.

Black dwarf

The final stage of solar evolution will be the black dwarf stage. The white dwarf will emit yellow light and then red light in the course of its evolution, drawing from the star's reservoir of thermal energy. Its nuclei will be packed as tightly as physically possible and no further collapse will be possible. The body is progressively cooling down and finally becomes as cold as the interstellar space around it, emitting no light at all. As a carbon–oxygen-rich black dwarf it will continue its journey through the galaxy (Milky Way) and may eventually encounter another giant gas cloud to become involved in the birth of a new star.

Hans. J. Haubold and A.M. Mathai

Bibliography

Allen, C.W. (1973) *Astrophysical Quantities*. London: Athlone Press.

Anselmann, P., Hampel, W., Heusser, G. *et al.* (1992a) Solar neutrinos observed by GALLEX at Gran Sasso. *Phys. Lett.*, **B285**, 376–89.

Anselmann, P., Hampel, W., Heusser, G. *et al.* (1992b) Implications of the GALLEX determination of the solar neutrino flux. *Phys. Lett.*, **B285**, 390–7.

Bahcall, J.N. and Davis, R., Jr (1982) An account of the development of the solar neutrino problem, in *Essays in Nuclear Astrophysics* (eds C.A. Barnes, D.D. Clayton and D.N. Schramm). Cambridge: Cambridge University Press, pp. 243–85.

Bahcall, J.N. (1989) *Neutrino Astrophysics*. Cambridge: Cambridge University Press.

Beer, J. (1987) Radioisotopes in natural archives: information about the history of the solar–terrestrial system, in *Solar–Terrestrial Relationships and the Earth Environment in the Last Millennia, Proc. International School of Physics 'Enrico Fermi', Course XCV*, (ed. G. Cini). Amsterdam: North Holland, pp. 183–98.

Bracewell, R.N. (1989) The solar cycle: a central-source wave theory. *Proc. Astron. Soc. Aust.*, **8**(2), 145–7.

Chandrasekhar, S. (1967) *An Introduction to the Study of Stellar Structure*. New York: Dover Publications.

Deubner, F.-L. and Gough, D. (1984) Helioseismology: oscillations as a diagnostic of the solar interior. *Ann. Rev. Astron. Astrophys.*, **22**, 593–619.

Dicke, R.H. and Goldenberg, H.M. (1967) Solar oblateness and general relativity. *Phys. Rev. Lett.*, **18**, 313–6.

Dicke, R.H., Kuhn, J.R. and Libbrecht, K.G. (1987) Is the solar oblateness variable? Measurements of 1985. *Astrophys. J.*, **318**, 451–8.

Fairbridge, R.W. (1987a) Sunspots, in *The Encyclopedia of Climatology* (eds J.E. Oliver and R.W. Fairbridge). New York: Van Nostrand Reinhold Company, pp. 815–23.

Fairbridge, R.W. (1987b) Climatic variation, historical record, in *The Encyclopedia of Climatology* (eds J.E. Oliver and R.W. Fairbridge). New York: Van Nostrand Reinhold Company. pp. 305–23.

Gibson, E.G. (1973) *The Quiet Sun*. NASA SP-303.

Guenther, D.B., Demarque, P., Kim, Y.-C. and Pinsonneault, M.H. (1992) Standard solar model. *Astrophys. J.*, **387**, 372–93.

Haubold, H.J. and Mathai, A.M. (1987) Analytical results connecting stellar structure parameters and neutrino fluxes. *Annal. Phys. (Leipzig)*, **44**(2), 103–16.

Haubold, H.J. and Mathai, A.M. (1992) Analytic stellar structure. *Astrophys. Space Sci.*, **197**, 153–61.

Herman, J.R. and Goldberg, R.A. (1978) *Sun, Weather, and Climate*. NASA SP-426.

Hill, H.A. and Stebbins, R.T. (1975) The intrinsic visual oblateness of the Sun. *Astrophys. J.*, **200**, 471–83.

Hill, H.A. and Kroll, R.J. (1992) Long-term solar variability and solar seismology: I, in *Basic Space Science* (eds H.J. Haubold and R.K. Khanna). New York: American Institute of Physics Conference Proceedings Vol. 245, pp. 170–80.

Hill, H.A., Oglesby, P. and Gu, Y.-M. (1992) Long-term solar variability and solar seismology: II, in *Basic Space Science* (eds H.J. Haubold and R.K. Khanna). New York: American Institute of Physics Conference Proceedings Vol. 245, pp. 181–92.

Hirata, K.S., Inoue, K., Ishida, T. *et al.* (1991) Real-time, directional measurement of ^{8}B solar neutrinos in the Kamiokande-II detector. *Phys. Rev.*, **D44**, 2241–60.

Iben, I., Jr (1969) The Cl^{37} solar neutrino experiment and the solar helium abundance. *Annal. Phys.*, **54**(1), 164–203.

Kavanagh, R.W. (1972) Reaction rates in the proton–proton chain, in *Cosmology, Fusion and Other Matters* (eds F. Reines). Boulder: Colorado Associated University Press, pp. 169–85.

Kourganoff, V. (1973) *Introduction to the Physics of Stellar Interiors*. Dordrecht: D. Reidel Publishing Company.

Kundt, W. (1992) The 22-year cycle of the Sun. *Astrophys. Space Sci.*, **187**; 75–85.

Lande, K. (1989) Status of solar neutrino observations and prospects for future experiments. *Annal. New York Acad. Sci.*, **571**, 553–60.

Lang, K.R. (1980) *Astrophysical Formulae*. New York: Springer-Verlag.

Longair, M. (1989) The new astrophysics, in *The New Physics*, (ed. P. Davies). Cambridge: Cambridge University Press, pp. 94–208.

Mathai, A.M. and Haubold, H.J. (1986) *Modern Problems in Nuclear and Neutrino Astrophysics*. Berlin: Akademie-Verlag.

Sackmann, I.-J., Boothrayd, A.I. and Fowler, W.A. (1990) Our Sun. I. The standard model: successes and failures. *Astrophys. J.*, **360**, 727–36.

Schatten, K.H. and Arking, A. (eds) (1990) *Climate Impact of Solar Variability*. NASA CP-3086.

Schou, J., Christensen-Dalsgaard, J. and Thompson, M.J. (1992) The resolving power of current helioseismic inversions for the Sun's internal rotation. *Astrophys. J.*, **385**, L59–62.

Schwarzschild, M. (1958) *Structure and Evolution of the Stars*. New York: Dover Publications.

Sears, R.L. (1964) Helium content and neutrino fluxes. *Astrophys. J.*, **140**, 477–84.

Sofia, S. (ed). (1981) *Variations of the solar Constant*. NASA CP-2191.

Sonett, C.P. (1984) Very long solar periods and the radiocarbon record. *Rev. Geophys. Space Phys.*, **22**(3), 239–54.

Stix, M. (1989) *The Sun*. Berlin: Springer-Verlag.

Weiss, W.W. and Schneider, H. (1991) Astroseismology. *The Messenger*, **66** (December), 36–40.

Zirin, H, (1988) *Astrophysics of the Sun*. Cambridge: Cambridge University Press.

Cross references

Galactic cycle
Solar activity
Solar motion
Solar neutrino
Solar photosphere
Solar system: origin
Ulysses mission

SURFACE GRAVITY

Surface gravity is the local gravitational acceleration (g) experienced by a freely falling body at the surface of a planet, moon or other celestial object. It may be calculated from Newton's law of gravitation:

$$g = \frac{GM}{R^2}$$

where M is the mass and R the radius of a spherical body, and G is the constant of gravitation (6.67×10^{-11} kg m^3 s^{-2}).

Knowledge of the mass and radius of an approximately spheroidal body permits the calculation of surface gravity for the body. The surface gravity of the Moon, for instance, with $M = 7.36 \times 10^{22}$ kg and $R = 1738$ km, is 1.67 m s^{-2}. This value is about one-fifth of that of the Earth. The gravitational acceleration at the surface of the Sun is on the other hand about 274 m s^{-2}.

The adopted value of g for the Earth is 9.80665 m s^{-2}. The rotational distortion of the Earth the (polar radius is 21 km less than the equatorial radius) gives rise to differences in measured values of g that vary with latitude. Theoretical gravity values are referred to the geoid, and corrections for elevation and mass intervening between the site of observation and the nominal geoid are applied. When such factors are accounted for, the remaining acceleration is known as the Bouguer anomaly. Gravity anomalies can provide information about the structures and the distribution of mass within the Earth and other planets.

James H. Shirley

Bibliography

Wilcox, L.E. (1989) Gravity anomalies: interpretation, in *Encyclopedia of Solid Earth Geophysics*, (ed. D.E. James). New York: Van Nostrand Reinhold, pp. 603–17.

Cross references

Geoid
Mars: gravity
Moon: gravity
Planetary geodesy
Venus: gravity

SURFACE PRESSURE

Pressure is the ratio of force to area. Surface pressure, or atmospheric pressure, is the force acting upon a unit area due to the weight of the overlying atmosphere. The surface pressure is a fluid force which exerts a compressive stress upon the surface.

Pressure may be measured with a number of scales. A common measure is the bar, equivalent to 100 kilopascals, 1 kg cm^{-2}, or 14.7 lb in^{-2}. Millibars are an often-employed measure of atmospheric pressure. Another common unit used in terrestrial meteorology is the standard atmospheric pressure ('atmospheres'), equivalent to 1.01325 bar or 760 column mm of mercury.

On Earth the surface pressure of about 1000 mbar corresponds to a mass of about 1 kg on each square centimeter of the planetary surface. For comparison, the surface pressure of the predominantly CO_2 atmosphere on Venus is about 90 000 mbar. Mars has a surface pressure of about 6 mbar, while Mercury's surface pressure is of the order of 10^{-9} mbar.

James H. Shirley

Bibliography

Considine, D.M. and Considine, G.D. (1989) Pressure, in *The Van Nostrand Reinhold Scientific Encyclopedia*, 7th edn. New York: Van Nostrand Reinhold.

Haman, S.D. (1957) *Physico-Chemical Effects of Pressure*. London: Butterworths.

Malone, T.F. (ed.) (1951) *Compendium of Meteorology*. Boston: American Meteorological Society.

Cross references

Atmosphere
Planet

SURFACE PROCESSES

Any natural (non-human) event or action that alters the shape of a planetary surface is referred to as a 'surface process.' All landforms are caused by one or more of the following types of surface processes: volcanism, tectonism, impact or gradation. The first two are derived from internal forces within a planet, the latter two are caused by agents external to the planet's surface.

Volcanism

Volcanism produces both constructional and destructional landforms. Many volcanic constructs are quite familiar; they have a wide range of morphologies depending primarily on the chemical composition and gas content of the material being extruded. Some lavas are very fluid (low silica content) and are erupted relatively quietly (molten rock is called lava on the surface and magma beneath the surface). Small eruptions produce isolated lava flows; larger eruptions can produce large volcanic shields like the Hawaiian Islands or the large Tharsis volcanoes on Mars, or even flat sheets that cover hundreds of square kilometers, like the Deccan Traps of India or the maria on the Moon. Magmas with a higher gas and/or silica content erupt more violently, producing cinder cones, spatter cones and other related features. Very silicic eruptions are explosive and produce little lava but large quantities of volcanic dust and ash. They form a combination of various constructional and destructional features; Mt St Helens in Washington state and Mt Pinatubo on the Philippines are examples. In other cases internal structures produced by volcanic intrusions become revealed by subsequent differential erosion. Examples include dikes, sills, and laccoliths. (See Volcanism in the solar system).

Tectonism

Tectonism, the fracturing or bending of the Earth's crust by internal forces, rarely acts alone to produce landforms. The fault scarp, or surface expression of relative movement along an earthquake fault, is the most notable example. Many tectonically produced internal structures are revealed by differential erosion, including fault-line scarps, where erosion has reversed the original topography caused by faulting, monoclines, or step-like bends, arch-like upfolds called anticlines, and synclines, or trough-like downfolds. Fractures along which no motion has occurred, called joints, can affect local erosion patterns.

Impact cratering

Impact craters are the dominant landform on planetary surfaces that have not undergone large-scale resurfacing for several billion years. Over 300 impact sites on Earth have been identified, but modification by other surface processes has largely buried, eroded or otherwise obliterated all but the very youngest; the best example being the famous Meteor Crater in Arizona. (See Impact cratering).

Gradation

The fourth general type of surface process, gradation, includes any natural process, erosional and/or depositional, that causes the alteration of a planetary surface due to the action of gravity alone or with the assistance of a fluid or gaseous transport agent. On Earth, gradation dominates the other surface processes because our atmosphere and hydrosphere are very active. Gradation also includes some biological (non-human) activity.

Mass wasting

The simplest type of gradation is termed mass wasting, in which there is no transport medium involved, only the downslope motion of material under the influence of gravity.

No internal deformation

In some types of mass wasting there is little or no internal deformation of the material in motion. Examples include rock fall, rock slide, debris fall, debris slide, and rotational slumps. The first four terms are self-explanatory; a rotational slump is the downslope movement of debris along a curved detachment zone. A tree or other vertically oriented object atop the slump would end up leaning uphill after the slump occurred. Groundwater can initiate and facilitate downslope movement for all five by acting as a lubricant.

Internal deformation

Another type of mass wasting involves the downslope movement of a mass of material that undergoes internal deformation during motion. Water, ice or even air can contribute significantly to the downslope movement, which can occur over a wide range of speeds. The slowest movement, soil creep, occurs at a rate of only a few millimeters per year and is caused by volumetric expansions and contractions of the uppermost soil layer due to temperature changes, freezing and thawing, or wetting and drying. Downslope motion of up to approximately 1 m per year can occur by solifluction or gelifluction if the ground is saturated either by water or ice respectively. In a rock glacier the bearing strength of coarse debris may be reduced by the presence of interstitial ice, resulting in downward movement that would not occur without the ice. Containing more water, and hence capable of flowing farther, are debris flows, comprised of 20–80% particles coarser than sand (~ 1 mm), and mudflows, in which ~ 80% of the particles are sand-sized and smaller.

Avalanche

The last class of mass wasting is avalanche, in which downhill speeds can exceed 50 m s^{-1}. The horizontal component of avalanche transport is much larger than the vertical drop traveled, implying that some sort of lubrication was present. Trapped air, snow and ice and ground water have all been proposed as lubricants for terrestrial avalanches. However, similar avalanche deposits have been found on the Moon and on Mars, which tends to discount the necessity of the lubricating agents mentioned, at least in those cases. Other possible explanations of the observed runouts include the avalanche behaving as a macroscopic gas due to elastic impacts and hence being very fluid, or that fluidization of the avalanche material occurs because large amounts of acoustic energy are released during avalanche motion.

Transport media

Much of the modification of the surface of the Earth by gradation is directly or indirectly caused by the flow of a transport medium, either over the surface or beneath it.

Subsurface water and ice

Periglacial surface features are caused by the subsurface migration of water and the formation of ground ice. One example are ice-cored mounds called pingos, which can form in two ways. If an enclosed basin of water-saturated silt freezes, the migration of the freezing front from the top, bottom and sides of the basin and expansion of the ice during freezing can concentrate the segregating ice into a large mass that heaves the surface above it. If the area of freezing is not confined, a pingo may still form because water tends to migrate to a zone of freezing, with the resulting pressure increase causing surface disruption and pingo formation. Another example of periglacial surface modification are features collectively termed thermokarst, in which the melting of segregated masses of ground ice undermines and disrupts the surface. Patterned ground is a third example, referring to a set of fractures, surface sorting and other features caused by repeated temperature-induced expansion, contraction and fracturing of the surface and attendant subsurface ice formation.

Subsurface water

A variety of surface features can result from the internal movement of ground water. Sapping is the undermining of an impermeable surface layer by the release of ground water from a permeable zone beneath it causing basal erosion. Sapping channels have distinctively flat floors and stubby tributaries that have ampitheater-shaped terminations. They are common in the American Southwest and have also been observed on Mars. Another surface expression of subsurface water flow is known as karst topography. Karst is caused by the dissolution of limestone by slightly acidic ground water, which forms dolines, a term for a variety of sinkholes, disrupted drainage patterns and other surface features with many different names. Dissolution of limestone can also form tower karst landscapes, typified by sharp, isolated pinnacles and steep mountains. A notable example of tower karst is found in the Guangxi Province of China.

Overland flow of ice

Glaciers produce a wide variety of distinctive erosional and depositional landforms, depending on the type of glacier in question, its rate of flow and the nature of the underlying topography. The grinding action and debris removal of some glaciers alters the shape of valley through which they flow to have a U-shaped cross-section; Yosemite Valley is a famous example. Erosion at the head of glaciers and their tributaries produces bowl-shaped indentations called cirques. As erosion continues from all sides of a pre-existing group of mountains, cirques tend to come together, forming mountain passes called cols, knife-edge ridges called aretes, and isolated peaks called horns. The European Alps contain many examples of each, including the famous Matterhorn. On a smaller scale, glaciers erode the surface over which they flow by plucking, where portions of the ground are frozen fast in the base of the glacier and are ripped out by its motion, and by glacial abrasion, where rocks imbedded in the base of the glacier gouge out grooves and striations as they move over the underlying rock.

Glaciers also produce many distinctive deposits. Ablation of ice at the margins of the glacier leaves debris, termed glacial drift or till, in piles called moraines. A moving glacier can pile up drift and underlying material into large streamlined hills called drumlins. Large quantities of drift can be washed into fractures in the disintegrating ice, forming kames. If a block of glacial ice is left behind in a moraine, it can form a small basin called a kettle when it melts. Meltwater flowing beneath a glacier often eroded a sinuous channel in the ice and deposits drift in a feature called an esker. The large quantity of material removed from beneath a glacier by water flowing beneath it can create a large, flat area in front of the glacier called an outwash plain.

Overland flow of water

One of the most effective types of terrestrial gradation is that caused by running water. Different erosional and depositional landforms occur in different local environments, depending on the relative quantities of material to be moved and the amount of water available, pre-existing landforms and the local topographic gradient. For example, alluvial fans and cones are formed where the amount of material to be moved greatly exceeds the capacity of the water to carry it. They are the result of the debouching of high-gradient streams, capable of carrying large quantities of coarse sediment, onto flatter surfaces where the water slows and dissipates. Only a small portion of an alluvial fan may have water flowing on its surface. The coarsest sediments are deposited first, closest to the head of the fan. Progressively finer sediments drop out farther down the fan, with only muds and dissolved salts reaching the distal ends of the fan.

If the topographic gradient is uniformly relatively low, conventional rivers and streams and their associated landforms result. Erosional features include, on the small scale, steep-side rills, gullies and, arroyos. Larger erosional valleys have a characteristic, V-shaped cross-section. Rivers form meanders when the sediment carrying capacity of the water is not exceeded. The meanders migrate over time, forming a variety of features, including point-bar deposits at the inside bends of the meanders, cut banks on the outside bends of the meanders, natural levees formed by overbank flooding, and cut-off meander loops, called oxbow lakes because of their shape. If the river is underfit for its sediment load, the river channel is braided rather than meandering. The Mississippi is a good example of a meandering river; the Platte is a good example of a braided river.

Standing water

When a river or stream enters a body of standing water, its flow speed and sediment carrying capacity decrease significantly, causing the formation of a depositional feature called a delta. As in the case of the alluvial fan, coarser material is deposited nearer the mouth of the stream, and progressively finer material farther from the mouth. Deep ocean basins are usually too far from river sediment sources to allow much deposition of clastic sediments; in those locations the fallout of carbonate shell material from foraminifera, single-celled organisms that live near the sea surface, ultimately leads to the formation of limestone, rather than the shales, siltstones and sandstones that form closer to shore. Submarine landslides can set bottom sediments into motion. Those sediments and the water in which they are suspended have a higher density than the surrounding water, and they can flow great distances down relatively shallow slopes. Such flows are called turbidity currents, and they can deliver coarse sediments to much greater depths than they would normally be deposited, producing distinctive features called turbidites. An excellent example of a turbidite sequence can be found on the Marin Headlands immediately north of the Golden Gate Bridge in San Francisco.

Large lakes and oceans can have currents and surface waves of sufficient strength to cause substantial shore erosion. Wave action can form wave-cut benches upon which beaches are built, wave-cut notches in the base of cliffs on shore, natural sea arches and isolated pinnacles called sea stacks. Lateral drift of sand and other material down the coastline can form sand spits and bars, connect offshore islands to the coast by sand bars called tombolos, and sculpt the coast into zetaform shapes, large-scale features that resemble a heart cut in half.

Wind (eolian) processes

The wind is capable of creating distinctive depositional and erosional landforms. A wide variety of types of dunes and other eolian deposits have been identified; their morphology is a function of particle size, the direction of prevailing wind(s), vegetation, topographic setting and other factors. The wind is also a powerful agent of erosion, producing deflation hollows, large streamlined hills called yardangs, and ventifacts, stones sculpted by the blasting action of sand-laden winds. Dunes, yardangs and other eolian features have also been found on Mars, Venus and Triton, the largest satellite of Neptune. (See Eolian transport)

Biological activity

A final type of surface modification is due to the action of (non-human) biological agents. The most significant are organic reefs, masses of biologically produced limestone that can be extremely large, produced by the action of colonial corals and associated calcareous algae. Reefs can extend seaward from a coast (fringing reef) or be separated form the coast by a lagoon (barrier reef) and can take on a variety of morphologies. Reef production is inhibited or does not occur in deep, cold or fresh water.

Steven H. Williams

Bibliography

Bloom, A.L. (1991) *Geomorphology: A Systematic Analysis of Late Cenozoic Landforms*, 2nd edn. Englewood Cliffs: Prentice Hall.
Cooke, R.U. and Warren, A. (1973) *Geomorphology in Deserts*. Berkeley: University of California Press.
French, H.M. (1976) *The Periglacial Environment*. London: Longman Group.
Greeley, R. (1985) *Planetary Landscapes*. London: Allen and Unwin.
Greeley, R. and Iversen, J.D. (1985) *Wind as a Geologic Process*. Cambridge: Cambridge University Press.
Press, F. and Siever, R. (1978) *Earth*, 2nd edn. San Francisco: Freeman and Co.
Thomas, D.S.G. (ed.) (1989) *Arid Zone Geomorphology*. London: Belhaven Press.

Cross references

Eolian transport
Impact cratering
Plate tectonics
Volcanism in the solar system

SURVEYOR MISSIONS

The Surveyor program was one of NASA's two robotic spacecraft programs mounted in the late 1960s to provide the information necessary for planning the landings of humans on the Moon. The other program comprised the Lunar Orbiters (q.v.).

The Surveyor program was directed by the Jet Propulsion Laboratory. Its objective was to achieve soft landings on the Moon by spacecraft capable of transmitting scientific and engineering measurements from the lunar surface (US Surveyor Program Office, 1969). Five of the seven Surveyors launched landed successfully, returning a total of more than 76 000 high-resolution television images of their surroundings, which provided the data for the selection of Apollo landing sites.

Surveyor 1 landed on the Moon on 2 June 1966 in the southwest portion of Oceanus Procellarum. It returned data on the bearing strength, temperatures and radar reflectivity of the lunar surface, along with 11 240 photographs. Surveyor 3 carried a remote-controlled scoop that could dig into the surface. This spacecraft was visited 30 months later by the Apollo 12 astronauts. Surveyor 5 carried an alpha-particle backscattering instrument that analyzed the chemical composition of the surface material and several small magnets to which the material was found to adhere. Surveyor 6 performed a short, controlled rocket burn to move the spacecraft about 8 ft (2.5 m) and observe the effects on the regolith. Surveyor 7 landed on 9 January 1968 in a highland area north of the crater Tycho, a location chosen for its scientific interest but not a candidate Apollo landing site.

Conway W. Snyder and James H. Shirley

Bibliography

US Surveyor Program Office (1969) *Surveyor Program Results*. Washington DC: NASA.

Cross references

Apollo missions
Lunar Orbiter missions
NASA

SYNERGETIC TIDAL FORCE

The word 'synergetic' is from the Greek *syn* (together) + *ergon* (work), hence *synergetikos* 'working together, cooperative.' A slight distinction exists between the word 'synergetic' and 'synergistic' – a term also used scientifically in medicine, anatomy, physiology and pharmacology. The latter use implies a net gain through the combined action of two muscles, hormones, chemical agents, etc., that is greater than the sum of the constituent elements acting independently. 'Synergetic,' while connoting the simultaneous, motion-impelling action of multiple forces, makes allowance for the vector nature of forces in the physical sciences. The expression 'synergetic tidal forces' implies the concurrent action of the gravitational forces of the Moon and Sun exerted upon the Earth's hydrosphere. Tidal changes occur therein which are usually the result of the mutual enhancement of codirectional force components. The net, resultant force may be either amplified or reduced with respect to the average value present.

Spring and neap tides

Two times during each month, at new Moon (conjunction) and full Moon (opposition), the Earth, Moon and Sun come into direct alignment in celestial longitude. Either position is known as syzygy (Figure S56a). In the additive combination of both the lunar and solar gravitational forces, enhanced tide-raising forces result, producing spring tides, usually twice each month. The Moon and Sun produce counteracting forces at times of lunar quadrature, resulting in neap tides.

Perigean spring tides

Since the Moon's orbit around the Earth is an ellipse, and the Earth is at an off-center focus, once each revolution the Moon attains its closest monthly approach to the Earth, a position known as perigee. The passage of the Moon through perigee and the alignment of Moon, Earth and Sun at new moon or full moon ordinarily do not take place at the same time. Commensurable relationships between the lengths of the synodic and anomalistic months do, however, make this circumstance possible from time to time. At approximately 6-month intervals, perigee and syzygy occur within 1½ days of each other (Figure S56b), producing perigean spring tides.

Proxigean and extreme proxigean spring tides

Exceptionally, such alignments between perigee and syzygy may occur within a few hours. At such times, such as when the Sun is aligned with the Moon's line of apsides in the lunar evectional cycle (Table S10), the eccentricity of the lunar orbit is increased, as are the lunar parallax and the Moon's orbital velocity. At times, solar-induced perturbations of the lunar orbit also reduce the Moon's perigee distance from the Earth (Figure S56b); the amount fluctuates

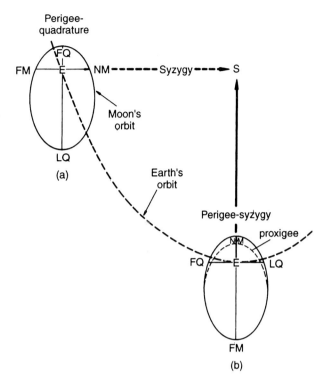

Figure S56 (a) The astronomical relationships producing ordinary spring tides, by a direct alignment in celestial longitude (termed 'syzygy') between Sun, Earth and either new or full Moon. The reinforcement of the Moon's gravitational force by that of the Sun increases the tides by 20% above the average high waters. The lunar perigee (position of closest monthly approach of the Moon, revolving in an elliptical orbit around the Earth) occurs nearer to either the first- or last-quarter lunar phase. (b) The astronomical relations in perigean and proxigean spring tides: these augmented tides likewise occur at syzygy, but involve a simultaneous alignment between the position of perigee and that of either new or full Moon. This dual alignment of perigee-syzygy raises high waters 40% above their mean values. A very close lunar line of apsides and solar force alignment, indicated by a very small perigee-syzygy separation interval, may also alter the shape of the lunar orbit, reduce the Moon's perigee distance from Earth and lift the high waters to exceptional levels, in the phenomenon of proxigee-syzygy.

throughout the year and is always greater the closer is the alignment between perigee and syzygy. The tide-raising force (q.v.) varies inversely as the cube of the distance between the Earth and the Moon (or Sun). This is because tides occur at the surface of the Earth rather than its center, the point designated in Newton's inverse square law of gravitation. On certain occasions, lunar passage through perigee involves a particularly close approach of the Moon to the Earth. These cases are called proxigee, causing tides of considerably increased amplitude and range, and even, where conditions justify, extreme proxigean spring tides. The forces may be further enhanced, in synergetic combination, by the Earth's annual proximity to the Sun at perihelion, as well as at coplanar alignment with the Sun and Moon at a solar or lunar eclipse (Wood, 1991).

Apogean neap tides

A situation opposite to that created by the additional tide-augmenting effects of perigee-syzygy and proxigee-syzygy occurs when the Moon is simultaneously at quadrature (first- or third-quarter phase) and at apogee. At this time the gravitational forces of the Sun and Moon combine vectorially to decrease the resultant force, and apogean neap tides are produced.

Approximately 2 weeks before or after the Moon reaches its closest possible approach to Earth (at extreme proxigee) in an orbit of

Table S10 Significant short- and long-period relationships between (1) the lunar evectional cycle (controlled by near-coincidence of the Sun with the lunar line of apsides) and (2) the occurrence of proxigee-syzygy and extreme proxigee-syzygy alignments associated with strongly enhanced luni-solar tidal forces upon the Earth

Decimal and calendar dates of proxigee-syzygy (UT)	Evectional cycle spans (years)	Days from perihelion	Mean anomaly (A) of Moon (deg)	Mean elongation (E) of Moon (deg)	Sun's angle from lunar line of apsides[a]	Lunar distance[b] at proxigee (km)	Parallax[c] value; class (min/sec)	Current alignment less 93.020 years	Matching proxigee-syzygy alignment date (UT)	Parallax[c] value; class (min/sec)
Series 1 (FM)										
CF 1972.890[a] 21 November, 0000		−43	358.87	178.96	(360−A)+(E−180) 0.09	356 441	61 30.1 ExPrS[f]	1879.870	1879.868 14 November, 0400	61 27.8 PrS
CF 1974.021 8 January, 1100	1.131	+4	0.42	179.07	(180+A)−E 1.35	356 450	61 30.0 ExPrS	1881.001	1880.998 31 December, 1500	61 28.0 PrS
1975.151 25 February, 2200	1.130	+54	2.80	179.95	2.85	356 431	61 30.2 ExPrS	1882.131	1882.132 18 February, 0100	61 28.7 PrS
Series 2 (NM)										
1977.939 10 December, 2300		−24	356.03	2.32	(360−A)+E 6.28	356 634	61 28.1 PrS	1884.919	1884.923 3 December, 0300	61 28.8 PrS
CF 1979.075 28 January, 1000	1.136	+24	358.40	3.19	4.79	356 653	61 27.9 PrS	1886.055	1886.054 20 January, 1300	61 29.4 ExPrS
Series 3 (FM)										
CF 1980.807[e] 23 October 1400		−70	3.24	173.65	(180+A)−E 9.59	356 673	61 27.7 PrS	1887.787	1887.791 16 October, 1800	61 26.0 PeS
CF 1981.939 11 December, 0000	1.132	−25	4.80	173.77	11.03	356 692	61 27.5 PrS	1888.919	1888.923 3 December, 0400	61 26.7 PrS
Series 4 (FM)										
CF 1990.918 2 December 1100		−32	356.56	180.37	(360−A)+(E−180) 2.81	356 441	61 30.1 ExPrS	1897.898	1897.897 24 November, 1500	62 27.5 PrS
1992.052[e] 19 January, 2200	1.134	+16	359.93	181.25	1.32	256 450	61 30.0 ExPrS	1899.032	1899.030 12 January, 0200	61 27.7 PrS
1993.181 8 March, 0800	1.129	+64	0.67	180.60	(180+A)−E 0.07	356 441	61 30.1 ExPrS	1900.161	1900.163 1 March, 1200	61 28.2 PrS

[a] Lunar orbital inclination to ecliptic is only 5°9′.
[b] Mean distance of Moon from Earth = 384 400 km.
[c] Mean lunar parallax = 57′ 02.608″.
[d] CF = Coastal flooding occurred.
[e] Leap year.
[f] Ex = extreme; PrS = proxigee-syzygy.

maximized eccentricity (Table S10), the Moon's greatest orbital separation from the Earth, known as exogee, is reached at the position of exogee-syzygy.

Short-period cycles

In Table S10 the importance of the short-period lunar evectional cycle (cycle A) to the tide-augmenting circumstances of a proxigee-syzygy alignment is indicated. Ten representative instances (in 1972–1993) possessing strong lunar evectional influences typify four different sequences of lunar evectional cycles. The average of these cycles is 1.134 years (column 2). Detailed explanations of the phenomenon of lunar evection may be found in Wood (1986, p. 154, 1990, 1991), together with additional examples of the type of tidal reinforcement under consideration.

Out of eight cases of proxigee-syzygy and extreme proxigee-syzygy alignments during 1979–1993, seven have been associated with strong, persistent onshore winds, and accompanied by tidal flooding on either the east or west (or both) coasts of the United States. In the 1979 example, ocean-floor current measurements obtained simultaneously in Scripps submarine canyon revealed current velocities seven times greater than ever before experienced, even at spring tides (Shepard, Sullivan and Wood, 1981). More than 100 other examples of coastal flooding accompanying perigean and proxigean spring tides, occurring worldwide over a 358-year period of record, have been investigated (Wood, 1986).

All the examples in Table S10 are instances involving either proxigee-syzygy or extreme proxigee-syzygy alignments. These result, in each case, in approaches of the Moon to the Earth to less than 356 692 km (the mean distance is 384 400 km), with lunar parallaxes greater than 61′ 27.5″. The contribution provided by the Sun to the total tidal force in these instances is shown as a result of the Earth's proximity to perihelion to be in no case in excess of 70 days. The differences between (1) the Moon's mean anomaly (its angular distance in orbit from perigee) and (2) the Moon's mean elongation (its angular distance in celestial longitude from the Sun) are indicated. These differences, projected into the correct trigonometric quadrant, and accounting for the existing phase of new moon or full moon, yield the Sun's angle (column 2) with respect to the lunar line of apsides. It is the closeness of this angle to 0° that is of importance to the tide-raising potential. A direct alignment between the Sun and the lunar line of apsides increases the eccentricity of the lunar orbit and results in considerably reduced distances of the Moon from Earth, the functional characteristics of proxigean spring and extreme proxigean spring tides. In Table S10 there is no single case when this angle was greater than 11.03°, and generally it was much smaller.

Long-period cycles

The pertinent short- and intermediate-period cycles that are responsible for the astronomical alignments producing greatly augmented tidal forces and potential tidal flooding – and which, significantly, are contained as individual or multiple components in all subsequent long-period cycles – are (1) the previously mentioned lunar evectional cycle, $A = 1.132$ years (average); and (2) the principal perigee-syzygy cycle, $B = 18.030$ years $= 16A - \frac{1}{14}A$. A typical example of the latter cycle exists in the exact 18.030-year interval between the major tidal floodings of 29 December 1959 (1959.991) and 9 January 1978 (1978.021). Each of these floodings occurred on both the east and west coasts of the United States within 24 h of a common epoch. A dynamically meaningful 3 days after tidal flooding struck the US East Coast, major flooding occurred along the coasts of England, Belgium, France and the Netherlands (Wylie, 1979, p. 71).

The first significant combination of the A and B cycles (that having the smallest period) is $C = 10.5A + 1.5B = 31.007$ years. In this case the 0.5 portion of each coefficient indicates that a syzygy phase shift is involved, from new Moon at the beginning of the cycle to full Moon at the end, or vice versa. In a limited sense, the 31.007-year cycle (and its integral multiples) might also be defined as a perihelion cycle.

Numerous examples exist linking alignments of proxigee-syzygy involving this period. Notable tidal flooding accompanied the instances of 4 March 1931 (1931.170) and 6 March 1962 (1962.175). In the 1962 instance the total damage from South Carolina to Maine was estimated at $0.5 billion. The damage occurred over a 62-h period, throughout five successive high tides, coupled with strong onshore winds (O'Brien and Johnson, 1963).

The significance of the 31.007-year cycle was predicted (Wood, 1978, pp. 326, 488) in connection with the date 8 March 1993 – following the tidal flooding events of 4 March 1931 and 6 March 1962. The 1993 alignment is noteworthy because of (1) the large lunar parallax (61′ 30.0″ – only 0.9″ below the maximum theoretically possible), (2) the close coplanar alignment in declination between the Sun and Moon (within 0.4°) and (3) the close separation interval between proxigee and syzygy (-2 h) – all factors conducive to greatly amplified astronomical tides. The 3× multiple of cycle C gives $C3 = 93.021 = 10.5A + 4.5B$, again with half-value coefficients and an alternation of syzygy phase. This meaningful cycle combines multiples of both the synodic and anomalistic months (1150.5 cycles of 29.530589 days = 93.02031024 years; 1233 cycles of 27.554551 days = 93.01981396 years). It also possesses a close agreement with five times the lunar nodical cycle of 18.6134 years (= 93.0670 years). The relationship between this cycle and the occurrence of both proxigee-syzygy and extreme proxigee-syzygy alignments is apparent from column 11 of Table S10.

In summary, therefore, it is evident that astronomically meaningful, cyclical relationships have been established between such synergetically amplified tides at locations widely separated in latitude and longitude, in totally different tidal regimes and over centuries of time. Measurements of local sea level, however, are entirely relative, being gauged with respect to coastal land surfaces which themselves may be subject to either subsidence or uplift from natural and artificial causes (Harris, 1981; Pugh, 1987; Emery and Aubrey, 1991).

Fergus J. Wood

Bibliography

Dolan, R., Inman, D.L. and Hayden, B. (1990) The Atlantic coast storm of March 1989. *J. Coastal Res.*, **6**, 721–5.

Emery, K.O. and Aubrey, D.G. (1991) *Sea levels, Land Levels, and Tide Gauges*. New York: Springer-Verlag, 237 pp.

Harris, D.L. (1981) *Tides and Tidal Datums in the United States*. US Army Corps of Engineers, Spec. Rep. No. 7, February 1981. Washington, DC: US Government Printing Office, 382 pp.

O'Brien, M.P. and Johnson, J.W. (1963) The March 1962 storm on the Atlantic coast of the United States, in *Proceedings, VIIIth Conference on Coastal Engineering*, Council on Wave Research. Richmond, VA: The Engineering Foundation.

Pugh, D.T. (1987) *Tides, Surges and Mean Sea Level*. New York: Wiley, 472 pp.

Shepard, F.P., Sullivan, G.G. and Wood, F.J. (1981) Greatly accelerated currents in submarine canyon head during optimum astronomical tide-producing conditions. *Shore and Beach*, **49**(1), 32–4.

US Weather Bureau (now National Weather Service, NOAA) (1931) *Monthly Weather Rev.*, **59**(3), 127.

US Weather Bureau (now National Weather Service, NOAA) (1959) *Monthly Weather Rev.*, **87**(12), 457; *Storm Data*, **1**(12), 120–1.

Wood, F.J. (1978) The strategic role of perigean spring tides, in *Nautical History and North American Coastal Flooding, 1635–1976*. Washington, DC: US Government Printing Office, 538 pp.

Wood, F.J. (1981) Astronomical and tidal analyses of unusual currents in a submarine canyon during proxigee-syzygy alignment. *Shore and Beach*, **49**(1), 35–6.

Wood, F.J. 1986. *Tidal Dynamics–Coastal Flooding, and Cycles of Gravitational Force*. Dordrecht and Boston: D. Reidel, 712 pp.

Wood, F.J. (1990) Cyclical astronomical 'supertides' – a coastal flooding caveat. *Cycles*, **41**(6), 305–15.

Wood, F.J. (1991) Short- and long-period cycles in proxigean spring tides–their role in prognostication of coastal flooding. *Cycles*, **42**(2), 68–83.

Wylie, F.E. (1979) High winds, high tides, in *Tides and the Pull of the Moon*. Brattleboro, Vt: Stephen Greene Press, 246 pp., Chapter 5.

Cross references

Earth–Moon system: dynamics
Hydrosphere
Tide-raising force

SYNODIC PERIOD

The synodic period of a planet is the interval between two consecutive oppositions (for a superior planet) or inferior conjunctions (for an inferior planet) with the Sun.

The synodic period P_{syn} of a planet in years is related to its sidereal period P_{sid} in years by

$$\frac{1}{P_{syn}} = \pm \left(\frac{1}{P_{sid}} - 1\right)$$

where the + sign is to be used for inferior planets, and the − sign for superior planets. Mars, at 780 days, has the longest synodic period of all the planets. The synodic period of Pluto is hardly more than 1 year.

Likewise the synodic period of the Moon (from full Moon to full Moon, 29.5 days) is distinguished from its sidereal period (with respect to the stars) of 27.3 days;

Again, although the sidereal period of rotation of the Sun is taken to be 25.38 days at its equator, for the purposes of calculating heliographic coordinates (see Coordinate systems), its synodic period (i.e. as seen from the Earth, which is in orbit around the Sun) is 27.28 days at its equator.

Jeremy B. Tatum

Bibliography

Green, R.M. (1985) *Spherical Astronomy*. New York: Cambridge University Press.
Woolard, E.W. and Clemence, G.M. (1966) *Spherical Astronomy*. New York: Academic Press.

SYZYGY

The Sun, Earth and a planet or the Moon are said to be in syzygy when they are roughly in a straight line, as when a planet is in opposition to or in conjunction with the Sun, or the Moon is new or full. The word is used more often by crossword compilers than by astronomers (but see Synergetic tidal force).

Jeremy B. Tatum

Bibliography

Green, R.M. (1985) *Spherical Astronomy*. New York: Cambridge University Press.
Wood, F.J. (1986) *Tidal Dynamics – Coastal Flooding, and Cycles of Gravitational Force*. Dordrecht and Boston: D. Reidel, 712 pp.
Woolard, E.W. and Clemence, G.M. (1966) *Spherical Astronomy*. New York: Academic Press.

Cross reference

Synergetic tidal force

T

TECTONICS

A subdiscipline of geology dealing with the deformation of the crust of any celestial body possessing a 'solid' (elasticobrittle) crust. Sometimes used as a synonym for structural geology, the collective processes involved are called tectonism. These terms have the same classical roots as 'architecture', from Greek *tektonikos* (pertaining to a builder), and Latin *tectonicus*. On planet Earth the overall system of crustal deformation is in the form of plate tectonics (q.v.), but that appears to be unique to this planet (Plate 10). The inner planets Mercury and Mars seem to be 'one-plate' bodies, as also is the Moon. On Mars both faulting and folding are observed, as well as a tectonic history (Carr, 1974; Scott and Dohm, 1990).

The root term is also employed with classical prefixes and suffixes, thus megatectonics, (giant scale), geotectonics (global or large scale), microtectonics (small scale, < 1 m, or requiring microscopic study), morphotectonics (structure reflected by surface morphology, or geomorphology), neotectonics (deformation due to 'contemporary' crustal dynamics, spanning the last 1 to 10 million years), seismotectonics (structures, particularly faults, associated with earthquakes and contemporary seismic activity), tectonophysics (structures and their deforming forces analyzed purely in terms of geophysics), tectonosphere (sometimes used for the solid layer above the asthenosphere (q.v.), tectonostratigraphy (the correlation or description of rock strata by structural criteria) and tectonoeustasy (global sea-level change controlled by deformation of ocean basins or sea-floor spreading).

A term for tectonic processes in general is diastrophism, commonly employed for global-scale deformation. The expression 'non-diastrophic structures' is used for small-scale intraformational folds and brecciation due to gravitational sliding or volume changes during the sedimentation process (Fairbridge, 1946).

Three major categories are recognized in tectonics, based on the dynamic geometry (James, 1989, p. 1224): compressional (in a horizontal sense), up- or downwarping (in a vertical sense), and fracturing and faulting (in extensional, or strike-slip sense), as follows.

1. Orogeny (Gilbert, 1890; Dennis, 1982), compressional folding (Cadell, 1890), literally the process of mountain building (Greek: *oros*, mountain), with orogenesis for the collective concept. A sediment-draped continental margin is the site of future orogeny and called an orogen or a geosyncline. Superficial folding is called 'thin-skinned' (detachment or décollement), in contrast to deep-seated deformation (Chapple, 1978).
2. Epeirogeny (collectively, epeirogenesis), proposed by Gilbert (1890) for vertical movements of the Earth's crust (from the Greek, *epeiros*, continent). A type example is the Colorado Plateau, which consists mainly of flat-lying marine sediments, vertically uplifted.
3. Taphrogeny (adjective, taphrogenic; collective noun, taphrogenesis; Krenkel, 1925) used for large-scale fracturing and rifting (graben,) such as in the Jordan/Red Sea–East African rift system (Illies, 1981; Quennell, 1982, 1985). The west coast of North America from the Gulf of California (Mexico) to Cape Mendocino is described as a 'taphrogenic coast', characterized by innumerable block-faulted sectors (Fairbridge, 1992). Major transposition of crustal slivers create 'exotic terranes' and rotation creates anomalous orientations (as in the 'boot' of Italy). Widespread patterns of fracture (called 'lineaments' on Earth; Hills, 1963) are also evident on Mars, the Moon and other natural satellites.

In Earth history tectonic activity is marked by crescendos at discrete times that reflect accelerations in plate tectonic activity (sea-floor spreading and subduction). They have received chronologic names in a hierarchy of terms: 'revolution' for the major, long drawn-out orogenic intervals; 'phase' for the more localized and time-constrained events (Bucher, 1933). For example: Alpine ('Alpide', 10–50 Ma), Variscan or Hercynian ('Variscide' or 'Hercynide', 260–300 Ma): Caledonian ('Caledonide', 400–440 Ma), Baikalian, Assyntic or Pan-African ('Baikalide', 560–600 Ma). Names have also been allotted to the individual orogenic phases (Stille's 'orogenic chronology law'), which are useful for regional studies (e.g. Bucher, 1993).

Rhodes W. Fairbridge

Bibliography

Bucher, W.H. (1933) *Deformation of the Earth's Crust*. Princeton: Princeton University Press, 518 pp.
Cadell, H.M. (1890) Experimental researches in mountain building. *Roy. Soc. Edinburgh, Trans.*, **35**, 337–57.
Carr, M.H. (1974) Tectonism and volcanism of the Tharsis region of Mars. *J. Geophys. Res.*, **79**, 3943–9 (reprinted in Garfunkel, Z., 1985, *Mantle Flow and Plate Theory*, New York: Van Nostrand Reinhold, pp. 168–74).
Chapple, W.M. (1978) Mechanics of thin-skinned fold-and-thrust belts. *Geol. Soc. Am. Bull.*, **89**, 1189–98.
Dennis, J.G. (ed.) (1982) *Orogeny*. Stroudsburg Pa.: Hutchinson Ross Publ. Co. (Benchmark Papers in Geology, Vol. 62), 379 pp.
Dennis, J.G. and Atwater, T.M. (1974) Terminology of geodynamics. *Am. Assoc. Petroleum Geol. Bull.*, **58**, 1030–6.
Dewey, J. (1972) Plate tectonics. *Sci. Am.*, **226**(5), 56–68.
Fairbridge, R.W. (1946) Submarine slumping and location of oil bodies. *Am. Assoc. Petroleum Geol. Bull.*, **30**, 84–92.
Fairbridge, R.W. (1992) Holocene marine coastal evolution of the United States. Tulsa: Society of Economic and Petroleum Geologists, Sp. Publ. 48.
Gilbert, G.K. (1890) Lake Bonneville. US Geological Survey, Monograph 1, 438 pp.
Hills, E.S. (1963) *Elements of Structural Geology*, New York: J. Wiley and Sons, 483 pp.
Illies, J.H. (ed.) (1981) Mechanism of Graben Formation. Amsterdam: Elsevier Sci., 266 pp. (reprinted from *Tectonophysics*, **73**(1–3)).

James, D.E. (ed.) (1989) *The Encyclopedia of Solid Earth Geophysics*. New York: Van Nostrand Reinhold, 1328 pp.

Krenkel, E. (1925) *Geologie Afrikas*, Vol. 1. Berlin: Gebr. Borntraeger Verlag.

Lister, G.S., Etheridge, M.A. and Symonds, P.A. (1986) Detachment faulting and the evolution of passive continental margins. *Geology*, **14**, 246–50.

Meissner, R. (1986) *The Continental Crust*. Orlando, FL: Academic Press.

Nierenberg, W.A. (1991) *Encyclopedia of Earth System Science*. San Diego: Academic Press, 4 vols.

Quennell, A.M. (ed.) (1982) *Rift Valleys: Afro-Arabian*. Stroudsburg, PA: Hutchinson Ross (Benchmark Papers in Geology, Vol. 60), 419 pp.

Quennell, A.M. (ed.) (1985) *Continental Rifts*. New York: Van Nostrand Reinhold (Benchmark Papers in Geology, Vol. 90), 349 pp.

Scott, D.H. and Dohm, J.M. (1990) Chronology and global distribution of fault and ridge systems on Mars in *Proc. Lunar Planet. Sci. Conf.*, **20**, 487–501.

Seyfert, C.K. (1987) *The Encyclopedia of Structural Geology and Plate Tectonics*. New York: Van Nostrand Reinhold, 896 pp.

Cross references

Asthenosphere
Lithosphere
Plate tectonics

TEKTITE

Tektites are naturally occurring glass objects formed by melting of sediment or rock during the impact of an extraterrestrial body.

Table T1 Tektite strewn fields

Strewn field	Geographic occurrence	Name	Age (Ma)
Australasian	Australia	Australites	0.78
	Java	Javanites	
	Indochina	Indochites	
	Philippines	Philippinites	
	Malaysia	Malaysianites	
	Borneo		
	Belitung Island	Billitonites	
	Southern China	Lei-gong-mo	
Ivory Coast	Ivory Coast of Africa		1.1
Czechoslovakian	Czechoslovakia	Moldavites	15
	Austria	Moldavites	
North America	Texas	Bediasites	35
	Georgia	Georgiaites	

Superficially they resemble silica-rich volcanic glass called obsidian. Tektites are found scattered over regions of the Earth's surface called strewn fields. Four major strewn fields are recognized: Australasian, Ivory Coast, Czechoslovakian and North American (Figure T1; Table T1). Tektites from a given strewn field can generally be recognized by their appearance, chemical composition or age. Tektites are often named after the areas where they are found: thus, for example, tektites found in Australia are called australites and those found in the Philippines are called philippinites (Table T1). An exception to this rule is the tektites found in Texas which are called bediasites after a local tribe of Indians – the Bedias Indians.

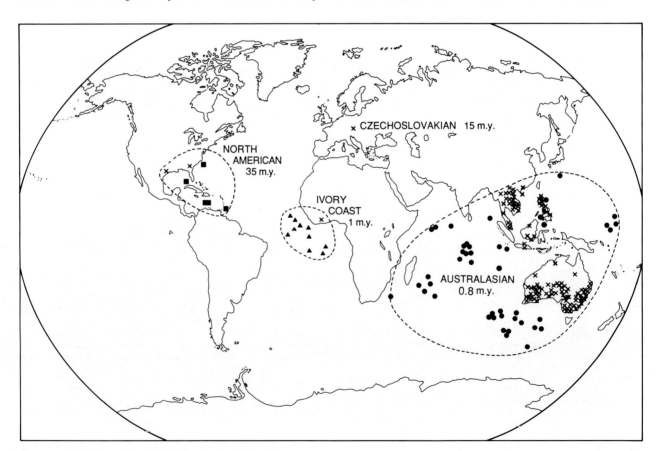

Figure T1 World map showing location and age of the four major tektite strewn fields. Crosses indicate tektite locations on land. Solid circles, solid triangles and solid squares indicate where Australasian, Ivory Coast and North American microtektites have been found in deep sea sediments. Dashed lines indicate approximate boundaries of the strewn fields.

The first written reference to tektites was by Liu Sun, about 950 AD, who wrote about tektites found in China where they are called *lei-gong-mo*. In 1844 Charles Darwin described Australian tektites which he thought were volcanic bombs. The first description of tektites in scientific literature appears to have been by J. Mayer in 1788. The word tektite was coined by F.E. Suess in 1900 from a Greek word meaning molten. Suess thought tektites were a type of glassy meteorite. L.J. Spencer was the first to suggest that tektites were formed by meteorite impact (Spencer, 1933), an hypothesis that is widely accepted today.

Tektites were used by prehistoric people as amulets and sacred objects. They were also used as raw material from which tools were made. In more modern times tektites have been used as semiprecious stones for making jewelry; especially the transparent green moldavite tektites from Czechoslovakia.

Description

Four major types or forms of tektites are recognized: splash forms, ablated forms, Muong Nong-type and microtektites (Figure T2). Splash forms have shapes that seemed to have been formed by rotating liquid drops. Most are spheroidal in shape, but dumbbells, teardrops and disc shapes are also common. The ablated forms are splash forms which have undergone a second period of melting. Ablated forms have lost varying amounts of their original mass when melted material was stripped from one side. In many cases some of the melted material flowed back from the melted surface to form a flange around the perimeter of the tektite (Figure T2). Muong Nong-type tektites are layered tektites which often have a blocky shape. They are named after a region in Laos where they were first recognized. Microtektites are splash forms which are generally less than 1 mm in size. Microtektites belonging to the Australasian, Ivory Coast and North American strewn fields have been found in deep sea sediments. Microtektites have also been reported from Devonian deposits (Wang, 1992).

A group of spherules, related to microtektites, are microscopic silicate glass beads of impact origin called microkrystites. These objects differ from microtektites in that they contain primary crystallites (microscopic crystals that crystallized from the melt). An example is the spherules that occur in Upper Eocene deposits in the Caribbean Sea, Gulf of Mexico, equatorial Pacific Ocean and eastern Indian Ocean. These spherules contain microscopic crystals of clinopyroxene and have thus been called clinopyroxene-bearing spherules or cpx spherules. The Upper Eocene clinopyroxene-bearing spherule layer is associated with an iridium anomaly. Some spherules found at the Cretaceous–Tertiary boundary and Precambrian rocks from South Africa and Australia may be diagenetically altered microkrystites.

Ablated tektites are generally only a few centimeters in size. Splash-form tektites can be up to 10 cm or so in size. Muong Nong-type tektites can be up to tens of centimeters in size. Most tektites appear to be opaque black, but in a thin edge they are seen to be green or brown. Moldavite tektites from Czechoslovakia are transparent green or brown and North American tektites from Georgia are an olive green color. Microtektites range from transparent to opaque and are generally some shade of green, brown or yellow, although some are black or colorless.

Tektites are composed of dense, fairly homogeneous glass, generally without any crystalline material. Most have a few bubble cavities, and bubble cavities are abundant in Muong Nong-type tektites. All tektites contain lechatelierite particles which were formed by melting of quartz grains. Although almost all tektites are free of crystalline material, some Muong Nong-type tektites contain a few relict mineral grains such as quartz (SiO_2), zircon ($ZrSiO_4$) and rutile (TiO_2). These mineral grains appear to be relicts of the rock that was melted to produce the tektites.

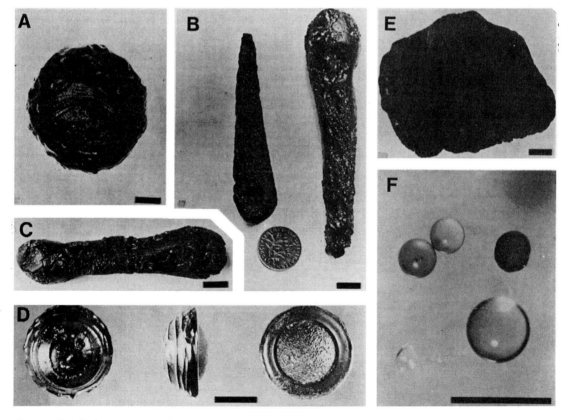

Figure T2 Photographs of tektites and microtektites. (A–C) Indochina splash form tektites. (D) Ablated australite. Anterior surface showing flow ridges is on the left. In the center is a side view. On the right is a posterior view showing posteria surface and flange. (E) Muong Nong-type (layered) tektite from Indochina. (F) Microtektites from the Australasian strewn field. Scale bars for A–E equal 1 cm. Scale bar for F equals 0.1 cm.

Table T2 Average major oxide composition of tektites

Oxide	Australites[a]	Ivory Coast[a]	Czechoslovakian[a]	North American[a]	Cretaceous/Tertiary[b]
SiO_2	73.45	68.0	78.82	76.37	63.09
TiO_2	0.69	0.57	0.35	0.76	0.67
Al_2O_3	11.53	16.3	10.62	13.78	15.21
Fe_2O_3	0.58	0.57	0.25	0.19	5.44[c]
FeO	4.05	5.99	1.61	3.81	
MgO	2.03	3.32	1.84	0.63	2.74
CaO	3.50	1.12	2.08	0.65	7.26
Na_2O	1.28	2.06	0.56	1.54	3.63
K_2O	2.28	1.89	2.61	2.08	1.59

[a] From Glass (1990).
[b] From Sigurdsson et al. (1991).
[c] Iron given as FeO.

The Muong Nong-type tektites that contain the relict mineral grains sometimes also contain grains that were altered or produced by high temperature or high pressure, such as coesite (SiO_2) and grains of corundum (Al_2O_3) in a matrix of lechatelierite (glass with an SiO_2 composition). In addition, some North American and Czechoslovakian tektites contain grains of baddeleyite (ZrO_2), in a groundmass of lechatelierite, which appear originally to have been zircon crystals. Some Australasian tektites contain metallic beads composed of iron with a few percent nickel. These metallic beads were originally thought to be of meteoritic origin, but Ganapathy and Larimer (1983) demonstrated that they have a terrestrial composition.

Tektites are composed of silicate glass (Table T2). They are made primarily of silica (SiO_2) and alumina (Al_2O_3) with smaller amounts of iron, calcium, magnesium, potassium, sodium and titanium oxides. Silica contents generally range between 65 and 85%. In general, Czechoslovakian and North American tektites have the highest silica contents, Ivory Coast tektites have the lowest and Australasian tektites have intermediate silica contents. The contents of the other major oxide vary inversely with silica content: an exception is the potassium oxide content, which is proportionate to the silica content below about 75% SiO_2. The major oxide contents of tektites are sufficient to distinguish tektites from different strewn fields (Table T2).

Tektites have lower water contents and higher FeO/Fe_2O_3 ratios than most terrestrial rocks or sediments. The water content of most tektites is less than 0.02%. The iron in tektites is mostly FeO, rather than Fe_2O_3, indicating that the glass was formed in a reducing rather than an oxidizing environment. Muong Nong-type tektites have somewhat higher water contents and lower FeO/Fe_2O_3 ratios than splash-form tektites, but otherwise have similar compositions.

Major and trace element compositions of tektites are similar to those of terrestrial sedimentary deposits (Koeberl, 1990), while the rare earth element compositions of tektites are similar to those of post-Archean upper crustal sediments. Rubidium–strontium and samarium–neodymium isotopic ratios are also similar to those of terrestrial sedimentary deposits (Shaw and Wasserburg, 1982). Tektites contain small quantities of cosmic ray-produced isotopes such as ^{10}Be, ^{26}Al and ^{56}Mn. The abundances and ratios of these cosmogenic isotopes are similar to those of near-surface sedimentary deposits.

Time of formation

The time of formation of the tektites can be determined by a number of methods involving radioactive decay. One method involves the decay of ^{40}K to ^{40}Ar. This method is used to determine the time since the tektites were heated to high enough temperatures to drive out all the argon. Presumably this is the time of formation of the tektites. A variation of this method is the $^{40}Ar/^{39}Ar$ method. These methods indicate that the four main tektite strewn fields were all formed between 0.78 and 35 million years ago (Table T1).

Tektites can also be dated by the fission-track method. This method involves the radioactive decay of uranium. In general this method gives the same ages as the K–Ar and $^{40}Ar/^{39}Ar$ methods, but it is much more sensitive to later reheating which can erase the fission tracks without removing the argon. Thus this method often gives ages that are somewhat younger than the K–Ar or $^{40}Ar/^{39}Ar$ method on the same tektites.

The formation ages of the tektites are almost always different from the ages of the deposits in which the tektites are found. For example, the australites are often found in sediments that were apparently deposited less than 20 000 years ago. This has led some authors to suggest that the time of formation, based on radiometric dating, is incorrect or that the australites fell about 0.77 million years after they were formed. A more likely explanation is that the tektites fell immediately after they were formed and that their occurrence in younger deposits is due to erosion and redeposition. This explanation is supported by two lines of evidence. First, the fission-track ages of the flanges of the australites, which were formed as the australite fell down through the atmosphere, have the same ages as the cores of the australites, which were formed when the tektites were first formed. Second, the Australasian microtektites are found in sediments with deposition ages the same as the formation ages of the Australasian tektites.

Origin

Numerous hypotheses have been proposed for the origin of tektites. Czechoslovakian tektites were at one time thought to be artificial glasses, but their formation age of about 15 million years rules out such a possibility. A meteoritic origin is excluded because of the low abundance and ratios between cosmogenic isotopes. In addition, no glass meteorites have been observed to fall and no glass meteorites have been found among the thousands of meteorites recovered from Antarctica. A terrestrial volcanic origin is precluded because of the very low water contents and high FeO/Fe_2O_3 ratios, the lack of microlites (microscopic minerals crystallized from the tektite melt) and the ubiquitous presence of lechatelierite particles.

Data obtained from aerodynamic studies of ablated australites have been interpreted as indicating that tektites must have entered the atmosphere at velocities greater than the Earth's escape velocity and that, therefore, tektites could not have originated on the Earth (e.g. Chapman and Larson, 1963). The low abundance of cosmogenic isotopes in the tektites indicate that they could not have originated outside the Earth–Moon system. Thus Chapman and Larson (1963) concluded that tektites were formed by meteorite impact on the Moon and were ejected to the Earth. However, studies of lunar samples have shown that appropriate parent material for tektites does not occur on the lunar surface. Furthermore, later studies by Melnik (1991) indicate that the ablation patterns of australites are consistent with a terrestrial origin.

O'Keefe (1976) accepted Chapman's argument that tektites must have originated on the Moon and, realizing that lunar surface material is not an appropriate parent material for tektites, proposed that tektites were formed by lunar volcanism – an idea that was originally put forth by Verbeek in 1897. However, this hypothesis does not explain the terrestrial composition of the tektites, nor does it explain the evidence for shock metamorphism. Furthermore, two tektite strewn fields are apparently associated with known terrestrial impact craters.

Most authors now accept the hypothesis that tektites were formed by meteorite impact on the Earth. According to this hypothesis tektites are terrestrial surface deposits that were melted and ejected by the impact of an extraterrestrial object. A terrestrial sedimentary parent material for the tektites is indicated by their elemental and isotopic compositions. The kinds of relict minerals found in some Muong Nong-type tektites also indicate a sedimentary parent material. An impact origin for tektites is indicated by the evidence for shock metamorphism and the possible association of two tektite strewn fields with known impact craters. Shock metamorphism refers to the changes that take place in rocks as a result of the high pressure and temperature produced during a high-velocity impact event. The relict mineral grains found in some Muong Nong-type tektites show evidence of shock metamorphism. Lechatelierite particles formed by melting of quartz grains are ubiquitous in tektites. Lechatelierite is common in impact glasses but absent in volcanic glasses. Other shock metamorphic minerals found in tektites include coesite and baddeleyite. Coesite is a high-pressure form of SiO_2 and baddeleyite in tektites was formed by high-temperature breakdown of zircon.

The 10.5 km diameter Bosumtwi crater in Ghana appears to have the same age as the Ivory Coast tektites and the 24 km diameter Ries crater in Germany appears to have the same age as the Czechoslovakian tektites. It has been proposed that the Bosumtwi and Ries craters are the source craters for the Ivory Coast and Czechoslovakian tektite strewn fields, respectively. The source craters for the Australasian and North American strewn fields have not been identified, but impact ejecta have been found associated with both the Australasian and North American microtektite layers. Thus all four major tektite events have directly or indirectly been associated with impact events.

Additional evidence for the terrestrial impact origin for tektites is the occurrence of tektite-like glass objects, called irghizites, associated with the Zhamanshin impact crater in southern Siberia. Irghizites are similar to tektites in many respects, but have somewhat higher water contents and somewhat lower FeO/Fe_2O_3 ratios.

Remaining problems

Although most authors agree that tektites were formed by impact, there still remain some unanswered questions. How were dense, relatively homogeneous, dry glasses produced in a short-lived impact event? How were tektites thrown over such large areas of the Earth's surface? Where are the source craters for the Australasian and North American strewn fields?

The formation of tektite glass required melting of rock material, homogenizing and refining of the melt and rapid cooling to prevent crystallization (refining refers to elimination of bubbles). A major impact event produces high enough temperatures to melt and even vaporize rock and the ejected melt will cool quickly enough to form glass unless it is buried deep in the ejecta blanket. However, O'Keefe (1976) has argued that according to glass technology theory there is not enough time to homogenize and refine the melt. This is one of the main reasons that he supports a lunar volcanic origin for tektites.

The Australasian tektite strewn field covers about 10% of the Earth's surface and contains approximately 100 million metric tons of glass. Furthermore, the australites appear to have been thrown out of the atmosphere and then were partly remelted and ablated when they re-entered the atmosphere. The closest Czechoslovakian tektites to the Ries crater, which is believed to be their source, are about 250 km and they are found as far away as 450 km. How the tektites were thrown great distances across the Earth's surface without being torn to a fine spray by the atmosphere is still not well understood.

The source craters for the North American and Australasian tektite strewn fields have not been identified. It is perhaps not surprising that the source crater for the North American strewn field has not been found since it is the oldest of the tektite strewn fields. On the other hand, the Australasian strewn field is the youngest and the largest strewn field and several lines of evidence suggest that the source crater should be in or adjacent to Indochina (Schnetzler, 1992).

About 65 million years ago nearly 75% of the species present on the Earth became extinct. This so-called mass extinction took place over a very short period of time. The last of the dinosaurs became extinct about this time. This extinction event marks the boundary between the Cretaceous and Tertiary Periods of geologic time. It was during the Tertiary Period that mammals became the dominant life form on Earth. According to Alvarez and colleagues the boundary between the Cretaceous and Tertiary or K–T boundary was marked by a major impact event which was responsible for the mass extinctions at that time (Alvarez et al., 1980). Microscopic beads found at the K–T boundary have been interpreted as altered microtektites. This conclusion has been supported by the discovery of tektite-like glass bodies at the K–T boundary in Haiti (Sigurdsson et al., 1991).

Some authors (e.g. Muller and Morris, 1986) have suggested that major impact events might trigger reversals of the Earth's magnetic field (i.e. cause the north and south magnetic poles to switch places). The major evidence for this is the close association of three of four tektite events with reversals of the Earth's magnetic field (Glass, Swincki and Zwart, 1979). However, Schneider and Kent (1990) have shown that in the case of the Ivory Coast tektite event, the impact occurred after the reversal and that, therefore, the impact that produced the tektites could not possibly have triggered the reversal. Debate about this issue continues.

Summary

Tektites are naturally occurring silicate glass objects formed by the impact of an extraterrestrial body on Earth (Taylor, 1973; Koeberl, 1990). Most tektites are black or green in color, a few centimeters in size and spheroidal, teardrop, dumbbell or disk shaped. Some tektites show evidence of high-velocity flight (approximately 10 km s^{-1}) through the atmosphere. Others are large, blocky, layered chunks that can weigh up to at least 10 kg (Muong Nong-type). Tektites have compositions similar to terrestrial sedimentary deposits and some of the Muong Nong-type tektites contain mineral grains showing evidence of shock metamorphism. The distribution of tektites is not random; they occur scattered over large areas of the Earth's surface called strewn fields. Four major tektite strewn fields are generally recognized: Australasian, Ivory Coast, Czechoslovakian and North American. The tektites in these strewn fields were formed about 0.77, 1, 15 and 35 million years ago respectively. The Ivory Coast and Czechoslovakian strewn fields appear to be associated with the Bosumtwi crater in Ghana and the Ries crater in Germany, respectively. The source craters for the Australasian (the largest and youngest) and the North American (the oldest) strewn fields have not been found. Tektite-like glasses have been found in 65 million year old rocks at the boundary between the Cretaceous and Tertiary geologic periods. This may be the largest and oldest recognized tektite strewn field.

Billy P. Glass

Bibliography

Alvarez, L.W., Alvarez, W., Asaro, F. and Michel, H.V. et al. (1980) Extraterrestrial cause for the Cretaceous–Tertiary extinction. *Science*, **208**, 1095–108.

Chapman, D.R. and Larson, H.K. (1963) On the lunar origin of tektites. *J. Geophys. Res.*, **68**, 4305–58.

Ganapathy, R. and Larimer, J.W. (1983) Nickel–iron spherules in tektites: non-meteoritic origin. *Earth Planet. Sci. Lett.*, **65**, 225–8.

Glass, B.P., Swincki, M.B. and Zwart, P.A. (1979) Australasian, Ivory Coast and North American tektite strewn fields: size, mass, and correlation with geomagnetic reversals and other Earth events. *Proc. Lunar Planet. Sci. Conf.*, **10**, Geochim. Cosmochim. Acta, Suppl. **11**, 2535–45.

Koeberl, C. (1990) The geochemistry of tektites: an overview. *Tectonophysics*, **171**, 405–22.

Melnik, W.L. (1991) Ablation of Australian tektites, supportive of a terrestrial origin? *Meteoritics*, **26**, 371.

Muller, E. and Morris, D. (1986) Geomagnetic reversals driven by abrupt sea level changes. *Geophys. Res. Lett.*, **13**, 1177–80.

O'Keefe, J.A. (1976) *Tektites and Their Origin*. Amsterdam: Elsevier, 245 pp.

Schneider, D.A. and Kent, D.V. (1990) Ivory Coast microtektites and geomagnetic reversals. *Geophys. Res. Lett.*, **17**, 163–6.

Schnetzler, C.C. (1992) Mechanism of Muong Nong-type tektite formation and speculation on the source of Australasian tektites. *Meteorites*, **27**, 154–65.

Shaw, H.F. and Wasserburg, G.J. (1982) Age and provenance of the target materials for tektites and possible impactites as inferred from Sm–Nd and Rb – Sr systematics. *Earth Planet. Sci. Lett.*, **60**, 155–77.

Sigurdsson, H., D'Hondt, S., Arthur, M.A. et al. (1991) Glass from the Cretaceous/Tertiary boundary in Haiti. *Nature*, **349**, 482–7.

Spencer, L.J. (1933) Origin of tektites. *Nature*, **131**, 117–8.
Taylor, S.R. (1973) Tektites: a post-Apollo view. *Earth Sci. Rev.*, **9**, 101–23.
Wang, K. (1992) Glassy microspherules (microtektites) from an Upper Devonian limestone. *Science*, **256**, 1547–50.

Cross references

Cretaceous–Tertiary boundary
Ejecta
Impact cratering
Meteorite

TEMPERATURE

Temperature is a concept linked originally with the measurment of the warmth or heat of objects. Heat is physically identified with the state of atomic and molecular motion of a substance. The condition of atomic and molecular motion in turn depends upon the energy of the object in question, in the sense that molecular motions in a hot object are more rapid and more energetic than in a cold one. Scales of temperature are measures of the warmth of the object in question.

Standard measures of temperature are the Kelvin, Celsius and Fahrenheit scales. Temperature may also refer to measures of radiation emitted from volumes or regions (see Blackbody radiation). Rather astonishingly, while astronomy had become quite sophisticated in classical Greece and the Middle Ages, no one appears to have even suggested the idea of a thermometer, until a first model was constructed by Galileo Galilei (q.v.) in the late 16th century. Mercury-bulb thermometers were developed in Florence, Italy, in the mid-17th century.

Lord Kelvin is credited with the development of the concept of an absolute scale of temperatures. The zero of the Kelvin scale is termed 'absolute zero', a state wherein the body in question gives up no heat. This state has been approached (but not attained) in laboratory experiments. Water melts at 273 K and boils at 373 K. No upper limit of the scale is defined. Among solar system surfaces, that of the Sun is warmest at 6000 K. The interior of the Sun is believed to reach 1.5×10^7 K. At the other extreme, the measured surface temperature of Neptune's moon Triton is a chilly 38 K.

The Celsius (or 'centigrade') scale is usually more convenient for general purposes, with freezing and boiling temperatures of water at 0° and 100° respectively. Anders Celsius described this scale in 1742. The Fahrenheit scale, proposed by Gabriel Daniel Fahrenheit in 1714, was extensively used in the English-speaking world. It is employed in the United States, although generally abandoned elsewhere; here freezing and melting occur at 32° and 212° respectively, and absolute zero has a value of −460°. The zero value of this scale corresponds to the freezing point of a saturated salt solution.

Conversion factors:

$1°C = 1.8°F$; $n°C = 5/9 (°F - 32)$; $n°F = 9/5 (°C + 32)$

James H. Shirley and Rhodes W. Fairbridge

Bibliography

Wolf, H.C. (ed.) (1955) Temperature, in *Temperature, Its Measurement and Control in Science and Industry*, Vol. 2. New York: Reinhold (American Institute of Physics).
Zemansky, M.W. (1964, 1981) *Temperatures Very Low and Very High*. New York: Van Nostrand Reinhold.

Cross references

Blackbody radiation
Emissivity

TERRESTRIAL PLANETS

The set of relatively small, rocky, 'Earth-like' planets nearest the Sun, including Mercury (Plate 5), Venus (Plate 6), Earth (Plates 9, 10) and Mars (Plate 16). They are often called the 'inner planets.' Comparative properties of the terrestrial planets are listed in Table T3. The densities of the terrestrial planets (3–6 g cm^{-3}) are high in comparison with those of the giant (or Jovian) planets (< 1.8 g cm^{-3}). The terrestrial planets have few or no satellites and lack rings. They have solid surfaces and atmospheres (though tenuous in some cases). Some authors include Pluto as a terrestrial planet (Kopal, 1979). The differences between the terrestrial planets and the giant planets may relate to their respective distances from the Sun during the early development and evolution of the solar system.

James H. Shirley

Bibliography

Encrenaz, T. and Bibring, J.-P. (1990) *The Solar System*. New York: Springer-Verlag.
Kopal, Z. (1979) *Realm of the Terrestrial Planets*. New York: John Wiley and Sons.
Murray, B., Malin, M.C. and Greeley, R. (1981) *Earthlike Planets*. San Francisco: W.H. Freeman and Company.

Cross references

Accretion
Earth
Earth–Moon system: origin
Mars
Mercury
Moon
Planet
Solar nebula
Solar system: origin
Venus

Table T3 Planetary data

	Mercury	Venus	Earth	Mars
Mean distance from Sun (AU)	0.387099	0.723332	1	1.523691
Sidereal period (d)	87.969	224.701	365.256	686.98
Orbital eccentricity	0.2056	0.0068	0.0167	0.0934
Orbital inclination (deg)	7	3.39	0	1.85
Obliquity (deg)	0	177	23.45	23.98
Mass (kg)	3.30×10^{23}	4.87×10^{24}	5.98×10^{24}	6.42×10^{23}
Density (g cm^{-3})	5.42	5.25	5.52	3.94
Equatorial radius (km)	2439	6051.5	6378	3398
Sidereal rotation period (h)	1403.75	5832.24	23.9345	24.6229
Surface gravity (m/s^{-2})	3.78	8.6	9.78	3.72
Surface temperature (K)	700[a]	730	288	220

[a] Approximate maximum temperature at subsolar point.

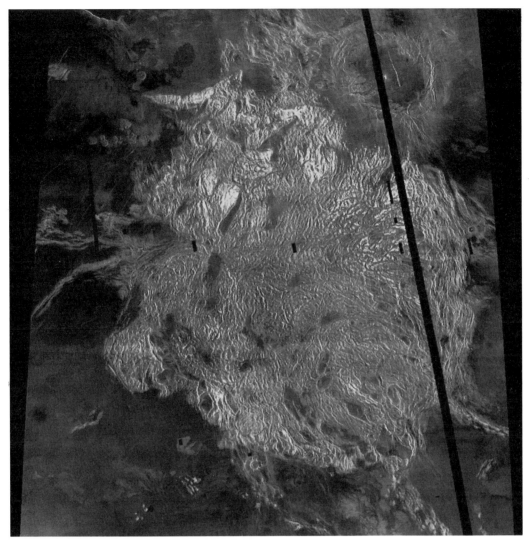

Figure T3 A low-resolution Magellan mosaic showing Alpha Regio, a large region of tesserae on Venus. Note the pervasive deformation and the many directions of lineations (image is approximately 3000 × 3000 km).

TESSERA

The expression 'tesserae' (singular, tessera), from the word for a mosaic of small squares, is used in a geotectonic sense for large-scale crustal blocks on planet Earth that are bounded by sutures, fracture zones or lineaments to form a polygonal or mosaic pattern. It appears to have been first used by the geologist B.B. Brock (1972, p. 61) to characterize the taphrogenically defined structural blocks of the African craton. The 'polygons' may be geometrically represented as triangles, rectangles, hexagons, octagons and so on. The simplest small-scale natural analog on planet Earth is the hexagonal shrinkage pattern developed on some volcanic terrains, e.g. at the Devil's Causeway in the Antrim basalts of Northern Ireland. The mechanism is a very slow homogeneous cooling beneath the insulating cover of a rapidly cooled crust.

Originally referred to as 'parquet' terrain, tesserae was first identified as a terrain type on Venus from synthetic aperture radar images from the Soviet Venera 15 and 16 missions (Sukhanov, 1986). Tesserae are areally extensive regions on the surface of Venus that are pervasively deformed by two or more intersecting sets of ridges and grooves. Tessera terrain covers about 15% of the planet's surface and, although each patch of tesserae is distinct, most of the tesserae areas share several properties. Tessera deformation occurs on scales of tens of kilometers to the limits of resolution of current imaging data (about 75 m). Tessera terrain is rough, and usually elevated above surrounding areas, and its low free-air gravity anomaly indicates that this extra elevation is supported by a low-density crustal root. Filling and embayment of low areas by radar-dark material within and on the margins of tesserae is common.

At present the origin of Venusian tessera terrain is unknown, but several hypotheses exist for its formation. These include the following: tesserae are remnants of paleocontinents; tessera are massive flood basalts, formed as part of mantle plume initiation (see Hot spot tectonics); tesserae are accreted crustal blocks over mantle downwellings; tessera are pre-existing underformed crustal blocks deformed by gravity sliding or basin-and-swell deformation. Although tessera is a terrain type sometimes considered unique to Venus, it does not necessarily represent a tectonic process unique to Venus, as similar landforms on Earth or Mars exist although their appearance has been altered by erosion and sedimentation.

Robert R. Herrick

Bibliography

Basilevsky, A.T., Pronin, A.A., Ronca, L.B. *et al.* (1986) Styles of tectonic deformations on Venus: analysis of Venera 15 and 16 data. *J. Geophys. Res.*, **91**, D399–411.

Bindschadler, D.L. and Head, J.W. (1991) Tessera terrain, Venus: characterization and models for origin and evolution. *J. Geophys. Res.*, **96**, 5889–907.

Nikishin, A.M. (1990) Tectonics of Venus: a review. *Earth, Moon, and Planets*, **50/51**, 101–25.

Solomon, S.C., Head, J.W., Kaula, W.M. *et al.* (1991) Venus tectonics: initial analysis from Magellan. *Science*, **252**, 297–312.

Sukhanov, A.L. (1986) Parquet: regions of areal plastic dislocations. *Geotectonics*, **20**, 294–305.

Cross references

Tectonics
Venus: geology and geophysics

THALES OF MILETUS (624–548 BC)

A learned and widely traveled merchant of Miletus (or Miletos; modern: Palatia), a coastal city-state located near the mouth of the rich Meander Valley in Ionia (part of Asia Minor), Thales is sometimes credited as the 'Father of Greek astronomy'. He was certainly one of the great natural philosophers of the day. In mathematics he was able to apply the principle of similar triangles to useful purposes such as measuring the distance of a ship at sea.

Thales is credited with being one of those who recognized the spherical nature of the Earth; a traveler at sea could hardly escape the observation of the 'disappearing ships'. He recognized and measured the obliquity of the ecliptic (q.v.), which made possible the prediction of eclipses (q.v.). Success in prediction is said to be the ultimate glory of science, and Thales exactly predicted a solar eclipse for 28 May, in 585 BC. He had studied astronomy and mathematics in Egypt and Mesopotamia, and learned about the 18-year Saros cycle (q.v.) from the Babylonian records. The eclipse coincided with a famous battle between the Medes and Lydians, and so overawed the combatants that they declared a truce and eventually made peace.

Thales is also respected in the history of geology because of his observations on the role of water in modeling the Earth's surface. Water, indeed, in liquid form is the unique characteristic of planet Earth that distinguishes it from all other planets in our solar system. He was impressed on the one hand by wave action and its erosive nature, and on the other by the annual build-up of sediments brought down by floods to the deltas of the Nile and the Meander. In fact, it was the progressive siltation of the harbors around the Meander delta that eventually caused many of the towns there to be shifted or abandoned.

A student of Thales in Miletus, Anaximander (611–547 BC) introduced, from Egypt or Mesopotamia, the idea of a sundial in the form of a vertical rod ('gnomon'), the shadow of which determined the hour and, over the year, specified the longest and shortest days, i.e. the solstices, and also the equinoxes. He also experimented with map making, working on the principle of a flat Earth within a geocentric universe with rotating celestial bodies. Whereas Thales viewed the universe as supported by water, Anaximander was emphatic that the 'primordial substance' was neither water nor one of the other elements, but *apeiron*, i.e. chaos. Another contemporary philosopher in Miletus was Anaximenes (born. c. 570 BC) who believed that the primordial substance was closer to what we know as air, i.e. atmospheric gases, and subject to change in temperature, volume and state. Of long-term importance is the concept developed by Thales and the 'Ionian philosophers' that in the physical world everything is in continuous motion. In leading the search for physical explanations for terrestrial and astronomical phenomena (in place of the Babylonian and Egyptian theistic approach), Thales led the way, recognizing the underlying unity of our universe and has accordingly been described as 'the first true scientist'.

Rhodes W. Fairbridge

Bibliography

Fenton, C.L. and Fenton, M.S. (1952) *Giants of Geology*. Garden City, N.Y.: Doubleday, 333 pp.

Longrigg, J. (1976) Thales. *Dict. Sci. Biogr.*, pp. 295–8.
Singer, C. (1959) *A Short History of Scientific Ideas to 1900*. Oxford: Clarendon Press, 525 pp.
Taton, R. (1957). *La Science Antique et Mediévale*. Paris: Presses Univ. de France [trans. by A.J. Promerans (1963) *Ancient and Medieval Science*. New York: Basic Books, 551 pp.]

THERMAL EVOLUTION OF PLANETS AND SATELLITES

The thermal history of a planet is the record of how its internal temperature has changed with time, from the planet's origin to the present. Basically a planet cools with time as it loses its interior heat to space through radiation from its surface or atmosphere, although this general cooling trend can be punctuated by periods of time in which its interior heats up. A certain amount of heat is trapped within a planet at the time of its formation and additional heat is produced later through the decay of radioactive elements. The heat is brought to the surface mainly by mantle convection (q.v.). The thermal evolution of a planet is then simply a consequence of the competition between internal energy sources producing heat and mantle convection removing it. A quantitative description of this history can be achieved through the application of the principle of energy conservation.

We apply this principle to a very simple model of a planet in order to illustrate the basic cooling history generally believed typical of the Earth and the other terrestrial planets. Suppose that a planet has internal temperature T and mass M. Conservation of energy for the planet is

$$MC\partial T/\partial t = MH - Aq \tag{T1}$$

where C is the specific heat of the planet, H is the rate of internal energy release per unit mass due to radioactivity, A is the planet's surface area, q is the heat flux at the surface and t is time. Surface heat flux is the rate of energy flow across the surface per unit area, with the energy brought to the surface from below by mantle convection. Specific heat is the energy per unit mass extracted from the planetary material when its temperature decreases by 1 K. The energy balance equates the time rate of change of internal thermal energy to the difference between the heat production rate throughout the planet and the rate of heat loss through the surface. Integration of the energy equation in time provides the thermal history $T(t)$ of the planet.

Before the energy equation can be integrated we must describe how H and q vary with time. The specific radiogenic heat production rate H decreases with time according to the exponential decay law

$$H = H_0\exp(-\lambda t) \tag{T2}$$

where H_0 is the specific heat production rate at $t = 0$ and λ is the decay constant. Substitution of (T2) into (T1) gives

$$MC\partial T/\partial t = MH_0\exp(-\lambda t) - Aq \tag{T3}$$

The surface heat flux is more difficult to specify because it depends on how mantle convection brings the heat to the surface, a process that is controlled largely by the planet's temperature T. Based on the results of a large number of laboratory and numerical experiments, it is possible to represent the convective heat flux by

$$q = \frac{k}{d}(T - T_s)\left(\frac{Ra}{Ra_{cr}}\right)^\beta \tag{T4}$$

where k is the thermal conductivity, d is the thickness of the mantle, T_s is the surface temperature, Ra is the Rayleigh number given by

$$Ra = \frac{g\alpha(T - T_s)d^3}{\kappa\nu} \tag{T5}$$

Ra_{cr} is the critical value of the Rayleigh number for the onset of convection in the mantle, and β is a constant. In (T5) g is the acceleration of gravity, α is the thermal expansivity, κ is the thermal diffusivity and ν is the kinematic viscosity. All quantities except for ν are assumed constant. The viscosity ν is a function of temperature, however, a dependence that controls the thermal evolution.

Relation (T4) is the so-called Nusselt number Nu – Rayleigh number relation $[Nu = qd/k(T - T_s)]$ of convection theory. The critical Rayleigh number for convection onset Ra_{cr} typically has a value of order 10^3. The power law exponent β generally has a value of about 0.3 according to boundary layer theory and a large number of experiments. The specific form (T4) of the surface heat flow parameterization is suggested by boundary layer theory and experiments on convection of a constant viscosity, Boussinesq fluid in a plane layer heated from below. Its applicability to other situations is surprisingly robust and has been discussed in detail by Schubert, Cassen and Young (1979) and Schubert, Stevenson and Cassen (1980). The Nusselt–Rayleigh number relations of other heating modes (e.g. internal heating) and geometries (e.g. spherical geometry) can all be written in the form of (T4) by appropriate definitions and identifications of Ra_{cr} and β. The use of (T4) for convection with temperature-dependent viscosity is generally appropriate if T is identified with the characteristic temperature of the convecting part of the fluid, Ra_{cr} and β are properly interpreted and T_s is chosen as either the surface temperature or the temperature near the base of any stagnant lid that forms over the convecting system (Schubert, Cassen and Young, 1979; Schubert, Stevenson and Cassen, 1980). For the parameterization of mantle convection in the Earth, T_s is properly taken as the surface temperature since the plates are mobile and do not form a stagnant cap over the convecting system.

The use of (T4) in planetary thermal history models to account for heat transfer by mantle convection has come to be known as the method of parameterized convection, since the spatially and temporally complex convective motions are not calculated in detail. The method has made possible essentially analytic models of planetary thermal history (Sharpe and Peltier, 1978, 1979; Schubert, 1979; Schubert, Cassen and Young, 1979; Stevenson and Turner, 1979; Turcotte, Cooke and Willeman, 1979; Turcotte, 1980; Davies, 1980).

The strong dependence of mantle viscosity on temperature exerts a controlling influence on the thermal evolution of a planet. Mantle kinematic viscosity ν depends on temperature according to

$$\nu = \bar{\nu} \exp\left(\frac{\overline{A}}{T}\right) \quad (T6)$$

where $\bar{\nu}$ and $\overline{A}$ are constants (e.g. Weertman and Weertman, 1975; Carter, 1976; Poirier, 1985). The parameter $\overline{A}$ is an activation temperature related to the activation energy E^* of the subsolidus creep deformation by $\overline{A} = E^*/R$, where R is the universal gas constant. The temperature dependence of mantle viscosity acts as a thermostat to regulate the mantle temperature (Tozer, 1967; Schubert, 1979). Initially, when a planet is hot, mantle viscosity is low, and extremely vigorous convection rapidly cools the planet. Later in its evolution, when the planet is cooler, its mantle viscosity is higher and more modest convection cools the planet at a reduced rate. Self-regulation tends to bring the viscosity of the mantle to a value that facilitates efficient removal of the heat generated in the mantle by convection. The temperature of the mantle adjusts to maintain or reach this preferred value of viscosity. If the mantle is excessively hot to start with, e.g. because of accretional heating and the heat released by core formation, it will rapidly cool to bring its viscosity in line with the value preferred by its internal heat generation. The farther the mantle is from the preferred viscosity, the more rapid is the adjustment. Thus the specific value of the initial temperature $T(0)$ chosen for modeling the thermal history is unimportant, though it should be high, since the adjustment by self-regulation rapidly rids the mantle of excess heat. Though not realistic, even an initially cold mantle would heat by radioactivity until the self-regulated viscosity was reached, a process that would have a billion-year timescale. Self-regulation ensures that the final state of the convecting mantle has no memory of initial conditions, a fortunate circumstance for thermal evolution models. As mantle radiogenic heat sources decay with time, convection needs to transfer less heat and the preferred mantle viscosity gradually increases, causing the mantle to undergo a secular cooling. The gradual decrease of mantle temperature with time is a fundamental aspect of terrestrial planetary evolution and requires that secular cooling contributes to the heat flow through a planet's surface (Schubert, Stevenson and Cassen, 1980; Davies, 1980). The relative contributions of secular cooling and mantle radioactivity to surface heat flow in the case of the Earth is uncertain, though it is generally accepted that q is dominated by radiogenic heating.

Upon combining equations (T3)–(T6) we obtain a single, first-order, nonlinear, ordinary differential equation for T that explicitly reveals the appearance of T in all terms of the equation

$$\frac{\partial T}{\partial t} = \frac{H_0}{C} e^{-\lambda t} - \frac{Ak}{MCd}\left(\frac{\alpha g d^3}{\kappa \bar{\nu} Ra_{cr}}\right)^{\beta}(T - T_s)^{1+\beta}\exp\left(\frac{-\beta \overline{A}}{T}\right) \quad (T7)$$

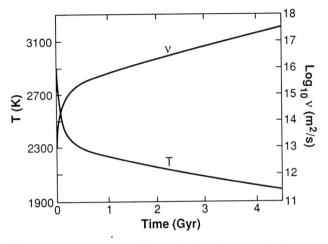

Figure T4 Temperature T and kinematic viscosity versus time for a model of the thermal evolution of the Earth's mantle. Parameter values for this example are $T(0) = 3273$ K, $T_s = 273$ K, $\beta = 0.3$, $\lambda = 1.42 \times 10^{-17}$ s^{-1}, $A = 7 \times 10^4$ K, $H_0/C = 4.317 \times 10^{-14}$ K s^{-1}, $k = 4.18$ W m^{-1} K^{-1}, $\kappa = 10^{-6}$ m^2 s^{-1}, $\alpha = 3 \times 10^{-5}$ K^{-1}, $d = 2.8 \times 10^6$ m, $g = 10$ m s^{-2}, $\bar{\nu} = 1.65 \times 10^2$ m^2 s^{-1}, $Ra_{cr} = 1100$ and $A/MC = 1.377 \times 10^{-13}$ m^2 K J^{-1}. (After Schubert, Stevenson and Cassen, 1980.)

Equation (T7) can be solved subject to the initial condition $T = T(0)$ at $t = 0$.

Equation (T7) describes the evolution of a planet subsequent to the violent early stages of its formation. The earliest stages of planetary evolution involve the accumulation of the planet from accreting planetesimals, large-body impacts during the late stages of accretion (Kaula, 1979; Wetherill, 1985) and major readjustments of the interior structure such as differentiation of a core. All of these events occur relatively quickly over a timespan of a few hundred million years. In addition, these events release enormous amounts of energy sufficient to heat a planet the size of the Earth by many thousand kelvin and differentiate it into a molten iron core and a mantle close to the solidus temperature (Stevenson, Spohn and Schubert, 1983; Stevenson, 1989). Given the retention of enough accretional energy inside a planet to initiate core–mantle differentiation, the formation of the core alone releases sufficient gravitational potential energy to heat an Earth-size planet about 2000 K (Birch, 1965; Tozer, 1965; Flasar and Birch, 1973; Shaw, 1978). Even radioactivity could contribute to the heating of planets during the first few hundred million years of their evolution if certain extinct radionuclides with very short half-lives such as ^{26}Al were incorporated into the planets. There is a surfeit of energy to ensure that the terrestrial planets Venus, Earth and Mars, were hot and fully differentiated at the end of accretion some 4.5 billion years ago. There is geochemical evidence for both Earth and Mars to support this view. Thus the initial state for the secular cooling history of a planet, to which (T3) applies, is a high, near-solidus temperature $T(0)$. As noted above, the self-regulation of internal temperature by mantle convection makes precise knowledge of $T(0)$ unnecessary, since adjustments to initial temperatures that are too high or too low are quite rapid. Though less eventful than the early stages of formation, the secular cooling evolution of a planet occupies almost all of geologic time since planet formation.

The result of a typical solution of (7) for an Earth-like thermal history is shown in Figure T4 (Schubert, Stevenson and Cassen, 1980). Viscosity increases monotonically from 3.2×10^{11} m^2 s^{-1} at the start of the model thermal history to 3.4×10^{17} m^2 s^{-1} after 4.5 Ga. Temperature undergoes a corresponding monotonic decrease of less than 50% because of the very strong temperature dependence of the viscosity. The large drop in temperature and the enormous increase in viscosity during the first few hundred million years of model thermal evolution is a consequence of the self-regulation discussed above. The mantle rapidly adjusts by early vigorous convection to a viscosity (temperature) that is higher (lower) than the viscosity (temperature) of its initial state. At the end of this early adjustment phase, the mantle has lost most of its initial excess heat, and it has come into a state in which temperature and viscosity are ideally suited to the convective removal of the remaining 'primordial' heat and the energy produced by radioactivity. During the rest of geologic time the model mantle undergoes a more gradual secular cooling with its attendant viscosity increase. The surface heat flow (not shown in the figure) declines throughout most of the evolution, tracking the decay in the total radiogenic heat production per unit surface area, but always remaining in excess of the internal heat release. The difference between the two heat flows is due to the loss of primordial heat (or heat produced earlier by previous radioactive decay). The loss of primordial heat, or secular cooling, contributes 25% of the surface heat flow in the model of Figure T4. At any time in the model thermal history, surface heat loss exceeds internal heat production.

That secular cooling contributes significantly to planetary surface heat flow is an important point because it had been widely believed that mantle convection was efficient enough to establish a balance between surface heat flow and radiogenic heat production. However, when internal heat sources decay with time, as is the case for mantle radiogenic heat sources, the surface heat loss and convection must also decline with time, and the system must cool. The secular decline in internal thermal energy must, by energy conservation, contribute to the flow of heat through the surface. No matter how efficiently convection transports heat through the mantle, the decay with time in the rate of internal heat production together with energy conservation, insure that secular cooling contributes to surface heat loss. The analyses of Schubert, Stevenson and Cassen, (1980) and Davies (1980) show that the magnitude of the contribution is substantial; about 25% of the Earth's surface heat flow could be due to cooling of the Earth. This conclusion is a robust result of numerous calculations designed to test the consequences of wide variations in the values of parameters that enter thermal history models (Schubert, Stevenson and Cassen, 1980).

The inequality between surface heat flow and interior heat production has been quantified in terms of the Urey ratio

$$Ur = \frac{MH}{Aq} \quad (T8)$$

defined as the ratio of the heat production term to the heat loss term on the right-hand side of (T1). A Urey ratio of less than unity implies a net loss of heat and a temperature decrease in the mantle given by

$$\frac{\partial T}{\partial t} = \frac{-Aq}{MC}(1 - Ur) \quad (T9)$$

The present value of the Urey ratio is 0.75 in the example of Figure T4. This value of the Urey ratio, together with (T9), $q = 60$ mW m^{-2} (mantle heat flux estimated by Turcotte and Schubert, 1982) and $A/MC = 1.38 \times 10^{-13}$ m^2 K J^{-1} give a present Earth mantle cooling rate of about 200 K Ga^{-1}.

The model of planetary thermal history discussed above is distinguished by its simplicity and ability to account for what is probably the most important effect in planetary evolution, namely the self-regulation of mantle convection by the strong Arrhenius-type dependence of mantle viscosity on temperature. Nevertheless, it is still important to build even more realism into models of planetary thermal history, though the models become more complex as a result. A first step in this regard is to incorporate more realistic internal structures by accounting for the core–mantle nature of the terrestrial planets. This requires separate energy equations for the core and mantle. The energy equation for the core should account for inner core solidification as the core cools with its attendant release of latent heat and gravitational potential energy as the light alloying element in the core is concentrated into the liquid outer core (Stevenson, Spohn and Schubert, 1983). The energy equation for the mantle should incorporate as a heat source the heat flux from the core which is also a heat sink for the core. The predictions of such a more realistic thermal history model include the decrease of core and mantle temperature with time, the decay of surface heat flux with time, the variation in heat flow from the core with time, and the onset time and growth history of a solid inner core.

Results of this type are shown in Figure T5 for thermal history models of Earth and Venus (Stevenson, Spohn and Schubert, 1983). Heat flow from the core initially decreases very rapidly with time during the period when early vigorous mantle convection removes heat very efficiently from the core. Inner core solidification begins at $t \approx 2.3$ Ga in the Earth model when the core has cooled sufficiently that its temperature drops to the core melting temperature at the center of the Earth model. Inner core solidification and growth

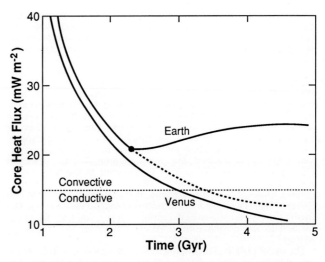

Figure T5 Core heat flux versus time for thermal history models of Earth and Venus. The horizontal line is the value of heat flux conducted upward along a core adiabat. The core is superadiabatic and convection is possible if core heat flux exceeds this value, but it is subadiabatic and non-convecting if core heat flux falls below this value. The filled circle marks the onset of core solidification in the Earth model. There is no core solidification in the Venus model. The dashed curve indicates how the Earth model would continue to cool without inner core freeze-out. Venus' core becomes subadiabatic and non-convective after 3 Ga in the model. Parameter values for these models and a more detailed discussion of them can be found in Stevenson, Spohn and Schubert (1983).

continues until the present, when the inner core in the model has a radius of 1207 km.

The monotonic decrease in core heat flux q_c for the Earth model would continue through geologic time (dashed curve in Figure T5) if not for inner core freezing and eventually q_c would fall below the value necessary to supply the conductive heat flow along the core adiabat (estimated at 15 mW m^{-2}, horizontal dashed line in Figure T5). Thermal convection in the core is not possible if q_c falls below the heat flux conducted along the core adiabat; a thermally driven geodynamo would also not be possible were q_c to drop below the conductive heat flux along the adiabat. (It is assumed that the temperature in the liquid outer core lies along an adiabat due to efficient convection in the liquid.) Figure T5 shows that thermal convection and thermal forcing of a core dynamo magnetic field would have ceased at about 3.2 Ga in the Earth model if not for inner core solidification. However, the core in the Earth model does begin to solidify at about 2.3 Ga and the decrease in core heat flow with time is arrested by the event. Once core freezing begins, the release of latent heat and gravitational energy contributes to the heat flow from the core, which is maintained above the conductive heat flow along the core adiabat for the rest of geologic time (Figure T5). Convection in the core of the Earth model is driven both thermally and compositionally subsequent to inner core freezing, which is seen to be essential for the maintenance of core convection and dynamo generation of a magnetic field at present. Gravitational energy release is probably more important in driving the dynamo than latent heat release since the mechanical energy is almost entirely available for dynamo generation while the thermal energy is subject to the Carnot efficiency factor (Gubbins, 1977).

The core thermal history of the Earth model thus has interesting implications for the Earth's magnetic field. The model shows onset of inner core freezing relatively late in Earth's thermal history, about 2 Ga ago. Since the Earth's magnetic field is at least 3.5 Ga old (McElhinny and Senanayake, 1980), the mode of powering the dynamo may have changed during the Earth's evolution. Early in Earth's thermal history, the magnetic field was probably powered by thermal convection with the heat derived from secular cooling of the fluid core. After initiation of inner core growth, the dominant source of energy for the dynamo became gravitational energy release upon concentration of the light element into the liquid outer core and latent heat release upon inner core solidification.

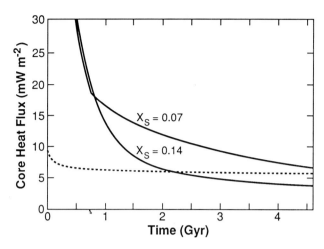

Figure T6 The decrease with time in heat flow from the core of the Martian thermal history model of Schubert and Spohn (1990) for two values of initial weight percent sulfur in the core $x_S = 7$ and 14%. The dashed curve is the heat flux conducted along the core adiabat. The sharp bend in the heat flux curve for $x_S = 0.07$ at $t \cong 750$ Ma marks the onset of inner core freeze-out.

In the Venus model, core heat flux decreases monotonically with time as the planet cools and, unlike the Earth model, there is no core solidification. The Venus core in the model of Figure T5 is entirely liquid at present. Venus' core heat flux dropped below the value necessary to sustain thermal convection in the core about 1.5 Ga ago. A thermally driven dynamo that might have been operative in Venus' core at earlier times would have ceased to function at 1.5 Ga ago. In this model Venus' core has been subadiabatic, non-convective and unable to sustain dynamo action for the last 1.5 Ga. Venus does not nucleate an inner core in the model mainly because of the lower pressure at the center of Venus compared with Earth's central pressure. Because of the lower central pressure in the Venus model, the freezing temperature at the center of Venus is lower than at the center of Earth and the Venus model needs to cool longer than the Earth model for solidification to occur (first at the planet's center).

According to the Venus model, the planet had an intrinsic magnetic field until about 1.5 Ga ago. This points to the possibility that the real planet had a magnetic field for several billion years of its evolution, a possibility that must be accounted for in studies involving the solar wind interaction with Venus and the evolution of Venus' atmosphere. Equally intriguing, from a speculative point of view, is the possibility suggested by the model that Venus may be on the verge of core solidification and rejuvenation of its magnetic field by compositionally driven convection in the core. It must be kept in mind that Venus thermal history calculations with a variety of outcomes, including inner core solidification with present magnetic field generation and nearly complete solidification of the core, are possible with changes in the values of model parameters (see Stevenson, Spohn and Schubert, 1983, for additional Venus thermal history calculations; also Solomatov, 1993). However, the Venus models that can be constructed with partially frozen cores and present magnetic fields are incompatible with observations of Venus that show it lacks a magnetic field. The Venus models that have nearly completely frozen cores are possible models for the real Venus, but they require very low core concentrations of light alloying elements compared to Earth and are therefore considered unlikely. It is significant that application of an Earth-like thermal history model to Venus predicts a different Venus core structure at present and provides a natural explanation for the lack of a Venus magnetic field.

The coupled core–mantle thermal history model used in the calculations of Figure T5 can also be applied to Mars (Schubert and Spohn, 1990; Schubert et al., 1992). Mars either has no magnetic field at present or at best a very weak one (Luhmann et al., 1992), an observation that places a major constraint on the planet's thermal history just as the lack of a present magnetic field strongly constrains the thermal history of Venus. The thermal history of the Martian core depends strongly on its sulfur content x_S (weight percent) as shown by the models of Figure T6. The amount of light alloying element (sulfur) in the core is important because of the dependence of core melting (solidification) temperature on x_S. Core melting temperature decreases with increasing concentration of the light constituent. Since the light element in the core is excluded from the solidifying inner core, its concentration in the liquid outer core increases with time as the inner core freezes. The melting temperature of the outer core accordingly decreases with time, thereby retarding inner core growth. Accordingly, it is difficult to freeze completely a planetary core that contains an alloying element, making complete core solidification an unlikely explanation, in general, for lack of a magnetic field.

The Mars model in Figure T6 with $x_S = 0.07$ undergoes a sudden change in the rate of core cooling at about 750 Ma, which marks the onset of inner core solidification. With $x_S = 0.14$ there is no freeze-out of an inner core. When inner core solidification occurs, the rate of decline in heat flow from the core is reduced because of latent heat and gravitational potential energy release, the latter occurring as a consequence of the exclusion of the light element S from the growing inner core and the concentration of S in the shrinking liquid outer core.

The dashed curve in Figure T6 represents the heat flux conducted along the core adiabat. When core heat flux falls below this curve, thermal convection can no longer be sustained within the core. This occurs in the model with $x_S = 0.14$ at about a time $t = 2.2$ Ga. In this model there would be no thermal convection in the core and no dynamo action and magnetic field generation after about 2.2 Ga. Thermal convection in the core of the model with $x_S = 0.07$ is possible beyond $t = 4.6$ Ga. In this model, core convection after $t = 750$ Ma is compositional as well as thermal. Indeed, the compositionally driven convection (i.e. convection driven by the release of latent heat and gravitational potential energy on inner core solidification) is responsible for the relatively slow cooling of the core and the maintenance of the core heat flux above the critical value associated with conduction along the adiabat. Core convection in the $x_S = 0.07$ model could occur even after the core heat flux curve falls below the dashed curve (at some time $t > 4.6$ Ga) as a consequence of chemical compositional buoyancy. Thermal convective transport during that time would actually be downward in the core, but compositional buoyancy would be more than adequate to offset the slightly stable state. Core convection, either thermal or compositional, could produce a magnetic field throughout the history of the model with $x_S = 0.07$. However, once the core heat flux falls below the conductive value along the adiabat, compositional convection driven by inner core freezing would be necessary for dynamo action and magnetic field generation. The Martian core thermal history with $x_S = 0.07$ is similar to the thermal evolution of the Earth's core in the model of Figure T5. With $x_S = 0.14$, the Martian core evolves similarly to the core in the Venus model of Figure T5.

According to the models of Figure T6, a weak or absent Martian magnetic field could be explained by the absence of thermal convection in a completely liquid core or the inefficient operation of a compositionally driven dynamo in the presence of a small solid inner core. If x_S in the Martian core is larger than about 10 wt%, the lack of a present Martian magnetic field would probably be explained by the absence of thermal convection in an entirely liquid core, similar to the probable cause of the lack of a magnetic field on Venus. If Mars actually has a small magnetic field and $x_S \gtrsim 10\%$, the magnetic field could be generated by weak compositional convection in the presence of a small solid inner core or by weak thermal convection in a liquid core if the estimate in Figure T6 of the conductive heat flux along the core adiabat is too large.

If x_S in the Martian core is less than about 10%, the absence of a Martian magnetic field would require non-operation of a compositionally driven dynamo in a partially to nearly solidified core (Young and Schubert, 1974). Such values of sulfur content in the Martian core would be easier to reconcile with the existence of a weak Martian magnetic field.

Though our knowledge of Mars (in particular x_S) is inadequate to distinguish among possible thermal history models of the planet, it is clear from these models that determination of the existence or nonexistence of a Martian magnetic field and the size of the Martian core and its sulfur content has much to teach us about planetary physics and dynamo theory. Future exploration of Mars should have the determination of these properties as a primary objective.

The thermal history of Mercury (Schubert et al., 1988), like the histories of Mars and Venus, is also strongly constrained by planetary magnetism but, unlike Mars and Venus, Mercury has a magnetic field

(Connerney and Ness, 1988). Like the other terrestrial planets, Mercury probably formed hot with early differentiation of an iron core. If this core is pure iron, it would have frozen very quickly (Cassen et al., 1976). However, volatile-bearing planetesimals most likely contributed to the accretion of the planet, adding a small amount of sulfur. Under these conditions the strongly depressed Fe–S eutectic has prevented the Mercurian core from freezing completely. Detailed thermal history models (Schubert et al., 1988) incorporating subsolidus convection of the Mercurian mantle indicate that, to date, it may be possible to maintain convection in the outer fluid core of Mercury, perhaps allowing for dynamo generation of the observed magnetic field. A total sulfur abundance of around 2 to 3% by mass relative to iron allows for both the rapid growth of an inner core prior to cessation of early bombardment (to satisfy the geologic constraint of little planetary contraction over geologic time), and the possibility of ongoing outer core convection. Permanent magnetism is at best a marginal explanation of the Mercurian magnetic field.

Additional realism in the modeling of planetary thermal history can be achieved by including effects of crustal differentiation in the models (Spohn, 1991; Schubert et al., 1992). Crustal formation is an important influence on planetary thermal history because the process of forming the crust by magmatism and volcanism removes heat-producing radiogenic elements from the mantle and concentrates them in the crust. The reduction in mantle heat sources allows the mantle to cool more rapidly and increases the lithosphere thickness. This in turn decreases mantle magmatism and crustal production, processes which occur proportionately with mantle convective vigor. Another effect of crustal formation is the thermal blanketing provided by thick crust. Heat escaping from the interior of a planet must be conducted through a surface crustal layer. Crustal rocks are not particularly good conductors of heat and temperatures tend to be higher beneath a crust. The precise way in which crustal formation is incorporated into a planetary thermal history model depends on whether the crust is recycled into the mantle. Planets like Mars and the Moon may be covered by a global basaltic crust which is not recycled. The basaltic oceanic crust of the Earth is largely recycled by plate tectonics, while the Earth's continental crust is somewhat more permanent but still undergoes recycling through the subduction of sediments.

For Earth, and perhaps Venus, the volatile dependence of mantle viscosity can play an important role in mantle self-regulation and thermal history. Mantle viscosity is not only a strong function of temperature, but it depends sensitively on mantle volatile content as well. Dissolved volatiles in the mantle lower the creep activation energy and thus reduce the viscosity at a given temperature. Thus a loss of volatiles from the mantle (degassing or outgassing) would stiffen the mantle and require an increase in mantle temperature to maintain a requisite vigor of convection. Conversely, volatile recharging of the mantle (regassing) by tectonic processes such as subduction, overthrusting and delamination would soften the mantle and require a decrease in mantle temperature to maintain a certain convective vigor. The dependence of mantle viscosity on both temperature and volatile content produces a strong coupling between the evolution of the mantle and that of the atmosphere–hydrosphere (Schubert et al., 1989). Thermal history models that include effects of both volatile- and temperature-dependent mantle viscosity have been developed by McGovern and Schubert (1989).

Many of the satellites in the outer solar system involve special circumstances in their evolution. Among these are Jupiter's satellite Io and Neptune's moon Triton. For both these bodies there is a strong coupling between their orbital and thermal evolutions. Io has a large surface heat loss of about 60 TW (Nash et al., 1986), about 1.7 times larger than the total heat flow through the Earth's surface of about 35.5 TW (Turcotte and Schubert, 1982), even though Earth's radius is 3.5 times larger than Io's radius. Io's surface heat flux of about 1500 mW m^{-2} is approximately 20 times larger than Earth's surface heat flux of about 70 mW m^{-2} (Turcotte and Schubert, 1982). The heat flow from the Moon, based on the mean of two lunar heat flux measurements by Apollo 15 and 17 (18 mW m^{-2}; Langseth, Keihm and Peters, 1976), is about 0.7 TW, only about 1% of Io's heat loss. Clearly, no cosmochemically reasonable concentration of radiogenic elements in Io's interior can account for the enormous amount of heat that must be generated inside the satellite. Io undoubtedly possesses an Earth-like inventory of radiogenic elements which would contribute at most about 1 TW to its surface heat flow, similar to the lunar heat flow. However, most of Io's surface heat loss originates from an energy source that is unimportant for the Earth and the other terrestrial planets. It is this energy source that distinguishes Io from other solar system bodies and is of particular interest here in connection with planetary thermal history.

The major source of Io's internally generated heat is widely accepted to be tidal dissipation (Peale, Cassen and Reynolds, 1979; Yoder, 1979; Yoder and Peale, 1981; Schubert, Stevenson and Ellsworth, 1981; Schubert, Spohn and Reynolds, 1986; Cassen, Peale and Reynolds, 1982; Ross and Schubert, 1985, 1986; Nash et al., 1986; Peale, 1986; Burns, 1986; Segatz et al., 1988). Although Io is rotating around Jupiter synchronously with its orbital period, its relatively large eccentricity, forced by an orbital resonance with Europa and Ganymede (known as the Laplace resonance), causes the tide-raising potential of Jupiter on Io's surface to oscillate. The periodic distortion of Io in response to the time-varying tidal potential gives rise to tidal dissipation and heating in its interior. Tidal friction in Io tends to reduce its orbital eccentricity, but circularization of Io's orbit is prevented by the orbital resonance. The energy dissipated as heat in Io's interior derives from orbital energy, the ultimate source being the rotational energy of Jupiter. Tides raised on Jupiter by Io produce torques which decelerate the planet's spin.

There is too much tidal dissipation in Io for its interior to be entirely solid with material parameters similar to those of the Earth or Moon. Peale, Cassen and Reynolds (1979) and Peale and Cassen (1978) have suggested that Io is a solid shell surrounding a liquid interior, a structure formed as a consequence of runaway melting due to tidal heating. Runaway melting could occur if melting was initiated near the center of Io by tidal dissipation and heat transfer in the surrounding solid part of Io was unable to remove the tidally generated heat because tidal dissipation would increase with the radius of the molten region. However, Schubert, Stevenson and Ellsworth (1981) have pointed out that the thermostat or self-regulation effect of subsolidus mantle convection should prevent runaway melting. Even if runaway melting occurred, efficient liquid state convection would solidify the satellite in a few hundred million years, leaving a differentiated satellite with a possibly molten iron-rich core, a solid mantle and a partially molten asthenosphere (Schubert, Stevenson and Ellsworth, 1981; Schubert, Spohn and Reynolds, 1986). The partially molten asthenosphere would exist because tidal dissipation in a silicate partial melt layer increases as the layer thickness decreases, thereby preventing complete solidification of the asthenosphere. While a largely molten Io cannot be entirely ruled out, it seems most plausible that Io has a hot, near-solidus mantle and a partially molten asthenosphere. The major uncertainty in Io's present structure concerns the existence and nature of the asthenosphere.

Coupled thermal and orbital evolution models of Io have been presented by Ojakangas and Stevenson (1986) and Fischer and Spohn (1990). Tidal dissipation couples Io's thermal and orbital evolution because it is a sink of energy in the orbital energy balance and a source of energy in the thermal balance. The tidal dissipation rate depends on the orbital parameters (mainly eccentricity and mean motion) and Io's internal temperature through the temperature dependence of the rheological parameters. The orbital evolution in the model of Fischer and Spohn (1990) is based on the theory of the Laplace resonance of Yoder and Peale (1981) and includes the effects of dissipation in both Io and Europa. The orbital evolution theory provides an ordinary differential equation for the secular variation of Io's mean motion, the integration of which, together with other dynamical relations, leads to the determination of Io's eccentricity. The thermal evolution in the model of Fischer and Spohn (1990) incorporates parameterized convective mantle heat transfer and the secular variation in Io's temperature, following the approach of Schubert, Cassen and Young (1979) with the radiogenic heating term replaced by tidal heating.

Triton has a retrograde motion, small eccentricity and large inclination, properties that suggest the satellite was captured by Neptune from heliocentric orbit (Farinella et al., 1980; McKinnon, 1984; Cruikshank and Brown, 1986; Burns, 1986; Goldreich et al., 1989). The capture event transforms orbital energy into a heat source in Triton's interior through tidal dissipation. Ross and Schubert (1990) have presented a coupled orbital–thermal model of this event. They show that the tidal heating is so enormous as to overwhelm the ability of solid state convection to remove it from Triton's interior, thereby resulting in global melting and differentiation of the interior into a silicate core and an ice mantle. Also, the temperature dependence of ice rheology can cause the dramatic changes in Triton's orbital and internal states to occur in a relatively brief period of time (100 Ma) long after (1 Ga) capture into Neptune orbit.

Still other outer solar system satellites (including the planet Pluto) may have experienced unique events in their evolutions. Many of these bodies, like Triton, are composed partly of ice and rock, and aside from special events, cooling histories of these bodies can be developed (Schubert, Spohn and Reynolds, 1986) similar to the models of secular cooling of the terrestrial planets discussed above. The rheology of ice governs subsolidus convection in these rock–ice satellites and the viscosity of ice, like the viscosity of rock, has an Arrhenius dependence on temperature. One of the major questions about the rock–ice satellites is the extent to which their interiors are differentiated, i.e. the extent to which rock and ice are separated. Accretional energy controls the extent of early differentiation with larger satellites being heated more. Accretional heating may have been large enough to have differentiated the Galilean satellite Ganymede which has an endogenically modified surface, but the Galilean satellite Callisto, about the same size as Ganymede, is probably undifferentiated, as attested to by its heavily cratered primordial surface. Accretional heating is probably insufficient to have differentiated the interiors of the small icy Saturnian satellites Mimas, Tethys, Dione, Rhea and Iapetus. Radiogenic heating in the rock part of a rock–ice mixture can lead to differentiation after accretion, but the process is mitigated against by subsolidus convective heat transport by the ice component. Tidal dissipation can also differentiate a rock–ice satellite under special circumstance, as we have seen probably happened with Triton.

While the rheology of water ice probably controls the thermal evolution of rock–ice satellites, preventing temperatures from rising above 273 K and melting the interiors, the presence of low melting point constituents, such as ammonia hydrate, may be crucial to their resurfacing. Internal melting of these components and their upward migration and extrusion onto the surface could occur even with a satellite in a relatively cold subsolidus state with regard to its major water-ice constituent. Some of the smaller rock–ice satellites such as Dione and Rhea may have been resurfaced in this way.

The major theme of this article, that most of the inner planets and rock–ice satellites of the outer planets have simply cooled from hot initial states in a gradual way over most of geologic time, is probably true as a first-order approximation to the thermal evolution of the solar system. However, there may be individual exceptions to this general picture, as we have already noted in our discussion of Triton and Io. Furthermore, even when a planet undergoes general cooling over the long term, it may also experience episodes of enhanced thermal activity or heating. Variability in hot spot and continental flood basalt activity on Earth (Larson and Olson, 1991) is one example of this. The superposition of episodic thermal activity on the long-term secular cooling of a planet could be expected on basic fluid dynamical grounds since mantle convection is a chaotic process (Schubert, 1992) subject to random oscillations in mantle temperature and heat flux. This aspect of mantle convection and other physical processes in the mantle that might lead to sudden overturnings, for example, are not included in the parameterized model of self-regulated planetary cooling presented in this article.

Gerald Schubert

Bibliography

Burns, J.A. (1986) The evolution of satellite orbits, in *Satellites* (eds J.A. Burns and M.S. Matthews). Tucson: University of Arizona Press, pp. 117–58.

Birch, F. (1965) Energetics of core formation. *J. Geophys. Res.*, **70**, 6217–221.

Cameron, A.G.W. (1984) Formation of the prelunar accretion disk (abstract), in *Papers Presented to the Conference on the Origin of the Moon*. Houston: Lunar and Planetary Institute, p. 58.

Cameron, A.G.W. (1986) The impact theory for origin of the Moon, in *Origin of the Moon* (eds W.K. Hartmann, R.J. Phillips and G.J. Taylor). Houston: Lunar and Planetary Institute, pp. 609–16.

Cameron, A.G.W. and Benz, W. (1991) The origin of the Moon and the single impact hypothesis IV. *Icarus*, **92**, 204–16.

Cameron, A.G.W. and Ward, W.R. (1975) The origin of the Moon (abstract), in *Lunar Sci. VII*. Houston: The Lunar Science Institute, pp. 120–2.

Carter, N.L. (1976) Steady state flow of rocks. *Rev. Geophys. Space Phys.*, **14**, 301–60.

Cassen, P.M., Peale, S.J. and Reynolds, R.T. (1982) Structure and thermal evolution of the Galilean satellites, in *Satellites of Jupiter* (ed. D. Morrison). Tucson: University of Arizona Press, pp. 93–128.

Cassen, P., Young, R.E., Schubert, G. and Reynolds, R.T. (1976) Implications of an internal dynamo for the thermal history of Mercury. *Icarus*, **28**, 501–8.

Cassen, P., Reynolds, R.T., Graziani, F. et al. (1979) Convection and lunar thermal history. *Phys. Earth Planet. Inter.*, **19**, 183–96.

Connerney, J.E. and Ness, N.F. (1988) Mercury's magnetic field and interior, in *Mercury*, (eds F. Villas, C.R. Chapman and M.S. Matthews). Tucson: University of Arizona Press, pp. 494–513.

Cruikshank, D. and Brown, R. (1986) Satellites of Uranus and Neptune and the Pluto–Charon system, in *Satellites* (eds J.A. Burns and M.S. Matthews). Tucson: University of Arizona Press, pp. 836–73.

Davies, G.F. (1980) Thermal histories of convective Earth models and constraints on radiogenic heat production in the Earth. *J. Geophys. Res.*, **85**, 2517–30.

Farinella, P., Milani, A., Nobili, A. and Valsecchi, V. (1980) Some remarks on the capture of Triton and the origin of Pluto. *Icarus*, **44**, 810–2.

Ferrari, A.J., Sinclair, W.S., Sjogren, W.L. et al. (1980) Geophysical parameters of the Earth–Moon system. *J. Geophys. Res.*, **85**, 3939–51.

Fischer, H.-J. and Spohn, T. (1990) Thermal–orbital histories of viscoelastic models of Io (J1). *Icarus*, **83**, 39–65.

Flasar, F.M. and Birch, F.J. (1973) Energetics of core formation: a correction. *J. Geophys. Res.*, **78**, 6101–3.

Goldreich, P., Murray, N., Longaretti, P.Y. and Banfield, D. (1989) Neptune's story. *Science*, **245**, 500–4.

Gubbins, D. (1977) Energetics of the Earth's core. *J. Geophys.*, **43**, 453–64.

Hartmann, W.K. (1986) The impact-trigger hypothesis, in *Origin of the Moon* (eds W.K. Hartmann, R.J. Phillips and G.J. Taylor). Houston: Lunar and Planetary Institute, pp. 579–608.

Hartmann, W.K. and Davis, D.R. (1975) Satellite-sized planetesimals and lunar origin. *Icarus*, **24**, 504–15.

Hood, L.L. (1986) Geophysical constraints on the lunar interior, in *Origin of the Moon* (eds W.K. Hartmann, R.J. Phillips and G.J. Taylor). Houston: Lunar and Planetary Institute, pp. 361–410.

Kaula, W.M. (1979) Thermal evolution of Earth and Moon growing by planetesimal impacts. *J. Geophys. Res.*, **84** 999–1008.

Langseth, M.G., Keihm, S.J. and Peters, K. (1976) Revised lunar heat-flow values, in *Proc. Lunar Planet. Sci. Conf.*, **7**, pp. 3143–71.

Larson, R.L. and Olson, P. (1991) Mantle plumes control magnetic reversal frequency. *Earth Planet. Sci. Lett.*, **107**, 437–47.

Luhmann, J.G., Russell, C.T., Brace, L.H. and Vaisberg, O.L. (1992) The intrinsic magnetic field and solar-wind interaction of Mars, in *Mars* (eds H.H. Kieffer, B.M. Jakosky, C.W. Snyder, and M.S. Matthews). Tucson: University of Arizona Press, pp. 1090–1134.

McElhinny, M.W. and Senanayake, W.E. (1980) Paleomagnetic evidence for the existence of the geomagnetic field 3.5 Ga ago. *J. Geophys. Res.*, **85**, 3523–8.

McGovern, P. and Schubert, G. (1989) Thermal evolution of the Earth: effects of volatile exchange between atmosphere and interior. *Earth Planet. Sci. Lett.*, **96** 27–37.

McKinnon, W.B. (1984) On the origin of Triton and Pluto. *Nature*, **311**, 355–8.

Nash, D.B., Carr, M.H., Gradie, J. et al. (1986) Io, in *Satellites* (eds J.A. Burns and M.S. Matthews). Tucson: The University of Arizona Press, pp. 629–88.

Ojakangas, G.W. and Stevenson, D.J. (1986) Episodic volcanics of tidally heated satellites with application to Io. *Icarus*, **66**, 341–58.

Peale, S.J. (1986) Orbital resonances, unusual configurations, and exotic rotation states among the planetary satellites, in *Satellites* (eds J.A. Burns and M.S. Matthews). Tucson: The University of Arizona Press, pp. 159–223.

Peale, S.J. and Cassen, P. (1978) Contribution of tidal dissipation to lunar thermal history. *Icarus*, **36**, 245–69.

Peale, S.J., Cassen, P. and Reynolds, R.T. (1979) Melting of Io by tidal dissipation, *Science*, **203**, 892–4.
Poirier, J.-P. (1985) *Creep of Crystals*. Cambridge: Cambridge University Press.
Ross, M. and Schubert, G. (1985) Tidally forced viscous heating in a partially molten Io. *Icarus*, **64**, 391–400.
Ross, M. and Schubert, G. (1986) Tidal dissipation in a viscoelastic planet, in *Proc. Lunar Planet. Sci. Conf.*, **16**. *J. Geophys. Res.*, **91**, D447–52.
Ross M. and Schubert, G. (1990) The coupled orbital and thermal evolution of Triton. *Geophys. Res. Lett.*, **17**, (10), 1749–52.
Runcorn, S.K. (1983) Lunar magnetism, polar displacements and primeval satellites in the Earth–Moon system. *Nature*, **304**, 589–96.
Russell, C.T., Coleman, P.J., Jr and Goldstein, B.E. (1981) Measurements of the lunar induced magnetic moment in the geomagnetic tail: evidence for a lunar core, in *Proc. Lunar Planet. Sci. Conf.* **12B**, pp. 831–6.
Schubert, G. (1979) Subsolidus convection in the mantles of terrestrial planets. *Ann. Rev. Earth Planet. Sci.*, **7**, 289–342.
Schubert, G. (1992) Numerical models of mantle convection. *Ann. Rev. Fluid Mech.*, **24**, 359–94.
Schubert, G. and Spohn, T. (1990) Thermal history of Mars and the sulfur content of its core. *J. Geophys. Res.*, **95**, 14095–104.
Schubert, G., Cassen, P. and Young, R.E. (1979) Subsolidus convective cooling histories of terrestrial planets. *Icarus*, **38**, 192–211.
Schubert, G., Spohn, T. and Reynolds, R.T. (1986) Thermal histories, compositions, and internal structures of the moons of the solar system, in *Satellites* (eds J.A. Burns and M.S. Matthews). Tucson: University of Arizona Press, pp. 224–92.
Schubert, G., Stevenson, D. and Cassen, P. (1980) Whole planet cooling and the radiogenic heat source contents of the Earth and moon. *J. Geophys. Res.*, **85**, 2531–8.
Schubert, G., Stevenson, D.J. and Ellsworth, K. (1981) Internal structures of the Galilean satellites. *Icarus*, **47**, 46–59.
Schubert, G., Ross, M.N., Stevenson, D.J. and Spohn, T. (1988) Mercury's thermal history and the generation of its magnetic field, in *Mercury*, (eds F. Villas, C.R. Chapman and M.S. Matthews). Tucson: University of Arizona Press, pp. 429–60.
Schubert, G., Turcotte, D.L., Solomon, S.C. and Sleep, N. (1989) Coupled evolution of the atmospheres and interiors of planets and satellites, in *Origin and Evolution of Planetary and Satellite Atmospheres* (eds S.K. Atreya, J.B. Pollack and M.S. Matthews). Tucson: University of Arizona Press, pp. 450–83.
Schubert, G., Solomon, S.C., Turcotte, D.L. *et al.* (1992) Origin and thermal evolution of Mars, in *Mars* (eds H.H. Kieffer, B.M. Jakosky, C.W. Snyder, and M.S. Matthews). Tucson: University of Arizona Press, pp. 147–83.
Segatz, M., Spohn, T., Ross, M.N. and Schubert, G. (1988) Tidal dissipation, surface heat flow, and figure of viscoelastic models of Io. *Icarus*, **75**, 187–206.
Sharpe, N.H. and Peltier, W.R. (1978) Parameterized mantle convection and the Earth's thermal history. *Geophys. Res. Lett.*, **5** 737–40.
Sharpe, N.H. and Peltier, W.R. (1979) A thermal history for the Earth with parameterized convection. *Geophys. J. Roy. Astron. Soc.*, **59**, 171–203.
Shaw, G.H. (1978) Effects of core formation. *Phys. Earth Planet. Inter.*, **16**, 361–9.
Smith, J.V., Anderson, A.T., Newton, R.C. *et al.* (1970) Petrologic history of the Moon inferred from petrography, mineralogy, and petrogenesis of Apollo 11 rocks, in *Proc. Apollo 11 Lunar Sci. Conf.*, pp. 897–925.
Solomatov, V.S. (1993) Parameterization of temperature- and stress-dependent viscosity convection and the thermal evolution of Venus, in *Creep in the Solar System: Observations, Modelling, and Theory* (eds D.B. Stone and S.K. Runcorn). Dordrecht: Kluwer, pp. 131–45.
Solomon, S.C. (1986) On the early thermal state of the Moon, in *Origin of the Moon* (eds W.K. Hartmann, R.J. Phillips and G.J. Taylor). Houston: Lunar and Planetary Institute, pp. 435–52.
Solomon, S.C. and Chaiken, J. (1976) Thermal expansion and thermal stress in the Moon and terrestrial planets: clues to early thermal history, in *Proc. Lunar Planet. Sci. Conf.*, **7**, pp. 3229–43.
Spohn, T. (1991) Mantle differentiation and thermal evolution of Mars, Mercury, and Venus. *Icarus*, **90**, 222–36.

Stevenson, D.J. (1984) Lunar origin from impact on the Earth: is it possible (abstract), in *Papers presented to the Conference on the Origin of the Moon*. Houston: Lunar and Planetary Institute, p. 60.
Stevenson, D.J. (1987) Origin of the Moon – the collision hypothesis. *Ann. Rev. Earth Planet. Sci.*, **15**, 271–315.
Stevenson, D.J. (1989) Formation and early evolution of the Earth, in *Mantle Convection*, (eds W.R. Peltier). New York: Gordon and Breach, pp. 817–73.
Stevenson, D.J. and Turner, J.S. (1979) Fluid models of mantle convection, in *The Earth, Its Origin, Evolution and Structure* (ed. M.W. McElhinny). New York: Academic Press. pp. 227–63.
Stevenson, D.J., Spohn, T. and Schubert, G. (1983) Magnetism and thermal evolution of the terrestrial planets. *Icarus*, **54**, 466–89.
Taylor, S.R. (1986) The origin of the Moon: geochemical considerations, in *Origin of the Moon* (eds W.K. Hartmann, R.J. Phillips and G.J. Taylor). Houston: Lunar and Planetary Institute, pp. 124–43.
Thompson, A.C. and Stevenson, D.J. (1983) Two-phase gravitational instabilities in thin disks with application to the origin of the Moon (abstract), in *Lunar Planet. Sci. XIV*. Houston: Lunar and Planetary Institute, pp. 787–8.
Tozer, D.C. (1965) Thermal history of the Earth: 1. The formation of the core. *Geophys. J. Roy. Astron. Soc.*, **9**, 95–112.
Tozer, D.C. (1967) Towards a theory of thermal convection in the mantle, in *The Earth's Mantle* (ed. T.F. Gaskell). London: Academic Press, pp. 325–53.
Turcotte, D.L. (1980) On the thermal evolution of the Earth. *Earth Planet. Sci. Lett.*, **48**, 53–8.
Turcotte, D.L. and Schubert, G. (1982) *Geodynamics*. New York: John Wiley.
Turcotte, D.L., Cooke, F.A. and Willeman, R.J. (1979) Parameterized convection within the moon and the terrestrial planets, in *Proc. Lunar Planet. Sci. Conf.*, **10**, pp. 2375–92.
Warren, P.H. (1985) The magma ocean concept and lunar evolution. *Ann. Rev. Earth Planet. Sci.*, **13**, 201–40.
Weertman, J. and Weertman, J.R. (1975) High temperature creep of rock and mantle viscosity. *Ann. Rev. Earth Planet. Sci.*, **3**, 293–315.
Wetherill, G.W. (1985) Occurrence of giant impacts during the growth of terrestrial planets. *Science*, **228**, 877–9.
Wood, J.A. (1986) Moon over Mauna Loa: a review of hypotheses of formation of Earth's Moon, in *Origin of the Moon* (eds W.K. Hartmann, R.J. Phillips and G.J. Taylor). Houston: Lunar and Planetary Institute. pp. 17–55.
Wood, J.A., Dickey, J.S., Marvin, U.B. and Powell, B.N. (1970) Lunar anorthosites and a geophysical model of the Moon, in *Proc. Apollo 11 Lunar Sci. Conf.*, pp. 965–88.
Yoder, C.F. (1979) How tidal heating in Io drives the Galilean orbital resonance locks. *Nature*, **279**, 767–70.
Yoder, C.F. (1981) The free librations of a dissipative Moon. *Phil. Trans. Roy. Soc. London*, **A303**, 327–38.
Yoder, C.F. and Peale, S.J. (1981) The tides of Io. *Icarus*, **47**, 1–35.
Young R. and Schubert, G. (1974) Temperatures inside Mars: is the core liquid or solid? *Geophys. Res. Lett.*, **1**, 157–60.

Cross references

Core
Earth–Moon system: origin
Mantle
Mantle convection
Moon: origin

THERMAL PLASMA INSTRUMENTATION

The incidence of solar photons with energies greater than the ionization potentials of planetary gases is the dominant ionization source for the dayside ionospheres of the inner planets Earth, Venus and Mars. Long-range coulomb forces increase the charged particle collision rates so that they can have a temperature that differs from that of the neutral atmosphere even where the atmosphere is only 0.1% ionized. In the upper F region of the ionosphere (q.v.), the vastly different electron and ion masses further allow these gases also to have distinct temperatures, which can differ by a factor of more than two. Elastic collisions of the photoelectrons ejected by solar

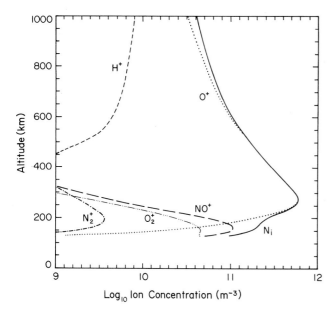

Figure T7 Altitude profile of Earth's ionosphere at middle latitudes on the dayside.

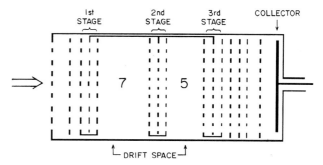

Figure T8 The Bennett tube ion mass spectrometer

ultraviolet radiation first heat the ambient electrons, which pass the heat to the colder ions, which in turn heat the neutral atmosphere (i.e. $T_e > T_i > T_n$). At high latitudes high-speed plasma convection (from externally applied electric fields) through the neutral atmosphere can heat the ions above the electron temperature. This section deals with instruments that have been used to diagnose properties of the thermal plasmas in these inner planetary ionospheres, which have temperatures in the range from a few hundred up to tens of thousands of kelvin. The sensors are usually moving supersonically through the ions, but not supersonically through the electrons. The sensors also have an electric potential that is different from that of the ambient plasma. Of interest are the ion composition and concentration, ion bulk motions, the ion and electron temperatures and spatial gradients in these quantities. There have been no *in situ* measurements made in the outer planetary ionospheres, where plasma conditions are quite different and where spacecraft velocities are much higher.

Before the end of World War II all ionospheric measurements were made using ground-based radio techniques, but the advent of sounding rockets, satellites and space probes has led to a substantial increase in our knowledge of planetary ionospheres. Figure T7 provides a representative altitude profile of Earth's mid-latitude ionosphere showing the range of ion concentrations and composition typically encountered by rocket- and satellite-borne instrumentation in the daytime. At night the ion concentrations below 200 km decrease by a factor of ten to a hundred. The high-latitude ionosphere, which is subjected to bombardment by energetic auroral particles, is much more structured, and vertical escape of the light ions H^+ and He^+ causes their concentration to be much less than at mid-latitudes. In spite of their different distances from the Sun, and the fact that the surface atmospheric pressure ratio of Mars : Earth : Venus is $\sim 1 : 100 : 10\,000$, the daytime electron concentration profiles versus altitude on Mars (Hanson, Sanatani and Zuccaro, 1977) and Venus (Taylor *et al.* 1980a) are quite similar to that shown in Figure T7, even though their ion composition is quite different.

Ion mass spectrometers

Ion mass spectrometers (IMS) have been used on rockets, satellites and space probes to measure the ion concentrations in planetary ionospheres. Many different kinds of mass analysers have been used for this purpose, but the Bennett tube, 90° magnetic sector and quadrupole devices have been the most widely employed.

The Bennett tube

The first IMS to be used in flight by both the US (Johnson, Meadows and Holmes, 1958) and the USSR (Istomin and Pokhunkov, 1963) was the Bennett tube, an ion velocity-selecting device operated at radio frequencies. Bennett (1950) showed that ions with velocity, v, would gain a maximum energy on traversing a set of three equally spaced grids with voltage $V_g = V_{rf} \cos(\omega t + \theta)$ applied to the center grid when

$$v = \frac{s\omega}{2.3311} \qquad (T10)$$

where s is the grid separation and $\omega = 2\pi f$ is the angular frequency. The maximum energy gain of ~ 1.45 ($e\,V_{rf}$) occurs when the phase angle θ upon ion arrival at the first grid is given by

$$\theta = \pi - \frac{s\omega}{v} \qquad (T11)$$

For a given energy the mass of the ion (in amu) having this resonant velocity is given by

$$m = \frac{0.266\,E}{s^2 f^2} \qquad (T12)$$

where E is the ion energy in electron volts, s is in centimeters, and f is in megahertz. The time required for a resonant ion to traverse a grid triplet is approximately 3/4 of a wave period, τ. The usual Bennett IMS flight configuration, shown in Fig. T8, uses three sets of grid triplets separated by $7v\tau$ and $5v\tau$, the so-called 7–5 cycle tube. Ions, entering from the left, are pre-accelerated into the tube to energy E by a sawtooth voltage waveform, and the mass of the resonant ions increases linearly with this voltage according to equation (T12), provided that the velocity change through the grids is small compared to v. A stopping potential is applied to a grid in front of the collector to select only resonant ions, which gained the most energy in the tube. Bennett IMSs have also been used in Earth orbit (Brinton *et al.*, 1973) and at Venus (Taylor *et al.*, 1980b).

Magnetic sector mass spectrometers

These devices utilize a uniform magnetic deflection field B to separate ions of different charge-to-mass ratio. The ions are pre-accelerated to energy qV, where q is the ionic charge and V is the accelerating voltage, so that their radius of curvature in the magnetic analyzer is given by

$$R = 1/B(2mV/q)^{1/2} \qquad (T13)$$

where m is the ion mass.

Many different geometries have been employed in these instruments, but one that was designed specifically for Earth's ionosphere (Hoffman *et al.*, 1973) is sketched in Figure T9. There are three exit apertures, each with an electron multiplier detector, that accept ions with the same energy but with mass-to-charge ratios that differ by $1 : 4 : 16$. Thus H^+, He^+ and O^+, the three dominant ions in the upper F region, can be detected simultaneously (D^+, O^{++} and O_2^+ are another pertinent combination). The entire mass range of interest (1 to 56 amu), including overlap at masses He^+ and O^+ where different channel sensitivities can be compared, is covered with a fourfold change in V.

Quadrupole mass filter

This device ideally uses four hyperbolic pole pieces, but has usually been constructed with four circular rods arranged as shown in Figure

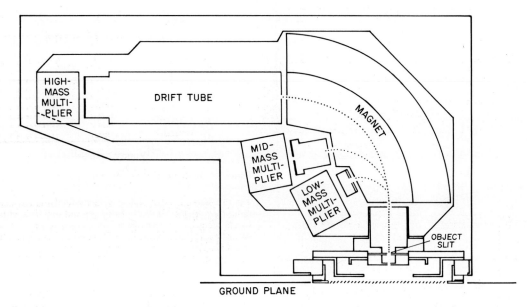

Figure T9 Magnetic sector mass spectrometer.

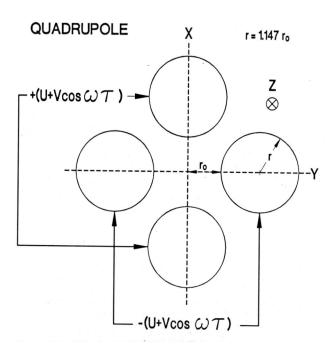

Figure T10 Quadrupole mass-filter spectrometer.

T10, with voltages applied to them as indicated. The pole pieces are extended in the Z-direction. Ions of either charge can be directed down the Z-axis and if they have the proper charge to mass ratio they will be stably trapped in the X–Y directions and be transmitted through the device to a detector. A ratio of (U/V) less than 0.168 is required for stable trapping, and the mass resolution $(m/\Delta m)$ increase and the transmission decreases as this limit is approached. Singly charged ions of mass m (amu) are selected when $(U/V) \leq 0.168$ where

$$m = \frac{V}{7.219 r_0^2 f^2} \qquad (T14)$$

and r_0 is in cm, f is in MHz and V is in volts. A mass scan can be obtained at a fixed frequency by sweeping V and U, keeping their ratio constant. The instrument is described in detail by Paul, Reinhard and von Zahn (1965). These devices have been utilized principally for low-altitude rocket measurements (Narcisi and Bailey, 1971; Goldberg and Aikin, 1971) for both positive and negative ions. For this purpose they have a small entrance aperture and are strongly pumped to reduce the gas pressure in the analyser.

Planar retarding potential analyzer

Planar retarding potential analyzers (RPAs) have been utilized on rockets and satellites for at least three decades (e.g. Hanson et al., 1973, 1981; Spenner and Dumbs, 1974; Knudsen et al. 1980). They usually consist of an axially symmetric planar sensor that has several transparent grids in front of a solid collector as shown in the cross-section in Figure T11. One or more of these grids, called the retarding or sweep grids, have a voltage waveform applied to them that limits the charged particles that can reach the collector according to their normal energy component. The devices have been used to measure the energy distribution of both ions and electrons, but their principal application has been on low-altitude satellites to measure parameters of the ion gas. In low Earth orbit, ambient ions have a bulk velocity of ~ 7500 m s^{-1} in the satellite frame of reference, which amounts to $\sim 1/3$ eV per amu. Figures T11b and T11c show typical collector current versus retarding voltage curves recorded in the ionospheric F layer, where O$^+$ and the molecular ions N$_2^+$, NO$^+$ and O$_2^+$ are present (T11b) and in the protonosphere boundary layer, where H$^+$, He$^+$ and O$^+$ are the major ion constituents (T11c). The lines drawn through the measured values are least-squares fits of a theoretical expression to the data, from which the absolute ion concentrations, the ion temperature and the bulk ion drift velocity component toward the satellite can be derived. The latter is obtained by subtracting the known vehicle velocity from the measured ram velocity.

Ion drift meter

The ion drift meter (IDM) is a device that has been used in low Earth orbit to determine the transverse bulk velocity components of the F-region ions (e.g. Hanson et al., 1973; Hanson and Heelis, 1975; Heelis et al., 1981). It does this by measuring the arrival angles of the ambient ions in the spacecraft frame of reference. This directional information, together with the longitudinal (ram) velocity component measured by an RPA, can be utilized to determine the bulk ion vector velocity if the spacecraft attitude is known. Attitude knowledge is typically available to a tenth or a few tenths of a degree, and an attitude uncertainty of 1° corresponds to an ion velocity uncertainty of ~ 135 m s^{-1}.

The sketch in Figure T12 shows the principle of operation of the IDM. A square aperture collimates the ions entering the internally gridded sensor at angle α in the X–Z plane, where V and v are the satellite and the ion velocity vectors; these are related by

$$\tan \alpha_z = (-V_z + v_z)/(-V_x + v_x) \qquad (T15)$$

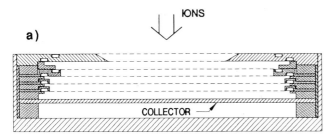

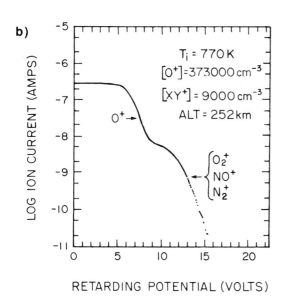

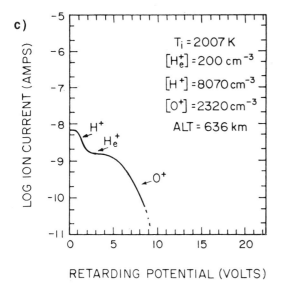

Figure T11 (a) Planar retarding potential analyzer. (b) F layer collector current versus retarding voltage. (c) Protonosphere boundary layer collector current versus retarding voltage.

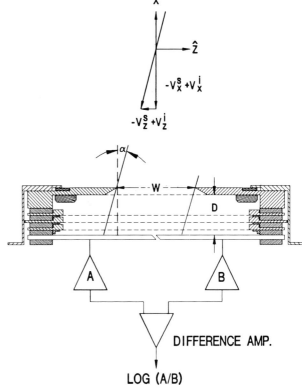

Figure T12 Ion drift meter.

After traversing a field-free 'drift space' the ions pass through a negative electron suppressor grid and strike four symmetric pie-shaped collector segments. Measurement of the ratios of the currents to the different segments determine the arrival angles of the ions. The log of the ratio of the currents to the logarithmic amplifiers A and B is measured with a difference amplifier. Vertical and horizontal arrival angles are obtained, depending on which collector segments are used for comparison. The corresponding velocity component, v, is then obtained from

$$v_{z,y} = v(\text{RPA}) \tan \alpha_{z,y} \qquad (T16)$$

where $v(\text{RPA})$ is the ram component $[v - V] \cdot \hat{x}$ measured by the RPA. If no RPA is available the satellite velocity alone $(-V \cdot \hat{x})$ can be used instead in equation (T16) with only a few percent error at invariant latitudes less than 60°. At high latitude this error would typically be $\sim 10\%$, but it could range up to $\sim 50\%$ because of the very strong convection sometimes observed there.

In-flight calibration of the IDM can be obtained from a spinning spacecraft, where the arrival angles vary in a known fashion, or from observing the corotation velocity of the ionosphere near the equator (~ 450 m s^{-1}) from a high-inclination satellite. The device works very well with the heavy ions in the F region, which have relatively large momentum in the satellite frame of reference, but it is unreliable when H$^+$ is a dominant ion.

Hyperbolic retarding potential analyzer

The hyperbolic retarding potential analyzer (HARP) is an electrostatic particle energy analyzer that uses hyperbolic shaped electrodes. Its axially symmetric quadrupole electric field is established by the body electrode (3) and the two end cap electrodes (1 and 2) shown in Figure T13. The parts of the end caps around the Z-axis are annular slits. The entrance and exit apertures are small openings on the Z-axis near the foci of the end caps. There is axial symmetry about the vertical Z-axis. Inside the cavity the trajectories for particles of energy E and charge q depends upon qU_b/E. The energy of the particles passing through the entrance aperture that can reach the exit

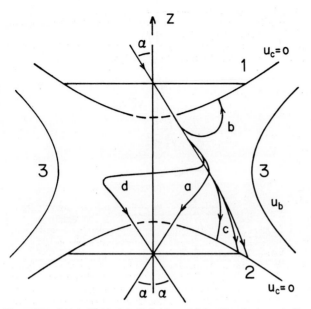

Figure T13 Quadrupole electric field established by the hyperbolic retarding potential analyzer (discussion in text).

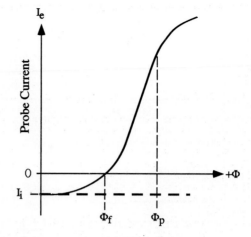

Figure T14 Langmuir probe current–voltage curve.

aperture (paths a and d) is slightly less than the potential of the body electrode (U_b) with respect to the grounded end caps. Either positive or negative particles can be differentially analyzed depending on the sign of the retarding potential applied to the body electrode. Particles with too low an energy will not be able to overcome the potential barrier at the central plane perpendicular to the Z-axis and will describe paths like the one labeled b. Particles with too high an energy cannot arrive at the exit aperture on the Z-axis and follow the paths labeled c. Those with the selected energy will exit along paths a or d. The location, size and geometry of the exit apertures define the elevation angle α and the azimuthal angular acceptance around the Z-axis.

The HARP was developed for flight by the University of Michigan rocket program (Shyn, Sharp and Hays, 1976), and a more elaborate version was included on the Phobos mission to Mars (Szucs *et al.*, 1990).

Langmuir probe

The electron temperature, T_e, is the principal parameter derived from a Langmuir probe (LP), though the electron and ion concentrations and the absolute potential difference between probe ground reference and plasma are also often obtained. These probes rely on the fact that the gas concentration of a Maxwellian gas in a potential field decreases exponentially with the increase in the ratio of the particle potential energy to the gas temperature. Thus the electron concentration n_e at a surface with an electric potential ϕ with respect to the ambient plasma is given by

$$n_e = n_o \exp - \frac{e\phi}{kT_e} \tag{T17}$$

where n_o is the ambient plasma concentration, e is the electron charge and k is the Boltzmann constant. The electron flux to such a surface is given by $n_e \bar{v}_e/4$ (where $\bar{v}_e$ is the average electron velocity) so that the electron current to a biased probe also varies as $\exp - (e\phi/kT)$.

Many different sensor geometries and waveforms, $\phi(t)$, have been employed to determine the electron temperature. These schemes can be divided into two broad classes depending on whether they utilize any shielding grids. Unshielded probes have a current collector directly exposed to the plasma, whereas shielded probes have the collector shielded by one or more grids. Unshielded probes measure the sum of the electron and ion currents, whereas gridded probes may measure only the electron current, and thus can see farther into the high-energy Boltzmann tail of the electron distribution. The unshielded probes can be spheres (Sagalyn, Smiddy and Bhargava, 1965), plates (Wrenn, 1969), or cylinders (Brace, Theis and Dalgarno, 1973), and they usually sweep some kind of 'guard' surface in addition to the active surface to eliminate edge or end effects and to improve the electron collection geometry.

The schematic in Figure T14 shows a typical unshielded LP current–voltage (I–ϕ) characteristic curve. The current is positive at the far left where the probe potential ϕ is very negative with respect to the plasma potential (ϕ_p) and goes to zero at the floating potential (ϕ_f) where the net current to the probe is zero. ϕ_f is often the reference potential (spacecraft ground) and defined to be zero. The electron current increases exponentially up to the plasma potential (ϕ_p), above which it is less sensitive to ϕ. The electron temperature is obtained from the shape of the I–ϕ curve in the so-called retarding region below ϕ_p. The shape of the curve above ϕ_p depends on the probe geometry. The ion current, represented by the heavy dashed line, is relatively insensitive to ϕ because the ion energy in the satellite frame is usually $\gg kT_i$.

Floating rectification probe

Problems have arisen on many flight LPs because of surface contamination, which causes hysteresis in the characteristic I–ϕ curves and results in derived temperatures that are too high. Putting a high (~ 100 V) potential on the sensor in the ionosphere results in particle bombardment that can clean the probe surface in orbit, but a rectification probe provides an alternative solution.

The rectification electron temperature probe operates in a high frequency range, greater than the ion plasma frequency but less than the electron plasma frequency, where the contamination layer does not distort the I–ϕ curves. Two symmetrically mounted sensors are used in this technique; one of them is electronically isolated and thus is maintained at the floating potential. The other has a sinusoidal voltage applied to it, but it is also electronically isolated so that it draws no net current. Because of the nonlinear exponential dependence of the electron current on the sensor potential, a greater electron current is gathered during the positive half of the sine wave than is suppressed during the negative half cycle. In order for the average electron current to remain the same (equal to the ion current), a negative shift takes place in the (average) floating potential. The difference in the floating potentials of the two sensors with and without the sine wave, $\Delta \phi$, can be shown to be (Hirao and Oyama, 1970)

$$\Delta\phi = -\frac{kT_e}{e} \ln I_o \left(\frac{ea}{kT_e}\right) \tag{T18}$$

where a is the amplitude of the sine wave and I_o is the modified Bessel function of zero order. There is a concern that the T_e measured in this manner is characteristic of the velocity distribution several kT_e out on the tail of the Boltzmann distribution, and this could be a problem if energy is being cascaded into the electron gas from a high-energy tail. As a guard against this, measurements at more than one amplitude (e.g. a and $2a$) are compared for consistency. These devices have been flown on rockets and several satellites (Oyama, 1991) with good results.

W.B. Hanson and R.A. Heelis

Bibliography

Bennett, W.H. (1950) Radio frequency mass spectrometer. *J. Appl. Phys.*, **21**, 143–9.
Brace, L.H., Theis, R.G. and Dalgarno, A. (1973) The cylindrical electrostatic probes for Atmosphere Explorer-C, -D, and -E. *Radio Sci.*, **8**, 341–8.
Brinton, H.C., Scott, L.R., Pharo, M.W. III and Coulson, J.T.C. (1973) The Bennett ion-mass spectrometer on Atmosphere Explorer-C and -E. *Radio Sci.*, **8**, 323–32.
Goldberg, R.A. and Aikin, A.C. (1971) Studies of positive-ion composition in the equatorial D region ionosphere. *J. Geophys. Res.*, **76**, 8352–64.
Hanson, W.B. and Heelis, R.A. (1975) Techniques for measuring bulk gas motions from satellites. *Space Sci. Instrum.*, **1**, 493–524.
Hanson, W.B., Sanatani, S. and Zuccaro, D.R. (1977) The Martian ionosphere as observed by the Viking retarding potential analyzers. *J. Geophys. Res.*, **82**, 4351–63.
Hanson, W.B., Zuccaro, D.R., Lippincott, C.R. and Sanatani, S. (1973) The retarding-potential analyzer on Atmosphere Explorer. *Radio Sci.*, **8**, 333–9.
Hanson, W.B., Heelis, R.A., Power, R.A. et al. (1981) The retarding potential analyzer for Dynamics Explorer-B. *Space Sci. Instrum.*, **5**, 503–10.
Heelis, R.A., Hanson, W.B., Lippincott, C.R. et al. (1981) The ion drift meter for Dynamics Explorer-B. *Space Sci. Instrum.*, **5**, 511–21.
Hirao, K. and Oyama, K.-I. (1970) An improved type of electron temperature probe. *J. Geomag. Geoelect.*, **22**, 393–402.
Hoffman, J.H., Hanson, W.B., Lippincott, C.R. and Ferguson, E.E. (1973) The magnetic ion-mass spectrometer on Atmosphere Explorer. *Radio Sci.*, **8**, 315–22.
Istomin, V.G. and Pokhunkov, A.A. (1963) Mass-spectrometer measurements of atmospheric composition in the USSR, in *Space Research III* (ed. W. Priester). Amsterdam: North-Holland Publishing Co., pp. 132–42.
Johnson C.Y., Meadows, E.B. and Holmes, J.C. (1958) Ion composition of the Arctic ionosphere. *J. Geophys. Res.*, **63**, 443–4.
Knudsen, W.C., Spenner, K., Bakke, J. and Novak, V. (1980) Pioneer Venus orbiter planar retarding potential analyzer plasma experiment. *IEEE Trans. Geosci. Remote Sensing*, **1**, 54–9.
Krehbiel, J.P., Brace, L.H., Theis, R.F. et al. (1980) Pioneer Venus orbiter electron temperature probe. *IEEE Trans. Geosci. Remote Sensing*, **1**, 49–54.
Krehbiel, J.P., Brace, L.H., Theis, R.F. et al. (1981) The Dynamics Explorer Langmuir probe instrument. *Space Sci. Instrum.*, **5**, 493–502.
Narcisi, R.S. and Bailey, A.D. (1965) Mass spectrometric measurements of positive ions at altitudes from 64 to 112 kilometers. *J. Geophys. Res.*, **70**, 3687–700.
Oyama, K.-I. (1991) Electron temperature measurements carried out by Japanese scientific satellites. *Adv. Space Res.*, **11**, 149–58.
Paul, W., Reinhard, H.P. and von Zahn, U. (1958) Das Elektrische Massenfilter als Massenspektrometer und Isotopentrenner. *Z. Phys.*, **152**, 143–82.
Sagalyn, R.C., Smiddy, M. and Bhargava, Y.N. (1965) Satellite measurements of the diurnal variation of electron temperatures in the F region. *Space Research V*, Amsterdam: North-Holland Publishing Co., pp. 189–206.
Shyn, T.W., Sharp, W.E. and Hays, P.B. (1976) Gridless retarding potential analyzer for use in very-low-energy charged particle detection. *Rev. Sci. Instrum.*, **47**, 1005–15.
Spenner, K. and Dumbs, A. (1974) The retarding potential analyzer on Aeros-B. *J. Geophys. Res.*, **40**, 585–92.
Szucs, I.T., Szemerey, I., Kiraly, P. et al. (1990) The HARP electron and ion sensor on the PHOBOS mission. *Nuclear Instrum. Meth. Phys. Res.*, **290**, 228–36.
Taylor, H.A., Jr, Brinton, H.C., Bauer, S.J. et al. (1980a) Global observations of the composition and dynamics of the ionosphere of Venus: implications for the solar wind interaction. *J. Geophys. Res.*, **85**, 7765–77.
Taylor, H.A., Jr, Brinton, H.C., Wagner, T.C.G. et al. (1980b) Bennett ion mass spectrometers on the Pioneer Venus bus and orbiter. *IEEE Trans. Geosci. Remote Sensing*, **1**, 44–49.
Wrenn, G.L. (1969) The Langmuir plate and spherical ion probe experiments aboard Explorer XXXI. *Proc. IEEE*, **57**, 1072–7.

Cross references

Ionosphere
Plasma
Plasma wave
Shock waves
Solar wind

THERMOSPHERE

The most fundamental variation of temperature within an atmosphere occurs with respect to altitude. Planetary atmospheres are thus conveniently divided into layers based on the vertical structure of the temperature profile. The names assigned to terrestrial atmospheric layers are traditionally given to comparable atmospheric layers found in other planets or moons. The thermosphere is that rarefied portion of a planetary atmosphere which extends from the top of the middle atmosphere (mesosphere) to the beginning of outer space (exosphere; Figure T15). On most planets the thermosphere is characterized by temperatures which increase dramatically with altitude due to various solar and non-solar heating mechanisms. The temperature asymptotically approaches a maximum somewhere in the exosphere. The thermosphere is also the topmost bound layer. Above, in the exosphere, molecules have mean free paths of greater than a scale height and temperatures become isothermal; this means that collisions are rare, and atoms or molecules may escape to space. The level of minimum temperature at the base of the thermosphere is termed the mesopause (Banks and Kockarts, 1973).

The thermosphere is also distinguished by a transition region called the homopause, below which atomic and molecular constituents are well mixed by winds and dissipative turbulence. The term homosphere is applied to this lower region where the rate of mixing is sufficiently rapid to produce a uniform relative composition for the major atmospheric constituents. This turbulent mixing process, providing net gas transport, is called eddy diffusion. Above, in the heterosphere, individual species begin to diffusively separate according to their unique masses. This net transport of gases, resulting from

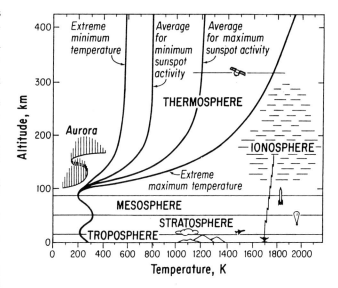

Figure T15 Temperature varies more in the thermosphere than in lower regions of the atmosphere.

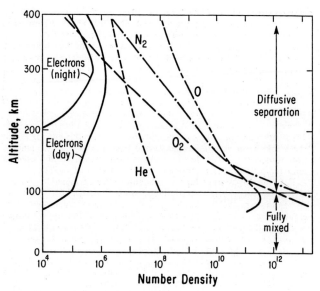

Figure T16 Heavier gases predominate in the lower thermosphere, lighter gases above.

molecular motions alone, is called molecular diffusion. To a first approximation, the heterospheric constituents adjust themselves to the planetary gravitational field according to their molecular weights. This decrease of a particular species number density (number per cm^3) with increasing altitude is usually described in terms of its scale height, that altitude interval over which a density drops by e^{-1}. The heavier molecular species are thus the most abundant in the lower thermosphere, leaving the lighter molecular and atomic species to dominate at the base of the exosphere (Figure T16). Number and mass (g cm^{-3}) densities decrease exponentially with increasing altitude throughout the entire atmosphere.

Embedded within the thermosphere is the ionosphere (q.v.), a weakly ionized plasma. The charged particles that make up a planetary ionosphere are formed by various ionization mechanisms that operate on the local neutral species that comprise the thermospheric composition. These charged particles are strongly influenced by the electric and magnetic fields that may exist about a planet. The absence of a significant intrinsic planetary magnetic field (e.g. for Venus and Mars) has some unique implications for the thermospheric heat budget.

The simple remote sensing instruments typically flown on flyby spacecraft (see Observations section below) are suited to thermospheric investigations. Orbiting spacecraft about Venus and Mars have also flown inside the thermosphere–ionosphere region to make *in situ* measurements. However, satellite drag due to the atmospheric mass becomes prohibitive at altitudes below the mesopause.

Planetary thermospheric processes

Why do thermospheres exist and what processes control their behavior? The thermosphere and ionosphere change dramatically because they absorb and are controlled by the highly variable components of the Sun's energy output. Solar energy reaches a planet mainly as (1) electromagnetic radiation and (2) in the form of the solar wind (q.v.), a flow of charged particles (plasma) from the Sun. Most of the Sun's electromagnetic energy is in the visible portion of the spectrum; it passes directly through the tenuous upper atmosphere and is either absorbed lower down or reflected back to space. Solar radiation in the ultraviolet (UV), at wavelengths shorter than 200 nm, strongly interacts with the gases in a planetary thermosphere to establish its basic chemical, thermal and dynamic structure. The amount of this ultraviolet radiation reaching a planet's upper atmosphere varies significantly with the Sun–planet (heliocentric) distance, the planet's obliquity or tilt which determines the local season, and the changing solar radiation itself. The Sun is observed to undergo a cyclic variation in its electromagnetic and solar wind output that spans a period of roughly 11 Earth years. The extreme ultraviolet (EUV) portion of the solar spectrum (5.0–110.0 nm) can vary by a factor of two to ten over this solar cycle period (Lean, 1991).

Large amounts of solar wind plasma and energy are also deposited in the high-latitude regions of the thermospheres of those planets having a significant intrinsic magnetic field, like the Earth and Jupiter. Solar wind particles are processed by the interaction of the magnetic cavity surrounding a planet (magnetosphere) and the solar wind. These charged particles ultimately are channeled by a planet's magnetic field into the high-latitude ionosphere where they collide with neutral gases and ionize them, dissociate molecular species and heat or excite atoms and molecules to higher energy states. These excited atoms and molecules subsequently relax back to their ground states by the emission of photons resulting in the auroral glow we see. The heat supplied locally at the Earth sometimes exceeds the absorbed solar electromagnetic energy. These polar auroral heating processes dominate the thermal balances at high latitudes, and to a lessor extent at lower latitudes due to global wind effects. However, auroral heating always exceeds solar heating on Jupiter, thereby driving its thermospheric winds.

What general processes typically maintain a vertical temperature profile in a planetary thermosphere? The large vertical temperature gradient results from the absorption of solar ultraviolet or solar wind energy and the lack of any effective means of thermal emission in the altitude region. For instance, these heat sources vary considerably over the solar cycle, causing the Earth's vertical temperature structure to change dramatically in the thermosphere and very little at lower altitudes (Figure T15). Heat sources and sinks eventually become negligible and temperatures become isothermal within the exosphere. Molecular conduction cools the upper thermosphere by transferring the heat down-gradient toward the mesopause, where it is effectively radiated to space by an infrared (IR) active cooling agent. The height and temperature of the mesopause is controlled by the effectiveness of the IR cooling process(es). For the Earth, Venus, and Mars this IR cooling results primarily from CO_2 emission at 15 μm. The IR agent must have an emitting 'window to space' at a wavelength that corresponds to the radiating temperature of that region. In this way, a thermostatic control is maintained which regulates lower thermosphere temperatures. This same thermostat may or may not be important at higher altitudes. A rough balance between solar UV heating, auroral heating and cooling by molecular thermal conduction occurs for the maintenance of observed terrestrial global mean temperatures (Bougher and Roble, 1991). However, the relative abundance of IR active gases may be large enough to permit IR cooling to be very important far above the mesopause (e.g. at Venus and Titan).

The local rate of absorption or heating depends on the product of the species absorption cross-section and the total abundance of the absorbing species along the path through the atmosphere. This product is known as the optical depth (q.v.); where it becomes unity, maximum absorption occurs. If the cross-section is large, solar energy is primarily absorbed high in the atmosphere; if it is small, absorption takes place lower down. For example, Figure T16 shows the vertical distribution of the Earth's major thermospheric gases, as sustained by molecular diffusion and eddy mixing processes. These are the UV absorbers that are important for controlling thermospheric heating. Figure T17 illustrates where the peak absorption of solar UV radiation occurs in the Earth's atmosphere as a function of these principal constituent absorbers (Roble, 1986; Paxton and Anderson, 1992). Such UV absorption within the Earth's atmosphere eventually results in an oxygen molecule being photodissociated into two O atoms. For these O atoms to recombine chemically to form O_2, they must collide with a third body ($O + O + M = O_2 + M$) below 100 km.

A thermospheric system is also subject to large-scale winds. These winds are driven by differential solar and non-solar heating which undergoes a daily, seasonal or orbital variation (Roble, 1986). Dayside heating causes the atmosphere to expand; the loss of heat at night causes it to contract. This heating pattern creates pressure differences that drive a global thermospheric circulation, transporting heat from the warm dayside to the cool nightside. The fluid motions are governed by the same equations as those used by meteorologists studying weather systems in the lower atmosphere. However, the thermosphere is more stable since the temperature increases with altitude. Also, the upper atmosphere is generally more responsive to external forcings; i.e. thermospheric heat and momentum sources are more effective per unit mass.

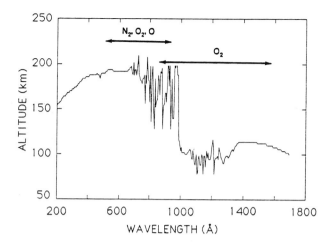

Figure T17 Solar UV radiation is absorbed by atmospheric constituents at various depths. Adapted from Paxton and Anderson (Venus and Mars: Atmosphere, Ionospheres, and Solar Wind Interactions, Geophysical Monograph 66, 113–189, 1992. Copyright 1992 by the American Geophysical Union.)

Table T4 Terrestrial planet parameters

Parameter	Earth	Venus	Mars
Gravity (cm s^{-2})	982	888	373
Heliocentric distance (AU)	1.0	0.72	1.38–1.67
Radius (km)	6371	6050	3388
Ω (rad s^{-1})	7.3 (−5)	3.0 (−7)	7.1 (−5)
Intrinsic B-field (with respect to Earth)	1.0	1.0 (−5)	≤ 1.0 (−5)?
Obliquity (deg)	23.5	1–3	24.0

From Bougher and Roble, *J. Geophys. Res.*, **96**, 11045–55, 1991. Copyright 1991 by the American Geophysical Union.

Comparison of thermospheres: data and models

Strategy

Thirty years of observational and theoretical advance prompted by spacecraft measurements have shown that the geocentric way of looking at planetary atmospheres is sometimes dangerous and misleading; it has typically been assumed that the terrestrial case is the best paradigm or reference for developing new theories of planet thermospheres. However, the geocentric perspective provides a framework for initial investigation that has proven helpful in many cases. The terrestrial planets (Venus, Earth and Mars) are often grouped together for discussion of planetary thermospheric features, since their composition, temperatures, winds and external parameters (Tables T4 and T5) are somewhat similar. The Jovian planet thermospheres (Jupiter, Saturn, Uranus and Neptune) are also grouped together for similar reasons (Table T6). Finally, recent work on the thermosphere of Titan suggests that it is Venus-like in some respects (Yelle, 1991). We will examine these three groupings in turn.

Observations: a brief historical review

Planetary spacecraft have surveyed all of the thermospheres, except that of Pluto. Early reconnaissance missions have been followed by probes and orbiters to Venus and Mars to examine the detailed structure and longer-term responses to solar cycle and solar wind variations (Chamberlain and Hunten, 1987). Venus has been visited by several flyby spacecraft (Venera 4, 6, 11 and 12, Mariner 5 and 10 and Galileo) that have remotely sensed the upper atmosphere structure using photometers and spectrometers. Orbiters at Venus include Venera 9 and 10 and Pioneer Venus. These missions have made remote and global *in situ* measurements of the composition, airglow and temperatures, giving insight into the thermospheric wind system at Venus. Neutral and ion mass spectrometers, drag experiments and visible and UV spectrometers were the key aeronomic instruments on these missions. Four Pioneer Venus probes have also made *in situ* measurements upon descent to the surface (Fox and Bougher, 1991). Current data for the Mars thermosphere is much less complete in its global coverage. Remote observations from flyby spacecraft instruments include those from Mariner 6 and 7 in 1969 and Mars 4 and 5 in 1973. Orbiters with thermospheric remote sensing capability include Mariner 9 in 1971, and Viking 1 and 2 in 1976 and subsequent years. The only *in situ* measurements made of the Mars thermosphere were obtained from the two Viking landers upon descent; i.e. two density profiles were measured and used to infer temperatures (Krasnopolsky, 1986; Barth *et al.*, 1992).

The thermospheres of Jupiter and Saturn were examined by Pioneer 10 and 11, and Voyager 1 and 2 flyby spacecraft using photometers and spectrometers. Stellar and solar UV occultation techniques were exploited to infer densities and temperatures in the upper atmospheres of these planets (Smith and Hunten, 1990). Only Voyager 2 continued on to probe remotely the thermospheres of Uranus and Neptune with the same techniques. Thus far, each of these Jovian planets has only been surveyed; no comprehensive picture of their global thermospheric structure or dynamics exists. Finally, Titan and Triton's thermospheres have been examined using the same UV occultation techniques aboard Voyager 1 and 2 respectively.

Data and one-dimensional modeling

Computer models are useful to examine the mean structure of a planetary thermosphere, and to discover the elementary processes and balances that determine the observed features. However, the lack of a self-consistent dynamical formulation typically limits these models to average conditions over the globe. If such averages are taken, the vertical transport of heat and constituents is reduced to a single dimension, and corresponding models are called 'one-dimensional' or 1-D. Many excellent 1-D model studies have been conducted over the past 30 years to interpret spacecraft data of various planetary thermospheres. Recent studies have compared the global mean features and processes of several planetary thermospheres (Bougher and Roble, 1991; Yelle, 1991; Stevens and Strobel, 1988).

Typical global mean density profiles found in the thermospheres of Venus and Mars are illustrated in Figures T18a and T18b. Venus atomic oxygen is seen to overtake CO_2 above 155 km; for Mars, O becomes the dominant constituent at much higher altitudes (190–220 km). The level of peak UV absorption and heating per unit volume by these constituents is shown in Figures T19a and T19b. Both thermospheres are controlled by solar EUV heating driven primarily by CO_2 absorption. Solar fluxes shortward of 227 nm dissociate CO_2 into CO and O; fluxes shortward of 90.0 nm ionize CO_2. Because of ion–neutral reactions involving atomic oxygen, O_2^+ is the major ion rather than CO_2^+. O and CO are transported to altitudes below 100 km where recombination processes reform CO_2 again.

Why is the Venus thermosphere relatively cold and unresponsive to solar flux changes when compared to the Earth and Mars, while Venus is the closest of these planets to the Sun? Figures T15, T20a, and T20b illustrate calculated thermospheric temperature profiles for global mean conditions that closely match those observed on all three planets; solar minimum (SMIN) and maximum (SMAX) conditions are assumed (Bougher and Roble, 1991). Their corresponding heat budgets reveal that the role of CO_2 15-μm cooling is distinctly different for each planet. The effect of this cooling is most strongly felt in the Venus thermosphere, where it serves as an efficient thermostat regulating temperatures. A rather weak global mean exospheric temperature variation (172–248 K) over the solar cycle is obtained. Venus CO_2 15-μm emission appears to be an effective cooling agent far above the mesopause. By comparison, the tenfold smaller O abundance on Mars relative to Venus renders CO_2 cooling less effective for moderating solar flux changes on Mars, resulting in a relatively large solar cycle variation of exospheric temperatures (180–290 K). Molecular conduction is most important in controlling the Mars exospheric temperatures. Lastly, the terrestrial exospheric temperatures are calculated to range from 737 to 1255 K over the solar cycle, much larger than the response obtained for either Venus or Mars. Warmer overall temperatures are expected for Earth since

Table T5 Implications of terrestrial planet parameters

Effect	Earth	Venus	Mars
Scale heights (km)	10–50	4–12	8–22
Peak EUV heating (km)	150–200	~ 140	120–130
O abundance (ion peak)	$\sim 40\%$	$\sim 10\text{–}20\%$	$\sim 1\text{–}3\%$
CO_2 15-μm cooling (km)	100–130	~ 140	≤ 120
Dayside thermostat	Winds + conduction	CO_2 cooling	Winds + conduction
Mean solar cycle T	737–1255 K	172–248 K	180–290 K
Rotational forces	Strong	Weak	Strong
Cryosphere	No	Yes	No
Auroral heating	Yes	No	No
Seasons	Yes	No	Yes

From Bougher and Roble, *J. Geophys. Res.*, **96**, 11045–55, 1991. Copyright 1991 by the American Geophysical Union.

Table T6 Comparison of Jovian planet thermospheres

	Heliocentric distance (AU)	T_∞ from Solar UV/EUV (K)	Voyager inferred T_∞ (K)
Jupiter	5.2	195	1000
Saturn	9.5	130	420
Uranus	19.2	113	850 ± 100
Neptune	30.1	141	750 ± 150

From Stevens and Strobel (1988).

its EUV heating efficiency is much larger than that for Venus and Mars (Bougher and Roble, 1991).

Fundamental planetary parameters such as gravity, heliocentric distance and intrinsic magnetic field are at the root of the basic differences in the heat budgets that are responsible for these thermal structures (Tables T4 and T5). In particular, Venus's proximity to the Sun permits strong photolysis of CO_2 and the production of large O abundances in the Venus upper thermosphere. Mars, on the other hand, is more distant from the Sun, and is therefore depleted in thermospheric atomic oxygen compared to Venus. This is important because collisions between O and CO_2 are known to be effective in enhancing CO_2 15-μm cooling in the thermospheres of the terrestrial planets (Bougher and Roble, 1991). The Mars 11-year temperature response is thus larger than that of Venus since molecular conduction is less effective as a thermostat than 15-μm cooling. By contrast, the Earth is subject to auroral heating processes that are not important for Venus and Mars, due to the presence of a relatively strong terrestrial intrinsic magnetic field. Also, the Earth's IR radiators (CO_2 and NO) are minor constituents which provide significant cooling only near the bottom of the Earth's thermosphere. This permits exospheric temperatures to be primarily regulated by molecular conduction, much like Mars.

Information on the thermospheres of the Jovian planets is sparse and primarily limited to that gleaned from the successful UV solar and stellar occultation experiments aboard the Voyager spacecraft (Atreya, 1986; Chamberlain and Hunten, 1987). The Jovian thermospheres are composed primarily of H_2 and H. Helium is estimated from model calculations. Methane (CH_4) and hydrocarbon species (e.g. C_2H_2 – acetylene, C_2H_6 – ethane) were found to be important at the nearly coincident mesopause–homopause levels of these planets. Uranian hydrocarbon densities (Figure T21) seem to be depleted with respect to those of Jupiter and Saturn. This implies relatively weak vertical mixing and eddy diffusion for the stratosphere of Uranus. Atomic hydrogen is the end result of ion chemistry in these H_2-dominated thermospheres. Chemical removal of H requires high pressures within the hydrocarbon regime where recombination can occur.

The inferred exospheric temperatures for the Jovian planets are far higher than expected for solar UV heating alone (Table T6). Clearly, much stronger sources of heat than the solar ionizing radiation must be present. High-latitude auroral heating, and its redistribution to lower altitudes by strong thermospheric winds, is likely to be most important for heating the thermosphere of Jupiter. The intrinsic magnetic field of Jupiter is the largest of all the planets, consistent with this predominance of auroral heating. Another mechanism may be responsible for heating the thermosphere of Saturn. Overall, molecular conduction appears to regulate the exospheric temperatures on all the giant planets. All four Jovian planets also have approximately the same mesospheric temperature (120–150 K); the spectral location of the hydrocarbon IR radiating gases (7.7–13.7-μm) enables them to be very effective thermostats which can accommodate a wide range of heat inputs. The measured mesosphere and exosphere temperatures are connected by models to obtain the thermospheric profiles available (e.g. Figure T21). One-dimensional models have only begun to examine the heat sources and their vertical distributions needed for maintaining observed temperatures of the Jovian planet thermospheres (Table T6; Stevens and Strobel, 1988; Strobel *et al.*, 1991).

Our knowledge of the Titan thermosphere is also very limited. However, this moon is particularly intriguing because its atmosphere is somewhat 'Earth-like' in composition and surface pressure, while its heat budget and slow rotation are reminiscent of Venus. Titan's thermospheric composition is predominantly N_2, with traces of IR active gases like CH_4, C_2H_2, C_2H_6 and HCN. The exospheric temperature is roughly 186 ± 20 K; to date the mesopause temperature has not been measured. Estimates are made based on known lower atmosphere temperatures and detailed one-dimensional models (Yelle, 1991). The observed exospheric temperature can be reproduced by a balance of UV heating and HCN cooling, with very little impact from molecular conduction. In this regard the thermospheres of Venus and Titan are similar since molecular thermal conduction plays little role in determining the exospheric temperatures of these bodies. Modeling is continuing in preparation for a future probe visit to Titan in connection with the Cassini mission to Saturn.

Dynamics and multidimensional modeling

As the understanding of planetary thermospheres becomes more mature, modeling approaches are evolving from simple one-dimensional to more realistic three-dimensional systems. These multidimensional models can address the feedbacks inherent in coupled dynamical, energetic and diffusion processes. The availability of global thermospheric temperature, density, wind and airglow data will drive the development of such complex models. A common thermospheric modeling framework is highly desirable to facilitate the systematic comparison of several planetary thermosphere systems. The National Center for Atmospheric Research (NCAR) thermospheric general circulation model (TGCM) is such a modeling tool. It has been successfully developed and exercised over the last decade to study individually and contrast the three-dimensional circulation, composition and temperature structure of the three terrestrial planet thermospheres (Dickinson, Ridley and Roble, 1984; Bougher *et al.*, 1988a,b; Roble *et al.*, 1988; Bougher *et al.*, 1990). Many other 3-D models have also successfully examined the thermospheres of Venus and Earth individually (Fox and Bougher, 1991; Roble *et al.*, 1988).

The key features of the dynamics of the Earth, Venus and Mars thermospheres can be understood by examining the implications of their fundamental planetary parameters (Tables T4 and T5). Venus is essentially a non-rotating planet with a very small intrinsic magnetic field. Its axial obliquity is also quite small, yielding little in the way of seasonal effects. The resulting circulation pattern, sketched in Figure T22a, is largely symmetric about the subsolar (SS) and antisolar (AS) points. The observed large day-to-night thermospheric temperature

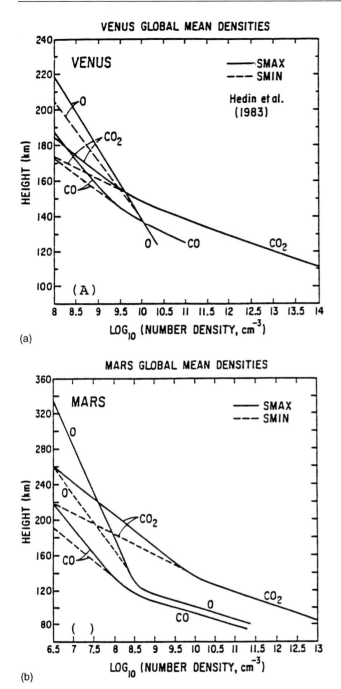

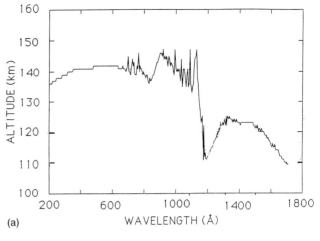

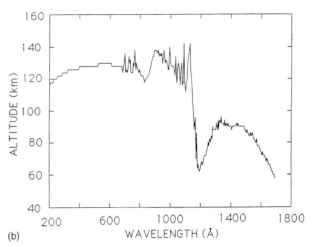

Figure T19 Solar UV radiation is absorbed by atmospheric constituents at various depths. (a) Venus, and (b) Mars. (From Paxton and Anderson, *Venus and Mars: Atmosphere, Ionospheres, and Solar Wind Interactions*, Geophysical Monograph 66: 113–89, 1992. Copyright 1992 by the American Geophysical Union.)

Figure T18 One-dimensional global mean SMIN/SMAX density profiles for (a) Venus and (b) Mars. (From Bougher and Roble, *J. Geophys. Res.*, **96**, 11045–55, 1991. Copyright 1991 by the American Geophysical Union.)

and density variations can be simulated by the Venus TGCM provided that the solar driven winds are mechanically slowed by turbulence-induced friction (Bougher *et al.*, 1988a; Fox and Bougher, 1991). This permits an even stronger isolation of the day and nightsides than that afforded by the slow planetary rotation. As a result, calculated nightside temperatures are quite cold, unlike any other upper atmosphere seen in the solar system. This region of the Venus nightside above 100 km is called the 'cryosphere'. Measured Pioneer Venus distributions of night airglow and helium densities provide excellent tracers of the thermospheric circulation, confirming an asymmetric zonal flow that modifies the predominant symmetric SS–AS circulation pattern. The physical mechanisms that generate this zonal flow and the frictional drag at thermospheric heights are not understood.

Mars also has a very weak intrinsic magnetic field, like Venus, suggesting the absence of high-latitude, aurorally driven winds. However, unlike Venus, Mars rotates nearly as fast as the Earth and has an 'Earth-like' seasonal cycle. The resulting circulation pattern for equinox conditions, sketched in Figure T22b, has features common to both Earth and Venus. The Mars rotation rate induces an asymmetric day–night temperature variation similar to that of the geomagnetically quiet (weak auroral period) terrestrial thermosphere. There is no predicted nightside cryosphere for Mars, unlike Venus. The Mars planetary rotation seems to preclude any effective isolation of the day and night sides. The Mars thermospheric wind speeds are similar to those of Venus, yet with an asymmetric circulation pattern; i.e. winds diverge at the warm subsolar point and converge near the cool morning terminator. Dayside temperatures also vary greatly near the exobase in response to changing solar EUV heating, molecular conduction and upwelling winds. Oxygen atoms are redistributed about the globe by the strong Mars wind system; this was also predicted for Venus. However, the observed Martian oxygen distribution clearly implies a significant forcing of the thermosphere from lower altitudes. Upward-propagating gravity waves or global-scale tides generated by solar heating of airborne dust may be important for modifying the thermosphere during Martian dust storm periods.

The terrestrial thermosphere is subject to the added complication of a relatively strong intrinsic magnetic field (Roble, 1986). This means that high-latitude auroral heating will usually be important in

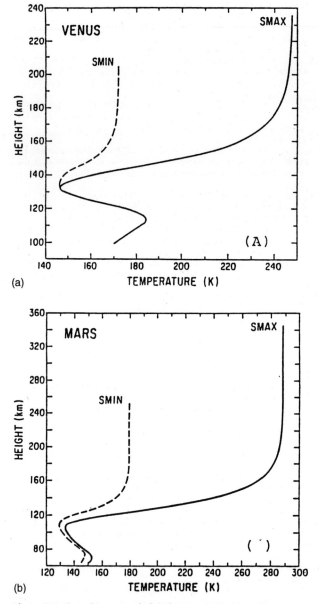

Figure T20 One-dimensional global mean SMIN/SMAX temperature profiles for (a) Venus and (b) Mars. (From Bougher and Roble, *J. Geophys. Res..*, **96**, 11045–55, 1991. Copyright 1991 by the American Geophysical Union.)

driving winds. Also, momentum forcing due to ion drag and magnetospheric induced ion convection modifies the global wind system from that otherwise expected. The resulting circulation pattern for equinox conditions, sketched in Figure T22c, is much different from that of Venus (Roble *et al.*, 1988). The winds are particularly strong near the poles, especially at night when ion drag is reduced. The day-to-night temperature variation and winds follow an asymmetric pattern, again due to the rotation of the planet. The helium distribution is strongly influenced by the global wind system, showing a winter hemisphere bulge. The terrestrial thermosphere is also influenced by tidal forcing; however, the lower atmosphere forcing agent is solar absorption by O_3 and water, not dust as on Mars. Finally, the terrestrial thermosphere is significantly perturbed by auroral storms which occur in response to the changing solar wind. During such times, the global wind system is dominated by a circulation pattern which rises at high latitudes and sinks near the equator.

The TGCM code is a useful tool for comparing the thermospheric structure and circulation of the three terrestrial planets Earth, Venus

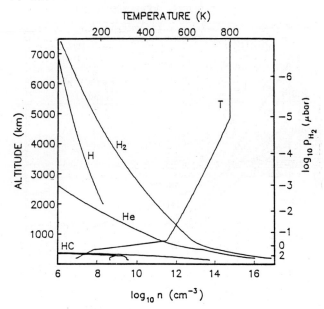

Figure T21 Jovian planet vertical temperature and density structures: Uranus, as an example. (From Herbert *et al.*, *J. Geophys. Res..*, **92**, 15093–109, 1987. Copyright 1987 by the American Geophysical Union.)

and Mars. Key parameters and processes (Tables T4 and T5) that distinguish the thermospheric structure and circulation of these three planets have been identified. The next adaptation of the TGCM should address the Jupiter circulation and structure. The very fast Jupiter planetary rotation plus its overwhelming auroral forcing suggest that the Jovian circulation pattern might be analogous to that for the Earth during a geomagnetic storm.

Summary

In our study of comparative terrestrial planet thermospheres, we hope to obtain a better understanding of the aeronomic processes responsible for the basic structure and dynamics of each of the planets and their response to solar activity. In particular, we hope to gain insight into the overall role of CO_2 as a cooling mechanism on each of the terrestrial planets. This, we believe, should enable us to understand better the effect that increasing CO_2 concentrations in the Earth's atmosphere, caused by human activities, will have in altering the basic structure and energy balance of our own planetary atmosphere.

The Jovian planets appear to provide an extreme example of the auroral effects on thermospheric structure and dynamics for comparison to the Earth. In particular, EUV heating is never sufficient to maintain the observed exospheric temperatures on the Jovian planets. Auroral and/or other sources need to be quantified and their vertical and horizontal distributions estimated. The Jupiter thermospheric circulation may permit high-latitude heating and aurorally produced hydrogen to be redistributed over the globe by strong winds, thereby contributing to the maintenance of observed low-latitude hydrogen and temperature distributions.

Those planets or moons with an appreciable intrinsic magnetic field (especially Earth and Jupiter) exhibit the warmest exospheric temperatures due to the significant role of auroral or other high-latitude heating mechanisms. Those without (e.g. Venus, Mars and Titan) are much cooler, since EUV heating is dominant. Most planet exospheric temperatures are regulated by molecular thermal conduction, such that heat deposited is conducted downward toward the mesopause to be radiated effectively to space. However, two thermospheres (Venus and Titan) are known to have IR radiative cooling mechanisms in operation far above the mesopause. Such strong IR cooling serves as a thermostat that very effectively moderates temperature changes due to solar flux variations.

Several spacecraft are either being planned or are in transit to the planets over the next 15 years. Those that will address the thermospheres of their target bodies include Galileo to Jupiter, Cassini to

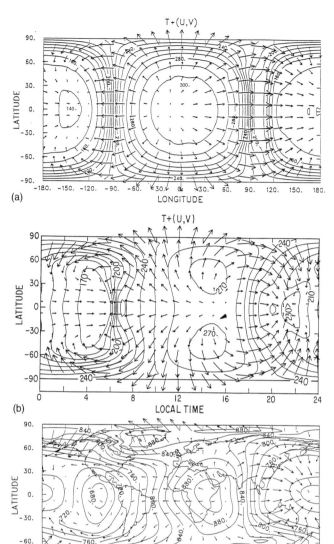

Figure T22 Contour plots of the basic thermospheric circulation patterns for (a) Venus (solar maximum), (b) Mars (near solar maximum) and (c) Earth (solar minimum). Contours represent exospheric temperatures in degrees K; strong gradients are visible from the day to the night sides of these planets. Arrows represent the magnitude and direction of the large-scale horizontal winds; circulation 'cells' can be seen. [From (a) Bougher et al., Icarus, **73**, 545–575, 1988a, copyright 1988 by Academic Press; (b) Bougher et al., Geophys. Res. Lett., **15**, 1511–4, 1988b, copyright 1988 by the American Geophysical Union; and (c) Roble et al., Geophys. Res. Lett., **15**, 1325–8, 1988. Copyright 1988 by the American Geophysical Union.]

Saturn, the Huygens probe to Titan, the Russian Mars 96 Mission to Mars and a Japanese Planet-B mission to Mars.

Stephen W. Bougher and Raymond G. Roble

Bibliography

Atreya, S.K. (1986) *Atmospheres and Ionospheres of the Outer Planets and their Satellites*. New York: Springer Verlag.

Banks, P.M. and Kockarts, G. (1973) *Aeronomy*, parts A and B. New York: Academic Press.

Barth, C.A., Stewart, A.I.F., Bougher, S.W. et al. (1992) Aeronomy of the current Martian atmosphere, in *Mars* (eds H.H. Kieffer, B.M. Jakovsky, C.W. Snyder and M.S. Matthews). Tucson: University of Arizona Press, pp. 1054–89.

Bougher, S.W. and Roble, R.G. (1991) Comparative terrestrial planet thermospheres: I. Solar cycle variation of global mean temperatures. *J. Geophys. Res.*, **96**, 11045–55.

Bougher, S.W., Dickinson, R.E., Ridley, E.C. and Roble, R.G. (1988a) Venus mesosphere and thermosphere: III. Three-dimensional general circulation with coupled dynamics and composition. *Icarus*, **73**, 545–73.

Bougher, S.W., Dickinson, R.E., Roble, R.G. and Ridley, E.C. (1988b) Mars thermospheric general circulation model: calculations for the arrival of Phobos at Mars. *Geophys. Res. Lett.*, **15**, 1511–4.

Bougher, S.W., Roble, R.G., Ridley, E.C. and Dickinson, R.E. (1990) The Mars thermosphere: II. General circulation with coupled dynamics and composition. *J. Geophys. Res.*, **95**, 14811–27.

Chamberlain, J.W. and Hunten, D.M. (1987) *Theory of Planetary Atmospheres: An Introduction to their Physics and Chemistry*, International Geophysics Series, No. 36. Academic Press.

Dickinson, R.E., Ridley, E.C. and Roble, R.G. (1984) Thermospheric general circulation with coupled dynamics and composition. *J. Atmos. Sci.*, **41**, 205–19.

Fox, J.L. and Bougher, S.W. (1991) Structure, luminosity and dynamics of the Venus thermosphere. *Space Sci. Rev.*, **55**, 357–489.

Herbert, F.L., Sandel, B.R., Broadfoot, A.L. et al. (1992) The upper atmosphere of Uranus: EUV occultations observed by Voyager 2. *J. Geophys. Res.*, **92**, 15093–109.

Krasnopolsky, V.A. (1986) *Photochemistry of the Atmospheres of Venus and Mars*. New York: Springer Verlag.

Lean, J. (1991) Variations in the Sun's radiative output. *Rev. Geophys.*, **29**, 505–35.

Paxton, L.J. and Anderson, D.E. (1992) Far ultraviolet sensing of Venus and Mars, in *Venus and Mars: Atmospheres, Ionospheres and Solar Wind Interactions*, Washington, DC: American Geophysical Union, Geophysical Monograph 66, pp. 113–89.

Roble, R.G. (1986) Chemistry in the thermosphere and ionosphere. *Chem. and Eng. News*, **64**(24), 23–38.

Roble, R.G., Ridley, E.C., Richmond, A.D. and Dickinson, R.E. (1988) A coupled thermosphere/ionosphere general circulation model. *Geophys. Res. Lett.*, **15**, 1325–8.

Smith, G.R. and Hunten, D.M. (1990) Study of planetary atmospheres by absorptive occultations, *Rev. Geophys.*, **28**, 117–43.

Stevens, M.H. and Strobel, D.F. (1988) Heat sources in the giant planet thermospheres, *Bull. Am. Astron. Soc.*, **20**, 1124.

Strobel, D.F., Yelle, R.V., Shemansky, D.E. and Atreya, S.K. (1991) The upper atmosphere of Uranus, in *Uranus* (eds J.T. Bergstralh, E.D. Miner and M.S. Matthews). Tucson: University of Arizona Press, pp. 65–109.

Yelle, R.V. (1991) Non-LTE models of Titan's upper atmosphere. *Ap. J.*, **383**, 380–400.

Cross references

Atmosphere
Atmospheric thermal structure
Exosphere
Mariner missions
Mesosphere
Pioneer 10 and 11 missions
Stratosphere
Venera missions
Venus: atmosphere
Viking mission
Voyager missions

TIDAL FRICTION

Tidal friction occurs when energy is dissipated inside a body distorted by tides. While ocean tides are the best-known example of tides, tides also occur in the solid Earth. The concept of tidal friction is in fact most easily illustrated by banishing the oceans, atmosphere and liquid core of the Earth and temporarily assuming the planet to be an

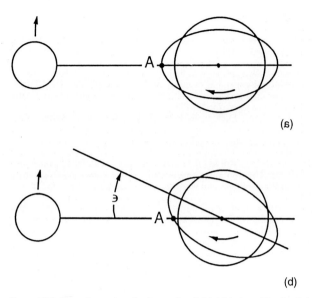

Figure T23 Two-dimensional schematic of tidal friction. The Earth is at left and the Moon is at right; A is the sub-lunar point. (a) In the absence of tidal friction high tide occurs directly beneath the Moon at A. (b) When tidal friction is present the bulge is rotated from A and leads the Moon by an angle ϵ.

elastic solid. Suppose, as in Figure T23a, that the Moon is the tide-raising body in a circular orbit in the Earth's equatorial plane. The viewpoint of the figure is from space, looking down on the North Pole. As the Earth rotates an observer stationed at a fixed point A on the equator will see the Moon pass overhead from east to west. If the Earth is completely elastic, the tidal bulge also sweeps over the observer's position from east to west, with high tide occurring directly beneath the Moon. However, if anelasticity is present, as shown in Figure 23b, the internal friction delays the bulge, so that high tide as seen from A occurs after the Moon passes overhead. Thus the axis of the bulge is to the east of the line which joins the Earth and Moon. As seen from space, the bulge leads the Moon, as shown in the lower part of the figure. Internal friction also decreases slightly the size of the bulge.

The tidal bulge on the Earth exerts a gravitational force on the Moon. In the case of no friction, the effect of this force is inconsequential: the force lies along the Earth–Moon line, merely altering slightly the central attraction of the Earth for the Moon. However, when friction is present, the force points at an angle to the line. The component perpendicular to the line torques the lunar orbit, increasing the orbital angular momentum. This causes the semimajor axis of the lunar orbit to increase, so that the Moon recedes from the Earth. Likewise, the Moon acts gravitationally on the tidal bulge, torquing the Earth. This torque must slow the Earth's rotation, in order to conserve the total angular momentum of the Earth–Moon system. These effects of tidal friction are small but secular, continuing century after century. Over time they build up to give substantial changes in the lunar orbit and the rotation of the Earth. The important result is that the present semimajor axis of the Moon's orbit and the current rotation rate of the Earth are not primordial. This is likewise true for other satellite systems (see below). While most tidal evolution occurs within satellite systems, tidal friction can also affect the rotation of the planets close to the Sun. The tidal evolution of planetary orbits about the Sun is negligible.

Earth–Moon system

The simplistic picture presented above must be modified when dealing with the actual Earth–Moon system. The bulk of the tidal dissipation occurs not in the solid Earth, but in the oceans. The sea has its own tidal bulge. There are also gravitational tides in the atmosphere, but these are small and are overwhelmed by atmospheric 'tides' caused by solar heating. Further, the Sun also raises tides on the Earth, slowing the planet's rotation. Both lunar and solar torques affect the Earth's obliquity as well as its rotation rate. The Moon is in an eccentric orbit inclined to the equator, and tidal friction affects not just the semimajor axis but all of the orbital elements. Moreover, the Earth and Sun raise tidal bulges on the Moon. These complications must be considered and dealt with in quantitative investigations.

The effects of tidal friction are observed in the Earth–Moon system. The most obvious consequence of tidal friction is visible to the naked eye: the Moon always presents the same face to the Earth. Since formation of the Moon directly in synchronous rotation seems improbable, it is likely that tidal friction in the Moon altered its primordial rotation until it achieved its present state. Likewise, tidal friction is believed to be responsible for the observed synchronism of other satellites (see Resonance in the solar system). Interestingly, triaxial satellites must go through a time of tumbling chaotically before achieving synchronism (Wisdom, 1987). An exception is Saturn's moon Hyperion, which may be permanently locked into chaotic tumbling (see Chaotic dynamics in the solar system).

Tidal friction is still acting in the Earth–Moon system. By Kepler's third law the mean motion n of a body decreases as its semimajor axis increases. Laser ranging from the Earth to the retroreflectors on the Moon reveals the acceleration $\dot{n}$ is presently $\dot{n} = -24.9 \pm 1.0$ arcsec century^{-2}, with the Moon retreating from the Earth at a rate of 4 cm year^{-1} (Newhall, Williams and Dickey, 1986). Artificial satellites gravitationally sense the Earth's tides; the satellite data can be extrapolated up to the Moon, giving $\dot{n} = -25.27 \pm 0.61$ arcsec century^{-2}, in good agreement with the lunar laser ranging data (Christodolidis et al., 1988). Modern methods are also used to observe accurately the present changes in the Earth's length of day (LOD). Direct observations of the secular slowing of the Earth are made difficult by the much larger decadal fluctuations in the LOD (see Earth rotation).

The amount of internal friction in the Earth can be expressed in terms of the quality factor $Q = 2\pi E/\Delta E$, where ΔE is the change in the tidal energy E over one cycle of flexure; the lower the value of Q, the more energy is dissipated. The lag angle ϵ (in radians) is of the order Q^{-1}. For the Earth's mantle Q is thought to be ~ 100 at tidal frequencies. From the observed value of $\dot{n}$, though, ϵ is a few degrees and the whole-Earth Q is ~ 13. This low value of Q is doubtless due to the oceans, but precisely where most of the energy is dissipated is not known. Bottom friction in the shallow seas, waves breaking on coastlines, the flexing of shelf ice and the scattering of internal density waves in the deep ocean have all been suggested as possible mechanisms, but none has been convincingly identified (Ray, 1994). The values of Q for the Moon necessary to explain its orbital evolution are surprisingly low, in the range 10 to 30, perhaps due to viscous core–mantle coupling. (The values of Q of the other large satellites in the solar system, both rocky and icy, are often taken to be ~ 100 in the absence of other information. The values of Q of the fluid outer planets are large, on the order of 10 000.)

Historical astronomical records of lunar occultations of stars, measurements of longitudes of the Sun, transits of Mercury and Venus across the Sun's face and observations of ancient solar eclipses have been used to estimate both the secular rotational acceleration of the Earth and the lunar recession. Historical records have the advantage of averaging over long time scales and the disadvantage of being increasingly unreliable the older they are. The recovered $\dot{n}$ is usually in the range -20 to -30 arcsec century^{-2}, in reasonable agreement with the modern value. Assuming the modern value of $\dot{n}$ holds throughout historical time and that ω is the angular rotational speed of the Earth, recovering the secular acceleration $\dot{\omega}$ from observations of ancient eclipses gives $\dot{\omega} \cong -1100$ arcsec century^{-2}. This value is only in rough accord with the tidal slowing expected from the observed value of $\dot{n}$ and angular momentum conservation, even after the solar tidal torques are taken into account. The Earth's secular rotational variations appear to be contaminated significantly by non-tidal mechanisms. Part of the discrepancy is probably due to postglacial rebound: as the polar regions rise the equatorial regions move closer to the rotation axis. Like spinning skaters pulling in their arms, the Earth speeds up, causing a secular acceleration of about $+200$ arcsec century^{-2}. Other mechanisms besides postglacial rebound, such as electromagnetic core–mantle coupling, may also be at work. The data suggest that the non-tidal mechanisms may not be constant in time (Stephenson and Morrison, 1984).

Fossil and geological evidence is thought to give information about the Earth's rotation rate in prehistoric times. The rhythmic banding seen in corals, bivalves, stromatalites and sedimentary layers is believed to reflect monthly and yearly cycles, indicating more days in the year than at present; the Earth may have rotated more than 400 times per year in the Precambrian. Since there is no reason to suspect that the Earth's orbit has changed significantly over geologic time, days must have been shorter in the past, i.e. the Earth was rotating faster. However, the data remain controversial (see Earth: rotational history).

Extrapolating the present rate of orbital evolution into the past indicates that the Moon closely approached the Earth about 2 billion years ago. Because there is no geological evidence on either the Earth or the Moon to indicate that this occurred, the inference is that tidal friction was smaller in the past than it is now. Models of the ocean tides appear to bear this out: for the present configuration and depth of the oceans the important M_2 tide is near resonance, so that the present amount of dissipation is probably anomalously high compared to the past. Plate tectonics, by changing the ocean basins, seems to have played an important role in the rate of evolution of the Earth–Moon system.

Tidal friction as the cause of the lunar recession was first proposed by the philosopher Immanuel Kant in the 18th century. Tidal friction was investigated mathematically by George Howard Darwin (q.v.; a son of famed naturalist Charles Darwin) in the 19th century, and by many investigators in the 20th century. The primary aim of this research was to help elucidate the origin of the Moon. By integrating the lunar orbit backwards in time for 4.5 billion years, it was hoped that a compelling dynamical state would be found which would favor a particular hypothesis of origin. Unfortunately, no compelling state was forthcoming. Tidal friction sheds little light on the Moon's origin (q.v.), whether it be by fission from the Earth, capture from heliocentric orbit or accretion in Earth orbit (the currently popular giant impact theory being a variant of accretion).

As for the future, billions of years from now the Moon will recede and the Earth will slow down until synchronism is reached: the day will equal the month in length. But solar tides will continue to brake the Earth, so that the day will become longer than the month. When that happens the bulge on the Earth raised by the Moon will lag behind the Moon (as the bulge on Mars lags Phobos; see below), dragging the Moon back towards the Earth, with attendant catastrophic consequences. This scenario assumes that the Earth is not first engulfed by the Sun when it becomes a red giant billions of years hence.

Mars

Earth-based and spacecraft observations of Phobos, the innermost moon of Mars, indicate that the satellite is spiraling inwards towards the Martian surface at the rate of $\ddot{n} = +12°$ century^{-2} and will impact about 40 million years hence (Sinclair, 1989). While some bizarre explanations have been proposed for this behavior (such as the suggestion of atmospheric drag on a hollow, artificial Phobos), tidal friction is almost certainly the culprit. Because Phobos revolves about Mars faster than the planet rotates (in contrast to Figure T23), the tidal bulge lags behind Phobos, causing the orbit to shrink, as observed. Reasonable guesses as to Mars' internal properties, such as $Q \cong 100$, can give quantitative agreement with the rate of decay. No orbital evolution of Mars' other satellite Deimos has been convincingly observed. This is expected in the tidal friction scenario: Deimos is too small and too close to geosynchronous rotation for its orbit to change much.

The gas giants

The effects of tidal friction in the satellite systems of Jupiter, Saturn, Uranus and Neptune are currently under active investigation; some of what follows below may soon become out of date.

The fact that Io (q.v.), the innermost Galilean satellite of Jupiter, is still relatively close to the planet indicates that Jupiter has a large Q; the lag angle of Jupiter's bulge is too small to have pushed Io very far away. A lower limit of $\sim 60\,000$ can be placed on Jupiter's Q, assuming the satellite system is primordial (or close to it) and that Io was originally near the planet.

Not only do the satellites raise tides on Jupiter, but also Jupiter raises tides on the satellites. These tidal bulges are not frozen in place on the synchronously rotating satellites because of the eccentricity of their orbits. An observer on Io, for example, will see Jupiter approach and recede, changing the size of the bulge; more importantly, the observer will also see Jupiter wander slightly in the sky, since the angular speed of Io in its orbit is not constant. The bulge will wander too, flexing the satellite. The effect of these tides is to try to circularize Io's orbit; but the eccentricity is maintained by the perturbations of the other satellites (see Resonance in the solar system).

Anyone who has bent a paperclip rapidly back and forth and then touched the joint knows that flexing can generate heat. In the case of Io, with its spectacular volcanoes, tidal dissipation makes it the most geologically active body in the solar system. Other sources, such as radiogenic or electromagnetic heating, are not thought to be sufficient to explain the volcanism. [That Io should undergo intense tidal heating was proposed before the arrival of Voyager 1 at Jupiter (Peale et al., 1979); this ranks as one of the finest accomplishments of planetary science.] Tidal heating (q.v.) may have been, or may still be, important for Europa (q.v.). Ganymede (q.v.) may have suffered early severe tidal heating, while Callisto (q.v.) experienced very little throughout its history.

Tidal friction is believed to have been important in the establishment and maintenance of the resonances observed between the Galilean satellites. While research is still ongoing, the elucidation of the complicated interplay of tides and orbital perturbations in the Jovian system (principally by Charles Yoder) has been hailed as one of the most outstanding feats in celestial mechanics in the last two decades (Burns and Matthews, 1986, p. 205).

Tidal friction presumably played a role in establishing and maintaining the resonances observed in the Saturnian satellite system (q.v.). The closeness of the small moons yields $Q \geq 14\,000$ for Saturn. Only the surface of Enceladus shows signs of recent heating, possibly due to tidal dissipation. Saturn's large moon Titan (q.v.) is still very much a mystery. Titan is now known to rotate synchronously but its orbital eccentricity of 0.0292 is unexplained.

The satellites of Uranus are currently not in a state of resonance, but may have been so in the past and evolved out of it, due to the interplay of tidal friction and chaotic orbits. If so, the value of Q of Uranus could be constrained to the range $11\,000 < Q < 39\,000$ (Tittlemore and Wisdom, 1990). Tidal dissipation may be responsible for the late resurfacing of Ariel and the strange features seen on Miranda (q.v.).

The story of Neptune's satellites (q.v.) may be dramatic. It has been proposed that Triton (q.v.) was captured from heliocentric orbit, perhaps by gas drag or impact with a pre-existing satellite, into an eccentric, retrograde orbit about Neptune. Triton's present small orbital eccentricity can easily be accounted for by tides on Triton circularizing its orbit. Triton may also have thrown Nereid (q.v.) into its peculiar orbit and destroyed any primordial satellites outside of five Neptune radii on its many passages through Neptune's equatorial plane. Satellites inside five radii may have been forced into chaotic orbits and collided with each other, with the present small satellite system being reconstituted from the debris. From these speculations the limits $12\,000 < Q < 330\,000$ are estimated for Neptune (Banfield and Murray, 1992). Tidal dissipation may have been responsible for early heating inside Triton, but cannot explain its present activity; solar heating may instead be responsible.

Pluto–Charon system

Tidal friction seems to have run its course with Pluto (q.v.) and Charon (q.v.): both objects appear to be locked into synchronous rotation, so that each always presents the same face to the other (Stern, 1992). In contrast to the Earth–Moon system, solar tides are much too weak to subtract much angular momentum from the Pluto–Charon system on the geologic timescale. Significant disturbances away from the present state probably come about only by impacts; the disturbances are then damped out by strong tidal friction.

Solar tidal torques

The solar tidal torque probably slowed Mercury's rotation until capture into the present state of spin–orbit coupling, in which the planet spins three times for every two trips around the Sun. Interestingly, it was not captured into synchronous rotation, as are many of the planetary satellites. Venus' retrograde rotation may be primordial; it is hard to see how solar tides could have reversed its spin, though they probably slowed its original rotation significantly. The Earth has been somewhat slowed by solar tides. The other planets are too far away from the Sun to have their rotation rates affected very much.

Conclusion

Tidal friction gives every indication of having played an important role in the evolution of satellite orbits and rotation states. Also, tides are probably responsible for the spin–orbit coupling of Mercury. Further, tidal friction is ongoing in the solar system, as evidenced by the recession of the Moon, the orbital decay of Phobos and the intense heating of Io.

David P. Rubincam

Bibliography

Banfield, D. and Murray, N. (1992) A dynamical history of the inner Neptunian satellites. *Icarus*, **99**, 390–401.
Brosche, P. and Sundermann, J. (eds) (1978) *Tidal Friction and the Earth's Rotation*. Berlin: Springer Verlag.
Brosche, P. and Sundermann, J. (eds) (1982) *Tidal Friction and the Earth's Rotation II*. Berlin: Springer-Verlag.
Burns, J.A. and Matthews, M.S. (eds) (1986) *Satellites*. Tucson: University of Arizona.
Christodoulidis, D.C., Smith, D.E., Williamson, R.G. and Klosko, S.M. (1988) Observed tidal braking in the Earth/Moon/Sun system. *J. Geophys. Res.*, **93**, 6216–36.
Goldreich, P. (1966) History of the lunar orbit. *Rev. Geophys.*, **4**, 411–39.
Lambeck, K. (1980) *The Earth's Variable Rotation*. Cambridge: Cambridge University.
Lambeck, K. (1988) *Geophysical Geodesy*. Oxford: Clarendon.
Newhall, X.X., Williams, J.G. and Dickey, J.O. (1986) Earth rotation from lunar laser ranging. Pasadena: Jet Propulsion Laboratory, December, JPL Geod. Geophys. Preprint 153.
Peale, S.J., Cassen, P.M. and Reynolds, R.T. (1979) Melting of Io by tidal dissipation. *Science*, **203**, 892–4.
Ray, R.D. (1994) Tidal energy dissipation: observations from astronomy, geodesy, and oceanography, in *The Oceans: Physical-Chemical Dynamics and Human Transport* (eds S.K. Majumdar, E.W. Miller, G.S. Forbes *et al.*). Easton, Pa.: Pennsylvania Academy of Science, pp. 171–85.
Sinclair, A.T. (1989) The orbits of the satellites of Mars determined from Earth-based and spacecraft observations. *Astron. Astrophys.*, **220**, 321–8.
Stephenson, F.R. and Morrison, L.V. (1984) Long-term changes in the rotation of the Earth: 700 B.C. to A.D. 1980. *Phil. Trans. Roy. Soc. London*, **A313**, 47–70.
Stern, S.A. (1992) The Pluto–Charon system, in *Ann. Rev. Astron. Astrophys.* (eds G. Burbidge, D. Layzer and A. Sandage). **30**, 185–233. Palo Alto: Annual Reviews Inc.
Tittlemore, W.C. and Wisdom, J. (1990) Tidal evolution of the Uranian satellites. III. *Icarus*, **85**, 394–443.
Wisdom, J. (1987) Urey Prize Lecture: chaotic dynamics in the solar system. *Icarus*, **72**, 241–75.

Cross references

Chaotic dynamics in the solar system
Commensurability
Earth–Moon system: dynamics
Earth rotation
Earth: rotational history
Planetary rotation
Resonance in the solar system
Satellite, natural

TIDAL HEATING

Tidal heating is the increase in internal thermal content of a planet or moon associated with the differential gravitational force (or tide) between two bodies in orbit about their common center of mass. Because the force of gravity depends upon distance, the gravitational force between two objects of finite size is not uniform within the interiors of the objects. Variation in distance between two bodies, for example a satellite in an eccentric orbit around a planet, will therefore lead to variation in the gravitational force felt within the bodies over the period of one orbit. This variation in force leads to distortion of the material structure, and hence movement of elements relative to each other with consequent frictional heating of the interior of both bodies. Most commonly, in the case of the moons of the giant planets, the fluid internal structure of the giant planet is less dissipative than the solid or partially molten interior of the rocky or icy moon. Much of the tidal heating, then, is expressed in the orbiting satellite.

Tidal heating has important consequences for both the evolution of the orbits of satellites, and for their thermal histories. Heating implies dissipation of the energy of the orbit within the interior of the satellite and parent planet; hence, orbits of satellites tend to circularize over time unless other mechanisms for keeping the eccentricity high are in play. Jupiter's satellites Io, Europa and Ganymede are locked in an orbital pattern called a resonance wherein the eccentricities are maintained at non-zero values over the age of the solar system, leading to intensive tidal heating and melting of Io (with consequent volcanism) and modest heating of icy Europa. In the Saturn system, Enceladus' youthful surface may be a result of tidal heating sometime in the past. The orbits of Miranda, Ariel and Umbriel around Uranus may have been such in the past as to heat those satellites enough to cause melting and volcanic resurfacing. Neptune's moon Triton was probably captured into an eccentric orbit around Neptune early on, leading to intensive tidal heating, with consequent outgassing to form an atmosphere.

Jonathan I. Lunine

Bibliography

Lunine, J.I. and Tittemore, W.C. (1992) Origin of outer planet satellites, in *Protostars and Planets III* (eds E.H. Levy and J.I. Lunine). Tucson: University of Arizona Press, pp. 1149–76.
Malhotra, R. (1991) Tidal origin of the Laplace resonance and the resurfacing of Ganymede. *Icarus*, **94**, 339–412.
Yoder, C.F. (1979) How tidal heating of Io drives the Galilean orbital resonance locks. *Nature*, **279**, 767–70.

Cross references

Europa
Io
Resonance in the solar system
Volcanism in the solar system

TIDE-RAISING FORCE

The tide-raising force (or tide-producing force, or tide-generating force) is not a single force in terms of its cause. Instead, it is a differential force. For a two-body system each unit particle except at the center of the celestial body concerned is subject to the tide-raising force resulting from an imbalance between the gravitational attraction force by another celestial body and the centrifugal force due to the orbital movements in which both of those bodies are involved. For each unit particle of the body in question, the tide-raising force can be determined from the vector difference between the direct gravitational attraction force exerted by the second body and the centrifugal force (with the same magnitude as the centripetal force but in an opposite direction) linked with the orbital motion of the bodies about their common center of mass or barycenter (q.v.). In other words, the tide-raising force is that part of the attractive force of the second body that does not affect the movement of the body in its orbital motion.

The celestial bodies on which tidal forces act may be satellites, planets or stars, as long as they are involved in orbital motion. For example, for the planet Earth in the Earth–Moon system, the lunar tidal force is the difference between the Moon's gravitational attraction and the centrifugal force produced by the revolution of the Earth's center of mass around the barycenter. Henceforth, 'tidal force' will be used in this context, which is applicable to any point on the Earth (including its lithosphere, mantle, oceans and atmosphere). As a result, there are lunar (and solar) tides in the solid Earth (body tides), in the hydrosphere and in the atmosphere because of the Moon's attraction. Conversely, there are also Moon tides felt on the lunar surface due to the tidal force exerted by the Earth (and the Sun). However, because there is neither ocean nor atmosphere on the Moon, only body tides are developed there. Since the Earth–

Moon system orbits the Sun, there are simultaneously solar tidal forces on both the Earth and the Moon.

Although it has long been known that there exist relationships between the positions and phases of the Moon, the heights of the ocean tides and the local time of day, it was impossible to formulate a tide-raising force until the 17th century. Isaac Newton, in his *Principia*, laid the foundations for the modern theory of the tides when he discovered the mathematical principles associated with gravitation, now known as the universal law of gravitation. This law enables the construction of the formulae of the tide-raising force. A full understanding of the tide-raising force and its effects on the solid Earth, oceans and atmosphere has only been achieved through a series of works by Love (body tides; 1911), Euler (fluid dynamics), Laplace (ocean tides) and many others. Laplace (1778/1779) first introduced the spherical harmonic function to expand the tide-raising force in a form that makes it easier to understand the nature of tidal phenomena. In order to simplify this discussion of how the tide-raising force is mathematically formulated, only the lunar tidal force on the Earth will be used for the examples through most of the following development.

On account of the distance between the Earth and the Moon, it is sufficiently accurate to take the Moon as a point mass located at its center of mass when considering the tides for the solid Earth, the oceans and the atmosphere. By the gravitational law, every unit particle of the solid Earth, oceans and atmosphere is attracted by the Moon. Such forces are all directed toward the center of mass of the Moon. The magnitudes of these forces are directly proportional to the mass of the Moon, and inversely proportional to the square of the distance between each unit particle and the Moon. Therefore all these forces are slightly different in direction and magnitude as shown in Figure T24a (with great exaggeration). In the meantime, because all the particles act as a whole and have an orbital movement around the center of mass (barycenter) of the Earth–Moon system, they are subject to the same centripetal force in a direction from the Earth's center toward the Moon's center (Figure T24b). The centripetal or 'orbital force' is actually the average of the gravitational attraction force over the whole Earth exerted by the Moon, which is equal to the attraction force at the Earth's center by the Moon. Therefore, at any time, the two forces are perfectly balanced at the center of the Earth. But this is not the case elsewhere. For other parts of the Earth at any instant, the vector difference between the spatially variable gravitational attraction force of the Moon and the spatially constant 'average' orbital force is thus the tide-raising force. Alternatively, it can also be said that the tide-raising force on the Earth is the vector sum of the gravitational attraction force and the centrifugal force, which is equal to the centripetal force in magnitude but in the opposite direction. Since the orbital force is equal to the lunar attraction force at the Earth's center, it is evident that on the side of the Earth facing toward the Moon the vector difference, i.e. the tide-raising force, would be directed toward the Moon and that, on the opposite side of the Earth, it would be away from the Moon. All positions around the circumference of the Earth along a great circle perpendicular to the line joining the centers of the Earth and the Moon would have tide-raising forces toward the Earth's center. Therefore the effect is symmetrical about the line joining them, tending to pull the Earth (with its hydrosphere and atmosphere) into an ellipsoidal shape, a tidal force 'envelope' with its major axis along the line connecting the Earth and Moon. Following Darwin (1898), Figure T24c shows the lunar tidal force at the surface of the Earth and Figure T24d shows its horizontal component schematically.

Because the gravitational force is a gradient of the gravitational potential, which is a scalar (i.e. a real number), and it is more convenient to deal with a scalar than a vector, it is a generally adopted procedure to first write the gravitational potential V for any point A in the solid Earth, oceans or atmosphere as

$$V = \frac{GM}{\rho} = GM \sum_{n=0}^{\infty} \frac{r^n}{d^{n+1}} P_n(\cos\psi) \quad (T19)$$

Where G is the gravitational constant, M is the Moon's mass, ρ is the distance between the point A and the Moon's center, r is radius at A, i.e. the distance between the Earth's center and A, d is the distance between the Moon's and the Earth's centers, $P_n(\cos\psi)$ are Legendre polynomials of nth degree, and ψ is the angle between the directions of r and d (Figure T25). The degree $n = 0$ term in equation (T19) is a constant and is physically meaningless, so it can be dropped. The $n = 1$ term is actually equal to the orbit potential (centripetal potential),

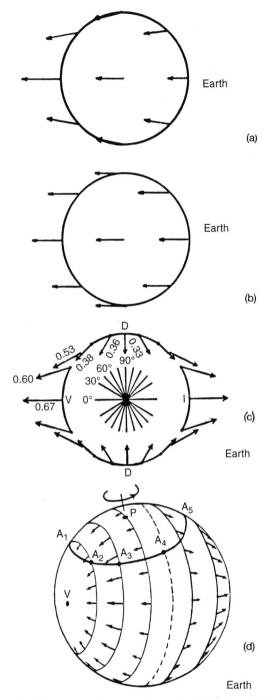

Figure T24 Schematic explanation of the lunar tidal force on the Earth. (a) The direct gravitational attraction force on each unit particle of the Earth by the Moon, which varies from point to point. (b) The orbit force or centripetal force, which is a constant vector equal to the gravitational attraction force at the center of the Earth by the Moon along the direction from the Earth's center to the Moon's center. (c) The vector tide-raising force, which is the vector difference of forces of (a) and (b). For an observer on the Earth, the Moon is at zenith, horizon and nadir when the observer is at points V, D and I respectively. The relative magnitudes of the force are also given by the numbers shown at a few representative points (after Darwin, 1898). (d) The horizontal tide-raising force (after Darwin, 1898). The great circle of D (shown by the dashed line) is the set of points where there is no lunar horizontal tidal force. P is the North Pole of the Earth. Circle A1 through A5 is a parallel of latitude (e.g. the latitude of London), where an observer would experience semi-diurnal tides during a day.

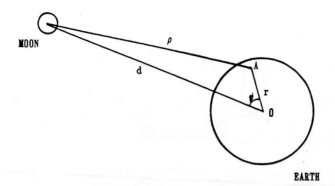

Figure T25 As explained in the text, point A is where the tide-raising potential and force are formulated.

which should be subtracted from the gravitational potential V during the derivation of the tide-raising potential. Consequently, the tide-raising potential is

$$V = GM \sum_{n=2}^{\infty} \frac{r^n}{d^{n+1}} P_n(\cos\psi) \tag{T20}$$

The tide-raising force F, which is a vector, then can be expressed by the gradient of V:

$$F = \text{grad } V = \text{grad} \left[GM \sum_{n=2}^{\infty} \frac{r^n}{d^{n+1}} P_n(\cos\psi) \right] \tag{T21}$$

Equation (T21) is the general expression of the tide-raising force.

If a spherical coordinate system (r, θ, λ) is chosen in such a way that its center is the Earth's center with its z-axis directed toward the north pole and x-axis along the Greenwich meridian, using a spherical harmonic expansion, equation (T20) can be further expressed as

$$V = \sum_{n=2}^{\infty} \frac{GMr^n}{d^{n+1}} \sum_{m=0}^{n} W_{nm} P_n^m(\sin\delta) P_n^m(\cos\theta) \cos[m(\omega t' - \alpha + \lambda)] \tag{T22}$$

where

$$W_{nm} = \frac{2(n-m)!}{(n+m)!}$$

if $m \neq 0$ and $W_{nm} = 1$ if $m = 0$. P_n^m is the associated Legendre polynomial of degree n and order m; δ is the declination of the Moon; θ is the polar angle, the colatitude of A; ω is the sideral rate of the Earth's rotation; t' is universal time (UT); α is the right ascension of the Moon; and λ is the azimuth angle, the eastward longitude of A.

The gradient of V of (T22) gives the three components of the tide-raising force along the spherical coordinate directions of r (outward from Earth's center), θ (southward along the meridian line) and λ (eastward), respectively.

All the above equations (T19) to (T22) can be applied to any celestial body if the Earth is replaced by another body whose tidal movements are being considered, and the Moon by a celestial body that exerts the tide-raising force on the former. If more than two heavenly bodies are involved, the total tidal force can be obtained by summing up all the tidal forces exerted by all the individual bodies, e.g. the total tidal force anywhere on the Earth is actually equal to superposition of all the individual tidal forces caused by the Moon, the Sun and other planets (and their associated orbital forces).

From equation (T22) it is clear that the tide-raising force is dependent not only on the geometric position and dynamic properties of the celestial body (Moon) that causes the tides, but also on those of the celestial body (Earth) that is under the action of the tide-raising force. In the same example of the Earth–Moon system, the magnitude of the tide-raising force is mainly determined by the first term ($n = 2$) of the right-hand side of (T22) since the second term ($n = 3$) is about two orders smaller (by a factor of r/d). Therefore, the lunar tidal force on the Earth is to a major degree proportional to the Moon's mass (M) and the reciprocal of the cube of the Earth–Moon distance (d). Furthermore, it also varies with the position of the Moon (δ and α) as well as the terrestrial position A (r, θ and λ) and the rotation rate of the Earth (ω).

As the main contribution of (T22) are the three terms of order $m = 0$, 1 and 2 of degree $n = 2$, we write down them as follows:

$$m = 0, \quad V = \frac{3GMr^2}{4d^3} 3(\tfrac{1}{3} - \sin^2\delta)(\tfrac{1}{3} - \cos^2\theta) \tag{T23}$$

$$m = 1, \quad V = \frac{3GMr^2}{4d^3} \sin 2\delta \sin 2\theta \cos(\omega t' - \alpha + \lambda) \tag{T24}$$

$$m = 2, \quad V = \frac{3GMr^2}{4d^3} \cos^2\delta \sin^2\theta \cos[2(\omega t' - \alpha + \lambda)] \tag{T25}$$

The three terms on the right-hand sides of (T23), (T24) and (T25) are three families of spherical harmonic functions which, following Laplace, are called respectively zonal, tesseral and sectorial functions. They represent long-period, diurnal and semi-diurnal spectra and, therefore, are responsible for the three corresponding species of tides respectively. Figure T26 (after Melchior, 1983) shows the three species of the spherical harmonic functions as functions of θ and λ (shaded areas are in opposite signs). Figure T26A represents the zonal function ($m = 0$). Because V is independent of λ, ω, α and t', it only varies with δ, which implicitly is a function of time t'. The principal periods are 14 days for the Moon and 6 months for the Sun. For $m = 1$ and 2, the main spectral lines are determined by ω and 2ω, i.e. diurnal and semidiurnal. Because δ, α and d also vary slowly with time, these main spectral lines are further split up into many spectral lines close to them. As a result, the tide-raising force in Figures T26b and T26c varies mainly with the rotation of the Earth with diurnal and semidiurnal periods respectively, and with small long-period variations.

When the common factor $3GMr/4d^3$ is dropped from the three equations, the gradients of (T23), (T24) and (T25) give the vertical components,

$$m = 0, \quad 6(\tfrac{1}{3} - \sin^2\delta)(\tfrac{1}{3} - \cos^2\theta) \tag{T26}$$

$$m = 1, \quad 2\sin 2\delta \sin 2\theta \cos(\omega t' - \alpha + \lambda) \tag{T27}$$

$$m = 2, \quad 2\cos^2\delta \sin^2\theta \cos[2(\omega t' - \alpha + \lambda)] \tag{T28}$$

the horizontal southward components,

$$m = 0, \quad 3(\tfrac{1}{3} - \sin^2\delta) \sin 2\theta \tag{T29}$$

$$m = 1, \quad 2\sin 2\delta \cos 2\theta \cos(\omega t' - \alpha + \lambda) \tag{T30}$$

$$m = 2, \quad \cos^2\delta \sin 2\theta \cos[2(\omega t' - \alpha + \lambda)] \tag{T31}$$

and the horizontal eastward components of the tide-raising force of $n = 2$,

$$m = 0, \quad 0 \tag{T32}$$

$$m = 1, \quad -2\sin 2\delta \cos\theta \sin(\omega t' - \alpha + \lambda) \tag{T33}$$

$$m = 2, \quad -2\cos^2\delta \sin\theta \sin[2(\omega t' - \alpha + \lambda)] \tag{T34}$$

For degree n higher than 2, there will be harmonic functions of higher degrees and orders for the tide-raising force, e.g. for $n = 3$ there will be also ter-diurnal components of the tide-raising force in addition to the three spectra. However, for the tide-raising force on the Earth, $n = 2$ and 3 for the Moon and $n = 2$ for the Sun are generally sufficiently accurate in terms of the resolution that may be obtained by modern tidal instruments.

All the terms of (T23) to (T34) can be separated into purely sinusoidal waves with arguments which are linear functions of time. Doodson (1921) performed such a harmonic expansion for the tide-raising force on the Earth by the Moon and Sun. His development leads to 386 components having an amplitude coefficient greater than or equal to 0.0001 D, where Doodson's constant D is given by

$$D = \tfrac{3}{4} GM \frac{\bar{r}^2}{\bar{d}^3}$$

$\bar{r}$ is the radius of the sphere of volume equal to that of the Earth, $\bar{d}$ is the mean distance between the Earth's center and the Moon's or the Sun's center, and M is the mass of the Moon or the Sun. For degree $n = 2$, D of the Sun is about 0.46 D of the Moon, so that on average the solar tidal force is about 0.46 of the lunar tidal force. Later on, with the aid of modern computers, Cartwright and Tayler (1971) extended Doodson's work and published more accurate harmonic expansion tables of the lunar–solar tidal potential.

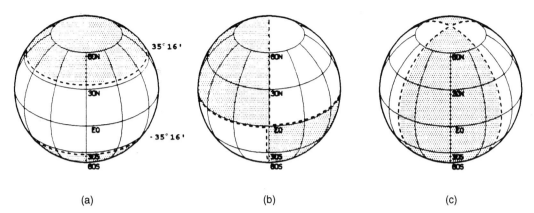

Figure T26 Three species of spherical hamonics of degree $n = 2$ as functions of θ and λ for the Earth: (a) zonal function ($m = 0$), (b) tesseral function ($m = 1$) and (c) sectorial function ($m = 2$). (After Melchior, 1983.)

The tide-raising force has various and complicated effects on the Earth. The main effects are fluid motions (tidal currents) in oceans and, to a much lesser degree, in the atmosphere, together with deformation in the solid Earth. Tidal torques change the direction of the rotation axis and rate of the rotation of the Earth (see Precession and nutation). Tidal disruption of a satellite or comet, as first proposed by E.A. Roche, is one of the three major hypotheses of the formation of planetary rings. According to that hypothesis, the rings could form when a single celestial body that comes too close to a planet (within the Roche limit; q.v.) and may be fragmented into a myriad of pieces. The body forming the rings may be a comet or a small satellite. In any case, fragmentation would have resulted from tidal disruption (Hamblin and Christiansen, 1990).

Yuan-Chong Zhang

Bibliography

Cartwright, D.E. and Tayler, R.J. (1971) New computations of the tide-generating potential. *Geophys. J. Roy. Astron. Soc.*, **23**, 45–74.
Darwin, G.H. (1898) *The Tides and Kindred Phenomena*. San Francisco: W.H. Freeman, 1962, 378 pp.
Doodson, A.T. (1921) The harmonic development of the tide-generating potential, *Proc. Roy. Soc. London, Ser. A.*, **100**, 305–29.
Godin, G. (1972) *The Analysis of Tides*, Toronto: University of Toronto Press, 294 pp.
Hamblin, W.K. and Christiansen, E.H. (1990) *Exploring the Planets*. New York: Macmillan Publishing Co., 451 pp.
Laplace, P.S. (1778/1779) Recherches sur plusieurs points du système du monde. *Mem. Acad. Roy. Sci. Paris*, for 1775, 75–182 (publ. 1778); for 1776, 117–267, 525–52 (publ. 1779).
Love, A.E.H. (1911) *Some Problems of Geodynamics*. New York: Dover Publ., 180 pp.
Melchior, P. (1983) *The Tides of the Planet Earth*, 2nd edn, Oxford: Pergamon Press.

Cross references

Earth–Moon system: dynamics
Gravitation
Synergetic tidal force

TITAN

Titan is the largest satellite in our solar system and in many ways it is more like a planet than some planets are (Plate 29). Its atmosphere was first detected by Kuiper (1944, 1952), but most of what we know about it has been learned since 1971. Titan's low mean density requires an interior model much richer in ices than is the case for the terrestrial planets, despite the other similarities between these bodies and Titan (Lewis, 1971). Titan's size and average density are bracketed by those for Jupiter's Galilean satellites, Ganymede and Callisto. The Galilean satellites are believed to consist of rock (silicates and iron compounds) and water ice (25–50% by mass). Candidate ice compounds for Titan's interior are different from Ganymede and Callisto, however, with $NH_3 \cdot H_2O$ and $CH_4 \cdot H_2O$ being the most likely alternatives to water ice. Titan's interior is probably made up of equal parts of rocky and icy material ($\sim 52 : 48$ by mass).

Titan is the only planetary moon with a significant atmosphere, and this atmosphere is denser than that of all the terrestrial planets, except Venus. Strong UV emissions measured by the Voyager 1 UV experiment (Strobel and Shemansky, 1982) showed that nitrogen was the atmosphere's main component. Other important constituents include methane and various other hydrocarbons. Voyager 1 images of Titan show that its surface is totally obscured by thick, dark orange or brown, stratospheric haze layers. The hazes are formed from complex organic polymers that are believed to be the end product of methane photochemistry. Titan has a small greenhouse warming caused primarily by pressure-induced opacity of N_2, CH_4 and H_2 (McKay, Pollack and Courtin, 1991).

Titan's photochemistry is a function of two processes: the photo-dissociation of methane and the dissociation of nitrogen (into N^+) by energetic particles from Saturn's magnetosphere. The N^+ ion then dissociates the methane into nitriles and higher-chain HCN polymers.

Titan has a hydrogen torus which extends $5 R_S$ from Titan's center. It is probably the result of dissociation of methane into carbon and hydrogen. Titan's weak gravitational acceleration, its extended atmosphere and warm thermosphere (~ 160 K) ensure the rapid escape of H and H_2.

Some basic Titan statistics are summarized in Table T7.

Titan's atmosphere

Titan's atmospheric temperature falls from a value of 94 K at the surface (pressure 1.5 bar) to a minimum of 71 K at the tropopause ($P = 128$ mbar; height = 42 km). It then rises rapidly to 160–170 K until levels of about 200 km are reached, at which point it levels off to a value of about 186 ± 20 K at the exobase (1600 km). The value of the surface temperature is somewhat controversial and could be as high as 101 K (Lellouch *et al.*, 1989).

The tropospheric composition, based on an inferred mean molecular weight of nearly 28 atomic mass units (amu), from Voyager IRIS and radio occultation measurements, is predominantly nitrogen or carbon monoxide. UV spectra from the Voyager ultraviolet spectrometer (UVS) support nitrogen as the dominant gas.

Initial estimates of the mean molecular weight ranged from 27.8–29.4. The mean value (28.6) is higher than that for pure nitrogen. Investigators need to suggest the presence of other gases in order to

Table T7 Characteristics of Titan

Property	Value	Notes
Surface radius (m)	2.575×10^6	⅖ of Earth's radius
Mass (kg)	1.35×10^{23}	$0.022 \times$ Earth's
Mean density (kg m^{-3})	1881	
Surface gravity (m s^{-2})	1.35	
Distance from Saturn (m)	5.0×10^8	$= 20\ R_S$
Distance from Sun (AU)	9.546	Same as Saturn
Orbital period around Saturn (d)	15.95	
Orbital period around Sun (years)	29.458	
Temperature (K)		
Surface	94	
Effective	86	
Surface pressure (Pa)	$149.6 \pm 0.2 \times 10^3$	

account for the difference. This additional gas must be significantly heavier than pure nitrogen, and spectroscopically inactive. Primordial (non-radiogenic) argon is the best candidate. Estimates of required mixing ratios give reasonable amounts of argon but this depends on the mixing that is assumed. The question of whether or not argon exists on Titan is still open.

Titan's highly reducing (or non-oxidizing) atmosphere suggests a striking analogy with what might have been the early Earth atmosphere. The existence and formation of life are generally believed to require the presence of water. Chemical reactions resulting in complex organic molecules occur easily in water. Water appears to be absent on Titan. The discovery of complex organics in Titan's atmosphere, however, reinforced the analogy that Titan's atmosphere may be a natural laboratory for studying prebiotic evolution.

Since temperatures in Titan's troposphere are no closer than 5 K to the condensation temperature of N_2 it is unlikely that nitrogen clouds can form. The condensation of clouds composed of CH_4 may be possible, but this would necessarily depend on the mixing ratio of methane assumed. Methane abundance is uncertain but it is very important for understanding thermal structure and possible surface states. Several detailed studies relating to a variety of feasible scenarios for the state of the Titan atmosphere and how they affect the methane abundance are given in Lindal et al. (1983), Eshlemann, Lindal and Tyler (1983) and Flasar (1983). Eshlemann, Lindal and Tyler derive an upper limit to the methane surface abundance of 3%, with condensation of methane clouds occurring at 15 km. Flasar's analysis concludes with a surface pressure for methane of 0.11 bar, and with methane clouds forming at about 3 km above the surface. IRIS data (Lindal et al., 1983) place a limit on the methane mixing ratio at the tropopause to between 1 and 2%. Each of these studies is extremely model dependent and leads to a different supposed thermal structure and a different model for the composition and state of the surface (Lunine, Atreya and Pollack, 1989). An interesting characteristic of Titan's atmosphere is its permanent, non-cyclic evolution. Methane is permanently dissociated (the heavy molecule dissociation products are believed to (1) form in the stratosphere, (2) fall down to the temperature minimum which acts as a cold trap where they condense and form aerosols, and then (3) fall to the surface). There must be a source or reservoir of methane at or below the surface which resupplies it to the atmosphere.

Other important constituents include molecular hydrogen and hydrocarbons such as ethane (C_2H_6), propane (C_3H_8) and acetylene (C_2H_2) in order of abundance. Carbon monoxide (CO) and carbon dioxide (CO_2) have been detected from Earth at levels of 60–150 ppm and 1.5 ppb respectively. Some less abundant organic constituents, such as methylacetylene (C_3H_4), diacetylene (C_4H_2) and hydrogen cyanide (HCN), were found by Voyager IRIS measurements to be more abundant in the north polar region. A steady southerly decline in their abundance was measured from the pole down to $-60°$ latitude. On the other hand, CO_2 was the only gas found to decrease near the north pole. These latitudinal variations of the gases have not been fully explained (Coustenis, 1990). A possible explanation for the nitrile (HCN, HC_3N and C_2N_2) enhancement in the northern regions is given by Yung (1987). Yung suggested that the nitrile accumulation occurred during the winter when the north polar region was in shadow. At the time of the Voyager flyby the polar region was just coming out of the winter season and thus had maximum nitrile abundances.

Titan's upper atmosphere was measured by the Voyager 1 UVS instrument. The temperature (186 K) is sufficient to assure the escape of hydrogen and helium. This process is rate limited by diffusion from below. The observed mixing ratio for H_2 is compatible with this rate and with its proposed source, namely photolysis of methane. The escaping gas is likely to be a mixture of H and H_2. It goes into orbit about Saturn to form a torus. The torus ranges from 8 to 25 R_S (Saturn radii), in the Saturn equatorial plane, with a vertical extent of 14 R_S. It is believed to enclose Titan's orbit at 20 R_S.

Titan's surface: are there oceans?

The composition of a liquid ocean on Titan's surface is highly model dependent (uncertainty in the surface temperature makes the physical state of the methane difficult to determine). Under the conditions of temperature and pressure presumed for Titan's surface, methane could be liquid. However, the temperature profile derived from the radio occultation experiment corresponds to a dry adiabatic gradient, which makes the existence of a pure methane ocean unlikely. Lunine, Stevenson and Yung (1983) suggested an ethane ocean with significant amounts of methane and other trace constituents. For example, different estimates of the surface temperature can lead to extremely different ocean depths (Lellouch et al., 1989). Titan might be covered, or partly covered, by an ocean composed of methane and ethane.

The only direct probes to the surface make use of microwave radio emission and radar frequencies. The results from these measurements are uncertain (Wagener et al., 1988; Muhleman et al., 1990). Muhleman et al. (1990) claim that their recent radar measurements preclude the possibility of a deep methane or methane–ethane ocean on the surface. Uncertainties in the thermal profile and surface temperature provide great uncertainty in the amount of methane present in the atmosphere near the surface. This further complicates the derived depth and composition of the ocean, if it indeed exists.

Although photochemical models (Yung, Allen and Pinto, 1984) are constrained to produce stratospheric column abundances of hydrocarbons that are consistent with Voyager IRIS measurements, there is uncertainty in our understanding of both aerosols and simpler hydrocarbons. One cannot rule out a model, therefore, in which most of the products of methane photolysis reach the surface as aerosols (Lunine, Atreya and Pollack, 1989) where they may have been accumulating since Titan's origin.

Recent investigations

Reanalysis of Voyager data (Owen and Gautier, 1989; McKay, Pollack and Courtin, 1991; Lellouch, 1990; Coustenis, 1991) has dealt mainly with improvements in abundance determinations of minor atmospheric constituents, greenhouse effects and reanalysis of the thermal profile with investigations into its latitudinal variations respectively. Recent results from Hubble Space Telescope (HST) images of Titan suggest evidence for seasonal changes in Titan's atmosphere (Caldwell et al., 1992). Caldwell et al. obtained three Titan images, at 439, 547 and 889 nm with HST's WF/PC camera on 26 August 1990. Comparison of these images with Voyager 1 and Voyager 2 images obtained 10 and 9 years earlier shows that the seasonal hemispheric brightness asymmetry has reversed at wavelengths near 440 nm and 550 nm with the northern hemisphere now being brighter. An additional, noisy HST image at 889 nm wavelength, for which there are no analogous Voyager data, suggests that the southern hemisphere may have been brighter than the northern hemisphere at that wavelength in 1990. Since this particular filter is centered on a very strong methane absorption, this result suggests that the upper cloud structure may have a different hemispheric dependency.

The work by Owen and Gautier (1989) discusses both primordial and evolutionary sources for CO in Titan's atmosphere in order to reconcile its abundance with available observations. McKay, Pollack and Courtin (1991) report that they have determined that an anti-greenhouse effect (as well as a greenhouse effect) is occurring on Titan. This anti-greenhouse effect results from the presence of the high-altitude haze layer that absorbs at solar wavelengths but which is transparent in the infrared. Lellouch (1990) reviews the information relating to Titan's thermal profile and discusses how the uncertainties in the temperature, density and composition are being used to define Cassini's instrument package. Coustenis (1990) discusses the latitudinal distribution of some of the minor constituents in Titan's atmosphere and reviews possible photochemical and seasonal models derived to explain these distributions.

NASA-ESA Cassini mission

Titan's atmosphere and surface will be studied in more detail by the Cassini mission (q.v.) which is scheduled to arrive in the early part of the 21st century (Lunine, 1990). This mission will obtain gravity assists from Venus, Earth and Jupiter in order to reach its intended target, Saturn. The mission is designed to orbit Saturn, and will also carry a probe that will be dropped into the Titan atmosphere. The Huygens probe (q.v.) is being supplied by the European Space Agency (ESA). Continuous sampling of the particle composition and size, temperature, pressure and composition of the gases, and the wind velocity and shear, will be some of the primary functions of the probe during is 3-h descent. If the probe survives the landing then it will continue to transmit valuable data about the surface for, approximately, another 30 min. Cassini, the orbiter, will carry a Titan Radar Mapper which it will use to map the surface of Titan during its numerous flybys of the moon. A more complete summary of a preliminary payload for the probe is given by Lunine (1990).

Cindy C. Cunningham

Bibliography

Barbato, J.P. and Ayer, E.A. (1981) *Atmospheres. A View of the Gaseous Envelopes Surrounding Members of our Solar System.* New York: Pergamon Press.
Caldwell, J., Cunningham, C.C., Anthony, D. *et al.* (1992) Titan: evidence for seasonal change. A comparison of Hubble Space Telescope and Voyager images. *Icarus*, **97**, 1–9.
Coustenis, A. (1990) Spatial variations of temperature and composition in Titan's atmosphere: recent results. *Annales Geophysicae*, **8**, 645–52.
Eshlemann, V.R., Lindal, G.F. and Tyler, G.L. (1983) Is Titan wet or dry? *Science*, **221**, 53–5.
Flasar, F.M. (1983) Oceans on Titan? *Science*, **221**, 55–7.
Hunten, D.M., Tomasko, M.G., Flasar, F.M. *et al.* (1984) Titan, in *Saturn* (eds T. Gehrels and M.S. Matthews). Tucson: University of Arizona Press, pp. 671–759.
Hunten, D.M. (1977) Titan's atmosphere and surface, in *Planetary Satellites* (ed. J.A. Burns). Tucson: University of Arizona Press, pp. 420–37.
Kuiper, G.P. (1944) Titan: a satellite with an atmosphere. *Astrophys. J.*, **62**, 245.
Kuiper, G.P. (1952) Planetary atmospheres and their origin, in *The Atmospheres of the Earth and Planets* (ed. G.P. Kuiper). Chicago: University of Chicago Press, pp. 306–405.
Lellouch, E., Coustenis, A., Raulin, F. *et al.* (1989) Titan atmosphere temperature profile: a reanalysis of Voyager 1 radio-occultation and IRIS 7.7 micron data. *Icarus*, **79**, 328–49.
Lellouch, E. (1990) Atmospheric models of Titan and Triton. *Annales Geophysicae*, **8**, 653–60.
Lewis, J.S. (1971) Satellites of the outer planets: their physical and chemical nature. *Icarus*, **15**, 174–85.
Lindal, G.F, Wood, G.E., Hotz, H.B. *et al.* (1983) The atmosphere of Titan: an analysis of the Voyager 1 radio occultation measurements. *Icarus*, **53**, 348–63.
Lunine, J.I., Stevenson, D.J. and Yung, Y.L. (1983) Ethane ocean on Titan. *Science*, **222**, 73.
Lunine, J.I., Atreya, S.K. and Pollack, J.B. (1989) Present state and chemical evolution of the atmospheres of Titan, Triton and Pluto, in *Origin and Evolution of Planetary Satellite Atmospheres* (ed. S.K. Atreya, J.B. Pollack and M.S. Matthews). Tuscon: University of Arizona Press, pp. 605–65.
Lunine, J.I. (1990) Titan. *Adv. Space. Res.*, **10**, 1137–44.
McKay, C.P., Pollack, J.B. and Courtin, R. (1991) The greenhouse and anti-greenhouse effects on Titan. *Science*, **253**, 1118–21.
Muhleman, D.O., Grossman, A.W., Butler, B.J. and Slade, M.A. (1990) Radar reflectivity of Titan. *Science*, **248**, 975–80.
Owen, T. and Gautier, D. (1989) Titan: some new results. *Adv. Space Res.*, **9**, 273–78.
Strobel, D.F. and Shemansky, D.E. (1982) EUV emission from Titan's upper atmosphere – Voyager 1 encounter. *J. Geophys. Res.*, **87**, 1361–8.
Wagener, R., Owen, T., Jaffe, W. and Caldwell. J. (1989) The surface emissivity of Titan at 2 cm. *Bull. Am. Astron. Soc.*, **20**, 843.
Yung, Y.L., Allen, M. and Pinto, J.P. (1984) Photochemistry of the atmosphere of Titan: comparison between model and observations. *Astrophys. J. Suppl.*, **55**, 465–506.
Yung, Y.L. (1987) An update of nitrile photochemistry on Titan. *Icarus*, **72**, 468–72.

Cross references

Saturn: satellite system
Satellite, natural

TITANIA

One of the moons of Uranus, Titania was discovered telescopically by William Herschel in 1789, 7 years after his discovery of Uranus. It is 1578 km in diameter and has a mass of 34.9×10^{20} kg, making it the largest of the Uranian moons.

The Voyager 2 spacecraft returned images of the surface of Titania in late January 1986 which showed it to be a heavily cratered body. There are a few large craters formed by high-energy impacts, including multiring craters such as Gertrude, which is 300 km in diameter, and Ursula, 155 km in diameter. However, the cratering record on Titania is dominated by smaller craters. The distribution of craters is non-uniform with some areas of the moon having far fewer craters than others, indicating that these areas have been resurfaced at some point in Titania's geologic history.

Titania also has a network of extensional fractures but these are less well developed than those on the Uranian moon Ariel. These fractures have scarps from 2 to 5 km high; they cut through the older large craters and do not appear to have been modified by the younger small craters. Thus they are among the youngest features on the moon. The largest fracture is Messina Chasmata, which extends for approximately 1500 km and varies from 50 to 100 km wide. Cross-cutting relationships among the canyons and the presence of multiple

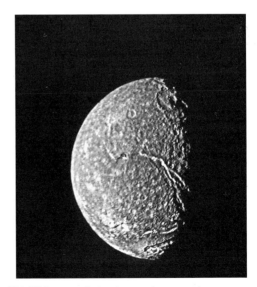

Figure T27 Highest-resolution image of Titania. This is a composite of two images taken by the Voyager 2 spacecraft from a distance of 369 000 km on 24 January 1986. The smallest features visible on the surface are approximately 13 km across. Gertrude is the large multiring crater at the upper left while Ursula is at the upper right. Messina Chasmata is the large fracture system between them. The south pole of the moon is near the lower center of the image. (NASA image P-29522.)

stepped faults in portions of Messina Chasmata indicate that there were probably several separate episodes of expansion on Titania (Croft and Soderblom, 1991). In addition Titania has a few ridges located between crater Gertrude and Messina Chasmata that appear to be of compressional origin, although they have also been interpreted as degraded crater rims. If they are compressional, they are unique in the Uranian system and indicate a period of horizontal crustal shortening early in Titania's geologic history.

Daniel M. Janes

Bibliography

Bergstralh, J., Miner, E. and Matthews, M. (1991) *Uranus*. Tucson: University of Arizona Press.
Croft, S.K. and Soderblom, L.A. (1991) Geology of the Uranian satellites, in *Uranus* (eds J. Bergstralh, E. Miner and M. Matthews). Tucson: University of Arizona Press, pp. 561–628.
Miner, E. (1990) *Uranus: The Planet, Rings and Satellites*. New York: Ellis Horwood.
Smith, B.A., Soderblom, L.A., Beebe, R. *et al.* (1986) Voyager 2 in the Uranian system: imaging science results. *Science*, **233**, 43–64.

Cross references

Satellite, natural
Uranus
Uranus: satellite system

TITIUS–BODE LAW

An empirical principle describing the distribution of planets within the solar system, which is found to conform to an approximately harmonic series with reference to their mean distance (orbital radius) from the Sun. A universally agreed cosmologic explanation for this relationship has not yet been presented.

J.B. Titius (1729–1796) and Johan Elert Bode (1747–1826) worked out the concept in a joint correspondence. The relationship can be formulated in several ways, the simplest being to take the numerical series 0, 3, 6, 16, 24, 48, 96, 192; then add 4 to each number and divide by 10. If the Earth's distance is taken as 1.0, by definition the astronomical unit (1.4959787066×10^8 km), the whole series, with theoretical and observed values (in AU), is given in Table T8. Only Earth (by definition) and Jupiter conform precisely, while Neptune and Pluto depart grossly. This may in fact eventually help to explain the anomalous interrelationships between those bodies; for example, Lyttleton suggested that Pluto was a former Moon-like satellite of Neptune that has spun off. The 'Law', as announced by Bode in 1772, was strongly reinforced by William Herschel's discovery of Uranus in 1781, although its position was slightly off. It also spurred a search for the 2.8-AU slot, which led to the discovery of Ceres by Piazzi in 1801. A third prediction, albeit somewhat misleading, eventually led to a calculation of Neptune's position by J.C. Adams (q.v.) and U.J.J. Leverrier (q.v.) and its actual discovery in 1846 by Galle and D'Arrest.

C.M. Stacey (1963, 1967) pointed out a fundamental weakness in the Titius–Bode law and proposed a modification. Using periodicities, not distances, and instead of choosing a minor planet (Earth) for unity, he selected the beat frequency between two giant planets, Saturn and Jupiter, which together possess 86% of the angular momentum of the entire solar system and set the tone for a harmonic resonating series comprising the entire system. In his words (Stacey, 1967, p. 745), 'the inner and outer planets oscillate in and out from the Sun along nodes that are binary fractions or multiples of the fundamental tone set by Jupiter each time it gains a lap on Saturn', which is every 19.859307 years. 'The interval between each pair of orbits, counting the planetoids as one, appear to be harmonic to a close degree of approximation.' Nieto (1972) has subsequently found that plotted logarithmically the planetary distribution corresponds very closely to a straight line. Departures may be explained perhaps by mass changes due to early asteroid impacts.

Rhodes W. Fairbridge

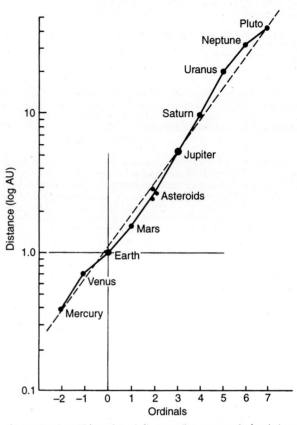

Figure T28 A semi-log plot of distance (in astronomical units) versus ordinal numbers (with Earth = 0), based on Nieto (1972, Figure 7.1). The straight-line fit has a slope of 1.7275, which is the Blagg–Richardson geometric progression ratio. Only Jupiter and Mercury are directly on the line; it may be postulated that during the very early stage of planetary formation, small displacements would result from major planetesimal–asteroid impacts.

Table T8 Relative mean distances of the planets and asteroid belt from the Sun (AU)

	Theoretical	Observed
Mercury	0.4	0.39
Venus	0.7	0.72
Earth	1.0	1.0
Mars	1.6	1.52
Asteroid belt	2.8	2.65
Jupiter	5.2	5.2
Saturn	10.0	9.54
Uranus	19.6	19.2
Neptune	38.2	30.1
Pluto	38.8	39.5

Bibliography

Nieto, M.M. (1972) *The Titius-Bode Law of Planetary Distances: Its History and Theory*. New York: Pergamon Press.
Stacey, C.M. (1963) Cyclical measures: some tidal aspects concerning equinoctial years. *New York Acad. Sci. Annals*, **105**(7), 421–60.
Stacey, C.M. (1967) Planetary intervals – the Titius–Bode rule and modification, in *The Encyclopedia of Atmospheric Sciences and Astrogeology* (ed. R.W. Fairbridge). New York: Van Nostrand Reinhold, pp. 745–6.

TOMBAUGH, CLYDE WILLIAM (1906–)

Tombaugh will always be remembered as the discoverer of the planet Pluto in 1930. He was born at a farm in Streator, Illinois, and grew up there; at an early age he borrowed a telescope and began to make observations of the planets, particularly Mars and Jupiter. He sent his drawings to the Lowell Observatory at Flagstaff, Arizona, and the Director, V.M. Slipher, was impressed. It had been decided to recommence the search for the trans-Neptunian planet which had been predicted on mathematical grounds by Percival Lowell, and Tombaugh was engaged specifically for this purpose. Using a 13-inch (33 cm) refracting telescope he began a photographic survey, and on 18 February 1930 he identified Pluto on plates which had been exposed on 23 and 29 January.

For the following years he continued to search, and also undertook an unsuccessful hunt for minor Earth satellites. After World War II he went to the White Sands proving ground to work on methods of optical tracking of rockets, and then moved to the New Mexico State University to set up a program of planetary observation. In addition to research, he was very active as a teacher of astronomy, and as the author of both popular and technical papers. He retired officially in 1973, but two decades later found him still active.

Patrick Moore

Bibliography

Levy, D.H. (1991) *Clyde Tombaugh*. Tucson: University of Arizona Press.
Tombaugh, C.W. and Moore, P. (1980) *Out of the Darkness: the Planet Pluto*. Harrisburg, PA: Stackpole Books.

TORQUE

A twisting or rotary motion in mechanics (from the Latin *torquere*, to twist). When a force is applied to a body at a point displaced from some reference origin or axis, a torque is produced. While the force applied at the origin determines the body's translational motion, its torque component or moment of force affects the body's rotational motion. The torque L of a system of particles is defined as $L = \Sigma\, r_i \times f_i$, where r_i is the position vector of the ith mass point with respect to some origin of a reference frame (often chosen to be the center of mass), f_i is the external force acting on the same mass point, and the summation is over all the particles of the system under consideration (for a continuous body, the summation should be replaced by integration). In engineering mechanics, a torque often refers to the torsional or twisting moment of a couple tending to twist a rigidly fixed object, such as a shaft about a rotational axis. Torque determines the time rate of change of the angular momentum (q.v.) of the system as expressed in Euler's equation of motion. Because both torque and angular momentum are vectors, torque affects both the direction and rotational rate of the system as a whole.

In our daily lives, every motion of our arms, legs, head and torso involves one or more torques within our bodies, and when we carry out our daily activities, whether it is opening a door, unscrewing a cover of a jar or steering a car, we are applying torques.

Torque plays an important role on rotational motions of the planets (q.v.) and the Sun (q.v.). The Earth always experiences forced precession (q.v.), nutation (q.v.) and changes of length of day caused by a net external torque on the Earth (with respect to its center of mass) due to the gravitational attractions (mainly by the Moon and the Sun) acting on Earth's equatorial bulge. It is concluded that our seasons will be reversed every hemicycle of about 13 000 years because of the precession of the Earth. The rotation rate of Earth's crust exhibits complicated fluctuations at an order of several milliseconds that may be roughly divided into three categories: (1) an overall linear increase from tidal dissipation, (2) irregular large variations on a timescale of decades arising primarily from exchanges of torque or angular momentum between the Earth's core and mantle and (3) the smaller higher-frequency changes: interannual, seasonal and interseasonal, driven primarily by exchanges of angular momentum with the atmosphere (Lambeck, 1980; Dickey, 1993; see Angular momentum cycle in planet Earth).

For the Earth–Moon system, the tidal dissipation causes a lag that displaces the Earth's tidal bulge slightly from the Earth–Moon axis. The bulge gives asymmetry to the geopotential. Consequently, the Earth's bulge exerts a torque on the Moon that tends to accelerate its orbital motion, while the Moon exerts a reciprocal torque on the Earth that slows its spin. By this mechanism, angular momentum is transferred from the rotation of the Earth to the lunar orbital motion. Assuming conservation of the total angular momentum in the Earth–Moon system, the orbital angular velocity of the system and the distance between the Moon and the Earth are related by Kepler's third law (see Kepler's laws). As a result, the acceleration of the orbital angular velocity causes the Moon to recede from the Earth (e.g. Jeffreys, 1976; Stacey, 1977). The rate of lunar retreat has been estimated to range from about 1 cm year^{-1} 2 billion years ago by paleogeologic studies to about 3–4 cm year^{-1} at present by recent measurements using lunar laser ranging (Williams, 1990).

Although the Sun possesses almost 1000 times as much mass as the nine known planets, it has only $\frac{1}{180}$ of the angular momentum of the whole solar system. Jupiter alone holds 71% of the total mass of the planets and 61% of the total angular momentum of the system. With such an uneven distribution of the angular momentum, and by the law of angular momentum conservation, different spatial arrangements of planets can cause variation of the Sun's orbital momentum with respect to the center of mass of the solar system (CM) from -0.1×10^{47} to 4.3×10^{47} g cm^2 s^{-1}, which is a difference of a factor of more than 40. A highly speculative coupling mechanism between the Sun's spin and its revolution about the CM may cause a variation of up to 5% of the Sun's equatorial rotational velocity, which is actually observed. The so–called strong impulses of the torque (IOT) in the Sun's rotation may be a cause of variation in solar activity (Landscheidt, 1988, 1989).

Yuan-Chong Zhang

Bibliography

Dickey, J.O. (1993) Angular momentum exchange between atmosphere and Earth. *EOS*, **74**(2), 17–22.
Jeffreys, H. (1976) *The Earth: Its Origin, History and Physical Constitution*, 6th edn. Cambridge: Cambridge University Press.
Lambeck, K. (1980) *The Earth's Variable Rotation: Geophysical Causes and Consequences*. Cambridge: Cambridge University Press.
Landscheidt, T. (1988) Solar rotation, impulses of the torque in the Sun's motion, and climatic variation. *Climate Change*, **12**, 265–95.
Landscheidt, T. (1989) *Sun–Earth–Man, A Mesh of Cosmic Oscillations*. London: Urania Trust.
Stacey, F.D. (1977) *Physics of the Earth*, 2nd edn. New York: John Wiley and Sons.
Williams, G.E. (1990) Tidal rhythmites: key to the history of the Earth's rotation and lunar orbit. *J. Phys. Earth*, **38**, 475–91.

TRANSIT

When a star in its daily apparent motion passes across the observer's meridian, it is said to transit across the meridian (see Coordinate systems). A specialized telescope mounted on an east–west axis and thereby restricted to move only in the meridian, and designed for the purpose of measuring the exact instant of meridian transit of the stars, is called a transit instrument.

When an inferior planet (i.e. Mercury or Venus) passes between the Earth and the Sun, it is said to transit across the Sun. Transits do not occur at every inferior conjunction of a planet, because the orbits of Mercury and Venus are inclined (at 7°.0 and 3°.4 respectively) to the plane of the ecliptic. Transits can happen only when conjunction occurs near a node (see Orbit); otherwise the planet passes above or below the Sun. Transits of Mercury occur about 13 times per century, always in May or November. Transits of Venus are exceedingly rare, and occur in pairs separated by 8 years, followed by a gap of either 105.0 or 121.5 years. They occur in June or November. None have occurred or will occur during the 20th century.

Mercury will just graze the limb of the Sun in 1999. Otherwise there will be no transit of either planet before the end of the 20th century. Thereafter, the next few transits are given in Table T9.

Table T9 Transits of Mercury and Venus across the Sun, 2000–2020

Date	Planet
7 May 2003	Mercury
8 June 2004	Venus
8 November 2006	Mercury
6 June 2012	Venus
9 May 2016	Mercury
11 November 2019	Mercury

Historically, transits of Venus were of considerable interest for determining the scale of the solar system. As pointed out by Edmond Halley, if a transit of Venus is observed from different places on the Earth's surface, the duration of the transit will be different for different observers, and the distance to the planet Venus (and hence the scale of the solar system) can be deduced from the observations. Expeditions to observe the 1761 and 1769 transits of Venus gave tolerably good values for the solar parallax, but difficulties in measuring the exact instants of ingress and egress limit the accuracy of the method, which is now of historical interest only.

Jeremy B. Tatum

Bibliography

Smart, W.M. (1962) *A Text-Book on Spherical Astronomy*. Cambridge: Cambridge University Press.
United States Naval Observatory (1961) Explanatory Supplement to the American Ephemeris and Nautical Almanac. United States Naval Observatory.

TRITON

Triton is the largest satellite of Neptune, and one of the most unusual worlds in the solar system (Plate 32). It was discovered by William Lassell, an English brewer and amateur astronomer, on 10 October 1846, less than 3 weeks after the discovery of Neptune. Triton's unusual retrograde orbit was quickly established, but due to its great distance (~ 30 AU), its mass, radius, albedo and geology remained unknown until 25 August 1989, when Voyager 2 revealed an active, diverse, geologically young and extremely cold world (Stone and Miner, 1989).

Fundamental characteristics

Triton's orbit is nearly circular with a semimajor axis of 14.33 Neptune radii (R_N) (Table T10; Jacobson, Riedel and Taylor, 1991). Its orbit is inclined 157° relative to Neptune's equatorial plane and is retrograde. Triton thus revolves about Neptune in a direction opposite that which the planet spins. Triton is one of only six retrograde moons in the solar system (four orbit Jupiter and one Saturn); among these, it orbits much closer to its primary, and it is by far the largest. Triton's circular and retrograde orbit suggests it

Table T10 Characteristics of Triton

Physical properties	
Mass (kg)	2.14×10^{22}
Radius (km)	1350
Density (g cm^{-3})	2.08
Gravity (m s^{-2})	0.78 (=0.08 g_{Earth})
Orbit	
Semimajor axis	14.33 R_N (=3.558×10^5 km)
Eccentricity	1.6×10^{-5}
Inclination (deg)	157
Period (days)	5.88

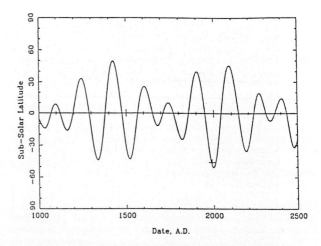

Figure T29 Subsolar point on Triton between 1000 and 2500 AD. A cross marks the latitude during the Voyager 2 encounter in 1989. (From Soderblom *et al.*, 1990, modified from Cruikshank and Brown, 1986, and Harris, 1984.)

experienced significant tidal interactions with Neptune that reduced its semimajor axis and eccentricity to their present values, and also suggests that it is a captured satellite (McCord, 1966; McKinnon, 1984). Triton's origin and evolution are probably significantly different from those of other large satellites of the outer planets.

Triton's orbit is synchronous; i.e. its orbital and rotational periods are identical, and one hemisphere continually faces Neptune. The gravitational effect of Neptune's oblateness causes Triton's orbital plane to precess with a period of ~ 688 years. Neptune's obliquity is $\sim 29°$ relative to the ecliptic, and it orbits the Sun once every 164.8 years. Due to Neptune's obliquity and period, and the precession rate and inclination of Triton's orbit, the subsolar point (at which the Sun appears directly overhead) oscillates in a complex pattern across Triton's surface between +52° and −52° of latitude (Figure T29; Harris, 1984). Seasons on Triton thus vary considerably in intensity and length and are among the most extreme in the solar system (with the possible exceptions of those of Uranus and Pluto–Charon).

With a radius of 1350 km, Triton ranks seventh among the largest moons in the solar system (after Ganymede, Titan, Callisto, Io, the Moon and Europa), and is slightly larger than Pluto. Triton's bulk density of 2.08 g cm^{-3} is consistent with a composition by mass of $\sim 70\%$ rock and $\sim 30\%$ ice (primarily H_2O, with minor amounts of methane and other volatile species; McKinnon and Mueller, 1989; Smith *et al.*, 1989). Triton is significantly more dense and rock-rich than many outer solar system satellites (e.g. Ganymede, Titan and the medium-sized moons of Saturn and Uranus), but quite similar in density and composition to Pluto.

As viewed from Earth, Triton has a maximum visual magnitude of only 13.5, and is thus a faint telescopic object. Its maximum angular separation from Neptune is only ~ 17 arcsec; consequently, reliable spectroscopic measurements of Triton's composition are difficult to obtain. Prior to Voyager 2, only methane (CH_4), in both gaseous and solid form, and molecular nitrogen (N_2) were detected (Cruikshank and Brown, 1986). We now know that abundant nitrogen resides both in Triton's thin atmosphere and as ice or frost on the surface. The Voyager ultraviolet spectrometer determined that N_2 is orders of magnitude more abundant than CH_4 as the lower atmospheric CH_4/N_2 mixing ratio is only $\sim 10^{-4}$ (Tyler *et al.*, 1989). More recently carbon monoxide (CO) and abundant carbon dioxide (CO_2) ices have been detected, discoveries that have important implications for Triton's origin (Cruikshank *et al.*, 1993). H_2O, which is expected to comprise the majority by mass of Triton's ices, has not been detected (this is also true for Pluto). This may be because H_2O bands are spectrally broad and hidden by CH_4 bands, or perhaps because the water ice acts as a bedrock that is obscured by surface volatile deposits.

Surface features

Although faint as viewed from Earth, Triton is actually among the most reflective objects in the solar system. Its global geometric

Figure T30 High-resolution mosaic of Triton. The sub-Neptune point at (0°, 0°) occurs slightly to the left of center. North is up. (Image processed by Alfred McEwen, US Geological Survey, Flagstaff.)

albedo (percentage of light reflected at zero phase angle) averages ~ 0.7 at visible wavelengths, and its bolometric Bond albedo (reflectance integrated over all wavelengths, radiated in all directions) is ~ 0.85 (Smith *et al.*, 1989; Hillier *et al.*, 1991). Voyager and ground-based measurements revealed that Triton's magnitude varies slightly as it rotates, with the leading hemisphere being slightly brighter. Triton's overall visual color is a very pale yellow-orange.

Voyager 2 imaged most of Triton south of ~ 20° latitude. The highest-resolution images covered ~ 40% of the surface centered on the hemisphere facing Neptune (Smith *et al.*, 1989). Triton's north polar region was not illuminated by sunlight during the encounter, thus surface features there remain unknown. Voyager 2 images revealed a number of albedo patterns and geologic units that only partially coincide.

Particularly prominent is a bright polar cap that dominates most of the southern hemisphere (Figure T30). The cap is asymmetric about the pole, which may account for Triton's slight lightcurve amplitude. The latitude of the northern edge varies between ~ −30° and ~ −10° in a pattern of concave 'scallops' at 10° to 30° longitudinal intervals (McEwen, 1990). A bright fringe outlines and appears to overlay the margin of the cap, from which diffuse bright swirls extend up to several hundred kilometers (generally toward the northeast). The cap also contains many relatively dark streaks, preferentially oriented to the northeast; these streaks, along with the bright rays, probably represent material transported by near-surface, Coriolis-displaced winds (Sagan and Chyba, 1990; Hansen *et al.*, 1990). The streaks are tens to hundreds of kilometers long, and resemble certain types of wind streaks on Mars. Of particular importance was the discovery of at least four relatively dark plumes rising from Triton's surface (Smith *et al.*, 1989, Soderblom *et al.*, 1990). The source of the plumes is the subject of much debate, although they may be related to the dark surface streaks. They are described below.

Portions of the southern polar cap and bright fringe have normal visual albedos close to 0.9 and are among Triton's brightest features (Smith *et al.*, 1989). The spectral characteristics and relationship to local topography suggest these are fresh deposits of frost or snow. Additional irregular and relatively dark spots (normal albedos ~ 0.8) occur within the cap near the edge, and may represent gaps in the frost. In general, the underlying geomorphology of the southern cap is difficult to discern due to the region's brightness, its mottled appearance and an unfavorable imaging geometry.

North of the polar cap, Triton's surface is generally darker and the surface morphology is more easily discerned. West of 0° longitude and slightly north of the equator is a large, topographically rugged region that resembles the surface of a cantaloupe. This terrain consists of many adjacent depressions that range in size from ~ 5 km to ~ 40 km, with an average size of ~ 25 km (Smith *et al.*, 1989). The origin of this 'cantaloupe terrain' is enigmatic. Early explanations that attributed the terrain to highly modified volcanic features, impact craters or interfering structural trends are now regarded as unsatisfactory. The most detailed and accepted explanation is that the terrain represents the surface expression of compositionally or thermally buoyant ice diapirs, analogous morphologically to terrestrial salt diapirs such as those in the Great Kavir of Iran (Schenk and Jackson, 1993).

The cantaloupe terrain is cross-cut by a pattern of intersecting ridges that appear to represent material erupted into extensional faults or graben. The faults extend south into the polar cap and east toward a smoother plains unit, where they are partially overlain by subsequent deposits. The wide geographic distribution of the faults suggests that they represent an episode of large-scale extension that may have been the last major tectonic event on Triton. Smaller faults have also been observed, including a few that suggest strike-slip movement (Smith *et al.*, 1989). The origin of the faults is not well understood.

East of the cantaloupe terrain is a large region of subdued topography that is dominated by a number of large, walled plains, the floors of which may have been volcanically emplaced. Near the northern terminator are two scarp-enclosed plains several hundred kilometers across that contain numerous pits and hills within their interiors. The pits and hills have been interpreted as either volcanic vents or as the result of ground collapse and scarp retreat due to sublimation and loss of interstitial material (Moore and Spencer, 1990). Slightly farther to the east is a structure that resembles a terrestrial volcanic caldera. Several long, wide, flat-topped ridges emanate radially for distances of up to several hundred kilometers from the caldera-like structure. The ridges are thought to be solidified flows of volcanically erupted ice.

Impact craters are relatively uncommon on Triton, indicating that the surface does not preserve a record of the late heavy bombardment represented by heavily cratered terrains on other satellites and on Mercury, the Moon and Mars (Smith *et al.*, 1989). Between 150 and 170 craters have been positively identified (Strom, Croft and Boyce, 1990; Schenk, 1992). Crater diameters range from a maximum of ~ 27 km down to the limit of Voyager resolution, or ~ 1.5 km. Craters smaller than ~ 11 km in diameter are structurally simple and bowl shaped; larger craters generally have central peaks and flat floors. The diameter at which the transition from simple to complex craters occurs is similar to that on other icy satellites (accounting for differences in surface gravity) for which H_2O is the dominant crustal component. None of the craters has observable ejecta or rays, perhaps because Triton's intrinsic brightness requires a phase angle close to zero for their detection.

The most heavily cratered region is located near the equator on the leading hemisphere, near Triton's eastern limb on the mosaic of the highest-resolution images. The number of craters per unit area in that region is similar to crater densities on the ~ 3-Ga lunar maria. The populations of crater-producing impactors at Triton and the Moon may be different, however, so an absolute age comparison may not be meaningful. Crater densities in the plains region encompassing the large volcanic features to the west and at the edge of the cap to the south are smaller by about a factor of two. Triton's leading hemisphere accumulates craters preferentially because the satellite is tidally locked in a synchronous orbit. Even when this effect is taken into account, the plains region to the west and the northern edge of the polar cap must still be about half the age of the most heavily cratered region. Comparisons with the crater densities of the cantaloupe terrain and most of the polar cap are not possible because craters are difficult to identify in those regions.

Although crater densities do not constrain the relative age of the cantaloupe terrain, stratigraphically it appears to be one of the older surface units. The graben are superimposed upon the cantaloupe terrain and clearly postdate it. In places, volcanic flows associated with the caldera-like structure obscure the graben and the edges of the cantaloupe terrain, and are therefore younger than both. Frost migration, wind streaks and active plumes discovered on the polar cap indicate that the cap surface is the youngest region on Triton (although the largely unknown underlying landforms are clearly older).

Atmosphere–surface interactions

Triton is one of only three solar system satellites that has an appreciable atmosphere (Titan and Io are the others). The surface temperature is a frigid ~ 38 K, the coldest measured on any solar system body, and is largely a consequence of Triton's highly reflective surface (Conrath *et al.*, 1989; Tyler *et al.*, 1989). Triton's tenuous atmosphere has a surface pressure of only ~ 16 μbar (16 millionths that at Earth's surface), consistent with buffering by solid N_2, which is present on the surface (Broadfoot *et al.*, 1989; Tyler *et al.*, 1989). The atmosphere is composed primarily of N_2 (scale height ≈ 15 km), with minor amounts of CH_4 that increase in abundance at lower altitudes.

The vertical temperature structure of Triton's atmosphere is unique in the solar system (Broadfoot *et al.*, 1989; Tyler *et al.*, 1989; Yelle, Lunine and Hunten, 1991). Near the surface, turbulence caused by horizontal atmospheric motion dominates vertical heat transfer, so the temperature decreases with increasing altitude along an N_2 adiabat, forming a troposphere. An inversion occurs near 8 km at a tropopause, above which the temperature rises conductively in a thermosphere to an altitude of 400 km. Above 400 km Triton's N_2 is ionized by solar radiation, forming an isothermal ionosphere with a temperature of ~ 95 K. The atmosphere of Triton is the only one known to possess an adjacent troposphere and thermosphere, without an intervening radiative stratosphere. Radiative effects within the atmosphere are negligible because of the low concentration of Triton's major greenhouse gas CH_4 and because of the low temperature.

The Voyager flyby occurred during late spring in Triton's southern hemisphere, when the subsolar point, at −45°, was at the most southerly latitude in recent centuries. Triton's southern hemisphere is presently receiving nearly maximum insolation, which is causing the polar cap to sublimate into thin clouds of condensing nitrogen at low atmospheric altitudes over the southern hemisphere. The sublimated material subsequently moves north, where frost deposition has been observed in a bright band of thin material at the edge of the polar cap and is strongly suspected to occur near Triton's dark north pole.

The subliming frost cap generates Ekman boundary layer winds (within ~ 3 km of the surface) that move north toward the equator at ~ 5–15 ms^{-1} (Ingersoll, 1990). Triton's 5.88-day rotation period induces a Coriolis force that deflects the winds to the east forming a polar anticyclone, which may explain the dark polar streaks if they result from wind-blown dust or inactive low-altitude plumes. At higher altitudes the atmosphere above the equator is slightly warmer than above the south pole because of surface albedo differences. The resulting north–south temperature gradient generates high-altitude winds that blow west, which are indicated by the active plume trails above ~ 8 km.

Triton's large-scale circulation pattern is thus one of seasonal condensation flow between the southern and northern hemispheres (Trafton, 1984; Ingersoll, 1990). Changes in Triton's spectrum since 1977, an interval in which the subsolar point has migrated ~ 10° south, are also consistent with seasonal volatile migration (Smith *et al.*, 1989). A similar circulation and volatile transport cycle occurs on Mars, where CO_2 takes the role of N_2, and H_2O that of CH_4. Changes in surface pressure of ~ 25% occur during seasonal cycles on Mars; even larger changes, perhaps by an order of magnitude or more compared to the present surface pressure, may occur on Triton. Atmosphere–surface interactions on Triton may also resemble those on Pluto, which experiences extreme seasonal variations and probably possesses a surface and atmosphere composition similar to Triton's (Owen *et al.*, 1993).

It is unusual for an object with a surface gravity as weak as Triton's to have an atmosphere. Triton's atmosphere exists because its large heliocentric distance and cold temperature inhibit thermally driven atmospheric escape, and because nitrogen in equilibrium at its surface regulates the pressure.

The discovery of active plumes on Triton was an unexpected and exciting result of the Voyager encounter (Figure T31) (Smith *et al.*, 1989). Some regions of Triton were imaged in stereo, which led to the discovery of the plumes (Soderblom *et al.*, 1990). The plumes originate at dark surface spots, rise ~ 8 km in narrow columns less than ~ 1 km in diameter, and trail abruptly to the west for up to 150 km. The plumes are ~ 10% darker than the surrounding surface, and two cast noticeable shadows. All four plumes are located between −49° and −60°, south of the subsolar point, in a region of continuous solar illumination at the time of the encounter (Hansen *et al.*, 1990).

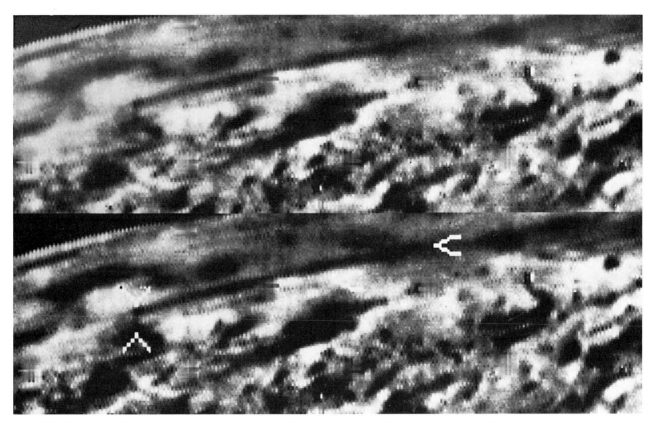

Figure T31 Highly magnified and contrast-enhanced image of the west plume on Triton. Arrows in the bottom frame indicate the plume's source (left) and trail (right). South is up. (Figure courtesy of NASA.)

Several explanations for the plumes have been proposed, including active volcanism, large dust devils and solar-driven geysers (Smith et al., 1989; Ingersoll and Tryka, 1990; Kirk, Brown and Soderblom, 1990). Plume locations, however, are a strong clue that solar power delivered to the surface is involved, perhaps as a solid state subsurface ice greenhouse. While details of the ice greenhouse models vary, each invokes the heating of translucent subsurface ices by a few degrees, substantially increasing the subsurface N_2 vapor pressure, and eruptively venting gas and dust into the atmosphere. Momentum and thermal buoyancy propel plume material to maximum altitudes of about 8 km, where prevailing winds change the plume direction to the west. As the driving mechanism weakens with time, the maximum elevation attained may diminish so that plume material is deflected by near-surface winds to the northeast, producing the observed dark surface streaks (Soderblom et al., 1990). The lifetimes of the plumes are not known, but the correlation of their locations with regions of near-maximum solar illumination suggests that Triton's plumes are seasonal phenomena and are probably not continuously active over geologic timescales.

Origin and evolution

The retrograde, inclined motion of Triton has inspired some imaginative speculation regarding its origin. Lyttleton (1936) hypothesized that Triton and Pluto originated as adjacent, prograde satellites of Neptune and were cast into their present orbits by a close gravitational interaction, a scenario now considered virtually impossible. Original satellite orbits should be prograde, nearly circular and lie in the plane of the planet's equator. The inclined and retrograde orbit of Triton strongly suggests that it is not an original satellite but is instead a captured object. Triton's status as the sole major moon of Neptune is also unusual, considering that the other Jovian planets have ordered satellite systems.

Triton's composition provides another clue to its origin. Icy outer solar system bodies formed in an environment that strongly affected their compositions and densities. Pluto-type objects formed in the cool outer solar nebula, have rock/(rock + ice) ratios higher than 'original' icy satellites and have densities of about 2.0 g cm^{-3} (McKinnon and Mueller, 1989; Prinn and Fegley, 1989). Triton's density exceeds that of all other icy satellites except Europa, but is similar to that of Pluto, suggesting that Pluto and Triton have similar compositions and formed at similar heliocentric distances. Furthermore, CO ice, which should be abundant in icy planetesimals but not in original satellites, has been detected on both Pluto and Triton (Owen et al., 1993).

Capture solely by a planet's gravity is at best temporary because no net orbital energy change relative to the planet occurs during a flyby. Therefore, the key to successful capture was for Triton to dissipate sufficient orbital energy to bind it to Neptune. The two most plausible capture mechanisms require at least one close flyby of Neptune and involve either gas drag within a protoplanetary nebula from which Neptune formed, or a collision with a previous satellite (Pollack et al., 1979; Goldreich et al., 1989; McKinnon and Leith, 1995). In either case it is likely that capture occurred billions of years ago, although collisional capture could have occurred more recently. The probability of capture is quite small, thus the existence of Triton suggests that it (and Pluto) represent the vestiges of a once-numerous outer solar system planetesimal population (Stern, 1991).

Subsequent to capture, Triton's inclined and eccentric orbit contracted due to tidal friction with Neptune, over a period of several hundred million years or less (McKinnon 1984; Goldreich et al., 1989). During this period, Neptune raised tides in Triton that were viscously dissipated, a process which converted Triton's orbital energy into internal heat and melted the satellite almost entirely. Consequently, Triton probably experienced an episode of extensive volcanism and other geologic activity, accompanied by significant internal structural changes and mineralogical alteration (McKinnon and Benner, 1990; Shock and McKinnon, 1993). The surface temperature may have been up to ~ 100 K warmer during this period, and a thick atmosphere may have been present (Lunine and Nolan, 1992). Geologic activity in response to tidal heating may be responsible for Triton's geologically youthful surface, although enough potassium, thorium and uranium probably exists in its rock-rich core to have driven surface activity through radiogenic heating (Stevenson and

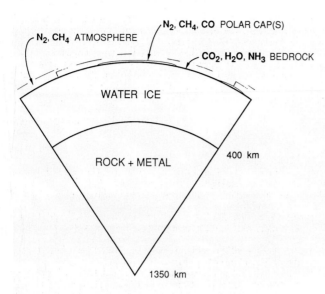

Figure T32 Possible present internal structure of Triton. (Modified from Shock and McKinnon, 1993.)

Gandhi, 1990). Triton's interior has probably differentiated into a rock and iron core ~ 950 km in radius and a ~ 400 km thick mantle consisting mostly of water ice (Figure T32) (Shock and McKinnon, 1993). During orbital contraction Triton probably made numerous close planetary flybys that could have gravitationally disrupted the satellites that may have been present about Neptune, and may have ejected Nereid into its present inclined, distant and eccentric orbit if Nereid is an original satellite (Goldreich et al., 1989). After tidal heating subsided, Triton slowly evolved to its present state.

There is, in a sense, a melodramatic aspect to Triton's continued existence. Tidal dynamics impose an inexorable orbital deterioration upon all retrograde satellites. Consequently, Triton journeys along an inward spiral, gradually encroaching upon Neptune and impending destruction. The fate of Triton is not a predestined collision with the planet, however. The satellite will be fragmented by the powerful gravitational field of Neptune. The remains of Triton will constitute the particles of a ring system more than 500 times as massive as the Saturnian rings. Although such an event is not anticipated for billions of years, it will represent the final metamorphosis of this most peculiar world.

Lance A.M. Benner

Bibliography

Broadfoot, A.L., Atreya, S.K., Bertaux, J.L. et al. (1989) Ultraviolet spectrometer observations of Neptune and Triton. *Science*, **246**, 1459–66.
Conrath, B., Flaser, F.M., Hanel, R. et al. (1989) Infrared observations of the Neptunian system. *Science*, **246**, 1454–9.
Cruikshank, D.P. and Brown, R.H. (1986) Satellites of Uranus and Neptune, and the Pluto–Charon system, in *Satellites*, (eds J.A. Burns and M.S. Matthews). Tucson: University of Arizona Press, pp. 836–73.
Cruikshank, D.P., Roush, T.L., Owen, T.C. et al. (1993) Ices on the surface of Triton. *Science*, **261**, 742–5.
Goldreich, P., Murray, N., Longaretti, P.Y. and Banfield, D. (1989) Neptune's story. *Science*, **245**, 500–4.
Hansen, C.J., McEwen, A.S., Ingersoll, A.P. and Terrile, R.J. (1990) Surface and airborne evidence for plumes and winds on Triton. *Science*, **250**, 421–4.
Harris, A.W. (1984) Physical properties of Neptune and Triton inferred from the orbit of Triton, in *Uranus and Neptune*, (ed. J.T. Bergstralh). NASA CP-2330, pp. 357–73.
Hillier, J., Helfenstein, P., Verbiscer, A. and Veverka, J. (1991) Voyager photometry of Triton: haze and surface photometric properties. *J. Geophys. Res. Suppl.*, **96**, 19203–19.
Ingersoll, A.P. (1990) Dynamics of Triton's atmosphere. *Nature*, **344**, 315–7.
Ingersoll, A.P. and Tryka, K.A. (1990) Triton's plumes: the dust devil hypothesis. *Science*, **250**, 435–7.
Jacobson, R.A., Riedel, J.E., and Taylor, A.H. (1991) The orbits of Triton and Nereid from spacecraft and Earthbased observations. *Astron. Astrophys*, **247**, 565–75.
Kirk, R.L., Brown, R.H., and Soderblom, L.A. (1990) Subsurface energy storage and transport for solar-powered geysers on Triton. *Science*, **250**, 424–9.
Lunine, J.I. and Nolan, M.C. (1992) A massive early atmosphere on Triton. *Icarus*, **100**, 221–34.
Lyttleton, R.A. (1936) On the possible results of an encounter of Pluto with the Neptunian system. *Mon. Not. Royal Astron. Soc.*, **97**, 108–15.
McCord, T.B. (1966) Dynamical evolution of the Neptunian system. *Astron. J.*, **71**, 585–90.
McEwen, A.S. (1990) Global color and albedo variations on Triton. *Geophys. Res. Lett.*, **17**, 1765–8.
McKinnon, W.B. (1984) On the origin of Triton and Pluto. *Nature*, **311**, 355–8.
McKinnon, W.B. and Benner, L.A.M. (1990) Triton's post-capture thermal history, in *Lunar Planet. Sci. Conf. XXI*, pp. 777–8.
McKinnon, W.B. and Leith, A.C. (1995) Gas drag and the orbital evolution of a captured Triton. *Icarus*, **118**, 392–413.
McKinnon, W.B. and Mueller, S. (1989) The density of Triton: a prediction. *Geophys. Res. Lett.*, **16**, 591–4.
Moore, J.M. and Spencer, J.R. (1990) Koyaanismuuyaw: the hypothesis of a perenially dichotomous Triton. *Geophys. Res. Lett.*, **17**, 1757–60.
Owen, T.C., Roush, T.L., Cruikshank, D.P. et al. (1993) Surface ices and the atmospheric composition of Pluto. *Science*, **261**, 745–8.
Pollack, J.B., Burns, J.A. and Tauber, M.E. (1979) Gas drag in primordial circumplanetary envelopes: a mechanism for satellite capture. *Icarus*, **37**, 587–611.
Prinn, R.G. and Fegley, B., Jr (1989) Solar nebula chemistry: origin of planetary, satellite and cometary volatiles, in *Origin and Evolution of Planetary and Satellite Atmospheres* (eds S.K. Atreya, J.B. Pollack and M.S. Matthews). Tucson: University of Arizona Press. pp. 78–136.
Sagan, C. and Chyba, C. (1990) Triton's streaks as windblown dust. *Nature*, **346**, 546–8.
Schenk, P. (1992) Volcanism, cratering and the geology of Triton. *Neptune and Triton Conf.*, p. 74.
Schenk, P. and Jackson, M.P.A. (1993) Diapirism on Triton: a record of crustal layering and instability. *Geology*, **21**, 299–302.
Shock, E.L. and McKinnon, W.B. (1993) Hydrothermal processing of cometary volatiles – applications to Triton. *Icarus*, **106**, 464–77.
Smith, B.A., Soderblom, L.A., Banfield, D. et al. (1989) Voyager 2 at Neptune: imaging science results. *Science*, **246**, 1422–49.
Soderblom, L.A., Kieffer, S.W., Becker, T.L. et al. (1990) Triton's geyser-like plumes: discovery and basic characterization. *Science*, **250**, 410–5.
Stern, S.A. (1991) On the number of planets in the outer solar system: evidence of a substantial population of 1000-km bodies. *Icarus*, **90**, 271–81.
Stevenson, D.J. and Gandhi, A.S. (1990) Puzzles of Triton, in *Lunar Planet. Sci. Conf. XXI*, pp. 1202–3.
Stone, E.C. and Miner, E.D. (1989) The Voyager 2 encounter with the Neptunian system. *Science*, **246**, 1417–21.
Strom, R.G., Croft, S.K. and Boyce, J.M. (1990) The impact cratering record on Triton. *Science*, **250** 437–9.
Trafton, L. (1984) Large seasonal variations in Triton's atmosphere. *Icarus*, **58**, 312–24.
Tyler, G.L., Sweetram, D.N., Anderson, J.D. et al, (1989) Voyager radio science observations of Neptune and Triton. *Science*, **246**, 1466–73.
Yelle, R.V., Lunine, J.I. and Hunten, D.M. (1991) Energy balance and plume dynamics in Triton's lower atmosphere. *Icarus*, **89**, 347–58.

Cross references

Neptune: satellite system
Volcanism in the solar system
Voyager missions

TROJAN ASTEROIDS

In 1906 an unusually slow-moving asteroid was discovered by Max Wolf, the founder and director of the Heidelberg Observatory. Orbital analysis showed that the object was librating about the Sun–Jupiter L4 Lagrange point, 60° ahead of Jupiter in its orbit (see Figure T33). The discovery confirmed, spectacularly, the theoretical prediction of J.L. Lagrange more than a century earlier that regions of stability exist in the triangular equilibrium points in the restricted three-body problem. The new asteroid was named 588 Achilles, beginning a tradition of naming these asteroids after heroes of the Trojan War. Objects in the L4 region are named after Greek heroes, while those in L5 are named after Trojans. For simplicity the two groups together are referred to as Trojans.

The Trojans do not move in rigid formation within a Sun–Jupiter reference frame, but rather librate around the Lagrangian points. Thus at any one time Trojans can be found along a wide range of ecliptic longitudes. Although the semimajor axes and eccentricities of Trojan orbits are strongly constrained, their inclinations range from nearly zero to greater than 40°. The combination of large libration amplitudes and a wide range of inclinations means that the total area over which Trojans may be found is many thousands of square degrees. Surveys prior to the 1980s were limited to regions near the ecliptic, so that many high-inclination Trojans may remain undiscovered. Approximately 200 Trojans with well-determined orbits are known to date, $\approx$ 120 in the L4 group and $\approx$ 80 in the L5 region. The larger number in the L4 region is probably due to observing bias, as L4 has been more thoroughly studied. The total Trojan population with diameters greater than 15 km may be $\approx$ 2300 (Shoemaker, Shoemaker and Wolfe, 1989).

The Trojans are different from main-belt asteroids both in shape and composition. Lightcurve studies show the larger Trojans (diameters > 150 km) to be more elongated on average than main-belt asteroids in the same size range (Binzel and Sauter, 1991). The largest known Trojan, 624 Hektor, is one of the most elongated asteroids known, with an axial ratio of 3 : 1. The elongated shapes presumably indicate a different collisional history for Trojans and for main-belt asteroids. Trojans are predominantly of compositional class D, with some P asteroids as well. No known meteorites in terrestrial collections match these spectral types. The best spectral analogs are mixtures of low-temperature condensate hydrocarbons with dark organic material (Gradie and Veverka, 1980; Vilas and Smith, 1985; French *et al.*, 1989). The D and P class asteroids have been called 'ultraprimitive' because they are believed to be unprocessed by reheating since the formation of the objects (Bell *et al.*, 1989). A further difference between Trojans and main-belt asteroids is seen in their phase curves, the variation of an asteroid's light with solar phase angle. The vast majority of dark asteroids in the main belt show a strong opposition effect, or brightening at low solar phase angle, as shadows between particles disappear. The two Trojans for which phase curves have been observed show no opposition effect, indicating either that their surfaces are less rough than those of main-belt asteroids or that the material making up the asteroids has a different single particle scattering function from the rocky material making up main-belt asteroids (French, 1987).

Linda M. French

Bibliography

Bell, J.F., Davis, D.R., Hartmann, W.K. and Gaffey, M.J. (1989) Asteroids: the big picture, in *Asteroids II*, (eds R.P. Binzel, T. Gehrels and M.S. Matthews). Tucson, University of Arizona Press, pp. 921–45.

Binzel, R.P. and Sauter, L. (1992) Trojan, Hilda, and Cybele asteroids: new lightcurve observations and analysis. *Icarus*, **95**, 222–38.

French, L.M. (1987) Rotation properties of four L5 Trojan asteroids from CCD photometry. *Icarus*, **72**, 325–41.

French, L.M., Vilas, F., Hartmann, W.K. and Tholen, D.J. (1989) Distant asteroids and Chiron, in *Asteroids II*, (eds R.P. Binzel, T. Gehrels and M.S. Matthews). Tucson: University of Arizona Press, pp. 468–86.

Gradie, J. and Veverka, J. (1980) The composition of the Trojan asteroids. *Nature*, **283**, 840–2.

Shoemaker, E.M., Shoemaker, C.S. and Wolfe, R.F. (1989) Trojan asteroids: populations, dynamical structure, and origin of the L4 and L5 swarms, in *Asteroids II* (eds R.P. Binzel, T. Gehrels and M.S. Matthews). Tucson: University of Arizona Press, pp. 487–523.

Vilas, F. and Smith, B.A. (1985) Reflectance spectrophotometry (0.5–1.0 μm) of outer-belt asteroids: implications for primitive, organic solar system material. *Icarus*, **64**, 503–16.

Cross references

Asteroid
Asteroid: lightcurve
Asteroid: resonance
Lagrangian point

TROPOSPHERE

The lowest layer of an atmosphere, adjacent to a planet's surface, is called the troposphere (from the Greek *tropos*-, to turn or change, plus *-sphaira*, globe). The troposphere is characterized by temperature decreasing with height. On Earth, the depth over which this occurs is 12–15 km in the tropics, but only 8 km near the poles. At the top of the troposphere, temperature becomes constant with height. This level, called the tropopause, marks the boundary between the troposphere and overlying stratosphere. On Earth 80–85% of the mass of the atmosphere is contained within the troposphere.

Temperature decreases with height in the troposphere because of heating by incident sunlight absorbed at the surface, combined with peak densities of infrared-absorbing gases at low altitude. The atmosphere is thus heated primarily from below and, much like a pot of water on a stove, the result is vertical motion and mixing. The mixing is accomplished primarily by small-scale convective overturning. Near the surface, convective turbulence mixes either dry or moist air to a height of 0.5–3 km, depending on location, season and time of day. If near-surface air is warm and moist, rising air can also lead to cloud formation, latent heat release and thunderstorms which mix air throughout the depth of the troposphere. Additional mixing is provided by gentle, large-scale rising and sinking motions driven by latitudinal gradients in solar heating. These large-scale motions exist primarily to redistribute heat from the tropics to the polar regions. All familiar weather phenomena have their origin in the troposphere.

Because the troposphere is so well-mixed, its composition varies little from place to place. The only exception is water vapor, which is highly variable in the presence of surface sources (ocean evaporation) and atmospheric sinks (condensation and precipitation). As a result,

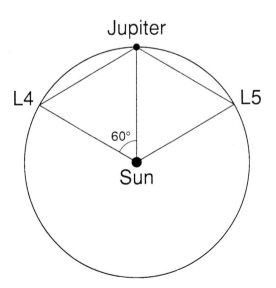

Figure T33 Geometry of the Trojan asteroids. L4 and L5 are the stable triangular Lagrangian points in Jupiter's orbit.

water vapor concentration decreases sharply with height. Almost all atmospheric water, whether in the vapor, liquid or ice phase, is contained within the troposphere. The troposphere also contains a substantial concentration of aerosols, derived from windblown dust, sea salt, marine phytoplankton gas emissions and anthropogenic combustion products. These aerosols are continually scavenged by precipitation, however, and individual aerosol particles reside in the troposphere for no more than a few days or weeks.

Anthony D. Del Genio

Bibliography

Goody, R.M. and Walker, J.C.G. (1972) *Atmospheres*. Englewood Cliffs: Prentice-Hall.
U.S Standard Atmosphere (1976) National Oceanic and Atmospheric Administration Report NOAA-S/T 76–1562. Washington, DC: US Govt Printing Office.

Cross references

Atmosphere
Earth: atmosphere
Stratosphere

TROUVELOT, ÉTIENNE LÉOPOLD (1827–1895)

Trouvelot was born at Guyencourt, in France, on 26 December 1827 and was educated at Paris. His interest in astronomy began at a very early age, but it was not for some time that he became a professional astronomer; this took place after he emigrated to America, in 1858, to become an assistant at the Harvard Observatory. Here he began systematic observations of the Sun and planets. His planetary drawings were excellent, and many of them are to be found in the *Harvard College Observatory Annals*; he concentrated largely upon Jupiter and Saturn, but also paid attention to the Moon. His main work, however, was in connection with the Sun, and in particular the solar prominences. Some 1000 sunspot drawings of his are in the Observatory library. He became a skilled spectroscopist, and began systematic studies. He eventually returned to France and in 1882 Jules Janssen, the Director of the Meudon Observatory near Paris, invited him to take up a post there. At Meudon he continued with his prominence work. He measured the height of one eruptive prominence, seen on 26 June 1885, as 300 000 miles (480 000 km). He also made a long series of observations of the planet Venus, and recorded diffuse detail; his drawings were among the best of their time, and in 1892 the Société Astronomique de France decided to print them in a special publication.

Trouvelot took part in several eclipse expeditions, and in 1883 spent the brief period of totality searching for the intra-Mercurian planet which had been predicted by U.J.J.Leverrier on the basis of irregularities in the movements of Mercury.

Trouvelot's work was widely recognized in France and elsewhere; he was a Founder Member of the Société Astronomique de France. He died in Paris on 22 April 1895.

Patrick Moore

Bibliography

Hoffleit, E.D. (1976) Trouvelot, Étienne Léopold. *Dict. Sci. Biogr.*, Vol. 13, 472–3.
Trouvelot, É.L. (1875) *Am. J. Sci.*, **11**, 169.
Trouvelot, E.L. (1892) Observations de Mercure et Vénus. *Soc. Astron. de France*.

TSIOLKOVSKY, KONSTANTIN EDUARDOVICH (1857–1935)

Generally regarded as the 'father of astronautics', Tsiolkovsky was born at Izhevsk, in the district of Kazan, Russia. The son of a forester, he became deaf at the age of 9 and was largely self-taught using his father's library. Following further self-training in Moscow in natural sciences, physics and mathematics, he became a high-school teacher and had time for research. He developed the reaction principle, which explained how a rocket in interplanetary space could operate (theoretically) in an environment approaching total vacuum. A fundamental law was established, defining the final speed of the rocket as depending directly on the ejection speed of the exhaust. Unfortunately, although the manuscript dealing with this discovery was sent to the *Technical Review* in Moscow in 1903, it was not actually published until 1923.

Tsiolkovsky also carried out the basic calculations required for putting artificial satellites into orbit, and thus eventually opened the way for interplanetary missions. Although his pioneer ideas were not greeted with any enthusiasm in Tsarist Russia, he was encouraged and supported by the Soviet regime until the time of his death. He wrote: 'The main motive of my life has been to . . . move humanity forward, if only slightly.'

Rhodes W. Fairbridge

Bibliography

Grigorian, A.T. (1976) Tsiolkovsky, Konstantin Eduardovich. *Dict. Sci. Biogr.* Vol. 13, pp. 482–4.
Tsiolkovsky, K.E. (1951–1967). Collected Works. *Sobranie Sochineny*, 5 vols, Moscow.

U

ULTRAVIOLET RADIATION

The ultraviolet (UV) portion of the electromagnetic spectrum begins at the upper limit of the range of visible light, at a wavelength of about 4000 angstroms (Å; 4.0×10^{-7} m). Its range extends to an upper limit beyond about 300 Å (3×10^{-8} m). X-rays and gamma rays have yet shorter wavelengths, but there is some overlap between the x-ray and ultraviolet ranges of the electromagnetic spectrum. The ultraviolet range is often divided into the 'near ultraviolet' (4000–3000 Å), the 'far ultraviolet' (2000–1050 Å), and the 'extreme ultraviolet' (~ 1050–100 Å).

The Sun is the principal source of ultraviolet radiation in the solar system. It is the principal cause of sunburn in humans. Most solar ultraviolet radiation is absorbed by gaseous molecules in the upper atmosphere. Reactions between UV radiation and molecular oxygen occur high in the stratosphere, generating an 'ozone layer' for the Earth. Prior to the development of an oxygen-rich atmosphere on Earth, unfiltered UV radiation at the Earth's surface may have inhibited the evolution of multicellular life.

The atmospheres of other planets likewise generally absorb incident ultraviolet radiation before it reaches the planet's surface (the exceptions are Mercury and Mars). Ultraviolet observations generally relate to questions of planetary atmospheric composition and structure; such observations must be made from space, since the Earth's atmosphere tends to block the radiation in question. Ultraviolet spectroscopy of planetary atmospheres is discussed in the following article.

Extreme-ultraviolet and far-ultraviolet spectra of Jupiter, Saturn and Uranus show bright lines originating in the excitation of hydrogen molecules (Yelle, 1988). These emissions are termed 'electroglow'.

James H. Shirley

Bibliography

Giese, A.C. (1945) Ultraviolet radiations and life. *Physiol. Zool.*, **18**, 223–50.
Koller, L.R. (1965) *Ultraviolet Radiation*. New York: John Wiley and Sons.
Yelle, R.V. (1988) H_2 emissions from the outer planets. *Geophys. Res. Lett.*, **15**, 1145–8.

Cross references

Atmosphere
Comet
Spectroscopy: atmosphere

ULTRAVIOLET SPECTROSCOPY

Ultraviolet spectroscopy is a versatile and effective tool for investigations of the atmospheres of planets and their satellites, of comets and of the interstellar medium. The technique has been used for remote sensing of atmospheres of all the planets of our solar system. Because Earth's atmosphere absorbs light at wavelengths less than about 3000 Å, the study of extra terrestrial bodies by UV spectroscopy requires access to space. Rocket experiments afforded early opportunities for pioneering work on Venus, Mars, Jupiter and Saturn. Now space missions and Earth-orbiting platforms are most productive. Flyby spacecraft equipped with UV spectrometers have visited all the planets except Pluto, and orbiting spacecraft have investigated Venus and Mars by UV spectroscopy.

The atmosphere of each planet except Mercury and Mars absorbs UV light before it reaches the planet's surface, and so UV spectroscopy addresses issues mainly related to the atmosphere. Typical topics amenable to study by UV spectroscopy include atmospheric composition, temperature structure and excitation mechanisms. Often parameters of physical interest must be derived from the observed spectrum by means of models. Thus UV spectroscopic investigations rely on and encourage advances in atomic and molecular physics, radiative transfer theory and deconvolution techniques.

The UV region of the spectrum is usually subdivided into wavelength bands according to differences in experimental approaches required. The names of these bands are fairly standard, but the boundaries between bands are not. The names used here are (1) ultraviolet (UV), 4000 to 300 Å, i.e. from shortward of visible wavelengths to the soft x-ray region; (2) far ultraviolet (FUV), 2000 to 1050 Å (the shortest wavelength at which transmitting optics can be used); and (3) extreme ultraviolet (EUV, sometimes XUV), 1050 to 100 Å.

UV spectroscopic techniques

Emission spectroscopy

A variety of mechanisms can produce UV emissions from a gas. The most important for planetary science include resonant scattering and fluorescent scattering of sunlight, photoionization and photodissociation reactions that give a product in an excited state, radiative recombination, chemiluminescence, and collisional excitation by precipitating charged particles. In addition, Rayleigh scattering of sunlight is measurable from the outer planets. Several of these mechanisms may act on a number of species simultaneously, so the UV emission spectrum of a planet may be a complex combination of atomic emission lines and molecular bands.

Resonant scattering occurs when an atom or molecule absorbs a quantum of frequency v in a transition of energy hv and re-emits (scatters) it at the same frequency. Scattering at a lower frequency (longer wavelength) is called fluorescent scattering. Resonant scattering of sunlight by an optically thin medium is easily treated analytically using the photon scattering coefficient g (often called the g-value, in photons s^{-1} atom^{-1}) given by

$$g = \frac{\pi e^2}{mc} f_{12} \pi F \tag{U1}$$

Here πF is the solar flux at the appropriate frequency ν, f_{12} is the oscillator strength for the transition, e and m are the charge and mass of the electron, and c is the speed of light. The quantity g is the rate at which an atom illuminated by the Sun scatters photons of frequency ν. For a given species and transition (or frequency ν) it depends only on the heliocentric distance through the solar flux, and is generally tabulated for a distance of 1 AU. Under optically thin conditions, a measured emission rate implies directly the column density of the emitting species, according to

$$4\pi I = g \int_0^\infty N(r) \mathrm{d}r \text{ photons cm}^{-2} \text{ s}^{-1} \tag{U2}$$

where $N(r)$ is the number density (cm^{-3}) of the emitter at r and the integration is over the line of sight. The column-integrated emission rate is usually expressed using the unit rayleigh (R), an emission rate of 10^6 photons cm^{-2} s^{-1}. If the column-integrated emission rate is expressed in units of 10^6 photons cm^{-2} s^{-1} sr^{-1}, then $4\pi I$ is in rayleighs. Many interesting examples of resonant scattering involve column densities so large that multiple scattering is important. Then radiative transfer modeling is required to relate measured emission rates to column densities. Tractable models include simplifying approximations; examples are discussed by Meier (1991).

Collisional excitation transfers part of the energy of particles colliding with atmospheric atoms or molecules to the target particle, thereby exciting internal states that radiate their energy in the UV. The same process may create new species, also in excited states, by ionization or dissociation. Important agents of collisional excitation are photoelectrons produced in the atmosphere and charged particles precipitating from the magnetosphere. Similarly, EUV solar photons are energetic enough to photodissociate molecules and photoionize atoms and molecules. The products are often in excited states, and radiate the extra energy in returning to the ground state. Radiative recombination of ions and electrons leads to an excited neutral that radiates and, in chemiluminescence species, recombines to an excited product. Emission processes are discussed by Chamberlain (1961), Chamberlain and Hunten (1987) and Meier (1991).

The emission spectrum recorded by a spectrometer or spectrograph represents an integral along the line of sight of the source for each wavelength. At wavelengths for which the atmosphere is optically thin, the integral can extend all the way through the atmosphere (for viewing above the limb) or to the surface (for nadir viewing). Then, if the excitation mechanism is understood, profiles of intensity as a function of altitude above the limb can be inverted directly to give the altitude profile of the emitting species, and nadir intensities give its column density. Practical situations are often more complicated than this ideal. Absorption by the same or different species may be important, and a combination of excitation mechanisms may be active. Nevertheless, by modeling of the atmosphere and excitation mechanisms, it is usually possible to derive a wealth of information about the composition and structure of the atmosphere.

Occultation spectroscopy

By observing the occultation of a light source by a planetary atmosphere, the transmission of the atmosphere as a function of altitude and wavelength can be determined. Most atmospheric gases have strong absorption cross-sections with characteristic structure in the FUV and EUV. Occultation measurements can therefore be inverted to retrieve the composition of the atmosphere and the vertical distribution of its constituents. For occultations observed at FUV and EUV wavelengths, the light from the source is diminished by absorption. These occultations therefore differ in a fundamental way from the stellar occultations by planets observed from Earth at visible wavelengths, in which refraction is the dominant process dimming the light. This discussion of occultation investigations is based on the review by Smith and Hunten (1991).

A thorough understanding of the absorption of UV light by atmospheric species is required to interpret occultation measurements. Atomic species can absorb light in their ionization continua, and molecules absorb in dissociation continua as well. Absorption in molecular bands leads to complex systems of discrete lines that are determined by vibrational and rotational energy states of the

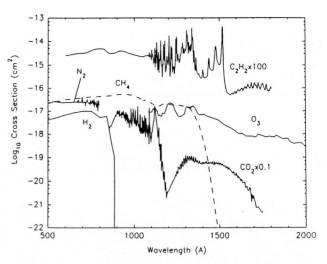

Figure U1 Photoabsorption cross-sections for several gases found in the atmospheres of the planets. In addition to its structure below 2000 Å the O_3 cross-section has a broad maximum in the 2400–2600 Å range. N_2 and H_2 have important band absorption features, not shown in the figure, longward of their continuum absorptions. The unique shapes of these cross sections make occultation spectroscopy practical; sometimes these absorption signatures are discernable in emission spectra as well.

molecule, and hence depend on the temperature of the absorber. Modeling these band systems is discussed in detail by Goody and Yung (1989), and H_2 band absorption, a case of particular interest for the outer planets, has been treated by Yelle (1988). Photoabsorption cross-sections for a variety of atmospheric gases are shown in Figure U1. Most species have cross-sections of about 10^{-17} cm^2 somewhere within the EUV or FUV range. The pronounced differences in the wavelength dependence of the cross-sections give occultation spectroscopy its power. This wavelength dependence serves as a signature of each absorbing species, even for a mixture of gases. Gases such as H_2 whose effective cross-section changes by orders of magnitude with wavelength can be measured over a correspondingly large range of column density, and hence a great range of altitude. When the cross-sections of two candidate species are similar within the resolution of the measurement, it has been possible to use emission spectroscopy to resolve the ambiguity.

Several methods for retrieving altitude profiles of the atmosphere from occultation lightcurves are available. For UV occultations refraction can be neglected, as can scattering into the beam from the source. Under these conditions the attenuation of light is given by an integral along the optical path from the source to the spectrometer,

$$\ln \left(\frac{I(\lambda, r)}{I(\lambda, \infty)} \right) = -\sum_i \int \sigma_i(\lambda) N_i(s) \mathrm{d}s \tag{U3}$$

where $I(\lambda, \infty)$ is the unattenuated intensity at wavelength λ, $I(\lambda, r)$ is the attenuated intensity measured at tangent ray height r, and N_i and $\sigma_i(\lambda)$ are the number density and the total extinction cross-section of the ith constituent. If the cross-sections σ_i are known, equation (U3) can be solved for N_i. For the simplest case of absorption by a single species, equation (U3) can be written as

$$\ln \left(\frac{I(\lambda, r)}{I(\lambda, \infty)} \right) = -2\sigma(\lambda) \int_{r_0}^\infty \frac{r N(r)}{(r^2 - r_0^2)^{1/2}} \mathrm{d}r \tag{U4}$$

By applying an Abel transform (Hays and Roble, 1968), we obtain

$$N(r) = \frac{-1}{\pi \sigma(\lambda)} \int_r^\infty \frac{\mathrm{d}}{\mathrm{d}r_0} \left[\ln \left(\frac{I(\lambda, r)}{I(\lambda, \infty)} \right) \right] (r_0^2 - r^2)^{-1/2} \mathrm{d}r_0 \tag{U5}$$

When the atmosphere includes several constituents with cross-sections that overlap in wavelength, and when the data are noisy, specialized techniques may be required. Specific examples arising from the analysis of solar occultations by Uranus and Triton are discussed by Herbert et al. (1987) and by Herbert and Sandel (1991).

Possible light sources for occultations include the Sun and hot stars. The size of the source projected on the atmosphere limits the potential spatial resolution, so stars are preferred to the Sun from this point of view. However, little starlight shortward of the Lyman limit at 912 Å is available for occultation measurements because it is absorbed by interstellar hydrogen. Several gases of interest in the study of planetary atmospheres absorb in their continua only at shorter wavelengths. Sunlight includes all UV wavelengths, and it is very bright, permitting measurements with high signal-to-noise ratios. Solar occultations are limited to probing the atmosphere near the terminator, but stars can be used at any local time. In practice, solar and stellar occultations are somewhat complementary and their results can be combined with good effect.

A productive occultation experiment requires a spectrometer with adequate time resolution and full spectral coverage. The requirement on time resolution is set by the need for at least two samples per scale height. For conditions typical of a planetary encounter or orbiting spacecraft, this requires a rate of several samples per second. Because the best determination of the optical depth τ is made when $\tau \sim 1$, measurements at a range of cross-sections (and hence at several wavelengths) are required to determine the number-density profile over a range of altitudes. Simultaneous detection of many wavelengths also gives the best chance of unraveling the effects of several absorbers having cross-sections that overlap in wavelength.

An important advantage of the occultation technique is its independence of the absolute calibration of the spectrometer; it is free of the effects of long-term drifts in instrument sensitivity. This is because the unattenuated and attenuated spectra of the source are measured by the same instrument, at nearly the same time, and only their ratio is needed to recover the atmospheric transmission. Thus comparisons between results of observations at very different times, and even between measurements by different instruments, can command an unusual degree of confidence.

Reflection spectroscopy

The surfaces of planets and satellites not obscured by an atmosphere may be studied by reflection spectroscopy in the UV. Many surface materials impose a characteristic absorption signature on the spectrum of sunlight reflected from the surface. Thus the spectrum of reflected sunlight can be used to investigate surface composition. This technique is particularly effective when combined with similar measurements at visible and infrared wavelengths. Reflection spectroscopy of the Galilean satellites is discussed by Nelson et al. (1987).

Spectroscopic instrumentation

Several factors add to the challenge of designing UV spectrometers. Traditional optical materials such as glass and quartz absorb light at wavelengths less than ~ 2200 and 1600 Å respectively, and hence are useless in transmitting optical systems for the FUV. Lenses, filters and windows of MgF_2 and LiF are used instead, but even these absorb wavelengths shorter than about 1150 and 1050 Å respectively. Thus only reflecting optics are suitable for the EUV. Reflectivity in the EUV is low as well, so for maximum throughput a single reflection is preferred. For these reasons the concave diffraction grating, a single optical element that can both focus the light and disperse it in wavelength, finds wide application in EUV spectrometry.

The theory of concave gratings, and the techniques for making them, originated in the late 19th century. This and many other topics related to UV spectroscopy are discussed in detail in the book by Samson (1967), from which much of the material of this section is drawn. The focusing and dispersion properties of the concave grating may be treated independently. Figure U2 illustrates the imaging properties. Light from point S strikes a concave grating having radius of curvature R, and is diffracted to point I. The incidence and diffracted angles are α and β, and the points S and I lie at distances r and r' from the center of the grating. If a horizontal image is formed at I, then

$$\frac{\cos^2\alpha}{r} - \frac{\cos\alpha}{R} + \frac{\cos^2\beta}{r'} - \frac{\cos\beta}{R} = 0 \quad (U6)$$

An important solution to this equation is

$$r = R\cos\alpha, \quad r' = R\cos\beta \quad (U7)$$

However, the concave mirror used in this way forms astigmatic images; the vertical foci satisfy

$$\frac{1}{r} - \frac{\cos\alpha}{R} + \frac{1}{r'} - \frac{\cos\beta}{R} = 0 \quad (U8)$$

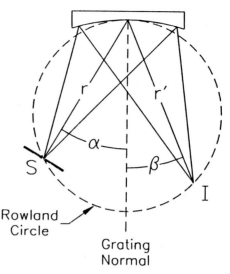

Figure U2 Imaging properties of a concave diffraction grating. A point at S on the Rowland circle comes to a horizontal focus at point I, also on the circle.

for which one solution is

$$r = \frac{R}{\cos\alpha}, \quad r' = \frac{R}{\cos\beta} \quad (U9)$$

According to equation (U7), horizontal focus is achieved with entrance and exit slits located on a circle of diameter R (called the Rowland circle) passing through the grating surface normal to the rulings, whose center lies on the grating normal. According to equation (U9), the vertical foci lie on the line tangent to the Rowland circle at its intersection with the grating normal.

Another solution to equation (U6) of particular importance to space flight instrumentation is found by setting $r = \infty$, which corresponds to illuminating the grating with collimated light. Then the location of horizontal focus is given by

$$r' = \frac{R\cos^2\beta}{\cos\alpha + \cos\beta} \quad (U10)$$

and the vertical focus by

$$r' = \frac{R}{\cos\alpha + \cos\beta} \quad (U11)$$

If $\cos\beta \sim 1$, the two expressions for r' are the same, and the image is stigmatic. A grating used under these circumstances is said to be in the Wadsworth mount. Stigmatic imaging is of particular importance now that 2-D array detectors are widely available. Stigmatic imaging with such a detector forms an 'imaging spectrograph' that can record spatial variations in source intensity along the entrance aperture in the direction perpendicular to the dispersion.

The dispersion of the concave grating spectrograph is described by

$$\pm m\lambda = \delta(\sin\alpha + \sin\beta) \quad (U12)$$

as for the plane grating. Here m is the order of the diffraction and δ is the separation between grooves. The angular dispersion is given by

$$\frac{d\beta}{d\lambda} = \frac{m}{\delta\cos\beta} \quad (U13)$$

and the linear dispersion is

$$\frac{d\lambda}{dl} = \frac{\delta\cos\beta}{mR} \quad (U14)$$

Ordinary gratings distribute the diffracted light through a number of diffracted orders, leading to low efficiency in each order. For maximum efficiency, a grating may be blazed for a particular wavelength diffracted in a particular order. Blazed gratings have grooves of controlled shape; at the blaze wavelength and order, the diffracted beam leaves the grating at the same angle as the specular reflection from the grating facets.

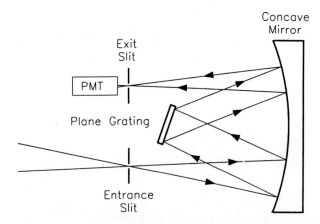

Figure U3 Ebert–Fastie spectrometer layout. Off-axis aberrations are introduced by the first reflection from the concave mirror, but they are canceled by the second reflection from the same mirror.

At wavelengths longer than about 1200 Å, reflective efficiencies are large enough to permit several reflections. A number of planetary missions have used an Ebert–Fastie spectrometer, a three-reflection design with a plane grating (Figure U3). The grating rotates to cover the spectral range, and the detector behind the exit slit is a photomultiplier tube. The ability of this design to accommodate long slits makes it useful for extended sources of emission such as airglow, and the throughput at a particular wavelength and resolution are high. However, astigmatism prevents its use as an imaging spectrograph, and the need to scan for wavelength coverage excludes occultation observations.

It is useful to distinguish the spectrometer from the spectrograph. Early spectroscopy used spectrographs, which recorded a range of wavelengths on photographic film. The development of the photomultiplier, which offered photoelectron counting sensitivity, digital readout and wide dynamic range, but lacked spatial resolution, led to widespread use of spectrometers. Typified by the Ebert–Fastie design, these were really scanning monochromators that measured only a narrow wavelength interval at one time. Modern array detectors combine the advantages of photomultipliers with the spatial resolution of film. Such detectors permit a return to the spectrograph configuration. Using special optical designs to take advantage of recent advances in holographic manufacturing techniques for gratings, it is possible to construct imaging spectrographs of excellent quality. These record their full spectral range in a single exposure, and at the same time record spatial variations in source intensity along the slit. They have not yet been used in planetary missions, but they have been demonstrated in space flight aboard the Shuttle (Broadfoot et al., 1992).

Knowledge of the widths of resonance-scattered emission lines such as H Lyα is important because the line width is related to the velocity distribution of the source atoms. Measuring the line shape, although possible with the International Ultraviolet Explorer (IUE), requires greater spectral resolution than is available from conventional general-purpose spectrographs that can be accommodated on interplanetary spacecraft. The H absorption cell is an alternative that can give the H Lyα line profile and is compact enough for space missions. The H cell is a vessel containing H_2 gas. Windows at each end permit H Lyα radiation to pass through. To perform a line width measurement, part of the H_2 is dissociated into H by heating filaments in the vessel. The column density of H, and hence the optical depth at H Lyα, can be controlled by adjusting the heating. As the optical depth increases, the fraction of H Lyα transmitted by the cell decreases because of increasing absorption in the Doppler wings of the H Lyα line. Thus measuring the fraction of the light from the source that is transmitted by the cell for several values of the H column abundance in the cell determines the profile of the H Lyα line. Bertaux and Lallement (1984) describe this device and its applications.

Photocathodes are the primary light sensing element in FUV photomultiplier tubes and image intensifiers. By the photoelectric effect, they produce the electrons needed to activate the multiplication process. A variety of photocathode materials are available for UV wavelengths, and the proper choice among them is important for optimizing a particular measurement. For the UV spectroscopist, the materials differ mainly in the long-wavelength cut-off in their response. Materials with high work functions, such as CsI and KBr, are often called 'solar blind.' They are not sensitive to wavelengths longer than about 1800 Å. The solar spectral irradiance increases rapidly toward longer wavelengths beginning at about 2000 Å. Without solar-blind photocathodes, scattering of this long-wavelength light could dominate the much weaker target emissions at shorter wavelengths. Photocathodes may be deposited on PMT and IIT windows or, for windowless detectors, on a mesh or directly on the microchannel plate.

UV spectroscopy in the solar system

Table U1 summarizes the UV investigations by planetary probes. Also shown in the table are an early Earth-orbiting satellite, Copernicus (Orbiting Astrophysical Observatory 2), and the two primary Earth-orbiting platforms, the International Ultraviolet Explorer (IUE), and Hubble Space Telescope (HST). HST includes two spectrographs, two cameras and a high-speed photometer, all with UV capability longward of 1050 Å. Passband filters and transmission-dispersive elements give the cameras spectral resolution. The planetary probes offer spatial resolution, viewing perspectives and, in many cases, wavelength coverage that is not presently available from Earth orbit, while the Earth orbiters offer the possibility of a long timebase of observations, and better spectral resolution. Important capabilities not shown in the table include Earth satellites equipped to measure H Lyα of the interstellar wind and recent Shuttle investigations.

Inner solar system

Mercury

The only investigations of Mercury's atmosphere by UV spectrometry were carried out by means of instruments on Mariner 10. The UV airglow spectrometer detected resonance scattered emissions at H Lyα, He (584 Å), and O (1304 Å). The atmosphere is extremely tenuous and is technically an exosphere. In Mercury's case, this means that atmospheric species collide mainly with the surface rather than with each other. The height profile of H Lyα implies a distribution of H having components at two temperatures. The surface abundances for the hot and cold components are 23 and 230 cm^{-3}. It has been argued that these two distributions correspond to the temperatures of the day and night sides of the planet, but the presence of the cold component is difficult to understand.

Mariner 10 measured profiles of He (584 Å) across the terminator and above the bright limb. The inferred He abundance at the surface near local noon is about 6000 cm^{-3}. Although the distribution of species within an exosphere is in general well understood, Mercury's exosphere is unusual in that interactions with the surface play an important role. Energy exchange between the atmosphere and the surface, and the related phenomena of transport between light (hot) and dark (cold) parts of the surface, are poorly understood. Owing to this difficulty, the He distribution has been only approximately modeled. The O (1304 Å) brightness implies a surface abundance of $\sim 4.4 \times 10^4$ cm^{-3}, but the signal-to-noise ratio for this measurement was not good enough to infer a scale height.

The Mariner 10 solar occultation experiment found no measurable absorption by the atmosphere, and placed stringent upper limits on a possible atmosphere consisting of species not detectable by resonance scattering in the UV. The upper limits depend on the assumed species, but generally fall in the range $(1–5) \times 10^7$ cm^{-3}. Measurements of Mercury's atmosphere are reviewed by Hunten, Morgan and Shemansky (1988).

Venus

A variety of experiments have recorded the UV spectrum of dayglow and nightglow from Venus. The Venera 4 and Mariner 5 flybys detected H Lyα from the exosphere and an extensive corona. A sounding rocket investigation revealed O (1304 Å) and O (1356 Å) emission and confirmed the presence of H inferred earlier. Later rocket flights revealed H, C, O and N emissions, as well as CO fourth-positive bands. SO_2 was first identified by its signature at 3210 Å found in ground-based work, and confirmed by the IUE, which measured the absorption signature at ~ 2150 Å. IUE spectra also furnished the first evidence for NO in Venus's atmosphere. Mariner 10 measured the H Lyα and O (1304 Å) emissions, and

Table U1 Space missions with UV capability

Spacecraft	Planet or satellite	Date of observation	Ultraviolet instrument	Wavelength range (Å)	Wavelength resolution (Å)
Venera 4	Venus	1967	Photometer	1050–1340	~ 150
Mariner 5	Venus	1967	Photometer	1050–2200	~ 150
Venera 6	Venus	1969		1050–1340	~ 130
Venera 9 and 10	Venus	1975 (orbiter)	H and D cells	–	–
Pioneer Venus	Venus	1978 (orbiter)	Spectrometer	1100–3400	13
Venera 11 and 12	Venus	1978	Spectrophotometer	304–1657	–
Mariner 6 and 7	Mars	1969	Spectrometer	1100–4300	10, 20
Mars 3	Mars	1971 (orbiter)	Photometer	H Lyα, O (1304 Å)	–
Mariner 9	Mars	1971 (orbiter)	Spectrometer	1100–3500	15
Mars 5	Mars	1974 (orbiter)	H and D cells	–	–
Phobos	Mars	1989	Spectrograph	2150–3280	12–24
Mariner 10	Venus	1974	Airglow, spectrophotometer, occultation, spectrophotometer	304–1657 470–890	20 75
	Mercury	1974			
Pioneer 10 and 11	Jupiter	1973, 1974	Photometer	H Lyα, He (584 Å)	–
Pioneer 11	Saturn	1979			
Voyager 1 and 2	Jupiter, Saturn and Titan	1979 1980, 1981	Spectrograph	500–1700	35
Voyager 2 only	Uranus, Neptune and Triton	1986 1989			
Galileo	Venus	1990	Spectrometer, Spectrograph	1130–4320 540–1280	7–13 35
	Jupiter	1995 (orbiter)			
Earth satellites					
Copernicus		1968 (launch)	Two spectrometers:		
			Short λ	1150–2000	10, 100
			Long λ	1800–3800	20, 200
International Ultraviolet Explorer		1978 (launch)	Two spectrographs:		
			Short λ	1150–2000	0.1–0.2; ~ 7
			Long λ	1900–3200	0.2–0.3; ~ 7
Hubble Space Telescope		1990 (launch)	Spectrographs, cameras	1150–visible	to 0.01

discovered the He (584 Å) and C (1657 Å) emission, demonstrating for the first time the presence of He and C in the atmosphere of Venus. Limb profiles of the He (584 Å) emission implied a He mixing ratio of 1×10^{-5} at the homopause, and a temperature of 375 ± 105 K for the dayside exosphere.

Mariner 5 and 10 measurements of H Lyα altitude profiles (Figure U4) suggested the presence of two scale heights differing by a factor of ~ 5, and this picture was confirmed by Venera 11 and 12. A number of processes were proposed to account for the observed structure, but now it seems certain that the two emission scale heights represent two populations of H characterized by different temperatures. The lower of these, ~ 250 K, corresponds to the thermospheric temperature. The higher scale height implies a temperature of ~ 1200 K. The source of this energetic hydrogen population has been attributed to several chemical reactions involving neutral and ionized hydrogen and oxygen, but the relative importance of these candidate reactions is presently uncertain. The question is an important one, because it is related to the escape of H and the observed enhancement in the D/H ratio. Venera 9 and 10 observations of the H Lyα line width made by means of a hydrogen absorption cell showed a second non-thermal component in the altitude range 3000–4500 km.

The Pioneer Venus Orbiter UVS carried out an extensive program of observations, including global mapping of NO, O and O_2 emissions. The bright CO fourth-positive bands are excited mainly by fluorescence scattering of solar H Lyα. Bands of this system are close enough in wavelength to O (1304 Å) and O (1356 Å) emissions to blend with them at the resolution of the Pioneer Venus Orbiter instrument, and therefore must be taken into account in the analysis of the oxygen emissions. An optically thin band near 1400 Å proved useful for determining the CO column abundance.

NO δ band emissions observed by Pioneer Venus Orbiter serve as tracers of atmospheric transport. O and N circulate from the dayside to the nightside. As they descend into the nightside atmosphere, they radiatively recombine to NO, with emission in the δ bands. Maps of

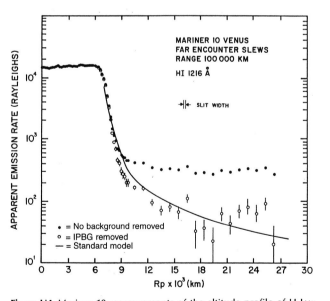

Figure U4 Mariner 10 measurements of the altitude profile of H Lyα emission at Venus. The rapidly decreasing component near the planet comes from H at a temperature of 275 K, while the slower decrease at higher altitudes corresponds to 1250 K. R_p is the minimum distance from the line of sight to the center of Venus. IPBG means 'interplanetary background.' (From Takacs, Broadfoot and Smith (1980), with permission from Pergamon Press Ltd.)

this emission have been used to study variations in the circulation pattern. From limb profiles of the same emission and a model of the atmosphere, it is possible to infer the downward fluxes of O and N, as well as the eddy diffusion coefficient.

Pioneer Venus Orbiter observations have shown that SO_2 plays an important role in determining the cloud contrasts and planetary albedo at wavelengths shorter than 3200 Å. The 20-R brightness of O (1304 Å) emission observed by Pioneer Venus Orbiter from the nightside of Venus has been interpreted in terms of excitation by electrons of energy < 300 eV. However, the required flux of electrons exceeds the measured flux by a factor of two to three, and a search for additional excitation mechanisms is in progress.

The EUV channel of Galileo provided the first EUV spectrum of Venus. (Earlier investigations were by spectrophotometers, which measured the brightness in several narrow and separate wavelength bands.) In addition to the earlier-known emissions of H, He and O^+, the spectrum shows a blend of features between 900 and 1200 Å. The blend probably includes O (989 Å), H Lyβ and perhaps other oxygen and nitrogen features (Hord et al, 1991). More details of Venus observations are given by Esposito et al. (1983), von Zahn et al. (1983) and Fox (1986).

Mars

The atmosphere of Mars resembles that of Venus in that it is composed mainly of CO_2. Most of the prominent UV airglow emissions arise from the ions or dissociation products of CO_2, and many of the airglow emission features prominent at Venus are present at Mars. Most of the UV airglow measurements come from the Mariner 6 and 7 flybys and the Mariner 9 orbiter.

The brightest UV emission features are the CO Cameron bands, which are in the 1900–2800 Å region (Figure U5). The main excitation mechanisms are dissociative excitation of CO_2 by electrons and photons. Dissociative recombination of CO_2^+ and, to a lesser extent, electron impact on CO also play roles. At altitudes above its peak, the emission rate is directly proportional to the CO_2 density. Therefore measurements of the altitude profile of the Cameron band intensity can be used to infer the scale height of CO_2, and thence the temperature of the atmosphere. The same mechanisms can also excite emissions in the CO fourth-positive bands. However, as at Venus, fluorescent scattering of solar H Lyα seems to be the dominant source of the CO fourth-positive emissions. A combination of photoionization and fluorescent scattering can account for the observed intensities in two prominent CO_2^+ band systems.

H Lyα and O (1304 Å) emissions have been measured by Mariner 6, 7 and 9, as well as by Copernicus and Soviet missions. Their brightness is directly related to the H and O abundances because the dominant excitation mechanism for both is resonant scattering of sunlight. The H Lyα profile implies an exospheric temperature of 350 K, a density of 3×10^4 cm^{-3} at the exobase, and an escape flux of about 1.8×10^8 cm^{-2} s^{-1} (Anderson and Hord, 1971). Analysis of the O (1304 Å) emission indicates an O mixing ratio of 1 to 3% at an altitude of 135 Km.

Because of the lower N_2 mixing ratio at Mars (2.5% at the homopause) compared to Venus, the spectral signatures of N-bearing species that on Venus served as tracers of atmospheric processes are too weak on Mars to compete with the brighter emissions at nearby wavelengths. A possible exception is a band in the N_2 Vegard–Kaplan system, which may appear in Mariner 9 spectra.

Mariner 9 mapped the relative surface pressure by nadir observations near 3050 Å (Hord et al., 1972). The technique is based on the fact that the UV albedo of the surface is low, so that the disk reflectance near 3050 Å is dominated by Rayleigh scattering in the atmosphere. For pure Rayleigh scattering, the measured backscatter is proportional to the column density between the orbiter and the surface, and hence to the surface pressure. Hord et al. used the pressure altitude so determined to investigate the topography of Mars. They used a similar approach to measure ozone absorption in the reflected spectrum, and thereby infer ozone column densities.

UV solar occultation observations by Phobos (Blamont et al., 1991) were compromised by an error in the software that directed the field of view of the spectrograph. Useful UV occultation data were obtained only above an altitude of 30 km. These data constrain the ozone abundance above 30 km and thereby imply rather weak mixing ($K \sim 10^6$ cm^2 s^{-1}). Barth (1985) discusses the excitation of UV emissions on Mars and the related photochemistry. Additional details are given by Fox (1986).

Outer solar system

The upper atmospheres of the outer planets are generally similar to one another in composition. Because of their large gravitational attraction, the outer planets have retained most of their primordial hydrogen and helium, and these gases dominate their upper atmospheres. Methane and its photochemical products are present near the lower boundary of the region probed by UV spectroscopy. Solar and stellar occultations have measured the exospheric temperature and determined the hydrocarbon distributions, and hence the strength of vertical mixing, for all the outer planets except Pluto.

The principal atomic emission features on the daysides are the H Lyα and He (584 Å) lines. The latter is excited by resonance scattering of the solar line, while the former is excited by that and other mechanisms. Resonance scattering of the sky background H Lyα leads to a nocturnal source for this emission. Molecular hydrogen emissions in the Lyman and Werner band systems have been detected from the daysides of all the outer planets except Pluto. Voyager found the brightness of these emissions to be inversely proportional to the square of the heliocentric distance of the planet, suggesting a strong connection with the solar flux. The H_2 band emissions probably arise from a combination of solar fluorescence and excitation by locally produced electrons, but the relative importance of these mechanisms is controversial. Figure U6 compares the dayglow spectra of Jupiter, Saturn and Uranus.

Jupiter

A unique feature of Jupiter is the H Lyα bulge, an enhancement in the equatorial H Lyα brightness. Voyager and IUE measurements show that the enhancement of ~ 30% is fixed in magnetic longitude near $\lambda_{III} = 100°$ and follows the magnetic equator rather than the spin equator (Dessler, Sandel and Atreya, 1981). It has been present continuously since at least 1979, but recent IUE observations show a marked change in shape (McGrath, 1991). Although a number of mechanisms have been proposed to account for the H Lyα bulge, none has succeeded in accounting for all its characteristics, and its origin is presently uncertain. Occultations of the Sun and the star Regulus imply a thermospheric temperature of 1000 K and an eddy diffusion coefficient $K \sim 10^7$ cm^2 s^{-1}. The exospheric temperature is much higher than can be accounted for by solar EUV heating alone, implying the presence of an additional energy source in the upper atmosphere. This is true of Saturn, Uranus and Neptune as well.

Jupiter's auroral displays are by far the most energetic in the solar system. About 10^{14} W are required to drive the EUV and FUV auroral emissions. The Jovian aurora has been intensively studied by Voyager, and the 12-year timebase of observations by IUE characterizes its long-term behavior (Figure U7). IUE measurements of the shape of the auroral H Lyα line show a blue shift corresponding to upward motion of ~ 50 km s^{-1}. The H Lyα probably comes from excited H atoms that have formed from upward-moving protons. This

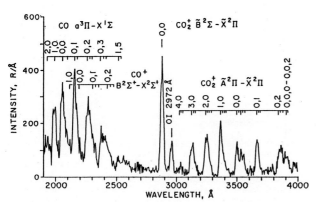

Figure U5 Spectrum from the 160–180 km altitude region at Mars. The CO Cameron bands and the CO_2^+ Fox–Duffenback–Barker bands are marked. (From Barth et al., 1971, copyright by the American Geophysical Union.)

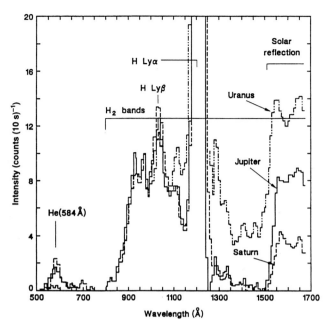

Figure U6 Comparison of the dayglow spectra of Jupiter, Saturn and Uranus measured by the Voyager UVS. The spectra have been scaled by the inverse square of the heliocentric distance of the planet. The similarity of the levels of the H_2 bands between 900 and 1200 Å means that their brightness closely follows the solar flux. Uranus is brighter longward of 1500 Å because reduced hydrocarbon abundances in its upper atmosphere permit greater reflectivity to the solar flux. (Adapted from Yelle et al., 1987.)

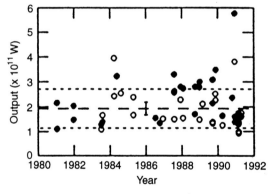

Figure U7 Power radiated in the 1557–1619 Å band by the Jovian aurora. These IUE measurements span a period of 12 years. At a particular time, the power from the north auroral region (filled circles) and the south auroral region (open circles) are the same within the sample error. Although there are strong temporal variations, there is no significant secular or solar-cycle trend. The dashed line shows the average and the dotted lines show ± 1 sample standard deviation about the average. (From Livengood et al., 1992).

particle population may be a significant source of magnetospheric plasma (Clarke, Trauger and Waite, 1989).

A torus-shaped cloud of plasma encircles Jupiter near the orbit of the innermost Galilean satellite Io. The plasma is made up of sulfur and oxygen lost from Io, ionized and trapped by Jupiter's magnetic field. These ions are excited by collisions with the plasma electrons, and radiate $\sim 3 \times 10^{12}$ W in the EUV. This radiation is the principal means of energy loss by the plasma. Voyager observations in the EUV and IUE observations in the FUV have played important roles in study of the composition and energy budget of this complex system (Brown, Pilcher and Strobel, 1983).

The source of the S and O is probably SO_2 from Io's volcanoes. UV reflection spectroscopy has been used to map the distribution of SO_2 frost on the surface of Io. The same technique has given evidence that Europa sweeps sulfur from the magnetosphere. A broad absorption band near 2800 Å has been interpreted as arising from sulfur atoms embedded in water ice on Europa's surface. Nelson et al. (1987) discuss these and other results from reflection spectroscopy of the Galilean satellites.

Saturn

No zonal variations analogous to Jupiter's H Lyα bulge have been detected, probably owing to the symmetry of the magnetic field. Solar occultations imply a thermospheric temperature of 420 ± 30 K and an eddy diffusion coefficient $K \sim 10^7$ cm^2 s^{-1}. The absence of H_2 band emissions in the shadow cast by Saturn's rings on the atmosphere emphasizes the close relationship between the excitation of the bands and the solar flux. Saturn's aurora has been observed only by the Voyager UVS. In marked contrast to Jupiter, Saturn's aurora seems Earth-like: it is found near the edges of the polar caps, and it is probably powered by the solar wind. The efficiency of converting solar wind power to auroral light is nearly the same as at Earth. Variations in the brightness of the aurora are correlated with kilometer-wavelength radio emissions from Saturn's high latitudes.

A cloud of H encircles Saturn, extending out to about 20 R_S (near the orbit of Titan) and above and below Saturn's equatorial plane by about 8 R_s. It is detectable in the FUV because it resonantly scatters the solar H Lyα line, which leads to a brightness of 10–20 R. The origin of the hydrogen is escape from the atmospheres of Titan and Saturn.

Uranus

The atmosphere of Uranus is unique among the outer planets in its weak eddy mixing. The solar occultation observed by Voyager implies $K \sim 10^4$ cm^2 s^{-1}, the lowest of the outer planets by about two orders of magnitude. This relatively stagnant atmosphere probably reflects Uranus's comparatively weak internal energy source. The low value of K limits mixing of hydrocarbons into the upper atmosphere, and thus permits FUV sunlight to penetrate deeply, to H_2 column densities great enough that Raman–Rayleigh scattering is important. Hydrocarbon absorption signatures in this scattered sunlight also imply a low value of K. Comparison between the hydrocarbon abundances derived from the occultations (near the equator) and from the spectrum of reflected sunlight (which refers mainly to the subsolar atmosphere near the pole) suggests an equator-to-pole decrease. To form a bright He (584 Å) line, He must be strongly mixed into the upper atmosphere, so that its concentration is significant at altitudes reached by solar He (584 Å) radiation, which is strongly absorbed by H_2. Therefore the observed absence of He (584 Å) emission from Uranus is also consistent with the low value of K. Because of the large tilt of its magnetic axis and the orientation of its spin axis, the form of Uranus' aurora is unique. H_2 bands are excited by precipitating particles in the vicinity of both magnetic poles, one of which is near the planet's spin equator. About 4×10^{10} W are required to power the aurora.

An extensive corona of H surrounding Uranus is detectable by its scattering of solar H Lyα. The H has escaped from the upper atmosphere under the influence of the 800-K thermospheric temperature and possibly an interaction with magnetospheric plasma. Its concentration is ~ 100 cm^{-3} at 2 R_U, high enough to reduce the lifetime of small ring particles by gas drag. More details of UV observations of Uranus are given by Atreya, Sandel and Romani (1991) and by Broadfoot et al. (1986).

Neptune

Neptune and Triton are so faint in the UV that only the Voyager flyby has provided detailed information by means of UV spectroscopy. Solar occultation observations probed an altitude range of 2000 Km in Neptune's atmosphere, corresponding to seven decades in pressure. Based on these data, the thermospheric temperature was assigned a provisional value of 750 ± 150 K and, as at Uranus, temperatures were inferred from H_2 band absorption and Rayleigh scattering at intermediate altitudes. The CH_4 distribution is consistent with $K \sim 10^7$–10^8 cm^2 s^{-1}, as was the measured He (584 Å) brightness of ~ 0.35 R. The relatively weak H Lyα brightness of

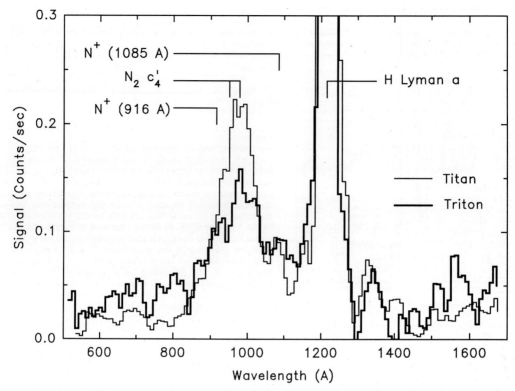

Figure U8 Comparison of the emission spectra of Triton and Titan. The two spectra have been normalized to equal levels at the N^+ (1085 Å) feature. The marked nitrogen features near 1000 Å were the first compelling evidence for the presence of that gas at both satellites. The H Lyα line goes off scale around 1216 Å.

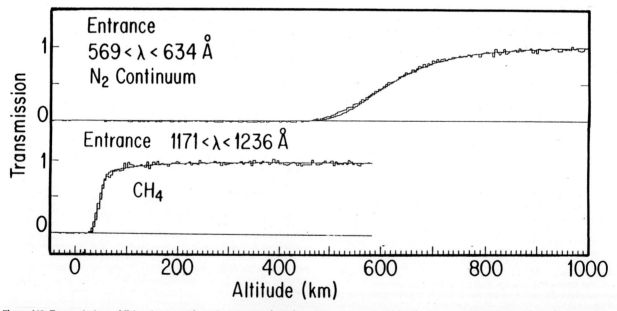

Figure U9 Transmission of Triton's atmosphere in two wavelength ranges, as measured by the Voyager UVS during the entrance solar occultations. The top panel shows continuum absorption by N_2 beginning about 750 km above the surface. The lower panel shows continuum absorption by CH_4 from a few tens of kilometers above the surface to the surface. Profiles of the temperature and the number densities of N_2 and CH_4 have been inferred from such absorption profiles.

340 R, and the low brightness in the 1300–1500 Å region, in comparison to Uranus, is understood in terms of Neptune's greater stratospheric CH_4 abundance. Nightside emissions include H Lyα (85 R) and weak emissions, probably H_2 bands, from two separate regions. The first region extends over more than 100° in latitude; the morphology and brightness seem consistent with excitation by photoelectrons from the magnetically conjugate dayside. The second region is more confined and may represent a classical aurora.

Titan and Triton

Both these satellites possess substantial atmospheres of N_2 with small amounts of CH_4. Their compositions have been inferred from the combination of emission (Figure U8) and occultation spectroscopy by the Voyager UVS, and the temperature structure comes from the occultation measurements (Figure U9). The atmospheric absorption inferred from the occultation measurements is consistent with the dominant species being either N_2 or Ar, but emission spectroscopy revealed strong features characteristic of the c'_4 bands of N_2, and other N and N^+ features, thus confirming N_2 as the main constituent. Nitrogen emissions are excited by a combination of solar fluorescence, photoelectrons and precipitating particles. Thermospheric temperatures are ~ 180 K at Titan and ~ 95 K at Triton. Occultation lightcurves define the CH_4 distributions on both satellites, and have been used to characterize the hazes found a few tens of kilometers above Triton's surface. For more details about Triton see Broadfoot et al. (1989); for Titan see Strobel and Shemansky (1982).

Comets

In the coma of a comet, resonance scattering at visible wavelengths is overpowered by the solar continuum scattered from dust. However, in the UV the solar flux is much weaker and scattering from dust is low compared with the intensity of resonance scattering, so UV spectroscopy of resonance emissions is feasible. The first UV investigations of comets, made by the Copernicus and OGO 5 Earth satellites at H Lyα, showed H clouds extending several million kilometers around the comet. Further investigations of the H distribution, combined with O (1304 Å) and the OH (0–0) band near 3085 Å, gave compelling support to the idea that comet nuclei consist mainly of water ice. Through models, the distributions of these emissions can be related to the production rate of water from the nucleus, a parameter not easily determined by other means. This rate, determined for a number of comets by IUE observations, generally falls in the range $1–30 \times 10^{28}$ molecules s^{-1}. Other species identified by their FUV signatures include C, S and neutral molecules and ions of these atoms combined with the products of water photolysis. The emissions are excited by resonance scattering of sunlight, and in interpreting their brightness it is necessary to take into account the Doppler shift caused by the comet's (typically) large radial velocity relative to the Sun. Feldman (1982) discusses UV spectroscopy of comets.

Interplanetary medium

Because the solar system is moving relative to the local interstellar medium (ISM), neutral H and He of the ISM enter the heliosphere, where they move under the influence of solar radiation pressure and gravity. The temperature and density of the ISM, the ionization lifetimes of H and He and some information about latitudinal variations in the solar flux and solar wind can be derived by mapping the brightness of He (584 Å) and H Lyα from the Sun that are resonantly scattered by the ISM. Both intensity maps made by means of conventional spectrometers and spectrographs (Lallement et al., 1991) and line shape measurements by means of absorption cells (Bertaux and Lallement, 1984) have been fruitful.

Bill R. Sandel

Bibliography

Anderson, D.E. and Hord, C.W. (1971) Mariner 6 and 7 ultraviolet spectrometer experiment: analysis of hydrogen Lyman-alpha data. *J. Geophys. Res.*, **76**, 6666–73.

Atreya, S.K., Sandel, B.R. and Romani, P.N. (1991) Photochemistry and vertical mixing, in *Uranus* (eds J.T. Bergstralh, E.D. Miner and M.S. Matthews), Tucson: University of Arizona Press.

Barth, C.A., Hord, C.W., Pearce, J.B. et al. (1971) Mariner 6 and 7 ultraviolet spectrometer experiment: upper atmosphere data. *J. Geophys. Res.*, **76**, 2213–27.

Barth, C.A. (1985) Mars, in *The Photochemistry of Atmospheres* (ed. J.S. Levine). New York: Academic Press, pp. 337–92.

Bertaux, J.L., Blamont, G., Marcelin, M. et al. (1978) Lyman-alpha observations of Venera-9 and 10. I. The non-thermal hydrogen population in the exosphere of Venus. *Planet. Space Sci.*, **26**, 817–31.

Bertaux, J.L. and Lallement, R. (1984) Analysis of interplanetary Lyman-alpha line profile with a hydrogen absorption cell: theory of the Doppler angular spectral scanning method. *Astron. Astrophys.*, **140**, 230–42.

Blamont, J.E., Chassefiere, E., Goutail, J.P. et al. (1991) Vertical profiles of dust and ozone in the Martian atmosphere deduced from solar occultation measurements. *Planet. Space Sci.*, **39**, 175–87.

Broadfoot, A.L., Herbert, F., Holberg, J.B. et al. (1986) Ultraviolet spectrometer observations of Uranus. *Science*, **233**, 74–9.

Broadfoot, A.L., Atreya, S.K., Bertaux, J.L. et al. (1989) Ultraviolet spectrometer observations of Neptune and Triton. *Science*, **246**, 1459–66.

Broadfoot, A.L., Sandel, B.R., Knecht, D. et al., (1992) Panchromatic spectrograph with supporting monochromatic imagers. *Appl. Optics*, **31**, 3083–6.

Brown, R.A., Pilcher, C.B. and Strobel, D.F. (1983) Spectrophotometric studies of the Io torus, in *Physics of the Jovian Magnetosphere* (ed. A.J. Dessler). London: Cambridge University Press. pp. 197–225.

Chamberlain, J.W. (1961) *Physics of the Aurora and Airglow*. New York: Academic Press.

Chamberlain, J.W. and Hunten, D.M. (1987) *Theory of Planetary Atmospheres*. New York: Academic Press.

Clarke, J.T., Trauger, J. and Waite, J.H., Jr (1989) Doppler shifted H Lyα emission from Jupiter's aurora. *Geophys. Res. Lett.*, **16**, 587–90.

Dessler, A.J., Sandel, B.R. and Atreya, S.K. (1981) The Jovian hydrogen bulge: evidence for corotating magnetospheric convection. *Planet. Space Sci.*, **29**, 215–24.

Esposito, L.W., Knollenberg, J.G., Ya, M. et al. (1983) The clouds and hazes of Venus, in *Venus*, (eds D.M. Hunten, L. Colin, T.M. Dohahue and V.I. Moroz). Tucson: University of Arizona, pp. 484–564.

Feldman, P.D. (1982) Ultraviolet spectroscopy of comae, in *Comets* (ed. L.L. Wilkening). Tucson: University of Arizona Press, pp. 461–79.

Fox, J.L. (1986) Models for aurora and airglow emissions from other planetary atmospheres. *Can. J. Phys.*, **64**, 1631–56.

Goody, R.M. and Yung, Y.L. (1989) *Atmospheric Radiation*. New York: Oxford University Press.

Hays, P.B. and Roble, R.G. (1986) Stellar spectra and atmospheric composition. *J. Atmos. Sci.*, **25**, 1141–53.

Herbert, F. and Sandel, B.R. (1991) CH_4 and haze in Triton's atmosphere. *J. Geophys. Res.*, **96** 19241–52.

Herbert, F., Sandel B.R., Broadfoot, A.L. et al. (1987) The upper atmosphere of Uranus: EUV occultations observed by Voyager 2. *J. Geophys. Res.*, **92** 15093–109.

Hord, C.W., Barth, C.A., Stewart, A.I. et al. Mariner 9 ultraviolet spectrometer experiment: photometry and topography of Mars. *Icarus*, **17**, 443–56.

Hord, C.W. et al. (1991) Galileo Ultraviolet Spectrometer experiment: initial Venus and interplanetary cruise results. *Science*, **253**, 1548–50.

Hunten, D.M., Morgan, T.H. and Shemansky, D.E. 1988. Mercury atmosphere, in *Mercury* (eds F. Vilas, C.R., Chapman, and M.S. Matthews). Tucson: University of Arizona Press, pp. 562–612.

Lallement, R., Bertaux, J.L., Chassefiere, E. and Sandel, B.R. (1991) Interplanetary Lyman observations with UVS on Voyager: direct measurements of the ionization lifetime. *Astron. Astrophys.*, **252**, 385–401.

Livengood, T.A., Moos, H.W., Ballester, G.E. and Prangé, R.M. (1992) Jovian ultraviolet auroral activity, 1981–1991. *Icarus*, **97**, 26–45.

McGrath, M.A. (1991) An unusual change in the Jovian Lyman-alpha bulge. *Geophys. Res. Lett.*, **18**, 1931–4.

Meier, R.R. (1991) Ultraviolet spectroscopy and remote sensing of the upper atmosphere. *Space Sci. Rev.*, **58**, 1–185.

Nelson, R.M., Lane, A.L., Matson, D.L. et al. (1987) Spectral geometric albedos of the Galilean satellites from 0.24 to 0.34 micrometers: observations with the International Ultraviolet Explorer. *Icarus*, **72**, 358–80.

Parkinson, C.D., McConnell, J.C., Sandel, B.R. et al. (1990) He 584 Å dayglow at Neptune. *Geophys. Res. Lett.*, **17**, 1709–12.

Samson, J.A.R. (1967) *Techniques of Vacuum Ultraviolet Spectroscopy*. Lincoln, Nebraska: Pied Publications.

Sandel, B.R., Herbert, F., Dessler, A.J. and Hill, T.W. (1990) Aurora and airglow on the night side of Neptune. *Geophys. Res. Lett.*, **17**, 1693–6.

Smith, G.R. and Hunten, D.M. (1990) Study of planetary atmospheres by absorptive occultations. *Revi. Geophys.*, **28**, 117–43.

Strobel, D.F. and Shemansky, D.E. (1982) EUV emission from Titan's upper atmosphere: Voyager 1 encounter. *J. Geophys. Res.*, **87**, 1361–8.

Takacs, P.Z., Broadfoot, A.L., and Smith, G.R. (1980) Mariner 10 observations of hydrogen Lyman alpha emission from the Venus exosphere: evidence of complex structure. *Planet. Space Sci.*, **28**, 687–701.

von Zahn, U., Kumar, S., Niemann, H. and Prinn, R. (1983) Composition of the Venus atmosphere, in *Venus*, (eds D.M. Hunten, L. Colin, T.M. Donahue and V.I. Moroz). Tucson: University of Arizona Press, pp. 299–430.

Yelle, R.V. (1988) H_2 emissions from the outer planets. *Geophys. Res. Lett.*, **15**, 1145–8.

Yelle, R.V., McConnell, J.C., Sandel, B.R. and Broadfoot, A.L. (1987) The dependence of the electroglow on the solar flux. *J. Geophys. Res.*, **92**, 15110–24.

Yelle, R.V., McConnell, J.C., Strobel, D.F. and Doose, L.R. (1989) The far ultraviolet reflection spectrum of Uranus: results from the Voyager encounter. *Icarus*, **77**, 439–56.

Cross references

Absorption spectrum
Comet: observation
Mariner missions
Radiative transfer
Spectroscopy: atmosphere
Voyager missions

ULYSSES MISSION

Ulysses, launched on 6 October 1990, is the first spacecraft to investigate the unexplored regions over the Sun's poles. The project is a joint venture of the European Space Agency (ESA) and the United States National Aeronautics and Space Administration (NASA).

The name Ulysses was suggested by the Italian scientist, Bruno Bertotti. He had been reading in Dante's *Inferno* how the adventurer in Greek legend exhorted his comrades 'to venture the unchartered distances . . . of the uninhabited world behind the Sun . . . to follow after knowledge and excellence'.

History

In the course of a year the Earth moves 7° north and then 7° south of the Sun's equator. (The Sun's rotation axis is tilted at 7° to the ecliptic plane). It was long ago realized that this narrow slice of heliolatitude scanned by the Earth, or by spacecraft in Earth orbit, was unlikely to be representative of the solar atmosphere as a whole. A round-table discussion of a possible journey out of the ecliptic plane was reported early in the space age (J.A. Simpson, B. Rossi, A.R. Hibbs, R. Jastrow, F.L. Whipple, T. Gold, E. Parker, N. Christofilos, J.A. Van Allen, 1959). At that time no launch vehicle was capable of imparting the necessary energy to a spacecraft. A craft launched from the moving platform that is Earth cannot avoid carrying the 30 km s^{-1} velocity of the Earth in the ecliptic plane. Even with the 11.4 km s^{-1} that modern launchers could give to Ulysses it would not have been possible to reach ecliptic inclinations greater than 23° without help. That help came from Jupiter. The spacecraft trajectory included a Jupiter flyby. This made use of the gravity of the giant planet in order to pull Ulysses out of the ecliptic plane.

Studies of an Out-of-Ecliptic (OOE) mission were carried out by ESA in the early 1970s. NASA joined these studies, and in 1978 approval was obtained for a joint ESA/NASA mission called ISPM (International Solar Polar mission). Each agency was to provide one spacecraft. These were to be launched as a tandem pair, one being directed over Jupiter's north pole en route to a journey south of the ecliptic plane, while the other would pass Jupiter's south pole and travel on through the Sun's north polar regions. Budget questions and other difficulties led NASA to cancel its spacecraft in 1981, but NASA did agree to supply the launcher, the tracking and data recovery, the Radioisotope Thermoelectric Generator (RTG) power supply, and about half of the scientific instruments for the ESA built spacecraft, by this time re-named Ulysses. More than 120 scientists from about 35 institutions are involved in the mission. About half are from Europe.

Scientific aims

Observation of flares and sunspots on the Sun long ago provided an indication that conditions were different at high latitudes. At the beginning of an 11-year solar cycle a few spots appear at latitudes around 35°. When the number of spots is at a maximum, most are located around 20°, and toward the end of the cycle a few spots are found below 10°. Sunspot activity is limited to this belt below about 40°.

Geomagnetic observations and spacecraft measurements in the ecliptic plane indicate that signatures of features on the solar surface propagate outward almost radially in the solar atmosphere. Consequently, it was reasonable to expect at solar latitudes above 40°, a regime different from anything encountered in the ecliptic.

From spacecraft measurements made in or near the ecliptic plane, it was known that a plasma (the solar wind) blows continuously from the Sun, with velocities averaging 450 km s^{-1}. This collisionless, superconducting plasma carries with it the solar magnetic field, out to at least 50 astronomical units (AU). Because the Sun rotates, the magnetic field formed in the ecliptic plane in interplanetary space is wound into a spiral. When the field-bearing solar wind is slow, the field lines are tightly coiled and when the solar wind is fast the magnetic field lines become more radial. When a fast solar wind stream overtakes a slow stream the picture becomes very confused. With increasing solar latitude, rotation plays a smaller and smaller role, until at the poles there are no rotation effects at all. High latitude studies therefore permit a fresh look at the solar wind flow patterns in a simpler and more easily understood regime.

The solar wind is believed to originate on the Sun in regions called 'coronal holes'. Magnetic field lines based in these 'holes' are carried out into interplanetary space and do not return to the Sun. At solar minimum, when all is quiet on the Sun, and its magnetic field is close to dipolar, these coronal holes occupy vast regions over the solar poles. A major Ulysses goal was therefore to get directly over the coronal holes, and examine the solar wind emerging without the hindrance of magnetic barriers. (At solar maximum the polar coronal holes shrink; smaller holes are seen here and there, and the solar magnetic structure becomes much more complicated.).

The heliosphere (q.v.), carved out in the interstellar space by the solar wind and the magnetic field which it carries, obstructs or 'modulates' the arrival of cosmic radiation to the inner solar system. The northern hemisphere of the heliosphere carries interplanetary magnetic field of one polarity while the southern hemisphere carries the opposite polarity. (This polarity arrangement reverses every 11 years with the change in solar magnetic field.) The two hemispheres are separated by a sheet of current which in quiet times lies close to the solar equator. At disturbed times this becomes rippled, like a ballerina skirt. It is then often referred to as the 'wavy current sheet'. The cosmic radiation appears to be modulated both by disturbances in the solar wind and by the wavy current sheet. Another basic Ulysses goal was to examine the arriving pristine cosmic radiation which had not yet been disturbed by the current sheet. It was thought that the charged particles could approach with relative ease, being guided along smooth and radial magnetic field lines emanating from the Sun's poles.

Most of the scientific instruments on the Ulysses are designed to study features of the high latitude heliosphere described above, i.e. the solar wind, the interplanetary field and the mechanisms by which these modulate the arriving cosmic radiation. Several other investigations are carried out on the spacecraft.

A novel instrument measures directly for the first time the arrival of interstellar neutral helium in the solar system. Our Sun rotates about the galactic center in such a way that it has a velocity in the region of 20 km s^{-1} relative to the local interstellar medium. As a consequence neutral particles, at rest in the local interstellar medium, appear to arrive in the solar system with a velocity around 20 km s^{-1}. Helium provides a more reliable measure than does the more plentiful hydrogen, because it is less subject to ionization and because its greater mass means that solar gravity dominates over radiation pressure. Analyzing the neutral helium trajectories about the Sun, in the three-dimensional regime provided by the Ulysses mission, provides outstanding accuracy for the measurement of the velocity of the solar system through interstellar space.

A sophisticated Unified Radio and Plasma Wave Detector System carries out local plasma diagnostics and analyses radio waves arriving from the Sun, from interplanetary space and from Jupiter. The ability

Table U2 Ulysses scientific investigations

Investigation	Principal investigator	Measurement	Instrumentation
Magnetic field	A. Balogh, Imperial College, London	Spatial and temporal variations of the heliospheric and Jovian magnetic field between 0.01 and 44000 nT	Triaxial vector helium and fluxgate magnetometers
Solar wind plasma	S. J. Bame, Los Alamos National Laboratory	Solar-wind ions between 260 eV/Q and 35 keV Q^{-1}; solar-wind electrons between 1 and 900 eV	Two electrostatic analyzers with channel electron multipliers
Solar wind ion composition	G. Gloeckler, University of Maryland/J. Geiss, University of Bern	Elemental and ionic-charge composition, temperature and mean velocity of solar wind ions for speeds from 145 km s^{-1} (H^+) to 1350 km s^{-1} (Fe^{+8})	Electrostatic analyzer with time-of-flight and energy measurement
Low-energy ions and electrons	L. Lanzerotti, Bell Laboratories	Energetic ions from 50 keV to 5 MeV; electrons from 30 keV to 300 keV	Two sensor heads with five solid state detector telescopes
Energetic particle composition and interstellar gas	E. Keppler, Max Planck Institute, Lindau, Germany	Composition of energetic ions from 80 keV to 15 MeV nuc^{-1}; interstellar neutral helium atoms	Four solid state detector telescopes; LiF-coated conversion plates with channel electron multipliers
Cosmic rays and solar particles	J. A. Simpson, University of Chicago	Cosmic rays and energetic solar particles in the range 0.3–600 MeV nuc^{-1}; electrons in the range 4–2000 MeV	Five solid state detector telescopes, one double Cerenkov and semiconductor telescope for electrons
Unified radio and plasma waves	R. G. Stone, Goddard Space Flight Center	Plasma waves, solar radio bursts and electron density; electric field plasma waves: 0–60 kHz; radio receiver: 1–940 kHz; magnetic field: 10–500 Hz	72-m radial dipole antenna, 8-m axial monopole antenna and two-axis search coil
Solar x-rays and cosmic gamma ray bursts	K. Hurley, University of California–Berkeley/M. Sommer, Max Planck Institute, Garching	Solar-flare x-rays and cosmic gamma-ray bursts in the energy range 5–150 keV	Two Si solid state detectors; two CsI scintillation crystals
Cosmic dust	E. Grün, Max Planck Institute, Heidelberg	Direct measurement of particulate matter in mass range 10^{-16}–10^{-7} g	Multi-coincidence impact detector with channeltron
Coronal sounding	M. Bird, University of Bonn	Density, velocity and turbulence spectra in the solar corona and solar wind	Spacecraft transponder
Gravitational waves	B. Bertotti, University of Pavia	Doppler shifts in radio signal received at Earth due to passage of wave	Spacecraft transponder

to find the arrival direction of the radio signals is a significant advance over earlier space measurements.

Another instrument takes advantage of the three-dimensional nature of the orbit to study the arrival direction of dust particles in order to determine whether these particles arrive from asteroids, comets, planets or from interstellar space.

There are also detectors which search for solar x-rays and for cosmic gamma-ray bursts. The radio transmitters on the spacecraft are used to probe the electron content of the heliosphere and to search for gravitational waves.

Scientific payload

The instruments carried in order to achieve the scientific aims described above are listed in Table U2.

Launch and trajectory

Ulysses was launched on 6 October 1990 from the NASA shuttle Discovery with the assistance of an inertial upper stage and a payload assist module. The trajectory followed by the spacecraft is illustrated in Figure U10. Closest approach to Jupiter, at 6.3 Jovian radii, was on 8 February 1992. High southern solar latitudes were reached in mid-1994 and high northern latitudes one year later. Altogether, 235 days were spent at latitudes about 70° and the maximum solar latitude achieved was just above 80°. Spacecraft operations are conducted by a joint ESA/NASA team at the Jet Propulsion Laboratory, Pasadena, California.

The spacecraft

The main features of the spacecraft are illustrated in Figure U11. The body is about 3 m wide and 2 m high, the magnetometers boom is 5.6 m long and the radial plasma wave dipole antennas are 72.5 m from tip to tip. Power is supplied by a radioactive thermoelectric generator (RTG) which at launch provided 283 W. Stability is achieved by spinning the spacecraft at 5 rpm, and Earth-pointing of the X-band high gain antenna directed along the spin axis is maintained to better than 0.4°. Earth pointing adjustments are required every 2 or 3 days.

A range of data rates is possible. In the usual operating mode one ground station receives interleaved real-time and recorded data for 8 h each day. The other 16 h are recorded on tape so that 24 h coverage is achieved. The usual record rate is 512 bit s^{-1} and playback or real time transmission 1024 bit s^{-1}.

Jupiter encounter

Ulysses flew closest to Jupiter at 6.3 planetary radii (1 R_J = 71 398 km) on 8 February 1992 and survived the hazardous radiation environment without apparent damage. Scientific instrumentation on board, although not optimised for the Jupiter magnetosphere, made novel measurements. The four earlier spacecraft that flew by Jupiter (Pioneer 10 and 11 and Voyager 1 and 2) did so close to the equator and exited on the morning side of the planet. Ulysses approached Jupiter on the morning side and exited on the dusk side, reaching magnetic latitudes above 40° both on the way in and the way out.

At magnetic latitudes just above 40° the spacecraft appeared to leave the trapped radiation region and enter a regime similar to that seen over the polar caps of the Earth. On the duskside departure from the planet, very large current layers were found parallel to the magnetic field. It was also found that the planetary field lines were swept back by the solar wind more than had been predicted. An overall conclusion is that the strong planetary rotation and magnetic field play a lesser role in the Jovian magnetosphere than was earlier believed, while the solar wind plays a more significant role.

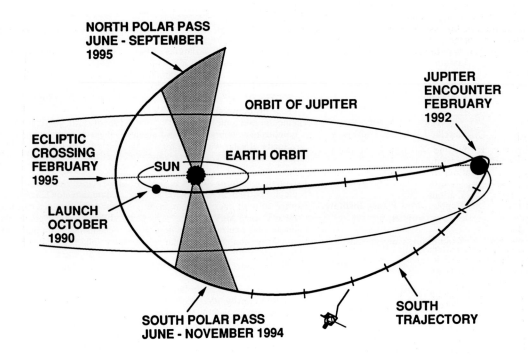

Figure U10 The Ulysses trajectory.

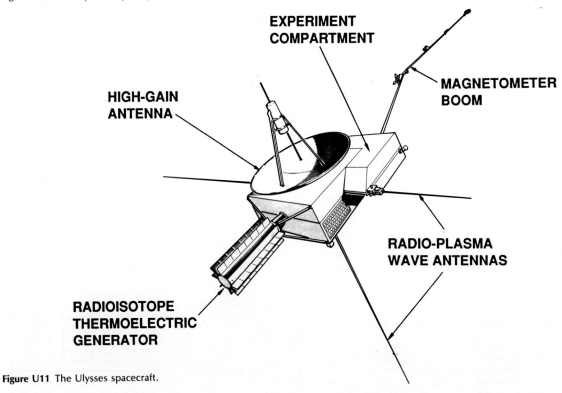

Figure U11 The Ulysses spacecraft.

Scientific discoveries

By January 1996 more than 400 papers based on Ulysses data had appeared in the scientific literature. It has to be remembered that the features seen during the high latitude passages (1993–1996) were those existing at a time when the Sun was quiet. In years of maximum solar activity the situation is likely to be very different.

Solar wind

There are two distinct and fundamentally different solar wind regimes and the boundary between the two is sharply delineated. Within about 35° of the solar equator the velocity averages around 450 km s^{-1} and is variable. At higher latitudes the wind, presumably from coronal holes, blows at all times in a remarkably steady fashion at around 750 km s^{-1}. The ions in the slow solar wind have charge states (number of electrons missing) different from those in the fast wind. The composition of the fast wind is close to that believed to exist in the solar photosphere but the slow wind composition is dependent on the ionization potential of the particular constituent.

Magnetic field

Measurements made from Earth indicate that a dipole magnetic field exists at the Sun's surface. A dipole has a concentration of magnetic flux at its poles. No such concentration was found at Ulysses altitudes (approximately 2 AU) over the solar poles. It has to be concluded that the conducting solar wind plasma smoothed out magnetic pressure differences, so that the radial magnetic field measured at Ulysses was independent of latitude.

Large amplitude fluctuations in magnetic field direction were seen over the solar poles. This is probably significant in explaining the surprising cosmic ray observations.

Cosmic rays

The long postulated increase in cosmic ray flux over the solar poles was not found. Only a small increase was seen from equator to pole, and as a result, theories of cosmic ray modulation by the heliosphere must be re-examined. The large magnetic field fluctuations, seen almost continuously in the polar regions, invalidated the concept of easy charged particle access along smooth radial field lines.

In all of the above areas – solar wind, magnetic field and cosmic rays – the configurations found over the north polar regions essentially repeated those first found over the south polar regions (with magnetic polarities reversed, of course). Differences of a few percent only were found in cosmic ray intensity and in solar wind momentum flux from south to north.

Interstellar neutrals

Interstellar neutral helium was directly measured for the first time. It arrives in the solar system at 25.3 ± 0.4 km s^{-1} and from a direction defined by ecliptic longitude 73.9 ± 0.8°, ecliptic latitude −5.6 ± 0.4°. Its temperature is 7000 ± 600 K. Its density is less certainly determined, but is believed to be in the range $1.4 - 1.7 \times 10^{-2}$ cm^{-3}.

Interstellar neutral nitrogen, oxygen, neon and carbon are difficult to ionize and therefore, like helium, can penetrate deep into the solar system before they become charged and are picked up by the magnetic field carried in the solar wind. The solar wind ion instrument on Ulysses has confirmed the pick-up process, and has for the first time, detected all of the above-mentioned interstellar neutrals in the form of pick-up ions. Pick-up hydrogen has also been identified.

Dust

Within 1 AU of Jupiter, bursts of submicron-sized dust grains were encountered at intervals corresponding approximately to the rotation period of the Sun.

A flow of micron-sized grains was detected at distances of around 5 AU from the Sun. These grains arrive on retrograde orbits at velocities greater than 26 km s^{-1}, and are therefore probably of interstellar origin.

Future mission goals

Ulysses passed over the south solar pole in the second half of 1994 and over the north pole 1 year later. This was a period of low solar activity. The spacecraft has now (January 1996) embarked on its second 6.2-year orbit. Aphelion (where Jupiter was at encounter in February 1992) will be reached in April 1998, the south solar pole in late 2000 and the north solar pole at the end of 2001. These polar passages will occur at a time when the Sun is active. The two orbits, of approximately 6 years each, allow coverage of an 11-year sunspot cycle.

D.Edgar Page

Bibliography

Astronomy and Astrophysics (1992) Special issue on Ulysses Instruments, **92**, 207–440.
Geophysical Research Letters (1995) Special issue on Ulysses Observations from Pole to Pole, **22**, 3297–432.
Page, D.E. (1975) Exploratory journey out of the ecliptic plane. *Science*, **190**, 845–50.
Planetary and Space Science (1993) Special issue on Ulysses at Jupiter, **41**, 797–1108.
Simpson, J.A., Rossi, B., Hibbs, A.R. *et al.* (1959) *J.Geophys. Res.*, **64**, 1691–93.
Smith, E.J., Page, D.E. and Wenzel, K.P. (1991) Ulysses: a journey above the Sun's poles. *EOS*, **72**, 241.
Space Science Reviews (1995) Special issue on the High Latitude Heliosphere (ed. R.G. Marsden), **72**, 1–498.
Wenzel, K.P., Marsden R.G. and Battrick, B. (1983) The International Solar Polar Mission, European Space Agency, SP-1050.

Cross references

Astronomical unit
Coronal mass ejection
Heliosphere
Plasma wave
Solar activity
Solar wind
Sun

UMBRIEL

One of the moons of Uranus, Umbriel was discovered telescopically by William Lassel in 1851. Umbriel orbits Uranus in approximately 4.1 days at a distance of 266 000 km. It has a diameter of 1169 km and a mass of 12.7×10^{20} kg.

Figure U12 Highest-resolution image of Umbriel taken by the Voyager 2 spacecraft from a distance of 557 000 km on 24 January 1986. The smallest features visible on the surface are approximately 10 km across. The bright feature at left is the crater Wunda. Crater Vuver, to the upper right of Wunda, shows a bright central peak. The south pole of the moon is near the lower center of the moon in this image. (NASA image P-29521.)

The surface of Umbriel is known from images returned by the Voyager 2 spacecraft in late January 1986. It has both the lowest average albedo and the least albedo variation of the major moons and shows the least evidence for geologic activity. Its surface shows many subdued large craters, and the major brightness variations on the moon are associated with them. Crater Wunda is approximately 30 km in diameter and has a ring of bright material in its interior while crater Vuver is slightly smaller and has a bright central peak. The bright markings associated with the large craters suggest that the dark surface of Umbriel is composed of a relatively thin coating of material and that bright ice lies below. However, no craters have bright rays emanating from them.

There are also a large number of smaller craters and a general lack of tectonic features on the moon, indicating that its surface is among the oldest of the major moons, formed approximately 4 billion years ago. The surface of Umbriel is not devoid of all tectonic activity, however. There are at least two sets of extensional canyons, but they are generally smaller and lack the prominent smooth, resurfaced floors seen on the other Uranian moons, such as Ariel.

Daniel M. Janes

Bibliography

Bergstralh, J. Miner, E. and Matthews. M. (1991) *Uranus*. Tucson: University of Arizona Press.
Miner, E. (1990) *Uranus: The Planet, Rings and Satellites*. New York: Ellis Horwood.
Smith, B.A. *et al.* (1986) Voyager 2 in the Uranian system: imaging science results. *Science*, **233**, 43–64.

Cross references

Uranus
Uranus: satellite system

URANUS

Uranus, one of the four giant planets, is the fourth most massive planet in the solar system (Plate 30). Its orbit has a period about the Sun equal to 30 685 Earth days, a semimajor axis of 2.87 billion km, and lies between the orbits of Saturn and Neptune. Uranus' orbit plane is inclined to the Earth's orbit plane by only 0.774°, and thus lies close to the orbit planes of most other planets. Uranus' orbital motion is prograde; i.e. the planet moves counterclockwise in its orbit when viewed from above the north pole of the Sun, in the same sense as other planetary orbits. Uranus' orbital eccentricity $e = 0.0472$, which is similar to the small values found for the other giant planets.

The one remarkable and unique dynamical characteristic of Uranus is its obliquity, or angle of inclination between the spin axis and the normal to the orbital plane. Uranus spins with an interior rotation period of 17.24 h and an obliquity of 98°, much larger than the obliquity of any other planet. It is currently believed that this large obliquity represents the effect of the impact of a single planet-sized body into the primordial Uranus at the time that Uranus was still accumulating from the early solar nebula. The fact that Uranus' compact and regular satellite system closely shares this large obliquity indicates that the planet's unusual spin vector was imparted before or during the time that the satellite system was forming, some 4.5×10^9 years ago.

Uranus presents a faint bluish disk to a terrestrial telescopic observer, with an apparent visual magnitude of about 6 and an apparent diameter of about 3.5 arcsec. Thus, although an acute unaided observer can just see Uranus in a dark sky, a telescope is needed to discern that it is a planet rather than a faint star. And because of Uranus' faintness, the planet's motion was not apparent to pretelescopic observers. Thus Uranus was the first planet to be discovered in historical times: it was accidentally discovered by the British astronomer William Herschel in 1781. Little was known about the planet, other than its mass and approximate size, for nearly 200 years afterward. In the 1970s and 1980s advances in ground-based planetary astronomy provided important new spectroscopic data about Uranus' atmospheric structure and composition, and about its interior structure from the motions of the rings and from heat flow measurements. A large increment in information resulted from the 1986 encounter of the Voyager 2 spacecraft with Uranus, so far the only completed or projected spacecraft mission to the planet. Table U3 presents basic data about Uranus.

Table U3 Uranus data

Mass M (Earth masses)	14.53
Equatorial radius a at 1 bar pressure (km)	25 559
Interior rotation period P (h)	17.24
Mean mass density (g cm^{-3})	1.27
First even zonal harmonic J_2	0.003516
Second even zonal harmonic J_4	−0.000032
Dynamical oblateness ϵ	0.01982
Dimensionless axial moment of inertia C/Ma^2	0.23
Internal energy flux at surface (W m^{-3})	< 0.09

Because Uranus has no discernible solid surface, its radius is defined at an arbitrary pressure level in its atmosphere, a level which is accessible to direct observation. Moreover, Uranus is non-spherical because of its substantial rotation rate (Uranus spins more rapidly than the Earth), and its polar radius b at the 1-bar pressure level is about 500 km smaller than its equatorial radius a. This difference is expressed in terms of the dynamical oblateness $\epsilon = (a - b)/a$.

Uranus' deviation from spherical symmetry can be measured in another way, through the departure of its gravitational potential V from the strict inverse dependence on radial distance r from the center of the planet. For a point mass M or for a spherically symmetric body of total mass M, the value of V external to the mass is given by the simple expression $V = GM/r$. When the body becomes oblate due to axial rotation, a correction term appears in V with a relative size characterized by the dimensionless zonal harmonic J_2. Knowing J_2 and the rotation period, it is possible to infer the body's moment of inertia about its rotation axis, C. For a liquid, rapidly spinning planet such as Uranus, only even zonal harmonics should be detectable in the external potential, as is observed for all four giant planets. The next even zonal harmonic J_4 represents a nonlinear response of the planet to rotation, and contains additional information about the distribution of mass in its interior.

Although Uranus' atmosphere is composed mostly of hydrogen, the atmosphere cannot extend to great depths nor comprise a major fraction of the planet's total mass. Because hydrogen is a highly volatile and compressible substance, a mass of hydrogen comparable to the total mass of Uranus would have a radius roughly twice the observed radius of Uranus, with the precise value depending sensitively on the temperature distribution in the hydrogen. On the other hand, Uranus' relatively low mean density indicates that dense rocky materials composed of magnesium silicates and iron do not predominate either. The mean density alone could be matched with a hydrogen envelope over a rocky core, but the gravitational moments require that the planet must be much less differentiated in density than this.

Current models of the interior of Uranus are based upon the assumption that the planet primarily accreted out of ice-rich planetesimals, similar in composition to the small icy bodies still present at great orbital distances from the Sun; bodies such as Chiron, Pluto and Charon, and the small icy satellites of Saturn, Uranus and Neptune. Although the precise composition of these bodies is not known, water ice (H_2O) appears to be a major component, as would also be expected from the high cosmic abundance of oxygen, and from nebular condensation models. Other probable chemical constituents of Uranus' interior are more difficult to identify. Carbon and nitrogen have high cosmic abundances and readily combine with hydrogen to form methane (CH_4) and ammonia (NH_3) respectively in a suitable hydrogen-rich environment. Plausibly, the 'ice' component of Uranus could be about two-thirds H_2O by mass, with CH_4 comprising most of the remaining mass. If rock is also present in accordance with cosmic abundances, it would have a mass equal to about one-third of the ice component.

Although the term 'ice' is used to refer to the mixture of water, methane and ammonia described above, the physical state of this material in Uranus is undoubtedly not solid. Although there are no direct data on interior temperatures in Uranus, calculations of compressional heating during planetary accumulation indicate that interior temperatures should be on the order of several thousand kelvin at pressures of several million bars. Under these circumstances, the 'ice' component would actually be a hot molecular fluid, and some molecular dissociation may occur as well.

Dynamic high-pressure experiments on chemical mixtures corresponding to putative Uranus interior material, so-called 'synthetic Uranus', have been performed in the laboratory to determine whether the compression curve of the material is compatible with Uranus' zonal harmonics, and whether such material has sufficient electrical conductivity to support Uranus' observed planetary magnetic field. The synthetic material is initially a liquid solution of water, ammonia and isopropanol, with molar abundances of H, O, C and N approximately equal to that of a solar-composition mixture of H_2O, CH_4 and NH_3. Single and double shock wave experiments on 'synthetic Uranus' have been conducted up to 2.2 million bar, and to temperatures in excess of 4000 K. The results closely match the compression curve required to yield Uranus' zonal harmonics, and thus provide indirect evidence of the bulk composition of the planet.

Interior models of Uranus have the following general characteristics. The outermost layers of the planet, roughly 20% of the radius (about 5000 km), are composed primarily of molecular hydrogen. The hydrogen is gaseous in the regions near the planet's 1-bar level, and gradually and continuously transforms to a hot molecular liquid as pressure and depth increase. At the bottom of the hydrogen-rich outer layer, pressures are about 0.1 million bars (100 kbar), and the density reaches about 0.2 g cm^{-3}. The total mass in this hydrogen-rich outer layer does not exceed about three Earth masses. At deeper layers there is an ice-rich region which appears, at the crude resolution available from model fitting, to be chemically homogeneous. The ice-rich region appears to extend virtually to the center of the planet, and there is no evidence that a denser rocky core has formed at the center of Uranus. Indeed, the lack of a rocky core is indicative that little gravitational separation has occurred in Uranus, with the magnesium silicate fraction instead suspended throughout the icy interior.

A small fraction of Uranus' hydrogen-rich outer layer may have originated from decomposition of the hydrogen-bearing ices in its interior. However, the outer layer is probably too massive for this contribution to be significant. Clear evidence for a nebular origin of most of the hydrogen layer comes from its helium abundance, which is virtually identical to the value in the solar atmosphere: there is about one helium atom for every 11 hydrogen atoms. At the same time, the atmosphere shows some evidence for the presence of an enormous ice reservoir beneath it. The abundance of methane in the atmosphere, relative to hydrogen, is enhanced by a factor of 20 to 30 over the value that would be expected for a solar ratio of carbon to hydrogen. As methane is the most volatile of the three gases, water, methane and ammonia, methane can penetrate well above the main Uranus cloud deck before suffering significant depletion due to condensation in Uranus' cold atmosphere. Thus it is in its full, enhanced, mixing ratio relative to hydrogen throughout Neptune's troposphere. In fact it is largely this phenomenon, together with methane's strong absorption of red photons, which gives Uranus its characteristic color.

In contrast to methane, little ammonia is observed in Uranus' atmosphere. However, this result is attributable to the lesser volatility of ammonia, its tendency to become incorporated in deeper cloud layers, and possibly to chemical reactions with other species, rather than a true depletion of ammonia in Uranus' deeper atmosphere. The same remarks apply to water, which has not been directly detected in Uranus, and which would require a very deep entry probe for its detection.

Observations of the heavy hydrogen isotope, deuterium, provide further evidence that Uranus' atmospheric methane abundance may be derived from the deep ice layer. The deuterated methane molecule, CH_3D, can be spectroscopically detected in the atmosphere and its abundance relative to ordinary methane measured. From this relative abundance, the deuterium to hydrogen ratio (D/H) in Uranus can be inferred. The ratio is about 7×10^{-5}, or about three to five times larger than the deuterium to hydrogen ratio in Jupiter and Saturn; the latter is about 2×10^{-5}. The ratio D/H in Jupiter and Saturn cannot have been altered by fractionation processes, because both bodies are mainly composed of hydrogen, and thus the D/H in Jupiter and Saturn should reflect the value in the hydrogen nebula from which both planets accumulated. Evidently there has been substantial enrichment of deuterium in Uranus with respect to the value for the two larger giant planets. It is plausible that the enrichment reflects an overall concentration of deuterium in Uranus' massive ice layers.

Observations of small ice-rich objects with gaseous envelopes in the outer solar system, such as Halley's comet and Saturn's moon Titan, indicate that deuterium also tends to be enhanced with respect to hydrogen in these objects. There are two possibilities for enhanced D/H in small icy objects. Since these have weak gravitational fields, light isotopes can preferentially escape their atmospheres with respect to more massive isotopes, which may lead to enhancement of D/H. However, it is also possible that D/H is initially enhanced in the ice component by disequilibrium chemical reactions in the interstellar medium; that is, primitive icy objects in the outer solar system may have a significant component of unprocessed interstellar ice grains, enriched in D/H. Since Uranus is so cold and massive that very little hydrogen is likely to have escaped from its atmosphere over the age of the solar system, its D/H ratio must be primordial, and may possibly reflect interstellar ice grain chemistry.

Uranus' internal energy flux is unusually small compared with the values for the other three giant planets. The internal energy flux for Neptune is 0.3 W m^{-2}, and Jupiter and Saturn emit several watts per square meter from their interiors. In contrast, as Table U3 shows, only an upper limit has been measured for Uranus' internal heat flux. Uranus radiates into space a total amount of infrared heat very nearly equal to that radiated by Neptune, but the two planets differ in that a large fraction of Neptune's heat flux comes from its interior, while nearly all of Uranus' heat flux is derived from the conversion of sunlight into heat in Uranus' atmosphere. Calculations of the cooling of Uranus, from a presumed initial hot state which resulted from initial accumulation of the planet's material, show that a present-day internal energy flux of as much as 1 W m^{-2} could be readily accounted for. Estimates of 'backwarming' of Uranus' atmosphere by absorbed sunlight show that this phenomenon may have a significant impact on Uranus' thermal evolution, and that the present-day heat flux may be affected by it. However, this effect cannot reduce the predicted heat flux to as low a value as the observed upper limit.

Theoretical estimates of the bulk composition of Uranus, as stated above, suggest that it may include as much as about four Earth masses of rocky material. If this rocky component generates heat due to radioactivity at the same rate as terrestrial rocky material, one might expect an equilibrium surface heat flux as large as 0.02 W m^{-2}, still well below the observed upper limit. Measurements of the heat flow to this level of precision will require careful monitoring of the planet's infrared and visible wavelength radiation balance over long time periods, possibly from a Uranus orbiter.

At present there is no consensus on the explanation for Uranus' abnormally low internal energy flux. It has been suggested that chemical composition gradients, such as the one apparently existing at a depth of 5000 km, may suppress convective currents and cut off the planet's hot deep interior from energy transport to the surface. Other, less pronounced gradients may exist at deeper layers and have a similar effect. However, this explanation has limited predictive power and has been unable to account for the much larger Neptune heat flux except in an a posteriori fashion.

William B. Hubbard

Bibliography

Atreya, S.K., Pollack, J.B. and Matthews, M.S. (1989) *Origin and Evolution of Planetary and Satellite Atmospheres*, part IV, *Outer Planets*. Tucson: University of Arizona Press.
Bergstrahl, J.T., Miner, E.D. and Matthews, M.S. (1991) *Uranus*. Tucson: University of Arizona.
Hubbard, W.B. (1984) *Planetary Interiors*. New York: Van Nostrand Reinhold.
Hubbard, W.B. (1990) Interiors of the giant planets, in *The New Solar System*, 3rd edn (eds J.K. Beatty and A. Chaikin). New York: Cambridge University Press. pp. 131–8.
J. Geophys. Res. (1987) Voyager 2 at Uranus issue, **92**, 14837–5375.
Stevenson, D.J. (1982) Interiors of the giant planets. *Ann. Rev. Earth Planet. Sci.*, **10**, 257–95.

Cross references

Jupiter: interior structure
Neptune
Saturn: interior structure

URANUS: ATMOSPHERE

Uranus is the smallest of the group of large, low-density, hydrogen-rich planets known as the gas giants, the others being Jupiter, Saturn and Neptune. Two salient characteristics set it apart from the other gas giants, including Neptune which has a similar mass, radius, density and rotation rate: (1) a relatively small internal heat flux and (2) a high obliquity (i.e. a large inclination of the equator to the orbital plane). For the other giant planets the emitted heat flux is greater than the absorbed solar flux, so that atmospheric temperatures are significantly greater than the solar equilibrium temperature. (Indeed Neptune, which receives less than half the solar flux of Uranus, is warmed to almost Uranian temperatures by its large internal heat.) For Uranus the internal heat flux is less than 14% of the solar input (Pollack *et al.*, 1986; Pearl *et al.*, 1990). The high obliquity of 98° means that Uranus essentially rotates on its side, so that each pole and the equator alternately receive direct sunlight during its 88-year orbital period.

Both of these factors contribute to the unusual calmness of the Uranian atmosphere, illustrated in Plate 30 and Figure U13. The weak internal heat flux results in relatively low spatial gradients in atmospheric temperatures and densities (which typically power planetary meteorology). The high obliquity causes the atmosphere to be warmed relatively uniformly, thus suppressing the need for meridional transport of gases from the equator to the pole, such as occurs on the Earth. Thus, in contrast to the other gas giants, all of which exhibit large-scale storm systems (e.g. the Great Red Spot on Jupiter, the Great Dark Spot on Neptune), the Uranian atmosphere is exceedingly bland, both visually and dynamically.

The Uranian atmosphere extends from space down to approximately 6000 km below the 1-bar level, where pressures exceed $\sim$ 1 Mbar, or 1 million times the atmospheric pressure of the Earth. There, due to this enormous pressure and concurrent high temperature (in excess of 2000 K), gases turn into superheated liquids, forming a deep, hot ocean comprised mainly of water, methane and ammonia. The upper 100 bars of the atmosphere have been remotely probed by ground-based and Earth-orbiting instruments, as well as instruments onboard the Voyager 2 spacecraft. The atmosphere below that level is hidden from view, owing to the overlying obscuring effects of an array of light-scattering and light-absorbing clouds and gases. We must therefore resort to thermodynamical and chemical theories to gain an understanding of the deep atmosphere.

Our direct knowledge is principally derived from measurements made over a large portion of the electromagnetic spectrum, from the far ultraviolet near 500 Å (0.000005 cm) to radio wavelengths near 20 cm. In the ultraviolet, from $\sim$ 1300 to 1600 Å, molecular Rayleigh scattering limits the view to the homopause near 50 μbar, above which gases are not well mixed but rather are largely segregated by molecular weight. Strong atomic hydrogen line emissions originating in this atmospheric region occur from $\sim$ 500 to 1300 Å, the strongest being the Lyman α line at 1216 Å (see Spectroscopy: atmosphere).

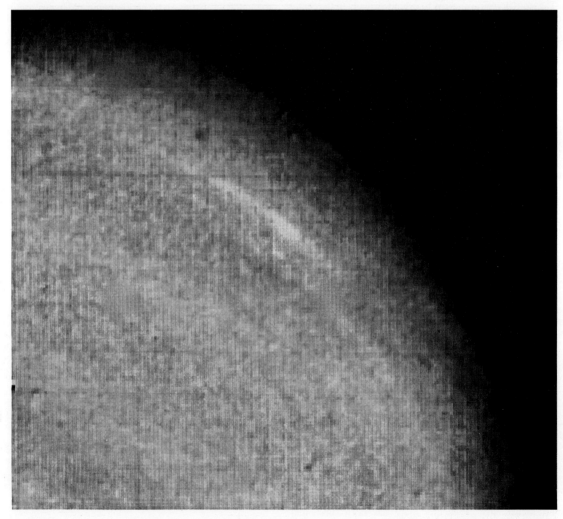

Figure U13 Voyager image of a plume feature at 35° latitude, observed just above the noise level in an extremely contrast-enhanced image. Only a few such features were observed in the relatively quiescent Uranian atmosphere, compared to hundreds seen in the atmospheres of the other Jovian planets, Jupiter, Saturn and Neptune. (NASA/JPL image P-29467.)

These yield clues to the temperature and energetics of the uppermost atmosphere. In the medium UV, 1700–2400 Å, solar radiation penetrates to ~ 1 to 80 mbar (depending monotonically on wavelength), well into the stratosphere, that stably stratified region above approximately 100 mbar where temperature increases with altitude.

In the visible, between approximately 3000 and 7000 Å, sunlight can for the most part penetrate through several thin haze layers down to an optically thick cloud deck at 2.7 bar (Baines and Bergstralh, 1986), well into the convective troposphere underlying the stratosphere. However, several methane-absorbing wavelength bands exist which restrict the view to the stratosphere and upper troposphere, allowing the properties of upper tropospheric hazes to be ascertained from ground-based observations. Special color filters onboard the Voyager 2 spacecraft have been used to image the planet's limb at close range to determine stratospheric aerosol properties (Pollack et al., 1987).

In the near infrared, from 0.7 μm to ~ 3 μm, methane bands become progressively stronger, correspondingly limiting the view to successively higher stratospheric regions. Beyond 3 μm, however, the predominant radiation is not reflected sunlight but rather indigenous thermal radiation. Between approximately 7 and 14 μm this radiation comes from thermally radiating molecules in the stratosphere (e.g. hydrocarbons such as methane, acetylene and ethane).

Due largely to wavelength-dependent hydrogen gas opacity, various other atmospheric levels produce the predominant radiation seen at other thermal wavelengths, extending out to the submillimeter and millimeter range, thus yielding the thermal profile from the high stratosphere down to 8 bar (Orton et al., 1986, 1987; Flasar et al., 1987). At radio wavelengths of 0.5–20 cm the minor constituent ammonia is the predominant opacity source, allowing views to the 100-bar level (Gulkis, Jannsen and Olsen, 1978).

Additional information about the thermal structure and composition of the Uranian ionosphere, thermosphere, stratosphere and upper troposphere comes from the Voyager radio occultation experiment (Tyler et al., 1986). Using the spacecraft's ultrastable radio transmitter, the radio occultation experiment probed various atmospheric levels with well-calibrated radio signals as the spacecraft, from the perspective of Earth-based receiving antennae, was occulted by the planet. Variations in signal strength and phase were then used to determine atmospheric and ionospheric mass and ion number densities.

Composition

Atmospheric composition relevant to planetary formation and evolutionary processes is determined primarily from examination of the deep troposphere below the condensation levels of the most prominent volatiles. In the stratosphere, ongoing condensation and complex photochemical processes result in significant depletions or enhancements of most minor gas constituents. In particular, photochemical processing of methane leads to a surprisingly rich set of molecular species.

Tropospheric abundances

The relative abundances of hydrogen, helium, methane and deuterium are particularly diagnostic of formation processes. Expressed in mole fractions, that is, by the fractional number of molecules of each species, the abundances are approximately 0.83, 0.15, 0.03 ± 0.01 and $7.5 \pm 2.5 \times 10^{-5}$ respectively. The helium mole fraction of 0.15 (Conrath et al., 1987) is consistent with that for the protosolar nebula (Fegley et al., 1991). This suggests that the He/H$_2$ ratio has not been enhanced by preferential depletion of H$_2$ through the process known as Jeans escape. This mechanism, wherein lightweight atoms and molecules – particularly hydrogen – are literally knocked out of the atmosphere by collisions with solar photons, explains, for example, why helium is depleted in Earth's atmosphere. The ineffectiveness of the mechanism's in Uranus is consistent with the planet's large orbital distance and large gravity well. The observed near-solar He/H$_2$ ratio also indicates that hydrogen has not been enriched by the dissociation of methane into its basic constituents, carbon and hydrogen, within the hot deep interior. It also indicates that rain-out of helium, due to the immiscibility of He in metallic H, does not occur, as has been suggested to explain the relatively low values of 0.06 ± 0.05 in Saturn's atmosphere. This is consistent with current interior models which indicate that metallic hydrogen is much more prevalent in Saturn than in Uranus.

The tropospheric mixing ratio CH$_4$/H$_2$ = 0.03 ± 0.01 (Baines and Bergstralh, 1986; Lindal et al., 1987), corresponds to a 30–60-fold enrichment. This result agrees with recent theories of planetary formation which hypothesize that CH$_4$ ice-rich protoplanetesimals were a major source of Uranian material.

For deuterium the observed D/H ratio of $7.5 \pm 2.5 \times 10^{-5}$ (Trafton and Ramsay, 1980) is somewhat greater than the solar nebula value, D/H ~ 2×10^{-5}. This suggests that a significant fraction of these icy protoplanetesimals were formed at the low temperatures characteristic of the outer solar nebulae since, under thermodynamic equilibrium conditions, low-temperature ices contain between 10 and 1000 times more deuterium than warmer ices (Podolak, Hubbard and Stevenson, 1991). However, since the deuterium enhancement is not very large, considering the large amount of protoplanetesimal material indicated by the high methane abundance, a significant fraction of the icy protoplanetesimal component may have originally formed at higher temperatures. Alternatively, the colder planetesimals may not have reached thermodynamic equilibrium prior to colliding with the planet.

Ground-based radio measurements indicate that ammonia (NH$_3$) is depleted by two orders of magnitude compared with the solar abundance ratio, when measured deep ($P > 100$ bar) within the atmosphere below its ~ 8-bar condensation level (Gulkis and de Pater, 1984). This is somewhat surprising, given the large methane abundance noted above, since both ammonia and methane ices were presumably present in supersolar abundances in the progenitor protoplanetesimals, and since ammonia is present in near-solar abundances on less ice-rich Jupiter and Saturn. The result suggests chemical processing of NH$_3$ within Uranus. Specifically, sequestration of ammonia through either (1) reaction with H$_2$S to form NH$_4$SH or (2) by aqueous solution in massive H$_2$O condensate clouds has been suggested. Alternatively, Lewis and Prinn (1980) suggest that the lack of ammonia is not due to its sequestration within clouds, but rather to the lack of accumulation of nitrogen within the planet during its formation.

Water has not been detected in Uranus because it condenses at atmospheric levels deeper than can be probed by remote sensing instruments. However, oxygen is cosmically abundant, so it is expected to be a major constituent of the Uranian atmosphere. For a solar oxygen abundance, water clouds would form near the 120-bar, 325-K level. For a 3% H$_2$O mole fraction, similar to the mole fraction of methane, the water cloudbase would be at about 550 bar pressure and a temperature of 470 K.

Hydrogen, the dominant constituent of the Uranian atmosphere, exists in two states, 'para' and 'ortho', as determined by the relative orientation of the molecule's two nuclear spin vectors (the ortho and para states denote, respectively, parallel and antiparallel nuclear spin vectors). The relative abundance of the ortho and para states of hydrogen is an indicator of vertical mixing (e.g. Conrath and Gierasch, 1984). While the equilibrium ratio of ortho to para states is a strong function of temperature in the cold Uranian atmosphere, equilibration itself takes several years in the absence of a catalyst, a time which may not be available for a parcel of circulating air. Thus the degree to which the observed ortho/para ratio is not in agreement with thermodynamic equilibrium indicates the timescale for vertical transport. The fraction of hydrogen observed in the equilibrium ortho/para state is 0.79 ± 0.16 (Baines and Bergstralh, 1986), indicating that a significant fraction of the H$_2$ population is in thermodynamic disequilibrium. Given a 3-year time constant for ortho/para conversion, this implies a maximum timescale for vertical mixing of 3–10 years. Vertical mixing may be substantially more energetic if catalytic processes (e.g. Massie and Hunten, 1982; Conrath and Gierasch, 1984) are involved.

Stratospheric abundances

In the upper troposphere above approximately the 1.6-bar level, the concentration of gaseous methane falls rapidly with height as the temperature decreases with height below the CH$_4$ condensation temperature (approximately 80 K for 2% methane mole fraction near 1.6 bar). At the minimum temperature of 53 K, located at the tropopause near 100 mbar, a methane mole fraction of 10^{-5} just saturates the atmosphere. Above this level atmospheric temperatures increase in the region of the atmosphere known as the stratosphere, so that the atmosphere could support a higher CH$_4$ concentration there. However, the nearest source of methane, the condensate cloud at about 1.0–1.6 bar, is virtually unreachable: convective processes are not powerful enough to carry cloud material upward over ~ 100 km from the cloudbase to the tropopause, as has been postulated, for example, for Neptune (see Neptune: atmosphere). Thus in the lower stratosphere the methane concentration is that allowed by the tropopause, or 'cold trap', temperature, i.e. 10^{-5}. Molecular diffusion is the primary mechanism which equilibrates the lower stratospheric concentration with that at the cold trap.

In the stratosphere, extending upward from approximately 100 mbar to 0.1 mbar, chemical composition is determined largely by methane photochemistry. Ethane (C_2H_6), acetylene (C_2H_2), ethylene (C_2H_4) and polyacetylenes ($C_{2n}H_2$, $n = 2,3,4...$) are the most abundant stable photolysis products generated from methane (Atreya, 1983) near the 0.1-mbar level. Voyager ultraviolet spectrometer (UVS) measurements indicate an ethane mixing ratio of several times 10^{-8} (Herbert et al., 1987), while ground-based thermal infrared observations impose an upper limit of 2×10^{-8} (Orton et al., 1987). IUE, Voyager UVS and ground-based thermal infrared emission measurements all indicate an acetylene mixing ratio of ~ 1×10^{-8} near 0.3 mbar (cf. Caldwell, Wagener and Fricke, 1988). The UVS also determined a methane mixing ratio of $1-3 \times 10^{-7}$ in this region (Yelle et al., 1989). This is the lowest concentration found in the upper stratosphere of any Jovian planet, thus indicating that vertical mixing on Uranus is the weakest among the giant planets.

Abundances in the upper atmosphere

Above the ~ 50-μbar level hydrocarbons are virtually absent, as empirically determined by the presence of the Raman-scattered Lyman α feature at 1280 Å (Yelle et al., 1987a), consistent with the sluggish vertical transport mentioned above. Consequently, hydrogen photochemistry dictates the composition. Indeed, the rate of atomic hydrogen production (~ 3×10^7 cm^{-2} s^{-1}, with an uncertainty factor of two) is consistent with the rate predicted by UV photochemistry (Strobel et al., 1991). Above the 5-μbar level the relative abundance of helium falls rapidly, decreasing to 1% of the atmosphere near 0.1 μbar and 0.1% near 0.005 ±bar (Figure U14).

Atomic hydrogen is the major constituent above the ~ 7500-km, 10^{-6} μbar level, although substantial production of H occurs down to ~ 10 μbar. Generally, atomic hydrogen is the end result of ion chemistry in an H_2 atmosphere, and is formed principally by solar extreme ultraviolet (EUV) ionization of H_2 and He followed by ion chemistry and plasma recombination (Strobel et al., 1991).

Atomic hydrogen is also subject to ionization by ultraviolet light, resulting in a large population of free electrons. This results in an extended ionosphere ranging upward to 10 000 km above the 1-bar pressure level with electron densities in excess of 100 cm^{-3}, as detected by the Voyager Radio Occultation Experiment (Tyler et al., 1986). Sharp ionization layers with electron concentrations up to 10^5 cm^{-3} are superimposed on the main ionospheric profile.

Voyager solar and stellar occultation experiments revealed that the thermosphere near the 10^{-5} μbar level some 5000 km above the 1-bar level is surprisingly warm (T ~ 750 K; Herbert et al., 1987), indicating a total heating rate, peaking near the 3-μbar level, of 0.4 erg cm^{-2} s^{-1}, orders of magnitude larger than the solar extreme ultraviolet (10^{-3} erg cm^{-2} s^{-1}) and auroral (2×10^{-2} erg cm^{-2} s^{-1}) inputs. Concurrently, the upper atmosphere emits copious amounts of UV light in the Lyman alpha line of atomic hydrogen at about 1216 Å. Energy sources responsible for both the UV emission and the anomalously warm upper atmospheric temperatures are uncertain, although gravity wave dissipation (Hinson and Eshleman, 1988), solar fluorescence (Yelle et al., 1987b) and a magnetic dynamo mechanism (Clarke, Hudson and Yung, 1987) have been suggested.

Cloud structure

Many of the trace gases found in the Uranian stratosphere and upper troposphere condense to form hazes and clouds (Figure U15). Phase equilibrium calculations (e.g. Atreya, 1983) indicate that ethane, diacetylene and acetylene condense to form discrete layers of aerosols as they diffuse downward in the temperature-inverted stratosphere. Spectroscopic (Trafton, 1987), spectrophotometric (cf. Smith et al., 1986) and limb imaging (Pollack et al., 1987) data indicate that aerosol hazes are indeed present in the lower stratosphere. The hazes, however, are not the white ices expected for pure hydrocarbon condensates. Instead, they are rather dark, exhibiting a visual imaginary index of refraction of ~ 0.01 (Pollack et al., 1987). One conjecture is that ultraviolet-induced polymerization produces dark particles, a mechanism which appears to be consistent with the dark particles which are also found on Neptune (Baines and Smith, 1990). These submicron-sized aerosols particles precipitate out of the stratosphere in 10–100 years.

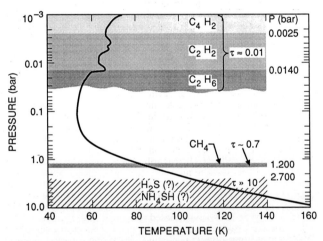

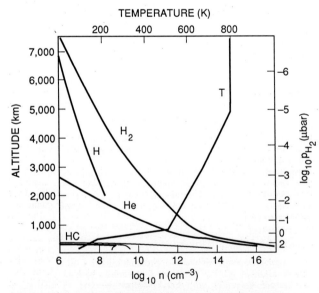

Figure U14 Number density and temperature profiles of the extreme upper atmosphere derived from the UVS occultation measurements as a function of altitude. Altitudes are referenced to the 1-bar level, located a distance of 25 550 km from the planet's center. HC denotes hydrocarbons in order of increasing abundance, C_2H_2, C_2H_6 and CH_4. The He profile is an extrapolation of the IRIS measurements from the tropopause (Hanel et al., 1986). (Figure from Strobel et al., 1991, after Herbert et al. 1987.)

Figure U15 Thermal and aerosol structure of the Uranian stratosphere and upper troposphere. Hydrocarbon condensates in the stratosphere are the end products of methane photochemistry occurring near the microbar level. Diffusion transports photochemically generated hydrocarbon gases downward to the colder environment at 1–10 mbar pressure where condensation for a number of individual species occurs, as shown. (The condensation pressure level for diacetylene, not shown, occurs near 0.15 μbar.) These hydrocarbon aerosols precipitate downward throughout the stratosphere and upper troposphere, evaporating near the 1-bar level. Methane condensation forms a physically and optically thin cloud near 1.2 bar. An optically thick cloud, most probably comprised of condensed hydrogen sulfide (H_2S), occurs near the 2.7-bar level. The thermal profile reported by the Voyager Radio Subsystem (RSS) occultation experiment (Lindal et al., 1987) is shown with an adiabatic extrapolation into the deep atmosphere.

In the troposphere, volatile species (CH_4, NH_4SH, H_2S and H_2O) are expected to form discrete condensation cloud layers as they are transported upward nearly adiabatically and cooled. Hydrocarbon aerosols formed in the stratosphere sediment down to this region as well. As they descend downward to sufficiently high temperatures, they are pyrolized back to CH_4.

Phase equilibrium calculations suggest that a cloud should form near $P = 1.2$ bar as CH_4, transported upward from the deeper troposphere by convection, cools and condenses. Yet, as observed from the ground and by the Voyager Radio Science Subsystem, this cloud is surprisingly thin, both optically and physically (Baines and Bergstralh, 1986; Lindal et al., 1987), given that methane comprises ~ 15% of the mass of the atmosphere just below the condensation level. An optical depth $0.4 < \tau < 1.0$ is found by Baines and Bergstralh (1986), more than an order of magnitude less than for terrestrial stratus clouds. This might be further evidence of weak vertical mixing: condensation and precipitation remove methane more rapidly than it can be replenished by upward transport.

The deepest detectable visible cloud is near $P = 2.7$ bar, as determined by the profiles of the H_2 4–0 and 3–0 quadrupole lines and a methane line at 6818 Å (cf. Baines and Bergstralh, 1986). H_2S is a thermodynamically plausible candidate consistent with the sluggish vertical mixing and the observed lack of NH_3 vapor (which would otherwise react with H_2S to form NH_4SH), but no positive identification has been made. The aerosol is bright in the blue–green portion of the spectrum, but darkens significantly at wavelengths longward of 0.6 μm.

Circulation and dynamics

Horizontal circulation

Two principal energy sources drive dynamics on the Jovian planets: the Sun and internally generated heat. Radiant energy from the Sun is absorbed by atmospheric gases (particularly CH_4) and light-absorbing aerosols in the stratosphere and upper troposphere, thereby heating the atmosphere. Heat released from the deep interior by a variety of processes (ongoing gravitational collapse, helium 'rain', radiogenic heating) may be absorbed at upper atmospheric levels as well. For most of the Jovian planets, the internal heat input dominates the external solar input throughout the atmosphere. Thus dynamics for Jupiter, Saturn and Neptune are largely governed by the internal heat source.

The Sun and internal heat drive dynamics by causing the atmospheric temperature to vary in latitude and longitude, thereby producing spatial gradients in atmospheric pressure forcing the air to move. Temperature variations may be caused by uneven solar heating, as on Earth, or by uneven internal heating. Since the mean internal heat flux dominates the external solar flux in most of the Jovian planets, spatial variations in that flux are likely to be the primary dynamics driver. However, little is known about this spatial distribution. In particular, the location and intensity of high-flux 'hot spots', if any, are unknown. Thus, dynamics on most of the Jovian planets is thought to be influenced primarily by a somewhat complicated internal heat flux distribution.

Uranus, however, is a different, simpler case. With a mean global internal heat flux of at most 14% of the mean solar flux at Uranus, the atmospheric circulation is dominated by the external solar flux. Indeed, Voyager and ground-based observations are consistent with the complete absence of internal heat flux (Pollack et al., 1986; Pearl et al., 1990). Thus the dominant atmospheric circulation on Uranus is that which is forced by the uneven distribution of sunlight from pole to equator.

Uranus' large obliquity produces an unusual distribution of sunlight. Rotating virtually on its side in relation to the plane of the ecliptic, Uranus experiences nearly direct sunlight over both poles as well as the equator during a Uranian year. Indeed, even though both poles experience a night half a Uranian year long, they nevertheless receive annually more than 50% more sunlight per unit area than the equator.

The dominant circulation in Uranus redistributes relatively warm polar air over the planet. This can be inferred directly from the Voyager IRIS (Infrared Interferometer Spectrometer) observations, which determined temperatures at several atmospheric levels over a wide range of latitudes. If the planet could be treated as a black sphere in equilibrium with the received insolation, then without this

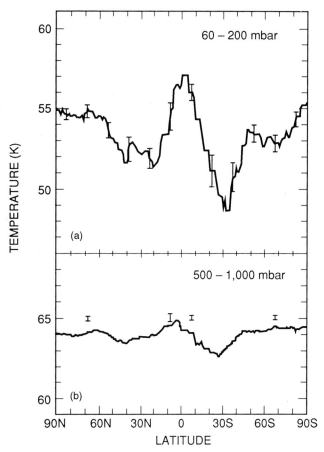

Figure U16 Temperature versus latitude inferred from Voyager 2 IRIS measurements. 15° latitude running mean temperatures for (a) a layer between approximately 60 and 200 mbar, and (b) temperatures for the 500–1000 mbar layer. The vertical bars show examples of 1-σ statistical error estimates. Temperatures are unusually uniform in the Uranian troposphere (b), attesting to the relatively uniform mean annual solar illumination over the planet and to the efficient transport of air away from the more greatly illuminated poles. (From Conrath et al., 1991.)

redistribution the planet would have an estimated equator-to-pole temperature gradient

$$\Delta T/T_e = [(\pi/2)^{1/4} - 1] = 0.12$$

or $\Delta T = 7°$ for $T_e = 59$ K, where T_e is the solar equilibrium temperature. Yet, as shown in Figure U16, Voyager IRIS found almost no equator-to-pole contrast in the 0.5–1.0 bar region where most of the visible sunlight is deposited in methane gas, and indeed found that in the stratosphere, from 60 to 200 mbar, equatorial temperatures are warmer than at the poles (Hanel et al., 1986). Thus air is effectively circulated over the globe of Uranus, allowing the upper troposphere in particular to reach the nearly isothermal condition to which all planetary atmospheres strive.

The temperature field measured by Voyager IRIS was used to determine quantitatively the magnitude of Uranus's circulation system through the thermal wind equation (Hanel et al., 1986). This equation relates the measured meridional temperature gradient (that is, equator-to-pole variations in the longitudinally averaged mean temperature) to vertical variations in the mean zonal winds (that is, the mean variation in altitude of the east–west component of wind speed). As depicted in Figure U17, Hanel et al. (1986) found that this equation implies the existence of a system of large-scale zonal jets blowing in the prograde direction at mid-latitudes (similar to the mid-latitude jets observed on Jupiter and Saturn) but strongly retrograde at low latitudes (opposite to the equatorial jets observed on Jupiter and Saturn). Such a distribution of winds was observed directly by

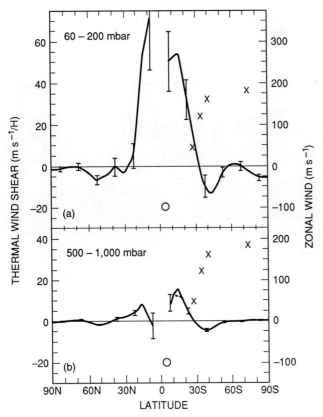

Figure U17 Mean zonal wind speeds and thermal wind shear derived from Voyager 2 imaging and infrared spectroscopy. Zonal winds of Smith *et al.* (1986) are depicted by crosses and refer to the scale on the right-hand side of the figure. The open circle represents a wind estimate from Voyager radio occultation measurements. The wind shear (solid curves with 1 σ error bars) are derived from application of the thermal wind equation to the latitudinal temperature profiles displayed in the previous figure. (From Conrath *et al.*, 1991.)

Voyager's cameras (the Imaging Subsystem, or ISS), which tracked various wind-blown cloud features (Smith *et al.*, 1986). Zonal winds reach peak prograde velocities of about 200 m s^{-1} between 50 and 60° S latitude, decline to zero at 20° S latitude, and reach a peak retrograde velocity of 100 m s^{-1} at the equator.

The thermal wind equation also implies that the vertical wind shear opposes the zonal wind, damping it more strongly with increasing altitude. Near the region of maximum zonal velocities at a latitude of about 45°, Flasar *et al.* (1987) find a vertical scale for decay to zero wind speed (as measured relative to the angular rotational speed of the planetary interior) of 10 scale heights or about 300 km.

Vertical dynamics

For the Jovian planets, quantitative constraints on the vertical circulation are inferred largely from (1) the distribution of disequilibrium chemical species, and (2) the vertical temperature structure. The presence in the upper atmosphere of disequilibrium species (that is, species which are chemically generated at high pressures and temperatures, typically 1000 bar and 1000 K, but which cannot sustain themselves at low pressures and temperatures for significant periods of time) is a direct indicator of vertical upwelling. In particular, the presence of phosphine (PH_3), germane (GeH_4) and arsine (AsH_3) in the upper tropospheres of Jupiter and Saturn, in concentrations some 10^4–10^{12} times greater than those predicted by equilibrium chemistry, is a direct indication of active vertical dynamics in these planets. However, this technique is inapplicable to Uranus due to the planet's extremely cold environment, which prevents detection of the infrared signatures of these molecules.

As noted earlier, the two primary states of hydrogen, the *ortho* and *para* states, allow hydrogen to be treated as a disequilibrium gas tracer of vertical dynamics. The relative abundance of the two states, in particular, the departure from the expected equilibrium value, indicates the timescale for vertical transport. Baines and Bergstrahl (1986) find a slightly non-equilibrium abundance which constrains vertical timescales to about 10 years. On the other hand, this spectroscopically deduced *ortho/para* ratio is inconsistent with the observed temperature lapse rate. (The lapse rate depends on the specific heat of hydrogen, C_p, which depends on the mixture of *ortho* and *para* components, each of which has its own C_p). One explanation, developed by Gierasch and Conrath (1987), is that the atmosphere is comprised of numerous superposed thin strata, which exchange heat but not material between them. The residence time of a parcel of gas within a particular stratum is long compared to the *ortho/para* conversion time, so that, on average each stratum is nearly in *ortho/para* equilibrium. However, the convective overturn time within a stratum is short compared with the rate of *ortho–para* conversion, so that the lapse rate within the stratum is not that predicted by an equilibrium *ortho/para* distribution spanning the stratum's range of temperatures. Such a stratified stability could be promoted by condensation of methane, a plentiful constituent with a mean molecular weight significantly heavier than the uncondensed constituents (i.e. hydrogen and helium). The condensation and subsequent precipitation of methane reduces the mean molecular weight with altitude in the CH_4-condensing region of the atmosphere, leading to a stepwise stratification of mean molecular weight that increases stability. This situation is similar to Earth's oceans, where stability is enhanced by an increase in salinity – and hence mean molecular weight – with depth.

Kevin H. Baines

Bibliography

Allison, M., Beebe, R.F., Conrath, B.J. *et al.* (1991) Uranus atmospheric dynamics and circulation, in *Uranus* (eds J.T. Bergstralh, E.D. Miner and M.S. Matthews). Tucson: University of Arizona Press, pp. 253–95.

Atreya, S.K. (1983) Photochemistry and condensation in the Uranus atmosphere. *Bull. Am. Astron. Soc.*, **15**, 859.

Atreya, S.K., Sandel, B.R. and Romani, P.N. (1991) Photochemistry and vertical mixing, in *Uranus* (eds J.T. Bergstralh, E.D. Miner and M.S. Matthews). Tucson: The University of Arizona Press, pp. 110–46.

Baines, K.H. and Bergstralh, J.T. (1986) The structure of the Uranian atmosphere: constraints from the geometric albedo spectrum and H_2 and CH_4 line profiles. *Icarus*, **65**, 406–41.

Baines, K.H. and Smith W.H. (1990) The atmospheric structure and dynamical properties of Neptune derived from ground-based and IUE spectrophotometry. *Icarus*, **85**, 65–108.

Broadfoot, A.L. Atreya, S.K., Bertaux, J.L. *et al.* (1989) Ultraviolet spectrometer observations of Neptune and Triton. *Science*, **246**, 1459–66.

Caldwell, J.T., Wagener, R. and Fricke, K. (1988) Observations of Neptune and Uranus below 2000 Å with the IUE. *Icarus*, **74**, 133–40.

Clarke, J.T., Hudson, M. and Yung, Y.L. (1987) The excitation for the far ultraviolet electroglow emissions on Uranus, Saturn and Jupiter. *J. Geophys. Res.*, **92**, 15138–47.

Conrath, B.J. and Gierasch, P.J. (1984) Global variations of the *para* hydrogen fraction in Jupiter's atmosphere and implications for dynamics in the Jovian planets. *Icarus*, **57**, 184–204.

Conrath, B.J., Gautier, D., Hanel, R.A. *et al.* (1987) The helium abundance of Uranus from Voyager measurements. *J. Geophys. Res.*, **92**, 15003–10.

Conrath, B.J., Pearl, J.C., Appleby, J.F. *et al.* (1991) Thermal structure and energy balance of Uranus, in *Uranus*, (eds J.T. Bergstralh, E.D. Miner and M.S. Matthews). Tucson: University of Arizona Press, pp. 204–52.

de Pater, I. and Gulkis, S. (1988) VLA observations of Uranus at 1.3–20 cm. *Icarus*, **75**, 306–23.

Fegley, B., Jr, Gautier, D., Owen, T. and Prinn, R.G. (1991) Spectroscopy and chemistry of the atmosphere of Uranus, in *Uranus* (eds J.T. Bergstralh, E.D. Miner and M.S. Matthews). Tuscon: University of Arizona Press, pp. 147–203.

Flasar, F.M. Conrath, B.J., Gierasch, P.J. and Pirraglia, J.A. et al. (1987) Voyager infrared observations of Uranus' atmosphere: thermal structure and dynamics. *J. Geophys. Res.*, **92**, 15011–8.
Gierasch, P.J., and Conrath, B.J. (1987) Vertical temperature gradients on Uranus: implications for layered convection. *J. Geophys. Res.*, **92**, 15019–29.
Gulkis, S. and de Pater, I. (1984) A review of the millimeter and centimeter observations of Uranus, in *Uranus and Neptune* (ed. J.T. Bergstralh) NASA CP-2330, pp. 225–62.
Gulkis, S., Jannsen, M.J. and Olsen, E.T. (1978) Evidence for depletion of ammonia in the Uranus atmosphere. *Icarus*, **34**, 10–9.
Hanel, R.A., Conrath, B.J., Flasar, F.M. et al. (1986) Infrared observations of the Uranian system. *Science*, **233**, 70–4.
Herbert, F.L., Sandel, B.R., Broadfoot, A.L. et al. (1987) The upper atmosphere of Uranus: EUV occultations observed by Voyager 2. *J. Geophys. Res.*, **92**, 15093–109.
Hinson, D.P. and Eshleman, V.R. (1988) Evidence of inertia-gravity waves in the stratosphere of Uranus derived from Voyager 2 radio occultation data. *Bull. Am. Astron. Soc.*, **20**, 822.
Lewis, J.S. and Prinn, R.G. (1980) Kinetic inhibition of CO and N_2 reduction in the solar nebula. *Astrophys. J.*, **238**, 357–64.
Lindal, G.F., Lyons, J.R., Sweetnam, D.M. et al. (1987) The atmosphere of Uranus: results of radio occultation measurements with Voyager 2. *J. Geophys. Res.*, **92**, 14987–5001.
Massie, S.T. and Hunten, D.M. (1982) Conversion of *para* and *ortho* hydrogen in the Jovian planets. *Icarus*, **49**, 213–26.
Orton, G.S., Griffin, M.J., Ade, P.A.R. et al. (1986) Submillimeter and millimeter observations of Uranus and Neptune. *Icarus*, **67**, 239–304.
Orton, G.S. Aitken, D.K., Smith, C. et al. (1987) The spectra of Uranus and Neptune at 8–14 and 17–23 μm. *Icarus*, **70**, 1–12.
Pearl, J.C., Conrath, B.J., Hanel, R.A. et al. (1990) The albedo, effective temperature, and energy balance of Uranus, as determined from Voyager IRIS data. *Icarus*, **84**, 12–28.
Podolak, M., Hubbard, W.B. and Stevenson, D.J. (1991) Models of Uranus' interior and magnetic field, in *Uranus* (eds J.T. Bergstralh, E.D. Miner, and M.S. Matthews. Tucson: University of Arizona Press. pp. 29–61.
Pollack, J.B., Rages K., Baines, K.H. et al. (1986) Estimates of the bolometric albedos and radiation balance of Uranus and Neptune. *Icarus*, **65**, 442–66.
Pollack, J.B. et al. (1987) Nature of stratospheric haze on Uranus: evidence for condensed hydrocarbons. *J. Geophys. Res.*, **92**, 15037–65.
Smith, B.A., Soderblom, L.A., Banfield, D. et al. (1986) Voyager 2 in the Uranian system: imaging science results. *Science*, **233**, 43–64.
Strobel, D.F., Yelle, R.V., Shemansky, D.E. et al. (1991) The upper atmosphere of Uranus, in *Uranus*, (eds J.T. Bergstralh, E.D. Miner and M.S. Matthews). Tucson: University of Arizona Press, pp. 65–109.
Trafton, L.M. (1987) Uranus' (3–0) H_2 quadrupole line profiles. *Icarus*, **70**, 13–30.
Trafton, L.M. and Ramsay, D.A. (1980) The D/H ratio in the atmosphere of Uranus: detection of the R5(1) line of HD. *Icarus*, **41**, 423–9.
Tyler, G.L., Sweetnam, D.N., Anderson, J.D. et al. (1986) Voyager 2 radio science observations of the Uranian system: atmospheres, rings, satellites. *Science*, **233**, 79–84.
Yelle, R.V., Doose, L.R., Tomasko, M.G. and Strobel, D.F. et al. (1987a) Analysis of Raman scattered Ly-α emissions from the atmosphere of Uranus. *Geophys. Res. Lett.*, **14**, 483–6.
Yelle, R.V., McConnell, J.C., Sandel, B.R. and Broadfoot, A. (1987b) The dependence of electroglow on the solar flux. *J. Geophys. Res.*, **92**, 15110–24.
Yelle, R.V., McConnell, J.C., Strobel, D.F. and Doose, L.R. et al. (1989) The far-ultraviolet spectrum of Uranus: results from the Voyager encounter. *Icarus*, **77**, 439–56.

Cross references

Atmosphere
Jupiter: atmosphere
Neptune: atmosphere
Radiative transfer
Radio astronomy
Radio science
Saturn: atmosphere
Spectroscopy: atmosphere

URANUS: MAGNETIC FIELD AND MAGNETOSPHERE

Prior to the Voyager encounter in January 1986, Uranus was expected to possess an unusual magnetosphere quite unlike that of the Earth, because the spin axis of Uranus is nearly in the plane in which Uranus orbits the Sun (Siscoe, 1975). If the intrinsic magnetic field of Uranus had been nearly aligned with the rotational axis, as the planets previously visited were, the polar axis of the magnetosphere, or the polar cusp as it is called, would have been aligned with the solar wind flow as Voyager flew by the planet. Ironically, the magnetic axis of the intrinsic magnetic field of Uranus was far from spin axis-aligned, so that the solar wind blew nearly perpendicular to the magnetic axis, as it does at Mercury, Earth, Jupiter and Saturn. Thus, while Uranus has an unusual intrinsic magnetic field, the resulting magnetosphere was found to be very Earth-like.

Planet and interior

Uranus has an equatorial radius of 25 600 km, less than half that of Saturn. It has a mean density of 1.25 g cm^{-3} and rotates with a period of 17.2 h. Its rotational axis is inclined at an angle of 97.9° to the pole of its orbital plane, in such a direction that, during the Voyager 2 flyby, the rotational axis was pointing nearly toward the Sun. The Uranian rings and satellites therefore orbit in planes almost orthogonal to the ecliptic plane. The interior is believed to consist of three principal layers, a rocky core, a water-ice and ammonia layer, and a gaseous envelope of hydrogen and helium, comprising the outer 30% of the planet.

Magnetic field

The magnetic dipole moment of Uranus is tilted at an angle of 59° to the spin axis of the planet. The extrapolated near-equatorial surface field is 23 μT, corresponding to a magnetic moment (equatorial surface field times radius cubed) of 3.9×10^{17} T m^3, close to 50 times greater than the terrestrial magnetic moment. The contribution of the quadrupole moment to the surface magnetic field is large, almost comparable to the contribution from the dipole moment. This relative contribution is in fact only exceeded by that of Neptune. The higher moments of the field are essentially unconstrained by the Voyager observations.

Magnetosphere

The Uranian magnetosphere is very similar to the terrestrial magnetosphere. There is a bow shock that deflects the supersonic flow of the solar wind in front of the magnetospheric cavity and a magnetic tail extending far downstream. The forward part of the magnetosphere extends to approximately 25 planetary radii and the bow shock to about 33 planetary radii. Evidence from the energetic particles fluxes and wave amplitudes in the magnetosphere indicate that it is somewhat less active than that of the Earth. However, there is some evidence for magnetospheric dynamics, or a magnetospheric 'substorm,' in the energetic particle data (Mauk et al., 1987; Sittler, Ogilvie and Selesniek, 1987).

Christopher T. Russell and Janet G. Luhmann

Bibliography

Connerney, J.E.P. and Ness, N.F. (1987) The magnetic field of Uranus. *J. Geophys. Res.*, **92**, 15329–36.
Mauk, B.H., Krimigis, S.J., Keath, E.P. et al. (1987) The hot plasma and radiation environment of the Uranian magnetosphere. *J. Geophys. Res.*, **92**, 15283–308.
Russell, C.T. (1987) Planetary magnetism, in *Geomagnetism*, Vol. 2 (ed. J.A. Jacobs). New York: Academic Press, 457–23.
Siscoe, G.L. (1975) Particle and field environment of Uranus. *Icarus*, **24**, 311.

Sittler, E.C., Ogilvie, K.W. and Selesniek, R. (1987) Survey of electrons in the Uranian magnetosphere: Voyager 2 observations. *J. Geophys. Res.*, **92**, 15263–81.

Acknowledgement

This work was supported in part by the National Aeronautics and Space Administration under research grant NAGW-2573.

Cross references

Magnetism
Magnetospheres of the outer planets

URANUS: RING SYSTEM

The Uranian ring system was discovered by accident in 1977, during the observation of a stellar occultation (the passage of Uranus in front of a star). A symmetric pattern of sharp dips in the stellar signal on either side of the planet revealed the presence of at least five narrow, almost opaque rings (Elliot, Dunham and Mink, 1977; Millis, Wasserman and Birch, 1977). Subsequent occultations observed in the late 1970s and 1980s, both from Earth-based telescopes and with instruments on the Voyager spacecraft, led to the discovery of five additional rings, while providing an abundance of data on the radii, widths and transparencies of the rings (Elliot and Nicholson, 1984; Holberg et al., 1987; Gresh et al., 1989; Colwell et al., 1990). In order of increasing distance from Uranus, the ten rings are designated 6, 5, 4, α, β, η, γ, δ, λ and ε, the somewhat curious nomenclature reflecting the naming of the original five rings (α–ε) as well as later discoveries.

Images of the Uranian rings were obtained by the Voyager spacecraft in 1986 and are now routinely made with near-infrared cameras on Earth-based telescopes, although the latter cannot separately resolve the individual rings. In addition to the ten narrow rings, the spacecraft images revealed several broader, less opaque ring structures (discussed below). Elliot and Nicholson (1984), French et al. (1991) and Esposito et al. (1991) review various aspects of the Uranian ring system.

Dimensions and orbits

One of the first of many surprises provided by the Uranian rings was that they are not quite circular and do not lie exactly in Uranus' equatorial plane. The occultation data for individual rings are, in

Figure U18 Complete ring system of Uranus. The difficult-to-discern lambda ring is inside the bright outer epsilon ring.

Table U4 Uranian ring parameters

Ring	Semimajor axis a (km)	Eccentricity e	Inclination i (deg)	Width (km)	Optical depth τ
6	41837.2	0.00101	0.062	~ 1.6	~ 0.3
5	42234.8	0.00190	0.054	1.6→2.8	0.6→0.4
4	42570.9	0.00106	0.032	~ 2.4	~ 0.3
α	44718.5	0.00076	0.015	3.6→10.7	0.9→0.3
β	45661.0	0.00044	0.005	5.1→11.2	0.4→0.2
η	47175.9	< 0.00001	< 0.002	~ 1.5	~ 0.3
γ	47626.9	0.00011[a]	< 0.003	0.6→3.8	> 3→1.1
δ	48300.1	< 0.00001[a]	0.001	2.7→7.0	0.9→0.3
λ	50032.9			~ 2.0	~ 0.13
ε	51149.3	0.00794	< 0.001	20→96	2.2→0.5

[a] Ring also exhibits a normal mode oscillation (see text).

most cases, fitted to a high degree of accuracy by Keplerian ellipses, i.e. the shapes of the rings are similar to orbits followed by satellites about Uranus' center of mass. Table U4 lists the semimajor axis a, eccentricity e and inclination to Uranus' equatorial plane i determined for each ring from 11 years of occultation observations (French et al., 1988). The eccentricities range from zero (rings η and λ) to a maximum of 0.0079 for the ε ring, while the inclinations range from essentially zero to 0.062°. The deviations of the observed ring radii from these simple models, except for γ and δ, are 300–600 m, or about one part in 10^5.

Occultation data also provide estimates of the widths and optical depths of the rings, and of the variations in these two quantities around each ring. Widths range from less than 1 km (γ ring) to a maximum of 96 km for the widest part of the ε ring. The optical depth τ of a ring is defined by the equation

$$I = I_0 e^{-\tau/\mu} \qquad (U15)$$

where I_0 is the intensity of starlight incident at an angle θ from the normal to the ring plane, I is the emergent intensity and $\mu = \cos \theta$. It is thus a measure of the fractional transmission of starlight through the ring at normal incidence. Table U4 also gives the observed range of widths and τ for each ring; in general a ring is most opaque at its narrowest point and least opaque at its widest, as is to be expected if the total amount of ring material per unit length of ring is constant. In the case of the α, β and ε rings, the width increases linearly with radius from a minimum at the ring's periapse to a maximum at apoapse, a result which implies that the inner and outer edges of these three rings are aligned, confocal Keplerian ellipses with slightly different eccentricities (Nicholson, Matthews and Goldreich, 1982). The η ring, in addition to the narrow feature listed in the table, has a 55 km wide exterior component with an optical depth of ~ 0.015. The δ ring shows a similar, but narrower, interior component (French et al., 1991).

The Voyager occultation observations revealed considerable internal structure within all ten rings, down to resolution limits of 10–100 m (Gresh et al, 1989; Colwell et al., 1990). Several of the ring edges were found to be at most 50 m wide, while vertical thicknesses are inferred from dynamical and other arguments to be in the range 15–30 m. Although most rings show substantial variations in their internal structure with longitude, the radial optical depth profile of the ε ring seems to expand and contract uniformly between its 20-km minimum width at periapse to the 96-km width at apoapse, behaving rather like an accordion.

A single long-exposure Voyager image taken at a phase angle (Sun–target–observer angle) of 172.5° revealed that the 'gaps' between the main rings are in fact not completely empty: they contain a complex series of narrow and broad lanes of widely dispersed fine material. The inferred maximum optical depth of the dust bands is 1.5×10^{-5}, making them quite undetectable in occultation observations. Although some of the dust structures are associated with the ten previously known rings, most notably the λ ring, many are not (Esposito et al., 1991). Interior to ring 6, a badly smeared image at 90° phase angle revealed the presence of a 2500 km wide ring of intermediate optical depth (~ 10^{-3}), which has been designated 1986U2R.

Dynamics

If Uranus were a perfectly spherical body, an elliptical orbit (or ring) would retain a fixed orientation in space. In reality the planet's oblate figure, induced by its rapid rotation, instead causes both a slow precession of a ring's apsidal line (the line connecting periapse and apoapse of an elliptical orbit) and a slow regression of the plane of the ring relative to the planet's equator. Both effects are readily observed in the Uranian rings' orbits, the precession of the ϵ ring's apsidal line amounting to $1.36325°$ day^{-1} while that of ring 6 is $2.7616°$ day^{-1}. The measured precession rates have led to an accurate determination of the quadrupole (J_2) and octupole (J_4) terms in Uranus' gravity field, which are important in constraining models of the planet's interior.

Precession also poses a problem for the continued existence of any eccentric or inclined ring, as the tendency of the inner parts of such a ring to precess (or regress) faster than the outer parts must rapidly circularize the ring unless counteracted by some other force. The timescale for the circularization of the ϵ ring by this process is less than 200 years. Although it has been shown (Goldreich and Tremaine, 1979) that gravitational attraction between the ring particles is capable of preventing differential precession, given an appropriate ring mass, this so-called 'self-gravity' model seems to be in conflict with several other lines of evidence (French *et al.*, 1991). It is possible that interparticle collisions could also act to counteract differential precession, but a successful quantitative model has yet to be constructed on this basis.

A more fundamental effect of interparticle collisions is a slow radial spreading of any initially narrow ring. Absent an external confining force, this spreading is an inevitable consequence of angular momentum conservation and the dissipation of kinetic energy in inelastic collisions. For the ϵ ring, the timescale for a doubling in mean width from the present 60 km to 120 km is at most 10^6 years. A proposal that a pair of small satellites orbiting on either side of the ring could provide the necessary confining forces (Goldreich and Tremaine, 1979) was elegantly confirmed by the discovery in Voyager images of just such a pair of 'shepherd' satellites, Cordelia and Ophelia, straddling the ϵ ring. Closer investigation of the orbits of these bodies showed that the precise locations of the ring's edges are determined by orbital resonances with the satellites, and also suggested that the same satellites may play roles in confining the δ and γ rings (Porco and Goldreich, 1987). It is suspected that additional shepherds exist among the Uranian rings, but are too faint to have been captured in the spacecraft images.

Two of the Uranian rings have been found to exhibit distortions from circular orbits which do not fit simple Keplerian ellipses. The γ ring's mean radius oscillates with an amplitude of ± 5 km and a period of 7.5421 h, while the δ ring's shape is that of a central ellipse (ie. the planet is at the center rather than at one focus) which rotates with a period of 15.3595 h (French *et al.*, 1988). These apparently bizarre motions in fact represent 'normal modes' of oscillation for a narrow ring, and arise from a careful phasing of the underlying Keplerian orbits followed by individual particles. It is likely that such modes arise spontaneously from gravitational instabilities in narrow, dense rings, but it is not known why a particular mode is favored by a particular ring, or what limits its amplitude.

Particle properties

Relatively little is known at present of the sizes and especially the composition of the particles which comprise the Uranian rings. The self-gravity model implies average particle diameters of 50 cm in the ϵ ring, 5 cm in the α and β rings and 15 cm in the δ ring, though the validity of this model is now in some doubt, as discussed above. Comparisons of the ring optical depths at microwave wavelengths (3 and 13 cm) with those in the near infrared and ultraviolet suggest mean particle diameters of 1.5 m or larger (Esposito *et al.*, 1991), although the precise value depends on the assumed steepness of the size distribution. (Typically, planetary rings are believed to contain a wide range of particle sizes, with the number of particles per unit volume, $n(d)$ increasing with decreasing diameter d roughly as d^{-3} in a state of collisional equilibrium.) In nine of the rings the fraction by surface area of tiny, dust-sized particles is ≤ 0.002; the exception is the very narrow λ ring, which may be composed almost entirely of particles in the micrometer size range. The thin inter-ring dust lanes are also likely to consist largely of such tiny particles, based on their light-scattering behavior (Ockert *et al.*, 1987).

Figure U19 Backlit, long-exposure view shows the distribution of fine particles throughout the ring system. Discontinuous short arcs are elongated star images produced by the motion of the spacecraft.

The reflectance spectrum of the Uranian rings between the violet (wavelength 0.41 μm) and near infrared (2.0–4.0 μm) appears to be flat and featureless, with an uncommonly low particle single-scattering albedo (i.e. the reflectivity integrated over all scattering directions) of about 0.015 (Elliot and Nicholson, 1984; Porco *et al.*, 1987). In the absence of any diagnostic spectral features, it is impossible to infer a unique composition, but the low albedo and flat spectrum are reminiscent of carbonaceous chondrite meteorites and C-type asteroids. Although carbon-rich materials of some kind seem the most likely surface materials, it has been suggested that the interior of a ring particle is likely to be an intimate mixture of H_2O, NH_3 and CH_4 ices, probably in clathrate form. When exposed at the particle's surface to bombardment by energetic atomic particles in the Uranian magnetosphere, the methane ice will be dissociated to form a layer composed primarily of elemental carbon (Cheng and Lanzerotti, 1978).

Origin and evolution

A common feature of the ring systems of all four giant planets is their close association with retinues of small (10–100 km radius) satellites. In many cases, such as the ϵ ring shepherds, there is obviously a close dynamical link between the rings and these nearby satellites, and in the case of Saturn's Encke gap we find a satellite (Pan) actually embedded within a broad ring. This association has led naturally to the hypothesis that the rings and inner satellites are also genetically related, although it is not clear whether rings represent primordial material which failed to accrete into satellites, or if they are the remnants of satellites destroyed by collisions with asteroids or comets (Harris, 1984). In the case of the Uranian system, nine small satellites ranging in radius from 13 km (Cordelia) to 77 km (Puck) orbit within or immediately exterior to the rings.

Several lines of evidence suggest that the Uranian ring system is currently evolving, and that it may indeed have taken on its present form only in the last 100 Ma or so (Esposito *et al.*, 1991). Voyager observations revealed that Uranus's upper atmosphere, or exosphere, is unexpectedly hot, with its tenuous outer parts extending out to the radii of the rings. Aerodynamic drag from this extended atmosphere on the inner rings makes it problematic whether (unseen) shepherd satellites can confine the α and β rings for more than about 50 Ma. The same drag limits the lifetimes of the tiny particles in the dust belts to ~ 1000 years, suggesting that they are continually replenished by micrometeoroid bombardment on unseen parent bodies orbiting between the narrow rings. On a somewhat longer timescale, collisional energy dissipation within the ϵ ring slowly forces the shepherd satellites to recede from the ring, giving up orbital energy. When augmented by atmospheric drag, which also depletes the ring's orbital energy and angular momentum, it is found that Cordelia (the inner shepherd) has receded to its current distance of 1400 km in at most 600 Ma. These arguments have led to a model (Esposito and Colwell, 1989) in which the rings and satellites are seen as elements in an evolving, coupled system, along with many presently undetected bodies in the 100 m to 10 km size range. Gradually, impacts break up the satellites into smaller bodies, which in turn serve as sources for the relatively short-lived rings and dust belts, and the whole system is seen as a 'collisional remnant' of a few larger bodies. At any given moment in this history, several rings may be present, shepherded by the larger remaining satellites.

Unsolved problems

Several outstanding gaps in our knowledge or understanding of the Uranian rings have been mentioned above. These include difficulties confronting the self-gravity model for maintenance of eccentric rings (and the absence of a viable alternative model); imperfect knowledge of the size distribution and especially the composition of the ring particles; and the long-term evolutionary history of the ring–satellite system. Although future theoretical work may throw some light on these problems, it is likely that some will remain unsolved until a future spacecraft returns to Uranus.

Philip D. Nicholson

Bibliography

Cheng, A.F. and Lanzerotti, L.J. (1978) Ice sputtering by radiation belt protons and the rings of Saturn and Uranus. *J. Geophys. Res.*, **83**, 2597–602.
Colwell, J.E. *et al.* (1990) Voyager polarimeter observations of Uranian ring occultations. *Icarus*, **83**, 102–25.
Elliot, J.L. and Nicholson, P.D. (1984) The rings of Uranus, in *Planetary Rings* (eds R. Greenberg and A. Brahic). Tucson: University of Arizona Press, pp. 25–72.
Elliot, J.L., Dunham, E.W. and Mink, D.J. (1977) The rings of Uranus. *Nature*, **267**, 328–30.
Esposito, L.W. and Colwell, J.E. (1989) Creation of the Uranus rings and dust bands. *Nature*, **339**, 637–40.
Esposito, L.W., Brahic, A., Burns, J.A. and Marouf, E.A. (1991) Particle properties and processes in Uranus' rings, in *Uranus* (eds J.T. Bergstralh, E.D. Miner and M.S. Matthews). Tucson: University of Arizona Press, pp. 410–65.
French, R.G., Nicholson, P.D., Porco, C.C. and Marouf, E.A. (1991) Dynamics and structure of the Uranian rings, in *Uranus* (eds J.T. Bergstralh, E.D. Miner and M.S. Matthews). Tucson: University of Arizona Press, pp. 327–409.
French, R.G. *et al.* (1988) Uranian ring orbits from Earth-based and Voyager occultation observations. *Icarus*, **73**, 349–78.
Goldreich, P. and Tremaine, S. (1979) Towards a theory for the Uranian rings. *Nature*, **277**, 97–9.
Gresh, D.L *et al.* (1989) Voyager radio occultation by Uranus's rings. I. Observational results. *Icarus*, **78**, 131–68.
Harris, A.W. (1984) The origin and evolution of planetary rings, in *Planetary Rings* (eds R. Greenberg and A. Brahic). Tucson: University of Arizona Press, pp. 641–59.
Holberg, J.B., Nicholson, P.D., French, R.G. and Elliot, J.L. (1987) Stellar occultation probes of the Uranian rings at 0.1 and 2.2 µm: a comparison of Voyager UVS and Earth-based results. *Astron. J.*, **94**, 178–88.
Millis, R.L., Wasserman, L.H. and Birch, P. (1977) Detection of rings around Uranus. *Nature*, **267**, 330–1.
Nicholson, P.D., Matthews, K. and Goldreich, P. (1982) Radial widths, optical depths and eccentricities of the Uranian rings. *Astron. J.*, **87**, 433–47.
Ockert, M.E., Cuzzi, J.N., Porco, C.C. and Johnson, T.V. (1987) Uranian ring photometry: results from Voyager 2. *J. Geophys. Res.*, **92**, 14969–78.
Porco, C.C. and Goldreich, P. (1987) Shepherding of the Uranian rings. I. Kinematics. *Astron. J.*, **93**, 724–9.
Porco, C.C., Cuzzi, J.N., Ockert, M.E. and Terrile, R.J. (1987) The color of the Uranian rings. *Icarus*, **72**, 69–78.

Cross references

Jupiter: ring system
Neptune: ring system
Planetary ring
Saturn: ring system
Shepherd satellite

URANUS: SATELLITE SYSTEM

The Uranian satellite system consists of 15 known moons. Earth-based telescopic observers discovered the five largest moons, while the rest were revealed by the Voyager 2 spacecraft. William Herschel (q.v.), who discovered Uranus in 1781, first observed the two outermost Uranian moons in 1789. Sixty-five years later his son named them Titania and Oberon after the king and queen of fairies in Shakespeare's *A Midsummer Night's Dream*. The English astronomer William Lassel discovered two more moons in 1851, naming them Ariel and Umbriel after characters in Pope's *Rape of the Lock*. The smallest and innermost of the moons observed from Earth was discovered by the Dutch-American Gerard Kuiper (q.v.) on photographic plates of the Uranian system in 1948. Kuiper continued the now-established tradition of naming Uranian moons after fairy characters in English drama, choosing the name Miranda from Shakespeare's *The Tempest*.

The family of Uranian moons was greatly expanded when, in late 1985 and early 1986, the Voyager 2 spacecraft discovered ten small, previously unknown satellites during its flyby of Uranus. Only one of these, discovered on 30 December 1985 and named Puck, was large enough to reveal any surface detail in the images obtained. The rest appeared only as bright points, a few pixels in diameter, whose motion from one image to another allowed their orbits about Uranus to be determined (Table U5). Names for these moons continued to be drawn from characters in English drama while features discovered on the larger moons were named after both characters and localities from the same source.

All the moons of Uranus orbit in nearly circular paths, in the same direction that Uranus rotates and, with the exception of Miranda, within 1° of its equatorial plane. All are spin–orbit coupled, i.e. their rotational and orbital periods are equal so that they keep the same face constantly toward Uranus. Most of the moons orbit outside the planetary ring system, but the two innermost small moons, Cordelia (1986U7) and Ophelia (1986U8), orbit just interior and exterior

Figure U20 The five largest moons of Uranus, demonstrating their relative sizes and reflectivities. Compiled from images returned by the Voyager 2 spacecraft 4 days before its closest approach to the Uranian system, the moons are shown in order of increasing distance from Uranus: Miranda, Ariel, Umbriel, Titania, and Oberon. (NASA image P-29464.)

Table U5 Physical and orbital data for the 15 known moons of Uranus; masses for the smaller moons have not been determined independently

	Diameter (km)	Mass ($\times 10^{20}$ kg)	Orbit radius ($\times 1000$ km)	Period (days)	Inclination (deg)	Eccentricity
Cordelia (1986U7)	52		49.75	0.335	0.14	0.0005
Ophelia (1986U8)	60		53.76	0.375	0.09	0.0101
Bianca (1986U9)	44		59.17	0.435	0.16	0.0009
Juliet (1986U3)	62		61.77	0.464	0.04	0.0002
Desdemona (1986U6)	54		62.66	0.474	0.16	0.0002
Rosalind (1986U2)	84		64.36	0.493	0.06	0.0006
Portia (1986U1)	104		66.10	0.513	0.09	0.0002
Cressida (1886U4)	54		69.93	0.558	0.28	0.0001
Belinda (1986U5)	66		75.26	0.624	0.03	0.0001
Puck (1985U1)	144		86.00	0.762	0.31	0.0001
Miranda (U5)	472	0.6	129.8	1.41	4.2	0.027
Ariel (U1)	1158	12.7	190.9	2.52	0.3	0.0034
Umbriel (U2)	1169	12.7	266.0	4.14	0.36	0.0050
Titania (U3)	1578	34.9	436.3	8.71	0.14	0.0022
Oberon (U4)	1523	30.3	583.4	13.46	0.10	0.0008

respectively to the outermost epsilon (ϵ) ring and help to keep it confined by gravitational shepherding (see Shepherd satellite).

The entire Uranian system has an axial tilt of 97.9° so that the north rotational and orbital poles of the moons lie slightly below the plane of the solar system. The fact that the moon's orbits share the axial tilt of Uranus strongly implies that the moons formed from a planetary nebula around Uranus after the development of the system's tilt. The orbital motion of the Uranian system around the Sun, combined with the system's axial tilt, produces a marked variation in the view of the system seen from the Sun as well as the areas of the planet and moons receiving sunlight. When the Uranian system is oriented with its rotational axis aligned perpendicular to the Sun–Uranus line, the moons appear to move linearly back and forth across Uranus when viewed from Earth. However, since the moons are in equatorial orbit around a tilted Uranus, this apparent motion is in a N–S direction rather than E–W as is the case for the moons of Jupiter and Saturn. In this orientation the moons experience 'normal' days with any given point on the surface seeing the Sun above the horizon for half the rotational period. At the other extreme, which occurs 21 years later when Uranus has moved approximately one-quarter of its orbital path around the Sun, the rotational axes of the system are aligned with the Sun. In this orientation the moons' orbits appear as circles when viewed from Earth, forming a 'bull's-eye' pattern, and the hemisphere facing the Sun is in constant sunlight while the opposite hemisphere is in constant darkness. Uranus was in this latter configuration during the Voyager 2 encounter and thus only the sunlit southern hemisphere of each moon could be imaged during the flyby.

Prior to the Voyager 2 encounter, the size of the Uranian moons could only be estimated from the total amount of light they reflected and assumptions about their albedo or reflectivity. Imaging and photometric experiments carried out during the encounter allowed the size and albedo of the larger satellites to be determined independently. Diameters were determined from modeling of the shape and size of a moon in several images and knowledge of the spacecraft's position and camera optics. Diameters measured in this manner range from 144 km for the newly discovered small moon Puck to 1578 km for Titania (q.v.), the largest of the Uranian moons. Geometric albedos (reflectivity at 0° phase angle) were computed from disk-integrated brightnesses of the five large moons measured at UV and IR wavelengths and various phase angles (Lane et al., 1986) and in the visual range using the Voyager 2 camera and filters (Smith et al., 1986), and proved to be lower than those of most of the large satellites of Jupiter and Saturn. In the visual range there is little variation in reflectivity at different wavelengths, and the moons are remarkably gray compared to the Jovian and Saturnian satellites, which generally show increased reflectivity at longer wavelengths.

By carefully tracking the motion of the Voyager 2 spacecraft through the Uranian system, monitoring Doppler shifts in its radio carrier wave and observing the mutual perturbations of the moons on each other's orbits, it was possible to determine masses for the larger moons (Table U5). Combined with the size of the moons, these masses yield densities for the four largest moons which are consistent with the 'solar mix' of components expected at Uranus' distance from the Sun during planetary accretion (Tyler et al., 1986). This 'solar mix' is approximately 39% water ice, 32% rock, 8% methane clathrate and 21% ammonia dihydrate. Earth-based infrared spectra confirm only the presence of water ice. Miranda has a slightly lower density than the four larger moons, indicating a larger fraction of ices in its composition.

Voyager 2 also carried an infrared spectrometer which measured thermal emissions from the surfaces of Ariel and Miranda at wavelengths between 25 and 50 µm. These emissions show no variation with wavelength and yield brightness temperatures of 84 K for Ariel and 86 K for Miranda (Hanel et al., 1986).

The moons of Uranus orbit either entirely or partially within the planetary magnetosphere, which serves to trap charged particles. The Voyager 2 cosmic ray system showed that the density of electrons with energies greater than 1.1 MeV rises above the galactic background level at approximately the location (in magnetic field coordinates) of Titania and that this density increases with decreasing distance from Uranus. The electron density shows distinct drops at the locations of Miranda, Ariel and Umbriel, indicating that the satellites are absorbing part of the electron flux (Stone et al., 1986). This absorption of charged particles may be partially responsible for the relatively low albedos of the Uranian moons. High-energy protons within the magnetosphere bombard the surface and can strip hydrogen atoms from CH_4 molecules, leaving a residue of carbon. This process would be particularly effective on the newly discovered small, inner moons which are presumed to be undifferentiated and orbit entirely within the magnetosphere. Puck, for example, has a reflectivity of only 7%.

The cratering record of the Uranian satellites demonstrates a dichotomy in size–frequency distribution, indicating that there were probably two distinct sources for impactors (Smith et al., 1986). One set of craters is characterized by relatively large numbers of closely spaced craters with diameters as great as 100 km. These are most likely to be the scars left by the sweep-up of objects remaining in solar orbit at the end of planetary accretion and are called population I impacts. The second set of craters are fewer in number and generally smaller in size with diameters less than 50 km. Their small size indicates lower impact energies and they are probably the result of impactors which were in orbit around Uranus. These are called population II impacts. The larger population I craters are most evident on the outer moons, Oberon, Titania and especially Umbriel. Their absence on the inner moons Ariel and Miranda indicates that these moons underwent extensive resurfacing episodes or disruption and re-accretion after the completion of the sweep-up of most of the population I impactors.

Prior to the Voyager 2 encounter it was expected that, due to their relatively small size and low temperature, the Uranian moons would be geologically inactive, heavily cratered iceballs. However, the five larger moons displayed a surprising measure of geologic activity and diversity, including both tectonic fracturing and resurfacing by flowing ice (cryovolcanism). Surface fractures could result from expansion of a thin frozen surface either as interiors are heated by tidal forces and radiogenic decay of unstable isotopes in the silicate fractions, or as the interior water fraction expands upon freezing. At the low surface temperatures of the Uranian system, however, the melting and flow of water ice onto the surface is unlikely. The presence of small amounts of ammonia mixed with the water would, however, lower the melting point of the mixture to 176 K.

Miranda (q.v.), the smallest of the moons known from Earth-based observations, shows extensive tectonic alteration, including a 20 km high scarp, three trapezoid to ovoid features called coronae, and some evidence for ice flows. Ariel is the least heavily cratered of the large moons and appears to have been extensively resurfaced by cryovolcanic flows. It also exhibits a complex set of extensional graben, most of which show smooth, resurfaced floors. This relatively fresh icy surface gives Ariel the highest albedo of the Uranian moons, but it nonetheless retains a few large impact basins. Umbriel (q.v.), by contrast, has a low albedo, many population I craters and appears to be the least altered of the major moons. Crater densities and populations on Titania (q.v.) show large variability from area to area, indicating that resurfacing has occurred over portions of this moon. While it also has a system of extensional fractures, they are less well developed than Ariel's. Oberon (q.v.) has a large number of population I craters, some of which have dark floors or bright rays, and a few extensional fractures.

Despite differing appearances, the moons of Uranus share certain common aspects of their geologic histories, albeit with significant variations in the timing and extent of the processes involved. This generalized geologic history begins with accretion of the moons from material in orbit around Uranus and a period of intense impact cratering as the remnants of the material in solar orbit were swept up. During accretion, high-energy collisions between large impactors and moons forming deep in Uranus' gravitational well may have catastrophically disrupted these moons, with Miranda and perhaps Ariel forming from the resulting fragments. There then followed a period of various degrees of warming, with the more intensely heated moons undergoing more extensive resurfacing as ice erupted and recoated the surfaces of the moons. The satellites then cooled and, with the expansion of the interior water ice upon freezing, the cold, brittle crusts fractured, producing graben and fault systems. Throughout these later episodes small bodies still in orbit around Uranus continually produced small craters, ultimately resulting in the surfaces we see today.

Daniel M. Janes

Bibliography

Bergstralh, J., Miner, E. and Matthews, M. (1991) *Uranus*. Tucson: University of Arizona Press.
Hanel, R., Conrath, B., Flasar, F.M. et al. (1986) Infrared observations of the Uranian system. *Science*, **233**, 70–4.
Lane, A.L., Hord, C.W., West, R.A. et al. (1986) Photometry from Voyager 2: initial results from the Uranian atmosphere, satellites, and rings. *Science*, **233**, 65–70.
Miner, E. (1990) *Uranus: The Planet Rings and Satellites*. New York: Ellis Horwood.
Smith, B.A., Soderblom, L.A., Beebe, R. et al. (1986) Voyager 2 in the Uranian system: imaging science results. *Science*, **233**, 43–64.
Stone, E.C., Cooper, J.F., Cummings, A.C. et al. (1986) Energetic charged particles in the Uranian magnetosphere. *Science*, **233**, 93–7.
Tyler, G.L., Sweetnam, D.N., Anderson, J.D. et al. (1986) Voyager 2 radio science observations of the Uranian system: atmosphere, rings, and satellites. *Science*, **233**, 79–84.

Cross references

Miranda
Oberon
Satellite, natural
Shepherd satellite
Small satellite
Titania
Umbriel

UREILITE METEORITES

Ureilites are carbonaceous ultramafic rocks that represent the second largest group of achondrite meteorites with 40 known specimens (Table U6). Only eucrites (q.v.) are more numerous. They are primarily composed of olivine and pyroxene (mostly pigeonite; some contain orthopyroxene and/or augite), with variable quantities of carbon minerals (mostly graphite and diamond), nickel–iron metal and sulfides (Plate 15). The origin of ureilites is controversial owing to mineralogical and textural features suggesting complex igneous/metamorphic differentiation, contrasting with geochemical properties pointing to a less-processed 'primitive' origin. No single hypothesis has yet been proposed that fully integrates the many diverse features of ureilites into a fully self-consistent model of origin.

The name 'ureilite' comes from Novo Urei, a meteorite that fell on 4 September 1886 on the territory of the present Gorkovski province, near the village of Karamzinka in the former USSR. (Vdovykin, 1970). Three individual specimens fell, but only one weighing 1.9 kg was preserved. Of the remaining two, one was lost in a bog, and the other was destroyed by superstitious peasants who reportedly ate some of the meteorite (McSween, 1987). Novo Urei lends its name to ureilites by virtue of being the first well-studied ureilite, but it is not the first known specimen. That distinction belongs to Goalpara, found in 1868 in the province of Assam in India. Dyalpur, the second known ureilite, fell in Sultanpur state, India, in 1872; Novo Urei is the third known ureilite.

In 1965 only seven ureilites existed in world collections (Table U6), and they were considered a rare meteorite type. Mostly because of recoveries during Antarctic collecting expeditions initiated in the mid 1970s, ureilites are no longer considered rare (Table U6). The large number of unpaired ureilites now available reveal that in spite of showing many common features, ureilites can be subdivided into subtypes based on texture and mineralogy (Goodrich, 1992).

The first scientific description of a ureilite is an article on Goalpara by G. Tschermak in 1870 (Vdovykin, 1970). In 1888 two Russian scientists, M.V. Jerofeev and P.A. Lachinov, discovered diamonds in Novo Urei. Later studies on ureilite diamonds and other petrologic features offered explanations for their origin. For example, Ringwood (1960) verified the existence of ureilite diamonds and attributed their origin – and that of ureilites – to high-pressure metamorphism and recrystallization of chondrite material in the interior of a 'moon-sized' body. Carter and Kennedy (1964) and Carter, Raleigh and de Carli (1968) agreed with a high static pressure origin, but found no evidence for recrystallization in Novo Urei. However, the discovery

Table U6 Ureilite data[a]

Ureilite	Year recovered	[b]Olivine composition	Pyroxene type(s)
Goalpara, India	Found 1868	78.6	Pigeonite
Dyalpur, India	Fell 1872	84.3	Pigeonite
Novo Urei, Russia	Fell 1886	78.9	Pigeonite
Lahrauli, India	Fell 1955	79	Pigeonite
Hajma, Oman	Found 1958	84.5	Pigeonite
North Haig, Australia	Found 1961	Breccia (76–92)	Pigeonite
Dingo Pup Donga, Australia	Found 1965	83.6	Pigeonite
Havero, Finland	Fell 1971	79.3	Clinoenstatite
Kenna, NM, USA	Found 1972	78.9	Pigeonite
Nilpena, Australia	Found 1975	Breccia (77–81)	Pigeonite–augite–hypersthene
RC027, NM, USA	Found 1984	79.5	Pigeonite
Nova 01, Mexico	Found 1991	78	Pigeonite–hypersthene
Antarctic ureilites			
Y74123	Found 1974	78.4	Pigeonite
Y74130	Found 1974	77.1	Pigeonite–augite–hypersthene
Y74659	Found 1974	91.1	Pigeonite–hypersthene
Y790981	Found 1979	79.8	Pigeonite–hypersthene
Y74154	Found 1974	84.6	Pigeonite
Y82100	Found 1982	81.4	Pigeonite
Y791538	Found 1979	91.2	Pigeonite–hypersthene
Y791839	Found 1979	74.6	Pigeonite
Y8448	Found 1984	77.3	Pigeonite
ALHA77257	Found 1977	85.1	Pigeonite
ALHA78019	Found 1979	76.7	Pigeonite
ALHA78262	Found 1978	78.0	Pigeonite
ALH82130–106–84136	[c]Paired; 1982, 1984	95	Pigeonite–augite
ALH83014	Found 1983	82	Pigeonite
ALHA81101	Found 1981	78.5	Pigeonite
EET83309	Found 1983	Breccia (77–95)	Pigeonite–augite–hypersthene
EET83225	Found 1983	87	Pigeonite
EET87517	Found 1987	92	Hypersthene
EET87511–523–717	Paired; 1987	84.8	Hypersthene
EET87720	Found 1987	Breccia (79–87)	
RKPA80239	Found 1980	84	Pigeonite
PCA82506	Found 1982	79	Pigeonite
META78008	Found 1978	77	Pigeonite–augite–hypersthene
LEW86216	Found 1986	80.6	Hypersthene
LEW88006	Found 1988	82	Pigeonite
LEW87165	Found 1987	85	Pigeonite
LEW85328	Found 1985	80	Pigeonite
LEW88201–012–281 LEW85440	Paired; 1988, 1985	91–92	Hypersthene–augite

[a] Data from Berkley (1986) and Goodrich (1992).
[b] In terms of mole % forsterite (Mg_2SiO_4) content.
[c] 'Paired' specimens are recovered as separate stones, but are fragments of the same meteorite.

of the hexagonal polymorph of carbon, lonsdaleite, in three ureilites by Vdovykin (1970) suggested that dynamic pressure (impact shock) alone could produce diamonds in ureilites. Lonsdaleite has only been created experimentally in intensely shocked materials. Besides, meteorite parent bodies (asteroids) (q.v.) are considered too small to create requisite static pressures to make diamonds.

A new twist to the ureilite controversy arose with the discovery of the Kenna, New Mexico, ureilite in 1972. Berkley et al. (1976) reported pronounced lineated and foliated silicate textures in Kenna (Plate 15b; Figure U21) which they interpreted as igneous cumulate textures. They emphatically rejected the idea that ureilites are merely little-modified, recrystallized chondrites. At the same time, however, Wasson et al. (1976) argued on the basis of siderophile element abundance ratios that ureilites represent residues from partial melting events. Boynton, Starzk and Schmitt (1976) came to the same conclusion on the basis of rare earth element distribution patterns. More recent studies have continued to advocate one of these processes (igneous cumulate versus partial melt residue) or the other (or a combination of both) to explain ureilite origins, although some of the plethora of current models employ entirely different means to produce ureilites. Some of these models are discussed later.

Textures and mineral chemistry

Olivine and pyroxene

Most ureilites are monomict (single rock type) but four of the 40 known ureilites are impact breccias containing fragments of many ureilite compositions as well as non-ureilite lithic clasts and mineral fragments (Table U6). Silicate textures in monomict ureilites come in two distinct modes, 'typical' and 'poikilitic'. Typical textures consist of straight or curved olivine and pyroxene (~ 1–2 mm diameter) grain boundaries meeting at triple-point junctures (Figure U21). Many of these ureilites also show mineral elongation lineations (Plate 15b) first described for Kenna, Poikilitic ureilites have pyroxene partially or completely surrounding rounded olivine grains. The pyroxene in these ureilites is commonly orthopyroxene or orthopyroxene plus augite, whereas 'typical ureilites' commonly have pigeonite as the sole or dominant pyroxene. Ureilites showing both types of textures in a single specimen are known, and some ureilites display textures intermediate between 'typical' and 'poikilitic' (Berkley, 1986). In general ureilites show textural evidence of varying degrees of textural annealing (metamorphism).

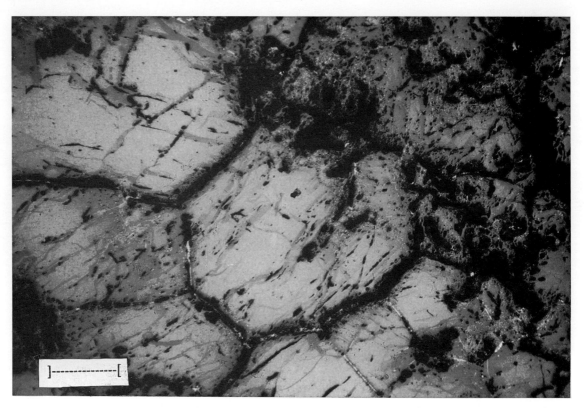

Figure U21 Reflected light photomicrograph of Kenna showing typical 'triple-juncture' texture. Dark carbonaceous material fills intergranular interstices between silicate grains (light gray) and invades some fractures. Bright spots are Ni–Fe metal grains. Scale bar equals 0.25 mm.

Olivine (ol) tends to dominate volumetrically over pyroxene (px) in most ureilites. The ol/px ratio commonly varies within any given sample, but 3 : 1 ol/px is probably a realistic ureilite average. Two ureilites, Y74123 and LEW86216, contain less than 1% pyroxene by volume and, as terrestrial rocks, would be considered dunites. Another dunite, occurring as a lithic fragment, was discovered in the EET83309 breccia. Olivine in ureilites tends toward the Mg-rich variety, ranging from about Fo_{75} to Fo_{95} (Table U6). Olivine typically shows very homogeneous (well-equilibrated) core compositions. Compared to olivine typical of terrestrial rocks, ureilite olivine is high in Ca and Cr.

Most ureilites contain well-equilibrated uninverted pigeonite as the sole or dominant pyroxene (Figure U22), although orthopyroxene and high-Ca clinopyroxene (augite) occur in some specimens. ALH82130 (Plate 15c) is the only ureilite showing obvious unmixing of pyroxene; in this case ropy augite blebs occur in a pigeonite host (Takeda, Mori and Ogata, 1989). Ureilite pyroxene is typically high in Cr and Al compared to comparable terrestrial pyroxenes.

Carbon minerals

All ureilites contain carbon-rich minerals, mostly finely granulated graphite concentrated along grain boundaries, intruding silicate grain fractures, and as isolated inclusions in silicates (Figure U21). Intergranular 'carbonaceous matrix' areas also contain much of the Ni–Fe metal and sulfide minerals found in ureilites. However, Berkley and Jones (1982) described graphite occurring in ALH78019 as prismatic graphite crystals along grain margins and occurring as inclusions in silicates (Plate 15d). Most ureilites also contain minute diamonds (diameter ~ 1 μm less) of the normal isometric variety, in addition to lonsdaleite (see above) and chaoite ('liquid' carbon), and Vdovykin (1970) identified traces of organic compounds in some ureilites. The absence of diamonds and lonsdaleite in ALH78019 suggests a very low-shock origin which may also explain the pristine state of graphite crystals. ALH83014 (Goodrich, 1992) and Nuevo Mercurio (b) (renamed Nova 001; Treiman and Berkley, 1992) also show the presence

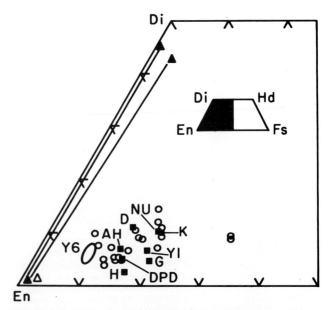

Figure U22 Magnesium-rich portion of the pyroxene composition quadrilateral (inset shows full diagram: Di = diopside; Hd = hedenbergite; Fs = ferrosillite; En = enstatite). CaO increases upward; MgO increases to left compared to FeO. Plotted compositions reflect the range of ureilite compositions. NU = Novo Urei; K = Kenna; G = Goalpara; Y1 = Yamato 74123; DPD = Dingo Pup Donga; H = Havero; Y6 = Yamato 74659 range; AH = Allan Hills 77257. Small unfilled circles and triangles at far left are pyroxenes in the North Haig Breccia.

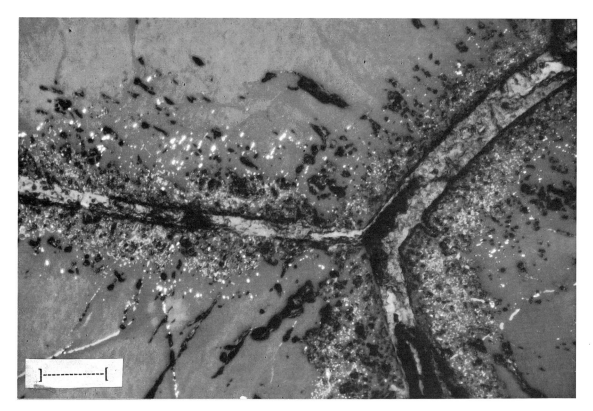

Figure U23 Reflected light photomicrograph of Novo Urei showing intergranular triple juncture filled with carbonaceous matrix (black, mostly plucked out). Bright granular spots are low-Ni metal reduced from silicates by reaction with graphite. Associated silicate rim material, thus depleted in iron, is Mg-rich compared to core silicate areas. Scale bar equals 50 μm.

of well-preserved graphite crystals suggesting that all ureilites contained such crystals prior to shock disruption.

Where they contact silicate material (particularly olivine), graphite has reacted with Fe–Mg silicates to produce more Mg-rich silicate plus low-Ni iron metal (Figure U23; Plate 15d) according to this or similar reaction:

$$2C + Mg_2SiO_4 + Fe_2SiO_4 \leftrightharpoons 2Fe° + Mg_2SiO_4 + SiO_2 + 2CO$$

2 graphite + forsterite + fayalite $\leftrightharpoons$ 2 iron metal + forsterite + silica + 2 carbon monoxide This reduction reaction is similar to that occurring in blast furnaces where 'coke' (devolatilized coal) is added to oxide iron ore at high temperatures to produce metallic iron. The result in ureilites is the production of a secondary generation of Ni–Fe metal that contains little or no Ni compared to 'matrix' metal (Figure U21; Plate 15d), which commonly contains about 2–7% Ni.

Another carbon mineral observed in at least five ureilites (Goodrich and Berkley, 1986) is cohenite (Fe_3C), a mineral also found in commercial steel (cementite). It occurs in metallic spherules (Figure U24) along with Ni–Fe metal and sulfide. Cohenite crystals in these spherules show euhedral (well-shaped) forms and otherwise display textures expected of eutectic crystallization in the Fe–C system.

Shock features

Even the most 'pristine' ureilites like ALH78019 show some evidence of shock deformation. In silicates shock deformation ranges from minor fracturing to kink banding, development of polysynthetic twins (pyroxene only) or, in the most extreme cases, 'mosaicism' (Lipschutz, 1964) as shown in Figure U24, Plate 15e. Silicate grain mosaicism occurs at very high shock pressures and is expressed as pervasively shattered crystals, especially olivine (Plate 15e). Shock also produces the pulverized mixture of graphite–metal–sulfide that constitutes the carbonaceous matrix of many ureilites, and results in the production of diamond and other C polymorphs. As noted by Matsuda, Fukunaga and Ito (1988, 1991), some diamond in ureilites may have formed by vapor growth from the solar nebula, later being incorporated into ureilites during accretion.

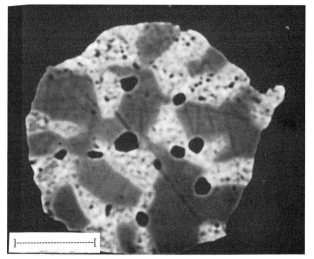

Figure U24 Backscattered electron picture of cohenite (Fe_3C)-bearing spherule in ALHA78262 Antarctic ureilite. Cohenite (iron carbide) crystals are dark; lighter areas are fine-grained intergrowths of cohenite plus Ni–Fe metal. These spherules are of definite igneous origin and also show that ureilite magmas were carbon rich. Scale bar equals 30 μm.

Bulk geochemical characteristics

Ureilites are highly scrutinized meteorites and most applicable geochemical techniques have been applied to their study. The brief summary below includes the most important of these in terms of ureilite formation modeling; however, data are available in other areas as well. A comprehensive discussion of all relevant parameters is given in Goodrich (1992).

Lithophile and volatile elements

Compared to CI chondrites, ureilites are enriched in Mn, Cr, V, Mg, Ti and Sc; they are highly depleted in the volatile alkali elements Rb, Na and K (e.g. Boynton, Starzyk and Schmitt, 1976; Wiik 1972; Wasson et al., 1976; Goodrich, Jones and Berkley, 1987; Spitz and Boynton, 1991). These trends reflect the lack of a feldspar component in ureilites and attest to their highly ultramafic character. Volatile siderophile elements are also depleted in ureilites (Wasson et al., 1976; see below). Ca/Al ratios are also high in ureilites, necessitating that they are derived from feldspar-depleted parent materials (Goodrich, Jones and Berkley, 1987).

Rare earth elements

Chondrite-normalized rare earth (REE; La through Lu) elements generally show V-shaped patterns with both light element (LREE) and heavy element (HREE) enrichments and maximum depletion at Eu (e.g. Boynton, Starzyk and Schmitt, 1976; Warren and Kallemeyn, 1989; Spitz and Boynton, 1991). Absolute abundances of REE are low in ureilites compared to chondrites. Acid leaching experiments have succeeded in separating a LREE-enriched leachate from a LREE-depleted residue, suggesting that ureilites may contain a yet-unidentified acid-soluble LREE-enriched phase (cf. Goodrich et al., 1991; Spitz and Boynton, 1991). Because no known natural process is capable of producing V-shaped patterns in rocks, the ureilite REE pattern probably results from the introduction of LREE-enriched material into HREE-enriched (LREE-depleted) ultramafic material.

Noble gases

Unlike other volatile elements, noble gas (He, Ne, Ar, Kr, Xe) abundances in ureilites are similar to those in chondrites (Weber, Begemann and Hintenberger, 1971; Gobel, Ott and Begemann, 1978). In addition, noble gas abundance patterns in ureilites are of the 'planetary' type, like those in chondrites, as opposed to solar or cosmic type (like that in the solar nebula and Sun). Diamond is the principle receptacle for trapped noble gases, graphite being nearly gas free (Gobel, Ott and Begemann, 1978). Wacker (1986) found that the typically high noble gas contents in the diamond-free ureilite ALH78019 were contained in fine-grained carbon species of unknown structure.

Siderophile elements

Siderophile trace element (e.g. Re, Os, Ir, Au, Ge) abundance patterns lie within ~ 0.07–$2.2 \times$ CI chondrites (e.g. Boynton, Starzyk and Schmitt, 1976; Higuchi et al., 1976; Janssens et al., 1987; Spitz 1992). Abundances of refractory siderophiles (Re, Os, W and Ir) are enriched over volatile siderophiles (Ni, Au, Ga, Co and Ge), and ureilite siderophile abundances in general are highly enriched over those in terrestrial, lunar and SNC parent body (Mars?) rocks. These trends suggest either that no major core-forming process occurred on the ureilite parent body, or that metal in ureilites is a late admixture (Goodrich, 1992).

Oxygen isotopes

Clayton and Mayeda (1988) expanded earlier studies on limited numbers of ureilites (Clayton et al., 1976; Goodrich et al., 1987) to 23 individual monomict samples and found that ureilites plot on the slope = 1 C2–C3 chondrite line on standard $\Delta^{18}O$‰ versus $\Delta^{17}O$‰ plots. This supports the contention of Vdovykin (1970) and Mueller (1969) that ureilites are related to carbonaceous chondrites in some manner. However, these data were a blow to those who advocated that ureilites are related to one another by mainly igneous processes (cf. Berkley et al., 1980; Goodrich, Jones and Berkley, 1987) because ureilites as a group do not fall on a slope = ½ mass-dependent fractionation trend (typical of igneous differentiates). On the other hand, mass fractionation-dependent trends were recognized within the FeO-rich, 'group 1' (Berkley et al., 1980) subgroups based on olivine mg (atomic Mg/Mg+Fe) number (Clayton and Mayeda, 1988). Also, ureilite olivine mg correlates well with $\Delta^{17}O$, a feature probably acquired in the solar nebula. Altogether oxygen isotope trends show that ureilites, in spite of other evidence to the contrary, cannot be highly differentiated rocks, but represent materials little altered compared to C2–C3 chondrites. Mass-dependent trends in the FeO-rich subgroup, however, shows that some limited differentiation and O isotope equilibration occurred within isolated areas of the ureilite parent body. Nevertheless, the data strongly preclude a ureilite origin by differentiation from a common magma reservoir.

Nd–Sm and Rb–Sr isotopic systematics and ages

Goodrich et al. (1991) performed a detailed study on Sm–Nd and Rb–Sr systematics in a suite of ureilites, augmenting earlier studies by Gobel, Ott and Begemann (1978). Nd–Sm systematics show that ureilites probably initially formed in 4.55 Ga. Rb–Sr data obtained by Takahashi and Matsuda (1990) also give a model age of 4.55 Ga and internal isochron age of about 4.0 Ga for META78008. Nd–Sm data also indicate younger isochron ages for various ureilites at 4.23 and 3.74 Ga. Thus ureilites appear to have formed initially at 4.55 Ga and were later affected by metamorphism, probably accompanied by injection of new material at various times. This material may have been the LREE-enriched component mentioned above (Goodrich and Lugmair 1991; Goodrich et al., 1991).

Petrogenesis models

Many models have been proposed to explain the varied features of ureilites; none of them succeed in explaining all aspects of ureilites without confronting insurmountable inconsistencies. Most current models can be placed in one of two general categories: (1) 'magmatic/high processing' models involving considerable differentiation from an original chondrite parent, and (2) 'primitive/low processing' models in which differentiation is kept to a minimum, mostly to preserve pre-planetary geochemical imprints (noble gas abundances, O isotope trends, siderophile patterns, etc.). A sample of current models is presented below to show the range of thought on the subject, but this list is not comprehensive. A detailed review is given in Goodrich (1992).

Multistage igneous cumulate model

This is the premier model in category 1 above. Proposed by Goodrich, Jones and Berkley (1987), it employs igneous processes to explain silicate textures and requires multiple processing to derive ureilites from plagioclase feldspar-depleted source rocks (required by high Ca/Al ratios). In simple terms the model requires (1) that chondritic material undergoes 10–30% partial melting, producing a plagioclase-depleted residue and basalt-rich crust; and (2) that remelting of the residue or mafic cumulates in the crust produces magmas that later crystallize as ureilites. This model successfully reproduces HREE-enriched patterns, critical lithophile ratios such as Ca/Al, 'igneous-like' textures, siderophile element patterns and noble gas abundances in present ureilites. It suffers in not adequately explaining O isotope data, and in making no allowance for the absence of complimentary crustal differentiates in world meteorite collections, among other considerations.

Explosive volcanism model

This model also belongs to category 1 (magmatic) above. Variations have been proposed by Scott, Keil and Taylor (1992) and by Warren and Kallemeyn (1992). It is similar to the 'igneous cumulate model' above, except that ureilites are partial melt residues, not igneous cumulates. The problem of the missing basaltic crust is resolved by removing it by explosive volcanism propelled by CO/CO_2 (generated at depth by C–silicate reactions). Wilson and Keil (1991) argue that volatiles moving in basaltic liquids through asteroids of < 100 km radius disrupt the melt into droplets capable of escaping the planetoid's gravity field. This model satisfies most constraints posed by ureilite geochemical data, but falls short of explaining lineated/foliated textures and fails to account for high noble gas contents in ureilites. Most noble gases would probably be lost during the explosive volcanism postulated by the model.

Nebular sedimentation model

This model belongs to category 2 (primitive) above. Its advocates (Takeda and Yanai, 1978; Takeda et al., 1980; Takeda 1989) propose that ureilites are not derived from chondritic parent materials, but represent a special kind of primitive material derived by nebular condensation. During solar nebular condensation, olivine and pyroxene were preferentially drawn toward the equator of the nebula, accreting with preferred lineation/foliation fabrics. Carbonaceous materials condensed and were added later. This assemblage (C + mafic minerals) later broke up into planetesimals, compacted and metamorphosed to explain annealed ureilite textures. Collision with other bodies caused sudden cooling, explaining the lack of pigeonite exsolution and low-pressure C–silicate reduction reactions (Figure U23). Although many aspects of ureilites can be explained by this model, it poses several problems. Foremost of these is whether olivine and pyroxene would, in fact, preferentially segregate to one locality of the solar nebula. The model also fails to account for ureilite siderophile patterns (and some other geochemical features). Finally, it is not certain that ureilite textures would result from this process. The elongation lineations (Plate 15b) probably require fluid flow specifically precluded in this nearly exclusively solid state model.

The three models above are examples of current approaches to the 'ureilite problem'. They are not the only proposals for ureilite origin, nor are they necessarily the best models. Other models worthy of consideration are the partially disruptive impact model (Warren and Kallemeyn, 1989), the planetesimal scale collision model (Takeda, Mori and Ogata, 1988), the unified primitive meteorite model (Kurat, 1988), and the impact melt model (Rubin, 1988).

Summary

Ureilites are carbonaceous olivine–pyroxene achondrites that, based on their ever-expanding numbers in meteorite collections, were important players in the formation of one or more asteroids in the solar system. Their mineralogy, textures and chemistry suggest a very complicated history. No generally accepted model of their formation has been advanced to date.

John L. Berkley

Bibliography

Berkley, J.L. (1986) Four Antarctic ureilites: petrology and observations on ureilite petrogenesis. *Meteoritics*, 21(2), 169–89.

Berkley, J.L. and Jones, J.H. (1982) Primary igneous carbon in ureilites: petrological implications. *Proc. Lunar Planet. Sci. Conf.*, 13; *J. Geophys. Res.*, 87, A353–64.

Berkley, J.L., Brown, H.G., IV, Keil, K. *et al.* (1976) The Kenna ureilite: an ultramafic rock with evidence for igneous, metamorphic, and shock origin. *Geochim. Cosmochim. Acta*, 40, 1429–37.

Berkley, J.L., Taylor, G.J., Keil, K. *et al.* (1980) The nature and origin of ureilites. *Geochim. Cosmochim. Acta*, 44, 1579–97.

Boynton, W.V., Starzk, P.M. and Schmitt, R.A. (1976) Chemical evidence for the genesis of ureilites, the achondrite Chasigny and nakhlites. *Geochim. Cosmochim. Acta*, 40, 1439–47.

Carter, N.L. and Kennedy, G.C. (1964) Origin of diamonds in the Canyon Diablo and Novo Urei meteorites. *J. Geophys. Res.*, 69; 2403–21.

Carter, N.L., Raleigh, C.B. and De Carli, P. (1968) Deformation of olivine in stony meteorites. *J. Geophys. Res.*, 73, 2403–61.

Clayton, R.N. and Mayeda, T. (1988) Formation of ureilites by nebular processes. *Geochim. Cosmochim. Acta*, 52, 1313–8.

Clayton, R.N., Mayeda, T.K., Onuma, N. and Shearer, J. (1976) Oxygen isotopic composition of minerals in the Kenna ureilite. *Geochim. Cosmochim. Act*, 40, 1475–6.

Gobel, R., Ott, U. and Begemann, F. (1978) On trapped noble gases in ureilites. *J. Geophys. Res*, 83, 855–67.

Goodrich, C.A. (1992) Ureilites: a critical review. *Meteoritics*, 27(4), 327–52.

Goodrich, C.A. and Berkley, J.L. (1986) Primary magmatic carbon in ureilites: evidence from cohenite-bearing metallic spherules. *Geochim. Cosmochim. Acta*, 50, 681–91.

Goodrich, C.A. and Berkley, J.L. (1986) Primary magmatic carbon in ureilites: evidence from cohenite-bearing metallic spherules. *Geochim. Cosmochim. Acta*, 50, 681–91.

Goodrich, C.A. and Lugmair, G.W. (1991) PCA82506: a ureilite with LREE-enriched component and a whole-rock Sm–Nd model age of 4.55 Ga, in *Lunar Planet. Sci.* XXII. Houston: Lunar and Planetary Institute, pp. 467–8.

Goodrich, C.A. and Lugmair, G.W. (1992) Addition of LREE-enriched material to a ureilite at 4.23 Ga: evidence for episodic metasomatism? in *Lunar Planet. Sci. XXIII*. Houston: Lunar and Planetary Institute, pp. 429–30.

Goodrich, C.A., Jones, J.H. and Berkley, J.L. (1987) Origin and evolution of the ureilite parent magmas: multi-stage igneous activity on a large parent body. *Geochim. Cosmochim. Acta*, 51, 2255–73.

Goodrich, C.A., Patchett, P.J., Lugmair, G.W. and Drake, M.J. (1991) Sm–Nd and Rb–Sr isotopic systematics of ureilites. *Geochim. Cosmochim. Acta*, 55, 829–48.

Goodrich, C.A., Keil, K., Berkley, J.L. *et al.* (1987) Roosevelt County 027: a low-shock ureilite with trapped silicate liquid and high noble gas concentrations. *Meteoritics*, 22, 191–218.

Goodrich, C.A., Keil, K., Berkley, J.L. *et al.* (1987) Roosevelt County 027: a low-shock ureilite with trapped silicate liquid and high noble gas concentrations. *Meteoritics*, 22, 191–218.

Higuchi, H., Morgan, J.W., Ganapathy, R. and Anders, E. (1976) Chemical variations in meteorites – X. Ureilites. *Geochim. Cosmochim. Acta*, 40, 1563–71.

Janssens, M.-J., Hertogen, J., Wolf, R. (1987) Ureilites: trace element clues to their origin. *Geochim. Cosmochim. Acta*, 51, 2275–83.

Kurat, G. (1988) Primitive meteorites: an attempt towards unification. *Phil. Trans. Roy. Soc. London*, A325, 459–82.

Lipschutz, M.E. (1964) Origin of diamonds in ureilites. *Science*, 143, 1431–4.

Matsuda, J.-I., Fukunaga, K. and Ito, K. (1988) On the vapor-growth diamonds formation in the solar nebula, in *Lunar Planet. Sci. XXIX*. Houston: Lunar and Planetary Institute, pp. 736–7.

Matsuda, J.-I., Fukunaga, K. and Ito, K. (1991) Noble gas studies in vapor-growth diamonds: comparison with shock produced diamonds and the origin of diamonds in ureilites. *Geochim. Cosmochim. Acta*, 55, 2011–23.

McSween, H.Y. (1987) *Meteorites and Their Parent Planets*. Cambridge: Cambridge University Press.

Mueller, G. (1969) Genetical interrelations between ureilites and carbonaceous chondrites, in *Meteorite Research* (ed. P.M. Millman). Hingham, Massachusetts: D. Reidel. pp. 505–17, 535–7.

Ringwood, A.E. (1960) The Novo Urei meteorite. *Geochim. Cosmochim. Acta*, 20, 1–4.

Rubin, A.E. (1988) Formation of ureilites by impact-melting of carbonaceous chondritic material. *Meteoritics*, 23, 333–7.

Scott, E.R.D., Keil, K. and Taylor, G.J. (1992) Origin of ureilites by partial melting and explosive volcanism on carbon-rich asteroids, in *Lunar Planet. Sci. XXIII*. Houston: Lunar and Planetary Institute, pp. 1253–4.

Spitz, A.H. (1992) ICP-MS trace element analysis of ureilites: Evidence for mixing of distinctive components, in *Lunar Planet. Sci. XXIII*. Houston: Lunar and Planetary Institute, 1339–40.

Spitz, A.H. and Boynton, W.V. (1991) Trace element analysis of ureilites: new constraints on their petrogenesis. *Geochim. Cosmochim. Acta*, 55, 3417–30.

Takahashi, K. and Matsuda, A. (1990) The Rb–Sr and Sm–Nd dating and REE measurements of ureilites (abstract). *Meteoritics*, 25, 413.

Takeda, H. (1989) Mineralogy of coexisting pyroxenes in magnesian ureilites and their formation conditions. *Earth Planet. Sci. Lett.*, 81, 358–70.

Takeda, H. and Yanai, K. (1978) A thought on the ureilite parent body as inferred from pyroxenes in Yamato-74659, in *Proc. 11th Lunar and Planet. Symp. Tokyo Inst. Space Aeronaut. Sci. Univ. Tokyo*, pp. 189–94.

Takeda, H., Mori, H., and Ogata, H. (1988) On the pairing of Antarctic ureilites with reference to their parent body, in *Proc. NIPR Symp. Antarctic Meteorites*, 1, pp. 145–172.

Takeda, H., Mori, H. and Ogata, H. (1989) Mineralogy of augite-bearing ureilites and the origin of their chemical trends. *Meteoritics*, 24, 73–81.

Takeda, H., Mori, H., Yanai, K. and Shiraishi, K. (1980) Mineralogical examination of the Allan Hills achondrites and their bearing on the parent bodies, in *Proc. 5th Symp. on Antarctic Meteorites, Natl Inst. Polar Research, Spec. Issue*, 17, pp. 119–44.

Treiman, A.H. and Berkley, J.L. (1992) Preliminary petrography of Nuevo Mercurio (B), a new ureilite, in *Lunar Planet. Sci. XXIII*. Houston: Lunar and Planetary Institute, pp. 1449–50.

Vdovykin, G.P. (1970) Ureilites. *Space Sci. Rev.*, 10, 483–510.

Wacker, J. (1986) Noble gases in the diamond-free ureilite, ALHA78019: the roles of shock and nebular processes. *Geochim. Cosmochim. Acta*, **50**, 633–42.

Warren, P.H. and Kallemeyn, G.W. (1989) Geochemistry of polymict ureilite EET83309, and a partially-disruptive impact model for ureilite origin. *Meteoritics*, **24**, 233–46.

Warren, P.H. and Kallemeyn, G.W. (1992) Ureilites: the graphite $f(O_2)$ buffer, explosive, and the gross dissipation of basalt from the parent asteroid(s), in *Lunar Planet. Sci. XXIII*. Houston: Lunar and Planetary Institute, 1497–8.

Wasson, J.T., Chou, C-L., Bild, R.W. and Baedecker, P.A. (1976) Classification of and elemental fractionation among ureilites. *Geochim. Cosmochim. Acta*, **40**, 1449–58.

Weber, H.W., Begemann, F. and Hintenberger, H. (1971) Noble gases in the Havero ureilite. *Earth Planet. Sci. Lett.*, **13**, 205–9.

Wiik, H.B. (1972) The chemical composition of the Havero meteorite and the genesis of the ureilites. *Meteoritics*, **7**, 553–557.

Wilson, L. and Keil, K. (1991) Consequences of explosive eruptions on small solar system bodies: the case of the missing basalts on the aubrite parent body. *Earth Planet Sci. Lett.*, **104**, 505–12.

Cross references

Achondrite meteorites
Differentiation
Igneous rock
Meteorite
Shock metamorphism

UREY, HAROLD CLAYTON (1893–1981)

Born in Walkerton, Indiana, Harold Urey earned a Nobel prize for chemistry in 1934 for his discovery of deuterium, the 'heavy hydrogen' isotope and rare constituent of 'heavy water'. His work on isotopic separation helped make possible the first atomic bomb during World War II. He coedited a volume entitled *Production of Heavy Water* (New York: McGraw-Hill, 1955).

After high school in Indiana, Urey taught in country schools for 3 years and then took a BS in zoology (minor in chemistry) at Montana State University (Missoula). During World War I he worked in industrial chemistry in Philadelphia and in 1919 returned to instruct in chemistry at Missoula. He then went to the University of California in Berkeley, obtaining a PhD (1923) with a thesis described as being 'on the calculation of diatomic gases from molecular-spectra data and with the equilibrium distribution of hydrogen atoms among the steady states'. After a postgraduate year in Denmark studying under Niels Bohr, he continued in research at Johns Hopkins University. From 1929 to 1945 he was professor of chemistry at Columbia University in New York City and 1945–1958 at the University of Chicago. From then on he was 'professor-at-large' at the University of California in La Jolla.

Following World War II Urey protested vigorously against misuse of nuclear energy and turned away from traditional chemistry to develop an abiding interest in planetary and especially lunar geology, becoming one of the founders of cosmochemistry (along with Libby, Suess, Harrison, Brown, Wasserburg and others). Together with Stanley Miller, he inspired the first laboratory experiment, utilizing simple basic chemicals believed to exist at the surface of the primordial Earth around 4 billion years ago, to create at room temperature a chemical 'soup' containing amino acids known to constitute the building blocks of organic life (see Miller–Urey experiment). Energy was supplied by an electric spark simulating lightning discharges. To this day, however, in spite of innumerable attempts, experimental production of self-reproducing life has not been successful. Following the appearance of photosynthesis at an early stage in Earth history, the atmospheric oxygen concentration only evolved very slowly, being self-regulated (the 'Urey effect') by UV photodissociation of H_2O, an idea earlier proposed by S. Arrhenius.

While at Chicago, Urey became more and more interested in geology, partly under the influence of Heinz Lowenstam, the paleoecologist. This opened up a totally new world for him. Here he devised the $^{16}O/^{18}O$ 'oxygen isotope paleothermometer' using the mass spectrometer to obtain the isotopic ratio in a geological sample, and thereby deducing the environmental temperature that existed at the time when certain carbonate-shelled organisms were living. Using this new technique, one of his students Cesare Emiliani (while at the University of Miami) was able to construct a temperature scale and glacial/interglacial cycle model for the last ice age in the Atlantic. This essentially marked the beginning of isotope geochemistry. Together with age determinations obtained with the aid of the ^{14}C isotope ('radiocarbon'), a technique pioneered by Libby, it now became possible quantitatively to reconstruct ancient environments, defined by age and climate, back to about 50 000 years.

In the meantime other isotopic studies, on a totally different timescale, were pushing back the history of the planet Earth, and indeed of the solar system, to about 4.6 billion years; Urey used meteorites as the chemical messengers of that history. His revolutionary book on *The Planets: Their Origin and Development* was published in 1952 and in 1960 he presented, at a Columbia University lecture, his new theory on the formation of meteorites. He served as a consultant to NASA on the Apollo and Viking missions.

Following World War II, C.F. von Weizsäcker and others had revived the classical nebular hypothesis (q.v.) for the origin of the solar system. Urey's geochemical studies showed that a cold accretion was the only acceptable hypothesis. The Moon, for its part, was thought to have been accreted separately from the Earth, and Urey rejected the old lunar fission theory completely (see however Moon: origin). He collaborated at first with Gerard Kuiper in lunar studies, but later they fell out. Convinced that the Moon was the 'Rosetta Stone' of the solar system, he persuaded Robert Jastrow at NASA to give top priority to its exploration.

In the 1950s Urey essentially 'invented' a new branch of science, meteoritics, which utilized the chemical make-up of meteorites as a basis for appraising and developing models for the birth and early evolution of the solar system.

Urey reasoned that chondrites represented average samples of solid (non-volatile) solar system material. Jointly with Hans Suess at La Jolla, Urey developed the idea of a cosmic abundance table for the solid chemical elements, with extrapolations to cover the volatiles, a table which has provided a basis for the theory of nuclear reactions within the Sun and kindred stars (see Cosmochemistry).

Urey's great contribution to planetary studies was in the foundations laid for an observationally based science of cosmochemistry (ter Haar and Cameron, 1967).

Rhodes W. Fairbridge

Bibliography

Aller, L.H. (1961) *The Abundance of the Elements*. New York: John Wiley (Interscience), 283 pp.

Anonymous (1960) Urey, Harold C(layton). *Current Biography Yearbook*, (New York: H.W. Wilson Co.).

Berkner, L.V. and Marshall, L.C. (1972) Oxygen: evolution in the Earth's atmosphere, in *The Encyclopedia of Geochemistry* (ed. R.W. Fairbridge). New York: Van Nostrand Reinhold, pp. 849–61.

Jastrow, R. and Cameron, A.G.W. (eds) (1963) *Origin of the Solar System*. New York: Academic Press, 176 pp.

Runcorn, S.K. (1985) Harold Clayton Urey. *Quart. J. Roy. Aston. Soc.*, **26**, 575–8.

Tatarewicz, J.N. (1990) Urey, Harold Clayton. *Dict. Sci. Biogr.*, Vol. 18, pp. 943–8.

ter Harr, D. and Cameron, A.G.W. (1967) Solar system: review of theories, in *The Encyclopedia of Atmospheric Sciences and Astrogeology* (ed. R.W. Fairbridge). New York: Reinhold Publ. Co., pp. 890–9.

Urey, H.C. (1952a) The planets: their origin and development. (Silliman Memorial Lecture). New Haven: Yale University Press.

Urey, H.C. (1952b) The abundance of the elements. *Phys. Rev.*, **88**, pp. 248–52.

Urey, H.C. (1959) The atmospheres of the planets, in *Handbuch der Physik*, Vol. 52: *Astrophysics III, the Solar System* Flügge. Berlin: Springer-Verlag, pp. 363–418.

V

VÄISÄLÄ ORBIT

The Gaussian method of orbit determination requires the use of three (or more) observations and, if they are accurate, well distributed and cover a reasonable arc, it will be possible to derive a meaningful orbit without making any assumptions about it. If there are only two observations, it is not possible to derive a general elliptical orbit. The Väisälä method of orbit determination requires only two observations, but makes certain assumptions.

Circular orbits may be calculated from two observations, but such orbits rarely mirror reality very closely. Also, it is not always possible to represent two observations by a circular orbit.

An elliptical orbit requires six elements to describe its size, shape and orientation in space. The usual form of these six elements are: T, the time of perihelion passage; ω, the argument of perihelion; Ω, the longitude of the ascending node; i, the orbital inclination; e, the orbital eccentricity; and a, the semimajor axis.

An elliptical orbit may be calculated from two observations, but it is necessary to assume two of the elements. In Väisälä's method, the assumptions are:

1. that the object is at perihelion (or aphelion) at the time of one of the observations;
2. that the geocentric distance at the moment of that observation is assumed, perhaps based on the object's apparent motion.

Assumption 1 is reasonable for observations of unknown minor planets that are brighter than photographic magnitude $B \sim 17$, as such objects will tend to be near very favorable oppositions, and hence near perihelion. For observations of fainter minor planets, reaching down to $B \sim 20$, this assumption is not so valid, but experience shows that Väisälä orbits are still useful in such situations. Assumption 2 allows a range of Väisälä orbits to be calculated for a given object for a variety of geocentric distances.

The Minor Planet Center, located at the Harvard–Smithsonian Center for Astrophysics in Cambridge, Massachusetts, is the central clearing house for all minor-planet and comet astrometry. In an average month, several hundred new minor planets with observations on at least two nights will be reported to the Minor Planet Center, and it is the Center's policy to compute Väisälä orbits for these discoveries.

Väisälä orbits are a very useful tool for following up discoveries of minor planets. This may be demonstrated by trying to link observations of the same minor planet. The following observations of 1991RJ$_{11}$ made by E. Elst, Uccle, at the European Southern Observatory (ESO) in September 1991, were published on *Minor Planet Circular* 19241:

Date	UT	α h m s	δ ° ′ ″	Obs
1991 09	04.02361	20 28 24.83	−20 56 54.9	809
1991 09	07.03611	20 27 27.57	−21 00 54.2	809

where the date is given in YYYY MM DD.ddddd form, the topocentric right ascension (α) and declination (δ) are J2000.0 positions and Obs is a code to indicate that the observations were made at ESO. The magnitude was given as $B = 19.0$. The following representative Väisälä orbit was derived: $T = 1991$ Sept. 7.027; $\omega = 213°.23$; $\Omega = 105°.27$; $i = 2°.25$; $a = 2.9225$ AU; and $e = 0.1090$. (The apparent slight discrepancy between the date of the second observation and the date of perihelion is mainly due to the light-time correction: the light that reached the earth at time t left the minor planet at time $t - 0.0057756 \Delta$, where Δ is the topocentric distance of the minor planet in AU and 0.0057756 is the reciprocal of the speed of light in AU day^{-1}.)

Some months later the same observer reported positions of objects from plates taken in August 1991. It was clear that 1991 RJ$_{11}$ should be on these plates, but the observer had not identified it. Using the above orbit the following predicted positions for 1991 RJ$_{11}$ were computed:

Date	UT	α h m	δ ° ′
1991 08	02.12	20 50.33	−19 16.7
1991 08	14.19	20 40.57	−20 03.8

A candidate was found close to these predicted positions as follows:

Date	UT	α h m s	δ ° ′ ″	Obs
1991 08	02.11806	20 49 24.37	−19 21 19.9	809
1991 08	14.18819	20 40 05.53	−20 05 55.9	809

The computed motion of 1991 RJ$_{11}$ between the two August observations was −9.76 min in right ascension and −47′.1 in declination. As a check, the observed motion of the candidate was −9.32 min in right ascension and −44′.6 in declination. The computed and observed motions are consistent, and this consistency and the reasonable agreement of the computed and observed positions indicated that the candidate object was indeed 1991RJ$_{11}$. Combining the August and September observations led to the following general orbit: $T = 1991$ May 25.952; $\omega = 188°.18$; $\Omega = 105°.99$; $i = 2°.44$; $a = 3.1299$ AU; $e = 0.1140$. There are broad similarities between this general orbit and the Väisälä orbit calculated from the September observations; in particular, the plane of the planet's orbit (given by Ω and i) was very well determined by the Väisälä orbit. The general orbit enabled 1991RJ$_{11}$ to be identified with 1989CT$_5$, giving the linked orbit: $T = 1991$ June 14.644 TT; $\omega = 192°.72$; $\Omega = 105°.84$; $i = 2°.42$; $a = 3.1296$ AU; $e = 0.1149$.

Gareth V. Williams

Bibliography

Bowell, E., Chernykh, N.S. and Marsden, B.G. (1989) Discovery and follow up of asteroids, in *Asteroids II* (ed. R.P. Binzel, T.

Gehrels and M.S. Matthews). Tucson: University of Arizona Press, pp. 21–38.
Bowell, E., Skiff, B.A., Wassermann, L.H. and Russell, K.S. (1990) Orbital information from asteroid motion vectors, in *Asteroids, Comets, Meteors III* (eds C.-I. Lagerkvist, H. Rickman, B.A. Lindblad, and M. Lindgren). Uppsala University, pp. 19–24.
Dubyago, A.D. (1949) *The Determination of Orbits*. Translated from the Russian by R.D. Burke *et al.* Published 1961 by Macmillan.
Elst, E.W. (1991) *Minor Planet Circular* 19241.
Elst, E.W. (1992) *Minor Planet Circular* 20599.
Marsden, B.G. (1991) The computation of orbits in indeterminate and uncertain cases. *Astron. J.*, **102**, 1539.
Väisälä, Y. (1939) Eine einfache Methode der Bahnbestimmung. *Annales Academiæ Scientiarum Fennicæ. Ser. A*, **52**(2).
Williams, G.V. (1992) *Minor Planet Circular* 20640.

Cross references

Asteroid
Orbit

VAN ALLEN, JAMES ALFRED (1914–)

A pioneer in space research, Van Allen was born in Mount Pleasant Iowa on 7 September 1914, the second of four children of Alfred Morris and Alma E. (Olney) Van Allen. His father was a lawyer as was his older brother George. The three younger sons all developed an interest in science; James and William received degrees in physics and Maurice became a neurosurgeon. James attended Iowa Wesleyan College in Mount Pleasant, graduating summa cum laude in 1935 with a BS in physics. The following year he received an MS from the State University of Iowa in Iowa City, Iowa (now called the University of Iowa). He graduated with his PhD in nuclear physics in 1939 from the same school and became a research fellow and later a physicist in nuclear physics at Carnegie Institution of Washington, DC. In 1941 he developed an interest in cosmic ray research and switched to the Department of Terrestrial Magnetism at the same institution.

In 1942 he became a physicist at the Applied Physics Laboratory (APL) of Johns Hopkins University and also lieutenant in the US Navy. During the war he worked on various military projects, including the proximity fuse. On 13 October 1945 he married Abigail Fithian Halsey, and subsequently had five children, three daughters and two sons. Leaving the Navy as a Lt Commander USNR, he return to research at APL where he was the supervisor of the high-altitude research group. This group was engaged in the measurement of cosmic rays, the Earth's magnetic field, atmospheric ozone and UV spectroscopy of the Sun at high altitudes using balloons. He pioneered the use of V-2 rockets for this work, and later supervised the development of the Aerobee rocket which replaced the V-2s in high-altitude research. He also led several scientific expeditions to measure cosmic rays at high altitudes using rockets launched from ships.

In January of 1951 Van Allen became Professor and Head of the Department of Physics and later (1959) Astronomy. In 1972 he was named the Carver Professor of Physics. He continued his high-altitude work at Iowa, developing small rockets that were launched from balloons.

Taking a leave of absence for the academic year 1953/4, he was a research associate at Princeton University working on Project Matterhorn, the early experimental work on controlled thermal-nuclear reactions. Feeling that there were years of work ahead in this area without much hope of success, he returned to Iowa and resumed his cosmic ray studies.

As part of the International Geophysical Year (IGY), his group at Iowa began preparing a cosmic ray instrument for the first US satellite. With the failure of the Vanguard launch vehicle, his group quickly modified the instrumentation for flight on the Jupiter-C rocket. This resulted in the first US satellite, Explorer 1. The next successful flight was Explorer 3, which included the same instrumentation along with a tape recorder. The data from these satellites were very puzzling. After several months of working with the data, Van Allen and his students came to the conclusion that they had detected energetic charged particles trapped in the Earth's magnetic field (the Van Allen belts). He and his former students have continued their research on the radiation belts of the Earth. He has had instrumentation on spacecraft that have made the first flybys past Venus, Mars, Jupiter and Saturn. The last two planets have very large natural radiation belts.

He officially retired in 1985 as the Head of the Physics and Astronomy Department, but has continued his research. His present research is concerned with the study of cosmic rays at great distances from the Sun using data from the Pioneer 10 and 11 spacecraft, the search for the boundary between the Sun and the interstellar medium, as well as the study of the natural radiation belts of the Earth, Jupiter and Saturn.

He is a member of numerous organizations, including the American Physical Society, American Geophysical Union and the National Academy of Sciences.

His numerous awards and medals include the National Medal of Science, the Nansen Medal and Prize and the Crafoord Prize.

Bruce Randall

Bibliography

Van Allen, J.A. (1959) Radiation belts around the Earth. *Sci. Am.*, **200**(3), 39–47.
Van Allen, J.A. (1975) Interplanetary particles and fields, *Sci. Am.*, **233**(3), 160–72.
Van Allen, J.A. (1983) *The Origins of Magnetospheric Physics*. Washington, DC: Smithsonian Institution Press.

Cross references

Earth: magnetic field and magnetosphere
Radiation belt

VEGA MISSION

The name of the Vega mission comes from a contraction of the Russian words *Venera* (Venus) and *Gallei* (Halley). The unique opportunity to combine missions to these celestial bodies was realized in 1984, when twin spacecraft Vega 1 and Vega 2 were launched by Proton rockets from the Baikonur cosmodrome on 15 and 21 December respectively. The Vega project was an international one. Though the spacecraft were controlled by the Soviet Union, the scientific program and payload were coordinated by the International Science and Technical Committee (ISTC), representing scientific institutions from nine countries. ISTC designed the Vega mission to be complementary to the European Giotto and Japanese Suisei cometary missions (See also Giotto mission; Sakigate and Suisei missions).

Each Vega spacecraft weighed about 4500 kg at launch including the 2000 kg Venus descent module. Half a year after the launch, on 11 and 15 June respectively, the spacecraft delivered the first balloons into the Venusian atmosphere (each carrying four scientific experiments) as well as delivering landers with nine experiments to the surface of the planet. The results of this part of the mission and more details on the complete project can be found in the papers of the scientific leadership of the mission (Sagdeev *et al.*, 1986a,b), and in papers describing preliminary results of separate scientific instruments.

After a gravitational maneuver near Venus the flyby modules were targeted to intercept Halley's comet (astronomical name P/Halley 1986 III). The interplanetary orbit of Vega 1 from Venus to Halley flyby is shown in Figure V1. The flight operation center was located in Evpatoria (Crimea), equipped with a 70-m main antenna, but most investigators were in Moscow at the Space Research Institute of the Russian Academy of Sciences (acronym IKI), where they were able to receive telemetry from deep space antennas in both Evpatoria and Medvezy Ozera (near Moscow, with a 64-m dish).

Vega 1 first encountered comet Halley on 6 March 1986 at 7 : 20 : 06 UT with a relative velocity of 79.2 km s^{-1}. Vega 2 passed near the comet nucleus 3 days later on 9 March 1986 at 7 : 20 : 00 UT with a velocity of 76.8 km s^{-1}. Their closest approaches to the nucleus were 8890 km and 8030 km respectively. The encounter trajectories of Vega 1 and 2 as well as those of Giotto and Suisei are presented in Figure V2.

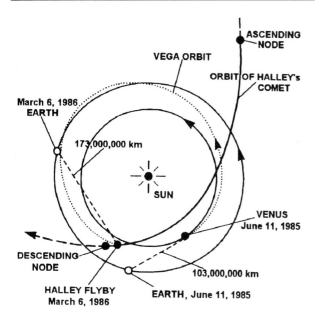

Figure V1 The path of Vega 1 probe from Venus to Halley. The projection of the inner portion of Halley's comet on the plane of the ecliptic is also shown.

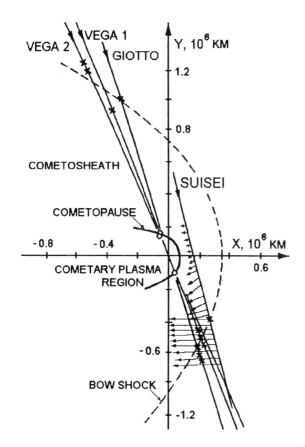

Figure V2 General overview of the spacecraft flyby trajectories and locations of the bow shock (crosses), cometopause (circles), cometosheath and cometary plasma region as identified from *in situ* plasma observations. Arrows on the Suisei trajectory show the direction of the plasma flow; their lengths are proportional to the plasma velocity.

Each Vega spacecraft, with a 'wingspan' of ~ 10 m, was three-axis stabilized and carried 14 experiments (Table V1). The three optical experiments. (TVS television system, IKS infrared spectrometer and TKS three-channel spectrometer) required a steerable platform (ASP-G), which could be automatically pointed at the cometary nucleus with an accuracy of 2.5 arcmin. It could scan an angular sector of 110° in the ecliptic plane and 60° in a plane perpendicular to the ecliptic.

The twin Vega spacecraft reached the comet a few days before the Giotto spacecraft. Optical observations of the cometary nucleus by Vega 1 and 2 TVS cameras, combined with a very accurate determination of trajectories using very long baseline interferometry, permitted the successful completion of the cooperative project 'Pathfinder.' As a result the uncertainty of Giotto's flyby distance was reduced from ± 1500 km to about ± 40 km, thus enabling the successful completion of its scientific program (Münch, Sagdeev and Jordan, 1986).

Of the three missions to comet Halley, the scientific package of Vega 1 and 2 (Table V1) could support the most composite studies. The payload of Giotto, though capable of more detailed plasma and neutral gas measurements, did not include optic and plasma wave experiments, while the Suisei payload consisted of only two scientific instruments.

During the journey of Vega 1 to comet Halley, the data returned from the scientific instruments were displayed in IKI in real time. It was first time that it was possible to observe a cometary nucleus. The existence of the solid, single nucleus, as predicted by F.L. Whipple, was demonstrated. Here the most obvious quantum jump in our knowledge was achieved. The nucleus turned out to be an irregular peanut-shaped body of about $14 \times 7.5 \times 7.5$ km in size. Its rotation period of 53 ± 3 h was obtained from a comparison of TVS images obtained from Vega 1 and Vega 2. First estimates based on measured brightness gave a geometrical albedo of $0.04^{+0.02}_{-0.01}$, which places Halley's nucleus among the darkest objects in the solar system.

The temperature of the nucleus (300–400 K as measured by the IKS spectrometer onboard Vega 1 at 0.8 AU) was much higher than the equilibrium temperature of subliming water ice of about 200 K. It was therefore reasonably assumed that what we were observing was not a bare icy nucleus but rather a nucleus covered by a layer of dark, warm dust.

The total gas production rate of Halley's nucleus was $Q \approx 1.3 \times 10^{30}$ molecules s^{-1} ≈ 40 metric ton s^{-1} (if the gas is H_2O). This was determined by a PLASMAG radial profile of the overall neutral gas density $n_n(r)$ of the cometary coma. Those data shown in Figure V3 were presented to the scientific community meeting in IKI on the day after the Vega 1 flyby. The evaluations of Q and ionization scale length (λ, 2×10^6 km) resulted from fitting observational data using a curve of the form.

$$n_n(r) = Q/(4\pi r^2 v_g) \exp(-r/\lambda)$$

where $v_g \approx 1$ km s^{-1} is the expansion velocity of the spherically expanding neutral gas with a specific ionization lifetime of λ/v_g.

In the internal part of the coma (< 500 km from the nucleus), the most abundant molecules (H_2O and CO_2) were identified by their infrared spectra as measured by the IKS and TKS experiments. Many secondary neutral species, such as OH, C_2, CH, CN and NH, were revealed by numerous spectra obtained by TKS. These provide information concerning the detailed chemistry of the inner coma.

Though the first dust particles were detected as far as $\sim 3 \times 10^5$ km from the nucleus, the sharp increase in the number densities of small dust particles of 10^{-15}–10^{-16} g took place at a distance of $\sim 1.5 \times 10^5$ km. This was interpreted as the inbound crossing of the dust paraboloid produced by the light pressure effect with an apex of $\sim 4.53 \times 10^4$ km. For larger masses of 10^{-12}–10^{-14} g, the boundary of the dust coma moves closer to the comet as a result of enhancement of the light pressure for such particles due to Mie resonances. Total dust production rate was estimated as 5–10 metric ton s^{-1} in the mass range 10^{-6}–10^{-16} g. The chemical composition of dust particles has been analyzed by the dust mass spectrometer PUMA. More than 1000 spectra have been classified into three broad classes: (1) particles mainly of low-z elements (C, H, O, N), (2) particles reminiscent of C1 carbonaceous chondrites but enriched in C, and (3) similar to class 2 particles but more enriched in hydrogen.

The outermost signatures of comet Halley were detected by the energetic particle telescope TÜNDE-M as far as 10 million km from the nucleus. How can this observation be linked with the presence of a comet? The charged particles originating from ionization of the cometary neutral gas have energies of only a few eV, which should

Table V1 Scientific experiments aboard Vega spacecraft

Acronym	Experiment	Goal and instrument parameters
TVS	Television system	Inner coma and nucleus imaging. Two CCD cameras, fields of view $0.43° \times 0.57°$ and $3.5° \times 5.3°$
IKS	Infrared spectrometer	Detection of infrared emissions of coma and thermal radiation of nucleus with $2.5 < \lambda < 12$ m
TKS	Three-channel spectrometer	Spectral mapping of coma emission in the range $0.12 < \lambda < 1.9$ m
PUMA	Dust mass spectrometer	Dust particle and elemental composition
SP-1	Dust particle counter	Dust particle flux and mass spectrum for $m > 10^{-16}$ g
SP-2	Dust particle counter	Dust particle flux and mass spectrum for $m > 10^{-16}$ g
DUCMA	Dust particle detector	Dust particle flux and mass spectrum for $m > 1.5 \times 10^{-13}$ g
FOTON	Shield penetration detector	Large dust particle detection (under anti-dust shield)
PLASMA G	Plasma energy spectrometer	Integral fluxes and energy spectra of cometary (15–3500 eV) and solar wind (0.05–25 keV) ions, electron spectra (3–10000 eV), neutral gas density ≥ 1 cm^{-3}
TÜNDE-M	Energetic particle telescope	Energy and flux of accelerated cometary ions, 0.04–13 MeV
ING	Neutral gas experiment	Neutral gas composition
MISHA	Magnetometer	Magnetic field ± 100 nT
APV-N	Wave and plasma analyzer	Plasma waves, 0.01–1000 Hz, plasma ion flux fluctuations
APV-V	Wave and plasma analyzer	Plasma waves, 0–30 kHz, plasma density and temperature

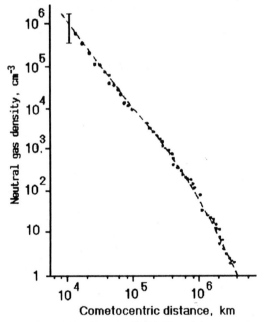

Figure V3 Comparison of the neutral gas density profile determined from the Ram Faraday Cup sensor of the PLASMAG experiment with that one predicted by theory (dashed line).

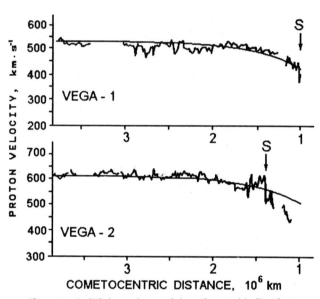

Figure V4 Radial dependence of the solar wind bulk velocity, as observed by the PLASMAG experiment at Vega 1 and 2 inbound trajectories upstream of the bow shock S of comet Halley. Solid line is the theoretical expectation for the cometary neutral gas production rate Q (1.3×10^{30} molecules s^{-1}). (From Gringauz and Verigin, 1990.)

be insufficient to register on this instrument. The explanation comes from the fact that the cometary nucleus moves not in an empty space but in the highly supersonic stream of solar plasma – 'solar wind'. The self-consistent electric and magnetic fields of the solar wind accelerate fresh cometary ions up to velocities less than twice that of the solar wind (~ 1000 km s^{-1}) for about one cyclotron period. This mechanism alone is still insufficient to explain TÜNDE-M observations, so it was concluded that the second-order Fermi process of turbulent acceleration must be effective even at great distances from the nucleus, and that this process is responsible for the acceleration of charged particles up to energies > 100 keV.

The general increase of the energetic particle fluxes was accompanied by deep and striking quasi-periodic variations. This effect was linked with the variation of the neutral gas production rate by the rotating cometary nucleus. The existence of periodic structures in the outer coma was not known from earlier distant Lyman alpha measurements.

Other effects of the increasing loading of the supersonic solar wind by fresh charged particles are the deceleration of the plasma flow on approach to the nucleus (Figure V4) and the formation of the cometary bow shock at $\sim 10^6$ km (Figure V2). Both effects were revealed by Vega's *in situ* measurements. The formation of the cometary bow shock is not caused by solar wind compression and heating due to interaction of the supersonic plasma flow with a sufficiently rigid obstacle (as in the case of the solar wind flow around the near-Earth magnetic obstacle or around the non-magnetized ionosphere plasma confined by the strong gravitational field of Venus). In this case the bow shock forms due to 'overloading' of the solar plasma by picked-up ions of cometary origin (Galeev, 1987). In addition to the plasma flow heating and deceleration, crossing of the cometary bow shock was accompanied by a rapid increase of the magnetic field (MISCHA magnetometer) and of plasma waves intensity with frequencies less than lower hybrid resonance (APV-N experiment).

It was proposed by the Vega plasma team that the plasma region downstream of the cometary bow shock should be called 'cometo-

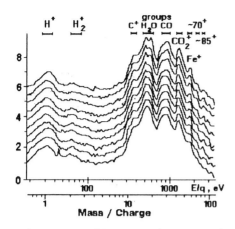

Figure V5 The sequence of ion energy/charge (mass/charge) spectra, as observed on board Vega 2 in the stagnating part of the cometary plasma region at cometocentric distance of (1.4–1.7) × 10^4 km.

sheath' (Gringauz et al., 1986) since the energy distribution of ions in this region is unique compared with similar regions near solar system planets, for example, the magnetosheath near Earth or the ionosheath near Venus. One of the differences is that the three different branches of ions are present in the ion energy distribution; the ratio of intensities of these branches changes with cometocentric distance. This feature of the cometosheath is associated with the above-noted principal difference in the bow shock formation process near planets and comets. Another characteristic feature of the cometosheath revealed by Vega measurements is cooling of the electron plasma component on approach to the comet. The decrease of electron temperature was explained by inelastic collisions between thermal electrons and cometary neutral gas (Gringauz and Verigin, 1990).

The Vega 2 spacecraft recorded a sharp (~ 10^4 km along the trajectory) change of the proton distribution function at a distance of ~ 1.6×10^5 km from the nucleus. This boundary, called the 'cometopause' (Gringauz, et al., 1986), was not predicted theoretically; it separates the cometosheath itself from the inner cometary plasma region (Figure V2). In the vicinity of the cometopause the dominance of solar wind protons changes to the dominance of heavy ions of cometary origin. The analysis of the physical processes at the cometopause based on PLASMAG, MISCHA, APV-V and APV-N data led to a conclusion that it is not one of the MHD discontinuities, and that it is the firehose instability that initiates the rapid change of the proton distribution function (Galeev et al., 1988).

Well inside the cometary plasma region Vega 2 encountered a region where both thermal and bulk velocities of cometary ions become small compared with the spacecraft relative velocity of 76.8 km s. Under these conditions energy per charge (E/q) spectra measured by the ram analyzer of the PLASMAG experiment can easily be interpreted as the mass spectra (Figure V5). These were the first in situ measurements of the mass composition of the cometary plasma. At $r \sim 1.5 \times 10^4$ km the relative abundance of water-group ions was 70–80%, with about 15–20% CO group ions and about 2–5% ions with m/q 44 (CO_2^+). A somewhat unexpected result was the observation of a well-defined peak at m/q 56 (Fe^+?). No ions of iron were identified earlier in optical spectra of Halley's comet; in optical spectra of other comets the metallic ions (usually light metals) have been identified at much smaller heliocentric distances compared with those where Vega 2 intercepted Halley's comet.

M.I. Verigin

Bibliography

Galeev, A.A., Gringauz, K.I., Klimov, S.I. et al. (1988) Physical processes in the vicinity of the cometopause interpreted on the basis of plasma, magnetic field and plasma wave data measured on board the Vega 2 spacecraft. *J. Geophys. Res.*, **93**(A7), 7527–31.

Galeev, A.A. (1987) Encounter with comets: discoveries and puzzles in cometary plasma physics. *Astron. Astrophys.*, **187**, 12–20.

Gringauz, K.I., Gombosi, T.I., Remizov, A.P. et al. (1986) First in situ plasma and neutral gas measurements at comet Halley. *Nature*, **321**, 282–5.

Gringauz, K.I. and Verigin, M.I. (1990) Some results of neutral and charged particle measurements in the vicinity of comet P/Halley aboard Vega 1, 2 spacecraft, in *Comet Halley Investigations, Results, Interpretations. Volume 1: Organization, Plasma, Gas* (ed. J. Mason). Chichester, England; Ellis Horwood, pp. 147–68.

Münch, R.E., Sagdeev, R.Z. and Jordan, J.F. (1986) Pathfinder: accuracy improvement of comet Halley trajectory for Giotto navigation. *Nature*, **321**(6067), 318–20.

Sagdeev, R.Z., Linkin, V.M., Blamont, J.E. and Preston, R.A. (1986a) The VEGTA Venus balloon experiment. *Science*, **231**, 1407–8.

Sagdeev, R.Z., Blamont, J., Galeev, A.A. et al. (1986b) Vega spacecraft encounters with comet Halley. *Nature*, **321**(6067), 259–62.

Cross references

Comet: observation
Giotto mission
Halley, Edmond, and Halley's comet
Sakigake and Suisei missions

VENERA MISSIONS

Venera 9 and 10: first satellites of Venus

Venera 9 and 10 were the first artificial satellites to orbit Venus. Their objectives included the study of the surface and atmosphere of Venus and its environment. Landers that separated from the mother spacecraft 2 days before orbit insertion provided the first panoramas of the rocky Venusian surface. Plasma and magnetic field packages on the two orbiters provided the first detailed study of solar wind–Venus interactions.

Orbit

The interplanetary probe Venera 9 was launched from the Baikonur Cosmodrome on 8 June 1975, and Venera 10 was launched on 14 June 1975. After the landers separated from the spacecraft (two days before insertion), final corrections of the trajectories were made to ensure the passage of the spacecraft through the planned pericenter regions. After braking maneuvers near Venus, on 22 and 25 October 1975 respectively, they entered Venus orbit. Initial parameters of the satellites' orbits are given in Table V2 (Abramovich et al., 1976).

Spacecraft

The Venera spacecraft is shown in Figure V6. It consists of an orbital module and lander. The orbiter is a combination of a cylinder that provides the main structure of the spacecraft and support subsystems of the spacecraft, and a pressurized torus for the electronics boxes of spacecraft systems and of scientific instrumentation.

A three-axis stabilization mode of the spacecraft allowed the high-gain antenna to be directed to the Earth, and the solar panels were oriented nearly perpendicular to the Sun direction. As Venus moved around the Sun the angle between the solar panels' normal and the

Table V2 Orbital parameters of Venera spacecraft

	Venera 9	Venera 10
Apocenter height (km)	118 190	119 930
Pericenter height (km)	1545	1665
Inclination	34°10′	29°30′
Period of revolution	48h 18min	49h 23min
Longitude of ascending node	76°25′	107°05′
Time of pericenter passage (UT)	03h 01min 17s	03h 02min 19s

Figure V6 Venera 9 and 10 spacecraft configuration with lander attached.

Sun direction changed in order to keep the high-gain antenna directed to the Earth.

The rate with which the Sun-oriented axis of the spacecraft changed its direction in the ecliptic plane was about 0.45° day^{-1}. Three-axis stabilization in other orientations was also used for pointing instruments at the planet. A spin-stabilized mode with the axis approximately directed to the Sun and with solar panels illuminated was also used on some orbits.

Landers

The Venera 9 lander, after descending through the atmosphere of Venus, landed on 22 October 1975 at 05 : 13 UT at latitude 31°42', longitude 290°50' and zenith angle 56°.6 (Abramovich et al., 1976). On the descent and for 53 min after landing, the instrumentation of the lander performed measurements of atmospheric parameters and of the soil properties. The Venera 10 lander touched down on 25 October 1975 at 05 : 17 UT at latitude 16°02', longitude 291°00' and zenith angle 62.5°. Measurements continued for 65 minutes after landing.

Instrumentation

Several optical instruments were installed on the orbiters for investigations of the atmosphere of Venus. An infrared spectrometer with circular wedge interference filter for the wavelength range 1.6–2.6 μm with resolution $\lambda/\Delta\lambda = 20$ was used to study the upper boundary of the cloud layer (Gnedykh et al., 1976). An infrared radiometer with two wave bands at 8–13 μm and 18–28 μm provided data on the temperature of the cloud layer (Ksanfomaliti et al., 1976a). The scattering of solar radiation by the atmosphere was studied with photometers with interference filters centered on 352 nm (Venera 9) and 345 nm (Venera 10), and with a photopolarimeter with eight filters in the wavelength range from 335 to 800 nm (Ksanfomaliti et al., 1976b). A grating spectrometer in the wavelength range 300–800 nm with a photomultiplier was used for measurements of the night sky emission of Venus (Krasnopol'skii et al., 1976). The hydrogen corona of Venus was measured with a Lyman α photometer with hydrogen and deuterium absorbing cells and a Geiger–Müller photon counter (Bertaux et al., 1976).

Two plasma instruments and a magnetometer were installed on Venera 9 and Venera 10 for investigation of the solar wind–Venus interaction and the planetary environment. This group of instruments was also used for measurements of the solar wind during the cruise phase. Dual radio occultation measurements were used for investigations of the Venusian ionosphere.

The plasma spectrometer, D-127, consisted of two wide-angle Faraday cups, one for ions and one for electrons, and was a modification of instruments previously used on Mars 2, 3 and 5 (Gringauz et al., 1974; See also Soviet Mars missions). The ion Faraday cup was directed towards the Sun and measured the differential energy flux of ions in eight energy bands in the energy range from spacecraft potential to 4400 eV Q^{-1}. The full width of the angular response was 90°. Only the Venera 9 ion detector was operated as the Venera 10 detector failed during the cruise phase. It recovered in April 1976, just before the termination of the mission. The electron Faraday cup was directed antisunward and measured the integral flux of electrons above each of eight energy steps from spacecraft potential to 300 eV. The full width of the angular response was 90°. In the basic operation mode the ion and electron fluxes on the collectors were measured every second, and the exposure time at every energy step was 20 s, so 20 measurements were made at every energy step. The total cycle of measurements was 160 s.

The plasma spectrometer, RIEP, was a modification of an instrument flown on Mars 2, 3 and 5 (Vaisberg et al., 1971). It consisted of seven narrow-angle cylindrical electrostatic analyzers for measurement of energy per charge spectra of ions in the energy range from 50 eV Q^{-1} to 19 keV Q^{-1} and one analyzer for measurement of electrons. Different ion analyzers performed measurements in different energy ranges in each of eight energy intervals. They were oriented in four different directions in one plane over an angle of 45°. This plane approximately coincided with the ecliptic plane. Channel electron multipliers were used for counting particles. Measurements

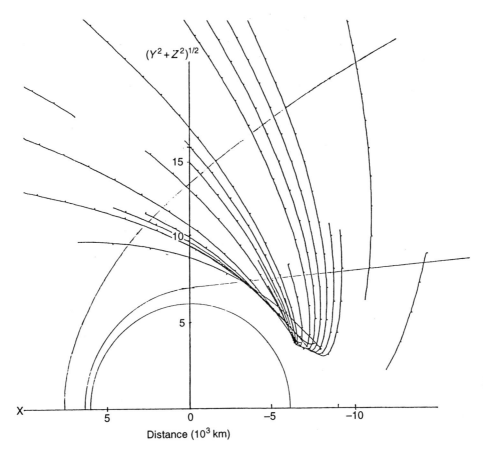

Figure V7 Parts of Venera 10 orbits on which plasma and magnetic field measurements were performed. Coordinate system is a cylindrical coordinate system with its X-axis oriented in the average aberrated solar wind direction (5° west of the Sun).

of the counting rate of every electrostatic analyzer were performed once per second. In the basic operating mode the energy steps of all analyzers were changed every 20 s, so the total cycle of measurements occupied 160 s. Analyzers for the measurement of electrons failed during the cruise phase, and the instrument on Venera 9 failed after performing measurements on first orbit around Venus.

The center of the field of view of the ion Faraday cup of the D-127 instrument and the center of the angular sector covered by the ion analyzers of RIEP were oriented towards the Sun at the beginning of the mission. They systematically moved out of the solar direction as time went on, as the orientation of the spacecraft changed to keep the high-gain antenna in the direction of the Earth.

A three-axis fluxgate magnetometer was used for magnetic field measurements (Dolginov et al., 1978). Its sensor was mounted on a boom attached to the solar panel. The precision of the measurements of the magnetic field was 1 nT.

Dual-frequency occultation measurements were made with coherent signals in the decimeter ($l = 32$ cm) and centimeter ($l = 8$ cm) radio bands (Alexandrov et al., 1976a,b; Yakovlev et al., 1976). A two-channel interferometer was used. The measured differential frequency shift was used for calculating the ionospheric density assuming spherical symmetry of ionosphere.

A transmitter with wavelength $\lambda = 32$ cm was also used for bistatic radar observations of the surface of Venus (Kucheryavenkov et al., 1976). Two solar energetic particle telescopes with silicon semiconductor detectors were used to study protons with energies of 0.075 to 25 MeV and electrons with energies of 50 to 300 keV during the cruise phase and in near-Venus orbit (Alekseev et al., 1976). Cosmic ray measurements were performed with a set of instruments including a semiconductor detector for measurements of 4–12 MeV protons, a gas discharge counter for protons with energy > 30 MeV and electrons with energy > 3 MeV, and a Cherenkov counter for protons with energies > 500 MeV and electrons with energies > 7 MeV (Vernov et al., 1976).

Telemetry and measurement modes of the plasma complex

Plasma measurements on Venera 9 and 10 were performed on selected orbits. Altogether plasma and magnetic field measurements were made on 22 orbits of Venera 9 within the time interval October 1975–January 1976 and on 20 orbits of Venera 10 from October 1975 to April 1976. The orbit of the spacecraft allowed one to perform measurements within the tail starting at approximately 0.5 R_V behind the terminator to about 5 R_V. Plasma group instruments had a lower priority compared to instruments for investigation of the neutral atmosphere, and measurements usually started 10 min after periapsis passage and continued for 90 min. The sampling rate for the plasma spectrometers and the magnetometer was 1 Hz. Two other regimes that were used occasionally provided a 0.25-Hz sampling rate for a 64-s time interval with a repetition rate of one measurement cycle every 2 min or 10 min.

Figure V7 shows the coverage of the solar wind–Venus interaction region, with the exception of two deep-tail passes, by the magnetic and plasma measurements on Venera 10 in a cylindrical coordinate system with its principal axis directed towards the Sun and with the plasma boundaries approximately indicated. It shows a quite reasonable sampling of the close tail region, ionosheath and the bow shock. Venera 9 had a similar coverage of the Venusian environment.

Lander equipment

Landers included panoramic television cameras (scanning telephotometers) that provided the first panoramas from the Venus surface (Selivanov et al., 1976a). Temperature and pressure during descent and on the surface were measured with a resistance thermometer and with six manometers (Avduevskii et al., 1976a). Measurements of winds in the atmosphere on descent were performed by an analysis of Doppler measurements of velocity of the landers (Antsibor et al., 1976). Measurements of the wind velocity on the surface of Venus were performed with cup anemometers (Avduevskii et al., 1976b). Clouds and aerosols were investigated with a nephelometer (Marov et al., 1976) and with a narrow-band photometer in the wavelength range 0.80–0.87 μm (Moroz et al., 1976). Illumination in the Venus atmosphere was measured with detectors having five bands in the wavelength range from 0.44 to 1.16 μm (Avduevskii, et al., 1976c). Finally, the density of the surface and the content of radioactive elements in Venusian rock were measured with a γ-radiation densimeter and γ-ray spectrometer (Surkov et al., 1976a,b).

Scientific results of the plasma complex

Venera 9 and 10 made measurements in the period close to solar minimum. They provided the first systematic mapping of plasma regions and boundaries, some of which were not known before. Figure V8 shows two sequences of ion spectra obtained with RIEP on two Venera 10 passes from the shadow to terminator of the planet through the ionosheath and farther into the undisturbed solar wind. Three distinct plasma domains are easily separated along the orbit, from top to bottom: the plasma tail, ionosheath and the solar wind. Low-energy ions with energies from tens to hundreds of electron volts were observed within the tail. Significantly hotter and faster ion flow was observed within the ionosheath (magnetosheath), the region behind the bow shock. The solar wind shows two peaks of cold protons and alpha particles.

Figure V9 summarizes the geometry of the plasma flow near Venus as deduced from observations of ion properties on Venera 9 and 10 with the RIEP plasma analyzer. In addition to the shock, three more boundaries separated different plasma regimes (Vaisberg et al., 1976a; Romanov et al., 1978). One boundary limits the boundary layer, or rarefaction region, within ionosheath where the ion spectra differ from those in the surrounding flow by larger fluctuations of ion flux and ion spectra, by modification of the ion spectra, especially by the disappearance of the higher-energy tail, by a decrease of the ion flux, or of the number density, and by the appearance of a second peak in the lower-energy part of the ion energy spectra. These phenomena were interpreted as consequences of the mass loading of the part of the external flow that is close to the gaseous obstacle (Vaisberg et al., 1976b).

The next boundary separates the boundary layer from the plasma region, where the energy of convective motion and ion temperature are significantly less than the one within the external flow. This boundary was initially called the 'ionopause' according to the hydrodynamic model of Spreiter (1976) that was used extensively for comparison with the observations. Because the solar wind plasma flow is terminated at this boundary, it is the boundary of the obstacle at the flank of the plasma tail of Venus (Vaisberg et al., 1976b).

The innermost boundary is the one below which no measurable fluxes of ions with energies > 50 eV were observed. The region behind the planet that is terminated by this boundary was called the cavity. Gringauz et al. (1976), reported observations of sporadic ion fluxes over a wide energy range and called this region the particulate shadow. The downstream extension of this region is unknown, as at $X \sim 5\, R_V$ the tail plasma fills most of the observed part of the tail (Vaisberg et al., 1976b), but it seems that it extends downstream, although its transverse dimension diminishes.

Bow shock

The bow shock was mapped both by the sharp changes of the plasma flow parameters and by the jump of the magnetic field. Figure V10 shows shock crossings on the dayside and on the flank (Smirnov et al., 1981). Different approximations to the observed crossings give the planetocentric distance of the shock at the subsolar point for the period of observations to be 6500 ± 200 km. It was also found that the shock is asymmetric in a $V \times B$ coordinate system (Romanov et al., 1978). The terminator projection of the shock is about 5% farther from the planet in the direction perpendicular to the IMF component that is transverse to solar wind flow (Smirnov et al., 1981).

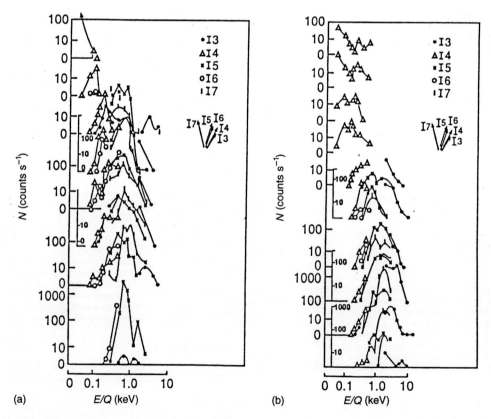

Figure V8 Two sets of ion spectra taken on parts of orbits passing (from top to bottom) through the plasma flow in the tail, ionosheath and to the solar wind: (a) at low magnetic latitudes and (b) high magnetic latitudes in a $V \times B$ coordinate system (Vaisberg and Zelenyi, 1984). Two upper spectra in (a) and four upper spectra in (b) show the difference in the tail ion populations at different magnetic latitudes.

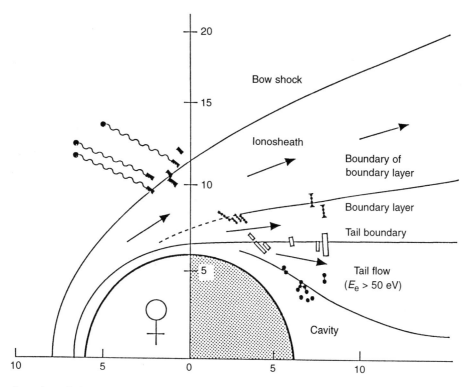

Figure V9 The plasma boundaries behind the terminator observed on Venera 9 and 10 (Vaisberg et al., 1976).

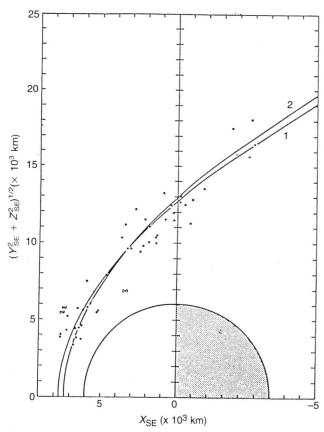

Figure V10 Venera 9 and 10 bow shock crossings and their approximations with conic sections (Smirnov et al., 1981).

Boundary layer

The boundary layer in the flow around Venus is characterized not only by the modification of ion spectra and by a decrease of number density, but is also seen in the velocity and temperature profiles. Figure V11 shows the boundary layer observed at larger distances from planet. Plasma parameters calculated from the observed ion spectra under the assumption that all ions are protons show the typical profile of velocity and temperature variations across a gas-dynamic boundary layer. These profiles support the explanation of the boundary layer in terms of mass loading of shocked solar wind flow by planetary ions. The addition of heavy ions leads to the decrease of velocity and to the increase of the width of ion spectra, interpreted as an increase of ion temperature.

Tail

Venus has a plasma tail, determined by the antisunward flow of plasma with significantly less convective energy and temperature than in the ionosheath (Vaisberg et al., 1976a). The flow of low-energy ions fills almost all the tail, and is separated from the shocked solar plasma flow by a distinct and thin (sometimes as thin as ~ 100 km) boundary. Accelerated ions with energies from 2.0 to 20 keV were observed almost at every crossing of the tail boundary. The dimension of the Venus tail of about 1.3 R_V was determined from plasma observations as the radius of the cavity formed by Venus in the solar wind flow (Vaisberg et al., 1976a). No evidence of solar wind plasma penetrating the tail has been found at close downstream distances.

Yeroshenko (1979) found that the tail magnetic field at distances close to Venus has a two-lobe structure with the polarity of the radial component controlled by the transverse component of the interplanetary magnetic field (shown draping around the planet in Figure V12). These lobes rotate as both the IMF transverse component and the current sheet separating these two tail lobes rotate. The boundary of the magnetic tail nearly coincides with the boundary of the plasma tail (Vaisberg et al., 1976b).

Plasma flow in the tail at relatively small distances behind the planet has a significant converging component, as if this plasma flow is filling the tail. It also suggests that the source region of this plasma is located near the terminator. The inflow of the plasma in the tail is

884 VENERA MISSIONS

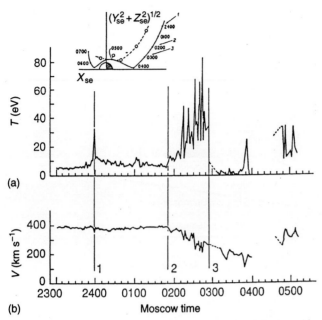

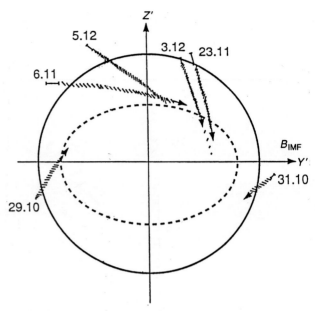

Figure V11 (a) Temperature and (b) ion velocity profiles along Venera 10 orbit (shown above) on 18/19 April 1976 according to RIEP spectrometer data (ion velocity is calculated assuming that the plasma everywhere consists of protons). Crossings of the bow shock (1) external boundary of the boundary layer (2) and the tail boundary (3) are indicated. (From Romanov, Smirnov and Vaisberg, 1978).

Figure V13 The parts of the Venera 10 orbits in the $Y'Z'$ plane of the $V \times B$ coordinate system where the tail plasma flow was observed (Vaisberg, 1980). This shows that the inflow of low-energy plasma in the tail is controlled by the IMF.

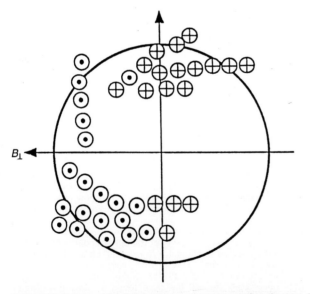

Figure V12 Polarity of B_x component in the Venus tail in the $V \times B$ coordinate system (solar wind velocity V is perpendicular to the plane of the figure, and the IMF transverse component is rotated to horizontal direction) demonstrates a two-lobe structure controlled by the IMF (Eroshenko, 1979).

asymmetric in the $V \times B$ coordinate system (Vaisberg, 1980). It can be seen from Figure V13 that at 'high magnetic latitude', i.e. far from V–B plane, the layer of plasma inflow is thick, while at 'low magnetic latitudes' the plasma is streaming in the tail within a thin layer.

The typical number flux of ions in the tail is about 2×10^6 cm^{-2} s^{-1}. At a distance $\sim 1\ R_V$ behind the planet the energy of convective motion is ~ 150 eV, the ion temperature is ~ 3 eV. To calculate the number density and velocity we must assume a plasma composition. There is evidence that the main ion of the tail flow is O$^+$. Then the typical velocity is 40 km s^{-1} and the number density is 0.5 cm^{-3}. The typical magnetic field magnitude in the tail is ~ 12 nT.

Plasma and magnetic field measurements at intermediate (2.5 to 5 R_V downstream) distances in the tail of Venus were made on two passes of Venera 10. Some measurements on one of these passes are shown on Figure VII. Figure V14 shows the dynamic spectrum of ions as measured within the tail on 26 March 1976 by RIEP, along with the magnetic field component B_x. This pass also shows that the Venusian tail is nearly permanently populated with plasma, though there are short time intervals when no ions have been observed. The mean energy of the ions is larger than at distances closer to Venus. Ions with significantly higher energy are observed in one-to-one correspondence when currents are seen in variations of the B_x component. Variation of the ion convective energy within the tail with the distance from the planet gives an average acceleration between $X = -1\ R_V$ and $X = -5\ R_V$ of $\sim 4 \times 10^4$ cm s^{-2} if the plasma consists of protons, and $\sim 2.5 \times 10^3$ cm s^{-2} if the tail flow consists primarily of singly ionized oxygen ions (Vaisberg et al., 1989).

Figure V15 is a composite of a distant Venera 10 pass on 18/19 April 1976 and a Venera 9 pass on 1 November 1975 at closer distances, as observed with the D-127 plasma spectrometer (Gringauz et al., 1976; Verigin et al., 1978). This gives an overview of plasma changes along the tail. With these measurements Verigin et al. (1978) separated corpuscular umbra and penumbra, characterized by lower (compared to the ionosheath) ion flux densities and velocities. As the major maximum of the ion spectrum systematically decreased as the satellite moved deeper into the ionosheath and penumbra regions, Verigin et al. (1978) concluded that the major maximum is determined by solar wind protons. The ion spectra measured at the magnetic field reversal within the tail showed high-energy ions with calculated ion temperatures up to 1 keV (Gringauz et al., 1977).

Using plasma measurements on Venera 9 and 10, Vaisberg and Zelenyi (1984) proposed a model of the plasma and magnetic tail as a result of the acceleration of pick-up photoions within the magnetic barrier. The magnetic barrier is formed above the dayside ionosphere due to the MHD deceleration of the flow at the obstacle and due to the depletion of the solar wind plasma from the field lines through escape along the field lines (Zwan and Wolfe, 1976). This magnetic barrier was observed by Pioneer Venus (q.v.), launched during the following opportunity in 1978 (Russell, Elphic and Slavin, 1979). Depleted field lines convect around the planet and are filled with the

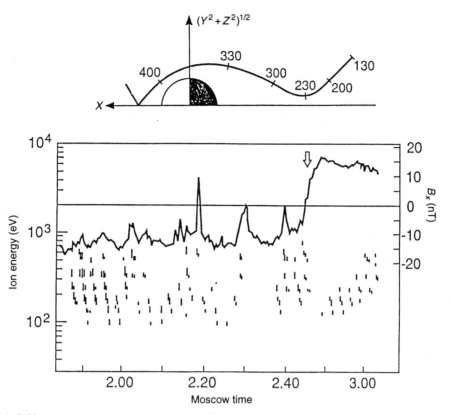

Figure V14 B_x magnetic field component and dynamic spectrum of ions as measured by the RIEP spectrometer on a Venera 10 pass within the tail on 26 March 1976. The Venera 10 orbit is shown above in a cylindrical coordinate system. The length of the bar shows the logarithm of the ion counting rate at a specific energy E/Q (scale is on the left). The B_x reversal that seems to be the main current sheet is indicated by an arrow.

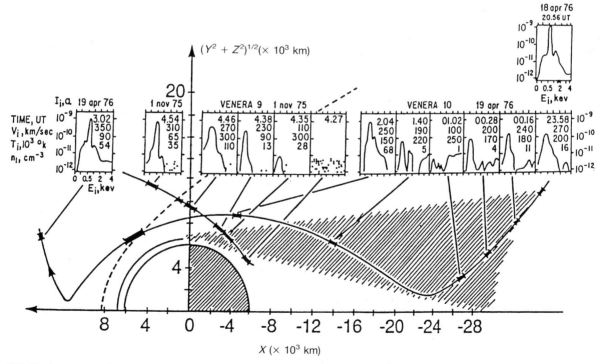

Figure V15 The ion spectra change in the downstream region from the wide-angle Faraday cup analyzer on the Venera 10 pass on 18/19 April 1976 and the Venera 9 pass on 1 November 1975. Particulate shadow behind the planet and penumbra (shaded) are indicated. (Verigin et al., 1978. Copyright American Geophysical Union.)

ions that are formed by photoionization of the upper atmospheric neutrals. The boundary between solar wind-dominated flow and the photoion-dominated flow forms at the level where the integrated mass flux of photoions equals the upstream mass flux of the solar wind. The loaded field lines are accelerated from the subsolar region to the terminator by the magnetic pressure gradient within the magnetic barrier. Then loaded magnetic field lines pass the terminator and fill the tail.

Ionosphere

A number of ionospheric profiles were obtained from dual radio occultation measurements (Alexandrov et al., 1976a,b). Gringauz et al. (1979) measured energetic electrons in the tail and concluded that they may be the main source of maintenance of the nightside ionosphere.

Atmosphere and surface of Venus

Unique panoramas of the Venus surface revealed rough rocky structure of the surface at landing sites and showed the existence of strong contrasts (Selivanov et al., 1976b). Bistatic radar measurements revealed that the surface of the Venus, on average, is smoother than the surface of the Moon (Kucheryavenko et al., 1976).

Measurements of Compton scattering of γ-radiation showed that the density of surface rock is 2.8 ± 0.1 g cm^{-3} which corresponds to massive textured basalts with low porosity (Surkov et al., 1976a). Gamma-ray spectrometer results confirmed this conclusion as the concentration of natural radioactive elements is close to terrestrial rocks of the basalt type (Surkov et al., 1976b).

Measurements of temperature and pressure profiles down from 63 km confirmed the results of earlier probes (Avduevskii et al., 1976a). Wind velocities in the atmosphere of Venus above ~ 35 km exceed 50 m s^{-1} (Antsibor et al., 1976), while near the surface it is, on average, within 0.5–1.0 m/s (Avduevskii et al., 1976b). Nephelometric measurements revealed that the lower boundary of the clouds is at 49 km, and that the clouds consist of particles of diameter 0.5–1. μm with a number density between 50 and 500 cm^{-3} (Marov et al., 1976). An aerosol component was also found below 49 km. Photometric measurements confirmed the number density of clouds, ~ 200 cm^{-3}, and found that the clouds are structured (Moroz et al., 1976). The relative abundance of water, H_2O/CO_2, was measured as $\sim 10^{-3}$ in the 25–45 km range. Measurements of lighting conditions showed the integral flux on surface to be ~ 100 W m^{-2} (Avduevskii et al., 1976c).

The upper boundary of the cloud layer was determined to range from 65 to 68 km from measurements by the infrared spectrometer (Gnedykh et al., 1976). Infrared radiometry showed the day–night asymmetry of temperature at the upper boundary of cloud layer to be about 10 K (Ksanfomaliti et al., 1976a). Thus the higher temperatures observed on the nightside suggest that at night the upper boundary of the cloud layer is several kilometers lower than during the day. UV photometry of Venus showed that the 6–9 km thick scattering layer of finely dispersed particles exists above the upper boundary of the principal clouds (Ksanfomaliti et al., 1976b).

Strong emission bands between ~ 4000 and ~ 7000 Å were observed in the nightside upper atmosphere at heights of 90 ± 25 km (Krasnopol'skii et al., 1976). They were interpreted as emission from CO molecules excited either by fast electron impact or by direct chemical reaction between C and O atoms with participation of a third particle (Slysh, 1976).

Measurements of Lyman α scattering by hydrogen in the upper atmosphere of Venus showed the existence of an extended hydrogen corona, with a temperature at the critical level of the atmosphere of 45 ± 50 K (Bertaux et al., 1976).

Conclusions

There are many results that were obtained for the first time with Venera 9 and Venera 10 experiments, and they contributed substantially to our present understanding of Venus and its environment. They provided a first close-up look at the Venusian surface, and led to better knowledge of the atmosphere, cloud layer and upper atmosphere of the planet. Plasma and magnetic field experiments on Venera 9 and 10 provided the first detailed picture of the induced magnetic field and pick-up processes in the solar wind–planetary atmosphere interaction.

Oleg L. Vaisberg

Bibliography

Abramovich, S.K., Ageeva, G.D., Akim, E.L. et al. (1976) Ballistics and navigation in the control of flight of automatic interplanetary stations Venera 9 and Venera 10. Cosmic Res., **14**, 586.

Aleksandrov, Yu.N., Vasil'ev, M.B., Vyshlov, A.S. et al. (1976a) Preliminary results of two-frequency radioscopy of the daytime ionosphere of Venus from Venera 9 and Venera 10. Cosmic Res., **14**, 703–6.

Aleksandrov, Yu.N., Vasil'ev, M.B., Vyshlov, A.S. et al. (1976b) Nighttime ionosphere of Venus from the results of two-frequency radioscopy from Venera 9 and Venera 10. Cosmic Res., **14**, 706.

Alekseev, N.V., Vakulov, P.V., Vologdin, N.I. et al. (1976) Measurements of interplanetary background due to low-energy particles on the Venera 9 and Venera 10. Cosmic Res., **14**, 728.

Antsibor, N.M., Bakit'ko, R.V., Ginzburg, A.L. et al. (1976) Estimates of wind velocity and turbulence from relayed Doppler measurements of the velocity of instruments dropped from Venera 9 and Venera 10. Cosmic Res., **14**, 625.

Avduevskii, V.S., Borodin, N.F., Burtsev, V.P. et al. (1976a) Automatic stations Venera 9 and Venera 10 – Functioning of descent vehicles and measurements of atmospheric parameters. Cosmic Res., **14**, 577.

Avduevskii, V.S., Vishnevetskii, S.L., Golov, I.A. et al. (1976b) Measurement of wind velocity on the surface of Venus during the operation of stations Venera 9 and Venera 10. Cosmic Res., **14**, 622.

Avduevskii, V.S., Golovin, Yu.M., Zavelevich, F.S. et al. (1976c) Preliminary results of an investigation of the lightning conditions in the atmosphere. Cosmic Res., **14**, 643.

Bertaux, J.L., Blamont, J.E., Dzyubenko, A.I. et al. (1976) Investigation of scattered L_α radiation in the vicinity of Venus. Cosmic Res., **14**, 695.

Dolginov, Sh.Sh., Zhuzgov, L.N., Sharova, V.A. and Buzin, V.B. (1978) Magnetic field and magnetosphere of the planet Venus. Cosmic Res., **16**, 657–87.

Dolginov, Sh.Sh., Dubinin, E.M., Eroshenko, E.G. et al. (1981) Field configuration in the magnetic tail of Venus. Cosmic Res., **19**, 434.

Eroshenko, E.G. (1979) Unipolar induction effects in the magnetic tail of Venus. Kosmich. Issled., **17**, 604.

Gnedykh, V.I., Zhegulev, V.S., Zasova, L.V. et al. (1976) Preliminary results of investigations of the infrared spectrum of Venus on the Venera 9 and Venera 10 spacecraft. Cosmic Res., **14**, 662.

Gringauz, K.I., Verigin, M.I., Breus, T.K. and Gombosi, T. (1979) The interaction of electrons in the optical umbra of Venus with the planetary atmosphere – the origin of nightside ionosphere. J. Geophys. Res., **84**, 2123.

Gringauz, K.I., Bezrukikh, V.V., Volkov, G.I. et al. (1974) Study of solar plasma near Mars and on the Earth–Mars route using charged-particle traps on Soviet spacecraft in 1971–1973, methods and instruments. Cosmic Res., **12**, 394

Gringauz, K.I. Bezrukikh, V.V., Breus, T.K. et al. (1976) Plasma observations near Venus onboard the Venera 9 and 10 satellites by means of wide angle plasma detectors, in Physics of Solar Planetary Environment (ed. D.J. Williams). Boulder: American Geophysical Union, pp. 918–32.

Krasnopol'skii, V.A., Krys'ko, A.A., Rogachev, V.N. and Parshev, V.A. (1976) Spectroscopy of the night-sky luminescence of Venus from the interplanetary spacecraft Venera 9 and Venera 10. Cosmic Res., **14**, 687.

Ksanfomaliti, L.V., Dedova, E.V., Obykhova, L.F. et al. (1976a) Infrared radiation of the clouds of Venus. Cosmic Res., **14**, 670.

Ksanfomaliti, L.V., Dedova, E.V., Zolotukhin, V.G. et al. (1976b) UV photometry of Venus. Scattering layer over absorbing clouds. Cosmic Res., **14**, 677.

Kucheryavenkov, A.I., Yakovlev, O.I., Pavelev, A.G. et al. (1976) Relief of Venus from the results of bistatic radar observations with Venera 9 and Venera 10. *Cosmic Res.*, **14**, 609.

Marov, M.Ya., Byvshev, B.V., Manuilov, K.N. et al. (1976) Nephelometric measurements by the Venera 9 and Venera 10 spacecraft. *Cosmic Res.*, **14**, 637.

Moroz, V.I., Parfent'ev, N.A., San'ko, N.F. et al. (1976) Preliminary results of narrow-band photometric probing of the cloud layer of Venus in the 0.80–0.87-μ spectral region of the Venera 9 and Venera 10 descent vehicles. *Cosmic Res.*, **14**, 649.

Romanov, S.A., Smirnov, V.N. and Vaisberg, O.L. (1978) On the nature of solar wind–Venus interaction. *Kosmich. Issled.*, **16**, 746.

Russell, C.T., Elphic, R.C., and Slavin, J.A. (1979) Initial Pioneer Venus magnetic field results: dayside observations. *Science*, **203**, 745–8.

Selivanov, A.S., Chemodanov, V.P., Naraeva, M.K. et al. (1976a) A television experiment on the surface of Venus. *Cosmic Res.*, **14**, 593.

Selivanov, A.S., Panfilov, A.S., Naraeva, M.K. et al. (1976b) Photometric analysis of panoramas on the surface of Venus. *Cosmic Res.*, **14**, 596.

Slysh, V.I. (1976) Identification of the CO molecule in the radiation spectrum of the Venusian nighttime sky. *Cosmic Res.*, **14**, 693.

Smirnov, V.N., Vaisberg, O.L., Romanov, S.A. et al. (1981) Three-dimentional shape and position of the shock wave at Venus. *Cosmic Res.*, **19**, 425.

Spreiter, J.R. (1976) Magnetohydroynamic and gasdynamic aspects of solar wind flow around terrestrial planets: a critical review, in *Solar-Wind Interaction with the Planets Mercury, Venus, and Mars* (ed. N.F. Ness). Washington: NASA Publ. NASA SP-397, pp. 135–50.

Surkov, Yu.A., Kirnozov, F.F., Khristianov, V.K. et al. (1976a) Density of surface rock on Venus from data obtained by the Venera 10 automatic interplanetary station. *Cosmic Res.*, **14**, 612.

Surkov, Yu.A., Kirnozov, F.F., Glazov, V.N. et al. (1976b) of natural radiactive elements in Venusian rock as determined by Venera 9 and Venera 10. *Cosmic Res.*, **14**, 618.

Vaisberg, O.L. (1980) On the asymmetry of the internal flow in the Venus wake. *Kosmich. Issled.*, **18**, 809.

Vaisberg, O.L. and Zelenyi, L.M. (1984) Formation of the plasma mantle in the Venusian magnetosphere. *Icarus*, **58**, 412.

Vaisberg, O.L., Smirnov, V.N., Zastenker, G.N. and Fedorov, A.O. (1989) Experimental data on the plasma envelopes of Mars, Venus, and Halley's and Giacobini-Zinner's comets: comparison of loading effects. *Cosmic Res.*, **27**, 638.

Vaisberg, O.L., Zhurina, L.S., Kovalenko, V.G. et al. (1971) Multichannel modular spectrometer of low energy electrons and ions. *Pribori i Technika Experimenta*, No. **6**, pp. 42–4.

Vaisberg, O.L., Romanov, S.A., Smirnov, V.N. et al. (1976a) Structure of the region of interaction of the solar wind with Venus inferred from measurements of ion-flux characteristics of Venera 9 and Venera 10. *Cosmic Res.*, **14**, 709.

Vaisberg, O.L., Smirnov, V.N., Karpinsky, I.P. et al. (1976b) Ion flux parameters in the solar-wind–Venus interaction region, in *Physics of Solar Planetary Environment* (ed. D.J. Williams). Boulder: American Geophysic Union, pp. 904–17.

Verigin, M.I., Gringauz, K.I., Gombosi, T. et al. (1978) Plasma near Venus from Venera 9 and 10 wide-angle analyzers data. *J. Geophys. Res.*, **83**, 3721.

Vernov, S.N., Tverskoi, B.A., Volga, V.I. et al. (1976) Cosmic-ray measurements with Venera 9 and Venera 10 probes. *Cosmic Res.*, **14**, 732.

Yakovlev, O.I., Efimov, A.I., Timofeeva, T.S. et al. (1976) Preliminary radio transmission data and the Venusian atmosphere from Venera 9 and Venera 10. *Cosmic Res.*, **14**, 632.

Yeroshenko, Ye.G. (1979) Unipolar induction effects in the magnetic fail of Venus. *Kosmich. Issled.*, **20**, 604.

Zwan, B.G. and Wolf, R.A. (1976) Depletion of solar wind plasma near a planetary boundary. *J. Geophys. Res.*, **81**, 1636.

Cross references

Plasma
Plasma wave
Solar wind
Venus
Venus: atmosphere
Venus: geology and geophysics

VENUS

Venus, along with the inner planets Mercury, Earth and Mars, is classified as a terrestrial planet (q.v.), one of the class of planets made up of solid, rocky material like the Earth, and distinct from the outer, gaseous planets Jupiter, Saturn, Uranus and Neptune. Pluto may be more like one of the icy Jovian satellites. Among the planets in the terrestrial class, Venus is the most Earth-like. Venus and Earth have very similar mean density, bulk composition and size.

Venus as an astronomical object

Venus is the second planet from the Sun in our solar system (Plate 6). It moves in a nearly circular orbit at a mean distance of 0.723 AU and a period of 224.7 Earth days. As seen from Earth, it is the brightest object in the sky, after the Moon. It appears as a 'morning star' before superior conjunction and as an 'evening star' setting later than the Sun after superior conjunction. The greatest angular distance from the Sun or maximum elongation angle is about 47°. Venus is most distant from Earth at superior conjunction, about 260 million km, where the full phase disk (0° phase angle) covers 10 arcsec. At the closest point to Earth, inferior conjunction, about 41.4 million km, the darkened disk subtends about 64 arcsec. The maximum brightness occurs at an elongation angle of approximately 39°. Venus has a retrograde rotation, clockwise to an observer looking down from the direction of the north celestial pole, of period 243.025 ± 0.001 days, as determined from Magellan mapping data. At closest approach to Earth, the same side of the planet faces us. This is almost, but not quite, exact. For synchronicity the spin would have to be 243.16 days, well outside the error bounds of the present determination. The mean equatorial radius of Venus is 6051.9 km. The current best estimate for the product of the universal gravitational constant (G) and the mass of the planet (M) is $GM_V = 324\,858.76561687$ km^3 s^{-2}. The uncertainty is probably about 0.8 km^3 s^{-2}.

Atmosphere

Venus has a dense atmosphere composed of 96.5% carbon dioxide and 3.5% nitrogen. Other gases account for no more than about 0.1% of the total, mostly water vapor, O_2, SO and SO_2, together with some argon, neon and helium and a trace of krypton. The surface temperature at the mean radius of 6051.9 km is 735 K at a pressure of

Table V3 Venus planetary parameters

Mean distance from Sun (AU)	0.723
(km)	108 159 000
Orbit eccentricity	0.0167
Sidereal period (days)	224.7
Rotation period, retrograde (days)	243.025 ± 0.001
Right ascension of pole α_o, J2000	$272.74 \pm 0.08°$
Declination of pole δ_o	$67.15 \pm 0.02°$
Product of universal gravitational constant and mass of planet, $GM_\venus$ (km^3 s^{-1})	324 858.7656169
Mass (kg)	4.87×10^{24}
Mean radius (km)	6051.9
Mean density (g cm^{-3})	5.25
Surface temperature	735 K at 6051.9 km radius
Surface pressure	92 bar at 6051.9 km
Pressure scale height at surface (km)	15.75
Lapse rate of lower atmosphere (K km^{-1})	7.8
Adiabatic lapse rate at surface (K km^{-1})	8.86
Albedo	0.71
Surface equatorial magnetic field	< 1/15 000 Earth field
Satellites	None

92 bar. The lapse rate near the surface is about 7.8 K km^{-1}. The adiabatic lapse rate for an ideal atmosphere with 96.5% carbon dioxide and 3.5% nitrogen is 8.86 K km^{-1}. Thus the atmosphere near the surface is thermally stable. Sulfur compounds are very important in the chemical and physical character of the Venus atmosphere. The dense clouds are made up of concentrated sulfuric acid droplets. The base of the clouds, at a height of about 50 km, represents the lower elevation limit of stability of sulfuric acid droplets. Sulfur dioxide is an important trace constituent in the upper atmosphere. The interactions of these compounds between the surface, the atmosphere and the clouds is little understood, but holds important consequences for the evolution of the surface and atmosphere. The lower atmosphere of Venus, below about 20 km, contains about 80% of the atmospheric mass, yet this region is the least understood and is largely unexplored. Venus's atmosphere differs from Earth's in several other ways. The surface temperature on Venus is virtually invariant, with no day or night variations or variations with latitude. The temperature is thus a simple function of altitude.

Another significant difference between the atmospheres of Venus and Earth is in their general circulation patterns. The primary circulation of the Venusian atmosphere is a zonal retrograde super-rotation of the atmosphere at heights of between about 10 km and more than 100 km. Zonal, east to west, wind speeds range from 1–2 m s^{-1} at the surface, to about 10 m s^{-1} at 10 km and 100 m s^{-1} at 50–100 km. Pioneer Venus probe data suggest that wind speeds are highest at about 60° latitude. Meridional winds with a Hadley cell component are probably required to redistribute heat from the equatorial regions toward the poles. Observational support for this circulation pattern is seen from mapping of wind streaks in the Magellan images which show a mean equatorward surface wind in both hemispheres (see Venus: atmosphere).

Surface

The surface of Venus has been imaged from Earth using large radio telescopes at Goldstone, California and Arecibo, Puerto Rico. This technique has yielded images with 1–2 km resolution, revealing impact craters, mountain belts and various types of volcanic features. Unfortunately, the Earth-based imaging can only be carried out near inferior conjunction, when Venus and Earth are close. Because of the nearly synchronous coupling of the spin of Venus and the orbital periods of the two planets, the same face of Venus appears at each inferior conjunction, and only about a quarter of the planet can be mapped from Earth. The Soviet Venera 15 and 16 spacecraft mapped the northern hemisphere of Venus from about 30° north latitude up to the north pole at 1–2 km resolution.

Venus has been explored by more spacecraft than any other planet. Thirty-one (nine of them unsuccessful) US and Soviet spacecraft have been sent to Venus. The Soviet Venera and Vega series of spacecraft and the US Mariners and Pioneers carried instruments to determine the composition of the surface and atmosphere, and the dynamics of the upper atmosphere. Several Venera landers transmitted images from the surface of Venus. The surface of Venus was seen, generally, to be rocky with a sprinkling of granular material.

The US Magellan mapping mission (q.v.) began mapping operations on 15 September 1990, and by the end of 1992 had mapped nearly 98% of the surface. The Magellan Venus mapping mission marks the end of the reconnaissance phase of NASA's exploration of the planets and the beginning of a new era of specifically focused missions to address scientific questions. Magellan was approved in 1983. It was launched from Cape Canaveral aboard the shuttle Atlantis on 4 May 1989, and placed in Venus orbit 15 months later, on 10 August 1990. Magellan has one instrument, a synthetic aperture radar capable of imaging the surface and mapping surface topography as an altimeter. The radar also measures radio emissions from the surface, revealing the electrical properties of the surface material. The synthetic aperture radar, or SAR, resolves features measuring about 120 m through the thick clouds that perpetually hide the planet. The altimeter measures surface elevations accurate to about 30 m, and the main antenna, when used as a radiometer, records the natural thermal emissions from the surface to help determine the surface composition. The primary mission was 243 Earth days, one Venus rotation beneath the orbiting spacecraft. The first cycle mission objective was to acquire imaging data from at least 70% of the planet. Magellan, in fact, mapped 84% of the planet's surface during the first 8 months. The mapping orbit had a periapsis (closest approach to the planet) altitude of 300 km and a period of 3 h 15 min.

Approximately 820 impact craters have been identified in the Magellan images. The smallest impact craters are about 2 km in diameter. This shows the efficiency of the dense atmosphere for shielding the surface from asteroid bombardment. The impact crater population indicates an average surface age of about 500 million years. Venus has no ancient cratered terrains such as occur on the Moon and Mars. Most craters appear to be unmodified by erosion or any other process, although a few are modified by more recent volcanic flows and tectonic fracturing. End-member models based on the crater record that are currently under consideration are catastrophic global volcanic resurfacing 500 Ma ago and continuum resurfacing of smaller areas such that the mean age is 500 Ma (see Venus: geology and geophysics).

Eolian features mapped by Magellan provide significant surface constraints on circulation of the atmosphere near the surface. Abundant bright and dark wind streaks have been mapped (Figure V17). Most streaks are associated with topographic barriers. Streaks are more abundant near impact craters, indicating that the impact

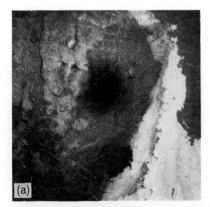
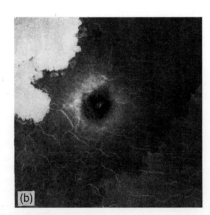
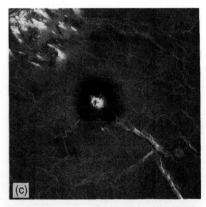

Figure V16 These dark, diffuse and circular patterns imaged by the Magellan spacecraft are thought to be effects generated by the impact of asteroids or comets into Venus' dense atmosphere. The three patterns, located in Sedna Planitia (near 45°N and 350°E), may represent a sequence in which (1) the impactor was large enough to penetrate the atmosphere largely intact and striking the surface producing a bright, rough impact crater and ejecta blanket (c); (2) the impactor was slowed down and broken up with only scattered fragments striking and scarring the surface (b); or (3) the impacting body was so small that it was disrupted and slowed down by the atmosphere creating no surface impact or scar. The dark circular regions (30–60 km in diameter) surrounding these possible impact features are thought to result from enormous shock waves transmitted to the surface by the atmosphere which are strong enough to crush surface materials into fine debris, which are dark (non-scattering) in the radar images. In some cases, notably (b), a brighter region also formed outside the dark zone. This may be a case where the shock was only strong enough to produce coarse fractures in volcanic flows and in this case increasing the ability to scatter radar signals. (P-37831.)

Figure V17 This Magellan image is centered at 17° north latitude and 62.5° longitude in the northwestern Ovda Region of Venus. The image, which is 100 km in width and 70 km in length, is a mosaic of orbits 829–834. The image shows some spectacular wind streaks 500 km northeast of the crater Mead. Mead is the largest impact crater known to exist on Venus with a diameter of 275 km. The large explosion associated with the impact spread debris as much as 500 km away from Mead onto the surrounding plains. The streaks in this image most likely represent debris from the impact that has been modified by surface winds. It is difficult to determine whether there are bright streaks on top of darker terrain or dark streaks on top of brighter terrain. If the streaks represent radar-bright debris, the wind is blowing from the northwest to the southeast. If the streaks are composed of radar-dark material, the wind direction is reversed. (P-38160.)

process produces either the winds or the sand and dust particles that are moved by subsequent winds. Typically one direction is indicated. A few streaks indicate multidirection winds. The mapped orientations indicate a significant equatorward component in each hemisphere providing evidence for Hadley circulation (q.v.).

The mean radius of Venus is 6051.9 km, with the lowest point at about 6048.0 km and the highest point, Maxwell Montes, at 6062.57 km. Thus the variation of topography on Venus is only about 14.6 km compared to about 20 km (19 881 m) for Earth. The temperature of the surface of Venus ranges from 653 K at the summit of Maxwell Montes to 766 K at the bottom of the Diana Chasma trench. The pressure varies from 45 bar at the highest elevations to 119 bar at the lowest points.

Interior

Magellan has revealed a unique global volcanic and tectonic style on Venus. Regional tectonism is evident in the widespread horizontal compression and stretching of the surface material. Large deformed regions are obvious results of crustal motion in the planet's geologic past. Venus apparently has or had a dynamic mantle that drives ongoing crustal warping. But, while various regions of the planet show evidence of motion, there has been no evidence found of Earth-like plate tectonics. Long narrow troughs are seen in several areas, showing where the crust has apparently ruptured; these linear rift zones are associated with extensive rises, or upward slopes, and large areas covered by lava that flowed from volcanic vents. Unlike on Earth, volcanism on Venus is global (Figure V18).

The only information about the structure of the interior of Venus is provided by gravity data. The gravity field of a planet, combined with topography, reveals variations in the density of the interior. Gravitational acceleration is determined by small changes in the frequency of the spacecraft radio signal, or Doppler shift, as it speeds up and slows down in its orbit around the planet. The line-of-sight accelerations of the spacecraft can be converted to variations in the planetary gravity field. Gravity information has provided one of the most fundamental differences between the interiors of Earth and Venus (see Venus: gravity). Long-wavelength gravity variations of Earth show no correlation with the global topography, and are thought to be produced by variations in mantle densities. The long wavelength gravity anomalies of Venus are correlated with topography. This is particularly puzzling and indicates a link between deep mantle density variations and surface topography. The observed correlation also indicates that Venus does not have a low-viscosity asthenosphere (q.v.) as on Earth. This zone on Earth is thought to be necessary to allow the motions that result in plate tectonics. On Venus the absence of such a zone would allow stresses to be transmitted from the surface topography to deep into the mantle.

R. Stephen Saunders

Bibliography

Saunders, R.S., Pettengill, G.H., Arvidson, R.E. *et al.* (1990) The Magellan Venus Radar mapping mission. *J. Geophys. Res.*, 95(B6), 8339–55.

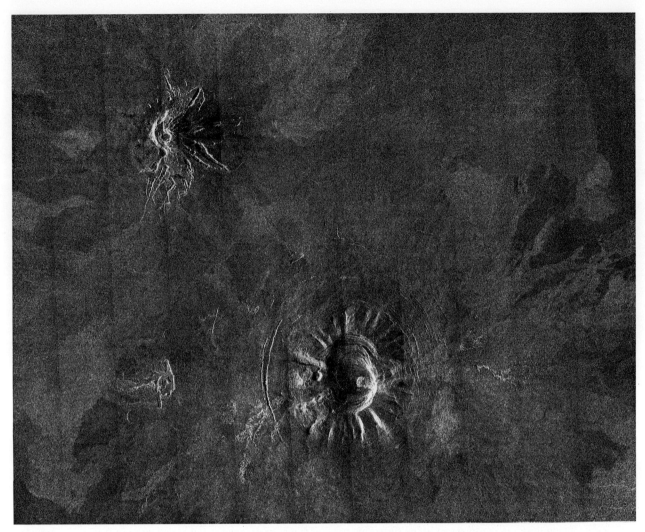

Figure V18 Two unusual volcanic domes are shown in this Magellan full-resolution mosaic. The image covers an area 180 by 240 km centered at 18° north latitude, 303.5° east longitude, just east of Beta Regio. The dome in the south center of the image is about 45 km across, with a 20-km caldera, or volcanic collapse crater, in the center. The dome in the northwest corner of the image is about 30 km across with a small (5-km summit crater. The very bright radar return from the western flank of this dome indicates that it has steep slopes. The flanks of the volcanoes display prominent gullies which may have been formed by slumping of surface material or thermal erosion by lava flows. Variations in the brightness with the surrounding plains show the extent of lava flows which originated at these volcanoes. Curved or bowed fractures surrounding the southern structure indicate that there has been subsidence, or sinking, of the dome following eruptive activity. (P-38811.)

Saunders, R.S. and Pettengill, G.H. (1991) Magellan: mission summary. *Science*, **252**, 247–9.
Saunders, R.S., Arvidson, R.E., Head, J.W., III *et al.* (1991) An overview of Venus geology. *Science*, **252**, 249–52.

Cross references

Magellan mission
Mariner missions
Pioneer Venus mission
Surface processes
Venera missions

VENUS: ATMOSPHERE

The solid planet Venus is very similar to the Earth in size and density (Table V4) and so it is initially surprising to find that the two atmospheres are quite different. Venus has a very thick atmosphere consisting largely of carbon dioxide, with nearly 100 times the base pressure of the Earth; this atmosphere is much hotter at the surface, with temperatures of nearly 750 K. The two planets are in nearly circular orbits at distances of 108.2 and 149.6 million km from the Sun, respectively, but this alone cannot account for the relatively very high surface temperature of Venus, and feedback processes involving the atmosphere have to be invoked. Venus has a much slower solid-body rotation rate than the Earth, and is possibly more active volcanically. Both of the factors contribute to the ubiquitous and highly reflective cloud cover which reflects 76% of the incoming solar flux and which results in a smaller net solar constant for Venus than for Earth. However, the cloud also plays a role in maintaining a very powerful greenhouse effect (q.v.), which accounts for the high equilibrium values of the observed surface and lower atmospheric temperatures.

The inclination of the planetary rotation axis, and the eccentricity of the orbit, are both very small for Venus, and so only very small Sun-driven seasonal changes would be expected, and none has been conclusively detected. The time taken for the planet to orbit the Sun is 224.7 Earth days, while it rotates on its axis once in 243 Earth days; the Venusian day is therefore longer than the Venusian year. The solar day, i.e. the time for the Sun to go from noon to noon as seen

Table V4 Physical constants for the terrestrial planets

	Venus	Earth	Mars
Orbital and rotational data			
Mean distance from Sun (km)	1.082×10^8	1.496×10^8	2.279×10^8
Eccentricity	0.0068	0.0167	0.0934
Obliquity (deg)	177	23.45	23.98
Sidereal period (days)	224.701	365.256	686.980
Rotational period (h)	5832.24	23.9345	24.6229
Solar day (days)	117	1	1.0287
Solar constant (kW m^{-2})	2.62	1.38	0.594
Net heat input (kW m^{-2})	0.367	0.842	0.499
Solid body data			
Mass (kg)	4.870×10^{24}	5.976×10^{24}	6.421×10^{23}
Radius (km)	6051.5	6378 to 6357	3398
Surface gravity (m s^{-2})	8.60	9.78	3.72
Atmospheric data			
Composition	See Table V5		
Mean molecular weight	43.44	28.98 (dry)	43.49
Mean surface temperature (K)	730	288	220
Mean surface pressure (N m^{-2})	92	1	0.007
Mass (kg)	4.77×10^{20}	5.30×10^{18}	$\sim 10^{16}$

Table V5 Composition of the terrestrial planet atmospheres[a]

	Venus	Earth	Mars
Carbon dioxide	0.96	0.0003	0.95
Nitrogen	0.035	0.770	0.027
Argon	0.00007	0.0093	0.016
Water vapor	~ 0.0001(?)	~ 0.01	~ 0.0003
Oxygen	~ 0	0.21	0.0013
Sulfur dioxide	0.00015	0.2 ppb	
Carbon monoxide	0.00004	0.12 ppm	0.0007
Neon	5 ppm	18 ppm	2.5 ppm

[a] Values are given as fractional abundances except where ppm (parts per million) or ppb (parts per billion) is stated.

from the surface of Venus, is about 117 Earth days. The atmosphere rotates much faster than this at some levels, for example completing a revolution once every 4 to 5 days near the cloud tops. As already noted, the fact that the surface of Venus rotates slowly has a profound effect on its atmospheric circulation and its surface climate.

Venus is too hot for liquid water to exist and, although the atmosphere contains large amounts of water as vapor and bound up with sulfur compounds in the clouds, the planet is overall very dry compared to the Earth. The main constituents, apart from CO_2 and H_2O, are inert gases, particularly nitrogen and argon (Table V5).

Structure of Venus' atmosphere

Composition

It is thought that the primary atmosphere of Venus, i.e. that which was originally formed with the solid body, in common with those of the other terrestrial planets, has been lost in the distant past, and that the present atmosphere is secondary, produced by outgassing from the crust and by influx of cometary and meteoritic material. The commonest gases which could accrue in this way are carbon dioxide, water vapor, methane and ammonia. The amounts of each remaining today, and the admixture of other gases, depends on the initial composition of the planet and its subsequent evolution, particularly its thermal and photochemical history. Moderately large amounts of some gases, such as argon, are produced by the decay of radioactive materials, and form a non-negligible component of the present-day atmosphere. Most of our knowledge of the abundances is derived from Earth-based spectroscopy and from mass spectrometer and gas chromatograph measurements made on the Pioneer Venus and Venera descent probes (von Zahn et al., 1983). With the notable exception of water vapor, they are broadly consistent with theoretical expectations as outlined above (Pollack and Yung, 1980).

Venus has between ten and 100 000 times less water in its atmosphere than exists in the oceans and atmosphere of the Earth (Donahue and Pollack, 1983). The fact that, at the same time, deuterium is about 100 times more abundant on Venus than Earth suggests that Venus had much more water initially, but that most of it has been lost. The loss processes involve dissociation to form hydrogen and oxygen followed by escape from the planet of hydrogen, a process which depends strongly on the abundance of water in the middle atmosphere. According to Kasting, Pollack and Ackerman (1984), Venus could have lost an ocean of present-day terrestrial proportions in only a few hundred million years. These authors also suggest a reason why the D/H ratio on Venus is only greater by 100 times that on Earth. It would be much larger if all the deuterium in the primordial Venusian ocean had been retained. However, deuterium as well as hydrogen can escape from the atmosphere when there is free water on the surface, if the UV heating of the upper atmosphere is sufficiently intense. Once the free water is all gone, the mixing ratio of vapour in the upper atmosphere falls and fractionation of the two isotopes becomes more pronounced. In the model of Kasting, Pollack and Ackerman, with the simplifying assumption that all the deuterium is lost until the last of the ocean evaporates and then none thereafter, the predicted enhancement is almost exactly that observed. These authors further point out that an extensive ocean on Venus would facilitate the disposal of the oxygen produced by water vapour dissociation. This cannot escape, and must be bound chemically within the crust. Most of the weathering processes by which this occurs on Earth involve liquid water.

Thermal structure

Enough solar energy diffuses through the cloud cover on Venus to provide about 17 W cm^{-2} of surface insolation on average, about 12% of the total absorbed by the planet as a whole (i.e. including the atmosphere). Thus heated, the surface warms the lower atmosphere, which responds by forming a deep convective region, the troposphere (Figure V19). Within the troposphere none of the atmosphere cools significantly by radiation to space, because the opacity of the overlying layers is too large. The basic process, whereby short-wavelength solar radiation penetrates to the lower atmosphere more easily than the longer thermal wavelengths can escape is usually known as the greenhouse effect, and its result is to raise the surface temperature significantly above that which would prevail on an airless planet. The effect is particularly extreme for Venus, where the surface temperature must rise to 730 K in order to force enough infrared cooling to balance the incoming sunlight. An airless body with the same albedo and heliocentric distance as Venus would reach equilibrium for a mean surface temperature of only about 230 K. This 500 K green-

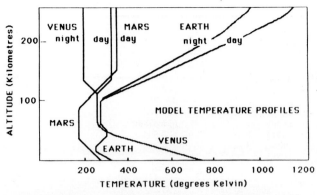

Figure V19 Mean temperature profiles for the atmospheres of the terrestrial planets.

house enhancement of the surface temperature compares with only about 30 K on Earth and 10 K on Mars. Radiative transfer models (Pollack, Toon and Boese, 1980) can account for the high surface temperatures by incorporating weak as well as strong bands of CO_2 and H_2O, plus those of the minor constituents CO, HCl and SO_2, and in particular the correct scattering and absorbing properties of the clouds (as measured by Pioneer Venus; Esposito et al., 1983).

Convection in the troposphere carries energy upwards to the base of the stratosphere, where strong radiative cooling to space can occur. On Venus this level (the tropopause) occurs about 40 km above the surface. Above the troposphere lies the middle atmosphere or mesosphere, a deep layer where the temperature tends to be constant with height, because the atmosphere here is optically thin and, to a first approximation, each layer tends to find the same equilibrium temperature. This is determined by the balance between the absorption of upwelling infrared from the surface and troposphere and cooling to space, if no significant absorption of direct solar energy takes place. On Venus the basically isothermal nature of the middle atmosphere is modified by absorption of moderate amounts of solar and thermal energy in the near-infrared bands of water vapor and carbon dioxide, and by dynamical effects, especially wave motions. Above the mesopause begins a low-density region called the thermosphere, which on Venus is very cold at night, around 100 K, and the transition from the day to night side values of temperature show surprisingly steep gradients (Keating et al., 1979). The implication is that the dynamics of Venus's thermosphere is such that the flow of air in response to the temperature gradient is inhibited, probably by large-scale eddies.

Clouds

The deep Venusian atmosphere of nearly pure CO_2 is completely enshrouded by a layer of sulfur-bearing clouds over 50 km deep (Table V6). As noted above, the H_2SO_4 clouds on Venus have properties which contribute very significantly to the high surface temperatures, by scattering conservatively at short (solar) wavelengths while strongly absorbing long-wavelength (planetary) radiation. Changes in the optical properties or depth of the cloud layers on timescales of years or longer would affect climatic variables such as surface temperature and the general circulation regime. At present the radiative, dynamical and chemical processes appear to be in balance, but the stability of the currently observed state is debatable.

The main cloud deck extends from about 45 to about 65 km above the surface, with haze layers above and below. Within this overall structure, detailed layering occurs and particles of different sizes congregate at different height levels. The particles range in diameter from less than 1 to over 30 μm and tend to a trimodal size distribution, with the commonest diameters falling towards the ends of the overall range and in the 2–3 μm region. It is these intermediate size or 'mode 2' droplets which are visible from outside Venus, and for which spectroscopic, polarimetric and other evidence yields a composition of 75% H_2SO_4 and 25% H_2O. The composition of the smaller, 'mode 1' drops is unknown; these form an aerosol haze extending throughout the cloud layer. Most of the mass of the clouds is in the big 'mode 3' drops, which may be crystalline sulfur or possibly more evolved sulfuric acid drops which have coagulated. The formation of the cloud droplets can be explained by a model in which H_2O and SO_2 (the latter possibly of volcanic origin) combine photochemically near the cloud top level. It is more difficult to explain the size distribution, particularly the existence of more than one mode. Compositional contrasts and dynamical effects may be at work but at present the observations which would elucidate these are lacking.

The detection of what was probably a CO_2 cloud on Venus at high altitudes over the dawn terminator was reported by the Pioneer Venus Orbiter infrared radiometer team (Taylor et al., 1980).

Dynamics

Dynamical measurements

The most straightforward way to obtain measurements of the winds at the surface is to use a simple cup anemometer, i.e. a rotating vane device similar to those seen on most terrestrial meteorological stations. The Russian landers Venera 9 and 10 (Avduevsky et al., 1976) made such measurements and found velocities of ≤ 1 m s^{-1}.

It is also possible to track the drift of descent probes as they pass through the atmosphere, and so to obtain vertical coverage and directional data, as well as wind strength. The Venera landers were tracked by the measurement of the Doppler shift in the radio signal from the spacecraft (Kerzhanovich et al., 1979), while the Pioneer Venus probes used an interferometric technique involving more than one receiving station (Councelman et al., 1980). Some results of both methods are summarized in Figures V20 and V21. They confirm that the temperature gradients in the lower atmosphere are close to adiabatic in the vertical, and close to zero in the horizontal, as would be expected theoretically from the high opacity and high density.

Global data have been obtained in the upper atmosphere by tracking cloud features seen in the ultraviolet images obtained by the Pioneer Venus Orbiter (Rossow et al., 1980) and other spacecraft. The main limitation of these is that the vertical levels being probed are not well defined; it is thought that the velocities derived in this way apply to the region where the pressure is of the order of, or somewhat less than, 1 bar. The measurement of Doppler shifted emission lines from atmospheric gases has been accomplished so far only by Earth-based observers (Traub and Carlton, 1975). The spatial resolution at this range is of course poor, and the main value of the existing measurements has been to verify that the cloud-tracked winds do, in fact, apply to mass motions, rather than the phase speed of waves as had been suggested.

Motions in the deeper atmosphere were observed by near-infrared imaging carried out by the Galileo probe in 1990 (Carlson et al., 1991). The features observed on the nightside of the planet at wavelengths from 1 to 3.5 μm originate in the main cloud deck, illuminated from behind by the hot lower atmosphere. The typical velocities inferred near the equator were about half as fast as those from UV markings, which is consistent with the vertical profiles of wind and cloud opacity measured by the Pioneer and Venera probes. The winds measured during the Galileo feature a zonal jet of more than 100 m s^{-1} at middle latitudes, and equator-to-pole drifts of a few meters per second.

The measurement of temperature by remote sensing, followed by a model-dependent analysis, is still the only means for obtaining information about the global dynamics in three dimensions (Taylor et al., 1980). The main limitations of this method are difficulties with the parameterization of viscosity, particularly that due to eddies, and the fact that the method fails near the equator, where assumptions of cyclostrophic balance break down. Also, of course, it is restricted to the region above the clouds, unless microwaves are used, which has not been the case to date. In the case of Pioneer Venus, five infrared bands near 15 μm were employed to cover the vertical range from 60 to 105 km with a mean vertical resolution of about 10 km. Radiance measurements were made every 0.2 s on a spacecraft spinning at 12 rpm; the net effect was coverage of the planet with a mean horizontal spatial resolution of the order of a few tens of kilometers. Time-averaged global maps of the Venusian temperature field obtained in this way are shown in Figure V22. These show several features clearly related to the general circulation, including (1) 'polar warming', a tendency for the temperature over a broad altitude range to increase from pole to equator, in spite of the fact that the trend in radiative heating is in the opposite direction; (2) the 'polar collar', an intense ribbon of cold air surrounding the pole at about 65° latitude;

Table V6 Properties of clouds and dust in the terrestrial planet atmospheres

	Venus	Earth	Mars
Fractional coverage	1.00	0.40	0.05 (cloud); 0–1.0 (dust)
Typical optical depth	25–40	5–7	0.01–1.0; 0.2–6 (dust)
Composition	$H_2SO_4.H_2O$	H_2O	H_2O, CO_2; magnetite etc. (dust)
Number density, liquid (cm^{-3})	50–300	100–1000	0
Number density, solid (cm^{-3})	10–50	0.1–50	30–1000 (near surface)
Typical mass loading (g m^{-3})	0.01–0.1	0.1–10	0.0002–0.1
Main production process	Chemistry	Condensation	Condensation; windblown (dust)
Equivalent depth[a] (mm)	0.1–0.2	0.03–0.05	1–100
Effective radius[b] (μm)	2–4	10	0.4–2.5 (dust)
Main forms	Stratiform	Stratiform, cumulus	Stratiform, mixed (dust)
Temporal variability	Slight	High	High
Dominant heat exchange process	Radiation	Latent heat	Radiation

[a] The equivalent depth is the estimated thickness of the cloud material if it were deposited on the surface.
[b] The effective radius is the radius of the spherical particles having most nearly the same scattering properties as the cloud at visible wavelengths.
After Esposito et al., 1983, with changes and additions.

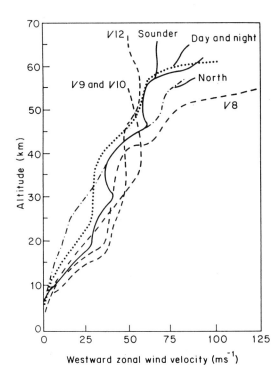

Figure V20 Profiles of the zonal (east to west) wind on Venus as measured by tracking the Pioneer Venus and the Venera 8, 9, 10 and 12 entry probes.

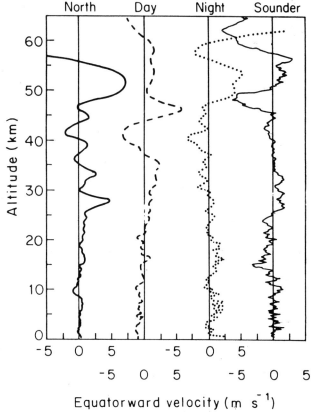

Figure V21 Meridional winds on Venus as measured by tracking the Pioneer Venus descent probes (Councelman et al., 1980). Note the suggestion of global layering, consisting perhaps of stacked 'Hadley' circulation cells of considerable latitudinal extent.

and (3) the solar tide, or variation with local time of day of the air temperature.

Zonal wind field

A striking feature of the zonal flow is the global 'superrotation', which manifests itself in cloud structure which moves rapidly around the planet in a direction parallel to the equator. The cloud markings, which appear with high contrast through an ultraviolet filter, have their origin at heights of 50 or 60 km above the surface (where the pressure is of the order of 100 mbar) and travel around the equator in 4 to 5 days, corresponding to speeds of about 100 m s^{-1}. This is more than 50 times faster than the rotation rate of the surface below. Measurements of the winds below the clouds, and calculations (from temperature data) of the winds above the cloud tops, show that the zonal wind speed declines at higher and lower levels, reaching values near zero at about 100 km and near the surface respectively.

Attempts have been made to explain these high zonal wind speeds on Venus by several mechanisms, all of which fall into one of three main categories, namely (1) the gravitational interaction of the Sun with the atmospheric tides, (2) the overhead motion of the Sun in the sky (the 'moving flame' mechanism), and (3) the upward transport of momentum from the surface. Currently prevailing opinion favors a version of mechanism 3, in which momentum from the solid planet is transported by waves whose interaction with the main flow is complex and in which the mean meridional circulation plays an important role. With suitable parameterizations, Young and Pollack (1977) were able

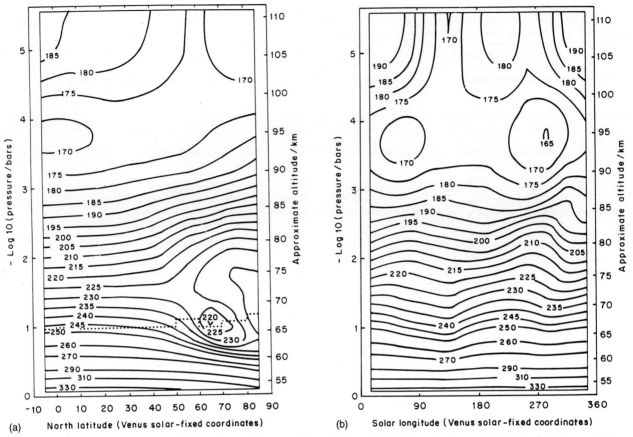

Figure V22 Time averaged temperature fields in the middle atmosphere of Venus (Schofield and Taylor, 1983). (a) The zonal mean field, and (b) the variations around a latitude belt from 0 to 30°N, both plotted against pressure and approximate height. The horizontal stepped line represents the retrieved mean cloud top height.

to to produce large zonal velocities in their three-dimensional spectral model of the Venusian atmosphere.

Meridional wind field

The cloud motions which trace the zonal winds also reveal the pattern of the meridional circulation on Venus. Hadley cells exist in each hemisphere, i.e. global-scale circulation cells characterized by rising motion all around one constant-latitude belt and descending motion at another. Each cell extends to higher latitudes than on Earth, in part a consequence of the slower zonal rotation speeds. Near the poles on Venus, a complex instability develops, resulting in dramatic long-lived wave structures. The polar collar takes the form of a ribbon of very cold air, some 10 km deep and 1000 km in radius, centered on the pole. Inside the collar, temperatures are some 40 K cooler than outside the feature (Figure V22). Poleward of the inner edge of the collar lies the polar dipole, a wavenumber 2 feature consisting of two well-defined warm regions circulating around the pole. Both the dipole and the collar have so far resisted attempts to model them as normal modes of the atmosphere.

The cloud-tracked winds obtained from Pioneer Venus and Mariner 10 both show equator-to-pole velocities of about 5 m s^{-1} in each hemisphere. Tracking of the Pioneer Venus probes shows winds of this magnitude at about 50–60 km altitude, with a very complicated vertical structure (Figure V21). One currently favored interpretation is that the alternations in the direction, as well as the magnitude, of the meridional wind, marks the passage of the probe through the different components of a stack of Hadley cells, each extending from the equator to high latitudes. This notion is supported by the fact that the Hadley cell seen at the cloud tops appears to be thermally indirect, that is to say, carries heat from the equator to the pole against the observed temperature gradient. It is probably driven by a stronger, direct cell underneath. The layered eddy sources and sinks which could drive the zonal superrotation may be related to the cell interfaces.

Tides

A particularly important form of wave motion is the solar tide, that induced by the apparent motion of the Sun overhead. This contains a whole spectrum of Fourier components, because the forcing is non-sinusoidal; the actual atmospheric response depends on the mean wind and the interference between the various components. The solar tide on Venus as measured by the Pioneer Venus Orbiter has been analyzed by Fels, Schofield and Crisp (1984), who find that experiment and classical tidal theory can be reconciled provided that a realistic representation of the zonal wind is incorporated.

Conclusion

For Venus, measurement programs of comparable sophistication to the terrestrial ones are needed, to add an understanding of the dynamical regimes present to our knowledge of the basic structure of the Venusian atmosphere. Details of basic things like the general circulation and the photochemistry of the clouds are still incomplete, although recent observations of the surface morphology by the Magellan orbiter may have clarified the scale of volcanism on Venus. Volcanoes may be important for maintaining the cloud layers and hence the extreme climate of Venus. The surface relief may shed some light on the question of whether or not Venus once had oceans, like the Earth.

A comprehensive understanding of the origin and maintenance of the zonal superrotation is still lacking. The thermal tides, on the other hand, although initially puzzling when first observed, now appear to fit a fairly simple model quite well. The biggest mysteries, which so far have defied any kind of explanation at all, are the spectacular quasi-permanent wave modes in the high-latitude regions, in particular the phenomena known as the polar collar and the polar dipole. An attempt has been made by Valdes (1984) to model these as trapped normal modes of the atmosphere, but without success. Thus, there is scope for more sophisticated models to explain some observed dynamical features on Venus, as well as for more data to allow the models of low-latitude phenomena to be further developed.

Fred W. Taylor

Bibliography

Avduevsky, V.S., Vishnevetskii, S.L., Golov, I.A. *et al.* (1976) Measurements of the wind velocity on the surface of Venus during the operation of stations Venera 9 and Venera 10. *Cosmic Res.*, **14**, 622–5.

Carlson, R.W., Baines, K.H., Encrenaz, Th. *et al.* (1991) Galileo infrared imaging spectroscopy measurements at Venus. *Science*, **253**, 1541–8.

Councelman, C.C., III, Gourevich, S.A., King, R.W. and Loriot, G.B. (1980) Zonal and meridional circulation of the lower atmosphere of Venus determined by radio interferometry. *J. Geophys. Res.*, **85**, 8026–30.

Donahue, T.M. and Pollack, J.B. (1983) Origin and evolution of the atmosphere of Venus in *Venus* (eds D.M. Hunten, T.M. Donahue and V. Moroz). Tucson: University of Arizona Press, pp. 1003–36.

Esposito, L.W., Knollenberg, R.G., Marov, M.Ya. *et al.* (1983) The clouds and hazes of Venus, in *Venus* (eds D.M. Hunten, L. Colin, T.M. Donahue and V.I. Moroz). Tucson: University of Arizona Press, pp. 484–564.

Fels, S.B., Schofield, J.T. and Crisp, D. (1984) Observations and theory of the solar semidiurnal tide in the mesosphere of Venus. *Nature*, **312**, 431–4.

Hou, A.Y. and Goody, R. (1985) Diagnostic requirements for the superrotation on Venus. *J. Atmos. Sci.*, **45**, 413–32.

Kasting, J.F., Pollack, J.B. and Ackerman, T. (1984) Response of Earth's atmosphere to increases in solar flux and implications for loss of water from Venus. *Icarus*, **57**, 335–54.

Keating, G.M., Taylor, F.W., Nicholson, J.Y. and Hinson, E.W. (1979) Short term cyclic variations and diurnal variations in the Venus upper atmosphere. *Science*, **205**, 62–4.

Kerzhanovich, V.V., Makarov, Yu.F., Marov, M.Ya. *et al.* (1979) An estimate of the wind velocity and turbulence in the atmosphere of Venus on the basis of reciprocal Doppler measurements by the Venera 11 and 12 spacecraft. *Cosmic Res.*, **17**, 565–79.

Pollack, J.B. (1981) Atmospheres of the terrestrial planets, in *The New Solar System* (eds J.K. Beatty, B. O'Leary and A. Chaikin). Cambridge University Press, pp. 57–70.

Pollack, J.B. and Yung, Y.L. (1980) Origin and evolution of planetary atmospheres. *Ann. Rev. Earth Planet. Sci.*, **8**, 425–87.

Pollack, J.B., Toon, O.B. and Boese, R. (1980) Greenhouse models of Venus' high surface temperature, as constrained by Pioneer Venus measurements. *J. Geophys. Res.*, **85**, 8223–31.

Rossow, W.B., Del Genio, A.D., Limaye, S.S. *et al.* (1980) Cloud morphology and motions from Pioneer Venus images. *J. Geophys. Res.*, **85**, 8107–28.

Schofield, J.T. and Taylor, F.W. (1983) Measurements of the mean, solar fixed temperature and cloud structure of the middle atmosphere of Venus. *Quart. J. Roy. Meteorol. Soc.*, **109**, 57–80.

Schubert, G. (1983) General circulation and the dynamical state of Venus' atmosphere, in *Venus* (eds D.M. Hunten, L. Colin, T.M. Donahue and V.I. Moroz). Tucson: University of Arizona Press, pp. 681–765.

Taylor, F.W., Beer, R., Chahine, M.T. *et al.* (1980) Structure and meteorology of the middle atmosphere of Venus: remote sensing from the Pioneer Venus Orbiter, *J. Geophys. Res.*, **85**, 7963–8006.

Traub, W.A. and Carleton, N.P. (1975) Spectroscopic observations of winds on Venus. *J. Atmos. Sci.*, **32**, 1045–59.

Valdes, P. (1984) Large scale waves in the atmosphere of Venus. D. Phil. Thesis, University of Oxford.

von Zahn, U., Kumar. S., Niemann, H. and Prinn, R. (1983) Composition of the Venus atmosphere, in *Venus* (eds D.M. Hunten, L. Colin, T.M. Donahue and V.I. Moroz). Tucson: University of Arizona Press, pp. 299–430.

Young, R.E. and Pollack, J.B. (1977) A three-dimensional model of dynamical processes in the Venus atmosphere, *J. Atmos. Sci.*, **34**, 1315–51.

Cross references

Atmosphere
Atmospheric thermal structure
Greenhouse effect
Mesosphere
Spectroscopy: atmosphere
Thermosphere
Troposphere
Ultraviolet spectroscopy

VENUS: GEOLOGY AND GEOPHYSICS

Venus shares many tectonic and volcanic characteristics with both Earth and the smaller terrestrial planetary bodies. Structures such as ridge belts and rift zones on Venus are, at least qualitatively, similar to features observed on Mars, Mercury and the Moon, as well as Earth. Linear mountain belts occur only on Venus and Earth. Venus also displays some structures not seen elsewhere, such as coronae (q.v.) and tesserae (q.v.), although these features are thought to be products of such familiar processes as faulting, folding and magmatism; and analogy with terrestrial landforms may be rendered difficult by the absence of significant weathering and erosion on Venus. Venus lacks global plate tectonics, in contrast to the Earth, but the relationship between gravity and topography on Venus provides strong evidence for active mantle convection (q.v.). Centers of volcanism on Venus are often associated with regions of apparent mantle upwelling, as on Earth, but volcanic plains are widespread, as on the smaller terrestrial planets. The average age of the Venus surface is about 500 Ma, comparable to that of the Earth but much less than that of the smaller terrestrial planets. Unlike the Earth, however, the age of as much as 80% of the Venus surface is indistinguishable from the global average value. The interior processes responsible for the evidently strongly time-variable history of resurfacing on Venus are topics of ongoing study.

Venus is the planet most similar to Earth in terms of radius, mass and probable bulk composition. Our nearest neighbor in the planetary system, Venus has been the object of intense and continuing spacecraft exploration by the United States and Soviet Union since 1962, when Mariner 2 became the first spacecraft to encounter and to return data successfully from another planet. During the 1970s and 1980s the Soviet Union sent a series of landers to Venus that returned images and chemical analyses of soil samples. The global geological exploration of Venus lagged behind that of the other inner planets, however, because Venus is shrouded by a global cloud layer and the surface is thus not discernible at visible wavelengths from Earth or from orbit. A series of Earth-based experiments nonetheless demonstrated that the cloud layer is transparent to radar and that radar mapping techniques are suitable for imaging geological structures on the surface. Radar images of the Venus surface made from Earth-based observatories and from spacecraft in planetary orbit have progressively improved in both coverage and resolution. The culmination of the radar exploration of Venus was the Magellan mission (q.v.). Inserted into Venus orbit in August 1990, the Magellan spacecraft has yielded images of 98% of the surface at a radar resolution of a few hundred meters or better, determined the topographic shape of the planet to wavelengths as short as 10 km, and measured the gravitational field of the planet to wavelengths as short as 200 km (see Plates 6 and 7 and Venus: gravity).

Global geological structures and processes

The geological evolution of Venus as revealed in its surface structures and units has been varied and complex, but a number of generalizations may be made. Elevations on Venus have a unimodal distribution, in contrast to the bimodal distribution of the Earth; 80% of the Venus surface consists of plains at elevations within 1 km of the modal value. In global map view the dominant topographic features

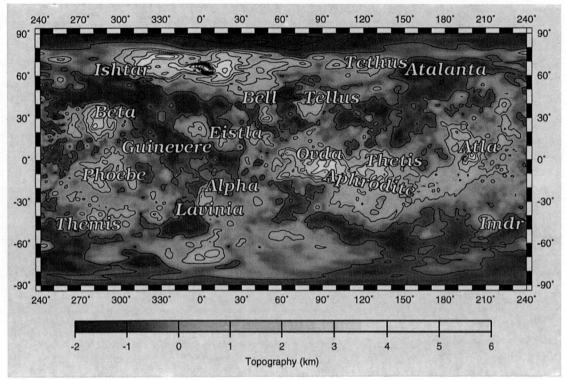

Figure V23 Topographic map of Venus. Prominent highland and lowland regions are indicated (cylindrical equidistant projection). Elevation, relative to mean planetary radius, is shown both by a continuous gray scale and by contours at 1-km intervals. Elevations in excess of 6 km (restricted to the mountain belts of Ishtar Terra) are shown without shading.

are the equatorial highlands, flanking mid-latitude lows and high-latitude uplands (Figure V23). The range in elevations relative to a spherical datum is 14 km, comparable to the relief on Earth (Plate 10).

The density of impact craters and basins on Venus (Figure V24) is considerably less than on the smaller terrestrial planets, Moon, Mars and Mercury, which have surfaces formed largely in the first half of solar system history. Impact craters smaller than about 30 km in diameter are deficient relative to larger craters on Venus because of the severe decrease in the kinetic energy of small meteoroids during transit through the dense Venus atmosphere. The areal density of craters larger than 30 km in diameter, together with estimates of the cratering rate scaled from the Earth and Moon or taken from the known population of Venus-crossing asteroids, indicate an average surface age of about 500 Ma, or 10% of the age of the solar system (Schaber et al., 1992). The largest multiringed impact basin has a diameter of 280 km, consistent with such an age. The average age of the surface of Venus appears, on this basis, to be much younger than those of the smaller terrestrial planetary bodies, older than the Earth's ocean basins, and generally younger than the Earth's continental crust.

While resurfacing processes must have removed craters significantly older than 500 Ma, the characteristics of craters on Venus are not what one would expect for a steady balance between crater formation and such crater removal processes as volcanic burial and tectonic disruption. Craters of all sizes are indistinguishable from a random population, and most of the craters (e.g. Figure V24) have not been significantly modified by tectonic deformation or by volcanic flows external to the crater rim (Phillips et al., 1992; Schaber et al., 1992). The leading interpretation of these characteristics is that much of the surface dates from the end of a global resurfacing event that ceased about 500 Ma ago, and that the small fraction of craters volcanically embayed or modified by deformation indicate that volcanic and tectonic activity subsequent to that time has been at much lower levels (Schaber et al., 1992). It is also important to acknowledge, however, that the Venus surface exhibits a spectrum of ages (Phillips et al., 1992). While the paucity of small craters prevents the use of crater density to determine the relative ages of individual geological units, as has been done for the solid planets and satellites lacking a significant atmosphere, estimates of the relative average ages of types of terrains are possible (Ivanov and Basilevsly, 1993; Namiki and Solomon, 1994; Price and Suppe, 1994).

The formation of sediment and its erosion, transport, and redeposition are much less important on Venus than on Earth or even the other terrestrial planets. As noted above, the dense, dominantly CO_2 atmosphere prevents small meteoroids from impacting the Venus surface; on the Moon the impact of such objects has fragmented surface material and led to global regolith formation. As a result of an atmospheric greenhouse and the global cloud cover, the lower atmosphere of Venus is extremely dry, the temperature at a given height is nearly independent of latitude or time of day, and the surface temperature at mean planetary radius is 740 K, about 450 K greater than on Earth. The lack of water and thermal cycling on Venus severely limits chemical and physical weathering and subsequent transport of surface materials. Geological processes such as the formation of large impact craters, tectonic activity and possibly pyroclastic volcanism provide local sources of sediment (Arvidson et al., 1992), and eolian activity is capable of sediment transport and forming such features as wind streaks (Greeley et al., 1992), but the fraction of the surface that contains significant thicknesses of soil deposits appears to be small.

Volcanism has been widespread on Venus; recognizable volcanic deposits constitute about 85% of the surface. Regional plains units interpreted to be of volcanic origin on the basis of their topography and smooth surface, associated flows and nearly ubiquitous small shields form the most abundant unit. Localized centers of volcanism have produced edifices spanning a range of morphologies and sizes (Figure V25). Nearly 300 volcanic centers between 20 and 100 km in diameter and approximately 150 shield volcanoes, large calderas and distinctive volcanic centers greater than 100 km in diameter and with relief as great as 6 km have been mapped from Magellan images (Head et al., 1992). The impact crater density on the latter group of large volcanoes is approximately half the global average value (Namiki and Solomon, 1994), and stratigraphic relations suggest that flows from some large volcanoes are among the youngest features on the planet (Basilevsky, 1993).

Coronae (q.v.), circular to oval structures typically several hundred kilometers in diameter with a generally elevated center and a narrow

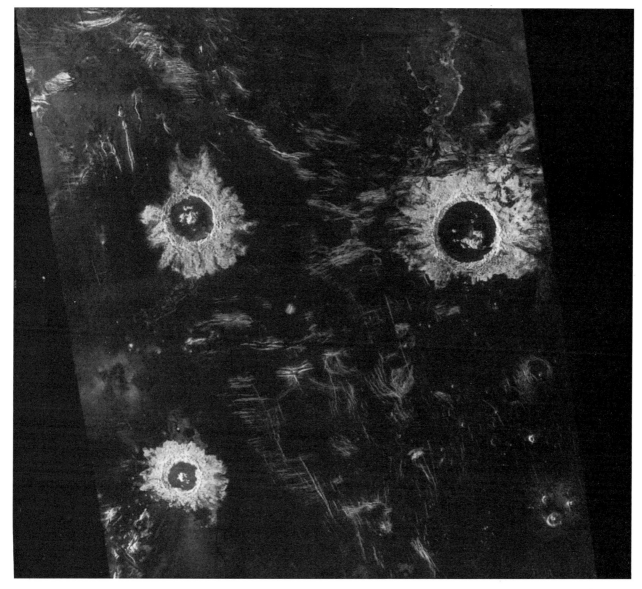

Figure V24 Impact craters. This image, obtained by the Magellan spacecraft (q.v.), shows three large impact craters (Saskia, Danilova and Aglaonice), with diameters ranging from 37 to 65 km respectively. The image corresponds to a location in the Lavinia Planitia region of Venus at 29°S, 339°E. The craters show features characteristic of impact structures, such as rough (bright) ejecta material outward of the rim, terraced inner walls and central peaks. (Magellan P-36711, courtesy of NASA.)

deformed annulus of concentric ridges (Figure V26), have been interpreted as products of local cylindrical upwelling of the upper mantle on the basis of their general characteristics, their approximately circular symmetry and the presence of volcanic sources and deposits in their interiors. Many coronae are clustered, as around Themis Regio and east and west of Ishtar Terra. More than 350 coronae have been identified and characterized (Stofan et al., 1992). The crater density on some types of coronae is not significantly different from the global average, but coronae with extensive associated volcanic deposits have lower crater densities and younger inferred average ages (Namiki and Solomon, 1994).

Volcanic centers tend to be concentrated on or near broad topographic rises 1000 to 3000 km in diameter, predominantly in portions of the equatorial highlands, such as Beta, Bell, Eistla and Atla Regiones. On the basis of their large apparent depths of compensation of long-wavelength relief and detailed radar images of selected areas, these rises are thought to be sites of mantle upwelling and associated volcanism at a larger scale than at coronae. The relative contributions to topographic relief, however, from thermal uplift, mantle convective tractions, volcanism and igneous intrusion are not known.

Tectonic features of a wide variety of styles and spatial scales are present over most of the Venus surface (Solomon et al., 1992). Deformation is manifested both in areally distributed strain of modest magnitude, such as families of faults and folds at spacing of a few to a few tens of kilometers in many volcanic plains, as well as in zones of more concentrated extension and shortening. Patterns of areally distributed strain are commonly coherent over distances of hundreds of kilometers or more. Features formed by both horizontal shortening and horizontal stretching of the crust are common. Few large-offset strike-slip faults, such as the San Andreas fault on Earth, are observed on Venus, but limited local horizontal shear has been accommodated across many zones of crustal stretching or shortening. Several large-scale tectonic features have topographic slopes in excess of 20–30° over a 10-km horizontal scale.

The most prominent extensional features are highland rift zones 100 to 300 km wide, thousands of kilometers long and marked by

Figure V25 Volcanic landforms on Venus. (a) Maat Mons, a 6 km high volcanic edifice centered at 1°N, 194.5°E, is seen in a perspective view from Magellan radar image data. Topographic data from the Magellan altimeter and radarclinometry provide the digital elevation model, shown here with a vertical exaggeration of 10 to 1. The view is toward the south. High-backscatter lava flows from the Maat Mons complex extend into the fractured plains in the foreground and partially embay ejecta from a 23 km diameter impact crater. (Magellan P-40175.)

linear troughs with flanking highs and as much as 5 km of topographic relief. The largest rifts are in the equatorial highlands and Beta-Phoebe Regiones; two prominent tectonic junctions occur at Beta and Atla Regiones. Large volcanic centers occur near the flanks of many rifts, but significant lengths of individual rift valleys appear to be free of volcanic deposits. On the basis of geometrical arguments, the rift systems in Beta Regio have been interpreted as sites of only limited extension (tens of kilometers) and may be analogous to intracontinental rift zones on Earth such as the East African or Rio Grande Rifts.

In contrast to these extensional features, linear mountain belts in western Ishtar Terra having elevations as great as 11 km above mean planetary radius, local relief of 3 to 7 km and a cumulative length of about 4000 km are inferred to have been sites of orogeny and associated underthrusting and shortening and thickening of the crust. These mountain belts are comparable in relief and horizontal dimensions to those on Earth, and like terrestrial mountains they show widespread evidence for lateral extension both during and following active crustal compression (Figure V27), a consequence of the tendency for elevated terrain to spread in response to gravitational stresses. The Ishtar Terra region, including the mountain belts, has an impact crater density indistinguishable from the global average. Outstanding questions for mountain belts on Venus include the mechanism and magnitude of the implied horizontal motions, the extent of recycling of crust into the mantle during episodes of active convergence, and the evolution of orogenic belts once convergence and underthrusting cease.

Smaller but probably analogous features in plains regions are the ridge belts, linear deformational features up to several hundred kilometers wide and hundreds to thousands of kilometers in length. Ridge belts form in two distinctive patterns: parallel to subparallel networks and fans within lowlands (Lavinia and Atalanta Planitiae areas) and more nearly orthogonal patterns adjacent or parallel to upland blocks. These features are characterized by narrow ridges and arches and generally rise up to several hundred meters above the surrounding plains. Their similarity to mare ridges and arches on the Moon, their topographic forms and their spacings support the view that the formation of ridge belts most commonly involved crustal shortening.

In addition to the linear deformational zones dominated by both extension (such as rifts) and compression (such as mountain belts), there are more equidimensional regions of areally distributed deformation known as tesserae (q.v.), which occupy about 8% of the planetary surface. Tesserae commonly are more elevated than adjacent plains and are characterized by at least two intersecting families of linear to arcuate deformational features (Figure V28). The patterns can be relatively simple and orthogonal, or they can be more variable and complex. The largest areas of tessera terrain are in western Aphrodite Terra and Ishtar Terra, but many tessera blocks occur as small inliers within younger plains units. While the density of impact

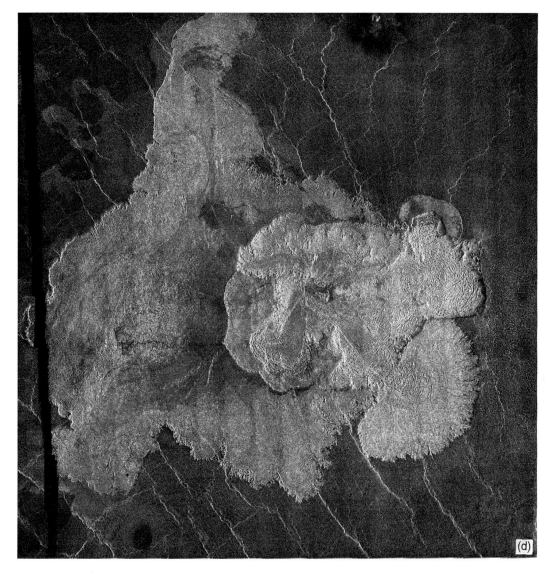

Figure V25 (continued) Volcanic landforms on Venus. (b) This unusual Venusian volcano is located on the plains between Artemis Chasma and Imdr Regio at 37.5°S, 164.5°E. The domical structure with lava channels radiating from the volcanic center is about 100 km across. Altimetry indicates that the relief of the domical structure is 0.5–1 km. This feature has been interpreted as the product of lava flows more viscous than typical of Venus. (Magellan P-39916.)

craters on tesserae is similar to the global average value (Schaber *et al.*, 1992), tesserae are stratigraphically older than most of the abutting plains and the average density of craters larger than 16 km in diameter is greater than that for the plains (Ivanov and Basilevsky, 1993). A number of formational and modificational processes have been proposed to account for the elevated topography and deformational patterns of tesserae, including crustal shortening and thickening in response to mantle downwelling and gravitational relaxation of elevated and thickened crust. On the basis of their diversity and complexity, it is likely that many tessera blocks are the result of multiple stages of strain of diverse origin and geometry.

Prior to the Magellan mission, competing hypotheses for the global geophysical workings of Venus tended to be strongly geomorphic, variants of either plate tectonics or a widespread system of mantle plumes. Further, most models were essentially time-stationary, i.e. it was supposed that observed features were the products of processes still operative somewhere on the planet. Images and gravity field measurements obtained by Magellan, however, have forced all prior models for global tectonics to be discarded. Instead, the data point to at least two distinct eras in the geological history of the planet, with neither of the two platelike or plumelike. In the earlier of the two eras, the lithosphere was able to deform so pervasively as to render the surface a nearly unreadable complex of faults and folds, a portion of which is preserved in tessera terrain, and plains volcanism was widespread. In the younger and present era, in striking contrast, the lithosphere shows signs of great strength, on the basis of gravity–topography relations and the support of the relatively young volcanic edifices, and deformation and volcanism are localized to a few regions constituting a small fraction of the surface area.

Interior heat budget

The heat budget of the Venus interior governs the planetary thermal history, the dynamics of the mantle and the mechanical properties of the lithosphere. Cosmochemical considerations and surface measurements of U, Th and K suggest that radiogenic heat production in Venus is broadly similar to that in the Earth. The atmospheric abundance of ^{40}Ar per planet mass is about one-fourth that of the Earth. This lower amount can be attributed to differences in the extent or timing of outgassing or in the bulk K abundance. The nearly terrestrial values for the K/U ratio in the crust (Barsukov, 1992), however, support the inference that differences in outgassing have dominated, along with

Figure V26 Fotla Corona. This Magellan radar image shows a corona structure approximately 150 km in diameter, centered at 59°S, 164°E in the Aino Planitia region. On and within the corona are three steep-sided volcanic domes (20–38 km in diameter). Such domes are thought to be the product of eruption of viscous lavas from a central vent. Fracturing and embayment of the domes and partial collapse of the eastern dome indicate that tectonic and volcanic activity within the corona has postdated dome formation. (Magellan P-38340.)

some contribution from a lesser degree of fractionation of K into the crust and a significantly lower rate of crustal erosion than on Earth.

If Venus loses heat at the same rate per mass as the Earth, the mean heat flux would be about 70 mW m^{-2} (Solomon and Head, 1991), with an uncertainty of perhaps 30% contributed by possible differences in the K abundance and in the fraction of heat loss contributed by secular cooling of the planet and by possible non-monotonic variations in heat loss for a large terrestrial planet. This heat flux, if derived mostly from beneath the crust, is sufficient to fuel vigorous mantle convection and is equivalent to an average conductive thermal gradient in the lithosphere of 15 to 30 K km^{-1}, depending on the crustal thickness and thermal conductivity structure. Implicit in this estimate is that Venus is similar to the Earth not only in radiogenic heat production but also in the fraction of global heat loss contributed by secular cooling of the interior. On Earth the fraction of heat loss contributed by interior cooling may be as high as 50%. That fraction may be smaller for Venus if its interior has cooled more than that of the Earth; if so, then the average heat flux and lithospheric thermal gradients could be less than the above values.

Although the prospects for direct measurement of surface heat flow on Venus are remote, the thermal gradient may be inferred indirectly from the surface temperature and the thickness of the elastic lithosphere estimated from the flexural response to lithospheric loads. Estimates of the elastic lithosphere thickness for Venus range from 10 to as much as 40 km (Sandwell and Schubert, 1992; Johnson and Sandwell, 1994). The smallest of these values of elastic lithosphere thickness are permissive of thermal gradients as high as 25 K km^{-1}; the largest values are comparable to those of oceanic lithosphere at deep sea trenches on Earth. These large values are surprising because the base of the mechanically strong lithosphere is thought to be limited by the temperature marking the onset of significant ductile flow over geological timescales (about 1000 K on Earth); because of the much hotter surface such a temperature would normally be expected to occur at shallower levels on Venus. One possibility is that the topographic profiles have been interpreted incorrectly; ductile flow accompanying gravitational relaxation of relief, for instance, can deform the surface in a manner that mimics elastic plate flexure. If real, a stronger than expected lithosphere on Venus could be the result of a lesser rate of heat loss per planet mass than on Earth, or it could indicate that under extremely anhydrous conditions crustal and mantle rocks are significantly more resistant to flow than in the Earth's crust and mantle. Recent measurements of strain rate in anhydrous diabase (Mackwell et al., 1995) have, in fact, documented a greater resistance to creep than if small amounts of water are present; but even these new measurements, if applied to the elastic lithosphere thicknesses derived from flexural analyses, require heat flow on Venus to be generally less than the Earth-scaled value (Johnson and Sandwell, 1994).

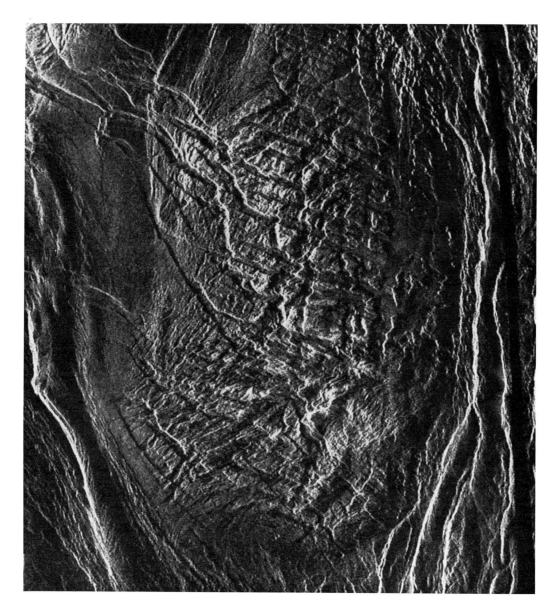

Figure V27 Multiply deformed mountainous terrain. This image, approximately 125 km wide and centered at 72°N, 342°E, shows the southern limit of the eastern limb of Freyja Montes in Ishtar Terra. The image is dominated by a highly fractured dome cut by two intersecting sets of graben, interpreted to be the product of gravitational spreading of a segment of the mountain belt following orogeny. The north–south–trending ridge belts to the east and west of the dome may be the result of the crustal shortening that accompanied mountain building, or may be, in part, a response to lateral spreading of the fractured dome. (Magellan P-37138.)

Crustal thickness and mantle dynamics

The relation between variations in gravity and topography on Venus provide important information on the interior structure and dynamics of the planet. Generally, that relationship depends on both spatial position and horizontal scale, or wavelength, with the longest wavelengths sensitive to the deepest interior structure. One class of models for gravity–topography relations are given by the principle of isostasy, namely that at some interior depth of compensation the mass per area in every overlying rock column is a constant. A familiar form of isostatic model is when variations in the thickness of a low-density crust compensate variations in topography. In the absence of other knowledge of structure, a gravity–topography relation can be cast in terms of an apparent depth of compensation. If that apparent depth is greater than the depth at which rocky material is expected to display long-term strength, or if that apparent depth varies strongly with horizontal scale, however, then the simple isostatic model must be discarded in favor of alternative models such as regional compensation by a lithosphere of finite strength or dynamic compensation by mantle convective tractions at the base of the lithosphere.

The relationship between gravity and topography on Venus differs between the largest tessera blocks and the plains, lowlands and rises that make up most of the surface. For the large tessera blocks, including western Aphrodite Terra and Ishtar Terra, the gravity–topography relation (at horizontal wavelengths less than the dimensions of the respective block, or about 2000 km for the largest such blocks) is consistent with istostatic compensation by variations in crustal thickness about a local average value of 25–40 km (Simons, Hager and Solomon, 1994). Neither lithospheric strength nor mantle convective tractions are required to match the observations for these regions. If the isostatic model is approximately representative of the actual structure beneath these regions, it may be inferred that tessera blocks

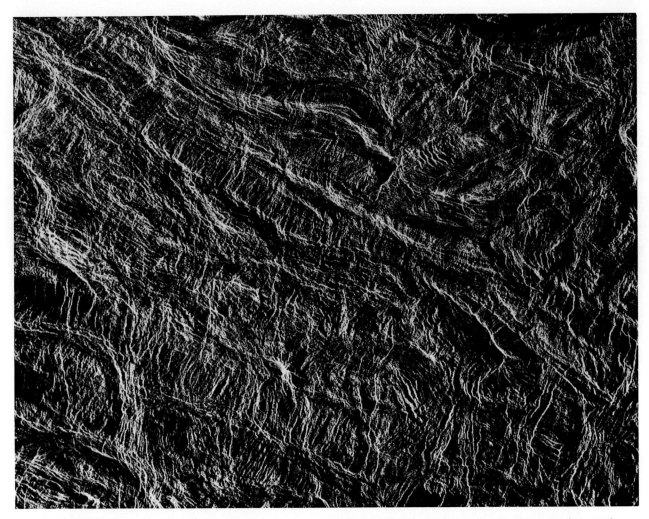

Figure V28 Tessera terrain. This radar image is of tessera terrain in Alpha Regio, centered at 24°N, 2°E, and shows an area about 150 km wide. The terrain has experienced intense folding, faulting, shearing, compression and extension. Here the terrain is dominated by WNW-trending troughs and ridges, spaced 20 to 50 km apart, that generally cross-cut a more closely spaced set of NNE-trending graben and normal faults. Multiple episodes of deformation of distinct geometry are common in tessera terrain. (Magellan P-37322.)

formed at a time when the local lithospheric strength was modest to negligible and that such blocks are now at sites where mantle convective activity does not dominate the topography at horizontal wavelengths of 2000 km and less.

For the plains, lowlands and broad rises, however, the relationship between gravity and topography is quite different. The ratio of gravitational to topographic variations decreases with decreasing wavelength in a manner inconsistent with a single depth of isostatic compensation (Simons, Hager and Solomon, 1994). Further, that ratio at the longest wavelengths is very large; if cast in terms of an apparent depth of compensation, that depth would be several hundred kilometers (Smrekar and Phillips, 1991). On Earth the apparent compensation depths are more typically several tens of kilometers, a result taken to indicate that the stresses associated with convective upwelling and downwelling in the sublithospheric upper mantle do not significantly deform the overlying lithosphere, i.e. they are decoupled by a low-viscosity zone at the base of the lithosphere. The large apparent depths of compensation on Venus indicate both that mantle convection is active and that the normal component of convective tractions couples strongly to the overlying lithosphere. In particular, the broad rises with associated rift zones and stratigraphically young volcanic flows have a gravity signature consistent with mantle upwelling, and the lowlands can be interpreted as sites of present mantle downwelling (Phillips, Grimm and Malin, 1991).

Crustal formational processes

The formation and evolution of the Venus crust are closely tied to the thermal and dynamical evolution of the interior. Planetary crustal material can be divided into three broad categories (Taylor, 1989): primary, the result of accretional heating (such as the lunar highland crust); secondary, the result of partial melting of the mantle (such as the lunar maria and the terrestrial oceanic crust); and tertiary, formed by the reprocessing of secondary crustal material (such as terrestrial continental crust). Although Venus may have once had a primary crust, the comparatively young age of the surface and the evidence that surface rocks sampled by landed spacecraft are basaltic in chemistry and likely to be products of partial melting of the mantle (Barsukov, 1992) support the premise that little or none of any primary crust is present at the Venus surface. Tertiary crustal material requires the remelting of secondary crust, either by basal melting of a thick crustal column or by melting of crustal material recycled by subduction or foundering into the mantle; such remelting on Earth is generally enabled by water, present at much lower abundances in the crust of Venus than in that of Earth. While the U, Th and K abundances at the Venera 8 landing site (Barsukov, 1992) and the morphology of some volcanic centers (Head et al., 1992) are consistent with magmas more silica rich than basalts, the extent of such material in the Venus crust is unknown. It should be remembered that all surface chemical measurements to date have been on volcanic plains, and the possibility that highlands or tesserae consist of compositionally distinct material remains open.

The simplest hypothesis is that most of the crust on Venus is secondary. Under this hypothesis the volume of the present crust provides a basis for estimating a lower bound on the accumulated volume of magma generated by partial melting of the Venus mantle. As noted above, gravity–topography relations in the highland blocks dominated by tessera terrain, including western Aphrodite Terra and Ishtar Terra, are consistent with average crustal thicknesses of 25–40 km in those regions. While the crustal thickness beneath plains and lowlands cannot be similarly estimated at the present resolution of the gravity field, simple isostatic considerations yield an average thickness of 10–20 km beneath such areas. If so, then the total crustal volume on Venus is about 10^{10} km^3 (Grimm and Solomon, 1988). This value is comparable to the present volume of crust on Earth and is about one order of magnitude less than the time-integrated volume of secondary crust produced over all of Earth's history. If little or no crust has been recycled into the mantle on Venus, average crustal production rates on Venus must be much lower than on Earth. If, on the other hand, crustal production rates have been comparable on the two planets, some type of crustal recycling must have occurred on Venus in the geologic past.

A perspective on these alternatives comes from the estimated rate of addition of new material to the crust and inferences as to the temporal variation in that rate. On the basis of the small fraction of large impact craters embayed by volcanic flows, the average rate of volcanic resurfacing over the last 500 Ma must be 0.2 km^2 year^{-1} or less (Phillips et al., 1992, Namiki and Solomon, 1994). This figure may be converted to a volumetric flux from the observation that the fraction of craters greater than 30 km in diameter that are partially embayed is greater than for smaller craters; the expected rim height of a 30 km diameter crater then yields an upper bound on the thickness of flows capable of embaying impact craters and gives an upper bound to the volumetric flux of 0.4 km^3 year^{-1} (Namiki and Solomon, 1994). This flux, if operative over the entire history of the planet, would generate only about one-fifth of the present crustal volume. Even if the ratio of intrusive to extrusive volumes is as great as 5 to 1, this upper bound on the magmatic flux would take 4.5 Ga to generate a volume of crust equal to the present volume. It must therefore be concluded that crustal formation rates were higher prior to 500 Ma ago than since that time.

Resurfacing processes

Hypotheses advanced to explain the tectonics of Venus, and in particular the resurfacing history, fall into two categories: catastrophic and evolutionary. Catastrophic resurfacing scenarios are motivated by the interpretation that the small fraction of impact craters modified by exterior volcanism or significant deformation implies that a global-scale resurfacing event ended about 500 Ma ago (Schaber et al., 1992). In one class of these scenarios, catastrophic resurfacing occurs because of an instability in the lithosphere. Turcotte (1993) has proposed that global lithospheric overturn, a version of plate tectonics, operates episodically on Venus, and that for the last 500 Ma the lithosphere has been cooling and mechanically stable. He argues that a cool and thick lithosphere is in better agreement (than one in steady-state conductive equilibrium with the long-term average heat flow from the interior) with the large values obtained for the flexural rigidity of the lithosphere and the large ratios of long-wavelength geoid anomaly to long-wavelength topographic relief. If episodic plate tectonics operated on Venus, then crustal production would have been rapid during episodes of lithospheric recycling but would occur at much lower rates during stable periods, such as the last 500 Ma. Parmentier and Hess (1992) have suggested that lithospheric stabilization may occur on Venus because of a decrease in the density of lithospheric mantle following melt extraction, but that the cooling of such a layer may subsequently increase the density sufficiently for the lithosphere later to become unstable. Global overturn of this unstable layer in their scenario would be followed by widespread partial melting of the upper mantle, global volcanic resurfacing and the gradual development of a new buoyant layer of lithospheric mantle that gives rise to another extended period of lithospheric stability. In one-dimensional models of this process, the time between lithospheric instability events is 300–500 Ma. Even if such instability mechanisms are operative, however, the assumption of one-dimensionality can be questioned. It would seem at least as likely that different parts of the planet at any given time would be at different stages in the stabilization and destabilization sequence, and that global-scale temporal variations will be smoothed out by such regional differences.

In other catastrophic scenarios it is time-variable mantle convection rather than lithospheric instability that serves as the mechanism for global resurfacing. An early effort to simulate time-variable, three-dimensional mantle convection on Venus by Arkani-Hamed and Toksöz (1984) led to models in which the characteristic flow speed and mantle heat flux showed large oscillations (by factors of two to ten) at intervals of 100–200 Ma. In improved calculations with better spatial resolution, mantle convection, while still time varying, shows significantly smaller fluctuations of several tens of percent. Steinbach and Yuen (1992) have drawn attention to the potential role of upper mantle phase transitions in governing the radial character of mantle convection in the large terrestrial planets. They argue that such phase transitions promote the formation of distinct convecting layers in the upper and lower mantle but that as the Rayleigh number in the mantle decreases in response to cooling of the central metallic core, whole-mantle convection tends to become favored over layered convection. If Venus cooled more rapidly than Earth early in its history, it may have a lesser mantle Rayleigh number at present and have gone through a transition to whole-mantle convection while the Earth may still be characterized by layered mantle flow. Such a transition would have been accompanied by overturn of the upper mantle, the upward transport of significant heat, the generation of substantial melt and probably global volcanic resurfacing.

The second category of resurfacing scenarios involve only a gradual lessening of volcanic or tectonic activity rather than one or more global catastrophes. There are several motivations for such evolutionary scenarios. While the known volcanic histories of Earth and Mars include pulses of greater than average activity and presumably therefore time-variable mantle convective flux, no other terrestrial planet displays evidence for rapid and complete global resurfacing during the past 4 Ga. Further, there are important differences in average surface age of a number of types of geological units (Ivanov and Basilevsky, 1993; Namiki and Solomon, 1994; Price and Suppe, 1994). Phillips et al. (1992) have described a model in which volcanic resurfacing occurs episodically in small patches a few hundred kilometers in extent, with a characteristic time between events of the order of 0.1 Ma. They argued that such a model, involving a gradual decline in the volcanic flux of the planet, is as consistent with the characteristics of impact craters on Venus as are the catastrophic scenarios.

This author has advanced the hypothesis that, prior to 500 Ma ago, the primary resurfacing mechanism on Venus may have been tectonic deformation rather than volcanism (Solomon, 1993). The reason for this view is as follows. An important difference between Venus and all the other terrestrial planets is its high surface temperature. Characteristic timescales for ductile deformation of crustal and mantle material are known to vary exponentially with reciprocal temperature, so for a given thermal gradient and stress field, high rates of ductile flow are expected to be reached at much shallower levels on Venus than on other terrestrial planets. Direct coupling of mantle convective stresses should give rise to lithospheric strains that are broadly coherent over large regions and, for a sufficiently weak crust, high rates of crustal deformation and consequently of surface strain.

Prior to the era of Venus history now preserved, therefore, if the surface temperature was comparable to that at present, the higher heat flow associated with early planetary cooling and enhanced levels of radiogenic heat production and a mantle convective vigor at least that of the present may have enabled geologically rapid rates of crustal deformation over most, if not all, of the surface. Such an era would have been characterized by a nearly global extent of intensely deformed terrain (such as tesserae) and few impact craters sufficiently undeformed as to be recognizable from surface images. At some point in the evolution of Venus, however, heat flow would decline to levels sufficiently low that the ductile strength of the lower crust would undergo a transition from high rates of deformation to rates which are significantly lower on geological timescales. Following such a transition, which may appear to have been relatively abrupt in the geological record because of the exponential dependence of strain rate on temperature, both volcanic deposits and impact craters would persist for long intervals with at most modest deformation. The characteristics of the impact crater population on Venus are consistent with this hypothesis if this transition from rapid to modest rates of surface strain occurred approximately 500 Ma ago.

This tectonic resurfacing hypothesis leads to some simple predictions that bear on the resurfacing controversy. If Venus was laterally uniform in both crustal thickness and heat flow, then the transition in surface strain rates would occur with global synchroneity; that is, there would be a rapid change on a planetary scale from high rates of resurfacing to low rates, as is called for by the catastrophic resurfacing model (Schaber et al., 1992), although no true catastrophe is involved. While the assumptions of uniform crustal thickness and uniform heat flow are not strictly correct, more than 80% of the Venus surface stands at elevations within 1 km of the mean value. To the extent that regions at similar elevations have similar crustal thicknesses and thermal structures, i.e. to the extent that the principle of isostasy applies, much of the Venus surface may not depart greatly from these assumptions and an apparently 'catastrophic,' nearly global change in tectonic resurfacing rates is not a bad first approximation. Departures from global synchroneity are to be expected, however. In particular, broad highland rises, which owe their elevations to greater than average temperatures and convective tractions associated with mantle upwelling, should persist as regions of concentrated strain long after the rate of deformation in areas of lower heat flow has dropped to modest levels.

There are several problems faced by this tectonic resurfacing scenario as a complete explanation for the resurfacing history of Venus. First, an explanation must be sought for the emplacement of volcanic plains constrained by the low fraction of embayed impact craters to have occurred over a geologically short time interval. Second, the change in heat flow into the base of the crust necessary to reduce crustal strain rates by one to two orders of magnitude is approximately 10–20%, too high a figure to occur over a geologically short time interval if the causative time dependence is secular global cooling of the planet. Finally, this scenario predicts a nearly steady balance at present between interior heat production and heat flow; thus the low thermal gradients inferred from flexural studies imply either a lower heat production per mass or a lower fraction of heat loss from global cooling on Venus than on Earth.

The resurfacing history of Venus probably reflects elements of both the catastrophic and the evolutionary models. The physical processes responsible for the apparently catastrophic component remain uncertain, as does the history of heat loss and resurfacing prior to the time of the oldest preserved terrain. Further studies of the sequence and timing of geological events on both regional and global bases should sharpen the constraints on the recorded history of deformation and magmatism, and improved models for interior dynamical processes should permit predictions to be tested with fewer restrictive assumptions and spanning a fuller range of observations than has been possible to date. Progress on both fronts can be expected as the enormous data set collected by the Magellan mission continues to be distilled.

Sean C. Solomon

Bibliography

Arkani-Hamed, J. and Toksöz, M.N. (1984) Thermal evolution of Venus. *Phys. Earth Planet. Inter.*, **34**(4), 232–50.
Arvidson, R.E., Greeley, R., Malin, M.C. et al. (1992) Surface modification of Venus as inferred from Magellan observations of plains. *J. Geophys. Res.*, **97**(E8), 13303–17.
Barsukov, V.L. (1992) Venusian igneous rocks, in *Venus Geology, Geochemistry, and Geophysics* (ed. V.L. Barsukov, A.T. Basilevsky, V.P. Volkov and V.N. Zharkov). Tucson: University of Arizona Press. pp. 165–76.
Basilevsky, A.T. (1993) Age of rifting and associated volcanism in Atla Regio, Venus. *Geophys. Res. Lett.*, **20**(10), 883–6.
Greeley, R., Arvidson, R.E., Elachi, C. et al. (1992) Aeolian features on Venus: preliminary Magellan results. *J. Geophys. Res.*, **97**(E8), 13319–45.
Grimm, R.E. and Solomon, S.C. (1988) Viscous relaxation of impact crater relief on Venus: constraints on crustal thickness and thermal gradient. *J. Geophys. Res.*, **93**(B10), 11911–29.
Head, J.W., Crumpler, L.S., Aubele, J.C. et al. (1992) Venus volcanism: classification of volcanic features and structures, associations, and global distribution from Magellan data. *J. Geophys. Res.*, **97**(E8), 13153–97.
Ivanov, M.A. and Basilevsky, A.T. (1993) Density and morphology of impact craters on tessera terrain, Venus. *Geophys. Res. Lett.*, **20**(23), 2579–82.
Johnson, C.L. and Sandwell, D.T. (1994) Lithospheric flexure on Venus. *Geophys. J. Int.*, **119**(2), 627–47.
Mackwell, S.J., Zimmerman, M.E., Kohlstedt, D.L. and Scherber, D.S. (1995) Experimental deformation of dry Columbia diabase: implications for tectonics on Venus, in *Proc. Rock Mechanics Symp.* 35 (eds J.J.K. Daemon and R.A. Schultz), pp. 207–14.
Namiki, N. and Solomon, S.C. (1994) Impact crater densities on volcanoes and coronae on Venus: implications for volcanic resurfacing. *Science*, **265**(5174), 929–33.
Parmentier, E.M. and Hess, P.C. (1992) Chemical differentiation of a convecting planetary interior: consequences for a one plate planet such as Venus. *Geophys. Res. Lett.*, **19**(20), 2015–8.
Phillips, R.J., Grimm, R.E. and Malin, M.C. (1991) Hot-spot evolution and the global tectonics of Venus. *Science*, **252**(5006), 651–8.
Phillips, R.J., Robertas, R.F., Arvidson, R.E. et al. (1992) Impact craters and Venus resurfacing history. *J. Geophys. Res.*, **97**(E10), 15923–48.
Price, M. and Suppe, J. (1994) Mean age of rifting and volcanism on Venus deduced from impact crater densities. *Nature*, **372**(6508), 756–9.
Sandwell, D.T. and Schubert, G. (1992) Flexural ridges, trenches, and outer rises around coronae on Venus. *J. Geophys. Res.*, **97**(E10), 16069–83.
Schaber, G.G., Strom, R.G., Moore, H.J. et al. (1992) Geology and distribution of impact craters on Venus: what are they telling us? *J. Geophys. Res.*, **97**(E8), 13257–301.
Simons, M., Hager, B.H. and Solomon, S.C. (1994) Global variations in the geoid/topography admittance of Venus. *Science*, **264**(5160), 798–803.
Smrekar, S.E. and Phillips, R.J. (1991) Venusian highlands: geoid to topography ratios and their implications. *Earth Planet. Sci. Lett.*, **107**(3/4), 582–97.
Solomon, S.C. (1993) The geophysics of Venus. *Physics Today*, **46**(7), 49–55.
Solomon, S.C. and Head, J.W. (1991) Fundamental issues in the geology and geophysics of Venus. *Science*, **252**(5003), 252–60.
Solomon, S.C., Smrekar, S.E., Bindschadler, D.L., et al. (1992) Venus tectonics: an overview of Magellan observations. *J. Geophys. Res.*, **97**(E8), 13199–255.
Steinbach, V. and Yuen, D.A. (1992) The effects of multiple phase transitions on Venusian mantle convection. *Geophys. Res. Lett.*, **19**(22), 2243–6.
Stofan, E.R., Sharpton, V.L., Schubert, G. et al. (1992) Global distribution and characteristics of coronae and related features on Venus: implications for origin and relation to mantle processes. *J. Geophys. Res.*, **97**(E8), 13347–78.
Taylor, S.R. (1989) Growth of planetary crusts. *Tectonophysics*, **161**(3/4), 147–56.
Turcotte, D.L. (1993) An episodic hypothesis for venusian tectonics. *J. Geophys. Res.*, **98**(E9), 17061–8.

Cross references

Corona (Venus)
Impact cratering
Isostasy
Magellan mission
Mantle convection
Surface processes
Tessera
Thermal evolution of planets and satellites
Volcanism in the solar system

VENUS: GRAVITY

Since Venus is approximately the same size as the Earth (6051 km versus 6378 km radius for the Earth) and its mass is only 23% smaller, its surface gravity is 887 cm s^{-2}. It is the variations about this mean gravitational value that interest the navigation engineer and the geophysicist, for the gravitational irregularity will greatly influence the lifetime of an orbiting spacecraft and the internal structure models that can be inferred. These variations are commonly called the Venus gravitational field. The gravitational field is measured from

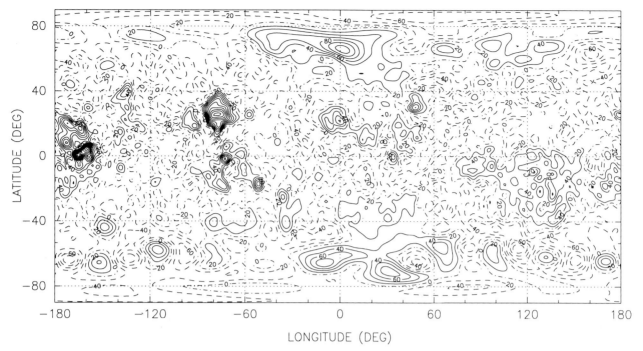

Figure V29 Venus gravity at the surface in milligals. (From A.S. Konopliv, *et al.*, 1993. Copyright American Geophysical Union.)

the reduction of radio Doppler data from the orbiting spacecraft Pioneer Venus Orbiter (1979–1992) and Magellan (1990–1994).

The contours of gravity, shown on Figure V29, are in milligals (1000 milligals = 1 cm s^{-2}). There is a very high correlation of gravity highs with topographic highs and similarly with the lows (Plate 7). This correlation is the largest of any observed terrestrial body (Earth, Mars and the Moon) and is the primary difference between the gravity fields of Venus and the Earth. The amplitudes (the peak values) are approximately the same, but on Earth there are gravity highs over the oceans (topographic lows) and gravity lows over the continents. These low-frequency variations on Earth are due to deep convection cells, causing crustal plate motions. Since these uncorrelated gravity/topography effects are not seen on Venus, geophysicists are concluding that, possibly, internal convection is not active.

The fact that the amplitudes of the gravity highs and lows on Earth and Venus are much smaller than on Mars and the Moon is due to their less rigid interior; this can accommodate relaxation or isostatic adjustment for the topographic loads (much like icebergs in the ocean). On Venus, Aphrodite Terra (a mountain range near the equator from 60°E to 160°E longitude) has a gravity value which can be explained with a relaxation or isostatic compensation depth of 65 km. This seems reasonable even for Earth models. However when Atla (200° longitude) and Beta (280° longitude) are similarly analyzed, depths of several hundred kilometers are obtained. This seems unrealistic and therefore models requiring dynamic support such as mantle plumes have been proposed.

William L. Sjogren

Bibliography

Konopliv, A.S., Borderies, N.J., Chodas, P.W. *et al.* (1993) Venus gravity and topography: 60th degree and order model. *Geophys. Res. Lett.*, **20**(21), 2403–6.

McNamee, J.B., Borderies, N.J. and Sjogren, W.L. (1993) Venus: global gravity and topography. *J. Geophys. Res.*, **98**(E5), 9113–28.

Mottinger, N.A., Sjogren, W.L. and Bills, B.G. (1985) Venus gravity: a harmonic analysis and geophysical implications, *J. Geophys. Res.*, **90**, C739–56.

Reasenberg, R.D., Goldberg, Z.M., MacNeil, P.E. and Shapiro, I.I. Venus gravity: a high resolution map, *J. Geophys. Res.*, **86**, 7173–9.

Sjogren, W.L., Bills, B.G., Birkland, P.W. *et al.* (1983) Venus gravity anomalies and their correlation with topography. *J. Geophys. Res.*, **88**, 1119–28.

Cross references

Geoid
Gravity fields of the terrestrial planets
Isostasy
Mars: gravity
Moon: gravity

VENUS: MAGNETIC FIELD AND MAGNETOSPHERE

Venus is sometimes characterized as Earth's 'twin' because of its close proximity in solar system location (~ 0.72 AU heliocentric distance compared to 1.0 AU) and its similar size (~ 6053 km radius compared to ~ 6371 km radius), but other close resemblances are few. Besides the more obvious atmospheric composition and pressure differences, and the related extreme temperatures at the surface described elsewhere in this volume, events in the history and evolution of the interior of Venus have left that planet with practically no intrinsic magnetic field. The consequences for the space environment and atmosphere are numerous, ranging from the presence of an 'induced' magnetotail in the wake, to an ionosphere and upper atmosphere that are constantly being scavenged by the passing solar wind.

Interior

Venus, like the other terrestrial planets, was presumably accreted from iron and silicate-bearing planetesimals some 4.5 billion years ago. These new planets are all likely to have differentiated in a similar manner, so that they have the common feature of a molten iron-rich core of about half the planet's radius, covered by a crust of the remaining (mainly silicate) material. Only indirect information is available about these cores, but seismic measurements on the surface of Earth tell us that a solid inner core, with a size depending on the size of the planet and on its thermal history, may also be a common feature. Since no seismic measurements have been obtained on the surface of Venus, we cannot be as certain about its interior; however, the large value of the mean density of ~ 5.25 g cm^{-3} derived from satellite orbits suggests that Venus contains an Earth-like core. Essentially, all other deductions about the interior of Venus are based on models of Earth-like planets with internal temperatures and pressures adjusted for the slightly different radius and possible compositional differences. One of these models has led to the hypothesis that the core of Venus may be completely solid or 'frozen' today, while others propose that core solidification has not yet commenced or has stopped at some time in the past (e.g. Stevenson, Spohn and Schubert, 1983). In all cases, evidence cited in support of these hypotheses always includes the known weakness of the intrinsic magnetic field.

Magnetic field

When Mariner 2 flew by Venus in 1962 at a distance of 6.6 planetary radii (R_V), it did not detect any evidence of an Earth-size magnetosphere. Mariner 5, passing within 1.4 R_V in 1967, detected the signatures in the solar wind of deflection around an 'obstacle' at Venus. The small inferred size of that obstacle placed an upper limit on the magnetic dipole moment of Venus of $\sim 10^{-3}$ that of Earth. Later Venera 4 made magnetic measurements down to 200 km altitude, still detecting no planetary field but providing data that reduced this estimate by about an order of magnitude. In a 1974 flyby, Mariner 10 merely confirmed the existence of a small, nearly planet-size obstacle. Venera 9 and 10 were put into orbit around Venus in 1975, but did not approach Venus closer than ~ 1500 km. Nevertheless, the data that these spacecraft obtained in the wake of the planet provided the first evidence that an Earth-like magnetotail was absent, and that instead a structure related to the interplanetary magnetic field occupied that region of space. The most definitive measurements of the magnetic moment of Venus were obtained during the Pioneer Venus Orbiter mission in its first years of operation (1979–1981). Repeated low-altitude (~ 150 km) passes by that spacecraft over the antisolar region, coupled with dayside observations to the same altitude, proved the insignificance of a field of internal origin in near-Venus space. The observed fields for the most part could be explained as solar wind interaction-induced features, to be described below. The new upper limit on the dipole moment obtained from the Pioneer Venus Orbiter wake measurements placed the Venus intrinsic magnetic field at $\sim 10^{-5}$ times that of Earth.

Of course, the weakness of the present measurement does not imply that Venus has always been bereft of an intrinsic field. Theories of the dynamos operating in the liquid cores of the newly accreted terrestrial planets suggest that there was a magnetic moment of Venus of the same order as Earth's for about the first billion years of Venus' life. During that time, thermal convection from the heat left over from accretion drove the dynamo. However, after that energy source diminished, there was apparently no source to replace it. While solid core formation in Earth's interior maintains its dynamo to this day by virtue of the related 'stirring' of the molten core around it, Venus appears to either lack the necessary internal ingredients (chemical or physical) for solid core formation, or to have ceased such processes at an earlier time if they resulted in complete core solidification or arrested core solidification. It is important to note that, contrary to popular belief, dynamo theory does not credit the smallness of the magnetic moment to the slow rotation of Venus (a Venus day of ~ 243 Earth days is almost equal to the length of its year of ~ 224 days, and its sense of rotation is retrograde). It is also notable that Venus would not have maintained any remanent crustal magnetic fields from its proposed early period of dynamo activity because the temperatures in the crust are expected to be above the Curie point (below which such fields could persist in rocky materials).

Solar wind interaction

The 'magnetosphere' of Venus that was detected by spacecraft is now known to be an example of an 'induced' magnetosphere. In an induced magnetosphere, the solar wind interacts directly with the planetary ionosphere. The fields and plasmas that are observed are generally of solar wind or ionospheric origin. There are no belts of trapped radiation such as Earth's Van Allen belts, and there is no 'magnetotail' composed of fields of planetary origin. The basic features of an induced magnetosphere are shown in Figure V30. The ionospheric obstacle to the solar wind is defined by a surface called the ionopause. At the ionopause, pressure balance exists between the solar wind dynamic pressure on the outside and the thermal pressure of the ionospheric ions and electrons on the inside. Outside of the ionopause the solar wind interaction has all of the features characteristic to a planetary magnetosphere. A bow shock forms upstream of the obstacle. An interesting feature of the Venus bow shock is that it appears to have a location that varies with the solar cycle. The 'nose', or subsolar position, of the bow shock at sunspot maximum is near 1.5 R_V, but the terminator location moves in to ~ 2.1 R_V. Inside of the bow shock the solar wind plasma is deflected around the obstacle in a magnetosheath region, which is sometimes referred to as an ionosheath, since the obstacle is an ionosphere. The embedded interplanetary magnetic field is compressed and draped around the obstacle in the magnetosheath region in the usual way.

Inside of the ionopause the plasma changes from solar wind-dominated to ionospheric in origin. During the primary Pioneer Venus mission, which occurred at a time of high solar activity when the planetary ionospheres are densest, this boundary between the

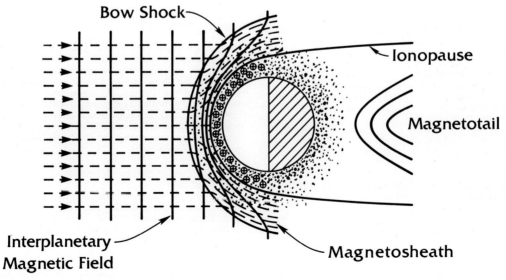

Figure V30 Illustration of the major features of the solar wind interaction with the ionosphere of Venus. The solid dots represent the neutral atmosphere, while the circled plus symbols represent ionized atmosphere. The ionized atmosphere above the ionopause is removed by the solar wind.

solar wind and ionosphere proper occurred at an average altitude of about 300 km, flaring to ~ 800 km average altitude near the terminator. The boundary, which moves up and down in response to changing external (solar wind) pressure, was typically thin at a few tens of kilometers, although it increases in thickness as its altitude decreases. The observations indicated that the magnetic fields of interplanetary origin in the magnetosheath generally remain confined above the ionosphere proper, although small-scale (dimensions of a few kilometers) field increases (to ~ 100 nT) of still-unknown origin were observed. These field intrusions appeared to have twisted internal structures and so were dubbed 'flux ropes'. The exception to this behavior occurred on the rare occasions (about 15% of the time) when the solar wind pressure was high enough to drive the pressure-balance boundary to altitudes of ~ 250 km or less (the ionospheric thermal pressure increases as altitude decreases down to about 190 km altitude). At these times it appeared as if the interplanetary magnetic field in the magnetosheath penetrated the ionosphere to at least the spacecraft minimum altitude of ~ 150 km. Its magnitude in the ionosphere can reach ~150 nT.

The nightside solar wind interaction features at altitudes below several hundred kilometers also show a dichotomy with solar wind pressure. When the conditions for the 'unmagnetized' dayside ionospheres prevail, the nightside ionosphere is supplied by planetary plasma flowing across the terminator from the dayside. The observed nightside ionospheric magnetic fields are fluctuating and weak (≤ 10 nT) at these times, and do not appear to be twisted like the dayside flux ropes. Near midnight, however, steady, almost vertical magnetic fields of a few tens of nanotesla were observed in conjunction with ionospheric density depletions called 'holes' by their discoverers (Brace, et al., 1982). These features are up to a significant fraction (~ ¼) of the planetary radius in horizontal scale, and they appear to have a field 'polarity' (e.g. sunward or antisunward) that depends on the interplanetary magnetic field orientation and the associated draped field in the magnetosheath. Their origin and nature remain controversial. The 'holes' disappear when the solar wind pressure is high. The nightside counterpart of the large-scale magnetosheath field penetration into the dayside ionosphere appears to be a large-scale horizontal field of somewhat smaller magnitude (tens of nanotesla) throughout the nightside. Its relationship to the dayside field is still poorly understood. It should be noted that the high solar wind pressure scenario is expected to be common during solar minimum, when the ionospheric pressure is always weaker than at solar maximum.

The high-altitude wake of Venus is permeated with structured magnetic fields that generally point sunward or antisunward and often exhibit a 'double-lobed' structure like an intrinsic planetary magnetotail. However, examination of the polarities of the fields in the lobes shows them to be coupled closely to the interplanetary field and resulting draped magnetosheath field orientations. As shown in Figure V30, this 'induced' magnetotail can be pictured as an extension of the magnetosheath, with the draped interplanetary fields sinking into the ionospheric obstacles' wake. The draping of the field in the induced magnetotail is observed to be enhanced beyond that in the surrounding magnetosheath. This enhancement has been attributed to the 'mass loading' of those interplanetary flux tubes that pass closest to the ionopause and form the magnetotail by virtue of heavy ionospheric ion production on those passing flux tubes. In this sense, Venus can be likened to a comet, which has an induced magnetotail of similar origin.

Janet G. Luhmann and Christopher T. Russell

Bibliography

Brace, L.H., Theis, R.F., Mayr, H.G. et al. (1982) Holes in the nightside ionosphere of Venus. *J. Geophys. Res.*, **87**, 199.
Hunten, D.M., Colin, L., Donahue, T.M. and Moroz, V.I. (eds) (1983) *Venus*. Tucson: University of Arizona Press.
Luhmann, J.G. (1986) The solar wind interaction with Venus. *Space Sci. Rev.*, **44**, 241.
Russell, C.T. (ed.) (1991) *Venus Aeronomy. Space Sci. Rev.*, **55**, London: Kluwer Academic Publishers.
Russell, C.T. (1987) Planetary magnetism in *Geomagnetism*, Vol. 2 (ed. J.A. Jacobs) London: Academic Press, pp. 457–523.
Stevenson, D.J., Spohn, T. and Schubert, G. (1983) Magnetism and thermal evolution of the terrestrial planets. *Icarus*, **54**, 466.

Acknowledgements

The authors are supported for work on this subject by NASA grant NAGW 2–501 through the Pioneer Venus project.

Cross references

Interplanetary magnetic field
Ionosphere
Magnetism
Magnetometry
Mariner missions
Pioneer Venus missions
Solar wind
Venera missions

VESTA

Vesta is one of the largest of the asteroids, and may be the parent body of the basaltic achondrite meteorites. It has a diameter of about 500 km (Drummond, Eckart and Hege, 1988), comparable in size to 2 Pallas and larger than any other asteroid except 1 Ceres. It is thus one of the easiest targets for many telescopic techniques. It also has a high visual albedo (0.38), which makes it the brightest of the main-belt asteroids.

Vesta's spectral reflectance signature is unique among large asteroids. It has strong bands caused by the mineral pyroxene, and weaker bands from the mineral feldspar, which are interpreted as components of the basaltic crust of a differentiated asteroid (Gaffey, 1983). Furthermore, there are spectral variations during Vesta's rotation, implying a heterogeneous surface. The only other asteroids with basaltic spectra are several small (5 km) main-belt asteroids in orbits very similar to Vesta's (Binzel and Xu, 1992) and a few small Earth-crossing asteroids (Cruikshank et al., 1991). The former are almost certainly chips off Vesta, and the latter might be as well. Vesta's spectrum matches that of the eucrite or, in some places, diogenite meteorites (see Basaltic achondrite meteorites), whose parent body must either be Vesta or some now-shattered asteroid that once resembled Vesta (Drake, 1979). The fact that Vesta is still covered by its basaltic crust has been used to set limits on the extent of collisional disruption in the main belt (Davis, 1985).

Vesta was discovered on 29 March 1807 by H.W. Olbers in Bremen. It was the fourth asteroid discovered. However, after those four were discovered in a period of slightly more than 6 years, it was nearly 40 years before another was found.

Timothy D. Swindle

Bibliography

Binzel, R.P. and Xu, S. (1992) Chips off of Vesta (abstract). *Bull. Am. Astron. Soc.*, **24**, 939.
Cruikshank, D.P., Tholen, D.J., Hartmann, W.K. et al. (1991) Three basaltic Earth-approaching asteroids and the source of the basaltic meteorites. *Icarus*, **89**, 1–13.
Davis, D.R. (1985) Collisional history of asteroids: evidence from Vesta on the Hirayama families. *Icarus*, **62**, 30–53.
Drake M.J. (1979) Geochemical evolution of the eucrite parent body: possible nature and evolution of asteroid 4 Vesta? in *Asteroids* (ed. T. Gehrels). Tucson: University of Arizona Press, pp. 765–82.
Drummond, J.D., Eckart, A. and Hege, E.K. (1988) Speckle interferometry of asteroids. IV. Reconstructed images of 4 Vesta. *Icarus*, **73**, 1–14.
Gaffey, M.J. (1983) The asteroid (4) Vesta: rotational spectral variations, surface material heterogeneity, and implications for the origin of the basaltic achondrites (abstract), in *Lunar Planet. Sci. XIV*. Houston: Lunar and Planetary Science Institute, pp. 231–2.

Cross references

Asteroid
Asteroid: compositional structure and taxonomy

Basaltic achondrite meteorites
Meteorite parent bodies

VIKING MISSION

The Viking mission to Mars, a continuation of NASA's Mariner program to explore the inner planets, accomplished the first successful landings of spacecraft on another planet (Plate 16). It was initiated in 1968 with the principal responsibility assigned to the Langley Research Center (LaRC) in Hampton, Virginia, assisted by the Jet Propulsion Laboratory (JPL), which built the orbiting spacecraft and conducted the mission operations. The landers were built on a contract by the Martin Marietta Aerospace Corporation in Denver.

The spacecraft

Each of the two Viking spacecraft consisted of two parts – an orbiter (Figure V31) sometimes called a 'bus' because it carried a passenger, and a landing capsule. As launched by the Titan–Centaur launch system, it weighed 3527 kg; the orbiter weighed 2328 kg, of which the propellant required to inject the spacecraft into Martian orbit weighed 1426 kg. The landing capsule, weighing 1168 kg, contained the lander (Figure V32), its parachute and the 153 kg of propellant that was required to ease the lander down onto the surface after its parachute was jettisoned. On the surface the lander weighed about 600 kg, including 103 kg for the nuclear-thermoelectric power system and 91 kg for the science instruments.

The mission

Originally planned for 1973, to follow immediately after Mariner 9, the mission was postponed 2 years because of funding problems. The two identical spacecraft, Viking 1 and 2, were launched on 20 August and 9 September 1975 and were inserted into orbit around Mars on 19 June and 7 August 1976. Each orbital mission began with several weeks of surface observations by the Orbiter Imaging Systems to locate a landing site that appeared to be safe. The landings were accomplished on 20 July and 3 September, with the first lander at 22°N latitude, 48°W longitude in Chryse Planitia and the second at 44°N latitude, 226°W longitude in Utopia.

The primary mission duration of the landers was specified to be 90 days, and they had been designed to this specification, but as all four spacecraft continued to operate virtually faultlessly, the Viking mission was repeatedly extended by NASA as long as data were being obtained. The final days of operation were: Orbiter 2, 25 July 1978; Lander 2, 11 April 1980, Orbiter 1, 7 August 1980; and Lander 1, 13 November 1982. Thus the total duration of the mission at Mars was 3.40 Martian years or 6.40 terrestrial years. In January 1991 NASA formally named Lander 1 the Thomas A. Mutch Memorial Station in memory of the leader of the Lander Imaging Team.

Throughout the Viking mission, the four spacecraft responded to 270 000 commands from the Deep Space Network, made hundreds of millions of scientific measurements and returned more than 500 billion bits of information to Earth.

The spacecraft and their scientific mission were the most complex of the unmanned space program up to that time. The primary scientific objective was to search for evidence of current or past living organisms on the Martian surface. Other objectives were to acquire close-up photographs of the entire surface of the planet and its moons, to measure the physical and chemical composition of the atmosphere and surface, to monitor the weather and to detect marsquakes.

Thirteen investigations were conducted by thirteen teams, which included 83 scientists from 24 universities, nine government laboratories and seven industrial or independent laboratories (Table V7). The Entry Science investigation, made by instruments on the descent capsule, is classed as an orbiter experiment because the data were relayed to the orbiters. The Radio Science Team analyzed data from both orbiters and landers.

Entry science

After its release from the orbiter, each landing capsule reoriented itself at 300 km altitude to prepare for its aerodynamic entry, and was slowed by atmospheric drag to 6 km altitude, where its aeroshell was jettisoned and its parachute deployed. At 1.5 km three retrorockets were fired for the final descent to the surface. Meanwhile, instruments made measurements on the atmosphere from 200 km altitude to the surface. The mass spectrometers measured the concentrations of CO_2, O_2, N_2, CO, NO and Ar atoms in the upper atmosphere. The retarding potential analyzers monitored the concentration of ions in the ionosphere. Data from pressure, temperature and acceleration

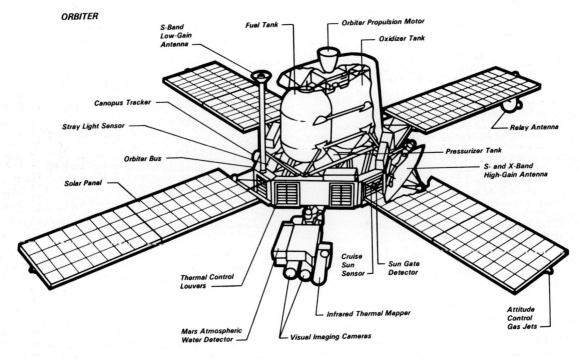

Figure V31 The Viking orbiter.

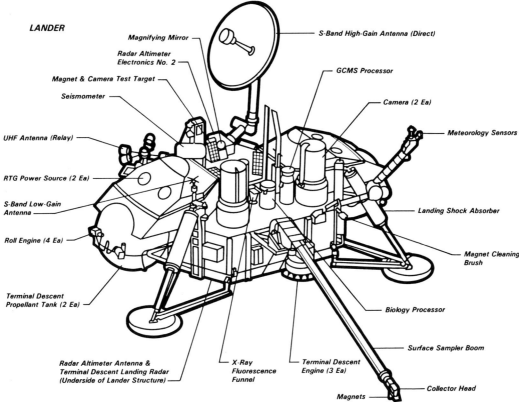

Figure V32 The Viking lander.

sensors, combined with those from radar altimeters and gyroscopes, determined the structure of the lower atmosphere and the winds.

Orbiter science

The cameras on the orbiters returned 52 603 pictures, which covered the entire planet at all Martian seasons with resolutions, near the end of the mission, as high as 7.5 m – about 150 times better than Mariner 4. Many pictures were in full color. They also obtained close-up pictures of both Martian moons and, at one time, photographed the shadow of Phobos at the instant that it crossed Lander 1, thus determining the precise position of the lander. The data obtained made possible much more detailed maps and established more than 9000 mapping control points on Mars with accuracies of 6 km or better.

The Mars Atmospheric Water Detectors (MAWD) continuously monitored the quantity of water vapor below the orbiters, thus documenting the spatial variation of vapor and its seasonal migrations. They produced nearly complete global maps of water vapor about every 30 days throughout the Martian year and showed that it is well mixed in the atmosphere at least up to 10 km altitude. The maximum column abundance observed was about 100 precipitable microns over the dark surface surrounding the northern polar cap in summer when the cap was releasing its frozen water.

The Infrared Thermal Mapper (IRTM) instruments continuously monitored the brightness temperature of the surface or clouds below them, thus producing numerous global maps of the temperature, albedo and thermal inertia of the surface. They detected many local dust storms and conclusively demonstrated (along with MAWD) that the northern polar cap in summer was water ice; the residual south polar cap was covered with carbon dioxide frost. Surface temperatures as high as 300 K (27°C) and as low as 150 K were measured.

Lander science

Within seconds of landing, the cameras on the landers, which also had color capability, began to send back pictures of the surrounding terrain: rock-strewn desert landscapes of reddish brown color. In the two missions 4587 pictures were received. With two cameras, stereoscopic photography made possible the production of very detailed maps of the landing areas. They recorded the trenches dug by the surface sampler to investigate the physical and magnetic properties of the surface material and observed the variations in the opacity of the atmosphere as dust storms came and went. The Lander 2 camera detected a coating of frost on the ground during one winter.

Each lander had a soil sampler arm that could move in three dimensions to furnish samples to the instruments for the Biology, Molecular Analysis, and Inorganic Analysis investigations.

The Biology Investigation included three separate experiments to search for evidence of microbial organisms in the Martian surface material. (1) The Carbon Assimilation Experiment was designed to detect the photosynthesis of organic compounds from atmospheric CO or CO_2. (2) The Labeled Release Experiment sought to detect metabolism, by monitoring the radioactive gas evolution following the addition of radioactive nutrients to the samples. (3) The Gas Exchange Experiment employed a gas chromatograph to detect the gaseous products of metabolism without using radioactivity. This experiment created great excitement when several atmospheric gases were evolved as soon as the sample was moistened. However, it soon became clear that the source was chemical reactions with the soil and not biological ones. Although there is still some controversy about the implications of the results of experiment 2, the concensus is that no evidence of Martian life was detected.

This conclusion was bolstered by the Molecular Analysis Experiment with its gas chromatograph–mass spectrometer, which was capable of detecting some organic compounds at the part-per-billion level, but found none. The instrument also analyzed the atmosphere and found, in essential agreement with the entry data, that the principal components were CO_2 (95.32%), N_2 (2.7%) Ar (1.6%) and O_2 (0.13%). These values apply to the early days of the mission, corresponding to spring in the northern hemisphere; the proportion of CO_2 is variable, in as much as the meteorology experiment found that the atmospheric pressure varied by about 30% over the year because atmospheric carbon dioxide was frozen out on the polar caps

Table V7 Science investigations and instruments

Investigations	Instruments
Orbiter	
Orbiter Imaging	Two vidicon cameras
Water Vapor Mapping	Infrared spectrometer
Thermal Mapping	Infrared radiometer
Entry Science	
Ionospheric properties	Retarding potential analyzer
Atmospheric composition	Mass spectrometer
Atmospheric structure	Pressure, temperature and acceleration sensors
Lander	
Lander Imaging	Two facsimile cameras
Biology[a]	Three analyses for metabolism, growth or photosynthesis
Molecular Analysis[a]	Gas chromatograph–mass spectrometer
Inorganic Analysis[a]	X-ray fluorescence spectrometer
Meteorology	Pressure, temperature, and wind velocity sensors
Seismology	Three-axis seismometer
Magnetic Properties	Magnet on sampler observed by camera
Physical Properties	Various engineering sensors
Investigations without onboard instruments	
Radio Science	Orbiter and lander radar and radio systems
Celestial mechanics, atmospheric properties and test of general relativity	

[a] Three investigations analyzed material from the soil sampler.

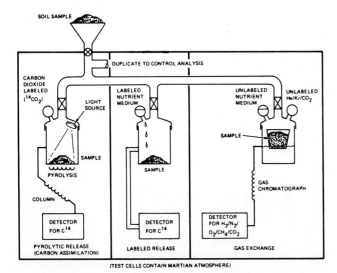

Figure V33 Operation procedures for the three Viking lander biology experiments.

and then released. The isotopic composition of the major components was also measured, and produced important implications about the source and history of the atmosphere (see Mars: atmosphere).

The x-ray fluorescence spectrometers of the Inorganic Analysis Experiments measured the concentrations in the surface material of thirteen elements with atomic numbers from 12 to 40, finding the most abundant to be Si (21%), Fe (13%), Mg (5%), Ca (4%), Al (3%), and S (3%); other elements not detectable constituted 50%.

The instrumentation of the Meteorology Investigation provided daily weather reports at the two locations on Mars for three martian years. Observed maximum daily temperatures of the atmosphere ranged from 160 K to 250 K and minima from 150 K to 190 K – all well below the freezing point of water (273 K). The atmospheric pressure at Lander 1 varied between 6.85 and 8.90 mbar; at Lander 2 the values were about 10% higher because of its lower altitude. The winds were usually very calm. The Orbiter Imaging Systems supplemented the meteorology data by observing a complex configuration of clouds, storms and weather fronts.

The Seismology Experiment was degraded by the failure of the seismometer on Lander 1. The Lander 2 instrument recorded the shaking of the lander by the wind but detected no signals that could be unequivocally attributed to seismic events. It was concluded that Mars is probably seismically less active than Earth and that much more sensitive seismometers would be required for future use on Mars missions.

The Physical Properties Experiment utilized the surface sampler arm to investigate the density, strength and cohesiveness of the surface material by digging trenches, building piles and pushing rocks, with the results documented by photography. Permanent magnets on the top of the landers and on the samplers attracted some components of the soil, permitting the Magnetic Properties Team to conclude that it contains a few percent of magnetic material.

Radio science

The radio science investigations of Viking were essentially identical with those of earlier Mariner spacecraft, but with the very significant advantages of dual frequencies, a longer time base, lower periapsis altitudes for the orbiters and, especially, transponders fixed on the planet surface; as a result the Viking measurements were greater in quantity and higher in precision. They provided information about the shape of the planet and its gravitational field, its period of rotation, the masses of asteroids and other planets, and the gravitational retardation of the velocity of light.

Analysis of the vast quantity of data from the four spacecraft is still adding to our understanding of the most earth-like planet.

Conway W. Snyder

Bibliography

Numerous papers on the scientific results of Viking were published in *Science*, *Journal of Geophysical Research* and *Icarus* beginning in August 1976.

Anonymous (1984) *Viking: The Exploration of Mars*. Washington, D.C.: NASA EP-208.
Biemann, H.-P. (1977) *The Vikings of '76*. Westford, MA: Murray Printing Company.
Carr, M.H. (1981) *The Surface of Mars*. New Haven, Conn.: Yale University Press.
Carr, M.H., and Evans, N. (1980) *Images of Mars: The Viking Extended Mission*. Washington, D.C.: NASA SP-444.
Corliss, W.R. (1974) *The Viking Mission to Mars*. Washington, D.C.: NASA SP-344.
Ezell, E.C., and Ezell, L.N. (1984) *On Mars: Exploration of the Red Planet*. Washington, D.C.: NASA SP-4212.
French, B. (1977) *Mars: The Viking Discoveries*. Washington, D.C.: NASA.
Lee, B.G. (1976) Mission operations strategy for Viking. *Science*, **194**, 59–62.
Liebes, S., Jr (1982) *Viking Lander Atlas of Mars*. NASA Contractor Report 3568.
Masursky, H., and Crabill, N.L. (1976a) The Viking landing sites: selection and certification. *Science*, **193**, 809–12.
Masursky, H. and Crabill, N.L. (1976b) Search for the Viking 2 landing site. *Science*, **193**, 62–8.
Snyder, C.W. (1977) The missions of the Viking orbiters. *J. Geophys. Res.*, **82**(28), 3971–83.
Snyder, C.W. (1979) The extended mission of Viking. *J. Geophys. Res.*, **84**(B14), 7917–33.
Snyder, C.W. (1979) The planet Mars as seen at the end of the Viking mission. *J. Geophys. Res.*, **84**(B4), 8487–519.

Snyder, C.W. and Evans, N. (1981) The final phases of the Viking mission to Mars. *Icarus*, **45**, 2–24.
Snyder, C.W. and Moroz, V.I. (1992) Spacecraft exploration of Mars, in *Mars* (eds H.H. Kieffer, B.M. Jakosky, C.W. Snyder, and M.S. Matthew). Tucson: University of Arizona Press, pp. 71–119.
Soffen, G.A. and Young, A.T. (1972) The Viking missions to Mars. *Icarus*, **16**, 1–16.
Soffen, G.A. and Snyder, C.W. (1976) The first Viking mission to Mars. *Science*, **193**, 759–66.
Soffen, G.A. (1976) Status of the Viking missions. *Science*, **194**, 57–9.
Soffen, G.A. (1976) Scientific results of the Viking missions. *Science*, **194**, 1274–6.
Soffen, G.A. (1977) The Viking project. *J. Geophys. Res.*, **82**(28), 3959–70.
Tyner, R.L. and Carroll, R.D. (1983) *A Catalog of Selected Viking Orbiter Images*. Washington, D.C: NASA Reference Publication 1093.
Viking Lander Imaging Team (1978) *The Martian Landscape*. Washington, DC: NASA SP-425.
Viking Orbiter Imaging Team (1980) *Viking Orbiter Views of Mars*. Washington, DC: NASA SP-441.

Cross references

Life: origin
Mariner missions
Mars
Mars: geology
Radio science

VISIBLE AND NEAR-INFRARED SPECTROSCOPY

Spectroscopy is the study of the interaction of matter and electromagnetic radiation. This article provides an introduction to spectroscopy in the visible (VIS ≈ 300–700 nm) and near-infrared (NIR ≈ 700–5000 nm) regions of the electromagnetic spectrum and outlines its applicability in terrestrial and planetary research. Laboratory and telescopic spectroscopy measurements in the VIS–NIR are the primary observational data with which to infer the composition and/or mineralogy of solar system bodies. More detailed background and discussions of this topic can be found in many of the textbooks and review papers discussed below and cited in the references (e.g. Wendlandt and Hecht, 1966; Kortüm, 1969; Burns, 1970; Hunt, 1980; Rossman, 1988a).

The useful wavelength limits of the VIS–NIR spectral region are primarily defined by the shape of the Sun's spectrum and the wavelength-dependent attenuation of light by the Earth's atmosphere (Figure V34), because sunlight provides the light source for most VIS–NIR spectroscopic measurements in planetary science and, until recently, most planetary spectral measurements have been obtained by ground-based methods.

The key variables used to define spectroscopic data are the desired wavelength coverage (λ), spectral bandwidth ($\Delta\lambda$), spectral resolution (R) and spectral sampling (S). $\Delta\lambda$ depends on the light dispersal method used during the measurement; for example, if an interference filter is used, $\Delta\lambda$ is usually defined as the full width at half maximum of the filter transmission profile. R is defined at any wavelength as $\lambda/\Delta\lambda$ and is a measure of the width of the narrowest spectral feature that can be detected. S is the number of spectral channels (or the number of 'colors' measured) per wavelength unit. Spectra are also frequently measured in energy units (wavenumbers, $v = \lambda^{-1}$, usually in units of cm^{-1}), and appropriately modified definitions of λ, $\Delta\lambda$, R and S apply. The key variables used in the analysis of spectral data are the positions, intensities and widths of the observed absorption or emission features.

VIS–NIR spectroscopy is used to address a broad range of questions in planetary science. The most common measurements are of the reflectance of planetary bodies or of laboratory samples. Reflectance is the ratio of the radiation reflected from a body to the radiation incident on it. This parameter is a strong function of viewing geometry (incidence, emission and phase angle), and also depends on particle size, temperature and atmospheric pressure. Reflectance measurements are often corrected for these effects and are presented in calibrated units such as radiance factor or albedo (Harris, 1961; Hapke, 1981). Reflectance measurements at low to moderate spectral resolution ($R = 10–1000$) have been used for the identification of mineralogic absorption features because spectra of solids typically

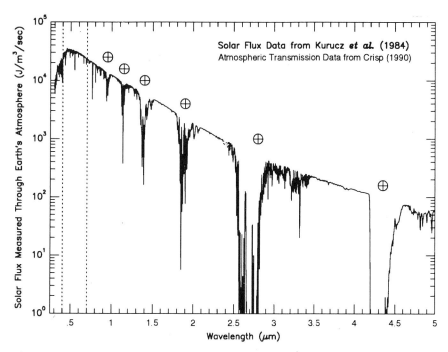

Figure V34 VIS to NIR solar spectrum times the Earth's atmospheric transmission. Within a distance-dependent constant, this is the 'baseline' signal upon which the spectral signature of a solar system object observed from Earth is superposed. The most easily measured spectral regions correspond to the atmospheric 'windows' between the Earth's strong H_2O absorption features (marked '⊕'). The wavelength region visible to the human eye occurs within the dashed lines. Note that beyond 3 μm the solar flux is from 100 to 500 times less than at visible wavelengths. (Solar flux data from Kurucz *et al.*, 1984; atmospheric transmission data from Crisp, 1990.)

do not exhibit the fine structure seen in spectra of liquids and gases (e.g. Burns, 1970; Adams, 1975). Reflectance measurements at higher resolution ($R=10^3–10^6$) have been used to identify gaseous and ionic absorptions and emissions in planetary and cometary atmospheres because absorptions due to these species are typically much narrower than those in most minerals. Recently it has been recognized that the reflectance spectra of certain minerals, such as phyllosilicates, contain very narrow, diagnostic absorptions that are characterized only at high R (e.g. Clark et al., 1990; Gaffey, McFadden and Nash, 1993). Spectroscopic experiments can also measure the transmittance of materials, or the ratio of the radiation passing through a substance to the incident radiation. In laboratory experiments, transmittance measurements are useful in the derivation of the optical properties (indices of refraction) and origin of colors of rocks and minerals (e.g. Rossman, 1988a and references therein).

Theory

Matter and radiation interact to produce observable spectral absorption features in the VIS-NIR via either electronic or vibrational processes.

Electronic transitions

The atomic orbitals surrounding each covalently-bonded atom in a molecule overlap, forming molecular orbitals in which electrons are shared by many atoms. There are four major types of absorption features caused by the resulting inter- and intra-atomic electron sharing. These electronic transitions are discussed in much greater detail by Burns (1970, 1993), Rossman (1988a), and Gaffey, McFadden and Nash (1993).

Crystal field (CF) transitions
Rearrangement of valence shell electrons within the d- and f-orbitals of transition elements such as Cr, Mn, Fe and Ti occurs as a result of absorption of VIS-NIR energy. The specific strengths and positions of these features depend on the local environment or crystal structure surrounding the cation. Specifically, the strength of this local crystal field depends on the types of cations and the types of ligands coordinated to these cations, the cation–oxygen interatomic distance, temperature (T) and pressure (P). The intensities of CF spectra are functions of T, P and of the number of unpaired electrons present in the cation. This latter dependency is often expressed through quantum mechanical selection rules, which are analytical expressions of the probabilities of electrons moving between different d- and f-orbital energy levels. The mathematics and spectroscopic notations for CF spectra are discussed in detail by Burns (1993) and some examples of CF transitions are presented in Figure V35a.

Charge transfer transitions: metal → metal
Displacement of electrons between cations in different oxidation states produces metal–metal or cation–cation intervalence charge transfer (IVCT) transitions in the VIS–NIR. These transitions may be either homonuclear, involving the same cation in different oxidation states (e.g. $Fe^{3+} \rightarrow Fe^{2+}$, $Ti^{4+} \rightarrow Ti^{3+}$) or they may be heteronuclear, involving different cations (e.g. $Ti^{4+} \rightarrow Fe^{2+}$). An example of an IVCT absorption feature is shown in Figure V35b. More detailed information on the character and origins of IVCT features can be found in Burns (1981, 1993).

Charge transfer transitions: ligand → metal
Displacement of electrons between an anion and a cation, most commonly between $O^{2-} \rightarrow Fe^{3+}$, produces oxygen–metal or ligand–metal charge transfer (LMCT) transitions. It requires more energy to produce an LMCT feature than a CF band, and thus the LMCT features are centered in the ultraviolet (UV, < 300 nm). However, LMCT features are typically $10^3–10^4$ times more intense than CF bands, and thus for several cations such as Fe^{3+}, Ti^{4+} and Cr^{6+}, the long-wavelength wing extends into the visible and contributes to the yellow-brown color of, for example, several iron and titanium oxide minerals. Figure V35c provides an example of an LMCT absorption. More details and references can be found in Sherman and Waite (1985) and Burns (1993).

Valence band transitions
Any photons with an energy greater than the difference between the valence and conduction bands in an atom will be absorbed by that atom. For most silicates this absorption edge is located at high energies, far into the UV. For some minerals, however, such as sulfides, this edge may be located in the VIS, resulting in the yellow-red color of minerals such as sulfur (Figure V35d). Additional details on valence band transitions may be found in Hunt (1980) and Rossman (1988a) and references therein.

Vibrational transitions

A molecule containing N atoms has $3N$ degrees of freedom and thus can vibrate in $3N$ fundamental modes. Molecular vibrations cause a change in the charge distribution of the molecule and, as with electronic transitions above, various selection rules govern whether or not a particular vibrational mode will absorb incident radiation. The theory behind vibrational transitions is discussed in great detail by Hunt (1980), Rossman (1988b) and Gaffey, McFadden and Nash (1993). Most fundamental vibrational transitions in molecules of interest in planetary science occur in the mid- to far infrared (> 5000 nm). However, a few fundamental modes and many overtones and combinations of these fundamental modes occur in the VIS–NIR region (Figure V36). Most common are vibrational fundamentals and overtones of OH and H_2O stretching and bending modes in minerals such as micas and amphiboles (Rossman, 1988b) and cation–OH vibrations in clay silicate minerals (Hunt, 1979; Clark et al., 1990).

Experimental methods

Wavelength dispersion in spectroscopic measurements is primarily accomplished using either discrete broad- or narrowband single-wavelength interference filters, multiwavelength variable filters, prisms, gratings or interferometers. Discrete filter spectroscopy is most often used for general spectral classification of objects such as asteroids (e.g. Gaffey, Bell and Cruikshank, 1989) or for examination of particular spectral regions that are characteristic of, for example, vegetation (e.g. Sabins, 1987). High spectral resolution measurements by this technique are impractical and often impossible given current limits on filter fabrication. An example of a variable filter is a CVF, which is a continuously variable series of narrow band filters mounted into a circular or linear-shaped holder. CVFs usually obtain data in the R = 80–150 regime, which is adequate for the spectral identification of many materials. Their main advantage is the ability to scan a wide wavelength range in a short period of time at moderate spectral resolution. Higher spectral resolution data ($R \approx 300$–3000) can be obtained with grating- or prism-based spectrographs (e.g. Walker, 1987; Clark, et al., 1990). Even higher values of R (upwards of 10^6), necessary for the detailed characterization of gaseous absorption features, can be obtained using instruments such as the Fabrey–Perot interferometer or specially adapted gratings such as the Échelle (e.g. Walker, 1987).

VIS–NIR spectroscopy is utilizing continually advancing technology to meet evolving scientific needs. For example, pioneering efforts in imaging spectroscopy, the simultaneous acquisition of high spatial and spectral resolution data, have been made over the past 10 years in both terrestrial (e.g. Goetz et al., 1985; Adams, Smith and Gillespie, 1993) and planetary (e.g. Bell, McCord and Lucey, 1990; Bibring et al., 1990; Carlson et al., 1991; Bell and Crisp, 1993) remote sensing applications. Other recent technological advances include tunable acousto-optical filters and imaging Fourier transform spectrometers.

Data resources

There have been a great number of VIS–NIR spectroscopic measurements of terrestrial materials and astronomical objects of interest to planetary scientists. While it is beyond the scope of this entry to review them all, this section provides a starting point at which the geologically oriented student or researcher can begin a more detailed literature search. Additional information regarding spectroscopy of gaseous species and planetary atmospheres can be found in a separate encyclopedia entry (Spectroscopy: atmosphere).

Laboratory spectra of rocks, minerals and ices.

Adams and Filice (1967) conducted one of the earliest systematic studies of the reflectance spectra of silicate rock powders with applications to remote sensing. A more detailed study of a larger

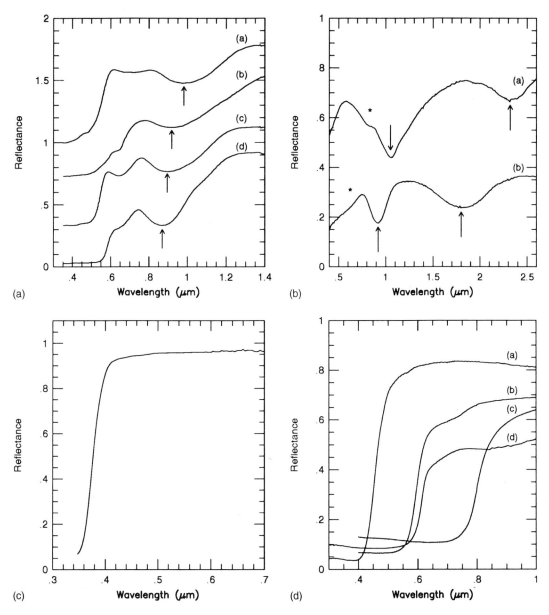

Figure V35 (a) Terrestrial iron oxide reflectance spectra from Morris et al. (1985). These spectra show several Fe^{3+} CF bands, and the variations in wavelength position of the band centers (indicated by arrows) can be used to identify the particular mineralogy. Spectra: (a) lepidocrocite, γ-FeOOH; (b) maghemite, γ-Fe_2O_3; (c) goethite, α-FeOOH; (d) hematite, α-Fe_2O_3. For clarity, each spectrum is offset by 0.3 reflectance units from the one below. (b) Terrestrial pyroxene reflectance spectra from the USGS spectral library (Clark et al., 1993). Spectra: (a) diopside (clinopyroxene), $CaMgSi_2O_6$; (b) hypersthene (orthopyroxene), (Mg, Fe)SiO_3. The absorptions marked with asterisks are Fe^{2+}–Fe^{3+} IVCT features resulting from small amounts of Fe^{3+} in these samples. Strong Fe^{2+} CF bands in these pyroxene spectra are marked with arrows. Determining the positions of these bands provides detailed information on pyroxene mineralogy and composition. (c) Laboratory TiO_2 spectrum in the visible showing the long-wavelength wing of the near-UV ligand-to-metal ($O^{2-} \rightarrow Ti^{4+}$) charge transfer feature. Since there are no free d-orbital electrons in Ti^{4+}, the origin of this absorption edge as an LMCT band is unambiguous. (Data from R.V. Morris, personal communication, 1993.) (d) Valence band absorption edge reflectance spectra for several sulfur-bearing minerals. Spectra: (a) elemental sulfur, S; (b) realgar, AsS; (c) stibnite, Sb_2S_3; (d) cinnabar, HgS. (Data from Clark et al., 1993, and Grove, Hook and Paylor, 1992.)

variety of samples was conducted in the early 1970s by Hunt, Salisbury, and Lenhoff in a series of 12 papers published in *Modern Geology* (e.g. Hunt and Salisbury, 1970, 1971; Hunt, Salisbury and Lenhoff, 1971). Specific studies of alteration minerals (oxides, hydroxides, clay silicates, etc.) and rock-forming minerals (olivine, pyroxene, etc.) were carried out by Hunt (1977, 1979), Adams (1974), Hunt and Ashley (1979), and Sherman and Waite (1985). Many of the samples studied by the above authors have been re-examined using modern instrumentation at higher spectral resolution by Morris et al. (1985), Clark et al. (1990, 1993), Grove, Hook and Paylor (1992) and Gaffey, McFadden and Nash (1993). Spectral studies of H_2O and CO_2 frosts and ices can be found in Kieffer (1970), Warren (1982), Fink and Sill (1982), and Warren, Wiscombe and Firestone (1990).

Telescopic and spacecraft spectra of solar system bodies

Harris (1961) provides a review of the historical multispectral observations of planets and satellites. Much of the more recent data can be found in review papers or book compendia such as the University of Arizona *Space Science Series*. For example, Mars spectral data are reviewed by *Soderblom* (1992) and Roush, Blaney and Singer (1993). Spectral data of the Moon and of lunar samples are summarized in Adams and Jones (1970), McCord and Adams,

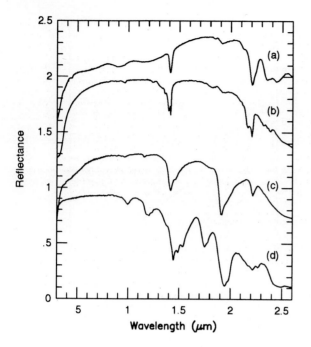

Figure V36 Sharp fundamental and overtone vibrational absorption features in terrestrial phyllosilicate and sulfate reflectance spectra. Most of the sharp spectral features in these spectra are caused by overtones and combinations of OH and H_2O stretching and bending modes. Features near 2.2 to 2.3 μm are due to cation–OH stretching modes and are of particular interest in planetary research because these features occur in a very transparent 'window' in the Earth's atmosphere (Figure V34). Spectra: (a) muscovite; (b) kaolinite; (c) montmorillonite; (d) gypsum. For clarity, each spectrum is offset by 0.5 reflectance units from the one below. (Data from Clark et al., 1993.)

(1973), McCord et al. (1981), and Heiken, Vaniman and French (1991). The very limited data set on Mercury is reviewed by Vilas (1988). Clark, Fanale and Gaffey (1986) provide a review of spectral data on planetary satellites, and reviews of asteroid and comet spectra are provided by Gaffey, Bell and Cruikshank (1989) and A'Hearn (1982).

James F. Bell III

Bibliography

Adams, J.B. (1974) Visible and near-infrared diffuse reflectance spectra of pyroxenes as applied to remote sensing of solid objects in the solar system. *J. Geophys. Res.*, **79**, 4829–36.

Adams, J.B. (1975) Interpretation of visible and near-infrared diffuse reflectance spectra of pyroxenes and other rock-forming minerals, in *Infrared and Raman Spectroscopy of Lunar and Terrestrial Minerals* (ed. C. Karr). New York: Academic Press, pp. 91–116.

Adams, J.B. and Filice, A.L. (1967) Spectral reflectance 0.4 to 2.0 μm of silicate rock powders. *J. Geophys. Res.*, **72**, 5705–15.

Adams, J.B. and Jones, R.L. (1970) Spectral reflectivity of lunar samples. *Science*, **167**, 737–9.

Adams, J.B., Smith, M.O. and Gillespie, A.R. (1993) Imaging spectroscopy: interpretation based on spectral mixture analysis, in *Remote Geochemical Analysis: Elemental and Mineralogical Composition* (eds C. Pieters and P. Englert). Cambridge: Cambridge University Press. pp. 145–66.

A'Hearn, T. (1982) Spectrophotometry of comets at optical wavelengths, in *Comets* (ed. L.L. Wilkening). Tucson: University of Arizona Press, pp. 433–60.

Bell, J.F., III and Crisp, D. (1993) Groundbased imaging spectroscopy of Mars in the near-infrared: preliminary results. *Icarus*, **104**, 2–19.

Bell, J.F., III, McCord, T.B. and Lucey, P.G. (1990) Imaging spectroscopy of Mars (0.4–1.1μm) during the 1988 opposition, in *Proc. Lunar Planet. Sci. Conf.*, **20**, 479–86.

Bibring, J.-P. and 16 others (1990) ISM observations of Mars and Phobos: first results, in *Proc. Lunar Planet Sci. Conf.*, **20**, 461–71.

Burns, R.G. (1970) *Mineralogical Applications of Crystal Field Theory*. Cambridge: Cambridge University Press.

Burns, R.G. (1981) Intervalence transitions in mixed-valence minerals of iron and titanium. *Ann. Rev. Earth Planet. Sci.*, **9**, 345–83.

Burns, R.G. (1993) Origin of electronic spectra of minerals in the visible-near infrared region, in *Remote Geochemical Analysis: Elemental and Mineralogical Composition* (eds C. Pieters and P. Englert). Cambridge: Cambridge University Press, pp. 3–29.

Carlson, R.W. and 20 others (1991) Galileo infrared imaging spectroscopy measurements at Venus. *Science*, **253**, 1541–8.

Clark, R.N., Fanale, F.P. and Gaffey, M.J. (1986) Surface composition of natural satellites, in *Satellites* (eds J.A. Burns and M.S. Matthews). Tucson: University of Arizona Press, pp. 437–91.

Clark, R.N., King, T.V.V., Klejwa, M. et al. (1990) High spectral resolution reflectance spectroscopy of minerals. *J. Geophys. Res.*, **95**, 12653–80.

Clark, R.N., Swayze, G.A., Gallagher, A.J. et al. (1993) The U.S.G.S. Digital Spectral Library: Version 1: 0.2 to 3.0 μm. US Geol. Surv. Open File Report 93–592.

Crisp, D. (1990) Infrared radiative transfer in the dust-free Martian atmosphere. *J. Geophys. Res.*, **95**, 14577–88.

Fink, U. and Sill, G.T. (1982) The infrared spectral properties of frozen volatiles, in *Comets*, (ed. L.L. Wilkening). Tucson: University of Arizona Press, pp. 164–202.

Gaffey, M.J., Bell, J.F. and Cruikshank, D.P. (1989) Reflectance spectroscopy and asteroid surface mineralogy, in *Asteroids II* (eds R.P. Binzel, T. Gehrels and M.S. Matthews). Tucson: University of Arizona Press, pp. 98–127.

Gaffey, S.J., McFadden, L.A. and Nash, D.B. (1993) Ultraviolet, visible, and near-infrared reflectance spectroscopy: laboratory spectra of geologic materials, in *Remote Geochemical Analysis: Elemental and Mineralogical Composition* (eds C. Pieters and P. Englert). Cambridge: Cambridge University Press, pp. 43–71.

Goetz, A.F.H., Vane, G. Solomon, J.E. and Rock, B.N. (1985) Imaging spectrometry for Earth remote sensing. *Science*, **228**, 1147–53.

Grove, C.I., Hook, S.J. and Paylor, E.D., II (1992) Laboratory reflectance spectra of 160 minerals, 0.4 to 2.5 μm. Pasadena: Jet Propulsion Laboratory, JPL Publication 92–2.

Hapke, B. (1981) Bidirectional reflectance spectroscopy. 1. Theory. *J. Geophys. Res.*, **86**, 3039–54.

Harris, D.L. (1961) Photometry and colorimetry of planets and satellites, in *The Solar System. III. Planets and Satellites* (eds G.P. Kuiper and B. Middlehurst). Chicago: University of Chicago Press. pp. 272–342.

Heiken, G.H., Vaniman, D.T. and French, B.M. (1991) *Lunar Sourcebook: A User's Guide to the Moon*. Cambridge: Cambridge University Press.

Hunt, G.R. (1977) Spectral signatures of particulate minerals in the visible and near-infrared. *Geophysics*, 42; 501–13.

Hunt, G.R. (1979) Near-infrared (1.3–2.4 μm) spectra of alteration minerals: potential for use in remote sensing. *Geophysics*, **44**, 1974–86.

Hunt, G.R. (1980) Electromagnetic radiation: the communication link in remote sensing, in *Remote Sensing in Geology* (eds B.S. Siegal and A.R. Gillespie). New York: John Wiley, pp. 5–45.

Hunt, G.R. and Ashley, R.P. (1979) Spectra of altered rocks in the visible and near-infrared. *Econ. Geol.*, **74**, 1613–29.

Hunt, G.R. and Salisbury, J.W. (1970) Visible and near-infrared spectra of minerals and rocks: I. Silicate minerals. *Mod. Geol.*, **1**, 283–300.

Hunt, G.R. and Salisbury, J.W. (1971) Visible and near-infrared spectra of minerals and rocks: II. Carbonates. *Mod Geol.*, **1**, 23–30.

Hunt, G.R., Salisbury, J.W. and Lenhoff, C.J. (1971) Visible and near-infrared spectra of minerals and rocks: III. Oxides and hydroxides. *Mod. Geol.*, **2**, 195–205.

Kieffer, H.H. (1970) Spectral reflectance of CO_2–H_2O frosts. *J. Geophys. Res.*, **75**, 501–9.

Kortüm, G. (1969) *Reflectance Spectroscopy*. New York: Springer.

Kurucz, R.L., Furenlid, I., Brault, J. and Testerman, L. (1984) *National Solar Observatory Solar Flux Atlas*. Cambridge: Harvard University Press.

McCord, T.B. and Adams, J.B. (1973) Progress in remote optical analysis of lunar surface composition. *The Moon*, **7**, 453–74.

McCord, T.B. Clark, R.N., Hawke, B.R. *et al.* (1981) Moon: near-infrared spectral reflectance, a first good look. *J. Geophys. Res.*, **86**, 10883–92.

Morris, R.V., Lauer, H.V., Jr, Lawson, C.A. *et al.* (1985) Spectral and other physicochemical properties of submicron powders of hematite (α-Fe_2O_3), maghemite (γFe_2O_3), magnetite (Fe_3O_4), goethite (α-FeOOH), and lepidocrocite (γ-FeOOH). *J. Geophys. Res.*, **90**, 3126–44.

Rossman, G.R. (1988a) Optical spectroscopy in *Spectroscopic Methods in Mineralogy and Geology* (ed. F.C. Hawthorne). Mineralogical Society of America Reviews in Mineralogy Series, Vol. 18, pp. 207–54.

Rossman, G.R. (1988b) Vibrational spectroscopy of hydrous components, in *Spectroscopic Methods in Mineralogy and Geology* (ed. F.C. Hawthorne). Mineralogical Society of America Reviews in Mineralogy Series, Vol. 18, pp. 193–206.

Roush, T.L., Blaney, D.L. and Singer, R.B. (1993) The surface composition of Mars as inferred from spectroscopic observations, in *Remote Geochemical Analysis: Elemental and Mineralogical Composition* (ed. C. Pieters and P. Englert). Cambridge: Cambridge University Press, pp. 367–93.

Sabins, F.F. (1987) *Remote Sensing: Principles and Interpretation.* New York: W.H. Freeman.

Sherman, D.M. and Waite, T.D. (1985) Electronic spectra of Fe^{3+} oxides and oxide hydroxides in the near-IR to near-UV. *Am. Mineral.*, **70**, 1262–9.

Soderblom, L.A. (1992) The composition and mineralogy of the martian surface from spectroscopic observations: 0.3–50 μm, in *Mars* (eds H. Kieffer, B.M. Jakowsky, C.W. Snyder and M.S. Matthews). Tucson: University of Arizona Press, pp. 557–93.

Vilas, F. (1988) Surface composition of Mercury from reflectance spectrophotometry, in *Mercury* (eds F. Vilas, C.R. Chapman and M.S. Matthews). Tucson: University of Arizona Press, pp. 59–76.

Walker, S.G. (1987) *Astronomical Observations: An Optical Perspective.* Cambridge: Cambridge University Press.

Warren, S.G. (1982) Optical properties of snow. *Rev. Geophys.*, **20**, 67–89.

Warren, S.G. Wiscombe, W.J. and Firestone, J.F. (1990) Spectral albedo and emissivity of CO_2 in Martian polar caps: model results. *J. Geophys. Res.*, **95**, 14717–41.

Wendlandt, W.W. and Hecht, H.G. (1966) *Reflectance Spectroscopy.* New York: John Wiley.

Cross references

Absorption, absorption spectrum
Infrared spectroscopy
Microwave spectroscopy
Reflectance spectroscopy
Spectroscopy: atmosphere
Ultraviolet spectroscopy

VOLCANISM IN THE SOLAR SYSTEM

Volcanism is one of the expressions of heat transfer from a planetary interior to the surface. It involves the movement of material – solid, liquid, gas or multiphase mixtures – and is one of the most important processes involved in chemical differentiation of planetary bodies and their resurfacing. Heat is generated by radioactive decay, tidal forces, electromagnetic induction, bombardment and phase changes within a planetary body. These mechanisms act in concert and their relative importance may change with time, so that volcanism on a planetary body is intimately related to its thermal and chemical evolution.

Planetary bodies lose internal heat in three ways:

1. by conduction from the interior to the surface, then off the surface into space;
2. by advection where melts (and other fluids) are generated in the interior and migrate to the surface or near-surface; and
3. by convection involving gravity-induced turn-over by plastic flow of the planetary interior.

Volcanism is a surface manifestation of advection and/or convection. On the Earth volcanism is recognized by the landforms produced, by the composition and petrogenesis of its products, by its tectonic setting and by its relationship to other geophysical phenomena. Furthermore, volcanism can be placed in a spatial and temporal context. To date, chemical analyses are only available for volcanic rocks from the Earth and Moon, with partial analyses available from the Viking samples from Mars and the Venera samples from Venus. Detailed geophysical data and the results of radiometric dating are only available for the Moon. On all other planetary bodies analysis of landforms is the only direct evidence for volcanism, unless volcanic activity has been directly observed by visiting probes. To date, volcanic landforms have been observed on Mercury, the Moon, Mars, Venus, the Galilean satellites and Triton, while active volcanism has only been observed on Io and Triton, and inferred on Venus. What follows is an analysis of volcanic landforms and processes on the Earth in the manner in which such features are visible on other planetary bodies, followed by summaries for the Moon and Mars, and an overview for Venus, Mercury and non-silicate bodies.

Volcanism on the Earth

Conduction, advection and convection all operate as heat-transfer mechanisms on the Earth, and areas where one or more process dominates are discrete terrains (Figure V37). Heat loss by conduction occurs everywhere, but only dominates through old and tectonically stable surfaces. On the Earth such terrains have low crater counts because no surfaces appreciably older than 3000 Ma have survived. No rocks older than 3950 Ma are known, though detrital minerals as old as 4000 Ma have been documented. The Earth has had an active surface throughout its history, and the rocks within the oldest terrains contain evidence of extensive volcanism and related igneous activity. They also contain water-lain metasediments.

Convective overturning of a plastic silicate mantle is and has been the driving force behind plate tectonic processes that have shaped the Earth's surface since at least 2000 Ma ago (Figure V38a). Resurfacing of the Earth by the upwelling of basaltic magma at the crests of spreading ridges in the ocean basins is the most distinctive feature of the Earth compared to the Moon, Mars and Venus. These basalts represent a 10–15% melt extracted from peridotite (spinel or garnet lherzolite), the remainder forming a depleted residue accreted beneath the oceanic crust. Volcanism is largely restricted to graben along the ridge axes, where it produces lava domes, flows, lava lakes and, where water depth permits, pyroclastic cones. Landforms are typically small, but in total, voluminous. Eruptions originate from central vents or along fissures. Typically, the ridge axis is repeatedly offset along fractures running perpendicular to the axis.

A pattern of spreading ridges similar in scale to the present appears to have existed for the last 2000 Ma. Prior to this, opinion diverges, though the present consensus is for some form of spreading-ridge magmatism to have been operative through most of Earth history. At present, some 70% of the planet's surface consists of terrain generated in this way, with rates of generation and destruction such that no in situ basaltic crust (oceanic crust) is currently older than 250 Ma.

Creation of new oceanic crust at spreading ridges is balanced by the recycling of this crust in subduction zones. Complex melting behaviour of subducted basaltic crust, overlying mantle and older crust, produces linear volcanic belts dominated by basaltic and andesitic lavas, alongside volumetrically minor amounts of dacite and rhyolite (highly viscous, siliceous, volatile-rich lavas). Eruptions typically occur from central vents, and volcanic landforms are dominated by stratified simple or complex cones up to 3000 m high, with aspect ratios (height to radius) greater than 0.3 (Figure V39).

Both convection and advection are involved with volcanism associated with the creation and destruction of oceanic crust. Magma generation at spreading ridges produces basalt and an ultramafic residue. At destructive subduction zones, complex melts ranging from basaltic to rhyolitic are evolved, and an ultramafic residue returns to the upper mantle. An upper mantle layer is being progressively depleted, while a progressively more felsic and siliceous layer is accreted to the older, stable continental terrains. In the past, a hotter and less depleted upper mantle may have produced more mafic melts than basalt, and between 3500 and 2500 Ma ago, ultramagnesian komatiites and magnesian basalts (picrites) are abundant. Few have been erupted since 2500 Ma ago (Figure V38a).

Advection associated with spreading ridges and subduction zones only involves the upper 90–120 km of the lithosphere and mantle. Another pattern of volcanism is superimposed on the spreading ridge–subduction zone pattern, relating to deeper seated melting, and is termed hot spot volcanism. This volcanism relates to deep

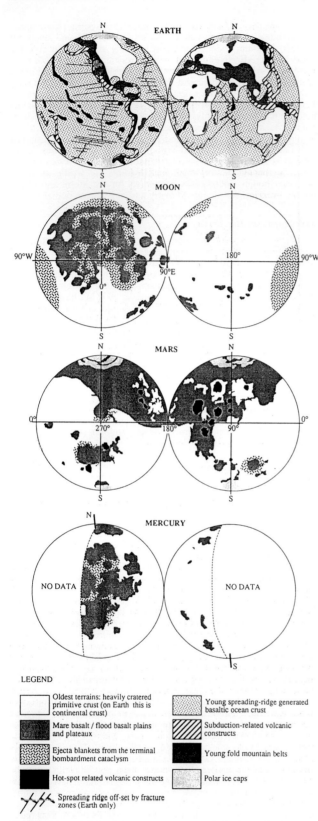

Figure V37 Terrain maps of the Earth, Moon, Mars and Mercury showing the distribution of the major types of terrain and the principal volcanic constructs on each body. Sedimentary basins and platforms have been omitted from the Earth and Mars, and many small 'hot spot' volcanoes are not shown on the Earth. (After Beatty and Chaikin, 1990.)

convection cells that appear stationary with respect to the moving lithospheric plates, and hot spot volcanic landforms often define linear volcanic chains on both oceanic and continental terrains. The magmas produced typically involve 5% or less melting of a more or less chondritic mantle peridotite (garnet lherzolite) at depths in excess of 150 km. Basalts are the predominant eruption product, building huge lava shields with aspect ratios well below 0.5 (Figure V39). Such shields typically have a summit caldera, and eruptions take place both from the central vent and from lateral rift zones. Volumetrically minor amounts of felsic trachytes, phonolites and alkaline rhyolites produce domes on the shields, and late-stage nephelinites are erupted from diatremes.

Nephelinites involve very small-scale melting of a mantle peridotite rich in volatiles. Mantle peridotites of near-chondritic composition, rich in water, F and CO_2 also yield gas-rich magmas. Typically, such material originates between 180 and 220 km below the surface, erupting as diatremes of kimberlite and lamproite (ultramafic igneous rocks rich in incompatible elements, carrying mantle xenoliths like diamonds). Differences in subcontinental and suboceanic mantle are reflected by the distribution of kimberlite and lamproite diatremes largely restricted to continents, while nephelinites are largely found in oceanic terrains.

Continental terrains need not remain permanently stable. Rifting leading to fragmentation, is evident in a number of places, notably east Africa and the Red Sea (Figure V40a). Basalt plateau formation on continental terrains predated rifting and continental break-up around the North and South Atlantic, in southern Australia and Antarctica, and around the Red Sea–Gulf of Yemen, while rift arms that did not spread are associated with basalt plateau and later alkaline activity throughout east and central Africa, and in the continental hinterlands on both sides of the Atlantic basin.

Overall the Earth is dominated by young, basaltic terrain underlying the ocean basins. Both spreading ridges and subduction zones produce planetwide patterns where tectonic activity is reflected by the distribution of volcanic landforms. Advection from deep within the mantle produces a pattern of hot spot volcanism, independently distributed with respect to the old, stable continental terrains, and young basaltic terrains (Figure V37). How old these patterns are is questionable, but the distinctive products of spreading ridge, hot spot and destructive margin volcanism are identified in geological formations at least 3000 Ma old (Figure V38a).

Volcanism on the Moon

The Earth's surface is dominated by material produced during the last 3000 Ma; by contrast, very little of the Moon's surface is younger than 3500 Ma. Three types of terrain dominate the surface of the Moon: highly cratered pale highland terrain, darker 'mare' terrain with lower crater counts, and ejecta blankets related to a small number of large ring basins (Figure V38b).

Highland terrain represents the oldest part of the lunar surface; it also represents the one component of the surface not directly sampled by the Apollo missions, as all highland material recovered came from ejecta blankets excavated by large impacts. The highland terrain consists largely of brecciated anorthosite, gabbro-anorthosite and troctolite with minor amounts of ultramafic cumulate (pyroxenite, dunite), potassium, rare earth element and phosphorus rich basalt (KREEP), and very high alumina basalt (VHA). Geophysical data suggest that this terrain is contiguous with the first 500 km of the lunar interior, which grades from anorthosite at the top to more peridotitic deeper levels. Brecciation by repeated impacts affects much of the upper 50 km.

Mare terrain overlies the highland material and consists of a diverse collection of basaltic lavas, ranging from undersaturated olivine types, through hypersthene to quartz-bearing types. They display a wider range of Cr and Ti contents than terrestrial basalts. Most of the basalts seem to have originated from fissure eruptions, but landforms resembling pyroclastic cones and collapsed lava tubes are common. Individual flows are often over 100 km long, over very shallow gradients, implying lower viscosities than are seen in terrestrial basalts.

Ejecta blankets are of two types, differing only in scale. Blankets covering large areas of the lunar surface, associated with geometric fracture patterns of global extent and ablation-sculpted landforms, can be related to a series of large ring basins like Imbrium and Orientale. These ejecta blankets predate the mare basalts. Smaller ejecta blankets postdate the mare basalts and occur as coherent

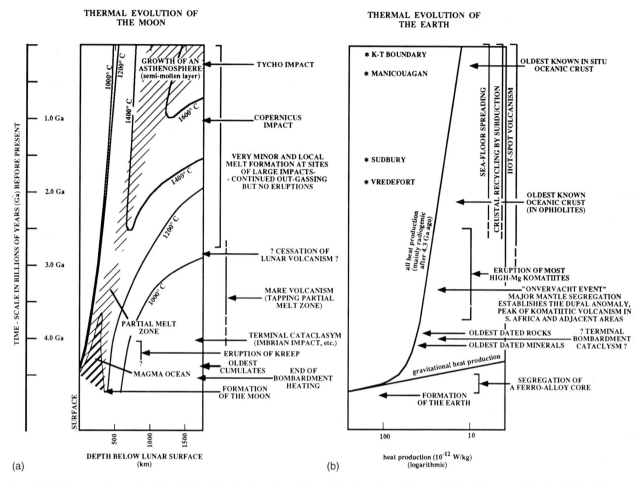

Figure V38 Diagrams illustrating the relationship of volcanic activity to the thermal evolution of (a) the Moon and (b) the Earth. (After Beatty and Chaikin, 1990.)

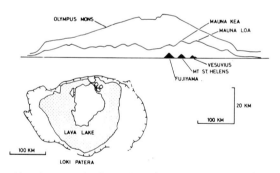

Figure V39 Aspect ratios for major volcanic constructs on the Earth and Mars, with a plan of Loki Patera (Io) at the same horizontal scale for comparison. Note the vertical exaggeration on the profiles. Olympus Mons and the Hawaiian chain are primarily basaltic constructs, Fujiyama, Mt St Helens and Vesuvius are more silicic/felsic.

sheets or high-albedo rays of global extent emanating from such late craters as Copernicus and Tycho.

The history of volcanism of the Moon directly reflects the thermal history of the planet (Figure V38b). Early bombardment of the lunar surface and possible capture of the Moon by the Earth were the principal heating mechanisms operating prior to 4.5 Ga ago. Consequently, the first 300–500 km below the surface became far hotter than the deeper levels. Complete melting produced a global magma ocean of approximately chondritic composition. During cooling, this vigorously convecting layer underwent crystal fractionation, during which a buoyant anorthosite accumulated at the top and ultramafic cumulates were deposited beneath. A middle basaltic layer became enriched in incompatible elements such as K, REEs (except Eu), P, Y, U and Th (KREEP and KREEPY) as a residual liquid (Figure V41). This tripartite primitive crust was still subjected to decreasing frequency of bombardment culminating in the Imbrian or 'terminal lunar cataclysm' around 4.0 Ga. These events produced the highland terrains and the large ejecta blankets. Volcanic products from this pre-Imbrian stage are ambiguous, but KREEP basalts in highland breccias may represent residual reservoirs of the magma ocean tapped by pre-Imbrian impacts. Alternatively, KREEP and VHA basalts may represent impact melts generated by pre-Imbrian events.

The terminal lunar cataclysm excavated the large ring basins of which Imbrium and Orientale are the most conspicuous. It is these basins that were subsequently flooded by the mare basalts. By this time in lunar history radiogenic heating from the interior was the principal heat source and an orthodox thermal gradient was established. The solidus isotherm had subsided into the cumulate layer at the base of the primitive crust, and the mare basalts all lie along the cotectic for this material. The composition suggests that the mare basalts represent a 5–10% partial melt of the cumulate layer.

Mare basalt eruptions continued until at least 3.0 Ga, at which point the solidus isotherm had subsided well below the cumulate layer (Figure V38b). Subsequently, the only melts generated that would have a chance of reaching the surface would be local impact melts. Volcanism effectively ceased on the Moon at or around 3.0 Ga, and though there is evidence for subsequent and continuing outgassing from the deep interior, no volcanic activity appears to be associated with this.

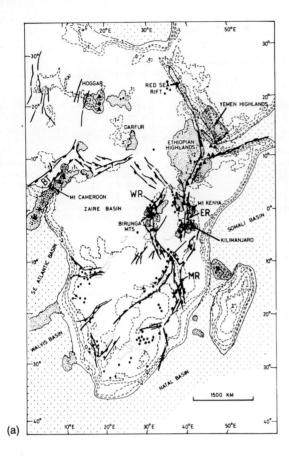

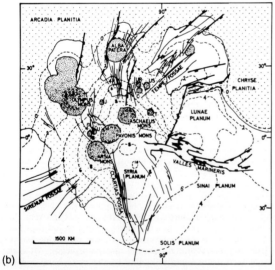

Figure V40 Volcanism associated with updomed and rifted areas on the Earth and Mars: (a) eastern and central Africa, and (b) the Tharsis Ridge on Mars. The maps are at the same scale, contours are in kilometers (above and below mean sea level in (a) and above mean altitude in (b).) Initials on (a): BT = Benue Trough, WR = Western Rift Valley, ER = Eastern Rift Valley, MR = Malawi Rift Valley; initials on (b): B = Biblis Patera, U = Ulysses Patera, J = Jovis Tholus, UR = Uranius Tholus, US = Uranius Patera, C = Ceraunius Tholus, T = Tharsis Tholus. Legend: no stipple = continental crust (a), older highland terrain (b); coarse stipple = basaltic ocean crust (a), basaltic lowland plains (b); fine stipple = major volcanic constructs (a and b); spots = kimberlite diatremes (a only); black triangles = recent volcanoes (a only). Major faults have downthrown side indicated.

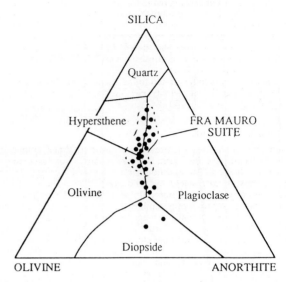

Figure V41 The low-pressure SiO_2–Mg_2SiO_4–$CaAl_2Si_2O_6$ anhydrous silicate system illustrating the near cotectic composition of KREEP, indicating its origin as either a residual liquid from the magma ocean cumulates or a low-volume partial melt(s) from the same material. (From Cadogan, 1981.)

An age for the onset of mare volcanism is more contentious. The oldest basalts sampled during the Apollo programme are 3.95 Ga old, but remote mapping suggests older flows exist, and supporters of the KREEP volcanic model would suggest that the residual reservoirs of the magma ocean produced extensive flood basalts prior to the Imbrian impacts.

Volcanism on Mars

Intermediate in size between the Moon and the Earth, with density and angular momentum that suggest an Earth-like differentiated interior, it might be expected that Mars would be Moon-like in retaining ancient cratered terrain, but Earth-like in having evidence of more recent volcanic activity (Plates 16, 17). Geophysical experiments on the Viking landers failed, so no seismic structure for the planetary interior has been established; however, the distribution of volcanic landforms on the surface implies a body intermediate between the Moon and the Earth (Figure V37).

Much of the southern hemisphere of the planet consists of old, heavily cratered highland terrain, while much of the northern hemisphere consists of younger basaltic lowland plains. Superimposed on both these areas are a number of enormous basaltic shields, including Olympus Mons, the largest volcanic edifice identified to date in the solar system (Figure V39 and V40b). Tectonic activity is identified in global fault systems, including the 4500 km long graben system that includes Valles Marineris. An exception to the bipartite highland–lowland division is the Tharsis ridge in the northern hemisphere. Standing 10 km above the mean level, it is crowned by a group of shield volcanoes a further 15 km high (Figure 40b).

Whether the duality of old, cratered terrain and basaltic plains reflects an early history comparable to the Moon remains an open question. Less contentious is the significance of the young basalt shields. The Tharsis ridge is criss-crossed with faults and graben structures implying inflation and extension (Figure V40b). In terrestrial terms the shields are hot spot volcanoes, rather than volcanoes related to the margins of lithospheric plates. Magmas were probably generated by advection deep in the mantle, and erupted in long-lived, non-migratory centres. The Tharsis area contains near pristine edifices like Olympus Mons, and older, deeply eroded and cratered volcanic centres like Alba Patera.

The age of this hot spot volcanism is contentious. Crater counts suggest ages for Olympus Mons anywhere between 2.5 and 1.5 Ga, while estimates for the duration of eruptive activity range from 100 to 1000 Ma. All the volcanic edifices have impact craters scattered over them, implying that volcanism ceased completely around 1.5 Ga ago, though some authorities maintain volcanism has continued almost to the present day. How evolved the volcanism became is hard to gauge. As in the case with terrestrial hot spot volcanoes such as Hawaii, basalts account for the major volume, however, structures with smaller aspect ratios exist (such as the 'tholus' landforms), and edifices like Tyrrhena Patera appear to consist largely of pyroclastic material, implying more viscous lavas and explosive eruptions. This hot spot volcanicity appears to be related to the decay of a radiogenic thermal gradient, with volcanic activity ceasing once the solidus isotherm subsided below the first 200–300 km.

Volcanism on Venus

Given the similarities of size, density and perhaps composition between Venus and the Earth, this planet should have a similar rate of heat production. The question is how the planet loses its internal heat. If heat is lost solely by conduction, the surface should be dominated by old, heavily cratered terrain, perhaps with 'mare'-style basaltic plains representing similar phases in Venusian and lunar history. On the other hand, perhaps the heat loss mechanisms are as complex as those on the Earth, and fully fledged plate tectonic processes operate. Rather than elucidate matters, the data from the Venera and Magellan probes has open up entirely new problems.

First, heavily cratered terrain does exist, but it is alongside smooth (basaltic?) plains with lower crater counts, plateau areas with crowning shield volcanoes and calderas, smaller domes with summit craters and aspect ratios higher than those of basalt shields, and possible ejecta plumes suggesting explosive eruptions. Second, transient heat sources have been recorded during infrared surveys. The Venera landers have also recorded transient acidic sulfur aerosols in the atmosphere, and areas of radar-bright terrain resemble recent terrestrial flow fields, suggest continuing volcanic activity. Third, though the volcanic edifices so far identified do not appear to have patterns expected along active along tectonic plate margins, there are features on the surface that resemble spreading ridges and fold mountain belts. Fourth, terrain types on Venus include at least two, tesserae and domed highlands, that resemble nothing seen on the Earth, the Moon or Mars.

The largest volcanic feature identified to date lies high in the northern hemisphere: Lakshmi Planum in the southwestern corner of Ishtar Terra – a candidate for a 'continent' comparable in size to Australia. Much of Ishtar Terra consists of the enigmatic tessera terrain (q.v.), but Lakshmi Planum is an extensive basaltic (?) lava plateau centred around two large calderas, Sacajawea and Colette. This plateau is surrounded by lineated (fold?) mountain belts, including Maxwell Montes, the highest elevation on Venus. Scattered across Lakshmi Planum are many smaller volcanic landforms including steep-sided cones, representing either pyroclastic cones or a more felsic and explosive volcanic activity than that represented by the larger calderas.

The overall distribution of caldera and radar-bright flow fields is suggestive of hot spot volcanism as seen on the Earth and Mars. Indeed the area of flood basalt terrain on Venus appears to exceed that seen on the other terrestrial planets. Extremely low crater counts imply that these areas are young, and possibly still active. Crater counts generally suggest a surface no older than 0.5 Ga.

Somewhat more contentious is the evidence for tectonic controls on volcanic activity. Aphrodite Terra is an elongate highland region running around much of the planet's equator. North–south profiles suggest an extremely symmetrical topography, with the possibility of an axial ridge graben (Figure V42). The ridge axis is offset by perpendicular fractures. Both these features are characteristic of spreading ridges on the Earth and have no counterpart on either the Moon or Mars.

This juxtaposition of old, heavily cratered terrain (lunar highland-like) and a potential spreading ridge (terrestrial ocean basin-like) implies the existence of a plate tectonic response to mantle convection, although on Venus this is not a process that has led to complete resurfacing of the planet. The reason for this may lie in the lack of liquid water on Venus, and its presence on the Earth. In terrestrial subduction zones, not only is a residue of partially melted oceanic crust returned to the upper mantle, but with it goes water in the partially hydrated old oceanic crust. The upper mantle on the Earth still contains a small amount of water, despite the general degassing of the mantle. This is sufficient to lower the solidus temperature for much of the peridotite in this zone. Indeed this may be one reason for the existence of the weak asthenosphere beneath the relatively rigid lithosphere. This mechanism for returning water to the upper mantle is unlikely to exist on Venus; consequently, a thinner asthenosphere or no asthenosphere at all may be expected. The planet may be losing heat primarily by advection through hot spots, rather than by convection-driven plate tectonic processes (Phillips, Grimm and Malin, 1991).

Volcanism on Mercury

Moon-like in size and density, but apparently with a ferro-alloy core, Mercury presents a very Moon-like surface. The hemisphere photographed by Mariner 10 consists of old, heavily cratered terrain, partially overlain by younger basaltic plains (Figure V37). Large ring basins are superimposed on the older terrain, and partly flooded by basalts. Crater counts suggest a history similar to the Moon, with an early bombardment culminating in a terminal cataclysm excavating basins like Caloris. Subsequently, 'mare'-type basaltic floods resurfaced upwards of 30% of the planet. The lunar analogy would imply that this activity ceased billions of years ago.

Volcanism in the outer solar system

The terrestrial planets are all silicate bodies with or without ferro-alloy cores. Volcanism is the surface expression of heat loss on bodies where radiogenic heat rapidly took over from surface impact melting as the principle source of magma. Where radiogenic decay heat could not maintain isotherms above the solidus for the outer mantle regions, volcanic activity ceased early in their histories. Advection from the deeper mantle and partial melting associated with convection may occur in dry peridotite, but on the Earth melting is crucially mediated by remnant volatiles, notably water. Silicate melts at or near the cotectic or eutectic temperatures for wet peridotite; wet basalt and wet granite lie in the range 900–1400 K.

Bodies that retain ices (H_2O, CO_2, CH_4, NH_3 and H_2) are expected to behave somewhat differently, whether or not they have silicate cores. They are unlikely to contain the same levels of the largely lithophile long-lived radiogenic isotopes of K, U and Th so important to heat production in silicate bodies. Consequently other heat production mechanisms will take on a new importance. Furthermore, on ice-dominated bodies the very nature and definition of volcanism may well be different. One of the surprises of the Voyager encounters was the direct observation of active volcanism on Io (a Moon-like silicate body; Plate 23) and Triton (a largely ice body). Io (q.v.) may be the most volcanically active body in the solar system.

The density and angular momentum of the Galilean satellites of Jupiter (Io, Ganymede, Callisto and Europa) suggest that they are layered ice–silicate bodies, with Io and Europa being most like the Moon and Ganymede and Callisto being significantly less dense (Figure V43a). In all four bodies models suggest that rocky cores are covered by an ice crust and a liquid or solid ice mantle. This ice layer on Io is debatable or of negligible thickness. On all the other Galilean satellites, it is the eruption of water from this mantle that appears to be the primary resurfacing process, producing lineated terrain, possibly analogous to that produced at spreading ridges on the Earth. On Europa no old cratered terrain remains.

Io, the densest Jovian satellite, is regarded as being largely rocky, with a thin, complex crust of ices and sulfur (Figure V43a). The volcanism, so evident on this body, involves sulfur and sulfur compounds, but the extent of silicate involvement remains unknown, though heat measurements from Loki Patera in excess of 900 K imply silicate lava in this caldera. The plumes from the volcanic eruptions witnessed by the Voyager probes are dominated by sulfur and SO_2, while the lava flows and lava lakes on the surface have colors dominated by sulfur compounds and sulfur allotropes. Whether this implies that these volcanic products are pure sulfur, or silicates with sulfur crusts, remains a controversial point. Nevertheless, aspect ratios of the volcanic edifices on Io are shield like, and the length and gradients of individual lava flows suggest the eruption of lavas with very low viscosities. Whether these are sulfur flows, basalts with sulfur crusts or secondary molten sulfur features mobilized by the extrusion or shallow intrusion of basalt remains to be seen (Figure V43b).

Voyager 1 observed eight volcanic plumes, of which seven remained active at the time of the Voyager 2 encounter. The morphology and compositions of the plumes implies that the volcanism on Io is propelled by explosive phase changes, with SO_2 and elemental sulfur the most important constituents. In this respect, volcanism on Io is more akin to geyser activity on the Earth. Two categories of plume

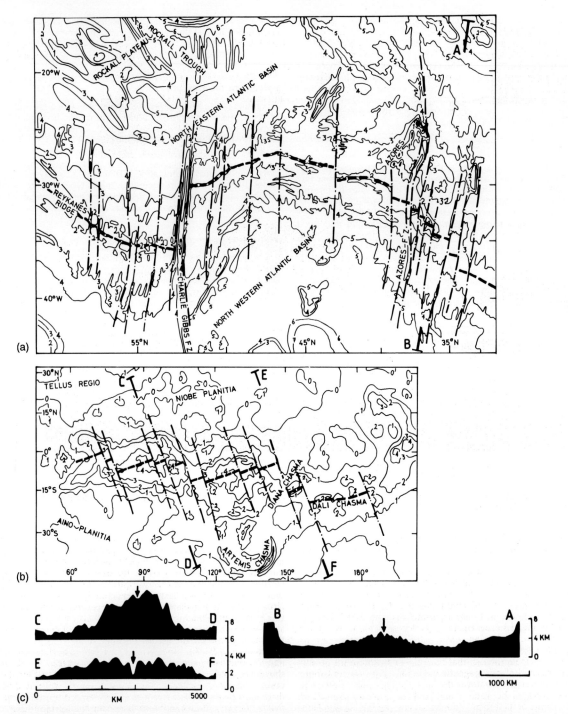

Figure V42 Symmetrical ridges with axial rifts and offsetting fractures on the Earth and Venus. (a) North Atlantic Ridge, a sea-floor spreading plate margin. Contours in kilometers below mean sea level indicating depth to ocean crust. (b) Aphrodite Terra on Venus, from Venera radar maps, with interpretation from Head and Crumpler (1987). Contours in kilometers above mean altitude. (c) Profiles across Aphrodite Terra and the North Atlantic Ridge – note the different vertical scales involved. The arrow in each case indicates the position of an axial rift valley.

have been recognized, suggesting two types of activity: (1) relatively low-velocity but long-lasting eruption powered by SO_2 – Prometheus type, and (2) higher-velocity, shorter-lived events powered by elemental sulfur – Pele type. A third form of eruption, involving altogether lower energy than either of the above may be associated with features resembling fault erosion scarps (Figure 43b). This would involve SO_2 degassing along fractures tapping shallow sulfur-rich reservoirs or 'aquifers' in Io's crust.

The largest volcanic edifices differ considerably in summit morphology. Ra Patera and Pele Patera (the largest volcano) have central vents surrounded by extensive flow fields; Loki Patera, on the other hand, consists of a huge caldera, 250 km across, occupied by a lava lake. The Loki caldera rim is not appreciably elevated above the mean surface (Fig. 39).

Silicate mountains poke through the sulfur-dominated surface, and some may well represent silicate volcanoes. Figure V43b shows a

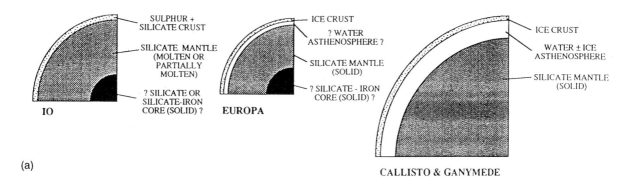

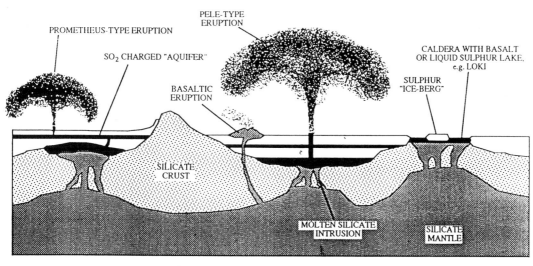

Figure V43 (a) Schematic sections through the Galilean satellites. (b) Schematic section through the crust of Io illustrating possible modes of volcanism. (After Beatty and Chaikin, 1990.)

schematic relationship between the different types of activity, illustrating a model in which basaltic magmas from a silicate mantle play a major role.

As a Moon-like body Io might be expected to have had a Moon-like thermal history, in which radiogenic heat was insufficient to drive volcanic activity after the first 2 billion years eons had passed. In order to explain present and continuing activity a different heat source(s) must be invoked. Io has an orbital resonance with Europa that imparts an unusual eccentricity. Tidal forces induced by Jupiter generate heat through friction, ensuring that the interior remains partially molten (see Tidal heating). A second potential heating mechanism arises from the fact that Io orbits deep within the Jovian magnetosphere, and electromagnetic induction may well contribute to heating Io's interior. Whatever the heat source(s), heat flow over Io's surface averages between 1 and 3 $W\,m^{-2}$, compared with a terrestrial average of $0.06\,W\,m^{-2}$ (thermal areas like Wairakei in New Zealand have heat flows of $1.7\,W\,m^{-2}$).

Unlike the Galilean satellites, Triton (q.v.) appears to be an ice body with a small rocky core. The volcanism observed by Voyager 2 falls into two categories: (1) eruption plumes and (2) resurfaced terrains with evidence of extrusion from central vents or from fissures. The observed plumes were geyserlike features representing explosive phase changes in a shallow reservoir. Possible sources for these plumes include vents associated with dark streaks on the surface. These dark streaks may consist of nitrogen frost and hydrocarbons vented during explosive eruptions. Re surfaced terrain dominates Triton. Lineated terrain representing upwelling of water ice or water-ammonia ice along dilatant fissures is common. Several impact craters also show signs of flooding with similar material.

Volcanism on Triton seems to reflect the interaction between a surface dominated by water ice and a shallow reservoir of liquid nitrogen with minor amounts of water, ammonia and hydrocarbons (Plate 32). This reservoir may be tapped by impacts or tidal stress-induced fracturing. Even the slight surface temperature changes between night and day may create the stresses necessary to crack the surface. The massive resurfacing apparent on Triton may reflect an unique event in the satellite's history, namely its capture into a near-circular but retrograde orbit by Neptune. At initial capture the satellites orbit would have been highly eccentric, and the tidal stresses during evolution into a more circular orbit may have ensured that much of the interior remained molten.

Comets may be the most abundant icy bodies in the solar system, and the only comet to have been closely observed, comet Halley, showed signs of volcanic activity on its surface. Such small bodies are unlikely to have much radiogenic material within them, and heat to drive the activity seen on comet Halley comes entirely from the Sun, making this volcanism a response to an exogenous process. Nevertheless, this is an important process in the evolution of these bodies. The results from the Giotto mission to comet Halley confirmed the 'dirty snowball' model of Whipple for cometary nuclei. Basically a mixture of volatile ices and chondritic material, each solar approach causes sublimation of ices into the coma, leaving behind a dark residue on the comet's surface. In comet Halley this dark crust appeared to cover the whole body, and degassing from the icy interior appeared to be largely concentrated at a small number of vents – volcanos in the broadest sense of the term.

Volcanism as a facet of evolution

Radiogenic decay in planetary interiors remains the most important long-term heat-generating mechanism, and as most long-lived radio-isotopes are strongly lithophile, radiogenic heat will be more important in rocky silicate bodies than in icy bodies. The volume of the planetary body will be a crucial factor in determining whether or not relatively shallow regions of their mantles are maintained at temperatures above the solidus for silicate, silicate + volatiles or ice mixtures. Earth and Venus appear to be the only bodies in the solar system where this heating process has provided long-term conditions conducive to continued volcanism. Primordial bombardment of planetary surfaces in the billion years following accretion, and events like orbital capture will heat up the shallow regions of a planetary body generating magma oceans and/or continuing volcanism. However, these heat sources are transient and the volcanism reflecting loss of heat from these sources will die out rapidly as the thermal structure of the body evolves. Whether or not volcanism continues will then depend on how much radiogenic heat is provided, or whether or not extraordinary factors intervene. Mercury and the Moon represent bodies where volcanism ceased shortly after the terminal bombardment cataclysm, while Mars represents a case where radiogenic heating of the interior allowed advection to drive volcanic activity perhaps for another billion years or more.

The Galilean satellites of Jupiter represent a case where tidal friction maintained a heat source for volcanic activity long after the terminal cataclysm, and where radiogenic heat alone would not have sufficed. On Io tidal heating and possibly electromagnetic induction continue to provide enough heat to sustain a very high level of surface activity. Triton appears to represent a case where capture provided heat sufficient to resurface much of the body, and tidal and/or thermal stresses continue to tap a molten or semimolten nitrogenous layer at shallow levels in the crust or upper mantle.

On icy bodies volcanism can continue because the driving force involves the behavior of solid–liquid–gas phases around eutectics or cotectics as low as 90–100 K, as oppose to the 800–1400 K eutectics and cotectics in silicate systems. On such bodies, external heating may become as important as internal heat loss as a driving mechanism for volcanic activity, and indeed, the very definition of volcanism may become blurred. The extreme case appears to be represented by comets, where external heating is probably the sole driving mechanism behind explosive degassing from vents through a residual dark (chondritic?) crust. Such activity would only occur during a brief interval before and after aphelion.

Adrian F. Park

Bibliography

Beatty, J.K. and Chaikin, A. (1990) *The New Solar System*. Cambridge, MA: Sky Publishing.
Burns, J. and Matthews, M.S. (eds) (1986) *Satellites*. Tucson: University of Arizona Press.
Cadogan, P. (1981) *The Moon – Our Sister Planet*. Cambridge, England: Cambridge University Press.
Greeley, R. (1987) *Planetary Landscapes*. New York: Allen & Unwin.
Hartmann, W.K. (1993) *Moons and Planets*, (3rd edn). New York: Wadsworth.
Head, J.W. and Crumpler, L.S. (1987) Evidence for divergent plate boundary characteristics and crustal spreading on Venus. *Science*, **238**, 1380–5.
Morrison, D. (ed.) (1982) *Satellites of Jupiter*. Tucson: University of Arizona Press.
Phillips, R.J., Grimm, R.E. and Malin, M.C. (1991) Hot-spot evolution and the global tectonics of Venus. *Science*, **252**, 651–4.
Wilhelms, D.E. (1987) *The Geologic History of the Moon*. US Geological Survey Professional Paper 1348.
Williams, H. and McBirney, A.R. (1979) *Volcanology*. San Francisco: Freeman-Cooper.

Cross references

Basalt
Callisto
Differentiation
Earth: geology, tectonics and seismicity
Ejecta
Europa
Ganymede
Hot spot tectonics
Igneous rock
Io
Mantle convection
Mars: structural geology and tectonics
Mercury: geology
Moon: geology
Plate tectonics
Tessera
Thermal evolution of planets and satellites
Triton
Venus: geology and geophysics

VOYAGER MISSIONS

NASA's Voyager 1 and 2 spacecraft have completed what by many scientists and other observers is acknowledged as the most successful exploratory mission to the planets ever flown (Plates 22–32). Two Mariner-type spacecraft were launched in late summer of 1977 (20 August and 5 September 1977) from Cape Canaveral, Florida. Voyager 1, although launched two weeks later than its twin, was on a faster trajectory, and flew by Jupiter on 5 March 1979 and by Saturn on 12 November 1980. Voyager 2's more leisurely pace brought it to Jupiter on 9 July 1979, to Saturn on 25 August 1981, to Uranus on 24 January 1986 and to Neptune on 25 August 1989. At the time of its Neptune flyby, Voyager 2 was more than 4.4 billion km (4.1 light hours) away from Earth. Each of the two spacecraft continue to operate nominally, and will measure the charged particle and magnetic field environment of the outer solar system until about the year 2015, transmitting their data back to Earth on a daily basis. Their goal is to detect the heliopause, the outer edge of the Sun's magnetic field, a boundary estimated to lie about 100 times the Earth's distance from the Sun. Beyond the heliopause the two hardy spacecraft will have the opportunity for the first time to measure the particles and magnetic fields in true interstellar space.

The extraordinary successes of the Voyager missions to the giant outer planets of the solar system cannot be attributed entirely to chance. A spacecraft design based on the proven Mariner series of planetary probes contributed greatly to Voyager's hardiness. The Titan–Centaur rockets which hoisted Voyager 2 and Voyager 1 from their Cape Canaveral launch pad were also tested and reliable. The onboard computers and those at NASA's Jet Propulsion Laboratory had been programmed to detect and correct many spacecraft problems aboard the two spacecraft. However, the success of the Voyager mission is most directly attributable to the dedication and expertise of the individuals, administrators, scientists, engineers and their associated staff members, who dedicated themselves to assuring that the two spacecraft operated as efficiently as possible and collected all the scientific data they were capable of collecting.

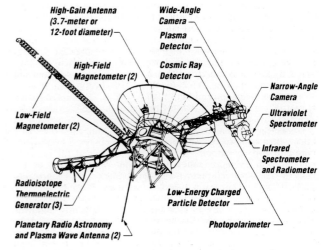

Figure V44 Schematic of the Voyager spacecraft showing the instruments carried during the missions to the planets Jupiter, Saturn, Uranus and Neptune. (Source: NASA).

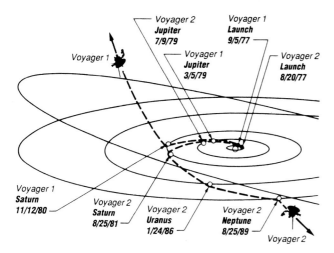

Figure V45 Schematic of the Voyager 1 and Voyager 2 spacecraft trajectories. The innermost circle represents the Sun, with the inner orbit shown representing the orbit of the Earth. The orbits of the planets Jupiter, Saturn, Uranus and Neptune lie in the ecliptic plane; the highly inclined plane represents the orbit of Pluto (which was not visited by the Voyager spacecraft).

Learning to use Voyager

Hundreds of unexpected problems on the two Voyager spacecraft had to be detected, analyzed and corrected by project personnel before they could return their magnificent findings. Most were minor, often as a result of human error; the complexity of the two Voyagers meant that project personnel had to go through a lengthy learning cycle before their capabilities could be used efficiently and safely. As each problem occurred, the causes were investigated and procedures and software were developed to prevent its recurrence or minimize its impact. Two potentially catastrophic problems on Voyager 2 are good examples. The first was the failure of the spacecraft's primary radio receiver; the second was the seizure of its steerable scan platform.

Receiver failure

Well before launch it was realized that a back-up receiver was useless unless the spacecraft had some means of sensing the failure of its primary receiver and automatically switching to the back-up. Two of the six spacecraft computers aboard each Voyager were therefore programmed to reset an internal counter each time a command from Earth was received. If a (typically) 7 day period passed without command detection, Voyager was to assume that its primary receiver had failed and automatically revert to use of the back-up.

When the immediate post-launch flurry of activity ceased, the frequency of 'real' commands to Voyager 2 slowed, and there was soon a period of 7 days when no commands were required. Voyager engineers, unaccustomed to the lower pace of activity, temporarily forgot to send a command within a 7-day period, and command inactivity caused Voyager 2 to switch to its back-up receiver. Recognition of the cause of the switch was swift, and the appropriate commands were quickly generated and transmitted to switch back to the primary receiver. During the switching, the primary receiver failed, possibly due to a momentary power surge. Another 7 days passed before Voyager 2 again switched to its back-up receiver, which has been in use since that time.

It was soon discovered that a feedback mechanism in the back-up receiver had failed. The receiver normally locks onto the frequency received from Earth, but Voyager 2's receiver would respond only to commands sent within a very narrow band of frequencies. The problem was traced to a shorted capacitor. The receiver frequency range was so narrow that the radio frequency of commands transmitted to Voyager 2 had to be adjusted to account for Doppler shifts due to the rotational and orbital motions of Earth as well as to the radial velocity of the spacecraft. To complicate matters, turning spacecraft instruments or heaters on or off would change the receiver's temperature and shift its sensitive frequency. The time required for the frequency to stabilize after power switching ranged up to 72 h. After each such command, Voyager engineers would be required to redetermine the 'best lock frequency' of Voyager's tone-deaf receiver. Until that frequency was determined, the project was faced with a moratorium on transmitting commands. In a relatively short period of time, Voyager engineers and personnel at the Deep Space Network tracking stations devised methods for predicting and transmitting at the correct frequencies, and the faulty receiver did little to hinder data collection during any of the four planetary encounters.

Scan platform problems

Instrumentation for four of Voyager's 11 scientific investigations is mounted on an articulated platform with two degrees of freedom – azimuth and elevation. About 100 min after Voyager 2's closest approach to Saturn, during its passage through the planet's shadow, the platform seized in its azimuth axis and would not respond to further commands from its computers. After about an hour of no response, the spacecraft computers automatically turned off power to the platform motors. Extensive analysis has led to the conclusion that the failure probably resulted from too frequent use of the scan platform at its highest rate, $1°\ s^{-1}$. The high level of activity may have driven lubricant from one of the motor gear shafts, and frictional wear of the shaft material caused the seizure. Subsequent heating and cooling of the motor and gear train freed the platform sufficiently to permit it to be pointed at the receding planet about 3 days after closest approach, and at Saturn's outermost satellite, Phoebe, 10 days after closest approach. Other than for a brief series of engineering tests, no platform motion was permitted for about 16 months thereafter. Since use of the platform resumed in February 1983, all slewing has been at rates of $0.083°\ s^{-1}$ or slower, except for brief periods of medium-rate slewing ($0.33°\ s^{-1}$) during near-encounter periods at Uranus and Neptune.

The spacecraft engineers also devised a method for checking whether the platform was beginning to experience excessive drag. In this procedure, known as torque margin testing, the duration of the motor drive pulses was reduced from the normal 200 ms to as low as 3 ms. The healthy actuators (as well as the Voyager 2 azimuth actuator) were found to drive at full rate with pulses of 6 ms or longer. Varying degrees of slowing occurred in all actuators with 5-ms pulses. Testing of identical actuators in the laboratory showed that when seizure was imminent, slowing occurred even when pulse durations were longer than 6 ms. Torque margin tests on both azimuth and elevation actuators on Voyagers 1 and 2 have shown that all four actuators continued to be healthy and in little danger of binding until they were shut off permanently in early 1993.

Image smear problems

Another problem for Voyager 2 was both predictable and solvable. The spacecraft were initially designed only for encounters with Jupiter and Saturn. NASA later approved extension of the mission to allow Voyager 2 to go on to Uranus and Neptune. At the distance of Neptune, light levels are reduced by a factor of 900 from those at Earth's distance from the Sun. Voyager 2's computers were reprogrammed to permit longer imaging exposures, but image smear then became a problem. Smear in the recorded images was reduced by automatic attitude-jet firings to compensate for the torque caused by starting and stopping the tape recorder. The normal quiescent attitude-control angular rates were also slowed by reducing jet pulses from 10 ms to 4 ms each. Methods were also devised to permit the cameras and other remote sensing instruments to track their targets more precisely, reducing image smear during times when the spacecraft was relatively close to its target.

Scientific instrumentation

Each Voyager spacecraft carried instrumentation for 11 investigations (see Table V8 for a list of the investigations and principal investigators.) These investigations included four remote sensing investigations, three charged-particle studies, magnetic field monitoring, plasma measurement and radio wave detection. The 11th investigation, radio science, used Earth-based instrumentation to analyze changes in the radio signals between the time of their transmission from Voyager and their receipt at tracking stations in California, Australia or Spain. Some additional radio science information was

Table V8 Voyager investigations and investigators

Investigation	Principal investigator (for which planets)[a] and affiliation
Imaging (ISS)	Bradford, A. Smith (J + S + U + N), University of Arizona, Tucson, AZ
Photopolarimetry (PPS)	Charles F. Lillie (J1) and Charles W. Hord (J2), University of Colorado, Boulder, CO; A. Lonne Lane (S + U + N), Jet Propulsion Laboratory, California Institute of Technology, Pasadena, CA
Infrared Spectroscopy (IRIS)	Rudolf A. Hanel (J + S + U) and Barney J. Conrath (N), Goddard Space Flight Center, Greenbelt, MD
Ultraviolet Spectroscopy (UVS)	A. Lyle Broadfoot (J + S + U + N), University of Arizona, Tucson, AZ
Radio Science (RSS)	Von R. Eshleman (J) and G. Leonard Tyler (S + U + N), Stanford University, Stanford, CA
Magnetometry (MAG)	Norman F. Ness (J + S + U + N), Bartol Research Institute, University of Delaware, Newark, DE
Plasma (PLS)	Herbert S. Bridge (J + S + U) and John W. Belcher (N), Massachusetts Institute of Technology, Cambridge, MA
Low-Energy Charged Particles (LECP)	Stamatios M. Krimigis (J + S + U + N), Applied Physics Laboratory, Johns Hopkins University, Laurel, MD
Cosmic Rays (CRS)	Rochus E. Vogt (J + S) and Edward C. Stone (U + N), California Institute of Technology, Pasadena, CA
Planetary Radio Astronomy (PRA)	James W. Warwick (J + S + U + N), Radiophysics Inc., Boulder, CO
Plasma Waves (PWS)	Frederick L. Scarf (J + S + U), TRW Defense and Space Systems Group, Redondo Beach, CA; Donald A. Gurnett (N), University of Iowa, Iowa City, IA

[a] J = Jupiter, S = Saturn, U = Uranus, N = Neptune, J1 = Voyager 1 Jupiter encounter, J2 = Voyager 2 Jupiter encounter.

collected during the Neptune encounter by the Parkes radiotelescope in Australia and the Usuda tracking station in Japan; the Very Large Array in New Mexico assisted in collecting high-rate telemetry from Voyager 2. Coordinated observations made it possible for the total science return from Voyager to exceed the sum of the results of the individual investigations.

The instruments for all 11 scientific investigations on Voyager 2 were still operating through the Neptune encounter periods. Voyager 1's photopolarimeter failed during the Jupiter encounter; its plasma instrument suffered a catastrophic loss of sensitivity shortly after the Saturn encounter. The other nine Voyager 1 investigations continued to function normally until after Voyager 2's encounter with Neptune. By early 1990 the imaging, photopolarimetry and infrared spectroscopy investigations had been permanently shut off on both spacecraft, and no more radio science observations were being made. Ultraviolet stellar astronomy continued at a low level until the end of 1992. Charged particle, radio wave and magnetic field investigations (except for the Voyager 1 plasma study) are engaged in a continuing search for the heliopause, the outer boundary of the Sun's magnetic field. Both Voyagers are expected to collect useful data well into the 21st century.

Scientific findings

During the planetary encounters data were gathered on:

- physical properties, dynamics and compositions of atmospheres;
- thermal properties, total radiated energy and total energy absorbed from the Sun (the latter two for comparison);
- charged particles and electromagnetic environments;
- ring systems;
- satellite surface features;
- periods of rotation, radii, shapes and other body properties;
- masses and gravitational fields.

Many of the scientific results from the Voyager missions are covered elsewhere in this encyclopedia. The results given below are but a brief summary of some of the major findings in the seven areas. Some of these findings are summarized in Tables V9 and V10.

Atmospheres

The atmospheres of the four giant planets above their cloud tops are composed mostly of hydrogen and helium. For every 100 g of atmospheric hydrogen, Jupiter has 18 ± 4 g of atmospheric helium, Saturn has 6 ± 5 g, Uranus has 26 ± 5 g and Neptune has 32 ± 5 g. The corresponding mole fractions of helium are 0.10 ± 0.03, 0.04 ± 0.02, 0.15 ± 0.03 and 0.19 ± 0.03 respectively. The Voyager data show that these giant bodies have undergone significant changes since their formation. The upper atmospheres of both Jupiter and Saturn have apparently been depleted of much of their original helium, presumably through gravitational settling of the heavier helium in the interiors of these planets. The upper atmospheres of Uranus and Neptune, on the other hand, seem to have somewhat larger fractions of helium than is estimated for the primordial solar nebula; it is possible that substantial amounts of hydrogen have escaped their lower gravitational fields.

The visible clouds of Jupiter and Saturn are composed predominantly of ammonia ice. Trace impurities give these clouds their colorations, such as the reddish tints of Jupiter's Great Red Spot, but their overall color closely replicates the yellow of the Sun. Methane is about ten times as abundant in the atmospheres of Uranus and Neptune as in the atmospheres of Jupiter and Saturn; the red-absorbing characteristics of methane gas give these planets their characteristic blue color. The uppermost clouds in the atmospheres of Uranus and Neptune consist of methane ice. Near the base of these methane clouds the measured abundance of methane gas implies that the abundance of carbon in the atmospheres of Uranus and Neptune is at least ten times that in the solar atmosphere. Either these planets have lost major fractions of their original hydrogen and helium, or a much larger fraction of their total mass is the result of the capture of carbon-rich asteroids or comets than is the case for Jupiter or Saturn.

Two of the satellites have substantial atmospheres. Both Saturn's Titan and Neptune's Triton have atmospheres dominated by nitrogen and containing substantially smaller amounts of methane, but there the similarity ends. Thick layers of haze in Titan's atmosphere hide the surface in visible light; Triton's surface is unobscured. The near-surface atmospheric pressure on Titan is 1.6 bar; Triton's is only 1.6×10^{-5} bar. Much of that difference is a result of the different temperatures of these two satellites.

Thermal properties

The atmospheres of all the giant planets have absolute temperature minima near the 0.1-bar pressure level. Jupiter's minimum temperature is about 110 K; Saturn's, 80 K; Uranus', 52 K; and Neptune's, 50 K. Their respective effective temperatures (temperatures of a black body which would radiate the same amount of energy per unit area) are 124.4 K, 95.0 K, 59.3 K and 59.3 K. Each of the planets, with the possible exception of Uranus, radiates more thermal energy than it receives from the Sun. The ratios of total radiated to total absorbed energy (also known as 'energy balance') are 1.67 ± 0.08 for Jupiter, 1.79 ± 0.10 for Saturn, less than 1.14 for Uranus and 2.7 ± 0.3 for Neptune.

Triton's surface temperature of 38 ± 3 K makes it the coldest body measured by Voyager. Titan has a surface temperature of 95 ± 1 K. Jupiter's Io has typical subsolar surface temperatures of about 135 K, but hot spots associated with active volcanism can reach temperatures of 650 K or more. Jupiter's Europa, Ganymede and Callisto have typical subsolar surface temperatures near 125 K, 156 K and 168 K. Saturn's satellites Rhea, Tethys and Enceladus have subsolar surface

Table V9 Outer planet data

	Earth	Jupiter	Saturn	Uranus	Neptune
Mean distance from Sun (10^9 km)	0.1496	0.7783	1.4294	2.8750	4.5043
Sidereal period of orbit (years)	1.0000	11.8623	29.458	84.01	164.79
Mean orbital velocity (km s^{-1})	29.79	13.06	9.64	6.81	5.43
Orbital eccentricity	0.0167	0.0485	0.0556	0.0472	0.0086
Inclination to ecliptic (deg)	0.00	1.30	2.49	0.77	1.77
Equatorial radius at 1 bar (km)	6378	71492	60268	25559	24764
Polar radius at 1 bar (km)	6357	66854	54364	24973	24340
Ellipticity of planet disk	0.00335	0.06487	0.09796	0.02293	0.0171
Volume of planet (Earth = 1)	1.0000	1321.3	763.6	63.1	57.7
Mass of planet (Earth = 1)	1.0000	317.892	95.184	14.536	17.148
Mass of planet (10^{27} kg)	0.00598	1.8997	0.5688	0.08686	0.10247
Mean density (g cm^{-3})	5.518	1.327	0.688	1.272	1.640
Body rotation period (hours)	23.9345	9.9249	10.6562	17.24	16.11
Tilt of equator or orbit (deg)	23.45	3.08	26.73	97.92	28.8
Effective temperature (K)	287	124.4	95.0	59.1	59.3
Atmospheric temperature at 1 bar (K)	287	165	134	76	74
Number of observed satellites	1	16	20	15	8
Number of observed rings	0	1[a]	7[a]	10[a]	4[a]
Magnetic dipole moment (gauss R_x^3)	0.308	4.28	0.218	0.228	0.133
Magnetic dipole moment (10^{27} gauss cm^{-3})	0.00798	1560	47.0	3.83	2.02
Magnetic dipole tilt (deg)	11.4	9.6	-0.0	58.6	46.8
Magnetic dipole offset (R_x)	0.0725	–	0.04	0.3	0.55

[a] No name designations have yet been made for Jupiter's ring, the most recently discovered ring of Uranus (1986U1R), nor the broadest of the four rings of Neptune (1989N4R). Saturn's seven rings, in order of distance from the planet, are D, C, B, A, F, G and E; those of Uranus are 6, 5, 4, α, β, η, γ, δ, (1986U1R) and ε; those of Neptune are Galle, Leverrier (1989N4R) and Adams.

temperatures of 100 ± 2 K, 93 ± 4 K and 75 ± 3 K respectively. The temperature differences among the various satellites arise from differences in surface reflectivity and in distance from the Sun. The satellites with the highest reflectivities generally possess surfaces that have been altered in recent geologic time, most often as a result of processes that heat and partially melt materials near the surface.

Electromagnetic environments

Each of the four giant planets possesses an intrinsic magnetic field and a trapped radiation field. Because the atmospheres rotate at different rates at different latitudes, periodic radio emissions (caused by the interaction between the planetary magnetic field and the solar wind) are the best means for determining the body rotation period. Jupiter's rotation period was found from Earth to be 9.9249 h. Saturn, Uranus and Neptune were discovered by Voyager to complete one rotation in 10.6562, 17.24 and 16.11 h respectively. Both pulsed (like an omnidirectional strobe light) and beamed (like a rotating lighthouse light) radio emissions have been detected.

Saturn's magnetic field may be represented by a dipole magnet aligned with the planet's rotation axis but offset to the north approximately 4% of the planet's radius. Jupiter's magnetic field is tilted by about 10° from the rotation axis. Uranus and Neptune each have magnetic fields highly tilted with respect to their rotation axes and offset from their centers by large fractions of the planetary radii. These offsets possibly result from internal dynamos (circulating electrical currents) that are much closer to the cloud tops on Uranus and Neptune than they are on the nearby large planets or Earth. Perhaps the deeper parts of the cores of Uranus and Neptune are not efficient electrical conductors.

The magnetospheres of the four giant planets are each populated with trapped charged particles. As Voyager 2 traveled outward in the solar system, it found that each successive planet is less populated with charged particles than its predecessor, and each is influenced less by the satellites which orbit within it. Radio emissions from lightning were detected both at Jupiter and Saturn, but not at Uranus or Neptune (see Planetary lightning).

Ring systems

Until 5 months before the 1977 launches of Voyagers 1 and 2, Saturn was the only planet known to have rings. Now Voyager 2 has relayed detailed data about ring systems around each of the giant planets. Each system is found to have unique characteristics. Jupiter's faint ring is composed primarily of tiny dust particles, probably the result of meteoroid bombardment of the satellites Metis and Adrastea. Saturn's ring has far more detail than had been supposed prior to Voyager; it is very bright, composed mainly of water ice, has a wide distribution of particle sizes and forms an extensive sheet of material closely confined to Saturn's equatorial plane. Self-gravity and the gravitational effects of nearby satellites combine to create a wonderfully complex structure including tightly wound spiral formations, gaps, narrow ringlets and sharp ring edges.

The Uranus and Neptune ring systems consist primarily of a series of narrow rings, most probably constrained from spreading by the gravitational action of both seen and unseen satellites. The extremely dark color of these ring systems is either a result of coating by carbonaceous material or is due to the action of high-energy protons on methane trapped in the ice in the particles. Strong azimuthal concentrations of ring particles in Neptune's outer ring, first detected from Earth, led to the conclusion that Neptune possessed only partial rings. These arclike segments within the unbroken ring would normally be expected to spread to azimuthal uniformity within a few years.

An unexpected outcome of the Voyager findings is the realization that planetary rings are apparently rapidly evolving structures. Their complex interactions include self-gravity effects, magnetic field interactions and atmospheric drag, in addition to 'shepherding' by nearby satellites. It now appears likely that a continual process of break-up of larger bodies into progressively smaller bodies is inherent in the creation and maintenance of ring systems (see Planetary ring).

Satellite surfaces

Voyager 1 and 2 were responsible for the discovery of at least 21 new satellites, for determination of the sizes of all but 12 of the solar system's 60 known satellites and for taking the only detailed images of the surfaces of the satellites of the giant planets. Surface maps from Voyager data now exist for Jupiter's Amalthea, Io, Europa, Ganymede and Callisto; for Saturn's Mimas, Enceladus, Tethys, Dione, Rhea, Hyperion and Iapetus; for Uranus' Puck, Miranda, Ariel, Umbriel, Titania and Oberon; and for Neptune's Triton (see also Cartography; Regional Planetary Image Facilities). Ancient cratered surfaces were expected to exist on all the satellites, but Voyager also disclosed an enormous variety of unexpected surface features. These range from the active volcanism (on Io and Triton) to evidence for flows of crustal materials (on Ganymede, Enceladus, Ariel and others) to unexplained landforms (on Miranda and others)

Table V10 Satellite data

Satellite name	Radius (km)	Density (g cm^{-3})	Distance (10^3 km)	Inclination (degrees)	Eccentricity	Period (days)
Moon (E)	1738	3.34	384.40	23.43	0.055	27.3217
Metis (J16)	20	–	127.96	0	0.0	0.2048
Adrastea (J15)	10	–	128.98	0	0.0	0.2983
Amalthea (J5)	94	–	181.3	0.40	0.003	0.4981
Thebe (J14)	50	–	221.90	0.8	0.015	0.6745
Io (J1)	1815	3.55	421.6	0.040	0.0041	1.7691
Europa (J2)	1569	3.01	670.9	0.470	0.0101	3.5512
Ganymede (J3)	2631	1.95	1070	0.195	0.0015	7.1546
Callisto (J4)	2400	1.86	1883	0.281	0.007	16.6890
Leda (J13)	8	–	11 094	26.07	0.148	238.72
Himalia (J6)	93	–	11 480	27.63	0.158	250.566
Lysithea (J10)	18	–	11 720	29.02	0.107	259.22
Elara (J7)	38	–	11 737	24.77	0.207	259.653
Ananke (J12)	15	–	21 200	147	0.169	631R[a]
Carme (J11)	20	–	22 600	163	0.207	692R
Pasiphae (J8)	25	–	23 500	148	0.378	735R
Sinope (J9)	18	–	23 700	153	0.275	758R
Pan (S18)	10	–	133.58	0.0	0.0	0.5750
Atlas (S15)	15	–	137.67	0.0	0.003	0.6019
Prometheus (S16)	50	–	139.35	0.0	0.0024	0.6130
Pandora (S17)	45	–	141.70	0.0	0.0042	0.6285
Epimetheus (S11)	60	–	151.42	0.34	0.009	0.6943
Janus (S10)	95	–	151.47	0.14	0.007	0.6947
Mimas (S1)	196	1.44	185.52	1.53	0.0202	0.9424
Enceladus (S2)	250	1.13	238.02	0.02	0.0045	1.3702
Tethys (S3)	530	1.21	294.66	1.09	0.000	1.8878
Telesto (S13)	15	–	294.66	0.0	0.0	1.8878
Calypso (S14)	13	–	294.66	0.0	0.0	1.8878
Dione (S4)	560	1.43	377.40	0.02	0.0022	2.7369
Helene (S12)	16	–	377.40	0.2	0.005	2.7369
Rhea (S5)	765	1.33	527.04	0.35	0.0010	4.5175
Titan (S6)	2575	1.88	1221.83	0.33	0.0292	15.9454
Hyperion (S7)	145	–	1481.1	0.43	0.1042	21.2766
Iapetus (S8)	730	1.15	3561.3	7.52	0.0283	79.3302
Phoebe (S9)	110	–	12 952	175.3	0.1633	550.48R
Cordelia (U13)	13	–	49.75	0.1	0.000	0.3350
Ophelia (U14)	16	–	53.77	0.1	0.001	0.3764
Bianca (U15)	22	–	59.16	0.2	0.001	0.4346
Cressida (U9)	33	–	61.77	0.0	0.000	0.4636
Desdemona (U12)	29	–	62.65	0.2	0.000	0.4737
Juliet (U8)	42	–	64.63	0.1	0.001	0.4931
Portia (U7)	55	–	66.10	0.1	0.000	0.5132
Rosalind (U10)	27	–	69.93	0.3	0.000	0.5585
Belinda (U11)	34	–	75.25	0.0	0.000	0.6235
Puck (U6)	77	–	86.00	0.3	0.000	0.7618
Miranda (U5)	236	1.25	129.8	4.22	0.0027	1.4135
Ariel (U1)	579	1.55	191.2	0.31	0.0034	2.5204
Umbriel (U2)	586	1.58	266.0	0.36	0.0050	4.1442
Titania (U3)	790	1.68	435.8	0.14	0.0022	8.7059
Oberon (U4)	762	1.64	582.6	0.10	0.0008	13.4632
Naiad (N8)	29	–	48.23	4.74	0.0003	0.2944
Thalassa (N7)	40	–	50.07	0.21	0.0002	0.3115
Despina (N5)	74	–	52.53	0.07	0.0001	0.3347
Galatea (N6)	79	–	61.95	0.05	0.0001	0.4287
Larissa (N4)	96	–	73.55	0.20	0.0014	0.5547
Proteus (N3)	208	–	117.64	0.04	0.0004	1.1223
Triton (N1)	1350	2.07	354.8	156.8	0.000	5.8768R
Nereid (N2)	170	–	5509.1	27.6	0.753	360.129

[a] R = retrograde.

to haze-enshrouded lakes of liquid ethane (on Titan). It is obvious that many processes due to internally generated heat have occurred since the early cratering epochs to alter the surfaces of these satellites.

Body properties

Most of the satellites of the solar system are in locked rotation, keeping one face toward their planet at all times. Noteworthy exceptions are Jupiter's outer eight satellites, Saturn's Hyperion and Phoebe, and probably Neptune's Nereid. Jupiter's outer eight satellites are all thought to be captured asteroids; Voyager did not study them. Phoebe makes a full rotation in approximately 9 h and orbits Saturn with a period of 550 days. Hyperion tumbles chaotically due to frequent gravitational interactions with nearby Titan. Voyager did not detect Nereid's rotation, but this satellite's highly elliptical orbit makes it an unlikely candidate for synchronous rotation.

Voyager found several satellites to be significantly non-spherical. These include Adrastea, Amalthea and Thebe at Jupiter; Atlas, Prometheus, Pandora, Epimetheus, Janus, Telesto, Calypso, Helene and Hyperion at Saturn; and Proteus and Larissa at Neptune. Voyager also measured differences between equatorial and polar radii for several other satellites and for the four planets. Surface reflectivities of the satellites range from almost 100% (Enceladus) to about 5% (Metis, Adrastea, Amalthea, Thebe, Phoebe and the dark face of Iapetus).

Bulk properties

The masses of the four planets were already known to reasonable accuracy from Earth-based observations. Voyager refined the densities, primarily by providing better estimates of sizes and shapes. Knowledge of the internal mass distribution was also improved by measuring body rotation periods and non-spherical gravity forces on the spacecraft. Models in vogue prior to Voyager are now seen to show Uranus and Neptune as too centrally condensed. Voyager tracking data led to the conclusion that much larger percentages of the planetary mass lie well away from the centers of the planets than was expected.

Voyager data allowed determination of the masses and densities of 17 satellites. Saturn's satellites all seem to have densities between 1.1 and 1.5 g cm^{-3}, except for Titan, whose density is 1.88 g cm^{-3}. Neptune's Triton is 2.05 g cm^{-3} in density. Uranus' four largest satellites range from 1.5 to 1.7 g cm^{-3}. These low densities all indicate large amounts of water ice. Jupiter's satellites, by contrast, range from a density of 1.86 g cm^{-3} for Callisto (the outermost of the Galilean satellites) to 3.55 g cm^{-3} for Io (the innermost of the Galilean satellites). This wide variation implies that heat from Jupiter in its formative stages was sufficient to drive water from its inner satellites.

To the heliopause and beyond

The two Voyager spacecraft were designed to have a high probability of surviving the 4 years between launch and Voyager 2's Saturn encounter. After more than 16 years of continuous operation they continue to operate well and to collect useful scientific data about the outer solar system environment. By 2015 both Voyager 1 and Voyager 2 will be more than three times the distance of Neptune (and Pluto) from the Sun; their respective outward velocities will be 16.6 and 14.9 km s^{-1}. The Sun's magnetic field and the solar wind are believed not to extend to those great distances, and so the Voyagers will probably become the first artificial spacecraft to exit the heliosphere and make direct measurements of the interstellar environment. Such measurements would represent yet another in a long line of firsts for the highly successful Voyager mission.

Ellis D. Miner

Bibliography

Detailed descriptions of the eleven scientific investigations are contained in *Space Science Reviews* (1977) **21**, 103–376.
A description of the Voyager 2 Neptune encounter and preliminary results is contained in *Science,* **246**, 1417–1501. See page 1421 of the overview paper for references to detailed Voyager findings at Jupiter, Saturn and Uranus. A popularized account of the Voyager mission and its findings at Jupiter, Saturn and Uranus is contained in Miner, (1990). More detailed information on Neptune is contained in *Journal of Geophysical Research*, **96**, 18903–19268 (1991).
Bergstralh, J.T., Miner, E.D. and Matthews, M.S. (eds) (1991) *Uranus*: Tucson: University of Arizona Press.
Cruikshank, D.P. and Matthews, M.S. (eds) (1995) *Neptune and Triton*. Tucson: University of Arizona Press.
Miner, E.D. (1990) *Uranus: The Planet, Rings and Satellites*. Chichester, UK: Ellis Horwood, 334 pp.

Cross references

Heliosphere
Jupiter
Neptune
Planetary lightning
Planetary ring
Plasma wave
Radio science
Regional Planetary Image Facilities
Saturn
Uranus

VOYAGER PLANETARY RADIO ASTRONOMY

The first two missions to the outer solar system were flown by two nearly identical spacecraft, Pioneer 10 and 11 (q.v.). These missions initially were known as the Galactic Jupiter Probes. Their principal target was interstellar space, where magnetic fields and cosmic-rays could be measured. To reach space in a timely manner required a gravitational boost from Jupiter, the largest planet. To radio astronomers, it appeared that mission planners understood rather belatedly the importance of Jupiter itself, a dynamic source of energetic particles and magnetic fields. Ground-based radio astronomical data demonstrating these phenomena had been available for more than one decade, but planning was too far advanced to include a radio telescope aboard the Pioneers.

The next missions to the outer solar system were initially known as the 'Grand Tours' which eventually became the missions executed by Voyager 1 and 2 (q.v.). From the outset these were organized primarily around planetary objectives and always contained a significant planetary radio astronomy capability. They have extended knowledge of Jupiter as a radio planet and established the radio emission characteristics of the other giant planets as well.

The planetary radio astronomy experiment (PRA) aboard the two Voyager spacecraft measures radio spectra of planetary and solar emissions in the range 1.2 kHz to 40.2 MHz. These emissions result from wave–particle–plasma interactions in the magnetospheres and ionospheres of the planets and in the solar corona. Prior to these missions Earth-based radio telescopes had routinely observed such emissions from Jupiter since 1955 (even since 1950 in a famous prediscovery series of data), but only ephemeral data had been obtained for Saturn, and no data at all for Uranus and Neptune. The frequency range of previously known emissions were extended by the two Voyagers downward to frequencies in the range of several kilohertz from Earth's ionospheric cut-offs (typically in the range 5 to 10 MHz, and in special circumstances as low as 1.5 MHz).

Non-thermal emissions from Uranus and Neptune, previously unknown, were observed during Voyager 2's closest approach to those planets. At Saturn the two Voyagers discovered impulsive bursts similar in many respects to radio emissions from Earth's lightning. At Saturn, Uranus and Neptune, Voyager observed impulsive bursts of radio energy as the spacecraft passed through the planets' equatorial planes. These bursts have been identified as the signature within PRA and Voyager's Plasma Wave Subsystem (PWS) of hypervelocity impacts by micron-sized dust on the spacecraft.

The design of the PRA receiver was based on known radio emissions from Jupiter and Earth as a model for all the planets. Table V11 summarizes these emissions from Jupiter. The brightness temperature T_B, is the temperature an optically thick black body would have to have if it were to reproduce the emissions observed in a narrow band centered on a given frequency, f. The table also gives

Table V11 Jupiter brightness temperatures

f (MHz)	30 000	3000	300	30
T_B (K)	140	6000	10^5	10^{18}
E_{av} (eV)	0.01	0.05	9	9×10^{13}

the average particle energy, $E_{av} = kT_B$, in this equivalent black body (k is the Boltzmann constant).

In the range from 10 000 MHz to about 40 MHz Jupiter's radio emissions are non-thermal, but would correspond to temperatures much hotter than the planet's atmosphere or exosphere. Their sources are energetic electrons whose powers add to form the observed signals. The lower-frequency emissions are also non-thermal. The most energetic electrons (10^8 eV) measured near Jupiter fail by a factor of 10^8 from being able to produce the 30-MHz emissions by incoherent processes such as in a black body.

Hardware

The PRA experiment includes two 10-m antenna booms and a receiver. The two boom tips and the receiver lie at the vertices of a right triangle whose short sides are defined by the booms themselves. The plane defined by the booms is orthogonal to Voyager's 13 m long magnetometer (MAG). The MAG boom is tilted 50° to the Earth-directed axis of the Voyager telemetry antenna (the negative z-axis of the spacecraft). The Voyager scan platform boom (length 5 m) and the radio isotope thermoelectric generator boom (length 3 m) are approximately collinear and located on opposite sides of the spacecraft. The trihedral angle formed by the magnetometer boom and the two PRA booms contains the RTG boom (Figure V46). Boom A, boom B and the MAG boom form a right-handed orthogonal set of axes, like Cartesian x, y and z axes.

This configuration of the PRA antennas was supposed to minimize their electrical coupling with the MAG boom. However, the remaining parts of the spacecraft also couple to the antennas and turn the idealized response of the PRA instrument into a complicated function of direction of arrival and of wave polarization.

The signals from the two antennas are separately amplified within the receiver and sent to a quadrature combiner (called a 'hybrid'). The combiner produces two outputs, C and D, which are appropriately phased sums of the antenna signals, $(2)^{-1/2} (V_A - jV_B)$ and $(2)^{-1/2} (V_B - jV_A)$, respectively. Ideally, if purely right-hand or purely left-hand planetary emissions were propagating toward the spacecraft along its positive z-axis only C or D would contain signal. In practice, this is virtually never the case because the spacecraft is a complicated electromagnetic structure that interacts with the booms in a manner that depends on wave polarization, angle of arrival and radio frequency. This interaction must be calculated numerically.

The frequency layout of the receiver is also complicated. The receiver scans the spectrum from 40.2 MHz to 1.2 kHz in 198 discrete steps, spaced 19.2 kHz apart from 1.2 to 1326 kHz and 307.2 kHz apart from 1228.8 kHz to 40 243.2 kHz. The spacecraft power supply produces a 2.4 kHz, 50 V square wave to which most Voyager subsystems are locked. The electromagnetic interference (EMI) resulting from the harmonics of the power supply would normally be bothersome to an investigation of this type. PRA's 70-step low-frequency scan is executed with a passband of 1.0 kHz centered between successive power supply harmonics. The PRA receiver's local oscillator is phase locked to the spacecraft power supply switch cycle. Therefore, unless the 1.0 kHz crystal filter drifts in frequency, EMI is largely avoided. The 128-step high-frequency scan is executed with a passband of 200 kHz. At each step in this part of the scan there are approximately 80 harmonics of the Voyager power supply within the passband. These are generated by all spacecraft subsystems; their power is the major signal within the high-frequency steps except when the spacecraft observes strong sources such as solar bursts, Jupiter's decametric emissions (DAM), or Saturn's electrostatic discharges (SED).

This passband arrangement successfully reduced EMI in the lowband region where it is most prominent. It was intended to increase system sensitivity in the high-band region where decreasing EMI, it was believed, would permit greater bandwidth. However, the receivers, now operating in space in the Voyager environment, suffer severe EMI throughout the high-band, typically 20 to 30 decibels above the cosmic noise background and occasionally even in the low band, especially the three lowest steps.

The gains of the receiver in the two frequency ranges are adjusted so that a broad band signal (not spacecraft interference) produces constant output across the transition between the low-and high-frequency portions of the scan. Because of the difference in bandwidth in these two portions, this requires a 23-dB increment to the gain of the LF scan.

The scan of the instrument is from high frequency to low frequency with a 30-ms dwell time per step; a scan therefore requires 5.94 s ($= 198 \times 0.03$). An additional pair of 30-millisecond steps is used for transmission of instrument status. The total time to acquire a PRA 'frame' is therefore exactly 6 s. Analog-to-digital conversion is carried to 8 bits of precision; the two steps allocated to receiver status each take 8 bits (16-bit status word). Therefore the bit rate of the receiver in full-up operation in its basic mode (called 'POLLO') is $200 \times 8/6 = 266 2/3$ bits per second (bps).

Consecutive steps are recorded from the C and D ports respectively. Emissions that are ideally polarized would appear in alternate steps across the spectrum. The first step of POLLO scans alternates between ports C and D from scan to scan. A partially polarized broadband planetary emission therefore produces a checkerboard pattern in the dynamic spectrum. (The spectrum as a function of time is called a 'dynamic spectrum'.) Since polarization response of the entire PRA system is a complicated function of frequency, direction of arrival and emission polarization, direct interpretation of a given checkerboard pattern may be difficult.

The instrument has two IF strips or channels (called 'upper' and 'lower'). They are respectively connected to ports C and D when a toggle switch called 'RH Cupper' is set to binary one. Ideal signals in the upper channel are then right-handed. In high band the upper channel processes signals in a 200-kHz passband 307.2 kHz lower in frequency than those in the lower channel. In low band, both upper and lower channels process signals in a 1-kHz passband at the same frequency. Whatever the position of RH Cupper, in both bands the two channels represent opposite states of polarization.

At the time of the design of the instrument it was recognized that an important objective of a planetary radio astronomy experiment carried to the giant planets should be the detection of lightning flashes, on Earth known to be prolific sources of radio emissions throughout the spectrum from VLF (very low frequencies, 3 to 30 kHz) to microwave frequencies (above 1000 MHz). It was also known that the exceedingly high Voyager telemetry rate to be used for the television imaging experiment (115.2 kbps) would, if it were in use full time, produce so many imaging data as to 'saturate' the image processing facilities on the ground. The window of opportunity for a high-rate PRA data stream was thereby opened. The PRA had all along intended to utilize its wide bandwidth in the high-frequency band for a very short but rapid series of data at fixed frequencies (so-called 'POLHI' mode). But now it became possible to divert the PRA data output in the high-rate mode to the television data bus aboard the Voyagers. This enabled the output of PRA data at 115.2 kbps to continue for a time equal to that of a television frame, 48 s. Data taken in this mode demonstrated the fine time structure of Jupiter's millisecond bursts and pulse bursts, SED and dust impacts on the spacecraft.

One year after the selection of the PRA experiment it became possible to add another wave instrument to Voyager's suite of experiments. The new plasma wave instrument (PWS) was able to take particularly effective advantage of the high-rate window of opportunity to execute waveform observations in the spectral range from 10 Hz to 12 kHz. PWS is mounted atop PRA and uses the same pair of booms; unlike PRA, PWS uses them as a balanced pair that forms a V-shaped short dipole antenna. The parallel operation of PRA and PWS is practical because both receivers have high-impedance radio frequency (RF) amplifiers at their inputs.

The sample rate for PHIEX, 8-bit samples taken at 14.4 kHz rate (note that 8 bits $\times$ 14 400 = 115 200), severely undersamples the detected signals in the 200-kHz bandwidth of the PRA receiver in its high-frequency range. The same rate in the 1-kHz bandwidth of the receiver in its low-frequency range would seriously oversample the signals. The possibility of using PHIEX to perform waveform analysis of PRA signals is ruled out, in any case, by the fact that the receiver acts as a normal radio astronomical radiometer, detecting the signals from its intermediate frequency amplifier.

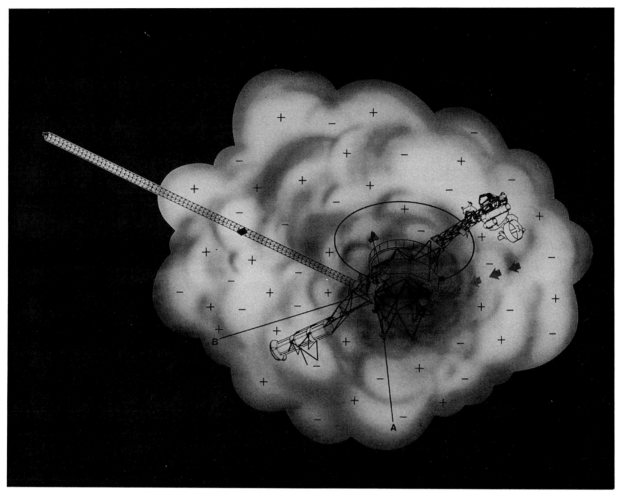

Figure V46 The Voyager magnetometer and two PRA booms (A and B).

PHIEX samples both upper and lower channels simultaneously. A single data word, 8 bits long, is presented to the imaging system data bus each $(14\,400)^{-1}$ s = 69.4 μs. The two simultaneous signals are centered in 200-kHz passbands separated by 307.2 kHz and in opposite states of circular polarization; as before, the upper channel carries the lower-frequency data. While one channel's data are being sent to the data bus, the other channel's data, sampled at the same time, are being held. The ultimate time resolution is therefore a 69.4-μs sample in one channel, separated in time by a further 69.4 μs from the next sample in the same channel. This sequential complexity represented a means of enriching the data base without great complexity of hardware.

There is just one analog-to-digital converter (ADC) on the PRA instrument. It is placed after the square-law detectors at the ends of the two IF strips. A toggle switch 'ADC-upper', set to binary one, allows signals from the upper channel to be converted; set to zero, it allows signals from the lower channel to be converted. This switch can be toggled either at 14.4 kHz or at 33⅓ Hz, for POLLO. This toggling is called 'channel toggle'. RH Cupper can also be toggled. Many distinct modes of operation of the receiver are therefore possible. One in particular which has proved very useful is 'FIXLO', the operation of the receiver on just one step (out of the many available ones). In this mode as finally utilized, RH Cupper is toggled while ADC-upper remains fixed. Polarization alternates between successive 30-μs samples.

The controls of the instrument are in the 'hands' of the Voyager flight data system (FDS). A 'minor frame' of FDS data is 48 s long. That comprises eight POLLO scans, or eight FIXLO scans. At the same time that the FDS commands PRA to perform FIXLO, the FDS sends it a 9-bit word designating a particular frequency step.

During a minor frame the FDS moves through a 'frequency look-up table' containing many separate frequencies and which allow different scans in a minor frame to be made at different frequencies.

The instrument determines its frequency of operation through a phase-lock loop controlled by the stable Voyager clock (which also controls the spacecraft power supply frequency). Its 9-bit digital registers allow for 512 rather than the 198 frequency steps described above. Only 198 of these are 'legal'. However, it is possible to operate the receiver in 'illegal' high band (200-kHz bandwidth) as high as 47 MHz (at which point RF filters in the input amplifiers cut off effective response at higher frequencies) or as low as 300 kHz (where spacecraft interference is a limiting factor except when exceedingly strong signals are present). It is also possible to operate in 'illegal' low band (1-kHz bandwidth) as high as 2500 kHz, above which, again, input filters limit the response. In actual operation during 14 years of flight (as of September 1991) all of these capabilities of the PRA receiver have been used to good advantage in one or more specific observational programs.

Scientific results

Earth

On departure from Earth, both Voyagers observed strong signals from Earth in the frequency range from 100 to 300 kHz. These were classic terrestrial kilometric radiation (TKR), produced in regions over the auroral zones during precipitation of energetic electrons towards Earth's surface (see Solar wind; Aurora, planetary). The new information provided by PRA was TKR's polarization; it corresponds to extraordinary mode radiation. This inference had previously been drawn from spectral data from other spacecraft.

The rotation of Earth's magnetic field manifests itself in modulation of TKR in a 24-h periodicity. For Earth, this period is locked to local solar time; that is, the sources of TKR are controlled by the magnetic field as it rotates past the solar meridian, and were fairly well understood prior to the Voyager missions.

Jupiter

Jupiter emissions, in the kilometric to hectometric range (100 to 500 kHz), were detected by PRA almost immediately after spacecraft launch. These of course increased in strength continuously over the subsequent 18 months until Voyager 1 encounter with Jupiter. Because of the high interference levels in high band, the observations of decametric radiation (DAM), familiar to ground-based radio astronomers, were possible only from December 1978 to November 1979, when DAM exceeded the interference in at least some high-band spectral ranges.

From the middle of 1978, both Voyager 1 and 2 in FIXLO and PHIEX detected what are now called 'pulse bursts' (ticking like a clock) from Jupiter in the hectometric range (80 to 1500 kHz). These powerful bursts occur nearly periodically at intervals from 0.3 to 3 s. Their source is unknown.

The first DAM dynamic spectra showed that unmodulated Jupiter emissions (seen from Earth, DAM is strongly modulated by both terrestrial and interplanetary scintillations) were arc shaped in the spectrum–time domain. 'Arc' is intended to suggest emission that is narrowband and curved like a parenthesis. Vertex-early arcs first appear at an intermediary frequency then develop along smoothly curving paths (in the frequency–time domain) towards both lower and higher frequencies. Vertex-late arcs appear simultaneously at both high and low frequencies then develop along curving paths to a vertex at an intermediary frequency. The occurrence of arcs is a strong function of longitude on Jupiter, vertex-early arcs appearing in data observed somewhat before sub-spacecraft longitude, $\lambda_{III} = 200°$ and vertex-late, after. (Radio longitudes are based on the radio rotation rate, known as 'system III', whose period is 9 h 55 min 29.71 s). This period was determined from ground-based decametric data taken long before the Voyager program. The significance of 200° longitude is that the northern end of Jupiter's magnetic dipole, tilted at about 10° to the rotation axis, lies in that meridian.

In general the PRA experiments confirmed the nature of DAM as recorded in ground-based observations. In particular they demonstrated that its emission patterns rotate like the beam of a lighthouse. This contrasts with the rotation of TKR.

The arcs stimulated many interpretive and theoretical discussions on the nature of the DAM source. All interpretations are based on the assumption that the DAM source lies near the planet's surface. In one family of interpretations it is related to the Io 'flux tube', which is the term for the instantaneous magnetic field lines that arch outward and ultimately thread through Jupiter's innermost Galilean satellite. Some data, however, suggested broadly that DAM originates from many other points (always close to the planet's ionosphere) rather than just the small set of points associated with the Io flux tube. Nevertheless, if the flux tube source is provisionally accepted, the arcs in DAM could be explained by emission into a broadly opened 'cuff' centered on the putative magnetic field direction within the source. In this description, arcs correspond to a cuff opening angle that is frequency dependent.

All researchers have assumed that DAM occurs close to the electron cyclotron frequency in the source. The Voyagers detected no DAM above 39.5 MHz frequency, its upper bound in ground-based data. The significance of this frequency is that it is very close to the maximum cyclotron frequency (14.4 gauss × 2.80 MHz per gauss = 40.3 MHz) predicted, by extrapolation of Pioneer 11 MAG data for the surface of Jupiter in the northern hemisphere. The DAM polarization tends to confirm the role of the electron cyclotron frequency. Emissions seen before and after longitude 200° are right-hand polarized, the extraordinary mode for decametric emissions outward along the inferred magnetic field in Jupiter's northern polar regions. That this may be only a curious coincidence is suggested, however, by the fact that DAM from the opposite hemisphere, seen 180° later in longitude, is not strongly left-handed nor does it even approach the maximum frequency (10.0 gauss, or 28.0 MHz) predicted for the surface of Jupiter in its southern hemisphere.

Voyager 1 and 2 detected complex patterns of hectometric emission whose interpretations appear to involve wave–particle interactions in regions of space far above Jupiter's DAM sources. These interpretations require ray paths consistent with inferred plasma distributions throughout the inner magnetosphere of the planet. In particular, the plasma torus surrounding the orbit of Io contains electron densities of up to several thousand per cubic centimeter which refractively modify emission patterns from sources closer to Jupiter. The plasma torus itself was discovered first by the Voyager ultraviolet spectrometer (UVS) and immediately thereafter by *in situ* measurements of plasma density directly by the Voyager plasma experiment (PLS) and indirectly by PRA, which detected the *in situ* plasma resonances associated with the upper hybrid frequency. At kilometric wavelengths (30 to 300 kHz) the Voyagers observed radio sources outside the region of the plasma torus. The rotation of these sources was slightly slower than that of DAM and was interpreted as a lagging of magnetospheric rotation beginning at radial distances near nine or ten Jupiter radii.

PRA measurements in the PHIEX mode confirmed ground-based data about millisecond-duration bursts in DAM. Despite a rich data base, the high-rate observations gave no indication of the presence of lightning on Jupiter. This result is somewhat surprising because PWS detected whistlers and the television system detected lights on Jupiter's nightside that were interpreted as very intense lightning storms.

Saturn

Saturn emissions in the kilometric range (SKR) may have been detected from an Earth-orbiting satellite (Radio Astronomy Explorer) as early as 1972. The first unambiguous detection of SKR was by the Voyagers in December 1979, only three months after the first Saturn flyby, accomplished by Pioneer 11 early in September 1979. Within one month, the PRA data established the rotation period of Saturn's magnetic field, 10.657 h.

The magnetic field of each of the giant planets rotates more rapidly than does Earth's. The radio emissions from these planets provide accurate measurements of the rotation periods. The explanation may be as follows.

For some planetary emissions, emission is highly directive; it is polarized; and its frequency is closely related to the electron cyclotron frequency in the emission sources. For these planets, we imagine that the planetary magnetic field is a scaffold upon which hang the various radio sources, whose energy propagates into well-defined beams. As the scaffold rotates, the beam pattern rotates in space. Consecutive passages of this beam across the spacecraft permit the measurement of the rotation period with respect to the spacecraft. This is a useful model for Jupiter's decametric emissions. DAM is not obviously correlated with solar activity.

For other planetary emissions, for example, kilometric radiation from Earth and Saturn, the emission is broadly directed and is intrinsically time dependent. The emission occurs when a special structure (for example, the meridian plane containing the dipole) crosses the solar noon meridian on the planet. Detection of successive emission events measures the time between successive presentations of the dipole to the Sun. For both Earth and Saturn, the kilometric radiation correlates with solar activity.

In the Voyager missions these different types of rotation have been observed for Uranus and Neptune. They are distinguishable because the emissions fixed in angle relative to the solar noon meridian on the planet remain spaced regularly in time as the spacecraft orbits from planet dayside to planet nightside. Emissions that are narrowly beamed and fixed in a rotating system, such as those defined by the magnetic field asymmetries, are not strictly periodic as seen from the spacecraft. They jump in phase as the spacecraft orbits the planet.

SKR has a very strong rotational pattern which indicates that its sources are strongly asymmetric. However, all three MAG experiments flown past Saturn have determined that the fields even near Saturn's surface are rotationally symmetric about an axis congruent to the planetary axis of rotation. It should therefore have been impossible to measure the rotation of the planetary magnetic field by radio emissions or in MAG data. To explain this paradox we require that there be small-scale complexity in the magnetic fields of the SKR sources. This complexity neither exists in nor is implied by the MAG data.

Saturn produced impulsive radio emissions throughout almost the entire spectral range of PRA, at a rate of about one per 5 s (on

Voyager 1; they were about one order of magnitude less abundant on Voyager 2) with a total of about 18 000 seen by Voyager 1. Their duration was typically several tens of milliseconds and their spectrum was very broad.

These events were called 'Saturn electrostatic discharges' ('SED') rather than 'Saturn lightning'. It is true that in many respects they resemble VHF emission from terrestrial lightning. Nevertheless, 'SED' seems more appropriate for reasons to be explained.

SED occurred in episodes lasting a few hours and recurred with a period of about 10 h 10 min (data from Voyager 1 only). They rotated with Saturn, but in the Jupiter (lighthouse) mode rather than the SKR mode. Their total energy was many orders of magnitude (10^6) greater than that of terrestrial lightning flashes and their spectrum, although very broad, appeared to peak in the range 10 to 20 MHz, with a strong drift toward lower frequencies in episodes during and after Voyager 1 closest approach.

Two very different physical sources have been suggested for SED. The earliest was that SED were produced in Saturn's rings. This explanation implies that the source orbits Saturn at about 1.8! Saturn radii, corresponding to 10 h 10 min episodic period, in the optically deepest part of the B ring. This source location is exotic but might be more acceptable if one remembers that other exotic phenomena ('spokes'), widely believed to be formed by electrostatic levitation of micron-sized dust, occur in very nearly the same region of the rings. The second explanation is that SED manifest Saturn lightning, produced in the planet's atmosphere in a longitudinally limited, narrow strip at very low latitude (within a few degrees of the planetary equator, where the clouds are known to rotate at the episodic period) containing a convective cloud complex. This rather ordinary location might seem less realistic if one cannot answer why Saturn's convective clouds are restricted to a narrow region near its equator, why the total power in the discharges is so large (at least 10^{15} W), and why the putative length of the discharges is so short (a few meters). Furthermore, no optical analog for the SED source was seen in the atmosphere at the time of the Voyager encounters.

Whatever may be the final explanation of SED, it was an extraordinary phenomenon, revealing Saturn to be an electrostatic planet on an unprecedented scale. This is an outstanding result of the Voyager missions.

PRA and PWS first detected the effect of particle impact on the spacecraft when they passed through the equatorial plane of Saturn. The combined orbital velocity of a typical particle and the velocity of the spacecraft produced a relative velocity of impact above 15 km s^{-1} in magnitude, far above the speed of sound in the materials of which the various surfaces of the spacecraft were constructed. Every single micron-sized dust particle thereby excavated a tiny crater in the surface of the spacecraft. The crater's material volatilized, ionized and expanded to engulf the spacecraft, the two PRA antenna booms in particular. PRA's detection was in the POLLO mode at Saturn and showed that the impact signature extended as high as one MHz. Neither PWS nor PRA had high time resolution ring-plane data from Voyager. PRA also had none from Voyager. PWS's high-rate, wideband (10 Hz to 12 kHz) data during Voyager 2's Saturn ring-plane crossing showed hundreds of such impacts per second, continuing for the several minutes required by Voyager to cross the ring plane. For each impact the electrical signature began rather abruptly and lasted for several milliseconds.

Uranus

Only Voyager 2 flew on to Uranus (and later Neptune) after its Saturn encounter. Radio emission from Uranus was not detected until five days before closest approach and then only in the frequency range below about 100 kHz. Uranus' positive rotation pole pointed nearly at Sun at the time of Voyager 2 encounter; the nightside was thus dark for a period of many years even though the planetary rotation period was 17.24 h (as measured from the radio emissions). The rotation was Jupiter-like. As Voyager came abreast of Uranus and passed over onto its nightside, emission suddenly increased in flux density by two orders of magnitude, and expanded upwards to occupy the spectrum to about 1100 kHz.

The dynamic spectrum of Uranus thus differed dramatically from dayside to nightside, although emissions from both sides were locked to the planetary rotation. On the dayside the emission showed a single peak, but for the nightside it had a double peak, separated by a longitude range containing little emission. Uranus also generated very strong emission in dramatic broadband bursts seen typically twice per rotation, when the planetary dipole axis presented itself 'sidewise' to the spacecraft.

Uranus appears to generate electrostatic discharges, but hundreds of times less abundantly than Saturn.

The Uranian magnetic field, as determined by MAG, does not even approximate that of a dipole either over the surface of the planet or in surrounding regions of space where radio emissions are typically generated. Furthermore, MAG data, extrapolated to the surface of the planet, suggested that the Sun-side rotational pole was a region of strong magnetic anomaly. Ten hours before closest approach, PRA detected weak emissions that had not been seen in the previous days of approach, but in the opposite polarization state from the other approach emissions. These anomalous emissions occurred precisely when Voyager lay above the polar magnetic anomaly.

As Voyager left Uranus' magnetosphere, PRA detected a curvilinear broadband modulation of Uranus' kilometric radiation in the range below 400 kHz. This modulation was produced by large-scale waves on the surface of the magnetopause. From the point of view of Voyager, the magnetopause was behaving as a 'scintillation screen'. The situation is analogous to Earth's ionosphere, which causes radio sources at great distance to scintillate, that is, to fluctuate rapidly in brightness as they are observed from ground stations, below the ionosphere.

Neptune

Neptune's radio emissions were detected many days before closest approach. Lying in the range from 500 to 1100 kHz, they were very intense, narrowband and shortlived. They showed a planetary rotation of 16.11 h and ultimately were understood to be of the Sun-locked, or Saturnian, rotation type. Within a few days of closest approach a different kind of emission appeared: broadband, from below 100 kHz to as high as 850 kHz, and also strongly controlled by Neptune's rotation. The kilometric emission was linked to rotation in the same manner as Jupiter's emission, thus opposite to the Sun-locked bursts seen earlier. Within an hour Voyager's passage (at 5000 km height, very close to the north pole of Neptune), additional emissions were seen which switched in polarization in a manner that suggested occultation of a lower-latitude emission source by local strong anomalies in Neptune's magnetic field.

MAG determined that Neptune's magnetic field was comparable in both strength and complexity to that of Uranus. The PRA observation of polarization changes in emission observed near the north pole connects with MAG's observation of an unexpected 2 : 1 enhancement of field in the space traversed just 1 h before Voyager 2's closest approach. This enhancement is computed relative to a best-fit dipole field for the overall planet.

At Neptune, for the first time in the Voyager missions, PHIEX (at 300, 600 and 900 kHz) was employed during dust impacts. The impact signature at these high frequencies unexpectedly persisted as long as at low frequencies (in the range 10 Hz to 12 kHz, in the observations obtained by PWS). This duration now shows that throughout the plasma cloud expanding through the antenna system there occur localized high-frequency broadband oscillations in electrical charge density (Figure V46, schematically indicated by pluses and minuses throughout the impact plasma cloud). As a consequence it is difficult to interpret either PRA or PWS impact data directly in terms of charge collection by the PRA booms.

Voyager interstellar missions

For PRA, Voyagers' last objective lies in the future. It is the detection of the boundary between the plasma of planetary space and that of interstellar space. A reliable estimate of the times of arrival of the spacecraft at this surface is difficult or impossible, inasmuch as it depends on the plasma conditions on the galactic side of the boundary as well as those on the planetary side. The spacecraft have perhaps 25 more years of life (as of 1991). Since they are now about 40 AU from the Sun and move at about 3 AU each year, this suggests that detection will be achieved if the heliopause lies within 115 AU of the Sun.

Two modes of detection seem possible. One is the detection of radio emissions from electron plasma oscillations near the heliopause. From 1983 to 1986 PWS may have detected these in the waveform mode at frequencies of 2 to 3 kHz. There has been no subsequent detection despite diligent observations. If the plasma density in the heliopause 'normally' is 2 electrons cm^{-3}, the associated emissions (at the plasma frequency) might lie above the detection range of PWS, and below that of the 199th PRA channel (20.4 kHz). This density more nearly corresponds to early estimates of interstellar electron density, 1 electron cm^{-3}, than the value implied by the heliopause detection at 3 kHz, that is, 0.1 electron cm^{-3}.

A second way of detecting the shock might be through the continued observation of solar radio emissions, which are still robust even at 40 AU and will remain so during all of the Voyager interstellar mission (VIM). The plan is to use the solar signals as a probe, like that provided by the Uranus kilometric emission with which PRA observed waves on the magnetopause of Uranus.

In VIM, the PRA receiver operates only in FIXLO. The FIXLO signals are summed in separate bins for RH and LH samples. In each polarization 99 30-ms samples of signals at a given frequency are gathered by the FDS and sent to the telemetry stream. The resulting PRA bit rate is approximately the normal PRA rate, 266⅔ bps, divided by the number of samples compressed into one pair, i.e. 198. Although the bit rate of PRA decreases to 1.3 bps paradoxically the sensitivity of the data increases by about 10 dB, to its highest-ever value. The price we pay for this advantage is an extremely long scan time, 384 s to cover just 64 frequency steps. The timing of data in the lowest-frequency channel, 1.2 kHz, is arranged so as to avoid periodic spikes of spacecraft EMI; system sensitivity improves even further.

James W. Warwick

Bibliography

Lang, G.J. and Peltzer, R.G. (1977) Planetary radio astronomy receiver, *IEEE Trans.*, **13**, 466.

Warwick, J.W., Pearce, J.B., Peltzer, R.G. and Riddle, A.C. (1977) Planetary radio astronomy experiment for Voyager missions. *Space Sci. Rev.*, **21**, 309–27.

The PRA Team:
J. Geophys. Res., (1981) (Jupiter Issue), **86** (A10).
J. Geophys. Res., (1983) (Saturn Issue), **88** (All).
J. Geophys. Res., (1987) (Uranus Issue), **92** (A13).
J. Geophys. Res., (1991) (Neptune Issue), **96**, 18903–9268.
Science, (1979) (Voyager 1 Jupiter Issue), **204**, 995.
Science, (1979) (Voyager 2 Jupiter Issue), **206**, 991.
Science, (1981) (Voyager 1 Saturn Issue), **212**, 239.
Science, (1982) (Voyager 2 Saturn Issue), **215**, 582.
Science, (1986) (Voyager Uranus Issue), **233**, 102.
Science, (1989) (Voyager Neptune Issue), **246**, 1498.

Cross references

Magnetospheres of the outer planets
Radio astronomy
Solar wind

W

WATER

On planet Earth the substance water, in its three physical states, solid (ice and snow), liquid (water and as droplets in clouds) and gaseous (water vapor, steam), is probably the most critical component in the planet's geochemistry, although the surficial water inventory only makes up 0.02% of the planet's mass (Plates 9, 10). Inasmuch as its liquid phase is one of the essentials for self-reproducing life, it is a substance of supreme interest to humankind, for survival, for the environment and for the economy (Kuenen, 1955).

Water in one form or another appears to be present in most of the planets and their satellites except for Mercury and the Moon, but on none does it play such a significant role as it does on Earth. Its unique place in terrestrial biogeochemistry is directly related to the Earth's distance of approximately 1.5×10^8 km from the Sun. A little closer, and a Venusian hothouse would result (no liquid phase), and a little farther away, a Martian cold world (ice house phase) would exist. Over approximately 4 billion years, planet Earth has enjoyed a global mean temperature in the range of $20 \pm 5°C$, even during the brief negative departures of the various ice ages (q.v.). Evidence for this generalization is based on (1) sedimentary products of running water (former sands, gravels); (2) biological evolutionary continuity and varied products such as fossil bacteria and algal mats (stromatolites which have present-day equivalents); and (3) geochemical equilibria (e.g. crystal forms of aragonite versus calcite; isotopic ratios).

This uniformitarian deduction leads to the assumption that the liquid-water phase has been continuously dominant on the Earth's surface since about 4 billion years ago, prior to which geological evidence is lacking. Traces of repeated ice ages during Phanerozoic time (the last 560 Ma) have never been so widespread that glaciers reached the equator and, in spite of evidence of periodic biological crises, no major phyla have ever become extinct. Even in Precambrian times, evolution of water-demanding organisms persisted without a break, although there is good evidence of an important shift in obliquity (see Obliquity: terrestrial record).

From the planetary science viewpoint the above statements would imply that the Earth/Moon orbital position must have remained very close to the present for at least 4 billion years. Although solar physics suggests that the Sun's luminosity was possibly as much as 30% less than today in early Precambrian time, the terrestrial insolation may have been buffered by a far higher CO_2 content and greenhouse effect in the atmosphere, so that surface temperatures have not materially increased or decreased, yielding important cosmological constraints.

The substance water, in its pure form as 'distilled water' is colorless, odorless and tasteless. This oxide of hydrogen, H_2O, freezes at 0°C and boils at 100°C (at standard atmospheric pressure; at lower pressure, as on mountain tops, the boiling point is lower). It has a high specific heat, is a poor conductor of electricity and is very slightly compressible. Its maximum density is reached at 4°C, so that in natural environments as on lakes, ice forms at the surface while the bottom water often remains warm.

As a natural solvent on planet Earth water is unequaled. It has a weak tendency to become ionized as H^+ and OH^- and these ions then form complexes with solute species. CO_2 is readily soluble in water creating a weak acid, which is generated during rainfall initiating a weathering process in common minerals of the rocks at the Earth's surface, e.g. $CaCO_3$ in limestones, or feldspars in granite or basalt. In contrast, in semiarid regions soils and ephemeral lake waters become enriched by alkaline products, notably the carbonates and sulfates of calcium and sodium. In these high-pH environments, silica (SiO_2, the principal component of quartz sand, sandstone and granite) hydrates to form a weak acid H_4SiO_4, represented by H^+ ions plus anionic species such as $H_3SiO_4^-$ and $H_2SiO_4^{2-}$ (or $O_4^=$). In this way the regional climates, through water's solvent behavior, are responsible for a large amount of geochemical rearrangement at the Earth's surface. Thanks to stratigraphic burial and the geologists' ability to date ancient deposits, a paleoclimatic record is thus established which contains built-in environmental data spanning almost the whole of geologic time.

Water in planet Earth is distributed in the successive spheres: atmosphere, hydrosphere and lithosphere. In the atmosphere its mass is 1.3×10^{13} metric tons (or 1.3×10^{19} g); in the hydrosphere 1.4×10^{18} metric tons; and in the crust, or outer lithosphere, 2.2 to 2.6×10^{18} metric tons.

In the atmosphere, spanning the Earth's surface area of 510×10^6 km², it is largely in the vapor phase except during intense convective lifting (such as tropical cumulus build-up when ice crystals form) and during rainfall when the liquid phase is created. Most of the water vapor is concentrated in the troposphere with very little in the stratosphere; 90% is in the lowest 5 km.

Water in the atmosphere plays a key role in the modulation of the Earth's climate on a day-to-day basis. The principal dynamic focus is the equatorial/intertropical zone where under solar radiation water is evaporated from the sea surface and humid forest lands to be convected up in pulses up to the lower stratosphere, and then redistributed polewards (a Hadley circulation, q.v.) Climatic trends of monthly to decadal significance are constrained, not by the atmosphere (which has a 'short memory'), but by the hydrosphere, where the ocean plays the role of 'the giant flywheel of the atmospheric heat engine'.

The distribution of water in the hydrosphere is 96.5%, in the oceans (1.34×10^{24} g or 10^{18} metric tons; a volume of 1.338×10^9 km³ or 1.34×10^{21} L); ~ 24×10^6 km³, in glacial ice; ~ 360 000 km³ in lakes and rivers; and 23×10^6 km³, in ground water and sedimentary rocks. During the last 2 million years the successive glacial/interglacial cycles (about 100 000 years) have left a eustatic signal of sea-level fluctuation, indicating expansion and contraction of the ocean by about 3% (by volume) with a corresponding vertical rise and fall of 100–135 m. If all the world's present glacier ice melted, sea level would rise 66 m.

Fresh water at the Earth's surface is thus severely restricted (only about 2.52% of the total water). The main difference of the rest is that the ocean is salty, on average about 3.5% or 35‰, with a usual range of 33–38‰ (Fegley, 1993). The salts consist largely of the chlorides and sulfates of sodium, magnesium and calcium, but including up to about 100 minor constituents. The corresponding mass of the dissolved solids is 4.9×10^{22} g (10^{16} metric tons). Because of this salinity, sea water is a good electrolyte (in contrast to fresh water) and conducts electricity at about 4 Ω m^{-1}; its conductivity increases with temperature.

Biologically important components of sea water include dissolved CO_2 and H_2CO_3 + Ca^{2+} (calcium bicarbonate) which provide respectively the basis for photosynthesis and the fixation of carbonate skeletal materials. Marine populations tend to be constrained by the supply of silicon, phosphorus and vanadium (each averaging ~3 mg L^{-1}). Dissolved oxygen has slowly increased through geologic time ($> 10^9$ years), while CO_2 has progressively decreased, much of it going into long-term storage as $CaCO_3$ (limestone) or incorporated into the minerals of igneous rocks (during subduction and seafloor spreading).

Through geologic time salts have been constantly added to the ocean through weathering on land and transport via the hydrologic cycle. Through evaporation and biologic extraction a long-term balance is maintained, however, and it seems likely that ocean salinity has been maintained approximately at the present level at least for the last billion years.

Average sea water has a density of 1.025 g cm^{-2}, and in an estuary the river's fresh water tends to 'float' above the sea water wedge below it. Hydrostatic pressure rises with depth at a rate of about 1 atmosphere per 10.1 m (= 1 dbar m^{-1} or 1×10^6 N m^2 per 100 m). At 5000 m (mean depth for 27% of the Pacific Ocean) the pressure is ~49×10^6 N m^{-2}.

There is also water in the Earth's lithosphere, in the soil, in porous sedimentary formations and chemically combined in various minerals. The total water reserves of the three outer spheres of planet Earth amount to about 1386 million km^3, of which 3.4 million km^3 are classified as ground water.

Other water present in the lithosphere is sometimes called 'pellicular', being chemically bound in minerals such as gypsum ($CaSO_4 \cdot 2H_2O$) which under slightly elevated temperature and/or pressure dehydrate to anhydrite ($CaSO_4$), liberating this 'bound water'. The reaction is reversible, so that buried anhydrite when exposed to weathering hydrates to gypsum. Hydration–dehydration reactions are common among the evaporite minerals (halides), the clays and micas (Fairbridge, 1983). Hot aqueous fluids triggered by igneous activity often reach the surface in volcanic regions, creating hot springs and geysers which liberate large quantities of steam to the atmosphere. On average, the rocks of the lithosphere have been estimated to contain 3.56% combined water, which would suggest that the whole lithosphere contained some 842 million km^3 of water (USSR Committee, 1978).

Besides the atmosphere, hydrosphere and lithosphere, there are two special types of 'sphere' that are superimposed on all three of the above. First, biosphere (q.v.) is the overall habitat of organized life. Second, the cryosphere is the boundary between average freezing and thawing, and is also one of the critical constraints for most inhabitants of the biosphere. The cryosphere boundary is defined in its relation to mean sea level; poleward of about 70° latitude it is generally below mean sea level (MSL) and the ocean is regularly frozen over (shifting seasonally), while near the equator it is at around 6000 m elevation. During former glacial periods (e.g. about 20 000 years ago) the cryosphere boundary fell dramatically; in the northern hemisphere it dropped below MSL at about 45°N and in mountainous regions of low latitudes it was about 1000 m lower (in exceptional spots like New Guinea near 5°S, glaciers almost reached present MSL). On planet Mars the cryosphere boundary undergoes a dramatic seasonal (polar) reversal and, with its major orbital variations, also long-term variations.

Water in the solar system

Evidence of water from outside of planet Earth is not lacking. Ordinary chondritic meteorites carry 10–1000 ppm water, and carbonaceous chondrites rather less. If these indicators are taken as representing the material of the proto-Earth planet, starting with a mass of 5.98×10^{27} g, if its water content were 0.5%, we would have 2.99×10^{25} g, or approximately 30 billion km^3. If today there is 1400×10^{21} g in the hydrosphere and 600×10^{22} g in the crust, somewhat less than 2.8×10^{25} g would remain in the mantle. Through geologic time, some water is lost to outer space by photodissociation of H_2O vapor in the outer atmosphere.

The other planets display (in varied ways) some evidence of water, spectroscopically or through its physiographic behavior. On the Moon or Mercury, water could only exist in the mineral-bound form, but none has ever been found. On Venus it is identified only as H_2O and as $H_2SO_4 \cdot xH_2O$ in cloud droplets. From the D to H ratio, Donahue and Hodges (1992) believe that Venus was formerly richer in H_2O.

On Mars there is geomorphological evidence of liquid, running water that at one time (or periodically) had been concentrated in sufficient amounts to erode dendritic drainage patterns and carve out giant canyons. Because of long-term orbital changes, Mars is probably subject to long-term climate cycles which mobilize water that is at present stored in permafrost form beneath the surface regolith (see Mars).

In the outer planets the atmospheres consist predominantly of hydrogen and helium. Water is only indicated spectroscopically on Jupiter; the others have not yet yielded any evidence. However, H_2O is present on some of the outer planets' satellites. Water was also identified on Halley's comet and the OH ion is found on some asteroids (e.g. Ceres).

Rhodes W. Fairbridge

Bibliography

Anders, E. and Grevesse, N. (1989) Abundances of the elements: meteoritic and solar. *Geochim. Cosmochin. Acta*, **53**, 197–214.

Back, W. and Freeze, R.A. (eds) (1983) *Chemical Hydrogeology* (Benchmark Papers in Geology, Vol. 73). Stroudsburg: Hutchinson Ross Publ. Co., 416 pp.

Donahue, T.M. and Hodges, R.R., Jr (1992) Past and present water budget of Venus. *J. Geophys. Res.*, **97**, 6083–91.

Fairbridge, R.W. (1967) Carbonate rocks and paleoclimatology in the biogeochemical history of the Earth, in *Carbonate Rocks* (eds G.V. Chilinger et al.). Amsterdam: Elsevier, pp. 399–432.

Fairbridge, R.W. (ed.) (1972) *The Encyclopedia of Geochemistry and Environmental Sciences*. New York: Van Nostrand Reinhold Co., 1321 pp.

Fairbridge, R.W. (1983) Syndiagenesis–anadiagenesis–epidiagenesis: phases in lithogenesis, in *Diagenesis in Sediments and Sedimentary Rocks*, Vol. 2 (eds G. Larsen and G.V. Chilinger). Amsterdam; Elsevier, pp. 17–113.

Fegley, B., Jr (1993) Properties of composition of terrestrial oceans and of the atmospheres of the Earth and other planets, in *AGU Handbook of Physical Constants*. Washington, DC: American Geophysical Union.

Holland, H.D. (1972) The geologic history of sea water – an attempt to solve the problem. *Geochim. Cosmochim. Acta*, **36**, 637–51.

Holland, H.D. (1984) *The Chemical Evolution of the Atmosphere and Oceans*. Princeton: Princeton University Press, 582 pp.

Kuenen, P.H. (1955) *Realms of Water*, New York: John Wiley and Sons, 327 pp.

Pepin, R.O. (1991) On the origin and early evolution of terrestrial planet atmospheres and meteoritic volatiles. *Icarus*, **92**, 2–79.

Short, N.M. (1975) *Planetary Geology*. Englewood Cliffs: Prentice-Hall, 361 pp.

Turekian, K.K. (1969) The oceans, streams, and atmosphere, in *Handbook of Geochemistry* Vol. 1 (ed. K. Wedepohl) 297–323.

USSR Committee for IHD (1978) *World Water Balance and Water Resources of the Earth*. Paris: UNESCO, 663 pp.

Wyllie, P.J. (1971) Role of water in magma generation and initiation of diapiric uprise in the mantle. *J. Geophys. Res.*, **76**, 1328–38. [Reprinted in Garfunkel, Z. (1985) *Mantle Flow and Plate Theory*. New York: Van Nostrand Reinhold, pp. 130–40].

Cross references

Callisto
Enceladus
Europa
Ganymede
Ice age

Life: origin
Mars: atmosphere
Polar cap

WEATHERING

Any geological, destructive process that tends to weaken or disintegrate the solid materials of a planetary surface (see encyclopedia entries in Fairbridge, 1968; Yatsu, 1988). It is a strongly climate-dependent process (Büdel, 1981). While first recognized on planet Earth, the process is also to be expected on all hard planetary surfaces at varies levels of intensity depending on the local conditions. Those variables include such factors as the Sun's radiation, atmospheric chemistry, and wind velocities.

There are two fundamental categories of weathering, (1) physical and (2) chemical weathering.

Physical weathering

Physical weathering is mainly thermal or abrasive. Thermal action results in brittle fracture and is usual diurnal or seasonal in its attack. It involves rock fracture and is thus called thermoclastic. It is well displayed on planet Earth in desert regions, both hot ones such as the Sahara or cold ones such as the 'dry valleys' of East Antarctica. Heating of an outer layer of rock is achieved under solar radiation, causing a scale-like surface layer to expand ('exfoliation'); abrupt cooling at night causes the scale to crack and split ('thermoclastic weathering'). In the Sahara some areas covering many hundreds of square kilometers are covered by sharp rock fragments ('thermoclasts') measuring about 5–15 cm across. The optimum size is physically determined by the lineal expansion coefficients of brittle solids.

On the Earth, in high latitudes and in high mountains, a wide variety of weathering phenomena are observed in 'geocryology'. A standard work on the subject (Washburn, 1980) cites no less than 2200 references to its various manifestations. The freezing process is known as cryofracture (from the Greek), or gelifracture or congelifraction (from the French, *gel* for frost). Anglo-Saxon terms for this process are 'frost riving' or 'frost splitting'.

Abrasion (or corrosion) is in contrast to abrupt fracturing, a gradual wearing down, scratching, scraping or scouring induced by friction which creates microfractures, comparable to sandpaper on a plank of wood. The energy for abrasion is provided by wind (eolian abrasion), moving water (fluvial or wave action) or moving ice (glacial action). Each requires the 'armament' of an abrasive agent such as boulders, sand grains or dust. The finest particles are easily dissolved, and can be detected by changes in glacial stream chemistry as an 'abrasion pH'.

On Mars (q.v.) wind action and eolian abrasion are widespread. Both thermoclastic and cryoclastic weathering are to be expected. Dust and larger particles, the armament for eolian abrasion, are abundant on the Martian surface and are presumptive evidence for chemical weathering (see below).

Chemical weathering

Chemical weathering is carried out in the presence of water or moist air, principally in the form of rain or ground water, by a number of chemical reactions. These include hydrolysis, hydration, oxidation, carbonation, ion exchange and solution (Barshad, 1972). Most are favored by humid climates and rising temperature (except for carbonate solution). For the various minerals involved a 'weathering index' has been constructed.

In contrast, the hot deserts have almost negligible ground water and chemical weathering is mostly limited to dew. Due to diurnal accumulation and reprecipitation, dew is a major cause of the 'desert varnish' or patina that forms as a thin veneer on desert thermoclasts; their color is dark red to black due to the concentration of iron and manganese oxides derived from dust.

In modern, hot deserts, apart from areas of sand dunes there are pervasive red colors in the local rocks and soils. These are iron oxides generated by seasonal rains (generally monsoonal), or may be inherited from a more humid period (Fairbridge, 1967).

In cold deserts in polar and subpolar regions chemical weathering is normally inhibited, but where daytime temperatures rise above freezing, it is for 24-h days when snow and permafrost melt. The meltwater, usually acidified by tundra vegetation, provides a solvent for most minerals except quartz (SiO_2); as a result, tundra soils are usually podsols (leached soils, predominantly quartz), with a strong organic component.

In certain places water dissolves caverns in the ice of the upper permafrost beneath a carpet of soil and peat, and are known as 'thermokarst'. Such caverns were lethal traps for late glacial-age mammoths. On planet Mars the strongly seasonal climate, with long-term cycles, is conducive to widespread thermokarst.

Chemical weathering also takes place in the marine realm as 'submarine weathering' or 'halmyrolysis' (Fairbridge, 1983, p. 50).

On Earth a special weathering form, unique to this planet, is recognized: biological weathering, which may modify each of the two principal categories. The physical geological process may be by the splitting of rocks and boulders by the slow expansion of a growing root or by the action of animals. The chemical biological (or biochemical) process is more pervasive, as at the microbial level or at the root hairs of a growing plant, in both cases due to the production of CO_2 and its weak acid H_2CO_3, which has a slow but highly destructive effect on many rocks, notably limestones ($CaCO_3$), locally dissolving channels and caverns (known as 'karst').

Rhodes W. Fairbridge

Bibliography

Barshad, I. (1972) Weathering – chemical. In *The Encyclopedia of Geochemistry and Environmental Sciences* (ed. R.W. Fairbridge). New York: Van Nostrand Reinhold, pp. 1264–9.
Birkeland, P. (1974) *Pedology, Weathering and Geomorphological Research*. New York, London: Oxford University Press, 285 pp.
Blackwelder, E.B. (1933) The insolation hypothesis of rock weathering: *Am. J. Sci.*, **226**, 97–113.
Büdel, J. (1981) *Klima-Geomorphologie*, 2nd edn. Berlin, Stuttgart: Gebr. Borntraeger, 304 pp.
Carroll, D. (1962) Rainwater as a chemical agent of geological processes – a review. *US Geological Survey*, Water-Supply Paper 1535-G, 18 pp.
Carroll, D. (1970) Rock weathering. New York: Plenum Press, 203 pp.
Cooke, R.U. and Smalley, I.J. (1968) Salt weathering in deserts: *Nature*, **220**, pp. 1226–7.
Fairbridge, R.W. (1968) *The Encyclopedia of Geomorphology*. New York: Van Nostrand Reinhold, 1295 pp.
Fairbridge, R.W. (1976) Effects of Holocene climate change on some tropical geomorphic processes. *Quaternary Res.*, **16**, 529–56.
Fairbridge, R.W. (1983) Syndiagenesis–anadiagenesis–epidiagenesis: phases in lithogenesis, in *Diagenesis in Sediments and Sedimentary Rocks*, Vol. 2 (eds G. Larsen and G.V. Chilanger). Amsterdam: Elsevier, pp. 17–113.
Fairbridge, R.W. and Finkl, C.W. Jr (eds) (1979) *The Encyclopedia of Soil Science*, Part 1. Stroudsburg: Dowden, Hutchinson & Ross, 646 pp.
Grant, W.H. (1969) Abrasion pH, an index of weathering. *Clays and Clay Minerals*, **17**, 151–5.
Loughnan, F.C. (1969) *Chemical Weathering of the Silicate Minerals*. Amsterdam: Elsevier, 154 pp.
Ollier, C.D. (1969) *Weathering*. Oliver and Boyd, Edinburgh, 304 pp.
Polynov, B.B. (1937) *The Cycle of Weathering*. New York: Nordmann Publ. Co. (transl. from Russian by A. Muir).
Walker, T.R. (1967) Formation of red beds in modern and ancient deserts. *Geol. Soc. Am. Bull.*, **78**, 353–68.
Washburn, A.L. (1980) *Geocryology: a Survey of Periglacial Processes and Environments*. New York: Halsted Press, 406 pp.
Yatsu, E. (1988). *The Nature of Weathering*. Tokyo: Sozo-Sha Publ., 624 p.

Cross references

Eolian transport
Regolith
Surface processes

WHISTLER

Whistlers are electromagnetic signals which are excited by atmospheric lightning and which propagate in the magnetized plasma of planetary ionospheres and magnetospheres at frequencies below both the electron cyclotron frequency f_c and the electron plasma frequency f_p. They were first detected on Earth possibly as early as 1894 (Preece, 1894) but certainly by Barkhausen in 1919. These were serendipitous detections made with early telephone and radio equipment which utilized long wires and audio frequency amplifiers used to eavesdrop on enemy communications during World War I. The peculiar frequency–time signature of whistlers (Fig. W1) is a rapidly decreasing tone which, when converted to an audible signal with an amplifier and speaker, sounds like a whistle.

When a lightning stroke occurs in the atmosphere, a short broadband burst of electromagnetic energy ranging from about 1 kHz to visible light frequencies is emitted. The lower frequency portion of this spectrum propagates in the Earth ionosphere waveguide and the higher-frequency portion (above a few megahertz) freely propagates away from the source through the ionosphere. A fraction of the very low-frequency radiation, however, can couple into the magnetized plasma in the ionosphere and magnetosphere. Once in the ionosphere, the waves propagate in what is known as the whistler mode (Stix, 1962). For the whistler mode the plasma medium is dispersive; that is, waves of different frequencies propagate at different speeds. Below a few kilohertz the lower frequency waves propagate slower and arrive at an observer after higher frequencies have arrived. The arrival time t as a function of frequency f is approximated by

$$t = D/f^{1/2} + t_0 \qquad (\text{W1})$$

where D is the dispersion constant and t_0 is the time of the lightning stroke (Eckersley, 1935). In addition, whistlers generally propagate along the magnetic field lines of the planet, often bouncing back and forth from one hemisphere to the other. Whistlers can propagate no more than about 19° from the field line. The amount of dispersion observed depends on the whistler's pathlength and the plasma density along the path such that longer pathlengths and higher densities increase the dispersion. For low frequencies, high densities and propagation parallel to the magnetic field Eckersley showed that the dispersion can be approximated by

$$D = \frac{1}{2c} \int \frac{f_p}{f_c^{1/2}} \, ds \qquad (\text{W2})$$

where c is the speed of light and the integral is along the path of the whistler. This characteristic has allowed whistlers at Earth and Jupiter to be used to determine the plasma density within limited regions of the respective magnetospheres. Sometimes whistlers are observed extending to higher frequencies for which there is a frequency of minimum delay and higher frequencies arrive later; these are called nose whistlers and their dispersion requires an extension to the Eckersley theory. The theory for whistlers was developed by Storey (1953) and is reviewed by Helliwell (1965).

The connection between whistlers and lightning on Earth is well established. The observations of these remarkable waves at other planets, then, can be used as reasonable evidence for extraterrestrial lightning. Figure W1 shows a frequency–time spectrogram of two whistlers observed by Voyager 1 near Jupiter. The characteristic decreasing frequency tones are very apparent in these two examples. The whistlers in Figure W1 and others like them were used by Gurnett and colleagues to confirm the presence of lightning in the Jovian atmosphere (which was also observed optically by time exposures of the nightside atmosphere). Whistlers were also observed by Voyager 2 near Neptune, again providing evidence for lightning. The high-frequency portion of the radio emissions of lightning called atmospherics (also called 'spherics') have been used to detect lightning at Venus, Saturn, Uranus and Neptune.

William S. Kurth

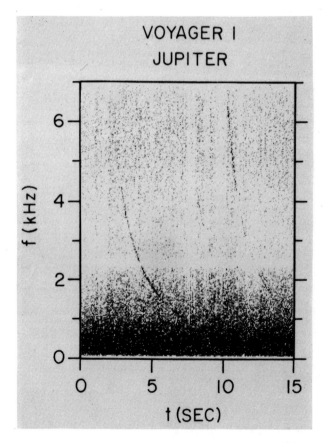

Figure W1 Two examples of whistlers detected near Jupiter by Voyager 1. Increasingly intense waves are represented by increasingly darker shading in the spectrogram.

Bibliography

Barkhausen, H. (1919) Zwei mit Hilfe der neuen Verstärker entdeckte Erscheinungen. *Physik. Z.*, **20**, 401–3.
Eckersley, T.L. (1935) Musical atmospherics. *Nature*, **135**, 104–5.
Gurnett, D.A., Shaw, R.R., Anderson, R.R. and Kurth, W.S. (1979) Whistlers observed by Voyager 1; detection of lightning on Jupiter. *Geophys. Res. Lett.*, **6**(6), 511–4.
Gurnett, D.A., Kurth, W.S., Cairns, I.H. and Granroth, L.J. (1990) Whistlers in Neptune's magnetosphere: evidence of atmospheric lightning. *J. Geophys. Res.*, **95**(A12), 20967–76.
Helliwell, R.A. (1965) *Whistlers and Related Ionospheric Phenomena*. Stanford: Stanford University Press.
Preece, W.H. (1894) Earth currents. *Nature*, **49**(1276), 554.
Stix, T.H. (1962) *The Theory of Plasma Waves*. New York: McGraw-Hill.
Storey, LRO (1953) An investigation of whistling atmospherics. *Phil. Trans. Roy. Soc. London*, A246, 113–41.

Acknowledgement

This research was supported by the National Aeronautics and Space Administration through Contracts 959193 and 958779 with the Jet Propulsion Laboratory.

Cross references

Ionosphere
Magnetospheres of the outer planets
Planetary lighting
Plasma wave
Radiation belt

WOLF, MAX (1863–1932)

A pioneer in astronomical photography, Max (Maximilian Franz Joseph Cornelius) Wolf first employed this new technique in 1891, using an exposure of several hours so that celestial objects appeared as arcs indicating any planet's movement relative to the stars in this period of time. In this way Wolf discovered the minor planet (asteroid) Svea in 1892. By the end of 1897 some 432 minor planets had been identified by various observers, and by the end of his career Wolf had discovered 582.

Fortunate in being the son of a wealthy physician, who built a private observatory for him, Wolf eventually became a professor of astronomy at the University of Heidelberg, Germany. He also turned his photographic skills to the exploration of galaxies, discovering more than 5000, for which he was honored by the gold medal of the Royal Astronomical Society.

Rhodes W. Fairbridge

Bibliography

Freiesleben, H.C. (1976) Wolf, Maximilian Franz Joseph Cornelius. *Dict. Sci. Biogr.*, Vol. 14, 481–2.
Schaifers, K. (1985) Max Wolf, 1863–1932, in *Semper apertus: Sechshundert Jahre Ruprecht-Karls-Universität Heidelberg 1386–1986*, Vol. 3 (ed. W. Doerr). Berlin: Springer-Verlag, pp. 97–113.
Wolf, M. (1893) Photographic observations of minor planets. *Astron. Astrophys.*, **12**, 779.

WOLF, RUDOLF (1816–1893), AND WOLF NUMBER

A Swiss astronomer notable for his long devotion to the sunspot problem, Johan Rudolf Wolf spent most of his life in Zürich, built the first astronomical observatory there and became professor of astronomy at the Eldgenossische Technische Hohschule (ETH) and director of the observatory. He was also a distinguished science historian. Going back to the earliest telescopic observations, by Galileo in 1610, he worked out the probable epochs of the solar cycle maxima and minima to establish that the mean period is 11.1 years for the years 1700–1848. (It was once known as the 'Schwabe cycle' after the pioneer who first identified the cycle in 1843.) One of Wolf's papers is available in translation (in Schove, 1983, pp. 83–9).

Wolf is best remembered for his development of a qualitative definition for sunspot number and its application to the existing historical data to see if Schwabe's cyclical variation could be found in longer spans of time. The Wolf sunspot number R is defined as

$$R = k(10\ g + f)$$

where f is the total number of spots regardless of size, g is the number of spot groups and k normalizes the counts from different observatories. It is evident that much greater weight is given to the groups than to individual sunspots. Daily means are averaged to monthly and then annual values. The long-term length of the cycle is now $11.1212\ (\pm 6)$ years and annual mean amplitude range is < 5 to > 200. Prior to the 'telescopic period' (i.e. since Galileo) the sunspot periods have been reconstructed by proxies, back to 649 BC by Schove (1955; reprint in 1983) with its spectrum determined by Jelbring (1995). The Wolf sunspot number has also been slightly refined subsequently. The 'Zurich relative sunspot number' was denoted by R_Z or lately as the 'international sunspot number'.

Wolf also observed that on planet Earth the geomagnetic cycle has the same period, a discovery that was made simultaneously and independently by Sir Edward Sabine (q.v.) in Britain and Alfred Gautier (1793–1881) in Switzerland.

When Wolf derived his values for the sunspot number, he did not have access to William Herschel's (1738–1822) notebooks. The period when Herschel was most active (1799–1806) was a time when few other observers recorded information about the solar surface. Those notebooks more than doubled the solar observations available to Wolf (Hoyt and Schatten, 1992).

Rhodes W. Fairbridge

Bibliography

Burckhardt, J.J. (1976) Wolf, Rudolf. *Dict. Sci. Biogr.*, Vol. 14, pp. 480–1.
Hoyt, D.V. and Schatten, K.H. (1992) New information on solar activity, 1779–1818, from Sir William Herschel's unpublished notebooks. *Astrophys. J.*, **384**, 361–84.
Jelbring, H. (1995) Analysis of sunspot phase variations – based on D. Justin Schove's proxy data. *J. Coastal Res.*, Special Issue **17**, 363–70.
Schove, D.J. (ed.) (1983) *Sunspot Cycles* (Benchmark Papers in Geology, Vol. 68). Stroudsburg: Hutchinson Ross Publ. Co., 392 pp.
Waldmeier, M. (1961) *The Sunspot Activity in the Years 1610–1960*. Zürich: Schulthess Co.
Wolf, R. (1858) Mitteilungen über die Sonnenflecken, no. 6, *Astron. Mitt.*, 127–43.

Y

YARKOVSKY EFFECT

The Yarkovsky effect is a force and a torque on a rotating body due to temperature differences on the body between the morning and evening hemispheres. The Yarkovsky effect was apparently discovered by one civil engineer Yarkovsky around 1900. The original reference has been lost, and the earliest known reference is that Ernst J. Öpik (1951, p. 194) 'read the pamphlet about 1909 and can refer to it only from memory.' Spacecraft engineers independently rediscovered the effect, as it explained the spontaneous despinning of spacecraft in Earth orbit, so it is only through the fortune of Öpik's memory that the effect retains Yarkovsky's name.

The Yarkovsky effect occurs when sunlight strikes a rotating body, warming its surface. The portion of the surface that has passed through the subsolar latitude will be warmer than the portion that has not yet reached it. Both sides receive equal solar irradiation, but emit blackbody radiation with a flux that is proportional to T^4 (ignoring other effects such as wavelength-dependent emissivity). Since one side is warmer than the other, it emits more radiation. This results in a net force on the body, perpendicular to the solar direction. It also results in a torque on the body, which slows the spin rate. For a body orbiting the Sun, the net result is either an increase or a decrease in the orbital energy (and semimajor axis), depending upon whether the spin is in the same or opposite direction as the orbit, respectively.

The sizes of objects for which the Yarkovsky effect is most important are in the range of centimeters to meters. For smaller bodies, Poynting–Robertson drag (q.v.) has a larger effect, and the temperature difference required to produce the Yarkovsky effect tends to be eliminated by conduction. For larger bodies the effect tends to be too small to be important. This size range of centimeters to meters is extremely important, however, as it includes the bodies which form meteors and meteorites, and also includes spacecraft. The principal effect on spacecraft is to lower their spin rates, requiring the expenditure of fuel to maintain the correct spin rate. Spacecraft design can reduce (or utilize) this effect, once recognized (e.g. Peterson, 1976). The effects of the Yarkovsky effect on the meteorite and meteor populations (and the populations of small bodies in general) have not been well examined, and are topics of current research. The effects on an individual body may be substantial, but it is not clear whether the effects on a population are important, as the Yarkovsky effect can both increase and decrease orbital energy, with (a priori) equal probability, and may leave the average properties of the population unchanged.

Michael C. Nolan

Bibliography

Öpik, E.J. (1951) Collision probabilities with the planets and the distribution of interplanetary matter. *Proc. Roy. Ir. Acad.*, **54A**, 165–9. Also as *Contributions from the Armagh Observatory*, **6**, 165–99.

Peterson, C. (1976) A source mechanism for meteorites controlled by the Yarkovsky effect. *Icarus*, **29**, 91–111.

YEAR

In the case of planet Earth, the year (a word of Scandinavian and Teutonic origin) is the interval of time that encompasses the four seasons (winter, spring, summer and autumn – or fall), i.e. the tropical or solar year (365.24219 days or 365 d 5 h 48 min 45 s), which is the time taken for the planet to travel from equinox to equinox or the duration between successive passages across the 'First Point of Aries'.

For human beings and most biological organisms, metabolism is intimately linked to the day/night (light/dark) rhythm of approximately 24 h. This 'circadian' rhythm (from Latin: *circa* + *dies*, about a day) is reflected by body temperature and chemistry, waking, sleeping and so on. Accordingly it is convenient to divide the year into an exact number of days. On this division is based the civil or calendar year of 365 days to which is added one day every fourth return, to give the 'leap year' of 366 days. This creates an artificial year of 365.25 d, which is a little too long for the tropical year by about 0.01 d, and which in turn is accommodated every four centuries by skipping a leap year.

For astronomical purposes an allowance must be made for the gyroscopic precession of the equinoxes, whereby the semimajor axis of the Earth's orbit drifts annually about 50 sec from west to east (see Precession and nutation), and therefore the year with reference to the 'fixed stars' is a little longer. This is the sidereal year (based on the Greek word for star) of 365.25636 days (365 d 6 h 9 min 10 s). The sidereal year is about 20 min longer than the tropical year. It is marked by the time of the heliacal rising of the star Sirius (just before dawn). This event, the rising of Sothis (the ancient Egyptian name for Sirius) came just about the time of the summer flood of the Nile, at the time of the summer solstice, and thus acquired a religious significance, and because it initiated the agricultural year. The ancient Egyptians used a 365-day civil year, as we do, and after 365 × 4 days (1460 d), the so-called Sothic cycle, they added one day, creating what we call a 'leap year'. This solar calendar appears to have been adopted in Egypt about 2773 BC, in preference to a lunar system.

The Egyptian system was adopted also by the Romans, with an edict by Julius Caesar in 45 BC. It thus became known as the 'Julian year'. Even so, the calendar got slightly out of phase with the seasons, and a correction was made on the authority of Pope Gregory XIII in 1582 AD, omitting 10 days and establishing the Gregorian year of 365.2422 days. Inasmuch as this is 0.0078 d shorter than the Julian year, an error of 3.14 d builds up every years, which is accommodated by establishing that those of every four centennial years has no leap year. This establishes the 'calendar year' which is 365.24 d (or 365 d 5 h 49 min 12 s).

A useful measure of annual time for long-term astronomical purposes is the *anomalistic year* (365.26 d or 365 d 6 h 13 min 53 s), which is the interval between one perihelion passage and the next. The latter drifts about 11 s each year in a retrograde (clockwise) motion.

In 1956 the International Commission on Weights and Measures (Paris) defined the second as 1/31 556 925.9747 of the tropical year at 12.00 h ephemeris time on 31 December 1899 (which was 'January 0' 1900 in astronomical usage). Subsequently, however, in 1967 the standard second' was redefined in terms of nuclear disintegration in the 'cesium clock' as 9 192 631 770 cycles of the unperturbed microwave transition between the two hyperfine levels of the ground state of ^{133}Cs. Added problems arise when applying Earth-bound time to the solar system (Moyer, 1981).

Among ancient primitive peoples, for agricultural purposes the Moon provides a more obvious timekeeper than the Sun and the lunar synodic month (or lunation) of 29.531 d is still widely employed in establishing the religious calendars, especially among Jews, Christians and Moslems. A lunar year would then be 12 synodic months or 354.372 d, some 10.87 d short of the tropical year. It became known in ancient Egypt that after 19 tropical years (6939.75 d) the solar seasons came almost into phase once more with the lunar (235 lunations). As worked out by the Egyptian astronomer Meton, a period of 6740 d (known as the Metonic Cycle) was established consisting of 125 × 30 d 'full' months and 110 × 29 d 'hollow' months.

An eclipse year is one revolution of the Earth relative to the same node of the lunar orbit, thus 346.62003 d. Nineteen eclipse years comprise 6585.78 d, thus almost the same as the period of the Saros (see Saros cycle).

The term 'year' is also applied to the orbital period (i.e. the sidereal period) for any planet with reference to the 'fixed' stars; thus we have a 'Venusian (or Cytherian) year', 'Martian year', 'Jovian year', etc. The expression 'sidereal period' is also used for a satellite in orbit around any planet.

The Earth completes 360° with reference to the fixed stars in a sidereal year; 360° − 50.26 arcsec in a tropical year; and 360° + 11.25 arcsec in an anomalistic year.

The light year, employed in stellar distances, is the distance travelled by light in one sidereal year, i.e. 9.4593 + 10^{12} km.

Rhodes W. Fairbridge

Bibliography

Moyer, T.D. (1981) Transformation from proper time on Earth to coordinate time in solar system barycentric space–time frame of reference. *Celest. Mech.*, **23**, 33–56.
Parker, R.A. (1950) *The Calendar of Ancient Egypt*. Chicago.
Rohan, C.A. *et al.* (1974) Calendar, in *Encyclopedia Britannica, Macropedia*, Vol. 3, 15th edn, Chicago, Encyclopedia Britannica Incorporated, p. 595.

Cross references

Orbital commensurability and resonance
Solar day and time

Z

ZEEMAN EFFECT

As early as 1845, Faraday was convinced of the reality of a tight relationship between optical phenomena and magnetism. In 1862 he performed an experiment to demonstrate the influence of the magnetic field on radiation which failed due to several technical difficulties. In 1896 these experiments were attempted again, this time with success, by Zeeman.

From theoretical considerations, Lorentz had predicted spectral line splitting into an odd number of lines. This is exactly what Zeeman observed in his first experiments on cadmium and zinc. However, he discovered an 'abnormal' Zeeman effect showing an even line splitting, as in the case of sodium.

This 'abnormal' line splitting remained unexplained for more than 25 years until the concept of the electronic spin was introduced by Pauli in 1925. The very existence of the spin was thus confirmed by this 'abnormal' Zeeman effect.

By introducing an additional term (related to the interaction of angular momentum with an external magnetic field) to the Hamiltonian describing the total energy of a given atom, it is possible to calculate the splitting of the different energy levels and then predict the spectral line separation to be expected as a function of the magnetic field strength. Due to the conservation of angular momentum, some of the lines produced show linear polarization, while some others show left-handed or right-handed circular polarizations.

These different properties of the Zeeman effect allows its detection in an astrophysical context, permitting remote evaluation of magnetic field strength.

Use of the Zeeman effect in solar physics

As early as 1892, Young and then Mitchell noticed that in sunspot spectra some of the iron lines, which are single in the solar spectrum, were double. The first photographic record of these line splittings was completed by Hale (1902). The proof that this was due to the Zeeman effect was given when the circular polarization was directly observed to be opposite in the two spectrally separated components (Hale, 1908). This opened the possibility of directly measuring magnetic fields in astrophysical objects.

Hale (1933) attempted to measure the much weaker quiet Sun magnetic field through Zeeman effect, but due to weakness of the Sun's field and to technical difficulties, Hale and his collaborators were unable to achieve a positive detection.

Twenty years later Babcock (1953) finally succeeded. He measured the Solar magnetic field in the range 1–20 gauss with a technique that was used on a daily basis. The method is as follows. At the focus of a spectrograph, an exit slit is placed on a wing of a Fraunhofer solar line selected for its sensitivity to the Zeeman effect. An analyzer of circular polarization will then produce a slight shift of the line position when changing from left-handed to right-handed position and, because of the sloping of the line wing, will induce a change in the recorded intensity.

For a magnetic field of 1 gauss, the line shift to be expected in the green region of the spectrum is only 8×10^{-6} nm even in the strongest cases. However due to the high signal recorded, the noise level corresponded to about 0.1 gauss and thus allowed excellent detection of the solar magnetic field.

Alfred Vidal-Madjar

Bibliography

Babcock, H.W. (1953) The solar magnetograph. *Astrophys. J.*, **118**, 387–96.
Hale, G.E. (1902) Solar research at the Yerkes Observatory. *Astrophys. J.*, **16**, 211–33.
Hale, G.E. (1908) On the probable existence of a magnetic field in sun-spots, *Astrophys. J.*, **28**, 315–43.
Hale, G.E. (1933) *Annual Report of the Director, Mount Wilson Observatory*. Carnegie Institution of Washington Yearbook, No. 32, pp. 143–69.

Cross references

Absorption, absorption spectrum
Sun

ZODIAC

The set of constellations that the Sun passes through in its annual journey around the ecliptic (q.v.). Traditionally there are twelve zodiacal constellations: Aries, Taurus, Gemini, Cancer, Leo, Virgo, Libra, Scorpius, Sagittarius, Capricornus, Aquarius and Pisces. However, the ecliptic also passes through the constellation Ophiuchus. The zodiacal constellations have little in common with the 'signs of the zodiac' beloved of astrologers.

J.B. Tatum

Bibliography

Green, R.M. (1985) *Spherical Astronomy*. New York: Cambridge University Press.
Woolard, E.W. and Clemence, G.M. (1966) *Spherical Astronomy*. New York: Academic Press.

Cross references

Archeoastronomy
Ecliptic

ZODIACAL LIGHT

The zodiacal light is a faint, cone-shaped glow in the sky, best viewed following sunset (to the west) or before sunrise (to the east); it is most easily seen in the absence of the Moon and any light pollution, in tropical latitudes of the Earth. The zodiacal light arises due to the presence of interplanetary dust in the inner solar system (see Dust). The light is generated by two mechanisms: the scattering of incident sunlight in the visible range, and the emission of thermal radiation in the infrared range of the spectrum. The zodiacal dust cloud is so named because the material is concentrated near the orbital planes of the planets (see Ecliptic; Zodiac). The dust cloud is relatively uniform in density and structure (though infrared observations show some distinct dust trails due to comets and certain asteroid families; see Dust).

J.D. Cassini (q.v.) was the first to study the zodiacal light in detail (Dumont, 1991); he ascribed the phenomenon to a cloud flattened in the plane of the solar equator. There is a small brightening of the zodiacal light in the antisolar direction, due to backscattering of the solar radiation; this was named the gegenschein ('counterglow') by A. von Humboldt in the 19th century. It has been difficult to study the phenomenon in isolation, due to the Earth's airglow (q.v.), but satellite observation capabilities have removed this difficulty.

Remote sensing observations of the zodiacal light demonstrate that the spectrum is similar to that of the Sun, indicating that the particles are larger than the wavelengths of visible light (Levasseur-Regourd, 1991). Doppler shifts indicate that they orbit in the same direction as the planets. Typical sizes are in the range from 10 to 100 μm; the mean temperature of the dust is about 260 K and the mean albedo is about 0.08 (Dumont, 1991). Observations at infrared wavelengths (5–25 μm) by the Infrared Astronomical Satellite (IRAS) revealed considerable previously unknown structure in the zodiacal cloud (Dermott and Nicholson, 1989). Relatively bright bands of dust were found to be associated with the orbits of the Koronis and Themis families of asteroids (see Asteroid: families). These particles are generated by collisions of the asteroids. Poynting–Robertson drag (q.v.) causes the dust particles to spiral inward toward the Sun. Comets represent another source of interplanetary dust; a typical short-period comet may lose 10^6 g s^{-1} of material during passage through the inner solar system. Dermott et al. (1992) summarize theories of the origin and evolution of the zodiacal dust cloud.

The zodiacal dust cloud interferes with some Earth-based observations, particularly in the infrared range, as fainter objects are targeted. It is less troublesome at visible wavelengths, but must be considered in observations of faint objects with the Hubble Space Telescope (Dumont, 1991).

James H. Shirley

Bibliography

Dermott, S.F. and Nicholson, P.D. (1989) IRAS dustbands and the origin of the zodiacal cloud. *Highlights Astron.*, **8**, 259–66.
Dermott, S.F., Durda, D.D., Gustafson, B.A.S. et al. (1992) Origin and evolution of the zodiacal dust cloud, in *Asteroids, Comets, Meteors*, Houston: Lunar and Planetary Institute, pp. 153–6.
Dumont, R. (1991) Zodiacal light and gegenschein, in *The Astronomy and Astrophysics Encyclopedia* (ed. S.P. Maran). New York: Van Nostrand Reinhold, pp. 969–71.
Levasseur-Regourd, A.-C. (1991) Interplanetary dust, remote sensing, in *The Astronomy and Astrophysics Encyclopedia* (ed. S.P. Maran). New York: Van Nostrand Reinhold, pp. 326–8.

Cross references

Asteroid: families
Dust
Ecliptic
Meteor, meteoroid
Poynting–Robertson drag
Zodiac

ZOND

Zond (Cyrillic spelling ЭОНД) was the name given to 11 robotic spacecraft launched by the USSR between 1964 and 1970. It is the Russian word for 'probe', and it was used because its non-specificity would not reveal the purpose of the mission.

The first three were launched in 1964 and 1965 from Tyuratam by the SL-6 launch system. Weighing less than 1000 kg, they were interplanetary probes. One was targeted for Venus and one for Mars, but neither returned any planetary data. The objective of Zond 3 was never announced, but it flew by the Moon and recorded and retransmitted photographs of the lunar surface.

Beginning in November 1967, eight Zonds were launched from Tyuratam by the SL-12 launch system; each weighed more than 5000 kg. All were apparently precursors to the intended manned lunar mission. Three failed to get out of Earth orbit and were not given numbers. The trajectory of Zond 4 is not known. Zond 5 to 8 each flew around the Moon and were recovered on the seventh day, all landing in the USSR except Zond 5, which landed in the Indian Ocean.

The series was apparently terminated when it became clear that the Soviets could not beat the Apollo program in landing on the Moon.

Conway W. Snyder

Bibliography

Thompson, T.D. (ed.) (1991) *TRW Space Log*. Redondo Beach, CA: Space Technology Group of TRW.

ZWICKY, FRITZ (1898–1974)

Born in Varna, Bulgaria, of Swiss and Czech parents, Zwicky made his name in astrophysics and rocketry, having a penchant for speculation 'outside the confines of prevailing theory'. He had the reputation of an 'irascible maverick' according to his biographer, Hufbauer (1990). Raised in Switzerland, he attended the ETH in Zürich, obtaining a doctorate on the theory of ionic crystals. In 1925 he went to the California Institute of Technology in Pasadena, serving as a faculty member from 1927 to 1968. Besides his scientific work he devoted time to the rebuilding of European war-damaged libraries and the problems of orphans.

In 1933, with W. Baade, he proposed the name 'supernovas' for those exceptionally bright phenomena that appear, perhaps once every millennium, in typical galaxies, converting ordinary stars to 'neutron stars' and thus generating cosmic rays. It was not however until the 1960s that radio astronomers were able to confirm these ideas. Meanwhile, he was also compiling catalogs of galaxies (issued 1961–1968, 1971).

During World War II he became interested in rocketry and in 1942 helped found the Aerojet Engineering Corporation. After the war he was sent to Germany and Japan to help appraise their rocket establishments; in 1949 he received the Presidential Medal of Freedom from Harry Truman.

Rhodes W. Fairbridge

Bibliography

Hufbauer, K. (1990) Zwicky, Fritz. *Dict. Sci. Biogr.*, Vol. 18, pp. 1011–3.
Müller, R. (1986) *Fritz Zwicky: Leben und Werk des Glarner Astrophysikers, Rakatenforschers und Morphologen*. Glarus, Switzerland.
Payne-Gaposchkin, C. (1974) A special kind of astronomer. *Sky and Telescope*, **47**, 311–3.
Zwicky, F. and Baade, W. (1934) On super-novae; and cosmic rays from super-novae. *Proc. Natl. Acad. Sci.*, **20**, 254–63.

Appendix A
List of entries by subject

This appendix consists of 15 tables. The tables contain main articles and partial listings of related entries for the following general subject areas: asteroids; astronomy; atmospheres; biographies; celestial mechanics and gravitation; comets; geology and geophysics; magnetic fields and the interplanetary environment; meteorites; missions to planets and the Moon; planets; rings; satellites; the Sun and solar physics; and techniques.

Table AA1 Asteroids and related articles

Amor object
Apollo object
Asteroid
Asteroid: compositional structure and taxonomy
Asteroid: families
Asteroid: lightcurve
Asteroid: photometry
Asteroid: resonance
Asteroid: thermal infrared studies
Aten object
Ceres
Chaotic dynamics in the solar system
Collision
Dust
Gaspra
Hilda asteroids
Ida
Mass extinction
Meteorite parent bodies
Near-Earth object
Planetesimal
Resonance in the solar system
Trojan asteroids
Väisälä orbit
Vesta

Table AA2 Astronomy and related articles

Absorption, absorption spectrum
Aphelion
Apsis, apsides
Archeoastronomy
Astrometric observation
Astronomical constants
Astronomical unit
Cassini, Jean-Dominique, and Cassini's laws
Charge-coupled device
Color
Coordinate systems
Eclipse
Ecliptic
Emissivity
Infrared radiation
Libration
Lyman alpha
Obliquity
Occultation
Opposition
Opposition effect
Orbit
Periapse, perihelion, perigee, peribac
Planetary dynamical astronomy
Precession and nutation
Reflectivity
Remote sensing
Sidereal period
Spacewatch
Synodic period
Syzygy
Transit
Year
Zodiac

Table AA3 Atmospheres

Aerosol
Angular momentum cycle in planet Earth
Atmosphere
Atmospheric thermal structure
Carbon, carbon dioxide
Coriolis force, geostrophic motion
Cyclone, anticyclone
Diurnal variation
Earth: atmosphere
Eolian transport
Greenhouse effect
Hadley circulation
Hydrosphere
Ice age
Insolation
Jupiter: atmosphere
Mars: atmosphere
Mercury: atmosphere
Mesosphere
Moon: atmosphere
Neptune: atmosphere
Optical depth
Polar cap
Radiative transfer
Saturn: atmosphere
Spectroscopy: atmosphere
Stratosphere
Surface pressure
Thermosphere
Troposphere
Uranus: atmosphere
Venus: atmosphere

Table AA5 Celestial mechanics and gravitation

Angular momentum
Barycenter
Capture mechanisms
Celestial mechanics
Chaotic dynamics in the solar system
Comet: dynamics
Commensurability
Earth–Moon system: dynamics
Ephemeris
Euler, Leonhard, and Eulerian motion
Gravitation
Gravity-assist navigation
Hohman transfer orbit
Inertia, inertial frame
Kepler's laws
Lagrangian point
Libration
Milankovitch, Milutin, and Milankovitch theory
Nemesis
Newton, Sir Isaac, and Newton's laws
Orbit
Planetary dynamical astronomy
Planetary rotation
Precession and nutation
Resonance in Saturn's rings
Resonance in the solar system
Roche limit
Solar motion
Solar system: stability
Tidal heating
Tidal friction
Torque
Väisälä orbit

Table AA4 Biographies

Adams, John Couch
Alfvén, Hannes, and Alfvén wave
Antoniadi, Eugendos
Aristarchus
Barnard, Edward Emerson
Bradley, James
Brahe, Tycho
Brouwer, Dirk
Campbell, William Wallace
Carrington, Richard Christopher
Cassini, Jean-Dominique, and Cassini's laws
Chamberlin, Thomas Chrowder, and planetesimal hypothesis
Copernicus, Nicolaus
Croll, James
Darwin, George Howard
Dawes, William Rutter
Denning, William Frederick
Encke, Johann Franz
Eratosthenes
Euler, Leonhard, and Eulerian motion
Galileo Galilei
Gauss, Carl Friedrich
Gilbert, William
Goddard, Robert Hutchings
Hale, George Ellery, and Hale cycle
Hall, Asaph
Halley, Edmond, and Halley's comet
Herschel, William
Hipparchus of Nicaea
Huygens, Christian
Jones, Harold Spencer
Kepler, Johannes
Kirchoff, Gustav Robert, and Kirchoff's laws
Kirkwood, Daniel
Kuiper, Gerard Peter
Lagrange, Joseph Louis
Laplace, Pierre Simon de
Leverrier, Urbain Jean Joseph
Lomonsov, Mikhael Vasilyevich
Lowell, Percival
Maraldi, Giacomo Filippo
Maunder, Edward Walter, and Maunder minimum
Milankovitch, Milutin, and Milankovitch theory
Newcomb, Simon
Newton, Sir Isaac, and Newton's laws
Olbers, Heinrich Wilhelm Matthäus
Oort, Jan Hendrik, and Oort cloud
Öpik, Ernst Julius
Piazzi, Guiseppe
Pickering, William Henry
Ptolemy
Roche, Édouard Albert
Sabine, Edward
Schiaparelli, Giovanni Virginio
Sharonov, Vsevolod Vasilievich
Slipher, Earl Carl
Thales of Miletus
Tombaugh, Clyde William
Trouvelot, Étienne Léopold
Tsiolkovsky, Konstantin Eduardovich
Urey, Harold Clayton
Van Allen, James Alfred
Wolf, Max
Wolf, Rudolf, and Wolf number
Zwicky, Fritz

Table AA6 Comets

Chiron
Comet
Comet: dynamics
Comet: historical perspective
Comet: impacts on Jupiter
Comet: impacts on Earth
Comet: observation
Comet: origin and evolution
Comet: structure and composition
Giotto mission
Halley, Edmond, and Halley's comet
Interstellar grains
Kuiper belt
Meteor, meteoroid
Meteor shower, meteoroid stream
Oort, Jan Hendrik, and Oort cloud
Sakigake and Suisei missions
Solar system: origin
Vega missions

Table AA7 Geology and geophysics

Asthenosphere
Astrogeology
Basalt
Breccia
Cartography
Chandler wobble
Chemical element
Core, terrestrial planetary
Corona (Venus)
Cosmochemistry
Crater
Cretaceous–Tertiary boundary
Crust
Dating methods
Differentiation
Dome
Dynamo theory
Earth rotation
Earth: geology, tectonics and seismicity
Earth: rotational history
Ejecta
Ellipsoid
Eolian transport
Erosion
Euler, Leonhard, and Eulerian motion
Flood basalt
Fracture, fault
Geoid
Geomagnetic polarity reversals and the geological record
Geomagnetic storm
Global Positioning System
Gravimetry
Gravity fields of the terrestrial planets
Hot spot tectonics
Hydrosphere
Ice age
Igneous rock
Impact cratering
Iridium anomaly
Iron
Isostasy
Length of day
Lithosphere
Mantle
Mantle convection
Map projections
Mars: geology
Mars: gravity
Mars: structural geology and tectonics
Mascon
Mercury: geology
Moon: geology
Moon: gravity
Moon: seismicity
Obliquity: terrestrial record
Paleomagnetism
Planetary geodesy
Planetary ice
Planetary rotation
Plate tectonics
Polar cap
Polarity reversals
Regolith
Seismicity
Shock metamorphism
Silica, silicate
Surface gravity
Surface processes
Tectonics
Tessera
Thermal evolution of planets and satellites
Tidal friction
Tidal heating
Tide-raising force
Venus: geology and geophysics
Venus: gravity
Volcanism in the solar system
Water
Weathering

Table AA8 Magnetic fields and the interplanetary environment

Airglow
Alfvén, Hannes, and Alfvén wave
Aurora, historical record
Aurora, planetary
Charged particle observation
Corpuscular radiation
Cosmic rays
Earth: magnetic field and magnetosphere
Eccentric dipole
Electromagnetic radiation
Exosphere
Forbush decrease
Forbush effect
Geomagnetic polarity reversals and the geological record
Geomagnetic storm
Heliosphere
Interplanetary magnetic field
Ionosphere
Jupiter: magnetic field and magnetosphere
Magnetism
Magnetometry
Magnetospheres of the outer planets
Mars: magnetic field and magnetosphere
Maxwell theory
Mercury: magnetic field and magnetosphere
Moon: magnetism and interior
Neptune: magnetic field and magnetosphere
Planetary lightning
Planetary torus
Plasma
Plasma wave
Polarity reversals
Radiation belt
Saturn: magnetic field and magnetosphere
Shock waves
Solar wind
Thermal plasma instrumentation
Uranus: magnetic field and magnetosphere
Venus: magnetic field and magnetosphere
Whistler

Table AA9 Meteorites

Achondrite meteorites
Antarctic meteorites
Basaltic achondrite meteorites
Carbonaceous chondrite
Chondrites, ordinary
Chronology: meteorite
Cosmic ray exposure age
Crater
Dating methods
Differentiation
Dust
Ejecta
Enstatite meteorites
Eucrite meteorites
Impact cratering
Interstellar grains
Iron meteorites
Lunar meteorites
Meteor, meteoroid
Meteor shower, meteoroid stream
Meteorite
Meteorite parent bodies
Micrometeorite
Noble gas
Poynting–Robertson drag
Shock metamorphism
SNC meteorites
Stony iron meteorites
Tektites
Ureilite meteorites

Table AA10 Missions to planets and the moon

Apollo missions
Cassini mission
Earth Observing System
Galileo mission
Giotto mission
History of planetary science II: space age
Huygens Titan atmospheric probe
Luna missions
Lunar Orbiter missions
Magellan mission
Mariner missions
Mars Observer mission
Phobos mission
Pioneer 10 and 11 missions
Pioneer Venus mission
Sakigake and Suisei missions
Soviet Mars missions
Surveyor missions
Ulysses mission
Vega mission
Venera missions
Viking mission
Voyager missions

Table AA11 Planets

Earth
Earth: atmosphere
Earth: geology, tectonics and seismicity
Earth: magnetic field and magnetosphere
Earth rotation
Earth: rotational history
Jupiter
Jupiter: interior structure
Jupiter: atmospheric circulation
Jupiter: magnetic field and magnetosphere
Jupiter: ring system
Jupiter: satellite system
Mars
Mars: atmosphere
Mars: geology
Mars: gravity
Mars: impact cratering
Mars: magnetic field and magnetosphere
Mars: remote sensing
Mars: structural geology and tectonics
Mercury
Mercury: atmosphere
Mercury: geology
Mercury: magnetic field and magnetosphere
Neptune
Neptune: atmosphere
Neptune: magnetic field and magnetosphere
Neptune: ring system
Neptune: satellite system
Planet
Planet: extrasolar
Planet X
Pluto
Saturn
Saturn: atmosphere
Saturn: interior structure
Saturn: magnetic field and magnetosphere
Saturn: ring system
Saturn: satellite system
Solar system
Solar system: origin
Solar system: stability
Terrestrial planets
Uranus
Uranus: atmosphere
Uranus: magnetic field and magnetosphere
Uranus: ring system
Uranus: satellite system
Venus
Venus: atmosphere
Venus: geology and geophysics
Venus: gravity
Venus: magnetic field and magnetosphere

Table AA12 Rings

Imaging science
Jupiter: ring system
Neptune: ring system
Planetary ring
Radio science
Resonance in Saturn's rings
Roche limit
Saturn: ring system
Shepherd satellite
Uranus: ring system

Table AA13 Satellites

Callisto
Charon
Deimos
Enceladus
Europa
Galilean satellites
Ganymede
Io
Jupiter: satellite system
Miranda
Moon (Earth's moon)
Moon: atmosphere
Moon: geology
Moon: gravity
Moon: magnetism and interior
Moon: origin
Moon: seismicity
Neptune: satellite system
Phobos
Satellite, natural
Saturn: satellite system
Shepherd satellite
Small satellite
Titan
Titania
Triton
Umbriel
Uranus: satellite system

Table AA14 The Sun, solar physics and related articles

Coronal mass ejection
Corpuscular radiation
Dynamo theory
Fraunhofer line
Hale, George Ellery, and Hale cycle
Insolation
Interplanetary magnetic field
Lyman alpha
Maunder, Edward Walter, and Maunder minimum
Solar activity
Solar constant
Solar corona
Solar flare
Solar luminosity
Solar motion
Solar neutrino
Solar particle event
Solar photosphere
Solar wind
Sun
Wolf, Rudolf, and Wolf number
Zeeman effect

Table AA15 Techniques

Cartography
Charge-coupled device
Comet: observation
Global Positioning System
Gravimetry
Gravity-assist navigation
Imaging science
Infrared spectroscopy
Ion and neutral mass spectrometry
Laser ranging
Magnetometry
Map projections
Microwave spectroscopy
Photoclinometry
Photogrammetry, radargrammetry and planetary topographic mapping
Photometry
Planetary sampling: *in situ* analysis
Polarimetry
Radar astronomy
Radio astronomy
Radiometry
Radio science
Reflectance spectroscopy
Remote sensing
Spectrophotometry
Spectroscopy: atmosphere
Thermal plasma instrumentation
Ultraviolet spectroscopy
Visible and near-infrared spectroscopy
Voyager planetary radio astronomy

Appendix B
The international system of units

The international system of units, abbreviated SI for Système International (d'Unités), was defined and given official status by the 11th General Conference on Weights and Measures, 1960. The complete listing of SI units and various other supplementary units may be found in *The International System of Units* (SI) (National Buteau of Standards Special Publication 330, 1981 edition) and in *The International System of Units: Physical Constants and Conversion Factors*, second revision, 1973, by E.A. Mechtly (National Aeronautics and Space Administration Publication SP-7012). Both publications are published by and available through the US Government Printing Office (Washington, DC 20402, USA).

The SI system is comprised of seven *base* units as well as a number of *derived* and *supplementary* units. The following compilation includes tables of the SI units (Tables B1–B7) as well as tables of certain important units that are outside the SI system (Tables B8–B11) but are still in common use. The tabular data are from NBS Special Publication 330, 1981 edition.

Table AB1 SI base units

Quantity	SI unit Name	Symbol	Quantity	SI unit Name	Symbol
Length	meter[a]	m	Thermodynamic temperature	kelvin[e]	K
Mass	kilogram[b]	kg	Amount of substance	mole[f]	mol
Time	second[c]	s	Luminous intensity	candela[g]	cd
Electric current	ampere[d]	A			

[a] A meter is the length equal to 1 650 763.73 wavelengths in vacuum of the radiation corresponding to the transition between the levels $2p_{10}$ and $5d_5$ of the ^{86}Kr atom.
[b] A kilogram is equal to the mass of the international prototype of the kilogram.
[c] A second is the duration of 9 192 631 770 periods of the radiation corresponding to the transition between the two hyperfine levels of the ground state of the ^{133}Cs atom.
[d] An ampere is that constant current which, if maintained in two straight parallel conductors of infinite length, of negligible circular cross section, and placed 1 meter apart in vacuum, would produce between these conductors a force equal to 2×10^{-7} newton per meter of length.
[e] A kelvin is the fraction 1/273.16 of the thermodynamic temperature of the triple point of water.
[f] A mole is the amount of substance of a system that contains as many elementary entities as there are atoms in 0.012 kilogram of ^{12}C.
[g] A candela is the luminous intensity, in a given direction, of a source that emits monochromatic radiation of frequency 540×10^{12} hertz and that has a radiant intensity in that direction of (1/683) watt per steradian.

Table AB2 Examples of SI derived units expressed in terms of base units

	SI unit	
Quantity	Name	Symbol
Area	square meter	m^2
Volume	cubic meter	m^3
Speed, velocity	meter per second	$m\ s^{-1}$
Acceleration	meter per second squared	$m\ s^{-2}$
Wave number	1 per meter	m^{-1}
Density, mass density	kilogram per cubic meter	$kg\ m^{-3}$
Specific volume	cubic meter per kilogram	$m^3\ kg^{-1}$
Current density	ampere per square meter	$A\ m^{-2}$
Magnetic field strength	ampere per meter	$A\ m^{-1}$
Concentration (of amount of substance)	mole per cubic meter	$mol\ m^{-3}$
Luminance	candela per square meter	$cd\ m^{-2}$

Table AB3 SI derived units with special names

	SI unit			
Quantity	Name	Symbol	Expression in terms of other units	Expression in terms of SI base units
Frequency	hertz	Hz		s^{-1}
Force	newton	N		$m\ kg\ s^{-2}$
Pressure, stress	pascal	Pa	$N\ m^{-2}$	$m^{-1}\ kg\ s^{-2}$
Energy, work, quantity of heat	joule	J	$N\ m$	$m^2\ kg\ s^{-2}$
Power, radiant flux	watt	W	$J\ s^{-1}$	$m^2\ kg\ s^{-3}$
Electric charge, quantity of electricity	coulomb	C		$s\ A$
Electric potential, potential difference, electromotive force	volt	V	$W\ A^{-1}$	$m^2\ kg\ s^{-3}\ A^{-1}$
Capacitance	farad	F	$C\ V^{-1}$	$m^{-2}\ kg^{-1}\ s^4\ A^2$
Electric resistance	ohm	Ω	$V\ A^{-1}$	$m^2\ kg\ s^{-3}\ A^{-2}$
Electric conductance	siemens	S	$A\ V^{-1}$	$m^{-2}\ kg^{-1}\ s^3\ A^2$
Magnetic flux	weber	Wb	$V\ s^{-1}$	$m^2\ kg\ s^{-2}\ A^{-1}$
Magnetic flux density	tesla	T	$Wb\ m^{-2}$	$kg\ s^{-2}\ A^{-1}$
Inductance	henry	H	$Wb\ A^{-1}$	$m^2\ kg\ s^{-2}\ A^{-2}$
Celsius temperature	degree Celsius	°C		K
Luminous flux	lumen	lm		$cd\ sr$
Illuminance	lux	lx	$lm\ m^{-2}$	$m^{-2}\ cd\ sr$

Table AB4 Examples of SI derived units expressed by means of special names

	SI unit		
Quantity	Name	Symbol	Expression in terms of SI base units
Dynamic viscosity	pascal second	$Pa\ s$	$m^{-1}\ kg\ s^{-1}$
Moment of force	newton meter	$N\ m$	$m^2\ kg\ s^{-1}$
Surface tension	newton per meter	$N\ m^{-1}$	$kg\ s^{-2}$
Heat flux density, irradiance	watt per square meter	$W\ m^{-2}$	$kg\ s^{-3}$
Heat capacity, entropy	joule per kelvin	$J\ K^{-1}$	$m^2\ kg\ s^{-2}\ K^{-1}$
Specific heat capacity, specific entropy	joule per kilogram kelvin	$J\ (kg\ K)^{-1}$	$m^2\ s^{-2}\ K^{-1}$
Specific energy	joule per kilogram	$J\ kg^{-1}$	$m^2\ s^{-2}$
Thermal conductivity	watt per meter kelvin	$W\ (m\ K)^{-1}$	$m\ kg\ s^{-3}\ K^{-1}$
Energy density	joule per cubic meter	$J\ m^{-3}$	$m^{-1}\ kg\ s^{-2}$
Electric field strength	volt per meter	$V\ m$	$m\ kg\ s^{-3}\ A^{-1}$
Electric charge density	coulomb per cubic meter	$C\ m^{-3}$	$m^{-3}\ s\ A$
Electric flux density	coulomb per square meter	$C\ m^{-2}$	$m^{-2}\ s\ A$
Permittivity	farad per meter	$F\ m^{-1}$	$m^{-3}\ kg^{-1}\ s^4\ A^2$
Permeability	henry per meter	$H\ m^{-1}$	$m\ kg\ s^{-2}\ A^{-2}$
Molar energy	joule per mole	$J\ mol^{-1}$	$m^2\ kg\ s^{-2}\ mol^{-1}$
Molar entropy, molar heat capacity	joule per mole kelvin	$J\ (mol\ K)^{-1}$	$m^2\ kg\ s^{-2}\ K^{-1}\ mol^{-1}$
Exposure (x and γ rays)	coulomb per kilogram	$C\ kg^{-1}$	$kg^{-1}\ s\ A$
Absorbed dose rate	gray per second	$Gy\ s^{-1}$	$m^2\ s^{-3}$

APPENDIX B THE INTERNATIONAL SYSTEM OF UNITS

Table AB5 SI supplementary units

Quantity	SI unit Name	Symbol
Plane angle	radian	rad
Solid angle	steradian	sr

Table AB6 Examples of SI derived units formed by using supplementary units

Quantity	SI unit Name	Symbol
Angular velocity	radian per second	rad s^{-1}
Angular acceleration	radian per second squared	rad s^{-2}
Radiant intensity	watt per steradian	W sr^{-1}
Radiance	watt per square meter per steradian	W m^{-2} sr^{-1}

Table AB7 SI prefixes

Factor	Prefix	Symbol	Factor	Prefix	Symbol
10^{18}	exa	E	10^{-1}	deci	d
10^{15}	peta	P	10^{-2}	centi	c
10^{12}	tera	T	10^{-3}	milli	m
10^{9}	giga	G	10^{-6}	micro	μ
10^{6}	mega	M	10^{-9}	nano	n
10^{3}	kilo	k	10^{-12}	pico	p
10^{2}	hecto	h	10^{-15}	femto	f
10^{1}	deka	da	10^{-18}	atto	a

Note: NBS Special Publication 330 (p. 12) states 'Among the base units of the International System, the unit of mass is the only one whose name, for historical reasons, contains a prefix. Names of decimal multiples and sub-multiples of the unit of mass are formed by attaching prefixes to the word "gram".' Thus, for example, 10^{-6} kg = 1 milligram, (not 1 microkilogram).

Table AB8 Units in use with the international system

Name	Symbol	Value in SI unit
Minute	min	1 min = 60 s
Hour	h	1 h = 60 min = 3600 s
Day	d	1 d = 24 h = 86 400 s
Degree	°	1° = (π/180) rad
Minute	'	1' = (1/60)° = (π/10 800) rad
Second	"	1" = (1/60)' = (π/648 000) rad
Liter	l, L	1 L = 1 dm^3 = 10^{-3} m^3
Metric ton	t	1 t = 10^3 kg

Table AB9 Units in use temporarily with the international system

Name	Symbol	Value in SI unit
Nautical mile		1 nautical mile = 1852 m
Knot		1 nautical mile per hour = (1852/3600) m s^{-1}
Ångström	Å	1 Å = 0.1 nm = 10^{-10} m
Are	a	1 a = 1 dam^2 = 10^2 m^2
Hectare	ha	1 ha = 1 hm^2 = 10^4 m^2
Barn	b	1 b = 100 fm^2 = 10^{-28} m^2
Bar	bar	1 bar = 0.1 MPa = 10^5 Pa
Gal	Gal	1 Gal = 1 cm s^{-2} = 10^{-2} m s^{-2}
Curie	Ci	1 Ci = 3.7 × 10^{10} Bq
Roentgen	R	1 R = 2.58 × 10^{-4} C kg^{-1}
Rad	rad	1 rad = 1 cGy = 10^{-2} Gy
Rem	rem	1 rem = 1 cSv = 10^{-2} Sv

Table AB10 CGS units with special names

Name	Symbol	Value in SI unit
Erg	erg	1 erg = 10^{-7} J
Dyne	dyn	1 dyn = 10^{-5} N
Poise	P	1 P = 1 dyn s cm^{-2} = 0.1 Pa s
Stokes	St	1 St = 1 cm^2 s^{-1} = 10^{-4} m^2 s^{-1}
Gauss	Gs, G	1 Gs corresponds to 10^{-4} T
Oersted	Oe	1 Oe corresponds to $\frac{1000}{4\pi}$ A m^{-1}
Maxwell	Mx	1 Mx corresponds to 10^{-8} Wb
Stilb	sb	1 sb = 1 cd cm^{-2} = 10^4 cd m^{-2}
Phot	ph	1 ph = 10^4 lx

Table AB11 Other units generally deprecated

Name	Value in SI unit
Fermi	1 fermi = 1 fm = 10^{-15} m
Metric carat	1 metric carat = 200 mg = 2 × 10^{-4} kg
Torr	1 torr = $\frac{101\ 325}{760}$ Pa
Standard atmosphere (am)	1 atm = 101 325 Pa
Kilogram-force (kgf)	1 kgf = 9.806 65 N
Calorie (cal)	
Micron (μ)	1 μ = 1 μm = 10^{-6} m
X unit	
Stere (st)	1 st = 1 m^3
Gamma (γ)	1 γ = 1 nT = 10^{-9} T
γ	1 γ = 1 μg = 10^{-9} kg
λ	1 λ = 1 μL = 10^{-6} L = 10^{-9} m^3

Author index

A'Hearn, M. 914
A'Hearn, M. and Feldman, P.D. 93
A'Hearn, M.F., Hoban, S., Birch, P.V. et al. 153
A'Hearn, M.F., Feldman, P.D. and Schleicher, D.G. 153
Abazov, A.I. et al. 754
Abbot, C.G. 756
Abbott, B.J. 302
Abe, K. 730
Abe, K. and Kanamori, H. 630
Abetti, G. 23, 579, 726
Abramovich, S.K., Ageeva, G.D., Akim, E.L. et al. 886
Abrams, J.W. 76
Ackley, S. 186
Acuña, M.H. 406, 410
Acuña, M.H., Connerney, J.F.P., Wasilewski, P. et al. 459
Acuña, M.H., Ness, N.F. 630
Acuña, M.H., Scearce, C.S., Seek, J.B. and Scheifele, J. 410
Adams, F.D. 207
Adams, J.B. 914
Adams, J.B. and Filice, A.L. 914
Adams, J.B. and Jones, R.L. 914
Adams, J.B. Smith, M.O. and Gillespie, A.R. 914
Adlam, J.H. and Allen, J.E. 736
Adler, I. 615
Afiattalab, F. and Wasson, J.T. 109
Agresti, D.G., Morris, R.V., Wills, E.L. et al. 615
Ahrens, L.H. 103
Ahrens, L.H. (ed.) 103
Ahrens, T.J. and O'Keefe, J. 114
Aikin, A.C. 153
Airy, G. 50
Akasofu, S.-I. and Kamide, Y. 61
Albarez, W. and Muller, R.A. 467
Albee, A.L., Arvidson, R.E. and Palluconi, F.D. 459
Albee, A.L. and Palluconi, F.D. 456
Aleksandrov, Yu.N., Vasil'ev, M.G., Vyshlov, A.S. et al. 886
Alekseev, N.V., Vakulov, P.V., Vologdin, N.I. et al. 886
Alexander, A.F.O'D. 70, 183, 607, 698, 722
Alexander, C.M.O., Hutchison, R. and Barber, D.J. 109

Alfvén, H. 11, 119, 153, 703
Alfvén, H. and Arrhenius, G. 13
Alfvén, H. and Arrhenius, G. (eds.) 11
Allcock, G.McK. 640
Allen, C. 716
Allen, C.W. 144, 793
Allen, D.A. 45
Allen, J., Sauer, H., Frank, L. and Reiff, P. 744
Aller, L.H. 874
Allison, M. 370
Allison, M. and Travis, I.D. 54
Allison, M., Beebe, R.F., Conrath, B.J. et al. 559, 862
Alvarez, L.W. 466
Alvarez, L.W., Alvarez, W., Asaro, F. and Michel, H.V. et al. 131, 467, 805
Alvarez, W. and Asaro, F. 131, 178
Alvarez, W., Asaro, F. and Montanari, A. 360
Alvarez, W. and Muller, R.A. 124, 256
Ananda, M.P., Sjogren, W.L., Phillips, R.J. et al. 681
Anders, E. and Albee, A.L. (eds.) 394
Anders, E. 79, 109, 344, 486
Anders, E. and Grevesse, N. 103, 176, 934
Anders, E. and Ebihara, M. 176
Anders, F. 131
Anders, F. and Owen, T. 131
Anders, F. and Zinner, E. 112
Anderson, A.J. and Cazenave, A. (eds.) 286
Anderson, D.E. and Hord, C.W. 851
Anderson, D.L. 360
Anderson, D.L., Miller, W.F., Latham, G.V. et al. 730
Anderson, H. 644
Anderson, J. 590
Anderson, J.D. 283, 287, 302, 520, 590, 596
Anderson, J.D. and Standish, E.M. Jr. 144
Anderson, J.D. and Hubbard, W.G. 287
Anderson, J.D., Armstrong, J.W., Campbell, J.K. et al. 287
Anderson, J.D., Hubbard, W.B. and Slattery, W.L. 287
Anderson, J.D., Colombo, G., Esposito, P.B. 287
Anderson, O.L. 161
Anderson, O.L., Oda, H. and Isaak, D. 421

Anderson, R.W. 237
Andrews, H.C. and Hunt, B.R. 326
Andrews, J.T. (ed.) 364
Anon. 79, 392, 560, 701, 774, 874, 910
Anselmann, P., Hampel, Kw., Heusser, G. et al. 793
Anselmann, P. et al. 754
Antoniadi, E.M. 23, 302, 552
Antonucci, E. and Somov, B.V. (eds.) 747
Antsibor, N.M., Bakit'ko, R.V., Ginzburg, A.L. et al. 886
Aoki, S. and Kakuta, C. 558
Apostolov, E.M. 744
Appleby, J.F. 531
Applegate, J.H., Douglas, M.R., Gursel, Y. et al. 594
Applegate, J.H., Douglas, M.R., Gursel, Y., Sussman, G.J. & Wisdom, J. 765
Apt, J., Clark, R. and Singer, R. 531
Araki, S. 722
Archibald, R.C. 540
Archimedes 27
Arkani-Hamed, J. and Toksöv, M.N. 904
Armi, L., Herbert, D., Oakey, N. et al. 370
Armitage, A. 297, 299
Armstrong, R.I. 207
Arnold, J.R. 37
Arnold, V.I. 765
Aronson, J.R. and Emslie, A.G. 460
Arpigny, C., Dossin, F., Woszcyk, A. et al. 140
Artem'ev, A.V. and Radzievskii, V.V. 612
Arthur, D.G.W., Pellicori, R.H. and Wood, C.A. 552
Arthur, D.G.W., Agnieray, A.P., Horvath, R.A. et al. 552
Arthur, D.G.W., Agnieray, A.P., Pellicori, R.H. et al. 552
Artyushkov, E.V. and Baer, M.A. 197
Arvidson, R.E. 590
Arvidson, R.E., Phillips, R.J. and Izenberg, N. 400
Arvidson, R.E., Greeley, R., Malin, M.C. et al. 904
Ash, M.F., Shapiro, I.I. and Smith, W. 50
Ashby, N. and Bertotti, B. 287
Astronomical Almanac 51, 159
Astronomy and Astrophysics 855
Atkinson, R.D'E. 365
Atreya, S. 716

Atreya, J.B., Pollack and M.S. Matthews 57
Atreya, S.K., Waite, J.H., Donahue, T.M. et al. 8
Atreya, S.K. 359, 825, 862
Atreya, S.K., Owen, T.C., Gautier, D. and Marten, A. 531
Atreya, S.K., Pollack, J.B. and Matthews, M.S. 372, 523 713, 857
Atreya, S.K., Sandel, B.R. and Romani, P.N. 851, 862
Attolini, M.R. et al. 295
Attolini, M.R., Cecchini, S., Gastagnoli, G.C. et al. 744
Aumann, H.H., Gillett, F.C., Beichman, C.A. et al. 589, 751
Avanesov, G., Zhukov, B., Ziman, Ya. et al. 575
Avanesov, G.A., Boney, B.I., Kemp, F. et al. 575
Avduvskii, V.S., Borodin, N.F., Burtsev, V.P. et al. 886
Avduvskii, V.S., Golovin, Yu.M., Zavelevich, F.S. et al. 886
Avduevsky, V.S., Vishnevetskii, S.L., Golov, I.A. et al. 886. 895
Aveni, A. 27
Ayres, I.D. and Thurston, P.C. 207
Ayres, I.D., Thurston, P.C., Card, K.D. and Weber, W. 207
Azimov, I. 79

Babcock, H.W. 192, 940
Babcock. H.E. 295
Back, W. and Freeze, R.A. (eds.) 934
Bagenal, F. 622, 613, 630
Bagenal, F. and Sullivan, J.D. 61, 622
Bagenal, F., Belcher, J.W., Sittler, E.C. and Lepping, R.P. 630
Bagenal, F., Shemansky, D.E., McNutt, R.L., Jr. et al. 622
Bagnold, R.A. 238
Bahcall, J.N. 124, 144, 336, 748, 754, 793
Bahcall, J.N. and Bahcall, S. 256
Bahcall, J.N. and Davis, R., Jr. 793
Bahcall, J.N. and Soneira, R.M. 124
Bai, T. and Sturrock, P.A. 252, 578, 744, 747
Bailey, M.E. 124, 126, 381, 560
Bailey, M., Clube, S. and Napier, W.M. 761
Bailey, M.E. and Stagg, C.R. 144
Bailey, M.E., Clube, S.V.M. and Napier, W.M. 256
Baines, K.H. 523, 858
Baines, K.H. and Bergstralh, J.T. 862
Baines, K.H. and Hammel, H.B. 531
Baines, K.H., Mickelson, M.F., Larsopn, L.E. and Ferguson, D.W. 531
Baines, K.H. and Smith, W.H. 531, 862
Baker, V.R. 431, 440, 449
Baker, V.R. (ed.) 449
Baker, V.R., Strom, R.G., Gulick, V.C. et al. 432, 449
Baldwin, R.B. 508
Ballester, G.E., Strobel, D.F., Moos, H.W. and Feldman, P.D. 350
Balmino, G., Maynot, B. and Vales, N. 450, 681
Balmino, G.B., Moynot, B. and Vales, N. 292
Balsiger, H., Eberhardt, P., Geiss, J. et al. 353
Bame, S.J., Asbridge, J.R., Feldman, W.C. et al. 772
Bame, S.J., Martin, R.H., Comas, D.J. et al. 630

Banerdt, W.B., Golombek, M.P. and Tanaka, K.L. 463, 730
Banfield, D. and Murray, N. 538, 828
Banks, P.M. and Kockarts, G. 359, 825
Baños, A. and Vernon, A.R. 736
Bar-nun, A., Bar-nun, N., Bauer, S.H. and Sagan, C. 131
Barat, C., Atteia, J.L., Jourdain, E. et al. 575
Barbato, J.P. and Ayer, E.A. 833
Barbosa, D.D. 61, 622
Barbosa, D.D. and Eviatar, A. 622
Barbosa, D.D. and Kivelson, M.G. 622
Barbosa, D.D. and Kurth, W.S. 640
Barker, E.S., Cazes, S., Emerich, C. et al. 622
Barkhausen, H. 641, 936
Barlow, N.G. 193, 430, 451, 453, 737
Barlow, N.G. and Hetherington, S. 393
Barlow, N.G. and Bradley, T.L. 453
Barnard, E.E. 62
Barnes, J.R., Pollack, J.B., Haberle, R.M. et al. 440
Barnes, J.R. 440
Barnet, D.C., Westphal, J.A., Beebe, R.F. and Huber, L.F. 716
Barrell, J. 45, 364, 392
Barrington, R.E. and Belrose, J.S. 641
Barron, E.J. 558
Barshad, I. 935
Barsukov, V.L., Basileusky, A.T., Volkov, V.P. and Zharkov, V.N. 161
Barsukov, V.L. 904
Barth, C.A. 8, 440, 851
Barth, C.A., Hord, C.W., Pearce, J.B. et al. 10, 851
Barth, C.A., Hord, C.W., Stewart, A.I. and Lane, A.I. 10
Barth, C.A., Fastie, W.G., Hord, C.W. et al. 10
Barth, C.A., Pearce, J.B., Kelly, K.K. et al. 8
Barth, C.A., Stewart, A.I.F., Bougher, S.W. et al. 835
Barton, C.E. 211
Barucci, M.A., Capria, M.T., Coradini, A. and Fulchignoni, M. 35
Barucci, M.A., Cellino, A., Desanctis, C. et al. 41
Baruskov, V.L., Basilersky, A.T., Burba, G.A. et al. 164
Basaltic Volcanism Study Project 131, 242, 318
Basilevsky, A.T. 904
Basilevsky, A.T., Pronin, A.A., Ronca, I.B. et al. 164, 807
Batson, R.M. 79, 83
Battin, R.H. 310
Battrick, B. and Buyenne, D. (eds.) 279
Baum, R.M. 387
Baur, F. 72
Bavassano, B., Mariani, F. and Ness, N.F. 736
Beatty, J.K. and Chaikin, A. 519, 922
Beatty, J.K. and Chaikin, A. (eds.) 500, 588
Beatty, J.K., O'Leary, B. and Chaikin, A. (eds.) 708
Becker, G.F. 688
Becker, R.H. and Pepin, R.O. 740
Becklin, E.E. and Zuckerman, B. 589
Beebe, R. 714
Beebe, R.F. et al. 159
Beebe, R.F. and Hockey, T.A. 370
Beebe, R.F., Barnet, D.C., Sada, P.V. and Murrell, A.S. 713, 716
Beer, A. 379

Beer, A. (ed.) 160
Beer, J., Blinov, A., Bonani, G. et al. 744
Beer, J. 793
Beery, A. 160
Behannon, K.W., Acuña, M.H., Burlaga, L.F., et al. 410
Bekov, G.I., Letokhov, V.S., Radaev, V.N. et al. 360
Belcher, J.W., Bridge, H.S., Bagenal, F. et al. 532
Belcher, J.W. 630, 722
Bell, A.E. 311
Bell, J.F. 299, 911
Bell, J.F., Davis, D.R., Hartmann, W.K. and Gaffey, M.J. 31, 35
Bell, J.F., Cruikshank, D.P. and Gaffey, M.J. 31
Bell, J.F., III, Morris, R.V. and Adams, J.B. 460
Bell, J.F., III, McCord, T.B. and Owensby, P.D. 449
Bell, J.F., Davis, D.R., Hartmann, W.K. and Gaffey, M.J. 488, 841
Bell, J.F., III, McCord, T.B. and Lucey, P.G. 914
Bell, J.F. III and Crisp, D. 914
Beloussov, V.V. 197
Belton, M.J.S. 140
Belton, M.S., West, R.A., Rahe, J. and Pereyda, M. (eds.) 366
Bender, P.I., Currie, D.G., Dicke, R.H. et al. 386
Bendjoya, Ph., Slézak, E. and Froeschlé, C. 37
Benner, L.A.M. 836
Bennett, W.H. 353, 819
Benoit, P.H. and Sears, D.W.G. 109
Benton, M.J. 467
Benz, W., Slattery, W. and Cameron, A. 513
Berger, A. 339
Berger, A., Imbrie, J., Hays, J.D. et al. 315
Berger, A., Imbrie, J., Hays, J. et al. (eds.) 495, 570
Berger, A., Loutre, M.F. and Laskar, J. 495, 570
Berger, A.L. 495, 570
Berger, A.L. and Loutre, M.F. 495
Bergerson, J. (ed.) 342
Bergstralh, J. and Miner, E. (eds.) 708
Bergstralh, J., Miner, E. and Matthews, M. 554, 588, 834, 856, 857, 868
Bergstralh, J.T., Miner, E.D. and Mathews, M.S. (eds.) 498, 927
Berkley, J.L. 868
Berkley, J.L. and Jones, J.H. 873
Berkley, J.L., Brown, H.G., IV, Keil, K. et al. 873
Berkley, J.L., Taylor, G.J., Keil, K. et al. 873
Berkner, L.V. and Marshall, L.C. 77
Berkner, L.V. and Marshall, L.C. 874
Beroza, G.C. and Jordan, T.H. 730
Berry, A. 3, 70, 83, 243, 258, 273, 300, 302, 384, 659
Bertaux, J.L., Blamont, J.E. and Festou, M. 398
Bertaux, J.L. and Blamont, J.E. 398
Bertaux, J.L., Blamont, G., Marcelin, M. et al. 851
Bertaux, J.L. and Lallement, R. 851
Bertaux, J.L., Blamont, J.E., Dzyubenko, A.I. et al. 886
Bessell, M.S. 578
Best, M.G. 208
Bevan, A.W.R. and Binns, R.A. 22, 109

AUTHOR INDEX

Bézard, B., Marten, A. and Paubert, G. 493
Bézard, B., Romani, P.N., Conrath, B.J. and Maguire, W.C. 531
Bhandari, N., Bonino, G., Callegari, E. et al. 480
Bibring, J.-P., Erard, S., Gondet, B. et al. 575
Bibring, J.-P. and 16 Others 914
Bibring, J.P., Combes, M., Langevin, Y. et al. 449, 460
Biemann, H.-P. 910
Biemann, K., Oro, J., Toulmin, P., III et al. 391, 449, 615
Biémont, E., Baudoux, M., Kurucz, R.L. et al. 176
Biermann, L. 124, 119, 153, 630
Biermann, L., Huebner, W.F. and Lüst, R. 124
Biermann, L. and Michel, K.W. 144
Biermann, L. and Trefftz, E. 119, 153
Bigg, E.K. 415
Bild, R.W. 3
Billings, D. 747
Billings, M.P. 253
Billingsley, F.C., Goetz, A.F.H. and Lindsley, J.N. 326
Bills, B.G. 287, 440, 594, 596
Bills, B.G. and Ferrari, A.J. 508
Bills, B.G. and Synnott, S.P. 596
Bilo, E.H. and Van De Hulst, H.C. 144
Binder, A.B. and Cruikshank, D.P. 350
Bindschadler, D.I. and Head, J.W. 808
Binzel, R.P. 3, 517, 708, 761
Binzel, R.P., Barucci, M.A. and Fulchignoni, M. 519
Binzel, R.P., Farinella, P., Zappala, V. and Cellino, A. 31, 41
Binzel, R.P., Gehrels, T. and Matthews, M.S. (eds.) 519
Binzel, R.P. and Sauter, L. 299
Binzel, R.P., Tholen, D.J. et al. 103
Binzel, R.P., Tholen, D.J., Tedesco, E.F. et al. 649
Binzel, R.P. and Sauter, L. 841
Binzel, R.P. and Xu, S. 67, 907
Binzel, R.P., Xu, S., Bus, S.J. et al. 25, 32, 51
Binzel, R.P. Xu, S., Bus, S.J. and Bowell, F. 12
Bioxham, J. and Gubbins, D. 406
Biraud, F., Bourgois, G., and Crovisier, J. 493
Birch, F. 421
Birch, F. 813
Birck, J.L. and Allégre, C.J. 67
Bird, J.M. 208
Bird, M.K., Asmar, S.W., Brenkle, J.P. et al. 681
Birkeland, P. 935
Bischoff, A. and Palme, H. 176
Bishop, J., Atreya, S.K., Romani, P.N. et al. 531
Biskup, M. 160
Biswas, A.K. 297
Biswas, S., Walsh, T., Bart, G. and Lipschutz, M.F. 236
Black, D.C. and Matthews, M.S. (eds.) 588
Black, D.C. 589
Blackshear, W.T., Daniels, E.F. and Anderson, S.G. 508
Blackwelder, E.B. 935
Blackwell, B.H. 578
Blackwell, H.R. 326
Blagg, M.A. and Müller, K. 552
Blamont, J.E., Chassefiere, E., Goutail, J.P. et al. 575, 851

Blanco, V.M. and McCuskey, S.W. 383, 650
Blewitt, G. 282
Blizard, J.B. 12
Bloom, A.I. 238, 796
Blouke, M.M. et al. 101
Blum, J.D., Wasserburg, G.J., Hutcheon, I.D. et al. 176
Blunck, J. 552
Bockelée-Morvan, D., Crovisier, J., Colom, P. et al. 140
Bockelée-Morvan, D., Colom, P., Crovisier, J. et al. 493
Bode, J. 761
Bodenheimer, P. and Pollack, J.B. 589
Bodenheimer, P., Ruzmaikina, T. and Matthiew, R. 761
Boehler, R. and Chopelas, A. 421
Boer, G.J. 18
Bogard, D. and Johnson, P. 64
Bogard, D.D. (ed.) 396
Bogard, D.D., Nyquist, L.E. and Johnson, P. 22
Bogard, D.D., Keil, K., Taylor, G.J. et al. 67
Bogard, D.D. and Garrison, D.H. 486
Bogard, D.D. and Johnson, P. 740
Bogard, D.D., Nyquist, L.E. and Johnson, P. 740
Bogard, D.D., Garrison, D.H., Jordan, J.L. and Mittlefehldt, D.W. 785
Bohren, C. and Huffman, D. 685
Boice, D.C. 381, 704
Boice, D.C. and Huebner, W.F. 114, 145
Boice, D.C. and Fairbridge, R.W. 295, 559
Bolt, B.A. 161
Bomford, D. 269
Bondi, H. and Hoyle, F. 2
Bonnet, R.M., Lemaire, P., Vial, J.C. et al. 398
Booker, H.G., Dessler, A.J., Eshleman, V.R. et al. 681
Borderies, N., Goldreich, P. and Tremaine, S. 607, 722
Born, M. 187
Born, M. and Wolf, E. 696
Borucki, W.J. 598
Borucki, W.J. and Magalhaes, J.A. 601
Borucki, W.J. and Chameides, W.L. 601
Borwn. E.W. 92
Bos, H.J.M. 311
Boss, A.P. 76, 589
Boss, A.P., Cameron, A.G.W. and Benz, W. 228
Bougher, S.W. and Roble, R.G. 819, 825
Bougher, S.W., Dickinson, R.E., Ridley, E.C. and Roble, R.G. 825
Bougher, S.W., Dickinson, R.E., Roble, R.G. and Ridley, E.C. 825
Bougher, S.W., Roble, R.G., Ridley, E.C. and Dickinson, R.E. 825
Bowell, E., Hapke, B., Domingue, D. et al. 41
Bowell, E. and Zeliner, B. 653
Bowell, E., Chernykh, N.S. and Marsden, B.G. 875
Bowell, E., Skiff, B.A., Wassermann, L.H. and Russell, K.S. 876
Bowell, F., Chapman, C.R., Gradie, J.C. et al. 32
Bowen, H.J.M. 77
Bowes, D.R. (ed.) 45, 319
Boxx, A.P. 223
Boxx, A.P. and Peale, S.J. 76
Boyce, J.M. 538
Boyer, C.B. 243

Boynston, W.V., Trombka, J.I., Feldman, W.C. et al. 459
Boynton, W.V., Starzk, P.M. and Schmitt, R.A. 873
Brace, L.H., Theis, R.G. and Dalgarno, A. 819
Brace, L.H., Theis, R.F., Mayr, H.G. et al. 907
Bracewell, R.N. 793
Brahic, A. and Hubbard, W. 534
Brahic, A. 722
Brandt, J.C. and Chapman, R.D. 119
Brandt, J.C. and Hodge, P.W. 415, 756
Bray, J.R. 57
Bray, R.J., Loughead, R.E. and Tappere, E.J. 756
Bredthauer, R.A. et al. 101
Breed, C.S., Grolier, M.J. and McCauley, J.F. 449
Breneman, H.H. 267
Breneman, H.H. and Stone, E.C. 754
Bretagnon, P. 596
Brett, R. and Keil, K. 236
Brice, N. 641
Bridges, E.M. 688
Brinton, H.C., Scott, L.R., Pharo, M.W. III and Coulson, J.T.C. 819
Bristow, J.W. and Saggerson, E.P. 251
Bristow, J.W., Allsopp, H.I., Erlank, A.J. et al. 251
Britt, D.T. and Lebofsky, L.A. 32, 486
Broadfoot, A.L. et al. 622
Broadfoot, A.L., Atreya, S.K., Bertaux, J.L. et al. 10, 61, 840, 851, 862
Broadfoot, A.L., Herbert, F., Holberg, J.B. et al. 10, 61, 851
Broadfoot, A.L., Sandel, B.R., Knecht, D. et al. 851
Broadfoot, A.L., Sandel, B.R., Shemansky, D.F. et al. 10, 398
Broadfoot, A.L., Shemansky, D.E. and Kumar, J.J. 473
Brock, B.B. 808
Brode, H.L. 736
Broecker, W.S. 495
Bronshten, V.A. and Zotkin, J.T. 131
Brooks, C.E.P. 72
Brosche, P. and Sündermann, J. (eds.) 221, 828
Brouwer, D. 37, 71, 92, 300
Brouwer, D., Eckert, W.J. and Clemence, G.M. 71
Brouwer, D. and Clemence, G.M. 92, 594
Browell, E.V. 696
Brower, D. and Clemence, G.M. 51
Brown, E.W. 223
Brown, E.W. and Shook, C.A. 92, 383
Brown, H. 176
Brown, R.A. and Yung, Y.L. 350
Brown, R.A. 622
Brown, R.A., Pilcher, C.B. and Strobel, D.F. 851
Brown, R.H. 598
Brownelee, D.E. 191
Brownlee, D. 489
Brownlee, D. and Kissel, J. 489
Brownlee, D.F., Wheelock, M.M., Temple, S. et al. 32
Brownlow, A.H. 103
Brückner, E. 71
Brumber, V.A. and Kovalevsky, J. 92
Brumberg, V.A. 92
Bruns, A.V. and Shumoko, S.M. 575
Brush, S.G. 94, 161, 228
Bucha, V. 57, 315
Bucher, W.H. 801
Buchwald, V.F. 236

Büdel, J. 935
Buffon, G.L. 761
Buie, M., Cruikshank, D.P., Lebofsky, L.A. and Tedesco, E.F. 649
Buie, M., Tholen, D.J. and Horne, K. 649
Bullen, K.E. and Bolt, B.A. 161
Bunbury, E.H. 241
Buratti, B.J. 75, 578, 598, 700, 705
Buratti, B.J., Nelson, R.M. and Lane, A.L. 246
Burbidge, E.M., Burbidge, G.R., Fowler, W.A. and Hoyle, F. 176
Burch, J.L. 61
Burckhardt, J.J. 937
Burke, B.F. and Franklin, K.L. 373
Burke, B.F., Rahe, J.H., Beebe, R.F. et al. 589
Burke, B.F. and Franklin, K.L. 675
Burke, J.G. 481, 486
Burlaga, L.F., Schwenn, R. and Rosenbauer, H. 722
Burnett, D.A. and Frank, L.A. 641
Burnett, D.S. and Wasserburg, G.J. 112
Burnett, D.S. and Woolum, D.S. 169
Burns, J. 761
Burns, J. and Matthews, M.D. (eds.) 708, 922
Burns, J.A. 185, 258, 573, 607, 713, 722
Burns, J.A. and Matthews, M.S. (eds.) 828
Burns, J.A., Showalter, M.R. and Morfill, G.E. 375
Burns, J.A., Schaffer, L.E., Greenberg, R.J. and Showalters, M.R. 375, 607
Burns, J.A., Showalter, M.R., Cuzzi, J.N. and Durisen, R.H. 601
Burns, J.A., Showalter, M.R. and Morfill, G.E. 607, 722
Burns, J.A. 813
Burns, R.G. 914
Burrows, A. and Liebert, J. 589
Burstyn, H.L. 179, 297
Burt, D.M. 360
Bus, E.S. and Larson, S. 540
Bus, S.J., A'Hearn, M.F., Schleicher, D.G. and Bowell, E. 105
Bus, S.J., Bowell, E., Harris, A.W. and Hewitt, A.V. 105
Buseck, P.R. and Goldstein, J.I. 785
Buseck, P.R. 785
Buseck, P.R. and Holdsworth, E. 785
Busse, F.H. 192, 406
Butcher, E.C. 744
Buxbaum, K. 258
BVSP [Basaltic Volcanism Study Project] 64, 251
Byl, J. 124
Byram, E.T., Chubb, T.A., Freidman, H. and Kupperian, J. E. 398

Cadell, H.M. 801
Cadogan, P. 922
Cahill, L.J. 410
Caho, J.K. and Olbert, S. 736
Cairns-Smith, A.G. 291
Cajori, F. 273
Calcutt, S., Taylor, F., Ade, P. et al. 87
Caldwell, J., Cunningham, C.C., Anthony, D. et al. 833
Caldwell, J.T., Wagener, R. and Fricke, K. 862
Calrson, R.W. 776
Calvin, W.M. and Clark, R.N. 75
Camberon, A.G.W., Colgate, S.A. and Grossman, L. 176
Camberson, A.G.W. 145
Cameron, A. 761
Cameron, A.G. and Ward, W.R. 513

Cameron, A.G.K. 119
Cameron, A.G.W. 144, 176, 589, 813
Cameron, A.G.W. and Benz, W. 228, 813
Cameron, A.G.W. and Ward, W.R. 813
Campbell, B., Walker, G.A.H. and Yang, S. 589
Campbell, D.B., Chandler, J.F., Ostro, S.J. et al. 663
Campbell, I.H. and Griffiths, R.W. 64, 251
Campbell, J.K. and Synnot, S.P. 287
Campbell, J.K. and Anderson, J.D. 287, 681
Campbell, W.W. 76
Cande, S.C. and Kent, D.V. 271
Card, K.D. 208
Card, K.D., Church, W.R., Franklin, J.M. et al. 208
Carlson, B.E., Rossow, W.B. and Orton, G.S. 531
Carlson, R.W., Lugmair, G.W. and Macdougall, J.D. 251
Carlson, R.W., Weissman, P.R., Smythe, W.D., et al. 316
Carlson, R.W., Baines, K.H., Encrenaz, Th. et al. 895
Carlson, R.W. and 20 others 914
Carovillano, R.L. and Forbes, J.M. (eds.) 359
Carozzi, A.V. 315
Carpino, M., Milani, A. and Nobili, A.M. 765
Carr, M.H. 314, 319, 391, 432, 440, 449, 801, 910
Carr, M.H. (ed.) 588
Carr, M.H., Kuzmin, R.A. and Masson, P.L. 319
Carr, M.H., Crumpler, L.S., Cutts, J.A. et al. 432, 453
Carr, M.H. and Evans 910
Carr, T.D., Desch, M.D. and Alexander, J.K. 675
Carroll, D. 935
Carter, N.L. 813
Carter, N.L. and Kennedy, G.C. 873
Carusi, A. 579
Carusi, A. and Massaro, E. 37
Carusi, A. and Valsecchi, G.B. 37, 124
Caspar, M. 379
Cassen, P. and Moosman, A. 761
Cassen, P., Reynolds, R.T., Graziani, F. et al. 813
Cassen, P., Young, R.E., Schubert, G. and Reynolds, R.T. 813
Cassen, P.M., Peale, S.J. and Reynolds, R.T. 421, 813
Cassidy, W.A. and Harvey, R.P. 22
Cassini, G.D. 83
Cassirer, E. 187
Castleman, K.R. 326
Cathles, L.M., III 421
Cattermole, P. 449
Cernicharo, J., Paubert, G., Encrenaz, T. et al. 493
Cess, R.D. et al. 339
Chabrier, G., Saumon, D., Hubbard, W.B. and Luniine, J.I. 713
Chamberlain, J.W. 851
Chamberlain, J.W. and Hunten, D.M. 479, 675, 716, 782, 825, 851
Chamberlain, T.E., and Hunten, D.M. 440
Chamberlin, R.T. 94
Chamberlin, T.C. 94, 612, 623
Chamberlin, T.C. and Chamberlin, R.T. 131
Chamberlin, T.C. and Moulton, F.R. 623
Chandrasekhar, S. 562, 685, 667, 793

Chang, Y., Benoit, P.H. and Sears, D.W.G. 236
Chao, B.F. 94, 215, 386
Chao, B.F. and Yu, A.Y. 386
Chao, B.F. and Gross, R.S. 730
Chao, E.C.T., Boreman, J.A., Minkin, J.A. et al. 71
Chapman, C.R. 32, 599
Chapman, C.R., Morrison, D. and Zellner, G. 32, 35
Chapman, C.R., Paolicchi, P., Zappala, V. et al. 32, 37
Chapman, C.R. and McKinnon, W.B. 75, 193, 266
Chapman, C.R. and Jones, K.L. 183, 453
Chapman, D.R. and Larson, H.K. 805
Chapman, G.A. 748
Chapman, G.A., Herzon, A.D. and Lawrence, J.K. 744
Chapman, S. 57, 297
Chapman, S. and Bartels, J. 211, 228, 704
Chapman, S. and Ferraro, V.C.A. 630
Chappell, B.W. and White, A.J.R. 319
Chapple, W.M. 801
Charney, J.G. 180
Charvátóva, I. 57, 63, 570, 744, 748, 750
Charvátóva, I. and Stréštik, J. 315, 570, 744, 750
Charvátóva-Jakabcová, Stréštik, J. and Křivsky, L. 57
Chen, J.H. and Wasserburg, G.J. 112
Cheng, A.F. 61, 102, 350, 410
Cheng, A.F. and Johnson, R.E. 415
Cheng, A.F., Haff, P.K., Johnson, R.E. and Lanzerotti, L.J. 415
Cheng, A.F., Johnson, R.E. 630
Cheng, A.F., Johnson, R.E., Krimingis, S.M. and Lanzerotti, L.J. 473
Cheng, A.F. and Lanzerotti, L.J. 866
Cheng, C.H. and Toksöz, M.N. 516
Chepil, W.S. and Woodruff, N.P. 238
Chi, X., Dahanayake, C., Wolowczyk, J. and Wolfendale, A.W. 169
Chirikov, B.V. and Vecheslavov, V.V. 98
Choi, D.R., Vasilyev, B.I. and Bhat, M.I. 208
Chopelas, A. and Boehler, R. 421
Christensen, N.I. and Salisbury, M.H. 208
Christensen, P.R., Anderson, D.I., Case, S.C. et al. 459
Christodoulidis, D.C., Smith, D.E., Williamson, R.G. and Klosko, S.M. 386, 828
Christon, S.P. 473
Christy, J.W. and Harrington, R.S. 594
Chupp, E.L. 748, 754
Chvostikov, I.A. 479
Chyba, C. and Sagan, C. 131, 391, 495
Chyba, C.F. 125, 131
Chyba, C.F., Thomas, P.J. and Zahnle, K.J. 131
Chyba, C.F., Owen, T.C. and Ip, W.-H. 131
Chyba, C.F., Brookshaw, T. and Sagan, C. 131
Chyba, C.F., Thomas, P.J. and Zahnie, K.J. 119, 335
Cisowski, S.M. 406
Cisowski, S.M., Collinson, D.W., Runcorn, S.K. et al. 512
Clacys, P., Casier, J.-G. and Margolis, S.V. 467
Clancy, R.T., Muhleman, D.O. and Berge, G.L. 493
Clancy, R.T., Grossman, A. and Muhleman, D.O. 493
Clancy, R.T. and Muhleman, D.O. 493

AUTHOR INDEX

Clark, B.C. 613
Clark, B.C., Baird, A.K., Weldon, R.D. et al. 391
Clark, B.C., III, Baird, A.K., Rose, H.J., Jr. et al. 449, 616
Clark, H. and Stephenson, F. 27
Clark, J.A., Farrell, W.E. and Peltier, W.R. 364
Clark, R.N., Brown, R.H., Owensby, P.D. and Steele, A. 598
Clark, R.N., Fanale, F.P. and Gaffey, M.J. 246, 266
Clark, R.N., McCord, T.B. 258, 713, 914
Clark, R.N., Swayze, G.A., Gallagher, A.J. et al. 914
Clark, R.N., Swayze, G.A., Singer, R.B. and Pollack, J.B. 460
Clark, R.N., King, T.V.V., Klejwa, M. et al. 914
Clarke, F.W. 176
Clarke, J.T., Weaver, H.A., Feldman, P.D. et al. 398
Clarke, J.T., Hudson, M. and Yung, Y.L. 862
Clarke, J.T., Trauger, J. and Waite, J.H., Jr. 851
Clarke, R.S. 22
Clayton, R.N. 242
Clayton, R.N. and Mayeda, T.K. 3, 64, 67
Clayton, R.N. Onuma, N. and Mayeda, T.K. 67
Clayton, R.N., Grossman, L. and Mayeda, T.K. 109, 486
Clayton, R.N., Mayeda, T.K., Goswami, J.N. and Olsen, E.J. 109
Clayton, R.N., Mayeda, T.K. and Rubin, A.F. 236
Clayton, R.N., Onuma, N. and Mayeda, T.K. 486
Clayton, R.N., Mayeda, T.K. and Nagahara, H. 486
Clayton, R.N. and Mayeda, T.K. 741, 785, 873
Clayton, R.N., Mayeda, T.K., Onuma, N. and Shearer, J. 873
Clemence, G.M. 48, 594
Clemett, S.J., Macching, C.R., Zare, R.N. et al. 191
Clerke, A.M. 295, 302
Cloos, M. 392
Clube, S.V.M. and Napier, W.M. 145
Clube, S.V.M. 256
Coates, A.J., Alsop, C., Coker, D.R. et al. 87
Cochran, W.D., Hatzes, A.P. and Hancock, R.J. 101
Cochran, W.D. and Hatzes, A.P. 589
Code, A.D., Houck, T.E. and Lillie, C.F. 119
Coffeen, D.L. and Hansen, J.E. 653
Cohen, I.B. 83, 336
Cohen, T.J. and Lintz, P.R. 744
Colburn, D.S., Pollack, J.B. and Haberle, R.M. 440
Colburn, D.S. and Sonett, C.P. 736
Coles, W.A. and Rickett, B.J. 772
Collis, R.T.H. and Russell, P.B. 696
Colombo, G. 83
Colombo, G. and Shapiro, I.I. 612
Colwell, J.E. and Esposito, L.W. 607
Colwell, J.E. et al. 607, 866
Colwell, J.F. and Esposito, L.W. 534
Colwell, R.N. (ed.) 696
Combes, M., Cara, C., Drossart, P. et al. 575
Condie, K.C. 208

Conley, J.M., Bradley, J.G., Giffin, C.E. et al. 616
Connerney, E.P., and Acuña, M.H. 228
Connerney, J.E.P. 61, 373, 630
Connerney, J.E.P. and Ness, N.F. 478, 813, 863
Connerney, J.E.P., Acuña, M.H. and Ness, N.F. 410, 532, 630, 719
Connes, P., Noxon, J.F., Traub, W.A. and Carleton, N.P. 10
Conrad, M. 187
Conrath, B., Flasar, F.M., Hanel, R. et al. 538, 840
Conrath, B.J., Hanel, R.A. and Samuelson, R.E. 56
Conrath, B.J., Flasar, F.M. and Gierasch, P.J. 531
Conrath, B.J., Gautier, D., Owen, T.C. and Samuelson, R.E. 531
Conrath, B.J. and Pirraglia, J.A. 716
Conrath, B.J. and Gierasch, P.J. 862
Conrath, B.J., Pearl, J.C., Appleby, J.F. et al. 862
Conrath, B.J., Gautier, D., Hanel, R.A. et al. 862
Considine, D.M. and Considine, G.D. 794
Consolmagno, G. 350
Constable, C. 656
Cook, A.H. 161, 287, 415
Cooke, R.U. and Warren, A. 238, 796
Cooke, R.U. and Smalley, I.J. 935
Cooper, H.J., Garstang, M. and Simpson, J. 189
Cooper, M.R. 208
Corliss, W.R. 910
Councelman, C.C. III, Gourevich, S.A., King. R.W. and Loriot, G.B. 895
Courant, R. and Friedrichs, K.O. 736
Courtillot, V., Lemouël, J.L., Ducruix, J. and Cazenave, A. 18
Courtillot, V. and Lemouël, J.L. 57
Courtillot, V., Besse, J., Vandamme, D. et al. 251
Courtillot, V. and Le Mouël, J.-L. 406
Courtright, E.M. (ed.) 24
Coustenis, A. 833
Couzens, D.A. and King, J.H. 772
Coveney, P. and Highfield, R. 188
Cowen, R. 519
Cowley, C.R. and Cowley, A.P. 254
Cowling, T.G. 192
Cox, A. 271, 644, 656
Cox, A. (ed.) 208
Cox, A. and Hart, R.B. 572, 644
Cox, A.V. 271
Cox, D.P. 345
Cox, K.G. 251
Crabb, J. and Schultz, I. 109
Craddock, R.A. and Maxwell, T.A. 453
Craig, C.H. and Sandwell, D.T. 208
Craig, R.A. and Willett, H.C. 71
Crain, I.K. and Crain, P.L. 256
Crain, I.K., Crain, P.L. and Plauut, M.G. 256
Cravens, T.E. 354, 359
Crawford, D. and Schultz, P. 512
Crisp, D. 914
Croft, S.K. 538, 540
Croft, S.K., Kargel, J.S., Kink, R.I. et al. 538
Croft, S.K. and Soderblom, L.A. 834
Croll, J. 179
Cronin, J.R., Pizzarello, S. and Cruikshank, D.P. 79
Cronk, G.W. 234
Crovisier, J. 132, 153
Crovisier, J. and Schloerb, F.P. 140, 493

Crovisier, J., Despois, D., Bockelée-Morvan, D. et al. 493
Cruikshank, D.P. 381, 508, 531
Cruikshank, D.P. et al. 538
Cruikshank, D.P. and Brown, R.H. 813, 840
Cruikshank, D.P., Brown, R.H., Tokunaga, A.T. et al. 598
Cruikshank, D., Morrison, D. and Pilcher, C. 598
Cruikshank, D.P. and Matthews, M.S. (eds.) 927
Cruikshank, D.P., Degewij, J. and Zellner, B. 377
Cruikshank, D.P. and Matthews, M.S. 523
Cruikshank, D.P., Pilcher, C.B. and Morrison, D. 649
Cruikshank, D.P., Roush, T.L., Owen, T.C. et al. 840
Cruikshank, D.P., Tholen, D.J., Hartmann, W.K. et al. 907
Cunningham, C.C. 519, 828
Curisen, R.H. et al. 607
Currie, R.G. 295, 570
Cutts, J.A., Blasius, K.R. and Roberts, W.J. 449
Cuzzi, J.-N. et al. 698
Cuzzi, J.C. 607
Cuzzi, J.N., Lissauer, J.J. and Esposito, L.W. 607, 681, 713
Cuzzi, J.N. and Burns, J.A. 607, 722
Cuzzi, J.N. and Scargle, J.D. 722
Cuzzi, J.N. and Durisen, R.H. 722
Cuzzi, J.N., Lissauer, J.J., Esposito, L.W. et al. 722

D'Amario, L.A., Bright, L. and Wolf, A.A. 289, 310
D'Amario, L.A., Bright, I.E., Byrnes, D.V. et al. 263
D'Hendecourt, L.B. and Jourdain De Muizon, M. 153
Dalrymple, G.B. and Ryder, G. 508
Daly, R.A. 188
Damon, .E. and Linick, T.W. 744
Damon, P. and Sonett, C.P. 570
Damon, P.D. and Linick, T. 750
Dana, J.D. 179
Danby, J.M.A. 71, 92, 221, 223, 594
Darwin, G.H. 181, 747
Davidson, A. 208
Davies, G.F. 813
Davies, G.L. 179
Davies, J.K., Green, S.F. and Geballe, T.R. 140
Davies, M.E. et al. 159
Davies, M.E., Seidelman, P.K., Standish, E.M. and Tholen, D.J. 197
Davies, M.E., Batson, R. and Wu, S.S.C. 578
Davies, M.E., Bursa, M., Abalakin, V.K. et al. 596
Davies, P. 188
Davies, P.C.W. 287
Davies, R.E. 391
Davies, R.E. and Koch, R.H. 391
Davis, A.M. and Olsen, E.J. 785
Davis, D.R. 498, 907
Davis, D.R., Chapman, C.R., Weidenschilling, S.J. and Greenberg, R. 32
Davis, D.R., Weidenschilling, S.J., Farinella, P. et al. 32
Davis, D.R. and Ryan, F.V. 114
Davis, D.R., Champman, C., Greenberg, R. et al. 114

Davis, J.L., Prescott, W.H., Svare, J.L. and Wendt, K. 282
Davis, J.L. Herring, T.A., Shapiro, I.I. 282
Davis, L., Jr., Lüst, R. and Schlüter, A. 736
Davis, M., Hut, P. and Muller, R.A. 521
Davis, M.S. 88
Davis, R., Jr. et al. 754
Davis, R., Jr., Mann, A.K. and Wolfenstein, L. 754
Dawes, W.R. 183
Dearborn, D.S.P. 26
Debergh, C., Lutz, B.I., Owen, T., et al. 531
Deer, W.A., Howie, R.A. and Zussman, J. 319
Degewij, J. and Van Houten, C.J. 377
Degewij, J., Andersson, L.E. and Zellner, B. 377
Dehoffman, F. and Teller, E. 736
Dehon, R.A. 432
Del Genio, A.D. and Suozzo, R.S. 293
Del Genio, A.D. 293
Delambre, J.B.J. 23
Delaney, J.S., Prinz, M., and Takeda, H. 67
Delaney, J.S., Takeda, H., Prinz, M. et al. 67
Delaney, J.S., Nehru, C.E., Prinz, M. and Harlow, G.E. 785
Delaney, J.W. 396
Delgenio, A.D. 51, 179, 162, 188, 786, 841
Delitsky, M.L., Eviatar, A. and Richardson, J.D. 622
Delsemme, A.H. 124, 131, 145, 598
Delsemme, A.H. and Swings, P. 153
Demets, C. 644
Demets, C., Gordon, R.G., Argus, D.F. and Stein, S. 282, 644
Deming, D., Mumma, M.J., Espenak, F. et al. 370
Dennis, J.D. et al. 179
Dennis, J.G. (ed.) 801
Dennis, J.G. et al. 253
Dennis, J.G. et al. (eds.) 392
Dennis, J.G. and Atwater, T.M. 392, 801
Dennison, J., Lingner, D. and Lipschutz, M. 22
Depater, I. 667, 675
Depater, I., Palmer, P. and Snyder, L.E. 493
Depater, I., Romani, P.N. and Atreya, S. 531
Depater, I. and Gulkis, S. 676, 862
Depater, I. and Geortz, C.K. 676
Depater, I. and Dickel, J.R. 676
Depater, I. and Mitchell, D.L. 676
Depater, I., Schloerb, F.P. and Rudolph, A. 676
Depater, I., Palmer, P. and Snyder, L.E. 676
Depater, I., Brown, R.A. and Dickel, J.R. 676
Derrbyshire, E., Gregory, K.J. and Haik, J.R. 241
Dermott, S.F. 570, 607
Dermott, S.F. and Murray, C.D. 98, 738
Dermott, S.F. and Gold, T. 145
Dermott, S.F., Nicholson, P.D., Burns, J.A. and Houck, J.A. 191
Dermott, S.F., Durda, D.D., Gustafson, B.A.S. et al. 941
Dermott, S.F. and Nicholson, P.D. 941
Descartes, R. 761
Desch, M.D. 61
Desitter, W. 594

Despois, D., Crovisier, J., Bockelée-Morvan, D. et al. 493
Dessler, A.J. (ed.) 373, 630
Dessler, A.J., Sandel, B.R. and Atreya, S.K. 851
Deubner, F.L. 221
Deubner, F.-L. and Gough, D. 793
Devaucouleurs, G. 596
Devries, D. and Wahr, J.M. 558
Dewey, J. 801
Dicke, R.H. 744
Dicke, R.H. and Goldenberg, H.M. 756, 793
Dicke, R.H., Kuhn, J.R. and Libbrecht, K.G. 793
Dickey, J.O. 835
Dickey, J.O., Williams, J.G. and Yoder, C.F. 386
Dickey, J.O., Williams, J.G. and Newhall, X.X. 221
Dickey, J.O., Bender, P.I., Faller, J.E. et al. 386
Dickinson, R.E. 756
Dickinson, R.E., Ridley, E.C. and Roble, R.G. 825
Dicks, D.R. 241
Dieke, S.H. 71
Dietz, R.S. 208
Dixon, T.H. 279, 282
Dixon, T.H., Gonzalez, G., Lichten, S. and Katsigris, E. 282
Dixon, T.H. and Kornreich, W.S. 282
Djorgovsky, S. 101
Dobrolovsky, O.V., Kiselev, N.N. and Chernova, G.P. 653
Dobrovolskis, A.R. 375, 439, 548, 607, 712, 649
Dobson, J.F. and Brodetsky, S. 160
Dodd, R.T. 79
Dodson, M.H. 183
Dolan, R., Inman, D.L. and Hayden, B. 799
Dole, S. 13
Dolfus, A. 650
Dolginov, Sh.Sh., Zhuzgov, L.N. and Shkolnikova, S.I. 410
Dolginov, Sh.Sh., Zhuzgov, L.N. and Selyutin, V.A. 410
Dolginov, Sh.Sh. and Pushkov, N.V. 410
Dolginov, Sh.Sh., Zhuzgov, L.N. and Pushkov, N.V. 410
Dolginov, Sh.Sh., Dubinin, E.M., Eroshenko, E.G. et al. 886
Dollfus, A. 688, 653, 654
Dollfus, A., Bastien, P., Leborgne, J.-F. et al. 654
Dollfus, A. and Bowell, E. 654
Dollfus, A., Deschamps, M. and Ksanfomaliti, L.V. 654
Dollfus, A., Deschamps, M. and Zimbelman, J.R. 654
Dollfus, A. and Geake, J.E. 654
Dollfus, A., Mandeville, J-C. and Duseaux, M. 654
Dollfus, A., Wolff, M., Geake, J.E. et al. 654
Dollfus, A. and Zellner, B. 654
Domingue, D.L. 257
Domingue, D.L., Hapke, B.W., Lockwood, G.W. and Thompson, D.T. 246
Donahue, T.M. 398
Donahue, T.M. and Pollack, J.B. 54, 895
Donahue, T.M. and Hodges, R.R., Jr. 934
Dones, L. and Tremaine, S. 558, 612
Dones, L. 607, 719, 722

Dones, L., Cuzzi, J.N. and Showalter, M.R. 722
Dong, D. and Bock, Y. 282
Donn, B. 145
Donn, B. and Urey, C.H. 153
Donnellan, A. 282
Dorman, I.I. 169, 252
Dowling, T.E. 367, 370
Dowling, T.E. and Ingersoll, A.P. 370
Doyle, L.R., Dones, L. and Cuzzi, J.N. 722
Drake, M. 761
Drake, M.J. 907
Drake, S. 258
Dreibus, G., Kruse, H., Spette, B. and Wänke, H. 67
Dreyer, J.L.E. 27, 70, 302
Drummond, J.D. 12, 486
Drummond, J.D., Eckart, A. and Hege, E.K. 907
Dubinin, E.M., Pissarenko, N.F., Barabash, S.V. et al. 575
Dubyago, A.D. 876
Duchaufour, P. 688
Duchesne, J.C. 64
Duckworth, H.F., Barber, R.C. and Venkatasubramanian, V.S. 353
Duennebier, F. and Sutton, G.H. 516
Duke, M.B. and Silver, L.T. 67
Dumont, R. 941
Duncan, M., Quinn, T. and Tremaine, S. 98, 124, 145, 381
Duncan, M.J. and Quinn, T. 119, 560
Duncan, R.A. and Pyle, G. 251
Dunn, J.H. and Howard, D.D. 696
Dunnington, G.W. 268
Duquennoy, A. and Mayor, M. 589
Duriez, L. and Vienne, A. 240
Durisen, R.H. 607
Durrance, S.T., Barth, C.A. and Stewart, A.I.F. 10
Durrance, S.T., Feldman, P. and Weaver, H. 350
Durrant, C.J. 747
Duxbury, T.C. 575
Dyal, P., Parkin, C.W. and Daily, W.D. 512
Dzurisin, D. and Blasius, K.R. 449

Earth Observing System 214
Earth Systems Sciences Committee 215
Eather, R.H. 61
Ebel, A. 479
Eberhart, J. 101
Eberhardt, P., Krankowsky, D., Schulte, W. et al. 153
Ebisawa, S. and Dollfus, A. 654
Eckersley, T.L. 936
Eckert, W.J., Brouwer, D. and Clemence, G.M. 594
Eddington, A.S. 689
Eddy, J.A. 744
Efelber, H.-J. 594
Einstein, A. 689
Eisenstaedt, J. 336
Ekeland, I. 188
Ekholm, N. and Arrhenius, S. 57, 570
El Goresey, A., Nagel, K. and Ramdohr, P. 176
Elachi, C. 69, 232, 233, 696
Elachi, C., Im, E., Roth, L.E. and Werner, C.L. 663
Eldredge, N. and Bould, S.J. 467
Elliot, J.L., Dunham, E.W., Bosh, A.S. 649
Elliot, J.L., Dunham, E.W., Bosh, A.S. et al. 101, 681

AUTHOR INDEX

Elliot, J.L., Dunham, E.W. and Mink, D.J. 607, 866
Elliot, J.L. and Nicholson, P.D. 601, 607, 866
Elliot, J.L. and Young, L.A. 649
Elst, E.W. 876
Elston, W.E. 463
Eluszkiewicz, J. 598
Emanuel, K.A. 180
Emergy, K.O. and Aubrey, D.G. 799
Emerich, C. and Cazes, S. 398
Emiliani, C. 495, 644
Emsley, J. 360
Encke, J.F. 234
Encounters With Comet Halley, The First Results. *Nature*. 279
Encrenaz, T. 490
Encrenaz, T., Lellouch, E., Paubert, G. and Gulkis, S. 493
Encrenaz, T., Lellouch, E., Cernicharo, J. et al. 493
Encrenaz, T., Bibring, J.-P. and Blanc, M. 588
Encrenaz, T. and Bibring, J.-P. 756, 886
Encrenaz, T., Lellouch, E., Rosenqvist, J. et al. 493
Engle, I.M. 373
Eppersom, P.M., Sweedler, J.V., Denton, M.B. et al. 101
Eroshenko, E.G. 886
Erwin, D.H., Valentine, J.W. and Sepkoski, J.J., Jr. 467
Esa Annual Report 248
Esa Bulletin 248
Esa Report To Cospar 248
Esa's Journals In The Fields 248
Eshleman, Tyler, G.L., Anderson, J.D. et al. 681
Eshleman, V.R. 681
Eshleman, V.R., Tyler, G.L., Wood, G.E. et al. 681
Eshleman, V.R. and Tyler, G.L. 681
Eshleman, V.R., Hinson, D.P., Lindal, G.F. and Tyler, G.L. 681
Eshlemann, V.R., Lindal, G.F. and Tyler, G.L. 833
Esposito, L.W. 722
Esposito, L.W. et al. 607, 698
Esposito, L.W., Brahic, A., Burns, J.A. and Marouf, E.A. 607, 866
Esposito, L.W. and Colwell, J.F. 534, 607, 866
Esposito, L.W., Harris, C.C. and Simmons, K.E. 607
Esposito, L.W., Cuzzi, J.N., Holberg, J.B. et al. 681, 722
Esposito, L.W., Knollenberg, R.G., Marov, M.Ya, et al. 895
Esposito, L.W., Knollenberg, J.G., Ya, M. et al. 851
Eugster, O. 396
Evans, J.V. and Hagfors, T. 663
Evans, M.E. 297
Evans, N.J., Gregoire, D.C., Goodfellow, W.D. et al. 360
Evans, J.V. and Hagfors, T. 686
Everhart, E. 124, 145
Eviatar, A., Goldstein, R., Young, D.T. et al. 153
Eviatar, A., Podolak, M. and Richardson, J.D. 622
Eviatar, A., Kennel, C.F. and Neugebauer, M. 622
Explanatory Supp. To The Amer. Ephemeris & Nautical Almanac 240, 555
Ezell, E.C. and Ezell, L.N. 429, 910

Fahleson, U.V. 641
Fahrig, W.F. and West, T.D. 208
Fairbridge, R.W. 10, 11, 26, 45, 57, 68, 71, 79, 178, 179, 229, 240, 241, 251, 267, 273, 283, 293, 295, 298, 299, 313, 315, 361, 365, 380, 392, 393, 495, 559, 562, 579, 623, 687, 658, 650, 701, 704, 705, 725, 732, 737, 744, 747, 748, 793, 801, 808, 834, 8
Fairbridge, R.W. (ed.) 189, 314, 703, 688
Fairbridge, R.W. and Finkl, C.W. Jr. (eds.) 935
Fairbridge, R.W. and Hameed, S. 57, 570, 744
Fairbridge, R.W. and Haubold, H.J. 83, 159, 310, 382, 383, 380
Fairbridge, R.W. and Hetherington, N.S. 93
Fairbridge, R.W. and Marshall, C.P. 76
Fairbridge, R.W. and Sanders, J.E. 26, 63, 315, 570, 744, 784
Fairbridge, R.W. and Shirley, J.H. 57, 63, 570, 744, 750
Fairchild, H.L. 623
Fan, C.Y., Chen, T.M., Yung, S.X. and Dai, K.M. 252
Fanale, F.P., Banerdt, W.B., Elson, I.S. et al. 350
Fanale, P.F., Salvail, J.R., Zent, A.P. and Postawko, S.E. 449
Farinella, P., Milani, A., Nobili, A. and Valsecchi, V. 813
Farmer, C.B., Davies, D.W. and Laportc, D.D. 460
Farthing, W.H. and Folz, W.C. 410
Faure, G. 176, 183
Fauvel, J.K., Flood, R., Shortland, M. and Wilson, R. (eds.) 188
Faye, H. 612
Fegley, B. 169
Fegley, B., Jr. 176, 934
Fegley, B., Jr. and Plame, H. 176
Fegley, B., Jr. and Lewis, J.S. 176
Fegley, B., Jr. and Kornacki, A.S. 176
Fegley, B., Jr. and Prinn, R.G. 176
Fegley, B., Jr. and Ireland, T.R. 176
Fegley, B., Jr., Gautier, D., Owen, T. and Prinn, R.G. 531, 862
Feigl, K.I., King, R.W. and Jordan, T.H. 282
Feldman, P.D. 140, 851
Feldman, W.C., Asbridge, J.R., Bame, S.J. and Montgomery, J.D. 772
Fels, S.B., Schofield, J.T. and Crisp, D. 895
Fenton, C.L. and Fenton, M.A. 94
Fenton, C.L. and Fenton, M.S. 808
Ferlet, R. 253
Fernández, J.A. 124, 145, 381
Fernández, J.A. and Ip, W.-H. 119, 124131, 145, 560
Ferrari, A.J., Sinclair, W.S., Sjogren, W.L. 283
Ferrari, A.J. and Bills, B.G. 596
Ferrari, A.J., Sinclair, W.S., Sjogren, W.L. et al. 813
Ferrari, C. 532
Ferrari, C. and Brahic, A. 534
Festou, M.C., Rickman, H. and West, R.M. 140
Festou, M.C. 140
Feynman, J. 295
Feynman, J. and Fougere, P.F. 295
Feynman, J., Armstrong, T.P. Dao-Gibner, L. and Silverman, S. 754
Feynman, J. and Gabriel, S.B. 744
Feynmann, R.P. et al. 69

Field, J. 379
Fimmel, R.O., Colin, I. and Burgess, E. 57
Fink, U., Hoffman, M., Grundy, W. et al. 32
Fink, U. and Larson, H.P. 531
Fink, U. and Sill, G.T. 598, 914
Fink, U. and Disanti, M.A. 649
Finson, M.L. and Probstein, R.F. 119
Fischbach, E. and Talmadge, C. 287
Fischer, A.G. 256
Fischer, A.G. and Arthur, M.A. 467
Fischer, H.-J. and Spohn, T. 813
Fisher, O. 392
Fitch, F.J. and Miller, J.A. 251
Fitzpatrick, P.M. 92
Fjeldbo, G., Kliore, A.J., Seidel, B. et al. 681
Fjeldbo, G. and Eshleman, V.R. 681
Flam, F. 344
Flammarion, C. 76, 183, 427
Flammer, K.R. 736
Flasar, F.M. 833
Flasar, F.M. and Gierasch, P.J. 370
Flasar, F.M. and Birch, F.J. 813
Flasar, F.M., Conrath, B.J., Gierasch, P.J. and Pirraglla, J.A. Et Al 863
Fleming, J.R. 94
Fletcher, C.H., Fairbridge, R.W., Møller, J.J. and Long, A.J. 364
Flint, R.F. 315
Floran, R.J., Prinz, M., Hlava, P.F. et al. 64
Floran, R.J., Caulfield, J.B.D., Harlow, G.E. and Prinz, M. 785
Floran, R.J. 785
Flückiger, E.O. 754
Flynn, B., Mendillo, M. and Baumgardner, J. 622
Flynn, G. and Sutton, S. 489
Fogel, R.A., Hess, P.C. and Rutherford, M.J. 236
Folland, C.K. 295
Fonkal, P.V. 11
Forbes, E.G. 79
Forbes, W.T.M. 256
Forbush, S.E. 169, 252, 298
Forslund, D.W. 772
Fossat, E. 756
Foukal, P. 744, 745
Foukal, P. and Lean, J. 339, 745, 746
Fowler, C.M.R. 644
Fox, J.I. 4, 10
Fox, J.L. and Dalgarno, A. 10
Fox, J.L. and Stewart, A.L.F. 61
Fox, J.L. and Bougher, S.W. 10, 825
Frakes, I.A. 315
Fram, M.S. and Lesher, C.E. 251
Frank, J., King, A.R. and Raine, D.J. 2
Franklin, F. et al. 698
Franz, O.G. and Millis, R.L. 350
Franz, O.G., Missis, R.L. and Pettauer, T.V. 598
Fraser, R.S. and Curran, R.J. 696
Fraunhofer, J. 254
Fredricks, R.W., Kennel, C.F., Scarf, F.L. et al. 641
Fredriksson, K., Nelen, J. and Fredriksson, B.J. 109
Freiesleben, H.C. 234, 937
French, B. 910
French, B.M. and Short, N.M. 178
French, B.M. and Short, N.M. (eds.) 733
French, H.M. 796
French, L.M. 27, 32, 183, 299, 841
French, L.M., Vilas, F., Hartmann, W.K. and Tholen, D.J. 32
French, L.M. and Binzel, R.P. 101

French, L.M., Vilas, F., Hartmann, W.K. and Tholen, D.J. 299, 841
French, R.G., Elliot, J.L., French, L.M. et al. 287
French, R.G., Nicholson, P.D., Porco, C.C. and Marouf, E.A. 559, 607
French, R.G. et al. 607, 866
French, R.G., Nicholson, P.D., Porco, C.C. and Marouf, E.A. 681
French, R.G., Nicholson, P.D., Porco, C.C. and Marouf, E.A. 866
Frey, H.V. and Schultz, R.A. 449, 453, 463
Friedman, M.P. 736
Frisch, J. (ed.) 379
Friss-Christensen, E. and Lassen, K. 745
Fritz, H. 57
Froeschlé, C. and Gonzci, R. 98
Froeschlé, C. and Scholl, H. 98
Frölich, C. et al. 748
Fuchs, L. and Blander, M. 176
Fujiwara, A., Kamimoto, G. and Tsukamoto, A. 114
Fukao, Y., Obayashi, M., Inoue, H. and Nenbal, M. 421
Fuller, M. and Cisowski, S. 512
Fuller, M., Williams, I. and Hoffman, K.A. 656
Fuller, M.D. 400

Gadsden, M. and Schröder, W. 479
Gaffey, M.J. 32, 109, 907
Gaffey, M.J., Bell, J.F. and Cruikshank, D.P. 32, 35, 242, 488, 914
Gaffey, M.J., Bell, J.F., Brown, R.H. et al. 35
Gaffey, M.J., Reed, K.L. and Kelley, M.S. 486
Gaffey, M.J., Bell, J.F., Brown, R.H. et al. 488
Gaffey, S.J., McFadden, L.A. and Nash, D.B. 914
Galeev, A.A., Gringauz, K.I., Klimov, S.I. et al. 879
Galeev, A.A. 879
Galileo Earth/Moon Encounter 500
Gallardo, T. and Fernández, J.A. 124
Gallino, R., Busso, M., Picchio, G. and Raiteri, C.M. 344
Gamow, G. 258
Ganapathy, R. 360
Ganapathy, R. and Larimer, J.W. 805
Gantier, D., Conrath, B.I., Owan, T. et al. 676
Gapeynski, J.P., Tolson, R. and Michael, W.H., Jr. 450
Garcia, R.R. 479
Garfinkel, B. 570
Garflunkel, Z. (ed.) 45
Garfunkel, A. 416
Garland, G.D. 713
Gatewood, G.D. 589
Gault, D.E., Guest, J.E., Murray, J.B. et al. 476
Gauss, C.F. 51
Gauss, K.F. 268
Gautier, D. and Owen, T. 713
Geake, J.E., Geake, M. and Zellner, B. 654
Geake, J.E. and Dollfus, A. 654
Geary, J.C. 101
Gegout, P. and Cazenave, A. 292
Gehrels, N. and Stone, E.C. 61, 622
Gehrels, T. 32, 48. 101, 377, 654, 774, 775
Gehrels, T. et al. 607
Gehrels, T. (ed.) 366, 372, 708
Gehrels, T. and Matthews, M.S. (eds.) 588, 698, 718

Gissler, P.E., Singer, R.B. and Luccitta, B.K. 449
Geophysical Research Letters 523, 855
Georgi, J. 704
Gérard, E., Bockelée-Morvan, D., Bourgois, G. et al. 493
Gerstenkorn, H. 703
Geyger, W.A. 410
Gibson, E.G. 745, 793
Gidon, P. 256
Gierasch, P.J. and Goody, R.M. 440
Gierasch, P.J. and Conrath, B.J. 863
Giese, A.C. 843
Gigerenzer, G., Swijtink, Z., Porter, T. et al. 188
Gilbert, G.K. 508, 801
Gilbert, W. 273
Gill, A.E. 370
Gill, D. 51
Gill, R. 103
Gillett, F.C. and Houck, J.R. 336
Gillett, F.C. and Ricke, G.H. 531
Gilluly, J. 644
Gilman, P.A. and Howard, R. 745
Gilvarry, J.J. and Hochstim, A.R. 131
Gingerich, O. 295, 379, 570
Gingerich, O. (ed.) 160
Giuli, R.T. 612
Gladman, B. and Duncan, M. 98, 124
Glanz, J. 192
Glass, B.P., Swincki, M.B. and Zwart, P.A. 805
Glass, B.P. 802
Glatzmaier, G.A., Schubert, G. and Bercovici, D. 421
Gleick, J. 92, 188
Gleissberg, W. 57
Gloeckler, G. et al. 103
Gloeckler, G. 353
Gnedykh, V.I., Zhegulev, V.S., Zasova, L.V. et al. 886
Gobel, R., Ott, U. and Bergemann, F. 873
Goddard, R.H. 283
Godfrey, D.A. and Moore, V. 159
Goertz, C. and Morfill, G. 722
Goertz, C.K. 350, 607, 622, 736
Goertz, C.K. et al. 607
Goertz, C.K. and Ip, W.H. 622
Goetz, A.F.H., Vane, G., Solomon, J.E. and Rock, B.N. 914
Goins, N.R., Dainty, A.M. and Toksöz, M.N. 516, 730
Gold, T. 558, 623, 630, 736
Goldberg, B.A., Garneau, G.W. and Lavoie, S.K. 350
Goldberg, R.A. and Aikin, A.C. 819
Golde, R.H. (ed.) 601
Golden, D.C., Morris, R.V., Ming, D.W. et al. 460
Goldreich, P. 154, 558, 700
Goldreich, P. and Tremaine, S. 607, 722, 733, 866
Goldreich, P., Murray, N., Longaretti, P.Y. and Banfield, D. et al. 538, 540
Goldreich, P., Tremaine, S. and Borderies, N. 534, 607
Goldreich, P. and Ward, W.R. 623
Goldreich, P., Murray, N., Longaretti, P.Y. and Banfield, D. 813, 840
Goldschmidt, V.M. 103, 176, 314
Goldsmith, D. 256
Goldstein, B.D., Suess, S.T. and Walker, R.J. 473
Goldstein, H. 92
Goldstein, R.M., Green, R.R., Pettengill, G.H. and Campbell, D.B. 663

Golombek, M., Plescia, J.G. and Franklin, B.J. 463
Gombosi, T.L. 622
Gonzalez, R.C. and Wintz, P. 326
Gooding, J.L. 616, 739, 741
Gooding, J.L. and Keil, K. 109
Goodrich, C.A. 486, 873
Goodrich, C.A. and Berkley, J.L. 873
Goodrich, C.A., Jones, J.H. and Berkley, J.L. 873
Goodrich, C.A., Keil, K., Berkley, J.L. et al. 873
Goodrich, C.A. and Lugmair, G.W. 873
Goodrich, C.A., Patchett, P.J., Lugmair, G.W. and Drake, M.J. 873
Goodwin, A.M. 208
Goody, R.M. and Walker, J.C.B. 54, 786, 842
Goody, R.M. and Yung, Y.L. 667, 782, 851
Göpel, C., Manhes, G. and Allegre, C.J. 110
Gordon, R.G. and Stein, S. 386
Gordon, R.W. 62
Gornitz, V. 441
Gornitz, V. (ed.) 726
Gosling, J.T. 164, 165, 343
Gosling, J.T. and McComas, D.J. 343
Gough, D. and Toomre, J. 748
Gould, S.J. 467
Gradie, J. and Tedesco, F.F. 32, 35
Gradie, J., Thomas, P. and Veverka, J. 375
Gradie, J. and Veverka, J. 32, 350, 841
Graham, A.L. and Annexstad, J.O. 22
Graham, A.L., Bevan, A.W.R. and Hutchison, R. 22
Grant, A.C. 208
Grant, J.A. and Schultz, P.H. 453
Grant, R. 302, 384
Grant, W.H. 935
Grard, R., Nairn, C., Pedersen, A. et al. 575, 641
Grasshoff, G. 659
Grattan-Guiness, I. 385
Gredel, R., Vandishoeck, F.F. and Black, J.H. 140
Greeley, R. 189, 796, 922
Greeley, R., Arvidson, R.E., Elachi, C. et al. 238, 904
Greeley, R. and Batson, R.M. 159, 233, 427
Greeley, R. and Batson, R.M. (eds.) 83
Greeley, R., Williams, S.H., White, B.R. et al. 238
Greeley, R. and Iversen, J.D. 238, 440, 449
Greeley, R., Leach, R.N., Williams, S.H. et al. 238
Greeley, R. and Guest, J.E. 449, 463
Greeley, R. and Schneid, B. 449
Greeley, R. and Crown, D.A. 449
Greeley, R. and Spudis, P.D. 432
Greeley, R., Fink, J., Gault, D.E. et al. 453
Greeley, R. and Iversen, J.D. 796
Green, D.H., Ware, N.G., Hibberson, W.O. and Major, A. 64
Green, H.W. 730
Green, J.C., Cash, W., Cook, T.A. and Stern, S.A. 543
Green, R.M. 159, 561, 800, 940
Greenberg, J.M. and Hage, J.I. 344, 623
Greenberg, R. 623
Greenberg, R. 698, 725
Greenberg, R. et al. 113
Greenberg, R. and Brahic, A. (eds.) 698

AUTHOR INDEX

Greenberg, R., Weidenschilling, S.J., Chapman, C.R. and Davis, D.R. 145
Greenberg, R., Wacker, J., Hartmann, W. and Chapman, C. 761
Greenstein, J.L. 153
Gregor, C.B. 197
Grene, M. 188
Gresh, D.L., Rosen, P.A., Tyler, G.L. and Lissauer, J.J. 607
Gresh, D.L., Marouf, E.A., Tyler, G.L. et al. 681
Gresh, D.L. et al. 866
Grevesse, N., Lambert, D.L., Sauval, A.J. et al. 176
Grewing, M., Praderie, F. and Reinhard, R. (eds.) 140, 257
Grieve, R.A., Stöffler, D. and Deutsch, A. 335
Grieve, R.A.F. 177, 178, 231, 733
Grigorian, A.T. 842
Grimm, R.E. and Solomon, S.C. 904
Gringauz, K.I., Bezrukikh, V.V., Volkov, G.I. et al. 886
Gringauz, K.I., Bezrukikh, V.V., Breus, T.K. et al. 886
Gringauz, K.I., Verigin, M.I., Breus, T.K. and Gombosi, T. 886
Gringauz, K.I. and Verigin, M.I. 879
Gringauz, K.T., Gombosi, T.I., Remizov, A.P. et al. 879
Grinspoon, D.H. 198
Grosser, M. 3, 387, 594
Grossman, L. and Larimer, J.W. 177
Grove, C.I., Hook, S.J. and Paylor, E.D. Ii 914
Grove, T.L. and Kinzler, R.J. 64
Grun, E. et al. 192
Grun, E., Et Al., Zook, H.A., Baghul, M. 344
Grün, E. and Morfill, G.F. 375
Grün, E., Morfill, G.E. and Mendis, D.A. 607
Grün, E., Garneau, G.W., Terrile, R.J. et al. 607
Gubbins, D. 192, 656, 813
Gudehus, D.H. and Hegyi, D.J. 101
Guenther, D.B., Demarque, P., Kim, Y.-C. and Pinsonneault, M.H. 793
Guest, J.E. and O'Donnell, W.P. 476
Guimon, R.K., Keck, B.D. and Sears, D.W.G. 110
Gulick, V.C. and Baker, V.R. 449
Gulkis, S., Kakar, R.K., Klein, M.J. et al. 493
Gulkis, S. and Depater, I. 863
Gulkis, S., Jannsen, M.J. and Olsen, E.T. 863
Gupta, H.K. 730
Gurevich, L. and Lebedinsky, A. 761
Gurnett, D.A. 630, 641
Gurnett, D.A., Kurth, W.S., Allendorf, S.C. and Poynter, R.L. 298
Gurnett, D.A., Kurth, W.S., Poynter, R.L. et al. 641
Gurnett, D.A., Kurth, W.S. and Scarf, F.L. 641
Gurnett, D.A., Kurth, W.S., Cairns, I.H. and Granroth, L.J. 936
Gurnett, D.A., Kurth, W.S., Scarf, F.L. and Pynter, R.L. 641
Gurnett, D.A. and O'Brien, B.J. 641
Gurnett, D.A. and Shaw, R.R. 641
Gurnett, D.A., Shaw, R.R., Anderson, R.R. et al. 641
Gurnett, D.A., Shaw, R.R., Anderson, R.R. and Kurth, W.S. 936
Gurnett, et al. 534

Gutenberg, B. 392
Gutenberg, B. and Richter, C.F. 644, 738

Haberle, R.M. 440
Haberle, R.M., Pollack, J.B, Barnes, J.R. et al. 440
Haberle, R.M., Leovy, C.B. and Pollack, J.B. 440
Haberle, R.M. 432
Hadingham, E. 27
Hadley, G. 293
Haff, P.K., Watson, C.C. and Yung, Y.I. 350
Hager, B.H. and Richards, M.A. 421
Hager, G.H., King, R.W. and Murray, M.H. 282
Hagihara, Y. 92
Hahn, G. and Rickman, H. 105, 124
Hahn, G. and Bailey, M.E. 131
Haines, K. 180, 370
Hakano, T. 589
Hale, G.E. 295, 756, 940
Hale, G.E. and Nicholson, S.B. 295
Hale, W.S. and Head, J.W. 453
Hall, A. 295
Hall, R.C. 684
Halley, E. 297
Halliday, D. and Resnick, R. 696
Halliday, I., Blackwell, A.T. and Griffin, A.A. 486
Halliday, I. 481
Hallion, R.P. and Crouch, T.D. (eds.) 24
Hallion, R.P. and Crouch, T.D. 397
Haman, S.D. 794
Hamblin, W.K. and Christiansen, E.H. 161
Hamilton, D.C., Gloecker, G., Iparich, F.M. et al. 353
Hamilton, D.P. and Burns, J.A. 534, 722
Hammel, H.B. 531
Hammel, H.B., Beebe, R.F., Dejong, E.M. et al. 531
Hammel, H.B., Lawson, S.L., Harrington, J. et al. 531
Hammel, H.B., Young, I.A., Hackwell, J. et al. 531
Hanel, R., Conrath, B., Hovis, W. et al. 449
Hanel, R., Conrath, B., Flasar, F.M. et al. 868
Hanel, R.A., Conrath, B.J., Jennings, D.J. and Samuelson, R.F. 57
Hanel, R.A. 338
Hanel, R.A., Conrath, B.J., Kunde, V.G. et al. 338, 370, 531 716
Hanel, R.A., Conrath, B., Hovis, W. et al. 338
Hanel, R.A., Conrath, B.J., Herath, L.W. Et Al 370, 531
Hanel, R.A., Conrath, B., Flaser, F.M. et al. 716, 863
Hanner, M.S. and Tokunaga, A.T. 140
Hansen, C.J., McEwen, A.S., Ingersoll, A.P. and Terrile, R.J. 840
Hansen, J.E. and Hovenier, J.W. 654
Hanson, W.B., Sanatani, S. and Zuccaro, D.R. 359
Hanson, W.B., Heelis, R.A., Power, R.A. et al. 819
Hanson, W.B., Sanatani, S. and Zuccaro, D.R. 819
Hanson, W.B. and Heelis, R.A. 814, 819
Hapke, B. 246, 590, 605, 914
Hapke, B. and Blewett, D. 663
Hapke, B.W. 561, 578, 684, 776
Hapke, B.W., Nelson, R.M. and Smythe, W.D. 561

Hardesty, R.M., Post, M.J. and Banta, R.M. 696
Hardie, R.H. 62
Hargraves, R.B. 208
Harland, W.B., Armstrong, R.I., Cox, A.V. et al. 271, 644
Harmon, J.K., Campbell, D.B., Hine, A.A. et al. 140
Harmon, J.K. and Slade, M.A. 470
Harmon, J.K., Campbell, D.B., Hine, A.A. et al. 663
Harmon, J.K., Sulzer, M.P., Perillat, P.J. and Chandler, J.F. 663
Harmon, J.K. and Slade, M.A. 663
Harper, J.F. 572, 641, 644
Harris, A. 722
Harris, A.I. 493
Harris, A.L. and Burns, J.A. 32
Harris, A.W. 32, 38, 41, 607, 840, 866
Harris, A.W. and Lupishko, D.F. 38, 41, 578
Harris, A.W. and Ward, W.R. 612
Harris, D.L. 350, 799, 914
Harris, R.A. 26
Harrison, R.A. et al. 165
Hartmann, W. 761
Hartmann, W.K. 113, 114, 201, 512, 513, 708, 813, 922
Hartmann, W.K., Phillips, R.J. and Taylor, G.J. (eds.) 228, 501, 558
Hartmann, W.K. and Tholen, D.J. 114
Hartmann, W.K., Tholen, D.J., Meech, K.J. and Cruikshank, D.P. 105
Hartmann, W.K. and Wood, C.A. 463
Hartmann, W.K., Phillips, R.J., Taylor, G.J. and Jeffrey, G. 421
Hartmann, W.K. and Vail, S.M. 558
Hartmann, W.K. and Davis, D.R. 813
Harvey, O.L. 366
Harwit, M. 287
Hassanzadeh, J., Rubin, A.E. and Wasson, J.T. 785
Hasselmann, K. 18
Hatfield, C.B. and Camp, M.J. 256
Hatzes, A., Wenkert, D.D., Ingersoll, A.P. and Danielson, G.F. 180, 370
Hauber, E., Regner, P., Schmidt, K. et al. 575
Haubold, H.J. 3, 70, 258, 378, 379, 751, 786
Haubold, H.J. and Chao, B.F. 242
Haubold, H.J. and Mathai, A.M. 793
Hawkes, R.I. 481
Hawkesworth, C.J., Mantovani, M.S.M., Taylor, P.N. and Palacz, Z. 251
Hawkesworth, C.J., Marsh, J.S., Duncan, A.R. et al. 251
Haxton, W.C. 754
Hayashi-Smith, J. and Mouginis-Mark, P.J. 453
Hays, J.D., Imbrie, J. and Shackleton, N.J. 495
Hays, P.B. and Roble, R.G. 851
Head, J.W., Crumpler, L.S., Aubele, J.C. et al. 904
Head, J.W. and Crumpler, L.S. 922
Heelis, R.A., Hanson, W.B., Lippincott, C.R. et al. 819
Heiken, G.H., Vaniman, D.T. and French, B.M. 25, 501, 914
Heiskanen, W.A. and Vening Meinesz, F.A. 292
Heiskanen, W.A. and Moritz, H. 283
Heisler, J. 124, 145
Heisler, J. and Tremaine, S. 124, 145
Heisler, J., Tremain, S. and Alcock, C. 124

Kovalevsky, J. and Brumberg, V.A. (eds.) 287
Kovalevsky, J., Mueller, I.I. and Kolaczek, B. 159
Kowal, C.T. 124
Koyré, A. 188
Kozi, Y. 37
Krall, N.A. and Trivelpiece, A.W. 641
Krasnopolsky, V.A. 10, 601, 825
Krasnopolsky, V.A. and Tomashova, 10
Krasnopolsky, V.A., Sandel, B.R. and Herbert, F. 538
Krasnpol'skii, V.A., Krys'ko, A.A., Rogachev, V.N. and Parshev, V.A. 886
Kraus, J.D. 232, 676
Krehbiel, J.P., Brace, L.H., Theis, R.F. et al. 819
Krenkel, E. 802
Kresak, L. 32, 124, 131
Krimigis, S.M. et al. 103
Krimigis, S.M., Carbary, J.F., Keath, E.P. et al. 350
Krisher, T.P. 287
Krisnamurthy, P. and Cox, K.G. 251
Kristian, J. and Blouke, M. 102
Krogdahl, W.S. 26
Kroner, A. 208
Kropotkin, P.N. 257
Krupp, E.C. 27
Kruse, P., McGlaughin, L. and McQuistan, R. 696
Ksanfomaliti, I.V., Dedova, E.V., Obykhova, L.F. et al. 886
Ksanfomaliti, I.V., Dedova, E.V., Zolotukhin, V.G. et al. 886
Ksanfomality, L., Murchie, S., Britt, D. et al. 575
Ksanfomality, L.V., Scarf, F.L. and Taylor, W.W.I. 601
Ksanfomality, L.V., Zubkova, V.M., Morozov, N.A. and Petrova, E.V. 730
Kucheryavenkov, A.I., Yakovlev, O.I., Pavelev, A.G. et al. 887
Kuenen, P.H. 934
Kuhn, J.R., Libbrecht, K.G. and Dicke, R.H. 745
Kuhn, T.S. 160, 302, 379
Kuiper, G.P. 145, 338, 381, 540, 833
Kukla, G. and Fairbridge, R.W. 495
Kukla, G. and Gavin, J. 495
Kukla, G.J. 315, 495
Kulčar, L. and Leftus, V. 570
Kulikov, K.A. 51, 658
Kulikovsky, P.G. 732
Kullmer, C.J. 295
Kundt, W. 794
Kupo, I., Mekler, Yu. and Eviatar, A. 350, 622
Kupperian, J.E., Byram, E.T., Chubb, T.A. and Freidman, H. 398
Kurat, G. 873
Kurt, V.G., Dostavalow, S.B. and Scheffer, E.K. 10
Kurth, W.S. 61, 641, 663, 936
Kurth, W.S. and Gurnett, D.A. 298, 641
Kurucz, R.L., Furenlid, I., Brault, J. and Testerman, L. 914
Kutzbach, J.F. 495
Kvenvolden, K.A. 392
Kyte, F.T., Smit, J. and Wasson, J.T. 360
Kyte, F.T. and Wasson, J.T. 360

Labitzke, K. and Van Loon, H. 570, 745
Lachenbruch, A.H. 421
Laclare, F. 756
Lagerkvist, C.-I., Harris, A.W. and Zappala, V. 32
Lagrange, J.L. 383
Laj, C., Mazaud, A., Fuller, M. and Herrero-Bervera, E. 271, 656
Lal, D. 169
Lallement, R. 345
Lallement, R., Bertaux, J.I., Chasselière, F. and Sandel, B. 345, 851
Lallement, R., Bertaux, J.L. and Kurt, V.G. 772
Lambeck, K. 217, 221, 268, 269, 289, 292, 364, 828, 835
Lammelein, D.R. et al. 516
Lammlein, D.R. 730
Lancaster, N. 238
Lancaster, N. and Greeley, R. 450
Lande, K. 794
Landgrebe, D.A. 696
Landscheidt, T. 252, 570, 835
Lane, A.I., West, R.A., Hord, C.W. et al. 531
Lane, A.L., Nelson, R.M. and Matson, D.L. 350, 598
Lane, A.L. et al. 607
Lane, A.L., Hord, C.W., West, R.A. et al. 868
Lang, G.J. and Peltzer, R.G. 932
Lang, K.R. 794
Langacker, P. and Mann, A.K. 287
Langel, R.A. 406
Langereis, C.G., Vanhoof, A.A.M. and Rochette, P. 656
Langseth, M.G., Keihm, S. and Peters, K. 512, 813
Lanzerotti, L.J. and Brown, W.L. 350
Lanzerotti, L.J., Brown, W.L., Augustyniak, W.M. et al. 350
Laplace, P.S. 23, 612, 761
Larimer, J.W. 177
Larmor, J. 192
Larson, K. 283
Larson, R. 761
Larson, R.L. and Olson, O. 656, 813
Larson, S.M., Edberg, S.J. and Levy, D.H. 140
Laskar, J. 94, 98, 229, 762, 765
Laskar, J., Joutel, F. and Robutel, P. 98, 765
Laskar, J., Quinn, T. and Tremaine, S. 761, 7655
Laskar, J. and Robutel, P. 98, 558, 765
Latham, D.W., Mazah, T., Stefanik, R.P. et al. 589
Lazarus, A.J. and McNutt, R.I., Jr. 622
Leach, R.W. 102
Lean, J. 339, 745, 746, 825
Lebas, M.J., Lemaitre, R.W. and Woolley, A.R. 319
Leberl, F.W. 578
Lebertre, T. and Zellner, B. 654
Lebofsky, L.A. 42, 93, 598
Lebofsky, L.A., Feierberg, M.A., Tokunaga, A.T. et al. 32, 93
Lebofsky, L.A., Greenberg, R., Tedesco, E.F. and Veeder, G.J. 45
Lebofsky, L.A., Jones, T.D., Owensby, P.D. et al. 488
Lebofsky, L.A. and Spencer, J.R. 32
Lebofsky, L.A. and Spencer, J.R. 45
Lebofsky, L.A., Sykes, M.V., Tedesco, E.F. et al. 45
Lebretin J.-P. 87, 311, 133
Lebreton, J.-P. and Matson, D.L. 87
Lee, B.G. 910
Lee, P., Ebisawa, S. and Dollfus, A. 654
Lee, T. 177

Leer, E. and Holzer, T.E. 772
Lefort, P. 319
Legg, M.P. and Westfold, K.C. 676
Lehninger, A.I. 392
Leick, A. 283
Lellouch, E. 833
Lellouch, E., Belton, M.J.S., Depater, I., et al. 493, 676
Lellouch, E., Coustenis, A., Raulin, F. et al. 833
Lellouch, E., Encrenaz, T., Phillips, T. et al. 493
Lellouch, E., Goldstein, J.J., Bougher, S.W. et al. 493
Lellouch, E., Paubert, G. and Encrenaz, T. 493
Lemaire, J. and Scherer, M. 772
Lemaitre, R.W., Lebas, M.J., Sabine, P.A. et al. 319
Lenogle, J. (ed.) 667
Leovy, C.B. 440
Lepichon, X. 645
Lerbekmo, J.F. and St Louis, R.M. 360
Levasseur-Regourd, A.-C. 941
Leverrier, V.J.J. 594
Levin, B. 761
Levin, G.V. 392
Levin, G.V. and Straat, P.A. 392
Levin, Z., Borucki, W.J. and Toon, O.B. 601
Levine, J. 761
Levison, H.F. and Duncan, M.J. 145
Levison, H.F. 124
Levison, H.F. and Duncan, M.J. 124
Levitus, S. and Oort, A.H. 18
Levy, D.H. 835
Levy, E. and Sonnett, C. 761
Levy, E., Ruzmaikin, A. and Ruzmaikina, T. 761
Levy, E.H. and Lunine, J.I. 588
Levy, E.H., Gatewood, G.D., Stein, J.W. and McMillan, R.S. 589
Levy, E.H., McMillan, R.S., Gatewood, G.D. et al. 589
Levy, J.R. 387, 701, 703
Lewis, J. 751
Lewis, J.S. 177, 350, 833
Lewis, J.S., Barshary, S.S. and Noyes, B. 177
Lewis, J.S. and Lewis, R.A. 394
Lewis, J.S. and Prinn, R.G. 177, 201, 531, 863
Lewis, R.S., Amari, S. and Anders, E. 112, 344
Lichten, S.M., Marcus, S. and Dickey, J.O. 283
Liebes, S., Jr. 910
Lieske, J.H. 240
Lightfoot, P. and Hawkesworth, C.J. 251
Likin, O.B., Golubkova, M.A., Dyachokov, A.V. et al. 575
Limaye, S.S. 370
Limaye, S.S. and Stromovsky, L.A. 531
Lin, C.C. and Shu, E.H. 608
Lin, D. and Papaloizou, J. 761
Lin, D.N.C., Papaloizou, J.C.B. and Ruden, S.P. 534
Lin, R.P., Anderson, K.A. and Hood, L.L. 512
Lin, Y.T., Nagel, H.-J., Lundberg, L.L. and El Goresy, A. 236
Lindal, G.F. 531, 602
Lindal, G.F., Wood, G.E., Hotz, H.B. et al. 833
Lindal, G.F., Hotz, H.B., Sweetnam, D.N. et al. 682

Lindal, G.F., Lyons, J.R., Sweetman, D.N. *et al.* 531, 863
Lindblad, B.A. and Southworth, R.B. 37
Lindqwister, U.J., Freedman, A. and Blewitt, G. 283
Lindsay, R.B. and Margenau, H. 92
Lindstrom, M.M. 19, 395
Lindstrom, M.M., Schwarz, C., Score, R. and Mason, B. 396
Liou, K.-N. 667
Lipa, B.J. and Tyler, G.L. 682
Lipschutz, M.E., Gaffey, M.J. and Pellas, P. 32
Lipschutz, M.E. and Samuels, S.M. 486
Lipschutz, M.E. 873
Lissauer, J. 534, 608, 623, 722, 761
Lissauer, J.J. and Kary, D.M. 612
Lissauer, J.J. and Safronov, V.S. 612
Lissauer, J.J., Shu, F.H. and Cuzzi, J.N. 722
Lissauer, J.J., Squyres, S.W. and Hartmann, W.K. 608
Lister, G.S., Etheridge, M.A. and Symonds, P.A. 802
Liu, A.S. and Laing, P. 509
Liu, H.-S. 495
Livengood, T.A., Moos, H.W., Ballester, G.E. and Prangé, R.M. 851
Livi, S. and Marsch, E. 772
Lockwood, G.W. 746
Lockwood, G.W. and Thompson, D.T. 531, 746
Lodge, J.P. 4
Lofgren, G.E. 110
Longair, M. 794
Longaretti, P.-Y. 608, 722
Longhi, J. 63, 64, 241, 249
Longhi, J. and Pan, V. 64, 67, 242
Longrigg, J. 808
Loper, D.E. 656
Lorell, J. 509
Lorenz, E.N. 18
Loughnan, F.C. 935
Love, S.G. and Brownlee, D.E. 192
Lovelock, J. 188, 201
Lovelock, J. and Whitfield, M. 201
Lovelock, J.E. 197
Lowell, A.L. 394
Lowell, P. 394, 594
Lowenstam, H.A. 406
Lowman, P.D. 703
Lowman, P.D., Jr. 25, 197, 194, 201, 208
Lowman, P.D., Jr., Allenby, R.J. and Frey, H.V. 208
Lowrie, W. 269, 654
Lowrie, W. and Alvarez, W. 271
Lubbock, C.A. 299
Lucchitta, B.K. and Soderblom, L.A. 246
Lucchitta, B.K., McEwen, A.S., Clow, G.D. *et al.* 463
Lucchitta, B.K. 450
Lugmair, G.W. and Galer, S.J.G. 112, 486
Luhmann, G. and Russell, C.T. 454
Luhmann, J.G. 61, 410, 630, 907
Luhmann, J.G. and Brace, L.H. 410, 455
Luhmann, J.G., Russell, C.T., Brace, L.H. and Vaisberg, O.L. 630, 813
Luhmann, J.G. and Russell, C.T. 905
Lunine, J.I. 177, 233, 708, 713, 828, 833
Lunine, J.I., Atreya, S.K. and Pollack, J.B. 543, 713, 833
Lunine, J.I. and Hunten, D.M. 531
Lunine, J.I. and Nolan, M.C. 538, 840
Lunine, J.I. and Stevenson, D.J. 598
Lunine, J.I., Stevenson, D.J. and Yung, Y.L. 713, 833
Lunine, J.I. and Tittemore, W.C. 828

Luu, J. 377
Luu, J.X. and Jewitt, D. 124, 145
Lynden-Bell, D. and Pringle, J.E. 722
Lyot, B. 654, 688
Lyttleton, R.A. 145, 783, 840

Macdonald, G.J.F. 223, 703
Macdonald, F.B., Teegarden, B.J., Trainor, J.H. and Webber, W.R. 298
Macdonald, G.A. 319
Macdougall, J.D., Lugmair, G.W. and Kerridge, J.F. 112
Macdougall, J.D. 251
Mach, E. 92
Mach, F. 336
Mackay, C. 102
Mackin, R.J. and Neugebauer, M. (eds.) 772
Mackinnon, D.J. and Tanaka, K.I. 450
Mackinnon, I. and Rietmeijer, F. 489
Mackwell, S.J., Zimmerman, M.E., Kohlstedt, D.L. and Scherber, D.S. 904
Maclow, M.M. and Ingersoll, A.P. 370
Macmillan, W.D. 94
Macpherson, G.J., Wark, D.A. and Armstrong, J.T. 177
Macpike, F.F. 297
Macy, W. and Trafton, L. 350
Macy, W. and Sinton, W. 531
Magalhaes, J.A., Weir, A.L., Conrath, B.J. *et al.* 370
Magnusson, P. 41
Magnusson, P., Barucci, M.A., Drummond, J.D. *et al.* 32, 41
Magnusson, P., Barucci, M.A., Binzel, R.P. *et al.* 41
Maher, K.A. and Stevenson, D.J. 131
Malaise, D.J. 153
Malcuit, R.J., Mehringer, D.M. and Winters, R.R. 76
Malcuit, R.J., Mehringer, D.M. and Winters, D.R. 538
Malhotra, R. 75, 266, 700, 828
Malhotra, R. and Dermott, S. 498, 701
Malin, M.C. and Pieri, D.C. 246
Malin, M.C., Danielson, G.E., Ingersoll, A.P. *et al.* 459
Maling, D.H. 427
Malone, T.F. (ed.) 794
Mansinha, L., Smylie, D.E. and Chapman, C.H. 730
Maquire, W. 531
Maraldi, G. 427
Marcialis, R.L., Rieke, G.H. and Lebofsky, L.A. 649
Marcy, G.W. and Butler, R.P. 102, 509
Mariolopoulos, E.G. 71
Markham, B.L. and Barker, J.L. 696
Marochnik, L.S., Mukhin, I.M. and Sagdeev, R.Z. 124, 145
Marouf, E.A., *et al.* 608
Marouf, E.A. and Tyler, G.L. 682
Marouf, E.A., Tyler, G.L. and Rosen, P.A. 562, 608, 682, 722
Marouf, E.A., Tyler, G.L. and Eshleman, V.R. 682
Marouf, E.A., Tyler, G.L., Zebker, H.A. *et al.* 722
Marov, M.Ya., Bryvshev, B.V., Manuilov, K.N. *et al.* 887
Marsch, E., Mühlhäser, K.-H., Rosenbauer, H. 772
Marsch, E., Mühlhäser, K.-H., Schwenn, R. *et al.* 772
Marsch, E. and Schwenn, R. (eds.) 630, 772

Marsden, B.G. 51, 124, 140, 145, 339, 342, 380, 394, 540, 876
Marsden, B.G., Sekanina, Z. and Everhart, E. 124, 560
Marsden, B.G. and Cameron, A.G.W. 83
Marsden, G.B. 380
Marsden, R.G., Wenzel, K.-P., Afonin, V.V. *et al.* 575
Marsh, B.C. 64
Marsh, J.G., Lerch, F.J., Putney, B.H. *et al.* 292
Marshack, A. 27
Marshall, C.P. 103
Marshall, J.R. and Oberbeck, V.R. 257
Marten, A., Gautier, D., Owen, T. *et al.* 493, 531
Marten, A., Gautier, D., Tanguy, L. *et al.* 493
Marti, K. and Graf, T. 169, 486
Martin, D.H. and Puplett, E. 338
Martin, P.M. and Mason, B. 177
Martin, T.Z. 460
Marvin, U.B. 22, 45, 486
Mason, B. 22, 67, 177, 242
Mason, D.L. 84
Mason J. (ed.) 140
Mason, J.W. (ed.) 279
Massey, P. and Jacoby, G.H. 102
Massie, S.T. and Hunten, D.M. 531, 863
Masursky, H. *et al.* 552
Masursky, H. and Crabill, N.L. 910
Matese, J.J. and Whitman, P.G. 257
Matese, J.J., Whitman, P.G., Innanen, K.A. and Valtonen, M.J. 257
Mathai, A.M. and Haubold, H.J. 754, 794
Mather, K.F. 94
Matson, D.L. 87
Matson, D.L., Veeder, G.J., Tedesco, E.F. and Lebofsky, L.A. 32
Matson, D.L. and Nash, D.B. 350
Matson, D.L., Goldberg, B.A., Johnson, T.V. and Carlson, 622
Matsuda, J-I., Fukunaga, K. and Ito, K. 873
Matthews, R. 12, 51
Mauk, B.H., Krimigis, S.J., Keath, E.P. *et al.* 863
Maxwell, J.C. 392
Maxwell, T.A. 463
May, K.O. 268
Mayer, C.H., McCullough, T.P. and Sloanaker, R.M. 492
Mayor, M. and Queloz, D. 589
Mazeh, T., Latham, D.W. and Stefanik, R.P. 589
Mazurs, E.G. 103
McCall, G.J.H. 236
McCarthy, D.D. and Babcock, A.K. 18
McCarthy, D.D., Seidelmann, P.K. and Vanflandern, T.C. 658
McCartney, E.J. 667
McCauley, J.F. 476
McCauley, J.F., Smith, B.A. and Soderblom, L.A. 350
McCauley, J.F., Guest, J.E., Schaber, G.G. *et al.* 476
McClain, J.S. and Orcutt, J.A. 179
McClatchey, R.A., Fenn, R.W., Selby, J.E.A. *et al.* 696
McCleese, D.J., Haskins, R.D., Schofield, J.T. *et al.* 459
McCluskey, S.C. 27
McComas, D.J. and Phillips, J.L. 165
McComas, D.J. and Nordholt, J.F. 353
McCord, T.B. 840
McCord, T.B. and Adams, J.B. 914

McCord, T.B., Adams, J.B. and Johnson, T.V. 67
McCord, T.B., Clark, R.N., Hawke, B.R. et al. 915
McCoy, T.J., Keil, K., Mayeda, T.K. and Clayton, R.N. 3
McCrea, W.H. 145, 257
McCulloch, M.T. 208
McCuskey, S.W. 92
McDonnell, J.A.M. (ed.) 489
McDonnell, J.A.M., Lamy, P.I. and Pankiewicz, G.S. 119
McDonnell, J.A.M., Sullivan, K., Stevenson, T.J. and Niblell, D.H. 192, 480
McDonough, T.R. and Brice, N.M. 415, 622
McDougall, I. 543
McElhinny, M.W. 572
McElhinny, M.W. and Senanayake, W.E. 813
McElroy, M.B., Kong, T.Y. and Yung. 440
McEntire, R.W. et al. 103
McEwen, A.S. 246, 576
McEwen, A.S. and Soderblom, L.A. 350
McFadden, L.-A., Tholen, D.J. and Veeder, G.J. 12, 25, 51
McFadden, L.A. 460
McFadden, P.I. and Merrill, R.T. 656
McGhee, G.R., Jr. 467
McGill, G.E. and Dimitriou, A.M. 450
McGlynn, T.A. and Chapman, R.D. 145
McGovern, P. and Schubert, G. 813
McGrath, M.A. 851
McGrath, M.A., Johnson, R.E. and Lanzerotti, L.J. 473
McIntyre, M.E. and Shepherd, T.G. 370
McIntyre, M.E. 479
McKay, C.P. 387, 392, 495
McKay, C.P., Pollack, J.B. and Courtin, R. 292, 833
McKay, C.P. and Davis, W.I. 392
McKay, C.P. and Stoker, C.R. 392
McKay, C.P., Toon, O.B. and Kasting, J.F. 392
McKay, D.S., Heiken, G., Basu, A. et al. 512
McKee, C.F. 345
McKee, E.D. (ed.) 238
McKenna-Lawlor, S.M.P., Afonin, V.V., Gringauz, K.I. et al. 575
McKinney, M.L. 467
McKinnon, W.B. 243, 649, 813, 840
McKinnon, W.B. and Melosh, H.J. 75
McKinnon, W.B. and Parmentier, E.M. 75, 266
McKinnon, W.B. and Leith, A.C. 538, 840
McKinnon, W.B. and Mueller, S. 649, 650, 840
McKinnon, W.B. and Benner, L.A.M. 840
McLaren, D.J. and Goodfellow, W.D. 197
McMahon, S. and Dveck, S. 591
McMillan, R.S. 98
McMillan, R.S., Moore, T.I., Perry, M.L. et al. 102, 589
McMillan, R.S., Moore, T.L., Perry, M.L. and Smith, P.H. 775
McMillan, R.S., Smith, P.H., Moore, T.L. et al. 589
McMillan, R.S., Smith, P.H., Frecker, J.E. et al. 589
McMillan, R.W. 588
McMullin, E. 160
McNally, D. (ed.) 342
McNamee, J.B., Borderies, N.J. and Sjogren, W.L. 905
McNish, A.G. and Lincoln, J.V. 745

McNutt, M.K. 364
McNutt, R.I., Jr. 61, 622
McPherron, R.L. 211
McSween, H.Y. 65, 432, 873
McSween, H.Y., Jr. 79, 486, 741
McSween, H.Y., Sears, D.W.G. and Dodd, R.T. 110
McVittie, G.C. 689
Measures, R.M. 696
Mechikunnel, A.T., Lee, R.B. III, Kyle, H.L. and Major, E.R. 746, 748
Meech, K.J. and Belton, M.J.S. 105
Mees, C.E.K. 326
Mees, C.E.K. and James, T.H. 326
Meeus, J. 223
Meier, R.R. 10, 851
Meissner, R. 802
Melnik, W.L. 805
Melosh, H.J. 761
Melosh, H.J. 79, 114, 197, 326, 395
Melosh, H.J. and McKinnon, W.B. 476
Melosh, H.J. and Schenk, P. 75
Melosh, H.J. and Sonett, C.P. 513
Melosh, H.J. and Vickery, A.M. 131, 335, 432, 453
Melrosse, D.B. 641
Memmesheimer, M. and Blum, P.W. 479
Mendell, W.W. (ed.) 508
Mendillo, M., Baumgardner, J. and Flynn, B. 501
Mendillo, M., Baumgardner, J., Flynn, B. and Hughes, J. 630
Mendis, D.A., Houpis, H.L.F. and Marconi, M.L. 354
Mendis, D.A., Hill, J.R., Ip, W.-H. 608
Menshutkin, B.V. 393
Menzel, D.H. 254
Meriwether, J.W. 10
Merrill, G.P. 688
Merrill, R.B. (ed.) 395
Merrill, R.T. and McElhinny, M.W. 211, 406
Message, P.J. 92
Metzger, A.E., Gilman, D.A., Luthey, J. and Hurley, K.C. 415
Metzger, A.F., Gilman, D.A., Luthey, J.L. et al. 61
Meyer, D.M. and Roth, K.C. 153
Meyer, J.P. 756
Meyerhoff, A.A. 257
Meyerhoff, A.A., Taner, I., Morris, A.F.I. et al. 208
Michelson, A.A. 338
Middlehurst, B.M. and Kuiper, G.P. (eds.) 71
Mignard, F. 540, 558
Mihalas, D. 254, 667
Mihalov, J.D., Sonett, C.P., Binsack, J.H. and Mitsoulas, M.D. 736
Mikereit, B., Forsyth, D.A., Green, A.G. et al. 208
Miki, S. 613
Milani, A. and Knezevic, Z. 38
Milani, A. and Nobili, A.M. 570
Milankovitch, M. 339, 495, 570
Miller, S.I. 392, 495
Miller, S.L., Smythe, W.D. 598
Millis, R.I. and Schleicher, D.G. 153
Millis, R.L. and Dunham, D.W. 32
Millis, R.L., Wasserman, L.H. and Birch, P. 866
Millis, R.L., Wasserman, L.H., Franz, O.G. et al. 93
Millman, P.M. 197
Milton, D.J. 46, 598
Milton, S. (ed.) 11, 480
Miner, E.D. 554, 834, 856, 868, 922, 927

Minster, J.-F., Ricard, L.-P. and Allegre, C.J. 236
Mischenko, M.I. 561, 654
Mishchenko, M.I. and Dlugash, J.M. 41
Misner, C.W., Thorne, K.S. and Wheeler, J.A. 287, 380
Mitchell, D.G., Roelof, E.C., Feldman, W.C. et al. 772
Mitchell, D.L. and Depater, I. 676
Mitrofanov, I., Chernenko, A., Dolidze, V. et al. 575
Mittlefehldt, D.W. 2, 67, 783, 785, 786
Mittlefehldt, D.W. and Lindstrom, M.M. 3, 68
Mittlefehldt, D.W., Chou, C.-L. and Wasson, J.T. 786
Mittlefihidt, D.W. and Longhi, J. 65
Mizuno, H. 613
Moberg, A. 295
Mogi, K. 731
Möhlmann, D., Riedler, W., Rustenbach, J. et al. 576
Moik, J.G. 326
Molchanov, A.M. 570
Molnar, P. and Tapponnier, P. 645
Monod, J. 188
Montanari, A., Asaro, F., Michel, H.V. and Kennett, J.P. 360
Moore, B. and Dozier, J. (eds.) 215
Moore, C.F. 254
Moore, J.G., Clague, D.A., Holcomb, R.T. et al. 450
Moore, J.M. and Malin, M.C. 266
Moore, J.M. and McKinnon, W.B. 263
Moore, J.M. and Spencer, J.R. 840
Moore, M.H. 350
Moore, P. 22, 27, 71, 75, 181, 183, 186, 233, 234, 295, 300
365, 386, 387, 427, 540, 560, 579, 835, 842
Moorison, D. and Owens, T. 197
Moos, H.W., Fastie, W.G. and Bottema, M. 10, 398
Morabito, L.A., Synnott, S.P., Kupferman, P.N. and Collins, S.A. 598
Morbidelli, A. and Moons, M. 701
Moreno, F., Molina, A. and Ortiz, J.I. 102
Morgan, J.S. 622
Morgan, P. and Phillips, R.J. 310
Morgan, T.H. 501
Morgan, T.H., Zook, H.A. and Potter, A.E. 473
Morgan, T.H. and Shemansky, D.E. 501
Morgan, W.J. 251, 310
Moriyasu, K. 287
Mörner, N.A. 364
Moroscow, A.A. 393
Moroz, V.I. 350, 731
Moroz, V.I., Spankuch, D., Kinkin, V.M. et al. 338
Moroz, V.I., Petrova, E.V., Ksanfomality, L.V. et al. 576
Moroz, V.I., Parfent'ev, N.A., San'ko, N.F. et al. 887
Morris, R.V., Agresti, D.G., Lauer, H.V., Jr. et al. 460
Morris, R.V., Gooding, J.L., Lauer, H.V., Jr. and Singer, R.B. 460
Morris, R.V., Lauer, H.V., Jr., Lawson, C.A. et al. 915
Morrison, D. 690, 718
Morrison, D. (ed.) 366, 708, 922
Morrison, D., Chapman, C.P., and Slovic, P. 131
Morrison, D., Johnson, T.V., Shoemaker, E.M. et al. 725
Morrison, D., Owen, T. and Soderblom, L.A. 233, 713, 725

Morrison, D. and Telesco, C.M. 350
Morrison, L.V. 594
Mörth, H.T. and Schlamminger, L. 570
Moses, J.I., Yung, Y.L. and Allen, M. 531
Moses, S.L., Coroniti, F.V., Kennel, C.F. et al. 630
Mottinger, N.A., Sjogren, W.L. and Bills, B.G. 364, 905
Motz, L. and Weaver, J.H. 336
Motz, L. and Duveen, A. 380
Mouginis-Mark, P.J., McCoy, T.J., Taylor, G.J. and Keil, K. 453
Moulton, F.R. 23, 92, 94, 229, 380, 383, 613
Moutsoulas, M.D. 387
Moyer, T.D. 287, 939
Mueller, B.E.A., Tholen, D.J., Hartmann, W.K. and Cruikshank, D.P. 32
Mueller, G. 873
Mueller, I.I. 658
Mueller, S., Talor, G.J. and Phillips, R.J. 512
Mueller, S. and McKinnon, W.B. 645
Muhleman, D.L., Grossman, A.W., Butler, B.J. and Slade, M.A. 676
Muhleman, D.O., Holdrige, D.B. and Block, N. 51
Muhleman, D.O., Berge, G.L. and Clancy, R.T. 493
Muhleman, D.O., Grossman, A.W., Butler, B.J. and Slade, M.A. 663, 833
Muller, E. and Morris, D. 805
Muller, P.M. and Sjogren, W.L. 283, 509, 682
Müller, R. 941
Muller, R.A. and Morris, D.E. 656
Multauf, L.S. 559
Mumma, M.J., Weaver, H.A., Larson, H.P. et al. 140
Münch, R.E., Sagdeev, R.Z. and Jordan, J.F. 879
Munitz, M.K. (ed.) 520
Munk, W.H. and Macdonald, G.J.F. 83, 217, 243, 747
Murchie, S., Mustard, J., Bishop, J. et al. 460
Murchie, S.L., Head, J.W. and Plescia, J.B. 75
Murchie, S.L., Britt, D.T., Head, J.W. et al. 185, 573
Murcray, F.H., Murcray, D.G. and Williams, W.J. 45
Murray, B. and Burgess, E. 476
Murray, B., Naraeva, M.K., Selivanov, A.S. et al. 576
Murray, B., Malin, M.C. and Greeley, R. 806
Murray, C.A. 92
Murray, C.D. 87, 732, 738
Murray, C.D. and Thompson, R.P. 733
Mustard, J.F., Erard, S., Bibring, J.-P. et al. 460
Mutch, T.A., Arvidson, R.E., Head, J.W., III et al. 432
Mutter, J.C., Buck, W.R. and Zehnder, C.M 251
Muursepp, P. and Preem, R. 560

Nakajima, T., Oppenheimer, B.R., Kulkarni, S.R. et al. 589
Nakamura, N., Unruh, D.M., Tatsumoto, M. and Hutchinson, R. 64
Nakamura, T. and Yasuda, N. 377
Nakamura, Y. 513, 516, 731
Nakamura, Y. et al. 516
Nakamura, Y., Latham, G.V. and Dorman, H.J. 512, 516

Namiki, N. and Solomon, S.C. 904
Napier, W.M. 257
Napier, W.M. and Staniucha, M. 124
Napp, G.R. 345
Narain, U. and Ulmschneider, P. 772
Narcisi, R.S. and Bailey, A.D. 819
Nash, D.B., Carr, M.H., Gradie, J. et al. 258, 350, 813
Nash, D.B., Fanale, F.P. and Nelson, R.M. 350
Nash, D.B. and Nelson, R.M. 350
Natenzon, M.Y., Neishtadt, A.I., Sagdeev, R.Z. et al. 98
National Aeronautics and Space Administration 215, 366
National Research Council 215
Nature. 616
Naval Observatory 48
Naylor, R.H. 258
Nedell, S.S., Squyres, S.W. and Andersen, D.W. 450
Needham, J. 188
Neel, I. 406
Negi, J.G. and Tiwari, R.K. 257
Nehru, C.E., Prinz, M., Weisberg, M.K. et al. 3
Nehru, C.E., Prinz, M., Weisberg, M.K. and Delaney, J.S. 236
Nelson, R.M. 561, 598, 775
Nelson, R.M., Smythe, W.D., Hapke, B.W. and Cohen, A.J. 351
Nelson, R.M. and Lane, A.L. 351
Nelson, R.M. and Smythe, W.D. 350
Nelson, R.M., Pieri, D.C., Baloga, S.M. et al. 350
Nelson, R.M. and Hapke, B.W. 350
Nelson, R.M., Hapke, B.W., Smythe, W.D. et al. 561
Nelson, R.M., Lane, A.L., Matson, D.I. et al. 350, 598, 851
Nerem, R.S., Chao, B.F., Chan, J.C. et al. 386
Ness, N.F., Acuña, M.H., Behannon, K.W. et al. 228
Ness, N.F. 410
Ness, N.F., Acuña, M.H., Behannon, K.W. et al. 410
Ness, N.F., Behannon, K.W., Lepping, R.P. and Shatten, K.H. 410
Neubauer, F.M., Luttgen, A. and Ness, N.F. 630
Neugebauer, O. 27, 300
Neugebauer, M. (ed.) 772
Neukum, G. and Hiller, K. 453
Neukum, G. and Wise, D.U. 453
Newburn, R.I., Jr., Neugebauer, M. and Rahe, J. (eds.) 119, 153, 297, 359
Newcomb, S. 51, 540, 594, 658
Newell, H.E. 684
Newhall, X.X. 238, 240, 838
Newhall, X.X., Standish, E.M., Jr. and Williams, J.G. 240, 287
Newkirk, G., Jr. and Frazier, K. 745
Newman, J.R. (ed.) 311, 382
Newsom, H.E. and Jones, J.H. (eds.) 188, 228
Newsom, H.E. and Taylor S.R. 228
Newsom, H.E. and Sims, K.W.W. 228
Newton, C.W. 18
Newton, H.A. 145
Newton, I. 188
Newton, R.R. 659
Nicholson, P.D. 864
Nicholson, P.D. et al. 608
Nicholson, P.D. and Dones, L. 608, 722
Nicholson, P.D., Hamilton, D.P., Matthews, K. and Yoder, C.F. 722

Nicholson, P.D. and Matthews, K. 375
Nicholson, P.D., Matthews, K. and Goldreich, P. 608, 866
Nicolis, G. and Prigogine, I. 188
Nier, A.J. and McElroy, M.B. 395
Nier, A.O. 354
Nierenberg, W.A. 802
Nieto, M.M. 570, 834
Nikishin, A.M. 808
Nisbet, E.G. 197, 208
Nishizumi, K., Arnold, J.R., Caffee, M.W. et al. 396
Nobili, A. 299
Nobili, A.M., Milani, A. and Carpino, M. 765
Noens, J.-C. 746
Nolan, M.C. 656, 938
Nordlund, A. 756
North, F.J. 315
North, G.R. et al. 339
North, J.D. 520
Northrup, T.G., Opp, A.G. and Wolfe, J.H. 713
Noxon, J.I., Traub, W.A., Carleton, N.P. and Connes, P. 10
Nozette, S. et al. 113
Nyquist, L.E. 453, 741
Nyquist, I.F., Takeda, H., Bansal, B.M. et al. 68
Nyquist, L.E., Lugmair, G., Signer, P. et al. 616

O'Brien, M.P. and Johnson, J.W. 799
O'Keefe, J.A. 805
O'Leary, B. and Ververka, J. 598
O'Neil, W.J. 263
O'Neil, W.J., Ausman, N.E., Johnson, T.V. and Landano, M.R. 263
Oberbeck, V.R. 231
O'Reilly, W. 273
Oberbeck, V.R. and Aggerwal, H. 131
Oberbye, D. 342
Oberst, J. 516
Oberst, J. and Nakamura, Y. 516
Ockert, M.E., Cuzzi, J.N., Porco, C.C. and Johnson, T.V. 866
Ogilvie, K.W., Kittredge, R.I. and Wilkerson, T.D. 354
Ohtsuki, K. 722
Ojakangas, G.W. and Stevenson, D.J. 246, 813
Okada, A., Keil, K., Taylor, G.J. and Newsom, H. 3
Oliver, J.E. and Fairbridge, R.W. (eds.) 4, 316, 650
Olivero, J.J. and Homas, G.E. 479
Oliverson, R.J., Scherb, F. and Poesler, F.L. 351
Ollier, C.D. 935
Olsen, E.E., Bunch, T.E., Jarosewich, E. 236
Oort, A.H. 13, 18
Oort, A.H. and Peixoto, J.P. 18, 201
Oort, J., Miller, S. and Lazcano, A. 761
Oort, J. 119, 124, 145, 153 560, 761
Oparin, A. 761
Öpik, E.J. 114, 119, 131, 560, 656, 938
Oppenheimer, M. 153
Oro, J., Miller, S.J. and Lazcanpo, A. 197
Oro, J. 131
Orrall, F.Q. 745, 748
Orth, C.J. 467
Orth, C.J., Attrep, M., Jr., and Quintana, L.R. 360
Orton, D.C. and Widing, K.G. 298
Orton, G.S., Aitken, D.K., Smith, C. et al. 863

Orton, G.S., Griffin, M.J., Ade, P.A.R. et al. 531, 863
Orton, G.S., Lacy, J.H., Achtermann, J.M. et al. 531
Osborn, W.H., A'Hearn, M.F., Carsenty, U. et al. 140
Osterbrock, D.F. 345
Ostro, S.J. 32, 663, 676
Ostro, S.J., Campbell, D.B. and Shapiro, I.I. 32
Ostro, S.J., Campbell, D.B., Simpson, R.A. et al. 75, 246
Ostro, S.J., Campbell, D.B., Chandler, J.F. et al. 663
Ostro, S.J., Chandler, J.F., Hine, A.A. et al. 663
Owen, T., Maillard, J.P., Debergh, C. and Lutz, B.L. 398
Owen, T. and Sagan, C. 381
Owen, T., Biemann, K., Rushneck, D. et al. 616
Owen, T. and Gautier, D. 833
Owen, T.C., Roush, T.L., Cruikshank, D.P. 650
Owen, T.C., Roush, T.L., Cruikshank, D.P. et al. 840
Owen, W.M., Jr., Vaughan, R.M. and Synnott, S.P. 534
Oyama, K.-I. 819
Oyama, V.I. and Berdahl, B.J. 392
Oyama, V.L., Carle, G.C. and Berdahl, B.J. 616
Ozima, M. and Podosek, F.A. 543

Pacheco, J.F., Scholz, C.H. and Sykes, L.R. 731
Page, D.E. 851, 855
Paige, D.A., Crisp, D. and Santec, M.L. 460
Paige, D.A. and Ingersoll, A.P. 440
Pais, A. 188
Palme, H. and Wlotzka, F. 177
Palme, H. and Fegley, G., Jr. 177
Pang, K.D., Voge, C.C., Rhoads, J. and Ajello, J. 415
Pannekock, A. 70, 302, 659
Pannella, G. 221
Panofsky, H.A. 163
Pap, J. 746
Papanastassiou, D.A., Rajan, R.J., Huneke, J.C. and Wasserburg, G.J. 112
Park, A.F. 189, 915
Park, C.G. 601
Parker, E.N. 192, 298, 343, 736, 772
Parker, G.D. 272
Parker, R.A. 939
Parker, T.J., Saunders, R.S. and Schneeberger, D.M. 450
Parkinson, C.D., McConnell, J.C., Sandel, B.R. et al. 851
Parkinson, T.D. and Hunten, D.M. 440
Parks, G.K. 630
Parmentier, E.M. and Hess, P.C. 904
Parry, A. 393
Parthasarathy, R. and King, J.H. 298
Pasachoff, J.M. 231, 480
Pascu, D., Panossian, S.P. and Schmidt, R.E. 377
Pascu, D. and Schmidt, R.E. 594
Pasquill, F. 4
Passey, Q.R. and Shoemaker, E.M. 75, 266
Paterson, W.S.B. 650
Patterson, C. 486
Paubert, G., Marten, A., Rosolen, C. et al. 493

Paul, W., Reinhard, H.P. and Vonzahn, U. 819
Paxton, L.J. and Anderson, D.E. 825
Payne-Gaposchkin, C. 480, 941
Peale, S.J. 83, 154, 558, 701, 813
Peale, S.J. and Cassen, P. 813
Peale, S.J., Cassen, P. and Reynolds, R.T. 351, 421, 814, 828
Pearl, J.C. and Conrath, B.J. 531
Pearl, J.C., Conrath, B.J., Harel, R.A. et al. 863
Pearl, J.C., Conrath, B.J., Hanel, R.A. and Pirraglia, J.A. 370
Pearl, J.C., Hanel, R. and Kunde, V. 351
Pearl, J.C. and Sinton, W.M. 351
Peck, B.M. 370
Pecker, J.C. 747
Pedersen, et al. 534
Peixoto, J.P. and Oort, A.H. 339
Pelayo, A.M. and Wiens, D.A. 645
Pellas, P. and Störzer, D. 112, 118, 183
Pellas, P. 486
Peltier, W.R. 645
Pepin, R. 761
Pepin, R.O. 543, 934
Percival, J.A. and Card, K.D. 208
Perez De Tejeda, H. 736
Perley, R.A., Schwab, F.R. and Bridle, A.H. 676
Peters, K.J. 663
Peterson, C. 938
Petersson, O. 26
Petrosky, T.Y. 98
Pettengil, G.H. 663
Pettengill, G.H., Ford, P.G. and Wilt, R.J. 233
Pettersson, O. 570
Philbrick, C., Fiare, A.C. and Fryklund, D.H. 479
Phillips, J.L., Stewart, A.I.F. and Luhmann, J.G. 10, 61
Phillips, R.J., Grimm, R.E. and Malon, M.C. 189
Phillips, R.J. 463
Phillips, R.J., Arvidson, R.E., Boyce, J.M. et al. 335
Phillips, R.J., Grimm, R.E. and Malon, M.C. 310, 904, 922
Phillips, R.J. and Lambeck, K. 292
Phillips, R.J. and Malin, M.C. 310
Phillips, R.J., Robertas, R.F., Arvidson, R.E. et al. 904
Philpotts, A.R. and Martello, A. 251
Pickering, W.H. 579, 594
Pieri, D.C., Baloga, S.M., Nelson, R.M. and Sagan, C. 351
Pieri, L.J. and Rudnyk, M.E. 686, 687
Pieters, C. and Englert, P. (eds.) 686
Pieters, C.M., Head, J.W., Adams, J.B. et al. 65
Pike, R.J. 453, 476
Pilcher, C.B., Ridgeway, S.T. and McCord, T.B. 598
Pilcher, C.B. and Morgan, J.S. 622
Pilipp, W.G., Miggenrieder, H., Montgomery, M.D. et al. 772
Pimentel, G.C., Forney, P.B. and Herr, K.C. 598
Piotrowski, S. 114
Pizzo, V.J. 772
Pizzo, V.J., Holzer, T.E. and Sime, D.G. (eds.) 630, 772
Plafker, G. and Savage, J.C. 731
Planetary and Space Science 855
Plaut, J.J., Kahn, R., Guinness, E.A. and Arvidson, R.E. 450
Plaxton, I.J. and Anderson, D.E. 10

Plescia, J.B. and Golombek, M.P. 463
Pleskot, L.K. 598
Podolak, M. and Reynolds, R.T. 287
Podolak, M., Hubbard, W.B. and Stevenson, D.J. 287, 531, 863
Podosek, F.A. 110, 131
Podosek, F.A., Zinner, E.K., Macpherson, G.J. et al. 183
Podosek, F.A. and Swindle, T.D. 183, 543
Poincaré, H. 98
Poirer, J.-P. 814
Pollack, J.B. 54, 201, 293, 713, 895
Pollack, J.B. et al. 863
Pollack, J.B. and Bodenheimer, P. 713
Pollack, J.B., Burns, J.A. and Tauber, M.E. 840
Pollack, J.B. and Fanale, F. 246, 377
Pollack, J.B., Kasting, J.F. and Poliakof, K. 432
Pollack, J.G., Kasting, J.F., Richardson, S.M. and Polikoff, K. 440, 450, 453
Pollack, J.B., Lunine, J.I. and Tittemore, W.C. 538, 713
Pollack, J.B., Podolak, M., Bodenheimer, P. and Christofferson, B. 531
Pollack, J.B., Rages, K., Baires, K.H. et al. 863
Pollack, J.B., Rages, K., Pope, S.K. et al. 531
Pollack, J.B., Roush, T., Witteborn, F. et al. 450, 460
Pollack, J.B., Toon, O.B. and Boese, R. 895
Pollack, J.B., Witteborn, F.C., Erickson, E..F. 258, 351
Pollack, J.B., Witteborn, F.C., Erickson, E.F. et al. 598
Pollard, H. 92
Polynov, B.B. 935
Porco, C. 534, 733, 738
Porco, C. and Goldreich, P. 733
Porco, C.C. 608, 722
Porco, C.C. et al. 608
Porco, C.C. and Eliason, P.T. 319
Porco, C.C., Nicholson, P.D., Cuzzi, J.N. et al. 608
Porco, C.C. and Goldreich, P. 866
Porco, C.C., Cuzzi, J.N., Ockert, M.E. and Terrile, R.J. 866
Posner, E., Rauch, L. and Madsen, B. 184
Potter, A.E. and Morgan, T.H. 473, 501
Potts, D.L. 459
Powell, B.N. 786
Poynting, J.H. 656
Prangé, R. 616, 622
Pratt, W.K. 326
Pravec, P., Tichy, M., Ticha, J. et al. 48
Preece, W.H. 936
Prentice, A.J.R. 385
Prentice, J.P.M. and Phillips, T.E.R. 186
Press, F. and Siever, R. 796
Prévot, M., Mankinen, E.A., Grommé, C.S. and Coe, R.S. 656
Price, M. and Suppe, J. 904
Price, M.J. and Franz, O.G. 531
Priest, E. (ed.) 630
Priest, E.R. 747
Prigogine, I. and Stengers, I. 188
Prinn, R.G. and Fegley, B., Jr. 177
Prinn, R.G., Larson, H.P., Caldwell, J. and Gautier, D. 716
Prinn, R.G. and Fegley, B., Jr. 840
Prinz, M., Nehru, C.E. and Delaney, J.S. 786
Prinz, M., Nehru, C.E. and Delaney, J.S., et al. 786

AUTHOR INDEX

Prinz, M., Nehru, C.E., Weisberg, M.K. and Delaney, J.S. 236
Prior, G.T. 786
Proctor, R. 761
Proctor, R.A. 183
Pryor, W.R., West, R.A., Simmons, K.E. and Delitsky, M. 531
Puffer, J.H., Hurtubise, D.O., Geiger, F.J. and Lechler, P. 251
Pugh, D.T. 799
Pumfrey, S. 273
Purcell, J.D. and Tousey, R. 398

Quennell, A.M. (ed.) 802
Quinn, T.R., Tremaine, S. and Duncan, M. 612, 765

Rabe, E. 51, 383
Rabin, D., Wilson, M.R. and Moore, R.L. 750
Rabinowitz, D.L. 102, 755
Raloff, J. 102
Rampino, M.R. 255, 257, 463, 467
Rampino, M.R. and Caldeira, K. 257, 467
Rampino, M.R. and Haggerty, B.M. 467
Rampino, M.R. and Stothers, R.B. 251, 257, 467
Randall, B. 876
Rasio, F.A. et al. 589
Rasool, S.I. 248
Ratcliffe, R.A. 359
Ratcliff, P.R., McDonnell, J.A.M., Firth, J.G. and Gruen, E. 87
Raulin, F., Mourey, D. and Toupance, G. 392
Raup, D.M. 257, 467
Raup, D.M. and Sepkoski, J.J., Jr. 257, 467
Ray, R.D. 828
Raynaud, D., Jouzel, J., Barnola, J.M. et al. 201
Razdan, H., Colburn, D.S. and Sonett, C.P. 736
Reasenberg, R.D. 450, 682
Reasenberg, R.D. and Goldberg, Z.M. 292
Reasenberg, R.D., Goldberg, Z.M., Macneil, P.E. and Shapiro, I.I. 905
Redman, R.O., Feldman, P.A., Mathers, H.E. et al. 45
Reedy, R.C. 169, 754
Reedy, R.C. and Marti, K. 754
Rees, M.H. 10, 359
Reiner, M.J., Fainberg, J., Stone, R.G. et al. 622
Reingold, N. 704
Reinhard, R. 246, 274
Reinhard, R. and Battrick, B. 616
Renne, P.R., Ernesto, M., Pacca, I.G. et al. 251
Rense, W.A. 398
Renzetti, N.A. 183
Reynolds, R.T., McKay, C.P. and Kasting, J.F. 246, 392
Reynolds, R.T., Squyres, S.W., Colburn, D.S. and McKay, C.P. 392
Ribes, E. et al. 748
Rice, A. and Fairbridge, R.W. 310
Richards, M.A., Duncan, R.A. and Courtillot, V.E. 251
Richardson, J.D., Belcher, J.W., Zhang, M. and McNutt, R.L., Jr. 622
Richardson, J.D. and Eviatar, A. 622
Richardson, J.D., Eviatar, A. and Siscoe, G.L. 622
Richardson, R.W. 738
Richardus, P. and Adler, R.K. 427

Richmond, A.D. 359
Richter, C.F. 731
Rickman, H. 119, 124, 297
Riedler, W., Möhlmann, D., Oraevsky, V.N. et al. 410
Riedler, W., Schwingenschuh, K., Lichtenegger, H. et al. 576
Rigaud, S.P. 70
Righter, K., Arculus, R.J., Delano, J.W. and Paslick, C. 786
Riley, J.P. and Skirrow, G. (eds.) 314
Ringwood, A.E. 181, 201, 228, 873
Rinnert, K. 601
Rizk, B., Wells, W.K., Hunten, D.M. et al. 102
Robertson, H.P. 656
Robinson, L.B., Brown, W. and Eilmore, K. 102
Robinson, M.S. and Tanaka, K.L. 450
Roble, R.G. 61, 825
Roble, R.G., Ridley, E.C., Richmond, A.D. and Dickinson, R.E. 825
Roddy, R.J., Pepin, R.O. and Merrill, R.B. (eds.) 178, 734
Rohan, C.A. et al. 939
Romani, P.N. and Atreya, S.K. 531
Romani, P.N., Bishop, J., Bézard, B. and Atreya, S. 531
Romani, P.N., Lellouch, F., Rosenqvist, J. et al. 532
Romanov, S.A., Smirnov, V.N. and Vaisberg, O.L. 887
Romanowicz, B. 416, 421, 731
Ronan, C.A. 297
Ronan, C.A. et al. 366
Ronca, L.B. 46
Rosen, E. 160
Rosen, P.A., Tyler, G.L. and Marouf, E.A. 682
Rosen, P.A., Tyler, G.L., Marouf, E.A. and Lissauer, J.J. 682, 722
Rosen, R. 188
Rosen, R.D. 18
Rosen, R.D., Salstein, D.A. and Wood, T.M. 18
Rosenbauer, H. (ed.) 772
Rosenberg, G.D. and Runcorn, S.K. (ed.) 221
Rosenfeld, A. and Kak, A.C. 326
Rosenfeld, L. 380
Rosenqvist, J., Lellouch, E., Romani, P.N. et al. 493, 532
Ross, M. and Schubert, G. 814
Ross, M.N. and Schubert, G. 421
Ross, M.N., Schubert, G., Gaskell, G.R.W. and Spohn, T. 421
Rossi, B. 169, 252, 326
Rossman, G.R. 915
Rossow, W.B. 54
Rossow, W.B., Delgenio, A.D., Limaye, S.S. et al. 895
Rossow, W.B., Zhang, Y.-C. and Lacis, A.A. 339
Rostoker, G. and Falthammar, C.-G. 343
Rothery, D.A. 233, 713
Rottman, G.J. and Moos, H.W. 10
Rotto, S.L. and Tanaka, K.L. 463
Roush, T.I. 459, 460
Roush, T.I. and Blake, D. 460
Roush, T.I., Blaney, D.I. and Singer, R.B. 461, 915
Roush, T.L., Pollack, J.B., Witteborn, F.C. et al. 75
Roux, A., Perraut, S., Rauch, J.L. et al. 641
Roy, A.E. 310
Roy, A.F. 92

Roy, A.E. and Ovendon, M.W. 154
Royal Society *Catalogue Of Scientific Papers* 579
Rozelot, J.P. 165, 745, 754
Rubey, W.W. 314, 598
Rubin, A.E. 873
Rubin, A.E. and Kallemeyn, G.W. 110
Rubin, A.E., Fegley, B., Jr. and Brett, R. 177
Rubin, A.F. and Mittlefehldt, D.W. 68, 786
Rubincam, D.P. 825
Ruden, S. and Pollack, J. 761
Rufener, F. 102
Ruggles, C.L.N. 27
Ruggles, C.L.N. and Whittle, A.W.R. 27
Rumaikina, T.V. and Wuchterl, B. 1
Runcorn, S.K. 161, 512, 572, 814, 874
Russell, C.T. 61, 373, 406, 410, 455, 478, 601, 719, 863, 907
Russell, C.T. (ed.) 263, 359, 772, 987
Russell, C.T., Baker, D.N. and Slavin, J.A. 478, 630
Russell, C.T., Coleman, P.J., Jr. and Goldstein, B.E. 512, 814
Russell, C.T., Elphic, R.C. and Slavin, J.A. 887
Russell, C.T., Hoppe, M.M. and Livesey, W.A. 630
Russell, C.T. and Luhmann, J.G. 208, 373, 476, 532, 718, 863
Russell, C.T. and McPherron, R.L. 211
Russell, C.T. and Vaisberg, O. 359
Russell, H.N. 42, 561, 776
Russell, J.F. and Batson, R.M. 543
Rust, D. 748
Ruzmaikina, T.V. 751, 756, 761
Ryabov, Y. 287
Ryan, E., Hartmann, W. and Davis, D. 114
Ryan, J.A., Hess, S.L., Henry, R.M. et al. 440
Ryder, G. 193, 501
Ryder, G., Bogard, D. and Garrison, D. 508

Sabine, P.A. 317, 319, 360
Sabins, F.F. 915
Sackmann, I.-J., Boothrayd, A.I. and Fowler, W.A. 794
Sadler, D.H. 747
Safronov, V. 513, 761
Safronov, V.S. 32, 113, 119, 145, 560, 612, 623
Sagalyn, R.C., Smiddy, M. and Bhargava, Y.N. 819
Sagan, C. 131, 351, 381, 392
Sagan, C. and Chyba, C. 840
Sagan, C. and Dermott, S.F. 713
Sagan, C. and Fox, P. 726
Sagan, C., Toon, O.B. and Gierasch, P.J. 453
Sagdeev, R.Z. and Zaslavsky, G.M. 98
Sagdeev, R.Z. and Zakharov, A.V. 576
Sagdeev, R.Z., Blamont, J., Galeev, A.A. et al. 879
Sagdeev, R.Z., Linkin, V.M., Blamont, J.E. and Preston, R.A. 879
Saikumar, V. and Goldstein, J.I. 786
Salisbury, J. 686
Salo, H. 722
Salstein, D.A. and Rosen, R.D. 18
Salters, V.J.M. and Hart, S.R. 65
Sampson, R.A. 240
Samson, J.A.R. 851
Samuelson, R.F. 54, 57, 292
Sanchez-Lavega, A. 716

Sanchez-Lavega, A., Colas, F., Lecacheur, J. et al. 717
Sandel, B.R. 843
Sandel, B.R. et al. 10
Sandel, B.R. and Broadfoot, A.L. 61
Sandel, B.R., Broadfoot, A.L. and Stroble, D.F. 398
Sandel, B.R., Herbert, F., Dessler, A.J. and Hill, T.W. 61, 851
Sanders, J.E. 179, 241, 495, 570
Sandford, M.C.W. 87
Sandwell, D.T. and Schubert, G. 645, 731, 904
Santer, R. and Dollfus, A. 654
Sargent, A. and Berkwith, S. 751
Sargent, A.I. and Beckwith, S.V.W. 589
Sarton, G. 382, 659
Saunder, R.S. 399, 887
Saunders, R.W., Arvidson, R.E., Head, J.W. III et al. 890
Saunders, R.S. and Carr, M.H. 319
Saunders, R.S. and Pettengill, G.H. 400, 663, 890
Saunders, R.S., Pettengill, G.H., Arvidson, R.E. et al. 663, 696, 889
Saunders, R.S., Spear, A.J., Allin, P.C. et al. 400
Savage, B.D. and Caldwell, J. 532
Scarf, F.L., Fredricks, R.W., Frank, L.A., et al. 641
Scarf, F.L., Fredricks, R.W., Frank, L.A. and Neugebauer, M. 641
Scarf, F.L., Gurnett, D.A. and Kurth, W.S. 641
Scarf, F.L., Gurnett, D.A., Kurth, W.S. and Poynter, R.L. 641
Scarf, F.L., Taylor, W.W.L. and Green, I.M. 641
Schaber, G.G., Boyce, J.M. and Trask, N.J. 476
Schaber, G.G., Strom, R.G., Moore, H.J. et al. 904
Schaefer, M.W. and Schaefer, B.E. 540
Schaffer, L. and Burns, J.A. 375
Schaffer, S. 297
Schaifers, K. 937
Schardt, A.W. and Goertz, C.K. 663
Schatten, K.H. and Arking, A. (eds.) 748, 746, 794
Schatten, K.H. and Sofia, S. 745
Schatzman, E. 762
Scheidegger, A. 364
Scheidegger, A.E. 241
Schenk, P. 498, 495, 840
Schenk, P. and Jackson, M.P.A. 840
Schenk, P.M. 75, 266, 267
Schenk, P.M. and McKinnon, W.B. 73, 75, 246, 267
Schiaparelli, G. 726
Schidlowski, M. 392
Schlesinger, M.E. 339
Schloerb, F.P. 676
Schloerb, F.P., Kinzel, W.M., Swade, D.A. and Irvine, W.M. 493
Schloerb, F.P., Robinson, S.E. and Irvine, W.M. 493
Schmidt, H.U., Wegmann, R., Huebner, W.F. and Boice, D.C. 153
Schmidt, O. 762
Schmidt, P.W., Williams, G.E. and Embleton, B.J.J. 558
Schmidt, W.K.H., Rosenbauer, H., Shelley, E.G. and Geiss, J. 772
Schmitt, H.H. 319
Schneider, D.A. and Kent, D.V. 656, 805
Schneider, N.M., Smyth, W.H. and McGrath, M.A. 622

Schneider, S.H. and Boston, P.J. (eds.) 201
Schneider, S.H. and Dickinson, R.E. 339
Schnetzler, C.C. 805
Schofield, J.T. and Taylor, F.W. 895
Scholl, H., Schmadel, L.D. and Roser, S. 32
Scholl, H. 105
Scholl, H. and Froeschlé, C. 701
Schopf, J.W. (ed.) 392
Schou, J., Christensen-Dalsgaard, J. and Thompson, M.J. 794
Schove, D.J. 57, 570, 750
Schove, D.J. (ed.) 27, 295, 937
Schröder, W. 478, 479
Schröder, W. and Treder, H.-J. 688
Schubart, J. 299
Schubart, J. and Matson, D.L. 32
Schubert, G. 416, 421, 808, 814, 895
Schubert, G., Cassen, P. and Young, R.E. 814
Schubert, G., Ross, M.N., Stevenson, D.J. and Spohn, T. 814
Schubert, G. and Spohn, T. 455, 814
Schubert, G., Solomon, S.C., Turcotte, D.I. et al. 421
Schubert, G., Spohn, T. and Reynolds, R.T. 188, 351, 421, 814
Schubert, G., Stevenson, D. and Cassen, P. 814
Schubert, G., Stevenson, D.J. and Ellsworth, K. 421, 814
Schubert, G., Solomon, S.C., Turcotte, D.L. et al. 814
Schubert, G., Turcotte, D.L., Solomon, S.C. and Sleep, N. 814
Schubert, J. and Bien, R. 383
Schultz, P.H. and Gault, D.E. 432, 453, 476
Schultz, P.H. and Lutz-Garihan, A.B. 454
Schultz, P.H., Schultz, R.A. and Rogers, J. 454
Schultz, P.H. and Srnka, L.J. 512
Schultz, P.H. 731
Schultz, R.A. 463
Schunk, R.W. 359
Schuster, A. 71, 570
Schutz, W. 393
Schwartz, R.D. and James, P.B. 257
Schwarz, J.H. 287
Schwarzbach, M. 316
Schwarzschilld, M. 794
Schwenn, R. and Marsch, E. (eds.) 772
Schwerdtner, W.M. and Lumbers, S.B. 208
Sciama, D.W. 336
Science 263, 366, 927, 932
Scott, D.H. 463
Scott, D.H. and Carr, M.H. 463
Scott, D.H. and Dohm, J.M. 461, 463, 802
Scott, D.H. and King, J.S. 463
Scott, D.H. and Tanaka, K.L. 463, 450
Scott, E.R.D. 22, 486, 786
Scott, E.R.D., Clayton, R.N. and Mayeda, T.K. 110
Scott, E.R.D., Keil, K. and Taylor, G.J. 873
Scott, E.R.D. and Rajan, R.S. 183
Scott, E.R.D., Rubin, A.E., Taylor, G.J. and Keil, K. 110
Scott, E.R.D., Taylor, G.J., Newsom, H.E. et al. 79
Scotti, J.V., Rabinowitz, D.L. and Marsden, B.G. 102
Scoville, N.Z. 345
Scrutton, C.T. 221
Scudder, J.D. and Olbert, S. 772

Scudder, J.D., Mangeney, A., Lacome, C. et al. 641
Sears, D.W. 177, 236
Sears, D.W. and Axon, H.J. 110
Sears, D.W., Grossman, J.N., Melcher, C.L. et al. 110
Sears, D.W., Kallemeyn, G.W. and Wasson, J.T. 236
Sears, D.W.G. 105, 110, 234
Sears, D.W.G., Jie, L., Benoit, P.H. et al. 110
Sears, D.W.G., Lu Jie, Keck, B.D. and Batchelor, D.J. 110
Sears, R.L. 794
Segatz, M., Spohn, T., Ross, M.N. and Schubert, G. 421, 814
Segrè, E. 243
Segura, M. 316
Seidelmann, P.K. 92, 591, 594
Seidelmann, P.K., and Harrington, R.W. 590
Seiff, A., Kirk, D.B., Young, R.E. et al. 189
Sekanina, Z. 131, 145
Selivanov, A.S., Panfilov, A.S., Naraeva, M.K. et al. 887
Selivanov, A.S., Chemodanov, V.P., Naraeva, M.K. et al. 887
Sellers, W.D. 189
Sepkoski, J.J., Jr. 467
Serret, J.A. (ed.) 382
Seyfert, C.K. 208, 802
Shackleford, J. 70
Shah, K.P., Muhleman, D.O. and Berge, G.L. 493
Shams, F.A. (ed.) 319
Shannon, C. and Weaver, W. 188
Shapiro, I.I. 51, 287
Shapiro, I.I., Chandler, J.F., Campbell, D.B. et al. 663
Sharonov, V.V. 393, 732
Sharp, R.P. 238, 450
Sharp, R.P. and Malin, M.C. 450
Sharpe, N.H. and Peltier, W.R. 814
Sharpley, H. 302
Sharpton, V.L. 197
Sharpton, V.L., Dalrymple, G.B., Marin, et al. 131
Sharpton, V.L., Dalrymple, G.B., Marin, L.E. 251
Sharpton, V.L. and Ward, P.D. (eds.) 178
Shaw, D.M. 208
Shaw, G.H. 814
Shaw, H.F. and Wasserburg, G.J. 805
Shaw, R.R. and Gurnett, D.A. 641
Shea, M.S. and Smart, D.F. 754
Shearer, P.M. and Masters, T.G. 421
Shearer, P.M. 421
Sheehan, W. 432
Shelley, E.G. and Ghielmetti, A.G. 354
Shemansky, D.E. and Sandel, B.R. 351
Shemansky, D.E. and Broadfoot, A.L. 473
Shemansky, D.E., Matheson, P., Hall, D.T. et al. 622
Shemansky, D.E. and Hall, D.T. 622
Shepard, F.P., Sullivan, G.G. and Wood, F.J. 799
Sherman, D.M. and Waite, T.D. 915
Shinagawa, H. and Cravens, T.E. 359
Shirley, J.H. 11, 23, 25, 48, 51, 103, 112, 153, 189, 192, 248, 336, 343, 366, 385, 415, 479, 480, 571, 578, 782, 683, 726, 731, 741, 747, 756, 794, 806, 843 and 941
Shirley, J.H. and Fairbridge, R.W. 62
Shirley, J.H., Sperber, K.R. and Fairbridge, R.W. 571, 745, 746

AUTHOR INDEX

Shklovskii, L.S. 747
Shkuratov, Yu. G. and Opanasenko, N.V. 654
Shmidt, O. 113
Shock, E.L. and McKinnon, W.B. 840
Shoemaker, E.M. 71, 232
Shoemaker, E.E. and Hackman, R.J. 508
Shoemaker, E.M., Lucchitta, B.K., Wilhems, D.E. et al. 267
Shoemaker, E.M. and Shoemaker, C. 450
Shoemaker, E.M., Shoemaker, C.S. and Wolfe, R.F. 383, 841
Shoemaker, E.M., Williams, J.G., Helin, E.F. and Wolfe, R.F. 131
Shoemaker, E.M. and Wolfe, R.F. 131, 145
Shoemaker, E.M., Wolfe, R.E. and Shoemaker, C.S. 131
Shoemaker, E.M., Williams, J.G., Helin, E.F. and Wolfe, R.F. 454
Short, N.M. 71, 688, 934
Showalter, M.R. 373, 608, 733, 738
Showalter, M.R., Burns, J.A., Cuzzi, J.N. and Pollack, J.B. 375, 608
Showalter, M.R. and Cuzzi, J.N. 534, 722
Showalter, M.R., Cuzzi, J.N. and Larson, S.M. 608, 722
Showalter, M.R., Cuzzi, J.N., Marouf, E.A. and Esposito, L.W. 608, 722
Showalter, M.R. and Nicholson, P.D. 722
Showalter, M.R., Pollack, J.B., Ockert, M.E. et al. 723
Shu, E.H. 608
Shu, F.H., Adams, F.C. and Lizano, S. 590, 762
Shu, E.H., Cuzzi, J.N. and Lissauer, J.J. 608
Shukolyukov, A. and Lugmair, G.W. 183
Shyn, T.W., Sharp, W.E. and Hays, P.B. 819
Sicardy, B., Roques, F. and Brahic, A. 608
Sicardy, B. and Lissauer, J.J. 534
Sidgwick, J.B. 299
Sieveka, E.M. and Johnson, R.E. 351
Sigurdsson, H., D'Hondt, S., Arthur, M.A. et al. 805
Sill, G.T. and Wilkening, I.L. 177
Silver, L.T. and Schultz, P.H. (eds.) 178
Silverman, S.M. 57, 745
Silverman, S.M. and Shapiro, R. 295
Simon, M. and Axford, W.I. 736
Simonelli, D.P., Pollack, J.B., McKay, C.P. et al. 650
Simonelli, D.P. and Veverka, J. 351
Simons, M., Hager, B.H. and Solomon, S.C. 904
Simpson, J.A., Rossi, B., Hibbs, A.R. et al. 855
Simpson, R.A. 232, 233, 558, 561, 682, 686
Simpson, R.A. and Ostro, S.J. 660
Simpson, R.A. and Tyler, G.L. 663
Simpson, R.A., Harmon, J.K., Zisk, S.H. et al. 663
Simpson, R.W., Russell, C.T. and Fairbridge, R.W. 11
Sinclair, A.T. 240, 828
Singer, C. 273, 297, 382, 385, 659, 808
Singer, R.B., McCord, T.B. and Clark, R.N. 450
Singer, R.B. and McSween, H.Y., Jr. 461
Singer, R.B. and Miller, J.S. 461
Singer, R.B., Owensby, P.D. and Clark, R.N. 461
Sinton, W.M. 351, 461
Siscoe, E.C. and Summers, D. 622
Siscoe, G.L. 57, 571, 630, 745, 863

Siscoe, G.L., Ness, N.F. and Yeates, C.M. 61, 473
Siscoe, G.L. and Verosub, K.I. 57
Sittler, E.C., Ogilvie, K.W. and Scutter, J.D. 622, 864
Sjogren, W.L. 283, 450, 451, 508, 596, 681. 0-4
Sjogren, W.L., Bills, B.G., Birkland, P.W. et al. 905
Sjogren, W.L., Phillips, R.J., Birkeland, P.W. and Wimberly, R.N. 682
Skinner, B.J. and Luce, F.D. 236
Skrutskie, M.F., Dutkevih, D., Strom, S.E. et al. 590
Skrutskie, M.F., Forrest, W.J. and Shure, M. 590
Sky, B.M. and Farmer, C.B. 440
Slade, M.A., Butler, B.J. and Mulheman, D.O. 470, 598, 650
Slade, M.A., Butler, B., Muhleman, D. and Jurgens, R. 663
Slavin, J.A., Smith, E.J., Spreiter, J.R. and Starhara, S.S. 630
Slavsky, D.B. and Smiith, H.J. 532
Sleep, N.H. 416, 421
Sleep, N.H. and Phillips, R.J. 189, 463
Sleep, N.H., Zahnle, K.F., Kasting, J.F. and Morowitz, H.J. 131
Slocum, R.E. 410
Slocum, R.E., Cabiness, P.C. and Blevins, S.L. 410
Slocum, R.E. and Reilly, F.N. 410
Slysh, V.I. 887
Smart, W.M. 92, 300, 555, 658, 836
Smirnov, V.N., Vaisberg, O.L., Romanov, S.A. et al. 887
Smith, B. and 42 Others 498
Smith, B.A. et al. 159, 538, 608, 698, 856
Smith, B.A., Soderblom, I.A., Banfield, D. et al. 180, 371, 532, 534, 540, 834, and 848
Smith, B.A., Soderblom, L.A., Batson, R.F. 717
Smith, B.A., Soderblom, L.A., Beebe, R. et al. 371, 554, 713, 717, 834, and 863, 868
Smith, B.A., Soderblom, L.A., Johnson, T.V. et al. 180
Smith, B.A. and Terrile, T.J. 590
Smith, D.E. 386
Smith, D.E. and Turcotte, D.L. (eds.) 208
Smith, D.E., Lerch, F.J., Nerem, R.S. et al. 451
Smith, E.J. and Gulkis, S. 373
Smith, E.J. and Wolfe, J.H. 772
Smith, E.J., Page, D.E. and Wentzel, K.P. 298, 855
Smith, G.R. and Hunten, D.M. 825, 851
Smith, G.R., Shemansky, D.F., Broadfoot, A.L. and Wallace, L. 473
Smith, J.E. and Sonnett, C.P. 410
Smith, J.R. 241
Smith, J.V. 319
Smith, J.V., Anderson, A.T., Newton, R.C. et al. 814
Smith, R.A. 351
Smith, R.A., Bagenal, F., Cheng, A.F. and Strobel, D. 415, 622
Smith, R.A. and Strobel, D. 622
Smrekar, S.E. and Phillips, R.J. 904
Smyth, W.H. 473
Smyth, W.H. and Combi, M.R. 622
Smythe, W.D., Nelson, R.M. and Nash, D.B. 351, 598
Smythe, W.D. 596
Snyder, C.W. 228, 397, 427, 573, 683, 772, 908, 910, 941

Snyder, C.W. and Moroz, V.I. 429, 774, 911
Snyder, C.W. and Evans, N. 911
Snyder, J.P. 232, 421, 427
Snyder, J.P. and Voxland, P.M. 427
Snyder, L.E. 493
Snyder, L.E., Palmer, P. and Depater, I. 493
Soderblom, L.A. 461, 915
Soderblom, L. and Johnson, T.V. 725
Soderblom, L.A., Kieffer, S.W., Becker, T.L. 598
Soderblom, L.A., Kieffer, S.W., Becker, T.L. et al. 538, 840
Soffen, G.A. 911
Soffen, G.A. and Snyder, C.W. 911
Soffen, G.A. and Young, A.T. 911
Sofia, S. (ed.) 794
Solomatov, V.S. 814
Solomon, C.S. and Head, J.W. 463
Solomon, S. 201, 215
Solomon, S.C. 10, 197, 364, 645, 814, 895, 904
Solomon, S.C., Bindschadler, D.L., Smrekar, S.E. et al. 904
Solomon, S.C. and Chaiken, J. 814
Solomon, S.C. and Head, J.W. 904
Solomon, S.C., Head, J.W., Kaula, W.M. et al. 364, 731, 808
Sonder, R.A. 253
Sonett, C.P. 734, 736, 745, 794
Sonett, C.P., Colburn, D.S., Davis, L., Jr. et al. 736
Sonett, C.P., Coleman, P.J. and Wilcox, J.M. (eds.) 772
Sonett, C.P. and Mihalov, J.D. 736
Sonett, C.P., Morfill, G.E. and Jokipii, J.R. 736
Sorensen, H. 319
Southwood, D.J. 87
Space Science Reviews 855, 927
Spaudis, P.D. 473
Spaudis, P.D. and Guest, J.E. 476
Spencer, E.W. 208
Spencer, I.J. 805
Spencer, J.R. 45, 75
Spencer, J.R., Buie, M.W. and Bjoraker, G.I. 538
Spencer, J.R. and Fanale, F.P. 450
Spencer Jones, A. 3
Spencer Jones, H. 51, 297
Spencer Jones, H. and Halm, J. 51
Spenner, K. and Dumbs, A. 819
Spilker, J.J. 283
Spitz, A.H. 873
Spitz, A.H. and Boynton, W.V. 873
Splencer, J.R., Lebofsky, L.A. and Sykes, M.V. 45
Spohn, T. 814
Sprague, A.I. 45, 471, 473
Sprague, A.I., Kozlowski, R.W.H. and Hunten, D.M. 473
Sprague, A.L., Kozlowski, R.W.J., Hunten, D.M. et al. 102, 501
Sprague, A.I., Kozlowski, R.W.H., Hunten, D.M. and Grosse, F.A. 473
Spreiter, J.R. 887
Spruit, H.D. 748
Spudis, P.D. 197
Squyres, S.W. 314, 432, 450
Sridhar, S. and Tremaine, S. 723
Srnka, L.J. 512
Sromovsky, L.A. 532
Sromovsky, L.A., Revercomb, H.E., Krauss, R.J. and Suomi, V.E. 371
Stacey, C.M. 571, 834
Stacey, F.D. 658, 835

Stacey, R.D. 211
Stagg, C.P. and Bailey, M.E. 124, 145
Stahl, W.H. 27
Standish, E.M. 32, 594, 650
Starr, V.P. 18, 19
Stauffer, M.R. (ed.) 253
Steel, D. 131
Steele, I.M. and Smith, J.V. 68
Steffes, P.G. and Eshleman, V.R. 1
Steffes, P.G. 676
Steigmann, G.A. 654
Stein, S. 208
Steinbach, V. and Yuen, D.A. 904
Steiner, J. 257
Steiner, J. and Grillmair, F. 257
Stephenson, B. 379
Stephenson, F.R. 705
Stephenson, F.R. and Wolfendale, A.W. (eds.) 57, 745
Stephenson, F.R., Yau, K.K.C. and Hunger, H. 98
Stephenson, F.R. and Morrison, L.V. 828
Stern, A.C. 4
Stern, D.P. 268
Stern, S.A. 103, 124, 145, 185, 623, 600, 828, 840
Stern, S.A., Shull, M.J. and Brandt, J.C. 145
Stern, S.A., Stocke, J. and Weissman, P.R. 145
Stetson, P. 102
Stevens, M.H. and Stroble, D.F. 825
Stevenson, D.J. 161, 228, 372, 421, 523, 713, 713, 814, 857
Stevenson, D.J. and Gandhi, A.S. 538, 840
Stevenson, D.J., Harris, A.W. and Lunine, J.I. 713
Stevenson, D.J. and Salpeter, E.E. 713
Stevenson, D.J., Spohn, T. and Schubert, G. 421, 814, 907
Stevenson, D.J. and Turner, J.S. 814
Stewart, A.I.F. and Barth, C.A. 10
Stewart, G.R., Lin, D.N.C. and Bodenheimer, P. 608
Stix, M. 748, 794
Stix, T.H. 641, 936
Stockwell, C.H., McGlynn, J.C., Emslie, R.F. et al. 208
Stofan, E.R. and Head, J.W. 164
Stofan, E.R., Sharpton, V.L., Schubert, G. et al. 904
Stöffler, D. 734
Stöffler, D., Bischoff, A., Buchwald, V. and Rubin, A.E. 762
Stöffler, D., Keil, K. and Scott, E.R.D. 110
Stoker, C.R. and Toon, O.B. 532
Stolper, E.M. 68, 242
Stone, E. 708
Stone, E.C. and Miner, E.D. 682, 840
Stone, E.C., Cooper, J.F., Cummings, A.C. et al. 868
Stone, E.T. 51
Stone, R.G., Pedersen, B.M., Harvey, C.C. et al. 641
Stooke, P.J. 159
Stooke, P.J. and Tatum, J.B. 154
Stooke, P.J. and Keller, C.P. 427
Storey, L.R.O. 641, 936
Störmer, C. 664
Stothers, R. 57, 571, 745, 750
Stothers, R.B. 257, 495
Strangway, D.W., Sharpe, H., Gose, W. and Pearch, G. 512
Stratton, F.J.M. 181
Streckeisen, A. 319
Streckeisen, A.L. 65

Streiff, A. 71
Strobel, D.F. 622
Strobell, M.E. and Masursky, H. 552
Strobel, D.F., Meier, R.R., Summers, M.E. and Strickland, D.L. 10
Strom, R.G and Schneider, N.M. 351
Strobel, D.F. and Shemansky, D.E. 622, 633, 851
Strobel, D.F., Yelle, R.V., Shemansky, D.E. et al. 863
Strobel, D.F., Yelle, R.V., Shemansky, D.E. and Atreya, S.K. 10, 825
Strom, R.G. 319, 361, 463, 476, 478
Strom, R.G. and Croft, S.K. 538
Strom, R.G., Croft, S.K. and Barlow, N.G. 454
Strom, R.G., Croft, S.K. and Boyce, J.M. 840
Strom, R.G., Kargel, J.S., Johnson, N. and Knight, C. 450
Strom, R.G., Trask, N.J. and Guest, J.E. 476
Strom, S., Edwards, S. and Skrutskie, M. 751
Struve, O. 559
Stuiver, M. and Braziunas, T.F. 571
Stuiver, M., Braziunas, T.F., Becker, B. and Kromer, B. 571
Suess, H.E. 177
Suess, H.E. and Urey, H.C. 177, 183
Sukhanov, A.L. 808
Sun, S.-S., Nesbitt, R.W. and Sharaskin, A.Y. 251
Sundquist, E.T. 201
Suomi, V.E., Limaye, S.S. and Johnson, D. 532
Suppe, F. 188
Suppe, J. and Connors, C. 197
Surkov, Y.A. 161
Surkov, Yu. A., Barsukov, V.L., Moskaleva, L.P. et al. 576
Surkov, Yu.A., Kirnozov, F.F., Glazov, V.N. et al. 887
Surkov, Yu.A., Kirnozov, F.F., Khristianov, V.K. et al. 887
Surkhov, Y.A., Schcheglov, O., Moskalyeva, L. et al. 616
Sussman, G.J. and Wisdom, J. 98, 229, 650, 766
Swain, J. 380
Swenson, I.S., Jr. 283
Swerdlow, N. and Neugebauer, O. 160
Swinbank, R. 19
Swindle, T.D. 188, 394, 481, 540, 907
Swindle, T.D. and Podosek, F.A. 110, 112, 183
Swings, P. 153
Swisher, D.D., Grajales-Nishimura, J.M. Montanari, A. et al. 178
Sykes, M.V. 192
Sykes, M.V., Greenberg, R., Dermott, S.F. et al. 192
Syler, G.L., Eshleman, V.R., Anderson, J.D. et al. 717
Synthesis Group Report 432
Szebehely, V. 92, 540
Szebehely, V. and McKenzie, R. 540
Szebehely, V.G. 594
Szucs, I.T., Szemerey, I., Kiraly, P. et al. 819

Tackley, P.J., Stevenson, D.J., Glatzmaier, G.A. and Schubert, G. 421
Tagger, M. and Henriksen, R.N. 608
Tagger, M., Henriksen, R.N. and Pellat, R. 723

Tagliaferri, E., Spalding, R., Jacobs, C. et al. 132
Takacs, P.Z., Broadfoot, A.L. and Smith, G.R. 851
Takahashi, K. and Matsuda, A. 873
Takeda, H. 873
Takeda, H. and Graham, H.I. 68
Takeda, H. and Yanai, K. 873
Takeda, H., Mori, H. and Ogata, H. 873
Takeda, H., Mori, H. and Yanai, K. 68
Takeda, H., Mori, H., Yanai, K. and Shiraishi, K. 873
Talbott, R.J. and Newman, M.J. 124
Tamaki, K. and Honza, E. 645
Tamrazyan, G.P. 257
Tanaka, K.I. 450, 454, 463
Tanaka, K.L. and Davis, P.A. 463
Tanaka, K.L., Isbell, N.K. and Scott, D.H. 463
Tanaka, K.I., Isbell, N.K., Scott, D.H. et al. 450, 454
Tanaka, K.I. and Scott, D.H. 450
Tanaka, K.L. and Scott, D.H. 463
Tanberg-Hanssen, E. 747
Tancredi, G. and Lindgren, M. 124
Tanguy, L., Bézard, B., Marten, A. et al. 493
Tarcotte, D.L., Haxby, J.F. and Ockendon, J.R. 645
Tatarewicz, N.J. 381, 874
Taton, R. 84, 241, 302, 427, 659, 808
Taton, R. and Wilson, C. (eds.) 385
Tatum, J.B. 46, 48, 307, 554, 737, 808, 835, 940
Taub, I. 160
Tauber, G.E. 25, 258
Taylor, F.W. 890
Taylor, F.W., Beer, R., Chahine, M.T. et al. 895
Taylor, G.J. 188
Taylor, G.J., Maggiore, P., Scott, E.R.D. et al. 110
Taylor, H.A., Jr., Brinton, H.C., Wagner, T.C.G. et al. 819
Taylor, H.A., Jr., Mayr, H.G. and Kramer, I. 19
Taylor, M.J. and Hill, M.J. 10
Taylor, S.R. 65, 193, 396, 512, 508, 590, 805, 814, 904
Taylor, S.R. and Norman, M.D. 361
Taylor, W.W.L., Parady, B.K. and Cahill, L.J.Jr. 641
Tedesco, E.F., William, J.G., Matson, D.L. et al. 35
ter Haar, D. and Cameron, A.G.W. 11, 302, 623, 762, 874
Terrile, R.J. and Smith, B.A. 532
Thalamas, A. 241
Thatcher, W. 731
Theon, J.S., Nordberg, W. and Smith, W.S. 479
Theon, J.S. and Smith, W.S. 479
Tholen, D.J. 35, 93, 488
Tholen, D. and Stern, A.S. 588
Tholen, D.J. and Barucci, M.A. 32
Tholen, D.J., Buie, M., Binzel, R. and Frueh, M. 32
Tholen, D.J. and Zellner, B. 377
Thom, R. 188
Thomas, B.T. and Smith, E.J. 343
Thomas, D.S.G. (ed.) 796
Thomas, G.E. 398
Thomas, G.E. and Krassa, R.F. 398
Thomas, N. 622
Thomas, P. 185, 540, 573
Thomas, P., Dermott, S. and Veverka, J. 540

AUTHOR INDEX

Thomas, P. and Veverka, J. 377, 450, 538, 540
Thomas, P., Veverka, J. and Dermott, S.F. 738
Thomas, P., Veverka, J., Lee, S. and Bloom, A. 238
Thomas, P., Veverka, J. and Helfenstein, P. 539, 540
Thomas, P.C. 159, 184, 572, 723
Thomas, P.J., Mackay, C.P. and Chyba, C.F. (eds.) 132
Thomas, P.J. and Squyres, S.W. 75, 267
Thompson, A.C. and Stevenson, D.J. 513, 814
Thompson, A.R., Moran, J.M. and Swenson, G.W., Jr. 676
Thompson, M.M. and Gruner, H. (eds.) 578
Thompson, R. and Sagan, C. 713
Thompson, T.D. (ed.) 941
Thomson, D.J. 26, 571
Thoren, V.E. 70
Thorne, A.P. 783
Thorne, K.S. 287
Thorne, R.M. 622
Thorne, R.M. and Moses, J. 641
Thorne, R.M. and Scarf, F.L. 641
Tikhomirov, V.V. 393
Tilton, G.R. 110, 112
Tisserand, F. 701
Tittlemore, W.C. and Wisdom, J. 701, 828
Tobin, W. 102
Tombaugh, C.W. 594
Tombaugh, C.W. and Moore, P. 835
Tonks, W.B. and Melosh, H.J. 161, 188, 623
Toomer, G.J. 229, 308, 659
Toomre, A. 558, 762
Toon, O.B., Pollack, J.B., Ward, W. et al. 440
Torbett, M.V. 145
Torbett, M.V. and Smoluchovski, R. 98, 124
Torge, W. 269
Toulmin, P., III, Baird, A.K., Clark, B.C. et al. 450
Touma, J. and Wisdom, J. 98, 558, 766
Townes, C.H. and Schawlon, A.L. 1, 783
Tozer, D.C. 421, 814
Trafton, L.M. 351, 650, 840, 863
Trafton, L.M. and Ramsay, D.A. 863
Traub, W.A. and Carleton, N.P. 895
Treiman, A.H. 741
Treiman, A.H. and Berkley, J.L. 873
Tremaine, S. 521
Tric, E., Laj, C., Jéhanno, C. et al. 656
Tric, E., Valet, J.-P., Trucholka, P. et al. 736
Trouvelot, E.L. 842
Trowell, N.F. and Johns, G.W. 208
Tsiolkovsky, K.E. 842
Tsoar, H. 238
Tsoar, H., Greeley, R. and Peterfreund, A.R. 238, 450
Tucker, C.J., Townshend, J.R.G. and Goff, T.E. 696
Tum, J.B. 560
Tupman, G.L. 51
Turcotte, D.L. 814, 904
Turcotte, D.L., Cooke, F.A. and Willeman, R.J. 814
Turcotte, D.L. and Schubert, G. 421, 814
Turekian, K.K. 934
Turkevich, A.L., Patterson, J.H. and Franzgrote, E.J. 616
Turncotte, D.T., Cisne, J.L. and Nordmann, J.C. 223

Turner, F.J. and Verhoogen, J. 208
Turner, G. 112
Tyler, G.L. 663, 676, 682
Tyler, G.I. et al. 608, 682
Tyler, G.L., Balmino, G., Hinson, D.P. et al. 459, 682
Tyler, G.L., Eshleman, V.R. and Anderson, J.D. et al. 682
Tyler, A.L., Kozlowski, R.W.H. and Lebofsky, L.A. 473
Tyler, G.L., Marouf, E.A., Simpson, R.A. et al. 682
Tyler, G.L., Marouf, E.A. and Wood, G.E. 682
Tyler, G.I., Sweetman, D.N., Anderson, J.D. et al. 359, 532, 682, 863, 868, 840
Tyner, R.L. and Carroll, R.D. 911
Tyson, J.A. 102

Ulaby, F.T. et al. 69
Ulaby, F.T., Moore, R.K. and Fung, A.K. 232, 233, 562, 686, 696
Ulrych, T. 257
Uman, M.A. 601
Umbgrove, J.H.F. 392
United States Naval Observatory 555, 836
Urey, H.C. 77, 113, 177, 508, 598, 623, 786, 874
Urey, H.C. and Craig, H. 110
US Geological Survey 552, 553, 578
US Standard Atmosphere 786, 842
US Weather Bureau (now National Weather Service, NOAA) 799
USSR Committee for IHD 934

Väisälä, Y. 876
Vaisberg, O.L. 879, 887
Vaisberg, O.L., Romanov, S.A., Smirnov, V.N. et al. 887
Vaisberg, O.L., Smirnov, V.N., Zastenker, G.N. and Fedorov, A.O. 887
Vaisberg, O.L., Smirnov, V.N., Karpinsky, I.P. et al. 887
Vaisberg, O.L. and Zelenyi, L.M. 887
Vaisberg, O.L., Zhurina, L.S., Kovalenko, V.G. et al. 887
Valdes, P. 895
Valentine, J.W. and Moores, E.M. 467
Valet, J.-P., Tucholka, P., Courtillot, V. and Mevnadier, L. 272, 656
Valsecchi, G.B. 35
Valsecchi, G.B., Carusi, A., Knezevic, Z. et al. 38
Valtonen, M., Zheng, J.-Q. and Mikkola, S. 257
Valtonen, M.J. 145
Valtonen, M.J. and Innanen, K.A. 145
van Allen, J.A. 608, 630, 664, 713, 876
van Allen, J.A. et al. 608
van De Hulst, H.C. 562
Van Schmus, W.R. 110
Van Schmus, W.R. and Wood, J.A. 110
vanDishoeck, E.F. and Dalgarno, A. 153
VanFlandern, T.C. and Pulkinen, K.F. 223
VanHelden, A. 725
VanSchmus, W.R. and Wood, J.A. 486
vanWoerkom, A.F.F. 124, 145
Varela, F. 188
Vasyliunas, V.M. 415, 622
Vdovykin, G.P. 873
Veillet, C. 540
Veizer, J. 198
Venkatesan, D. 298, 397
Venkatesan, D. and Krimigis, S.M. 298
Verigin, M.I. 876
Verigin, M.I., Gringauz, K.I., Gombosi, T. et al. 887

Verigin, M.I., Shutte, N.M., Galeev, A.A. et al. 576
Verma, S.D. 571
Vernazza, J.E., Avrett, E.H. and Looser, R. 756
Vernov, S.N., Tverskoi, B.A., Volga, V.I. et al. 887
Verpaelst, P., Brooks, C. and Franconi, A. 208
Vessot, R.F.C. and twelve others. 287
Veverka, J. 11, 540, 561, 776
Veverka, J., Belton, M., Chapman, C. and the Galileo Imaging Team 102
Veverka, J. and Burns, J. 185, 573
Veverka, J.P., Simonelli, D.P., Thomas, P. et al. 351
Vickery, A.M. and Melosh, H.J. 741
Vidal-Madjar, A. 397, 398, 940
Vidal-Madjar, A., Encrenaz, T., Ferlet, R. et al. 398
Vienne, A. and Duriez, L. 240
Viking Lander Imaging Team 911
Viking Orbiter Imaging Team 911
Vilas, F., Chapman, C.R. and Matthews, M.S. 161
Vilas, F., Chapman, C.R. and Matthews, M.S. (eds.) 470, 588
Vilas, F. 467, 915
Vilas, F. and Smith, B.A. 841
Vincent, M.A. 287
Vine, F.J. and Matthews, D.H. 645
Virga, A., Wopenka, B. Amari, S. et al. 177
Vogt, S., Herzog, G.F. and Reedy, R.C. 169
Voight, B. (ed.) 253
Von Braun, W. and Ordway, F.I., III 283
Von Waltershausen, S. 268
vonBertalanffy, L. 188
vonFoerster, H. 188
vonNeumann, J. 188
vonWeizsäcker, C. 762
vonZahn, U., Kumar, S., Niemann, H. and Prinn, R. 493, 851, 895
Voshage, H., Feldman, H. and Braun, O. 169
Vsekhsvyatskii, S.K. 145
Vucinich, A. 393

Wacker, J. 874
Waff, C.B. 184
Wagener, R., Owen, T., Jaffe, W. and Caldwell, J. 833
Wahr, J.M. 217
Wahr, J.M. and Oort, A.H. 19
Wai, C.M. and Wasson, J.T. 177
Waite, J.H., Jr. 58
Waite, J.H., Jr., Clarke, J.T., Cravens, T.E. and Hammond, C.M. 61
Waite, J.H., Jr., Cravens, T.E., Kizyn, J.U. et al. 61
Wald, R.M. 336
Waldmeier, M. 937
Waldrop, M.M. 381
Walker, G.A.H., Walker, A.R., Irwin, A.W. et al. 590
Walker, J.C.G., Hays, P.B. and Kasting, J.F. 201
Walker, M.F. and Hardie, R. 650
Walker, S.G. 915
Walker, T.R. 935
Walker. A. 102
Wallace, J.M. and Hobbs, P.V. 163
Wallace, M.W., Gostin, V.A. and Keays, R.R. 360

Subject index

Encyclopedia article titles are listed in capital letters, and the page number of the corresponding article is in bold type.

A-class asteroids 488
Abbot, C.G. 745, 754
Ablation patterns 804
Abrasion 935
ABSORPTION **1**, 561
Absorption bands 684
Absorption features 777
Absorption lines 54
Acapulco (Mexico) 483
ACCRETION **1**, 223, 708
Accretion simulation 142
Accretion sequence 522
Accretional stage 759
Accretionary process 649
ACHONDRITE METEORITES **2**, 19, 483, 686, 907
Acropolis 71
Acryogenic 315
Active Cavity Radiometer Irradiance Monitor (ACRIM) 745
Active plumes 838
Active regions 755
Active volcanoes 257, 346
Active volcanism on Venus 584
Adams ring 532, 606
ADAMS, J.C. **3**, 301, 521
Adhémar 178, 494
Adiabatic temperature gradient 418
Adrastea 706
Advanced Very High Resolution Radiometer (AVHRR) 695
Advection 915
Aerial photography 690, 694
Aerobee rocket 876
Aerodynamic braking 312
Aeronautics 517
AEROSOL **4**, 575, 842
Aerosol (Venus) 886
Agassiz, L. 315
Agglutinates 652
Agricultural soil 688
AIRGLOW **4**, 941
Airy model 364
Al-Ma'mun 300
Alba patera 441, 918
ALBEDO **10**, 33, 233, 494, 578, 596, 662, 684, 685, 693, 711, 738, 775, 856, 907, 911
Albedo (Mars) 433
Albers conic projection 426

ALFVÉN, H.O.G. **11**
ALFVÉN WAVE **11**, 272, 624, 704, 769
Alfvénic 624
Alien civilizations 390
Allende (Mexico) 482
Almagest 659
Alpha backscatter spectrometer 613
Alpha-particle backscattering instrument 796
Alphonsine tables 300
Altazimuth coordinates 154
Altimeter 399
Altimetry 279
Aluminous crust 509
Amalthea 706
Amazonian 461
Amazonian (Mars) 441
American space missions 427
Amino acids 495
Ammonia 388, 495, 724
Ammonia (Neptune) 524
Ammonia (Uranus) 857
Ammonia ices 716
AMOR OBJECT **11**, 518
Ampere's law 468
Ananke 706
Anaximander 808
Andesitic volcanoes 641
Angrites 3
ANGULAR MOMENTUM 2, **12**, 123, 216, 226, 273, 288, 384, 513, 556, 565, 609, 610, 749, 759, 826, 834, 835
ANGULAR MOMENTUM CYCLE IN PLANET EARTH **13**
Annular eclipse 230
Annulus of concentric ridges (see Coronae) 897
Anomalistic month 222, 572, 799
Anomalistic year 939
ANTARCTIC METEORITES **19**, 482
Antenna reflector 580
Antennas 632
Anthropogenic activity 11
Anthropogenic radiation 232
Anti-greenhouse effect 832
anticyclone 179
Anticyclones (Jupiter) 369
Antipode zones 511
ANTONIADI, E. **22**, 469
APHELION **23**, 26, 566, 572, 747

Aphelion/perihelion drift 566
Aphrodite Terra 919
Apoapse 572
Apobac 26, 466, 572
Apogalactic position 255
Apogean neap tides 798
Apogee 26, 572
Apollo astronaut missions 613
Apollo 12 astronauts 796
APOLLO MISSIONS **23**
APOLLO OBJECT **25**
Apollo photographs 694
Apollo program 500, 501
Apollo–Soyuz 690
Apollo Soyuz test 517
Apparent annual path of the sun 231
Applied gravitation 285
Apsides 23, 567, 799
APSIS, APSIDES **26**
Aquarids 480
Aquarius 940
Arabic translations 241
Arcadia planitia 441
Archean crust 201
ARCHEOASTRONOMY **26**
Archimedean Spiral 343
Archimedes 92
Arecibo observatory 660
Argon 542
Argyre 441
Ariane program 247
Ariel 706
Aries 940
ARISTARCHUS **27**
Aristotelian orthodoxy 258
Aristotle 186, 284
Artificial satellites 91, 842
Ascending node 562
Ascensions 231
ASTEROID 12, 25, **27**, 114, 178, 193, 299, 316, 739, 774, 907
Asteroid belt 141, 189, 518, 579, 587, 751
Asteroid rotation 38
ASTEROID: COMPOSITIONAL STRUCTURE AND TAXONOMY **33**
ASTEROID: LIGHTCURVES **38**
ASTEROID: FAMILIES **35**
ASTEROID: PHOTOMETRY **38**
ASTEROID: RESONANCES **42**, 380

CRETACEOUS–TERTIARY (K–T)
  BOUNDARY **178**, 251, 335, 360, 520
CROLL, J. **178**
CRUST **179**, 801
Crustal convergence 201
Crustal dichotomy (Mars) 461
Crustal evolution 197
Crustal magnetization 509
Crustal rocks 416
Crustal thickness (Venus) 901
Crustal warping (Venus) 889
Cryofracture 935
Cryogenic cooling 337
Cryosphere 314, 650, 823, 934
Cryovolcanism 868
Curie point 404, 643
Cyclic changes in geologic record 255
CYCLONE **179**
Cyclones (Jupiter) 370
Cyclotron resonance 638
Cyclotron waves 704
Cytherian year 939

D'Arrest, H. 301
Dalton minimum 704
Dark matter 559
Dark plains 506
Dark terrain 266
Darkening 511, 597
DARWIN, G.H. **181**, 217, 226, 302, 747
Darwin–deSitter–Wise model 702
Data archive 591
Data center in Sioux Falls 694
Data publishing 592
DATING METHODS **181**
Dating of rocks 542
DAWES, W.R. **183**
Day/night rhythms 938
Day–night temperature variation 822
De Magnete 209, 273
Dead body 502
Dead worlds 711
Debye length 624
Decametric radiation (DAM) 672, 930
Decay of radioactive elements 808
Decay of radioactive isotopes 416
Decay time 400
Deccan traps 249
Decimetric radiation 366
Deep sea drilling project 642
DEEP SPACE NETWORK (DSN) **183**, 354, 400, 574, 660, 677
Deep winds (Jupiter) 368
Deep-focus earthquakes 727
Deferent 300
Deflation 237
Degassing 314, 415
Degradation (Mars) 452
Degree of order 564
DEIMOS **184**, 295, 431, 550
Delta T. 594
Deltas 808
Dendrochronology 570
DENNING, W.F. **186**
Density 496, 595
Density (Mercury) 477
Density (Phobos) 573
Density (Saturn rings) 721
Density of rock 363
Depressions 326
Descartes, R. 540
Descending node 562
Descent modules 773
Desdemona 706
Desert dust 4
Desert varnish or patina 935
Despina 534, 606, 706

Despina, Thalass, Naiad 532
Despinning 938
Detectors 321
Determination of distances 385
DETERMINISM **186**, 763
Deuterium 791
Devil's Causeway 807
Diamagnet 404
Diamond 543
Diapir 163, 310
Diapiric flow 536
Diastrophism 835
Dielectric materials 233
Differential Absorption lidar (DIAL) 695
Differential rotation of Sun 741
Differentiated meteorites 111
DIFFERENTIATION **188**, 415, 809
Diffraction 691
Diffraction gratings 253
Diffuse reflectance 684
Digital image processing 322, 694
Digital Number or Data Number (DN) 694
Dinosaurs 196
Diogenites 3, 65, 241
Dione 551, 706, 724
Dipolar Magnetic Field 625, 663
Dipolar magnetic field of Sun 741
Dipole tilt angles 629
Dipole component 654
Dipole moment 401
Dipole term 209
Dirac's hypothesis 689
Dirty ice 597
Dirty ice-ball 126
Discipline nodes 591
Disk accretion 2
Dissipative force 698
Distortion 421
Disturbing function 540
Diurnal libration 387
Diurnal periods 830
DIURNAL VARIATION **188**
Diurnal windows 303
Diversity (biotic) 465
DNA 197, 387
DOME **188**
Domes, highlands (Venus) 919
Doppler effect 593, 676, 781
Doppler shift 5, 283, 490, 588, 665, 695, 691
Doppler tracking 215, 428, 450
Doppler Wind Experiment (DWE) 86
Dorsa 550
Double planet 498
Double sunspot cycle 293, 569
Drifting continents 642
Dry ice 650
Dual-frequency occultation experiment 428
Ductile flow (Venus) 900
Dumbbell, Dumbbells 748, 803
Dunes 237
Dunite 3
DUST **189**, 751, 756, 759
Dust bands 606
Dust clouds 653
Dust condensations 751
Dust devils 536
Dust grains 374, 605, 653
Dust jets 277
Dust (Mars) 432
Dust mass spectrometer PUMA 877
Dust particles (IDPs) 169, 277, 344
Dust storms (Mars) 436, 600
Dust tail 148
Dust-cloud hypothesis 623
Dusty rings 719
Dutton, C.E. 363

Dynamical astronomy 88, 592
Dynamo 810
Dynamo mechanism 791
DYNAMO THEORY **192**
Dynamo theory (Mars) 454

EARLY BOMBARDMENT **193**
Early bombardment (Mercury) 919
EARTH 162, **194**, 387, 401, 584
EARTH: ATMOSPHERE **198**
EARTH: GEOLOGY, TECTONICS AND SEISMICITY **201**
Earth history 495
EARTH: MAGNETIC FIELD AND MAGNETOSPHERE 208
Earth/Moon body 181
EARTH OBSERVING SYSTEM **211**, 690
Earth (Planetary properties) 608
Earth Radiation Budget (ERB) 745
Earth resources technology satellite 690
Earth Rotation Parameters (ERP) 215
EARTH ROTATION 4, 13, 71, **215**, 365, 385, 826
EARTH: ROTATIONAL HISTORY **217**
Earth Sciences Data and Information System (ESDIS) 214
Earth terrain camera 694
Earth's axis of rotation 181, 657
Earth's core 295
Earth's crust 179
Earth's equator 657
Earth's hydrosphere 797
Earth's magnetic declination 295
Earth's magnetic field 296, 570, 704, 876
Earth's moon 708
Earth's obliquity 765
Earth's perihelion 572
Earth's rotation axis 608
Earth's spin 226
Earth-crossing orbits 127, 487
Earth-like celestial bodies 392
Earth-like seasonal cycle 823
Earth–Moon system 76, 217, 556, 826
Earth–Moon tides 499
EARTH–MOON SYSTEM: DYNAMICS **221**
EARTH–MOON SYSTEM: ORIGIN **223**
Earth-orbiting satellites 211
Earth–Sun distance 540
Earthquakes 642
ECCENTRIC DIPOLE **228**
Eccentric orbit 610, 839
ECCENTRICITY 221, **228**, 384, 494, 521, 562, 566, 646
Eccentricity (Nereid) 539
Eccentricity of the lunar orbit 799
Echoes from the Moon 660
Eckersley theory 936
ECLIPSE **229**, 741, 747
Eclipse timing 610
Eclipse year 221, 939
ECLIPTIC **229**, **231**, 498, 554, 657, 940
Ecliptic coordinates 155
Ecliptic plane 593, 852
Economic statistics 299
Eddington–Lemaitre model 689
Eddy diffusion 819
Einstein's general relativistic gravitation 688
Einstein's theory of general relativity 284, 688
Einstein–de sitter cosmos 689
EJECTA 178, **231**
Ejecta blanket (Mars) 451
Ejecta blanket (Mercury) 473
Ejecta blankets 507, 573
Ejecta blankets (Moon) 916

SUBJECT INDEX

Ejecta deposits 496
El Niño (ENSO) 569
El Niño/Southern oscillation 216, 286
Elara 706
Electrical circuits 380
Electrical conductivity 400
Electro-glow 843
Electromagnetic environments (giant planets) 925
Electromagnetic force 468
Electromagnetic ion cyclotron waves 640
ELECTROMAGNETIC RADIATION 232, 664, 775, 820
Electromagnetic resonances 604
Electromagnetic signals 936
Electromagnetic sounding 509
Electromagnetic spectrum 336
Electromagnetic waves 630, 691
Electron cyclotron 636
Electron cyclotron frequency 631
Electron microscope 613
Electron plasma 150
Electron plasma oscillations 633
Electronic circuitry 307
Electronic excitation states 665
Electronic spectra: atoms 778
Electronic spectra: molecules 779
Electronic transitions (spectroscopy) 912
Electrons 624, 766
Electrostatic ion cyclotron waves 639
Electrostatic waves 630
Element 360
Elements 170
Elevation compensation 363
Ellipse(s) 378, 494, 562
ELLIPSOID 232
Ellipsoid 268
Elliptical orbit 562
Ellipticity 494
Elysium mons 441
Embryo(s) 2, 760
Embryo disk formation 759
Emel: Earth–Mercury lap 565
Emiliani, C. 494
Emission 777
Emission spectroscopy 843
EMISSIVITY 233
Emitted thermal infrared radiation 664
Emitted energy 233
ENCELADUS 233, 551, 605, 712, 706
Encke gap 604, 733, 738
ENCKE, J.F. 233, 518
Endothermic phase change 420
Energetic charged particles 663
Energetic eruptions (Sun) 747
Energetic particles 411, 575, 667, 708, 754
Energetic particle fluxes 878
Energetic particle telescope TÜNDE-M 877
Energy 787
Energy emission 709
Energy balance 52
Energy spectrum 166
ENSTATITE METEORITES 105, 234, 483
Entropy 187
Entry and descent scenario 313
Environmental trauma(s) 128, 520
Eolian activity (Venus) 896
Eolian features (Mars) 446
Eolian features (Venus) 888
Eolian blanket 688
Eolian features 431
EOLIAN TRANSPORT 237
EOSAT 690
Eötvös, R. 283
Epeirogeny 801
EPHEMERIS 238, 342, 563, 592, 764
Ephemeris time 594

Epicycle(s) 23, 300, 762
Epimetheus 551, 706
Episodicity 310
Epitrochoids 566
Equal areas 378, 562
Equal-area map 422
Equation of mass conservation 787
Equation of state 787
Equatorial bulge 215, 610, 657, 835
Equatorial coordinates 154
Equatorial stratospheric wind reversal 569
Equilibrium 187
Equilibrium figure 363
Equinox 593, 808
Equipotential surface(s) 268, 290
ERATOSTHENES 240
Eratosthenian system (period) 505
Ergs, or sand seas 237
EROSION 241
Erosion 116, 128
Erratic stars 584
EUCRITE METEORITES 3, 64, 65, 241
EULER, L. 222, 242, 829
Euler's equation of motion 835
EULERIAN MOTION 242
EUROPA 243, 257, 551, 660, 706
Europa (Torus) 620
EUROPEAN SPACE AGENCY (ESA) 246
EUV heating 824
EUV solar photons 844
Evolution by catastrophe 466
Evolution of Mars 456
Evolution of solar system 486
Evolutionary processes 464
Excitation 1
Excursion 269
Exfoliation 935
Exobase (Titan) 831
EXOSPHERE 248, 819
Exospheric 770
Expanding gases 231
Explorer 1 302, 876
Exposure ages 169
Extensional canyons 856
Extensional features (Venus) 898
Extensional fractures 833, 860
External forcings 820
Extinct comet nuclei 519
Extraterrestrial agencies 495
Extraterrestrial body 326
Extraterrestrial lightning (Venus) 599
Extraterrestrial magnetic fields 294
Extraterrestrial matter 479
Extraterrestrial rocks 361
Extreme ultraviolet (EUV) 843

Faculae 741, 745, 747, 789
Fahrenheit 806
Fahrenheit, G.D. 806
Faint young Sun paradox 748
Fair-weather electric field 598
Faraday's law 468
Farside of the Moon 395, 513
Father of astronautics 842
Fault motion 279
Faults 838
Feedback 187, 494
Feeding zone 760
Feldspar 733
Ferromagnesian silicates 77
Ferromagnetism 273
Field regeneration 401
Figure of the Earth 541
Filigree 755
Fireballs 480, 482
First Point of Aries 155, 562, 938

Fission hypothesis 181, 226, 556
Fissure 253
Fissure eruptions (Moon) 916
Fixed stars 300, 747
Flares (Sun) 790
Flares and plages 789
Flatness of the solar system 751
Flattened protoplanetary disk 757
Flattening 232, 268
Flexus 551
Floating rectification probe 818
Flood of Noah 541
FLOOD BASALT 249
Flow and fast flows (solar wind) 766
Flowing water 430
Fluctus 550
Fluidized debris flows (Mars) 443
Fluorescence 777, 780
Fluvial landforms 195
Flux ropes (Venus) 907
Flux of planetary ions 575
Fluxgate magnetometer 408, 704
Flyby 287, 303, 308, 496, 821, 838, 843, 852, 876
Flyby mission 428
Flybys 297, 394, 604, 614, 709, 876
Flywheel 933
Focus of gravitational attraction 572
Folded mountains 206
Folklore 26
FORBUSH DECREASE 251
FORBUSH EFFECT 165, 252, 298
Forced precession 835
Formation of solar system 110
Fossae 551
Fourier series 592
Fra Mauro formation 505, 510
Fracture systems 74
FRACTURE, FAULT 252
Frames of reference 336
Franck–Condon factor 780
Fraunhofer (F) Corona 746
FRAUNHOFER LINE 253, 380, 576, 940
Free nutation 557
Fresh water 313, 934
Frictional torque 386
Frost deposition 838
Frost riving or frost splitting 935
Frozen field effect 400
Frozen volatiles 705
Fundamental tone 566, 834
Fusion reactor 788
Future of planetary exploration 308

G-modes 756
Gabriel cycle 567
Gaia 194
Gaia hypothesis 194, 200, 390
Galactic center 749
GALACTIC CYCLE 255
Galactic cosmic rays 543
Galactic tidal force 256
Galactic tidal perturbations 559
Galactic tide 520
Galactic year 559
Galatea 532, 534, 606, 706
Galaxies 757
GALILEAN SATELLITES 231, 239, 243, 257, 263, 345, 366, 389, 596, 662, 674, 699, 707, 813, 831, 919
GALILEO 186, 258, 284, 300, 366, 498, 708, 755, 806
GALILEO MISSION 258
Galileo probe 368
Galle, J.G. 301, 521
Galle ring (Neptune) 532, 606
GALLEX experiment 792

Kant 757
KANT–LAPLACE **384**, 519
Karoo 249
Karst 935
Karst-like subsidence (Mars) 443
Karst topography 795
Kayser 777
Keeler gap 604
Kelvin 806
Kelvin–Helmholtz mechanism 717
KEPLER, J. 186, 228, 284, **378**, 562, 592
KEPLER'S LAWS **379**
Keplerian orbits 309
Keplerian shear 609
Keweenawan Volcanics 251
Kind–Hele cycle 566
Kinetic energy 630
KIRCHHOFF, G.R. **380**
Kirchoff's laws 380, 665
KIRKWOOD, D. **380**, 609
Kirkwood gaps 91, 95, 154, 190, 486, 519, 698, 699
Kirkwood's theory 42
KOCMOC (cosmos) 773
Koronis family 316
KREEP (Moon) 64, 916
Krypton 542
KUIPER BELT 117, 123, 126, **381**, 518, 560, 623
KUIPER, G.P. **381**

L'Aigle, France 482
Laboratory spectra of rocks, minerals and ices (spectroscopy) 912
Lacaille 23
Lacus, Maria, Paludes, Sinus 550
LAGRANGE, J.L. **382**
Lagrange's equation 542
LAGRANGIAN POINT 42, **382**, 841
Lakshmi Planum 919
Lambert 422
Lambert conic projection 425
Lambertian surface 693
Lamina cycles 219
Landenberg–Reiche function 782
Lander science (Mars) 909
Landers 773
Landers (Venera) 880
Landsat 690
Landsat thematic mapper 694
Langley, S.P. 745
Langmuir probe 818
Lap period 565
Laplace 222, 757, 762, 829
LAPLACE, P.S. DE **383**, 520, 609
Laplace relation 699
Laplace resonance 812
Lapse rate 52
Lapse rates (Mars) 434
Larissa 534, 706
Laser altimetry 385
Laser Geodynamics Satellite (LAGEOS) 385
LASER RANGING **385**
Laser remote sensing atmosphere 695
Lassell, W. 301, 855
Launch opportunities 303
Laurasia 643
Lava tubes 506
Lava tubes (Moon) 916
Law of action and reaction 541
Law of biologic continuity 68
Law of differential variation of terrestrial density 701
Law of superposition 503
Law of universal gravitation 762
Laws of nature 186

Layer of no strength 45
Le Verrier Ring (Neptune) 532
Leading spots 741
Leaning tower of Pisa 258
Leap year 365, 938
Leda 706
Leibniz, G.W. 540
LENGTH OF DAY (LOD) 13, 215, **386**, 499, 826, 835
Leo 940
Leonid shower 480
Leonids 559
Le Verrier ring 606
LE VERRIER, U.J.J. 301, **386**, 521, 763
Libra 940
Library at Alexandria 241
LIBRATION 91, **387**, 841
Librations (Nereid) 540
Lidar ('light detection and ranging') 695
Lidar systems 694
Life 197, 200
Life detection system 614
Life on Earth 191
LIFE: ORIGIN **387**
Light 320
Light flux 578
Light rare earth elements (LREE) 785
Lightcurve amplitude 299
Lightcurves 38, 578
Lightning 734, 936
Lightning discharges 598, 874
Lightning flashes 598
Lightning (Jupiter) 600
Limestones 933
Lineae 550
Lineament 253, 801
Lineament structures 470
Linear adiabatic perturbation theory 790
Linear depressions 573
Liquid hydrocarbons 712
Liquid water 194, 387
Liquid-water phase 933
Liquid water (Venus) 891
Lithophile elements (Ureilite) 104, 172, 872
LITHOSPHERE 179, 201, **392**, 415, 417, 644
Lithosphere (Earth) 933
Lithosphere (Moon) 509
Lithosphere (Venus) 899
Little Ice Age 369, 468, 566, 744, 790
Loading 363
Local Thermodynamic Equilibrium (LTE) 665
LOD 216
Lodestone 273
Lodran (Pakistan) 483
LOMOSOMOV, M.V. 301, **393**
Lomosomov ridge 393
Long Duration Exposure Facility (LDEF) 190, 480, 489
Long-term geologic cycles 255
Long-term sea level change 282
Longitude of the perihelion 494
Lord Kelvin 806
Lorentz expression 665
Lorentz Force 468
Lorentz profile 782
Lorentz resonances 604
Lorentz, H.A. 284
LOWELL, P. 301, **393**, 430, 579, 590, 645
Lowland plains (Mars) 442
Luminosity 4, 745
Luminosity spectrum 569
Luminous energy 754
Luna I 500
LUNA MISSIONS **394**
Lunakhod 500

Lunar apsides cycle 220
Lunar atlases 381
Lunar core 509
Lunar craters 498, 560
Lunar eclipses 229, 705
Lunar ephemeris 238
Lunar evectional cycles 798
Lunar exosphere 501
Lunar fission theory 874
Lunar formation 226
Lunar geology 874
Lunar gravitational field 508
Lunar highlands 499
Lunar highlands meteorites 395
Lunar interior 509
Lunar landforms 502
Lunar Laser Ranging (LLR) 385
Lunar mare basalts 318
LUNAR METEORITES 20, **387**, 939
Lunar month 217, 220
Lunar nomenclature 381
Lunar orbit 221
LUNAR ORBITER MISSIONS **397**, 500, 939
Lunar Orbiters 796
Lunar paleomagnetism 512
Lunar periodicities 567
Lunar recession 826
Lunar regolith 507, 669
Lunar retreat 835
Lunar samples 500
Lunar 'secular acceleration' 385
Lunar seismic activity **513**, 730
Lunar soil 732
Lunar stratigraphy 503
Lunar swirls 511
Lunar theory 91
Lunar tide cycle 495
Lunar (and solar) tides 828
Lunar year 939
Lunation 387, 939
Luni-solar precession 657
Lunokhod 613
Lunokhods 394
LYMAN ALPHA **397**
Lyman alpha in comets 397
Lyman alpha scattering 886
Lyman series 779
Lyrids 480
Lysithea 706

Mach's principle 336
Maculae 551
Magellan mapping 887, 888
MAGELLAN MISSION **399**
Magellanic clouds 256
Magma 317
Magnetic anomalies 643
Magnetic declination 572
Magnetic deflection mass spectrometers 352
Magnetic field 167, 192, 208, 272, 499, 940
Magnetic field (giant planets) 930
Magnetic field (Jupiter) 372
Magnetic field (Neptune) 532
Magnetic field (Saturn) 581
Magnetic field (Solar) 852
Magnetic field (Uranus) 863
Magnetic field lines 210, 936
Magnetic field and magnetosphere 710
Magnetic field units 406
Magnetic fields and interplanetary environment, **Table AA8** 944
Magnetic inclination 572
Magnetic polarity 343
Magnetic poles 269
Magnetic sector mass spectrometers 815
Magnetic solar cycle 790

SUBJECT INDEX

Magnetic storms 268, 747, 790
MAGNETISM **400**
Magnetism (Lunar) 510
Magnetohydrodynamic dynamo 523, 655, 718
Magnetohydrodynamic processes 400
Magnetohydrodynamic Theory (MHD) 626
Magnetohydrodynamics 791
Magnetometer(s) 457, 575, 580
MAGNETOMETRY **406**
Magnetopause 273
Magnetosheath (Venus) 906
Magnetosphere (Saturn) 718
Magnetosphere (Uranus) 863, 867
Magnetosphere(s) 208, 210, 820
Magnetospheres 624
MAGNETOSPHERES OF THE OUTER PLANETS **410**
Magnetospheric plasmas 629
Magnetostratigraphy 269
Magnetotactic bacteria 405
Magnetotail 211
Magnetotail (Venus) 905, 906
Main-belt asteroids 841
Main-sequence star 786
Manganese nodules 206
MANTLE 162, 400, **415**, 641
Mantle (Venus) 889
MANTLE CONVECTION 163, **416**, 808
Mantle convection (Venus) 895
Mantle dynamics (Venus) 901
Mantle plumes 310, 646
Mantle viscosity 644, 809
Map 79
MAP PROJECTIONS **421**
Mare 498
Mare imbrium 502
Mare meteorites 395
Mare terrain (Moon) 916
Mare-type basaltic floods (Mercury) 919
Marginal basins 644
Maria (sing. mare) 502
Marine compass 273
Marine Observation Satellite (MOS) 690
Mariner Venus–Mercury mission 289
Mariner 847
Mariner 4 451
Mariner 7 573
Mariner 6 and 7 451
Mariner 9 451, 573, 650
Mariner 10 469, 477
MARINER MISSIONS **427**
MARS 5, 11, 23, 76, 158, 162, 180, 183, 189, 198, 232, 308, 390, 398, 401, 429, **430**, 489, 490, 546, 560, 564, 568, 579, 584, 592, 671, 699, 725
MARS: ATMOSPHERE 430, **432**
MARS: GEOLOGY **441**
MARS: GRAVITY **450**
MARS: IMPACT CRATERING **451**
Mars ionosphere 455
MARS: MAGNETIC FIELD AND MAGNETOSPHERE **454**
Mars (named features) 550
Mars (named features) [map] 545
MARS OBSERVER MISSION **456**
Mars Observer Camera (MOC) 457
Mars Observer Laser Altimeter (MOLA) 386, 458
Mars (planetary properties) 608
Mars program (Soviet) 772
Mars, question of life 430
MARS: REMOTE SENSING **459**
Mars (sampling) 613
MARS: STRUCTURAL GEOLOGY AND TECTONICS **461**
Mars, surface geology 430

Mars (tidal friction) 827
Mars (UV) 848
Mars (Viking Mission) 908
Mars-sized bodies 226
Martian 237
Martian atmosphere 456
Martian 'canals' 737
Martian chronologies 453
Martian cold world 933
Martian core 811
Martian crustal dichotomy 442
Martian dust storms 237
Martian gravity field 292
Martian magnetotail 455
Martian meteorites 20
Martian moons 184, 573
Martian nomenclature 543
Martian origin of SNC meteorites 739
Martian polar caps 447
Martian samples 740
Martian satellites 295
Martian soil 360, 431
Martian terrains 652
Martian year 939
MASCON **463**
Maskelynite 241
Mass 366, 508, 541
Mass spectrometry 351
MASS EXTINCTION 129, 178, 255, **464**, 518, 520
Mass wasting (Earth) 795
Mass–luminosity relation 381
Matter 787
MAUNDER MINIMUM **467**, 744, 790
MAUNDER, E.W. **467**
Maunder's 'Butterfly Diagram' 467
Max Wolf 841
Maxwell Montes 919
MAXWELL THEORY **468**
Maxwell–Boltzmann distribution 665
Mayan 27
Mean solar day 215, 608
Mean temperature of planet Earth 315
Mécanique Céleste 186
Mechanique analytique 382
Megacycles 203
Megalithic 26
Megatectonics 801
Melting 224
Mensae 551
Mercator 422
Mercator projection 424
Mercurian core 812
Mercurian crust 476
Mercurian magnetic field 812
MERCURY 12, 162, 287, 360, 401, 429, **469**, 593, 610, 764, 584, 662, 669, 699
MERCURY: ATMOSPHERE **471**
Mercury flights 517
MERCURY: GEOLOGY **473**
MERCURY: MAGNETIC FIELD AND MAGNETOSPHERE **476**
Mercury (named features) 550
Mercury (planetary properties) 608
Mercury (UV) 846
Mercury's rotation 660
Mercury's orbital motion 284
Meridian 156
Meridian transit 835
Meridians of longitude 421
Meridional wind field (Venus) 894
Mesogranulation 755
Mesopause 478, 819
Mesosiderites 784
MESOSPHERE **478**
Mesospheric temperature 822
Messina Chasmata 833

Metabolic processes 68
Metal 360
Metal–silicate mixing 107, 197, 235, 484, 785
Meteor Crater, Arizona 327
METEOR SHOWER, METEOROID STREAM **480**
Meteor showers 31, 118
Meteor swarms 725
METEOR, METEOROID **479**
Meteor-slag theory 732
METEORITE 169, 479, **481**, 874
Meteorite grains 344
METEORITE PARENT BODIES **486**
Meteorites, **Table AA9** 945
Meteoritic impact 804
Meteoritic bombardment 652, 708
Meteoritics 874
Meteoroid impacts 510
Meteorological network 387
Meteor(s) 342, 656
Methane (Neptune) 524
Methane (Triton) 536
Methane (Uranus) 857
Methane 195, 388, 495, 523, 589, 596, 712, 715, 786, 822, 831, 846
Methane ice 724
Methane + Nitrogen ice 596
Method of multiple resonance-commensurability construction 565
Metis 706
Metonic Cycle 939
Metric system 384
Mevnadier, L. 272
Microbial life 387, 430
Microkrystites 803
MICROMETEORITE **489**, 507
Micrometeoroid bombardment 604
Microtekites 803
Microwave opacity (Venus' atmosphere) 670
Microwave radiometer 428
MICROWAVE SPECTROSCOPY **490**
Microwaves 691
Mid-ocean Ridge Basalts (MORB) 250
Mid-ocean ridges 641
Mie theory 653, 666, 685
Milankovitch cycles 216
MILANKOVITCH, M. **494**
Milankovitch orbital cycles 567
MILANKOVITCH THEORY 300, **494**, 610
Milky Way 519, 559, 756, 786
MILLER–UREY EXPERIMENT 389, **495**, 874
Mimas 551, 723, 706
Minor planets 875
Minovitch, M.A. 288
MIR space station 489
MIRANDA **495**
Miranda 551, 699, 706
Missions to planets and the moon, **Table AA10** 945
Mimas 697
Mohorovicic discontinuity 179, 392
Moldavite 803
Molecular Analysis (Mars) 909
Molecular conduction 820
Molecular rotational transitions 777
Molecular hydrogen 759
Molten core (Mars) 454
Molten ejecta 510
Moment of inertia 12, 13, 610
Momentum 541
Monostatic radars 662
Montes 550, 551
Monthly cycles 498, 826

Precambrian 933
PRECESSION 155, 494, 562, **656**
Precession rate (Triton) 836
Precession of the equinoxes 179, 241, 300, 610, 656
Precession of Moon's orbit 215
Precession of nodes 763
Precession of the perihelion 763
Precession-inducing force 657
Precursors 273
Predictability 186
Prediction of solar activity 749
Prehistoric humans 361
Preplanetary dust grains 113
Presolar evolution stages 792
Pressure gradients 293
Presolar materials 112
Pressure modes (p-modes) 756, 790
Pressure Modulator Infrared Radiometer (PMIRR) 458
Prime Meridian 156
Primordial Earth 495
Primordial heat sources 416
Primordial substance 808
Primordial Venusian ocean 891
Principia Mathematica 541
Probabilism 187
Probability theory 383
Probe 313
Proctor, R.A. 302
Prograde rotation 608, 611
Project Longstop 763
Prolonged minima of sunspots 749
Prometheus 605, 706
Prominences (Sun) 230, 741, 746, 790
Proportionality of integer values 564
Proteins 495
Proterozoic crust 203
Proterozoic glacial environment 555
Proteus 551, 534, 706
Proto-Earth 556, 783, 934
Proto-Mercury 224
Proto-Moon 226
Proto-Sun 757
Proton density 766
Proton fluxes 754
Protons 168, 624, 766
Protoplanetary bodies 623
Protoplanetary disk 609, 751
Protosolar nebula 649, 707
Protostar 792
Proxigean spring tides 797
Proxigee 26, 572
Ptolemaic system 88
PTOLEMY 23, 88, 228, 231, 300, **658**, 659
Puck 551, 706, 865, 867
Pulse of the solar system 565, 749
Punctuated equilibrium 466
Pyroclastic cones (Moon) 916
Pyroclastic volcanism (Venus) 896
Pyrolytic release experiment 430

Quadrupole mass filter 815
Quadrupole model for solar magnetic field 791
Quantitative topography 576
Quantum theory 187
Quartz 317, 737
Quasi-biennial period 743
Quasi-biennial oscillation(s) 216, 386, 479
Quasi-periodic climate change (Mars) 438
Quasi-Periodic Variations 36
Quasi-resonant term 762

Radar 137, 399
Radar altimeters 290
RADAR ASTRONOMY **660**

Radar images (Venus) 895
Radar ranging 593
Radar signal 691
Radar-projection lines 576
RADARGRAMMETRY **576**
Radiant heat 232
Radiation 55, 754
RADIATION BELT **663**
Radiation pressure 489
Radiation theory 494
Radiative zone (Sun) 789
Radiative transfer 692
RADIATIVE TRANSFER IN PLANETARY ATMOSPHERES **664**
Radio 137
RADIO ASTRONOMY **667**
Radio bursts 711
Radio frequency mass spectrometers 352
Radio frequency transmitters 279
Radio occultation 605, 677
Radio occultation experiment 428
Radio occultation measurements 432
Radio occultation technique 359
RADIO SCIENCE **676**
Radio science (Mars) 910
Radio telemetry 307
Radio telescope(s) 167, 660, 888
Radio waves 354, 691
Radioactive decay 181
Radiocarbon fluctuations 743
Radiogenic decay 921
Radiogenic heat production 416
Radiogenic heat production (Venus) 899
Radiogenic heat sources 809
Radiogenic heating 839
Radiogenic heating (Moon) 917
Radiogenic isotopes 169
Radioisotope Thermoelectric Generators (RTGs) 580
Radiometric albedo 43
RADIOMETRY **682**
Radionuclides 169, 182
Radon 542
Raman scattering 695
Range gate 385
Ranger program 381, 500
RANGER MISSIONS **683**
Rayleigh number 808
Rayleigh scattering 695
Rayleigh scattering theory 692
Reaction principle 842
Recession of the moon 500, 828
Red giant 793
Reducing atmosphere 495, 832
Reflectance 10, 693
REFLECTANCE SPECTROSCOPY **684**
Reflection spectroscopy (UV) 1, 845
Reflecting surface 578
Reflecting telescope 540
Reflective objects 836
REFLECTIVITY **686**
Reflector array 385
Refraction 393
Refractive index 253, 684
Refractory elements 172
REGIONAL PLANETARY IMAGE FACILITIES (RPIFs) **686**
Regiones 550
REGOLITH 266, 573, 652, **687**, 732, 776
Regolith breccias 395, 483
Regression of lunar nodes 567
Regression of nodes 554
Relative age 269
RELATIVISTIC COSMOLOGY **688**
Relativity 240, 688
Remnant planetesimal 649
REMOTE SENSING 232, **689**

Remote sensing of ionosphere 695
Resonance 299, 380, 498, 564, 573, 605, 724, 764, 828
Resonance gaps 42
Resonance 'ladders' 569
Resonance origin 697
Resonance scattering 777, 780, 843
RESONANCE IN THE SOLAR SYSTEM **698**, 826
RESONANCE IN SATURN'S RINGS **696**
Resonance-induced variations 610
Resurfacing processes (Venus) 896, 903
Retrograde 587
Retrograde motion 566
Retrograde orbit (Triton) 536
Retrograde orbit 836
Retrograde rotation of Uranus 513
Retrograde rotation (Venus) 887
Retrograde satellites 840
Revolution 378
Rhea 551, 706, 724
Rhyolites 249
Rhythm 566
Rhythmic banding 826
Riccioli 300
Richter scale 726
Ridge belts (Venus) 898
Ridged plains (Mars) 441
Ries crater 805
Ries Kessel, Germany 327
RIFMA 613
Rilles 506
Ring basins (Moon) 917
Ring bombardment 710
Ring moons 696
Ring nebula 344
Ring occultation 680
Ring particles 603
Ring particles (Saturn) 718
Ring system 500
Ring system (Uranus) 864
Ring systems (giant planets) 925
Ringed features 551
Ringed maria basins 464
Ringlets 696
Ringlets (Saturn) 721
Rings, **Table AA12** 945
Rings of Uranus 680, 681
Ripples 237
Rising of Sothis 938
Robotic spacecraft 517, 796
Robotic spacecraft programs 397
ROCHE, E.A. **701**
ROCHE LIMIT 76, 217, 226, 500, 534, **701**, 721, 831
Roche lobe 703
Rock magnetism 404
Rock flour 688
Rocket equation 306
Rocket technology 307
Rocket trajectories 309
Rocketry 283, 941
Rocks 542, 651
Rocks from space 481
Rocky core 857
Roemer, O. 345, 366
Rosalind 706
Rose–Tschermak–Brezina classification 2
Rossby Number 53
Rosseland mean opacity 789
Rotation 587, 594, 608
Rotation (Mercury) 477
Rotation axis 268
Rotation cloud 759
Rotation period 267
Rotation rate 280, 609
Rotation states 828

Rotation of the Earth 593, 747
Rotation of the galaxy 560
Rotation of the Moon 387
Rotation of the Sun 742
Rotational angular momentum 611
Rotational fission 226
Rotational motion 835
Rotational periods 587
Rovers 308
Royal Observatory, Greenwich 156
Rudolphine tables 379
Runaway growth 760
Running water (Earth) 796
Rupes 550

SABINE, E. **704**
Sabine Minimum 704, 749
Safronov, V.S. 512, 611, 758, 760
Sagittarius 940
SAKIGAKE AND SUISEI MISSIONS **704**
Salinity-density-driven circulation 314
Saltation 237
San Andreas fault 727
Sand, Sandstone 737
Sand dunes 431
Saros 230, 567
SAROS CYCLE **705**, 808
Satellite 345
Satellite data 926
Satellite (largest, Titan) 831
Satellite Laser Ranging (SLR) 280, 385
SATELLITE, NATURAL **705**
Satellite surfaces (giant planets) 925
Satellite system(s) 153, 375
Satellites of planets (list) 585
Satellites, **Table AA13** 946
SATURN 59, 183, 186, 228, 233, 295, 402, 412, **708** 504, 579, 673
Saturn and Jupiter 834
SATURN: ATMOSPHERE **714**
Saturn electrostatic discharges (SED) 931
Saturn emissions kilometric range (SKR) 930
SATURN: INTERIOR STRUCTURE **717**
Saturn lightning 931
SATURN: MAGNETIC FIELD AND MAGNETOSPHERE **718**
Saturn occultation experiments 676
Saturn (planetary properties) 608
SATURN: RING SYSTEM 604, 696, **719**
SATURN: SATELLITE SYSTEM **723**
Saturn satellites (names) 551
Saturn (UV) 849
Saturn's rings 310, 653, 660, 711
Saturn's ring system 181, 581
Saturn-Jupiter resonance 153
Saturnian satellites 596, 813
Scalar magnetometers 407
Scaliger, J. 365
Scanner cameras 576
Scaphe 241
Scarp retreat 838
Scattering 561, 666, 685, 770
SCHIAPARELLI, G.V. 301, 393, 430, 469, 480, **725**
Schiaparelli's system of names 543
Schimper, K. 314
Schröter, J.H. 301
Schuster, A. 570
Schwabe cycle 937
Scientific philosophy 757
Scoop 796
Scorpius 940
Sea level 268, 279
Sea of Tranquility 684
Sea salts 4, 314
Sea water 313, 934

Seafloor sediments 190
Searches for planets 589
SEASAT 290, 695
Seasonal cycling of CO_2 (Mars) 435
Seasonal cycles 650
Seasonal insolation 555
Seasons on Triton 836
Seasonal variation(s) 490, 838
Seasons 555
Sector areas 564
Secular acceleration of the Moon 222, 384
Secular cooling 810
Secular cooling (Venus) 900
Secular resonances 700
Secular variation 210, 403, 572, 646
Sedimentation 241
Segregation 188
Seismic energy release 726, 729
Seismic gap 729
Seismic tomography 418
SEISMICITY 201, **726**
Seismicity of the Earth 726
Seismicity of Moon 727
Seismicity of Venus 728
Seismicity of Mars 728
Seismographs 161
Seismometer(s) 24, 726
Seismotectonics 801
Self-reproducing life 933
Semidiurnal periods 830
Separatrices 698
Shadow hiding hypothesis 561
Shakespeare 866
Shakespearean characters 496
Shallow earthquakes 726
Shallow moonquakes 514
SHARONOV, V.V. **732**
Shatter cones 331, 733
SHEPHERD SATELLITE 700, **732**
Shepherds 696
Shergottites 3
Shield volcanoes (Mars) 442
Shock features (Ureilite) 871
Shock heating 108
SHOCK METAMORPHISM **733**, 804
SHOCK WAVES 177, 231, 598, **734**, 766
Shocked rocks 733
Shoemaker-Levy 9 **132**
Shooting stars 480
Short-period comets 381
Shuttle imaging radars 694
Shuttle Atlantis 399
Sidereal period of rotation of the Sun 800
Sidereal day 608
Sidereal month 222
Sidereal year 300, 930
Siderophile elements (Ureilite) 872
Siderophile 188
Siderophile element 104, 172, 869
SILICA, SILICATE **737**
Silicate absorption (Mars) 460
Silicate mantle 416
Silicate planets 194
Silicate rocks 499
Silicates 783
Silicon oxide ($SiO2$) 737
Silicon carbide 543
Single-scattering albedo 776
Sinope 706
SJ:TLC = Triad Length Cycle of SJL 565
SJL 565
Skylab 517, 690
SLIPHER, E.C. **737**
Slowing of the Earth's rotation 747
SMALL SATELLITE **738**
Smog 714
SNC METEORITES 454, 483, **739**

Snell's law of refraction 692
Snow 933
Sodium 616, 940
Soft landers 308, 394
Soil (Lunar) 669
Soil analysis, Viking 390
Soil physics 651
SOLAR ACTIVITY 164, 272, **741**, 749, 789
Solar activity and weather on Earth 743
Solar activity cycle(s) 252, 566, 663, 742
Solar backscatter 766
SOLAR CONSTANT **745**
SOLAR CORONA 630, **746**
Solar cycle (Neptune) 528
Solar cycle(s) 523, 569, 479, 766
Solar cycle period 820
SOLAR DAY AND TIME **747**
Solar eclipses 229
SOLAR FLARES 165, 543, 569, 742, **747**, 754
Solar flux 844
Solar flux (Mars) 433
Solar flux (Venus) 890
Solar gravity 852
Solar hour angle 730
Solar inertial motion 745
Solar irradiance 338
Solar jerk 569
Solar linkage 704
SOLAR LUMINOSITY 700, 745, **748**
Solar Lyman alpha 397
Solar magnetic field(s) 342, 741, 791, 940
Solar maximum (SMAX) 821
Solar maximum mission satellite 748
Solar minimum (SMIN) 821
Solar magnetic reversal 294
SOLAR MOTION **748**
SOLAR NEBULA 2, 223, 484, **751**
SOLAR NEUTRINO **751**
Solar panels 429
Solar parallax 50, 365
solar particle event 754
Solar photons 753
SOLAR PHOTOSPHERE **754**
Solar prominences 842
Solar proton bombardment 652
Solar radiation 336
Solar radiation pressure 118, 656, 719
Solar radius 748
Solar rotation 790
SOLAR SYSTEM 88, 255, 381, 592, **756**, 786, 707
Solar system barycenter 62
SOLAR SYSTEM: ORIGIN **756**
SOLAR SYSTEM: STABILITY **762**
Solar tides 610
Solar torque period 252
Solar ultraviolet (Mars) 434
Solar ultraviolet flux 745
Solar ultraviolet radiation 200
Solar waves 756
SOLAR WIND 147, 164, 210, 251, 242, 272, 297, 342, 411, 470, 501, 575, 616, 624, 663, 704, 719, 741, **766**, 852, 854, 710, 746, 820
Solar wind convection 627
Solar wind heat flux 769
Solar wind interaction (Venus) 906
Solar wind: ionosphere (Venus) 583
Solar wind scavenging 454
Solar wind speed 629
Solfatara 314
Solstices 808
SOP: Sun's Barycentric Inertial Orbital Progression 565
Sosigenes 365

Sothic cycle 938
South Pole: Aitken Basin 113
Soviet Union 394
SOVIET MARS MISSIONS **772**
Space shuttle 598
Space missions 613
Space, space vehicles 517
Space techniques 302
Space shuttle 517
Spaceborne photography 694
Spacecraft 84, 283, 489
Spacecraft cameras 577
Spacecraft measurements 593
Spacecraft missions 591
Spaceprobes 704
SPACEWATCH 578, **774**
Spectral line 665
Spectral strengths 781
SPECTROPHOTOMETRY 135, **775**
Spectroscope 253
Spectroscopic instrumentation (UV) 845
Spectroscopic remote sensing (Mars) 459
Spectroscopy 135, 253, 682
SPECTROSCOPY: ATMOSPHERES **776**
Spectroscopy (UV) 843
Spherical harmonics 384, 402, 595
Spherics 936
Spherules 803
Spicules (Sun) 789
Spin axes 611
Spin rate 253
Spin-orbit coupling 828
Spin-orbit 1:1 resonance 699
Spinel: Perovskite transition 420
Spiral arm 256, 559
Spiral density waves 605
Spontaneous fission 183, 512
Spörer's Law 79, 294
SPOT 690
Spreading centers 201
Spreading ridges (Earth) 915
Spring and neap tides 797
Sputnik 394, 500, 517, 772
Sputnik 1 302
Sputtering 374, 473, 489, 501, 626, 708, 710
SQ-1 through 12 = Solar Quadrature Arrangements 565
SSC: Mean Sunspot Cycle 565
Stability of the solar system 384, 762
Stability of systems 88
Stability of volatiles 596
Standing water (Earth) 796
Star of Bethlehem 274, 296
Stars 584
Stefan–Boltzmann constant 787
Stellar evolution 759
Stellar occultations 676
Stereo terrain analysis 694
Stereophotographs 576
Stereoplanigraph 577
Stickney impact 431
Stishovite 327, 737
Stochastic mechanisms 791
Stochastic processes 612
Stony meteorites 2
STONY IRON METEORITES 483, **783**
Strahl 769
Stratigraphic timescale 46
Stratopause 786
STRATOSPHERE **786**
Stratospheric abundances (Uranus) 859
Stratospheric hazes (Neptune) 525
Stratospheric inversions 56
Streaks (Venus) 889
Streamers 599
Strewn fields 802
Strike-slip (horizontal) motion 727

Strike-slip faulting 253
Stromatolites 217, 387
Structural geology 801
Subduction 727
Subduction zones 201, 641
Subduction zones (Earth) 915
Sublimation 536
Submarine weathering 935
Substrate 68
Subsurface water (Earth, Mars) 795
Sudbury 206
Sulci 551
Sulfur 412
Sulfur volcanoes 346
Sulfur-bearing gases 786
Sulfur-metabolizing organism 388
Sulfuric Acid Clouds (Venus) 892
SUN 159, 164, 165, 253, 401, 520, 756, **786**
Sun rotation 79
Sun's brightness 468
Sun's circum-galactic trajectory 786
Sun's diameter 229
Sun's epicyclic motion 256
Sun's magnetic field reversal 569
Sun's mean spin rate 569
Sun's orbit 88
Sun's rotation axis 852
Sun–barycenter separation distance 567
Sunburn 843
Sundial 808
Sunlight 293
Sunrise 941
Sunset 941
Sunspot cycle(s) 299, 704, 789
Sunspot dearth 467
Sunspot number 252, 937
Sunspots 79, 258, 338, 568, 741, 745, 747, 754, 755, 789
Super-Alfvén speeds 766
Superchrons 270
Superconducting Quantum Interference Device (SQUID) 401
Supergranulation 755
Supernovas 941
Supersonic flight 734
SURFACE GRAVITY **794**
SURFACE PRESSURE **794**
SURFACE PROCESSES **794**
Surveyor landers 397
SURVEYOR MISSIONS 500, **796**
Survivorship 464
Swarms of earthquakes 727
Swedenborg, E. 541
Swell 310
Synchronism 826
Synchronous corotation 535
Synchronous rotation 498, 609, 707
Synchrotron radiation 663, 667, 672
SYNERGETIC TIDAL FORCE **797**
Synodic conjunctions 564
Synodic months 221
SYNODIC PERIOD **800**
Synthetic Aperture Radar (SAR) 695, 888
Syrtis Major 441
SYZYGY 26, 572, **800**

Taphrogenic coast 801
Taphrogeny 801
Tasman sea 290
Taurus molecular cloud 759
Taurus 940
Techniques, **Table AA15** 946
Tectonic boundaries 201
Tectonic features (Mercury) 475
Tectonic fracturing 868
Tectonic lineaments 536
Tectonic plates 417

Tectonic style (Venus) 889
TECTONICS **801**
Tectonism (Earth) 795
Tectonoeustasy 801
Tectonophysics 801
Tectonosphere 801
Tectonostratigraphy 801
TEKTITE **802**
Telescope 258, 300, 321
Telescopic and spacecraft spectra of solar system bodies (spectroscopy) 913
Telescopic period 937
Telesto 706
Television images 796
Temperate latitudes 555
TEMPERATURE **806**
Temperature profile 820
Tension 727
Terminal shock 736
Terminator 44
Terrae (lands) 502, 550
Terrain information 576
Terrestrial Dynamical Time (TDT) 594
Terrestrial geology 194
Terrestrial ionosphere 356
Terrestrial kilometric radiation (TKR) 929
Terrestrial magnetism 294
TERRESTRIAL PLANETS 180, 223, 586, **806**, 887
Tessera deformation 807
TESSERA (terrain) 310, 550, **807**
Tesserae (Venus) 898, 919
Tethys 551, 724, 706
Textured terrain 505
Thalassa 534, 706
THALES OF MILETUS 705, **808**
Tharsis 431, 441
Tharsis bulge 431
Tharsis montes 461
Tharsis ridge 918
The Sun, solar physics, **Table AA14** 946
Thebe 706
Theory of universal gravitation 540
Thermal convection 417, 642, 757
Thermal emission 667, 672
THERMAL EVOLUTION OF PLANETS AND SATELLITES **808**
Thermal excitation 777
Thermal fusion processes 753
Thermal histories of satellites 828
Thermal history 809
Thermal infrared wavelengths 691
Thermal lithosphere 392
Thermal moonquakes 514
THERMAL PLASMA INSTRUMENTATION **814**
Thermal properties (giant planets) 924
Thermal radio emission 673
Thermal runaway mechanism 310
Thermal winds (Jupiter) 368
Thermoclastic and cryoclastic weathering 935
Thermoclasts 935
Thermodynamical equilibrium (TE) 490
Thermokarst (Earth) 795, 935
Thermonuclear energy 787
Thermosphere 354
THERMOSPHERE **819**
Thermospheric general circulation model (TGCM) 822
Thermostatic control 820
Tholi 551
Three-body problem 243, 382, 383, 763
Thunderstorms 598
Tidal braking of the Earth's rotation 386
Tidal bulge(s) 610, 826, 835
Tidal changes 797

Tidal disruption 226, 831
Tidal dissipation 556, 812, 835
Tidal dynamics 840
Tidal equilibrium state 611
Tidal flexure 587
Tidal flooding 799
Tidal force 'envelope' 829
Tidal forces 738
TIDAL FRICTION 14, 153, 181, 555, 610, 747, **825**, 839
Tidal friction theory of George Darwin 702
TIDAL HEATING 266, 498, 712, **828**
Tidal rhythmites 217
Tidal torque 215
Tidal periods 215
Tidal torques 573, 831
Tidal years 220
TIDE-RAISING FORCE **828**
Tides 498
Tides (Venus) 894
Tillites 256
Time 747
Time-of-flight mass spectrometers 353
Timescales 593
TITAN 7, 237, 311, 412, 490, 706, 707, 708, 709, 712, 713, 723, **831**
Titan Torus 620
Titan (UV) 850
TITANIA 551, 706, **833**, 867
Titius, J.B. 834
TITIUS-BODE LAW 27, 267, 564, 579, 587, 758, **834**
TOMBAUGH, C.W. **835**
Top of the Atmosphere (TOA) 338
Topex/Poseidon mission 282
Topography 594
Topography of Venus 583
Tornado 187
Toroidal field 401
TORQUE 17, 84, 387, 566, 610, 646, 657, 733, 812, 826, **834**, 938
Torus 349, **616**
Tracking 303
Tracking and Data Relay Satellites (TDRSS) 694
Trade gases 598
Trade winds 293
Trans-Neptunian planet 737
Transform faulting 253
Transform faults 642
Transform motion 727
TRANSIT 300, 747, 826, **835**
Transit circle observations 593
Transit of Mercury 295
Transitional paths 655
Transits of Venus 836
Transmittance 692
Transport 241
Transport media (Earth) 795
Trapped continuum radiation 634
Tree rings 252, 743
Trenches 201, 641
Triangulation 267
Triaxial ellipsoid(s) 233, 269
TRITON 7, 237, 441, 534, 551, 646, 649, 786, 812, **836**
Triton (named features) [map] 549
Triton Torus 621
Triton (UV) 850
Triton (Volcanism) 921
Triton year 536
Troilite 172, 783
Troilite (FeS) 234
TROJAN ASTEROIDS 28, 383, **841**
Tropical month 222
Tropical year 300

Tropopause 841
Tropopause (Titan) 831
Tropopause (Triton) 838
TROPOSPHERE **841**
Tropospheric abundances (Uranus) 859
Tropospheric clouds and hazes (Neptune) 525
Tropospheric delay 280
TROUVELOT, É. L. **842**
TSIOLKOVSKY, K.E. 309, **842**
Tsunami earthquakes 727
Tunguska 130
Turbidity currents (Earth) 796
Twin-planet system 62
Two-body problem and orbit determination 88

Ultramafic rocks 317
Ultraviolet (UV) 138, 784, 820
Ultraviolet emissions 415
Ultraviolet flux 471
Ultraviolet Imaging Spectrograph (UVIS) 86
Ultraviolet photodissociation 786
Ultraviolet photons 708
ULTRAVIOLET RADIATION 630, 712, **843**
ULTRAVIOLET SPECTROSCOPY **843**
Ultraviolet spectrometer experiment 429
Ultraviolet wavelengths 691
ULYSSES MISSION 298, **852**
Umbra 229, 741
UMBRIEL 551, 706, **855**
Unexplained orbital motions 590
Uniformitarian 933
Uniformitarian philosophy 314
Universal Time (UT) 215, 386
Universe 215, 273, 386, 762
Uranian atmosphere 858
Uranian magnetic field 931
Uranian moons 833
Uranian satellites 596
URANUS 59, 228, 387, 402, 414, 495, 504, 593, 834, **856**
Uranus (UV) 849
URANUS: ATMOSPHERE **858**
URANUS: MAGNETIC FIELD AND MAGNETOSPHERE **863**
Uranus (Planetary properties) 608
URANUS: RING SYSTEM 605, **864**
URANUS: SATELLITE SYSTEM **866**
UREILITE METEORITES 3, **868**
Urey effect 874
Urey ratio 810
UREY, H.C. **874**
US Geological Survey's Earth Resources Observation Systems (EROS) 694
UTC or Coordinated Universal Time 747
UV emissions 831
UV heating 820
UV heating (Venus) 891
UV spectroscopy **843**, 876

VÄISÄLÄ ORBIT **875**
Validation 214
Valles 550
Valles Marineris 429, 431, 441, 450, 918
Valley network (Mars) 444
VAN ALLEN, J.A. **876**
Van Allen belts 411, 663
Variable star 741
Vastitas borealis 441
Vector magnetometers 408
VEGA MISSION **876**
Velocity of light 366
Venera (Venus) 876
VENERA MISSIONS 847, **879**

VENUS 5, 51, 59, 158, 162, 163, 198, 310, 393, 401, 429, 490, 582, 584, 599, 660, 669, **887**
VENUS: ATMOSPHERE 658, **890**
Venus core 811
VENUS: GEOLOGY AND GEOPHYSICS **895**
VENUS: GRAVITY **904**
Venus gravity assist 260
VENUS: MAGNETIC FIELD AND MAGNETOSPHERE **905**
Venus (named features) 550
Venus (named features) [map] 547
Venus orbit 399
Venus (planetary properties) 608
Venus (sampling) 614
Venus surface 886
Venus transit 296
Venus (UV) 846
Venus-crossing asteroids 896
Venusian atmosphere 658, 888
Venusian hothouse 933
Venusian mesosphere 490
Venusian surface 879
Venusian tessera 807
Venusian wind streaks 237
Verona rupes 496
Vertical dynamics (Neptune) 530
Very large array (VLA) 593, 660
Very Long Baseline Interferometry (VLBI) 280
VESTA 29, 242, 559, **907**
VHA (Moon) 916
Vibration-rotation spectra 778
Vidicon cameras 428, 576
VIKING MISSION 390, 694, **908**
Viking orbiters 573
Viking 1 and 2 451
Virgo 940
Virtual Geomagnetic Pole (VGP) 403, 572, 654
Viscosity 809
Viscous fluid 417
VISIBLE AND NEAR-INFRARED SPECTROSCOPY (VIS-NIR) **911**
Visible and Infrared Mapping Spectrometer (VMS) 337
Visible light 232, 336
Visible wavelengths 691
Volatile elements 175
Volatile loss 649
Volatile recharging (regassing) 812
Volcanic ash 688
Volcanic center 310
Volcanic eruptions 335, 786
Volcanic exhalation 314
Volcanic features (Mars) 442
Volcanic flows (Venus) 896
Volcanic haze 786
Volcanic overplating 196
Volcanic plains (Mars) 443
Volcanic resurfacing 828
Volcanic rocks 317
Volcanic tremor 727
Volcanic vents (Venus) 889
Volcanically active body 919
Volcanically erupted ice 838
Volcanism 233, 400, 496
Volcanism (Earth) 795, 915
VOLCANISM IN THE SOLAR SYSTEM **915**
Volcanism (Mars) 918
Volcanism (Mercury) 919
Volcanism (Moon) 916
Volcanism (Venus) 895, 919
Volcanotectonic landform 163
von Mädler, J.H. 301

von Humboldt, A. 941
Voyager 694, 838
Voyager images 858
VOYAGER MISSIONS 775, **922**
Voyager radio occultation experiment 524
Voyager spacecraft 367
Voyager infrared (IRIS) 368
Voyager 1 297, 373, 587, 712, 718, 832
Voyager 2 297, 414, 496, 521, 523, 536, 718, 856
VOYAGER PLANETARY RADIO ASTRONOMY **927**

Wanderers 584
Wandering stars 300
WATER 194, 197, 199, 587, **933**
Water ice 176, 243, 257, 496, 534, 573, 575, 596, 650, 707, 718, 724, 738, 813, 868
Water in solar system 934
Water mass redistribution 217
Water reservoirs on Mars 445
Water vapor 841, 933
Water vapor (Mars) 432, 491
Water-ice clouds 715
Water-ice particles 605
Water-level experiments 541

Wave of darkening 430
Wave theory of light 310
Weather radar 695
WEATHERING 241, 688, **935**
Weathering index 935
Wegener 642
Westward drift 295
WHISTLER 600, 638, **936**
White dwarf 793
White light 540
White spot on Saturn 295
Whole-number relations 384
Wind (eolian) processes (Earth) 796
Wind streaks (Venus) 896
Window to space 820
Wind velocities (Venus) 886
Wobble 94, 187
WOLF, M. **936**
WOLF NUMBER **937**
Wolf number 741
WOLF, R. 704, 741, **937**
Wolf's quantity 790
Working Group for Planetary System Nomenclature (WGPSN) 543
Wrinkle ridges (Mars) 441, 461

X Rays 232, 691
X-ray Fluorescence Spectrometers (XRFS) 613
Xenon 542

Yardangs 431
YARKOVSKY EFFECT **938**
YEAR **938**
Yearly cycles 826

Z Oscillation 255
ZEEMAN EFFECT 253, 294, **940**
Zero electric charge 751
Zero rest mass 751
Zeuner, F. 494
Zhamanshin impact crater 805
ZODIAC **940**
Zodiacal dust cloud 941
ZODIACAL LIGHT 656, **941**
Zonal flow 718
Zonal wind field (Venus) 893
Zonal winds (Jupiter) 367
Zones of instability 95
Zurich sunspot number 790
ZWICKY, F. **941**

NOV 1 3 2018

ST. MARY'S COLLEGE OF MARYLAND LIBRARY

3 3127 00166 4640

QB 600.2 .E53 1997

Encyclopedia of planetary sciences